WZMWq   BMBCR   P9-AQD-041

QGKWH   FCFMG

Intellectual Property
in the
New Technological Age

Intellectual Property in the New Technological Age

Fourth Edition

Robert P. Merges
> Wilson, Sonsini, Goodrich, and Rosati
> Professor of Law and Technology
> University of California at Berkeley

Peter S. Menell
> Professor of Law
> Director, Berkeley Center for Law and Technology
> University of California at Berkeley

Mark A. Lemley
> William H. Neukom Professor of Law
> Director, Stanford Program in Law, Science and
> Technology
> Stanford University

ASPEN

PUBLISHERS

76 Ninth Avenue, New York, NY 10011
http://lawschool.aspenpublishers.com

Aspen Publishers
Attn: Permissions Department
76 Ninth Avenue, 7th Floor
New York, NY 10011-5201

Printed in the United States of America.

1 2 3 4 5 6 7 8 9 0

ISBN 0-7355-5866-3

Library of Congress Cataloging-in-Publication Data

Merges, Robert P.
    Intellectual property in the new technological age / Robert P. Merges, Peter S. Menell, Mark A. Lemley. — 4th ed.
        p. cm.
    ISBN 0-7355-5866-3
        1. Intellectual property — United States. 2. Technological innovations — Law and legislation — United States. I. Menell, Peter Seth. II. Lemley, Mark A., 1966- III. Title.
    KF2979.I432 2006
    346.7304'8 — dc22

                                                                        2006006972

About Aspen Publishers

Aspen Publishers, headquartered in New York City, is a leading information provider for attorneys, business professionals, and law students. Written by preeminent authorities, our products consist of analytical and practical information covering both U.S. and international topics. We publish in the full range of formats, including updated manuals, books, periodicals, CDs, and online products.

Our proprietary content is complemented by 2,500 legal databases, containing over 11 million documents, available through our Loislaw division. Aspen Publishers also offers a wide range of topical legal and business databases linked to Loislaw's primary material. Our mission is to provide accurate, timely, and authoritative content in easily accessible formats, supported by unmatched customer care.

To order any Aspen Publishers title, go to *http://lawschool.aspenpublishers.com* or call 1-800-638-8437.

To reinstate your manual update service, call 1-800-638-8437.

For more information on Loislaw products, go to *www.loislaw.com* or call 1-800-364-2512.

For Customer Care issues, e-mail *CustomerCare@aspenpublishers.com*; call 1-800-234-1660; or fax 1-800-901-9075.

Aspen Publishers
a Wolters Kluwer business

Summary of Contents

Contents

3 ══ *Patent Law* — *117*

4   *Copyright Law*   367

5   *Trademark Law*   *617*

6   State Intellectual Property Law and Federal Preemption   835

8   *Intellectual Property and Competition Policy*   *1097*

Preface

Rapid advances in digital and life sciences technology continue to spur the evolution of intellectual property law. As professors and practitioners in this field know all too well, Congress and the courts continue to develop intellectual property law and jurisprudence at a rapid pace. For that reason, we have significantly augmented and revised our text yet again.

Here is a synopsis of the prinicipal changes in this fourth edition:

Chapter 1. Introduction. We have expanded coverage of theories of intellectual property, including discussion of the open source movement.

Chapter 2. Trade Secret Protection. We have substituted new cases on reverse engineering and remedies.

Chapter 3. Patent Law. We have reorganized the validity materials, moving the materials on written description before novelty and non-obviousness. This chapter contains expanded coverage of patenting of DNA-related inventions (with the addition of *In re Fisher*). We have substantially rewritten and updated the materials on claim construction to take into consideration the Federal Circuit's *en banc* decision in *Phillips v. AWH Corporation*. This chapter also includes new material on extraterritorial infringement and remedies.

Chapter 4. Copyright Law. The past several years have demonstrated the extraoridinary potential of the Internet as a distribution medium, and with that, tremendous dynamism in the field of copyright protection. In order to present all of the new developments coherently, we have substantially expanded our coverage of digital copyright law. We have expanded our coverage of indirect liability (*MGM v. Grokster*) and added material on the application of fair use to search engines (*Kelly v. Arriba Soft. Corp.*), as well as added several new problems relating to Internet copyright issues. We have also expanded coverage of international copyright law.

Chapter 5. Trademark Law. The past three years have seen tremendous growth in what might be called cyber-trademark law. We have added the leading case on the requirement of trademark use (*1-800 Contacts, Inc. v. WhenU.com, Inc.*). We have also updated the materials on trademark dilution.

Chapter 6. State Intellectual Property Law and Federal Preemption. This chapter includes expanded coverage of the right of publicity, trespass to chattels, and preemption of state law.

Chapter 7. Protection of Computer Software. We have substantially expanded coverage of open source licensing, patent protection for software-related and business method inventions, and the Semiconductor Chip Protection Act.

Chapter 8. Intellectual Property and Competition Policy. We have updated coverage of the Microsoft antitrust litigation and pharmaceutical patent settlement cases.

Acknowledgments

We are indebted to a great many people who have helped us since this project began in 1991. We would like to thank our many colleagues who reviewed earlier drafts of the book and provided helpful guidance. While many of these reviews were anonymous, we have also benefitted from the advice of Lynn Baker, Paul Heald, and Pam Samuelson, each of whom read several different drafts of the book as it made its way through the editorial process. We gratefully acknowledge the reserach assistance of Evelyn Findeis, Edwin Flores, Ryan Garcia, Shari Heino, Toni Moore Knudson, Christopher Leslie, and Barbara Parvis. We would also like to thank Michele Co for exceptional secretarial and administrative assistance in completing this text.

We are grateful to many colleagues for providing suggestions for improving this book. In particular, we would like to thank John Allison, Julie Cohen, Ken Dam, Robin Feldman, Terry Fisher, Marshall Leaffer, Glynn Lunney, Ron Mann, David Nimmer, Ruth Okediji, Malla Pollack, Peggy Radin, Jerry Reichmann, Paul Schwartz, Lon Sobel, and several anonymous reviewers for their comments and suggestions in preparing the second, third, and fourth editions. We have also benefitted greatly from the research assistance and proofreading of Brian Carver, Colleen Chien, Will Devries, Tom Fletcher, Ryan Garcia, Ines Gonzalez, David Grady, Victoria H. Kane, Michelle A. Marzahn, Selena R. Medlen, Pilar Ossorio, Ryan Owens, Stephanie N.-P. Pham-Quang, Laura Quilter, John Sasson, Helaine Schweitzer, Shannon Scott, Laurence Trask, Allison Watkins, Emily Wohl, and Tarra Zynda.

Finally, we acknowledge the authors of the following images and excerpts used in this volume with their permission:

Bowman, Ward, Jr., Patent and Antitrust Law: A Legal and Economic Appraisal (1973). Reprinted courtesy of University of Chicago Press.

Carver, Brian W., Share and Share Alike: Understanding and Enforcing Open Source and Free Software Licences, 20 Berkeley Technology Law Journal 443 (2005).

Kellner, Lauren Fisher, Trade Dress Protection for Computer User Interface "Look and Feel," 61 U. Chi. L. Rev 1011 (1994). Reprinted by permission of the University of Chicago Law Review.

Menell, Peter S., The Challenges of Reforming Intellectual Property Protection for Computer Software, 94 Colum. L. Rev. 2644, 2651-54 (1994), reprinted by permission of the Columbia Law Review.

Menell, Peter S., Tailoring Legal Protection for the Computer Software, 39 Stan. L. Rev 1329 (1987), © 1987 by the Board of Trustees of the Leland Stanford Junior University.

Radin, Margaret Jane, Property and Personhood (1982), as revised in Margaret Jane Radin, Reinterpreting Property (University of Chicago Press, 1993). Reprinted courtesy of Professor Radin.

Samuelson, Pamela, Randall Davis, Mitchell Kapor, and J.H. Reichman, A Manifesto Concerning the Legal Protection of Computer Programs. This article orginally appeared at 94 Colum. L. Rev. 2308 (1994). Reprinted by permission of Pamela Samuelson and the Columbia Law Review.

Steinberg, Saul, A View of the World from 9th Avenue, 1976. © 2002 The Saul Steinberg Foundation/Artists Rights Society (ARS), New York. Cover reprinted with permission of The New Yorker. All rights reserved.

Xerox advertisement courtesy of Xerox Corporation.

Note: We have selectively omitted citations and footnotes from cases without the uses of ellipses or other indications. All footnotes are numbered consecutively within each chapter, except that footnotes in cases and other excerpts correspond to the actual footnote numbers in the published reports.

Many of the problems in this text are taken from actual cases. However, in many instances we have altered the facts of the case. In most cases we have also altered the names of the parties involved. In a few cases, however, particularly in the trademark and antitrust chapters, we felt that it was important to the problem to use the name of a product or company with which the reader would be familiar. Readers should understand that the problems are hypothetical in nature and that we do not intend them to represent the actual facts of any case or situation.

1 *Introduction*

The concept of property is well understood in Western society. It is among the oldest institutions of human civilization. It is widely recognized that people may own real property and tangible objects. The common law and the criminal law protect private property from interference by others. The Fifth Amendment to the U.S. Constitution protects private property against takings by the government without just compensation. The philosophical bases for protection of private property are well entrenched in our culture. Private property has been viewed as resulting when labor is applied to nature, as an incentive for discovery, as an essential part of personhood, and as a foundation for an ordered economic system.

Ideas, by definition, are less tangible. They exist in the mind and work of humans. Legal protection for intellectual work evolved much later in the development of human society than did protection for tangible property. The protection of such "intellectual property" raises complex philosophical questions. Should the first person to discover a way of performing an important task—for example, a procedure for closing a wound—be entitled to prevent others from using this procedure? Should the first person to pen a phrase or hum a melody be entitled to prevent others from copying such words or singing the song? Should such "intellectual property rights" be more limited than traditional property rights (i.e., the fee simple)? This book explores the legal institutions and rules that have developed to protect intellectual property.

This chapter has two principal purposes. It first explores the principal philosophical foundations for the protection of intellectual property. Understanding the reasons why we protect intellectual property—and how those reasons differ from the justifications for real property—will help the reader grasp the details of the many legal rules that will follow in this book. The second section provides a comparative overview of the principal modes of intellectual property protection: patent, copyright, trademark/trade dress, and trade secret. Understanding the intellectual property landscape requires thinking about each form of intellectual property not just in isolation but as it interacts with all the others. The remainder of this book will explore these areas in detail, highlighting their logic and interplay in promoting progress in technology and the arts.

A. PHILOSOPHICAL PERSPECTIVES

All justifications for intellectual property protection, whether based in economics or morality, must contend with a fundamental difference between ideas and tangible property. Tangible property, whether land or chattels, is composed of atoms, physical things that can occupy only one place at any given time. This means that possession of a physical thing is necessarily "exclusive" — if I have it, you don't. Indeed, the core of the Western concept of property lies in the right granted to the "owner" of a thing or a piece of land to exclude others from certain uses of it. Settled ownership rights in land and goods are thought to prevent both disputes over who can use the property for what purpose, and the overuse of property that would result if everyone had common access to it.

Ideas, though, do not have this characteristic of excludability. If I know a particular piece of information, and I tell it to you, you have not deprived me of it. Rather, we both possess it. The fact that the possession and use of ideas is largely "nonrivalrous" is critical to intellectual property theory because it means that the traditional economic justification for tangible property does not fit intellectual property. In the state of nature, there is no danger of overusing or overdistributing an idea, and no danger of fighting over who gets to use it. Everyone can use the idea without diminishing its value. See generally Peter S. Menell, Intellectual Property: General Theories, Encyclopedia of Law and Economics (B. Bouckaert & G. De Geest, eds. 2000).

Theorists have therefore turned elsewhere to justify exclusive rights in ideas. Over the course of human history, numerous theories have been put forth to explain intellectual property protection. The principal basis for such protection in the United States is the utilitarian or economic incentive framework. Nonetheless, other theories — most notably the natural rights and personhood justifications — have been important in understanding the development and scope of intellectual property law, both here and abroad. The two excerpts that follow are written as justifications for tangible property. Consider how well they apply to ideas.

1. The Natural Rights Perspective

John Locke, Two Treatises on Government
Third Edition, 1698

Though the earth and all inferior creatures be common to all men, yet every man has a "property" in his own "person." This nobody has any right to but himself. The "labour" of his body and the "work" of his hands, we may say, are properly his. Whatsoever, then, he removes out of the state that Nature hath provided and left it in, he hath mixed his labor with it, and joined to it something that is his own, and thereby makes it his property. It being by him removed from the common state Nature placed it in, it hath by this labour something annexed to it that excludes the common right of other men. For this "labour" being the unquestionable property of the labourer, no man but he can have a right to what that is once joined to, at least where there is enough, and as good left in common for others.

He that is nourished by the acorns he picked up under an oak, or the apples he gathered from the trees in the wood, has certainly appropriated them to himself. . . .

That labour put a distinction between them and common.... And will any one say he had no right to those acorns or apples he thus appropriated because he had not the consent of all mankind to make them his? Was it a robbery thus to assume to himself what belonged to all in common? If such a consent as that was necessary, man had starved, notwithstanding the plenty God had given him. We see in commons, which remain so by compact, that it is the taking any part of what is common, and removing it out of the state Nature leaves it in, which begins the property, without which the common is of no use. And the taking of this or that part does not depend on the express consent of all the commoners....

It will, perhaps, be objected to this, that if gathering the acorns or other fruits, of the earth, etc., makes a right to them, then any one may engross as much as he will. To which I answer, Not so. The same law of Nature that does by this means give us property, does also bound that property too.... As much as any one can make use of to any advantage of life before it spoils, so much he may by his labor fix his property in. Whatever is beyond this is more than his share, and belongs to others....

As much land as a man tills, plants, improves, cultivates, and can use the product of, so much is his property. He by his labor does, as it were, enclose it from the common.... Nor was this appropriation of any parcel of land, by improving it, any prejudice to any other man, since there was still enough and as good left, and more than the yet unprovided could use. So that, in effect, there was never the less left for others because of his enclosure for himself. For he that leaves as much as another can make use of does as good as take nothing at all.

COMMENTS AND QUESTIONS

1. How do Locke's theories of real property apply to intellectual property? Should we treat the two as the same for ownership purposes? What would Locke say about the exclusive rights granted by the patent laws, to prevent others from using the claimed invention for up to 20 years, whether or not they discovered the invention on their own? Surely if Locke considered the working of land and raw materials to be "labor" that justified ownership of the resulting product, he would have considered labor toward the creation of a new *idea* — the "sweat of the brow" — to be equally deserving of protection. Or do the differences between real and intellectual property mean that Locke's arguments shouldn't apply to intellectual property?

For illumination on these and related points, see Edwin C. Hettinger, Justifying Intellectual Property, 18 Phil. & Pub. Aff. 31 (1989). Hettinger critiques the major theories of intellectual property rights. As to Lockean labor theory, Hettinger makes these useful observations:

- [A]ssuming that labor's fruits are valuable, and that laboring gives the laborer a property right in this value, this would entitle the laborer only to the value she added, and not to the *total* value of the resulting product. Though exceedingly difficult to measure, these two components of value (that attributable to the object labored on and that attributable to the labor) need to be distinguished. [p. 37]
- Property rights in the thing produced are ... not a fitting reward if the value of these rights is disproportional to the effort expended by the laborer. "Effort"

includes (1) how hard someone tries to achieve a result, (2) the amount of risk voluntarily incurred in seeking this result, and (3) the degree to which moral considerations played a role in choosing the result intended. The harder one tries, the more one is willing to sacrifice, and the worthier the goal, the greater are one's deserts. [pp. 41-42]

The philosopher Robert Nozick made the first point by means of *reductio ad absurdum:* he asks whether the owner of a can of tomato juice who dumps it into the ocean can thereafter claim ownership of all the high seas. See Robert Nozick, Anarchy, State and Utopia 175 (1984).

2. An application of the Lockean approach with an especially detailed consideration of the Lockean "proviso" (i.e., that "as much and as good" be left for others after appropriation) can be found in Wendy J. Gordon, A Property Right in Self-Expression: Equality and Individualism in the Natural Law of Intellectual Property, 102 Yale L.J. 1533 (1993). The proviso poses a problem we will see throughout the course: how to delimit the rights of a creator in the face of claims by consumers and other members of the public at large.

More particularly, Gordon challenges the extreme view taken by some commentators that a creator's rights should be absolute. The absolutist view proceeds from the idea that since the creator is solely responsible for the creation, no one is harmed if the creation is withheld from the public entirely. And since the creator can withhold it entirely, he or she can naturally restrict its availability in any manner, including a high price or conditions on its purchase. See, e.g., John Stuart Mill, Principles of Political Economy 142 (1872); Steven N. S. Cheung, Property Rights and Invention, in 8 Research in Law and Economics: The Economics of Patents and Copyrights 5, 6 (John Palmer & Richard O. Zerbe, Jr., eds., 1986). Through the examination of various examples — such as the Church of Scientology's efforts to restrain books critical of its teachings, West Publishing's suit to enjoin Lexis from offering page numbers corresponding to West's case reports, and Disney's efforts to prevent bawdy parodies of its images that ridicule sanitized popular culture — Gordon suggests that, sometimes at least, the public can be worse off if a creation is offered and then limited in its use than it would have been had the creation never been made.

Should it matter whether Locke's hypothetical creator was the only one likely to come up with his particular invention or discovery? If others would have discovered the same phenomenon a few years (or a few weeks) later, does Locke's argument for property rights lose its force?

For detailed discussions of the "Lockean proviso," see Jeremy Waldron, Enough and as Good Left for Others, 29 Phil. Q. 319 (1979); Robert Nozick, Anarchy, State and Utopia 175-182 (1984); David Gauthier, Morals by Agreement 190-232 (1986). For more detailed treatment of another problem addressed by Gordon — the follow-on creator who builds on a preexisting work — see Lawrence C. Becker, Deserving to Own Intellectual Property, 68 Chicago-Kent L. Rev. 609 (1993).

3. Nozick's book (cited above) offers a different philosophical perspective on intellectual property, one rooted in the libertarian tradition. It is not clear how libertarians should think of intellectual property rights. On the one hand, ownership of property seems necessarily to underlie the market exchange that is at the heart of the libertarian model of society. On the other hand, one might view the free flow of information unfettered by property rights as the norm, and view government-enforced intellectual property rights as an unnecessary aberration. For a libertarian

approach that is decidedly hostile to intellectual property, see John Perry Barlow, The Economy of Ideas, 2.03 Wired 84 (Mar. 1994).

4. Natural rights are strongly emphasized in the continental European justifications for intellectual property. Those justifications to some extent parallel Locke's arguments, but there are important differences. Continental scholars emphasize the importance of reputation and noneconomic aspects of intellectual property, factors that lead them to support moral rights in copyright law. Professor Alfred Yen presents a thorough account of the role of natural rights in American copyright law. See Alfred Yen, Restoring the Natural Law: Copyright as Labor and Possession, 51 Ohio St. L.J. 517 (1990).

PROBLEM

Problem 1-1. You are a botanist exploring a remote region of a small tropical country. You stumble across a field of strange flowers that you have never encountered before. The tribespeople tell you that they use the flower to heal various ailments by rubbing its petals on the skin and chanting a healing prayer. You pluck one of the flowers and when you return to your campsite that night, you show it to a fellow explorer who is an expert in biochemistry. The biochemist smells the flower and says that it is vaguely reminiscent of substance P. Substance P is a medicine widely used to treat a variety of serious diseases. She tells you that Substance P is easy to detect: It turns bright yellow when exposed to intense heat. That evening, you put the flower over the campfire and, sure enough, it turns bright yellow. When you return home, you work for months to isolate the active ingredient in the flower. It is not substance P, but a close structural analog. In chemical experiments, the extract shows great promise for fighting many of the diseases that Substance P can treat without Substance P's dreaded side effects.

a) What rights, if any, should you have in this discovery and research? Should those rights prevent anyone else from going back to the tropical country, finding the flower, and isolating the chemical you have discovered?

b) You make a profit selling your medicine throughout the world, including to the native tribespeople of the tropical country. John, a chemist in your company, is angered by this policy. Your formula is a carefully guarded secret, but John publishes it in the New England Journal of Medicine. A nonprofit organization begins producing the medicine and selling your product to the native tribespeople at a discount. The organization advertises that it is selling your medicine for pennies a year. Should you be able to stop them from selling your medicine? Should you be able to stop them from using the name of your medicine in their ads?

c) You have named your medicine "Tropicurical." To advertise the product, one of the employees in your advertising department writes a song based on the very distinctive sounds of the wind in the inland coves of the country mixed with the native birds' calls. The tribespeople of the tropical country have a song that sounds remarkably similar to your advertising tune. The song is important to their tribal identity, and they argue that it is inappropriate to use their tribal identity song to commercialize your product. They ask you to stop using it. Should they be able to stop you from using the name or song?

d) When you refuse to stop using the Tropicurical song to advertise your product, the tribespeople write a new version of their song entitled "Tropicursical" with angry lyrics claiming that your product destroys culture. Should you be able to stop them from singing the song? National Geographic records the tribespeople singing "Tropicursical." The recording plays as part of the news report on their television show. It is also available for download on National Geographic's website. Individuals download copies of the song and e-mail it to their friends. It spreads like wildfire over the Internet and eventually popular radio stations begin playing it over the air. Should you be able to stop these copies and performances?

2. The Personhood Perspective

Margaret Jane Radin, Property and Personhood
34 Stan. L. Rev. 957 (1982)

This article explores the relationship between property and personhood, a relationship that has commonly been both ignored and taken for granted in legal thought. The premise underlying the personhood perspective is that to achieve proper self-development — to be *a person* — an individual needs some control over resources in the external environment. The necessary assurances of control take the form of property rights. Although explicit elaboration of this perspective is wanting in modern writing on property, the personhood perspective is often implicit in the connections that courts and commentators find between property and privacy or between property and liberty. In addition to its power to explain certain aspects of existing schemes of property entitlement, the personhood perspective can also serve as an explicit source of values for making moral distinctions in property disputes and hence for either justifying or criticizing current law. . . .

In what follows I shall discuss the personhood perspective as Hegel developed it in *Philosophy of Right*, trace some of its later permutations and entanglements with other perspectives on property, and try to develop a contemporary view useful in the context of the American legal system. . . .

I. Property for Personhood: An Intuitive View

Most people possess certain objects they feel are almost part of themselves. These objects are closely bound up with personhood because they are part of the way we constitute ourselves as continuing personal entities in the world. They may be as different as people are different, but some common examples might be a wedding ring, a portrait, an heirloom, or a house.

One may gauge the strength or significance of someone's relationship with an object by the kind of pain that would be occasioned by its loss. On this view, an object is closely related to one's personhood if its loss causes pain that cannot be relieved by the object's replacement. If so, that particular object is bound up with the holder. For instance, if a wedding ring is stolen from a jeweler, insurance proceeds can reimburse the jeweler, but if a wedding ring is stolen from a loving wearer, the price of a replacement will not restore the status quo — perhaps no amount of money can do so.

The opposite of holding an object that has become a part of oneself is holding an object that is perfectly replaceable with other goods of equal market value. One holds such an object for purely instrumental reasons. The archetype of such a good is, of course, money, which is almost always held only to buy other things. A dollar is worth no more than what one chooses to buy with it, and one dollar bill is as good as another. Other examples are the wedding ring in the hands of the jeweler, the automobile in the hands of the dealer, the land in the hands of the developer, or the apartment in the hands of the commercial landlord. I shall call these theoretical opposites — property that is bound up with a person and property that is held purely instrumentally — personal property and fungible property respectively. . . .

III. Hegel, Property, and Personhood

A. *Hegel's Philosophy of Right* . . .

Because the person in Hegel's conception is merely an abstract unit of free will or autonomy, it has no concrete existence until that will acts on the external world. . . .

Hegel concludes that the person becomes a real self only by engaging in a property relationship with something external. Such a relationship is the goal of the person. In perhaps the best-known passage from this book, Hegel says:

> The person has for its substantive end the right of placing its will in any and every thing, which thing is thereby mine; [and] because that thing has no such end in itself, its destiny and soul take on my will. [This constitutes] mankind's absolute right of appropriation over all things.

Hence, "property is the first embodiment of freedom and so is in itself a substantive end." . . .

Hegel seems to make property "private" on the same level as the unit of autonomy that is embodying its will by holding it. He argues that property is private to individuals when discussing it in the context of the autonomous individual will and that it is essentially common within a family, when discussing it in the context of the autonomous family unit. He does not make the leap to state property, however, even though his theory of the state might suggest it. For Hegel, the properly developed state (in contrast to civil society) is an organic moral entity . . . and individuals within the state are subsumed into its community morality. . . .

B. *Hegel and Property for Personhood*

[A] theory of personal property can build upon some of Hegel's insights. First, the notion that the will is embodied in things suggests that the entity we know as a person cannot come to exist without both differentiating itself from the physical environment and yet maintaining relationships with portions of that environment.

The idea of embodied will, cut loose from Hegel's grand scheme of absolute mind, reminds us that people and things have ongoing relationships which have their own ebb and flow, and that these relationships can be very close to a person's center and sanity. If these relationships justify ownership, or at least contribute to its justification, Hegel's notion that ownership requires continuous embodiment of the will is appealing.

Second, Hegel's incompletely developed notion that property is held by the unit to which one attributes autonomy has powerful implications for the concept of group development and group rights. Hegel thought that freedom (rational self-determination) was only possible in the context of a group (the properly organized and fully developed state). Without accepting this role for the state, one may still conclude that in a given social context certain groups are likely to be constitutive of their members in the sense that the members find self-determination only within the groups. This might have political consequences for claims of the group on certain resources of the external world (i.e., property).

Third, there may be an echo of Hegel's notion of an objective community morality in the intuition that certain kinds of property relationships can be presumed to bear close bonds to personhood. If property in one's body is not too close to personhood to be considered property at all, then it is the clearest case of property for personhood. The property/privacy nexus of the home is also a relatively clear case in our particular history and culture....

[T]he personhood theory helps us understand the nature of the right dictating that discrete units [i.e., an undivided, individual asset] ought to be protected.

An argument that discrete units are more important than total assets takes the following form. A person cannot be fully a person without a sense of continuity of self over time. To maintain that sense of continuity over time and to exercise one's liberty or autonomy, one must have an ongoing relationship with the external environment, consisting of both "things" and other people. One perceives the ongoing relationship to the environment as a set of individual relationships, corresponding to the way our perception separates the world into distinct "things." Some things must remain stationary if anything is to move; some points of reference must be constant or thought and action is not possible. In order to lead a normal life, there must be some continuity in relating to "things." One's expectations crystallize around certain "things," the loss of which causes more disruption and disorientation than does a simple decrease in aggregate wealth. For example, if someone returns home to find her sofa has disappeared, that is more disorienting than to discover that her house has decreased in market value by 5%. If, by magic, her white sofa were instantly replaced by a blue one of equal market value, it would cause no loss in net worth but would still cause some disruption in her life.

This argument assumes that all discrete units one owns and perceives as part of her continuing environment are to some degree personal. If the white sofa were totally fungible, then magically replacing it with a blue one would cause no disruption. In fact, neither would replacing it with money....

But the theory of personal property suggests that not all object-loss is equally important. Some objects may approach the fungible end of the continuum so that the justification for protecting them as specially related to persons disappears. They might just as well be treated by whatever general moral rules govern wealth-loss at the hands of the government. If the moral rules governing wealth-loss correspond to Michelman's utilitarian suggestion—government may take whatever wealth is necessary to generate higher welfare in which the individual can confidently expect to share — then the government could take some fungible items without compensation. In general, the moral inquiry for whether fungible property could be taken would be the same as the moral inquiry for whether it is fair to impose a tax on this particular person.

On the other hand, a few objects may be so close to the personal end of the continuum that no compensation could be "just." That is, hypothetically, if some object were so bound up with me that I would cease to be "myself" if it were taken, then a government that must respect persons ought not to take it. If my kidney may be called my property, it is not property subject to condemnation for the general public welfare. Hence, in the context of a legal system, one might expect to find the characteristic use of standards of review and burdens of proof designed to shift risk of error away from protected interests in personal property. For instance, if there were reason to suspect that some object were close to the personal end of the continuum, there might be a prima facie case against taking it. That prima facie case might be rebutted if the government could show that the object is not personal, or perhaps that the object is not "too" personal compared with the importance to the government of acquiring that particular object for social purposes.

COMMENTS AND QUESTIONS

1. How well does Professor Radin's theory of real property apply to intellectual property? Can an individual be so "bound up" in her own inventions or works of authorship that their loss would occasion more than economic damage? Does it affect your answer that intellectual property can be used simultaneously by many people without depleting its functional value to any one — so that an author's "loss" is not the physical deprivation of stolen chattels, but the less personal fact that someone else has copied her work?

It may be that the investment of "personhood" in intellectual property varies greatly, both with the type of intellectual property at issue and with the time and effort the owner put into developing it. For example, a novel on which one has worked for several years may have more personal value than a massive software program for navigating a jumbo jet or a company's customer list. Should the law take account of these differences, giving greater protection to more personal works?

For applications of Radin's "personhood" ideas in the context of intellectual property, see Neil W. Netanel, Copyright Alienability Restrictions and the Enhancement of Author Autonomy: A Normative Evaluation, 24 Rutgers L.J. 347 (1993); Steven Cherensky, Comment, A Penny for Their Thoughts: Employee-Inventors, Pre-invention Assignment Agreements, Property, and Personhood, 81 Cal. L. Rev. 595, 641 (1993). Cherensky argues, using personhood theory and the related idea of market-inalienability, that employee-inventors should retain greater property interests in their inventions than they typically do under conventional employee assignment contracts. On market-inalienability, i.e., things which should not be subject to market exchange at all, see Margaret Jane Radin, Market-Inalienability, 100 Harv. L. Rev. 1849 (1987). Cf. Justin Hughes, The Philosophy of Intellectual Property, 77 Georgetown L.J. 287, 350-53 (1988) (suggesting various strains of the personhood justification in American copyright law). Professor Roberta Kwall offers a related "moral" basis for legal protection emphasizing the dignity, honor, self-worth, and autonomy of the author. See Roberta Kwall, "Author-Stories": Narrative's Implications for Moral Rights and Copyright's Joint Authorship Doctrine, 75 S. Cal. L. Rev. 1 (2001).

2. For a critique of Radin's overall theory, see Stephen J. Schnably, Property and Pragmatism: A Critique of Radin's Theory of Property and Personhood, 45 Stan. L. Rev. 347 (1993) (challenging Radin's appeal to consensus and arguing that this focus obscures issues of power and the like). See also Margaret Jane Radin, Reinterpreting Property (1993) (a collection of related essays); A. John Simmons, The Lockean Theory of Rights (1992); Jeremy Waldron, The Right to Private Property (1991) (comparing Lockean and Hegelian property rights theories); Symposium, Property Rights, 11 Soc. Phil. & Policy 1-286 (1994).

3. A radical critique of some bedrock notions implicit in intellectual property — most notably, the concept of "*the* author" herself — has grown up in recent years, fueled by a general deconstructive trend in literary criticism. While not aimed directly at Radin's personhood approach to certain forms of property, these ideas do pose a challenge to the application of Radin's approach in the intellectual property context. They suggest that the concept of authorship is so malleable, contingent, and "socially constructed" that we should be wary about identifying a creative work too closely with a particular author, let alone her personality. In this view, all creations are largely a product of communal forces. Dividing the stream of intellectual discourse into discrete units — each owned by and closely associated with a particular author — is therefore a logically incoherent exercise subject more to the political force of asserted authors' groups than to recognition of inherent claims of "personhood." See, e.g., Martha Woodmansee & Peter Jaszi, eds., The Construction of Authorship: Textual Appropriation in Law and Literature (1994); Peter Jaszi, Toward a Theory of Copyright: The Metamorphoses of "Authorship," 1991 Duke L.J. 455; James Boyle, A Theory of Law and Information: Copyright, Spleens, Blackmail and Insider Trading, 80 Cal. L. Rev. 1413 (1992).

If authorship is an inchoate concept, is there any role at all for copyright law? How can one protect the rights (natural, moral, or economic) of the author, if in fact there is no author? Does the literary critique answer the charge that authors will not create in the absence of economic reward? Or is it directed only at personality-based theories of intellectual property?

4. Consider the observations of philosopher Lawrence Becker:

> So if property-as-personality [à la Hegel] again turns out to be a dead end, perhaps we should dispense with the search for a deep justification for property rights (from metaphysics, moral psychology, sociobiology, or whatever) and focus on the behavioral surface: the observed, persistent, robust behavioral connections between various property arrangements and human well-being, broadly conceived. This may provide a foundation for egalitarian arguments that is more secure than speculative metaphysics, and a foundation for private property that is more stable than a pluralistic account of the standard array of bedrock justifications for it.

Lawrence Becker, Too Much Property, 21 Phil. & Pub. Aff. 196, 206 (1992).

3. The Utilitarian/Economic Incentive Perspective

Utilitarian theory, and the economic framework built upon it, has long provided the dominant paradigm for analyzing and justifying the various forms of intellectual property protection. See generally Peter S. Menell & Suzanne Scotchmer, Intellectual

Property, in Handbook of Law and Economics (A. Mitchell Polinsky & Steven Shavell eds., 2006); William M. Landes & Richard A. Posner, The Economic Structure of Intellectual Property Law (2003). For purposes of exploring the economic dimensions of the intellectual property field, it is important to distinguish between two quite distinct functions. The principal objective of much of intellectual property law is the promotion of new and improved works — whether technological or expressive. This purpose encompasses patent, copyright, and trade secret law, as well as several more specialized protection systems (for mask works (semiconductor chip layouts), databases, and designs). Trademark and related bodies of unfair competition law focus primarily on a very different economic problem — ensuring the integrity of the marketplace.

a. Promoting Innovation and Creativity

Both the United States Constitution and judicial decisions emphasize incentive theory in justifying intellectual property. The Constitution expressly conditions the grant of power in the patent and copyright clause on a particular end, namely "to Promote the Progress of Science and useful Arts." U.S. Const., art. I, cl. 8. As the Supreme Court explained in Mazer v. Stein, 347 U.S. 201 (1954):

> "The copyright law, like the patent statutes, makes reward to the owner a secondary consideration." United States v. Paramount Pictures, 334 U.S. 131, 158. However, it is "intended definitely to grant valuable, enforceable rights to authors, publishers, etc., without burdensome requirements: 'to afford greater encouragement to the production of literary [or artistic] works of lasting benefit to the world.'" Washington Pub. Co. v. Pearson, 306 U.S. 30.
>
> The economic philosophy behind the clause empowering Congress to grant patents and copyrights is the conviction that it is the best way to advance public welfare through the talents of authors and inventors in "Science and useful Arts." Sacrificial days devoted to such creative activities deserve rewards commensurate with the services rendered.

To understand why the Framers thought exclusive rights in inventions and creations would promote the public welfare, consider what would happen absent any sort of intellectual property protection. Invention and creation require the investment of resources — the time of an author or inventor, and often expenditures on facilities, prototypes, supplies, etc. In a private market economy, individuals will not invest in invention or creation unless the expected return from doing so exceeds the cost of doing so — that is, unless they can reasonably expect to make a profit from the endeavor. To profit from a new idea or work of authorship, the creator must be able either to sell it to others or to put it to some use that provides her with a comparative advantage in a market, such as by reducing the cost of producing goods.

But ideas (and writings, for that matter) are notoriously hard to control. Even if the idea is one that the creator can use herself, for example to boost productivity in her business, she will reap a reward from that idea only to the extent that her competitors don't find out about it. A creator who depends on secrecy for value therefore lives in constant peril of discovery and disclosure. Competitors may steal the idea or learn of it from an ex-employee. They may be able to figure it out by watching the creator's production process or by examining the products she sells. Finally, they may come upon the idea on their own or discover it in the published literature. In all of these cases, the secrecy value of the idea will be irretrievably lost.

The creator who wants to sell her idea is in an even more difficult position. Selling information requires disclosing it to others. Once the information has been disclosed outside a small group, however, it is extremely difficult to control. Information has the characteristics of what economists call a "public good" — it may be "consumed" by many people without depletion, and it is difficult to identify those who will not pay and prevent them from using the information. See Kenneth J. Arrow, Economic Welfare and the Allocation of Resources for Invention, in The Rate and Direction of Inventive Activity: Economic and Social Factors 609, 614-616 (Natl. Bureau of Economic Research ed., 1962). Once the idea of the intermittent windshield wiper is disclosed, others can imitate its design relatively easily. Once a book is published, others can copy it at low cost. It is difficult to exclude nonpurchasers from sharing in the benefits of the idea. Ideas and information can also be used by many without depleting the enjoyment of others. Unlike an ice cream cone, a good story or the concept of intermittent windshield wipers can be enjoyed by many without diminishing the enjoyment of these creations by others.[1]

If we assume that it is nearly costless to distribute information to others — an assumption that was once unrealistic but now much more reasonable with advances in digital technology (including the Internet) — it will prove virtually impossible to charge for information over the medium run in the absence of effective intellectual property or some other means of protection (such as technological protection measures). If the author of a book charges more than the cost of distribution, hoping to recover some of her expenditures in writing the work, competitors will quickly jump in to offer the book at a lower price. Competition will drive the price of the book toward its marginal cost — in this case the cost of producing and distributing one additional copy. In such a competitive market, the author will be unable to recoup the fixed cost of writing the book. More to the point, if this holds true generally, authors may be expected to leave the profession in droves, since they cannot make any money at it. The result, according to economic theory, is an underproduction of books and of other works of invention and creation with similar public goods characteristics.[2]

Information is not the only example of a public good. Economists generally offer lighthouses and national defense as examples of public goods, since it is virtually impossible to provide the benefits of either one only to paying clients. It is impossible, for example, to exclude some ships and not others from the benefits of a lighthouse. Furthermore, the use of the lighthouse by one ship does not deplete the value of its hazard warning to others. As a result, it would be inefficient to exclude nonpayers from using the lighthouse's warning system even if we could, since consumption of this good is "nonrivalrous" (meaning that everyone can benefit from it once it is produced). For these reasons, the market will in theory undersupply such goods

---

1. To some extent this statement oversimplifies the problem by ignoring possible second-order distorting effects. In practice, if you taught several hundred million people to fish, the result might be depletion of a physical resource (fish) that would otherwise not have occurred. Similarly, wide dissemination of information may have particular effects on secondary markets, depending on the information.

2. See, e.g., F. M. Scherer, Industrial Market Structure and Economic Performance 444 (2d ed. 1980) ("If pure and perfect competition in the strictest sense prevailed continuously . . . incentives for invention and innovation would be fatally defective without a patent system or some equivalent substitute."). Professor Scherer goes on to note, however, that natural market imperfections may give advantages to first movers, reducing the need for intellectual property protection. Id. at 444-45.

because producers cannot reap the marginal (incremental) value of their investment in providing them.[3]

Can you see why broadcast television signals, beautiful gardens on a public street, and national defense are also public goods?

By contrast, markets for pure private goods, such as ice cream cones, feature exclusivity and rivalrous competition—the ice cream vendor provides the good only to those who pay the price, and the consumer certainly depletes the amount of the good available to others. Thus the market system provides adequate incentives for the creation of ice cream cones: sellers can exact their cost of production, and the value of the product is fully enjoyed by the purchaser.

In the case of national defense (and most lighthouses), we avoid the underproduction that would result from leaving it to the market by having the government step in and pay for the public good. For a variety of good reasons, discussed in note 3 following this section, we have not gone that route with many forms of information. Instead, government has created intellectual property rights in an effort to give authors and inventors control over the use and distribution of their ideas, thereby encouraging them to invest in the production of new ideas and works of authorship. Thus the economic justification for intellectual property lies not in rewarding creators for their labor but in ensuring that they (and other creators) have appropriate incentives to engage in creative activities.

Unfortunately, this approach comes at a cost. Granting authors and inventors the right to exclude others from using their ideas necessarily limits the diffusion of those ideas and so prevents many people from benefiting from them. In economic terms, intellectual property rights prevent competition in the sale of the particular work or invention covered by those rights, and therefore may allow the intellectual property owner to raise the price of that work above the marginal cost of reproducing it. This means that in many cases fewer people will buy the work than if it were distributed on a competitive basis, and they will pay more for the privilege. A fundamental principle of our economic system is the proposition that free market competition will ensure an efficient allocation of resources, absent market failures. In fact, the principal thrust of the antitrust laws serves precisely this goal. In this limited sense, then, intellectual property rights appear to run counter to free market competition: they limit the ability of competitors to copy or otherwise imitate the intellectual efforts of the first person to develop an idea. These rights enable those possessing intellectual property rights to charge monopoly prices or to otherwise limit competition, such as by controlling the use of the intellectual work in subsequent products.

Because intellectual property rights impose social costs on the public, the intellectual property laws can be justified by the public goods argument only to the extent that they do on balance encourage enough creation and dissemination of new works to offset those costs. One of the reasons that intellectual property rights are limited in scope, duration, and effect is precisely in order to balance these costs and benefits. For example, the limited term of intellectual property rights ensures that inventions will be freely available after that fixed term. The key to economic efficiency lies in balancing the social benefit of providing economic incentives for creation and the social costs of limiting the diffusion of knowledge. We will encounter this critical trade-off throughout

---

3. Ronald Coase has offered evidence casting doubt on the economic assumption that lighthouses must be publicly provided. See Ronald H. Coase, The Lighthouse in Economics, in The Firm, the Market, and the Law (1988).

our study of intellectual property. The two examples below highlight some of the major issues.

Economic Incentive Benefit. Intellectual property protection is necessary to encourage inventors, authors, and artists to invest in the process of creation. Without such protection, others could copy or otherwise imitate the intellectual work without incurring the costs and effort of creation, thereby inhibiting the original creators from reaping a reasonable return on their investment. Consider the following example:

> After years of effort and substantial expense, Earnest Inventor develops the Mouso-matic, a significantly better mousetrap. Not only does it catch mice better than the competition's trap, but it also neatly packages the dead mice in sanitary disposable bags. Consumers are willing to pay substantially more for this product than for the competition's. The Mousomatic catches the attention of Gizmo Gadget Incorporated. Gizmo copies the basic design of the Mousomatic and offers its version of the Mouso-matic at a substantial discount. (Gizmo can still earn a profit at this lower price because it had minimal research and development expense.) In order to stay in business, Earnest is forced to lower his price. Market competition pushes the price down to the cost of production and distribution. In the end, Earnest is unable to recover his cost of research and development and suffers a loss. Although he has numerous other interesting ideas, he decides that they are not worth pursuing because Gizmo, or some other company, will simply copy them if they turn out well.

The existence of intellectual property rights encourages Earnest and other inventors to pursue their creative efforts. If Earnest can obtain the right to prevent others from copying his inventions, then he stands a much better chance to reap a profit. Hence he will be much more inclined to make the initial investment in research and development. In the end, not only will Earnest be wealthier, but the public will be enriched by the new and better products brought forth by intellectual property protection.

Costs of Limiting Diffusion. Legal protection for ideas and their expression prevents others from using those works to develop similar works that build upon them. Knowledge in society is cumulative. In the words of Sir Isaac Newton, "If I have seen further [than others], it is by standing on the shoulders of giants." Hence society at large can be harmed by intellectual property protection to the extent that it unnecessarily raises the cost of acquiring a product (through monopoly pricing by the right holder) and limits others from making further advances. Consider the following scenario:

> Professor Lee conducts research on drug treatments at University College. Her laboratory is generously funded by grants from the federal government. For the past decade, Professor Lee has competed with colleagues at other laboratories to discover the cure for a prevalent form of cancer. It is likely that the first person to discover the cure will win a Nobel prize, as well as numerous other financial and professional rewards. In early 1995, Professor Lee hits upon the Alpha drug, which cures the disease. She files for and receives a patent. Professor Hu, a researcher at another research institution, independently discovers the identical cure a few months later. With patent in hand, Professor Lee starts a company to sell Alpha and charges a price 100 times the cost of production. Because Alpha is a life-saving cure, those stricken with the disease who can

afford the treatment are more than willing to pay the price. Then, to relieve the suffering of millions, Professor Hu begins selling Alpha at the cost of production. Moreover, she has developed an improvement on the Alpha drug, Alpha+, that reduces the side effects of the treatment. Professor Lee quickly obtains an injunction preventing Hu from selling either version of Alpha for the life of the patent.

This example raises serious questions about whether intellectual property protection is desirable, at least for this class of invention. Professor Lee does not bear significant risk in pursuing the invention because her research is generously supported by the government and the university. Furthermore, the potential for a Nobel prize, expanded research funding, and professional recognition provide substantial encouragement for Professor Lee to pursue a cure whether or not she gains financially from sales of Alpha. Moreover, other researchers were poised to make the same discovery at about the time that Professor Lee made her discovery. Yet she has the right not only to block sales of Alpha by competitors, but also to block sales of improvements such as Alpha+. Does such a system benefit society? One must also consider that without the financial incentive of a patent, there would perhaps have been less competition to discover and market any cure for cancer. Problems such as this one have led some scholars to question the economic efficiency of the patent system in particular circumstances. See F. M. Scherer, Industrial Market Structure and Economic Performance 445-55 (2d ed. 1980).

Similar problems can arise in the copyright domain. In 1936, Margaret Mitchell's epic novel *Gone with the Wind* was published. The plot revolves around Scarlett O'Hara, a hard-working and ambitious Southern woman who lives through the American Civil War and Reconstruction. *Gone with the Wind* quickly became one of the most popular books ever written, eventually selling more than 28 million copies throughout the world. Ms. Mitchell would receive a Pulitzer Prize, and the novel would be become the basis for one of the most successful movies ever.

Sixty-five years later, Alice Randall, an African-American historian, set out to tell the history of the Civil War-era South from another viewpoint. She invented the character of Cynara, a mulatto slave on Scarlett's plantation who also happened to be Scarlett's half-sister. Although Cynara was not depicted in Mitchell's novel, many of the other characters in Randall's novel mimic characters in the original. Randall's *The Wind Done Gone* depicts the mistreatment and suffering of slaves in their own vernacular dialect. Randall believed that *Gone with the Wind*, as the lens through which millions viewed the deep South during the Civil War era, provided a unique vehicle for communicating the racial injustice of American history. Margaret Mitchell's estate sued Randall for copying *Gone with the Wind* without authorization, obtaining a preliminary injunction. Although Randall eventually prevailed on appeal under the fair use doctrine, Suntrust Bank v. Houghton Mifflin Co. 268 F.3d 1257 (11th Cir. 2001), this controversy illustrates how strong copyright protection can potentially inhibit cumulative creativity.

In applying a utilitarian framework, the economic incentive benefits of intellectual property rights must be balanced against the costs of limiting diffusion of knowledge. A critical issue in assessing the need for intellectual property protection is whether innovators have sufficient means to appropriate an adequate return on investment in research and development. In this regard, the market itself often provides means by which inventors can realize sufficient rewards to pursue innovation without formal intellectual property rights beyond contract law. The first to introduce a product can in many contexts earn substantial and long-lived advantages in the market. In many markets, the costs or time required to imitate a product (for example, to reverse engineer a complex machine) are so great that the first to market a product has substantial opportunity for profit. Moreover, as we will see in Chapter 2, inventors can often prevent imitation through contractual means, such as trade secrecy and licensing agreements with customers. Where the invention relates to a manufacturing process, the innovator may be able to maintain protection through secrecy even after the product is on the market. Alternatively, a producer may be able to bundle products with essential services and contracts for updates of the product. In addition, the producer may be able to spread the costs of research and development among a group of firms through research joint ventures.

In those areas in which economic incentives for innovation are inadequate and the creation of intellectual property rights is the most efficacious way of encouraging progress, society must determine the appropriate requirements for, duration and scope of, and set of rights afforded intellectual property. Over the past several decades, economists have developed and refined models to assess the appropriate trade-off between the social benefits of providing economic incentives for innovation through intellectual property rights and the social costs of limiting diffusion of knowledge. Professor William Nordhaus developed the first formal model analyzing the optimal duration of intellectual property. His model of the innovative process assumed that investments in research produced a single independent innovation. W. Nordhaus, Invention, Growth, and Welfare: A Theoretical Treatment of Technological Change 3-7 (1969). The principal policy implication of this model is that the term of intellectual property protection should be calibrated to balance the incentive benefits of protection against the deadweight loss of monopoly pricing and the resulting limitations on dissemination.

Since Nordhaus's important early work, economic historians and economic theorists have greatly enriched our understanding of the innovative process and the implications for public policy. Historical and industry studies of the innovation process find that inventions are highly interdependent: "Technologies . . . undergo a gradual, evolutionary development which is intimately bound up with the course of their diffusion."[4] In fact, "secondary inventions" — including essential design improvements, refinements, and adaptations to a variety of uses — are often as crucial to the generation of social benefits as the initial discovery.[5] Economic theorists have more recently developed models of the innovative process incorporating concepts of rivalrous and cumulative innovation,[6] uncovering a range of important effects. Most

4. Paul David, New Technology, Diffusion, Public Policy, and Industrial Competitiveness 20 (Center for Economic Policy Research, Publication No. 46, Apr. 1985). See generally, Richard Nelson & Steven Winter, An Evolutionary Theory of Economic Change (1982); Nathan Rosenberg, Factors Affecting the Diffusion of Technology, 10 Explorations Econ. Hist. 3 (1972).

5. See, e.g., Enos, A Measure of the Rate of Technological Progress in the Petroleum Refining Industry, 6 J. Indus. Econ. 180, 189 (1958); Mak & Walton, Steamboats and the Great Productivity Surge in River Transportation, 32 J. Econ. Hist. 619, 625 (1972).

6. See Suzanne Scotchmer, Standing on the Shoulders of Giants: Cumulative Research and the Patent Law, 5 J. Econ. Perspectives 29 (1991); Robert P. Merges & Richard Nelson, On the Complex Economics of Patent Scope, 90 Colum. L. Rev. 839, 843, 868-879 (1990); M. Kamien & N. Schwartz, Market Structure and Innovation (1982).

notably, excessive protection for first-generation innovation can impede later innovations if licensing is costly. More generally, these models cast doubt on the notion that society can perfectly calibrate intellectual property rewards for each innovation.

As is increasingly evident, the range of innovative activity and creative expression in our society is vast and ever changing. As the materials in this book will highlight, the intellectual property institutions and rules that have evolved to promote technology and the arts are intricate. It will be the challenge of future generations of policymakers, judges, and lawyers to refine the ability of the intellectual property system to enhance the public welfare.

COMMENTS AND QUESTIONS

1. One significant difference between the natural rights perspective and the utilitarian perspective relates to who is entitled to the fruits of productive labor. In the natural rights framework, the inventor or author is entitled to all the social benefits produced by his or her efforts. In the utilitarian framework, reward to the inventor or author is a secondary consideration; the principal objective is to enrich the public at large. Which view is more compelling? Consider in this regard the optimal division of benefits from the invention of Alpha among Professor Lee, Professor Hu, and the public at large. Is Professor Lee entitled to all or even a lion's share of the benefits?

2. In 1966, the Report of the President's Commission on the Patent System identified four major economic justifications for the patent laws.

First, a patent system provides an incentive to invent by offering the possibility of reward to the inventor and to those who support him. This prospect encourages the expenditure of time and private risk capital in research and development efforts.

Second, and complementary to the first, a patent system stimulates the investment of additional capital needed for the further development and marketing of the invention. In return, the patent owner is given the right, for a limited period, to exclude others from making, using, or selling the invented product or process.

Third, by affording protection, a patent system encourages early public disclosure of technological information, some of which might otherwise be kept secret. Early disclosure reduces the likelihood of duplication of effort by others and provides a basis for further advances in the technology involved.

Fourth, a patent system promotes the beneficial exchange of products, services, and technological information across national boundaries by providing protection for industrial property of foreign nationals.

While directed specifically at the patent system, many of these arguments have application to all forms of intellectual property.

Are these incentives necessary to invention and creation? Using cost and other data from publishing companies, Professor (now Justice) Stephen Breyer contended that lead time advantages and the threat of retaliation reduce the cost advantages of copiers, thus obviating if not eliminating the need for copyright protection for books. See Stephen Breyer, The Uneasy Case for Copyright: A Study in Copyright of Books, Photocopies and Computer Programs, 84 Harv. L. Rev. 281 (1970). Cf. Barry W. Tyerman, The Economic Rationale for Copyright Protection for Published Books: A Reply to Professor Breyer, 18 UCLA L. Rev. 1100 (1971); Stephen Breyer,

Copyright: A Rejoinder, 20 UCLA L. Rev. 75 (1972). This debate took place more than a quarter-century ago. Have advances in technology strengthened or weakened Breyer's argument?

3. *Comparative Institutional Analysis.* Numerous institutional mechanisms exist for addressing the public goods problem inherent in the production of ideas and information — direct government funding of research, government research subsidies, promotion of joint ventures, and prizes. The case for intellectual property rights ideally compares all of these options. Intellectual property rights have the advantage of limiting the government's role in allocating resources to a finite set of decentralized decisions: whether particular inventions are worthy of a fixed period of protection. The market then serves as the principal engine of progress. Decentralized consumers generate demand for products and competing decentralized sellers produce them. By contrast, most other incentive systems, especially large-scale research funding, require central planning on a mass scale. Most economists place more confidence in the former means of allocating resources. The case for intellectual property rights, then, is based more on a generalized perception of institutional choice than on strong direct evidence of the superiority of intellectual property rights relative to the alternatives.

4. *The Open Source Movement.* The emergence of cooperative working environments for the development of software has raised questions about the core precept underlying the utilitarian/economic perspective: that exclusive property rights represent the most effective means for promoting creative enterprise. See generally Steven Weber, The Success of Open Source (2004); Yochai Benkler, Coase's Penguin, or Linux and the Nature of the Firm, 112 Yale L.J. 369 (2002); David McGowan, The Legal Implications of Open Source Software, 2001 U. Ill. L. Rev. 241. Open source software traces its origins to the early 1970s and the culture of collaborative research on computer software that existed in many software research environments. In an effort to perpetuate that model in the face of increasingly proprietary software, Richard Stallman, a former researcher in MIT's Artificial Intelligence Laboratory, established the Free Software Foundation (FSF) to promote users' rights to use, study, copy, modify, and redistribute computer programs. Such rights obviously conflict with the default bundle of rights of copyright law. For that reason, FSF developed the GNU General Public License (GPL), a complex licensing agreement designed to prevent programmers from building proprietary limitations into "free" software. Stallman set forth a task list for the development of a viable UNIX-compatible open source operating system. Many programmers from throughout the world contributed to this effort on a voluntary basis, and by the late 1980s most of the components had been assembled. The project gained substantial momentum in 1991 when Linus Torvalds developed a UNIX-compatible kernel, which he called "Linux." Torvalds structured the evolution of his component on the GNU GPL "open source" model. The integration of the GNU and Linux components resulted in a UNIX-compatible open source program (referred to as GNU/Linux) and has since become widely used throughout the computing world. In the process, it has spawned a large community of computer programmers and service organizations committed to the principles of open source development. The growth and success of Linux has brought the open source movement into the mainstream computer software industry. Today, a variety of vendors, such as Red Hat, Caldera, and Debian, distribute open source software, and it has tens of millions of users worldwide. Does this experience refute the logic

underlying the property rights paradigm or merely broaden the range of viable governance structures?

5. Drawing upon Thomas Jefferson's natural rights insight that "ideas should freely spread from one to another over the globe, for the moral and mutual instruction of man, and improvement of his condition, seems to have been peculiarly and benevolently designed by nature,"[7] John Perry Barlow's essay "The Economy of Ideas" has emerged as a manifesto for a new libertarianism that resists intellectual property protection in cyberspace. John Perry Barlow, The Economy of Ideas, Wired 2.03 (Mar. 1994). Barlow questions whether a right of property can or should exist in a medium (digital networks) lacking physical structure or any significant cost of distribution. This approach resonates with many in the computer hacker community who believe that "information wants to be free." Professor Lawrence Lessig explores and expands upon this perspective in his books Free Culture: How Big Media Uses Technology and the Law to Lock Down Culture and Control Creativity (2004) and The Future of Ideas: The Fate of the Commons in a Connected World (2001), although his focus is not on eliminating intellectual property but on preserving spaces free of such property.

What is likely to happen if we abolish the concept of ownership of information on the Internet? In general? Consider the implications for particular works of authorship: musical compositions, sound recordings, software, movies, databases. Will people stop producing these works? Or will other types of incentives and appropriation mechanisms (e.g., encryption, secrecy) continue to encourage invention and creativity? What are the advantages and disadvantages of these alternative mechanisms?

PROBLEM

Problem 1-2. Building on your lifelong interest in music and computers, you have just spent ten years creating a computer-based Encyclopedia of Music. The encyclopedia, based on a CD-ROM, includes text, snippets of recorded music (no longer than seven seconds each), pictures, and other graphics, all of which can be called up by the user. It is thus an instance of the new generation of "multimedia" computer products.

Like many other multimedia products, your encyclopedia allows the user to select a topic of interest and see (or hear) more about it. For example, from the opening menu a user can select classical composers, then select Mozart, and then choose from a biographical sketch (with text, pictures, and samples of sheet music) or musical samples that actually play on the computer's speakers.

To make it easier for computer users to access all the information in your encyclopedia, you have used an "access interface" similar to the widely used "MediaMate" interface written and made popular by multigazillionaire computer impresario Gil Bates. (Bates adapted the "MediaMate" interface from a publicly available interface designed by the U.S. Army to teach electronics and other technical subjects to new recruits.) Although MediaMate has now become

---

7. See Andrew A. Lipscomb and Albert Ellery Bergh (eds.) The Writings of Thomas Jefferson, vol. 13: 333-35 (Writings (document 12): letter from Thomas Jefferson to Isaac McPherson, 13 Aug. 1813) (1905).

the standard interface program, Bates, because of his awesome programming talent, wrote the program in his spare time over several weeks while watching soap operas. According to his autobiography, he "never broke a sweat" while writing it; "in fact, it was a breeze."

Since Bates's company, MacroLoft, also owns many of the legal rights needed to play music from such musicians as Jimi Hendrix and Aretha Franklin, you have decided it is time to negotiate directly with Bates prior to placing your encyclopedia on the market. Note that although Bates has received valid licenses in the relevant copyrights for these songs, it is widely known that some of the musicians who created the songs are not pleased with the use of short snippets on inadequate computer sound systems, some of which sound "tinny."

You are aware that negotiating with Bates is far different from standard business practice. He is known to do all negotiating standing up in an unheated room in a mountain cabin, with no lawyers or assistants of any kind. Most unusually, he is not interested at all in formal legal rights. Reflecting his training as a serious student of philosophy, he insists on negotiating from what he calls "first principles." This usually involves long discussions of who should have which rights in which situations, and why.

You would like to convince Bates to let you market your encyclopedia without paying any royalties to him for songs, the access interface, or anything else. Failing this, you would like to keep these royalties to a minimum. Draw up a negotiating strategy that will convince him.

b. Ensuring Integrity of the Marketplace

Quite unlike patent and copyright, trademark law does not protect innovation or creativity directly. Rather, it aims to protect the integrity of the marketplace by prohibiting the use of marks associated with particular manufacturers in ways that would cause confusion as to the sources of the goods. In so doing, trademark law reduces consumer confusion and enhances incentives for firms to invest in activities (including R&D) that improve brand reputation. This function, however, is part of a larger framework of laws and institutions that regulate the quality of information in the marketplace. See generally Peter S. Menell & Suzanne Scotchmer, Intellectual Property, in Handbook of Law and Economics (A. Mitchell Polinsky & Steven Shavell eds., 2006).

The efficiency of the marketplace depends critically upon the quality of information available to consumers. In markets in which the quality of goods is uniform or easily inspected at the time of purchase, consumers can determine the attributes themselves and no information problem arises. In many markets, however — such as used automobiles, computers, watches, as well as designer handbags — an information asymmetry exists: sellers typically have better information about the products or services being offered than buyers can uncover without buying the product. See George A. Ackerlof, The Market for "Lemons": Quality Uncertainty and the Market Mechanism, 84 Quarterly Journal of Economics 488 (1970). Unscrupulous sellers will be tempted to make false or misleading product claims or copy the trademark of a rival producer known for superior quality. It is often easier to copy a trademark than to duplicate production techniques, quality assurance programs, and the like. For example, two watches that look the same on the outside may have very different mechanical features, manufacturing quality, and composition of materials used.

Proliferation of unreliable information in the marketplace increases consumers' costs of search and distorts the provision of goods. Consumers will have to spend more time and effort inspecting goods, researching the product market, and actually testing products. Manufacturers will have less incentive to produce quality goods as others will be able to free-ride on such reputations. In markets for products where quality is costly to observe, high-quality manufacturers might not be able to survive without effective mechanisms for policing the source of products and the accuracy of claims relating to unobservable product characteristics.

Trademarks, as concise and unequivocal indicators of the source (e.g., Intel) and nature (e.g., Pentium) of particular goods, counteract the "market for lemons" problem by communicating to consumers the enterprise responsible for the goods and, in some cases, the specifications of the goods. The brand name Coca-Cola, for example, informs the consumer of the maker of the soft drink beverage as well as the taste that they can expect. If the product lives up to or exceeds expectations, then the trademark owner gains a loyal customer who will be willing to pay a premium in future transactions; if the product disappoints, then the trademark owner will have more difficulty making future sales to that consumer (or will have to offer a discount to attract their business). In this way, trademarks implicitly communicate unobservable characteristics about the quality of branded products, thereby fostering incentives for firms to invest in product quality, even when such attributes are not directly observable prior to a purchasing decision. Sellers who enter the high-quality segment of the market must initially invest in building a strong reputation. Only after consumers become acquainted with the attributes of their brand can they recoup these costs. As this process unfolds, high-quality items can sell for a premium above their costs of production, since consumers will expect them to be of high quality. Trademarks also facilitate efficient new business models, such as franchising, which generate economies of scale and scope in marketing and facilitate rapid business diffusion across vast geographic areas.

The marking of products also creates incentives for disreputable sellers to pass off their own wares as the goods of better-respected manufacturers. Trademark law (as well as false advertising and unfair competition laws more generally) harnesses the incentives of sellers in the marketplace to police the use of marks and advertising claims of competitors. Sellers often have the best information about the quality of products in the marketplace; they also have a direct stake in preventing competitors from free riding on their brand, reputation, and consumer loyalty. By creating private causes of action, trademark and false advertising law take advantage of this informational base and incentive structure as well as the vast decentralized enforcement resources of trademark owners to regulate the informational marketplace, effectively in the name of consumers.

As with patent and copyright law, the creation of intellectual property rights in words, phrases, logos, and other identifying product features can entail several types of costs. Protection of descriptive terms as trademarks can increase search costs and impair competition by raising the marketing costs of competitors. For example, if a cookie manufacturer were to obtain a trademark on the word "cookie," then other companies interested in selling cookies would have a much more difficult time communicating the nature of their goods to consumers. If, however, the trademark was to "Mrs. Fields Cookies" and any protection for "cookies" was disclaimed, then potential competitors would be able to describe their products in the most easily recognized manner and would be able to develop their own marks—such as "ACME Cookies." At a minimum, trademark protection for descriptive terms significantly reduces the effective range of terms that may be used others.

More generally, trademark protection for descriptive terms can impede competition. Gaining control over the most effective term for describing a product raises the costs of potential competitors seeking to sell in that marketplace. By not being able to use a term or means of communication most easily understood by the consuming public, the entrant must bear higher marketing costs. Limitations on the use of trademarked terms for purposes of comparative advertising would also impede vigorous competition.

Trademark protection can also interfere with both communicative and creative expression. Broad exclusive trademark rights would limit the ability of others (including non-competitors) to comment on and poke fun at trademarks and their owners. As we will see, various doctrines limit such adverse effects. But as trademark protection has expanded beyond the traditional core—for example, to encompass a broad conception of connection to, sponsorship of, and affiliation with a trademark owner—it becomes more difficult to assess the boundaries, leading film and television production companies, for example, to tread carefully (and increasingly incur the costs of licensing transactions) in the use of trademarks in their works.

COMMENTS AND QUESTIONS

1. *Comparative Institutional Analysis.* In addition to private rights of action for trademark infringement and false advertising, several other mechanisms are available to provide and regulate market information, such as deceit and fraud common law causes of action and privately enforced consumer protection statutes, public regulation and public enforcement of unfair competition laws, and industry self-regulation and certification organizations. How do these institutions compare with trademark protection? Should they supplement or substitute for trademark protection?

2. *Status Goods.* Some trademarks also serve a more ambiguous function: signaling status or identity for some consumers. Some have referred to such commodities as "Veblen" goods, reflecting Thorstein Veblen's theory of conspicuous consumption. See Thorstein Veblen, The Theory of the Leisure Class: An Economic Study of Institutions (1899). This theory posits that unlike normal goods, demand for status goods actually increases with their price. Purchasers of such goods may be interested in being associated with a particular brand—such as a Rolex watch, a t-shirt with the name and colors of a particular university, or a corporate brand—possibly apart from whether it is authentic or the quality associated with the authentic good. Some purchasers of such goods may well prefer a less expensive, counterfeit version. They presumably would not be confused when purchasing such goods (e.g., a Rolex watch sold on a street corner for $10).

The marketing of less expensive, lower quality imitations of status goods creates the possibility of separate harm to the sellers and purchasers of authentic goods. The availability of counterfeit articles could well divert some consumers who would otherwise purchase the authentic article, although this effect is likely to be relatively small due to the large price differential and the availability of the authentic goods for those who are interested. The lower quality of the counterfeit goods could, however, erode the goodwill associated with the authentic manufacturer through post-sale confusion — on-lookers who mistake the shoddier counterfeit good for the authentic good and are thereby less inclined to purchase the authentic version, thereby reducing sales by the trademark owner. In addition, due to the proliferation of non-easily recognized "fakes," prior and potential purchasers of the authentic "status" goods may be less interested in owning a much less rare commodity. The value of ownership may be sullied. In essence, status goods exhibit a negative network externality, whereby proliferation of such goods erodes the value to prior purchasers. Should trademark law be concerned with this effect?

3. *Anti-Dilution Protection.* In 1996, Congress expanded trademark protection beyond the prevention of consumer confusion to safeguard famous marks — such as Coca-Cola, Nike, Rolls Royce, Disney, and Tiffany's — against blurring (loss of distinctiveness) and tarnishment of brand identity. As consumers develop their mental lexicon of brands, they associate both specific products and general attributes with particular trademarks. For example, Rolls Royce connotes both the source of a luxury automobile as well as a brand of uncompromising quality and ornate styling (as well as high cost). If another company were to introduce Rolls Royce candy bars, it is unlikely that many (if any) consumers would believe that the automobile manufacturer was the source. Whether intended or not, the candy company might benefit from the particular general attributes that the consuming public associates with the Rolls Royce brand. They might also gain some "status" equity to the extent that consumers value the signal associated with a mark. Thus, adopting the Rolls Royce name enables the newcomer some ability to free-ride on the general brand reputation of the famous trademark owner.

Such use, however, might impose some costs on consumers and the owner of the famous trademark. As this new use of the Rolls Royce term gained popularity, the association between the mark and a particular source would become blurred. Furthermore, as more companies in unrelated markets adopt this moniker — Rolls Royce tennis racquets, Rolls Royce landscaping, Rolls Royce tacos — the distinctive quality of the mark would become further eroded. Over time, consumers would lose the non-product specific identity (i.e., Rolls Royce as a brand of uncompromising quality and ornate styling) that the original Rolls Royce mark once evoked. Arguably this raises consumers' search costs: consumers' mental lexicon has become more difficult to parse. See Stacey L. Dogan & Mark A. Lemley, The Merchandising Right: Fragile Theory or Fait Accompli?, 54 Emory L.J. 461 (2005).

As another example, Disney has developed a strong reputation for wholesome, family entertainment. If a pornographic film maker were to brand its films Disney Smut Productions, it is unlikely that many consumers would be confused as to the source of such films. Nonetheless, consumers' shopping lexicon would arguably be distorted because the Disney name would trigger associations with both family-oriented content and pornography. Such a negative association could well injure the Disney Corporation's brand equity.

Anti-dilution protection prevents this erosion of the distinctive quality of a mark by prohibiting famous marks from being used by others — even in unrelated product markets and in non-confusing ways. This preserves distinctive brands and affords the owners exclusive rights to carry their brand names into wholly new markets (or not). We see examples of such brand migration in many markets. Sony Corporation, for example, which honed its reputation in the consumer electronics marketplace, has now developed products in the sound recording and motion picture marketplaces. Cross-branding, such as the marketing of a Barbie doll adorned with Coca-Cola's logo and a distinctive red ensemble, is also increasingly common.

What are the costs of anti-dilution protection? Is the harm to consumers of non-confusing, diluting uses of famous marks likely to be significant? Would scaling back or eliminating anti-dilution protection significantly weaken the incentives of firms to invest in and maintain their brand equity?

B. OVERVIEW OF INTELLECTUAL PROPERTY

Intellectual property law has traditionally been taught along doctrinal lines. Separate courses have covered patent, copyright, and trademark law, with trade secrets often lost between the gaps. Yet the practice of intellectual property law increasingly cuts across these lines. Computer technology companies, for example, frequently require lawyers to address trade secret, copyright, patent, trademark, and antitrust issues simultaneously. Moreover, from a purely practical standpoint, clients are ultimately interested in appropriating a return from their investments, not in how many patents, copyrights, trademarks, or trade secrets their lawyers can obtain. Thus intellectual property lawyers must possess an integrated understanding of these various fields in order to provide sound advice.

With this objective in mind, our book integrates the various modes of intellectual property in a functional manner. Nonetheless, it is still necessary to devote significant time to mastering each of the distinct fields. Therefore, the next five chapters survey the principal modes of intellectual property — trade secret, patent, copyright, trademark/trade dress, and related state law doctrines — while emphasizing the overlaps and interactions of the various bodies of law. Chapter 7 integrates these fields directly in exploring legal protection for computer technology. Chapter 8 surveys the interplay between intellectual property law and antitrust law.

Before we begin this more detailed study, however, a brief survey of the overall landscape of intellectual property is in order. The following section sketches the elements of each of the principal modes of intellectual property protection in a comparative framework. These elements are summarized in Table 1-1. As an initial exploration, we conclude the chapter with a problem highlighting the integrated nature of intellectual property law and the challenges of applying its many branches to a real-world problem.

1. Trade Secret

Trade secret laws are state law doctrines that protect against the misappropriation of certain confidential information. As such, they are more akin to traditional tort and contract law than to patent or copyright law. While protection for trade secrets has long been a part of the common law, most states today protect trade secrets by statute. The basic purpose behind protecting trade secrets is to prevent "theft" of information by unfair or commercially unreasonable means. In essence, trade secret law is a form of

TABLE 1-1
Principal Modes of Legal Protection for Intellectual Work

	Trade secret	Patent	Copyright	Trademark/dress
Underlying Theory	Freedom of contract; protection against unfair means of competition	Limited monopoly to encourage production of utilitarian works in exchange for immediate disclosure and ultimate enrichment of the public domain	Limited (although relatively long-lived) monopoly to encourage the authorship of expressive works; developed initially as a means of promoting publishing	Perpetual protection for distinctive nonfunctional names and dress in order to improve the quality of information in the market place
Source of Law	State statute (e.g., Uniform Trade Secrets Act); common law	Patent Act (federal)	Copyright Act (federal) common law (limited)	Lanham Act (federal); common law (unfair competition)
Subject Matter	Formula, pattern, compilation, program, device, method, technique, process	Process, machine, manufacture, or composition of matter; plants (asexually reproducing); designs — *excluding:* laws of nature, natural substances, printed matter (forms), mental steps	Literary, musical, choreographic, dramatic and artistic works as well as computer software and aesthetic elements of useful articles *limited by* idea/expression dichotomy *(no protection for ideas, systems, methods, procedures)*; no protection for facts/research	Trademarks; service marks; certification marks (e.g., Good Housekeeping); collective marks (e.g., Toy Manufacturers of America); trade dress (§43(a)); *no protection for* functional features, descriptive terms, geographic names, misleading aspects, or "generic" names (e.g., thermos)
Standard for Protection	Information not generally known or available; reasonable efforts to maintain secrecy; commercial value	Novelty; non-obviousness; and utility (distinctiveness for plant patents; ornamentality for design patents)	Originality; authorship; fixation in a tangible medium	Distinctiveness; secondary meaning (for descriptive and geographic marks); use in commerce (minimal); famous mark (for dilution protection)

25

TABLE 1-1
(Continued)

	Trade secret	Patent	Copyright	Trademark/dress
Scope of Protection	Protection against misappropriation — acquisition by improper means or authorized disclosure	Exclusive rights to make, use, sell innovation as limited by contribution to art; extends to "equivalents"	Rights of performance, display, reproduction, derivative works; limited protection for attribution and integrity afforded some works of visual art; protection against circumvention of technical protection measures	Exclusive rights in U.S.; likelihood of confusion; false designation of origin (§ 43(a)); dilution (for famous marks)
Period of Protection	Until becomes public knowledge	20 years from filing (utility); extensions up to 5 years for drugs, medical devices and additives; 14 years (design)	Life of author + 70 years; "works for hire": minimum of 95 years after publication or 120 years after creation	Perpetual, subject to abandonment
Disclosure	Loss of protection (unless sub rosa)	Right to patent lost if inventor delays too long after publishing before filing application; full disclosure is required as part of application; notice of patent required for damages	© notice and publication no longer required, but confer certain benefits	® notice optional; establishes prima facie evidence of validity, constructive knowledge of registration, confers federal jurisdiction, becomes incontestable after 5 years of continuous use, authorizes treble damages and attorney fees, and right to bar imports bearing infringing mark

Rights of Others	Independent discovery; reverse engineering	Only if licensed; can request reexamination of patent by Patent Office	Fair use; compulsory licensing for musical compositions, cable TV, et al.; independent creation	Truthful reflection of source of product; fair and collateral use (e.g., comment)
Costs of Protection	Security expenses; personnel dissatisfaction; litigation costs	Filing, issue, and maintenance fees; litigation costs	None (protection attaches upon fixation); publication requires notice; suit requires registration; litigation costs	Registration search; marking product (optional—see above); litigation costs
Licensing and Assignment	Complicated by inherent nature of bargaining (seller wants guarantee before disclosure; buyer wants to know what is offered)	Encouraged by completeness of property rights, subject to antitrust constraints	Assignor has termination right between 36th and 41st years following transfer	No naked licenses (owner must monitor licensee); no sales of trademark "in gross"
Remedies	Civil suit for misappropriation; conversion, unjust enrichment, breach of contract; damages (potentially treble) and injunctive relief; criminal prosecution for theft	Injunctive relief and damages (potentially treble); attorney fees (in exceptional cases)	Injunction against further infringement; destruction of infringing articles; damages (actual or profits); statutory ($200–$150,000 damages within court's discretion); attorney fees (within court's discretion); criminal prosecution	Injunction; accounting for profits; damages (potentially treble); attorney fees (in exceptional cases); seizure and destruction of infringing goods; criminal prosecution for trafficking in counterfeit goods or services

private intellectual property law under which creators establish contractual limitations or build legal "fences" that afford protection from misappropriation.

The definition of subject matter eligible for protection is quite broad: business or technical information of any sort. To benefit from trade secret protection, the information must be a secret. However, only relative and not absolute secrecy is required. In addition, the owner of a trade secret must take reasonable steps to maintain its secrecy. Trade secrets have no definite term of protection but may be protected only as long as they are secret. Once a trade secret is disclosed, protection is lost.

There is no state agency in charge of "issuing" (or even registering) secrets. Rather, any information that meets the above criteria can be protected. Courts will find misappropriation of trade secrets in two circumstances: where the secrets were obtained by theft or other improper means, or where they were used or disclosed by the defendant in violation of a confidential relationship. However, trade secret laws do not protect against independent discovery or invention. Nor do they prevent competitors from "reverse engineering" a legally obtained product in order to determine the secrets contained inside. Violations of trade secret law entitle the owner to damages and in some cases injunctions against use or further disclosure.

2. Patent

Patent law is the classic example of an intellectual property regime modeled on the utilitarian framework. Following the constitutional authorization, patent law offers the possibility of a limited period of exclusive rights to encourage research and development aimed at discovering new processes, machines, and compositions of matter, and improvements thereof. The public benefits directly through the spur to innovation and disclosure of new technology. After the term of the patent expires, the innovation becomes part of the public domain, freely available to all.

To obtain a utility patent, an inventor must submit an application to the Patent and Trademark Office (PTO) that meets five requirements: patentable subject matter, usefulness, novelty, non-obviousness, and disclosure sufficient to enable others skilled in the art to make and use the invention. While the threshold for usefulness is low, the novelty and non-obviousness standards are substantial, and the PTO conducts an independent review of the application to ensure that it meets these requirements. If the PTO grants the patent, the inventor obtains exclusive rights to make, use, and sell the innovation for a term of up to 20 years. The patent grant is nearly absolute, barring even those who independently develop the invention from practicing its art. Infringement will be found where the accused device, composition, or process embodies all of the elements of a valid patent claim (or accomplishes substantially the same function in substantially the same way to achieve the same result).

The PTO also issues plant patents for distinctive plants and design patents for ornamental designs for articles of manufacture. Design patents have a term of 14 years.

3. Copyright

Although the copyright and patent laws flow from the same constitutional basis and share the same general approach—statutorily created monopolies to foster progress—they feature different elements and rights, reflecting the very different fields of creativity that they seek to encourage. In general, copyrights are easier to secure and last substantially longer than patents, although the scope of protection afforded copyrights is narrower and less absolute than that given to patents.

Copyright law covers the broad range of literary and artistic expression — including books, poetry, song, dance, dramatic works, computer programs, movies, sculpture, and paintings. Ideas themselves are not copyrightable, but the author's particular expression of an idea is protectable. A work must exhibit a modicum of originality and be fixed in a "tangible medium of expression" to receive protection. Copyright protection attaches as soon as a work is fixed. There is no examination by a governmental authority, although the Copyright Office registers copyrightable works. Such registration is no longer required for validity, but U.S. authors must register their works prior to filing an infringement suit. A copyright lasts for the life of the author plus 70 years, or a total of 95 years in the case of entity authors.

The breadth and ease of acquisition of copyright protection are balanced by the more limited rights that copyright law confers. Ownership of a valid copyright protects a copyright holder from unauthorized copying, public performance, and display, and it entitles the holder to make derivative works and to control sale and distribution of the work. These rights, however, are limited in a number of ways. Others may make "fair use" of the material in certain circumstances. The Copyright Act also establishes compulsory licensing for musical compositions and cable television. A limited set of moral rights protects against misattribution or destruction of a narrow class of works of visual art.

Copyright law protects only against *copying* of protected expression. Independent creation of a copyrighted work does not violate the Copyright Act, nor does copying the unprotected elements of a work. Therefore, copyright law must have some mechanism for determining when a work has been copied illegally. While in rare cases direct proof of copying may be available, usually it is not. In its place, courts infer copying from proof that the defendant has had *access* to the plaintiff's work combined with evidence that the two works are *similar*. Even if copying is established, it must be further shown that the defendant's work is *substantially similar* to protected elements (e.g., excluding ideas) of the plaintiff's work.

With the proliferation of digital technology, Congress has augmented traditional copyright protection by prohibiting the circumvention of technical protection measures intended to prevent unauthorized use and distribution of copyrighted works and alteration of copyright management information. These new rights are subject to various exceptions and limitations.

4. Trademark/Trade Dress

Trademarks are also protected by federal statute, although the source of constitutional authority is different from that of the Patent and Copyright Acts. Rather than deriving from a specific grant of constitutional power, federal power to regulate trademarks and unfair competition derives from the Commerce Clause of the Constitution, which authorizes Congress to regulate interstate commerce. Unlike patent and copyright protection, trademark law did not evolve from a desire to stimulate particular types of economic activity. Rather, its original purpose was to protect consumers in a world of mass merchandising from unscrupulous sellers attempting to fly under the banner of someone else's well-known logo or identifying symbol. Only in recent years has trademark law begun to embrace the incentive, personhood, and natural rights rationales. The Lanham Act (the federal trademark statute) protects words, symbols, and other attributes that serve to identify the nature and source of goods or services. Examples of marks protectable under the Lanham Act include corporate and product names, symbols, logos, slogans, pictures and designs, product configurations, colors, and even smells. Not all such marks are protectable, however. To receive trademark protection, an identifying mark need not be new or previously unused, but it must represent to consumers the source of the good or service identified. It cannot be merely a description of the good itself or a generic term for the class of goods or services offered. Further, the identifying mark may not be a functional element of the product itself but must serve a purely identifying purpose. Since 1996, famous marks also receive federal protection against "dilution" by blurring or tarnishment. Finally, trademark protection is directly tied to the use of the mark to identify goods in commerce. Trademarks do not expire on any particular date but continue in force until they are "abandoned" by their owner or become unprotectable.

The PTO examines trademark applications and issues trademark registrations that confer significant benefits upon the registrants, including prima facie evidence of validity; constructive notice to others of the claim of ownership; federal subject matter jurisdiction; incontestability after five years, which confers exclusive right to use the mark; authorization to seek treble damages and attorney fees; and the right to bar importation of goods bearing the infringing mark. Federal trademark registration, however, is not necessary to obtain trademark protection. A trademark owner who believes that another is using the same or a similar mark to identify competing goods can bring suit for trademark infringement. Infringement turns on whether consumers are likely to be confused as to the origin of the goods or services. If so, the trademark owner is entitled to an injunction against the confusing use, damages for past infringement, and in some cases the seizure and destruction of infringing goods.

PROBLEM

Problem 1-3.
MEMORANDUM
To: Associate
From: Senior Partner
Re: HEALTHWARE Inc.

Janet Peterson called me yesterday about a new venture that she plans to try to get off the ground. As you may know, Janet is a computer programmer and a registered nurse. She has an interesting idea for a new venture and would like our advice on how she might structure the business to have the best potential for success.

The proposed venture will be called Healthware. Janet believes that she can tap into the current diet/health/environmental/personal computer craze by developing a user-friendly computer program that would monitor the user's diet and fitness activity. The user would input information on his or her health (e.g., age, weight, medical history, dietary restrictions). Each day, the user would input information on diet and physical exercise. The program would have simple "pull down" menus for making this quick and easy; nutritional information on all foods would be stored in the program. The computer would periodically provide an analysis of the user's health, as well as suggestions for achieving the user's goals, whether weight reduction, better fitness, or general health. In addition, the program would compile a record of the user's activities which could be brought to annual physicals. Other subroutines would be available for pregnant and lactating women, children, the elderly, diabetics, vegetarians, triatheletes, etc.

Janet thinks that she could put together the diverse people necessary to pull this project off: programmers, a nutritionist, a physician, a fitness consultant. She is concerned, however, that any one of these people could, after they are familiar with the product, develop a competing program.

What are the options for structuring Healthware? What problems do you foresee in structuring this venture? Assuming that the product is popular, what are the major risks to Healthware's success? How can we structure Healthware so as to overcome these problems?

2

Trade Secret Protection

A. INTRODUCTION

1. History

The idea that information should be protected against "theft" (which may include the physical taking of tangible goods containing information or simply the copying or memorization of data) is a venerable one in the law. One scholar traces the earliest legal protection against "misappropriation of trade secrets" to the Roman empire.[1] The Roman courts created a cause of action called "actio servi corrupti"—literally, an action for corrupting a slave. According to Schiller, the actio servi corrupti was used to protect slave owners from third parties who would "corrupt" slaves (by bribery or intimidation) into disclosing their owners' confidential business information. The law made such third parties liable to the slave owner for twice the damages he suffered as a result of the disclosure.

While recent scholarship has cast some doubt on the enforcement of trade secret protection in the Roman empire,[2] the concept that so-called business or "trade secrets" were entitled to legal protection spread rapidly throughout the world. As

1. Arguably, trade secrets existed before this time, albeit in unusual forms. Consider Mark C. Suchman, Invention and Ritual: Notes on the Interrelation of Magic and Intellectual Property in Preliterate Societies, 89 Colum. L. Rev. 1264, 1274 (1989):

> [L]et us imagine a hypothetical preliterate inventor who, through diligence or good fortune, discovers that her maize crop is larger when she plants a small fish next to each kernel of corn.... Clearly, this technique has economic value and could garner its creator material and social rewards if she could monopolize it and license it for a fee. Unfortunately, the odds of keeping such a discovery secret are slight.... [T]he procedure is so simple that even a casual observer could replicate the process.... Magic, however, provides a way out of the dilemma. By claiming, for example, that the power of the fish is activated by a talisman that she alone possesses, the inventor can remove her idea from the public domain.... [T]he magicked process is far easier to monopolize than the simple technology alone.

2. For a discussion of the Roman law, see A. Arthur Schiller, Trade Secrets and the Roman Law: The Actio Servi Corrupti, 30 Colum. L. Rev. 837 (1930). For a dissenting view on the role of Roman law in protecting trade secrets, see Alan Watson, Trade Secrets and Roman Law: The Myth Exploded, 11 Tul. Eur. & Civ. L.F. 19 (1996).

33

early as the Renaissance, most European nation-states had laws that protected businesses (notably, the Guild cartels) from those who used their secret processes and ideas without permission. These early laws were translated during the Industrial Revolution into statutes that protected "industrial secrets." Many of these statutes are still in force today, albeit in modified form.

The roots of trade secrecy in slavery law were further evident in the treatment of employees in the centuries before the Industrial Revolution. Both commerce and foreign policy included a strong dose of "mercantilism." Governments and private guilds attempted to keep "their" intellectual property within their grasp, using a combination of rewards to inventors and rules that reduced employee mobility. These developments are well traced in Carlo M. Cipolla, Before the Industrial Revolution: European Economy and Society 1000-1700 (2d ed. 1980); David J. Jeremy, Transatlantic Industrial Revolution: The Diffusion of Textile Technologies Between Britain and America, 1790-1830's, at 185-189 (MIT Press 1981). The authors provide such examples as restrictive British secrecy laws, city rewards to woollen craftsmen in thirteenth-century Bologna and, on the other side, the kidnapping of skilled Swedish ironworkers by France in 1660.

This patchwork of public and private regulation arguably served as a substitute for strong intellectual property protection by protecting a country's or a corporation's "human capital" and so encouraging investment in employee skills and training. The guild system of apprenticeship and mandatory periods of servitude operated to prevent excessive "spillovers" of guild knowledge to competitors and hence may have encouraged innovation. On this point, compare Paul A. David, Intellectual Property and the Panda's Thumb: Patents, Copyrights, and Trade Secrets in Economic Theory and History, in Global Dimensions of Intellectual Property Rights in Science and Technology 19, 45 (Natl. Academy Press 1993) (guild system encouraged investment in innovation) with Cipolla, *supra*, at 261 (guilds restricted competition and therefore restrained business and technological progress). In effect, guilds and countries solved the public goods problem by doing their best to privatize the good — keeping information secret from prying eyes.

Unfortunately, this form of private protection comes at a price. Companies (or countries) that must rely on current and former employees to keep secrets run into a host of problems. Companies may not disclose to employees all the information they need to know. They may take inefficient physical security precautions. They may hire employees based on their loyalty rather than their productivity. And finally, they may simply choose to prey on other companies rather than investing in innovation themselves. Robert Sherwood makes the case that many of these problems occurred in the recent past in Mexico and Brazil as a result of inadequate trade secret protection. See Robert M. Sherwood, Intellectual Property and Economic Development 113 (1990).

Obviously, it would be preferable to have a legal system that could protect an employer's secrets while allowing employee mobility and the dissemination of information through licensing. And such rules have generally developed as legal systems and norms became stable enough to support them. European countries, for example, developed trade secret protection quite early.[3]

---

3. Interestingly, Japan has only recently enacted a trade secret protection statute. See Hideo Nakoshi, New Japanese Trade Secret Act, 75 J. Pat. & Trademark Off. Society 631 (1993). Nakoshi provides some interesting thoughts on why Japan considered such a trade secret statute unnecessary for so long. He argues

By contrast, Anglo-American jurisprudence was a relative latecomer to the protection of trade secrets. English and American courts first recognized a cause of action for damages for misappropriation of trade secrets in 1817 and 1837, respectively;[4] injunctive relief against actual or threatened misappropriation came still later.[5] These early decisions concerned issues that are still debated in trade secret cases today: the circumstances in which an employee may continue her business after departing her employer, the circumstances in which a competitor may copy another's publicly sold product, and whether courts will enforce a contract requiring that business information be kept confidential.

Trade secret litigation is particularly important to small companies. A study of 530 manufacturing firms in Massachusetts found that small firms spent a greater proportion of their litigation resources on trade secret cases than did large firms, perhaps because of the enormous cost associated with patent litigation or because of the delay in obtaining patents. Forty-three percent of the intellectual property disputes involving these firms had a trade secret component. Josh Lerner, The Importance of Trade Secrecy: Evidence from Civil Litigation, Harv. Bus. School working paper #95-043 (Dec. 1994). Lerner's data indicate that trade secrets, though important to all firms, are most important for the small companies that drive innovation in many developing fields. Since the time of his study, however, patent litigation has been growing in significance relative to trade secrecy.

2. Overview of Trade Secret Protection

Today, every one of the United States protects trade secrets in some form or another. Improper use or disclosure of a trade secret was traditionally a common law tort. In the twentieth century, the principal document setting forth the "law" of trade secrets was the Restatement of Torts, published in 1939. Sections 757 and 758 of the Restatement set forth basic principles of trade secret law that were adopted by courts in most states in the country. National protection for trade secrets is also compelled by United States adherence to the intellectual property agreement in the Uruguay Round of the General Agreement on Tariffs and Trade (GATT) Trade-Related Aspects of Intellectual Property Rights (TRIPs) in 1994. GATT TRIPs article 39 requires member nations to protect trade secrets.

The Restatement protected as a trade secret any information "used in one's business" that gives its owner "an opportunity to obtain an advantage over competitors who do not know or use it," so long as the information was in fact a secret. Restatement § 757, Comment *b*.

When the Restatement (Second) of Torts was published in 1979, the authors decided not to include sections 757 and 758, on the grounds that the law of trade

---

that cultural norms of assumed trust made it socially uncomfortable to insist on formal confidentiality agreements, and that long-term or lifetime employment eliminated many trade secret issues associated with employee mobility.

4. See Newberry v. James, 35 Eng. Rep. 1011, 1013 (Ct. Ch. 1817); Vickey v. Welch, 36 Mass. 523, 527 (1837).

5. See Yovett v. Winyard, 37 Eng. Rep. 525 (Ct. Ch. 1820); Taylor v. Blanchard, 95 Mass. 370 (1866).

secrets had developed into an independent body of law that no longer relied on general principles of tort law. The influence of the original Restatement remained, however, because it had been adopted by so many courts.[6]

Beginning in 1979, the National Conference of Commissioners on Uniform State Laws promulgated a model state statute, the Uniform Trade Secrets Act, which differed in some respects from the common law. The Uniform Trade Secrets Act (UTSA) has now been enacted (in one form or another) by 40 states and the District of Columbia. See Restatement (Third) of Unfair Competition § 39, at 437-438 (listing state statutes). Because of the Uniform Act's importance, we reproduce its primary provisions here.

Uniform Trade Secrets Act, with 1985 Amendments

§ 1. Definitions

As used in this [Act], unless the context requires otherwise:

(1) "Improper means" includes theft, bribery, misrepresentation, breach or inducement of a breach of a duty to maintain secrecy, or espionage through electronic or other means;

(2) "Misappropriation" means:

(i) acquisition of a trade secret of another by a person who knows or has reason to know that the trade secret was acquired by improper means; or

(ii) disclosure or use of a trade secret of another without express or implied consent by a person who

(A) used improper means to acquire knowledge of the trade secret; or

(B) at the time of disclosure or use, knew or had reason to know that his knowledge of the trade secret was

(I) derived from or through a person who had utilized improper means to acquire it;

(II) acquired under circumstances giving rise to a duty to maintain its secrecy or limit its use; or

(III) derived from or through a person who owed a duty to the person seeking relief to maintain its secrecy or limit its use; or

(C) before a material change of his [or her] position, knew or had reason to know that it was a trade secret and that knowledge of it had been acquired by accident or mistake....

(4) "Trade secret" means information, including a formula, pattern, compilation, program, device, method, technique, or process, that:

(i) derives independent economic value, actual or potential, from not being generally known to, and not being readily ascertainable by proper means by, other persons who can obtain economic value from its disclosure or use, and

(ii) is the subject of efforts that are reasonable under the circumstances to maintain its secrecy.

---

6. The authors of the Restatement (Third) of Unfair Competition, published in 1994, once again included a section on the law of trade secrets. The section is organized to follow the Uniform Act, and many of its provisions are consistent with the act. Nonetheless, there are some differences between the act and the Restatement. They are discussed below.

A trade secret claim can be broken down into three essential elements. First, the subject matter involved must qualify for trade secret protection; it must be the type of knowledge or information that trade secret law was meant to protect, and it must not be generally known to all. On eligible subject matter, the current trend, exemplified once again by the UTSA, is to protect as a trade secret *any* valuable information. So long as the information is capable of adding economic value to the plaintiff, it can be protected by trade secret law. The requirement that the information not be generally known follows from the label *secret*. The requirement is meant to ensure that no one claims intellectual property protection for information commonly known in a trade or industry.

The second element to be established by the plaintiff in a trade secret case is that the plaintiff, holder of the trade secret, took *reasonable precautions* under the circumstances to prevent its disclosure. Courts have shown some confusion over the rationale for this requirement. Some see in it evidence that the trade secret is valuable enough to bother litigating; others argue that where reasonable precautions are taken, chances are that a defendant acquired the trade secret wrongfully. Whatever the justification, it is clear that no one may let information about products and operations flow freely to competitors at one time and then later claim that competitors have wrongfully acquired valuable trade secrets. To establish the right to sue later, one must be consistently diligent in protecting information. As always, however, the presence of the term "reasonable" ensures close cases and difficult line-drawing for courts, a theme reflected in several of the cases that follow.

Finally, a trade secret plaintiff also must prove that the defendant acquired the information wrongfully—in a word, that the defendant *misappropriated* the trade secret. Just because a person's information is valuable does not make it wrong for another to use it or disclose it. But use or disclosure is wrong, in the eyes of trade secret law, when the information is acquired through deception, skulduggery, or outright theft. As we will see in the cases that follow, close cases abound in this area because of the creativity of competitors in rooting out information about their rivals' businesses and products.

In many cases a defendant's use or disclosure is wrongful because of a preexisting obligation to the plaintiff not to disclose or appropriate the trade secret. Such an obligation can arise in either of two ways: explicitly, by contract; and implicitly, because of an *implied duty*. A classic example of an implied duty is the case of an employee. Even in the absence of an explicit contract, most employees are held to have a duty to protect their employers' interests in the employers' secret practices, information, and the like. Even where the duty arises by explicit contract, however, public policy limitations on the scope and duration of the agreement will often come into play, in some cases resulting in substantial judicial modification of the explicit obligations laid out in the contract.

3. Theory of Trade Secrets

Legal protection for trade secrets is premised primarily on two theories that are only partly complementary. The first is utilitarian. Under this view, protecting against the theft of proprietary information encourages investment in such information. This idea is sometimes associated with the view that trade secrets are a form of property. The second theory emphasizes deterrence of wrongful acts and is therefore sometimes described as a tort theory. Here the aim of trade secret law is to punish and prevent illicit behavior, and even to uphold reasonable standards of commercial behavior. Cf. Kim Lane Scheppele, Legal Secrets: Equality and Efficiency in the Common Law (1988) (arguing that cases involving legal secrets — including trade secrets cases — are better explained in terms of principles all would be willing to agree to rather than in the efficiency terms of law and economics). Although under the tort theory trade secret protection is not explicitly about encouraging investments, it is plain that one consequence of deterring wrongful behavior would be to encourage investment in trade secrets. Hence, despite their conceptual differences, the tort and property/incentive approaches to trade secrets may well push in the same direction.

Property Rights. Some jurists have conceptualized "intellectual property" as a species of the broader concept of "property." The Supreme Court adopted this view of trade secret law in Ruckelshaus v. Monsanto Co., 467 U.S. 986, 1001-1004 (1984). There the Court faced the question (considered in more detail below) of whether a federal law that required Monsanto to publicly disclose its trade secrets was a "taking of private property" for which the Fifth Amendment required compensation. The Court, in finding that trade secrets were "property," reasoned in part that "[t]rade secrets have many of the characteristics of more tangible forms of property. A trade secret is assignable. A trade secret can form the res of a trust, and it passes to a trustee in bankruptcy." Id. at 1002-1004. Treatment of trade secrets as property rights vested in the trade secret "owner" is consistent with a view of trade secrets law as providing an additional incentive to innovate, beyond those provided in patent law. The Supreme Court has offered some support for this incentive view in cases such as Kewanee Oil Co. v. Bicron Corp., 416 U.S. 470, 481-485 (1974).[7]

Tort Law. An alternate explanation for much of trade secrets law is what might be described as a "duty-based" theory, or what Melvin Jager calls "the maintenance of commercial morality." 1 Melvin Jager, Trade Secrets Law § 1.03, at 1-4. The Supreme Court adopted this view in a famous early decision:

> The word "property" as applied to trademarks and trade secrets is an unanalyzed expression of certain secondary consequences of the primary fact that the law makes some rudimentary requirements of good faith. Whether the plaintiffs have any valuable secret or not, the defendant knows the facts, whatever they are, through a special confidence that he accepted. The property may be denied, but the confidence cannot be. Therefore

---

7. In *Kewanee,* the Court held that:

> Certainly the patent policy of encouraging invention is not disturbed by the existence of another form of incentive to invention. In this respect the two systems are not and never would be in conflict. . . .
> Trade secret law will encourage invention in areas where patent law does not reach, and will prompt the independent innovator to proceed with the discovery and exploitation of his invention. Competition is fostered and the public is not deprived of the use of valuable, if not quite patentable, invention.

the starting point for the present matter is not property or due process of law, but that the defendant stood in confidential relations with the plaintiffs. . . .

E.I. du Pont & Co. v. Masland, 244 U.S. 100, 102 (1917).[8] Closely related to *Masland*'s theory of "breach of confidence" is the contract basis for trade secret law. While not always applicable, many trade secret cases arise out of a "duty" explicitly stated in a contract, such as a technology license or an employment agreement. The tort-based theory of breach of duty merges in those cases with a standard common law action for breach of contract.

For an argument that there is no theoretical justification for trade secret law, and that other contract and tort doctrines can do just fine in protecting confidential information, see Robert G. Bone, A New Look at Trade Secret Law: Doctrine in Search of Justification, 86 Calif. L. Rev. 241 (1998).

The different theories of trade secret law reflect different views of the nature of intellectual property law as a whole, discussed in Chapter 1. These debates recur throughout this book, because the reason that we protect intellectual property turns out to matter a great deal in deciding whether and to what extent we protect it.

Keep in mind the different rationales for trade secret protection as you read the material that follows. Consider the extent to which each of these theories can explain the law of trade secrets.

B. SUBJECT MATTER

1. Defining Trade Secrets

Metallurgical Industries Inc. v. Fourtek, Inc.
United States Court of Appeals for the Fifth Circuit
790 F.2d 1195 (5th Cir. 1986)

GEE, Circuit Judge:

Today's case requires us to review Texas law on the misappropriation of trade secrets. Having done so, we conclude that the district court misconceived the nature and elements of this cause of action, a misconception that led it to direct a verdict erroneously in favor of appellee Bielefeldt. We also conclude that the court abused its discretion in excluding certain evidence. Accordingly, we affirm in part, reverse in part, and remand the case for a new trial.

key word

8. The *Monsanto* Court attempted to distinguish *Masland* in a footnote, claiming that "Justice Holmes did not deny the existence of a property interest; he simply deemed determination of the existence of that interest irrelevant to the resolution of the case." *Monsanto*, 467 U.S. 1004 n.9.

I. Facts of the Case

We commence with a brief description of the scientific process concerned. Tungsten carbide is a metallic compound of great value in certain industrial processes. Combined with the metal cobalt, it forms an extremely hard alloy known as "cemented tungsten carbide" used in oil drills, tools for manufacturing metals, and wear-resistant coatings. Because of its great value, reclamation of carbide from scrap metals is feasible. For a long time, however, the alloy's extreme resistance to machining made reclamation difficult. In the late 1960's and early 1970's, a new solution — known as the zinc recovery process — was devised, a solution based on carbide's reaction with zinc at high temperatures. In the crucibles of a furnace, molten zinc will react with the cobalt in the carbide to cause swelling and cracking of the scrap metal. After this has occurred, the zinc is distilled from the crucible, leaving the scrap in a more brittle state. The carbide is then ground into a powder, usable in new products as an alternative to virgin carbide. This process is the generally recognized modern method of carbide reclamation.

Metallurgical Industries has been in the business of reclaiming carbide since 1976, using the more primitive "cold-stream process." In the mid-1970's, Metallurgical began to consider using the zinc recovery process. In that connection, it came to know appellee Irvin Bielefeldt, a representative of Therm-O-Vac Engineering & Manufacturing Company (Therm-O-Vac). Negotiations led to a contract authorizing Therm-O-Vac to design and construct two zinc recovery furnaces, the purchase order for the first being executed in July 1976.

The furnace arrived in April 1977. Dissatisfied with its performance, Metallurgical modified it extensively. First, it inserted chill plates in one part of the furnace to create a better temperature differential for distilling the zinc. Second, Metallurgical replaced the one large crucible then in place with several smaller crucibles to prevent the zinc from dispersing in the furnace. Third, it replaced segmented heating elements which had caused electric arcing with unitary graphite heating elements. Last, it installed a filter in the furnace's vacuum-pumps, which zinc particles had continually clogged. These efforts proved successful and the modified furnace soon began commercial operation.

...[M]etallurgical returned to Therm-O-Vac for its second furnace. A purchase order was signed in January 1979, and the furnace arrived that July. Further modifications again had to be made, but commercial production was allegedly achieved in January 1980.

In 1980, after Therm-O-Vac went bankrupt, Bielefeldt and three other former Therm-O-Vac employees — Norman Montesino, Gary Boehm, and Michael Sarvadi — formed Fourtek, Incorporated. Soon thereafter, Fourtek agreed to build a zinc recovery furnace for appellee Smith International, Incorporated (Smith)....

Metallurgical ... brought a diversity action against Smith, Bielefeldt, Montesino, Boehm, and Sarvadi in November 1981. In its complaint, Metallurgical charged the defendants with misappropriating its trade secrets....[Trial] testimony indicated Metallurgical's frequent notices to Bielefeldt that the process was a secret and that the disclosures to him were made in confidence. Another witness recounted meetings in which the modifications were agreed to. Bielefeldt was allegedly unconvinced about the efficacy of these changes and contributed little to the discussion. Metallurgical also presented evidence that it had expended considerable time, effort, and money to modify the furnaces.

[The district court nonetheless granted defendants' motions for directed verdicts.] The principal reason advanced was the court's conclusion that no trade secret is involved. At trial, Metallurgical acknowledged that the individual changes, by themselves, are not secrets; chill plates and pump filters, for example, are well-known. Metallurgical's position instead was that the process, taken as a whole, is a trade secret in the carbide business. The court, however, refused to recognize any protection Texas law provides to a modification process. It also concluded that the information Bielefeldt obtained from working with Metallurgical is too general to be legally protected. Finally, it ruled that "negative know-how" — the knowledge of what not to do — is unprotected. . . .

III.   Defining a "Trade Secret"

We begin by reviewing the legal definition of a trade secret. Of course, to qualify as one, the subject matter involved must, in fact, be a secret; "matters of general knowledge in an industry cannot be appropriated by one as his secret." Wissman v. Boucher, 150 Tex. 326, 240 S.W.2d 278, 280 (Tex. 1951); see also Zoecon Industries v. American Stockman Tag Co., 713 F.2d 1174, 1179 (5th Cir. 1983) ("a customer list of readily ascertainable names and addresses will not be protected as a trade secret"). Smith emphasizes the absence of any secret because the basic zinc recovery process has been publicized in the trade. Acknowledging the publicity of the zinc recovery process, however, we nevertheless conclude that Metallurgical's particular modification efforts can be as yet unknown to the industry. A general description of the zinc recovery process reveals nothing about the benefits unitary heating elements and vacuum pump filters can provide to that procedure. That the scientific principles involved are generally known does not necessarily refute Metallurgical's claim of trade secrets.

Metallurgical, furthermore, presented evidence to back up its claim. One of its main witnesses was Arnold Blum, a consultant very influential in the decisions to modify the furnaces. Blum testified as to his belief that Metallurgical's changes were unknown in the carbide reclamation industry. The evidence also shows Metallurgical's efforts to keep secret its modifications. Blum testified that he noted security measures taken to conceal the furnaces from all but authorized personnel. The furnaces were in areas hidden from public view, while signs warned all about restricted access. Company policy, moreover, required everyone authorized to see the furnace to sign a non-disclosure agreement. These measures constitute evidence probative of the existence of secrets. One's subjective belief of a secret's existence suggests that the secret exists. Security measures, after all, cost money; a manufacturer therefore presumably would not incur these costs if it believed its competitors already knew about the information involved. In University Computing Co. v. Lykes-Youngstown Corp., 504 F.2d 518, 535 (5th Cir. 1974), we regarded subjective belief as a factor to consider in determining whether secrecy exists. Because evidence of security measures is relevant, that shown here helps us conclude that a reasonable jury could have found the existence of the requisite secrecy.

Smith argues, however, that Metallurgical's disclosure to other parties vitiated the secrecy required to obtain legal protection. Metallurgical revealed its information to Consarc Corporation in 1978; it also disclosed information in 1980 to La Floridienne, its European licensee of carbide reclamation technology. Because both these disclosures occurred before Bielefeldt allegedly misappropriated the knowledge of modifications, others knew of the information when the Smith furnace was built. This being so, Smith argues, no trade secret in fact existed.

Although the law requires secrecy, it need not be absolute. Public revelation would, of course, dispel all secrecy, but the holder of a secret need not remain totally silent:

> He may, without losing his protection, communicate to employees involved in its use. He may likewise communicate it to others pledged to secrecy. . . . Nevertheless, a substantial element of secrecy must exist, so that except by the use of improper means, there would be difficulty in acquiring the information.

Restatement of Torts, §757 Comment b (1939). We conclude that a holder may divulge his information to a limited extent without destroying its status as a trade secret. To hold otherwise would greatly limit the holder's ability to profit from his secret. If disclosure to others is made to further the holder's economic interests, it should, in appropriate circumstances, be considered a limited disclosure that does not destroy the requisite secrecy. The only question is whether we are dealing with a limited disclosure here. . . .

Looking . . . to the policy considerations involved, we glean two reasons why Metallurgical's disclosures to others are limited and therefore insufficient to extinguish the secrecy Metallurgical's other evidence has suggested. First, the disclosures were not public announcements; rather, Metallurgical divulged its information to only two businesses with whom it was dealing. This case thus differs from Luccous v. J. C. Kinley Co., 376 S.W.2d 336 (Tex. 1964), in which the court concluded that the design of a device could not be a trade secret because it had been patented — and thus revealed to all the world — before any dealing between the parties. Second, the disclosures were made to further Metallurgical's economic interests. Disclosure to Consarc was made with the hope that Consarc could build the second furnace. A long-standing agreement gave La Floridienne the right, as a licensee, to the information in exchange for royalty payments. Metallurgical therefore revealed its discoveries as part of business transactions by which it expected to profit.

Metallurgical's case would have been stronger had it also presented evidence of confidential relationships with these two companies, but we are unwilling to regard this failure as conclusively disproving the limited nature of the disclosures. Smith correctly points out that Metallurgical bears the burden of showing the existence of confidential relationships. Contrary to Smith's assertion, however, confidentiality is not a requisite; it is only a factor to consider. Whether a disclosure is limited is an issue the resolution of which depends on weighing many facts. The inference from those facts, construed favorably to Metallurgical, is that it wished only to profit from [disclosure of] its secrets to the public. We therefore are unpersuaded by Smith's argument.

Existing law, however, emphasizes other requisites for legal recognition of a trade secret. In Huffines, 314 S.W.2d 763, a seminal case of trade secret law, Texas adopted the widely-recognized pronouncements of the American Law Institute's Restatement of the Law. The Texas Supreme Court quoted the Restatement's definition of a trade secret:

> A trade secret may consist of any formula, pattern, device, or compilation of information which is used in one's business, and which gives him an opportunity to obtain an advantage over competitors who do not know or use it. It may be a chemical compound,

a process of manufacturing, treating or preserving materials, a pattern for a machine or other device or a list of customers.

Id. at 776, quoting Restatement of Torts, §757 Comment b (1939). From this the criterion of value to the holder of the alleged secret arises. . . .

Metallurgical met the burden of showing the value of its modifications. Lawrence Lorman, the company's vice president, testified that the zinc recovery process gave Metallurgical an advantage over its two competitors by aiding in the production of the highest quality reclaimed carbide powder. The quality of the powder, in fact, makes it an alternative to the more costly virgin carbide. Lorman testified that customers regarded Metallurgical's zinc reclaimed powder as a better product than that reclaimed by the coldstream process used by others. This evidence clearly indicates that the modifications that led to the commercial operation of the zinc recovery furnace provided a clear advantage over the competition.

Another requisite is the cost of developing the secret device or process. In Huffines' companion case, K&G Oil, Tool & Service Co. v. G&G Fishing Tool Service, 158 Tex. 594, 314 S.W.2d 782, 790, 117 U.S.P.Q. (BNA) 471 (Tex. 1958), the court recognized the cost involved in developing the device in question; "the record shows . . . that much work and ingenuity have been applied to the development of a practical and successful device." See also Zoecon Industries, 713 F.2d at 1179 ("even if the names and addresses were readily ascertainable through trade journals as the defendants allege, the other information could be compiled only at considerable expense"). No question exists that Metallurgical expended much time, effort, and money to make the necessary changes. It clearly has met the burden of demonstrating the effort involved in making a complex manufacturing process work.

That the cost of devising the secret and the value the secret provides are criteria in the legal formulation of a trade secret shows the equitable underpinnings of this area of the law. It seems only fair that one should be able to keep and enjoy the fruits of his labor. If a businessman has worked hard, has used his imagination, and has taken bold steps to gain an advantage over his competitors, he should be able to profit from his efforts. Because a commercial advantage can vanish once the competition learns of it, the law should protect the businessman's efforts to keep his achievements secret. As is discussed below, this is an area of law in which simple fairness still plays a large role.

We do not say, however, that all these factors need exist in every case. Because each case must turn on its own facts, no standard formula for weighing the factors can be devised. Secrecy is always required, of course, but beyond that there are no universal requirements. In a future case, for example, should the defendant's breach of confidence be particularly egregious, the injured party might still seek redress in court despite the possibility that the subject matter was discovered at little or no cost or that the object of secrecy is not of great value to him. The definition of "trade secret" will therefore be determined by weighing all equitable considerations. It is easy to recognize the possibility of a trade secret here, however, because Metallurgical presented evidence of all three factors discussed above. . . .

COMMENTS AND QUESTIONS

1. Sales of goods in mass retail markets are made to further the seller's economic interests. Under the rationale in *Fourtek*, why do such sales destroy the value of the secret?

2. If the defendants had acquired information about Metallurgical's process from Consarc or La Floridienne — the Metallurgical licensees — would the case have come out differently? Should it matter whether they acquired the information with knowledge that Metallurgical still considered it a trade secret?

3. The categories of information *eligible* for protection as trade secrets are quite expansive. As *Fourtek* makes clear, they include secret combinations of items which by themselves are publicly known. They also include both scientific and technical information and business information, such as customer lists and business plans. Can you think of any type of information that should be ineligible for trade secret protection?

Note that eligibility for protection is only the first of many hurdles required to establish the existence of a trade secret. Many alleged trade secrets that are *eligible* for protection do not in fact receive protection because they do not meet one of the other standards described in this section. Courts have frequently held certain basic ideas or concepts incapable of protection as secrets on the grounds that they were too well known to derive value from secrecy. For example, in Buffets, Inc. v. Klinke, 73 F.3d 765 (9th Cir. 1996), the court held that the plaintiff could not claim its relatively straightforward recipes for barbequed chicken and macaroni and cheese as trade secrets. On the other hand, in Camp Creek Hospitality Inns v. Sheraton Franchise Corp., 139 F.3d 1396 (11th Cir. 1997), the court held that a hotel could protect information about its prices, discounts, and occupancy levels as a trade secret where it was closely guarded information in the industry.

4. Courts have made it clear that strict novelty is not required for trade secret protection. The idea may have occurred to someone before; it may even be in use by another. But if it is not generally known or readily ascertainable to the competitors in an industry, it may still qualify for trade secret protection. One widely cited decision described the standard for protectable ideas as follows:

✳ [U]niqueness in the patent law sense is not an essential element of a trade secret, for the patent laws are designed to encourage invention, whereas trade secret law is designed to protect against a breach of faith. However, the trade secret must "possess at least that modicum of originality which will separate it from everyday knowledge." Cataphote Corporation v. Hudson, 444 F.2d 1313, 1315 (5th Cir. 1971). As stated in an authoritative treatise on this subject:

> As distinguished from a patent, a trade secret need not be essentially new, novel or unique; therefore, prior art is a less effective defense in a trade secret case than it is in a patent infringement case. The idea need not be complicated; it may be intrinsically simple and nevertheless qualify as a secret, unless it is in common knowledge and, therefore, within the public domain.

2 Callman, Unfair Competition, Trademarks and Monopolies § 52.1 (3d ed. 1968).

Forest Laboratories v. The Pillsbury Co., 425 F.2d 621, 624 (7th Cir. 1971). Some courts have gone even further, suggesting that "[a] trade secret may be no more than 'merely a mechanical improvement that a good mechanic can make.'" SI Handling Systems, Inc. v. Heisley, 753 F.2d 1244, 1256 (3d Cir. 1985).

Why not require novelty in order to protect a trade secret? That is, why should the law protect the "secrecy" of a piece of information if others have already discovered it?

5. The Restatement of Torts offers further guidance in determining whether information constitutes a trade secret. It lists six factors to be considered:

- The extent to which the information is known outside the claimant's business.
- The extent to which it is known by employees and others involved in the business.
- The extent of measures taken by the claimant to guard the secrecy of the information.
- The value of the information to the business and its competitors.
- The amount of effort or money expended by the business in developing the information.
- The ease or difficulty with which the information could be properly acquired or duplicated by others.

Each of these factors is open-ended. Together, they provide the basis for an inquiry into how secret the information really is. This is also the point of *Forest Laboratories'* truncated inquiry into novelty. To be protectable, information must not be "generally known" or "readily ascertainable" by competitors in an industry. See Burbank Grease Servs. v. Sokolowski, 693 N.W.2d 89 (Wis. App. 2005), *review granted* 700 N.W.2d 271 (Wis. 2005) (Burbank's list of potential customers was readily ascertainable from the Internet, trade associations, and by asking customers who to contact).

6. Why should we bother to protect secrets that were stumbled upon with little or no investment in research but that happen to have "value"? Does the economic rationale for intellectual property suggest that such secrets will be underproduced absent protection?

7. Why is secrecy required at all? Trade secrets are not misappropriated unless information is taken by improper means or from a confidential relationship. Why aren't those tortious elements enough? It is certainly possible to envision a "misappropriation" tort that punishes diversion of information, whether or not it is secret.[9] Indeed, a number of controversial cases discussed in Chapter 6 have created just such a common law tort. In addition to the cases cited there, see United States Sporting Products v. Johnny Stewart Game Calls, 865 S.W.2d 214 (Tex. Ct. App. 1993) (publicly sold uncopyrighted recordings protectable under a "labor theory"); Note, The "Genetic Message" from the Cornfields of Iowa: Expanding the Law of Trade Secrets, 38 Drake L. Rev. 631 (1989) (describing a similar case involving publicly sold grain).

One possible objection to such a scheme is that it may chill the legitimate acquisition of information from competitors. But the only information protected by a misappropriation tort that is not also protected by trade secret law is public information. Since it is public, the need for competitors to acquire it directly from another company or through dubious means is presumably low.[10]

---

9. For suggestions along these lines, see Dennis Karjala, Misappropriation as a Third Intellectual Property Paradigm, 94 Colum. L. Rev. 2594 (1994); J. H. Reichman, Legal Hybrids Between the Patent and Copyright Paradigms, 94 Colum. L. Rev. 2432 (1994).

10. Another reason for limiting common law protection to secret information may be concern over preemption of state common law by the federal intellectual property laws. While the course of the law is not completely clear, federal courts have generally held that state laws that create property rights in public

But there may be a wide gulf between "secret" information and "public" information. If three large companies all use the same process but guard it closely, is it a secret? If a company guards a process closely as a secret, but an account of the process is available in an obscure published source, does the company have a protectable trade secret? Does it matter how obscure the published source is, if the defendant in fact steals the information from the plaintiff rather than going to the public source? What theory of trade secrets would support a finding of liability in such a case?

In Rohm & Haas Co. v. Adco Chemical Co., 689 F.2d 424 (3d Cir. 1982), defendant Harvey was a former Rohm & Haas employee who was hired by Adco to duplicate a process for producing "paint delivery vehicles," the chemicals added to paint that allow it to be applied to surfaces easily. There seems no question in the case that Harvey did in fact memorize the plaintiff's formula and take it to Adco. In their defense, the defendants offered evidence that Rohm & Haas's "secret" process was in fact disclosed in a series of prior publications. The court nonetheless concluded that it was a protectable trade secret, in part because the defendants *did not in fact* obtain the information from those publications. Why should this matter? Certainly the defendants' conduct proves that they took the information from the plaintiff, but that should be irrelevant if the plaintiff does not have a protectable secret. Has the *Rohm & Haas* court in effect abolished the requirement of secrecy by requiring only that the defendants in fact obtain the information from the plaintiff?

Rohm & Haas is at the center of a critical debate in trade secret law. At issue is whether information must actually be "known" to competitors or merely be "knowable" for the court to conclude that it is not secret. Many courts following the Restatement of Torts have taken the former view, as *Rohm & Haas* does. This view gives broad scope to trade secret law, because it allows a plaintiff to protect information that could have been acquired properly but in fact was not. It also underscores the unfair competition rationale for trade secret protection — the problem is not that the defendant acquired the information at all, but the way in which it was acquired.

In a significant break with the old Restatement rule, the Uniform Trade Secrets Act provides that information is not a trade secret if it is "generally known" or "readily ascertainable by proper means."[11] Under this view, once a secret is readily available through public sources, it loses all trade secret protection. At this point, the defendant is free to obtain the information from the public source *or from the plaintiff herself*. See Restatement (Third) of Unfair Competition § 39 Comment f, at 433 ("When the information is readily ascertainable from such [public] sources, however, actual resort to the public domain is a formality that should not determine liability.").

---

information are preempted by the patent and copyright laws. The policy being served by preemption is to protect the balance struck by federal intellectual property laws. See, e.g., Synercom Technology v. University Computing Co., 474 F. Supp. 37 (N.D. Tex. 1979) (state unfair competition laws cannot prevent copying of public information).

There are two rationales for such preemption. First, it may be that the federal laws reflect a judgment that unpatentable inventions ought to belong to the public. On this rationale, see Bonito Boats v. Thunder Craft Boats, 489 U.S. 141 (1989). Second, preemption may serve to erect barriers between different types of intellectual property protection, "channeling" inventions into one or another form of protection. We discuss federal preemption in more detail in Chapter 6.

11. Some states, however, refused to accede to this change and have modified the language of the act to remove the reference to "readily ascertainable" information. California is one prominent example. See Cal. Civ. Code § 3426.1(d)(1); ABBA Rubber Co. v. Seaquist, 286 Cal. Rptr. 518 (Ct. App. 1991) (customer list qualified for trade secret protection even though information was available in trade directories). California law does permit proof of ready ascertainability as a defense, however.

Even jurisdictions that follow the Restatement of Torts view place some limit on what can qualify for trade secret protection. If information is generally known to the public, or even within a specialized industry, it does not qualify for protection. No company can claim that "$E = mc^2$" is a trade secret, for example, even if it keeps the formula under lock and key, and even if the defendant steals it from the company rather than obtaining it elsewhere. See Spring Indus. v. Nicolozakes, 58 U.S.P.Q.2d 1794 (Ohio App. 2000) (information on gravel mining not a trade secret despite efforts to keep it confidential). We consider the rather different issue of whether two parties could *agree* to treat the formula as a secret in section D, below.

8. What relevance do the competing theories of trade secret protection have for the known vs. knowable debate? If the purpose of trade secret law is to promote innovation, why should a company be entitled to protect an idea that already exists in the literature and can readily be found there? On the other hand, doesn't the tort theory compel the conclusion that a company should be able to protect an idea against being stolen *from it*, even if the idea is commonly known elsewhere?

9. The Restatement factors used in *Forest Labs* seem to focus on how widely the information is known, suggesting that information used by many people cannot be a secret. The secrecy requirement, therefore, may reflect a policy judgment in favor of the distribution of information once it has reached a certain "critical mass." *Metallurgical Industries*, by contrast, focuses on attempts to prevent disclosure of the information. This approach suggests that even widely known information (or published information, as that in *Metallurgical Industries* was) may receive trade secret protection.

Compare the definition of a trade secret offered in the new Restatement (Third) of Unfair Competition:

> A trade secret is any information that can be used in the operation of a business or other enterprise and that is sufficiently valuable and secret to afford an actual or potential economic advantage over others.

Restatement (Third) of Unfair Competition § 39. Under the new Restatement, many different companies can possess the same information and each protect it as a secret. Id. at illustration 1.[12] In a comment interpreting this provision, the drafters note that "[t]he concept of a trade secret as defined in this Section is intended to be consistent with the definition of 'trade secret' in § 1(4) of the [Uniform Trade Secrets] Act." Is it?

---

12. The Restatement does note, however, that "[w]hen information is no longer sufficiently secret to qualify for protection as a trade secret, its use should not serve as a basis for the imposition of liability," id. at 433 Comment *f*, suggesting that at some point information is not secret even though companies protect it as if it were.

PROBLEMS

Problem 2-1. Company *X* possesses a valuable piece of information about the process for making its product. That information is not known at all outside company *X*. Suppose *X* discloses the information to two companies, *A* and *B*. *A* receives the information in confidence, and subject to a written agreement that it will not use or disclose the information outside the bounds of the relationship. *B* receives the information without any restrictions whatever on its use. Does *X* have a protectable trade secret that it can assert against *A*? Against *B*? Against *C*, who steals the information from *X*'s computer network?

Does your answer change if *X*, *A*, *B*, and *C* are the only companies in the industry?

Problem 2-2. StartUp, Inc., is the only participant in a new market. The market is based on a product for which StartUp has a nonexclusive license from the inventor; that is, StartUp cannot prevent others from obtaining a similar license. Nonetheless, StartUp has exhaustively researched the demand for the product, has concluded that a market exists, and has worked to stimulate demand. As a result, it has both "made" a market for the product and developed a comprehensive list of customers.

Thaddeus, a sales representative for StartUp, leaves to found his own company. He gets a license to make the product from its inventor. He takes with him from StartUp the customer list he worked with as an employee, his personal knowledge of and contacts with specific customers, and StartUp's knowledge of the market. StartUp sues Thaddeus for misappropriation of trade secrets. What result?

Problem 2-3. Research Co. is a major pharmaceutical company working on a cure for certain types of cancer. Derek is a molecular biologist employed by Research. After several years on the job, Derek leaves Research for Conglomerate, Inc., another pharmaceutical company, which has decided to work on the same cancer cure. At the time Derek leaves, Research has not been successful in finding a cancer cure. However, as a result of his work at Research, Derek is able to help Conglomerate avoid several unproductive avenues of research. Aided in part by this knowledge, Conglomerate (using scientists other than Derek) develops a cancer cure before Research. Research sues Conglomerate, alleging misappropriation of trade secrets. Does Research have a case?

Problem 2-4. The Church of True Belief is a religious group founded around a set of closely guarded scriptural materials supposedly handed down to Church's elders from Church's deity. After a bitter theological dispute, a group of adherents leaves the church to form the House of Absolute Belief. They take with them a copy of Church's confidential scriptures, which they rely on in gaining adherents and founding the new House. Church sues House for misappropriation of trade secrets.

At trial, the issue is whether the scriptures qualify as a trade secret. The evidence indicates that the scriptures had never before been removed from the confines of Church, that both Church and House are tax-exempt nonprofit organizations which rely on donations for their funding, and that Church (but not House) rations access to the scriptures in proportion to the size of an adherent's donation.

Can the scriptures qualify as a trade secret? Does your answer depend on whether the governing law is the UTSA, the Restatement of Torts, or the Restatement (Third) of Unfair Competition?

2. Reasonable Efforts to Maintain Secrecy

Besides the existence of a trade secret, plaintiffs must show under the Uniform Act that they have taken "reasonable measures" to protect the secrecy of their idea. Certainly, a plaintiff cannot publicly disclose the secret and still expect to protect it. But precautions must go further than that. Generally, they must include certain efforts to prevent theft or use of the idea by former employees.

≡ ### *Rockwell Graphic Systems, Inc. v. DEV Industries, Inc.*
United States Court of Appeals for the Seventh Circuit
925 F.2d 174 (7th Cir. 1991)

POSNER, Circuit Judge:

This is a suit for misappropriation of trade secrets. Rockwell Graphic Systems, a manufacturer of printing presses used by newspapers, and of parts for those presses, brought the suit against DEV Industries, a competing manufacturer, and against the president of DEV, who used to be employed by Rockwell. . . .

When we said that Rockwell manufactures both printing presses and replacement parts for its presses — "wear parts" or "piece parts," they are called — we were speaking approximately. Rockwell does not always manufacture the parts itself. Sometimes when an owner of one of Rockwell's presses needs a particular part, or when Rockwell anticipates demand for the part, it will subcontract the manufacture of it to an independent machine shop, called a "vendor" by the parties. When it does this it must give the vendor a "piece part drawing" indicating materials, dimensions, tolerances, and methods of manufacture. Without that information the vendor could not manufacture the part. Rockwell has not tried to patent the piece parts. It believes that the purchaser cannot, either by inspection or by "reverse engineering" (taking something apart in an effort to figure out how it was made), discover how to manufacture the part; to do that you need the piece part drawing, which contains much information concerning methods of manufacture, alloys, tolerances, etc. that cannot be gleaned from the part itself. So Rockwell tries — whether hard enough is the central issue in the case — to keep the piece part drawings secret, though not of course from the vendors; they could not manufacture the parts for Rockwell without the drawings. DEV points out that some of the parts are for presses that Rockwell no longer manufactures. But as long as the presses are in service — which can be a very long time — there is a demand for replacement parts.

Rockwell employed Fleck and Peloso in responsible positions that gave them access to piece part drawings. Fleck left Rockwell in 1975 and three years later joined DEV as its president. Peloso joined DEV the following year after being fired by Rockwell when a security guard caught him removing piece part drawings from Rockwell's plant. This suit was brought in 1984, and pretrial discovery by Rockwell turned up 600 piece part drawings in DEV's possession, of which 100 were Rockwell's. DEV claimed to have obtained them lawfully, either from customers of Rockwell or from Rockwell vendors, contrary to Rockwell's claim that either Fleck and Peloso stole them when they were employed by it or DEV obtained them in some other unlawful manner, perhaps from a vendor who violated his confidentiality agreement with Rockwell. Thus far in the litigation DEV has not been able to show which customers or vendors lawfully supplied it with Rockwell's piece part drawings.

The defendants persuaded the magistrate and the district judge that the piece part drawings weren't really trade secrets at all, because Rockwell made only perfunctory efforts to keep them secret. Not only were there thousands of drawings in the hands of the vendors; there were thousands more in the hands of owners of Rockwell presses, the customers for piece parts. The drawings held by customers, however, are not relevant. They are not piece part drawings, but assembly drawings. . . . An assembly drawing shows how the parts of a printing press fit together for installation and also how to integrate the press with the printer's other equipment. Whenever Rockwell sells a printing press it gives the buyer assembly drawings as well. These are the equivalent of instructions for assembling a piece of furniture. Rockwell does not claim that they contain trade secrets. It admits having supplied a few piece part drawings to customers, but they were piece part drawings of obsolete parts that Rockwell has no interest in manufacturing and of a safety device that was not part of the press as originally delivered but that its customers were clamoring for; more to the point, none of these drawings is among those that Rockwell claims DEV misappropriated.

. . . DEV's main argument is that Rockwell was impermissibly sloppy in its efforts to keep the piece part drawings secret.

On this, the critical, issue, the record shows the following. (Because summary judgment was granted to DEV, we must construe the facts as favorably to Rockwell as is reasonable to do.) Rockwell keeps all its engineering drawings, including both piece part and assembly drawings, in a vault. Access not only to the vault, but also to the building in which it is located, is limited to authorized employees who display identification. These are mainly engineers, of whom Rockwell employs 200. They are required to sign agreements not to disseminate the drawings, or disclose their contents, other than as authorized by the company. An authorized employee who needs a drawing must sign it out from the vault and return it when he has finished with it. But he is permitted to make copies, which he is to destroy when he no longer needs them in his work. The only outsiders allowed to see piece part drawings are the vendors (who are given copies, not originals). They too are required to sign confidentiality agreements, and in addition each drawing is stamped with a legend stating that it contains proprietary material. Vendors, like Rockwell's own engineers, are allowed to make copies for internal working purposes, and although the confidentiality agreement that they sign requires the vendor to return the drawing when the order has been filled, Rockwell does not enforce this requirement. The rationale for not enforcing it is that the vendor will need the drawing if Rockwell reorders the part. Rockwell even permits unsuccessful bidders for a piece part contract to keep the drawings, on the theory that the high bidder this round may be the low bidder the next. But it does consider the ethical standards of a machine shop before making it a vendor, and so far as appears no shop has ever abused the confidence reposed in it.

The mere fact that Rockwell gave piece part drawings to vendors — that is, disclosed its trade secrets to "a limited number of outsiders for a particular purpose" — did not forfeit trade secret protection. On the contrary, such disclosure, which is often necessary to the efficient exploitation of a trade secret, imposes a duty of confidentiality on the part of the person to whom the disclosure is made. But with 200 engineers checking out piece part drawings and making copies of them to work from, and numerous vendors receiving copies of piece part drawings and copying them, tens of thousands of copies of these drawings are floating around outside Rockwell's vault, and many of these outside the company altogether. Although the magistrate and the district judge based their conclusion that Rockwell had not made adequate efforts to maintain secrecy in part at least on the irrelevant fact that it took no efforts at all to keep its assembly drawings secret, DEV in defending the judgment that it obtained in the district court argues that Rockwell failed to take adequate measures to keep even the piece part drawings secret. Not only did Rockwell not limit copying of those drawings or insist that copies be returned; it did not segregate the piece part drawings from the assembly drawings and institute more secure procedures for the former. So Rockwell could have done more to maintain the confidentiality of its piece part drawings than it did, and we must decide whether its failure to do more was so plain a breach of the obligation of a trade secret owner to make reasonable efforts to maintain secrecy as to justify the entry of summary judgment for the defendants.

The requirement of reasonable efforts has both evidentiary and remedial significance, and this regardless of which of the two different conceptions of trade secret protection prevails. . . . [T]he two different conceptions of trade secret protection are better described as different emphases. The first emphasizes the desirability of deterring efforts that have as their sole purpose and effect the redistribution of wealth from one firm to another. The second emphasizes the desirability of encouraging inventive activity by protecting its fruits from efforts at appropriation that are, indeed, sterile wealth-redistributive — not productive — activities. The approaches differ, if at all, only in that the second does not limit the class of improper means to those that fit a preexisting pigeonhole in the law of tort or contract or fiduciary duty — and it is by no means clear that the first approach assumes a closed class of wrongful acts, either.

Under the first approach, at least if narrowly interpreted so that it does not merge with the second, the plaintiff must prove that the defendant obtained the plaintiff's trade secret by a wrongful act, illustrated here by the alleged acts of Fleck and Peloso in removing piece part drawings from Rockwell's premises without authorization, in violation of their employment contracts and confidentiality agreements, and using them in competition with Rockwell. Rockwell is unable to prove directly that the 100 piece part drawings it got from DEV in discovery were stolen by Fleck and Peloso or obtained by other improper means. But if it can show that the probability that DEV could have obtained them otherwise — that is, without engaging in wrongdoing — is slight, then it will have taken a giant step toward proving what it must prove in order to recover under the first theory of trade secret protection. The greater the precautions that Rockwell took to maintain the secrecy of the piece part drawings, the lower the probability that DEV obtained them properly and the higher the probability that it obtained them through a wrongful act; the owner had taken pains to prevent them from being obtained otherwise.

Under the second theory of trade secret protection, the owner's precautions still have evidentiary significance, but now primarily as evidence that the secret has real value. For the precise means by which the defendant acquired it is less important under the second theory, though not completely unimportant; remember that even the second theory allows the unmasking of a trade secret by some means, such as reverse engineering. If Rockwell expended only paltry resources on preventing its piece part drawings from falling into the hands of competitors such as DEV, why should the law, whose machinery is far from costless, bother to provide Rockwell with a remedy? The information contained in the drawings cannot have been worth much if Rockwell did not think it worthwhile to make serious efforts to keep the information secret.

The remedial significance of such efforts lies in the fact that if the plaintiff has allowed his trade secret to fall into the public domain, he would enjoy a windfall if permitted to recover damages merely because the defendant took the secret from him, rather than from the public domain as it could have done with impunity. Brunswick Corp. v. Outboard Marine Corp., supra, 79 Ill. 2d at 479, 404 N.E.2d at 207; Van Products Co. v. General Welding & Fabricating Co., 419 Pa. 248, 267-68, 213 A.2d 769, 779-80 (1965) (repudiating the interpretation of Pennsylvania law that this court had adopted in Smith v. Dravo Corp., 203 F.2d 369, 374-75 (7th Cir. 1953)). It would be like punishing a person for stealing property that he believes is owned by another but that actually is abandoned property. If it were true, as apparently it is not, that Rockwell had given the piece part drawings at issue to customers, and it had done so without requiring the customers to hold them in confidence, DEV could have obtained the drawings from the customers without committing any wrong. The harm to Rockwell would have been the same as if DEV had stolen the drawings from it, but it would have had no remedy, having parted with its rights to the trade secret. This is true whether the trade secret is regarded as property protected only against wrongdoers or (the logical extreme of the second conception, although no case — not even *Christopher* — has yet embraced it and the patent statute may preempt it) as property protected against the world. In the first case, a defendant is perfectly entitled to obtain the property by lawful conduct if he can, and he can if the property is in the hands of persons who themselves committed no wrong to get it. In the second case the defendant is perfectly entitled to obtain the property if the plaintiff has abandoned it by giving it away without restrictions.

It is easy to understand therefore why the law of trade secrets requires a plaintiff to show that he took reasonable precautions to keep the secret a secret. If analogies are needed, one that springs to mind is the duty of the holder of a trademark to take reasonable efforts to police infringements of his mark, failing which the mark is likely to be deemed abandoned, or to become generic or descriptive (and in either event be unprotectable). The trademark owner who fails to police his mark both shows that he doesn't really value it very much and creates a situation in which an infringer may have been unaware that he was using a proprietary mark because the mark had drifted into the public domain, much as DEV contends Rockwell's piece part drawings have done.

But only in an extreme case can what is a "reasonable" precaution be determined on a motion for summary judgment, because the answer depends on a balancing of costs and benefits that will vary from case to case and so require estimation and measurement by persons knowledgeable in the particular field of endeavor involved. On the one hand, the more the owner of the trade secret spends on preventing the secret from leaking out, the more he demonstrates that the secret has real value deserving of legal protection, that he really was hurt as a result of the misappropriation of it, and that there really was misappropriation. On the other hand, the more he spends, the higher his costs. The costs can be indirect as well as direct. The more Rockwell restricts access to its drawings, either by its engineers or by the vendors, the harder it will be for either group to do the work expected of it. Suppose Rockwell forbids any copying of its drawings. Then a team of engineers would have to share a single drawing, perhaps by passing it around or by working in the same room, huddled over the drawing. And how would a vendor be able to make a piece part — would Rockwell have to bring all that work in house? Such reconfigurations of patterns of work and production are far from costless; and therefore perfect security is not optimum security.

There are contested factual issues here, bearing in mind that what is reasonable is itself a fact for purposes of Rule 56 of the civil rules. Obviously Rockwell took some precautions, both physical (the vault security, the security guards — one of whom apprehended Peloso in flagrante delicto) and contractual, to maintain the confidentiality of its piece part drawings. Obviously it could have taken more precautions. But at a cost, and the question is whether the additional benefit in security would have exceeded that cost. We do not suggest that the question can be answered with the same precision with which it can be posed, but neither can we say that no reasonable jury could find that Rockwell had done enough and could then go on to infer misappropriation from a combination of the precautions Rockwell took and DEV's inability to establish the existence of a lawful source of the Rockwell piece part drawings in its possession.

This is an important case because trade secret protection is an important part of intellectual property, a form of property that is of growing importance to the competitiveness of American industry. Patent protection is at once costly and temporary, and therefore cannot be regarded as a perfect substitute. If trade secrets are protected only if their owners take extravagant, productivity-impairing measures to maintain their secrecy, the incentive to invest resources in discovering more efficient methods of production will be reduced, and with it the amount of invention. . . .

Reversed and remanded.

COMMENTS AND QUESTIONS

1. There is an intuitive relationship between the existence of a secret and reasonable efforts to protect a secret. After all, if something is not a secret, there would not seem to be any point to protecting it. And the fact that an idea is well protected may be evidence that it is in fact a secret. Nonetheless, the requirements are conceptually distinct. Information in the public domain cannot be turned into a secret merely by treating it like a secret, a point that lawyers and even courts sometimes forget. This distinction is made clear in the Uniform Trade Secrets Act, which defines a trade secret as information that is both "not generally known" *and* the subject of reasonable efforts to maintain secrecy. UTSA § 1(4).

Consider whether the opinion in *Rockwell* conflates these two into a single requirement. The court seems to emphasize the evidentiary significance of the precautions Rockwell took in proving misappropriation. Since it was clear (to the court, anyway) that the DEV employees did in fact take the information from Rockwell, the court did not consider the precautions to be that important. This approach was apparently followed in the Restatement (Third) of Unfair Competition. Unlike the Uniform Act, the new Restatement does not contain a requirement that plaintiffs take reasonable precautions to protect their secrets. The comment to section 39 takes the position that, while "[p]recautions taken to maintain the secrecy of information are relevant in determining whether the information qualifies for protection as a trade secret," "if the value and secrecy of the information are clear, evidence of specific precautions taken by the trade secret owner may be unnecessary." Restatement (Third) of Unfair Competition § 39, Comment *g*, at 435-436.

Contrast *Rockwell* with the Minnesota Supreme Court's decision in Electro-Craft Corp. v. Controlled Motion, Inc., 332 N.W.2d 890 (Minn. 1983), a case that also involved information taken by former employees and used in starting a competing company. The court found that the information the employees took was not generally known or readily ascertainable in the industry. However, it found that the information did not constitute a trade secret:

> (c) Reasonable efforts to maintain secrecy. It is this element upon which [plaintiff Electro-Craft Corp., or "ECC"]'s claim founders. The district court found that, even though ECC had no "meaningful security provisions," ECC showed an intention to keep its data and processes secret. This finding does not bear upon the statutory requirement that ECC use "efforts that are reasonable under the circumstances to maintain . . . secrecy." Minn. Stat. § 325C.01, subd. 5(ii). . . . [E]ven under the common law, more than an "intention" was required — the plaintiff was required to show that it had manifested that intention by making some effort to keep the information secret.
>
> This element of trade secret law does not require maintenance of absolute secrecy; only partial or qualified secrecy has been required under the common law. What is actually required is conduct which will allow a court acting in equity to enforce plaintiff's rights. . . .
>
> In the present case, even viewing the evidence most favorably to the findings below, we hold that ECC did not meet its burden of proving that it used reasonable efforts to maintain secrecy as to [the subject matter of the suit, a product called the ECC 1125]. We acknowledge that ECC took minimal precautions in screening its Handbook and publications for confidential information and by requiring some of its employees to sign a confidentiality agreement, but these were not enough.
>
> First, ECC's physical security measures did not demonstrate any effort to maintain secrecy. By "security" we mean the protection of information from discovery by outsiders. Security was lax in this case. For example, the main plant had a few guarded entrances, but seven unlocked entrances existed without signs warning of limited access. Employees were at one time required to wear badges, but that system was abandoned by the time of the events giving rise to this case. The same was generally true of the Amery, Wisconsin plant where ECC 1125 and brushless motors were manufactured. One sign was posted at each plant, however, marking the research and development lab at Hopkins and the machine shop at Amery as restricted to "authorized personnel." Discarded drawings and plans for motors were simply thrown away, not destroyed. Documents such as motor drawings were not kept in a central or locked location, although some design notebooks were kept locked.
>
> The relaxed security by itself, however, does not preclude a finding of reasonable efforts by ECC to maintain secrecy. Other evidence did not indicate that industrial espionage is a major problem in the servo motor industry. Therefore, "security" measures may not have been needed, and the trial court could have found trade secrets if ECC had taken other reasonable measures to preserve secrecy.

However, ECC's "confidentiality" procedures were also fatally lax, and the district court was clearly in error in finding ECC's efforts to be reasonable. By "confidentiality" in this case we mean the procedures by which the employer signals to its employees and to others that certain information is secret and should not be disclosed. Confidentiality was important in this case, for testimony demonstrated that employees in the servo motor business frequently leave their employers in order to produce similar or identical devices for new employers. ECC has hired many employees from other corporations manufacturing similar products.[13] If ECC wanted to prevent its employees from doing the same thing, it had an obligation to inform its employees that certain information was secret.

ECC's efforts were especially inadequate because of the non-intuitive nature of ECC's claimed secrets here. The dimensions, etc., of ECC's motors are not trade secrets in as obvious a way as a "secret formula" might be. ECC should have let its employees know in no uncertain terms that those features were secret.

Instead, ECC treated its information as if it were not secret. None of its technical documents were marked "Confidential," and drawings, dimensions and parts were sent to customers and vendors without special marking. Employee access to documents was not restricted. ECC never issued a policy statement outlining what it considered to be secret. Many informal tours were given to vendors and customers without warnings as to confidential information. Further, two plants each had an "open house" at which the public was invited to observe manufacturing processes. . . .

In summary, ECC has not met its burden of proof in establishing the existence of any trade secrets. The evidence does not show that ECC was ever consistent in treating the information here as secret.

In a part of the opinion not reprinted here, the court noted that the information in question was not in fact known at all outside ECC. Given that, why shouldn't the company be able to prevent its employees from using the information they acquired there? Should the laxity of ECC's precautions matter if no one other than the defendants in fact took advantage of it?

2. In *Rockwell*, the court apparently assumed that the manufacturer was in a confidential relationship with the subcontractors that it sent drawings to. Is it reasonable to assume that there was an implied relationship of confidentiality in the absence of an express agreement? Two decisions by the Ninth Circuit set the parameters of this debate. In Entertainment Research Group v. Genesis Creative Group, 122 F.3d 1211 (9th Cir. 1997), the court found no confidential relationship between a manufacturer of inflatable costumes and the marketing firm hired to distribute them. By contrast, in IMAX Corp. v. Cinema Technologies, 152 F.3d 1161 (9th Cir. 1998), the court concluded that the proprietor of an IMAX movie theater had a duty to IMAX to protect the specifications of the projector as trade secrets from a competitor who was attempting to reverse engineer the IMAX technology.

3. How much effort should be required of trade secret owners? Obviously, the best way to protect a secret is not to tell anyone at all. In the modern commercial world, however, this is normally impractical. Companies with trade secrets must tell the secret to their employees, their business partners, and often their distributors and customers as well. But the risk of *inadvertent* use or disclosure can be reduced in

---

13. One ECC employee actually prided himself on the information he had brought with him from his former employer. One day, just before that employee left ECC to join another company, the president of ECC found him copying documents after hours. ECC never questioned the employee or warned him or his new employer that certain information was confidential.

a number of ways: for example, by requiring employees, licensees, and even customers to sign confidentiality agreements; by investing in physical security measures against theft, such as fences, safes, and guards; and by designing products themselves so that they do not reveal their secrets upon casual (or even detailed) inspection.

4. Will reasonable precautions *always* be a question of fact? Or are certain activities (publishing a secret formula, for example) so inconsistent with trade secret protection that they automatically preclude a successful trade secret suit? See section 3, discussing disclosure of trade secrets.

One issue that arises frequently is whether companies which sell products on the open market that embody their trade secret have disclosed the secret. As we shall see, customers who buy a product on the open market are entitled to break it apart to see how it works. This process is called "reverse engineering" the product. Trade secret law does not protect owners against legitimate purchasers who discover the secret through reverse engineering, absent a valid nondisclosure agreement. But does the possibility that a product might be reverse-engineered foreclose *any* trade secret protection, even against people who have not actually reverse-engineered the product? At least one court has said no. In Data General Corp. v. Grumman Systems Support Corp., 825 F. Supp. 340, 359 (D. Mass. 1993), the court upheld a jury's verdict that Grumman had misappropriated trade secrets contained in object code form in Data General's computer program, despite the fact that many copies of the program had been sold on the open market. The court reasoned:"With the exception of those who lawfully licensed or unlawfully misappropriated MV/ADEX, Data General enjoyed the exclusive use of MV/ADEX. Even those who obtained MV/ADEX and were able to *use* MV/ADEX were unable to discover its trade secrets because MV/ADEX was distributed only in its object code form, which is essentially unintelligible to humans."

Data General suggests that reasonable efforts to protect the secrecy of an idea contained in a commercial product — such as locks, black boxes, or the use of unreadable code — may suffice to maintain trade secret protection even after the product itself is widely circulated. Does this result make sense?

On the basis of the reasoning in *Data General* — that the source code, which contained the secrets, was not widely disclosed — would the distribution of the Data General software in object code form to the general public constitute misappropriation of those trade secrets?

5. If the idea is to encourage investment in trade secrets, why require any degree of "reasonable precautions" at all? One sometimes hears in this regard that *all* "fencing" expenditures are inefficient. Cf. Edmund W. Kitch, The Law and Economics of Rights in Valuable Information, 9 J. Legal Stud. 683 (1980) (arguing that reasonable precautions make sense only as evidence of the existence of a trade secret). Why not simply require explicit notice — large neon signs, stamps on all documents, or publication of a secrecy policy — in place of physical precautions? For a suggestion that proceeds along these lines to some extent, see J. H. Reichman, Legal Hybrids Between the Patent and Copyright Paradigms, 94 Colum. L. Rev. 2432 (1994).

Is it appropriate to punish the creator of a valuable secret who has not invested in "reasonable precautions"? What value is there in a legal rule that requires investment in precautions up to the level that would be rational *in the absence of* the legal rule? Are prospective trade secret thieves actually encouraged by the reasonable precautions argument to steal ideas when they observe a lapse in security, and does this rule give them an incentive to search out such lapses? Professor Kitch asks the related question of why these expenses should be required in addition to the expense of bringing a trade secret lawsuit.

David Friedman, William Landes & Richard Posner, Some Economics of Trade Secret Law, 5 J. Econ. Perspectives 61, 67 (1991), makes the related point that a trade secret cause of action which yields a legal remedy ought to be available when it is cheaper than the physical precautions that would be necessary to protect a piece of information. They also state:

> Where on the contrary the social costs of enforcing secrecy through the legal system would be high, the benefits of shared information are likely to exceed the net benefits of legal protection.

Should it be a defense to a trade secret action that the plaintiff could more easily have protected the secret through physical precautions?

Kitch notes by way of analogy that we do not prohibit criminal complaints for larceny just because a property owner was careless. (On the other hand, many states do reduce recovery in tort suits on the basis of "comparative negligence.") He also suggests that reasonable precautions are required only to put prospective infringers on notice about the existence of a right and to serve as evidence of the fact that the secret is worth protecting legally. The fencing thus serves a notice function, akin to "marking" products with patent numbers, copyright symbols, or trademark symbols. Kitch, *supra*, at 698.

The Court in duPont & Co. v. Christopher, 431 F.2d 1012 (5th Cir. 1970), *cert. denied*, 400 U.S. 1024 (1971), pays significant attention to the role of fencing costs in trade secrets suits. We will return to that case when we consider misappropriation of trade secrets in section C.

One might imagine other rationales for requiring reasonable protection. For example, one might treat the requirement of reasonable precautions as serving a gatekeeper function that helps to weed out frivolous trade secret claims by requiring evidence of investment by the plaintiff in protecting the secret.

PROBLEMS

Problem 2-5. Smith, a bar owner in rural Alabama, develops by accident one night the relatively simple formula for a new alcoholic beverage. The drink is simply a mixture of three common ingredients. Smith begins selling the drink, which he calls "Mobile Mud," in his bar. However, he instructs his bartenders not to reveal the formula to anyone and has them premix "Mud" in the back of the bar, out of sight of customers. Smith is outraged when he learns that Jimmy Dean, an international distributor of alcoholic beverages, has copied his formula and is marketing it under a different name. At trial, Dean employees and independent experts unanimously testify that it is possible for someone with experience in the beverage industry to determine the formula for Mud by looking at, smelling and tasting the drink.

Has Smith taken reasonable precautions? What more could he have done to protect the "secret formula" of Mobile Mud? Is the secret so obvious to consumers that selling the product on the open market destroys protection? Does your opinion of the case change if you learn that Dean's representative went to Smith's bar and bribed a bartender to disclose the formula?

Problem 2-6. MidContinent is a small manufacturer of signs and decals. It has only five employees, two of whom are father and son and two more of whom are family friends. The company describes itself as having a "relaxed, congenial" working atmosphere. In order to avoid what the president considers excessive formality, the employees have never been required to sign confidentiality agreements, and documents kept within the company aren't stamped confidential. The company has never conducted "exit interviews" or instructed its employees about trade secrecy. According to the company president, "we trust our employees, and that trust has never been misplaced." On the other hand, the company does take certain steps to keep outsiders from accessing its customer lists and its adhesive manufacturing process. And there is little history of economic espionage in the decal-manufacturing business.

Has MidContinent taken reasonable efforts to protect its secrets? Should it matter whether the party accused of stealing those secrets is an employee or an outsider?

3. Disclosure of Trade Secrets

It is axiomatic that public disclosure of a trade secret destroys the "secret," and therefore ends protection forever. The corollary to this rule is that as long as a trade secret remains secret, it is protectable. Thus, trade secrets do not last for a specific term of years but continue indefinitely until the occurence of a particular event — the public disclosure of the secret.

Disclosure of a once-protected trade secret can occur in a number of ways:

- First, a trade secret owner may publish the secret, whether in an academic journal or any other forum. In that case, secrecy is lost, at least provided the publication is accessible to those interested in the subject matter. This loss might reasonably be considered a substantial disincentive to publication of scientific or technical advances. But publication of secret information regularly occurs, either because the inventors have not thought through the consequences of their actions or because the value or prestige of first publication is deemed to outweigh the potential loss of commercial trade secret protection.

  One common form of disclosure is the publication of an issued patent. Because (as we shall see) patent law requires the public disclosure of an invention with sufficient specificity to enable one of ordinary skill in the art to make it, obtaining a patent on an invention destroys trade secret protection. See Ferroline Corp. v. General Aniline & Film Corp., 207 F.2d 912 (7th Cir. 1953), *cert. denied,* 347 U.S. 953 (1954). Thus an inventor must "elect" either patent or trade secret protection, for the two cannot protect the same invention simultaneously.

The Federal Circuit applied a notable exception to this seemingly absolute rule in Rhone-Poulenc Agro v. DeKalb Genetics Corp., 272 F.3d 1335 (Fed. Cir. 2001). In that case, the defendant had stolen the plaintiff's trade secrets and published them in its own patent application. The court concluded that in those circumstances the trade secret owner hadn't had the opportunity to "elect" to give up trade secret protection, and so ruled that the publication of the defendant's patent hadn't disclosed plaintiff's trade secrets. This result seems equitable to the plaintiff. But does it really comport with the principle that information must be secret to be protected? For a contrary rule, see Evans v. General Motors, 125 F.3d 1448 (Fed. Cir. 1997) (holding that publication by a thief started the one-year clock running for patenting an invention, and reasoning that because the trade secret owner knew of the theft he could have acted to seek patent protection within a reasonable period of time).

- Second, a trade secret owner may in some cases disclose the secret by selling a commercial product that embodies the secret. As one court explained, both the Restatement of Torts and the Uniform Trade Secrets Act "necessarily compel the conclusion that a trade secret is protectable only so long as it is kept secret by the party creating it. If a so-called trade secret is fully disclosed by the products produced by use of the secret then the right to protection is lost." Vacco Indus. v. Van den Berg, 6 Cal. Rptr. 2d 602, 611 (Ct. App. 1992) (citations omitted). Further, disclosure may occur even without sale of the product itself, if the secret is disclosed freely and without restriction during the manufacturing or development processes.

  However, sales of a product to the public do not necessarily disclose a trade secret simply because the product embodies the trade secret. Rather, the question is whether the secret is apparent from the product itself. Secrets that are apparent to the buyers of a product are considered disclosed by the product, but secrets contained in undecipherable form within the product (such as object code in a computer program) are considered secret even when the product is sold.

  Of what relevance is the motivation behind the disclosure of a secret? Recall that in *Metallurgical Industries,* the court found the fact that Metallurgical had disclosed its secrets only for profit to weigh in favor of trade secret status. Why should this be the case? On the one hand, licensing is evidence that a secret has value and is worth protecting. On the other hand, one could argue that the fact that a secret holder has sold its information for profit suggests that it is not trying to keep this information secret at all but rather is attempting to profit from its disclosure. Which of these arguments you find persuasive may depend on the view you take of the reasons for trade secret protection.

- Third, trade secrets may be publicly disclosed (through publication or the sale of a product) by someone other than the trade secret owner. Commonly, this occurs when someone other than the trade secret owner has independently developed or discovered the secret. Call the first trade secret "owner" *A,* and the independent developer *B. A* has no control over what *B* does with her independent discovery; if she chooses to publish the secret, she defeats not only her rights to trade secret protection, but *A*'s rights as well. Suppose *B* did not develop the secret independently of *A* but in fact stole it from *A.* What happens if *B* publishes the secret? Can *A* still protect it? If so, what happens to *C,* who began using the secret after reading *B*'s publication? This issue was addressed in Religious Technology Center v. Lerma, 908 F. Supp. 1362 (E.D. Va. 1995). In that case, the Church of Scientology sued (among others) the Washington Post, which had quoted from part of its confidential scriptures. The court concluded that the fact that the scriptures were posted on a Usenet newsgroup for ten days defeated any claim of trade secrecy:

> [For ten days, the documents] remained potentially available to the millions of Internet users around the world.
> As other courts who have dealt with similar issues have observed,"posting works to the Internet makes them generally known" at least to the relevant people interested in the news group. Once a trade secret is posted on the Internet, it is effectively part of the public domain, impossible to retrieve. Although the person who originally posted a trade secret on the Internet may be liable for trade secret misappropriation, the party who merely down loads Internet information cannot be liable for misappropriation because there is no misconduct involved in interacting with the Internet.

908 F. Supp. at 1368; accord American Red Cross v. Palm Beach Blood Bank Inc., 143 F.3d 1407 (11th Cir. 1998) (Red Cross donor list lost trade secret status because it was posted on a publicly accessible computer bulletin board).

Should the obscurity of the Web site matter? What if it is not indexed in a search engine? In DVD Copy Control Ass'n v. Bunner, 116 Cal. App. 4th 241, 10 Cal. Rptr. 3d 185 (Ct. App. 2004), the court found disclosure of a secret on the Internet only because it was "quickly and widely republished to an eager audience," and cautioned that it did not assume that the secrets were lost merely because they were put on the Internet.

Because of the risk of loss of trade secrecy through Internet posting, companies have been more aggressive in suing individuals who post information they consider confidential. See Apple Computer, Inc. v. Doe 1, 2005 WL 578641, *4, 74 U.S.P.Q.2d 1191, 1191, 33 Media L. Rep. 1449, 1449 (Cal. Superior Mar 11, 2005) (NO. 1-04-CV-032178)); Ford v. Lane (67 F. Supp. 2d 745 (E.D. Mich. 1999)). In some of these cases, notably *Apple,* the company knows only that the information has been disclosed, and not who has done so. Do such lawsuits present First Amendment issues? Several have involved claims by the defendant that they are being sued for reporting legitimate news. See Franklin B. Goldberg, Ford Motor Co. v. Lane, 16 Berkeley Tech. L.J. 271 (2001).

• Fourth, trade secrets may be disclosed inadvertently (for example, by being left on a train or elsewhere in public view). While the case law on this issue is sparse, it seems reasonable to argue that a truly accidental disclosure should not defeat trade secret protection if reasonable precautions have been taken. On the other hand, if the inadvertent disclosure is widespread, it would seem unfair (as well as impracticable) to require the public as a whole to "give back" the secret. Note that section 1(2)(ii)(C) of the Uniform Trade Secrets Act provides that it is misappropriation for someone to disclose a secret that they have reason to know has been acquired "by accident or mistake." The Restatement (Third) of Unfair Competition § 40(b)(4) takes the same position,"unless the [accidental] acquisition was the result of the [trade secret owner]'s failure to take reasonable precautions to maintain the secrecy of the information." See also Williams v. Curtis-Wright Corp., 681 F.2d 161 (3d Cir. 1982) (user of secrets disclosed by mistake was liable for misappropriation because he had constructive notice of the secrecy of the information). But in DVD Copy Control Ass'n v. Bunner, 116 Cal. App. 4th 241, 10 Cal. Rptr. 3d 185 (Ct. App. 2004), the court rejected the idea that "once the information became *publicly* available everyone else would be liable under the trade secret laws simply because they knew about its unethical origins." "This," the court said, "is not what trade secret law is designed to do."

- Finally, government agencies sometimes require the disclosure of trade secrets by private parties in order to serve some other social purpose. See Corn Products Refining Co. v. Eddy, 249 U.S. 427 (1919) (requiring a food manufacturer to label its product with an accurate list of ingredients). Health and environmental concerns are a very common reason for the government to require disclosure of product contents. For example, the Federal Insecticide, Fungicide and Rodenticide Act (FIFRA), 7 U.S.C. § 136 et seq., requires disclosure of the contents of pesticides as well as a great deal of other information. FIFRA makes two concessions to trade secret protection, however. First, it limits public disclosure of information concerning manufacturing processes and inert (as opposed to active) contents. Second, it provides for compensation to be paid to the inventors of trade secrets which the government appropriates by public disclosure. See also Ruckelshaus v. Monsanto Co., 467 U.S. 986 (1984) (federal requirement that private parties disclose trade secrets may constitute a taking under the Fifth Amendment); Phillip Morris Inc. v. Reilly, 312 F.3d 24 (1st Cir. 2002) (Massachusetts law requiring labeling of cigarette ingredients was a taking of tobacco companies' trade secrets).

COMMENTS AND QUESTIONS

1. Under what circumstances does the sale of a commercial product embodying a trade secret destroy the secret? Is the answer different for a commercially available product produced by a secret manufacturing process?

2. Until 1999, patent applications were kept secret by the U.S. Patent Office unless and until the patent issued. See 35 U.S.C. § 122. If a patent application was not actively prosecuted, or if the patent did not issue, it was declared abandoned by the Patent Office. See 37 C.F.R. § 114 (1995). Abandoned applications were *not* available to the public. In fact, the application itself was destroyed after 20 years. Id.

In Europe, Japan, and elsewhere, patent applications are published 18 months after they are filed, regardless of whether the patent has yet issued. Normally, however, an applicant is notified before the application is published, allowing him or her to "elect" whether to continue to pursue a patent (and have the application be disclosed to the world) or to forgo prosecuting the patent (and retain the secrecy of the application). In 1999, Congress changed U.S. law to require that some (but not all) patent applications be published after 18 months. Since it takes almost three years on average for a patent to issue, many applicants will face an election not between patent and trade secret protection, but between trade secret protection and the prospect of future patent protection.

The notion of an "election" between trade secret and patent protection assumes that the patent application actually describes all the details of an invention. For more on this issue — known as the "enablement" requirement in patent law — see 35 U.S.C. § 112, and Chapter 3 in this book. Certainly there is evidence that firms sometimes make precisely such an election. For an interesting historical example, see Henry Petroski, The Pencil 114-115 (1990) (describing how the family of Henry David Thoreau kept its pencil-making technology secret rather than disclose it by obtaining a patent).Most patent licenses also allow the licensee to use trade secrets and know-how related to or associated with the patent. If both a patent and its associated secrets are licensed, should there be any limits on the duration of the license? Do these licenses suggest that an inventor can avoid the "election" doctrine and "have it both ways" in some respects? See United States v. Pilkington plc, No. 94-345-TUC-WDB (D. Ariz. May 25, 1994) (government antitrust action against a company that sought to enforce trade secrets against competitors after a related patent had expired).

3. To what extent does the incorporation of a secret in a public governmental record preclude trade secret protection? See Frazee v. U.S. Forest Service, 97 F.3d 367 (9th Cir. 1996) (information was not a trade secret because it could be obtained from the government under the Freedom of Information Act); Weygand v. CBS, Inc., 43 U.S.P.Q.2d 1120 (C.D. Cal. 1997) (depositing a work with the U.S. Copyright Office destroys trade secrecy).

C. MISAPPROPRIATION OF TRADE SECRETS

Not all uses of another's trade secrets constitute misappropriation. Acquisition or use of a trade secret is illegal only in two basic situations: where it is done through improper means, or where it involves a breach of confidence. Uniform Trade Secrets Act § 1; Restatement (Third) of Unfair Competition § 40.

1. Improper Means

E. I. duPont deNemours & Co. v. Rolfe Christopher et al.
United States Court of Appeals for the Fifth Circuit
431 F.2d 1012 (5th Cir. 1970)

GOLDBERG, Circuit Judge:

This is a case of industrial espionage in which an airplane is the cloak and a camera the dagger. The defendants-appellants, Rolfe and Gary Christopher, are photographers in Beaumont, Texas. The Christophers were hired by an unknown third party to take aerial photographs of new construction at the Beaumont plant of E. I. duPont deNemours & Company, Inc. Sixteen photographs of the DuPont facility were taken from the air on March 19, 1969, and these photographs were later developed and delivered to the third party.

DuPont employees apparently noticed the airplane on March 19 and immediately began an investigation to determine why the craft was circling over the plant. By that afternoon the investigation had disclosed that the craft was involved in a photographic expedition and that the Christophers were the photographers. DuPont contacted the

Christophers that same afternoon and asked them to reveal the name of the person or corporation requesting the photographs. The Christophers refused to disclose this information, giving as their reason the client's desire to remain anonymous.

Having reached a dead end in the investigation, DuPont subsequently filed suit against the Christophers, alleging that the Christophers had wrongfully obtained photographs revealing DuPont's trade secrets which they then sold to the undisclosed third party. DuPont contended that it had developed a highly secret but unpatented process for producing methanol, a process which gave DuPont a competitive advantage over other producers. This process, DuPont alleged, was a trade secret developed after much expensive and time-consuming research, and a secret which the company had taken special precautions to safeguard. The area photographed by the Christophers was the plant designed to produce methanol by this secret process, and because the plant was still under construction parts of the process were exposed to view from directly above the construction area. Photographs of that area, DuPont alleged, would enable a skilled person to deduce the secret process for making methanol. DuPont thus contended that the Christophers had wrongfully appropriated DuPont trade secrets by taking the photographs and delivering them to the undisclosed third party. In its suit DuPont asked for damages to cover the loss it had already sustained as a result of the wrongful disclosure of the trade secret and sought temporary and permanent injunctions prohibiting any further circulation of the photographs already taken and prohibiting any additional photographing of the methanol plant. . . .

. . . [T]he Christophers argue that for an appropriation of trade secrets to be wrongful there must be a trespass, other illegal conduct, or breach of a confidential relationship. We disagree.

It is true, as the Christophers assert, that the previous trade secret cases have contained one or more of these elements. However, we do not think that the Texas courts would limit the trade secret protection exclusively to these elements. On the contrary, in Hyde Corporation v. Huffines, 1958, 158 Tex. 566, 314 S.W.2d 763, the Texas Supreme Court specifically adopted the rule found in the Restatement of Torts which provides:

> One who discloses or uses another's trade secret, without a privilege to do so, is liable to the other if
>
> (a) he discovered the secret by improper means, or
> (b) his disclosure or use constitutes a breach of confidence reposed in him by the other in disclosing the secret to him. . . .

Restatement of Torts § 757 (1939). Thus, although the previous cases have dealt with a breach of a confidential relationship, a trespass, or other illegal conduct, the rule is much broader than the cases heretofore encountered. Not limiting itself to specific wrongs, Texas adopted subsection (a) of the Restatement which recognizes a cause of action for the discovery of a trade secret by any "improper" means. . . .

The question remaining, therefore, is whether aerial photography of plant construction is an improper means of obtaining another's trade secret. We conclude that it is and that the Texas courts would so hold. The Supreme Court of that state has declared that "the undoubted tendency of the law has been to recognize and enforce higher standards of commercial morality in the business world." Hyde Corporation v.

Huffines, supra 314 S.W.2d at 773. That court has quoted with approval articles indicating that the proper means of gaining possession of a competitor's secret process is "through inspection and analysis" of the product in order to create a duplicate. K & G Tool & Service Co. v. G & G Fishing Tool Service, 1958, 158 Tex. 594, 314 S.W.2d 782, 783, 788. Later another Texas court explained:

> The means by which the discovery is made may be obvious, and the experimentation leading from known factors to presently unknown results may be simple and lying in the public domain. But these facts do not destroy the value of the discovery and will not advantage a competitor who by unfair means obtains the knowledge without paying the price expended by the discoverer.

Brown v. Fowler, Tex. Civ. App. 1958, 316 S.W.2d 111, 114, *writ ref'd n.r.e.*

We think, therefore, that the Texas rule is clear. One may use his competitor's secret process if he discovers the process by reverse engineering applied to the finished product; one may use a competitor's process if he discovers it by his own independent research; but one may not avoid these labors by taking the process from the discoverer without his permission at a time when he is taking reasonable precautions to maintain its secrecy. To obtain knowledge of a process without spending the time and money to discover it independently is improper unless the holder voluntarily discloses it or fails to take reasonable precautions to ensure its secrecy.

In the instant case the Christophers deliberately flew over the DuPont plant to get pictures of a process which DuPont had attempted to keep secret. The Christophers delivered their pictures to a third party who was certainly aware of the means by which they had been acquired and who may be planning to use the information contained therein to manufacture methanol by the DuPont process. The third party has a right to use this process only if he obtains this knowledge through his own research efforts, but thus far all information indicates that the third party has gained this knowledge solely by taking it from DuPont at a time when DuPont was making reasonable efforts to preserve its secrecy. In such a situation DuPont has a valid cause of action to prohibit the Christophers from improperly discovering its trade secret and to prohibit the undisclosed third party from using the improperly obtained information.

We note that this view is in perfect accord with the position taken by the authors of the Restatement. In commenting on improper means of discovery the savants of the Restatement said:

> f. Improper means of discovery. The discovery of another's trade secret by improper means subjects the actor to liability independently of the harm to the interest in the secret. Thus, if one uses physical force to take a secret formula from another's pocket, or breaks into another's office to steal the formula, his conduct is wrongful and subjects him to liability apart from the rule stated in this Section. Such conduct is also an improper means of procuring the secret under this rule. But means may be improper under this rule even though they do not cause any other harm than that to the interest in the trade secret. Examples of such means are fraudulent misrepresentations to induce disclosure, tapping of telephone wires, eavesdropping or other espionage. A complete catalogue of improper means is not possible. In general they are means which fall below the generally accepted standards of commercial morality and reasonable conduct.

Restatement of Torts §757, Comment *f* at 10 (1939).

In taking this position we realize that industrial espionage of the sort here perpetrated has become a popular sport in some segments of our industrial community. However, our devotion to free-wheeling industrial competition must not force us into accepting the law of the jungle as the standard of morality expected in our commercial relations. Our tolerance of the espionage game must cease when the protections required to prevent another's spying cost so much that the spirit of inventiveness is dampened. Commercial privacy must be protected from espionage which could not have been reasonably anticipated or prevented. We do not mean to imply, however, that everything not in plain view is within the protected vale, nor that all information obtained through every extra optical extension is forbidden. Indeed, for our industrial competition to remain healthy there must be breathing room for observing a competing industrialist. A competitor can and must shop his competition for pricing and examine his products for quality, components, and methods of manufacture. Perhaps ordinary fences and roofs must be built to shut out incursive eyes, but we need not require the discoverer of a trade secret to guard against the unanticipated, the undetectable, or the unpreventable methods of espionage now available.

In the instant case DuPont was in the midst of constructing a plant. Although after construction the finished plant would have protected much of the process from view, during the period of construction the trade secret was exposed to view from the air. To require DuPont to put a roof over the unfinished plant to guard its secret would impose an enormous expense to prevent nothing more than a school boy's trick. We introduce here no new or radical ethic since our ethos has never given moral sanction to piracy. The market place must not deviate far from our mores. We should not require a person or corporation to take unreasonable precautions to prevent another from doing that which he ought not do in the first place. Reasonable precautions against predatory eyes we may require, but an impenetrable fortress is an unreasonable requirement, and we are not disposed to burden industrial inventors with such a duty in order to protect the fruits of their efforts. "Improper" will always be a word of many nuances, determined by time, place, and circumstances. We therefore need not proclaim a catalogue of commercial improprieties. Clearly, however, one of its commandments does say "thou shall not appropriate a trade secret through deviousness under circumstances in which countervailing defenses are not reasonably available." . . .

COMMENTS AND QUESTIONS

1. Improper means has a substantial basis in other torts. See Comment *f* to the Restatement, cited in *duPont*, and the Restatement (Third) of Unfair Competition §43 (defining improper means as including "theft, fraud, unauthorized interception of communications, inducement of or knowing participation in a breach of confidence, and other means either wrongful in themselves or wrongful under the circumstances of the case.").

DuPont itself provides an example of conduct prohibited by trade secret law that is probably not otherwise tortious. Should otherwise legitimate conduct be prohibited because it will disclose a trade secret? Most people would probably say yes in the context of the *duPont* case. As Judge Posner noted in *Rockwell,* misappropriation of trade secrets is largely redundant if it does not reach any further than other torts. Nonetheless, the reach of the case is troubling. Consider the last sentence of the *duPont* opinion. What is wrong with "deviousness"? Is reverse engineering (which is legal under both the Restatement and the Uniform Act) any less devious than aerial photography? And why should "countervailing defenses" enter the picture?

2. There is general agreement that "reverse engineering" — that is, buying a product and taking it apart to see how it works — is not a misappropriation of a trade secret.[14] But while it is easy to see how to reverse engineer a product, how does one reverse engineer a process? Are there realistic alternatives to the sort of espionage condemned in *duPont* for those who seek to reproduce a process? Should there be some legally protected way of discovering a competitor's process?

3. Note how the decision in *duPont* dovetails with the discussion of precautions in section B.2 above. DuPont could have protected itself against aerial photography by building a roof over its plant area before beginning internal construction. The Fifth Circuit rejected this alternative because it would "impose an enormous expense to prevent nothing more than a school boy's trick." On the other hand, the courts are clearly willing to put duPont to *some* expense to protect its secrets from prying eyes. If duPont had allowed its engineers to leave copies of the plant blueprints on subways, the result of the case might be very different.

How much expense is duPont required to incur to protect itself? The uniform answer of the courts is that only "reasonable" precautions must be taken. While this is not a terribly helpful answer, it may be a fairly practical one in any given industry, where companies can protect themselves by taking those precautions that are customary in that industry.

On the other hand, consider the Second Circuit's statement in Franke v. Wiltschek, 209 F.2d 493, 495 (2d Cir. 1953):

> It matters not that the defendants could have gained their knowledge from a study of the expired patent and plaintiff's publicly marketed product. The fact is they did not. Instead they gained it from plaintiffs via their confidential relationship, and in doing so incurred a duty not to use it to plaintiff's detriment. This duty they have breached.

Is this consistent with the UTSA?

4. An interesting case involving only circumstantial evidence of misappropriation is Pioneer Hi-Bred International v. Holden Foundation Seeds, Inc., 35 F.3d 1226, 31 U.S.P.Q.2d 1385 (8th Cir. 1994). Plaintiff Pioneer could not establish a specific act of misappropriation, but it did show (through the use of three sophisticated scientific tests) that it was highly unlikely that defendant's hybrid seeds had been developed independently of plaintiff's seeds. Instead, due to genetic similarities, Pioneer showed that it was much more likely that those seeds were "derived from" a popular Pioneer hybrid. Pioneer was found to have maintained reasonable measures to guard against disclosure, including putting experimental seeds in bags marked with a secret code, allowing seeds to be grown only under strict nondisclosure arrangements, and leaving unmarked the fields in which the seeds had been planted. Pioneer was awarded $46.7 million in damages.

---

14. As we shall see, however, reverse engineering of a patented invention is generally not permissible.

PROBLEM

Problem 2-7. Suppose that, at trial in the *duPont* case, the Christophers proved that they could have discovered the secrets contained in the layout of the duPont plant from another source (such as a copy of the blueprints on file with the Environmental Protection Agency and available under the Freedom of Information Act) that placed no restrictions on their use of the information. Are the Christophers then relieved from liability for misappropriation of trade secrets? Would your answer change if they had obtained the photos by sneaking onto duPont property on the ground?

---

2. Confidential Relationship

The previous section concerned misappropriation of trade secrets by improperly obtaining them from their owner. But secrets that have been properly obtained may still be misappropriated if they are improperly used or disclosed. Most often, this occurs when the secrets are used or disclosed in violation of a confidential relationship.

What is a confidential relationship? Obviously, the easiest way to create a confidential relationship is to sign a contract to that effect. Agreements to keep certain information confidential are generally enforceable, at least if the information meets the definition of a trade secret. See Restatement (Third) of Unfair Competition § 39 Comment *d*, at 430. But confidential relationships may also arise without any express agreement. Consider the following case.

≡ *Smith v. Dravo Corp.*
United States Court of Appeals for the Seventh Circuit
203 F.2d 369 (7th Cir. 1953)

LINDLEY, J.

Plaintiffs appeal from a judgment for defendant entered at the close of a trial by the court without a jury. The complaint is in four counts. 1 and 2 charge an unlawful appropriation by defendant of plaintiffs' trade secrets relating to the design and construction and selling and leasing of freight containers; 3 and 4 aver infringement of plaintiffs' patents Nos. 2,457,841 and 2,457,842. . . .

In the early 1940s Leathem D. Smith, now deceased, began toying with an idea which, he believed, would greatly facilitate the ship and shore handling and transportation of cargoes. As he was primarily engaged in the shipbuilding business, it was quite natural that his thinking was chiefly concerned with water transportation and dock handling. Nevertheless his overall plan encompassed rail shipping as well. He envisioned construction of ships especially designed to carry their cargo in uniformly sized steel freight containers. These devices (which, it appears, were the crux of his idea) were: equipped with high doors at one end; large enough for a man to enter easily; weather and pilfer proof; and bore collapsible legs, which (1) served to lock them (a) to the deck of the ship by fitting into recesses in the deck, or (b) to each other, when stacked, by reason of receiving sockets located in the upper four corners of each container, and (2) allowed sufficient clearance between deck and container or container and container for the facile insertion of a fork of a lift tractor, and (3) were equipped with lifting eyelets, which, together with a specially designed hoist, made possible placement of the containers upon or removal from a ship, railroad car or truck, while filled with cargo. The outer dimensions of the devices were such that they would fit compactly in standard gauge North American railroad cars, or narrow gauge South American trains, and in the holds of most water vessels.

[At the end of World War II, Smith's company — Safeway Containers — had some success building and selling such containers.]

On June 23, 1946, Smith died in a sailing accident. The need for cash for inheritance tax purposes prompted his estate to survey his holdings for disposable assets. It was decided that the container business should be sold. Devices in process were completed but no work on new ones was started.

Defendant was interested in the Safeway container, primarily, it appears, for use by its subsidiary, the Union Barge Lines. In October 1946 it contacted Agwilines [one of Smith's customers] seeking information. It watched a loading operation in which Agwilines used the box. At approximately the same time, defendant approached the shipbuilding company and inquired as to the possibility of purchase of a number of the containers. It was told to communicate with Cowan, plaintiffs' eastern representative. This it did, and, on October 29, 1946, in Pittsburgh, Cowan met with defendant's officials to discuss the proposed sale of [containers]. But, as negotiations progressed, defendant demonstrated an interest in the entire container development. Thus, what started as a meeting to discuss the purchase of individual containers ended in the possible foundation for a sale of the entire business.

Based upon this display of interest, Cowan sent detailed information to defendant concerning the business. This included: (1) patent applications for both the 'knock-down' and 'rigid' crates; (2) blue prints of both designs; (3) a miniature Safeway container; (4) letters of inquiry from possible users; (5) further correspondence with prospective users. In addition, defendant's representatives journeyed to Sturgeon Bay, Wisconsin, the home of the shipbuilding company, and viewed the physical plant, inventory and manufacturing operation.

Plaintiffs quoted a price of $150,000 plus a royalty of $10 per unit. This was rejected. Subsequent offers of $100,000 and $75,000 with royalties of $10 per container were also rejected. Negotiations continued until January 30, 1947, at which time defendant finally rejected plaintiffs' offer.

On January 31, 1947 defendant announced to Agwilines that it "intended to design and produce a shipping container of the widest possible utility" for "coastal steamship application . . . [and] use . . . on the inland rivers and . . . connecting highway and rail carriers." Development of the project moved rapidly, so that by February 5, 1947 defendant had set up a stock order for a freight container which was designed, by use of plaintiffs' patent applications, so as to avoid any claim of infringement. One differing feature was the use of skids and recesses running the length of the container, rather than legs and sockets as employed in plaintiffs' design. However, Agwilines rejected this design, insisting on an adaptation of plaintiffs' idea. In short defendant's final product incorporated many, if not all, of the features of plaintiffs' design. So conceived, it was accepted by the trade to the extent that, by March 1948, defendant had sold some 500 containers. Many of these sales were made to firms who had shown considerable prior interest in plaintiffs' design and had been included in the prospective users disclosed to defendant.

One particular feature of defendant's container differed from plaintiffs: its width was four inches less. As a result plaintiffs' product became obsolete. Their container could not be used interchangeably with defendant's; they ceased production. Consequently the prospects of disposing of the entire operation vanished.

The foregoing is the essence of plaintiffs' cause of action. Stripped of surplusage, the averment is that defendant obtained, through a confidential relationship, knowledge of plaintiffs' secret designs, plans and prospective customers, and then wrongfully breached that confidence by using the information to its own advantage and plaintiffs' detriment.

[The court found that, notwithstanding certain disclosures of information during the operation of Safeway, plaintiffs' information about how to design its containers remained a trade secret.]

(3) Was Defendant in a Position of Trust and Confidence at the Time of the Disclosure?

Mr. Justice Holmes once said that the existence of the confidential relationship is the "starting point" in a cause of action such as this. E. I. DuPont de Nemours Powder Co. v. Masland, 244 U.S. 100, 102, 37 S. Ct. 575, 61 L. Ed. 1016. While we take a slightly different tack, there is no doubt as to the importance of this element of plaintiffs' case.

Certain it is that a non-confidential disclosure will not supply the basis for a lawsuit. Plaintiffs' information is afforded protection because it is secret. Promiscuous disclosures quite naturally destroy the secrecy and the corresponding protection. But this is not true where a confidence has been reposed carrying with it communication of an idea.

It is clear that no express promise of trust was exacted from defendant. There is, however, the further question of whether one was implied from the relationship of the parties. Pennsylvania has not provided us with a decision precisely in point but Pressed Steel Car Co. v. Standard Car Co., 210 Pa. 464, 60 A. 4, furnishes abundant guideposts. There plaintiff delivered its blue prints to customers in order that they might acquaint themselves more thoroughly with the railroad cars they were purchasing; from these customers, defendant obtained the drawings. In holding that the customers held the plans as a result of a confidence reposed in them by plaintiff, and that the confidence was breached by delivery of the blue prints to defendant, the court said, 60 A. at page 10: "While there was no expressed restriction placed on the ownership of the prints, or any expressed limitation as to the use to which they were to be put, it is clear . . . that the purpose for which they were delivered by the plaintiff was understood by all parties. . . ."

The quoted language is applicable and determinative. Here plaintiffs disclosed their design for one purpose, to enable defendant to appraise it with a view in mind of purchasing the business. There can be no question that defendant knew and understood this limited purpose. Trust was reposed in it by plaintiffs that the information thus transmitted would be accepted subject to that limitation. "[T]he first thing to be made sure of is that the defendant shall not fraudulently abuse the trust reposed in him. It is the usual incident of confidential relations. If there is any disadvantage in the fact that he knew the plaintiffs' secrets, he must take the burden with the good." E. I. DuPont de Nemours Powder Co. v. Masland, 244 U.S. 100 at page 102, 37 S. Ct. 575, at page 576, 61 L. Ed. 1016.

Nor is it an adequate answer for defendant to say that the transactions with plaintiffs were at arms length. So, too, were the overall dealings between plaintiffs and defendants in Booth v. Stutz Motor Car Co., 7 Cir., 56 F.2d 962; Allen-Qualley Co. v. Shellmar Products Co., D.C., 31 F.2d 293, affirmed, 7 Cir., 36 F.2d 623 and Schavoir v. American Rebonded Leather Co., 104 Conn. 472, 133 A. 582. That fact does not detract from the conclusion that but for those very transactions defendant would not have learned, from plaintiffs, of the container design. The implied limitation on the use to be made of the information had its roots in the "arms-length" transaction.

(4) The Improper Use by Defendant of the Secret Information

Defendant's own evidence discloses that it did not begin to design its container until after it had access to plaintiffs' plans. Defendant's engineers admittedly referred to plaintiffs' patent applications, as they said, to avoid infringement. It is not disputed that, at the urging of Agwilines, defendant revised its proposed design to incorporate the folding leg and socket principles of plaintiffs' containers. These evidentiary facts, together with the striking similarity between defendant's and plaintiffs' finished product, were more than enough to convict defendant of the improper use of the structural information obtained from plaintiffs.

COMMENTS AND QUESTIONS

1. Since the shipping containers were available on the open market, couldn't Dravo have argued that they were not a trade secret? Did Smith's agent, Cowan, disclose anything that was *not* readily apparent from inspection of the containers?

2. In Van Prod. Co. v. General Welding & Fabricating Co., 213 A.2d 769, 779-780 (Pa. 1965), the Pennsylvania Supreme Court criticized *Smith* for focusing on the existence of a confidential relationship to the exclusion of whether there was a trade secret at all. The court said: "The starting point in every case of this sort is not whether there was a confidential relationship, but whether, in fact, there was a trade secret to be misappropriated."

3. Compare Smith v. Dravo with Omnitech Intl. v. Clorox Co., 29 U.S.P.Q.2d 1665 (5th Cir. 1994), *cert. denied*, 115 S. Ct. 71 (1994), where the court held that it was not an actionable "use" of a trade secret for the defendant to *evaluate* it in the course of trying to decide whether to (a) acquire the company or (b) take a license to use the trade secret. A finding of no liability here makes sense, because it allows the potential licensee to make an informed judgment and therefore promotes efficient licensing. But how far does it extend? Is the potential licensee entitled to replicate the research or build models in an effort to evaluate it? Are there special limits that should be placed on companies engaged in a "make or buy" decision — that is, who are considering *either* licensing the plaintiff's technology *or* entering the market themselves?

Smith and *Omnitech* both involved disclosures in the course of licensing negotiations. Cases such as these form part of the amorphous law of "precontractual liability." See E. Allen Farnsworth, Precontractual Liability and Preliminary Agreements: Fair Dealing and Failed Negotiations, 87 Colum. L. Rev. 217 (1987). Both the *Smith* and *Omnitech* cases present the problem of Arrow's Information Paradox. What the plaintiff has to sell to the defendant is information that is valuable only because it is secret. If there is no legal protection and if the plaintiff discloses the secret to the defendant, its value will be lost. But the defendant cannot be expected to pay for information unless it can see the information to determine its value. Thus, absent some form of legal protection for confidential disclosures, sale or licensing of trade secrets may not occur. Note that this problem does not arise in other areas of intellectual property, such as patent law, since the inventions being licensed are already publicly disclosed.

4. The Restatement (Third) of Unfair Competition § 41 holds that a confidential relationship is established in the following circumstances:

> (a) the person made an express promise of confidentiality prior to the disclosure of the trade secret; or
> (b) the trade secret was disclosed to the person under circumstances in which the relationship between the parties to the disclosure or the other facts surrounding the disclosure justify the conclusions that, at the time of the disclosure,
> > (1) the person knew or had reason to know that the disclosure was intended to be in confidence, and
> > (2) the other party to the disclosure was reasonable in inferring that the person consented to an obligation of confidentiality.

On the issue of knowledge, compare this "had reason to know" standard with the Eleventh Circuit's decision in Bateman v. Mnemonics, Inc., 79 F.3d 1532 (11th Cir. 1996). In that case, the court indicated that it was "wary" of trade secret claims based on implied confidential relationships because they were subject to abuse. The court rejected Bateman's allegation that such a relationship existed because Bateman had not "made it clear to the parties involved that there was an expectation and obligation of confidentiality." Thus the court seemed to create a standard of actual knowledge on the part of the recipient of confidential information that the discloser of such information intended the disclosure to be confidential. At the other extreme, the Fifth Circuit in Phillips v. Frey, 20 F.3d 623, 631-632 (5th Cir. 1994) found an implied confidential relationship to exist in the course of negotiations over the sale of a business, despite the fact that the disclosing party never even requested that the information remain confidential.

Which approach makes more sense? Who is in the best position to clarify the question, the discloser or the recipient?

5. Is there any way for a potential licensee to prevent the formation of a confidential relationship in such a situation?

6. The defendant in *Smith* actually sold a device that was not identical to the plaintiff's. Under what circumstances can a defendant be liable for misappropriation without literally copying the trade secret? In Mangren Res. & Dev. Corp. v. National Chem. Co., 87 F.3d 1339 (7th Cir. 1996), the court defined improper "use" broadly, stating that "the user of another's trade secret is liable even if he uses it with modifications or improvements upon it effected by his own efforts, so long as the substance of the process used by the actor is derived from the other's secret. . . . If trade secret law were not flexible enough to encompass modified or even new products that are substantially derived from the trade secrets of another, the protections that the law provides would be hollow indeed." To similar effect as *Mangren* is Texas Tanks Inc. v. Owens-Corning Fiberglas Corp., 99 F.3d 740 (5th Cir. 1997). In *Texas Tanks,* the Fifth Circuit held that any improper "exercise of control and domination" over a

secret constituted a commercial use of that secret. It rejected the defendant's argument that it could not be liable for taking a secret unless the secret was actually incorporated in a commercial product. The court noted that Owens Corning's awareness of the secret would likely influence the development of its own competing product, and that this was enough to demonstrate improper "use" of the secret.

PROBLEMS

Problem 2-8. Solomon, a regular customer of ToolCo's products, comes up with an idea for a new tool. He sends the idea to ToolCo, suggesting that they manufacture it. ToolCo does in fact produce and sell the new tool. Is Solomon entitled to compensation for ToolCo's use of his idea? Does it make any difference if (1) ToolCo actively solicited the idea from Solomon? (2) Solomon sent the suggestion to ToolCo along with a letter saying he wanted to open negotiations over a possible licensing arrangement to use the idea? (3) ToolCo has in the past had an informal, unwritten policy of compensating inventors who submit good ideas?

Problem 2-9. VenCo, a venture capitalist in the business of financing start-up companies, investigates TechCo in an effort to decide whether or not to finance it. To aid in its investigation, VenCo asks for and receives confidential information about TechCo's products and market position. VenCo eventually decides not to finance TechCo because of concerns about its management, but it does finance a start-up competitor of TechCo in the same field. Are VenCo or the start-up liable to TechCo? What obligations, if any, does VenCo undertake as a result of its exposure to TechCo's secrets?

Problem 2-10. MegaGene (MG) is a genomics company: a company whose business is to isolate and identify individual human genes. Founded in 1996, MG was originally conceived as a company that would accumulate patents on various genes and partial gene sequences. The idea was that as researchers — either within MG or outside of it — figured out what particular proteins were produced ("coded for") by individual genes, some of those genes would be very valuable. Some might code for important proteins that some humans needed as disease therapies; others might contain errors that caused diseases, and thereby be useful in diagnosing or treating the diseases.

MG's original strategy was to obtain patents on as many genes and partial gene sequences as it could. Today it has over 100 such patents, almost exclusively on full gene sequences. These patents were acquired because MG was able to establish the actual function of each of the genes. Legal changes — most notably a tighter interpretation of patentable "utility" (see Chapter 3) — precluded MG from acquiring any patents on partial gene sequences, however.

Although some of the full-gene patents may someday be valuable, none are as yet. So to try to generate cash in the short term, MG has been pursuing another approach to the genomics business: it charges users for access to its database of gene sequences and related information. The database has been painstakingly assembled not only from information in MG's own 100 patents, but also from various public domain sources. It includes a great deal of information from the publicly funded Human Genome Project. For example, it includes information on many "markers" that help researchers determine which individual chromosome certain sequences come from. It also contains a great deal of information about individual genes: when they are expressed or repressed, and what they code for in various tissue samples and patient populations. This information helps researchers identify which genetic sequences are part of genes that may be important — because they are associated with a disease, for example.

IceGene is a company based in Iceland that has designed a highly promising approach to genomics. IceGene has combined a vast genetic database with detailed information on disease incidence in specific people. This is made possible by the fact that for many years detailed genealogical records have been kept for the highly homogenous Icelandic population. Now, IceGene has compiled a large database of tissue samples from individual Icelanders. Its business goal is to identify, from genealogical records, specific individuals whose family histories show they may be at risk for certain diseases and other conditions. Then, with access to the genetic code of the individual in question, IceGene can locate specific genetic variations and mutations that might be associated with the disease or condition. Once a correlation or association is established, specific genes and genetic problems can be identified for future research. (Much of this information is derived with reference to public data, including the Human Genome Project data.) IceGene plans to pursue some of these research opportunities itself, but there will be too many for the firm to research all of them. So it too plans to establish a database of genes and individual genetic variations that will be accessible (for a fee) to other companies doing genomics research.

Both MG and IceGene have established web-based access to their databases. The web pages are accessible only by subscribers of the firms' respective services, but customers can subscribe online using an interactive signup menu. The signup procedure requires users to agree to a "Database and Trade Secrecy Access Agreement," which includes these two provisions:

> [1] Customer, upon accepting this Access Agreement, agrees that the Genetic Data accessible hereon is and will be a Trade Secret of Company, and further agrees not to disclose, publish, or otherwise disseminate said Genetic Data to any other company, person, or institution. Customer also agrees that payment under the Payment provision herein will continue as provided for in this Agreement, notwithstanding the subsequent disclosure, publication, or other dissemination of any portion of said Genetic Data by a third party not subject to this Agreement. . . .
>
> [2] Customer agrees to pay to the Company a royalty of 5 per cent of the gross selling price of any diagnostic kit, therapeutic drug, or other commercial product developed as a result of Customer's access to or use of the Genetic Data contained herein.

MG becomes the dominant genomics database firm, with over 8,000 subscribers. It is widely known in the industry that virtually every serious researcher uses MG data in his or her genomics research. Meanwhile, IceGene has become an accepted source for information about the relationship between certain gene mutations and actual human diseases. Gene O'Mick, a researcher based in

Europe, is a subscriber to both the MG and the IceGene databases. In the course of his research, Gene discovers a gene, called "Cobb1," that he believes is linked to a rare genetic disorder called Ty Cobb syndrome. (Sufferers are mentally unstable and often kleptomaniacs.) Although Gene is an MG subscriber, he contends that he did not use the database in locating the Cobb1 gene; his published research on the subject reveals that he used the publicly available Human Genome Project database to first home in on the gene's location and then sequence it. Gene now wants to confirm the link between the Cobb1 gene and Ty Cobb syndrome, by running an analysis using the IceGene database. He balks, however, when he comes to clause [2] above in the IceGene Access Agreement. Gene believes that the rarity of Ty Cobb's disease will make it very difficult to interest pharmaceutical firms in developing a therapy to deal with it. (This could for example take the form of a chemical compound that inhibits expression of the Cobb1 gene, effectively suppressing the disorder.) Because of this, Gene wants to keep the cost of any Cobb1 therapy as low as possible. So he objects to the "reach through" or "pass through" royalty called for in Clause [2].

Gene has asked you to argue his position with MG and IceGene. What arguments would you marshal on his behalf?

3. Reverse Engineering

Not all use or disclosure of someone's secret is actionable misappropriation. Rather, as the commissioners who drafted the Uniform Trade Secrets Act noted, there are several categories of "proper means" of obtaining a trade secret. We classify these "proper means" as defenses, because they do not directly deny the existence of a trade secret or the defendant's use of that secret. Rather, they are *legitimate* uses of trade secrets by a competitor.

The commissioners' Comment to section 1 of the act had this to say on the subject of defenses:

> One of the broadly stated policies behind trade secret law is "the maintenance of standards of commercial ethics." The Restatement of Torts, Section 757, Comment (f), notes: "A complete catalogue of improper means is not possible," but Section 1(1) includes a partial listing.
>   Proper means include:
>   1. Discovery by independent invention;
>   2. Discovery by "reverse engineering," that is, by starting with the known product and working backward to find the method by which it was developed. The acquisition of the known product must, of course, also be by a fair and honest means, such as purchase of the item on the open market, for reverse engineering to be lawful; . . .
>   4. Observation of the item in public use or on public display;
>   5. Obtaining the trade secret from published literature.

See also Restatement (Third) of Unfair Competition §43 ("Independent discovery and analysis of publicly available products or information are not improper means of acquisition.").

Kadant, Inc. v. Seeley Machine, Inc.
United States District Court for the Northern District of New York
244 F. Supp. 2d 19 (N.D.N.Y. 2003)

HURD, District Judge.

. . .

II. FACTS

Kadant is a publicly traded corporation with yearly sales eclipsing $100 million. Kadant AES ("AES") is a division of Kadant located in Queensbury, New York. For twenty-eight years, AES has manufactured and sold to customers worldwide products that clean and condition papermaking machines and filter water used in the paper-making process. "There are three main products areas of the business of AES: shower and spray devices and nozzles; foil blades to remove water from the paper during the papermaking process; and structures that hold foils and filters for straining water utilized in the papermaking process so that it can be reused." . . . Plaintiff occupies a dominant place in the national papermaking market.

Corlew was hired by AES in July of 1995 as a machinist. Nearly three years later, in April of 1998, he was promoted to a position in the engineering department. In that capacity, Corlew would be provided with a refined customer order, made by an AES engineer and a customer, and it was his job to create a "manufacturing drawing (with instructions and a bill of materials — i.e., the 'recipe') for the order." In order to create such a drawing, Corlew used a computer assisted drawing machine. The computer assisted drawing machine contained "the recipes for the AES products and generate[d] drawings and bills of materials." Plaintiff contends that AES took steps to protect the secrecy of the information contained in the computer assisted drawing machine. In June of 1999, Corlew was promoted again and now had the responsibility to assist the engineers in designing customer orders. Corlew was terminated in the summer of 2001.

During his tenure at AES, Corlew had access to plaintiff's "recipes" ("design specifications") for its products and to plaintiff's computerized database of prospective customers, which includes "names, addresses, and e-mails for all potential customers in the papermaking industry, including the names of individuals key to these companies' purchasing orders." Plaintiff contends that one of Corlew's final assignments while at AES was to coordinate this database with the database containing information on all current AES customers and their purchasing histories. Corlew had access to AES's entire computer system, which apparently included both databases as well as the computer assisted drawing machine containing the design specifications. At the end of his employment, Corlew's access to AES's computer system was terminated, and his laptop was wiped clean of AES information, and/or access thereto. Throughout his employment with AES, Corlew was subject to a signed confidentiality agreement, in which he agreed not to disclose or use to his benefit any confidential information, including information about AES customers.

Following his termination, Corlew formed APE "to outsource the manufacture of his own line of products for sale to the pulp and papermaking industry." According to defendants, APE was not a successful venture, forcing Corlew to begin working for Cambridge Valley Machine, Inc. ("CVM"). In April or May of 2002, Corlew resigned from CVM. Defendants contended in their moving papers that, as a condition of his resignation, Corlew was required to relinquish the rights to the name APE to CVM.

Near the end of April of 2002, Corlew began working for Seeley, "developing and marketing a new line of Seeley products for sale to the pulp and papermaking industry." According to defendants, "[t]he products comprising this new line were reverse-engineered from existing products, freely available in the public domain and unprotected by published patent applications, in-force patents or trade secrets." According to plaintiff, the only way defendants could have developed and put out for sale this new line of products in so short of time is not by reverse engineering, which it alleges is time consuming, expensive, and requiring technical skill, but by Corlew's theft of AES's trade secrets–its design specifications and the customer databases — and infringement of its trademark.

Specifically, claims plaintiff, using as a frame of reference defendants' own expert, it would take defendants 1.7 years to reverse engineer all of plaintiff's nozzles. Defendants maintain, however, that only a small fraction, not all, of plaintiff's products were reverse engineered. The parties also dispute how much time would be associated with the manufacturing process. . . .

III. DISCUSSION

Plaintiff has moved for preliminary injunctive relief pursuant to Fed.R.Civ.P. 52(a). . . .

In order for plaintiff to successfully move for a preliminary injunction, it must demonstrate: 1) a likelihood of irreparable harm if the injunction is not granted; and 2) either a likelihood of success on the merits of its claims, or the existence of serious questions going to the merits of its claims plus a balance of the hardships tipping decidedly in its favor. Bery v. City of New York, 97 F.3d 689, 693-94 (2d Cir.1996). Because the issuance of a preliminary injunction is a drastic remedy, plaintiff is required to make a "clear showing" of these requirements. *See* Mazurek v. Armstrong, 520 U.S. 968, 972, 117 S.Ct. 1865, 138 L.Ed.2d 162 (1997). . . .

B. TRADE SECRET THEFT/MISAPPROPRIATION

. . .

1. *Likelihood of success on the merits*

To establish misappropriation of a trade secret, a plaintiff must prove 1) that "it possessed a trade secret; and 2) that defendants are using that trade secret in breach of an agreement, confidence, or duty, or as a result of discovery by improper means." Integrated Cash Management Services, Inc. v. Digital Transactions, Inc., 920 F.2d 171, 173 (2d Cir.1990); *see also* Carpetmaster of Latham v. Dupont Flooring Systems, 12 F.Supp.2d 257, 261 (N.D.N.Y.1998).

. . .

b. design specifications

Secrecy again takes center stage and is dispositive when determining whether plaintiff's product design specifications are entitled to trade secret protection for the purposes of obtaining preliminary injunctive relief. If secrecy is lost when a product is placed on the market, there is no trade secret protection. *See LinkCo,* 230 F.Supp.2d at 498-99 (collecting cases). The primary issue with respect to this alleged trade secret is whether plaintiff's products could be reverse engineered in the time span between Corlew's hiring at Seeley and defendants' marketing and putting out their products for sale.[11] As will be shown, *infra,* the parties disagree about every material fact that goes toward resolving this debate.

"Trade secret law . . . does not offer protection against discovery by . . . so-called reverse engineering[.]" Kewanee Oil Co. v. Bicron Corp., 416 U.S. 470, 476, 94 S.Ct. 1879, 40 L.Ed.2d 315 (1974). However, "the term 'reverse engineering' is not a talisman that may immunize the theft of trade secrets." Telerate Systems, Inc. v. Caro, 689 F.Supp. 221, 233 (S.D.N.Y.1988). The relevant inquiry is whether the means to obtain the alleged trade secret were proper or "honest," as opposed to being obtained by virtue of a confidential relationship with an employer. *See* Franke v. Wiltschek, 209 F.2d 493, 495 (2d Cir.1953); *Telerate,* 689 F.Supp. at 233. Reverse engineering a product to determine its design specifications is therefore permissible so long as the means used to get the information necessary to reverse engineer is in the public domain, and not through the confidential relationship established with the maker or owner of the product.

One court, not satisfied with this distinction, has held that even where a product is out in the public domain, and was thus subject to being reverse engineered by a purchaser, trade secret status remains intact because the defendant's former employment with the plaintiff was the only basis for the defendants being "able to select particular items from a vast sea of public information." *See* Monovis, Inc. v. Aquino, 905 F.Supp. 1205, 1228 (W.D.N.Y.1994). This view of the law would effectively eviscerate any benefit reverse engineering would provide in the preliminary injunction analysis as applied to trade secrets, forestall healthy notions of commercial competitiveness, and heavily contribute to an inert marketplace where products can only be developed and sold under an impenetrable cloak of originality. It is therefore rejected.

Plaintiff has presented no evidence that the means used by defendants to obtain the alleged trade secret were improper or dishonest. In short, it has no evidence Corlew actually stole the design specifications. It instead necessarily relies upon an inference — that the only way defendants could develop, market, and sell their products in so short of time is if Corlew stole the design specification information — that is, as far as the evidence to this point shows, is unjustified. Plaintiff does not seem to argue that reverse engineering is impossible, just that it would take a great deal of time, skill, and expense, and that the lack thereof demonstrates that the design specifications must have been stolen. Defendants have argued that the plaintiff's products were simple, consisting of non-technical and few parts, that reverse engineering would take little time, and that, in any event, they only reverse engineered a small fraction, not all, of plaintiff's products. Plaintiff has not sufficiently rebutted these contentions.

---

11. "Reverse engineering is the process by which an engineer takes an already existing product and works backward to re-create its design and/or manufacturing process." *United Technologies Corp. by Pratt & Whitney v. F.A.A.,* 102 F.3d 688, 690 n. 1 (2d Cir.1996).

Thus, because plaintiff has failed to make a clear showing that defendants improperly obtained and reverse engineered its products, trade secret protection at this stage of the litigation is improper. *See, e.g.,* Bridge C.A.T. Scan Associates v. Technicare Corp., 710 F.2d 940, 946-47 (2d Cir.1983) (lack of evidence that means to receive information that plaintiff claimed was trade secret were improper mandated denying preliminary injunctive relief).

COMMENTS AND QUESTIONS

1. Why is reverse engineering lawful? If, as the Commissioners suggested, one purpose of trade secret law is to promote standards of commercial ethics, doesn't there seem to be something wrong with taking apart a competitor's product in order to figure out how to copy it? Does reverse engineering benefit only those competitors who are not smart enough to develop ideas or products for themselves?

In Chicago Lock Co. v. Fanberg, 676 F.2d 400 (9th Cir. 1982), the plaintiff, a company that made locks, sued a locksmith who compiled a list of master key codes for the plaintiff's locks by soliciting information from those who had picked the locks on behalf of their clients. The court rejected the claim, holding that the owners of the individual locks had the right to open them, and therefore to authorize others to reverse engineer the key codes:

> A lock purchaser's own reverse-engineering of his own lock, and subsequent publication of the serial number-key code correlation, is an example of the independent invention and reverse engineering expressly allowed by trade secret doctrine.[15] Imposing an obligation of nondisclosure on lock owners here would frustrate the intent of California courts to disallow protection to trade secrets discovered through "fair and honest means." See id. Further, such an implied obligation upon the lock owners in this case would, in effect, convert the Company's trade secret into a state-conferred monopoly akin to the absolute protection that a federal patent affords. Such an extension of California trade secrets law would certainly be preempted by the federal scheme of patent regulation.
>
> Appellants, therefore, cannot be said to have procured the individual locksmiths to breach a duty of nondisclosure they owed to the Company, for the locksmiths owed no such duty.

2. Does reverse engineering by one or a few firms mean that the plaintiff's information is no longer secret? See Barr-Mullin, Inc. v. Browning, 442 S.E.2d 226 (Ct. App. N.C. 1993). In Reingold v. Swiftships, 126 F.3d 645 (5th Cir. 1997), the court rejected a claim that a boat hull mold could not be a trade secret because it could readily be reverse engineered. While reverse engineering may protect one who engages in it, the court held, it does not protect those who actually acquire the secret by improper means. Should it matter whether the reverse engineering is very difficult or easy enough that any competitor could do it?

---

15. If a group of lock owners, for their own convenience, together published a listing of their own key codes for use by locksmiths, the owners would not have breached any duty owed to the Company. Indeed, the Company concedes that a lock owner's reverse engineering of his own lock is not "improper means."

Reverse engineering may be explained as a legal rule designed to weaken trade secret protection relative to patent protection. Can you think of reasons why we would want to weaken trade secret protection? See Kewanee Oil Co. v. Bicron Corp., 416 U.S. 470 (1974) (patent law does not preempt trade secret law, in part because the reverse engineering rule "weakens" trade secret law).

3. Consider the recurring question of whether the parties can agree to override trade secret law. Suppose that the owner of a trade secret includes in a license or sale contract a provision prohibiting the buyer from reverse engineering the product. Is that contractual provision enforceable? The courts are split. Compare K & G Oil & Tool Service Co. v. G & G Fishing Tool Service, 158 Tex. 94, 314 S.W.2d 782, 785-86 (1958) (contract preventing disassembly of tools to protect trade secrets was enforceable) with Vault Corp. v. Quaid Software Ltd., 847 F.2d 255, 265-267 (5th Cir. 1988) (contract provision prohibiting reverse engineering of software void as against public policy).

4. If a company is looking for a piece of information — say, the solution to a given technical problem — society is very much concerned that the company be allowed free access to information in the public domain. The reason is that minimizing the search costs of the company is thought to be a social good, consistent with some level of protection for the prior investment of others. In a similar vein, suppose the costs of obtaining that information from the public domain are very high, but the costs (to the company) of stealing it from a competitor are very low. Should society punish — and therefore deter — the theft? Wouldn't that simply create an inefficiency, since the company could get the information but would have to incur greater search costs? Cf. David Friedman, William Landes & Richard Posner, Some Economics of Trade Secret Law, 5 J. Econ. Perspectives 61, 62 (1991) (arguing that trade secret law prohibits only costly means of obtaining competitors' information, while encouraging cheaper forms of obtaining information such as reverse engineering). Are there other considerations that militate in favor of requiring such a search?

PROBLEMS

Problem 2-11. Atech and Alpha both manufacture complex medical devices used in diagnosing a variety of ailments. These devices are sold almost exclusively to hospitals, since they cost in excess of $100,000 each. There are several hundred Atech devices currently in use in hospitals throughout the country. Atech, which claims a trade secret in the internal workings of its device, carefully monitors the purchasers of its device. Alpha pays a third party to buy a device from Atech without disclosing that it will be given to Alpha. Once it has obtained the device, Alpha disassembles it and studies it in order to compare it to Alpha's own device. In the course of opening it up, Alpha's engineers pick two internal padlocks on the Atech device. When Atech discovers that Alpha has obtained the device, it demands the unit's return, offering to refund the purchase price. Alpha refuses, and Atech sues for misappropriation of trade secrets.

a) Assume that Atech's trade secrets were worth $5 million. Assume further that the padlocks cost $5 each, and that it costs $100 per lock to pick these padlocks. Has Atech taken reasonable precautions to preserve its secrets? What other security measures must Atech take—and at what cost—both to deter Alpha (and others) from reverse engineering and to preserve its secrecy? Does Atech's sale of the products on the open market automatically preclude a finding of secrecy?

b) Assume Atech has presented the buyer of the machine with a contract that licenses (rather than sells) the machine, subject to the following restrictions: (a) the buyer is prohibited from disclosing anything she learns during the course of using the machine; (b) the buyer is prohibited from reselling the machine; and (c) the buyer is liable for Atech's damages in the event that any third party learns of Atech's secrets from the buyer or the buyer's machine. Is such an agreement enforceable? Would you advise a client thinking of licensing an Atech machine to sign this contract? How would you redraft the agreement to protect the buyer?

c) Assume that after Alpha's engineers picked the padlocks, they gained access to the inner workings of the machine. Assume further that they discovered numerous flaws in the imaging mechanism that caused potentially serious defects in the images (and hence the diagnoses) stemming from the Atech machine. Finally, assume that Alpha's engineers not only fixed these problems but significantly improved on Atech's design and hence the reliability of the machine. Should these facts affect Alpha's liability? Atech's remedy?

Problem 2-12. Bonnie Bluenote, a world-famous blues guitarist, is noted for her distinctive sound, which she gets by tuning her guitar specially every time before she plays. Although the guitar itself is a standard professional model, Bonnie's adjustments of settings on the guitar, amplifier, and sound system combine to produce a distinctive sound. Because the sound is so important to her image, Bonnie guards it carefully. While she is tuning her guitar, only band members and close associates are allowed in the room. When she records in a new studio, she has the sound engineers sign nondisclosure agreements.

One day Freddie Fender-Rhodes, a big fan of Bonnie's and a budding bluesman himself, is hanging around outside the studio where Bonnie is recording her newest album. He happens to see an ID tag, worn by all guests in the studio, in the wastebasket. He fills in his name and walks into the studio. The band members and recording engineers, seeing the tag, let Freddie stay. He observes how Bonnie tunes her guitar, sees the sound board settings, and makes extensive mental notes.

Five months later, Bonnie is shocked to see in a record store a CD by Freddie titled "The Bonnie Bluenote Sound." Then she discovers that Freddie is planning to publish an article in Blues Guitar magazine revealing the secrets to Bonnie's sound. Does she have any recourse against Freddie? What additional precautions should she have taken?

4. The Special Case of Departing Employees

Many of the thorniest issues in trade secret (and contract) law arise when employees leave a company in order either to start their own business or to take a job elsewhere. Such cases present a fundamental clash of rights between an employee and an employer, as Judge Adams suggests in the following excerpt from SI Handling Systems v. Heisley, 753 F.2d 1244, 1266-1269 (3d Cir. 1985) (Adams, J., concurring):

> When deciding the equitable issues surrounding the request for a trade secret injunction, it would seem that a court cannot act as a pure engineer or scientist, assessing the technical import of the information in question. Rather, the court must also consider economic factors since the very definition of "trade secret" requires an assessment of the competitive advantage a particular item of information affords to a business. Similarly, among the elements to be weighed in determining trade secret status are the value of the information to its owner and to competitors, and the ease or difficulty with which the information may be properly acquired or duplicated.
>
> While the majority may be correct in suggesting that the trial court need not always "engage in extended analysis of the public interest," the court on occasion must apply the elements of sociology. This is so since trade secret cases frequently implicate the important countervailing policies served on one hand by protecting a business person from unfair competition stemming from the usurpation of trade secrets, and on the other by permitting an individual to pursue unhampered the occupation for which he or she is best suited. "Trade secrets are not . . . so important to society that the interests of employees, competitors and competition should automatically be relegated to a lower position whenever trade secrets are proved to exist." Robison, The Confidence Game: An Approach to the Law About Trade Secrets, 25 Ariz. L. Rev. 347, 382 (1983).
>
> These observations take on more force, I believe, when a case such as the present one involves the concept of "know-how." Under Pennsylvania law an employee's general knowledge, skill, and experience are not trade secrets. Thus in theory an employer generally may not inhibit the manner in which an employee uses his or her knowledge, skill, and experience — even if these were acquired during employment. When these attributes of the employee are inextricably related to the information or process that constitutes an employer's competitive advantage — as increasingly seems to be the case in newer, high-technology industries — the legal questions confronting the court necessarily become bound up with competing public policies.
>
> It is noteworthy that in such cases the balance struck by the Pennsylvania courts apparently has favored greater freedom for employees to pursue a chosen profession. The courts have recognized that someone who has worked in a particular field cannot be expected to forego the accumulated skills, knowledge, and experience gained before the employee changes jobs. Such qualifications are obviously very valuable to an employee seeking to sell his services in the marketplace. A person leaving one employer and going into the marketplace will seek to compete in the area of his or her greatest aptitude. In light of the highly mobile nature of our society, and as the economy becomes increasingly comprised of highly skilled or high-tech jobs, the individual's economic interests will more and more be buffeted by competitive advantage. Courts must be cautious not to strike a balance that unduly disadvantages the individual worker. . .
>
> In my view a proper injunction necessarily would impose the minimum restraint upon the free utilization of employee skill consistent with denying unfaithful employees an advantage from misappropriation of information. Thus, as I see it, the district court, on remand, should fashion an injunction that extends only so long as is essential to negate any unfair advantage that may have been gained by the appellants.

The majority opinion in *SI Handling* partially upheld a finding that two former employees had misappropriated trade secrets, but it vacated an injunction against them in order for the district court to reconsider its scope.

a. Employee Trade Secrets

Most companies require their employees to sign some sort of employment agreement, either when they are hired or at some point during their tenure. Employment agreements generally fall into one or more of three categories: confidentiality agreements, invention assignments, and noncompetition agreements. Confidentiality agreements, with which we are concerned here, generally recite that the employee will receive confidential information during her employment, and that she undertakes to keep it secret and not to use it for anyone other than the employer. Invention assignments give the employer the right to intellectual property created by the employee while she was employed. Noncompetition agreements limit the circumstances in which former employees can compete for customers with their former employers.

Presumably, employers may require their employees to agree not to use or disclose trade secrets to which they have access during employment. Virtually all employers do so. The problem comes when employers attempt to limit an employee's use of information that does not constitute a trade secret under the legal standards described above. Can employers and employees "agree" that the employee will not use *any* information obtained during employment, whether it is public or not? This question parallels the problem of restrictive license provisions — are employers limited by the intellectual property laws to protecting only trade secrets that meet the statutory requirements, or are they free to impose additional restrictions on their employees so long as they do so by agreement? Courts struggle with this issue, with the majority concluding that "reasonable" contract restrictions on employees are enforceable even in the absence of a protectable trade secret. See Bernier v. Merrill Air Engineers, 770 A.2d 97 (Me. 2001); 12 Roger Milgrim, Milgrim on Trade Secrets § 3.05[1][a], at 3-209 to 3-210.

An employee's obligation to refrain from using an employer's trade secrets looks more onerous when it is the employee herself who came up with the secret. Consider the following case.

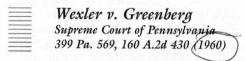

Wexler v. Greenberg
Supreme Court of Pennsylvania
399 Pa. 569, 160 A.2d 430 (1960)

Buckingham Wax Company is engaged in the manufacture, compounding and blending of sanitation and maintenance chemicals. In March, 1949, appellant Greenberg, a qualified chemist in the sanitation and maintenance field, entered the employ of Buckingham as its chief chemist and continued there until April 28, 1957. In the performance of his duties, Greenberg consumed half of his working time in Buckingham's laboratory where he would analyze and duplicate competitors' products and then use the resulting information to develop various new formulas. He would change or modify these formulas for color, odor or viscosity in order that greater commercial use could be made of Buckingham's products. . . . As a result of his activities Greenberg was not only familiar with Buckingham's formulas, he was also fully conversant with the costs of the products and the most efficient method of producing them. . . .

... In August, 1957, Greenberg left Buckingham and went to work for Brite [, who up until that time had purchased its chemicals exclusively from Buckingham]. At no time during Greenberg's employment with Buckingham did there exist between them a written or oral contract of employment or any restrictive agreement.

Prior to Greenberg's association with Brite, the corporation's business consisted solely of selling a complete line of maintenance and sanitation chemicals, including liquid soap cleaners, wax base cleaners, disinfectants, and floor finishes. Upon Greenberg's arrival, however, the corporation purchased equipment and machinery and, under the guidance and supervision of Greenberg, embarked on a full-scale program for the manufacture of a cleaner, floor finish and disinfectant, products previously purchased from Buckingham. The formulas in issue in this litigation are the formulas for each of these respective products. The appellants dispute the chancellor's findings as to the identity of their formulas with those of Buckingham, but there was evidence that a spectrophotometer examination of the respective products of the parties revealed that the formulas used in making these products are substantially identical. Appellants cannot deny that they thought the products sufficiently similar as to continue delivery of their own products to their customers in the same cans and drums and with the same labels attached which they had previously used in distributing the products manufactured by Buckingham, and to continue using the identical promotional advertising material. Appellees' formulas had been developed during the tenure of Greenberg as chief chemist and are unquestionably known to him.

The Chancellor found that Greenberg did not develop the formulas for Brite's products after he left Buckingham, but rather that he had appropriated them by carrying over the knowledge of them which he had acquired in Buckingham's employ. The Chancellor went on to find that the formulas constituted trade secrets and that their appropriation was in violation of the duty that Greenberg owed to Buckingham by virtue of his employment and the trust reposed in him. Accordingly, the relief outlined above was ordered [enjoining Greenberg and Brite from disclosing the secret formulas or making or selling any products using the formulas].

We are initially concerned with the fact that the final formulations claimed to be trade secrets were not disclosed to Greenberg by the appellees during his service or because of his position. Rather, the fact is that these formulas had been developed by Greenberg himself, while in the pursuit of his duties as Buckingham's chief chemist, or under Greenberg's direct supervision. We are thus faced with the problem of determining the extent to which a former employer, without the aid of any express covenant, can restrict his ex-employee, a highly skilled chemist, in the uses to which this employee can put his knowledge of formulas and methods he himself developed during the course of his former employment because this employer claims these same formulas, as against the rest of the world, as his trade secrets. This problem becomes particularly significant when one recognizes that Greenberg's situation is not uncommon. In this era of electronic, chemical, missile and atomic development, many skilled technicians and expert employees are currently in the process of developing potential trade secrets. Competition for personnel of this caliber is exceptionally keen, and the interchange of employment is commonplace. ... We must therefore be particularly mindful of any effect our decision in this case might have in disrupting this pattern of employee mobility, both in view of possible restraints upon an individual in the pursuit of his livelihood and the harm to the public in general in forestalling, to any extent, widespread technological advances.

The principles outlining this area of the law are clear. A court of equity will protect an employer from the unlicensed disclosure or use of his trade secrets by an ex-employee provided the employee entered into an enforceable covenant so restricting his use, Fralich v. Despar, 165 Pa. 24, 30 Atl. 521 (1894), or was bound to secrecy by virtue of a confidential relationship existing between the employer and employee, Pittsburgh Cut Wire Co. v. Sufrin, 350 Pa. 31, 38 A.2d 33 (1944). Where, however, an employer has no legally protectible trade secret, an employee's "aptitude, his skill, his dexterity, his manual and mental ability, and such other subjective knowledge as he obtains while in the course of his employment, are not the property of his employer and the right to use and expand these powers remains his property unless curtailed through some restrictive covenant entered into with the employer." Id. at 35. The employer thus has the burden of showing two things: (1) a legally protectable trade secret; and (2) a legal basis, either a covenant or a confidential relationship, upon which to predicate relief. . . .

The usual situation involving misappropriation of trade secrets in violation of a confidential relationship is one in which an employer *discloses to his employee* a pre-existing trade secret (one already developed or formulated) so that the employee may duly perform his work. In such a case, the trust and confidence upon which legal relief is predicated stems from the instance of the employer's *turning over to the employee* the pre-existing trade secret. It is then that a pledge of secrecy is impliedly extracted from the employee, a pledge which he carries with him even beyond the ties of his employment relationship. Since it is conceptually impossible, however, to elicit an implied pledge of secrecy from the sole act of an employee turning over to his employer a trade secret which he, the employee, has developed, as occurred in the present case, the appellees must show a different manner in which the present circumstances support the permanent cloak of confidence cast upon Greenberg by the chancellor. The only avenue open to the appellees is to show that the nature of the employment relationship itself gave rise to a duty of nondisclosure.

The burden the appellees must thus meet brings to the fore a problem of accommodating competing policies in our law: the right of a businessman to be protected against unfair competition stemming from the usurpation of his trade secrets and the right of an individual to the unhampered pursuit of the occupations and livelihoods for which he is best suited. There are cogent socio-economic arguments in favor of either position. Society as a whole greatly benefits from technological improvements. Without some means of post-employment protection to assure that valuable developments or improvements are exclusively those of the employer, the businessman could not afford to subsidize research or improve current methods. In addition, it must be recognized that modern economic growth and development has pushed the business venture beyond the size of the one-man firm, forcing the businessman to a much greater degree to entrust confidential business information relating to technological development to appropriate employees. While recognizing the utility in the dispersion of responsibilities in larger firms, the optimum amount of "entrusting" will not occur unless the risk of loss to the businessman through a breach of trust can be held to a minimum.

On the other hand, any form of post-employment restraint reduces the economic mobility of employees and limits their personal freedom to pursue a preferred course of livelihood. The employee's bargaining position is weakened because he is potentially shackled by the acquisition of alleged trade secrets; and thus, paradoxically, he is restrained, because of his increased expertise, from advancing further in the industry in which he is most productive. Moreover, as previously mentioned, society suffers because competition is diminished by slackening the dissemination of ideas, processes and methods.

Were we to measure the sentiment of the law by the weight of both English and American decisions in order to determine whether it favors protecting a businessman from certain forms of competition or protecting an individual in his unrestricted pursuit of a livelihood, the balance would heavily favor the latter....

[The court examined cases from other jurisdictions that found a confidential relationship to exist when the employer disclosed trade secrets to the employee.] Upon our examination of the record here, however, we find that the instant circumstances fall far short of such a relationship. The Chancellor's finding that Greenberg, while in the employ of Buckingham, never engaged in research nor conducted any experiments nor created or invented any formula was undisputed. There is nothing in the record to indicate that the formulas in issue were specific projects of great concern and concentration by Buckingham; instead it appears that they were merely the result of Greenberg's routine work of changing and modifying formulas derived from competitors. Since there was no experimentation or research, the developments by change and modification were fruits of Greenberg's own skill as a chemist without any appreciable assistance by way of information or great expense or supervision by Buckingham....

COMMENTS AND QUESTIONS

1. *Wexler* is explicitly based on the absence of a written agreement between employer and employee.[16] Suppose that Greenberg had signed a contract when he was hired which stated that "all discoveries, inventions, or intellectual property rights stemming from employee's work at Buckingham are the exclusive property of Buckingham." Would such a contract be enforceable? Does it matter whether Greenberg's "discoveries" were new formulas or ways to copy competitors' formulas?

An example of the more common situation, in which the court finds either an express or implied agreement to assign an employee's own inventions, is Winston Research Corp. v. 3M Co., 350 F.2d 134 (9th Cir. 1965). There, the court dismissed an argument akin to Greenberg's as follows:

> Winston argues that information is protected from disclosure only if communicated to the employee by the employer who is seeking protection, and that the information involved in this case was not disclosed by [3M] to the employees subsequently hired by Winston, but rather was developed by these employees themselves, albeit while employed by [3M].
>
> We need not examine the soundness of the rule for which Winston contends, or its applicability to a case such as this in which a group of specialists engaged in related facets

16. The absence of such an agreement can be significant even in situations in which it might be assumed that the "employer" had a reasonable claim to the employee's invention. Cf. Lariscey v. United States, 20 U.S.P.Q.2d 1845 (Fed. Cir. 1991) (prison inmate owns rights to a secret process he developed while working in a prison shop).

of a single development project change their employer. . . . [A]n obligation not to disclose may arise from circumstances other than communication in confidence by the employer. It may also rest upon an express or implied agreement. In the present case, an agreement not to disclose might be implied from [3M]'s elaborate efforts to maintain the secrecy of its development program, and the employees' knowledge of those efforts and participation in them. In any event, [3M] and its employees entered into express written agreements binding the latter not to disclose confidential information, and these agreements did not exclude information which the employee himself contributed.

Id. at 140.

In addition, even in Pennsylvania *Wexler* is not always followed. See Healthcare Affiliated Services v. Lippancy, 701 F. Supp. 1142, 1155 (W.D. Pa. 1988) (rejecting *Wexler* analysis, emphasizing that although defendant developed inventions on his own, he did so using knowledge and information made available by the plaintiff employer). But see Fidelity Fund v. DiSanto, 500 A.2d 431 (Sup. Ct. Pa. 1985) (denying recovery against ex-employee salesman partly on the basis that he developed client contacts himself during employment).

2. Does the *Wexler* court confuse an employee's *general* knowledge and skills with specific inventions developed by the employee while at work *based* on that knowledge and skills? If an employee's job involves two tasks, working on established products (covered by trade secrets) and developing new ones, what effect will the *Wexler* rule have on the employee's allocation of time between these tasks?

3. A variant on the departing employee who wishes to take her own inventions is the departing sales representative who wishes to "take" a list of customers (either a written list or one that they have memorized) in order to call on those customers for a competitor.[17] Customer lists are generally protectable as trade secrets, but enjoining employees from calling on customers with whom they have had long-standing relationships raises serious concerns about employee mobility. Two California cases frame the dispute. In Moss, Adams & Co. v. Shilling, 179 Cal. App. 3d 124 (1986), the court drew a line between an employee announcing her departure to start a competing company and actively soliciting old clients to follow her. The former was permissible, but the latter was not. This compromise was thrown into doubt by the decision in Morlife Inc. v. Perry, 66 Cal. Rptr. 2d 731 (Ct. App. 1997), where the court held that the passage of the Uniform Trade Secrets Act in California superseded *Moss, Adams* and that continuing personal relationships with former customers could constitute misappropriation of a trade secret.

Where should the line be drawn between permissible work and impermissible use of "secret" lists of customers? To whom does the value inherent in personal relationships belong?

---

17. Merely memorizing a trade secret rather than taking physical documents will not preclude a finding of misappropriation, though it may make misappropriation harder to detect. See Ed Nowogroski Ins. Inc. v. Rucker, 971 P. 2d 936 (Wash. 1999). In Stampede Tool Warehouse Inc. v. May, 651 N.E.2d 209 (Ill. Ct. App. 1995), for example, the plaintiff alleged that former employees who started a competing business had misappropriated its trade secrets by memorizing a list of plaintiff's customers and soliciting those customers. The court found the defendants liable, reasoning that "memorization is one method of misappropriation." But if the case involved merely a former employee contacting those she knew from experience to be potential customers, rather than explicitly attempting to memorize a list of customers, it is hard to fault the employee's conduct.

Note on the Common Law Obligation to Assign Inventions

One important dimension of the *Wexler* case is that none of Greenberg's research yielded a patentable invention. If it had, ownership of the patent rights would have been determined according to Greenberg's status under a long line of common law employee invention cases. In general, employees such as Greenberg fall into one of three categories: (1) employees "hired to invent," which results in employer ownership of the invention; (2) employees who invent on the employer's time or using its resources, which results in a limited, nonexclusive "shop right" on the part of the employer to practice the invention; and (3) an employee's "independent invention," in which case the employee owns the invention. See generally United States v. Dubilier Condenser Corp., 289 U.S. 178 (1933); John C. Stedman, Employer-Employee Relations, in Fredrik Neumeyer, The Employed Inventor in the United States 30, 40-41 (1971).

The first category is relatively straightforward. It seems logical to extend this treatment to consultants and others who are not "employees" in the strict sense. Cf. McElmurry v. Arkansas Power & Light Co., 27 U.S.P.Q.2d 1129, 1135 n.15 (Fed. Cir. 1993) (upholding shop right in employer where inventor/patentee was a consultant); Robert P. Merges, Intellectual Property and the Costs of Commercial Exchange: A Review Essay, 93 Mich. L. Rev. 1570 (1995) (highlighting the role of intellectual property in structuring non-employment-based organizations, such as consulting companies and joint ventures). Category (2) reflects situations where employers have less than a complete claim to the invention, but where the employer's facilities or resources are combined with the inventor's talent and industry to produce the invention. The employee owns it, but the employer is compensated by receiving a limited right to practice the invention. Category (3) covers cases where the employee invents on his or her own time, outside the field of employment. See *Dubilier, supra.*

All the above rules apply only in the absence of an agreement. Of course, most companies generally require their employees to sign an employment agreement, usually on the company's terms. One state has even gone so far as to enact a statute that automatically assigns inventions to an employer provided they were developed in the course of employment and relate to the scope of the employee's work. Nev. Rev. Stat. § 600.500. On the other hand, in an attempt to alleviate the disparity in bargaining power between the employer and the inventor/employee, several states have adopted "freedom to create" statutes restricting the instances in which employers may compel assignment through contract. See Minn. Stat. Ann. § 181.78 (1980); N.C. Gen. Stat. § 66-57.1 to 57.2 (1981); Wash. Rev. Code Ann. § 49.44.140 (1987); Cal. Labor Code § 2870 (West 1987).

These statutes provide that any employee invention assignment agreement that purports to give employers greater rights than they have under the statute is against public policy and unenforceable. The California statute is a good example. It provides that contracts may not require assignment of "invention[s] that the employee developed entirely on his or her own time without using the employer's equipment, supplies, facilities, or trade secret information" unless the invention relates to the employer's current or demonstrably anticipated business. Cal. Labor Code § 2870.

One commentator has argued that the United States is slipping in technological innovation because this country's inventors (the vast majority of whom are employed) have lost their incentive to create partially due to assignment agreements. After labeling the state legislation described above as "a step forward," the author writes: "Comprehensive federal legislation is necessary to provide employed inventors greater control over, and interests in, their inventions in order to create the psychological and financial incentives required to increase technological innovation in the United States.... Federal legislation would obviate the need for state reform and would bring uniformity that would solve the potential problem of inventors employed by multistate corporations." Henrik D. Parker, Reform for Rights of Employed Inventors, 57 S. Cal. L. Rev. 603 (1984). But see Robert P. Merges, The Law and Economics of Employee Inventions, 13 Harv. J. L. & Tech. 1 (1999).

PROBLEM

Problem 2-13. Helen, an employee of CarTech, is assigned to work on solving a particular problem in car design. She works for two years on that problem for CarTech, without success. Two months after Helen leaves CarTech to start her own company, she puts out a product that incorporates a solution to the same problem. CarTech sues Helen for misappropriation of trade secrets. At trial, CarTech proves that Helen came up with her system while at CarTech and that she decided to start her own company to exploit her invention rather than disclose it to CarTech. As a result, CarTech had no knowledge of the idea until Helen came out with her product. Should CarTech prevail in its trade secret suit, assuming that there was no employment or invention contract between the parties?

Note on Trailer Clauses

The common law rules just described apply in the absence of contracts, as stated; but they also have some relevance even when written agreements are in place. This is so because, under principles similar to those applied in the law of noncompetition agreements, the law limits the ability of employers to claim ownership by contract of all employee inventions no matter how tenuously related to the employment relationship. A case study in the limits of invention assignment agreements involves the "trailer clause," a contractual provision that requires employees to assign their rights not only in inventions made during the period of employment but also for a certain time thereafter. Trailer clauses developed in response to a number of cases which held that, absent contractual provisions to the contrary, an employee's ideas did not belong to the employer unless they were written down in tangible form (such as a patent disclosure statement) before the employee left work. See Jamesbury Corp. v. Worcester Valve Co., 443 F.2d 205 (1st Cir. 1971).

Of course, even if an employer uses a trailer clause, there is always the risk that the former employee will simply wait out the duration of the term and then conveniently make the discovery upon the trailer clause's expiration. Such a strategy was attempted by an inventor in at least one case. Fortunately for the employer, the court did not believe the defendant inventor's story that the conception date for his flow meter was five days after the expiration of the six months specified in the trailer clause:

The perfection of a flow meter proved to be a painstakingly intricate process involving extensive testing. It is therefore difficult to believe that after a long and distinguished career with Plaintiff, Mr. Halmi in his musing five days after the trailer clause expired for the first time came up with the idea for the NTV. Although the word "Eureka!" has allegedly been uttered by more than one inventor over the years, the concept at issue does not lend itself to such sudden discovery. The court finds that the concept of the '434 patent must have existed in Mr. Halmi's mind before his employment with GSC ended. Mr. Halmi therefore violated his agreement with GSC.

General Signal Corp. v. Primary Flow Signal, Inc., 1987 U.S. Dist. LEXIS 6929 at 10 (D.R.I. Jul. 27, 1987).

Generally, trailer clauses are enforceable only to the extent that they are "reasonable." Although trailer clauses are generally enforceable, clauses of particularly long or indefinite duration may run afoul of the antitrust laws as well as being unenforceable. See United Shoe Machinery Co. v. La Chapelle, 212 Mass. 467, 99 N.E. 289 (1912). One court expressed the requirement of reasonableness as follows:

> Hold-over clauses are simply a recognition of the fact of business life that employees sometimes carry with them to new employers inventions or ideas so related to work done for a former employer that in equity and good conscience the fruits of that work should belong to that former employer. In construing and applying hold-over clauses, the courts have held that they must be limited to *reasonable times*... and to *subject matter* which an employee worked on or had knowledge of during his employment.... Unless expressly agreed otherwise, an employer has no right under a holdover clause to inventions made outside the scope of the employee's former activities, and made on and with a subsequent employer's time and funds.

Dorr-Oliver, Inc. v. United States, 432 F.2d 447, 452 (Ct. Cl. 1970). Corporations may have trouble enforcing restrictions that are broader in scope. This is particularly true of large conglomerates that attempt to require the assignment of any invention related to their (diverse) fields of business. See Ingersoll-Rand Co. v. Ciavatta, 110 N.J. 609, 542 A.2d 879, 896 n.6 (1988).

Several proposals designed to improve the rights of the employed inventor have been introduced into Congress. In 1981, for example, Rep. Robert Kastenmeier introduced two bills, one of which would have totally eliminated trailer clauses. Do you see any problems with doing so? Who should bear the burden of proof with respect to inventions arguably made during the period of employment?

b. Noncompetition Agreements

Employees inevitably learn a great deal about a business on the job. They may learn the trade secrets of their employer, but they may also learn such basic information as how many people it takes to make a product, who is likely to buy the product, where to advertise, etc. Further, employees develop personal relationships with vendors, customers, and others that are immensely useful in business. For this reason, many employers would like to prevent their employees from competing against them at all. Such employers often ask their employees to sign "noncompetition agreements," which prevent the employee from competing with his former employer for customers for a set period of time. Should such agreements be enforceable?

In *Comprehensive Technologies Intl. v. Software Artisans, Inc.*, 3 F.3d 730 (4th Cir. 1993), CTI brought suit for copyright infringement and misappropriation of trade secrets against a group of former employees who left the company to form a competing company which shortly thereafter came out with a new product. The court concluded that the departing employees had not infringed CTI's copyrights or misappropriated any CTI trade secrets. Nonetheless, the court enforced an agreement signed by one of the employees, Dean Hawkes. The agreement provided that for a period of twelve months after he left CTI, Hawkes would not

> engage directly or indirectly in any business within the United States (financially as an investor or lender or as an employee, director, officer, partner, independent contractor, consultant or owner or in any other capacity calling for the rendition of personal services or acts of management, operation or control) which is in competition with the business of CTI. For purposes of this Agreement, the "business of CTI" shall be defined as the design, development, marketing, and sales of CLAIMS EXPRESS) and EDI LINK) type PC-based software with the same functionality and methodology. . . .

The court stated the general legal standard governing covenants not to compete:

> Virginia has established a three-part test for assessing the reasonableness of restrictive employment covenants. Under the test, the court must ask the following questions:
>
> (1) Is the restraint, from the standpoint of the employer, reasonable in the sense that it is no greater than is necessary to protect the employer in some legitimate business interest?
> (2) From the standpoint of the employee, is the restraint reasonable in the sense that it is not unduly harsh and oppressive in curtailing his legitimate efforts to earn a livelihood?
> (3) Is the restraint reasonable from the standpoint of a sound public policy?
>
> *Blue Ridge Anesthesia & Critical Care, Inc. v. Gidick*, 239 Va. 369, 389 S.E.2d 467, 469 (Va. 1990). If a covenant not to compete meets each of these standards of reasonableness, it must be enforced. As a general rule, however, the Virginia courts do not look favorably upon covenants not to compete, and will strictly construe them against the employer. The employer bears the burden of demonstrating that the restraint is reasonable.

The court found that Hawkes's agreement, which prevented him from competing with CTI anywhere in the United States, was reasonable because CTI had offices, clients, or prospects in many (though not all) states throughout the country. Further, the court noted:

> . . . As the individual primarily responsible for the design, development, marketing and sale of CTI's software, Hawkes became intimately familiar with every aspect of CTI's operation, and necessarily acquired information that he could use to compete with CTI in the marketplace. When an employee has access to confidential and trade secret information crucial to the success of the employer's business, the employer has a strong interest in enforcing a covenant not to compete because other legal remedies often prove inadequate. It will often be difficult, if not impossible, to prove that a competing employee has misappropriated trade secret information belonging to his former employer. On the facts of this case, we conclude that the scope of the employment restrictions is no broader than necessary to protect CTI's legitimate business interests.

Most courts similarly apply an overarching requirement of "reasonableness" to covenants not to compete. See Mich. Comp. Laws §445.774a (noncompetition agreements enforceable if the agreement is "reasonable as to its duration, geographical area, and type of employment or line of business"). They may disagree, however, on what restrictions are reasonable. In Gateway 2000 Inc. v. Kelley, 9 F. Supp. 2d 790 (E.D. Mich. 1998), the court invalidated an agreement that was similar to the one upheld in CTI. The court relied in part on the fact that the company had later adopted a less restrictive noncompetition provision, suggesting that the older, broader provision was not necessary to protect its interests. And the Virginia Supreme Court has held unreasonable a noncompetition agreement that prevented the defendant from working for a competitor in any capacity, rather than specifying particular positions the defendant could not take. Modern Environments, Inc. v. Stinnett, 561 S.E.2d 694 (Va. 2002).

COMMENTS AND QUESTIONS

1. There seems to be no question in the court's mind that none of the defendants misappropriated any CTI trade secrets, infringed any copyrights, or otherwise "took" anything belonging to CTI in starting Software Artisans. Why doesn't that dispose of the case? What social purpose is served by enjoining former employees from pursuing their livelihood? Shouldn't the mobility and liberty of individuals be the paramount consideration, as California courts have suggested? See Diodes, Inc. v. Franzen, 260 Cal. App. 2d 244 (1968). One author has suggested that there is a more practical economic motivation for precluding such noncompetition agreements. She argues that the relative success of California's Silicon Valley compared to Boston's Route 128 is directly attributable to the prevalence of noncompetition agreements in Route 128 companies, which prevented the free movement of employees and therefore discouraged start-up companies. See Annalee Saxenian, Regional Advantage: Culture and Competition in Silicon Valley and Route 128 (1994).

Some courts have limited the enforcement of noncompetition agreements to situations where trade secrets are likely to be used or disclosed if an employee is allowed to compete. The New York Court of Appeals, for example, took the following view:

> Undoubtedly judicial disfavor of these covenants is provoked by "powerful considerations of public policy which militate against sanctioning the loss of a man's livelihood" (Purchasing Assoc. v. Weitz. . . .) Indeed, our economy is premised on the competition engendered by the uninhibited flow of services, talent and ideas. Therefore, no restrictions should fetter an employee's right to apply to his own best advantage the skills and knowledge acquired by the overall experience of his previous employment. This includes those techniques which are but "skillful variations of general processes known to the particular trade" (Restatement, Agency 2d, §396 Comment *b*).
>
> Of course, the courts must also recognize the legitimate interest an employer has in safeguarding that which has made his business successful and to protect himself against deliberate surreptitious commercial piracy. Thus restrictive covenants will be enforceable to the extent necessary to prevent the disclosure or use of trade secrets or confidential customer information. In addition injunctive relief may be available where

an employee's services are unique or extraordinary and the covenant is reasonable. This latter principle has been interpreted to reach agreements between members of the learned professions.

Reed Roberts Assoc. v. Strauman, 40 N.Y.2d 303, 353 N.E.2d 590 (Ct. App. 1976). Does the *Reed Roberts* approach in essence hold noncompetition agreements unenforceable, since it allows them to operate only when trade secret laws also provide relief? Are there sound reasons to enforce an employer-employee agreement that prevents the employee from competing after termination? State statutes which address the issue have generally been interpreted to allow such "reasonable" employee agreements, regardless of how the statutes themselves are worded. Of course, what agreements are "reasonable" is far from clear in this context, and is the subject of considerable litigation. Many states have invalidated noncompetition agreements on the ground that they were unreasonable. See, e.g., Mutual Service Casualty Ins. Co. v. Brass, 625 N.W.2d 648 (Wisc. App. 2001); Brentlinger Ents. v. Curran, 752 N.E.2d 994 (Ohio App. 2001); Harvey Barnett, Inc. v. Shidler, 143 F. Supp. 2d 1247 (D. Colo. 2001); Mertz v. Pharmacists Mutual Ins., 625 N.W.2d 197 (Neb. 2001); City Slickers, Inc. v. Douglas, 40 S.W.3d 805 (Ark. App. 2001).

2. Other states, notably California, have an even more restrictive rule. Cal. Bus. & Prof. Code § 16600 provides that "every contract by which anyone is restrained from engaging in a lawful profession . . . is to that extent void." California courts have interpreted this statute to bar noncompetition agreements altogether in employee contracts but to permit such agreements if they are ancillary to the sale of a business, so long as the terms of the agreement are "reasonable." See Monogram Indus., Inc. v. SAR Indus., Inc., 64 Cal. App. 3d 692, 134 Cal. Rptr. 714, 718 (1976). Further, while California courts will not enforce a noncompetition agreement, they will prevent departing employees from using or disclosing their former employer's trade secrets. See State Farm Mutual Automobile Ins. Co. v. Dempster, 344 P.2d 821 (Cal. App. 1959); Gordon v. Landau, 49 Cal. 2d 690, 321 P.2d 456 (Cal. 1958). Other states, including Alabama, Florida, Louisiana, Montana, and North Dakota, also forbid noncompetition agreements. Still other states, including Colorado, Delaware, Massachusetts, and Tennessee, forbid them in professional settings. See, e.g., Murfreesboro Med. Clinic v. Udom, 166 S.W.3d 674 (Tenn. 2005).

The strength of California's commitment to the free movement of employees was demonstrated in The Application Group, Inc. v. The Hunter Group, Inc., 72 Cal. Rptr. 2d 73 (Ct. App. 1998). There, the California Court of Appeals held that § 16600 precluded the enforcement of a noncompetition agreement entered into in Maryland between a Maryland employer and employee, where the employee subsequently left to take a job telecommuting from Maryland for a California company. Despite the fact that Maryland courts would enforce the agreement, the California court concluded that California's interests were "materially stronger" than Maryland's in this case. See also D'Sa v. Playhut, Inc., 85 Cal. App. 4th 927, 102 Cal. Rptr. 2d 495 (2000) (disregarding choice of law provision in holding a noncompete agreement unenforceable).

California's strong public policy has led to conflicts with other states. The most notable example is Advanced Bionics v. Medtronic, 87 Cal. App. 4th 1235 (2001), in which both California and Minnesota courts asserted that their law should control, with the Minnesota court enjoining the departing employee and the California court enjoining the employer from proceeding with the suit. The California Supreme Court ultimately reversed, not because it didn't consider the policy of employee mobility important but because it thought the specter of conflicting judgments unseemly. Advanced Bionics v. Medtronic, 29 Cal. 4th 697, 128 Cal. Rptr. 2d 172 (2002). But doing so may simply subjugate California's policy to the law of any other state that would enforce a noncompete agreement, even if the employee doesn't work in that state. See IBM Corp. v. Bajorek, 191 F.3d 1033 (9th Cir. 1999) (applying New York law to enjoin competition against New York company by employee in California, and disregarding California policy to the contrary). For a contrary ruling giving nation-wide effect to a refusal to enforce a noncompete agreement under Georgia law, see Palmer & Cay v. Marsh & McLennan, 404 F.3d 1297 (11th Cir. 2005).

Is there a reasoned way to resolve such conflicts in public policy? Or will the inevitable result be a "race to the courthouse"?

3. What are the competing policy interests at stake in noncompetition clauses? On the one hand, such restrictions seem onerous burdens to impose on employees. Imagine how you would feel as an attorney if you left a firm only to find that you were prevented from practicing law in the same field or geographic region for the next two years. (In this regard, it is significant that even *Reed Roberts* expressed the view that the "learned professions" were properly subject to noncompetition agreements.) In addition, it is not completely clear that such provisions benefit companies in the long run. Strauman, the defendant in *Reed Roberts*, came to Reed Roberts after having worked for a competitor for four years. He was hired in part because of his valuable experience in the industry. What if Strauman's former employer had required him to sign an enforceable noncompete agreement?

On the other hand, it seems unfair to employers to simply allow their employees to do whatever they want upon leaving. Particularly where the employees were in positions of importance, their knowledge of the employer's trade secrets may leave the former employer at a competitive disadvantage. In a competitive industry, preventing the disclosure of trade secrets is far preferable to suing for misappropriation after they have already been disclosed. A noncompetition agreement may be a reasonable way for an employer to prevent a problem — and a lawsuit — before it starts.

4. When a firm requires a new employee to sign an employment agreement containing a covenant not to compete, the employee is giving up something substantial. What is the employer giving up? Some cases have raised the issue of consideration (in the contract law sense) in such an agreement on the part of the employer; they generally conclude that there is consideration, on one theory or another. See, e.g., Central Adjustment Bureau v. Ingram, 678 S.W.2d 28 (Tenn. 1984) (consideration in the form of continuous employment over a long period of time). But see Light v. Centel Cellular Co., 883 S.W.2d 642 (Tex. 1994) (holding that an at-will employment agreement did not provide adequate consideration for a noncompete clause, since the employee could be fired at any time).

Why not require, out of fairness, that an employer who insists on such a covenant must pay the employee's salary during the term of the noncompete provision?[18] How would this change the behavior of an ex-employer vis-à-vis the newly departed

18. German law requires just this. See Abschn. Handlungsgehilfen and Handlungslehrlinge §74, in Handelsgesetzbuch 27 (2001). In the United States, some companies have adopted such a rule voluntarily. See Marcam Corp. v. Orchard, 885 F. Supp. 294 (D. Mass. 1995) (enforcing a contractual provision preventing Orchard from working for any competitor in the country for one year, provided that Marcam paid 110 percent of the salary offered by the competitor).

employee? Would you expect the rate of employee mobility to be higher under such a system?

5. Suppose that a California employer offered an employee stock options, but required that the employee not go to work for a competitor for six months after exercising those options. Would such a restriction violate Bus. & Prof. Code § 16600? The Ninth Circuit said no in IBM Corp. v. Bajorek, 191 F.3d 1033 (9th Cir. 1999). The court's reasoning is suspect, however, because it interpreted § 16600 as prohibiting only contracts that "completely restrain" an employee from competing. California state court decisions have been less sympathetic to such partial restraints.

6. Can an employer avoid state laws restricting noncompetition agreements by requiring the employee to sign an agreement that does not forbid employment, but calls for the payment of a "liquidated damage" penalty if the employee goes to work for a competitor? Is such a monetary penalty effectively the same as enforcing a non-compete?

Note on the "Inevitable Disclosure" of Trade Secrets

Are there circumstances in which an employee's use or disclosure of trade secrets is "inevitable"? Consider the position of a former head of research and development at a computer company. Is there any way that that employee can "keep separate" the ideas and projects he was working on for his old employer from the ideas and projects he will be asked to develop for his new employer? If not, should the employer be entitled to prevent the employee from competing even if it cannot show that the employee *intends* to use its trade secrets? IBM made just this argument when one of its disk drive specialists, Peter Bonyhard, left IBM to work for Seagate Technology in the same field. A preliminary injunction granted by the district court against Bonyhard's employment was reversed by the Eighth Circuit in an unpublished decision. IBM v. Bonyhard, 962 F.2d 12 (8th Cir. 1992). Several other decisions have also rejected inevitable disclosure as a basis for an injunction against employment. Campbell Soup Co. v. Giles, 47 F.3d 467 (1st Cir. 1995); FMC Corp. v. Cyprus Foote Mineral Co., 899 F. Supp. 1477 (W.D.N.C. 1995). In *Campbell,* the court grounded its rejection of inevitable disclosure theory in the public policy that favors employee mobility. It quoted the district court opinion in the case, holding that public policy "counsels against unilateral conversion of non-disclosure agreements into non-competitive agreements. If Campbell wanted to protect itself against the competition of former employees, it should have done so by contract. This court will not afford such protection after the fact." See also Carolina Chem. Equip. Co. v. Muckenfuss, 471 S.E.2d 721 (S.C. App. 1996) (agreement entitled "Covenant Not to Divulge Trade Secrets" was actually an overly broad and unenforceable covenant not to compete, because the definition of trade secrets in the agreement was so broad as to effectively prevent any competition).

On the other hand, the Seventh Circuit ordered the issuance of a preliminary injunction preventing Quaker Oats Co. from employing Redmond, a former general manager for PepsiCo North America. Redmond was general manager of PepsiCo's California business unit for ten years until, in 1994, he accepted Quaker's offer to become the chief operating officer of its Gatorade and Snapple Co. divisions. The court held that Redmond would inevitably be forced to use PepsiCo trade secrets for his new employer:

> PepsiCo asserts that Redmond cannot help but rely on PCNA [PepsiCo North America] trade secrets as he helps plot Gatorade and Snapple's new course, and that these secrets will enable Quaker to achieve a substantial advantage by knowing exactly how PCNA will price, distribute, and market its sports drinks and new age drinks and being able to respond strategically. This type of trade secret problem may arise less often, but it nevertheless falls within the realm of trade secret protection under the present circumstance.
>
> Quaker and Redmond assert that they have not and do not intend to use whatever confidential information Redmond has by virtue of his former employment. They point out that Redmond has already signed an agreement with Quaker not to disclose any trade secrets or confidential information gleaned from his earlier employment. They also note with regard to distribution systems that even if Quaker wanted to steal information about PCNA's distribution plans, they would be completely useless in attempting to integrate the Gatorade and Snapple beverage lines.
>
> The defendants' arguments fall somewhat short of the mark. Again, the danger of misappropriation in the present case is not that Quaker threatens to use PCNA's secrets to create distribution systems or coopt PCNA's advertising and marketing ideas. Rather, PepsiCo believes that Quaker, unfairly armed with knowledge of PCNA's plans, will be able to anticipate its distribution, packaging, pricing, and marketing moves. Redmond and Quaker even concede that Redmond might be faced with a decision that could be influenced by certain confidential information that he obtained while at PepsiCo. In other words, PepsiCo finds itself in the position of a coach, one of whose players has left, playbook in hand, to join the opposing team before the big game. Quaker and Redmond's protestations that their distribution systems and plans are entirely different from PCNA's are thus not really responsive. . . .
>
> Quaker and Redmond do not assert that the confidentiality agreement is invalid; such agreements are enforceable when supported by adequate consideration.[10] Rather, they argue that "inevitable" breaches of these contracts may not be enjoined. The case on which they rely, however, R. R. Donnelley & Sons Co. v. Fagan, 767 F. Supp. 1259 (S.D.N.Y. 1991) (applying Illinois law), says nothing of the sort. The *R. R. Donnelley* court merely found that the plaintiffs had failed to prove the existence of any confidential information or any indication that the defendant would ever use it. Id. at 1267. The threat of misappropriation that drives our holding with regard to trade secrets dictates the same result here.

PepsiCo, Inc. v. Redmond, 54 F.3d 1262, 35 U.S.P.Q.2d 1010 (7th Cir. 1995). The court distinguished two prior cases refusing to find that an ex-employee's disclosure of trade secrets was inevitable:

> In *Teradyne,* Teradyne alleged that a competitor, Clear Communications, had lured employees away from Teradyne and intended to employ them in the same field. In an insightful opinion, Judge Zagel observed that "threatened misappropriation can be enjoined under Illinois law" where there is a "high degree of probability of inevitable

---

10. The confidentiality agreement is also not invalid for want of a time limitation. See 765 ILCS 1065/8(b)(1) ("[A] contractual or other duty to maintain secrecy or limit use of a trade secret shall not be deemed to be void or unenforceable solely for lack of durational or geographic limitation on the duty."). Nor is there any question that the confidentiality agreement covers much of the information PepsiCo fears Redmond will necessarily use in his new employment with Quaker.

and immediate . . . use of . . . trade secrets." *Teradyne,* 707 F. Supp. at 356. Judge Zagel held, however, that Teradyne's complaint failed to state a claim because Teradyne did not allege "that defendants have in fact threatened to use Teradyne's secrets or that they will inevitably do so." [The *Teradyne* court held]:

> the defendants' claimed acts, working for Teradyne, knowing its business, leaving its business, hiring employees from Teradyne and entering the same field (though in a market not yet serviced by Teradyne) do not state a claim of threatened misappropriation. All that is alleged, at bottom, is that defendants could misuse plaintiff's secrets, and plaintiffs fear they will. This is not enough. It may be that little more is needed, but falling a little short is still falling short.

Id. at 357.

In *AMP* we affirmed the denial of a preliminary injunction on the grounds that the plaintiff AMP had failed to show either the existence of any trade secrets or the likelihood that defendant Fleischhacker, a former AMP employee, would compromise those secrets or any other confidential business information. AMP, which produced electrical and electronic connection devices, argued that Fleishhacker's new position at AMP's competitor would inevitably lead him to compromise AMP's trade secrets regarding the manufacture of connectors. *AMP,* 823 F.2d at 1207. In rejecting that argument, we emphasized that the mere fact that a person assumed a similar position at a competitor does not, without more, make it "inevitable that he will use or disclose . . . trade secret information" so as to "demonstrate irreparable injury." Id.

The ITSA, *Teradyne,* and *AMP* lead to the same conclusion: a plaintiff may prove a claim of trade secret misappropriation by demonstrating that defendant's new employment will inevitably lead him to rely on the plaintiff's trade secrets. See also 1 Jager, . . . § 7.02[2][a] at 7-20 (noting claims where "the allegation is based on the fact that the disclosure of trade secrets in the new employment is inevitable, whether or not the former employee acts consciously or unconsciously"). . . .

PepsiCo presented substantial evidence at the preliminary injunction hearing that Redmond possessed extensive and intimate knowledge about PCNA's strategic goals for 1995 in sports drinks and new age drinks. The district court concluded on the basis of that presentation that unless Redmond possessed an uncanny ability to compartmentalize information, he would necessarily be making decisions about Gatorade and Snapple by relying on his knowledge of PCNA trade secrets. It is not the "general skills and knowledge acquired during his tenure with" PepsiCo that PepsiCo seeks to keep from falling into Quaker's hands, but rather "the particularized plans or processes developed by [PCNA] and disclosed to him while the employer-employee relationship existed, which are unknown to others in the industry and which give the employer an advantage over his competitors." *AMP,* 823 F.2d at 1202. The Teradyne and AMP plaintiffs could do nothing more than assert that skilled employees were taking their skills elsewhere; PepsiCo has done much more.

Id.

COMMENTS AND QUESTIONS

1. What Pepsi trade secrets are threatened by Redmond's "defection" to Quaker? The information Redmond possesses includes (1) new flavor and product packaging information; (2) pricing strategies; (3) Pepsi's "attack plans" for specific markets; and

(4) Pepsi's new distribution plan, being pilot tested in California. What advantages would Quaker obtain by knowing this information? How long would it take Quaker to find out about each in the absence of inside knowledge from Redmond? If all these items would become readily apparent the moment Pepsi's plans were implemented, does this circumstance suggest a limit on the appropriate remedy?

Assuming that Quaker will learn of Pepsi's strategies from Redmond unless enjoined, how expensive would it be for Pepsi to develop a new marketing strategy? Could the new strategy take advantage of the fact that Quaker *thinks* it knows what Pepsi will do? How would this possibility affect Quaker's use of the information?

2. Assume that the market for sports and new age drinks is increasingly concentrated in the hands of two companies, Pepsi and Quaker. Should this affect the outcome of the case? Given the court's decision, what can one predict about future salary and benefits negotiations in this industry for employees like Redmond?

3. Is it fair to preclude former employees from doing any work for a competitor simply because they would be incapable of not using the information they obtained from their former employer? Note that the Seventh Circuit upheld an injunction against Redmond's employment for only six months. At the same time, the court issued a permanent injunction against the disclosure of Pepsi's trade secrets. Does this result make sense? If the theory of inevitable disclosure is that Redmond *must* use Pepsi's secrets in his employment, could he go to work for Quaker at the end of six months without violating the permanent injunction? Perhaps the court was implicitly seeking to balance Redmond's interests in employment mobility, reasoning that the business secrets would be less important after that time.

4. The judicial debate over the appropriateness of the inevitable disclosure doctrine has continued. Several recent cases have followed *PepsiCo* and enjoined employment absent either proof of trade secret misappropriation or an enforceable noncompetition agreement. See Uncle B's Bakery v. O'Rourke, 920 F. Supp. 1405 (N.D. Iowa 1996) (enjoining former plant manager at a bagel manufacturer from working for any competing business within a 500-mile radius); National Starch & Chem. Corp. v. Parker Chem. Corp., 530 A.2d 31 (N.J. App. 1997) (citing *PepsiCo* with approval). By contrast, one court has adopted what might be called a "partial inevitable disclosure" injunction. In Merck & Co. v. Lyon, 941 F. Supp. 1443 (M.D.N.C. 1996), the court enjoined a pharmaceutical marketing director from discussing his former employer's products or pricing for a period of two years, but refused to enjoin him from competing employment altogether absent a "showing of bad faith." Another court has accepted an inevitable disclosure theory, but issued an injunction limited to nine months, reasoning that the secrets likely to be disclosed would turn stale over time. Novell Inc. v. Timpanogos Research Group, 46 U.S.P.Q.2d 1197 (D. Utah 1998). And other courts, particularly those in California, continue to reject inevitable disclosure altogether. See, e.g., GlobeSpan, Inc. v. O'Neill, 151 F. Supp. 2d 1229 (C.D. Cal. 2001); Danjaq, LLC v. Sony Corp., 50 U.S.P.Q.2d 1638 (C.D. Cal. 1999), aff'd on other grounds 263 F.3d 942 (9th Cir. 2001); Bayer Corp. v. Roche Molecular Sys., 72 F. Supp. 2d 1111, 1120 (N.D. Cal. 1999); Cardinal Health Staffing Network v. Bowen, 106 S.W.3d 452 (Tex. App. 2004); Del Monte Fresh Produce Co. v. Dole Food Co., 148 F. Supp. 2d 1326, 1337-39 (S.D. Fla. 2001); LeJeune v. Coin Acceptors, 849 A.2d 451 (Md. App. 2004); Whyte v. Schlage Lock Co., 101 Cal. App. 4th 1443, 125 Cal. Rptr. 2d 277 (Ct. App. 2002).

Note on Nonsolicitation Agreements

If the departing employees can avoid using the employer's trade secrets when they leave, the question becomes a rather different one — is it acceptable to hire away a group of employees? In Diodes, Inc. v. Franzen, 260 Cal. App. 2d 244, 67 Cal. Rptr. 19 (1968), the president and vice-president of Diodes left to form a competing company, called Semtech. Before they left, the officers solicited a number of Diodes employees to join them. Diodes sued the departing employees, alleging a number of claims centering on unfair competition and breach of fiduciary duty. The court dismissed the complaint, stating:

> As a general principle, one who unjustifiably interferes with an advantageous business relationship to another's damage may be held liable therefor. The product is bottled under a variety of labels, including unfair competition, interference with advantageous relations, contract interference, and inducing breach of contract.
>
> Even though the relationship between an employer and his employee is an advantageous one, no actionable wrong is committed by a competitor who solicits his competitor's employees or who hires away one or more of his competitor's employees who are not under contract, so long as the inducement to leave is not accompanied by unlawful action. In the employee situation the courts are concerned not solely with the interests of the competing employers, but also with the employee's interest. The interests of the employee in his own mobility and betterment are deemed paramount to the competitive business interests of the employers, where neither the employee nor his new employer has committed any illegal act accompanying the employment change.

67 Cal. Rptr. at 25-26 (citations omitted).

Can employers change this result by forcing their employees to sign "non-solicitation agreements" that prevent a departing employee from soliciting other employees to join him? What if such agreements also prohibit departing employees from soliciting their former customers? In a case that has been heavily criticized on other grounds, the Ninth Circuit has upheld (albeit without discussion) an injunction against solicitation of employees or customers based on such a nonsolicitation agreement. See MAI Systems Corp. v. Peak Computing, Inc., 991 F.2d 511, 523 (9th Cir. 1993). Other courts and commentators take a more guarded approach, prohibiting solicitation only where there is active inducement or where departing employees have taken a customer list which is itself a trade secret. Cf. James H. A. Pooley, Restrictive Employee Covenants in California, 4 Santa Clara Computer & High Tech. L.J. 251, 259 (1988).

Should former employees be allowed to compete for the business of their old customers? Should they be allowed to hire away their coworkers? Does your answer depend on whether the employees are using trade secrets in the solicitation? On whether the employee has signed a nonsolicitation agreement?

Should the dominance of the employer in the industry matter? In Wood v. Acordia of West Virginia, 2005 WL 1567337 (W. Va. 2005), the court upheld a two-year agreement preventing insurance agents from soliciting any of Acordia's current, former, or prospective customers. In dissent, Justice Starcher pointed out that Acordia was so dominant in the West Virginia insurance market that "every prospect in the market has been spoken for by an Acordia salesman." The agreement at issue thus effectively prevents all competition."

PROBLEM

Problem 2-14. You have been offered a position with a high-technology start-up company. They ask you to sign the following agreement. Do you sign it? Is it enforceable?

EMPLOYMENT, CONFIDENTIAL INFORMATION, AND INVENTION ASSIGNMENT AGREEMENT

As a condition of my employment with Science Company, its subsidiaries, affiliates, successors, or assigns (together the "Company"), and in consideration of my employment with the Company and my receipt of the compensation now and hereafter paid to me by the Company, I agree to the following:

1. Confidential Information

(a) *Company Information.* I agree at all times during the term of my employment and thereafter, to hold in strictest confidence, and not to use, except for the benefit of the Company, or to disclose to any person, firm, or corporation without written authorization of the Board of Directors of the Company, any Confidential Information of the Company. I understand that **"Confidential Information"** means any Company proprietary information, technical data, trade secrets or know-how, including, but not limited to, research, product plans, products, services, customer lists and customers (including, but not limited to, customers of the Company on whom I called or with whom I became acquainted during the term of my employment), markets, software, developments, inventions, processes, formulas, technology, designs, drawings, engineering, hardware configuration information, marketing, finances, or other business information disclosed to me by the Company either directly or indirectly in writing, orally, or by drawings or observation of parts or equipment. I further understand that Confidential Information does not include any of the foregoing items which has become publicly known and made generally available through no wrongful act of mine or of others who were under confidentiality obligations as to the item or items involved.

(b) *Third Party Information.* I recognize that the Company has received and in the future will receive from third parties their confidential or proprietary information subject to a duty on the Company's part to maintain the confidentiality of such information and to use it only for certain limited purposes. I agree to hold all such confidential or proprietary information in the strictest confidence and not to disclose it to any person, firm, or corporation or to use it except as necessary in carrying out my work for the Company consistent with the Company's agreement with such third party.

2. Inventions

I agree that I will promptly make full written disclosure to the Company, will hold in trust for the sole right and benefit of the Company, and hereby assign to the Company, or its designee, all my right, title, and interest in and to any and all inventions, original works of authorship, developments, concepts, improvements or trade secrets, whether or not patentable or registrable under copyright or similar laws, which I may solely or jointly conceive or develop or reduce to practice, or cause to be conceived or developed or reduced to practice (collectively referred to as "Inventions"), during the period of time I am in the employ of the Company and for three months thereafter, except as provided below. I further acknowledge that all original works of authorship which are made by me (solely or jointly with others) within the scope of and during the period of my employment with the Company and which are protectable by copyright are "works made for hire," as that term is defined in the United States Copyright Act.

3. Conflicting Employment

I agree that, during the term of my employment with the Company and for a period of one year thereafter, I will not engage in any other employment, occupation, consulting, or other business activity in competition with or directly related to the business in which the Company is now involved or becomes involved during the term of my employment, nor will I engage in any other activities that conflict with my obligations to the Company.

4. Returning Company Documents

I agree that at the time of leaving the employ of the Company, I will deliver to the Company (and will not keep in my possession, recreate or deliver to anyone else) any and all devices, records, data, notes, reports, proposals, lists, correspondence, specifications, drawings, blueprints, sketches, materials, equipment, other documents or property, or reproductions of any aforementioned items developed by me pursuant to my employment with the Company or otherwise belonging to the Company, its successors or assigns.

5. Notification to New Employer

In the event that I leave the employ of the Company, I hereby grant consent to notification by the Company to my new employer about my rights and obligations under this Agreement.

6. Solicitation of Employees

I agree that for a period of twelve (12) months immediately following the termination of my relationship with the Company for any reason, whether with or without cause, I shall not either directly or indirectly solicit, induce, recruit or encourage any of the Company's employees to leave their employment, or take away such employees, or attempt to solicit, induce, recruit, encourage, or take away employees of the Company, either for myself or for any other person or entity.

D. AGREEMENTS TO KEEP SECRETS

As we have seen, trade secret law imposes certain limitations on the owner of a trade secret. To qualify for trade secret protection, information must not be generally known, must be valuable, and must not be disclosed. But can an owner of information avoid those restrictions by requiring others to *agree* to keep the information secret, whether or not the information meets the requirements for protection? The question is fundamental in intellectual property law.

Licenses are generally considered good from an economic standpoint because they promote efficiency. The company (or individual) that develops a new product may not be in the best position to market it, particularly if the invention has uses in several different fields. Absent licenses, that company would either have to sell the product outright, use it incompletely or inefficiently, enter a new field itself, or not use it at all. Private contract remedies allow the market to reorder itself efficiently and still determine the appropriate reward for invention. But we cannot always rely on the market to determine an efficient outcome. Indeed, intellectual property in general is an example of pervasive market failure, a failure that may color the role of contract law.

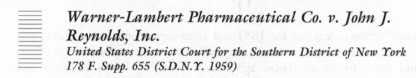

Warner-Lambert Pharmaceutical Co. v. John J. Reynolds, Inc.
United States District Court for the Southern District of New York
178 F. Supp. 655 (S.D.N.Y. 1959)

BRYAN, District Judge.

Plaintiff sues under the Federal Declaratory Judgment Act, 28 U.S.C. §§ 2201 and 2202, for a judgment declaring that it is no longer obligated to make periodic payments to defendants based on its manufacture or sale of the well known product "Listerine," under agreements made between Dr. J. J. Lawrence and J. W. Lambert in 1881, and between Dr. Lawrence and Lambert Pharmacal Company in 1885. Plaintiff also seeks to recover the payments made to defendants pursuant to these agreements since the commencement of the action.

Plaintiff is a Delaware corporation which manufactures and sells Listerine, among other pharmaceutical products. It is thesuccessor in interest to Lambert and Lambert Pharmacal Company which acquired the formula for Listerine from Dr. Lawrence under the agreements in question. Defendants are the successors in interest to Dr. Lawrence.

Jurisdiction is based on diversity of citizenship.

For some seventy-five years plaintiff and its predecessors have been making the periodic payments based on the quantity of Listerine manufactured or sold which are called for by the agreements in suit. The payments have totalled more than twenty-two million dollars and are presently in excess of one million five hundred thousand dollars yearly....

[J. J. Lawrence developed the formula for Listerine in 1880. He licensed the secret formula exclusively to Lambert (later Warner-Lambert) in 1881 under a contract which provided that

I, Jordan Lambert, hereby agree for myself, my heirs, executors and assigns to pay monthly to Dr. Lawrence, his heirs, executors or assigns, the sum of twenty dollars for each and every gross of said Listerine hereafter sold by myself, my heirs, executors or assigns.

The amount was reduced by subsequent agreement to $6.00 per gross.]

The agreements between the parties contemplated, it is alleged, "the periodic payment of royalties to Lawrence for the use of a trade secret, to wit, the secret formula for" Listerine. After some modifications made with Lawrence's knowledge and approval, the formula was introduced on the market. The composition of the compound has remained the same since then and it is still being manufactured and sold by the plaintiff.

It is then alleged that the "trade secret" (the formula for Listerine) has gradually become a matter of publicknowledge through the years following 1881 and prior to 1949, and has been published in the United States Pharmacopo[e]ia, the National Formulary and the Journal of the American Medical Association, and also as a result of proceedingsbrought against plaintiff's predecessor by the Federal Trade Commission. Such publications were not the fault of plaintiff or its predecessors. . . .

(1)

The plaintiff seems to feel that the 1881 and 1885 agreements are indefinite and unclear, at least as to the length of time during which they would continue in effect. I do not find them to be so. These agreements seem to me to be plain and unambiguous. . . .

The obligation to pay on each and every gross of Listerine continues as long as this preparation is manufactured or sold by Lambert and his successors. It comes to an end when they cease to manufacture or sell the preparation. . . . The plain meaning of the language used in these agreements is simply that Lambert's obligation to pay is co-extensive with manufacture or sale of Listerine by him and his successors.

(3)

However, plaintiff urges with vigor that the agreement must be differently construed because it involved the conveyance of a secret formula. The main thrust of its argument is that despite the language which the parties used the court must imply a limitation upon Lambert's obligation to pay measured by the length of time that the Listerine formula remained secret.

To sustain this theory plaintiff relies upon a number of cases involving the obligations of licensees of copyrights or patents to make continuing payments to the owner or licensor, and argues that these cases are controlling here. . . .

. . . [A]ll [these cases hold] is that when parties agree upon a license under a patent or copyright the court will assume, in the absence of express language to the contrary, that their actual intention as to the term is measured by the definite term of the underlying grant fixed by statute.

It is quite plain that were it not for the patent and copyright features of such license agreements the term would be measured by use. . . .

In the patent and copyright cases the parties are dealing with a fixed statutory term and the monopoly granted by that term. This monopoly, created by Congress, is designed to preserve exclusivity in the grantee during the statutory term and to release the patented or copyrighted material to the general public for general use thereafter. This is the public policy of the statutes in reference to which such contracts are made and it is against this background that the parties to patent and copyright license agreements contract.

Here, however, there is no such public policy. The parties are free to contract with respect to a secret formula or trade secret in any manner which they determine for their own best interests. A secret formula or trade secret may remain secret indefinitely. It may be discovered by someone else almost immediately after the agreement is entered into. Whoever discovers it for himself by legitimate means is entitled to its use.

But that does not mean that one who acquires a secret formula or a trade secret through a valid and binding contract is then enabled to escape from an obligation to which he bound himself simply because the secret is discovered by a third party or by the general public. I see no reason why the court should imply such a term or condition in a contract providing on its face that payment shall be co-extensive with use. To do so here would be to rewrite the contract for the parties without any indication that they intended such a result. . . .

One who acquires a trade secret or secret formula takes it subject to the risk that there be a disclosure. The inventor makes no representation that the secret is non-discoverable. All the inventor does is to convey the knowledge of the formula or process which is unknown to the purchaser and which in so far as both parties then know is unknown to any one else. The terms upon which they contract with reference to this subject matter are purely up to them and are governed by what the contract they enter into provides.

If they desire the payments or royalties should continue only until the secret is disclosed to the public it is easy enough for them to say so. But there is no justification for implying such a provision if the parties do not include it in their contract, particularly where the language which they use by fair intendment provides otherwise. . . .

COMMENTS AND QUESTIONS

1. The scope of trade secret protection is limited, not by a fixed term of years but by the length of time the information remains secret. Indeed, trade secrets become a part of the public domain once secrecy is lost, and their owner cannot prevent even direct copying. It seems odd that contract law would require royalty payments from Warner-Lambert for eighty years (or presumably 120 or more years, as Listerine is still sold commercially), while every other competitor can copy the formula for free. Are there sound reasons for such a rule?

One such reason may be the economic value of freedom of contract. If a party chooses to contract to pay royalties for as long as it uses a product, it is free to do so. Presumably the amount of the royalty payments as well as their duration will reflect the value of the secret to the buyer, discounted by the likelihood of public disclosure. But doesn't this argument also apply to extensions of the patent or copyright term as well? Compare Meehan v. PPG Indus., 802 F.2d 881, 886 (7th Cir. 1986), *cert. denied*, 479 U.S. 1091 (1987) and Boggild v. Kenner Prods., 776 F.2d 15, 1320-1321 (6th Cir. 1985), *cert. denied*, 477 U.S. 908 (1986), which hold that it is illegal patent or copyright misuse to agree to extend a patent or copyright beyond its term. Why does contract law appear to override trade secret law but not patent or copyright law?

Could a licensee evade its royalty obligations by acquiring the now-public secret from an independent source? The answer seems to depend on the way in which the particular license is written. But cf. Aronson v. Quick Point Pencil Co., 440 U.S. 257 (1979) (holding that the contract obligated licensee to continue paying royalties on an invention even though licensor's patent application had been rejected).

2. The *Warner-Lambert* result is controversial. The Restatement (Third) of Unfair Competition takes the position that nondisclosure agreements which purport to protect information in the public domain may be unenforceable as an unreasonable restraint on trade. See § 41, Comment *d*, at 472. Further, the Restatement notes that "because of the public interest in preserving access to information that is in the public domain, such an agreement will not ordinarily estop a defendant from contesting the existence of a trade secret." § 39, Comment *d* at 430. A number of cases support this view, which seems at odds with *Warner-Lambert*. See,e.g, Gary Van Zeeland Talent, Inc. v. Sandas, 267 N.W.2d 242 (Wisc. 1978); Sarkes Tarzian, Inc. v. Audio Devices, Inc., 166 F. Supp. 250 (S.D. Cal. 1958), aff'd, 283 F.2d 695 (9th Cir. 1960), *cert. denied*, 364 U.S. 869 (1961).

On the other hand, the Federal Circuit has seemingly endorsed the *Warner-Lambert* approach, holding in a recent case that the issuance of a patent did not extinguish the confidentiality obligation imposed by a nondisclosure agreement, even though the issuance of the patent destroyed the trade secret that was the basis for the agreement. Celeritas Technologies v. Rockwell Intl. Corp., 150 F.3d 1354 (Fed. Cir. 1998). The equities in that case might be thought to favor the plaintiff: the patent was held invalid, and so offered the plaintiff no relief against a theft of its technology. But isn't that the risk a trade secret owner takes in deciding to patent (and therefore disclose) her invention? The Seventh Circuit recently held that courts had no power to review business nondisclosure agreements for reasonableness, even though employee agreements were subject to a reasonableness requirement. IDX Systems v. Epic Systems, 285 F.3d 581 (7th Cir. 2002). And in Bernier v. Merrill Air Engineers, 770 A.2d 97 (Me. 2001), the Maine Supreme Court held that an employee violated a nondisclosure agreement by disclosing information that did not qualify as a trade secret.

3. Should such restrictions be governed by intellectual property law — making them unenforceable — or by contract law, which presumably would enforce them? The answer to that question will determine whether intellectual property laws (including trade secret laws) constitute binding governmental rules balancing competing interests or merely "default" rules that parties may opt to change. In practice, the courts have walked a hazily defined middle line, refusing to hold that intellectual property statues preempt contracts which alter their terms, but also refusing to enforce certain contracts that go "too far" in upsetting the balance the intellectual property laws have struck.

4. Many licensing agreements are between competitors in an industry, since firms in the same industry are generally those who will be interested in a particular invention. Agreements between competitors, though, are generally considered suspect under the antitrust laws. So too are common license restrictions that limit the use the licensee can make of the technology. We consider the restrictions that antitrust law places on private agreements of this sort in Chapter 8.

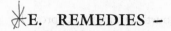

E. REMEDIES –

As might be expected given the fluidity of doctrine in trade secret law, there are disparate remedies for trade secret misappropriation stemming from different remedial concerns. The Uniform Trade Secrets Act sets forth the following remedies for misappropriation:

Section 2

(a) Actual or threatened misappropriation may be enjoined. Upon application to the court, an injunction shall be terminated when the trade secret has ceased to exist, but the injunction may be continued for an additional reasonable period of time in order to eliminate commercial advantage that otherwise would be derived from the misappropriation.

(b) If the court determines that it would be unreasonable to prohibit future use, an injunction may condition future use upon payment of a reasonable royalty for no longer than the period of time the use could have been prohibited.

(c) In appropriate circumstances, affirmative acts to protect a trade secret may be compelled by court order.

Section 3

(a) In addition to or in lieu of injunctive relief, a complainant may recover damages for the actual loss caused by misappropriation. A complainant also may recover for the unjust enrichment caused by misappropriation that is not taken into account in computing damages for actual loss.

(b) If willful and malicious misappropriation exists, the court may award exemplary damages in an amount not exceeding twice any award made under subsection (a).

Section 4

If (i) a claim of misappropriation is made in bad faith, (ii) a motion to terminate an injunction is made or resisted in bad faith, or (iii) willful and malicious misappropriation exists, the court may award reasonable attorney's fees to the prevailing party.

What motivates this hodgepodge of remedial measures? As we shall see in the chapters to come, most intellectual property statutes operate on the basis of "property rules." That is, as in cases involving real property, the owner of the intellectual property right is entitled to judicial assistance in protecting the right from future interference. Normally this assistance comes in the form of injunctive relief. By contrast, most tort and contract cases do not involve injunctive relief but rather damages designed to make the plaintiff "whole" in the sense of restoring her to the position she occupied before the tort, or to the position she expected to occupy if the contract had been performed.

Section 2 of the Uniform Trade Secrets Act seems to entitle trade secret plaintiffs to property-like protection, as least so long as their secret remains a secret. But section 2(b) holds open the possibility that courts may refuse to grant such an injunction, settling instead for a reasonable "royalty" (presumably a court's attempt to approximate what the parties might have agreed to pay in a licensing transaction). This provision casts some doubt on the "property entitlement" a trade secret owner might expect.

Similar doubt pervades the provisions on damages. Concepts like "reasonable royalty," "lost profits," and limited-time injunctions designed to "eliminate commercial advantage" all sound like restitutionary measures, aimed at making the plaintiff whole after a loss without necessarily punishing or deterring the defendant. But further provisions permit trade secret plaintiffs to recover for "unjust enrichment" on the part of defendants, and to recover treble damages and attorney fees in the case of willful misappropriation. And in some circumstances misappropriation of trade secrets can be a criminal offense, an idea that is certainly more consistent with a property entitlement rule than a tort or contract rule.

Generally, injunctions are available as a remedy in trade secret cases. Since injunctions offer only prospective relief, however, damages for preinjunction activities may also be collected. Since, unlike patents and copyrights, trade secrets have no definite term, the length of the injunction is often a difficult issue. The following case illustrates the use of one important measure of trade secret injunctions, the "head start" theory.

Winston Research Corp. v. 3M Corp.,
United States Court of Appeals for the Ninth Circuit
350 F.2d 134 (9th Cir. 1965)

BROWNING, Circuit Judge:

For some uses of precision tape recorder/reproducers, the time interval between coded signals must be recorded and reproduced with great accuracy. To accomplish this, the tape must move at as constant a speed as possible during both recording and reproduction, and any changes in tape speed during recording must be duplicated as nearly as possible during reproduction. The degree to which a particular tape recorder/reproducer accomplishes, this result is measured by its "time-displacement error."

An electronic device known as a "servo" system is commonly used to reduce time-displacement error by detecting fluctuations in tape speed and immediately adjusting the speed of the motor. Machines prior to the Mincom machine employed a flywheel to inhibit fluctuation in tape speed by increasing the inertia of the system. However, the flywheel reduced the effectiveness of the servo system since the increased inertia prevented rapid adjustments in the speed of the motor.

The effectiveness of the servo system in prior machines was also reduced by resonances created by the moving parts. The range of sensitivity of the servo system was limited to exclude the frequencies of the interfering resonances. This had the disadvantage of limiting the capacity of the servo system to respond to a full range of variations in the speed of the tape.

To solve these problems Mincom eliminated the flywheel and reduced the mass of all other rotating parts. This reduced the inertia of the tape transport system, permitting rapid adjustments in tape speed. Interfering resonances were eliminated by mechanical means. This permitted use of a servo system sensitive to a wide range of frequencies, and hence capable of rapid response to a wide range of variations in tape speed. After four years of research and development based upon this approach, Mincom produced a successful machine with an unusually low time-displacement error.

In May 1962, when Mincom had substantially completed the research phase of its program and was beginning the development of a production prototype, Johnson, who was in charge of Mincom's program, left Mincom's employment. He joined Tobias, who had previously been discharged as Mincom's sales manager, in forming Winston Research Corporation. In late 1962, Winston contracted with the government to develop a precision tape reproducer. Winston hired many of the technicians who had participated in the development of the Mincom machine to work on the design and development of the Winston machine.

In approximately fourteen months, Winston completed a machine having the same low time-displacement error as the Mincom machine.

Conflicting policy considerations come into play in deciding what limitations should be imposed upon an employee in the use and disclosure of information acquired in the course of a terminated employment relationship — or, conversely, what protection should be extended to the former employer against use and disclosure of such information.

On the one hand, restrictions upon the use and disclosure of such information limit the employee's employment opportunities, tie him to a particular employer, and weaken his bargaining power with that employer. Such restrictions interfere with the employee's movement to the job in which he may most effectively use his skills. They inhibit an employee from either setting up his own business or from adding his strength to a competitor of his employer, and thus they diminish potential competition. Such restrictions impede the dissemination of ideas and skills throughout industry. The burdens that they impose upon the employee and society increase in proportion to the significance of the employee's accomplishments, and the degree of his specialization.

On the other hand, restrictions upon an employee's disclosure of information that was developed as a result of the employer's initiative and investment, and which was entrusted to the employee in confidence, are necessary to the maintenance of decent standards of morality in the business community. Unless protection is given against unauthorized disclosure of confidential business information by employees, employee-employer relationships will be demoralized; employers will be compelled to limit communication among employees with a consequent loss in efficiency; and business, espionage, deceit, and fraud among employers will be encouraged. . . .

The federal patent statutes require full disclosure of the invention as a condition to the grant of monopoly so that at the end of the period of monopoly the development may be freely available to all. Thus, the federal patent statutes would seem to reflect a congressional determination that any individual or social interests that may be served by secrecy are outweighed by those served by full disclosure. And it would seem to follow that state law providing protection for trade secrets cannot be applied to serve a premise that the balance of interests favors secrecy. Thus, state law protecting trade secrets cannot be based "on a policy of rewarding or otherwise encouraging the development of secret processes or devices. The protection is merely against breach of faith and reprehensible means of learning another's secret." Restatement, Torts § 757, comment b.

The district court found, and Winston concedes, that Johnson and the other former Mincom employees based Winston's development program upon the same approach to the problem of achieving a low time-displacement error as they had pursued in developing the Mincom machine. The district court further found that this general approach was not a trade secret of Mincom's. Finally, the district court found that the particular embodiment of these general concepts in the Mincom machine was Mincom's trade secret, and had been improperly utilized by the former Mincom employees in developing the Winston machine.

Mincom contends that the court defined Mincom's trade secrets too narrowly; Winston, that the court's definition was too broad.

In describing Mincom's trade secrets in the judgment, the district court first outlined the general approach which both Mincom and Winston followed in the development of their machines, closing with the phrase, "as accomplished and adopted by" Mincom. The court then set out the detailed specifications of the various elements of the Mincom machine. Winston reads the court's initial outline of the general approach as a determination that this was itself a Mincom trade secret, and then vigorously attacks this supposed determination.

We read the initial description of the general approach as merely introductory to, and limited by, the detailed specifications that follow. We think the district court rejected Mincom's contention that the general approach itself was a Mincom trade secret — and properly so. There was expert testimony that if a technician in the field were asked to design a machine having the time — displacement error achieved by the Mincom machine he would adopt the general approach of reducing the inertia of rotating parts and utilizing a wide band servo system — that this approach was dictated by well known principles of physics. It was therefore not protectible under accepted trade secret doctrine. It was not "secret," for it consisted essentially of general engineering principles in the public domain and part of the intellectual equipment of technical employees. Its disclosure could not be treated as betrayal of a confidence placed by Mincom in its technical employees.

Winston points out that the basic mechanical elements of the Mincom machine were found in various publicly known recorder-reproducers, and that all of them were disclosed in a patent issued to Johnson and assigned to Mincom in 1961. But there was substantial evidence that the specifications of these basic mechanical elements and their relationship to each other embodied in the Mincom machine and detailed in the district court's judgment were not publicly known, and were arrived at by Mincom only after painstaking research and extensive trial and error.

As we read the findings and judgment, it was these specifications and relationships that the court found to constitute Mincom's trade secret, and not the general approach itself or the basic mechanical elements as such. We think this determination was correct.

Winston also suggests that the specifications were "mere choices of engineering design," and not of critical importance. Even if this were so, they were choices made first by Mincom, and gave Mincom product having superior characteristics that Winston achieved only by utilizing elements with substantially the same specifications.

Winston argues that information is protected from disclosure only if communicated to the employee by the employer who is seeking protection, and that the information involved in this case was not disclosed by Mincom to the employees subsequently hired by Winston, but rather was developed by these employees themselves, albeit while employed by Mincom.

We need not examine the soundness of [this argument], or its applicability to a case such as this, in which a group of specialists engaged in related facets of a single development project change their employer. The argument is apparently based upon the notion that unless the first employer conveys the information to the employee, subsequent disclosure by the employee cannot be a breach of a duty of confidence owed that employer. Futurecraft Corp. v. Clary Corp., 23 Cal.Rptr. 198. As the court in Futurecraft recognized, an obligation not to disclose may arise from circumstances other than communication in confidence by the employer. It may also rest upon an express or implied agreement. In the present case, an agreement not to disclose might be implied from Mincom's elaborate efforts to maintain the secrecy of its development program, and the employees' knowledge of these efforts and participation in them. In any event, Mincom and its employees entered into express written agreements binding the latter not to disclose confidential information, and these agreements did not exclude information which the employee himself contributed.

The district court enjoined Winston Research Corporation, Johnson, and Tobias from disclosing or using Mincom's trade secrets in any manner for a period of two years from the date of judgment. The court also required the assignment of certain patent applications to Mincom. No damages were awarded.... Mincom argues that the injunction should have been permanent, or at least for a substantially longer period. Winston contends that no injunctive relief was appropriate.

Mincom was, of course, entitled to protection of its trade secrets for as long as they remained secret. The district court's decision to limit the duration of injunctive relief was necessarily premised upon a determination that Mincom's trade secrets would shortly be fully disclosed, through no fault of Winston, as a result of public announcements, demonstrations, and sales and deliveries of Mincom machines. Mincom has not seriously challenged this implicit finding, and we think the record fully supports it....

We think the district court's approach was sound. A permanent injunction would subvert the public's interest in allowing technical employees to make full use of their knowledge and skill and in fostering research and development. On the other hand, denial of any injunction at all would leave the faithless employee unpunished where, as here, no damages were awarded; and he and his new employer would retain the benefit of a headstart over legitimate competitors who did not have access to the trade secrets until they were publicly disclosed. By enjoining use of the trade secrets for the approximate period it would require a legitimate Mincom competitor to develop a successful machine after public disclosure of the secret information, the district court denied the employees any advantage from their faithlessness, placed Mincom in the position it would have occupied if the breach of confidence had not occurred prior to the public disclosure, and imposed the minimum restraint consistent with the realization of these objectives upon the utilization of the employees' skills....

Winston also challenges the district court's determination that "knowledge of the reasons for" the particular specifications of the Mincom machine, and "knowledge of what not to do . . . and how not to make the same mistakes" as Mincom had made in arriving at these specifications, were Mincom trade secrets. [Although we agree with the district court's conclusion that such "negative know-how" is a trade secret, in the circumstances of this case we can see no way to prohibit Mincom's former employees from using such knowledge without prohibiting them from using their general knowledge and experience at the same time. In an appropriate case, this kind of knowledge can be protected by an injunction or even an award of damages, but this is not such a case.]....

Mincom argues that the district court should have awarded money damages as well as injunctive relief. We think the district court acted well within its discretion in declining to do so. Since Winston sold none of its machines, it had no profits to disgorge. The evidence as to possible future profits was at best highly speculative. To enjoin future sales and at the same time make an award based on future profits from the prohibited sales would result in duplicative and inconsistent relief, and the choice that the district court made between these mutually exclusive alternatives was not an unreasonable one. There was evidence that Winston would probably sell its machine and realize profits after the injunction expired, but these sales and profits, as we have seen, would not be tainted by breach of confidence, since Winston could by that time have developed its machine from publicly disclosed information. . . .

Finally, Mincom's employment contracts required its employees to assign to Mincom inventions conceived during employment, and inventions conceived within one year of termination of employment that were "based upon" confidential information. Mincom sought to require the assignment of one patent and a number of patent applications. The district court received evidence as to when each invention was conceived and its relationship to Mincom's secrets.

The district court found that three patent applications involved inventions conceived during the inventors' employment by Mincom, and ordered assignment on that basis. The court also found that a patent and some other patent applications involving inventions conceived after the employees left Mincom and went to Winston were not based on Mincom's confidential trade secrets. [The mere fact that they were conceived after the employees moved to Winston was not the reason for this finding — the reason was that these later inventions simply were not based on Mincom's confidential information.] We think these findings have support in the record and are not clearly erroneous. [Judgment affirmed.]

COMMENTS AND QUESTIONS

1. "Head-start injunctions" are available to plaintiffs who have published or otherwise disclosed their secret at some point after it was misappropriated. Suppose Anne possesses a secret that she is in the process of commercializing. Suppose further that it takes Anne two years after developing the secret to bring the product to market, at which point the secret is disclosed. If Benjamin steals Anne's idea during the development process (say, at month 12), Benjamin will be able to get to market one year earlier than if he had waited until the information became public. In such a case, courts will issue a "head-start" injunction for a period of one year, putting Benjamin in the same position he would have been in without the secret.

Even if such an injunction is impossible (for example, because Benjamin has already entered the market), courts may allow him to continue using the former secret but require him to pay a "reasonable royalty" to Anne. The reasonable royalty is presumably set by the court, in an effort to approximate the royalty Anne might have charged Benjamin in a voluntary transaction.

Is a reasonable royalty a fair solution in such a situation? See Mid-Michigan Comp. Sys. v. Marc Glassman Inc., 416 F.3d 505 (6th Cir. 2005) (basing damages for misappropriation of computer software on a reasonable royalty). At least one commentator has suggested that such a remedy "is peculiarly inappropriate to redress a situation where injunctive relief ought to be applied." See 12 Roger Milgrim, Milgrim on Trade Secrets § 1.01[2][a], at 1-36 n.20 (citing a district court decision concluding that limiting relief to a reasonable royalty invites misappropriation).

2. In Kewanee Oil Co. v. Bicron Corp., 416 U.S. 470, 473 (1974), the former employees had signed confidentiality agreements. The Court upheld the district court's granting of a permanent injunction against the disclosure or use by respondents of 20 of the 40 claimed trade secrets until such time as the trade secrets had been released to the public, had otherwise generally become available to the public, or had been obtained by respondents from sources having the legal right to convey the information. Id. at 473-74. See also Henry Hope X-Ray Prods., Inc. v. Marron Carrel, Inc., 674 F.2d 1336, 1342 (9th Cir. 1982) (the limitation on confidential information contains the implicit temporal limitation that information may be disclosed when it ceases to be confidential). But see Howard Schultz & Assoc. v. Broniec, 239 Ga. 181, 236 S.E.2d 265, 270 (1977) ("The nondisclosure covenant here contains no time limitation and hence it is unenforceable"); Gary Van Zeeland Talent, Inc. v. Sandas, 54 Wis. 2d 202, 267 N.W.2d 242, 250 (1978) (unlimited duration of agreement not to disclose trade secret customer list makes the agreement per se void).

The Eighth Circuit in Sigma Chemical Co. v. Harris, 794 F.2d 371, 375 (8th Cir. 1986) explained, in the course of rejecting a "temporally unlimited" injunction, the rationale for limiting injunctions in time:

> [E]xtending the injunction beyond the time needed for independent development would give the employer "a windfall protection and would subvert the public interest in fostering competition and in allowing employees to make full use of their knowledge and ability."
>
> We believe the part of the injunction prohibiting disclosure of trade secrets must be limited in duration and, accordingly, reverse in part and remand the case to the district court for consideration of the time it would take a "legitimate competitor" to independently reproduce the information contained in the product and vendor files. On remand, the district court should also modify the language of the injunction to expressly state that Harris may use that information which is already in the public domain.

3. Should there be a presumption that injunctive relief is appropriate? Such a presumption is typical of cases involving real property, but no such rule exists in most tort cases, and certainly not in typical contract cases. Are damages sufficient to protect trade secret owners? Perhaps not, since trade secrets are often hard to value. On the other hand, perhaps the parties could contract for injunctive relief, at least in employment cases, using noncompetition or confidentiality agreements. Should the law impose obligations the parties have not undertaken voluntarily? Should it refuse to enforce contract terms that the parties have agreed to? See East v. Aqua Gaming Inc., 805 So. 2d 932 (Fla. App. 2001) (affirming injunction against use of trade secrets, but vacating injunction that prevented competition in the absence of an enforceable noncompete agreement).

One problem with such "automatic" injunctive relief is that it is difficult — and costly — to enforce. Is it an appropriate use of judicial resources to supervise employer-employee relationships on an ongoing basis? Is there a more cost-effective alternative to such supervision?

Who should be enjoined? Several courts have refused to hold the new employer liable for an employee's malfeasance where they were unaware of it. See Infinity Prods. v. Quandt, 810 N.E.2d 1028 (Ind. 2004) (no respondeat superior liability for trade secret law); BEA Sys. v. WebMethods, Inc., 595 S.E.2d 87 (Ga. App. 2004).

Note on Criminal Trade Secret Statutes

Misappropriation of trade secrets is not only a tort; in some circumstances, it is a crime. See generally Eli Lederman, Criminal Liability for Breach of Confidential Commercial Information, 38 Emory L.J. 921 (1989) (summarizing criminal laws governing misappropriation of confidential information). A series of well-publicized criminal prosecutions of computer executives accused of taking trade secrets to their new employers have raised the consciousness of industry professionals about trade secrets. The prosecutions have also raised ethical and political questions about the propriety of trade secret prosecutions "engineered" by the real parties in interest, often major companies such as Intel or Borland.

Criminal trade secret cases differ from civil ones in several respects. Obviously, the complaining party is the government, rather than the injured company. However, the injured companies are the "real parties in interest" and usually have some presence in the case. Even though they are not parties to the criminal proceeding, they at least supply a significant number of the witnesses and enjoy a high level of communication with the district attorney.

Just as obviously, the burden of proof is higher than in a civil case. Some cases that could be won by the plaintiffs as civil cases will be lost in criminal court. This situation is even more likely because several states have definitions of trade secrets in their criminal laws that are more limited than their civil counterparts. For example, Cal. Penal Code §499c, which governs theft of trade secrets, historically limited the definition of a trade secret to "scientific or technical" information. (The constitutionality of this definition of trade secrets was upheld against a vagueness attack in People v. Serrata, 133 Cal. Rptr. 144 (Ct. App. 1976).) But in 1996, the California legislature amended its criminal trade secret statute to be coextensive with the definition of trade secrets in the Uniform Trade Secrets Act.

Defendants accused of stealing trade secrets may be charged with other offenses as well. For example, defendants who acquire a secret through improper means, as opposed to acquisition in a confidential relationship, may be guilty of larceny, receiving stolen property, or a host of similar crimes. See People v. Gopal, 217 Cal. Rptr. 487, 493-94 (Ct. App. 1985), cert. denied, 476 U.S. 1105 (1986). Further, the growth of computer technology has expanded the federal role in prosecuting theft of trade secrets, since data taken over a computer network is considered to cross state lines. See United States v. Riggs, 739 F. Supp. 414 (N.D. Ill. 1990) (allowing indictment of computer hackers who published data from a Bell South computer text file for wire fraud and interstate transportation of stolen property).

The prosecution of a criminal (rather than civil) trade secret case has other effects on the parties involved. First, criminal trade secret courtrooms are the scene of constant battles over the publication of information. The real parties in interest will naturally oppose the disclosure in a public courtroom of the very secrets the defendant is accused of stealing. This concern runs headlong into the defendant's constitutional right to a public trial.[19] Second, civil cases are generally stayed pending the outcome

---

19. There is no such right in civil cases. To avoid the very real danger that a misappropriation action will result in disclosure of the very secrets the plaintiff seeks to protect, civil trade secret actions will almost invariably include protective orders limiting the disclosure of information produced in discovery. Such orders are usually agreed to by the parties but may sometimes be imposed by the court. They will sometimes go so far as to prevent the parties themselves (as opposed to the attorneys and hired experts) from reviewing the other side's documents. In such a case, should in-house counsel be given access to discovery documents? See Brown Bag Software v. Symantec Corp., 960 F.2d 1465, 1470 (9th Cir. 1992).

of a criminal prosecution. Thus a criminal prosecution may actually delay injunctive relief—the kind of remedy a civil plaintiff is often most interested in.

COMMENTS AND QUESTIONS

1. Given the stay imposed on a parallel civil action and the higher burden of proof in a criminal case, why would a civil plaintiff ever seek a criminal prosecution?

2. Should theft of trade secrets be a criminal offense? Does the presence of criminal sanctions have any effect on the optimal level of deterrence provided in damages suits, and therefore on the damages that should be awarded in a civil suit?

Do you see any problems with California's recent inclusion of business information in the criminal trade secret statute? Are there reasons to treat theft of scientific information more harshly than theft of business information?

Note on Federal Criminal Liability for Trade Secret Misappropriation[20]

Prior to the passage of the Economic Espionage Act of 1996, there was only a single, very limited federal statute that directly prohibited the misappropriation of a trade secret. When confronted with an allegation that an individual had misappropriated a valuable trade secret, federal prosecutors were forced to turn to a number of federal statutes that were clearly not designed to cover trade secrets. In particular, prosecutors attempted to use the depression-era Interstate Transportation of Stolen Property Act (ITSP) and the wire fraud and mail fraud statutes. However a recent court decision severely curtails the use of ITSP, and the use of the mail and wire fraud statutes has always been limited because the theft often does not involve the use of mail or wire. In addition, since a thief usually merely copies information and does not necessarily "defraud" the victim permanently of the data, prosecutions are further limited. These shortcomings of federal law, coupled with testimony about the threat of economic espionage sponsored by foreign governments, underlie Congress's decision to enact the Economic Espionage Act of 1996 (EEA), 18 U.S.C. § 1831 et seq.

The structure of the EEA reflects its rather disparate origins and the confusion that surrounded major changes made in the last days of the 1996 Congressional session. The EEA contains a definition of trade secrets that is taken, with only minor modifications, from the definition in the Uniform Trade Secrets Act. At the same time, the sections defining misappropriation are entirely new, and have no parallels in existing trade secret law.

Sections 1831(a) and 1832(a) contain identical language regarding acts of misappropriation, punishing a specified individual who:

(1) steals, or without authorization appropriates, takes, carries away, or conceals, or by fraud, artifice or deception obtains a trade secret;

(2) without authorization copies, duplicates, sketches, draws, photographs, downloads, uploads, alters, destroys, photocopies, replicates, transmits, delivers, sends, mails, communicates, or conveys a trade secret;

---

20. This section is adapted from James H.A. Pooley et al., Understanding the Economic Espionage Act of 1996, 5 Tex. Intell. Prop. L.J. 177 (1997).

(3) receives, buys, or possesses a trade secret, knowing the same to have been stolen or appropriated, obtained, or converted without authorization . . .

These provisions are significantly broader in some respects than corresponding civil trade secret laws such as the Uniform Act.

Section (a)(1) lists a number of means of improper acquisition of a trade secret. In general, the provisions in this section seem to track the "improper means" of acquisition punished in civil cases. In particular, the references to stealing, concealment, fraud, artifice, and deception are reminiscent of the civil law's prohibition against both illegal and immoral business conduct. However, the provision also makes it a crime to "appropriate" or "take" a secret without authorization from the trade secret owner. These terms might be thought to encompass the sort of lawful business espionage that has long been permitted by civil trade secret law — conduct on the order of observing a competitor's property from across the street. There is some suggestion in the legislative history that the EEA is not intended to inhibit robust competition.[21] While this might be enough to declare broad categories of competitive intelligence gathering proper and therefore lawful, the ambiguity of some of the terms in section (a)(1) is troubling.

Broader still is section (2), which gives the trade secret owner the right to control a whole host of activities on the order of duplication, transportation, or destruction of a trade secret. This provision encompasses more than its civil counterparts. Section (2) is not by its terms limited to secrets acquired by "improper means" like those listed in section (1). Even legally acquired secrets can be misappropriated under the EEA if they are analyzed or duplicated in one of the ways listed in section (2). The civil law does prohibit some such uses of a lawfully acquired trade secret, but limits its reach to disclosure or use of a secret in violation of a confidential relationship. By contrast, there is no requirement of such a confidential relationship in the EEA. So long as the requirements of secrecy are satisfied, the trade secret owner is apparently entitled to seek the aid of the Justice Department in preventing the world from engaging in any of the enumerated acts in section (2), even if the secret is contained in a product that the defendant lawfully acquired.[22] As a result, the EEA effectively implies a confidential relationship between the trade secret owner and the world at large, something trade secret law has never before attempted and an indication that Congress believes trade secrets law is based in property theory, not tort or contract. Further, some of the specific provisions in section (2), such as the prohibition against "altering" or "destroying" a trade secret, are outside the normal reach of trade secret law.[23]

The most troubling thing about 18 U.S.C. § 1831(a)(2) is that it arguably prohibits many forms of heretofore lawful reverse engineering activity. While reverse engineering is not expressly prohibited under this section, neither is it expressly permitted. Rather, its legality appears to be judged according to whether the reverse engineer engages in any of the prohibited acts.[24] To be sure, some types of reverse

---

21. S. Rep. No. 142-12212 ("Other companies can and must have the ability to determine the elements of a trade secret through their own inventiveness, creativity and hard work. . . . [P]arallel development of a trade secret cannot and should not constitute a violation of this statute.").

22. See H.R. Rep. No. 104-788 ("The concept of control also includes the mere possession of the information, regardless of the manner by which the non-owner gained possession of the information.").

23. "Destroying" a secret might mean "destroying its value by publishing it to the world," in which case civil law does provide a remedy. Alternatively, it might mean the physical destruction of valuable information, which is not covered by the Uniform Act.

24. See S. Rep. No. 142-12212.

engineering, such as looking at or tasting a lawfully acquired product in order to determine what is in it, will not be illegal under the EEA. On the other hand, several of the restrictions in section (2) will, if read literally, encompass some forms of reverse engineering. For example, reverse engineering of computer software by "decompilation" almost always involves the making of a prohibited "copy" of the program. Reverse engineering of mechanical devices and computer hardware may well involve prohibited "sketching, drawing, or photographing" of the trade secret contained in the publicly sold device. It is even possible that the prohibition against "altering" a trade secret will be interpreted to prevent chemical analysis of a trade secret product if such analysis involves the use of chemical reactants that bond to secret chemicals or that precipitate out certain elements from the formula.

The legislative history of the EEA is not encouraging to reverse engineers. The Senate report suggests that "the important thing is to focus on whether the accused has committed one of the prohibited acts of this statute rather than whether he or she has 'reverse engineered.' If someone has lawfully gained access to a trade secret and can replicate it without violating copyright, patent *or this law,* then that form of 'reverse engineering' should be fine."[25] The report goes on to suggest that observing a lawfully purchased product, or drinking a Coca-Cola, are legitimate activities.[26] On the other hand, there appears to be no reverse engineering defense protecting any of the forms of analysis suggested above. A computer programmer has the right to decompile a software program in certain circumstances under the Uniform Act, copyright law, and the common law without fear of civil liability. Under the EEA, though, that programmer has arguably committed a felony.

The Justice Department is actively prosecuting cases under the EEA. As of this writing, over three dozen cases had been filed. One potentially significant problem with the Act, and criminal trade secret statutes in general, is that the Act's provisions for protecting the confidentiality of the trade secrets during trial may be in conflict with the constitutional rights of criminal defendants, particularly the right to a public trial.

In United States v. Hsu, 155 F.3d 189 (3d Cir. 1998), the first appellate decision interpreting the Act, the court concluded that the defendants (who were the subject of a sting operation) could be found guilty of conspiring and attempting to misappropriate information that was not in fact a trade secret, because the crime of attempt did not require proof that the information the defendants tried to steal was in fact a secret.

---

25. S. Rep. No. 142-12212, at 12-13 (emphasis added).
26. Id. at 12213. Cf. Mason v. Jack Daniels Distillery, 518 So. 2d 130 (Ala. Ct. App. 1987).

3

Patent Law

A. INTRODUCTION

This chapter introduces the basics of patent law. After a brief survey of the historical origins of our modern patent system and an overview of the elements of patentability, we turn to a substantive treatment of each of those elements. Then we discuss the rules and doctrines concerning patent infringement. Finally we consider remedies in patent cases. [1]

1. Historical Background

The first recorded reference to patents seems to be in Aristotle's *Politics,* composed in the fourth century B.C. In the course of a discussion of rival descriptions of a good constitution, Aristotle mentions a proposal by one Hippodamus of Miletos. According to Aristotle, Hippodamus called for a system of rewards to those who discover things useful to the state. Aristotle condemned the proposal as likely to lead to instability. After this isolated classical reference, the history of patents skips several historical epochs. Although the Middle Ages are known to have produced important innovations, [2] this era does not appear to have been conducive to the idea of patents, at least in the West. [3] Individual credit and gain for inventions was

1. The Statutory Supplement to this book contains a brief introduction to the science of biotechnology. Those without a background in the science may wish to review that material at this time.

2. See Donald Hill, A History of Engineering in Classical and Medieval Times (1984).

3. One historian, however, claims that the modern patent system has its origins in the Byzantine Empire:

> It is likely that the modern monopoly originated in Byzantium, and became the invention patent at the Renaissance when numbers of inventions appeared. A traveller of the twelfth century, Benjamin of Tudela, mentions an exclusive privilege for dyeing cloth in the semi-Byzantine kingdom of Jerusalem.

Frumkin, Early History of Patents for Invention, 26 Trans. Newcomen Soc. 47, 47 (1947). Frumkin notes several other isolated instances of protective grants: in Bordeaux, in 1236, a fifteen-year monopoly for the manufacture of cloth; and in 1331, by King Edward III of Great Britain, a nonexclusive grant to export woollen cloth. Id. at 48.

inconsistent with prevailing social mores.[4] Whatever the reason, one must look to the early Renaissance to find the first references to a real patent system.

The first regular administrative apparatus for granting patents — the first real patent "system" — arose in Venice in the late fifteenth century. Isolated grants in Venice and elsewhere were made earlier: in Venice in the early fourteenth century (for corn mill designs), and in Florence to the celebrated architect Brunelleschi for his 1421 invention of a barge with a hoist for transporting marble. The term *patent* — from the Latin *patere* (to be open), referring to an open letter of privilege from the sovereign — originated in this period. But not until the Venetian Senate's 1474 Act was the practice of granting patents regularized:

> Be it enacted that, by the authority of this Council, every person who shall build any new and ingenious device in this City, not previously made in this Commonwealth, shall give notice of it to the office of our General Welfare Board when it has been reduced to perfection so that it can be used and operated. It being forbidden to every other person in any of our territories and towns to make any further device conforming with and similar to said one, without the consent and license of the author, for the term of 10 years. And if anybody builds it in violation hereof, the aforesaid author and inventor shall be entitled to have him summoned before any Magistrate of this City, by which Magistrate the said infringer shall be constrained to pay him one hundred ducats; and the device shall be destroyed at once.[5]

The Venetian Act lays out all the essential features of a modern patent statute. It covers "devices"; states that they must be registered with a specific administrative agency; says that they must be "new and ingenious," "reduced to perfection," and "not previously made in this Commonwealth"; provides a fixed term of ten years; and sets forth a procedure to determine infringement, as well as a remedy. Interestingly, the Venetian Act reserved to the Republic the right to use any invention without compensating the inventor.[6] This is an early attempt to reconcile individual interest with the good of the community, a recurring problem in patent law.

The opening of trade in Europe ensured that the new Venetian concept would spread. As Italian craftsmen — particularly glassworkers — fanned out across Europe, they brought with them the idea of legal protection for inventions.

Patents came to Great Britain by this route, sometime in the middle of the sixteenth century. The chief minister under Elizabeth I, William Cecil (Lord Burghley), used patent grants to induce foreign artisans to introduce continental technologies into England.[7] Thus what later became the Anglo-American patent system originated

---

4. Yet the culture of "collective invention" common to craft guilds did originate many important technologies. On the guilds generally, see Carlo M. Cipolla, Before the Industrial Revolution 256 et seq. (2d ed. 1980). It is also interesting that Venice, home to the first patent statute, was populated by a number of vigorous guilds. Indeed, guilds and patents were seen as alternative mechanisms to achieve the same end — new technologies for the benefit of the Venetian Republic. See Pamela Long, Openness, Secrecy, Authorship: Technical Arts and the Culture of Knowledge from Antiquity to the Renaissance (2000).

5. Mandich, Venetian Patents (1450-1550), 30 J. Pat. & Trademk. Off. Society 166, 177 (1948).

6. Id.

7. For a detailed discussion of the early development of patent law in England and on the continent, see Edward C. Walterscheid, The Early Evolution of the United States Patent Law: Antecedents, 76 J. Pat. & Trademk. Off. Society 697 (1994) (Part I); 76 J. Pat. & Trademk. Off. Society 849 (1994) (Part II); 77 J. Pat. & Trademk. Off. Society 771 and 847 (1995) (Part III); 78 J. Pat. & Trademk. Off. Society 77 (1996) (Part IV).

as a mercantilist instrument—what today would be called a "strategic international trade" policy. The idea was to lure immigrants who had desirable skills and know-how with the promise of an exclusive privilege. Faint glimmers of this early policy survive in certain odd corners of today's patent laws.[8] Note that this policy reflects another attempt to balance the rights of the community against the individual interests of inventors. By luring skilled artisans with the reward of exclusive rights, it was hoped the community would gain the fruits of this skilled labor.[9] Ironically, by the mid-eighteenth century, Britain began to show concern over the reverse problem—leakage of its technical prowess to overseas rivals, including the American colonies.[10]

With the accession of James I in England in the early seventeenth century, patents became less an incentive for inventors of new arts and more a royal favor dispensed to well-placed courtiers. Under this rubric, "patents" were granted on such enterprises as running ale-houses. Parliament, whose members represented many trades injured by these special privileges, was displeased. Thus arose the Statute of Monopolies of 1624, which forbade all grants of exclusive privilege except those described in the famous Section 6:

> [B]e it declared and enacted that any declaration before mentioned shall not extend to any letters patent and grants of privilege for the term of fourteen years or under, hereafter to be made, of the sole working or making of any manner of new manufactures within this realm, to the true and first inventor and inventors of such manufacture, which others at the time of making such letters patent shall not use, so as also they be not contrary to law, nor mischievous to the State, by raising prices of commodities at home, or hurt of trade, or generally inconvenient; the said fourteen years to be accounted from the date of the first letters patents, or grant of such privilege hereafter to be made....[11]

This statute called on the common law courts to review all privileges granted by the crown and outlawed all but those based on true inventions. The Statute of Monopolies, with its general ban on exclusive rights to manufacture and sell goods, and its limited exception for the purpose of fostering new inventions, is an early example of both antitrust laws and the complex economic interaction between a desire for competition and a desire for new inventions.

Even with the Statute of Monopolies in effect, the British patent system remained a largely informal administrative apparatus. After the vicissitudes of the Civil War period, during which the Cromwell government "called in" all extant patents and privileges, the *status quo ante* was for the most part restored. Court influence was still helpful in obtaining a patent until the latter part of the eighteenth century. Patent applications were registered rather than examined. And, most tellingly, very few patents were granted.

---

8. See Donald Chisum, Foreign Activity: Its Effect on Patentability Under United States Law, 11 Intl. Rev. of Indus. Prop. & Copyright L. 26 (1980). One such provision, which favored U.S. inventors in priority disputes with foreigners, was partially abolished in 1995 when the United States ratified GATT. See *infra* section B.3.e.

9. Thus the first patents were described as "passports," which allowed their holders to move about freely and practice their trade. Walton H. Hamilton, The Politics of Industry 68 (1957).

10. See, e.g., David Jeremy, Transatlantic Industrial Revolution 36-49 (1981) (describing the key role of English emigrants in transfer of technology to the colonies, and the failure of the prohibitory English export laws due to the fact that it was not necessary to transfer physical embodiments of technology when people could simply memorize the necessary information).

11. Great Britain, Statutes at Large, 21 Jac. I, c. 3 (1624).

But as the Industrial Revolution picked up steam (so to speak), attention began to focus on patents once again. An important change at this time was the increasingly stringent requirement that the applicant for a patent describe his or her invention clearly and completely, a development most often associated with the 1778 opinion of the well-respected Judge Mansfield in *Liardet v. Johnson.*[12] The importance of the specification requirement is that it reflected a changed perception about what the inventor was contributing to society in exchange for the patent grant. Under the original patent systems, society's benefit was the introduction of a new art or technology into the country. By the late eighteenth century, the primary benefit was seen as the technological know-how behind the inventor's patent. The beneficiaries of this view were not just the public at large, but also others skilled in the technical arts who could learn something from the patentee's invention. This was a major change in the economic role of patents, for it shifted the emphasis from the introduction of finished products into commerce to the introduction of new and useful *information* to the technical arts. While it is difficult to estimate the significance of this transition, it does seem to address a complaint voiced by Lord Burghley over the original patent system — its dismal success rate in introducing new industries to the country. Perhaps paradoxically, the emphasis on technical specifications, while recognizing that not every invention will lead to a new industry, may have more efficiently fostered the growth of industry as a whole, by ensuring that up-to-date technical information was disseminated rapidly after its creation.

Although the overall contribution of the patent system to the Industrial Revolution has been a matter of debate in historical circles, it seems no coincidence that the patent system matured alongside the early industrial technologies. One historian, H. I. Dutton, noted that the British patent system of this period was less than watertight from the inventor's point of view. But Dutton argues that this actually redounded to the benefit of the economy as a whole, since "leaks" in the grant to one inventor benefitted other inventors.[13] In this, too, the early British experience foreshadows problems that courts struggle with today — how to encourage invention through the use of exclusive rights, without at the same time stifling the creativity of inventors other than the patentee.

---

12. There is no official report of this case, but knowledge survives from contemporary accounts. See E. Walterscheid, Part III, *supra*, at 793-797.

13. In his thorough review of the role of the patent system in the "first" Industrial Revolution in Great Britain, Dutton concludes that the system was instrumental in fostering almost all of the key technologies of the era. In addition, in chapters on "Trade in Invention" and "Investment in Patents," he documents the historical connections between patents and the financing of invention, thus illustrating that the early patent system did not reward innovation directly but instead played much the same role it does today, i.e., fostering *invention* (creation of new technology) and thus indirectly encouraging innovation (introduction of new products embodying that technology on the market). H. I. Dutton, The Patent System and Inventive Activity During the Industrial Revolution 1750-1852, 103-48 (1984). And in his conclusion Dutton argues that the patent system's inefficiencies actually made it close to an ideal system, since it encouraged invention but did not protect new technology too much from those who would try to improve it. Id. at 204-205. See also Christine MacLeod, Accident or Design? George Ravenscroft's Patent and the Invention of Lead-Crystal Glass, 28 Tech. & Culture 776-780 (1987) (describing long time lag between invention of lead crystal glass and introduction of final product with "bugs" all worked out); F. M. Scherer, Invention and Innovation in the Watt-Boulton Steam-Engine Venture, 6 Tech. & Culture 184 (1965) (role of the patent system in Watt's seminal steam engine invention). But cf. Joel Mokyr, The Industrial Revolution and the New Economic History, in The Economics of the Industrial Revolution 1, 28 (Joel Mokyr, ed. 1989) (arguing that "[p]roperty rights in new techniques were protected, albeit imperfectly by British patent law," yet "[t]he cumulative effect of small improvements made by mostly anonymous workers and technicians was often more important than most of the great inventions.").

Patents were among the many British legal concepts introduced to the American colonies between 1640 and 1776. State patents were granted in most of the original thirteen colonies, beginning with a Massachusetts patent in 1641. Even after the Revolution, under the Articles of Confederation, the individual states continued to issue patents.

Perhaps inevitably, however, conflicts began to arise between the states — most notably over steamboat patents, which were issued to two different inventors during this period. This led to a great deal of confusion over who was actually the inventor of the steamboat, which created an obstacle to the successful operation of interstate steam lines. With this problem (among others) in mind, the Constitutional Convention of 1789 resolved to create a national patent system rooted in the Constitution itself. Thus the provision of Article I, Section 8, authorizing Congress to award exclusive rights for a limited time to authors and inventors "for their respective writings and discoveries." One historical footnote is worth mentioning in this connection: An early draft of this provision, set out in James Madison's notes to the Convention, called for both exclusive rights *and* outright subsidies for new inventions. But this was rejected in favor of exclusive rights only.[14] In any event the first U.S. patent statute was passed in 1790,[15] in the very early days of the first Congress (reflecting the importance of this matter), and the first patent was issued shortly thereafter — to Samuel Hopkins of Pittsford, Vermont, for a process for making potash from wood ashes.

The story of Thomas Jefferson's involvement in the early national patent system has often been told; he was a significant contributor to the original statute, and he helped to administer the patent system established in 1790.[16] But while the patent system got on its feet under Jefferson,[17] it did not grow to its full stature until the 1836 revision, when a formal system of examination, with professional examiners, was substituted for the pro forma registration system of the 1793 Act, which had itself been substituted for the original (1790) procedure involving three high-level government officials (including Jefferson as Secretary of State).

Since 1836 the patent system has grown dramatically by any standard — number of patents issued, number of cases litigated, number of significant inventions patented, and so on. As greater demands were placed on it, the patent system developed new rules. For example, the requirement that an invention be more than novel, that it reveal

14. Bruce Bugbee, The Genesis of American Patent and Copyright Law 126, 143 (1967) (describing Madison and Pinckney proposals, both of which included some form of subsidies; and noting that the Senate proposed a compulsory licensing provision in the first Patent Act of 1790, modeled on similar provisions in state copyright acts, but it was rejected by the House).

15. Patent Act of 1790, Ch. 7, 1 Stat. 109-112 (April 10, 1790).

16. For a corrective to the traditional view that Jefferson single-handedly created U.S. patent law, see Edward C. Walterscheid, The Use and Abuse of History: The Supreme Court's Interpretation of Thomas Jefferson's Influence on the Patent Law, 39 IDEA 195 (1999).

17. There is some evidence that the early patent acts were a significant stimulus to invention in the new nation. See, e.g., Michael B. Folsom & Steven D. Lubar, Introduction, in The Philosophy of Manufacturers: Early Debates over Industrialization in the United States xxvii-xxviii (Folsom & Lubar eds. 1982) ("Given the fact that the primary inducement and reward for industrial development was money, new industrial interests exerted pressure to establish safeguards for the investment of time and capital in technological innovation. The first great American industrial corporation, the Boston Manufacturing Company, made a significant early profit selling other textile companies rights to the machine patents it held."). See also Zorina Khan, The Democratization of Invention: Patents and Copyrights in American Economic Development, 1790-1920 (2005).

Whether the patent laws spurred invention or not, a patent bar was quick to form to help inventors deal with the new act. The first patent treatise, by the eminent Fessenden, appeared in Boston very early in the nineteenth century. See Thomas Fessenden, An Essay on the Law of Patents for Inventions (1st ed. 1810).

an "inventive leap," or what is now called nonobviousness, developed in the mid-nineteenth century to limit the number of patents that were being issued. Late in the nineteenth century the bureaucratic structure of the Patent Office as we know it began to take shape as well.

In Europe, the nineteenth century was a time when a new generation of analytical economists questioned the economic foundations of the patent system.[18] Indeed, Switzerland and the Netherlands had no patent systems for more than fifty years during this period.[19] Despite the period of transition and questioning, the patent system was a well-accepted feature of the economic landscape by the beginning of the twentieth century. Key patents on the lightbulb, the telephone system, the basic design of the automobile, and the first airplanes symbolized the technical virtuosity and dynamism of the age. As the scale of industry grew, research and development departments began to appear in the larger companies.[20] Patents were not only a valuable output of these departments; they helped measure the productivity of the departments and served to justify their importance.

The history of the U.S. patent system in the 20th century reflects swings between greater and lesser protection. By the 1920s and 1930s, a number of people began to believe that large companies with patent portfolios were a little *too* powerful. Spurred in part by a series of anticompetitive acts by large companies whose patents dominated their respective industries, courts became less willing to enforce patent rights and more willing to punish patentees for exceeding the scope of their patent grant. The pendulum swung back towards patentability during the 1940s. As the nation threw all available resources into the war effort, the armed forces called on engineers and scientists to perfect a vast array of new technologies in short order. Companies during this era had neither the time nor the ability to try to exclude their competitors: government-directed industrial development and "mandatory cooperation" replaced exclusionary acts. By the time the war was over, there was a consensus in Congress in favor of a strong patent system. In fact the 1952 Patent Act, the first major revision of the patent code since the nineteenth century, marked a return to the principles of that century in many respects.[21]

The patent system by general consensus reached a low-water mark during the 1960s, in the wake of this period of strong protection. The patent office issued patents rather freely, without particularly rigorous examination in many circumstances. On the other hand, it was difficult to get a patent upheld in many federal circuit courts, and the circuits diverged widely both as to doctrine and basic attitudes toward patents. As a consequence, industry downplayed the significance of patents.

---

18. See Fritz Machlup & Edith Penrose, The Patent Controversy in the Nineteenth Century, 10 J. Econ. Hist. 1 (1950); Fritz Machlup, Patents, in 2 Intl. Encyclopedia of the Social Sciences 461 (1968).

19. E. Schiff, Industrialization Without National Patents: The Netherlands 1869-1912; Switzerland 1850-1907 (1971). Schiff's book is rather inconclusive on the effect of this non-patent era on the economic development of the two countries. He states in regard to the Netherlands, for instance, that "it seems unlikely that the overall rate of progress in industry would have been markedly different if a patent system ... had been in operation." Id. at 40. On the other hand, he notes statistical evidence "that the reintroduction of a patent system in 1912 has given an extra spur to Dutch inventive activity." Id.

20. See Thomas P. Hughes, American Genesis: A Century of Invention and Technological Enthusiasm 1870-1970, 150-180 (1989).

21. For an overview of the period 1900-2000, see Robert P. Merges, One Hundred Years of Solicitude: Intellectual Property Law, 1900-2000, 88 Cal. L. Rev. 2187 (2000).

In 1982 Congress passed the Federal Courts Improvement Act, creating the new Court of Appeals for the Federal Circuit.[22] The Federal Circuit handles several important types of cases, but from the beginning one of its primary functions has been to hear all appeals involving patents. While the Federal Circuit was ostensibly formed strictly to unify patent doctrine, it was no doubt hoped by some (and expected by others) that the new court would make subtle alterations in the doctrinal fabric, with an eye toward expanding the scope of patent protection.

To judge by results, that is exactly what happened. Overall, patents are more likely to be held valid now than in previous decades.[23] It is generally easier to get an injunction against an infringer. (Although this is subject to pressure now that companies have emerged that specialize in acquiring and asserting patents against large manufacturers — see below, section F.1, on injunctions.) In general, patent cases lead to higher damage awards now, both on average and in the highest-visibility cases. Whether intentional or not, the creation of the Federal Circuit will surely be seen as a watershed event by future historians of the patent system.[24] Not that the court is completely "pro-patentee": far from it, as recent developments in claim scope (particularly the "doctrine of equivalents"; see section C.3.a below) demonstrate.

Overall, the perceived value of patents has increased since the court's founding. The sheer number of patents being granted has increased roughly threefold in the past 20 years, and the number of litigated cases has roughly tripled as well. At the same time, the environment in which the Federal Circuit operates has been changing. Exports have grown considerably as a source of national wealth. As a result, policymakers in the government now pay much closer attention to the legal and economic features of our major trading partners. This naturally includes much closer scrutiny of the intellectual property systems in these other countries. At the same time, it has become clear that other countries have identified technology-intensive industries as keys to economic growth in the future. This awareness has led to a new focus on economic policy instruments that can be used to foster these industries — including protection of intellectual property. Finally, because intellectual property legislation has no direct, immediate cost to the government, it seems to many to be a relatively cheap aid to industry. The result of these factors has been a growing trend in world trade policy toward exporting or "internationalizing" patent law. A number of countries (including the European Union) have created a reciprocal system in which applicants need not file for protection separately in every country. International treaties contain similar arrangements.

---

22. See Rochelle Dreyfuss, The Federal Circuit: A Case Study in Specialized Courts, 64 N.Y.U. L. Rev. 1, 25-26 (1989); Jordan, Specialized Courts: A Choice?, 76 Nw. U. L. Rev. 745 (1981) (describing competing arguments over specialized courts).

23. Several recent studies have provided statistical support for the proposition (widely accepted in the patent bar) that the Federal Circuit is at least moderately "pro-patent." See John R. Allison & Mark A. Lemley, Empirical Evidence on the Validity of Litigated Patents, 26 Am. Intell. Prop. L. Assn. Q.J. 185 (1998); Donald R. Dunner, J. Michael Jakes, and Jeffrey D. Kerceski, A Statistical Look at the Federal Circuit's Patent Decisions: 1982-1994, 5 Fed. Circuit B.J. 151, 154-155 (1995). See Robert P. Merges, Commercial Success and Patent Standards: Economic Perspectives on Innovation, 76 Cal. L. Rev. 803, 820-821 (1988) (comparing pre- and post-Federal Circuit era statistics); Kimberly A. Moore, Judges, Juries and Patent Cases: An Empirical Peek Inside the Black Box, 99 Mich. L. Rev. 365 (2000).

24. A 2001 study shows that Federal Circuit judges with patent experience prior to appointment to the bench have a disproportionate impact, based on number of judicial opinions written. See John R. Allison & Mark A. Lemley, How Federal Circuit Judges Vote in Patent Validity Cases, 10 Fed. Cir. B.J. 435 (2001). For a history of the formation of the court, and the policy studies justifying its creation, see Judge Pauline Newman, The Federal Circuit in Perspective, 54 Am. U. L. Rev. 821 (2005).

If recent changes wrought by the Uruguay Round of GATT, GATT-TRIPS, are any indication, the patent laws of most major countries will come into ever-closer harmony.

2. An Overview of the Patent Laws

It is important to appreciate some basic attributes of the patent system before reading the following cases. This is a complex body of law with its own terminology and tradition, and one must become acquainted with the rudiments before diving into the details.

a. Requirements for Patentability

The Patent and Trademark Office (PTO) reviews each patent application to see if it meets five requirements: an invention fits one of the general categories of patentable subject matter; it has not been preceded in identical form in the public prior art; it is useful; it represents a nontrivial extension of what was known; and it is disclosed and described by the applicant in such a way as to enable others to make and use the invention. 35 U.S.C. §§ 101 (utility), 102 (novelty), 103 (nonobviousness), and 112 (enablement) (1982). See generally Graham v. John Deere, 383 U.S. 1 (1966).

Patent lawyers call the no-identical-prior-invention requirement "novelty." In practice, novelty is established by applying a set of technical rules to determine if a patent applicant was really the first to make the invention she is claiming. The novelty test determines whether the claimed invention is unpatentable because it was made before, sold more than a year before a patent application was filed, or otherwise disqualified by prior use or knowledge.

Utility, the second requirement, has devolved over the years into a rather minimal obstacle to obtaining a patent. Section 101 is the source of this requirement. 35 U.S.C. § 101 (1982) ("Whoever invents any new *and useful* process, machine, manufacture, or composition of matter, may obtain a patent therefor. . . ." (emphasis added)). Today, a patent will not be withheld even though the invention works only in an experimental setting and has no proven use in the field or factory. Only if an invention has absolutely no "practical utility" will a patent be denied. The only exception is inventions pertaining to pharmaceuticals, where some cases question whether laboratory promise is enough to establish utility in treating human patients.

The next requirement, <u>nontriviality</u>, is known to patent lawyers as "<u>non-obviousness</u>." 35 U.S.C § 103. This is the <u>most important requirement</u>; it has been called "<u>the ultimate condition of patentability</u>." The reason is that nonobviousness attempts to measure an even more abstract quality than novelty or utility: the *technical accomplishment* reflected in an invention. This requirement asks whether an invention is a big enough technical advance over the prior art. Even if an invention is new and useful, it will still not merit a patent if it represents merely a trivial step forward in the art.

Finally, 35 U.S.C. § 112 requires a patentee to give a sufficiently good description of her invention that "one of ordinary skill in the art" would be able to make and use the invention. The concern here is not so much with whether the inventor has developed something worth patenting as it is with the benefit the public obtains from the patent "bargain." The disclosure and "enablement" requirements of section 112 ensure that those "skilled in the art" of the invention can read and understand the inventor's contribution, and that after the patent expires they will be able to make and use the invention themselves.

b. Rights Conferred by a Patent

"Claims" are the heart of patent law. Claims define the boundaries of the property right that the patent confers. (Innumerable patent cases therefore analogize claims to the "metes and bounds" of a real property deed.) Claims come at the end of the written description of the invention. Most patents also have one or more drawings.

The specification describes the invention. It names all the parts or components of the invention, describes how they work, and illustrates how they work together to perform the invention's function. Only at the end of the specification does the inventor (or, more usually, her patent lawyer) state the precise legal definition of the invention. These are the claims. Here are a few (fanciful) examples of claims:

1. Element 95.
2. A composition comprising
   (a) a solid selected from the group consisting of
       (1) sodium chloride,
       (2) potassium chloride and
       (3) lithium chloride;
   (b) a liquid selected from the group consisting of
       (1) sulfuric acid,
       (2) nitric acid....[25]
3. The material wrought tungsten, having a specific gravity of approximately 19 or greater, and capable of being forged and worked.
4. A windmill comprising a wind-catching device, directed to face the oncoming wind force, said device turning a shaft, said shaft acting on gears or another device to change the direction of said wind force, so as to operate a pump that pumps water.
5. A windmill according to claim 4 wherein the force-changing device is a set of gears.
6. A method for treating baldness comprising applying to the scalp an aqueous solution of the compound minoxidil.
7. An apparatus for playing record albums comprising a cartridge or stylus made from at least 40 percent graphite by weight, a tone arm on which

---

25. This is an example of a claim to a "genus" that includes several "species," a common construct in patent law. A special form of genus claim, termed a "Markush expression," is often encountered. Such claims take the form of an expression such as "An X selected from the group consisting of A, B, and C," where X describes the functional class to which A, B, and C belong. These claims are often, but certainly not always, used in patents to chemical compositions. There is, for example, a famous patent claiming a cigarette filter made of cheese, whose claim 2 reads: "A cigarette filter according to claim 1, in which the cheese comprises grated particles of cheese selected from the group consisting of Parmesan, Romano, Swiss and cheddar cheeses." U.S. Patent No. 3,234,948, to Stebbings, cited in Robert C. Faber, Landis on Mechanics of Patent Claim Drafting §50 (3rd ed. 1990).

said cartridge or stylus is mounted, a turntable on which said record albums are placed for playing, and means for turning said turntable at appropriate speeds for the playing of said record albums.

Note that all the claims except claim 5 are examples of *independent claims;* they do not refer to any other claim or claims. This is in contrast to a dependent claim, such as claim 5 above. You can tell it is a dependent claim because it begins with the phrase "A windmill according to claim 4 . . . ." A dependent claim incorporates all the limitations of the independent claim on which it depends.

Note also that claim 6 is a claim to a process or method rather than a device, and that the last element of claim 7 is stated in "means plus function" format: the element is not described in detail but is merely listed as "a means" for accomplishing some goal. The importance of this specialized claim format will be discussed in section C.5. A final note: the phrase "consisting of" is known as an "open" transition: claims using this terminology normally cover devices that include all the listed elements *plus* any additional elements. The phrase "consisting of," by contrast, is said to be a closed transition: it does not cover devices that include additional elements. *See, e.g.,* Gillette Co. v. Energizer Holdings, Inc., 405 F.3d 1367, 1373 (Fed. Cir. 2005) (claim to safety razor "comprising . . . a group of first, second and third blades . . . ." covered an accused device with four blades).

A patent confers the right to exclude others from making, using, selling, offering for sale, or importing the claimed invention for a specific term of years. Until 1995, that "patent term" was 17 years from the date the patent issued. With United States adherence to GATT-TRIPS, the patent term was changed to extend for 20 years from the date the patent application was filed, rather than the old term of 17 years from the date it was issued by the Patent Office. The length of a patent term under the new law therefore varies from case to case — a patent is in force for 20 years minus the amount of time the patent spent in the application (or "prosecution") process.[26]

The exclusionary right is in a sense a negative right, for two reasons. First, a patent does not automatically grant an affirmative right to do anything; patented pharmaceuticals, for instance, must still pass regulatory review at the Food and Drug Administration to be sold legally. Second, a patented invention may itself be covered by a preexisting patent. For instance, a broad "pioneering" patent on a product or process may cover later-developed inventions, themselves patented as

---

26. Uruguay Round Agreements Act of 1994, P.L. 103-465, 108 Stat. 4809 (1994), at §532, codified at 35 U.S.C. §154(a)(2). For an analysis of the likely effects of this change in patent term, see Mark A. Lemley, An Empirical Study of the Twenty Year Patent Term, 22 AIPLA Q.J. 369 (1994).

The history of the patent term in the United States prior to the 1994 Act is interesting in its own right. The first patent act, Act of April 10, 1790, ch. 7, 1 Stat. 109, adopted the same patent term as the British Statute of Monopolies of 1624, 21 Jac. 1, ch. 3 — namely, 14 years. This was based on the notion that a patent should protect an inventor for a period equal to two terms of a standard British trade apprenticeship of seven years. See White, Why a Seventeen Year Patent?, 38 J. Pat. & Trademk. Off. Society 839 (1956). In the patent act of 1836 — which once again required patents to be examined and approved by government officials, as under the 1790 Act — a seven-year renewal term was added, making the total possible term 21 years. The 17-year term that prevailed for many years was the result of a compromise between the House, which wanted to retain the 14 plus 7 term, and the Senate, which wanted to eliminate the renewal term. Economists have long noted that a uniform term sometimes rewards inventors too much and sometimes gives too little incentive; some have proposed to "craft" individual terms to fit individual inventions. See, e.g., William D. Nordhaus, Invention, Growth, and Welfare (1969) (economic model deriving optimal patent term under different circumstances).

improvements. In such a case the holder of an improvement patent has the right to exclude everyone from her improvement—including the holder of the broad patent—while at the same time being barred from use of the improvement herself unless the holder of the broad patent authorizes such use. Patents so related are said to be "blocking patents."

3. Theories of Patent Law

By contrast with trade secret law, which draws on a number of different (and sometimes contradictory) theoretical bases, the central theory behind patent law is relatively straightforward. This theory posits that inventions are public goods that are costly to make and that are difficult to control once they are released into the world. As a result, absent patent protection inventors will not have sufficient incentive to invest in creating, developing, and marketing new products. Patent law provides a market-driven incentive to invest in innovation, by allowing the inventor to appropriate the full economic rewards of her invention.

Other theories often advanced to explain intellectual property law—natural rights and personhood theories, for example—play a much less significant role in patent law than in other areas of intellectual property.[27] In part, this is because of the broad nature of the patent grant. Patents give the inventor the right to sue not only those who "steal" his invention, but those who reverse engineer it and even those who develop the same invention independently. The broad nature of this grant makes it difficult to speak of a "moral entitlement" to a patent. And in part the focus on utilitarian theory mirrors the subject matter of patents, which revolves around mechanical devices, chemical formulae, and the like. It would be anomalous to most to speak of the "personality" invested in stamping machinery or pesticides.

Although the economic incentive story is straightforward, it does not tell us very much about how to design a patent system to provide optimal incentives. For the reasons described in Chapter 1, overprotecting is as bad as underprotecting in many ways. Thus designing the proper economic incentive requires the policymaker to balance the length of the patent term, the appropriate standard of invention, and the nature of the rights granted to patentees. Resolving these conflicts has occupied courts and Congress since the passage of the first patent statute in 1790. We discuss them in some detail below.

---

27. That is not to say such theories do not exist. For a discussion of reward-based and even natural law theories of scientific invention, see, e.g., Adam Mossoff, Rethinking the Development of Patents: An Intellectual History, 1550-1800, 52 Hastings L.J. 1255 (2001) (reviewing case that natural rights theories contributed to development of patent system); A. Samuel Oddi, Un-Unified Economic Theories of Patents—the Not-Quite-Holy Grail, 71 Notre Dame L. Rev. 267, 274-277 (1996); Kevin Rhodes, Comment, The Federal Circuit's Patent Nonobviousness Standards: Theoretical Perspectives on Recent Doctrinal Changes, 85 Nw. U.L. Rev. 1051 (1991); cf. Lawrence C. Becker, Deserving to Own Intellectual Property, 68 Chicago-Kent L. Rev. 609 (1993) (arguing that entitlement-based rationales for patent law are intuitively appealing, but do not necessarily justify the scope of current patent doctrine). An alternative to classical incentive theory is the prospect theory of patents, advanced by Edmund Kitch in an important article in 1977. See Edmund Kitch, The Nature and Function of the Patent System, 30 J.L. & Econ. 265 (1977). Kitch offers a property-based vision of patents as entitlements to innovate within a particular field, granted to those who have already started such innovation. For a refinement of Kitch's approach that takes account of rent-seeking, see Mark F. Grady & Jay I. Alexander, Patent Law and Rent Dissipation, 78 Va. L. Rev. 305 (1992). For a challenging extension of the theory, see John F. Duffy, Rethinking the Prospect Theory of Patents, 71 U. Chi. L. Rev. 439 (2004).

B. THE ELEMENTS OF PATENTABILITY

very important

In the sections that follow we investigate the five primary requirements of patentability: (1) patentable subject matter, (2) novelty, (3) utility, (4) nonobviousness, and (5) enablement.

1. Patentable Subject Matter

This section deals with what is known as patentable subject matter under § 101 of the patent code: that is, the issue of which *types* of inventions will be considered for patent protection. Section 101 determines, for instance, whether living things can be patented (as in the case that follows), or whether mathematical algorithms are proper patent subject matter. (Cases specifically dealing with computer-related patents are discussed in Chapter 7.) We are not concerned here with the technical requirements of patentability (novelty, utility, nonobviousness, etc.); these requirements are the subject of subsequent sections.

Section 101 of the statute defines the categories of patentable invention broadly: "any ... process, machine, manufacture, ... composition of matter, or ... improvement thereof...." Patent lawyers often employ a working distinction between process claims and "product" claims, which are simply claims to any of the other three types of invention, i.e, machines, manufactures, or compositions of matter.

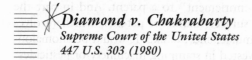

Diamond v. Chakrabarty
Supreme Court of the United States
447 U.S. 303 (1980)

BURGER, C.J.

We granted certiorari to determine whether a live, human-made micro-organism is patentable subject matter under 35 U.S.C. § 101.

I

In 1972, respondent Chakrabarty, a microbiologist, filed a patent application, assigned to the General Electric Co. The application asserted 36 claims related to Chakrabarty's invention of "a bacterium from the genus *Pseudomonas* containing therein at least two stable energy-generating plasmids, each of said plasmids providing a separate hydrocarbon degradative pathway."[1] This human-made, genetically engineered bacterium is capable of breaking down multiple components of crude oil. Because of this property, which is possessed by no naturally occurring bacteria,

---

1. Plasmids are hereditary units physically separate from the chromosomes of the cell. In prior research, Chakrabarty and an associate discovered that plasmids control the oil degradation abilities of certain bacteria. In particular, the two researchers discovered plasmids capable of degrading camphor and octane, two components of crude oil. In the work represented by the patent application at issue here, Chakrabarty discovered a process by which four different plasmids, capable of degrading four different oil components, could be transferred to and maintained stably in a single *Pseudomonas* bacterium, which itself has no capacity for degrading oil.

Chakrabarty's invention is believed to have significant value for the treatment of oil spills.

Chakrabarty's patent claims were of three types: first, process claims for the method of producing the bacteria; second, claims for an inoculum comprised of a carrier material floating on water, such as straw, and the new bacteria; and third, claims to the bacteria themselves. The patent examiner allowed the claims falling into the first two categories, but rejected claims for the bacteria. His decision rested on two grounds: (1) that micro-organisms are "products of nature," and (2) that as living things they are not patentable subject matter under 35 U.S.C. § 101.

Chakrabarty appealed the rejection of these claims to the Patent Office Board of Appeals, and the Board affirmed the examiner on the second ground.[3] Relying on the legislative history of the 1930 Plant Patent Act, in which Congress extended patent protection to certain asexually reproduced plants, the Board concluded that § 101 was not intended to cover living things such as these laboratory created micro-organisms. [The Court of Appeals reversed.]

The Commissioner of Patents and Trademarks again sought certiorari, and we granted the writ. . . .

II

The Constitution grants Congress broad power to legislate to "promote the Progress of Science and useful Arts." . . . The patent laws promote this progress by offering inventors exclusive rights for a limited period as an incentive for their inventiveness and research efforts. The authority of Congress is exercised in the hope that "[t]he productive effort thereby fostered will have a positive effect on society through the introduction of new products and processes of manufacture into the economy, and the emanations by way of increased employment and better lives for our citizens."

The question before us in this case is a narrow one of statutory interpretation requiring us to construe 35 U.S.C. § 101. Specifically, we must determine whether respondent's micro-organism constitutes a "manufacture" or "composition of matter" within the meaning of the statute.

III

[T]his Court has read the term "manufacture" in § 101 in accordance with its dictionary definition to mean "the production of articles for use from raw or prepared materials by giving to these materials new forms, qualities, properties, or combinations, whether by hand-labor or by machinery." American Fruit Growers, Inc. v. Brogdex Co., 283 U.S. 1, 11 (1931). Similarly, "composition of matter" has been construed consistent with its common usage to include "all compositions of two or more substances and . . . all composite articles, whether they be the results of chemical union, or of mechanical mixture, or whether they be gases, fluids, powders or

---

3. The Board concluded that the new bacteria were not "products of nature," because *Pseudomonas* bacteria containing two or more different energy-generating plasmids are not naturally occurring.

solids." Shell Development Co. v. Watson, 149 F. Supp. 279, 280 (D.C. 1957). In choosing such expansive terms as "manufacture" and "composition of matter," modified by the comprehensive "any," Congress plainly contemplated that the patent laws would be given wide scope.

The relevant legislative history also supports a broad construction. The Patent Act of 1793, authored by Thomas Jefferson, defined statutory subject matter as "any new and useful art, machine, manufacture, or composition of matter, or any new or useful improvement [thereof]." Act of Feb. 21, 1793, §1, 1 Stat. 319. The Act embodied Jefferson's philosophy that "ingenuity should receive a liberal encouragement." 5 Writings of Thomas Jefferson 75-76 (Washington ed. 1871). Subsequent patent statutes in 1836, 1870, and 1874 employed this same broad language. In 1952, when the patent laws were recodified, Congress replaced the word "art" with "process," but otherwise left Jefferson's language intact. The Committee Reports accompanying the 1952 Act inform us that Congress intended statutory subject matter to "include anything under the sun that is made by man." S. Rep. No. 1979, 82d Cong., 2d Sess., 5 (1952); H.R. Rep. No. 1923, 82d Cong., 2d Sess., 6 (1952).

This is not to suggest that § 101 has no limits or that it embraces every discovery. *The laws of nature, physical phenomena, and abstract ideas have been held not patentable.* Thus, a new mineral discovered in the earth or a new plant found in the wild is not patentable subject matter. Likewise, Einstein could not patent his celebrated law that $E = mc^2$; nor could Newton have patented the law of gravity. Such discoveries are "manifestations of... nature, free to all men and reserved exclusively to none."

Judged in this light, respondent's micro-organism plainly qualifies as patentable subject matter. His claim is not to a hitherto unknown natural phenomenon, but to a nonnaturally occurring manufacture or composition of matter—a product of human ingenuity "having a distinctive name, character [and] use." The point is underscored dramatically by comparison of the invention here with that in *Funk* [Bros. Seed Co. v. Kalo Inoculant Co., 333 U.S. 127 (1948)]. There, the patentee had discovered that there existed in nature certain species of root-nodule bacteria which did not exert a mutually inhibitive effect on each other. He used that discovery to produce a mixed culture capable of inoculating the seeds of leguminous plants. Concluding that the patentee had discovered "only some of the handiwork of nature," the Court ruled the product nonpatentable:

> Each of the species of root-nodule bacteria contained in the package infects the same group of leguminous plants which it always infected. No species acquires a different use. The combination of species produces no new bacteria, no change in the six species of bacteria, and no enlargement of the range of their utility. Each species has the same effect it always had. The bacteria perform in their natural way. Their use in combination does not improve in any way their natural functioning. They serve the ends nature originally provided and act quite independently of any effort of the patentee.

333 U.S., at 131.

Here, by contrast, the patentee has produced a new bacterium with markedly different characteristics from any found in nature and one having the potential for significant utility. His discovery is not nature's handiwork, but his own; accordingly it is patentable subject matter under § 101.

IV

Two contrary arguments are advanced, neither of which we find persuasive.

The petitioner's first argument rests on the enactment of the 1930 Plant Patent Act, which afforded patent protection to certain asexually reproduced plants, and the 1970 Plant Variety Protection Act, which authorized protection for certain sexually reproduced plants but excluded bacteria from its protection.[7] In the petitioner's view, the passage of these Acts evidences congressional understanding that the terms "manufacture" or "composition of matter" do not include living things; if they did, the petitioner argues, neither Act would have been necessary.

We reject this argument. Prior to 1930, two factors were thought to remove plants from patent protection. The first was the belief that plants, even those artificially bred, were products of nature for purposes of the patent law.... The second obstacle to patent protection for plants was the fact that plants were thought not amenable to the "written description" requirement of the patent law. See 35 U.S.C. § 112. Because new plants may differ from old only in color or perfume, differentiation by written description was often impossible.

In enacting the Plant Patent Act, Congress addressed both of these concerns.... Moreover, there is language in the House and Senate Committee Reports suggesting that to the extent Congress considered the matter it found the Secretary's dichotomy unpersuasive. The Reports observe:

> There is a clear and logical distinction *between the discovery of a new variety of plant and of certain inanimate things,* such, for example, as a new and useful natural mineral. The mineral is created wholly by nature unassisted by man.... On the other hand, a plant discovery resulting from cultivation is unique, isolated, and is not repeated by nature, nor can it be reproduced by nature unaided by man....

S. Rep. No. 315, supra, at 6; H.R. Rep. No. 1129, supra, at 7 (emphasis added).

Congress thus recognized that the relevant distinction was not between living and inanimate things, but between products of nature, whether living or not, and human-made inventions. Here, respondent's micro-organism is the result of human ingenuity and research. Hence, the passage of the Plant Patent Act affords the Government no support.

Nor does the passage of the 1970 Plant Variety Protection Act support the Government's position. As the Government acknowledges, sexually reproduced plants were not included under the 1930 Act because new varieties could not be reproduced true-to-type through seedlings. By 1970, however, it was generally recognized that true-to-type reproduction was possible and that plant patent protection was therefore

---

7. The Plant Patent Act of 1930, 35 U.S.C. § 161, provides in relevant part:

Whoever invents or discovers and asexually reproduces any distinct and new variety of plant, including cultivated sports, mutants, hybrids, and newly found seedlings, other than a tuber propagated plant or a plant found in an uncultivated state, may obtain a patent therefor....

The Plant Variety Protection Act of 1970, provides in relevant part:

The breeder of any novel variety of sexually reproduced plant (other than fungi, bacteria, or first generation hybrids) who has so reproduced the variety, or his successor in interest, shall be entitled to plant variety protection therefor....

84 Stat. 1547, 7 U.S.C. § 2402(a). See generally, 3 A. Deller, Walker on Patents, ch. IX (2d ed. 1964); R. Allyn, The First Plant Patents (1934).

appropriate. The 1970 Act extended that protection. There is nothing in its language or history to suggest that it was enacted because § 101 did not include living things....

The petitioner's second argument is that micro-organisms cannot qualify as patentable subject matter until Congress expressly authorizes such protection. His position rests on the fact that genetic technology was unforeseen when Congress enacted § 101. From this it is argued that resolution of the patentability of inventions such as respondent's should be left to Congress. The legislative process, the petitioner argues, is best equipped to weigh the competing economic, social, and scientific considerations involved, and to determine whether living organisms produced by genetic engineering should receive patent protection. In support of this position, the petitioner relies on our recent holding in Parker v. Flook, 437 U.S. 584 (1978), and the statement that the judiciary "must proceed cautiously when... asked to extend patent rights into areas wholly unforeseen by Congress." Id., at 596.

It is, of course, correct that Congress, not the courts, must define the limits of patentability; but it is equally true that once Congress has spoken it is "the province and duty of the judicial department to say what the law is." Marbury v. Madison, 1 Cranch 137, 177 (1803). Congress has performed its constitutional role in defining patentable subject matter in § 101; we perform ours in construing the language Congress has employed. In so doing, our obligation is to take statutes as we find them, guided, if ambiguity appears, by the legislative history and statutory purpose. Here, we perceive no ambiguity. The subject-matter provisions of the patent law have been cast in broad terms to fulfill the constitutional and statutory goal of promoting "the Progress of Science and the useful Arts" with all that means for the social and economic benefits envisioned by Jefferson. Broad general language is not necessarily ambiguous when congressional objectives require broad terms.

Nothing in *Flook* is to the contrary. The Court carefully scrutinized the claim at issue to determine whether it was precluded from patent protection under "the principles underlying the prohibition against patents for 'ideas' or phenomena of nature." Id., at 593. We have done that here. *Flook* did not announce a new principle that inventions in areas not contemplated by Congress when the patent laws were enacted are unpatentable per se.

To read that concept into *Flook* would frustrate the purposes of the patent law. This is especially true in the field of patent law. A rule that unanticipated inventions are without protection would conflict with the core concept of the patent law that anticipation undermines patentability. Mr. Justice Douglas reminded that the inventions most benefiting mankind are those that "push back the frontiers of chemistry, physics, and the like." Great A.&P. Tea Co. v. Supermarket Corp., 340 U.S. 147, 154 (1950) (concurring opinion). Congress employed broad general language in drafting § 101 precisely because such inventions are often unforeseeable.[10]

To buttress his argument, the petitioner, with the support of *amicus*, points to grave risks that may be generated by research endeavors such as respondent's. The briefs present a gruesome parade of horribles. Scientists, among them Nobel laureates, are quoted suggesting that genetic research may pose a serious threat to the human race, or, at the very least, that the dangers are far too substantial to permit such research to proceed apace at this time. We are told that genetic research and

---

10. Even an abbreviated list of patented inventions underscores the point: telegraph (Morse, No. 1,647); telephone (Bell, No. 174,465); electric lamp (Edison, No. 223,898); airplane (the Wrights, No. 821,393); transistor (Bardeen & Brattain, No. 2,524,035); neutronic reactor (Fermi & Szilard, No. 2,708,656); laser (Schawlow & Townes, No. 2,929,922). See generally Revolutionary Ideas, Patents & Progress in America, United States Patent and Trademark Office (1976).

related technological developments may spread pollution and disease, that it may result in a loss of genetic diversity, and that its practice may tend to depreciate the value of human life. These arguments are forcefully, even passionately, presented; they remind us that, at times, human ingenuity seems unable to control fully the forces it creates — that, with Hamlet, it is sometimes better "to bear those ills we have than fly to others that we know not of."

It is argued that this Court should weigh these potential hazards in considering whether respondent's invention is patentable subject matter under § 101. We disagree. The grant or denial of patents on micro-organisms is not likely to put an end to genetic research or to its attendant risks. The large amount of research that has already occurred when no researcher had sure knowledge that patent protection would be available suggests that legislative or judicial fiat as to patentability will not deter the scientific mind from probing into the unknown any more than Canute could command the tides. Whether respondent's claims are patentable may determine whether research efforts are accelerated by the hope of reward or slowed by want of incentives, but that is all.

What is more important is that we are without competence to entertain these arguments — either to brush them aside as fantasies generated by fear of the unknown, or to act on them. The choice we are urged to make is a matter of high policy for resolution within the legislative process after the kind of investigation, examination, and study that legislative bodies can provide and courts cannot. That process involves the balancing of competing values and interests, which in our democratic system is the business of elected representatives. Whatever their validity, the contentions now pressed on us should be addressed to the political branches of the Government, the Congress and the Executive, and not to the courts.

Accordingly, the judgment of the Court of Customs and Patent Appeals is [a]ffirmed.

Mr. Justice BRENNAN, with whom Mr. Justice WHITE, Mr. Justice MARSHALL, and Mr. Justice POWELL join, dissenting.

I agree with the Court that the question before us is a narrow one. Neither the future of scientific research, nor even the ability of respondent Chakrabarty to reap some monopoly profits from his pioneering work, is at stake. Patents on the processes by which he has produced and employed the new living organism are not contested. The only question we need decide is whether Congress intended that he be able to secure a monopoly on the living organism itself, no matter how produced or how used. Because I believe the Court has misread the applicable legislation, I dissent....

The patent laws attempt to reconcile th[e] Nation's deep-seated antipathy to monopolies with the need to encourage progress. Given the complexity and legislative nature of this delicate task, we must be careful to extend patent protection no further than Congress has provided. In particular, were there an absence of legislative direction, the courts should leave to Congress the decisions whether and how far to extend the patent privilege into areas where the common understanding has been that patents are not available.

In this case, however, we do not confront a complete legislative vacuum. The sweeping language of the Patent Act of 1793, as re-enacted in 1952, is not the last pronouncement Congress has made in this area. In 1930 Congress enacted the Plant Patent Act affording patent protection to developers of certain asexually reproduced plants. In 1970 Congress enacted the Plant Variety Protection Act to extend protection to certain new plant varieties capable of sexual reproduction. Thus, we are not dealing — as the Court would have it — with the routine problem of "unanticipated inventions." In these two Acts Congress has addressed the general problem of patenting animate inventions and has chosen carefully limited language granting protection to some kinds of discoveries, but specifically excluding others. These Acts strongly evidence a congressional limitation that excludes bacteria from patentability. [2]

First, the Acts evidence Congress' understanding, at least since 1930, that § 101 does not include living organisms. If newly developed living organisms not naturally occurring had been patentable under § 101, the plants included in the scope of the 1930 and 1970 Acts could have been patented without new legislation. Those plants, like the bacteria involved in this case, were new varieties not naturally occurring. [3] Although the Court rejects this line of argument, it does not explain why the Acts were necessary unless to correct a pre-existing situation. I cannot share the Court's implicit assumption that Congress was engaged in either idle exercises or mere correction of the public record when it enacted the 1930 and 1970 Acts. And Congress certainly thought it was doing something significant. The Committee Reports contain expansive prose about the previously unavailable benefits to be derived from extending patent protection to plants. [5] H.R. Rep. No. 91-1605, pp. 1-3 (1970); S. Rep. No. 315, 71st Cong., 2d Sess., 1-3 (1930). Because Congress thought it had to legislate in order to make agricultural "human-made inventions" patentable and because the legislation Congress enacted is limited, it follows that Congress never meant to make items outside the scope of the legislation patentable....

The Court protests that its holding today is dictated by the broad language of § 101, which cannot "be confined to the 'particular application[s]... contemplated by the legislators.' But as I have shown, the Court's decision does not follow the unavoidable implications of the statute. Rather, it extends the patent system to cover living material even though Congress plainly has legislated in the belief that § 101 does not encompass living organisms. It is the role of Congress, not this Court, to broaden or narrow the reach of the patent laws. This is especially true where, as here, the composition sought to be patented uniquely implicates matters of public concern.

2. But even if I agreed with the Court that the 1930 and 1970 Acts were not dispositive, I would dissent. This case presents cogent reasons not to extend the patent monopoly in the face of uncertainty. At the very least, these Acts are signs of legislative attention to the problems of patenting living organisms, but they give no affirmative indication of congressional intent that bacteria be patentable. The caveat of Parker v. Flook, 437 U.S. 584, 596 (1978), an admonition to "proceed cautiously when we are asked to extend patent rights into areas wholly unforeseen by Congress," therefore becomes pertinent. I should think the necessity for caution is that much greater when we are asked to extend patent rights into areas Congress has foreseen and considered but has not resolved.

3. The Court refers to the logic employed by Congress in choosing not to perpetuate the "dichotomy" suggested by Secretary Hyde. But by this logic the bacteria at issue here are distinguishable from a "mineral ... created wholly by nature" in exactly the same way as were the new varieties of plants. If a new Act was needed to provide patent protection for the plants, it was equally necessary for bacteria. Yet Congress provided for patents on plants but not on these bacteria. In short, Congress decided to make only a subset of animate "human-made inventions" patentable.

5. Secretary Hyde's letter was not the only explicit indication in the legislative history of these Acts that Congress was acting on the assumption that legislation was necessary to make living organisms patentable. The Senate Judiciary Committee Report on the 1970 Act states the Committee's understanding that patent protection extended no further than the explicit provisions of these Acts:

> Under the patent law, patent protection is limited to those varieties of plants which reproduce asexually, that is, by such methods as grafting or budding. No protection is available to those varieties of plants which reproduce sexually, that is, generally by seeds. S. Rep. No. 91-1246, p. 3 (1970).

Similarly, Representative Poage, speaking for the 1970 Act, after noting the protection accorded asexually developed plants, stated that "for plants produced from seed, there has been no such protection." 116 Cong. Rec. 40295 (1970).

COMMENTS AND QUESTIONS

1. The Court revisited the legislative history behind the 1930 Plant Patent Act and the 1970 Plant Variety Protection Act in holding that neither preempted the applicability of the Patent Act of 1952 — i.e., the general utility patent statute studied in this chapter — to plant-related subject matter. J.E.M. Ag-Supply, Inc. v. Pioneer Hi-Bred Intern. Inc., 534 U.S. 124 (2002).

2. The Court cites Funk Brothers Seed Co. v. Kalo Inoculant Co., 333 U.S. 127 (1948), which centered around a patent for bacteria that assists in the process of nitrogen fixation in the roots of plants. The invention underlying the patent was said to be the realization that, contrary to accepted wisdom, varieties from different species of bacteria could be mixed together and sold in a single root-treatment formula. Experts had previously believed that this was not possible because different species would inhibit each other if mixed, and had adopted the practice of selling individual bacterial species. The Supreme Court held that combining the bacteria in this way was a discovery, not invention: "[it] is no more than the discovery of some of the handiwork of nature and hence is not patentable."

What distinguishes *Chakrabarty* from *Funk Brothers?* One possible answer is that the Court is drawing a line between *discovery* and *invention.* In *Chakrabarty,* the Court emphasizes the transformations that the inventor makes on the admittedly natural raw materials of the invention. But in *Funk Bros.,* the Court emphasizes that the bacteria, though combined in a novel way, still perform their same old natural function. Can the cases be reconciled on this ground? Consider 35 U.S.C. § 100, which provides that "invention" as the term is used in the Patent Act means both "invention and discovery."

Does *Funk Bros.* suggest that new combinations of known elements can never be patentable? At a minimum, does the case require some sort of "synergy" resulting from a combination — that the elements when combined must do more than they each would have done alone? Would such a requirement make sense?

3. Should human intervention be the touchstone for patentability in this area? Consider the following case.

Parke-Davis & Co. v. H. K. Mulford Co.
United States Circuit Court for the Southern District of New York
189 F. 95 (C.C.S.D.N.Y. 1911)

L. HAND, District Judge:

[Jokichi Takamine, inventor of the two patents in issue, discovered how to isolate a purified substance of significant medical use from the suprarenal glands of animals. Parke-Davis, assignee of the patents, called the product Adrenalin. Takamine's adrenalin was medically superior to the older isolates of this gland that had been in use, because those products were not nearly as pure, which resulted in the injection into humans of a good deal of extraneous gland tissue left over from incomplete purification techniques. Defendant made and sold a similar product, which it called Adrin.]

The patentee originally attempted to claim the active principle itself. This was in his first application where he claimed process and product; but the examiner would not allow these claims, basing his rejection upon his interpretation of American

Wood Paper Co. v. Fibre Disintegrating Co., 23 Wall. 566, 23 L. Ed. 31, that no product is patentable, however it be of the process, which is merely separated by the patentee from its surrounding materials and remains unchanged.... In [an] amendment..., which was about two months after his first application, he changed all the claims so that they...were not limited to the active principle. I think that this effected a substantial change in meaning, and that...the claims are now broader than a mere claim for the chemically free base, or active principle, and that they cover any substance which possesses the physiological characteristics of the glands and is substantially pure....[28]

[After deciding that defendant's product infringed several claims of both Takamine patents, Judge Hand turned to the issue of whether the Takamine patents were invalid.]

...Nor is the patent only for a degree of purity, and therefore not for a new "composition of matter." As I have already shown, it does not include a salt, and no one had ever isolated a substance which was not in salt form, and which was anything like Takamine's.... But, even if it were merely an extracted product without change, there is no rule that such products are not patentable.... Takamine was the first to make it available for any use by removing it from the other gland-tissue in which it was found, and, while it is of course possible logically to call this a purification of the principle, it became for every practical purpose a new thing commercially and therapeutically. That was a good ground for a patent. That the change here resulted in ample practical differences is fully proved. Everyone, not already saturated with scholastic distinctions, would recognize that Takamine's crystals were not merely the old dried glands in a purer state, nor would his opinion change if he learned that the crystals were obtained from the glands by a process of eliminating the inactive organic substances. The line between different substances and degrees of the same substance is to be drawn rather from the common usages of men than from nice considerations of dialectic....

COMMENTS AND QUESTIONS

1. To Judge Hand, it is of paramount importance that the patentee has created "for every practical purpose a new thing commercially and therapeutically." But if commercial and therapeutic novelty were enough, wouldn't it be possible to obtain a patent on the unprocessed bark of a rare tree which turns out to be an effective therapy for cancer, and therefore a commercially successful product? Can you think of any reasons why society should be reluctant to grant such a patent on the bark itself to the first person who discovers the therapeutic effects of the bark? For criticism that the "purification exception" to the "product of nature" doctrine has gone too far, see Linda J. Demaine & Aaron Xavier Fellmeth, Reinventing the Double Helix: A Novel and Nonobvious Reconceptualization of the Biotechnology Patent, 55 Stan. L. Rev. 303 (2002) (highlighting the distinction between

---

28. Claim 1 in U.S. patent 730,176, one of the patents in issue in the case, reads:

1. A substance possessing the herein-described physiological characteristics and reactions of the suprarenal glands in a stable and concentrate form, and practically free from inert and associated gland-tissue. — EDS.

impure and purified natural products, and introducing a more rigorous "substantial transformation test" for patentable subject matter related to genetic material).

2. A parallel line of cases involves metal compounds found in nature. See, e.g., General Electric Co. v. DeForest Radio Co., 28 F.2d 641 (3d Cir. 1928), *cert. denied,* 278 U.S. 656 (1929) (affirming trial court finding of invalidity of claims to "substantially pure tungsten"). In the *DeForest* case, the court stated:

> What [the patentee] produced by his process was natural tungsten in substantially pure form. What he discovered were natural qualities of pure tungsten. Manifestly he did not create pure tungsten, nor did he create its characteristics. These were created by nature....

How was the adrenalin in the *Parke-Davis* case different from the tungsten in *DeForest*? What is the difference between the "discovery" of an "inherent property" and the "invention" of a "new property or use" of a natural product?

3. By contrast with *Parke-Davis,* see Ex parte Latimer, 1889 Dec. Commr. Pat. 123, 46 Pat. Off. Gazz. 1638 (Commr. Patents 1889), where the Commissioner of Patents affirmed two holdings of the Patent Office: (1) the issuance of process claims to a method of extracting weavable fibers from the *Pinus australis* tree; and (2) the final rejection of the applicant's product claim to the same subject matter, viz., "the cellular tissues of the *Pinus australis* eliminated in full lengths from the silicious, resinous, and pulpy parts of the pine needles and subdivided into long, pliant filaments adapted to be spun and woven," because:

> the mere ascertaining of the character or quality of trees that grow in the forest and the construction of the woody fiber and tissue of which they are composed is not a patentable invention, recognized by the statute, any more than to find a new gem or jewel in the earth would entitle the discoverer to patent all gems which should be subsequently found....

See also In re Bergy, 563 F.2d 1031, 1036 (C.C.P.A. 1977) (distinguishing the patentable, humanly-transformed bacteria claimed in the case from a prior case where "[w]e were thinking of something preexisting and merely plucked from the earth and claimed as such, a far cry from a biologically pure culture produced by great labor in a laboratory and so claimed." (citing In re Mancy, 499 F.2d 1289 (C.C.P.A. 1974)).

4. Consider the following passage written by the noted libertarian philosopher Robert Nozick, in which he sets forth the reasons why one who discovers a new plant is entitled to assert property rights over it:

> He does not worsen the situation of others; if he did not stumble upon the substance no one else would have, and the others would remain without it. However, as time passes, the likelihood increases that others would have come across the substance; upon this fact might be based a limit to his property right in the substance so that others are not below their baseline position; for example, its bequest might be limited.

Robert Nozick, Anarchy, State and Utopia 181 (1974).[29] Does this view imply any particular *content* to the right that follows from discovery? Would trade secret

---

29. The passage comes in the midst of a section discussing John Locke's theory of property; thus the emphasis on not worsening anyone else's position, one of the "Lockean provisos" that must be met under this theory for property rights to be defensible. See *supra* Chapter 1.

protection — a minimal right protecting the discoverer's information concerning the source of the natural product, for instance — be sufficient? Or is a patent called for, with its protection against independent discovery of the product and its prohibition of any competing makers or sellers of the product?

5. It is possible to discern from this patchwork of cases a set of rules for the patenting of living things. First, there is generally no objection to patent claims covering a *process* for extracting a natural product, as opposed to claims to the product itself. See, e.g., Merck & Co. v. Olin Mathieson Chem. Corp., 253 F.2d 156 (4th Cir. 1958). What practical difference would there be between the two types of claims? Consider two cases: (a) a natural product identical to the patentee's but made from a different process, and (b) a natural product identical to the patentee's made from a process whose details are unknown to the patentee.

Should the first person to purify a natural product to a certain degree be entitled to a patent on the purified product itself, however produced? Or only on purified products produced using her method? For a discussion of these issues, see Eileen M. Kane, *Splitting the Gene: DNA Patents and the Genetic Code*, 71 Tenn. L. Rev. 707, 707 (2004) (arguing that patenting genes amounts to "constructive preemption of the genetic code," and ought to be found contrary to the Supreme Court's dictate that the laws of nature are not patentable).

Second, Nozick's suggestion to the contrary notwithstanding, it does not appear to be enough to discover a previously unknown plant or animal (or, for that matter, a previously unknown mineral). On the other hand, human creation of an entirely new bacterium does qualify for patent protection under *Chakrabarty*. Subsequent decisions have uniformly extended this rule to allow the patenting of "new" plants and animals, including mammals. See, e.g., Ex Parte Hibberd, 227 U.S.P.Q. 443 (Bd. Pat. App. & Int. 1985) (section 101 did not bar the issuance of a utility patent for seeds and plants developed by sexual reproduction in a way not found in nature). See also Ex Parte Allen, 2 U.S.P.Q.2d 1425 (Bd. Pat. App. & Int. 1987) (a genetically developed Pacific Oyster was patentable subject matter but was found "obvious" in view of the prior art); U.S. Patent No. 4,736,866 (Apr. 12, 1988) (issued to Philip Leder and Harvard University for "Transgenic Non-Human Mammals," specifically mice). And patenting human DNA sequences — though not humans themselves, of course — is a booming industry. Under *Parke-Davis, supra*, these patents are considered valid, because they typically claim *isolated* and *purified* gene sequences. The same logic applies to the patenting of stem cells — progenitor cells that can produce other cells. See, e.g., U.S. Patent 6,200,806, issued Mar. 13, 2001.

While a number of groups have raised concerns about the patenting of living things, particularly as the living things move up the scale of complexity from bacteria towards humans, these arguments about the morality of patenting life have not found a warm reception in the courts. Patentable subject matter appears to be "morally neutral," at least after *Chakrabarty*. There is a doctrine of moral *utility* that may apply in these cases, however. We consider that doctrine in the next section.

6. Apart from the morality of "patenting life" itself, is there ever a valid concern about allocating exclusive ownership rights to certain types of inventions? Consider the problem of a doctor who has come up with a radically new surgical technique. Should the doctor be allowed to patent his technique? Recall that a patent would allow him not only to prevent others from copying his method, but from using the same method even if they developed it independently. Legislation passed by

Congress in 1996 precludes owners of patents on medical and surgical procedures from enforcing those patents, thus carving out a rare legislative exception to the scope of patentable subject matter. 35 U.S.C. § 287(c). Is the doctor's situation any different than that of a pharmaceutical company that wants to patent a new drug it has developed? One type of patent attempts to claim a way of treating disease starting with an observation (or diagnosis) that certain disease-indicating conditions are present. A case along these lines was argued in the Supreme Court in March 2006. See Metabolite Laboratories, Inc. v. Laboratory Corp. of America Holdings, 370 F 3d 1354 (Feb. Cir. 2004), cert. granted 126 S. Ct. 1317 (2006).

PROBLEM

Problem 3-1. Inventor *A* discovers a technique for purifying Hormone *Z* from human tissue and applies for a patent on the purified form of the hormone. Inventor *B* discovers the gene coding for the hormone, clones it, and obtains expression in a mammalian cell culture environment. Are these inventions/discoveries patentable?

Note on Patenting "Abstract Ideas"

While general subject matter limitations are disfavored under section 101 (the "products of nature" rule being a notable exception), there are a number of judicially developed doctrines that limit the subject matter of patents. Of these, one of the most important is the rule against patenting "abstract ideas." This rule appears to derive from the traditional idea that patents are intended to cover "devices" or physical things in the useful arts, not more esoteric matters. As the Supreme Court put it in Rubber-Tip Pencil Co. v. Howard, 87 U.S. (20 Wall.) 498, 507 (1874), "[a]n idea of itself is not patentable, but a new device by which it may be made ✳ practically useful is."

In O'Reilly v. Morse, 56 U.S. (15 How.) 62 (1853), the telegraph pioneer Morse (of "Morse code" fame) was allowed a broad patent for a process of using electromagnetism to produce discernible signals over telegraph wires. But the Court denied Morse's famous eighth claim, in which Morse claimed the use of "electro magnetism, however developed for marking or printing intelligible characters, signs, or letters, at any distances." Id. at 112. The Court in disallowing that claim said,

> If this claim can be maintained, it matters not by what process or machinery the result is accomplished. For aught that we now know, some future inventor, in the onward march of science, may discover a mode of writing or printing at a distance by means of the electric or galvanic current, without using any part of the process or combination set forth in the plaintiff's specification. His invention may be less complicated — less liable to get out of order — less expensive in construction, and in its operation. But yet, if it is covered by this patent, the inventor could not use it, nor the public have the benefit of it, without the permission of this patentee.

Id. at 113. Is this truly an objection that Morse's claim does not extend to patentable subject matter, or is the Court really expressing the concern that Morse's claim is *too broad?* That the Court's concern is the latter is suggested by the fact that it allowed Morse's other claims to particular uses of electricity in communication (i.e., the telegraph). It is instructive in this regard to consider The Telephone Cases, 126 U.S. 1, 534 (1887), in which the Court upheld a patent to Alexander Graham Bell for his telephone. The Court explained the *Morse* case as follows:

> The effect of that decision was, therefore, that the use of magnetism as a motive power, without regard to the particular process with which it was connected in the patent, could not be claimed, but that its use in that connection could. Bell's invention was the use of electric current to transmit vocal or other sounds. The claim was not "for the use of a current of electricity in its natural state as it comes from the battery, but for putting a continuous current in a closed circuit into a certain specified condition suited to the transmission of vocal and other sounds, and using it in that condition for that purpose." The claim, in other words, was not "one for the use of electricity distinct from the particular process with which it is connected in his patent." The patent was for that use of electricity "both for the magneto and variable resistance *methods.*"

Bell's claim, in other words, was not one for all telephonic use of electricity.

But there are certainly circumstances in which the new idea itself is so abstract that it may run afoul of this doctrine. Consider the hypothetical claim advanced by the Court in *Chakrabarty*—an attempt by Einstein to patent the formula $E = mc^2$. This claim is not overbroad—Einstein has given the world precisely what he claims to own. But there is still an intuitive sense in which this idea is too abstract, too fundamental for society to countenance its ownership by a private individual. Perhaps section 101 is designed in part to exclude from patentability discoveries and inventions so basic and fundamental to future advances that patenting them would unduly burden future inventors.

For an interesting account of a movement from the early part of the twentieth century to protect the findings of basic scientific research with a patent-like property right, see C. J. Hamson, Patent Rights for Scientific Discoveries (1930).

COMMENTS AND QUESTIONS

1. Typically, there is a good deal of development work necessary to turn an invention into a viable commercial product. This first *commercial* version of an invention is usually termed an innovation. The patent systems of today reward such innovations only indirectly, through the granting of patents on inventions. Some observers have argued that this is a fundamental flaw in the patent system. They contend that the original function of patents was to reward innovation directly. See generally F. Scott Kieff, Property Rights and Property Rules for Commercializing Inventions, 85 Minn. L. Rev. 697 (2001). Some who take this view propose a new patent regime, that would provide *direct* protection of innovation by means of exclusive rights to market newly introduced products. See Direct Protection of Innovation (W. Kingston, ed. 1987). In this book, Kingston proposes a system of property rights to come into effect only when a new product is actually introduced on the market. See id. at 1-34. For example, under such a system, the first entrepreneur who introduced an overnight package delivery service, or any other new business practice, service, or product, would receive an exclusive property right over that concept for a limited period of time.

Can you think of any potential problems with such a system — e.g., administrative or other practical difficulties? Does such a proposal shed any light on why the current patent system focuses on invention rather than innovation? Could you defend the current system, perhaps by arguing that technology, rather than business practices or services, contributes something unique to the economy, or otherwise merits special encouragement? Cf. H. I. Dutton, The Patent System and Inventive Activity During the Industrial Revolution 1750-1852 (1984) (arguing that the early patent system did not reward innovation directly but instead played much the same role it does today). Robert P. Merges, Uncertainty and the Standard of Patentability, 7 High Tech. L. Rev. 1 (1993) (model demonstrating that prospect of receiving patent can have dramatic effect on incentive to develop a basic invention).

Note on Patenting Business Methods and "Printed Matter"

Other artifacts of the traditional focus of patent law on physical machines and devices in the useful arts are the twin rules against patenting "printed matter" and "business methods," the latter now defunct. Both rules originally developed from a series of cases in which the "inventor" of new printed business forms sought to patent those forms. Prior to the Federal Circuit decision in State Street Bank & Trust Co. v. Signature Financial Group, 149 F. 3d 1368 (Fed. Cir. 1998), courts uniformly rejected both the idea that a printed piece of paper could be a patentable invention and that the new system of conducting business embodied in the paper was patentable. United States Credit System Co. v. American Credit Indemnity Co., 59 F. 139, 143 (2d Cir. 1893), for example, involved a patent for a means of credit insurance, which the court struck down on section 101 grounds as a business method.

In *State Street Bank*, the court announced that it had decided "to lay this ill-conceived exception to rest." It further opined that the business-methods exception had been without legal force since the 1952 Patent Act, and the court distinguished cases that arguably invoked the exception since that time. In so doing it bolstered the idea that a "process" can be statutory subject matter despite the fact that it doesn't act on anything tangible.

In the wake of *State Street*, the Patent office has issued hundreds of patents covering methods of doing business.

At the same time, courts have held that a system of transacting business was not patentable unless what was claimed were the physical means for carrying out the system rather than merely new text. See, e.g., Hotel Security Checking Co. v. Lorraine Co., 160 F. 467, 469 (2d Cir. 1908). The result of these two doctrines in combination has been that most business and financial innovations were not traditionally considered patentable. The printed matter rule has been extended to deny patents for such physical items as games, in which the physical elements used (a board, dice, cards) are not new, but what is printed on those elements was new. See Ex parte Gwinn, 112 U.S.P.Q. 439 (Bd. Pat. App. & Int. 1955). The printed matter rule retains its force. See, e.g., In re Ngai, 367 F.3d 1336 (Fed. Cir. 2004) (claim to known kits for normalizing and amplifying RNA population by simply attaching new set of instructions to an existing product were unpatentable); In re Shanahan, 2005 WL 191069 (Bd. Pat. App. & Int.) (n.d.) (unpub.) (fortune cookie card game unpatentable because only difference between claimed game and prior art fortune cookies is the material printed on the paper strips inside the cookie).

Because business method patents have been so controversial, they have engendered a good deal of discussion about issues of "patent quality" generally. See, e.g., Robert P. Merges, As Many as Six Impossible Patents for Breakfast: Property Right for Business Concepts and Patent System Reform, 14 Berkeley Tech. L. J. 579 (1999); Mark A. Lemley, Rational Ignorance at the Patent Office, 95 Nw. UL Rev 1491 (2001). See also John R. Allison and Emerson H. Tiller, The Business Method Patent Myth, 18 Berkeley Tech. L.J. 987 (2003) (describing Patent Office efforts to improve "patent quality" for business method patents, including "second pair of eyes" review, and using statistical evidence to argue that business method patents are no more suspect than other patents).

Printed matter may be part of a patentable invention, however, if the invention as a whole claims a new and useful physical structure, or if the relationship between the printed matter and the physical structure is a new and nonobvious one. How this exception interacts with the printed matter rejection is complex and not always clear. For example, in In re Lowry, 32 F.3d 1579 (Fed. Cir. 1994), the Federal Circuit held that a claim for a novel "data structure" contained entirely within an ordinary computer memory was not printed matter because the data structure necessarily reorganized the electronic components of the computer memory in a way that was not intelligible to humans, but only to the machine itself. The court also noted that broad application of the printed matter doctrine was "disfavored."

COMMENTS AND QUESTIONS

1. One justification for the printed matter rule is that it is necessary to "channel" certain creations into the realm of patent law, and other creations (notably those in written form) into copyright law. Does this channeling justification make sense? Does it explain the *Lowry* decision?

2. Commentators have been quite critical of the *State Street* decision. See, e.g., John R. Thomas, The Patenting of the Liberal Professions, 40 B.C. L. Rev. 1139 (1999); Alan Durham, "Useful Arts" in the Information Age, 1999 BYU L. Rev. 1419. These commentators point to the likely expansion of *State Street* to permit patenting of advances in economics, law, architecture, and sports, and question the need for the change in patent law rules.

PROBLEMS

Problem 3-2. In 1986, an American scientist by the name of John Anderson perfected a process that yielded a new man-made chemical element. The element, named Litigacium (Li), was assigned atomic number 116. Anderson's method of synthesizing the element involves using a neutronic reactor at relative high power (about 200 kilowatts) for approximately 100 days. Anderson's patent application noted that a suitable reactor for producing Litigacium is described in a 1984 patent of Dr. Enrago Firmi.

Although Anderson's process claim met with no resistance from the Patent and Trademark Office, the claim relating to the product Litigacium raised more difficult issues. An examiner notified Anderson that his claimed element was not patentable because it preexists in nature, in some form, since a chemical element cannot truly be "man-made."

Dr. Anderson has asked you to represent him in his appeal to the Federal Circuit. The appeal presents the following issues:

a) The Patent Office admits that the level of Litigacium produced under natural circumstances — e.g., in the intense heat and pressure that accompanies the collapse of certain large stars — would be miniscule. For example, the examiner states that using well-known theoretical calculations, star collapse would result in an infinitesimally small amount of the element spread throughout an enormous amount of other matter under tremendous heat and pressure. How can the mere possibility that a star would create an undetectable amount of Litigacium permit rejection of your client's patent application?

b) The examiner concedes that it would not be practical given current technology to "harvest" Litigacium from natural sources such as stars. Should this have a bearing on the outcome of the case? Is it the function of section 101's patentable subject matter requirement to find patentable anything that is cheaper (i.e., more efficient) for humans to make than to harvest in raw form from nature? Or is section 101 an expression of a deeply felt aversion to permitting humans to claim "authorship" over products given to us by nature's bounty? Is there a viable distinction between things we "find" in nature and things we ourselves "invent"? Does the language in the patent clause of the Constitution — i.e., that patents shall be conferred to inventors for their "discoveries" — shed any light on these matters?

Problem 3-3. You are a newly elected U.S. senator. As part of your duties on the legislative committee assigned to introduce intellectual property legislation, you have been asked to co-sponsor a bill that seeks to rewrite section 101 of the patent code. According to the sponsor of the bill, the "archaic" language of the current section 101 was drafted during the era of "quill pens and buggy whips." She claims that it is therefore "out of step in an era of computers and space shuttles."

Several lobbyists have seized on the occasion to introduce their pet revisions into section 101. Environmentalists have asked that "newly discovered natural products" be added to the categories of patentable subject matter; they believe this addition will create an incentive to identify and preserve rare species of potential benefit, e.g., as ingredients in medicines or as sources of raw materials. In addition, academic scientists have testified that they feel "slighted" by the exclusion of natural laws and scientific principles from the coverage of the current patent act. They have stated that they fail to see why the patent system should discriminate against basic research, especially in an era when federal funding of science is shrinking. Even the social scientists have gotten into the act. A famous economist who pioneered a revolutionary technique for measuring the volume of money in circulation testified in hearings on the bill that he would have continued thinking up pioneering ideas like this but "the current system gave me no incentive to do so, at least after I made tenure." And some sports stars have sent letters to your committee, claiming that they wish they had received some form of protection for what everyone refers to as their "patented" backhand or putting style or hurdling technique.

How can you evaluate these claims? Is there any justification in refusing to change the current language — e.g., because it has shown enough flexibility to adapt to computer technology and biotechnology and the like? Is there anything special about traditional technology—machines, manufactures, etc.—that makes it more valuable, or more in need of protection, than other "innovations" such as newly discovered natural products, new "pure" science results, new academic theories, or new techniques for doing business (such as the leveraged buyout, or the multidivisional corporate structure)? Why not grant property rights over all sorts of new things, and let the market figure out which ones are truly worthwhile?

2. Utility

The patent code protects all inventions that are novel, *useful,* and nonobvious. This section concentrates on this second requirement, known as the utility requirement.

At first glance it might seem as though this is a simple requirement to apply. After all, whether or not something is useful is normally easy to determine. In a sense patent law reflects this; utility is a relatively rare issue in the Patent Office, or in an infringement suit. But in another sense there is more to utility. Both conceptually and as borne out in the cases, subtle issues lurk within the waters of utility.

Chemists often synthesize compounds that they believe might be useful someday for something but for which no particular use is known. When they apply for patents on these compounds, they sometimes run headlong into the utility requirement, as the following case demonstrates.

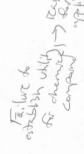

Brenner v. Manson
Supreme Court of the United States
383 U.S. 519 (1966)

Mr. Justice FORTAS delivered the opinion of the Court.

A Patent Office examiner denied Manson's application, and the denial was affirmed by the Board of Appeals within the Patent Office. The ground for rejection was the failure "to disclose any utility for" the chemical compound produced by the process. This omission was not cured, in the opinion of the Patent Office, by Manson's reference to an article in the November 1956 issue of the Journal of Organic Chemistry, 21 J. Org. Chem. 1333-1335, which revealed that steroids of a class which included the compound in question were undergoing screening for possible tumor-inhibiting effects in mice, and that a homologue[3] adjacent to Manson's steroid had proven effective in that role. Said the Board of Appeals, "It is our view that the statutory requirement of usefulness of a product cannot be presumed merely because it happens to be closely related to another compound which is known to be useful."

The Court of Customs and Patent Appeals (hereinafter CCPA) reversed[, stating] "where a claimed process produces a known product it is not necessary to show utility for the product," so long as the product "is not alleged to be detrimental to the public interest."

3. "A homologous series is a family of chemically related compounds, the composition of which varies from member to member by CH2 (one atom of carbon and two atoms of hydrogen).... Chemists knowing the properties of one member of a series would in general know what to expect in adjacent members."

Our starting point is the proposition, neither disputed nor disputable, that one may patent only that which is "useful." [T]he concept of utility has maintained a central place in all of our patent legislation, beginning with the first patent law in 1790 and culminating in the present [§ 101]....

Respondent does not — at least in the first instance — rest upon the extreme proposition, advanced by the court below, that a novel chemical process is patentable so long as it yields the intended product and so long as the product is not itself "detrimental." Nor does he commit the outcome of his claim to the slightly more conventional proposition that any process is "useful" within the meaning of § 101 if it produces a compound whose potential usefulness is under investigation by serious scientific researchers, although he urges this position, too, as an alternative basis for affirming the decision. Rather, he begins with the much more orthodox argument that his process has a specific utility which would entitle him to a declaration of interference even under the Patent Office's reading of § 101. The claim is that the supporting affidavits, by reference to Ringold's 1956 article, reveal that an adjacent homologue of the steroid yielded by his process has been demonstrated to have tumor-inhibiting effects in mice, and that this discloses the requisite utility. We do not accept any of these theories as an adequate basis for overriding the determination of the Patent Office that the "utility" requirement has not been met.

Even on the assumption that the process would be patentable were respondent to show that the steroid produced had a tumor-inhibiting effect in mice, we would not overrule the Patent Office finding that respondent has not made such a showing. The Patent Office held that, despite the reference to the adjacent homologue, respondent's papers did not disclose a sufficient likelihood that the steroid yielded by his process would have similar tumor-inhibiting characteristics. Indeed, respondent himself recognized that the presumption that adjacent homologues have the same utility has been challenged in the steroid field because of "a greater known unpredictability of compounds in that field." In these circumstances and in this technical area, we would not overturn the finding of the Primary Examiner, affirmed by the Board of Appeals and not challenged by the CCPA.

The second and third points of respondent's argument present issues of much importance. Is a chemical process "useful" within the meaning of § 101 either (1) because it works — i.e., produces the intended product? or (2) because the compound yielded belongs to a class of compounds now the subject of serious scientific investigation? These contentions present the basic problem for our adjudication. Since we find no specific assistance in the legislative materials underlying § 101, we are remitted to an analysis of the problem in light of the general intent of Congress, the purposes of the patent system, and the implications of a decision one way or the other.

In support of his plea that we attenuate the requirement of "utility," respondent relies upon Justice Story's well-known statement that a "useful" invention is one "which may be applied to a beneficial use in society, in contradistinction to an invention injurious to the morals, health, or good order of society, or frivolous and insignificant" [20] — and upon the assertion that to do so would encourage inventors of new processes to publicize the event for the benefit of the entire scientific

---

20. Note on the Patent Laws, 3 Wheat. App. 13, 24. See also Justice Story's decisions on circuit in Lowell v. Lewis, 15 Fed. Cas. 1018 (No. 8568) (C.C.D. Mass.), and Bedford v. Hunt, 3 Fed. Cas. 37 (No. 1217) (C.C.D. Mass.).

community, thus widening the search for uses and increasing the fund of scientific knowledge. Justice Story's language sheds little light on our subject. Narrowly read, it does no more than compel us to decide whether the invention in question is "frivolous and insignificant" — a query no easier of application than the one built into the statute. Read more broadly, so as to allow the patenting of any invention not positively harmful to society, it places such a special meaning on the word "useful" that we cannot accept it in the absence of evidence that Congress so intended. There are, after all, many things in this world which may not be considered "useful" but which, nevertheless, are totally without a capacity for harm.

Whatever weight is attached to the value of encouraging disclosure and of inhibiting secrecy, we believe a more compelling consideration is that a process patent in the chemical field, which has not been developed and pointed to the degree of specific utility, creates a monopoly of knowledge which should be granted only if clearly commanded by the statute. Until the process claim has been reduced to production of a product shown to be useful, the metes and bounds of that monopoly are not capable of precise delineation. It may engross a vast, unknown, and perhaps unknowable area. Such a patent may confer power to block off whole areas of scientific development, without compensating benefit to the public. The basic *quid pro quo* contemplated by the Constitution and the Congress for granting a patent monopoly is the benefit derived by the public from an invention with substantial utility. Unless and until a process is refined and developed to this point — where specific benefit exists in currently available form — there is insufficient justification for permitting an applicant to engross what may prove to be a broad field.

These arguments for and against the patentability of a process which either has no known use or is useful only in the sense that it may be an object of scientific research would apply equally to the patenting of the product produced by the process. Respondent appears to concede that with respect to a product, as opposed to a process, Congress has struck the balance on the side of nonpatentability unless "utility" is shown. Indeed, the decisions of the CCPA are in accord with the view that a product may not be patented absent a showing of utility greater than any adduced in the present case. We find absolutely no warrant for the proposition that although Congress intended that no patent be granted on a chemical compound whose sole "utility" consists of its potential role as an object of use-testing, a different set of rules was meant to apply to the process which yielded the unpatentable product. That proposition seems to us little more than an attempt to evade the impact of the rules which concededly govern patentability of the product itself.

This is not to say that we mean to disparage the importance of contributions to the fund of scientific information short of the invention of something "useful," or that we are blind to the prospect that what now seems without "use" may tomorrow command the grateful attention of the public. But a patent is not a hunting license. It is not a reward for the search, but compensation for its successful conclusion. "[A] patent system must be related to the world of commerce rather than to the realm of philosophy...."

The judgment of the CCPA is Reversed.

Mr. Justice HARLAN, concurring in part and dissenting in part:

...Because I believe that the Court's policy arguments are not convincing and that past practice favors the respondent, I would reject the narrow definition of "useful" and uphold the judgment of the CCPA.

The Court's opinion sets out about half a dozen reasons in support of its interpretation. Several of these arguments seem to me to have almost no force. For instance, it is suggested that "[u]ntil the process claim has been reduced to production of a product shown to be useful, the metes and bounds of that monopoly are not capable of precise delineation" and "[i]t may engross a vast, unknown, and perhaps unknowable area." I fail to see the relevance of these assertions; process claims are not disallowed because the products they produce may be of "vast" importance nor, in any event, does advance knowledge of a specific product use provide much safeguard on this score or fix "metes and bounds" precisely since a hundred more uses may be found after a patent is granted and greatly enhance its value....

More to the point, I think, are the Court's remaining, prudential arguments against patentability: namely, that disclosure induced by allowing a patent is partly undercut by patent-application drafting techniques, that disclosure may occur without granting a patent, and that a patent will discourage others from inventing uses for the product. How far opaque drafting may lessen the public benefits resulting from the issuance of a patent is not shown by any evidence in this case but, more important, the argument operates against all patents and gives no reason for singling out the class involved here. The thought that these inventions may be more likely than most to be disclosed even if patents are not allowed may have more force; but while empirical study of the industry might reveal that chemical researchers would behave in this fashion, the abstractly logical choice for them seems to me to [be to] maintain secrecy until a product use can be discovered. As to discouraging the search by others for product uses, there is no doubt this risk exists but the price paid for any patent is that research on other uses or improvements may be hampered because the original patentee will reap much of the reward. From the standpoint of the public interest the Constitution seems to have resolved that choice in favor of patentability.

What I find most troubling about the result reached by the Court is the impact it may have on chemical research. Chemistry is a highly interrelated field and a tangible benefit for society may be the outcome of a number of different discoveries, one discovery building upon the next. To encourage one chemist or research facility to invent and disseminate new processes and products may be vital to progress, although the product or process be without "utility" as the Court defines the term, because that discovery permits someone else to take a further but perhaps less difficult step leading to a commercially useful item. In my view, our awareness in this age of the importance of achieving and publicizing basic research should lead this Court to resolve uncertainties in its favor and uphold the respondent's position in this case.

COMMENTS AND QUESTIONS

1. What is the basis of the majority's claim that until an invention is shown to be useful, the inventor has not supplied the *quid pro quo* for a patent? What harm is there in granting a patent on an invention later shown to be useful? In granting a patent on a useless device? Who would suffer from the grant of such a patent? Would granting the patent help the inventor attract investment, which could be used to establish utility?

2. Would the existence of a market demand for the "intermediate" products at issue in *Brenner* change the result in the case? Companies certainly exist that make money by selling both raw materials and research tools to laboratories. Should

we treat patents on these products differently than we do "end" products because of their intended use in the laboratory?

3. The concept of a "use patent" is described in Robert P. Merges & Richard Nelson, On the Complex Economics of Patent Scope, 90 Colum. L. Rev. 839 (1990), where the authors discuss an anomaly alluded to in Justice Harlan's dissent in *Brenner*: an inventor who obtains a patent for a product, e.g., a particular molecule, has the right to exclude all others from making, using or selling that product for *any* and all purposes, including purposes that the inventor did not herself discover or invent. For example, a patented compound created for its use as a leather tanning agent might turn out to be an effective anti-AIDS drug. Cf. In re Thuau, 135 F.2d 344 (C.C.P.A. 1943). If so, the patentee would have the right to prevent all others from selling the drug as an AIDS treatment — including the person who discovered that the leather tanning compound had anti-AIDS properties. Note that the utility requirement is met so long as the patentee shows *any* specific utility for the chemical when the patent is first filed — in our example, when the leather tanning property of the compound is discovered. Some have responded to the anomaly by calling for a special type of patent in these situations, a "new use" patent. See id.

New use patents already exist in a limited way. In our example, the one who discovers the anti-AIDS property of the leather tanning agent can obtain a *process* patent for "the process of using [the leather tanning compound] to treat AIDS." See Rohm & Haas v. Roberts Chem. Co., 245 F.2d 693 (4th Cir. 1957) (upholding defendant's patent on a new use of a well-known product as a fungicide); 1 Donald Chisum, Patents § 1.03[8] (1978 & Supp. 2005(collecting other cases on this point)). This is in essence only an improvement patent; the discoverer would still have to obtain a license from the patentee to use the compound for treatment of AIDS. But the reverse is also true; unlike the example outlined above, if the one who discovered the leather tanning agent's anti-AIDS properties obtained a process patent, the patentee would have to obtain a license from the improver to have the right to use the compound to treat AIDS.

4. Utility for pharmaceutical products can generally be established by animal testing. See U.S. Patent and Trademark Office, Manual of Patent Examining Procedures § 2107.01 (8th ed. 2001). In the past, the biotechnology industry sometimes had trouble overcoming utility rejections without clinical data proving that a particular drug was effective *in humans,* despite Federal Circuit precedent holding that a showing of efficacy in a laboratory experiment was sufficient to establish the utility of a new drug. Cross v. Iizuka, 753 F.2d 1040, 1051 (Fed. Cir. 1985).

The Federal Circuit made it clear that human clinical trials are not necessary to prove utility of a drug in In re Brana, 51 F.3d 1560 (Fed. Cir. 1995). The court held that the results of in vivo tests in mice were sufficiently probative of efficacy in humans to pass the utility threshold:

> The Commissioner, as did the Board, confuses the requirements under the law for obtaining a patent with the requirements for obtaining government approval to market a particular drug for human consumption. . . . FDA approval is not a prerequisite for finding a compound useful within the meaning of the patent laws. . . . Usefulness in patent law, and in particular in the context of pharmaceutical inventions, necessarily includes the expectation of further research and development. . . . Were we to require Phase II testing in order to prove utility, the associated costs would prevent many companies from obtaining patent protection on promising new inventions, thereby eliminating an incentive to pursue, through research and development, potential cures in many crucial areas such as the treatment of cancer.

Brana, 51 F.3d at 1567. On the other hand, in In re Ziegler, 992 F.2d 1197, 1203 (Fed. Cir. 1993), the Federal Circuit held that a patent applicant was not entitled to claim priority to a 1954 application for "polypropylene," since at the time of the 1954 application "at best, Ziegler was on the way to discovering a practical utility for polypropylene...but in that application Ziegler had not yet gotten there." Are these two cases distinguishable?

5. *Brenner* in many ways represents the "high-water mark" of the utility doctrine. Most applications of the doctrine have been quite limited in the hurdles they place before inventors. Certainly any number of "frivolous" ideas are patented, suggesting that the burden of showing utility cannot be all that high. Thus the reader should not assume that *Brenner* can readily be extended by analogy.

In re Fisher
421 F.3d 1365 (Fed. Cir. 2005)

MICHEL, Chief Judge.

Because we conclude that substantial evidence supports the Board's findings that the claimed invention lacks a specific and substantial utility and that the '643 application does not enable one of ordinary skill in the art to use the invention, we affirm [the decision of the Board of Appeals and Interferences].

The claimed invention relates to five purified nucleic acid sequences that encode proteins and protein fragments in maize plants. The claimed sequences are commonly referred to as "expressed sequence tags" or "ESTs." When a gene is expressed in a cell, the relevant double-stranded DNA sequence is transcribed into a single strand of messenger ribonucleic acid ("mRNA"). Messenger RNA contains three of the same bases as DNA (A, G, and C), but contains uracil ("U") instead of thymine. mRNA is released from the nucleus of a cell and used by ribosomes found in the cytoplasm to produce proteins. [For more background, see Appendix A, "Biotechnology: An Introduction to the Technology," in the Case and Statutory Supplement.]

Complementary DNA ("cDNA") is produced synthetically by reverse transcribing mRNA. cDNA, like naturally occurring DNA, is composed of nucleotides containing the four nitrogenous bases, A, T, G, and C. Scientists routinely compile cDNA into libraries to study the kinds of genes expressed in a certain tissue at a particular point in time. One of the goals of this research is to learn what genes and downstream proteins are expressed in a cell so as to regulate gene expression and control protein synthesis.

An EST is a short nucleotide sequence that represents a fragment of a cDNA clone. It is typically generated by isolating a cDNA clone and sequencing a small number of nucleotides located at the end of one of the two cDNA strands. When an EST is introduced into a sample containing a mixture of DNA, the EST may hybridize with a portion of DNA. Such binding shows that the gene corresponding to the EST was being expressed at the time of mRNA extraction.

Claim 1 of the '643 application recites:

A substantially purified nucleic acid molecule that encodes a maize protein or fragment thereof comprising a nucleic acid sequence selected from the group consisting of SEQ ID NO: 1 through SEQ ID NO: 5.

The ESTs set forth in SEQ ID NO: 1 through SEQ ID NO: 5 are obtained from [a particular] cDNA library. SEQ ID NO:1 through SEQ ID NO:5 consist of 429, 423, 365, 411, and 331 nucleotides, respectively. When Fisher filed the '643 application, he .... did not know the precise structure or function of either the genes or the proteins encoded for by those genes.

The '643 application generally discloses that the five claimed ESTs may be used in a variety of ways, including: (1) serving as a molecular marker for mapping the entire maize genome, which consists of ten chromosomes that collectively encompass roughly 50,000 genes; (2) measuring the level of mRNA in a tissue sample...; (3) providing a source for primers for use in the polymerase chain reaction ("PCR") process to enable rapid and inexpensive duplication of specific genes; (4) identifying the presence or absence of a polymorphism; (5) isolating promoters via chromosome walking; (6) controlling protein expression; and (7) locating genetic molecules of other plants and organisms. [The Board of Appeals and Interferences affirmed a final rejection by the examiner, who found that none of the recited uses for the ESTs satisfied the "substantial utility" standard required for patentability.]

Under the correct application of the law, Fisher argues, the record shows that the claimed ESTs provide seven specific and substantial uses, regardless whether the functions of the genes corresponding to the claimed ESTs are known.... Fisher likewise argues that the general commercial success of ESTs in the marketplace confirms the utility of the claimed ESTs. Hence, Fisher avers that the Board's decision was not supported by substantial evidence and should be reversed.

The government contends that.... Fisher's alleged uses are so general as to be meaningless. What is more, the government asserts that the same generic uses could apply not only to the five claimed ESTs but also to any EST derived from any organism. It thus argues that the seven utilities alleged by Fisher are merely starting points for further research, not the end point of any research effort....

Regarding the seven uses asserted by Fisher, we observe that each claimed EST uniquely corresponds to the single gene from which it was transcribed ("underlying gene"). As of the filing date of the 643 application, Fisher admits that the underlying genes have no known functions. Fisher, nevertheless, claims that this fact is irrelevant because the seven asserted uses are not related to the functions of the underlying genes. We are not convinced by this contention. Essentially, the claimed ESTs act as no more than research intermediates that may help scientists to isolate the particular underlying protein-encoding genes and conduct further experimentation on those genes. The overall goal of such experimentation is presumably to understand the maize genome — the functions of the underlying genes, the identity of the encoded proteins, the role those proteins play during anthesis [i.e., flowering], whether polymorphisms exist, the identity of promoters that trigger protein expression, whether protein expression may be controlled, etc. Accordingly, the claimed ESTs are, in words of the Supreme Court, mere "object[s] of use-testing," to wit, objects upon which scientific research could be performed with no assurance that anything useful will be discovered in the end. Brenner, 383 U.S. at 535.

Fisher compares the claimed ESTs to certain other patentable research tools, such as a microscope. Although this comparison may, on first blush, be appealing in that both a microscope and one of the claimed ESTs can be used to generate scientific data about a sample having unknown properties, Fisher's analogy is flawed. As the government points out, a microscope has the specific benefit of optically magnifying an object to immediately reveal its structure. One of the claimed ESTs, by contrast, can only be used to detect the presence of genetic material having the same structure as the EST itself. It is unable to provide any information about the overall structure let alone the function of the underlying gene. Accordingly, while a microscope can offer an immediate, real world benefit in a variety of applications, the same cannot be said for the claimed ESTs. Fisher's proposed analogy is thus inapt. Hence, we conclude that Fisher's asserted uses are insufficient to meet the standard for a "substantial" utility under § 101.

We agree with the Board that the facts here are similar to those in *Brenner*. . . .

Here, granting a patent to Fisher for its five claimed ESTs would amount to a hunting license [under *Brenner*] because the claimed ESTs can be used only to gain further information about the underlying genes and the proteins encoded for by those genes. The claimed ESTs themselves are not an end of Fisher's research effort, but only tools to be used along the way in the search for a practical utility. Thus, while Fisher's claimed ESTs may add a noteworthy contribution to biotechnology research, our precedent dictates that the '643 application does not meet the utility requirement of § 101 because Fisher does not identify the function for the underlying protein-encoding genes. Absent such identification, we hold that the claimed ESTs have not been researched and understood to the point of providing an immediate, well-defined, real world benefit to the public meriting the grant of a patent.

We conclude that substantial evidence supports the Board's findings that each of the five claimed ESTs lacks a specific and substantial utility and that they are not enabled. Accordingly, the Board's decision affirming the final rejection of claim 1 of the '643 patent for lack of utility under § 101 and lack of enablement under § 112, first paragraph, is affirmed.

COMMENTS AND QUESTIONS

1. For an argument that ESTs really are much like a microscope, and hence satisfy the utility standard, see the dissent in this case by Judge Rader, 421 F.3d at 1379. Judge Rader states:

> These research tools are similar to a microscope; both take a researcher one step closer to identifying and understanding a previously unknown and invisible structure. Both supply information about a molecular structure. Both advance research and bring scientists closer to unlocking the secrets of the corn genome to provide better food production for the hungry world. If a microscope has § 101 utility, so too do these ESTs.

421 F.3d at 1380 (Rader, J., dissenting).

Are you convinced by the majority's argument that a microscope delivers immediate information about a specimen's structure, while ESTs only point to a specimen of interest? For an early discussion of the EST issue, arguing in favor of the position adopted by the majority in *In re Fisher*, see Rebecca S. Eisenberg and Robert P. Merges, Opinion Letter as to the Patentability of Certain Inventions Associated with the Identification of Partial cDNA Sequences, 23 Am. Intell. Prop. L. Ass'n Q.J. 1 (1995).

2. In an omitted part of his opinion, Judge Rader argues in dissent that:

> In truth, I have some sympathy with the Patent Office's dilemma. The Office needs some tool to reject inventions that may advance the "useful arts" but not sufficiently to

warrant the valuable exclusive right of a patent. The Patent Office has seized upon this utility requirement to reject these research tools as contributing "insubstantially" to the advance of the useful arts. The utility requirement is ill suited to that task, however, because it lacks any standard for assessing the state of the prior art and the contributions of the claimed advance. The proper tool for assessing sufficient contribution to the useful arts is the obviousness requirement of 35 U.S.C. § 103. Unfortunately this court has deprived the Patent Office of the obviousness requirement for genomic inventions. See In re Deuel, 51 F.3d 1552 (Fed.Cir.1995); Martin J. Adelman et al., Patent Law, 517 (West Group 1998) (commenting that scholars have been critical of Deuel, which "overly favored patent applicants in biotech by adopting an overly lax nonobviousness standard." (citing Anita Varma & David Abraham, DNA Is Different: Legal Obviousness and the Balance Between Biotech Inventors and the Market, 9 Harv. J.L. & Tech. 53 (1996))); Philippe Ducor, The Federal Circuit and In re Deuel: Does § 103 apply to Naturally Occurring DNA?, 77 J. Pat. & Trademark Off. Soc'y 871, 883 (Nov.1995) ("The Court of Appeals for the Federal Circuit could have formulated its opinion in only one sentence: '35 U.S.C. § 103 does not apply to newly retrieved natural DNA sequences.'"); Philippe Ducor, Recombinant Products and Nonobviousness: A Typology, 13 Santa Clara Computer and High Tech. L.J. 1, 44-45 (Feb.1997) ("This amounts to a practical elimination of the requirement for nonobviousness for these products, even when all the information necessary to discover them is previously available."); see also over fifty additional articles critical of Deuel in the "Citing References" tab for Deuel on Westlaw. Nonetheless, rather than distort the utility test, the Patent Office should seek ways to apply the correct test, the test used world wide for such assessments (other than in the United States), namely inventive step or obviousness.

421 F.3d at 1379, 1381-82 (Rader, J., dissenting).

3. Will there be incentive to identify ESTs without patent protection for them? Is an intermediate solution possible, such as granting patents on ESTs but limiting the patent to its use as a research tool rather than extending to any use of the DNA sequence?

Note on the Patent Office Utility Guidelines

The Patent and Trademark Office (PTO) has promulgated guidelines for determining the specific utility of an invention. See Utility Examination Guidelines, 66 Fed. Reg. 1092 (Jan. 5, 2001), available at *http://www.uspto.gov/web/offices/ pac/mpep/documents/2100_2107.htm*. The PTO guidelines require that an asserted utility be "specific, credible, and substantial." The credibility element was well known; it is the basis of utility rejections for farfetched inventions such as perpetual motion machines. The novel aspects of the guidelines were (1) the definition of a "specific" utility, and (2) the addition of a new requirement of "substantial" utility. On the issue of specific utility, the USPTO training materials based on the guidelines give as an example an asserted utility for a gene sequence, claimed to be useful to identify a disease. The Patent Office would now consider this not specific enough. More detail — such as the *specific type* of disease sought to be identified — would be required.

Closely related is the notion of "substantial" utility. On this issue the guidelines state:

2.(a)(1) A claimed invention must have a specific and substantial utility. This requirement excludes "throw-away," "insubstantial," or "nonspecific" utilities, such as the use of a complex invention as landfill, as a way of satisfying the utility requirement of 35 U.S.C. 101....

(b) If no assertion of specific and substantial utility for the claimed invention made by the applicant is credible, and the claimed invention does not have a readily apparent well-established utility, reject the claim(s) under § 101 on the grounds that the invention as claimed lacks utility....

As examples of "insubstantial" utility, training materials accompanying the guidelines include: (1) using an expensive transgenic mouse as snake food; and (2) use of a laboratory-derived protein as a shampoo ingredient. See Revised Interim Utility Guidelines Training Materials, available at *www.uspto.gov/web/menu/utility.pdf.*

The utility guidelines are a good example of how large policy issues are often played out in the details of patent rules and doctrine. There has been growing concern in the academic literature, particularly from scientists and economists, regarding the fragmentation of individual ownership claims. This theme is discussed in depth in an important article from the journal Science:

> In 1980, in an effort to promote commercial development of new technologies, Congress began encouraging universities and other institutions to patent discoveries arising from federally supported research and development and to transfer their technology to the private sector. Supporters applaud the resulting increase in patent filings and private investment, whereas critics fear deterioration in the culture of upstream research. Building on [Michael] Heller's theory of anticommons property, this article identifies an unintended and paradoxical consequence of biomedical privatization: A proliferation of intellectual property rights upstream may be stifling life-saving innovations further downstream in the course of research and product development.

Michael A. Heller and Rebecca Eisenberg, Can Patents Deter Innovation? The Anticommons in Biomedical Research, 280 Science 698 (1998) (footnotes omitted). See also Michael A. Heller, The Tragedy of the Anticommons: Property in the Transition from Marx to Markets, 111 Harv. L. Rev. 622 (1998). Recent empirical evidence finds little support for the anticommons idea so far. See John P. Walsh, Charlene Cho, and Wesley M. Cohen, The View from the Bench: Patents and Material Transfers, 309 Science 2002 (23 Sept. 2005). For a cautiously contrary view of the anticommons theory from a more theoretical perspective, see Richard A. Epstein, "Steady the Course: Property Rights in Genetic Material," in Perspectives on Properties of the Human Genome Project (F. Scott Kieff, ed., 2003) (noting that patent-related "holdouts" are not considered a problem by practitioners and businesspeople in the field).

While they do not take explicit notice of the academic debate, the guidelines do reflect a similar concern. It is clear that the *reason* that "throwaway" utilities are not acceptable is that they permit patents to issue on inventions whose greatest value has yet to be realized. This is particularly acute in the area of gene sequences. The raw sequence information is of only nominal value in itself. The real value lies in identifying the whole genes that include each sequence, and even more importantly, identifying what those genes code for. Then of course proteins can be mass produced, diseases can be identified, and gene therapies of one type or another may even be devised. These are the very "downstream" uses that Heller and Eisenberg are at pains to protect from patent blockages, i.e., "anticommons."

For commentary on the utility guidelines, see Nathan Machin, Comment, Prospective Utility: A New Interpretation of the Utility Requirement of Section 101 of the Patent Act, 87 Cal. L. Rev. 421 (1999); Andrew T. Kight, Note, Pregnant with Ambiguity: Credibility and the PTO Utility Guidelines in Light of Brenner, 73 Ind. L. J. 997 (1998).

PROBLEMS

Problem 3-4. Acid Look, Inc. (ALI) is a fabric design company that specializes in designing jeans for the high-end fashion market. ALI develops a process for producing a "random faded effect" on the fabric of new jeans, catering to the market for new jeans that appear to be pre-worn. ALI seeks a patent not only on the process of treating the jeans to produce a random faded effect, but on jeans treated by this process. The patent examiner rejects the claim to pre-faded jeans on the ground that they lack utility. Is this rejection proper? What test of utility is being applied here?

Problem 3-5. As described in Problem 3-2, physicist John Anderson has isolated and produced a new element in the laboratory—Litigacium, element 116. Although Anderson is the first person to have produced Litigacium in any appreciable amount, and the first to have isolated it, mathematical calculations based on prior experiments show that Dr. Firmi's nuclear reactor inherently produces 3/1,000,000,000 of one gram of Litigacium every year that it operates. Further, this minute amount of Litigacium is spread among 40 tons of highly radioactive uranium fuel, making it impossible to recover or isolate. Nonetheless, Dr. Firmi and his coworkers were aware at the time they ran the reactor that it was producing trace quantities of Litigacium.

Dr. Anderson claims "element 116" in a patent application. Is this claim anticipated by Firmi's prior work?

Problem 3-6. In late 1967 the Navy encountered a problem with its new fighter plane. The Navy's aircraft—stationed on aircraft carriers—were more likely to suffer structural damage than Air Force planes of the same type that were positioned on land. Navy scientists concluded that the widespread problem of fuselage cracking, known as "stress corrosion," was attributable to the harsh marine environment. The Navy presented its problem to the aluminum industry, and Algol Corp. was awarded a research contract to develop a new alloy that would be more resistant to the phenomenon.

Pursuant to the terms of the defense contract, Algol would periodically report on its research results in "progress letters." The Navy drew up a distribution list of 33 designees to receive Algol's progress letters, and each letter bore the following "export control" notice: "This Document Is Subject to Special Export Controls and Each Transmittal to Foreign Governments or Foreign Nationals May Be Made Only with Prior Approval of the Naval Air Systems Command."

The designees included aluminum producers (including Algol's competitors), aircraft manufacturers, government agencies, branches of the military, and academic researchers. Several progress letters were mailed, the last on

April 5, 1969. As of the last progress letter, Algol had succeeded in developing a high-strength, stress-corrosion-resistant alloy that met the Navy's needs. Algol successfully applied for a patent covering its process for aging and treating its new alloy on February 5, 1971.

When Algol refused to license the patent to its competitor Richards, the latter began to manufacture and sell a high-strength alloy using Algol's patented technology. Algol filed suit alleging infringement.

You are Senior Counsel for Intellectual Property Matters at Richards Aluminum. Your defense strategy is limited by the fact that if the Algol patent is valid, your company is undeniably engaged in infringement. Algol gathers evidence that the 33 designees who received Algol's progress reports treated them as highly confidential although they contained no express limitation on access other than the export control notice. Depositions of Richards' own executives reveal that your company kept its copies of the letters in a protected area, screened even from some of its own employees. Such procedures were typical in defense industry circles when a report bearing an export control notice was received. Nevertheless, the absence of any other access restrictions on the letters supports Richards' argument that Richards could have shared the report's contents with every American citizen.

Is the patent anticipated by Algol's prior "publication"? If so, under section 102(a) or under section 102(b)?

Note on Different Types of Utility

Patent law covers more than one type of utility. We identify three separate types of utility arguments that have been used by the courts in rejecting patent applications. The first type centers on whether an invention is operable or capable of any use. The inquiry here is whether the invention as claimed can really *do* anything. The second major issue is whether the invention works to solve the problem it is designed to solve. The focus here is on the operability of the invention to serve its intended purpose. The third issue, in some ways the most interesting, is whether the intended purpose of the invention has some minimum social benefit, or at least is not completely harmful or deleterious. That is, if the invention does what it is supposed to, is that something that society wants done? The first problem may be thought of as general utility; the second, specific utility, and the third, beneficial or moral utility.

General Utility

In his landmark 1890 treatise, Professor William Robinson wrote that to be patentable, an invention must be more than "a mere curiosity, a scientific process exciting wonder yet not producing physical results, or [a] frivolous or trifling article or operation not aiding in the progress nor increasing the possession of the human race." 1 W. Robinson, Treatise on the Law of Patents for Useful Inventions 463 (1890). In applying this approach, one might perhaps imagine a "machine" with working parts that did not really do anything; perhaps it just spins around, or oscillates back and forth for no particular purpose. Such a machine would fail the test of utility under section 101 of the patent code. Note that machines that serve only to amuse or entertain *are* deemed useful under the patent code, however, so the limitation does not appear to be a very strict one.

Specific Utility

Even if an invention is directed toward a certain function, it must actually perform that function. Otherwise it is not "useful" for achieving that function. A good example of this principle is the case of Newman v. Quigg, 877 F.2d 1575 (Fed. Cir. 1989), *cert. denied,* 495 U.S. 932 (1990), where patent applicant Newman claimed an "Energy Generation System Having Higher Energy Output Than Input"—i.e., a "perpetual motion" machine. The Federal Circuit, in upholding the denial of Newman's patent, noted that the applicant had not rebutted data from tests performed by the National Bureau of Standards showing that, as feared, the device did not function perpetually. In short, it had no utility because it did not work.

For an argument that the utility requirement diminishes incentives for "revolutionary" inventions, see Samuel Oddi, Beyond Obviousness: Invention Protection in the Twenty-First Century, 38 Am. U.L. Rev. 1097, 1127 (1989) (arguing that utility requirement and statutory classifications under section 101 [i.e., process, machine, manufacture, etc.] discriminate against revolutionary inventions, and calling for special protection of revolutionary inventions to offset these factors).

Beneficial or Moral Utility

The concept of immoral subject matter is thought to have originated in dictum from a Joseph Story opinion. Lowell v. Lewis, 1 Mason 182, 15 F. Cas. 1018 (No. 8568) (C.C.D. Mass. 1817). As examples of nonuseful inventions, he cited patents to "poison people, or to promote debauchery, or to facilitate private assassination." 15 F. Cas. 1018, 1019.

This doctrine was often invoked in the late nineteenth century to deny patents on gambling devices. Interestingly, it was a successful bar to patentability even where inventions appeared to be useful for things other than gambling. See, e.g., Schultz v. Holtz, 82 F. 448 (N.D. Cal. 1897) (patent on coin return device for coin-operated machines denied because it had application to slot machines); Nat'l Automatic Device Corp. v. Lloyd, 40 F. 89, 90 (N.D. Ill. 1889) (patent on toy horse race course denied on evidence that toy course was used in bars for betting purposes). Patents were struck down on this basis well into the twentieth century; see, e.g., Meyer v. Buckley Mfg. Co., 15 F. Supp. 640, 641 (N.D. Ill. 1936) (patent denied on "game of chance" vending machine, where user inserted coin and tried to manipulate miniature steam shovel to scoop up a toy), and even as late as 1941, in a pinball machine patent case, the Seventh Circuit was careful to note the distinction between playing pinball and gambling. Chicago Patent v. Genco, 124 F.2d 725, 728 (7th Cir. 1941) (upholding patent on pinball machine). By the 1970s, however, the courts were regularly upholding patents on gambling devices — both because gambling was no longer seen as a major moral issue and because courts had become more wary of denying patents on the basis of an indeterminate moral standard. See, e.g., Ex parte Murphy, 200 U.S.P.Q. (BNA) 801, 803 (Bd. Pat. App. & Int. 1977) (upholding claim for "one-armed bandit").

The fight against immoral inventions was not limited to patents for gambling devices. Another line of cases denied patents for inventions that could be used only to defraud. These cases are reviewed and rejected in the more recent case of Juicy Whip, Inc. v. Orange Bang, Inc., 185 F.3d 1364 (Fed. Cir. 1999). The case involved U.S. Patent 5,575,405, drawn to a juice dispensing system. The invention is depicted in Fig. 3-1. The glass bowl on top only *appears* to circulate fresh juice. In actuality, it circulates an undrinkable liquid. The actual juice is dispensed from the tanks hidden underneath the glass bowl display. In reversing a district court opinion holding the patent invalid for lack of utility, the Federal Circuit stated:

> In holding the patent in this case invalid for lack of utility, the district court relied on two Second Circuit cases dating from the early years of this century, Rickard v. Du Bon, 103 F. 868 (2d Cir.1900), and Scott & Williams v. Aristo Hosiery Co., 7 F.2d 1003 (2d Cir.1925). In the *Rickard* case, the court held invalid a patent on a process for treating tobacco plants to make their leaves appear spotted. At the time of the invention, according to the court, cigar smokers considered cigars with spotted wrappers to be of superior quality, and the invention was designed to make unspotted tobacco leaves appear to be of the spotted — and thus more desirable — type. The court noted that the invention did not promote the burning quality of the leaf or improve its quality in any way; "the only effect, if not the only object, of such treatment, is to spot the tobacco, and counterfeit the leaf spotted by natural causes." Id. at 869.

The *Aristo Hosiery* case concerned a patent claiming a seamless stocking with a structure on the back of the stocking that imitated a seamed stocking. The imitation was commercially useful because at the time of the invention many consumers regarded seams in stockings as an indication of higher quality. The court noted that the imitation seam did not "change or improve the structure or the utility of

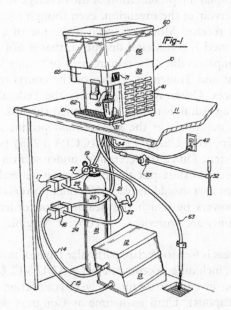

FIGURE 3-1
Juicy Whip patent drawing.

the article," and that the record in the case justified the conclusion that true seamed stockings were superior to the seamless stockings that were the subject of the patent. See *Aristo Hosiery,* 7 F.2d at 1004. "At best," the court stated, "the seamless stocking has imitation marks for the purposes of deception, and the idea prevails that with such imitation the article is more salable." Id. That was not enough, the court concluded, to render the invention patentable.

We decline to follow *Rickard* and *Aristo Hosiery,* as we do not regard them as representing the correct view of the doctrine of utility under the Patent Act of 1952. The fact that one product can be altered to make it look like another is in itself a specific benefit sufficient to satisfy the statutory requirement of utility.

It is not at all unusual for a product to be designed to appear to viewers to be something it is not. For example, cubic zirconium is designed to simulate a diamond, imitation gold leaf is designed to imitate real gold leaf, synthetic fabrics are designed to simulate expensive natural fabrics, and imitation leather is designed to look like real leather. In each case, the invention of the product or process that makes such imitation possible has "utility" within the meaning of the patent statute, and indeed there are numerous patents directed toward making one product imitate another. See, e.g., U.S. Pat. No. 5,762,968 (method for producing imitation grill marks on food without using heat); U.S. Pat. No. 5,899,038 (laminated flooring imitating wood); U.S. Pat. No. 5,571,545 (imitation hamburger). Much of the value of such products resides in the fact that they appear to be something they are not. Thus, in this case the claimed post-mix dispenser meets the statutory requirement of utility by embodying the features of a post-mix dispenser while imitating the visual appearance of a pre-mix dispenser.

The fact that customers may believe they are receiving fluid directly from the display tank does not deprive the invention of utility. Orange Bang has not argued that it is unlawful to display a representation of the beverage in the manner that fluid is displayed in the reservoir of the invention, even though the fluid is not what the customer will actually receive. Moreover, even if the use of a reservoir containing fluid that is not dispensed is considered deceptive, that is not by itself sufficient to render the invention unpatentable. The requirement of "utility" in patent law is not a directive to the Patent and Trademark Office or the courts to serve as arbiters of deceptive trade practices. Other agencies, such as the Federal Trade Commission and the Food and Drug Administration, are assigned the task of protecting consumers from fraud and deception in the sale of food products. Cf. In re Watson, 517 F.2d 465, 474-76, 186 USPQ 11, 19 (CCPA 1975) (stating that it is not the province of the Patent Office to determine, under section 101, whether drugs are safe). As the Supreme Court put the point more generally, "Congress never intended that the patent laws should displace the police powers of the States, meaning by that term those powers by which the health, good order, peace and general welfare of the community are promoted." Webber v. Virginia, 103 U.S. (13 Otto) 344, 347-48 (1880).

Of course, Congress is free to declare particular types of inventions unpatentable for a variety of reasons, including deceptiveness. Cf. 42 U.S.C. § 2181(a) (exempting from patent protection inventions useful solely in connection with special nuclear material or atomic weapons). Until such time as Congress does so, however, we find no basis in section 101 to hold that inventions can be ruled unpatentable for lack of utility simply because they have the capacity to fool some members of the public. The district court therefore erred in holding that the invention of the '405 patent lacks utility because it deceives the public through imitation in a manner that is designed to increase product sales.

COMMENTS AND QUESTIONS

1. What conclusions can be drawn from the attempts of the courts to enforce moral norms by denying patents? First, as in the case of gambling devices, moral norms — or at least the courts' perceptions of them — change over time. Gambling is not perceived (by most) as a pernicious social ill in the current age, perhaps because other problems now seem so much worse. And other technologies once thought very wrong are now accepted as commonplace, e.g., birth control.

Even conceding that a particular new technology is analogous to gambling or selling fake medicines, another problem remains: what are the limits of the immorality test? How far into the future can the patent challenger look for the immoral effects of an invention, and what consensus version of morality can the courts rely on?

For example, historians and sociologists have long noted the profound social changes that accompanied the invention of the automobile. Some of these changes had unquestionable moral dimensions, such as the impact of automobiles on the incidence of pollution and premarital sex. If these changes could have been foreseen, immoral use might have been raised as a reason not to enforce the patent. A host of other technologies can be thought of in this vein: e.g., cattle prods (sometimes used in torture) and abortion-inducing drugs (safer than other procedures but considered immoral by some). On these points, see Robert Merges, The Patentability of Higher Life Forms: Intellectual Property Rights and Controversial Technologies, 47 Md. L. Rev. 1051 (1988).

3. Describing and Enabling the Invention

A patent can be a potent property right. In exchange for this grant from the government, an inventor must disclose the workings of his or her invention in enough detail to be informative to other people working in the same field. The sufficiency of the patentee's disclosure, and its relationship to the claims of a patent, are the topics we consider in this section. To begin, we must understand some basics about how patents are obtained.

a. Procedures for Obtaining a Patent

The process of obtaining a patent from the Patent Office is known as "prosecution." The time and effort required to prosecute a patent varies immensely from case to case. A few applications are reviewed quickly and issued within a year of the application date. Others languish in the PTO for years and, especially when several inventors claim they were the first to produce a particular invention, even decades. The "average" prosecution takes approximately 2.77 years. See John R. Allison and Mark A. Lemley, Who's Patenting What? An Empirical Exploration of Patent Prosecution,

53 Vand L. Rev. 2099 (2000). And it has been estimated that over the course of prosecution, the typical application receives only eighteen total hours of attention.[30]

Prosecution begins with the filing of a patent application. Although an inventor or his or her lawyer[31] may file a temporary "placeholder" application, called a provisional application, it must be converted into a standard, full-blown application within one year if it is ever to issue as a patent. See 35 U.S.C. § 111(b). The basics of an application are: (1) the specification, including a summary of the invention and drawings in most cases; (2) one or more claims, at the end of the specification; (3) an oath, declaring the inventor(s) actually invented what is described in the specification; and (4) applicable filing fees.

Immediately after an application is received, it is assigned to one of 17 main examining groups, then to a specific "art unit" that specializes in the relevant technology, and ultimately to one of the roughly 3,400 patent examiners employed at the Patent Office.[32] It is normally 12 months or more until a patent examiner actually picks up an initial application; applications are handled in the order they are received. During this period, applicants may file additional papers, such as an information disclosure statement (IDS), describing prior art known to the applicant at the time of filing, 37 C.F.R. §§ 1.97, 1.98 (2002), or preliminary amendments, which are changes made to the application before the examiner's first response or "office action." Examiners usually make an initial review of the application, which includes a search of the prior art. The most common office action following this initial review is a rejection of most of the claims in the application, often on grounds of lack of novelty or nonobviousness. The examiner must state the reasons for each rejection, together with information and references to aid the applicant in judging the desirability of continuing prosecution or overcoming the rejection. 35 U.S.C. § 132. The examiner has the burden to show why an application should not issue as a patent. In re Lee, 277 F.3d 1338 (Fed. Cir. 2002). In response, the applicant can either contest (or "traverse," as patent lawyers say) the rejection or acquiesce in it.

Aside from an outright "rejection," an examiner can issue what is known as an "objection," which notes problems of form rather than substance pertaining to the application. The most common objection is called a "restriction requirement," which is used when the examiner finds that the applicant has claimed more than one distinct invention in a single application, a violation of section 121 of the Patent Act. In response, an applicant can "elect" some claims from the initial application and continue prosecuting them. Other claims (and supporting disclosure) can be split off into a separate application, known as a "divisional application." This application retains the priority date of the initial filing, but becomes a separate application. (Prior to 1995, and the advent of the 20-year-from-filing patent term, applicants such as pharmaceutical companies used this device to extend patent terms on variants of profitable drug molecules. Subject matter was split off from a basic application

30. See U.S. Patent and Trademark office, 2000 Fiscal Year Report, avail. at *www.uspto.gov/web/offices/com/annual/2000*; John R. Allison and Mark A. Lemley, The Growing Complexity of the United States Patent System, 82 B.U.L. Rev. 77, 135 (2002).

31. To be precise, one need not be a lawyer to represent inventors before the Patent Office. Non-lawyer "patent agents" may do so as well, as long as they have passed the Patent Bar Exam administered by the U.S. Patent Office.

32. There is currently a lively debate over the resources to be devoted to the patent examination system, and the best way to deploy them. For a good summary, see Berkeley Symposium on Patent System Reform, 17 Berkeley Tech. L. J. 623 et seq. (2002).

and left dormant via amendments and continuations until deep into the patent term of the first "elected" invention.)

In general applications can be amended quite liberally during prosecution.[33] The most common reason for an amendment is to respond to a rejection or other office action. But applicants can also correct mistakes, add or change drawings, and even update the disclosure portion of the specification (though for amendments of this latter type newly disclosed material may not enjoy the benefit of the initial filing date). Standard amendments are usually permitted up to the time when an examiner issues a final rejection, which may be as early as the second office action.

The label "final rejection" is a misnomer if ever there was one. An applicant can respond to a final rejection in a number of ways, most typically via a continuation or amendment after final rejection under 37 C.F.R. §1.116. (If the applicant feels continued prosecution would not be worthwhile, e.g., because there are clear disagreements with the examiner, he or she may file an appeal to the Board of Patent Appeals and Interferences, an administrative tribunal within the Patent Office.) A continuation is basically a new version of the application, though continuations come in several flavors.[34] Filing a continuation resets the examination process, without having to wait a year for a response (since continuations receive priority in processing). A simple continuation application retains the benefit of the initial application's filing date; it therefore may serve much the same role as an amendment. A continuation-in-part ("CIP"), by contrast, adds new matter to the specification. This new matter does not retain the benefit of the earlier filing date, though any information carried over from the original filing does.

Applicants may employ a variation on the continuation theme when the examiner has decided to allow some claims but has issued a final rejection as to others. Typical prosecution strategy is to take the bird in the hand and fight over the contested claims separately. To achieve this, the applicant can file an amendment after final rejection canceling the rejected claims. This puts the remaining claims in condition for allowance. Meanwhile, the rejected claims may be carried over into a separate continuation application. Then the battle over these claims can be joined again; in the meantime, the acceptable claims will become enforceable as soon as the original, slimmed-down application is allowed to issue as a patent.

Continuations come in handy for other objectives as well. For important inventions, patent lawyers often file a continuation application just prior to the issuance of a patent, so that prosecution based on the original disclosure may continue. This

---

33. Only recently has the Federal Circuit approved of a doctrine limiting excessive abuse of applicant's latitude in prosecuting applications. See Symbol Technologies, Inc. v. Lemelson Med., Educ. & Res. Fdn., Ltd., 277 F.3d 1361 (Fed. Cir. 2002) (announcing doctrine of "prosecution history laches"); M. Scott Carey, Note, Ford Motor Co. v. Lemelson, 16 Berkeley J. L. & Tech. 219, 221 (1998) ("[T]wenty-two of the claims asserted against Ford appeared thirty-three years after the original disclosure. All the claims asserted against Ford claimed the benefit of the 1954 and 1956 [original applications] for their priority."). For subsequent opinions in *Lemelson* after remand, see 301 F.Supp.2d 1147 (D. Nev. 2004), aff'd 422 F.3d 1378 (Fed. Cir. 2005).

34. Technically, there are two types of continuations: (1) a "standard" one, which may be thought of as a new branch of the patent; and (2) one that completely replaces the prior application and therefore renders it abandoned. Continuations in this second category filed after May 29, 2000 must be filed under a request for continuing examination (RCE). Note that an application prosecuted after an RCE is not technically a new application. See 37 C.F.R. §1.114 (2002). An RCE is not to be confused with the pre-2000 continued prosecution application (CPA) procedure. See 37 C.F.R. §1.53(d). A "standard" continuation does not cause the parent application to be abandoned, which is why this is used to split off contested subject matter. The parent application, now minus the contested claims, carries on—and hence can issue as a patent.

is valuable where a competitor may attempt to design around a patent by adopting minor variants. In that event, it may be possible to revise the continuation application claims to cover the competitor's new variant, considerably enhancing the effective scope of the patent. Ideally, this maneuver can be accomplished without sacrificing the original priority date. But even if some new matter is added, making the filing a CIP (see above), this still has advantages over filing a completely separate invention. For one thing, the earlier versions of the application will not generally be prior art as against the material in the CIP. And examination is faster as well.

In addition to these various filings, applicants can communicate with patent examiners by phone or in person, through what is called an examiner interview. Any notes made during these communications form part of the prosecution history of the patent.

Prior to 2001, U.S. patent applications were not published. They were held in secrecy until the issue date. Long-pending applications, such as those described in the *Lemelson* case *infra* section III.D.2 (notes after *Kingsdown*), were referred to as "submarine patents" for this reason. These abuses, together with the felt need to harmonize U.S. practice with that of Japan and Europe, led to the publication of patent applications under the American Inventors Protection Act of 1999 ("AIPA"), 35 U.S.C. § 122(b). Under the AIPA, all applications that cover inventions that will be the subject of foreign applications are published eighteen months after their filing date.[35] Thus where an applicant is interested only in a U.S. patent, the application will not be published. An interesting feature of the AIPA is that after publication but before the patent issues, the applicant has certain limited legal rights in the invention. So long as the infringer has actual notice of the published patent application, and so long as the claims in the published application are "substantially identical" to the claims in the patent when issued, the applicant/patentee can recover a reasonable royalty from the infringer. Unlike an issued patent, rights in a published application cannot be enforced with an injunction. 35 U.S.C. § 154(d).

Interested third parties — in particular, competitors of the applicant — may read published patent applications carefully. Regulations put in place after the AIPA permit third parties to submit prior art references to be included in the prosecution file of the application. While third parties have always been permitted to submit prior art during prosecution[36] or after a patent issues,[37] the availability of published applications on the World Wide Web may lead to a large increase in the volume of third-party submissions. The U.S. may thus be said to have taken a small step in the direction of a full-blown third party "opposition system" such as those available in Japan and Europe.[38] But it is only a small step; third parties who submit prior art informally to the PTO cannot participate in the examination of the patent.

Often the claims section of a patent begins with the broadest claim, which is then "qualified" in a series of dependent claims. This is followed perhaps by a

---

35. There are exceptions for design patents, inventions subject to government secrecy orders due to national security, and provisional patent applications. See 35 U.S.C. § 12(b)(2) (2002).

36. Under the "public protest" provisions of 37 C.F.R. § 1.291 (2002). Note that prior to the publication of applications, third parties had to find out some other way that an application was pending (presumably usually through some disclosure by the applicant him- or herself).

37. 37 C.F.R. § 1.501 (2002).

38. Commentators are split over whether we have enough information to decide whether that would be a good thing. See, e.g., Mark A. Lemley, Rational Ignorance at the Patent Office, 95 Nw. U. L. Rev. 1495, 1524-25 (2001) (agnostic on oppositions); Robert P. Merges, As Many as Six Impossible Patents before Breakfast: Property Rights for Business Concepts and Patent System Reform, 14 Berkeley Tech. L.J. 577 (1999) (enthusiastic about oppositions).

narrower independent claim, which may itself be qualified by a series of dependent claims. In this way, the general structure of a patent often resembles an inverted pyramid: the broadest claims are first, the narrowest last, and the scope of the claims generally "tapers" from the first to the last.

Prosecution ends when the patent is granted or the application is abandoned. Issued patents are summarized in the Patent Gazette and made available to the public. Prosecution can in a sense continue after a patent has issued. A patentee who comes to believe that her patent claims are either too broad or too narrow can seek a reissue of the patent, so long as the deficiency in the original patent is the result of a bona fide error or omission. Reissues to broaden the scope of claims must be initiated within two years of the original issuance, however.

In a related proceeding called a reexamination, *anyone* (including the patentee) can seek a reexamination of a patent if a substantial new basis for questioning the patentability of the invention arises after issuance. Although the AIPA of 1999 gave the party requesting a reexamination broader powers to participate in the proceeding,[39] lawyers do not expect much of an increase in the traditionally low volume of reexamination requests. This is primarily because issues raised in a participatory (or *inter partes*) reexamination may not be raised again in later litigation — a form of *res judicata* that patent lawyers are not generally willing to risk.

Once a patent is issued, a patentee may bring a lawsuit against someone accused of infringing the patent. There are typically two defenses to such a suit: first, the accused infringer will argue that the patentee's patent is invalid[40] for any number of reasons; an invalid patent, of course, cannot be enforced. Second, the accused infringer will usually argue that even if the patent is valid, the products being made or sold by the accused do not infringe. Because invalidity is often argued as a defense in an infringement case, all the issues of patent validity dealt with by the Patent Office in granting a patent may also be considered by a court. As we shall see, the standard of review of the Patent Office's decision to issue a patent is a contentious question. See, e.g., Dickinson v. Zurko, 527 U.S. 150 (1999) (adopting the deferential "substantial evidence" standard for judicial review of PTO decisions). But there is no doubt that a patent is "born valid": the Patent Act includes a "presumption of validity" that accompanies any issued patent. 35 U.S.C. § 282. Since 1982, appeals in patent infringement cases have gone automatically to the Court of Appeals for the Federal Circuit, regardless of the district in which the case was filed.

Three other forms of patent litigation deserve brief mention here.

- A unique form of patent proceeding is a priority dispute between two or more inventors, all of whom claim to have been the first inventor of a particular invention. These are known as interference proceedings. They are an outgrowth of the fact that the United States awards a patent to the first inventor, unlike almost every other country, which awards the patent to the first to file a

---

39. Formally, the AIPA created a new "optional" *inter partes* reexamination procedure that parallels the existing procedure giving very limited rights of ongoing participation to the party requesting the reexamination. See 35 U.S.C. § 311 et seq. (2002).

40. Technically, some defenses in patent cases result in a finding that a patent is unenforceable, rather than invalid. This is the case with the defense of inequitable conduct, for instance, which is concerned with improper disclosure or nondisclosure by a patent attorney during prosecution of the patent application. The antitrust-related defense of patent misuse is another example; when patent misuse is proven the affected patents are found unenforceable (at least until the misuse can be "purged" or cured).

patent application. Interference proceedings are handled by the PTO internally through the Board of Patent Appeals and Interferences, again with a right of appeal to the CAFC.

➤ • People who reasonably fear being sued for patent infringement (for example, because the patentee has threatened them with suit) can file suit themselves for a declaratory judgment that the patent at issue is invalid, or that their conduct does not infringe the patent. Such declaratory judgment actions are typically met with counterclaims charging infringement; these proceed much as normal patent infringement suits would.

• The International Trade Commission has jurisdiction to block the importation of products into the United States if they infringe U.S. patents and to block the importation of products made abroad by processes that are patented in the United States. Patent owners can complain to the ITC, which will bring "section 337" actions against likely infringers. These actions have their own set of procedural rules that operate parallel to district court actions for patent infringement.

Finally, a recent Supreme Court case opens the door to regional courts of appeals to consider patent issues in lawsuits when the original complaint does not relate to the Patent Act, but the counterclaim does. Holmes Group, Inc. v. Vornado Air Circulation Systems, Inc., 535 U.S.826 (2002).

b. Disclosure Doctrines: Enablement and Written Description

The overall goal when drafting patent claims is to make them as broad as the Patent Office will allow. There are essentially two constraints on the breadth of the claims you can draft: (a) the mass of publicly available information on your problem — what patent practitioners call "the prior art"; and (b) the actual work the inventor has done, in the sense that you may not claim anything beyond what the inventor has discovered, i.e., beyond the limits of the principle of the invention. To use a famous example: the inventor of the telegraph, Samuel Morse, was not permitted to claim "all forms of communicating at a distance" using electromagnetic waves, since he had only discovered one form — the telegraph. It would be unfair to permit Morse to claim, e.g., microwave communications, since he did not actually discover this. In patent parlance, these "embodiments" were not "enabled" by Morse, and so he may not claim them.

i. Enablement

Some inventions are capable of being manifested in a very wide array of embodiments. Think of the Velcro fastener, for instance, present on everything from shoes to spacesuits to huge industrial storage sacks. The point to keep in mind is that all these embodiments share the same inventive principle, i.e., are all instances of the same underlying invention. Thus it is reasonable to give the person who invented Velcro fasteners a patent on the entire concept of the new fastener, in whatever context it might be applied. But the patentee must be in possession of the inventive principle. The reach of this doctrine is explored in the following foundational case.

The Incandescent Lamp Patent → *unique to Patent*
Supreme Court of the United States
159 U.S. 465 (1895)
law this case.

This was a bill in equity, filed by the consolidated Electric Light Company against the McKeesport Light Company, to recover damages for the infringement of letters patent No. 317,076, issued May 12, 1885, to the Electro-Dynamic Light Company, assignee of Sawyer and Man, for an electric light. The defendants justified [their actions] under certain patents to Thomas A. Edison, particularly No. 223,898, issued January 27, 1880; denied the novelty and utility of the complainants' patent, and averred that the same had been fraudulently and illegally procured. The real defendant was the Edison Electric Light Company, and the case involved a contest between what are known as the Sawyer and Man and the Edison systems of electric lighting.

In their application, Sawyer and Man stated that their invention related to "that class of electric lamps employing an incandescent conductor enclosed in a transparent, hermetically-sealed vessel or chamber, from which oxygen is excluded, and . . . more especially to the incandescing conductor, its substance, its form, and its combination with the other elements composing the lamp. Its object is to secure a cheap and effective apparatus; and our improvement consists, first, of the combination, in a lamp chamber, composed wholly of glass, . . . of an incandescing conductor of carbon made from a vegetable fibrous material, in contradistinction to a similar conductor made from mineral or gas carbon, and also in the form of such conductor so made from such vegetable carbon, and combined in the lighting circuit with the exhausted chamber of the lamp."

The specification further stated that:

In the practice of our invention we have made use of carbonized paper, and also wood carbon. We have also used such conductors or burners of various shapes, such as pieces with their lower ends secured to their respective supports, and having their upper ends united so as to form an inverted V-shaped burner. We have also used conductors of varying contours — that is, with rectangular bends instead of curvilinear ones; but we prefer the arch shape.

No especial description of making the illuminating carbon conductors, described in this specification and making the subject-matter of this improvement, is thought necessary, as any of the ordinary methods of forming the material to be carbonized to the desired shape and size, and carbonizing it according to the methods in practice before the date of this improvement, may be adopted in the practice thereof by any one skilled in the arts appertaining to the making of carbons for electric lighting or for other use in the arts. The advantages resulting from the manufacture of the carbon from vegetable fibrous or textile material instead of mineral or gas carbon are many. Among them may be mentioned the convenience afforded for cutting and making the conductor in the desired form and size, the purity and equality of the carbon obtained, its susceptibility to tempering, both as to hardness and resistance, and its toughness and durability.

The claims were as follows:

1. An incandescing conductor for an electric lamp, of carbonized fibrous or textile material and of an arch or horseshoe shape, substantially as hereinbefore set forth.
2. The combination, substantially as hereinbefore set forth, of an electric circuit and an incandescing conductor of carbonized fibrous material, included in and forming part of said circuit, and a transparent hermetically sealed chamber in which the conductor is enclosed.
3. The incandescing conductor for an electric lamp, formed of carbonized paper, substantially as described.

The commercial Edison lamp used by the appellee is composed of a burner made of carbonized bamboo of a peculiar quality discovered by Mr. Edison to be highly useful for the purpose, and having a length of about six inches, a diameter of about five one thousandths of an inch, and an electrical resistance of upwards of 100 ohms. This filament of carbon is bent into the form of a loop, and its ends are secured by good electrical and mechanical connections to two fine platinum wires....

Upon a hearing in the Circuit Court before Mr. Justice Bradley upon pleadings and proofs, the court held the patent to be invalid, and dismissed the bill. 40 Fed. Rep. 21 [C.C.W.D.P.A. (1889)]. Thereupon complainant appealed to this court.

Mr. Justice BROWN, after stating the case as above reported, delivered the opinion of the court.

In order to obtain a complete understanding of the scope of the Sawyer and Man patent, it is desirable to consider briefly the state of the art at the time the application was originally made, which was in January, 1880.

...The form of illumination...known as the incandescent system...consists generally in the passage of a current of electricity through a continuous strip or piece of refractory material, which is a conductor of electricity, but a poor conductor—in other words, a conductor offering a considerable resistance to the flow of the current through it. It was discovered early in this century that various substances might be heated to a white heat by passing a sufficiently strong current of electricity through them....

For many years prior to 1880, experiments had been made by a large number of persons, in various countries, with a view to the production of an incandescent light which could be made available for domestic purposes, and could compete with gas in the matter of expense. Owing partly to a failure to find a proper material, which should burn but not consume, partly to the difficulty of obtaining a perfect vacuum in the globe in which the light was suspended, and partly to a misapprehension of the true principle of incandescent lighting, these experiments had not been attended with success; although it had been demonstrated as early as 1845 that, whatever material was used, the conductor must be enclosed in an air-tight bulb [i.e., vacuum], to prevent it from being consumed by the oxygen in the atmosphere. The chief difficulty was that the carbon burners were subject to a rapid disintegration or evaporation, which electricians assumed was due to the disrupting action of the electric current, and, hence, the conclusion was reached that carbon contained in itself the elements of its own destruction, and was not a suitable material for the burner of an incandescent lamp.

It is admitted that the lamp described in the Sawyer and Man patent is no longer in use, and was never a commercial success; that it does not embody the principle of high resistance with a small illuminating surface; that it does not have the filament burner of the modern incandescent lamp; that the lamp chamber is defective, and that the lamp manufactured by the complainant and put upon the market is substantially the Edison lamp; but it is said that, in the conductor used by Edison (a particular part of the stem of the bamboo lying directly beneath the silicious cuticle, the peculiar fitness for which purpose was undoubtedly discovered by him), he made use of a fibrous or textile material, covered by the patent to Sawyer and Man, and is, therefore, an infringer. It was admitted, however, that the third claim — for a conductor of carbonized paper — was not infringed.

The two main defences to this patent are (1) that it is defective upon its face, in attempting to monopolize the use of all fibrous and textile materials for the purpose of electric illumination; and (2) that Sawyer and Man were not in fact the first to discover that these were better adapted than mineral carbons to such purposes.

Is the complainant entitled to a monopoly of all fibrous and textile materials for incandescent conductors? If the patentees had discovered in fibrous and textile substances a quality common to them all, or to them generally, as distinguishing them from other materials, such as minerals, etc., and such quality or characteristic adapted them peculiarly to incandescent conductors, such claim might not be too broad. If, for instance, minerals or porcelains had always been used for a particular purpose, and a person should take out a patent for a similar article of wood, and woods generally were adapted to that purpose, the claim might not be too broad, though defendant used wood of a different kind from that of the patentee. But if woods generally were not adapted to the purpose, and yet the patentee had discovered a wood possessing certain qualities, which gave it a peculiar fitness for such purpose, it would not constitute an infringement for another to discover and use a different kind of wood, which was found to contain similar or superior qualities. The present case is an apt illustration of this principle. Sawyer and Man supposed they had discovered in carbonized paper the best material for an incandescent conductor. Instead of confining themselves to carbonized paper, as they might properly have done, and in fact did in their third claim, they made a broad claim for every fibrous or textile material, when in fact an examination of over six thousand vegetable growths showed that none of them possessed the peculiar qualities that fitted them for that purpose. Was everybody then precluded by this broad claim from making further investigation? We think not.

The injustice of so holding is manifest in view of the experiments made, and continued for several months, by Mr. Edison and his assistants, among the different species of vegetable growth, for the purpose of ascertaining the one best adapted to an incandescent conductor. Of these he found suitable for his purpose only about three species of bamboo, one species of cane from the Valley of the Amazon, impossible to be procured in quantities on account of the climate, and one or two species of fibres from the agave family. Of the special bamboo, the walls of which have a thickness of about three-eighths of an inch, he used only about twenty-thousandths of an inch in thickness. In this portion of the bamboo the fibres are more nearly parallel, the cell walls are apparently smallest, and the pithy matter between the fibres is at its minimum. It seems that carbon filaments cannot be made of wood — that is, exogenous vegetable growth — because the fibres are not parallel and the longitudinal fibres are intercepted by radial fibres. The cells composing the fibres are all so large that the resulting carbon is very porous and friable. Lamps made of this material proved of no commercial value. After trying as many as thirty or forty different woods of exogenous growth, he gave them up as hopeless. But finally, while experimenting with a bamboo strip which formed the edge of a palmleaf fan, cut into filaments, he obtained surprising results. After microscopic examination of the material, he despatched a man to Japan to make arrangements for securing the bamboo in quantities. It seems that the characteristic of the bamboo which makes it particularly suitable is, that the fibres run more nearly parallel than in other species of wood. Owing to this, it can be cut up into filaments having parallel fibres, running throughout their length, and producing a homogeneous carbon. There is no generic quality, however, in vegetable fibres, because they are fibrous, which adapts them to the purpose. Indeed, the fibres are rather a disadvantage. If the bamboo grew solid without fibres, but had its peculiar cellular formation, it would be a perfect material, and incandescent lamps would last at least six times as long as at present. All vegetable fibrous growths do not have a suitable cellular structure. In some the cells are so large that they are valueless for that purpose. No exogenous, and very few endogenous, growths are suitable. The messenger whom he dispatched to different parts of Japan and China sent him about forty different kinds of bamboo, in such quantities as to enable him to make a number of lamps, and from a test of these different species he ascertained which was best for the purpose. From this it appears very clearly that there is no such quality common to fibrous and textile substances generally as makes them suitable for an incandescent conductor, and that the bamboo which was finally pitched upon, and is now generally used, was not selected because it was of vegetable growth, but because it contained certain peculiarities in its fibrous structure which distinguished it from every other fibrous substance. The question really is whether the imperfectly successful experiments of Sawyer and Man, with carbonized paper and wood carbon, conceding all that is claimed for them, authorize them to put under tribute the results of the brilliant discoveries made by others.

It is required by Rev. Stat. §4888 that the application shall contain a written description of the device "and of the manner and process of making, constructing, compounding, and using it in such full, clear, concise, and exact terms as to enable any person, skilled in the art or science to which it appertains or with which it is most nearly connected, to make, construct, compound, and use the same." The object of this is to apprise the public of what the patentee claims as his own, the courts of what they are called upon to construe, and competing manufacturers and dealers of exactly what they are bound to avoid. Grant v. Raymond, 6 Pet. [31 U.S.] 218, 247 [1832]. If the description be so vague and uncertain that no one can tell, except by independent experiments, how to construct the patented device, the patent is void.

It was said by Mr. Chief Justice Taney in Wood v. Underhill, 5 How. [46 U.S.] 1, 5 [1857], with respect to a patented compound for the purpose of making brick or tile, which did not give the relative proportions of the different ingredients:

> But when the specification of a new composition of matter gives only the names of the substances which are to be mixed together, without stating any relative proportion, undoubtedly it would be the duty of the court to declare the patent void. And the same rule would prevail where it was apparent that the proportions were stated ambiguously and vaguely. For in such cases it would be evident, on the face of the specification, that no one could use the invention without first ascertaining, by experiment, the exact proportion of the different ingredients required to produce the result intended to be

obtained. . . . And if, from the nature and character of the ingredients to be used, they are not susceptible of such exact description, the inventor is not entitled to a patent.

So in Tyler v. Boston, 7 Wall. [74 U.S.] 327, 330 [1868], wherein the plaintiff professed to have discovered a combination of fuel oil with the mineral and earthy oils, constituting a burning fluid, the patentee stated that the exact quantity of fuel oil, which is necessary to produce the most desirable compound, must be determined by experiment. And the court observed: "Where a patent is claimed for such a discovery it should state the component parts of the new manufacture claimed with clearness and precision, and not leave a person attempting to use the discovery to find it out 'by experiment.'" See also Bene v. Jeantet, 129 U.S. 683 [(1889)]; Howard v. Detroit Stove Works, 150 U.S. 164, 167 [(1893)]; Schneider v. Lovell, 10 Fed. Rep. 666 [C.C.S.D.N.Y. 1882]; Welling v. Crane, 14 Fed. Rep. 571 [C.C.D.N.J. 1882].

If Sawyer and Man had discovered that a certain carbonized paper would answer the purpose, their claim to all carbonized paper would, perhaps, not be extravagant; but the fact that paper happens to belong to the fibrous kingdom did not invest them with sovereignty over this entire kingdom, and thereby practically limit other experimenters to the domain of minerals.

In fact, such a construction of this patent as would exclude competitors from making use of any fibrous or textile material would probably defeat itself, since, if the patent were infringed by the use of any such material, it would be anticipated by proof of the prior use of any such material. In this connection it would appear, not only that wood charcoal had been constantly used since the days of Sir Humphry Davy for arc lighting, but that in the English patent to Greener and Staite of 1846, for an incandescent light, "charcoal, reduced to a state of powder," was one of the materials employed. So also, in the English patent of 1841 to De Moleyns, "a finely pulverized boxwood charcoal or plumbago" was used for an incandescent electric lamp. Indeed, in the experiments of Sir Humphry Davy, early in the century, pieces of well-burned charcoal were heated to a vivid whiteness by the electric current, and other experiments were made which evidently contemplated the use of charcoal heated to the point of incandescence. Mr. Broadnax, the attorney who prepared the application, it seems, was also of opinion that a broad claim for vegetable carbons could not be sustained because charcoal had been used before in incandescent lighting.

We are all agreed that the claims of this patent, with the exception of the third, are too indefinite to be the subject of a valid monopoly. For the reasons above stated the decree of the Circuit Court is Affirmed.

Patent office tried to prevent someone from having a monopoly.

COMMENTS AND QUESTIONS

1. The modern equivalent to old Rev. Stat. §4888 is 35 U.S.C. §112 (1986). The first paragraph of §112 has been interpreted in such a way that three distinct requirements are now said to spring from it: (1) the written description requirement; (2) the best mode requirement; and (3) the enablement requirement. The second paragraph adds the separate requirement of claim definiteness. Which of these requirements was at issue in *Incandescent Lamp*? More than one?

2. The Court puts forth several rationales for the enablement requirement. For example, the Court asks: "Was everybody then precluded by this broad claim from making further investigation? We think not." What assumption lies behind this statement? Why should the patent system be concerned with "further investigation"? What purpose behind section 112 is being served here?

3. Were *all* claims of the Sawyer and Man patent invalidated? Of what relevance, if any, is it that Edison's research resulted in several patents of his own? Does the fact that inventor *A* received a patent suggest anything about whether *A*'s research results are enabled by the specification of a prior patent issued to *B*?

4. Edison's discovery that only a certain type of bamboo plant would work as a filament is an example of his exhaustive research efforts. Of course, the bamboo itself is not patentable, since it is a product of nature. After isolating the precise type of bamboo that would work, Edison acted with characteristic speed on a characteristically grand scale: he tried to lock up as many acres of production of the bamboo as he could. See A. Millard, Edison and the Business of Innovation (1990). Cf. Hirshleifer, The Private and Social Value of Information and the Reward to Inventive Activity, 61 Am. Econ. Rev. 561 (1971) (positing reduced need for intellectual property protection, since inventors have "inside information" about their inventions, so they can reap gains by investing in assets that their inventions will make more valuable and selling short assets that their inventions will make less valuable).

If the court was right to invalidate the entire patent, how much research should Sawyer and Man have done to render their broad claims valid? Must they test *every* possible variant known to them?

For more on the early history of the lighting industry, see W. Bright, The Electric-Lamp Industry (1949), and W. MacLaren, The Rise of the Electrical Industry During the Nineteenth Century (1943); A. Millard, *supra*.

5. How much would Sawyer and Man have had to disclose in order to justify a patent on all vegetable and fibrous filaments? What if they had tested 100 filaments, and they had all worked about as well? See In re Wands, 858 F.2d 731 (Fed. Cir. 1988) (enablement is tested by asking whether the person having ordinary skill in the art would be able to make and use all the species covered by the patent without "undue experimentation").

Note on "Analog" Claims in Chemical and Biotechnology Patents: An Exploration of Patent Breadth

Many chemicals have "analogs." Two different molecules with almost identical chemical structure will normally behave in approximately the same way. So, too, similar DNA sequences normally have similar effects. The prevalence of analogs creates problems for those who try to patent new chemical or biological products. If they claim only the precise DNA sequence or chemical structure they have identified, it is relatively easy for others to "design around" their invention by varying an insignificant part of the total structure.

To avoid this problem, many patentees claim not only the precise structure they have produced, but its analogs as well. (This is akin to the claims of Sawyer & Man, in the Incandescent Lamp case, to fibrous and textile materials other than the paper they had actually experimented with.) To what extent a claim should be allowed to preempt such analogs is a difficult problem in patent law. In Amgen, Inc. v. Chugai Pharmaceutical Co., Ltd., 927 F.2d 1200 (Fed. Cir.), *cert. denied,* 502 U.S. 856 (1991), the plaintiff and defendant each held patents on technology related to the production of erythropoietin (EPO), a critical biological protein that stimulates production of red blood cells and is therefore effective in combating anemia and related conditions. Plaintiff Amgen held a patent on a recombinant DNA version of EPO, while defendant Chugai held a license from codefendant Genetics Institute under a product patent for purified EPO made by concentrating trace amounts of the protein from natural sources. The trial court had held valid certain claims in defendant's patent, and it ruled that plaintiff infringed; it also held that certain claims in both patents were invalid for failure to enable.

Plaintiff's recombinant EPO patent included key claim 7, which reads: "7. A purified and isolated DNA sequence consisting essentially of a DNA sequence encoding a polypeptide having an amino acid sequence sufficiently duplicative of that of erythropoietin to allow possession of the biological property of causing bone marrow cells to increase production of reticulocytes and red blood cells, and to increase hemoglobin synthesis or iron uptake." The "biological property" language, together with the descriptions of the two key functions — blood cell production and iron uptake — was intended to broaden the claim so as to cover any functional substitute or "analog" for the natural EPO protein. The following excerpt comes from the Federal Circuit's discussion of the validity of claim 7 and related claims.

> The essential question here is whether the scope of enablement of claim 7 is as broad as the scope of the claim. See generally *In re Fisher.* . . .
> The specification of [Amgen's] patent provides that:
>
>> one may readily design and manufacture genes coding for microbial expression of polypeptides having primary conformations [i.e., proteins with a basic shape] which differ from that herein specified for mature EPO in terms of the identity or location of one or more residues [i.e., amino acids] (e.g., substitutions, terminal and intermediate additions and deletions). . . .
>>   DNA sequences provided by the present invention are thus seen to comprehend all DNA sequences suitable for use in securing expression in a procaryotic or eucaryotic host cell of a polypeptide product having at least a part of the primary structural conformation and one or more of the biological properties of erythropoietin. . . .
>
> The district court found that over 3,600 different EPO analogs can be made by substituting at only a single amino acid position [in the entire protein], and over a million different analogs can be made by substituting three amino acids. The patent indicates that it embraces means for preparation of "numerous" polypeptide analogs of EPO. Thus, the number of claimed DNA encoding sequences that can produce an EPO-like product is potentially enormous.
>
> In a deposition, Dr. Elliott, who was head of Amgen's EPO analog program, testified that he did not know whether the fifty to eighty EPO analogs Amgen had made "had the biological property of causing bone marrow cells to increase production of reticulocytes and red blood cells, and to increase hemoglobin synthesis or iron uptake" [as required by some of the claims]. Based on this evidence, the trial court [found a lack of enablement]. In making this determination, the court relied in particular on the lack of predictability in the art. . . . After five years of experimentation, the court noted, "Amgen is still unable to specify which analogs have the biological properties set forth in claim 7."
>
> [Although] it is not necessary that a patent applicant test all the embodiments of his invention, In re Angstadt, 537 F.2d 498, 502 (C.C.P.A. 1976)[,] what is necessary is that he provide a disclosure sufficient to enable one skilled in the art to carry out the invention commensurate with the scope of his claims. For DNA sequences, that means disclosing

how to make and use enough sequences to justify grant of the claims sought. Amgen has not done that here. It is well established that a patent applicant is entitled to claim his invention generically, when he describes it sufficiently to meet the requirements of Section 112. See Utter v. Hiraga, 845 F.2d 993, 998 (Fed. Cir. 1988) ("A specification may, within the meaning of 35 U.S.C. § 112 ¶1, contain a written description of a broadly claimed invention without describing all species that claim encompasses."). Here, however, despite extensive statements in the specification concerning all the analogs of the EPO gene that can be made, there is little enabling disclosure of particular analogs and how to make them. Details for preparing only a few EPO analog genes are disclosed. Amgen argues that this is sufficient to support its claims; we disagree. This "disclosure" might well justify a generic claim encompassing these and similar analogs, but it represents inadequate support for Amgen's desire to claim all EPO gene analogs. There may be many other genetic sequences that code for EPO-type products. Amgen has told how to make and use only a few of them and is therefore not entitled to claim all of them.

In affirming the district court's [invalidation of these] claims, we do not intend to imply that generic claims to genetic sequences cannot be valid where they are of a scope appropriate to the invention disclosed by an applicant. That is not the case here, where Amgen has claimed every possible analog of a gene containing about 4,000 nucleotides, with a disclosure only of how to make EPO and a very few analogs.

Considering the structural complexity of the EPO gene, the manifold possibilities for change in its structure, with attendant uncertainty as to what utility will be possessed by these analogs, we consider that more is needed concerning identifying the various analogs that are within the scope of the claim, methods for making them, and structural requirements for producing compounds with EPO-like activity. It is not sufficient, having made the gene and a handful of analogs whose activity has not been clearly ascertained, to claim all possible genetic sequences that have EPO-like activity. Under the circumstances, we find no error in the court's conclusion that the generic DNA sequence claims are invalid under Section 112.

COMMENTS AND QUESTIONS

1. To similar effect as *Amgen* is In re Goodman, 11 F.3d 1046, 1052 (Fed. Cir. 1993), where the court held that disclosure of a single working example could not support a broad claim to a method for producing mammalian proteins in the cells of plants.

Consider the interaction between section 103 and section 112 of the Act. When we considered nonobviousness in the context of biotechnology, we saw that the courts found the isolation of a human DNA sequence from the protein it creates to be nonobvious, on the grounds that it would take a great deal of effort to identify the relevant DNA sequence. There, the difficulty of working in the field benefited the patentee. But that same conclusion may make it hard for inventors to enable broad claims in the biotechnology field, since they would have to prove that *the same* "person skilled in the art" would be able to produce other examples covered by the claim without undue experimentation.

2. In addition to the types of broad claims at issue in *Amgen*, recent patents illustrate another expansive claim format. This characteristically reads: "[I claim] the nucleotide sequence in Figure 1, and functionally equivalent sequences having at least *x* percent sequence homology with said sequence."

Homologous regions are regions with identical nucleotide sequences. Thus the "percent homology" is a measure of the similarity between two strands of DNA;

100 percent homology means the two strands are identical. Homology is normally measured by hybridization, which takes advantage of the complementarity between the two strands of a DNA molecule. One of the two strands from one segment of DNA is joined with the complementary strand from another DNA segment; this forms a *hybrid* segment. If all the nucleotides "match up," or form complementary base pairs, there will be no unmatched regions on the hybrid, and one may conclude that the two original segments are 100 percent homologous. But if there are any mismatches, i.e., if any nucleotide on one strand is not matched with its complementary nucleotide on the other, the hybrid strands will not be bound together at that point.

Does the requirement that the sequence be "functionally equivalent" *and* similar in substantial part provide adequate protection against the concern in *Amgen* that the patentee will control numerous gene sequences that he has not produced or investigated and that may have value in themselves in addition to their use as analogs?

3. To what extent must the invention itself be fully described in the patent specification? In In re Wands, 858 F.2d 731 (Fed. Cir. 1988), the patentee claimed the use of particular monoclonal antibodies to perform a particular immunoassay.[41] The PTO challenged the application under the enablement requirement of 35 U.S.C. §112. The Federal Circuit reversed, finding the patent valid. The court noted that complete description was not always required:

> Appellants contend that their written specification fully enables the practice of their claimed invention because the monoclonal antibodies needed to perform the immunoassays can be made from readily available starting materials using methods that are well known in the monoclonal antibody art. There is no challenge to their contention that the starting materials (i.e., mice, HBsAg antigen, and myeloma cells) are available to the public. The PTO concedes that the methods used to prepare hybridomas and to screen them for high-affinity IgM antibodies against HBsAg were either well known in the monoclonal antibody art or adequately disclosed in the '145 patent and in the current application. The sole issue is whether, in this particular case, it would require undue experimentation to produce high-affinity IgM monoclonal antibodies.
>
> Enablement is not precluded by the necessity for some experimentation such as routine screening. However, experimentation needed to practice the invention must not be undue experimentation. "The key word is 'undue,' not 'experimentation.'" [In re Angstadt, 537 F.2d 498, 504 (C.C.P.A. 1976).]
>
> The determination of what constitutes undue experimentation in a given case requires the application of a standard of reasonableness, having due regard for the nature of the invention and the state of the art.
>
> The test is not merely quantitative, since a considerable amount of experimentation is permissible, if it is merely routine, or if the specification in question provides a reasonable amount of guidance with respect to the direction in which the experimentation should proceed....
>
> Factors to be considered in determining whether a disclosure would require undue experimentation have been summarized by the board in In re Forman. They include (1) the quantity of experimentation necessary, (2) the amount of direction or guidance presented, (3) the presence or absence of working examples, (4) the nature of the invention, (5) the state of the prior art, (6) the relative skill of those in the art, (7) the

---

41. On monoclonal antibody technology, see the Appendix on biotechnology in the statutory supplement.

predictability or unpredictability of the art, and (8) the breadth of the claims. [In re Forman, 230 U.S.P.Q. (BNA) at 547.]

4. In Atlas Powder Co. v. E. I. du Pont De Nemours & Co., 750 F.2d 1569 (Fed. Cir. 1984), the court affirmed a finding that du Pont, the accused infringer, had not proved lack of enablement on the part of the patentee, Atlas Powder. The patent, for explosive compounds, listed in its specification numerous salts, fuels, and emulsifiers that could form thousands of emulsions, but it gave no commensurate information as to which combinations would work. Du Pont had argued that its tests showed a 40 percent failure rate in constructing various embodiments of the claimed invention. The court rejected this "inoperable species" argument:

> Of course, if the number of inoperative combinations becomes significant, and in effect forces one of ordinary skill in the art to experiment unduly in order to practice the claimed invention, the claims might indeed be invalid. See, e.g., In re Cook, 439 F.2d 730, 735 (1971). That, however, has not been shown to be the case here.... The district court also found that one skilled in the art would know how to modify slightly many of [the experimental] "failures" to form a better emulsion.

Id. at 1576-1577.

5. Enablement cannot be defined without first determining *who* must be enabled. As one court has noted:

> A patent specification is not addressed to judges or lawyers, but to those skilled in the art; it must be comprehensible to them, even though the unskilled may not be able to gather from it how to use the invention, and even if it is 'all Greek' to the unskilled.
>
> Gould v. Mossinghoff, 229 U.S.P.Q. 1 (D.D.C. 1985), *aff'd in part, vacated in part sub nom.* Gould v. Quigg, 822 F.2d 1074 (Fed. Cir. 1987).

ii. The Written Description Requirement

Section 112 of the Patent Act reads:

> The specification shall contain a written description of the invention, and of the manner and process of making and using it, in such full, clear, concise and exact terms as to enable any person skilled in the art to which it pertains .... to make and use the same, and shall set forth the best mode contemplated by the inventor of carrying out his invention.

The "description ... sufficient to enable" language has long been the source of the enablement requirement described in connection with the *Incandescent Lamp Patent* case, above. In addition, a series of cases later developed the notion that § 112 added another requirement: the "written description" requirement. See, e.g., In re Ruschig, 379 F.2d 990 (C.C.P.A. 1967); Vas-Cath v. Mahurkar, 935 F.2d 1555 (Fed. Cir. 1991). (On the history of the doctrine, see Salima Merani, Written Description: Hyatt v. Boone, 14 Berkeley Tech. L.J. 137 (1999).) Originally seen as a doctrine that applied only when claims were amended during prosecution, it has been expanded in recent years and now forms an integral component of the law of patent disclosure. The basic standard is stated clearly enough in the cases: as of the application filing date, the applicant must show he or she was "in possession" of the invention as later claimed. Courts compare the original disclosure with the final claims of an issued patent, and ask whether the applicant signified that the specific features and embodiments later claimed were in fact important aspects of the invention. As it has evolved, the doctrine has become a way to prevent applicants from very generally describing a broad class of embodiments but then claiming a narrow subset. The "possession" test therefore means, in effect, that if you are going to focus on a feature or aspect of the disclosed technology in the claims, you need to indicate that in the specification itself.

It is important to see that the written description doctrine is more restrictive than enablement. That is, claims may meet the enablement test but fail the written description test. See, e.g., In re Barker, 559 F.2d 588 (C.C.P.A. 1977); Amgen Inc. v. Hoechst Marion Roussel, Inc., 314 F.3d 1313, 1334 ("The enablement requirement is often more indulgent than the written description requirement. The specification need not explicitly teach those in the art to make and use the invention; the requirement is satisfied if, given what they already know, the specification teaches those in the art enough that they can make and use the invention without 'undue experimentation.'"). For example, a patent's disclosure may teach about a certain group of embodiments sufficiently that they can be made without "undue experimentation" under *The Incandescent Lamp Patent*, and yet not single out a specific feature that later appears in a claim. That patent would be invalid under the written description requirement. See, e.g., In re Barker, *supra* (disclosure of pre-fabricated shingle panels having eight and sixteen shingles across enabled, did not adequately describe, later claim to "at least six shingles"). Under the current state of the law, it is possible to enable, yet not possess. To see how this is possible, consider the following case:

The Gentry Gallery, Inc. v. The Berkline Corp.
United States Court of Appeals for the Federal Circuit
134 F.3d 1473 (Fed. Cir. 1998)

LOURIE, Circuit Judge.

The Gentry Gallery appeals from the judgment of the United States District Court for the District of Massachusetts holding that the Berkline Corporation does not infringe U.S. Patent 5,064,244, and declining to award attorney fees for Gentry's defense to Berkline's assertion that the patent was unenforceable.... Berkline cross-appeals from the decision that the patent was not shown to be invalid.... [B]ecause the court clearly erred in finding that the written description portion of the specification supported certain of the broader claims asserted by Gentry, we reverse the decision that those claims are not invalid under 35 U.S.C. § 112, ¶1 (1994).

Background

Gentry owns the '244 patent, which is directed to a unit of a sectional sofa in which two independent reclining seats ("recliners") face in the same direction.

Sectional sofas are typically organized in an L-shape with "arms" at the exposed ends of the linear sections. According to the patent specification, because recliners usually have had adjustment controls on their arms, sectional sofas were able to contain two recliners only if they were located at the exposed ends of the linear sections. Due to the typical L-shaped configuration of sectional sofas, the recliners therefore faced in different directions. See '244 patent; col. 1, ll. 15-19. Such an arrangement was "not usually comfortable when the occupants are watching television because one or both occupants must turn their heads to watch the same [television] set. Furthermore, the separation of the two reclining seats at opposite ends of a sectional sofa is not comfortable or conducive to intimate conversation." Id. at col. 1, ll. 19-25.

The invention of the patent solved this supposed dilemma by, *inter alia,* placing a "console" between two recliners which face in the same direction. This console "accommodates the controls for both reclining seats," thus eliminating the need to position each recliner at an exposed end of a linear section. Id. at col. 1, ll. 36-37. Accordingly, both recliners can then be located on the same linear section allowing two people to recline while watching television and facing in the same direction. Claim 1, which is the broadest claim of the patent, reads in relevant part:

> A sectional sofa comprising:
>
> a pair of reclining seats disposed in parallel relationship with one another in a double reclining seat sectional sofa section being without an arm at one end . . . ,
> each of said reclining seats having a backrest and seat cushions and movable between upright and reclined positions . . . ,
> *a fixed console* disposed in the double reclining seat sofa section between the pair of reclining seats and with the console and reclining seats together comprising a unitary structure,
> said console including an armrest portion for each of the reclining seats; said arm rests remaining fixed when the reclining seats move from one to another of their positions,
> and *a pair of control means,* one for each reclining seat; *mounted on the double reclining seat sofa section.* . . .

Id. at col. 4, line 68 to col. 5, ll. 1-27 (emphasis added to most relevant claim language). Claims 9, 10, 12-15, and 19-21 are directed to a sectional sofa in which the control means are specifically located on the console.

In 1991, Gentry filed suit . . . alleging that Berkline infringed the patent by manufacturing and selling sectional sofas having two recliners facing in the same direction. In the allegedly infringing sofas, the recliners were separated by a seat which has a back cushion that may be pivoted down onto the seat, so that the seat back may serve as a tabletop between the recliners. . . . The district court granted Berkline's motion for summary judgment of non-infringement, but denied its motions for summary judgment of invalidity and unenforceability. In construing the language "fixed console," the court relied on, *inter alia,* a statement made by the inventor named in the patent, James Sproule, in a Petition to Make Special (PTMS). See 37 C.F.R. § 1.102 (1997). Sproule had attempted to distinguish his invention from a prior art reference by arguing that the reference, U.S. Patent 3,877,747 to Brennan *et al.* ("Brennan"), "shows a complete center seat with a tray in its back." *Gentry I,* 30 U.S.P.Q.2d at 1137. Based on Sproule's argument, the court concluded that, as a matter of law, Berkline's sofas "contain[] a drop-down tray identical to the one employed by the Brennan product" and therefore did not have a "fixed console" and did not literally infringe the patent. Id. The court held that Gentry was also "precluded from recovery" under the doctrine of equivalents. Id. at 1138.

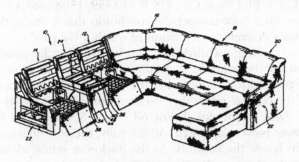

FIGURE 3-2
Patent drawing for the Gentry '244 patent.

Invalidity . . .

Berkline . . . argues that claims 1-8, 11, and 16-18 are invalid because they are directed to sectional sofas in which the location of the recliner controls is not limited to the console. According to Berkline, because the patent only describes sofas having controls on the console and an object of the invention is to provide a sectional sofa "with a console . . . that accommodates the controls for both the reclining seats," '244 patent, col. 1, ll. 35-37, the claimed sofas are not described within the meaning of § 112, ¶1. Berkline also relies on Sproule's testimony that "locating the controls on the console is definitely the way we solved it [the problem of building sectional sofa with parallel recliners] on the original group [of sofas]." Gentry responds that the disclosure represents only Sproule's preferred embodiment, in which the controls are on the console, and therefore supports claims directed to a sofa in which the controls may be located elsewhere. Gentry relies on Ethicon Endo-Surgery, Inc. v. United States Surgical Corp., 93 F.3d 1572, 1582 n.7 (Fed. Cir. 1993), and In re Rasmussen, 650 F.2d 1212, 1214 (CCPA 1981), for the proposition that an applicant need not describe more than one embodiment of a broad claim to adequately support that claim.

We agree with Berkline that the patent's disclosure does not support claims in which the location of the recliner controls is other than on the console. Whether a specification complies with the written description requirement of § 112, ¶1, is a question of fact, which we review for clear error on appeal from a bench trial. See Vas-Cath Inc. v. Mahurkar, 935 F.2d 1555, 1563 (Fed. Cir. 1991). To fulfill the written description requirement, the patent specification "must clearly allow persons of ordinary skill in the art to recognize that [the inventor] invented what is claimed." In re Gosteli, 872 F.2d 1008, 1012 (Fed. Cir. 1989). An applicant complies with the written description requirement "by describing *the invention,* with all its claimed limitations." Lockwood v. American Airlines, Inc., 107 F.3d 1565, 1572 (Fed. Cir. 1997).

It is a truism that a claim need not be limited to a preferred embodiment. However, in a given case, the scope of the right to exclude may be limited by a narrow disclosure. For example, as we have recently held, a disclosure of a television set with a keypad, connected to a central computer with a video disk player did not support claims directed to "an individual terminal containing a video disk player." See id.

(stating that claims directed to a "distinct invention from that disclosed in the specification" do not satisfy the written description requirement); see also Regents of the Univ. of Cal. v. Eli Lilly & Co., 119 F.3d 1559, 1568 (Fed. Cir. 1997) (stating that the case law does "not compel the conclusion that a description of a species always constitutes a description of a genus of which it is a part").

In this case, the original disclosure clearly identifies the console as the only possible location for the controls. It provides for only the most minor variation in the location of the controls, noting that the control "may be mounted on top or side surfaces of the console rather than on the front wall . . . without departing from this invention." '244 patent, col. 2, line 68 to col. 3, line 3. No similar variation beyond the console is even suggested. Additionally, the only discernible purpose for the console is to house the controls. As the disclosure states, identifying the only purpose relevant to the console, "[a]nother object of the present invention is to provide . . . a console positioned between [the reclining seats] that accommodates the controls for both of the reclining seats." Id. at col. 1, 11. 33-37. Thus, locating the controls anywhere but on the console is outside the stated purpose of the invention. Moreover, consistent with this disclosure, Sproule's broadest original claim was directed to a sofa comprising, *inter alia*, "control means located upon the center console to enable each of the pair of reclining seats to move separately between the reclined and upright positions." Finally, although not dispositive, because one can add claims to a pending application directed to adequately described subject matter, Sproule admitted at trial that he did not consider placing the controls outside the console until he became aware that some of Gentry's competitors were so locating the recliner controls. Accordingly, when viewed in its entirety, the disclosure is limited to sofas in which the recliner control is located on the console.

Gentry's reliance on *Ethicon* is misplaced. It is true, as Gentry observes, that we noted that "an applicant . . . is generally allowed claims, when the art permits, which cover more than the specific embodiment shown." *Ethicon*, 93 F.3d at 1582 n.7 (quoting In re Vickers, 141 F.2d 522, 525 (CCPA 1944)). However, we were also careful to point out in that opinion that the applicant "was free to draft claim[s] broadly (within the limits imposed by the prior art) to exclude the lockout precise location as a limitation of the claimed invention" only because he "did not consider the precise location of the lockout to be an element of his invention." Id. Here, as indicated above, it is clear that Sproule considered the location of the recliner controls on the console to be an essential element of his invention. Accordingly, his original disclosure serves to limit the permissible breadth of his later-drafted claims.

Similarly, In re Rasmussen does not support Gentry's position. In that case, our predecessor court restated the uncontroversial proposition that "a claim may be broader than the specific embodiment disclosed in a specification." 650 F.2d at 1215. However, the court also made clear that "[a]n applicant is entitled to claims as broad as the prior art *and his disclosure* will allow." Id. at 1214 (emphasis added). The claims at issue in *Rasmussen*, which were limited to the generic step of "adheringly applying" one layer to an adjacent layer, satisfied the written description requirement only because "one skilled in the art who read [the] specification would understand that it is unimportant how the layers are adhered, so long as they are adhered." Here, on the contrary, one skilled in the art would clearly understand that it was not only important, but essential to Sproule's invention, for the controls to be on the console.

In sum, the cases on which Gentry relies do not stand for the proposition that an applicant can broaden his claims to the extent that they are effectively bounded only by the prior art. Rather, they make clear that claims may be no broader than the supporting disclosure, and therefore that a narrow disclosure will limit claim breadth. Here, Sproule's disclosure unambiguously limited the location of the controls to the console. Accordingly, the district court clearly erred in finding that he was entitled to claims in which the recliner controls are not located on the console. We therefore reverse the judgment that claims 1-8, 11, and 16-18, were not shown to be invalid.

COMMENTS AND QUESTIONS

1. Berkline's sofa recliners have the reclining controls on arm rests, seat edges, etc. See *www.berkline.com*. The district court stated that Gentry Gallery's amended claims were "'broad' but not unsupported." 939 F. Supp. 98, 105 (D. Mass. 1996). The district court stated that the invention had solved the "control quandary" of removing controls from arm rests and underneath seat cushions, and that this location of the controls on this central console was only one way of implementing the invention.

2. *Gentry Gallery* has been described as a case about an "omitted element," Reiffen v. Microsoft Corp., 48 U.S.P.Q.2d 1274 (N.D. Cal. 1998), but the Federal Circuit has rejected that characterization, see 213 F.3d 1342 (Fed. Cir. 2000). To those who favor this terminology, the case centers on the *omission* of an originally disclosed element from one or more final claims. See Benjamin Hattenbach, On Illuminating Black Holes in Patent Disclosures: Toward a Structured Approach to Identifying Omitted Elements under the Written Description Requirement of Patent Law, 38 Hous. L. Rev. 1195 (2001). There are two things to consider in this analysis. First, who suffers from the inventor's change of heart? In cases such as *Gentry Gallery,* if the amended claims had been allowed and found valid, the Berkline Corporation (the patentee's competitor, and the defendant in the case) would have suffered. Berkline's product (a variant on the sofa design disclosed in Gentry's initial application) would be bounced from the market, perhaps to be replaced by an identical product from Gentry. This would amount to an appropriation of Berkline's variation by the original applicant, Gentry. Before you conclude that this is unfair, however, recall that it is only a trivial variant on Gentry's bona fide invention. Put another way, Gentry had on file in the U.S. Patent Office a document that taught (enabled) skilled artisans enough to arrive at this variant with minimal effort. Keeping this in mind may temper your sense of the wrongfulness of Gentry's appropriation. As a final consideration, recall that as far as Berkline is concerned, it was purely fortuitous in the actual case that Gentry had not "adequately described" Berkline's variation. There is no way that Berkline could have found out what precisely constituted Gentry's initial disclosure. (Why not?) Thus there is nothing akin to a "reliance" argument in favor of Berkline. But just because there is nothing that Berkline could have done to avoid Gentry's appropriation does not mean that the appropriation is rightful — does it?

3. Does it make sense to have both an enablement requirement and a written description requirement? If the inventor must teach one of ordinary skill in the art how to make and use the invention, why require them to write down the exact

invention as well? Cf. Moba v. Diamond Automation, Inc., 325 F.3d 1306, 1323 (Fed. Cir. 2003) (Rader, J., dissenting) ("Under Federal Circuit case law, [the plaintiff in this case] asked the jury to decide that the patent's disclosure can enable a skilled artisan to make and practice the entire invention, but still not inform that same artisan that the inventor was in possession of the invention. Perplexing.").

4. The *Gentry Gallery* case has been robustly criticized. See, e.g., Harris A. Pitlick, The Mutation on the Description Requirement Gene, 80 J. Pat. & Trademark Off. Soc'y 209, 222 (1998) (recent cases are "an unmitigated disaster"); Laurence H. Pretty, The Recline and Fall of Mechanical Genus Claim Scope Under "Written Description" in the Sofa Case, 80 J. Pat. & Trademark Off. Soc'y 469 (1998).

Note on "Written Description" and Biotechnology

The application of the written description requirement to biotechnology claims has been a particularly controversial issue. A number of cases beginning in the late 1990s required inventors to disclose specific gene sequences in order to claim them, even when functional properties of a gene (such as the protein it codes for) were already known. See Janice M. Mueller, The Evolving Application of the Written Description Requirement to Biotechnological Inventions, 13 Berkeley Tech. L.J. 615 (1998). For example, consider Fiers v. Revel, 984 F.2d 1164, 1170-71 (Fed. Cir. 1993). Fiers, Revel, and Sugano all claimed to be the first to have isolated the DNA sequence coding for human β-interferon, a protein with potential use in the treatment of cancer. The Federal Circuit had to determine which of the three had the first claim on the invention, since only that party could obtain a patent.[42] The court rejected Revel's claim to priority on the grounds that he had not listed the actual human DNA sequence in his patent application:

> An adequate written description of a DNA requires more than a mere statement that it is part of the invention and reference to a potential method for isolating it; what is required is a description of the DNA itself. Revel's specification does not do that.... A bare reference to a DNA with a statement that it can be obtained by reverse transcription is not a description; it does not indicate that Revel was in possession of the DNA.... [S]uch a disclosure just represents a wish, or arguably a plan, for obtaining the DNA.

Perhaps the most controversial in this line of cases was Regents of the University of California v. Eli Lilly & Co., 119 F.3d 1559 (Fed. Cir. 1997). In the *Eli Lilly* case, University of California researchers filed a patent application disclosing the DNA sequence for rat insulin. The final version of the patent claimed rat, mammalian, vertebrate, and human insulin. These are very closely related variants; human insulin differs from the rat version by only twelve (out of more than 300) DNA base pairs that code for the protein. As a result, scientists could quite easily use the rat insulin DNA sequence to construct a gene probe that would enable them to find human insulin. Cf. Michael Delmas Plimier, Genentech, Inc. v. Novo Nordisk & University

---

42. Indeed, that party would also obtain the right to exclude the other two from practicing the invention, even though all three were researching in the area contemporaneously. See *infra* section C.

of California v. Eli Lilly and Co., 13 Berkeley Tech. L.J. 149 (1998). Nonetheless, the court invalidated the claim to human insulin on the grounds that the specification did not meet the written description requirement. The court cited contemporary cases holding that disclosure of an amino acid sequence for a protein, together with well-known techniques for identifying a DNA sequence from the amino acids, did not render the resulting sequence obvious under § 103. Because even obvious variants of disclosed material might not be within the written description of a patent specification, it was clear to the court that nonobvious variants (under these cases) were also not adequately described. For more background and analysis, see the Plimier article.

Some defended the decision in *Eli Lilly* on the grounds that it prevented anticipatory patenting. See, e.g., Mark J. Stewart, Note, The Written Description Requirement of 35 U.S.C. § 112(1): The Standard After Regents of the University of California v. Eli Lilly & Co., 32 Ind. L. Rev. 537, 563-64 (1999):

> [I]t would seem that the holding in *Lilly*... avoided a disaster that would have crippled the biotechnology industry.... [T]he Federal Circuit has decided that the uniqueness of biotechnology inventions claiming DNA sequences requires the application of a stringent written description requirement to protect the public from inventors seeking to slow the pace of research by preempting future developments before they arrive.

And it has influenced a number of important subsequent cases. See, e.g., Noelle v. Lederman, 355 F.3d 1343, 1349 (Fed. Cir. 2004) (finding that the human antibody there at issue was not adequately described by the structure and function of the mouse antigen that had been described by the patentee); University of Rochester v. G.D. Searle & Co., 358 F.3d 916, 925-26 (Fed. Cir. 2004) (holding that the patentee's description of a new enzyme, the COX-2 enzyme, did not serve to describe unknown compounds capable of selectively inhibiting the enzyme; and therefore invalidating broad claims to any compound that could selectively inhibit the described enzyme).

From the outset, however, *Lilly* and progeny have had their critics. See, e.g., Mueller, The Evolving Application of the Written Description Requirement to Biotechnological Inventions, *supra*; Mark Janis, On Courts Herding Cats: Contending with the "Written Description Requirement" (And Other Unruly Patent Disclosure Doctrines), 2000 Wash. U. J. L. & Pol'y 55 (2000). To judge from more recent cases, some of those critics now sit on the Federal Circuit itself. See Moba v. Diamond Automation, Inc., 325 F.3d 1306, 1323-24 (Fed. Cir. 2003) (Rader, J., dissenting) (questioning the statutory basis and need for the written description requirement separate from enablement). The court has since backed off considerably from its early insistence that inventors obtain detailed characterizations of biological materials before claiming them. One important step along the way was the case of Enzo Biochem, Inc. v. Gen-Probe Inc., 323 F.3d 956 (Fed. Cir. 2002), vacating earlier opinion at 285 F.3d 1013 (Fed. Cir. 2002). In *Enzo*, a patentee claimed nucleic acid probes that selectively hybridize to the bacteria that cause gonorrhea. At the time the application was filed, the patentee had placed on deposit with a public depository several versions of the nucleotides claimed, but had not obtained the actual sequence information of those nucleotides. The Federal Circuit, after an initial opinion finding the claims invalid, issued a revised opinion holding that the claims did satisfy the written description requirement, by virtue of the public deposit. Harold C. Wegner, The Disclosure Requirements of the 1952 Patent Act: Looking Back and a New Statute for the Next Fifty Years, 37 Akron L. Rev. 243, 248 (2004) (criticizing written description case, and *Enzo* in particular as "judicial legislation" with no foundation in the history of the 1952 Patent Act).

The *Enzo* opinion was followed by Amgen Inc. v. Hoechst Marion Roussel, Inc. 314 F.3d 1313 (Fed. Cir. 2003), in which the court refused to invalidate broad claims to the production of erythropoietin or EPO. Although Amgen had produced EPO using hamster and monkey cell lines, it claimed protein expression in all vertebrate and mammalian cells. Nonetheless, the court found the written description adequate to support the claims. It distinguished *Eli Lilly* on the ground that what was claimed there was a broad genus of new, previously uncharacterized DNA sequences, whereas Amgen claimed a broad class cell types used in the production of a novel protein (EPO). Id., at 1332.

Then came Capon v. Eshhar, 418 F.3d 1349 (Fed. Cir. 2005). In *Capon* a patent interference was declared between two groups of inventors claiming a new type of chimeric genes: genes designed to enhance the human body's immune response by providing cells with specific cell-surface antibodies in a form that can penetrate diseased sites, such as solid tumors. The invention claimed by both groups combined (1) DNA coding for cell surface receptors with (2) DNA for certain proteins that are highly effective at penetrating and destroying tumors, called lymphocyte signaling proteins. The Board of Appeals and Patent Interferences held that neither party's specification listed the structure, formula, chemical name, or physical properties of the DNA claimed. Hence under cases such as *Eli Lilly*, the Board found that neither of the two parties to the interference had satisfied the written description requirement. The Federal Circuit reversed, further distancing written description doctrine from the *Eli Lilly* case:

> For the chimeric genes of the Capon and Eshhar inventions, the law must take cognizance of the scientific facts. The Board erred in refusing to consider the state of the scientific knowledge, as explained by both parties, and in declining to consider the separate scope of each of the claims. None of the cases to which the Board attributes the requirement of total DNA re-analysis, i.e., *Regents v. Lilly, Fiers v. Revel, Amgen,* or *Enzo Biochem,* require a re-description of what was already known. In *Lilly,* the cDNA for human insulin had never been characterized. Similarly in *Fiers,* much of the DNA sought to be claimed was of unknown structure, whereby this court viewed the breadth of the claims as embracing a "wish" or research "plan." In *Amgen,* the court explained that a novel gene was not adequately characterized by its biological function alone because such a description would represent a mere "wish to know the identity" of the novel material. In *Enzo Biochem,* this court reaffirmed that deposit of a physical sample may replace words when description is beyond present scientific capability....The "written description" requirement states that the patentee must describe the invention; it does not state that every invention must be described in the same way. As each field evolves, the balance also evolves between what is known and what is added by each inventive contribution.... The Board erred in holding that the specifications do not meet the written description requirement because they do not reiterate the structure or formula or chemical name for the nucleotide sequences of the claimed chimeric genes.

Slip op. 14-15.

NOTES AND COMMENTS

1. To what extent do these arguments result in a consideration of the relative merits of (and incentives facing) original "pioneer" inventors and subsequent improvers? Cf. Mark A. Lemley, The Economics of Improvement in Intellectual Property Law, 75 Tex. L. Rev. 989 (1997); Jerry R. Green & Suzanne Scotchmer, On the Division of Profit in Sequential Innovation, 26 Rand J. Econ. 20 (1995); Robert P. Merges & Richard R. Nelson, On the Complex Economics of Patent Scope, 90 Colum. L. Rev. 839 (1990).

2. One possible benefit of the written description requirement is to preclude patent owners from later claiming what they did not think of at the time they filed their applications. This is a particularly significant issue when the patent does not issue until long after the application is filed. In Hyatt v. Boone, 47 U.S.P.Q.2d 1128 (Fed. Cir. 1998), the court held that Hyatt's patent on a microprocessor did not deserve the benefit of its earliest filing date because the original application did not contain a written description of the invention ultimately claimed. Compare In re Hogan, 559 F.2d 595, 606 (C.C.P.A. 1977) (refusing to find defect in written description of first-filed application where art continued to develop and applicant updated disclosure accordingly):

> The PTO has not challenged appellants' assertion that their 1953 application enabled those skilled in the art in 1953 to make and use "a solid polymer" as described in claim 13. Appellants disclosed, as the only then existing way to make such a polymer, a method of making the crystalline form. To now say that appellants should have disclosed in 1953 the amorphous form which on this record did not exist until 1962, would be to impose an impossible burden on inventors and thus on the patent system. There cannot, in an effective patent system, be such a burden placed on the right to broad claims. To restrict appellants to the crystalline form disclosed, under such circumstances, would be a poor way to stimulate invention, and particularly to encourage its early disclosure. To demand such restriction is merely to state a policy against broad protection for pioneer inventions, a policy both shortsighted and unsound from the standpoint of promoting progress in the useful arts, the constitutional purpose of the patent laws.

Note on the Best Mode Requirement

In addition to the written description and enablement requirements, section 112 also mandates that the patent disclose the "best mode" of carrying out the invention contemplated by the inventor. This "best mode" requirement is designed to prevent a patentee from "holding back" knowledge from the public, in effect maintaining part of the invention as a trade secret while protecting the whole under patent law. Indeed, one opinion describes the best mode requirement as effectuating "a statutory bargained-for-exchange by which a patentee obtains the right to exclude others from practicing the claimed invention for a certain time period, and the public receives knowledge of the preferred embodiments for practicing the claimed invention." Eli Lilly & Co. v. Barr Laboratories, Inc. 251 F.3d 955, 963 (Fed. Cir. 2001).

The Federal Circuit described the best mode cases in detail in Bayer AG v. Schein Pharmaceuticals, Inc., 301 F.3d 1306 (Fed. Cir. 2002). *Bayer* reviews prior case law and concludes that best mode violations are found where there is either a "failure to disclose a preferred embodiment, or else failure to disclose a preference that materially affected making or using the invention." As the *Bayer* case notes, there are two essential elements to the doctrine: (1) The factfinder must determine whether, at the time the patent application was filed, the inventor *had* a best mode of practicing the claimed invention—whether there was a preferred way to build or use the claimed device or process. If the inventor him or herself had no preferred way, there cannot be a best mode violation. (This is an entirely subjective inquiry, turning on the inventor's belief in the existence of a best mode.) The actual extent of information that an inventor must disclose depends on the scope of the claimed invention. (2) Once it is established that the inventor actually contemplated a best mode, a court will inquire whether "the disclosure [in the patent specification] is adequate to enable one of ordinary skill in the art to practice the best mode of the invention. This inquiry is objective and depends upon the scope of the claimed invention and the level of skill in the relevant art." Id. at 10, quoting from N. Telecom Ltd. v. Samsung Elec. Co., 215 F.3d 1281, 1286 (Fed. Cir. 2000).

One important issue that is sometimes litigated is the question of whether an inventor must update his or her best mode disclosure during patent prosecution. In general, if the material in a continuation application is "common subject matter" with that of the original application, the inventor need not have updated his best mode disclosure in the continuation application. See Transco v. Performance Contracting, Inc., 38 F.3d 551 (Fed. Cir. 1994). An inventor has the obligation to disclose a best mode in a continuation application only if the claim feature associated with that best mode first appeared or first received adequate written description in the continuation.

COMMENTS AND QUESTIONS

1. On the sometimes difficult decisions practitioners are called upon to make regarding best mode disclosures, see William F. Heinze, A Risk-Balancing Approach to Best Mode Disclosure in Software Patent Applications, 84 J. Pat. & Trademark Off. Soc'y 40, 41 (2002) (hypothetical question from a client: "Isn't there some way to include the essence of the code without making it so easy for my competitors to copy?").

PROBLEM

Problem 3-7. Scientists involved in the field of spectroscopy were being haunted by ghosts. Spectroscopy is the study and analysis of materials to determine their components and molecular structure. One form of spectroscopy, magnetic resonance spectroscopy, works by observing a material's reaction to imposed radiation. A spectroscope bombards the sample material (usually contained in a test tube) with pulses of radio frequency radiation and then records and analyzes the material's response. If the signal response given by the material was weak, the spectroscope would often confuse interfering noise and static for the material's response. Scientists attempted to solve the problem of unwanted frequencies (or "ghost" frequencies) by various mathematical manipulations.

One such scientist was Daniel Murray, a doctoral candidate studying at Oxford under the supervision of Sir William Akeroyd. In the course of his experiments, Murray arrived at a means of "ghostbusting" through shifting the phase of the frequency transmitter at 90 degree angles between pulses, thus avoiding sophisticated manipulations of data. A patent application embodying the Murray invention was filed in the United Kingdom on April 8, 1996, and in the United States on December 21, 1998. Murray secured patent protection in both countries by 2001.

In the spring of 1995, Akeroyd attended a scientific meeting in Boulder, Colorado. On a bus ride during the conference, Akeroyd engaged Dr. Christopher Squire in an informal conversation. Squire headed a research team at the U.S.-based Monsanto Company that developed related technology for use in the company's chemical analyses. Akeroyd informed Squire of the promising results of Murray's preliminary research on the phase shifting spectroscope. By mid-1997, Squire had updated his spectrometer at Monsanto to include a similar phase shifting transmitter.

Squire immediately applied his improved spectrometer technology to assess whether a particular Monsanto herbicide was safe for release into the environment. This was a project Squire's team had been working on for a full year prior to the improvement in the lab's spectrometer. Excited by the improved results, Squire invited several of his superiors and fellow employees into the lab to observe his data.

Is Squire's use a public use under section 102(b)? Does your analysis depend on any of the following hypothetical facts:

- Dr. Squire was the only person who was permitted access to the Monsanto lab where the updated spectrometer was located.
- Anyone *could* have entered the lab, but in fact Squire was the only one who did.
- The only person who set eyes upon the updated spectrometer besides Squire was a Monsanto custodian.
- Squire left the door unlocked regularly at night.
- The phase shifting transmitter is located deep inside the apparatus, and cannot be seen without dismantling the spectrometer.

4. Novelty and Statutory Bars

Section 102 of the Patent Act embodies the principle that only truly new inventions deserve patents. In the technical analysis that has been developed to interpret and apply this principle, two distinct requirements are discernible: novelty and statutory bars. Novelty means "new compared to the prior art"; it states the requirement that, to be patentable, an invention must be somehow different from all published articles, known techniques, and marketed products. Statutory bar means "a bar to patentability based on too long a delay in seeking patent protection." In most cases, the bar arises because of something the inventor herself does. The classic example is publication of a scientific article. If an inventor fails to file a patent application within a year of the article's publication, she is barred from receiving a patent. She has suffered a "loss of right," in the words of the statute.

One key difference between novelty and statutory bars is that generally speaking an inventor's own work cannot destroy the novelty of her invention, but an inventor can create a statutory bar by her own actions (e.g., publication of an article). This distinction follows from the fact that novelty is measured from the date of *invention* whereas statutory bars are a function of the date a patent application is *filed*. It would be anomalous to say that an inventor did not invent first because she herself invented earlier; it is perfectly logical to say that an inventor is barred from receiving a patent because she herself filed an application more than a year after she published her invention.

Major portions of the text of section 102 are printed below, with bracketed insertions. Note that section 102(a) is concerned with novelty, while the statutory bars are set forth in section 102(b).

35 U.S.C. § 102. Conditions for Patentability; Novelty and Loss of Right to Patent

A person shall be entitled to a patent unless —

(a) [**Novelty**] the invention was known or used by others in this country, or patented or described in a printed publication in this or a foreign country, before the invention thereof by the applicant for patent, or

(b) [**Statutory Bars**] the invention was patented or described in a printed publication in this or a foreign country or in public use or on sale in this country, more than one year prior to the date of the application for patent in the United States, or

[(c) and (d) omitted]

(e) [**Secret Prior Art: Previously-filed applications**] The invention was described in (1) [a published patent application] by another filed in the United States before the invention by the applicant for patent . . . or (2) a patent granted on an application for patent by another filed in the United States before the invention thereof by the applicant for patent . . . or

(f) [**Derivation**] he did not himself invent the subject matter sought to be patented, or

(g)(1) during the course of an interference . . . , another inventor involved therein establishes, to the extent permitted in section 104 . . . that before such person's invention thereof the invention was made by such other inventor and not abandoned, suppressed, or concealed, or (2) before such person's invention thereof, the invention was made in this country by another inventor who had not abandoned, suppressed, or concealed it. In determining priority of invention under this subsection, there shall be considered not only the respective dates of conception and reduction to practice of the invention, but also the reasonable diligence of one who was first to conceive and last to reduce to practice, from a time prior to conception by the other.

a. The Nature of Novelty

Rosaire v. National Lead Co.
United States Court of Appeals for the Fifth Circuit
218 F.2d 72 (5th Cir. 1955), cert. denied, 349 U.S. 916 (1955)

TUTTLE, Circuit Judge.

In this suit for patent infringement there is presented to us for determination the correctness of the judgment of the trial court, based on findings of fact and conclusions of law, holding that the two patents involved in the litigation were invalid and void and that furthermore there had been no infringement by defendant.

The Rosaire and Horvitz patents relate to methods of prospecting for oil or other hydrocarbons. The inventions are based upon the assumption that gases have emanated from deposits of hydrocarbons which have been trapped in the earth and that these emanations have modified the surrounding rock. The methods claimed involve the steps of taking a number of samples of soil from formations which are not themselves productive of hydrocarbons, either over a horizontal area or vertically down a well bore, treating each sample, as by grinding and heating in a closed vessel, to cause entrained or absorbed hydrocarbons therein to evolve as a gas, quantitatively measuring the amount of hydrocarbon gas so evolved from each sample, and correlating the measurements with the locations from which the samples were taken.

Plaintiff claims that in 1936 he and Horvitz invented this new method of prospecting for oil. In due course the two patents in suit, Nos. 2,192,525 and 2,324,085, were issued thereon. Horvitz assigned his interest to Rosaire.

In view of the fact that the trial court's judgment that the patents were invalid, would of course dispose of [this infringement suit] if correct, we turn our attention to this issue. [Appellee argues] that work carried on by one Teplitz for the Gulf Oil Corporation invalidated both patents by reason of the relevant provisions of the patent laws which state that an invention is not patentable if it "was known or used by others in this country" before the patentee's invention thereof, 35 U.S.C.A. § 102(a). Appellee contends that Teplitz and his coworkers knew and extensively used in the field the same alleged inventions before any date asserted by Rosaire and Horvitz.

On this point appellant himself in his brief admits that "Teplitz conceived of the idea of extracting and quantitatively measuring entrained or absorbed gas from the samples of rock, rather than relying upon the free gas in the samples. We do not deny that Teplitz conceived of the methods of the patents in suit." And further appellant makes the following admission: "We admit that the Teplitz-Gulf work was done before Rosaire and Horvitz conceived of the inventions. We will show, however, that Gulf did not apply for patent until 1939, did not publish Teplitz's ideas, and did not otherwise give the public the benefit of the experimental work."

The question as to whether the work of Teplitz was "an unsuccessful experiment," as claimed by appellant, or was a successful trial of the method in question and a reduction of that method to actual practice, as contended by appellee, is, of course, a question of fact. On this point the trial court made the following finding of fact:

> I find as a fact that Abraham J. Teplitz and his coworkers with Gulf Oil Corporation and its Research Department during 1935 and early 1936, before any date claimed by

Rosaire, spent more than a year in the oil fields and adjacent territory around Palestine, Texas, taking and analyzing samples both over an area and down drill holes, exactly as called for in the claims of the patents which Rosaire and Horvitz subsequently applied for and which are here in suit. This Teplitz work was a successful and adequate field trial of the prospecting method involved and a reduction to practice of that method. The work was performed in the field under ordinary conditions without any deliberate attempt at concealment or effort to exclude the public and without any instructions of secrecy to the employees performing the work.

As we view it, if the court's findings of fact are correct then under the statute as construed by the courts, we must affirm the finding of the trial court that appellee's patents were invalid.

[T]here was sufficient evidence to sustain the finding of the trial court that there was more here than an unsuccessful or incomplete experiment. It is clear that the work was not carried forward, but that appears to be a result of two things: (1) that the geographical area did not lend itself properly to the test, and (2) that the "entire gas prospecting program was therefore suspended in September of 1936, in order that the accumulated information might be thoroughly reviewed." It will be noted that the program was not suspended to test the worth of the method but to examine the data that was produced by use of the method involved.

With respect to the argument advanced by appellant that the lack of publication of Teplitz's work deprived an alleged infringer of the defense of prior use, we find no case which constrains us to hold that where such work was done openly and in the ordinary course of the activities of the employer, a large producing company in the oil industry, the statute is to be so modified by construction as to require some affirmative act to bring the work to the attention of the public at large.

While there is authority for the proposition that one of the basic principles underlying the patent laws is the enrichment of the art, and that a patent is given to encourage disclosure of inventions, no case we have found requires a holding that, under the circumstances that attended the work of Teplitz, the fact of public knowledge must be shown before it can be urged to invalidate a subsequent patent.

COMMENTS AND QUESTIONS

1. In New Railhead Mfg., L.L.C. v. Vermeer Mfg. Co., 298 F.3d 1290 (Fed. Cir. 2002), the Federal Circuit considered a similar fact situation. Patentee New Railhead had tested a drill bit in underground drilling which occurred before the critical date for its patent. In response to Vermeer's argument that its use was either non-public or experimental, the court followed the reasoning of *Rosaire*, and invalidated the patent. Judge Dyk took issue in a dissent. See 298 F.3d 1290, 1299, at 1300 (Dyk. J., dissenting) ("The use actually took place under public land, hidden from view, and there has been no showing whatsoever that the use was anything but confidential. In order to understand the method of using the drill bit a person at the job site would have to view the drill bit or see it in operation, and this was impossible to do while the drill bit was underground."). In the earlier case of W. L. Gore & Assocs., Inc. v. Garlock, Inc., 721 F.2d 1540, 1548 (Fed. Cir. 1983), *cert. denied*, 469 U.S. 851 (1984), the Federal Circuit had this to say about "secret" prior use,

where patentee Gore argued that prior use by another of a machine conforming to the elements of Gore's claim was nonpublic and therefore nonanticipatory:

> The nonsecret use of a claimed process in the usual course of producing articles for commercial purposes is a public use. Electric Storage Battery Co. v. Shimadzu, 307 U.S. 5, 20 (1939).... Thus it cannot be said that the district court erred in determining that the invention set forth in claim 1 of the '566 patent was known or used by others under §102(a), as evidenced by...operation of the '401 machine before Dr. Gore's asserted date of that invention.

2. In Woodland Trust v. Flowertree Nursery, Inc., 148 F.3d 1368 (Fed. Cir. 1998), defendant Flowertree argued the invalidity of the patent owned by Woodland based on prior use of a nursery watering system identical to that claimed in the patent:

> Under §102(a), Flowertree stated that the method was previously known and used by each of Joseph Burke and William Hawkins (an owner of Flowertree) in the 1960s and 1970s at their nurseries in Florida, and was then discontinued by these users in 1976 and 1978. The district court found that Hawkins' system at Flowertree was reconstructed in 1988.
>
> Four witnesses testified in support of the defense of prior knowledge and use: Mark Hawkins, Joseph Burke, Charles Hudson, and John Kaufmann. Mark Hawkins is the son of William Hawkins; he testified that his father's system, on which he worked as a child, was destroyed by a tornado in 1978, and was not reconstructed until 1988. Joseph Burke is a nursery owner who has known William Hawkins since the 1960s; he testified that he used the same system as shown in the patent, but tore it down in 1976 and did not rebuild it. Charles Hudson is a nursery owner who had worked for Joseph Burke, and John Kaufmann is a life-long friend of William Hawkins; they testified that they observed the patented system at the Burke or Hawkins nursery, before its use was discontinued.

148 F.3d at 1369. The Federal Circuit reviewed Supreme Court precedent laying down a stringent corroboration requirement, and then concluded:

> This guidance, applied to this case, reinforces the heavy burden when establishing prior public knowledge and use based on long-past events. The Supreme Court's view of human nature as well as human recollection, whether deemed cynical or realistic, retains its cogency. This view is reinforced, in modern times, by the ubiquitous paper trail of virtually all commercial activity. It is rare indeed that some physical record (e.g., a written document such as notes, letters, invoices, notebooks, or a sketch or drawing or photograph showing the device, a model, or some other contemporaneous record) does not exist.
>
> In this case, despite the asserted many years of commercial and public use, we take note of the absence of any physical record to support the oral evidence. The asserted prior knowledge and use by Hawkins and Burke was said to have begun approximately thirty years ago and to have continued for about a decade. Hawkins testified that his prior use was terminated in 1978, and the district court found that Hawkins' system was not reconstructed until 1988. The relationship of the witnesses and the fact that the asserted prior uses ended twenty years before the trial, and were abandoned until the defendant reportedly learned of the patentee's practices, underscore the failure of this oral evidence to provide clear and convincing evidence of prior knowledge and use. The district court did not rely on the two undated photographs, and indeed their lack of detail and clarity can not have provided documentary support.

With the guidance of precedent, whose cautions stressed the frailty of memory of things long past and the temptation to remember facts favorable to the cause of one's relative or friend, we conclude that this oral evidence, standing alone, did not provide the clear and convincing evidence necessary to invalidate a patent on the ground of prior knowledge and use under § 102(a).

Id. at 1373.

Do you agree with the Federal Circuit's implication: that in the absence of a strong corroboration requirement for prior public use evidence, there would be a temptation for competitors of a patentee to lie? With such a strong requirement of documentary evidence, is there much difference between public use evidence and proof that an invention appeared in a prior publication or patent?

3. The very expansive view taken in *Rosaire* of what it means for a disclosure to be "public" was criticized by one scholar:

> The term "public" ... seems merely to mean "not secret." It is unnecessary to show that the previous discovery was ever used commercially, that it was in fact observable by the public if such a process or device would not normally be so viewed, or that it was known to more than a few persons. This construction of the term "public" seems questionable, since it may result in the denial of a patent even though the subsequent inventor has conferred a benefit by filing the invention with the public records.

Comment, Prior Art in the Patent Law, 73 Harv. L. Rev. 369, 373 (1959). The author goes on to propose a "higher standard of knowledge or use," arguing that

> an invention should be considered "known or used" only if it was so widely known or used that an ordinary skilled worker exercising reasonable diligence to learn the state of the art would have discovered, recognized, and been able to construct the invention.

Id. at 373. In effect, this proposal would apply a trade secret standard of novelty to the patent law — an invention could be "new" even though it had been made before, as long as it was not in general knowledge or use. Are there any problems with applying this weaker standard of novelty to determine whether a patent should be issued? Can it be reconciled with the broad language of section 102(a), precluding patents on inventions that were previously "known or used by others in this country"?

In Hall v. MacNeale, 107 U.S. (17 Otto) 90 (1883), Hall had in 1866 received a patent on an improved design for the door and walls of burglar-proof safes. The question before the Court was whether the use of certain earlier-model safes before the critical date was *public* enough to create a statutory bar. The Court held:

> The construction and arrangement and purpose and mode of operation and use of the [hidden feature] in the safes were necessarily known to the workmen who put them in. They were, it is true, hidden from view, after the safes were completed, and it required a destruction of the safe to bring them into view. But this was no concealment of them or use of them in secret. They had no more concealment than was inseparable from any legitimate use of them.

4. One issue in *Rosaire* was the patentee's allegation that the Teplitz work was not prior art because it was incomplete, a mere "abandoned experiment." On this

issue, consider Picard v. United Aircraft Corp., 128 F.2d 632 (2d Cir. 1942), *cert. denied*, 317 U.S. 651 (1942) (L. Hand, J.):

> It is true that another's experiment, imperfect and never perfected, will not serve either as an anticipation or as part of the prior art, for it has not served to enrich it. The patented invention does not become "known" by such a use or sale, or by anything of which the art cannot take hold and make use as it stands. But the mere fact that an earlier "machine" or "manufacture," sold or used, was an experiment does not prevent its becoming an anticipation or a part of the prior art, provided it was perfected and thereafter became publicly known. Whether it does become so depends upon how far it becomes a part of the stock of knowledge of the art in question. Judged by that standard, the Curtiss engine [prior art reference] was not an "abandoned experiment"; it had been perfected; it had withstood a severer test than was necessary in use; it had been sold; it remained permanently accessible to the art, a contribution to the sum of knowledge so far as it went.

This issue would likely arise today under a different rubric: was the prior art reference "enabled," or capable of teaching one skilled in the art its details. See, e.g., In re Paulsen, 30 F.3d 1475 (Fed. Cir. 1994).

Note on the Inherency Doctrine

Section 102(a) provides that an applicant is not entitled to a patent if her invention was "known or used by others" prior to the date of the applicant's invention. The meaning of this phrase has been called into question in a series of cases involving unintended, "accidental" anticipation of an invention. Most of these cases involve the inherent, unintended production of a particular physical product. When an inventor later *intentionally* makes the product, presumably because she has some use for it, the prior unintended production of the product may be raised as prior art to the invention.

On the one hand, a venerable line of cases holds that where the first, accidental producer was not aware of the product and did not attempt to produce it, the first production did not bar a patent on the "invention" of the product. Thus, in Tilghman v. Proctor, 102 U.S. 707 (1880), the Supreme Court held that the accidental separation of fat acids from tallow during operation of a steam engine lubricated by tallow did not anticipate Tilghman's patent for a similar separation process. Of the prior production, the Court said:

> They revealed no process for the manufacture of fat acids. If the acids were accidentally and unwittingly produced, whilst the operators were in pursuit of other and different results, without exciting attention and without it even being known what was done or how it had been done, it would be absurd to say that this was an anticipation of Tilghman's discovery.

Id. at 711. See also Abbott Labs v. Geneva Pharmaceuticals, Inc., 182 F.3d 1315 (Fed. Cir. 1999) (distinguishing *Tilghman* in case where prior art product was appreciated and sold, even though party did not appreciate qualities of product). On the other hand, an equally venerable line of cases holds that if a product is known in the art already, an inventor cannot obtain a patent *on the product* merely by putting it to a new use, "even if the new result had not before been contemplated." Ansonia

Brass & Copper Co. v. Elec. Supply Co., 144 U.S. 11 (1892). The inventor could obtain a patent on the new process using that product, however. For more on this distinction, see Atlas Powder Co. v. Ireco, Inc., 190 F.3d 1342 (Fed. Cir. 1999).

In recent years, Federal Circuit cases have focused on whether an invention was present in the prior art in a way that provided a public benefit. The knowledge of those in the art has been de-emphasized as a factor. In Schering Corp. v. Geneva Pharms., Inc., 339 F.3d 1373, 1377 (Fed. Cir. 2003), the court invalidated a claim to a compound that is necessarily produced in the human body whenever a person ingests the allergy medicine Claritin — even though the patentee Schering-Plough argued that no one knew about the claimed product until Schering characterized it. Another case is Prima Tek II, L.L.C. v. Polypap, S.A.R.L., 412 F.3d 1284 (Fed. Cir. 2005), where the court invalidated a claim to a "potless" floral arrangement holder. The claim included, among other things, the elements of (1) a band to hold the floral arrangement; (2) a sheet of decorative material covering the floral arrangement; and (3) "means for forming a crimped portion in the sheet of material." 412 F.3d 1284, 1287 n. 1. Polypap, the accused infringer, argued that a published French patent application, the "Charrin reference," inherently anticipated the claim. The patentee Prima Tek countered with the argument that Charrin did not disclose the "crimping" limitation. The court agreed with Polypap, noting that in the drawing below Charrin clearly discloses a string along the top of the floral arrangement holder which necessarily creates crimping in the covering material:

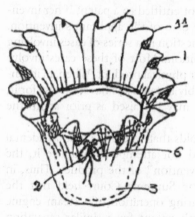

In other words, the invention was already present in the prior art, because the decorative material would be crimped whenever the Charrin device was used, whether or not people recognized or intended that result.See Dan L. Burk & Mark A. Lemley, Inherency, 47 Wm. & Mary L. Rev. (2005) (endorsing the move to public benefit, and making the point that if the PHOSITA knows about the prior use there is no need for an inherency doctrine at all).

b. Statutory Bars: Publications

Recall from the introduction to this section that one difference between statutory bars (which we cover in this subsection) and novelty is that an inventor can create a statutory bar by her own actions (e.g., publication of an article), whereas she cannot destroy the novelty of her own inventions. In short, an inventor's own work cannot be cited against her (in general) under section 102(a), but it is "fair game" under section 102(b).

Notwithstanding this fact, for many purposes the two provisions are not appreciably different. For example, the definition of "publication"—a term you will recall occurs in both sections 102(a) and (b)—is the same for both. Thus although the following case involves a section 102(b) reference, the reasoning is equally applicable to cases decided under section 102(a).

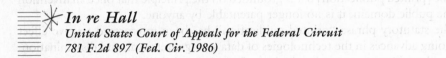

In re Hall
United States Court of Appeals for the Federal Circuit
781 F.2d 897 (Fed. Cir. 1986)

BALDWIN, Circuit Judge.

This is an appeal from the decision of the U.S. Patent and Trademark Office's (PTO) former Board of Appeals, adhered to on reconsideration by the Board of Patent Appeals and Interferences (board), sustaining the final rejection of claims 1-25 of [Hall's] reissue [a]pplication, based principally on a "printed publication" bar under 35 U.S.C. § 102(b). The reference is a doctoral thesis. Because appellant concedes that his claims are unpatentable if the thesis is available as a "printed publication" more than one year prior to the application's effective filing date of February 27, 1979, the only issue is whether the thesis is available as such a printed publication. On the record before us, we affirm the board's decision.

A protest was filed during prosecution of appellant's reissue application which included in an appendix a copy of the dissertation "1,4-α-Glucanglukohydrolase ein amylotylisches Enzym..." by Peter Foldi (Foldi thesis or dissertation). The record indicates that in September 1977, Foldi submitted his dissertation to the Department of Chemistry and Pharmacy at Freiburg University in the Federal Republic of Germany, and that Foldi was awarded a doctorate degree on November 2, 1977....

The examiner made a final rejection of the application claims. He said: "On the basis of the instant record it is reasonable to assume that the Foldi thesis was available (accessible) prior to February 27, 1979."[43]

By letter, the PTO's Scientific Library asked Dr. Will whether the Foldi dissertation was made available to the public by being cataloged and placed in the main collection. Dr. Will replied in an October 20, 1983 letter, as translated: "Our dissertations, thus also the Foldi dissertation, are indexed in a special dissertations catalogue, which is part of the general users' catalogue. In the stacks they are likewise set apart in a special dissertation section, which is part of the general stacks."

In response to a further inquiry by the PTO's Scientific Library requesting (1) the exact date of indexing and cataloging of the Foldi dissertation or (2) "the time such procedures normally take," Dr. Will replied in a June 18, 1984 letter: "The Library copies of the Foldi dissertation were sent to us by the faculty on November 4, 1977. Accordingly, the dissertation most probably was available for general use toward the beginning of the month of December, 1977."

43. The court presumably means February 27, 1978. — EDS.

The board held that the unrebutted evidence of record was sufficient to conclude that the Foldi dissertation had an effective date as prior art more than one year prior to the filing date of the appellant's initial application.

On appeal, appellant raises two arguments: (1) the § 102(b) "printed publication" bar requires that the publication be accessible to the interested public, but there is no evidence that the dissertation was properly indexed in the library catalog prior to the critical date; and (2) even if the Foldi thesis were cataloged prior to the critical date, the presence of a single cataloged thesis in one university library does not constitute sufficient accessibility of the publication's teachings to those interested in the art exercising reasonable diligence.

The [printed publication] bar is grounded on the principle that once an invention is in the public domain, it is no longer patentable by anyone.

The statutory phrase "printed publication" has been interpreted to give effect to ongoing advances in the technologies of data storage, retrieval, and dissemination. Because there are many ways in which a reference may be disseminated to the interested public, "public accessibility" has been called the touchstone in determining whether a reference constitutes a "printed publication" bar under 35 U.S.C. § 102(b). The § 102 publication bar is a legal determination based on underlying fact issues, and therefore must be approached on a case-by-case basis. The proponent of the publication bar must show that prior to the critical date the reference was sufficiently accessible, at least to the public interested in the art, so that such a one by examining the reference could make the claimed invention without further research or experimentation.

[A]ppellant argues that the Foldi thesis was not shown to be accessible because Dr. Will's affidavits do not say when the thesis was indexed in the library catalog and do not chronicle the procedures for receiving and processing a thesis in the library.

[A]ppellant would have it that accessibility can only be shown by evidence establishing a *specific* date of cataloging and shelving before the critical date. While such evidence would be desirable, in lending greater certainty to the accessibility determination, the realities of routine business practice counsel against requiring such evidence. The probative value of routine business practice to show the performance of a specific act has long been recognized. See, e.g., 1 Wigmore, Evidence § 92 (1940); rule 406, Fed. R. Evid. Therefore, we conclude that competent evidence of the general library practice may be relied upon to establish an approximate time when a thesis became accessible.

We agree with the board that the evidence of record consisting of Dr. Will's affidavits establishes a prima facie case for unpatentability of the claims under the § 102(b) publication bar. It is a case which stands unrebutted.

Accordingly, the board's decision sustaining the rejection of appellant's claims is affirmed.

COMMENTS AND QUESTIONS

1. Note that the issue in this case could arise under either section 102(a) or 102(b). If the publication had occurred before the *invention* of the subject matter by the applicant, section 102(a) would apply. Section 102(b) applied in this case because the publication occurred more than one year before the applicant *filed* for a patent.

2. Compare the "publicness" standard in *Hall* to the rule in trade secret law (information must be "generally known or readily ascertainable" in an industry). Why is it so much easier to lose patent rights than trade secrets?

3. In general, a publication becomes public when it becomes available to at least one member of "the general public." Thus a magazine or technical journal is effective as of its date of publication, i.e., when someone first receives it, rather than the date a manuscript was sent to the publisher or the date the journal or magazine was mailed. See In re Schlittler, 234 F.2d 882 (C.C.P.A. 1956) ("[T]he mere placing of a manuscript in the hands of a publisher does not necessarily make it available to the public within the meaning of [the] authorities."). After *Schlittler,* mere receipt of a manuscript by a publisher does not constitute "publication" under section 102(a) or (b). Actual publication—i.e., receipt by subscribers—is required. See, e.g., Protein Found. v. Brenner, 147 U.S.P.Q. (BNA) 429 (1965); U.S. Patent and Trademark Office, Manual of Patent Examining Procedures § 706.02 (5th ed. rev. 1991); Ex parte Hudson, 18 U.S.P.Q.2d (BNA) 1322 (Bd. App. & Int. 1990).

Is a grant proposal sent to a limited number of expert reviewers a publication under § 102(a)? See E. I. du Pont de Nemours & Co. v. Cetus Corp., 19 U.S.P.Q.2d (BNA) 1174 (N.D. Cal. 1990) (no).

PROBLEM

Problem 3-8. Slice-O-Rama is a company that makes food preparation items, including cheese slicers. It owns a U.S. patent covering a cheese slicer with a novel twist: the cutting element is a wire, which can be stretched between an end point and a handle. The slicer is well-received by the public and is even the subject of a national design award. One element in the claim calls for "a cutting wire" stretched between a handle and a fixed point on the base of the cutting element arm (see Figure 3-3).

OmniCorp is a large conglomerate with numerous divisions in various industries. One division, BigChem, Inc., is in the business of developing and manufacturing plastics. In the wake of research on new plastics, BigChem has come up with a plastic that is capable of holding a sharp cutting edge. Deemed "sharpylene," it can be manufactured in a variety of thicknesses, including a very thin version that looks like a wire or piece of string. As part of an internal initiative to encourage various divisions of OmniCorp to work more closely, Fran Fromage, product manager for the food processor and blender product lines in OmniCorp's FoodMulch division, comes across a sample of the Big-Chem plastic. Being familiar with the popularity of the Slice-O-Rama cheese slicer, she hits on the idea of substituting BigChem's new plastic in place of the metal wire. She believes that this cheese slicer would be an important addition to the FoodMulch product line. She is aware of the Slice-O-Rama patent, but as she puts it, "how can that patent stop us from selling blades made from plastic that *we invented*?! Those folks over at Slice-O-Rama have never even heard of sharpylene!"

You are a member of the patent department at OmniCorp. Assuming that the FoodMulch cheese slicer will meet all limitations other than the "cutting wire" element in the Slice-O-Rama patent, how would you advise Ms. Fromage?

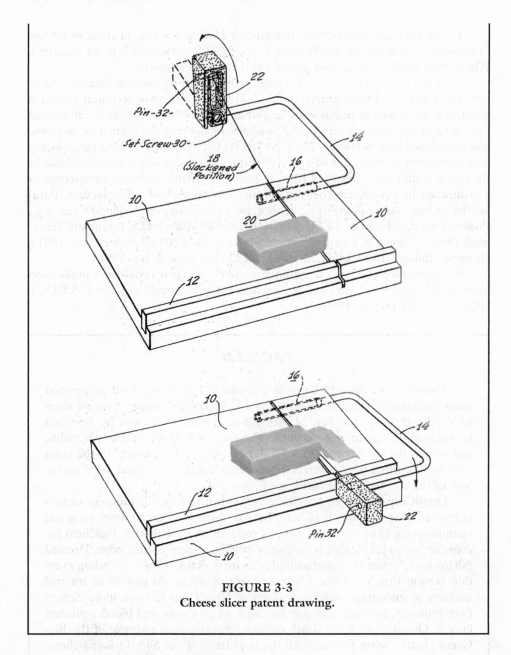

FIGURE 3-3
Cheese slicer patent drawing.

c. Statutory Bars: Public Use

Egbert v. Lippmann
Supreme Court of the United States
104 U.S. 333 (1881)

Mr. Justice WOODS delivered the opinion of the court.

This suit was brought for an alleged infringement of the complainant's reissued letters-patent, No. 5216, dated Jan. 7, 1873, for an improvement in corset-springs.

The original letters were issued to Samuel H. Barnes. The reissue was made to the complainant, under her then name, Frances Lee Barnes, executrix of the original patentee.

The specification for the reissue declares: —

This invention consists in forming the springs of corsets of two or more metallic plates, placed one upon another, and so connected as to prevent them from sliding off each other laterally or edgewise, and at the same time admit of their playing or sliding upon each other, in the direction of their length or longitudinally, whereby their flexibility and elasticity are greatly increased, while at the same time much strength is obtained.

The bill alleges that Barnes was the original and first inventor of the improvement covered by the reissued letters-patent, and that it had not, at the time of his application for the original letters, been for more than two years in public use or on sale, with his consent or allowance.

The answer takes issue on this averment and also denies infringement. On a final hearing the court dismissed the bill, and the complainant appealed.

We have to consider whether the defence that the patented invention had, with the consent of the inventor, been publicly used for more than two years prior to his application for the original letters, is sustained by the testimony in the record.

[The patent statute] render[s] letters-patent invalid if the invention which they cover was in public use, with the consent and allowance of the inventor, for more than two years prior to his application.[44] Since the passage of the act of 1839 it has been strenuously contended that the public use of an invention for more than two years before such application, even without his consent and allowance, renders the letters-patent therefor void.

It is unnecessary in this case to decide this question, for the alleged use of the invention covered by the letters-patent to Barnes is conceded to have been with his express consent.

The evidence on which the defendants rely to establish a prior public use of the invention consists mainly of the testimony of the complainant.

She testifies that Barnes invented the improvement covered by his patent between January and May, 1855; that between the dates named the witness and her friend Miss Cugier were complaining of the breaking of their corset-steels. Barnes, who was present, and was an intimate friend of the witness, said he thought he could make her a pair that would not break. At their next interview he presented her with a pair of corset-steels which he himself had made. The witness wore these steels a long time. In 1858 Barnes made and presented to her another pair, which she also wore a long time. When the corsets in which these steels were used wore out, the witness ripped them open and took out the steels and put them in new corsets. This was done several times.

It is admitted, and, in fact, is asserted, by complainant, that these steels embodied the invention afterwards patented by Barnes and covered by the reissued letters-patent on which this suit is brought.

44. This has since been changed to one year. — EDS.

Joseph H. Sturgis, another witness for complainant, testifies that in 1863 Barnes spoke to him about two inventions made by himself, one of which was a corset-steel, and that he went to the house of Barnes to see them. Before this time, and after the transactions testified to by the complainant, Barnes and [the complainant] had intermarried. Barnes said his wife had a pair of steels made according to his invention in the corsets which she was then wearing, and if she would take them off he would show them to [Sturgis]. Mrs. Barnes went out, and returned with a pair of corsets and a pair of scissors, and ripped the corsets open and took out the steels. Barnes then explained to the witness how they were made and used.

The question for our decision is, whether this testimony shows a public use within the meaning of the statute.

We observe, in the first place, that to constitute the public use of an invention it is not necessary that more than one of the patented articles should be publicly used. The use of a great number may tend to strengthen the proof, but one well-defined case of such use is just as effectual to annul the patent as many. For instance, if the inventor of a mower, a printing press, or a railway-car makes and sells only one of the articles invented by him, and allows the vendee to use it for two years, without restriction or limitation, the use is just as public as if he had sold and allowed the use of a great number.

We remark, secondly, that, whether the use of an invention is public or private does not necessarily depend upon the number of persons to whom its use is known. If an inventor, having made his device, gives or sells it to another, to be used by the donee or vendee, without limitation or restriction, or injunction of secrecy, and it is so used, such use is public, even though the use and knowledge of the use may be confined to one person.

We say, thirdly, that some inventions are by their very character only capable of being used where they cannot be seen or observed by the public eye. An invention may consist of a lever or spring, hidden in the running gear of a watch, or of a rachet, shaft, or cog-wheel covered from view in the recesses of a machine for spinning or weaving. Nevertheless, if its inventor sells a machine of which his invention forms a part, and allows it to be used without restriction of any kind, the use is a public one. So, on the other hand, a use necessarily open to public view, if made in good faith solely to test the qualities of the invention, and for the purpose of experiment, is not a public use within the meaning of the statute.

Tested by these principles, we think the evidence of the complainant herself shows that for more than two years before the application for the original letters there was, by the consent and allowance of Barnes, a public use of the invention, covered by them. He made and gave to her two pairs of corset-steels, constructed according to his device, one in 1855 and one in 1858. They were presented to her for use. He imposed no obligation of secrecy, nor any condition or restriction whatever. They were not presented for the purpose of experiment, nor to test their qualities. No such claim is set up in her testimony. The invention was at the time complete, and there is no evidence that it was afterwards changed or improved. The donee of the steels used them for years for the purpose and in the manner designed by the inventor. They were not capable of any other use. She might have exhibited them to any person, or made other steels of the same kind, and used or sold them without violating any condition or restriction imposed on her by the inventor.

According to the testimony of the complainant, the invention was completed and put to use in 1855. The inventor slept on his rights for eleven years. Letters-patent were not applied for till March, 1866. In the mean time, the invention had found its way into general, and almost universal, use. A great part of the record is taken up with the testimony of the manufacturers and venders [sic] of corset-steels, showing that before he applied for letters the principle of his device was almost universally used in the manufacture of corset-steels. It is fair to presume that having learned from this general use that there was some value in his invention, he attempted to resume, by his application, what by his acts he had clearly dedicated to the public.

We are of opinion that the defence of two years' public use, by the consent and allowance of the inventor, before he made application for letters-patent, is satisfactorily established by the evidence.

Decree affirmed.

Mr. Justice MILLER dissenting.

A private use with consent, which could lead to no copy or reproduction of the machine, which taught the nature of the invention to no one but the party to whom such consent was given, which left the public at large as ignorant of this as it was before the author's discovery, was no abandonment to the public, and did not defeat his claim for a patent. If the little steel spring inserted in a single pair of corsets, and used by only one woman, covered by her outer-clothing, and in a position always withheld from public observation, is a *public* use of that piece of steel, I am at a loss to know the line between a private and a public use.

COMMENTS AND QUESTIONS

1. In what sense was Frances Lee (later Barnes) using the corset steels (i.e., the springs) "publicly"? Why doesn't it matter how many people use the invention, for purposes of deciding if the use was public? Compare this holding to that of the court in *Rosaire* above.

2. The Court emphasizes that Ms. Barnes never entered into a confidentiality agreement with the inventor. But the Court also relates that the two later married. Does Barnes have a reasonable argument that although no *express* "injunction of secrecy" was made, nonetheless an *implied* requirement of secrecy could be inferred from the surrounding circumstances: i.e., an invention of an intimate nature, by a boyfriend (later husband), disclosed to few people? Justice Miller, in dissent, added: "It may well be imagined that a prohibition to the party so permitted [to use the springs] against her use of the steel spring to public observation, would have been supposed to be a piece of irony." 104 U.S. (14 Otto), at 339. Perhaps the embarrassment of asking was enough to keep Mr. Barnes from requesting that Frances Lee not show the invention to anyone.

3. Subsequent cases have not abandoned *Egbert,* but they have perhaps not read it as broadly as they could. In particular, the Federal Circuit has focused on the nature and purpose of the use in deciding whether it was "public" or not. Use for a commercial purpose is generally a public use, even if it is secret and even if it only occurs once. However, use for personal interest or enjoyment will generally not be considered a "public" use.

A similar rule has developed with respect to the "on sale" prong of the section 102(b) statutory bar. A single sale or offer to sell a product will start the clock running, whether or not it was made in secret. On the other hand, discussions that do not rise to the level of a definite offer to sell the invention will not bar a later patent on that invention.

4. Obviously, the policy behind the statutory bars in section 102(b) is to encourage inventors to file patent applications early. The inventor in *Egbert* did not do so. Was it reasonable to bar him from receiving a patent on his invention for this reason? Is anyone entitled to a patent on this invention?

5. While "public use" and "on sale" bars frequently result from the inventor's own conduct, they do not have to. Indeed, one recent case held that sales of the invention by a third party precluded the inventor from later obtaining a patent, even though the third party had stolen the invention from the inventor. Evans Cooling Sys., Inc. v. General Motors, 125 F.3d 1448 (Fed. Cir. 1997). The case arose out of inventor John Evans' demonstration of his aqueous reverse flow cooling system to GM engineers in the spring of 1989. GM, as early as June 13, 1991, received a binding order for sale of a Corvette model car incorporating the aqueous reverse flow cooling system. The district court agreed with GM that this sale created a statutory bar for Evans' subsequent patent application filed July 1, 1992. Evans appealed on two grounds: that no enforceable contract could have been formed in June of 1991 because GM had yet to clear certain requisite regulatory hurdles routinely required for sale of a new car model and that the theft of a trade secret ought to be considered an exception to the general statutory bar rule. The court disposed of the first ground by stating simply that "the mere fact that the offer for sale was illegal or ineffective does not remove it from the purview of the section 102(b) bar." Id. at 1452.

The court found the second ground—the "theft exception"—equally unpersuasive in light of existing case law. It wrote:

> While such a result may not seem fair, Evans is not without recourse if GM in fact misappropriated his invention. Evans would have an appropriate remedy in state court for misappropriation of a trade secret. We note as well that the facts Evans alleges in support of its misappropriation claim demonstrate that Evans knew GM stole the invention at the very time it was allegedly stolen because during the demonstration GM employees allegedly told Mr. Evans they intended to steal the invention and a sealed room was unsealed during the night between the tests. Evans' patent rights would have nevertheless been protected if Mr. Evans had filed a patent application no more than one year from the date of the demonstration. This he did not do; instead Mr. Evans waited for more than two years after the demonstration and some six years after it was reduced to practice.

Id. at 1453-54

6. Must an invention actually be completed and built before it can be sold or "offered for sale?" The Supreme Court said no in a recent case. Pfaff v. Wells Elec., Inc., 525 U.S. 55 (1998). The Court reasoned that "invention" occurred when the inventor conceived of the product and did not require proof that the invention has actually been built. Indeed, it noted, many famous inventions (including Alexander Graham Bell's telephone) were patented before a prototype was ever built. Id. at 308-09. However, the Court did require evidence that the inventive concept itself was "complete, rather than merely . . . substantially complete":

[T]he on-sale bar applies when two conditions are satisfied before the critical date. First, the product must be the subject of a commercial offer for sale....

Second, the invention must be ready for patenting. That condition may be satisfied in at least two ways: by proof of reduction to practice before the critical date; or by proof that prior to the critical date the inventor had prepared drawings or other descriptions of the invention that were sufficiently specific to enable a person skilled in the art to practice the invention. In this case the second condition of the on-sale bar is satisfied because the drawings Pfaff sent to the manufacturer before the critical date fully disclosed the invention.

Id. at 67. See also Weatherchem Corp. v. J.L. Clark, Inc., 163 F.3d 1326 (Fed. Cir. 1998) (applying *Pfaff* to case where detailed drawings had been made in preparation of an offer for sale). Contrast *Pfaff* with Micro Chem., Inc. v. Great Plains Chem. Co., 103 F.3d 1538 (Fed. Cir. 1997) ("Because Pratt was not close to completion of the invention at the time of the alleged offer and had not demonstrated a high likelihood that the invention would work for its intended purpose upon completion, his December 1984 'offer' could not trigger the on-sale bar.").

Does it make sense to use an enablement standard in determining whether an invention has been made, or should there be some requirement that the invention be physically constructed before it can be placed on sale? Does the 1994 addition of the "offer for sale" language to the list of exclusive rights in 35 U.S.C. §271(a) suggest that completing the invention should not be required?

PROBLEM

Problem 3-9. Mike Molar, a production engineer for Tasty Toothpaste Corp. (Tasty) hit upon a way to make a sanitary, cheap, and small disposable toothbrush. The problem he had been running into was how to keep the toothbrush small while still providing a feature he thought necessary to make it attractive — a built-in supply of toothpaste. The solution: a small reservoir for holding toothpaste in the handle of the toothbrush. Then the user could squeeze the toothpaste from the reservoir to the brush bristles.

He quickly built a prototype of the invention. To build it, he bought a fountain pen, pulled out the ink cartridge and other parts from the interior, tore the bristles off a toothbrush and glued them onto a hollow tube, poked two holes in the top of the tube between two rows of bristles, fitted up a plastic plunger that slid in and out at the back of the fountain pen body, pulled the plunger all the way out, filled the tube with toothpaste, glued the tube inside the fountain pen body, and pushed the plunger. Voila! Toothpaste came out of the holes more or less onto the bristles, ready for brushing. It worked! He drafted a patent specification that included this passage:

> It is one important object of the present invention to provide a portable toothbrush which is easily carried in the pocket and which is thus available whenever it is required for brushing teeth.

The open end of the device includes a reduced section portion which carries the bristles. The cylindrical body shaft includes an interior passage that extends into the reduced section end (i.e., the one with the bristles). The passage includes two termination openings which are in the area at the base of the toothbrush bristles. An open space for placing toothpaste is provided within the interior passage, and a movable plunger is provided for forcing the toothpaste from this space through the narrow part of the passage and out through the termination openings into the toothpaste bristles for use.

In the preferred embodiment, the body shaft is made of plastic. The invention may be made so as to be disposable. Various means for filling the space with toothpaste are envisioned, including pressure-injection through a small hole in the top of the body, which hole can then be sealed. This would make the toothbrush usable only once; it would then be disposed.

He concluded with the following claim.

I claim:

1. A pocket toothbrush having an exterior structure resembling a traditional fountain pen case comprising

a. a removable cylindrical end cap cover,

b. a main cylindrical body shaft over at least one end of which said end cap cover fits and having means for engaging the interior of said end cap cover to retain said end cap cover,

c. said cylindrical body shaft having one end which contains toothbrush bristles [the "bristle end"] extending transversely and capable of being confined within said end cap,

d. said cylindrical body shaft including an interior passage extending into said bristle end and having at least one termination opening in the area at the base of said bristles,

e. a movable plunger extending into said cylindrical body shaft in said main cylindrical body shaft,

f. said body shaft including an interior space for the accommodation of a charge of toothpaste to be fed to said bristles by the operation of said movable plunger, said space being at least big enough to hold a charge for a single application of toothpaste.

Once the patent issues, Tasty begins selling a disposable, portable toothbrush that garners a loyal following. Soon competitors begin entering the market. One, KopyCat Industries, Inc. (KCI), begins selling a portable toothbrush that includes a replaceable toothpaste cartridge so that the brush can be used over and over if the user wishes. The cartridges are designed to have a weak plastic closure that easily breaks when the plunger is pushed against the cartridge. This keeps the toothpaste from hardening in the openings to the bristles. Also, instead of a cap, the KCI design has a telescoping retractable cover that remains attached to the nonbristle end of the brush. The cover is collapsed down, the brush used, and then the cover is pulled back into place. The retractable cover is attached very firmly with two tiny screws. The screws can be taken out and the cover removed, but it takes a tiny jeweler's screwdriver and is difficult.

Tasty has threatened to sue KCI for infringement of the Molar patent. KCI has come to you for advice. Focusing on the *claim language*—and using the specification only to interpret that language — determine (1) whether KCI runs a serious risk of being found liable for infringing the Molar patent; and (2) what changes KCI might make in its product to avoid a future infringement action by Tasty. In particular, pay attention to these issues: Does the KCI product include a "removable end cap cover"? Is it relevant that KCI improved the design of the replaceable toothpaste cartridge?

Several cases have now addressed the relationship between contract law and the "on sale" bar. In Group One, Ltd. v. Hallmark Cards, Inc., 254 F.3d 1041 (Fed. Cir. 2001), the patentee and a third party had conducted correspondence and engaged in other interactions in connection with a later-patented technology. The Federal Circuit refused to find that this exploratory placed the invention "on sale" under § 102(b). Only a transaction which "rises to the level of a commercial offer for sale" under the Uniform Commercial Code triggers the on-sale bar, according to the court. 254 F.3d at 1047.

In re Kollar, 286 F.3d 1326 (Fed. Cir. 2002), concerned a licensing transaction involving a chemical process. The Federal Circuit refused to find that the transaction triggered the on-sale bar: "[L]icensing the invention," the court stated, "under which development of the claimed process would have to occur before the process is successfully commercialized, is not . . . a sale." Id., at 1333. Compare Elan Corp. v. Andrx Pharmaceuticals, Inc., 366 F.3d 1336 (Fed. Cir. 2004) (offer to supply potential licensee with limited quantities of bulk tablets not enough to constitute on-sale activity in what otherwise appeared to be a proposed licensing transaction) with Minton v. National Ass'n. of Securities Dealers, Inc., 336 F.3d 1373, 1378 (Fed. Cir. 2003) (Inventor's conveyance to lessee of fully operational computer program implementing and thus embodying method subsequently claimed in patent for interactive securities trading system, along with warranty of workability, enabled lessee to practice invention, and thus was "offer for sale" within meaning of on-sale bar).

Does this distinction make sense? Why should those who license a process invention be able to avoid the on-sale bar when those who sell a patented product cannot?

d. The Experimental Use Exception

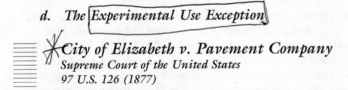

City of Elizabeth v. Pavement Company
Supreme Court of the United States
97 U.S. 126 (1877)

Mr. Justice BRADLEY delivered the opinion of the court.

This suit was brought by the American Nicholson Pavement Company against the city of Elizabeth, N. J., upon a patent issued to Samuel Nicholson, dated Aug. 20, 1867, for a new and improved wooden pavement. [I]n the specification, it is declared that the nature and object of the invention consists in providing a process or mode of constructing wooden block pavements upon a foundation along a street or roadway with facility, cheapness, and accuracy, and also in the creation and construction of such a wooden pavement as shall be comparatively permanent and durable, by so uniting and combining all its parts, both superstructure and foundation, as to provide against the slipping of the horses' feet, against noise, against unequal wear, and against rot and consequent sinking away from below.

The bill charges that the defendants infringed this patent by laying down wooden pavements in the city of Elizabeth, N.J., constructed in substantial conformity with the process patented, and prays an account of profits, and an injunction.

The defendants . . . averred that the alleged invention of Nicholson was in public use, with his consent and allowance, for six years before he applied for a patent, on a certain avenue in Boston called the Mill-dam; and contended that said public use worked an abandonment of the pretended invention. . . .

The next question to be considered is, whether Nicholson's invention was in public use or on sale, with his consent and allowance, for more than two years prior to his application for a patent, within the meaning of . . . the acts in force in 1854, when he obtained his patent. It is contended by the appellants that the pavement which Nicholson put down by way of experiment, on Mill-dam Avenue in Boston, in 1848, was publicly used for the space of six years before his application for a patent, and that this was a public use within the meaning of the law.

To determine this question, it is necessary to examine the circumstances under which this pavement was put down, and the object and purpose that Nicholson had in view. It is perfectly clear from the evidence that he did not intend to abandon his right to a patent. He had filed a *caveat* in August, 1847, and he constructed the pavement in question by way of experiment, for the purpose of testing its qualities. The road in which it was put down, though a public road, belonged to the Boston and Roxbury Mill Corporation, which received toll for its use; and Nicholson was a stockholder and treasurer of the corporation. The pavement in question was about seventy-five feet in length, and was laid adjoining to the toll-gate and in front of the toll-house. It was constructed by Nicholson at his own expense, and was placed by him where it was, in order to see the effect upon it of heavily loaded wagons, and of varied and constant use; and also to ascertain its durability, and liability to decay. Joseph L. Lang, who was toll-collector for many years, commencing in 1849, familiar with the road before that time, and with this pavement from the time of its origin, testified as follows:

> Mr. Nicholson was there almost daily, and when he came he would examine the pavement, would often walk over it, cane in hand, striking it with his cane, and making particular examination of its condition. He asked me very often how people liked it, and asked me a great many questions about it. I have heard him say a number of times that this was his first experiment with this pavement, and he thought that it was wearing very well. The circumstances that made this locality desirable for the purpose of obtaining a satisfactory test of the durability and value of the pavement were: that there would be a better chance to lay it there; he would have more room and a better chance than in the city; and, besides, it was a place where most everybody went over it, rich and poor. It was a great thoroughfare out of Boston. It was frequently travelled by teams having a load of five or six tons, and some larger. As these teams usually stopped at the toll-house, and started again, the stopping and starting would make as severe a trial to the pavement as it could be put to.

This evidence is corroborated by that of several other witnesses in the cause; the result of the whole being that Nicholson merely intended this piece of pavement as an experiment, to test its usefulness and durability. Was this a public use, within the meaning of the law?

An abandonment of an invention to the public may be evinced by the conduct of the inventor at any time, even within the two years named in the law. The effect of the law is, that no such consequence will necessarily follow from the invention being in public use or on sale, with the inventor's consent and allowance, at any time within two years before his application; but that, if the invention is in public use or on sale prior to that time, it will be conclusive evidence of abandonment, and the patent will be void.

But, in this case, it becomes important to inquire what is such a public use as will have the effect referred to. That the use of the pavement in question was public in one sense cannot be disputed. But can it be said that the invention was in public use? The use of an invention by the inventor himself, or of any other person under his direction, by way of experiment, and in order to bring the invention to perfection, has never been regarded as such a use. Curtis, Patents, § 381; Shaw v. Cooper, [32 U.S.] 7 Pet. 292 [1833].

Now, the nature of a street pavement is such that it cannot be experimented upon satisfactorily except on a highway, which is always public.

When the subject of invention is a machine, it may be tested and tried in a building, either with or without closed doors. In either case, such use is not a public use, within the meaning of the statute, so long as the inventor is engaged, in good faith, in testing its operation. He may see cause to alter it and improve it, or not. His experiments will reveal the fact whether any and what alterations may be necessary. If durability is one of the qualities to be attained, a long period, perhaps years, may be necessary to enable the inventor to discover whether his purpose is accomplished. And though, during all that period, he may not find that any changes are necessary, yet he may be justly said to be using his machine only by way of experiment; and no one would say that such a use, pursued with a *bona fide* intent of testing the qualities of the machine, would be a public use, within the meaning of the statute. So long as he does not voluntarily allow others to make it and use it, and so long as it is not on sale for general use, he keeps the invention under his own control, and does not lose his title to a patent.

It would not be necessary, in such a case, that the machine should be put up and used only in the inventor's own shop or premises. He may have it put up and used in the premises of another, and the use may inure to the benefit of the owner of the establishment. Still, if used under the surveillance of the inventor, and for the purpose of enabling him to test the machine, and ascertain whether it will answer the purpose intended, and make such alterations and improvements as experience demonstrates to be necessary, it will still be a mere experimental use, and not a public use, within the meaning of the statute.

Whilst the supposed machine is in such experimental use, the public may be incidentally deriving a benefit from it. If it be a grist-mill, or a carding-machine, customers from the surrounding country may enjoy the use of it by having their grain made into flour, or their wool into rolls, and still it will not be in public use, within the meaning of the law.

But if the inventor allows his machine to be used by other persons generally, either with or without compensation, or if it is, with his consent, put on sale for such use, then it will be in public use and on public sale, within the meaning of the law.

If, now, we apply the same principles to this case, the analogy will be seen at once. Nicholson wished to experiment on his pavement. He believed it to be a good thing, but he was not sure; and the only mode in which he could test it was to place a specimen of it in a public roadway. He did this at his own expense, and with the consent of the owners of the road. Durability was one of the qualities to be attained. He wanted to know whether his pavement would stand, and whether it would resist decay. Its character for durability could not be ascertained without its being subjected to use for a considerable time. He subjected it to such use, in good faith, for the simple purpose of ascertaining whether it was what he claimed it to be. Did he do any thing more than the inventor of the supposed machine might do, in testing his invention? The public had the incidental use of the pavement, it is true; but was the invention in public use, within the meaning of the statute? We think not. The proprietors of the road alone used the invention, and used it at Nicholson's request, by way of experiment. The only way in which they could use it was by allowing the public to pass over the pavement.

Had the city of Boston, or other parties, used the invention, by laying down the pavement in other streets and places, with Nicholson's consent and allowance, then, indeed, the invention itself would have been in public use, within the meaning of the law; but this was not the case. Nicholson did not sell it, nor allow others to use it or sell it. He did not let it go beyond his control. He did nothing that indicated any intent to do so. He kept it under his own eyes, and never for a moment abandoned the intent to obtain a patent for it. . . .

It is sometimes said that an inventor acquires an undue advantage over the public by delaying to take out a patent, inasmuch as he thereby preserves the monopoly to himself for a longer period than is allowed by the policy of the law; but this cannot be said with justice when the delay is occasioned by a *bona fide* effort to bring his invention to perfection, or to ascertain whether it will answer the purpose intended. His monopoly only continues for the allotted period, in any event; and it is the interest of the public, as well as himself, that the invention should be perfect and properly tested, before a patent is granted for it. Any attempt to use it for a profit, and not by way of experiment, for a longer period than two years before the application, would deprive the inventor of his right to a patent.

COMMENTS AND QUESTIONS

1. Compare the facts in the preceding case to those in Egbert v. Lippmann, *supra*, the "corset case." How did Nicholson's actions in this case differ from those of Barnes in the *Egbert* case? What evidence did Nicholson have regarding his six-year· prefiling period that Barnes did not have for his comparably long period?

2. Could Nicholson simply have filed a patent application at the end of year one and prosecuted it while conducting his continued test? Would a patent have been granted on such an application? Note that because Nicholson was able to delay filing his application for six years, he received a patent that expired later and was therefore arguably more valuable to him than the patent he would have received if he had applied at the end of one year.

3. A 2005 Federal Circuit case applied the *City of Elizabeth* principles to a case involving "experimental sales." Electromotive Div. of General Motors Corp. v. Transportation Systems Div. of General Elec. Co., 417 F.3d 1203 (Fed. Cir. 2005).

In this case, the patentee (GM) claimed planetary compressor bearings for use in turbochargers for diesel train engines. Pursuant to its standard development and testing procedure, GM used prototypes of its new bearing design in numerous train engines before the critical date. It also agreed to supply some bearings covered by one of its patents as spares to one of its customers if it needed them. Defendant GE argued that the sale of the train engines containing the new bearings, and the agreement to supply them as spares, constituted "on sale" activity under § 102(b). GM disagreed, and argued that because the bearings were part of its development and testing program, the customer's use of the engines with the bearings was covered by the "experimental use" of *City of Elizabeth*. The Federal Circuit sided with defendant GE:

> When sales are made in an ordinary commercial environment and the goods are placed outside the inventor's control, an inventor's secretly held subjective intent to "experiment," even if true, is unavailing without objective evidence to support the contention. Under such circumstances, the customer at a minimum must be made aware of the experimentation:
>
> > We have generally looked to objective evidence to show that a pre-critical date sale was primarily for experimentation.... The length of the test period is merely a piece of evidence to add to the evidentiary scale. The same is true with respect to whether payment is made for the device, whether a user agreed to use secretly, whether records were kept of progress, whether persons other than the inventor conducted the asserted experiments, how many tests were conducted, how long the testing period was in relationship to tests of other similar devices.
>
> [T.P. Laboratories, Inc. v. Professional Positioners, Inc., 724 F.2d 965, 972 (Fed.Cir.1984)], at 971-2.
>
> We agree... that a customer's awareness of the purported testing in the context of a sale is a critical attribute of experimentation. If an inventor fails to communicate to a customer that the sale of the invention was made in pursuit of experimentation, then the customer, as well as the general public, can only view the sale as a normal commercial transaction.... Accordingly, we hold not only that customer awareness is among the experimentation factors, but also that it is critical.
>
> Our precedent has treated control and customer awareness of the testing as especially important to experimentation. Indeed, this court has effectively made control and customer awareness dispositive. Accordingly, we conclude that control and customer awareness ordinarily must be proven if experimentation is to be found.

417 F.3d 1203, at 1212-1215 (citations omitted). See also Allen Eng'g Corp. v. Bartell Indus., Inc., 299 F.3d 1336, 1353 (Fed. Cir. 2002) (listing 13 factors, collected from previous cases, relevant to finding of experimental use).

e. *Priority Rules and the First to Invent*

Having dealt with novelty, we turn now to the closely related topic of *priority*.

35 U.S.C. § 102 Novelty and Loss of Right

An inventor shall be entitled to a patent unless —
(g)(1) during the course of an interference..., another inventor involved therein

establishes, to the extent permitted in section 104 . . . that before such person's invention thereof the invention was made by such other inventor and not abandoned, suppressed, or concealed, or (2) before such person's invention thereof, the invention was made in this country by another inventor who had not abandoned, suppressed, or concealed it. In determining priority of invention under this subsection, there shall be considered not only the respective dates of conception and reduction to practice of the invention, but also the reasonable diligence of one who was first to conceive and last to reduce to practice, from a time prior to conception by the other.

Although there are a number of fine points, this section states some basic rules for determining *priority* of invention. Priority arises in two different contexts, treated in the two separate subsections of § 102(g). Subsection (1) covers *interference proceedings*—formal priority contests between rival claimants to the same patentable subject matter. Subsection (2) covers the use of prior inventions as a source of prior art. One may invoke subsection (2) outside the interference context; for example, a defendant in an infringement suit might point to a third party's prior invention as a prior art reference to defeat the validity of the plaintiff's patent.

Besides the different contexts in which they arise, there is a major substantive difference between (g)(1) and (g)(2): their territorial scope. The key phrases are "to the extent permitted in section 104" in (g)(1) and "in this country" in (g)(2). Section 104 allows proof of prior inventions in any country that is a signatory to the World Trade Organization (WTO). Under (g)(1), then, inventors from any WTO country may introduce evidence of prior inventions. Contrast this with the language of (g)(2): only evidence of prior inventions made *in the U.S.* may be introduced outside the interference context.

Putting aside the difference in territorial scope, the basic rules of priority are the same. They are stated in the third sentence. The courts have fleshed out the meaning of this priority rule over many years; the case and notes that follow serve as an introduction. In brief, priority generally goes to the first inventor to (1) reduce an invention to practice, without (2) abandoning the invention. Note in this respect that filing a valid, enabling patent application constitutes a constructive reduction to practice. An exception to this rule is where an inventor is the first to conceive, but the last to reduce to practice: priority may be retained, *if* the first conceiver was diligent in reducing to practice, with diligence being measured from a time just prior to the second conceiver's conception date. (Got that? Now you may understand why some famous interferences take a decade or more to work through!)

The following case centers on the last sentence of section 102(g), particularly the "reasonable diligence of one who was first to conceive but last to reduce to practice." For variations on the facts that present other issues under section 102(g), see Note 1 after the case. Also, notice that one of the rival inventors in this priority contest, Kanamaru, introduced only evidence of a patent *filing date* but no evidence regarding dates of conception and reduction to practice. In such cases the Patent Office effectively collapses the entire sequence of inventive events into the single date of patent filing; it is as if Kanamaru conceived, reduced to practice, and filed a patent application on the invention all on the same day. For a likely explanation of *why* Kanamaru relied only on this date, see Note 2 after the case.

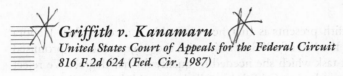

Griffith v. Kanamaru
United States Court of Appeals for the Federal Circuit
816 F.2d 624 (Fed. Cir. 1987)

NICHOLS, Senior Circuit Judge.

Owen W. Griffith (Griffith) appeals the decision of the Board of Patent Appeals and Interferences (board) that Griffith failed to establish a prima facie case that he is entitled to an award of priority against the filing date of Tsuneo Kanamaru, et al. (Kanamaru) for a patent on aminocarnitine compounds. We affirm.

Background

This patent interference case involves the application of Griffith, an Associate Professor in the Department of Biochemistry at Cornell University Medical College, for a patent on an aminocarnitine compound, useful in the treatment of diabetes, and a patent issued for the same invention to Kanamaru, an employee of Takeda Chemical Industries. The inventors assigned their rights to the inventions to the Cornell Research Foundation, Inc. (Cornell) and to Takeda Chemical Industries respectively.

Griffith had established conception by June 30, 1981, and reduction to practice on January 11, 1984. Kanamaru filed for a United States patent on November 17, 1982. The board found, however, that Griffith failed to establish reasonable diligence for a prima facie case of prior invention....

The board ... decided that Griffith failed to establish a prima facie case for priority against Kanamaru's filing date. This result was based on the board's conclusion that Griffith's explanation for inactivity between June 15, 1983, and September 13, 1983, failed to provide a legally sufficient excuse to satisfy the "reasonable diligence" requirement of 35 U.S.C. § 102(g). Griffith appeals on the issue of reasonable diligence.

Analysis

This is a case of first impression and presents the novel circumstances of a university suggesting that it is reasonable for the public to wait for disclosure until the most satisfactory funding arrangements are made. The applicable law is the "reasonable diligence" standard contained in 35 U.S.C. § 102(g) and we must determine the appropriate role of the courts in construing this exception to the ordinary first-in-time rule. As a preliminary matter we note that, although the board focused on the June 1983 to September 1983 lapse in work, and Griffith's reasons for this lapse, Griffith is burdened with establishing a prima facie case of reasonable diligence from immediately before Kanamaru's filing date of November 17, 1982, until Griffith's reduction to practice on January 11, 1984. 35 U.S.C. § 102(g).

On appeal, Griffith presents two grounds intended to justify his inactivity on the aminocarnitine project between June 15, 1983, and September 13, 1983. The first is that ... it is reasonable, and as a policy matter desirable, for Cornell to require Griffith and other research scientists to obtain funding from outside the university.

The second reason Griffith presents is that he reasonably waited for Ms. Debora Jenkins to matriculate in the Fall of 1983 to assist with the project. He had promised her she should have that task which she needed to qualify for her degree. We reject these arguments and conclude that Griffith has failed to establish grounds to excuse his inactivity prior to reduction to practice.

The reasonable diligence standard balances the interest in rewarding and encouraging invention with the public's interest in the earliest possible disclosure of innovation. Griffith must account for the entire period from just before Kanamaru's filing date until his [i.e., Griffith's] reduction to practice....

The board in this case was, but not properly, asked to pass judgment on the reasonableness of Cornell's policy regarding outside funding of research. The correct inquiry is rather whether it is reasonable for Cornell to require the public to wait for the innovation, given the well settled policy in favor of early disclosure.... A review of caselaw on excuses for inactivity in reduction to practice reveals a common thread that courts may consider the reasonable everyday problems and limitations encountered by an inventor. See, e.g.,... Reed v. Tornqvist, 436 F.2d 501 (C.C.P.A. 1971) (concluding it is not unreasonable for inventor to delay completing a patent application until after returning from a three week vacation in Sweden, extended by illness of inventor's father).... De Wallace v. Scott, 15 App. D.C. 157 (1899) (where applicant made bona fide attempts to perfect his invention, applicant's poor health, responsibility to feed his family, and daily job demands excused his delay in reducing his invention to practice).

...We first note that, in regard to waiting for a graduate student, Griffith does not even suggest that he faced a genuine shortage of personnel. He does not suggest that Ms. Jenkins was the only person capable of carrying on with the amino-carnitine experiment. We can see no application of precedent to suggest that the convenience of the timing of the semester schedule justifies a three-month delay for the purpose of reasonable diligence. Neither do we believe that this excuse, absent even a suggestion by Griffith that Jenkins was uniquely qualified to do his research, is reasonable.

Griffith's second contention that it was reasonable for Cornell to require outside funding, therefore causing a delay in order to apply for such funds, is also insufficient to excuse his inactivity. The crux of Griffith's argument is that outside funding is desirable as a form of peer review, or monitoring of the worthiness of a given project. He also suggests that, as a policy matter, universities should not be treated as businesses, which ultimately would detract from scholarly inquiry. Griffith states that these considerations, if accepted as valid, would fit within the scope of the caselaw excusing inactivity for "reasonable" delays in reduction to practice and filing.

Griffith's excuses sound more in the nature of commercial development, not accepted as an excuse for delay, than the "hardship" cases most commonly found and discussed supra. Delays in reduction to practice caused by an inventor's efforts to refine an invention to the most marketable and profitable form have not been accepted as sufficient excuses for inactivity.

...[I]t seems evident that Cornell has consciously chosen to assume the risk that priority in the invention might be lost to an outside inventor, yet, having chosen a noncommercial policy, it asks us to save it the property that would have inured to it if it had acted in single-minded pursuit of gain.

Although we agree with the board's conclusion, it is appropriate to go further and consider other circumstances as they apply to the reasonable diligence analysis of 35 U.S.C. § 102(g). The record reveals that from the relevant period of November 17, 1982 (Kanamaru's filing date), to September 13, 1983 (when Griffith renewed his efforts towards reduction to practice), Griffith interrupted and often put aside the aminocarnitine project to work on other experiments. Between June 1982 and June 1983 Griffith admits that, at the request of the chairman of his department, he was primarily engaged in an unrelated research project.... Griffith also put aside the aminocarnitine experiment to work on a grant proposal on an unrelated project.... Griffith made only minimal efforts to secure funding directly for the aminocarnitine project.

The conclusion we reach from the record is that the aminocarnitine project was second and often third priority in laboratory research as well as the solicitation of funds. We agree that Griffith failed to establish a prima facie case of reasonable diligence or a legally sufficient excuse for inactivity to establish priority over Kanamaru.

COMMENTS AND QUESTIONS

1. What would the result in the interference have been if Professor Griffith had reduced the invention to practice on November 16, 1982? (Recall that Kanamaru's filing date—and therefore, in this case, his effective conception and reduction to practice dates—was November 17, 1982.)

If Griffith had reduced to practice on November 16, 1982, would it matter whether he was diligent between his conception date (June 30, 1981) and his reduction to practice? If not, why not?

Assume that Kanamaru introduced the following evidence: a conception date of January 1, 1982, and reduction to practice on November 1, 1982. Assume also, as actually happened, that Griffith's reduction to practice came after Kanamaru's. With Griffith the first to conceive of the invention, is *Kanamaru's* diligence an issue? *Should* it be?

2. Since Takeda Chemical, Kanamaru's assignee, is based primarily in Japan, it is reasonable to conclude that Kanamaru did the research leading to his patent application in Japan. This would explain the exclusive reliance on the patent filing date in this priority contest. Before 1995, and thus at the time this case was decided, foreign inventors could not introduce evidence of foreign inventive activity (e.g., conception and reduction to practice). In 1994, however, as part of the legislative package implementing the Trade Related Aspects of Intellectual Property (TRIPs) portion of the Uruguay Round negotiations under the GATT, Congress changed 35 U.S.C. § 104 to permit evidence of inventive activity taking place in any country that is a member of the World Trade Organization (the successor organization to the GATT) beginning in 1996. See Uruguay Round Agreements Act of 1994, P.L. 103-465, 108 Stat. 4809 (1994), at § 531, codified at 35 U.S.C. § 104.

3. Most countries in the world award patents not to the first person to invent the subject matter but to the first person to file a patent application covering the subject. Who would have won if priority were determined not by a first to invent rule but by a first to *file* rule? Would the outcome in that case be fair? In deciding,

keep in mind that such a rule would have saved the cost of the entire interference proceeding, including the appeal that yielded the decision reproduced above. Is the extra cost of this additional "due process" worth it? For many years, it was thought that most interferences were won by senior parties anyway. Then some scholars thought to test that conventional wisdom. See Mark A. Lemley and Colleen V. Chien, Are the U.S. Priority Rules Really Necessary?, 54 Hastings L.J. 1299 (2003) ("Of the 100 cases in our population that have final outcomes, junior parties [i.e., those second to file] won 33 (or 33%). More significantly, in the 76 cases that are actually resolved on priority grounds, junior parties won 33 times (or 43%). Thus, it seems that when priority is actually adjudicated, the first to invent is quite frequently not the first to file."). Does this sway your thinking about the desirability of giving up the traditional U.S. "first to invent" standard? For a recent legislative proposal that would do just that, see Section 3 of H.R. 2795, 109th Cong., 1st Sess., introduced by Representative Lamar Smith, summarized at 70 Pat. Tm. & Copyrt. J. 142 (June 10, 2005), which would have amended 35 U.S.C. to provide for "first to file" priority.

4. Griffith and Kanamaru were well on their way to the same invention at roughly the same time. Under the Federal Circuit's decision, only Kanamaru gets a patent on the invention. What happens to Griffith? Not only is he not entitled to the exclusionary power of a patent on his idea, but Kanamaru can exclude *him* from using his own, independently developed idea. This result seems harsh, particularly where (as here) Griffith was working on his version of the invention long before the patent ever issued to Kanamaru.

To ameliorate this problem, a number of bills have been proposed in recent years that would grant "prior user rights" to non-patentees, if they could show that they were using the invention before it was patented by someone else. See H.R. 2795, discussed supra. And in fact, limited prior user rights are now in place in the area of business method patents. See American Inventors Protection Act, Pub. L. 106-113, 113 Stat. 1501, at §43, codified at 35 U.S.C §273 (Supp. V 1999). This proposal is similar to the rule in many European nations, which give prior user rights to continue doing whatever you were doing before the patent issued. Note that these rights are limited, however, because they do not allow the prior user to expand its sales or improve its product if doing so would infringe on the patent.

Are prior user rights a good idea? Do they unfairly weaken the patent grant? If prior user rights are to be enacted, should the patentee be compensated somehow—say, by receiving a compulsory licensing fee from the prior user? Should such rights be transferable? If so, would a patentee worry that its most aggressive competitor could buy a prior user right from someone else? Stephen Maurer and Suzanne Scotchmer, The Independent Inventor Defense in Intellectual Property, 69 Economica 535 (2002). Note that prior user rights for business method inventions are not generally transferable. See 35 U.S.C. §273(3)(C)(6) (Supp. V 1999).

Note on Recent Changes to Patent Law:
International Harmonization

As part of its obligations under the Uruguay Round of trade negotiations under the General Agreement on Trade and Tariffs (GATT), concluded in 1994, the United States made the following changes in domestic U.S. patent law. They were the result of the 1994 GATT agreement known as "Trade-Related Aspects of Intellectual Property," or TRIPs.

- The term of a patent is now 20 years, measured from the date the patent application was filed. It had been 17 years, measured from the date the patent was issued by the Patent Office. 35 U.S.C. § 154. Under a number of circumstances, such as interferences and appealed rejections, this term may be extended for up to five years. § 154(b).
- Members of the World Trade Organization (WTO) may introduce evidence of pre-patent filing inventive acts in their home country for purposes of establishing entitlement to a patent under the U.S. "first to invent" system. See 35 U.S.C. § 104. Prior to this amendment, only U.S. inventors were permitted to introduce such evidence, i.e., evidence of conception and reduction to practice; foreign inventors applying for a U.S. patent could only rely on inventive activity *in the U.S.* or on their first patent filing date in the United States or elsewhere. This was an obvious disadvantage to foreign applicants, and in keeping with GATT-TRIPs, it was eliminated. (The same treatment had been extended earlier to members of the North American Free Trade Agreement, or NAFTA, viz., Canada and Mexico.)
- The definition of infringement was expanded to include the acts of unauthorized offering for sale and importing. The old definition specified only making, using, and selling. 35 U.S.C. § 271.
- A new section of the patent statute permitting "provisional applications," 35 U.S.C. § 111, was added. While a provisional application must be fully enabling under § 112, it does not have to include any claims. A brief description and drawing suffice under this new section to establish the applicant's priority, provided that a more complete application is filed within one year. Importantly, filing a provisional application does not begin the 20-year clock for the applicant's patent term; only the filing of a full-blown application, with a claim or claims, can do so.

Some additional steps toward harmonization were taken with the American Inventors Protection Act of 1999 (AIPA). The AIPA made a number of substantive changes to U.S. patent law, the most important of which were:

- Prior user rights for business method inventions only, protecting prior inventors from a later, valid patent (35 U.S.C. § 273);
- Publication of most U.S. patent applications 18 months after filing; applications without any foreign counterpart, i.e., where only U.S. protection is sought for an invention, are exempt (35 U.S.C. § 122(b));
- Expanded patent reexamination procedures permitting more third party participation — but with a very large caveat: full *res judicata* effect for all issues covered in the reexamination (which is expected to make this promising change effectively useless) (35 U.S.C. § 311 et seq.).

In general, these changes are a far cry from full international harmonization. In Germany, Britain, France, and Japan, for example, prior user rights are available for all patented inventions, not just business methods. In Europe and Japan, all patent applications—not just those with foreign counterparts—are published. And in Europe and Japan, robust opposition systems allow competitors to challenge patents in an administrative proceeding for limited periods after issuance.

At the same time, these are major deviations from the U.S. baseline of only a few years ago. Hence they may represent an opening wedge for even more systematic harmonization to come.

5. Nonobviousness

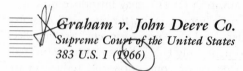

Graham v. John Deere Co.
Supreme Court of the United States
383 U.S. 1 (1966)

Mr. Justice CLARK delivered the opinion of the Court.

After a lapse of 15 years, the Court again focuses its attention on the patentability of inventions under the standard of Art. I, §8, cl. 8, of the Constitution and under the conditions prescribed by the laws of the United States. Since our last expression on patent validity, A. & P. Tea Co. v. Supermarket Corp., 340 U.S. 147 (1950), the Congress has for the first time expressly added a third statutory dimension to the two requirements of novelty and utility that had been the sole statutory test since the Patent Act of 1793. This is the test of obviousness, i.e., whether "the subject matter sought to be patented and the prior art are such that the subject matter as a whole would have been obvious at the time the invention was made to a person having ordinary skill in the art to which said subject matter pertains. Patentability shall not be negatived by the manner in which the invention was made." §103 of the Patent Act of 1952, 35 U.S.C. §103 (1964 ed.).

The questions, involved in each of the companion cases before us, are what effect the 1952 Act had upon traditional statutory and judicial tests of patentability and what definitive tests are now required. We have concluded that the 1952 Act was intended to codify judicial precedents embracing the principle long ago announced by this Court in Hotchkiss v. Greenwood, 11 How. 248 [52 U.S.] (1851), and that, while the clear language of §103 places emphasis on an inquiry into obviousness, the general level of innovation necessary to sustain patentability remains the same.

I

(a). Graham v. John Deere Co., an infringement suit by petitioners, presents a conflict between two Circuits over the validity of a single patent on a "Clamp for vibrating Shank Plows." The invention, a combination of old mechanical elements, involves a device designed to absorb shock from plow shanks as they plow through rocky soil and thus to prevent damage to the plow. We granted certiorari. Although we have determined that neither Circuit applied the correct test, we conclude that the patent is invalid under §103....

II

At the outset it must be remembered that the federal patent power stems from a specific constitutional provision which authorizes the Congress "To promote the Progress of . . . useful Arts, by securing for limited Times to . . . Inventors the exclusive Right to their . . . Discoveries." Art. I, § 8, cl. 8. The clause is both a grant of power and a limitation. This qualified authority, unlike the power often exercised in the sixteenth and seventeenth centuries by the English Crown, is limited to the promotion of advances in the "useful arts." It was written against the backdrop of the practices — eventually curtailed by the Statute of Monopolies — of the Crown in granting monopolies to court favorites in goods or businesses which had long before been enjoyed by the public. See Meinhardt, Inventions, Patents and Monopoly, pp. 30-35 (London, 1946). The Congress in the exercise of the patent power may not overreach the restraints imposed by the stated constitutional purpose. Nor may it enlarge the patent monopoly without regard to the innovation, advancement or social benefit gained thereby. Moreover, Congress may not authorize the issuance of patents whose effects are to remove existent knowledge from the public domain, or to restrict free access to materials already available. Innovation, advancement, and things which add to the sum of useful knowledge are inherent requisites in a patent system which by constitutional command must "promote the Progress of . . . useful Arts." This is the standard expressed in the Constitution and it may not be ignored. And it is in this light that patent validity "requires reference to a standard written into the Constitution." A. & P. Tea Co. v. Supermarket Corp., supra, at 154 (concurring opinion).

Within the limits of the constitutional grant, the Congress may, of course, implement the stated purpose of the Framers by selecting the policy which in its judgment best effectuates the constitutional aim. This is but a corollary to the grant to Congress of any Article I power. Gibbons v. Ogden, 9 Wheat. [22 U.S.] 1 [1824]. Within the scope established by the Constitution, Congress may set out conditions and tests for patentability. McClurg v. Kingsland, 1 How. [42 U.S.] 202, 206 [1843]. It is the duty of the Commissioner of Patents and of the courts in the administration of the patent system to give effect to the constitutional standard by appropriate application, in each case, of the statutory scheme of the Congress.

Congress quickly responded to the bidding of the Constitution by enacting the Patent Act of 1790 during the second session of the First Congress. It created an agency in the Department of State headed by the Secretary of State, the Secretary of the Department of War and the Attorney General, any two of whom could issue a patent for a period not exceeding 14 years to any petitioner that "hath . . . invented or discovered any useful art, manufacture, . . . or device, or any improvement therein not before known or used" if the board found that "the invention or discovery [was] sufficiently useful and important. . . ." 1 Stat. 110. This group, whose members administered the patent system along with their other public duties, was known by its own designation as "Commissioners for the Promotion of Useful Arts."

Thomas Jefferson, who as Secretary of State was a member of the group, was its moving spirit and might well be called the "first administrator of our patent system." See Federico, Operation of the Patent Act of 1790, 18 J. Pat. Off. Soc. 237, 238 (1936). He was not only an administrator of the patent system under the

1790 Act, but was also the author of the 1793 Patent Act. In addition, Jefferson was himself an inventor of great note. His unpatented improvements on plows, to mention but one line of his inventions, won acclaim and recognition on both sides of the Atlantic. Because of his active interest and influence in the early development of the patent system, Jefferson's views on the general nature of the limited patent monopoly under the Constitution, as well as his conclusions as to conditions for patentability under the statutory scheme, are worthy of note.

Jefferson, like other Americans, had an instinctive aversion to monopolies. It was a monopoly on tea that sparked the Revolution and Jefferson certainly did not favor an equivalent form of monopoly under the new government. His abhorrence of monopoly extended initially to patents as well. From France, he wrote to Madison (July 1788) urging a Bill of Rights provision restricting monopoly, and as against the argument that limited monopoly might serve to incite "ingenuity," he argued forcefully that "the benefit even of limited monopolies is too doubtful to be opposed to that of their general suppression," V Writings of Thomas Jefferson, at 47 (Ford ed., 1895).

His views ripened, however, and in another letter to Madison (Aug. 1789) after the drafting of the Bill of Rights, Jefferson stated that he would have been pleased by an express provision in this form: "Art. 9. Monopolies may be allowed to persons for their own productions in literature & their own inventions in the arts, for a term not exceeding—years but for no longer term & no other purpose." Id., at 113. And he later wrote: "Certainly an inventor ought to be allowed a right to the benefit of his invention for some certain time.... Nobody wishes more than I do that ingenuity should receive a liberal encouragement." Letter to Oliver Evans (May 1807), V Writings of Thomas Jefferson, at 75-76 (Washington ed.).

Jefferson's philosophy on the nature and purpose of the patent monopoly is expressed in a letter to Isaac McPherson (Aug. 1813), a portion of which we set out in the margin. [2] He rejected a natural-rights theory in intellectual property rights and clearly recognized the social and economic rationale of the patent system. The patent monopoly was not designed to secure to the inventor his natural right in his discoveries. Rather, it was a reward, an inducement, to bring forth new knowledge. The grant of an exclusive right to an invention was the creation of society — at odds with the inherent free nature of disclosed ideas — and was not to be freely given. Only inventions and discoveries which furthered human knowledge, and were new and

---

2. Stable ownership is the gift of social law, and is given late in the progress of society. It would be curious then, if an idea, the fugitive fermentation of an individual brain, could, of natural right, be claimed in exclusive and stable property. If nature has made any one thing less susceptible than all others of exclusive property, it is the action of the thinking power called an idea, which an individual may exclusively possess as long as he keeps it to himself; but the moment it is divulged, it forces itself into the possession of every one, and the receiver cannot dispossess himself of it. Its peculiar character, too, is that no one possesses the less, because every other possesses the whole of it. He who receives an idea from me, receives instruction himself without lessening mine; as he who lights his taper at mine, receives light without darkening me. That ideas should freely spread from one to another over the globe, for the moral and mutual instruction of man, and improvement of his condition, seems to have been peculiarly and benevolently designed by nature, when she made them, like fire, expansible over all space, without lessening their density in any point, and like the air in which we breathe, move, and have our physical being, incapable of confinement or exclusive appropriation. Inventions then cannot, in nature, be a subject of property. Society may give an exclusive right to the profits arising from them, as an encouragement to men to pursue ideas which may produce utility, but this may or may not be done, according to the will and convenience of the society, without claim or complaint from any body.

VI Writings of Thomas Jefferson 180-81 (H. A. Washington ed.).

useful, justified the special inducement of a limited private monopoly. Jefferson did not believe in granting patents for small details, obvious improvements, or frivolous devices. His writings evidence his insistence upon a high level of patentability.

As a member of the patent board for several years, Jefferson saw clearly the difficulty in "drawing a line between the things which are worth to the public the embarrassment of an exclusive patent, and those which are not." The board on which he served sought to draw such a line and formulated several rules which are preserved in Jefferson's correspondence.[3] Despite the board's efforts, Jefferson saw "with what slow progress a system of general rules could be matured." Because of the "abundance" of cases and the fact that the investigations occupied "more time of the members of the board than they could spare from higher duties, the whole was turned over to the judiciary, to be matured into a system, under which every one might know when his actions were safe and lawful." Letter to McPherson, supra, at 181, 182. Apparently Congress agreed with Jefferson and the board that the courts should develop additional conditions for patentability. Although the Patent Act was amended, revised or codified some 50 times between 1790 and 1950, Congress steered clear of a statutory set of requirements other than the bare novelty and utility tests reformulated in Jefferson's draft of the 1793 Patent Act.

III

The difficulty of formulating conditions for patentability was heightened by the generality of the constitutional grant and the statutes implementing it, together with the underlying policy of the patent system that "the things which are worth to the public the embarrassment of an exclusive patent," as Jefferson put it, must outweigh the restrictive effect of the limited patent monopoly. The inherent problem was to develop some means of weeding out those inventions which would not be disclosed or devised but for the inducement of a patent.

This Court formulated a general condition of patentability in 1851 in Hotchkiss v. Greenwood, 11 How. 248 [52 U.S. (1851)]. The patent involved a mere substitution of materials — porcelain or clay for wood or metal in doorknobs — and the Court condemned it, holding:

> [Unless] more ingenuity and skill ... were required ... than were possessed by an ordinary mechanic acquainted with the business, there was an absence of that degree of skill and ingenuity which constitute essential elements of every invention. In other words, the improvement is the work of the skilful mechanic, not that of the inventor. At p. 267.

Hotchkiss test of patentability

---

3. "[A] machine of which we are possessed, might be applied by every man to any use of which it is susceptible." Letter to Isaac McPherson, id. at 181.

"[A] change of material should not give title to a patent. As the making a ploughshare of cast rather than of wrought iron; a comb of iron instead of horn or of ivory ..." Id.

"[A] mere change of form should give no right to a patent, as a high-quartered shoe instead of a low one; a round hat instead of a three-square; or a square bucket instead of a round one." Id. at 181-82.

"[A combined use of old implements.] A man has a right to use a saw, an axe, a plane separately; may he not combine their uses on the same piece of wood?" Letter to Oliver Evans (Jan. 1814), id. at 298.

Hotchkiss, by positing the condition that a patentable invention evidence more ingenuity and skill than that possessed by an ordinary mechanic acquainted with the business, merely distinguished between new and useful innovations that were capable of sustaining a patent and those that were not. The *Hotchkiss* test laid the cornerstone of the judicial evolution suggested by Jefferson and left to the courts by Congress. The language in the case, and in those which followed, gave birth to "invention" as a word of legal art signifying patentable inventions. Yet, as this Court has observed, "[t]he truth is the word [invention] cannot be defined in such manner as to afford any substantial aid in determining whether a particular device involves an exercise of the inventive faculty or not." McClain v. Ortmayer, 141 U.S. 419, 427 (1891); A. & P. Tea Co. v. Supermarket Corp., supra, at 151. Its use as a label brought about a large variety of opinions as to its meaning both in the Patent Office, in the courts, and at the bar. The *Hotchkiss* formulation, however, lies not in any label, but in its functional approach to questions of patentability. In practice, *Hotchkiss* has required a comparison between the subject matter of the patent, or patent application, and the background skill of the calling. It has been from this comparison that patentability was in each case determined.

IV. The 1952 Patent Act

The Act sets out the conditions of patentability in three sections. An analysis of the structure of these three sections indicates that patentability is dependent upon three explicit conditions: novelty and utility as articulated and defined in § 101 and § 102, and non-obviousness, the new statutory formulation, as set out in § 103. The first two sections, which trace closely the 1874 codification, express the "new and useful" tests which have always existed in the statutory scheme and, for our purposes here, need no clarification. The pivotal section around which the present controversy centers is § 103. It provides:

§ 103. Conditions for patentability; non-obvious subject matter

> A patent may not be obtained though the invention is not identically disclosed or described as set forth in section 102 of this title, if the differences between the subject matter sought to be patented and the prior art are such that the subject matter as a whole would have been obvious at the time the invention was made to a person having ordinary skill in the art to which said subject matter pertains. Patentability shall not be negatived by the manner in which the invention was made.

The section is cast in relatively unambiguous terms. Patentability is to depend, in addition to novelty and utility, upon the "non-obvious" nature of the "subject matter sought to be patented" to a person having ordinary skill in the pertinent art....

It is undisputed that this section was, for the first time, a statutory expression of an additional requirement for patentability, originally expressed in *Hotchkiss.* It also seems apparent that Congress intended by the last sentence of § 103 to abolish the test it believed this Court announced in the controversial phrase "flash of creative genius," used in Cuno Corp. v. Automatic Devices Corp., 314 U.S. 84 (1941).[7] ...

V ...

While the ultimate question of patent validity is one of law, A. & P. Tea Co. v. Supermarket Corp., supra, at 155, the § 103 condition, which is but one of three conditions, each of which must be satisfied, lends itself to several basic factual inquiries. Under § 103, the scope and content of the prior art are to be determined; differences between the prior art and the claims at issue are to be ascertained; and the level of ordinary skill in the pertinent art resolved. Against this background the obviousness or nonobviousness of the subject matter is determined. Such secondary considerations as commercial success, long felt but unsolved needs, failure of others, etc., might be utilized to give light to the circumstances surrounding the origin of the subject matter sought to be patented. As indicia of obviousness or nonobviousness, these inquiries may have relevancy. See Note, Subtests of "Nonobviousness": A Nontechnical Approach to Patent Validity, 112 U. Pa. L. Rev. 1169 (1964).

This is not to say, however, that there will not be difficulties in applying the nonobviousness test. What is obvious is not a question upon which there is likely to be uniformity of thought in every given factual context. The difficulties, however, are comparable to those encountered daily by the courts in such frames of reference as negligence and scienter, and should be amenable to a case-by-case development. We believe that strict observance of the requirements laid down here will result in that uniformity and definiteness which Congress called for in the 1952 Act.

While we have focused attention on the appropriate standard to be applied by the courts, it must be remembered that the primary responsibility for sifting out unpatentable material lies in the Patent Office. To await litigation is — for all practical purposes — to debilitate the patent system. We have observed a notorious difference between the standards applied by the Patent Office and by the courts. While many reasons can be adduced to explain the discrepancy, one may well be the free rein often exercised by Examiners in their use of the concept of "invention." In this connection we note that the Patent Office is confronted with a most difficult task. Almost

---

7. The sentence in which the phrase occurs reads: "[T]he new device, however useful it may be, must reveal the flash of creative genius, not merely the skill of the calling." At p. 91. Although some writers and lower courts found in the language connotations as to the frame of mind of the inventors, none were so intended. The opinion approved *Hotchkiss* specifically, and the reference to "flash of creative genius" was but a rhetorical embellishment of language going back to 1833. Cf. "exercise of genius," Shaw v. Cooper, 7 Pet. 292; "inventive genius," Reckendorfer v. Faber, 92 U.S. 347 (1876); Concrete Appliances Co. v. Gomery, 269 U.S. 177; "flash of thought," Densmore v. Scofield, 102 U.S. 375 (1880); "intuitive genius," Potts v. Creager, 155 U.S. 597 (1895). Rather than establishing a more exacting standard, *Cuno* merely rhetorically restated the requirement that the subject matter sought to be patented must be beyond the skill of the calling. It was the device, not the invention, that had to reveal the "flash of creative genius." See Boyajian, The Flash of Creative Genius, An Alternative Interpretation, 25 J. Pat. Off. Society 776, 780, 781 (1943); Pacific Contact Laboratories, Inc. v. Solex Laboratories, Inc., 209 F.2d 529, 533; Brown & Sharpe Mfg. Co. v. Kar Engineering Co., 154 F.2d 48, 51-52; In re Shortell, 31 C.C.P.A. (Pat.) 1062, 1069, 142 F.2d 292, 295-96.

100,000 applications for patents are filed each year. Of these, about 50,000 are granted and the backlog now runs well over 200,000. 1965 Annual Report of the Commissioner of Patents 13-14. This is itself a compelling reason for the Commissioner to strictly adhere to the 1952 Act as interpreted here. This would, we believe, not only expedite disposition but bring about a closer concurrence between administrative and judicial precedent.

We have been urged to find in § 103 a relaxed standard, supposedly a congressional reaction to the "increased standard" applied by this Court in its decisions over the last 20 or 30 years. The standard has remained invariable in this Court. Technology, however, has advanced — and with remarkable rapidity in the last 50 years. Moreover, the ambit of applicable art in given fields of science has widened by disciplines unheard of a half century ago. It is but an evenhanded application to require that those persons granted the benefit of a patent monopoly be charged with an awareness of these changed conditions. The same is true of the less technical, but still useful arts. He who seeks to build a better mousetrap today has a long path to tread before reaching the Patent Office.

VI

We now turn to the application of the conditions found necessary for patentability to the cases involved here:

A. *The Patent in Issue in Graham v. John Deere Co.*

This patent, No. 2,627,798 (hereinafter called the '798 patent) relates to a spring clamp which permits plow shanks to be pushed upward when they hit obstructions in the soil, and then springs the shanks back into normal position when the obstruction is passed over. The device . . . is fixed to the plow frame as a unit. The mechanism around which the controversy centers is basically a hinge. The top half of it, known as the upper plate, is a heavy metal piece clamped to the plow frame and is stationary relative to the plow frame. The lower half of the hinge, known as the hinge plate, is connected to the rear of the upper plate by a hinge pin and rotates downward with respect to it. The shank, which is bolted to the forward end of the hinge plate, runs beneath the plate and parallel to it for about nine inches, passes through a stirrup, and then continues backward for several feet curving down toward the ground. The chisel, which does the actual plowing, is attached to the rear end of the shank. As the plow frame is pulled forward, the chisel rips through the soil, thereby plowing it. In the normal position, the hinge plate and the shank are kept tight against the upper plate by a spring, which is atop the upper plate. A rod runs through the center of the spring, extending down through holes in both plates and the shank. Its upper end is bolted to the top of the spring while its lower end is hooked against the underside of the shank [see Figure 3-4].

When the chisel hits a rock or other obstruction in the soil, the obstruction forces the chisel and the rear portion of the shank to move upward. The shank is pivoted against the rear of the hinge plate and pries open the hinge against the closing tendency of the spring. This closing tendency is caused by the fact that, as the hinge

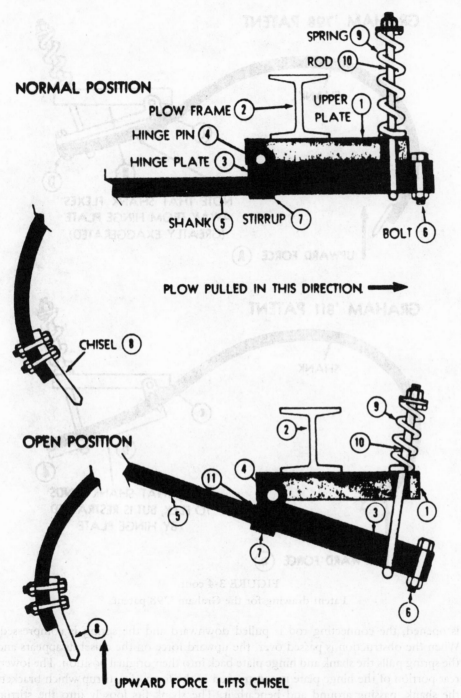

NORMAL POSITION

SPRING ⑨
ROD ⑩
PLOW FRAME ②
UPPER PLATE ①
HINGE PIN ④
HINGE PLATE ③
SHANK ⑤ STIRRUP ⑦
BOLT ⑥

PLOW PULLED IN THIS DIRECTION ➡

CHISEL ⑧

OPEN POSITION

UPWARD FORCE LIFTS CHISEL

FIGURE 3-4
Patent drawing for the Graham '798 patent.

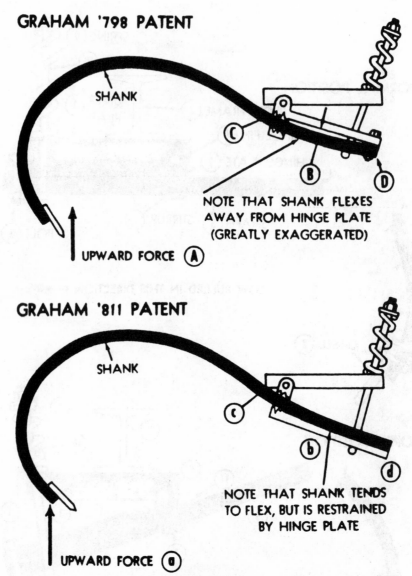

FIGURE 3-4 cont.
Patent drawing for the Graham '798 patent.

is opened, the connecting rod is pulled downward and the spring is compressed. When the obstruction is passed over, the upward force on the chisel disappears and the spring pulls the shank and hinge plate back into their original position. The lower, rear portion of the hinge plate is constructed in the form of a stirrup which brackets the shank, passing around and beneath it. The shank fits loosely into the stirrup (permitting a slight up and down play). The stirrup is designed to prevent the shank from recoiling away from the hinge plate, and thus prevents excessive strain on the shank near its bolted connection. The stirrup also girds the shank, preventing it from fishtailing from side to side.

In practical use, a number of spring-hinge-shank combinations are clamped to a plow frame, forming a set of ground-working chisels capable of withstanding the shock of rocks and other obstructions in the soil without breaking the shanks.

Background of the Patent

Chisel plows, as they are called, were developed for plowing in areas where the ground is relatively free from rocks or stones. Originally, the shanks were rigidly attached to the plow frames. When such plows were used in the rocky, glacial soils of some of the Northern States, they were found to have serious defects. As the chisels hit buried rocks, a vibratory motion was set up and tremendous forces were transmitted to the shank near its connection to the frame. The shanks would break. Graham, one of the petitioners, sought to meet that problem, and in 1950 obtained a patent, U.S. No. 2,493,811 (hereinafter '811), on a spring clamp which solved some of the difficulties. Graham and his companies manufactured and sold the '811 clamps. In 1950, Graham modified the '811 structure and filed for a patent. That patent, the one in issue, was granted in 1953. This suit against competing plow manufacturers resulted from charges by petitioners that several of respondents' devices infringed the '798 patent.

The Prior Art

Five prior patents indicating the state of the art were cited by the Patent Office in the prosecution of the '798 application. Four of these patents, 10 other United States patents and two prior-use spring-clamp arrangements not of record in the '798 file wrapper were relied upon by respondents as revealing the prior art. The District Court and the Court of Appeals found that the prior art "as a whole in one form or another contains all of the mechanical elements of the '798 Patent." One of the prior-use clamp devices not before the Patent Examiner—Glencoe—was found to have "all of the elements."

We confine our discussion to the prior patent of Graham, '811, and to the Glencoe clamp device, both among the references asserted by respondents. The Graham '811 and '798 patent devices are similar in all elements, save two: (1) the stirrup and the bolted connection of the shank to the hinge plate do not appear in '811; and (2) the position of the shank is reversed, being placed in patent '811 above the hinge plate, sandwiched between it and the upper plate. The shank is held in place by the spring rod which is hooked against the bottom of the hinge plate passing through a slot in the shank. Other differences are of no consequence to our examination. In practice the '811 patent arrangement permitted the shank to wobble or fishtail because it was not rigidly fixed to the hinge plate; moreover, as the hinge plate was below the shank, the latter caused wear on the upper plate, a member difficult to repair or replace.

Graham's '798 patent application contained 12 claims. All were rejected as not distinguished from the Graham '811 patent. The inverted position of the shank was specifically rejected as was the bolting of the shank to the hinge plate. The Patent Office examiner found these to be "matters of design well within the expected skill of the art and devoid of invention." Graham withdrew the original claims and substituted the two new ones which are substantially those in issue here. His contention was that wear was reduced in patent '798 between the shank and the heel or rear of the upper plate.[11] He also emphasized several new features, the relevant one here being that the bolt used to connect the hinge plate and shank maintained the upper face of the shank in continuing and constant contact with the underface of the hinge plate.

Graham did not urge before the Patent Office the greater "flexing" qualities of the '798 patent arrangement which he so heavily relied on in the courts. The sole element in patent '798 which petitioners argue before us is the interchanging of the shank and hinge plate and the consequences flowing from this arrangement. The contention is that this arrangement—which petitioners claim is not disclosed in the prior art—permits the shank to flex under stress for its *entire* length. [W]hen the chisel hits an obstruction the resultant force (A) pushes the rear of the shank upward and the shank pivots against the rear of the hinge plate at (C). The natural tendency is for that portion of the shank between the pivot point and the bolted connection (i.e., between C and D) to bow downward and away from the hinge plate. The maximum distance (B) that the shank moves away from the plate is slight—for emphasis, greatly exaggerated in the sketches. This is so because of the strength of the shank and the short—nine inches or so—length of that portion of the shank between (C) and (D). On the contrary, in patent '811 the pivot point is the upper plate at point (c); and while the tendency for the shank to bow between points (c) and (d) is the same as in '798, the shank is restricted because of the under-lying hinge plate and cannot flex as freely. In practical effect, the shank flexes only between points (a) and (c), and not along the entire length of the shank, as in '798. Petitioners say that this difference in flex, though small, effectively absorbs the tremendous forces of the shock of obstructions whereas prior art arrangements failed.

The Obviousness of the Differences

We cannot agree with petitioners. We assume that the prior art does not disclose such an arrangement as petitioners claim in patent '798. Still we do not believe that the argument on which petitioners' contention is bottomed supports the validity of the patent. The tendency of the shank to flex is the same in all cases. If free-flexing, as petitioners now argue, is the crucial difference above the prior art, then it appears evident that the desired result would be obtainable by not boxing the shank within the confines of the hinge. The only other effective place available in the arrangement was to attach it below the hinge plate and run it through a stirrup or bracket that would not disturb its flexing qualities. Certainly a person having ordinary skill in the prior art, given the fact that the flex in the shank could be utilized more effectively if allowed to run the entire length of the shank, would immediately see that the thing to do was what Graham did, i.e., invert the shank and the hinge plate.

11. In '811, where the shank was above the hinge plate, an upward movement of the chisel forced the shank up against the underside of the rear of the upper plate. The upper plate thus provided the fulcrum about which the hinge was pried open. Because of this, as well as the location of the hinge pin, the shank rubbed against the heel of the upper plate causing wear both to the plate and to the shank. By relocating the hinge pin and by placing the hinge plate between the shank and the upper plate, as in '798, the rubbing was eliminated and the wear point was changed to the hinge plate, a member more easily removed or replaced for repair.

Petitioners' argument basing validity on the free-flex theory raised for the first time on appeal is reminiscent of Lincoln Engineering Co. v. Stewart-Warner Corp., 303 U.S. 545 (1938), where the Court called such an effort "an afterthought. No such function...is hinted at in the specifications of the patent. If this were so vital an element in the functioning of the apparatus it is strange that all mention of it was omitted." At p. 550. No "flexing" argument was raised in the Patent Office. Indeed, the trial judge specifically found that "flexing is not a claim of the patent in suit..." and would not permit interrogation as to flexing in the accused devices. Moreover, the clear testimony of petitioners' experts shows that the flexing advantages flowing from the '798 arrangement are not, in fact, a significant feature in the patent.

We find no nonobvious facets in the '798 arrangement. The wear and repair claims were sufficient to overcome the patent examiner's original conclusions as to the validity of the patent. However, some of the prior art, notably Glencoe, was not before him. There the hinge plate is below the shank but, as the courts below found, all of the elements in the '798 patent are present in the Glencoe structure. Furthermore, even though the position of the shank and hinge plate appears reversed in Glencoe, the mechanical operation is identical. The shank there pivots about the underside of the stirrup, which in Glencoe is *above* the shank. In other words, the stirrup in Glencoe serves exactly the same function as the heel of the hinge plate in '798. The mere shifting of the wear point to the heel of the '798 hinge plate from the stirrup of Glencoe—itself a part of the hinge plate—presents no operative mechanical distinctions, much less nonobvious differences....

The judgment of the Court of Appeals in [Graham v. John Deere] is affirmed....

COMMENTS AND QUESTIONS

1. For extensive background on the *Graham* case, see John F. Duffy and Robert P. Merges, "The Standard of Creativity and the *Graham* Cases," in Intellectual Property Stories (Jane C. Ginsburg and Rochelle Cooper Dreyfuss, eds., 2006).

2. The Court states that it is proper in interpreting the Graham '798 claims to refer to the prosecution history (or "file wrapper") of the patent. We will see that this is an important tool in the law of infringement later in this chapter. For now, what is important is that the validity of the patent claims is only one part of a larger picture — "claim interpretation." Patentees not only want to have their patents upheld over the prior art, but they also want those patents to be interpreted broadly, to cover a wide range of potential infringements. Because these goals are in tension, it is important to keep them both in mind whenever you interpret the claims of a patent.

3. Without question, the plow design in *Graham* was new, i.e., novel under section 102 of the patent code. What policy is served by the detailed inquiry into whether it is "new enough" to deserve a patent, i.e., nonobvious? What would be the effect of a patent system that only required novelty? For one view of the matter, see Robert P. Merges, Uncertainty and the Standard of Patentability, 7 High Tech. L.J. 1 (1993). Arguably, section 103 requires an "inventive leap" of some degree over what has been done before as a counterbalance to the strong rights given patentholders.

The test established in 35 U.S.C. §103—whether the invention as a whole would be "obvious to one of ordinary skill in the art"—does not itself tell courts very much about how to decide what is obvious. Courts have developed a number of rules to assist in this determination. Two of the most important rules are discussed below.

a. Combining References

To anticipate a patent application under section 102, a single prior art reference must disclose every element of what the patentee claims as his invention. If a prior art reference does not disclose all the parts of an invention, it does not "anticipate" the application. Under section 103, however, a single reference need not disclose the entire invention to bar a patent. Thus section 103 asks whether a researcher who is aware of all the prior art would think to create the claimed invention. In deciding the question of obviousness, it is sometimes permissible to analyze a combination of ideas from different sources of prior art (known as prior art "references").

In re Vaeck
United States Court of Appeals for the Federal Circuit
947 F.2d 488 (Fed. Cir. 1991)

RICH, J.

This appeal is from the September 12, 1990 decision of the Patent and Trademark Office (PTO) Board of Patent Appeals and Interferences (Board), affirming the examiner's rejection of [almost all] claims...of [appellant's] application..., filed March 4, 1987,...as unpatentable under 35 U.S.C. 103.... We reverse the §103 rejection.

Background

A. The Invention

The claimed invention is directed to the use of genetic engineering techniques for production of proteins that are toxic to insects such as larvae of mosquitos and black flies. These swamp-dwelling pests are the source of numerous human health problems, including malaria. It is known that certain species of the naturally-occurring Bacillus genus of bacteria produce proteins ("endotoxins") that are toxic to these insects. Prior art methods of combatting the insects involved spreading or spraying crystalline spores of the insecticidal Bacillus proteins over swamps. The spores were environmentally unstable, however, and would often sink to the bottom of a swamp before being consumed, thus rendering this method prohibitively expensive. Hence the need for a lower-cost method of producing the insecticidal Bacillus proteins in high volume, with application in a more stable vehicle.

As described by appellants, the claimed subject matter meets this need by providing for the production of the insecticidal Bacillus proteins within host cyanobacteria. Although both cyanobacteria and bacteria are members of the procaryote kingdom [i.e., they both are organisms lacking a distinct cellular nucleus, as opposed to eukaryotes], the cyanobacteria (which in the past have been referred to as "blue-green algae") are unique among procaryotes in that the cyanobacteria are capable of oxygenic photosynthesis. The cyanobacteria grow on top of swamps where they are consumed by mosquitos and black flies. Thus, when Bacillus proteins are produced within transformed cyanobacterial hosts [i.e., when the Bacillus genes for those proteins have been successfully taken up by the foreign cyanobacteria and the genetic material has been made a permanent part of that host organism, to be replicated when it reproduces] according to the claimed invention, the presence of the insecticide in the food of the targeted insects advantageously guarantees direct uptake by the insects.

More particularly, the subject matter of the application on appeal includes a chimeric (i.e., hybrid) gene comprising (1) a gene derived from a bacterium of the Bacillus genus whose product is an insecticidal protein, united with (2) a DNA promoter effective for expressing the Bacillus gene in a host cyanobacterium, so as to produce the desired insecticidal protein....

D. The Grounds of Rejection

1. The § 103 Rejections

Claims 1-6, 16-21, 33-38, 47-48 and 52 (which include all independent claims in the application) were rejected as unpatentable under 35 U.S.C. 103 based upon [the] Dzelzkalns [reference] in view of [the] Sekar I or Sekar II and Ganesan [references]. The examiner stated that Dzelzkalns discloses a chimeric gene capable of being highly expressed in a cyanobacterium, said gene comprising a promoter region effective for expression in a cyanobacterium operably linked to a structural gene encoding [a protein called chloramphenicol acetyl transferase, abbreviated "CAT"]. The examiner acknowledged [the differences between the proteins, but] pointed out [that] Sekar I, Sekar II, and Ganesan teach genes encoding insecticidally active proteins produced by Bacillus, and the advantages of expressing such genes in... hosts [from another species] to obtain larger quantities of the protein. The examiner contended that it would have been obvious to one of ordinary skill in the art to substitute the Bacillus genes taught by Sekar I, Sekar II, and Ganesan for the CAT gene in the... Dzelzkalns [reference] in order to obtain high level expression of the Bacillus genes in the transformed cyanobacteria. The examiner further contended that it would have been obvious to use cyanobacteria as [a host] for expression of the claimed genes....

Opinion

A. Obviousness

We first address whether the PTO erred in rejecting the claims on appeal as prima facie obvious within the meaning of 35 U.S.C. § 103. Obviousness is a legal question which this court independently reviews, though based upon underlying factual findings which we review under the clearly erroneous standard.

Where claimed subject matter has been rejected as obvious in view of a combination of prior art references, a proper analysis under § 103 requires, inter alia, consideration of two factors: (1) whether the prior art would have suggested to those of ordinary skill in the art that they should make the claimed composition or device, or carry out the claimed process; and (2) whether the prior art would also have revealed that in so making or carrying out, those of ordinary skill would have a reasonable expectation of success. See In re Dow Chemical Co., 837 F.2d 469, 473 (Fed. Cir. 1988). Both the suggestion and the reasonable expectation of success must be founded in the prior art, not in the applicant's disclosure. Id.

We agree with appellants that the PTO has not established the prima facie obviousness of the claimed subject matter. The prior art simply does not disclose or suggest the expression in cyanobacteria of a chimeric gene encoding an insecticidally active protein, or convey to those of ordinary skill a reasonable expectation of success in doing so. More particularly, there is no suggestion in Dzelzkalns, the primary reference cited against all claims, of substituting in the disclosed plasmid a structural gene encoding Bacillus insecticidal proteins for the CAT gene utilized for selection purposes. The expression of antibiotic resistance-conferring genes in cyanobacteria, without more, does not render obvious the expression of unrelated genes in cyanobacteria for unrelated purposes.

The PTO argues that the substitution of insecticidal Bacillus genes for CAT marker genes in cyanobacteria is suggested by the secondary references Sekar I, Sekar II, and Ganesan, which collectively disclose expression of genes encoding Bacillus insecticidal proteins in two species of host Bacillus bacteria (B. megaterium and B. subtilis) as well as in the bacterium E. coli. While these references disclose expression of Bacillus genes encoding insecticidal proteins in certain transformed bacterial hosts, nowhere do these references disclose or suggest expression of such genes in transformed cyanobacterial hosts.

To remedy this deficiency, the PTO emphasizes similarity between bacteria and cyanobacteria, namely, that these are both procaryotic organisms, and argues that this fact would suggest to those of ordinary skill the use of cyanobacteria as hosts for expression of the claimed chimeric genes. While it is true that bacteria and cyanobacteria are now both classified as procaryotes, that fact alone is not sufficient to motivate the art worker as the PTO contends. As the PTO concedes, cyanobacteria and bacteria are not identical; they are classified as two separate divisions of the kingdom Procaryotae. Moreover, it is only in recent years that the biology of cyanobacteria has been clarified, as evidenced by references in the prior art to "blue-green algae." Such evidence of recent uncertainty regarding the biology of cyanobacteria tends to rebut, rather than support, the PTO's position that one would consider the cyanobacteria effectively interchangeable with bacteria as hosts for expression of the claimed gene....

The PTO asks us to agree that the prior art would lead those of ordinary skill to conclude that cyanobacteria are attractive hosts for expression of any and all heterologous genes. Again, we can not. The relevant prior art does indicate that cyanobacteria are attractive hosts for expression of both native and heterologous [i.e., foreign] genes involved in photosynthesis (not surprisingly, for the capability of undergoing oxygenic photosynthesis is what makes the cyanobacteria unique among procaryotes). However, these references do not suggest that cyanobacteria would be equally attractive hosts for expression of unrelated heterologous genes, such as the claimed genes encoding Bacillus insecticidal proteins.

In [In re] O'Farrell [853 F.2d 894 (Fed. Cir. 1988)], this court affirmed an obviousness rejection of a claim to a method for producing a "predetermined protein in a stable form" in a transformed bacterial host. 853 F.2d at 895.... The main difference between the prior art and the claim at issue was that in [the] Polisky [reference], the heterologous gene was a gene for ribosomal RNA, while the claimed invention substituted a gene coding for a predetermined protein. Id. at 901. Although, as the appellants therein pointed out, the ribosomal RNA gene is not normally translated into protein, Polisky mentioned preliminary evidence that the transcript of the ribosomal RNA gene was translated into protein, and further predicted that if a gene coding for a protein were to be substituted, extensive translation might result. Id. We thus affirmed, explaining that

> the prior art explicitly suggested the substitution that is the difference between the claimed invention and the prior art, and presented preliminary evidence suggesting that the [claimed] method could be used to make proteins....
> ...Polisky contained detailed enabling methodology for practicing the claimed invention, a suggestion to modify the prior art to practice the claimed invention, and evidence suggesting that it would be successful.

Id. at 901-02.

In contrast with the situation in O'Farrell, the prior art in this case offers no suggestion, explicit or implicit, of the substitution that is the difference between the claimed invention and the prior art. Moreover, the "reasonable expectation of success" that was present in O'Farrell is not present here. Accordingly, we reverse the § 103 rejections.

In re Dembiczak
United States Court of Appeals for the Federal Circuit
175 F.3d 994 (Fed. Cir. 1999)

CLEVENGER, Circuit Judge.

Anita Dembiczak and Benson Zinbarg appeal the rejection, upheld by the Board of Patent Appeals and Interferences, of all pending claims in their Application. Because the Board erred in sustaining rejections of the pending claims as obvious under 35 U.S.C. § 103(a) (Supp.1998), we reverse.

I

The invention at issue in this case is, generally speaking, a large trash bag made of orange plastic and decorated with lines and facial features, allowing the bag, when filled with trash or leaves, to resemble a Halloween-style pumpkin, or jack-o'-lantern. [See Figure 3-5] As the inventors, Anita Dembiczak and Benson Zinbarg (collectively, "Dembiczak") note, the invention solves the long-standing problem of unsightly trash bags placed on the curbs of America, and, by fortuitous happenstance, allows users to express their whimsical or festive nature while properly storing garbage, leaves, or other household debris awaiting collection. Embodiments of the invention — sold under a variety of names, including Giant Stuff-A-Pumpkin,

FIGURE 3-5
Dembiczak Pumpkin.

Funkins, Jack Sak, and Bag-O-Fun — have undisputedly been well-received by consumers, who bought more than seven million units in 1990 alone. Indeed, in 1990, the popularity of the pumpkin bags engendered a rash of thefts around Houston, Texas, leading some owners to resort to preventative measures, such as greasing the bags with petroleum jelly and tying them to trees.

The road to profits has proved much easier than the path to patentability, however. In July 1989, Dembiczak filed a utility patent application generally directed to the pumpkin bags. [After it was rejected, the Patent Office Board of Appeals affirmed the rejection of the patent application on grounds of obviousness, and Dembiczak appealed.]

A

The patent application at issue includes claims directed to various embodiments of the pumpkin bag. Claims 37, 49, 51, 52, 58 through 64, 66 through 69, and 72 through 81 are at issue in this appeal. Though the claims vary, independent claim 74 is perhaps most representative:

> 74. A decorative bag for use by a user with trash filling material, the bag simulating the general outer appearance of an outer surface of a pumpkin having facial indicia thereon, comprising:
> a flexible waterproof plastic trash or leaf bag having
> an outer surface which is premanufactured orange in color for the user to simulate the general appearance of the outer skin of a pumpkin, and having
> facial indicia including at least two of an eye, a nose and a mouth on the orange color outer surface for forming a face pattern on said orange color outer surface to simulate the general outer appearance of a decorative pumpkin with a face thereon,

said trash or leaf bag having first and second opposite ends, at least said second end having an opening extending substantially across the full width of said trash or leaf bag for receiving the trash filling material,

wherein when said trash or leaf bag is filled with trash filling material and closed, said trash or leaf bag takes the form and general appearance of a pumpkin with a face thereon.

All of the independent claims on appeal, namely 37, 52, 72, and 74, contain limitations that the bag must be "premanufactured orange in color," have "facial indicia," have openings suitable for filling with trash material, and that when filled, the bag must have a generally rounded appearance, like a pumpkin. Independent claims 37, 52, and 72 add the limitation that the bag's height must be at least 36 inches. Claim 72 requires that the bag be made of a "weatherproof material," and claim 74, as shown above, requires that the bag be "waterproof." Claim 52 recites a "method of assembling" a bag with the general characteristics of apparatus claim 37.

B

The prior art cited by the Board includes:

(1) pages 24-25 of a book entitled "A Handbook for Teachers of Elementary Art," by Holiday Art Activities ("Holiday"), describing how to teach children to make a "Crepe Paper Jack-O-Lantern" out of a strip of orange crepe paper, construction paper cut-outs in the shape of facial features, and "wadded newspapers" as filling;

(2) page 73 of a book entitled "The Everything Book for Teachers of Young Children," by Martha Shapiro and Valerie Indenbaum ("Shapiro"), describing a method of making a "paper bag pumpkin" by stuffing a bag with newspapers, painting it orange, and then painting on facial features with black paint;

(3) U.S. Patent No. 3,349,991 to Leonard Kessler, entitled "Flexible Container" ("Kessler"), describing a bag apparatus wherein the bag closure is accomplished by the use of folds or gussets in the bag material;

(4) Prior art "conventional" plastic lawn or trash bags ("the conventional trash bags").

[handwritten margin note: prior art]

Using this art, the Board affirmed the Examiner's final rejection of all the independent claims (37, 52, 72, 74) under 35 U.S.C. § 103, holding that they would have been obvious in light of the conventional trash bags in view of the Holiday and Shapiro references. The Board determined that, in its view of the prior art, "the only difference between the invention presently defined in the independent claims on appeal and the orange plastic trash bags of the prior art and the use of such bags resides in the application of the facial indicia to the outer surface of the bag." The Board further held that the missing facial indicia elements were provided by the Holiday and Shapiro references' description of painting jack-o'-lantern faces on paper bags. Dependent claims 49 and 79, which include a "gussets" limitation, were considered obvious under similar reasoning, except that the references cited against them included Kessler.

[handwritten margin note: Anticipated by prior art]

II ...

A

Our analysis begins in the text of section 103...with the phrase "at the time the invention was made." For it is this phrase that guards against entry into the "tempting but forbidden zone of hindsight," when analyzing the patentability of claims pursuant to that section. Measuring a claimed invention against the standard established by section 103 requires the oft-difficult but critical step of casting the mind back to the time of invention, to consider the thinking of one of ordinary skill in the art, guided only by the prior art references and the then-accepted wisdom in the field. See, e.g., W.L. Gore & Assocs., Inc. v. Garlock, Inc., 721 F.2d 1540, 1553, 220 UPSQ 303, 313 (Fed.Cir.1983). Close adherence to this methodology is especially important in the case of less technologically complex inventions, where the very ease with which the invention can be understood may prompt one "to fall victim to the insidious effect of a hindsight syndrome wherein that which only the inventor taught is used against its teacher." Id.

Our case law makes clear that the best defense against the subtle but powerful attraction of a hindsight-based obviousness analysis is rigorous application of the requirement for a showing of the teaching or motivation to combine prior art references. See, e.g., C.R. Bard, Inc. v. M3 Sys., Inc., 157 F.3d 1340, 1352, 48 USPQ2d 1225, 1232 (Fed.Cir.1998) (describing "teaching or suggestion or motivation [to combine]" as an "essential evidentiary component of an obviousness holding").

We have noted that evidence of a suggestion, teaching, or motivation to combine may flow from the prior art references themselves, the knowledge of one of ordinary skill in the art, or, in some cases, from the nature of the problem to be solved. The range of sources available, however, does not diminish the requirement for actual evidence. That is, the showing must be clear and particular. Broad conclusory statements regarding the teaching of multiple references, standing alone, are not "evidence."

All the obviousness rejections affirmed by the Board resulted from a combination of prior art references, e.g., the conventional trash or yard bags, and the Holiday and Shapiro publications teaching the construction of decorated paper bags. To justify this combination, the Board simply stated that "the Holiday and Shapiro references would have suggested the application of...facial indicia to the prior art plastic trash bags." However, rather than pointing to specific information in Holiday or Shapiro that suggest the combination with the conventional bags, the Board instead described in detail the similarities between the Holiday and Shapiro references and the claimed invention, noting that one reference or the other — in combination with each other and the conventional trash bags — described all of the limitations of the pending claims. Nowhere does the Board particularly identify any suggestion, teaching, or motivation to combine the children's art references (Holiday and Shapiro) with the conventional trash or lawn bag references, nor does the Board make specific — or even inferential — findings concerning the identification of the relevant art, the level of ordinary skill in the art, the nature of the problem to be solved, or any other factual findings that might serve to support a proper obviousness analysis.

To the contrary, the obviousness analysis in the Board's decision is limited to a discussion of the ways that the multiple prior art references can be combined to read on the claimed invention. For example, the Board finds that the Holiday bag reference depicts a "premanufactured orange" bag material, finds that Shapiro teaches the use of paper bags in various sizes, including "large", and concludes that the substitution of orange plastic for the crepe paper of Holiday and the paper bags of Shapiro would be an obvious design choice. Yet this reference-by-reference, limitation-by-limitation analysis fails to demonstrate how the Holiday and Shapiro references teach or suggest their combination with the conventional trash or lawn bags to yield the claimed invention. Because we do not discern any finding by the Board that there was a suggestion, teaching, or motivation to combine the prior art references cited against the pending claims, the Board's conclusion of obviousness, as a matter of law, cannot stand.

B

Because the Board has not established a prima facie case of obviousness, we therefore reverse the obviousness rejections, and have no need to address the parties' arguments with respect to secondary factors.

COMMENTS AND QUESTIONS

1. A recent turn toward administrative law principles in Federal Circuit review of the PTO bears closely on the suggestion/motivation issue. In In re Gartside, 203 F.3d 1305 (Fed. Cir. 2000), the Federal Circuit held that, contrary to an omitted part of the *Dembiczak* opinion, the court would review the Patent Office Board of Appeals' fact-findings under the "substantial evidence" standard of the Administrative Procedure Act (APA). This in turn followed in the wake of a Supreme Court decision, Dickinson v. Zurko, 527 U.S. 150 (1999), holding that the standards governing judicial review of findings of fact made by federal administrative agencies apply when the Federal Circuit reviews findings of fact made by the Patent and Trademark Office (PTO).

The impact of this "administrative law" turn in PTO review cases has been widely felt. For example, when the *Zurko* case reached the Federal Circuit on remand from the Supreme Court, the Federal Circuit once again reversed the Patent Office Board of Appeals and Interferences. The court stated that the Board's reliance on basic knowledge or common sense was inadequate under the newly-imposed standard of review. In re Zurko, 258 F.3d 1379 (Fed. Cir. 2001). The court continued the theme in In re Lee, 277 F.3d 1338 (Fed. Cir. 2002), holding that the Board's reliance on "common knowledge and common sense" as the motive to combine references did not fulfill the agency's obligation to cite references to support its conclusions:

> Deferential judicial review under the Administrative Procedure Act does not relieve the agency of its obligation to develop an evidentiary basis for its findings.... In its decision on Lee's patent application, the Board rejected the need for "any specific hint or suggestion in a particular reference" to support the combination of the ... references. Omission of a relevant factor required by precedent is both legal error and arbitrary agency

action.... 'Common knowledge and common sense,' even if assumed to derive from the agency's expertise, do not substitute for authority when the law requires authority.

277 F.3d at 1344-45. See also In re Thrift, 298 F.3d 1357 (Fed. Cir. 2002) (overturning obviousness rejection to claim in speech recognition software invention, which had been premised on ground that "[t]he use of [the claimed element] is old and well known in the art of speech recognition as a means of optimization which is highly desirable."). For a critique of requirements that patent examiners fully document obviousness rejections, see Arti K. Rai, Addressing the Patent Gold Rush: The Role of Deference to PTO Patent Denials, 2 Wash. U.. L.& Pol'y 199 (2000).

2. As in so many areas of nonobviousness doctrine, the suggestion/motivation test sometimes seems less like a rule, and more like a dialectic conversation. One case illustrating the flip side of *Dembiczak* is SIBIA Neurosciences, Inc. v. Cadus Pharmaceutical Corp., 225 F.3d 1349 (Fed. Cir. 2000). The case involved a patent on a biotechnology invention: a cell-based screening method useful for the identification of potentially useful pharmaceutical compounds. The claimed screening method takes advantage of cellular activity in response to the presence of compounds on the surface of the cell. After interacting with the compound binding to it, the surface proteins send signals to the interior of the cell, and genes within the cell are activated. The claimed method measures the level of activation of the genes. Compounds producing high activity demonstrate a good ability to stimulate a cellular response, and hence are good candidates to study further. The Federal Circuit in *SIBIA* overturned a jury verdict that had found the patent valid. In reversing, and holding the patent invalid for obviousness, Judge Gajarsa discussed the fact that obviousness here is based primarily on a single reference—an article by Stumpo. The Stumpo reference described a technique for generating cells with surface proteins similar to those employed in the claimed method. Stumpo, however, was aimed not at drug compound screening, but at precisely studying how insulin and other molecules stimulate intracellular activity after binding to the surface of a cell. A second reference, an article by Lester, described with understated excitement (as befits the prestigious journal Science) the prospect of using cell surface proteins to screen large numbers of drug compound candidates. Judge Gajarsa, after noting the absence of any express combination-type statements in either Stumpo or Lester, nevertheless found the invention obvious:

> Lester ... describes [how] drug screening methods utilizing the expression of excitability molecules (i.e., cell surface receptors) can overcome the "highly empirical approach to the design of drugs" and the lack of functional assays for determining which compounds act on which cell surface receptors. Id. at 1062. These are the identical problems that were being addressed by the [SIBIA] patent.... Th[e] suggestion or motivation may be derived from the prior art reference itself, from the knowledge of one of ordinary skill in the art, or from the nature of the problem to be solved ... [SIBIA's argument] that the Stumpo reference itself does not teach the modification is not substantial evidence of no motivation to modify, given the express teaching of the prior art.... [This] ignores the possibility that the motivation to modify Stumpo can be found outside the reference itself. Thus, while Stumpo does not expressly suggest that the cells described therein could be used in drug screening methods, the knowledge of those skilled in the art, in particular as embodied in the Lester review article, suggests this modification.

225 F.3d at 1357-58.

To similar effect is Ruiz v. A.B. Chance Co., 357 F.3d 1270 (Fed. Cir. 2004). The claim in *Ruiz* covered a device for shoring up sagging foundations. It detailed two elements, among others: (1) a "screw anchor," twisted into the ground to provide support; and (2) a metal bracket for holding the screw anchor onto the foundation. One prior art reference disclosed a metal bracket, used with a different anchoring device. Another disclosed a screw anchor minus the bracket. The court held that "the nature of the problem to be solved" could provide the requisite motivation to combine; therefore the claim was obvious.

Note on Nonobviousness and Biotechnology Inventions

Biotechnology has proved to be a fertile area for application of the nonobviousness standard. For example, consider the important case of In re Bell, 991 F.2d 781 (Fed. Cir. 1993).

Bell involved a patent application claiming human gene sequences that code for insulin-like growth factors I and II (IGF-I and II). The examiner rejected the claims on grounds of obviousness, citing two publications by Rinderknecht disclosing amino acid sequences for IGF-I and IGF-II, and a patent to Weissman on a "Method for Cloning Genes." After the Board upheld the examiner, the applicants appealed. The Federal Circuit reversed.

The court held that although Rinderknecht provided the structure of the protein, there were an extraordinary number of possible nucleotide sequences that might code for it. Given the large number of possibilities suggested by the prior art, and the failure of the cited prior art to suggest which of those possibilities is the correct human sequence, the claimed sequences would not have been obvious. Further, the court stated, combining Rinderknecht with Weissman does not make the claimed sequences obvious, since Weissmann does not expressly teach nor fairly suggest that its general method for isolating genes should be combined with the disclosed protein of the Rinderknecht references. The Board clearly erred, the court continued, when it held that Weissman teaches toward, rather than away from, the claimed sequences. Therefore, the requisite teaching or suggestion to combine the teachings of the cited prior art references is absent. In the words of the court:

> It may be true that, knowing the structure of the protein, one can use the genetic code to hypothesize possible structures for the corresponding gene and that one thus has the potential for obtaining that gene.

However, because of the degeneracy of the genetic code, there are a vast number of nucleotide sequences that might code for a specific protein. In the case of IGF, Bell has argued without contradiction that the Rinderknecht amino acid sequences could be coded for by more than 10^{36} different nucleotide sequences, only a few of which are the human sequences that Bell now claims. Therefore, given the nearly infinite number of possibilities suggested by the prior art, and the failure of the cited prior art to suggest which of those possibilities is the human sequence, the claimed sequences would not have been obvious.

Bell does not claim all of the 10^{36} nucleic acids that might potentially code for IGF. Neither does Bell claim all nucleic acids coding for a protein having the biological activity of IGF. Rather, Bell claims only the human nucleic acid sequences coding for IGF. Absent anything in the cited prior art suggesting which of the 10^{36} possible sequences suggested by Rinderknecht corresponds to the IGF gene, the PTO has not met its burden of establishing that the prior art would have suggested the claimed sequences.

Accord In re Deuel, 51 F.3d 1552, 1559 (Fed. Cir. 1995). Does the result in *Bell* and *Deuel* make sense? Does it matter how difficult it would be to test all of the 10^{36} possible nucleic acids to see if they code for IGF? Does it matter whether there is some reason to prefer the actual code used in the human body, as opposed to codons producing equivalent proteins?

For an argument that the court's treatment of nonobviousness in the biotechnology area is fundamentally flawed, see Dan L. Burk and Mark A. Lemley, Biotechnology's Uncertainty Principle, 54 Case W. Res. L. Rev. 691 (2004). For an argument that rewarding discovery in the face of uncertainty is appropriate, and that the cost of overcoming the uncertainty is as important as the degree of uncertainty, see Robert P. Merges, Uncertainty and the Standard of Patentability, 7 High Tech. L.J. 1 (1993).

PROBLEM

Problem 3-10. You have drafted a patent application for a client claiming a lollipop in the shape of a human thumb. An examiner rejected the claims as obvious over a combination of numerous prior art references. You must now prepare an argument trying to overturn the examiner's decision by appeal to the Board of Patent Appeals and Interferences. The claimed invention consists of a lollipop filled with a plug of gum, chocolate or food-grade wax. A thumb-shaped elastomeric mold served as the product's internal wrapper, and after it was peeled from the candy, the user could wear the mold on his or her own thumb. The examiner relied on the following prior art references in rejecting claims to the thumb-shaped lollipop invention:

a) Siciliano shows ice cream in a mold with a stick inserted. The removable mold also serves as the product's wrapper.

b) Copeman shows candy lollipops in elastomeric molds taking "varying shapes, such as fruit or animals." The molds may be used as toy balloons after being removed.

c) Harris shows a hollow, thumb-shaped lollipop into which the user's thumb is inserted.

d) Webster shows a chewing gum entirely enclosing a liquid syrup product. This patent also suggests the greater appeal to consumers of providing two different components in the same confection.

> Although some of the references cite at least one other reference, no reference explicitly suggests combining its teaching with that of any other reference. What is your basis for arguing that the invention is patentable?

b. *"Secondary" Considerations*

In *Graham*, the Supreme Court stated that the "secondary factors" of commercial success, long-felt need, and so on "may have relevancy." 383 U.S. at 18. The Federal Circuit, by contrast, routinely speaks of these factors — under the rubric "objective evidence" — as a *required* fourth element in the § 103 analysis. See, e.g., Greenwood v. Haitori Seiko Co., Ltd., 900 F.2d 238, 241 (Fed. Cir. 1990):

[C]ertain factual predicates are required before the legal conclusion of obviousness or nonobviousness can be reached . . . . The underlying factual determinations to be made are (1) the scope and content of the prior art, (2) the differences between the claimed invention and the prior art, (3) the level of ordinary skill in the art, and (4) *objective evidence of non-obviousness, such as commercial success, long-felt but unsolved need, failure of others, copying, and unexpected results.*

These "secondary considerations" may work as "plus factors," tipping the balance of obviousness one way or the other in a particular case. But the significance of each of these factors has been hotly debated. Consider the following argument from Rochelle Dreyfuss, The Federal Circuit: A Case Study in Specialized Courts, 64 N.Y.U. L. Rev. 1 (1989):

Th[e] use of secondary considerations is not new to the CAFC. Rather, these considerations were previously accorded little weight because their appearance can sometimes be attributed to factors other than nonobviousness. For instance, commercial success may be due to the dominant market position of the patentee before the introduction of the new invention; the sudden ability to meet long felt need could derive from other technological advances, unrelated to the inventor's contribution; acquiescence may be attributed to the relative cost of obtaining a license, as opposed to challenging the patent. Rather than reject these considerations entirely, the CAFC has recognized their importance in making the law precise and instead has sought to minimize the extent to which they can be misused. Thus, the court has elaborated a "nexus" requirement, which requires that before secondary considerations can be used to demonstrate nonobviousness, a showing must be made that their appearance is attributable to the inventive characteristics of the discovery as claimed in the patent. Secondary considerations do not constitute a complete answer to the problem posed by obviousness. It is, for instance, possible for a nonobvious invention to fail to present secondary considerations. Nonetheless, it is now less probable that a lower court will declare invalid the patent on an invention that, because of the insight of its inventor, met long felt need, enjoyed commercial success, or displayed other objective indicia of having made an important social contribution. Since it is likely that the inconsistent treatment of such inventions was the most destabilizing element of the system, the CAFC has, in this area, made strides in achieving the appearance of precision.

For an economically oriented critique of one particular factor, evidence of commercial success, see Robert P. Merges, Economic Perspectives on Innovation: Patent Standards and Commercial Success, 76 Cal. L. Rev. 803 (1988):

Commercial success is a poor indicator of patentability because it depends for its effectiveness on a long chain of inferences, and because the links in the chain are often subject to doubt. This was one of the central insights of a seminal article on patentability, Graham v. John Deere Co.: New Standards for Patents, written by Edmund Kitch in 1966.[45] In it Kitch argued that commercial success was an unreliable indicator of nonobviousness. To make his point, Kitch identified four inferences a judge must make to work backward from evidence of market success to a conclusion of patentable invention:

> First, that the commercial success is due to the innovation. Second, that...potential commercial success was perceived before its development. Third, the potential commercial success having been perceived, it is likely that efforts were made [by a number of firms] to develop the improvement. Fourth, the efforts having been made by men of skill in the art, they failed because the patentee was the first to reduce his development to practice.

With only this last event as a starting point, a court is asked to reconstruct a long series of events, and, more importantly, to decide how much of the final success is attributable to each factor introduced along the way. Each inference is weak, because there are almost always several explanations why a product was successful or why other firms missed a market opportunity. Only the *last* piece of the puzzle is indisputably established; the goal of the exercise is reached through a series of inferences that only begins with this last piece. It is an altogether extraordinary job of factual reconstruction, one that reveals the falsity of the term "objective evidence," which is often used by proponents of the secondary considerations.

Merges goes on to argue in favor of another "secondary consideration" — failure of others:

> Unlike commercial success, the failure of others to make an invention proves *directly* that parallel research efforts were under way at a number of firms, and that one firm (the patentee) won the race to a common goal. So long as the race was long enough, and so long as there was a clear winner, it is difficult to find fault with such evidence as proof of patentability.[242] In fact, since the failure of others is often one of the inferential steps underlying the commercial success doctrine, it makes sense for courts to adopt a rule of thumb requiring the patentee in most cases to prove failure of others before commercial success will be given substantial weight.

In Merck & Co., Inc. v. Teva Pharmaceuticals USA, Inc., 395 F.3d 1364 Fed. Cir. 2005), the Federal Circuit adopted views consistent with Merges's article. The court noted that the commercial success of the product in question was based at least in part on the fact that the sale of substitute products was barred by a blocking patent, not on the value of the litigated patent. In dissent from a ruling denying rehearing en banc, Judge Lourie took issue, arguing that "[s]uccess is success." 405 F.3d 1338, 1339 (2005).

---

45. 1966 Sup. Ct. Rev. 293. — EDS.

242. Many judges have sung the praises of long felt need. Justice William R. Day of the Supreme Court said: "It may be safely said that if those skilled in the mechanical arts are working in a given field and have failed after repeated efforts to discover a certain new and useful improvement, that he who first makes the discovery ... is entitled to protection as an inventor." Expanded Metal Co. v. Bradford, 214 U.S. 366, 381 (1908). See also Krementz v. S. Cottle Co., 148 U.S. 556, 560 (1892). Recently, the Federal Circuit has shown a willingness to consider such evidence, but has at times appeared to relax one of the two elements conventionally required to establish it — actual parallel research.

C. INFRINGEMENT

1. Claim Interpretation

Patent claims define a patent owner's property right; they have been analogized from time immemorial to the "metes and bounds" of a real property deed. Word meanings determine the precise boundaries of claims. And these boundaries can be crucial: Linguistically minor variations in phraseology and meaning are the difference between a finding of infringement (and thus exclusion of the competitor, and perhaps damages) and noninfringement (and thus open entry in at least part of the patentee's market). To the hardheaded businessperson, claim interpretation defines the "shelf space" that belongs exclusively to a patentee. To a patent lawyer, there is deep fascination in the almost scholastic debate over the difference between "therein" and "thereon" in a certain patent claim (see the *Larami* case below). The average businessperson looks straight to the bottom line: can the patentee exclude a competitor (expanding the reach of its exclusive legal franchise) or will it have to share shelf space and profits?

Courts employ a number of rules and procedures to mark off this space, most of which share a common spirit with general canons of legal interpretation applied to statutes, contracts, and the like.

a. The Proper Role of Judge and Jury in Patent Cases

In patent cases, as elsewhere, the distinction between questions of law and fact is an important issue. As we have seen so far, patent cases often center on scientific or technical details such as how an invention differs from the prior art, or how it works compared to a competitor's product. Whether a judge or a jury decides these key issues may make a significant difference in the outcome of a patent infringement case.

In Markman v. Westview Instruments, 52 F.3d 967 (Fed. Cir. 1995) (en banc), the Federal Circuit put judges squarely in the driver's seat when it comes to interpreting patent claims. Upholding a district judge who had overturned a jury verdict based on improper claim construction, the court ruled (over a vigorous dissent) that claim construction is a matter of law. In addition, the Federal Circuit has held that it will review district court findings de novo, notwithstanding such factors as the trial court's proximity to expert witnesses. See Cybor Corp. v. FAS Technologies, Inc., 138 F.3d 1448 (Fed. Cir. 1998).

The Supreme Court affirmed in a unanimous opinion. Markman v. Westview Instruments, 517 U.S. 370 (1996). The Court evaluated Markman's claim that the Seventh Amendment required that a jury interpret the language of the patent claims by turning to the historical treatment of claim interpretation under English patent law. While the Court noted that "there is no dispute that infringement cases today must be tried to a jury, as their predecessors were more than two centuries ago," it could find no equivalent rule governing the construction of patent claims, since claims per se did not exist in early U.S. patent cases. Those cases the Court did consider did not unambiguously establish that terms of art in patents were historically interpreted by juries. The Court then turned to the policy considerations, which it found to support leaving claim construction in the hands of judges:

The construction of written instruments is one of those things that judges often do and are likely to do better than jurors unburdened by training in exegesis. Patent construction in particular "is a special occupation, requiring, like all others, special training and practice. The judge, from his training and discipline, is more likely to give a proper interpretation to such instruments than a jury; and he is, therefore, more likely to be right, in performing such a duty, than a jury can be expected to be." Parker v. Hulme, 18 F. Cas. at 1140. Such was the understanding nearly a century and a half ago, and there is no reason to weigh the respective strengths of judge and jury differently in relation to the modern claim; quite the contrary, for "the claims of patents have become highly technical in many respects as the result of special doctrines relating to the proper form and scope of claims that have been developed by the courts and the Patent Office." Woodward, Definiteness and Particularity in Patent Claims, 46 Mich. L. Rev. 755, 765 (1948).

Markman would trump these considerations with his argument that a jury should decide a question of meaning peculiar to a trade or profession simply because the question is a subject of testimony requiring credibility determinations, which are the jury's forte. It is, of course, true that credibility judgments have to be made about the experts who testify in patent cases, and in theory there could be a case in which a simple credibility judgment would suffice to choose between experts whose testimony was equally consistent with a patent's internal logic. But our own experience with document construction leaves us doubtful that trial courts will run into many cases like that. In the main, we expect, any credibility determinations will be subsumed within the necessarily sophisticated analysis of the whole document, required by the standard construction rule that a term can be defined only in a way that comports with the instrument as a whole. Thus, in these cases a jury's capabilities to evaluate demeanor, to sense the "mainsprings of human conduct," or to reflect community standards, are much less significant than a trained ability to evaluate the testimony in relation to the overall structure of the patent. The decisionmaker vested with the task of construing the patent is in the better position to ascertain whether an expert's proposed definition fully comports with the specification and claims and so will preserve the patent's internal coherence. We accordingly think there is sufficient reason to treat construction of terms of art like many other responsibilities that we cede to a judge in the normal course of trial, notwithstanding its evidentiary underpinnings....

Finally, we see the importance of uniformity in the treatment of a given patent as an independent reason to allocate all issues of construction to the court. As we noted in General Elec. Co. v. Wabash Appliance Corp., 304 U.S. 364, 369, 82 L. Ed. 1402, 58 S. Ct. 899 (1938), "the limits of a patent must be known for the protection of the patentee, the encouragement of the inventive genius of others and the assurance that the subject of the patent will be dedicated ultimately to the public." Otherwise, a "zone of uncertainty which enterprise and experimentation may enter only at the risk of infringement claims would discourage invention only a little less than unequivocal foreclosure of the field," United Carbon Co. v. Binney & Smith Co., 317 U.S. 228, 236, 87 L. Ed. 232, 63 S. Ct. 165 (1942), and "the public [would] be deprived of rights supposed to belong to it, without being clearly told what it is that limits these rights." Merrill v. Yeomans, 94 U.S. 568, 573, 24 L. Ed. 235 (1877). It was just for the sake of such desirable uniformity that Congress created the Court of Appeals for the Federal Circuit as an exclusive appellate court for patent cases, observing that increased uniformity would "strengthen the United States patent system in such a way as to foster technological growth and industrial innovation."

On the practical side, *Markman* raises a number of potentially challenging issues for the patent system. Most importantly, it has permanently changed a number of routine practices in patent cases. For example, patent cases are usually appealed only after a full trial. The trial will very often include a claim interpretation issue but will of course also cover a host of other issues. *Markman* raises the concern that a full trial on many subordinate issues may become moot depending on how the Federal Circuit interprets a claim. For example, if the claim is interpreted as excluding all of the accused infringer's products, there is no need to consider defenses, such as lack of enablement, failure to disclose best mode, and inequitable conduct. Indeed, to ensure that time and money are not spent on pointless issues, the parties in patent cases increasingly seek to obtain a trial court's legal interpretation of claims via summary judgment, and then appeal to the Federal Circuit.[46] This practice, together with the de novo standard of review for claim interpretation at the Federal Circuit, see Cybor Corp. v. FAS Technologies, Inc., 138 F.3d 1448 (Fed. Cir. 1998), contributes to a significant reversal rate by the Federal Circuit in claim construction cases. See, e.g., Kimberly A. Moore, *Markman Eight Years Later: Is Claim Construction More Predictable?*, 9 Lewis & Clark L. Rev. 231, 239 (2005) (claim construction becoming less, not more, predictable; study of claim construction cases from 1996 to 2003 shows "the Federal Circuit held at least one term was wrongly construed [by the district court] in 37.5% of the cases."). For a description of current *Markman* hearing practices, and a suggestion that district court judges be given more discretion in this area, see Edward Brunet, *Markman* Hearings, Summary Judgment, and Judicial Discretion, 9 Lewis & Clark L. Rev. 93 (2005).

Phillips v. AWH Corporation
United States Court of Appeals for the Federal Circuit
415 F.3d 1303 (Fed. Cir. 2005) (en banc)

BRYSON, Circuit Judge [for the court en banc]:

Edward H. Phillips invented modular, steel-shell panels that can be welded together to form vandalism-resistant walls. The panels are especially useful in building prisons because they are load-bearing and impact-resistant, while also insulating against fire and noise. Mr. Phillips obtained a patent on the invention, U.S. Patent No. 4,677,798 ("the '798 patent") . . . .

In February 1997, Mr. Phillips brought suit in the United States District Court for the District of Colorado charging AWH with . . . infringement . . .

[In deciding]the patent infringement issue, the district court focused on the language of claim 1, which recites "further means disposed inside the shell for increasing its load bearing capacity comprising internal steel baffles extending inwardly from the steel shell walls." . . . [Finding that the accused product did not contain "baffles" as that term is used in claim 1,] the district court granted summary judgment of non-infringement.

Mr. Phillips appealed [and a divided] panel of this court affirmed. Phillips v. AWH Corp., 363 F.3d 1207 (Fed. Cir. 2004). The majority [Judge Lourie,

---

46. Note, however, that the Federal Circuit has taken a firm stand against interlocutory appeals for patent claim interpretation issues. See *Cybor Corp., supra*, 138 F.3d 1448, 1479 (Newman, J., additional views).

joined by Judge Newman] sustained the district court's summary judgment of noninfringement, although on different grounds. The dissenting judge [Judge Dyk] would have reversed the summary judgment of noninfringement....

[T]he [original Federal Circuit] panel concluded that the patent uses the term "baffles" in a restrictive manner. Based on the patent's written description, the panel held that the claim term "baffles" excludes structures that extend at a 90 degree angle from the walls. The panel noted that the specification repeatedly refers to the ability of the claimed baffles to deflect projectiles and that it describes the baffles as being "disposed at such angles that bullets which might penetrate the outer steel panels are deflected." '798 patent, col. 2, ll. 13-15; see also id. at col. 5, ll. 17-19 (baffles are "disposed at angles which tend to deflect the bullets"). In addition, the panel observed that nowhere in the patent is there any disclosure of a baffle projecting from the wall at a right angle and that baffles oriented at 90 degrees to the wall were found in the prior art. Based on "the specification's explicit descriptions," the panel concluded "that the patentee regarded his invention as panels providing impact or projectile resistance and that the baffles must be oriented at angles other than 90 [degrees]." Phillips, 363 F.3d at 1213. The panel added that the patent specification "is intended to support and inform the claims, and here it makes it unmistakably clear that the invention involves baffles angled at other than 90 [degrees]." Id. at 1214. The panel therefore upheld the district court's summary judgment of noninfringement.

The dissenting judge argued that the panel had improperly limited the claims to the particular embodiment of the invention disclosed in the specification, rather than adopting the "plain meaning" of the term "baffles."... [T]he dissenting judge contended, the specification "merely identifies impact resistance as one of several objectives of the invention." Id. at 1217. In sum, the dissent concluded that "there is no reason to supplement the plain meaning of the claim language with a limitation from the preferred embodiment." Id. at 1218. Consequently, the dissenting judge argued that the court should have adopted the general purpose dictionary definition of the term baffle, i.e., "something for deflecting, checking, or otherwise regulating flow," id., and therefore should have reversed the summary judgment of noninfringement.

This court agreed to rehear the appeal en banc and vacated the judgment of the panel. We now . . . reverse the portion of the court's judgment addressed to the issue of infringement.

I

Claim 1 of the '798 patent is representative of the asserted claims with respect to the use of the term "baffles." It recites:

> Building modules adapted to fit together for construction of fire, sound and impact resistant security barriers and rooms for use in securing records and persons, comprising in combination, an outer shell . . . , sealant means . . . and further means disposed inside the shell for increasing its load bearing capacity comprising internal steel baffles extending inwardly from the steel shell walls. . . .

II

The first paragraph of section 112 of the Patent Act, 35 U.S.C. § 112, states that the specification

> shall contain a written description of the invention, and of the manner and process of making and using it, in such full, clear, concise, and exact terms as to enable any person skilled in the art to which it pertains . . . to make and use the same. . . .

The second paragraph of section 112 provides that the specification

> shall conclude with one or more claims particularly pointing out and distinctly claiming the subject matter which the applicant regards as his invention.

Those two paragraphs of section 112 frame the issue of claim interpretation for us. The second paragraph requires us to look to the language of the claims to determine what "the applicant regards as his invention." On the other hand, the first paragraph requires that the specification describe the invention set forth in the claims. The principal question that this case presents to us is the extent to which we should resort to and rely on a patent's specification in seeking to ascertain the proper scope of its claims. . . .

A

It is a "bedrock principle" of patent law that "the claims of a patent define the invention to which the patentee is entitled the right to exclude." Innova [/Pure Water, Inc. v. Safari Water Filtration Systems, Inc., 381 F.3d 1111 (Fed. Cir. 2004)], at 1115. That principle has been recognized since at least 1836, when Congress first required that the specification include a portion in which the inventor "shall particularly specify and point out the part, improvement, or combination, which he claims as his own invention or discovery." Act of July 4, 1836, ch. 357, § 6, 5 Stat. 117, 119. In the following years, the Supreme Court made clear that the claims are "of primary importance, in the effort to ascertain precisely what it is that is patented." Merrill v. Yeomans, 94 U.S. 568, 570 (1876). . . . We have frequently stated that the words of a claim "are generally given their ordinary and customary meaning." Vitronics [Corp. v. Conceptronic, Inc., 90 F.3d 1576 (Fed. Cir. 1996)], at 1582. We have made clear, moreover, that the ordinary and customary meaning of a claim term is the meaning that the term would have to a person of ordinary skill in the art in question at the time of the invention, i.e., as of the effective filing date of the patent application.

The inquiry into how a person of ordinary skill in the art understands a claim term provides an objective baseline from which to begin claim interpretation. . . . That starting point is based on the well-settled understanding that inventors are typically persons skilled in the field of the invention and that patents are addressed to and intended to be read by others of skill in the pertinent art.

Importantly, the person of ordinary skill in the art is deemed to read the claim term not only in the context of the particular claim in which the disputed term appears, but in the context of the entire patent, including the specification.

B

In some cases, the ordinary meaning of claim language as understood by a person of skill in the art may be readily apparent even to lay judges, and claim construction in such cases involves little more than the application of the widely accepted meaning of commonly understood words. See Brown v. 3M, 265 F.3d 1349, 1352 (Fed Cir. 2001) (holding that the claims did "not require elaborate interpretation"). In such circumstances, general purpose dictionaries may be helpful. In many cases that give rise to litigation, however, determining the ordinary and customary meaning of the claim requires examination of terms that have a particular meaning in a field of art. Because the meaning of a claim term as understood by persons of skill in the art is often not immediately apparent, and because patentees frequently use terms idiosyncratically, the court looks to "those sources available to the public that show what a person of skill in the art would have understood disputed claim language to mean." Innova, 381 F.3d at 1116. Those sources include "the words of the claims themselves, the remainder of the specification, the prosecution history, and extrinsic evidence concerning relevant scientific principles, the meaning of technical terms, and the state of the art." Id. . . .

1

Quite apart from the written description and the prosecution history, the claims themselves provide substantial guidance as to the meaning of particular claim terms. See Vitronics, 90 F.3d at 1582.

To begin with, the context in which a term is used in the asserted claim can be highly instructive. To take a simple example, the claim in this case refers to "steel baffles," which strongly implies that the term "baffles" does not inherently mean objects made of steel. . . .

Other claims of the patent in question, both asserted and unasserted, can also be valuable sources of enlightenment as to the meaning of a claim term. *Vitronics, 90 F.3d at 1582*. Because claim terms are normally used consistently throughout the patent, the usage of a term in one claim can often illuminate the meaning of the same term in other claims. . . . Differences among claims can also be a useful guide in understanding the meaning of particular claim terms. . . . For example, the presence of a dependent claim that adds a particular limitation gives rise to a presumption that the limitation in question is not present in the independent claim. See Liebel-Flarsheim Co. v. Medrad, Inc., 358 F.3d 898, 910 (Fed. Cir. 2004).

2

The claims, of course, do not stand alone. Rather, they are part of "a fully integrated written instrument," Markman, 52 F.3d at 978, consisting principally of a specification that concludes with the claims. For that reason, claims "must be read in view of the specification, of which they are a part." Id. at 979. As we stated in *Vitronics*, the specification "is always highly relevant to the claim construction analysis. Usually, it is dispositive; it is the single best guide to the meaning of a disputed term." 90 F.3d at 1582.

This court and its predecessors have long emphasized the importance of the specification in claim construction. In Autogiro Co. of America v. United States, 384 F.2d 391, 397-98 (Ct. Cl. 1967), the Court of Claims characterized the specification as "a concordance for the claims," based on the statutory requirement that the specification "describe the manner and process of making and using" the patented invention.... That principle has a long pedigree in Supreme Court decisions as well. [Citations omitted.]

3

In addition to consulting the specification, we have held that a court "should also consider the patent's prosecution history, if it is in evidence." Markman, 52 F.3d at 980. Like the specification, the prosecution history provides evidence of how the PTO and the inventor understood the patent. Furthermore, like the specification, the prosecution history was created by the patentee in attempting to explain and obtain the patent. Yet because the prosecution history represents an ongoing negotiation between the PTO and the applicant, rather than the final product of that negotiation, it often lacks the clarity of the specification and thus is less useful for claim construction purposes. Nonetheless, the prosecution history can often inform the meaning of the claim language by demonstrating how the inventor understood the invention and whether the inventor limited the invention in the course of prosecution, making the claim scope narrower than it would otherwise be.

C

Although we have emphasized the importance of intrinsic evidence in claim construction, we have also authorized district courts to rely on extrinsic evidence, which "consists of all evidence external to the patent and prosecution history, including expert and inventor testimony, dictionaries, and learned treatises." Markman, 52 F.3d at 980, citing Seymour v. Osborne, 78 U.S. (11 Wall.) 516, 546 (1870). Within the class of extrinsic evidence, the court has observed that dictionaries and treatises can be useful in claim construction.... We have especially noted the help that technical dictionaries may provide to a court "to better understand the underlying technology" and the way in which one of skill in the art might use the claim terms. Vitronics, 90 F.3d at 1584 n.6. Because dictionaries, and especially technical dictionaries, endeavor to collect the accepted meanings of terms used in various fields of science and technology, those resources have been properly recognized as among the many tools that can assist the court in determining the meaning of particular terminology to those of skill in the art of the invention.

We have also held that extrinsic evidence in the form of expert testimony can be useful to a court for a variety of purposes, such as to provide background on the technology at issue, to explain how an invention works, to ensure that the court's understanding of the technical aspects of the patent is consistent with that of a person of skill in the art, or to establish that a particular term in the patent or the prior art has a particular meaning in the pertinent field.... However, conclusory, unsupported assertions by experts as to the definition of a claim term are not useful to a court....

[U]ndue reliance on extrinsic evidence poses the risk that it will be used to change the meaning of claims in derogation of the "indisputable public records consisting of the claims, the specification and the prosecution history," thereby undermining the public notice function of patents. Southwall Techs.[, Inc. v. Cardinal IG Co.,] 54 F.3d [1570] at 1578 [Fed. Cir. 1995].

III

Although the principles outlined above have been articulated on numerous occasions, some of this court's cases have suggested a somewhat different approach to claim construction, in which the court has given greater emphasis to dictionary definitions of claim terms and has assigned a less prominent role to the specification and the prosecution history. The leading case in this line is Texas Digital Systems, Inc. v. Telegenix, Inc., 308 F.3d 1193 (Fed. Cir. 2002) [written by Linn, J., joined by Michel and Schall, JJ.].

A

In *Texas Digital,* the court noted that "dictionaries, encyclopedias and treatises are particularly useful resources to assist the court in determining the ordinary and customary meanings of claim terms." 308 F.3d at 1202. Those texts, the court explained, are "objective resources that serve as reliable sources of information on the established meanings that would have been attributed to the terms of the claims by those of skill in the art," and they "deserve no less fealty in the context of claim construction" than in any other area of law. Id. at 1203. The court added that because words often have multiple dictionary meanings, the intrinsic record must be consulted to determine which of the different possible dictionary meanings is most consistent with the use of the term in question by the inventor. If more than one dictionary definition is consistent with the use of the words in the intrinsic record, the court stated, "the claim terms may be construed to encompass all such consistent meanings." Id.

The Texas Digital court further explained that the patent's specification and prosecution history must be consulted to determine if the patentee has used "the words [of the claim] in a manner clearly inconsistent with the ordinary meaning reflected, for example, in a dictionary definition." 308 F.3d at 1204. The court identified two circumstances in which such an inconsistency may be found. First, the court stated, "the presumption in favor of a dictionary definition will be overcome where the patentee, acting as his or her own lexicographer, has clearly set forth an explicit definition of the term different from its ordinary meaning." Id. Second, "the presumption also will be rebutted if the inventor has disavowed or disclaimed scope of coverage, by using words or expressions of manifest exclusion or restriction, representing a clear disavowal of claim scope." Id.

The court concluded that it is improper to consult "the written description and prosecution history as a threshold step in the claim construction process, before any effort is made to discern the ordinary and customary meanings attributed to the words themselves." Texas Digital, 308 F.3d at 1204. To do so, the court reasoned, "invites a violation of our precedent counseling against importing limitations into the claims." Id.

B

Although the concern expressed by the court in *Texas Digital* was valid, the methodology it adopted placed too much reliance on extrinsic sources such as dictionaries, treatises, and encyclopedias and too little on intrinsic sources, in particular the specification and prosecution history. In effect, the Texas Digital approach limits the role of the specification in claim construction to serving as a check on the dictionary meaning of a claim term. . . . That approach, in our view, improperly restricts the role of the specification in claim construction.

Assigning such a limited role to the specification, and in particular requiring that any definition of claim language in the specification be express, is inconsistent with our rulings that the specification is "the single best guide to the meaning of a disputed term," and that the specification "acts as a dictionary when it expressly defines terms used in the claims or when it defines terms by implication." Vitronics, 90 F.3d at 1582.

The main problem with elevating the dictionary to such prominence is that it focuses the inquiry on the abstract meaning of words rather than on the meaning of claim terms within the context of the patent. Properly viewed, the "ordinary meaning" of a claim term is its meaning to the ordinary artisan after reading the entire patent. Yet heavy reliance on the dictionary divorced from the intrinsic evidence risks transforming the meaning of the claim term to the artisan into the meaning of the term in the abstract, out of its particular context, which is the specification. Thus, there may be a disconnect between the patentee's responsibility to describe and claim his invention, and the dictionary editors' objective of aggregating all possible definitions for particular words.

The problem is that if the district court starts with the broad dictionary definition in every case and fails to fully appreciate how the specification implicitly limits that definition, the error will systematically cause the construction of the claim to be unduly expansive. The risk of systematic overbreadth is greatly reduced if the court instead focuses at the outset on how the patentee used the claim term in the claims, specification, and prosecution history, rather than starting with a broad definition and whittling it down.

Thus, the use of the dictionary may extend patent protection beyond what should properly be afforded by the inventor's patent. . . .

Even technical dictionaries or treatises, under certain circumstances, may suffer from some of these deficiencies. There is no guarantee that a term is used in the same way in a treatise as it would be by the patentee. In fact, discrepancies between the patent and treatises are apt to be common because the patent by its nature describes something novel.

Moreover, different dictionaries may contain somewhat different sets of definitions for the same words. A claim should not rise or fall based upon the preferences of a particular dictionary editor, or the court's independent decision, uninformed by the specification, to rely on one dictionary rather than another. Finally, the authors of dictionaries or treatises may simplify ideas to communicate them most effectively to the public and may thus choose a meaning that is not pertinent to the understanding of particular claim language. The resulting definitions therefore do not necessarily reflect the inventor's goal of distinctly setting forth his invention as a person of ordinary skill in that particular art would understand it.

As we have noted above, however, we do not intend to preclude the appropriate use of dictionaries. Dictionaries or comparable sources are often useful to assist in understanding the commonly understood meaning of words and have been used both by our court and the Supreme Court in claim interpretation.

To avoid importing limitations from the specification into the claims, it is important to keep in mind that the purposes of the specification are to teach and enable those of skill in the art to make and use the invention and to provide a best mode for doing so.... One of the best ways to teach a person of ordinary skill in the art how to make and use the invention is to provide an example of how to practice the invention in a particular case. Much of the time, upon reading the specification in that context, it will become clear whether the patentee is setting out specific examples of the invention to accomplish those goals, or whether the patentee instead intends for the claims and the embodiments in the specification to be strictly coextensive.... The manner in which the patentee uses a term within the specification and claims usually will make the distinction apparent....

In the end, there will still remain some cases in which it will be hard to determine whether a person of skill in the art would understand the embodiments to define the outer limits of the claim term or merely to be exemplary in nature. While that task may present difficulties in some cases, we nonetheless believe that attempting to resolve that problem in the context of the particular patent is likely to capture the scope of the actual invention more accurately than either strictly limiting the scope of the claims to the embodiments disclosed in the specification or divorcing the claim language from the specification.

In *Vitronics*, we did not attempt to provide a rigid algorithm for claim construction, but simply attempted to explain why, in general, certain types of evidence are more valuable than others. Today, we adhere to that approach and reaffirm the approach to claim construction outlined in that case, in *Markman*, and in *Innova*. We now turn to the application of those principles to the case at bar.

IV

A

The critical language of claim 1 of the '798 patent — "further means disposed inside the shell for increasing its load bearing capacity comprising internal steel baffles extending inwardly from the steel shell walls" — imposes three clear requirements with respect to the baffles. First, the baffles must be made of steel. Second, they must be part of the load-bearing means for the wall section. Third, they must be pointed inward from the walls. Both parties, stipulating to a dictionary definition, also conceded that the term "baffles" refers to objects that check, impede, or obstruct the flow of something. The intrinsic evidence confirms that a person of skill in the art would understand that the term "baffles," as used in the '798 patent, would have that generic meaning.

The other claims of the '798 patent specify particular functions to be served by the baffles. For example, dependent claim 2 states that the baffles may be "oriented with the panel sections disposed at angles for deflecting projectiles such as bullets able to penetrate the steel plates." The inclusion of such a specific limitation on the term "baffles" in claim 2 makes it likely that the patentee did not contemplate that the term "baffles" already contained that limitation. See Dow Chem. Co. v. United States, 226 F.3d 1334, 1341-42 (Fed. Cir. 2000) (concluding that an independent claim should be given broader scope than a dependent claim to avoid rendering the dependent claim redundant). Independent claim 17 further supports that proposition. It states that baffles are placed "projecting inwardly from the outer shell at angles tending to deflect projectiles that penetrate the outer shell." That limitation would be unnecessary if persons of skill in the art understood that the baffles inherently served such a function.... Dependent claim 6 provides an additional requirement for the baffles, stating that "the internal baffles of both outer panel sections overlap and interlock at angles providing deflector panels extending from one end of the module to the other." If the baffles recited in claim 1 were inherently placed at specific angles, or interlocked to form an intermediate barrier, claim 6 would be redundant.

The specification further supports the conclusion that persons of ordinary skill in the art would understand the baffles recited in the '798 patent to be load-bearing objects that serve to check, impede, or obstruct flow. At several points, the specification discusses positioning the baffles so as to deflect projectiles. See '798 patent, col. 2, ll. 13-15; id., col. 5, ll. 17-19. The patent states that one advantage of the invention over the prior art is that "there have not been effective ways of dealing with these powerful impact weapons with inexpensive housing." Id., col. 3, ll. 28-30. While that statement makes clear the invention envisions baffles that serve that function, it does not imply that in order to qualify as baffles within the meaning of the claims, the internal support structures must serve the projectile-deflecting function in all the embodiments of all the claims. The specification must teach and enable all the claims, and the section of the written description discussing the use of baffles to deflect projectiles serves that purpose for claims 2, 6, 17, and 23, which specifically claim baffles that deflect projectiles....

The specification discusses several other purposes served by the baffles. For example, the baffles are described as providing structural support. The patent states that one way to increase load-bearing capacity is to use "at least in part inwardly directed steel baffles 15, 16." '798 patent, col. 4, ll. 14-15. The baffle 16 is described as a "strengthening triangular baffle." Id., col. 4, line 37. Importantly, Figures 4 and 6 do not show the baffles as part of an "intermediate interlocking, but not solid, internal barrier." In those figures, the baffle 16 simply provides structural support for one of the walls, as depicted below:

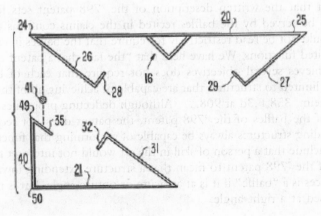

FIG. 4.

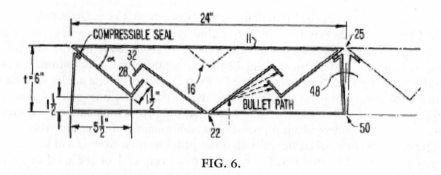

FIG. 6.

Other uses for the baffles are listed in the specification as well. In Figure 7, the overlapping flanges "provide for overlapping and interlocking the baffles to produce substantially an intermediate barrier wall between the opposite [wall] faces":

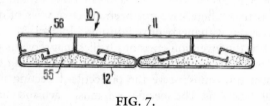

FIG. 7.

'798 patent, col. 5, ll. 26-29. Those baffles thus create small compartments that can be filled with either sound and thermal insulation or rock and gravel to stop projectiles. Id., col. 5, ll. 29-34. By separating the interwall area into compartments (see, e.g., compartment 55 in Figure 7), the user of the modules can choose different types of material for each compartment, so that the module can be "easily custom tailored for the specific needs of each installation." Id., col. 5, ll. 36-37. When material is placed into the wall during installation, the baffles obstruct the flow of material from one compartment to another so that this "custom tailoring" is possible.

The fact that the written description of the '798 patent sets forth multiple objectives to be served by the baffles recited in the claims confirms that the term "baffles" should not be read restrictively to require that the baffles in each case serve all of the recited functions. We have held that "the fact that a patent asserts that an invention achieves several objectives does not require that each of the claims be construed as limited to structures that are capable of achieving all of the objectives." Liebel-Flarsheim, 358 F.3d at 908. . . . Although deflecting projectiles is one of the advantages of the baffles of the '798 patent, the patent does not require that the inward extending structures always be capable of performing that function. Accordingly, we conclude that a person of skill in the art would not interpret the disclosure and claims of the '798 patent to mean that a structure extending inward from one of the wall faces is a "baffle" if it is at an acute or obtuse angle, but is not a "baffle" if it is disposed at a right angle.

B

Invoking the principle that "claims should be so construed, if possible, as to sustain their validity," Rhine v. Casio, Inc., 183 F.3d 1342, 1345 (Fed Cir. 1999), argues that the term "baffles" should be given a restrictive meaning because if the term is not construed restrictively, the asserted claims would be invalid.

While we have acknowledged the maxim that claims should be construed to preserve their validity, we have not applied that principle broadly, and we have certainly not endorsed a regime in which validity analysis is a regular component of claim construction.... Instead, we have limited the maxim to cases in which "the court concludes, after applying all the available tools of claim construction, that the claim is still ambiguous." Liebel-Flarsheim, 358 F.3d at 911. In such cases, we have looked to whether it is reasonable to infer that the PTO would not have issued an invalid patent, and that the ambiguity in the claim language should therefore be resolved in a manner that would preserve the patent's validity.

In this case ... the claim term at issue is not ambiguous. Thus, it can be construed without the need to consider whether one possible construction would render the claim invalid while the other would not. The doctrine of construing claims to preserve their validity, a doctrine of limited utility in any event, therefore has no applicability here.

In sum, we reject AWH's arguments in favor of a restrictive definition of the term "baffles." Because we disagree with the district court's claim construction, we reverse the summary judgment of noninfringement. In light of our decision on claim construction, it is necessary to remand the infringement claims to the district court for further proceedings.

* * *

[Remanded.]

MAYER, Circuit Judge, with whom NEWMAN, Circuit Judge, joins, dissenting.

Now more than ever I am convinced of the futility, indeed the absurdity, of this court's persistence in adhering to the falsehood that claim construction is a matter of law devoid of any factual component. Because any attempt to fashion a coherent standard under this regime is pointless, as illustrated by our many failed attempts to do so, I dissent.

This court was created for the purpose of bringing consistency to the patent field. See H.R. Rep. No. 312, 97th Cong., 1st Sess. 20-23 (1981). Instead, we have taken this noble mandate, to reinvigorate the patent and introduce predictability to the field, and focused inappropriate power in this court. In our quest to elevate our importance, we have, however, disregarded our role as an appellate court; the resulting mayhem has seriously undermined the legitimacy of the process, if not the integrity of the institution.

In the name of uniformity, Cybor Corp. v. FAS Technologies, Inc., 138 F.3d 1448 (Fed. Cir. 1998) (en banc), held that claim construction does not involve subsidiary or underlying questions of fact and that we are, therefore, unbridled by either the expertise or efforts of the district court. What we have wrought, instead, is the substitution of a black box, as it so pejoratively has been said of the jury, with the black hole of this court. Out of this void we emit "legal" pronouncements by way of "interpretive necromancy"; these rulings resemble reality, if at all, only by chance. Regardless, and with a blind eye to the consequences, we continue to struggle under this irrational and reckless regime, trying every alternative — dictionaries first, dictionaries second, never dictionaries, etc., etc., etc.

Again today we vainly attempt to establish standards by which this court will interpret claims. But after proposing no fewer than seven questions, receiving more than thirty amici curiae briefs, and whipping the bar into a frenzy of expectation, we say nothing new, but merely restate what has become the practice over the last ten years — that we will decide cases according to whatever mode or method results in the outcome we desire, or at least allows us a seemingly plausible way out of the case. I am not surprised by this. Indeed, there can be no workable standards by which this court will interpret claims so long as we are blind to the factual component of the task.

While this court may persist in the delusion that claim construction is a purely legal determination, unaffected by underlying facts, it is plainly not the case. Claim construction is, or should be, made in context: a claim should be interpreted both from the perspective of one of ordinary skill in the art and in view of the state of the art at the time of invention. . . . These questions, which are critical to the correct interpretation of a claim, are inherently factual. They are hotly contested by the parties, not by resort to case law as one would expect for legal issues, but based on testimony and documentary evidence. During so called *Markman* "hearings," which are often longer than jury trials, parties battle over experts offering conflicting evidence regarding who qualifies as one of ordinary skill in the art; the meaning of patent terms to that person; the state of the art at the time of the invention; contradictory dictionary definitions and which would be consulted by the skilled artisan; the scope of specialized terms; the problem a patent was solving; what is related or pertinent art; whether a construction was disallowed during prosecution; how one of skill in the art would understand statements during prosecution; and on and on. In order to reconcile the parties' inconsistent submissions and arrive at a sound interpretation, the district court is required to sift through and weigh volumes of evidence. While this court treats the district court as an intake clerk, whose only role is to collect, shuffle and collate evidence, the reality, as revealed by conventional practice, is far different.

* * * *

If we persist in deciding the subsidiary factual components of claim construction without deference, there is no reason why litigants should be required to parade their evidence before the district courts or for district courts to waste time and resources evaluating such evidence. . . . If the proceedings before the district court are merely a "tryout on the road," [citation omitted] . . . as they are under our current regimen, it is wasteful to require such proceedings at all. Instead, all patent cases could be filed in this court; we would determine whether claim construction is necessary, and, if so, the meaning of the claims. Those few cases in which claim construction is not dispositive can be remanded to the district court for trial. In this way, we would at least eliminate the time and expense of the charade currently played out before the district court.

Eloquent words can mask much mischief. The court's opinion today is akin to rearranging the deck chairs on the Titanic — the orchestra is playing as if nothing is amiss, but the ship is still heading for Davey Jones' locker.

NOTES AND COMMENTS

1. *The Death of a Rigid Dictionary "Methodology"?* The *Texas Digital* interpretive framework described and critiqued in the *Phillips* majority opinion had offered the hope of objectivity to some observers, while to others it invited only more difficulties.

See, e.g., Joseph Scott Miller, The Proven Key: Roles and Rules for Dictionaries at the Patent Office and the Courts, 54 Am U. L. Rev. 829 (2005) (noting the prevalence of resort to dictionaries in all courts, including the Federal Circuit, and advocating that the Patent Office require patent applicants to indicate in their applications which general purpose and technical dictionaries are to be used to interpret the patent); Ben Hattenbach, Chickens, Eggs and Other Impediments to Escalating Reliance on Dictionaries in Patent Claim Construction, 85 J. Pat. & Trademark Off. Soc'y 181, 182 (2003) ("*Telegenix*' suggestion that courts should look to dictionaries first, before considering the specifications they are used to interpret, seems conceptually irreconcilable with the need to consider specifications before deciding which dictionaries would be pertinent aids to interpretation in a given case. The *Telegenix* approach also arguably introduces unnecessary inefficiency and inaccuracy into the claim construction process.") After *Phillips*, courts may use dictionaries in conjunction with the specification, but the dictionary definitions enjoy no presumptive position of prominence. For more on different theories of claim construction, see Craig Allen Nard, A Theory of Claim Interpretation, 14 Harv. L. & Tech. Rev. 1 (2000) (describing patent claims as writings produced by a technological community and emphasizing the importance of resorting to the specification as a source of meaning for claims). See also Robert P. Merges and Jane Ginsburg, Foundations of Intellectual Property 122-128 (2004) (excerpting the Nard article and exploring some of its suppositions).

2. *How Much Justice Can We Afford?* The opinions in *Phillips* seem to assume that the goal of patent claim interpretation is to obtain the most accurate interpretation possible, taking into account the inventor's contribution and the intended "audience" for the patent, those in the same art or field. The opinion says nothing about the cost of applying the chosen methodology. For an appeal to a different approach, one which is intentionally more "formulaic," see R. Polk Wagner and Lee Petherbridge, Is the Federal Circuit Succeeding? An Empirical Assessment of Judicial Performance, 152 U. Pa. L. Rev. 1105 (2004). The authors of this article identify two primary interpretive approaches, the "holistic," drawing on the specification and diverse other sources, and "proceduralist," a reproducible approach based on dictionaries and plain meaning. The authors note that many competitors and artisans need to be able to predict how a court will interpret a claim, and therefore a more predictable procedure is desirable. Cf. David Charny, The New Formalism in Contract, 66 U. Chi. L. Rev. 842 (1999) (speculating that formalism and technical expertise, as with tax and patent courts, may go hand-in-hand). Do you believe there is something of a tradeoff between accuracy and efficiency in claim interpretation methodologies? Cf. Louis Kaplow, Rules Versus Standards: An Economic Analysis, 42 Duke L.J. 557 (1992-1993) (standards are more difficult for individuals to apply ex ante, but more flexible because they can be tailored by courts ex post). Even if so, do you believe that a "proceduralist" approach is in fact more reproducible, leading to more predictable results, than a "holistic" approach? Cf. Alan Schwartz, Incomplete Contracts, in 2 The New Palgrave Dictionary of Economics and the Law (Peter Newman ed., 1998), at 277, 280 (noting that the relative costs and benefits of interpretive approaches is an empirical issue for which we have little data). Does the existence of dozens of different dictionaries shed any light on whether a dictionary focus would increase predictability? Federal Circuit cases have had to decide plausible disagreements over the meanings of the words "a," (North American Vaccine, Inc. v. American Cyanamid Co., 7 F.3d 1571, 1581 (Fed. Cir. 1993), *cert. denied,* 511 U.S. 1069 (1994)); "or," Kustom Signals, Inc. v. Applied Concepts, Inc., 264 F.3d 1326 (Fed. Cir. 2001) "to," Corp. v.

FAS Technologies, Inc., 138 F.3d 1448, 1459 (Fed. Cir. 1998); "including," (Toro Co. v. White Consolidated Industries, Inc., 199 F.3d 1295 (Fed. Cir. 1999)); and "through," (Sage Products, Inc. v. Devon Industries, Inc., 126 F.3d 1420 (Fed. Cir. 1997)), to name but a few, suggesting that even words we think have clear meanings may be subject to interpretation.

3. *Judge Mayer's Views on Dodging the Question of Deference.* If the Federal Circuit did defer to trial court claim constructions, would that solve the interpretative problems discussed in *Phillips?* Or would deference merely remove the controversy from the Federal Circuit while providing less guidance to trial courts? Is there anything to be said for deference on efficiency grounds — for example, that some degree of interpretive disagreement is inherent in the process of claim interpretation and cannot be eliminated by appellate rules, and therefore trial courts' decisions should be left largely alone on the theory that additional scrutiny will only add uncertainty and expense?

4. *The Post-*Phillips *Landscape.* Several opinions in the wake of *Phillips* show that the opinion has begun to work significant changes in the law of claim interpretation. For example, in Nystrom v. Trex Co., Inc., 424 F.3d 1136 (Fed.Cir. 2005), superseding 374 F.3d 1105 (Fed.Cir. 2004), the Federal Circuit withdrew an earlier opinion centering around the interpretation of the word "board" in a claim covering decking construction in which the boards are designed to shed water more effectively. The accused infringer, Trex, sold a decking system in which the boards were made of composite wood and recycled plastic material. The patentee Nystrom argued, among other things, that certain dictionary definitions of the word "board" covered the Trex decking materials. The Federal Circuit stated:

> In this case, both parties acknowledge the ordinary meaning of "board" as "a piece of sawed lumber," [i.e., wood]. Nystrom, however, seeks to broaden the term "board" to encompass relatively obscure definitions that are not supported by the written description or prosecution history. Broadening of the ordinary meaning of a term in the absence of support in the intrinsic record indicating that such a broad meaning was intended violates the principles articulated in *Phillips.* Therefore, we affirm the district court's construction of the term "board."

374 F.3d 1136, 1145-46.

As *Phillips* makes clear, however, dictionaries are not a dead letter in Federal Circuit claim construction. Consider for example Free Motion Fitness, Inc. v. Cybex International, Inc., 423 F.3d 1343 (Fed. Cir. 2005), where the court confronted a claim to an exercise machine in which one part was "adjacent to" another. The court held that where there was no intrinsic evidence in the specification giving the term "adjacent to" a specialized meaning, the lower court properly looked to dictionaries for definitions for assistance in determining the term's meaning to one skilled in the art. But the court also cautioned against a return to the pre-*Phillips* reliance on dictionaries alone, with little reference to the specification:

> The court must ensure that any reliance on dictionaries accords with the intrinsic evidence: the claims themselves, the specification, and the prosecution history.... Under *Phillips,* the rule that "a court will give a claim term the full range of its ordinary meaning," Rexnord Corp. v. Laitram Corp., 274 F.3d 1336, 1342 (Fed.Cir.2001), does not mean that the term will presumptively receive its broadest dictionary definition or the

aggregate of multiple dictionary definitions.... Rather, in those circumstances where reference to dictionaries is appropriate, the task is to scrutinize the intrinsic evidence in order to determine the most appropriate definition.

423 F.3d 1343, 1348-49.

5. *Parsing the Specification: Of the "Lexicographer" Rule and "Disavowal."* A longstanding claim interpretation rule is that a patentee may "act as his own lexicographer," defining terms as he or she sees fit. An idiosyncratically defined term that differs from its ordinary meaning is of course given the meaning ascribed by the patentee. See, e.g., Johnson Worldwide Associates, Inc. v. Zebco Corp., 175 F.3d 985, 990 (Fed. Cir. 1999). In a related vein, some pre-*Phillips* Federal Circuit opinions introduced the notion of "claim scope disavowal"—statements in a patent specification that explicitly disclaim coverage of certain embodiments. These cases build on the "patentee as lexicographer" cases, in that they use statements in the specification to limit or interpret terms in the claim.

A good example of the emerging "disavowal" doctrine comes from the case of *Astrazeneca AB, Inc. v. Mutual*, 384 F.3d 1333 (Fed. Cir. 2004) (Michel, J. writing for Archer, J. and Bryson, J.). The key claim in the key patent at issue reads as follows:

> 1. A solid preparation providing extended release of an active compound with very low solubility in water comprising
> a solution or dispersion of an effective amount of the active compound in a semi-solid or liquid nonionic solubilizer, wherein the amount by weight of the solubilizer is at least equal to the amount by weight of the active compound, and a release controlling system to provide extended release.

Claim construction in *Asrazeneca* centered on the term "solubilizer." According to the Federal Circuit opinion, "[t]he parties agree that as a general matter, artisans would understand the term "solubilizer" to embrace three distinct types of chemicals: (1) surface active agents (also known as "surfactants"), (2) co-solvents, and (3) complexation agents." 384 F.3d 1333, 1336 (footnote omitted).

Relying on a dictionary definition, the district court held that the "ordinary meaning" of "solubilizer" embraced the three types of chemicals noted above. The Federal Circuit reversed on two grounds. First, the patentee acted as its own lexicographer by implicitly providing a specialized definition of "solubilizer" in the specification.

Second—and here is where the new doctrinal twist shows up—the court said that the patentee had specifically "disavowed" co-solvents and complexation agents in the definition of "solubilizer." Addressing the "lexicographer" issue, the court wrote:

> [W]e hold that the inventors deliberately acted as their own lexicographers. The 'Description of the Invention' states that '[t]he solubilizers suitable according to the invention *are defined below*' (emphasis added), and two paragraphs later, states that '[t]he solubilizers suitable for the preparations according to the invention are semi-solid or liquid non-ionic *surface active agents*' (emphasis added). Astrazeneca maintains that these statements simply refer to preferred embodiments of 'suitable' solubilizers. We might agree if the specification stated, for example, '*a solubilizer* suitable for the preparations according to the invention,' but in fact, the specification definitively states '*the solubilizers* suitable for the preparations according to the invention' (emphasis added). Astrazeneca seems to suggest that lexicography requires a statement in the form "I define _____ to mean _____," but such rigid formalism is

not required. See, e.g., Bell Atl. Network Servs., Inc. [v. Covad Comm. Group, Inc., 262 F.3d 1258 (Fed. Cir. 2001)] at 1268 ("[A] claim term may be clearly redefined without an explicit statement of redefinition.... [T]he specification may define claim terms 'by implication' such that the meaning may be 'found in or ascertained by a reading of the patent documents.'"(citation omitted)). Certainly the '081 specification's statement that "[t]he solubilizers suitable according to the invention are defined below" provides a strong signal of lexicography.

384 F.3d 1333, 1340. The "disavowal" issue was next:

> Second, we hold the specification clearly disavows nonsurfactant solubilizers. The inventors' lexicography alone works an implicit disavowal of nonsurfactant solubilizers, but the rest of the specification goes further. The "Description of the Invention" twice describes micelle structures as a feature of the novel formulation structure conceived by the inventors.... It is undisputed that surfactants are the only solubilizers believed to form micelle structures in watery environments. Indeed, immediately after the reference to the "micelle-structure formed by the solubilizer" of the invention, the specification criticizes other types of solubilizers—and specifically co-solvents—as leading to undesirable precipitation.... Astrazeneca contends that these statements in the specification simply address the features of preferred embodiments. Astrazeneca seems to suggest that clear disavowal requires an "expression of manifest exclusion or restriction" in the form of "my invention does not include _____." But again, such rigid formalism is not required. Where the general summary or description of the invention describes a feature of the invention (here, micelles formed by the solubilizer) and criticizes other products (here, other solubilizers, including co-solvents) that lack that same feature, this operates as a clear disavowal of these other products (and processes using these products).

384 F.3d 1333, 1340. The disavowal idea appears in other opinions as well. See, e.g., Gaus v. Conair Corp., 363 F.3d 1284, 1291 (Fed. Cir. 2004); Omega Engineering, Inc, v. Raytek Corp., 334 F.3d 1314 (Fed. Cir. 2003).

After *Phillips*'s endorsement of a more specification-intensive form of claim interpretation, the disavowal notion could increase in importance.

6. *Meaning as of When?* A recent case sharpens an issue that has lurked below the surface of claim interpretation cases for some time: at what point in time should a court consider the meaning of a claim term—when the claim was filed in a patent application, at the time the infringement took place, or some other time? PC Connector Solutions LLC v. SmartDisk Corp., 406 F.3d 1359 (Fed. Cir. 2005) holds that the claim words "conventional," "traditional," and "standard" should be interpreted as being temporally limited to the conventions, traditions, and standards that existed at the time when the patent was filed. The patent at issue covered a device for connecting peripheral devices to a computer via the computer's diskette drive. Claim 1 of the patent is representative:

> 1. In combination a computer having a diskette drive, an end user computer peripheral device having an input/output port normally connectible to a *conventional* computer input/output port, and a coupler which couples the computer with the end user computer peripheral device without using a *conventional* computer input/output port:
>
> said coupler being sized and shaped for insertion within the diskette drive of the computer... whereby data is transferred from said computer to said end user computer peripheral device via said read/write head of said computer, said coupler, and said input/output port of said end user computer peripheral device.

The accused devices in the case were couplers to link peripheral devices through a computer disk drive, but the couplers were designed to handle a particular class of peripheral devices — flash memories and smart cards that have flat, planar surface contact electrodes as their normal connecting elements (i.e., connectors radically different from the sort of multi-pronged connectors conventional in 1988). The court held that the accused devices could not infringe the patent because the relevant claim language limited the invention to an adapter for peripheral devices having 1988-style connectors:

> We find nothing in the written description that amounts to a clear attempt by the patentee to impart any special meaning to the words "normally," "conventional," "traditionally," or "standard." Likewise, in the prosecution history, we do not read the patentability arguments that were made specifically to distinguish a test fixture as having effected the particular redefinition now advanced on appeal. As a consequence, the terms "normally," "conventional," "traditionally," and "standard" are governed by their ordinary and customary meanings, and that, in view of their implicit time-dependence, the district court did not err in construing the literal scope of the claim limitations qualified by those terms as being limited to technologies existing at the time of the invention.
>
> PC Connector's argument that the dictionary definitions of "normally," "conventional," "traditionally," and "standard" contain no explicit reference to a time limitation is not persuasive, as their time-related significance is implicit from their ordinary usage — just as other words are implicitly not time-related. A comparison of the words "conventional" and "dedicated" is instructive in this regard, as PC Connector's briefs appear to treat them as interchangeable and coterminous in claim scope, when, in fact, they are not. To illustrate, a present-day USB port may be described as a "dedicated" I/O port within the ordinary meaning of "dedicated" as that word would be used to characterize the I/O ports found on a computer built in 1988, yet it would not be considered "conventional" back then, even though it is "conventional" today. (Martin Aff. P11; J.A. 264). Thus, unlike the word "dedicated," the word "conventional" necessarily has a meaning specific to the time of filing.

406 F.3d at 1363-64. Do these words have "implicit" time-related significance? Why did the patentee invite this discussion by limiting the claim to an adapter for peripherals having "conventional" connectors? For more on this important issue, see Mark Lemley, The Changing Meaning of Patent Claim Terms, 104 Mich. L. Rev. 101 (2005) (arguing that patent claim terms should have a fixed meaning throughout time, and that that meaning should be fixed at the time the patent application is first filed).

2. Literal Infringement

35 U.S.C. § 271 gives the patentee an infringement cause of action against anyone who makes, uses, sells, offers for sale, or imports the invention described in the claims of the patent. This right runs from the day the patent is issued until the end of the patent term (formerly 17 years from the day of issue, now 20 years from the day the patent application is filed). Because a patent is defined in terms of its claims, a patent infringement lawsuit is resolved by comparing the claims of the patent to the accused product.

Larami Corp. v. Amron

United States District Court for the Eastern District of Pennsylvania
27 U.S.P.Q.2d 1280 (E.D. Pa. 1993)

REED, J.

This is a patent case concerning toy water guns manufactured by plaintiff Larami Corporation ("Larami"). Currently before me is Larami's motion for partial summary judgment of noninfringement of United States Patent No. 4,239,129 ("the '129 patent")....

For the reasons discussed below, the motion will be granted.

I. Background

Larami manufactures a line of toy water guns called "SUPER SOAKERS." This line includes five models: SUPER SOAKER 20, SUPER SOAKER 30, SUPER SOAKER 50, SUPER SOAKER 100, and SUPER SOAKER 200. All use a hand-operated air pump to pressurize water and a "pinch trigger" valve mechanism for controlling the ejection of the pressurized water. All feature detachable water reservoirs prominently situated outside and above the barrel of the gun. The United States Patent and Trademark Office has issued patents covering four of these models. Larami does not claim to have a patent which covers SUPER SOAKER 20.

Defendants Alan Amron and Talk To Me Products, Inc. (hereinafter referred to collectively as "TTMP") claim that the SUPER SOAKER guns infringe on the '129 patent which TTMP obtained by assignment from Gary Esposito ("Esposito"), the inventor. The '129 patent covers a water gun which, like the SUPER SOAKERS, operates by pressurizing water housed in a tank with an air pump. In the '129 patent, the pressure enables the water to travel out of the tank through a trigger-operated valve into an outlet tube and to squirt through a nozzle. Unlike the SUPER SOAKERS, the '129 patent also contains various electrical features to illuminate the water stream and create noises. Also, the water tank in the '129 patent is not detachable, but is contained within a housing in the body of the water gun.

The "Background of the Invention" contained in the '129 patent reads as follows:

> Children of all ages, especially boys, through the years have exhibited a fascination for water, lights and noise and the subject invention deals with these factors embodied in a toy simulating a pistol.
>
> An appreciable number of U.S. patents have been issued which are directed to water pistols but none appear to disclose a unique assembly of components which can be utilized to simultaneously produce a jet or stream of water, means for illuminating the stream and a noise, or if so desired, one which can be operated without employing the noise and stream illuminating means. A reciprocal pump is employed to obtain sufficient pressure whereby the pistol can eject a stream an appreciable distance in the neighborhood of thirty feet and this stream can be illuminated to more or less simulate a lazer [sic] beam....

Larami has moved for partial summary judgment of noninfringement of the '129 patent ... and for partial summary judgment on TTMP's counterclaim for infringement of the '129 patent.

II. Discussion...

B. Infringement and Claim Interpretation

A patent owner's right to exclude others from making, using or selling the patented invention is defined and limited by the language in that patent's claims. Corning Glass Works v. Sumitomo Electric U.S.A., Inc., 868 F.2d 1251, 1257 (Fed. Cir. 1989). Thus, establishing infringement requires the interpretation of the "elements" or "limitations" of the claim and a comparison of the accused product with those elements as so interpreted. Because claim interpretation is a question of law, it is amenable to summary judgment.

The words in a claim should be given their "ordinary or accustomed" meaning. Senmed, Inc. v. Richard-Allan Medical Industries, Inc., 888 F.2d 815, 819 & n.8 (Fed. Cir. 1989). An inventor's interpretations of words in a claim that are proffered after the patent has issued for purposes of litigation are given no weight....

A patent holder can seek to establish patent infringement in either of two ways: by demonstrating that every element of a claim (1) is literally infringed or (2) is infringed under the doctrine of equivalents. To put it a different way, because every element of a claim is essential and material to that claim, a patent owner must, to meet the burden of establishing infringement, "show the presence of every element or its substantial equivalent in the accused device." If even one element of a patent's claim is missing from the accused product, then "[t]here can be no infringement as a matter of law."... London v. Carson Pirie Scott & Co., 946 F.2d 1534, 1538-39 (Fed. Cir. 1991).

Larami contends, and TTMP does not dispute, that twenty-eight (28) of the thirty-five (35) claims in the '129 patent are directed to the electrical components that create the light and noise. Larami's SUPER SOAKER water guns have no light or noise components. Larami also contends, again with no rebuttal from TTMP, that claim 28 relates to a "poppet valve" mechanism for controlling the flow of water that is entirely different from Larami's "pinch trigger" mechanism. Thus, according to Larami, the six remaining claims (claims 1, 5, 10, 11, 12 and 16) are the only ones in dispute. Larami admits that these six claims address the one thing that the SUPER SOAKERS and the '129 patent have in common — the use of air pressure created by a hand pump to dispense liquid. Larami argues, however, that the SUPER SOAKERS and the '129 patent go about this task in such fundamentally different ways that no claim of patent infringement is sustainable as a matter of law.

1. Literal Infringement of Claim 1

TTMP claims that SUPER SOAKER 20 literally infringes claim 1 of the '129 patent. Claim 1 describes the water gun as:

[a] toy comprising an elongated housing [case] having a chamber therein for a liquid [tank], a pump including a piston having an exposed rod [piston rod] and extending rearwardly of said toy facilitating manual operation for building up an appreciable

amount of pressure in said chamber for ejecting a stream of liquid therefrom an appreciable distance substantially forwardly of said toy, and means for controlling the ejection.

U.S. Patent No. 4,239,129 (bracketed words supplied; see Diagram A, the '129 patent, attached hereto [Figure 3-6]).

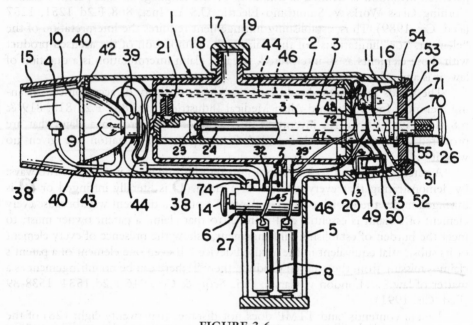

FIGURE 3-6
Patent drawing showing water pistol and/or flash light structure.

Claim 1 requires, among other things, that the toy gun have "an elongated housing having a chamber therein for a liquid." The SUPER SOAKER 20 water gun, in contrast, has an external water reservoir (chamber) that is detachable from the gun housing, and not contained within the housing. TTMP argues that SUPER SOAKER 20 contains a "chamber therein for a liquid" as well as a detachable water reservoir. It is difficult to discern from TTMP's memorandum of law exactly where it contends the "chamber therein" is located in SUPER SOAKER 20. Furthermore, after having examined SUPER SOAKER 20 . . . , I find that it is plain that there is no "chamber" for liquid contained within the housing of the water gun. The only element of SUPER SOAKER 20 which could be described as a "chamber" for liquid is the external water reservoir located atop the housing. Indeed, liquid is located within the housing only when the trigger causes the liquid to pass from the external water reservoir through the tubing in the housing and out of the nozzle at the front end of the barrel. SUPER SOAKER 20 itself shows that such a transitory avenue for the release of liquid is clearly not a "chamber therein for liquid." Therefore, because the absence of even one element of a patent's claim from the accused product means there can be no finding of literal infringement, London, 946 F.2d at 1538-39, I find that SUPER SOAKER 20 does not infringe claim 1 of the '129 patent as a matter of law. . . .

Accordingly, I conclude that the SUPER SOAKER 20 water gun does not literally infringe claim 1 of the '129 patent.

2. Infringement by Equivalents of Claim 10

[The court further found that defendants TTMP failed to produce evidence which would support a finding that there was a genuine issue of material fact as to whether the Super Soaker guns infringed claim 10 under the doctrine of equivalents.

The court began its analysis by noting that successful use of the doctrine to show infringement requires the patent owner to prove that the accused product has the "substantial equivalent" of every limitation or element of a patent claim. TTMP argued that claim 10 teaches an arrangement of the tank, air pump and outlet nozzle along the same axis. The court, citing a previous decision regarding this same claim, denied that claim 10 required the three components to be located on the same axis; thus the axial placement of these components on the Super Soakers "cannot infringe claim 10 of the '129 patent because there is nothing in the language of claim 10 to which it could be substantially equivalent."

Additionally, Super Soakers' use of an external, detachable water reservoir was found to be such a dramatic improvement over the traditional design — benefiting both the manufacturer and user — that it could not be held to be the "substantial equivalent" of the claim 10 requirement of "a tank in the barrel for a liquid."]

III. Conclusion

In patent cases, summary judgment is appropriate where the accused product does not literally infringe the patent and where the patent owner does not muster evidence that is "sufficient to satisfy the legal standard for infringement under the doctrine of equivalents." London, 946 F.2d at 1538. Thus, and for the foregoing reasons, Larami's motion for partial summary judgment of noninfringement of the '129 patent will be granted.

COMMENTS AND QUESTIONS

1. Suppose TTMP had drafted claim 1 using the following phrase: "an elongated housing having a chamber for a liquid." Further suppose that the dictionary defines conjoining as "attached to; connected to." If the specification failed to attach any special meaning to this term, would the court's holding regarding literal infringement of claim 1 have been affected? See Casler v. U.S., 9 U.S.P.Q.2d 1753, 1772 (Ct. Cl. 1988); Universal Oil Products Co. v. Globe Oil & Refining Co., 40 F. Supp. 575, 582 (N.D. Ill. 1941), aff'd, 137 F.2d 3 (7th Cir. 1943), aff'd, 322 U.S. 471 (1944).

2. As in question 1 (above), assume the specification did not discuss further the phrase "having a chamber for a gun." But suppose that plaintiff Larami discovers claim 1 was originally rejected by an examiner because a prior art reference showed a squirt gun with an oversized chamber on top of the body of the gun. The patent applicant successfully overcame this rejection by pointing out that in the invention the chamber

is inside the housing of the gun. Recall that the face of the '129 patent would not give a clue to potential infringers such as Larami as to the potentially narrower definition of the claim language conceded during the prosecution history. To find out what the inventor argued during prosecution, one would need to obtain the prosecution history, or "file wrapper," for the patent. See Jonsson v. Stanley Works, 903 F.2d 812, 818 (Fed. Cir. 1990).

3. The elements of a patent claim are of considerable importance in determining its scope. For an accused product to literally infringe a patent, *every* element contained in the patent claim must also be present in the accused product or device. If a claimed apparatus has five parts, or "elements," and the allegedly infringing apparatus has only four of those five, it does not literally infringe. This is true even though the defendant may have copied the four elements exactly, and regardless of how significant or insignificant the missing element is.

What happens if an infringer *adds* elements? The outcome then depends on the wording of the patentee's claim. If the patentee has drafted an "open" claim, usually indicated by the term "comprising," adding elements does not help the infringer. See, e.g., Genentech, Inc. v. Chiron Corp., 112 F.3d 495, 501 (Fed.Cir.1997) ("'Comprising' is a term of art used in claim language which means that the named elements are essential, but other elements may be added and still form a construct within the scope of the claim."). See also Mars, Inc. v. H.J. Heinz Co., L.P., 377 F.3d 1369, 1375 (Fed. Cir. 2004) ("containing" and "mixture" are open, like "comprising"). But if the patentee uses the term "consisting of," the result will likely be the opposite: adding elements makes the accused product non-infringing. See Norian Corp. v. Stryker Corp., 363 F.3d 1321 (Fed. Cir. 2004) (claims to dental repair kit "consisting of" certain chemicals not infringed by infringer's kit, which also included mixing spatula). If an accused device has all the elements of a patent claim, it will generally be found to infringe. In most cases, that is, it does not matter that the defendant has *added* several new elements — adding new features cannot help a defendant escape infringement. The exact outcome of such a case is determined by the precise claim language and the nature and significance of the "added elements."

This rule has an important consequence for the process of innovation. Patentees who have properly claimed a fundamental technology can assert their patent against anyone who uses that technology, even if the defendants have improved it or put it to different use. A broad basic patent therefore gives its owner a great deal of control not only over potential direct competitors, but over a number of derivative or ancillary markets during the term of the patent.

For a discussion and critique of the way courts determine the "elements" of an invention, see Dan A. Burk and Mark Lemley, Quantum Patent Mechanics, 9 Lewis & Clark L. Rev. 29, 29 (2005):

> There are no hard and fast standards in the law by which to make the "right" decision as to either the size of the textual element or the level of abstraction at which it will be evaluated. Indeed, the indeterminacy is so acute that courts generally don't even acknowledge that they are engaging in either inquiry. They define an element almost arbitrarily, and even when judges disagree as to the proper definition they offer no principled basis for doing so.

To reduce the indeterminacy, the authors advocate that a court "start with the patentee's invention itself, construing patent claims narrowly and in light of the actual invention when the claim terms are ambiguous. Courts could then supplement this narrower claim construction with a doctrine of equivalents analysis, which would permit them to decide how broadly to apply the principle of the invention." Id., at 30. They caution that such an approach would require a more functional, and less formalistic, approach to that doctrine, however. We turn next to a discussion of this doctrine of equivalents.

3. The Doctrine of Equivalents

a. Basic Issues

As in other areas of the law, literal interpretation can result in injustice. If competitors could circumvent patents through insubstantial changes in the design of a product, then many patents would lose their value, and patent drafters would expend unreasonable efforts trying to include every possible variation. Unlike geophysical measurement, the English language lacks numerical precision. So courts have developed the doctrine of equivalents. The rationale for the doctrine was explained by the Supreme Court half a century ago in Graver Tank & Mfg. Co. v. Linde Air Prods., Inc., 339 U.S. 605 (1950):

> In determining whether an accused device or composition infringes a valid patent, resort must be had in the first instance to the words of the claim. If accused matter falls clearly within the claim, infringement is made out and that is the end of it.
>
> But courts have also recognized that to permit imitation of a patented invention which does not copy every literal detail would be to convert the protection of the patent grant into a hollow and useless thing. Such a limitation would leave room for — indeed encourage — the unscrupulous copyist to make unimportant and insubstantial changes and substitutions in the patent which, though adding nothing, would be enough to take the copied matter outside the claim, and hence outside the reach of law. One who seeks to pirate an invention, like one who seeks to pirate a copyrighted book or play, may be expected to introduce minor variations to conceal and shelter the piracy. Outright and forthright duplication is a dull and very rare type of infringement. To prohibit no other would place the inventor at the mercy of verbalism and would be subordinating substance to form. It would deprive him of the benefit of his invention and would foster concealment rather than disclosure of inventions, which is one of the primary purposes of the patent system.
>
> The doctrine of equivalents evolved in response to this experience. The essence of the doctrine is that one may not practice a fraud on a patent. Originating almost a century ago in the case of *Winans v. Denmead*, [56 U.S.] 15 How. 330 [1853], it has been consistently applied by this Court and the lower federal courts, and continues today ready and available for utilization when the proper circumstances for its application arise. "To temper unsparing logic and prevent an infringer from stealing the benefit of the invention" a patentee may invoke this doctrine to proceed against the producer of a device "if it performs substantially the same function in substantially the same way to obtain the same result." *Sanitary Refrigerator Co. v. Winters*, 280 U.S. 30, 42 (1929). The theory on which it is founded is that "if two devices do the same work in substantially the same way, and accomplish substantially the same result, they are the same, even though they differ in name, form or shape." *Union Paper-Bag Machine Co. v. Murphy*, 97 U.S. 120, 125 [1877]....
>
> What constitutes equivalency must be determined against the context of the patent, the prior art, and the particular circumstances of the case. Equivalence, in the patent law,

is not the prisoner of a formula and is not an absolute to be considered in a vacuum. It does not require complete identity for every purpose and in every respect. In determining equivalents, things equal to the same thing may not be equal to each other and, by the same token, things for most purposes different may sometimes be equivalents. Consideration must be given to the purpose for which an ingredient is used in a patent, the qualities it has when combined with the other ingredients, and the function which it is intended to perform. An important factor is whether persons reasonably skilled in the art would have known of the interchangeability of an ingredient not contained in the patent with one that was.

In Warner-Jenkinson v. Hilton Davis Chemical Co., 520 U.S. 17 (1997), the Supreme Court revisited the doctrine of equivalents for the first time since *Graver Tank, supra. Warner-Jenkinson* involved a claim to a process for purifying dyes. The claim had three key limitations: (1) the pore diameter of the filtration screen; (2) a pressure range under which the filtration took place; and (3) a requirement that the filtration take place in an aqueous solution having a pH range "from approximately 6.0 to 9.0." Limitation (3), the pH range, was added by the patentee during the prosecution of the patent. The accused infringer, Warner-Jenkinson, used a filtration process that met the first two limitations, but operated at a pH of 5.0. A classic doctrine of equivalents case.

The Supreme Court first affirmed the continued viability of the doctrine of equivalents, as it had in *Graver Tank.* Next, it laid down some basic principles regarding the doctrine. The Court did not reject the venerable "triple identity" test of *Graver Tank* – i.e., that equivalence may be found where the accused product produces substantially the same function, in substantially the same way, to achieve substantially the same result as the device of the patent claim. But it noted that this test might be more suitable to mechanical inventions than to chemical processes such as the one in this case. On the ultimate question of the proper standard in a doctrine of equivalents case, the Court held:

> In our view, the particular linguistic framework used is less important than whether the test is probative of the essential inquiry: Does the accused product or process contain elements identical or equivalent to each claimed element of the patented invention? Different linguistic frameworks may be more suitable to different cases, depending on their particular facts. A focus on individual elements and a special vigilance against allowing the concept of equivalence to eliminate completely any such elements should reduce considerably the imprecision of whatever language is used. An analysis of the role played by each element in the context of the specific patent claim will thus inform the inquiry as to whether a substitute element matches the function, way, and result of the claimed element, or whether the substitute element plays a role substantially different from the claimed element.

520 U.S. at 40. This approach is known as the "all elements rule."

But *Warner-Jenkinson* left open one crucial question: what exactly is an element? (For a detailed discussion of this issue, see Dan A. Burk and Mark Lemley, Quantum Patent Mechanics, 9 Lewis & Clark L. Rev. 29, 29 (2005).) And how precisely must the accused product track each element to infringe under the doctrine of equivalents?

It did not take long for these issues to present themselves in the wake of *Warner-Jenkinson.* After deciding that case, the Supreme Court vacated the Federal Circuit's decision in Hughes Aircraft Co. v. United States—further complicating an already torturous procedural history. On remand, the Federal Circuit was obliged to consider the impact of *Warner-Jenkinson* on its prior ruling finding infringement under the doctrine of equivalents. See 140 F.3d 1470 (Fed. Cir. 1998). In its opinion, the Federal Circuit revisited the equivalents issue and in particular the application of the Supreme Court's "all elements" test. The Federal Circuit began with a brief description of the satellite technology at issue in the case:

> The Williams patent relates to an apparatus for control over the orientation, or attitude, of a spacecraft using commands from a ground control station. The relevant limitations of claim 1 of the Williams patent read:
>
>> (e) means disposed on said body for providing an *indication to a location external to said body of the instantaneous spin angle position* of said body about said axis and the orientation of said axis with reference to a fixed external coordinate system;
>>
>> (f) and means disposed on said body for *receiving from said location control signals synchronized* with said indication;
>>
>> (g) said valve being coupled to said last-named means and responsive to said control signals for applying fluid to said fluid expulsion means *in synchronism therewith* for precessing said body to orient said axis into a predetermined desired relationship with said fixed external coordinate system.

(Emphasis added). In order to correct the attitude of the spacecraft, the ground crew must be able to calculate the instantaneous spin angle (ISA) position. The ISA position is the angle between two specific planes. The first plane, the rotating plane, is defined by the location of the precessing jet and the satellite's axis of rotation. The second plane, the reference plane, is defined by a fixed reference point in an external coordinate system (such as the sun or another star) and the spin axis. The angle between these planes at a given moment in time is the ISA position with reference to a fixed external coordinate system. Thus, the ISA position generally measures the location of the precessing jet in its rotational cycle relative to the reference plane.

Two pieces of information are needed to calculate the ISA position: the spin rate of the satellite and the instant in time at which the rotating plane passes by the fixed point in the fixed external coordinate system and at which the jet is closest to the fixed reference point. The invention uses onboard sensors to collect this data and then transmits this information to earth to allow the ground crew to determine the satellite's existing and desired orientations. After making the necessary calculations, the ground crew pulses the attitude jet by radio signal commands to precess, or tip, the spin axis of the satellite to the desired position.

In the accused "store and execute" (S/E) craft, the satellite retrieves the same raw data but calculates the ISA position onboard. The spin rate and information to determine the orientation of the satellite is transmitted to the ground. In most of the S/E craft, the satellite does not provide information sufficient to calculate the ISA position. After receiving the spin rate, the ground crew performs the necessary calculations to adjust the attitude of the craft. This information is then sent to and stored in the satellite. The precession does not occur, however, until the ground crew sends an execute command to the satellite.

In 1982, the then-Court of Claims originally determined, inter alia, that the accused S/E devices did not infringe the patent literally or under the doctrine of equivalents. See Hughes Aircraft Co. v. United States, 215 U.S.P.Q. 787, 812 (Ct.Cl. Trial Div.1982) (Hughes VI). On appeal, this court reversed the noninfringement judgment, holding that the S/E devices infringe under the doctrine of equivalents, and remanded for a determination of just compensation. See Hughes VII, 717 F.2d at 1366. After the decision by the Court of Federal Claims on remand, Hughes appealed, challenging the assessment of damages, and the government cross-appealed, again challenging the liability determination of Hughes VII in light of this court's in banc decision in Pennwalt Corp. v. Durand-Wayland, Inc., 833 F.2d 931 (Fed. Cir. 1987) (in banc). See Hughes XIII, 86 F.3d at 1566 (Fed. Cir. 1996). This court affirmed the damages determination and refused, under the doctrine of law of the case, to reconsider the Hughes VII decision. See id. at 1576, 86 F.3d 1566. The Supreme Court, however, granted certiorari, vacated the judgment, and remanded the case (GVR order) to this court for reconsideration in light of its decision in Warner-Jenkinson. See Hughes XIV, [520 U.S. 1183]. . . .

Turning to the merits, we first address the effect of the "all-elements" rule (sometimes referred to as the "all-limitations" rule) enunciated in Warner-Jenkinson on the Hughes VII decision. The Supreme Court clarified the doctrine of equivalents by noting that:

> [e]ach element contained in a patent claim is deemed material to defining the scope of the patented invention, and thus the doctrine of equivalents must be applied to individual elements of the claim, not to the invention as a whole. It is important to ensure that the application of the doctrine, even as to an individual element, is not allowed such broad play as to effectively eliminate that element in its entirety.

520 U.S. at 29. Thus, the test for equivalence is to be applied to the individual claim limitations:

> An analysis of the role played by each element in the context of the specific patent claim will thus inform the inquiry as to whether a substitute element matches the function, way, and result of the claimed element, or whether the substitute element plays a role substantially different from the claimed element.

Id. at 40; see Pennwalt, 833 F.2d at 935 ("'[T]he plaintiff must show the presence of every element or its substantial equivalent in the accused device.' To be a 'substantial equivalent,' the element substituted in the accused device for the element set forth in the claim must not be such as would substantially change the way in which the function of the claimed invention is performed." (citations omitted) (quoting Perkin-Elmer Corp. v. Westinghouse Elec. Corp., 822 F.2d 1528, 1532-33 (Fed. Cir. 1987))).

The government argues that the all-elements rule demands that we depart from the reasoning in Hughes VII, in which the court stated that the trial court erred in not "apply[ing] the doctrine of equivalents to the claimed invention as a whole." Hughes VII, 717 F.3d at 1364. According to the government, to conclude that the claim limitations in paragraphs (e), (f), and (g) are met equivalently by elements in the accused devices would vitiate those claim limitations. The government additionally urges that the arguably corresponding elements of the S/E system differ substantially from the claim limitations by storing the ISA position value onboard in lieu of transmitting an indication of the ISA position to the ground, by not acting in synchronism with the control signals, and by not firing the precession jet within a fixed period of time after receiving the command signal.

Hughes responds that there is no reason to depart from the conclusion reached in Hughes VII because Warner-Jenkinson did not significantly alter the all-elements rule as stated in Pennwalt. Moreover, the Hughes VII court, in Hughes' opinion, did perform the required element-by-element analysis mandated by Warner-Jenkinson.

We conclude that the analysis performed in Hughes VII satisfies the all-elements rule as stated in Warner-Jenkinson. Regarding claim paragraph (e), the court in Hughes VII concluded that the transmission to the ground crew of the spin rate and information sufficient to calculate the sun angle in the S/E vehicles "is the modern day equivalent to providing an indication of the ISA to the ground. . . ." 717 F.2d at 1365. This information, while insufficient to calculate the ISA position, was sufficient to enable the ground crew to control the satellite, which is substantially the same function performed and the identical result achieved by transmitting the indication of the ISA position to the ground. See id.

Furthermore, although the information sent is insufficient to calculate the ISA position, this does not demonstrate a substantial difference in the way the element functions. This is a case in which a "subsequent change in the state of the art, such as later-developed technology, obfuscated the significance of [the] limitation at the time of its incorporation into the claim." Sage Prods., Inc. v. Devon Indus., Inc., 126 F.3d 1420, 1425 (Fed. Cir. 1997); cf. Warner-Jenkinson, 520 U.S. at 37 ("Insofar as the question under the doctrine of equivalents is whether an accused element is equivalent to a claimed element, the proper time for evaluating equivalency . . . is at the time of infringement, not at the time the patent was issued."); Pennwalt, 833 F.2d at 938 ("[T]he facts here do not involve later developed computer technology which should be deemed within the scope of the claims to avoid the pirating of an invention."). The court in Hughes VII determined that the change in the S/E devices was the result of a technological advance not available until after the patent issued. See Hughes VII, 717 F.2d at 1365. Relying on testimony of one of skill in the art at the time of infringement, the court in Hughes VII concluded that this advance resulted in an insubstantial change in the way the element performed its function. See id. (citing testimony that an engineer would realize that transmission of the ISA position was no longer necessary as a result of the change in technology).

The court in Hughes VII also concluded that the "synchronism" limitations in paragraphs (f) and (g) were also equivalently met by the accused devices. Again, as a result of an advance in technology, the satellite system was able at the time of infringement to store the precession information and to wait to precess the satellite until receipt of the execute command. Thus, the synchronism in the accused device is coordinated by the computer instead of by real-time execution of the command from the ground. As recognized in Hughes VII, "[t]he difference between operation by retention and operation by sending is achieved by relocating the function, making no change in the function performed, or in the basic manner of operation, or in the result obtained." Id. at 1366, 717 F.2d 1351. The court in Hughes VII correctly performed an analysis of the function, way, and result of the individual elements in the accused devices and concluded that these elements equivalently met the claim limitations at issue.

Accordingly, we conclude that Warner-Jenkinson provides no basis to alter the decision in Hughes VII because the court properly applied the all-elements rule.

Id. at 1472-75.

COMMENTS AND QUESTIONS

1. *An Essential Tension.* In the *Hughes Satellite* case, the Federal Circuit struggled to reconcile two difficult aspects of the *Warner-Jenkinson* opinion: (1) the adoption of the "element by element" approach to equivalence and (2) the preservation of the rule that equivalence is measured as of the time of infringement, and not as of the time of invention. The difficulty here — best illustrated by the facts in the *Hughes* case itself — is that "after arising" technologies (those developed after the date of invention) will often challenge the definition of an element in the original invention. Recall that in *Hughes,* the accused embodiments incorporated "on board" computers that were not in existence at the time of the invention of the Hughes satellite. The Federal Circuit, in the original opinion finding infringement under the doctrine of equivalents, found that these after-developed satellites were the equivalent of those claimed in the original patent. This was certainly defensible under prevailing doctrine of equivalents analysis at the time of this opinion. However, the Federal Circuit's subsequent deci-

sion in *Pennwalt,* which was endorsed by the Supreme Court in the *Warner-Jenkinson* case, called this conclusion into doubt. This is so because the original claims in the Hughes patent include two crucial elements or limitations that are arguably not present in the later-developed satellites: (1) "means for providing... [an] *indication to a location external to said body of the instantaneous spin angle position"*; and (2) "means disposed on said body *for receiving from said location control signals synchronized* with said indication." [emphasis added].

2. *Hard Questions.* Ultimately, the court in *Hughes* concludes that subsequent developments in the technology "*obfuscate[d] the significance of [the] limitation...,*" (emphasis added), quoting *Sage Products.* What good is the "all-elements rule" if it is ignored in the important and common case of after-developed technologies, or applied in such a way that it is difficult to tell what an "element" is? How does the temporal aspect of the post–*Warner-Jenkinson* test, i.e., measuring equivalence at the time of infringement, square with the Supreme Court's emphasis on the "notice to competitors" function of patent claims?

b. Prosecution History Estoppel

An additional issue central to *Warner-Jenkinson* was the fact that the pH limit in the claim was added during prosecution of the patent. This triggered application of the doctrine of "prosecution history estoppel." In this case, the infringer argued, that doctrine prevented the patentee from arguing infringement under the doctrine of equivalents, because the patentee had surrendered coverage for any pH under 6.0 when the "6.0 to 9.0" range was added to the claim during prosecution. The Court in *Warner-Jenkinson* created a presumption under these circumstances that the claim had been amended to avoid prior art, and thus that prosecution history estoppel did indeed apply. 520 U.S. at 33. But it permitted the patentee to rebut this presumption with evidence that the claim had been amended for some other purpose, unrelated to avoiding the prior art. The status of this presumption, and the doctrine of prosecution history estoppel generally, was the key issue in the *Festo* case decided by the Supreme Court five years later.

Festo Corp. v. Shoketsu Kinzoku Kogyo Kabushiki Co., Ltd.
Supreme Court of the United States
535 U.S. 722 (2002)

Justice KENNEDY delivered the opinion of the Court.

This case requires us to address once again the relation between two patent law concepts, the doctrine of equivalents and the rule of prosecution history estoppel. The Court considered the same concepts in Warner-Jenkinson Co. v. Hilton Davis Chemical Co., 520 U.S. 17 (1997), and reaffirmed that a patent protects its holder against efforts of copyists to evade liability for infringement by making only insubstantial changes to a patented invention. At the same time, we appreciated that by extending protection beyond the literal terms in a patent the doctrine of equivalents can create substantial uncertainty about where the patent monopoly ends. Id., at 29. If the range of equivalents is unclear, competitors may be unable to determine what is a permitted alternative to a patented invention and what is an infringing equivalent.

To reduce the uncertainty, Warner-Jenkinson acknowledged that competitors may rely on the prosecution history, the public record of the patent proceedings. In some cases the Patent and Trademark Office (PTO) may have rejected an earlier version of the patent application on the ground that a claim does not meet a statutory requirement for patentability. 35 U.S.C. § 132 (1994 ed., Supp. V). When the patentee responds to the rejection by narrowing his claims, this prosecution history estops him from later arguing that the subject matter covered by the original, broader claim was nothing more than an equivalent. Competitors may rely on the estoppel to ensure that their own devices will not be found to infringe by equivalence.

In the decision now under review the Court of Appeals for the Federal Circuit held that by narrowing a claim to obtain a patent, the patentee surrenders all equivalents to the amended claim element. Petitioner asserts this holding departs from past precedent in two respects. First, it applies estoppel to every amendment made to satisfy the requirements of the Patent Act and not just to amendments made to avoid preemption by an earlier invention, i.e., the prior art. Second, it holds that when estoppel arises, it bars suit against every equivalent to the amended claim element. The Court of Appeals acknowledged that this holding departed from its own cases, which applied a flexible bar when considering what claims of equivalence were estopped by the prosecution history. Petitioner argues that by replacing the flexible bar with a complete bar the Court of Appeals cast doubt on many existing patents that were amended during the application process when the law, as it then stood, did not apply so rigorous a standard.

We granted certiorari to consider these questions.

I

Petitioner Festo Corporation owns two patents for an improved magnetic rodless cylinder, a piston-driven device that relies on magnets to move objects in a conveying system. The device has many industrial uses and has been employed in machinery as diverse as sewing equipment and the Thunder Mountain ride at Disney World. Although the precise details of the cylinder's operation are not essential here, the prosecution history must be considered.

Petitioner's patent applications, as often occurs, were amended during the prosecution proceedings. The application for the first patent, the Stoll Patent (U.S. Patent No. 4,354,125), was amended after the patent examiner rejected the initial application because the exact method of operation was unclear and some claims were made in an impermissible way. (They were multiply dependent.) 35 U.S.C. § 112 (1994 ed.). The inventor, Dr. Stoll, submitted a new application designed to meet the examiner's objections and also added certain references to prior art. 37 CFR § 1.56 (2000). The second patent, the Carroll Patent (U.S. Patent No. 3,779,401), was also amended during a reexamination proceeding. The prior art references were added to this amended application as well. Both amended patents added a new limitation — that the inventions contain a pair of sealing rings, each having a lip on one side, which would prevent impurities from getting on the piston assembly. The amended Stoll Patent added the further limitation that the outer shell of the device, the sleeve, be made of a magnetizable material.

After Festo began selling its rodless cylinder, respondents (whom we refer to as SMC) entered the market with a device similar, but not identical, to the ones disclosed by Festo's patents. SMC's cylinder, rather than using two one-way sealing rings, employs a single sealing ring with a two-way lip. Furthermore, SMC's sleeve is made of a nonmagnetizable alloy. SMC's device does not fall within the literal claims of either patent, but petitioner contends that it is so similar that it infringes under the doctrine of equivalents.

SMC contends that Festo is estopped from making this argument because of the prosecution history of its patents. The sealing rings and the magnetized alloy in the Festo product were both disclosed for the first time in the amended applications. In SMC's view, these amendments narrowed the earlier applications, surrendering alternatives that are the very points of difference in the competing devices — the sealing rings and the type of alloy used to make the sleeve. As Festo narrowed its claims in these ways in order to obtain the patents, says SMC, Festo is now estopped from saying that these features are immaterial and that SMC's device is an equivalent of its own.

The United States District Court for the District of Massachusetts disagreed. It held that Festo's amendments were not made to avoid prior art, and therefore the amendments were not the kind that give rise to estoppel. A panel of the Court of Appeals for the Federal Circuit affirmed. 72 F. 3d 857 (1995). We granted certiorari, vacated, and remanded in light of our intervening decision in Warner-Jenkinson v. Hilton Davis Chemical Co., 520 U.S. 17 (1997). After a decision by the original panel on remand, 172 F. 3d 1361 (1999), the Court of Appeals ordered rehearing en banc to address questions that had divided its judges since our decision in *Warner-Jenkinson*. 187 F. 3d 1381 (1999).

The en banc court reversed, holding that prosecution history estoppel barred Festo from asserting that the accused device infringed its patents under the doctrine of equivalents. 234 F. 3d 558 (2000). The court held, with only one judge dissenting, that estoppel arises from any amendment that narrows a claim to comply with the Patent Act, not only from amendments made to avoid prior art. Id., at 566. More controversial in the Court of Appeals was its further holding: When estoppel applies, it stands as a complete bar against any claim of equivalence for the element that was amended. Id., at 574-575. The court acknowledged that its own prior case law did not go so far. Previous decisions had held that prosecution history estoppel constituted a flexible bar, foreclosing some, but not all, claims of equivalence, depending on the purpose of the amendment and the alterations in the text. The court concluded, however, that its precedents applying the flexible-bar rule should be overruled because this case-by-case approach has proved unworkable. In the court's view a complete-bar rule, under which estoppel bars all claims of equivalence to the narrowed element, would promote certainty in the determination of infringement cases.

Four judges dissented from the decision to adopt a complete bar. Id., at 562. In four separate opinions, the dissenters argued that the majority's decision to overrule precedent was contrary to *Warner-Jenkinson* and would unsettle the expectations of many existing patentees. Judge Michel, in his dissent, described in detail how the complete bar required the Court of Appeals to disregard 8 older decisions of this Court, as well as more than 50 of its own cases. 234 F. 3d, at 601-616.

We granted certiorari. 533 U.S. 915 (2001).

II

The patent laws "promote the Progress of Science and useful Arts" by rewarding innovation with a temporary monopoly. U.S. Const., Art. I, § 8, cl. 8. The monopoly is a property right; and like any property right, its boundaries should be clear. This clarity is essential to promote progress, because it enables efficient investment in innovation. A patent holder should know what he owns, and the public should know what he does not. For this reason, the patent laws require inventors to describe their work in "full, clear, concise, and exact terms," 35 U.S.C. § 112, as part of the delicate balance the law attempts to maintain between inventors, who rely on the promise of the law to bring the invention forth, and the public, which should be encouraged to pursue innovations, creations, and new ideas beyond the inventor's exclusive rights. Bonito Boats, Inc. v. Thunder Craft Boats, Inc., 489 U.S. 141, 150 (1989).

Unfortunately, the nature of language makes it impossible to capture the essence of a thing in a patent application. The inventor who chooses to patent an invention and disclose it to the public, rather than exploit it in secret, bears the risk that others will devote their efforts toward exploiting the limits of the patent's language:

> An invention exists most importantly as a tangible structure or a series of drawings. A verbal portrayal is usually an afterthought written to satisfy the requirements of patent law. This conversion of machine to words allows for unintended idea gaps which cannot be satisfactorily filled. Often the invention is novel and words do not exist to describe it. The dictionary does not always keep abreast of the inventor. It cannot. Things are not made for the sake of words, but words for things.

Autogiro Co. of America v. United States, 384 F. 2d 391, 397 (Ct. Cl. 1967).

The language in the patent claims may not capture every nuance of the invention or describe with complete precision the range of its novelty. If patents were always interpreted by their literal terms, their value would be greatly diminished. Unimportant and insubstantial substitutes for certain elements could defeat the patent, and its value to inventors could be destroyed by simple acts of copying. For this reason, the clearest rule of patent interpretation, literalism, may conserve judicial resources but is not necessarily the most efficient rule. The scope of a patent is not limited to its literal terms but instead embraces all equivalents to the claims described. See Winans v. Denmead, 15 How. 330, 347 (1854).

It is true that the doctrine of equivalents renders the scope of patents less certain. It may be difficult to determine what is, or is not, an equivalent to a particular element of an invention. If competitors cannot be certain about a patent's extent, they may be deterred from engaging in legitimate manufactures outside its limits, or they may invest by mistake in competing products that the patent secures. In addition the uncertainty may lead to wasteful litigation between competitors, suits that a rule of literalism might avoid. These concerns with the doctrine of equivalents, however, are not new. Each time the Court has considered the doctrine, it has acknowledged this uncertainty as the price of ensuring the appropriate incentives for innovation, and it has affirmed the doctrine over dissents that urged a more certain rule. When the Court in Winans v. Denmead, supra, first adopted what has become the doctrine of equivalents, it stated that "[t]he exclusive right to the thing patented is not secured, if the public are at liberty to make substantial copies of it, varying its form or proportions." Id., at 343. The dissent argued that the Court had sacrificed the objective of "[f]ul[l]-

ness, clearness, exactness, preciseness, and particularity, in the description of the invention." Id., at 347 (opinion of Campbell, J.).

The debate continued in Graver Tank & Mfg. Co. v. Linde Air Products Co., 339 U.S. 605 (1950), where the Court reaffirmed the doctrine. *Graver Tank* held that patent claims must protect the inventor not only from those who produce devices falling within the literal claims of the patent but also from copyists who "make unimportant and insubstantial changes and substitutions in the patent which, though adding nothing, would be enough to take the copied matter outside the claim, and hence outside the reach of law." Id., at 607. Justice Black, in dissent, objected that under the doctrine of equivalents a competitor "cannot rely on what the language of a patent claims. He must be able, at the peril of heavy infringement damages, to forecast how far a court relatively un-versed in a particular technological field will expand the claim's language. . . ." Id., at 617.

Most recently, in *Warner-Jenkinson,* the Court reaffirmed that equivalents remain a firmly entrenched part of the settled rights protected by the patent. A unanimous opinion concluded that if the doctrine is to be discarded, it is Congress and not the Court that should do so:

> [T]he lengthy history of the doctrine of equivalents strongly supports adherence to our refusal in *Graver Tank* to find that the Patent Act conflicts with that doctrine. Congress can legislate the doctrine of equivalents out of existence any time it chooses. The various policy arguments now made by both sides are thus best addressed to Congress, not this Court. 520 U.S., at 28.

III

Prosecution history estoppel requires that the claims of a patent be interpreted in light of the proceedings in the PTO during the application process. Estoppel is a "rule of patent construction" that ensures that claims are interpreted by reference to those "that have been cancelled or rejected." Schriber-Schroth Co. v. Cleveland Trust Co., 311 U.S. 211, 220-221 (1940). The doctrine of equivalents allows the patentee to claim those insubstantial alterations that were not captured in drafting the original patent claim but which could be created through trivial changes. When, however, the patentee originally claimed the subject matter alleged to infringe but then narrowed the claim in response to a rejection, he may not argue that the surrendered territory comprised unforeseen subject matter that should be deemed equivalent to the literal claims of the issued patent. On the contrary, "[b]y the amendment [the patentee] recognized and emphasized the difference between the two phrases[,] . . . and [t]he difference which [the patentee] thus disclaimed must be regarded as material." Exhibit Supply Co. v. Ace Patents Corp., 315 U.S. 126, 136 C137 (1942).

A rejection indicates that the patent examiner does not believe the original claim could be patented. While the patentee has the right to appeal, his decision to forgo an appeal and submit an amended claim is taken as a concession that the invention as patented does not reach as far as the original claim. See Goodyear Dental Vulcanite Co. v. Davis, 102 U.S. 222, 228 (1880) ("In view of [the amendment] there can be no doubt of what [the patentee] understood he had patented, and that both he and the commissioner regarded the patent to be for a manufacture made exclusively of vulcanites by the detailed process"); Wang Laboratories, Inc. v. Mitsubishi Electronics

America, Inc., 103 F. 3d 1571, 1577-1578 (CA Fed. 1997) ("Prosecution history estoppel...preclud[es] a patentee from re-gaining, through litigation, coverage of subject matter relinquished during prosecution of the application for the patent"). Were it otherwise, the inventor might avoid the PTO's gatekeeping role and seek to recapture in an infringement action the very subject matter surrendered as a condition of receiving the patent.

Prosecution history estoppel ensures that the doctrine of equivalents remains tied to its underlying purpose. Where the original application once embraced the purported equivalent but the patentee narrowed his claims to obtain the patent or to protect its validity, the patentee cannot assert that he lacked the words to describe the subject matter in question. The doctrine of equivalents is premised on language's inability to capture the essence of innovation, but a prior application describing the precise element at issue undercuts that premise. In that instance the prosecution history has established that the inventor turned his attention to the subject matter in question, knew the words for both the broader and narrower claim, and affirmatively chose the latter.

A

The first question in this case concerns the kinds of amendments that may give rise to estoppel. Petitioner argues that estoppel should arise when amendments are intended to narrow the subject matter of the patented invention, for instance, amendments to avoid prior art, but not when the amendments are made to comply with requirements concerning the form of the patent application. In *Warner-Jenkinson* we recognized that prosecution history estoppel does not arise in every instance when a patent application is amended. Our "prior cases have consistently applied prosecution history estoppel only where claims have been amended for a limited set of reasons," such as "to avoid the prior art, or otherwise to address a specific concern — such as obviousness — that arguably would have rendered the claimed subject matter unpatentable." 520 U.S., at 30..C32. While we made clear that estoppel applies to amendments made for a "substantial reason related to patentability," id., at 33, we did not purport to define that term or to catalog every reason that might raise an estoppel. Indeed, we stated that even if the amendment's purpose were unrelated to patentability, the court might consider whether it was the kind of reason that nonetheless might require resort to the estoppel doctrine. Id., at 40-41.

Petitioner is correct that estoppel has been discussed most often in the context of amendments made to avoid the prior art. See *Exhibit Supply Co.*, supra, at 137; Keystone Driller Co. v. Northwest Engineering Corp., 294 U.S. 42, 48 (1935). Amendment to accommodate prior art was the emphasis, too, of our decision in *Warner-Jenkinson*, supra, at 30. It does not follow, however, that amendments for other purposes will not give rise to estoppel. Prosecution history may rebut the inference that a thing not described was indescribable. That rationale does not cease simply because the narrowing amendment, submitted to secure a patent, was for some purpose other than avoiding prior art.

We agree with the Court of Appeals that a narrowing amendment made to satisfy any requirement of the Patent Act may give rise to an estoppel. As that court explained, a number of statutory requirements must be satisfied before a patent can issue. The claimed subject matter must be useful, novel, and not obvious.

35 U.S.C. §§ 101-103 (1994 ed. and Supp. V). In addition, the patent application must describe, enable, and set forth the best mode of carrying out the invention. § 112 (1994 ed.). These latter requirements must be satisfied before issuance of the patent, for exclusive patent rights are given in exchange for disclosing the invention to the public. See *Bonito Boats,* 489 U.S., at 150-151. What is claimed by the patent application must be the same as what is disclosed in the specification; otherwise the patent should not issue. The patent also should not issue if the other requirements of § 112 are not satisfied, and an applicant's failure to meet these requirements could lead to the issued patent being held invalid in later litigation.

Petitioner contends that amendments made to comply with § 112 concern the form of the application and not the subject matter of the invention. The PTO might require the applicant to clarify an ambiguous term, to improve the translation of a foreign word, or to rewrite a dependent claim as an independent one. In these cases, petitioner argues, the applicant has no intention of surrendering subject matter and should not be estopped from challenging equivalent devices. While this may be true in some cases, petitioner's argument conflates the patentee's reason for making the amendment with the impact the amendment has on the subject matter.

Estoppel arises when an amendment is made to secure the patent and the amendment narrows the patent's scope. If a § 112 amendment is truly cosmetic, then it would not narrow the patent's scope or raise an estoppel. On the other hand, if a § 112 amendment is necessary and narrows the patent's scope — even if only for the purpose of better description — estoppel may apply. A patentee who narrows a claim as a condition for obtaining a patent disavows his claim to the broader subject matter, whether the amendment was made to avoid the prior art or to comply with § 112. We must regard the patentee as having conceded an inability to claim the broader subject matter or at least as having abandoned his right to appeal a rejection. In either case estoppel may apply.

B

Petitioner concedes that the limitations at issue — the sealing rings and the composition of the sleeve — were made for reasons related to § 112, if not also to avoid the prior art. Our conclusion that prosecution history estoppel arises when a claim is narrowed to comply with § 112 gives rise to the second question presented: Does the estoppel bar the inventor from asserting infringement against any equivalent to the narrowed element or might some equivalents still infringe? The Court of Appeals held that prosecution history estoppel is a complete bar, and so the narrowed element must be limited to its strict literal terms. Based upon its experience the Court of Appeals decided that the flexible-bar rule is unworkable because it leads to excessive uncertainty and burdens legitimate innovation. For the reasons that follow, we disagree with the decision to adopt the complete bar.

Though prosecution history estoppel can bar challenges to a wide range of equivalents, its reach requires an examination of the subject matter surrendered by the narrowing amendment. The complete bar avoids this inquiry by establishing a per se rule; but that approach is inconsistent with the purpose of applying the estoppel in the first place — to hold the inventor to the representations made during the application process and to the inferences that may reasonably be drawn from the amendment. By amending the application, the inventor is deemed to concede that the patent does not extend as far as the original claim. It does not follow, however, that the amended claim becomes so perfect in its description that no one could devise an equivalent. After amendment, as before, language remains an imperfect fit for invention. The narrowing amendment may demonstrate what the claim is not; but it may still fail to capture precisely what the claim is. There is no reason why a narrowing amendment should be deemed to relinquish equivalents unforeseeable at the time of the amendment and beyond a fair interpretation of what was surrendered. Nor is there any call to foreclose claims of equivalence for aspects of the invention that have only a peripheral relation to the reason the amendment was submitted. The amendment does not show that the inventor suddenly had more foresight in the drafting of claims than an inventor whose application was granted without amendments having been submitted. It shows only that he was familiar with the broader text and with the difference between the two. As a result, there is no more reason for holding the patentee to the literal terms of an amended claim than there is for abolishing the doctrine of equivalents altogether and holding every patentee to the literal terms of the patent.

This view of prosecution history estoppel is consistent with our precedents and respectful of the real practice before the PTO. While this Court has not weighed the merits of the complete bar against the flexible bar in its prior cases, we have consistently applied the doctrine in a flexible way, not a rigid one. We have considered what equivalents were surrendered during the prosecution of the patent, rather than imposing a complete bar that resorts to the very literalism the equivalents rule is designed to overcome. E.g., *Goodyear Dental Vulcanite Co.*, 102 U.S., at 230; Hurlbut v. Schillinger, 130 U.S. 456, 465 (1889).

The Court of Appeals ignored the guidance of *Warner-Jenkinson*, which instructed that courts must be cautious before adopting changes that disrupt the settled expectations of the inventing community. See 520 U.S., at 28. In that case we made it clear that the doctrine of equivalents and the rule of prosecution history estoppel are settled law. The responsibility for changing them rests with Congress. Ibid. Fundamental alterations in these rules risk destroying the legitimate expectations of inventors in their property. The petitioner in *Warner-Jenkinson* requested another bright-line rule that would have provided more certainty in determining when estoppel applies but at the cost of disrupting the expectations of countless existing patent holders. We rejected that approach: "To change so substantially the rules of the game now could very well subvert the various balances the PTO sought to strike when issuing the numerous patents which have not yet expired and which would be affected by our decision." Id., at 32, n. 6; see also id., at 41 (Ginsburg, J., concurring) ("The new presumption, if applied woodenly, might in some instances unfairly discount the expectations of a patentee who had no notice at the time of patent prosecution that such a presumption would apply"). As *Warner-Jenkinson* recognized, patent prosecution occurs in the light of our case law. Inventors who amended their claims under the previous regime had no reason to believe they were conceding all equivalents. If they had known, they might have appealed the rejection instead. There is no justification for applying a new and more robust estoppel to those who relied on prior doctrine.

In *Warner-Jenkinson* we struck the appropriate balance by placing the burden on the patentee to show that an amendment was not for purposes of patentability:

> Where no explanation is established, however, the court should presume that the patent application had a substantial reason related to patentability for including the limiting element added by amendment. In those circumstances, prosecution history estoppel would bar the application of the doctrine of equivalents as to that element. Id., at 33.

When the patentee is unable to explain the reason for amendment, estoppel not only applies but also "bar[s] the application of the doctrine of equivalents as to that element." Ibid. These words do not mandate a complete bar; they are limited to the circumstance where "no explanation is established." They do provide, however, that when the court is unable to determine the purpose underlying a narrowing amendment — and hence a rationale for limiting the estoppel to the surrender of particular equivalents — the court should presume that the patentee surrendered all subject matter between the broader and the narrower language.

Just as *Warner-Jenkinson* held that the patentee bears the burden of proving that an amendment was not made for a reason that would give rise to estoppel, we hold here that the patentee should bear the burden of showing that the amendment does not surrender the particular equivalent in question. This is the approach advocated by the United States, and we regard it to be sound. The patentee, as the author of the claim language, may be expected to draft claims encompassing readily known equivalents. A patentee's decision to narrow his claims through amendment may be presumed to be a general disclaimer of the territory between the original claim and the amended claim. *Exhibit Supply,* 315 U.S., at 136-137 ("By the amendment [the patentee] recognized and emphasized the difference between the two phrases and proclaimed his abandonment of all that is embraced in that difference"). There are some cases, however, where the amendment cannot reasonably be viewed as surrendering a particular equivalent. The equivalent may have been unforeseeable at the time of the application; the rationale underlying the amendment may bear no more than a tangential relation to the equivalent in question; or there may be some other reason suggesting that the patentee could not reasonably be expected to have described the insubstantial substitute in question. In those cases the patentee can overcome the presumption that prosecution history estoppel bars a finding of equivalence.

This presumption is not, then, just the complete bar by another name. Rather, it reflects the fact that the interpretation of the patent must begin with its literal claims, and the prosecution history is relevant to construing those claims. When the patentee has chosen to narrow a claim, courts may presume the amended text was composed with awareness of this rule and that the territory surrendered is not an equivalent of the territory claimed. In those instances, however, the patentee still might rebut the presumption that estoppel bars a claim of equivalence. The patentee must show that at the time of the amendment one skilled in the art could not reasonably be expected to have drafted a claim that would have literally encompassed the alleged equivalent.

IV

On the record before us, we cannot say petitioner has rebutted the presumptions that estoppel applies and that the equivalents at issue have been surrendered. Petitioner concedes that the limitations at issue — the sealing rings and the composition of the sleeve — were made in response to a rejection for reasons under § 112, if not also because of the prior art references. As the amendments were made for a reason relating to patentability, the question is not whether estoppel applies but what territory the amendments surrendered. While estoppel does not effect a complete bar, the question remains whether petitioner can demonstrate that the narrowing amendments did not surrender the particular equivalents at issue. On these questions, respondents may well prevail, for the sealing rings and the composition of the sleeve both were noted expressly in the prosecution history. These matters, however, should be determined in the first instance by further proceedings in the Court of Appeals or the District Court.

The judgment of the Federal Circuit is vacated, and the case is remanded for further proceedings consistent with this opinion.

COMMENTS AND QUESTIONS

1. *In the Wake of* Festo. At the doctrinal level, the Supreme Court states succinctly its disagreement with the Federal Circuit's "absolute bar," and adopts a more flexible test. At the same time, however, the court also rejects the idea that patentees who have amended their claims are nonetheless free to seek protection under the doctrine of equivalents. The Court finds intermediate ground in what might be called a "foreseeable bar": a patentee is barred from seeking doctrine of equivalents protection for any matter it should have known it was giving up by amending its claims. By contrast, if a reasonable patentee could not have foreseen the limiting effect of an amendment, the doctrine of equivalents remains available. The foreseeability approach is foreshadowed in the Federal Circuit's opinion in Sage Products v. Devon, 126 F.3d 1420 (Fed. Cir. 1997). It is most fully articulated in Matthew Conigliaro, Andrew Greenberg, and Mark A. Lemley, Foreseeability in Patent Law, 16 Berkeley Tech. L.J. 1045 (2001) (urging the court to adopt the foreseeability approach).

In a key passage, the court fleshes out the foreseeability approach:

> There are some cases, however, where the amendment cannot reasonably be viewed as surrendering a particular equivalent. The equivalent may have been unforeseeable at the time of the application; the rationale underlying the amendment may bear no more than a tangential relation to the equivalent in question; or there may be some other reason suggesting that the patentee could not reasonably be expected to have described the insubstantial substitute in question. In those cases the patentee can overcome the presumption that prosecution history estoppel bars a finding of equivalence.

535 U.S. 722, 740-741.

This passage suggests three different scenarios under which the absolute bar is inappropriate: (1) unforeseeable equivalents, which would seem to be at least primarily cases of "after-arising" technologies; (2) amendments made for reasons "tangential" to the equivalent in question; and (3) a residual category where "for some other reason" the patentee could not be expected to have described the equivalent. Note that in all cases, the patentee bears the burden of overcoming the *Hilton-Davis* presumption that amendments relate to patentability and bar all equivalents.

The first *Festo* category, "after-arising technology," is discussed in the following "Note on the Problem of After-Arising Technologies." The second category emphasizes the *rationale* for a claim amendment. Here the Court suggests that some reasons for an amendment should not fairly trigger the surrender of equivalents.

What would these be? The Court gives no examples, but Conigliaro et al. point to Lockheed v. Space Systems/Loral, 249 F.3d 1314 (Fed. Cir. 2001), *vacated*, 535 U.S. 1109 (2002), as an example of a case that might be decided differently under the foreseeable bar:

> There, the original patent claim involved a method of orienting a satellite using a wheel whose rotation "varied sinusoidally" [i.e., in the shape of a sine wave] around an axis as the satellite deviated from an equatorial orbit. In response to prior art rejections, the claim was limited to sinusoidal variations with the same frequency as the orbital frequency of the satellite. The accused satellite did not literally infringe because the nature of the variation was not quite sinusoidal. The Federal Circuit held that the patentee could have no range of equivalents for the entire phrase, since the phrase had been changed to add the orbital frequency limitation. We think the result in this case would be different under the foreseeable bar. Lockheed could not reasonably have foreseen that an amendment related to orbital frequency would bar it from using the doctrine of equivalents to expand the definition of "sinusoidally," a term that was present in the original patent claim. Lockheed is an example of a case in which the patentee made an amendment intended to disclaim a totally different type of change than the one the defendant made.

Conigliaro et al., *supra*, at 1070-71.

Other Federal Circuit cases decided under the "flexible bar" may provide some hints. See, e.g., Insituform Technologies, Inc. v. CAT Contracting, Inc., 156 F.3d 1199 (Fed. Cir. 1998), *rev'd after remand*, 10 Fed.Appx. 871 (Fed. Cir. 2001) (unpub.), *vacated in light of Supreme Court decision in* Festo, 535 U.S. 1108 (2002), *rev'd* 385 F.3d 1360 (Fed. Cir. 2004). In its 1998 opinion in *Insituform*, the Federal Circuit permitted application of the doctrine of equivalents to a claim to a method for relining old sewer pipes. The relining method involved putting a liner on the inside of old pipes. The liner is made by coating a felt tube with resin. The patentee's method involved placing a suction device in the felt tube and using it to pull resin down the length of the tube. When the resin dries, the tube liner is rigid and is placed inside the old sewer pipe. The patentee's claim was amended during prosecution in light of prior art—the "Everson" reference—showing a single large vacuum device at the end of the pipe. The amended claim limited the patentee to use of a smaller suction device, which is moved down the length of the felt tube to pull the resin along. An important aspect of the amendment was the claim limitation to a single suction cup. The accused infringer, apparently taking advantage of this limitation, introduced a competing process that used multiple small suction devices to pull the resin down the tube. The Federal Circuit in its 1998 decision noted that the reason for the "single cup" amendment was to distinguish the "large suction device" prior art, and thus held that the accused process did indeed infringe under the doctrine of equivalents:

> The stated reason, therefore, for Insituform's amendment to overcome the Everson reference was to avoid the need to use a large compressor when the vacuum is created a significant distance from the resin source.... Therefore, we hold that Insituform did explain the reason for its changes in claim 1, and those reasons do not prohibit application of the doctrine of equivalents completely, but only with respect to the subject matter surrender of a large, single vacuum source placed a significant distance from the resin front.

156 F.3d 1199, 1203 (Fed. Cir. 1998). Notice the interplay of three factors in cases such as *Insituform:* (a) the contribution of the claimed technology; (b) the nature of the technology of the reference leading to the amendment; and (c) the nature of the accused infringer's technology. Do you think the *Festo* approach balances these three factors appropriately? What incentives does *Festo* create for patent applicants wishing to preserve some scope of equivalents in the wake of a claim amendment? How would you characterize the "scope" of a reference such as Everson in the *Insituform* case when making an amendment to a patent application—narrow or broad?

2. *On Remand.* The Federal Circuit applied the Court's new prosecution history estoppel test in a follow-up opinion in *Festo* reported at 344 F.3d 1359 (Fed. Cir. 2003). After holding that prosecution history estoppel was an issue of law for the judge to decide, the court held there were open factual issues regarding whether an ordinarily skilled artisan at the time of the claim amendment would have thought an aluminum sleeve was an objectively unforeseeable equivalent of the magnetizable sleeve, or whether the single two-way sealing ring was unforeseeable in light of the two sealing rings. These issues it remanded to the district court. The Federal Circuit also held that neither the aluminum sleeve nor double sealing ring amendments satisfied either of the other criteria—the "tangential" and "some other reason" exceptions. On remand, the district court found that both the single sealing ring and aluminum sleeve material were known in the prior art at the time of the amendment, and therefore were objectively foreseeable under *Festo.* 75 U.S.P.Q.2d 1830, 1836, 1837 (D. Mass. 2005). For more on "foreseeability," see Pioneer Magnetics, Inc. v. Micro Linear Corp., 330 F.3d 1352 (Fed Cir. 2003).

3. *Estoppel: Threshold Issues.* Several post-*Festo* cases state two threshold rules for reaching the estoppel issue: (1) claim amendments must surrender claim scope to trigger operation of the doctrine; and (2) arguments to examiners are treated differently than actual amendments. On (1), see Business Objects S.A. v. MicroStrategy, Inc., 393 F.3d 1366 (Fed. Cir. 2005) (amendment made claim broader, not narrower, so no prosecution history estoppel in this case); Boston Scientific Scimed, Inc. v. Cordis Corp., 2005 WL 1322974 (D. Del., Jun 3, 2005), at 3 ("The court is not convinced that the patentees narrowed their claims when they amended claim one to more accurately describe a coating comprised of two separate coats. The amended claim merely better described what the patentees had been claiming all along.")— subsequent proceedings at 392 F.Supp. 2d 676 (D.Del. 2005), 2006 WL 760714 (D.Del. Mar. 21, 2006). On issue (2), see Pharmacia & Upjohn Co. v. Mylan Pharm., Inc., 170 F.3d 1373, 1376-77 (Fed. Cir. 1999) ("The doctrine of prosecution history estoppel limits the doctrine of equivalents when an applicant makes a narrowing amendment for purposes of patentability, or clearly and unmistakably surrenders subject matter by arguments made to an examiner."). This notion was applied post-*Festo* in Aquatex Indus., Inc. v. Techniche Solutions, 419 F.3d 1374, 1382 (Fed. Cir. 2005) (confirming differing standards for "argument-based" and "amendment-based" estoppel; finding that in this case, "We do not see the clear and unmistakable surrender of subject matter required to invoke argument-based prosecution history estoppel.").

4. *Amendments for "Tangential" and "Other Reasons."* Many of the post-*Festo* cases have focused on the "foreseeability" of equivalents at the time a claim is amended. But the "tangential reasons" exception to the prosecution history estoppel rule of *Festo* has also factored into some cases. See, e.g., Insituform Technologies, Inc. v. CAT Contracting, Inc., 385 F.3d 1360, 1368 (Fed. Cir. 2004) (amendment made to distinguish one aspect of claimed invention from prior art

does not bar doctrine of equivalents argument where accused device differs substantially from that prior art); Engineered Products Co. v. Donaldson Co., Inc., 313 F.Supp.2d 951, 974 (N.D.Iowa 2004) (because amendment was made to avoid an element in the prior art that was merely similar to the element in the accused product alleged to be equivalent, that amendment was tangential and patentee could proceed with doctrine of equivalents argument). The "some other reason" exception has not been developed in the cases yet, however. Cf. Biagro W. Sales, Inc. v. Grow More, Inc., 423 F.3d 1296, 1307 (Fed. Cir. 2005) ("Biagro's ["some other reason"] contention should be rejected as merely an attempt to reargue the claim construction issue.").

5. *Law Lords Say No to the Doctrine.* In Kirin-Amgen Inc. v. Hoechst Marion Roussel Ltd., [2004] UKHL 46 (Oct. 21, 2004), Lord Hoffmann, a notable jurist on the Appellate Committee of the House of Lords (i.e, the British "Supreme Court"), considered the question whether Britain, operating under the European Patent Convention, must recognize a doctrine of equivalents. In passing, the opinion mentions both the "pith and marrow" doctrine, essentially a "gist of the invention" doctrine, and Article 69 of the European Patent Convention:

> There is often discussion about whether we have a European doctrine of equivalents and, if not, whether we should. It seems to me that both the doctrine of equivalents in the United States and the pith and marrow doctrine in the United Kingdom were born of despair. The courts felt unable to escape from interpretations which "unsparing logic" appeared to require and which prevented them from according the patentee the full extent of the monopoly which the person skilled in the art would reasonably have thought he was claiming. The background was the tendency to literalism which then characterised the approach of the courts to the interpretation of documents generally and the fact that patents are likely to attract the skills of lawyers seeking to exploit literalism to find loopholes in the monopoly they create. (Similar skills are devoted to revenue statutes).
>
> If literalism stands in the way of construing patent claims so as to give fair protection to the patentee, there are two things that you can do. One is to adhere to literalism in construing the claims and evolve a doctrine which supplements the claims by extending protection to equivalents. That is what the Americans have done. The other is to abandon literalism. That is what the House of Lords did in the *Catnic* case (at [1982] RPC 183, 242).... The solution, said Lord Diplock, was to adopt a principle of construction which actually gave effect to what the person skilled in the art would have understood the patentee to be claiming.
>
> Since the *Catnic* case we have article 69 which, as it seems to me, firmly shuts the door on any doctrine which extends protection outside the claims. I cannot say that I am sorry because the Festo litigation suggests, with all respect to the courts of the United States, that American patent litigants pay dearly for results which are no more just or predictable than could be achieved by simply reading the claims.

c. Subject Matter "Disclosed but Not Claimed"

Even before the Supreme Court issued its opinion in *Festo,* the Federal Circuit decided another *en banc* case that could significantly restrict the scope of the doctrine of equivalents: the doctrine of dedication to the public domain. Along with *Festo,* the *Johnson and Johnston* case represents an another important inroad on the doctrine. As you read the case, consider where patentees and infringers are left in the wake of these two decisions.

Johnson & Johnston Associates Inc. v. R.E. Service Co., Inc.

United States Court of Appeals for the Federal Circuit
285 F.3d 1046 (Fed. Cir. 2002) (en banc)

PER CURIAM.

Johnson and Johnston Associates (Johnston) asserted United States Patent No. 5,153,050 (the '050 patent) against R.E. Service Co. (RES). A jury found that RES willfully infringed claims 1 and 2 of the patent under the doctrine of equivalents and awarded Johnston $1,138,764 in damages. Upon entry of judgment, the United States District Court for the Northern District of California further granted Johnston enhanced damages, attorney fees, and expenses. Johnson & Johnston Assocs. v. R.E. Serv. Co., No. C97-04382CRB (N.D.Cal.1998). After a hearing before a three-judge panel on December 7, 1999, this court ordered *en banc* rehearing of the doctrine of equivalents issue, Johnson & Johnston Assocs. v. R.E. Serv. Co., 238 F.3d 1347 (Fed. Cir. 2001), which occurred on October 3, 2001. Because this court concludes that RES, as a matter of law, could not have infringed the '050 patent under the doctrine of equivalents, this court reverses the district court's judgment of infringement under the doctrine of equivalents, willfulness, damages, attorneys fees, and expenses.

I.

The '050 patent, which issued October 6, 1992, relates to the manufacture of printed circuit boards. Printed circuit boards are composed of extremely thin sheets of conductive copper foil joined to sheets of a dielectric (nonconductive) resin-impregnated material called "prepreg." The process for making multi-layered printed circuit boards stacks sheets of copper foil and prepreg in a press, heats them to melt the resin in the prepreg, and thereby bonds the layers.

In creating these circuit boards, workers manually handle the thin sheets of copper foil during the layering process. Without the invention claimed in the '050 patent, stacking by hand can damage or contaminate the fragile foil, causing discontinuities in the etched copper circuits. The '050 patent claims an assembly that prevents most damage during manual handling. The invention adheres the fragile copper foil to a stiffer substrate sheet of aluminum. With the aluminum substrate for protection, workers can handle the assembly without damaging the fragile copper foil. After the pressing and heating steps, workers can remove and even recycle the aluminum substrate.... [The specification explains:]

> The use of the adhered substrate, regardless of what material it is made of, makes the consumer's (manufacturer's) objective of using thinner and thinner foils and ultimately automating the procedure more realistic since the foil, by use of the invention, is no longer without the much needed physical support.

'050 patent, col. 8, ll. 21-30. The specification further describes the composition of the substrate sheet:

> While aluminum is currently the preferred material for the substrate, other metals, such as stainless steel or nickel alloys, may be used. In some instances ... polypropelene [sic] can be used.

'050 patent, col. 5, ll. 5-8.

As noted, the jury found infringement of claims 1 and 2:

Claim 1. A component for use in manufacturing articles such as printed circuit boards comprising:

a laminate constructed of a sheet of copper foil which, in a finished printed circuit board, constitutes a functional element and a sheet of *aluminum* which constitutes a discardable element;

one surface of each of the copper sheet and the *aluminum* sheet being essentially uncontaminated and engageable with each other at an interface,

a band of flexible adhesive joining the uncontaminated surfaces of the sheets together at their borders and defining a substantially uncontaminated central zone inwardly of the edges of the sheets and unjoined at the interface.

'050 patent, Claim 1, col. 8, ll. 47-60 (emphasis supplied). Claim 2 defines a similar laminate having sheets of copper foil adhered to both sides of the aluminum sheet. . . .

In 1997, RES began making new laminates for manufacture of printed circuit boards. The RES products, designated "SC2" and "SC3," joined copper foil to a sheet of steel as the substrate instead of a sheet of aluminum. Johnston filed a suit for infringement. In this case, the district court granted RES's motion for summary judgment of no literal infringement. With respect to the doctrine of equivalents, RES argued, citing Maxwell v. J. Baker, Inc., 86 F.3d 1098, 39 USPQ2d 1001 (Fed. Cir. 1996), that the '050 specification, which disclosed a steel substrate but did not claim it, constituted a dedication of the steel substrate to the public. Johnston argued that the steel substrate was not dedicated to the public, citing YBM Magnex, Inc. v. Int'l Trade Comm'n, 145 F.3d 1317 (Fed.Cir. 1998). On cross-motions for summary judgment, the district court ruled that the '050 patent did not dedicate the steel substrate to the public, and set the question of infringement by equivalents for trial, along with the issues of damages and willful infringement.

The jury found RES liable for willful infringement under the doctrine of equivalents and awarded Johnston $1,138,764 in damages. . . .

II.

On appeal, RES does not challenge the jury's factual finding of equivalency between the copper-steel and copper-aluminum laminates. Instead, citing *Maxwell,* RES argues that Johnston did not claim steel substrates, but limited its patent scope to aluminum substrates, thus dedicating to the public this unclaimed subject matter. On this ground, RES challenges the district court's denial of its motion for summary judgment that RES's copper-steel laminates are not equivalent, as a matter of law, to the claimed copper-aluminum laminates. Johnston responds that the steel substrates are not dedicated to the public, citing *YBM Magnex.* In other words, the two parties dispute whether *Maxwell* or *YBM Magnex* applies in this case with regard to infringement under the doctrine of equivalents.

In *Maxwell,* the patent claimed a system for attaching together a mated pair of shoes. 86 F.3d at 1101-02. Maxwell claimed fastening tabs between the inner and outer soles of the attached shoes. Maxwell disclosed in the specification, but did not claim, fastening tabs that could be "stitched into a lining seam of the shoes." U.S. Patent No. 4,624,060, col. 2, l. 42. Based on the "well-established rule that 'subject matter disclosed but not claimed in a patent application is dedicated to the public,'" this court held that Baker could not, as a matter of law, infringe under the doctrine of equivalents by using the disclosed but unclaimed shoe attachment system. *Maxwell*, 86 F.3d at 1106 (quoting Unique Concepts, Inc. v. Brown, 939 F.2d 1558, 1562-63 (Fed.Cir. 1991)) This court stated further:

> By [Maxwell's failure] to claim these alternatives, the Patent and Trademark Office was deprived of the opportunity to consider whether these alternatives were patentable. A person of ordinary skill in the shoe industry, reading the specification and prosecution history, and interpreting the claims, would conclude that Maxwell, by failing to claim the alternate shoe attachment systems in which the tabs were attached to the inside shoe lining, dedicated the use of such systems to the public.

Maxwell, 86 F.3d at 1108.

In *YBM Magnex*, the patent claimed a permanent magnet alloy comprising certain elements, including "6,000 to 35,000 ppm oxygen." U.S. Patent No. 4,588,439 (the '439 patent), col. 3, l. 12. The accused infringer used similar magnet alloys with an oxygen content between 5,450 and 6,000 ppm (parts per million), which was allegedly disclosed but not claimed in the '439 patent. In *YBM Magnex*, this court stated that *Maxwell* did not create a new rule of law that the doctrine of equivalents could never encompass subject matter disclosed in the specification but not claimed. 45 F.3d at 1321. Distinguishing *Maxwell*, this court noted:

> Maxwell avoided examination of the unclaimed alternative, which was distinct from the claimed alternative. In view of the distinctness of the two embodiments, both of which were fully described in the specification, the Federal Circuit denied Maxwell the opportunity to enforce the unclaimed embodiment as an equivalent of the one that was claimed.

Id. at 1320. In other words, this court in *YBM Magnex* purported to limit *Maxwell* to situations where a patent discloses an unclaimed alternative distinct from the claimed invention. Thus, this court must decide whether a patentee can apply the doctrine of equivalents to cover unclaimed subject matter disclosed in the specification.

III.

Both the Supreme Court and this court have adhered to the fundamental principle that claims define the scope of patent protection. See, e.g., Aro Mfg. v. Convertible Top Replacement Co., 365 U.S. 336, 339 (1961) ("[T]he claims made in the patent are the sole measure of the grant. . . . "); Cont'l Paper Bag Co. v. E. Paper Bag Co., 210 U.S. 405, 419 (1908) ("[T]he claims measure the invention."); Atl. Thermoplastics Co. v. Faytex Corp., 974 F.2d 1299 (Fed.Cir.1992) ("The claims alone define the patent right."); SRI Int'l v. Matsushita Elec. Corp., 775 F.2d 1107, 1121 (Fed.Cir.1985) ("It is the *claims* that measure the invention."). The claims thus give notice of the scope of patent protection. See, e.g., Mahn v. Harwood, 112 U.S. 354, 361 (1884) ("The public is notified and informed by the most solemn act on the part of the patentee, that his claim to invention is for such and such an element or combination, and for nothing more."). The claims give notice both to the examiner at the U.S. Patent and Trademark Office during prosecution, and to the public at large, including potential competitors, after the patent has issued.

Consistent with its scope definition and notice functions, the claim requirement presupposes that a patent applicant defines his invention in the claims, not in the specification. After all, the claims, not the specification, provide the measure of the patentee's right to exclude. . . .

Moreover, the law of infringement compares the accused product with the claims as construed by the court. Infringement, either literally or under the doctrine of equivalents, does not arise by comparing the accused product "with a preferred embodiment described in the specification, or with a commercialized embodiment of the patentee." *SRI Int'l,* 775 F.2d at 1121.

Even as early as the 1880s, the Supreme Court emphasized the predominant role of claims. For example, in Miller v. Bridgeport Brass Co., a case addressing a reissue patent filed fifteen years after the original patent, the Supreme Court broadly stated: "[T]he claim of a specific device or combination, and an omission to claim other devices or combinations apparent on the face of the patent, are, in law, a dedication to the public of that which is not claimed." 104 U.S. 350, 352 (1881). Just a few years later, the Court repeated that sentiment in another reissue patent case: "[T]he claim actually made operates in law as a disclaimer of what is not claimed; and of all this the law charges the patentee with the fullest notice." *Mahn,* 112 U.S. at 361. The Court explained further:

> Of course, what is not claimed is public property. The presumption is, and such is generally the fact, that what is not claimed was not invented by the patentee, but was known and used before he made his invention. But, whether so or not, his own act has made it public property if it was not so before. The patent itself, as soon as it is issued, is the evidence of this. The public has the undoubted right to use, and it is to be presumed does use, what is not specifically claimed in the patent.

Id. at 361.

[The Court reviewed *Graver Tank* and *Warner-Jenkinson*].

IV.

As stated in *Maxwell,* when a patent drafter discloses but declines to claim subject matter, as in this case, this action dedicates that unclaimed subject matter to the public. Application of the doctrine of equivalents to recapture subject matter deliberately left unclaimed would "conflict with the primacy of the claims in defining the scope of the patentee's exclusive right." Sage Prods. Inc. v. Devon Indus., Inc., 126 F.3d 1420, 1424 (Fed.Cir. 1997) (citing *Warner-Jenkinson,* 520 U.S. at 29). . . .

Moreover, a patentee cannot narrowly claim an invention to avoid prosecution scrutiny by the PTO, and then, after patent issuance, use the doctrine of equivalents to establish infringement because the specification discloses equivalents. "Such a result would merely encourage a patent applicant to present a broad disclosure in the specification of the application and file narrow claims, avoiding examination of broader claims that the applicant could have filed consistent with the specification." *Maxwell,*

86 F.3d at 1107 (citing Genentech, Inc. v. Wellcome Found. Ltd., 29 F.3d 1555, 1564 (Fed.Cir. 1994)). By enforcing the *Maxwell* rule, the courts avoid the problem of extending the coverage of an exclusive right to encompass more than that properly examined by the PTO. Keystone Bridge Co. v. Phoenix Iron Co., 95 U.S. 274, 278 (1877) ("[T]he courts have no right to enlarge a patent beyond the scope of its claim as allowed by the Patent Office, or the appellate tribunal to which contested applications are referred.").

V.

In this case, Johnston's '050 patent specifically limited the claims to "a sheet of aluminum" and "the aluminum sheet." The specification of the '050 patent, however, reads: "While aluminum is currently the preferred material for the substrate, other metals, such as stainless steel or nickel alloys may be used." Col. 5, ll. 5-10. Having disclosed without claiming the steel substrates, Johnston cannot now invoke the doctrine of equivalents to extend its aluminum limitation to encompass steel. Thus, Johnston cannot assert the doctrine of equivalents to cover the disclosed but unclaimed steel substrate. To the extent that *YBM Magnex* conflicts with this holding, this *en banc* court now overrules that case.

A patentee who inadvertently fails to claim disclosed subject matter, however, is not left without remedy. Within two years from the grant of the original patent, a patentee may file a reissue application and attempt to enlarge the scope of the original claims to include the disclosed but previously unclaimed subject matter. 35 U.S.C. § 251 (2000). In addition, a patentee can file a separate application claiming the disclosed subject matter under 35 U.S.C. § 120 (2000) (allowing filing as a continuation application if filed before all applications in the chain issue). Notably, Johnston took advantage of the latter of the two options by filing two continuation applications that literally claim the relevant subject matter.

Conclusion

For the reasons stated above, the district court erred as a matter of law in concluding that RES infringed the '050 patent under the doctrine of equivalents by using a steel substrate. Consequently, this court reverses the district court's judgment of infringement under the doctrine of equivalents . . . REVERSED.

RADER, Circuit Judge, with whom MAYER, Chief Judge, joins, concurring.

While endorsing the results and reasoning of the court, I would offer an alternative reasoning. This alternative would also help reconcile the preeminent notice function of patent claims with the protective function of the doctrine of equivalents. This reconciling principle is simple: the doctrine of equivalents does not capture subject matter that the patent drafter reasonably could have foreseen during the application process and included in the claims. This principle enhances the notice function of claims by making them the sole definition of invention scope in all foreseeable circumstances. This principle also protects patentees against copyists who employ insubstantial variations to expropriate the claimed invention in some unforeseeable circumstances.

Few problems have vexed this court more than articulating discernible standards for non-textual infringement. On the one hand, the Supreme Court has recognized that the doctrine of equivalents provides essential protection for inventions: "[T]o permit imitation of a patented invention which does not copy every literal detail would be to convert the protection of the patent grant into a hollow and useless thing...leav[ing] room for indeed encourag[ing] the unscrupulous copyist to make unimportant and insubstantial changes." Graver Tank & Mfg. Co. v. Linde Air Prods. Co., 339 U.S. 605, 607 (1950). The protective function of non-textual infringement, however, has a price. Recently, the Supreme Court acknowledged that a broad doctrine of equivalents can threaten the notice function of claims: "There can be no denying that the doctrine of equivalents, when applied broadly, conflicts with the definitional and public-notice functions of the statutory claiming requirement." Warner-Jenkinson Co. v. Hilton Davis Chem. Co., 520 U.S. 17, 29 (1997). These competing policies make it difficult to set a standard that protects the patentee against insubstantial changes while simultaneously providing the public with adequate notice of potentially infringing behavior.

In general, the Supreme Court and this court have attempted to deal with these competing principles by placing limits on non-textual infringement. Thus, in further-ance of the notice objective, *Pennwalt* and *Warner-Jenkinson* require an equivalent for each and every element of a claim (applying the doctrine of equivalents to the claim as a whole gives too much room to enforce the claim beyond its notifying limitations). Similarly, to enhance notice, *Festo* and *Warner-Jenkinson* propose to bar patentees from expanding their claim to embrace subject matter surrendered during the patent acquisition process. Finally, *Wilson Sporting Goods* prevents the doctrine of equivalents from expanding claim scope to embrace prior art.

Perhaps more than each of these other restraints on non-textual infringement, a foreseeability bar would concurrently serve both the predominant notice function of the claims and the protective function of the doctrine of equivalents. When one of ordinary skill in the relevant art would foresee coverage of an invention, a patent drafter has an obligation to claim those foreseeable limits. This rule enhances the notice function of claims by making them the sole definition of invention scope in all foreseeable circumstances. When the skilled artisan cannot have foreseen a variation that copyists employ to evade the literal text of the claims, the rule permits the patentee to attempt to prove that an "insubstantial variation" warrants a finding of non-textual infringement. In either event, the claims themselves and the prior art erect a foreseeability bar that circumscribes the protective function of non-textual infringe-ment. Thus, foreseeability sets an objective standard for assessing when to apply the doctrine of equivalents.

A foreseeability bar thus places a premium on claim drafting and enhances the notice function of claims. To restate, if one of ordinary skill in the relevant art would reasonably anticipate ways to evade the literal claim language, the patent applicant has an obligation to cast its claims to provide notice of that coverage. In other words, the patentee has an obligation to draft claims that capture all reasonably foreseeable ways to practice the invention. The doctrine of equivalents would not rescue a claim drafter who does not provide such notice. Foreseeability thus places a premium on notice while reserving a limited role for the protective function of the doctrine of equivalents.

[Judge Rader cites other cases adopting a foreseeability approach]

In this case, Johnston's '050 patent claimed only a "sheet of aluminum" and "the aluminum sheet" twice specifying the aluminum limitation. The patent specification then expressly mentioned other potential substrate metals, including stainless steel. col. 5, [lines] 5-10. Johnston's patent disclosure expressly admits that it foresaw other metals serving as substrates. Yet the patent did not claim anything beyond aluminum. Foreseeability bars Johnston from recapturing as an equivalent subject matter not claimed but disclosed. In *Sage* terms, "as between [Johnston] who had a clear opportunity to negotiate broader claims but did not do so, and the public at large, it is [Johnston] who must bear the cost of its failure to seek protection for this foreseeable alteration of its claimed structure." 126 F.3d at 1425.

Foreseeability relegates non-textual infringement to its appropriate exceptional place in patent policy. The doctrine of equivalents should not rescue claim drafters who fail to give accurate notice of an invention's scope in the claims. The Patent Act supplies a correction process for applicants who have claimed "more or less than [they] had a right to claim in the patent." 35 U.S.C. §§ 251, 252 (2001). The doctrine of equivalents need not duplicate the statute's means of correcting claiming errors.

NEWMAN, Circuit Judge, dissenting.

Instead of deciding this appeal on the basis on which it reaches us — that is, whether to sustain the jury verdict that stainless steel and aluminum are equivalent substrates for copper foil laminates — my colleagues launch yet another assault on the doctrine of equivalents. The court today holds that there is no access to equivalency for any subject matter that is disclosed in a patent specification but not claimed. Thus the court establishes a new absolute bar to equivalency, a bar that applies when there is no prosecution history estoppel, no prior art, no disclaimer, no abandonment.

The court overrules not only its own decisions but also those of the Supreme Court, and reaches out to create a new, unnecessary and often unjust, *per se* rule. This decision jettisons even the possibility of relief when relief is warranted, and further distorts the long-established balance of policies that undergird patent-supported industrial innovation. It is self-evident that the placement of an increasing number of pitfalls in the path of patentees serves only as a deterrent to innovation. Before taking so deliberate a step, the court should at least consider the consequences. For example, the adverse effect on the disclosure of information in patents — an immediate and foreseeable result of this ruling — has not been recognized by any of my colleagues, despite the fervid warnings of the bar.

Even were this court expert in its understanding of technology policy and innovation economics, we have no authority to change the precedent that binds us. This *en banc* court has placed itself in egregious conflict with the Supreme Court's decisions in Graver Tank & Mfg. Co. v. Linde Air Prods. Co., 339 U.S. 605 (1950) and Warner-Jenkinson Co. v. Hilton Davis Chemical Co., 520 U.S. 17 (1997). The Federal Circuit is no less subject to *stare decisis* than is any other court of appeals. See Rodriguez de Quijas v. Shearson/American Express, Inc., 490 U.S. 477, 484 (1989) (the courts of appeals should "leav[e] to this Court the prerogative of overruling its own decisions"). Shrugging off this obligation, my intrepid colleagues adopt the position of the dissenters in *Graver Tank*. I must, respectfully, dissent....

The public interest in fostering innovation and technological advance is not served by a judicial decision that imposes legal obstacles to the disclosure of scientific and technologic information. Information dissemination is a critical purpose of the patent system. By penalizing the inclusion of information in the specification the patent becomes less useful as a source of knowledge, and more a guarded legal contract.

No patentee deliberately chooses the doctrine of equivalents to protect commercial investment. Yet every patentee must guard against infringement at the edges of the invention. After today, whenever a patentee draws a line in a disclosed continuum, the copier who simply crosses the line can avoid even the charge of equivalency; a safe and cheap way to garner the successes of another. Each new pitfall for inventors simply diminishes the value of the patent incentive, and ultimately inhibits technological innovation.

COMMENTS AND QUESTIONS

1. The court emphasizes the "notice function" of claims, a theme also prevalent in the *en banc* Federal Circuit opinion in *Festo*, reversed by the Supreme Court in the opinion earlier in this section. Why is it not sufficient notice to mention a substitute material in the disclosure portion of the patent specification? As between an inventor who hits upon an idea, claiming only a portion, and a later copyist who reads through the inventor's specification for information on unclaimed variants, why should the latter prevail? Recall in this respect that the jury found RES's product to be an equivalent of the claimed technology. Do you agree with Judge Newman's argument in dissent that no patentee intentionally relies upon the doctrine of equivalents?

For an argument that patents don't actually serve a notice function very well, see John R. Thomas, Claim Re-Construction: The Doctrine of Equivalents in the Post-Markman Era, 9 Lewis & Clark L. Rev. 153 (2005).

2. It can be argued that the "dedication to the public" rule fits well with the "foreseeability" test of *Festo*. After all, it seems difficult for an inventor to argue that a particular variant of the claimed subject matter in his or her patent is unforeseeable when the variant is mentioned right in the specification. In this regard, see Scott R. Boalick, Note, The Dedication Rule and the Doctrine of Equivalents: A Proposal for Reconciliation, 87 Geo. L. J. 2363 (1999).

3. While it may be foreseeable that a certain variant will work, it may not be foreseeable that the variant will be commercially viable. What if circumstances change after the issuance of the patent, making an unclaimed variant a more viable alternative to those recited in the claims. Who should reap the benefit of this change?

4. The court's opinion describes two techniques for avoiding the effect of its holding: (1) broadening reissues and (2) continuation applications. (Recall that a patentee must file the continuation before all related applications have issued or been abandoned.) As the court notes, the patentee in *Johnson and Johnston* made use of the latter technique by filing two continuations based on the parent application that led to the '050 patent at issue in the case. One, U.S. Patent 5,725,937, is discussed in Judge Newman's dissent; she uses it to demonstrate that the applicant did not intend to dedicate the equivalents in question to the public via his disclosure. The other, U.S. Patent 5,942,315, issued Aug. 24, 1999 (during the *Johnson & Johnston* litigation), appears to claim explicitly what the patentee tried to cover under the doctrine of equivalents. See claims 1-5 (generically claiming "metal sheets" for holding the thin copper sheet; and specifically disclosing aluminum as the substrate sheet). If the court had ruled in favor of the patentee in *Johnson and Johnston*, it would have found in effect that patent A (the '050 patent in the case) covered subject matter under the doctrine of equivalents that was later claimed explicitly in patent B (the '305 patent issued in 1999). Is this second patent good evidence that the court reached the

right result in this case? Or do you agree with Judge Newman that the two additional patents show no intent to dedicate disclosed but unclaimed product variants to the public? Alternatively, might you argue that the issuance of patent B here shows that the patentee in fact invented (as proven by the issued patent) the product variant sold by the infringer — and hence that the patentee deserved to win the case after all?

5. *Johnson & Johnston* has been followed and applied in a number of cases. See Toro Co. v. White Consolidated Industries, Inc., 383 F.3d 1326, 1333 (Fed. Cir. 2004) (patentee's intent irrelevant to question of whether embodiments were dedicated to the public in the specification; also, disclosure need not meet standards of §112 to dedicate subject matter to the public). But see PSC Computer Products, Inc. v. Foxconn Intern., Inc., 355 F.3d 1353, 1359 (Fed. Cir. 2004) (considering standard for determining if embodiments are disclosed but not claimed, and holding that "if one of ordinary skill in the art can understand the unclaimed disclosed teaching upon reading the written description, the alternative matter disclosed has been dedicated to the public.").

d. After-Arising Technologies

Courts have determined how broadly they see "equivalents" based on the degree of advance over the art the original patent represents. When the patent is on a "mere improvement" the courts tend not to consider as "equivalent" a product or process that is even a modest distance beyond the literal terms of the claims. Brill v. Washington Elec. & Ry. Co., 215 U.S. 527 (1910); Kinzenbaw v. Deere & Co., 741 F.2d 383 (Fed. Cir. 1984), *cert. denied*, 470 U.S. 1004 (1985). On the other hand, a patent representing a "pioneer invention" — which the Supreme Court has defined as "a patent concerning a function never before performed, a wholly novel device, or one of such novelty and importance as to make a distinct step in the progress in the art," Boyden Power-Brake Co. v. Westinghouse, 170 U.S. 537, 569 (1898) — is "entitled to a broad range of equivalents." 4 Donald Chisum, Patents §18.04[2] (1998). That is, when a pioneer patent is involved, a court will stretch to find infringement even by a product whose characteristics lie considerably outside the boundaries of the literal claims. See generally John R. Thomas, The Question Concerning Patent Law and Pioneer Inventions, 10 High Tech. L.J. 35 (1995).

The question of infringement also turns on the precise characteristics of the allegedly infringing device. Following the test laid down by the Supreme Court in *Graver Tank,* courts confronted with a device accused of infringing inquire whether it performs the same function and achieves the same result as the invention in the claims, and whether it does so in the same way. Where the accused device shows only minor or "insubstantial" variations in one of these elements — such as the small movement of one part or a minor change in structure — infringement will be found even if the patentee's invention is a "mere improvement." See, e.g., Tigrett Indus., Inc. v. Standard Indus., Inc., 162 U.S.P.Q. (BNA) 32, 36 (W.D. Tenn. 1967), *aff'd*, 411 F.2d 1218 (6th Cir. 1969), *aff'd by an equally divided court,* 397 U.S. 586 (1970) (claim for playpen calling for "a pair of spaced openings" for two converging drawstrings to adjust side webbing infringed by device with one hole for drawstrings). And even a pioneer patent is not infringed by a device that achieves a different result, or achieves it in a different way. See, e.g., Mead Digital Sys., Inc. v. A. B. Dick Co., 723 F.2d 455, 464 (6th Cir. 1983) (finding that ink-jet printer patent, though a "quantum leap" in the art, was not infringed).

One important set of cases under this doctrine has grappled with the question of whether new technologies, unforeseen at the time the patent was issued, can constitute equivalents. This issue arises when a subsequent device that uses new technology is accused of infringing the original patent. The early cases were split, but the prevailing view now is that new technology can be equivalent. This is true despite the statement in *Graver Tank* that an important determinant in the equivalents inquiry is whether "persons reasonably skilled in the art would have known of the interchangeability of an ingredient not contained in the patent with one that was." Cf. Marty Adelman & Gary Francione, The Doctrine of Equivalents in Patent Law: Questions Pennwalt Did Not Answer, 137 U. Pa. L. Rev. 673, 696 n.103, 697 (1989) (arguing that "this factor [interchangeability] should be used to reject rather than support the application of the doctrine of equivalents," because it signifies that a patentee could have, but mistakenly or intentionally did not, include these interchangeable elements in her original claims).

Notwithstanding the "interchangeability" language in the leading Supreme Court case on the subject, a device performing the same function and achieving the same result in the same way as a patented invention can be found to infringe even if it uses technology developed after the patent was issued.

In a partial dissent from the Federal Circuit's *en banc* decision in *Festo*, later reversed by the Supreme Court, Judge Rader focused directly on the problem of after-arising technologies:

> A primary justification for the doctrine of equivalents is to accommodate after-arising technology. Without a doctrine of equivalents, any claim drafted in current technological terms could be easily circumvented after the advent of an advance in technology. A claim using the terms "anode" and "cathode" from tube technology would lack the "collectors" and "emitters" of transistor technology that emerged in 1948. Thus, without a doctrine of equivalents, infringers in 1949 would have unfettered license to appropriate all patented technology using the out-dated terms "cathode" and "anode." Fortunately, the doctrine of equivalents accommodates that unforeseeable dilemma for claim drafters. Indeed, in Warner-Jenkinson Co., Inc. v. Hilton Davis Chemical Co., 520 U.S. 17, 37 (1997), the Supreme Court acknowledged the doctrine's role in accommodating after-arising technology. Unfortunately, by barring all application of the doctrine of equivalents for amended claims, this court does not account at all for the primary role of the doctrine. All patent protection for amended claims is lost when it comes to after-arising technology, while the doctrine of equivalents will continue to accommodate after-arising technology in unamended claims.

234 F. 3d 558, 619-20 (2000) (Rader, J., concurring in part, dissenting in part). For more on this point, see Anthony H. Azure, Note, *Festo*'s Effect on After-Arising Technology and the Doctrine of Equivalents, 76 Wash, L. Rev. 1153 (2001) (criticizing Federal Circuit for unfairly penalizing patentees who amend claims and then face after-arising equivalents).

The after-arising technology problem is hardly new. Consider B.G. Corp. v. Walter Kiddie & Co., 79 F.2d 20 (2d Cir. 1935) (L. Hand, J.), in which the inventor, Paulson, had perfected a spark plug with a head that resisted deformation at high temperatures, and a shank that conducted heat away from the head. (This is the ancestor of the still-common ceramic-shrouded aluminum sparkplug of today.) Judge Learned Hand, in one of his many eminently reasoned and pithily worded patent opinions, said this about the infringer's argument that Paulson had not foreseen the adaptation of his design to airplane engines:

> It is true that Paulson did not foresee the particular adaptability of his plug to the airplane; indeed, we may assume that he did not even know the especial needs of its engine. Nevertheless he did not shoot in the dark; he laid down with perfect certainty what he wished to accomplish and how.... The unsuspected value of this flexibility of heat control contributes to establish his discovery as an invention; he is not charged with a prophetic understanding of the entire field of its usefulness.

79 F.2d 20, at 22.

For a more recent case, consider Hughes Aircraft Co. v. United States, 717 F.2d 1351 (Fed. Cir. 1983). In its 1983 opinion, the Federal Circuit stated: "Advanced computers and digital communications techniques developed since [the] Williams [patent] permit doing on-board a *part* of what Williams taught as done on the ground." The court concluded: "[P]artial variation in technique, an embellishment made possible by post-Williams technology, does not allow the accused spacecraft to escape the 'web of infringement.'" Id. at 1365 (emphasis in original) (citation omitted). Another case found a patented method for laying pipe, calling for a beam of light to align pipe segments, infringed by the use of later-developed laser beam technology. Laser Alignment, Inc. v. Woodruff & Sons, Inc., 491 F.2d 866 (7th Cir. 1974), *cert. denied,* 419 U.S. 874 (1974).

On the other hand, in Texas Instruments, Inc. v. U.S. Intern. Trade Commn., 805 F.2d 1558 (Fed. Cir. 1986), the Federal Circuit held that major improvements in all the essential elements of hand-held calculators rendered the improved devices non-infringing. 805 F.2d at 1570:

> It is not appropriate in this case, where all of the claimed functions are performed in the accused devices by subsequently developed or improved means, to view each such change as if it were the only change from the disclosed embodiment of the invention. It is the entirety of the technology embodied in the accused devices that must be compared with the patent disclosure....

The specification supporting Texas Instruments' pioneer patent, for instance, described the use of integrated circuits containing bipolar transistors. The improvements all used integrated circuits having metal oxide semiconductor (MOS) transistors. This is an example of improvements in *materials.*

The improved calculators receive input via a device that scans the "matrix" under the keyboard at frequent intervals, whereas the original design had a conductive strip underneath the keypad. This is an example of an improvement that *reduced the number of components* in the invention. Also, the original Texas Instruments display was shown in its specification as a small thermal printer that printed dots on a tape in response to output signals from the processor. The accused devices all use liquid crystal displays (LCDs), the familiar lighted display that does not produce a paper copy, an example of an improvement that *increases the efficiency of an individual component.*

Finally, the internal processing elements of the original calculator were manufactured as discrete components, that were electrically interconnected only in the final design. The newer calculators, in contrast, have all their logic on one integrated circuit, eliminating the necessity for many electrical interconnections. This is an example of enhanced *overall design*. For an argument that changes of this nature militate against a finding of infringement under the doctrine of equivalents, see Robert Merges & Richard Nelson, On the Complex Economics of Patent Scope, 90 Colum. L. Rev. 839 (1990).

PROBLEM

Problem 3-11. Nichols, a scientist who enjoys puzzles, designs a "rotating cube" puzzle in which each face of the cube is composed of a number of smaller cubes, each face is initially of a different color, and the object of the puzzle is to restore the original color scheme once it has been disturbed. Nichols obtains a patent on a method of solving this puzzle, but not on the physical puzzle itself.

Rubik builds and sells puzzles similar to the ones Nichols has designed. Has Rubik infringed the Nichols patent, either directly or indirectly? Does it matter whether Nichols's patent covers the only known solution to the puzzle, or only one among many possible solutions? [See Figure 3-7.] What if Rubik's product includes a "cheat" sheet advising buyers how to solve the puzzle using a number of methods, including Nichols's? What if Rubik includes a copy of the Nichols patent with each cube sold, ostensibly to advise users how to avoid infringement, but arguably with the intent of encouraging them to use the Nichols method?

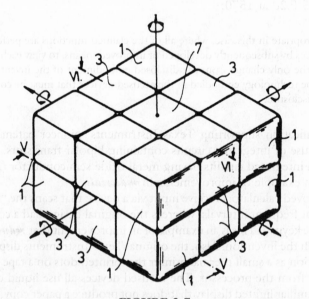

FIGURE 3-7
Patent drawing for Rubik's cube, to which Nichols patented the solutions.

4. The "Reverse" Doctrine of Equivalents

Courts have long recognized that, "[c]arried to an extreme, the doctrine of equivalents could undermine the entire patent system." Scope could be enlarged so far beyond the literal language of claims that patents would take on unlimited power. To check the potentially destructive impact of the doctrine and to preserve symmetry in the rules on infringement, in 1898 the Supreme Court ruled that while

> a charge of infringement is sometimes made out, though the letter of the claims be avoided . . . [t]he converse is equally true. The patentee may bring the defendant within the letter of his claims, but if the latter has so far changed the principle of the device that the claims of the patent, literally construed, have ceased to represent his actual invention, he is as little subject to be adjudged an infringer as one who has violated the letter of a statute has to be convicted, when he has done nothing in conflict with its spirit and intent.

Boyden Power-Brake Co. v. Westinghouse, 170 U.S. 537, 562 (1898). An example drawn from this case may help to illuminate the doctrine. In 1869 George Westinghouse invented a train brake that used a central reservoir of compressed air for stopping power. Further advances in his design, primarily the addition of an air reservoir in each brake cylinder, resulted in a brake that was patented in 1887. An improvement on this 1887 brake, invented by George Boyden, added an ingenious mechanism for pushing compressed air into the brake piston both from the central reservoir *and* from a local reservoir in each brake cylinder. (Westinghouse's brake required a complicated series of passageways to supply air from the two sources.) With the added stopping power of the Boyden brake, engineers could safely operate the increasingly long trains of the late nineteenth century.

The Westinghouse patent included a claim for "the combination of a main air-pipe, an auxiliary reservoir, a brake-cylinder, a triple valve [the device that coordinated the airflows from the main reservoir and the individual brake reservoir] and an auxiliary-valve device, actuated by the piston of the triple-valve . . . for admitting air in the application of the brake." The Court noted that the literal wording of the Westinghouse patent could be read to cover Boyden's brake, since it included what could be described as a "triple valve." But it refused to find infringement, on the ground that Boyden's was a significant contribution that took the invention outside the equitable bounds of the patent:

> We are induced to look with more favor upon this device, not only because it is a novel one and a manifest departure from the principle of the Westinghouse patent, but because it solved at once in the simplest manner the problem of quick [braking] action, whereas the Westinghouse patent did not prove to be a success until certain additional members had been incorporated in it.[47]

---

47. 170 U.S. 537 at 572. On the application of this standard to specific cases, see Jacoby-Bender, Inc. v. Foster Metal Products, Inc., 152 F. Supp. 289 (D. Mass. 1957), *aff'd,* 255 F.2d 869 (1st Cir. 1958) ("I am disposed to regard [the accused] device as . . . an equivalent unless what it accomplished was a marked improvement. . . . In such event it would be appropriate to judge equivalency by the extent of the improvement — the significance of the departure in relation to the remaining basic concept."); Piggott, *Equivalents in Reverse,* 48 J. Pat. and Trademk. Off. Society 291, 295-299 (1966).

The doctrine was recognized — but not applied — once again by the Supreme Court in *Graver Tank:*

> [W]here a device is so far changed in principle from a patented article that it performs the same or a similar function in a substantially different way, but nevertheless falls within the literal words of the claim, the doctrine of equivalents may be used to restrict the claim and defeat the patentee's action for infringement.

Graver Tank & Mfg. Co. v. Linde Air Products Co., 339 U.S. 605, 608 (1950). But since 1898, no case has squarely applied the doctrine to excuse infringement. According to the Federal Circuit, "because products on which patent claims are readable word for word often are in fact the same, perform the same function in the same way, and achieve the same result, as the claimed invention, a defense based on the reverse doctrine of equivalents is rarely offered." SRI Int'l v. Matsushita Electric Corp. of America, 775 F.2d 1107, 1123, n.19 (Fed. Cir. 1985). See, e.g., Monsanto Co. v. Mycogen Plant Science, Inc., 61 F. Supp. 2d 133, 187-188 (D.Del. 1999) (granting JMOL motion overturning jury's exoneration of accused infringer under reverse doctrine of equivalents).

Indeed, the recent Federal Circuit opinion in Tate Access Floors, Inc. v. Interface Architectural Resources, Inc., 279 F.3d 1357 (Fed. Cir. 2002) comes very close to sounding the death knell for the doctrine. *Tate Access* was directly concerned with an infringement suit where the accused infringer, Interface, argued that it could not be held liable for infringement because the design for its product was drawn from the prior art predating the patentee's claimed invention. Interface claimed that this "practicing the prior art defense" was akin to the reverse doctrine of equivalents, and asked the court to consider the defense notwithstanding the finding of literal infringement. The Federal Circuit declined in no uncertain terms:

> Not once has this court affirmed a decision finding noninfringement based on the reverse doctrine of equivalents. And with good reason: when Congress enacted 35 U.S.C. § 112, after the decision in *Graver Tank,* it imposed requirements for the written description, enablement, definiteness, and means-plus-function claims that are co-extensive with the broadest possible reach of the reverse doctrine of equivalents. . . . cf. Hilton Davis Chem. Co. v. Warner-Jenkinson Co., 62 F.3d 1512, 1569 (Fed. Cir. 1995) (en banc) (Nies, J., dissenting) (noting that the reverse doctrine of equivalents was originally used by the courts to reduce the scope of broad "means" claims to "cover only what the inventor discloses and equivalents thereof"), *rev'd and remanded,* 520 U.S. 17 (1997). Even were this court likely ever to affirm a defense to literal infringement based on the reverse doctrine of equivalents, the presence of one anachronistic exception, long mentioned but rarely applied, is hardly reason to create another. We therefore decline the invitation to adopt Interface's "practicing the prior art" defense on these grounds.

COMMENTS AND QUESTIONS

1. Does the court's reasoning in *Tate Access* make sense? What exactly is wrong with defending against an infringement claim by proving that you were practicing the prior art? If the defendant is correct that they were following the prior art, what does that say about the validity of the patent?

2. The statement in *Tate Access* that the Federal Circuit had never affirmed a judgment on the basis of the reverse doctrine is true but highly misleading. The court

has applied the doctrine to reverse a finding of infringement in Scripps Clinic & Research Foundation v. Genentech, Inc., 927 F.2d 1565 (Fed. Cir. 1991), a case typical of an early wave of biotechnology patent actions. Genentech invented and patented the recombinant DNA form of the blood protein Factor VIII:C, a blood clotting agent made by the body and useful in treating patients with clotting disorders. Scripps had previously obtained a patent on purified Factor VIII:C, which it made by isolating and purifying the protein from raw human blood. Scripps sued Genentech for infringement of its product patent, citing the conventional rule that a product patent covers the product no matter how it is made. After attempting to distinguish its recombinant version from Scripps' purified natural protein, Genentech ultimately relied on a pragmatic defense: that the recombinant version was by far cheaper to make, and therefore ought not to be deemed an infringement. The Federal Circuit remanded the case for a determination whether the reverse doctrine of equivalents applied in these circumstances:

> The so-called "reverse doctrine of equivalents" is an equitable doctrine invoked in applying properly construed claims to an accused device. Just as the purpose of the "doctrine of equivalents" is to prevent "pirating" of the patentee's invention, *Graver Tank*, 339 U.S. 605, 607, 608 (1950), so the purpose of the "reverse" doctrine is to prevent unwarranted extension of the claims beyond a fair scope of the patentee's invention.
>
> ...Application of the doctrine requires that facts specific to the accused device be determined and weighed against the equitable scope of the claims, which in turn is determined in light of the specification, the prosecution history, and the prior act.
>
> The record contained evidence of the properties of plasma-derived and recombinantly produced VIII:C, which was presented primarily by Scripps in connection with its proofs of infringement. There was deposition testimony that there were differences between VIII:C from plasma and VIII:C obtained by recombinant techniques; a Scripps' witness described the products as "apples and oranges," referring specifically to stability and formulations. The parties disputed, in connection with the summary judgment motions, the capabilities of the respective processes in terms of the purity and specific activities that were enabled for the respective products. The record on this point is extensive.
>
> Genentech argues that its product is equitable seen as changed "in principle," particularly when viewed in the context of the prior art. Genentech asserts that the specific activities and purity that are obtainable by recombinant technology exceed those available by the Scripps process; an assertion disputed by Scripps, but which if found to be correct could provide — depending on the specific facts of similarities and differences — sufficient ground for invoking the reverse doctrine. These aspects were not discussed by the district court.
>
> The principles of patent law must be applied in accordance with the statutory purpose, and this issues raised by new technologies require considered analysis. Genentech has raised questions of scientific and evidentiary fact that are material to the issue of infringement. Consideration of extrinsic evidence is required, and summary judgment is inappropriate. The grant of summary judgment of infringement of [the product] claims is reversed. The issue requires trial.

927 F.2d at 1581.

The *Genentech* case was settled before the district court reached a decision on the reverse doctrine of equivalents issue on remand.

Since *Tate Access*, the Federal Circuit has gone out of its way on several occasions to make it clear that the reverse doctrine of equivalents, while rare, does still have viability. See, e.g., Amgen, Inc. v. Hoechst Marion Roussel, 314 F.3d 1313, 1351 (Fed. Cir. 2003); Plant Genetic Systems, N.V. v. DeKalb Genetics Corp., 315 F.3d

1335, 1341 (Fed. Cir. 2003) ("[T]he judicially-developed 'reverse doctrine of equiva-lents,' requiring interpretation of claims in light of the specification, may be safely relied upon to preclude improper enforcement against later developers.") See also Amgen, Inc. v. Hoechst Marion Roussel, Inc., 339 F.Supp.2d 202, 300-301 (D. Mass. 2004) (thoroughly reviewing the doctrine, and concluding that it almost – but not quite – was applicable). In addition, some observers have argued that con-temporary disclosure doctrines have obviated the need for the reverse doctrine. See Joseph Yang & Roxana H. Yang, An Update on the Doctrine of Equivalents — Where Do We Stand in 2003?, 749 Practicing L. Inst. 521, 561 (2003) (concluding that the reverse doctrine is dead due to the enactment of section 112 in 1952).

5. Equivalents for Means-Plus-Function Claims

The trick in drafting patent claims, as we have seen, is to make them as broad as the prior art and other patent doctrines will allow. Because of this, and because drafting claims is time-consuming and expensive, patent drafters are always on the lookout for convenient ways to broadly cover the elements of inventions. So, for example, assume you have invented a new type of "animated corsage," containing artificial flowers together with flashing lights and a miniature speaker for playing music stored on a tiny device inside the corsage. You may want to include a way to attach the corsage to clothing as an element in one or more claims. But how to do so as broadly as possible? One way is to simply draft a separate claim for every conceivable way of attaching things to clothing — buttons, clips, pins, Velcro, etc. That would quickly become tedious, however, not to mention expensive (the Patent Office charges more for applications that exceed a modest number of claims). One solution to your problem would be to use a "catch-all" phrase that covers all conventional ways of attaching things to cloth-ing. This approach is officially sanctioned by § 112 of the Patent Act, paragraph six. This provision allows the claim drafter to use what are known as "means plus function" claims. In our corsage example, a claim using this approach would read: "A corsage, comprising a main body, said main body comprising [element 1, element 2, etc.], and [3] *means for attaching* said main body to an article of clothing."

Prior to the Patent Act of 1952, courts invalidated many such broad claims on the grounds that they covered entire "functions" rather than specific machines. See, e.g., Halliburton Oil Well Cementing Co. v. Walker, 329 U.S. 1 (1946). Section 112, ¶6 of the 1952 Act reflected the widespread sentiment in favor of some mechanism to claim broad functional features of inventions. Section 112 ¶6 falls far short of permit-ting claims to an entire function, e.g., "broadcasting" or "cutting," but it does allow functional language in claims to cover claim *elements*.

Section 112, ¶6 reads in full as follows:

> An element in a claim for a combination may be expressed as a means or step for per-forming a specified function without the recital of structure, material, or acts in support thereof, and such claim shall be construed to cover the corresponding structure, material, or acts described in the specification and equivalents thereof.

35 U.S.C. § 112. Under this section, the phrase "means for" is a signal to the PTO and the courts to turn to the specification in order to define the invention. The patentee's claim does not capture *all* means for achieving the specified function — only those

actually discussed in the specification, and equivalents thereof. Returning to our corsage example, you might list various clothes-attaching means in the part of the specification where you discuss how to attach the corsage to clothing, in a passage something like this:

> The corsage described herein may be attached to clothing in a variety of ways: using buttons, pins, clips, Velcro, snaps, or any other mechanism suitable for the purpose.

Means-plus-function claims present a number of unique issues. This section briefly considers two: (1) Which claims are considered means-plus-function claims? and (2) What is the relationship between means-plus-function claims and the doctrine of equivalents?

Which Claims Are Means-Plus-Function Claims?

Is every claim that uses the phrase "means for..." considered a means-plus-function claim, subject to treatment under § 112 ¶6? No. This rule can be helpful for patentees. Recall that § 112 ¶6 ties claim scope to the structure(s) disclosed in the patent's specification. Escaping the strictures of § 112 ¶6 allows a patentee to argue that an accused device infringes despite the fact that the device differs in some respects from that disclosed in the patentee's specification.

So when does "means for" language invoke § 112 ¶6? The use of the word "means" "triggers *a presumption* that the inventor used this term advisedly to invoke the statutory mandate for means-plus-function clauses." York Prods., Inc. v. Cent. Tractor Farm & Family Ctr., 99 F.3d 1568, 1574 (Fed. Cir. 1996) (emphasis added). This presumption may be overcome in two ways. First, "a claim element that uses the word 'means' but recites no function corresponding to the means does not invoke § 112, ¶6." Rodime PLC v. Seagate Tech., Inc., 174 F.3d 1294, 1302 (Fed. Cir. 1999). Second, "even if the claim element specifies a function, if it also recites sufficient structure or material for performing that function, § 112, ¶6 does not apply." Id.; see Cole v. Kimberly-Clark Corp., 102 F.3d 524, 531 (Fed. Cir. 1996) ("To invoke [§ 112, paragraph 6], the alleged means-plus-function claim element must not recite a definite structure which performs the described function."). A claim term recites sufficient structure if "the 'term, as the name for structure, has a reasonably well understood meaning in the art.'" Watts v. XL Sys., Inc., 232 F.3d 877, 880-81 (Fed. Cir. 2000) (quoting Greenberg v. Ethicon Endo-Surgery, Inc., 91 F.3d 1580, 1583 (Fed. Cir. 1996)). The mere use of the word "means" does not suffice to make that limitation a means-plus-function limitation. *Cole, supra*, 102 F.3d at 531. Thus in *Rodime*, which dealt with a patent to computer hard disk control technology, the Federal Circuit held that a number of claim elements that included the word "means" were not means-plus-function claims subject to § 112 ¶6. The reason was the detailed recitation of structure following the "means for" language. For example, claim 3 of the patent at issue included a "positioning means" element. But the claim as a whole was held to recite too much structure to be considered a means-plus-function claim. The element read more fully:

> Positioning means for moving said transducer means between the concentrically adjacent tracks on said micro hard-disk, said positioning means including: two support arms... a pivot shaft... a positioning arm... a bearing assembly... a stepper motor... means for operating said stepper motor... and a tensioned steel band....

In *Cole*, the court reached a similar conclusion. The court noted that "[t]he claim describes not only the structure that supports the [claimed] function, but also its location...and extent.... An element with such a detailed recitation of its structure, as opposed to its function, cannot meet the requirements of the statute." *Cole*, 102 F.3d at 531. In contrast, in Sage Products, Inc. v. Devon Industries, Inc., 126 F.3d 1420 (Fed. Cir. 1997), the Federal Circuit held that a claim element to "closure means...for controlling access" was properly construed as a means-plus-function limitation, since a function was recited for the means and the claim did not "explicitly recite[ ] the structure, material, or acts needed to perform [the function]. *Sage*, 126 F.3d at 1428.

§ 112 ¶6 Equivalents and the Doctrine of Equivalents

Can a means-plus-function claim that is not literally infringed be infringed under the doctrine of equivalents? The cases say yes, but only in two limited circumstances. Specifically, the Federal Circuit has found separate ground for the doctrine of equivalents in section 112 ¶6 cases where the *function* (as opposed to the corresponding structure) was equivalent but not identical, and where the accused device contained after-arising technology that was equivalent to the patented structure, but was not known at the time the patent application was filed.

Consider for example WMS Gaming, Inc. v. International Game Technology, 184 F.3d 1339 (Fed. Cir. 1999). The patent in the case, issued to International Game Technology (IGT), was to an improved slot machine. Specifically, the claimed slot machine achieved a desired result: lowering the probability of a winning pull of the slot machine lever, thereby increasing the potential payoff to players of the machine. It did this without a disadvantageous side effect: increasing the number of reels required on the machine or otherwise increasing the size of the machine. (The reels are the spinning elements of the machine, which usually carry on their outer face familiar symbols such as cherries, lemons, and the number "7." Each of these symbols represents a "stop position" on the reel.) The claimed invention used a microprocessor to generate random numbers and assign them to positions on the reels of the machine. The stop positions of the reels are determined first and then the payoff is calculated based on the stop positions. In the accused device sold by WMS Gaming, the payoff is calculated first and then stop positions that represent that payoff are chosen. The court found that the WMS device did not literally infringe the IGT patent. The IGT claim included the phrase "means for assigning" random numbers to stop positions. While the accused device did have a *structure* corresponding to this element in IGT's specification—a microprocessor programmed to assign a single number to each stop position—the court concluded that the structure did not perform the identical *function* that it did in IGT's patent.

Nonetheless, the court found infringement under the doctrine of equivalents. The accused microprocessor algorithm performed a *function* that was equivalent, but not identical to, that in the claimed invention: to assign numerical values to stop positions on the slot machine reels. And the difference between the assigning numerical values singly and in a group was insubstantial. Thus, there was infringement under the doctrine of equivalents. 184 F.3d at 1583.

The *WMS* court went on to distinguish another Federal Circuit case, *Chiuminatta Concrete*. That case held that the structural equivalence could not be found under the doctrine of equivalents in a § 112, ¶6 claim unless the infringing structure did not exist at the time of the claim:

Recently, in Chiuminatta Concrete Concepts, Inc. v. Cardinal Industries, Inc., we stated:

> Both § 112, ¶6, and the doctrine of equivalents protect the substance of a patentee's right to exclude by preventing mere colorable differences or slight improvements from escaping infringement, the former, by incorporating equivalents of disclosed structures into the literal scope of a functional claim limitation, and the latter, by holding as infringements equivalents that are beyond the literal scope of the claim. They do so by applying similar analyses of insubstantiality of the differences.

145 F.3d at 1310, 46 USPQ2d at 1758. We went on to point out in *Chiuminatta* that a "lack of equivalent structure under a means-plus-function limitation may preclude a finding of equivalence under the doctrine of equivalents." Id. We stated that such would be the case unless a variant that was accused of infringement—but that did not literally infringe a means-plus-function limitation—was due to technological advances developed after the patent was granted and "constitute[d] so insubstantial a change from what [was] claimed in the patent that it should be held to be an infringement." Id.

As just seen, our holding that the WMS 400 slot machine does not literally infringe claim 1 of the Telnaes patent is not based on a finding that the accused device lacks structure equivalent to that disclosed in the patent. On the contrary, we have sustained the district court's finding that the WMS 400 slot machine has equivalent structure. However, we have reversed the district court's holding of literal infringement based on a lack of identity of function. Consequently, unlike *Chiuminatta*, the accused device in this case may still infringe under the doctrine of equivalents. See Al-Site Corp. v. VSI Int'l, Inc., 174 F.3d 1308, 1320-21 (Fed. Cir. 1999) (an accused device can infringe under the doctrine of equivalents without infringing literally under 35 U.S.C. §112, ¶6 because the doctrine only requires substantially the same function, not identicality of function as in section 112, ¶6). *Al-Site* also points out that only the doctrine of equivalence, and not §112, ¶6, could form the basis for patent infringement in the case of an "after-arising" improvement. See also Medtronic Minimed Inc. v. Smiths Medical MD Inc., 373 F.Supp.2d 466, 470 (D. Del. 2005) (same).

COMMENTS AND QUESTIONS

1. Is section 112, ¶6 consistent with the point of having patent claims in the first place? If a potential infringer reads a patent claim that includes an element phrased in "means for *x*" language, how can he be sure he is not infringing on the patent? What if the accused infringer invents a new "means for" performing an old function—given the importance of "after-arising" technology under the doctrine of equivalents, how can he or she determine whether the new means is covered by the old patent?

2. Is the scope of a means-plus-function claim element a question of law, like any other claim construction, or a question of fact, like any other equivalents determination?

6. Contributory Infringement

C. R. Bard, Inc. v. Advanced Cardiovascular Systems, Inc.

United States Court of Appeals for the Federal Circuit
911 F.2d 670 (Fed. Cir. 1990)

This is a case of claimed infringement of a method patent for a medical treatment. Defendant-Appellant Advanced Cardiovascular Systems, Inc. (ACS) was marketing [a] perfusion catheter for use in coronary angioplasty. Plaintiff-Appellee C.R. Bard, Inc. (Bard) sued ACS for alleged infringement of U.S. Patent No. 4,581,017 ('017), which Bard had purchased all rights to as of December 31, 1986. The '017 patent relates to a method for using a catheter in coronary angioplasty. The district court granted plaintiff Bard summary judgment against ACS, finding infringement of claim 1 of the '017 patent. We reverse the grant of summary judgment and remand the case for further proceedings.

Plaintiff Bard alleges that the ACS catheter is especially adapted for use by a surgeon in the course of administering a coronary angioplasty in a manner that infringes claim 1 of the '017 patent, that therefore ACS is a contributory infringer, and that ACS actively induces infringement. Of course, a finding of induced or contributory infringement must be predicated on a direct infringement of claim 1 by the users of the ACS catheter.

For purposes of this case, the statute requires that ACS sell a catheter for use in practicing the '017 process, which use constitutes a material part of the invention, knowing that the catheter is especially made or adapted for use in infringing the patent, and that the catheter is not a staple article or commodity of commerce suitable for substantial noninfringing use.

In asserting ACS's contributory infringement of claim 1, Bard seeks to establish the requisite direct infringement by arguing that there is no evidence that any angioplasty procedures using the ACS catheter would be noninfringing. Testing this assertion requires a two step analysis. First is a determination of the scope of the claim at issue. Second is an examination of the evidence before the court to ascertain whether, under §271(c), use of the ACS catheter would infringe the claim as interpreted.

Bard argues that [a] prior art patent teaches the use of the catheter with the inlets (side openings) where the blood enters the tube placed only in the aorta, whereas the '017 method in suit involves insertion of the catheter into the coronary artery in such a manner that the openings "immediately adjacent [the] balloon fluidly connect locations within [the] coronary artery surrounding [the] proximal and distal portions of [the] tube." Thus, Bard argues, a surgeon, inserting the ACS catheter into a coronary artery to a point where an inlet at the catheter's proximal end draws blood from the artery, infringes the '017 patent.

[I]t is important to note that the ACS catheter has a series of ten openings in the tube near, and at the proximal end of, the balloon. The first of these openings — the one closest to the balloon — is approximately six millimeters (less than 1/4 inch) from the edge of the proximal end of the balloon. The remainder are located along the main lumen at intervals, the furthest from the balloon being 6.3 centimeters (approximately 2 1/2 inches) away.

It would appear that three possible fact patterns may arise in the course of using the ACS catheter. The first pattern involves positioning the catheter such that all of its side openings are located only in the aorta. This is clearly contemplated by the prior art '725 patent cited by the examiner. In the second of the possible fact patterns, all of the side openings are located within the coronary artery. This situation appears to have been contemplated by the '017 patent, the method patent at issue. In the third fact pattern, some of the side openings are located in the aorta and some are located in the artery.

There is evidence in the record that 40 to 60 percent of the stenoses that require angioplasty are located less than three centimeters from the entrance to the coronary artery. ACS argues that therefore the ACS catheter may be used in such a way that all of the openings are located in the aorta. Even assuming that the trial judge's conclusion is correct that claim 1 is applicable to the third of the fact patterns, it remains true that on this record a reasonable jury could find that, pursuant to the procedure described in the first of the fact patterns (a noninfringing procedure), there are substantial noninfringing uses for the ACS catheter.

Whether the ACS catheter "has no use except through practice of the patented method," Dawson Chemical Co. v. Rohm & Haas Co., is thus a critical issue to be decided in this case. As the Supreme Court recently noted, "[w]hen a charge of contributory infringement is predicated entirely on the sale of an article of commerce that is used by the purchaser [allegedly] to infringe a patent, the public interest in access to that article of commerce is necessarily implicated." Sony Corp. v. Universal City Studios, Inc., 464 U.S. 417, 440 (198[4]) [declining to find contributory copyright infringement in sale of video cassette recorders]. Viewing the evidence in this case in a light most favorable to the nonmoving party, and resolving reasonable inferences in ACS's favor, it cannot be said that Bard is entitled to judgment as a matter of law. The grant of summary judgment finding ACS a contributory infringer under § 271(c) is not appropriate.

A person induces infringement under § 271(b) by actively and knowingly aiding and abetting another's direct infringement. Bard argues that ACS induced infringement under § 271(b) by: 1) providing detailed instructions and other literature on how to use its catheter in a manner which would infringe claim 1; and 2) having positioned the inlets near the balloon's proximal end so as to allow a user of the ACS catheter to infringe claim 1. Because a genuine issue of material fact exists, a grant of summary judgment finding ACS induced infringement is also not appropriate.

COMMENTS AND QUESTIONS

1. While Bard sued ACS, a competing manufacturer of catheters, the "real" infringer in this case is the doctor who completed the catheterization under circumstances that violated Bard's patent. What should doctors do in such a situation? Refuse to perform procedures that infringe a patent? Seek a license? In 1996, Congress amended the patent laws to exempt doctors who perform medical processes from liability for infringement, rendering this problem moot. However, Congress apparently intended to leave device manufacturers liable for contributory infringement of such patents. See 35 U.S.C. § 287.

2. The Federal Circuit has held that section 271(c) requires a showing that the alleged contributory infringer knew that the combination for which her component was especially designed was both patented and infringing. Trell v. Marlee Elec. Corp., 912 F.2d 1443 (Fed. Cir. 1990); Cybiotronics, Ltd. v. Golden Source Elec., Inc., 130 F. Supp. 2d 1152 (C.D. Cal. 2001). Does this requirement make sense? Is it consistent with the rule for direct infringement, which does not require evidence of knowledge or intent?

If an unpatented product has no other use except in conjunction with the patented machine, process or product, is it reasonable automatically to presume that making, using or selling the *unpatented* product is contributory infringement? Or should companies be allowed to compete in the market for nonstaple products that work with the patented invention? For a discussion of these issues, see Dawson Chem. Co. v. Rohm & Haas, Inc., 448 U.S. 176 (1980); Husky Injection Molding Systems, Ltd. v. R&D Tool Engineering Co., 291 F.3d 780 (Fed. Cir. 2002).

Note on Inducement

The concepts of contributory infringement and inducement evolved to address infringing activity that somehow lacked the element of a direct making, using, or selling of the patented invention. As we have seen, contributory infringement sweeps into the net of infringement the making, use, or sale of *less than* the entire patented device. Inducement, on the other hand, involves behavior that omits any making, using, or selling but that nevertheless amounts to an attempt to appropriate the value of an invention. It is often described as activity that "aids and abets" infringement. Although inducing infringement commonly involves instructing another to violate a patent, this branch of liability is broad enough to ensnare a host of diverse activities.

A good example of inducement at work is Water Technologies Corp. v. Calco, Ltd., 850 F.2d 660, 669 (Fed. Cir.), *cert. denied*, 488 U.S. 968 (1988). In this case Gartner, one of the defendants, was hired as a consultant by Calco to design a portable water purification system to compete with that sold by the plaintiff/patentee. (The patentee's system involved an advanced purification resin.) Gartner complied by supplying plans for an infringing device. The court called it a classic case of inducement:

> Although section 271(b) does not use the word "knowing," the case law and legislative history uniformly assert such a requirement. Gartner argues that no proof of a specific, knowing intent to induce infringement exists. While proof of intent is necessary, direct evidence is not required; rather, circumstantial evidence may suffice. The requisite intent to induce infringement may be inferred from all of the circumstances. Gartner's activities provide sufficient circumstantial evidence for this court to affirm the district court's finding that he intentionally induced Calco's and the public's direct infringement. Under the facts here, although Gartner's liability as a direct infringer may be de minimis, we see no reason to hold him liable for less than all damages attributable to Calco's infringing sales on the basis of his inducement of direct infringement.

Id. at 668-669.

Water Technologies brings out a number of themes common to cases involving inducement to infringe. First is the importance of intent. As the court states, though the statute does not mention intent, the cases all require it. Notice, however, what evidence the court recites in support of its conclusion that Gartner (one of the defendants in the case) did intend to infringe. Why is extensive circumstantial evidence — whether called intent or something else — required to establish infringement in such a case? Is it relevant that Gartner did not himself make, use, or sell the item that infringed the patent? Why would more extensive evidence be required in such a case?

Second, notice that one item of evidence centers on the existence of a patent. Part of the requisite intent, in other words, is that the one accused of inducing infringement must know of the patent. Beyond this, it is also generally required that the accused infringer know that his or her activities will lead to infringement of the patent. See, e.g., Hewlett-Packard Co. v. Bausch & Lomb, Inc., 909 F.2d 1464 (Fed. Cir. 1990). But see Manville Sales Corp. v. Paramount Systems, Inc., 917 F.2d 544 (Fed. Cir. 1990) (setting forth a divergent level of intent required to find inducement). See generally, Mark A. Lemley, Inducing Patent Infringement, 39 U.C. Davis L. Rev. 225 (2005) (discussing this split in the law on requisite intent and proposing a sliding scale to resolve it). Since knowledge of infringement depends in part on the accused infringer's understanding of the meaning of the patentee's claims, a number of cases turn on the existence and credibility of a patent attorney's opinion letter regarding whether the accused activities will in fact result in patent infringement. See Micro Chem., Inc. v. Great Plains Chem. Co., 194 F.3d 1250 (Fed. Cir. 1999) (refusing to find inducement to infringe, partially due to an opinion letter obtained by the accused infringer).

COMMENTS AND QUESTIONS

1. A classic example of inducement/contributory infringement is the sale of device components or "replacement parts" for a patented device, even though the parts themselves constitute less than the complete device and hence are not direct infringement. Contributory infringement is often used to attack sellers of parts where the actual ultimate infringers are end-users who replace parts or assemble components. See, e.g., Aro Mfg. Corp. v. Convertible Top Co., 377 U.S. 176 (1964); Husky Injection Molding Systems, Ltd. v. R&D Tool Engineering Co., 291 F.3d 780 (Fed. Cir. 2002). Another example is the sale of products with instructions that assist buyers in employing a process that infringes a process patent. See Dawson Chem. Co. v. Rohm & Haas, Inc., 448 U.S. 176 (1980).

2. For more, see Tegal Corp. v. Tokyo Electron Corp., 248 F.3d 1376 (Fed. Cir. 2001); Case Note, Patent Law-Active Inducement of Infringement, 115 Harv. L. Rev. 1246 (2002). Note that the Supreme Court "borrowed" patent law notions of inducement in deciding the *Grokster* copyright case in 2005. See Chapter 4, section H.

7. Infringement Involving Foreign Activities

Contemporary technology often involves the manufacture of different components in different locations, increasingly in other countries. And in a "networked world," the *use* of many technologies can be spread across different countries as well. The software and wireless Internet industries are at the vanguard of these developments. As many have noted, the technologies underlying these industries are increasingly straining the legal fabric of patent law, in which the law of each country reaches only strictly domestic activities. The cases in this section show the legal system's efforts to keep up with these new realities. One set of issues centers on when a product is "used" in the United States for purposes of 35 U.S.C. §271(a). Another concerns §271(f)(1) of the Patent Act, which was enacted in the wake of Deepsouth Packing Co. v. Laitram Corp., 406 U.S. 518 (1972) (finding no infringement under U.S. law when components of product were shipped separately for assembly abroad).

One important case involved the popular "Blackberry" handheld e-mail device. NTP, Inc. v. Research In Motion, Ltd., 418 F.3d 1282 (Fed. Cir. 2005). NTP, Inc. holds a number of patents on wireless e-mail technology. NTP sued Research in Motion, Ltd. (RIM), maker of the "Blackberry" device. A jury awarded over $53 million in damages to NTP. NTP, Inc. v. Research in Motion, Ltd., 2003 WL 23325540 (E.D.Va. Aug.5, 2003), and RIM appealed. Aside from issues of claim construction, an important issue on appeal was the design of RIM's e-mail system, which routes all messages through a central hub in Canada. See Figure 3-8.

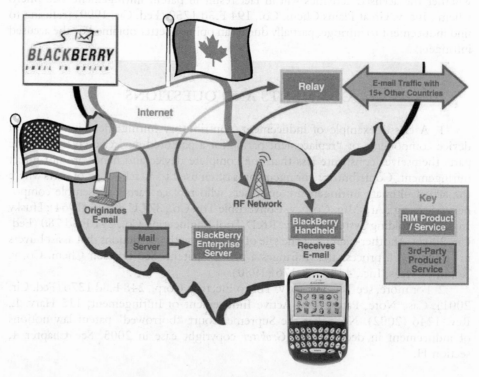

FIGURE 3-8
Wireless e-mail system.

Source: RIM: Federal Circuit Brief.

Claim 1 of [U.S. Patent 5,436,960, the '960 patent] covers a wireless e-mail system that uses radio frequency ("RF") transmissions to tell a portable RF device (such as a Blackberry handheld device) that someone has sent an e-mail. The full content of the e-mail is sent over a "wireline," i.e., it is not transmitted directly to the RF device. The claim reads as follows:

1. A system for transmitting originated information in an *electronic mail system* [from an originating "processor" or computer to at least one other "destination processor," or computer]...comprising:
[1] at least one gateway switch [for receiving and storing the originated information]...;
[2] a [radio frequency, or RF] information transmission network for transmitting the originated information to at least one RF receiver which transfers the originated information to...at least one of the...destination processors;
[3] at least one interface switch,...connecting...[one or more] gateway switch to the RF information transmission network and transmitting the originated information received from the gateway switch to the RF information transmission network; and wherein
[4] the originated information is transmitted to the...interface switch by the...gateway switch in response to an address of the...interface switch added to the originated information at the...originating processors or by the electronic mail system and the originated information is transmitted from the...interface switch to the RF information transmission network with an address of...one of the plurality of destination processors to receive the originated information....; and
[5] the electronic mail system transmits other originated information from one of the...originating processors...to at least one of the...destination processors in the electronic mail system through a wireline without transmission using the RF information transmission network.

'960 patent, col. 49, ll. 2-45 (emphasis added in Federal Circuit opinion). Claim 18, a representative process or method claim, claimed "[a] method for transmitting originated information...in an electronic mail system to at least one of a plurality of destination processors in the electronic mail system," including the same essential elements listed in the product claim, expressed as steps in the method instead of components in the system. (This is a common claim drafting technique.)

The Federal Circuit addressed the question of whether RIM's system for handling Blackberry e-mails infringed NTP's U.S. patent:

The use of a claimed system under section 271(a) is the place at which the system as a whole is put into service, i.e., the place where control of the system is exercised and beneficial use of the system obtained. See Decca [Ltd. v. United States, 544 F.2d 1070 (1976)], at 1083. Based on this interpretation of section 271(a), it was proper for the jury to have found that use of NTP's asserted system [i.e., product] claims occurred within the United States. RIM's customers located within the United States controlled the transmission of the originated information and also benefited from such an exchange of information. Thus, the location of the Relay in Canada did not, as a matter of law, preclude infringement of the asserted system claims in this case.

418 F.3d 1282, 1318.
Another line of cases brings into play § 271(f)(1) of the Patent Act. It reads:

Whoever without authority supplies or causes to be supplied in or from the United States all or a substantial portion of the components of a patented invention, where such components are uncombined in whole or in part, in such a manner as to actively induce the combination of such components outside the United States in a manner that would infringe the patent if such combination occurred within the United States shall be liable as an infringer.

35 U.S.C. § 271(f)(1) (2000). This provision was enacted in the wake of the *Deepsouth Packing* case, mentioned earlier. It was applied in the following case to a novel set of facts drawn from the software industry.

Eolas Technologies Inc. v. Microsoft Corp.
United States Court of Appeals for the Federal Circuit
399 F.3d 1325 (Fed. Cir. 2005)

[Microsoft appealed a trial court ruling that it had infringed claims 1 and 6 of U.S. Patent No. 5,838,906 (the '906 patent), and had actively induced United States users of Internet Explorer to infringe claim 1. The district court also invoked 35 U.S.C. § 271(f) to include foreign sales of Internet Explorer in the royalty awarded to Eolas.]

[T]his court affirms the district court's holding that "components," according to section 271(f)(1), includes software code on golden master disks.

The '906 patent carries the title "distributed hypermedia method for automatically invoking external application providing interaction and display of embedded objects within a hypermedia document." Eolas alleged that certain aspects of Microsoft's Internet Explorer (IE) product incorporate its invention. Microsoft denied infringement and asserted that the claims were invalid and unenforceable.

Essentially, the claimed invention allows a user to use a web browser in a fully interactive environment. For example, the invention enables a user to view news clips or play games across the Internet. The '906 claims require a web browser with certain properties. Specifically, the invention calls for a browser located in a "distributed hypermedia environment."... The inventors have consistently maintained that "[t]his invention was the first instance where interactive applications were embedded in Web pages."

* * * *

Eolas claimed royalty damages for both foreign and domestic sales of Windows with IE. Microsoft moved, in limine, to prevent Eolas from seeking damages based on foreign sales under section 271(f).

Microsoft exports a limited number of golden master disks containing the software code for the Windows operating system to Original Equipment Manufacturers (OEMs) abroad who use that disk to replicate the code onto computer hard drives for sale outside of the United States. The golden master disk itself does not end up as a physical part of an infringing product. The district court denied Microsoft's motion, finding that source code is the legal equivalent of a piece of computer hardware and that "in a legal sense, a [sic] source code is a made part of a computer product." The district court determined that the code on the golden master disks constitutes "components" of an infringing product for combination outside of the United States under section 271(f).

This court must ... decide whether software code made in the United States and exported abroad is a "component[ ] of a patented invention" under section 271(f).

Section 271(f) refers to "components of a patented invention." This statutory language uses the broad and inclusive term "patented invention." Title 35, in the definitions section, defines "invention" to mean "invention or discovery"—again broad and inclusive terminology. 35 U.S.C. § 100(a) (2000). This statutory language did not limit section 271(f) to patented "machines" or patented "physical structures."

Rather every form of invention eligible for patenting falls within the protection of section 271(f). By the same token, the statute did not limit section 271(f) to "machine" components or "structural or physical" components. Rather every component of every form of invention deserves the protection of section 271(f).

In examining the statutory language of section 271(f), this court must next examine whether the software code on the golden master disk is a "component" of the computer program invention. A "component" of a process invention would encompass method steps or acts. See, e.g., 35 U.S.C. § 112, ¶6 (2000). A "component" of an article of manufacture invention would encompass a part of that construct. Because a computer program product is a patented invention within the meaning of Title 35, then the "computer readable program code" claimed in claim 6 of the '906 patent is a part or component of that patented invention.

Exact duplicates of the software code on the golden master disk are incorporated as an operating element of the ultimate device. This part of the software code is much more than a prototype, mold, or detailed set of instructions. This operating element in effect drives the "functional nucleus of the finished computer product." Imagexpo, L.L.C. v. Microsoft, Corp., 299 F.Supp.2d 550, 553 (E.D.Va.2003). Without this aspect of the patented invention, the invention would not work at all and thus would not even qualify as new and "useful." Thus, the software code on the golden master disk is not only a component, it is probably the key part of this patented invention. Therefore, the language of section 271(f) in the context of Title 35 shows that this part of the claimed computer product is a "component of a patented invention."

Sound patent policy also supports the meaning of section 271(f). In the first place, this court accords the same treatment to all forms of invention. See, e.g., TRIPS Agreement, Part II, Section 5 (1994) ("[P]atents shall be available and patent rights enjoyable without discrimination as to the place of invention[ ][and] the field of technology. . . . "). This court cannot construct a principled reason for treating process inventions different than structural products. Moreover, as the district court pointed out, process and product — software and hardware — are practically interchangeable in the field of computer technology. Eolas Techs., Inc. v. Microsoft Corp., 274 F.Supp.2d 972, 974 (N.D.Ill.2003). On a functioning computer, software morphs into hardware and vice versa at the touch of a button. In other words, software converts its functioning code into hardware and vice versa. Thus in the context of this patented invention, the computer transforms the code on the golden disk into a machine component in operation. Thus, sound policy again counsels against varying the definition of "component of a patented invention" according to the particular form of the part under consideration, particularly when those parts change form during operation of the invention as occurs with software code.
[Affirmed.]

COMMENTS AND QUESTIONS

1. Another 2005 case applied § 271(f) to a chemical patent claim. Union Carbide Chemicals & Plastics Technology Corp. v. Shell Oil Co., 425 F.3d 1366 (Fed. Cir. 2005), reh'g denied 434 F.3d 1357 (Fed. Cir. 2006), centered on Union Carbide's patent claiming improved silver catalysts for the commercial production of ethylene oxide, an intermediate chemical compound used ultimately for production of polyester fiber, resin, and film. After a finding that the patent was infringed, Union Carbide

attempted to submit evidence of Shell's profits from the sale of silver catalysts overseas. Shell argued in turn that because Union Carbide claimed the production of the silver catalysts in *process* form, overseas sales of the catalyst were not infringing acts under § 271(f). The Federal Circuit disagreed, relying in part on Eolas v. Microsoft. As in *Eolas*, the court stressed the broad wording of the statute ("components of a patented invention"). The court also distinguished a portion of the Blackberry (NTP v. RIM case), excluded from the excerpt of that case earlier in this chapter, in which the Federal Circuit found no § 271(f) liability for users who purchased their Blackberries in the U.S. but took them overseas. That situation, which the court called "the domestic sale of a component being used, in part, outside the United States," was different from Union Carbide's wholesale exporting of catalysts to foreign affiliates. Union Carbide, 425 F.3d at 1380.

2. Are these interpretations of the Patent Act merely stopgap measures until true international patent protection becomes available? In the meantime, how far should courts stretch to provide effective (quasi-international) coverage for innovators operating in a global competitive environment? For a detailed discussion, see Mark A. Lemley et al., Divided Infringement Claims, 33 AIPLA Q.J.

3. How can patent claims be drafted so that infringement can occur within a single jurisdiction? For an example of the problem, see U.S. Patent 5,586,937, "Interactive, Computerised Gaming System with Remote Terminals," issued Dec. 24, 1996, assigned to Julian Mehashe. Claim 1 reads: "A gaming system including a host computer, a plurality of terminal computers...remote from the host computer...." As written, no one infringes this claim if the host and the terminals are in different countries or owned by different users. How might this be re-drafted to avoid a potential infringement issue? Consider: "A gaming terminal having a connection to a computer network, said terminal comprising software downloaded from said network, said software capable of...." Since the terminal and the "network connection" are in the same jurisdiction, this might solve the problem.

D. DEFENSES

1. The "Experimental Use" Defense

Despite the absolute language of section 271, there is a judicially created exception to infringement that is commonly known as the "experimental use" exception. The exception allows for the unlicensed construction and use of a patented invention for purposes of pure scientific inquiry. Though questioned in a concurring opinion in one recent case, the doctrine is generally thought to be viable. See Embrex, Inc. v. Service Engineering Corp., 216 F.3d 1343, 1352 (Fed. Cir. 2000) (Rader, J., concurring) ("[T]he Patent Act leaves no room for any *de minimis* or experimental use exception for infringement.").

The doctrine had its origins in Justice Story's opinion in Whittemore v. Cutter, 29 F. Cas. 1120 (C.C.D. Mass. 1813) (No. 17,600). In this case, the defendant appealed a jury instruction to the effect that the "making of a machine... with a design to use it for profit" constituted infringement. Justice Story upheld the trial judge's instruction and stated that "it could never have been the intention of the legislature to punish a man, who constructed such a machine merely for philosophical experiments, or for the purpose of ascertaining the sufficiency of the machine to produce its described effects." 29 F. Cas. at 1121. Other cases followed, generally limiting the exception to these quite narrow grounds. See Note, Experimental Use as Patent Infringement: The Impropriety of a Broad Exception, 100 Yale L.J. 2169 (1991) (stating that the experimental use exception "should be applied as it has been in the past: in a very restrictive manner, consistent with the purpose and function of the patent system.").

In Roche Products, Inc. v. Bolar Pharmaceutical Co., 733 F.2d 858 (Fed. Cir.), *cert. denied*, 469 U.S. 856 (1984), the Federal Circuit considered the experimental use defense for the first time. Here the defendant (Bolar) engaged in infringing acts prior to expiration of plaintiff's patent in order to facilitate FDA testing, so as to be ready to market the drug as soon as the patent expired. The Federal Circuit overruled the district court's finding of noninfringement, holding the experimental use exception did not include "the limited use of a patented drug for testing and investigation strictly related to FDA drug approval requirements. . . . " 733 F.2d at 861.

> Bolar's intended "experimental" use is solely for business reasons and not for amusement, to satisfy idle curiosity, or for strictly philosophical inquiry. . . . Bolar may intend to perform "experiments," but unlicensed experiments conducted with a view to the adaption of the patented invention to the experimentor's business are a violation of the rights of the patentee to exclude others from using his patented invention.

Id. at 863.

In 2002, the Federal Circuit construed *Roche* narrowly in the case of Madey v. Duke University, 307 F.3d 1351 (Fed. Cir. 2002). In 1989 Madey, a laser researcher, had moved himself and his research lab to Duke from Stanford University. During his time at Stanford, Madey had obtained sole ownership of two patents related to his research. Because of a dispute over the administration of his lab, Madey resigned from Duke in 1998. Other researchers continued to use the Madey-designed equipment in the lab, however — which led to an infringement suit by Madey. Duke defended on the grounds that the research was covered by the experimental use defense. The district court ruled for Duke on summary judgment, partly on the strength of this defense.

The Federal Circuit reversed and remanded, in an opinion giving narrow scope to the experimental use defense. After summarizing its cases in the area, the court turned to the heart of the district court's ruling, which was premised on the nonprofit, research orientation of Duke University:

> Our precedent clearly does not immunize use that is in any way commercial in nature. Similarly, our precedent does not immunize any conduct that is in keeping with the alleged infringer's legitimate business, regardless of commercial implications. For example, major research universities, such as Duke, often sanction and fund research projects with arguably no commercial application whatsoever. However, these projects unmistakably further the institution's legitimate business objectives, including educating and enlightening students and faculty participating in these projects. These projects also serve, for example, to increase the status of the institution and lure lucrative research grants, students and faculty.
>
> In short, regardless of whether a particular institution or entity is engaged in an endeavor for commercial gain, so long as the act is in furtherance of the alleged infringer's legitimate business and is not solely for amusement, to satisfy idle curiosity, or for strictly

philosophical inquiry, the act does not qualify for the very narrow and strictly limited experimental use defense. Moreover, the profit or non-profit status of the user is not determinative.

In the present case, the district court attached too great a weight to the non-profit, educational status of Duke, effectively suppressing the fact that Duke's acts appear to be in accordance with any reasonable interpretation of Duke's legitimate business objectives. On remand, the district court will have to significantly narrow and limit its conception of the experimental use defense. The correct focus should not be on the non-profit status of Duke but on the legitimate business Duke is involved in and whether or not the use was solely for amusement, to satisfy idle curiosity, or for strictly philosophical inquiry.

While *Madey* essentially eliminates the common law experimental use doctrine, there is also a limited statutory experimental use defense for FDA-related experimentation. 35 U.S.C. § 271(e)(1). In 2005, the Supreme Court interpreted that statute. Merck KGaA v. Integra Life Sciences I Ltd., _____ U.S. _____, 125 S.Ct. 2372 (2005). As the Court stated the issue:

> It is generally an act of patent infringement to "mak[e], us[e], offe[r] to sell, or sel[l] any patented invention . . . during the term of the patent therefor." [35 U.S.C.]§ 271(a). In 1984, Congress enacted an exemption to this general rule, see Drug Price Competition and Patent Term Restoration Act of 1984, § 202, 98 Stat. 1585, as amended, 35 U.S.C. § 271(e)(1), which provides:
>
> > "It shall not be an act of infringement to make, use, offer to sell, or sell within the United States or import into the United States a patented invention (other than a new animal drug or veterinary biological product (as those terms are used in the Federal Food, Drug, and Cosmetic Act and the Act of March 4, 1913) . . .) solely for uses reasonably related to the development and submission of information under a Federal law which regulates the manufacture, use, or sale of drugs. . . . "

125 S.Ct. at 2376-77.

Integra and the other initial plaintiffs owned five patents to "RGD peptides," small proteins that attach tightly to certain binding sites on the surfaces of blood vessel cells. Researchers working in cooperation with Merck discovered that RGD peptides could be used to block receptors active during the proliferation of blood vessel cells. Because this has the effect of slowing down blood vessel growth, it is an effective way to inhibit the growth of tumors. Merck and its research collaborators used RGD peptides — later found to infringe Integra's patents — in two different ways: (1) they tested them directly as tumor-inhibiting agents; and (2) they used them as "benchmarks" against which to measure other compounds they were developing for tumor inhibition.

At trial, Merck argued that its research was exempt from patent infringement for two reasons: the "common law" research exemption discussed above (e.g., Duke v. Madey); and the special statutory "safe harbor" of § 271(e)(1). A jury found that neither exemption applied, and awarded damages to Integra in the amount of $15 million. On appeal, the Federal Circuit affirmed. It rejected Merck's arguments based on both the common law exemption and the statutory safe harbor, concluding with respect to the latter that "the Scripps work sponsored by [petitioner] was not clinical testing to supply information to the FDA, but only general biomedical research to identify new pharmaceutical compounds."

Integra Life Sciences I, Ltd. v. Merck KGaA, 331 F.3d 860, 866 (Fed. Cir. 2003).

The Supreme Court reversed:

> [T]o construe § 271(e)(1), as the Court of Appeals did, not to protect research conducted on patented compounds for which an IND is not ultimately filed is effectively to limit assurance of exemption to the activities necessary to seek approval of a generic drug: One can know at the outset that a particular compound will be the subject of an eventual application to the FDA only if the active ingredient in the drug being tested is identical to that in a drug that has already been approved.
>
> The statutory text does not require such a result. Congress did not limit § 271(e)(1)'s safe harbor to the development of information for inclusion in a submission to the FDA; nor did it create an exemption applicable only to the research relevant to filing an ANDA for approval of a generic drug. Rather, it exempted from infringement all uses of patented compounds "reasonably related" to the process of developing information for submission under any federal law regulating the manufacture, use, or distribution of drugs. See Eli Lilly, 496 U.S., at 674. We decline to read the "reasonable relation" requirement so narrowly as to render § 271(e)(1)'s stated protection of activities leading to FDA approval for all drugs illusory. Properly construed, § 271(e)(1) leaves adequate space for experimentation and failure on the road to regulatory approval: At least where a drugmaker has a reasonable basis for believing that a patented compound may work, through a particular biological process, to produce a particular physiological effect, and uses the compound in research that, if successful, would be appropriate to include in a submission to the FDA, that use is "reasonably related" to the "development and submission of information under . . . Federal law."

125 S.Ct. at 2383.

While rejecting the narrow view of the safe harbor espoused by the Federal Circuit, the Supreme Court did not reach two other issues of widespread interest to researchers and patent lawyers: the scope of the "common law" exemption and the application of § 271(e)(1) to "research tools" such as genetic isolation and cloning techniques. Tools such as these are used primarily to facilitate ongoing research, so a liberal research exemption under § 271(e)(1) could have a deleterious effect on the development and deployment of such tools.

COMMENTS AND QUESTIONS

1. *Some Dissent Over* Merck. Defenders of the Federal Circuit opinion in *Merck* were particularly concerned that the opposite result might disproportionately harm the market for research tools. See for example Lawrence B. Ebert, In Favor of the Federal Circuit Position in Merck v. Integra, 87 J. Pat. & Trademark Off. Soc'y 321, 339-340 (2005):

> The difficulty with [a liberal approach to § 271(e)(1)] . . . is that it leaves the people in the shoes of the Integras of the world with no incentive to disclose their work to the public. If the work of the Integras can be used as a stepping stone for later commercializations by others, without payment of royalties to the Integras, there won't be any Integras. Although it may be that most issued patents are not commercialized or otherwise licensed, that does not mean we want to remove incentives for people to seek patents,

and, through that effort, to enrich the base of public knowledge. If players, especially larger ones, know that they can free-ride under the guise of an experimental use exemption, it will have a detrimental effect on the willingness of many to seek patents. Rather than expressing the argument in terms of infringement vs. an infringement "safe harbor," the better approach for cases involving the use of "stepping stone" patents may be to acknowledge infringement, where it exists, and to monetize the value of that infringement. This is true both for the present case (which is not a research tool case) as well as for cases that might involve research tools. As in the case of the "de minimis" infringement debate, infringement should be recognized and the "amount" of infringement fairly compensated in the analysis of damages.

2. *Is the Ivory Tower Just Another Office Building?* In an omitted footnote in the *Madey* case, the court said:

> Duke's patent and licensing policy may support its primary function as an educational institution. See Duke University Policy on Inventions, Patents, and Technology Transfer (1996), available at *http://www.ors.duke.edu/policies/patpol.htm* (last visited Oct. 3, 2002). Duke, however, like other major research institutions of higher learning, is not shy in pursuing an aggressive patent licensing program from which it derives a not insubstantial revenue stream. See id.

Do the aggressive patent acquisition, licensing, and enforcement activities of universities undermine the legitimacy of their claims to "special treatment" under the research exemption? For a general discussion, see Derek Bok, Universities in the Marketplace: The Commercialization of Higher Education 201 (2003) ("Universities may not yet be willing to trade all of their values for money, but they have proceeded much further down that road than they are generally willing to acknowledge.").

3. In an excellent article, Rebecca Eisenberg thoroughly reviews the history and rationale behind the doctrine and makes the following recommendations concerning its scope:

> (1) Research use of a patented invention to check the adequacy of the specification and the validity of the patent holder's claims about the invention should be exempt from infringement liability.
> (2) Research use of a patented invention with a primary or significant market among research users should not be exempt from infringement liability when the research user is an ordinary consumer of the patented invention.
> (3) A patent holder should not be entitled to enjoin the use of a patented invention in subsequent research in the field of the invention, which could potentially lead to improvements in the patented technology or to the development of alternative means of achieving the same purpose. However, it might be appropriate in some cases to award a reasonable royalty after the fact to be sure that the patent holder receives an adequate return on the initial investment in developing the patented invention.

Rebecca Eisenberg, Patents and the Progress of Science: Exclusive Rights and Experimental Use, 56 U. Chi. L. Rev. 1017 (1989).

How would you go about setting the reasonable royalty mentioned by Professor Eisenberg in a case where a subsequent improver infringed a patent in the course of developing a superior alternative that destroys the patentee's market?

2. Inequitable Conduct

Kingsdown Medical Consultants, Ltd. v. Hollister Inc.

United States Court of Appeals for the Federal Circuit
863 F.2d 867 (Fed. Cir. 1988), cert. denied, 490 U.S. 1067 (1989)

Kingsdown Medical Consultants, Ltd. and E. R. Squibb & Sons, Inc., (Kingsdown) appeal from a judgment of the United States District Court for the Northern District of Illinois, holding U.S. Patent No. 4,460,363 ('363) unenforceable because of inequitable conduct before the United States Patent and Trademark Office (PTO). We reverse and remand.

Background

Kingsdown sued Hollister Incorporated (Hollister) for infringement of . . . Kingsdown's '363 patent. The district court held the patent unenforceable because of Kingsdown's conduct in respect of claim 9 and reached no other issue.

The invention claimed in the '363 patent is a two-piece ostomy appliance for use by patients with openings in their abdominal walls for release of waste.

The two pieces of the appliance are a pad and a detachable pouch. The pad is secured to the patient's body encircling the abdominal wall opening. Matching coupling rings are attached to the pad and to the pouch. When engaged, the rings provide a water tight seal. Disengaging the rings allows for removal of the pouch.

The Prosecution History

Kingsdown filed its original patent application in February 1978. The '363 patent issued July 17, 1984. The intervening period of more than six-and-a-half years saw a complex prosecution, involving the submission, rejection, amendment, re-numbering, etc., of 118 claims, a continuation application, an appeal, a petition to make special, and citation and discussion of 44 references. . . .

[Certain claims in Kingsdown's original application were allowed by the examiner; others were rejected. Kingsdown appealed the rejection of the latter claims.] While Kingsdown's appeal of other rejected claims was pending, Kingsdown's patent attorney saw a two-piece ostomy appliance manufactured by Hollister. Kingsdown engaged an outside counsel to file a continuation application and withdrew the appeal.

[The continuation application] indicated, incorrectly, that claim 43 in the continuation application corresponded to allowed claim 50 in the parent application. Claim 43 actually corresponded to [one of the other claims] that had been rejected for indefiniteness under § 112. Claim 43 was [allowed by the Examiner and] renumbered as the present claim 9 in the '363 patent. . . .

Having examined the prosecution history, the district court found that the examiner could have relied on the representation that claim 43 corresponded to allowable claim 50 and rejected Kingsdown's suggestion that the examiner must have made an independent examination of claim 43. . . .

The district court stated that the narrower language of amended claim 50 gave Hollister, [the defendant accused of infringement here], a possible defense, i.e., that Hollister's coupling member does not encircle the intersection of the aperture and the pad surface because it has an intervening "floating flange" member. The court inferred motive to deceive the PTO because Kingsdown's patent attorney viewed the Hollister appliance after he had amended claim 50 and before the continuation application was filed. The court expressly declined to make any finding on whether the accused device would or would not infringe any claims, but stated that Kingsdown's patent attorney must have perceived that Hollister would have a defense against infringement of the amended version of claim 50 that it would not have against the unamended version.

Issue

Whether the district court's finding of intent to deceive was clearly erroneous, rendering its determination that inequitable conduct occurred an abuse of discretion.[5]

Opinion

We confront a case of first impression, in which inequitable conduct has been held to reside in an incorrect inclusion in a continuation application of a claim that contained allowable subject matter, but had been rejected as indefinite in the parent application.

 Inequitable conduct resides in failure to disclose material information, or submission of false material information, with an intent to deceive, and those two elements, materiality and intent, must be proven by clear and convincing evidence. J. P. Stevens & Co., Inc. v. Lex Tex Ltd., Inc., 747 F.2d 1553, 1559 (Fed. Cir. 1984), *cert. denied*, 474 U.S. 822 (1985). The findings on materiality and intent are subject to the clearly erroneous standard of Rule 52(a) Fed. R. Civ. P. and are not to be disturbed unless this court has a definite and firm conviction that a mistake has been committed. J. P. Stevens, 747 F.2d at 1562.

"To be guilty of inequitable conduct, one must have intended to act inequitably." FMC Corp. v. Manitowoc Co., Inc., 835 F.2d 1411, 1415 (Fed. Cir. 1987). Kingsdown's attorney testified that he was not aware of the error until Hollister mentioned it in March 1987, and the experts for both parties testified that they saw no evidence of deceptive intent. As above indicated, the district court's finding of Kingsdown's intent to mislead is based on the alternative grounds of: (a) gross negligence; and (b) acts indicating an intent to deceive. Neither ground, however, supports a finding of intent in this case.

a. Negligence

The district court inferred intent based on what it perceived to be Kingsdown's gross negligence. Whether the intent element of inequitable conduct is present

---

5. Because of our decision on intent, it is unnecessary to discuss materiality. Allen Archery, Inc. v. Browning Mfg. Co., 819 F.2d 1087, 1094 (Fed. Cir. 1987).

cannot always be inferred from a pattern of conduct that may be described as gross negligence. That conduct must be sufficient to require a finding of deceitful intent in the light of all the circumstances. We are not convinced that deceitful intent was present in Kingsdown's negligent filing of its continuation application or, in fact, that its conduct even rises to a level that would warrant the description "gross negligence."

It is well to be reminded of what actually occurred in this case — a ministerial act involving two claims, which, because both claims contained allowable subject matter, did not result in the patenting of anything anticipated or rendered obvious by anything in the prior art and thus took nothing from the public domain. In preparing and filing the continuation application, a newly-hired counsel for Kingsdown had two versions of "claim 50" in the parent application, an unamended rejected version and an amended allowed version. As is common, counsel renumbered and transferred into the continuation all (here, 22) claims "previously allowed." In filing its claim 43, it copied the "wrong," i.e., the rejected, version of claim 50. That error led to the incorrect listing of claim 43 as corresponding to allowed claim 50 and to incorporation of claim 43 as claim 9 in the patent. In approving the continuation for filing, Kingsdown's regular attorney did not, as the district court said, "catch" the mistake.

In view of the relative ease with which others also overlooked the differences in the claims, Kingsdown's failure to notice that claim 43 did not correspond to the amended and allowed version of claim 50 is insufficient to warrant a finding of an intent to deceive the PTO. Undisputed facts indicating that relative ease are: (1) the similarity in language of the two claims; (2) the use of the same claim number, 50, for the amended and unamended claims; (3) the multiplicity of claims involved in the prosecution of both applications; (4) the examiner's failure to reject claims using "encircled" in the parent application's first and second office actions, making its presence in claim 43 something less than a glaring error; (5) the two-year interval between the rejection/amendment of claim 50 and the filing of the continuation; (6) failure of the examiner to reject claim 43 under §112 or to notice the differences between claim 43 and amended claim 50 during what must be presumed, absent contrary evidence, to have been an examination of the continuation; and (7) the failure of Hollister to notice the lack of correspondence between claim 43 and the amended version of claim 50 during three years of discovery and until after it had carefully and critically reviewed the file history 10 to 15 times with an eye toward litigation. That Kingsdown did not notice its mistake during more than one opportunity of doing so, does not in this case, and in view of Hollister's frequent and focused opportunities, establish that Kingsdown intended to deceive the PTO.

... The district court correctly noted that an examiner has a right to expect candor from counsel. Its indication that examiners "must" rely on counsel's candor would be applicable, however, only when the examiner does not have the involved documents or information before him, as the examiner did here. Blind reliance on presumed candor would render examination unnecessary, and nothing in the statute or Manual of Patent Examining Procedure would justify reliance on counsel's candor as a substitute for an examiner's duty to examine the claims.

Thus the first basis for the district court's finding of deceitful intent (what it viewed as "gross negligence") cannot stand.

b. Acts

The district court also based its finding of deceitful intent on the separate and alternative inferences it drew from Kingsdown's acts in viewing the Hollister device, in desiring to obtain a patent that would "cover" that device, and in failing to disclaim or reissue after Hollister charged it with inequitable conduct. The district court limited its analysis here to claim 9 and amended claim 50.

It should be made clear at the outset of the present discussion that there is nothing improper, illegal or inequitable in filing a patent application for the purpose of obtaining a right to exclude a known competitor's product from the market; nor is it in any manner improper to amend or insert claims intended to cover a competitor's product the applicant's attorney has learned about during the prosecution of a patent application. Any such amendment or insertion must comply with all statutes and regulations, of course, but, if it does, its genesis in the marketplace is simply irrelevant and cannot of itself evidence deceitful intent....

Faced with Hollister's assertion that an experienced patent attorney would knowingly and intentionally transfer into a continuing application a claim earlier rejected for indefiniteness, without rearguing that the claim was not indefinite, the district court stated that "how an experienced patent attorney could allow such conduct to take place" gave it "the greatest difficulty." A knowing failure to disclose and knowingly false statements are always difficult to understand. However, a transfer of numerous claims en masse from a parent to a continuing application, as the district court stated, is a ministerial act. As such, it is more vulnerable to errors which by definition result from inattention, and is less likely to result from the scienter involved in the more egregious acts of omission and commission that have been seen as reflecting the deceitful intent element of inequitable conduct in our cases....

Resolution of Conflicting Precedent

"Gross Negligence" and the Intent Element of Inequitable Conduct

Some of our opinions have suggested that a finding of gross negligence compels a finding of an intent to deceive. Others have indicated that gross negligence alone does not mandate a finding of intent to deceive.

"Gross negligence" has been used as a label for various patterns of conduct. It is definable, however, only in terms of a particular act or acts viewed in light of all the circumstances. We adopt the view that a finding that particular conduct amounts to "gross negligence" does not of itself justify an inference of intent to deceive; the involved conduct, viewed in light of all the evidence, including evidence indicative of good faith, must indicate sufficient culpability to require a finding of intent to deceive....

Effect of Inequitable Conduct

When a court has finally determined that inequitable conduct occurred in relation to one or more claims during prosecution of the patent application, the entire patent is rendered unenforceable. We, in banc, reaffirm that rule as set forth in J. P. Stevens & Co. v. Lex Tex Ltd.

Conclusion

Having determined that the district court's finding of intent is clearly erroneous, the panel reverses the judgment based on a conclusion of inequitable conduct before the PTO and remands the case for such further proceedings as the district court may deem appropriate.

COMMENTS AND QUESTIONS

1. Why is the penalty for inequitable conduct so harsh—rendering the patent completely unenforceable? Who has more information about the inventor's work and the precise prior art it relates to—the examiner or the inventor? Who has more to gain from the issuance (or rejection) of a patent?

2. Many inequitable conduct cases involve suppressing a key prior art reference. See, e.g., *J. P. Stevens,* cited in the *Kingsdown* case. It is important to note, however, that only prior art that was *known* by the inventor or the patent lawyer gives rise to a duty of disclosure. See, e.g., Elk Corp. of Dallas v. GAF Bldg. Materials Corp., 168 F.3d 28 (Fed. Cir. 1999). This rule is embodied not only in the doctrine of inequitable conduct in the courts, but also in Rule 56 of the rules of the patent practice, pertaining to the duty of patent applicants during prosecution. See 37 C.F.R. § 1.56 (2002), at 1.56(a) ("Each individual associated with the filing and prosecution of a patent application has a duty of candor and good faith in dealing with the [patent] office, which includes a duty to disclose to the office all information *known to that individual* to be material to patentability...") (emphasis added). Note that this means that patentees do *not* have to conduct a prior art search. But if they know about prior art, it must be disclosed. Do you think it should matter whether (1) the reference was in a class or group that the Examiner *should* have searched?; (2) that it was disclosed, but in a long "string cite" of less relevant prior art?; (3) that the patentee didn't know of the key reference, but *should have?*

The Federal Circuit continues to find inequitable conduct under a wide variety of factual situations. See, e.g., Purdue Pharma L.P. v. Endo Pharms., Inc., 410 F.3d 690 (Fed. Cir. 2005) (finding patent unenforceable because applicant characterized "prophetic" examples as actual experimental results); Dayco Products, Inc. v. Total Containment, Inc., 329 F.3d 1358 (Fed. Cir. 2003) (noting in dicta that failure to disclose a co-pending but related application to an examiner could be inequitable conduct; no inequitable conduct found in the case, however); Pharmacia Corp. v. Par Pharmaceuticals, Inc., 417 F.3d 1369 (Fed. Cir. 2005) (patent applicant's submission of inaccurate and misleading declaration in order to overcome obviousness objection was inequitable conduct, especially given that the declaration conflicted with several undisclosed references, including one co-authored by the inventor); Frazier v. Roessel Cine Photo Tech Inc., 417 F.3d 1230 (Fed. Cir. 2005) (inequitable conduct to submit video purportedly created with patented camera lens that in fact was not).

3. The practice of "submarine patenting"—deliberately keeping a secret patent application pending for years in the PTO, only to spring it on an unsuspecting

industry—has periodically been challenged under a variety of patent law doctrines. Submarine patenting has survived charges of inequitable conduct in some notable district court cases. See Ford Motor Co. v. Lemelson, 42 U.S.P.Q.2d 1706 (D. Nev. 1997); Advanced Cardiovascular Sys. v. Medtronic Inc., 1996 WL 467293 (N.D. Cal. 1996) (both rejecting inequitable conduct claims). But one court has held that allegations of submarine patenting may state an *antitrust* claim against the patentee. See Discovision Assoc. v. Disc Mfg. Inc., 42 U.S.P.Q.2d 1749 (D. Del. 1997).

In Symbol Technologies v. Lemelson, 277 F.3d 1361 (Fed. Cir. 2002), the Federal Circuit adopted a theory of "continuing application laches" under which excessive delay in prosecuting a patent could preclude its enforcement against an unsuspecting industry. *Symbol Technologies* involved notorious submarine patentee Jerome Lemelson, who had amassed well over 1,000 paper patents and had kept some applications pending for over 40 years. The Federal Circuit concluded that unreasonable and unexplained delay in prosecuting a patent could be grounds to bar enforcement of that patent. For subsequent opinions after remand, see 301 F.Supp.2d 1147 (D. Nev. 2004), aff'd 422 F.3d 1378 (Fed. Cir. 2005). Judge Newman, dissenting in the original opinion, argued that because Lemelson had followed the rules for filing continuation applications set out in the patent statute, he should not be barred from enforcing his long-pending patents. See also In re Bogese, 303 F.3d 1362 (Fed. Cir. 2002) (upholding examiner's final rejection of patent due to applicant's unreasonable delay). On the problem of successive continuations in patent applications, see Mark A. Lemley and Kimberly Moore, Ending Abuse of Patent Continuations, 84 B.U. L. Rev. 63 (2004).

Does *Symbol Technologies* solve the problem of submarine patents by effectively discouraging strategic refiling? Would a bright-line rule serve this purpose better than a reasonableness standard? How can the courts distinguish between delay caused by the patentee and delay that is the fault of the PTO?

3. Patent Misuse

Patent misuse is a judicially created doctrine that bars patentees from enforcing their patent against infringers when they have "misused" the patent. It was first adopted by the Supreme Court in 1917.

Motion Picture Patents Company v. Universal Film Manufacturing Company et al.
Supreme Court of the United States
243 U.S. 502 (1917)

Mr. Justice CLARKE delivered the opinion of the court.

In this suit relief is sought against three defendant corporations as joint infringers of claim number seven of United States letters patent No. 707,934 granted to Woodville Latham, assignor, on August 26, 1902, for improvements in Projecting-Kinetoscopes. It is sufficient description of the patent to say that it covers a part of the mechanism used in motion picture exhibiting machines for feeding a film through the machine with a regular, uniform and accurate movement and so as not to expose the film to excessive strain or wear.

The defendants in a joint answer do not dispute the title of the plaintiff to the patent but they deny the validity of it, deny infringement, and claim an implied license to use the patented machine.

Evidence which is undisputed shows that the plaintiff on June 20, 1912, in a paper styled "License Agreement" granted to The Precision Machine Company a right and license to manufacture and sell machines embodying the inventions described and claimed in the patent in suit, and in other patents, throughout the United States, its territories and possessions. This agreement contains a covenant on the part of the grantee that every machine sold by it, except those for export, shall be sold "under the restriction and condition that such exhibiting or projecting machines shall be used solely for exhibiting or projecting motion pictures containing the inventions of reissued letters patent No. 12,192, leased by a licensee of the licensor while it owns said patents, and upon other terms to be fixed by the licensor and complied with by the user while the said machine is in use and while the licensor owns said patents (which other terms shall only be the payment of a royalty or rental to the licensor while in use)." . . .

The agreement further provides that the grantee shall not sell any machine at less than the plaintiff's list price, except to jobbers and others for purposes of resale and that it will require such jobbers and others to sell at not less than plaintiff's list price. . . .

It was admitted at the bar that 40,000 of the plaintiff's machines are now in use in this country and that the mechanism covered by the patent in suit is the only one with which motion picture films can be used successfully.

This state of facts presents two questions for decision:

. . .

Second. May the assignee of a patent, which has licensed another to make and sell the machine covered by it, by a mere notice attached to such machine, limit the use of it by the purchaser or by the purchaser's lessee to terms not stated in the notice but which are to be fixed, after sale, by such assignee in its discretion? . . .

Since Pennock v. Dialogue, 2 Pet. 1, was decided in 1829 this court has consistently held that the primary purpose of our patent laws is not the creation of private fortunes for the owners of patents but is "to promote the progress of science and useful arts" (Constitution, Art. I, § 8). . . .

Plainly, this language of the statute and the established rules to which we have referred restrict the patent granted on a machine, such as we have in this case, to the mechanism described in the patent as necessary to produce the described results. It is not concerned with and has nothing to do with the materials with which or on which the machine operates. The grant is of the exclusive right to use the mechanism to produce the result with any appropriate material, and the materials with which the machine is operated are no part of the patented machine or of the combination which produces the patented result. The difference is clear and vital between the exclusive right to use the machine which the law gives to the inventor and the right to use it exclusively with prescribed materials to which such a license notice as we have here seeks to restrict it. . . . Both in form and in substance the notice attempts a restriction upon the use of the supplies only and it cannot with any regard to propriety in the use of language be termed a restriction upon the use of the machine itself.

Whatever right the owner may have to control by restriction the materials to be used in operating the machine must be derived through the general law from the ownership of the property in the machine and it cannot be derived from or protected by the patent law....

This construction gives to the inventor the exclusive use of just what his inventive genius has discovered. It is all that the statute provides shall be given to him and it is all that he should receive, for it is the fair as well as the statutory measure of his reward for his contribution to the public stock of knowledge. If his discovery is an important one his reward under such a construction of the law will be large, as experience has abundantly proved, and if it be unimportant he should not be permitted by legal devices to impose an unjust charge upon the public in return for the use of it. For more than a century this plain meaning of the statute was accepted as its technical meaning, and that it afforded ample incentive to exertion by inventive genius is proved by the fact that under it the greatest inventions of our time, teeming with inventions, were made....

The construction of the patent law which justifies as valid the restriction of patented machines, by notice, to use with unpatented supplies necessary in the operation of them, but which are no part of them, is believed to have originated in Heaton-Peninsular Button-Fastener Co. v. Eureka Specialty Co., 77 Fed. Rep. 288 (which has come to be widely referred to as the *Button-Fastener Case*), decided by the Circuit Court of Appeals of the Sixth Circuit in 1896. In this case the court, recognizing the pioneer character of the decision it was rendering, speaks of the "novel restrictions" which it is considering and says that it is called upon "to mark another boundary line around the patentee's monopoly, which will debar him from engrossing the market for an article not the subject of a patent," which it declined to do.

This decision proceeds upon the argument that, since the patentee may withhold his patent altogether from public use he must logically and necessarily be permitted to impose any conditions which he chooses upon any use which he may allow of it. The defect in this thinking springs from the substituting of inference and argument for the language of the statute and from failure to distinguish between the rights which are given to the inventor by the patent law and which he may assert against all the world through an infringement proceeding and rights which he may create for himself by private contract which, however, are subject to the rules of general as distinguished from those of the patent law. While it is true that under the statutes as they were (and now are) a patentee might withhold his patented machine from public use, yet if he consented to use it himself or through others, such use immediately fell within the terms of the statute and as we have seen he is thereby restricted to the use of the invention as it is described in the claims of his patent and not as it may be expanded by limitations as to materials and supplies necessary to the operation of it imposed by mere notice to the public.

...The perfect instrument of favoritism and oppression which such a system of doing business, if valid, would put into the control of the owner of such a patent should make courts astute, if need be, to defeat its operation. If these restrictions were sustained plainly the plaintiff might, for its own profit or that of its favorites, by the obviously simple expedient of varying its royalty charge, ruin anyone unfortunate enough to be dependent upon its confessedly important improvements for the doing of business.

...[F]ollowing the decision of the *Button-Fastener Case*, it was widely contended as obviously sound, that the right existed in the owner of a patent to fix a price at which the patented article might be sold and resold under penalty of patent infringement. But this court, when the question came before it in Bauer v. O'Donnell, 229

U.S. 1, . . . decided that the owner of a patent is not authorized by either the letter or the purpose of the law to fix, by notice, the price at which a patented article must be sold after the first sale of it, declaring that the right to vend is exhausted by a single, unconditional sale, the article sold being thereby carried outside the monopoly of the patent law and rendered free of every restriction which the vendor may attempt to put upon it. The statutory authority to grant the exclusive right to "use" a patented machine is not greater, indeed it is precisely the same, as the authority to grant the exclusive right to "vend," and, looking to that authority, for the reasons stated in this opinion we are convinced that the exclusive right granted in every patent must be limited to the invention described in the claims of the patent and that it is not competent for the owner of a patent by notice attached to its machine to, in effect, extend the scope of its patent monopoly by restricting the use of it to materials necessary in its operation but which are no part of the patented invention, or to send its machines forth into the channels of trade of the country subject to conditions as to use or royalty to be paid to be imposed thereafter at the discretion of such patent owner. The patent law furnishes no warrant for such a practice and the cost, inconvenience and annoyance to the public which the opposite conclusion would occasion forbid it.

It is argued as a merit of this system of sale under a license notice that the public is benefitted by the sale of the machine at what is practically its cost and by the fact that the owner of the patent makes its entire profit from the sale of the supplies with which it is operated. This fact, if it be a fact, instead of commending, is the clearest possible condemnation of, the practice adopted, for it proves that under color of its patent the owner intends to and does derive its profit, not from the invention on which the law gives it a monopoly but from the unpatented supplies with which it is used and which are wholly without the scope of the patent monopoly, thus in effect extending the power to the owner of the patent to fix the price to the public of the unpatented supplies as effectively as he may fix the price on the patented machine. . . .

Coming now to the terms of the notice attached to the machine sold to the Seventy-second Street Amusement Company under the license of the plaintiff and to the first question as we have stated it.

This notice first provides that the machine, which was sold to and paid for by the Amusement Company may be used only with moving picture films containing the invention of reissued patent No. 12,192, so long as the plaintiff continues to own this reissued patent.

Such a restriction is invalid because such a film is obviously not any part of the invention of the patent in suit; because it is an attempt, without statutory warrant, to continue the patent monopoly in this particular character of film after it has expired, and because to enforce it would be to create a monopoly in the manufacture and use of moving picture films, wholly outside of the patent in suit and of the patent law as we have interpreted it. . . .

A restriction which would give to the plaintiff such a potential power for evil over an industry which must be recognized as an important element in the amusement life of the nation, under the conclusions we have stated in this opinion, is plainly void, because wholly without the scope and purpose of our patent laws and because, if sustained, it would be gravely injurious to that public interest, which we have seen is more a favorite of the law than is the promotion of private fortunes. . . .

Mr. Justice HOLMES, dissenting.

I suppose that a patentee has no less property in his patented machine than any other owner, and that in addition to keeping the machine to himself the patent gives him the further right to forbid the rest of the world from making others like it. In short, for whatever motive, he may keep his device wholly out of use. Continental Paper Bag Co. v. Eastern Paper Bag Co., 210 U.S. 405, 422. So much being undisputed, I cannot understand why he may not keep it out of use unless the licensee, or, for the matter of that, the buyer, will use some unpatented thing in connection with it. Generally speaking the measure of a condition is the consequence of a breach, and if that consequence is one that the owner may impose unconditionally, he may impose it conditionally upon a certain event. Ashley v. Ryan, 153 U.S. 436, 443. Lloyd v. Dollison, 194 U.S. 445, 449....

No doubt this principle might be limited or excluded in cases where the condition tends to bring about a state of things that there is a predominant public interest to prevent. But there is no predominant public interest to prevent a patented tea pot or film feeder from being kept from the public, because, as I have said, the patentee may keep them tied up at will while his patent lasts. Neither is there any such interest to prevent the purchase of the tea or films, that is made the condition of the use of the machine. The supposed contravention of public interest sometimes is stated as an attempt to extend the patent law to unpatented articles, which of course it is not, and more accurately as a possible domination to be established by such means. But the domination is one only to the extent of the desire for the tea pot or film feeder, and if the owner prefers to keep the pot or the feeder unless you will buy his tea or films, I cannot see in allowing him the right to do so anything more than an ordinary incident of ownership, or at most, a consequence of the *Paper Bag Case,* on which, as it seems to me, this case ought to turn. See Grant v. Raymond, 6 Pet. 218, 242....

COMMENTS AND QUESTIONS

1. The asserted justification for the patent misuse doctrine lies in the public policy underpinning the patent laws. It is no coincidence, however, that the doctrine first developed around 1900, shortly after the antitrust laws were passed. Patent misuse and antitrust are strikingly similar in the sorts of conduct they prohibit. For example, in *Motion Picture,* the conduct at issue was (among other things) resale price maintenance, which is per se illegal under the antitrust laws. Patent misuse, as originally conceived, was broader than the antitrust laws, however. Neither market power nor actual effect on competition need be proven to show patent misuse, as they would to prove an antitrust violation. We will return to the scope of the doctrine vis-à-vis the antitrust laws shortly.

2. The Court continually expanded the scope of the patent misuse doctrine in the decades that followed *Motion Picture.* That expansion culminated in the cases of Mercoid Corp. v. Mid-Continent Inv. Co., 320 U.S. 661 (1944) (*Mercoid I*) and Carbice Corp. v. American Patents Dev. Corp., 283 U.S. 27 (1931). *Mercoid* held that it was patent misuse to tie a patented product to a nonstaple product.[48] *Carbice* held that a showing of patent misuse (which does not require proof of market power or impact) was prima facie evidence of an antitrust violation.

Both of these rules were short-lived. Congress amended the patent laws in 1952 by adding 35 U.S.C. §271(d)(1)-(3).

---

48. A staple product is one which has an existing market, beyond use with the patent. A non-staple product has no substantial commercial use except in connection with the patent.

35 U.S.C. § 271. Infringement of Patent

. . .

(d) No patent owner otherwise entitled to relief for infringement or contributory infringement of a patent shall be denied relief or deemed guilty of misuse or illegal extension of the patent right by reason of his having done one or more of the following: (1) derived revenue from acts which if performed by another without his consent would constitute contributory infringement of the patent; (2) licensed or authorized another to perform acts which if performed without his consent would constitute contributory infringement of the patent; (3) sought to enforce his patent rights against infringement or contributory infringement; . . .

The import of this statutory limit on the judicially created doctrine of misuse was described in detail by the Supreme Court in the case of Dawson Chem. Co. v. Rohm & Haas Co., 448 U.S. 176 (1980). The Court stated:

Section 271(c) identifies the basic dividing line between contributory infringement and patent misuse. It adopts a restrictive definition of contributory infringement that distinguishes between staple and nonstaple articles of commerce. It also defines the class of nonstaple items narrowly. In essence, this provision places materials like the dry ice of the Carbice case outside the scope of the contributory infringement doctrine. As a result, it is no longer necessary to resort to the doctrine of patent misuse in order to deny patentees control over staple goods used in their inventions.

The limitations on contributory infringement written into §271(c) are counterbalanced by limitations on patent misuse in §271(d). Three species of conduct by patentees are expressly excluded from characterization as misuse. First, the patentee may "deriv[e] revenue" from acts that "would constitute contributory infringement" if "performed by another without his consent." This provision clearly signifies that a patentee may make and sell nonstaple goods used in connection with his invention. Second, the patentee may "licens[e] or authoriz[e] another to perform acts" which without such authorization would constitute contributory infringement. This provision's use in the disjunctive of the term "authoriz[e]" suggests that more than explicit licensing agreements is contemplated. Finally, the patentee may "enforce his patent rights against . . . contributory infringement." This provision plainly means that the patentee may bring suit without fear that his doing so will be regarded as an unlawful attempt to suppress competition. The statute explicitly states that a patentee may do "one or more" of these permitted acts, and it does not state that he must do any of them.

In our view, the provisions of §271(d) effectively confer upon the patentee, as a lawful adjunct of his patent rights, a limited power to exclude others from competition in nonstaple goods. A patentee may sell a nonstaple article himself while enjoining others from marketing that same good without his authorization. By doing so, he is able to eliminate competitors and thereby to control the market for that product. Moreover, his power to demand royalties from others for the privilege of selling the nonstaple item itself implies that the patentee may control the market for the nonstaple good; otherwise, his "right" to sell licenses for the marketing of the nonstaple good would be meaningless, since no one would be willing to pay him for a superfluous authorization. See Note, 70 Yale. L.J. 649, 659 (1961).

3. Much of the debate over the wisdom of the patent misuse doctrine parallels similar debates over the wisdom of particular antitrust rules. We discuss these rules in the context of intellectual property in more detail in Chapter 8.

4. The patent misuse doctrine is closely connected to the idea of *exhaustion* of patent rights by a first sale in commerce. The exhaustion doctrine provides that once a patent owner has sold or licensed goods that embody the patented product, it cannot exercise further control over those goods. The exhaustion doctrine is venerable and universally accepted.

Recent Federal Circuit precedent casts serious doubt on the modern efficacy of the exhaustion doctrine, however. First, in Mallinckrodt v. MediPart, 976 F.2d 700 (Fed. Cir. 1992), the court was asked to review a district court's grant of summary judgment in favor of an accused infringer. The patentee had attempted to prevent buyers from reusing a patented medical device by putting a "label license" on the device indicating that it was authorized for a single use only. The district court did not rule on the enforceability of the "label license"; it simply ruled that such a license was *per se* unenforceable under the Patent Act. The Federal Circuit reversed, holding that the restriction did not *per se* violate either the patent misuse doctrine or antitrust law. 976 F.2d at 701. The Federal Circuit extended *Mallinckrodt* in B. Braun Medical, Inc. v. Abbott Labs., Inc., 124 F.3d 1419 (Fed. Cir. 1997), but also noted the limitations that patent misuse places on such restrictive licenses:

> The resolution of this issue is governed by our precedent in *Mallinckrodt*. . . . In that case, we canvassed precedent concerning the legality of restrictions placed upon the post-sale use of patented goods. As a general matter, we explained that an unconditional sale of a patented device exhausts the patentee's right to control the purchaser's use of the device thereafter. . . . The theory behind this rule is that in such a transaction, the patentee has bargained for, and received, an amount equal to the full value of the goods. This exhaustion doctrine, however, does not apply to an expressly conditional sale or license. In such a transaction, it is more reasonable to infer that the parties negotiated a price that reflects only the value of the "use" rights conferred by the patentee. As a result, express conditions accompanying the sale or license of a patented product are generally upheld. . . . Accordingly, conditions that violate some law or equitable consideration are unenforceable. On the other hand, violation of valid conditions entitles the patentee to a remedy for either patent infringement or breach of contract. . . . This, then, is the general framework.
>
> In *Mallinckrodt*, we also outlined the framework for evaluating whether an express condition on the post-sale use of a patented product constitutes patent misuse. The patent misuse doctrine, born from the equitable doctrine of unclean hands, is a method of limiting abuse of patent rights separate from the antitrust laws. The key inquiry under this fact-intensive doctrine is whether, by imposing the condition, the patentee has "impermissibly broadened the 'physical or temporal scope' of the patent grant with anticompetitive effect." . . . Two common examples of such impermissible broadening are using a patent which enjoys market power in the relevant market, see 35 U.S.C. § 271(d)(5) (1994), to restrain competition in an unpatented product or employing the patent beyond its 17-year term. In contrast, field of use restrictions (such as those at issue in the present case) are generally upheld, see *General Talking Pictures*, 305 U.S. at 127, and any anticompetitive effects they may cause are reviewed in accordance with the rule of reason. See *Mallinckrodt*, 976 F.2d at 708.

Id. at 1426-1427. Is this consistent with the *Motion Picture* patents case? Is it good policy? On the growing use of restrictions on personal property, see Glen O. Robinson, Personal Property Servitudes, 71 U. Chi. L. Rev. 1449 (2004).

Note that limited-use contracts run counter to a bedrock assumption lying behind the first-sale doctrine in all areas of intellectual property law: that at least some of the property rights held by an intellectual property owner are extinguished when an item embodying those rights is sold. To the extent that *Mallinckrodt* and *Braun* assume a different status for tangible items incorporating an intellectual property right, they undercut this basic assumption. See, e.g., Richard H. Stern, Post-Sale Patent Restrictions After *Mallinckrodt*—An Idea in Search of Definition, 5 Alb. L.J. Sci. & Tech. 1 (1994).

For a discussion of the relationship between price discrimination and the exhaustion doctrine, albeit in the copyright context, see Michael Meurer, Price Discrimination, Personal Use and Piracy: Copyright Protection of Digital Works, 45 Buff. L. Rev. 845 (1997).

PROBLEM

Problem 3-12. Big Soya, Inc. is a large agricultural chemicals firm that sells a widely-used herbicide named WEEDAWAY. Beginning in the 1980s, Big Soya began investing heavily in agricultural biotechnology research, with an eye toward developing advanced herbicides and related products. One of the problems that Big Soya faces, like all herbicide firms, is "selectivity": designing products that kill weeds, but do little or no damage to farm crops. In the mid-1990s, Big Soya researchers hit on the idea of developing a genetically engineered strain of soybeans that is particularly immune to the effects of the popular WEEDAWAY herbicide. By the late 1990s, they had perfected—and patented—special strains of soybean seeds with just these properties, which they called "WEEDAWAY-PROOF" seeds. As part of its strategy to protect the market for its proprietary seeds, Big Soya asks farmers using the seeds to sign a contract restricting the farmers' activities in a number of ways. Specifically, the farmers agree: (1) To use the seed containing Big Soya gene technologies for planting a commercial crop only in a single season; (2) Not to supply any of the seed to any other person or entity for planting, and to not save any crop produced from this seed for replanting, or supply saved seed to anyone for replanting; (3) Not to use the seed or provide it to anyone for crop breeding, research, generation of herbicide registration data, or seed production. The farmers' agreement also contains a clause specifying damages in the event of breach by the farmer:

> In the event that the Grower saves, supplies, sells or acquires seed for replant in violation of this Agreement and license restriction, in addition to other remedies available to the technology provider(s), the Grower agrees that damages will include a claim for liquidated damages, which will be based on 120 times the applicable Technology Fee.

In a lawsuit brought against a farmer, Betty Overall, Big Soya alleges that Betty saved seeds from a batch of WEEDAWAY-PROOF soybeans, and used them again the next season. Betty argues that she bought the seeds outright, despite having signed the agreement (which she claims neither to have read nor understood); that farmers have traditionally had the right to plant "saved seed" from the previous year's crops, and that the Plant Variety Protection Act (see below) even enshrines that tradition in a "saved seed" exemption. Betty argues that Big Soya's farmers' agreement and the assertion of its patents against farmers such as

> Betty is patent misuse. Do you agree? How would you argue the case, based on *Motion Picture* and the other materials in this chapter?

Note on the Scope of the Patent Misuse Doctrine

Congress has shown continued dissatisfaction with the scope of the patent misuse doctrine. In 1988, it passed the Patent Misuse Reform Act (PMRA), which provides:

35 U.S.C. § 271. Infringement of Patent

. . .

(d) No patent owner otherwise entitled to relief for infringement or contributory infringement of a patent shall be denied relief or deemed guilty of misuse or illegal extension of the patent right by reason of his having done one or more of the following: . . .

(4) refused to license or use any rights to the patent; or (5) conditioned the license of any rights to the patent or the sale of the patented product on the acquisition of a license to rights in another patent or purchase of a separate product, unless, in view of the circumstances, the patent owner has market power in the relevant market for the patent or patented product on which the license or sale is conditioned.

There is obviously a great deal of overlap between the policies of the patent misuse doctrine and those underlying the antitrust laws. A similar overlap existed in scope between the two doctrines, even before (and certainly after) passage of the Patent Misuse Reform Act. In fact, Judge Posner argues that patent misuse and antitrust are actually coextensive — that an act cannot be misuse unless it also violates the antitrust laws. See USM Corp. v. SPS Tech., 694 F.2d 505 (7th Cir. 1982), *cert. denied*, 462 U.S. 107 (1983):

> The [patent misuse] doctrine arose before there was any significant body of federal antitrust law, and reached maturity long before that law (a product very largely of free interpretation of unclear statutory language) attained its present broad scope. Since the antitrust laws as currently interpreted reach every practice that could impair competition substantially, it is not easy to define a separate role for a doctrine also designed to prevent an anticompetitive practice — the abuse of a patent monopoly. One possibility is that the doctrine of patent misuse, unlike antitrust law, condemns any patent licensing practice that is even trivially anticompetitive, at least if it has no socially beneficial effects. . . .
>
> If misuse claims are not tested by conventional antitrust principles, by what principles shall they be tested? Our law is not rich in alternative concepts of monopolistic abuse; and it is rather late in the day to try to develop one without in the process subjecting the rights of patent holders to debilitating uncertainty. Cf. Hensley Equipment Co. v. Esco Corp., 383 F.2d 252, 261-262 n.19, *amended*, 386 F.2d 442 (5th Cir. 1967).

Contrary to Judge Posner's suggestion, however, there are several differences in scope between misuse and conventional antitrust doctrines. Patent misuse extends (or may extend) to a number of practices that do not fall within the antitrust laws at all. Many of these are practices that arise exclusively or primarily in the context of patents. Several examples follow.

Nonmetered Licenses. Patentees are entitled to charge royalties on the patents they license to others. What form may those license payments take? Hazeltine Research, a consumer electronics patent consortium, charged a royalty based on a flat percentage of a purchaser's sales, regardless of how many patents the purchaser needed or how important they were to the end product. The Court upheld this arrangement in 1950 in Automatic Radio Mfg. Co. v. Hazeltine Research, Inc., 339 U.S. 827, 833-34 (1950). But by 1969 it was of a different view. In Zenith Radio Corp. v. Hazeltine Research, Inc., 395 U.S. 100 (1969), the Court struck down the royalty provision:

> We also think patent misuse inheres in a patentee's insistence on a percentage-of-sales royalty, regardless of use, and his rejection of licensee proposals to pay only for actual use. Unquestionably, a licensee must pay if he uses the patent. Equally, however, he may insist upon paying only for use, and not on the basis of total sales, including products in which he may use a competing patent or in which no patented ideas are used at all. There is nothing in the right granted the patentee to keep others from using, selling, or manufacturing his invention which empowers him to insist on payment not only for use but also for producing products which do not employ his discoveries at all.

What is wrong with nonmetered royalties? Here, Bork's and Posner's arguments in the leveraging debate seem to have some force. See Robert Bork, The Antitrust Paradox (1978); Richard A. Posner, Antitrust Law: An Economic Perspective (1976). If a licensee is only willing to pay a fixed amount for a license, should it matter how that amount is collected? Certainly the "percentage of sales" formula is not itself anticompetitive, and it may be far more convenient than more exact measures of how much a licensee uses a patent. Should the transaction costs of royalty accounting come into play in deciding whether a practice amounts to misuse?

Should it matter whether the licensee voluntarily agrees to the restriction? See Engel Indus., Inc. v. Lockformer Co., 96 F.3d 1398 (Fed. Cir. 1996). In *Engel*, the accused infringer had signed a licensing agreement with the patentee for use of the patented system for connecting sheet metal ducts together. The agreement called for calculation of royalties based in part on the number of "corner connectors" used by the licensee; these components were not covered by the patent. The court pointed out that this was a convenient way to calculate royalties, and noted that the licensee "voluntarily agreed to the royalties provisions," and that the licensor had not "used its market power to shoehorn [the licensee] into license agreement that forced [it] to purchase unpatented items..." 96 F.3d at 1408.

One problem with agreements requiring licensees to pay a percentage of all sales is that they also provide a perfect source of information for policing a cartel. A licensee subject to a percentage of sales license must turn its sales and output information over to the licensor. With this information, the licensor can detect attempts to cheat on the cartel by expanding output. This danger is particularly acute when a participant in a market licenses his patent horizontally — that is, to its competitors in the same market. There is therefore an antitrust danger associated with permitting the use of these agreements.

Grantback Clauses. As a condition to a license agreement, a patentee will sometimes require the licensee to grant him rights to any "improvement patents" the licensee is issued while using the licensed patent. Courts have been fairly lenient with respect to such "grantback clauses." See Transparent-Wrap Mach. Corp. v. Stokes & Smith Co., 329 U.S. 637 (1947). But grantbacks may sometimes run afoul of the patent misuse doctrine.

Field-of-Use Restrictions. Most patent licenses contain some sort of restriction on how the patent may be used by the licensee. As we discuss in Chapter 8, restrictions that attempt to control the price at which the licensee sells products made using the patented process or equipment are illegal per se under Sherman Act section 1. Other restrictions, however, control not the price but the geographic or product market in which the licensee sells. These agreements have generally been upheld, unless they are part of a tying arrangement. See, e.g., U.S. v. Studiengesellschaft Kohle, 670 F.2d 1122 (D.C. Cir. 1981) (upholding field of use restriction in licenses for industrial catalysts).

Patent Suppression. For a variety of reasons, many (indeed, most) patents are never commercialized. The reasons can include lack of adequate financing, absence of commercial value, or simply bad business judgment on the part of an inventor. But there are less benign reasons as well. Inventions may be deliberately patented as a matter of course, in case they turn out to be useful. Many biotechnology companies follow this practice. Inventions may be deliberately patented and then not used in order to deny competitors the opportunity that a new technological advance might present, even though the patentee does not intend to alter its equipment or production process to take advantage of the new invention. Cf. Special Equip. Co. v. Coe, 324 U.S. 370 (1945) (Douglas, J., dissenting) ("The right of suppression of a patent came into the law over a century after the first patent act was passed. . . . I think it is time to be rid of that rule. It is inconsistent with the Constitution and the patent legislation which Congress has enacted.").

Commentators have long argued over whether suppression actually occurs, and whether it is an economically rational strategy. See, e.g., Kurt M. Saunders, Patent Nonuse and the Role of Public Interest as a Deterrent to Technology Suppression, 15 Harv. J.L. & Tech. 389, 392 (2002) ("There is evidence to suggest that products as obscure as artificial caviar and as important as photocopiers have been strategically shelved."); Mark Clark, Suppressing Innovation: Bell Laboratories and Magnetic Recording, 34 Tech. & Cult. 516, 532 (1993) (documenting suppression of Hickman patents covering magnetic recording, a policy that the author argues was pursued by AT&T to preserve markets and for "ideological reasons," i.e., the threat to privacy from recorded phone conversations). Nonuse of patents for benign reasons certainly occurs on a daily basis. Whether true suppression occurs has never been resolved. Advocates of compulsory licensing point to several instances in which companies are supposed to have designed or purchased a patented invention for the sole purpose of denying it to others. Automobile manufacturers are charged with suppressing new auto anti-pollution technology because they feared that they would be forced by government regulators to install the new devices. See Saunders, *supra*, 15 Harv. J.L. & Tech. at 393. And according to a rumor, manufacturers of panty hose are supposed to have suppressed their design for "no-run" hose in order to maintain the lucrative market for disposable hose. For other examples, see Charles Allen Black, The Cure for Deadly Patent Practices: Preventing Technology Suppression and Patent Shelving in the Life Sciences, 14 Alb. L.J. Sci. & Tech. 397 (2004).

In all these cases (and numerous others), patents for the allegedly suppressed products exist. What is uncertain is *why* the products were never commercialized. The companies involved normally claim that the products do not, in fact, work as promised. Those claims are bolstered by the argument of economists that suppression simply makes no sense. If a new invention is truly superior to current products, they argue, the patentee could sell that invention and more than make up for any losses it might sustain in the market for its current products. Indeed, economists continue, that is the definition of a superior product.

This argument has substantial force when the allegedly suppressed invention is in the same market as the patentee's current products. Thus panty hose manufacturers could presumably switch from selling disposable hose to selling no-run hose and, assuming the no-run hose was really a better product, charge prices high enough to make up for the fact that they would sell fewer pairs of hose. This argument is not completely convincing, however. If the process for producing no-run hose requires different machinery from that which the manufacturers currently use, there will be substantial fixed costs associated with the switch. Manufacturers may prefer to delay introducing the new product until they have to replace their machines anyway; they will suppress the invention until then. By contrast, if a new entrant into the market could use the patented process, he would choose to build the new machines immediately. By suppressing the patent, therefore, the patentee causes society to lose the benefits of the immediate production of the new process.[49] An example of this occurred in the rubber industry, when Standard Oil admitted in a 1942 consent decree to inhibiting the introduction of synthetic rubber. The reason was clearly to preserve its extensive investment in natural rubber production processes. See D. R. B. Ross, Patents and Bureaucrats: U.S. Synthetic Rubber Development Before Pearl Harbor, in Business and Government 119, 120 (J. R. Frese & J. Judd, eds., 1985). See generally Dunford, Suppression of Technology, 32 Admin. Sci. Q. 512 (1987).

This "retooling" problem is even more serious when the patentee is not in the same market as the patented invention but rather in an upstream or downstream market that would be affected by the patent. Consider the rumor that Exxon purchased and buried the design for the "momentum engine," which would tremendously increase automobile engine efficiency (and therefore tremendously decrease the demand for gasoline). It could produce and sell the momentum engine, using the revenues from those sales to offset its loss in gasoline revenues. However, Exxon is not in the engine business and is likely to be less efficient at that business than it is at refining and selling gasoline. Its profit-maximizing course may therefore be to conceal the invention, so that no one else can use it, and to continue to sell gasoline.[50] On the other hand, there is evidence of a more robust corporate market for patent licenses in

---

49. Of course, the patentee could always license the patent to the new entrant. However, innovation is associated with strong first-mover advantages, so the first company to manufacture and sell a product is likely to maintain a dominant position even after the patent expires and after further inventions supersede the original one. Thus licensing to competitors may not be an attractive option for many patentees.

50. For reasons similar to those discussed above, licensing to an existing engine manufacturer may not be an attractive option in this situation. Although Exxon would not be in direct competition with the licensee, its business would be adversely affected by licensing the engine. And there is often doubt that a licensor can extract the full value of its invention in the licensing agreement. Exxon would have changed from a continuing enterprise to one whose major asset was licensing revenue for the next 20 years, and which could then expect to go out of business. This option, even if economically rational, is not attractive to most major corporations.

the contemporary economy — suggesting that perhaps Exxon would license such a technology today. See Ashish Arora, et al., Markets for Technology: The Economics of Innovation and Corporate Strategy (2001).

In any event, Congress — and the courts before it — deemed that the costs of determining when nonuse or suppression is actionable, together with potential strategic uses of any remedies for it, outweigh the benefits of rooting out actual cases of suppression. See 35 U.S.C. §271(d)(4), amended in 1988 to protect nonuse.

E. INTERNATIONAL PATENT LAW

To obtain international protection for patented technology, lawyers and businesspeople must ultimately rely on the domestic patent law of each country. In two respects, however, international treaties do come into play. First, a number of international conventions streamline the procedures under which patents originating outside the host country are prosecuted. Second, certain international treaties provide a uniform substantive floor of protection, a minimum set of rights below which treaty signatories must not fall in conferring patents. Chief among these is the Trade-Related Aspects of Intellectual Property (TRIPs) Agreement, part of the 1994 Uruguay Round of GATT revisions, which are discussed later in this section.

1. Procedural Rules

The most important set of international procedural rules for patentees involve filing and priority dates. To understand them, however, one must first recognize that in practice they interact with elements of each country's domestic law. Recall, in this regard, the crucial divergence between U.S. and foreign priority rules. The United States is virtually the world's only "first-to-invent" country. All other countries, with trivial exceptions, establish priority based on a "first-to-file" system. If an inventor has an interest in securing patent protection overseas — whether in addition to U.S. protection or in place of it — he or she must file there *as soon after invention as possible.* Unlike the situation in the United States, where one's filing date is not the only significant event, in foreign countries a late filing (relative to a rival) can be fatal. There is no opportunity to prove that although a rival filed first, you in fact invented earlier. The filing date is the *only* relevant event in most foreign countries.

One additional rule in place in many foreign countries is the notion of "absolute priority." This means that an application must be filed before the invention is described or used in a publicly available forum. In other words, there is no one-year statutory grace period for filing in foreign systems. An application must be filed *before* publication of an article describing the same invention, or before any public use or sale — as opposed to within one year of these events as under 35 U.S.C. §102(b).

Thus even if one has solid proof of the date of invention, one must still make haste to the patent office, since any activity — by the inventor or a third party — prior to the date of filing can disqualify the invention from patentability. Contrast this with the statutory bars, whose critical date of one year prior to filing gives the patentee one year from the date of certain activities to file for a patent.

Another difference in domestic patent laws is also relevant. In many foreign countries, most notably those in Europe, "prior user" rights protect first inventors/later filers. Prior inventors are allowed to continue their use of the invention after a patent issues to the first filer/second inventor. See Brownlee, Trade Secret Use of Patentable Inventions, Prior User Rights and Patent Law Harmonization, 72 J. Pat. & Trademark Off. Soc. 523 (1990). Would this rule make sense for the United States also? See, e.g., Franklin Pierce Law Center: Third Patent System Major Problems Conference, 32 Idea 1, 41-64 (1991) (discussing the need for prior user rights system in U.S.). Such a bill was pending in Congress at this writing. H.R. 3460, 104th Cong., 2d Sess. Note that at a minimum a system of prior user rights mitigates somewhat the harshness of the first-to-file system. Be aware, however, that this is true only to a limited extent; the prior user right allows the prior inventor to practice the invention only in the manner and to the extent that she was using it when the patent issued to the other inventor. This limitation will often keep the first inventor/prior user right holder from commercializing the patented invention, unless of course efforts to commercialize were already under way when the patent issued. See Brownlee, *supra*.

Coordinating International Prosecution

Patent lawyers face two problems in coordinating the prosecution of a series of national patents. First, a common priority date must be obtained to ensure that protection will be uniform and unaffected by prior art published (or otherwise having an effective date) before one or more of the national patent applications. Also, a common date will ensure that prosecution of a patent in country *A* does not somehow compromise the patentability of the invention in Country *B*. Second, the patent lawyer has to deal with the logistics of international protection; she must oversee multiple filings in diverse languages in numerous countries. The wide variations in national practices and the high cost of conducting a large-scale application barrage make multiple filings one of the more challenging professional tasks in patent law.

Fortunately, two international agreements make these tasks a bit more tolerable. First is the Paris Convention, a longstanding international organization created by treaty in 1886 whose primary function is to guarantee a uniform worldwide priority date across all member countries. An applicant may file in any member country of the Convention up to one year after an initial (typically home-country) filing, without losing the priority date of the initial filing. (This treatment is provided under United States law in 35 U.S.C. §119.) The second international agreement is the Patent Cooperation Treaty (PCT), which streamlines the filing of multiple national patent applications. Each agreement in its own way is an indispensable tool of the patent trade. Although detailed discussion of the agreements would take up too much room for this volume, a few words about the essential features of each is in order.

a) The Paris Convention

The Paris Convention was signed in 1883, a product of the first true "internationalization" wave in the field of patent law. Paris Convention for the Protection of

Industrial Property, as last revised, July 14, 1967, 21 U.S.T. 1583, T.I.A.S. No. 6295, 828 U.N.T.S. 305 (the last revision, sometimes referred to as the "Stockholm" revision, entered into force April 26, 1970). Its primary function is to define a common priority date so that one may file an application in one member state and have the benefit of that same filing date when filing later in another member state. One purpose of this is to prevent interlopers from copying patents applied for or issued in one state and claiming them as their own in another, before the legitimate owner has time to file in the other country.

The key provision in the Convention as regards priority is Article 4. The relevant portions read as follows:

Article 4

A(1) Any person who has duly filed an application for a patent, or for the registration of a utility model, or of an industrial design, or of a trademark, in one of the countries of the Union, or his successor in title, shall enjoy, for the purpose of filing in the other countries, a right of priority in the periods hereinafter fixed.

A(2) Any filing that is equivalent to a regular national filing under the domestic legislation of any country of the Union or under bilateral or multilateral treaties concluded between countries of the Union shall be recognized as giving rise to the right of priority.

A(3) By a regular national filing is meant any filing that is adequate to establish the date on which the application was filed in the country concerned, whatever may be the subsequent fate of the application.

B. Consequently, any subsequent filing in any of the other countries of the Union before the expiration of the periods referred to above shall not be invalidated by reason of any acts accomplished in the interval, in particular, another filing, the publication or exploitation of the invention, the putting on sale of copies of the design, or the use of the mark, and such acts cannot give rise to any third-party right or any right of personal possession. Rights acquired by third parties before the date of the first application that serves as the basis for the right of priority are reserved in accordance with the domestic legislation of each country of the Union.

C(1) The periods of priority referred to above shall be twelve months for patents and utility models, and six months for industrial designs and trademarks.

Thus filing in one country that is a signatory to the Paris convention gives an applicant some "breathing room" — 12 months in which to prepare to file in other signatory nations.

b) The Patent Cooperation Treaty (PCT)

The PCT was signed in 1970. The Patent Cooperation Treaty, opened for signature June 19, 1970, 28 U.S.T. 7645, T.I.A.S. No. 8733 (entered into force Jan. 24, 1978). Its major purpose is to streamline the early prosecution stages of patent applications filed in numerous countries. It is often described as a clearinghouse for international patent applications. As a practical matter, its major advantage is that it gives an inventor (and her patent lawyer) more time, a precious commodity in the prosecution of an application destined for many countries. The signatories to the PCT have agreed to permit an applicant to wait for up to 30 months after the initial filing of a patent application in one country to begin the in-depth prosecution of the application in other countries. This allows the inventor more time, compared to non-PCT prosecution, in which to test the product, decide which countries' protection is worthwhile, and pay the patent office filing fees in the various countries.

There are two main parts of the PCT. Chapter 1 provides that an applicant who files in a national patent office may elect within 12 months to add a PCT filing. The PCT filing is simply an additional filing in any national patent office designated in the PCT. In this case, the applicant has up to 20 months from the initial filing to request that the PCT preliminary prosecution procedure be initiated. At that time, the applicant must also select the PCT member nations in which the applicant wishes to prosecute the patent. Note that Chapter 1 preserves the applicant's priority date (in PCT member countries), without having to begin active prosecution, for eight months longer than the simple Paris Convention priority period.

Chapter 2 of the PCT extends the election period to 30 months. To qualify under Chapter 2, the applicant must make her PCT filing at most five months after the first national filing. Chapter 2 gives an inventor 18 extra months, compared to the Paris Convention, to select countries for coverage and initiate multiple national prosecutions. In other words, so-called "Chapter Two" PCT filings give the inventor up to 30 months to make his or her "national elections."

The extra time is a substantial advantage. Besides simply delaying the expenditure of filing and examination fees, the PCT allows an inventor a significant extra period to assess the technical merits and commercial potential of the invention. This extra time helps the inventor save wasted filing fees for inventions that fail to blossom; for those that show great promise, the various patent applications that grow out of the PCT filing can be tailored to reflect the commercially significant embodiments that have emerged from the extensive testing.

2. Substantive Harmonization and GATT-TRIPs

As noted earlier, the lasting impact of the Paris Convention was primarily procedural, especially with respect to the uniform, worldwide priority date it made possible. The Paris Convention did, however, contain some minimum substantive standards of protection, as described earlier. Even so, further efforts at harmonizing worldwide standards of patentability, patentees' rights, and infringement bore little fruit for most of the twentieth century.

The slow pace of harmonization was in part due to the fact that the international business community seemed little interested in investing time in the esoteric issues of detailed harmonization. Although it was cumbersome to deal with the divergent standards held by the world's many individual domestic patent jurisdictions, some level of effective protection was available in most of the commercially important countries. Furthermore, patent law harmonization was the domain of a highly specialized affiliate of the United Nations, the World Intellectual Property Organization (WIPO), a forum not thought to be particularly friendly to Western business interests.

Things were changing by the mid-1980s, however. For one thing, the perceived value of intellectual property was increasing; it was beginning to take on a more central role in business planning and strategy. Of special concern was the increasing importance to U.S. businesses of overseas markets in developing countries—

countries that had traditionally opposed strong intellectual property protection. Businesses in the United States and Europe, aware both of the increasing importance of intellectual property and of WIPO's slow progress in harmonization, were thus on the lookout for an alternative forum in which to pursue harmonization. The search ended with the announcement of the Uruguay Round of negotiations to revise the main international trade agreement/organization, the General Agreement on Tariffs and Trade (GATT).

In addition to proposed reforms to the core function and structure of the GATT, the early Uruguay Round agenda soon grew to encompass negotiations on the "Trade-Related Aspects of Intellectual Property," or TRIPs, as it became known. The reference in the title to "Trade-Related" issues was a concession to those who doubted the relevance of intellectual property to the basic GATT mission; it soon became clear, however, that most of the basic elements of intellectual property protection would be up for discussion and potential harmonization in the TRIPs negotiations. By the time the GATT round ended in late 1993, TRIPs had become one of the principal components in the overall package of changes.

Although the post-TRIPs amendments to U.S. law are important, they pale in contrast to the revolutionary changes the agreement makes to the intellectual property regimes of many developing countries. To summarize the highlights, all signatories of the Uruguay Round treaty (who, under the agreement, become members of the newly created World Trade Organization (WTO)) must now:

- Include virtually all important commercial fields within the ambit of patentable subject matter, a major change for countries that, for example, have traditionally refused to enforce pharmaceutical patents on public health/access grounds.
- Test patent applications for (a) the presence of an inventive step, which is defined as precisely synonymous with nonobviousness under section 103 of the U.S. Patent Act, and for (b) "industrial application," similarly defined as coextensive with the U.S. utility requirement.
- Include in the patentees' bundle of exclusive rights the right to control the market for imports of the patented products.
- Eliminate or severely curtail the practice of granting compulsory licenses for patented technology.

See Final Act Embodying the Results of the Uruguay Round of Multilateral Trade Negotiations, April 15, 1994 2-3 (GATT Secretariat 1994); Annex 1C: Agreement on Trade-Related Aspects of Intellectual Property Rights, id. at 6-19, 365-403. For implementation in the United States, see Uruguay Round Agreements Act, Pub. L. No. 103-465 (H.R. 5110), Dec. 8, 1994. See also J. H. Reichman, Universal Minimum Standards of Intellectual Property Protection Under the TRIPs Component of the WTO Agreement, 29 Int'l Law. 345 (1995) (able summary of provisions and open questions).

We have discussed GATT-related changes to U.S. law earlier in this chapter, in the section on priority, novelty, and statutory bars. By way of summary, the most important provisions:

- Changed the U.S. patent term to 20 years, measured from the date the patent application is filed, rather than 17 years from the date the patent was issued by the Patent Office. 35 U.S.C. §154. (Under certain circumstances, such as interferences and appealed rejections, this term may be extended for up to five years. Id.)

- Opened up the U.S. "first-to-invent" system by allowing members of the WTO to introduce evidence of inventive acts in their home country for purposes of establishing priority. See 35 U.S.C. §104.
- Expanded the definition of infringement to include acts of unauthorized offering for sale and importing. 35 U.S.C. §271.
- Added a new procedure for filing "provisional applications," 35 U.S.C. §111, which must satisfy section 112 but need not include claims. Such an application does not begin the 20-year clock for the applicant's patent term.
- Require publication of U.S. patent applications covering inventions also claimed in foreign applications. 35 U.S.C. §122(b).

These changes furthered the substantive harmonization of world patent law by bringing the United States into line with the rest of the world on a number of important issues. So long as patent applicants must file and prosecute (and pay fees!) in each country in which they require protection, though, the impact of substantive harmonization will be limited.

F. REMEDIES

As we have seen in the cases in this chapter, the rights conferred by a patent are different from those that grow out of other property rights. On the other hand, there are a number of similarities between patent law and real property law. Like real property, intellectual property is freely alienable, for example. One of the most important similarities comes in the area of remedies. With patents, as with most other legal entitlements that share the label "property," the rule is that the owner of the property right may obtain an injunction to prevent ongoing "trespasses." Indeed, in one important conceptual framework used to describe many types of legal entitlements, it is the injunctive remedy that distinguishes the property right from other entitlements.[51]

The holder of a legal entitlement other than a property right can obtain compensation when the right is violated but cannot prevent violations *ex ante*. Because those who violate such a right must pay damages, this form of entitlement is referred

---

51. Guido Calabresi & A. Douglas Melamed, Property Rules, Liability Rules, and Inalienability: One View of the Cathedral, 85 Harv. L. Rev. 1089 (1972). According to Calabresi and Melamed, the extent and nature of transaction costs in a particular case dictate whether one of the parties to the Coasian bargain ought to have an absolute property right or simply the right to collect damages caused by the other party's encroachment (i.e., a "liability rule"). This doctrine holds that several factors point toward a property rule: few parties, difficult valuation problems, and otherwise low transaction costs. Other factors indicate that a liability rule might better effectuate the bargain: many parties (especially where any one has the power to "hold up" the whole enterprise), likelihood of strategic bargaining, and otherwise high transaction costs. Note that the authors refer to entitlements with injunctive remedies as property *rules,* to facilitate discussion of unconventional entitlements not normally thought of as property rights.

to as a *liability rule*. Classic examples of liability rule entitlements (which we will call "liability rights") are the rights of tort victims and the rights of parties to a contract in the face of a breach. In the latter case, for example, it is generally understood that the normal remedy for breach of contract is damages—a monetary payment—rather than compelling the breaching party to perform the contract. Contracting parties cannot usually obtain injunctions to prevent breach. ("Specific performance" is an exception to this rule.)

Comparing intellectual property rights to the rights of a party to a contract provides insight into why the standard remedies are different in these two areas. Consider first the paradigmatic contract, say for delivery of 100 bushels of wheat at $5.00 per bushel. If the seller breaches, what is the measure of the buyer's injury? Under basic contract law, it is simply the difference between the contract price and the market price at the time of breach. And of course, since wheat is a standard commodity, the market price can be determined easily—found in a newspaper, for example.

Now consider the infringement of a patented invention. What is the measure of the patentee's injury? At first blush, this seems easy to establish: just subtract the patentee's profits after infringement from those she would have made absent the infringement, and extrapolate for the full 15- to 20-year patent term. A moment's reflection should reveal a number of tricky issues, however. What would have happened to the patentee's operations without the infringement? What could competitors have done if the patentee had controlled the market for the patented component by herself? In the field of technology in which the patent exists, what will happen over the next 15 to 20 years?

In most, if not all, cases, these questions will be very difficult to answer. It is hard to put a precise dollar value on a particular piece of patented technology, especially over a protracted period. This difficulty—which we refer to as the *valuation problem*—is the key to understanding why the injunction, and not damages, is the standard remedy in a patent case. (Of course, with respect to infringement that takes place *prior to* litigation, damages are the only remedy available, so valuation problems cannot be avoided entirely.)

As an aside, you should be aware that the subject matter of copyrights, trademarks, trade secrets, and other forms of intellectual property suffer from similar valuation problems. Indeed, you might say that the various requirements of "uniqueness" in intellectual property law (e.g., novelty and nonobviousness in patent law) almost guarantee that its subject matter will not be amenable to treatment as a commodity. As we explore the remedies issues presented below, and in later chapters, keep this in mind. It explains much about the policies behind intellectual property remedies, and it also helps explain why, when they are called for (e.g., in assessing past damages for infringement), intellectual property remedies of the "liability right" variety seem so difficult to quantify and defend.

The purpose of injunctive remedies goes beyond allowing the rightholder to prevent activities of the infringer. To the extent that a rightholder will consider negotiating a license with the infringer, the threat of an injunction will heavily influence the *terms* of the license. Specifically, it allows the rightholder to set her own price for the injury. In intellectual property cases, it allows the *rightholder,* and not a court, to set the terms of a license agreement settling the infringement litigation. This is assumed to be the efficient result, as a court called on to set the terms of the exchange would have a difficult time doing so quickly and cheaply, given the specialized nature of the assets and the varied and complex business environments in which they are deployed.[52] Hence the parties are left to make their own deal.

On the other hand, a property right is a potent weapon. Sometimes this causes problems. Especially where courts may have a difficult time determining how much of a composite product's value is attributable to a (perhaps small) patented component, the availability of property right may lead to what economists call a "holdup problem"—an entitlement holder who uses his rights strategically to extract more than the fair market value of his or her asset. A classic example is the owner of the last of many parcels being assembled for a large real estate project. Should injunctions be readily available when holdup is a potential problem? Cf. MercExchange, LLC v. eBay, Inc., 401 F.3d 1323 (Fed. Cir. 2005) (injunction granted in patent infringement case arguably presenting this scenario).

COMMENTS AND QUESTIONS

1. Is it true that giving a property entitlement entirely to one party to a transaction and allowing both parties to bargain will produce the efficient price? There are at least three potential problems with this approach. First, the parties themselves may not know the value of the technology over which they are negotiating. See David J. Teece, The Market for Know-How and the Efficient International Transfer of Technology, Annals of the Am. Assn. Pol. & Soc. Sci., Nov. 1981, at 81, 86 ("[N]either [the] buyer nor seller of technology seems to have a clear idea of the value of the commodity in which they are trading," which means that "royalty rates may simply be a function of [the] negotiating skills of the parties involved."). Second, the transaction costs associated with such licensing arrangements can be significant. See David J. Teece, The Multinational Corporation and the Resource Cost of International Technology Transfer 36 (1976) (transaction costs average 20 percent of total value of technology exchanges studied); Farok J. Contractor, International Technology Licensing: Compensation, Costs, and Negotiation 105 (1981) (transaction costs averaged more than $100,000 for licensing deals studied). Finally, strategic behavior may cause parties to forgo some deals that would be in the best interests of both sides. See Robert Cooter, The Cost of Coase, 11 J. Legal Stud. 1, 19-23 (1982). For a suggestion that the cumulative effect of these problems is that efficient licensing of improvements will not always occur, see Mark A. Lemley, The Economics of Improvement in Intellectual Property Law, 75 Tex. L. Rev. 989 (1997).

Even if you think that bargaining is likely to be inefficient for one of the foregoing reasons, what alternatives are there? Are they any better?

2. As an example of strategic behavior, consider the following: A pioneering inventor patents a basic discovery which leads to a product that yields $100 in profits. An improver comes up with a radical extension of the technology that builds on the insight of the pioneer and produces a product yielding much higher profits: say $1,000. The improver may offer to split the cooperative surplus—$1,000 (the $1,000 that the improver could make selling the improved product, plus the $100

---

52. This follows from the fact that the parties are participants in the same industry and therefore have more knowledge of the technology at work in their industry. Courts, by contrast, usually have very little technical sophistication; even if they have some, it is of the most generalized variety.

the inventor can continue to make, less the $100 profit the inventor would make absent a bargain) — available in the joint profit-maximizing scenario under which the patentee licenses the basic patent to the improver. That would leave the patentee with $600: her $100 profits from selling her product, plus the royalty of $500, which again is half the cooperative surplus of $1000. Can we expect the patentee to agree to this bargain? Recall that the patentee has a wonderful threat: the right to banish the improver from the market altogether, i.e., to exercise the "property right" conferred by the patent, in the form of a permanent injunction. Will the patentee therefore be inclined to ask for more than half the surplus? At what point will the improver give up on the whole deal? Does this scenario justify replacing the property right with a liability right, at least in some cases? How would you determine which cases would be subject to this modified rule?

3. What role (if any) should culpability play in determining the appropriate remedy for patent infringement? Is it reasonable to treat a willful infringer more harshly than an innocent one? How would such differential treatment square with the rationale for patent remedies discussed above?

1. Injunctions

H. H. Robertson Co. v. United Steel Deck, Inc.
United States Court of Appeals for the Federal Circuit
820 F.2d 384 (Fed. Cir. 1987)

The district court applied to Robertson's motion the Third Circuit standard:

> An applicant for a preliminary injunction against patent infringement must show: . . . (1) a reasonable probability of eventual success in the litigation and (2) that the movant will be irreparably injured pendente lite if relief is not granted. . . . Moreover, while the burden rests upon the moving party to make these two requisite showings, the district court "should take into account, when they are relevant, (3) the possibility of harm to other interested persons from the grant or denial of the injunction, and (4) the public interest."

This is substantially the same standard enunciated by this court. See, for example, Roper Corp. v. Litton Systems, Inc., 757 F.2d 1266, 1270-73 (Fed. Cir. 1985), and Atlas Powder, 773 F.2d at 1231-34.

In matters involving patent rights, irreparable harm has been presumed when a clear showing has been made of patent validity and infringement. Smith International, 718 F.2d at 1581. This presumption derives in part from the finite term of the patent grant, for patent expiration is not suspended during litigation, and the passage of time can work irremediable harm. The opportunity to practice an invention during the notoriously lengthy course of patent litigation may itself tempt infringers.

The nature of the patent grant thus weighs against holding that monetary damages will always suffice to make the patentee whole, for the principal value of a patent is its statutory right to exclude. The presumption of irreparable harm in patent cases is analogous to that applicable to other forms of intellectual property, as discussed in Roper Corp., 757 F.2d at 1271-72.

The district court held that irreparable injury was presumed because Robertson had established a "strong likelihood of success in establishing validity and infringement." Such presumption of injury was not, however, irrebuttable. During oral argument, in response to an inquiry from the bench, USD and Bouras urged that money damages are an adequate remedy. In response, Robertson emphasized the few remaining years of patent life.

Even when irreparable injury is presumed and not rebutted, it is still necessary to consider the balance of hardships. The magnitude of the threatened injury to the patent owner is weighed, in the light of the strength of the showing of likelihood of success on the merits, against the injury to the accused infringer if the preliminary decision is in error. Results of other litigation involving the same patent may be taken into account, and the public interest is considered. No one element controls the result.

When the movant has shown the likelihood that the acts complained of are unlawful, the preliminary injunction "preserves the status quo if it prevents future trespasses but does not undertake to assess the pecuniary or other consequences of past trespasses." Atlas Powder Co., 773 F.2d at 1232. The court in *Atlas Powder* thus distinguished between remedies for past infringement, where there is no possibility of other than monetary relief, and prospective infringement, "which may have market effects never fully compensable in money." Id. The cautionary corollary is that a preliminary injunction improvidently granted may impart undeserved value to an unworthy patent.

COMMENTS AND QUESTIONS

1. The Federal Circuit has modified its approach in preliminary injunction cases since the *H.H. Robertson* decision. See, e.g., Amazon.com v. Barnesandnoble.com, 239 F.3d 1343 (Fed. Cir. 2001) (reversing trial court grant of preliminary injunction in case involving Amazon.com's "one click shopping" patent); Anton/Bauer, Inc. v. PAG, Ltd., 329 F.3d 1343 (Fed. Cir. 2003) (reversing district court grant of preliminary injunction). But see Oakley, Inc. v. Sunglass Hut Intern., 316 F.3d 1331 (Fed. Cir. 2003) (upholding grant of preliminary injunction).

2. The district court in its opinion in *H.H. Robertson* referred to the USD/Bouras portrayal of disruption, loss of business, and loss of jobs, and to Robertson's business needs and patent rights. Observing that "[t]his patent does not have many more years to run," the court held "the equities weigh heavily against the wrongdoer." Id. (Why?) The court stated that the "protection of patents furthers a strong public policy . . . advanced by granting preliminary injunctive relief when it appears that, absent such relief, patent rights will be flagrantly violated."

To the same effect is the case of Polaroid v. Kodak, where the court was asked to consider the "public interest" in deciding whether to grant a stay of an injunction. The response:

> Not only will an injunction injure Kodak customers and goodwill, it will also, according to Kodak, cause a "major disruption" of business. If and when Kodak is forced to shut down its instant camera production, 800 full-time and 3700 part-time employees will lose their jobs, and the company will lose its $200 million investment in plant and equipment.
>
> I am not unmindful of the hardship an injunction will cause — particularly to Kodak customers and employees. It is worth noting, however, that the harm Kodak will suffer simply mirrors the success it has enjoyed in the field of instant photography. To the extent

Kodak has purchased that success at Polaroid's expense, it has taken a "calculated risk" that it might infringe existing patents....

Kodak's characterization of the public interest . . . misconstrues the very concept of public benefit. The public policy at issue in patent cases is the 'protection of rights secured by valid patents.' Courts grant — or refuse to stay — injunctions in order to safeguard that policy, even if those injunctions discommode business and the consuming public.

Polaroid Corp. v. *Eastman Kodak Co.*, Civ. No. 76-1634-Z, D. Mass., October 11, 1985, Slip Op. at 3, 4.

3. Keep in mind that *H. H. Robertson* and *Kodak* were decisions involving *preliminary* injunctions. When a full trial concludes with a finding of infringement, there is almost never any doubt that a permanent injunction will issue. The Federal Circuit reiterated these principles in a 2005 opinion in the case of MercExchange, LLC v. eBay, Inc., 401 F.3d 1323 (Fed. Cir. 2005), *cert. granted*, 126 S. Ct. 733 (2005). The district court had denied a permanent injunction in the case, pending appeal to the Federal Circuit, on numerous grounds, including concerns over the validity and wisdom of "methods of doing business" patents such as the one MercExchange had asserted against defendant eBay, Inc. The Federal Circuit reversed on this issue, in definitive fashion:

> In its post-trial order, the district court stated that the public interest favors denial of a permanent injunction in view of "a growing concern over the issuance of business-method patents, which forced the PTO to implement a second level review policy and cause legislation to be introduced in Congress to eliminate the presumption of validity for such patents." A general concern regarding business-method patents, however, is not the type of important public need that justifies the unusual step of denying injunctive relief. Another reason the court gave for denying a permanent injunction was that the litigation in this case had been contentious and that if a permanent injunction were granted, the defendants would attempt to design around it. That strategy, in turn, would result in "contempt hearing after contempt hearing requiring the court to essentially conduct separate infringement trials to determine if the changes to the defendants' systems violates the injunction" and in "extraordinary costs to the parties, as well as considerable judicial resources." The court's concern about the likelihood of continuing disputes over whether the defendants' subsequent actions would violate MercExchange's rights is not a sufficient basis for denying a permanent injunction. A continuing dispute of that sort is not unusual in a patent case, and even absent an injunction, such a dispute would be likely to continue in the form of successive infringement actions if the patentee believed the defendants' conduct continued to violate its rights.
>
> The trial court also noted that MercExchange had made public statements regarding its willingness to license its patents, and the court justified its denial of a permanent injunction based in part on those statements. The fact that MercExchange may have expressed willingness to license its patents should not, however, deprive it of the right to an injunction to which it would otherwise be entitled. Injunctions are not reserved for patentees who intend to practice their patents, as opposed to those who choose to license. The statutory right to exclude is equally available to both groups, and the right to an adequate remedy to enforce that right should be equally available to both as well. If the injunction gives the patentee additional leverage in licensing, that is a natural consequence of the right to exclude and not an inappropriate reward to a party that does not intend to compete in the marketplace with potential infringers.
>
> Finally, we do not agree with the court that MercExchange's failure to move for a preliminary injunction militates against its right to a permanent injunction. A preliminary injunction is extraordinary relief that is available only on a special showing of need for

relief pendente lite; a preliminary injunction and a permanent injunction "are distinct forms of equitable relief that have different prerequisites and serve entirely different purposes." Lermer Germany GmbH v. Lermer Corp., 94 F.3d 1575, 1577 (Fed. Cir. 1996). We therefore see no reason to depart from the general rule that courts will issue permanent injunctions against patent infringement absent exceptional circumstances. Accordingly, we reverse the district court's denial of MercExchange's motion for a permanent injunction.

401 F.3d 1323, 1339. For more on the genesis and effects of business method patents, see Robert P. Merges, The Uninvited Guest: Patents on Wall Street, Fed. Res. Bk. Atlanta Econ. Rev., 4th Quarter, 2003, at p. 1. Should injunctions be available to a patentee whether the patentee manufactures the patented product, or is simply in the business of licensing (and litigating) patents? What if the plaintiff in a case has as its business model the acquisition of third-party patents, strictly with an eye toward maximizing patent-related revenue by suing as many actual manufacturers as possible? As patents have become more valuable, these and related issues are finding their way into legal arguments regarding the desirability of injunctions and other remedies.

4. In very rare cases, however, the court may deny a permanent injunction even after a finding of infringement. In Foster v. American Mach. & Foundry Co., 492 F.2d 1317 (2d Cir. 1974), for example, the court upheld what amounted to a compulsory license: a reasonable royalty damage award (described later in this chapter), but no injunction. This is an example of a rare case in patent law: the treatment of a patent as a liability right. The appellate court opinion is concise on this point:

> We do not find any difficulty in agreeing that an injunction would be an inappropriate remedy in this case. An injunction to protect a patent against infringement, like any other injunction, is an equitable remedy to be determined by the circumstances. It is not intended as a club to be wielded by a patentee to enhance his negotiating stance. Here, as the District Court noted, the appellee manufactures a product; the appellant does not. In the assessment of relative equities, the court could properly conclude that to impose irreparable hardship on the infringer by injunction, without any concomitant benefit to the patentee, would be inequitable.
>
> Instead, the District Court avoided ordering a cessation of business to the benefit of neither party by compensating appellant in the form of a compulsory license with royalties. This Court has approved such a "flexible approach" in patent litigation. Here the compulsory license is a benefit to the patentee who has been unable to prevail in his quest for injunctive relief. To grant him a compulsory royalty is to give him half a loaf. In the circumstance of his utter failure to exploit the patent on his own, that seems fair.

492 F.2d at 1324. What difference, if any, should it make that the patentee (appellant in *Foster*) does not manufacture anything—i.e., is not "practicing" the patent? While in theory the patentee has the right not to practice the patent, other cases have taken this into account in setting remedies as well. See, e.g., E. I. du Pont de Nemours & Co. v. Phillips Petroleum Co., 835 F.2d 277, 278 (Fed. Cir. 1987) (upholding grant of stay of preliminary injunction where patentee du Pont (1) had licensed all who desired entry into the polyethylene business, the subject of the patent, and (2) planned to exit the market; "harm to duPont here is of a different nature than harm to a patentee who is practicing its invention and fully excluding others"); Vitamin Technologists, Inc. v. Wisconsin Alumni Research Found., 146 F.2d 941 (9th Cir. 1944),

cert. denied, 325 U.S. 876 (1945) (describing policy reasons why a permanent injunction should be denied in a case where the patentee had refused to license a health-related patent). For some fresh, insightful thinking about injunctions and patent cases, emphasizing a connection between innovation, manufacturing, and the right to obtain an injunction, see Michelle Armond, Note, Introducing the Defense of Independent Invention to Motions for Preliminary Injunctions in Patent Infringement Lawsuits, 91 Cal. L. Rev. 117 (2003); Julie S. Turner, Comment, The Nonmanufacturing Patent Owner: Toward a Theory of Efficient Infringement, 86 Calif. L. Rev. 179 (1998). Cf. MercExchange, LLC v. eBay, Inc., 401 F.3d 1323 (Fed. Cir. 2005) (rejecting arguments that patentees ought to actually produce a patented invention to merit an injunction).

2. Damages: Reasonable Royalty and Lost Profits

Courts have struggled with the problem of setting appropriate measures of damages for past infringement. In Panduit Corp. v. Stahlin Bros. Fibre Works, Inc., 575 F.2d 1152, 1158 n.5 (6th Cir. 1978), the court adopted an oft-used four-factor test for determining lost profits:

> To obtain as damages the profits on sales he would have made absent the infringement, i.e., the sales made by the infringer, a patent owner must prove: (1) demand for the patented product, (2) absence of acceptable noninfringing substitutes, (3) his manufacturing and marketing capability to exploit the demand, and (4) the amount of the profit he would have made.
>
> When actual damages, e.g., lost profits, cannot be proved, the patent owner is entitled to a reasonable royalty. A reasonable royalty is an amount "which a person, desiring to manufacture and sell a patented article, as a business proposition, would be willing to pay as a royalty and yet be able to make and sell the patented article, in the market, at a reasonable profit." Goodyear Tire and Rubber Co. v. Overman Cushion Tire Co., 95 F.2d 978 at 984 (6th Cir. 1937) (citing Rockwood v. General Fire Extinguisher Co., 37 F.2d 62 at 66, (2d Cir. 1930)), *appeal dismissed*, 306 U.S. 665 (1938).

In Polaroid Corp v. Eastman Kodak Co., 16 U.S.P.Q.2d (BNA) 1481 (D. Mass. 1990), the court ultimately assessed total damages of $873,158,971 (corrected from the original figure of $909,457,567). In its discussion of damages, the court said:

> At various times during the trial Polaroid seemed to argue that all evidence about conventional photography should be excluded because conventional photography was not an acceptable substitute for instant photography. Kodak argued that conventional photography was an "economic substitute" for instant. Kodak did not attempt to quantify the number of Kodak instant purchasers who would have turned to conventional products in Kodak's absence; Kodak simply urged that the relative price of instant and conventional photography was a significant variable in the demand formula for instant products and therefore must be considered in any assessment of the market. Instant photography occupied a unique niche in the overall photography market during the infringement period. Consumers sought the emotional "instant experience" of having a picture develop immediately, usually in the presence of the subject. Although instant photography was unique, it did not exist in a vacuum; it also competed with conventional photography for the consumer's photographic dollar. Those who purchased infringing Kodak instant products at Kodak prices would not have considered conventional products as an

acceptable substitute at the time of purchase, but the relationship between the relative advantages of conventional and instant photography was an integral part of each consumer's decision. If Polaroid were simply claiming that it could have made Kodak's sales at the same time at similar prices, there would be no question in my mind that consumers would have made the same choice vis-à-vis conventional photography. However, Polaroid's claim is not so simple.

Polaroid's experts spun a scenario in which the prices Polaroid would have charged [if Kodak had not been in the market] were substantially higher and in which, through a complicated massaging of the demand curve, the great bulk of sales would have occurred later than they did historically. In this "but for" scenario, the effect of conventional photography on the instant photography market must be considered. The evidence of competition between instant and conventional products is overwhelming. Even Polaroid's econometric expert attempted to include the effect of competition from conventional photography in his computation. The relative values of instant and conventional changed throughout the period of infringement. On the facts before me, I must consider that relationship when deciding whether Polaroid's scenario, which differs so substantially from the historical world, is feasible. *Contrary to Polaroid's urging, the law does not require this Court to ignore the effect of competitive forces on the market for patented goods just because there are no non-infringing alternatives. Indeed, the law requires a careful assessment of all market influences when determining lost profit or reasonable royalty damages.* I find that conventional photography was not an acceptable substitute for instant photography during the period of infringement. Competition between conventional and instant photography (which changed throughout the ten years of infringement) did, however, affect the price that consumers were willing to pay for instant photography. Therefore, I have assessed what effect the competition would have had on Polaroid's ability to charge more for its instant products and on the profitability of delaying the introduction of certain lower-priced products. (Emphasis added.)

[In another section of the opinion, the court concludes that in fact this competition had no effect on Polaroid's profit margins or product-introduction-related profits.]

Under *Panduit*, lost profits are available if there are "no adequate substitutes." Since some Federal Circuit lost profits cases suggested that a patent almost ensures that there are no adequate substitutes (recall our discussion of "uniqueness" in the introduction to this section), the *Polaroid* court might have simply refused to consider evidence of conventional camera sales altogether. At the other extreme, recent antitrust thinking comes close to arguing that a patent almost never confers "market power" in an economic sense. Under this view, the existence of the patents on elements of instant photography would have been irrelevant in *Polaroid*. Only evidence and testimony on cross-elasticities and the like could establish the absence of substitute products.

Polaroid steers an interesting middle course. While conceding that there are no substitutes in the market for patented products, the court goes on to consider the effect of economically significant products on the *profit* the patentee would have made absent infringement, even though those products are not close substitutes in a strict sense. It admits, in other words, that the patented product is unique, but it refuses to equate uniqueness with total insulation from market forces. In a word, there may be no exact substitutes, but at some price (and for some consumers) there are *economic* substitutes. These exert price discipline on the patentee's pricing.

One way to state this in terms of conventional antitrust analysis is to say that although the cross-elasticities of demand for the patented item and related items are not high enough to call them identical products, the availability of some (perhaps imperfect) substitutes, coupled with at least some price elasticity in the demand for the patented good, means that the patentee would have faced some price discipline even if she could have made the infringer's sales. One can even imagine a scenario where, if the infringer had not been in the patentee's market, he would have moved into a parallel market from which he could have exerted greater price discipline on the patentee.

COMMENTS AND QUESTIONS

1. Some scholarship in the area of antitrust market definition is of special relevance to a consideration of the noninfringing substitute element of the *Panduit* test. In a paper by R. S. Hartman, D. J. Teece, W. Mitchell and T. M. Jorde, Assessing Market Power in Regimes of Rapid Technological Change, 2 Indus. & Corp. Change 3 (1993), the authors argue that traditional market definition techniques are seriously flawed when applied to markets for highly innovative and technology-intensive products. Specifically, they point out that current Department of Justice Guidelines assume a much too static view, which "biases market definition downwards in industries experiencing rapid technological change, where competition often takes place on performance attributes, and not price." "The more innovative the new product or process," the authors argue, "the greater the conventional market power will appear, because price changes will have little or no influence on demand for a truly innovative product." Id. at 6-7. The authors propose an approach whereby product attributes are included in the assessment of demand and supply. See also Thomas M. Jorde & David J. Teece, Rule of Reason Analysis of Horizontal Arrangements: Agreements Designed to Advance Innovation and Commercialize Technology, 61 Antitrust L.J. 579 (1993).

Views such as these may get a hearing at the Supreme Court. In Independent Ink, Inc. v. Illinois Tool Works, Inc., 396 F.3d 1342, 1348-49 (Fed. Cir. 2005), which involved a claim by an ink manufacturer that the defendant Illinois Tool was illegally tying the sale of its patented ink jet printheads to its unpatented ink, the court held that this tying claim could go forward even if the plaintiff could not prove that Illinois Tool had market power in the market for ink jet printheads. The court relied on older Supreme Court cases to hold that market power need not be proven in such a case:

> [T]he Supreme Court cases in this area squarely establish that patent and copyright tying, unlike other tying cases, do not require an affirmative demonstration of market power. Rather, International Salt [v. United States, 332 U.S. 392 (1947)] and [United States v.] Loew's [371 U.S. 38 (1962),] make clear that the necessary market power to establish a section 1 violation is presumed. The continued validity of International Salt and Loew's as binding authority, and the distinction between patent tying and other tying cases that was articulated in Loew's, have been consistently reaffirmed by the Court ever since.

Illinois Tool sought and obtained certiorari from the Supreme Court to address whether, in an action under Section 1 of the Sherman Act, 15 U.S.C. § 1, plaintiff must prove that the defendant possessed market power in the relevant market for the tying product. The Supreme Court should decide the case by the spring of 2006.

a. Reasonable Royalty

Upon a finding of infringement, the plaintiff is entitled to "damages adequate to compensate for the infringement, but in no event less than a reasonable royalty for the use made of the invention by the infringer." 35 U.S.C. § 284 (2001). In *Aro Manufacturing*, the Supreme Court stated that the statutory measure of damages represents "the difference between [the patent owner's] pecuniary condition after the infringement, and what his condition would have been if the infringement had not occurred." Aro Mfg. Co. v. Convertible Top Replacement Co., 377 U.S. 476, 507 (1964). In practice, the statute has been interpreted to provide for a "floor" of compensation, equal to the reasonable royalty the parties would have agreed to prior to infringement. To go above this floor, and recover higher "lost profits" damages, the patentee must pass a four-part test explained below. Here, we begin with the reasonable royalty "floor."

According to Federal Circuit Judge Rader,

> The statute guarantees patentees a reasonable royalty even when they are unable to prove entitlement to lost profits or an established royalty rate. A "reasonable royalty" contemplates a hypothetical negotiation between the patentee and the infringer at a time before the infringement began. [citation omitted.]...[T]his analysis necessarily involves some approximation of the market as it would have hypothetically developed absent infringement. This analysis, in turn, requires sound economic and factual predicates.

Riles v. Shell Exploration and Production Co., 298 F.3d 1302, 1311 (Fed. Cir. 2002) (Rader, J.).

The artificiality of the hypothetical royalty negotiation has received a good deal of attention. "In a normal negotiation, the potential licensee has three basic choices: forego all use of the invention; pay an agreed royalty; infringe the patent and risk litigation." Fromson v. Western Litho Plate and Supply Co., 853 F.2d 1568, 1576 (Fed.Cir.1988). In a hypothetical negotiation, however, the factfinder "presumes that the licensee has made the second choice, when in fact it made the third." Id. Commentators — and some courts — have noted that this standard could be applied in a way unfair to patentees, if a court were "to pretend that the infringement never happened." TWM Mfg. Co. v. Dura Corp., 789 F.2d 895, 900 (Fed.Cir.1986). As a consequence, courts keep in mind that the hypothetical negotiation approach "must be flexibly applied as a device in the aid of justice." Id. (internal quotations omitted). In practice, this means that courts walk a fine line when awarding reasonable royalties: between on the one hand undercompensating the patentee, by resort to a purely fictional negotiation which takes place outside the shadow of any actual infringement, and on the other awarding damages far removed from industry standards, the negotiating positions of the parties, and other real world constraints. In addition, although the cases sometimes hint at a quasi-punitive component in a reasonable royalty calculation,[53] the statute provides for an explicit "multiplier" *only* in the case

---

53. See, e.g., Panduit Corp. v. Stahlin Brothers Fibre Works, 575 F.2d 1152 (6th Cir. 1978), where the court pointed out that if courts calculated reasonable royalties without any cognizance of actual infringement, potential infringers would have no incentive to approach patentees prior to infringement — they would be no worse off having waited, and if the infringement were never detected, they would get away without paying anything. For a sophisticated analysis of these issues, see Roger D. Blair and Thomas F. Cotter, Intellectual Property: Economic and Legal Dimensions of Rights and Remedies (2004).

of willful infringement (considered below). Nevertheless, courts have been unsympathetic to infringers who argue that the patentee's proposed royalty would have meant that the infringer would have lost money on sales of the product. See, e.g., Golight, Inc. v. Wal-Mart Stores, Inc., 355 F.3d 1327, 1338 (Fed. Cir. 2004); Monsanto Co. v. Ralph, 382 F.3d 1374, 1384 (Fed. Cir. 2004) ("[T]he law does not require that an infringer be permitted to make a profit. And, where, as here, a patentee is unwilling to grant an unlimited license, the hypothetical negotiation process has its limits.").

In practice, then, setting reasonable royalties entails a detailed review of the technology at issue, the bargaining positions of the parties, and other factors that might have been relevant if the hypothetical negotiation had actually taken place. A rich and varied list of factors was laid out in Georgia-Pacific v. U.S. Plywood Corp., 318 F. Supp. 1116, 1120 (S.D.N.Y. 1970), and has been referred to ever since. The *Georgia-Pacific* factors are:

1. The royalties received by the patentee for the licensing of the patent in suit, proving or tending to prove an established royalty.
2. The rates paid by the licensee for the use of other patents comparable to the patent in suit.
3. The nature and scope of the license, as exclusive or non-exclusive; or as restricted or non-restricted in terms of territory or with respect to whom the manufactured product may be sold.
4. The licensor's established policy and marketing program to maintain his patent monopoly by not licensing others to use the invention or by granting licenses under special conditions designed to preserve that monopoly.
5. The commercial relationship between the licensor and licensee, such as, whether they are competitors in the same territory in the same line of business; or whether they are inventor and promoter.
6. The effect of selling the patented specialty in promoting sales of other products of the licensee; that existing value of the invention to the licensor as a generator of sales of his non-patented items; and the extent of such derivative or convoyed sales.
7. The duration of the patent and the term of the license.
8. The established profitability of the product made under the patent, its commercial success, and its current popularity.
9. The utility and advantages of the patent property over the old modes or devices, if any, that had been used for working out similar results.
10. The nature of the patented invention, the character of the commercial embodiment of it as owned and produced by the licensor, and the benefits to those who have used the invention.
11. The extent to which the infringer has made use of the invention, and any evidence probative of the value of that use.
12. The portion of the profit or of the selling price that may be customary in the particular business or in comparable businesses to allow for the use of the invention or analogous inventions.
13. The portion of the realizable profit that should be credited to the invention as distinguished from non-patented elements, the manufacturing process, business risks, or significant features or improvements added by the infringer.

14. The opinion testimony of qualified experts.
15. The amount that a licensor (such as the patentee) and a licensee (such as the infringer) would have agreed upon (at the time the infringement began) if both had been reasonably and voluntarily trying to reach an agreement; that is, the amount which a prudent licensee — who desired, as a business proposition, to obtain a license to manufacture and sell a particular article embodying the patented invention — would have been willing to pay as a royalty and yet be able to make a reasonable profit and which amount would have been acceptable by a prudent patentee who was willing to grant a license.

Is the reasonable royalty approach circular? While it seems to make sense to base a royalty on what other parties were negotiating for similar rights, those license negotiations themselves will be driven by what the parties think will happen if they don't come to terms and the dispute goes to court.

b. Lost Profits Damages

Although the statute establishes the "reasonable royalty" as the floor for patent damages, much of the action in recent years has been in the law of lost damages. In particular, four novel theories have been used to increase damage awards in the past several years.

i. Price Erosion

In a market with only two suppliers, it is obvious that sales made by the infringer would have been made by the patentee if there had been no infringement. Thus it is easy to establish the *number* of units the patentee would have sold without the infringement: the infringer's sales plus the patentee's actual sales. But at what *price* could the patentee have sold? Economic theory advises that a monopolist will charge a higher price than a duopolist, i.e., a seller in a two-firm market. As a result, some patentees have successfully argued that they would have sold the infringer's units plus their own, all at a higher price than they actually charged, given the presence of the infringer. See, e.g., Lam, Inc. v. Johns-Manville Corp., 718 F.2d 1056 (Fed. Cir. 1983).

What assumption does this theory make about the demand for the patentee's product? Wouldn't you expect the number of units sold to decline as the price increases?

An important element of symmetry was introduced in this area with the Federal Circuit decision in Grain Processing Corp. v. American Maize Products Co., 185 F.3d 1341 (Fed. Cir. 1999). Plaintiff Grain Processing was the holder of a patent on a form of maltodextrin, a food additive used in frostings, syrups, drinks, cereals, and frozen foods. After finding infringement, Judge Frank Easterbrook of the Seventh Circuit (sitting by designation on the district court) denied the patentee lost profits damages. Judge Easterbrook found that although the infringer did not in fact sell a non-infringing alternative during the damages period, it easily *could have*. The Federal Circuit affirmed, opening the way for accused infringers to argue more freely about hypothetical scenarios that might have played out if they had known they were infringing.

In summary, this court requires reliable economic proof of the market that establishes an accurate context to project the likely results "but for" the infringement. The availability of substitutes invariably will influence the market forces defining this "but for" marketplace, as it did in this case. Moreover, a substitute need not be openly on sale to exert this influence. Thus, with proper economic proof of availability, as American Maize provided the district court in this case, an acceptable substitute not on the market during the infringement may nonetheless become part of the lost profits calculus and therefore limit or preclude those damages.

185 F.3d at 1356.

It should be noted that the facts in *Grain Processing* made it particularly easy to make a persuasive case that the noninfringing substitute would very likely have been employed by the infringer. The technology was well-understood — so much so that the infringer switched to a noninfringing version of the product within two weeks of being ordered to do so by the district court; buyers and consumers of maltodextrin did not value the features of the claimed product in a way that set it apart from other versions, making other versions good substitutes; and the accused infringer had especially convincing proof that it could easily have ramped up production of the noninfringing alternative.

The Federal Circuit reaffirmed and applied *Grain Processing* in Crystal Semiconductor Corp. v. Tritech Microelectronics Int'l, Inc., 246 F.3d 1336, 1358-60 (Fed. Cir. 2001) (affirming trial court findings and emphasizing need to analyze market reaction to patented technology). But the court also made clear in a subsequent case that it was serious about the need for the infringer to prove that that the noninfringing substitute was *readily* available. See Micro Chemical, Inc. v. Lextron, Inc., 318 F.3d 1119, 1123 (Fed. Cir. 2003) (reversing a trial court finding of ready availability under *Grain Processing*). On the economic soundness of the overall approach, see John W. Schlicher, Measuring Patent Damages by the Market Value of Inventions — The *Grain Processing, Rite-Hite,* and *Aro* Rules, 82 J. Pat. & Trademark Off. Soc'y 503 (2000).

ii. The "Market Share" Rule

When more than two sellers share a market and at least one seller is a noninfringing competitor of the patentee, it would appear that one element in the *Panduit* test is missing: the absence of noninfringing substitutes. However, some patentees have overcome this deficiency with a clever argument regarding what would have happened if the infringer had not been in the market. Under this theory, the court is asked to assume that the patentee's market share *relative to the noninfringer* would have remained the same in the absence of the infringer. Consequently, it is assumed that the patentee would have made the same percentage of the infringer's sales as the patentee made in the overall market. See, e.g., State Indus. v. Mor-Flo, Inc., 883 F.2d 1573 (Fed. Cir. 1988), *cert. denied,* 493 U.S. 1022 (1990). With this theory, in other words, the presence or absence of noninfringing substitutes does not prevent the patentee from laying claim to at least some of the infringing sales. Id. at 1578. Some have argued, however, that the "market share" rule of *State Industries* must be adjusted to take account of the fact that in the absence of (at least some) competition, the price for a product would have been higher, and therefore that demand for the product would have been lower — undercutting an implicit assumption in *State Industries* that sales volume in the hypothetical world without the infringer would have been the same as the actual sales volume. See Roy J. Epstein, The Market Share Rule with Price Erosion: Patent Infringement Lost Profits Damages After *Crystal*, 31 Am. Intell. Prop. L.Q.J. 1 (2003).

Does it make sense that the patentee would have captured the sales actually made by the infringer? Is it possible that the infringer's advertising, customer base, or other assets were responsible for the sales, and that no one else — patentee or competitor — would have made them? Cf. Litton Sys. v. Honeywell Inc., 87 F.3d 1559 (Fed. Cir. 1996) ($1.2 billion jury verdict vacated because plaintiff's expert improperly assumed without evidence that all government contracts would have been awarded to plaintiff if the patented component had not been included in defendant's product). In deciding whether the infringer's customers will switch to the patentee or to another competitor, is the relative price of the infringer's goods relevant? What should a court do if the remaining competitors in the industry also sell infringing products, but were not parties to the suit?

iii. Lost Sales of Unpatented Components or Products

In yet a third line of cases, patentees have argued that since the infringer's presence cost it sales of unpatented components and products, these losses are compensable. The components cases are analyzed under the "entire market value" rule, which directs the court to ascertain whether the unpatentable components at issue are functionally integral to an overall product whose consumer demand is based on the patented component. See, e.g., TWM Mfg. Co. v. Dura Corp., 789 F.2d 895, 900 (Fed. Cir.), *cert. denied*, 479 U.S. 852 (1986). The idea is that the unpatented component would naturally and normally be sold along with the patented one, and thus the patentee deserves compensation for the lost sales of both. Does this amount to stretching the boundary of the claimed subject matter?[54] Would the patentee price patented components differently if the rule were otherwise?

A related theory was adopted by the Federal Circuit in the 1995 case of Rite-Hite Corp. v. Kelley Co., 56 F.3d 1538 (Fed. Cir. 1995) (en banc), *cert. denied*, 116 S. Ct. 184 (1995). In this case, the patentee, Rite-Hite, argued that the infringer's presence in the market cost it not only sales of its patented product but also sales of an *unpatented* competing product as well. The court accepted this argument, reasoning that "[i]f a particular injury was or should have been reasonably foreseeable by an infringing competitor in the relevant market, broadly defined, that injury is generally compensable absent a persuasive reason to the contrary." 56 F.3d at 1546.

Does this impermissibly shift the focus from the claimed invention to the infringer's behavior? If we accept a counterfactual world when the patentee is alone in the market for the patented item, why not consider an alternative where the infringer sells only the unpatented, competing product?

---

54. Compare the patent misuse doctrine, which distinguishes between patentees who attempt to control "staple" products and those who attempt to control "nonstaple" products associated with the invention.

Knorr-Bremse Systeme Fuer Nutzfahrzeuge GmbH v. Dana Corp., 133 F.Supp.2d 833 (E.D.Va.2001) (partial summary judgment) ("*Knorr-Bremse I*"); 133 F.Supp.2d 843 (E.D.Va.2001) (findings of fact and conclusions of law) ("*Knorr-Bremse II*"); Civ. A. No. 00-803-A (E.D.Va. Mar. 7, 2001) (final judgment); No. 00-803-A (E.D.Va. Apr. 9, 2001) (amended final judgment) ("*Knorr-Bremse III*").

iv. Post-Expiration Sales

It is accepted black-letter law that once a patent expires, others are free to use its teachings and to make, use, and sell competing products. But some ex-patentees have successfully asserted a claim for post-expiration damages on the theory that infringement during the term of the patent gave the infringer a "head start" on post-expiration sales, since the infringer did not have to start up production and marketing after the patent expired. Courts awarding or endorsing damages on such an "accelerated reentry" theory include TP Orthodontics, Inc. v. Professional Positioners, Inc., 17 U.S.P.Q.2d 1497, 1504-1506 (E.D. Wisc. 1990); Amsted Indus. v. National Castings, Inc., 16 U.S.P.Q.2d 1737 (N.D. Ill. 1990); BIC Leisure Prods. v. Windsurfing Int'l, 687 F. Supp. 134 (S.D.N.Y. 1988). In each case, post-expiration damages were part of an award that included damages for infringement during the term of the patent as well.

Compare these cases to the "head-start" injunction rule in trade secret law. If trade secret plaintiffs can get damages and injunctive relief after the secret has been disclosed, is there any reason to deny similar relief to patent plaintiffs?

COMMENTS AND QUESTIONS

The bottom line on patent damages—whether based on lost profits or reasonable royalty—is that they are growing. See William O. Kerr and Gauri Prakash-Canjels, Patent Damages and Royalty Awards: The Convergence of Economics and Law, LES Nouvelles, June, 2003 at 83 (collecting judgments and settlements of patent infringement cases accompanied by large damages). In this environment, one can expect even more novel arguments to be put forward.

3. Willful Infringement

Knorr-Bremse Systeme Fuer Nutzfahrzeuge Gmbh v. Dana Corporation
United States Court of Appeals for the Federal Circuit
383 F.3d 1337 (Fed. Cir. 2004) (in banc)

NEWMAN, Circuit Judge.

Knorr-Bremse Systeme Fuer Nutzfahrzeuge GmbH is the owner of United States Patent No. 5,927,445 (the '445 patent) entitled "Disk Brake For Vehicles Having Insertable Actuator," [and was] sued on July 27, 1999. At trial to the United States District Court for the Eastern District of Virginia, the appellants Dana Corporation, Haldex Brake Products Corporation, and Haldex Brake Products AB were found liable for infringement and willful infringement.[1] No damages were awarded, for

---

1. *Knorr-Bremse Systeme Fuer Nutzfahrzeuge GmbH v. Dana Corp.*, 133 F.Supp.2d 833 (E.D.Va.2001) (partial summary judgment) ("*Knorr-Bremse I*"); 133 F.Supp.2d 843 (E.D.Va.2001) (findings of fact and conclusions of law) ("*Knorr-Bremse II*"); Civ. A. No. 00-803-A (E.D.Va. Mar. 7, 2001) (final judgment); No. 00-803-A (E.D.Va. Apr. 9, 2001) (amended final judgment) ("*Knorr-Bremse III*").

there were no sales of the infringing brakes. Based on the finding of willful infringement the court awarded partial attorney fees under 35 U.S.C. § 285.

The appellants seek reversal of the finding of willful infringement, arguing that an adverse inference should not have been drawn from the withholding by Haldex of an opinion of counsel concerning the patent issues, and from the failure of Dana to obtain its own opinion of counsel. Applying our precedent, the district court inferred that the opinion of counsel withheld by Haldex was unfavorable to the defendants. After argument of the appeal we took this case en banc in order to reconsider our precedent with respect to these aspects. The parties were asked to submit additional briefing on four questions, and amicus curiae briefs were invited. We now hold that no adverse inference that an opinion of counsel was or would have been unfavorable flows from an alleged infringer's failure to obtain or produce an exculpatory opinion of counsel. Precedent to the contrary is overruled. We therefore vacate the judgment of willful infringement and remand for re-determination, on consideration of the totality of the circumstances but without the evidentiary contribution or presumptive weight of an adverse inference that any opinion of counsel was or would have been unfavorable.

Background

Knorr-Bremse, a German corporation, manufactures air disk brakes for use in heavy commercial vehicles, primarily Class 6-8 trucks known as eighteen wheelers, semis, or tractor-trailers. Knorr-Bremse states that air disk brake technology is superior to the previously dominant technology of hydraulically or pneumatically actuated drum brakes, and that air disk brakes have widely supplanted drum brakes for trucks in the European market.

Dana, an American corporation, and the Swedish company Haldex Brake Products AB and its United States affiliate, agreed to collaborate to sell in the United States an air disk brake manufactured by Haldex in Sweden. The appellants imported into the United States about 100 units of a Haldex brake designated the Mark II model. Between 1997 and 1999 the Mark II brake was installed in approximately eighteen trucks of Dana and various potential customers. The trucks were used in transport, and brake performance records were required to be kept and provided to Dana. Dana and Haldex advertised these brakes at trade shows and in industry media in the United States.

Knorr-Bremse in December 1998 orally notified Dana of patent disputes with Haldex in Europe involving the Mark II brake, and told the appellants that patent applications were pending in the United States. On August 31, 1999 Knorr-Bremse notified Dana in writing of infringement litigation against Haldex in Europe, and that Knorr-Bremse's United States '445 patent had issued on July 27, 1999. Knorr-Bremse filed this infringement suit on May 15, 2000. [The district court found that defendant's two different brake designs, "Mark II" and "Mark III," infringed the patent.]

On the issue of willful infringement, Haldex told the court that it had consulted European and United States counsel concerning Knorr-Bremse's patents, but declined to produce any legal opinion or to disclose the advice received, asserting the attorney-client privilege. Dana stated that it did not itself consult counsel, but relied on Haldex. Applying Federal Circuit precedent, the district court found: "It is reasonable to conclude that such opinions were unfavorable." The court discussed the evidence for and against willful infringement and concluded that "the totality of the circumstances compels the conclusion that defendants' use of the Mark II air disk brake, and indeed Dana's continued use of the Mark II air disk brake on various of its vehicles [after the judgment of infringement] amounts to willful infringement of the '445 patent." . . . Based on the finding of willful infringement the court found that the case was "exceptional" under 35 U.S.C. § 285, and awarded Knorr-Bremse its attorney fees for the portion of the litigation that related to the Mark II brake, but not the Mark III.

The appellants appeal only the issue of willfulness of the infringement and the ensuing award of attorney fees. Knorr-Bremse cross-appeals, seeking to enjoin the appellants from retaining and using the brake performance records and test data obtained through use of the Mark II brake.

I Willful Infringement

In discussing "willful" behavior and its consequences, the Supreme Court has observed that "[t]he word 'willful' is widely used in the law, and, although it has not by any means been given a perfectly consistent interpretation, it is generally understood to refer to conduct that is not merely negligent," *McLaughlin v. Richland Shoe Co.*, 486 U.S. 128, 133 (1988). . . . The concept of "willful infringement" is not simply a conduit for enhancement of damages; it is a statement that patent infringement, like other civil wrongs, is disfavored, and intentional disregard of legal rights warrants deterrence. Remedy for willful infringement is founded on 35 U.S.C. § 284 ("the court may increase the damages up to three times the amount found or assessed") and 35 U.S.C. § 285 ("the court in exceptional cases may award reasonable attorney fees to the prevailing party").

Determination of willfulness is made on consideration of the totality of the circumstances . . . and may include contributions of several factors, as compiled, *e.g.*, in *Rolls-Royce Ltd. v. GTE Valeron Corp.*, 800 F.2d 1101, 1110 (Fed.Cir.1986) and *Read Corp. v. Portec, Inc.*, 970 F.2d 816, 826-27 (Fed.Cir.1992). These contributions are evaluated and weighed by the trier of fact. . . .

Fundamental to determination of willful infringement is the duty to act in accordance with law. Reinforcement of this duty was a foundation of the formation of the Federal Circuit court, at a time when widespread disregard of patent rights was undermining the national innovation incentive. *See Advisory Committee On Industrial Innovation Final Report,* Dep't of Commerce (Sep.1979). Thus in *Underwater Devices, Inc. v. Morrison-Knudsen Co.*, 717 F.2d 1380 (Fed.Cir.1983) the court stressed the legal obligation to respect valid patent rights. The court's opinion quoted the infringer's attorney who, without obtaining review by patent counsel of the patents at issue, advised the client to "continue to refuse to even discuss the payment of a royalty." *Id.* at 1385. The attorney advised that "[c]ourts, in recent years, have — in patent infringement cases — found the patents claimed to be infringed upon invalid in approximately 80% of the cases," and that for this reason the patentee would probably not risk filing suit. *Id.* On this record of flagrant disregard of presumptively valid patents without analysis, the Federal Circuit ruled that "where, as here, a potential infringer has actual notice of another's patent rights, he has an affirmative duty to exercise due care to determine whether or not he is infringing," including "the duty to seek and obtain competent legal advice from counsel before the initiation of any possible infringing activity." *Id.* at 1389-90.

Underwater Devices did not raise any issue of attorney-client privilege . . . [but this issue] arose in *Kloster Speedsteel AB v. Crucible Inc.*, 793 F.2d 1565 (Fed.Cir.1986), where the Federal Circuit observed that the infringer "has not even asserted that it sought advice of counsel when notified of the allowed claims and [the patentee's] warning, or at any time before it began this litigation," and held that the infringer's "silence on the subject, in alleged reliance on the attorney-client privilege, would warrant the conclusion that it either obtained no advice of counsel or did so and was advised that its importation and sale of the accused products would be an infringement of valid U.S. patents." *Id.* at 1580. Thus arose the adverse inference, reinforced in *Fromson v. Western Litho Plate & Supply Co.*, 853 F.2d 1568 (Fed.Cir.1988), and establishing the general rule that "a court must be free to infer that either no opinion was obtained or, if an opinion were obtained, it was contrary to the infringer's desire to initiate or continue its use of the patentee's invention." *Id.* at 1572-73. Throughout this evolution the focus was not on attorney-client relationships, but on disrespect for law. However, implementation of this precedent has resulted in inappropriate burdens on the attorney-client relationship.

We took this case en banc to review this precedent. While judicial departure from stare decisis always requires "special justification," . . . the "conceptual underpinnings" of this precedent . . . have significantly diminished in force. The adverse inference that an opinion was or would have been unfavorable, flowing from the infringer's failure to obtain or produce an exculpatory opinion of counsel, is no longer warranted. Precedent authorizing such inference is overruled.

We answer the four questions presented for en banc review, as follows:

Question 1

When the attorney-client privilege and/or work-product privilege is invoked by a defendant in an infringement suit, is it appropriate for the trier of fact to draw an adverse inference with respect to willful infringement?

The answer is "no." Although the duty to respect the law is undiminished, no adverse inference shall arise from invocation of the attorney-client and/or work product privilege. . . .

Although this court has never suggested that opinions of counsel concerning patents are not privileged, the inference that withheld opinions are adverse to the client's actions can distort the attorney-client relationship, in derogation of the foundations of that relationship. We conclude that a special rule affecting attorney-client relationships in patent cases is not warranted. . . . There should be no risk of liability in disclosures to and from counsel in patent matters; such risk can intrude upon full communication and ultimately the public interest in encouraging open and confident relationships between client and attorney.

Question 2

When the defendant had not obtained legal advice, is it appropriate to draw an adverse inference with respect to willful infringement?

The answer, again, is "no." The issue here is not of privilege, but whether there is a legal duty upon a potential infringer to consult with counsel, such that failure to do so will provide an inference or evidentiary presumption that such opinion would have been negative.

Dana Corporation did not seek independent legal advice, upon notice by Knorr-Bremse of the pendency of the '445 application in the United States and of the issuance of the '445 patent, followed by the charge of infringement. In tandem with our holding that it is inappropriate to draw an adverse inference that undisclosed legal advice for which attorney-client privilege is claimed was unfavorable, we also hold that it is inappropriate to draw a similar adverse inference from failure to consult counsel. The amici curiae describe the burdens and costs of the requirement, as pressed in litigation, for early and full study by counsel of every potentially adverse patent of which the defendant had knowledge, citing cases such as *Johns Hopkins Univ. v. Cellpro, Inc.,* 152 F.3d 1342, 1364 (Fed.Cir.1998), wherein the court held that to avoid liability for willful infringement in that case, an exculpatory opinion of counsel must fully address all potential infringement and validity issues. Although other cases have imposed less rigorous criteria, the issue has occasioned extensive satellite litigation, distorting the "conceptual underpinnings" of *Underwater Devices* and *Kloster Speedsteel.* Although there continues to be "an affirmative duty of due care to avoid infringement of the known patent rights of others," *L.A. Gear, Inc. v. Thom McAn Shoe Co.,* 988 F.2d 1117, 1127 (Fed.Cir.1993), the failure to obtain an exculpatory opinion of counsel shall no longer provide an adverse inference or evidentiary presumption that such an opinion would have been unfavorable. * * *

Question 4

Should the existence of a substantial defense to infringement be sufficient to defeat liability for willful infringement even if no legal advice has been secured?

The answer is "no." Precedent includes this factor with others to be considered among the totality of circumstances, stressing the "theme of whether a prudent person would have sound reason to believe that the patent was not infringed or was invalid or unenforceable, and would be so held if litigated," *SRI Int'l, Inc. v. Advanced Tech. Labs. Inc.,* 127 F.3d 1462, 1465 (Fed.Cir.1997). However, precedent also authorizes the trier of fact to accord each factor the weight warranted by its strength in the particular case. We deem this approach preferable to abstracting any factor for per se treatment, for this greater flexibility enables the trier of fact to fit the decision to all of the circumstances. We thus decline to adopt a per se rule....

Summary

An adverse inference that a legal opinion was or would have been unfavorable shall not be drawn from invocation of the attorney-client and/or work product privileges or from failure to consult with counsel. Contrary holdings and suggestions of precedent are overruled.

This case is remanded for redetermination of the issue of willful infringement and any remedy therefor.

DYK, Circuit Judge, concurring-in-part and dissenting-in-part.

I join the majority opinion insofar as it eliminates an adverse inference (that an opinion of counsel would be unfavorable) from the infringer's failure to disclose or obtain an opinion of counsel. I do not join the majority opinion to the extent that it may be read as reaffirming that "where, as here, a potential infringer has actual notice of another's patent rights, he has an affirmative duty to exercise due care to determine whether or not he is infringing." Maj. Op., *ante,* at 1343 (quoting *Underwater Devices, Inc. v. Morrison-Knudsen Co.,* 717 F.2d 1380, 1389 (Fed.Cir.1983)).

There is a substantial question as to whether the due care requirement is consistent with the Supreme Court cases holding that punitive damages can only be awarded in situations where the conduct is reprehensible. *See, e.g., State Farm Mut. Auto. Ins. Co. v. Campbell,* 538 U.S. 408 (2003); *BMW of N. Am., Inc. v. Gore,* 517 U.S. 559 (1996) ("*Gore*"). While the majority properly refrains from addressing this constitutional issue, as it has not been briefed or argued by the parties or *amici* in this case, I write separately to note my view that enhancing damages for failure to comply with the due care requirement cannot be squared with those recent Supreme Court cases.

* * * *

COMMENTS AND QUESTIONS

1. The consensus response to *Knorr-Bremse* is that (1) it was wise for the Federal Circuit to do away with the negative inference concerning lack of opinion letters, but (2) the remaining "duty of care" premised on a "totality of the circumstances" still leaves plenty of room for infringement defendants to worry about willful infringement damage awards. See, e.g., Carol Johns, Comment, Knorr-Bremse Systeme Fuer Nutzfahrzeuge Gmbh v. Dana Corp.: A Step In The Right Direction For Willful Infringement, 20 Berkeley Tech. L.J. 69 (2005). This means that, as a practical matter, opinion letters may still be very helpful in avoiding willful infringement liability. See Mark A. Lemley & Ragesh K. Tangri, Ending Patent Law's Willfulness Game, 18 Berkeley Tech. L.J. 1085, 1115 (2003) (predicting even pre-*Knoerr-Bremse* that "while eliminating the adverse inference from failure to disclose an opinion of counsel may be a good idea, . . . [t]he defendant's best hope of avoiding liability for willfulness is [still] to rely on the advice of counsel.").

2. The Read v. Portec "totality of the circumstances" factors, mentioned in the *Knoerr-Bremse* opinion, are:

(1) [W]hether the infringer deliberately copied the ideas or design of another[.]
(2) [W]hether the infringer, when he knew of the other's patent protection, investigated the scope of the patent and formed a good-faith belief that it was invalid or that it was not infringed[.]
(3) [T]he infringer's behavior as a party to the litigation.
(4) Defendants' size and financial condition.
(5) Closeness of the case.
(6) Duration of defendant's misconduct.
(7) Remedial action by the defendant.
(8) Defendant's motivation for harm.
(9) [W]hether defendant attempted to conceal its misconduct.

Read, 970 F.2d at 827 (citations omitted). The open-ended nature of this list makes willfulness a difficult issue for lawyers to predict with any accuracy, meaning there is still plenty of ambiguity and nuance in the law of willfulness, even post-*Knoerr-Bremse*. See, e.g., Fuji Photo Film Co., Ltd. v. Jazz Photo Corp., 394 F.3d 1368, 1379 (Fed. Cir. 2005) (considering "totality of circumstances," defendant had actual notice of patent, so jury's willfulness finding upheld). As many have pointed out, this means that despite all the discussion of these issues in recent years, there is still reason for firms to worry about research employees gaining access to and reading patents issued to competitors. See, e.g., Note, The Disclosure Function of the Patent System (Or Lack Thereof), 118 Harv. L. Rev. 2007 (2005).

And in general, even post-*Knoerr-Bremse*, there is still reason to be concerned over the possibility of willful infringement liability. See, e.g., Liquid Dynamics Corp. v. Vaughan Co., Inc., 2005 WL 711993 (N.D.Ill. 2005), at *2 (Upholding district court finding of willful infringement, based in part on jury findings: "Presumably, the jury disregarded [defendant's in-house patent counsel's] testimony [regarding his opinion that plaintiff's patent was invalid and not infringed], resolved credibility determinations in [plaintiff] Liquid Dynamics' favor, and rejected [defendant] Vaughan's good faith defense.... The jury's determination was reasonable, given [defendant's patent lawyer's] bias and interest in the case's outcome, and was supported by sufficient evidence, given the evidence of deliberate copying [of plaintiff's patented design]."); Software AG v. Bea Systems, Inc., 2005 WL 859266, *2 (D. Del. 2005) (declining defendant's effort to introduce evidence of plaintiff's patent lawyer's involvement in pre-litigation consultations, because this might prejudice jury's thinking regarding defendant's intent to infringe, and thus plaintiff's prospects of receiving willful infringement damages).

3. Some recent proposals for patent reform further reduce the threat of willful infringement. See for example Section 6 of H.R. 2795, 109th Cong., 1st Sess., introduced by Representative Lamar Smith, summarized at 70 Pat. Tm. & Copyrt. J. 142 (June 10, 2005). The proposed amendment to 35 U.S.C. §284 limits willfulness to cases where the infringer (1) receives notice from the patentee; (2) intentionally copies an invention with knowledge of the patent; or (3) engages in further infringing activities after having been found liable for infringement in an earlier court proceeding. They also make it much harder for patentees to allege willfulness.

Do we need a doctrine of willful infringement? Is there a risk that without some sort of damages enhancement infringers will just copy patented inventions with impunity, secure in the knowledge that they will only have to pay a reasonable royalty? Or does the risk of being enjoined mid-production serve as an adequate deterrent? See generally Roger D. Blair and Thomas F. Cotter, Intellectual Property: Economic and Legal Dimensions of Rights and Remedies 42-66 (2005) (modeling compensation and deterrence in patent cases, and discussing doctrinal implications).

G. DESIGN AND PLANT PATENTS

Section 101 is the subject matter provision of what is referred to as the "utility" patent statute. Most patents issued in the United States are utility patents, and utility patents are what most people generally think of when they speak generically of "patent law." Several other provisions of the patent law cover related material, most importantly designs and certain types of plants. Here we review briefly these special statutes.

1. Design Patents

The aesthetic appearance of a product rather than its functional features are protected by the design patent. Design patents have been obtained for a wide range of products including shoes, hats, furniture, tools, packaging, televisions, automobile designs, and computer graphics. Design patent law overlaps both copyright law and trademark and unfair competition law in its coverage of the nonfunctional features of useful objects. 1 Donald S. Chisum, Patents, § 1.04 at 1-180 (1992). Due to a lengthy processing time, high application cost, strict requirements that are vague and difficult to apply, and a long history of judicial hostility, the design patent system has been criticized as ineffective and in need of reform. As the courts have recently become more receptive to design patents, interest has been renewed in this form of intellectual property protection.

a. Introduction

The original design patent law in the United States was enacted in 1842 to fill a gap that then existed between copyright protection for authors and patent protection for inventors in the mechanical arts. Chisum, *supra*, § 1.04[1] at 1-180. The intent of the statute, which extended patent protection to "new and original designs for articles of manufacture" was to "give encouragement to the decorative arts." Gorham Mfg. Co. v. White, 81 U.S. (14 Wall.) 511, 524-525 (1871). Appearance, rather than utility, is the crucial factor for consideration in design patent protection. A design may consist of surface ornamentation, configuration, or a combination of both. Id. The Design Patent Act, which was codified in 35 U.S.C. § 171, allows a design patent to be obtained for "any new, original and ornamental design for an article of manufacture" and provides that most provisions relating to patents for inventions also apply to design patents. 35 U.S.C. § 171 (1988). Design patents are issued for a 14-year term.

b. Requirements for Patentability

A design is patentable if it meets the requirements of novelty, originality, and nonobviousness, is ornamental, and is not dictated by functional considerations. Chisum, *supra*, § 1.04[2] at 1-184. The Patent and Trademark Office defines a design as "the visual characteristics or aspects displayed by the object. It is the appearance presented by the object which creates a visual impact upon the mind of the observer." U.S. Patent and Trademark Office, Manual of Patent Examining Procedure, § 1502 (5th ed. Rev. 8, 1988).

An "article of manufacture" has been broadly defined to include silverware, Gorham Mfg. Co. v. White, 81 U.S. (14 Wall.) 511 (1871); cement mixers, In re Koehring, 37 F.2d 421 (C.C.P.A. 1930); furniture, In re Rosen, 673 F.2d 388 (C.C.P.A. 1982); and containers for liquids, Unette Corp. v. Unit Pack Co., 785

F.2d 1026 (Fed. Cir. 1986). In In re Hruby, 373 F.2d 997, 1000 (C.C.P.A. 1967), the Court of Customs and Patent Appeals held that "a manufacture is anything 'made by the hands of man' from raw materials whether literally by hand or by machinery or by art" and found that a design created by the flow of water in a fountain was patentable. While the design must be embodied in an article of manufacture, a design patent may be obtained for only part of an article. In re Zahn, 617 F.2d 261 (C.C.P.A. 1980) (holding that the shank of a drill bit was patentable under section 171). Thus the subject matter of a design patent is not limited to designs for discrete articles.

i. Novelty

Novelty is established if no prior art shows exactly the same design. A design is novel if the "ordinary observer," viewing the new design as a whole, would consider it to be different from, rather than a modification of, an already existing design. See, e.g., Clark Equip. Co. v. Keller, 570 F.2d 778, 799 (8th Cir. 1978). Section 102(b) of the Patent Act of 1952 provides for a one-year novelty grace period, allowing a patent for a design as long as the design has not been published anywhere or sold or displayed within the United States more than one year before the application is filed with the Patent and Trademark Office. 35 U.S.C. § 102(b) (1988). This gives a designer the opportunity to "test market" a design before incurring the expense of a patent application. J.H. Reichman, Design Protection and the New Technologies: The United States Experience in a Transnational Perspective, 19 U. Balt. L. Rev. 6, 23 (1991).

ii. Nonobviousness

Because section 171 provides that the provisions of Title 35 relating to utility patents also apply to design patents, a design must be nonobvious, which requires the "exercise of the inventive or originative faculty." Smith v. Whitman, 148 U.S. 674, 679 (1893). The test for determining nonobviousness of a utility patent is whether the differences between the invention sought to be patented and the prior art would be obvious at the time of invention to a person with "ordinary skill in the pertinent art." Graham v. John Deere Co., 383 U.S. 1, 14 (1966). See also Chisum, *supra*, § 1.04[2] at 1-200. The *Graham* test requires the court to ascertain the scope and content of the prior art, the differences between the prior art and the claim at issue, and the level of ordinary skill in the art. *Graham*, 383 U.S. at 17. In addition to these subjective inquiries, secondary considerations such as commercial success and long-felt need in the industry are relevant to determine obviousness. Id. at 17-18.

Prior to the adoption of a uniform standard to determine nonobviousness in design patents, the courts were in conflict over whether to use an "ordinary designer" standard or that of an "ordinary intelligent man" when applying the *Graham* test. See In re Laverne, 356 F.2d 1003 (C.C.P.A. 1966). Determining the nonobviousness of a design patent, as opposed to a utility patent, is unpredictable, as it is an inherently subjective inquiry, depending largely on personal taste. In evaluating nonobviousness of utility patents, judges can measure the distance between the prior art and a new invention on the basis of uniform scientific criteria and technical data, while the evaluation of the distance between an appearance design and its predecessors necessarily involves "value judgments that are hard to quantify and unreliable at best." Reichman, *supra*, at 33 n.164. See In re Bartlett, 300 F.2d 942, 944 (C.C.P.A. 1962) (noting that the determination of patentability in design cases depends on the subjective conclusion of each judge). The nonobviousness requirement has been cited as a primary factor in limiting the availability of design protection in the United States from the 1920s on. Reichman, *supra*, at 24. Without a clear standard to follow, the appellate courts could apply their own stricter view of nonobviousness, and they routinely invalidated design patents, despite the presumption of validity accorded to a grant by the Patent and Trademark Office. Id.

In 1981, the Court of Customs and Patent Appeals held that the *Graham* test could apply to design patents, and nonobviousness would be measured in terms of a "designer of ordinary capability who designs articles of the type presented in the application." In re Nalbandian, 661 F.2d 1214, 1216 (C.C.P.A. 1981). The new standard allows for objective evidence of expert testimony from designers in the field to be used to prove nonobviousness. Id. at 1217. The Federal Circuit, which was created in 1982 to hear all patent appeals, subsequently adopted this standard. While this approach appears more evenhanded, and has led to more patents being upheld as valid, it may not solve the problem of unpredictability, because the opinions of different designers can vary considerably. William T. Fryer, III, Industrial Design Protection in the United States of America — Present Situation and Plans for Revision, 70 J. Pat. & Trademark Off. Soc'y 821, 829 (1988).

The Federal Circuit has emphasized the presumptive validity of a design patent, and placed the burden on the challenger to come forward with clear and convincing proof of nonobviousness. See, e.g., Trans-World Mfg. Corp. v. Al Nyman & Sons, Inc., 750 F.2d 1552, 1559-60 (Fed. Cir. 1984); Avia Group Intl., Inc. v. L.A. Gear Cal., 853 F.2d 1557, 1562 (Fed. Cir. 1988). The court also upheld the notion that to find obviousness, reference must be made to prior art with the same overall appearance of the patented design, rather than to a combination of features from several references. Litton Sys., Inc. v. Whirlpool Corp., 728 F.2d 1423, 1443 (Fed. Cir. 1984); see also In re Rosen, 673 F.2d 388, 390-91 (C.C.P.A. 1982). Most significantly, the Federal Circuit held that objective secondary considerations which apply to utility patents, such as commercial success and copying, are also relevant to determine nonobviousness of design patents. See, e.g., Litton, 728 F.2d at 1441; Avia, 853 F.2d at 1564. The theory supporting the consideration of commercial success is that the purpose of a design patent is to increase salability, so if a design has been a success it "must have been sufficiently novel and superior to attract attention." Robert W. Brown & Co. v. De Bell, 243 F.2d 200, 202 (9th Cir. 1957); see also Chisum, *supra*, at §1.04[2][f] at 1-208. Evidence of commercial success must be related to the patented design rather than to factors such as functional improvement or advertising. See, e.g., Litton, 728 F.2d at 1443; Avia, 853 F.2d at 1564. The approach adopted by the Federal Circuit with respect to designs challenged for obviousness has been cited as effectively lowering the invalidation rate to only 38 percent from a rate of 75 to 100 percent only a few years earlier. Reichman, *supra*, at 37.

iii. Ornamentality

A patentable design must be ornamental, creating a more pleasing appearance. To satisfy the requirement of ornamentality, a design "must be the product of aesthetic skill and artistic conception." Blisscraft of Hollywood v. United Plastics Co.,

294 F.2d 694, 696 (2d Cir. 1964). This requirement has been met by articles which are outside the realm of traditional "art." See In re Koehring, 37 F.2d 421, 422 (C.C.P.A. 1930) (determining a design for a cement mixer to be ornamental because it "possessed more grace and pleasing appearance" than prior art). A number of cases have denied patentability to designs which are concealed during the normal use of an object, on the basis that ornamentality requires the design to be visible while the object is in its normal and intended use. See Chisum, *supra*, § 1.04[2][c] at 1-190-91.

iv. Functionality

If a design is "primarily functional rather than ornamental," or is "dictated by functional considerations," it is not patentable. Power Controls Corp. v. Hybrinetics, Inc., 806 F.2d 234, 238 (Fed. Cir. 1986). The functionality rule furthers the purpose of the design patent statute, which is to promote the decorative arts. In addition, the rule prevents granting in essence a monopoly to functional features that do not meet the requirements of a utility patent. Chisum, *supra*, § 1.04[2][d] at 1-195. Recognizing that strict application of the functionality rule would invalidate the majority of modern designs, the Federal Circuit validated designs with a higher functionality factor than had been tolerated by the courts previously. Reichman, *supra*, at 40. This is evidenced in cases upholding design patents for an eyeglass display rack, Trans-World Mfg. Corp. v. Al Nyman & Sons, Inc., 750 F.2d 1552 (Fed. Cir. 1984); fiberglass camper shells, Fiberglass in Motion, Inc. v. Hindelang, No. 83-1266 (Fed. Cir. Apr. 19, 1984); and containers for dispensing liquids, Unette Corp v. Unit Pack Co., 785 F.2d 1026 (Fed. Cir. 1986). The Federal Circuit has held that a design may have functional components as long as the design does not embody a function that is necessary to compete in the market. Avia, 853 F.2d at 1563. If the functional aspect of a design may be achieved by other design techniques, then it is not primarily functional. Id. This more flexible approach reflects a recognition by the court that the majority of valuable industrial designs which should be granted protection in order to stimulate economic growth are a combination of functional and aesthetic features.

c. Claim Requirements and Procedure

Two major criticisms of the design patent system in the United States are that it is too expensive and that protection takes too long to obtain. See Fryer, *supra*, at 834; Reichman, *supra*, at 24 (procedural requirements make design protection in the United States much "slower and costlier to obtain" than in other countries); Perry J. Saidman, Design Patents — The Whipping Boy Bites Back, 73 J. Pat. & Trademark Off. Soc'y 859 (1991) (defending against these criticisms).

In 1988, the cost of a design patent application was estimated at $1000. Fryer, *supra*, at 835. A large part of this cost is the expense of preparing the drawings which constitute the claim. The drawings "must contain a sufficient number of views to constitute a complete disclosure of the appearance of the article." 37 CFR § 1.152. All that is required in writing is a very brief description of the drawings. The adequate disclosure and definiteness of the claim required by section 112 are accomplished by the drawings. Chisum, *supra*, § 1.04[3]. Only one claim can be included in a design application. An application may illustrate more than one embodiment of a design only if the embodiments involve a "single inventive concept" and can be protected by a single claim. In re Rubenfield, 270 F.2d 391, 396 (C.C.P.A. 1959), *cert. denied*, 362 U.S. 903 (1960).

The pendency time for a design patent is approximately two to three years. Saidman, *supra*, at 861. Reasons for this delay include budget and staffing restraints at the Patent and Trademark Office. Id. This leaves a design patent applicant without patent protection from copiers during this long waiting period, as opposed to copyright protection, which requires no initial procedural requirement and takes only a few months for registration to be issued. Fryer, *supra*, at 840, 835. Of course, it is still possible to protect such designs in appropriate cases under both copyright and trade dress principles. More on that in Chapters 4 and 5.

It has been noted that the current system is "unsuited to the fast-moving but short-lived product cycle characteristic of today's market for mass-produced consumer goods." Reichman, *supra*, at 24. While there is general agreement that a new form of protection for industrial design is needed, Congress has yet to adopt any of the proposed legislation. An example of proposed legislation that would afford better protection for designs is the Design Copyright Protection Act, which employs a modified copyright form of protection as an alternative to the design patent. See Fryer, *supra*, at 839-846. Alternatively, some current proposed legislation would protect functional industrial designs of every kind, including those that are neither aesthetically nor technically innovative. Reichman, *supra*, at 121-122.

d. Infringement

The standard for finding infringement of a design patent was defined in Gorham Mfg. Co. v. White, where the Supreme Court held that "if in the eye of an ordinary observer, giving such attention as a purchaser usually gives, two designs are substantially the same, if the resemblance is such as to deceive such an observer, inducing him to purchase one supposing it to be the other, the first one patented is infringed by the other." 81 U.S. (14 Wall.) 511, 528 (1872). This "eye of the ordinary observer" standard continues to be the rule followed by the Federal Circuit. See Oakley, Inc. v. Intern. Tropic-Cal, Inc., 923 F.2d 167, 169 (Fed. Cir. 1991). The ordinary observer is one who has "reasonable familiarity" with the object in question and is capable of making a comparison to other objects which have preceded it. Applied Arts Corp. v. Grand Rapids Metalcraft Corp., 67 F.2d 428, 430 (6th Cir. 1933). The key factor to determine infringement under the Gorham test is similarity, rather than consumer confusion. Unette, 785 F.2d at 1029 (holding that "likelihood of confusion as to the source of goods is not a necessary or appropriate factor for determining infringement of a design patent").

The second prong of the infringement analysis is the "point of novelty" test, which is distinct from the issue of similarity. Under the "point of novelty" test, the similarity found by the ordinary observer must be attributable to the novel elements of the patented design that distinguish it from prior art. Litton, 728 F.2d at 1444; Avia, 853 F.2d at 1565; FMC Corp. v. Hennessy Indus., Inc., 836 F.2d 521, 527 (Fed. Cir. 1987). Unless the accused design appropriates the novel features of the patented design, there has been no infringement. Avia, 853 F.2d at 1565. The scope of the patented claim and its points of novelty are determined by examining the field of the prior art. See Litton, 728 F.2d at 1444 (holding that where the field of prior art is "crowded," the scope of a claim will be construed narrowly).

After determining the scope of the patented claim, the infringement inquiry focuses only on the protectable aesthetic components of a patented design. See Lee v. Dayton-Hudson Corp., 838 F.2d 1186, 1188 (Fed. Cir. 1988) (holding that "it is the non-functional, design aspects that are pertinent to determinations of infringement"). Thus this test permits strong similarities to be excused if the defendant can prove that she borrowed "only commonplace or generic ideas, functional features, or other nonprotectable matter" while adding sufficient variation to protectable elements of the design. Reichman, *supra*, at 44.

Whether a design is infringed when it is used on an entirely different type of article than the patented one has not been settled. Chisum, *supra* at § 1.04[4] at 1-225. In Avia Group Intl. v. L.A. Gear Cal., where the patented design was for an adult athletic shoe and the accused design was for a children's shoe, the court found infringement and held that even in a situation where the patent holder has not put out a product or where the patented design is embodied in a product that does not compete with the patent holder's product, a finding of infringement is not precluded. 853 F.2d at 1565. This decision indicates that the Federal Circuit is willing to extend design patent protection beyond "literal infringement" and protect the design concept itself. Reichman, *supra*, at 53.

2. Plant Patents

Today, new plants can be protected with utility patents under the general standards discussed earlier in this chapter. J.E.M. Ag Supply, Inc. v. Pioneer Hi-Bred Int'l, Inc., 534 U.S. 124 (2001). In addition to such protection, plants can be protected under two acts specifically designed for them — the Plant Patent Act and the Plant Variety Protection Act.

a. The Plant Patent Act

Prior to passage of the Plant Patent Act of 1930, 35 U.S.C. §§ 161-164 (1988) (PPA), inventions in plants faced two obstacles to patentability. Like other living organisms, plants were believed to be ineligible for patent protection as they were products of nature. See, e.g., Ex parte Latimer, 1889 Commn. Dec. 123 (1889) (fiber from needle of evergreen tree unpatentable product of nature). The second barrier was the fact that "plants were not thought amenable to the 'written description' requirement" now found at § 112 of the Patent Act. Congress addressed these issues in enacting the Plant Patent Act, now embodied in sections 161-64 of Title 35 of the U.S. Code.

The PPA extends patent protection to inventors and discoverers of "any distinct and new" variety of asexually reproducing plant.[55] By judicial interpretation, the term

---

55. 35 U.S.C. § 161 (1988). Asexual reproduction is reproduction that does not involve the use of seeds. Examples of such methods are budding, grafting, cutting, division, or layering. This limitation was apparently premised on the perception that plants which reproduced other than asexually could not be reproduced reliably true to type. Nicholas O. Seay, Protecting the Seeds of Innovation: Patenting Plants, 16 AIPLA Q.J. 418 (1989).

plant is used in its lay rather than scientific sense and thus bacteria are not eligible subject matter. In re Arzberger, 112 F.2d 834 (C.C.P.A. 1940).

According to section 161 of the PPA, the provisions of Title 35 that apply to utility patents apply equally to plant patents "except as otherwise provided." However, the elements of plant patentability are not the same as for utility patents. For example, the PPA substitutes the requirement of distinctness for that of utility. Yoder Brothers, Inc. v. California-Florida Plant Corp., 537 F.2d 1347 (5th Cir. 1976), *cert. denied,* 429 U.S. 1094 (1977). Distinctness is measured by examining the characteristics that make the plant clearly distinguishable from other existing plants. As with utility patents, the characteristics which distinguish the plant from others do not have to be superior, merely different.

Although the nonobviousness requirement is theoretically applicable to plant patents, courts have admitted difficulty in applying it in this context. After rephrasing the traditional three-part obviousness test for the plant world, the courts are left with a standard reminiscent of the "distinctness" requirement discussed above. Thus nonobviousness is viewed as a requirement for "invention" resulting in substantial, rather than merely minor, distinctions in the plant.

Perhaps the most significant difference between prosecution of utility patents and plant patents lies in the exemption from the written description requirement (35 U.S.C. § 112) granted plant patent applicants by the PPA. 35 U.S.C. § 162. Noncompliance with § 112 is permitted provided the applicant's description is as complete as is reasonably possible.

The protection offered by plant patents is the exclusive right to reproduce the plant asexually. 35 U.S.C. § 163. Each plant patent application may include only one claim directed specifically to the plant shown and described. 35 U.S.C. § 162. Courts are split as to whether a plant patent covers independent derivation of a plant having the same varietal characteristics or only covers plant material actually derived from the patentee's plant. Cole Nursery Co. v. Youdath Perennial Gardens, Inc., 17 F. Supp. 159 (N.D. Ohio 1936); Yoder Bros., 537 F.2d at 1380; Pan-American Plant Co. v. Matsui, 433 F. Supp. 693, 694 (N.D. Cal. 1977). Regarding the doctrine of equivalents, courts are unanimous in holding that the doctrine is inappropriate in interpreting the claim of a plant patent.

While prosecution under the PPA may be less expensive than obtaining a utility patent, the applicant must be aware of the shortcomings of a plant patent. First, parts of plants are not protected as they would be under a utility patent. Additionally, there may be no doctrine of equivalents for plant patents. Lastly, there is no infringement in the sexual reproduction of patented plant material nor any biological material derived from the sexual reproduction of the plant.

b. The Plant Variety Protection Act

The second special statutory system to protect rights to plant properties is the Plant Variety Protection Act, 7 U.S.C. § 2321-2582 (PVPA). Enacted in 1970, the PVPA essentially parallels the PPA for protection of plant varieties that are sexually reproduced by seed. Successful applicants under the PVPA are awarded not a patent, but a "Certificate of Plant Variety Protection."

PVPA certificates are available to the breeder "of any novel variety of sexually reproduced plant," provided the variety possess "distinctness," "uniformity," and "stability." 7 U.S.C. § 2402. The "distinctness" requirement is most critical, requiring that the variety "clearly differs by one or more identifiable morphological, physiological or other characteristics. . . . " 7 U.S.C. § 2401. Just as the PPA has been interpreted to include "plants" only in the layman's sense, the PVPA explicitly denies protection to fungi, bacteria, or first-generation hybrid plants.[56] In re Bergy, 596 F.2d 952 (C.C.P.A. 1979), *vacated on other grounds,* 444 U.S. 1028 (1980).

Although the PVPA is an example of a registration system, it does share many procedural requirements with the utility patent scheme: a series of statutory bars, content requirements for the application, and a requirement for a seed deposit to be made with the Plant Variety Protection Office. Still, the PVPA features several provisions found in neither the utility patent nor plant patent systems. These provisions include a mandatory license, a series of statutory exemptions for saved seed, an exemption for sales by farmers, and a research exemption.

A PVPA certificate grants the breeder the right to exclude others from "selling the variety, or offering it for sale, or reproducing it, or importing it, or exporting it, or using it in producing . . . a hybrid or different variety therefrom. . . ." 7 U.S.C. § 2483(a). The protection lasts for 18 years. Although the PVPA does not specifically address the question of independent derivation as a potential defense to infringement, the inclusion of a provision for an interference proceeding suggests that independent derivation is not a defense. Recent technological advances enable scientists to determine a plant's ancestry from its molecular composition, thus making detection of infringement much easier. This development, coupled with the significant monetary investment required to develop novel seed varieties, may account for the recent willingness of firms to vigorously assert their legal claims to specific plant varieties. See A. Hagedorn, Suits Sprout over Rights to Seeds, Wall St. J., Mar. 5, 1990, B1, B8.

The "farmer's exemption" mentioned above permits individual farmers to sell protected varieties to other farmers without liability. By diluting the exclusive rights enjoyed by a PVPA certificate holder, this exemption appears to be at odds with the PVPA's primary purpose of establishing incentives for breeders to develop new strains. After struggling with this inherent contradiction in Delta and Pine Land Co. v. People Gin Co., 694 F.2d 1012 (5th Cir. 1983), the Fifth Circuit held that Congress intended the farmer's exemption to be read narrowly; thus the exemption requires that sales be made directly from farmer to farmer — the active participation of a middleman being prohibited.

More recently, defendant farmers accused of engaging in "brown bag sales" invoked the farmer's exemption as a defense to infringement liability. The Supreme Court held that the "farmer's exemption" for saved seed allowed farmers to sell only such seed as was saved to replant their own acreage and not to engage in the business of selling protected seeds. Asgrow Seed Co. v. Winterboer, 513 U.S. 179 (1995).

COMMENTS AND QUESTIONS

1. There is no question that intellectual property rights have affected the growth and structure of the seed industry. One article observes:

> [Since *Chakrabarty,*] [n]umerous companies have . . . filed patent applications that cover the genes, the processes of isolating the genes, and making the genetically modified plants

---

56. First-generation hybrid plants are those produced by mass-breeding two different inbred varieties. First-generation hybrids do not reproduce true to type.

and seeds themselves.... Although no one disputes that companies that have invested heavily in R&D to isolate, test, and commercialize genes are entitled to protection for their inventions, there is considerable debate within the seed industry concerning how much protection is deserved and what impact patents will have on the cooperative nature of the seed industry itself.

C. S. Gasser & R. T. Fraley, Genetically Engineering Plants for Crop Improvement, 244 Science 1293 (16 June 1989). The authors note that before the passage of the Plant Variety Protection Act in 1970, only three companies sold commercial soybean seeds; now there are "more than 40." Id. See also W. Lesser & R. Masson, An Economic Analysis of the Plant Variety Protection Act 123 (1985) (summarizing positive stimulus to industry from PVPA); Evenson, Intellectual Property Rights and Agribusiness Research and Development: Implications for the Public Agricultural Research System, 65 Am. J. Agric. Econ. 967 (1983) (same). For a less optimistic view, see Julian M. Alston and Raymond J. Venner, The Effects of the U.S. Plant Variety Protection Act on Wheat Genetic Improvement, 31 Res. Pol'y 527 (2002) (PVPA had no effect on experimental or commercial wheat yields).

A good case can be made that the expanded coverage of the conventional Patent Act (as opposed to the PVPA) will further spur plant-related research in this country. See Lesser, Patenting Seeds in the United States of America: What to Expect, Industrial Prop., Sept. 1986, at 360. There is an inevitable cost, however: more litigation. See, e.g., A. Hagedorn, Suits Sprout over Rights to Seeds, Wall St. J., Mar. 5, 1990, at 131, 138 (describing suit under PVPA over "Napolean" celery variety: "With companies spending millions of dollars yearly on biotechnology to create novel seed varieties, the costs of losing the seeds to competitors are greater than ever.").

2. Critics of intellectual property rights in the agriculture sector contend that they help accelerate undesirable trends such as centralization and the loss of economic power by small farmers. See, e.g., J. Kloppenburg, First the Seed: The Political Economy of Plant Biotechnology (1988). Kloppenburg notes that there was a great controversy in the agriculture world when in 1956 a researcher received a patent for hybrid plant breeding techniques. Id. at 113. Kloppenburg uses this as an example of a longstanding rift between the open and public-minded nature of federally funded agriculture research and the orientation toward private gain of private commercial researchers in this sector, whose growing prominence he says contributes to "the commodification of the seed." Id. at 282-284. His general thrust, echoed by others, is that agriculture is a special industry that is not always well-served by competition among private interests.

3. Recall the excerpt from the book by Robert Nozick at the beginning of this chapter. What is the difference between what is protected by plant patents and the protection of "discovered" plants alluded to in the excerpts? What arguments could you make that no special incentives are needed to discover plants, as opposed to the creation of plant-related "inventions" protectable under the patent code and the UPOV? Even if "discovered" plants were given some sort of protection, does that mean that all work on plant-related inventions would cease? See J. Brodovsky, The Mexican Pharmochemical and Pharmaceutical Industries, in The United States and Mexico: Face to Face with New Technology 198 (C. Thorup ed. 1987) (Mexican monopoly on barbasco, a plant that was a good source for making steroids, ended when purchasers developed alternative supply sources after Mexican government imposed higher price for barbasco).

4. Some countries have asserted that they "own" the genetic material from plants that grow inside their borders. Many are poor countries from tropical regions where a great variety of plant species grow. These countries insist that companies from developed countries pay "royalties" for the right to remove genetic material for research or the development of new products. Cf. Jorge Pina, Greenpeace Heads Global Campaign Against "Biopiracy," Inter-Press Service, Jun. 25, 2001, available at Westlaw "News" library (discussing a proposed UN treaty that would prevent patents related to the world's main food crops). Is this a form of intellectual property? Could the U.S. government make such a claim?

Others have contended that for humans to assert any "ownership" over inventions derived essentially from nature is sheer hubris. See Leon R. Kass, Patenting Life, 63 J. Pat. & Trademark Off. Soc'y 571, 599 (1981). The standard reply is to point out that incentives are needed to induce people to perform research in this socially valuable field, an argument that does not directly address the moral objection. Cf. Comment, In His Image: On Patenting Human-Based Bioproducts, 25 U.S.F. L. Rev. 583 (1991). An interesting middle ground is suggested in the following excerpt, which is drawn from a longer discussion of the extent to which property rights ought to be based on how hard someone works to create something — i.e., on a "labor" theory.

> [A]ssuming that labor's fruits are valuable, and that laboring gives the laborer a property right in this value, this would entitle the laborer only to the value she added, and not to the *total* value of the resulting product. Though exceedingly difficult to measure, these two components of value (that attributable to the object labored on and that attributable to the labor) need to be distinguished.

Edwin C. Hettinger, Justifying Intellectual Property, 18 Phil. & Pub. Aff. 31, 37 (1989). Imagine that a company from a developed country "prospects" for genetic material in a poor tropical country, takes some plant specimens back to the lab, and inserts a gene from the "prospected" material into an ordinary domestic plant, thereby producing a very valuable new plant. According to the approach taken in the excerpt, how should the rights be allocated between (1) the poor country's government, which claims ownership of the raw genetic material, and (2) the company that developed the new plant? How would one determine the "value added" by the company's researchers as opposed to the value contributed by the original genetic material?

5. The notion of granting intellectual property rights over the genetic material in native species may seem strange to some, but it might actually serve laudable social purposes. In addition to more fairly distributing the gains from recombinant genetic products based on those species, it would also give developing countries an incentive to protect rainforests and other genetically rich areas. In general, the granting of property rights over a resource can be expected to lead to more efficient use of the resource; at the very least, it will prevent over-exploitation of the resource due to its free (or "public good") quality. See, e.g., Harold Demsetz, Toward a Theory of Property Rights, in Ownership, Control, and the Firm 104 (1988); Charles, Fishery Socioeconomics: A Survey, 64 Land Econ. 276, 279-280 (1988) (describing allocation of fish catches via property rights).

Can a similar argument be made in favor of the attempts of some scientists to patent portions of the human genome? How do the legal and policy considerations differ?

4

Copyright Law

A. INTRODUCTION

This chapter explores the broad and expanding domain of copyright law, a principal means for protecting works of authorship. Although focused upon expressive (and non-functional) works, copyright has since its inception responded to advances in technologies for reproducing, disseminating, and storing information. Copyright laws emerged in the wake of the printing press and have evolved to encompass other methods of reproducing works of authorship, such as photography, motion pictures, and sound recordings. The development of broadcasting technology—enabling the performance of works at distant points—triggered a second wave of expansions and adjustments to copyright. The digital revolution represents a third distinct wave of technological innovation reshaping copyright law. By bringing about new modes of expression (such as computer programming, synthesized music, video games, and interactive multimedia works) and empowering anyone with a computer and an Internet connection to flawlessly, inexpensively, and instantaneously reproduce and distribute works of authorship, digital technology represents possibly the greatest set of challenges to copyright law. This latest wave is just cresting—as the Internet and digital technology have become widely diffused—and hence the future of copyright law is very much in flux.

We begin with a brief survey of the origins of copyright law, its philosophical underpinnings, and of its vast provisions. Building upon this introduction, we examine the traditional components of copyright as it has developed on "analog" platforms:[1] its subject matter, ownership structure, rights (including infringement

---

1. The term "analog" is used to signify that the medium uses an "analogy" to represent the phenomenon. For most of the history of copyright law, content storage and distribution innovations have centered around means of mechanically capturing and reproducing works of authorship—such as printing, phonographs, photographs, film, and photocopies—all of which record or, to use copyright law's rubric, "fix," works of authorship through some human or mechanical process of deforming a physical object (such as stone, paper, vinyl, film) in a manner that conveys an image (a letter, number, or graphic image) or signal varying in audio frequency (sound) or light or color intensity (film). Even broadcasting technology has been based upon analog propagation (wave forms) of analog encoded content (sound recordings etched in vinyl and later tape and audiovisual works fixed in film).

analysis), defenses, and the evolution of doctrines of indirect liability. We then explore the new and rapidly developing frontier of digital copyright law. (Chapter 7 examines a particular dimension of digital copyright law: legal protection for computer software.) The chapter concludes by surveying international dimensions of copyright law and remedies.

1. Brief History of Copyright Protection

The invention of the printing press in the West provided the impetus for the establishment of copyright protection.[2] Working from wine press technology from his native Rhine Valley, Johannes Gutenberg, a German goldsmith, developed a printing press with wood and metal movable type by the year 1440. Gutenberg experimented with this technology for the next decade with funding from a German businessman. An infusion of new funding around 1450 enabled Gutenberg to build a larger press, ultimately leading to the first printed version of the Bible in 1452. This extraordinary technological achievement encountered some resistance from nobles who refused to tarnish their libraries of handcopied manuscripts with printed books, the Catholic Church which sought to control technology of mass communication, and the Islamic world with its calligraphic traditions. Nonetheless, the printing press spread rapidly across Europe. With its highly developed guild system and fertile technological culture, Venice emerged as the "capital of printing" in the late fifteenth century.

Not surprisingly given the confluence then occurring in Venice, patent protection presaged the development of copyright. The Venetian Republic granted Johann Speyer, the first printer in the city, a patent in 1469 for the printing press, affording Speyer an exclusive right to print books in all Venetian territories for the next five years. See John Feather, Publishing, Piracy and Politics: An Historical Study of Copyright in Britain 10-11 (1994); Frank D. Prager, A History of Intellectual Property from 1545 to 1787, 26 J. Pat. Off. Soc'y 711, 718 (1944) (noting that it remains "contested whether [Speyer] was recognized as first importer of the whole art of typography, or as inventor of improvements"). Within a few decades, the Venetian Cabinet recognized for the first time exclusive rights in the printing of particular books (as distinguished from the technology of reproduction), awarding Daniele Barbaro a ten-year exclusive grant to publish a book by his late brother, Ermolao. See Christopher May, The Venetian Moment: New Technologies, Legal Innovation and the Institutional Origins of Intellectual Property, 20 Prometheus 159, 172 (2002). Ease of entry into printing and an oversupply of books ultimately led the Venetian Senate to restrict the printing privilege to "new and previously unprinted works." By the middle of the sixteenth century, a new decree organized all of Venice's printers and booksellers into a guild and provided a means for allowing the Church to suppress heretical works.

With the growth of international commerce and the rise of London, England became a focal point for the development of copyright law. The first "copyright" was granted in England by royal decree in 1556, not long after the introduction of the printing press in England.[3] See Elizabeth L. Eisenstein, The Printing Revolution

---

2. Printing technology developed in the Far East much earlier, resulting in the earliest dated printed book ("Diamond Sutra," a Buddhist scripture) using a block printing technology in China at least as early as the year 868. A printing device using movable clay type was invented in China in 1041 by Bi Sheng.

3. Prior to royal decrees, an author had the property right to physical possession of his manuscript at common law. Recent scholarship casts serious doubt on the existence of a common law right to prevent

in Early Modern Europe (1993). For political reasons, the Crown consolidated the new printing business in the hands of the Stationers' Company. It granted printers of this company — not to authors — the exclusive right to control the printing and sale of books, forever. Not incidentally, the government conferred these copyrights upon loyal publishers who would not publish books that the Crown considered politically or religiously objectionable, and indeed it subjected the printing business to the oversight of the Star Chamber.[4]

After the exclusive right of the Stationers' Company ended in 1694, members of the company faced for the first time substantial competition in the printing of books. They promptly sought assistance from Parliament. In 1710, Parliament responded by passing the Statute of Anne. The Statute of Anne vested in *authors* of books a monopoly over their works. Unlike the perpetual rights granted to publishers by decree, the statutory right was limited to only 14 years, renewable for an additional 14 years by the author. The statute contained a complex system of registration, notice, and deposit requirements. The Statute of Anne further turned its back on its monopolistic roots by providing that the government could set maximum prices for copyrighted books upon application by disgruntled consumers — what might be described as the first compulsory license in copyright. Coexisting with the English history of copyright is the continental approach, which treats an author's right in his works of authorship as a fundamental moral right. Justice Breyer's work on copyright quotes a sixteenth-century French lawyer who argued that "as the heavens and the earth belong to God, because they are the work of his word . . . so the author of a book is its complete master, and as such can dispose of it as he chooses." Stephen Breyer, The Uneasy Case for Copyright: A Study of Copyright in Books, Photocopies, and Computer Programs, 84 Harv. L. Rev. 281, 284 (1970) (quoting Marion). Copyright on the Continent, and particularly in France, accordingly developed parallel means of protecting authors: a property right similar to the English statutory right, based on a decree by the Revolutionary government in January 1791[5] and on a series of moral rights established in judicial decisions. These rights have only relatively recently been embodied in statutory form in the United States.

Most of the United States passed state copyright laws modeled on the Statute of Anne shortly after gaining their independence. Some of these state statutes went even further than the Statute of Anne in their "pro-consumer" orientation, allowing the government to regulate both the price charged for copyrighted works and the number of such works produced. These state acts also contained complex and often conflicting formal requirements. At the same time, early state statutes contained

---

copying of words or ideas. See Howard B. Abrams, The Historic Foundation of American Copyright Law: Exploding the Myth of Common Law Copyright, 29 Wayne L. Rev. 1119 (1983). In Donaldson v. Becket, 98 Eng. Rep. 257, 262 (H.L. 1774), the House of Lords concluded that a copyright had *never* existed at common law and therefore could not be said to survive the passage of the Statute of Anne, the first statute to afford protection against unauthorized copying.

4. The English Crown's desire to control the unlicensed printing of manuscripts manifested itself even earlier, in a 1529 edict directed at dissident speech during a tumultuous time. See David Lange, At Play in the Fields of the Word: Copyright and the Construction of Authorship in the Post-Literate Millennium, 55 L. & Contemp. Probs. 139 (1992).

5. Pre-Revolution French copyright took the form of royal grants of exclusive publishing monopolies, with rights sometimes granted to authors and sometimes to printers. See Jane C. Ginsburg, A Tale of Two Copyrights: Literary Property in Revolutionary France and America, 64 Tulane L. Rev. 991 (1990).

elements of the continental tradition favoring a natural right of authors. For example, the 1783 Massachusetts statute defined the purpose of copyright as *both* the utilitarian goal of producing new creative works, and the securing of "one of the natural rights of all men."

Problems with applying these conflicting state laws across state borders led to a general consensus that a national law was necessary, and the Constitution expressly granted power to the federal government to create both patents and copyrights. The patent and copyright clause passed the Convention and was ratified without significant debate.[6] One of the first acts of the new Congress was to pass the Copyright Act of 1790. That Act, like the Statute of Anne, granted authors protection for books, maps, and charts for 14 years, and allowed renewal for a second 14-year term. Indeed, this new act was similar in significant respects to the English law that preceded it. The 1790 Act allowed copyrights to be registered with the local district court and notice to be published in local newspapers. As technology for making and reproducing works of authorship expanded and the arts flourished, Congress amended the Copyright Act to extend to new media and means of exploitation. By the end of the nineteenth century, copyright protection extended to prints, musical compositions, dramatic works, photographs, graphic works, and sculpture. Since the advent of the printing press, advances in the technologies for creating and distributing works of authorship have played a critical role in shaping copyright law. See generally Peter S. Menell, Envisioning Copyright Law's Digital Future, 46 N.Y.L. Rev. 63 (2003); Paul Goldstein, Copyright's Highway: From Gutenberg to the Celestial Jukebox (2d ed. 2003); Jessica Litman, Copyright Legislation and Technological Change, 68 Or. L. Rev. 265 (1989).

The Copyright Act has undergone numerous changes during its history. Although these changes are embodied en masse within Title 17 of the U.S. Code, it will be helpful to put this immensely complex statute into historical perspective. See generally David Nimmer, Codifying Copyright Comprehensibly, 51 UCLA L. Rev. 1233 (2004). Here is a brief sketch of the principal milestones and motivating forces behind the evolution of U.S. copyright law over the past century:

1909 Act. The most significant overhaul of the Copyright Act since its founding occurred in 1909. Like the amendments that led up to it, the 1909 Act generally broadened the scope of copyright protection. Protection for literary works was expanded to include "all writings," reaching works in progress, and speeches, among other new matter. Copyright protection lasted for an initial term of 28 years, with an additional 28 years available upon renewal. Building upon the International Copyright Act of 1891 which grudgingly afforded protection for some works of foreign origin, the 1909 Act perpetuated policies that discriminated against foreign works, keeping the United States at odds with the growing number of nations joining the Berne Convention for the Protection of Literary and Artistic Works (promulgated in 1886).

1976 Act and Related Reforms. Advances in technology for creating and distributing works of authorship—most notably, sound recording and

---

6. For a discussion of the history of this clause, see Edward C. Walterscheid, The Nature of the Intellectual Property Clause: A Study in Historical Perspective (2002); Karl Fenning, The Origin of the Patent and Copyright Clause of the Constitution, 17 Geo. L.J. 109 (1929).

broadcasting—as well as anachronisms of the 1909 Act (such as the dual term of protection) periodically aroused interest in reforming the Copyright Act through the middle of the twentieth century, but none resulted in significant change. In 1955, Congress requested that the Copyright Office undertake a series of studies aimed at assessing the copyright system and set in motion an effort aimed at comprehensive reform of the statute through negotiation among the principal interest groups affected by copyright policy. The complex process bogged down over such issues as compulsory licenses for jukeboxes and cable television and before the omnibus project was completed, Congress approved a law providing copyright protection for sound recordings in 1971. After two decades of study, negotiation, and debate, Congress approved the 1976 Act, which continues to serve as the principal framework for copyright protection in the United States. The 1976 Act expanded both the scope and duration of protection. All written works became protected upon being "fixed in a tangible medium of expression," even if they were unpublished. The duration of copyright was expanded to the life of the author plus 50 years, or 75 years in the case of corporate "authors." Further, the formal notice and registration requirements were loosened, although not discarded. In other respects, the 1976 Act weakened intellectual property protection by establishing several compulsory licensing regimes, approving numerous exemptions from liability, codifying the fair use doctrine that had been developing through the courts, and preempting most state and common law protections that impinge upon federal copyright protection. In 1980, Congress expressly incorporated protection for computer programs into the Copyright Act. At the urging of the major copyright industries, Congress added an additional 20 years to the duration of copyright protection in 1998.

Berne Convention Accession. As the global content marketplace expanded to unprecedented levels in the 1980s and piracy of copyrighted works in many corners of the world increased, the United States government resolved to join the Berne Convention as a means for expanding protection for U.S. works throughout the world and enhancing U.S. influence on the direction of global copyright protection. As a result, Congress approved several amendments between 1988 and 1994 (scaling back of formalities, extending protection for moral rights and architectural works, restoration of copyright to certain foreign works under protection in the source country but in the public domain in the United States) needed to bring U.S. copyright law into compliance with the minimum standards set forth in the Berne Convention, which the United States joined in 1989.

The Digital Age. By the early 1990s, advances in digital technology were beginning to be felt in the major content marketplaces. The traditional content industries feared that widespread availability of technology for making low-cost, perfect copies of digital media could undermine their ability to enforce their rights. In response, Congress has passed several detailed amendments to the Copyright Act during the 1990s aimed at reforming copyright law for the digital age. The Audio Home Recording Act of 1992 regulates the design of digital audio tape technology and imposes a levy on the sale of devices and blank media intended to compensate copyright owners for losses from home copying. The Digital Performance Right in Sound Recording Act of 1995 affords creators and owners of sound recordings a basis for earning income on digital streams (webcasts) of their works. The No Electronic Theft (NET) Act of 1996 expands criminal enforcement for piracy over digital networks. The Digital

Millennium Copyright Act (DMCA) of 1998 affords to copyright owners rights against those who circumvent copy protection technologies and insulates online service providers from liability for infringing acts of their subscribers subject to various limitations.

2. An Overview of the Copyright Regime

Although the copyright and patent laws flow from the same constitutional basis and share the same general approach—statutorily created monopolies to foster progress—they feature different elements and rights, reflecting the very different fields of creativity that they seek to encourage. We sketch below the basic elements and rights of copyright law. As you review these features, contrast them to the analogous provisions of the patent law. How do you explain the differences?

A protectable copyright has the following elements:

Copyrightable Subject Matter. The subject matter protectable by copyright spans the broad range of literary and artistic expression—including literature, song, dance, sculpture, graphics, painting, photography, sound, movies, and computer programming. Ideas themselves are not copyrightable, but the author's particular expression of an idea is protectable.

Threshold for Protection. A work need only exhibit a modicum of originality and be fixed in a "tangible medium of expression."

Formalities. Notice of copyright is required on all works published prior to 1989. Registration of a copyright is not strictly required for its validity, but is required of U.S. authors prior to instituting an infringement suit. Deposit of copies of the work is required to obtain registration of copyright.

Authorship and Ownership. The work must have been created by the party bringing suit, or rights in the work must have been transferred by the author to the party bringing suit. In the case of works made "for hire," the employer and not the original creator is considered the author and the owner of the work.

Duration of Copyright. A copyright lasts for the life of the author plus 70 years, or 95 years from first publication in the case of entity authors (or 120 years from the year of creation, whichever occurs first).

Although the United States Copyright Office registers works, unlike the Patent Office it does not compare works to the prior art or make any assessment of validity (other than to ensure a modicum of creativity). The Copyright Office functions more like a title office for land. A copyright, like a trade secret, is protectable at the moment the work is created.

In relation to patents, the ease with which copyrights may be obtained and the duration for which they last are counter-balanced by the more limited rights accorded and the numerous and substantial exceptions and limitations to protection. Ownership of a valid copyright confers the following rights:

Copying. The owner has the exclusive right to make copies. She may sue a copier for infringement if the copying is "material" and "substantial," even if the copy is in a different form or is of only part of the whole.

ex. sequel (like Harry Potter's) not cool.
does not → parity of sequels.
A. *Introduction* 373

Derivative Works. The owner has the exclusive right to prepare derivative works, which are works based on the original but in different forms or otherwise altered (such as translations, movies based on books, etc.). These derivative works are themselves copyrightable, to the extent that they contain their own original expression. Note that the right to create derivative works overlaps with the right to make copies.

Distribution. The owner has the right to control the sale and distribution of the original and all copies or derivative works, including licensed copies. However, this right extends only to the first sale of such works. The owner does not have the right to limit resale by purchasers of her works (except in certain limited circumstances).

Performance and Display. The owner has the right to control the public (but not private) performance and display of her works, including both literary and performance-oriented works. This right extends to computer programs and other audiovisual works. The owner generally does not, however, have the right to prevent the display of a particular original or copy of a work of art in a public place.

Anticircumvention. The Copyright Act prohibits the circumvention of technological protection measures (such as encryption) designed to safeguard digitally-encoded works, subject to several exceptions and limitations.

Moral Rights. Visual artists possess the right to claim authorship in their works, prevent intentional distortion, mutilation, or other modification of their work, and block destruction of works of "recognized stature," subject to several limitations. *→ more visual things*

As with patents, copyrights are protected against both direct and indirect (contributory, vicarious, or inducement) infringement.

These rights are limited in several ways. The fair use doctrine, intended to create leeway for criticism, comment, news reporting, teaching, scholarship, and research, applies a balancing test to determine whether a use of copyrighted material goes too far. In addition, the Copyright Act establishes compulsory licensing for musical compositions, cable television, and webcasts, among others, and exempts some uses from liability. The Act also establishes a safe harbor that partially immunizes online service providers from liability for infringing acts of their subscribers.

There is a fundamental difference between the rights granted by copyright law and those granted by patent law. Copyrights do not give their owner the exclusive right to prevent others from making, using, or selling their creations. Rather, they give the author only the right to prevent unauthorized copying of their works, as well as the right to prevent some limited types of uses of those works (such as public performances) when derived from the copyright owner. The independent development of a similar or even identical work is perfectly legal. This means that copyright law must have some mechanism for determining when a work has been copied illegally. While in rare cases direct proof of copying may be available, in most cases courts determine whether copying has occurred on the basis of the defendant's access to the plaintiff's work and the extent which the two works are similar. If copying is proven — whether directly or by inference — then infringement will be found if the defendant's work is substantially similar to protected expression — in whole or substantial part — in the plaintiff's work.

3. Philosophical Perspectives on Copyright Protection

In the vast body of court decisions, legislation, and commentaries on copyright law, one can find references to a great many philosophical justifications for copyright protection. Lord Justice Mansfield, writing in the mid-eighteenth century, states: "From what source, then is the common law drawn, which is admitted to be so clear, in respect of the copy before publication? From this argument—because it is just, that an author should reap the pecuniary profits of his own ingenuity and labor." Millar v. Taylor, 4 Burr. 230, 238 (1769).[7]

Scholars have long debated the philosophical foundations of copyright law. See generally Alfred Yen, Restoring the Natural Law: Copyright as Labor and Possession, 51 Ohio St. L.J. 517 (1990) (tracing roots of natural law in American copyright law); Justin Hughes, The Philosophy of Intellectual Property, 77 Geo. L.J. 287, 350-353 (1988) (suggesting various strains of the personhood justification in American copyright law). Immanuel Kant spoke of the "natural obligation" to respect the author's ownership of his works. See Immanuel Kant, Of the Injustice of Counterfeiting Books, 1 Essays and Treatises on Moral, Political, and Various Philosophical Subjects 225, 229-230 (Richardson ed. 1798). And, as we have seen, the "natural right" of the author to control the use of his work and to be rewarded for it is one of the significant underpinnings of at least some American copyright jurisprudence. See Benjamin Kaplan, An Unhurried View of Copyright 79 (1967 ed.); James M. Treece, American Law Analogues of the Author's Moral Right, 16 Am. J. Comp. L. 487 (1968).

The predominant philosophical framework undergirding American copyright law, however, is utilitarian. The Constitution grants Congress the power to enact copyright laws in order to "promote the Progress of Science and useful Arts." Art. I, § 8, cl. 8. In the early decision of Wheaton v. Peters, 33 U.S. (8 Pet.) 591 (1834), the Court treated copyright as a statutory creation designed primarily to enhance the public interest and only secondarily to confer a reward upon authors. Id. at 661. More recently, Justice Stewart described the basic purpose of the Copyright Act as follows:

> The limited scope of the copyright holder's statutory monopoly, like the limited duration required by the Constitution, reflects a balance of competing claims upon the public interest: Creative work is to be encouraged and rewarded, but private motivation must ultimately serve the cause of promoting broad public availability of literature, music, and the other arts.[8] The immediate effect of our copyright law is to secure a fair return to an "author's" creative labor. But the ultimate aim is, by this incentive, to stimulate artistic creativity for the general public good. "The sole interest of the United States and the primary object in conferring the monopoly," this Court has said, "lie in the general benefits derived by the public from the labors of authors."

---

7. This ruling was overturned by the House of Lords five years later, in Donaldson v. Becket, 98 Eng. Rep. 257 (H.L. 1774).

8. Lord Mansfield's statement of the problem almost 200 years ago in Sayre v. Moore, quoted in a footnote to Cary v. Longman, 1 East *358, 362 n.(b), 102 Eng. Rep. 138, 150 n.(b) (1801), bears repeating:

> [W]e must take care to guard against two extremes equally prejudicial; the one, that men of ability, who have employed their time for the service of the community, may not be deprived of their just merits, and the reward of their ingenuity and labour; the other, that the world may not be deprived of improvements, nor the progress of the arts be retarded.

Twentieth Century Music Corp. v. Aiken, 422 U.S. 151, 156 (1975). See also Mazer v. Stein, 347 U.S. 201, 219 (1954). And Justice Stevens has commented that:

> The monopoly privileges that Congress may authorize are neither unlimited nor primarily designed to provide a special private benefit. Rather, the limited grant is a means by which an important public purpose may be achieved. It is intended to motivate the creative activity of authors and inventors by the provision of a special reward, and to allow the public access to the products of this genius after the limited period of exclusive control has expired.

Sony Corp. of America v. Universal City Studios, Inc., 464 U.S. 417, 429 (1984).

American copyright law can thus be seen as primarily striving to achieve an optimal balance between fostering incentives for the creation of literary and artistic works and the optimal use and dissemination of such works. Nonetheless, copyright law reflects the other philosophical perspectives as well. Society grants copyrights both because it wants to encourage creation and because it wants to reward authors for their work. Copyright also reflects the Lockean principle that authors deserve to own the works they have created. The law limits the duration and scope of copyrights because it wants to make sure that copyright protection does not unduly burden other creators or free expression, that works are widely disseminated, and that the next generation of authors can make use of ideas in creating still more works. As we will see later in this chapter, international pressure and appeals by artists have brought increased recognition of the moral rights of artists. These policies interact in complex ways. In many cases, there is still great controversy over which policy should prevail.

COMMENTS AND QUESTIONS

1. Does copyright law appear to strike the proper balance between fostering incentives for the creation of literary and artistic works and the optimal use and dissemination of such works? Does the same duration of protection for all copyrightable works — whether books, computer programs, songs, paintings, or choreographic works — make sense? Do other justifications beyond the utilitarian balance better explain copyright's structure and provisions? Cf. Peter S. Menell, Intellectual Property: General Theories, Encyclopedia of Law and Economics (B. Bouckaert and G. De Geest, eds., 2000); Alfred Yen, Restoring the Natural Law: Copyright as Labor and Possession, 51 Ohio St. L.J. 517 (1990); Wendy Gordon, An Inquiry into the Merits of Copyright: The Challenges of Consistency, Consent, and Encouragement Theory, 41 Stan. L. Rev. 1343 (1989); Justin Hughes, The Philosophy of Intellectual Property, 77 Geo. L.J. 287 (1988).

2. Contrast the way in which copyright law, trade secret law, and patent law vary along the following dimensions:

- threshold for protection
- duration of protection
- rights conferred
- treatment of independent creation
- defenses to infringing use

To what extent can the differences among these legal regimes be explained by differences in the subject of coverage (and the nature of the innovative process in these areas)? differences in the philosophical justifications for these modes of protection? other factors?

3. The term "copyright" reflects the underlying philosophy of the Anglo-American regime for protecting literary and artistic works — regulation of the right to make *copies* for the purpose of promoting progress in the arts and literature. The emphasis is on the benefit to the public, not the benefits or rights of authors. By contrast, the civil law analog to copyright has a different name and orientation. In France, the comparable body of law is *droit d'auteur,* which translates to "author's rights." The laws in Germany and Spain are similar — *Urheberrecht* and *derecho de autor.* This civil law tradition derives more from a Kantian (natural rights) or Hegelian (personhood) justification for legal entitlements, and thus focuses on the rights of *authors.* Thus, the civil law countries have long expressly protected the moral rights of authors — e.g., the right of an author to prevent the mutilation of his or her work after it is sold. See generally Jane C. Ginsburg, A Tale of Two Copyrights: Literary Property in Revolutionary France and America, 64 Tul. L. Rev. 991 (1990).

Of what significance is the underlying philosophical perspective — whether utilitarian, natural rights, or personhood — for the structure and content of copyright law? Which perspective is more appropriate as a matter of social justice? public policy? Can these perspectives be effectively harmonized without losing their coherence?

4. To what extent does the open source model of collaborative creativity dispel the principles underlying traditional copyright law? Does the Internet's essentially free distribution system call for radical changes in the structure of copyright law? Should copyright law be strengthened or weakened in the digital age? What considerations guide your analysis? Cf. Lawrence Lessig, Free Culture: How Big Media Uses Technology and the Law to Lock Down Culture and Control (2004); Peter S. Menell, Envisioning Copyright Law's Digital Future, 46 N.Y. L. Rev. 63 (2003); Paul Goldstein, Copyright's Highway: From Gutenberg to the Celestial Jukebox (2d ed. 2003); Lawrence Lessig, The Future of Ideas: The Fate of the Commons in a Connected World (2001); Jessica Litman, Digital Copyright (2001).

B. REQUIREMENTS

1. Original Works of Authorship

17 U.S.C. § 102. Subject Matter of Copyright: In General

(a) Copyright protection subsists, in accordance with this title, in original works of authorship fixed in any tangible medium of expression, now known or later developed, from which they can be perceived, reproduced, or otherwise communicated, either directly or with the aid of a machine or device....

The legislative history to the 1976 Copyright Act provides:

> The two fundamental criteria of copyright protection—originality and fixation in a tangible form—are restated in the first sentence of this cornerstone provision. The phrase "original works of authorship," which is purposively left undefined, is intended to incorporate without change the standard of originality established by the courts under the present copyright statute. This standard does not include requirements of novelty, ingenuity, or esthetic merit, and there is no intention to enlarge the standard of copyright protection to require them. . . .

As developed by the courts, originality entails *independent creation* of a work featuring a *modicum of creativity*. Independent creation requires only that the author not have copied the work from some other source. As the eminent copyright jurist Learned Hand eloquently observed,

> If by some magic a man who had never known it were to compose anew Keats's *Ode on a Grecian Urn,* he would be an "author" and, if he copyrighted it, others might not copy that poem, though they might of course copy Keats's.

Sheldon v. Metro-Goldwyn Pictures Corp., 81 F.2d 49, 54 (2d Cir. 1936). This highlights an important distinction between patent and copyright law.

> The alleged inventor is chargeable with full knowledge of all the prior art, although in fact he may be utterly ignorant of it. The "author" is entitled to a copyright if he independently contrived a work completely identical with what went before; similarly, although he obtains a valid copyright, he has no right to prevent another from publishing a work identical with his, if not copied from his.

Alfred Bell & Co. v. Catalda Fine Arts, Inc., 191 F.2d 99, 103 (2d Cir. 1951).
Courts have set the threshold of creativity necessary to satisfy the originality requirement quite low. Copyright law does not require that a work be

> strikingly unique or novel. . . . All that is needed to satisfy both the Constitution and the statute is that the "author" contributed something more than "merely trivial" variation, something recognizably "his own." Originality in this context "means little more than a prohibition of actual copying." No matter how poor artistically the "author's" addition, it is enough if it be his own.

Id. at 102-103.
Courts do not judge the artistic merit of a work:

> It would be a dangerous undertaking for persons trained only to the law to constitute themselves final judges of the worth of pictorial illustrations, outside of the narrowest and most obvious limits. At one extreme some works of genius would be sure to miss appreciation. Their very novelty would make them repulsive until the public had learned the new language in which their author spoke. It may be more than doubted, for instance, whether the etchings of Goya or the paintings of Manet would have been sure of protection when seen for the first time. At the other end, copyright would be denied to pictures which appealed to a public less educated than the judge.

Bleistein v. Donaldson Lithographing Co., 188 U.S. 239, 251-252 (1903) (finding a circus advertisement to be sufficiently original). Courts have rarely found literary or artistic works to fall below the *de minimis* originality threshold of copyright law. The few exceptions generally relate to simple slogans and exceedingly modest variations on another work. See 37 C.F.R. §202.1 Material Not Subject to Copyright ("(a) Words and short phrases such as names, titles, and slogans; familiar symbols or designs; mere variations of typographical ornamentation, letter or coloring; mere listing of ingredients or contents...(e) Type face as typeface.").

A more difficult problem arises when an author creates a work in a mechanical or functional manner. Such assemblage of information can be costly and time consuming (the "sweat of the brow"), but may lack creativity. Should copyright law protect such works?

John Locke proposed a theory of property in which labor over a previously unowned piece of property (intellectual or physical) can vest ownership rights in the laborer. See John Locke, Two Treatises of Government §§27-28. The sweat of the brow theory also commands as adherents a number of venerable court decisions and some modern commentators. See, e.g., International News Service v. Associated Press, 248 U.S. 215, 236 (1918); Blunt v. Patten, 3 F. Cas. 763, 765 (C.C.S.D.N.Y. 1828); Robert Denicola, Copyright in Collections of Facts: A Theory for the Protection of Nonfiction Literary Works, 81 Colum. L. Rev. 516 (1981). Proponents of the sweat of the brow theory advance two basic rationales. The first is economic: compilation is necessary and efficient, and unless it is rewarded there will be no incentive to compile or create factual works. The second rationale is based on fairness: it is unjust to permit one person to benefit from the hard work of another.

Opponents of the sweat of the brow approach see it as essentially at odds with the rationale of intellectual property protection, because it protects work without regard to creativity. If a list compiled at great effort is protectable, why not a ditch dug with similar effort? Further, they point to the dangers of monopoly (if no one can duplicate or chooses to duplicate the effort of the compiler) and of wasted resources (if several competing companies do the same fact-intensive work to produce the same product). See, e.g., Jerome H. Reichman and Pamela Samuelson, Intellectual Property Rights in Data?, 50 Vand. L. Rev. 51 (1997); Jane C. Ginsburg, Creation and Commercial Value: Copyright Protection of Works of Information, 90 Colum. L. Rev. 1865 (1990).

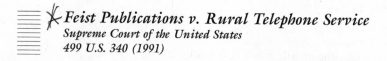

Feist Publications v. Rural Telephone Service
Supreme Court of the United States
499 U.S. 340 (1991)

Justice O'CONNOR delivered the opinion of the Court.
This case requires us to clarify the extent of copyright protection available to telephone directory white pages.

I

Rural Telephone Service Company is a certified public utility that provides telephone service to several communities in northwest Kansas. It is subject to a state regulation that requires all telephone companies operating in Kansas to issue annually an updated telephone directory. Accordingly, as a condition of its monopoly franchise, Rural publishes a typical telephone directory, consisting of white pages and yellow pages. The white pages list in alphabetical order the names of Rural's subscribers, together with their towns and telephone numbers. The yellow pages list Rural's business subscribers alphabetically by category and feature classified advertisements of various sizes. Rural distributes its directory free of charge to its subscribers, but earns revenue by selling yellow pages advertisements.

Feist Publications, Inc., is a publishing company that specializes in area-wide telephone directories. Unlike a typical directory, which covers only a particular calling area, Feist's area-wide directories cover a much larger geographical range, reducing the need to call directory assistance or consult multiple directories. The Feist directory that is the subject of this litigation covers 11 different telephone service areas in 15 counties and contains 46,878 white pages listings — compared to Rural's approximately 7,700 listings. Like Rural's directory, Feist's is distributed free of charge and includes both white pages and yellow pages. Feist and Rural compete vigorously for yellow pages advertising.

As the sole provider of telephone service in its service area, Rural obtains subscriber information quite easily. Persons desiring telephone service must apply to Rural and provide their names and addresses; Rural then assigns them a telephone number. Feist is not a telephone company, let alone one with monopoly status, and therefore lacks independent access to any subscriber information. To obtain white pages listings for its area-wide directory, Feist approached each of the 11 telephone companies operating in northwest Kansas and offered to pay for the right to use its white pages listings.

Of the 11 telephone companies, only Rural refused to license its listings to Feist. Rural's refusal created a problem for Feist, as omitting these listings would have left a gaping hole in its area-wide directory, rendering it less attractive to potential yellow pages advertisers. . . .

Unable to license Rural's white pages listings, Feist used them without Rural's consent. Feist began by removing several thousand listings that fell outside the geographic range of its area-wide directory, then hired personnel to investigate the 4,935 that remained. These employees verified the data reported by Rural and sought to obtain additional information. As a result, a typical Feist listing includes the individual's street address; most of Rural's listings do not. Notwithstanding these additions, however, 1,309 of the 46,878 listings in Feist's 1983 directory were identical to listings in Rural's 1982-1983 white pages. Four of these were fictitious listings that Rural had inserted into its directory to detect copying.

Rural sued for copyright infringement. . . .

II

A

This case concerns the interaction of two well-established propositions. The first is that facts are not copyrightable; the other, that compilations of facts generally are. Each of these propositions possesses an impeccable pedigree. That there can be no valid copyright in facts is universally understood. The most fundamental axiom of copyright law is that "no author may copyright his ideas or the facts he narrates."

Harper & Row, Publishers, Inc. v. Nation Enterprises, 471 U.S. 539, 556 (1985). Rural wisely concedes this point, noting in its brief that "facts and discoveries, of course, are not themselves subject to copyright protection." At the same time, however, it is beyond dispute that compilations of facts are within the subject matter of copyright. Compilations were expressly mentioned in the Copyright Act of 1909, and again in the Copyright Act of 1976.

There is an undeniable tension between these two propositions. Many compilations consist of nothing but raw data — i.e., wholly factual information not accompanied by any original written expression. On what basis may one claim a copyright in such a work? Common sense tells us that 100 uncopyrightable facts do not magically change their status when gathered together in one place. Yet copyright law seems to contemplate that compilations that consist exclusively of facts are potentially within its scope.

The key to resolving the tension lies in understanding why facts are not copyrightable. The *sine qua non* of copyright is originality. To qualify for copyright protection, a work must be original to the author. See *Harper & Row*, supra, at 547-549. Original, as the term is used in copyright, means only that the work was independently created by the author (as opposed to copied from other works), and that it possesses at least some minimal degree of creativity. 1 M. Nimmer & D. Nimmer, Copyright §§2.01[A], [B] (1990) (hereinafter Nimmer). To be sure, the requisite level of creativity is extremely low; even a slight amount will suffice. The vast majority of works make the grade quite easily, as they possess some creative spark, "no matter how crude, humble or obvious" it might be. Id., §1.08[C][1]. Originality does not signify novelty; a work may be original even though it closely resembles other works so long as the similarity is fortuitous, not the result of copying. . . .

Originality is a constitutional requirement. The source of Congress' power to enact copyright laws is Article I, §8, cl. 8, of the Constitution, which authorizes Congress to "secur[e] for limited Times to Authors . . . the exclusive Right to their respective Writings." In two decisions from the late 19th century — The Trade-Mark Cases, 100 U.S. 82 (1879); and Burrow-Giles Lithographic Co. v. Sarony, 111 U.S. 53 (1884) — this Court defined the crucial terms "authors" and "writings." In so doing, the Court made it unmistakably clear that these terms presuppose a degree of originality.

In The Trade-Mark Cases, the Court addressed the constitutional scope of "writings." For a particular work to be classified "under the head of writings of authors," the Court determined, "originality is required." 100 U.S., at 94. The Court explained that originality requires independent creation plus a modicum of creativity: "[W]hile the word *writings* may be liberally construed, as it has been, to include original designs for engraving, prints, &c., it is only such as are *original*, and are founded in the creative powers of the mind. The writings which are to be protected are *the fruits of intellectual labor*, embodied in the form of books, prints, engravings, and the like." Ibid. (emphasis in original).

In Burrow-Giles, the Court distilled the same requirement from the Constitution's use of the word "authors." The Court defined "author," in a constitutional sense, to mean "he to whom anything owes its origin; originator; maker." 111 U.S., at 58 (internal quotation marks omitted). As in The Trade-Mark Cases, the Court emphasized the creative component of originality. It described copyright as being limited to "original intellectual conceptions of the author," 111 U.S., at 58, and stressed the importance of requiring an author who accuses another of infringement to prove "the existence of those facts of originality, of intellectual production, of thought, and conception." Id., at 59-60.

The originality requirement articulated in The Trade-Mark Cases and Burrow-Giles remains the touchstone of copyright protection today. See Goldstein v. California, 412 U.S. 546, 561-562 (1973). It is the very "premise of copyright law." Miller v. Universal City Studios, Inc., 650 F.2d 1365, 1368 (CA5 1981). Leading scholars agree on this point. As one pair of commentators succinctly puts it: "The originality requirement is *constitutionally mandated* for all works." Patterson & Joyce, Monopolizing the Law: The Scope of Copyright Protection for Law Reports and Statutory Compilations, 36 UCLA L.Rev. 719, 763, n. 155 (1989) (emphasis in original) (hereinafter Patterson & Joyce). Accord, id., at 759-760, and n. 140; Nimmer § 1.06[A] ("[O]riginality is a statutory as well as a constitutional requirement"); id., § 1.08[C][1] ("[A] modicum of intellectual labor . . . clearly constitutes an essential constitutional element").

It is this bedrock principle of copyright that mandates the law's seemingly disparate treatment of facts and factual compilations. "No one may claim originality as to facts." Id., § 2.11[A], p. 2-157. This is because facts do not owe their origin to an act of authorship. The distinction is one between creation and discovery: the first person to find and report a particular fact has not created the fact; he or she has merely discovered its existence. . . .

Factual compilations, on the other hand, may possess the requisite originality. The compilation author typically chooses which facts to include, in what order to place them, and how to arrange the collected data so that they may be used effectively by readers. These choices as to selection and arrangement, so long as they are made independently by the compiler and entail a minimal degree of creativity, are sufficiently original that Congress may protect such compilations through the copyright laws. Nimmer §§ 2.11[D], 3.03; Denicola 523, n.38. Thus, even a directory that contains absolutely no protectable written expression, only facts, meets the constitutional minimum for copyright protection if it features an original selection or arrangement. See Harper & Row, 471 U.S., at 547. Accord Nimmer § 3.03.

This protection is subject to an important limitation. The mere fact that a work is copyrighted does not mean that every element of the work may be protected. Originality remains the sine qua non of copyright; accordingly, copyright protection may extend only to those components of a work that are original to the author. . . .

This inevitably means that the copyright in a factual compilation is thin. Notwithstanding a valid copyright, a subsequent compiler remains free to use the facts contained in an another's publication to aid in preparing a competing work, so long as the competing work does not feature the same selection and arrangement. As one commentator explains it: "[N]o matter how much original authorship the work displays, the facts and ideas it exposes are free for the taking. . . . [T]he very same facts and ideas may be divorced from the context imposed by the author, and restated or reshuffled by second comers, even if the author was the first to discover the facts or to propose the ideas." Ginsburg 186-8.

It may seem unfair that much of the fruit of the compiler's labor may be used by others without compensation. As Justice Brennan has correctly observed, however, this is not "some unforeseen byproduct of a statutory scheme." Harper & Row, 471 U.S., at 589 (dissenting opinion). It is, rather, "the essence of copyright," ibid., and a constitutional requirement. The primary objective of copyright is not to reward the labor of authors, but "to promote the Progress of Science and useful Arts." Art. I, § 8, cl. 8. Accord Twentieth Century Music Corp. v. Aiken, 422 U.S. 151, 156 (1975)....

III

There is no doubt that Feist took from the white pages of Rural's directory a substantial amount of factual information. At a minimum, Feist copied the names, towns, and telephone numbers of 1,309 of Rural's subscribers. Not all copying, however, is copyright infringement. To establish infringement, two elements must be proven: (1) ownership of a valid copyright, and (2) copying of constituent elements of the work that are original. See Harper & Row, 471 U.S., at 548. The first element is not at issue here; Feist appears to concede that Rural's directory, considered as a whole, is subject to a valid copyright because it contains some foreword text, as well as original material in its yellow pages advertisements.

The question is whether Rural has proved the second element. In other words, did Feist, by taking 1,309 names, towns, and telephone numbers from Rural's white pages, copy anything that was "original" to Rural? Certainly, the raw data does not satisfy the originality requirement. Rural may have been the first to discover and report the names, towns, and telephone numbers of its subscribers, but this data does not "'owe its origin'" to Rural. Burrow-Giles, 111 U.S., at 58. Rather, these bits of information are uncopyrightable facts; they existed before Rural reported them and would have continued to exist if Rural had never published a telephone directory. The originality requirement "rules out protecting...names, addresses, and telephone numbers of which the plaintiff by no stretch of the imagination could be called the author." Patterson & Joyce 776....

The question that remains is whether Rural selected, coordinated, or arranged these uncopyrightable facts in an original way. As mentioned, originality is not a stringent standard; it does not require that facts be presented in an innovative or surprising way. It is equally true, however, that the selection and arrangement of facts cannot be so mechanical or routine as to require no creativity whatsoever. The standard of originality is low, but it does exist. See Patterson & Joyce 760, n.144 ("While this requirement is sometimes characterized as modest, or a low threshold, it is not without effect"). As this Court has explained, the Constitution mandates some minimal degree of creativity, see The Trade-Mark Cases, 100 U.S., at 94; and an author who claims infringement must prove "the existence of...intellec-intellectual production, of thought, and conception." *Burrow-Giles*, supra, at 59-60.

The selection, coordination, and arrangement of Rural's white pages do not satisfy the minimum constitutional standards for copyright protection. As mentioned at the outset, Rural's white pages are entirely typical. Persons desiring telephone service in Rural's service area fill out an application and Rural issues them a telephone number. In preparing its white pages, Rural simply takes the data provided by its subscribers and lists it alphabetically by surname. The end product is a garden-variety white pages directory, devoid of even the slightest trace of creativity.

Rural's selection of listings could not be more obvious: it publishes the most basic information — name, town, and telephone number — about each person who applies to it for telephone service. This is "selection" of a sort, but it lacks the modicum of creativity necessary to transform mere selection into copyrightable expression.

Rural expended sufficient effort to make the white pages directory useful, but insufficient creativity to make it original.

We note in passing that the selection featured in Rural's white pages may also fail the originality requirement for another reason. Feist points out that Rural did not truly "select" to publish the names and telephone numbers of its subscribers; rather, it was required to do so by the Kansas Corporation Commission as part of its monopoly franchise. See 737 F. Supp., at 612. Accordingly, one could plausibly conclude that this selection was dictated by state law, not by Rural.

Nor can Rural claim originality in its coordination and arrangement of facts. The white pages do nothing more than list Rural's subscribers in alphabetical order. This arrangement may, technically speaking, owe its origin to Rural; no one disputes that Rural undertook the task of alphabetizing the names itself. But there is nothing remotely creative about arranging names alphabetically in a white pages directory. It is an age-old practice, firmly rooted in tradition and so commonplace that it has come to be expected as a matter of course. See Brief for Information Industry Association et al. as *Amici Curiae* 10 (alphabetical arrangement "is universally observed in directories published by local exchange telephone companies"). It is not only unoriginal, it is practically inevitable. This time-honored tradition does not possess the minimal creative spark required by the Copyright Act and the Constitution.

We conclude that the names, towns, and telephone numbers copied by Feist were not original to Rural and therefore were not protected by the copyright in Rural's combined white and yellow pages directory....

COMMENTS AND QUESTIONS

1. *Originality Intent?* The Court's conclusion turns critically upon the premise that the framers of the Constitution — through their use of the word "authors" and "writings" — intended originality as the touchstone and the bedrock for copyright protection. Does this square with the scope of protection for the 1790 Act — books, *maps*, and *charts* — approved by many of the framers during the first Congress? Cf. Burrow-Giles Lithographic Co. v. Sarony, 111 U.S. 53, 56 (1884) (expressing doubt that the constitutional language makes creativity or artistic accomplishment a prerequisite to copyright protection); Jane Ginsburg, Creation and Commercial Value: Copyright Protection of Works of Information, 90 Colum. L. Rev. 1865, 1873-93 (1990). Aren't maps and charts factual labor-intensive works valued not for their "creativity" but rather for their accuracy and precision? Why would the first Congress of a new nation bordering vast and largely unexplored territory seek to promote only "creative" rather than *accurate* maps and charts? Is the premise of the *Feist* decision — an originalist form of analysis — factually credible? Or is this a case of questionable history masquerading as post-hoc rationalization?

2. *Fictitious "Facts" and Copyright Estoppel.* Suppose that an author writes a manuscript that she holds out to the world as fact, even though the work is fictional. Should that author be able to prevent another from appropriating these fictitious "facts" in another work, such as a movie, on the basis of copyright infringement? The courts have developed a defense of copyright estoppel to bar such lawsuits: "equity and good morals will not permit one who asserts something as a fact which he insists his readers believe as the real foundation for its appeal to those who may buy and read the work, to change that position for profit in a law suit."

Oliver v. Saint Germain Foundation, 41 F.Supp. 296 (S.D. Cal. 1941) (dismissing lawsuit by alleged copyright owner who averred that his manuscript had been dictated by a spirit from another planet); Urantia Found. v. Maaherra, 114 F.3d 955 (9th Cir. 1997) (same); see also Arica Institute, Inc. v. Palmer, 970 F.2d 1067 (2d Cir. 1992).

3. *Is Anything Truly Original?* Drawing upon postmodern literary theory, some scholars question the very concept of originality as a romantic myth. See James Boyle, Shamans, Software, and Spleens: Law and the Construction of the Information Society (1996); Peter Jaszi, Toward a Theory of Copyright: The Metamorphoses of "Authorship," 41 Duke L.J. 455 (1991); David Lange, At Play in the Fields of the Word: Copyright and the Construction of Authorship in the Post-Literate Millennium, 55 L. & Contemp. Probs. 139 (1992). From this perspective, nothing is truly original; all authorship derives from the work of those who came before. Do you consider this argument persuasive? Does this approach suggest that the goal of identifying and rewarding "originality" is misguided? If so, what sorts of works *should* we protect?

4. *Database Protection Under* Feist. Is there any room for protection of factual compilations after *Feist*? Does *Feist* preclude copyright protection of automated databases, since such databases were built around user search capabilities rather than an original (and copyrightable) organizational structure? Would an electronic "phone book" that users could search by entering the name of the desired party be copyrightable under *Feist*? Does it matter how the database itself is organized (something that the user may not see)?

Cases decided after *Feist* have refused to protect compilations similar in nature to the telephone white pages. For example, in Warren Publishing v. Microdos Data Corp., 115 F.3d 1509 (11th Cir. 1997), the court held that a compilation of cable television companies serving each community in the nation was not copyrightable. The arrangement was unprotectable because the communities were listed alphabetically, and the selection was not original because every community in the U.S. served by a cable company was included. See also TransWestern Publishing Co. v. Multimedia Marketing Assoc., 133 F.3d 773 (10th Cir. 1998); cf. Southco, Inc. v. Kanebridge Corp., 258 F.3d 148 (3d Cir. 2001) (holding that a system of part numbers lacks sufficient originality for protection). Other cases have denied protection for codes used in parts catalogs. See ATC Distribution Group, Inc. v. Whatever it Takes Transmissions & Parts, Inc., 402 F.3d 700 (6th Cir. 2005); Southco, Inc. v. Kanebridge Corp. 390 F.3d 276 (3d Cir. 2004) (en banc) (part numbers used to identify and distinguish among types of screw fasteners not protectable); Toro Co. v. R & R Products Co., 787 F.2d 1208 (8th Cir.1986) (pre-*Feist* case denying protection for a numbering system for lawn care machine parts because the "random and arbitrary use of numbers...does not evince enough originality to distinguish authorship"). Still other cases, however, have found sufficient originality in other numbering systems. See American Dental Association v. Delta Dental Plans Association, 126 F.3d 977 (7th Cir. 1997) (a taxonomy of dental insurance billing codes for dental procedures was sufficiently original); CDN Inc. v. Kapes, 197 F.3d 1256 (9th Cir. 1999) (a list of wholesale prices for rare coins listed by publisher contained sufficient originality). Can you reconcile these decisions?

5. *Policy Analysis of Database Protection.* Putting aside the debatable historical analysis underlying the *Feist* decision, is the Court's determination sound from a larger policy perspective?

In the wake of *Feist*, owners of databases and other factual compilations have turned increasingly to contract law to protect their "sweat of the brow" investment. See ProCD, Inc. v. Zeidenberg, 86 F.3d 1447 (7th Cir. 1996). These cases allow database vendors to protect by contract what they cannot protect by copyright, so that someone who copies a database may be liable for breach of contract even if the database is composed entirely of unprotectable facts. (On this point, compare the discussion of protecting trade secrets by contract in Chapter 2.B.3.) As we will discuss in Chapter 6, these cases pose a difficult questions relating to federal Copyright Act preemption of state contract law and the enforceability of shrinkwrap and click-wrap licensing agreements.

An alternative to contract protection is the development of a new intellectual property regime designed specifically to protect "sweatworks." The European Community has adopted this approach. See Directive 96/9/EC of the European Parliament and of the Council of 11 March 1996 on the Legal Protection of Databases, 1996 O.J. (L77) 20; Xuqiong (Joanna) Wu, E.C. Database Directive, 17 Berkeley Tech. L.J. 571 (2002). The EC *sui generis* regime does not protect data directly, but rather the efforts of those who make a "substantial investment" in compiling "collections of independent data . . . arranged in a systematic . . . and individually accessible" manner. Others are free to collect the same data themselves. Such protection expires after 15 years, although substantial changes to the database (such as updating) restarts the clock. Some scholars worry that such protection will hinder the progress of science by causing unnecessary duplication of effort, raising the transaction and direct costs of scientific research, constraining fact checking, and complicating efforts to reformat and otherwise improve data collections. See Stephen M. Maurer and Suzanne Scotchmer, Database Protection: Is It Broken and Should We Fix It? 284 Science 1129 (May 14, 1999); Stephen M. Maurer, P. Bernt Hugenholtz & Harlan J. Onsrud, Europe's Database Experiment, 294 Science 789 (Oct. 26, 2001); James Boyle: A Natural Experiment; Thomas Hazlett: Changes Merit Symmetric Scrutiny; Richard Epstein: Why Rule Out Copyright Protection for All Databases? Financial Times (Nov. 22, 2004).

Several scholars favor a separate statute to protect some "sweatworks." See Jane C. Ginsburg, "No Sweat?" Copyright and Other Protection of Works of Information After Feist v. Rural Telephone, 92 Colum. L. Rev. 338 (1992); Wendy J. Gordon, On Owning Information: Intellectual Property and the Restitutionary Impulse, 78 Va. L. Rev. 149 (1992) (corrective justice approach justifies giving authors a right—albeit conditional and defeasible—to a reward for their efforts). Should Congress amend the Copyright Act to protect such "sweat of the brow" works? Is Congress constitutionally empowered to protect such works under the Intellectual Property Clause? Could such legislation be grounded on the Commerce Clause?

6. *Copyright Protection for Maps.* Recall that the first federal copyright statute expressly included maps and charts (e.g., nautical maps) within the scope of copyright protection. Under the present statute, copyright protection extends to maps under the definition of "pictorial, graphic and sculptural works." Can a map satisfy the originality requirement set forth in *Feist*? Isn't a map by its nature predominantly if not entirely driven by functional considerations (i.e., accurate representation of geographic information)?

In Mason v. Montgomery Data, Inc., 967 F.2d 135 (5th Cir. 1992), a case decided after *Feist*, the court held that a series of real estate maps of a Texas county were eligible for copyright protection. The court found the maps original in two distinct respects: (1) Mason had exercised "sufficient creativity in both the selection, coordination and arrangement of the facts that [the maps] depict"; and (2) the graphic artistry of the maps themselves was sufficiently original to qualify for protection. Are either of these elements truly original in the typical map? For example, what if the map includes every publicly accessible road? How many ways are there to accurately depict the United States and its 50 states? Should copyright depend on something as trivial as the colors chosen for each political subdivision? Cf. Dennis Karjala, Copyright in Electronic Maps, 35 Jurimetrics J. 395 (1995).

PROBLEM

Problem 4-1. Central Bell, the local telephone utility, distributes both a "white pages" telephone book and a "yellow pages," which lists businesses who have chosen to advertise there. Bell's yellow pages are organized alphabetically by subject matter of the business, and alphabetically within each subject. Bell itself created the subject headings with input from its advertisers. Bell also sells larger advertisements to certain companies for more money.

Christopher Publications decides to create and distribute its own yellow pages directory in competition with Bell. Christopher does this by taking a copy of Bell's yellow pages and calling every business that advertises there, asking each if it would like to advertise in Christopher's publication. Christopher places the resulting ads in its own subject matter listings (which do overlap somewhat with Bell's). Many of the advertisements themselves (which are submitted by the advertisers) are identical to those in the Bell directory. Bell sues Christopher for copyright infringement. Who should prevail?

---

2. Fixation in a Tangible Medium of Expression

H.R. Rep. No. 94-1476, 94th Cong., 2d Sess. 52-53 (1976)

As a basic condition of copyright protection, the bill [the Copyright Act of 1976] perpetuates the existing requirement that a work be fixed in a "tangible medium of expression," and adds that this medium may be one "now known or later developed," and that the fixation is sufficient if the work "can be perceived, reproduced, or otherwise communicated, either directly or with the aid of a machine or device." This broad language is intended to avoid the artificial and largely unjustifiable distinctions, derived from cases such as White-Smith [Music] Publishing Co. v. Apollo Co., 209 U.S. 1 (1908), under which statutory copyrightability in certain cases has been made to depend upon the form or medium in which the work is fixed. Under the bill it makes no difference what the form, manner, or medium of fixation may be whether it is in words, numbers, notes, sounds, pictures, or any other graphic or symbolic indicia, whether embodied in a physical object in written, printed, photographic, sculptural, punched, magnetic, or any other stable form, and whether it is capable of perception directly or by means of any machine or device "now known or later developed."

Under the bill, the concept of fixation is important since it not only determines whether the provisions of the statute apply to a work, but it also represents the dividing line between common law and statutory protection. As will be noted in more detail in connection with section 301, an unfixed work of authorship, such as an improvisation or an unrecorded choreographic work, performance, or broadcast, would continue to be subject to protection under State common law or statute, but would not be eligible for Federal statutory protection under section 102.

The bill seeks to resolve, through the definition of "fixation" in section 101, the status of live broadcasts — sports, news coverage, live performances of music, etc. — that are reaching the public in unfixed form but that are simultaneously being recorded. When a football game is being covered by four television cameras, with a director guiding the activities of the four cameramen and choosing which of their electronic images are sent out to the public and in what order, there is little doubt that what the cameramen and the director are doing constitutes "authorship." The further question to be considered is whether there has been a fixation. If the images and sounds to be broadcast are first recorded (on a video tape, film, etc.) and then transmitted, the recorded work would be considered a "motion picture" subject to statutory protection against unauthorized reproduction or retransmission of the broadcast. If the program content is transmitted live to the public while being recorded at the same time, the case would be treated the same; the copyright owner would not be forced to rely on common law rather than statutory rights in proceeding against an infringing user of the live broadcast.

Thus, assuming it is copyrightable — as a "motion picture" or "sound recording," for example — the content of a live transmission should be accorded statutory protection if it is being recorded simultaneously with its transmission. On the other hand, the definition of "fixation" would exclude from the concept purely evanescent or transient reproductions such as those projected briefly on a screen, shown electronically on a television or other cathode ray tube, or captured momentarily in the "memory" of a computer.

Under the first sentence of the definition of "fixed" in section 101, a work would be considered "fixed in a tangible medium of expression" if there has been an authorized embodiment in a copy or phonorecord and if that embodiment "is sufficiently permanent or stable" to permit the work "to be perceived, reproduced, or otherwise communicated for a period of more than transitory duration." The second sentence makes clear that, in the case of "a work consisting of sounds, images, or both, that are being transmitted," the work is regarded as "fixed" if a fixation is being made at the same time as the transmission.

Under this definition "copies" and "phonorecords" together will comprise all of the material objects in which copyrightable works are capable of being fixed. The definitions of these terms in section 101, together with their usage in section 102 and throughout the bill, reflect a fundamental distinction between the "original work" which is the product of "authorship" and the multitude of material objects in which it can be embodied. Thus, in the sense of the bill, a "book" is not a work of authorship, but is a particular kind of "copy." Instead, the author may write a literary "work," which in turn can be embodied in a wide range of "copies" and "phonorecords," including books, periodicals, computer punch cards, microfilm, tape recordings, and so forth. It is possible to have an "original work of authorship" without having a "copy" or "phonorecord" embodying it, and it is also possible to have a "copy" or "phonorecord" embodying something that does not qualify as an "original work of authorship." The two essential elements—original work and tangible object—must merge through fixation in order to produce subject matter copyrightable under the statute.

COMMENTS AND QUESTIONS

1. The fixation requirement arises in two separate portions of the Copyright Act. First, it is a requirement for copyright protection. Section 102(a) of the Copyright Act provides that "[c]opyright protection subsists . . . in original works of authorship *fixed in any tangible medium of expression,* now known or later developed, from which they can be perceived, reproduced, or otherwise communicated, either directly or with the aid of a machine or device" (emphasis added). Unless and until a work of authorship is so fixed, it does not qualify for copyright protection.

Fixation also plays a role in determining whether a defendant has infringed a copyright. Section 106(1) of the act provides that the copyright owner has the exclusive right to "reproduce the copyrighted work in copies or phonorecords." Section 101 of the act defines "copies" as "material objects . . . in which a work is *fixed.* . . ." (emphasis added). Thus a defendant does not infringe the right to reproduce unless she has reproduced the copyrighted work in fixed form. This rule has been at issue in a number of recent cases concerned with "copies" of computer programs or data made by the computer during its operation. Such "RAM copies" occur every time a computer is turned on or a file is opened; they are necessary for a computer user to view a program or file; and they disappear when the computer is turned off. Does opening a file make a "copy" for copyright purposes? See MAI v. Peak Computing, 991 F.2d 511 (9th Cir. 1993) (yes); see *infra* Chapter 7 (discussing the issue in detail).

2. *Constitutional Mandate?* As with the originality requirement, the Supreme Court has indicated that fixation is a constitutional requirement based on the Founders' use of the term "Writings" in Article I, Section 8, Clause 8. See Goldstein v. California, 412 U.S. 546, 561 (1973) (interpreting "Writings" to mean "physical rendering[s]" of expression). As a result, several courts have invalidated 1994 amendments to the Copyright Act, 17 U.S.C. § 1101, allowing performers to sue makers and distributors of bootleg recordings on the ground that such works were not "fixed." See United States v. Martignon, 346 F.Supp.2d 413 (S.D.N.Y. 2004); KISS Catalog v. Passport Intern. Productions, Inc., 350 F.Supp.2d 823 (C.D.Cal. 2004). Does this provision exceed Congress's power under the Intellectual Property Clause? Could it be sustained under the Commerce Clause? See United States v. Moghadam, 175 F.3d 1269 (11th Cir. 1999)(upholding anti-bootlegging under the Commerce Clause). What about Congress's power to enact legislation necessary and proper to implement an international treaty, such as the Agreement on Trade-Related Aspects of Intellectual Property Rights (TRIPs)? Should performers be able to prevent unauthorized distribution of their live performances?

3. *Policy Rationale(s) for a Fixation Requirement.* Why have a fixation requirement at all? One explanation for fixation lies in a view of copyright as intended to protect communication. See David G. Luettgen, Functional Usefulness vs. Communicative Usefulness: Thin Copyright Protection for the Nonliteral Elements of Computer Programs, 4 Tex. Intell. Prop. L.J. 233 (1996). Certainly, the original copyright laws—and the Constitution—speak of authors and writings, which we

associate with relatively direct communication between writer and audience. On this view, material that does not communicate (directly) to people is undeserving of copyright protection. Is there any necessary connection between whether a work is "fixed" and whether it communicates directly with an audience? Compare an extemporaneous public speech (unfixed, but communicative) with an individual's private diary (fixed, but private).

Another argument for the fixation requirement relates to the practical requirements of copyright litigation. Fixation helps in proving authorship. See Douglas Lichtman, Copyright as a Rule of Evidence, 52 Duke L.J. 683, 716-35 (2003). If any expression could be copyrighted, the law might face a large number of frivolous infringement suits that would be virtually impossible to verify — along the lines of "I gave them the idea (or rather the expression of the idea) for that book!" Analogous rules exist in other areas of law. Consider the statute of frauds in contract law and the best evidence rule, requiring the use of original documents as evidence in trials so as to ensure authenticity. But these rules are subject to various exceptions. Couldn't the authenticity function of the fixation rule be accomplished by a heightened evidentiary requirement (e.g., corroborating evidence, eye or ear witnesses, strong proof of access by the defendant) for unfixed works to be the basis for copyright infringement?

4. *State Law Protection for Unfixed Works.* Even when the fixation requirement bars federal copyright protection, common law protection may still be available. See, e.g., Estate of Hemingway v. Random House, 244 N.E.2d 250 (N.Y. 1968). The Copyright Act's preemption provision expressly retains state laws that protect unfixed expression. 17 U.S.C. § 301(b)(1).

3. Formalities

Copyright "formalities" are procedural requirements imposed on authors by the government that are necessary to obtain copyright protection but do not relate to the substance of the copyright. The United States has traditionally had four such formal requirements: notice of copyright, publication of the work, registration of the work with the Copyright Office, and deposit of a copy of the work with the Library of Congress. Over the past century, U.S. law has progressed from a regime in which failure to adhere to certain technical requirements resulted in forfeiture of copyright protection to the current regime in which formalities are largely voluntary and failure to comply does not risk forfeiture. The principal reason for this transformation was the decision by the United States to join the Berne Convention, which provides that copyright shall "not be subject to any formality." Nonetheless, copyright formalities have continuing relevance to the practice of copyright law, particularly for works created prior to January 1, 1978, the effective date of the 1976 Act.

a. Notice

U.S. copyright law has experienced three decreasingly restrictive notice regimes since the early part of the twentieth century:

1909 Act. Under the 1909 Act, federal copyright law protected only those works that contained the following information in the appropriate form and location: the year of first publication; the word "Copyright," the abbreviation "Copr." or the symbol ©; and the name of the copyright holder.[9] 17 U.S.C. §§ 19, 20. Failure to satisfy the precise technical requirements of the 1909 Act, such as placing notice on the "title page or the page immediately following" of books or other printed publications, typically resulted in forfeiture of copyright protection and the work's falling into the public domain. Section 21 of the 1909 Act excused omissions due to "accident or mistake," although this provision has been interpreted narrowly.

1976 Act/Pre-Ratification of Berne Convention (January 1, 1978–March 1, 1989). The 1976 Act provided that copyright protection begins upon the creation of a work, not upon publication. Congress maintained a notice requirement, although it liberalized the rules governing form and location of notice, see 17 U.S.C. §§ 401-03,[10] and it took much of the harshness out of the requirement, see 17 U.S.C. §§ 405-06. Failure to give notice on a small number of copies would not result in forfeiture; nor would even large-scale omissions, so long as they were inadvertent and the copyright holder registered the work within five years after publication and made reasonable efforts to give notice after the omission was discovered.

Post-Ratification of the Berne Convention (since March 1, 1989). The Berne Convention Implementation Act, Pub. L. No. 100-568, 102 Stat. 2853 (1988), eliminated the notice requirement of U.S. copyright law prospectively. Thus both the 1909 Act and 1976 Act (pre-Berne) regimes still apply to works publicly distributed without proper notice prior to March 1, 1989. Congress, however, *encouraged* voluntary notice by precluding an alleged infringer from claiming "innocent infringement" in mitigation of actual or statutory damages except where the copy to which she had access lacked proper notice. 17 U.S.C. § 401(d).

b. Publication

1909 Act. Federal copyright protection under the 1909 Act was triggered by the act of publishing a work. Unpublished works could be protected under state common law, or "constructively" published by registration with the Copyright Office. Despite the importance of "publication" to federal copyright protection, the 1909 Act did not specifically define the term. This gap spawned a rich and complex jurisprudence defining publication. Because of the grave consequences of publishing a work without proper notice — forfeiture of copyright protection — the courts developed a distinction between *divestive* publication, resulting in forfeiture (divestiture) of common law copyright protection, and *investive* publication, resulting in forfeiture of

---

9. Unpublished works were protectable at common law prior to the 1976 Act. See Chapter 6, discussing common law copyright and preemption.

10. The Copyright Office has issued regulations to accommodate different works by requiring only "reasonable" placement of the notice. In some cases, this may be on packaging, particularly where the user would not encounter the notice on the actual work (e.g., computer software object code) or where affixing the notice to the actual work would interfere with the work (e.g., sound recordings, some forms of visual art).

federal copyright protection if notice is inadequate. In a leading case, the Second Circuit observed that

> ...courts apply different tests of publication depending on whether plaintiff is claiming protection because he did not publish and hence has a common law claim of infringement—in which case the distribution must be quite large to constitute "publication"—or whether he is claiming under the copyright statute—in which case the requirements for publication are quite narrow. In each case the courts appear so to treat the concept of "publication" as to prevent piracy.

American Visuals Corp. v. Holland, 239 F.2d 740, 744 (2d Cir. 1956). Thus the extent of distribution required to divest common law copyright protection is substantially greater than that required to invest, i.e., require notice for, a federal copyrighted work. In White v. Kimmell, 193 F.2d 744, 746-47 (9th Cir. 1952), *cert. denied,* 343 U.S. 957 (1952), the court distinguished between "limited publication" whereby a distribution of copies to "a definitely selected group and for a limited purpose, and without the right of diffusion, reproduction, distribution or sale" did not constitute a publication for purposes of the 1909 Act, and "general publication," which operated to divest common law protection. In addition, courts generally found that public performance or display of a work did not constitute publication unless tangible copies of the work were distributed to the public. See Ferris v. Frohman, 223 U.S. 424 (1912) (applying pre-1909 Act law); McCarthy v. White, 259 F. 364 (S.D.N.Y. 1919) (public performance of musical work); Estate of Martin Luther King, Jr., Inc. v. CBS, Inc., 194 F.3d 1211 (11th Cir. 1999) (holding that King's delivery of his renowned "I Have a Dream" speech in August 1963 to a large audience, along with the fact that sponsors of event obtained live broadcasts on radio and television and extensive contemporary coverage of event in the news media, did not alone amount to general publication of the speech for copyright purposes and hence did not divest King of common law copyright protection for the speech); King v. Mister Maestro, Inc., 224 F.Supp. 101 (S.D.N.Y. 1963) (same). Furthermore, courts determined that unauthorized distribution did not constitute publication. In these ways, courts alleviated some of the harsh effects of failure to adhere to the notice requirements.

Is a television broadcast a "publication" within the meaning of the 1909 Act? Does it matter whether permanent copies are made at the time of the broadcast?

1976 Act/Pre-Ratification of Berne Convention (January 1, 1978–March 1, 1989). Under the 1976 Act, federal copyright protection is triggered by the act of creating a work fixed in a tangible medium of expression, and common law copyright is preempted. Hence publication no longer served to distinguish between statutory and common law copyright. Nonetheless, publication still served to determine when notice was required.

The 1976 Act clarified the copyright law by defining "publication" as

> ...the distribution of copies or phonorecords of a work to the lay public by sale or other transfer of ownership, or by rental, lease, or lending. The offering to distribute copies or phonorecords to a group of persons for purposes of further distribution, public performance, or public display, constitutes publication. A public performance or display of a work does not of itself constitute publication.

17 U.S.C. §101. This definition largely codified the principal considerations that evolved through judicial interpretation of the 1909 Act. See H.R. Rep. No. 1476, 94th Cong., 2d Sess., 138 (1976). Publication turns on physical transfer of copies to the public generally without disclosure restrictions or to a narrower group for purposes of distribution or dissemination. The 1976 Act definition of publication expressly excludes public performance or display.

Post-Ratification of the Berne Convention (since March 1, 1989). With the elimination of a mandatory notice requirement, the act of publication is no longer a factor in determining the validity of works created after March 1, 1989. Nonetheless, publication still has relevance for works created after March 1, 1989 in the following respects:

- *Deposit.* Deposit at the Library of Congress is mandatory only for published works. 17 U.S.C. §407.
- *Works of Foreign Authors.* Whereas all unpublished works are protected regardless of nationality or domicile of the author, *published* works of foreign authors are protected only under the conditions described in section 104(b).
- *Duration of Copyright Protection.* The term of protection for entity owners and works for hire is 95 years from the year of first publication and 120 years from creation for unpublished works. 17 U.S.C. §302(c); see also 17 U.S.C. §302(d), (e).
- The reproduction rights of libraries depend on whether the work has been published. See 17 U.S.C. §108(b), (c).
- *Termination of Transfers.* See 17 U.S.C. §203(a)(3).
- *Certain Performance Rights.* See 17 U.S.C. §§110(9), 118(b), (d).
- *Establishing Prima Facie Evidence of Validity of Copyright.* Registration of copyright must occur within five years of first publication. 17 U.S.C. §410(c).
- *Damages.* Statutory damages and attorney fees are available for published works only if registration preceded the infringement or if the work was registered within three months after publication. 17 U.S.C. §412.

PROBLEM

Problem 4-2. Which of the following acts would constitute publication under the 1909 Act? the 1976 Act?

a) Penelope Poet brings copies of her latest three poems to the monthly meeting of the Philadelphia Aspiring Poets Society. She distributes copies to the eight people in attendance that month prior to reading the poems.

b) Professor Edgar Edifice assembles a course reader consisting of public domain materials and excerpts he wrote himself. Students in his Advanced Copyright Theory seminar purchase the reader for the cost of reproduction from the school's reprographics service.

c) Arnold Author recently completed a draft of his first novel. He sends a copy to his agent, whom he authorizes to distribute copies to publishing houses for consideration.

---

c. Registration

Unlike the notice requirement, registration of a copyrighted work with the Copyright Office has always been "voluntary," although with a significant qualification with respect to works governed by the 1909 Act. Also unlike the notice requirement, the registration requirement remains in effect today with regard to works of U.S. origin, notwithstanding United States' adherence to the Berne Convention.[11]

1909 Act. The term of a copyright for a 1909 Act work initially ran for 28 years, with the author having the right to renew the copyright for an additional 28 years. The Act did not require registration in order to obtain a copyright (publication with proper notice triggered protection), but it did require registration by the 28th year in order to renew the copyright.[12] Further, registration of the work was a prerequisite to bringing an infringement action.

1976 Act/Pre-Ratification of Berne Convention (January 1, 1978–March 1, 1989). By shifting to a unitary term, the 1976 Act abolished renewal for works created after January 1, 1978, making copyright registration entirely optional for the maintenance of copyright protection. Duration of copyright protection no longer depends on registration. Nonetheless, Congress retained the registration system and encouraged its use through various incentives. First, successful registration constitutes *prima facie* evidence of the validity of the copyright. 17 U.S.C. §410. Second, a copyright holder must register the copyright before bringing an infringement action, 17 U.S.C. §411, so in practice an unregistered copyright is only a potential rather than an actual right.[13] Finally, there is also a powerful incentive for *early* registration: a copyright holder can obtain statutory damages and attorney fees only for infringements that occurred after registration (or which occurred after publication if the work was registered within three months after publication).

Post-Ratification of the Berne Convention (since March 1, 1989). To comply with the Berne Convention, Congress amended section 411 to eliminate the requirement that copyright owners whose country of origin[14] is another Berne member nation must register their works prior to instituting suits. However, Congress retained the requirement of registration prior to suit for domestic works, thereby imposing a greater burden on those who first publish in the United States.

---

11. As we will see in the international copyright law section (I), the Berne Convention does not mandate that nations jettison formalities entirely, only that they may not impose formalities as a prerequisite for copyright protection with respect to works of foreign origin. Nations remain free to discriminate *against* works of domestic origin.

12. Pursuant to the Copyright Renewal Act of 1992, Congress made renewal automatic. Therefore, works published with proper notice after January 1, 1964 (1992-28=1964) no longer need be registered to be protected for the full duration available under the statute. Nonetheless, Congress encouraged renewal registration by offering several benefits, such as automatic vesting, reversion rights in derivative works (17 U.S.C. §304(a)(4)(A), see *infra* section D(3)(c) (discussion of *Abend*-rights)), evidentiary weight (renewal registration constitutes *prima facie* evidence of the facts asserted in the renewal certificate), and enhanced remedies.

13. Interestingly, a copyright holder need only *file* for registration of copyright and obtain a response before bringing suit. The putative copyright owner is entitled to bring suit even if the Copyright Office rejects the registration application. See 17 U.S.C. §411(a).

14. A work's country of origin is the country in which it was first published, or in the case of unpublished works, the country in which it was created.

d. Deposit

Section 407 of the Copyright Act requires deposit of two copies of each work published in the United States[15] for which copyright is claimed within three months after publication. (Certain categories of works are exempted from the requirement.) The purpose of this requirement is to enhance the collection of the Library of Congress. Under the 1909 Act, the Register of Copyrights could demand compliance with this deposit requirement, and failure to comply could result in forfeiture of the copyright. Under the 1976 Act, however, the Library deposit requirement — while still mandatory — does not affect the validity of a copyright or the author's right to bring suit. 17 U.S.C. §407. Failure to comply gives rise to a fine, but no other penalty.[16]

A separate deposit with the Copyright Office is also required for copyright registration by section 408. However, this requirement may generally be satisfied by section 407's Library deposit. Unlike the Library of Congress deposit requirement, failure to comply with the section 408 deposit requirement results in a refusal to register the copyright.

The purpose behind registration deposit under section 408 is to enable the Copyright Office to know what it is registering. The section 407 Library of Congress deposit requirement promotes progress by making knowledge accessible.

Note on the Restoration of Foreign Copyrighted Works

The Berne Convention Implementation Act of 1988 brought U.S. copyright law formalities into compliance with the Berne Convention's minimum criteria on a prospective basis, but did not address the retrospective problem — that a number of foreign works that received copyright protection in their home country (and elsewhere in the world) entered the public domain in the United States because their authors failed to comply with then-applicable notice requirements of U.S. law. This problem was of more than technical or theoretical interest. Among the foreign works that were still under copyright protection in much of the rest of the world but in the public domain in the United States were many of the works of J.R.R. Tolkein, including *The Hobbit* and *The Fellowship of the Ring*. These works have been freely copyable in the United States but still received copyright protection elsewhere in the world.

Article 18 of the Berne Convention requires that new members protect all works of other member nations whose copyrights had not yet expired in their countries of origin. The United States eventually got around to complying with this provision in 1993 (with regard to works from Canada and Mexico as part of the North American Free Trade Agreement Act (NAFTA)) and in 1994 (with regard to all other Berne Convention nations as part of the Agreement on Trade-Related Aspects of Intellectual Property Rights (TRIPs) as part of the General Agreement

---

15. The provision limiting the Library deposit requirement to works published in the United States was added by the Berne implementing legislation in 1988.

16. In amending the 1976 Act to adhere to the Berne Convention, Congress concluded that retention of the deposit requirement was not inconsistent with the Berne Convention because failure to comply with the deposit requirement did not result in forfeiture of any copyright protection. H.R. Rep. No. 609, 100th Cong., 2d Sess. 44 (1988).

on Tariffs and Trade (GATT) Uruguay Round). To comply with TRIPs, Congress in 1994 added section 104A to the Copyright Act, restoring copyright protection for foreign works[17] from Berne Convention nations that had lost protection in the United States "due to noncompliance with formalities imposed...by United States copyright law." See Dam Things from Denmark, a/k/a Troll Company ApS, v. Russ Berrie & Co., Inc., 290 F.3d 548 (3rd Cir. 2002). In effect, these works are retrieved from the public domain in the United States and are treated in the same way as any other copyrighted works for purposes of duration, ownership, and so on. Thanks to this law, copyrights in *The Hobbit, The Fellowship of the Ring,* and various other works of J.R.R. Tolkien were restored in the United States in 1996. See Library of Congress, Copyright Office, Copyright Restoration of Works in Accordance With the Uruguay Round Agreements Act, 61 Fed. Reg. 46133 (Aug. 30, 1996).

Copyright restoration presents a number of potential problems, and efforts to deal with them have made section 104A a complex provision. Copyright protection in restored works is not retroactive. However, the copyright owner does have the right to "cut off" future uses of the work after a limited grace period by giving the Copyright Office notice of her intention to enforce the copyright. More complex problems are presented by those who have created "derivative works" in reliance on the U.S. public domain status of a work of foreign origin. Should such authors be prevented from exploiting their creations following restoration of the underlying work's copyright? The Act provides that the authors of such derivative works may continue to exploit them, provided they pay "reasonable compensation" to the original copyright owner. See 17 U.S.C. § 104A(d)(3)(B).

Does restoring copyright in works already in the public domain "promote the progress of science and the useful arts"? How? What additional incentive does such a right give to current authors? If it does not confer such an additional incentive, is it constitutional? See Luck's Music Library, Inc. v. Gonzales, 407 (D.C. Cir. 2005) (upholding the constitutionality of the copyright restoration provisions); Golan v. Ashcroft, 310 F.Supp.2d 1215 (D.Colo. 2004) (same).

C. COPYRIGHTABLE SUBJECT MATTER

1. Limitations on Copyrightability: Distinguishing Function and Expression

This section considers the fundamental doctrines that operate to channel protection for works between the patent and copyright regimes.

a. *The Idea-Expression Dichotomy*

The most significant doctrine limiting the copyrightability of works is the "idea-expression" dichotomy, which is codified in section 102(b):

---

17. This provision excludes works of foreign authors that were first published in the United States without proper notice.

17 U.S.C. § 102. Subject Matter of Copyright: In General

(a) Copyright protection subsists, in accordance with this title, in original works of authorship. . . .

(b) In no case does copyright protection for an original work of authorship extend to any idea, procedure, process, system, method of operation, concept, principle, or discovery regardless of the form in which it is described, explained, illustrated, or embodied in such work.

The division between protectable expression and unprotectable ideas was developed in the seminal case of *Baker v. Selden*.

Baker v. Selden
Supreme Court of the United States
101 U.S. 99 (1879)

Mr. Justice BRADLEY delivered the opinion of the court.

Charles Selden, the testator of the complainant in this case, in the year 1859 took the requisite steps for obtaining the copyright of a book, entitled "Selden's Condensed Ledger, or Book-keeping Simplified," the object of which was to exhibit and explain a peculiar system of book-keeping. In 1860 and 1861, he took the copyright of several other books, containing additions to and improvements upon the said system. The bill of complaint was filed against the defendant, Baker, for an alleged infringement of these copyrights. The latter, in his answer, denied that Selden was the author or designer of the books, and denied the infringement charged, and contends on the argument that the matter alleged to be infringed is not a lawful subject of copyright. . . .

The book or series of books of which the complainant claims the copyright consists of an introductory essay explaining the system of book-keeping referred to, to which are annexed certain forms or blanks, consisting of ruled lines, and headings, illustrating the system and showing how it is to be used and carried out in practice. This system effects the same results as book-keeping by double entry; but, by a peculiar arrangement of columns and headings, presents the entire operation, of a day, a week, or a month, on a single page, or on two pages facing each other, in an account-book. The defendant uses a similar plan so far as results are concerned; but makes a different arrangement of the columns, and uses different headings. If the complainant's testator had the exclusive right to the use of the system explained in his book, it would be difficult to contend that the defendant does not infringe it, notwithstanding the difference in his form of arrangement; but if it be assumed that the system is open to public use, it seems to be equally difficult to contend that the books made and sold by the defendant are a violation of the copyright of the complainant's book considered merely as a book explanatory of the system. Where the truths of a science or the methods of an art are the common property of the whole world, an author has the right to express the one, or explain and use the other, in his own way. As an author, Selden explained the system in a particular way. It may be conceded that Baker makes and uses account-books arranged on substantially the same system; but the proof fails to show that he has violated the copyright of Selden's book, regarding the latter merely as an explanatory work; or that he has infringed

Selden's right in any way, unless the latter became entitled to an exclusive right in the system.

✗ . . . [T]he question is, whether the exclusive property in a system of book-keeping can be claimed, under the law of copyright, by means of a book in which that system is explained? . . .

There is no doubt that a work on the subject of book-keeping, though only explanatory of well-known systems, may be the subject of a copyright; but, then, it is claimed only as a book. Such a book may be explanatory either of old systems, or of an entirely new system; and, considered as a book, as the work of an author, conveying information on the subject of book-keeping, and containing detailed explanations of the art, it may be a very valuable acquisition to the practical knowledge of the community. But there is a clear distinction between the book, as such, and the art which it is intended to illustrate. The mere statement of the proposition is so evident, that it requires hardly any argument to support it. The same distinction may be predicated of every other art as well as that of book-keeping. A treatise on the composition and use of medicines, be they old or new; on the construction and use of ploughs, or watches, or churns; or on the mixture and application of colors for painting or dyeing; or on the mode of drawing lines to produce the effect of perspective, — would be the subject of copyright; but no one would contend that the copyright of the treatise would give the exclusive right to the art or manufacture described therein. The copyright of the book, if not pirated from other works, would be valid without regard to the novelty, or want of novelty, of its subject-matter. The novelty of the art or thing described or explained has nothing to do with the validity of the copyright. To give to the author of the book an exclusive property in the art described therein, when no examination of its novelty has ever been officially made, would be a surprise and a fraud upon the public. That is the province of letters-patent, not of copyright. The claim to an invention or discovery of an art or manufacture must be subjected to the examination of the Patent Office before an exclusive right therein can be obtained; and it can only be secured by a patent from the government.

The difference between the two things, letters-patent and copyright, may be illustrated by reference to the subjects just enumerated. Take the case of medicines. Certain mixtures are found to be of great value in the healing art. If the discoverer writes and publishes a book on the subject (as regular physicians generally do), he gains no exclusive right to the manufacture and sale of the medicine; he gives that to the public. If he desires to acquire such exclusive right, he must obtain a patent for the mixture as a new art, manufacture, or composition of matter. He may copyright his book, if he pleases; but that only secures to him the exclusive right of printing and publishing his book. So of all other inventions or discoveries. . . .

Of course, these observations are not intended to apply to ornamental designs, or pictorial illustrations addressed to the taste. Of these it may be said, that their form is their essence, and their object, the production of pleasure in their contemplation. This is their final end. They are as much the product of genius and the result of composition, as are the lines of the poet or the historian's periods. On the other hand, the teachings of science and the rules and methods of useful art have their final end in application and use; and this application and use are what the public derive from the publication of a book which teaches them. But as embodied and taught in a literary composition or book, their essence consists only in their statement. This alone is what is secured by the copyright. The use by another of the same methods of statement, whether in words or illustrations, in a book published for teaching the art, would undoubtedly be an infringement of the copyright.

Returning to the case before us, we observe that Charles Selden, by his books, explained and described a peculiar system of book-keeping, and illustrated his method by means of ruled lines and blank columns, with proper headings on a page, or on successive pages. Now, whilst no one has a right to print or publish his book, or any material part thereof, as a book intended to convey instruction in the art, any person may practice and use the art itself which he has described and illustrated therein. The use of the art is a totally different thing from a publication of the book explaining it. The copyright of a book on book-keeping cannot secure the exclusive right to make, sell, and use account-books prepared upon the plan set forth in such book.

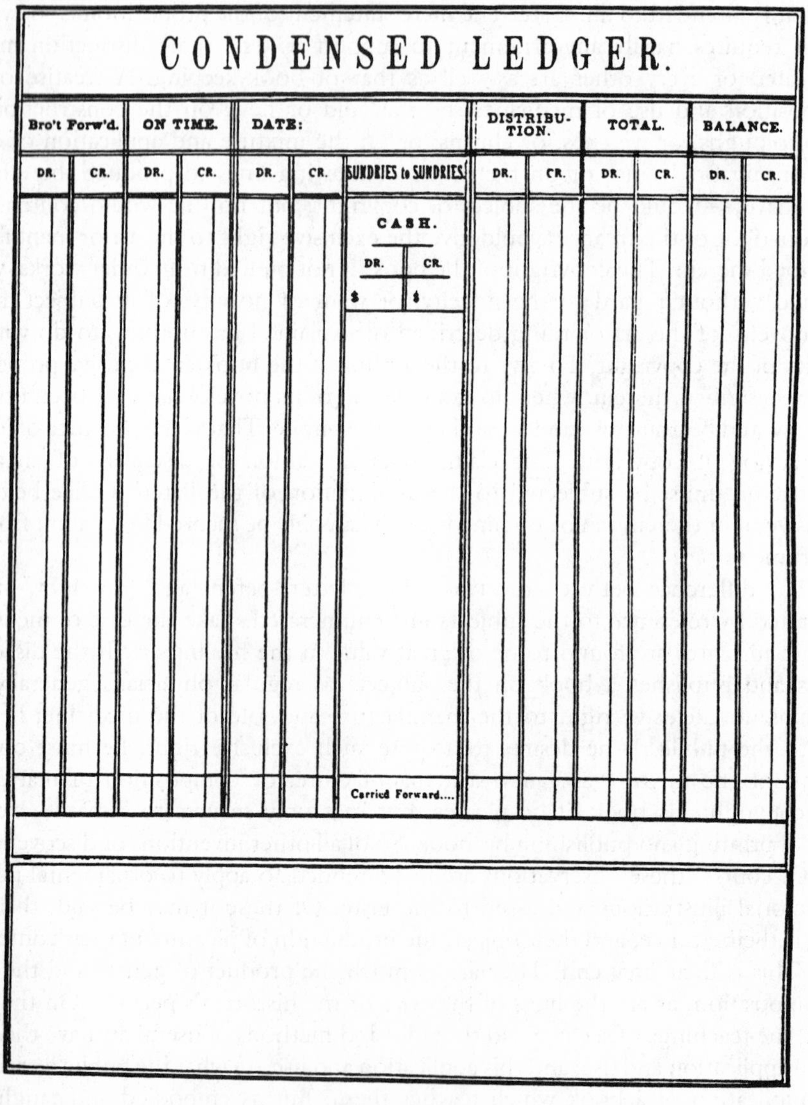

FIGURE 4-1
Selden's blank form for condensed ledger.

Whether the art might or might not have been patented, is a question which is not before us. It was not patented, and is open and free to the use of the public. And, of course, in using the art, the ruled lines and headings of accounts must necessarily be used as incident to it [see Figure 4-1].

The plausibility of the claim put forward by the complainant in this case arises from a confusion of ideas produced by the peculiar nature of the art described in the books which have been made the subject of copyright. In describing the art, the illustrations and diagrams employed happen to correspond more closely than usual with the actual work performed by the operator who uses the art. Those illustrations and diagrams consist of ruled lines and headings of accounts; and it is similar ruled lines and headings of accounts which, in the application of the art, the book-keeper makes with his pen, or the stationer with his press; whilst in most other cases the diagrams and illustrations can only be represented in concrete forms of wood, metal, stone, or some other physical embodiment. But the principle is the same in all. The description of the art in a book, though entitled to the benefit of copyright, lays no foundation for an exclusive claim to the art itself. The object of the one is explanation; the object of the other is use. The former may be secured by copyright. The latter can only be secured, if it can be secured at all, by letters-patent. . . .

The conclusion to which we have come is, that blank account books are not the subject of copyright; and that the mere copyright of Selden's book did not confer upon him the exclusive right to make and use account-books, ruled and arranged as designated by him and described and illustrated in said book.

The decree of the Circuit Court must be reversed, and the cause remanded with instructions to dismiss the complainant's bill; and it is

So ordered.

COMMENTS AND QUESTIONS

1. What is the Supreme Court's holding in this case? Did the Court rule that Selden's subject matter — accounting forms — are not copyrightable? Or that Selden's particular forms are not copyrightable? Alternatively, did the Court rule that Selden's forms are copyrightable but that the copyright does not prevent Baker's particular use of the forms, because such a result would in effect bestow upon Selden a monopoly over the system in question?

2. What does this case suggest about the relationship between copyright and patent protection? Does it imply that a particular work cannot be protected by both patent and copyright law? In this regard, consider whether Selden could prevent Baker from reproducing the book explaining the Selden system. Does the doctrine of Baker v. Selden — which establishes an idea-expression dichotomy — coherently channel intellectual property protection between the copyright and patent modes of protection? What differences between patentable subject matter (and the process of innovation) and copyrightable subject matter (and the process of artistic creativity) justify such a doctrine?

Consider the Seventh Circuit's discussion of copyright protection (or lack thereof) for recipes:

The recipes' directions for preparing the assorted dishes fall squarely within the class of subject matter specifically excluded from copyright protection by Section 102(b). . . . The recipes at issue here describe a procedure by which the reader may produce many dishes featuring a specific yogurt. As such, they are excluded from copyright protection as either a "procedure, process, or system" under Section 102(b).

Protection for ideas or processes is the purview of patent....

Nothing in our decision today runs counter to the proposition that recipes may be copyrightable. There are cookbooks in which the authors lace their directions for producing dishes with musings about the spiritual nature of cooking or reminiscences they associate with the wafting odors of certain dishes in various stages or preparation.... Other recipes may be accompanied by tales of their historical or ethnic origin.

Publications Intl. v. Meredith Corp., 88 F.3d 473 (7th Cir. 1996). What aspects of a cookbook are protectable?

3. *Distinguishing Idea from Expression.* The most challenging aspect of applying the idea-expression dichotomy is determining where to draw the line between idea and expression. Consider the work in question in *Baker v. Selden.* What was the "idea" behind Selden's book? Was it to describe a better system of accounting? Was it to describe a system of double-entry bookkeeping? Was it to describe the particular system he had invented? Or was it all of the above?

In the classic statement of how to draw the line between idea and expression, Judge Learned Hand stated the problem as follows:

> Upon any work, and especially upon a play, a great number of patterns of increasing generality will fit equally well, as more and more of the incident is left out. The last may perhaps be no more than the most general statement of what the play is about, and at times may consist only of its title; but there is a point in this series of abstractions where they are no longer protected, since otherwise the playwright could prevent the use of his "ideas."...

Nichols v. Universal Pictures Corp., 45 F.2d 119, 121 (2d Cir. 1930). Apply this reasoning to *Baker* itself. If the idea of Selden's book is to write a step-by-step guide explaining the use of his new forms (a low level of abstraction), even the detailed structure of the explanatory text may be unprotectable (because it is "necessary" to the idea). At a slightly higher level of abstraction, the Court considered the forms themselves unprotectable but gave Selden considerable latitude to protect his description of his system. At the highest levels of abstraction, had the Supreme Court determined that Selden's idea was to improve accounting, his double-entry bookkeeping forms might well have been considered simply one means of expressing that idea. In that case, Selden would presumably be entitled to copyright the forms.

How do courts determine the appropriate "level of abstraction" for distinguishing idea and expression? Paul Goldstein suggests that there are three categories of unprotectable ideas: the "animating concept" behind the work, the functional principles or "solutions" described or embodied in the work (such as Selden's forms), and the fundamental "building blocks" of creative expression (such as basic plot or character outlines in literary or dramatic works). 1 Paul Goldstein, Copyright § 2.3.1.1. In his view, courts engage in a rough sort of balancing between the dangers of overprotecting and underprotecting a particular work in determining on which side of the idea/expression line the work falls.

4. *Blank Forms.* The Copyright Office considers the following works not copyrightable and hence ineligible for registration: "Blank forms, such as time cards, graph paper, account books, diaries, bank checks, scorecards, address books, report

forms, order forms and the like, which are designed for recording information and do not in themselves convey information." 37 C.F.R. § 202(1)(c).

5. For a detailed history of the background behind *Baker*, see Pamela Samuelson, The Story of *Baker v. Selden*, in Intellectual Property Stories (Rochelle Dreyfuss and Jane Ginsburg, eds., 2005).

PROBLEM

Problem 4-3. Mediforms designs and sells health care forms to doctors, who submit them to insurance carriers. A sample form is shown in Figure 4-2.

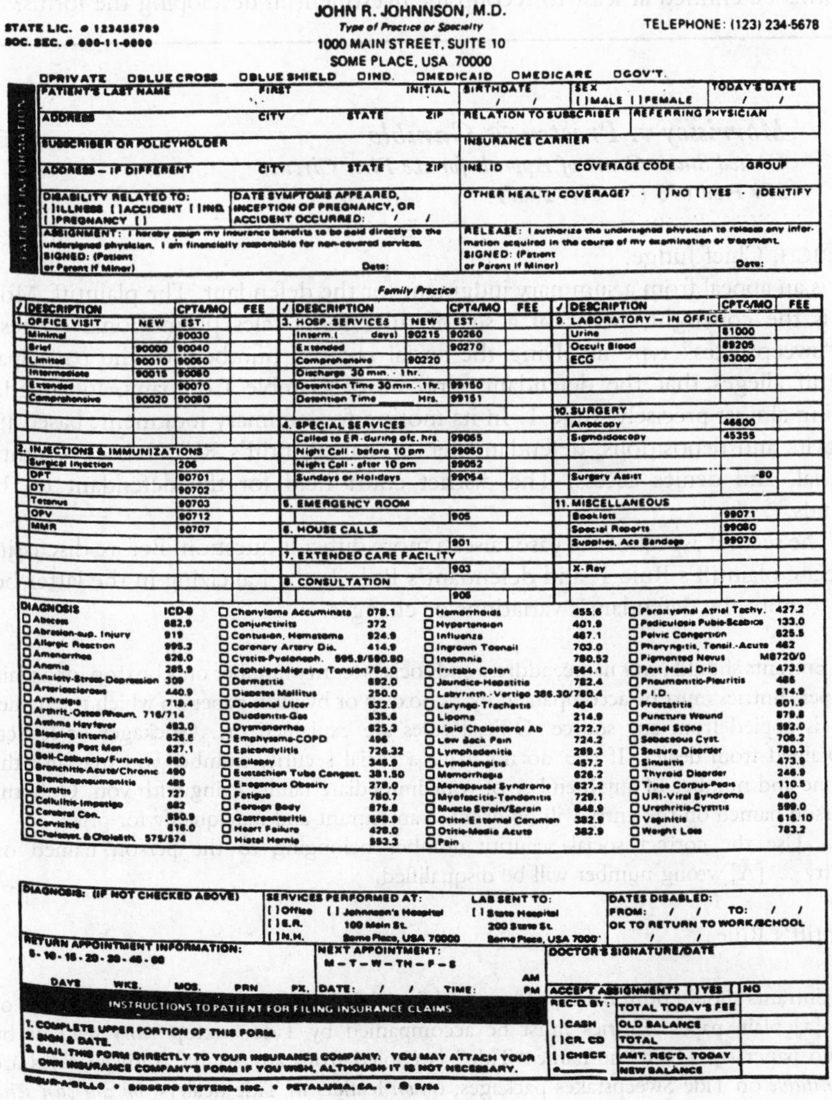

JOHN R. JOHNNSON, M.D.
Type of Practice or Specialty
1000 MAIN STREET, SUITE 10
SOME PLACE, USA 70000

STATE LIC. # 123456789
SOC. SEC. # 000-11-0000
TELEPHONE: (123) 234-5678

☐PRIVATE ☐BLUE CROSS ☐BLUE SHIELD ☐IND. ☐MEDICAID ☐MEDICARE ☐GOV'T.

PATIENT INFORMATION

PATIENT'S LAST NAME | FIRST | INITIAL | BIRTHDATE / / | SEX [ ]MALE [ ]FEMALE | TODAY'S DATE / /
ADDRESS | CITY | STATE | ZIP | RELATION TO SUBSCRIBER | REFERRING PHYSICIAN
SUBSCRIBER OR POLICYHOLDER | INSURANCE CARRIER
ADDRESS — IF DIFFERENT | CITY | STATE | ZIP | INS. ID | COVERAGE CODE | GROUP
DISABILITY RELATED TO: [ ]ILLNESS [ ]ACCIDENT [ ]IND. [ ]PREGNANCY [ ] | DATE SYMPTOMS APPEARED, INCEPTION OF PREGNANCY, OR ACCIDENT OCCURRED: / / | OTHER HEALTH COVERAGE? · [ ]NO [ ]YES · IDENTIFY
ASSIGNMENT: I hereby assign my insurance benefits to be paid directly to the undersigned physician. I am financially responsible for non-covered services. SIGNED: (Patient or Parent if Minor) Date:
RELEASE: I authorize the undersigned physician to release any information acquired in the course of my examination or treatment. SIGNED: (Patient or Parent if Minor) Date:

Family Practice

✓	DESCRIPTION		CPT4/MO	FEE	✓	DESCRIPTION		CPT4/MO	FEE	✓	DESCRIPTION	CPT4/MO	FEE
	1. OFFICE VISIT	NEW	EST.			3. HOSP. SERVICES	NEW	EST.			9. LABORATORY — IN OFFICE		
	Minimal		90030			Interm.(days)	90215	90260			Urine	81000	
	Brief	90000	90040			Extended		90270			Occult Blood	89205	
	Limited	90010	90050			Comprehensive	90220				ECG	93000	
	Intermediate	90015	90060			Discharge 30 min. - 1 hr.							
	Extended		90070			Detention Time 30 min. - 1 hr.		99150					
	Comprehensive	90020	90080			Detention Time ____ Hrs.		99151					
											10. SURGERY		
						4. SPECIAL SERVICES					Anoscopy	46600	
						Called to ER - during ofc. hrs.		99065			Sigmoidoscopy	45355	
	2. INJECTIONS & IMMUNIZATIONS					Night Call - before 10 pm		99050					
	Surgical Injection		206			Night Call - after 10 pm		99052					
	DPT		90701			Sundays or Holidays		99054			Surgery Assist	-80	
	DT		90702										
	Tetanus		90703			5. EMERGENCY ROOM					11. MISCELLANEOUS		
	OPV		90712					905			Booklets	99071	
	MMR		90707			6. HOUSE CALLS					Special Reports	99080	
								901			Supplies, Ace Bandage	99070	
						7. EXTENDED CARE FACILITY							
								903			X-Ray		
						8. CONSULTATION							
								906					

DIAGNOSIS	ICD-9						
☐ Abscess	682.9	☐ Chonyloma Accuminata	078.1	☐ Hemorrhoids	455.6	☐ Paroxysmal Atrial Tachy.	427.2
☐ Abrasion-sup. Injury	919	☐ Conjunctivitis	372	☐ Hypertension	401.9	☐ Pediculosis Pubis-Scabies	133.0
☐ Allergic Reaction	995.3	☐ Contusion, Hematoma	924.9	☐ Influenza	487.1	☐ Pelvic Congestion	625.5
☐ Amenorrhea	626.0	☐ Coronary Artery Dis.	414.9	☐ Ingrown Toenail	703.0	☐ Pharyngitis, Tonsil.-Acute	462
☐ Anemia	285.9	☐ Cystitis-Pyeloneph.	595.9/590.80	☐ Insomnia	780.52	☐ Pigmented Nevus	M8720/0
☐ Anxiety-Stress-Depression	309	☐ Cephalgia-Migraine Tension	784.0	☐ Irritable Colon	564.1	☐ Post Nasal Drip	473.9
☐ Arteriosclerosis	440.9	☐ Dermatitis	692.0	☐ Jaundice-Hepatitis	782.4	☐ Pneumonitis-Pleuritis	486
☐ Arthralgia	719.4	☐ Diabetes Mellitus	250.0	☐ Labyrinth.-Vertigo	386.30/780.4	☐ P.I.D.	614.9
☐ Arthrit.-Osteo Rheum.	716/714	☐ Duodenal Ulcer	532.9	☐ Laryngo-Tracheitis	464	☐ Prostatitis	601.9
☐ Asthma Hayfever	493.0	☐ Duodenitis-Gas	535.6	☐ Lipoma	214.9	☐ Puncture Wound	879.8
☐ Bleeding Internal	626.6	☐ Dysmenorrhea	625.3	☐ Lipid Cholesterol Ab	272.7	☐ Renal Stone	592.0
☐ Bleeding Post Men	627.1	☐ Emphysema-COPD	496	☐ Low Back Pain	724.2	☐ Sebaceous Cyst	706.2
☐ Boil-Carbuncle/Furuncle	680	☐ Epicondylitis	726.32	☐ Lymphadenitis	289.3	☐ Seizure Disorder	780.3
☐ Bronchitis-Acute/Chronic	490	☐ Epilepsy	345.9	☐ Lymphangitis	457.2	☐ Sinusitis	473.9
☐ Bronchopneumonitis	485	☐ Eustachian Tube Congest.	381.50	☐ Menorrhagia	626.2	☐ Thyroid Disorder	246.9
☐ Bursitis	727	☐ Exogenous Obesity	278.0	☐ Menopausal Syndrome	627.2	☐ Tinea Corpus-Pedis	110.5
☐ Cellulitis-Impetigo	682	☐ Fatigue	780.7	☐ Myofascitis-Tendonitis	729.1	☐ URI-Viral Syndrome	460
☐ Cerebral Con.	850.9	☐ Foreign Body	879.8	☐ Muscle Strain/Sprain	848.9	☐ Urethritis-Cystitis	599.0
☐ Cervicitis	616.0	☐ Gastroenteritis	558.9	☐ Otitis-External Cerumen	382.9	☐ Vaginitis	616.10
☐ Cholecyst. & Cholelith.	575/574	☐ Heart Failure	428.0	☐ Otitis-Media Acute	382.9	☐ Weight Loss	783.2
		☐ Histal Hernia	553.3	☐ Pain		☐	

DIAGNOSIS: (IF NOT CHECKED ABOVE)
SERVICES PERFORMED AT: [ ]Office [ ]E.R. [ ]N.H. [ ] Johnnson's Hospital 100 Main St. Some Place, USA 70000
LAB SENT TO: [ ] State Hospital 200 State St. Some Place, USA 7000
DATES DISABLED: FROM: / / TO: / / OK TO RETURN TO WORK/SCHOOL / /
RETURN APPOINTMENT INFORMATION: 5 - 10 - 15 - 20 - 30 - 45 - 60 DAYS WKS. MOS. PRN PX.
NEXT APPOINTMENT: M — T — W — TH — F — S DATE: / / TIME: AM PM
DOCTOR'S SIGNATURE/DATE
ACCEPT ASSIGNMENT? [ ]YES [ ]NO
REC'D. BY: [ ]CASH [ ]CR. CD [ ]CHECK #____
TOTAL TODAY'S FEE
OLD BALANCE
TOTAL
AMT. REC'D. TODAY
NEW BALANCE

INSTRUCTIONS TO PATIENT FOR FILING INSURANCE CLAIMS
1. COMPLETE UPPER PORTION OF THIS FORM.
2. SIGN & DATE.
3. MAIL THIS FORM DIRECTLY TO YOUR INSURANCE COMPANY. YOU MAY ATTACH YOUR OWN INSURANCE COMPANY'S FORM IF YOU WISH, ALTHOUGH IT IS NOT NECESSARY.

INSUR-A-BILL® • BIBBERO SYSTEMS, INC. • PETALUMA, CA. • © 8/84

FIGURE 4-2
Sample health care form.

The diagnosis checklist contains categories specified by the American Medical Association or government publications (including official code numbers). Mediforms offers a variety of forms, according to specialty, to reflect the illnesses and treatments most relevant to the particular doctor. The forms are personalized to include the doctor's name and address, the nature of the doctor's practice, and the hospitals or clinics at which the doctor performs services. Doctors may use the checklists provided or may customize their own checklists, which most doctors choose to do. The forms also contain brief instructions for filling in each blank and instructions explaining how to obtain insurance reimbursement. The forms were widely praised in the industry and were copied by competitors and individual doctors. Are the forms copyrightable? Should Mediforms be entitled at least to recoup its investment in developing the forms?

Morrissey v. Procter & Gamble
United States Court of Appeals for the First Circuit
379 F.2d 675 (1st Cir. 1967)

ALDRICH, Chief Judge.

This is an appeal from a summary judgment for the defendant. The plaintiff, Morrissey, is the copyright owner of a set of rules for a sales promotional contest of the "sweepstakes" type involving the social security numbers of the participants. Plaintiff alleges that the defendant, Procter & Gamble Company, infringed, by copying almost precisely, Rule 1. In its motion for summary judgment, based upon affidavits and depositions, defendant denies that plaintiff's Rule 1 is copyrightable material, and denies access. The district court held for the defendant on both grounds. . . .

The second aspect of the case raises a more difficult question. Before discussing it we recite plaintiff's Rule 1, and defendant's Rule 1, the italicizing in the latter being ours to note the defendant's variations or changes.

> 1. Entrants should print name, address and social security number on a boxtop, or a plain paper. Entries must be accompanied by . . . boxtop or by plain paper on which the name-
> . . . is copied from any source. Official rules are explained on . . . packages or leaflets obtained from dealer. If you do not have a social security number you may use the name and number of any member of your immediate family living with you. Only the person named on the entry will be deemed an entrant and may qualify for prize.
>
> Use the correct social security number belonging to the person named on entry. . . . [A] wrong number will be disqualified.

(Plaintiff's Rule)

> 1. Entrants should print name, address and Social Security number on a Tide boxtop, or *on* [a] plain paper. Entries must be accompanied by Tide boxtop *(any size)* or by plain paper on which the name "Tide" is copied from any source. Official rules are *available* on Tide Sweepstakes packages, or *on* leaflets *at* Tide dealers, *or you can send a stamped, self-addressed envelope to:* Tide "Shopping Fling" Sweepstakes, P.O. Box 4459, Chicago 77, Illinois.

If you do not have a Social Security number, you may use the name and number of any member of your immediate family living with you. Only the person named on the entry will be deemed an entrant and may qualify for a prize.

Use the correct Social Security number, belonging to the person named on *the* entry — wrong numbers will be disqualified.

(Defendant's Rule)

The district court, following an earlier decision, Gaye v. Gillis, D. Mass., 1958, 167 F. Supp. 416, took the position that since the substance of the contest was not copyrightable, which is unquestionably correct, Baker v. Selden, 1879, 101 U.S. 99, 25 L. Ed. 841; Affiliated Enterprises v. Gruber, 1 Cir., 1936, 86 F.2d 958; Chamberlin v. Uris Sales Corp., 2 Cir., 1945, 150 F.2d 512, and the substance was relatively simple, it must follow that plaintiff's rule sprung directly from the substance and "contains no original creative authorship." 262 F. Supp. at 738. This does not follow. Copyright attaches to form of expression, and defendant's own proof, introduced to deluge the court on the issue of access, itself established that there was more than one way of expressing even this simple substance. Nor, in view of the almost precise similarity of the two rules, could defendant successfully invoke the principle of a stringent standard for showing infringement which some courts apply when the subject matter involved admits of little variation in form of expression. . . .

Nonetheless, we must hold for the defendant. When the uncopyrightable subject matter is very narrow, so that "the topic necessarily requires," Sampson & Murdock Co. v. Seaver-Radford Co., 1 Cir., 1905, 140 F. 539, 541; cf. Kaplan, An Unhurried View of Copyright, 64-65 (1967), if not only one form of expression, at best only a limited number, to permit copyrighting would mean that a party or parties, by copyrighting a mere handful of forms, could exhaust all possibilities of future use of the substance. In such circumstances it does not seem accurate to say that any particular form of expression comes from the subject matter. However, it is necessary to say that the subject matter would be appropriated by permitting the copyrighting of its expression. We cannot recognize copyright as a game of chess in which the public can be checkmated. Cf. Baker v. Selden, supra.

Upon examination the matters embraced in Rule 1 are so straightforward and simple that we find this limiting principle to be applicable. Furthermore, its operation need not await an attempt to copyright all possible forms. It cannot be only the last form of expression which is to be condemned, as completing defendant's exclusion from the substance. Rather, in these circumstances, we hold that copyright does not extend to the subject matter at all, and plaintiff cannot complain even if his particular expression was deliberately adopted.
Affirmed.

COMMENTS AND QUESTIONS

1. *The Merger Doctrine.* The *Morrissey* case applies what has come to be known as the "merger" doctrine: When there is only one or but a few ways of expressing an idea, then courts will find that the idea behind the work *merges* with its expression and the work is not copyrightable. The doctrine is an extension of the basic rationale of *Baker v. Selden.*

2. *Objective or Pragmatic Standard.* Application of the merger doctrine involves the same "levels of abstraction" problem described earlier. Characterizing the "idea" of a work at a lower level of abstraction brings the idea in close alignment with the expression, rendering that expression unprotectable. Logically, the doctrine follows the form, "Categories (or genuses) are unprotectable; instances (or species) are protectable." The game then becomes to define what is the category or genus to which the work belongs. The more fine-grained and precise the definition of the category or genus, the more likely that the "expression" alleged to have been infringed is in fact an "idea."

Some courts have softened the harsh consequences of finding merger (i.e., no protection against copying) by finding that works with limited means of expression are copyrightable but that the scope of copyright protection for such works is "thin." In Continental Casualty Co. v. Beardsley, 253 F.2d 702 (2d Cir. 1958), *cert. denied,* 358 U.S. 816 (1958), the Second Circuit determined that the forms developed by the creator of new type of insurance covering lost securities were copyrightable. The forms in question included explanatory information. The court reasoned that

> in the fields of insurance and commerce the use of specific language in forms and documents may be so essential to accomplish a desired result and so integrated with the use of a legal or commercial conception that the proper standard of infringement is one which will protect as far as possible the copyrighted language and yet allow free use of the thought beneath the language.

Since the competitor copied only the forms and not the explanatory language, however, the court determined that no infringement had occurred. Does this approach better comport with the objectives and structure of copyright law? Or might it force subsequent authors to use awkward variations in expression in order to make use of an unprotectable idea?

3. *Historical Facts and Research.* Copyright law does not protect historical facts, on the ground that such information is not an original work of authorship. Some courts have extended this doctrine to deny copyright protection for historical research. In Miller v. Universal Studios, 650 F.2d 1365 (5th Cir. 1981), an investigative reporter spent more than 2,500 hours researching a bizarre kidnapping and rescue in which the victim had been buried alive in an underground coffin for 5 days. The researcher and the victim published a book describing the events. After efforts to obtain movie rights for the book failed, Universal Studios proceeded to produce a film based largely upon the book. The Fifth Circuit held "the valuable distinction in copyright between facts and the expression of facts cannot be maintained if research is held to be copyrightable." 650 F.2d at 1365.

Should copyright deny protection for the discoveries of historians? Is this consistent with the incentive basis for copyright protection? What arguments can be made in defense of the doctrine?

To what extent is the selection and arrangement of facts protectable under copyright? What about an original theory interpreting historical research? These questions have been the subject of extensive scholarly debate. See Jane C. Ginsburg, Sabotaging and Reconstructing History: A Comment on the Scope of Copyright Protection in Works of History After Hoehling v. Universal Studios, 29 J. Copyright Soc'y 647 (1982); Robert Denicola, Copyright in Collections of Facts: A Theory for the Protection of Nonfiction Literary Works, 81 Colum. L. Rev. 516 (1981); Robert Gorman, Copyright Protection for the Collection and Representation of Facts, 76 Harv. L. Rev. 1569 (1963).

4. *Scenes à Faire.* Copyright protection does not extend to the "incidents, characters or settings which are as a practical matter indispensable, or at least standard, in the treatment of a given topic." Atari, Inc. v. North American Phillips Consumer Electronics, 672 F.2d 607, 616 (7th Cir. 1982). To allow protection for such aspects of a work would unduly restrict subsequent authors in building their own works within general settings with which their audiences will relate.

5. *Computer Software.* The application of *Baker v. Selden* and the merger doctrine has become particularly important in computer software cases where subsequent computer manufacturers seek to develop computers offering compatible operating systems and subsequent programmers seek to develop application programs featuring the same menu command structure of popular programs. We study these cases in Chapter 7.

PROBLEMS

Problem 4-4. Insect Representations, Inc., sells a highly successful line of jewelry in the shape of various insects. One of its products is a stickpin in the shape of a bumblebee, made of gold veining and encrusted with a number of white gems. Animal Jewelry Corp. designs and markets a jewelled bee pin that also has gold veining and is encrusted with gems. However, Animal uses a different number of total gems, and some of its gems are colors other than white. When Insect sues Animal for copyright infringement, Animal defends on the grounds that it has taken merely the "idea" of a jewelled bee pin. Who should prevail?

Problem 4-5. Harry Historian had always been curious about the cause of the Hindenberg disaster, the explosion of a German zeppelin in 1938. After carefully investigating records and news accounts of the disaster and interviewing witnesses, Harry concluded that the disaster was caused by a disgruntled crew member who sabotaged the dirigible so as to embarrass the Nazi regime. He then wrote a book developing his hypothesis. The book contained rich descriptions of the events leading up to the disaster, including detailed accounts of German beer hall revelry and the passionate patriotism of German nationals (as expressed in their enthusiastic singing of the German national anthem). Without obtaining the movie rights to Harry's book, Capitalistic Studios produced a movie of the disaster which featured a crewman-saboteur and many of the richly detailed scenes in Harry's book. Does Harry have a valid copyright infringement claim? Should he have such a claim?

---

b. The Useful Article Doctrine

Section 101 of the Copyright Act defines "Pictorial, graphic, and sculptural [PGS] works," one of the categories of works protected under section 102, to include two-dimensional and three-dimensional works of fine, graphic, and applied art, photo-graphs, prints and art reproductions, maps, globes, charts, diagrams, models, and tech-nical drawings, including architectual plans. Such works shall include works of artistic craftsmanship insofar as their form but not their mechanical or utilitarian aspects are concerned; the design of a useful article, as defined in this section, shall be considered a pictorial, graphic, or sculptural work only if, and only to the extent that, such design incorporates pictorial, graphic, or sculptural features that can be identified separately from, and are capable of existing independently of, the utilitarian aspects of the article.

The Act defines a "useful article" as an "article having an intrinsic utilitarian function that is not merely to portray the appearance of the article or to convey information. An article that is normally a part of a useful article is considered a 'useful article.'" 17 U.S.C. § 101.

H.R. Rep. No. 94-1476, 94th Cong., 2d Sess., 47, 54-55 (1976)

[T]he definition of "pictorial, graphic, and sculptural works" carries with it no implied criterion of artistic taste, aesthetic value, or intrinsic quality. The term is intended to comprise not only "works of art" in the traditional sense but also . . . works of "applied art." . . .

In accordance with the Supreme Court's decision in Mazer v. Stein, 347 U.S. 201 (1954), works of "applied art" encompass all original pictorial, graphic, and sculptural works that are intended to be or have been embodied in useful articles, regardless of factors such as mass production, commercial exploitation, and the potential availability of design patent protection. . . .

The Committee has added language to the definition of "pictorial, graphic, and sculptural works" in an effort to make clearer the distinction between works of applied art protectable under the bill and industrial designs not subject to copyright protection. The declaration that "pictorial, graphic, and sculptural works" include "works of artistic craftsmanship insofar as their form but not their mechanical or utilitarian aspects are concerned" is classic language: it is drawn from Copyright Office regulations promulgated in the 1940s and was expressly endorsed by the Supreme Court in the *Mazer* case.

The second part of the amendment states that "the design of a useful arti-cle . . . shall be considered a pictorial, graphic, or sculptural work only if, and only to the extent that, such design incorporates pictorial, graphic, or sculptural features that can be identified separately from, and are capable of existing independently of, the utilitarian aspects of the article." . . .

In adopting this amendatory language, the Committee is seeking to draw as clear a line as possible between copyrightable works of applied art and uncopyrighted works of industrial design. A two-dimensional painting, drawing, or graphic work is still capable of being identified as such when it is printed on or applied to utilitarian articles such as textile fabrics, wallpaper, containers, and the like. The same is true when a statue or carving is used to embellish an industrial product or, as in the *Mazer* case, is incorporated into a product without losing its ability to exist independently as a work of art. On the other hand, although the shape of an industrial product may be aesthetically satisfying and valuable, the Committee's intention is not to offer it copyright protection under the bill. Unless the shape of an automobile, airplane, ladies' dress, food processor, television set, or any other industrial product contains some element that, physically or conceptually, can be identified as separable from the utilitarian aspects of that article, the design would not be copyrighted under the bill. The test of separability and independence from "the utilitarian aspects of the article" does not depend upon the nature of the design—that is, even if the appearance of an article is determined by aesthetic (as opposed to functional) considerations, only elements, if any, which can be identified separately from the useful article as such are copyrightable. And, even if the three-dimensional design contains some such element (for example, a carving on the back of a chair or a floral relief design on silver flatware), copyright protection would extend only to that element, and would not cover the over-all configuration of the utilitarian article as such.

Brandir International, Inc. v. Cascade Pacific Lumber Co.
United States Court of Appeals for the Second Circuit
834 F.2d 1142 (2d Cir. 1987)

OAKES, Circuit Judge:
In passing the Copyright Act of 1976 Congress attempted to distinguish between protectable "works of applied art" and "industrial designs not subject to copyright protection." See H.R. Rep. No. 1476, 94th Cong., 2d Sess. 54, reprinted in 1976 U.S. Code Cong. & Admin. News 5659, 5667 (hereinafter H.R. Rep. No. 1476). The courts, however, have had difficulty framing tests by which the fine line establishing what is and what is not copyrightable can be drawn. Once again we are called upon to draw such a line, this time in a case involving the "RIBBON Rack," a bicycle rack made of bent tubing that is said to have originated from a wire sculpture. [Figure 4-3]...The Register of Copyright, named as a third-party defendant under the statute, 17 U.S.C. §411, but electing not to appear, denied copyright-ability. In the subsequent suit brought in the United States District Court for the Southern District of New York, Charles S. Haight, Jr., Judge, the district court granted summary judgment on...the copyright claim....Against the history of copyright protection well set out in the majority opinion in Carol Barnhart Inc. v. Economy Cover Corp., 773 F.2d 411, 415-18 (2d Cir. 1985), and in Denicola, Applied Art and Industrial Design: A Suggested Approach to Copyright in Useful Articles, 67 Minn. L. Rev. 707, 709-17 (1983), Congress adopted the Copyright Act of 1976....

As courts and commentators have come to realize, however, the line Congress attempted to draw between copyrightable art and noncopyrightable design "was neither clear nor new." Denicola, supra, 67 Minn. L. Rev. at 720. One aspect of the distinction that has drawn considerable attention is the reference in the House Report to "physically *or conceptually*" (emphasis added) separable elements....

...As Judge Newman's dissent made clear, the *Carol Barnhart* majority did not dispute "that 'conceptual separability' is distinct from 'physical sepability' and, when present, entitles the creator of a useful article to a copyright on its design." 773 F.2d at 420.

"Conceptual separability" is thus alive and well, at least in this circuit. The problem, however, is determining exactly what it is and how it is to be applied.

Judge Newman's illuminating discussion in dissent in Carol Barnhart, see 773 F.2d at 419-24, proposed a test that aesthetic features are conceptually separable if "the article...stimulate[s] in the mind of the beholder a concept that is separate from the concept evoked by its utilitarian function." Id. at 422. This approach has received favorable endorsement by at least one commentator, W. Patry, Latman's The Copyright Law 43-45 (6th ed. 1986), who calls Judge Newman's test the "temporal displacement" test. It is to be distinguished from other possible ways in which conceptual separability can be tested, including whether the primary use is as a utilitarian article as opposed to an artistic work, whether the aesthetic aspects of the work can be said to be "primary," and whether the article is marketable as art, none of which is very satisfactory. But Judge Newman's test was rejected outright by the majority as "a standard so ethereal as to amount to a 'nontest' that would be extremely difficult, if not impossible, to administer or apply." 773 F.2d at 419 n.5.

Perhaps the differences between the majority and the dissent in *Carol Barnhart* might have been resolved had they had before them the Denicola article on

FIGURE 4-3
Brandir Ribbon bicycle rack.

Applied Art and Industrial Design: A Suggested Approach to Copyright in Useful Articles, supra. There, Professor Denicola points out that although the Copyright Act of 1976 was an effort "to draw as clear a line as possible," in truth "there is no line, but merely a spectrum of forms and shapes responsive in varying degrees to utilitarian concerns." 67 Minn. L. Rev. at 741. Denicola argues that "the statutory directive requires a distinction between works of industrial design and works whose origins lie outside the design process, despite the utilitarian environment in which they appear." He views the statutory limitation of copyrightability as "an attempt to identify elements whose form and appearance reflect the unconstrained perspective of the artist," such features not being the product of industrial design. Id. at 742. "Copyrightability, therefore, should turn on the relationship between the proffered work and the process of industrial design." Id. at 741. He suggests that "the dominant characteristic of industrial design is the influence of nonaesthetic, utilitarian concerns" and hence concludes that copyrightability "ultimately should depend on the extent to which the work reflects artistic expression uninhibited by functional considerations."[2] Id. To state the Denicola test in the language of conceptual separability, if design elements reflect a merger of aesthetic and functional considerations, the artistic aspects of a work cannot be said to be conceptually separable from the utilitarian elements. Conversely, where design elements can be identified as reflecting the designer's artistic judgment exercised independently of functional influences, conceptual separability exists.

We believe that Professor Denicola's approach provides the best test for conceptual separability and, accordingly, adopt it here for several reasons. First, the approach is consistent with the holdings of our previous cases. . . . Second, the test's emphasis on the influence of utilitarian concerns in the design process may help, as Denicola notes, to "alleviate the de facto discrimination against nonrepresentational art that has regrettably accompanied much of the current analysis." Id. at 745. Finally, and perhaps most importantly, we think Denicola's test will not be too difficult to administer in practice. The work itself will continue to give "mute testimony" of its origins. In addition, the parties will be required to present evidence relating to the design process and the nature of the work, with the trier of fact making the determination whether the aesthetic design elements are significantly influenced by functional considerations.

Turning now to the facts of this case, we note first that Brandir contends, and its chief owner David Levine testified, that the original design of the RIBBON Rack stemmed from wire sculptures that Levine had created, each formed from one continuous undulating piece of wire. These sculptures were, he said, created and displayed in his home as a means of personal expression, but apparently were never sold or displayed elsewhere. He also created a wire sculpture in the shape of a bicycle and states that he did not give any thought to the utilitarian application of any of his sculptures until he accidentally juxtaposed the bicycle sculpture with one of the self-standing wire sculptures. It was not until November 1978 that Levine seriously began pursuing the utilitarian application of his sculptures,

2. Professor Denicola rejects the exclusion of all works created with some utilitarian application in view, for that would not only overturn Mazer v. Stein, 347 U.S. 201 (1954), on which much of the legislation is based, but also "a host of other eminently sensible decisions, in favor of an intractable factual inquiry of questionable relevance." 67 Minn. L. Rev. at 741. He adds that "any such categorical approach would also undermine the legislative determination to preserve an artist's ability to exploit utilitarian markets." Id. (citing 17 U.S.C. §113(a) (1976)).

when a friend, G. Duff Bailey, a bicycle buff and author of numerous articles about urban cycling, was at Levine's home and informed him that the sculptures would make excellent bicycle racks, permitting bicycles to be parked under the overloops as well as on top of the underloops. Following this meeting, Levine met several times with Bailey and others, completing the designs for the RIBBON Rack by the use of a vacuum cleaner hose, and submitting his drawings to a fabricator complete with dimensions. The Brandir RIBBON Rack began being nationally advertised and promoted for sale in September 1979.

In November 1982 Levine discovered that another company, Cascade Pacific Lumber Co., was selling a similar product. Thereafter, beginning in December 1982, a copyright notice was placed on all RIBBON Racks before shipment and on December 10, 1982, five copyright applications for registration were submitted to the Copyright Office. The Copyright Office refused registration by letter, stating that the RIBBON Rack did not contain any element that was "capable of independent existence as a copyrightable pictorial, graphic or sculptural work apart from the shape of the useful article." An appeal to the Copyright Office was denied by letter dated March 23, 1983, refusing registration on the above ground and alternatively on the ground that the design lacked originality, consisting of "nothing more than a familiar public domain symbol." In February 1984, after the denial of the second appeal of the examiner's decision, Brandir sent letters to customers enclosing copyright notices to be placed on racks sold prior to December 1982.

Between September 1979 and August 1982 Brandir spent some $38,500 for advertising and promoting the RIBBON Rack, including some 85,000 pieces of promotional literature to architects and landscape architects. Additionally, since October 1982 Brandir has spent some $66,000, including full-, half-, and quarter-page advertisements in architectural magazines such as Landscape Architecture, Progressive Architecture, and Architectural Record, indeed winning an advertising award from Progressive Architecture in January 1983. The RIBBON Rack has been featured in Popular Science, Art and Architecture, and Design 384 magazines, and it won an Industrial Designers Society of America design award in the spring of 1980. In the spring of 1984 the RIBBON Rack was selected from 200 designs to be included among 77 of the designs exhibited at the Katonah Gallery in an exhibition entitled "The Product of Design: An Exploration of the Industrial Design Process," an exhibition that was written up in the New York Times.

Sales of the RIBBON Rack from September 1979 through January 1985 were in excess of $1,367,000. Prior to the time Cascade Pacific began offering for sale its bicycle rack in August 1982, Brandir's sales were $436,000. The price of the RIBBON Rack ranges from $395 up to $2,025 for a stainless steel model and generally depends on the size of the rack, one of the most popular being the RB-7, selling for $485.

Applying Professor Denicola's test to the RIBBON Rack, we find that the rack is not copyrightable. It seems clear that the form of the rack is influenced in significant measure by utilitarian concerns and thus any aesthetic elements cannot be said to be conceptually separable from the utilitarian elements. This is true even though the sculptures which inspired the RIBBON Rack may well have been — the issue of originality aside — copyrightable.

Brandir argues correctly that a copyrighted work of art does not lose its protected status merely because it subsequently is put to a functional use. The Supreme Court so held in Mazer v. Stein, 347 U.S. 201, 98 L. Ed. 630, 74 S. Ct. 460 (1954), and

Congress specifically intended to accept and codify *Mazer* in section 101 of the Copyright Act of 1976. See H.R. Rep. No. 1476 at 54-55. The district court thus erred in ruling that, whatever the RIBBON Rack's origins, Brandir's commercialization of the rack disposed of the issue of its copyrightability.

Had Brandir merely adopted one of the existing sculptures as a bicycle rack, neither the application to a utilitarian end nor commercialization of that use would have caused the object to forfeit its copyrighted status. Comparison of the RIBBON Rack with the earlier sculptures, however, reveals that while the rack may have been derived in part from one or more "works of art," it is in its final form essentially a product of industrial design. In creating the RIBBON Rack, the designer has clearly adapted the original aesthetic elements to accommodate and further a utilitarian purpose. These altered design features of the RIBBON Rack, including the space-saving, open design achieved by widening the upper loops to permit parking under as well as over the rack's curves, the straightened vertical elements that allow in- and above-ground installation of the rack, the ability to fit all types of bicycles and mopeds, and the heavy-gauged tubular construction of rustproof galvanized steel, are all features that combine to make for a safe, secure, and maintenance-free system of parking bicycles and mopeds. Its undulating shape is said in Progressive Architecture, January 1982, to permit double the storage of conventional bicycle racks. Moreover, the rack is manufactured from 2-3/8-inch standard steam pipe that is bent into form, the six-inch radius of the bends evidently resulting from bending the pipe according to a standard formula that yields bends having a radius equal to three times the nominal internal diameter of the pipe.

Brandir argues that its RIBBON Rack can and should be characteriz as a sculptural work of art within the minimalist art movement. Minimalist sculpture's most outstanding feature is said to be its clarity and simplicity, in that it often takes the form of geometric shapes, lines, and forms that are pure and free of ornamentation and void of association. As Brandir's expert put it, "The meaning is to be found in, within, around and outside the work of art, allowing the artistic sensation to be experienced as well as intellectualized." People who use Foley Square in New York City see in the form of minimalist art the "Tilted Arc," which is on the plaza at 26 Federal Plaza. Numerous museums have had exhibitions of such art, and the school of minimalist art has many admirers.

It is unnecessary to determine whether to the art world the RIBBON Rack properly would be considered an example of minimalist sculpture. The result under the copyright statute is not changed. Using the test we have adopted, it is not enough that, to paraphrase Judge Newman, the rack may stimulate in the mind of the reasonable observer a concept separate from the bicycle rack concept. While the RIBBON Rack may be worthy of admiration for its aesthetic qualities alone, it remains nonetheless the product of industrial design. Form and function are inextricably intertwined in the rack, its ultimate design being as much the result of utilitarian pressures as aesthetic choices. Indeed, the visually pleasing proportions and symmetricality of the rack represent design changes made in response to functional concerns. Judging from the awards the rack has received, it would seem in fact that Brandir has achieved with the RIBBON Rack the highest goal of modern industrial design, that is, the harmonious fusion of function and aesthetics. Thus there remains no artistic element of the RIBBON Rack that can be identified as separate and "capable of existing independently of, the utilitarian aspects of the article." Accordingly, we must affirm on the copyright claim. . . .

WINTER, Circuit Judge, concurring in part and dissenting in part:

...I respectfully dissent from the majority's discussion and disposition of the copyright claim.

My colleagues, applying an adaptation of Professor Denicola's test, hold that the aesthetic elements of the design of a useful article are not conceptually separable from its utilitarian aspects if "form and function are inextricably intertwined" in the article, and "its ultimate design [is] as much the result of utilitarian pressures as aesthetic choices." Applying that test to the instant matter, they observe that the dispositive fact is that "in creating the Ribbon Rack, [Levine] has clearly adapted the *original* aesthetic elements to accommodate and further a utilitarian purpose." (emphasis added). The grounds of my disagreement are that: (1) my colleagues' adaptation of Professor Denicola's test diminishes the statutory concept of "conceptual separability" to the vanishing point; and (2) their focus on the process or sequence followed by the particular designer makes copyright protection depend upon largely fortuitous circumstances concerning the creation of the design in issue....

[T]he relevant question is whether the design of a useful article, however intertwined with the article's utilitarian aspects, causes an ordinary reasonable observer to perceive an aesthetic concept not related to the article's use. The answer to this question is clear in the instant case because any reasonable observer would easily view the Ribbon Rack as an ornamental sculpture.[2] Indeed, there is evidence of actual confusion over whether it is strictly ornamental in the refusal of a building manager to accept delivery until assured by the buyer that the Ribbon Rack was in fact a bicycle rack. Moreover, Brandir has received a request to use the Ribbon Rack as environmental sculpture, and has offered testimony of art experts who claim that the Ribbon Rack may be valued solely for its artistic features. As one of those experts observed: "If one were to place a Ribbon Rack on an island without access, or in a park and surround the work with a barrier,...its status as a work of art would be beyond dispute."

My colleagues also allow too much to turn upon the process or sequence of design followed by the designer of the Ribbon Rack. They thus suggest that copyright protection would have been accorded "had Brandir merely adopted...as a bicycle rack" an enlarged version of one of David Levine's original sculptures rather than one that had wider upper loops and straightened vertical elements. I cannot agree that copyright protection for the Ribbon Rack turns on whether Levine serendipitously chose the final design of the Ribbon Rack during his initial sculptural musings or whether the original design had to be slightly modified to accommodate bicycles. Copyright protection, which is intended to generate incentives for designers by according property rights in their creations, should not turn on purely fortuitous events. For that reason, the Copyright Act expressly states that the legal test is how the final article is perceived, not how it was developed through various stages. It thus states in pertinent part: "[T]he design of a useful article...shall be considered a...sculptural work only if, and only to the extent that, such design incorporates... *sculptural features that can be identified separately from, and are capable of existing independently of, the utilitarian aspects of the article.*" 17 U.S.C. § 101 (1982) (emphasis added)....

---

2. The reasonable observer may be forgiven, however, if he or she does not recognize the Ribbon Rack as an example of minimalist art.

COMMENTS AND QUESTIONS

1. *Useful Article.* A threshold issue is whether a work is a useful article at all. According to section 101, a useful article is a work "having an intrinsic utilitarian function that is not merely to portray the appearance of the article or to convey information." Thus a drawing of a tablecloth design would not satisfy this definition, but the tablecloth itself would be a useful article. Nonetheless, the definition is difficult to apply as the degree of functionality wanes and the degree of fanciful and artistic expression rises. Consider the following examples:

- a distinctively decorated toy airplane
- an ornamental fireplace hearth that cannot burn wood
- pet rocks in various shapes

2. *Physical and/or Conceptual Separability.* In one of the first cases interpreting the useful article doctrine as embodied in the Copyright Act of 1976, the D.C. Circuit took the narrow view that the artistic elements of a useful article (an outdoor lamp) must be physically separable from its overall design to be protectable. The court did not apply any conceptual separability test. Esquire, Inc. v. Ringer, 591 F.2d 796 (D.C. Cir. 1978). Most courts and commentators have rejected such a narrow reading of the statute. As noted in *Brandir,* it is well established in the Second Circuit that physical *or* conceptual separability is sufficient to establish copyrightability.

The physical separability test is relatively straightforward to apply. An expressive element of a useful article is physically separable if it can stand alone from the article as a whole and if such separation does not impair the utility of the article.

3. *The Conceptual Separability Test.* As the *Brandir* case suggests, the conceptual separability test has been difficult to apply. In addition to the various tests discussed in *Brandir*—the temporal displacement test (Newman dissent in *Barnhart*, a case involving mannequin torsos), Denicola's test (majority in *Brandir*), Winter's test (dissent in *Brandir*)—Paul Goldstein has suggested the following formulation: "[A] pictorial, graphic or sculptural feature incorporated in the design of a useful article is conceptually separable if it can stand on its own as a work of art traditionally conceived, and if the useful article in which it is embodied would be equally useful without it." Paul Goldstein, Copyright § 2.5.3.1 (1989). Which test comports best with the statute and its legislative history? Which of these tests will produce the most predictable results? Of the various tests for conceptual separability, which test best promotes innovation in utilitarian works? artistic works?

Controversy persists over the proper formulation and application of the conceptual separability standard. See, e.g., Pivot Point Int'l, Inc. v. Charlene Prods., Inc., 372 F.3d 913 (7th Cir. 2004) (allowing protection for a mannequin head); Bonzali v. R.S.V.P. Int'l, Inc., 353 F.Supp.2d 218 (D.R.I. 2005) (heart-shaped measuring spoons). Recall the Supreme Court's warning in *Bleistein* of the dangers of embroiling the courts in artistic value judgments. Is there any way to decide the separability question without rendering such a value judgment? See Alfred C. Yen, Copyright Opinions and Aesthetic Theory, 71 S. Cal. L. Rev. 247 (1998).

4. *Binary Choice versus Sliding Scale.* Court decisions treat separability of expression from utility to be a binary choice—the expression is either separable or not, with that determination largely determining liability. Could the philosophical indeterminacies of this doctrine be at least partially defused by employing a sliding

scale — in which works lacking clear physical or conceptual separability but involving significant artistic creativity are subject to a higher standard for infringement (e.g., virtual identity) and narrow scope (so as not to interfere with competition for the functional elements of the article)? Would this approach jeopardize the supremacy of patent law in protecting utilitarian inventions?

5. *Relation to Patent and Design Patent Protection.* Contrast the useful article doctrine with the idea-expression dichotomy. As discussed in *Baker v. Selden, supra,* the idea-expression dichotomy channels protection for functional works toward the patent system, which applies a relatively high threshold for protection (novelty and non-obviousness), requires examination by a skilled examiner, and affords protection for only 20 years from the time an application is filed, thereby encouraging others to build upon patented advances following a relatively limited period of exclusive protection. Awarding protection for functional works through copyright law — with its low threshold for protection and much greater duration — would undermine the role of the patent system as the principal means for protecting utilitarian works and hinder the process of sequential innovation essential to technological progress. See Dennis S. Karjala and Peter S. Menell, Applying Fundamental Copyright Principles to Lotus Development Corp. v. Borland International, Inc., 10 High Tech. L.J. 177, 184-86 (1995). Does the useful article doctrine reflect a similar objective? Or does it show greater solicitude for the protection of artistic works? If so, what justifies the difference?

How is your analysis affected by Congress's statement in the legislative history to the Copyright Act of 1976 that copyright protection extends to works of "applied art" satisfying the separability requirement regardless of "the potential availability of design patent protection"? In re Yardley, 493 F.2d 1389 (C.C.P.A. 1974) (allowing a design patent for a copyrighted work). But cf. 37 C.F.R. § 202.10(a) (1989) (Copyright Office regulation disallowing copyright registration for works for which a design patent has issued). Contrast copyright and design patents, *supra* Chapter 3.G.1, in their protection for ornamental designs. What purpose is served by overlapping protection?

6. *Design Protection Legislation.* Congress has considered proposals to protect design by way of copyright at various times during the past century. Title II of the 1976 Copyright Revision Bill would have raised the threshold for originality (by excluding "staple or commonplace" designs) and required registration. Designs would have been protected for a term of ten years. The provision passed the Senate but failed to get out of committee in the House. Is such protection necessary to stimulate innovation in design given the availability of design patents? If so, what is the appropriate form of protection? See generally J. H. Reichman, Design Protection and the Legislative Agenda, 55 L. & Contemp. Probs. 281 (1992); Ralph Brown, Design Protection: An Overview, 34 U.C.L.A. L. Rev. 1341 (1987).

One example of such protection involves vessel hull designs. Through a plug-molding technique, boat builders can copy hull designs relatively easily. In 1989, the Supreme Court struck down a Florida law protecting boat hull designs as pre-empted by federal patent law. Bonito Boats, Inc. v. Thunder Craft Boats, Inc., 489 U.S. 141 (1989). Boat designers successfully brought their concerns to Congress. As part of the Digital Millennium Copyright Act of 1998, see *infra* Chapter 7, Congress afforded designers of original boat hulls ten years of protection. 17 U.S.C. §§ 1301-32. The Act excludes from such protection designs that are "dictated solely by a utilitarian function of the article that embodies it." § 1302(4). In order to secure protection, the designer must register the work within two years of making the design public. Like patent law, the Act affords the owner the exclusive right to make, sell, import, or distribute for sale or any commercial use any hull embodying the design. Like copyright law, § 1308 does not impose liability for independently created designs. Should specialized protection be available for other fields of design? Some see the Vessel Hull Design Protection Act as a template for a possible general design protection law. Cf. Maverick Boat Co. v. American Marine Holdings Inc., 418 F.3d 1186 (11th Cir. 2005) (boat manufacturer's minor changes to pre-existing design did not constitute "substantial revision" of design, and hence did not qualify for boat hull design protection).

7. *Other Utilitarian Works.* The utilitarian function exception to copyright protection applies only in the case of a particular type of copyrighted works — "pictorial, graphic and sculptural works." It has no counterpart in literary or musical works, for example. Does this limitation make sense? Do other doctrines, such as the idea-expression dichotomy, serve the same purpose in other contexts?

PROBLEMS

Problem 4-6. The Walt Disney Company markets a line of telephones in the shape of the cartoon characters Mickey and Minnie Mouse. The telephones resemble the characters standing up, with push buttons on the torso and the telephone receiver resting on the hand. Are these designs copyrightable?

Problem 4-7. Armond Artist designs a belt buckle cast in gold and silver that is generally recognized as a work of abstract art. The shape of the buckle is the key artistic component.

The buckles retail for $200 to $6,000 at high-fashion jewelry stores. Some of the designs have been made a part of the permanent collection of the Metropolitan Museum of Art. Is the buckle copyrightable, or is it merely a "useful article"?

Problem 4-8. Versace Fashion Display Inc. has developed a collection of original sculptural forms in the shape of the human torso. They are life-size and anatomically accurate, but without neck, arms, or a back. Versace supplies these forms to clothing retailers for the display of fashion clothing. These forms became popular in the market. Versace customers found the distinctive lines of the forms to be visually attractive and effective in selling blouses, shirts, and sweaters displayed on the forms. Discount Display, Inc., began to notice a decline in their mannequin sales after Versace entered the market. Thereafter, Discount Display expanded its catalog to include forms that it copied from Versace. In response, Versace registers its forms with the Copyright Office and sues Discount Display for copyright infringement.

How should a court resolve this dispute?

---

c. Government Works

The question of whether works of the government could be protected by copyright law arose early in the history of the American republic. Wheaton, one of the Supreme Court's "official" reporters, asserted protection for his annotated compilations of

Supreme Court opinions. Distinguishing between the reporter's work in annotating, editing, and condensing judicial opinions and the opinions themselves, the Supreme Court held that "no reporter has or can have any copyright in the written opinions delivered by this Court..." Wheaton v. Peters, 33 U.S. (8 Pet.) 591, 668 (1834). See also Banks v. Manchester, 128 U.S. 244, 253 (1888) (noting that the "whole work done by the judges constitutes the authentic exposition and interpretation of the law, which, binding every citizen, is free for publication to all, whether it is a declaration of unwritten law, or an interpretation of a constitution or statute."); Nash v. Lathrop, 142 Mass. 29, 6 N.E. 559 (1886) (observing that a legislature could not deny public access to statutes and that "the law" is in the public domain); see generally L. Ray Patterson & Craig Joyce, Monopolizing the Law: The Scope of Copyright Protection for Law Reports and Statutory Compilations, 36 UCLA L. Rev. 719, 751-58 (1989); 1 Melville B. Nimmer & David Nimmer, Nimmer on Copyright § 5.06[c] at 5-92 (2000) ("state statutes, no less than federal statutes, are regarded as being in the public domain.")

With regard to works of the federal government, Congress has extended the bar against copyright protection well beyond statutes and judicial decisions. Section 105 of the Copyright Act provides that copyright protection is not available "for any work of the United States Government, but the United States Government is not precluded from receiving and holding copyrights transferred to it by assignment, bequest, or otherwise." Section 101 defines "a work of the United States Government" as "a work prepared by an officer or employee of the United States Government as part of that person's official duties." The House Report notes that this definition should be construed in the same manner as the definition of "work made for hire." We focus on the "work for hire" doctrine in section D.1, *infra*.

A potential problem arises where a federal government agency commissions a work by an independent contractor. The legislative history states:

> The bill deliberately avoids making any sort of outright, unqualified prohibition against copyright in works prepared under Government contract or grant. There may well be cases where it would be in the public interest to deny copyright in the writings generated by Government research contracts and the like; it can be assumed that, where a government agency commissions a work for its own use merely as an alternative to having one of its own employees prepare the work, the right to secure a private copyright would be withheld. However, there are almost certainly many other cases where the denial of copyright protection would be unfair or would hamper the production and publication of important works. Where, under the particular circumstances, Congress or the agency involved finds that the need to have a work freely available outweighs the need of the private author to secure a copyright, the problem can be dealt with by specific legislation, agency regulations, or contractual restrictions.

H.R. Rep. No. 94-1476, 94th Cong., 2d Sess. 59 (1976).

The Copyright Act does not expressly limit the protectibility of works created by state government officers or employees in their official capacities. Nonetheless courts have held that certain types of government works, such as statutes, codes, and judicial opinions, are inherently part of the public domain. See, e.g., Veeck v. Southern Bldg. Code Congress Intern., Inc., 293 F.3d 791 (5th Cir. 2002) (en banc) (holding that model codes enter the public domain when they are enacted into law); Georgia v. The Harrison Co., 548 F.Supp. 110 (N.D.Ga. 1982), *vacated by agreement of the parties,* 559 F.Supp. 37 (N.D. Ga. 1983) (holding new statutory codification to be in public domain); Building Officials & Code Administrators Int'l v. Code Technology,

Inc., 628 F.2d 730 (1st Cir. 1980) (suggesting that building code written by private party and later enacted into law would be in the public domain). These decisions have emphasized the inherently public nature of laws and the sense in which the citizens of a state are owners of such works. Limitations on access to such works would undoubtedly raise due process concerns where failure to adhere to such laws or regulations threatens civil or criminal liability.

In a decision distinguishing tax maps from statutes and judicial opinions, the Second Circuit has held that a municipality may retain copyright in its tax maps to the extent that the entity or individual who created the work needs an economic incentive to create the work and the public does not need notice of this particular work to have notice of the law. See County of Suffolk v. First American Real Estate Solutions, 261 F.3d 179 (2d Cir. 2001).

COMMENTS AND QUESTIONS

1. *Incentives to Develop Model Laws.* The dissent in *Veeck* asserted that copyright protection for model codes provides valuable incentives for creation of laws and model codes:

Congress itself has provided the strongest support for the proposition that these privately created codes should be treated differently than other laws. Recognizing that the production of a comprehensive technical code requires a great deal of research, labor, time, and expertise, Congress in the National Technology and Transfer Act of 1995 (the "NTTA") expressly directs that "Federal agencies and departments shall use technical standards that are developed or adopted by voluntary consensus standards bodies...." The OMB [Office of Management and Budget], in its Circular A-119, which was designed to provide guidance to federal agencies in the wake of the NTTA, requires that "[i]f a voluntary standard is used and published in an agency document, your agency must observe and protect the rights of the copyright holder and any other similar obligations." These pronouncements by Congress and the OMB strongly evince a recognition that the privately created regulatory codes and standards differ greatly from either judicial opinions or a statutes. Technical codes are indispensable resources in today's increasingly complex, high-tech society, and they deserve authorship protections not afforded to other types of "THE law."

The policy concerns supporting the retention of at least some copyright protection [for those who create model codes] are more persuasive and probative. First and most importantly, unlike judges and legislators who are paid from public funds to issue opinions and draft laws, SBCCI is a private sector, not-for-profit organization which relies for its existence and continuing services, in significant part, on revenues from the sale of its model codes.[46] ...

The importance of affording organizations like SBCCI protection from attenuated third parties like Veeck–even when motives are pure and unfair financial competition is not the goal–is best underscored by verbalizing the natural consequence of reducing the revenues, and thus the creative incentives, for organizations like SBCCI. Without private code-creating entities, our smaller towns — and even some of our larger cities, states, and agencies of the federal government — would be forced to author their own regulatory

---

46. Approximately one third, or $3 million, of SBCCI's annual $9 million dollar revenue is generated by sales of model codes to contractors and other interested parties. The remaining revenue is mainly derived from the annual fees of voluntary members and member organizations. Voluntary members include scholars, builders, contractors, and governmental entities that have adopted the code.

codes. Such a task would inefficiently expend the time and resources of the legislative and executive bodies of these governmental entities, not to mention the question of available expertise. To create codes of appropriate detail, accuracy, and information, governmental bodies would have to enlist the aid of technical experts, undoubtedly at considerable cost. Finally, causing municipalities, states, and the federal agencies to engage in this activity could lead to innumerable variations of any given code, thereby undermining uniformity and, with it, safety and efficiency. For small towns like Anna and Savoy, such a result could be even more detrimental, as their limited resources well might be insufficient to absorb the costs of creating their own codes. Ultimately, taxpayers would end up paying for a service that is currently provided efficiently, expertly, and at no expense to them.

I hasten to add that, for my analysis to have force, SBCCI need not be put completely out of business. Continued maintenance of a revenue source from sales of codes to individual owners, architects, engineers, materials suppliers, builders and contractors as well as libraries and other more attenuated purchasers, all of whom buy copies of the codes directly from SBCCI, serves another public interest. I refer to the continuation of SBCCI's independence from the self interest of its dues-paying members, who otherwise might be in a position to command more influence were SBCCI forced to obtain too great a share of its revenue from such supporters. Clearly, SBCCI's receipts from sales of the codes substantially reduces the potential for greater dependence on its membership, presumably allowing SBCCI to operate without becoming entirely beholden for its existence to self-interested entities.

Veeck v. Southern Bldg. Code Congress Int'l Inc., 293 F.3d 791, 815-17 (5th Cir. 2002) (dissenting opinion). The majority countered that:

> SBCCI, like other code-writing organizations, has survived and grown over 60 years, yet no court has previously awarded copyright protection for the copying of an enacted building code under circumstances like these. Second, the success of voluntary code-writing groups is attributable to the technological complexity of modern life, which impels government entities to standardize their regulations. The entities would have to promulgate standards even if SBCCI did not exist, but the most fruitful approach for the public entities and the potentially regulated industries lies in mutual cooperation. The self-interest of the builders, engineers, designers and other relevant tradesmen should also not be overlooked in the calculus promoting uniform codes. As one commentator explained,
>
> > . . . it is difficult to imagine an area of creative endeavor in which the copyright incentive is needed less. Trade organizations have powerful reasons stemming from industry standardization, quality control, and self-regulation to produce these model codes; it is unlikely that, without copyright, they will cease producing them.
>
> 1 Goldstein § 2.5.2, at 2:51.
> Third, to enhance the market value of its model codes, SBCCI could easily publish them as do the compilers of statutes and judicial opinions, with "value-added" in the form of commentary, questions and answers, lists of adopting jurisdictions and other information valuable to a reader. The organization could also charge fees for the massive amount of interpretive information about the codes that it doles out. In short, we are unpersuaded that the removal of copyright protection from model codes only when and to the extent they are enacted into law disserves "the Progress of Science and useful Arts." U.S. Const. art. I. § 8, cl. 8.

Id. at 805-06. Who has the better of this argument? Is it appropriate for courts to be subjecting particular classes of works to this sort of open-ended analysis of whether copyright protection "promotes progress"? Doesn't the case turn on whether or not idea and expression merge in the wording of law? Should Congress amend the Copyright Act in order to resolve this issue one way or the other?

2. The dissent in *Veeck* suggests that the Copyright Act offers a middle ground for balancing the competing considerations at play in the protection of privately developed model codes that have been adopted as law:

> [D]enying the Veecks of the world unrestricted republication and dissemination rights does not obstruct reasonable and necessary usage of and compliance with the adopted codes. I remain confident that the copyright doctrines of fair use and implied license or waiver are more than adequate to preserve the ability of residents and construction industry participants to copy any portions of the code that they want or need to view....

Veeck v. Southern Bldg. Code Congress Intern., Inc., 293 F.3d 791, 817 (5th Cir. 2002) (dissenting opinion). Is this realistic? If the public is free to copy the code without paying, how exactly will SBCCI get paid? Surely if they are seeking copyright protection for enacted codes it is because they expect to be able to stop (or alternatively, to charge for) some copying that now occurs for free.

3. Over the course of several decades, the American Medical Association developed the Physician's Current Procedural Terminology ("CPT"), a comprehensive classification system for identifying more than 6,000 medical procedures comprising a five-digit code and brief description for each procedure. The AMA revises the CPT each year to reflect new developments in medical procedures. In 1977, Congress instructed the Health Care Financing Administration to establish a uniform code for identifying physicians' services for use in completing Medicare and Medicaid claim forms. Rather than creating a new code, HCFA contracted with the AMA. The AMA gave HCFA a "non-exclusive, royalty free, and irrevocable license to use, copy, publish and distribute" the CPT on the conditions that the HCFA not "use any other system of procedure nomenclature... for reporting physicians' services" and require use of the CPT in programs administered by HCFA, by its agents, and by other agencies whenever possible. HCFA published notices in the Federal Register incorporating the CPT in HCFA's Common Procedure Coding System and adopted regulations requiring applicants for Medicaid reimbursement to use the CPT. A publisher and distributor of medical books challenged the copyrightability of the CPT. Should this code be protectable? See Practice Management Info. Corp. v. American Medical Ass'n, 121 F.3d 516 (9th Cir. 1997) (upholding the AMA's copyright but finding that the AMA had misused its copyright rights), *amended by* 133 F.3d 1140, *cert. denied* 522 U.S. 933 (1998).

PROBLEM

Problem 4-9. To celebrate the bicentennial of the federal court system, the Administrative Office of the United States Courts commissioned KPBS, a public broadcasting station, to create a television series dramatizing famous early federal cases. The production contract assigned all copyright interests in the works to the federal government. The History Channel seeks to air these programs. Must it obtain a license from the federal government? Would your analysis differ if there were no assignment clause? What if a historian employed by the Administrative Office of the United States Courts served as a consultant to the television series? What if federal public officials — including Supreme Court Justices and the Solicitor General — provided commentaries about the famous cases?

2. The Domain and Scope of Copyright Protection

17 U.S.C. § 102. Subject Matter of Copyright: In General

(a) Copyright protection subsists, in accordance with this title, in original works of authorship fixed in any tangible medium of expression, now known or later developed, from which they can be perceived, reproduced, or otherwise communicated, either directly or with the aid of a machine or device. Works of authorship include the following categories:

(1) literary works;
(2) musical works, including any accompanying words;
(3) dramatic works, including any accompanying music;
(4) pantomimes and choreographic works;
(5) pictorial, graphic, and sculptural works;
(6) motion pictures and other audiovisual works;
(7) sound recordings; and
(8) architectural works.

H.R. Rep. No. 94-1476, 94th Cong., 2d Sess. (1976)

The second sentence of section 102 lists [eight][18] broad categories which the concept of "works of authorship" is said to "include." The use of "include," as defined in section 101, makes clear that the listing is "illustrative and not limitative," and that the [eight] catgories do not necessarily exhaust the scope of "original works of authorship" that the bill is intended to protect. Rather, the list sets out the general area of copyrightable subject matter, but with sufficient flexibility to free the courts from rigid and outmoded concepts of the scope of particular categories. The items are also overlapping in the sense that a work falling within one class may encompass works coming within some or all of the other categories....

a. Literary Works

Section 101 of the Copyright Act defines "literary works" as "works, other than audiovisual works, expressed in words, numbers, or other verbal or numerical symbols or indicia, regardless of the nature of the material objects, such as books, periodicals, manuscripts, phonorecords, films, tapes, disks, or cards, in which they are embodied."

The legislative history notes that "[t]he term 'literary works' does not connote any criterion of literary merit or qualitative value: it includes catalogs, directories, and similar factual reference, or instructional works and compilations of data. It also

---

18. The 1976 Act did not separately list architectural works, although the legislative history made clear that copyright extended to an architect's plans and drawings. H.R. Rep. at 55. The category of "architectural works" was added to section 102(a) by the Architectural Works Copyright Protection Act of 1990. — EDS.

includes computer data bases, and computer programs to the extent that they incorporate authorship in the programmer's expression of original ideas, as distinguished from the ideas themselves." Copyright Law Revision, H.R. Rep. No. 94-1476, 94th Cong., 2d Sess. (1976). Although the domain of literary works is quite broad and the originality threshold for protection is low, the Copyright Office regulations state that "words and short phrases such as names, titles, and slogans" are not subject to copyright. 37 C.F.R. § 202.1.[19]

The scope of copyright protection for literary works extends not only to the literal text but also to non-literal elements of a work such as its structure, sequence, and organization. Thus a second comer may not circumvent copyright law merely by paraphrasing an original text. As Judge Learned Hand has explained, were the rule otherwise, a "plagiarist would escape liability by immaterial variations" of the copyrighted work. Nichols v. Universal Pictures, 45 F.2d 119, 121 (2d Cir. 1930).

While the words of a story and other expressive elements of its text are clearly protectable, courts have struggled to delineate the scope of protection for other elements of literary works such as fictional characters. As we saw earlier, the scope of copyright protection is limited to expressive content and does not extend to the underlying ideas. To what extent, therefore, should an original literary description of a fictional character, such as James Bond or Superman, limit other authors' use of characters featuring similar attributes? In *Nichols,* Judge Learned Hand provided the following standard for determining the scope of protection for fictional characters:

> [W]e do not doubt that two plays may correspond in plot closely enough for infringement. How far that correspondence must go is another matter. Nor need we hold that the same may not be true as to the characters, quite independently of the "plot" proper, though, as far as we know, such a case has never arisen. If *Twelfth Night* were copyrighted, it is quite possible that a second comer might so closely imitate Sir Toby Belch or Malvolio as to infringe, but it would not be enough that for one of his characters he cast a riotous knight who kept wassail to the discomfort of the household, or a vain and foppish steward who became amorous of his mistress. These would be no more than Shakespeare's "ideas" in the play, as little capable of monopoly as Einstein's Doctrine of Relativity, or Darwin's theory of the Origin of Species. It follows that the less developed the characters, the less they can be copyrighted; that is the penalty an author must bear for marking them too indistinctly.

b. Pictorial, Graphic, and Sculptural Works

Section 101 of the Copyright Act defines "[p]ictorial, graphic, and sculptural works" to include "two-dimensional and three-dimensional works of fine, graphic, and applied art, photographs, prints and art reproductions, maps, globes, charts, diagrams, models, and technical drawings, including architectural plans. . . ." As with literary works, courts are not authorized to judge the artistic merit of the work in deciding whether pictorial, graphic, and sculptural works ought to qualify for copyright protection other than to determine that the threshold of original expression has been attained.

The most significant limitation placed on pictorial, graphic, and sculptural works is the utilitarian function exception discussed above. Such works are not protectable to the

---

19. Titles, names of characters, and phrases may, however, be protectable under unfair competition law and trademark law. See *infra* Chapters 5 and 6.

extent that they have a utilitarian rather than artistic function. Thus the useful article doctrine poses a significant limitation on the scope of protection for sculptural works.

As with literary works, the scope of protection for pictorial, graphic, and sculptural works depends on the degree to which the author has delineated the subjects of the work. In some cases, particularly photographs, drawings and maps, the limited range of expressive choices necessarily limits the scope of protection afforded by copyright law.

PROBLEM

Problem 4-10. The New York Arrows, a professional soccer team, hired Sports Images, an advertising firm, to develop a logo to advertise the team. Sports Images developed the accompanying logo (Figure 4-4).

FIGURE 4-4
Proposed logo for the New York Arrows.

After a dispute with the team about fees, Sports Images submitted an application to the Copyright Office to register the work. Is the work copyrightable?

---

c. Architectural Works

Under the 1976 Act, achitectural plans in the United States were protected as a species of pictorial, graphic, and sculptural works. The protection afforded architectural works — actual structures — was therefore limited by the useful article doctrine and the idea/expression dichotomy. As provided in the definition of pictorial, graphic, and sculptural works in § 101, "the design of a useful article, as defined in this section, shall be considered a pictorial, graphic, or sculptural work only if, and only to the extent that, such design incorporates pictorial, graphic, or sculptural features that can be identified separately from, and are capable of existing independently of, the utilitarian aspects of the article." Hence architectural structures, as opposed to the drawings for them, had little if any protection under copyright law. See Demetriades v. Kaufmann, 690 F. Supp. 658 (S.D.N.Y. 1988) (holding that traced architectural plans infringed the originals, but that the construction of an identical building would not violate a copyright in architectural plans under the principles of *Baker v. Selden*).

In order to bring the United States into compliance with the Berne Convention — which expressly requires protection for architectural works, see Article

2(1) — Congress passed the Architectural Works Copyright Protection Act of 1990. The Act specifically protects "the design of a building as embodied in any tangible medium of expression, including a building, architectural plans, or drawings. The work includes the overall form and elements in the design, but does not include individual standard features." 17 U.S.C. § 101. The House Report states:

> ... By creating a new category of protectable subject matter in new section 102(a)(8), and therefore, by deliberately not encompassing architectural works as pictorial, graphic, or sculptural works in existing section 102(a)(5), the copyrightability of architectural works shall not be evaluated under the separability test applicable to pictorial, graphic, or sculptural works embodied in useful articles. There is considerable scholarly and judicial disagreement over how to apply the separability test, and the principal reason for not treating architectural works as pictorial, graphic, or sculptural works is to avoid entangling architectural works in this disagreement.
>
> The Committee does not suggest, though, that in evaluating the copyrightability or scope of protection for architectural works, the Copyright Office or the courts should ignore functionality. A two-step analysis is envisioned. First, an architectural work should be examined to determine whether there are original design elements present, including overall shape and interior architecture. If such design elements are present, a second step is reached to examine whether the design elements are functionally required. If the design elements are not functionally required, the work is protectable without regard to physical or conceptual separability. As a consequence, contrary to the Committee's report accompanying the 1976 Copyright Act with respect to industrial products, the aesthetically pleasing overall shape of an architectual work would be protected under this bill.

H.R. Rep. No. 101-735, 101st Cong., 2d Sess. 20-21 (1990).

The effective date for the Architectural Works Copyright Protection Act of 1990 is December 1, 1990 and it does not apply retroactively. Therefore, architectural works produced before that time are governed by the standard for PGS works. In addition, the protection of section 102(a)(8) is subject to two important limitations set forth in section 120:

(a) **Pictorial representations permitted.** — The copyright in an architectural work that has been constructed does not include the right to prevent the making, distributing, or public display of pictures, paintings, photographs, or other pictorial representations of the work, if the building in which the work is embodied is located in or ordinarily visible from a public place.

(b) **Alterations to and destruction of buildings.** — Notwithstanding the provisions of section 106(2), the owners of a building embodying an architectural work may, without the consent of the author or copyright owner of the architectural work, make or authorize the making of alterations to such building, and destroy or authorize the destruction of such building.

COMMENTS AND QUESTIONS

1. Is the distinction that Congress seeks to make between the conceptual separability analysis required for pictorial, graphic, and sculptural works and the two-step functionality limitation for architectural works coherent? In what circumstances do they produce different outcomes? What aspects of a house design are protectable?

2. Relatively few architectural cases have been litigated, even after the 1990 amendments expanding protection. What might explain this phenomenon — a broad public domain for architectural features? broad limitations on the scope of protection? norms within the architectural field?

3. *Distinguishing Architectural Works and Sculptural Structures.* The artist who designed the streetwall and courtyard space for a building in downtown Los Angeles sued the producers of the film *Batman Forever,* which featured this structure in the film (the building was the Second Bank of Gotham), for infringing the artist's copyright in its "sculptural work." The Ninth Circuit held that the work more properly constituted part of a larger architectural work comprising the building towers and hence could be photographed without authorization of the copyright owner under § 120(a). See Leicester v. Warner Bros., 232 F.3d 1212 (9th Cir. 2000).

4. *Broadening the Pictorial Representation Limitation.* Section 120(a) makes a lot of sense in enabling other creators to film in the vicinity of publicly accessible architectural works. The transaction costs of having to negotiate licenses would be high, and it is difficult to imagine that architects' incentives to create would be much affected by not being able to control such uses. Couldn't a more general argument be made with regard to all publicly displayed sculpture, whether or not it is integrated into an architectural work? What about publicly visible artwork generally? Should it depend on the extent to which it is featured in the film or promotional advertising? Shouldn't film makers be allowed to capture that which is publicly viewable?

d. Musical Works and Sound Recordings

Musical creativity manifests in various ways, including individual musical compositions, sound recordings of such compositions, dramatic works (such as operas and musical plays) incorporating music, motion pictures (sound tracks, musical performances), and audiovisual works (theme music, sound effects). Copyright law distinguishes among these aspects of musical creativity. We begin here with the basic building blocks: musical works (music and lyrics) and sound recordings. Copyright law affords each of these different aspects of music protection in recognition of the distinctive skills and individuality involved in their creation. This distinction can have important ramifications. As we will discuss in more detail in the section on copyright owner rights, U.S. copyright law accords owners of musical composition the full complement of rights (including the right to perform in public), whereas sound recordings do not receive a traditional performance right (although they now have a digital performance right). Therefore, when you hear a sound recording on traditional "over the air" radio broadcasts (although not satellite radio), the owner of the copyright in the musical composition typically is paid a royalty whereas the performer is not.

Congress first extended copyright protection for musical compositions in 1831 and added a performance right in 1897. Sound recordings did not receive federal copyright protection until 1972. The more limited right structure reflects the lobbying clout of broadcasters. They successfully resisted copyright protection for sound recordings for many years. When it finally gained enough momentum to pass, they were able to extract as a compromise that such protection would not include a performance right, thereby avoiding the burden of obtaining additional licenses.

Musical works can be written on paper, pressed onto a phonorecord, recorded on audiotape, or otherwise fixed in a tangible medium of expression. The work must be original in its melody, harmony, or rhythm, individually or in combination. Sound recordings are magnetically or electronically recorded versions of a musical composition or any aural performance (such as spoken words, animal sounds, or sound effects). The definition of "sound recordings" excludes "the sounds accompanying a motion picture or other audiovisual work." 17 U.S.C. § 101. Such recordings are protected as part of the motion picture or audiovisual work.

e. Dramatic, Pantomime, and Choreographic Works

The protection afforded these three distinct copyrightable forms is similar, in that each extends to written or otherwise fixed instructions for performing a work of art. A dramatic work portrays a story by means of dialogue or acting. "It gives direction for performance or actually represents all or a substantial portion of the action as actually occurring rather than merely being narrated or described." U.S. Copyright Office, Compendium II of Copyright Office Practices § 431. Distinguishing between literary, musical, and dramatic works can be important in practice. Although the three types of works are often captured in the same form, i.e., written text, performance and display rights may vary depending on whether the work is dramatic or nondramatic. See, e.g., 17 U.S.C. §§ 110(2), 115.

Pantomime and choreographic works were first brought within copyright law by the 1976 Act. Copyright in such works inheres either in notation — such as Labanotation, a system of symbols for representing movements that can be related to a musical score — or (more commonly) in a film recording. Impromptu, unrecorded dancing is not a protectable work because it is not fixed in a tangible medium of expression.

Many of the rules governing dramatic, pantomime, and choreographic works parallel the law with regard to literary works. For example, copyright protection for choreographic works does not extend to simple dance steps, and protection for pantomimes does not extend to "conventional gestures." Both are akin to the "short phrases" denied protection among literary works. All three forms of work are entitled to protection not only against literal copying, but also against copying of their expressive elements, character, action, and dialogue. Should a series of yoga exercises be protectable? See Open Source Yogan Unity v. Choudhury, 74 U.S.P.Q.2d 1435 (N.D. Cal. 2005) (relating to a claim to copyright protection for "Bikram yoga," a sequence of 26 yoga positions and two breathing exercises).

PROBLEM

Problem 4-11. At the 1995 Pan-World Figure Skating Championships in Denver, Colorado, skater Kurt Klutzinovich performed the first series of quadruple lutzes ever seen in competition. ABC sports commentator Bill McKay proclaimed it the "klutz," "Kurt's patented move." At the 1996 Championships, Kurt's coach threatens to bring a copyright infringement action against any other skater who performs a "klutz." Is Kurt's series of moves protectable as a choreographic work? If so, would another skater's performance of a klutz infringe Kurt's copyright?

f. Motion Pictures and Other Audiovisual Works

The Copyright Act defines "audiovisual works" as

> works that consist of a series of related images which are intrinsically intended to be shown by the use of machines or devices such as projectors, viewers, or electronic equipment, together with accompanying sounds, if any, regardless of the nature of the material objects, such as films or tapes, in which the works are embodied.

17 U.S.C. § 101. "Motion pictures" are a subset of audiovisual works "consisting of a series of related images, which, when shown in succession, impart an impression of motion, together with accompanying sounds, if any." 17 U.S.C. § 101. Thus, sound tracks are treated as an integral part of the motion pictures that they accompany.

Protection for audiovisual works has taken on increased importance with the growth of computer software, particularly game and multimedia products. We discuss these aspects of audiovisual works in detail in Chapter 7.

g. Derivative Works and Compilations

Section 103 of the Copyright Act provides protection for derivative works and compilations. The copyright in a derivative work or compilation "extends only to the material contributed by the author of such work, as distinguished from the preexisting material employed in the work." § 103(b).

A "derivative work" is

> based upon one or more preexisting works, such as a translation, musical arrangement, dramatization, fictionalization, motion picture version, sound recording, art reproduction, abridgment, condensation, or any other form in which a work may be recast, transformed, or adapted. A work consisting of editorial revisions, annotations, elaborations, or other modifications which, as a whole, represent an original work of authorship, is a "derivative work."

§ 101. We explore this definition in detail in section E.2, *infra*, addressing the right to make derivative works.

Why should the copyright owner be entitled to a separate copyright in the derivative work, in addition to the copyright in the original work? There are at least two possible explanations for granting separate copyright protection to the original elements of a derivative work.

First, derivative works along a "chain" of related works may well entail significant new creativity beyond the original copyrighted work. For example, an author may produce a children's book, then a movie script based on the book, followed by a movie based on the script, and a series of stuffed animals based on characters from the movie. The stuffed animals at the end of this "chain" may bear little if any resemblance to identifiable characters in the original book. Allowing the derivative works to be copyrighted gives the copyright holder a stronger argument that each new step in the chain is in fact protectable as a derivative of the prior copyright. A second justification is that we may wish to protect new expression "derived" from works that are already in the public domain. See Paul Goldstein, Derivative Rights and Derivative Works in Copyright, 30 J. Copr. Soc'y 209 (1983).

In addition, because derivative works often capture different markets, the copyright owner in the original work may wish to license to others the right to produce derivative works. Just because someone is a successful writer, for example, does not mean that he will be particularly successful at manufacturing and selling plush toys. Separating the copyright in original elements of a derivative work may facilitate the division of ownership between the original author and his licensees.

A "compilation" is a work "formed by the collection and assembling of preexisting materials or of data that are selected, coordinated, or arranged in such a way that the resulting work as a whole constitutes an original work of authorship. The term 'compilation' includes collective works." §101. A "collective work" is a work, "such as a periodical issue, anthology, or encyclopedia, in which a number of contributions, constituting separate and independent works in themselves, are assembled into a collective whole." §101.

The level of originality required for a compilation to be copyrightable has been a contentious issue in copyright law. Many fact-based works, such as telephone directories, require tremendous time, effort, and expense to compile. As we saw in the *Feist* case, however, there must be "some minimal degree of creativity" to garner copyright protection. Since neither the underlying material being compiled (names, which are factual) nor the method of arrangement (alphabetical order) were original, the compilation thus created was not copyrightable, regardless of how much effort was involved in producing the directory.

Many controversial copyright cases turn on whether a collection of uncopyrightable elements becomes copyrightable as a compilation. What is the threshold for originality in a compilation?

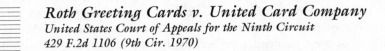

Roth Greeting Cards v. United Card Company
United States Court of Appeals for the Ninth Circuit
429 F.2d 1106 (9th Cir. 1970)

HAMLEY, Circuit Judge:

Roth Greeting Cards (Roth) and United Card Company (United), both corporations, are engaged in the greeting card business. Roth brought this suit against United to recover damages and obtain injunctive relief for copyright infringement of seven studio greeting cards. After a trial to the court without a jury, judgment was entered for defendant. Plaintiff appeals.

Roth's claim involves the production and distribution by United of seven greeting cards which bear a remarkable resemblance to seven of Roth's cards on which copyrights had been granted. Roth employed a writer to develop the textual material for its cards. When Roth's president determined that a textual idea was acceptable, he would integrate that text into a rough layout of a greeting card with his suggested design for the art work. He would then call in the company artist who would make a comprehensive layout of the card. If the card was approved, the artist would do a finished layout and the card would go into production.

During the period just prior to the alleged infringements, United did not have any writers on its payroll. Most of its greeting cards came into fruition primarily through the activities of United's president, Mr. Koenig, and its vice-president, Edward Letwenko.

The source of the art and text of the cards of United, here in question, is unclear. Letwenko was unable to recall the origin of the ideas for most of United's cards.

He speculated that the gags used may have come from plant personnel, persons in bars, friends at a party, Koenig, or someone else. He contended that the art work was his own. But he also stated that he visited greeting card stores and gift shows in order to observe what was going on in the greeting card business. Letwenko admitted that he may have seen the Roth cards during these visits or that the Roth cards may have been in his office prior to the time that he did his art work on the United cards....

[T]he trial court found that the art work in plaintiff's greeting cards was copyrightable, but not infringed by defendant. The trial court also found that, although copied by defendant, the wording or textual matter of each of the plaintiff's cards in question consist of common and ordinary English words and phrases which are not original with Roth and were in the public domain prior to first use by plaintiff.

Arguing that the trial court erred in ruling against it on merits, Roth agrees that the textual material involved in their greeting cards may have been in the public domain, but argues that this alone did not end the inquiry into the copyrightability of the entire card. Roth argued that "It is the arrangement of the words, their combination and plan, together with the appropriate art work..." which is original, the creation of Roth, and entitled to copyright protection.

In order to be copyrightable, the work must be the original work of the copyright claimant or of his predecessor in interest. M. Nimmer, Copyright (hereafter Nimmer), 10 at 32 (1970). But the originality necessary to support a copyright merely calls for independent creation, not novelty. Alfred Bell & Co., Ltd. v. Catalda Fine Arts, Inc., 191 F.2d 99, 102 (2d Cir. 1951). Cf. Baker v. Selden, 101 U.S. 99, 102-103 (1879).

United argues, and we agree, that there was substantial evidence to support the district court's finding that the textual matter of each card, considered apart from its arrangement on the cards and its association with artistic representations, was not original to Roth and therefore not copyrightable. However, proper analysis of the problem requires that all elements of each card, including text, arrangement of text, art work, and association between art work and text, be considered as a whole.

Considering all of these elements together, the Roth cards are, in our opinion, both original and copyrightable. In reaching this conclusion we recognize that copyright protection is not available for ideas, but only for the tangible expression of ideas. Mazer v. Stein, 347 U.S. 201, 217 (1954). We conclude that each of Roth's cards, considered as a whole, represents a tangible expression of an idea and that such expression was, in totality, created by Roth. See Dorsey v. Old Surety Life Ins. Co., 98 F.2d 872, 873 (10th Cir. 1938).

This brings us to the question of infringement. Greeting cards are protected.... They are the embodiment of humor, praise, regret or some other message in a pictorial and literary arrangement. As proper subjects of copyright, they are susceptible to infringement in violation of the Act. Detective Comics, Inc. v. Bruns Publications, Inc., 111 F.2d 432 (2d Cir. 1940).

To constitute an infringement under the Act there must be substantial similarity between the infringing work and the work copyrighted; and that similarity must have been caused by the defendant's having copied the copyright holder's creation. The protection is thus against copying—not against any possible infringement caused when an independently created work coincidentally duplicates copyrighted material. Sheldon v. Metro-Goldwyn Pictures Corp., 81 F.2d 49, 54 (2d Cir. 1936).

It appears to us that in total concept and feel the cards of United are the same as the copyrighted cards of Roth. With the possible exception of one United card (exhibit 6), the characters depicted in the art work, the mood they portrayed, the combination of art work conveying a particular mood with a particular message, and the arrangement of the words on the greeting card are substantially the same as in Roth's cards. In several instances the lettering is also very similar.

It is true, as the trial court found, that each of United's cards employed art work somewhat different from that used in the corresponding Roth cards. However, "The test of infringement is whether the work is recognizable by an ordinary observer as having been taken from the copyrighted source." White-Smith Music Pub. Co. v. Apollo Company, 209 U.S. 1, 17 (1907), Bradbury v. Columbia Broadcasting System, Inc., 287 F.2d 478, 485 (9th Cir. 1961). The remarkable similarity between the Roth and United cards in issue (with the possible exception of exhibits 5 and 6) is apparent to even a casual observer. For example, one Roth card (exhibit 9) has, on its front, a colored drawing of a cute moppet suppressing a smile and, on the inside, the words "i wuv you." With the exception of minor variations in color and style, defendant's card (exhibit 10) is identical. Likewise, Roth's card entitled "I miss you already," depicts a forlorn boy sitting on a curb weeping, with an inside message reading "...and You Haven't even Left..." (exhibit 7), is closely paralleled by United's card with the same caption, showing a forlorn and weeping man, and with the identical inside message (exhibit 8)....

KILKENNY, Circuit Judge (dissenting).

The majority agrees with a specific finding of the lower court that the words on the cards are not the subject of copyright. By strong implication, it likewise accepts the finding of the trial court that the art work on the cards, although subject to copyright, was not infringed. Thus far, I agree.

I cannot, however, follow the logic of the majority in holding that the uncopyrightable words and the imitated, but not copied art work, constitutes such total composition as to be subject to protection under the copyright laws. The majority concludes that in the overall arrangement of the text, the art work and the association of the art work to the text, the cards were copyrightable and the copyright infringed. This conclusion, as I view it, results in the whole becoming substantially greater than the sum total of its parts. With this conclusion, of course, I cannot agree....

Feeling, as I do, that the copyright act is a grant of limited monopoly to the authors of creative literature and art, I do not think that we should extend a 56-year monopoly in a situation where neither infringement of text, nor infringement of art work can be found. On these facts, we should adhere to our historic philosophy requiring freedom of competition. I would affirm.

COMMENTS AND QUESTIONS

1. Would there be sufficient incentive for authors and publishers to produce greeting cards without the type of protection recognized by the court in *Roth*? On the contrary, does this decision enable a greeting card company to monopolize most of the stock phrases and images simply by being first?

2. Is *Roth* still good law after *Feist*? *Feist* suggests that the copyright in compilations of facts or other uncopyrightable elements is "thin." As noted more recently by the Ninth Circuit in a case denying protection to a sculpture entombing a jellyfish in glass, "a combination of unprotectable elements is eligible for copyright protection only if those elements are numerous enough and their selection and arrangement

original enough that their combination constitutes an original work of authorship." Satava v. Lowry, 323 F.3d 805, 811 (9th Cir. 2003).

PROBLEMS

Problem 4-12. Assume that West Publishing Company, which has published both official and unofficial reports of federal and state court decisions for over 100 years, and Mead Data Central are the only competitors in the market for computer legal research databases. Because cases (especially in the federal courts) are cited by the West's volume and page number, Mead decides to copy West's pagination in its computer database. West sues, arguing that the arrangement of its cases in its reporters is copyrighted. Does West have a valid copyright?

Problem 4-13. The Bond News and Investor's World are both financial reporting services. Each provides to its subscribers a weekly update of all municipal bonds that have been "called" by the city, and several pieces of information about the bonds: the bond series, the call price, the date of the call, and the address and phone number of the calling agency. Bond obtains this information by having researchers cull through published notices in 250 newspapers nationwide each day. Bond, suspecting that Investor's is copying its data rather than conducting a similar search, plants false information in its updates. When Investor's publishes the same false information in its update, Bond sues. Investor's admits copying the data but claims that it had a right to do so because the data were not copyrightable. Who should prevail?

D. OWNERSHIP AND DURATION

Copyrights today are in many ways like other forms of property interests. As we will see in section E, a copyright includes numerous distinct rights. Once a copyright is acquired, the entire bundle of rights may be assigned to others. Alternatively, the owner may divide and transfer particular rights. Nonetheless, there are significant differences between the ownership rules governing copyrights and those governing other property interests. The most important difference is that copyrights have limited duration. As required by the U.S. Constitution, copyrights enter the public domain after a statutorily determined limited time. Furthermore, Congress has differentiated the ownership, transfer, and termination of these rights in other ways as well. As you work through these provisions, consider how copyright ownership rules differ from the ownership rules for other forms of intellectual (e.g., patents) and real (e.g., land, chattel) property. What justifies these differences? More specifically, how do these differences promote the objectives of the copyright law?

This section first surveys the law that determines who initially acquires a copyright. This law includes the rules governing works for hire and jointly produced works. We then outline the rules governing the duration of copyright protection. These rules are complicated by the renewal process under the 1909 Act. We conclude with a discussion of the rules governing the division, transfer, and termination of transfers of copyright interests.

1. Initial Ownership of Copyrights

As noted earlier, copyright law since the 1700s has vested copyright in authors and artists who create works satisfying the requirements of copyright law. In the case of an individual who writes, composes, or paints an original work of authorship on her or his own today, that person acquires the copyright upon creation. 17 U.S.C. § 201(a). Many works in our modern and increasingly interconnected economy, however, are produced in the employment context or commissioned by a production company or publisher. In addition, many works are the result of collaborations among numerous authors and artists.

a. Works Made for Hire

The 1909 Act recognized that "the word 'author' shall include an employer in the case of works made for hire," § 26, although the Act did not define the term "work made for hire." Courts took an expansive view of what constituted a work made for hire, presuming that any works created within the scope of employment or commissioned by independent contractors (at the "instance and expense" of the employer) vested in the employer. See May v. Morganelli-Heumann & Assoc., 618 F.2d 1363 (9th Cir. 1980); Scherr v. Universal Match Corp., 417 F.2d 497 (2d Cir. 1969), *cert. denied*, 397 U.S. 936 (1970). The presumption could be overcome by an express agreement to the contrary or other evidence suggesting an alternative intention of the parties, such as industry custom or lack of supervision or creative control by the employer. Works created prior to January 1, 1978 remain subject to this test. See Roth v. Pritikin, 710 F.2d 934 (2d Cir.), *cert. denied*, 464 U.S. 961 (1983).

In the legislative process leading up to the 1976 Act, authors' representatives expressed concern that freelance authors lacked the bargaining power to reject contractual clauses designating their works as "works made for hire." The motion picture industry and other producers and publishers of works involving many creative contributors were concerned about holdout problems complicating commercial exploitation if they did not own copyrights in works they commissioned outright and for the entire duration of copyright protection. The ultimate compromise, hashed out over more than a decade, defines "work made for hire" to include: (1) a work prepared by an employee within the scope of his or her employment; or (2) a work falling within one of nine enumerated categories—a contribution to a collective work, a part of a motion picture or other audiovisual work, a translation, a supplementary work, a compilation, a test, answer material for a test, or an atlas—and evidenced by a written agreement signed by both parties expressly stating that the work is intended to be a "work made for hire." § 101. Disputes quickly arose as to what Congress intended by the term "employee."

Community for Creative Non-Violence et al. v. Reid

Supreme Court of the United States
490 U.S. 730 (1989)

Justice MARSHALL delivered the opinion of the Court.

In this case, an artist and the organization that hired him to produce a sculpture contest the ownership of the copyright in that work. To resolve this dispute, we must construe the "work made for hire" provisions of the Copyright Act of 1976 (Act or 1976 Act), 17 U.S.C. §§ 101 and 201(b), and in particular, the provision in § 101, which defines as a "work made for hire" a "work prepared by an employee within the scope of his or her employment" (hereinafter § 101(1)).

Petitioners are the Community for Creative Non-Violence (CCNV), a nonprofit unincorporated association dedicated to eliminating homelessness in America, and Mitch Snyder, a member and trustee of CCNV. In the fall of 1985, CCNV decided to participate in the annual Christmastime Pageant of Peace in Washington, D.C., by sponsoring a display to dramatize the plight of the homeless. As the District Court recounted:

> Snyder and fellow CCNV members conceived the idea for the nature of the display: a sculpture of a modern Nativity scene in which, in lieu of the traditional Holy Family, the two adult figures and the infant would appear as contemporary homeless people huddled on a streetside steam grate. The family was to be black (most of the homeless in Washington being black); the figures were to be life-sized, and the steam grate would be positioned atop a platform "pedestal," or base, within which special-effects equipment would be enclosed to emit simulated "steam" through the grid to swirl about the figures. They also settled upon a title for the work — "Third World America" — and a legend for the pedestal: "and still there is no room at the inn." 652 F. Supp. 1453, 1454 (DC 1987).

Snyder made inquiries to locate an artist to produce the sculpture. He was referred to respondent James Earl Reid, a Baltimore, Maryland, sculptor. In the course of two telephone calls, Reid agreed to sculpt the three human figures. CCNV agreed to make the steam grate and pedestal for the statue. Reid proposed that the work be cast in bronze, at a total cost of approximately $100,000 and taking six to eight months to complete. Snyder rejected that proposal because CCNV did not have sufficient funds, and because the statue had to be completed by December 12 to be included in the pageant. Reid then suggested, and Snyder agreed, that the sculpture would be made of a material known as "Design Cast 62," a synthetic substance that could meet CCNV's monetary and time constraints, could be tinted to resemble bronze, and could withstand the elements. The parties agreed that the project would cost no more than $15,000, not including Reid's services, which he offered to donate. The parties did not sign a written agreement. Neither party mentioned copyright.

After Reid received an advance of $3,000, he made several sketches of figures in various poses. At Snyder's request, Reid sent CCNV a sketch of a proposed sculpture showing the family in a crechelike setting: the mother seated, cradling a baby in her lap; the father standing behind her, bending over her shoulder to touch the baby's foot. Reid testified that Snyder asked for the sketch to use in raising funds for the sculpture. Snyder testified that it was also for his approval. Reid sought a black family to serve as a model for the sculpture. Upon Snyder's suggestion, Reid visited a family living at CCNV's Washington shelter but decided that only their newly born child was a suitable model. While Reid was in Washington, Snyder took him to see homeless people living on the streets. Snyder pointed out that they tended to recline on steam grates, rather than sit or stand, in order to warm their bodies. From that time on, Reid's sketches contained only reclining figures.

Throughout November and the first two weeks of December 1985, Reid worked exclusively on the statue, assisted at various times by a dozen different people who were paid with funds provided in installments by CCNV. On a number of occasions,

CCNV members visited Reid to check on his progress and to coordinate CCNV's construction of the base. CCNV rejected Reid's proposal to use suitcases or shopping bags to hold the family's personal belongings, insisting instead on a shopping cart. Reid and CCNV members did not discuss copyright ownership on any of these visits.

On December 24, 1985, 12 days after the agreed-upon date, Reid delivered the completed statue to Washington. There it was joined to the steam grate and pedestal prepared by CCNV and placed on display near the site of the pageant. Snyder paid Reid the final installment of the $15,000. The statue remained on display for a month. In late January 1986, CCNV members returned it to Reid's studio in Baltimore for minor repairs. Several weeks later, Snyder began making plans to take the statue on a tour of several cities to raise money for the homeless. Reid objected, contending that the Design Cast 62 material was not strong enough to withstand the ambitious itinerary. He urged CCNV to cast the statue in bronze at a cost of $35,000, or to create a master mold at a cost of $5,000. Snyder declined to spend more of CCNV's money on the project.

In March 1986, Snyder asked Reid to return the sculpture. Reid refused. He then filed a certificate of copyright registration for "Third World America" in his name and announced plans to take the sculpture on a more modest tour than the one CCNV had proposed. Snyder, acting in his capacity as CCNV's trustee, immediately filed a competing certificate of copyright registration.

Snyder and CCNV then commenced this action against Reid and his photographer, Ronald Purtee, seeking return of the sculpture and a determination of copyright ownership. The District Court granted a preliminary injunction, ordering the sculpture's return. After a 2-day bench trial, the District Court declared that "Third World America" was a "work made for hire" under §101 of the Copyright Act and that Snyder, as trustee for CCNV, was the exclusive owner of the copyright in the sculpture. 652 F. Supp., at 1457. The court reasoned that Reid had been an "employee" of CCNV within the meaning of §101(1) because CCNV was the motivating force in the statue's production. Snyder and other CCNV members, the court explained, "conceived the idea of a contemporary Nativity scene to contrast with the national celebration of the season," and "directed enough of [Reid's] effort to assure that, in the end, he had produced what they, not he, wanted." Id., at 1456.

The Court of Appeals for the District of Columbia Circuit reversed and remanded.

II

A

The Copyright Act of 1976 provides that copyright ownership "vests initially in the author or authors of the work." 17 U.S.C. §201(a). As a general rule, the author is the party who actually creates the work, that is, the person who translates an idea into a fixed, tangible expression entitled to copyright protection. §102. The Act carves out an important exception, however, for "works made for hire." If the work is for hire, "the employer or other person for whom the work was prepared is considered the author" and owns the copyright, unless there is a written agreement to the contrary. §201(b). Classifying a work as "made for hire" determines not only the initial ownership of its copyright, but also the copyright's duration, §302(c), and the owners' renewal rights, §304(a), termination rights, §203(a), and right to import certain goods bearing the copyright, § 601(b)(1). See 1 M. Nimmer & D. Nimmer, Nimmer on Copyright § 5.03 [A], pp. 5-10 (1988). The contours of the work for hire doctrine therefore carry profound significance for freelance creators — including artists, writers, photographers, designers, composers, and computer programmers — and for the publishing, advertising, music, and other industries which commission their works.

Section 101 of the 1976 Act provides that a work is "for hire" under two sets of circumstances:

> (1) a work prepared by an employee within the scope of his or her employment; or
>
> (2) a work specially ordered or commissioned for use as a contribution to a collective work, as a part of a motion picture or other audiovisual work, as a translation, as a supplementary work, as a compilation, as an instructional text, as a test, as answer material for a test, or as an atlas, if the parties expressly agree in a written instrument signed by them that the work shall be considered a work made for hire.

The Petitioners do not claim that the statue satisfies the terms of § 101(2). Quite clearly, it does not. Sculpture does not fit within any of the nine categories of "specially ordered or commissioned" works enumerated in that subsection, and no written agreement between the parties establishes "Third World America" as a work for hire.

The dispositive inquiry in this case therefore is whether "Third World America" is "a work prepared by an employee within the scope of his or her employment" under § 101(1). The Act does not define these terms. In the absence of such guidance, four interpretations have emerged. The first holds that a work is prepared by an employee whenever the hiring party retains the right to control the product. See Peregrine v. Lauren Corp., 601 F. Supp. 828, 829 (Colo. 1985); Clarkstown v. Reeder, 566 F. Supp. 137, 142 (S.D.N.Y. 1983). Petitioners take this view. A second, and closely related, view is that a work is prepared by an employee under § 101(1) when the hiring party has actually wielded control with respect to the creation of a particular work. This approach was formulated by the Court of Appeals for the Second Circuit, Aldon Accessories Ltd. v. Spiegel, Inc., 738 F.2d 548, *cert. denied,* 469 U.S. 982 (1984), and adopted by the Fourth Circuit, Brunswick Beacon, Inc. v. Schock-Hopchas Publishing Co., 810 F.2d 410 (1987), the Seventh Circuit, Evans Newton, Inc. v. Chicago Systems Software, 793 F.2d 889, *cert. denied,* 479 U.S. 949 (1986), and, at times, by petitioners. A third view is that the term "employee" within § 101(1) carries its common-law agency law meaning. This view was endorsed by the Fifth Circuit in Easter Seal Society for Crippled Children and Adults of Louisiana, Inc. v. Playboy Enterprises, 815 F.2d 323 (1987), and by the Court of Appeals below. Finally, respondent and numerous amici curiae contend that the term "employee" only refers to "formal, salaried" employees. See, e.g., Brief for Respondent 23-24; Brief for Register of Copyrights as Amicus Curiae 7. The Court of Appeals for the Ninth Circuit recently adopted this view. See Dumas v. Gommerman, 865 F.2d 1093 (1989).

The starting point for our interpretation of a statute is always its language. Consumer Product Safety Comm'n v. GTE Sylvania, Inc., 447 U.S. 102, 108 (1980). The Act nowhere defines the terms "employee" or "scope of employment." It is, however, well established that "where Congress uses terms that have accumulated settled meaning under . . . the common law, a court must infer, unless the statute otherwise dictates, that Congress means to incorporate the established meaning of these terms." NLRB v. Amax Coal Co., 453 U.S. 322, 329 (1981). In the past, when Congress has used the term "employee" without defining it, we have concluded that Congress intended to describe the conventional master-servant relationship as understood by common-law agency doctrine. See, e.g., Kelley v. Southern Pacific Co., 419 U.S. 318, 322-323 (1974). Nothing in the text of the work for hire provisions indicates that Congress used the words "employee" and "employment" to describe anything other than "'the conventional relation of employer and employee.'" *Kelley,* supra, at 323, quoting *Robinson*; compare NLRB v. Hearst Publications, Inc., 322 U.S. 111, 124-132 (1944) (rejecting agency law conception of employee for purposes of the National Labor Relations Act where structure and context of statute indicated broader definition). On the contrary, Congress' intent to incorporate the agency law definition is suggested by § 101(1)'s use of the term, "scope of employment," a widely used term of art in agency law. See Restatement (Second) of Agency § 228 (1958) (hereinafter Restatement).

We thus agree with the Court of Appeals that the term "employee" should be understood in light of the general common law of agency.

In contrast, neither test proposed by petitioners is consistent with the text of the Act. The exclusive focus of the right to control the product test on the relationship between the hiring party and the product clashes with the language of § 101(1), which focuses on the relationship between the hired and hiring parties. The right to control the product test also would distort the meaning of the ensuing subsection, § 101(2). Section 101 plainly creates two distinct ways in which a work can be deemed for hire: one for works prepared by employees, the other for those specially ordered or commissioned works which fall within one of the nine enumerated categories and are the subject of a written agreement. The right to control the product test ignores this dichotomy by transforming into a work for hire under § 101(1) any "specially ordered or commissioned" work that is subject to the supervision and control of the hiring party. Because a party who hires a "specially ordered or commissioned" work by definition has a right to specify the characteristics of the product desired, at the time the commission is accepted, and frequently until it is completed, the right to control the product test would mean that many works that could satisfy § 101(2) would already have been deemed works for hire under § 101(1). Petitioners' interpretation is particularly hard to square with § 101(2)'s enumeration of the nine specific categories of specially ordered or commissioned works eligible to be works for hire, e.g., "a contribution to a collective work," "a part of a motion picture," and "answer material for a test." The unifying feature of these works is that they are usually prepared at the instance, direction, and risk of a publisher or producer. By their very nature, therefore, these types of works would be works by an employee under petitioners' right to control the product test.

We therefore conclude that the language and structure of § 101 of the Act do not support either the right to control the product or the actual control approaches.[8] The structure of § 101 indicates that a work for hire can arise through one of two mutually exclusive means, one for employees and one for independent contractors, and ordinary canons of statutory interpretation indicate that the classification of a particular hired party should be made with reference to agency law.

This reading of the undefined statutory terms finds considerable support in the Act's legislative history....

---

8. We also reject the suggestion of respondent and amici that the § 101(1) term "employee" refers only to formal, salaried employees. While there is some support for such a definition in the legislative history, see Varmer, Works Made for Hire 130, the language of § 101(1) cannot support it. The Act does not say "formal" or "salaried" employee, but simply "employee." ...

We turn, finally, to an application of § 101 to Reid's production of "Third World America." In determining whether a hired party is an employee under the general common law of agency, we consider the hiring party's right to control the manner and means by which the product is accomplished. Among the other factors relevant to this inquiry are the skill required; the source of the instrumentalities and tools; the location of the work; the duration of the relationship between the parties; whether the hiring party has the right to assign additional projects to the hired party; the extent of the hired party's discretion over when and how long to work; the method of payment; the hired party's role in hiring and paying assistants; whether the work is part of the regular business of the hiring party; whether the hiring party is in business; the provision of employee benefits; and the tax treatment of the hired party. See Restatement § 220(2) (setting forth a nonexhaustive list of factors relevant to determining whether a hired party is an employee). No one of these factors is determinative. See Ward, 362 U.S., at 400; Hilton Int'l Co. v. NLRB, 690 F.2d 318, 321 (CA2 1982).

Examining the circumstances of this case in light of these factors, we agree with the Court of Appeals that Reid was not an employee of CCNV but an independent contractor. 846 F.2d, at 1494, n.11. True, CCNV members directed enough of Reid's work to ensure that he produced a sculpture that met their specifications. 652 F. Supp., at 1456. But the extent of control the hiring party exercises over the details of the product is not dispositive. Indeed, all the other circumstances weigh heavily against finding an employment relationship. Reid is a sculptor, a skilled occupation. Reid supplied his own tools. He worked in his own studio in Baltimore, making daily supervision of his activities from Washington practicably impossible. Reid was retained for less than two months, a relatively short period of time. During and after this time, CCNV had no right to assign additional projects to Reid. Apart from the deadline for completing the sculpture, Reid had absolute freedom to decide when and how long to work. CCNV paid Reid $15,000, a sum dependent on "completion of a specific job, a method by which independent contractors are often compensated." Holt v. Winpisinger, 811 F.2d 1532, 1540 (1987). Reid had total discretion in hiring and paying assistants. "Creating sculptures was hardly 'regular business' for CCNV." 846 F.2d, at 1494, n.11. Indeed, CCNV is not a business at all. Finally, CCNV did not pay payroll or Social Security taxes, provide any employee benefits, or contribute to unemployment insurance or workers' compensation funds.

Because Reid was an independent contractor, whether "Third World America" is a work for hire depends on whether it satisfies the terms of § 101(2). This petitioners concede it cannot do. Thus, CCNV is not the author of "Third World America" by virtue of the work for hire provisions of the Act. However, as the Court of Appeals made clear, CCNV nevertheless may be a joint author of the sculpture if, on remand, the District Court determines that CCNV and Reid prepared the work "with the intention that their contributions be merged into inseparable or interdependent parts of a unitary whole." 17 U.S.C. § 101. In that case, CCNV and Reid would be co-owners of the copyright in the work. See § 201(a).

For the aforestated reasons, we affirm the judgment of the Court of Appeals for the District of Columbia Circuit.

It is so ordered.

COMMENTS AND QUESTIONS

1. What policies underlie the work made for hire doctrine? From which philosophical perspectives do these policies flow? Does the court's analysis and holding in *CCNV* comport with these policies?

We learned in the introduction to this chapter that copyright law, while originally designed to protect publishers, has since the early days of English law been intended to benefit the authors themselves. Does the "work made for hire" provision in the 1976 Act defeat this purpose? In the case of works made for hire, section 101 automatically vests copyright in employers, not the particular employees who author the work. This result is somewhat surprising, especially in light of patent law's approach of considering the individual inventor to be the patentee (at least nominally). Why is this so? Why are inventors always people, whereas authors can be companies? In view of assignability rules in patent law, does the distinction make a difference? Unlike copyright law, patent law has no termination of transfer rule or other provision safeguarding inventors against unremunerative transfers.

One justification for the work made for hire doctrine may be that it reduces transaction costs. A work made for hire is deemed to be a corporate creation, so it need not be assigned. The doctrine essentially "preassigns" a work to the employer. This has the important effect of eliminating the costs of negotiating and executing assignment agreements. In addition, as we will see in section 3 *infra*, the work made for hire doctrine avoids the inalienability of the termination of transfer right. The work belongs to the employer *ab initio*, i.e., from the moment of creation, rather than by assignment. Such treatment solves a potentially large "holdout" problem in compilations, multimedia, motion pictures, and other works involving numerous contributors. Are the works listed in the second part of section 101's definition of work made for hire the types of works for which holdout problems are most likely?

2. *Within the Scope of Employment.* Even if a work is created by an employee, it must also be "within the scope of his or her employment" in order to be deemed a work made for hire. From the Supreme Court's instruction to apply the common law of agency, most courts reaching this issue have adopted the test set forth in Restatement 2d of Agency. An employee's work is deemed within the scope of his employment if:

   a) it is within the kind he is employed to perform;
   b) it occurs substantially within the authorized time and space limits; [and]
   c) it is actuated, at least in part, by a purpose to serve the master.

Restatement 2d of Agency § 228 (1958). See Quinn v. City of Detroit, 988 F.Supp. 1044, 1049 (E.D. Mich. 1997).

3. *Copyright Shop Rights?* In the areas of trade secrets and patented inventions, courts have recognized a common law "shop right" which allows an employer to use an employee's invention to the extent necessary for its regular business where the employer contributed to the development of the invention, for example, by providing wages, tools, or a workplace. See McElmurry v. Arkansas Power & Light Co., 995 F.2d 1576, 1580 (Fed. Cir. 1993). In the negotiations leading up to the 1976 Act, screenwriters and film music composers advocated a comparable "shop right," whereby the employee would retain all other rights, subject to a covenant not to authorize competing uses. Congress rejected this proposal in favor of providing employers full copyright for works prepared within the scope of the worker's

employment. H.R.Rep. No. 94-1476, at 121, U.S. Code Cong. & Admin. News 1976, p. 5737. What about works that the employee completes on the employer's time and using the employer's resources but fall outside of the employee's scope of employment? Should the employer be able to assert a shop right in this circumstance?

4. *Works of the U.S. Government.* As we saw earlier, works of the U.S. government are not copyrightable. Section 101 defines U.S. government works as works "prepared by an officer or employee of the United States government as part of that person's official duties." The legislative history makes clear that Congress intended this provision to parallel the scope of the work made for hire doctrine.

5. *Teacher Exception.* There is a significant exception to the work made for hire doctrine that prevents universities from claiming to own professors' works under the works made for hire doctrine. This exception is venerable, but lacks textual support in the 1976 Act. Some courts and commentators have concluded that the Act did in fact abolish the "teacher exception," but most work hard to find a way to keep the doctrine alive. See Hays v. Sony Corp. of Am., 847 F.2d 412, 416 (7th Cir. 1988) (Judge Posner, a prolific former professor, justified the exception upon the "havoc that [a contrary] conclusion would wreak in the settled practices of academic institutions, the lack of fit between the policy of the work-for-hire doctrine and the conditions of academic production, and the absence of any indication that Congress meant to abolish" the exception.); Weinstein v. University of Illinois, 811 F.2d 1091, 1093-94 (7th Cir. 1987); Rochelle Cooper Dreyfuss, The Creative Employee and the Copyright Act of 1976, 54 U. Chi. L. Rev. 590, 597-98 (1987); but see University of Colorado Found. v. American Cyanamid, 880 F. Supp. 1387 (D. Colo. 1995) (assuming that an academic article written by university professors was a work made for hire). Can you think of a way for the exception to survive the passage of the 1976 Act? Should it?

6. *The Role of Contract Law.* Regardless of whether a work is created within the confines of the Copyright Act's "work made for hire" provisions, parties are free to assign copyright interests through contract. Use of such instruments is quite common in the production of many commercial copyrighted works. There is, however, at least one important respect in which contract cannot substitute for "work made for hire" status. As we will discuss further below, the 1976 Act creates an inalienable right in authors to terminate transfers of copyright in a 5-year window beginning 35 years after a transfer. See 17 U.S.C. § 203(a). The only way that assignees can avoid this termination is by establishing that the copyright vested in their name *ab initio.* A non-author can satisfy this requirement only if the work falls within the definition of a "work made for hire." Assigned rights also differ from works made for hire in certain other respects, such as the duration of copyright protection.

7. *The Evolving Nature of Employers.* The precise language of the work made for hire doctrine is becoming increasingly important in a world of "network organizations." With the proliferation of consultants, "outsourcing," "downsizing," and "strategic partnering," businesses are placing much more emphasis on quasi-integration through contracts. This in turn makes for some interesting problems in view of the fact that copyright law was designed not for a complex, "organizationally promiscuous" world but a simple binary (big company vs. solo artist/author) world. How should this shift affect the interpretation of the work made for hire doctrine?

b. Joint Works

Aalmuhammed v. Lee
United States Court of Appeals for the Ninth Circuit
202 F.3d 1227 (9th Cir. 2000)

KLEINFELD, Circuit Judge:

This is a copyright case involving a claim of coauthorship of the movie *Malcolm X.* We reject the "joint work" claim but remand for further proceedings on a quantum meruit claim.

I. Facts

In 1991, Warner Brothers contracted with Spike Lee and his production companies to make the movie *Malcolm X,* to be based on the book, *The Autobiography of Malcolm X.* Lee co-wrote the screenplay, directed, and co-produced the movie, which starred Denzel Washington as Malcolm X. Washington asked Jefri Aalmuhammed to assist him in his preparation for the starring role because Aalmuhammed knew a great deal about Malcolm X and Islam. Aalmuhammed, a devout Muslim, was particularly knowledgeable about the life of Malcolm X, having previously written, directed, and produced a documentary film about Malcolm X.

Aalmuhammed joined Washington on the movie set. The movie was filmed in the New York metropolitan area and Egypt. Aalmuhammed presented evidence that his involvement in making the movie was very extensive. He reviewed the shooting script for Spike Lee and Denzel Washington and suggested extensive script revisions. Some of his script revisions were included in the released version of the film; others were filmed but not included in the released version. Most of the revisions Aalmuhammed made were to ensure the religious and historical accuracy and authenticity of scenes depicting Malcolm X's religious conversion and pilgrimage to Mecca.

Aalmuhammed submitted evidence that he directed Denzel Washington and other actors while on the set, created at least two entire scenes with new characters, translated Arabic into English for subtitles, supplied his own voice for voice-overs, selected the proper prayers and religious practices for the characters, and edited parts of the movie during post production. Washington testified in his deposition that Aalmuhammed's contribution to the movie was "great" because he "helped to rewrite, to make more authentic." Once production ended, Aalmuhammed met with numerous Islamic organizations to persuade them that the movie was an accurate depiction of Malcolm X's life.

Aalmuhammed never had a written contract with Warner Brothers, Lee, or Lee's production companies, but he expected Lee to compensate him for his work. He did not intend to work and bear his expenses in New York and Egypt gratuitously. Aalmuhammed ultimately received a check for $25,000 from Lee, which he cashed, and a check for $100,000 from Washington, which he did not cash.

During the summer before *Malcolm X's* November 1992 release, Aalmuhammed asked for a writing credit as a co-writer of the film, but was turned down. When the film was released, it credited Aalmuhammed only as an "Islamic Technical Consultant," far down the list. In November 1995, Aalmuhammed applied for a copyright with the U.S. Copyright Office, claiming he was a co-creator, co-writer, and co-director of the movie. The Copyright Office issued him a "Certificate of Registration," but advised him in a letter that his "claims conflict with previous registrations" of the film.

On November 17, 1995, Aalmuhammed filed a complaint against Spike Lee, his production companies, and Warner Brothers, (collectively "Lee") . . . The suit sought declaratory relief and an accounting under the Copyright Act. . . .

II. Analysis

A. Copyright Claim

Aalmuhammed claimed that the movie *Malcolm X* was a "joint work" of which he was an author, thus making him a co-owner of the copyright. . . .

Aalmuhammed argues that he established a genuine issue of fact as to whether he was an author of a "joint work," *Malcolm X*. The Copyright Act does not define "author," but it does define "joint work":

> A "joint work" is a work prepared by two or more authors with the intention that their contributions be merged into inseparable or interdependent parts of a unitary whole.

"When interpreting a statute, we look first to the language." The statutory language establishes that for a work to be a "joint work" there must be (1) a copyrightable work, (2) two or more "authors," and (3) the authors must intend their contributions be merged into inseparable or interdependent parts of a unitary whole. A "joint work" in this circuit "requires each author to make an independently copyrightable contribution" to the disputed work. *Malcolm X* is a copyrightable work, and it is undisputed that the movie was intended by everyone involved with it to be a unitary whole. It is also undisputed that Aalmuhammed made substantial and valuable contributions to the movie, including technical help, such as speaking Arabic to the persons in charge of the mosque in Egypt, scholarly and creative help, such as teaching the actors how to pray properly as Muslims, and script changes to add verisimilitude to the religious aspects of the movie. Speaking Arabic to persons in charge of the mosque, however, does not result in a copyrightable contribution to the motion picture. Coaching of actors, to be copyrightable, must be turned into an expression in a form subject to copyright. The same may be said for many of Aalmuhammed's other activities. Aalmuhammed has, however, submitted evidence that he rewrote several specific passages of dialogue that appeared in *Malcolm X,* and that he wrote scenes relating to Malcolm X's Hajj pilgrimage that were enacted in the movie. If Aalmuhammed's evidence is accepted, as it must be on summary judgment, these items would have been independently copyrightable. Aalmuhammed, therefore, has presented a genuine issue of fact as to whether he made a copyrightable contribution. All persons involved intended that Aalmuhammed's contributions would be merged into interdependent parts of the movie as a unitary whole. Aalmuhammed maintains that he has shown a genuine issue of fact for each element of a "joint work."

But there is another element to a "joint work." A "joint work" includes "two or more authors." Aalmuhammed established that he contributed substantially to the film, but not that he was one of its "authors." We hold that authorship is required under the statutory definition of a joint work, and that authorship is not the same thing as making a valuable and copyrightable contribution. We recognize that a contributor of an expression may be deemed to be the "author" of that expression for purposes of determining whether it is independently copyrightable. The issue we deal with is a different and larger one: is the contributor an author of the joint work within the meaning of 17 U.S.C. § 101.

By statutory definition, a "joint work" requires "two or more authors." The word "author" is taken from the traditional activity of one person sitting at a desk with a pen and writing something for publication. It is relatively easy to apply the word "author" to a novel. It is also easy to apply the word to two people who work together in a fairly traditional pen-and-ink way, like, perhaps, Gilbert and Sullivan. In the song, "I Am the Very Model of a Modern Major General," Gilbert's words and Sullivan's tune are inseparable, and anyone who has heard the song knows that it owes its existence to both men, Sir William Gilbert and Sir Arthur Sullivan, as its creative originator. But as the number of contributors grows and the work itself becomes less the product of one or two individuals who create it without much help, the word is harder to apply.

Who, in the absence of contract, can be considered an author of a movie? The word is traditionally used to mean the originator or the person who causes something to come into being, or even the first cause, as when Chaucer refers to the "Author of Nature." For a movie, that might be the producer who raises the money. Eisenstein thought the author of a movie was the editor. The "auteur" theory suggests that it might be the director, at least if the director is able to impose his artistic judgments on the film. Traditionally, by analogy to books, the author was regarded as the person who writes the screenplay, but often a movie reflects the work of many screenwriters. Grenier suggests that the person with creative control tends to be the person in whose name the money is raised, perhaps a star, perhaps the director, perhaps the producer, with control gravitating to the star as the financial investment in scenes already shot grows. Where the visual aspect of the movie is especially important, the chief cinematographer might be regarded as the author. And for, say, a Disney animated movie like *The Jungle Book,* it might perhaps be the animators and the composers of the music.

The Supreme Court dealt with the problem of defining "author" in new media in Burrow-Giles Lithographic Co. v. Sarony. The question there was, who is the author of a photograph: the person who sets it up and snaps the shutter, or the person who makes the lithograph from it. Oscar Wilde, the person whose picture was at issue, doubtless offered some creative advice as well. The Court decided that the photographer was the author, quoting various English authorities: "the person who has superintended the arrangement, who has actually formed the picture by putting the persons in position, and arranging the place where the people are to be — the man who is the effective cause of that"; "'author' involves originating, making, producing, as the inventive or master mind, the thing which is to be protected"; "the man who really represents, creates, or gives effect to the idea, fancy, or imagination." The Court said that an "author," in the sense that the Founding Fathers used the term in the Constitution, was "'he to whom anything owes its origin; originator; maker; one who completes a work of science or literature.'"

Answering a different question, what is a copyrightable "work," as opposed to who is the "author," the Supreme Court held in *Feist Publications* that "some minimal level of creativity" or "originality" suffices. But that measure of a "work" would be too broad and indeterminate to be useful if applied to determine who are "authors" of a movie. So many people might qualify as an "author" if the question were limited to whether they made a substantial creative contribution that that test would not distinguish one from another. Everyone from the producer and director to casting director, costumer, hairstylist, and "best boy" gets listed in the movie credits because all of their creative contributions really do matter. It is striking in *Malcolm X* how much the person who controlled the hue of the lighting contributed, yet no one would use the word "author" to denote that individual's relationship to the movie. A creative contribution does not suffice to establish authorship of the movie.

Burrow-Giles, in defining "author," requires more than a minimal creative or original contribution to the work. *Burrow-Giles* is still good law, and was recently reaffirmed in *Feist Publications. Burrow-Giles* and *Feist Publications* answer two distinct questions; who is an author, and what is a copyrightable work. *Burrow-Giles* defines author as the person to whom the work owes its origin and who superintended the whole work, the "master mind." In a movie this definition, in the absence of a contract to the contrary, would generally limit authorship to someone at the top of the screen credits, sometimes the producer, sometimes the director, possibly the star, or the screenwriter — someone who has artistic control. After all, in *Burrow-Giles* the lithographer made a substantial copyrightable creative contribution, and so did the person who posed, Oscar Wilde, but the Court held that the photographer was the author.

The Second and Seventh Circuits have likewise concluded that contribution of independently copyrightable material to a work intended to be an inseparable whole will not suffice to establish authorship of a joint work.[24] Although the Second and Seventh Circuits do not base their decisions on the word "authors" in the statute, the practical results they reach are consistent with ours. These circuits have held that a person claiming to be an author of a joint work must prove that both parties intended each other to be joint authors. In determining whether the parties have the intent to be joint authors, the Second Circuit looks at who has decision making authority, how the parties bill themselves, and other evidence.

In Thomson v. Larson, an off-Broadway playwright had created a modern version of *La Boheme,* and had been adamant throughout its creation on being the sole author. He hired a drama professor for "dramaturgical assistance and research," agreeing to credit her as "dramaturg" but not author, but saying nothing about "joint work" or copyright. The playwright tragically died immediately after the final dress rehearsal, just before his play became the tremendous Broadway hit, *Rent.* The dramaturg then sued his estate for a declaratory judgment that she was an author of *Rent* as a "joint work," and for an accounting. The Second Circuit noted that the dramaturg had no decision making authority, had neither sought nor was billed as a coauthor, and that the defendant entered into contracts as the sole author. On this reasoning, the Second Circuit held that there was no intent to be joint authors by the putative parties and therefore it was not a joint work.

Considering *Burrow-Giles,* the recent cases on joint works (especially the thoughtful opinion in Thomson v. Larson), and the Gilbert and Sullivan example, several factors suggest themselves as among the criteria for joint authorship, in the absence of contract. First, an author "superintend[s]" the work by exercising control. This will likely be a person "who has actually formed the picture by putting the persons in position, and arranging the place where the people are to be — the man who is the effective cause of that," or "the inventive or master mind" who "creates, or gives effect to the idea." Second, putative coauthors make objective manifestations of a shared intent to be coauthors, as by denoting the authorship of *The Pirates of*

24. Thomson v. Larson, 147 F.3d 195, (2d Cir. 1998); Erickson v. Trinity Theatre, Inc., 13 F.3d 1061 (7th Cir. 1994); Childress v. Taylor, 945 F.2d 500 (2d Cir. 1991).

Penzance as "Gilbert and Sullivan." We say objective manifestations because, were the mutual intent to be determined by subjective intent, it could become an instrument of fraud, were one coauthor to hide from the other an intention to take sole credit for the work. Third, the audience appeal of the work turns on both contributions and "the share of each in its success cannot be appraised." Control in many cases will be the most important factor.

The best objective manifestation of a shared intent, of course, is a contract saying that the parties intend to be or not to be coauthors. In the absence of a contract, the inquiry must of necessity focus on the facts. The factors articulated in this decision and the Second and Seventh Circuit decisions cannot be reduced to a rigid formula, because the creative relationships to which they apply vary too much. Different people do creative work together in different ways, and even among the same people working together the relationship may change over time as the work proceeds.

Aalmuhammed did not at any time have superintendence of the work. Warner Brothers and Spike Lee controlled it. Aalmuhammed was not the person "who has actually formed the picture by putting the persons in position, and arranging the place. . . ." Spike Lee was, so far as we can tell from the record. Aalmuhammed, like Larson's dramaturg, could make extremely helpful recommendations, but Spike Lee was not bound to accept any of them, and the work would not benefit in the slightest unless Spike Lee chose to accept them. Aalmuhammed lacked control over the work, and absence of control is strong evidence of the absence of coauthorship. Also, neither Aalmuhammed, nor Spike Lee, nor Warner Brothers, made any objective manifestations of an intent to be coauthors. Warner Brothers required Spike Lee to sign a "work for hire" agreement, so that even Lee would not be a coauthor and co-owner with Warner Brothers. It would be illogical to conclude that Warner Brothers, while not wanting to permit Lee to own the copyright, intended to share ownership with individuals like Aalmuhammed who worked under Lee's control, especially ones who at the time had made known no claim to the role of coauthor. No one, including Aalmuhammed, made any indication to anyone prior to litigation that Aalmuhammed was intended to be a coauthor and co-owner.

Aalmuhammed offered no evidence that he was the "inventive or master mind" of the movie. He was the author of another less widely known documentary about Malcolm X, but was not the master of this one. What Aalmuhammed's evidence showed, and all it showed, was that, subject to Spike Lee's authority to accept them, he made very valuable contributions to the movie. That is not enough for coauthorship of a joint work. . . .

COMMENTS AND QUESTIONS

1. *Statutory Interpretation of Joint Authorship.* The legislative history of the joint authorship definition comments that:

> a work is "joint" if the authors collaborated with each other, or if each of the authors prepared his or her contribution with the knowledge and intention that it would be merged with the contributions of other authors as "inseparable or interdependent parts of a unitary whole." The touchstone here is the intention, at the time the writing is done, that the parts be absorbed or combined into an integrated unit. . . .

House Report at 120. Aalmuhammed would appear to have a strong claim under both the "collaboration" and the "parties' intent" criteria.[20] The Ninth Circuit found that he had made substantial and "very valuable contributions to the movie" and that all persons involved in the project intended that his contributions would be merged into the "unitary whole." Shouldn't that be enough? From where does the Ninth Circuit get the requirement of "superintending" the work?

2. *Originality and Joint Authorship.* As we saw in *Feist*, copyright law applies a low threshold for meeting the requirement of originality. Courts generally do not judge the "quality" of a contribution for purposes of determining whether the originality hurdle has been cleared. In the words of Justice Holmes, "It would be a dangerous undertaking for persons trained only to the law to constitute themselves final judges of the work of pictorial illustrations, outside the narrowest and most obvious limits." Bleistein v. Donaldson Lithographing Co., 188 U.S. 239, 251 (1903). By contrast, the courts have developed a much higher threshold and more searching inquiry (subjective intent of the putative authors) for determining joint authorship. Does this make sense? See Mary K. LaFrance, Authorship, Dominance, and the Captive Collaborator: Preserving the Rights of Joint Authors, 50 Emory L.J. 193 (2001). Can the complex collaborative projects made possible in today's increasingly interconnected world be realistically sorted out in this manner? See generally Rochelle Cooper Dreyfuss, Collaborative Research: Conflicts on Authorship, Ownership, and Accountability, 53 Vand. L. Rev. 1161 (2000).

3. *Requirement of Independently Copyrightable Expression.* The Ninth Circuit, as well as others, see, e.g., Childress v. Taylor, 945 F.2d 500, 506-07 (2d Cir. 1991), requires that each joint author must "make an independently copyrightable contribution" to a joint work. Writing for the Seventh Circuit, Judge Posner takes a less rigid approach:

> [W]here two or more people set out to create a character jointly in such mixed media as comic books and motion pictures and succeed in creating a copyrightable character, it would be paradoxical if though the result of their joint labors had more than enough originality and creativity to be copyrightable, no one could claim copyright. That would be peeling the onion until it disappeared. The decisions that say, rightly in the generality of cases, that each contributor to a joint work must make a contribution that if it stood alone would be copyrightable weren't thinking of the case in which it couldn't stand alone because of the nature of the particular creative process that had produced it.
>
> Here is a typical case from academe. One professor has brilliant ideas but can't write; another is an excellent writer, but his ideas are commonplace. So they collaborate on an academic article, one contributing the ideas, which are not copyrightable, and the other the prose envelope, and . . . they sign as coauthors. Their intent to be the joint owners of the copyright in the article would be plain, and that should be enough to constitute them joint authors within the meaning of 17 U.S.C. §201(a). This is the valid core of the Nimmers' heretical suggestion that "if authors A and B work in collaboration, but A's contribution is limited to plot ideas that standing alone would not be copyrightable, and

---

20. Judge Newman, in Childress v. Taylor, 945 F.2d 500, 505-06 (2d Cir. 1991), downplays the separateness of the two criteria in noting that "it is hard to imagine activity that would constitute meaningful 'collaboration' unaccompanied by the requisite intent on the part of both participants that their contributions be merged into a unitary whole, and the case law has read the statutory language literally so that the intent requirement applies to all works of joint authorship."

B weaves the ideas into a completed literary expression, it would seem that A and B are joint authors of the resulting work." 1 Nimmer & Nimmer §6.07.

See Gaiman v. McFarlane, 360 F.3d 644, 658-59 (7th Cir. 2004).

4. *Co-Ownership of Jointly Authored Works: Tenancy-in-Common.* The Copyright Act treats joint authors as tenants-in-common, which means that in the absence of an agreement to the contrary, each of the co-owners has an equal, undivided ownership interest in the entire work,[21] even where it is clear that their respective contributions to the joint work are not equal. Community for Creative Non-Violence v. Reid, 846 F.2d 1485, 1498 (D.C. Cir. 1988), *aff'd*, 490 U.S. 730 (1989). Each owner can use the work and license others to do so,[22] subject to a duty to account to the others for profits. Had Aalmuhammed been deemed a joint author of the film Malcolm X, he would have been entitled to half of all profits from the film. He also would have had freedom to license the work to another film distribution studio, theaters, television broadcasters, or DVD distributors, so long as he provided Warner Brothers with half of the profits. Might this help explain why the court came out the way it did?

5. *Implied License?* Notwithstanding the "work made for hire" device for working around the joint authorship problem in films and other collaborative works, there is still a risk that a contributor to a project falls outside of the "work made for hire" designation. As in the *Aalmuhammed* case, studio lawyers may fail to get all contributors to a project to sign a "work made for hire" agreement. Although such a disaster was avoided in the *Aalmuhammed* case through the application of a relatively high joint authorship hurdle, what should a court do in a case where a relatively modest contributor to a major production who has not signed "work made for hire" agreement clears the joint author-ship hurdle? Cf. F. Jay Dougherty, Not a Spike Lee Joint? Issues in the Authorship of Motion Pictures Under U.S. Copyright Law, 49 UCLA L. Rev. 225, 317-33 (2001) (arguing that courts should find an implied license enabling the film producer to exploit the work without liability for damages absent a showing by the author that the use exceeded that which was reasonably foreseeable).

c. Collective Works

Section 201(c) provides:

> Copyright in each separate contribution to a collective work is distinct from copyright in the collective work as a whole, and vests initially in the author of the contribution. In the absence of an express transfer of the copyright or of any rights under it, the owner of copyright in the collective work is presumed to have acquired only the privilege of reproducing and distributing the contribution as part of that particular collective work, any revision of that collective work, and a later collective work in the same series.

Section 101 defines a "collective work" as "a work, such as a periodical issue, anthology, or encyclopedia, in which a number of contributions, constituting separate and independent works in themselves, are assembled into a collective whole."

---

21. Upon the death of each joint owner, the decedent's respective share of the joint work goes to his heirs. Joint authors can, however, agree to a right of survivorship or some other ownership arrangement.

22. All joint owners must agree in writing to transfer of the copyright or an exclusive license. 17 U.S.C. §204(a).

Note on the Rights of Authors and Publishers in Electronic Compilations

Section 201(c) of the Copyright Act provides that when a copyrighted work is contributed to a collective work, the copyright in the collective work (held by its publisher) is separate from the copyright in the component works (held in the first instance by the authors). The statute provides that unless the parties agree otherwise, "the owner of the copyright in the collective work is presumed to have acquired only the privilege of reproducing and distributing the contribution as part of that particular collective work, any revision of that collective work, and any later collective work in the same series." 17 U.S.C. § 201(c).

What happens when a collective work such as a magazine or law review is made available on the Internet through a searchable database? At least until recently, most publication contracts did not address this possibility. Nonetheless, publishers have treated the electronic version of their publication as if it were the same as the print version. That is, they have acted as though they had the rights to authorize the online use or reproduction of articles from their magazine. Digital dissemination of previously licensed works has resulted in the principal area of dispute.

In New York Times Co., Inc. v. Tasini, 533 U.S. 483 (2001), freelance authors of articles previously published in the New York Times, Newsday, and Sports Illustrated sued to enjoin these publications (and LEXIS/NEXIS) from distributing their articles through online databases and CD-ROM products.[23] The litigation focused on a relatively technical issue of statutory interpretation: whether distribution of the freelancers' articles falls within section 201(c) of the Act, which affords publishers of collective works a limited privilege to publish any *revision* of that collective work and any later collective work in the same series without the permission of the authors. The Supreme Court found for the freelance authors, holding that removing articles from their context as part of a collective work and placing them into searchable databases as separate files went beyond the section 201(c) privilege. See also Faulkner v. National Geographic Society, 409 F.3d 26 (2d Cir. 2005). What advice would you offer the New York Times and National Geographic about how they could continue to offer the works of freelance contributors through online databases and CD-ROM products without obtaining consent? What advice would you provide about how they could avoid such problems in the future? Cf. Amy Terry, Note, *Tasini* Aftermath: The Consequence of the Freelancers' Victory, 14 DePaul-LCA J. Art & Ent. L. 231 (2004) (arguing that publishers have responded in various ways to undermine the freelancers' victory in *Tasini*). What are the larger policy implications with regard to incentives to create and equity to authors posed by digital technology and online distribution channels? Does new technology favor publishers or authors, or does it improve the lots of both groups?

PROBLEMS

Problem 4-14. Smith, a graphic artist, is employed full time by ADCO to design and flesh out illustrations for advertising campaigns. In his spare time, at

---

23. These authors operated under oral contracts with the publishers and hence the publishers cannot be deemed to be the owners of the works under the "work made for hire" doctrine. See 17 U.S.C. § 101 (definition of "works made for hire"); discussion *infra* section D(1).

home and using no materials taken from work, Smith designs an ad campaign for another company on a freelance basis. Who owns the copyright in Smith's freelance work?

Problem 4-15. Edwards, a playwright, wrote three short plays to be produced by a community theater company. The plays were written on a tight budget, and Edwards made a number of revisions to the script during rehearsals. Some of the changes, including the reconstruction of two scenes, were made at the suggestion of actors during rehearsals, and the new scenes were worked out largely by consensus.

After a creative disagreement, the theater company performs the plays without Edwards' permission. Edwards sues for copyright infringement, and the actors claim that they are joint authors with a right to perform the work. Who should prevail?

Problem 4-16. Bable comes up with an idea for a toy car with an integrated circuit that responds to commands as well as speaking and singing songs. Bable founds a company called Up and Running, Inc. to market her "talking car" concept. Bable finds several people to record some new material for the talking car.

• The first is Sally Singer, who agrees to record a children's song she has written called "Red Light Go, Green Light Stop — Whoops!" for use on the car. Bable has Sally sign an agreement giving Up and Running "all ownership in the song." The song becomes a hit, and Bable licenses the song to Warner Kids Records for inclusion on an album of children's songs. Sally protests, saying she had planned to release her own album featuring the song.

• The second is Telly Talker, a multilingual kids' storyteller who enters into a "long-term requirements employment" contract with Up and Running. Telly's job is to record translations of the songs and slogans that the toy car says into as many languages as Bable requires. The Japanese version of the toy car becomes a big hit, and Telly informs Bable that he is planning to license his recorded voice to a third party, Toyco, for use in their Japanese talking bear product.

• The third is Gary Guitar, a musician who records guitar music for Up and Running. His practice is to record a snippet of music in whatever genre (bluegrass, jazz, etc.) Bable requests. Bable then sends a check with a standard form legend saying "cashing this check confirms your employment relationship with Up and Running, Inc., and the latter's ownership of the copyright in the music paid for hereby."

Who owns what copyrights in the car's songs and slogans?

Problem 4-17. After graduating from law school and passing the bar examination in 1994, George Jones joined the law firm of Blaketree, Hickman, and Charles (BH&C). He signed a standard associate employment contract stating that he would provide legal services on behalf of BH&C in exchange for a compensation package (salary plus bonus tied to billable hours, bar dues, health benefits, and contribution to retirement plan). George rotated among a variety of practice areas during his first two years — IP litigation, trademark prosecution, and corporate transactions. He received solid evaluations. He particularly liked the trademark group and requested to specialize in that department, which already had four partners, six associates, and four paralegals. Glenda Elston, the lead attorney in the department, viewed George favorably and he joined the department full-time early in his third year with the firm.

Six months into his third year, George began to feel that the trademark prosecution work was becoming monotonous. He increasingly recognized that many aspects of his work — validity analysis, record keeping, correspondence, etc. — could be usefully automated through the use of computer macros. In addition, he felt that the efficiency of the office could be vastly improved by creating an integrated database that all members of the trademark department (and clients) could access and use. He mentioned this idea at a department lunch and received mixed reactions. A few members of the group liked the idea and wanted to hear more about it. Glenda noted that there was a tremendous work load and that she did not want this type of project to distract George from his prosecution responsibilities. Without discussing the matter further with his colleagues, George decided to pursue the project in his spare time. He had taken some computer classes while in college and had kept up with advances in microcomputers, network computing, web-based computing, JAVA, and HTML. He liked fooling around with computers and began chipping away at this project at home. Using his own computer and software he purchased with his own funds, George developed a prototype of the program over the next year. He worked around the clock, putting in regular hours at the office and then spending evenings and weekends in his home office on his software project. George occasionally accessed the law firm's website and files in developing the program and eventually ported a version of the program onto his office machine. He also tested the program with his office files. As the program components reached the operational stage, George began using (and refining) the program at the office. His productivity increased, as did his enjoyment of work. The challenge of automating his practice brought tremendous satisfaction. After more than a year of effort, George completed a prototype for what he called "TM Prosecution Toolkit." The program stores general information on clients and their applications, provides a checklist/expert system for assessing the applicability of § 2 bars to registration, identifies necessary actions for trademark prosecution, records information on actions taken, generates draft forms and correspondence for use in prosecution, calculates prosecution deadline dates, and alerts responsible individuals to impending deadlines.

George demonstrated the program to the trademark group on February 12, 2000, at a department-wide planning retreat. By that point in time, the department had grown to almost 20 attorneys. Everyone was duly impressed. Glenda recommended that the system be implemented throughout the department immediately. BH&C touted the use of the program on its website and promotional materials. The materials referred to the program as "BH&C's innovative TM Prosecution Toolkit^{TM}." Steven Roland, the webmaster at the firm, adapted aspects of George's program so that clients and attorneys could access the firm's trademark database through a password protected portal. The firm touted this new service as its TM Fileshare^{SM}.

George received a particularly large bonus the following December. He was initially quite pleased by the reaction of the firm and colleagues, but was chagrined when Glenda informed him that BH&C planned to register the copyright and trademark in the program under the firm's name and market the programs to other law firms and trademark consulting firms. She informed George that BH&C owned all rights in the programs and that the firm intended to exploit this market edge aggressively. She invited George to serve on a committee that she was forming to guide Steve Roland and the firm's computer staff on improving the products and services. Although there were some other programs of this general type available, the firm considered George's program to be the best available — its user interface (designed by George) and networking features (originally designed by George, but later augmented by Steve Roland) went well beyond the competing programs.

George has come to you for advice. (You work at another IP law firm in town.) A prominent trademark search firm, TMs r Us, has approached George about acquiring the program. In addition, Shiply and Elrod, another law firm in town specializing in IP is interested in bringing George on board in a lucrative "Of Counsel" capacity, which would allow George to develop a business around the TM Prosecution Toolkit. (Although George still enjoys practicing law, the past two years have surfaced a latent entrepreneurial streak in him. The Shiply and Elrod opportunity provides what he considers to be an ideal mix of legal practice and entrepreneurship. In addition, his income would go up significantly.) Shiply and Elrod would, however, want George to license the program (on a non-exclusive basis) to the firm and charge relatively high fees to any other firms (including BH&C, a principal competitor) using the software.

a) Who owns copyright in "TM Prosecution Toolkit"? "TM Fileshare"?

b) Could George prevent BH&C from further use of the programs?

2. Duration and Renewal

The duration of copyright protection has evolved significantly over the past century, generally moving in the direction of a longer term of protection.

1909 Act. The 1909 Act employed a dual term of protection, granting a first term of 28 years from the date of first publication (with proper notice) that could be renewed in the final year for a second term of 28 years. Failure to renew registration of copyright in that last year resulted in the work falling into the public domain.

1976 Act. The 1976 Act moved to a unitary term of protection lasting for the life of the author plus 50 years[24] (or, in the case of corporate, anonymous, or pseudonymous entities, or work made for hire, 75 years from publication or 100 years from creation, whichever occurred first). The 1976 Act also extended the renewal term for works for 1909 Act works to 47 years. For purposes of administrative convenience, the 1976 Act provided that copyright terms shall run until the end of the calendar year in which they would otherwise expire, thereby adding an additional period of up to one year. 1992 legislation made renewal registration optional on a prospective basis. As a result, works not yet in their renewal term (i.e., those published after 1964 (1992

---

24. In the case of joint works, the term of copyright protection is measured from the death of the last surviving author. 17 U.S.C. § 302(b).

minus 28 years)) no longer risked falling into the public domain prematurely. Section 106A(d) governs the duration of certain moral rights of visual artists.

Sonny Bono Copyright Term Extension Act of 1998. With copyrights from the 1920s and 1930s set to expire, heirs of music composers (such as George and Ira Gershwin) as well as major content companies (such as the Walt Disney Corporation, which feared the loss of protection for Mickey Mouse) lobbied Congress to extend copyright protection for an additional 20 years. The fact that the European Union had added 20 years just a few years earlier worked in their favor. The legislation was passed without organized opposition. It was named in memory of Representative Sonny Bono, a successful songwriter and recording artist from the 1960s who had died earlier that year in a skiing accident and who reportedly said that copyright should last forever.

Sections 302-05 of the Act govern copyright duration. Table 4-1 summarizes the principal features.

TABLE 4-1
Duration of Copyright Protection

Date Work Created or Published	Protected from	Term of Protection
Created January 1, 1978, or thereafter	When the work is fixed in a tangible medium of expression	Life of the author + 70 years (or, if work of corporate, anonymous, or pseudonymous entity, or works made for hire, 95 years from publication, or 120 years from creation, whichever is less) (through to the end of the calendar year)
Published between beginning of 1964 and end of 1977	When published with notice	28 years for first term; automatic extension of 67 years for second term (through to the end of the calendar year)
Published between 1950 and end of 1964	When published with notice	28 years for first term; could be renewed for 67 years (through to the end of the calendar year); if not so renewed, now in public domain
Published between 1923 and end of 1949	When published with notice	28 years for first term; could be renewed for 67 years; if not so renewed, now in public domain
Created before January 1, 1978 and published between January 1, 1978 and December 31, 2002	January 1, 1978, the effective date of the 1976 Act which eliminated common law copyright protection	Later of life of the author plus 70 years (through to the end of the calendar year) or December 31, 2047
Created before January 1, 1978 and not yet published	January 1, 1978, the effective date of the 1976 Act which eliminated common law copyright protection	Life of the author plus 70 years (through to the end of the calendar year) or December 31, 2002, whichever is later
Published before 1923		In public domain

Sound recordings created prior to February 15, 1972	Depends upon treatment under applicable state law. "Any rights or remedies [for such works] under the common law or statutes of any State shall not be annulled or limited by this title until February 15, 2067." 17 U.S.C. § 301(c). See Capitol Records, Inc. v. Naxos of Am., Inc., 4 N.Y.3d 540, 797 N.Y.S.2d 352, 830 N.E.2d 250 (2005) (all pre-1972 sound recordings protected in New York until 2067).

COMMENTS AND QUESTIONS

1. *Determining the Optimal Duration of Copyright Protection.* In Article 1, Section 8, Clause 8 of the U.S. Constitution, Congress is authorized to provide limited terms of protection in order to promote progress in science and the useful arts. Does the duration of copyright protection effectuate the appropriate balance between reward to authors and enrichment of the public domain so as to best promote progress in the arts? Why is the term so much longer than for patents? Determining the appropriate balance requires consideration of a broad range of variables, each of which is difficult to assess and measure.

The optimal duration of copyright protection from a utilitarian perspective requires a balancing of the costs and benefits of lengthening protection. Benefits presumably come in the form of an enhanced incentive for authors and artists to create, while the costs imposed are the limitations on the rights of subsequent creators to make use of copyrighted works in their creative efforts and the social cost from monopoly pricing. Unfortunately, there is no good empirical data on this trade-off. Using cost and other data from publishing companies, Professor (now Justice) Stephen Breyer has questioned the need for copyright protection of books in view of the lead time advantages and the threat of retaliation in the form of competitive pricing against later market entrants. See Stephen Breyer, The Uneasy Case for Copyright: A Study in Copyright of Books, Photocopies and Computer Programs, 84 Harv. L. Rev. 281 (1970). Although Professor Breyer's conclusion that any copyright protection for books may be too long seems to overstate what his data can support, cf. Barry W. Tyerman, The Economic Rationale for Copyright Protection for Published Books: A Reply to Professor Breyer, 18 UCLA L. Rev. 1100 (1971); Stephen Breyer, Copyright: A Rejoinder, 20 UCLA L. Rev. 75 (1972), he certainly raises serious qualms about lengthening the duration of copyright protection. A second source of data relating to the appropriate duration of copyright protection is the Copyright Office's records on renewal. Over the period 1927 through 1954, only about 9.5 percent of registered works were renewed prior to the end of their first term. The renewal rate was highest for "published music" (45 percent) and "motion picture photoplays" (43.7 percent). See Guinan, Duration of Copyright, Appendices A, B, Copyr. L., Revision Study No. 30 (1957), in 1 Studies in Copyr. 473 (1963).

Consider the following argument for significantly limiting copyright protection: At a qualitative level, there would seem little basis for protecting most copyrightable works beyond 10 or 15 years. This observation is supported by the renewal data discussed above. Casebooks, for example, are rarely marketable after 5 years unless they are revised. Similarly, the public's interest in many works of literature and art tends to follow popular waves of a decade or less. Moreover, one can argue that after 25 years, the main interest in most literary works is historical. The public would be served by allowing historians the ability to draw upon such works in creating new works of history and social commentary. With regard to those relatively few works that have enduring commercial value beyond a decade or two, there is little question that such works generate substantial revenue for their authors. Therefore the public would be best served by limiting copyright protection for literary works to 25 years. Are you persuaded by this argument? What counterarguments would you offer? What philosophical basis or bases underlie your arguments? See William M. Landes & Richard A. Posner, Indefinitely Renewable Copyright, 70 U. Chi. L. Rev. 471 (2003).

2. *The Political Economy of Copyright Term Extension.* Who benefits most from the extension of the copyright term? Does extension of the copyright term pose a significant threat to the public? What reasons might explain the lack of public concern about this type of legislation? See William Patry, The Failure of the American Copyright System: Protecting the Idle Rich, 72 Notre Dame L. Rev. 907 (1997).

3. *Constitutionality of Copyright Term Extensions.* Shortly after the passage of the Sonny Bono Copyright Term Extension Act of 1998 (CTEA), various entities seeking to republish and distribute works that would otherwise have fallen into the public domain challenged the CTEA on three constitutional grounds: (1) that it violates the First Amendment by unduly restraining speech; (2) that it violates the originality requirement of the Intellectual Property Clause, Art. I, §8, cl. 8, by conferring additional protection to works that already exist; and (3) that it exceeds the "limited Times" constraint upon Congress's authority to enact copyright legislation. In a 7-2 decision, the Supreme Court upheld the CTEA as within Congress's broad discretion to prescribe "limited Times" and not at odds with the First Amendment. See Eldred v. Ashcroft, 537 U.S. 186 (2003). Emphasizing nearly two centuries of evolution of intellectual property law, the Court determined that Congress could extend protection for extant works consistent with the Intellectual Property Clause: "Congress could rationally seek to 'promote...Progress' by including in every copyright statute an express guarantee that authors would receive the benefit of any later legislative extension of the copyright term." Not surprisingly, Justice Breyer dissented, reiterating the themes of his 1970 article. Given the Court's deferential standard, are there any constitutional limitations on the duration of copyright protection or is this a matter solely for the legislative branch?

4. *Political Reform?* The CTEA and the *Eldred* case have served to rally various entities opposed to the ever-expanding nature of intellectual property protection. Various proposals have surfaced, including the reintroduction of a copyright maintenance fee, aimed at accelerating growth of the public domain. See, e.g., Christopher Sprigman, Reform(aliz)ing Copyright, 57 Stan. L. Rev. 485 (2004); William M. Landes & Richard A. Posner, Indefinitely Renewable Copyright, 70 U. Chi. L. Rev. 471 (2003). The Public Domain Enhancement Act, H.R. 2601, 108th Cong. (2003), would require U.S. authors to renew their copyrights for the modest fee of $1 after 50 years and again at 10-year intervals until the copyright expired. Since relatively few works were renewed under the 1909 Act regime, this statute would mean that most older works would likely become available after 50 years and the registration system would make it easier to trace ownership for those works that remain protected. Should the U.S. adopt this reform? What reforms would you recommend?

PROBLEMS

Problem 4-18. Determine the duration of copyright in the following cases:
- Arnold Author completes his novel *You'll Be Mine 'Til the End of Time* on February 28, 1996. The next day, February 29th, Arnold is hit by a bus and dies instantly. On what day does his copyright expire?
- While working for the New Englander Magazine, Arnold Author writes a story entitled "You'll Always Be Mine." The story is finally published by The New Englander in 2010. When does the copyright expire?
- Arnold Author began work on his greatest novel, *Time Is on My Side,* in 1990. In 1991, he completes the first three chapters. In 1992, he writes the middle three chapters. In 1993, he completes the final three chapters. In 1995, he signs an agreement with Time/Life Books to publish the novel. The contract assigns all copyright interests to Time/Life Books in exchange for 20 percent royalties based on the wholesale price. The book is finally published on January 1, 1996. Arnold dies on February 28, 1996. On what day does the copyright expire? Does your answer change if you assume that Arnold entered into a contract with Time/Life Books before writing the novel?
- At the time of his death on February 29, 1996, Arnold Author has completed three-fourths of his novel *Time Lives On.* His will leaves all his property to his spouse, Angela Author, who is also a writer. She plans to complete the novel by 1999. Time/Life Books has agreed to publish the completed manuscript in the year 2000. Assuming that all goes according to plan, when will copyright in *Time Lives On* expire?

Problem 4-19. What is the duration of copyright in the following cases?
- Loretta Wrighter composed and sent a letter to her friend, Emily Johnson, in 1961. Emily has saved the letter in her correspondence file since that time. Loretta died in 1970. What is the term of protection for this work?
- Stephen Morris published his first novel, *Child's Play,* at the age of 15 in 1920. He is still alive today. What is the term of protection for this work?
- Anita Author published (with proper notice) her novel entitled *The Winds of Change* in 1970. She died three years later. What is the term of protection for this work?
- Penelope Painter painted her masterpiece entitled "Garden of Wildflowers" in 1953. She distributed copies with notice of copyright in 1955. She renewed the copyright in 1982. She died the next year. When does the copyright expire?

3. Division, Transfer, and Reclaiming of Copyrights

The preceding section has explored one important difference between copyright interests and the traditional fee simple absolute in real property law: the limited duration of protection. Copyrights, therefore, can be analogized to a hybrid of a term of years and life estate in that the owner has control of the rights of copyright for a defined and limited period of time (life plus 70 years), after which such rights fall to the public at large.

Another important aspect of copyright ownership is the distinction between ownership of the material object on which the work of authorship is fixed — the book manuscript or oil canvas — and ownership of the copyright interests themselves. Section 202 states:

> Ownership of a copyright, or of any of the exclusive rights under a copyright, is distinct from ownership of any material object in which the work is embodied. Transfer of ownership of any material object, including the copy or phonorecord in which the work is first fixed, does not of itself convey any rights in the copyrighted work embodied in the object; nor, in the absence of an agreement, does transfer of ownership of a copyright or of any exclusive rights under a copyright convey property rights in any material object.

Thus the author of a letter retains her copyright interests in the writing even though she sends the letter to the addressee. The addressee thereby obtains ownership of the material object but may not infringe the copyright interests of the author. The addressee may view the material object and he may show it to others, but he may not make copies, prepare derivative works, distribute the work, or perform or display the work publicly.

Two other elements of real property interests are the rights of property owners and the alienability of such rights. Part E (Rights of Copyright Holders) and Part F (Defenses, including fair use) explore the extent of and limitations on copyrights. This section discusses the division, transfer, and termination of transfer rights of copyright holders. In the domain of real property, the owner of a fee interest may freely divide and alienate the various rights of property ownership. For example, a property owner may divide her lot and sell a portion to another person. Alternatively, she may sell one particular right within the bundle of rights, such as the right to use a path running across the property (an easement). Moreover, such transfers are generally not terminable unless so specified in the transfer agreement. Thus a property owner who creates an easement across her land may not unilaterally terminate such right at a later time.

By contrast, copyright law restricts the alienability of the rights of copyright owners in certain ways. As you study these materials, scrutinize the reasons for restricting the alienability of copyright interests.

a. Division and Transfer of Copyright Interests Under the 1909 Act

Courts interpreted the 1909 Act to preclude the formal divisibility of the rights comprising a copyright. A copyright owner could assign the entire copyright to another, but a transfer of any lesser interest was considered a license. This doctrine of indivisibility simplified the notice requirement. As noted earlier, failure to provide proper notice could result in forfeiture of copyright protection. The "owner" of the copyright (or "proprietor"), whether the author or the assignee of the entire copyright, was the appropriate name to be included in the copyright notice. One consequence of the doctrine of indivisibility was that only proprietors had standing to bring suit to enforce the copyright. Thus a licensee would have to join the proprietor in order to protect his or her rights. In practice, however, courts' decisions interpreting the 1909 Act limited the effect of the indivisibility rule.

b. Division and Transfer of Copyright Interests Under the 1976 Act

The 1976 Act eliminated restrictions on the formal divisibility of copyright interests. Section 201(d) provides:

(1) The ownership of a copyright may be transferred in whole or in part by any means of conveyance or by operation of law, and may be bequeathed by will or pass as personal property by the applicable laws of intestate succession.

(2) Any of the exclusive rights comprised in a copyright, including any subdivision of any of the rights specified by section 106, may be transferred as provided by clause (1) and owned separately. The owner of any particular exclusive right is entitled, to the extent of that right, to all of the protection and remedies accorded to the copyright owner by this title.

Section 101 defines "transfer of copyright ownership" to include an assignment or an *exclusive* license of any of the exclusive rights comprised in a copyright. The divisibility of copyright ownership enables any owner of an exclusive right to bring an infringement suit without having to join the copyright proprietor.

Section 204 requires transfers of copyright ownership to be executed in writing and signed by the copyright owner. Section 205 provides for the voluntary recordation of transfers with the Copyright Office. It also sets forth priority rules for resolving cases of conflicting transfers.

c. Reclaiming Copyrights

Under the 1909 Act, authors could reclaim copyright interests that they had licensed at the time of renewal. This opportunity to reclaim copyright upon renewal proved unavailing in most commercial circumstances as book and music publishers routinely insisted upon assignment of the initial copyright term and the advance assignment of the renewal term in exchange for publishing a work. The Supreme Court long ago upheld the enforceability of such contingent (upon the author's survival until renewal) advance assignment clauses. See Fred Fisher Music Co. v. M. Witmark & Sons, 318 U.S. 643 (1943).

While it eliminated the renewal regime, the 1976 Act created a much more potent means for authors and their survivors to reclaim copyright interests at a later time: an inalienable right to terminate transfers of copyright between the thirty-fifth and fortieth year from the execution of the transfer of rights for works created after 1977. 17 U.S.C. § 203(a)(3). With respect to works in their second renewal term prior to 1978, § 304(c) allows authors and their families to terminate transfers between the fifty-sixth and sixty-first year of copyright protection for such works so as to allow them to profit from the 19 years of protection for such works added by the 1976 Act. A comparable right was bestowed upon authors in the Sonny Bono Copyright Term Extension Act of 1998, ensuring that they may reclaim the 20 years added to their copyrights. Termination of transfer rights may not be assigned in advance. Congress enacted these provisions to better ensure that authors and their families are able to reap a fair portion of the benefits of the author's creative efforts. Congress was concerned that authors had "unequal bargaining power" in negotiating rights with publishers and marketers "resulting in part from the impossibility of determining a work's value until it has been exploited." H.R. Rep. No. 1476, 94th Cong., 2d Sess. 124 (1976).

As noted in the discussion of the "works made for hire" doctrine, the only way for a transferee to prevent a termination of transfer is by establishing that the work was "made for hire" and therefore owned by the employer or commissioning party *ab initio* (from the outset). With regard to commissioned works (works prepared by independent contractors as opposed to "employees"), Congress limited this exception to the termination of transfer provision in two ways: by allowing only certain enumerated categories of works to be treated as works made for hire and by requiring that the parties specifically agree in writing that the work shall be treated as a "work made for hire." The film, magazine, newspaper, and textbook industries foresaw that the termination of transfer provision could seriously disrupt their operations and persuaded Congress to include "motion picture[s]" and "contribution[s] to collective works" among the enumerated categories.

COMMENTS AND QUESTIONS

1. *Renewal and Derivative Works: Abend Rights.* Cornell Woolrich's story, "It Had to Be Murder," was published by Dime Detective Magazine in 1942. Woolrich retained all other rights in the story and in 1945 assigned the motion picture rights — for both the initial term of copyright and the contingent renewal term — to DeSylva Productions. In 1953, DeSylva assigned the story rights to Jimmy Stewart and Alfred Hitchcock, who together produced the film *Rear Window*, in 1954. Just prior to the renewal period for the story, Woolrich died, leaving the property in trust. The trustee renewed the copyright and assigned the renewal term to Sheldon Abend, an enterprising author's representative, in exchange for $650 plus 10 percent of any proceeds from exploitation of the story. Abend then sued Stewart, Hitchcock, and the distributor of *Rear Window*, alleging that further exploitation of the film without his consent infringed his copyright in the underlying story. The Supreme Court agreed, resting its decision on two critical interpretations of the 1909 Act: (1) when an author dies before the renewal period vests, the renewal right passes to the author's statutory successors and any advance assignments of rights in the renewal term go "unfulfilled"; and (2) that during the renewal term, continued exploitation of derivative works made with permission of the owners of underlying works during the original term of copyright nonetheless requires continued authorization to utilize such underlying copy-

righted elements during the renewal term. See Stewart v. Abend, 495 U.S. 207 (1990). Shortly thereafter, the parties settled the dispute, enabling *Rear Window* to be exploited and Abend to license a remake for worldwide distribution. Note that Abend would not have been able to exploit a remake outside the United States without consent of the owners of the copyright in Hitchcock's version of the film. In this sense, the Supreme Court's decision created a blocking right structure.

When Congress amended the Copyright Act in 1992 to provide for automatic renewal, it held out the carrot of *Abend* rights as an incentive for renewal registration. Failure to file a renewal registration forfeited any claim to so-called *Abend* rights.

2. *Termination of Transfers and Derivative Works.* When Congress added 19 years to the term or pre-1978 works in the 1976 Act and an additional 20 years in the CTEA in 1998, it expressly authorized continued exploitation of derivative works. See 17 U.S.C. §§ 203(b)(1); 304(c)(6)(A).

3. *Sound Recordings and Works Made for Hire.* When the "work made for hire" provision of the 1976 Act was being hammered out in the early to mid 1960s, sound recordings were not yet a part of the Copyright Act. Record industry representatives were understandably focused on getting federal protection for their works. Furthermore, at that time many record labels considered recording artists to be employees. Therefore, there was not much attention focused on getting sound recordings included within the "work made for hire" exception to termination of transfer rule. Over the next two decades, the record industry underwent substantial changes. Recording artists became more independent, making the "employee" classification dubious.

Fearing that some of its prize assets might be vulnerable to notice of termination beginning in 2003 (Congress allowed for a 10-year window for notifying transferees) and actual termination in 2013 of works transferred in 1978, the sound recording industry surreptitiously persuaded Congress to insert "sound recordings" into the Act through a "technical amendment" buried within the Satellite Home Viewer Improvement Act of 1999. After this legislation was signed into law, recording artists protested the backroom deal. Soon thereafter, Congress held hearings at which the Register of Copyrights acknowledged that the change was more than a "technical amendment." The industry defended its actions on the ground that record albums constitute "collective works" and hence sound recordings contained therein are owned by record companies as "contributions to collective works," one of the enumerated categories. Do you agree with this reading of the Copyright Act?

In an effort to wash the slate clean, the recording industry and recording artist representatives drafted a compromise bill repealing the 1999 provision and restoring the *status quo ante.* See Work Made for Hire and Copyright Corrections Act of 2000, Pub. L. No. 106-379, sec. 1, 114 Stat. 1444 (Oct. 27, 2000); David Nimmer and Peter S. Menell, Sound Recordings, Works for Hire, and the Termination-of-Transfers Time Bomb, 49 J. Copyright Soc'y 387 (2001). Whether sound recordings should be deemed "works made for hire" awaits judicial resolution[25] or further legislation. How should sound recordings be handled? Will the exploitation of sound recordings become a legal morass due to a multiplicity of claimants? Who is eligible to terminate in the case of a typical popular record — record producer, arranger, featured artists, band members, background musicians, recording engineer,

---

25. Scattered district court rulings have reached the unremarkable conclusion that a work cannot qualify as commissioned if the basis is that it is a sound recording. See Ballas v. Tedesco, 41 F. Supp. 2d 531, 541 (D.N.J. 1999); Staggers v. Real Authentic Sound, 77 F. Supp. 2d 57 (D.D.C. 1999).

remixing engineer? Are they all joint authors? See generally Peter S. Menell and David Nimmer, Defusing the Termination of Transfers Time Bomb (manuscript 2006).

4. *Policy Analysis of the Termination of Transfers.* Will the termination of transfers provision alleviate the problem of "unremunerative transfers" that Congress sought to address through the creation of an inalienable termination right? Do you agree with Congress's premise that authors are at a serious bargaining disadvantage in negotiating the rights to their works? Is making the power of reverter inalienable necessary to address this concern? For a discussion of the inalienability provision from a moral rights perspective, see R. Anthony Reese, Reflections on the Intellectual Commons: Two Perspectives on Copyright Duration and Reversion, 47 Stan. L. Rev. 707 (1995).

What problems might arise as a result of the power to terminate copyright transfers? What alternative means might Congress have used to protect the interests of authors and their families short of an inalienable power of reverter? See generally Peter S. Menell and David Nimmer, Defusing the Termination of Transfers Time Bomb (manuscript 2006).

E. TRADITIONAL RIGHTS OF COPYRIGHT OWNERS

As noted in the introduction to this chapter, copyright law has for much of its history focused upon the protection of expressive works fixed in analog media—books, vinyl, tapes, and film, as well as over-the-air broadcasting. This section addresses the main bundle of rights that has grown up around these technology platforms. We will turn in section H to the new array of protections that has been put in place to address the challenges and opportunities posed by digital technology.

17 U.S.C. § 106. Exclusive Rights in Copyrighted Works

Subject to sections 107 through 118, the owner of copyright under this title has the exclusive rights to do and to authorize any of the following:

(1) to reproduce the copyrighted work in copies or phonorecords;

(2) to prepare derivative works based on the copyrighted work;

(3) to distribute copies or phonorecords of the copyrighted work to the public by sale or other transfer of ownership, or by rental, lease, or lending;

(4) in the case of literary, musical, dramatic, and choreographic works, pantomimes, and motion pictures and other audiovisual works, to perform the copyrighted work publicly; and

(5) in the case of literary, musical, dramatic, and choreographic works, pantomimes, and pictorial, graphic, or sculptural works, including the individual images of a motion picture or other audiovisual work, to display the copyrighted work publicly.

The 1976 Act states that anyone who "violates any of the exclusive rights of the copyright owner as provided by sections 106 through 121" is liable for copyright infringement. 17 U.S.C. § 501(a). Copyright infringement may occur by two distinct sets of actors: (1) those who directly infringe the rights of copyright holders; and (2) those who encourage or assist a third party to infringe.

1. The Right to Make Copies

There is a close relationship between the rights in subsections (1), (2), and (3) of section 106. The right to "reproduce" is, of course, the most fundamental of the rights granted to the copyright owner. It includes all rights to fix a work in a tangible medium of expression, including the right to make phonorecords.

The right to copy granted by section 106(1) is not limited to exact reproduction (such as "copying" a disk or making a photocopy). Rather, it is a broad grant of the right to prevent others from making exact or "substantially similar" reproductions by any means "now known or later developed, and from which the work can be perceived, reproduced, or otherwise communicated, either directly or with the aid of a machine or device," 17 U.S.C. § 101 (definition of "copies"). "Phonorecords" includes not only disks that may be played on record players — the technology that existed at the time of the 1976 Act — but all "material objects in which sounds, other than those accompanying a motion picture or other audiovisual works, are fixed by any method now known or later developed, and from which the sounds can be perceived, reproduced, or otherwise communicated, either directly or with the aid of a machine or device." 17 U.S.C. § 101.

The doctrines governing direct infringement of copyright address two distinct problems in determining copyright liability. The first relates to proving whether someone has actually copied the work of another. As you will recall, copyright law prohibits copying, but it does not prohibit independent creation. The fact that two composers claim similar musical compositions does not necessarily mean that one copied from the other. They may have independently composed similar songs. Ideally, a court would like to have direct proof of copying — e.g., eyewitness testimony, records indicating that one author obtained the work from another, videotape of direct copying, or distinctive flaws (such as unusual errors that are common to the two works). In many cases, however, such evidence is not available. Nonetheless, the circumstances surrounding the works may point inexorably toward a finding of copying, for example, where a work has been widely disseminated and a second author claims authorship of a nearly identical work.

The second problem that courts have confronted in assessing infringement arises even when the defendant acknowledges having developed his or her work with knowledge of the plaintiff's work. In these cases, the question is whether the second comer appropriated sufficient protected material to violate a copyright owner's rights. The legislative history to § 106 provides the following guidance:

> [A] copyrighted work would be infringed by reproducing it *in whole or in any substantial part*, and by duplicating it exactly or by imitation or simulation. Wide departures or variations from the copyrighted works would still be an infringement as long as the author's "expression" rather [than] merely the author's "ideas" are taken.

H.R. Rep. No. 94-1476, 94th Cong., 2d Sess. 61 (1976) (emphasis added). As suggested by this passage, the inquiry into whether the defendant has violated a plaintiff's copyright is often complicated by the fact that many copyrightable works intermingle original expression with public domain materials, ideas, facts, stock literary elements, scenes à faire, and other nonprotectable elements. Hence courts have had to develop an infringement filter that adequately protects the interests of copyright owners but at the same time does not extend the scope of copyright protection beyond its statutory limits.

As you will see, courts use the term "substantial similarity" in discussing both problems. This can be confusing. As you study these materials, pay special attention to the context in which the courts are applying this term. "Substantial similarity" has a different meaning depending on whether it is being used as an aid in determining proof of copying, on the one hand, or whether the appropriation of protectable material was improper, on the other.

a. Copying

Arnstein v. Porter
United States Court of Appeals for the Second Circuit
154 F.2d 464 (2d Cir. 1946)

FRANK, Circuit Judge.

[Ira B. Arnstein sued Cole Porter for infringement of copyrights in various of plaintiff's musical compositions. He sought a jury trial. Plaintiff alleged that the defendant's "Begin the Beguine" had been plagiarized from plaintiff's "The Lord Is My Shepherd" and "A Mother's Prayer" and that defendant's "My Heart Belongs to Daddy" had been plagiarized from "A Mother's Prayer." Plaintiff testified in deposition that both works had been published and that about 2,000 copies of "The Lord Is My Shepherd" and over a million copies of "A Mother's Prayer" had been sold. Plaintiff offered no direct proof that defendant saw or heard these compositions. Plaintiff further testified that defendant's "Night and Day" had been plagiarized from plaintiff's "I Love You Madly" and that although the latter composition had not been published, it had been performed publicly over the radio. In addition, plaintiff averred that a copy of the song had been stolen from his room. Plaintiff alleged that some other songs of the defendant had been plagiarized from the plaintiff's unpublished works. He suggested in deposition that the defendant had gained access to these songs either through publishers or a movie producer who were sent copies or through "stooges" who defendant had hired to follow, watch, and live with the plaintiff (and who may have been responsible for the ransacking of his room). When asked how he knew that defendant had anything to do with the "burglaries," plaintiff testified "I don't know that he had to do with it, but I only know that he could have." Defendant testified in depositions that he had never seen nor heard the plaintiff's compositions and that he did not have any connection to the alleged theft of such works.

The district court granted defendant's motion for summary judgment.]

... The principal question on this appeal is whether the lower court, under Rule 56, properly deprived plaintiff of a trial of his copyright infringement action. The answer depends on whether "there is the slightest doubt as to the facts." In applying that standard here, it is important to avoid confusing two separate elements essential to a plaintiff's case in such a suit: (a) that defendant copied from plaintiff's copyrighted work and (b) that the copying (assuming it to be proved) went so far as to constitute improper appropriation.

As to the first — copying — the evidence may consist (a) of defendant's admission that he copied or (b) of circumstantial evidence — usually evidence of access — from which the trier of the facts may reasonably infer copying. Of course, if there are no similarities, no amount of evidence of access will suffice to prove copying. If there is evidence of access and similarities exist, then the trier of the facts must determine whether the similarities are sufficient to prove copying. On this issue, analysis ("dissection") is relevant, and the testimony of experts may be received to aid the trier of the facts. If evidence of access is absent, the similarities must be so striking as to preclude the possibility that plaintiff and defendant independently arrived at the same result.

If copying is established, then only does there arise the second issue, that of illicit copying (unlawful appropriation). On that issue (as noted more in detail below) the test is the response of the ordinary lay hearer; accordingly, on that issue, "dissection" and expert testimony are irrelevant.

In some cases, the similarities between the plaintiff's and defendant's work are so extensive and striking as, without more, both to justify an inference of copying and to prove improper appropriation. But such double-purpose evidence is not required; that is, if copying is otherwise shown, proof of improper appropriation need not consist of similarities which, standing alone, would support an inference of copying.

Each of these two issues — copying and improper appropriation — is an issue of fact. If there is a trial, the conclusions on those issues of the trier of the facts — of the judge if he sat without a jury, or of the jury if there was a jury trial — bind this court on appeal, provided the evidence supports those findings, regardless of whether we would ourselves have reached the same conclusions. But a case could occur in which the similarities were so striking that we would reverse a finding of no access, despite weak evidence of access (or no evidence thereof other than the similarities); and similarly as to a finding of no illicit appropriation.

We turn first to the issue of copying. After listening to the compositions as played in the phonograph recordings submitted by defendant, we find similarities; but we hold that unquestionably, standing alone, they do not compel the conclusion, or permit the inference, that defendant copied. The similarities, however, are sufficient so that, if there is enough evidence of access to permit the case to go to the jury, the jury may properly infer that the similarities did not result from coincidence.

Summary judgment was, then, proper if indubitably defendant did not have access to plaintiff's compositions. Plainly that presents an issue of fact. On that issue, the district judge, who heard no oral testimony, had before him the depositions of plaintiff and defendant. The judge characterized plaintiff's story as "fantastic"; and, in the light of the references in his opinion to defendant's deposition, the judge obviously accepted defendant's denial of access and copying. Although part of plaintiff's testimony on deposition (as to "stooges" and the like) does seem "fantastic," yet plaintiff's credibility, even as to those improbabilities, should be left to the jury. If evidence is "of a kind that greatly taxes the credulity of the judge, he can say so, or, if he totally disbelieves it, he may announce that fact, leaving the jury free to believe it or not." If, said Winslow, J., "evidence is to be always disbelieved because the story told seems remarkable or impossible, then a party whose rights depend on the proof of some facts out of the usual course of events will always be denied justice simply because his story is improbable." We should not overlook the shrewd proverbial admonition that sometimes truth is stranger than fiction.

But even if we were to disregard the improbable aspects of plaintiff's story, there remain parts by no means "fantastic." On the record now before us, more than a million copies of one of his compositions were sold; copies of others were sold in smaller quantities or distributed to radio stations or band leaders or publishers, or the pieces were publicly performed. If, after hearing both parties testify, the jury disbelieves defendant's denials, it can, from such facts, reasonably infer access. It follows that, as credibility is unavoidably involved, a genuine issue of material fact presents itself. With credibility a vital factor, plaintiff is entitled to a trial where the jury can observe the witnesses while testifying. . . .

Assuming that adequate proof is made of copying, that is not enough; for there can be "permissible copying," copying which is not illicit. Whether (if he copied) defendant unlawfully appropriated presents, too, an issue of fact. The proper criterion on that issue is not an analytic or other comparison of the respective musical compositions as they appear on paper or in the judgment of trained musicians. The plaintiff's legally protected interest is not, as such, his reputation as a musician but his interest in the potential financial returns from his compositions which derive from the lay public's approbation of his efforts. The question, therefore, is whether defendant took from plaintiff's works so much of what is pleasing to the ears of lay listeners, who comprise the audience for whom such popular music is composed, that defendant wrongfully appropriated something which belongs to the plaintiff.

Surely, then, we have an issue of fact which a jury is peculiarly fitted to determine. Indeed, even if there were to be a trial before a judge, it would be desirable (although not necessary) for him to summon an advisory jury on this question.

We should not be taken as saying that a plagiarism case can never arise in which absence of similarities is so patent that a summary judgment for defendant would be correct. Thus suppose that Ravel's *Bolero* or Shostakovitch's *Fifth Symphony* were alleged to infringe "When Irish Eyes Are Smiling." But this is not such a case. For, after listening to the playing of the respective compositions, we are, at this time, unable to conclude that the likenesses are so trifling that, on the issue of misappropriation, a trial judge could legitimately direct a verdict for defendant.

At the trial, plaintiff may play, or cause to be played, the pieces in such manner that they may seem to a jury to be inexcusably alike, in terms of the way in which lay listeners of such music would be likely to react. The plaintiff may call witnesses whose testimony may aid the jury in reaching its conclusion as to the responses of such audiences. Expert testimony of musicians may also be received, but it will in no way be controlling on the issue of illicit copying, and should be utilized only to assist in determining the reactions of lay auditors. The impression made on the refined ears of musical experts or their views as to the musical excellence of plaintiff's or defendant's works are utterly immaterial on the issue of misappropriation; for the views of such persons are caviar to the general — and plaintiff's and defendant's compositions are not caviar. . . .

CLARK, Circuit Judge (dissenting).

[A]fter repeated hearings of the records, I could not find therein what my brothers found. The only thing definitely mentioned seemed to be the repetitive use of the note e, in certain places by both plaintiff and defendant, surely too simple and ordinary a device of composition to be significant. In our former musical plagiarism cases we have, naturally, relied on what seemed the total sound effect; but we have also analyzed the music enough to make sure of an intelligible and intellectual decision. Thus in Arnstein v. Edward B. Marks Music Corp., 2 Cir., 82 F.2d 275, 277, Judge L. Hand made quite an extended comparison of the songs, concluding, inter alia: "... the seven notes available do not admit of so many agreeable permutations that we need be amazed at the re-appearance of old themes, even though the identity extend through a sequence of twelve notes." See also the discussion in Marks v. Leo Feist, Inc., 2 Cir., 290 F. 959, and Darrell v. Joe Morris Music Co., 2 Cir., 113 F.2d 80, where the use of six similar bars and of an eight-note sequence frequently repeated were respectively held not to constitute infringement, and Wilkie v. Santly Bros., 2 Cir., 91 F.2d 978, *affirming* D.C.S.D.N.Y., 13 F. Supp. 136, *certiorari denied,* Santly Bros. v. Wilkie, 302 U.S. 735, 58 S. Ct. 120, 82 L. Ed. 568, where use of eight bars with other similarities amounting to over three-quarters of the significant parts was held infringement.

It is true that in Arnstein v. Broadcast Music, Inc., 2 Cir., 137 F.2d 410, 412, we considered "dissection" or "technical analysis" not the proper approach to support a finding of plagiarism, and said that it must be "more ingenuous, more like that of a spectator, who would rely upon the complex of his impressions." But in its context that seems to me clearly sound and in accord with what I have in mind. Thus one may look to the total impression to repulse the charge of plagiarism where a minute "dissection" might dredge up some points of similarity. Hence one cannot use a purely theoretical disquisition to supply a tonal resemblance which does not otherwise exist. Certainly, however, that does not suggest or compel the converse—that one must keep his brain in torpor for fear that otherwise it would make clear differences which do exist. Music is a matter of the intellect as well as the emotions; that is why eminent musical scholars insist upon the employment of the intellectual faculties for a just appreciation of music.

Consequently I do not think we should abolish the use of the intellect here even if we could. When, however, we start with an examination of the written and printed material supplied by the plaintiff in his complaint and exhibits, we find at once that he does not and cannot claim extensive copying, measure by measure, of his compositions. He therefore has resorted to a comparative analysis—the "dissection" found unpersuasive in the earlier cases—to support his claim of plagiarism of small detached portions here and there, the musical fillers between the better known parts of the melody. And plaintiff's compositions, as pointed out in the cases cited above, are of the simple and trite character where small repetitive sequences are not hard to discover. It is as though we found Shakespeare a plagiarist on the basis of his use of articles, pronouns, prepositions, and adjectives also used by others. The surprising thing, however, is to note the small amount of even this type of reproduction which plaintiff by dint of extreme dissection has been able to find.

... The usual claim seems to be rested upon a sequence of three, of four, or of five—never more than five—identical notes, usually of different rhythmical values. Nowhere is there anything approaching the twelve-note sequence of the *Marks* case ...

In the light of these utmost claims of the plaintiff, I do not see a legal basis for the claim of plagiarism. So far as I have been able to discover, no earlier case approaches the holding that a simple and trite sequence of this type, even if copying may seem indicated, constitutes proof either of access or of plagiarism. ...

COMMENTS AND QUESTIONS

1. *Infringement Analysis: Copying + Improper Appropriation.* The *Arnstein* majority articulates both elements needed to establish infringement: (a) copying and (b) improper (or unlawful) appropriation ("whether defendant took from plaintiff's works so much of what is pleasing to the ears of lay listeners, who comprise the audience for whom such popular music is composed, that defendant wrongfully appropriated something which belongs to the plaintiff"). We explore the former stage in these comments and problems and the latter stage in the next section.

2. *Circumstantial Proof of Copying: Access + Similarity.* The majority opinion in *Arnstein* articulates the leading test for establishing copying by way of circumstantial evidence. The court suggests a sliding scale: "[I]f there are no similarities, no amount of evidence of access will suffice to prove copying. If there is evidence of access and similarities exist, then the trier of the facts must determine whether the similarities are sufficient to prove copying. . . . If evidence of access is absent, the similarities must be so striking as to preclude the possibility that plaintiff and defendant independently arrived at the same result."

Courts often loosely use the term "substantial similarity" in referring to both the comparison of similarity in assessing circumstantial proof of copying and the comparison of similarity in assessing improper appropriation. For this reason, the application of the doctrine would be more coherent if courts uniformly referred to the type of similarity relevant to determining copying as a factual matter as "*probative similarity*" — i.e., tending to *prove* copying, irrespective of whether the copying infringed the plaintiff's copyright. See Johnson v. Gordon, 409 F.3d 12 (1st Cir. 2005); 4 Melville B. Nimmer & David Nimmer, Nimmer on Copyright § 13.03[A]. As reflected in *Arnstein,* the degree of similarity required to establish circumstantial proof of copying is gauged along a sliding scale and incorporates objective criteria.[26]

Courts have differed in their application of this sliding scale. In the Second Circuit, a plaintiff need not adduce any evidence of access "if a copyrighted work and an allegedly infringing work are strikingly similar." Gaste v. Kaiserman, 863 F.2d 1061 (2d Cir. 1988). Thus the trier of fact may infer from evidence of striking similarity that the defendant could not have independently created his or her work. In Repp v. Lloyd, 132 F.3d 882 (2d Cir. 1997), a songwriter accused Lord Andrew Lloyd Webber of copying one of his songs in "Phantom's Song," part of Webber's *Phantom of the Opera.* Webber submitted his own affidavit and several supporting affidavits indicating that he had never heard Repp's song and that he had written "Phantom's Song" without outside influence. In response, Repp submitted expert declarations that the two songs were so "strikingly similar" as to preclude the possibility of independent creation. The Second Circuit reversed summary judgment for Webber, holding that the expert testimony about similarity standing alone was sufficient to create a genuine issue of fact regarding access.

By contrast, the Seventh Circuit requires the plaintiff to "present sufficient evidence to support a reasonable possibility of access because the jury cannot draw an inference of access based upon speculation and conjecture alone." Selle v. Gibb, 741 F.2d 896, 901 (7th Cir. 1984). In that case, a member of a local Chicago-area

---

26. By contrast, the "substantial similarity" standard applied in assessing improper appropriation is subjective. The issue at this second stage is determining whether, in the eyes or ears of the trier of fact, the defendant's work reproduces so much of the plaintiff's protected expression as to infringe the copyright. See *infra,* subsection b.

band alleged that the Bee Gees' 1978 hit tune "How Deep Is Your Love" infringed Selle's 1975 composition "Let It End." The only evidence that the Bee Gees might have had access to Selle's song was that it had been played two or three times by Selle's band in the Chicago area and that Selle had sent a recording to eleven recording and publishing companies, eight of whom returned it and three of whom did not respond. The jury found for Selle, but the trial court judge entered a judgment notwithstanding the verdict in favor of the Bee Gees. On appeal, the Seventh Circuit affirmed on the grounds that there was insufficient proof of access and that the plaintiffs had failed to establish that the similarity could only have been the result of copying—i.e., not the result of independent creation or derivation from a common prior source.

In Ty, Inc. v. GMA Accessories, Inc., 132 F.3d 1167 (7th Cir. 1997), Judge Posner sought to harmonize the Seventh and Second Circuit tests for proving copying through circumstantial evidence by characterizing *Selle* as dealing with a situation in which "two works may be strikingly similar—may in fact be identical—not because one is copied from the other but because both are copies of the same thing in the public domain." Id. at 1170. "[T]he tension between *Gaste* and *Selle* can be resolved and the true relation between similarity and access expressed. Access (and copying) may be inferred when two works are so similar to each other and not to anything in the public domain that it is likely that the creator of the second work copied the first, but the inference can be rebutted by disproving access or otherwise showing independent creation." Id. at 1171.

Does it make sense to dispense with the need for *any* proof of access in some cases? Or would a burden-shifting presumption be more appropriate?

3. *Subconscious Copying of a Work*. Robert Mack composed the song "He's So Fine," which The Chiffons recorded in 1962. It enjoyed popular success, rising to No. 1 on the U.S. billboard charts for five weeks in 1963; it was among the top hits in England for about seven weeks in 1963 as well. In 1970, George Harrison, formerly of The Beatles, wrote the song "My Sweet Lord." A recording of the song by Billy Preston became a popular hit. Both songs consisted of four repetitions of a very short basic musical phrase, "sol-me-ri," followed by four repetitions of another short basic musical phrase, "sol-la-do-la-do." While neither phrase is novel (or uncommon), experts at trial agreed that the pattern of juxtaposing four repetitions of each phrase is highly unique. George Harrison testified at trial that he had composed the song while in Copenhagen, Denmark, while on a gig. He recalled "vamping" some guitar chords while singing "Hallelujah" and "Hare Krishna." The song developed further as he improvised with the other musicians in his entourage. In assessing copying, the court determined that George Harrison had not deliberately reproduced Mack's work. Nonetheless, the court found infringement on the grounds that " 'My Sweet Lord' is the very same song as 'He's So Fine' with different words, and Harrison had access to 'He's So Fine.' This is, under the law, infringement of copyright, and is no less so even though subconsciously accomplished." Bright Tunes Music Corp. v. Harrisongs Music, Ltd., 420 F. Supp. 177 (S.D.N.Y. 1976). See also Lipton v. Nature Co., 71 F.3d 464 (2d Cir. 1995) (finding infringement where a defendant had licensed a work from an author who in turn had copied it, even though the defendant had no way of knowing of the unlawful copying).

4. *Proof of Copying by Deliberate Error or Common Mistake*. Creators of fact works often deliberately plant minor errors in their works to trap copyists. We saw an example of this in *Feist, supra* section B.1, where a telephone company placed false names in a telephone directory. Cartographers also frequently bury erroneous

or misspelled place names in their maps for this reason. In addition, accidental misspellings and other mistakes common to both works have aided copyright owners in establishing copying.

5. *Techniques for Reducing the Risk of Infringement: Returning Unopened Submissions and Clean Rooms.* Potential copyright defendants sometimes go to great lengths to avoid having access to works that might influence them. For example, many movie studios and television producers routinely return unsolicited scripts unopened. In the computer industry, program developers occasionally prepare new programs in "clean rooms," in which procedures are established to regulate the entry and exit of material from the location where the work is being created and the creative process is carefully documented so as to be able to prove that a program was independently created without access to code written by others.

PROBLEM

Problem 4-20. Scooter, a ventriloquist, performs a traveling show with a dummy that vocalizes the catchphrase "You Got the Right One, Uh-Huh." Scooter has performed this show since 1984. His performances have primarily been at elementary schools and Job Corps camps, but he did have a pavilion at the 1984 World's Fair in which he used the phrase. Scooter also attempted to promote his show by mailing unsolicited information packets to corporate executives. Included in these packets were letters that referred to his catchphrase. He mailed one such packet to a Pepsi executive in Baltimore in 1988, but the executive cannot recall ever receiving it.

In 1991, Pepsi starts a massive advertising campaign using Ray Charles singing "You Got the Right One Baby, Uh-Huh" with similar voice inflections. Scooter sues for copyright infringement. Does he have a case?

---

b. Improper Appropriation

The second problem that arises in assessing infringement is determining whether the defendant has copied sufficient expression to violate the plaintiff's copyright interests. This requires careful assessment of the extent to which the defendant has copied "protected" expression. Judge Learned Hand's opinion in Nichols v. Universal Pictures, Corp., decided more than 70 years ago, remains the starting point for this inquiry. As you study this opinion, pay close attention to the manner in which the court distinguishes protected and unprotected expression and how it determines whether the defendant has improperly appropriated the plaintiff's work.

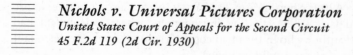

Nichols v. Universal Pictures Corporation
United States Court of Appeals for the Second Circuit
45 F.2d 119 (2d Cir. 1930)

L. HAND, Circuit Judge.

The plaintiff is the author of a play, "Abie's Irish Rose," which it may be assumed was properly copyrighted under section five, subdivision (d), of the Copyright Act, 17

USCA § 5(d). The defendant produced publicly a motion picture play, "The Cohens and The Kellys," which the plaintiff alleges was taken from it. As we think the defendant's play too unlike the plaintiff's to be an infringement, we may assume, arguendo, that in some details the defendant used the plaintiff's play, as will subsequently appear, though we do not so decide. It therefore becomes necessary to give an outline of the two plays.

"Abie's Irish Rose" presents a Jewish family living in prosperous circumstances in New York. The father, a widower, is in business as a merchant, in which his son and only child helps him. The boy has philandered with young women, who to his father's great disgust have always been Gentiles, for he is obsessed with a passion that his daughter-in-law shall be an orthodox Jewess. When the play opens the son, who has been courting a young Irish Catholic girl, has already married her secretly before a Protestant minister, and is concerned to soften the blow for his father, by securing a favorable impression of his bride, while concealing her faith and race. To accomplish this he introduces her to his father at his home as a Jewess, and lets it appear that he is interested in her, though he conceals the marriage. The girl somewhat reluctantly falls in with the plan; the father takes the bait, becomes infatuated with the girl, concludes that they must marry, and assumes that of course they will, if he so decides. He calls in a rabbi, and prepares for the wedding according to the Jewish rite.

Meanwhile the girl's father, also a widower, who lives in California, and is as intense in his own religious antagonism as the Jew, has been called to New York, supposing that his daughter is to marry an Irishman and a Catholic. Accompanied by a priest, he arrives at the house at the moment when the marriage is being celebrated, but too late to prevent it and the two fathers, each infuriated by the proposed union of his child to a heretic, fall into unseemly and grotesque antics. The priest and the rabbi become friendly, exchange trite sentiments about religion, and agree that the match is good. Apparently out of abundant caution, the priest celebrates the marriage for a third time, while the girl's father is inveigled away. The second act closes with each father, still outraged, seeking to find some way by which the union, thus trebly insured, may be dissolved.

The last act takes place about a year later, the young couple having meanwhile been abjured by each father, and left to their own resources. They have had twins, a boy and a girl, but their fathers know no more than that a child has been born. At Christmas each, led by his craving to see his grandchild, goes separately to the young folks' home, where they encounter each other, each laden with gifts, one for a boy, the other for a girl. After some slapstick comedy, depending upon the insistence of each that he is right about the sex of the grandchild, they become reconciled when they learn the truth, and that each child is to bear the given name of a grandparent. The curtain falls as the fathers are exchanging amenities, and the Jew giving evidence of an abatement in the strictness of his orthodoxy.

"The Cohens and The Kellys" presents two families, Jewish and Irish, living side by side in the poorer quarters of New York in a state of perpetual enmity. The wives in both cases are still living, and share in the mutual animosity, as do two small sons, and even the respective dogs. The Jews have a daughter, the Irish a son; the Jewish father is in the clothing business; the Irishman is a policeman. The children are in love with each other, and secretly marry, apparently after the play opens. The Jew, being in great financial straits, learns from a lawyer that he has fallen heir to a large fortune from a great-aunt, and moves into a great house, fitted luxuriously. Here he and his family live in vulgar ostentation, and here the Irish boy seeks out his Jewish bride, and is chased away by the angry father. The Jew then abuses the Irishman over the telephone, and both become hysterically excited. The extremity of his feelings make[s] the Jew sick, so that he must go to Florida for a rest, just before which the daughter discloses her marriage to her mother.

On his return the Jew finds that his daughter has borne a child; at first he suspects the lawyer, but eventually learns the truth and is overcome with anger at such a low alliance. Meanwhile, the Irish family who have been forbidden to see the grand-child, go to the Jew's house, and after a violent scene between the two fathers in which the Jew disowns his daughter, who decides to go back with her husband, the Irishman takes her back with her baby to his own poor lodgings. The lawyer, who had hoped to marry the Jew's daughter, seeing his plan foiled, tells the Jew that his fortune really belongs to the Irishman, who was also related to the dead woman, but offers to conceal his knowledge, if the Jew will share the loot. This the Jew repudiates, and, leaving the astonished lawyer, walks through the rain to his enemy's house to surrender the property. He arrives in great dejection, tells the truth, and abjectly turns to leave. A reconciliation ensues, the Irishman agreeing to share with him equally. The Jew shows some interest in his grandchild, though this is at most a minor motive in the reconciliation, and the curtain falls while the two are in their cups, the Jew insisting that in the firm name for the business, which they are to carry on jointly, his name shall stand first.

It is of course essential to any protection of literary property, whether at com-mon-law or under the statute, that the right cannot be limited literally to the text, else a plagiarist would escape by immaterial variations. That has never been the law, but, as soon as literal appropriation ceases to be the test, the whole matter is necessarily at large, so that, as was recently well said by a distinguished judge, the decisions can-not help much in a new case. Fendler v. Morosco, 253 N.Y. 281, 292, 171 N.E. 56. When plays are concerned, the plagiarist may excise a separate scene; or he may appropriate part of the dialogue. Then the question is whether the part so taken is "substantial," and therefore not a "fair use" of the copyrighted work; it is the same question as arises in the case of any other copyrighted work. But when the plagiarist does not take out a block in suit, but an abstract of the whole, decision is more troublesome. Upon any work, and especially upon a play, a great number of patterns of increasing generality will fit equally well, as more and more of the incident is left out. The last may perhaps be no more than the most general statement of what the play is about, and at times might consist only of its title; but there is a point in this series of abstractions where they are no longer protected, since otherwise the playwright could prevent the use of his "ideas," to which, apart from their expres-sion, his property is never extended. Nobody has ever been able to fix that boundary, and nobody ever can. In some cases the question has been treated as though it were analogous to lifting a portion out of the copyrighted work; but the analogy is not a good one, because, though the skeleton is a part of the body, it pervades and supports the whole. In such cases we are rather concerned with the line between expression and what is expressed. As respects plays, the controversy chiefly centers upon the characters and sequence of incident, these being the substance.

We did not in Dymow v. Bolton, 11 F.2d 690, hold that a plagiarist was never liable for stealing a plot; that would have been flatly against our ruling in Dam v. Kirk La Shelle Co., 175 F. 902, and Stodart v. Mutual Film Co., 249 F. 513, affirming my decision in (D.C.) 249 F. 507; neither of which we meant to overrule. We found the plot of the second play was too different to infringe, because the most detailed pattern, common to both, eliminated so much from each that its content went into the public domain; and for this reason we said, "this mere subsection of a plot was not susceptible of copyright." But we do not doubt that two plays may correspond in plot closely enough for infringement. How far that correspondence must go is another matter. Nor need we hold that the same may not be true as to the characters, quite independently of the "plot" proper, though, as far as we know such a case has never arisen. If Twelfth Night were copyrighted, it is quite possible that a second comer might so closely imitate Sir Toby Belch or Malvolio as to infringe, but it would not be enough that for one of his characters he cast a riotous knight who kept wassail to the discomfort of the household, or a vain and foppish steward who became amorous of his mistress. These would be no more than Shakespeare's "ideas" in the play, as little capable of monopoly as Einstein's Doctrine of Relativity, or Darwin's theory of the Origin of Species. It follows that the less developed the characters, the less they can be copyrighted; that is the penalty an author must bear for marking them too indistinctly.

In the two plays at bar we think both as to incident and character, the defendant took no more — assuming that it took anything at all — than the law allowed. The stories are quite different. One is of a religious zealot who insists upon his child's marrying no one outside his faith; opposed by another who is in this respect just like him, and is his foil. Their difference in race is merely an obligato to the main theme, religion. They sink their differences through grandparental pride and affection. In the other, zealotry is wholly absent; religion does not even appear. It is true that the parents are hostile to each other in part because they differ in race; but the marriage of their son to a Jew does no[t] apparently offend the Irish family at all, and it exacerbates the existing animosity of the Jew, principally because he has become rich, when he learns it. They are reconciled through the honesty of the Jew and the generosity of the Irishman; the grandchild has nothing whatever to do with it. The only matter common to the two is a quarrel between a Jewish and an Irish father, the marriage of their children, the birth of grandchildren and a reconciliation.

If the defendant took so much from the plaintiff, it may well have been because her amazing success seemed to prove that this was a subject of enduring popularity. Even so, granting that the plaintiff's play was wholly original, and assuming that novelty is not essential to a copyright, there is no monopoly in such a background. Though the plaintiff discovered the vein, she could not keep it to herself; so defined, the theme was too generalized an abstraction from what she wrote. It was only a part of her "ideas."

Nor does she fare better as to her characters. It is indeed scarcely credible that she should not have been aware of those stock figures, the low comedy Jew and Irishman. The defendant has not taken from her more than their prototypes have contained for many decades. If so, obviously so to generalize her copyright, would allow her to cover what was not original with her. But we need not hold this as matter of fact, much as we might be justified. Even though we take it that she devised her figures out of her brain de novo, still the defendant was within its rights.

There are but four characters common to both plays, the lovers and the fathers. The lovers are so faintly indicated as to be no more than stage properties. They are loving and fertile; that is really all that can be said of them, and anyone else is quite within his rights if he puts loving and fertile lovers in a play of his own, wherever he gets the cue. The plaintiff's Jew is quite unlike the defendant's. His obsession is his religion, on which depends such racial animosity as he has. He is affectionate, warm and patriarchal. None of these fit the defendant's Jew, who shows affection for his daughter only once, and who has none but the most superficial interest in his grandchild. He is tricky, ostentatious and vulgar, only by misfortune redeemed into honesty. Both are grotesque, extravagant and quarrelsome; both are fond of display; but these common qualities make up only a small part of their simple pictures, no more than any one might lift if he chose. The Irish fathers are even more unlike; the plaintiff's a mere symbol for religious fanaticism and patriarchal pride, scarcely a character at all. Neither quality appears in the defendant's, for while he goes to get his grandchild, it is rather out of a truculent determination not to be forbidden, than from pride in his progeny. For the rest he is only a grotesque hobbledehoy, used for low comedy of the most conventional sort, which any one might borrow, if he chanced not to know the exemplar.

The defendant argues that the case is controlled by my decision in Fisher v. Dillingham, (D.C.) 298 F. 145. Neither my brothers nor I wish to throw doubt upon the doctrine of that case, but it is not applicable here. We assume that the plaintiff's play is altogether original, even to an extent that in fact it is hard to believe. We assume further that, so far as it has been anticipated by earlier plays of which she knew nothing, that fact is immaterial. Still, as we have already said, her copyright did not cover everything that might be drawn from her play; its content went to some extent into the public domain. We have to decide how much, and while we are as aware as any one that the line, wherever it is drawn, will seem arbitrary, that is no excuse for not drawing it; it is a question such as courts must answer in nearly all cases. Whatever may be the difficulties a priori, we have no question on which side of the line this case falls. A comedy based upon conflicts between Irish and Jews, into which the marriage of their children enters, is no more susceptible of copyright than the outline of Romeo and Juliet.

The plaintiff has prepared an elaborate analysis of the two plays, showing a "quadrangle" of the common characters, in which each is represented by the emotions which he discovers. She presents the resulting parallelism as proof of infringement, but the adjectives employed are so general as to be quite useless. Take for example the attribute of "love" ascribed to both Jews. The plaintiff has depicted her father as deeply attached to his son, who is his hope and joy; not so, the defendant, whose father's conduct is throughout not actuated by any affection for his daughter, and who is merely once overcome for the moment by her distress when he has violently dismissed her lover. "Anger" covers emotions aroused by quite different occasions in each case; so do "anxiety," "despondency" and "disgust." It is unnecessary to go through the catalogue for emotions are too much colored by their causes to be a test when used so broadly. This is not the proper approach to a solution; it must be more ingenuous, more like that of a spectator, who would rely upon the complex of his impressions of each character. . . .

Decree affirmed.

COMMENTS AND QUESTIONS

1. *Test for Improper Appropriation.* Some decisions in the Ninth Circuit have bifurcated the analysis into an objective test and a subjective test. See, e.g., Shaw v. Lindheim, 919 F.2d 1353 (9th Cir. 1990). The objective test analytically dissects

the objective manifestations of creativity (plots, themes, dialogue, mood, setting, pace, sequence, characters) in the plaintiff's work in order to determine the elements that are protectable under copyright law. In the second stage of analysis the trier of fact, applying a purely subjective perspective, determines whether the defendant's work improperly appropriates the plaintiff's protected expression.

Does this bifurcated approach make sense? How can the trier of fact compare the two works without relying on the objective approach to determine what is protectable?

2. *Framing the Subjective Analysis Comparison.* A key issue in applying the intrinsic or subjective stage of analysis is delineating what the fact finder compares in deciding whether two works are substantially similar. Does the fact finder compare the two works as a whole or only those elements that are protectable? Courts have differed in their treatment of this critical issue. Some courts have held that the fact-finder shall compare the entirety of the two works, including the "unprotectable" elements. See, e.g., Roth Greeting Cards v. United Card Co., 429 F.2d 1106 (9th Cir. 1970); Sheldon v. Metro-Goldwyn Pictures Corp., 81 F.2d 49 (2d Cir. 1936). Other courts have excluded nonprotectable elements from the comparison. See, e.g., Hoehling v. Universal City Studios, Inc., 618 F.2d 972 (2d Cir.), *cert. denied,* 449 U.S. 841 (1980); Computer Associates International v. Altai, Inc., 982 F.2d 693 (2d Cir. 1992) (addressing the protection of computer code), *infra* Chapter 7. Which view comports best with copyright principles?

3. *How Much Must Be Taken to Constitute Improper Appropriation?* A copyright owner need not prove that all or nearly all of his or her work has been appropriated to establish infringement. Although the quantum necessary depends on the nature of the work, recall that the legislative history to § 106 provides that "a copyrighted work would be infringed by reproducing it *in whole or in any substantial part,* and by duplicating it exactly or by imitation or simulation. Wide departures or variations from the copyrighted works would still be an infringement as long as the author's 'expression' rather than merely the author's 'ideas' are taken." H.R. Rep. No. 94-1476, 94th Cong., 2d Sess. 61 (1976) (emphasis added). Thus courts have held that "[e]ven a small amount of the original, if it is qualitatively significant, may be sufficient to be an infringement. . . ." Horgan v. Macmillan, Inc., 789 F.2d 157, 162 (2d Cir. 1986).

Determining the threshold for infringement is particularly difficult in those cases in which a defendant has copied distinct literal elements of the plaintiff's work and incorporated them into a larger work of his or her own. This class of cases has been referred to as *fragmented literal similarity.* See Melville B. Nimmer & David Nimmer, Nimmer on Copyright § 13.03[A][2]. The Nimmer treatise states:

> The question in each case is whether the similarity relates to matter which constitutes a substantial portion of plaintiff's work—not whether such material constitutes a substantial portion of defendant's work. The quantitative relation of the similar material to the total material contained in plaintiff's work is certainly of importance. However, even if the similar material is quantitatively small, if it is qualitatively important the trier of fact may properly find substantial similarity. In such circumstances the defendant may not claim immunity on the grounds the infringement "is such a little one." If, however, the similarity is only as to nonessential matters, then a finding of no substantial similarity should result.

4. *The Sliding Scale and the Virtual Identity Test.* As courts have increasingly recognized, "more similarity is required when less protectable matter is at issue." Nimmer on Copyright § 13.03(A). Therefore, many courts now require "virtual identity" when dealing with works in which copyright protection is "thin" — i.e., works involving unprotectible elements and/or where the range of creative expression is limited. See, e.g., Incredible Technologies, Inc. v. Virtual Technologies, 400 F.3d 1007 (7th Cir. 2005) (screen displays for video golf game); Data East USA, Inc. v. Epyx, Inc., 862 F.2d 204 (9th Cir. 1988) (screen displays for video karate game); Satava v. Lowry, 323 F.3d 805, 811 (9th Cir. 2003) (glass sculptures encasing jellyfish); cf. Jacobsen v. Deseret Book Co., 287 F.3d 936 (10th Cir. 2002) (noting that "[b]ecause fact-based works differ as to the relative proportion of fact and fancy [ranging from 'sparsely embellished maps and directories' to 'elegantly written biography'], the quantum of similarity required to establish infringement differs in each case").

5. *The De Minimis Doctrine — In General.* Copyright recognizes the maxim *de minimis non curat* — the law does not concern itself with trifles. The cases applying this principle use it as a shorthand for lack of substantial similarity — where "the copying of the protected material is so trivial 'as to fall below the quantitative threshold of substantial similarity.'" Gordon v. Nextel Communications and Mullen Advertising, Inc., 345 F.3d 922, 924 (6th Cir. 2003) (quoting Ringgold v. Black Entertainment Television Inc., 126 F.3d 70, 74 (2d Cir. 1997)); Fisher v. Dees, 794 F.2d 432, 435 (9th Cir. 1986) (noting that *de minimis* copying "is so meager and fragmentary that the average audience would not recognize the appropriation"); Warner Bros., Inc. v. American Broadcasting Cos., 720 F.2d 231, 242 (2d Cir. 1983) (holding that the *de minimis* doctrine allows "literal copying of a small and usually insignificant portion of the plaintiff's work"); cf. Use of Certain Copyrighted Works in Connection with Noncommercial Educational Broadcasting, 37 C.F.R. § 253.8 (regulation issued by the Copyright Office for royalties to be paid by public broadcasting entities for the use of published pictorial and visual works pursuant to 17 U.S.C. § 118(b)). Courts will not apply the doctrine, however, without attention to qualitative considerations. See CyberMedia, Inc. v. Symantec Corp., 19 F.Supp.2d 1070, 1077 (N.D.Cal. 1998) (holding that "even if a copied portion be relatively small in proportion to the entire work, if qualitatively important, the finder of fact may properly find substantial similarity").

6. *De Minimis Copying and Digital Sampling.* The rap and hip-hop genres have built new compositions upon digital samples (literal copying) of existing sound recordings. Several early cases held that such copying infringed copyrights in the underlying musical compositions on the basis of fragmented literal similarity. See Grand Upright Music, Ltd. v. Warner Brothers Records, Inc., 780 F.Supp. 182 (S.D.N.Y. 1991); Jarvis v. A & M Records, 827 F.Supp. 282 (D.N.J. 1993). In the first case to squarely address digital samples of sound recordings, the Sixth Circuit ruled that the Copyright Act bars application of the *de minimis* doctrine in this class of works, with the result that even the copying of a single note could constitute copyright infringement. See Bridgeport Music, Inc. v. Dimension Films, 410 F.3d 792, 800-01 (6th Cir. 2005).

Section 114(b) provides that "[t]he exclusive right of the owner of copyright in a sound recording under clause (2) of section 106 is limited to the right to prepare a derivative work in which the actual sounds fixed in the sound recording are rearranged, remixed, or otherwise altered in sequence or quality." Further, the rights of sound recording copyright holders under clauses (1) and (2) of section 106 "do not extend to the making or duplication of another sound recording that consists *entirely* of an independent fixation of other sounds, even though such sounds imitate or simulate those in the copyrighted sound recording." 17 U.S.C. § 114(b) (emphasis added). The significance of this provision is amplified by the fact that the Copyright Act of 1976 added the word "entirely" to this language. Compare Sound Recording Act of 1971, Pub.L. 92-140, 85 Stat. 391 (Oct. 15, 1971) (adding subsection (f) to former 17 U.S.C. § 1) ("does not extend to the making or duplication of another sound recording that is an independent fixation of other sounds"). In other words, a sound recording owner has the exclusive right to "sample" his own recording.

In view of the fact that Congress could not have had digital sampling in mind when it drafted the Sound Recording Act of 1971 or the Copyright Act of 1976, does this strike you as a reasonable interpretation? Do you think that Congress intended to override the *de minimis* doctrine through this rather ambiguous language? The court bolstered its analysis on policy grounds, asserting that such a bright line rule ("Get a license or do not sample") would ease enforcement and would not stifle creativity because a well-functioning sampling market currently exists and because artists are free to record a *de minimis* "riff" in the studio. Do you agree? In any case, is it likely to matter much in practice? The *Bridgeport* case does not preclude a finding of fair use. See *id.* at 805.

A case involving the rap song "Pass the Mic" presents another variation on this theme. The Beastie Boys obtained a sampling license from ECM, the record label controlling rights to noted-jazz flutist James Newton's recording of his composition "Choir," to use a six-second clip from the song's opening as a backdrop for their sound recording. Newton sued, alleging that the Beastie Boys also needed a license to the underlying musical composition, for which he held the copyright. The court held that although the sound recording of the six-second sample may well have qualified for copyright protection due to the complexity of the performance, copying of the underlying musical composition — involving a three-note sequence sung above a finger-held C note to be played in a "largo/senza-misura" (slowly/without measure) tempo while overblowing the background C note — was not actionable under the *de minimis* doctrine. Newton v. Diamond, 349 F.3d 591 (9th Cir. 2003), amended 388 F.3d 1189 (9th Cir. 2004), *cert. denied*, 125 S.Ct. 2905 (2005). Note that this case does not contradict the *Bridgeport* ruling because it involves copying of the musical composition, and not the sound recording.

Does it make sense to treat the scope of musical composition and sound recording copyrights differently in this way?

7. *The Role of Expert Testimony in Determining Improper Appropriation.* The court in *Arnstein* held that expert opinion is "utterly immaterial" to the determination of improper appropriation. Does this limitation on evidence make sense with regard to all works? Expert testimony would seem essential in assessing appropriation with regard to technically complex material written for specialized audiences. The issue arises frequently in the context of computer software copyright cases. See *infra* Chapter 7. Is it desirable to assess similarities in two database programs from the standpoint of the ordinary person on the street rather than the ordinary user of database programs? Isn't expert testimony on the extent to which programming elements are common in the trade essential to determining improper appropriation?

8. *The Appropriate Perspective for Assessing Substantial Similarity: The Ordinary Observer.* The Second Circuit has defined "substantial similarity" as whether the "ordinary observer, unless he set out to detect the disparities [between two works], would be disposed to overlook them, and regard their aesthetic appeal as the same." Peter Pan Fabrics, Inc. v. Martin Weiner Corp., 274 F.2d 487, 489 (2d Cir. 1960).

A number of cases have narrowed the "ordinary observer" perspective by focusing on the impressions of the target audience for the work in question. For example, in Original Appalachian Artworks, Inc. v. Blue Box Factory (USA) Ltd., 577 F. Supp. 625 (S.D.N.Y. 1983), involving copyright protection for a popular line of dolls called "Cabbage Patch Kids," the court allowed expert evidence about how the works would be perceived by children. In Data East USA, Inc. v. Epyx, Inc., 862 F.2d 204 (9th Cir. 1988), the court assessed substantial similarity of two karate video games from the perspective of a "discerning 17.5 year-old boy," based on the district court's finding that "the average age of individuals purchasing 'Karate Champ' is 17.5 years, that the purchasers are predominantly male, and comprise a knowledgeable, critical, and discerning group."

Should the "ordinary observer" test be tailored to the target audience for the works? Is the "ordinary observer" perspective, even if tailored to reflect the target audience for the work, likely to distinguish between the protectable and nonprotectable elements of a work in assessing infringement?

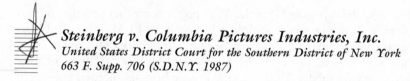

Steinberg v. Columbia Pictures Industries, Inc.
United States District Court for the Southern District of New York
663 F. Supp. 706 (S.D.N.Y. 1987)

STANTON, District Judge.

In these actions for copyright infringement, plaintiff Saul Steinberg is suing the producers, promoters, distributors and advertisers of the movie "Moscow on the Hudson" ("Moscow"). Steinberg is an artist whose fame derives in part from cartoons and illustrations he has drawn for *The New Yorker* magazine. Defendant Columbia Pictures Industries, Inc. (Columbia) is in the business of producing, promoting and distributing motion pictures, including "Moscow." . . .

Plaintiff alleges that defendants' promotional poster for "Moscow" infringes his copyright on an illustration that he drew for *The New Yorker* and that appeared on the cover of the March 29, 1976 issue of the magazine, in violation of 17 U.S.C. §§ 101-810. Defendants deny this allegation and assert the affirmative defenses of fair use as a parody, estoppel and laches. Defendants have moved, and plaintiff has cross-moved, for summary judgment. For the reasons set forth below, this court rejects defendants' asserted defenses and grants summary judgment on the issue of copying to plaintiff. . . .

II

The essential facts are not disputed by the parties despite their disagreements on nonessential matters. On March 29, 1976, *The New Yorker* published as a cover illustration the work at issue in this suit, widely known as a parochial New Yorker's view of the world. The magazine registered this illustration with the United States Copyright Office and subsequently assigned the copyright to Steinberg. Approximately three months later, plaintiff and *The New Yorker* entered into an agreement to print and sell a certain number of posters of the cover illustration.

It is undisputed that unauthorized duplications of the poster were made and distributed by unknown persons, although the parties disagree on the extent to which plaintiff attempted to prevent the distribution of those counterfeits. Plaintiff has also conceded that numerous posters have been created and published depicting other localities in the same manner that he depicted New York in his illustration. These facts, however, are irrelevant to the merits of this case, which concerns only the relationship between plaintiff's and defendants' illustrations.

Defendants' illustration was created to advertise the movie "Moscow on the Hudson," which recounts the adventures of a Muscovite who defects in New York. In designing this illustration, Columbia's executive art director, Kevin Nolan, has admitted that he specifically referred to Steinberg's poster, and indeed, that he purchased it and hung it, among others, in his office. Furthermore, Nolan explicitly directed the outside artist whom he retained to execute his design, Craig Nelson, to use Steinberg's poster to achieve a more recognizably New York look. Indeed, Nelson acknowledged having used the facade of one particular edifice, at Nolan's suggestion that it would render his drawing more "New York-ish." While the two buildings are not identical, they are so similar that it is impossible, especially in view of the artist's testimony, not to find that defendants' impermissibly copied plaintiff's.

To decide the issue of infringement, it is necessary to consider the posters themselves. Steinberg's illustration presents a bird's eye view across a portion of the western edge of Manhattan, past the Hudson River and a telescoped version of the rest of the United States and the Pacific Ocean, to a red strip of horizon, beneath which are three flat land masses labeled China, Japan and Russia. The name of the magazine, in *The New Yorker*'s usual typeface, occupies the top fifth of the poster, beneath a thin band of blue wash representing a stylized sky.

The parts of the poster beyond New York are minimalized, to symbolize a New Yorker's myopic view of the centrality of his city to the world. The entire United States west of the Hudson River, for example, is reduced to a brown strip labeled "Jersey," together with a light green trapezoid with a few rudimentary rock outcroppings and the names of only seven cities and two states scattered across it. The few blocks of Manhattan, by contrast, are depicted and colored in detail. The four square blocks of the city, which occupy the whole lower half of the poster, include numerous buildings, pedestrians and cars, as well as parking lots and lamp posts, with water towers atop a few of the buildings. The whimsical, sketchy style and spiky lettering are recognizable as Steinberg's.

The "Moscow" illustration depicts the three main characters of the film on the lower third of their poster, superimposed on a bird's eye view of New York City, and continues eastward across Manhattan and the Atlantic Ocean, past a rudimentary evocation of Europe, to a clump of recognizably Russian-styled buildings on the horizon, labeled "Moscow." The movie credits appear over the lower portion of the characters. The central part of the poster depicts approximately four New York city blocks, with fairly detailed buildings, pedestrians and vehicles, a parking lot, and some water towers and lamp posts. Columbia's artist added a few New York landmarks at apparently random places in his illustration, apparently to render the locale more easily recognizable. Beyond the blue strip labeled "Atlantic Ocean," Europe is

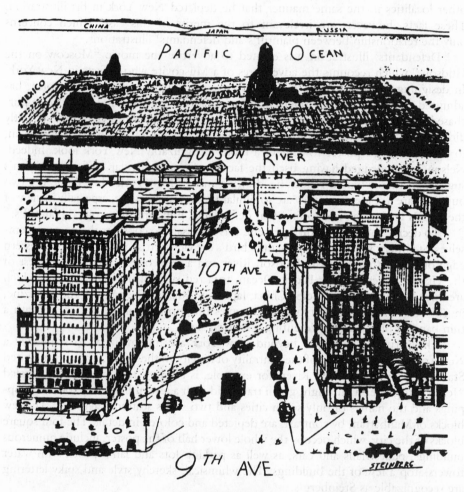

Mar. 29, 1976 · THE · Price 75 cents

NEW YORKER

FIGURE 4-5
Saul Steinberg. *A view of the world from 9th Avenue*, 1976. © 2002
The Saul Steinberg Foundation/Artists Rights Society (ARS), New York.

FIGURE 4-6
Moscow on the Hudson poster.

represented by London, Paris and Rome, each anchored by a single landmark (although the landmark used for Rome is the Leaning Tower of Pisa).

The horizon behind Moscow is delineated by a red crayoned strip, above which are the title of the movie and a brief textual introduction to the plot. The poster is crowned by a thin strip of blue wash, apparently a stylization of the sky. This poster is executed in a blend of styles: the three characters, whose likenesses were copied from a photograph, have realistic faces and somewhat sketchy clothing, and the city blocks are drawn in a fairly detailed but sketchy style. The lettering on the drawing is spiky, in block-printed handwritten capital letters substantially identical to plaintiff's, while the printed texts at the top and bottom of the poster are in the typeface commonly associated with *The New Yorker* magazine.[2]

III

To succeed in a copyright infringement action, a plaintiff must prove ownership of the copyright and copying by the defendant. Reyher v. Children's Television Workshop, 533 F.2d 87, 90 (2d Cir. 1976). There is no substantial dispute concerning plaintiff's ownership of a valid copyright in his illustration. Therefore, in order to prevail on liability, plaintiff need establish only the second element of the cause of action.

"Because of the inherent difficulty in obtaining direct evidence of copying, it is usually proved by circumstantial evidence of access to the copyrighted work and substantial similarities as to protectible material in the two works." *Reyher,* 533 F.2d at 90, citing Arnstein v. Porter, 154 F.2d 464, 468 (2d Cir. 1946). "Of course, if there are no similarities, no amount of evidence of access will suffice to prove copying." Arnstein v. Porter, 154 F.2d at 468.

Defendants' access to plaintiff's illustration is established beyond peradventure. Therefore, the sole issue remaining with respect to liability is whether there is such substantial similarity between the copyrighted and accused works as to establish a violation of plaintiff's copyright. The central issue of "substantial similarity," which can be considered a close question of fact, may also validly be decided as a question of law.

"Substantial similarity" is an elusive concept. This circuit has recently recognized that

> [t]he "substantial similarity" that supports an inference of copying sufficient to establish infringement of a copyright is not a concept familiar to the public at large. It is a term to be used in a courtroom to strike a delicate balance between the protection to which authors are entitled under an act of Congress and the freedom that exists for all others to create their works outside the area protected by infringement.

Warner Bros., 720 F.2d at 245.

---

2. The typeface is not a subject of copyright, but the similarity reinforces the impression that defendants copied plaintiff's illustration.

The definition of "substantial similarity" in this circuit is "whether an average lay observer would recognize the alleged copy as having been appropriated from the copyrighted work." Ideal Toy Corp. v. Fab-Lu Ltd., 360 F.2d 1021, 1022 (2d Cir. 1966); Silverman v. CBS, Inc., 632 F. Supp. at 1351-52. A plaintiff need no longer meet the severe "ordinary observer" test established by Judge Learned Hand in Peter Pan Fabrics, Inc. v. Martin Weiner Corp., 274 F.2d 487 (2d Cir. 1960). Uneeda Doll Co., Inc. v. Regent Baby Products Corp., 355 F.Supp. 438, 450 (E.D.N.Y. 1972). Under Judge Hand's formulation, there would be substantial similarity only where "the ordinary observer, unless he set out to detect the disparities, would be disposed to overlook them, and regard their aesthetic appeal as the same." 274 F.2d at 489.

Moreover, it is now recognized that "[t]he copying need not be of every detail so long as the copy is substantially similar to the copyrighted work." Comptone Co. v. Rayex Corp., 251 F.2d 487, 488 (2d Cir. 1958).

In determining whether there is substantial similarity between two works, it is crucial to distinguish between an idea and its expression. It is an axiom of copyright law, established in the case law and since codified at 17 U.S.C. § 102(b), that only the particular expression of an idea is protectible, while the idea itself is not.

> "The idea/expression distinction, although an imprecise tool, has not been abandoned because we have as yet discovered no better way to reconcile the two competing societal interests that provide the rationale for the granting of and restrictions on copyright protection," namely, both rewarding individual ingenuity, and nevertheless allowing progress and improvements based on the same subject matter by others than the original author.

Durham Industries, 630 F.2d at 912, quoting *Reyher,* 533 F.2d at 90.

There is no dispute that defendants cannot be held liable for using the idea of a map of the world from an egocentrically myopic perspective. No rigid principle has been developed, however, to ascertain when one has gone beyond the idea to the expression, and "[d]ecisions must therefore inevitably be ad hoc." Peter Pan Fabrics, Inc. v. Martin Weiner Corp., 274 F.2d 487, 489 (2d Cir. 1960) (L. Hand, J.). As Judge Frankel once observed, "Good eyes and common sense may be as useful as deep study of reported and unreported cases, which themselves are tied to highly particularized facts." Couleur International Ltd. v. Opulent Fabrics, Inc., 330 F. Supp. 152, 153 (S.D.N.Y. 1971).

Even at first glance, one can see the striking stylistic relationship between the posters, and since style is one ingredient of "expression," this relationship is significant. Defendants' illustration was executed in the sketchy, whimsical style that has become one of Steinberg's hallmarks. Both illustrations represent a bird's eye view across the edge of Manhattan and a river bordering New York City to the world beyond. Both depict approximately four city blocks in detail and become increasingly minimalist as the design recedes into the background. Both use the device of a narrow band of blue wash across the top of the poster to represent the sky, and both delineate the horizon with a band of primary red.

The strongest similarity is evident in the rendering of the New York City blocks. Both artists chose a vantage point that looks directly down a wide two-way cross street that intersects two avenues before reaching a river. Despite defendants' protestations, this is not an inevitable way of depicting blocks in a city with a grid-like street system, particularly since most New York City cross streets are one-way. Since even a photograph may be copyrighted because "no photograph, however simple, can be unaffected by the personal influence of the author," Time Inc. v. Bernard Geis Assoc., 293 F. Supp. 130, 141 (S.D.N.Y. 1968), quoting *Bleistein,* supra, one can hardly gainsay the right of an artist to protect his choice of perspective and lay-out in a drawing, especially in conjunction with the overall concept and individual details. Indeed, the fact that defendants changed the names of the streets while retaining the same graphic depiction weakens their case: had they intended their illustration realistically to depict the streets labeled on the poster, their four city blocks would not so closely resemble plaintiff's four city blocks. Moreover, their argument that they intended the jumble of streets and landmarks and buildings to symbolize their Muscovite protagonist's confusion in a new city does not detract from the strong similarity between their poster and Steinberg's.

While not all of the details are identical, many of them could be mistaken for one another; for example, the depiction of the water towers, and the cars, and the red sign above a parking lot, and even many of the individual buildings. The shapes, windows, and configurations of various edifices are substantially similar. The ornaments, facades and details of Steinberg's buildings appear in defendants', although occasionally at other locations. In this context, it is significant that Steinberg did not depict any buildings actually erected in New York; rather, he was inspired by the general appearance of the structures on the West Side of Manhattan to create his own New York-ish structures. Thus, the similarity between the buildings depicted in the "Moscow" and Steinberg posters cannot be explained by an assertion that the artists happened to choose the same buildings to draw. The close similarity can be explained only by the defendants' artist having copied the plaintiff's work. Similarly, the locations and size, the errors and anomalies of Steinberg's shadows and streetlight, are meticulously imitated.

In addition, the Columbia artist's use of the childlike, spiky block print that has become one of Steinberg's hallmarks to letter the names of the streets in the "Moscow" poster can be explained only as copying. There is no inherent justification for using this style of lettering to label New York City streets as it is associated with New York only through Steinberg's poster.

While defendants' poster shows the city of Moscow on the horizon in far greater detail than anything is depicted in the background of plaintiff's illustration, this fact alone cannot alter the conclusion. "Substantial similarity" does not require identity, and "duplication or near identity is not necessary to establish infringement." *Krofft,* 562 F.2d at 1167. Neither the depiction of Moscow, nor the eastward perspective, nor the presence of randomly scattered New York City landmarks in defendants' poster suffices to eliminate the substantial similarity between the posters. As Judge Learned Hand wrote, "no plagiarist can excuse the wrong by showing how much of his work he did not pirate." Sheldon v. Metro-Goldwyn Pictures Corp., 81 F.2d 49, 56 (2d Cir.), *cert. denied,* 298 U.S. 669, 56 S. Ct. 835, 80 L. Ed. 1392 (1936).

Defendants argue that their poster could not infringe plaintiff's copyright because only a small proportion of its design could possibly be considered similar. This argument is both factually and legally without merit. "[A] copyright infringement may occur by reason of a substantial similarity that involves only a small portion of each work." Burroughs v. Metro-Goldwyn-Mayer, Inc., 683 F.2d 610, 624 n.14 (2d Cir. 1982). Moreover, this case involves the entire protected work and an iconographically, as well as proportionately, significant portion of the allegedly infringing work.

The process by which defendants' poster was created also undermines this argument. The "map," that is, the portion about which plaintiff is complaining, was designed separately from the rest of the poster. The likenesses of the three main characters, which were copied from a photograph, and the blocks of text were superimposed on the completed map.

I also reject defendants' argument that any similarities between the works are unprotectible *scenes a faire,* or "incidents, characters or settings which, as a practical matter, are indispensable or standard in the treatment of a given topic." *Walker,* 615 F. Supp. at 436. See also *Reyher,* 533 F.2d at 92. It is undeniable that a drawing of New York City blocks could be expected to include buildings, pedestrians, vehicles, lampposts and water towers. Plaintiff, however, does not complain of defendants' mere use of these elements in their poster; rather, his complaint is that defendants copied his expression of those elements of a street scene.

While evidence of independent creation by the defendants would rebut plaintiff's prima facie case, "the absence of any countervailing evidence of creation independent of the copyrighted source may well render clearly erroneous a finding that there was not copying." Roth Greeting Cards v. United Card Co., 429 F.2d 1106, 1110 (9th Cir. 1970).

Moreover, it is generally recognized that ". . . since a very high degree of similarity is required in order to dispense with proof of access, it must logically follow that where proof of access is offered, the required degree of similarity may be somewhat less than would be necessary in the absence of such proof." 2 Nimmer § 143.4 at 634, quoted in *Krofft,* 562 F.2d at 1172. As defendants have conceded access to plaintiff's copyrighted illustration, a somewhat lesser degree of similarity suffices to establish a copyright infringement than might otherwise be required. Here, however, the demonstrable similarities are such that proof of access, although in fact conceded, is almost unnecessary. . . .

VI

For the reasons set out above, summary judgment is granted to plaintiffs as to copying. . . .

COMMENTS AND QUESTIONS

1. *Proof of Copying.* Many copyright decisions, like this one, do not clearly distinguish between the analysis of copying and improper appropriation. Was there sufficient evidence to prove copying?

2. *Improper Appropriation.* Do you agree with the court's assessment of improper appropriation? Which elements of Steinberg's work are protected under copyright law? Did Columbia Pictures copy those elements? Did they copy Steinberg's compilation of otherwise unprotectable elements? Suppose you were arguing this case on appeal. What arguments would you make on behalf of the defendant? What counterarguments would you make on behalf of the plaintiff?

PROBLEMS

Problem 4-21. Dinopets markets a line of stuffed animal toys for children. The line includes five popular dinosaurs with exaggerated facial features (e.g., large droopy eyes, long teeth, rounded noses), cheerful pastel colors (pink, lemon, lime), distinctive stitching, and a soft cuddly cotton texture. About a year after Dinopets were on the market, Gigatoys, Inc., a leading toy manufacturer, developed a line of stuffed dinosaur toys. Its line, the Dinomites, featured the five dinosaurs in the Dinopets line as well as three others. Dinomites are about 25 percent larger than comparable Dinopets. Dinomites feature cute facial features (including droopy eyes and long teeth) and come in earth-tone colors (light brown, clay, sand, and stone). They are made of a suede-like material (somewhat coarser than the Dinopets). Dinopets sues Gigatoys, alleging copyright infringement. How would the analysis be conducted? What result?

Problem 4-22. The Stunted Intellectuals, a psychedelic punk rap jazz fusion band, writes songs by taking digital "samples" from previous compositions of a number of different genres, juxtaposing or superimposing them, and laying an electronic beat under and continuous rap commentary over the mixture. Some of the "samples" are from works in the public domain; of the remainder, some are identifiable and unaltered, some are identifiable but altered, and some are unrecognizable as a prior song. No sample lasts for longer than four seconds in any song, but some are repeated within the same song. Have the Stunted Intellectuals infringed the copyright on any of the songs they sample? On all of them?

Problem 4-23. In preparing a biography of the reclusive author J. D. Salinger, Ian Hamilton gained access to letters Salinger wrote to a number of notable people that had been donated to university libraries. Through these letters and other sources, Hamilton constructed his biography of Salinger's life. Out of concern for copyright infringement, Hamilton quotes barely more than 200 words from the letters throughout the entire biography. Nonetheless, the letters are paraphrased or otherwise drawn upon in approximately 40 percent of the 192-page biography. To accurately describe events and emotions, impart some of Salinger's distinctive style, and avoid "pedestrian" reporting, Hamilton follows some of the passages from the letters closely.

The following examples illustrate Hamilton's use of the letters to present Salinger's life. In a 1943 letter to Whit Burnett, Salinger's friend, teacher, and editor at *Story* magazine, Salinger expressed his disapproval of the marriage of Oona O'Neil, with whom Salinger had been romantically involved, and Charlie Chaplin, the silent screen film star.

Salinger's Letter	*Hamilton's Biography*
I can see them at home evenings. Chaplin squatting grey and nude, atop his chiffonier, swinging his thyroid around his	At one point in a letter to Whit Burnett, he provides a pen portrait of the Happy Hour Chez Chaplin: the comedian,

head by his bamboo cane, like a dead rat. Oona in an aquamarine gown, applauding madly from the bathroom. Agnes (her mother) in a Jantzen bathing suit, passing between them with cocktails. I'm facetious, but I'm sorry. Sorry for anyone with a profile as young and lovely as Oona's.

ancient and unclothed, is brandishing his walking stick — attached to the stick, and horribly resembling a lifeless rodent, is one of Chaplin's vital organs. Oona claps her hands in appreciation and Agnes, togged out in a bathing suit, pours drinks. Salinger goes on to say he's sorry — sorry not for what he has just written, but for Oona: far too youthful and exquisite for such a dreadful fate.

In another letter to Burnett, Salinger expresses his disfavor of presidential candidate Wendell Wilkie.

Salinger's Letter	*Hamilton's Biography*
He looks to me like a guy who makes his wife keep a scrapbook for him.	[Salinger] had fingered [Wilkie] as the sort of fellow who makes his wife keep an album of his press cuttings.

In another letter describing Parisians' adulation of American soldiers at the liberation of Paris, Salinger writes that the Parisians would have said "What a charming custom!" if:

Salinger's Letter	*Hamilton's Biography*
we had stood on top of the jeep and taken a leak.	the conquerors had chosen to urinate from the roofs of their vehicles.

Has Hamilton infringed Salinger's copyright in his letters?

Limitations on the Exclusive Right to Copy. In general, the right to prevent copying does not turn on the intent of the copyist. The copyright laws do not apply only to copying for a commercial purpose or to large-scale copying, but to any copying. There are, however, several statutory exceptions to this general rule:

- *Archival Copies for Public Libraries.* Section 108 exempts from copyright liability a public "library or archives" which makes only one copy of a work at a time, assuming the copy is made for specified purposes. Copies may be made for the preservation and replacement of existing works, but only if the work cannot be replaced by purchase at a "fair price." Libraries can also make single copies for noncommercial users, as long as the library does not engage in the "systematic reproduction or distribution" of such copies. Finally, libraries are not liable for copyright violations by their patrons (even those using on-site photocopiers), as long as the library posts conspicuous warnings notifying users of the copyright laws.
- *Ephemeral Copies by Broadcasters.* Sections 112 and 118(d) permit broadcasters to make "ephemeral" or "ancillary" copies of certain performances and displays during the course of broadcasting. For example, broadcasters are permitted

to make a copy during the course of retransmitting a program. We discuss these exceptions in more detail when we consider performance and display rights.

- *Cover License.* Section 115 of the Act permits anyone to record a musical composition upon the payment of a royalty determined by a formula specified in the statute. However, artists cannot record another's musical compositions if they have not been recorded with permission at least once before. Further, section 115 does not permit new artists to change the "basic melody or fundamental character of the work." The copyright law also does not prohibit close imitations of an artist's sound recording, so-called "sound-alike" versions, see § 114(b), although such recordings can run afoul of the right of publicity and the Federal Trade Commission's unfair trade practice guidelines under some circumstances. See, e.g., Midler v. Ford Motor Co., 849 F.2d 460 (9th Cir. 1988) (discussed in Chapter 6); M. William Krasilovsky and Sidney Shemel, This Business of Music 417 (8th ed. 2000).

- *Non-commercial Copies of Musical Compositions and Sound Recordings.* Section 1008, added by the Audio Home Recording Act of 1992 (AHRA), authorizes "consumers" to make copies of sound recordings for "non-commercial use." The immunity for home taping is part of a broader compromise that resolved issues surrounding the use of "digital audio tape" (DAT) technology, which was thought in 1992 to be the wave of the future. In return for this immunity from suit, manufacturers of DAT decks and tapes must pay a royalty to the Copyright Office for distribution to copyright owners. Further, the Act outlaws the sale of DAT decks that can copy copies. We discuss the AHRA in greater detail in Section H.

In addition to these limitations, the fair use doctrine provides a defense in many instances of copying. We consider fair use in detail in Section F.

2. The Right to Prepare Derivative Works

In its early form, copyright law protected only against the production of substantially similar copies in the same medium. It did not protect against translations or dramatic renditions of a novel. Over time, copyright law has expanded to afford the copyright protection in a wide range of derivative media. No one has exploited this broad reach of protection more successfully than George Lucas, who built a multibillion dollar *Star Wars* empire on merchandise — including toys, commercial tie-ins, clothing, books, and games. Just as sports icons like Michael Jordan or Tiger Woods can earn more money off the playing field through endorsements than through actual salary or tournament winnings, many authors today — such as John Grisham (*The Firm, The Pelican Brief*), Michael Crichton (*Jurassic Park, Disclosure*), and Tom Clancy (*Patriot Games, The Hunt for Red October*) — can earn far greater returns from movie and commercial tie-ins than from the novels on which these works are based. Disney and Universal Studios have been particularly effective in leveraging their copyrights in characters and films to television series, toys, commercial tie-ins, and theme parks.

Building upon the foundation laid by the 1909 Act and subsequent court decisions expanding the uses and media protected by copyright law, the 1976 Act provides the copyright owner the exclusive right "to prepare derivative works based upon the copyrighted work." 17 U.S.C. § 106(2). The Act defines a "derivative work" as "a work based on one or more preexisting works, such as a translation, musical arrangement, dramatization, fictionalization, motion picture version, sound recording, art reproduction, abridgement, condensation, or any other form in which a work may be recast, transformed, or adapted." It also includes "a work consisting of editorial revisions, annotations, elaborations, or other modifications which, as a whole, represent an original work of authorship." As the following case reflects, this right structure can operate like a "prospect," cf. Edmund W. Kitch, The Nature and Function of the Patent System, 20 J. Law & Econ. 265 (1977); Mark A. Lemley, The Economics of Improvement in Intellectual Property Law, 75 Tex. L. Rev. 989 (1997), affording the original author a broad opportunity to exploit new markets and incremental improvements in his or her work. How far this right should extend remains a controversial issue.

Anderson v. Stallone
United States District Court for the Central District of California
11 USPQ2d 1161 (C.D. Cal. 1989)

KELLER, District Judge:

. . .

Factual Background

The movies *Rocky I, II,* and *III* were extremely successful motion pictures. Sylvester Stallone wrote each script and played the role of Rocky Balboa, the dominant character in each of the movies. In May of 1982, while on a promotional tour for the movie *Rocky III,* Stallone informed members of the press of his ideas for *Rocky IV.* Although Stallone's description of his ideas would vary slightly in each of the press conferences, he would generally describe his ideas as follows:

> I'd do it [*Rocky IV*] if Rocky himself could step out a bit. Maybe tackle world problems.... So what would happen, say, if Russia allowed her boxers to enter the professional ranks? Say Rocky is the United States' representative and the White House wants him to fight with the Russians before the Olympics. It's in Russia with everything against him. It's a giant stadium in Moscow and everything is Russian Red. It's a fight of astounding proportions with 50 monitors sent to 50 countries. It's the World Cup — a war between 2 countries.

Waco Tribune-Herald, May 28, 1982; Section D, pg. 1. In June of 1982, after viewing the movie *Rocky III,* Timothy Anderson wrote a thirty-one page treatment entitled "*Rocky IV*" that he hoped would be used by Stallone and MGM-UA Communications Co. (hereinafter "MGM") as a sequel to *Rocky III.* The treatment incorporated the characters created by Stallone in his prior movies and cited Stallone as a co-author.

In October of 1982, Mr. Anderson met with Art Linkletter, who was a member of MGM's board of directors. Mr. Linkletter set up a meeting on October 11, 1982, between Mr. Anderson and Mr. Fields, who was president of MGM at the time. Mr. Linkletter was also present at this October 11, 1982 meeting. During the meeting, the parties discussed the possibility that plaintiff's treatment would be used by defendants as the script for *Rocky IV*. At the suggestion of Mr. Fields, the plaintiff, who is a lawyer and was accompanied by a lawyer at the meeting, signed a release that purported to relieve MGM from liability stemming from use of the treatment. Plaintiff alleges that Mr. Fields told him and his attorney that "if they [MGM & Stallone] use his stuff [Anderson's treatment] it will be big money, big bucks for Tim."

On April 22, 1984, Anderson's attorney wrote MGM requesting compensation for the alleged use of his treatment in the forthcoming *Rocky IV* movie. On July 12, 1984, Stallone described his plans for the *Rocky IV* script on the Today Show before a national television audience. Anderson, in his deposition, states that his parents and friends called him to tell him that Stallone was telling "his story" on television. . . .

Stallone completed his *Rocky IV* script in October of 1984. *Rocky IV* was released in November of 1985. The complaint in this action was filed on January 29, 1987.

Conclusions of Law . . .

IV. Defendants Are Entitled to Summary Judgment on Anderson's Copyright Infringement Claims . . .

A. Defendants Are Entitled to Summary Judgment Because Anderson's Treatment Is an Infringing Work That Is Not Entitled to Copyright Protection

The Court finds that Anderson's treatment is not entitled to copyright protection. This finding is based upon the following determinations that will be delineated further below: (a) the Rocky characters developed in *Rocky I, II* and *III* constitute expression protected by copyright independent from the story in which they are contained; (b) Anderson's treatment appropriated these characters and created a derivative work based upon these characters without Stallone's permission in violation of Section 106(2); (c) no part of Anderson's treatment is entitled to copyright protection as his work is pervaded by the characters of the first three *Rocky* movies that are afforded copyright protection.

1. Visually Depicted Characters Can Be Granted Copyright Protection

The precise legal standard this Court should apply in determining when a character may be afforded copyright protection is fraught with uncertainty. The Second Circuit has followed Judge Learned Hand's opinion in Nichols v. Universal Pictures, 45 F.2d 119 (2d Cir. 1930), *cert. denied*, 282 U.S. 902 (1931). Judge Hand set forth a test, simple in theory but elusive in application, to determine when a character should be granted copyright protection. Essentially, under this test, copyright protection is granted to a character if it is developed with enough specificity so as to constitute protectable expression. Id. at 121.

This circuit originally created a more rigorous test for granting copyright protection to characters. In Warner Bros. Pictures, Inc. v. Columbia Broadcasting System, Inc., (hereinafter the "Sam Spade" opinion) this circuit held that the literary character Sam Spade was not copyrightable, opining that a character could not be granted copyright protection unless it "constituted the story being told." 216 F.2d 945, 950 (9th Cir. 1954). The Sam Spade case has not been explicitly overruled by this circuit and its requirement that a character "constitute the story being told" appears to greatly circumscribe the protection of characters in this circuit. Subsequent decisions in the Ninth Circuit cast doubt on the reasoning and implicitly limit the holding of the Sam Spade case. In Walt Disney Productions v. Air Pirates, this circuit held that several Disney comic characters were protected by copyright. 581 F.2d 751, 755 (9th Cir. 1978). In doing so the Court of Appeals reasoned that because "comic book characters ... are distinguishable from literary characters, the *Warner Bros.* language does not preclude protection of Disney's characters." Id. *Air Pirates* can be interpreted as either attempting to harmonize granting copyright protection to graphic characters with the "story being told" test enunciated in the Sam Spade case or narrowing the "story being told" test to characters in literary works.

[The court proceeded to determine that the characters and relationships in the *Rocky* series "are delineated so extensively that they are protected from bodily appropriation when taken as a group and transposed into a sequel by another author."]

3. Anderson's Work Is an Unauthorized Derivative Work

Under 17 U.S.C. section 106(2), the holder of a copyright has the exclusive right to prepare derivative works based upon his copyrighted work. In this circuit a work is derivative "*only if it would be considered an infringing work if* the material which it had derived from a prior work had been taken without the consent of the copyright proprietor of the prior work." Litchfield v. Spielberg, 736 F.2d 1352, 1354 (9th Cir. 1984) (emphasis in original), citing United States v. Taxe, 540 F.2d 961, 965 n. 2 (9th Cir. 1976). This Court must now examine whether Anderson's treatment is an unauthorized derivative work under this standard.

Usually a court would be required to undertake the extensive comparisons under the *Krofft* substantial similarity test to determine whether Anderson's work is a derivative work. *See* 1 M. Nimmer, §3.01 at 3-3. However, in this case, Anderson has bodily appropriated the *Rocky* characters in his treatment. This Court need not determine whether the characters in Anderson's treatment are substantially similar to Stallone's characters, as it is uncontroverted that the characters were lifted lock, stock, and barrel from the prior *Rocky* movies. Anderson retained the names, relationships and built on the experiences of these characters from the three prior *Rocky* movies. 1 M. Nimmer, §2.12 at 2-177 (copying names of characters is highly probative evidence of infringement). His characters are not merely substantially similar to Stallone's, they are Stallone's characters. As Professor Nimmer stated, "Where there is literal similarity. ... [i]t is not necessary to determine the level of abstraction at which similarity ceases to consist of an 'expression of ideas' since literal similarity by definition is always a similarity as to the expression of ideas." 3 M. Nimmer, §13.03[3], pg. 13-35. Anderson's bodily appropriation of these characters infringes upon the protected expression in the *Rocky* characters and renders his work an unauthorized derivative work. 1 Nimmer, § 2.12 at 2-171. By bodily appropriating the significant elements of protected expression in the *Rocky* characters, Anderson has copied protected expression and his treatment infringes on Stallone's copyrighted work.

4. Since Anderson's Work Is an Unauthorized Derivative Work, No Part of the Treatment Can Be Granted Copyright Protection

Stallone owns the copyrights for the first three *Rocky* movies. Under 17 U.S.C. section 106(2), he has the exclusive right to prepare derivative works based on these copyrighted works. This Court has determined that Anderson's treatment is an unauthorized derivative work. Thus, Anderson has infringed upon Stallone's copyright. See 17 U.S.C. section 501(a).

Nevertheless, plaintiff contends that his infringing work is entitled to copyright protection and he can sue Stallone for infringing upon his treatment. Plaintiff relies upon 17 U.S.C. section 103(a) as support for his position that he is entitled to copyright protection for the non-infringing portions of his treatment. 17 U.S.C. section 103(a) reads:

> The subject matter of copyright as specified by section 102 includes compilations and derivative works, but protection for a work employing preexisting material in which copyright subsists does not extend to any part of the work in which the material has been used unlawfully.

Plaintiff has not argued that section 103(a), on its face, requires that an infringer be granted copyright protection for the non-infringing portions of his work. He has not and cannot provide this Court with a single case that has held that an infringer of a copyright is entitled to sue a third party for infringing the original portions of his work. Nor can he provide a single case that stands for the extraordinary proposition he proposes here, namely, allowing a plaintiff to sue the party whose work he has infringed upon for infringement of his infringing derivative work.

Instead, Anderson alleges that the House Report on section 103(a) indicates that Congress intended protection for the non-infringing portions of derivative works such as his treatment. The House Report for section 103(a) first delineates the differences between compilations and derivative works. H.R. No. 1476, 94th Cong., 2d Sess. at 57-58 (1976). The House Report then reads as follows:

> The second part of the sentence that makes up section 103(a) deals with the status of a compilation or derivative work unlawfully employing preexisting copyrighted material. In providing that protection does not extend to "any part of the work in which such material has been used unlawfully," the bill prevents an infringer from benefiting, through copyright protection, from committing an unlawful act, but preserves protection for those parts of the work that do not employ the preexisting work. Thus, an unauthorized translation of a novel could not be copyrighted at all, but the owner of copyright in an anthology of poetry could sue someone who infringed the whole anthology, even though the infringer proves that publication of one of the poems was unauthorized.

The Court recognizes that the House Report language is muddled. It makes a general statement that non-infringing portions of a work should be granted protection if these portions do not employ the pre-existing work. The report then provides two examples: one involving a compilation where the non-infringing portion was deemed protected, and another involving a derivative work where no part of the work could be protected. The general statement, when taken in the context of the comparison of compilations and derivative works in the section and the two examples given, is best understood as applying only to compilations. Although it is not crystal clear, it appears that the Committee assumed that in a derivative work the underlying work is "employed" throughout.

Professor Nimmer also interprets the House Report language as generally denying copyright protection to any portion of an unauthorized derivative work. After setting forth some of the language from the House Report regarding section 103(a) he states,

> the effect [of section 103(a)] generally would be to deny copyright to derivative works, in which the pre-existing work tends to pervade the entire derivative work, but not to collective works, where the infringement arises from the copying of the selection and arrangement of a number of pre-existing works, and not per se from the reproduction of any particular prior work.

1 M. Nimmer, § 3.06, pgs. 3-22.3 thru 3-22.4. Like the House Report, Nimmer also preceded his conclusion that no part of derivative work unlawfully employing preexisting material should be copyrightable with a general statement that "only that portion of a derivative or collective work which employs the pre-existing work would be denied copyright". 1 M. Nimmer, § 3.06, pg. 3-22.3. At first blush, both Nimmer's and the Committee's language are internally inconsistent. Both start with a general proposition that only the portion of a work which unlawfully employs the prior work should be denied copyright protection. Both then appear to conclude that no part of an infringing derivative work should be granted copyright protection. Only if a derivative work is *assumed* to employ the infringing work throughout do these passages read coherently.

The case law interpreting section 103(a) also supports the conclusion that generally no part of an infringing derivative work should be granted copyright protection. In Eden Toys, Inc. v. Florelee Undergarment Co., the circuit court dealt primarily with the question of whether an authorized derivative work contained sufficient originality to gain copyright protection. 697 F.2d 27, 34-35 (2d Cir. 1982). However, in dicta the court opined on what result would be warranted if the derivative work had been made without the permission of the original author. The Court cited to the aforementioned passages from Professor Nimmer's treatise and the House Report and *assumed* without discussion that the "derivative copyrights would be invalid, since the preexisting illustration used without permission would tend to pervade the entire work." Id. at 34 n. 6. In Gracen v. Bradford, the Seventh Circuit also dealt primarily with whether plaintiff's derivative work had sufficient originality to comply with requirements of section 103. 798 F.2d 300, 302-303 (7th Cir. 1983). *Gracen* also discussed the issue of the copyrightability of an unauthorized derivative work. The Court stated "if Miss Gracen had no authority to make derivative works from the movie, she could not copyright the painting and drawings, and she infringed MGM's copyright by displaying them publicly." Id. at

303. Once again, the Circuit court *assumed* that no part of an unlawful derivative work could be copyrighted.

Plaintiff has written a treatment which is an unauthorized derivative work. This treatment infringes upon Stallone's copyrights and his exclusive right to prepare derivative works which are based upon these movies. 17 U.S.C. § 106(2). Section 103(a) was not intended to arm an infringer and limit the applicability of section 106(2) on unified derivative works. As the House Report and Professor Nimmer's treatise explain, 103(a) was not intended to apply to derivative works and most certainly was not an attempt to modify section 106(2). Section 103(a) allows an author whose authorship essentially is the arrangement or ordering of several independent works to keep the copyright for his arrangement even if one of the underlying works he arranged is found to be used unlawfully. The infringing portion would be easily severable and the scope of the compilation author's own work would be easily ascertainable. Even if this Court were to interpret section 103(a) as allowing an author of an infringing derivative work to sue third parties based on the non-infringing portions of his work, section 106(2) most certainly precludes the author of an unauthorized infringing derivative work from suing the author of the work which he has already infringed. Thus, the Court HOLDS that the defendants are entitled to summary judgment on plaintiff's copyright claims as the plaintiff cannot gain copyright protection for any portion of his work under section 103(a). In addition, Anderson is precluded by section 106(2) from bringing an action for copyright infringement against Stallone and the other defendants.

COMMENTS AND QUESTIONS

1. Why protect derivative works under section 106(2) at all? The Nimmer treatise refers to section 106(2)'s right to prepare derivative works as "completely superfluous." Melville B. Nimmer & David Nimmer, 2 Nimmer on Copyright § 8.09[A], at 8-114. It reasons that infringement of the right to prepare derivative works necessarily also infringes either the right to make copies or the right to perform works, so there is no reason to have both. If derivative works must be "substantially similar" to the underlying work to infringe the section 106(2) right, does it add anything to the protections granted elsewhere in the Copyright Act?

Paul Goldstein suggests that protecting derivative works is necessary in some instances to ensure that adequate incentives are given to copyright holders to develop new works. He argues that the author of a book should own the rights to a movie made out of that book, for example. See Paul Goldstein, Derivative Rights and Derivative Works in Copyright, 30 J. Copyr. Off. Soc'y 209 (1982). But why are derivative rights necessary to accomplish this? In most such situations, the movie (or sequel) will of necessity copy places and characters exactly from the original. This seems to be a clear case of direct infringement on protected expression. If, on the other hand, expression from the original is *not* copied, but only general ideas or themes, is copyright protection desirable?

2. Were Stallone to lose this case, what would prevent prospective authors from generating numerous potential plot permutations for future James Bond films, publicizing them on the Internet, and then waiting to sue when MGM eventually selects one of these story lines? What counterarguments could you offer to this scenario? Is this risk really significant? What could MGM do to protect itself against such potential

plaintiffs? On the other hand, does it seem fair that Stallone is free to take original work written by others without compensation?

3. *The Originality Requirement for Derivative Works.* In Gracen v. Bradford Exchange, 698 F.2d 300, 304 (7th Cir. 1983), MGM had licensed Bradford Exchange to use characters and scenes from the movie in a series of collectors' plates. Bradford invited artists to submit paintings of Dorothy as played by Judy Garland in *The Wizard of Oz*, with the understanding that the artist who submitted the best painting would be offered a contract for the entire series. Bradford supplied contestants with photographs from the movie and the following instruction: "We do want *your* interpretation of these images, but your interpretation must evoke all the warm feeling the people have for the film and its actors. So, *your* Judy/Dorothy must be very recognizable as everybody's Judy/Dorothy." Jorie Gracen submitted a painting of Dorothy skipping along the yellow brick road that closely resembled two still images from the film. It integrated and embellished on the original scenes by, for example, adding a rainbow in the background, placing Toto under Dorothy's arm, and altering the lighting in the scene. After Gracen won the contest, Bradford offered her a contract for the plate series. Gracen declined the offer, after which Bradford hired another artist to produce the series. Bradford provided the artist with a copy of Gracen's entry, from which he produced a similar representation. Gracen sued Bradford for copyright infringement. In finding against Gracen, Judge Posner reasoned that a derivative work must be substantially different from the underlying work to be copyrightable in order to provide courts with a plausible basis for determining whether a derivative work is based on the original or a prior derivative work. Cf. L. Batlin & Son, Inc. v. Snyder, 536 F.2d 486 (2d Cir. 1976) (en banc) (requiring substantial variation); Pickett v. Prince, 207 F.3d 402 (7th Cir. 2000). Should the threshold for originality be higher for derivative works?

4. *Economic Incentives.* Does a broad right to prepare derivative works comport with the economic incentive approach to intellectual property protection? Proponents of an incentive-based view of copyright might well challenge the assumption that authors should receive royalties from derivative works. Do authors really create under the assumption that their works will be translated into different forms? (Some do, certainly; Disney markets its animated films with an eye towards selling T-shirts and stuffed animals as well as movie tickets.) From a strict incentive perspective, should we reward authors in markets they did not originally enter? To what extent does your answer depend on assumptions about the capacity of authors and artists to develop other applications of their work? To what extent does your answer depend on assumptions about the transaction costs of licensing?

The economic rationale for derivative works may break down where the derivative right is used to preclude defendants from developing their own creative works in a market the plaintiff has not herself exploited, but which depends somehow on the plaintiff's work. Something of this sort may have happened in Castle Rock Entertainment v. Carol Publishing Group, 150 F.3d 132 (2d Cir. 1998). There the court enjoined the publication of a Seinfeld trivia book called the *Seinfeld Aptitude Test.* The court reasoned that the book infringed the copyright in the Seinfeld television series because it took numerous "facts" from the episodes created by the copyright owner. Because Seinfeld was fictional, the court concluded that these "facts" constituted copyrightable expression, and the defendants could not prevail on a fair use defense.

Do you find the result in the Seinfeld case problematic? What if the book in question had been an unauthorized biography of Jerry Seinfeld? A news report about a particularly controversial episode? What light, if any, does the case shed on the appropriate limits of the derivative works right?

5. *Comparison to Patent Scope.* Copyright law appears to afford the copyright proprietor broad control of all extensions of their original expression. Section 103(b) extends protection only to new expression, and not to preexisting material included in the derivative work. Only the original author or a licensee is entitled to a copyright in the derivative work. This means that if a filmmaker makes a movie out of a copyrighted book without authorization, adding substantial expression of her own in the process, *she is not entitled to a copyright in any portion of the movie in which infringing material appears.* In addition to *Anderson*, see Sobhani v. @Radical.Media Inc., 257 F.Supp.2d 1234 (C.D.Cal. 2003); Pickett v. Prince, 207 F.3d 402 (7th Cir. 2000).

Does this result make sense? If only new expression is copyrightable as part of a derivative work, why assign that new expression (as distinguished from the original expression) exclusively to the original author?

Compare this result with the "blocking patents" situation in patent law. As Merges explains, this doctrine permits a second inventor to obtain a patent on his improvement even though that improvement also infringes another patent. Robert P. Merges, Intellectual Property Rights and Bargaining Breakdown: The Case of Blocking Patents, 62 Tenn. L. Rev. 74 (1994). Historically this circumstance has led to enhanced bargaining, but the social costs of occasional bargaining breakdown justify a sort of "patent fair use" principle (the "reverse doctrine of equivalents"). Fortunately, although situations are serious given that they often involve significant new technologies, they are relatively rare. In the great run of cases, the ingenious institution of blocking patents balances the rights of original creators and subsequent improvers rather nicely.

No such institution exists in the law of copyrights. See Mark A. Lemley, The Economics of Improvement in Intellectual Property Law, 75 Tex. L. Rev. 989 (1997). Copyright doctrine prohibits a follow-on creator from appropriating and adding to the copyrighted material of an original creator. Goldstein has argued that the cases provide no compelling reason to deny the derivative artist a copyright in his additional creative efforts. See Paul Goldstein, Derivative Rights and Derivative Works in Copyright, 30 J. Copyr. Soc'y 209 (1982) ("The rule [denying copyrightability for unauthorized derivative works] is, however, hard to justify when applied to derivative works such as the motion picture in Sheldon v. MGM [309 U.S. 390 (1940)] in which the underlying work represents only a small part of the value of the derivative work but, because it underlies the whole, will defeat copyright protection for the entire derivative work. Just as an injunction against the motion picture gave plaintiff there a greater return than was needed to induce his investment in the underlying work, so depriving the motion picture owner of all protection against others will give it far less return than is needed to justify investment in the derivative work."); Wendy Gordon, Toward a Jurisprudence of Benefits: The Norms of Copyright and the Problem of Private Censorship, 57 U. Chi. L. Rev. 1009 (1990). Perhaps the best explanation for the lack of a doctrine of "blocking copyrights" is copyright law's policy favoring the reputational interest of authors: by requiring *ex ante* licensing of anyone who wishes to incorporate a copyrighted work into another work, the law ensures that the owner of the copyrighted work will completely control all manifestations of it. The law of derivative works is one way in which a personality or "moral rights" aspect creeps into U.S. copyright law. In effect, the strong protection given derivative works affords the original copyright owner control over alterations or "improvements" upon her work.

Because there is no blocking copyrights doctrine, copyright law is left with a vacuum in certain cases. What should be done with the hypothetical infringer who creates otherwise protectable new expression? Should that new expression be unprotectable because it derives from an infringement? Should it be in the public domain? Should it be deemed "captured" by the original copyright holder?

6. *Video Game "Enhancements."* A series of cases involving video games has addressed whether add-on devices and software designed to enhance the playing experience constitute derivative works. In Midway Mfg. Co. v. Arctic International, Inc., 704 F.2d 1009 (7th Cir. 1983), *cert. denied,* 464 U.S. 923 (1983), the defendant sold printed circuit boards that sped up the play of plaintiff's Galaxian and Pac-Man video games. Interpreting the definition of "derivative work" in section 101 of the Act, the court explained that

> [i]t is not obvious from this language whether a speeded-up video game is a derivative work. A speeded-up phonograph record probably is not. Cf. Shapiro, Bernstein & Co. v. Jerry Vogel Music Co., 73 F.Supp. 165, 167 (S.D.N.Y.1947) ("The change in time of the added chorus, and the slight variation in the base of the accompaniment, there being no change in the tune or lyrics, would not be 'new work'"); 1 Nimmer on Copyright § 3.03 (1982). But that is because the additional value to the copyright owner of having the right to market separately the speeded-up version of the recorded performance is too trivial to warrant legal protection for that right. A speeded-up video game is a substantially different product from the original game. As noted, it is more exciting to play and it requires some creative effort to produce. For that reason, the owner of the copyright on the game should be entitled to monopolize it on the same theory that he is entitled to monopolize the derivative works specifically listed in Section 101. The current rage for video games was not anticipated in 1976, and like any new technology the video game does not fit with complete ease the definition of derivative work in Section 101 of the 1976 Act. But the amount by which the language of Section 101 must be stretched to accommodate speeded-up video games is, we believe, within the limits within which Congress wanted the new Act to operate.

Midway Mfg. Co., 704 F.2d at 1014. Do you agree? We return in Chapter 7 to question of whether video game enhancement tools fall within the fair use privilege. See Lewis Galoob Toys., Inc. v. Nintendo of America, Inc., 964 F.2d 965 (9th Cir. 1992), *cert. denied,* 507 U.S. 985 (1993) (finding fair use); Micro Star v. Formgen, Inc., 154 F.3d 1107 (9th Cir. 1998) (rejecting fair use defense).

7. *Content Filtering.* In 2002, ClearPlay introduced a technology that allows consumers to activate film-specific filters to blackout violent and sexual scenes and to mute profanity for DVDs. In order to accomplish this functionality, ClearPlay wrote software masks that run in parallel with DVDs that instruct the player to skip over particular scenes and mute specific segments. Motion picture studios brought suit, alleging that such scripts constituted unauthorized derivative works. Do such scripts constitute derivative works? Surely a consumer can fast-forward through scenes and press the mute button without violating the rights of the copyright owner. Should the owner of the copyright in the work have the exclusive right to provide add-on technologies to assist the consumer in these efforts?

After several years of litigation, Congress settled this dispute by passing the Family Entertainment and Copyright Act of 2005, which immunizes:

> the making imperceptible, by or at the direction of a member of a private household, of limited portions of audio or video content of a motion picture, during a performance in or transmitted to that household for private home viewing, from an authorized copy of the motion picture, or the creation or provision of a computer program or other technology that enables such making imperceptible and that is designed and marketed to be used, at the direction of a member of a private household, for such making imperceptible, if no fixed copy of the altered version of the motion picture is created by such computer program or other technology.

17 U.S.C. § 110(11).

8. *Software Guides "for Dummies."* Are *Windows for Dummies, Excel for Dummies,* and *Word for Dummies* derivative works? Whereas Hollywood seems to go after derivative uses vigorously, the software industry has been much more welcoming of guides for software products. How would you explain this difference?

PROBLEMS

Problem 4-24. Garamon, a French author, wrote a successful novel in French. The novel is copyrighted in France in 1954. Garamon authorized an English translation of his novel but failed to comply with the formalities then required under U.S. law to obtain a U.S. copyright in the translation. Thus the translation fell into the public domain. Subsequently, Oaktree Press photocopies and distributes the English translation of the novel in the United States. Garamon sues for infringement, not of the translation, but of the copyright on the underlying French novel. Who should prevail?

Problem 4-25. A graphic artist for World Enquirer magazine is asked to produce a seamless integration of two photographs so that it appears that two figures from separate photographs were in the same picture. To accomplish this, he uses digital imaging technology. He feeds both photographs into a scanner, and stores both images separately in the computer. Using a commercially available graphics program, he merges the pictures.

Assume one picture was copyrighted, and the other was in the public domain. What rights do the photo owner and World Enquirer have in the resulting image?

---

3. The Distribution Right

Section 106(3) grants copyright owners the right to distribute, through sale or other means, either the original or subsequent copies of a copyrighted work. This right is closely allied with the right to copy, since copying is the usual means of commercial exploitation of works of authorship. Thus, copying and selling a copyright owner's work without authorization violates both the right to copy and the right to distribute. As a corollary, both the copier who never does anything with his or her copies and the unknowing distributor of unauthorized copies are liable for copyright infringement.

First Sale Doctrine. An important limitation on the exclusive right to distribute is the "first sale doctrine." 17 U.S.C. § 109(a). This doctrine provides that a copyright holder cannot restrict what a purchaser of a particular lawful copy does *with that copy*. The purchaser may not copy it, but may resell it without restriction or liability. In response to the availability of home copying technologies, Congress limited the first sale doctrine by prohibiting the rental of phonorecords and computer programs for profit. 17 U.S.C. § 109(b).

Importation Right. Manufacturers often sell goods in multiple national markets, sometimes pricing goods differently depending upon supply (e.g., cost of manufacturing locally, advertising expense, provision of repair or other services) and demand (e.g., income, tastes, availability of substitutes) conditions. In order to preserve this ability to maintain price differentials across national borders, section 602 of the Copyright Act affords copyright owners a right to prohibit commercial importation into the United States of copyrighted works acquired abroad. This prohibition contains various exceptions allowing importation of works for government use (but not for schools and excluding audiovisual works (except for archival purposes)), personal use (limited numbers of copies), and scholarly, educational, or religious purposes (also subject to specified limitations).

COMMENTS AND QUESTIONS

1. What is the point of having a "right to distribute" if it is limited by the first sale doctrine? Is there any right granted in section 106(3) that does not already exist in the other parts of section 106? One purpose of section 106(3) may be to allow authors to control the right of *first* publication of a work. For example, a copyright owner may be able to prevent a bookstore that acquires advance copies of the new Harry Potter book from "jumping the gun" by selling that book before its official release date. But if the copyright owner has not imposed such a restriction by contract, should copyright law give her the right to limit what the bookstore does with the work?

Granting copyright owners the right to control distribution may also make it easier for them to find and sue infringers. Distributors may be easier to identify than upstream copyists, and because they operate on a larger scale, there may be fewer to go after. Suing distributors may also be more palatable to copyright owners in some cases than suing actual infringers. For example, textbook publishers may be willing to sue commercial copy shops who sell class readers, but unwilling to sue the university professors who authorized making the copies. After all, most of their customers are the same university professors.

2. *Interaction of the First Sale Doctrine and the Right to Prepare Derivative Works: Art Tiles.* Can the purchaser of cards and books containing artwork mount and affix the pictures to ceramic tiles for later sale? The Ninth Circuit has held that the process of attaching art from a book onto ceramic tiles constituted the preparation of a derivative work. See Mirage Editions, Inc. v. Albuquerque A.R.T. Co., 856 F.2d 1341 (9th Cir. 1988). In rejecting a first sale defense, the court recognized that the defendant could purchase a copy of the plaintiff's book and subsequently alienate

its ownership in that book. "However, the right to transfer applies only to the particular copy of the book which appellant has purchased and nothing else. The mere sale of the book to the appellant without a specific transfer by the copyright holder of its exclusive right to prepare derivative works, does not transfer that right to appellant. The derivative works right, remains unimpaired and with the copyright proprietors." Id. at 1344. On nearly identical facts involving the same defendant, the Seventh Circuit reached a contrary result, holding that the mere mounting of pictures on tiles did not rise to the level of originality required to create a derivative work and that the first sale doctrine immunizes the defendants from liability for unauthorized distribution. See Lee v. A.R.T. Company, 125 F.3d 580 (7th Cir. 1997).

Do you agree with the reasoning of the Seventh Circuit or Ninth Circuit in the A.R.T. cases? Is it possible to envision circumstances under which framing a work of art involves original, creative expression? Even if it is, should the copyright owner be entitled to control how the work is framed once it has been sold?

3. *Software Marketing and "License versus Sale."* As a means of controlling post-transaction use of software programs, many software vendors have characterized the distribution of their products as "licenses" rather than "sales" and argued that the first sale doctrine accordingly does not apply. Should they be able to circumvent the first sale doctrine through semantics? In Softman Products Co., LLC v. Adobe Systems, Inc., 171 F. Supp. 2d 1075 (C.D. Cal. 2001), a software distributor had lawfully acquired a retail collection of Adobe software products and then unbundled them for later sale in violation of the terms of an End User Licensing Agreement (EULA), which prohibited the distribution of individual software titles that were originally distributed as part of a collection. Looking to the "economic realities of the exchange" by which Softman acquired the Adobe product, the court determined that "a single payment giving the buyer an unlimited period in which it has a right to possession . . . is a sale." As such, the court found that the first sale doctrine overrode the EULA. See David A. Rice, Licensing the Use of Computer Program Copies and the Copyright Act First Sale Doctrine, 30 Jurimetrics J. 157, 172 (1990) (noting the following factors in characterizing a software transaction as a sale rather than a license: temporally unlimited possession; absence of time limits on copy possession; pricing and payment schemes that are unitary, not serial; subsequent transfer is neither prohibited nor conditioned on obtaining the licensor's prior approval; and the principal purpose of the use restrictions is to protect intangible copyrightable subject matter and not to preserve property interests in individual program copies). We return to this topic in Chapter 7.

4. *Interaction of the First Sale Doctrine and the Right to Import: The Grey Goods/ Parallel Imports Problem.* The interrelationship of the "right to import" and the first sale doctrine conflict in situations in which companies legitimately acquire inexpensively priced copyrighted products in a region of the world with low prices and export them to the United States for resale. Such goods are referred to as "grey goods" or "parallel imports." Does the first sale doctrine provide a defense to the claim of unauthorized importation into the United States?

In Quality King Distributors, Inc. v. L'Anza Research International, Inc., 523 U.S. 135 (1998), a U.S. manufacturer of hair care products (affixed with copyrighted labels) sold some of its products at premium prices to domestic hair salons (where it had spent heavily on advertising and training) and others in foreign markets for 35-40 percent less. The defendant acquired these products abroad, reimported them into the

United States, and sold them through discount outlets for substantially less than the retail prices available in domestic hair salons. L'Anza sued for copyright infringement, claiming that the products were imported without its authorization; Quality King defended under the first sale doctrine. In a decision limited to these particular facts — most significantly, U.S. manufacturing of the product — the Supreme Court held that the first sale doctrine applied since the goods were "lawfully made under this title" under § 109.

The Supreme Court did not, however, address the more significant scenario — where the imported products are manufactured abroad with the authorization of the copyright owner, purchased lawfully there, and then imported into the United States. Several lower court cases hold that the first sale doctrine would not nullify the § 602 importation right in these circumstances. See, e.g., Columbia Broadcasting System v. Scorpio Music Distributors, Inc., 569 F.Supp. 47 (E.D. Pa. 1983), *aff'd mem.*, 738 F.2d 424 (3d Cir. 1984); Parfums Givenchy, Inc. v. Drug Emporium, Inc., 38 F.3d 477 (9th Cir. 1994); but see Cosmair, Inc. v. Dynamite Enters., Inc., 226 U.S.P.Q. 344 (S.D. Fla. 1985).

5. *Interpretation of Prior Distribution Deals in the Wake of New Distribution Platforms.* Whenever new technologies for distributing works of authorship emerge, the question inevitably arises whether older contracts extend to such new distribution channels. Rosetta Books, one of the first companies to distribute electronic books ("eBooks"), obtained permission from well-known authors Kurt Vonnegut, William Styron, and Robert B. Parker to distribute some of their classic titles (*Slaughterhouse-Five, Sophie's Choice*, and *Promised Land*) in eBook form. Their works became available for downloading on Rosetta Books' Web site shortly thereafter. Random House, the exclusive U.S. publisher of the print versions of these books, promptly sued, claiming that its book publishing agreements with these authors to "print, publish, and sell the work in book form" dating back to the 1960s extended to this new medium. This litigation is reminiscent of prior waves of cases following the development of other new distribution media. See Bartsch v. Metro-Goldwyn-Mayer, 391 F.2d 150 (2d Cir. 1950) (whether license to exhibit motion pictures extends to television broadcasts), *cert. denied*, 393 U.S. 826 (1968); Boosey & Hawkes Music Publishers v. Walt Disney, 145 F.3d 481 (2d Cir. 1998) (whether license to record musical composition for use in a motion picture extends to video cassettes); Bourne v. Walt Disney, 68 F.3d 621 (2d Cir. 1995) (same), *cert. denied*, 517 U.S. 1240 (1996). Applying a narrow interpretation of the contractual language, the district court denied Random House's motion for a preliminary injunction, and the Second Circuit affirmed. See Random House, Inc. v. Rosetta Books LLC, 150 F.Supp.2d 613 (S.D.N.Y. 2001), *aff'd*, 283 F.3d 490 (2d Cir. 2002). Should a court interpret the terms of such contracts narrowly (limiting the contract to the literal media described in the agreement) or more expansively, seeking to gauge the larger intent and bargaining positions of the parties and consider knowledge available at the time of interpretation?

PROBLEM

Problem 4-26. Lee, a law student, attends an expensive private school. After buying the books for his first-year classes, he underlines key passages in the books and takes some notes in the margins. At the end of the first year of law school, Lee sells his books back to the law school bookstore. The bookstore in turn sells them to incoming law students the next fall as "used books." Under what theory might Lee or the bookstore be liable for copyright infringement? Should they be?

4. Public Performance and Display Rights

In addition to the rights to make and distribute copies and adaptations of a work, copyright owners possess the exclusive right to perform or display their works publicly. (A 1909 Act requirement that the performance or display be made "for profit" was abolished in the 1976 Act.) 17 U.S.C. § 106(4), (5).

The performance and display rights roughly parallel each other, and most types of copyrighted works are covered by one or the other of the two rights. As a rough approximation, if it moves it's a performance, and if it stays still, it's a display. See 17 U.S.C. § 101 (definitions of "display" and "perform"). Thus paintings, sculptures, photographs, single photo frames from movies or video games, and physical copies of books are displayed. Plays, dances, movies, copyrighted combinations of still photographs (such as slide shows), and readings of books are performed publicly. Importantly, no public display right exists in architectural works, and only a limited public performance right (discussed below) exists in sound recordings.[27] This means, for instance, that it is not copyright infringement to show a house where people can see it, and playing music over the airwaves does not infringe the copyright *in the sound recordings played*.[28] Further, courts have narrowed the definition of a performance in some contexts. For example, one court has held that playing a board game in a national games tournament was not a "performance" of the game. The court restricted the word "play" in section 101 to the playing of sound recordings. *Allen v. Academic Games League of America*, 89 F.3d 614 (9th Cir. 1996).

The distinction between performances and displays is important because the scope of the exclusive protections is quite different. Any physical act taken to make a work perceivable to the viewer or listener, or cause a work to be reproduced (even in a transient form that does not create a "copy" under section 106(1)) is a performance. Playing a CD or a videotape is a performance. Reading a book aloud (but not silently) is a performance. Dancing a work of choreography is a performance. See 17 U.S.C. § 101. By contrast, the definition of a display is limited by section 109(c), which provides that the owner of a particular copy of a work is entitled to display it "to viewers present at the place where the copy is located." This limitation allows most common displays of a work—including those works fixed at a single location from which they are visible to the public. Showing artwork in a

---

27. This right covers only "digital" performance of sound recordings, for example performances transmitted over the Internet. The right was added by Congress in 1995. 17 U.S.C. § 106(6). It is discussed in section H.

28. There is also generally a copyright in the musical composition that is infringed by public performance unless a royalty is paid to the composer or the music publisher to whom the composer transferred the copyright (or the public performance right) to the musical composition. These are typically handled through blanket licenses—permission to publicly perform an song from a vast catalog in exchange for a fee based on the scale of the business activity—administered by the American Society of Composers, Authors and Publishers (ASCAP), Brodacast Music Inc. (BMI), or Society of European Stage Authors and Composers (SESAC).

gallery without authorization is not copyright infringement, but showing a movie in a theater is infringement. Showing the same piece of art through a television broadcast or over a computer network, by contrast, would infringe the display right. See R. Anthony Reese, The First Sale Doctrine in the Era of Digital Networks, 44 B.C. L. Rev. 577 (2003) (explaining the role of the display right in the online environment).

The definitions of both performance and display are quite broad. The limitation to "public" performances and displays, however, prevents many commonplace activities from infringing performance and display rights. The definition of a public performance or display is set forth in section 101:

> To perform or display a work "publicly" means—
> (1) to perform or display it at a place open to the public or at any place where a substantial number of persons outside of a normal circle of a family and its social acquaintances is gathered; or
> (2) to transmit or otherwise communicate a performance or display of the work to a place specified by clause (1) or to the public, by means of any device or process, whether the members of the public capable of receiving the performance or display receive it in the same place or in separate places and at the same time or at different times.

Under this definition, you can probably have a (small) party at which you play a CD or video. Large parties—if they go beyond the amorphous definition of your "social acquaintances"—may pose problems. Further, if the place of performance or display is "open to the public," it does not appear to matter how many people actually view the performance. Clause (2) appears to cover all broadcasts to the "public" even if members of the public do not view the copyrighted work at the same place or time. Thus a television broadcast is a performance, even though no one may watch it at all, or if people watch it only in the privacy of their own homes.

A number of courts have grappled with the problem of when a performance becomes public. In two separate cases, the Third Circuit has held that video rental stores cannot provide viewing rooms for customers,[29] because the performance of a rented movie in such a room is "public" (even though the room is rented only to one group at a time). The Third Circuit explained its rationale in terms that interpret the phrase "public performance" very broadly:

> The Copyright Act speaks of performances at a place open to the public. It does not require that the public place be actually crowded with people. A telephone booth, a taxi cab, and even a pay toilet are commonly regarded as "open to the public," even though they are usually occupied only by one party at a time.

Columbia Pictures v. Aveco, Inc., 800 F.2d 59, 63 (3d Cir. 1986); *accord* Columbia Pictures v. Redd Horne, 749 F.2d 154, 158 (3d Cir. 1984). The Ninth Circuit distinguished these cases from the situation in which hotels rent their guests video-cassettes and provide in-room videocassette players. "While the hotel may indeed be

---

29. In the early days of the VCR, when relatively few households had home video recorders, video rental stores provided separate viewing rooms where a renter could watch the movie they had rented.

'open to the public,' a guest's hotel room, once rented, is not." Columbia Pictures v. Professional Real Estate Investors, Inc., 866 F.2d 278, 281 (9th Cir. 1989).

Sound Recordings

As noted earlier, sound recordings did not come under the protection of federal copyright law until 1972. Even then, the political strength of broadcasters persuaded Congress to deny public performance rights to the owners of sound recordings. Broadcasters did not want to pay royalties to sound recording owners as well as musical composition owners. Musical composition copyright owners, and their collecting societies (ASCAP, BMI, and SESAC) were also concerned that public performance royalties paid to sound recording owners would cut into their share of licensing revenues paid by broadcasters.

The introduction of digital transmission technology in the 1990s altered this balance. Recording artists feared that transmission of near perfect quality recordings that could be recorded on digital audio devices posed a significant risk of piracy. Traditional broadcasters partially aligned with sound recording owners in favoring some protection for digital performance rights in sound recordings because of the competitive threat posed by new broadcasting entities such as digital audio subscription and interactive services. The resulting compromise became the Digital Performance Right in Sound Recordings Act (DPRSRA) of 1995, codified at 17 U.S.C. §§ 106(6), 114. This right is discussed in section H.

STATUTORY LIMITS ON PERFORMANCE AND DISPLAY RIGHTS

Besides the general fair use provision of section 107, which we discuss in section F, there are a number of specific statutory exceptions that limit the scope of the performance and display rights. We summarize each only briefly here; you are encouraged to study the statutory provisions at issue. The provisions are of two basic types.

Public Interest Exemptions

Section 110 of the Copyright Act exempts many "public interest" performances and displays from the reach of sections 106(4) and (5). Thus, most live educational performances and displays are exempt under section 110(1), as are distance learning broadcasts (including via webcasts) made by accredited, nonprofit educational institutions (subject to various conditions) § 110(2). Religious performances and displays are exempt from the act under section 110(3). Face-to-face performances of "nondramatic literary or musical works" for free or for charitable purposes are exempt, reviving in part the "for profit" requirement of the 1909 Act. § 110(4). Record stores may play records without charge to promote their sale under section 110(7), although the closely analogous performance of videos in video stores without permission appears to be prohibited.

After many years of complaints by small business and restaurant owners and performing rights societies over the collection of public performance royalties for broadcast and other recorded music played in these establishments, Congress passed the Fairness in Music Licensing Act in 1998. Due to the political strength of retailers and restaurant owners, Congress substantially broadened an exemption for home listening of transmitted performances to extend to small businesses (less than 2,000 square feet), restaurants (less than 3,750 square feet), and larger establishments conforming to limitations on the number of loudspeakers and television screen size. § 110(5). These establishments would still require a public performance license from the musical composition owner to host live or taped performances. § 513.

In May 2000, a World Trade Organization (WTO) panel found the Fairness in Music Licensing Act to be in violation of the Agreement on Trade Related Aspects of Intellectual Property Rights (TRIPS) and the Berne Convention for the Protection of Literary and Artistic Works requiring that member nations afford copyright owners minimum levels of protection. WTO Dispute Panel Report on Section 110(5) of the U.S. Copyright Act, WT/DS160/R § 7.1, at 69 (Jun. 15, 2000). Although these accords allow member nations to craft limited exceptions that "do not conflict with a normal exploitation of the work and do not unreasonably prejudice the legitimate interests of the right holder," the WTO panel concluded that § 110(5) operated on too large a scale. A Congressional Research Service showed that 65.2 percent of all establishments, 71.8 percent of all drinking establishments, and 27 percent of all retail establishments qualified for the exemption. Rather than appeal the WTO decision, the United States has indicated its intention to amend § 110(5) and has agreed to pay approximately $1.1 million in damages per year until it brings its law into compliance. At this writing, however, there had been no movement to change the law.

Given that the music being played in a bar or restaurant often comes over the radio, and the copyright owner has already been paid for the public performance over the radio, does section 110(5) really "conflict with a normal exploitation of the work"?

Compulsory Licenses

Cable Retransmission. Section 111 authorizes television broadcast relays, or "secondary transmissions," under a variety of circumstances where they are not for profit and are not content-controlled. For example, the owner of an apartment building with a single reception antenna may relay its signal to residents of the building without charge. Cable systems are also entitled to retransmit television broadcasts over their networks and charge a fee for the service, provided that the cable network registers its intent to do so and pays a royalty based on the revenues it receives from subscribers. The royalty rate, a percentage of the cable system's gross receipts (as set forth in the statute), is based on the cable system's retransmission of distant *non-network* programming. Congress determined that content owners could fully recover the value of retransmission of their shows through their contracts with the networks. After all, retransmission in the local market for network programming enhanced advertising revenue, which would be passed through to content owners through the market for shows. Any loss to content owners would come from retransmission to distant markets as such programming lost the opportunity to license their content into the local network media channels. Cable companies cannot delay or alter the programming they relay from local television broadcasts.

Until 1993, the royalties paid to content owners were collected, calculated, and distributed under the supervision of the Copyright Royalty Tribunal, a governmental entity. From 1993 through 2004, distribution disputes were resolved privately or through ad hoc arbitration panels (Copyright Arbitration Royalty Panels (CARPs)). A 2004 law replaced the CARP process with three Copyright Royalty Judges (CRJ), full-time employees of the Library of Congress appointed for six-year terms with an opportunity for reappointment.

Satellite Retransmission. Section 119 provides a similar right to a compulsory license for satellite transmission to "unserved households" — that is, households that do not receive the normal transmission signal from a particular network or other station, either through broadcast or cable. Satellites may broadcast the signal of those stations to subscribing recipients upon the payment of a royalty. Unlike section 111, however, section 119 does not specify the royalty rate. Rather, the rate is subject to voluntary negotiation or compulsory arbitration between the satellite owner and the individual stations or networks.

Jukeboxes. Section 116 authorizes owners of jukeboxes ("coin-operated pho-norecord players," in the words of the statute) to publicly perform the musical works contained in the jukebox subject to a compulsory license. This compulsory license, like that in section 111, is fixed by statute.

Public Broadcasting. Section 118 authorizes public broadcasting stations to transmit musical and artistic (but not literary or audiovisual) works upon payment of a compulsory license. This section does not, however, set royalty rates. Rather, it requires public broadcasters and the owners of such works to negotiate a rate every five years under the supervision of the Librarian of Congress.

Webcasting. Section 114(d)-(j) limits the newly created right of performance in digital sound recordings, both restricting the scope of the new right and creating a compulsory license mechanism to be invoked should the parties fail to agree on a royalty. We explore this compulsory license in section H.

PROBLEMS

Problem 4-27. Ralston Hotels, a national hotel chain, offers guests an "in-room video rental" service. A menu is displayed on the guest's interactive television screen, and the guest can select both a movie and a starting time by using his remote control. Portland Pictures, a major movie producer that owns the video rental rights to its movies, sues Ralston, alleging that each selected movie is a "public performance" and demanding royalties. Who should prevail?

Problem 4-28. Buford, the owner of the Pumpwell service station on a busy commercial street, receives approximately 20 calls each day from customers. Because of the volume of repair work that Pumpwell does, it frequently must put callers on hold. Buford decides to provide those callers with music while they are holding, so he patches his radio into the telephone. When a call is placed on hold, the caller hears a station selected by Buford until the call is picked up again. ASCAP sues Pumpwell for copyright infringement, claiming that Pumpwell is publicly performing its songs. Does ASCAP have a meritorious case?

5.   Moral Rights

Moral rights extend beyond ownership of economic control of works of authorship to encompass protections of the "personality" of the author. They include the right to have one's name associated with one's work (right of attribution) and the right to protect one's works from mutilation or distortion (right of integrity). Moral rights derive from the continental European "author's rights" tradition in intellectual property law, which differs in many ways from the Anglo-American tradition emphasizing the economic function of intellectual property.

The carrot of joining the Berne Convention ultimately led the United States to bring moral rights protection into the Copyright Act. Article 6^bis(1) of the Berne Convention requires that "even after the transfer of [copyrighted works], the author shall have the right to claim authorship of the work, and to object to any distortion, mutilation or other modification of, or other derogatory action in relation to the said work, which would be prejudicial to his honor or reputation." After initially taking the position that a patchwork of existing federal and state protections discharged its moral rights obligations under the Convention,[30] Congress passed the Visual Artists Rights Act of 1990 (VARA), which directly recognizes moral rights of visual artists. Compared to moral rights protections in Europe, U.S. protections apply quite narrowly (only to "work[s] of visual art") and are subject to significant exceptions:

Works Protected.   A "work of visual art" is defined to include paintings, drawings, prints, still photographs (signed by the authors), and sculptures, existing in a single copy or in limited edition (200 or fewer copies). 17 U.S.C. § 101.

Exclusions.   A work of visual art does not include (i) "any poster, map, globe, chart, technical drawing, diagram, model, applied art, motion picture or other audiovisual work, book, magazine, newspaper, periodical, data base, electronic information service, electronic publication, or similar publication"; (ii) "any merchandising item or advertising, promotional, descriptive, covering, or packaging material or container"; or (iii) any work made for hire. 17 U.S.C. § 101.

Rights.   An author of a work of visual art has a right to claim authorship of the work, to prevent the use of her name on works she did not create, and to prevent the use of her name on works that have been modified or distorted. Authors of works of "recognized stature" have the right to prevent destruction of such works. 17 U.S.C. § 106A(a)(3).

Duration.   For works created on or after June 1, 1991, moral rights subsist for the life of the author (or the life of the last surviving artist in the case of joint works). 17 U.S.C. § 106A(d)(3). As a result of legislative confusion about reconciling House and Senate versions of VARA, moral rights protection for works created prior to June

---

30. The legislative history of the Berne Convention Implementation Act of 1988 pointed to rights under section 106; section 115(a)(2) (relating to distortions of musical works under the compulsory license); section 203 (termination of transfers); section 43(a) of the Lanham Act (relating to false designations of origins and false descriptions); state and local laws relating to publicity, contract, fraud and misrepresentation, unfair competition, defamation, and invasion of privacy; and new state statutes in eight states protecting rights of integrity and paternity in certain works of art. H.R. Rep. No. 609, 100th Cong., 2d Sess. 33-34 (1988). We discuss these protections in Chapters 5 and 6. It was highly debatable, however, that this collection of protections fully discharged U.S. obligations under the Berne Convention.

1, 1991 for which the artist retained a copy or copies lasts for the life of the author plus 70 years. If the artist did not retain a copy, then no moral rights exist whatsoever.

Transfer and Waiver. Moral rights are personal to the artist and may be exercised only by the author. Such rights are not transferable, but may be waived (in a writing identifying the work and signed by the author). 17 U.S.C. § 106A(b),(e).

Exceptions. A work that changes as a result of the passage of time or conservation measures does not trigger liability. In the case of works of visual art incorporated into buildings in such a way that removing the work will cause harm to the work, VARA seeks to strike a balance between the economic interests of property owners and the moral rights of authors. The integrity right does not apply to the extent that the artist consented to the installation of the work in the building (and acknowledged that such installation "may subject the work to destruction, distortion, mutilation, or other modification, by reason of its removal"). 17 U.S.C. § 113(d).

Remedies. Violations of moral rights are subject to the full extent of civil copyright remedies, although not to criminal sanctions. Artists need not register their works in order to recover statutory damages or attorneys' fees.

COMMENTS AND QUESTIONS

1. *Comparison to Moral Rights Protection in Continental Europe.* Significant constraints on moral rights protection built into section 106A make it a much more limited than artist protection in Europe. Consider the case involving Picasso's painting "Trois Femmes." Two art investors cut the painting into one-inch squares, which they then marketed as "original Picassos." Daniel Grant, Before You Cut Up That Picasso . . . , World Monitor, Feb. 1992, at 58-59. This mutilation is illegal in France. Because Picasso was dead when this occurred, however, section 106A would not prohibit it in the United States. 17 U.S.C. § 106A(d)(1). In addition, section 106A does not extend to a host of works such as films that are protected under European moral rights regimes. Cf. John Huston — Asphalt Jungle Case, reported at 22 Int'l. Rev. Ind. Prop. & Copyrt. L. 121 (1991) (describing French court's decision to bar showing of "colorized" version of film made in black and white by director John Huston).

Moral rights in the continental tradition are normally thought to be "inalienable." Certainly, the Berne Convention speaks of an author retaining such rights even after relinquishing the copyright. And Sarraute refers to the "inalienable, unbarrable, and perpetual nature of the French moral right." Raymond Sarraute, Current Theory on the Moral Right of Authors and Artists Under French Law, 16 Am. J. Comp. L. 465, 485 (1968). But see Neil W. Netanel, Alienability Restrictions and the Enhancement of Author Autonomy in United States and Continental Copyright Law, 12 Cardozo Arts & Ent. L.J. 1 (1994) (suggesting that the extent of inalienability in continental moral rights law has been overstated). Should moral rights be subject to sale or waiver by contract? If so, what good are they?

2. *Works of "Recognized Stature."* Congress left this important category undefined. Courts require that an artist show: "(1) that the visual art in question has 'stature,' i.e. is viewed as meritorious, and (2) that this stature is 'recognized' by art experts, other members of the artistic community, or by some cross-section of

society." See Carter v. Helmsley-Spear, Inc., 861 F.Supp. 303, 325 (S.D.N.Y. 1994) (granting injunction to prevent destruction of a sculpture in a building lobby), *rev'd on other grounds*, 71 F.3d 77 (2d Cir. 1995) (overturning injunction upon finding that sculpture was a work made for hire, a category of works outside of moral rights protection); Martin v. City of Indianapolis, 982 F.Supp. 625 (S.D. Ind. 1997). Thus, application of the "recognized stature" standard will ordinarily require testimony by expert witnesses. The standard is intended to be objective and not whether the trier of fact considers the work to be aesthetically pleasing.

3. *Other Copyright-Based Moral Rights Protections.* In Gilliam v. American Broadcasting Companies, 538 F.2d 14 (2d Cir. 1976), the British comedy group Monty Python sued to prevent ABC from broadcasting edited versions of their television series "Monty Python's Flying Circus." On a motion for preliminary injunction, the Second Circuit observed that

> It also seems likely that appellants will succeed on the theory that, regardless of the right ABC had to broadcast an edited program, the cuts made constituted an actionable mutilation of Monty Python's work. This cause of action, which seeks redress for defamation of an artist's work, finds its roots in the continental concept of droit moral, or moral right, which may generally be summarized as including the right of the artist to have his work attributed to him in the form in which he created it.
>
> American copyright law, as presently written, does not recognize moral rights or provide a cause of action for their violation, since the law seeks to vindicate the economic, rather than the personal, rights of authors. Nevertheless, the economic incentive for artistic and intellectual creation that serves as the foundation for American copyright law cannot be reconciled with the inability of artists to obtain relief for mutilation or misrepresentation of their work to the public....

The court went on to conclude that this "mutilation" was actionable under the Lanham Act (discussed in Chapter 5):

> It is sufficient to violate the Act that a representation of a product, although technically true, creates a false impression of the product's origin. [citing cases]...
>
> These cases cannot be distinguished from the situation in which a television network broadcasts a program properly designated as having been written and performed by a group, but which has been edited, without the writer's consent, into a form that departs substantially from the original work.... Thus, an allegation that a defendant has presented to the public a "garbled," distorted version of plaintiff's work seeks to redress the very rights sought to be protected by the Lanham Act.

Id. at 23-24. Other cases have held that the publication of public domain fairy tales edited in a form substantially similar to plaintiff's books violated plaintiff's "right of attribution" under the Lanham Act, Waldman Publishing Corp. v. Landoll, Inc., 43 F.3d 775 (2d Cir. 1994), and that the failure to attribute quotes taken from the plaintiff's work to the plaintiff doomed a fair use defense under copyright law. Robinson v. Random House, 877 F. Supp. 830 (S.D.N.Y. 1995). The *Dastar* case, discussed in Chapter 5, may make the application of the Lanham Act problematic in these situations. See Jane Ginsburg, The Right to Claim Authorship in U.S. Copyright and Trademarks Law, 41 Hous. L. Rev. 263 (2004).

Recall Mirage Editions v. Albuquerque A.R.T., discussed *supra* section E.3. The court there held that altering a copyrighted work created an infringing derivative work, even though the alteration did not involve the making of a new copy. Does this rule serve to protect the moral rights of artists by allowing them to prevent alterations of their works even after the artists have sold the work? Does it matter that the copyright owner, rather than the original author, is the one with the power to exercise this right?

4. *Policy Analysis.* Moral rights remain controversial in the United States. They raise a host of difficult philosophical and implementation questions: Should U.S. law offer general protections to authors and artists against alteration or misattribution of their works? Do such rights interfere with the free licensing of works of intellectual property by giving the creator a continual "veto power" over editing and publication? Should moral rights extend beyond the first sale of a book or work of art, preventing its owner from (for example) destroying or mutilating that particular copy? Is it desirable for an artist to be able to control what the owner of a piece of art does with it? Even in the privacy of her own home? The statutes speak of "defacement, mutilation, or destruction" of works of art, but also prevent mere "alteration" and "modification." Who should make this determination? The artist? The courts? What might constitute actionable alteration of a work of art? Moving it? Cf. Pollara v. Seymour, 344 F.3d 265 (2d Cir. 2003).

F. DEFENSES

1. Fair Use

The fair use doctrine traces its origins to a decision by Justice Story involving the copying of the private letters of George Washington.

> Patents and copyrights approach, nearer than any other class of cases belonging to forensic discussions, to what may be called the metaphysics of the law, where the distinctions are, or at least may be, very subtile [sic] and refined, and, sometimes, almost evanescent. . . . [I]n cases of copyright, . . . the identity of the two works in substance, and the question of piracy, often depend upon a nice balance of the comparative use made in one of the materials of the other; the nature, extent, and value of the materials thus used; the objects of each work; and the degree to which each writer may be fairly presumed to have resorted to the same common sources of information, or to have exercised the same common diligence in the selection and arrangement of the materials. Thus, for example, no one can doubt that a reviewer may fairly cite largely from the original work, if his design be really and truly to use the passages for the purposes of fair and reasonable criticism. On the other hand, it is as clear, that if he thus cites the most important parts of the work, with a view, not to criticise, but to supersede the use of the original work, and substitute the review for it, such a use will be deemed in law a piracy. A wide interval might, of course, exist between these two extremes, calling for great caution and involving great difficulty . . .

Folsom v. Marsh, 9 F. Cas. 342 (C.C.D. Mass. 1841). Although ultimately determining that Washington's letters could not be reproduced in substantial part without authorization, the case provided the foundation for one of copyright law's most important safety valves for promoting cumulative creativity and free expression.

Numerous cases refined, elaborated, and applied this doctrine, although such efforts have failed to bring about predictable results. Thus, nearly a century after Justice Story conceptualized the fair use doctrine, the great copyright jurist Learned Hand characterized it as "the most troublesome in the whole law of copyright." Dellar v. Samuel Goldwyn, Inc., 104 F.2d 661, 662 (2d Cir. 1939).

Congress sought to bring some clarity to this doctrine in its comprehensive codification of copyright law in 1976:

17 U.S.C. § 107. Limitations on Exclusive Rights: Fair Use

Notwithstanding the provisions of section 106, the fair use of a copyrighted work, including such use by reproduction in copies or phonorecords or by any other means specified by that section, for purposes such as criticism, comment, news reporting, teaching (including multiple copies for classroom use), scholarship, or research, is not an infringement of copyright. In determining whether the use made of a work in any particular case is a fair use the factors to be considered shall include —

(1) the purpose and character of the use, including whether such use is of a commercial nature or is for nonprofit educational purposes;

(2) the nature of the copyrighted work;

(3) the amount and substantiality of the portion used in relation to the copyrighted work as a whole; and

(4) the effect of the use upon the potential market for or value of the copyrighted work.

The fact that a work is unpublished shall not itself bar a finding of fair use if such finding is made upon consideration of all the above factors.

In the accompanying report, Congress indicated that the list of factors in section 107 is illustrative rather than exhaustive and that it "intended to restate the present judicial doctrine of fair use, not to change, narrow, or enlarge it in any way." House Report No. 94-1476 at 66. Therefore, the many fair use decisions continue to provide the backdrop for applying this doctrine. Lawyers must often review dozens of opinions in order to discern the contours of this defense.

Harper & Row, Publishers, Inc., et al. v. Nation Enterprises et al.
Supreme Court of the United States
471 U.S. 539 (1985)

Justice O'CONNOR delivered the opinion of the Court.

This case requires us to consider to what extent the "fair use" provision of the Copyright Revision Act of 1976 (hereinafter the Copyright Act), 17 U.S.C. § 107, sanctions the unauthorized use of quotations from a public figure's unpublished manuscript. In March 1979, an undisclosed source provided The Nation Magazine with the unpublished manuscript of "A Time to Heal: The Autobiography of Gerald R. Ford." Working directly from the purloined manuscript, an editor of The Nation produced a short piece entitled "The Ford Memoirs—Behind the Nixon Pardon." The piece was timed to "scoop" an article scheduled shortly to appear in Time Magazine. Time had agreed to purchase the exclusive right to print prepublication excerpts from the copyright holders, Harper & Row Publishers, Inc. (hereinafter Harper & Row), and Reader's Digest Association, Inc. (hereinafter Reader's Digest). As a result of The Nation article, Time canceled its agreement. Petitioners brought a successful copyright action against The Nation. On appeal, the Second Circuit reversed the lower court's finding of infringement, holding that The Nation's act was sanctioned as a "fair use" of the copyrighted material. We granted certiorari, 467 U.S. 1214 (1984), and we now reverse.

. . . The [Time] issue featuring the excerpts was timed to appear approximately one week before shipment of the full length book version to bookstores. Exclusivity was an important consideration; Harper & Row instituted procedures designed to maintain the confidentiality of the manuscript, and Time retained the right to renegotiate the second payment should the material appear in print prior to its release of the excerpts.

Two to three weeks before the Time article's scheduled release, an unidentified person secretly brought a copy of the Ford manuscript to Victor Navasky, editor of The Nation, a political commentary magazine. Mr. Navasky knew that his possession of the manuscript was not authorized and that the manuscript must be returned quickly to his "source" to avoid discovery. He hastily put together what he believed was "a real hot news story" composed of quotes, paraphrases, and facts drawn exclusively from the manuscript. Ibid. Mr. Navasky attempted no independent commentary, research or criticism, in part because of the need for speed if he was to "make news" by "publish[ing] in advance of publication of the Ford book." The 2,250-word article, reprinted in the Appendix to this opinion, appeared on April 3, 1979. As a result of The Nation's article, Time canceled its piece and refused to pay the remaining $12,500. . . .

II

We agree with the Court of Appeals that copyright is intended to increase and not to impede the harvest of knowledge. But we believe the Second Circuit gave insufficient deference to the scheme established by the Copyright Act for fostering the original works that provide the seed and substance of this harvest. The rights conferred by copyright are designed to assure contributors to the store of knowledge a fair return for their labors. Twentieth Century Music Corp. v. Aiken, 422 U.S. 151, 156 (1975). . . .

III

A . . .

"[T]he author's consent to a reasonable use of his copyrighted works ha[d] always been implied by the courts as a necessary incident of the constitutional policy of promoting the progress of science and the useful arts, since a prohibition of such use would inhibit subsequent writers from attempting to improve upon prior works and thus...frustrate the very ends sought to be attained." Ball 260. Professor Latman, in a study of the doctrine of fair use commissioned by Congress for the revision effort, see Sony Corp. of America v. Universal City Studios, Inc., 464 U.S., at 462-463, n.9 (dissenting opinion), summarized prior law as turning on the "importance of the material copied or performed from the point of view of the reasonable copyright owner. In other words, would the reasonable copyright owner have consented to the use?"

Perhaps because the fair use doctrine was predicated on the author's implied consent to "reasonable and customary" use when he released his work for public consumption, fair use traditionally was not recognized as a defense to charges of copying from an author's as yet unpublished works. Under common-law copyright, "the property of the author...in his intellectual creation [was] absolute until he voluntarily part[ed] with the same." American Tobacco Co. v. Werckmeister, 207 U.S. 284, 299 (1907); 2 Nimmer § 8.23, at 8-273. This absolute rule, however, was tempered in practice by the equitable nature of the fair use doctrine. In a given case, factors such as implied consent through de facto publication or performance or dissemination of a work may tip the balance of equities in favor of prepublication use. See Copyright Law Revision — Part 2: Discussion and Comments on Report of the Register of Copyrights on General Revision of the U.S. Copyright Law, 88th Cong., 1st Sess., 27 (H.R. Comm. Print 1963) (discussion suggesting works disseminated to the public in a form not constituting a technical "publication" should nevertheless be subject to fair use); 3 Nimmer § 13.05, at 13-62, n.2. But it has never been seriously disputed that "the fact that the plaintiff's work is unpublished...is a factor tending to negate the defense of fair use." Ibid. Publication of an author's expression before he has authorized its dissemination seriously infringes the author's right to decide when and whether it will be made public, a factor not present in fair use of published works. Respondents contend, however, that Congress, in including first publication among the rights enumerated in § 106, which are expressly subject to fair use under § 107, intended that fair use would apply in pari materia to published and unpublished works. The Copyright Act does not support this proposition.

We also find unpersuasive respondents' argument that fair use may be made of a soon-to-be-published manuscript on the ground that the author has demonstrated he has no interest in nonpublication. This argument assumes that the unpublished nature of copyrighted material is only relevant to letters or other confidential writings not intended for dissemination. It is true that common-law copyright was often enlisted in the service of personal privacy. See Brandeis & Warren, The Right to Privacy, 4 Harv. L. Rev. 193, 198-199 (1890). In its commercial guise, however, an author's right to choose when he will publish is no less deserving of protection. The period encompassing the work's initiation, its preparation, and its grooming for public dissemination is a crucial one for any literary endeavor. The Copyright Act, which accords the copyright owner the "right to control the first public distribution" of his work, House Report, at 62, echos the common law's concern that the author or copyright owner retain control throughout this critical stage. See generally Comment, The Stage of Publication as a "Fair Use" Factor: Harper & Row, Publishers v. Nation Enterprises, 58 St. John's L. Rev. 597 (1984). The obvious benefit to author and public alike of assuring authors the leisure to develop their ideas free from fear of expropriation outweighs any short-term "news value" to be gained from premature publication of the author's expression. See Goldstein, Copyright and the First Amendment, 70 Colum. L. Rev. 983, 1004-1006 (1970) (The absolute protection the common law accorded to soon-to-be published works "[was] justified by [its] brevity and expedience"). The author's control of first public distribution implicates not only his personal interest in creative control but his property interest in exploitation of prepublication rights, which are valuable in themselves and serve as a valuable adjunct to publicity and marketing. See Belushi v. Woodward, 598 F. Supp. 36 (D.C. 1984) (successful marketing depends on coordination of serialization and release to public); Marks, Subsidiary Rights and Permissions, in What Happens in Book Publishing 230 (C. Grannis ed. 1967) (exploitation of subsidiary rights is necessary to financial success of new books). Under ordinary circumstances, the author's right to control the first public appearance of his undisseminated expression will outweigh a claim of fair use.

B.

Respondents, however, contend that First Amendment values require a different rule under the circumstances of this case. The thrust of the decision below is that "[t]he scope of [fair use] is undoubtedly wider when the information conveyed relates to matters of high public concern." Consumers Union of the United States, Inc. v. General Signal Corp., 724 F.2d 1044, 1050 (CA2 1983) (construing 723 F.2d 195 (CA2 1983) (case below) as allowing advertiser to quote Consumer Reports), cert. denied, 469 U.S. 823 (1984). Respondents advance the substantial public import of the subject matter of the Ford memoirs as grounds for excusing a use that would ordinarily not pass muster as a fair use — the piracy of verbatim quotations for the purpose of "scooping" the authorized first serialization. Respondents explain their copying of Mr. Ford's expression as essential to reporting the news story it claims the book itself represents. In respondents' view, not only the facts contained in Mr. Ford's memoirs, but "the precise manner in which [he] expressed himself [were] as news-worthy as what he had to say." Respondents argue that the public's interest in learning this news as fast as possible outweighs the right of the author to control its first publication.

The Second Circuit noted, correctly, that copyright's idea/expression dichotomy "strike[s] a definitional balance between the First Amendment and the Copyright Act by permitting free communication of facts while still protecting an author's expression." 723 F.2d, at 203. No author may copyright his ideas or the facts he narrates. 17 U.S.C. §§ 102(b). See, e.g., New York Times Co. v. United States, 403 U.S. 713, 726, n.*(Brennan, J., concurring) (Copyright laws are not restrictions on freedom of speech as copyright protects only form of expression and not the ideas expressed); 1 Nimmer § 1.10[B][2]. As this Court long ago observed: "[T]he news element — the information respecting current events contained in the literary produc-tion — is not the creation of the writer, but is a report of matters that ordinarily are publici juris; it is the history of the day." International News Service v. Associated Press, 248 U.S. 215, 234 (1918). But copyright assures those who write and publish factual narratives such as "A Time to Heal" that they may at least enjoy the right to market the original expression contained therein as just compensation for their investment. Cf. Zacchini v. Scripps-Howard Broadcasting Co., 433 U.S. 562, 575 (1977).

Respondents' theory, however, would expand fair use to effectively destroy any expectation of copyright protection in the work of a public figure. Absent such pro-tection, there would be little incentive to create or profit in financing such memoirs, and the public would be denied an important source of significant historical information. The promise of copyright would be an empty one if it could be avoided merely by dubbing the infringement a fair use "news report" of the book. See Wainwright Securities Inc. v. Wall Street Transcript Corp., 558 F.2d 91 (CA2 1977), *cert. denied*, 434 U.S. 1014 (1978).

Nor do respondents assert any actual necessity for circumventing the copyright scheme with respect to the types of works and users at issue here. Where an author and publisher have invested extensive resources in creating an original work and are poised to release it to the public, no legitimate aim is served by pre-empting the right of first publication. The fact that the words the author has chosen to clothe his narrative may of themselves be "newsworthy" is not an independent justification for unauthorized copying of the author's expression prior to publication....

In our haste to disseminate news, it should not be forgotten that the Framers intended copyright itself to be the engine of free expression. By establishing a marketable right to the use of one's expression, copyright supplies the economic incentive to create and disseminate ideas....

It is fundamentally at odds with the scheme of copyright to accord lesser rights in those works that are of greatest importance to the public. Such a notion ignores the major premise of copyright and injures author and public alike. "[T]o propose that fair use be imposed whenever the 'social value [of dissemination]...outweighs any detriment to the artist,' would be to propose depriving copyright owners of their right in the property precisely when they encounter those users who could afford to pay for it." Gordon, Fair Use as Market Failure: A Structural and Economic Analysis of the *Betamax* Case and its Predecessors, 82 Colum. L. Rev. 1600, 1615 (1982). And as one commentator has noted: "If every volume that was in the public interest could be pirated away by a competing publisher,...the public [soon] would have nothing worth reading." Sobel, Copyright and the First Amendment: A Gathering Storm?, 19 ASCAP Copyright Law Symposium 43, 78 (1971). See generally Comment, Copyright and the First Amendment; Where Lies the Public Interest?, 59 Tulane L. Rev. 135 (1984).

Moreover, freedom of thought and expression "includes both the right to speak freely and the right to refrain from speaking at all." Wooley v. Maynard, 430 U.S. 705, 714 (1977) (Burger, C.J.). We do not suggest this right not to speak would sanction abuse of the copyright owner's monopoly as an instrument to suppress facts. But in the words of New York's Chief Judge Fuld:

> The essential thrust of the First Amendment is to prohibit improper restraints on the *voluntary* public expression of ideas; it shields the man who wants to speak or publish when others wish him to be quiet. There is necessarily, and within suitably defined areas, a concomitant freedom *not* to speak publicly, one which serves the same ultimate end as freedom of speech in its affirmative aspect.

Estate of Hemingway v. Random House, Inc., 23 N.Y.2d 341, 348, 296 N.Y.S.2d 771, 776, 244 N.E.2d 250, 255 (1968). Courts and commentators have recognized that copyright, and the right of first publication in particular, serve this countervailing First Amendment value. See Schnapper v. Foley, 667 F.2d 102 (1981), *cert. denied*, 455 U.S. 948 (1982); 1 Nimmer §§1.10[B], at 1-70, n. 24; Patry 140-142.

In view of the First Amendment protections already embodied in the Copyright Act's distinction between copyrightable expression and uncopyrightable facts and ideas, and the latitude for scholarship and comment traditionally afforded by fair use, we see no warrant for expanding the doctrine of fair use to create what amounts to a public figure exception to copyright. Whether verbatim copying from a public figure's manuscript in a given case is or is not fair must be judged according to the traditional equities of fair use.

IV

...

The four factors identified by Congress as especially relevant in determining whether the use was fair are: (1) the purpose and character of the use; (2) the nature of the copyrighted work; (3) the substantiality of the portion used in relation to the copyrighted work as a whole; (4) the effect on the potential market for or value of the copyrighted work. We address each one separately.

Purpose of the Use. The Second Circuit correctly identified news reporting as the general purpose of The Nation's use. News reporting is one of the examples enumerated in § 107 to "give some idea of the sort of activities the courts might regard as fair use under the circumstances." Senate Report, at 61. This listing was not intended to be exhaustive, see ibid.; § 101 (definition of "including" and "such as"), or to single out any particular use as presumptively a "fair" use. The drafters resisted pressures from special interest groups to create presumptive categories of fair use, but structured the provision as an affirmative defense requiring a case-by-case analysis. See H.R. Rep. No. 83, 90th Cong., 1st Sess., 37 (1967); Patry 477, n.4. "[W]hether a use referred to in the first sentence of section 107 is a fair use in a particular case will depend upon the application of the determinative factors, including those mentioned in the second sentence." Senate Report, at 62. The fact that an article arguably is "news" and therefore a productive use is simply one factor in a fair use analysis.

We agree with the Second Circuit that the trial court erred in fixing on whether the information contained in the memoirs was actually new to the public. As Judge Meskill wisely noted, "[c]ourts should be chary of deciding what is and what is not news." 723 F.2d, at 215 (dissenting). Cf. Gertz v. Robert Welch, Inc., 418 U.S. 323, 345-346 (1974). "The issue is not what constitutes 'news,' but whether a claim of news reporting is a valid fair use defense to an infringement of copyrightable expression." Patry 119. The Nation has every right to seek to be the first to publish information. But The Nation went beyond simply reporting uncopyrightable information and actively sought to exploit the headline value of its infringement, making a "news event" out of its unauthorized first publication of a noted figure's copyrighted expression.

The fact that a publication was commercial as opposed to nonprofit is a separate factor that tends to weigh against a finding of fair use. "[E]very commercial use of copyrighted material is presumptively an unfair exploitation of the monopoly privilege that belongs to the owner of the copyright." Sony Corp. of America v. Universal City Studios, Inc., 464 U.S., at 451. In arguing that the purpose of news reporting is not purely commercial, The Nation misses the point entirely. The crux of the profit/nonprofit distinction is not whether the sole motive of the use is monetary gain but whether the user stands to profit from exploitation of the copyrighted material without paying the customary price. See Roy Export Co. Establishment v. Columbia

Broadcasting System, Inc., 503 F. Supp., at 1144; 3 Nimmer § 13.05[A][1], at 13-71, n.25.3.

In evaluating character and purpose we cannot ignore The Nation's stated purpose of scooping the forthcoming hardcover and Time abstracts. The Nation's use had not merely the incidental effect but the intended purpose of supplanting the copyright holder's commercially valuable right of first publication. See Meredith Corp. v. Harper & Row, Publishers, Inc., 378 F. Supp. 686, 690 (S.D.N.Y.) (purpose of text was to compete with original), *aff'd*, 500 F.2d 1221 (CA2 1974). Also relevant to the "character" of the use is "the propriety of the defendant's conduct." 3 Nimmer § 13.05[A], at 13-72. "Fair use presupposes 'good faith' and 'fair dealing.'" Time Inc. v. Bernard Geis Associates, 293 F. Supp. 130, 146 (S.D.N.Y. 1968), quoting Schulman, Fair Use and the Revision of the Copyright Act, 53 Iowa L. Rev. 832 (1968). The trial court found that The Nation knowingly exploited a purloined manuscript. Unlike the typical claim of fair use, The Nation cannot offer up even the fiction of consent as justification. Like its competitor newsweekly, it was free to bid for the right of abstracting excerpts from "A Time to Heal." Fair use "distinguishes between 'a true scholar and a chiseler who infringes a work for personal profit.'" Wainwright Securities Inc. v. Wall Street Transcript Corp., 558 F.2d, at 94, quoting from Hearings on Bills for the General Revision of the Copyright Law before the House Committee on the Judiciary, 89th Cong., 1st Sess., ser. 8, pt. 3, p. 1706 (1966) (statement of John Schulman).

Nature of the Copyrighted Work. Second, the Act directs attention to the nature of the copyrighted work. "A Time to Heal" may be characterized as an unpublished historical narrative or autobiography. The law generally recognizes a greater need to disseminate factual works than works of fiction or fantasy. See Gorman, Fact or Fancy? The Implications for Copyright, 29 J. Copyright Soc. 560, 561 (1982).

> [E]ven within the field of fact works, there are gradations as to the relative proportion of fact and fancy. One may move from sparsely embellished maps and directories to elegantly written biography. The extent to which one must permit expressive language to be copied, in order to assure dissemination of the underlying facts, will thus vary from case to case. Id., at 563.

Some of the briefer quotes from the memoirs are arguably necessary adequately to convey the facts; for example, Mr. Ford's characterization of the White House tapes as the "smoking gun" is perhaps so integral to the idea expressed as to be inseparable from it. Cf. 1 Nimmer § 1.10[C]. But The Nation did not stop at isolated phrases and instead excerpted subjective descriptions and portraits of public figures whose power lies in the author's individualized expression. Such use, focusing on the most expressive elements of the work, exceeds that necessary to disseminate the facts.

The fact that a work is unpublished is a critical element of its "nature." 3 Nimmer § 13.05[A]; Comment, 58 St. John's L. Rev., at 613. Our prior discussion establishes that the scope of fair use is narrower with respect to unpublished works. While even substantial quotations might qualify as fair use in a review of a published work or a news account of a speech that had been delivered to the public or disseminated to the press, see House Report, at 65, the author's right to control the first public appearance of his expression weighs against such use of the work before its release. The right of first publication encompasses not only the choice whether to publish at all, but also the choices of when, where, and in what form first to publish a work.

In the case of Mr. Ford's manuscript, the copyright holders' interest in confidentiality is irrefutable; the copyright holders had entered into a contractual undertaking to "keep the manuscript confidential" and required that all those to whom the manuscript was shown also "sign an agreement to keep the manuscript confidential." While the copyright holders' contract with Time required Time to submit its proposed article seven days before publication, The Nation's clandestine publication afforded no such opportunity for creative or quality control. It was hastily patched together and contained "a number of inaccuracies." A use that so clearly infringes the copyright holder's interests in confidentiality and creative control is difficult to characterize as "fair."

Amount and Substantiality of the Portion Used. Next, the Act directs us to examine the amount and substantiality of the portion used in relation to the copyrighted work as a whole. In absolute terms, the words actually quoted were an insubstantial portion of "A Time to Heal." The District Court, however, found that "[T]he Nation took what was essentially the heart of the book." 557 F. Supp., at 1072. We believe the Court of Appeals erred in overruling the District Judge's evaluation of the qualitative nature of the taking. See, e.g., Roy Export Co. Establishment v. Columbia Broadcasting System, Inc., 503 F. Supp., at 1145 (taking of 55 seconds out of 1 hour and 29-minute film deemed qualitatively substantial). A Time editor described the chapters on the pardon as "the most interesting and moving parts of the entire manuscript." The portions actually quoted were selected by Mr. Navasky as among the most powerful passages in those chapters. He testified that he used verbatim excerpts because simply reciting the information could not adequately convey the "absolute certainty with which [Ford] expressed himself," or show that "this comes from President Ford," or carry the "definitive quality" of the original. In short, he quoted these passages precisely because they qualitatively embodied Ford's distinctive expression.

As the statutory language indicates, a taking may not be excused merely because it is insubstantial with respect to the infringing work. As Judge Learned Hand cogently remarked, "no plagiarist can excuse the wrong by showing how much of his work he did not pirate." Sheldon v. Metro-Goldwyn Pictures Corp., 81 F.2d 49, 56 (CA2), *cert. denied*, 298 U.S. 669 (1936). Conversely, the fact that a substantial portion of the infringing work was copied verbatim is evidence of the qualitative value of the copied material, both to the originator and to the plagiarist who seeks to profit from marketing someone else's copyrighted expression.

Stripped to the verbatim quotes, the direct takings from the unpublished manuscript constitute at least 13% of the infringing article. See Meeropol v. Nizer, 560 F.2d 1061, 1071 (CA2 1977) (copyrighted letters constituted less than 1% of infringing work but were prominently featured). The Nation article is structured around the quoted excerpts which serve as its dramatic focal points. In view of the expressive value of the excerpts and their key role in the infringing work, we cannot agree with the Second Circuit that the "magazine took a meager, indeed an infinitesimal amount of Ford's original language." 723 F.2d, at 209.

Effect on the Market. Finally, the Act focuses on "the effect of the use upon the potential market for or value of the copyrighted work." This last factor is undoubtedly the single most important element of fair use.[9] See 3 Nimmer § 13.05[A], at 13-

---

9. Economists who have addressed the issue believe the fair use exception should come into play only in those situations in which the market fails or the price the copyright holder would ask is near zero. See,

76, and cases cited therein. "Fair use, when properly applied, is limited to copying by others which does not materially impair the marketability of the work which is copied." 1 Nimmer § 1.10[D], at 1-87. The trial court found not merely a potential but an actual effect on the market. Time's cancellation of its projected serialization and its refusal to pay the $12,500 were the direct effect of the infringement. . . .

More important, to negate fair use one need only show that if the challenged use "should become widespread, it would adversely affect the *potential* market for the copyrighted work." Sony Corp. of America v. Universal City Studios, Inc., 464 U.S., at 451 (emphasis added); id., at 484, and n. 36 (collecting cases) (dissenting opinion). This inquiry must take account not only of harm to the original but also of harm to the market for derivative works. See Iowa State University Research Foundation, Inc. v. American Broadcasting Cos., 621 F.2d 57 (CA2 1980); Meeropol v. Nizer, supra, at 1070; Roy Export v. Columbia Broadcasting System, Inc., 503 F. Supp., at 1146. "If the defendant's work adversely affects the value of any of the rights in the copyrighted work (in this case the adaptation [and serialization] right) the use is not fair." 3 Nimmer § 13.05[B], at 13-77 — 13-78 (footnote omitted).

It is undisputed that the factual material in the balance of The Nation's article, besides the verbatim quotes at issue here, was drawn exclusively from the chapters on the pardon. The excerpts were employed as featured episodes in a story about the Nixon pardon — precisely the use petitioners had licensed to Time. The borrowing of these verbatim quotes from the unpublished manuscript lent The Nation's piece a special air of authenticity — as Navasky expressed it, the reader would know it was Ford speaking and not The Nation. Thus it directly competed for a share of the market for prepublication excerpts. The Senate Report states:

> "With certain special exceptions . . . a use that supplants any part of the normal market for a copyrighted work would ordinarily be considered an infringement."

Senate Report, at 65. Placed in a broader perspective, a fair use doctrine that permits extensive prepublication quotations from an unreleased manuscript without the copyright owner's consent poses substantial potential for damage to the marketability of first serialization rights in general.

> "Isolated instances of minor infringements, when multiplied many times, become in the aggregate a major inroad on copyright that must be prevented."

Ibid.

COMMENTS AND QUESTIONS

1. A spirited dissent in *Harper & Row* (omitted here) argued that *The Nation* had taken no more than was necessary to report the story, and that most of what was taken

---

e.g., T. Brennan, Harper & Row v. The Nation, Copyrightability and Fair Use, Dept. of Justice Economic Policy Office Discussion Paper 13-17 (1984); Gordon, Fair Use as Market Failure: A Structural and Economic Analysis of the Betamax Case and its Predecessors, 82 Colum. L. Rev. 1600, 1615 (1982). As the facts here demonstrate, there is a fully functioning market that encourages the creation and dissemination of memoirs of public figures. In the economists' view, permitting "fair use" to displace normal copyright channels disrupts the copyright market without a commensurate public benefit.

reflected ideas rather than expression. Given the newsworthiness of this information, shouldn't such use fit within the fair use doctrine?

2. *Commercial Use.* One common category of "fair use" involves a minor, non-commercial use of a copyrighted work, often on behalf of a non-profit organization such as an educational institution. There are two important differences between the use in *Harper & Row* and the traditional "public interest" concept of fair use: the use in *Harper & Row* was for profit, and it had a direct market impact. *The Nation* took significant excerpts from Ford's book, and those excerpts constituted the bulk of the *Nation* article. These factors weigh against a finding of fair use. Is there a strong "public interest" in allowing *The Nation* to publish its story that should outweigh those factors?

3. *Unclean Hands.* As noted in *Harper & Row*, the fair use defense, as an equitable doctrine, involves consideration of the "propriety of [a] defendant's conduct." But how far should the defendant's bad faith be taken? Some have argued that this consideration conflicts with the policy interests underlying copyright and fair use protections. See Pierre N. Leval, Toward a Fair Use Standard, 103 Harv. L. Rev. 1105, 1126-28 (1990). The Federal Circuit, however, has held that "an individual must possess an authorized copy of a literary work" in order to invoke the fair use defense. Atari Games Corp. v. Nintendo of Am. Inc., 975 F.2d 832, 843 (Fed. Cir. 1992). Other courts take a more flexible stance. See NXIVM Corp. v. Ross Institute, 364 F.3d 471, 478 (2d Cir. 2004); Religious Tech. Ctr. v. Netcom On-Line Communication Servs., Inc., 923 F.Supp. 1231, 1244 n. 14 (N.D. Cal. 1995) (noting that "[n]othing in Harper & Row indicates that [the defendants'] bad faith [is] itself conclusive of the fair use question, or even of the first factor"); see generally Nimmer on Copyright § 13.05[A][1][d] (2003)(noting that "knowing use of a purloined manuscript militates against a fair use defense," but not suggesting that bad faith is an absolute bar to fair use).

4. *Copyright and Time.* Suppose *The Nation* had published its article two weeks *after Time*'s article. Would the court reach a different result? Should it? What about a biographer who quotes extensively from Ford's memoirs 30 years later? Several scholars have suggested that copyright can both ensure adequate primary incentives (appropriability) and better promote creativity by allowing greater use of copyrighted works as they age. See Justin Hughes, Fair Use Across Time, 50 UCLA L. Rev. 775 (2003); Joseph P. Liu, Copyright and Time: A Proposal, 101 Mich. L. Rev. 409 (2002); Peter S. Menell, Tailoring Legal Protection for Computer Software, 39 Stan. L. Rev. 1329 (1987) (showing that the economics of network markets favors affording greater access to software systems after they become *de facto* standards, a form of copyright genericide).

5. *Unpublished Works.* As with other intellectual property regimes, the basic theory behind copyright law is that protection will promote more creation and the public will therefore benefit. What does this suggest for copyrighted works that are kept private? Should they be entitled to more or less protection? The *Harper & Row* court seems to believe that such works are entitled to more protection because the privacy of the author is implicated. Can an argument be made to the contrary, particularly where as here the work is about to be released? Do we really want the monopoly provided to copyright holders to be extended to "monopoly reporting" of news events by the media?

In recognition of concern that strong protection for unpublished works could adversely affect historical research and other important values reflected in the fair use doctrine, Congress amended section 107 in 1992 to add the following caveat: "The fact that a work is unpublished shall not itself bar a finding of fair use if such finding is made upon consideration of all of [the section 107] factors." 102 Pub. L. No. 492, 106 Stat. 3145 (1992). The extent to which this provision re-equilibrates the balance in favor of the use of unpublished works remains unclear. The House Report accompanying this bill expressly approved the statement in *Harper & Row* that the unpublished nature of the work is a "key though not necessarily determinative factor tending to negate the defense of fair use." See H.R. No. 836, 102d Cong., 2d Sess. 9 (1992).

Are there circumstances in which the "public interest" is so strong that it will outweigh the privacy interests of those who have no intention of *ever* publishing a copyrighted work? Consider the Pentagon Papers case (New York Times v. United States, 403 U.S. 713 (1971)). In that case, the *New York Times* was prosecuted for appropriating and publishing written government secrets about the Vietnam War. The Supreme Court concluded that the government could not prevent the *New York Times* from publishing the Papers. But suppose instead that the author of the Pentagon Papers had sued for copyright infringement. [Put aside for the moment that works of the federal government are not eligible for copyright protection.] What result? Is publication of a document intended to be kept private a "fair use"? The *New York Times* acted "for profit" in the same way *The Nation* did, although the element of market loss is missing. Would the plaintiff in such a case be entitled to damages? To injunctive relief?

6. *Creative Control.* One of the factors the *Harper & Row* court cites as militating against allowing *The Nation* to publish excerpts from Ford's memoir is that the author of the book will lose "creative control" over the excerpts. Is this a valid consideration? Must fair use be use within the control or intent of the original author? Courts in general have said no and have allowed parody or negative criticism of a work to use excerpts from the work. We discuss parodies later in this section.

7. *Nature of the Infringed Work.* The copyrighted work in *Harper & Row* was an autobiography, which is a factual work. As we discovered earlier, the copyright protection afforded factual works is much "thinner" than for works of fiction, because the facts themselves cannot be protected. Should the Court have distinguished between factual and protectable material in determining the extent of copying by *The Nation*? Would such an analysis have changed the result? The dissent in the case argued that there could be no copyright infringement because *The Nation* took only ideas, and not protectable expression, from Ford's book.

8. *Percentage of the Potentially Infringing Work?* The Court refers to the four factors in section 107 as "nonexclusive." Can you think of other factors that are (or should be) relevant to the fair use inquiry?

The factor the Court relies on that is not identified in section 107 is the percentage of the *potentially infringing* work composed of copied material. (The statute mentions only the amount of the *copyrighted work* that is taken.) The Court thus makes much of the fact that copied material constituted 13 percent of the infringing work, even though it was well less than 1 percent of the copyrighted work. Should this be a factor in fair use analysis? If so, how should book reviews be analyzed for fair use purposes?

The percentage of the potentially infringing work is particularly noteworthy in the recent spate of suits against the makers of movies and television shows alleging that the movie or show infringes the copyright of a work that is shown briefly in the background of the film. In these cases, the copyrighted work (a chair, a sculpture, a poster, or a song) may be "taken" in its entirety, but it is generally a miniscule portion of the defendant's work. Courts dealing with these cases have come to different conclusions under the fair use doctrine. For example, in Woods v. Universal City Studios, 920 F. Supp. 62 (S.D.N.Y. 1996), the court enjoined the distribution of the film *12 Monkeys* because three scenes in the film featured a distinctive futuristic torture chamber modeled after a graphic work in the plaintiff's book. And in Ringgold v. Black Entertainment Television Inc., 126 F.3d 70 (2d Cir. 1997), the court held that the use of an artistic poster as part of the background scenery that appeared for less than thirty seconds of a television sitcom was not fair use. By contrast, the depiction of background art for sixty seconds was held to be fair use in Jackson v. Warner Bros., 993 F. Supp. 585 (E.D. Mich. 1997). Cf. Sandoval v. New Line Cinema Corp., 147 F.3d 215 (2d Cir. 1998) (holding unpublished photographs appearing in the background of the movie *Seven* for less than 30 seconds constituted only *de minimis* copying, obviating fair use analysis).

How would the court that decided *Harper & Row* approach such a case? Does it make sense to allow the copying of most of a work simply because the defendant has contributed so much noninfringing material? Cf. Mark A. Lemley, The Economics of Improvement in Intellectual Property Law, 75 Tex. L. Rev. 989 (1997) (suggesting that copyright should excuse copying in some such circumstances).

9. *Fair Use as a Solution to Market Failure.* Wendy Gordon argues that the fair use doctrine can and does serve to remedy market failures.

> ... Though the copyright law has provided a means for excluding nonpurchasers and thus has attempted to cure the public goods problem, and though it has provided mechanisms to facilitate consensual transfers, at times bargaining may be exceedingly expensive or it may be impractical to obtain enforcement against nonpurchasers, or other market flaws might preclude achievement of desirable consensual exchanges. In those cases, the market cannot be relied on to mediate public interests in dissemination and private interests in remuneration....
>
> Fair use should be awarded to the defendant in a copyright infringement action when (1) market failure is present; (2) transfer of the use to defendant is socially desirable; and (3) an award of fair use would not cause substantial injury to the incentives of the plaintiff copyright owner. The first element of this test ensures that market bypass will not be approved without good cause. The second element of the test ensures that the transfer of a license to use from the copyright holder to the unauthorized user effects a net gain in social value. The third element ensures that the grant of fair use will not undermine the incentive-creating purpose of the copyright law.

Wendy Gordon, Fair Use as Market Failure: A Structural and Economic Analysis of the Betamax Case and Its Predecessors, 82 Colum. L. Rev. 1600 (1982). In footnote 9, the Court appears to endorse such a framework, noting that "there is a fully functioning market that encourages the creation and dissemination of memoirs of public figures." But is that the right question? Is the relevant market here the one for memoirs, or for articles like *The Nation*'s? If the latter, is there a "fully functioning market" for excerpts of this nature? For critical reviews of memoirs?

Does a focus on market failure and licensing unreasonably limit the scope of the fair use doctrine? If the point of the fair use doctrine is merely to avoid transactions costs, is it ever likely to play a role in a case that is litigated, where the parties have already shown a willingness to pay hundreds of thousands of dollars? Should some copying be allowed without the copyist having to pay a royalty, even if the copyright owner would have demanded one? In a more recent article, Gordon argues that her paper has been read too narrowly. See Wendy J. Gordon, Excuse and Justification in the Law of Fair Use: Transactions Costs Have Always Been Only Part of the Story, 50 J. Copyr. Soc'y 149 (2003).

10. *Fair Use as a Cause of Market Failure.* Robert Merges has suggested that the fair use doctrine and other liability-type rules in intellectual property law might undermine economic efficiency by discouraging the formation of effective licensing institutions. See Robert P. Merges, Contracting into Liability Rules: Intellectual Property Rights and Collective Rights Organizations, 84 Cal. L. Rev. 1293 (1996). He points to the role of clear and relatively rigid property-type copyright rules in bringing about the creation of collective rights organizations, such as ASCAP and BMI, which have fostered creativity and dissemination. Because they draw upon experienced industry professionals, such institutions can deal with complex valuation problems more effectively than courts. And they are more sensitive to market forces than government institutions. With appropriate antitrust limitations, such institutions can ensure that authors get paid without undue transaction costs.

PROBLEMS

Problem 4-29. A home movie taken by a witness named Zales captures the shooting of President John F. Kennedy in Dallas in 1963. Only days after the shooting, Earth magazine buys the exclusive rights to the Zales film. It subsequently publishes some Zales frames in a magazine special on the assassination. The Zales frames are also appended to a government report by the Whitewash Commission on the assassination. Stone, a writer who is convinced that the Whitewash Report is flawed, unsuccessfully seeks permission from Earth to reprint the Zales pictures in his book alleging a conspiracy to kill the president. Undaunted by Earth's refusal, Stone breaks into Earth's offices and photocopies the pictures, which he then publishes in his book. Earth sues for copyright infringement. Stone defends on grounds of fair use, and offers to turn over all profits from his book to Earth. What result?

Problem 4-30. Garrison, a scholar who believes that Lee Harvey Oswald acted alone, is incensed by Stone's book. Garrison publishes a book that he entitles *A Rebuttal to Stone.* In it, Garrison follows Stone's organization in detail, presenting and refuting each of Stone's arguments. In doing so, Garrison quotes liberally from Stone's work. Garrison does not use the Zales pictures, however. Stone sues for copyright infringement. Is Garrison's work fair use?

Problem 4-31. Refer back to the Salinger-Hamilton case (problem 4-23 above). Can Hamilton successfully argue a fair use defense?

a. Videotaping

Sony Corporation of America v. Universal City Studios, Inc.

Supreme Court of the United States
464 U.S. 417 (1984)

Justice STEVENS delivered the opinion of the Court.

[The respondents, a group of movie studios, sued the makers of video cassette recorders ("VCRs," or "VTR's" in the opinion), alleging that they were liable for contributory copyright infringement because consumers bought VCRs and used them to tape movies and other programming broadcast by television stations. In order for there to be indirect liability, there must direct infringement. The record showed that most consumers used Sony's product for "time shifting" — recording shows for later viewing; relatively few consumers engaged in "archiving," recording and storing programs in a library. As summarized by the District Court, "[a]ccording to plaintiffs' survey, 75.4% of the VTR owners use their machines to record for time-shifting purposes half or most of the time. Defendants' survey showed that 96% of the Betamax owners had used the machine to record programs they otherwise would have missed." "When plaintiffs asked interviewees how many cassettes were in their library, 55.8% said there were 10 or fewer. In defendants' survey, of the total programs viewed by interviewees in the past month, 70.4% had been viewed only that one time and for 57.9%, there were no plans for further viewing." 480 F.Supp., at 438. Both activities constituted copying of protected expression. The critical question was therefore whether these activities fell within the fair use exception. If not, then Sony could not be held indirectly liable. We will return to this case in section G, discussing indirect liability.]

IV

The question is thus whether the Betamax is capable of commercially significant noninfringing uses. In order to resolve that question, we need not explore *all* the different potential uses of the machine and determine whether or not they would constitute infringement. Rather, we need only consider whether on the basis of the facts as found by the district court a significant number of them would be non-infringing. Moreover, in order to resolve this case we need not give precise content to the question of how much use is commercially significant. For one potential use of the Betamax plainly satisfies this standard, however it is understood: private, non-commercial time-shifting in the home. It does so both (A) because respondents have no right to prevent other copyright holders from authorizing it for their programs, and (B) because the District Court's factual findings reveal that even the unauthorized home time-shifting of respondents' programs is legitimate fair use.

A. Authorized Time Shifting

Each of the respondents owns a large inventory of valuable copyrights, but in the total spectrum of television programming their combined market share is small. The exact percentage is not specified, but it is well below 10%. If they were to prevail, the outcome of this litigation would have a significant impact on both the producers and the viewers of the remaining 90% of the programming in the Nation. No doubt, many other producers share respondents' concern about the possible consequences of unrestricted copying. Nevertheless the findings of the District Court make it clear that time-shifting may enlarge the total viewing audience and that many producers are willing to allow private time-shifting to continue, at least for an experimental time period.]

The District Court found:

> Even if it were deemed that home-use recording of copyrighted material constituted infringement, the Betamax could still legally be used to record noncopyrighted material or material whose owners consented to the copying. An injunction would deprive the public of the ability to use the Betamax for this noninfringing off-the-air recording.
>
> Defendants introduced considerable testimony at trial about the potential for such copying of sports, religious, educational and other programming. This included testimony from representatives of the Offices of the Commissioners of the National Football, Basketball, Baseball and Hockey Leagues and Associations, the Executive Director of National Religious Broadcasters and various educational communications agencies. . . .

If there are millions of owners of VTR's who make copies of televised sports events, religious broadcasts, and educational programs such as Mister Rogers' Neighborhood, and if the proprietors of those programs welcome the practice, the business of supplying the equipment that makes such copying feasible should not be stifled simply because the equipment is used by some individuals to make unauthorized reproductions of respondents' works. The respondents do not represent a class composed of all copyright holders. Yet a finding of contributory infringement would inevitably frustrate the interests of broadcasters in reaching the portion of their audience that is available only through time-shifting.

Of course, the fact that other copyright holders may welcome the practice of time-shifting does not mean that respondents should be deemed to have granted a license to copy their programs. Third party conduct would be wholly irrelevant in an action for direct infringement of respondents' copyrights. But in an action for contributory infringement against the seller of copying equipment, the copyright holder may not prevail unless the relief that he seeks affects only his programs, or unless he speaks for virtually all copyright holders with an interest in the outcome. In this case, the record makes it perfectly clear that there are many important producers of national and local television programs who find nothing objectionable about the enlargement in the size of the television audience that results from the practice of time-shifting for private home use.[28] In the context of television programming, some producers evidently believe that permitting home viewers to make copies of their works off the air actually enhances the value of their copyrights. Irrespective of their reasons for authorizing the practice, they do so, and in significant enough numbers to create a

---

28. It may be rare for large numbers of copyright owners to authorize duplication of their works without demanding a fee from the copier . . . The traditional method by which copyright owners capitalize upon the television medium — commercially sponsored free public broadcast over the public airwaves — is predicated upon the assumption that compensation for the value of displaying the works will be received in the form of advertising revenues.

substantial market for a non-infringing use of the Sony VTR's. No one could dispute the legitimacy of that market if the producers had authorized home taping of their programs in exchange for a license fee paid directly by the home user. The legitimacy of that market is not compromised simply because these producers have authorized home taping of their programs without demanding a fee from the home user.... The seller of the equipment that expands those producers' audiences cannot be a contributory infringer if, as is true in this case, it has had no direct involvement with any infringing activity.

B. *Unauthorized Time-Shifting*

Even unauthorized uses of a copyrighted work are not necessarily infringing. An unlicensed use of the copyright is not an infringement unless it conflicts with one of the specific exclusive rights conferred by the copyright statute. Twentieth Century Music Corp. v. Aiken, 422 U.S. 151, 154-155. Moreover, the definition of exclusive rights in § 106 of the present Act is prefaced by the words "subject to sections 107 through 118." Those sections describe a variety of uses of copyrighted material that "are not infringements of copyright notwithstanding the provisions of § 106." The most pertinent in this case is § 107, the legislative endorsement of the doctrine of "fair use."

That section identifies various factors that enable a Court to apply an "equitable rule of reason" analysis to particular claims of infringement. Although not conclusive, the first factor requires that "the commercial or nonprofit character of an activity" be weighed in any fair use decision. If the Betamax were used to make copies for a commercial or profit-making purpose, such use would presumptively be unfair. The contrary presumption is appropriate here, however, because the District Court's findings plainly establish that time-shifting for private home use must be characterized as a noncommercial, nonprofit activity. Moreover, when one considers the nature of a televised copyrighted audiovisual work, see 17 U.S.C. § 107(2), and that timeshifting merely enables a viewer to see such a work which he had been invited to witness in its entirety free of charge, the fact that the entire work is reproduced, see id., at § 107(3), does not have its ordinary effect of militating against a finding of fair use. [33]

This is not, however, the end of the inquiry because Congress has also directed us to consider "the effect of the use upon the potential market for or value of the copyrighted work." Id., at § 107(4). The purpose of copyright is to create incentives for creative effort. Even copying for noncommercial purposes may impair the copyright holder's ability to obtain the rewards that Congress intended him to have. But a use that has no demonstrable effect upon the potential market for, or the value of, the copyrighted work need not be prohibited in order to protect the author's incentive to create. The prohibition of such noncommercial uses would merely inhibit access to ideas without any countervailing benefit.

---

33. It has been suggested that "consumptive uses of copyrights by home VTR users are commercial even if the consumer does not sell the homemade tape because the consumer will not buy tapes separately sold by the copyrightholder." Home Recording of Copyrighted Works: Hearing before Subcommittee on Courts, Civil Liberties and the Administration of Justice of the House Committee on the Judiciary, 97th Congress, 2d Session, pt. 2, p. 1250 (1982) (memorandum of Prof. Laurence H. Tribe)... [T]he live viewer is no more likely to buy pre-recorded videotapes than is the timeshifter. Indeed, no live viewer would buy a pre-recorded videotape if he did not have access to a VTR.

Thus, although every commercial use of copyrighted material is presumptively an unfair exploitation of the monopoly privilege that belongs to the owner of the copyright, noncommercial uses are a different matter. A challenge to a noncommercial use of a copyrighted work requires proof either that the particular use is harmful, or that if it should become widespread, it would adversely affect the potential market for the copyrighted work. Actual present harm need not be shown; such a requirement would leave the copyright holder with no defense against predictable damage. Nor is it necessary to show with certainty that future harm will result. What is necessary is a showing by a preponderance of the evidence that some meaningful likelihood of future harm exists. If the intended use is for commercial gain, that likelihood may be presumed. But if it is for a noncommercial purpose, the likelihood must be demonstrated.

In this case, respondents failed to carry their burden with regard to home time-shifting. The District Court described respondents' evidence as follows:

> Plaintiffs experts admitted at several points in the trial that the time-shifting without librarying would result in 'not a great deal of harm.' Plaintiffs' greatest concern about time-shifting is with 'a point of important philosophy that transcends even commercial judgment.' They fear that with any Betamax usage, 'invisible boundaries' are passed: 'the copyright owner has lost control over his program.'

There was no need for the District Court to say much about past harm. "Plaintiffs have admitted that no actual harm to their copyrights has occurred to date."

On the question of potential future harm from time-shifting, the District Court offered a more detailed analysis of the evidence. It rejected respondents' "fear that persons 'watching' the original telecast of a program will not be measured in the live audience and the ratings and revenues will decrease," by observing that current measurement technology allows the Betamax audience to be reflected.[36] It rejected respondents' prediction "that live television or movie audiences will decrease as more people watch Betamax tapes as an alternative," with the observation that "[t]here is no factual basis for [the underlying] assumption." It rejected respondents' "fear that time-shifting will reduce audiences for telecast reruns," and concluded instead that "given current market practices, this should aid plaintiffs rather than harm them." And it declared that respondents' suggestion "that theater or film rental exhibition of a program will suffer because of time-shift recording of that program" "lacks merit."

---

36. "There was testimony at trial, however, that Nielsen Ratings has already developed the ability to measure when a Betamax in a sample home is recording the program. Thus, the Betamax will be measured as a part of the live audience. The later diary can augment that measurement with information about subsequent viewing." 480 F.Supp., at 466.

In a separate section, the District Court rejected plaintiffs' suggestion that the commercial attractiveness of television broadcasts would be diminished because Betamax owners would use the pause button or fast-forward control to avoid viewing advertisements:

> It must be remembered, however, that to omit commercials, Betamax owners must view the program, including the commercials, while recording. To avoid commercials during playback, the viewer must fast-forward and, for the most part, guess as to when the commercial has passed. For most recordings, either practice may be too tedious. As defendants] survey showed, 92% of the programs were recorded with commercials and only 25% of the owners fast-forward through them. Advertisers will have to make the same kinds of judgments they do now about whether persons viewing televised programs actually watch the advertisements which interrupt them.

After completing that review, the District Court restated its overall conclusion several times, in several different ways. "Harm from time-shifting is speculative and, at best, minimal." "The audience benefits from the time-shifting capability have already been discussed. It is not implausible that benefits could also accrue to plaintiffs, broadcasters, and advertisers, as the Betamax makes it possible for more persons to view their broadcasts." "No likelihood of harm was shown at trial, and plaintiffs admitted that there had been no actual harm to date." "Testimony at trial suggested that Betamax may require adjustments in marketing strategy, but it did not establish even a likelihood of harm." "Television production by plaintiffs today is more profitable than it has ever been, and, in five weeks of trial, there was no concrete evidence to suggest that the Betamax will change the studios' financial picture."

The District Court's conclusions are buttressed by the fact that to the extent time-shifting expands public access to freely broadcast television programs, it yields societal benefits. Earlier this year, in Community Television of Southern California v. Gottfried, 459 U.S. 498, 508, n.12 (1983), we acknowledged the public interest in making television broadcasting more available. Concededly, that interest is not unlimited. But it supports an interpretation of the concept of "fair use" that requires the copyright holder to demonstrate some likelihood of harm before he may condemn a private act of time-shifting as a violation of federal law.

When these factors are all weighed in the "equitable rule of reason" balance, we must conclude that this record amply supports the District Court's conclusion that home time-shifting is fair use....

In summary, the record and findings of the District Court lead us to two conclusions. First, Sony demonstrated a significant likelihood that substantial numbers of copyright holders who license their works for broadcast on free television would not object to having their broadcasts time-shifted by private viewers. And second, respondents failed to demonstrate that time-shifting would cause any likelihood of nonminimal harm to the potential market for, or the value of, their copyrighted works....

COMMENTS AND QUESTIONS

1. The Court never directly addresses whether consumers who do not merely time-shift but use their VCRs to archive — i.e., develop large collections of recorded movies and television shows — would also fall within the ambit of the fair use doctrine. Such a use is common today. Is it a fair use? Why or why not?

2. Would the fair use analysis of time-shifting have changed if the plaintiffs had presented evidence that most consumers fast forwarded through television commercials? What about the fact that the widespread availability of VCRs (and later DVDs) vastly expanded the marketplace for films and television content? Does the fact that a technology the copyright owners sought to ban now accounts for the majority of their revenue suggest that we should not trust copyright owners to decide what innovations should and should not be permissible?

3. *Sony* focused on the nature of the "public benefit" conferred by home taping. Public benefits are not expressly listed as a factor in section 107, although the "purpose and character" of the use is one of the four factors. Is there a public benefit to home taping? If so, what is it and how should it be factored into the analysis?

4. Judge Kozinski of the Ninth Circuit Court of Appeals has proposed eliminating the fair use doctrine and injunctive relief altogether in favor of liability rules — essentially a form of judicially imposed compulsory licensing. See Alex Kozinski and Christopher Newman, What's So Fair About Fair Use?, 46 J. Copyr. Soc. 513 (1999). Does this strike you as a fairer regime? Would it better serve the purposes that Congress has sought to effectuate through section 107? What would be the effects of such a regime on licensing activity, royalty rates, and courts' dockets?

5. The *Sony* decision is complicated by the fact that while it is the home tapers who are doing the copying, they are not the ones being sued. Rather, content industries sued the makers of VCRs and sought to ban them altogether on a theory of contributory infringement. The legality of copying by end users was relevant, because the court decided that Sony couldn't be liable for selling VCRs if consumers were using them for legal purposes. The question of whether the sale of products capable of both legal and illegal uses can contribute to or induce copyright infringement is one we take up in section G, below.

PROBLEM

Problem 4-32. ReplayTV has introduced a digital video recorder (DVR), a digital version of the VCR. Thanks to digital technology, the ReplayTV 4500 offers consumers the ability to instantly skip over commercials with the click of a 30-second advance fast forward button. In addition, the large storage capacity of the ReplayTV 4500 hard drive enables consumers to store up to 60 hours of content, making archiving of content much more convenient than on VCRs. A recent survey of DVR owners finds that 35 percent say they never watch commercials while nearly 60 percent say they watch them only occasionally. The television industry fears that these capabilities will make advertisers much less willing to support their broadcasts. They sue to enjoin sale of ReplayTV's DVR. How should the court rule on this case?

---

b. *Photocopying*

American Geophysical Union, et al. v. Texaco Inc.
United States Court of Appeals for the Second Circuit
60 F.3d 913 (2d Cir. 1994)

Jon O. NEWMAN, Chief Judge:

... Plaintiffs American Geophysical Union and 82 other publishers of scientific and technical journals (the "publishers") brought a class action claiming that Texaco's unauthorized photocopying of articles from their journals constituted copyright infringement. Among other defenses, Texaco claimed that its copying was fair use under section 107 of the Copyright Act....

Although Texaco employs 400 to 500 research scientists, of whom all or most presumably photocopy scientific journal articles to support their Texaco research, the parties stipulated — in order to spare the enormous expense of exploring the photocopying practices of each of them — that one scientist would be chosen at random as the representative of the entire group. The scientist chosen was Dr. Donald H. Chickering, II, a scientist at Texaco's research center in Beacon, New York. For consideration at trial, the publishers selected from Chickering's files photocopies of eight particular articles from the Journal of Catalysis. . . .

Chickering, a chemical engineer at the Beacon research facility, has worked for Texaco since 1981 conducting research in the field of catalysis, which concerns changes in the rates of chemical reactions. To keep abreast of developments in his field, Chickering must review works published in various scientific and technical journals related to his area of research. Texaco assists in this endeavor by having its library circulate current issues of relevant journals to Chickering when he places his name on the appropriate routing list.

The copies of the eight articles from Catalysis found in Chickering's files that the parties have made the exclusive focus of the fair use trial were photocopied in their entirety by Chickering or by other Texaco employees at Chickering's request. Chickering apparently believed that the material and data found within these articles would facilitate his current or future professional research. The evidence developed at trial indicated that Chickering did not generally use the Catalysis articles in his research immediately upon copying, but placed the photocopied articles in his files to have them available for later reference as needed. Chickering became aware of six of the photocopied articles when the original issues of Catalysis containing the articles were circulated to him. He learned of the other two articles upon seeing a reference to them in another published article. As it turned out, Chickering did not have occasion to make use of five of the articles that were copied. . . .

I. The Nature of the Dispute . . .

A. Fair Use and Photocopying . . .

As with the development of other easy and accessible means of mechanical reproduction of documents, the invention and widespread availability of photocopying technology threatens to disrupt the delicate balances established by the Copyright Act. As a leading commentator astutely notes, the advent of modern photocopying technology creates a pressing need for the law "to strike an appropriate balance between the authors' interest in preserving the integrity of copyright, and the public's right to enjoy the benefits that photocopying technology offers." 3 Nimmer on Copyright § 13.05[E][1], at 13-226.

Indeed, if the issue were open, we would seriously question whether the fair use analysis that has developed with respect to works of authorship alleged to use portions of copyrighted material is precisely applicable to copies produced by mechanical means. The traditional fair use analysis, now codified in section 107, developed in an effort to adjust the competing interests of authors—the author of the original copyrighted work and the author of the secondary work that "copies" a portion of the original work in the course of producing what is claimed to be a new work. Mechanical "copying" of an entire document, made readily feasible and economical by the advent of xerography, is obviously an activity entirely different from creating a work of authorship. Whatever social utility copying of this sort achieves, it is not concerned with creative authorship. . . .

II. The Enumerated Fair Use Factors of Section 107 . . .

A. First Factor: Purpose and Character of Use

The first factor listed in section 107 is "the purpose and character of the use, including whether such use is of a commercial nature or is for nonprofit educational purposes." 17 U.S.C. § 107(1). Especially pertinent to an assessment of the first fair use factor are the precise circumstances under which copies of the eight Catalysis articles were made. After noticing six of these articles when the original copy of the journal issue containing each of them was circulated to him, Chickering had them photocopied, at least initially, for the same basic purpose that one would normally seek to obtain the original — to have it available on his shelf for ready reference if and when he needed to look at it. The library circulated one copy and invited all the researchers to make their own photocopies. It is a reasonable inference that the library staff wanted each journal issue moved around the building quickly and returned to the library so that it would be available for others to look at. Making copies enabled all researchers who might one day be interested in examining the contents of an article in the issue to have the article readily available in their own offices. In Chickering's own words, the copies of the articles were made for "my personal convenience," since it is "far more convenient to have access in my office to a photocopy of an article than to have to go to the library each time I wanted to refer to it." Significantly, Chickering did not even have occasion to use five of the photocopied articles at all, further revealing that the photocopies of the eight Catalysis articles were primarily made just for "future retrieval and reference."

It is true that photocopying these articles also served other purposes. The most favorable for Texaco is the purpose of enabling Chickering, if the need should arise, to go into the lab with pieces of paper that (a) were not as bulky as the entire issue or a bound volume of a year's issues, and (b) presented no risk of damaging the original by exposure to chemicals. And these purposes might suffice to tilt the first fair use factor in favor of Texaco if these purposes were dominant. For example, if Chickering had asked the library to buy him a copy of the pertinent issue of Catalysis and had placed it on his shelf, and one day while reading it had noticed a chart, formula, or other material that he wanted to take right into the lab, it might be a fair use for him to make a photocopy, and use that copy in the lab (especially if he did not retain it and build up a mini-library of photocopied articles). This is the sort of "spontaneous" copying that is part of the test for permissible nonprofit classroom copying. See Agreement on Guidelines for Classroom Copying in Not-For-Profit Educational Institutions, quoted in Patry, The Fair Use Privilege, at 308. But that is not what happened here as to the six items copied from the circulated issues.

As to the other two articles, the circumstances are not quite as clear, but they too appear more to serve the purpose of being additions to Chickering's office "library" than to be spontaneous copying of a critical page that he was reading on his way to the lab. One was copied apparently when he saw a reference to it in another article, which was in an issue circulated to him. The most likely inference is that he decided that he ought to have copies of both items — again for placement on his shelf for later use if the need arose. The last article was copied, according to his affidavit, when he saw a reference to it "elsewhere." What is clear is that this item too was simply placed "on the shelf." As he testified, "I kept a copy to refer to in case I became more involved in support effects research."

The photocopying of these eight Catalysis articles may be characterized as "archival"—i.e., done for the primary purpose of providing numerous Texaco scientists (for whom Chickering served as an example) each with his or her own personal copy of each article without Texaco's having to purchase another original journal. The photocopying "merely 'supersedes the objects' of the original creation," Campbell, 114 S. Ct. at 1171 (quoting Folsom v. Marsh, 9 F. Cas. 342, 348 (C.C.D. Mass. 1841) (No. 4,901)), and tilts the first fair use factor against Texaco. We do not mean to suggest that no instance of archival copying would be fair use, but the first factor tilts against Texaco in this case because the making of copies to be placed on the shelf in Chickering's office is part of a systematic process of encouraging employee researchers to copy articles so as to multiply available copies while avoiding payment.

Texaco criticizes three aspects of the District Court's analysis of the first factor. . . . We consider these three lines of attack separately.

1. Commercial use. We generally agree with Texaco's contention that the District Court placed undue emphasis on the fact that Texaco is a for-profit corporation conducting research primarily for commercial gain. Since many, if not most, secondary users seek at least some measure of commercial gain from their use, unduly emphasizing the commercial motivation of a copier will lead to an overly restrictive view of fair use. . . .

We do not consider Texaco's status as a for-profit company irrelevant to the fair use analysis. Though Texaco properly contends that a court's focus should be on the use of the copyrighted material and not simply on the user, it is overly simplistic to suggest that the "purpose and character of the use" can be fully discerned without considering the nature and objectives of the user.

Ultimately, the somewhat cryptic suggestion in section 107(1) to consider whether the secondary use "is of a commercial nature or is for nonprofit educational purposes" connotes that a court should examine, among other factors, the value obtained by the secondary user from the use of the copyrighted material. See Rogers, 960 F.2d at 309 ("The first factor . . . asks whether the original was copied in good faith to benefit the public or primarily for the commercial interests of the infringer."); MCA, Inc. v. Wilson, 677 F.2d 180, 182 (2d Cir. 1981) (court is to consider "whether the alleged infringing use was primarily for public benefit or for private commercial gain"). The commercial/nonprofit dichotomy concerns the unfairness that arises when a secondary user makes unauthorized use of copyrighted material to capture significant revenues as a direct consequence of copying the original work. See Harper & Row, 471 U.S. at 562 ("The crux of the profit/nonprofit distinction is . . . whether the user stands to profit from exploitation of the copyrighted material without paying the customary price.").

Consistent with these principles, courts will not sustain a claimed defense of fair use when the secondary use can fairly be characterized as a form of "commercial exploitation," i.e., when the copier directly and exclusively acquires conspicuous financial rewards from its use of the copyrighted material. . . . Conversely, courts are more willing to find a secondary use fair when it produces a value that benefits the broader public interest. . . . The greater the private economic rewards reaped by the secondary user (to the exclusion of broader public benefits), the more likely the first factor will favor the copyright holder and the less likely the use will be considered fair.

As noted before, in this particular case the link between Texaco's commercial gain and its copying is somewhat attenuated: the copying, at most, merely facilitated Chickering's research that might have led to the production of commercially valuable products. Thus, it would not be accurate to conclude that Texaco's copying of eight particular Catalysis articles amounted to "commercial exploitation," especially since the immediate goal of Texaco's copying was to facilitate Chickering's research in the sciences, an objective that might well serve a broader public purpose. See Twin Peaks, 996 F.2d at 1375; Sega Enterprises, 977 F.2d at 1522. Still, we need not ignore the for-profit nature of Texaco's enterprise, especially since we can confidently conclude that Texaco reaps at least some indirect economic advantage from its photocopying. As the publishers emphasize, Texaco's photocopying for Chickering could be regarded simply as another "factor of production" utilized in Texaco's efforts to develop profitable products. Conceptualized in this way, it is not obvious why it is fair for Texaco to avoid having to pay at least some price to copyright holders for the right to photocopy the original articles.

2. *Transformative Use.* The District Court properly emphasized that Texaco's photocopying was not "transformative." After the District Court issued its opinion, the Supreme Court explicitly ruled that the concept of a "transformative use" is central to a proper analysis under the first factor, see Campbell, 114 S. Ct. at 1171-73. The Court explained that though a "transformative use is not absolutely necessary for a finding of fair use, . . . the more transformative the new work, the less will be the significance of other factors, like commercialism, that may weigh against a finding of fair use." Id. at 1171.

The "transformative use" concept is pertinent to a court's investigation under the first factor because it assesses the value generated by the secondary use and the means by which such value is generated. To the extent that the secondary use involves merely an untransformed duplication, the value generated by the secondary use is little or nothing more than the value that inheres in the original. Rather than making some contribution of new intellectual value and thereby fostering the advancement of the arts and sciences, an untransformed copy is likely to be used simply for the same intrinsic purpose as the original, thereby providing limited justification for a finding of fair use. See Weissmann v. Freeman, 868 F.2d 1313, 1324 (2d Cir.) (explaining that a use merely for the same "intrinsic purpose" as original "moves the balance of the calibration on the first factor against" secondary user and "seriously weakens a claimed fair use"), *cert. denied*, 493 U.S. 883, 107 L. Ed. 2d 172, 110 S. Ct. 219 (1989).

In contrast, to the extent that the secondary use "adds something new, with a further purpose or different character," the value generated goes beyond the value that inheres in the original and "the goal of copyright, to promote science and the arts, is generally furthered." Campbell, 114 S. Ct. at 1171; see also Pierre N. Leval, Toward a Fair Use Standard, 103 Harv. L. Rev. 1105, 1111 (1990) [hereinafter Leval, Toward a Fair Use Standard]. It is therefore not surprising that the "preferred" uses illustrated in the preamble to section 107, such as criticism and comment, generally involve some transformative use of the original work. See 3 Nimmer on Copyright § 13.05[A][1][b], at 13-160.

Texaco suggests that its conversion of the individual Catalysis articles through photocopying into a form more easily used in a laboratory might constitute a transformative use. However, Texaco's photocopying merely transforms the material object embodying the intangible article that is the copyrighted original work. See 17 U.S.C. §§ 101, 102 (explaining that copyright protection in literary works subsists in the original work of authorship "regardless of the nature of the material objects . . . in which they are embodied"). Texaco's making of copies cannot properly be regarded as a transformative use of the copyrighted material.

Even though Texaco's photocopying is not technically a transformative use of the copyrighted material, we should not overlook the significant independent value that can stem from conversion of original journal articles into a format different from their normal appearance. See generally Sony, 464 U.S. at 454, 455 n.40 (acknowledging possible benefits from copying that might otherwise seem to serve "no productive purpose"); Weinreb, Fair's Fair, at 1143 & n.29 (discussing potential value from nontransformative copying). As previously explained, Texaco's photocopying converts the individual Catalysis articles into a useful format. Before modern photocopying, Chickering probably would have converted the original article into a more serviceable form by taking notes, whether cursory or extended; today he can do so with a photocopying machine. Nevertheless, whatever independent value derives from the more usable format of the photocopy does not mean that every instance of photocopying wins on the first factor. In this case, the predominant archival purpose of the copying tips the first factor against the copier, despite the benefit of a more usable format. . . .

On balance, we agree with the District Court that the first factor favors the publishers, primarily because the dominant purpose of the use is "archival." . . .

B. *Second Factor: Nature of Copyrighted Work*

The second statutory fair use factor is "the nature of the copyrighted work." 17 U.S.C. § 107(2). . . . Ultimately the manifestly factual character of the eight articles precludes us from considering the articles as "within the core of the copyright's protective purposes," Campbell, 114 S. Ct. at 1175; see also Harper & Row, 471 U.S. at 563 ("The law generally recognizes a greater need to disseminate factual works than works of fiction or fantasy."). Thus, in agreement with the District Court, we conclude that the second factor favors Texaco.

C. *Third Factor: Amount and Substantiality of Portion Used*

The third statutory fair use factor is "the amount and substantiality of the portion used in relation to the copyrighted work as a whole." 17 U.S.C. § 107(3). The District Court concluded that this factor clearly favors the publishers because Texaco copied the eight articles from Catalysis in their entirety. [The Court of Appeals agreed].

D. *Fourth Factor: Effect Upon Potential Market or Value*

The fourth statutory fair use factor is "the effect of the use upon the potential market for or value of the copyrighted work." 17 U.S.C. § 107(4). Assessing this factor, the District Court detailed the range of procedures Texaco could use to obtain-authorized copies of the articles that it photocopied and found that "whatever combination of procedures Texaco used, the publishers' revenues would grow significantly." 802 F. Supp. at 19. The Court concluded that the publishers "powerfully demonstrated entitlement to prevail as to the fourth factor," since they had shown "a substantial harm to the value of their copyrights" as the consequence of Texaco's copying. See id. at 18-21. . . .

In analyzing the fourth factor, it is important (1) to bear in mind the precise copyrighted works, namely the eight journal articles, and (2) to recognize the distinctive nature and history of "the potential market for or value of" these particular works. Specifically, though there is a traditional market for, and hence a clearly defined value of, journal issues and volumes, in the form of per-issue purchases and journal subscriptions, there is neither a traditional market for, nor a clearly defined value of, individual journal articles. As a result, analysis of the fourth factor cannot proceed as simply as would have been the case if Texaco had copied a work that carries a stated or negotiated selling price in the market.

Like most authors, writers of journal articles do not directly seek to capture the potential financial rewards that stem from their copyrights by personally marketing copies of their writings. Rather, like other creators of literary works, the author of a journal article "commonly sells his rights to publishers who offer royalties in exchange for their services in producing and marketing the author's work." Harper & Row, 471 U.S. at 547. In the distinctive realm of academic and scientific articles, however, the only form of royalty paid by a publisher is often just the reward of being published, publication being a key to professional advancement and prestige for the author, see Weissmann, 868 F.2d at 1324 (noting that "in an academic setting, profit is ill-measured in dollars. Instead, what is valuable is recognition because it so often influences professional advancement and academic tenure."). The publishers in turn incur the costs and labor of producing and marketing authors' articles, driven by the prospect of capturing the economic value stemming from the copyrights in the original works, which the authors have transferred to them. Ultimately, the monopoly privileges conferred by copyright protection and the potential financial rewards therefrom are not directly serving to motivate authors to write individual articles; rather, they serve to motivate publishers to produce journals, which provide the conventional and often exclusive means for disseminating these individual articles. It is the prospect of such dissemination that contributes to the motivation of these authors. . . .

1. Sales of Additional Journal Subscriptions, Back Issues, and Back Volumes

Since we are concerned with the claim of fair use in copying the eight individual articles from Catalysis, the analysis under the fourth factor must focus on the effect of Texaco's photocopying upon the potential market for or value of these individual articles. Yet, in their respective discussions of the fourth statutory factor, the parties initially focus on the impact of Texaco's photocopying of individual journal articles upon the market for Catalysis journals through sales of Catalysis subscriptions, back issues, or back volumes.

As a general matter, examining the effect on the marketability of the composite work containing a particular individual copyrighted work serves as a useful means to gauge the impact of a secondary use "upon the potential market for or value of" that individual work, since the effect on the marketability of the composite work will frequently be directly relevant to the effect on the market for or value of that individual work. Quite significantly, though, in the unique world of academic and scientific articles, the effect on the marketability of the composite work in which individual articles appear is not obviously related to the effect on the market for or value of the individual articles. Since (1) articles are submitted unsolicited to journals, (2) publishers do not make any payment to authors for the right to publish their articles or to acquire their copyrights, and (3) there is no evidence in the record suggesting that publishers seek to reprint particular articles in new composite works, we cannot readily conclude that evidence concerning the effect of Texaco's use on the marketability of journals provides an effective means to appraise the effect of Texaco's use on the market for or value of individual journal articles.

These considerations persuade us that evidence concerning the effect of Texaco's photocopying of individual articles within Catalysis on the traditional market for Catalysis subscriptions is of somewhat limited significance in determining and evaluating the effect of Texaco's photocopying "upon the potential market for or value of" the individual articles. We do not mean to suggest that we believe the effect on the marketability of journal subscriptions is completely irrelevant to gauging the effect on the market for and value of individual articles. Were the publishers able to demonstrate that Texaco's type of photocopying, if widespread, would impair the marketability of journals, then they might have a strong claim under the fourth factor. Likewise, were Texaco able to demonstrate that its type of photocopying, even if widespread, would have virtually no effect on the marketability of journals, then it might have a strong claim under this fourth factor.

On this record, however, the evidence is not resounding for either side. The District Court specifically found that, in the absence of photocopying, (1) "Texaco would not ordinarily fill the need now being supplied by photocopies through the purchase of back issues or back volumes . . . [or] by enormously enlarging the number of its subscriptions," but (2) Texaco still "would increase the number of subscriptions somewhat." 802 F. Supp. at 19. This moderate conclusion concerning the actual effect on the marketability of journals, combined with the uncertain relationship between the market for journals and the market for and value of individual articles, leads us to conclude that the evidence concerning sales of additional journal subscriptions, back issues, and back volumes does not strongly support either side with regard to the fourth factor. Cf. Sony, 464 U.S. at 451-55 (rejecting various predictions of harm to value of copyrighted work based on speculation about possible consequences of secondary use). At best, the loss of a few journal subscriptions tips the fourth factor only slightly toward the publishers because evidence of such loss is weak evidence that the copied articles themselves have lost any value.

2. Licensing Revenues and Fees

The District Court, however, went beyond discussing the sales of additional journal subscriptions in holding that Texaco's photocopying affected the value of the publishers' copyrights. Specifically, the Court pointed out that, if Texaco's unauthorized photocopying was not permitted as fair use, the publishers' revenues would increase significantly since Texaco would (1) obtain articles from document delivery services (which pay royalties to publishers for the right to photocopy articles),

(2) negotiate photocopying licenses directly with individual publishers, and/or (3) acquire some form of photocopying license from the Copyright Clearance Center Inc. ("CCC"). See 802 F. Supp. at 19. Texaco claims that the District Court's reasoning is faulty because, in determining that the value of the publishers' copyrights was affected, the Court assumed that the publishers were entitled to demand and receive licensing royalties and fees for photocopying. Yet, continues Texaco, whether the publishers can demand a fee for permission to make photocopies is the very question that the fair use trial is supposed to answer.

It is indisputable that, as a general matter, a copyright holder is entitled to demand a royalty for licensing others to use its copyrighted work, see 17 U.S.C. § 106 (copyright owner has exclusive right "to authorize" certain uses), and that the impact on potential licensing revenues is a proper subject for consideration in assessing the fourth factor, see, e.g., Campbell, 114 S. Ct. at 1178; Harper & Row, 471 U.S. at 568-69; Twin Peaks, 996 F.2d at 1377; DC Comics Inc. v. Reel Fantasy, Inc., 696 F.2d 24, 28 (2d Cir. 1982); United Telephone Co. of Missouri v. Johnson Publishing Co., Inc., 855 F.2d 604, 610 (8th Cir. 1988).

However, not every effect on potential licensing revenues enters the analysis under the fourth factor. Specifically, courts have recognized limits on the concept of "potential licensing revenues" by considering only traditional, reasonable, or likely to be developed markets when examining and assessing a secondary use's "effect upon the potential market for or value of the copyrighted work." See Campbell, 114 S. Ct. at 1178 ("The market for potential derivative uses includes only those that creators of original works would in general develop or license others to develop."); Harper & Row, 471 U.S. at 568 (fourth factor concerned with "use that supplants any part of the normal market for a copyrighted work") (emphasis added) (quoting S. Rep. No. 473, 94th Cong., 1st Sess. 65 (1975)); see also Mathieson v. Associated Press, 23 U.S.P.Q.2d 1685, 1690-91 (S.D.N.Y. 1992) (refusing to find fourth factor in favor of copyright holder because secondary use did not affect any aspect of the normal market for copyrighted work).

...Texaco is correct, at least as a general matter, when it contends that it is not always appropriate for a court to be swayed on the fourth factor by the effects on potential licensing revenues. Only an impact on potential licensing revenues for traditional, reasonable, or likely to be developed markets should be legally cognizable when evaluating a secondary use's "effect upon the potential market for or value of the copyrighted work."

Though the publishers still have not established a conventional market for the direct sale and distribution of individual articles, they have created, primarily through the CCC, a workable market for institutional users to obtain licenses for the right to produce their own copies of individual articles via photocopying. The District Court found that many major corporations now subscribe to the CCC systems for photocopying licenses. 802 F. Supp. at 25. Indeed, it appears from the pleadings, especially Texaco's counterclaim, that Texaco itself has been paying royalties to the CCC. Since the Copyright Act explicitly provides that copyright holders have the "exclusive rights" to "reproduce" and "distribute copies" of their works, see 17 U.S.C. § 106(1) & (3), and since there currently exists a viable market for licensing these rights for individual journal articles, it is appropriate that potential licensing revenues for photocopying be considered in a fair use analysis.

Despite Texaco's claims to the contrary, it is not unsound to conclude that the right to seek payment for a particular use tends to become legally cognizable under the fourth fair use factor when the means for paying for such a use is made easier. This notion is not inherently troubling: it is sensible that a particular unauthorized use should be considered "more fair" when there is no ready market or means to pay for the use, while such an unauthorized use should be considered "less fair" when there is a ready market or means to pay for the use. The vice of circular reasoning arises only if the availability of payment is conclusive against fair use. Whatever the situation may have been previously, before the development of a market for institutional users to obtain licenses to photocopy articles, see Williams & Wilkins, 487 F.2d at 1357-59, it is now appropriate to consider the loss of licensing revenues in evaluating "the effect of the use upon the potential market for or value of" journal articles. It is especially appropriate to do so with respect to copying of articles from Catalysis, a publication as to which a photocopying license is now available. We do not decide how the fair use balance would be resolved if a photocopying license for Catalysis articles were not currently available. . . .

Primarily because of lost licensing revenue, and to a minor extent because of lost subscription revenue, we agree with the District Court that "the publishers have demonstrated a substantial harm to the value of their copyrights through [Texaco's] copying," 802 F. Supp. at 21, and thus conclude that the fourth statutory factor favors the publishers. . . .

The order of the District Court is affirmed.

JACOBS, Circuit Judge, dissenting:

The stipulated facts crisply present the fair use issues that govern the photocopying of entire journal articles for a scientist's own use, either in the laboratory or as part of a personal file assisting that scientist's particular inquiries. I agree with much in the majority's admirable review of the facts and the law. Specifically, I agree that, of the four nonexclusive considerations bearing on fair use enumerated in section 107, the second factor (the nature of the copyrighted work) tends to support a conclusion of fair use, and the third factor (the ratio of the copied portion to the whole copyrighted work) militates against it. I respectfully dissent, however, in respect of the first and fourth factors. As to the first factor: the purpose and character of Dr. Chickering's use is integral to transformative and productive ends of scientific research. As to the fourth factor: the adverse effect of Dr. Chickering's use upon the potential market for the work, or upon its value, is illusory. For these reasons, and in light of certain equitable considerations and the overarching purpose of the copyright laws, I conclude that Dr. Chickering's photocopying of the Catalysis articles was fair use.

A. Purpose and Character of the Use . . .

A use that is reasonable and customary is likely to be a fair one. See Harper & Row Publishers, Inc. v. Nation Enterprises, 471 U.S. 539, 550 (1985) ("the fair use doctrine was predicated on the author's implied consent to 'reasonable and customary' use"). The district court, the majority and I start from the same place in assessing whether Dr. Chickering's photocopying is a reasonable and customary use of the material: making single photocopies for research and scholarly purposes has been considered both reasonable and customary for as long as photocopying technology has been in existence. See Williams & Wilkins Co. v. United States, 487 F.2d 1345, 1355-56 (Ct. Cl. 1973), aff'd by an equally divided court, 420 U.S. 376 (1976). . . .

The majority emphasizes passim that the photocopying condemned here is "systematic" and "institutional". These terms furnish a ground for distinguishing this case from the case that the majority expressly does not reach: the copying of journal articles by an individual researcher outside an institutional framework. For all the reasons adduced above, I conclude that the institutional environment in which Dr. Chickering works does not alter the character of the copying done by him or at his instance, and that the selection by an individual scientist of the articles useful to that scientist's own inquiries is not systematic copying, and does not become systematic because some number of other scientists in the same institution — four hundred or four — are doing the same thing.

First, the majority's reliance on Texaco's institutional framework does not limit the potentially uncontrolled ramifications of the result. Research is largely an institutional endeavor nowadays, conducted by employees pursuing the overall goals of corporations, university laboratories, courts and law firms, governments and their agencies, think-tanks, publishers of newspapers and magazines, and other kinds of institutions. The majority's limitation of its holding to institutional environments may give comfort to inventors in bicycle shops, scientists in garage laboratories, freelance book reviewers, and solo conspiracy theorists, but it is not otherwise meaningful.

The majority's reliance on the systematic character of the photocopying here also seems to me erroneous. The majority deems Texaco's photocopying systematic because Texaco uses circulation lists to route a copy of each journal issue to the scientists interested in the field. The majority, however, ignores the one determinative issue: whether the decision to photocopy individual articles is made by the individual researcher, as Dr. Chickering did here. Journal issues may be systematically circulated to all scientists in a given group, rather than (say) at random, but the circulation of journal issues is not photocopying, systematic or otherwise. The journal issues circulated by Texaco are procured by subscription. Once Texaco receives the subscription copies from the publisher, Texaco is free to circulate them in-house so that they can be seen by as many scientists as can lay eyes on them. This circulation of copies allows individual scientists to select individual articles for copying. The majority opinion, which leaves open the idea that this practice may comport with copyright law if done by an individual scientist, does not explain why it is impermissible when done by more than one. . . .

B. *Effect Upon Potential Market or Value*

In gauging the effect of Dr. Chickering's photocopying on the potential market or value of the copyrighted work, the majority properly considers two separate means of marketing: (1) journal subscriptions and sales, and (2) licensing revenues and fees.

(1) Subscriptions and Sales . . .

As to the individual articles photocopied by Dr. Chickering, I agree with the majority — as I read the opinion — that one cannot put a finger on any loss suffered by the publisher in the value of the individual articles or in the traditional market for subscriptions and back issues. The district court found that Texaco would not purchase back-issues or back volumes in the numbers needed to supply individual copies of articles to individual scientists.

Finally, the circulation of Catalysis among a number of Texaco scientists can come as no surprise to the publisher of Catalysis, which charges double the normal subscription rate to institutional subscribers. The publisher must therefore assume that, unless they are reading Catalysis for pleasure or committing it to memory, the scientists will extract what they need and arrange to copy it for personal use before passing along the institutional copies.

(2) Licensing Revenues and Fees...

In this case the only harm to a market is to the supposed market in photocopy licenses. The CCC scheme is neither traditional nor reasonable; and its development into a real market is subject to substantial impediments. There is a circularity to the problem: the market will not crystallize unless courts reject the fair use argument that Texaco presents; but, under the statutory test, we cannot declare a use to be an infringement unless (assuming other factors also weigh in favor of the secondary user) there is a market to be harmed. At present, only a fraction of journal publishers have sought to exact these fees. I would hold that this fourth factor decisively weighs in favor of Texaco, because there is no normal market in photocopy licenses, and no real consensus among publishers that there ought to be one....

COMMENTS AND QUESTIONS

1. Is the majority's reasoning in *Texaco* circular? Isn't the issue to be resolved whether or not Texaco has to pay a license fee in order to photocopy journal articles? The answer to that question depends on whether Texaco's copies constitute a "fair use." But the court hinges the fair use inquiry on whether or not the copyright owners have lost licensing revenue from Texaco. Does this make sense?

The majority suggests that the "vice of circularity" can be avoided by considering "only traditional, reasonable, or likely to be developed markets" when considering a challenged use upon a potential market. See Ringgold v. Black Entertainment Television Inc., 126 F.3d 70, 81 (2d Cir. 1997). Do you agree?

2. *From Borrowed Passages to Mechanically Reproduced Works.* Judge Newman questions whether the fair use doctrine, which initially developed to allow use of portions of copyrighted material as part of new works, should have been extended to allow copying of entire documents through the use of xerography at all. Does this perspective look too narrowly at creativity, focusing on the output and overlooking the research process? Judge Leval, the trial court judge in the case (who now sits on the Second Circuit), noted that the effect of requiring permissions would be that Texaco would have to buy more copies of journals it wished to circulate (or at least would have to pay more money to the CCC for permission to make photocopies). Assuming a budget constraint, Texaco will purchase a narrower range of journals, since it must buy more copies of the more widely used journals. Does this promote "the progress of science and the useful arts"? Is it a result the authors of journal articles would applaud? At the least, isn't the loss of information transfer that results from these decisions a factor to be considered in determining the issue of fair use?

Alternatively, might Texaco allocate its research acquisition budget more broadly by allowing researchers to acquire articles on a pay-as-you-read/download basis? In the end, determining actual market effects can be quite difficult.

Most people today would agree that making a single photocopy of a single academic article for research purposes—whether by the NIH or by a commercial group—promotes the scholarly enterprise and falls within the fair use privilege. But at what point do such individual fair uses become a collective infringement of copyright? Would a well-functioning licensing market—such as that envisioned by the CCC and encouraged by Judge Newman—serve to both promote primary and cumulative creativity?

3. *Coursepacks.* Copyright owners have brought a number of cases against "copy shops" that reproduce course readers. In Basic Books v. Kinko's Graphics Corp., 758 F. Supp. 1522 (S.D.N.Y. 1991), the court rejected Kinko's fair use defense, largely on the grounds that the copy shop is a for-profit operation. See also Princeton University Press v. Michigan Document Servs., 99 F.3d 1381 (6th Cir. 1996) (en banc) (holding as a matter of law that the for-profit copying of academic readers (or "coursepacks") could not be a fair use).

Kinko's argued that it is the professors, not the copy shop, that have perpetrated and authorized the copyright infringement. The court dismissed the argument fairly summarily, but it is worthy of further consideration. Were the staff at Kinko's or MDS in a position to know whether professors who submitted packets had obtained the proper copyright permissions? Should all copy shops be required to verify the copyright status of any works copied under their auspices? Should such checking extend even to "self-serve" copiers? Isn't the truly responsible entity here the professor who took 110 pages from a work and put it into a reader? Could Kinko's absolve itself from liability by having the professors sign a statement that they take the responsibility for complying with copyright law?

As a practical matter, academic publishers are unlikely to sue college professors for copyright infringement. First, most of their authors are college professors. Second, college professors are responsible (directly or indirectly) for most of the purchases of academic books. Thus they have a powerful influence on the market. Finally, it is much more difficult to sue many individual college professors than a few large national copy shops.

It is noteworthy that public libraries are not subject to the requirements the *Kinko's* decision imposes on commercial shops. 17 U.S.C. § 108 exempts libraries from liability for copies made on their premises as long as they post a warning notice as specified in 37 C.F.R. § 201.14. Is there a justification for this difference in treatment? Does it stem from the presumed public, non-commercial nature of libraries?

4. *Copyright Law Driving Social Change and Market Formation.* Prior to the coursepack decisions, many university teachers and copy shops ignored copyright law. See Robert Ellickson, Order Without Law 258-264 (1991) (documenting how informal norms among professors "trump" the formal law of copyright as regards photocopying articles for course packets). In less than a decade, the *Texaco* and coursepack decisions have swung social practices 180 degrees. Most university bookstores will not distribute unauthorized course readers. As a result, the CCC has become well known and well used.

Does the existence of licensing transactions vindicate Professor Merges' conjecture about strong property rules producing effective licensing institutions? To the extent that licensing markets function well—i.e., producing efficient, low cost access to scholarly works—is fair use needed? Note that had the court found fair use, it would have deflated the CCC's sails by making private licensing of these copies unnecessary. Does this one-way ratchet suggest that courts should be cautious in finding fair use? Or has society lost something important by moving from a system under which people were free to make copies for academic research to one in which the ability to make those copies exists only at the collective sufferance of a host of copyright owners? The recent decision by some authors and publishers to sue Google over its project to index books and make them searchable online is to some an illustration of the dangers of assuming that copyright owners will always license a use merely because it is rational for them to do so.

The rise of the Internet has offered a new model for creating coursepacks. Some professors now use links on course websites to "distribute" course readings. In many cases, these works are freely available, although licensing sometimes plays a role. Cf. Robert P. Merges, A New Dynamism in the Public Domain, 71 U. Chi. L. Rev. 183 (2004) (noting how technology and market forces have spurred public accessibility of works in unanticipated ways). Should this electronic distribution be illegal absent a license from the CCC? Does *Texaco* compel that result?

5. *The Orphan Work Problem.* The CCC regime works reasonably well for licensing the many works available through this licensing clearinghouse. But what about works that are still under copyright for which the owners cannot be located? As universities and copy shops increasingly police the assemblage of coursepacks, it has become more difficult and in some cases impossible to get permissions to copy rare or out-of-print works. The scaling back of formalities in copyright law means that copyright owners have no obligation to maintain contact information in any central repository. Should the fair use standard be lower in such circumstances? What about where a licensor refuses to license or only on exorbitant terms? Cf. William F. Patry & Richard A. Posner, Fair Use and Statutory Reform in the Wake of *Eldred*, 92 Cal. L. Rev. 1639 (2004). The Copyright Office has initiated a project aimed at studying ways of alleviating the orphan work problem. See U.S. Copyright Office, 70 Federal Register 3739 (Jan. 26, 2005).

6. *Academic Journal Publishing.* The *Texaco* court focuses on the market effects of Texaco's practices on the publication of academic research. Yet the authors of such works—principally university professors and scientists—typically assign their copyrights in exchange for publication and do not receive royalties. Subscription (and licensing) fees go to the publisher. This makes sense within the traditional copyright system, since without print journals the spread of knowledge and research would suffer.

The Internet now offers an open source alternative, which is generating pressure on traditional print publishers whose prices have continued to rise, straining the budgets of their principal customers—university libraries. New online publishing markets—such as the Public Library of Science (PLoS), the California Digital Library, the Social Science Research Network (SSRN), the Berkeley Economics Press (BePress)—offer zero- or lower-cost alternatives for the distribution of academic research. Google Scholar now offers powerful search tools for identifying research online. Professors are increasingly seeking rights to post their published works in free and open online archives. Until now, most professors have preferred the higher prestige print journals, but such preference is waning as on line alternatives develop reputations and speed works into circulation. Will technology ultimately solve the problems raised by *Texaco* by bringing about fundamental change in the organization of academic publishing? Cf. Gabe Bloch, Transformation in Publishing: Modeling the Effect of New Media, 20 Berkeley Tech. L.J. 647 (2005). Does the existence of a new, low-cost means of distribution of ideas mean that there is less need for economic support of the publishing industry through copyright?

c. Parodies

One circumstance in which the fair use defense has repeatedly arisen is in the treatment of parodies. A parody makes fun of an original work by at once imitating and distorting the work. For the parody to be effective, the audience must recognize the connection between the parody and the original work. This necessarily involves some deliberate copying of the original. Not surprisingly, authors unhappy at being parodied have invoked copyright law as a means of gaining control of the uses of their works, or at least sharing of the profits of parodies. As the following case shows, copyright law affords parodists (but less so satirists) wide berth in drawing upon protected works.

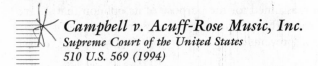

Campbell v. Acuff-Rose Music, Inc.
Supreme Court of the United States
510 U.S. 569 (1994)

Justice SOUTER delivered the opinion of the Court.

We are called upon to decide whether 2 Live Crew's commercial parody of Roy Orbison's song, "Oh, Pretty Woman," may be a fair use within the meaning of the Copyright Act of 1976, 17 U.S.C. § 107. Although the District Court granted summary judgment for 2 Live Crew, the Court of Appeals reversed, holding the defense of fair use barred by the song's commercial character and excessive borrowing. Because we hold that a parody's commercial character is only one element to be weighed in a fair use enquiry, and that insufficient consideration was given to the nature of parody in weighing the degree of copying, we reverse and remand.

I

In 1964, Roy Orbison and William Dees wrote a rock ballad called "Oh, Pretty Woman" and assigned their rights in it to respondent Acuff-Rose Music, Inc. Acuff-Rose registered the song for copyright protection.

Petitioners Luther R. Campbell, Christopher Wongwon, Mark Ross, and David Hobbs are collectively known as 2 Live Crew, a popular rap music group. In 1989, Campbell wrote a song entitled "Pretty Woman," which he later described in an affidavit as intended, "through comical lyrics, to satirize the original work...." On July 5, 1989, 2 Live Crew's manager informed Acuff-Rose that 2 Live Crew had written a parody of "Oh, Pretty Woman," that they would afford all credit for ownership and authorship of the original song to Acuff-Rose, Dees, and Orbison, and that they were willing to pay a fee for the use they wished to make of it. Enclosed with the letter were a copy of the lyrics and a recording of 2 Live Crew's song. Acuff-Rose's agent refused permission, stating that "I am aware of the success enjoyed by 'The

2 Live Crews,' but I must inform you that we cannot permit the use of a parody of 'Oh, Pretty Woman.'" Nonetheless, in June or July 1989, 2 Live Crew released records, cassette tapes, and compact discs of "Pretty Woman" in a collection of songs entitled "As Clean As They Wanna Be." The albums and compact discs identify the authors of "Pretty Woman" as Orbison and Dees and its publisher as Acuff-Rose.

Almost a year later, after nearly a quarter of a million copies of the recording had been sold, Acuff-Rose sued 2 Live Crew and its record company, Luke Skyywalker Records, for copyright infringement. The District Court granted summary judgment for 2 Live Crew, reasoning that the commercial purpose of 2 Live Crew's song was no bar to fair use; that 2 Live Crew's version was a parody, which "quickly degenerates into a play on words, substituting predictable lyrics with shocking ones" to show "how bland and banal the Orbison song" is; that 2 Live Crew had taken no more than was necessary to "conjure up" the original in order to parody it; and that it was "extremely unlikely that 2 Live Crew's song could adversely affect the market for the original." 754 F. Supp. 1150, 1154-1155, 1157-1158 (MD Tenn. 1991). The District Court weighed these factors and held that 2 Live Crew's song made fair use of Orbison's original. Id., at 1158-1159.

The Court of Appeals for the Sixth Circuit reversed and remanded. 972 F.2d 1429, 1439 (1992). Although it assumed for the purpose of its opinion that 2 Live Crew's song was a parody of the Orbison original, . . . the court concluded that its "blatantly commercial purpose . . . prevents this parody from being a fair use." Id., at 1439.

We granted certiorari, 507 U.S. 1003 (1993), to determine whether 2 Live Crew's commercial parody could be a fair use.

II

It is uncontested here that 2 Live Crew's song would be an infringement of Acuff-Rose's rights in "Oh, Pretty Woman," under the Copyright Act of 1976, 17 U.S.C. §106, but for a finding of fair use through parody. From the infancy of copyright protection, some opportunity for fair use of copyrighted materials has been thought necessary to fulfill copyright's very purpose, "to promote the Progress of Science and useful Arts. . . ." U.S. Const., Art. I, §8, cl. 8. For as Justice Story explained, "in truth, in literature, in science and in art, there are, and can be, few, if any, things, which in an abstract sense, are strictly new and original throughout. Every book in literature, science and art, borrows, and must necessarily borrow, and use much which was well known and used before." Emerson v. Davies, 8 F. Cas. 615, 619 (No. 4,436) (CCD Mass. 1845). Similarly, Lord Ellenborough expressed the inherent tension in the need simultaneously to protect copyrighted material and to allow others to build upon it when he wrote, "while I shall think myself bound to secure every man in the enjoyment of his copy-right, one must not put manacles upon science." Carey v. Kearsley, 4 Esp. 168, 170, 170 Eng. Rep. 679, 681 (K.B. 1803). In copyright cases brought under the Statute of Anne of 1710, English courts held that in some instances "fair abridgements" would not infringe an author's rights, see W. Patry, The Fair Use Privilege in Copyright Law 6-17 (1985) (hereinafter Patry); Leval, Toward a Fair Use Standard, 103 Harv. L. Rev. 1105, 1105 (1990) (hereinafter Leval), and although the First Congress enacted our initial copyright statute, Act of May 31, 1790, 1 Stat. 124, without any explicit reference to "fair use," as it later came to be known, the doctrine was recognized by the American courts nonetheless.... [10]

A

The first factor in a fair use enquiry is "the purpose and character of the use, including whether such use is of a commercial nature or is for nonprofit educational purposes." § 107(1). This factor draws on Justice Story's formulation, "the nature and objects of the selections made." Folsom v. Marsh, 9 F. Cas., at 348. The enquiry here may be guided by the examples given in the preamble to § 107, looking to whether the use is for criticism, or comment, or news reporting, and the like, see § 107. The central purpose of this investigation is to see, in Justice Story's words, whether the new work merely "supersedes the objects" of the original creation, Folsom v. Marsh, supra, at 348; accord, *Harper & Row,* supra, at 562 ("supplanting" the original), or instead adds something new, with a further purpose or different character, altering the first with new expression, meaning, or message; it asks, in other words, whether and to what extent the new work is "transformative." Leval 1111. Although such transformative use is not absolutely necessary for a finding of fair use, *Sony,* supra, at 455, n.40, [11] the goal of copyright, to promote science and the arts, is generally furthered by the creation of transformative works. Such works thus lie at the heart of the fair use doctrine's guarantee of breathing space within the confines of copyright, see, e.g., *Sony,* supra, at 478-480 (Blackmun, J., dissenting), and the more transformative the new work, the less will be the significance of other factors, like commercialism, that may weigh against a finding of fair use.

This Court has only once before even considered whether parody may be fair use, and that time issued no opinion because of the Court's equal division. Benny v. Loew's Inc., 239 F.2d 532 (CA9 1956), *aff'd sub nom.* Columbia Broadcasting System, Inc. v. Loew's Inc., 356 U.S. 43 (1958). Suffice it to say now that parody has an obvious claim to transformative value, as Acuff-Rose itself does not deny. Like less ostensibly humorous forms of criticism, it can provide social benefit, by shedding light on an earlier work, and, in the process, creating a new one. We thus line up with the courts that have held that parody, like other comment or criticism, may claim fair use under § 107.

The germ of parody lies in the definition of the Greek parodeia, quoted in Judge Nelson's Court of Appeals dissent, as "a song sung alongside another." 972 F.2d, at

10. Because the fair use inquiry often requires close questions of judgment as to the extent of permissible borrowing in cases involving parodies (or other critical works), courts may also wish to bear in mind that the goals of the copyright law, "to stimulate the creation and publication of edifying matter," Leval 1134, are not always best served by automatically granting injunctive relief when parodists are found to have gone beyond the bounds of fair use. See § 17 U.S.C. 502(a) (court "*may* ... grant ... injunctions on such terms as it may deem reasonable to prevent or restrain infringement") (emphasis added); Leval 1132 (while in the "vast majority of cases, [an injunctive] remedy is justified because most infringements are simple piracy," such cases are "worlds apart from many of those raising reasonable contentions of fair use" where "there may be a strong public interest in the publication of the secondary work [and] the copyright owner's interest may be adequately protected by an award of damages for whatever infringement is found"); Abend v. MCA, Inc., 863 F.2d 1465, 1479 (CA9 1988) (finding "special circumstances" that would cause "great injustice" to defendants and "public injury" were injunction to issue).

11. The obvious statutory exception to this focus on transformative uses is the straight reproduction of multiple copies for classroom distribution.

1440, quoting 7 Encyclopedia Britannica 768 (15th ed. 1975). Modern dictionaries accordingly describe a parody as a "literary or artistic work that imitates the characteristic style of an author or a work for comic effect or ridicule," or as a "composition in prose or verse in which the characteristic turns of thought and phrase in an author or class of authors are imitated in such a way as to make them appear ridiculous." For the purposes of copyright law, the nub of the definitions, and the heart of any parodist's claim to quote from existing material, is the use of some elements of a prior author's composition to create a new one that, at least in part, comments on that author's works. See, e.g., Fisher v. Dees, supra, at 437; MCA, Inc. v. Wilson, 677 F.2d 180, 185 (CA2 1981). If, on the contrary, the commentary has no critical bearing on the substance or style of the original composition, which the alleged infringer merely uses to get attention or to avoid the drudgery in working up something fresh, the claim to fairness in borrowing from another's work diminishes accordingly (if it does not vanish), and other factors, like the extent of its commerciality, loom larger.[14] Parody needs to mimic an original to make its point, and so has some claim to use the creation of its victim's (or collective victims') imagination, whereas satire can stand on its own two feet and so requires justification for the very act of borrowing. See ibid.; Bisceglia, Parody and Copyright Protection: Turning the Balancing Act Into a Juggling Act, in ASCAP, Copyright Law Symposium, No. 34, p. 25 (1987).

The fact that parody can claim legitimacy for some appropriation does not, of course, tell either parodist or judge much about where to draw the line. Like a book review quoting the copyrighted material criticized, parody may or may not be fair use, and petitioner's suggestion that any parodic use is presumptively fair has no more justification in law or fact than the equally hopeful claim that any use for news reporting should be presumed fair, see Harper & Row, 471 U.S., at 561. The Act has no hint of an evidentiary preference for parodists over their victims, and no workable presumption for parody could take account of the fact that parody often shades into satire when society is lampooned through its creative artifacts, or that a work may contain both parodic and non-parodic elements. Accordingly, parody, like any other use, has to work its way through the relevant factors, and be judged case by case, in light of the ends of the copyright law.

Here, the District Court held, and the Court of Appeals assumed, that 2 Live Crew's "Pretty Woman" contains parody, commenting on and criticizing the original work, whatever it may have to say about society at large. . . .

We have less difficulty in finding that critical element in 2 Live Crew's song than the Court of Appeals did, although having found it we will not take the further step of evaluating its quality. The threshold question when fair use is raised in defense of parody is whether a parodic character may reasonably be perceived.[16] Whether, going

---

14. A parody that more loosely targets an original than the parody presented here may still be sufficiently aimed at an original work to come within our analysis of parody. If a parody whose wide dissemination in the market runs the risk of serving as a substitute for the original or licensed derivatives (see infra, discussing factor four), it is more incumbent on one claiming fair use to establish the extent of transformation and the parody's critical relationship to the original. By contrast, when there is little or no risk of market substitution, whether because of the large extent of transformation of the earlier work, the new work's minimal distribution in the market, the small extent to which it borrows from an original, or other factors, taking parodic aim at an original is a less critical factor in the analysis, and looser forms of parody may be found to be fair use, as may satire with lesser justification for the borrowing than would otherwise be required.

16. The only further judgment, indeed, that a court may pass on a work goes to an assessment of whether the parodic element is slight or great, and the copying small or extensive in relation to the parodic element, for a work with slight parodic element and extensive copying will be more likely to merely "supersede the objects" of the original.

beyond that, parody is in good taste or bad does not and should not matter to fair use. As Justice Holmes explained, "it would be a dangerous undertaking for persons trained only to the law to constitute themselves final judges of the worth of [a work], outside of the narrowest and most obvious limits. At the one extreme some works of genius would be sure to miss appreciation. Their very novelty would make them repulsive until the public had learned the new language in which their author spoke." Bleistein v. Donaldson Lithographing Co., 188 U.S. 239, 251 (1903) (circus posters have copyright protection); cf. Yankee Publishing Inc. v. News America Publishing, Inc., 809 F. Supp. 267, 280 (S.D.N.Y. 1992) (Leval, J.) ("First Amendment protections do not apply only to those who speak clearly, whose jokes are funny, and whose parodies succeed") (trademark case).

While we might not assign a high rank to the parodic element here, we think it fair to say that 2 Live Crew's song reasonably could be perceived as commenting on the original or criticizing it, to some degree. 2 Live Crew juxtaposes the romantic musings of a man whose fantasy comes true, with degrading taunts, a bawdy demand for sex, and a sigh of relief from paternal responsibility. The later words can be taken as a comment on the naivete of the original of an earlier day, as a rejection of its sentiment that ignores the ugliness of street life and the debasement that it signifies. It is this joinder of reference and ridicule that marks off the author's choice of parody from the other types of comment and criticism that traditionally have had a claim to fair use protection as transformative works.[17]

The Court of Appeals, however, immediately cut short the enquiry into 2 Live Crew's fair use claim by confining its treatment of the first factor essentially to one relevant fact, the commercial nature of the use. The court then inflated the significance of this fact by applying a presumption ostensibly culled from Sony, that "every commercial use of copyrighted material is presumptively...unfair...." Sony, 464 U.S., at 451. In giving virtually dispositive weight to the commercial nature of the parody, the Court of Appeals erred.

The language of the statute makes clear that the commercial or non profit educational purpose of a work is only one element of the first factor enquiry into its purpose and character....

B

The second statutory factor, "the nature of the copyrighted work," § 107(2), draws on Justice Story's expression, the "value of the materials used." Folsom v. Marsh, 9 F. Cas., at 348. This factor calls for recognition that some works are closer to the core of intended copyright protection than others, with the consequence that fair use is more difficult to establish when the former works are copied. See, e.g., Stewart v. Abend, 495 U.S., at 237-238 (contrasting fictional short story with factual works); Harper & Row, 471 U.S., at 563-564 (contrasting soon-to-be-published memoir with published speech); Sony, 464 U.S., at 455, n.40 (contrasting motion

---

17. We note in passing that 2 Live Crew need not label its whole album, or even this song, a parody in order to claim fair use protection, nor should 2 Live Crew be penalized for this being its first parodic essay. Parody serves its goals whether labeled or not, and there is no reason to require parody to state the obvious (or even the reasonably perceived). See Patry & Perlmutter 716-717.

pictures with news broadcasts); Feist, 499 U.S., at 348-351 (contrasting creative works with bare factual compilations); 3 M. Nimmer & D. Nimmer, Nimmer on Copyright § 13.05[A][2] (1993) (hereinafter Nimmer); Leval 1116. We agree with both the District Court and the Court of Appeals that the Orbison original's creative expression for public dissemination falls within the core of the copyright's protective purposes. 754 F. Supp., at 1155-1156; 972 F.2d, at 1437. This fact, however, is not much help in this case, or ever likely to help much in separating the fair use sheep from the infringing goats in a parody case, since parodies almost invariably copy publicly known, expressive works.

C

The third factor asks whether "the amount and substantiality of the portion used in relation to the copyrighted work as a whole," § 107(3) (or, in Justice Story's words, "the quantity and value of the materials used," Folsom v. Marsh, supra, at 348) are reasonable in relation to the purpose of the copying. Here, attention turns to the persuasiveness of a parodist's justification for the particular copying done, and the enquiry will harken back to the first of the statutory factors, for, as in prior cases, we recognize that the extent of permissible copying varies with the purpose and character of the use. See Sony, 464 U.S., at 449-450 (reproduction of entire work "does not have its ordinary effect of militating against a finding of fair use" as to home videotaping of television programs); Harper & Row, 471 U.S., at 564 ("Even substantial quotations might qualify as fair use in a review of a published work or a news account of a speech" but not in a scoop of a soon-to-be-published memoir). The facts bearing on this factor will also tend to address the fourth, by revealing the degree to which the parody may serve as a market substitute for the original or potentially licensed derivatives. See Leval 1123.

The District Court considered the song's parodic purpose in finding that 2 Live Crew had not helped themselves overmuch. 754 F. Supp., at 1156-1157. The Court of Appeals disagreed, stating that "while it may not be inappropriate to find that no more was taken than necessary, the copying was qualitatively substantial. . . . We conclude that taking the heart of the original and making it the heart of a new work was to purloin a substantial portion of the essence of the original." 972 F.2d, at 1438. . . .

Where we part company with the court below is in applying [this factor] to parody, and in particular to parody in the song before us. Parody presents a difficult case. Parody's humor, or in any event its comment, necessarily springs from recognizable allusion to its object through distorted imitation. Its art lies in the tension between a known original and its parodic twin. When parody takes aim at a particular original work, the parody must be able to "conjure up" at least enough of that original to make the object of its critical wit recognizable. See, e.g., Elsmere Music, 623 F.2d, at 253, n.1; Fisher v. Dees, 794 F.2d, at 438-439. What makes for this recognition is quotation of the original's most distinctive or memorable features, which the parodist can be sure the audience will know. Once enough has been taken to assure identification, how much more is reasonable will depend, say, on the extent to which the song's overriding purpose and character is to parody the original or, in contrast, the likelihood that the parody may serve as a market substitute for the original. But using some characteristic features cannot be avoided.

We think the Court of Appeals was insufficiently appreciative of parody's need for the recognizable sight or sound when it ruled 2 Live Crew's use unreasonable as a matter of law. It is true, of course, that 2 Live Crew copied the characteristic opening bass riff (or musical phrase) of the original, and true that the words of the first line copy the Orbison lyrics. But if quotation of the opening riff and the first line may be said to go to the "heart" of the original, the heart is also what most readily conjures up the song for parody, and it is the heart at which parody takes aim. Copying does not become excessive in relation to parodic purpose merely because the portion taken was the original's heart. If 2 Live Crew had copied a significantly less memorable part of the original, it is difficult to see how its parodic character would have come through. See Fisher v. Dees, 794 F.2d, at 439.

This is not, of course, to say that anyone who calls himself a parodist can skim the cream and get away scot free. In parody, as in news reporting, see *Harper & Row*, supra, context is everything, and the question of fairness asks what else the parodist did besides go to the heart of the original. It is significant that 2 Live Crew not only copied the first line of the original, but thereafter departed markedly from the Orbison lyrics for its own ends. 2 Live Crew not only copied the bass riff and repeated it, but also produced otherwise distinctive sounds, interposing "scraper" noise, overlaying the music with solos in different keys, and altering the drum beat. See 754 F. Supp., at 1155. This is not a case, then, where "a substantial portion" of the parody itself is composed of a "verbatim" copying of the original. It is not, that is, a case where the parody is so insubstantial, as compared to the copying, that the third factor must be resolved as a matter of law against the parodists.

Suffice it to say here that, as to the lyrics, we think the Court of Appeals correctly suggested that "no more was taken than necessary," 972 F.2d, at 1438, but just for that reason, we fail to see how the copying can be excessive in relation to its parodic purpose, even if the portion taken is the original's "heart." As to the music, we express no opinion whether repetition of the bass riff is excessive copying, and we remand to permit evaluation of the amount taken, in light of the song's parodic purpose and character, its transformative elements, and considerations of the potential for market substitution sketched more fully below.

D

The fourth fair use factor is "the effect of the use upon the potential market for or value of the copyrighted work." § 107(4). It requires courts to consider not only the extent of market harm caused by the particular actions of the alleged infringer, but also "whether unrestricted and widespread conduct of the sort engaged in by the defendant . . . would result in a substantially adverse impact on the potential market" for the original. Nimmer § 13.05[A][4], p. 13-102.61 (footnote omitted); accord Harper & Row, 471 U.S., at 569; Senate Report, p. 65; Folsom v. Marsh, 9 F. Cas., at 349. The enquiry "must take account not only of harm to the original but also of harm to the market for derivative works." Harper & Row, supra, at 568.

Since fair use is an affirmative defense, its proponent would have difficulty carrying the burden of demonstrating fair use without favorable evidence about relevant markets. In moving for summary judgment, 2 Live Crew left themselves at just such a disadvantage when they failed to address the effect on the market for rap derivatives, and confined themselves to uncontroverted submissions that there was no likely effect on the market for the original. They did not, however, thereby subject themselves to the evidentiary presumption applied by the Court of Appeals. In assessing the likelihood of significant market harm, the Court of Appeals quoted from language in *Sony* that "'if the intended use is for commercial gain, that likelihood may be presumed. But if it is for a noncommercial purpose, the likelihood must be demonstrated.'" 972 F.2d, at 1438, quoting Sony, 464 U.S., at 451. The court reasoned that because "the use of the copyrighted work is wholly commercial, . . . we presume a likelihood of future harm to Acuff-Rose exists." 972 F.2d, at 1438. In so doing, the court resolved the fourth factor against 2 Live Crew, just as it had the first, by applying a presumption about the effect of commercial use, a presumption which as applied here we hold to be error.

No "presumption" or inference of market harm that might find support in *Sony* is applicable to a case involving something beyond mere duplication for commercial purposes. *Sony*'s discussion of a presumption contrasts a context of verbatim copying of the original in its entirety for commercial purposes, with the non-commercial context of *Sony* itself (home copying of television programming). In the former circumstances, what *Sony* said simply makes common sense: when a commercial use amounts to mere duplication of the entirety of an original, it clearly "supersedes the objects," Folsom v. Marsh, 9 F. Cas., at 348, of the original and serves as a market replacement for it, making it likely that cognizable market harm to the original will occur. Sony, 464 U.S., at 451. But when, on the contrary, the second use is transformative, market substitution is at least less certain, and market harm may not be so readily inferred. Indeed, as to parody pure and simple, it is more likely that the new work will not affect the market for the original in a way cognizable under this factor, that is, by acting as a substitute for it ("superseding [its] objects"). See Leval 1125; Patry & Perlmutter 692, 697-698. This is so because the parody and the original usually serve different market functions. Bisceglia, ASCAP, Copyright Law Symposium, No. 34, p. 23.

We do not, of course, suggest that a parody may not harm the market at all, but when a lethal parody, like a scathing theater review, kills demand for the original, it does not produce a harm cognizable under the Copyright Act. Because "parody may quite legitimately aim at garroting the original, destroying it commercially as well as artistically," B. Kaplan, An Unhurried View of Copyright 69 (1967), the role of the courts is to distinguish between "biting criticism [that merely] suppresses demand [and] copyright infringement[, which] usurps it." Fisher v. Dees, 794 F.2d, at 438.

This distinction between potentially remediable displacement and unremediable disparagement is reflected in the rule that there is no protectable derivative market for criticism. The market for potential derivative uses includes only those that creators of original works would in general develop or license others to develop. Yet the unlikelihood that creators of imaginative works will license critical reviews or lampoons of their own productions removes such uses from the very notion of a potential licensing market. "People ask . . . for criticism, but they only want praise." S. Maugham, Of Human Bondage 241 (Penguin ed. 1992). Thus, to the extent that the opinion below may be read to have considered harm to the market for parodies of "Oh, Pretty Woman," see 972 F.2d, at 1439, the court erred. Accord, Fisher v. Dees, 794 F.2d, at 437; Leval 1125; Patry & Perlmutter 688-691, n.22.

In explaining why the law recognizes no derivative market for critical works, including parody, we have, of course, been speaking of the later work as if it had nothing but a critical aspect (i.e., "parody pure and simple," *supra*, at 22). But the later work may have a more complex character, with effects not only in the arena of criticism but also in protectable markets for derivative works, too. In that sort of case, the law looks beyond the criticism to the other elements of the work, as it does here. 2 Live Crew's song comprises not only parody but also rap music, and the derivative market for rap music is a proper focus of enquiry, see Harper & Row, 471 U.S., at 568; Nimmer § 13.05[B]. Evidence of substantial harm to it would weigh against a finding of fair use, because the licensing of derivatives is an important economic incentive to the creation of originals. See 17 U.S.C. § 106(2) (copyright owner has rights to derivative works). Of course, the only harm to derivatives that need concern us, as discussed above, is the harm of market substitution. The fact that a parody may impair the market for derivative uses by the very effectiveness of its critical commentary is no more relevant under copyright than the like threat to the original market.

Although 2 Live Crew submitted uncontroverted affidavits on the question of market harm to the original, neither they, nor Acuff-Rose, introduced evidence or affidavits addressing the likely effect of 2 Live Crew's parodic rap song on the market for a non-parody, rap version of "Oh, Pretty Woman." . . . It is impossible to deal with the fourth factor except by recognizing that a silent record on an important factor bearing on fair use disentitled the proponent of the defense, 2 Live Crew, to summary judgment. The evidentiary hole will doubtless be plugged on remand.

III

It was error for the Court of Appeals to conclude that the commercial nature of 2 Live Crew's parody of "Oh, Pretty Woman" rendered it presumptively unfair. No such evidentiary presumption is available to address either the first factor, the character and purpose of the use, or the fourth, market harm, in determining whether a transformative use, such as parody, is a fair one. The court also erred in holding that 2 Live Crew had necessarily copied excessively from the Orbison original, considering the parodic purpose of the use. We therefore reverse the judgment of the Court of Appeals and remand for further proceedings consistent with this opinion.

It is so ordered.

Appendix A	*Appendix B*
"Oh, Pretty Woman" by Roy Orbison and William Dees	"Pretty Woman" as Recorded by 2 Live Crew
Pretty Woman, walking down the street,	Pretty woman walkin' down the street
Pretty Woman, the kind I like to meet,	Pretty woman girl you look so sweet
Pretty Woman, I don't believe you, you're not the truth,	Pretty woman you bring me down to that knee
No one could look as good as you Mercy	Pretty woman you make me wanna beg please Oh, pretty woman

Pretty Woman, won't you pardon me,
Pretty Woman, I couldn't help but see,
Pretty Woman, that you look lovely as
  can be
Are you lonely just like me?

Pretty Woman, stop a while,
Pretty Woman, talk a while,
Pretty Woman give your smile
  to me
Pretty Woman, yeah, yeah, yeah
Pretty Woman, look my way,
Pretty Woman, say you'll stay with me
'Cause I need you, I'll treat you right
Come to me baby, Be mine tonight

Pretty Woman, don't walk on by,
Pretty Woman, don't make me cry,
Pretty Woman, don't walk away,
Hey, O. K.
If that's the way it must be, O. K.
I guess I'll go on home, it's late
There'll be tomorrow night, but wait!

What do I see
Is she walking back to me?
Yeah, she's walking back to me!
Oh, Pretty Woman.

Big hairy woman you need to shave that stuff
Big hairy woman you know I bet it's tough
Big hairy woman all that hair it ain't legit
'Cause you look like 'Cousin It' Big hairy woman

Bald headed woman girl your hair won't grow
Bald headed woman you got a teeny weeny afro
Bald headed woman you know your hair could
  look nice
Bald headed woman first you got to roll it
  with rice
Bald headed woman here, let me get this hunk
  of biz for ya
Ya know what I'm saying you look better than
  rice a roni
Oh bald headed woman
Big hairy woman come on in
And don't forget your bald headed friend
Hey pretty woman let the boys
Jump in

Two timin' woman girl you know you ain't right
Two timin' woman you's out with my boy
  last night
Two timin' woman that takes a load off
  my mind
Oh, two timin' woman
Oh pretty woman

COMMENTS AND QUESTIONS

1. *Burden of Proof.* 2 Live Crew was successful in persuading the Supreme Court to reverse the Court of Appeals and reduce the significance of the presumption that commercial uses are unfair. But the Supreme Court did not simply affirm the district court's ruling that 2 Live Crew was entitled to summary judgment. Instead, the Court remanded the case for a determination of the parody's effect on the market for (non-parodic) rap derivatives of "Pretty Woman." Further, the Court obviously assumed that it was 2 Live Crew's burden to show that its parody had no such effect. Is 2 Live Crew—or any defendant—likely to be able to present such evidence and thus to prevail on summary judgment? Is it appropriate to assign the burden of proof to defendants on this issue?

2. *Scope of the Derivative Work Right.* The debate over parodies reflects a larger copyright issue that the Court does not directly address. Does the copyright owner have the right to prevent derivative works from being created, or only the right to receive royalties from derivative works produced by others? Traditionally, copyright law has assumed that the copyright owner has the right to prevent others from publishing a work at all without his or her permission. But the Court in *Campbell*

seems to assume the reverse — that copyright owners are expected to license their rights in order to promote the widespread distribution of information (a key policy underlying copyright law). In particular, see footnote 10 of the Court's opinion.

Are copyright owners likely to allow parodies voluntarily? Because many parodies lampoon the original, the authors of the original are often unwilling to permit a parody to be prepared at any price. Robert Merges has described this reluctance as the "bargaining breakdown" problem — noneconomic factors prevent the parties from agreeing to what might be an efficient license (in purely economic terms). See Robert Merges, Are You Making Fun of Me? Notes on Market Failure and the Parody Defense in Copyright, 21 Am. Intellectual Prop. L. Ass'n Q.J. 305 (1993). Cf. Richard Posner, When Is Parody Fair Use?, 21 J. Legal Stud. 79 (1992). In particular, in this case, Acuff-Rose refused to negotiate a mutually advantageous transfer because they didn't like the defendant's song. Does the breakdown of bargaining justify fair use treatment of parodies? Does the policy of encouraging dissemination do so? Note that some parodists like Weird Al Yankovic do succeed in getting licenses for their work, although the resulting parodies might be less critical than those produced without a license.

3. Must the alleged infringer's primary intent be to parody the work? Arguably, 2 Live Crew simply wanted to engage in satire. After all, their song is only obliquely directed at Orbison's. Can a work ever be a parody if the author did not intend it to be? (Note that 2 Live Crew originally claimed that their song was a *cover* of the original, and was therefore protected under section 115. This argument failed because 2 Live Crew substantially changed the song from its original version.)

4. *Parody or Satire?* The problems with the parody/satire distinction were put into sharp relief by the Ninth Circuit's decision in Dr. Seuss Enterprises v. Penguin Books, 109 F.3d 1394 (9th Cir. 1997). In that case, the defendant published a book making fun of the O.J. Simpson murder trial. Called "The Cat NOT in the Hat," the poem copied the graphic artistic style and lyric structure of Theodore Geisel's famous children's book *The Cat in the Hat*. The Ninth Circuit rejected the defendant's claim of fair use. It held that the defendant's work was satire, not parody, because his intent was to poke fun at the O.J. Simpson trial and not at the Dr. Seuss book, and therefore was not within the scope of fair use. The case is troubling because the defendant's work seems precisely the sort of transformative use that *Campbell* seemed to encourage and because there is no way to create such a work without relying on Geisel's preexisting work. See Tyler T. Ochoa, Dr. Seuss, the Juice and Fair Use: How the Grinch Silenced a Parody, 45 J. Copyright Society 546 (1998).

5. Margaret Mitchell's 1936 novel, *Gone with the Wind*, has become one of the best-selling books in the world, second in sales only to the Bible. The Mitchell Trust has actively managed the copyright, authorizing derivative works and a variety of commercial items. Alice Randall decided to retell this classic story of the pre-Civil War South from the perspective of Cyana, the illegitimate daughter of the plantation owner and the slave who cared for his children. Randall's book, *The Wind Done Gone*, contains many of the characters, settings, plot devices, and scenes from Mitchell's work, although viewed from a very different perspective. Randall's publisher, Houghton and Mifflin, no doubt familiar with the *Campbell*

case, billed the work as a "provocative literary parody that explodes the mythology perpetrated by a Southern classic." Suntrust, the Trustee of the Mitchell Trust, sued to halt publication of Randall's book. The trial court issued a preliminary injunction, finding that the plaintiff was likely to prevail on both substantial similarity and lack of fair use. Suntrust Bank v. Houghton Mifflin Co., 136 F.Supp.2d 1357, 1364 (N.D.Ga. 2001). On appeal, the Eleventh Circuit reversed the injunction. It agreed that the works were substantially similar, but held that Randall's book would likely prevail on its fair use defense. 268 F.3d 1257 (11th Cir. 2001). With respect to the first factor, the purpose and character of the work, the court found Randall's work to criticize and transform the original, stripping away the romantic portrait of the antebellum South painted by *Gone with the Wind* and reversing the race roles originally protrayed. Besides tipping the first factor in favor of fair use, the work's highly transformative nature meant that it would not compete with the original or harm the market for derivatives under the fourth factor. With respect to the third factor, the court acknowledged that *The Wind Done Gone* borrowed heavily from the original, but noted, citing *Campbell*, that parodies need to copy substantially in order to conjure up the original. It also noted that the second factor, nature of the copyrighted work, had little bearing since parodies nearly always copy expressive works. Thus, based principally on its analysis of factors one and four, coupled with First Amendment concerns about prior restraints on speech, the appellate court concluded that *The Wind Done Gone* would likely prevail on a fair use defense, and overturned the lower court's preliminary injunction. Cf. Jed Rubenfeld, The Freedom of Imagination: Copyright's Constitutionality. 112 Yale L. J. 1 (2002). The parties settled the case in a manner that allowed Randall's book to remain on the market.

PROBLEMS

Problem 4-33. Greenwich Systems is a software company that sells the world's most popular "screen saver." Greenwich's screen saver, called "flying toasters," shows pop-up toasters with wings flying across the screen.

Delaware, a small start-up company, licenses the popular "Bloom County/Outland" characters Opus and Bill for use in its own screen saver. Its program contains a number of humorous replacement screens, including one screen in which Opus shoots down flying toasters with a shotgun. The flying toasters look very similar to Greenwich's. Greenwich sues for copyright infringement, and Delaware defends on grounds of parody. Who should prevail?

Problem 4-34. Air Pirates, a publisher of counter-culture adult comic books, developed a series of comic books cynically depicting popular children's cartoon characters, including Mickey and Minnie Mouse, Donald Duck, and Goofy, as promiscuous, drug-dealing rogues. One of the magazine covers had the accompanying image (Figure 4-7).

Should Air Pirates' use be considered fair use?

FIGURE 4-7
Air Pirates cover.

2. Other Defenses

In addition to statutory defenses (17 U.S.C. §§ 108-118) and the fair use doctrine (17 U.S.C. § 107), there are a number of other defenses against a claim of copyright infringement. Many of these are straightforward, and the reader has no doubt encountered similar defenses elsewhere in the law. The most significant defenses include:

Independent Creation. A defendant may present evidence to prove that he or she created the work independently.

Consent/License. A defendant may defend a copyright action on the ground that he or she has the copyright owner's permission to make use of the protected material. Note that section 204 requires transfers of copyright ownership to be executed in writing and signed by the copyright owner. Nonexclusive licenses need not be in writing, however.

Inequitable Conduct. Closely related to copyright invalidity, the inequitable conduct doctrine parallels the patent law defense of inequitable conduct discussed in Chapter 3. Inequitable conduct occurs when a copyright owner obtains a copyright through fraud or other deceptive conduct on the Copyright Office — for example, by failing to disclose the owner's own plagiarism of a prior work. Circumstances that give rise to inequitable conduct generally also render copyrights invalid, but the defense is distinct because some of the consequences may be different.

Copyright Misuse. As copyright has expanded into the protection of technological products — most notably, computer software — courts have become increasingly sensitive to the ways in which copyright protection may be leveraged to undermine free competition. In 1990, a court held that a software licensing agreement that prohibited a licensee from developing any kind of computer-assisted die-making software for a term of 99 years constituted misuse, rendering the copyright unenforceable until the improper effects of the overreaching had been purged. See Lasercomb America, Inc. v. Reynolds, 911 F.2d 970 (4th Cir. 1990); see also DSC, Inc. v. DGI, Inc., 81 F.3d 597 (5th Cir. 1996); Practice Management Info. Corp. v. American Medical Ass'n, 121 F.3d 516 (9th Cir. 1997); Alcatel v. DGI Technologies, 166 F.3d 772 (5th Cir. 1999); Assessment Technologies of WI, LLC v. WIREdata, Inc. 350 F.3d 640 (7th Cir. 2003); but see Bellsouth Advertising & Pub. Corp. v. Donnelley Information Pub., Inc., 933 F.2d 952, 961 (11th Cir. 1991) (declining to follow *Lasercomb*). Copyright misuse is a blend of antitrust policies and copyright-specific policies against the improper extension of copyrights. We focus on the application of antitrust doctrine to intellectual property disputes in Chapter 8.

First Amendment. The argument is sometimes made that an accused infringer cannot be prevented from publishing an infringing work because to do so would violate her rights under the First Amendment to the U.S. Constitution. Courts generally reject this argument on the grounds that the Constitution also contains an explicit provision allowing Congress to protect copyrights and that this explicit provision has been squared with the First Amendment by the fair use and idea-expression limits on copyright. See Harper & Row, Publishers, Inc., et al. v. Nation Enterprises et al., 471 U.S. 539, 555-60 (1985); Eldred v. Ashcroft, 537 U.S. 186 (2003). Does this argument make sense? The Constitution also contains an explicit provision authorizing the government to regulate interstate commerce, but no one would argue that it prevents the application of the First Amendment to laws enacted under that power. If copyright is different because it can promote as well as interfere with speech, doesn't that suggest that the courts should be sensitive to whether copyright law strikes the right balance between these competing speech concerns?

Are certain statements on matters of public concern so important that a copyright owner should not be allowed to appropriate them to his own use? Do statements by political candidates fall in this category? Should there be a general constitutional protection against copyright infringement for the press? Cf. Melville Nimmer, Does Copyright Abridge the First Amendment Guarantees of Free Speech and Press?, 17 UCLA L. Rev. 1180, 1199 (1970) (suggesting a compulsory licensing scheme for "news photographs"). The Ninth Circuit has rejected this view, concluding that unedited videotapes of news events (in this case an airplane crash and a train wreck) were copyrightable. The court did, however, leave open the possibility that copyright owners would be forced to release their footage if there was no other avenue of access to the news contained in the film. See Los Angeles News Service v. Tullo, 24 U.S.P.Q.2d 1026 (9th Cir. 1992). Compare Los Angeles News Service v. KCAL-TV, 108 F.3d 1119 (9th Cir. 1997). There, the court held that LANS' videotape of the Reginald Denny beating could not be broadcast by a television news program that did not pay for a license. The court noted that in that case LANS was, in fact, offering to license the news footage to various news stations for a fee.

Several scholars have argued that copyright law causes broader and more fundamental tension with the First Amendment. See Jed Rubenfeld, The Freedom of Imagination: Copyright's Constitutionality, 112 Yale L.J. 1 (2002). Cf. Neil Weinstock Netanel, Locating Copyright Within the First Amendment Skein, 54 Stan. L. Rev. 1 (2001); Neil Weinstock Netanel, Copyright and a Democratic Civil Society, 106 Yale L.J. 283 (1996).

Immoral/Illegal/Obscene Works. Although some early cases refused to enforce copyrights in obscene or otherwise antisocial works, the modern trend rejects such a defense. See Mitchell Bros. Film Group v. Adult Cinema Theater, 604 F.2d 852 (9th Cir. 1980) (pornography), *cert. denied,* 445 U.S. 917 (1980); Belcher v. Tarbox, 486 F.2d 1087 (9th Cir. 1973) (racing forms).

Statute of Limitations. The statute of limitations for copyright infringement is three years "after the claim accrued." § 507. There is some question whether a continuing infringement gives rise to one or many claims and, if it is only a single claim, when it accrues. To see the difference, consider the position of an accused infringer who sold two copies of the same painting, one four years ago and one two years ago. If each sale constitutes a separate infringement, she can be held liable for the recent sale but not for the older sale. If selling the same painting constitutes a single act of infringement, however, whether she can be held liable at all depends on whether the infringement occurred two years ago or four years ago.

G. INDIRECT LIABILITY

As digital technology increasingly brings copyrighted works to wide audiences in forms that are easily copied, adapted, and distributed, the contours of liability for those who contribute to, induce, or profit from the infringing acts of others, or who merely sell products that others can use to infringe, has become the most critical question in the field. Before reaching the particular challenges posed by digital technology, addressed in section H, it is necessary to understand how indirect copyright liability emerged and developed. [32]

Copyright infringement standards developed from an austere statutory foundation. The 1790 Act provided that "any person or persons who shall print or publish any manuscript, without the consent and approbation of the author or proprietor thereof... shall be liable to suffer and pay to the said author or proprietor all damages occasioned by such injury." The Act did not provide a formal definition of infringement. The 1909 did not elucidate copyright's reach any further, stating simply that any person who "shall infringe the copyright in any work protected under the copyright laws of the United States... shall be liable" for various remedies. See 17 U.S.C. § 25 (1909 Act), recodified § 101 (1912 Act); see also H. Committee Print, 89th Cong., 1st Sess., Copyright Law Revision Part 6, Supplementary Report of the Register of Copyrights on the General Revision of the U.S. Copyright Law; 1965 Revision Bill (May 1965), chapter 7 (Copyright Infringement and Remedies) at p. 131 ("It seems strange, though not very serious, that the present law lacks any statement or definition of what constitutes an infringement.")

Against this bare legislative backdrop and drawing upon general principles of civil liability (tort law), courts recognized that copyright liability extends not just to those who infringe directly but also to those who contribute to or vicariously profit from the infringing acts of others. As noted more than a century ago, "[t]he evidence shows that the defendants bought the pictures from the complainants, furnished them to the photogravure company, ordered the copies made, and gave general directions as to how the work should be done. They are therefore liable as joint tort feasors." Fishel v. Lueckel, 53 F. 499 (S.D.N.Y. 1892); see also Kalem Co. v. Harper Brothers, 222 U.S. 55, 62-63 (1911) (observing that contributory liability is a principle "recognized in every part of the law").

Respondeat Superior. Courts readily recognized that employers should be liable for the infringing acts of their employees under traditional master-servant principles:

> Neither does the fact, if it is a fact, that young Williams, the operator of the player piano, borrowed this music without the direction, knowledge, or consent of the owner or manager of the theater affect the question. The rule of the common law applies, to wit, that the master is civilly liable in damages for the wrongful act of his servant in the transaction of the business which he was employed to do, although the particular act may have been done without express authority from the master, or even against his orders.

M. Witmark & Sons v. Calloway, 22 F.2d 412, 415 (D.Tenn. 1927).

Vicarious Liability. Even beyond the master-servant context, courts extended liability to those who profit from infringing activity where an enterprise has the right and ability to prevent infringement.

> [T]he owner of a dance hall at whose place copyrighted musical compositions are played in violation of the rights of the copyright holder is liable, if the playing be for the profit of the proprietor of the dance hall. And this is so even though the orchestra be employed under a contract that would ordinarily make it an independent contractor.

---

32. This section draws upon Peter S. Menell and David Nimmer, Direct Analysis of Indirect Copyright Liability (manuscript 2006).

Dreamland Ball Room v. Shapiro, Bernstein & Co., 36 F.2d 354 (7th Cir. 1929); see also Gershwin Publ'g Corp. v. Columbia Artists Mgt., 443 F.2d 1159 (2d Cir. 1971) ("When the right and ability to supervise coalesce with an obvious and direct financial interest in the exploitation of copyrighted material—even in the absence of actual knowledge . . . —the purposes of copyright law may be best effectuated by the imposition of liability upon the beneficiary of that exploitation.") By contrast, courts did not extend liability to landlords who leased premises to a direct infringer for a fixed rental and did not participate directly in organizing or soliciting the infringing activity. See Deutsch v. Arnold, 98 F.2d 686 (2d Cir. 1938). Cf. Fonovisa v. Cherry Auction, 76 F.3d 259 (9th Cir. 1996) (extending liability to the operator of a swap meet who repeatedly leased booth space to concessionaires selling infringing tapes).

Contributory Liability. "[O]ne who, with knowledge of the infringing activity, induces, causes, or materially contributes . . . may be held liable as a contributory infringer." Gershwin Publ'g Corp. v. Columbia Artists Mgt., 443 F.2d 1159 (2d Cir. 1971). Thus, in Elektra Records v. Gem Elec. Distribs., 360 F.Supp. 821 (E.D.N.Y. 1973), an electronics store which sold blank tapes and made available both pre-recorded tapes of copyrighted works and a high speed, coin-operated "Make-A-Tape" system was held contributorily liable for the infringing activities of its customers.

One of the studies commissioned for what ultimately became the 1976 Act reviewed the jurisprudence of indirect liability. See Alan Latman & William S. Tager, Study No. 25: Liability of Innocent Infringers of Copyrights (1958), reprinted in Subcomm. on Patents, Trademarks, and Copyrights, Senate Comm. on the Judiciary, 86th Cong., Copyright Law Revision: Studies 22-25, at 135 (Comm. Print 1960). The comments largely endorsed what the courts had done in extending copyright liability upstream. None of the many participants in the decade and a half of legislative hearings advocated change in the way such liability was addressed under the 1909 Act.

The principal reports accompanying the final version of the 1976 Act confirm that Congress intended to perpetuate indirect copyright liability. In explaining the general scope of copyright, the House Report recognizes contributory liability:

> The exclusive rights accorded to a copyright owner under section 106 are 'to do and to authorize' any of the activities specified in the five numbered clauses. Use of the phrase 'to authorize' is intended to avoid any questions as to the liability of *contributory* infringers. For example, a person who lawfully acquires an authorized copy of a motion picture would be an infringer if he or she engages in the business of renting it to others for purposes of unauthorized public performance.

H.R. Rep. No. 94-1478 at 61 (emphasis added). In discussing the infringement section, the House Report includes the following explanation:

> Vicarious liability for infringing performances
>
> The committee has considered and rejected an amendment to this section intended to exempt the proprietors of an establishment, such as a ballroom or night club, from liability for copyright infringement committed by an independent contractor, such as an orchestra leader. A well-established principle of copyright law is that a person who violates any of the exclusive rights of the copyright owner is an infringer, including persons who

can be considered related or vicarious infringers. To be held a related or vicarious infringer in the case of performing rights, a defendant must either actively operate or supervise the operation of the place wherein the performances occur, or control the content of the infringing program, and expect commercial gain from the operation and either direct or indirect benefit from the infringing performance. The committee has decided that no justification exists for changing existing law, and causing a significant erosion of the public performance right.

Id. at 159-60.

Not long after the passage of the 1976 Act, the scope of contributory infringement liability was tested in a case in which the copyright owners claimed that the sale of a recording device — the VCR — illegally contributed to infringement.

Sony Corporation of America v. Universal City Studios, Inc.

Supreme Court of the United States
464 U.S. 417 (1984)

Justice STEVENS delivered the opinion of the Court.

[The background to this case and the analysis of direct liability (fair use) can be found in section F.]

II

Article I, Sec. 8 of the Constitution provides that:

> The Congress shall have Power . . . to Promote the Progress of Science and useful Arts, by securing for limited Times to Authors and Inventors the exclusive Right to their respective Writings and Discoveries.

The monopoly privileges that Congress may authorize are neither unlimited nor primarily designed to provide a special private benefit. Rather, the limited grant is a means by which an important public purpose may be achieved. It is intended to motivate the creative activity of authors and inventors by the provision of a special reward, and to allow the public access to the products of their genius after the limited period of exclusive control has expired.

> The copyright law, like the patent statute, makes reward to the owner a secondary consideration. In Fox Film Corp. v. Doyal, 286 U.S. 123, 127, Chief Justice Hughes spoke as follows respecting the copyright monopoly granted by Congress, "The sole interest of the United States and the primary object in conferring the monopoly lie in the general benefits derived by the public from the labors of authors." It is said that reward to the author or artist serves to induce release to the public of the products of his creative genius.

United States v. Paramount Pictures, 334 U.S. 131, 158.

As the text of the Constitution makes plain, it is Congress that has been assigned the task of defining the scope of the limited monopoly that should be granted to authors or to inventors in order to give the public appropriate access to their work product. Because this task involves a difficult balance between the interests of authors and inventors in the control and exploitation of their writings and discoveries on the one hand, and society's competing interest in the free flow of ideas, information, and commerce on the other hand, our patent and copyright statutes have been amended repeatedly.[10]

From its beginning, the law of copyright has developed in response to significant changes in technology. Indeed, it was the invention of a new form of copying equipment — the printing press — that gave rise to the original need for copyright protection. Repeatedly, as new developments have occurred in this country, it has been the Congress that has fashioned the new rules that new technology made necessary. . . .

The judiciary's reluctance to expand the protections afforded by the copyright without explicit legislative guidance is a recurring theme. See, e.g., Teleprompter Corp. v. CBS, 415 U.S. 394 (1974); Fortnightly Corp. v. United Artists, 392 U.S. 390 (1968); White-Smith Music Publishing Co. v. Apollo Co., 209 U.S. 1 (1908); Williams and Wilkins v. United States, 487 F.2d 1345 (1973), affirmed by an equally divided court, 420 U.S. 376 (1975). Sound policy, as well as history, supports our consistent deference to Congress when major technological innovations alter the market for copyrighted materials. Congress has the constitutional authority and the institutional ability to accommodate fully the varied permutations of competing interests that are inevitably implicated by such new technology.

In a case like this, in which Congress has not plainly marked our course, we must be circumspect in construing the scope of rights created by a legislative enactment which never contemplated such a calculus of interests. In doing so, we are guided by Justice Stewart's exposition of the correct approach to ambiguities in the law of copyright:

> The limited scope of the copyright holder's statutory monopoly, like the limited copyright duration required by the Constitution, reflects a balance of competing claims upon the public interest: Creative work is to be encouraged and rewarded, but private motivation must ultimately serve the cause of promoting broad public availability of literature, music, and the other arts. The immediate effect of our copyright law is to secure a fair return for an "author's" creative labor. But the ultimate aim is, by this incentive, to stimulate artistic creativity for the general public good. "The sole interest of the United States and the primary object in conferring the monopoly," this Court has said, "lie in the general benefits derived by the public from the labors of authors." Fox Film Corp. v. Doyal, 286 U.S. 123, 127. When technological change has rendered its literal terms ambiguous, the Copyright Act must be construed in light of this basic purpose.

10. In its report accompanying the comprehensive revision of the Copyright Act in 1909, the Judiciary Committee of the House of Representatives explained this balance:

> The enactment of copyright legislation by Congress under the terms of the Constitution is not based upon any natural right that the author has in his writings, . . . but upon the ground that the welfare of the public will be served and progress of science and useful arts will be promoted by securing to authors for limited periods the exclusive rights to their writings.
> ****
> In enacting a copyright law Congress must consider . . . two questions: First, how much will the legislation stimulate the producer and so benefit the public, and, second, how much will the monopoly granted be detrimental to the public? The granting of such exclusive rights, under the proper terms and conditions, confers a benefit upon the public that outweighs the evils of the temporary monopoly.

H.R.Rep. No. 2222, 60th Cong., 2d Sess. 7 (1909).

Twentieth Century Music Corp. v. Aiken, 422 U.S. 151, 156 (footnotes omitted)....

The Copyright Act provides the owner of a copyright with a potent arsenal of remedies against an infringer of his work, including an injunction to restrain the infringer from violating his rights, the impoundment and destruction of all reproductions of his work made in violation of his rights, a recovery of his actual damages and any additional profits realized by the infringer or a recovery of statutory damages, and attorneys fees.

The two respondents in this case do not seek relief against the Betamax users who have allegedly infringed their copyrights. Moreover, this is not a class action on behalf of all copyright owners who license their works for television broadcast, and respondents have no right to invoke whatever rights other copyright holders may have to bring infringement actions based on Betamax copying of their works. As was made clear by their own evidence, the copying of the respondents' programs represents a small portion of the total use of VTR's. It is, however, the taping of respondents own copyrighted programs that provides them with standing to charge Sony with contributory infringement. To prevail, they have the burden of proving that users of the Betamax have infringed their copyrights and that Sony should be held responsible for that infringement.

III

The Copyright Act does not expressly render anyone liable for infringement committed by another. In contrast, the Patent Act expressly brands anyone who "actively induces infringement of a patent" as an infringer, 35 U.S.C. §271(b), and further imposes liability on certain individuals labeled "contributory" infringers, §271(c). The absence of such express language in the copyright statute does not preclude the imposition of liability for copyright infringements on certain parties who have not themselves engaged in the infringing activity. For vicarious liability is imposed in virtually all areas of the law, and the concept of contributory infringement is merely a species of the broader problem of identifying the circumstances in which it is just to hold one individual accountable for the actions of another.

We note the parties' statements that the questions of petitioners' liability under the "doctrines" of "direct infringement" and "vicarious liability" are not nominally before this Court. We also observe, however, that reasoned analysis of respondents' unprecedented contributory infringement claim necessarily entails consideration of arguments and case law which may also be forwarded under the other labels, and indeed the parties to a large extent rely upon such arguments and authority in support of their respective positions on the issue of contributory infringement.

Such circumstances were plainly present in Kalem Co. v. Harper Brothers, 222 U.S. 55 (1911), the copyright decision of this Court on which respondents place their principal reliance. In *Kalem*, the Court held that the producer of an unauthorized film dramatization of the copyrighted book Ben Hur was liable for his sale of the motion picture to jobbers, who in turn arranged for the commercial exhibition of the film....

Respondents argue that *Kalem* stands for the proposition that supplying the "means" to accomplish an infringing activity and encouraging that activity through advertisement are sufficient to establish liability for copyright infringement. This argument rests on a gross generalization that cannot withstand scrutiny. The producer in *Kalem* did not merely provide the "means" to accomplish an infringing activity; the producer supplied the work itself, albeit in a new medium of expression. Petitioners in the instant case do not supply Betamax consumers with respondents' works; respondents do. Petitioners supply a piece of equipment that is generally capable of copying the entire range of programs that may be televised: those that are uncopyrighted, those that are copyrighted but may be copied without objection from the copyright holder, and those that the copyright holder would prefer not to have copied. The Betamax can be used to make authorized or unauthorized uses of copyrighted works, but the range of its potential use is much broader than the particular infringing use of the film Ben Hur involved in *Kalem*. *Kalem* does not support respondents' novel theory of liability.... The only contact between Sony and the users of the Betamax that is disclosed by this record occurred at the moment of sale. The District Court expressly found that "no employee of Sony... had either direct involvement with the allegedly infringing activity or direct contact with purchasers of Betamax who recorded copyrighted works off-the-air." And it further found that "there was no evidence that any of the copies made by... individual witnesses in this suit were influenced or encouraged by [Sony's] advertisements."

If vicarious liability is to be imposed on petitioners in this case, it must rest on the fact that they have sold equipment with constructive knowledge of the fact that their customers may use that equipment to make unauthorized copies of copyrighted material. There is no precedent in the law of copyright for the imposition of vicarious liability on such a theory. The closest analogy is provided by the patent law cases to which it is appropriate to refer because of the historic kinship between patent law and copyright law. [19] ...

In the Patent Code both the concept of infringement and the concept of contributory infringement are expressly defined by statute. The prohibition against contributory infringement is confined to the knowing sale of a component especially made for use in connection with a particular patent. There is no suggestion in the statute that one patentee may object to the sale of a product that might be used in connection with other patents. Moreover, the Act expressly provides that the sale of a "staple article or commodity of commerce suitable for substantial noninfringing use" is not contributory infringement.

When a charge of contributory infringement is predicated entirely on the sale of an article of commerce that is used by the purchaser to infringe a patent, the public interest in access to that article of commerce is necessarily implicated. A finding of contributory infringement does not, of course, remove the article from the market altogether; it does, however, give the patentee effective control over the sale of that item. Indeed, a finding of contributory infringement is normally the functional equivalent of holding that the disputed article is within the monopoly granted to the patentee. [21]

19. United States v. Paramount Pictures, 334 U.S. 131, 158 (1948); Fox Film Corp. v. Doyal, 286 U.S. 123, 131 (1932); Wheaton and Donaldson v. Peters and Grigg, 33 U.S. (8 Pet.) 591, 657-658 (1834). The two areas of the law, naturally, are not identical twins, and we exercise the caution which we have expressed in the past in applying doctrine formulated in one area to the other. See generally, Mazer v. Stein, 347 U.S. 201, 217-218 (1954); Bobbs-Merrill Co. v. Straus, 210 U.S. 339, 345 (1908).

21. It seems extraordinary to suggest that the Copyright Act confers upon all copyright owners collectively, much less the two respondents in this case, the exclusive right to distribute VTR's simply because they may be used to infringe copyrights. That, however, is the logical implication of their claim. The request for an injunction below indicates that respondents seek, in effect, to declare VTR's contraband. Their suggestion in this Court that a continuing royalty pursuant to a judicially created compulsory license would be an acceptable remedy merely indicates that respondents, for their part, would be willing to license their claimed monopoly interest in VTR's to petitioners in return for a royalty.

For that reason, in contributory infringement cases arising under the patent laws the Court has always recognized the critical importance of not allowing the patentee to extend his monopoly beyond the limits of his specific grant. These cases deny the patentee any right to control the distribution of unpatented articles unless they are "unsuited for any commercial noninfringing use." Unless a commodity "has no use except through practice of the patented method," the patentee has no right to claim that its distribution constitutes contributory infringement. "To form the basis for contributory infringement the item must almost be uniquely suited as a component of the patented invention." P. Rosenberg, Patent Law Fundamentals § 17.02[2] (1982). "[A] sale of an article which though adapted to an infringing use is also adapted to other and lawful uses, is not enough to make the seller a contributory infringer. Such a rule would block the wheels of commerce." Henry v. A.B. Dick Co., 224 U.S. 1, 48 (1912), overruled on other grounds, Motion Picture Patents Co. v. Universal Film Mfg. Co., 243 U.S. 502, 517(1917).

We recognize there are substantial differences between the patent and copyright laws. But in both areas the contributory infringement doctrine is grounded on the recognition that adequate protection of a monopoly may require the courts to look beyond actual duplication of a device or publication to the products or activities that make such duplication possible. The staple article of commerce doctrine must strike a balance between a copyright holder's legitimate demand for effective–not merely symbolic–protection of the statutory monopoly, and the rights of others freely to engage in substantially unrelated areas of commerce. Accordingly, the sale of copying equipment, like the sale of other articles of commerce, does not constitute contributory infringement if the product is widely used for legitimate, unobjectionable purposes. Indeed, it need merely be capable of substantial noninfringing uses.

IV

[The court found that a substantial portion of the public's use of VCRs did not implicate copyright at all, and also that the most common use — time-shifting — was a fair use]. . . . The Betamax is, therefore, capable of substantial noninfringing uses. Sony's sale of such equipment to the general public does not constitute contributory infringement of respondent's copyrights.

V

"The direction of Art. I is that Congress shall have the power to promote the progress of science and the useful arts. When, as here, the Constitution is permissive, the sign of how far Congress has chosen to go can come only from Congress." Deepsouth Packing Co. v. Laitram Corp., 406 U.S. 518, 530 (1972).

One may search the Copyright Act in vain for any sign that the elected representatives of the millions of people who watch television every day have made it unlawful to copy a program for later viewing at home, or have enacted a flat prohibition against the sale of machines that make such copying possible.

It may well be that Congress will take a fresh look at this new technology, just as it so often has examined other innovations in the past. But it is not our job to apply laws that have not yet been written. Applying the copyright statute, as it now reads, to the facts as they have been developed in this case, the judgment of the Court of Appeals must be reversed.

It is so ordered.

COMMENTS AND QUESTIONS

1. *Statutory Interpretation or Judicial Legislation?* How would you characterize the methodology underlying the resolution of indirect liability in *Sony?* Given that Congress had comprehensively revised the Copyright Act just a few years earlier, was the Court justified in transplanting an express provision of the 1952 Patent Act into the Copyright Act? in applying the common law to create a tort of contributory infringement at all? How else might the Court have determined the appropriate liability standard in the absence of Congressional guidance?

2. The Court notes that "[t]he judiciary's reluctance to expand the protections afforded by the copyright without explicit legislative guidance is a recurring theme." Yet, it was the judiciary, and not Congress, that brought doctrines of indirect liability into copyright law, and Congress endorsed that approach in the 1976 Act. By contrast, the principal cases on which the Court relies for its comment about "judicial reluctance" addressed whether a statutory definition — created by Congress — applied to a new activity.

3. *Exploring the "Historic Kinship."* The Supreme Court justifies its transplantation of a categorical safe harbor from the patent statute into copyright law on the basis of a terse characterization of a rather complex historical relationship. Consider the following effort to delve more deeply into that relationship:

> While central to both patent and copyright law, technology plays very different roles in the two regimes. In patent law, technological innovation is the end to which the system is directed.... The staple article of commerce doctrine arose as a way of balancing contributory liability and patent misuse (antitrust-like limits on the leveraging of patent rights)....
>
> By contrast, in copyright law, technology serves as a means to the end of promoting creation and dissemination of works of authorship — art, music, literature, film, and other expressive works. Technology provides the platforms for instantiation, reproduction, and distribution on which creative expression flourishes and commerce occurs. When new technology platforms threaten the economic infrastructure supporting creative expression, copyright law seeks to protect the system that supports the creative arts.
>
> ...If, contrary to the Court's findings, VCRs did pose a serious threat to the "golden goose" of creative expression, then copyright law would have required a very different analytical perspective. Rather than look to patent law — which seeks to delineate the proper scope of exclusive rights in order to promote technological advance and freedom to use that which is not protected — the Court might have been better served by looking to statutory and common law regimes aimed at protecting interests threatened by technologies that can produce harmful side effects — such as tort law (nuisance, product defect) and environmental law. Thus, when a court enjoins a factory that spews noxious chemicals under nuisance or statutory environmental law, it would be misleading to characterize such a result as giving pollution victims "exclusive rights" over the factory's technology. A more apt characterization would be that society does not believe that the activity should be permitted in its current form. Such a perspective would not necessarily

mean that the factory should be shut down permanently. But it might mean that it would have to install filters to limit the adverse effects on neighbors.

Similarly, copyright law has long constrained technologies and business practices that jeopardize the system that supports creative expression. In *Jerome H. Remick & Co. v. General Electric Co.*, 16 F.2d 829 (S.D.N.Y. 1926), a lawsuit pitting music publishers against the newly-emerging radio industry, the court had little difficulty finding that the defendant's broadcast of plaintiff's copyrighted musical composition constituted copyright infringement, despite the fact that such a holding conferred a measure of "control" over the nascent radio broadcasting industry. What that case established was that radio broadcasters would have to obtain valid copyright licenses if they were going to build the popularity of their medium using copyrighted content.[3] This decision did not "shut down" the radio industry. Rather it led to the development of institutions for monitoring of broadcasts and compensation of artists — such as the ASCAP blanket license — which have fostered both commercial broadcasting and the creative arts.

Brief of Professors Peter S. Menell, David Nimmer, Robert P. Merges, and Justin Hughes, as Amici Curiae in Support of Petitioners, Metro-Goldwyn-Mayer Studios Inc., et al., v. Grokster, Ltd., et al., No. 04-480 (2005), reprinted in 20 Berkeley Tech. L.J. 511 (2005); see also Peter S. Menell and David Nimmer, Direct Analysis of Indirect Copyright Liability (manuscript 2006). Should the Court have looked to tort law or patent law as the default regime for determining the scope of indirect copyright liability? If it looked to tort law, should all services or technologies — from Internet service providers (ISPs) to general purpose computers and iPods — be subject to scrutiny (in the absence of a legislative safe harbor, as ISPs now have, discussed *infra* section H)? If it looked to patent law, should any technology regardless of its infringement consequences be immune from liability if it is merely capable of noninfringing use? Since we are dealing with indirect liability, should the economics of enforcement factor into the equation? Are you comforted by the brief's suggestion that copyright owners might not be entitled to block technologies that contribute to infringement altogether, but should be able to request that courts order design changes to counteract demonstrated piracy-causing effects? Cf. Universal Music Australia v. Sharman License Holdings ([2005] FCA 1242) (Australia) (requiring provider of file sharing software to implement a keywork filtering technology that excludes copyrighted music from search results).

4. "*Capable of Substantial Noninfringing Use.*" The Patent Act defines the staple article of commerce safe harbor as "suitable for substantial noninfringing use," 35 U.S.C. § 271(c), whereas the *Sony* decision speaks of "capable" of noninfringing use. What explains this difference? It appears that the Court wanted to get at both present and future uses. Are courts able to gauge such possibilities? Note in this regard that Hollywood was wrong to predict imminent disaster if Sony were allowed to continue making VCRs. Indeed, the movie industry has discovered ways to make a tidy profit from video rentals and sales. Does this suggest that courts should be cautious about using copyright law to screen markets for technology? On the other hand, might the technology sector and copyright owners have resolved any liability through a licensing arrangement, akin to ASCAP's blanket license? How many copyright owners would have had to agree to such a license in order for the Sony product to be legal? All of them?

3. The dance hall cases can also be characterized in this way. . . . Dance halls, like radio . . . can be used for infringing and non-infringing uses. The dance hall cases established that the proprietors of such facilities bore some responsibility to ensure that their clubs were not used for infringing uses. In the end, most clubs complied with the law by obtaining blanket licenses through ASCAP and BMI.

Both the patent and copyright law "staple article of commerce" standards restrict the safe harbor to "substantial" noninfringing uses. In *Sony*, the Court also defines it as a "commercially significant" noninfringing use. Is this a reasonable definition? Can you think of circumstances in which a noninfringing use might be substantial enough to justify marketing a product even though it is not "commercially significant"? Aren't noncommercial uses more likely to avoid infringement under the fair use doctrine?

Should the analysis focus on the entire product or be conducted on a feature-by-feature basis? Is there an argument that even if the VCR had substantial noninfringing uses, particular parts of its design didn't (e.g., fast-forward button)?

Note that the dissent would have held a defendant liable for contributory infringement unless use of its product was "predominantly" noninfringing. Legislation introduced in Congress in 1996, but which did not pass, would have adopted a version of the dissent's test for judging contributory infringement (whether the "primary purpose or effect" of a device was to encourage infringement), at least in the context of the Internet. Similar legislation passed in 1998 as part of the Digital Millenium Copyright Act, though with significant changes and limitations. With regard to tools for gaining access to a copyrighted work secured by a technological protection measure, Congress adopted the dissent's standard for what is a substantial noninfringing use. Therefore, *Sony* has been overruled in this respect. The Act is discussed in detail in section H. Is this a better standard? Why or why not?

5. *Dicta?* In view of the specific facts and other determinations in the *Sony* case — that time-shifting, the predominant use of the VCRs at issue, did not infringe under the fair use doctrine — it is not at all clear that the outcome of the case (victory for Sony) turned on the resolution of the standard for indirect liability. Even under the dissent's proposed standard for indirect liability — whether the primary use of technology is infringing — Sony would have prevailed given the majority's determination that the predominant use of the Betamax (time-shifting) constituted fair use. The correspondence among the justices reveals that the legality of "time shifting" of over-the-air television broadcasts was the critical point of disagreement between the majority and the dissent, not the standard for indirect liability. See Jonathan Band and Andrew J. McLaughlin, The Marshall Papers: A Peek Behind the Scenes at the Making of Sony v. Universal, 17 Colum.-VLA J.L. & the Arts 427 (1993). Therefore, but for the unresolved issue of archiving, it seems that there was no need for the Court to have even reached the question of indirect liability.

6. Should distributors, including retail stores, have a duty to inquire into the copyright status of the works they sell? How could such a duty be discharged? Should a party's status as a contributory infringer differ depending on whether an injunction or damages are sought? This issue arises with particular force in the case of ISPs. We will examine the online service provider safe harbor added to copyright law in 1998 in section H.

PROBLEM

Problem 4-35. Industrial Music Co. (IM) develops and produces a device known as a digital music filter for home and personal use. The filter allows individuals with access to audio tapes to "remix" those tapes themselves, and create new works in which sounds from existing works are speeded up, slowed down, added to or subtracted from, or combined with other preexisting works. IM markets its filter as a device for amateur musicians who want to edit their own works, but IM knows that approximately half of the filters it sells are used for "sampling" or otherwise copying copyrighted works without authorization. Is IM guilty of contributory infringement? If so, with every sale, or only with some sales? Can the sale of the filter be enjoined?

H. DIGITAL COPYRIGHT LAW[32]

Even though computer technology became a reality more than half a century ago, it is only in the past decade that it has begun to shake the foundations of the principal content industries — publishing, music, film, and television. The content industries' long-standing business models — selling books, newspapers, magazines, and records (and later tapes and compact discs, or CDs), exhibiting films (and later selling and renting home videos and DVDs), and broadcasting music and television shows — proved quite adaptable to the early generations of computer technology. The relatively late onset of the digital "piracy" threat can be attributed to the sheer informational magnitude of music and film and the inability, until recently, to bring to market affordable, high resolution means for perceiving (listening to and viewing) digital content. Even with the introduction and rapid popularity of digitally encoded CDs and the proliferation of microcomputers beginning in the early 1980s, the record industry did not appreciate the dramatic changes that would be wrought by a digital distribution platform. Available microprocessors, the low fidelity of computer peripherals, and limitations of memory storage capacity prevented music from being stored, perceived, and reproduced effectively on affordable computer devices until the mid-1990s.

With Moore's law continuing apace (predicting the doubling of storage capacity of semiconductor chips every 18 months) and related advances in information technology improving the capability and reducing the cost of computing, microcomputers became a viable platform for video games, multimedia content, and music by the late 1980s and early 1990s. The development of consumer versions of digital audio tape (DAT) technology around the same time set off the first alarm bells within the record industry. Concomitant with these developments, advances in network technology, eventually leading to the World Wide Web, data compression, a new wave of consumer electronics (including portable hard drives for storing music), the development of peer-to-peer architectures, and the deployment of broadband access for Internet home users drove the convergence of digital computers and traditional content. In so doing, the diffusion of digital technology set the stage for what has become an epic battle over the future of copyright law.

The content industries have increasingly focused their energies on forestalling and bracing for the blossoming of a digital platform. Even before the free flow of content in the post-Napster era, the content industries actively resisted the introduction of digital technologies and used the threat of such technologies as a basis for obtaining new

---

32. This section draws significantly upon Peter S. Menell, Envisioning Copyright Law's Digital Future, 46 N.Y.L. Sch. L. Rev. 63, 98-199 (2003).

legislation expanding rights and enforcement powers of copyright owners. This section summarizes the various amendments to copyright law during the 1990s and the efforts by the content industries to preclude and combat technologies that contribute to the unauthorized reproduction and distribution of copyrighted works. We then turn to the enforcement battles being waged on several fronts. The final section examines how the fair use doctrine is being applied in cyberspace.

1. Digital Copyright Legislation

a. *Prohibition on Commercial Record and Software Rental*

Even before the availability of home digital recording technology, the sound recording industry became concerned that home copying on widely available and improving analog cassette recorders threatened the retail market for sound recordings. See generally United States Office of Technology Assessment, Copyright and Home Copying: Technology Challenges Law (Oct. 1989). In 1984, the industry persuaded Congress to amend the first sale doctrine to prohibit the rental of sound recordings. See Record Rental Amendment of 1985 (codified at 17 U.S.C. § 109(b)). The software industry obtained comparable prohibitions on rentals of software in 1990. See Computer Software Rental Amendments of 1990 (codified at 17 U.S.C. § 109(b)). By contrast, it is legal to rent videos and books.

b. *Digital Audio Tape Devices*

As analog recording technology improved during the 1980s, the sound recording industry became particularly concerned about the inevitable arrival of digital recording technology. While listeners had been recording off the airwaves since the introduction of the audio cassette tape, copyright owners feared that digital equipment could produce the viral spread of high quality copies. By the mid-1980s, just a few years after the release of the record labels' catalogs in unencrypted digital format (on CDs), consumer electronics companies sought to introduce a host of new products that would enable consumers to make digital copies of audio recordings. These technologies, DAT and mini-disc (DCC), made it possible to produce identical copies of copyrighted works without any significant degradation of quality. As occurred with the introduction of video cassette recording technology in the early 1980s, copyright owners sued the principal manufacturer of this technology, the Sony Corporation. See Cahn v. Sony Corp., 90 Civ. 4537 (S.D.N.Y. filed July 9, 1990).

In the shadow of costly and uncertain litigation (and following Sony's acquisition of CBS Records, one of the leading record labels, in 1987), the various interests resolved their differences through negotiations which culminated in Congress's passage of the Audio Home Recording Rights Act of 1992 (codified at 17 U.S.C. §§ 1001-10). For the first time in the history of copyright, the government imposed technological design constraints on the manufacture of copying devices. This legislation also established a royalty on the sale of devices and blank recording media. Section 1002(a) prohibits the importation, manufacture, and distribution of any digital audio recording device that does not incorporate technological controls (Serial Copy Management System or functional equivalents) that block second-generation digital copying. This technology control allowed users to make copies directly from a compact disc, but not from digital copies made using this technology. In so doing, the AHRA sought to limit the viral spread of copies. Consumers could make first-generation copies, but no further copies could be made from those copies.

As a means to compensate copyright owners for the copying that could result from these new technologies, the Act requires manufacturers and importers of digital audio recording equipment and blank tapes, disks, or other storage media to pay a percentage of their transfer prices (2% for digital audio devices and 3% for storage media) into a royalty pool, which is distributed to owners of musical compositions (one-third) and sound recordings (two-thirds) based on prior year sales and air time. See 17 U.S.C. §§ 1003-07. This compensation mechanism is administered by the Register of Copyright, with provisions for arbitration of disputes. Section 1008 affords immunity for the manufacture, importation, and distribution of digital audio devices meeting the § 1002 design requirements, any analog audio recording devices, and any recording media. It also immunizes consumers from infringement liability for the noncommercial use of analog or qualifying digital devices for making copies. Violations of the AHRA are not copyright violations. Rather, the AHRA contains its own enforcement, remedy, and dispute resolution provisions. See 17 U.S.C. §§ 1009-10.

Digital audio tapes never took off. Does that fact suggest any lessons for legislation in this field? Might the AHRA itself have had something to do with that failure? Or was it simply the 8-track of its day, quickly supplanted by better technology (advances in computers)?

c. Webcasting

Sound recordings, as distinguished from the underlying musical compositions, did not receive federal copyright protection until 1972. By that point in time, radio broadcasters had sufficient political clout to exclude a public performance right from the rights accorded owners of sound recording copyrights. As a result, when a radio station broadcasts a post-1972 Frank Sinatra rendition of Cole Porter's "I've Got You Under My Skin," only Cole Porter receives a public performance royalty payment (typically through ASCAP's or BMI's blanket performance right license for musical compositions). This arrangement has always rankled record labels and recording artists. See Steven J. D'Onofrio, In Support of Performance Rights in Sound Recordings, 29 UCLA L. Rev. 169 (1981). When the Internet opened up a new distribution channel for sound recordings — what has come to be known as webcasting — record labels seized the opportunity to establish a performance right, even if only with respect to digital audio performances. See Lionel Sobel, A New Music Law for the Age of Digital Technology, 17 Ent. L.J. 3 (Nov. 1995). They voiced great concern that this new medium could seriously disrupt the market for sound recordings. If consumers could access and possibly even download high-quality recordings of their favorite songs over the Internet whenever they desired, they argued, there would be little demand for retail record (CD) sales or authorized online distribution outlets.

The prospect of webcasting and other online subscription services united traditional broadcasters and the sound recording industry in support of a digital performance right. Recording artists and record labels could at least partially rectify the omission of a public performance right in sound recordings, while traditional broadcasters could impose a new licensing requirement (and cost) upon potential new competitors. Since this new industry was not yet well developed, it lacked the political clout to block this new right, although the owners of musical composition copyrights (and their performing rights societies, ASCAP, BMI, and SESAC), who did not wish to empower another set of music licensing claimants, succeeded in constraining the reach of this new right. Furthermore, Congress sought to ensure that the new right would not unduly interfere with the development of new digital transmission business models.

The ultimate compromise amended sections 106 and 114 of the Copyright Act to establish an exclusive right to perform sound recordings "publicly by means of a digital audio transmission." See Digital Performance Right in Sound Recording Act of 1995 (DPRSRA), as amended by the DMCA. The DPRSRA created a three-tiered system: (1) "exempt transmissions," including nonsubscription digital broadcast transmissions, largely by traditional broadcasters, are exempt from the digital performance right;[33] (2) non-exempt transmissions such as non-interactive subscription digital transmissions meeting specified statutory criteria are subject to the new right but are granted a compulsory license; (3) non-exempt transmissions, including interactive or user-selected streaming of music, are ineligible for the compulsory license and hence require consent of or a license from the sound recording copyright owner. The Act sets procedures to determine the compulsory license rate.

COMMENTS AND QUESTIONS

1. *AM/FM Simulcasts.* Traditional broadcasters came to realize that they would want to simulcast their over-the-airwaves broadcasts as webcasts. They took the position that simultaneous Internet streaming by radio broadcasters of their AM/FM broadcast programming constituted a "nonsubscription broadcast transmission" that was exempt from the §106(6) digital audio transmission performance right. In a rulemaking proceeding implementing the DPRSRA, the Copyright Office interpreted the statutory exemption otherwise, thereby subjecting traditional AM/FM broadcasters to the compulsory license. See Bonneville Int'l Corp. v. Peters, 347 F.3d 485 (3d Cir. 2003) (upholding the Copyright Office's interpretation). In so doing, the Copyright Office made for a more level playing field in the webcasting market.

2. *Rate Setting and Small Webcaster Rules.* The statute provides for compulsory license rates to be set on the basis of "economic, competitive and programming information presented" by interested parties. The first such ratesetting adopted a rate of .07¢ per digital performance and a minimum fee of $500. See Copyright Office, Determination of Reasonable Rates and Terms for the Digital Performance of Sound Recordings and Ephemeral Recordings, 67 Fed. Reg. 45,240 (Jul. 8, 2002); Beethoven.com v. Librarian of Congress, 394 F.3d 939 (D.C. Cir. 2005) (upholding

---

33. Traditional television and radio broadcasters may continue to perform sound recordings without being subject to this new right, even if they convert their signal to digital form. See 17 U.S.C. §114(d)(1). In addition, various secondary transmissions of exempt primary transmissions and transmissions within business establishments (such as MUZAK) do not implicate the digital performance right.

rates). To put this into perspective, a webcaster eligible for the statutory license with an average base of 1,000 listeners would incur approximately $250 per day in licensing fees for digital performances of sound recordings. (1,000 listeners per day x $.0007 x 15 songs per hour x 24 hours = $252 per day). They would also need licenses from ASCAP, BMI, and/or SESAC for the performance rights in the underlying musical compositions. A special set of rates applies to "small webcasters." SoundExchange, an arm of the Recording Industry Association of America (RIAA), administers the licensing system for digital performance rights and allocates royalty payments to performing artists and sound recording copyright owners. Digital broadcasters complained loudly to Congress that the rates the Copyright Office had set were prohibitory.

d. Criminal Enforcement

Due to the large potential magnitude of harm from digital piracy, the difficulties of detecting illegal acts and holding those responsible accountable, and the distinctive profile of digital piracy (often involving "hacking" of digital protections without any clear commercial motivation), policymakers have expanded criminal penalties for digital piracy as a means of deterring such activities. In 1997, Congress enacted the No Electronic Theft (NET) Act in order to strengthen criminal prosecution and penalties against those who distribute copyrighted works without authorization. It specifically responded to the ruling in United States v. La Macchia, 871 F.Supp. 531 (D. Mass. 1994), in which the court held that a computer bulletin board operator providing users with unauthorized copies of copyrighted software without charge could not be prosecuted under federal criminal copyright law provisions because the government could not show that the operator benefited financially from the copyright infringement. The NET Act closed this loophole by criminalizing various intentional acts of copyright infringement without regard to whether the defendant received any financial benefit from such acts. It also stiffened the criminal penalties applicable to copyright infringement committed through electronic means. A person found guilty under this provision can receive three years in prison for a first offense and be forced to pay a substantial fine, even for illegally distributing sound recordings valued as little as $1,000. See 18 U.S.C. §2319.

In the first NET Act prosecution, completed in November 1999, federal prosecutors proceeded against a college student who had posted MP3 files, movie clips, and software on his website. Although a plea bargain kept the student out of jail, the case received substantial publicity. The more general problem of computer crime — fraud and the spreading of computer viruses — has led the United States Department of Justice to establish specialized cybercrime units throughout the nation. In December 2001, federal agents carried out raids in 27 cities as part of effort to break up a particularly notorious software piracy ring known by the name "DrinkorDie."

In response to the growing availablility of movies on peer-to-peer networks soon after (and in some cases, even before) their theatrical release, Congress passed the Artists' Rights and Theft Prevention Act of 2005 (ART Act). This law prohibits the unauthorized, knowing use or attempted use of a video camera or similar device to transmit or make a copy of a motion picture in movie theaters. The so-called "camcorder law" authorizes movie theater employees to detain those suspected of committing an offense in a reasonable manner and imposes imprisonment and stiff fines for violations. Several states also ban cameras and other recording devices such as image-capturing cell phones in theaters. The ART Act also establishes criminal penalties for willful copyright infringement by the distribution of a computer program, musical work, motion picture or other audiovisual work, or sound recording being prepared for commercial distribution by making it available on a computer network accessible to members of the public if the person knew or should have known that the work was intended for commercial distribution. On the civil side, the ART Act provides for preregistration of a work that is being prepared for commercial distribution so as to allow copyright owners to recover statutory damages and attorney fees for infringement of works in the production pipeline.

e. *Anticircumvention Prohibitions*

In the mid 1990s, content owners came to see encryption and other digital rights managements technologies as a means of "self-help" in the effort to discourage unauthorized distribution of their works. They recognized, however, that such technologies would be vulnerable to hacking—unauthorized circumvention of technological protection measures. As a result, they sought to expand copyright protection beyond its traditional prohibitions against infringement to include limits on the decrypting or circumventing of technological protection systems and the trafficking in such decryption tools. They argued to Congress that without such protection, they would be unwilling to release content onto the Internet, which in turn would hamper the adoption of broadband services. Various other interests—ranging from ISPs and telecommunications companies to consumer electronic manufacturers, library associations, computer scientists, and copyright professors—expressed concern about chilling effects of such an expansion of copyright law upon those who transmit content and wish to make "fair use" of copyrighted works. The resulting legislation—the Digital Millennium Copyright Act of 1998 (DMCA)—responded to the core concerns of the content owners by enacting anti-circumvention and anti-trafficking bans, while assuaging the fears of the most powerful opposing interest group coalition— ISPs and telecom companies—by creating a series of online service provider safe harbors.

Somewhat like the AHRA, Title I of the DMCA goes beyond traditional copyright protections in order to address the threat of unauthorized reproduction and distribution of copyrighted works in the digital age.[34] But rather than mandating specific technology controls,[35] the DMCA focuses on ensuring the efficacy of technological control measures put in place by copyright owners. With regard to technological measures controlling *access* to a work (e.g., encryption governing access to an eBook), Section 1201(a) prohibits both specific acts to circumvent the technological measure[36] and the manufacture, importation, trafficking in, and marketing of devices

---

34. Although codified as part of Title 17 of the U.S. Code, violations of the DMCA do not constitute copyright infringements. See 17 U.S.C. §§ 1203-04 (specifying civil and criminal remedies for violations of the DMCA's anticircumvention and anti-trafficking provisions).

35. While generally eschewing technology mandates, see 17 U.S.C. § 1201(c)(3) (the "no mandate" provision), the DMCA does impose limited technology controls on some videocassette recorders. See 17 U.S.C. § 1201(k) (requiring future analog VCRs to incorporate new anticopying technology).

36. 17 U.S.C. § 1201(a)(1). To circumvent a technological measure is defined as descrambling a scrambled work, decrypting an encrypted work, or "otherwise to avoid, bypass, remove, deactivate, or

that: (1) are primarily designed or produced for the purpose of circumventing a technological measure that effectively "controls access to" a copyrighted work; (2) have only limited commercially significant purpose or use other than to circumvent such technological protection measures; or (3) are marketed for use in circumventing such technological protection measures. 17 U.S.C. § 1201(a)(2). With regard to technological measures regulating *use* of a work where access has been lawfully obtained (e.g., through the purchase of a DVD), section 1201(b) prohibits not the act of circumvention but only trafficking in and marketing of circumvention devices. This more limited protection was purportedly designed so as not to impair users' ability to make fair use of content to which they have been given access.[37] This limitation, however, provides little solace to advocates of broad fair use standards because although it allows circumvention of use controls, the ban on trafficking of circumvention devices (including instructions) puts the means for such access beyond the reach of all but the most technically adept—those possessing the ability to decrypt restricted works. Section 1202 further bolsters encryption efforts by prohibiting the falsification, removal, or alteration of "copyright management information," such as digital watermarks and identifying information, when done with the intent to encourage or conceal infringement.

The DMCA addresses the many objections and concerns raised by various groups through a complex series of narrow exemptions.[38] In order to reduce adverse effects of Section 1201 upon fair use of copyrighted works, the DMCA authorizes the Librarian of Congress to exempt any classes of copyrighted works where persons making non-infringing uses are likely to be adversely affected by the anticircumvention ban. Perhaps of most significance, the DMCA authorizes the circumvention of technological protection measures for purposes of reverse engineering of computer programs for the "sole purpose of identifying and analyzing those elements of the program that are necessary to achieve interoperability of an independently created computer program." 17 U.S.C. § 1201(f)(1). Many observers consider the exemptions to be overly narrow, severely restricting the traditional fair use of copyrighted works. See Pamela Samuelson, Intellectual Property and the Digital Economy: Why the Anti-Circumvention Regulations Need to Be Revised, 14 Berkeley Tech. L.J. 519 (1999); David Nimmer, A Riff on Fair Use in the Digital Millennium Copyright Act, 148 U. Pa. L. Rev. 673 (2000).

COMMENTS AND QUESTIONS

1. *Stream Capture Technology.* One of the first tests of the anticircumvention prohibitions arose with regard to technology for streaming music and video over the Internet. RealNetworks developed technology that allows Internet users to access protected content encoded in its RealMedia formats using its RealPlayer software. The user cannot, however, store the content on their computer (unless the content provider activated the download capability). Streambox began offering its "VCR"

---

impair a technological measure, without the authority of the copyright owner." 17 U.S.C. § 1201(a)(3)(A).

37. See H.R. Rep. 105-551, pt. 1, at 18 (1998); Exemption to Prohibition on Circumvention of Copyright Protection Systems for Access Control Technologies, 65 Fed. Reg. 64,557 (2000) (codified at 37 C.F.R. § 201).

38. Detailed exemptions exist for law enforcement activities, radio and television broadcasters, libraries, encryption researchers, filtering of content to prevent access by minors, and protection for personally identifying information. See 17.U.S.C. §§ 1201(d), (e), (h), (i).

and "Ripper" technologies. The Streambox VCR product enables users to access and download copies of RealMedia files that are streamed over the Internet by mimicking the operation of RealPlayer software. It then circumvents the authentication procedure in order to gain access to streamed content. Unlike the RealPlayer, however, the Streambox VCR bypasses the copy switch so that users can download content, even if the content owner had intended that it only be streamed. Once downloaded, the content can then be accessed, copied, and distributed at the user's discretion. Streambox's Ripper technology enables users to convert files from RealMedia (.RMA) format to other formats such as .WAV (a format commonly used for music editing), .WMA (Windows Media Player), and MP3.

RealNetworks sued Streambox for violating the DMCA's anticircumvention prohibitions. On RealNetworks's motion for a preliminary injunction, the court held that aspects of the Streambox VCR were likely to violate the new law. See RealNetworks, Inc. v. Streambox, Inc., 2000 WL 127311 (W.D.Wash. 2000). In particular, the court found that the authentication process used to establish a handshake between the RealPlayer and a RealNetworks server constitutes a "technological measure" that "effectively controls access" to copyrighted works. The Streambox VCR's means of establishing access and then bypassing the copy switch circumvents the technological protection measures. The court further found that it had no significant commercial purpose other than to enable users to access and record protected content. The court rejected Streambox's defense that its software allows consumers to make "fair use" copies, such as to time or space shift access to content. It distinguished the *Sony* case on two grounds: (1) many of the copyright owners there authorized or would not have objected to having their content time-shifted whereas all of the content owners using the RealNetworks' technology to stream their works specifically chose not to authorize downloading; and (2) *Sony* did not address the new protections afforded by the DMCA.[39] The court declined to enjoin Streambox's Ripper software, raising doubts as to whether the .RMA format constituted a "technological protection measure" within the meaning of the DMCA and noting that it could serve significant legitimate purposes.

2. *Constitutionality of the DMCA's Anticircumvention Provisions.* In a series of high-profile cases, the content industries have pursued publishers of decryption code who have asserted as a defense that the DMCA interferes with their freedom of expression protected by the First Amendment. The courts have upheld the anticircumvention provisions under the intermediate scrutiny test applied to content-neutral constraints on speech. Under this standard, the courts determined that the ban on distributing decryption code were adequately justified by the substantial governmental interest in restraining unauthorized distribution of copyrighted works in the digital age, were not related to the suppression of free expression, and did not burden substantially more speech than necessary to further the interest in preventing

---

39. The court cited Nimmer on Copyright for the proposition that "those who manufacture equipment and products generally can no longer gauge their conduct as permitted or forbidden by reference to the *Sony* doctrine. For a given piece of machinery might qualify as a staple item of commerce, with a substantial noninfringing use, and hence be immune from attack under *Sony*'s construction of the Copyright Act—but nonetheless still be subject to suppression under Section 1201." 1 Nimmer on Copyright (1999 Supp.), § 12A.18[B]. As such, "[e]quipment manufacturers in the twenty-first century will need to vet their products for compliance with Section 1201 in order to avoid a circumvention claim, rather than under *Sony* to negate a copyright claim." Id.

piracy. See Universal City Studios, Inc. v. Corley, 273 F.3d 429 (2d Cir. 2001); see also United States v. Elcom, Ltd., 203 F. Supp. 2d 1111 (N.D. Cal. 2002); 321 Studios v. Metro-Goldwyn-Mayer, Inc., 307 F.Supp.2d 1085 (N.D. Cal. 2004).

3. *Circumventing Fair Use?* As noted above, violations of the DMCA are not acts of copyright infringement, but separate offenses. As a result, some courts have held that the defenses available under the Copyright Act, including fair use, simply don't apply to a DMCA claim. 321 Studios v. Metro-Goldwyn-Mayer, Inc., 307 F.Supp.2d 1085 (N.D. Cal. 2004) (holding liable a provider of software that decrypted a DVD but allowed owners of a DVD only to make a single backup copy, notwithstanding the fair use claim for the making of that copy). While section 1201(c)(1) of the DMCA provides that "nothing in this law" shall interfere with "fair use," among other defenses, the courts coming to this conclusion have reasoned that the DMCA doesn't interfere with fair use, but merely renders it irrelevant by allowing copyright owners to bring a non-copyright claim. On the other hand, in Chamberlain Group, Inc., v. Skylink Tech., Inc., 381 F.3d 1178 (Fed. Cir. 2004), the court engaged in a tortured interpretation of the statute in order to conclude that the fair use defense did apply to anticircumvention claims.

4. *Exemption Process.* As noted above, the anticircumvention provisions impinge upon fair use of copyrighted works. In addition to the narrow exceptions and limitations specified in the DMCA, Congress instructed the Copyright Office to conduct periodic rulemaking proceedings to consider whether additional categorical exemptions should be made. In the first such proceeding, conducted in 2000, the Copyright Office interpreted its authority narrowly. Its final rule exempted only two class of works from the anticircumvention ban: (1) literary works whose storage formats had malfunctioned or become obsolete; and (2) lists of filtering criteria employed by software vendors to screen undesirable Internet content. See U.S. Copyright Office, Exemption to Prohibition on Circumvention of Copyright Protection Systems for Access Control Technologies: Final Rule, 65 Fed. Reg. 64,556 (Oct. 27, 2000). The Copyright Office limited these exemptions still further when they came up for review after three years. See U.S. Copyright Office, Exemption to Prohibition on Circumvention of Copyright Protection Systems for Access Control Technologies: Final Rule, 68 Fed. Reg. 62,011 (Oct. 28, 2003). Do you think that this shift toward a regulatory model for copyright protection makes sense? Cf. Peter S. Menell, Envisioning Copyright Law's Digital Future, 46 N.Y.L. Sch. Rev. 63, 194-97 (2003); Joseph P. Liu, Regulatory Copyright, 83 N.C. L. Rev. 87 (2004). As a result, this rulemaking provision has provided little protection for users.

5. *Control of Aftermarket Products.* The general wording of the DMCA's anticircumvention provisions has produced an unanticipated wave of litigation involving the use of technological protection measures to exclude competitors from "aftermarkets" — goods or services supplied for a durable product (e.g., a printer) after its initial sale, such as replacement ink. Several companies embedded digital code into their products and aftermarket components that must interoperate in order to function as a means of exerting control over such aftermarkets. When competitors in these aftermarkets decrypted such digital code in order to manufacture their own components, these durable product manufacturers sued, alleging violation of the anticircumvention provisions of the DMCA. The courts declined to find liability, emphasizing that the careful balance that Congress sought to achieve between the "interests of content creators and information users," H.R.Rep. No. 105-551, pt. 1, at 26, would be upset if the anticircumvention prohibitions could be applied to activities that did

not facilitate copyright infringement. See Chamberlain Group, Inc., v. Skylink Tech., Inc., 381 F.3d 1178 (Fed. Cir. 2004) (holding that access to the copyrighted work was in fact authorized and that "section 1201 prohibits only forms of access that bear a reasonable relationship to the protections that the Copyright Act otherwise affords copyright owners"); Lexmark Int'l, Inc. v. Static Control Components, Inc., 387 F.3d 522 (6th Cir. 2005) (holding that the lock-out technology at issue did not effectively control access to a copyrighted work); see also Storage Technology Corp. v. Custom Hardware Engineering, 421 F.3d 1307 (Fed. Cir. 2005) (decryption (by third-party software repair entity) in order to perform software maintenance activities not actionable). Are these claims legitimate uses of the DMCA? Or are they are a form of "bootstrapping," alleging that the lockout code is itself the copyrighted work that the code is nominally designed to protect?

f. Online Service Provider Safe Harbors

As the Internet emerged as a communications and commercial medium, new commercial enterprises providing access to the Internet found themselves in a precarious situation by hosting websites posting content uploaded by their customers. Did these Internet businesses face liability for hosting websites containing infringing content and transmitting copyrighted works without authorization?

In Religious Technology Center v. Netcom On-Line Communication Services, Inc., 907 F.Supp. 1361 (N.D.Cal. 1995), a copyright owner sued an Internet access provider (Netcom) and its subscriber, a computer bulletin board service (BBS) hosting a Usenet newsgroup, for direct and indirect copyright liability on the basis of material posted to the newsgroup. Netcom employed an automated system for relaying messages on Usenet groups without any monitoring or control of content. The court determined that "[a]lthough copyright is a strict liability statute, there should still be some element of volition or causation which is lacking where a defendant's system is merely used to create a copy by a third party." To hold otherwise would "also result in liability for every single Usenet server in the worldwide link of computers transmitting [the poster's infringing] message to every other computer." With regard to contributory liability, the court held that when a BBS cannot reasonably verify a claim of infringement — because of "the copyright holder's failure to provide the necessary documentation to show that there is a likely infringement" or otherwise — then the knowledge element of contributory infringement should be deemed lacking. As to participation, the court concluded that "it is fair, assuming Netcom is able to take simple measures to prevent further damage to plaintiffs' copyrighted works, to hold Netcom liable for contributory infringement where Netcom has knowledge of [the poster's] infringing postings yet continues to aid in the accomplishment of [the infringer's] purpose of publicly distributing the postings." Other courts adopted the *Netcom* approach. See ALS Scan, Inc. v. RemarQ Cmtys., Inc., 239 F.3d 619 (4th Cir. 2001); Marobie-Fl, Inc. v. National Ass'n of Fire and Equipment Distributors, 983 F.Supp. 1167 (N.D.Ill. 1997).

Controversy over the scope of liability of online service providers (OSPs) fed directly into negotiations over the DMCA. OSPs, such as America Online (AOL) and Yahoo, warned that failure to incorporate immunity directly into the Copyright Act could severely impair their rapidly emerging industry and impede the growth of economic activity on the Internet. At their urging (and over the resistance of the content industries), Congress established a series of safe harbors insulating OSPs from liability for various acts, such as transmitting, storing, or linking to unauthorized content as Title II of the DMCA. These provisions codify the key aspects of the *Netcom* decision.

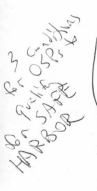

In order to qualify for safe harbor treatment, an OSP must meet three threshold conditions: (i) adopt, implement, and inform its subscribers of its policy for terminating service to of users who are repeat copyright infringers, 17 U.S.C. §512(i)(1)(A); (ii) adopt standard technical measures used by copyright owners to identify and protect copyrighted works, 17 U.S.C. §512(i)(1)(B); and (iii) designate an agent to receive notification of claimed infringement from copyright owners and register that agent with the Copyright Office. The Act imposes more specific criteria in order to qualify for the particular safe harbors: (1) transmission and routing — transmitting copyrighted material without authorization, see 17 U.S.C. §512(m) (requiring that the OSP act as a passive conduit, neither directing, initiating, selecting, or modifying the content being transmitted by third parties); (2) storage — storing such material on their servers, see 17 U.S.C. §512(c) (requiring that the OSP not have actual notice that infringing material resides on its servers and promptly remove or block access to infringing material upon learning of copyright violations);[40] (3) caching — making temporary copies on their systems, see 17 U.S.C. §512(b); and (4) linking — providing links to infringing material, see 17 U.S.C. §512(d). OSPs satisfying these requirements are shielded from monetary relief and most forms of equitable relief.

COMMENTS AND QUESTIONS

1. *Terminating Repeat Infringers.* The DMCA imposes responsibilities upon OSPs that seek to obtain the benefits of the specified safe harbors. In addition to meeting the three threshold requirements, they must reasonably implement a policy of terminating repeat infringers. Several cases explore the ambiguities of the statute, including the problematic question of who is a repeat infringer. See Ellison v. Robertson, 357 F.3d 1072 (9th Cir. 2004); Corbis Corp. v. Amazon.com, 351 F.Supp.2d 1090 (W.D. Wash. 2004); see generally David Nimmer, Repeat Infringers, 52 J. Copyright Soc'y 167 (2005).

2. *Erroneous Takedown Notices.* The DMCA has aroused concern that copyright owners might use overly aggressive tactics or misrepresentations to suppress legal activities. Courts apply a subjective standard for determining whether a takedown notice was propounded in "good faith." See Rossi v. Motion Picture Association of America, 391 F.3d 1000 (9th Cir. 2004); Online Policy Group v. Diebold, Inc., 337 F.Supp.2d 1195 (N.D.Cal. 2004) (finding "knowing misrepresentation" in takedown request and awarding damages and attorneys' fees pursuant to §512(f)).

---

40. In order to protect users' rights, however, the OSP must promptly notify users that material has been blocked or removed. The user may then provide a "counter notification" stating that the material may properly be stored, which the OSP must promptly pass along to the copyright owner. The OSP must replace or unblock the material within 10 to 14 business days of receiving the counter notification unless the copyright owner informs the OSP that it has filed a court action to restrain the infringement. 17 U.S.C. §512(g)(2).

PROBLEM

Problem 4-36. Real Estate Net offers local independent real estate agents an inexpensive means for posting real estate listings to a large searchable online database. In order to post a listing, subscribers fill out an online form and agree to "Terms and Conditions" that include a promise not to post copies of photographs without authorization. The subscriber then transmits the text of the listing and any applicable photographs to Real Estate Net. The text gets posted immediately, while the photograph goes into a temporary folder. Within 48 hours, a Real Estate Net employee reviews the photograph to determine whether the photograph in fact depicts commercial real estate and to identify any obvious evidence, such as a copyright notice, that the photograph may have been copyrighted by another. If the photograph fails either one of these criteria, the employee deletes the photograph and notifies the subscriber. Otherwise, the employee clicks an "accept" button that uploads the photograph to Real Estate Net's website.

National Realtors, Inc. (NRI) maintains a large online database of real estate information, including photographs to which it holds copyright. NRI's online customers agree not to post its photographs on other websites. Notwithstanding such agreements, NRI discovered approximately 100 of its copyrighted photographs among the nearly 50,000 photographs posted on Real Estate Net's website. With regard to 25 of the photographs in question, NRI has sent letters to Real Estate Net demanding that the specific works be removed, to which Real Estate Net has promptly complied. NRI has also made general allegations of infringement and mentioned the names of several real estate agents that have posted photographs without authorization.

Please advise Real Estate Net on any potential copyright liability based on the above facts. In addition, please identify any steps that Real Estate Net might take to reduce its exposure to liability in the future.

2. Enforcement

In addition to their various efforts to strengthen and reorient copyright protection to address the risks posed by digital technology, the Recording Industry Association of America (RIAA), which represents more than 500 companies engaged in the creation, manufacturing, and distribution of sound recordings, as well as the Motion Picture Association of America (MPAA), which represents the leading motion picture studios, have spearheaded an aggressive litigation campaign against the entire digital music distribution pipeline. In view of the vast reaches of digital technology, these associations initially focused their efforts on the most significant "leakage" points in order to have the greatest impact. The most prominent legal battles have focused on MP3 devices and peer-to-peer networks, although it has also stepped up enforcement efforts against other intermediaries in the pipeline, such as universities and businesses, and, most recently, users of peer-to-peer networks.

a. Digital Hardware Devices

In 1998, Diamond Multimedia introduced the Rio portable digital audio technology, a portable hard drive capable of storing approximately one hour of music compressed using the MP3 file format. This product dramatically increased consumer interest in downloading MP3 files over the Internet and extracting or "ripping" sound recording files from CDs to a computer hard drive and compressing them. Prior to the introduction of this product, the principal benefit that consumers could derive from downloading or ripping sound recordings was to listen to these files through headphones or speakers at their computers. The Rio rendered these files portable. In comparison to portable cassette players, the Rio 300 was more compact, easier to use, and more resistant to distortion from motion. The RIAA brought suit to enjoin the manufacture and distribution of the Rio, alleging that the Rio violated the requirements for digital audio recording devices under the Audio Home Recording Act of 1992 because it did not employ a Serial Copyright Management System and Diamond Multimedia failed to pay royalties on sales of a digital audio recording device. Concluding that the legislative bargain effectuated by the AHRA applied narrowly to digital audio recording devices (and not general computer technology), the Ninth Circuit held that the Rio device did not implicate the AHRA and dismissed the action. Recording Industry Ass'n of America v. Diamond Multimedia Systems, Inc., 180 F.3d 1072 (9th Cir. 1999). Echoing the Supreme Court's decision in the *Sony* case that "time shifting" fell within the fair use doctrine, the Ninth Circuit observed, in dicta, that "space shifting" was "paradigmatic noncommercial personal use."

In November 2001, television networks and production studios brought suit against ReplayTV, alleging that its features enabling consumers to skip commercials and transmit digital copies of television programming over the Internet to other ReplayTV owners violated copyright law. Although the recording capabilities of digital video recorders (DVRs) parallel those of the VCRs at issue in the *Sony* case, ReplayTV vastly simplified advertisement skipping through the use of a 30-second advance button and allowed consumers to set its latest model to skip commercials automatically. The litigation ultimately drove the maker of ReplayTV into bankruptcy before the legality of its devices could be evaluated. Note that the television industry has not sued the makers of the TiVo DVR. Unlike ReplayTV, TiVo does not offer an automated commercial skip feature. And unlike ReplayTV, TiVo has sought to work with the television industry and has garnered significant investments from that sector. TiVo has developed several new advertising vehicles in conjunction with television studios and has downplayed its commercial skipping features.

b. Search Engines, Services, and Software

The first legal skirmishes over the unauthorized distribution of copyrighted works over the Internet took place at the level of websites. The RIAA sent cease-and-desist letters to thousands of sites containing protected works without authorization. As the Internet grew and websites and services evolved to ease the search for and access to content, the record industry took aim at these businesses. Of greatest significance, the development of peer-to-peer networks vastly expanded the stakes and legal complexity surrounding online distribution of content. The legality of these various tools, software products, and service-based systems centers around the application of the fair use doctrine, the *Sony* decision, and doctrines of contributory and vicarious liability to decentralized distribution architectures in which the consumers engage in reproduction, uploading, and downloading of protected works.

Search Engines. In 1999, MP3board.com developed a generalized search engine that automatically scoured the Internet for MP3 files and provided links to such sites. In October of that year, the RIAA sent cease-and-desist letters to MP3board.com and its online service providers demanding that they halt operation of this service. In order to remove the liability cloud hanging over its business, MP3board.com filed an action in June 2000 seeking to declare that hypertext linking created by automated processes does not constitute copyright infringement even if the destination of a link is a website containing copyrighted material posted without authorization. The district court denied both parties' motions for summary judgment, concluding that the legality of the MP3 Board site depended on proof of the defendant's knowledge of infringement and the uses to which the service is put. After several years of litigation, the matter was settled with MP3board.com dropping its MP3 search engine.

Online Music "Lockers." In January 2000, MP3.com, a successful portal for recording artists to make their songs available for downloading and for general information about the music industry, launched its "MyMP3.com" service. This service enabled subscribers to develop a virtual online music locker from which they could access sound recordings copied and stored on MP3.com's servers from any Internet portal through a password protected user interface. The service was premised on the idea that the fair use doctrine authorizes consumers to "space shift" music that they have lawfully acquired, and it required users to load their CDs into a machine so the service could verify they had lawfully purchased a copy. Its manner of ascertaining whether consumers "owned" the CDs was, however, vulnerable to abuse.

The major record labels and music publishers sued MP3.com on the ground that MP3.com's initial copying of CDs onto its server and its distribution of such music to its subscribers over the Internet infringed their copyrights. As we discuss in subsection 3 (relating to fair use in cyberspace), the trial court had little trouble rejecting MP3.com fair use defense. See UMG Recordings, Inc. v. MP3.com, Inc., 92 F.Supp.2d 349 (S.D.N.Y. 2000). Facing the possibility of crushing liability, MP3.com settled the case with four out of the five major record labels for approximately $80 million. After the court assessed liability to Universal Music Group (UMG) at $25,000 per CD copied, the parties settled for another $53.4 million. MP3.com faced further exposure to independent record labels and music publishers. The various legal problems and licensing complexities eventually led MP3.com to abandon its efforts to establish a broad-based "music locker" service, limiting this venture to the files voluntarily loaded onto its website by independent artists and a few smaller record labels. Universal later acquired what was left of MP3.com and promptly sued MP3.com's outside law firm, alleging that it committed malpractice by advising MP3.com that its MyMP3.com service fell within the boundaries of fair use. The case eventually settled.

Peer-to-Peer Networks. The amount of content available over the Internet took a quantum leap in late 1999 with the introduction of Napster's peer-to-peer network technology. This technology vastly expanded the effective storage and exchange capacity of the Internet by enabling computer users running Napster's software to search the hard drives of thousands of other users for files encoded in the MP3 compression format commonly used for music files. Napster's server contained the labels of MP3 files, typically some combination of band and song titles, which could be searched by Napster users. Searches produced a list of Internet addresses of computers containing the search term. The Napster software would then form a connection through the Internet to the particular computer containing the file, establish a link, and quickly transfer the file to the searcher's hard drive. In essence, the Napster platform converted every computer running the software and connected to Napster into a "servent" — enabling it to function as both a server and a client. It became the fastest adopted software application in the history of computer technology, attaining 70 million users within its relatively brief period of operation.

The trial court and the Ninth Circuit ultimately concluded that Napster's direct knowledge of copyright infringement by users of its software and its ability to control such activities through the index of file names maintained on its central servers created a *responsibility* to remove links to infringing content and engage in efforts to police its network. See A&M Records, Inc., v. Napster, Inc., 239 F.3d 1004 (9th Cir. 2001). The burden of this responsibility and the prospect of crushing liability ultimately pushed Napster into bankruptcy. During the pendency of this litigation, a new generation of peer-to-peer software providers entered the market, prompting another legal battle.

MGM Studios Inc. v. Grokster, Ltd.
Supreme Court of the United States
125 S. Ct. 2764 (2005)

Justice SOUTER delivered the opinion of the Court.

The question is under what circumstances the distributor of a product capable of both lawful and unlawful use is liable for acts of copyright infringement by third parties using the product. We hold that one who distributes a device with the object of promoting its use to infringe copyright, as shown by clear expression or other affirmative steps taken to foster infringement, is liable for the resulting acts of infringement by third parties.

I

A

Respondents, Grokster, Ltd., and StreamCast Networks, Inc., defendants in the trial court, distribute free software products that allow computer users to share electronic files through peer-to-peer networks, so called because users' computers communicate directly with each other, not through central servers. The advantage of peer-to-peer networks over information networks of other types shows up in their substantial and growing popularity. Because they need no central computer server to mediate the exchange of information or files among users, the high-bandwidth communications capacity for a server may be dispensed with, and the need for costly server storage space is eliminated. Since copies of a file (particularly a popular one) are available on many users' computers, file requests and retrievals may be faster than on other types of networks, and since file exchanges do not travel through a server, communications can take place between any computers that remain connected to the network without risk that a glitch in the server will disable the network in its entirety. Given these benefits in security, cost, and efficiency, peer-to-peer networks are employed to store and distribute electronic files by universities, government agencies, corporations, and libraries, among others.

Other users of peer-to-peer networks include individual recipients of Grokster's and StreamCast's software, and although the networks that they enjoy through using the software can be used to share any type of digital file, they have prominently employed those networks in sharing copyrighted music and video files without authorization. A group of copyright holders (MGM for short, but including motion picture studios, recording companies, songwriters, and music publishers) sued Grokster and StreamCast for their users' copyright infringements, alleging that they knowingly and intentionally distributed their software to enable users to reproduce and distribute the copyrighted works in violation of the Copyright Act, 17 U.S.C. § 101 et seq. (2000 ed. and Supp. II). MGM sought damages and an injunction.

Discovery during the litigation revealed the way the software worked, the business aims of each defendant company, and the predilections of the users. Grokster's eponymous software employs what is known as FastTrack technology, a protocol developed by others and licensed to Grokster. StreamCast distributes a very similar product except that its software, called Morpheus, relies on what is known as Gnutella technology. A user who downloads and installs either software possesses the protocol to send requests for files directly to the computers of others using software compatible with FastTrack or Gnutella. On the FastTrack network opened by the Grokster software, the user's request goes to a computer given an indexing capacity by the software and designated a supernode, or to some other computer with comparable power and capacity to collect temporary indexes of the files available on the computers of users connected to it. The supernode (or indexing computer) searches its own index and may communicate the search request to other supernodes. If the file is found, the supernode discloses its location to the computer requesting it, and the requesting user can download the file directly from the computer located. The copied file is placed in a designated sharing folder on the requesting user's computer, where it is available for other users to download in turn, along with any other file in that folder.

In the Gnutella network made available by Morpheus, the process is mostly the same, except that in some versions of the Gnutella protocol there are no supernodes. In these versions, peer computers using the protocol communicate directly with each other. When a user enters a search request into the Morpheus software, it sends the request to computers connected with it, which in turn pass the request along to other connected peers. The search results are communicated to the requesting computer, and the user can download desired files directly from peers' computers. As this description indicates, Grokster and StreamCast use no servers to intercept the content of the search requests or to mediate the file transfers conducted by users of the software, there being no central point through which the substance of the communications passes in either direction.

Although Grokster and StreamCast do not therefore know when particular files are copied, a few searches using their software would show what is available on the networks the software reaches. MGM commissioned a statistician to conduct a systematic search, and his study showed that nearly 90% of the files available for download on the FastTrack system were copyrighted works. Grokster and StreamCast dispute this figure, raising methodological problems and arguing that free copying even of copyrighted works may be authorized by the rightholders. They also argue that potential noninfringing uses of their software are significant in kind, even if infrequent in practice. Some musical performers, for example, have gained new audiences by distributing their copyrighted works for free across peer-to-peer networks, and some distributors of unprotected content have used peer-to-peer networks to disseminate files, Shakespeare being an example. Indeed, StreamCast has given Morpheus users the opportunity to download the briefs in this very case, though their popularity has not been quantified.

As for quantification, the parties' anecdotal and statistical evidence entered thus far to show the content available on the FastTrack and Gnutella networks does not say much about which files are actually downloaded by users, and no one can say how often the software is used to obtain copies of unprotected material. But MGM's evidence gives reason to think that the vast majority of users' downloads are acts of infringement, and because well over 100 million copies of the software in question are known to have been downloaded, and billions of files are shared across the FastTrack and Gnutella networks each month, the probable scope of copyright infringement is staggering.

Grokster and StreamCast concede the infringement in most downloads, and it is uncontested that they are aware that users employ their software primarily to download copyrighted files, even if the decentralized FastTrack and Gnutella networks fail to reveal which files are being copied, and when. From time to time, moreover, the companies have learned about their users' infringement directly, as from users who have sent e-mail to each company with questions about playing copyrighted movies they had downloaded, to whom the companies have responded with guidance. And MGM notified the companies of 8 million copyrighted files that could be obtained using their software.

Grokster and StreamCast are not, however, merely passive recipients of information about infringing use. The record is replete with evidence that from the moment Grokster and StreamCast began to distribute their free software, each one clearly voiced the objective that recipients use it to download copyrighted works, and each took active steps to encourage infringement.

After the notorious file-sharing service, Napster, was sued by copyright holders for facilitation of copyright infringement, A & M Records, Inc. v. Napster, Inc., 114 F.Supp.2d 896 (N.D.Cal.2000), aff'd in part, rev'd in part, 239 F.3d 1004 (C.A.9 2001), StreamCast gave away a software program of a kind known as OpenNap, designed as compatible with the Napster program and open to Napster users for downloading files from other Napster and OpenNap users' computers. Evidence indicates that "[i]t was always [StreamCast's] intent to use [its OpenNap network] to be able to capture email addresses of [its] initial target market so that [it] could promote [its] StreamCast Morpheus interface to them," App. 861; indeed, the Open-Nap program was engineered "'to leverage Napster's 50 million user base,'" id., at 746.

StreamCast monitored both the number of users downloading its OpenNap program and the number of music files they downloaded. It also used the resulting OpenNap network to distribute copies of the Morpheus software and to encourage users to adopt it. Internal company documents indicate that StreamCast hoped to attract large numbers of former Napster users if that company was shut down by court order or otherwise, and that StreamCast planned to be the next Napster. A kit developed by StreamCast to be delivered to advertisers, for example, contained press articles about StreamCast's potential to capture former Napster users and it introduced itself to some potential advertisers as a company "which is similar to what Napster was." It broadcast banner advertisements to users of other Napster-compatible software, urging them to adopt its OpenNap. An internal e-mail from a company executive stated: "'We have put this network in place so that when Napster pulls the plug on their free service . . . or if the Court orders them shut down prior to that . . . we will be positioned to capture the flood of their 32 million users that will be actively looking for an alternative.'"

Thus, StreamCast developed promotional materials to market its service as the best Napster alternative. One proposed advertisement read: "Napster Inc. has announced that it will soon begin charging you a fee. That's if the courts don't order it shut down first. What will you do to get around it?" Id., at 897. Another proposed ad touted StreamCast's software as the "# 1 alternative to Napster" and asked "[w]hen the lights went off at Napster . . . where did the users go?" Id., at 836 (ellipsis in original).[7] StreamCast even planned to flaunt the illegal uses of its software; when it launched the OpenNap network, the chief technology officer of the company averred that "[t]he goal is to get in trouble with the law and get sued. It's the best way to get in the new[s]." Id., at 916.

The evidence that Grokster sought to capture the market of former Napster users is sparser but revealing, for Grokster launched its own OpenNap system called Swaptor and inserted digital codes into its Web site so that computer users using Web search engines to look for "Napster" or "[f]ree filesharing" would be directed to the Grokster Web site, where they could download the Grokster software. Id., at 992-993. And Grokster's name is an apparent derivative of Napster.

StreamCast's executives monitored the number of songs by certain commercial artists available on their networks, and an internal communication indicates they aimed to have a larger number of copyrighted songs available on their networks than other file-sharing networks. Id., at 868. The point, of course, would be to attract users of a mind to infringe, just as it would be with their promotional materials developed showing copyrighted songs as examples of the kinds of files available through Morpheus. Id., at 848. Morpheus in fact allowed users to search specifically for "Top 40" songs, id., at 735, which were inevitably copyrighted. Similarly, Grokster sent users a newsletter promoting its ability to provide particular, popular copyrighted materials.

In addition to this evidence of express promotion, marketing, and intent to promote further, the business models employed by Grokster and StreamCast confirm that their principal object was use of their software to download copyrighted works. Grokster and StreamCast receive no revenue from users, who obtain the software itself for nothing. Instead, both companies generate income by selling advertising space, and they stream the advertising to Grokster and Morpheus users while they are employing the programs. As the number of users of each program increases, advertising opportunities become worth more. Cf. App. 539, 804. While there is doubtless some demand for free Shakespeare, the evidence shows that substantive volume is a function of free access to copyrighted work. Users seeking Top 40 songs, for example, or the latest release by Modest Mouse, are certain to be far more numerous than those seeking a free Decameron, and Grokster and StreamCast translated that demand into dollars.

7. The record makes clear that StreamCast developed these promotional materials but not whether it released them to the public. Even if these advertisements were not released to the public and do not show encouragement to infringe, they illuminate StreamCast's purposes.

Finally, there is no evidence that either company made an effort to filter copyrighted material from users' downloads or otherwise impede the sharing of copyrighted files. Although Grokster appears to have sent e-mails warning users about infringing content when it received threatening notice from the copyright holders, it never blocked anyone from continuing to use its software to share copyrighted files. Id., at 75-76. StreamCast not only rejected another company's offer of help to monitor infringement, id., at 928-929, but blocked the Internet Protocol addresses of entities it believed were trying to engage in such monitoring on its networks, id., at 917-922.

B

After discovery, the parties on each side of the case cross-moved for summary judgment. The District Court limited its consideration to the asserted liability of Grokster and StreamCast for distributing the current versions of their software, leaving aside whether either was liable "for damages arising from past versions of their software, or from other past activities." The District Court held that those who used the Grokster and Morpheus software to download copyrighted media files directly infringed MGM's copyrights, a conclusion not contested on appeal, but the court nonetheless granted summary judgment in favor of Grokster and StreamCast as to any liability arising from distribution of the then current versions of their software. Distributing that software gave rise to no liability in the court's view, because its use did not provide the distributors with actual knowledge of specific acts of infringement.

The Court of Appeals affirmed. 380 F.3d 1154 (C.A.9 2004). In the court's analysis, a defendant was liable as a contributory infringer when it had knowledge of direct infringement and materially contributed to the infringement. But the court read Sony Corp. of America v. Universal City Studios, Inc., 464 U.S. 417, as holding that distribution of a commercial product capable of substantial noninfringing uses could not give rise to contributory liability for infringement unless the distributor had actual knowledge of specific instances of infringement and failed to act on that knowledge. The fact that the software was capable of substantial noninfringing uses in the Ninth Circuit's view meant that Grokster and StreamCast were not liable, because they had no such actual knowledge, owing to the decentralized architecture of their software....

The Ninth Circuit also considered whether Grokster and StreamCast could be liable under a theory of vicarious infringement. The court held against liability because the defendants did not monitor or control the use of the software, had no agreed-upon right or current ability to supervise its use, and had no independent duty to police infringement. We granted certiorari.

II

A

MGM and many of the *amici* fault the Court of Appeals's holding for upsetting a sound balance between the respective values of supporting creative pursuits through copyright protection and promoting innovation in new communication technologies by limiting the incidence of liability for copyright infringement. The more artistic protection is favored, the more technological innovation may be discouraged; the administration of copyright law is an exercise in managing the trade-off.

The tension between the two values is the subject of this case, with its claim that digital distribution of copyrighted material threatens copyright holders as never before, because every copy is identical to the original, copying is easy, and many people (especially the young) use file-sharing software to download copyrighted works. This very breadth of the software's use may well draw the public directly into the debate over copyright policy, and the indications are that the ease of copying songs or movies using software like Grokster's and Napster's is fostering disdain for copyright protection. As the case has been presented to us, these fears are said to be offset by the different concern that imposing liability, not only on infringers but on distributors of software based on its potential for unlawful use, could limit further development of beneficial technologies.

The argument for imposing indirect liability in this case is, however, a powerful one, given the number of infringing downloads that occur every day using Stream-Cast's and Grokster's software. When a widely shared service or product is used to commit infringement, it may be impossible to enforce rights in the protected work effectively against all direct infringers, the only practical alternative being to go against the distributor of the copying device for secondary liability on a theory of contributory or vicarious infringement. See In re Aimster Copyright Litigation, 334 F.3d 643, 645-646 (C.A.7 2003).

One infringes contributorily by intentionally inducing or encouraging direct infringement, see Gershwin Pub. Corp. v. Columbia Artists Management, Inc., 443 F.2d 1159, 1162 (C.A.2 1971), and infringes vicariously by profiting from direct infringement while declining to exercise a right to stop or limit it, Shapiro, Bernstein & Co. v. H.L. Green Co., 316 F.2d 304, 307 (C.A.2 1963). Although "[t]he Copyright Act does not expressly render anyone liable for infringement committed by another," Sony Corp. v. Universal City Studios, 464 U.S., at 434, these doctrines of secondary liability emerged from common law principles and are well established in the law.

In the present case MGM has argued a vicarious liability theory, which allows imposition of liability when the defendant profits directly from the infringement and has a right and ability to supervise the direct infringer, even if the defendant initially lacks knowledge of the infringement. Because we resolve the case based on an inducement theory, there is no need to analyze separately MGM's vicarious liability theory.

B

Despite the currency of these principles of secondary liability, this Court has dealt with secondary copyright infringement in only one recent case, and because MGM has tailored its principal claim to our opinion there, a look at our earlier holding is in order. In Sony Corp. v. Universal City Studios, this Court addressed a claim that secondary liability for infringement can arise from the very distribution of a commercial product. There, the product, novel at the time, was what we know today as the videocassette recorder or VCR. Copyright holders sued Sony as the manufacturer, claiming it was contributorily liable for infringement that occurred when VCR owners taped copyrighted programs because it supplied the means used to infringe, and it had constructive knowledge that infringement would occur. At the trial on the merits, the evidence showed that the principal use of the VCR was for "'time-shifting,'" or taping a program for later viewing at a more convenient time, which the Court found to be a fair, not an infringing, use. There was no evidence that Sony had expressed an object of bringing about taping in violation of copyright or had taken active steps to increase its profits from unlawful taping. Although Sony's advertisements urged consumers to buy the VCR to "'record favorite shows'" or "'build a library'" of recorded programs, neither of these uses was necessarily infringing.

On those facts, with no evidence of stated or indicated intent to promote infringing uses, the only conceivable basis for imposing liability was on a theory of contributory infringement arising from its sale of VCRs to consumers with knowledge that some would use them to infringe. But because the VCR was "capable of commercially significant noninfringing uses," we held the manufacturer could not be faulted solely on the basis of its distribution.

This analysis reflected patent law's traditional staple article of commerce doctrine, now codified, that distribution of a component of a patented device will not violate the patent if it is suitable for use in other ways. 35 U.S.C. §271(c); Aro Mfg. Co. v. Convertible Top Replacement Co., 377 U.S. 476, 485 (1964) (noting codification of cases). The doctrine was devised to identify instances in which it may be presumed from distribution of an article in commerce that the distributor intended the article to be used to infringe another's patent, and so may justly be held liable for that infringement. "One who makes and sells articles which are only adapted to be used in a patented combination will be presumed to intend the natural consequences of his acts; he will be presumed to intend that they shall be used in the combination of the patent." New York Scaffolding Co. v. Whitney, 224 F. 452, 459 (C.A.8 1915).

In sum, where an article is "good for nothing else" but infringement, there is no legitimate public interest in its unlicensed availability, and there is no injustice in presuming or imputing an intent to infringe, see Henry v. A.B. Dick Co., 224 U.S. 1, 48 (1912), overruled on other grounds, Motion Picture Patents Co. v. Universal Film Mfg. Co., 243 U.S. 502 (1917). Conversely, the doctrine absolves the equivocal conduct of selling an item with substantial lawful as well as unlawful uses, and limits liability to instances of more acute fault than the mere understanding that some of one's products will be misused. It leaves breathing room for innovation and a vigorous commerce. See Sony Corp. v. Universal City Studios, supra, at 442; Dawson Chemical Co. v. Rohm & Haas Co., 448 U.S. 176, 221 (1980); Henry v. A.B. Dick Co., supra, at 48.

The parties and many of the *amici* in this case think the key to resolving it is the *Sony* rule and, in particular, what it means for a product to be "capable of commercially significant noninfringing uses." MGM advances the argument that granting summary judgment to Grokster and StreamCast as to their current activities gave too much weight to the value of innovative technology, and too little to the copyrights infringed by users of their software, given that 90% of works available on one of the networks was shown to be copyrighted. Assuming the remaining 10% to be its noninfringing use, MGM says this should not qualify as "substantial," and the Court should quantify *Sony* to the extent of holding that a product used "principally" for infringement does not qualify. As mentioned before, Grokster and StreamCast reply by citing evidence that their software can be used to reproduce public domain works, and they point to copyright holders who actually encourage copying. Even if infringement is the principal practice with their software today, they argue, the noninfringing uses are significant and will grow.

We agree with MGM that the Court of Appeals misapplied *Sony*, which it read as limiting secondary liability quite beyond the circumstances to which the case applied. *Sony* barred secondary liability based on presuming or imputing intent to cause infringement solely from the design or distribution of a product capable of substantial lawful use, which the distributor knows is in fact used for infringement. The Ninth Circuit has read *Sony*'s limitation to mean that whenever a product is capable of substantial lawful use, the producer can never be held contributorily liable for third parties' infringing use of it; it read the rule as being this broad, even when an actual purpose to cause infringing use is shown by evidence independent of design and distribution of the product, unless the distributors had "specific knowledge of infringement at a time at which they contributed to the infringement, and failed to act upon that information." Because the Circuit found the StreamCast and Grokster software capable of substantial lawful use, it concluded on the basis of its reading of *Sony* that neither company could be held liable, since there was no showing that their software, being without any central server, afforded them knowledge of specific unlawful uses.

This view of *Sony*, however, was error, converting the case from one about liability resting on imputed intent to one about liability on any theory. Because *Sony* did not displace other theories of secondary liability, and because we find below that it was error to grant summary judgment to the companies on MGM's inducement claim, we do not revisit *Sony* further, as MGM requests, to add a more quantified description of the point of balance between protection and commerce when liability rests solely on distribution with knowledge that unlawful use will occur. It is enough to note that the Ninth Circuit's judgment rested on an erroneous understanding of *Sony* and to leave further consideration of the *Sony* rule for a day when that may be required.

C

Sony's rule limits imputing culpable intent as a matter of law from the characteristics or uses of a distributed product. But nothing in *Sony* requires courts to ignore evidence of intent if there is such evidence, and the case was never meant to foreclose rules of fault-based liability derived from the common law. Sony Corp. v. Universal City Studios, 464 U.S., at 439 ("If vicarious liability is to be imposed on Sony in this case, it must rest on the fact that it has sold equipment with constructive knowledge" of the potential for infringement). Thus, where evidence goes beyond a product's characteristics or the knowledge that it may be put to infringing uses, and shows statements or actions directed to promoting infringement, *Sony*'s staple-article rule will not preclude liability.

The classic case of direct evidence of unlawful purpose occurs when one induces commission of infringement by another, or "entic[es] or persuad[es] another" to infringe, Black's Law Dictionary 790 (8th ed. 2004), as by advertising. Thus at common law a copyright or patent defendant who "not only expected but invoked [infringing use] by advertisement" was liable for infringement "on principles recognized in every part of the law." Kalem Co. v. Harper Brothers, 222 U.S., at 62-63 (copyright infringement).

The rule on inducement of infringement as developed in the early cases is no different today. Evidence of "active steps . . . taken to encourage direct infringement," Oak Industries, Inc. v. Zenith Electronics Corp., 697 F.Supp. 988, 992 (N.D.Ill. 1988), such as advertising an infringing use or instructing how to engage in an infringing use, show an affirmative intent that the product be used to infringe, and a showing that infringement was encouraged overcomes the law's reluctance to find liability when a defendant merely sells a commercial product suitable for some lawful use, see, e.g., Water Technologies Corp. v. Calco, Ltd., 850 F.2d 660, 668 (C.A.Fed.1988) (liability for inducement where one "actively and knowingly aid[s] and abet[s] another's direct infringement" (emphasis omitted)); Fromberg, Inc. v. Thornhill, 315 F.2d 407, 412-413 (C.A.5 1963) (demonstrations by sales staff of infringing uses supported liability for inducement); Haworth Inc. v. Herman Miller Inc., 37 USPQ 2d 1080, 1090 (WD Mich.1994) (evidence that defendant "demon-strate[d] and recommend[ed] infringing configurations" of its product could support inducement liability); Sims v. Mack Trucks, Inc., 459 F.Supp. 1198, 1215 (E.D.Pa.1978) (finding inducement where the use "depicted by the defendant in its promotional film and brochures infringes the . . . patent"), overruled on other grounds, 608 F.2d 87 (C.A.3 1979). Cf. W. Keeton, D. Dobbs, R. Keeton, & D. Owen, Prosser and Keeton on Law of Torts 37 (5th ed. 1984) ("There is a definite tendency to impose greater responsibility upon a defendant whose conduct was intended to do harm, or was morally wrong").

For the same reasons that *Sony* took the staple-article doctrine of patent law as a model for its copyright safe-harbor rule, the inducement rule, too, is a sensible one for copyright. We adopt it here, holding that one who distributes a device with the object of promoting its use to infringe copyright, as shown by clear expression or other affirmative steps taken to foster infringement, is liable for the resulting acts of infringement by third parties. We are, of course, mindful of the need to keep from trenching on regular commerce or discouraging the development of technologies with lawful and unlawful potential. Accordingly, just as *Sony* did not find intentional inducement despite the knowledge of the VCR manufacturer that its device could be used to infringe, 464 U.S., at 439, n. 19, mere knowledge of infringing potential or of actual infringing uses would not be enough here to subject a distributor to liability. Nor would ordinary acts incident to product distribution, such as offering customers technical support or product updates, support liability in themselves. The inducement rule, instead, premises liability on purposeful, culpable expression and conduct, and thus does nothing to compromise legitimate commerce or discourage innovation having a lawful promise.

III

A

The only apparent question about treating MGM's evidence as sufficient to with-stand summary judgment under the theory of inducement goes to the need on MGM's part to adduce evidence that StreamCast and Grokster communicated an inducing message to their software users. The classic instance of inducement is by advertisement or solicitation that broadcasts a message designed to stimulate others to commit violations. MGM claims that such a message is shown here. It is undisputed that StreamCast beamed onto the computer screens of users of Napster-compatible programs ads urging the adoption of its OpenNap program, which was designed, as its name implied, to invite the custom of patrons of Napster, then under attack in the courts for facilitating massive infringement. Those who accepted StreamCast's Open-Nap program were offered software to perform the same services, which a factfinder could conclude would readily have been understood in the Napster market as the ability to download copyrighted music files. Grokster distributed an electronic newsletter containing links to articles promoting its software's ability to access popular copyrighted music. And anyone whose Napster or free file-sharing searches turned up a link to Grokster would have understood Grokster to be offering the same file-sharing ability as Napster, and to the same people who probably used Napster for infringing downloads; that would also have been the understanding of anyone offered Grokster's suggestively named Swaptor software, its version of OpenNap. And both companies communicated a clear message by responding affirmatively to requests for help in locating and playing copyrighted materials.

In StreamCast's case, of course, the evidence just described was supplemented by other unequivocal indications of unlawful purpose in the internal communications and advertising designs aimed at Napster users ("When the lights went off at Napster . . . where did the users go?" App. 836 (ellipsis in original)). Whether the messages were communicated is not to the point on this record. The function of the message in the theory of inducement is to prove by a defendant's own statements that his unlawful purpose disqualifies him from claiming protection (and incidentally to point to actual violators likely to be found among those who hear or read the message). Proving that a message was sent out, then, is the preeminent but not exclusive way of showing that active steps were taken with the purpose of bringing about infringing acts, and of showing that infringing acts took place by using the device distributed. Here, the summary judgment record is replete with other evidence that Grokster and StreamCast, unlike the manufacturer and distributor in *Sony*, acted with a purpose to cause copyright violations by use of software suitable for illegal use.

Three features of this evidence of intent are particularly notable. First, each company showed itself to be aiming to satisfy a known source of demand for copyright infringement, the market comprising former Napster users. StreamCast's internal documents made constant reference to Napster, it initially distributed its Morpheus software through an OpenNap program compatible with Napster, it advertised its OpenNap program to Napster users, and its Morpheus software functions as Napster did except that it could be used to distribute more kinds of files, including copyrighted movies and software programs. Grokster's name is apparently derived from Napster, it too initially offered an OpenNap program, its software's function is likewise comparable to Napster's, and it attempted to divert queries for Napster onto its own Web site. Grokster and StreamCast's efforts to supply services to former Napster users, deprived of a mechanism to copy and distribute what were overwhelmingly infringing files, indicate a principal, if not exclusive, intent on the part of each to bring about infringement.

Second, this evidence of unlawful objective is given added significance by MGM's showing that neither company attempted to develop filtering tools or other mechanisms to diminish the infringing activity using their software. While the Ninth Circuit treated the defendants' failure to develop such tools as irrelevant because they lacked an independent duty to monitor their users' activity, we think this evidence underscores Grokster's and StreamCast's intentional facilitation of their users' infringement.[12]

Third, there is a further complement to the direct evidence of unlawful objective. It is useful to recall that StreamCast and Grokster make money by selling advertising space, by directing ads to the screens of computers employing their software. As the record shows, the more the software is used, the more ads are sent out and the greater the advertising revenue becomes. Since the extent of the software's use determines the gain to the distributors, the commercial sense of their enterprise turns on high-volume use, which the record shows is infringing. This evidence alone would not justify an inference of unlawful intent, but viewed in the context of the entire record its import is clear.

The unlawful objective is unmistakable.

B

In addition to intent to bring about infringement and distribution of a device suitable for infringing use, the inducement theory of course requires evidence of actual infringement by recipients of the device, the software in this case. As the account of the facts indicates, there is evidence of infringement on a gigantic scale, and there is no serious issue of the adequacy of MGM's showing on this point in order to survive the companies' summary judgment requests. Although an exact calculation of infringing use, as a basis for a claim of damages, is subject to dispute, there is no question that the summary judgment evidence is at least adequate to entitle MGM to go forward with claims for damages and equitable relief.

* * *

In sum, this case is significantly different from *Sony* and reliance on that case to rule in favor of StreamCast and Grokster was error. *Sony* dealt with a claim of liability based solely on distributing a product with alternative lawful and unlawful uses, with knowledge that some users would follow the unlawful course. The case struck a balance between the interests of protection and innovation by holding that the product's capability of substantial lawful employment should bar the imputation of fault and consequent secondary liability for the unlawful acts of others.

MGM's evidence in this case most obviously addresses a different basis of liability for distributing a product open to alternative uses. Here, evidence of the distributors' words and deeds going beyond distribution as such shows a purpose to cause and profit from third-party acts of copyright infringement. If liability for inducing infringement is ultimately found, it will not be on the basis of presuming or imputing fault, but from inferring a patently illegal objective from statements and actions showing what that objective was.

---

12. Of course, in the absence of other evidence of intent, a court would be unable to find contributory infringement liability merely based on a failure to take affirmative steps to prevent infringement, if the device otherwise was capable of substantial noninfringing uses. Such a holding would tread too close to the *Sony* safe harbor.

There is substantial evidence in MGM's favor on all elements of inducement, and summary judgment in favor of Grokster and StreamCast was error. On remand, reconsideration of MGM's motion for summary judgment will be in order. . . .

Justice GINSBURG, with whom THE CHIEF JUSTICE and Justice KENNEDY join, concurring.

I concur in the Court's decision, which vacates in full the judgment of the Court of Appeals for the Ninth Circuit and write separately to clarify why I conclude that the Court of Appeals misperceived, and hence misapplied, our holding in *Sony*. . . .

This case differs markedly from *Sony*. Here, there has been no finding of any fair use and little beyond anecdotal evidence of noninfringing uses. . . .

[W]hen the record in this case was developed, there was evidence that Grokster's and StreamCast's products were, and had been for some time, overwhelmingly used to infringe and that this infringement was the overwhelming source of revenue from the products. Fairly appraised, the evidence was insufficient to demonstrate, beyond genuine debate, a reasonable prospect that substantial or commercially significant noninfringing uses were likely to develop over time. . . . [3]

If, on remand, the case is not resolved on summary judgment in favor of MGM based on Grokster and StreamCast actively inducing infringement, the Court of Appeals, I would emphasize, should reconsider, on a fuller record, its interpretation of *Sony*'s product distribution holding. . . .

Justice BREYER, with whom Justice STEVENS and Justice O'CONNOR join, concurring.

I agree with the Court that the distributor of a dual-use technology may be liable for the infringing activities of third parties where he or she actively seeks to advance the infringement. I further agree that, in light of our holding today, we need not now "revisit" *Sony*. Other Members of the Court, however, take up the *Sony* question: whether Grokster's product is "capable of 'substantial' or 'commercially significant' noninfringing uses." And they answer that question by stating that the Court of Appeals was wrong when it granted summary judgment on the issue in Grokster's favor. I write to explain why I disagree with them on this matter. . . .

When measured against *Sony*' s underlying evidence and analysis, the evidence now before us shows that Grokster passes *Sony*'s test—that is, whether the company's product is capable of substantial or commercially significant noninfringing uses. For one thing, petitioners' (hereinafter MGM) own expert declared that 75% of current files available on Grokster are infringing and 15% are "likely infringing." That leaves some number of files near 10% that apparently are noninfringing, a figure very similar to the 9% or so of authorized time-shifting uses of the VCR that the Court faced in *Sony*. . . .

---

3. Justice Breyer finds support for summary judgment in this motley collection of declarations and in a survey conducted by an expert retained by MGM. That survey identified 75% of the files available through Grokster as copyrighted works owned or controlled by the plaintiffs, and 15% of the files as works likely copyrighted. As to the remaining 10% of the files, "there was not enough information to form reasonable conclusions either as to what those files even consisted of, and/or whether they were infringing or non-infringing." Even assuming, as Justice Breyer does, that the *Sony* Court would have absolved Sony of contributory liability solely on the basis of the use of the Betamax for authorized time-shifting, summary judgment is not inevitably appropriate here. *Sony* stressed that the plaintiffs there owned "well below 10%" of copyrighted television programming, and found, based on trial testimony from representatives of the four major sports leagues and other individuals authorized to consent to home-recording of their copyrighted broadcasts, that a similar percentage of program copying was authorized. Here, the plaintiffs allegedly control copyrights for 70% or 75% of the material exchanged through the Grokster and StreamCast software and the District Court does not appear to have relied on comparable testimony about authorized copying from copyright holders.

Importantly, *Sony* also used the word "capable," asking whether the product is "capable of" substantial noninfringing uses. Its language and analysis suggest that a figure like 10%, if fixed for all time, might well prove insufficient, but that such a figure serves as an adequate foundation where there is a reasonable prospect of expanded legitimate uses over time. See ibid. (noting a "significant potential for future authorized copying"). And its language also indicates the appropriateness of looking to potential future uses of the product to determine its "capability."

Here the record reveals a significant future market for noninfringing uses of Grokster-type peer-to-peer software. . . .

And that is just what is happening. Such legitimate noninfringing uses are coming to include the swapping of: research information (the initial purpose of many peer-to-peer networks); public domain films (e.g., those owned by the Prelinger Archive); historical recordings and digital educational materials (e.g., those stored on the Internet Archive); digital photos (OurPictures, for example, is starting a P2P photo-swapping service); "shareware" and "freeware" (e.g., Linux and certain Windows software); secure licensed music and movie files (Intent MediaWorks, for example, protects licensed content sent across P2P networks); news broadcasts past and present (the BBC Creative Archive lets users "rip, mix and share the BBC"); user-created audio and video files (including "podcasts" that may be distributed through P2P software); and all manner of free "open content" works collected by Creative Commons (one can search for Creative Commons material on StreamCast). I can find nothing in the record that suggests that this course of events will not continue to flow naturally as a consequence of the character of the software taken together with the foreseeable development of the Internet and of information technology.

There may be other now-unforeseen noninfringing uses that develop for peer-to-peer software, just as the home-video rental industry (unmentioned in *Sony*) developed for the VCR. But the foreseeable development of such uses, when taken together with an estimated 10% noninfringing material, is sufficient to meet *Sony*'s standard. . . .

COMMENTS AND QUESTIONS

1. At first blush, the unanimous Supreme Court in *Grokster* would appear to be a significant victory for content providers over peer-to-peer companies. But what will be the long-term impact of this decision on the distribution of file-sharing technology and enforcement of copyrights on the Internet? Could a new entrant into the file-sharing marketplace offering the very same functionality as Grokster or Streamcast shield itself from liability by avoiding statements encouraging infringement?

2. What light does the *Grokster* decision shed on contributory or vicarious liability in copyright law? How would you advise a client developing peer-to-peer technology (or other technology that could be used for copyright infringement) about the scope and predictability of the *Sony* safe harbor? Has this case clarified the scope of contributory or vicarious liability to any significant degree (and if so, in what direction)? Do the concurrences enhance or detract from the clarity of the *Sony* standard? Does the *Sony* standard even matter anymore, now that copyright owners can bring a claim based on inducement even if the only conduct is selling a product capable of substantial noninfringing uses so long as they can allege bad intent?

3. Just as Grokster and Steamcast emerged during the litigation over Napster, BitTorrent, a powerful new generation of file-sharing technology, became available during the pendency of the *Grokster* litigation. BitTorrent breaks files into smaller pieces and provides users information about the location of the various pieces. This program allows rapid distribution of very large files and is used for full-length feature films, software, and music, as well as Linux upgrades and podcasts. BitTorrent is an open source program that is available for free (and does not earn any income from advertising). How would this technology fare under the *Grokster* decision?

4. *Policy Analysis.* As a venture capitalist remarked in a discussion with Napster's developers, "You've distributed more music [in your first year of operation] than the whole record industry since it came into existence." Joseph Menn, All the Rave 161 (2003). Doesn't it seem peculiar that the copyright system's treatment of the most dramatic advance in the means for reproducing and distributing works of authorship since the printing press is being decided not by a legislative policy judgment but reliance on a pre-digital age case focused on the legality of home taping of television shows on video cassette recorders? Various proposals have been floated on how to reform copyright law in the wake of peer-to-peer technology. Consider the following proposals:

- replace the staple article of commerce standard with a predominant use standard that considers: (1) whether non-infringing uses can be achieved for most consumers through other means without significant added expense, inconvenience, or loss of functionality; (2) the extent to which copyright owners can protect themselves against infringements without undue cost (e.g., through encryption); and (3) the cost and efficacy of enforcement against direct infringers.
- abolish indirect liability and make it easier to enforce copyrights against direct infringers.
- abolish liability for noncommercial file sharing and impose a levy on Internet users designed to compensate copyright owners for estimated losses due to unauthorized copying.

These and related proposals are discussed in Peter S. Menell and David Nimmer, Direct Analysis of Indirect Copyright Liability (manuscript 2006); Mark A. Lemley & R. Anthony Reese, Reducing Digital Copyright Infringement Without Restricting Innovation, 56 Stan. L. Rev. 1345 (2004); Robert P. Merges, Compulsory Licensing v. the Three "Golden Oldies": Property Rights, Contracts, and Markets, 508 Cato Policy Analysis (2004); William W. Fisher III, Promises to Keep: Technology, Law, and the Future of Entertainment (2004); Neil W. Netanel, Impose a Noncommercial Use Levy to Allow Free Peer-to-Peer File Sharing, 17 Harv. J.L. & Tech. 1 (2003).

PROBLEM

Problem 4-37. Return to problem 4-32 regarding digital video recorders and consider the following additional feature of the ReplayTV 4500: consumers can connect their device to the Internet and send files of television broadcasts to others owning ReplayTV devices. How should this feature be analyzed? Does it depend on whether any premium channel content (e.g., HBO) can be distributed in this manner?

c. End Users

Prior to the advent of peer-to-peer technology, the DMCA's notice and take-down provisions provided a relatively effective means of tamping down unauthorized distribution of copyrighted works over computer networks. Content companies could locate files just as easily as consumers by using the leading search engines. Following the passage of the DMCA, content companies began sending notice letters to online service providers hosting websites containing unauthorized content. These entities promptly complied so as to maintain their immunity under section 512.

The enforcement problem expanded exponentially following Napster's arrival in late 1999. Nonetheless, content industries initially resisted bringing enforcement actions directly against end users for several reasons. From a business standpoint, such lawsuits could trigger a consumer backlash. At a practical level, it was not clear whether suing individuals would produce desirable or cost-effective results. With the proliferation of file sharing on peer-to-peer networks, millions of Internet users—many in their teenage years—were engaging in copyright infringement. Yet developing a strategy for identifying and pursuing infringers without being seen as bullies required great care.

Before going down this path, the content industries attempted several other approaches to the problem including suing the distributors of file sharing software, consumer education (advertisements, "copyright awareness week," and movie trailers aimed at discouraging file sharing), and efforts to sabotage file sharing networks by seeding the peer-to-peer systems with fake and degraded files. They also introduced online distribution outlets, although the pricing structures, digital rights management constraints, gaps in available catalog, and the challenge of competing with freely available unrestricted content on Napster doomed these enterprises.

The recording industry's ultimate success in its lawsuit against Napster did little to quell file sharing. New, decentralized peer-to-peer networks supplanted Napster on an even larger scale before Napster's servers shut down. With sales volume slumping and with Grokster's victory in the district court under the "staple article of commerce doctrine" (ultimately reversed), the content industries decided to begin suing large-scale file sharers directly.

Even apart from the public relations concerns, such lawsuits entailed several complexities. Although content owners can determine the Internet Protocol (IP) addresses of file sharers through forensic efforts, they often require the assistance of online service providers to identify offenders. Several leading OSPs resisted requests to turn over the names of their customers on privacy and other grounds. Copyright owners sought to use the subpoena provisions of section 512(h) to unmask file sharers. After some initially favorable rulings on the scope of this power, the D.C. Circuit ruled that this provision applied narrowly only to website owners and did not extend to those distributing files on peer-to-peer networks. See RIAA v. Verizon Internet Services, Inc., 351 F.3d 1229 (D.C. Cir. 2003). Every other court to consider the issue since has concurred. As a result, content owners must now follow a time-consuming process of filing "John Doe" lawsuits before obtaining subpoenas. And even those suits have problems because there is no way to tell whether a particular John Doe lives in a particular judicial district (and therefore whether there is personal jurisdiction) until he is identified. Efforts to consolidate cases have met with only mixed success. Content owners have also had to deal with problems of wide-scale litigation efforts against often impecunious defendants. In order to avoid being seen as bullies, the plaintiffs have been willing to settle cases for relatively modest amounts (averaging $3,000 per case), which covers the costs of the lawsuits but doesn't generate returns to the copyright owners.

COMMENTS AND QUESTIONS

1. Does suing file sharers make business sense for content owners? What are the advantages? disadvantages? How would you advise a major record label or motion picture studio about this strategy? See Justin Hughes, On the Logic of Suing One's Customers and the Dilemma of Infringement-Based Business Models, 22 Cardozo Arts & Ent. L.J. 725 (2005); Kristina Groenings, Costs and Benefits of the Recording Industry's Litigation Against Individuals, 20 Berkeley Tech. L.J. 571 (2005).

2. *Class Defense.* Class actions have long been seen as an efficient mechanism for consolidating claims of many similarly situated plaintiffs. By analogy, should content owners be able to use consolidated claims involving similarly situated defendants — e.g., file sharers at a particular university — on similar grounds? See Assaf Hamdani and Alon Klement, The Class Defense, 93 Cal. L. Rev. 685 (2005).

3. *Investor Exposure.* As the legal exposure for digital distribution ventures became apparent after the Scour, MyMP3.com, and Napster cases, venture capitalists became increasingly wary of the legal costs, economic risk, and potential vicarious liability associated with investing in these enterprises. The recording industry has substantially raised the stakes for investors by bringing suits directly against officers, directors, and venture capitalists involved in Napster, alleging that they should be liable under what might be termed "tertiary liability" — contributing to the acts of Napster, who was in turn contributing to the acts of direct infringers. Such actions have chilled investment in new technology ventures relating to reproduction and distribution of content. How would you assess such litigation (and threats of litigation) — as unduly chilling investment or properly internalizing the external harms?

4. *Competing with Free.* Perhaps the most promising strategy that the record industry has pursued has been the encouragement of new online distribution models, such as Apple's iTunes Music Store. Music copyright owners have also sought to transition college students to legitimate music services by offering steep discounts for subscription music services to college students. Does enforcement against file sharers help or hurt these initiatives?

3. Fair Use in Cyberspace

Kelly v. Arriba Soft Corp.
United States Court of Appeals for the Ninth Circuit
336 F.3d 811 (9th Cir. 2003)

T.G. NELSON, Circuit Judge.

This case involves the application of copyright law to the vast world of the internet and internet search engines. The plaintiff, Leslie Kelly, is a professional photographer who has copyrighted many of his images of the American West. Some of these images are located on Kelly's web site or other web sites with which Kelly has a license agreement. The defendant, Arriba Soft Corp., operates an internet search engine that displays its results in the form of small pictures rather than the more usual form of text. . . .

I.

The search engine at issue in this case is unconventional in that it displays the results of a user's query as "thumbnail" images. When a user wants to search the internet for information on a certain topic, he or she types a search term into a search engine, which then produces a list of web sites that contain information relating to the search term. Normally, the list of results is in text format. The Arriba search engine, however, produces its list of results as small pictures.

To provide this service, Arriba developed a computer program that "crawls" the web looking for images to index. This crawler downloads full-sized copies of the images onto Arriba's server. The program then uses these copies to generate smaller, lower-resolution thumbnails of the images. Once the thumbnails are created, the program deletes the full-sized originals from the server. Although a user could copy these thumbnails to his computer or disk, he cannot increase the resolution of the thumbnail; any enlargement would result in a loss of clarity of the image. . . .

II.

. . .

A.

. . . As to the thumbnails, Arriba conceded that Kelly established a prima facie case of infringement of Kelly's reproduction rights.

A claim of copyright infringement is subject to certain statutory exceptions, including the fair use exception. This exception "permits courts to avoid rigid application of the copyright statute when, on occasion, it would stifle the very creativity which that law is designed to foster." The statute sets out four factors to consider in determining whether the use in a particular case is a fair use. We must balance these factors in light of the objectives of copyright law, rather than view them as definitive or determinative tests. We now turn to the four fair use factors.

1. Purpose and Character of the Use

The Supreme Court has rejected the proposition that a commercial use of the copyrighted material ends the inquiry under this factor. Instead,

[t]he central purpose of this investigation is to see . . . whether the new work merely supersede[s] the objects of the original creation, or instead adds something new, with a further purpose or different character, altering the first with new expression,

meaning, or message; it asks, in other words, whether and to what extent the new work is transformative.

The more transformative the new work, the less important the other factors, including commercialism, become.

There is no dispute that Arriba operates its web site for commercial purposes and that Kelly's images were part of Arriba's search engine database. As the district court found, while such use of Kelly's images was commercial, it was more incidental and less exploitative in nature than more traditional types of commercial use. Arriba was neither using Kelly's images to directly promote its web site nor trying to profit by selling Kelly's images. Instead, Kelly's images were among thousands of images in Arriba's search engine database. Because the use of Kelly's images was not highly exploitative, the commercial nature of the use weighs only slightly against a finding of fair use.

The second part of the inquiry as to this factor involves the transformative nature of the use. We must determine if Arriba's use of the images merely superseded the object of the originals or instead added a further purpose or different character. We find that Arriba's use of Kelly's images for its thumbnails was transformative.

Although Arriba made exact replications of Kelly's images, the thumbnails were much smaller, lower-resolution images that served an entirely different function than Kelly's original images. Kelly's images are artistic works intended to inform and to engage the viewer in an aesthetic experience. His images are used to portray scenes from the American West in an aesthetic manner. Arriba's use of Kelly's images in the thumbnails is unrelated to any aesthetic purpose. Arriba's search engine functions as a tool to help index and improve access to images on the internet and their related web sites. In fact, users are unlikely to enlarge the thumbnails and use them for artistic purposes because the thumbnails are of much lower-resolution than the originals; any enlargement results in a significant loss of clarity of the image, making them inappropriate as display material.

Kelly asserts that because Arriba reproduced his exact images and added nothing to them, Arriba's use cannot be transformative. Courts have been reluctant to find fair use when an original work is merely retransmitted in a different medium.[19] Those cases are inapposite, however, because the resulting use of the copyrighted work in those cases was the same as the original use. For instance, reproducing music CDs in computer MP3 format does not change the fact that both formats are used for entertainment purposes. Likewise, reproducing news footage into a different format does not change the ultimate purpose of informing the public about current affairs.

Even in Infinity Broadcast Corp. v. Kirkwood, where the retransmission of radio broadcasts over telephone lines was for the purpose of allowing advertisers and radio stations to check on the broadcast of commercials or on-air talent, there was nothing preventing listeners from subscribing to the service for entertainment purposes. Even though the intended purpose of the retransmission may have been different from the

---

19. See Infinity Broad. Corp. v. Kirkwood, 150 F.3d 104, 108 (2d Cir.1998) (concluding that retransmission of radio broadcast over telephone lines is not transformative); UMG Recordings, Inc. v. MP3.com, Inc., 92 F.Supp.2d 349, 351 (S.D.N.Y.2000) (finding that reproduction of audio CD into computer MP3 format does not transform the work); Los Angeles News Serv., 149 F.3d at 993 (finding that reproducing news footage without editing the footage "was not very transformative").

purpose of the original transmission, the result was that people could use both types of transmissions for the same purpose.

This case involves more than merely a retransmission of Kelly's images in a different medium. Arriba's use of the images serves a different function than Kelly's use — improving access to information on the internet versus artistic expression. Furthermore, it would be unlikely that anyone would use Arriba's thumbnails for illustrative or aesthetic purposes because enlarging them sacrifices their clarity. Because Arriba's use is not superseding Kelly's use but, rather, has created a different purpose for the images, Arriba's use is transformative. . . .

The Copyright Act was intended to promote creativity, thereby benefitting the artist and the public alike. To preserve the potential future use of artistic works for purposes of teaching, research, criticism, and news reporting, Congress created the fair use exception. Arriba's use of Kelly's images promotes the goals of the Copyright Act and the fair use exception. The thumbnails do not stifle artistic creativity because they are not used for illustrative or artistic purposes and therefore do not supplant the need for the originals. In addition, they benefit the public by enhancing information-gathering techniques on the internet.

In Sony Computer Entertainment America, Inc. v. Bleem, 214 F.3d 1022 (9th Cir. 2000), we held that when Bleem copied "screen shots" from Sony computer games and used them in its own advertising, it was a fair use. In finding that the first factor weighed in favor of Bleem, we noted that "comparative advertising redounds greatly to the purchasing public's benefit with very little corresponding loss to the integrity of Sony's copyrighted material." Similarly, this first factor weighs in favor of Arriba due to the public benefit of the search engine and the minimal loss of integrity to Kelly's images.

2. Nature of the Copyrighted Work

"Works that are creative in nature are closer to the core of intended copyright protection than are more fact-based works." Photographs that are meant to be viewed by the public for informative and aesthetic purposes, such as Kelly's, are generally creative in nature. The fact that a work is published or unpublished also is a critical element of its nature. Published works are more likely to qualify as fair use because the first appearance of the artist's expression has already occurred. Kelly's images appeared on the internet before Arriba used them in its search image. When considering both of these elements, we find that this factor weighs only slightly in favor of Kelly.

3. Amount and Substantiality of Portion Used

"While wholesale copying does not preclude fair use per se, copying an entire work militates against a finding of fair use." However, the extent of permissible copying varies with the purpose and character of the use. If the secondary user only copies as much as is necessary for his or her intended use, then this factor will not weigh against him or her.

This factor neither weighs for nor against either party because, although Arriba did copy each of Kelly's images as a whole, it was reasonable to do so in light of

Arriba's use of the images. It was necessary for Arriba to copy the entire image to allow users to recognize the image and decide whether to pursue more information about the image or the originating web site. If Arriba only copied part of the image, it would be more difficult to identify it, thereby reducing the usefulness of the visual search engine.

4. Effect of the Use upon the Potential Market for or Value of the Copyrighted Work

This last factor requires courts to consider "not only the extent of market harm caused by the particular actions of the alleged infringer, but also 'whether unrestricted and widespread conduct of the sort engaged in by the defendant . . . would result in a substantially adverse impact on the potential market for the original.'" A transformative work is less likely to have an adverse impact on the market of the original than a work that merely supersedes the copyrighted work.

Kelly's images are related to several potential markets. One purpose of the photographs is to attract internet users to his web site, where he sells advertising space as well as books and travel packages. In addition, Kelly could sell or license his photographs to other web sites or to a stock photo database, which then could offer the images to its customers.

Arriba's use of Kelly's images in its thumbnails does not harm the market for Kelly's images or the value of his images. By showing the thumbnails on its results page when users entered terms related to Kelly's images, the search engine would guide users to Kelly's web site rather than away from it. Even if users were more interested in the image itself rather than the information on the web page, they would still have to go to Kelly's site to see the full-sized image. The thumbnails would not be a substitute for the full-sized images because the thumbnails lose their clarity when enlarged. If a user wanted to view or download a quality image, he or she would have to visit Kelly's web site. [37] This would hold true whether the thumbnails are solely in Arriba's database or are more widespread and found in other search engine databases.

In addition, we note that in the unique context of photographic images, the quality of the reproduction may matter more than in other fields of creative endeavor. The appearance of photographic images accounts for virtually their entire aesthetic value.

Arriba's use of Kelly's images also would not harm Kelly's ability to sell or license his full-sized images. Arriba does not sell or license its thumbnails to other parties. Anyone who downloaded the thumbnails would not be successful selling full-sized images enlarged from the thumbnails because of the low resolution of the thumbnails. There would be no way to view, create, or sell a clear, full-sized image without going to Kelly's web sites. Therefore, Arriba's creation and use of the thumbnails does not harm the market for or value of Kelly's images. This factor weighs in favor of Arriba.

37. We do not suggest that the inferior display quality of a reproduction is in any way dispositive or will always assist an alleged infringer in demonstrating fair use. In this case, however, it is extremely unlikely that users would download thumbnails for display purposes, as the quality full-size versions are easily accessible from Kelly's Web sites.

In addition, we note that in the unique context of photographic images, the quality of the reproduction may matter more than in other fields of creative endeavor. The appearance of photographic images accounts for virtually their entire aesthetic value.

Having considered the four fair use factors and found that two weigh in favor of Arriba, one is neutral, and one weighs slightly in favor of Kelly, we conclude that Arriba's use of Kelly's images as thumbnails in its search engine is a fair use.

COMMENTS AND QUESTIONS

1. *Transformative Use or Transformative Technology.* This decision takes an expansive view of tranformative use, emphasizing the functionality of Arriba's technology. In that way, the case seems more akin to the *Sony* indirect liability rule (asking whether a product is capable of substantial non-infringing use) than it does to conventional fair use cases.

Would Kelly have had a stronger claim if he had focused on the full copy that Arriba's search engine necessarily made in its database in order to index the images, but which was never shown to the public? Is such a use "transformative"? Why or why not?

2. *Harm Standard/Functionality Ratio.* Also as in *Sony*, the plaintiff in *Arriba* did not have any evidence of economic harm and there was good reason to believe that the technology was offering valuable functionality to users. Such technology cases are quite different from the traditional fair use scenario in which an author or artist incorporates some of a copyright owner's work into their own. The real social trade-off is whether new functionality adversely affects markets for copyright owners. Might fair use be more predictable (and effective at promoting economic efficiency) if plaintiffs had to meet a threshold test of demonstrable harm when challenging new technologies that offer significant economic benefits to consumers? If harm is shown, should technology developers be able to prevail if they can show a favorable ratio of consumer benefit to harm? What if the harm could be offset by design changes in technology? Even if not, why shouldn't the technology still be required to compensate the copyright owner(s) — which brings us full circle to copyright's traditional schema?

3. *Burden of Protection and Technological Controls.* Perhaps the solution to this problem is for copyright owners not wishing their graphical images to be indexed to use technological protection measures. Alternatively, web search technologies could provide convenient means for copyright owners to block their images from being indexed. Who should bear the burden in the cyberspace context? Would a least cost avoider standard provide an efficient solution?

4. Notwithstanding *Kelly*, litigation against search engines continues. Perfect 10, a subscription service providing access to nude photographs, has sued Google and Amazon.com over the same kind of thumbnail images held legal in *Kelly*. See Perfect 10 v. Google, Inc., ___ F. Supp. 2d ___, 2006 WL 454354 (C.D. Cal. Feb. 17, 2006) (entering preliminary injunction against Google for copyright infringement).

5. *Online Music "Lockers."* The *Arriba* decision contrasts with an earlier decision relating to online music lockers. As noted above, MP3.com offered subscribers the ability to maintain a virtual online music locker from which they could access sound recordings from any Internet portal through a password-protected user interface. MP3.com purchased and uploaded thousands of CDs onto its servers. Subscribers to this service could establish access to particular CDs by purchasing the CD online through a cooperating online retailer (the "Instant Listening Service") or by loading a CD that the subscriber owned into his or her computer CD-ROM drive (the "Beam-it

Service"). Software on the computer could verify the presence of the particular CD, but not actual ownership. MyMP3.com provided access to the copy of the CD stored on MP3.com's server. Thus, subscribers did not in fact access their own copy but rather MP3.com's copy. The notion of an actual private locker was metaphorical. In fact, subscribers had differential access to the same "locker," MP3.com's servers.

Judge Rakoff rejected MP3.com's fair use defense on the basis of the following findings. He considered the service to be commercial in purpose and non-transformative in character, rejecting MP3.com's argument that "space shifting" of a copyrighted work transforms it in legally cognizable ways. The court instead applied a more literal test: whether the defendant added "new aesthetics, new insights and understandings" to the sound recordings. The second and third fair use factors — the nature of the copyrighted work and the amount taken — clearly favored the plaintiffs. MP3.com relied principally upon the fourth factor — the effect upon the potential market for or value of the work — arguing that its service promotes sales of CDs by providing a means to make them more readily available. The court concluded, however, that this service impinged upon the copyright owners' ability to develop their own online distribution channels.

Do you agree with Judge Rakoff's statement that "[t]he complex marvels of cyberspatial communication may create difficult legal issues; but not in this case"? How would you respond to his assessment of the fair use factors?

PROBLEM

Problem 4-38. At the beginning of the 3rd century BC, Ptolemy II set out to collect and house the world's knowledge in the Library of Alexandria. At its peak, the library is believed to have stored 400,000 to 700,000 parchment scrolls. Its ultimate destruction remains shrouded in mystery.

In late 2004, Google set out to accomplish a similar feat — "to organize the world's information and make it universally accessible and useful." In conjunction with the University of Michigan, Harvard University, Stanford University, The New York Public Library, and Oxford University, Google plans to scan all or portions of their collections and make those texts searchable on Google. For works that are in the public domain, users will be able to access the entire book. For books still under copyright, users will receive only eight lines surrounding their search term. The goal will be to help users discover books and provide information about where to obtain a complete copy if the book is not in the public domain. Google does, however, earn a return on this investment. It would deliver advertisements along with search results, in much the way that its search engine and e-mail products function.

Soon after the project was announced, the Association of American University Presses (AAUP) castigated Google' for posing a risk of "systematic infringement of copyright on a massive scale." The Authors' Guild, a group representing writers, has filed a class-action lawsuit to stop or modify the Google Print Library Project. Various publishers have also sued.

Please advise Google on any copyright exposure that it might face and what steps it might take to limit such exposure.

I. INTERNATIONAL COPYRIGHT LAW

Although copyright protection, like patent protection, is territorial (country-specific) in nature, the lack of formalities in copyright law today along with nearly universal adherence to international treaties affording authors of signatory countries national treatment has resulted in nearly worldwide protection for copyrighted works. Therefore, U.S. copyright owners can enforce their rights in the courts of most nations in the world without even registering their works. The effectiveness of such efforts in stemming piracy, however, varies significantly on the basis of local conditions. Similarly, most foreign authors of works first published just about anywhere in the world can enforce their copyrights against infringement in the United States in U.S. courts.

Such near global copyright protection has not always been the case. Our purpose in this section is to provide a concise history of U.S. involvement in the international copyright system and survey the principal contours of international copyright law. Students interested in delving more deeply into the subject should consult one of the comprehensive international copyright treatises. See, e.g., Nimmer on Copyright chapters 17, 18; Paul Goldstein, International Copyright: Principles, Law, and Practice (2001); Paul Edward Geller, International Copyright Law and Practice.

1. Evolution of the International Copyright System and U.S. Participation

Early copyright laws did not afford protection to the works of foreign nationals. As printing businesses grew and literacy expanded, nations came to recognize the common interest in affording protection to authors from other nations. Without protection for works of foreign authors, publishers of domestic authors would have to compete with publishers of foreign works without any royalty obligations. Furthermore, domestic authors lost out on foreign royalty streams unless they could obtain protection abroad. As a solution, by the mid-nineteenth century, several European countries entered into bilateral treaties or passed legislation affording the works of foreign nationals protection on a reciprocal basis (i.e., where works of that nation's authors were afforded protection in the courts of the foreign national's home country). This movement eventually led most of the nations of Western Europe, as well as a few other nations, to adopt the Berne Convention for the Protection of Literary and Artistic Works in 1886. In order to join the convention, a nation's copyright law had to meet specified minimum criteria for protection. Authors from signatory nations obtained "national treatment" — i.e., the same rights as domestic authors — in each member nation. Over the course of the next century, most nations of the world would join this treaty.

Somewhat surprisingly in view of the strength of its content industries throughout much of the twentieth century, the United States long took an isolationist position in the development of international copyright law. See generally Hamish R. Sandison, The Berne Convention and the Universal Copyright Convention: The American Experience, 11 Colum.-V.L.A. J.L. & Arts 89 (1986). This in part reflected the relatively undeveloped nature of the arts in the early republic and the protectionist political influences of American publishers. England produced many of the popular

English-language authors well into the nineteenth century, and U.S. publishers had much to gain from being able to copy the works of British novelists and poets. During a visit to the United States in 1842, Charles Dickens bemoaned "the exquisite [in]justice of never deriving sixpence from an enormous American sale of all my books" and castigated American publishers for undermining the flourishing of native literary talent. As aptly noted by a later commentator, "American readers were less inclined to read the novels of Cooper or Hawthorne for a dollar when they could buy a novel of Scott or Dickens for a quarter." Max Kampelman, The United States and International Copyright, 41 Am. J. Int'l L. 406, 413 (1947). To the disappointment of Dickens, as well as many foreign and U.S. authors at the time, domestic publishers successfully blocked protection for foreign works until the late nineteenth century.

The United States grudgingly yielded to protection for works of foreign authors with the passage of the International Copyright Act of 1891 (the Chace Act). The U.S. insisted upon compliance with U.S. notice, registration, and deposit requirements and erected a further protectionist measure: a requirement that any printed book or periodical in the English language, as well as any printed book or periodical of domestic origin in any language, had to be printed from type set in the United States. This manufacturing clause continued until July 1, 1986. By complying with these requirements, foreign authors whose nations provided reciprocal protection to American nationals could obtain protection for their works in the United States. Following the Chace Act, the United States also began to enter into bilateral copyright reciprocity agreements with a growing number of nations. As an alternative to Berne, from which the U.S. was barred by its formalities and protectionist policies, the United States became a charter member of the less demanding, but also less widely adopted, Universal Copyright Convention (U.C.C.) in 1955. For the next three decades, the U.C.C. and bilateral treaties afforded American authors the ability to enforce their copyrights in much of the world. Furthermore, U.S. copyright owners could obtain "back door" protection for their works under the Berne Convention by publishing their works simultaneously in the United States and a Berne signatory nation.

Nonetheless, by the mid-1980s, as the global content marketplace expanded to unprecedented levels and piracy of copyrighted works in many corners of the world increased, the U.S. government came to believe that Berne membership was essential if the U.S. was to persuade other nations to join the international copyright system and exert its influence on the reform of global copyright protection. Notwithstanding reluctance to jettison formalities and revise its domestic copyright system to satisfy Berne's minimum criteria, the United States joined the Berne Convention in 1989. (We have discussed throughout this chapter the myriad changes aimed at achieving Berne compliance — including eliminating formalities for foreign authors, creating moral rights, expanding protection for architectural works, and restoring copyrights for foreign works that were lost as a result of failure to comply with U.S. formalities). In so doing, U.S. copyright proprietors gained the ability to protect their works directly in two dozen additional nations and the United States gained a seat at the Berne negotiation table. The Berne Convention now includes over 150 nations.

Soon thereafter, the United States and other developed nations elevated the role of intellectual property on the global stage by placing the formation of a new, more readily enforceable treaty mandating minimum intellectual property standards on the agenda for the ongoing multilateral trade negotiations as part of the Uruguay Round of the General Agreement on Tariffs and Trade (GATT) (reconstituted in 1995 as the

World Trade Organization (WTO)).[41] These negotiations resulted in the Agreement on Trade-Related Aspects of Intellectual Property Rights (TRIPs), a treaty signed by more than 100 nations that entered into force on January 1, 1995. That treaty set minimum standards for copyright protection required of each member nation, and also for the first time set expectations regarding enforcement of those legal rules. Developed nations had one year to bring their domestic laws into compliance; developing nations and ex-communist states were afforded an additional grace period of up to 10 years.

Rapid advances in digital technology fueled continuing international negotiations over copyright standards. In 1996, under the auspices of the World Intellectual Property Organization (WIPO), an arm of the United Nations which administers the Berne Convention, national representatives reached agreement on two supplementary agreements — the WIPO Copyright Treaty and the WIPO Performances and Phonograms Treaty. These treaties strengthen copyright protection along several dimensions. Of most significance (and controversy) were provisions aimed at protecting copyrights in the digital age. It was pursuant to these treaties that the United States adopted the anticircumvention, protection of copyright management information, and online service provider safe harbor provisions of the Digital Millennium Copyright Act. These treaties entered into force in 2002, upon ratification by 30 nations, although their implementation awaits domestic action in many WIPO nations.

2. International Copyright Treaties

This section summarizes the main features of the Berne Convention and the TRIPs treaty, the principal international copyright conventions operating today.[42]

Berne Convention for the Protection of Literary and Artistic Works

The Berne Convention has undergone several revisions over its long history. The latest text, to which most Berne signatories adhere, was negotiated in Paris in 1971. The Berne Convention was structured so that members were not required to adhere to revisions as a condition of membership. As a result, several nations still adhere to prior versions, such as the Rome (1928) or Brussels (1948) text. With China's and the Russian Federation's accession in 1992 and 1995, respectively, the Berne Convention includes all of the most significant economies in the world.

The Berne Convention is built upon two pillars: (1) national treatment — member nations must afford works of nationals of other Berne member nations the same protections as works of domestic authors (Article 5(1)); and (2) minimum

---

41. In addition, the United States, Canada, and Mexico entered into the North American Free Trade Agreement (NAFTA) in 1993. This agreement mandates that each of the member nations afford minimum standards for protection, most notably the availability of preliminary injunctive relief, and provides an enforcement mechanism for addressing disputes among the member nations.

42. The Universal Copyright Convention and bilateral agreements have a continuing role with regard to the protection of U.S. works published prior to U.S. entry into Berne or TRIPs in countries that are not Berne signatories.

standards — the copyright laws of member nations must satisfy the following minimum criteria:

Works Covered (Article 2). The convention covers "literary and artistic works," which is defined broadly to include "every production in the literary, scientific and artistic domain, whatever may be the mode or form of expression." It also includes derivative works and collective works, but excludes newsworthy facts. Although initially considered outside of the copyright domain, recent developments such as the E.C. Software Directive, TRIPs, and the WIPO Copyright Treaty indicate that computer programs are to be protected as "literary works" within the meaning of the Berne Convention. By contrast, commentary and national practice suggest that the Berne Convention does not extend to phonograms (sound recordings), which are commonly accorded rights outside the United States under the rubric of "neighboring" rights. Several other treaties apply to these works.[43]

Limitations on Formalities (Article 5(2)). "The enjoyment and the exercise of [copyright] shall not be subject to any formality . . . other than in the country of origin of the work." Thus, Berne prohibits only those formalities that would operate to preclude copyright protection for works from other member states. A Berne member may impose formalities upon works of its own authors and may impose formalities as a condition to certain types of remedies (e.g., statutory damages, attorney's fees), compulsory licenses, or exemptions.

Duration (Article 7). Berne members must afford protection for no less than life of the author plus 50 years or 50 years from publication in the case of motion pictures and anonymous or pseudonymous works. Photographic works and applied art shall receive no less than 25 years of protection from the time of creation.

Exclusive Rights (Articles 8, 9, 11, 12, and 14). The Berne Convention requires that member nations afford exclusive rights to make and authorize translation, reproduction, public performance, and adaptation of their works. In contrast to U.S. law, the Berne Convention does not include display and distribution among the exclusive rights that must be accorded.

Moral Rights (Article 6[bis]). The Berne Convention provides:

(1) Independently of the author's economic rights, and even after the transfer of the said rights, the author shall have the right to claim authorship of the work and to object to any distortion, mutilation or other modification of, or other derogatory action in relation to, the said work, which would be prejudicial to his honor or reputation.

(2) The rights granted to the author in accordance with the preceding paragraph shall, after his death, be maintained, at least until the expiry of the economic rights . . .

(3) The means of redress for safeguarding the rights granted by this Article shall be governed by the legislation of the country where protection is claimed.

Exceptions (Articles 2[bis], 10, 10[bis]). Berne members may limit protection for political speeches, allow the press broader rights to reproduce public lectures, and carve out a fair use privilege.

Restoration of Rights (Article 18). The Convention applies to "all works which, at the moment of its coming into force, have not yet fallen into the public domain in the country of origin through the expiry of the term of protection."

43. See International Convention for the Protection of Performers, Producers of Phonograms and Broadcasting Organizations (1961) (Rome Convention); Convention for the Protection of Producers of Phonograms Against Unauthorized Duplication of Their Phonograms (1971) (Geneva Phonograms Convention); Convention Relating to the Distribution of Programme-Carrying Signals Transmitted by Satellite (1974) (Brussels Satellite Convention); WIPO Performances and Phonogram Treaty (1996).

Agreement on Trade-Related Aspects of Intellectual Property Rights (TRIPs)

The TRIPs Agreement incorporates and expands upon the Berne foundation. By its terms, all WTO Member Countries must enforce all requirements of the Berne Convention, save the moral rights provisions (Article 9). Furthermore, WTO Members must adhere to requirements set forth in the agreement relating to neighboring rights, but not extending to prior conventions (Article 14). TRIPs specifies more extensive civil and criminal enforcement obligations and incorporates the new WTO dispute-settlement process for resolving disputes among the member nations. It was pursuant to this procedure that the U.S. Fairness in Music Licensing Act was held to be in violation of TRIPs and the Berne Convention. See United States — Section 110(5) of the U.S. Copyright Act, Report of the Panel, World Trade Organization, WT/DS160/R Jun. 15, 2000. Subject to several exemptions, TRIPs (Article 4) also goes beyond the Berne framework by requiring that members afford all foreign authors the same protections as those offered to authors from the "most favored nation." In most cases, such treatment will be the same as national treatment — that which a nation affords its own authors.

TRIPs expands on the minimum criteria of the Berne Convention in several respects:

> *Works Covered* (Article 10). TRIPs requires that member nations afford protection for computer programs as literary works under the Berne Convention. Original selection or arrangement of databases must also be protected.
>
> *Exclusive Rights* (Article 11). Member nations must afford copyright owners the right to authorize or prohibit rental of computer programs and films. The same provision, however, authorizes nations to permit movie rental "unless such rental has led to widespread copying of such works which is materially impairing the exclusive right of reproduction conferred in that Member on authors and their successors in title."
>
> *Exceptions* (Article 13). TRIPs sought to limit exceptions to Berne's exclusive rights by requiring that Members "confine limitations or exceptions to exclusive rights to certain special cases which do not conflict with a normal exploitation of the work and do not unreasonably prejudice the legitimate interests of the right holder."

COMMENTS AND QUESTIONS

1. *Global Copyright Governance.* Certainly by comparison to the patent field, international copyright protection would seem to be a model of global governance. The nearly universal reach of the Berne Convention in combination with the World Trade Organization's formal dispute resolution mechanism have brought about a reasonably well coordinated system for the protection of copyrights throughout the world. Given the challenges facing copyright owners in the digital age, it could be said that these developments could not have come about at a more opportune time.

2. *Political Economy of International Copyright Law and Policy.* Political economists and advocates for the public domain tend to view recent international developments in the copyright arena somewhat more cynically. See Michael P. Ryan, Knowledge Diplomacy: Global Competition and the Politics of Intellectual Property

(1998); Pamela Samuelson, The Copyright Grab, Wired 4.01 (Jan. 1996). They see the TRIPs Agreement and the WIPO Copyright Treaties as the product of well-coordinated content industry lobbyists seeking ever stronger protection, quite possibly at the expense of lesser developed nations and the public at large. Are these industries using their superior organizational skills and political influence unjustly, or are they rightly concerned to be advocating for enhanced protections in the face of digital technologies that threaten to undermine the legal institutions supporting creative expression? Is it possible that both things are true? Have new interest groups formed in the global arena to counterbalance the protectionist bias?

3. Protection of U.S. Works Against Infringement Abroad

International copyright law incorporates the doctrine of "territoriality." This doctrine embodies three precepts. First, a nation's laws apply only within its territorial boundaries. Second, nations may exercise jurisdiction over those within its boundaries. And third, principles of comity caution against any nation state applying its laws in such a way as to interfere with the sovereignty interests of other nations. Stated another way, U.S. copyright law does not apply extraterritorially. See Subafilms, Ltd. v. MGM-Pathe Communications Co., 24 F.3d 1088, 1095 (9th Cir. 1994) (en banc) (noting "eighty years of consistent jurisprudence").

As a result, U.S. copyright owners must look to the law of the nation in which infringement occurs to obtain redress. This poses the question whether a U.S. copyright owner possesses enforceable rights in another nation, which turns on whether the nation is party to a multi-lateral treaty (such as TRIPs, Berne, or U.C.C.) or whether a bilateral agreement with the United States which confers protection upon U.S. copyright owners or the nation extends protection to foreign copyright owners on some other basis.[44] Just a few decades ago, analysis of the complexity of international accords arose with some frequency. The nearly universal reach of Berne today makes 1989, the year in which the U.S. joined this convention, an historic dividing point in the protection U.S. copyright interests enjoy abroad.

Due to the widespread adoption of Berne, U.S. works—both post-1989 and pre-1989[45]—receive national treatment in most nations of the world. That does not, however, ensure that an enforcement action by a U.S. copyright owner can proceed in a foreign court. A threshold question of standing arises: does the foreign country (commonly referred to as the "protecting country") recognize the party seeking to enforce a copyright interest as the owner of rights in the work? The proper test depends upon choice of law rules, which determine whether U.S. law or the law of the protecting country governs. United States courts apply the law of the state with the "most significant relationship" to the work to determine ownership. See Itar-Tass Russian News Agency v. Russian Kurrier, 153 F.3d 82 (2d Cir. 1998) (finding that some of the plaintiffs lacked standing because they were not owners of an exclusive right under Russian copyright law). Other nations may apply their own law. If the

---

44. Germany and France, for example, extend protection universally. U.S. copyright law protects all unpublished works irrespective of the author's nationality.

45. Under Article 18 of the Berne Convention, member states must afford protection for all U.S. works still under copyright protection in the United States.

putative U.S. copyright owner is not recognized as such by the protecting state, then it will not be permitted to proceed.

Once the standing hurdle is cleared, there remains the determination whether and to what extent the foreign country protects the work for which enforcement is sought. Although the Berne Convention (now augmented by TRIPs) establishes minimum standards for protection, U.S. copyright protection extends beyond these limits in several respects (and arguably falls short in other areas, such as protection of moral rights). For example, whereas the United States protects sound recordings under copyright law, many other countries treat such works under neighboring rights regimes that fall outside of the Berne Convention. Lastly, liability and remedies must be determined. Reflecting the territoriality doctrine, most nations, including the United States, apply the conflicts principle of *lex loci delicti* ("the law of the place where the tort or other wrong was committed") to infringement analysis and determination of remedies.

Furthermore, there is an important exception to national treatment relating to duration of copyright protection for works of foreign origin. Under Article 7(8) of the Berne Convention, the duration of protection may be the shorter of the duration as between the protecting country and the country of origin, unless the law of the protecting country chooses otherwise. The rule of the shorter term applies in most countries of the world. The E.C. Term Directive requires that member countries adhere to the rule of the shorter term.

COMMENTS AND QUESTIONS

1. *The Rule of the Shorter Term and Duration of U.S. Copyrights Under the 1909 Act.* Under the Berne convention, what is the duration of U.S. copyright protection for pre-1976 Act U.S. works which lost (or never received) copyright protection for failure to comply with required formalities (publication with proper notice)? What about those works for which renewal was not sought or successfully attained? In an "informal" advisory opinion, a WIPO official took the position that in the case of "those works, which, due to the non-compliance with formalities (such as the requirement of publication of notice), have never been protected, it is clear that they have not fallen into the public . . . through the expiry of the term of protection since there has been no term of protection applicable for them." For works that received only one term of protection under the 1909 Act, the WIPO official opined that the term should be ninety-five years from publication, the term which would had been available had renewal been successfully obtained (and applicable term extensions been added), because this is the duration that the owner would have enjoyed had formalities "been fulfilled (or would not have existed)." See Letter of Shozo Uemura, Deputy Director General, World Intellectual Property Organization, reprinted in 47 J. Copyr. Soc'y 91 (1999). It can certainly be argued, however, that U.S. works for which no renewal registration was made fell into the public domain through the "expiry of the term of protection."

2. *Protection of U.S. Works Outside of Berne/TRIPs Member Nations.* In one of the relatively few nations not party to Berne or TRIPs, a U.S. copyright owner can enforce its rights only if that nation extends protection to foreign copyright owners directly or if that nation is party to a bilateral agreement with the United States conferring copyright protection upon each nation's copyright owners. If a basis for

protection is identified, the U.S. copyright owner's standing to sue and substantive rights would be determined under the protecting nation's copyright and choice of law regime.

3. *Combating Foreign Infringing Activities in U.S. Courts.* Notwithstanding the territoriality doctrine, U.S. copyright owners can pursue remedies in U.S. courts for foreign infringement in several circumstances:

- Where an "act" of infringement occurs within the United States even though the infringement is completed abroad, the copyright owner can sometimes pursue remedies under U.S. copyright law. See Ortman v. Stanray Corp., 163 U.S.P.Q. 331 (N.D. Ill. 1969). Nonetheless, a defendant cannot be held liable under U.S. law for merely authorizing conduct that occurred overseas that would constitute copyright infringement if it occurred in the United States. See Subafilms, Ltd. v. M.G.M.-Pathe Communications, 24 F.3d 1088 (9th Cir. 1994) (en banc).
- Sections 602 and 603 of the U.S. Copyright Act prohibit the importation of infringing articles. See also 19 U.S.C. § 337 (providing for exclusion of infringing articles through an International Trade Commission proceeding).
- A U.S. copyright owner, or any copyright owner for that matter, may be able to sue for infringement occurring elsewhere in the world if personal jurisdiction over the defendant can be obtained in an American court under a theory that copyright infringement constitutes a transitory cause of action. Cf. Curtis Bradley, Territorial Intellectual Property Rights in an Age of Globalism, 37 Va. J. Int'l L. 505 (1997). 28 U.S.C. § 1332(a)(2) confers jurisdiction in federal court where there is diversity of citizenship (including citizens of a State and subjects of a foreign state) and at least $75,000 in dispute. The U.S. court would apply the law of the country in which the alleged infringement occurred. See London Film Prods. Ltd. v. Intercontinental Communications, Inc., 580 F. Supp. 47 (S.D.N.Y. 1984). There are, however, potential legal and practical impediments to such a course of action. The defendant may persuade the court to dismiss the action under the *forum non conveniens* doctrine. See, e.g., Murray v. British Broad. Corp., 81 F.3d 287 (2d Cir. 1996); but cf. Boosey & Hawkes Music Publishers, Ltd. v. Walt Disney Co., 145 F.3d 481 (2d Cir. 1998) (overturning dismissal of an action under this doctrine); Armstrong v. Virgin Records, Ltd., 91 F. Supp. 2d 628, 637-38 (S.D.N.Y. 2000). Furthermore, to the extent that the plaintiff seeks injunctive or other relief that is available only in the nation in which the infringement is occurring, then she will still have to pursue enforcement abroad.
- If copyright infringement occurs within the United States, the plaintiff can recover the defendant's profits from overseas sales or use of infringing works. See Los Angeles News Service v. Reuters Television, Int'l, 149 F.3d 987 (9th Cir. 1998), *cert. denied*, 525 U.S. 1141 (1999). Does this make sense? What if liability would differ under the laws of other jurisdictions? Note that in Computer Associates Int'l v. Altai, Inc., 126 F.3d 365 (2d Cir. 1997), the court held that a copyright owner that had lost its claim in the U.S. could bring the same claim in France under French law. Is it fair that if a copyright owner wins in the U.S. they can get worldwide damages, but if they lose in the U.S. they are free to try again elsewhere?

4. *Other Means for Combating Foreign Piracy of U.S. Copyrighted Works.* Under Section 301 of the Trade Act of 1974, as amended (including "Special 301" authority), the U.S. Trade Representative monitors the protection of U.S. intellectual property rights abroad and uses the threat of trade sanctions as a means of promoting greater enforcement by other nations. U.S. administrations have used this provision to put countries on a "priority watch list" if they do not adequately enforce intellectual property rights and to obtain a series of bilateral treaty concessions on intellectual property rights from those countries. Some scholars argue that Special 301 is an efficient means of enforcing intellectual property rights. See Tara Kalagher Giunta & Lily H. Shang, Ownership of Information in a Global Economy, 27 Geo. Wash. J. Int'l L. & Econ. 327 (1994). Many nations, however, have criticized the procedure as discriminatory and at odds with multilateral dispute resolution mechanisms such as the World Trade Organization's process. See Lina M. Montén, The Inconsistency Between Section 301 and TRIPs: Counterproductive with Respect to the Future of International Protection of Intellectual Property Rights?, 9 Marq. Intell. Prop. L. Rev. 387 (2005).

5. *The Territoriality Doctrine and the Internet.* Will the territoriality doctrine function effectively in cyberspace? For example, courts in several nations have come to diametrically opposed conclusions in resolving copyright disputes relating to file-sharing activities that cross international boundaries. In the United States and in Australia, distributors of peer-to-peer file sharing software have been held liable for copyright infringement, but in the Netherlands the same defendants have been held not to be liable. Which rule should govern? It seems difficult to say that a web site can be shut down in one country but can continue to operate in another. Especially given the possible res judicata ramifications, but cf. Computer Associates Int'l v. Altai, 126 F.3d 365 (2d Cir. 1997) (refusing to enjoin the French suit under principles of res judicata or collateral estoppel, reasoning that the application of French law to conduct occurring in France presented a separate issue from the one litigated in the U.S. case), cyberspace copyright disputes could fuel races to courthouses in several nations and produce friction among nations. See Peter P. Swire, Elephants and Mice Revisited: Law and Choice of Law on the Internet 153 U. Pa. L. Rev. 1975 (2005) (suggesting that such conflict has not yet manifested on a large scale due to technology's ability to trump law, lack of jurisdiction over defendants, harmonization of substantive law, and the existence of self-regulatory and other systems that suppress choice-of-law conflicts for transactions); see generally Symposium: Copyright's Long Arm: Enforcing U.S. Copyrights Abroad, 24 Loy. L.A. Ent. L. Rev. 45 (2004); Paul Schiff Berman, The Globalization of Jurisdiction, 151 U. Pa. L. Rev. 311 (2002). As the problem of unauthorized distribution of copyrighted works on the Internet escalates and nations' legal decisions and legislation diverge, will we see an increase in conflicts of law problems or will copyright owners direct their efforts toward enforcement within separate countries in combination with efforts to harmonize legal regimes through multi-lateral treaty negotiations?

4. Protection of Foreign Works Against Infringement in the United States

The framework for analyzing protection for foreign works in the United States mirrors the framework applied to the protection of U.S. works abroad. The United

States affords protection for all works of Berne and TRIPs members for which copyright subsists in their country of origin, irrespective of when such nations joined these treaties. See 17 U.S.C. §§ 104(a), 104A (restoration of copyright in works of foreign origin). The United States does not apply the rule of the shorter term, and therefore foreign copyright owners may obtain the full benefit of the U.S. term of copyright protection even if protection has expired in the country of origin. Works receiving the benefit of Berne protection that would otherwise have fallen into the public domain in the United States on the basis of a notice defect may be enforced.

J. ENFORCEMENT AND REMEDIES

Economists often classify legal rules into one of two types depending on the remedies available. The first type of legal regime is known as the "liability rule." Under a liability regime, such as contract law, one who violates the rights of another (by, for example, breaching a contractual responsibility) is liable for damages—a monetary measure of the harm caused to the plaintiff. The liability rule contrasts with a property rule, which entitles the right holder to prevent violations, not merely obtain monetary compensation. Thus, landowners can typically obtain injunctive relief preventing trespassory invasions (as well as recover for any harm caused by prior invasions). For a more detailed discussion of property and liability rules, see A. Mitchell Polinsky, An Introduction to Law and Economics (3d ed. 2003); Robert Cooter & Thomas Ulen, Law and Economics (4th ed. 2004); John P. Dwyer and Peter S. Menell, Property Law and Policy: A Comparative Institutional Perspective 281-287 (1998).

Reflecting the traditional treatment of intellectual property rights as a form of property interest, copyright law largely reflects the property rule framework. Copyright infringement is not made acceptable simply by the infringer's payment of damages. Indeed, intentional infringement of copyright for financial gain is a criminal offense. 17 U.S.C. § 506(a).[46] In civil cases, infringement can be—and routinely is—enjoined under 17 U.S.C. § 502. Further, courts will often order the seizure or impoundment of allegedly infringing articles while a copyright case is pending, under the authority of 17 U.S.C. § 503(a) (in civil cases) or § 509(a) (in criminal cases). Those articles may be destroyed or forfeited to the United States after judgment. Finally, in addition to the above remedies, the copyright holder may collect as damages either "actual damages" (defined as the loss to the copyright holder plus additional profits made by the infringer) or statutory damages of up to $30,000 per infringed work for unintentional infringement and $150,000 per infringed work for willful infringement. 17 U.S.C. § 504. Prevailing parties (plaintiffs *and* defendants) are also entitled to costs and attorney fees in "exceptional" cases. 17 U.S.C. § 505.

Not all aspects of copyright law follow the property rule model. The fair use doctrine, for example, has strong liability rule components. Unlike the traditional liability rule framework, however, the fair use doctrine actually absolves the defendant

---

46. Criminal penalties for copyright infringement are limited to those set forth in section 506. The Supreme Court has rejected an attempt by the government to prosecute copyright infringement as transportation of stolen property. The Court reasoned that the specific statutory scheme of section 506 was meant to replace, not supplement, general laws that were not written with intellectual property in mind. Dowling v. United States, 473 U.S. 207 (1985).

of any responsibility to compensate the copyright owner, although the amount of harm plays a role in determining whether the defense succeeds. As we have discussed, there are also a number of other provisions for compulsory licensing in copyright law. A further analogy to contract law can be found in the provision for statutory damages, which may be seen as a sort of liquidated damages clause designed to avoid market failure in cases in which injury to the copyright holder is hard to quantify—though unlike liquidated damages, statutory damages can be punitive.

As is clear from even cursory exposure to intellectual property right valuation, including licensing negotiations, these assets are exceedingly difficult to value. This is partly because, by statutory definition, each of these rights is in some sense unique. And when a tribunal comprised of people without much direct experience in the relevant industry or industries affected resolves the litigation, as is often the case in copyright disputes, the costs rise even more. Law and economics scholarship suggests that a property rule is superior to a liability rule when "the court lacks information about both damages and benefits." A. Mitchell Polinksy, Resolving Nuisance Disputes: The Simple Economics of Injunctive and Damage Remedies, 32 Stan. L. Rev. 1075, 1112 (1980); see also Robert P. Merges, Contracting into Liability Rules: Intellectual Property Rights and Collective Rights Organizations, 84 Cal. L. Rev. 1293 (1996). The following case illustrates the challenges of measuring damages in copyright cases.

≡≡≡ *Sheldon et al. v. Metro-Goldwyn*
≡≡≡ *Pictures Corp. et al.*
≡ *Supreme Court of the United States*
≡ *309 U.S. 390 (1940)*

Mr. Chief Justice HUGHES delivered the opinion of the Court.

The questions presented are whether, in computing an award of profits against an infringer of a copyright, there may be an apportionment so as to give to the owner of the copyright only that part of the profits found to be attributable to the use of the copyrighted material as distinguished from what the infringer himself has supplied, and, if so, whether the evidence affords a proper basis for the apportionment decreed in this case.

Petitioners' complaint charged infringement of their play "Dishonored Lady" by respondents' motion picture "Letty Lynton," and sought an injunction and an accounting of profits. The Circuit Court of Appeals, reversing the District Court, found and enjoined the infringement and directed an accounting. 81 F.2d 49. Thereupon the District Court confirmed with slight modifications the report of a special master which awarded to petitioners all the net profits made by respondents from their exhibitions of the motion picture, amounting to $587,604.37. 26 F. Supp. 134, 136. The Circuit Court of Appeals reversed, holding that there should be an apportionment and fixing petitioners' share of the net profits at one-fifth. 106 F.2d 45, 51. In view of the importance of the question, which appears to be one of first impression in the application of the copyright law, we granted certiorari. December 4, 1939.

Petitioners' play "Dishonored Lady" was based upon the trial in Scotland, in 1857, of Madeleine Smith for the murder of her lover,—a *cause celebre* included in the series of "Notable British Trials" which was published in 1927. The play was copyrighted as an unpublished work in 1930, and was produced here and abroad. Respondents took the title of their motion picture "Letty Lynton" from a novel of that name written by an English author, Mrs. Belloc Lowndes, and published in 1930. That novel was also based upon the story of Madeleine Smith and the motion picture rights were bought by respondents. There had been negotiations for the motion picture rights in petitioners' play, and the price had been fixed at $30,000, but these negotiations fell through.

As the Court of Appeals found, respondents in producing the motion picture in question worked over old material; "the general skeleton was already in the public demesne. A wanton girl kills her lover to free herself for a better match; she is brought to trial for the murder and escapes." [106 F.2d 50.] But not content with the mere use of that basic plot, respondents resorted to petitioners' copyrighted play. They were not innocent offenders. From comparison and analysis, the Court of Appeals concluded that they had "deliberately lifted the play"; their "borrowing was a deliberate plagiarism." It is from that standpoint that we approach the questions now raised.

Respondents contend that the material taken by infringement contributed in but a small measure to the production and success of the motion picture. They say that they themselves contributed the main factors in producing the large net profits; that is, the popular actors, the scenery, and the expert producers and directors. Both courts below have sustained this contention.

The District Court thought it "punitive and unjust" to award all the net profits to petitioners. The court said that, if that were done, petitioners would receive the profits that the "motion picture stars" had made for the picture "by their dramatic talent and the drawing power of their reputations." "The directors who supervised the production of the picture and the experts who filmed it also contributed in piling up these tremendous net profits." The court thought an allowance to petitioners of 25 percent of these profits "could be justly fixed as a limit beyond which complainants would be receiving profits in no way attributable to the use of their play in the production of the picture." But, though holding these views, the District Court awarded all the net profits to petitioners, feeling bound by the decision of the Court of Appeals in Dam v. Kirk La Shelle Co., 175 F. 902, 903, a decision which the Court of Appeals has now overruled.

The Court of Appeals was satisfied that but a small part of the net profits was attributable to the infringement, and, fully recognizing the difficulty in finding a satisfactory standard, the court decided that there should be an apportionment and that it could fairly be made. The court was resolved "to avoid the one certainly unjust course of giving the plaintiffs everything, because the defendants cannot with certainty compute their own share." The court would not deny "the one fact that stands undoubted," and, making the best estimate it could, it fixed petitioners' share at one-fifth of the net profits, considering that to be a figure "which will favor the plaintiffs in every reasonable chance of error."

Petitioners stress the provision for recovery of "all" the profits, but this is plainly qualified by the words "which the infringer shall have made from such infringement." This provision in purpose is cognate to that for the recovery of "such damages as the copyright proprietor may have suffered due to the infringement." The purpose is thus to provide just compensation for the wrong, not to impose a penalty by giving to the copyright proprietor profits which are not attributable to the infringement.

Prior to the Copyright Act of 1909, there had been no statutory provision for the recovery of profits, but that recovery had been allowed in equity both in copyright and patent cases as appropriate equitable relief incident to a decree for an injunction. Stevens v. Gladding, 17 How. 447, 455. That relief had been given in accordance with the principles governing equity jurisdiction, not to inflict punishment but to prevent an unjust enrichment by allowing injured complainants to claim "that which, ex aequo et bono, is theirs, and nothing beyond this." Livingston v. Woodworth, 15 How. 546, 560. See Root v. Railway Co., 105 U.S. 189, 194, 195. Statutory provision for the recovery of profits in patent cases was enacted in 1870. The principle which was applied both prior to this statute and later was thus stated in the leading case of Tilghman v. Proctor, 125 U.S. 136, 146:

> The infringer is liable for actual, not for possible gains. The profits, therefore, which he must account for, are not those which he might reasonably have made, but those which he did make, by the use of the plaintiff's invention; or, in other words, the fruits of the advantage which he derived from the use of that invention, over what he would have had in using other means then open to the public and adequate to enable him to obtain an equally beneficial result. If there was no such advantage in his use of the plaintiff's invention, there can be no decree for profits, and the plaintiff's only remedy is by an action at law for damages....

Petitioners stress the point that respondents have been found guilty of deliberate plagiarism, but we perceive no ground for saying that in awarding profits to the copyright proprietor as a means of compensation, the court may make an award of profits which have been shown not to be due to the infringement. That would be not to do equity but to inflict an unauthorized penalty. To call the infringer a trustee *ex maleficio* merely indicates "a mode of approach and an imperfect analogy by which the wrongdoer will be made to hand over the proceeds of his wrong." Larson Co. v. Wrigley Co., 277 U.S. 97, 99, 100. He is in the position of one who has confused his own gains with those which belong to another. Westinghouse Co. v. Wagner Co., supra, p. 618. He "must yield the gains begotten of his wrong." Duplate Corp. v. Triplex Co., 298 U.S. 448, 457. Where there is a commingling of gains, he must abide the consequences, unless he can make a separation of the profits so as to assure to the injured party all that justly belongs to him. When such an apportionment has been fairly made, the copyright proprietor receives all the profits which have been gained through the use of the infringing material and that is all that the statute authorizes and equity sanctions.

Both courts below have held in this case that but a small part of the profits were due to the infringement, and, accepting that fact and the principle that an apportionment may be had if the evidence justifies it, we pass to the consideration of the basis of the actual apportionment which has been allowed.

The controlling fact in the determination of the apportionment was that the profits had been derived, not from the mere performance of a copyrighted play, but from the exhibition of a motion picture which had its distinctive profit-making features, apart from the use of any infringing material, by reason of the expert and creative operations involved in its production and direction. In that aspect the case has a certain resemblance to that of a patent infringement, where the infringer has created profits by the addition of non-infringing and valuable improvements. And, in this instance, it plainly appeared that what respondents had contributed accounted for by far the larger part of their gains....

COMMENTS AND QUESTIONS

1. Does *Sheldon* unfairly prevent copyright owners from prohibiting access to their works? Consider the case of private diaries that are published. Might not the injury to the author in some cases exceed the profits of the publisher? 17 U.S.C. §504(a) attempts to take care of this problem by allowing the copyright owner to recover both his actual losses and any additional profits incurred by the infringer, so long as the copyright owner does not thereby obtain a "double recovery."

2. *Attorneys' Fees.* Section 505 of the Copyright Act authorizes the award of attorney fees in the discretion of the court to "prevailing parties." The Supreme Court interpreted this provision as allowing fees to be awarded to prevailing plaintiffs or defendants on an equal basis. Fogerty v. Fantasy, Inc., 510 U.S. 517 (1994). The Court's reasoning in rejecting a "dual standard" that favored plaintiffs is instructive on the issue of appropriate copyright incentives:

> [T]he policies served by the Copyright Act are more complex, more measured, than simply maximizing the number of meritorious suits for copyright infringement....
>
> Because copyright law ultimately serves the purpose of enriching the general public through access to creative works, it is peculiarly important that the boundaries of copyright law be demarcated as clearly as possible. To that end, defendants who seek to advance a variety of meritorious copyright defenses should be encouraged to litigate them to the same extent that plaintiffs are encouraged to litigate meritorious claims of infringement. In the case before us, the successful defense of "The Old Man Down the Road" increased public exposure to a musical work that could, as a result, lead to further creative pieces. Thus a successful defense of a copyright infringement action may further the policies of the Copyright Act every bit as much as a successful prosecution of an infringement claim by the holder of a copyright.

Id. at 526-27. Are statutory damages, injunctions and criminal penalties consistent with the Court's statement of copyright policy?

3. Is it reasonable to draw a distinction between "counterfeiting" a work — that is, copying it in its entirety for profit — and infringing the copyright through another means (say, by writing a substantially similar screenplay)? Should counterfeiting be punished more severely than other forms of infringement? In 1982, Congress passed the Piracy and Counterfeiting Amendments Act, 18 U.S.C. §§2318-19. The Act makes it a felony to "knowingly traffic" in counterfeit phonorecords or audiovisual works, and it sets fines of up to $250,000 and prison terms of up to five years.

4. *Jury Role in Determining Statutory Damages.* In Feltner v. Columbia Pictures Television, 523 U.S. 340 (1998), the Supreme Court held that 17 U.S.C. §504(c), which provides that statutory damages are to be awarded in the discretion of the Court, violated the Seventh Amendment right to a jury trial. The Court concluded that the Seventh Amendment required both that liability issues be tried before a jury even if the plaintiff seeks only statutory damages and that the amount of the statutory damages itself must be determined by a jury. How this will interact with the Copyright Act's provision allowing plaintiffs to elect statutory damages at any time before judgment is unclear. It is possible that if a copyright plaintiff is dissatisfied with a jury's award of actual damages, he can then reconvene the jury and ask them to determine statutory damages.

5. *Punitive Damages.* Almost all courts have held that punitive damages are not available for copyright infringement. Such a requirement goes beyond the

restitutionary nature of copyright law. See Bucklew v. Hawkins, Ash, Baptie & Co., 329 F.3d 923, 931-32 (7th Cir. 2003). "The public policy rationale for punitive damages of punishing and preventing malicious conduct can be properly accounted for in the provisions for increasing a maximum statutory damage award" in cases of willful infringement. Id. at 177. The one case to leave open the possibility of punitive damages, TVT Records v. Island Def Jam Music Group, 262 F. Supp. 2d 185, 187 (S.D.N.Y. 2003), rests on a questionable foundation. But in some cases the availability of statutory damages may have the same effect, since a plaintiff can recover up to $150,000 per work infringed even if the work in question was worth only a few dollars and was copied only once.

Note on Injunctive Relief

In addition to damage remedies, injunctive relief is freely granted in copyright cases, even when the relief is sought before trial on the merits. See, e.g., Concrete Machinery Co. v. Classic Lawn Ornaments Inc., 843 F.2d 600 (1st Cir. 1988); Wainwright Securities, Inc. v. Wall Street Transcript Corp., 558 F.2d 91, 94 (2d Cir. 1977); Conrad Fabrics, Inc. v. Marcus Bros. Textile Corp., 409 F.2d 1315, 1316-17 (2d Cir. 1969). These preliminary injunctions are generally granted as a matter of course where a plaintiff can convince the court that a finding of infringement is likely. Further, courts are often willing to presume irreparable harm to the copyright owner, on the theory that it is hard to "close the door" after an infringing work has been publicly distributed.

After trial, the "entitlement" to injunctive relief is even more firmly established. In some cases, however, particularly those involving derivative works, a permanent injunction against infringement may appear inequitable. For example, in Abend v. MCA, Inc., 863 F.2d 1465 (9th Cir. 1988), Abend, the licensee of the original owner of the copyright in the story *It Had to Be Murder,* sued the owners of the Alfred Hitchcock film *Rear Window,* which was based on the story. While the copyright in the story had been validly licensed at the time the movie was produced, the Ninth Circuit held that Abend owned the renewal right in the copyright, and that continued exploitation of the film *Rear Window* without permission constituted infringement.[47] However, the court denied Abend injunctive relief:

> ... We are mindful that this case presents compelling equitable considerations which should be taken into account by the district court in fashioning an appropriate remedy.... Defendants invested substantial money, effort, and talent in creating the "Rear Window" film. Clearly the tremendous success of that venture initially and upon re-release is attributable in significant measure to, inter alia, the outstanding performances of its stars — Grace Kelly and James Stewart — and the brilliant directing of Alfred Hitchcock. The district court must recognize this contribution in determining Abend's remedy.
>
> The district court may choose from several available remedies for the infringement. Abend seeks first an injunction against the continued exploitation of the "Rear Window" film. 17 U.S.C. sec. 502(a) provides that the court "*may...* grant temporary and final injunctions on such terms as it may deem reasonable to prevent or restrain infringement

47. The Supreme Court affirmed this holding. Stewart v. Abend, 495 U.S. 207 (1990).

of a copyright." Defendants argue...that a finding of infringement presumptively entitles the plaintiff to an injunction, citing Professor Nimmer. See 3 M. Nimmer, Nimmer on Copyright sec. 14.06[B] at 14-55 to 14-56.2 (1988). However, Professor Nimmer also states that "where great public injury would be worked by an injunction, the courts might...award damages or a continuing royalty instead of an injunction in such special circumstances."

We believe such special circumstances exist here. The "Rear Window" film resulted from the collaborative efforts of many talented individuals other than Cornell Woolrich, the author of the underlying story. The success of the movie resulted in large part from factors completely unrelated to the underlying story, "It Had to be Murder." It would cause a great injustice for the owners of the film if the court enjoined them from further exhibition of the movie. An injunction would also effectively foreclose defendants from enjoying legitimate profits derived from exploitation of the "new matter" comprising the derivative work, which is given express copyright protection by section 7 of the 1909 Act. Since defendants could not possibly separate out the "new matter" from the underlying work, their right to enjoy the renewal copyright *in the derivative work* would be rendered meaningless by the grant of an injunction. We also note that an injunction would cause public injury by denying the public the opportunity to view a classic film for many years to come.

This is not the first time that we have recognized that an injunction may be an inappropriate remedy for copyright infringement. In Universal City Studios v. Sony Corp., we stated that Professor Nimmer's suggestion of damages or a continuing royalty would constitute an acceptable resolution for infringement caused by in-home taping of television programs by VCR—"time-shifting."

As the district court pointed out in the *Sony* case, an injunction is a "harsh and drastic" discretionary remedy, never an absolute right. Abend argues nonetheless that defendants' attempts to interfere with his production of new derivative works can only be remedied by an injunction. We disagree. Abend has not shown irreparable injury which would justify imposing the severe remedy of an injunction on defendants. Abend can be compensated adequately for the infringement by monetary compensation. 17 U.S.C. sec. 504(b) provides that the copyright owner can recover actual damages and "any profits of the infringement that are *attributable to the infringement* and are not taken into account in computing the actual damages." (Emphasis added by court.)

The district court is capable of calculating damages caused to the fair market value of plaintiff's story by the re-release of the film....

COMMENTS AND QUESTIONS

1. Why shouldn't Abend be entitled to an injunction? Given the high public value of the *Rear Window* film, isn't it reasonable to expect that he would simply license MCA to continue showing the film? Should the court really be in the business of determining the relative value of Abend's contribution to the film?

2. Protecting the moral rights of the artists offers one possible justification for injunctive relief. Would the result in *Abend* be different if the author objected to the use that was made of his work? Should it?

3. The fair use doctrine may serve as a limit on the strong form of the property rule in copyright law. This is particularly true given the Supreme Court's distinction between superseding and transformative works. The copyright owner's theoretical right to "hold up" all progress based on the copyright will not prevent transformative (and therefore fair) uses, at least in some instances. Dictum in the Supreme Court's

most recent pronouncement on the issue supports this view (and cites the Ninth Circuit decision in *Abend* with approval):

> Because the fair use enquiry often requires close questions of judgment as to the extent of permissible borrowing in cases involving parodies (or other critical works), courts may also wish to bear in mind that the goals of the copyright law, to stimulate the creation and publication of edifying matter, are not always best served by automatically granting injunctive relief when parodists are found to have gone beyond the bounds of fair use.

Campbell v. Acuff-Rose Music Co., 510 U.S. 569, 578 n.10 (1994); see also New York Times Co., Inc. v. Tasini, 533 U.S. 483, 505 (2001) (suggesting that injunctive relief might not be warranted in a case involving the continued availability of online databases of copyrighted newspaper articles).

PROBLEM

Problem 4-39: Lebbeus, a writer and artist, writes a surrealist novel that features an elaborate, futuristic torture chamber with a distinctive chair attached to moving rails on a wall. He draws a picture of the chair and uses it to illustrate the cover of his novel. Pinnacle Entertainment, a major movie studio, releases *Seven Apes*, a science-fiction movie whose plot is entirely unlike Lebbeus's novel, but which in one 90-second scene features a chair on rails strikingly similar to the one on the book cover. Assume that the court determines that Pinnacle has copied the chair from Lebbeus, and that it has no legal defense. What is the appropriate remedy? Should Lebbeus be entitled to enjoin distribution of *Seven Apes*? How would a court determine the appropriate share of profits from the movie?

5

Trademark Law

A. INTRODUCTION

1. Background

Trademarks have existed for almost as long as trade itself. Once human economies progressed to the point where a merchant class specialized in making goods for others, the people who made and sold clothing or pottery began to "mark" their wares with a word or symbol to identify the maker. Such marks — often no more than the name of the maker — have been discovered on goods from China, India, Persia, Egypt, Rome, Greece, and elsewhere, and date back as much as 4000 years.[1]

These early marks served several purposes. First, they were a form of advertising, allowing makers to get their name in front of potential customers. Second, they may have been used to prove that the goods were sold by a particular merchant, thus helping to resolve ownership disputes. Third, the marks served as a guarantee of quality, since a merchant who identifies herself with her goods puts her reputation on the line.

These functions coalesced in modern practice, where trademarks are widely viewed as devices that help to reduce information and transaction costs by allowing customers to estimate the nature and quality of goods before purchase.

Consumers rely most on trademarks when it is difficult to inspect a product quickly and cheaply to determine its quality. Many products fit this description: cars, computers, electronic equipment, even food and toys. In precisely these cases, unscrupulous competitors may be tempted to copy the trademark of a rival producer known for superior quality. After all, it is easier to copy a trademark than to duplicate production techniques, quality assurance programs, and the like.

Even the earliest trademark law cases reflect an awareness of the need to provide a legal remedy against counterfeiting.[2] Under English common law, a party who used a

1. See, e.g., William H. Browne, A Treatise on the Law of Trademarks 1-14 (1885) (tracing the history of trademarks).
2. See Frank Schechter, The Historical Foundations of the Law Relating to Trademarks (1925) (reviewing medieval origins of modern trademark law).

617

trademark was entitled to prevent subsequent use of the same mark by others selling the same types of goods. See, e.g., Sykes v. Sykes, 107 Eng. Rep. 834 (1824).

In the United States, statutory trademark law appeared late on the scene, particularly given the early legislative interest in patents and copyrights.[3] This was no doubt a function of the low volume of trade in finished goods early in the history of the Republic. As Lawrence Friedman has written,

> [T]rademark law...was relatively undeveloped in [the early nineteenth century]. No trademark infringement case was decided in the United States before 1825. Joseph Story granted the first injunction for trademark infringement, in 1844, to protect the makers of "Taylor's Persian Thread." Congress provided neither guidance nor any machinery for registration. Legal protection for designers of trademarks had to be forged in the rough mills of the courts. The economy was still deeply rooted in land and its produce. Intellectual property, despite the name, was not valued for intellectual reasons at all, but because of mercantile and industrial applications. As such, this property was not a central concern of the law until the full-blown factory age.

Lawrence Friedman, A History of American Law 257 (2nd ed. 1985).

As Kenneth J. Vandevelde, points out in The New Property of the Nineteenth Century: The Development of the Modern Concept of Property, 29 Buffalo L. Rev. 325, 341 (1980), trademarks in the eighteenth century were protected only by the common law of fraud. Beginning with Millington v. Fox, 3 My. & Cr. 338, 40 Eng. Rep. 956 (Ch. 1838), where a tradesman was permanently enjoined from using another's mark, dicta began to appear suggesting that marks were a form of property. Further support for the notion that trademark law served primarily to prevent consumer fraud can be found in Mira Wilkins, The Neglected Intangible Asset: The Influence of the Trademark on the Rise of the Modern Corporation, 34 Bus. & Hist. 66, 72 (1992), where the author writes: "[I]n the American colonies laws passed to maintain the quality of manufactured articles came to form the basis of the country's subsequent trademark legislation.... [B]ut not until the late 1840s was the first state law passed to 'prevent fraud in the use of false stamps and labels.'..."

Trademarks were protected at common law in the United States until 1870, when Congress enacted the first federal trademark statute. That statute, which grounded protection for trademark rights in the patent and copyright clause of the Constitution, was struck down by the Supreme Court as beyond the powers of Congress. The Court reasoned that the patent and copyright clause of the Constitution could not support the statute, since it protected all marks regardless of any novelty or originality. Trade-Mark Cases, 100 U.S. 82, 94 (1879). Congress reenacted limited federal trademark protection in the Act of 1881, this time basing the statute in the Commerce Clause. The trademark statute was significantly modified in the Act of 1905 and further changed by subsequent amendment in 1920. Today, trademarks are protected by the Lanham Act, 15 U.S.C. § 1051 et seq., which was enacted in 1946.

With few exceptions, the history of these statutory changes has been one of expansion of the rights of trademark owners. The Act of 1905 eliminated the requirements of identity and intention to deceive, substituting instead the more fluid test of likelihood of confusion. The Lanham Act further liberalized trademark law by providing advantages to registration of trademarks and introducing a separate statutory

---

3. By contrast, one commentator has traced the earliest common law American trademark decision as far back as 1584, to the decision in *Sandforth's Case.*

prohibition against "unfair methods of competition" that afforded protection even to unregistered marks. 15 U.S.C. § 1125(a). The result is that a broad class of "marks" now qualify for Lanham Act protection.

2. A Brief Overview of Trademark Theory

Trademarks differ in fundamental ways from the other types of intellectual property protection we have studied so far. Patents, copyrights, and trade secrets are designed to protect and/or reward something new, inventive, or creative, whether it be an idea, a physical creation, or an expression. A trademark, by contrast, does not "depend upon novelty, invention, discovery, or any work of the brain. It requires no fancy or imagination, no genius, no laborious thought." Trade-Mark Cases, 100 U.S. at 94. Rather, trademark protection is awarded merely to those who were the first to use a distinctive mark in commerce. In trademark parlance, the senior (that is, first) user of a mark may prevent junior (subsequent) users from employing the same or a similar mark, where there is a "likelihood of confusion" between the two marks.

Traditionally, there has been nothing in trademark law analogous to the desire to encourage invention or creation that underlies (at least in part) patent and copyright law. There is no explicit federal policy to encourage the creation of more trademarks.[4] Rather, the fundamental principles of trademark law have essentially been ones of tort: unfair competition and the tort of deception of the consumer.[5] In this sense, trademarks may not be thought of as analogous to "property rights" at all. See, e.g., Hanover Star Milling Co. v. Metcalf, 240 U.S. 403 (1916) and cases cited therein. Rather, they are rights acquired with the use of a trademark in commerce, and they continue only so long as that use continues.[6]

Early trademark cases (and statutes) took a restrictive view of these rights. Trademark infringement originally was limited to the use of a name or mark identical to the trademark in the sale of identical goods, where the infringer's use was intended to deceive consumers. These cases were essentially an extension of common law misrepresentation principles. Normally, if retailer *R* misrepresents to buyer *B* that *R*'s goods are made by manufacturer *M*, then buyer *B* has a cause of action for misrepresentation against *R*. This would void any contract between *R* and *B*. *M* might also have a common law cause of action against *R*, but the *consumer's* interests can be directly protected at common law only through the misrepresentation cause of action.

---

4. Some commentators discern in federal trademark law a purpose to promote interbrand competition by strengthening brands, thereby providing information to consumers and potential competitors. See William P. Kratzke, Normative Economic Analysis of Trademark Law, 21 Memphis St. U.L. Rev. 199, 212-219 (1991). To the extent that this is true, trademarks may be thought of as comporting with federal competition policy, as embodied in the antitrust laws and elsewhere. See Continental T.V., Inc. v. GTE Sylvania, 433 U.S. 36 (1977) (policy of the antitrust laws is to promote brand strength in order to promote interbrand competition).

Further, many commentators believe that recent decisions are moving the Court in the direction of an incentive-based theory of trademark law. We discuss this trend *infra*.

5. These are two very different interests held by different parties, both protected by a trademark granted to one of the parties. This combination of producer and consumer interests in trademark protection is significant, and we will return to it later in this chapter.

6. The Trademark Law Revision Act of 1988 changed this general principle in an important respect. Under that act, it is now possible to register and protect a trademark based on an *intention* to use that mark in commerce at some time within the next three years. 15 U.S.C. § 1051(b) (1988). But the right remains inchoate until use actually occurs.

One reading of these old common law unfair competition cases is that they conflated *B*'s and *M*'s discrete harms into a single cause of action. They in some sense allowed *M* to stand in *B*'s shoes and sue *R* for misrepresentation.

One way to analyze this "propertization" trend is to say that it vested rights in competitors in order to increase the likelihood that these causes of action would be brought in the first place. Individual consumers are isolated, and each generally purchases only a few of a given trademarked item. If we also consider the United States' elaborate system of interstate transportation, an individual consumer has very little incentive to police trademark infringements. The difficulty of forming a class for purposes of class action remedies is simply too great.

Competitors — *M* in the story here — have a much greater incentive to police misuse of their marks. In this view, the legitimate trademark user's lower transaction costs in policing the mark are harnessed to the original, fundamental consumer protection rationale to obtain the modern trademark infringement suit. Trademark "ownership," in this view, essentially begins as something like a legal fiction that gives the trademark owner a cause of action he would not otherwise have, in order to benefit consumers and the competitive process.

Giving the originator of a mark the right to police counterfeiting also serves to protect three types of investment: (1) investment in the creation of the mark; (2) investment in advertising and promoting the product in association with the mark; and (3) product-related investments such as high-quality raw materials, production equipment, and quality assurance techniques.

3. The Basic Economics of Trademarks and Advertising

Economists have studied advertising and promotion extensively. Despite the fact that developed economies spend more than 2 percent of their GNPs on advertising, economists have not reached consensus regarding the economic function of trademarks. See Richard Schmalensee, Advertising and Market Structure, in New Developments in the Analysis of Market Structure 373 (Joseph Stiglitz and G. Frank Matathewson eds. 1991). In an earlier era, analysis centered on the notion that advertising "unnaturally" stimulated demand and perpetuated oligopoly through "artificial" product differentiation. See, e.g., Joan Robinson, The Economics of Imperfect Competition 89 (1933); Sherwin Rosen, Advertising, Information, and Product Differentiation, in Issues in Advertising: The Economics of Persuasion 161-91 (David G. Tuerck ed. 1978) (summarizing these arguments). See generally Beverly W. Pattishall, Trademarks and the Monopoly Phobia, 50 Mich. L. Rev. 967 (1952) (dismissing these arguments).

There is something to this concern. Certainly, advertising may have the effect of differentiating in the minds of consumers products that are in fact similar or identical. The result of this brand differentiation may be that the trademark owner obtains some power over price. An example is over-the-counter drugs, where brand-name drugs regularly sell for twice the price of their "generic" equivalents, even though the two drugs are chemically identical. Thus one could argue that advertising actually hurts rather than helps consumers. See Ralph S. Brown Jr., Advertising and the Public Interest: Legal Protection of Trade Symbols, 57 Yale L.J. 1165 (1948).

Starting in the 1960s and 1970s, however, economists have viewed advertising in a much more positive light. See George Stigler, The Economics of Information, 69 J. Pol. Econ. 213 (1961). The consensus view now is that advertising cheaply conveys information to consumers. See Phillip Nelson, Advertising as Information, 82 J. Pol. Econ. 729 (1974). Cf. Thomas T. Nagle, Do Advertising-Profitability Studies Really Show That Advertising Creates a Barrier to Entry?, 24 J.L. & Econ. 333 (1981) (brand loyalty demonstrates the costliness of information; advertising and brand promotion lower these costs somewhat).

Economists studying the informational role of advertising make an important distinction, one worth keeping in mind when studying trademark law. They distinguish between a product's simple "search characteristics" (price, color, shape, size, and product category) and its more complex "experience characteristics," such as taste or long-term durability. The former are aspects of product quality that consumers can verify by inspecting the product before they buy it. It is easy to see why firms would invest in advertising these qualities: they are basic aspects of a product, and consumers can easily verify whether the claims are true.

Experience characteristics, by contrast, cannot be cheaply investigated before purchase. They will be discovered only through use. The taste of fruit or wine is an obvious example, but many other goods share the same characteristic: motor oil, house paint, cough suppressant, and computer software, to name a few.

Since consumers cannot easily verify advertising about these aspects of a product, economists wondered why advertisers spend so much money on such advertising. They have come to believe that it serves a "signaling function." Jack Hirshleifer, Where Are We in the Theory of Information?, 63 Am. Econ. Rev. 31, 35-38 (1973) (interpreting work of Phillip Nelson, *supra*); Paul Milgrom and John Roberts, Price and Advertising Signals of Product Quality, 94 J. Pol. Econ. 796 (1986). That is, "[t]he primary information content of advertisements for experience goods is the information that the brand advertises." Phillip Nelson, *supra*, at 745. The idea is that advertising pays only when consumers become repeat purchasers; consumers know this about advertising; hence advertising signals a product that the producer believes will attract repeat buyers. Cf. Benjamin Klein and Keith Leffler, The Role of Market Forces in Assuring Contractual Performance, 89 J. Pol. Econ. 615, 630-631 (1981) (advertising signals high "firm-specific selling costs," and a "high premium," hence signals presence of high-quality goods). The basic idea has borne up under recent empirical investigation. See I. P. L. Png & David Reitman, Why Are Some Products Branded and Others Not?, 38 J.L. & Econ. 207 (1995) (summarizing empirical research on branded versus unbranded sales of gasoline and other products at gasoline stations, which shows branding is more common where goods are hard to inspect and where consumer search costs are higher); Lee Benham, The Effect of Advertising on the Price of Eyeglasses, 15 J.L. & Econ. 337, 344-345 (1972) (advertising correlates with lower prices). But see John A. Rizzo, Advertising and Competition in the Ethical Pharmaceutical Industry: The Case of Antihypertensive Drugs, 42 J.L. Econ. 89 (1999) (primary effect of advertising drugs is to reduce consumer price sensitivity). While there may be something to this idea, it is inherently self-limiting: If consumers take advertising as a signal of quality, makers of low-quality products will begin to advertise too, hoping to deceive consumers into purchasing their low-quality products.

Obviously, no amount of advertising makes sense unless consumers remember the name of the product. This is where trademarks come in: they are essential shorthand. When they are effective, consumers strongly associate the trademark with the producer's product. The trademark comes to embody all of the firm's informational investments.

Often, courts and commentators speak of the economic value of consumers' associations with a firm and its trademark as the firm's *goodwill*. Goodwill can be thought of as the residual benefit the firm receives from making the three types of investments described above. The direct benefit is the "good feeling" consumers have when they see, hear, or think of the firm and/or its trademark; in economic terms, it is the probability that, based on this good feeling, customers will come back in the future. For many firms — as for Wall Street — the value of goodwill is considerable indeed. By one estimate, for example, the value of Coca-Cola's trademark independent of any of its tangible assets is $24 billion. See Industry Calls for Stiffer Enforcement of Anti-Counterfeiting Laws Abroad, 44 Pat., Trademark & Copyright J. (BNA) 585, 586, Oct. 1, 1992. See generally Russell L. Parr, The Value of Trademarks, in ALI-ABA Trademarks, Copyrights, and Unfair Competition for the General Practitioner, April 14, 1994 (Pub. No. C913 ALI-ABA 229), at 235 (using Marlboro brand as case study for trademark valuation technique; estimated value at $65 billion); Jerre B. Swann & Theodore H. Davis, Jr., Dilution, An Idea Whose Time Has Gone: Brand Equity as Protectable Property, The New/Old Paradigm, 1 J. Intell. Prop. L. 219, 229 (1994) (documenting key role played by famous trademarks — Nabisco, Winston, Miracle Whip, Burger King — in raising acquisition prices up to five times book value in recent corporate mergers).

In general, then, the "consumer protection" and "producer incentive" theories of trademark law often seem to be flip sides of the same coin. Yet, as we will see, the two theories may lead to different outcomes in certain cases. You can get a preliminary sense of this by considering a simple question: What if a producer has spent a great deal of money creating and advertising a mark, but the public has not yet begun to associate it with the product? Can a competitor use the mark on identical or similar goods? What incentives are created by a rule either way?

What if instead an accused infringer could demonstrate conclusively in a particular case that (i) the trademark owner invested nothing in the creation of the mark; and (ii) she has spent next to nothing promoting the mark or the product to which it is attached. Her success, in other words, has been entirely through repeat purchases and word of mouth. If an incentive/reward theory is at the heart of trademark, what would we be rewarding if we enforced the trademark in this case — *choice* of an effective trademark? (Note that the trademark chosen would appear to be highly successful in one sense: it fosters repeat business with little promotional spending.)

COMMENTS AND QUESTIONS

1. Note how the two economic approaches to trademarks — the older product differentiation theory (trademarks bad) and the newer product information theory (trademarks good) — follow from differing views of the economic role of advertising and promotional expenditure in general. Do you think advertising (1) communicates valuable price/quality information; (2) artificially creates demand for nonessential product features and product "image"; or (3) some combination of the two?

2. Why is a legal remedy necessary for false representations about the origins of consumer goods? If you hire a carpenter to fix your roof and he does a bad job, you

will not hire him again. What are the differences between this scenario and products purchased less often from more diverse sources? See George A. Akerlof, The Market for Lemons: Quality Uncertainty and the Market Mechanism, 84 Q.J. Econ. 488 (1970), reprinted in An Economic Theorist's Book of Tales 7 (1984) (observing that under some conditions markets for goods such as used cars may not function effectively because buyers find it difficult to test the quality of the goods offered, giving sellers an incentive to sell poor quality items, with a resultant diminution of activity across the entire market). Akerlof concludes that one way to stop the "market for lemons" dynamic, where bad (low quality) sellers drive out good ones, is through the use of brand names.

> Brand names not only indicate quality but also give the consumer means of retaliation if the quality does not meet expectations. For the consumer will then curtail future purchases. Often too, new products are associated with old brand names. This ensures the prospective consumer of the quality of the product.

Id., in An Economic Theorist's Book of Tales, at 21. Akerlof identifies other institutions that serve the same purpose: guarantees; chain stores; and government licensing, as of doctors. Could all trademark law be eliminated by mandatory warranty terms? Would consumers necessarily trust them? Would such terms lower search costs as much as brand names? How about government certification programs — e.g., "U.S. Grade A Refrigerators"? If the idea is that consumers need certification of quality levels, why wouldn't they prefer independent, third-party certification? Under what circumstances do consumers demand just that? Do they do so, e.g., for airlines, doctors, lawyers? Why in some cases and not others? (Consider the costs of a bad product choice.)

Alternatively, private organizations might be expected to spring up to provide unbiased evaluation of products for a fee. Such organizations exist. Groups like Consumer's Union sell their evaluations of products, and their reputation depends on continued accuracy and integrity in product investigation. Other examples of private, third-party certification include Good Housekeeping magazine and Underwriters Laboratories, both of which give "seals of approval" to certain products and services that meet their standards for quality.

3. As you will see as you work your way through this chapter, trademark law has been expanding in recent years. Trade dress, antidilution protection, and other developments are all part of the trend. This expansion has moved trademark's conceptual center of gravity well beyond its traditional moorings in "consumer confusion." For cogent critiques of this movement — often called the "propertization" of trademark — see Mark A. Lemley, The Modern Lanham Act and the Death of Common Sense, 108 Yale L.J. 1687 (1999) and Glynn S. Lunney, Jr., Trademark Monopolies, 48 Emory L.J. 367 (1999). Lunney summarizes things this way:

> [M]any courts and commentators succumbed to "property mania" — the belief that expanded trademark protection was necessarily desirable so long as the result could be characterized as "property." The result has been a radical and ongoing expansion of trademark protection, both in terms of what can be owned as a trademark and in terms of what trademark ownership entails. This expansion, and its associated reinterpretation of trademark's underlying policies, presents a serious threat to social welfare and has placed at risk the competitive balance that deception-based trademark law originally established. Like deception-based trademark protection, property-based trademark

protection can enable a trademark owner to differentiate her product and exclude others from using the differentiating feature. It can thereby cede control over distinct product markets to individual producers and generate for a trademark owner the downward sloping demand curve of a monopolist. However, unlike deception-based trademark, property-based trademark has only a tenuous relationship to consumer deception, and therefore lacks the offsetting efficiency advantages associated with deception-based trademark's quality control and certification functions. As a result, property-based trademark appears presumptively anticompetitive — it generates market power and associated efficiency losses without the offsetting efficiency gains that are thought to justify deception-based trademark.

Id. at 372. Critics of expansive trademark law address one form of trademark expansion: the strengthening of the bundle of rights associated with a particular trademark. However, trademark owners often engage in another type of expansion: once a trademark is established for one type of product, they try to "leverage the brand" by selling other products under the same trademark. This has become especially popular in recent years as businesspeople and consultants awake to the economic power that accompanies established brands. See, e.g., Brands: The New Wealth Creators (Susannah Hart & John Murphy, eds. 1998).

B. WHAT CAN BE PROTECTED AS A TRADEMARK?

What exactly is a trademark? As noted above, the first trademarks were simply names or identifying symbols attached to goods. Names, symbols, and logos remain important trademarks, but they have been joined by a host of other sorts of trademarks. Company names now exist alongside product names. Slogans or phrases qualify for protection as trademarks. The design of a product itself or its packaging may be distinctive "trade dress" entitled to protection under the Lanham Act.

Of course, not all identifying marks are entitled to protection. If all a party needed to do to gain exclusive use of a mark or name was to use it, product comparisons would be virtually impossible because the very words that describe a product would be appropriated to a specific company. Suppose, for example, that Ford owned the exclusive right to describe its products as "cars" or "automobiles." Customers looking for automated means of ground transportation offered by other companies might encounter difficulty in knowing what to ask for. See Kenneth L. Port, Foreword: Symposium on Intellectual Property Law Theory, 68 Chicago-Kent L. Rev. 585, 596-598 (1993).

To avoid this problem, only certain trademarks are entitled to legal protection. Whether an identifying name or phrase may be trademarked at all, and the degree of protection accorded to it, both depend on the "strength" of the mark. This in turn depends on, among other things, the "classification" of the mark as either (1) arbitrary, (2) suggestive, (3) descriptive, or (4) generic. An arbitrary or fanciful mark is a word or phrase that bears no relationship whatsoever to the product it describes. "Exxon" is a good example of a fanciful mark. Arbitrary and fanciful marks are the strongest, because any value they possess in terms of name recognition obviously comes from the corporate use of the name, rather than the natural association in people's minds between a name and a product. The other three categories decrease in strength as they increase in natural association: "suggestive" marks suggest a product in people's minds; "descriptive" marks describe the product or service offered; "generic" marks are so associated with a particular product class that they have become the natural way to refer to that type of product.

1. Trademarks, Trade Names, and Service Marks

To the layperson, trademarks are often thought of as the public name of a producer or other business. In fact, however, the Lanham Act distinguishes in form between several different types of marks. "Trademarks" are the words, phrases, logos, and symbols that producers use to identify their goods:

> **Trademark.** The term "trademark" includes any word, name, symbol, or device, or any combination thereof—
> (1) used by a person, or
> (2) which a person has a bona fide intention to use in commerce and applies to register on the principal register established by this Act,
> to identify and distinguish his or her goods, including a unique product, from those manufactured or sold by others and to indicate the source of the goods, even if that source is unknown.

Lanham Act § 45, 15 U.S.C. § 1127.

The term *trademark* does not cover another closely associated type of business identifier, the service mark. Service marks serve the same purposes as trademarks, but they are used to identify services rather than goods. The Act defines service marks as follows:

> **Service Mark:** The term "service mark" means any word, name, symbol, or device, or any combination thereof—
> (1) used by a person, or
> (2) which a person has a bona fide intention to use in commerce and applies to register on the principal register established by this Act,
> to identify and distinguish the services of one person, including a unique service, from the services of others and to indicate the source of the services, even if that source is unknown. Titles, character names, and other distinctive features of radio and television programs may be registered as service marks notwithstanding that they, or the programs, may advertise the goods of the sponsor.

Lanham Act § 45, 15 U.S.C. § 1127. In general, service marks are subject to many of the same rules as trademarks, e.g., the rules on establishing priority of use. See, e.g., Martahus v. Video Duplication Services, Inc., 3 F.3d 417 (Fed. Cir. 1993) (cancellation of service mark on grounds that similar service mark had priority).

One issue that sometimes arises involves attempts to register service marks for services that are closely related to the sale of goods. In general, where the services are "expected or routine" in connection with the goods, such registrations are rejected. The theory is that, otherwise, closely related registrations will proliferate, clogging the register. See, e.g., In re Dr. Pepper Co., 836 F.2d 508 (Fed. Cir. 1987) (affirming trademark office rejection of service mark for conducting contests in connection with sale of soft drinks).

A third category of marks is trade names. Rather than goods or services, trade names identify the company itself. Unlike trademarks and service marks, trade names cannot be registered under the Lanham Act unless they actually function to identify the source of particular goods or services, rather than merely identifying a company. See Bell v. Streetwise Records, Ltd., 761 F.2d 67, 75 (1st Cir. 1985). However, trade names are generally registerable in state offices, and state and federal common law may provide protection against confusingly similar company names.

COMMENTS AND QUESTIONS

Although the Lanham Act defines both trade names and service marks, it allows registration of only the latter. Why?

2. Color, Fragrance, and Sounds

Qualitex Co. v. Jacobson Products Co., Inc.
Supreme Court of the United States
514 U.S. 159 (1995)

Justice BREYER delivered the opinion of the Court.

The question in this case is whether the Lanham Trademark Act... permits the registration of a trademark that consists, purely and simply, of a color. We conclude that, sometimes, a color will meet ordinary legal trademark requirements. And, when it does so, no special legal rule prevents color alone from serving as a trademark.

I

The case before us grows out of petitioner Qualitex Company's use (since the 1950's) of a special shade of green-gold color on the pads that it makes and sells to dry cleaning firms for use on dry cleaning presses. In 1989 respondent Jacobson Products (a Qualitex rival) began to sell its own press pads to dry cleaning firms; and it colored those pads a similar green-gold. In 1991 Qualitex registered the special green-gold color on press pads with the Patent and Trademark Office as a trademark. Registration No. 1,633,711 (Feb. 5, 1991). Qualitex subsequently added a trademark infringement count... in a lawsuit it had already filed challenging Jacobson's use of the green-gold color.

Qualitex won the lawsuit in the District Court. 21 U.S.P.Q.2d 1457, 1991 WL 318798 (CD Cal. 1991). But, the Court of Appeals for the Ninth Circuit set aside the judgment in Qualitex's favor on the trademark infringement claim because, in that Circuit's view, the Lanham Act does not permit Qualitex, or anyone else, to register "color alone" as a trademark. 13 F.3d 1297, 1300, 1302 (1994).

The courts of appeals have differed as to whether or not the law recognizes the use of color alone as a trademark. Compare NutraSweet Co. v. Stadt Corp., 917 F.2d 1024, 1028 (CA7 1990) (absolute prohibition against protection of color alone), with In re Owens-Corning Fiberglas Corp., 774 F.2d 1116, 1128 (CA Fed. 1985) (allowing registration of color pink for fiberglass insulation). . . . Therefore, this Court granted certiorari. . . . We now hold that there is no rule absolutely barring the use of color alone, and we reverse the judgment of the Ninth Circuit.

II

The Lanham Act gives a seller or producer the exclusive right to "register" a trademark . . . and to prevent his or her competitors from using that trademark. . . . Both the language of the Act and the basic underlying principles of trademark law would seem to include color within the universe of things that can qualify as a trademark. The language of the Lanham Act describes that universe in the broadest of terms. It says that trademarks "includ[e] any word, name, symbol, or device, or any combination thereof." §1127. Since human beings might use as a "symbol" or "device" almost anything at all that is capable of carrying meaning, this language, read literally, is not restrictive. The courts and the Patent and Trademark Office have authorized for use as a mark a particular shape (of a Coca-Cola bottle), a particular sound (of NBC's three chimes), and even a particular scent (of plumeria blossoms on sewing thread). See, e.g., Registration No. 696,147 (Apr. 12, 1960); Registration Nos. 523,616 (Apr. 4, 1950) and 916,522 (July 13, 1971); In re Clarke, 17 U.S.P.Q.2d 1238, 1240 (TTAB 1990). If a shape, a sound, and a fragrance can act as symbols why, one might ask, can a color not do the same?

. . . True, a product's color is unlike "fanciful," "arbitrary," or "suggestive" words or designs, which almost automatically tell a customer that they refer to a brand. . . . [S]ee Two Pesos, Inc. v. Taco Cabana, Inc., 112 S. Ct. 2753, 2757 (1992). The imaginary word "Suntost," or the words "Suntost Marmalade," on a jar of orange jam immediately would signal a brand or a product "source"; the jam's orange color does not do so. But, over time, customers may come to treat a particular color on a product or its packaging (say, a color that in context seems unusual, such as pink on a firm's insulating material or red on the head of a large industrial bolt) as signifying a brand. And, if so, that color would have come to identify and distinguish the goods—i.e. to "indicate" their "source"—much in the way that descriptive words on a product (say, "Trim" on nail clippers or "Car-Freshner" on deodorizer) can come to indicate a product's origin. . . . In this circumstance, trademark law says that the word (e.g., "Trim"), although not inherently distinctive, has developed "secondary meaning." See Inwood Laboratories, Inc. v. Ives Laboratories, Inc., 456 U.S. 844, 851, n.11, 102 S. Ct. 2182, 2187, n.11, 72 L. Ed. 2d 606 (1982) ("secondary meaning" is acquired when "in the minds of the public, the primary significance of a product feature . . . is to identify the source of the product rather than the product itself"). Again, one might ask, if trademark law permits a descriptive word with secondary meaning to act as a mark, why would it not permit a color, under similar circumstances, to do the same?

We cannot find in the basic objectives of trademark law any obvious theoretical objection to the use of color alone as a trademark, where that color has attained "secondary meaning" and therefore identifies and distinguishes a particular brand (and thus indicates its "source"). In principle, trademark law, by preventing others from copying a source-identifying mark, "reduce[s] the customer's costs of shopping and making purchasing decisions," 1 J. McCarthy, McCarthy on Trademarks and

Unfair Competition § 2.01[2], p. 2-3 (3d ed. 1994) (hereinafter McCarthy), for it quickly and easily assures a potential customer that this item — the item with this mark — is made by the same producer as other similarly marked items that he or she liked (or disliked) in the past. At the same time, the law helps assure a producer that it (and not an imitating competitor) will reap the financial, reputation-related rewards associated with a desirable product. The law thereby "encourage[s] the production of quality products," ibid., and simultaneously discourages those who hope to sell inferior products by capitalizing on a consumer's inability quickly to evaluate the quality of an item offered for sale.... It is the source-distinguishing ability of a mark — not its ontological status as color, shape, fragrance, word, or sign — that permits it to serve these basic purposes. See Landes & Posner, Trademark Law: An Economic Perspective, 30 J. Law & Econ. 265, 290 (1987). And, for that reason, it is difficult to find, in basic trademark objectives, a reason to disqualify absolutely the use of a color as a mark.

Neither can we find a principled objection to the use of color as a mark in the important "functionality" doctrine of trademark law. The functionality doctrine prevents trademark law, which seeks to promote competition by protecting a firm's reputation, from instead inhibiting legitimate competition by allowing a producer to control a useful product feature. It is the province of patent law, not trademark law, to encourage invention by granting inventors a monopoly over new product designs or functions for a limited time, 35 U.S.C. § § 154, 173, after which competitors are free to use the innovation. If a product's functional features could be used as trademarks, however, a monopoly over such features could be obtained without regard to whether they qualify as patents and could be extended forever (because trademarks may be renewed in perpetuity). See Kellogg Co. v. National Biscuit Co., 305 U.S. 111, 119-120, 59 S. Ct. 109, 113-114...(1938) (Brandeis, J.); Inwood Laboratories, Inc., supra, 456 U.S., at 863, 102 S. Ct., at 2193 (White, J., concurring in result) ("A functional characteristic is 'an important ingredient in the commercial success of the product,' and, after expiration of a patent, it is no more the property of the originator than the product itself") (citation omitted). Functionality doctrine therefore would require, to take an imaginary example, that even if customers have come to identify the special illumination-enhancing shape of a new patented light bulb with a particular manufacturer, the manufacturer may not use that shape as a trademark, for doing so, after the patent had expired, would impede competition — not by protecting the reputation of the original bulb maker, but by frustrating competitors' legitimate efforts to produce an equivalent illumination-enhancing bulb. See, e.g., Kellogg Co., supra, 305 U.S., at 119-120, 59 S. Ct., at 113-114 (trademark law cannot be used to extend monopoly over "pillow" shape of shredded wheat biscuit after the patent for that shape had expired). This Court consequently has explained that, "[i]n general terms, a product feature is functional," and cannot serve as a trademark, "if it is essential to the use or purpose of the article or if it affects the cost or quality of the article," that is, if exclusive use of the feature would put competitors at a significant non-reputation-related disadvantage. Inwood Laboratories, Inc., 456 U.S., at 850, n.10, 102 S. Ct., at 2186, n.10. Although sometimes color plays an important role (unrelated to source identification) in making a product more desirable, sometimes it does not. And, this latter fact — the fact that sometimes color is not essential to a product's use or purpose and does not affect cost or quality — indicates that the doctrine of "functionality" does not create an absolute bar to the use of color alone as a mark. See Owens-Corning, 774 F.2d, at 1123 (pink color of insulation in wall "performs no nontrademark function").

It would seem, then, that color alone, at least sometimes, can meet the basic legal requirements for use as a trademark. It can act as a symbol that distinguishes a firm's goods and identifies their source, without serving any other significant function. See U.S. Dept. of Commerce, Patent and Trademark Office, Trademark Manual of Examining Procedures 1202.04(e), p. 1202-13 (2d ed. May, 1993) (hereinafter PTO Manual) (approving trademark registration of color alone where it "has become distinctive of the applicant's goods in commerce," provided that "there is [no] competitive need for colors to remain available in the industry" and the color is not "functional"); see also 1 McCarthy §§ 3.01[1], 7.26 ("requirements for qualification of a word or symbol as a trademark" are that it be (1) a "symbol," (2) "use[d] . . . as a mark," (3) "to identify and distinguish the seller's goods from goods made or sold by others," but that it not be "functional"). Indeed, the District Court, in this case, entered findings (accepted by the Ninth Circuit) that show Qualitex's green-gold press pad color has met these requirements. The green-gold color acts as a symbol. Having developed secondary meaning (for customers identified the green-gold color as Qualitex's), it identifies the press pads' source. And, the green-gold color serves no other function. (Although it is important to use some color on press pads to avoid noticeable stains, the court found "no competitive need in the press pad industry for the green-gold color, since other colors are equally usable." 21 U.S.P.Q.2d, at 1460, 1991 WL 318798.) Accordingly, unless there is some special reason that convincingly militates against the use of color alone as a trademark, trademark law would protect Qualitex's use of the green-gold color on its press pads.

III

Respondent Jacobson Products says that there are four special reasons why the law should forbid the use of color alone as a trademark. We shall explain, in turn, why we, ultimately, find them unpersuasive.

First, Jacobson says that, if the law permits the use of color as a trademark, it will produce uncertainty and unresolvable court disputes about what shades of a color a competitor may lawfully use. Because lighting (morning sun, twilight mist) will affect perceptions of protected color, competitors and courts will suffer from "shade confusion" as they try to decide whether use of a similar color on a similar product does, or does not, confuse customers and thereby infringe a trademark. Jacobson adds that the "shade confusion" problem is "more difficult" and "far different from" the "determination of the similarity of words or symbols." . . .

We do not believe, however, that color, in this respect, is special. Courts traditionally decide quite difficult questions about whether two words or phrases or symbols are sufficiently similar, in context, to confuse buyers. They have had to compare, for example, such words as "Bonamine" and "Dramamine" (motion-sickness remedies); "Huggies" and "Dougies" (diapers); "Cheracol" and "Syrocol" (cough syrup); "Cyclone" and "Tornado" (wire fences); and "Mattres" and "1-800-Mattres" (mattress franchisor telephone numbers) . . . . Legal standards exist to guide courts in making such comparisons. See, e.g., 2 McCarthy § 15.08; 1 McCarthy §§ 11.24-11.25 ("[S]trong" marks, with greater secondary meaning, receive broader protection than "weak" marks). We do not see why courts could not apply those standards to a color, replicating, if necessary, lighting conditions under which a colored product is normally sold. . . .

Second, Jacobson argues, as have others, that colors are in limited supply. See, e.g., NutraSweet Co., 917 F.2d, at 1028; Campbell Soup Co. v. Armour & Co., 175 F.2d 795, 798 (CA3 1949). Jacobson claims that, if one of many competitors can appropriate a particular color for use as a trademark, and each competitor then tries to do the same, the supply of colors will soon be depleted. Put in its strongest form, this argument would concede that "[h]undreds of color pigments are manufactured and thousands of colors can be obtained by mixing." L. Cheskin, Colors: What They Can Do For You 47 (1947). But, it would add that, in the context of a particular product, only some colors are usable. By the time one discards colors that, say, for reasons of customer appeal, are not usable, and adds the shades that competitors cannot use lest they risk infringing a similar, registered shade, then one is left with only a handful of possible colors. And, under these circumstances, to permit one, or a few, producers to use colors as trademarks will "deplete" the supply of usable colors to the point where a competitor's inability to find a suitable color will put that competitor at a significant disadvantage.

This argument is unpersuasive, however, largely because it relies on an occasional problem to justify a blanket prohibition. When a color serves as a mark, normally alternative colors will likely be available for similar use by others. See, e.g., Owens-Corning, 774 F.2d, at 1121 (pink insulation). Moreover, if that is not so — if a "color depletion" or "color scarcity" problem does arise — the trademark doctrine of "functionality" normally would seem available to prevent the anticompetitive consequences that Jacobson's argument posits, thereby minimizing that argument's practical force.

The functionality doctrine, as we have said, forbids the use of a product's feature as a trademark where doing so will put a competitor at a significant disadvantage because the feature is "essential to the use or purpose of the article" or "affects [its] cost or quality." Inwood Laboratories, Inc., 456 U.S., at 850, n.10, 102 S. Ct., at 2186, n.10. The functionality doctrine thus protects competitors against a disadvantage (unrelated to recognition or reputation) that trademark protection might otherwise impose, namely their inability reasonably to replicate important non-reputation-related product features. For example, this Court has written that competitors might be free to copy the color of a medical pill where that color serves to identify the kind of medication (e.g., a type of blood medicine) in addition to its source. See id., at 853, 858, n.20, 102 S. Ct., at 2188, 2190, n.20 ("[S]ome patients commingle medications in a container and rely on color to differentiate one from another"); see also J. Ginsburg, D. Goldberg, & A. Greenbaum, Trademark and Unfair Competition Law 194-195 (1991) (noting that drug color cases "have more to do with public health policy" regarding generic drug substitution "than with trademark law"). And, the federal courts have demonstrated that they can apply this doctrine in a careful and reasoned manner, with sensitivity to the effect on competition. Although we need not comment on the merits of specific cases, we note that lower courts have permitted competitors to copy the green color of farm machinery (because customers wanted their farm equipment to match) and have barred the use of black as a trademark on outboard boat motors (because black has the special functional attributes of decreasing the apparent size of the motor and ensuring compatibility with many different boat colors).... The Restatement (Third) of Unfair Competition adds that, if a design's "aesthetic value" lies in its ability to "confe[r] a significant benefit that cannot practically be duplicated by the use of alternative designs," then the design is "functional." Restatement (Third) of Unfair Competition § 17, Comment c, pp. 175-176 (1995). The "ultimate test of aesthetic functionality," it explains, "is whether the recognition of trademark rights would significantly hinder competition." Id., at 176.

The upshot is that, where a color serves a significant nontrademark function—whether to distinguish a heart pill from a digestive medicine or to satisfy the "noble instinct for giving the right touch of beauty to common and necessary things," G.K. Chesterton, Simplicity and Tolstoy 61 (1912)—courts will examine whether its use as a mark would permit one competitor (or a group) to interfere with legitimate (nontrademark-related) competition through actual or potential exclusive use of an important product ingredient. That examination should not discourage firms from creating aesthetically pleasing mark designs, for it is open to their competitors to do the same. See, e.g., W. T. Rogers Co. v. Keene, 778 F.2d 334, 343 (CA7 1985) (Posner, J.). But, ordinarily, it should prevent the anticompetitive consequences of Jacobson's hypothetical "color depletion" argument, when, and if, the circumstances of a particular case threaten "color depletion."

Third, Jacobson points to many older cases—including Supreme Court cases—in support of its position. [The Court distinguishes these as arising before the Lanham Act of 1946.]

Fourth, Jacobson argues that there is no need to permit color alone to function as a trademark because a firm already may use color as part of a trademark, say, as a colored circle or colored letter or colored word, and may rely upon "trade dress" protection, under § 43(a) of the Lanham Act, if a competitor copies its color and thereby causes consumer confusion regarding the overall appearance of the competing products or their packaging, see 15 U.S.C. § 1125(a) (1988 ed., Supp. V). The first part of this argument begs the question. One can understand why a firm might find it difficult to place a usable symbol or word on a product (say, a large industrial bolt that customers normally see from a distance); and, in such instances, a firm might want to use color, pure and simple, instead of color as part of a design. Neither is the second portion of the argument convincing. Trademark law helps the holder of a mark in many ways that "trade dress" protection does not. See 15 U.S.C. § 1124 (ability to prevent importation of confusingly similar goods); § 1072 (constructive notice of ownership); § 1065 (incontestable status); § 1057(b) (prima facie evidence of validity and ownership). Thus, one can easily find reasons why the law might provide trademark protection in addition to trade dress protection.

IV

Having determined that a color may sometimes meet the basic legal requirements for use as a trademark and that respondent Jacobson's arguments do not justify a special legal rule preventing color alone from serving as a trademark (and, in light of the District Court's here undisputed findings that Qualitex's use of the green-gold color on its press pads meets the basic trademark requirements), we conclude that the Ninth Circuit erred in barring Qualitex's use of color as a trademark. For these reasons, the judgment of the Ninth Circuit is

Reversed.

COMMENTS AND QUESTIONS

1. The Court mentions "NBC's three chimes" as an example of a sound that is registered as a trademark under the Lanham Act. Another famous example is MGM's "lion's roar," usually heard at the beginning of an MGM film. See also Harley Wants Roar of Engine Protected by a Trademark, Sacramento Bee, Mar. 27, 1996, p. D6 (describing Harley-Davidson trademark application for engine sound opposed by competitors). Compare this with section 43(a) cases alleging imitation of famous voices, e.g., Midler v. Ford Motor Co., 849 F.2d 460 (9th Cir. 1988), *cert. denied*, 503 U.S. 951 (1992); see Chapter 6.

2. Justice Breyer's opinion for a unanimous Court in *Qualitex* is quite important not only for resolution of the color issue but also as a statement of current Court thinking on the cutting-edge issue of the functionality defense. For more on this topic, see *infra*.

3. Certification and Collective Marks

For the most part, the "source" identified by a trademark is a single company or individual. But for two special types of marks — certification and collective marks — this is not the case.

Lanham Act §45 (15 U.S.C. §1127) defines a certification mark as follows.

> The term "certification mark" means any word, name, symbol, or device, or any combination thereof—
> (1) used by a person other than its owner, or
> (2) which its owner has a bona fide intention to permit a person other than the owner to use in commerce and files an application to register on the principal register established by this chapter,

to certify regional or other origin, material, mode of manufacture, quality, accuracy, or other characteristics of such person's goods or services or that the work or labor on the goods or services was performed by members of a union or other organization.

Certification marks are generally used by trade associations or other commercial groups to identify a particular type of goods. For example, the city of Roquefort, France, holds a certification mark in "Roquefort" as a sheep's milk cheese cured in the limestone caves of Roquefort, France. See Community of Roquefort v. William Faehndrich, Inc., 303 F.2d 494 (2d Cir. 1962) (enjoining use of the term "Imported Roquefort Cheese" on cheese not made in Roquefort, France). Certification marks cannot be limited to a single producer; they must be open to anyone who meets the standards set forth for certification.

Certification marks serve to certify conformity with centralized standards. See, e.g., Levy v. Kosher Overseers Association of America Inc., 36 U.S.P.Q.2d 1724 (S.D.N.Y. 1995), rev'd 104 F.3d 38 (2d Cir. 1997) (case involving plaintiff, Organized Kashruth Laboratories, suing another for infringement of its well-known kosher certification mark, the "circle K" found on many kosher foods). Because trademarks are thought by many to have grown out of trade guilds, which had much the same quality-control function, it could be argued that certification marks were the first true modern trademarks. See Frank Schechter, The Historical Foundations of the Law Relating to Trade-marks 47 (1925).

Certification marks are meant to bear the "seal of approval" of a central organization, so they can be cancelled on the ground that the organization no longer exercises sufficient control over its members to ensure consistent product standards. See 15 U.S.C. § 1064 (providing that a certification mark may be cancelled if not policed effectively); see American Angus Association v. Sysco Corp., 865 F. Supp. 1180 (W.D.N.C. 1993) (denying cancellation standing to trademark infringement defendant who sought to cancel plaintiff's beef quality certification mark).

Collective marks are defined as follows in § 45 of the Lanham Act (15 U.S.C. § 1127):

> The term "collective mark" means a trademark or service mark —
>
> (1) used by the members of a cooperative, an association, or other collective group or organization, or
>
> (2) which such cooperative, association, or other collective group or organization has a bona fide intention to use in commerce and applies to register on the principal register established by this chapter, and includes marks indicating membership in a union, an association, or other organization.

See also Aloe Creme Laboratories, Inc. v. American Society for Aesthetic Plastic Surgery, Inc., 192 U.S.P.Q. 170, (TTAB 1976), a decision denying an opposition to registration of a collective mark:

> There are two basic types of collective marks. A collective trademark or collective service mark is a mark adopted by a "collective" (i.e., an association, union, cooperative, fraternal organization, or other organized collective group) for use only by its members, who in turn use the mark to identify their goods or services and distinguish them from those of nonmembers. The "collective" itself neither sells goods nor performs services under a collective trademark or collective service mark, but the collective may advertise or otherwise promote the goods or services sold or rendered by its members under the mark. A collective membership mark is a mark adopted for the purpose of indicating membership in an organized collective group, such as a union, an association, or other organization. Neither the collective nor its members uses the collective membership mark to identify and distinguish goods or services; rather, the sole function of such a mark is to indicate that the person displaying the mark is a member of the organized collective group. For example, if the collective group is a fraternal organization, members may display the mark by wearing pins or rings upon which the mark appears, by carrying membership cards bearing the mark, etc. Of course, a collective group may itself be engaged in the marketing of its own goods or services under a particular mark, in which case the mark is not a collective mark but is rather a trademark for the collective's goods or service mark for the collective's services.

192 U.S.P.Q. 170, 173-174.

Collective marks of the first type — those attached to goods and services — are useful in franchising and related arrangements where individual stores or outlets are at least somewhat independent from the central organization holding the collective mark. For the most part, these types of collective marks are treated the same as ordinary trademarks. See, e.g., Sebastian Intl. v. Long's Drug Stores Corp., 53 F.3d

1073 (9th Cir. 1995), *cert. denied,* 116 S. Ct. 302 (1995) (holding trademark "first sale" doctrine applicable to sales under a collective mark).

4. Trade Dress and Product Configurations

Words or phrases that serve to identify a product are not all that the Lanham Act protects. The act also protects "trade dress," the design and packaging of materials, and even the design and shape of a product itself, if the packaging or the product configuration serve the same source-identifying function as trademarks. It is possible to register both trade dress and product configurations as "trademarks" under the Lanham Act. (Indeed, such a registration was at issue in the *Qualitex* case discussed above.) However, because of their complex and changing nature, most trade dress and product configurations are protected without registration under section 43(a) of the Lanham Act, 15 U.S.C. § 1125(a).

§ 1125 [Lanham Act § 43]. False Designations of Origin and False Descriptions Forbidden

(a)(1) Any person who, on or in connection with any goods or services, or any container for goods, uses in commerce any word, term, name, symbol, or device, or any combination thereof, or any false designation of origin, false or misleading description of fact, or false or misleading representation of fact, which —

(A) Is likely to cause confusion, or to cause mistake, or to deceive as to the affiliation, connection, or association of such person with another person, or as to the origin, sponsorship, or approval of his or her goods, services, or commercial activities by another person, or

(B) in commercial advertising or promotion, misrepresents the nature, characteristics, qualities, or geographic origin of his or her or another person's goods, services, or commercial activities,

shall be liable in a civil action by any person who believes that he or she is or is likely to be damaged by such act.

Section 43(a) is commonly referred to as providing "federal common law" protection for trademarks and related source identifiers. In the next section, we discuss the requirements for protecting a trademark, either under the Lanham Act's registration procedures or under section 43(a).

C. ESTABLISHMENT OF TRADEMARK RIGHTS

1. Distinctiveness

a. *Classification of Marks and Requirements for Protection*

When a trademark is immediately capable of identifying a unique product source, rights to the mark are determined solely by priority of use. (See for example the

Zazu Designs case at the beginning of section 2.) Marks such as these are labeled "inherently distinctive," though for analytical completeness they are further subdivided into arbitrary, fanciful, and suggestive marks. For all other trademarks — those deemed not inherently distinctive — the Lanham Act requires proof of an additional element to secure trademark rights: secondary meaning.

The *Zatarain's* case that follows explores the hierarchy of trademark classifications and illustrates why classifications matter.

The most important type of word or symbol requiring proof of secondary meaning is the *descriptive* trademark. A descriptive mark is defined as "[a] word, picture, or other symbol that directly describes something about the goods or services in connection with which it is used as a mark." J. Thomas McCarthy, McCarthy's Desk Encyclopedia of Intellectual Property 119 (2d ed. 1995). Examples include: Tender Vittles for cat food, Arthriticare for arthritis treatment, and Investacorp for financial services.

In addition to descriptive marks, several other categories of marks require secondary meaning to acquire legal protection: most notably *geographic marks* (such as Nantucket soft drinks), and *personal name marks* (such as O'Malley's beer).[7]

Secondary meaning exists when buyers associate a descriptive term with a single source of products. Thus when consumers recognize the Tender Vittles brand of cat food — when they expect the can so labeled to be of that brand — this descriptive term is functioning as a trademark. To be sure, Tender Vittles retains its primary meaning as a product descriptor. But proof that it has acquired a secondary meaning as a source identifier elevates it to trademark status.

It is important to understand the nature of this secondary meaning. It does not mean that buyers need to know the *identity* of the source, only that the product or service comes from a *single* source. The phrase "single source" may thus be understood to mean "single though anonymous source." See A. J. Canfield Co. v. Honickman, 808 F.2d 291 (3d Cir. 1986).

The next case is a leading example of the application of these and related concepts.

Zatarain's, Inc. v. Oak Grove Smokehouse, Inc.
United States Court of Appeals for the Fifth Circuit
698 F.2d 786 (5th Cir. 1983)

GOLDBERG, Circuit Judge:

This appeal of a trademark dispute presents us with a menu of edible delights sure to tempt connoisseurs of fish and fowl alike. At issue is the alleged infringement of two trademarks, "Fish-Fri" and "Chick-Fri," held by appellant Zatarain's, Inc. ("Zatarain's"). The district court held that the alleged infringers had a "fair use" defense to any asserted infringement of the term "Fish-Fri" and that the registration of the term "Chick-Fri" should be cancelled. We affirm.

---

7. In addition, the following require proof of secondary meaning: titles of single literary works; descriptive titles of literary series; non-inherently distinctive designs and symbols; non-inherently distinctive trade dress and packaging; and product and container shapes. McCarthy, Trademarks and Unfair Competition § 15.01[2].

I. Facts and Proceedings Below

A. *The Tale of the Town Frier*

Zatarain's is the manufacturer and distributor of a line of over one hundred food products. Two of these products, "Fish-Fri" and "Chick-Fri," are coatings or batter mixes used to fry foods. These marks serve as the entree in the present litigation.

Zatarain's "Fish-Fri" consists of 100% corn flour and is used to fry fish and other seafood. "Fish-Fri" is packaged in rectangular cardboard boxes containing twelve or twenty-four ounces of coating mix. The legend "Wonderful FISH-FRI™" is displayed prominently on the front panel, along with the block Z used to identify all Zatarain's products. The term "Fish-Fri" has been used by Zatarain's or its predecessor since 1950 and has been registered as a trademark since 1962.

Zatarain's "Chick-Fri" is a seasoned corn flour batter mix used for frying chicken and other foods. The "Chick-Fri" package, which is very similar to that used for "Fish-Fri," is a rectangular cardboard container labelled "Wonderful CHICK-FRI." Zatarain's began to use the term "Chick-Fri" in 1968 and registered the term as a trademark in 1976.

Zatarain's products are not alone in the marketplace. At least four other companies market coatings for fried foods that are denominated "fish fry" or "chicken fry." Two of these competing companies are the appellees here, and therein hangs this fish tale.

Appellee Oak Grove Smokehouse, Inc. ("Oak Grove") began marketing a "fish fry" and a "chicken fry" in March 1979. Both products are packaged in clear glassine packets that contain a quantity of coating mix sufficient to fry enough food for one meal. The packets are labelled with Oak Grove's name and emblem, along with the words "FISH FRY" or "CHICKEN FRY." Oak Grove's "FISH FRY" has a corn flour base seasoned with various spices; Oak Grove's "CHICKEN FRY" is a seasoned coating with a wheat flour base.

B. *Out of the Frying Pan, Into the Fire*

Zatarain's first claimed foul play in its original complaint filed against Oak Grove on June 19, 1979, in the United States District Court for the Eastern District of Louisiana. The complaint alleged trademark infringement and unfair competition under the Lanham Act §§32(1), 43(a), 15 U.S.C. §§1114(1), 1125(a) (1976), and La. Rev. Stat. Ann. §51:1405(A) (West Supp. 1982).

The district court found that Zatarain's trademark "Fish-Fri" was a descriptive term with an established secondary meaning, but held that Oak Grove and Visko's had a "fair use" defense to their asserted infringement of the mark. The court further found that Zatarain's trademark "Chick-Fri" was a descriptive term that lacked secondary meaning, and accordingly ordered the trademark registration cancelled.

Battered, but not fried, Zatarain's appeals from the adverse judgment on several grounds. First, Zatarain's argues that its trademark "Fish-Fri" is a suggestive term and therefore not subject to the "fair use" defense. Second, Zatarain's asserts that even if the "fair use" defense is applicable in this case, appellees cannot invoke the doctrine because their use of Zatarain's trademarks is not a good faith attempt to describe their products. Third, Zatarain's urges that the district court erred in cancelling the trademark registration for the term "Chick-Fri" because Zatarain's presented sufficient evidence to establish a secondary meaning for the term. For these reasons, Zatarain's argues that the district court should be reversed....

III. The Trademark Claims

A. Basic Principles

1. Classifications of Marks

The threshold issue in any action for trademark infringement is whether the work or phrase is initially registerable or protectable. Vision Center v. Opticks, Inc., 596 F.2d 111, 115 (5th Cir. 1980); American Heritage Life Insurance Co. v. Heritage Life Insurance Co., 494 F.2d 3, 10 (5th Cir. 1974). Courts and commentators have traditionally divided potential trademarks into four categories. A potential trademark may be classified as (1) generic, (2) descriptive, (3) suggestive, or (4) arbitrary or fanciful. These categories, like the tones in a spectrum, tend to blur at the edges and merge together. The labels are more advisory than definitional, more like guidelines than pigeonholes. Not surprisingly, they are somewhat difficult to articulate and to apply. Soweco, Inc. v. Shell Oil Co., 617 F.2d 1178, 1183 (5th Cir. 1980); Vision Center, 596 F.2d at 115.

A generic term is "the name of a particular genus or class of which an individual article or service is but a member." Vision Center, 596 F.2d at 115; Abercrombie & Fitch Co. v. Hunting World, Inc., 537 F.2d 4, 9 (2d Cir. 1976). A generic term connotes the "basic nature of articles or services" rather than the more individualized characteristics of a particular product. American Heritage, 494 F.2d at 11. Generic terms can never attain trademark protection. William R. Warner & Co. v. Eli Lilly & Co., 265 U.S. 526, 528, 44 S. Ct. 615, 616, 68 L. Ed. 1161 (1924); Soweco, 617 F.2d at 1183; Vision Center, 596 F.2d at 115. Furthermore, if at any time a registered trademark becomes generic as to a particular product or service, the mark's registration is subject to cancellation. Lanham Act § 14, 15 U.S.C. § 1064(c) (1976). Such terms as aspirin and cellophane have been held generic and therefore unprotectable as trademarks. See Bayer Co. v. United Drug Co., 272 F. 505 (S.D.N.Y. 1921) (aspirin); DuPont Cellophane Co. v. Waxed Products Co., 85 F.2d 75 (2d Cir. 1936) (cellophane).

A descriptive term "identifies a characteristic or quality of an article or service," Vision Center, 596 F.2d at 115, such as its color, odor, function, dimensions, or ingredients. American Heritage, 494 F.2d at 11. Descriptive terms ordinarily are not protectable as trademarks, Lanham Act § 2(e)(1), 15 U.S.C. § 1052(e)(1) (1976); they may become valid marks, however, by acquiring a secondary meaning in the minds of the consuming public. See id. § 2(f), 15 U.S.C. § 1052(f). Examples of descriptive marks would include "Alo" with reference to products containing gel of the aloe vera plant, Aloe Creme Laboratories, Inc. v. Milsan, Inc., 423 F.2d 845 (5th Cir. 1970), and "Vision Center" in reference to a business offering optical goods and services, Vision Center, 596 F.2d at 117. As this court has often noted, the distinction between descriptive and generic terms is one of degree. Soweco, 617 F.2d at 1184; Vision Center, 596 F.2d at 115 n.11 (citing 3 R. Callman, The Law of Unfair

Competition, Trademarks and Monopolies §70.4 (3d ed. 1969)); American Heritage, 494 F.2d at 11. The distinction has important practical consequences, however; while a descriptive term may be elevated to trademark status with proof of secondary meaning, a generic term may never achieve trademark protection. Vision Center, 596 F.2d at 115 n.11.

A suggestive term suggests, rather than describes, some particular characteristic of the goods or services to which it applies and requires the consumer to exercise the imagination in order to draw a conclusion as to the nature of the goods and services. Soweco, 617 F.2d at 1184; Vision Center, 596 F.2d at 115-116. A suggestive mark is protected without the necessity for proof of secondary meaning. The term "Coppertone" has been held suggestive in regard to sun tanning products. See Douglas Laboratories, Inc. v. Copper Tan, Inc., 210 F.2d 453 (2d Cir. 1954).

Arbitrary or fanciful terms bear no relationship to the products or services to which they are applied. Like suggestive terms, arbitrary and fanciful marks are protectable without proof of secondary meaning. The term "Kodak" is properly classified as a fanciful term for photographic supplies, see Eastman Kodak Co. v. Weil, 137 Misc. 506, 243 N.Y.S. 319 (1930) ("Kodak"); "Ivory" is an arbitrary term as applied to soap. Abercrombie & Fitch, 537 F.2d at 9 n.6.

2. Secondary Meaning

As noted earlier, descriptive terms are ordinarily not protectable as trademarks. They may be protected, however, if they have acquired a secondary meaning for the consuming public. The concept of secondary meaning recognizes that words with an ordinary and primary meaning of their own "may [after] long use with a particular product, come to be known by the public as specifically designating that product." Volkswagenwerk Aktiengesellschaft v. Rickard, 492 F.2d 474, 477 (5th Cir. 1974). In order to establish a secondary meaning for a term, a plaintiff "must show that the primary significance of the term in the minds of the consuming public is not the product but the producer." Kellogg Co. v. National Biscuit Co., 305 U.S. 111, 118, 59 S. Ct. 109, 113, 83 L. Ed. 73 (1938). The burden of proof to establish secondary meaning rests at all times with the plaintiff; this burden is not an easy one to satisfy, for "[a] high degree of proof is necessary to establish secondary meaning for a descriptive term." Vision Center, 596 F.2d at 118 (quoting 3 R. Callman, supra, §77.3, at 359). Proof of secondary meaning is an issue only with respect to descriptive marks; suggestive and arbitrary or fanciful marks are automatically protected upon registration, and generic terms are unprotectable even if they have acquired secondary meaning. See Soweco, 617 F.2d at 1185 n.20.

3. The "Fair Use" Defense

Even when a descriptive term has acquired a secondary meaning sufficient to warrant trademark protection, others may be entitled to use the mark without incurring liability for trademark infringement. When the allegedly infringing term is "used fairly and in good faith only to describe to users the goods or services of [a] party, or their geographic origin," Lanham Act §33(b)(4), 15 U.S.C. §1115(b)(4) (1976), a defendant in a trademark infringement action may assert the "fair use" defense. The defense is available only in actions involving descriptive terms and only when the term is used in its descriptive sense rather than its trademark sense. Soweco, 617 F.2d at 1185; see Venetianaire Corp. v. A & P Import Co., 429 F.2d 1079, 1081-1082 (2d Cir. 1970). In essence, the fair use defense prevents a trademark registrant from appropriating a descriptive term for its own use to the exclusion of others, who may be prevented thereby from accurately describing their own goods. Soweco, 617 F.2d at 1185. The holder of a protectable descriptive mark has no legal claim to an exclusive right in the primary, descriptive meaning of the term; consequently, anyone is free to use the term in its primary, descriptive sense so long as such use does not lead to customer confusion as to the source of the goods or services. See 1 J. McCarthy, Trademarks and Unfair Competition § 11.17, at 379 (1973).

4. Cancellation of Trademarks

Section 37 of the Lanham Act, 15 U.S.C. § 1119 (1976), provides as follows:

> In any action involving a registered mark the court may determine the right to registration, order the cancellation of registrations, in whole or in part, restore cancelled registrations, and otherwise rectify the register with respect to the registrations of any party to the action. Decrees and orders shall be certified by the court to the Commissioner, who shall make appropriate entry upon the records of the Patent Office, and shall be controlled thereby.

This circuit has held that when a court determines that a mark is either a generic term or a descriptive term lacking secondary meaning, the purposes of the Lanham Act are well served by an order cancelling the mark's registration. American Heritage, 494 F.2d at 14.

We now turn to the facts of the instant case.

B. "FISH-FRI"[3]

1. Classification

Throughout this litigation, Zatarain's has maintained that the term "Fish-Fri" is a suggestive mark automatically protected from infringing uses by virtue of its registration in 1962. Oak Grove and Visko's assert that "fish fry" is a generic term identifying a class of foodstuffs used to fry fish; alternatively, Oak Grove and Visko's argue that "fish fry" is merely descriptive of the characteristics of the product. The district court found that "Fish-Fri" was a descriptive term identifying a function of the product being sold. Having reviewed this finding under the appropriate "clearly erroneous" standard, we affirm. See Vision Center, 596 F.2d at 113.

We are mindful that "[t]he concept of descriptiveness must be construed rather broadly." 3 R. Callman, supra, § 70.2. Whenever a word or phrase conveys an immediate idea of the qualities, characteristics, effect, purpose, or ingredients of a product or service, it is classified as descriptive and cannot be claimed as an exclusive trademark. Id. § 71.1; see Stix Products, Inc. v. United Merchants & Manufacturers,

---

3. We note at the outset that Zatarain's use of the phonetic equivalent of the words "fish fry" — that is, misspelling it — does not render the mark protectable. Soweco, 617 F.2d at 1186 n.24.

Inc., 295 F.Supp. 479, 488 (S.D.N.Y. 1968). Courts and commentators have formulated a number of tests to be used in classifying a mark as descriptive.

A suitable starting place is the dictionary, for "[t]he dictionary definition of the word is an appropriate and relevant indication 'of the ordinary significance and meaning of words' to the public." American Heritage, 494 F.2d at 11 n.5; see also Vision Center, 596 F.2d at 116. Webster's Third New International Dictionary 858 (1966) lists the following definitions for the term "fish fry": "1. a picnic at which fish are caught, fried, and eaten; . . . 2. fried fish." Thus, the basic dictionary definitions of the term refer to the preparation and consumption of fried fish. This is at least preliminary evidence that the term "Fish-Fri" is descriptive of Zatarain's product in the sense that the words naturally direct attention to the purpose or function of the product.

The "imagination test" is a second standard used by the courts to identify descriptive terms. This test seeks to measure the relationship between the actual words of the mark and the product to which they are applied. If a term "requires imagination, thought and perception to reach a conclusion as to the nature of goods," Stix Products, 295 F. Supp. at 488, it is considered a suggestive term. Alternatively, a term is descriptive if standing alone it conveys information as to the characteristics of the product. In this case, mere observation compels the conclusion that a product branded "Fish-Fri" is a prepackaged coating or batter mix applied to fish prior to cooking. The connection between this merchandise and its identifying terminology is so close and direct that even a consumer unfamiliar with the product would doubtless have an idea of its purpose or function. It simply does not require an exercise of the imagination to deduce that "Fish-Fri" is used to fry fish. See Vision Center, 596 F.2d at 116-17; Stix Products, 295 F. Supp. at 487-88. Accordingly, the term "Fish-Fri" must be considered descriptive when examined under the "imagination test."

A third test used by courts and commentators to classify descriptive marks is "whether competitors would be likely to need the terms used in the trademark in describing their products." Union Carbide Corp. v. Ever-Ready, Inc., 531 F.2d 366, 379 (7th Cir. 1976). A descriptive term generally relates so closely and directly to a product or service that other merchants marketing similar goods would find the term useful in identifying their own goods. Vision Center, 596 F.2d at 116-17; Stix Products, 295 F. Supp. at 488. Common sense indicates that in this case merchants other than Zatarain's might find the term "fish fry" useful in describing their own particular batter mixes. While Zatarain's has argued strenuously that Visko's and Oak Grove could have chosen from dozens of other possible terms in naming their coating mix, we find this position to be without merit. As this court has held, the fact that a term is not the only or even the most common name for a product is not determinative, for there is no legal foundation that a product can be described in only one fashion. Vision Center, 596 F.2d at 117 n.17. There are many edible fish in the sea, and as many ways to prepare them as there are varieties to be prepared. Even piscatorial gastronomes would agree, however, that frying is a form of preparation accepted virtually around the world, at restaurants starred and unstarred. The paucity of synonyms for the words "fish" and "fry" suggests that a merchant whose batter mix is specially spiced for frying fish is likely to find "fish fry" a useful term for describing his product.

A final barometer of the descriptiveness of a particular term examines the extent to which a term actually has been used by others marketing a similar service or product. Vision Center, 596 F.2d at 117; Shoe Corp. of America v. Juvenile Shoe Corp., 266 F.2d 793, 796 (C.C.P.A. 1959). This final test is closely related to the question whether competitors are likely to find a mark useful in describing their products. As noted above, a number of companies other than Zatarain's have chosen the word combination "fish fry" to identify their batter mixes. Arnaud's product, "Oyster Shrimp and Fish Fry," has been in competition with Zatarain's "Fish-Fri" for some ten to twenty years. When companies from A to Z, from Arnaud to Zatarain's, select the same term to describe their similar products, the term in question is most likely a descriptive one.

The correct categorization of a given term is a factual issue, Soweco, 617 F.2d at 1183 n.12; consequently, we review the district court's findings under the "clearly erroneous" standard of Fed. R. Civ. P. 52. See Vision Center, 596 F.2d at 113; Volkswagenwerk, 492 F.2d at 478. The district court in this case found that Zatarain's trademark "Fish-Fri" was descriptive of the function of the product being sold. Having applied the four prevailing tests of descriptiveness to the term "Fish-Fri," we are convinced that the district court's judgment in this matter is not only not clearly erroneous, but clearly correct....

2. Secondary Meaning

Descriptive terms are not protectable by trademark absent a showing of secondary meaning in the minds of the consuming public.[5] To prevail in its trademark infringement action, therefore, Zatarain's must prove that its mark "Fish-Fri" has acquired a secondary meaning and thus warrants trademark protection. The district court found that Zatarain's evidence established a secondary meaning for the term "Fish-Fri" in the New Orleans area. We affirm.

The existence of secondary meaning presents a question for the trier of fact, and a district court's finding on the issue will not be disturbed unless clearly erroneous. American Heritage, 494 F.2d at 13; Volkswagenwerk, 492 F.2d at 477. The burden of proof rests with the party seeking to establish legal protection for the mark — the plaintiff in an infringement suit. Vision Center, 596 F.2d at 118. The evidentiary burden necessary to establish secondary meaning for a descriptive term is substantial. Id.; American Heritage, 494 F.2d at 12; 3 R. Callman, supra, § 77.3, at 359.

In assessing a claim of secondary meaning, the major inquiry is the consumer's attitude toward the mark. The mark must denote to the consumer "a single thing coming from a single source," Coca-Cola Co. v. Koke Co., 254 U.S. 143, 146, 41 S. Ct. 113, 114, 65 L. Ed. 189 (1920); Aloe Creme Laboratories, 423 F.2d at 849, to support a finding of secondary meaning. Both direct and circumstantial evidence may be relevant and persuasive on the issue.

Factors such as amount and manner of advertising, volume of sales, and length and manner of use may serve as circumstantial evidence relevant to the issue of secondary meaning. See, e.g., Vision Center, 596 F.2d at 119; Union Carbide Corp., 531 F.2d at 380; Aloe Creme Laboratories, 423 F.2d at 849-50. While none of these factors alone will prove secondary meaning, in combination they may establish the

---

5. A mark that has become "incontestable" under section 15 of the Lanham Act, 15 U.S.C. § 1065 (1976), cannot be challenged as lacking secondary meaning, although it is subject to seven statutory defenses. See id. § 33(b), 15 U.S.C. § 1115(b). In order for a registrant's mark to be deemed "incontestable," the registrant must use the mark for five continuous years following the registration date and must file certain affidavits with the Commissioner of Patents. Id. § 15, 15 U.S.C. § 1065. No evidence in the record indicates that Zatarain's has satisfied the requirements of "incontestability"; consequently, we must determine whether proof of secondary meaning otherwise exists.

necessary link in the minds of consumers between a product and its source. It must be remembered, however, that "the question is not the extent of the promotional efforts, but their effectiveness in altering the meaning of [the term] to the consuming public." Aloe Creme Laboratories, 423 F.2d at 850.

Since 1950, Zatarain's and its predecessor have continuously used the term "Fish-Fri" to identify this particular batter mix. Through the expenditure of over $400,000 for advertising during the period from 1976 through 1981, Zatarain's has promoted its name and its product to the buying public. Sales of twelve-ounce boxes of "Fish-Fri" increased from 37,265 cases in 1969 to 59,439 cases in 1979. From 1964 through 1979, Zatarian's sold a total of 916,385 cases of "Fish-Fri." The district court considered this circumstantial evidence of secondary meaning to weigh heavily in Zatarain's favor.

In addition to these circumstantial factors, Zatarain's introduced at trial two surveys conducted by its expert witness, Allen Rosenzweig. In one survey, telephone interviewers questioned 100 women in the New Orleans area who fry fish or other seafood three or more times per month. Of the women surveyed, twenty-three percent specified Zatarain's "Fish-Fri" as a product they "would buy at the grocery to use as a coating" or a "product on the market that is especially made for frying fish." In a similar survey conducted in person at a New Orleans area mall, twenty-eight of the 100 respondents answered "Zatarain's 'Fish-Fri'" to the same questions....

The authorities are in agreement that survey evidence is the most direct and persuasive way of establishing secondary meaning. Vision Center, 596 F.2d at 119; Aloe Creme Laboratories, 423 F.2d at 849; 1 J. McCarthy, supra, § 15.12(D). The district court believed that the survey evidence produced by Zatarain's, when coupled with the circumstantial evidence of advertising and usage, tipped the scales in favor of a finding of secondary meaning. Were we considering the question of secondary meaning de novo, we might reach a different conclusion than did the district court, for the issue is close.

Mindful, however, that there is evidence in the record to support the finding below, we cannot say that the district court's conclusion was clearly erroneous. Accordingly, the finding of secondary meaning in the New Orleans area for Zatarain's descriptive term "Fish-Fri" must be affirmed.

3. The "Fair Use" Defense

Although Zatarain's term "Fish-Fri" has acquired a secondary meaning in the New Orleans geographical area, Zatarain's does not now prevail automatically on its trademark infringement claim, for it cannot prevent the fair use of the term by Oak Grove and Visko's. The "fair use" defense applies only to descriptive terms and requires that the term be "used fairly and in good faith only to describe to users the goods or services of such party, or their geographic origin." Lanham Act § 33(b), 15 U.S.C. § 1115(b)(4) (1976). The district court determined that Oak Grove and Visko's were entitled to fair use of the term "fish fry" to describe a characteristic of their goods; we affirm that conclusion.

Zatarain's term "Fish-Fri" is a descriptive term that has acquired a secondary meaning in the New Orleans area. Although the trademark is valid by virtue of having acquired a secondary meaning, only that penumbra or fringe of secondary meaning is given legal protection. Zatarain's has no legal claim to an exclusive right in the original, descriptive sense of the term; therefore, Oak Grove and Visko's are still free to use the words "fish fry" in their ordinary, descriptive sense, so long as such use will not tend to confuse customers as to the source of the goods. See 1 J. McCarthy, supra, § 11.17.

The record contains ample evidence to support the district court's determination that Oak Grove's and Visko's use of the words "fish fry" was fair and in good faith. Testimony at trial indicated that the appellees did not intend to use the term in a trademark sense and had never attempted to register the words as a trademark. Oak Grove and Visko's apparently believed "fish fry" was a generic name for the type of coating mix they manufactured. In addition, Oak Grove and Visko's consciously packaged and labelled their products in such a way as to minimize any potential confusion in the minds of consumers. The dissimilar trade dress of these products prompted the district court to observe that confusion at the point of purchase — the grocery shelves — would be virtually impossible. Our review of the record convinces us that the district court's determinations are correct. We hold, therefore, that Oak Grove and Visko's are entitled to fair use of the term "fish fry" to describe their products; accordingly, Zatarain's claim of trademark infringement must fail.

C. "CHICK-FRI"

1. Classification

Most of what has been said about "Fish-Fri" applies with equal force to Zatarain's other culinary concoction, "Chick-Fri." "Chick-Fri" is at least as descriptive of the act of frying chicken as "Fish-Fri" is descriptive of frying fish. It takes no effort of the imagination to associate the term "Chick-Fri" with Southern fried chicken. Other merchants are likely to want to use the words "chicken fry" to describe similar products, and others have in fact done so. Sufficient evidence exists to support the district court's finding that "Chick-Fri" is a descriptive term; accordingly, we affirm.

2. Secondary Meaning

The district court concluded that Zatarain's had failed to establish a secondary meaning for the term "Chick-Fri." We affirm this finding. The mark "Chick-Fri" has been in use only since 1968; it was registered even more recently, in 1976. In sharp contrast to its promotions with regard to "Fish-Fri," Zatarain's advertising expenditures for "Chick-Fri" were mere chickenfeed; in fact, Zatarain's conducted no direct advertising campaign to publicize the product. Thus the circumstantial evidence presented in support of a secondary meaning for the term "Chick-Fri" was paltry.

Allen Rosenzweig's survey evidence regarding a secondary meaning for "Chick-Fri" also "lays an egg." The initial survey question was a "qualifier": "Approximately how many times in an average month do you, yourself, fry fish or other seafood?" Only if respondents replied "three or more times a month" were they asked to continue the survey. This qualifier, which may have been perfectly adequate for purposes of the "Fish-Fri" questions, seems highly unlikely to provide an adequate sample of potential consumers of "Chick-Fri." This survey provides us with nothing more than some data regarding fish friers' perceptions about products used for frying chicken. As such, it is entitled to little evidentiary weight.[10]

It is well settled that Zatarain's, the original plaintiff in this trademark infringement action, has the burden of proof to establish secondary meaning for its term. Vision Center, 596 F.2d at 118; American Heritage, 494 F.2d at 12. This it has failed to do. The district court's finding that the term "Chick-Fri" lacks secondary meaning is affirmed.

COMMENTS AND QUESTIONS

1. Do not be misled by *Zatarain's* reference to the "fair use" of a trademark. The trademark "fair use" defense is a totally different and much more limited defense than its counterpart in copyright law. Fair use in the trademark context establishes that a junior user will not be liable for using a mark in its descriptive sense, as opposed to trading on the senior user's established trademark meaning. While many argue that "fair use" applies only to descriptive marks that have acquired secondary meaning, the leading trademark commentator has argued for a broader role for the doctrine. See 1 J. McCarthy, Trademarks and Unfair Competition § 11.17[1], at 11-80 (3d ed. 1992 & Supp. 1996). Professor McCarthy's position was endorsed by the Second Circuit in Car-Freshner Corp. v. S. C. Johnson & Son, 70 F.3d 267 (2d Cir. 1995).

An example of the application of trademark's fair use doctrine can be found in Cosmetically Sealed Industries, Inc. v. Chesebrough-Pond's USA, 125 F.3d 28 (2d Cir. 1997). There the plaintiff registered the trademark "Sealed With a Kiss" for a brand of long-lasting lipstick. Plaintiff sued after the defendant began an advertising campaign for its own brand of lipstick that encouraged users to place a lipstick "kiss" on a postcard and mail it to someone. The defendant's campaign used the phrase "Seal it with a Kiss!!" The court held that the phrase "sealed with a kiss" was in common use and that the defendant was not liable because it merely used that common phrase in its descriptive (rather than its trademark) sense.

The doctrine of fair use is an affirmative defense to a claim of trademark infringement. In KP Permanent Makeup v. Lasting Impression I, 125 S.Ct. 542 (2004), the Supreme Court confirmed that that defense is not negated by demonstrating a likelihood of consumer confusion. If fair use could be asserted only in the absence of consumer confusion, it would not serve as a defense at all, since as we will see, proof of consumer confusion is necessary to make out a claim of trademark infringement. While the Court held that fair uses were permissible even when they caused some confusion, it declined to specify how much confusion might render a use unfair. On remand, the Ninth Circuit held that while no longer determinative, consumer confusion was still relevant in deciding whether a use was fair. 408 F.3d 596 (9th Cir. 2005).

2. In the course of its opinion, the court states: "Oak Grove and Visko's apparently believed 'fish fry' was a generic name for the type of coating mix they

---

10. Even were we to accept the results of the survey as relevant, the result would not change. In the New Orleans area, only 11 of the 100 respondents in the telephone survey named "Chick-Fri," "chicken fry," or Zatarain's "Chick-Fri" as a product used as a coating for frying chicken. Rosenzweig himself testified that this number was inconclusive for sampling purposes. Thus the survey evidence cannot be said to establish a secondary meaning for the term "Chick-Fri."

manufactured." A generic term is a term that indicates what *type* or *category* a product belongs to, rather than indicating the producer or other source of the product. A generic term for a product "tell[s] the buyer what it is, not where it came from." 2 J. McCarthy, Trademarks and Unfair Competition § 12.01[1] (3d ed. 1992 & Supp. 1996). Examples would include "bed," "chair," and "pants." (Contrast these with "Eas-a-bed," "ComfortMax Chairs," and "Aunt's Pants"—all at least capable of indicating the source of a product.) A generic term has no distinctiveness vis-à-vis source, so it cannot serve as a trademark. For more on generic trademarks, including the process by which formerly distinctive terms become generic ("genericide"), see section E.1 below.

3. Note that in the case, the survey sample seems to be drawn from a very small area—the New Orleans metropolitan region. Should secondary meaning in this area permit a descriptive mark to be enforced anywhere, or only in this area? Is this in effect a geographic limitation on the scope of a trademark that is not apparent from the statute—which provides ostensibly national protection for registered marks? How does this geographic limitation interact with the doctrines governing the conflict between junior common law users of a mark and a senior, federally registered mark owner?

A recent case discusses geographical issues related to secondary meaning. Adray v. Adry-Mart, Inc., 76 F.3d 984 (9th Cir. 1996), *amending* 68 F.3d 362 (9th Cir. 1995), centered around rival operators of electronics stores in Southern California (Lou Adray and Adry-Mart), both of whom used the name "Adray's." In 1979 Adry-Mart acquired from Lou Adray's relatives the right to use the Adray's trademark in stores in Los Angeles County. Lou Adray continued to operate his Orange County store, as he had since 1968. In 1989, however, Adry-Mart began to expand to areas first near, and then inside, Orange County.

In this appeal from Lou Adray's trademark infringement suit, the Ninth Circuit held that the district court was correct in instructing the jury that for Lou to receive damages he must show secondary meaning in the market areas in which Adry-Mart operated its Orange County stores. The court observed that there was no possibility that Lou could show national secondary meaning. Also, the court noted, the alleged infringer had not adopted its mark in bad faith in this case. The court went on to hold, however, that it was clear error for the district court to find that Adry-Mart's market included all of Los Angeles County; the evidence established that in some parts of the county Lou Adray had a bigger market share than Adry-Mart. Given the overlap of sales and advertising in these areas, it is likely that neither Lou nor Adry-Mart had secondary meaning there. On remand, the court ordered, the district court should reexamine the issue of where each party had established secondary meaning.

To achieve registration and therefore nationwide protection, the Trademark Office has generally required applicants for federal registration (which confers presumptively nationwide protection) to show more than secondary meaning in a limited area. See Phillip Morris, Inc. v. Liggett & Myers Tobacco Co., 139 U.S.P.Q. 240 (TTAB 1963).

Compare *Adray* to another case, Fuddruckers, Inc. v. Doc's B.R. Others, Inc., 826 F.2d 837 (9th Cir. 1987). Here defendant Doc's negotiated with Fuddruckers for a Fuddruckers franchise, but the negotiations fell through and Fuddruckers found another franchisee. In mid-June of 1983, Fuddruckers announced plans to open a restaurant in Phoenix in December of that year. Defendants knew of Fuddruckers's plans but proceeded to open a restaurant in Phoenix with trade dress quite similar to that of Fuddruckers. As Fuddruckers's trade dress (restaurant décor) was not inherently distinctive, Fuddruckers was required to show that its trade dress had acquired secondary meaning in its action under §43(a) of the Lanham Act. The court ruled, however, that Fuddruckers did not have to show secondary meaning *in Arizona* prior to Doc's opening its restaurant there: "Fuddruckers is a national restaurant chain, and restaurant customers travel. Fuddruckers should be permitted to show that its trade dress had acquired secondary meaning among some substantial portion of consumers nationally." 826 F.2d at 844.

4. What is the proper role of "circumstantial evidence" of secondary meaning, such as advertising expenditures, the commercial success of the product, and attempts at imitation? Such evidence is generally allowed by courts in cases where secondary meaning is at issue. But should it be? If it is clear, for example, that advertising expenditures have been completely ineffective in swaying the public, is the fact of such expenditures relevant? The answer may depend on your views as to why we are protecting trademarks. If our goal is to provide incentives for businesses to invest in marks (and therefore in quality control), we may want to encourage such expenditures directly.

Note that some courts (particularly the Second Circuit) have explicitly done away with the need to prove secondary meaning where plaintiffs can point to some other equitable factor in their favor, such as intentional deception or "palming off" by the defendants. These courts reason that if the defendant is intentionally copying the plaintiff's mark, the defendant at least must think there is some accumulated goodwill to appropriate. This view is known as the New York rule, although there is some question as to its continued vitality in New York state courts. For a detailed discussion of the New York rule, see 2 J. Thomas McCarthy, Trademarks and Unfair Competition §15.04. What does the New York rule suggest about the rationale for trademark protection? Is it incentive-based, property based, or tort based?

The test of "secondary meaning" is very fact-specific and relies on the reactions of consumers to a mark, generally as tested through consumer surveys. We will see the use of consumer reaction as a test again when we consider trademark infringement. Does it make sense to rely on such empirical evidence to establish a party's intellectual property rights, particularly given the problems with most consumer surveys? Is there a better alternative?

5. *Foreign Descriptive Terms.* In some circumstances, it may not even be clear what a descriptive term is. Consider the problem of terms that are descriptive in a language other than English. The doctrine of foreign equivalents holds that foreign words must be translated into English for purposes of determining their protectability. But application of this doctrine has been uneven. Is "La Posada" (Spanish for "inn") descriptive of lodging services? Does it matter whether a substantial portion of the clientele speaks Spanish? See In re Pan Tex Hotel Corp., 190 U.S.P.Q. 109 (TTAB 1976) (La Posada is not descriptive because it is unlikely that consumers will translate the name into English); Palm Bay Imports v. Veuve Clicquot, 396 F.3d 1369 (Fed. Cir. 2005) ("Veuve Clicquot" not confusingly similar to "The Widow," since most American consumers won't know that "veuve" is French for "widow"). Compare In re Hag Aktiengesellschaft, 155 U.S.P.Q. 598 (TTAB 1967) ("Kaba," meaning coffee, is descriptive of coffee). Similarity of meaning in translation is important, but not determinative, in deciding the issue of descriptiveness. The relation in sight and sound between the English and foreign terms is also important; "a much closer approximation [between the meaning of the foreign term

and the English equivalent] is necessary to justify a refusal to register on that basis alone where the marks otherwise are totally dissimilar." In re Sarkli, Ltd., 220 U.S.P.Q. 111, 113 (Fed. Cir. 1983).

6. *Acronyms.* One possible way around a finding that a term is descriptive (or generic) is to alter the term, either by misspelling it or by using an acronym. This, and not an intrinsic aversion to proper spelling, explains the profusion of product names with words like "Evr," "EZ" and "Klear." Acronyms are occasionally, but not normally, effective in creating a distinctive mark. The relevant question for the courts is whether the misspelling or acronym has the same connotation as the original descriptive or generic mark. If it does, the acronym is not entitled to protection. Thus, ROM is generic because it conveys the same meaning to the listener as "read-only memory." Intel Corp. v. Radiation, Inc., 184 U.S.P.Q. 54 (TTAB 1984). By contrast, "L.A." was not merely descriptive of "low alcohol" beer. Anheuser-Busch, Inc. v. Stroh Brewery Co., 750 F.2d 631 (8th Cir. 1984) ("if some operation of the imagination is required to connect the initials with the product, the initials cannot be equated with the generic phrase, but are suggestive in nature, thereby rendering them protectable."). See generally McCarthy, Trademarks and Unfair Competition 12-72 to 12-75 (1993).

7. *Personal Names.* Like descriptive marks, personal names (e.g., "Jones Antiques") need to acquire secondary meaning in order to obtain trademark protection. But does the acquisition of trademark status of a personal name by one enterprise preclude a second comer from using his or her own name? Although early cases recognized an absolute right to use one's name in business, "the more recent trend is to forbid any use of the name as part of the proprietor's trademark, permitting use only in a subsidiary capacity, and . . . with the first name attached [in equal size] . . . . In either event, the junior user has almost uniformly been bound to display negative disclaimers." Basile, S.p.A. v. Basile, 899 F.2d 35, 38 (D.C. Cir. 1990). Thus, the courts have developed a limited fair use defense tailored to personal names along the general contours of the fair use defense for descriptive terms discussed in *Zatarain's.*

8. *Geographic Marks.* Geographic marks must also generally establish secondary meaning in order to receive trademark protection. The issue of what constitutes a geographic mark, however, raises various complications. For example, is Philadelphia Cream Cheese a geographic mark if the product is not made in or sold primarily in Philadelphia? The Trademark Office and the courts have developed an elaborate structure distinguishing between descriptive and misdescriptive geographic names. We will examine these doctrines in detail. See Section 3 (Trademark Office Procedures).

Even beyond the threshold for protection, can the establishment of trademark protection in a geographic mark through secondary meaning (e.g., "New York Times") prevent other companies from using geographic designations? As with descriptive names and personal names, the courts have developed a fair use doctrine for analyzing such uses, balancing a merchant's interest in accurately describing its location against the interest of the senior users and the consumer. The junior user must confine and adapt geographical usage so as to avoid likelihood of confusion. Courts will often accommodate these competing interests by requiring junior users to employ disclaimers, prefixes, suffixes, and other means of reducing confusion. See generally McCarthy, Trademarks and Unfair Competition § 14.07.

9. *Prefixes.* The Trademark Trial and Appeal Board has held that use of the letter "e" as a prefix in front of a commonly known and understood word — such as

"eFashion"—yields a descriptive term for an online retailer. See In re Styleclick, 57 USPQ2d 1445 (TTAB 2000). In a companion case, the TTAB held that use of the word "virtual" in front of a common word—as in "Virtual Fashion"—also results in a descriptive term. See In re Styleclick, 57 USPQ2d 1523 (TTAB 2000).

b. Distinctiveness of Trade Dress and Product Configuration

≡≡≡≡ **Two Pesos, Inc. v. Taco Cabana, Inc.**
Supreme Court of the United States
505 U.S. 763 (1992)

Justice WHITE delivered the opinion of the Court.

The issue in this case is whether the trade dress [1] of a restaurant may be protected under § 43(a) of the Trademark Act of 1946 (Lanham Act), 60 Stat. 441, 15 U.S.C. § 1125(a) (1982 ed.), based on a finding of inherent distinctiveness, without proof that the trade dress has secondary meaning.

I

Respondent Taco Cabana, Inc., operates a chain of fast-food restaurants in Texas. The restaurants serve Mexican food. The first Taco Cabana restaurant was opened in San Antonio in September 1978, and five more restaurants had been opened in San Antonio by 1985. Taco Cabana describes its Mexican trade dress as "a festive eating atmosphere having interior dining and patio areas decorated with artifacts, bright colors, paintings and murals. The patio includes interior and exterior areas with the interior patio capable of being sealed off from the outside patio by overhead garage doors. The stepped exterior of the building is a festive and vivid color scheme using top border paint and neon stripes. Bright awnings and umbrellas continue the theme." 932 F.2d 1113, 1117 (CA5 1991).

In December 1985, a Two Pesos, Inc., restaurant was opened in Houston. Two Pesos adopted a motif very similar to the foregoing description of Taco Cabana's trade dress. Two Pesos restaurants expanded rapidly in Houston and other markets, but did not enter San Antonio. In 1986, Taco Cabana entered the Houston and Austin markets and expanded into other Texas cities, including Dallas and El Paso where Two Pesos was also doing business.

In 1987, Taco Cabana sued Two Pesos in the United States District Court for the Southern District of Texas for trade dress infringement under § 43(a) of the Lanham Act, 15 U.S.C. § 1125(a) (1982 ed.), and for theft of trade secrets under Texas

---

1. The District Court instructed the jury: "'Trade dress' is the total image of the business. Taco Cabana's trade dress may include the shape and general appearance of the exterior of the restaurant, the identifying sign, the interior kitchen floor plan, the decor, the menu, the equipment used to serve food, the servers' uniforms and other features reflecting on the total image of the restaurant." The Court of Appeals accepted this definition and quoted from Blue Bell Bio-Medical v. Cin-Bad, Inc., 864 F.2d 1253, 1256 (CA5 1989): "The 'trade dress' of a product is essentially its total image and overall appearance." See 932 F.2d 1113, 1118 (CA5 1991). It "involves the total image of a product and may include features such as size, shape, color or color combinations, texture, graphics, or even particular sales techniques." John H. Harland Co. v. Clarke Checks, Inc., 711 F.2d 966, 980 (CA11 1983). Restatement (Third) of Unfair Competition § 16, Comment a (Tent. Draft No. 2, Mar. 23, 1990).

common law. The case was tried to a jury, which was instructed to return its verdict in the form of answers to five questions propounded by the trial judge. The jury's answers were: Taco Cabana has a trade dress; taken as a whole, the trade dress is nonfunctional; the trade dress is inherently distinctive; the trade dress has not acquired a secondary meaning in the Texas market; and the alleged infringement creates a likelihood of confusion on the part of ordinary customers as to the source or association of the restaurant's goods or services. Because, as the jury was told, Taco Cabana's trade dress was protected if it either was inherently distinctive or had acquired a secondary meaning, judgment was entered awarding damages to Taco Cabana. In the course of calculating damages, the trial court held that Two Pesos had intentionally and deliberately infringed Taco Cabana's trade dress.

The Court of Appeals ruled that the instructions adequately stated the applicable law and that the evidence supported the jury's findings. In particular, the Court of Appeals rejected petitioner's argument that a finding of no secondary meaning contradicted a finding of inherent distinctiveness.

II

The Lanham Act was intended to make "actionable the deceptive and misleading use of marks" and "to protect persons engaged in . . . commerce against unfair competition." §45, 15 U.S.C. §1127. Section 43(a) "prohibits a broader range of practices than does §32," which applies to registered marks, Inwood Laboratories, Inc. v. Ives Laboratories, Inc., 456 U.S. 844, 858 (1982), but it is common ground that §43(a) protects qualifying unregistered trademarks and that the general principles qualifying a mark for registration under §2 of the Lanham Act are for the most part applicable in determining whether an unregistered mark is entitled to protection under §43(a). See A. J. Canfield Co., v. Honickman, 808 F.2d 291, 299, n.9 (CA3 1986); Thompson Medical Co. v. Pfizer Inc., 753 F.2d 208, 215-216 (CA2 1985).

The Court of Appeals determined that the District Court's instructions were consistent with the foregoing principles and that the evidence supported the jury's verdict. Both courts thus ruled that Taco Cabana's trade dress was not descriptive but rather inherently distinctive, and that it was not functional. None of these rulings is before us in this case, and for present purposes we assume, without deciding, that each of them is correct. In going on to affirm the judgment for respondent, the Court of Appeals, following its prior decision in *Chevron,* held that Taco Cabana's inherently distinctive trade dress was entitled to protection despite the lack of proof of secondary meaning. It is this issue that is before us for decision, and we agree with its resolution by the Court of Appeals. There is no persuasive reason to apply to trade dress a general requirement of secondary meaning which is at odds with the principles generally applicable to infringement suits under §43(a). Petitioner devotes much of its briefing to arguing issues that are not before us, and we address only its arguments relevant to whether proof of secondary meaning is essential to qualify an inherently distinctive trade dress for protection under §43(a). Petitioner argues that the jury's finding that the trade dress has not acquired a secondary meaning shows conclusively that the trade dress is not inherently distinctive. The Court of Appeals' disposition of this issue was sound:

> Two Pesos' argument—that the jury finding of inherent distinctiveness contradicts its finding of no secondary meaning in the Texas market—ignores the law in this circuit.

While the necessarily imperfect (and often prohibitively difficult) methods for assessing secondary meaning address the empirical question of current consumer association, the legal recognition of an inherently distinctive trademark or trade dress acknowledges the owner's legitimate proprietary interest in its unique and valuable informational device, regardless of whether substantial consumer association yet bestows the additional empirical protection of secondary meaning.

932 F.2d, at 1120, n.7.

This brings us to the line of decisions by the Court of Appeals for the Second Circuit that would find protection for trade dress unavailable absent proof of secondary meaning, a position that petitioner concedes would have to be modified if the temporary protection that it suggests is to be recognized. In Vibrant Sales, Inc. v. New Body Boutique, Inc., 652 F.2d 299 (1981), the plaintiff claimed protection under § 43(a) for a product whose features the defendant had allegedly copied. The Court of Appeals held that unregistered marks did not enjoy the "presumptive source association" enjoyed by registered marks and hence could not qualify for protection under § 43(a) without proof of secondary meaning. Id. at 303, 304. The court's rationale seemingly denied protection for unregistered but inherently distinctive marks of all kinds, whether the claimed mark used distinctive words or symbols or distinctive product design. The court thus did not accept the arguments that an unregistered mark was capable of identifying a source and that copying such a mark could be making any kind of a false statement or representation under § 43(a).

This holding is in considerable tension with the provisions of the Act. If a verbal or symbolic mark or the features of a product design may be registered under § 2, it necessarily is a mark "by which the goods of the applicant may be distinguished from the goods of others," 60 Stat. 428, and must be registered unless otherwise disqualified. Since § 2 requires secondary meaning only as a condition to registering descriptive marks, there are plainly marks that are registrable without showing secondary meaning. These same marks, even if not registered, remain inherently capable of distinguishing the goods of the users of these marks. Furthermore, the copier of such a mark may be seen as falsely claiming that his products may for some reason be thought of as originating from the plaintiff.

Some years after *Vibrant,* the Second Circuit announced in Thompson Medical Co. v. Pfizer Inc., 753 F.2d 208 (CA2 1985), that in deciding whether an unregistered mark is eligible for protection under § 43(a), it would follow the classification of marks set out by Judge Friendly in Abercrombie & Fitch, 537 F.2d, at 9. Hence, if an unregistered mark is deemed merely descriptive, which the verbal mark before the court proved to be, proof of secondary meaning is required; however, "suggestive marks are eligible for protection without any proof of secondary meaning, since the connection between the mark and the source is presumed." 753 F.2d, at 216. The Second Circuit has nevertheless continued to deny protection for trade dress under § 43(a) absent proof of secondary meaning, despite the fact that § 43(a) provides no basis for distinguishing between trademark and trade dress. See, e.g., Stormy Clime Ltd. v. ProGroup, Inc., 809 F.2d, at 974; Union Mfg. Co. v. Han Baek Trading Co., 763 F.2d 42, 48 (1985); Le Sportsac, Inc. v. K Mart Corp., 754 F.2d 71, 75 (1985).

The Fifth Circuit was quite right in *Chevron,* and in this case, to follow the *Abercrombie* classifications consistently and to inquire whether trade dress for which protection is claimed under § 43(a) is inherently distinctive. If it is, it is capable of identifying products or services as coming from a specific source and secondary meaning is not required. This is the rule generally applicable to trademark, and the protection of trademarks and trade dress under § 43(a) serves the same statutory purpose of preventing deception and unfair competition. There is no persuasive reason to apply different analysis to the two. The "proposition that secondary meaning must be shown even if the trade dress is a distinctive, identifying mark, [is] wrong, for the reasons explained by Judge Rubin for the Fifth Circuit in *Chevron*." Blau Plumbing, Inc. v. S.O.S. Fix-it, Inc., 781 F.2d 604, 608 (CA7 1986). The Court of Appeals for the Eleventh Circuit also follows *Chevron*, Ambrit, Inc. v. Kraft, Inc., 805 F.2d 974, 979 (1986), and the Court of Appeals for the Ninth Circuit appears to think that proof of secondary meaning is superfluous if a trade dress is inherently distinctive. Fuddruckers, Inc. v. Doc's B. R. Others, Inc., 826 F.2d 837, 843 (1987).

It would be a different matter if there were textual basis in § 43(a) for treating inherently distinctive verbal or symbolic trademarks differently from inherently distinctive trade dress. But there is none. The section does not mention trademarks or trade dress, whether they be called generic, descriptive, suggestive, arbitrary, fanciful, or functional. Nor does the concept of secondary meaning appear in the text of § 43(a). Where secondary meaning does appear in the statute, 15 U.S.C. § 1052 (1982 ed.), it is a requirement that applies only to merely descriptive marks and not to inherently distinctive ones. We see no basis for requiring secondary meaning for inherently distinctive trade dress protection under § 43(a) but not for other distinctive words, symbols, or devices capable of identifying a producer's product.

Engrafting onto § 43(a) a requirement of secondary meaning for inherently distinctive trade dress also would undermine the purposes of the Lanham Act. Protection of trade dress, no less than of trademarks, serves the Act's purpose to "secure to the owner of the mark the goodwill of his business and to protect the ability of consumers to distinguish among competing producers. National protection of trademarks is desirable, Congress concluded, because trademarks foster competition and the maintenance of quality by securing to the producer the benefits of good reputation." *Park'N Fly*, 469 U.S., at 198, citing S. Rep. No. 1333, 79th Cong., 2d Sess., 3-5 (1946) (citations omitted). By making more difficult the identification of a producer with its product, a secondary meaning requirement for a nondescriptive trade dress would hinder improving or maintaining the producer's competitive position.

Suggestions that under the Fifth Circuit's law, the initial user of any shape or design would cut off competition from products of like design and shape are not persuasive. Only nonfunctional, distinctive trade dress is protected under § 43(a). The Fifth Circuit holds that a design is legally functional, and thus unprotectable, if it is one of a limited number of equally efficient options available to competitors and free competition would be unduly hindered by according the design trademark protection. See Sicilia Di R. Biebow & Co. v. Cox, 732 F.2d 417, 426 (CA5 1984). This serves to assure that competition will not be stifled by the exhaustion of a limited number of trade dresses.

On the other hand, adding a secondary meaning requirement could have anticompetitive effects, creating particular burdens on the start-up of small companies. It would present special difficulties for a business, such as respondent, that seeks to start a new product in a limited area and then expand into new markets. Denying protection for inherently distinctive nonfunctional trade dress until after secondary meaning has been established would allow a competitor, which has not adopted a distinctive trade dress of its own, to appropriate the originator's dress in other markets and to deter the originator from expanding into and competing in these areas.

As noted above, petitioner concedes that protecting an inherently distinctive trade dress from its inception may be critical to new entrants to the market and that withholding protection until secondary meaning has been established would be contrary to the goals of the Lanham Act. Petitioner specifically suggests, however, that the solution is to dispense with the requirement of secondary meaning for a reasonable, but brief period at the outset of the use of a trade dress. If §43(a) does not require secondary meaning at the outset of a business' adoption of trade dress, there is no basis in the statute to support the suggestion that such a requirement comes into being after some unspecified time.

III

We agree with the Court of Appeals that proof of secondary meaning is not required to prevail on a claim under §43(a) of the Lanham Act where the trade dress at issue is inherently distinctive, and accordingly the judgment of that court is affirmed.

COMMENTS AND QUESTIONS

1. *Two Pesos* reverses the prior rule of a number of circuit courts, which had held that secondary meaning was a required element of a section 43(a) action for trade dress infringement, even if the trade dress was inherently distinctive. See, e.g., Ferrari S.P.A. v. Roberts, 944 F.2d 1235, 1239 (6th Cir. 1991), *cert. denied*, 505 U.S. 1219 (1992). But see the case that follows these comments, throwing the issue into some doubt.

2. Focus for a moment on the policy justifications the Court offers in support of its ruling. Two such justifications appear: that protecting distinctive trade dress will enable companies to appropriate their own goodwill, and that small companies may be unable to protect their trade dress from larger infringers if they are required to establish secondary meaning. Are these justifications persuasive? Should the Court focus on the rights of the trademark holder to protect its goodwill, or on the rights of the public to accurate product information? Are these two goals likely to come into conflict here?

3. Does the Court adopt either the "incentive" or property-rights theory of trademark law? See Stephen L. Carter, Does It Matter Whether Intellectual Property Is Property?, 68 Chicago-Kent L. Rev. 715, 721-22 (1993) (arguing that *Two Pesos* is indicative of a trend to strengthen the "property" nature of intellectual property). Certainly, the Court expresses a concern that seems consistent with the incentive theory: "protecting an inherently distinctive trade dress from its inception may be critical to new entrants to the market." Early protection encourages the development of new marks, in the Court's view.

A moral argument can also be made in favor of protecting inherently distinctive trade dress, at least against intentional copying. Such an argument asserts that the infringer can make no valid moral claim to copy the trade dress, so a right to it ought to reside in its creator. This is different from saying that the originator of the trade dress has earned it; it says that he gets it by default, because no one else has a superior claim. See Lawrence Becker, Property Rights: Philosophic Foundations 41 (1977) ("[i]t is not so much that the producers *deserve* the produce of their labors. It is rather that no one else does...."); see also Robert C. Denicola, Institutional Publicity Rights: An Analysis of the Merchandising of Famous Trade Symbols, 62 N.C. L. Rev. 603, 640-41 (1984) (arguing no one has better claim to trademark's commercial value than its producer). This moral view — that knock-offs are bad — appears to be behind the "New York rule" discussed above. Arguably, though, this approach misses the point. Isn't the question not *who* should own the right to a mark, but whether *anyone* should have exclusive rights to it?

Judge Posner gives an economic answer to this question; he says that in some situations we declare property rights merely to prevent wasteful dissipation of the value of the asset covered by the right. He explicitly applies this to the right of publicity. See Richard Posner, Economic Analysis of Law 43 (4th ed. 1992):

> [W]hatever information value a celebrity's endorsement has to consumers would be lost if every advertiser can use the celebrity's name and picture.... The existence of a congestion externality provides an argument that rights of publicity should be perpetual and thus inheritable (a matter of legal controversy today). We don't want this form of information or expression to be in the public domain, because it will be less valuable there, whether the celebrity is dead or alive.

Is this such a situation? What waste would occur if other restaurants were permitted to duplicate the Taco Cabana décor? For a contrary argument, that the copying of product design should be permitted in a market economy unless society needs to subsidize the creation of such designs, see Robert C. Denicola, Freedom to Copy, 108 Yale L.J. 1661 (1999); Mark A. Lemley, Ex Ante Versus Ex Post Justifications for Intellectual Property, 71 U. Chi. L. Rev. 147 (2004).

4. Should trade dress protection extend to the décor of a restaurant? Does *Two Pesos* open the door to a wide variety of claims that a particular style of doing business is protectable "trade dress?" An extreme example of a trade dress claim allowed in the wake of *Two Pesos* is Toy Mfgrs. of America v. Helmsley-Spear, Inc., 960 F. Supp. 673 (S.D.N.Y. 1997), where the court held that the plaintiff's "unique" registration process, forms, and location for a toy fair (taken together) constituted protectable trade dress which the defendants had infringed.

Wal-Mart Stores, Inc. v. Samara Brothers, Inc.
Supreme Court of the United States
529 U.S. 205 (2000)

Justice SCALIA delivered the opinion of the Court.

In this case, we decide under what circumstances a product's design is distinctive, and therefore protectible, in an action for infringement of unregistered trade dress under § 43(a) of the Trademark Act of 1946 (Lanham Act), 60 Stat. 441, as amended, 15 U.S.C. § 1125(a).

I

Respondent Samara Brothers, Inc., designs and manufactures children's clothing. Its primary product is a line of spring/summer one-piece seersucker outfits decorated with appliques of hearts, flowers, fruits, and the like. A number of chain stores, including JCPenney, sell this line of clothing under contract with Samara.

Petitioner Wal-Mart Stores, Inc., is one of the nation's best known retailers, selling among other things children's clothing. In 1995, Wal-Mart contracted with one of its suppliers, Judy-Philippine, Inc., to manufacture a line of children's outfits for sale in the 1996 spring/summer season. Wal-Mart sent Judy-Philippine photographs of a number of garments from Samara's line, on which Judy-Philippine's garments were to be based; Judy-Philippine duly copied, with only minor modifications, 16 of Samara's garments, many of which contained copyrighted elements. In 1996, Wal-Mart briskly sold the so-called knockoffs, generating more than $1.15 million in gross profits....

After sending cease-and-desist letters, Samara brought this action in the United States District Court for the Southern District of New York against Wal-Mart, Judy-Philippine, Kmart, Caldor, Hills, and Goody's for copyright infringement under federal law, consumer fraud and unfair competition under New York law, and — most relevant for our purposes — infringement of unregistered trade dress under § 43(a) of the Lanham Act, 15 U.S.C. § 1125(a)....

After a weeklong trial, the jury found in favor of Samara on all of its claims.... The Second Circuit affirmed..., 165 F.3d 120 (1998), and we granted certiorari...

II

The Lanham Act provides for the registration of trademarks, which it defines in § 45 to include "any word, name, symbol, or device, or any combination thereof [used or intended to be used] to identify and distinguish [a producer's] goods ... from those manufactured or sold by others and to indicate the source of the goods...." 15 U.S.C. § 1127.... In addition to protecting registered marks, the Lanham Act, in § 43(a), gives a producer a cause of action for the use by any person of "any word, term, name, symbol, or device, or any combination thereof ... which ... is likely to cause confusion ... as to the origin, sponsorship, or approval of his or her goods...." 15 U.S.C. § 1125(a). It is the latter provision that is at issue in this case.

The breadth of the definition of marks registrable under § 2, and of the confusion-producing elements recited as actionable by § 43(a), has been held to embrace not just word marks, such as "Nike," and symbol marks, such as Nike's "swoosh" symbol, but also "trade dress" — a category that originally included only the packaging, or "dressing," of a product, but in recent years has been expanded by many courts of appeals to encompass the design of a product. See, e.g., Ashley Furniture Industries, Inc. v. Sangiacomo N. A., Ltd., 187 F.3d 363 (C.A.4 1999) (bedroom furniture); Knitwaves, Inc. v. Lollytogs, Ltd., 71 F.3d 996 (C.A.2 1995) (sweaters); Stuart Hall Co., Inc. v. Ampad Corp., 51 F.3d 780 (C.A.8 1995) (notebooks). These courts have assumed, often without discussion, that trade dress constitutes a "symbol" or "device" for purposes of the relevant sections, and we conclude likewise. "Since human beings might use as a 'symbol' or 'device' almost anything at all that is capable of carrying meaning, this language, read literally, is not restrictive." Qualitex Co. v. Jacobson Products Co., 514 U.S. 159, 162 (1995). This reading of § 2 and § 43(a) is buttressed by a recently added subsection of § 43(a), § 43(a)(3), which refers specifically to "civil action[s] for trade dress infringement under this chapter for trade dress not registered on the principal register." 15 U.S.C.A. § 1125(a)(3) (Oct.1999 Supp.).

The text of § 43(a) provides little guidance as to the circumstances under which unregistered trade dress may be protected. It does require that a producer show that the allegedly infringing feature is not "functional," see § 43(a)(3), and is likely to cause confusion with the product for which protection is sought, see § 43(a)(1)(A), 15 U.S.C. § 1125(a)(1)(A). Nothing in § 43(a) explicitly requires a producer to show that its trade dress is distinctive, but courts have universally imposed that requirement, since without distinctiveness the trade dress would not "cause confusion . . . as to the origin, sponsorship, or approval of [the] goods," as the section requires. Distinctiveness is, moreover, an explicit prerequisite for registration of trade dress under § 2, and "the general principles qualifying a mark for registration under § 2 of the Lanham Act are for the most part applicable in determining whether an unregistered mark is entitled to protection under § 43(a)." Two Pesos, Inc. v. Taco Cabana, Inc., 505 U.S. 763, 768(1992) (citations omitted).

In evaluating the distinctiveness of a mark under § 2 (and therefore, by analogy, under § 43(a)), courts have held that a mark can be distinctive in one of two ways. First, a mark is inherently distinctive if "[its] intrinsic nature serves to identify a particular source." Ibid. In the context of word marks, courts have applied the now-classic test originally formulated by Judge Friendly, in which word marks that are "arbitrary" ("Camel" cigarettes), "fanciful" ("Kodak" film), or "suggestive" ("Tide" laundry detergent) are held to be inherently distinctive. See Abercrombie & Fitch Co. v. Hunting World, Inc., 537 F.2d 4, 10-11 (C.A.2 1976). Second, a mark has acquired distinctiveness, even if it is not inherently distinctive, if it has developed secondary meaning, which occurs when, "in the minds of the public, the primary significance of a [mark] is to identify the source of the product rather than the product itself." Inwood Laboratories, Inc. v. Ives Laboratories, Inc., 456 U.S. 844, 851, n. 11 (1982).

The judicial differentiation between marks that are inherently distinctive and those that have developed secondary meaning has solid foundation in the statute itself. Section 2 requires that registration be granted to any trademark "by which the goods of the applicant may be distinguished from the goods of others" — subject to various limited exceptions. 15 U.S.C. § 1052. It also provides, again with limited exceptions, that "nothing in this chapter shall prevent the registration of a mark used by the applicant which has become distinctive of the applicant's goods in commerce" — that is, which is not inherently distinctive but has become so only through secondary meaning. § 2(f), 15 U.S.C. § 1052(f). Nothing in § 2, however, demands the conclusion that *every* category of mark necessarily includes some marks "by which the goods of the applicant may be distinguished from the goods of others" *without* secondary meaning — that in every category some marks are inherently distinctive.

Indeed, with respect to at least one category of mark — colors — we have held that no mark can ever be inherently distinctive. See *Qualitex,* 514 U.S., at 162-163. In *Qualitex,* petitioner manufactured and sold green-gold dry-cleaning press pads. After respondent began selling pads of a similar color, petitioner brought suit under § 43(a), then added a claim under § 32 after obtaining registration for the color of its pads. We held that a color could be protected as a trademark, but only upon a showing of secondary meaning. Reasoning by analogy to the *Abercrombie & Fitch* test developed for word marks, we noted that a product's color is unlike a "fanciful," "arbitrary," or "suggestive" mark, since it does not "almost *automatically* tell a customer that [it] refer[s] to a brand," ibid., and does not "immediately . . . signal a brand or a product 'source,'" id., at 163. However, we noted that, "over time, customers may come to treat a particular color on a product or its packaging . . . as signifying a brand." Id., at 162-163. Because a color, like a "descriptive" word mark, could eventually "come to indicate a product's origin," we concluded that it could be protected *upon a showing of secondary meaning.* Ibid.

It seems to us that design, like color, is not inherently distinctive. The attribution of inherent distinctiveness to certain categories of word marks and product packaging derives from the fact that the very purpose of attaching a particular word to a product, or encasing it in a distinctive packaging, is most often to identify the source of the product. Although the words and packaging can serve subsidiary functions — a suggestive word mark (such as "Tide" for laundry detergent), for instance, may invoke positive connotations in the consumer's mind, and a garish form of packaging (such as Tide's squat, brightly decorated plastic bottles for its liquid laundry detergent) may attract an otherwise indifferent consumer's attention on a crowded store shelf — their predominant function remains source identification. Consumers are therefore predisposed to regard those symbols as indication of the producer, which is why such symbols "almost *automatically* tell a customer that they refer to a brand," id., at 162-163, and "immediately . . . signal a brand or a product 'source,'" id., at 163. And where it is not reasonable to assume consumer predisposition to take an affixed word or packaging as indication of source — where, for example, the affixed word is descriptive of the product ("Tasty" bread) or of a geographic origin ("Georgia" peaches) — inherent distinctiveness will not be found. That is why the statute generally excludes, from those word marks that can be registered as inherently distinctive, words that are "merely descriptive" of the goods, § 2(e)(1), 15 U.S.C. § 1052(e)(1), or "primarily geographically descriptive of them," see § 2(e)(2), 15 U.S.C. § 1052(e)(2). In the case of product design, as in the case of color, we think consumer predisposition to equate the feature with the source does not exist. Consumers are aware of the reality that, almost invariably, even the most unusual of product designs — such as a cocktail shaker shaped like a penguin — is intended not to identify the source, but to render the product itself more useful or more appealing.

The fact that product design almost invariably serves purposes other than source identification not only renders inherent distinctiveness problematic; it also renders application of an inherent-distinctiveness principle more harmful to other consumer interests. Consumers should not be deprived of the benefits of competition with regard to the utilitarian and esthetic purposes that product design ordinarily serves by a rule of law that facilitates plausible threats of suit against new entrants based upon alleged inherent distinctiveness. How easy it is to mount a plausible suit depends, of course, upon the clarity of the test for inherent distinctiveness, and where product design is concerned we have little confidence that a reasonably clear test can be devised. Respondent and the United States as *amicus curiae* urge us to adopt for product design relevant portions of the test formulated by the Court of Customs and Patent Appeals for product packaging in Seabrook Foods, Inc. v. Bar-Well Foods, Ltd., 568 F.2d 1342 (1977). That opinion, in determining the inherent distinctiveness of a product's packaging, considered, among other things, "whether it was a 'common' basic shape or design, whether it was unique or unusual in a particular field, [and] whether it was a mere refinement of a commonly-adopted and well-known form of ornamentation for a particular class of goods viewed by the public as a dress or ornamentation for the goods." Id., at 1344 (footnotes omitted). Such a test would rarely provide the basis for summary disposition of an anticompetitive strike suit. Indeed, at oral argument, counsel for the United States quite understandably would not give a definitive answer as to whether the test was met in this very case, saying only that "[t]his is a very difficult case for that purpose."

It is true, of course, that the person seeking to exclude new entrants would have to establish the nonfunctionality of the design feature, see §43(a)(3), 15 U.S.C.A. §1125(a)(3) (Oct. 1999 Supp.)—a showing that may involve consideration of its esthetic appeal, see *Qualitex,* 514 U.S., at 170. Competition is deterred, however, not merely by successful suit but by the plausible threat of successful suit, and given the unlikelihood of inherently source-identifying design, the game of allowing suit based upon alleged inherent distinctiveness seems to us not worth the candle. That is especially so since the producer can ordinarily obtain protection for a design that *is* inherently source identifying (if any such exists), but that does not yet have secondary meaning, by securing a design patent or a copyright for the design—as, indeed, respondent did for certain elements of the designs in this case. The availability of these other protections greatly reduces any harm to the producer that might ensue from our conclusion that a product design cannot be protected under §43(a) without a showing of secondary meaning.

Respondent contends that our decision in *Two Pesos* forecloses a conclusion that product-design trade dress can never be inherently distinctive. In that case, we held that the trade dress of a chain of Mexican restaurants, which the plaintiff described as "a festive eating atmosphere having interior dining and patio areas decorated with artifacts, bright colors, paintings and murals," 505 U.S., at 765 (internal quotation marks and citation omitted), could be protected under §43(a) without a showing of secondary meaning, see id., at 776. *Two Pesos* unquestionably establishes the legal principle that trade dress can be inherently distinctive, see, e.g., id., at 773, but it does not establish that *product-design* trade dress can be. *Two Pesos* is inapposite to our holding here because the trade dress at issue, the decor of a restaurant, seems to us not to constitute product *design.* It was either product packaging—which, as we have discussed, normally *is* taken by the consumer to indicate origin—or else some tertium quid that is akin to product packaging and has no bearing on the present case.

Respondent replies that this manner of distinguishing *Two Pesos* will force courts to draw difficult lines between product-design and product-packaging trade dress. There will indeed be some hard cases at the margin: a classic glass Coca-Cola bottle, for instance, may constitute packaging for those consumers who drink the Coke and then discard the bottle, but may constitute the product itself for those consumers who are bottle collectors, or part of the product itself for those consumers who buy Coke in the classic glass bottle, rather than a can, because they think it more stylish to drink from the former. We believe, however, that the frequency and the difficulty of having to distinguish between product design and product packaging will be much less than the frequency and the difficulty of having to decide when a product design is inherently distinctive. To the extent there are close cases, we believe that courts should err on the side of caution and classify ambiguous trade dress as product design, thereby requiring secondary meaning. The very closeness will suggest the existence of relatively small utility in adopting an inherent-distinctiveness principle, and relatively great consumer benefit in requiring a demonstration of secondary meaning....

We hold that, in an action for infringement of unregistered trade dress under §43(a) of the Lanham Act, a product's design is distinctive, and therefore protectible, only upon a showing of secondary meaning. The judgment of the Second Circuit is reversed, and the case is remanded for further proceedings consistent with this opinion.

It is so ordered.

COMMENTS AND QUESTIONS

1. *Trade Dress Classification.* The Supreme Court establishes three categories for determining whether a product's trade dress is protectable: (1) product packaging; (2) product design; and (3) a "tertium quid" (defined in the Oxford English Dictionary as "[s]omething (indefinite or left undefined) related in some way to two (definite or known) things, but distinct from both."). Product packaging can be inherently distinctive while product design cannot. But how is one to distinguish the two? Should close cases (such as the Coca-Cola bottle) be characterized as product design and hence subject to the higher threshold of secondary meaning? Cf. Restatement (Third) of Unfair Competition § 16 (interpreting case authority to require proof of secondary meaning as a prerequisite for trade dress protection). Did the Court introduce needless confusion by not simply acknowledging that its finding in *Two Pesos* that the décor of a Mexican restaurant was inherently distinctive was ill-advised? Furthermore, the Court leaves unresolved who decides on the classification of trade dress. *Two Pesos* seems to suggest that this question is for the jury, whereas *Samara* emphasizes summary adjudication. Who should decide?

2. One recent court decision highlights the difficulty of applying a regime originally designed for word marks to product configurations. In Rock and Roll Hall of Fame and Museum, Inc. v. Gentile Productions, 134 F.3d 749 (6th Cir. 1998), the owner of the Rock and Roll Hall of Fame building in Cleveland, Ohio, sued a photographer who marketed pictures of the building set against the Cleveland skyline, alleging that the photographer was infringing the building's trade dress. The district court issued a preliminary injunction, but the Sixth Circuit reversed. The majority and the dissent battled over the proper classification of the museum. The majority agreed that the design of the museum was "fanciful" but denied that it was "fanciful in a trademark sense." It concluded that the design of the museum itself did not function as a trademark and that the public did not recognize it as such. Important to the court's determination was the trademark owner's inconsistent use of the museum's design as a source-identifying function:

> [A]lthough the Museum has used drawings or pictures of its building design on various goods, it has not done so with any consistency. As museum director Robert Bosak stated in his affidavit, "the Museum has used versions of the building shape trademark on . . . a wide variety of products." Several items marketed by the Museum display only the rear of the Museum's building, which looks dramatically different from the front. Drawings of the front of the Museum on the two T-shirts in the record are similar, but they are quite different from the photograph featured in the Museum's poster. And, although the photograph from the poster is also used on a postcard, another postcard displays various close-up photographs of the Museum which, individually and perhaps even collectively, are not even immediately recognizable as photographs of the Museum. In this regard, this case is similar to those in which a party has claimed trademark rights in a famous person's likeness. In Estate of Presley [v. Russen, 513 F.Supp. 1339, 1363-64 (D.N.J. 1981)], the court concluded that, although one particular image of [Elvis] Presley had been used consistently as a mark, "the available evidence [did] not support [the estate's] broad position" that all images of Presley served such a function.

134 F.3d 749, 754-755.

In another case, the Sixth Circuit had little difficulty in applying *Samara*'s classification scheme to furniture. See Herman Miller v. Palazzetti Imports and

Exports, Inc., 270 F.3d 298 (6th Cir. 2001) (finding genuine issue of material fact as to whether lounge chair and ottoman had acquired secondary meaning).

3. *Product Configurations and Overlapping Intellectual Property Rights.* An old rule of intellectual property law called the "principle of election" used to limit an author to one type of intellectual property protection. That doctrine was resoundingly rejected in In re Yardley, 493 F.2d 1389 (C.C.P.A. 1974). *Yardley* involved a design patent application for a watch which featured a caricature of Spiro Agnew. The patent examiner rejected the application on the ground that Yardley had already received several copyright registrations for the watch. The Court of Customs and Patent Appeals reversed, concluding that the principle of election was in direct conflict with the patent, copyright and trademark statutes. The court reasoned that because each statute strikes its own "coverage" balance, a work that qualifies for protection under more than one statute is entitled to the protection of each.

Nonetheless, where Congress has not expressly dictated the principles for channeling protection among the various modes of intellectual property, the courts have construed the scope of protection in order to effectuate the larger policies of federal intellectual property law. In general, utility patent protection trumps all other modes of intellectual property with regard to the functional features of products. The courts recognize that patent law imposes high thresholds for protection (novelty, non-obviousness, and disclosure) and an examination process and affords a relatively short duration of protection in order to balance the short-run hampering of free competition with the longer-term benefits of innovation. Were copyright law or trademark law to afford protection for utilitarian features of products, patent law would be unnecessary and competition would give way to monopolization by the first to design a product, regardless of the effort that went into the design. In order to minimize this threat, the Supreme Court in Baker v. Selden foreclosed competition under copyright law for accounting systems. (Congress reinforced this limitation in the 1976 Copyright Act, see 17 U.S.C. § 102(b).) The Supreme Court recently acknowledged a similarly high burden for establishing that a product feature or configuration that has been the subject of a utility patent can ever qualify for trade dress protection. See TrafFix Devices, Inc. v. Marketing Displays, Inc., 532 U.S. 23 (2001) excerpted, *infra*, section E.2 (functionality defense).

How do dress designs fit into this framework? Copyright law denies protection for dress (as opposed to fabric) designs, see Whimsicality, Inc. v. Rubie's Costume Co., Inc., 891 F.2d 452 (2d Cir. 1989); Fashion Originators Guild v. FTC, 114 F.2d 80, 84 (2d Cir. 1940) (L. Hand, J.) (holding that clothes, as useful articles, are not copyrightable), *aff'd*, 312 U.S. 457 (1941); see also 1 Nimmer on Copyright, *supra*, § 2.08[H]; H.R.Rep. No. 1476, 94th Cong., 2d Sess. 55 (1976), reprinted in 1976 U.S.Code Cong. & Admin. News 5659, 5668 (noting that the Copyright Act of 1976 did not affect the prior law in this regard), and design patent law affords protection for the ornamental features of articles of manufacture. Should trade dress law be permitted to enter this sphere? While not excluding trade dress protection for dress designs, the *Samara* decision reflects a concern for not unduly hindering free competition. The Court noted the fact that product design "almost invariably serves purposes other than source identification," such as making products more useful or appealing. The Court also stated that application of the inherently distinctive test in product design cases could deprive consumers of "the benefits of competition with regard to the utilitarian and esthetic purposes that product design ordinarily serves." See also Samara Bros., Inc. v. Wal-Mart Stores, Inc., 165 F.3d 120, 133 (2d Cir. 1999) (Newman, dissenting) (noting that "we must construe the Lanham Act 'in the light of a strong federal policy in favor of vigorously competitive markets'; "courts are to 'exercise[ ] particular "caution" when extending [trademark] protection to product designs'"; "'just as copyright law does not protect ideas but only their concrete expression, neither does trade dress law protect an idea, a concept, or a generalized type of appearance.'").

The Supreme Court took a sharply different tack from *Yardley* in its most recent treatment of the issue. In *Dastar Corp. v. Twentieth Century Fox Film Corp.*, 539 U.S. 23 (2003), defendant Dastar produced and sold its own copies of plaintiff's video series, in which the copyright had expired. Fox sued on the grounds that Dastar was misrepresenting the origin of the goods in violation of the Lanham Act. The Court rejected that argument, construing "origin" narrowly to mean only the physical source of the videotapes, not who had actually filmed or produced them. In so holding, the Court was expressly concerned to avoid reading the Lanham Act to conflict with copyright law, which would permit such copying. It warned "against misuse or over-extension of trademark and related protections into areas traditionally occupied by patent or copyright," and explained that "allowing a cause of action under § 43(a) for that representation would create a species of mutant copyright law that limits the public's federal right to copy and to use expired copyrights." Id. at 34. The Court seems quite concerned with channeling protection between the copyright and trademark regimes and unwilling to permit much overlap between the two.

4. Should "Tide's squat, brightly decorated plastic bottles for its liquid laundry detergent" be deemed inherently distinctive, as Justice Scalia seems to suggest? How would you analyze this example?

PROBLEM

Problem 5-1. Alice Richland, a gourmet chef world-renowned for her chocolate desserts, designs a new line of chocolate products. Called "Chocolate Shells," these products are made of dark chocolate flavored in special ways with a combination of ingredients Alice hit upon after months of work in her kitchens. The use of the special ingredients imparts a unique flavor to the Shells, and has the additional property of making the usually soft chocolate feel sandy or grainy to the touch. The Shells are made in the shapes of different seashells native to the Florida coast, and each shell is colored in a different, unusual pattern. On the advice of her lawyer, Alice files a trademark application seeking to register each of her shell designs. What aspects of the Shells are entitled to registration? To section 43(a) protection?

---

2. Priority

As in patent law, many trademark cases turn on issues of priority. Section 45(a) of the Lanham Act, which defines a trademark, requires that the mark either be (1) "used in commerce" or (2) registered with a bona fide intention to use it in commerce. Both at common law and under the traditional Lanham Act registration procedures, determining who owned a trademark meant determining who was first to use it to identify her goods.

The requirement of "use in commerce" is an historical result of the constitutional basis for the trademark laws, which (unlike the patent and copyright statutes) rely on the congressional power to regulate interstate commerce. This requirement also goes hand in hand with the basic trademark theory elaborated in the introduction to this chapter; recall that this theory emphasizes the protection of consumer associations of a brand with a particular product, which can arise only after a trademark is placed on goods sold in commerce.

But just what constitutes use of a trademark? And how much use is enough to secure legal protection for the mark? The case and notes that follow provide some guidance.

Zazu Designs v. L'Oreal, S.A.
United States Court of Appeals for the Seventh Circuit
979 F.2d 499 (7th Cir. 1992)

EASTERBROOK, Circuit Judge.

In 1985 Cosmair, Inc., concluded that young women craved pink and blue hair. To meet the anticipated demand, Cosmair developed a line of "hair cosmetics" — hair coloring that is easily washed out. These inexpensive products, under the name ZAZU, were sold in the cosmetic sections of mass merchandise stores. Apparently the teenagers of the late 1980s had better taste than Cosmair's marketing staff thought. The product flopped, but its name gave rise to this trademark suit. Cosmair is the United States licensee of L'Oreal, S.A., a French firm specializing in perfumes, beauty aids, and related products. Cosmair placed L'Oreal's marks on the bottles and ads....

L'Oreal hired Wordmark, a consulting firm, to help it find a name for the new line of hair cosmetics. After checking the United States Trademark Register for conflicts, Wordmark suggested 250 names. L'Oreal narrowed this field to three, including ZAZU, and investigated their availability. This investigation turned up one federal registration of ZAZU as a mark for clothing and two state service mark registrations including that word. One of these is Zazu Hair Designs; the other was defunct.

Zazu Hair Designs is a hair salon in Hinsdale, Illinois, a suburb of Chicago. We call it "ZHD" to avoid confusion with the ZAZU mark.... The salon is a partnership between Raymond R. Koubek and Salvatore J. Segretto, hairstylists who joined forces in 1979. ZHD registered ZAZU with Illinois in 1980 as a trade name for its salon. L'Oreal called the salon to find out if ZHD was selling its own products. The employee who answered reported that the salon was not but added, "we're working on it." L'Oreal called again; this time it was told that ZHD had no products available under the name ZAZU.

L'Oreal took the sole federal registration, held by Riviera Slacks, Inc., as a serious obstacle. Some apparel makers have migrated to cosmetics, and if Riviera were about to follow Ralph Lauren (which makes perfumes in addition to shirts and skirts) it might have a legitimate complaint against a competing use of the mark. Sands, Taylor & Wood Co. v. Quaker Oats Co., 24 U.S.P.Q.2d 1001, 1011 (7th Cir. 1992). Riviera charged L'Oreal $125,000 for a covenant not to sue if L'Oreal used the ZAZU mark on cosmetics. In April 1986, covenant in hand and satisfied that ZHD's state trade name did not prevent the introduction of a national product, L'Oreal made a small interstate shipment of hair cosmetics under the ZAZU name. It used this shipment as the basis of an application for federal registration, filed on June 12, 1986. By August L'Oreal had advertised and sold its products nationally.

Unknown to L'Oreal, Koubek and Segretto had for some time aspired to emulate Vidal Sassoon by marketing shampoos and conditioners under their salon's trade name. In 1985 Koubek began meeting with chemists to develop ZHD's products. Early efforts were unsuccessful; no one offered a product that satisfied ZHD. Eventually ZHD received acceptable samples from Gift Cosmetics, some of which Segretto sold to customers of the salon in plain bottles to which he taped the salon's business card. Between November 1985 and February 1986 ZHD made a few other sales. Koubek shipped two bottles to a friend in Texas, who paid $13. He also made two shipments to a hair stylist friend in Florida — 40 bottles of shampoo for $78.58. These were designed to interest the Floridian in the future marketing of the product line. These bottles could not have been sold to the public, because they lacked labels listing the ingredients and weight. See 21 U.S.C. § 362(b); 15 U.S.C. §§ 1452, 1453(a); 21 CFR §§ 701.3, 701.13(a). After L'Oreal's national marketing was under way, its representatives thrice visited ZHD and found that the salon still had no products for sale under the ZAZU name. Which is not to say that ZHD was supine. Late in 1985 ZHD had ordered 25,000 bottles silkscreened with the name ZAZU. Later it ordered stick-on labels listing the ingredients of its products. In September 1986 ZHD began to sell small quantities of shampoo in bottles filled (and labeled) by hand in the salon. After the turn of the year ZHD directed the supplier of the shampoo and conditioner to fill some bottles; the record does not reveal how many.

After a bench trial the district court held that ZHD's sales gave it an exclusive right to use the ZAZU name nationally for hair products. 9 U.S.P.Q.2d 1972 (N.D. Ill. 1988). The court enjoined L'Oreal from using the mark (a gesture, since the product had bombed and L'Oreal disclaimed any interest in using ZAZU again). It also awarded ZHD $100,000 in damages on account of lost profits and $1 million more to pay for corrective advertising to restore luster to the ZAZU mark. [The final judgment also included a $1 million punitive damage award for "oppressive and deceitful" tactics used in the litigation.]

Federal law permits the registration of trademarks and the enforcement of registered marks. Through § 43(a) of the Lanham Act, 15 U.S.C. § 1125(a), a provision addressed to deceit, it also indirectly allows the enforcement of unregistered marks. But until 1988 federal law did not specify how one acquired the rights that could be registered or enforced without registration. That subject fell into the domain of state law, plus federal common law elaborating on the word "use" in § 43(a).... At common law, "use" meant sales to the public of a product with the mark attached. Trade-Mark Cases, 100 U.S. 82, 94-95 (1879). See also Hanover Star Milling Co. v. Metcalf, 240 U.S. 403, 414 (1916); United Drug Co. v. Theodore Rectanus Co., 248 U.S. 90, 97 (1918).

"Use" is neither a glitch in the Lanham Act nor a historical relic. By insisting that firms use marks to obtain rights in them, the law prevents entrepreneurs from reserving brand names in order to make their rivals' marketing more costly. Public sales let others know that they should not invest resources to develop a mark similar to one already used in the trade. Blue Bell, Inc. v. Farah Manufacturing Co., 508 F.2d 1260, 1264-65 (5th Cir. 1975); see also William M. Landes and Richard A. Posner, Trademark Law: An Economic Perspective, 30 J.L. & Econ. 265, 281-84 (1987). Only active use allows consumers to associate a mark with particular goods and notifies other firms that the mark is so associated.

Under the common law, one must win the race to the marketplace to establish the exclusive right to a mark. Blue Bell v. Farah; La Societe Anonyme des Parfums LeGalion v. Jean Patou, Inc., 495 F.2d 1265, 1271-74 (2d Cir. 1974). Registration modifies this system slightly, allowing slight sales plus notice in the register to substitute for substantial sales without notice. 15 U.S.C. § 1051(a). (The legislation in 1988 modifies the use requirement further, but we disregard this.) ZHD's sales of its product are insufficient use to establish priority over L'Oreal. A few bottles sold over the counter in Hinsdale, and a few more mailed to friends in Texas and Florida, neither link the ZAZU mark with ZHD's product in the minds of consumers nor put other producers on notice. As a practical matter ZHD had no product, period, until months after L'Oreal had embarked on its doomed campaign.

In finding that ZHD's few sales secured rights against the world, the district court relied on cases such as Department of Justice v. Calspan Corp., 578 F.2d 295 (C.C.P.A. 1978), which hold that a single sale, combined with proof of intent to go on selling, permit the vendor to register the mark. See also Axton-Fisher Tobacco Co. v. Fortune Tobacco Co., 82 F.2d 295 (C.C.P.A. 1936); Maternally Yours, Inc. v. Your Maternity Shop, Inc., 234 F.2d 538, 542 (2d Cir. 1956)....
But use sufficient to register a mark that soon is widely distributed is not necessarily enough to acquire rights in the absence of registration. The Lanham Act allows only trademarks "used in commerce" to be registered. 15 U.S.C. § 1051(a). Courts have read "used" in a way that allows firms to seek protection for a mark before investing substantial sums in promotion. See Fort Howard Paper Co. v. Kimberly-Clark Corp., 390 F.2d 1015 (C.C.P.A. 1968); cf. Jim Dandy Co. v. Martha White Foods, Inc., 458 F.2d 1397, 1399 (C.C.P.A. 1972) (party may rely on advertising to show superior registration rights); but see Weight Watchers International, Inc. v. I. Rokeach & Sons, Inc., 211 U.S.P.Q. 700, 709 (T.M.T.A.B. 1981) (more than minimal use is required to register because the statute allows only "owner[s]" to register, and ownership of a mark depends on commercial use). Liberality in registering marks is not problematic, because the registration gives notice to latecomers, which token use alone does not. Firms need only search the register before embarking on development. Had ZHD registered ZAZU, the parties could have negotiated before L'Oreal committed large sums to marketing.

ZHD applied for registration of ZAZU after L'Oreal not only had applied to register the mark but also had put its product on the market nationwide. Efforts to register came too late. At oral argument ZHD suggested that L'Oreal's knowledge of ZHD's plan to enter the hair care market using ZAZU establishes ZHD's superior right to the name. Such an argument is unavailing. Intent to use a mark, like a naked registration, establishes no rights at all. Hydro-Dynamics, Inc. v. George Putnam & Co., 811 F.2d 1470, 1472 (Fed. Cir. 1987). Even under the 1988 amendments (see note), which allow registration in advance of contemplated use, an unregistered plan to use a mark creates no rights. Just as an intent to buy a choice parcel of land does not prevent a rival from closing the deal first, so an intent to use a mark creates no rights a competitor is bound to respect. A statute granting no rights in bare registrations cannot plausibly be understood to grant rights in "intents" divorced from either sales or registrations. Registration itself establishes only a rebuttable presumption of use as of the filing date. Rolley, Inc. v. Younghusband, 204 F.2d 209, 211 (9th Cir. 1953). ZHD made first use of ZAZU in connection with hair services in Illinois, but this does not translate to a protectable right to market hair products nationally. The district court construed L'Oreal's knowledge of ZHD's use of

ZAZU for salon services as knowledge "of [ZHD's] superior rights in the mark." 9 U.S.P.Q.2d at 1978. ZHD did not, however, have superior rights in the mark as applied to hair products, because it neither marketed such nor registered the mark before L'Oreal's use. Because the mark was not registered for use in conjunction with hair products, any knowledge L'Oreal may have had of ZHD's plans is irrelevant. Cf. Weiner King, Inc. v. Wiener King Corp., 615 F.2d 512 (C.C.P.A. 1980).

Imagine the consequences of ZHD's approach. Businesses that knew of an intended use would not be entitled to the mark even if they made the first significant use of it. Businesses with their heads in the sand, however, could stand on the actual date they introduced their products, and so would have priority over firms that intended to use a mark but had not done so. Ignorance would be rewarded—and knowledgeable firms might back off even though the rivals' "plans" or "intent" were unlikely to come to fruition. Yet investigations of the sort L'Oreal undertook prevent costly duplication in the development of trademarks and protect consumers from the confusion resulting from two products being sold under the same mark. See Natural Footwear Ltd. v. Hart, Shaffner & Marx, 760 F.2d 1383, 1395 (3d Cir. 1985). L'Oreal should not be worse off because it made inquiries and found that, although no one had yet used the mark for hair products, ZHD intended to do so. Nor should a potential user have to bide its time until it learns whether other firms are serious about marketing a product. The use requirement rewards those who act quickly in getting new products in the hands of consumers. Had L'Oreal discovered that ZHD had a product on the market under the ZAZU mark or that ZHD had registered ZAZU for hair products, L'Oreal could have chosen another mark before committing extensive marketing resources. Knowledge that ZHD planned to use the ZAZU mark in the future does not present an obstacle to L'Oreal's adopting it today. Selfway, Inc. v. Travelers Petroleum, Inc., 579 F.2d 75, 79 (C.C.P.A. 1978).

Occasionally courts suggest that "bad faith" adoption of a mark defeats a claim to priority. See California Cedar Products Co. v. Pine Mountain Corp., 724 F.2d 827, 830 (9th Cir. 1984); Stern Electronics, Inc. v. Kaufman, 669 F.2d 852, 857 (2d Cir. 1982); Blue Bell v. Farah, 508 F.2d at 1267. Although ZHD equates L'Oreal's knowledge of its impending use with "bad faith," the cases use the term differently. In each instance the court applied the label "bad faith" to transactions designed merely to reserve a mark, not to link the name to a product ready to be sold to the public. In California Cedar Products, for example, two firms sprinted to acquire the abandoned DURAFLAME mark. One shipped some of its goods in the abandoning company's wrapper with a new name pasted over it. Two days later the other commenced bona fide sales under the DURAFLAME mark. The court disregarded the first shipment, calling it "both premature and in bad faith," 724 F.2d at 830, and held that the first firm to make bona fide sales to customers was the prior user. "Bad faith" was no more than an epithet stapled to the basic conclusion: that reserving a mark is forbidden, so that the first producer to make genuine sales gets the rights. If these cases find a parallel in our dispute, ZHD occupies the place of the firm trying to reserve a mark for "intended" exploitation. ZHD doled out a few samples in bottles lacking labeling necessary for sale to the public. Such transactions are the sort of pre-marketing maneuvers that these cases hold insufficient to establish rights in a trademark.

The district court erred in equating a use sufficient to support registration with a use sufficient to generate nationwide rights in the absence of registration. Although whether ZHD's use is sufficient to grant it rights in the ZAZU mark is a question of fact on which appellate review is deferential, California Cedar Products, 724 F.2d at 830 . . . , the extent to which ZHD used the mark is not disputed. ZHD's sales of hair care products were insufficient as a matter of law to establish national trademark rights at the time L'Oreal put its electric hair colors on the market.

[In a forcefully stated section of the opinion, the court also reversed the punitive damages holding.]

Reversed and remanded.

CUDAHY, C.J., dissenting:

On the important issue of good faith, L'Oreal's conduct here merits a very hard look. In the case of Riviera, a men's clothing retailer, L'Oreal was careful to pay $125,000 for an agreement not to sue. Yet men's clothing and hair cosmetics marketed to women hardly seem related at all. On the other hand, a women's hair salon developing a line of hair care products is a purveyor of goods and services that seem closely related to hair cosmetics. Therefore, L'Oreal's knowledge of ZHD's use defeats any claim L'Oreal may have to priority.

One of the keys here seems to be the use of ZAZU as a service mark connected with the provision of salon services by ZHD. A service mark can be infringed by its use on a closely related product.[1] . . . [See] 2 J. Thomas McCarthy, Trademark and Unfair Competition § 24:6, at 71 (2d ed. 1984 & Supp. 1991) (stating that "[w]here the services consist of retail sales services, likelihood of confusion is found when another mark is used on goods which are commonly sold through such a retail outlet"). A service and a product are related if buyers are likely to assume a common source or sponsorship . . . . The salon services and hair products at issue in this case, which are nearly as kindred as a service and product can be, offer the paradigmatic illustration of things that are closely related. Thus the majority's disregard for ZHD's substantial use of ZAZU in connection with salon services is unfounded.

As the majority correctly notes, the standard for granting federal registration is somewhat less exacting than that for establishing common law trademark rights . . . . But that distinction is so slight as to be inconsequential here. While any bona fide transaction that is more than a "mere sham" will, when combined with intent to continue use, suffice to support federal registration, Avakoff v. Southern Pacific Co., 765 F.2d 1097, 1098 (Fed. Cir. 1985) . . . , any use greater than de minimis will still warrant trademark protection in the absence of registration. Thus, bona fide test marketing or small experimental sales — indeed, any use that is not nominal or token — can satisfy the test . . . .

In this case, ZHD's use of the ZAZU mark, both in its highly successful salon service business, which drew some out-of-state clients, and in its local and interstate product sales to customers and to a potential marketer, surely is more than de minimis. The extensive evidence of ZHD's intent to step up hair product sales — such as its

---

1. Our recent opinion in Sands, Taylor & Wood Co. v. Quaker Oats Co., 24 U.S.P.Q.2d 1001 (7th Cir. 1992), states explicitly that modern trademark law prohibits the use of a senior user's mark on products that are closely related to the senior user's, as well as those products in direct competition. See 24 U.S.P.Q.2d at 1010; International Kennel Club, Inc. v. Mighty Star, Inc., 846 F.2d 1079, 1089 (7th Cir. 1988). This "protects the owner's ability to enter product markets . . . into which it might reasonably be expected to expand in the future." Sands, Taylor & Wood, 24 U.S.P.Q.2d at 1011; 2 J. Thomas McCarthy, Trademark and Unfair Competition § 24:5, at 177 (2d ed. 1984).

order for 25,000 ZAZU-emblazoned bottles and its inquiry about advertising rates in a national magazine — bolsters this assessment. Even if ZHD did fail to demonstrate more than a de minimis market penetration nationally, at the very least it successfully established exclusive rights within its primary area of operation. The salon's substantial advertising, increasing revenue and staff and preliminary product sales indicate sufficient market penetration to afford trademark protection in that region. See Natural Footwear Ltd. v. Hart, Schaffner & Marx, 760 F.2d 1383 (3d Cir.) (senior user can establish common law rights in geographic areas where it achieved market penetration), *cert. denied,* 474 U.S. 920, 106 S. Ct. 249, 88 L. Ed. 2d 257 (1985)....

L'Oreal concedes that ZHD has exclusive rights to use ZAZU for salon services in the Hinsdale area. Those exclusive rights also preclude L'Oreal from using the mark on hair products in the local area because of the likelihood of confusion between those products and ZHD's salon services, even apart from any confusion between the two parties' products. Given the deferential standard of review on the factual question of use, therefore, I think it clear that ZHD has achieved market penetration and exclusive rights to the ZAZU mark at the very least in the Chicago area.

ZHD's contention that its rights in the ZAZU mark extend beyond the local area is enhanced by evidence that L'Oreal did not, as we have noted, act in good faith. The majority's consideration of the good faith issue minimizes the important role good faith plays in trademark disputes, particularly disputes involving unregistered marks.... See, e.g., A. J. Canfield Co. v. Honickman, 808 F.2d 291 (3d Cir. 1986) (stating the doctrine that a senior user "has enforceable rights against any junior user who adopted the mark with knowledge of its senior use").... Contrary to the majority's narrow characterization of bad faith as a concept employed solely to deter attempts to reserve marks prior to genuine sales, courts have examined junior users' good faith in a variety of contexts. In fact, this court has held that a good faith in a variety of contexts. In fact, this court has held that a good faith junior user is simply one that begins using a mark without knowledge that another party already is using it. The Money Store v. Harriscorp Finance, Inc., 689 F.2d 666, 674 (7th Cir. 1982); see 2 McCarthy, supra, § 26:4 at 292 (equating good faith to "the junior user's lack of knowledge"). And while such knowledge may not automatically negate good faith, only the most unusual situations encompass both knowledge and good faith....

COMMENTS AND QUESTIONS

1. The majority opinion mentions that Zazu Hair Design, "ZHD," had registered its "trade name" as a "service mark" in Illinois in 1980. But how does prior use of a trade name or "service mark" such as Zazu Hair Design, used on a hair salon, affect subsequent use of the same term for hair color products? See Malcolm Nicol & Co. v. Witco Corp., 881 F.2d 1063, 1065 (Fed. Cir. 1989) (a trade name, even one that lacks any independent trademark or service mark significance, may bar registration of a trademark or service mark that is confusingly similar to that trade name).

2. The *Zazu* case was decided under pre-1989 federal trademark law. As noted earlier, Congress in 1989 made several important changes in trademark law, including the adoption of "intent to use" registration under section 1 of the Lanham Act. See Trademark Law Revision Act of 1989 (hereinafter "TLRA"), Pub. L. No. 100-667, 102 Stat. 3935, codified at 15 U.S.C. § 1051. Under pre-1989 law, *actual* use in commerce, prior to application for registration, was a requirement for registration.

This requirement spurred prospective applicants to ship and/or sell a small batch of goods in order to secure trademark protection, a practice that came to be known as "token use."

Among other things, the TLRA amended section 1 of the Lanham Act, 15 U.S.C. §1051, to provide that "[a] person who has a bona fide intention, under circumstances showing the good faith of such person, to use a trademark in commerce may apply to register the trademark...on the principal register." Assuming an application based on an intention to use a mark is otherwise allowable, the Trademark Office will issue a "notice of allowance" to the trademark owner (rather than simply registering the mark on the Principal Register). 15 U.S.C. §1063(b)(2). After the notice of allowance is granted, the trademark owner has six months (extendable to one year automatically and to three years for good cause shown) to submit a verified statement that the trademark has in fact been used in commerce, at which point it is entered on the Principal Register. If the trademark owner does not submit such a statement, the trademark is considered abandoned. 15 U.S.C. §1051(d). Assuming that the intent-to-use registrant does eventually use the mark, however, the initial application will be considered "constructive use," entitling the registrant to nationwide priority from the date of the application. 15 U.S.C. §1057(c).

The TLRA was passed largely to bring United States law into harmony with the law in the rest of the world. No country other than the United States required actual use before a mark could be registered. Because international treaties required the United States to give full credit to foreign applicants who had registered marks outside the United States, the United States use requirement in practice meant that *only* U.S. citizens were required to show use in commerce in order to register a mark in the United States.

Is the intent-to-use provision constitutional? Recall that the original United States trademark statute was struck down as unsupported by the patent and copyright clause, and that subsequent trademark statutes (including the Lanham Act) were then enacted under the Commerce Clause. The old requirement in the Lanham Act that registered marks be used in interstate commerce was thought necessary to invoke federal jurisdiction under the Commerce Clause. Does an intent to use a product in interstate commerce confer federal jurisdiction? Compare Charles J. Vinicombe, The Constitutionality of an Intent to Use Amendment to the Lanham Act, 78 Trademark Rep. 361, 369-373 (1988) (courts likely to defer to Congress by assuming that an intention to use a mark means that it is part of the "flow of commerce") with Kenneth L. Port, Foreword: Symposium on Intellectual Property Law Theory, 68 Chicago-Kent L. Rev. 585, 595 n.63 (1993) (Trademark Law Revision Act "should fail constitutional review based on the Commerce Clause").

3. It is widely believed that the TLRA has eliminated "token use" by raising the standard for determining when a mark has been "use[d] in commerce." Consider for example the case of Paramount Pictures Corp. v. White, 31 U.S.P.Q.2d 1768 (TTAB 1994) aff'd 108 F.3d 1392 (Fed Cir. 1997) (unpub.), where Paramount opposed registration of "The Romulans" for a connect-the-dots game distributed by White, leader of a rock group called The Romulans. One ground for the opposition was that the mark had not been used in commerce; in particular, that the distribution of connect-the-dots games on promotional fliers for the band was not a statutory use in commerce justifying registration. In commenting on the magnitude of use, the Trademark Trial and Appeal Board stated: "The legislative history of the Trademark Law Revision Act reveals that the purpose of the amendment was to eliminate 'token

use' as a basis for registration, and that the new stricter standard contemplates instead commercial use of the type common to the particular industry in question." 31 U.S.P.Q.2d at 1774. Footnote 8 of the Board's opinion quotes from the Congressional Record of November 19, 1987, p. 196-197:

> Amendment of the definition of "use in commerce" is one of the most far-reaching changes the legislation contains. Revised to eliminate the commercially-transparent practice of token use, which becomes unnecessary with the legislation's provision for an intent-to-use application system, it will have a measurable effect on improving the accuracy of the register.... The committee intends that the revised definition of "use in commerce" be interpreted to mean commercial use which is typical in a particular industry.

Id. But cf. Allard Ent., Inc. v. Advanced Programming Resources, Inc., 146 F.3d 350 (6th Cir. 1998) (plaintiff's somewhat extensive "word-of-mouth" campaign to popularize "APR" mark for computer professional placement service established priority under post-TLRA law). For an example of the pre-1989 rule, see Fort Howard Paper v. Kimberly Clark Corp., 390 F.2d 1015 (C.C.P.A. 1968), *cert. denied*, 393 U.S. 831 (1968) (very limited use sufficient to establish priority). But cf. La Societe Anonyme des Parfums LeGalion v. Jean Patou, Inc., 495 F.2d 1265 (2d Cir. 1974) (token sales program not sufficient use to avoid abandonment); Procter and Gamble v. Johnson & Johnson, Inc., 485 F. Supp. 1185 (S.D.N.Y. 1979), *aff'd without opinion*, 636 F.2d 1203 (2d Cir. 1980) ("minor brands program" not sufficient use).

Could it be argued in the *Zazu* case that ZHD's shipments of hair products were only a "token use"? How might the TLRA have affected the dissent in *Zazu*? Would L'Oreal's good faith still be an issue under the TLRA?

4. Consider this passage from Judge Easterbrook's opinion: "Liberality in registering marks is not problematic, because the registration gives notice to latecomers, which token use alone does not. Firms need only search the register before embarking on development. Had ZHD registered ZAZU, the parties could have negotiated before L'Oreal committed large sums to marketing."

But ZHD *did* register Zazu — not as a federal trademark but as a business (trade name) in the State of Illinois. Why is this any different from the *national* registration contemplated by Easterbrook? Does it provide less opportunity for negotiating prior to large investments? Recall the evidence in the case, which established not only that L'Oreal did find the ZHD trade name in its pre-product introduction search, but also that L'Oreal did contact ZHD. Standard trademark search services generally find all state and federal registrations, together with many "common law" (i.e., nonregistered) uses. Given this evidence, why was state registration any less of a basis for negotiation than federal registration?

5. Judge Easterbrook says that through the use requirement, "the law prevents entrepreneurs from reserving brand names in order to make their rivals' marketing more costly." Could a rival reserve all the potential trademarks that would allow a firm to identify its products? Cf. Stephen L. Carter, The Trouble With Trademarks, 99 Yale L.J. 759, 760 (1990):

> The traditional economic justification for trademark law rests on the premise that the set of available marks is virtually infinite and, in consequence, that the actual mark chosen by a firm to represent its goods is irrelevant. If that assumption turns out to be false — if even before the public comes to associate a mark with any particular goods or services, some

marks are more desirable than others—then allowing protection of marks devoid of market significance may raise substantial barriers to entry by competitors. For this reason, I . . . argue, the theoretical case is clearer for the common law model than for the Federal scheme.

Note that another rationale for the use requirement stems not from rivals' costs, but from the desire not to encourage firms to specialize in identifying and registering potential trademarks. See the note below on "Priority and Trademark Theory."

6. Judge Easterbrook provides another explanation for the use requirement: "Public sales let others know that they should not invest resources to develop a mark similar to one already used in the trade." Is there a better and more accurate way to do this? After all, public sales may be quite limited. Does the new federal intent-to-use registration process serve this purpose more effectively?

7. An important case, Blue Bell, Inc. v. Farah Mfg. Co., 508 F.2d 1260 (5th Cir. 1975), teaches that although the race is to the swift in trademark law, the race must be run cleanly if a firm wants to prevail. In the case, Farah's management settled on the name "Time Out" for a new line of blue jeans on May 16. Sample tags using the new name were drawn up on June 27, and on July 3 Farah sent out 12 pair of jeans bearing the new mark to its regional sales managers. More extensive shipments occurred on July 11. Meanwhile Blue Bell management decided on the name "Time Out" for *its* new line of jeans on June 18. Blue Bell commissioned several hundred sample tags (bearing the new logo) that were attached over the top of existing tags on a large shipment of jeans sent out on July 5. By October both firms had received substantial orders for their respective new lines of jeans. The court ruled (1) that Blue Bell's "secondary" use of the new logo was in "bad faith" and therefore its July 5 shipment did not establish priority and (2) that Farah's minimal shipments on July 3 also did not establish priority. This left Farah's July 11 shipment as the first substantial use of the new mark in commerce—so Farah won.

When Blue Bell slapped new labels on old jeans, the court condemned it as "a bad faith attempt to reserve a mark." What if, subsequent to the relabelling but before Farah's first shipments, Blue Bell had made actual sales to consumers; would their prior bad faith deprive them of priority? Should it? Why did the court not characterize Farah's sales of pants to employees as an attempt to "reserve a mark"? For more on the myriad roles that bad faith plays in trademark cases, see below, sections D.2, Likelihood of Consumer Confusion, and E.3, Abandonment.

8. Both Blue Bell and Farah began advertising prior to their first sales to customers, yet ultimately the respective sales dates to consumers determined priority. Can advertising or other "promotional activities" ever form the basis for priority?

Maryland Stadium Authority v. Becker, 806 F. Supp. 1236 (D. Md. 1992), *aff'd*, 36 F.3d 1093 (4th Cir. 1994), is instructive on this point. As early as 1987, the Maryland Stadium Authority (MSA) used the name "Camden Yards" in promotional materials (widely distributed to the press and public) describing the new stadium for the Baltimore Orioles baseball team. A number of other names were under consideration at the time, however. In 1989, MSA distributed to the press and sold to the public photographic renditions of the proposed ballpark labeled with the name Camden Yards—again, before the name had been finalized. Defendant Becker, meanwhile, began selling baseball T-shirts under the "Camden Yards" mark in July 1991, shortly after the ballpark's name was made final and official but before the park opened. In an action by MSA against Becker for trademark infringement under

Maryland law, the court found that Becker had intentionally adopted MSA's mark, creating a presumption of the likelihood of confusion that Becker could not overcome. The totality of MSA's activities constituted adoption and use of the mark, generated secondary meaning, and was sufficient to establish priority over Becker. Compare Paramount Pictures Corp. v. White, 31 U.S.P.Q. 2d 1768 (TTAB 1994) (use of name The Romulans on a connect-the-dot game distributed by a rock band held mere "promotional use" and hence not use in commerce under § 1 of the Lanham Act). But cf. Societe de Developments et D'Innovations des Marches Agricoles et Aliminetaries-Sodima Union de Cooperatives Agricoles v. International Yogurt, 662 F. Supp. 839, 847 (D. Ore. 1987) ("An axiom of trademark law is: no trade, no trademark."). Note the emphasis on "bad faith" in the *Maryland Stadium* case. Does this distinguish it from cases such as *Paramount?* Or is it another instance of Judge Easterbrook's observation in the *Zazu* opinion that bad faith can be merely "an epithet stapled to...[a] conclusion"?

Another use-based limitation on trademark rights is highlighted by Buti v. Impressa Perosa S.R.L., 139 F.3d 98 (2d Cir. 1998). In that case, a restaurant located in Milan, Italy, named "Fashion Café" advertised its business in the United States just before Buti began using the same mark for restaurants in the United States. The court held that the Italian restaurant could not obtain United States trademark rights through its advertising because the services it was providing were not themselves offered in commerce in the United States.

Note on Geographic Limitations on Trademark Use

At common law (and today for unregistered marks), ownership of a trademark does not necessarily confer nationwide protection. Rather, common law trademarks are protected only in the areas where the marked products are sold or advertised. Thus the owner of an unregistered trademark for goods sold in Oregon and Washington, but not elsewhere, is entitled to prevent others from using that mark for similar goods only in Oregon and Washington. The rationale is that trademarks are not intended to confer a broad property right but merely to protect the goodwill the trademark owner has invested in the mark. Because no one outside Oregon or Washington could associate the mark with the owner, there is no reason to protect it elsewhere. Thus a seller of similar goods in New York can use the same name for the goods without conflict.

There are two exceptions to this common law rule, both based on concerns that the trademark owner's goodwill will be unfairly taken. First, a trademark owner is entitled to the exclusive use of her mark in any geographic area in which the mark's reputation has been established, even if the product is not sold in that geographic area. Such a broader geographic reputation might be established, for example, by national advertising or media coverage of a local business such as a restaurant. Further, the trademark owner is entitled to protect the mark in a territory which he is expected to reach in the normal expansion of his business, even if there is no current likelihood of confusion in that area. See Hanover Star Milling Co. v. Metcalf, 240 U.S. 403 (1916). Second, a trademark owner is entitled to prevent anyone from intentionally trading on her goodwill, even outside her established geographic area. Only innocent (or "good faith") use of the same mark is protected.

One of the principal advantages of trademark registration is that it automatically confers nationwide protection of the mark, retroactive to the date of the trademark application, even if the goods for which the mark is used are sold or advertised in only a small part of the country. Thus trademark registration is vital to protect businesses that plan to expand geographically, as well as those that fear a large national company might use the same name.

What happens, then, if two parties use the same mark for the same goods? If neither party registers its mark, then the common law rule applies. Each party is entitled to exclusive use of the mark in the areas where it has established goodwill. Should the two marks come into conflict in a particular geographic area, the conflict will be resolved in favor of the earliest user *in that area*. If one party registers her mark and the other does not, the registrant will generally be entitled to the exclusive right to use the mark throughout the country. However, the non-registering party may assert a "limited area" defense. This defense allows the non-registering party to claim priority in those geographic areas where he has made continuous use of the mark since *before* the registering party filed her application. The non-registering party is "frozen" in the use of his mark, however, and cannot expand it outside his existing territory or a natural "zone of expansion." See Weiner King, Inc. v. Wiener King Corp., 615 F.2d 512 (C.C.P.A. 1980).

Finally, if both parties file for registration of marks that were in use in different areas at the same time, the Trademark Office will generally declare an "interference" between the two applications. But if the parties agree, or if the Trademark Board determines that registration of both marks is unlikely to cause confusion, it is possible that both marks may be registered for "concurrent use."[8] If two or more marks are registered concurrently, however, the Trademark Office will impose whatever restrictions on the use of the marks are necessary to prevent confusion among consumers.

COMMENTS AND QUESTIONS

1. Compare the current U.S. trademark system to the various schemes for filing patent applications. You will recall that the United States has long used a "first to invent" rule for determining who is entitled to own a patent, while the rest of the world awards a patent to the "first to file." Is there a similar distinction in trademark law, between "first to use" and "first to file"? If so, does the Trademark Law Revision Act turn the United States into a "first to file" system?

2. Why should the user of an arbitrary or suggestive, but unregistered, trademark be limited to protection in a particular geographic area? The asserted justification is that the trademark owner has established goodwill only in that limited area. But why should that matter? Distinctive marks, unlike descriptive marks, are entitled to automatic protection under trademark law without proof that the public associates them with a particular product. See *infra* section C.1. Isn't it inconsistent to limit the scope of that protection to geographic areas in which the public has formed such an association?

3. Priority disputes are often resolved by the Trademark Trial and Appeal Board through oppositions or interferences. Section 13 of the Lanham Act, 15 U.S.C. § 1063, provides that "[a]ny person who believes that he would be damaged by the registration of a mark upon the principal register may...file an opposition in

---

8. Concurrent use applies only to two marks that are actually in use, not to applications based on an intent to use.

the Patent and Trademark Office, stating the grounds therefor." Further, 15 U.S.C. § 1062(a) expressly provides that trademark applications be published before issuance, so that interested parties may have the opportunity to search for and oppose potentially damaging applications. Applications may be opposed by showing that the mark is not entitled to registration, for example because others had made use of it before the applicant did.

Often the party objecting to an application will herself have filed an application for the same or a similar mark. In that circumstance, the Trademark Commissioner may declare an interference between the two applications. Under section 16 of the Lanham Act, 15 U.S.C. § 1066, interferences may also be declared between pending applications and registered trademarks (unless they have become "incontestable"). Interferences are heard by the Trademark Trial and Appeal Board, which may register neither, either, or both marks, and may impose restrictions or conditions on the registration of the mark. 15 U.S.C. §§ 1067, 1068. Both oppositions and interference decisions may be appealed to the Court of Appeals for the Federal Circuit.

Note on Priority and Trademark Theory

Do any of the tests for priority discussed so far make sense in terms of lowering consumer search costs? Consider: If a second user can freely appropriate a mark, then consumers who have begun to rely on the association between the mark and the first user's product will be thwarted. This situation not only destroys that particular association with the first user's product; it may also make consumers less likely to establish such associations in the future. The upshot is that unless we protect the rights of the first user, more consumers will spend more time searching for goods.

On the other hand, many priority cases involve very limited uses in commerce; these cases typically occur early in the life of a new product or marketing campaign. As a consequence, very few consumers will, at the time of the litigation, have come to associate the mark in question with any goods. Where this is so, we might consider issues other than absolute priority to be important. For example, we might ask whether the first or second user was better positioned to distribute the goods bearing the mark. If the second user was in a better position — e.g., was larger, had more money to spend on advertising, etc. — why not let it use the mark? If consumer search costs are the key, why not take into account the interests of *future* consumers, who may be better served by allowing a search-cost-reducing trademark to fall into the hands of a large company that can make best use of it?

Then again, if the second user can really make better use of the mark, wouldn't it buy the trademark from the first user? When this is so, an award of priority to the first user will give it a share of the value of the market that will ultimately be served by the second user. Perhaps the payment to the first user will compensate it to some extent for the effort expended in coming up with a mark that, in the second user's hands, has real commercial value. (Notice that if this is so, it is a far cry from the consumer protection theory of trademark protection; it comes closer to the property theory described in the Introduction. No one is paying the consumers whose expectations are dashed.)

Priority by Contract: Covenants Not to Sue and the Like. Notice in this connection the $125,000 paid by L'Oreal to Riviera Slacks, Inc. in *Zazu*. Recall that Riviera was an early user of the Zazu mark on related goods (clothes). (Note also that the court characterizes this agreement as a "covenant not to sue." For an explanation of this characterization — as opposed to "trademark license agreement" — see below, section E.3.a.) This payment is interesting for several reasons. First, it suggests that priority is not a once-and-for-all determination. Since the first user can license the trademark to the second-comer, an award of priority does not determine for all time who can use the mark. In some cases, it merely sets the scene for a subsequent contract. Where the second-comer is in a better position to exploit the mark, an award of priority to the first user may simply result in an exchange of money for the right to use the mark. On the other hand, where the first user can more efficiently exploit the mark, economic theory suggests that it will continue to do so.

Priority and the Prevention of Trademark Races. Why should the law grant rights to the first user of a mark? Consider the alternative: if two entities have to "race" to establish nationwide recognition for their mark, will they spend more money widely distributing their new product — and advertising it — than they would otherwise? Would such extensive, early promotional efforts be wasteful? Is the role of trademark priority to forestall such wasteful expenditures and instead promote a more rational product "rollout" nationwide? Should the parties ever be able to divide up the nation?

Some guidance may be gleaned from a prominent justification for secure title in real property holdings. It has been argued that without secure property rights, assigned in advance, those seeking use of a resource will make wasteful expenditures seeking to claim it. The idea is that the orderly, rational development of the resource will be distorted by the absence of property rights. For example, consider "land rushes," such as the famous Oklahoma Land Rush. See Terry L. Anderson & Peter J. Hill, The Race for Property Rights, 33 J.L. & Econ. 177 (1990); R. Taylor Dennen, Some Efficiency Effects of Nineteenth-Century Federal Land Policy: A Dynamic Analysis, 51 Agric. Hist. 718 (1977); David D. Haddock, First Possession Versus Optimal Timing: Limiting the Dissipation of Economic Value, 64 Wash. U. L.Q. 775 (1986). Economic historians have detected evidence of wasteful spending by prospective claimants to real property. See generally Gary D. Libecap, Contracting for Property Rights (1989). Instead of allocating early expenditures rationally — e.g., some money for land, some for seeds, fertilizer, and building materials — people put all their money into the pursuit of land claims. The upshot was that the land was not utilized in an efficient manner. (Compare this line of reasoning to the "prospect theory" of patents described in Chapter 3.)

If the analogy between land development and trademark investment makes sense, why not go all the way to a pure registration system, under which virtually no expenditures need be made to secure trademark rights?

As another branch of economic theory would predict, "pure" registration systems — those where broad rights can be acquired without actual use in commerce — have been known to give rise to the scattershot acquisition of numerous trademarks solely for their resale value to real prospective users. In short, pure registration also invites "rent seeking." Cf. William M. Landes & Richard A. Posner, Trademark Law: An Economic Perspective, 30 J.L. & Econ. 265, 275 (1987). Consider the case of Robert Aries, who in 1965 had the foresight to register over 100 valuable American trademarks, including Pan American, NBC, Texaco, Monsanto, and Goodrich, with the National Trademark Office of Monaco. After registering the trademarks, Mr. Aries forced the American companies to buy their own marks back from him.

Gerald D. O'Brien, The Madrid Agreement Adherence Question, 56 Trademark Rep. 326, 328 (1966). Similar practices were well known under French trademark law until recent reforms. See, e.g., Andre Armengaud, The New French Law on Trademarks, 56 Trademark Rep. 430, 435-36 (1966):

> In France, during the last few years before the enactment of the 1964 Act, trademark registrations were becoming more and more numerous. This was due to the fact that many persons to whom a fancy name would come to mind, would register the name with the ulterior motive of obtaining some financial return from a possible subsequent user should the occasion arise. As a result, the area of choice for marks for new products was becoming narrower every day. Another drawback was that a merchant, or a manufacturer engaged in a particular trade area like hosiery for instance, would register his mark in all thirty-four classes. The relatively low cost of trademark registration in France, negligible as compared to the high costs a large company usually bears for advertising, made such practices possible.... [T]hese two practices...were responsible for the tremendous volume of trademark registration.

On the end of these practices in France, see Gerard Dassas, Survey of Experience Under the French Trademark Law, 66 Trademark Rep. 485, 491 (1976).

A similar phenomenon has occurred more recently with Internet domain names. The registration of domain names was initially handled by Network Solutions, Inc. (NSI) and is now overseen by a variety of domain name registries operating under the auspices of a non-profit entity called the Internet Corporation for Assigned Names and Numbers (ICANN). These entities have allocated domain names on a first-come, first-served basis for a modest fee. During its initial period of operation, this system wreaked havoc among trademark holders as "cyber-pirates" rushed in to register well-known trademarks held by others. These registrants would then offer to sell them to the trademark owners at exorbitant prices. Some companies even registered the marks of their competitors and used the sites to put up comparative product information. See, e.g., Joshua Quittner, Making a Name on the Internet, Newsday, Oct. 7, 1994, at A4 ("It['s]...like a gold rush: Two thousand requests a month are coming in to stake claim to a name on the Internet, nearly 10 times as many as a year ago."). As we will discuss below, courts, ICANN, and legislatures developed rules and dispute resolution processes to curtail such extortionate rent seeking. The potential to profit from registering domain names closely related to the trademarks of others continues to occur, although the opportunities for extracting value from trademark owners have been greatly reduced and the penalties for crossing the line greatly enhanced. See, e.g., Shields v. Zuccarini, 254 F.3d 476 (3d Cir. 2001) (enjoining "typo-squatting" — registering misspelled versions of trademarks — and imposing a large fine under the Anticybersquatting Consumer Protection Act).

There seems to be good reason to allow some sort of early claiming system (i.e., registration with national effect after minimal use) without going all the way to "pure" registration (which invites equally wasteful rent seeking). An alternative would be to allow pure registration with lapse for nonuse after some period of time. Notice that this quite adequately describes the current system of "intent to use" registration.

Incentives to Create Trademarks. Recall the respective efforts of Zazu Hair Design (ZHD), the plaintiff in the case that begins this section, and Wordmark, the consulting firm that L'Oreal hired to help identify a catchy name for its new hair coloring products. Why should ZHD potentially receive rights to the trademark when it appears that Wordmark thought to use Zazu as a product name first? (ZHD had only been using it as a trade name.) In other words, why doesn't the law grant rights to firms that specialize in creating new trademarks? Why must a mark be "used in commerce," as the statute phrases it, to receive legal protection? An argument for abolishing this requirement was made by Stephen Carter in his article, Does It Matter Whether Intellectual Property Is Property?, 68 Chicago-Kent L. Rev. 715, 721 (1993): "But perhaps we should say, So what? How simple and elegant it would be to conclude that secondary meaning is unnecessary because the first to appropriate the mark owns it; owns it not because of its representational nature, but because it is a product of the mind."

Consider some counterarguments. In a seminal article written more than 50 years ago, Ralph S. Brown, Jr. highlighted the dual nature of trademark as serving both private and public interests. See Ralph S. Brown, Jr., Advertising and the Public Interest: Legal Protection of Trade Symbols, 57 Yale L.J. 1165, 1167 (1948). Expanding on this theme, Professor Wendy Gordon observes that:

> Promotional value, good will, popularity, and similar elements of value are joint products of both the public and the creator to a greater extent than are intellectual products themselves. True, even standard intellectual products — collocations of words, music, scenes — will be beneficial only if someone appreciates them; labor is never the only source of value, even for Locke. But with standard intellectual products the active role of the producer and the comparatively passive role of the public makes it easier to assign the resulting value primarily to the laborer. By contrast, with products such as popularity and "commercial magnetism," the chain of causality and responsibility is much harder to trace.... Usually a trademark is not purposely created for its own sake; the "benefit" purposely created is the good will of the owning entity (such as its reputation for manufacturing high-quality products), which the trademark merely happens to represent.

Wendy J. Gordon, A Property Right in Self-Expression: Equality and Individualism in the Natural Law of Intellectual Property, 102 Yale L. J. 1533, 1588 n.277 (1993); cf. Hal Morgan, Symbols of America (1987) (noting broad cultural impact of well-known trademarks). Note the skepticism Professor Gordon apparently feels regarding the claim that trademarks are brought about by legal incentives. This skepticism is shared by Rochelle Dreyfuss. See Rochelle C. Dreyfuss, Expressive Genericity: Trademarks as Language in the Pepsi Generation, 65 Notre Dame L. Rev. 397, 399 (1990) ("[T]here is little need to create economic incentives to encourage businesses to develop a vocabulary with which to conduct commerce....").

What do you think would happen if the law did give rights to firms such as Wordmark? Would it deplete the "stock" of potential trademarks available to actual manufacturers and sellers? Would it be burdensome if many firms who wanted to start a business had to go first to a firm like Wordmark to "shop for a trademark"? How would the parties agree on a fair price for a trademark?

Keep in mind when thinking about these issues that even though firms such as Wordmark are not granted a legal right under the Lanham Act, such firms exist anyway. Indeed, from published reports, it would appear that the image/identity industry is thriving. Is this a good argument against a system that granted rights to these firms?

Do you suppose ZHD would have a cause of action against Wordmark? How about L'Oreal, on the grounds that Wordmark chose a trademark that wound up in costly litigation? Perhaps in the contract specifying Wordmark's services, Wordmark expressly disclaims liability for subsequent litigation.

COMMENTS AND QUESTIONS

1. The *Zazu* case and its ilk center on what constitutes "use" for trademark purposes. But what happens when the "use" asserted by a plaintiff originates not with the seller of goods itself, but with the common parlance of consumers? Consider, for example, Volkswagenwerk A.G. v. Advanced Welding & Mfg. Co., 193 U.S.P.Q. 673 (TTAB 1976), which concerned trademark rights in the word "Bug." To Volkswagen (VW), the official moniker for its classic economy car was "Type I," or, later, "Beetle." But in common parlance, this car was universally referred to as a "Bug." Importantly, it was the consuming public, and not VW, that originated this usage. Should trademark law protect the association between VW and "Bug," even though "Bug" originated with consumers themselves? If so, it is a reflection of the "consumer protection" rationale for trademarks, as opposed to the incentive/property rationale. (Do you see why? How much did VW invest in creating the "Bug" mark?) Courts tend toward the view that consumer associations should be protected in this context. Id. See also National Cable Television Assn. v. American Cinema Editors, Inc., 937 F.2d 1572 (Fed. Cir. 1991) (ACE used in common parlance as an acronym for association of film (cinema) editors).

2. Difficult priority problems arise when a well-known trademark is abandoned by its original owner. The priority question involves a race among rivals to capture the mark. See California Cedar Products Co. v. Pine Mountain Corp., 724 F.2d 827, 830 (9th Cir. 1984) (bad faith of first rival to claim abandoned Duraflame mark for ersatz fireplace logs negates its claim, leaving second rival to claim the mark with priority). Are the issues the same as when firms race to obtain rights to a new mark? Under a strict consumer protection rationale for trademarks, is there an argument that abandoned marks should be off-limits to rival firms, at least for a number of years? Why would a firm abandon a mark when it has value to rivals — why not sell it? (See the discussion below on Assignments in Gross, in the section on Licensing, section E.3.b.)

Note on Secondary Meaning in the Making

Consider how the doctrine of priority interacts with the doctrine of secondary meaning. Under trademark priority rules, the trademark is presumptively owned by the first person to use it in commerce (barring a federal registration). But the secondary meaning doctrine provides that *descriptive* marks are not entitled to protection until their owner can prove secondary meaning in the minds of consumers. So when does a trademark owner obtain priority of use in a descriptive mark? When she first uses the mark? Or only after she can establish secondary meaning?

This issue has arisen in a number of cases where the defendant is accused of quickly adopting a plaintiff's descriptive mark before the plaintiff can establish secondary meaning. In Laureyssens v. Idea Group, Inc., 964 F.2d 131 (2d Cir. 1992), the court considered a trade dress infringement suit by the makers of "Happy Cube" 3-D puzzles against the makers of "Snafooz" puzzles:

> The district court found that there was no serious question whether actual secondary meaning exists in the HAPPY CUBE trade dress. We think this conclusion is sound given the weak sales of the HAPPY CUBE puzzles, low expenditures for advertising and promotion, minimal unsolicited media coverage, and the brief period of exclusive

use of the HAPPY CUBE trade dress.[4] And, while there was evidence of intentional imitation as to the puzzles themselves, there was no evidence of copying of the trade dress.

The district court concluded, however, that Laureyssens satisfied the requirement of secondary meaning by raising a serious question whether the flat-form, shrink-wrapped HAPPY CUBE trade dress should be protected under the doctrine of secondary meaning in the making.

In Metro Kane Imports, Ltd. v. Federated Dept. Stores, Inc., 625 F. Supp. 313, 316 (S.D.N.Y. 1985), *aff'd*, 800 F.2d 1128 (2d Cir. 1986), Judge Sweet explained that a trade dress will be "protected against intentional, deliberate attempts to capitalize on a distinctive product" where "secondary meaning is 'in the making' but not yet fully developed." See also Jolly Good Industries, Inc. v. Elegra, Inc., 690 F. Supp. 227, 230-31 (S.D.N.Y. 1988) (indicating that the theory has been "well-received by commentators" and citing as an example, 3 R. Callman, The Law of Unfair Competition, Trademarks and Monopolies §77.3, at 356 (3d ed. 1971)).[5] The supposed doctrine seeks to prevent pirates from intentionally siphoning off another's nascent consumer recognition and goodwill. See, e.g., Jolly Good, 690 F. Supp. at 230-31.

In this case, Judge Sweet found that although Idea Group offered evidence that the SNAFOOZ packaging was developed without prior knowledge of the HAPPY CUBE trade dress, "the evidence does not indicate when [Idea Group] developed its flat-form shrink wrapped package, and therefore Plaintiffs have barely established a serious question of [Idea Group's] 'intentional deliberate attempts' to copy their trade dress." 768 F. Supp. at 1048.

We are, then, squarely presented for the first time with the question whether the doctrine of secondary meaning in the making should be recognized under the Lanham Act....

... "The doctrine, if taken literally, is inimical to the purpose of the secondary meaning requirement." Restatement, supra, §13 reporter's note, comment *e* at 53. The secondary meaning requirement exists to insure that something worth protecting exists—an association that has developed in the purchasing public's mind between a distinctive trade dress and its producer—before trademark law applies to limit the freedom of a competitor to compete by copying. As the drafters of the Restatement, supra, §17 comment b at 104-05, explain:

> The freedom to copy product and packaging features is limited by the law of trademarks only when the copying is likely to confuse prospective purchasers as to the source or sponsorship of the goods. The imitation or even complete duplication of another's product or packaging creates no risk of confusion unless some aspect of the duplicated appearance is identified with a particular source. Unless a design is distinctive...and thus distinguishes the goods of one producer from those of others, it is ineligible for protection as a trademark.

See also Norwich Pharmacal Co. v. Sterling Drug, Inc., 271 F.2d 569, 572 (2d Cir. 1959) ("Absent confusion, imitation of certain successful features in another's product is not unlawful and to that extent a 'free ride' is permitted."), *cert. denied*, 362 U.S. 919, 4 L. Ed. 2d 739, 80 S. Ct. 671 (1960); Perfect Fit Industries, Inc. v. Acme Quilting Co., 618 F.2d 950, 952-53 (2d Cir. 1980). The so-called doctrine of secondary meaning in the making, by affording protection before prospective purchasers are likely to associate the trade dress with a particular sponsor, constrains unnecessarily the freedom to copy and compete.

---

4. The record indicates that the HAPPY CUBE trade dress was adopted in 1990, and for some of that time Idea Group was marketing its puzzles in the allegedly infringing trade dress.

5. The subsequent edition of Callman's treatise, however, contains no language expressing approval for the doctrine of secondary meaning in the making. See 3 R. Callman, The Law of Unfair Competition, Trademarks, and Monopolies §19.27 (4th ed. 1989).

The Eighth Circuit previously recognized the improper focus of the concept of secondary meaning in the making in Black & Decker Mfg. v. Ever-Ready Appliance Mfg., 684 F.2d 546, 550 (8th Cir. 1982): "Such a theory focuses solely upon the intent and actions of the seller of the product to the exclusion of the consuming public; but the very essence of secondary meaning is the association in the mind of the public of particular aspects of trade dress with a particular product and producer." See also Scagnelli, Dawn of a New Doctrine? — Trademark Protection for Incipient Secondary Meaning, 71 Trademark Rep. 527, 542-43 (1981).

The argument in favor of permitting development of a doctrine of secondary meaning in the making, offered by Laureyssens, rests principally on the supposition that, without such a doctrine, there will be strong incentives for pirates to capitalize on products that have not yet developed secondary meaning. This argument, however, underestimates the level of protection afforded under existing law to prevent piracy in the early stages of product development. See Scagnelli, supra, at 543-49. For example, intentional copying is "persuasive evidence" of secondary meaning. Coach Leatherware, 933 F.2d at 169 (noting, however, that "conscious replication alone does not establish secondary meaning"); see also Restatement, supra, § 17 comment b at 106. Furthermore, secondary meaning can develop quickly to preclude knock-off artists from infringing. See Maternally Yours, Inc. v. Your Maternity Shop, Inc., 234 F.2d 538, 541 (2d Cir. 1956) (secondary meaning acquired in mark MATERNALLY YOURS for maternity apparel store in the 11 months preceding defendant's opening of store named YOUR MATERNITY SHOP). Finally, under New York's common law of unfair competition, a producer's trade dress is protected without proof of secondary meaning against practices imbued with an odor of bad faith. See Saratoga Vichy Spring Co. v. Lehman, 625 F.2d 1037, 1044 (2d Cir. 1980). These practices include palming off, actual deception, appropriation of another's property, see Norwich Pharmacal, 271 F.2d at 570-71; Upjohn Co. v. Schwartz, 246 F.2d 254, 261-62 (2d Cir. 1957), or deliberate copying. See Morex S.P.A. v. Design Inst. of Am., Inc., 779 F.2d 799, 801-02 (2d Cir. 1985); Perfect Fit Indus., 618 F.2d at 952-54; Restatement, supra, § 16 reporter's note at 115. Therefore, true innovators, at least under New York law, have adequate means of recourse against free-riders.

For these reasons, we reject the doctrine of secondary meaning in the making under section 43(a) of the Lanham Act. Accordingly, we reverse the district court's decision to grant a preliminary injunction against Idea Group based on section 43(a) of the Lanham Act. Given our holding, we need not address the likelihood of confusion which may have been created by the SNAFOOZ trade dress.

COMMENTS AND QUESTIONS

1. Courts have been virtually unanimous in rejecting the idea of protecting "secondary meaning in the making" as inconsistent with the idea that descriptive marks are not protected. Indeed, many courts expressly require that a trademark plaintiff obtain secondary meaning before the defendant begins any use of the term. See McCarthy, Trademarks and Unfair Competition § 16.12, at 16-40 to 16-43 (collecting cases). Cf. Fuddruckers, Inc. v. Doc's B.R. Others, Inc., 826 F.2d 837 (9th Cir. 1987) (plaintiff was entitled to protect restaurant with trade dress nationally recognized among travelers against infringement by restaurant in Arizona, even though plaintiff had not established secondary meaning in Arizona directly).

What happens to trademarks that acquire secondary meaning too late? Are they wholly unprotectable? Or can they be asserted against a third party who uses the mark after secondary meaning is acquired? Note that the Trademark Office will register a descriptive mark that is first to acquire secondary meaning, regardless of when other users began use without secondary meaning. Presumably that registration confers some rights, at least against subsequent adopters. But does it make sense that earlier users are "grandfathered" into the market and can continue to use the mark in competition with its owner?

2. What is wrong with protecting secondary meaning "in the making"? Suppose that in *Laureyssens,* the plaintiff had proven that Idea Group copied its descriptive packaging exactly because it knew that Laureyssens was advertising its product heavily and that Idea Group intended to trade on Laureyssens' expected success in making its packaging distinctive. Is it really fair to permit Idea Group to borrow Laureyssens' future goodwill merely because Laureyssens hasn't yet succeeded in establishing that goodwill? The court acknowledges this argument but contends that the doctrine of secondary meaning adequately protects against "piracy in the early stages of product development." Is the court's argument persuasive?

PROBLEMS

Problem 5-2. Preco Industries began using the term "Porcelaincote" for its porcelain resurfacing material in 1966. Preco concedes that, at the time it began use, the "Porcelaincote" mark was unprotectable because it was descriptive. Preco continued to use the mark for a relatively minor product line. In 1977, Ceramco began to use the identical mark for identical goods. [Both companies sell their products on a nationwide basis.] In 1979, Ceramco began a nationwide advertising campaign using the "Porcelaincote" mark. Shortly thereafter, in 1980, Preco began a similar campaign. In 1981 the parties filed complaints against each other for trademark infringement. The court determined at trial that Ceramco established secondary meaning in 1979, and that Preco established it in 1980.

Assume that neither party ever registered its mark. Who should prevail in the lawsuits? Does either party have rights against a third company that began using the name in 1990? Does your answer to either of these questions change if one or both parties registered its mark on the Principal Register?

Problem 5-3. In May 1989, Shalom Children's Wear begins advertising and planning a line of clothes to be called "Body Gear." In November 1989, In-Wear Corp. files an intent-to-use application for the mark "Body Gear." Shalom files an intent-to-use application for the same mark in December 1989 and begins selling its products in February 1990. Is either party entitled to register the Body Gear mark? (Assume that the mark is otherwise protectable.) Does the result change if In-Wear fails to begin selling its clothes by November 1992?

---

3. Trademark Office Procedures

Administrative proceedings concerning trademarks in the United States are among the most stringent in the world. The U.S. Patent and Trademark Office (PTO) actively examines applications and — with the help of the courts — polices the trademark registers as well. In this section we review these administrative procedures.

a. *Principal vs. Supplemental Register*

Although registration is not a prerequisite to trademark protection, trademarks registered on the Principal Register enjoy a number of significant advantages. The primary advantages are: (1) nationwide constructive use and constructive notice, which cut off rights of other users of the same or similar marks, Lanham Act § 22 (15 U.S.C. § 1072) and Lanham Act § 7(c) (15 U.S.C. § 1057(c)), and (2) the possibility of achieving incontestable status after five years, which greatly enhances rights by eliminating a number of defenses, Lanham Act § 15 (15 U.S.C. § 1065).[9]

Trademark applications are maintained in an index at the PTO and made available for public scrutiny soon after filing. This procedure is different from patent applications, the contents of which are kept secret for 18 months after filing in most cases. See Chapter 3.

The Supplemental Register was established by the 1946 Lanham Act "to enable persons in this country to domestically register trademarks so that they might obtain registration under the laws of foreign countries." McCarthy, Trademarks and Unfair Competition, § 19.09[1] (1996), at pp. 19-68. Under the Paris Convention, foreign registration could not be granted in the absence of domestic registration. Because there are countries where trademark registration is granted to marks that would not qualify for the U.S. Principal Register, the Supplemental Register was created. Thus, even if a U.S. mark cannot gain the advantages of registration on the Principal Register, it may obtain protection in foreign countries.

To be eligible for the Supplemental Register a mark need only be capable of distinguishing goods or services. There is no need to prove that it actually functions in that capacity. The Supplemental Register is not available for clearly generic names, but it is available for the registration of trade dress.

Unlike the Principal Register, registration on the Supplemental Register confers no substantive trademark rights "beyond common law." See, e.g., Clairol, Inc. v. Gillette Co., 389 F.2d 264 (2d Cir. 1968). Registration on the Supplemental Register has no evidentiary effects, it does not provide constructive notice of ownership, the mark cannot become incontestable, and it cannot be used as a basis for the Treasury Department to prevent the importation of infringing goods. However, a mark on the Supplemental Register may be litigated in federal court, may be cited by the PTO against a later applicant, and may provide notice to others that the mark is in use. See In re Clorox Co., 578 F.2d 305 (C.C.P.A. 1978). Marks registered on the Supplemental Register are not subject to intent-to-use filings, interference proceedings, or opposition challenges, but may be canceled at any time by a court.

b. *Grounds for Refusing Registration*

Section 2 of the Lanham Act provides the basis for many of the grounds for refusing registration on the Principal Register:

> No trademark by which the goods of the applicant may be distinguished from the goods of others shall be refused registration on the principal register on account of its nature unless it —

---

9. Other advantages include: (1) the right to request customs officials to bar the importation of goods bearing infringing trademarks, Lanham Act § 42 (15 U.S.C. § 1124), and (2) provisions for treble damages, attorney fees, and certain other remedies in civil infringement actions, Lanham Act §§ 34-38 (15 U.S.C. §§ 1116-1120).

(a) Consists of or comprises immoral, deceptive, or scandalous matter; or matter which may disparage or falsely suggest a connection with persons, living or dead, institutions, beliefs, or national symbols, or bring them into contempt, or disrepute.

(b) Consists of or comprises the flag or coat of arms or other insignia of the United States, or of any State or municipality, or of any foreign nation, or any simulation thereof.

(c) Consists of or comprises a name, portrait, or signature identifying a particular living individual except by his written consent, or the name, signature, or portrait, of a deceased President of the United States during the life of his widow, if any, except by the written consent of the widow.

(d) Consists of or comprises a mark which so resembles a mark registered in the Patent and Trademark Office, or a mark or trade name previously used in the United States by another and not abandoned, as to be likely, when used on or in connection with the goods of the applicant, to cause confusion, or to cause mistake, or to deceive: *Provided,* That, if the Commissioner determines [accordingly, concurrent registrations may be possible; see below].

(e) Consists of a mark which, (1) when used on or in connection with the goods of the applicant is merely descriptive or deceptively misdescriptive of them, (2) when used on or in connection with the goods of the applicant is primarily geographically descriptive of them, except as indications of regional origin may be registrable under section 1054 of this title, (3) when used on or in connection with the goods of the applicant is primarily geographically deceptively misdescriptive of them, (4) is primarily merely a surname, or (5) comprises any matter that, as a whole, is functional.

Lanham Act § 2, 15 U.S.C. § 1052. Marks rejected under subsection (e)(1), (2), and (4) may be registered if the applicant can demonstrate secondary meaning, however.

i. Immoral or Scandalous Marks

Section 2(a) forbids the registration of "immoral" or "scandalous" matter, even if it is otherwise eligible for trademark protection. While this section has been used to deny protection on a number of occasions, the definition of what is scandalous has changed over time to reflect contemporary mores.

In re Old Glory Condom Corp., 26 U.S.P.Q.2d 1216 (TTAB 1993), is a good example of a case involving allegedly "immoral or scandalous matter" under § 2(a).

OLD GLORY CONDOM CORP

FIGURE 5-1
Logo for Old Glory Condom Corp.

The applicant was attempting to register the following mark to denote its condom product:

The Board reversed rejection of the application by the examining attorney: In the course of reviewing older cases finding scandalous marks such as "QUEEN MARY" for women's underwear and "BUBBY TRAP" for brassieres, the court said:

> What was once considered scandalous as a trademark or service mark twenty, thirty or fifty years ago may no longer be considered so. Marks one thought scandalous may now be thought merely humorous (or even quaint). The point to be made here is that, in deciding whether a mark is scandalous under Section 2(a), we must consider that mark in the context of contemporary attitudes. . . .

Id. at 1218-19.

While the Board relies on "contemporary attitudes" in *Old Glory*, in other cases it points to traditional disparaging meanings to find a term offensive. Which approach makes most sense?

Because section 2(a) applies only to registration, a cancellation or refusal to register would affect only the mark owner's ownership of federal registrations. What effect will it have on enforcement of the trademarks under Lanham Act § 43(a) or under state statutory and common law? Under federal antidilution protection? Later in this chapter, we explain that a mark must be found to be famous to be protected against dilution under the federal statute. Can a mark be disparaging yet still famous? Is it consistent with trademark law to protect consumer associations even though a mark has been declared disparaging?

Who can challenge a mark as scandalous? The short answer appears to be: virtually anyone. In Ritchie v. Simpson, 170 F.3d 1092 (Fed. Cir. 1999), William B. Ritchie opposed registration of various trademarks associated with O.J. Simpson. Ritchie argued that the trademarks recited immoral or scandalous material given the well-known past of the registrant (i.e., spousal abuse, accusation of murder, etc.). The Federal Circuit upheld Ritchie's standing to oppose the marks, holding that he had shown a sufficiently real and personal interest in the outcome of the opposition proceeding (in the form of a deep personal belief in non-abusive relationships) to defeat a motion to dismiss for lack of standing.

PROBLEM

Problem 5-4. The Washington Redskins is a professional football team that has used that name for decades. Harjo is a Native American who is offended by the use of the term "redskin," which he says is offensive to Native Americans. He petitions the PTO to cancel the Redskins mark.

In support, Harjo presents the testimony of its linguistics expert, Dr. Geoffrey Nunberg, regarding the usage of the word "redskin(s)." He contends that the primary denotation of "redskin(s)" is Native American people; that, only with the addition of the word "Washington," has "redskin(s)" acquired a secondary denotation in the sports world, denoting the NFL Football club; that the "offensive and disparaging qualities" of "redskin(s)" arise from its connotations; and that these negative connotations pertain to the word "redskin(s)" in the context of the team name "Washington Redskins." Regarding whether the negative connotations of "redskin(s)" are inherent or arise from the context of its usage, he argues that "redskin(s)" is inherently offensive and disparaging.

The NFL contends that the word "redskin(s) has throughout history, been a purely denotative term, used interchangeably with 'Indian'." It argues that "redskin(s)" is "an entirely neutral and ordinary term of reference" from the relevant time period to the present; and that, as such, "redskin(s)" is "[synonymous] with ethnic identifiers such as 'American Indian,' 'Indian,' and 'Native American'." It also states that, through its long and extensive use of "Redskins" in connection with professional football, the word has developed a meaning, "separate and distinct from the core, ethnic meaning" of the word "redskin(s)," denoting the "Washington Redskins" football team; and that such use by respondent "has absolutely no negative effect on the word's neutrality — and, indeed, serves to enhance the word's already positive associations — as football is neither of questionable morality nor per se offensive to or prohibited by American Indian religious or cultural practices."

Petitioners conducted a survey of 301 people, which showed that 46.6 percent of all respondents, and 36.6 percent of Native Americans, found "redskin" offensive.

Should the PTO cancel the mark?

FIGURE 5-2
Trademark Registration Number 986, 668.

ii. Geographic Marks

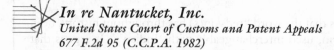

In re Nantucket, Inc.
United States Court of Customs and Patent Appeals
677 F.2d 95 (C.C.P.A. 1982)

MARKEY, Chief Judge.

Nantucket, Inc. (Nantucket) appeals from a decision of the Trademark Trial and Appeal Board (board) affirming a refusal to register the mark NANTUCKET for men's shirts on the ground that it is "primarily geographically deceptively misdescriptive." In re Nantucket, Inc., 209 U.S.P.Q. 868 (TTAB 1981). We reverse.

Background

... Nantucket, based in North Carolina, filed [an] application ... for registration of NANTUCKET for men's shirts on the principal register in the Patent and Trademark Office (PTO) ....

Refusal to register was based on § 2(e)(2) of the Lanham Act, 15 U.S.C. § 1052(e)(2), as interpreted by the board in In re Charles S. Loeb Pipes, Inc., 190 U.S.P.Q. 238 (TTAB 1975), and in § § 1208.02, 1208.05 and 1208.06 of the Trademark Manual of Examining Procedure (TMEP) ....

TMEP § 1208.02 indicates that a mark is primarily geographical, inter alia, if it "is the name of a place which has general renown to the public at large and which is a place from which goods and services are known to emanate as a result of commercial activity."

The examiner, citing a dictionary definition of "Nantucket" as an island in the Atlantic Ocean south of Massachusetts, concluded that the mark NANTUCKET was either primarily geographically descriptive or primarily geographically deceptively misdescriptive, depending upon whether Nantucket's shirts did or did not come from Nantucket Island.

Nantucket informed the PTO that its shirts "do not originate from Nantucket Island," and insisted that the mark would not be understood by purchasers as representing that the shirts were produced there because the island has no market place significance vis-à-vis men's shirts .... As applied to shirts, it was argued, NANTUCKET is arbitrary and nondescriptive, because there is no association in the public mind of men's shirts with Nantucket Island.

The examiner's final refusal was based on the view that, because the shirts did not come from Nantucket Island, NANTUCKET is "primarily geographically deceptively misdescriptive."

Before the board, Nantucket relied upon a number of cases, of which In re Circus Ices, Inc., 158 U.S.P.Q. 64 (TTAB 1968), is representative, for its asserted "public association" or "noted for" test. In that case, the board said:

> The term "HAWAIIAN," meaning of or pertaining to Hawaii or the Hawaiian Islands, possesses an obvious geographical connotation, but it does not necessarily follow therefrom that it is primarily geographically descriptive of applicant's product within the meaning of Section 2(e). In determining whether or not a geographical term is primarily geographically descriptive of a product, of primary consideration is whether or not there is an association in the public mind of the product with the particular geographical area, as for example perfumes and wines with France, potatoes with Idaho, rum with

Puerto Rico, and beef with Argentina.... In the present case, it has not been made to appear that Hawaii or the Hawaiian Islands are noted for flavored-ice products or that the term "HAWAIIAN" is used by anyone to denote the geographical origin of such products. [...]

In referring to *Amerise,* the board viewed the "noted for" test, mentioned in *Circus Ices,* as relevant only to whether a geographic term is deceptive under § 2(a) ["No trademark...shall be refused registration...unless it (a) consists of or comprises immoral, *deceptive,* or scandalous matter...." (emphasis added).] Regarding § 2(e)(2), the board said:

> (A)ny term which, when applied to the goods or services of the applicant, conveys to consumers primarily or immediately a geographical connotation is precluded from registration under Section 2(e)(2) notwithstanding the fact that the area or place named is not "noted for" goods or services of that type....

Id. at 871.

The board concluded that the "term NANTUCKET" has a readily recognizable geographic meaning, and...no alternative non-geographic significance...and hence falls within the proscription of Section 2(e)(2)." Id.

The Board's Test

The board correctly notes that its test for registrability of geographic terms is "easy to administer" and "minimizes subjective determinations by eliminating any need to make unnecessary inquiry into the nebulous question of whether the public associates particular goods with a particular geographical area in applying Section 2(e)(2)."...Ease-of-administration considerations aside, the board's approach does raise the question of whether public association of goods with an area must be considered in applying § 2(e)(2). That question is one of first impression in this court. We answer in the affirmative.

The board's test rests mechanistically on the one question of whether the mark is recognizable, at least to some large segment of the public, as the name of a geographical area. NANTUCKET is such. That ends the board's test. Once it is found that the mark is the name of a known place, i.e., that it has "a readily recognizable geographic meaning," the next question, whether applicant's goods do or do not come from that place, becomes irrelevant under the board's test, for if they do, the mark is "primarily geographically descriptive"; if they don't, the mark is "primarily geographically deceptively misdescriptive."[10]

The Statute

One flaw in the board's test resides in its factoring out the nature of applicant's goods, in contravention of § 2(e)(2)'s requirement that the mark be evaluated "when

10. Under the modern statute, primarily geographically descriptive marks can be registered if they have acquired secondary meaning. By contrast, geographically deceptively misdescriptive marks can never be registered. 15 U.S.C. § 1052(f). — EDS.

applied to the goods of the applicant," and that registration be denied only when the mark is geographically descriptive or deceptively misdescriptive "of them" (the goods).

Another flaw in the board's test lies in its failure to give appropriate weight to the presence of "deceptively" in § 2(e)(2). If the goods do not originate in the geographic area denoted by the mark, the mark might in a vacuum be characterized as geographically misdescriptive, but the statutory characterization required for denial of registration is "geographically *deceptively* misdescriptive." (Emphasis supplied.) Before that statutory characterization may be properly applied, there must be a reasonable basis for believing that purchasers are likely to be deceived.[5] . . .

Geographic terms are merely a specific kind of potential trademark, subject to characterization as having a particular kind of descriptiveness or misdescriptiveness. Registration of marks that would be perceived by potential purchasers as describing or deceptively misdescribing the goods themselves may be denied under § 2(e)(1). Registration of marks that would be perceived by potential purchasers as describing or deceptively misdescribing the geographic origin of the goods may be denied under § 2(e)(2). In either case, the mark must be judged on the basis of its role in the marketplace.

As the courts have made plain, geographically deceptive misdescriptiveness cannot be determined without considering whether the public associates the goods with the place which the mark names. If the goods do not come from the place named, and the public makes no goods-place association, the public is not deceived and the mark is accordingly not geographically deceptively misdescriptive. . . .

In National Lead Co. v. Wolfe, 223 F.2d 195, 199, (CA 9), *cert. denied,* 350 U.S. 883 (1955), the court held that neither DUTCH, nor DUTCH BOY, as applied to paint, was used "otherwise than in a fictitious, arbitrary and fanciful manner," and noted that "there is no likelihood that the use of the name 'Dutch' or 'Dutch Boy' in connection with the appellant's goods would be understood by purchasers as representing that the goods or their constituent materials were produced or processed in Holland or that they are of the same distinctive kind or quality as those produced, processed or used in that place."

. . . There is no indication that the purchasing public would expect men's shirts to have their origin in Nantucket when seen in the market place with NANTUCKET on them. Hence buyers are not likely to be deceived, and registration cannot be refused on the ground that the mark is "primarily geographically deceptively misdescriptive."

Accordingly, the decision of the board is reversed.

COMMENTS AND QUESTIONS

1. In some cases prospective trademark registrants may be trying to seize on the descriptiveness of a term of geographic origin; "Napa" for wine would be an example. Another motivation is to seize on the good feelings engendered by a place name. See, e.g., Singer Mfg. Co. v. Birginal-Bigsby Corp., 319 F.2d 273 (C.C.P.A. 1963) (American Beauty is primarily geographically deceptively misdescriptive when applied

---

5. Reasonable persons are unlikely to believe that bananas labeled ALASKA originated or were grown in Alaska. On the other hand, reasonable persons are quite likely to believe that salmon labeled ALASKA originated in the waters of that state. The board recognized this decisional parameter in its references to French wine, Idaho potatoes, etc., in *Circus Ices,* 158 U.S.P.Q. at 64.

to sewing machines of Japanese origin). For example, a trademark applicant might try to register "Cuban" for cigars made in New Hampshire; this would very likely lead to a finding of deceptiveness (section 2(a)), and the application would be barred completely.

On the other hand, the use of a geographic term is not always forbidden. As the *Nantucket* case demonstrates, the question is whether the public connects the geographic indication with the source of goods. For a questionable application of the deceptive misdescriptiveness doctrine, see In re Broyhill Furn. Indus., 60 U.S.P.Q.2d 1511 (T.T.A.B. 2001) ("Toscana" primarily geographically deceptively misdescriptive for furniture).

2. In many countries, geographic terms — usually for food or wine — serve also to certify composition, traditional preparation techniques, quality, and actual taste. In Europe, for example, so-called "appellations of origin" for wine — such as "Champagne," "Chablis," and "Chianti" — are protected by statute and administered by national authorities. See Institut National des Appellations D'Origine v. Vintners Intl. Co., 958 F.2d 1574 (Fed. Cir. 1992) (affirming denial of opposition by plaintiff, French national authority for establishing and policing appellations of origin, for registration of "Chablis with a Twist" for citrus-flavored wine drink that did not originate in France).

As part of the GATT amendments in 1994, Congress (as a concession to European interests) adopted limited recognition of the appellations of origin concept.

Lanham Act § 2(a) [15 U.S.C. § 1052(a)]

No trademark . . . shall be refused registration on the principal register on account of its nature unless it —

(a) Consists of or comprises . . . a geographical indication which, when used on or in connection with wines or spirits, identifies a place other than the origin of the goods and is first used on or in connection with wines or spirits by the applicant on or after one year after the date on which the WTO [World Trade Organization] Agreement (as defined in section 3501(9) of Title 19) enters into force with respect to the United States.

Section 2(e) has also long provided that appellations of origin registered under the Lanham Act § 4 (15 U.S.C § 1054) (collective and certification marks) are an exception to the prohibition on geographically descriptive trademarks. A certification mark is a guarantee of authenticity — "Belgian chocolates" must come from Belgium, for example. Certification marks may be registered without proof of secondary meaning but must be made available in a nondiscriminatory fashion to anyone who complies with the terms of the certification. See Community of Roquefort v. William Faehndrich, Inc., 303 F.2d 494 (2d Cir. 1962) (French city of Roquefort was entitled to prevent use of the term "Roquefort" to describe cheese made in Hungary, based on United States certification mark). Note that group certification marks serve much the same quality assurance function as individual trademarks, but (assuming collective action problems are overcome) they may effectively protect small businesses who could not establish a well-known mark on their own.

For a spirited debate on appellations of origin, see Jim Chen, A Sober Second Look at Appellations of Origin: How the United States Will Crash France's Wine and Cheese Party, 5 Minn. J. Global Trade 29 (1996) (arguing that appellations of origin are artificial market segmentation devices that detract from consumer welfare); Louis Lorvellec, You've Got to Fight for Your Right to Party, 5 Minn. J. Global Trade 65 (1996) (begging to differ, from French perspective).

Europe has a detailed set of rules regulating the use of geographically descriptive terms, particularly certification marks. Generally, these rules are more restrictive than United States law has been.

3. Marks that are "deceptively misdescriptive" cannot be registered even if they are not geographic in nature. See, e.g., Glendale Int'l Corp. v. PTO, 374 F. Supp. 2d 479 (E.D. Va. 2005) (the mark "Titanium" was deceptively misdescriptive for a recreational vehicle that contained no titanium).

iii. Marks Which Are "Primarily Merely a Surname"

Section 2 treats as "descriptive" marks that are "primarily merely a surname" in the minds of consumers, whether or not the mark actually represents the owner's surname. From In re Garan, Inc., 3 U.S.P.Q.2d 1537, 1539-1540 (TTAB 1987):

> That there are no other meanings of the name in the English language will not support refusal of registration of the surname under the "primarily merely a surname" statutory language unless the average member of the purchasing public would, upon seeing it used as a trademark, recognize it as a surname. This rule was first announced by the late Assistant Commissioner Leeds in her landmark decision, Ex parte Rivera Watch Corporation, 106 U.S.P.Q. 145, 149 (Comr. Pats. 1955), as follows:
>
>> There are some names which by their very nature have only a surname significance even though they are rare surnames. "Seidenberg," if rare, would be in this class. And there are others which have no meaning — well known or otherwisee — and are in fact surnames which do not, when applied to goods as trademarks, create the impression of being surnames. It seems to me that the test to be applied in the administration of this provision in the Act is not the rarity of the name, nor whether it is the applicant's name, nor whether it appears in one or more telephone directories, nor whether it is coupled with a baptismal name or initials. The test should be: What is the primary significance to the purchasing public? ...
>
> Some twenty years later, in In re Kahan & Weisz Jewelry Manufacturing Corp., 508 F.2d 831 (C.C.P.A. 1975), a case involving facts strikingly similar to those in the one now before us, the predecessor of the Court above expressly adopted the Rivera rule and it has since been consistently followed by the Board. See, e.g., In re Pillsbury Co., 174 U.S.P.Q. 318 (TTAB 1972)....
>
> Clearly, based on the above, we cannot stop with the directory listings and absence of other meanings, but must evaluate all of the relevant factors.
>
> The six (or seven) directory and NEXIS listings of "Garan" as a surname have limited persuasive impact in view of the fact these were the only ones found in an enormous NEXIS database and some 43 directories of major population centers. In this context, we conclude that "Garan" is an extremely rare surname. While rare surnames may nevertheless be considered to be "primarily merely" surnames, e.g., In re Possis Medical, Inc., 230 U.S.P.Q. 72, 74 (TTAB 1986), and cases cited therein, we agree with appellant that the degree of a surname's rareness should have material impact on the weight given the directory evidence. Here, since it appears that the directory and NEXIS evidence shows "Garan" to be an extremely rare surname, we conclude that the directory and NEXIS evidence only slightly supports the Office's position that GARAN is "primarily merely a surname."
>
> On analysis, we find all of the other factors either neutral or supportive of appellant's position that GARAN would be perceived by purchasers as an arbitrary term or as the trademark and trade name of appellant.

COMMENTS AND QUESTIONS

1. The statutory hesitance to register surnames springs in part from an old common law policy in favor of the right to use one's name, a policy that still has some life. Basile, S.p.A. v. Basile, 899 F.2d 35 (D.C. Cir. 1990), for example, is a case where an Italian watchmaker by the name of Basile sought to use that name on watches in the United States — against the wishes of an established company, well known for its watches sold under the same name. In his opinion reviewing the scope of the injunction granted by the district court in the case, Judge Williams of the D.C. Circuit summarized the common law rule and its more recent interpretation:

A seller's right to use his family name might have carried the day against a risk of buyer confusion in an era when the role of personal and localized reputation gave the right a more exalted status. In Burgess v. Burgess, 3 De G.M. & G. 896, 903-904, 43 Eng. Rep. 351, 354 (1853), the plaintiff's son had followed him into the trade of making anchovy paste; Lord Justice Bruce said:

All the Queen's subjects have a right, if they will, to manufacture and sell pickles and sauces, and not the less that their fathers have done so before them. All the Queen's subjects have a right to sell these articles in their own names, and not the less so that they bear the same name as their fathers.

See also Brown Chemical Co. v. Meyer, 139 U.S. 540, 544, 11 S. Ct. 625, 627, 35 L. Ed. 247 (1891) ("[a] man's name is his own property, and he has the same right to its use and enjoyment as he has to that of any other species of property"); Stix, Baer & Fuller Dry Goods Co. v. American Piano Co., 211 F. 271, 274 (8th Cir. 1913). But even quite old decisions have enjoined a second comer's use of his name where necessary to prevent confusion. See, e.g., Thaddeus Davids Co. v. Davids Manufacturing Co., 233 U.S. 461, 472, 34 S. Ct. 648, 652, 58 L. Ed. 1046 (1914); Hat Corp. of America v. D. L. Davis Corp., 4 F. Supp. 613, 623 (D. Conn. 1933). The commentators, moreover, have been scornful of protecting the second comer's right to use his name at the expense of customer confusion, anathematizing it as the "sacred rights" theory. See Milton Handler & Charles Pickett, Trade-Marks and Trade Names — An Analysis and Synthesis, 30 Colum. L. Rev. 168, 197-200 (1930).... Case Note, 38 Harv. L. Rev. 405, 406 (1925).

True, even recent decisions have invoked the right to use one's name, at least as an interest against which the senior user's are balanced. See Taylor Wine [Co., Inc. v. Bully Hill Vineyards, Inc., 569 F.2d 731, 734 (2d Cir. 1978)], ... at 735 .... But its weight has decidedly diminished. The courts are now consistent in imposing tighter restrictions on the second comer in the face of possible confusion ...; any residual protection of the second comer's use of his own name seems amply explained by the more general principle that an equitable remedy should be no broader than necessary to correct the wrong.... Where the second comer had established no reputation under his own name, one court did not even purport to "balance."

This trend in the law unsurprisingly reflects trends in the marketplace. In a world of primarily local trade, the goodwill of an anchovy paste seller may well have depended on his individual reputation within the community. Indeed, one elderly decision tells us that an entrepreneur's failure to use his family name once risked "the reproach of doing business under false colors." Hat Corp., 4 F. Supp. at 623. By contrast, the court went on, "[i]n an age when by corporate activity, mass production, and national distribution, the truly personal element has been so largely squeezed out of business, there is naturally less legitimate pecuniary value in a family name." Id.; see also Handler & Pickett, 30 Colum. L. Rev. at 199. That was in 1933 and the point is more obvious today. Other than understandable pride and sense of identity, the modern businessman loses nothing by losing the name. A junior user's right to use his name thus must yield to the extent its exercise causes confusion with the senior user's mark . . . .

Here Francesco Basile's interest in the use of his name is peculiarly weak. He has no reputation in the United States as a watchmaker . . . . There is no suggestion that the watch industry is one where an individual proprietor's personal presentation of his wares plays a key commercial role, as may still be true of high fashion designers such as Yves St. Laurent and Christian Dior. Nor is it a business that has remained largely local in scope. Although none of these conditions would allow a second comer to use his own name at a serious cost of customer confusion, their absence means that Francesco's interest in his own name here is scarcely greater than his interest in the name Bulova . . . . [W]e think the only plausible motivation for his fight here is a wish to free-ride on Basile's goodwill, precisely what the law means to suppress.

Id. at 39-40.

2. Should it matter whether the surname sought to be registered is common or rare? That the senior trademark user is well known or obscure? See S. Smith, Primarily Merely, 63 Trademark Rep. 24 (1973).

3. The scope of injunctions shows the remnants of the common law concern with the right to use one's name. Courts are more likely to permit the junior use if the user disclaims any connection with the more famous senior user. See Taylor Wine Co. v. Bully Hill Vineyards, 569 F.2d 731 (2d Cir. 1978).

4. Should the nature of the defendant's use matter? One court has held that the restrictions applied to registering personal names should not apply where the name in question ("Niles") was used for a plush toy camel, since unlike personal names there was no strong business interest in using one's own name for a camel. Peaceable Planet, Inc. v. Ty, Inc., 362 F.3d 986 (7th Cir. 2004).

PROBLEM

Problem 5-5. Ernest and Julio Gallo are the nation's largest seller of wine. They sell their wine under the "Gallo" label and have advertised that trademark heavily. Ernest and Julio's younger brother Joseph Gallo runs a company called Gallo Cattle Co. Gallo Cattle begins selling cheese and salami under the name "Gallo Cheese" and "Gallo Salami." If Ernest and Julio Gallo wish to stop their brother from marketing meat and cheese under the Gallo name, what must they be able to show at trial? If they win, how should a court craft the injunction granting relief? What should it forbid and permit?

---

c. Opposition

Section 13(a) of the Lanham Act reads in part:

(a) Any person who believes that he would be damaged by the registration of a mark upon the principal register may, upon payment of the prescribed fee, file an opposition in the Patent and Trademark Office, stating the grounds therefor, within thirty days after the publication...of the mark sought to be registered.

Because potential opposers must become aware of the contested mark's future registration, opposition is meaningful protection only for firms possessing the resources to conduct frequent searches of the Official Gazette, where prospective trademarks are published.

An opposer must plead and prove that: (1) it is likely to be damaged by registration of the applicant's mark (this is the standing requirement); and (2) that there are valid legal grounds why the applicant is not entitled to register its claimed mark. The opposer is in the plaintiff position, and the applicant is in the defendant position. The opposer has the burden of proving that the applicant has no right to register the contested mark. In general, there has been a trend toward liberalization of the standing requirement. 3 McCarthy § 20.02[1][a] (1996). "To establish standing to oppose, the opposer need only be something more than a gratuitous interloper or a vicarious avenger of someone else's rights." Id., at p. 20-16.

Once the standing threshold has been crossed, the opposer may rely on any legal ground that negates the applicant's right to registration. See, e.g., Estate of Biro v. Bic Corp., 18 U.S.P.Q.2d 1382 (TTAB 1991). For example, an opposer might argue that the applicant did not make sufficient use of the mark in interstate commerce to receive a use-based registration. Generally, opposers rely on one of the bars to registration described in Lanham Act § 2, with section 2(d) being the most common. Section 2(d) prohibits registration where the applicant's mark so resembles either (1) opposer's registered mark or (2) opposer's prior common law mark or trade name, as to be likely to cause confusion. The test for likelihood of confusion is the same used in general litigation. 3 McCarthy, *supra*, at p. 20-24. It is described in detail below in section D.1.

d. Cancellation

If opposition is the first "backstop" to ex parte examination of trademarks, cancellation may be thought of as the second. Even after a trademark examiner is satisfied that a mark meets the requirements for registration and the mark is in fact registered, it may still be challenged in an inter partes proceeding. The Lanham Act allows one who "believes that he is or will be damaged by registration" to petition for cancellation of marks on either the Principal or Supplemental Register. Lanham Act § 18. Cancellation petitions are heard by the Trademark Trial and Appeal Board (TTAB). In civil suits where a federally registered mark is at issue, such as suits under Lanham Act § 2, the court may order cancellation of the registration.

The standing requirement for cancellation proceedings is quite similar to that for opposition. In both cases the statute speaks in terms of the plaintiff's belief that he or she will be damaged.

Even after cancellation of the registration, a mark may still enjoy common law rights. See, e.g., National Trailways Bus System v. Trailway Van Lines, 269 F. Supp. 352 (E.D.N.Y. 1965).

e. Concurrent Registration

Concurrent use registration is provided for in Lanham Act § 2(d):

> No trademark . . . shall be refused registration on the principal register . . . unless it — . . .
>
>   (d) Consists of or comprises a mark which so resembles a mark . . . as to be likely . . . to cause confusion . . . : *Provided,* That if the Commissioner determines that confusion, mistake, or deception is not likely to result from the continued use by more than one person of the same or similar marks under conditions and limitations as to the mode or place of use of the marks or the goods on or in connection with which such marks are used, concurrent registrations may be issued to such persons when they have become entitled to use such marks as a result of their concurrent lawful use in commerce prior to (1) the earliest of the filing dates of the applications pending or of any registration issued under this Act . . . .

Lanham Act § 2(d), 15 U.S.C. § 1052(d). Section 2(d) goes on to say that prior use may be waived by agreement of the parties seeking concurrent registration, and that the commissioner may also issue concurrent use registrations when ordered to do so by a court. Id.

The most important condition for approval of concurrent use registration is that such registration cannot be likely to cause confusion of buyers or others. Application of Beatrice Foods Co., 429 F.2d 466 (C.C.P.A. 1970). In *Beatrice,* the senior user (Beatrice) had used a mark in 23 states and the junior user (Fairway) had used a mark in five states by the time the registration hearing commenced. Both had filed registration applications, and a concurrent use proceeding was instituted. The court established the general rule that the senior user can be awarded registration covering all parts of the United States except those regions where the subsequent (junior) user can establish existing rights in its actual area of use or zones of natural expansion. The junior user must show that confusion is not likely to result from the concurrent registration. The court then recognized three exceptions to this general rule: (1) the PTO is not required to grant registration contrary to an agreement between the parties that leaves some territory open; (2) where the junior user is the first to obtain federal registration, the junior user obtains nationwide rights subject only to the territorial limitations of the senior user (see, e.g., Weiner King, Inc. v. Weiner King Corp., 615 F.2d 512 (C.C.P.A. 1980) (junior user who registered first and expanded after discovering the senior user was entitled to registration covering the entire United States with the exception of a small enclave encompassing the senior user's territory); (3) areas of mutual nonuse may be maintained if the mark, goods, and territories are such that this is the only way to avoid the likelihood of confusion.

COMMENTS AND QUESTIONS

Should an agreement settling a dispute between parties seeking concurrent registration — hence carving up the nation into exclusive territories — be scrutinized for antitrust issues? See VMG Enterprises Inc. v. F. Quesada & Franco Inc., 788 F. Supp. 648 (D.P.R. 1992). Absent intellectual property rights, an agreement among competitiors to divide up markets would be illegal per se. See Chapter 8.

4. Incontestability

Park 'N Fly, Inc. v. Dollar Park and Fly, Inc.
Supreme Court of the United States
469 U.S. 189 (1985)

Justice O'CONNOR delivered the opinion of the Court.

In this case we consider whether an action to enjoin the infringement of an incontestable trade or service mark may be defended on the grounds that the mark is merely descriptive. We conclude that neither the language of the relevant statutes nor the legislative history supports such a defense.

I

Petitioner operates long-term parking lots near airports. After starting business in St. Louis in 1967, petitioner subsequently opened facilities in Cleveland, Houston, Boston, Memphis, and San Francisco. Petitioner applied in 1969 to the United States Patent and Trademark Office (Patent Office) to register a service mark consisting of the logo of an airplane and the words "Park 'N Fly." The registration issued in August 1971. Nearly six years later, petitioner filed an affidavit with the Patent Office to establish the incontestable status of the mark. As required by § 15 of the Trademark Act of 1946 (Lanham Act), 60 Stat. 433, as amended, 15 U.S.C. § 1065, the affidavit stated that the mark had been registered and in continuous use for five consecutive years, that there had been no final adverse decision to petitioner's claim of ownership or right to registration, and that no proceedings involving such rights were pending. Incontestable status provides, subject to the provisions of § 15 and § 33(b) of the Lanham Act, "conclusive evidence of the registrant's exclusive right to use the registered mark...." § 33(b), 15 U.S.C. § 1115(b).

Respondent also provides long-term airport parking services, but only has operations in Portland, Oregon. Respondent calls its business "Dollar Park and Fly." Petitioner filed this infringement action in 1978 in the United States District Court for the District of Oregon and requested the court permanently to enjoin respondent from using the words "Park and Fly" in connection with its business. Respondent counterclaimed and sought cancellation of petitioner's mark on the grounds that it is a generic term. See § 14(c), 15 U.S.C. § 1064(c). Respondent also argued that petitioner's mark is unenforceable because it is merely descriptive. See § 2(e), 15 U.S.C. § 1052(e). As two additional defenses, respondent maintained that it is in privity with a Seattle corporation that has used the expression "Park and Fly" since a date prior to the registration of petitioner's mark, see § 33(b)(5), 15 U.S.C. § 1115(b)(5), and that it has not infringed because there is no likelihood of confusion. See § 32(1), 15 U.S.C. § 1114(1)....

II ...

This case requires us to consider the effect of the incontestability provisions of the Lanham Act in the context of an infringement action defended on the grounds that the mark is merely descriptive. Statutory construction must begin with the language employed by Congress and the assumption that the ordinary meaning of that language accurately expresses the legislative purpose. See American Tobacco Co. v. Patterson, 456 U.S. 63, 68 (1982). With respect to incontestable trade or service marks, § 33(b) of the Lanham Act states that "registration shall be conclusive evidence of the registrant's exclusive right to use the registered mark" subject to the conditions of § 15 and certain enumerated defenses. Section 15 incorporates by reference subsections (c) and (e) of § 14, 15 U.S.C. § 1064. An incontestable mark that becomes generic may be canceled at any time pursuant to § 14(c). That section also allows cancellation of an incontestable mark at any time if it has been abandoned, if it is being used to misrepresent the source of the goods or services in connection with which it is used, or if it was obtained fraudulently or contrary to the provisions of § 4, 15 U.S.C. § 1054, or § § 2(a)-(c), 15 U.S.C. § § 1052(a)-(c)....

The language of the Lanham Act also refutes any conclusion that an incontestable mark may be challenged as merely descriptive. A mark that is merely descriptive of an applicant's goods or services is not registrable unless the mark has secondary meaning. Before a mark achieves incontestable status, registration provides prima facie evidence of the registrant's exclusive right to use the mark in commerce. § 33(a), 15 U.S.C. § 1115(a). The Lanham Act expressly provides that before a mark becomes incontestable an opposing party may prove any legal or equitable defense which might have been asserted if the mark had not been registered. Ibid. Thus, § 33(a) would have allowed respondent to challenge petitioner's mark as merely descriptive if the mark had not become incontestable. With respect to incontestable marks, however, § 33(b) provides that registration is conclusive evidence of the registrant's exclusive right to use the mark, subject to the conditions of § 15 and the seven defenses enumerated in § 33(b) itself. Mere descriptiveness is not recognized by either § 15 or § 33(b) as a basis for challenging an incontestable mark.

The statutory provisions that prohibit registration of a merely descriptive mark but do not allow an incontestable mark to be challenged on this ground cannot be attributed to inadvertence by Congress. The Conference Committee rejected an amendment that would have denied registration to any descriptive mark, and instead retained the provisions allowing registration of a merely descriptive mark that has acquired secondary meaning. See H.R. Conf. Rep. No. 2322, 79th Cong., 2d Sess., 4 (1946) (explanatory statement of House managers). The Conference Committee agreed to an amendment providing that no incontestable right can be acquired in a mark that is a common descriptive, i.e., generic, term. Id., at 5. Congress could easily have denied incontestability to merely descriptive marks as well as to generic marks had that been its intention.

The Court of Appeals in discussing the offensive/defensive distinction observed that incontestability protects a registrant against cancellation of his mark. 718 F.2d, at 331. This observation is incorrect with respect to marks that become generic or which otherwise may be canceled at any time pursuant to § § 14(c) and (e). Moreover, as applied to marks that are merely descriptive, the approach of the Court of Appeals makes incontestable status superfluous. Without regard to its incontestable status, a mark that has been registered five years is protected from cancellation except on the grounds stated in § § 14(c) and (e). Pursuant to § 14, a mark may be canceled on the grounds that it is merely descriptive only if the petition to cancel is filed within five years of the date of registration. § 14(a), 15 U.S.C. § 1064(a). The approach adopted by the Court of Appeals implies that incontestability adds nothing to the protections against cancellation already provided in § 14. The decision below not only lacks support in the words of the statute; it effectively emasculates § 33(b) under the circumstances of this case.

III

Nothing in the legislative history of the Lanham Act supports a departure from the plain language of the statutory provisions concerning incontestability. Indeed, a conclusion that incontestable status can provide the basis for enforcement of the registrant's exclusive right to use a trade or service mark promotes the goals of the statute. The Lanham Act provides national protection of trademarks in order to secure to the owner of the mark the goodwill of his business and to protect the ability of consumers to distinguish among competing producers. See S. Rep. No. 1333, at 3, 5. National protection of trademarks is desirable, Congress concluded, because trademarks foster competition and the maintenance of quality by securing to the producer the benefits of good reputation. Id., at 4. The incontestability provisions, as the proponents of the Lanham Act emphasized, provide a means for the registrant to quiet title in the ownership of his mark. See Hearings on H.R. 82 before the Subcommittee of the Senate Committee on Patents, 78th Cong., 2d Sess., 21 (1944) (remarks of Rep. Lanham); id., at 21, 113 (testimony of Daphne Robert, ABA Committee on Trade Mark Legislation); Hearings on H.R. 102 et al. before the Subcommittee on Trade-Marks of the House Committee on Patents, 77th Cong., 1st Sess., 73 (1941) (remarks of Rep. Lanham). The opportunity to obtain incontestable status by satisfying the requirements of § 15 thus encourages producers to cultivate the goodwill associated with a particular mark. This function of the incontestability provisions would be utterly frustrated if the holder of an incontestable mark could not enjoin infringement by others so long as they established that the mark would not be registrable but for its incontestable status.

Respondent argues, however, that enforcing petitioner's mark would conflict with the goals of the Lanham Act because the mark is merely descriptive and should never have been registered in the first place ....

The alternative of refusing to provide incontestable status for descriptive marks with secondary meaning was expressly noted in the hearings on the Lanham Act. Id., at 64, 69 (testimony of Robert Byerley, New York Patent Law Assn.); Hearings on S. 895 before the Subcommittee of the Senate Committee on Patents, 77th Cong., 2d Sess., 42 (1942) (testimony of Elliot Moyer, Special Assistant to the Attorney General). Also mentioned was the possibility of including as a defense to infringement of an incontestable mark the "fact that a mark is a descriptive, generic, or geographical term or device." Id., at 45, 47. Congress, however, did not adopt either of these alternatives. Instead, Congress expressly provided in §§ 33(b) and 15 that an incontestable mark could be challenged on specified grounds, and the grounds identified by Congress do not include mere descriptiveness.

The dissent echoes arguments made by opponents of the Lanham Act that the incontestable status of a descriptive mark might take from the public domain language that is merely descriptive. As we have explained, Congress has already addressed concerns to prevent the "commercial monopolization" of descriptive language. The Lanham Act allows a mark to be challenged at any time if it becomes generic, and, under certain circumstances, permits the nontrademark use of descriptive terms contained in an incontestable mark. Finally, if "monopolization" of an incontestable mark threatens economic competition, § 33(b)(7), 15 U.S.C. § 1115(b)(7), provides a defense on the grounds that the mark is being used to violate federal antitrust laws. At bottom, the dissent simply disagrees with the balance struck by Congress in determining the protection to be given to incontestable marks . . . .

VI

We conclude that the holder of a registered mark may rely on incontestability to enjoin infringement and that such an action may not be defended on the grounds that the mark is merely descriptive. Respondent urges that we nevertheless affirm the decision below based on the "prior use" defense recognized by § 33(b)(5) of the Lanham Act. Alternatively, respondent argues that there is no likelihood of confusion and therefore no infringement justifying injunctive relief. The District Court rejected each of these arguments, but they were not addressed by the Court of Appeals. 718 F.2d, at 331-332, n.4. That court may consider them on remand. The judgment of the Court of Appeals is reversed, and the case is remanded for further proceedings consistent with this opinion. It is so ordered.

Justice STEVENS, dissenting.

In trademark law, the term "incontestable" is itself somewhat confusing and misleading because the Lanham Act expressly identifies over 20 situations in which infringement of an allegedly incontestable mark is permitted.[1] Moreover, in § 37 of the Act, Congress unambiguously authorized judicial review of the validity of the registration "in any action involving a registered mark."[2] The problem in this case arises because of petitioner's attempt to enforce as "incontestable" a mark that Congress has plainly stated is inherently unregistrable.

The mark "Park 'N Fly" is at best merely descriptive in the context of airport parking.[3] Section 2 of the Lanham Act plainly prohibits the registration of such a mark unless the applicant proves to the Commissioner of the Patent and Trademark Office that the mark "has become distinctive of the applicant's goods in commerce," or to use the accepted shorthand, that it has acquired a "secondary meaning." See 15

1. Section 33(b) enumerates seven categories of defenses to an action to enforce an incontestable mark. See 15 U.S.C. § 1115(b), quoted ante, at 3, n.1. In addition, a defendant is free to argue that a mark should never have become incontestable for any of the four reasons enumerated in § 15. 15 U.S.C. § 1065. Moreover, § 15 expressly provides that an incontestable mark may be challenged on any of the grounds set forth in subsections (c) and (e) of § 14, 15 U.S.C. § 1064, and those sections, in turn, incorporate the objections to registrability that are defined in § § 2(a), 2(b), and 2(c) of the Act. 15 U.S.C. § § 1052(a), (b), and (c).

2. Section 37, in pertinent part, provides: "In any action involving a registered mark the court may determine the right to registration, order the cancellation of registrations in whole or in part, restore canceled registrations, and otherwise rectify the register with respect to registrations of any party to the action." 15 U.S.C. § 1119.

3. In the Court of Appeals petitioner argued that its mark was suggestive with respect to airport parking lots. The Court of Appeals responded: "We are unpersuaded. Given the clarity of its first word, Park 'N Fly's mark seen in context can be understood readily by consumers as an offering of airport parking — imagination, thought, or perception is not needed. Simply understood, 'park and fly' is a clear and concise description of a characteristic or ingredient of the service offered — the customer parks his car and flies from the airport. We conclude that Park 'N Fly's mark used in the context of airport parking is, at best, a merely descriptive mark." 718 F.2d 327, 331 (CA9 1983).

Although the Court appears to speculate that even though the mark is now merely descriptive it might not have been merely descriptive in 1971 when it was first registered . . . . I find such speculation totally unpersuasive. But even if the Court's speculation were valid, the entire rationale of its opinion is based on the assumption that the mark is in the "merely descriptive" category.

U.S.C. §§1052(e), (f). Petitioner never submitted any such proof to the Commissioner, or indeed to the District Court in this case. Thus, the registration plainly violated the Act.

The violation of the literal wording of the Act also contravened the central purpose of the entire legislative scheme. Statutory protection for trademarks was granted in order to safeguard the goodwill that is associated with particular enterprises. A mark must perform the function of distinguishing the producer or provider of a good or service in order to have any legitimate claim to protection. A merely descriptive mark that has not acquired secondary meaning does not perform that function because it simply "describes the qualities or characteristics of a good or service." No legislative purpose is served by granting anyone a monopoly in the use of such a mark....

The word "incontestable" is not defined in the Act. Nor, surprisingly, is the concept explained in the Committee Reports on the bill that was enacted in 1946. The word itself implies that it was intended to resolve potential contests between rival claimants to a particular mark. And, indeed, the testimony of the proponents of the concept in the Committee hearings that occurred from time to time during the period when this legislation was being considered reveals that they were primarily concerned with the problem that potential contests over the ownership of registrable marks might present. No one ever suggested that any public purpose would be served by granting incontestable status to a mark that should never have been accepted for registration in the first instance.

In those hearings the witnesses frequently referred to incontestability as comparable to a decree quieting title to real property. Such a decree forecloses any further contest over ownership of the property, but it cannot create the property itself. Similarly the incontestability of a trademark precludes any competitor from contesting the registrant's ownership, but cannot convert unregistrable subject matter into a valid mark. Such a claim would be clearly unenforceable....

... In light of this legislative history, it is apparent that Congress could not have intended that incontestability should preserve a merely descriptive trademark from challenge when the statutory procedure for establishing secondary meaning was not followed and when the record still contains no evidence that the mark has ever acquired a secondary meaning....

In sum, if petitioner had complied with §2(f) at the time of its initial registration, or if it had been able to prove secondary meaning in this case, I would agree with the Court's disposition. I cannot, however, subscribe to its conclusion that the holder of a mark which was registered in violation of an unambiguous statutory command "may rely on incontestability to enjoin infringement."... Accordingly, I respectfully dissent.

COMMENTS AND QUESTIONS

1. Why should marks be allowed to become incontestable? Does this privilege reflect a judgment that judicial review of trademarks is unnecessary after a certain time? That the Trademark Office can be trusted to make the right decision in an ex parte proceeding? Certainly, no similar right is afforded the holders of patents, even though they are in some sense a "stronger" intellectual property right. At least one commentator has suggested that incontestability ought to be abolished. Kenneth L. Port, The Illegitimacy of Trademark Incontestability, 26 Ind. L. Rev. 519 (1993). Port reasons

that trademark incontestability creates what is in effect a property right in trademarks, since 15 U.S.C. § 1115(b) speaks of a conclusive presumption of the trademark holder's "ownership of the mark." (Certainly, the Court's "quiet title" rationale smacks of real property law.) This property right, he asserts, is inconsistent with general principles of trademark law. He explains:

> Patent and copyright owners enjoy the "bundle of rights" notion of property. Their rights are divisible, freely alienable, and exclusive for the duration of statutory protection.
>
> Trademarks, on the other hand, enjoy none of the "bundle of rights" that other forms of property enjoy. Trademark holders possess only the right to exclude others from using that specific trademark on similar goods. Holders of marks possess the right to protect the sphere of interest in which they are using the mark by excluding others, but nothing more. Mark holders do not possess a property right in the mark itself, because trademarks are nothing when devoid of the goodwill they have come to represent or the product on which they are used.

Id. Read section 1115(b) carefully. Is this a fair criticism?

2. How does the dissent in *Park 'N Fly* deal with the majority's argument that "incontestability" would be worthless if a trademark holder could not use it to enforce his rights against an infringer? Is Justice Stevens's argument persuasive? If the owner of a descriptive mark improperly registered is not entitled to judicial enforcement of his "incontestable" trademark, why should he be entitled to Customs assistance in stopping the importation of goods? To registration of the invalid trademark in other countries? Further, why have incontestability at all? Wouldn't the legal system be better served by a rule that improperly issued trademarks could be cancelled at any time?

This debate was resurrected in Shakespeare Co. v. Silstar Corp., 9 F.3d 1091 (4th Cir. 1993). There, the District Court cancelled plaintiff's registration for a color scheme on a fishing rod on the grounds that it was functional (apparently because it helped to catch fish). The Fourth Circuit reversed, noting that the mark had become incontestable and holding that incontestable marks cannot be cancelled for functionality because functionality was not specifically enumerated in section 1064. The dissent argued that federal courts should not enforce the mark, even if they could not cancel it.

The functionality problem raises an interesting question about incontestability. What happens when trade dress which is initially not functional *becomes* functional over time? (This may happen because of widespread adoption of the owner's trade dress. For example, a computer graphical user interface (GUI) may become so familiar to so many users that it becomes a "standard" in the industry, and it is a commercial advantage to use the standard product.) Would trademark defendants lose *any* opportunity to defeat the mark on functionality grounds? Apparently concerned by the result in *Shakespeare,* Congress revised the Lanham Act in 1998 to provide that functionality may be asserted even against an "incontestable" mark. 15 U.S.C. § 1064(3).

3. Note that the presumptions afforded registered trademarks differ in several ways from those given to issued patents. Registered trademarks, while entitled to a presumption of validity, are not entitled to the benefit of the "clear and convincing evidence" rule applied in patent cases. At the same time, there is no provision for incontestability of patents after a certain number of years. Is there any reason for these differing presumptions?

4. Patent law gives inventors no rights at all unless the Patent Office issues them a patent. By contrast, copyright law gives full protection to unregistered works as soon

as they are created, and there are current legislative efforts to abolish the Copyright Office entirely. Trademark law appears to fall somewhere in between. Trademark owners are entitled to protection without registration, but there are still substantial benefits to registering a trademark. Can you think of any reason for the different administrative schemes? Which is preferable?

D. INFRINGEMENT

We now turn to the circumstances in which a defendant is liable for the use of a mark similar or identical to the plaintiff's mark. We begin with the threshold question of whether the defendant has "used" the mark at all.

1. The Requirement of Trademark Use

1-800 Contacts, Inc. v. WhenU.com, Inc.
United States Court of Appeals for the Second Circuit
414 F.3d 400 (2d Cir. 2005)

JOHN M. WALKER, JR., Chief Judge.

Defendant-appellant WhenU.com, Inc. ("WhenU") is an internet marketing company that uses a proprietary software called "SaveNow" to monitor a computer user's internet activity in order to provide the computer user ("C-user") with advertising, in the form of "pop-up ads," that is relevant to that activity. Plaintiff-appellee 1-800 Contacts, Inc. ("1-800") is a distributor that sells contact lenses and related products by mail, telephone, and internet website. . . .

1-800 filed a complaint alleging, *inter alia,* that WhenU was infringing 1-800's trademarks, in violation of the Lanham Act, 15 U.S.C. §§ 1114(1), 1125(a)(1), by causing pop-up ads of 1-800's competitors to appear on a C-user's desktop when the C-user has accessed 1-800's website. In an Opinion entered January 7, 2004, the district court granted 1-800's motion for a preliminary injunction as it related to 1-800's trademark claims, and enjoined WhenU from using or otherwise displaying 1-800's trademarks, or anything confusingly similar to such trademarks, in connection with WhenU's contextually relevant advertising. *1-800 Contacts, Inc. v. WhenU.com,* 309 F.Supp.2d 467 (S.D.N.Y.2003) ("*1-800 Contacts*"). WhenU has filed this interlocutory appeal.

We hold that, as a matter of law, WhenU does not "use" 1-800's trademarks within the meaning of the Lanham Act, 15 U.S.C. § 1127, when it (1) includes 1-800's website address, which is almost identical to 1-800's trademark, in an unpublished directory of terms that trigger delivery of WhenU's contextually relevant advertising to C-users; or (2) causes separate, branded pop-up ads to appear on a C-user's computer screen either above, below, or along the bottom edge of the 1-800 website window. Accordingly, we reverse the district court's entry of a preliminary injunction and remand with instructions to (1) dismiss with prejudice 1-800's trademark infringement claims against WhenU, and (2) proceed with 1-800's remaining claims.

BACKGROUND...

II.   The Challenged Conduct...

WhenU provides a proprietary software called "SaveNow" without charge to individual C-users, usually as part of a bundle of software that the C-user voluntarily downloads from the internet. "Once installed, the SaveNow software requires no action by the [C-user] to activate its operations; instead, the SaveNow software responds to a [C-user]'s 'in-the-moment' activities by generating pop-up advertisement windows" that are relevant to those specific activities. *1-800 Contacts*, 309 F.Supp.2d at 477. To deliver contextually relevant advertising to C-users, the SaveNow software employs an internal directory comprising "approximately 32,000 [website addresses] and [address] fragments, 29,000 search terms and 1,200 keyword algorithms," *Wells Fargo*, 293 F.Supp.2d at 743 ¶58, that correlate with particular consumer interests to screen the words a C-user types into a web browser or search engine or that appear within the internet sites a C-user visits.

When the SaveNow software recognizes a term, it randomly selects an advertisement from the corresponding product or service category to deliver to the C-user's computer screen at roughly the same time the website or search result sought by the C-user appears. As the district court explained,

> The SaveNow software generates at least three kinds of ads — an ad may be a small 'pop-up' . . . [that appears] in the bottom right-hand corner of a [C-user]'s screen; it may be a 'pop-under' advertisement that appears behind the webpage the [C-user] initially visited; or it may be a 'panoramic' ad[ ] that stretches across the bottom of the [C-user]'s computer screen.

1-800 Contacts, 309 F.Supp.2d at 478. Each type of ad appears in a window that is separate from the particular website or search-results page the C-user has accessed. *Id*. In addition, a label stating "A WhenU Offer — click? for info." appears in the window frame surrounding the ad, together with a button on the top right of the window frame marked "?" that, when clicked by the C-user, displays a new window containing information about WhenU and its ads,[7] as well as instructions for uninstalling the resident SaveNow software. *Id*. at 478 nn.22 & 23.

Usually there is a "few-second" delay between the moment a user accesses a website, and the point at which a SaveNow pop-up advertisement appears on the [C-user]'s screen.

> If a SaveNow user who has accessed the 1-800 Contacts website and has received a WhenU.com pop-up advertisement does not want to view the advertisement or the advertiser's website, the user can click on the visible portion of the [1-800] window . . . , [which will move] the 1-800 Contacts website . . . to the front of the screen display, with the pop-up ad moving behind the website window. Or, . . . the [C-user] can close

7. Specifically, C-users are informed that

"[t]his offer is brought to you by WhenU.com, through the SaveNow service. SaveNow alerts you to offers and services at the moment when they are most relevant to you. SaveNow does not collect any personal information or browsing history from its users. Your privacy is 100 percent protected. The offers shown to you by SaveNow are not affiliated with the site you are visiting. For more about SaveNow, click here or e-mail information at WhenU.com. At WhenU, we are committed to putting you in control of your Internet experience."

1-800 Contacts, 309 F.Supp.2d at 478 n. 22.

the pop-up website by clicking on its "X," or close, button. If the user clicks on the pop-up ad, the main browser window (containing the 1-800 Contacts website) will be navigated to the website of the advertiser that was featured inside the pop-up advertisement.

Id. at 476-77 (internal citations omitted).

In its complaint, 1-800 alleges that WhenU's conduct infringes 1-800's trademarks, in violation of Sections 32(1) and 43(a) of the Lanham Act, 15 U.S.C. §§ 1114(1), 1125(a), by delivering advertisements of 1-800's competitors (*e.g.*, Vision Direct, Inc.) to C-users who have intentionally accessed 1-800's website. Although somewhat difficult to discern from the complaint, the allegations that pertain specifically to 1-800's trademark claims appear to be as follows: (1) WhenU's pop-up ads appear "on," "over," or "on top of" the 1-800 website without 1-800's authorization, and change its appearance; (2) as a result, the ads impermissibly "appear to be an integral and fully authorized part of [1-800's] website"; (3) in addition, WhenU's unauthorized pop-up ads "interfere with and disrupt the carefully designed display of content" on the website, thereby altering and hindering a C-user's access to 1-800's website; (4) WhenU is thereby "free-riding" and "trad[ing] upon the goodwill and substantial customer recognition associated with the 1-800 Contacts marks"; and (5) WhenU is using 1-800's trademarks in a manner that creates a likelihood of confusion.

Following an evidentiary hearing on 1-800's motion for a preliminary injunction, the district court held that 1-800 had demonstrated a likelihood of success on its trademark infringement claims and issued a preliminary injunction prohibiting WhenU from utilizing 1-800's trademarks. *1-800 Contacts*, 309 F.Supp.2d at 510. WhenU appeals the district court's decision.

Discussion

WhenU challenges the district court's finding that WhenU "uses" 1-800's trademarks within the meaning of the Lanham Act, 15 U.S.C. § 1127. *See 1-800 Contacts*, 309 F.Supp.2d at 489. In the alternative, WhenU argues that the district court erred in finding that WhenU's pop-up ads create a likelihood of both source confusion and "initial interest confusion," as to whether WhenU is "somehow associated with [1-800] or that [1-800] has consented to [WhenU's] use of the pop-up ad[s]." *Id*. at 494; *see id*. at 503-04. Because we agree with WhenU that it does not "use" 1-800's trademarks, we need not and do not address the issue of likelihood of confusion.

I. *Legal Standards*...

B. *Lanham Act*

In order to prevail on a trademark infringement claim for registered trademarks, pursuant to 15 U.S.C. § 1114, or unregistered trademarks, pursuant to 15 U.S.C. § 1125(a)(1), a plaintiff must establish that (1) it has a valid mark that is entitled to protection under the Lanham Act; and that (2) the defendant used the mark, (3) in commerce, (4) "in connection with the sale . . . or advertising of goods or services," 15 U.S.C. § 1114(1)(a), (5), without the plaintiff's consent. See Time, Inc. v. Petersen Publ'g Co., 173 F.3d 113, 117 (2d Cir.1999); Genesee Brewing Co., Inc. v. Stroh Brewing Co., 124 F.3d 137, 142 (2d Cir.1997). In addition, the plaintiff must show that defendant's use of that mark "is likely to cause confusion . . . as to the affiliation, connection, or association of [defendant] with [plaintiff], or as to the origin, sponsorship, or approval of [the defendant's] goods, services, or commercial activities by [plaintiff]." 15 U.S.C. § 1125(a)(1)(A); *see also* Estee Lauder Inc. v. The Gap, Inc., 108 F.3d 1503, 1508-09 (2d Cir.1997); Gruner + Jahr USA Publ'g v. Meredith Corp., 991 F.2d 1072, 1075 (2d Cir.1993).

The only issue before us on appeal is whether the district court abused its discretion when it entered the preliminary injunction against WhenU; specifically, whether the district court erred in finding that 1-800 had demonstrated a likelihood of success on its trademark claims. As a result, the threshold of error required to reverse the district court's decision is higher than it would be were we reviewing a decision on 1-800's trademark claims themselves. That higher threshold is met in this case, however, because the district court erred as a matter of law in finding that WhenU "uses" 1-800's trademark. Because 1-800 cannot establish an essential element of its trademark claims, not only must the preliminary injunction be vacated, but 1-800's trademark infringement claims must be dismissed as well.

II. "Use" Under the Lanham Act . . .

In issuing the preliminary injunction, the district court held that WhenU use[s] [1-800]'s mark in two ways.

> First, in causing pop-up advertisements for Defendant Vision Direct to appear when SaveNow users have specifically attempted to access [1-800]'s website — on which Plaintiff's trademark appears — [WhenU is] displaying Plaintiff's mark "in the . . . advertising of" Defendant Vision Direct's services . . . [and, t]hus, . . . [is] "using" Plaintiff's marks that appear on Plaintiff's website.
>
> Second, Defendant WhenU.com includes Plaintiff's [website address], <www. 1800contacts.com>, [which incorporates 1-800's trademark,] in the proprietary WhenU.com directory of terms that triggers pop-up advertisements on SaveNow users' computers. In so doing, Defendant WhenU.com "uses" Plaintiff's mark . . . to advertise and publicize companies that are in direct competition with Plaintiff.

1-800 Contacts, 309 F.Supp.2d at 489.

Prior to the district court's decision, two other courts had addressed the issue of "use" as it applies to WhenU's specific activities and reached the opposite conclusion. In Wells Fargo & Co. v. WhenU.com, Inc., 293 F.Supp.2d 734 (E.D.Mich.2003), the district court denied Wells Fargo's motion for a preliminary injunction after finding that WhenU's inclusion of plaintiff Wells Fargo's trademarked website address in WhenU's proprietary directory of keywords was not "use" for purposes of the Lanham Act, and that WhenU did not alter or interfere with Wells Fargo's website in any manner. *Id.* at 757-61. The district court in U-Haul International, Inc. v. WhenU.com, Inc., 279 F.Supp.2d 723 (E.D.Va.2003), employing a very similar analysis, granted summary judgment in favor of WhenU after concluding that WhenU's inclusion of U-Haul's trademarked website address in the SaveNow directory was not actionable because it was for a "pure machine-linking function" that was not "use" under the Lanham Act. *Id.* at 728 (internal quotation marks omitted).

In the case before us, the district court's consideration of these two comprehensive decisions on the precise issue at hand was confined to a footnote in which it cited the cases, summarized their holdings in parentheticals, and concluded, without discussion, that it "disagree[d] with, and [was] not bound by these findings." *1-800 Contacts,* 309 F.Supp.2d at 490 n. 43. Unlike the district court, we find the thorough analyses set forth in both *U-Haul* and *Wells Fargo* to be persuasive and compelling.

A. *The SaveNow Directory*

The district court held that WhenU's inclusion of 1-800's website address in the SaveNow directory constitutes a prohibited "use" of 1-800's trademark. *Id.* at 489. We disagree.

At the outset, we note that WhenU does not "use" 1-800's trademark in the manner ordinarily at issue in an infringement claim: it does not "place" 1-800 trademarks on any goods or services in order to pass them off as emanating from or authorized by 1-800. See U-Haul, 279 F.Supp.2d at 728; cf. L.L. Bean, Inc. v. Drake Publishers, Inc., 811 F.2d 26, 32-34 (1st Cir.1987); Societe Comptoir de L'Industrie Cotonniere Etablissements Boussac v. Alexander's Dep't Stores, Inc., 299 F.2d 33, 37 (2d Cir.1962). The fact is that WhenU does not reproduce or display 1-800's trademarks at all, nor does it cause the trademarks to be displayed to a C-user. Rather, WhenU reproduces 1-800's website address, fwww.1800contacts.com.", which is similar, but not identical, to 1-800's 1-800CONTACTS trademark. *See 1-800 Contacts,* 309 F.Supp.2d at 478-79.

The district court found that the differences between 1-800's trademarks and the website address utilized by WhenU were insignificant because they were limited to the addition of the "www." and ".com" and the omission of the hyphen and a space. *See id.* We conclude that, to the contrary, the differences between the marks are quite significant because they transform 1-800's trademark—which is entitled to protection under the Lanham Act—into a word combination that functions more or less like a public key to 1-800's website.

Moreover, it is plain that WhenU is using 1-800's website address precisely because it is a website address, rather than because it bears any resemblance to 1-800's trademark, because the only place WhenU reproduces the address is in the SaveNow directory. Although the directory resides in the C-user's computer, it is inaccessible to both the C-user and the general public. *See id.* at 476 (noting that directory is scrambled to preclude access). Thus, the appearance of 1-800's website address in the directory does not create a possibility of visual confusion with 1-800's mark. More important, a WhenU pop-up ad cannot be triggered by a C-user's input of the 1-800 trademark or the appearance of that trademark on a webpage accessed by the c-user. Rather, in order for WhenU to capitalize on the fame and recognition of 1-800's trademark—the improper motivation both 1-800 and the district court ascribe to WhenU—it would have needed to put the actual trademark on the list.[11]

In contrast to some of its competitors, moreover, WhenU does not disclose the proprietary contents of the SaveNow directory to its advertising clients nor does it permit these clients to request or purchase specified keywords to add to the directory. See GEICO v. Google, Inc., 330 F.Supp.2d 700, 703-04 (E.D.Va.2004) (distinguishing

---

11. This observation, however, is not intended to suggest that inclusion of a trademark in the directory would necessarily be an infringing "use." We express no view on this distinct issue.

WhenU's conduct from defendants' practice of selling "keywords" to its advertising clients), *claim dism'd,* Order, Dec. 15, 2004 (dismissing Lanham Act claim following bench trial on finding no likelihood of confusion); see also U-Haul, 279 F.Supp.2d at 728 (discussing other practices).

A company's internal utilization of a trademark in a way that does not communicate it to the public is analogous to a individual's private thoughts about a trademark. Such conduct simply does not violate the Lanham Act, which is concerned with the use of trademarks in connection with the sale of goods or services in a manner likely to lead to consumer confusion as to the source of such goods or services. *See* 15 U.S.C. § 1127; *see also* Louis Altman, 4 Callmann on Unfair Competition, Trademarks and Monopolies § 22:25 n.1 (4th ed. 2004) ("A fortiori, a defendant who does not sell, but merely uses internally within his own company, the trademarked product of another, is not a trademark infringer or unfair competitor by virtue of such use.").

Accordingly, we conclude that WhenU's inclusion of the 1-800 website address in its SaveNow directory does not infringe on 1-800's trademark.

B. *The Pop-up Advertisements*

The primary issue to be resolved by this appeal is whether the placement of pop-up ads on a C-user's screen contemporaneously with either the 1-800 website or a list of search results obtained by the C-user's input of the 1-800 website address constitutes "use" under the Lanham Act, 15 U.S.C. §§ 1114(1), 1125(a). The district court reasoned that WhenU, by "causing pop-up advertisements for Defendant Vision Direct to appear when SaveNow users have specifically attempted to access [1-800]'s website, . . . [is] displaying [1-800]'s mark in the . . . advertising of . . . Vision Direct's services." *1-800 Contacts,* 309 F.Supp.2d at 489.

The fatal flaw with this holding is that WhenU's pop-up ads do *not* display the 1-800 trademark. The district court's holding, however, appears to have been based on the court's acceptance of 1-800's claim that WhenU's pop-up ads appear "on" and affect 1-800's website. *See, e.g., id.* at 479 (stating that WhenU has "no relationship with the companies *on* whose websites the pop-up advertisements appear") (emphasis omitted) (emphasis added). As we explained above, the WhenU pop-up ads appear in a separate window that is prominently branded with the WhenU mark; they have absolutely no tangible effect on the appearance or functionality of the 1-800 website.

More important, the appearance of WhenU's pop-up ad is not contingent upon or related to 1-800's trademark, the trademark's appearance on 1-800's website, or the mark's similarity to 1-800's website address. Rather, the contemporaneous display of the ads and trademarks is the result of the happenstance that 1-800 chose to use a mark similar to its trademark as the address to its web page and to place its trademark on its website. The pop-up ad, which is triggered by the C-user's input of 1-800's website address, would appear even if 1-800's trademarks were not displayed on its website. A pop-up ad could also appear if the C-user typed the 1-800 website address, not as an address, but as a search term in the browser's search engine, and then accessed 1-800's website by using the hyperlink that appeared in the list of search results.

In addition, 1-800's website address is not the only term in the SaveNow directory that could trigger a Vision Direct ad to "pop up" on 1-800's website. For example, an ad could be triggered if a C-user's searched for "contacts" or "eye care," both terms contained in the directory, and then clicked on the listed hyperlink to 1-800's website.

Exemplifying the conceptual difficulty that inheres in this issue, the district court's decision suggests that the crux of WhenU's wrongdoing — and the primary basis for the district court's finding of "use" — is WhenU's alleged effort to capitalize on a C-user's specific attempt to access the 1-800 website. As the court explained it,

> WhenU.com is doing far more than merely "displaying" Plaintiff's mark. WhenU's advertisements are delivered to a SaveNow user when the user directly accesses Plaintiff's website — thus allowing Defendant Vision Direct to profit from the goodwill and reputation in Plaintiff's website that led the user to access Plaintiff's website in the first place.

1-800 Contacts, 309 F.Supp.2d at 490. Absent improper use of 1-800's trademark, however, such conduct does not violate the Lanham Act. See TrafFix Devices, Inc. v. Mktg. Displays, Inc., 532 U.S. 23, 29, 121 S.Ct. 1255, 149 L.Ed.2d 164 (2001); Kellogg Co. v. Nat'l Biscuit Co., 305 U.S. 111, 122, 59 S.Ct. 109, 83 L.Ed. 73 (1938) (holding that Kellogg's sharing in the goodwill of the unprotected "Shredded Wheat" market was "not unfair"); see also William P. Kratzke, Normative Economic Analysis of Trademark Law, 21 Memphis St. U.L.Rev. 199, 223 (1991) (criticizing importation into trademark law of "unjust enrichment" and "free riding" theories based on a trademark holder's goodwill). Indeed, it is routine for vendors to seek specific "product placement" in retail stores precisely to capitalize on their competitors' name recognition. For example, a drug store typically places its own store-brand generic products next to the trademarked products they emulate in order to induce a customer who has specifically sought out the trademarked product to consider the store's less-expensive alternative. WhenU employs this same marketing strategy by informing C-users who have sought out a specific trademarked product about available coupons, discounts, or alternative products that may be of interest to them.

1-800 disputes this analogy by arguing that unlike a drugstore, only the 1-800 website is displayed when the pop-up ad appears. This response, however, ignores the fact that a C-user who has installed the SaveNow software receives WhenU pop-up ads in a myriad of contexts, the vast majority of which are unlikely to have anything to do with 1-800 or the C-user's input of the 1-800 website address.[14]

The cases relied on by 1-800 do not alter our analysis. As explained in detail by the court in *U-Haul,* they are all readily distinguishable because WhenU's conduct does not involve any of the activities those courts found to constitute "use." *U-Haul,* 279 F.Supp.2d at 728-29 (collecting cases). Significantly, WhenU's activities do not alter or affect 1-800's website in any way. Nor do they divert or misdirect C-users away from 1-800's website, or alter in any way the results a C-user will obtain when searching with the 1-800 trademark or website address. *Id.* at 728-29. Compare Playboy Enters., Inc. v. Netscape Communications Corp., 354 F.3d 1020, 1024 (9th Cir.2004) (holding that infringement could be based on defendant's insertion of unidentified banner ads on C-user's search-results page); Brookfield Communications v. West Coast Entm't Corp., 174 F.3d 1036 (9th Cir.1999) (holding that defendant's use of trademarks in "metatags," invisible text within websites that search engines use for ranking results, constituted a "use in commerce" under the Lanham Act); see

---

14. Indeed, although we do not address the district court's finding of a likelihood of confusion, we note that 1-800's claim that C-users will likely be confused into thinking that 1-800 has sponsored its competitor's pop-up ads is fairly incredulous given that C-users who have downloaded the SaveNow software receive numerous WhenU pop-up ads — each displaying the WhenU brand — in varying contexts and for a broad range of products.

generally Bihari v. Gross, 119 F.Supp.2d 309 (S.D.N.Y.2000) (discussing *Brookfield* and similar cases). [15]

In addition, unlike several other internet advertising companies, WhenU does not "sell" keyword trademarks to its customers or otherwise manipulate which category-related advertisement will pop up in response to any particular terms on the internal directory. *See, e.g., GEICO,* 330 F.Supp.2d at 703-04 (finding that Google's sale to advertisers of right to use specific trademarks as "keywords" to trigger their ads constituted "use in commerce"). In other words, WhenU does not link trademarks to any particular competitor's ads, and a customer cannot pay to have its pop-up ad appear on any specific website or in connection with any particular trademark. *See id.* at 704 (distinguishing WhenU's conduct on this basis). Instead, the SaveNow directory terms trigger categorical associations (e.g., www.1800Contacts.com might trigger the category of "eye care"), at which point, the software will randomly select one of the pop-up ads contained in the eye-care category to send to the C-user's desktop.

Perhaps because ultimately 1-800 is unable to explain precisely how WhenU "uses" its trademark, it resorts to bootstrapping a finding of "use" by alleging other elements of a trademark claim. For example, 1-800 invariably refers to WhenU's pop-up ads as "unauthorized" in an effort, it would seem, to establish by sheer force of repetition the element of unauthorized use of a trademark. Not surprisingly, 1-800 cites no legal authority for the proposition that advertisements, software applications, or any other visual image that can appear on a C-user's computer screen must be authorized by the owner of any website that will appear contemporaneously with that image. The fact is that WhenU does not need 1-800's authorization to display a separate window containing an ad any more than Corel would need authorization from Microsoft to display its WordPerfect word-processor in a window contemporaneously with a Word word-processing window. Moreover, contrary to 1-800's repeated admonitions, WhenU's pop-up ads *are* authorized—if unwittingly—by the C-user who has downloaded the SaveNow software.

1-800 also argues that WhenU's conduct is "use" because it is likely to confuse C-users as to the source of the ad. It buttresses this claim with a survey it submitted to the district court that purportedly demonstrates, *inter alia,* that (1) a majority of C-users believe that pop-up ads that appear *on* websites are sponsored by those websites, and (2) numerous C-users are unaware that they have downloaded the SaveNow software. 1-800 also relies on several cases in which the court seemingly based a finding of trademark "use" on the confusion such "use" was likely to cause. *See, e.g., Bihari,* 119 F.Supp.2d at 318 (holding that defendant's use of trademarks in metatags constituted a "use in commerce" under the Lanham Act in part because the hyperlinks "effectively act[ed] as a conduit, steering potential customers away from Bihari Interiors and toward its competitors"); *GEICO,* 330 F.Supp.2d at 703-04 (finding that Google's sale to advertisers of right to have specific trademarks trigger their ads was "use in commerce" because it created likelihood of confusion that Google had the trademark holder's authority to do so). Again, this rationale puts the cart before the horse. Not only are "use," "in commerce," and "likelihood of confusion" three distinct elements of a trademark infringement claim, but "use" must be decided as a threshold matter because, while any number of activities may

---

15. We note that in distinguishing cases such as *Brookfield, Playboy* and *Bihari,* we do not necessarily endorse their holdings. See Playboy, 354 F.3d at 1034-36 (Berzon, *C.J.,* concurring, noting disagreement with holding in *Brookfield*).

be "in commerce" or create a likelihood of confusion, no such activity is actionable under the Lanham Act absent the "use" of a trademark. 15 U.S.C. § 1114; see People for the Ethical Treatment of Animals v. Doughney, 263 F.3d 359, 364 (4th Cir. 2001). Because 1-800 has failed to establish such "use," its trademark infringement claims fail.

. . .

Conclusion

For the foregoing reasons, we reverse the district court's entry of a preliminary injunction and remand with instructions to (1) dismiss with prejudice 1-800's trademark infringement claims against WhenU . . . .

COMMENTS AND QUESTIONS

1. The trademark use doctrine has been recognized and applied by a number of courts. See, e.g., DaimlerChrysler AG v. Bloom, 315 F.3d 932 (8th Cir. 2003) (holding that a telecommunications company did not use the term "Mercedes" in a trademark sense merely by licensing a vanity phone number that spelled "1-800-MERCEDES" to Mercedes dealers); Interactive Prods. Corp. v. A2Z Mobile Office Solutions, Inc., 326 F.3d 687, 695 (6th Cir. 2003) ("If defendants are only using [plaintiff's] trademark in a 'non-trademark' way—that is, in a way that does not identify the source of a product—then trademark infringement and false designation of origin laws do not apply."); Lockheed Martin Corp. v. Network Solutions, Inc., 141 F. Supp. 2d 648, 656 (N.D. Tex. 2001) (holding that Network Solutions' "acceptance of registrations for domain names is not a use of a mark; nor does it reflect an intent to profit from the mark"); Acad. of Motion Picture Arts & Scis. v. Network Solutions, Inc., 989 F. Supp. 1276, 1280 (C.D. Cal. 1997) (concluding that Network Solutions' registering and cataloging of domain names did not constitute "commercial use" of the marks and thus is not liable as a direct trademark infringement); Lockheed Martin Corp. v. Network Solutions, Inc., 985 F. Supp. 949, 957 (C.D. Cal. 1997) ("[Network Solutions] merely uses domain names to designate host computers on the Internet. This is the type of purely 'nominative' function that is not prohibited by trademark law.").

But other courts have failed to apply the doctrine, often by simply ignoring it, see Playboy Enters., Inc. v. Netscape Communications Corp., 354 F.3d 1020, 1024 (9th Cir. 2004), but sometimes by concluding that the doctrine is coterminous with the jurisdictional requirement of "use in commerce" and therefore satisfied whenever interstate commerce is present. See, e.g., Planned Parenthood Federation of America Inc. v. Bucci, 42 U.S.P.Q.2d 1430, 1434–37 (S.D.N.Y. 1997). And the leading trademark treatise, McCarthy on Trademarks and Unfair Competition, contains no reference to the doctrine.

2. Dogan and Lemley argue that the trademark use doctrine serves an important limiting function. They explain:

The speech-oriented objectives of the trademark use doctrine protect more than just intermediaries; they prevent trademark holders from asserting a generalized right to control language, an interest that applies equally—and sometimes especially—when the speaker competes directly with the trademark holder. The trademark use doctrine has broad application—because of it, newspapers aren't liable for using a trademarked term in a headline, even if the use is confusing or misleading. Writers of movies and books aren't liable for using trademarked goods in their stories. Makers of telephone directories aren't liable for putting all the ads for taxi services together on the same page. Marketing surveyors aren't liable for asking people what they think of a competitor's brand-name product. Magazines aren't liable for selling advertisements that relates to the content of their special issues, even when that content involves trademark owners. Gas stations and restaurants aren't liable for locating across the street from an established competitor, trading on the attraction the established company has created or benefiting from the size of the sign the established company has put up. Individuals aren't liable for their use of a trademark in conversation, even in an inaccurate or misleading way (referring to a Puffs brand facial tissue as a "Kleenex," or a competing cola as a "Coke," for example). Generic drug manufacturers aren't liable for placing their drugs near their brand-name equivalents on drug store shelves, and the stores aren't liable for accepting the placement. They may be making money from their "uses" of the trademark, and the uses may be ones the trademark owner objects to, but they are not *trademark* uses and therefore are not within the ambit of the statute.

Stacey L. Dogan & Mark A. Lemley, Trademarks and Consumer Search Costs on the Internet, 41 Hous. L. Rev. 777 (2004).

The statute speaks only of a defendant's use "on or in connection with" the sale of goods and services. 15 U.S.C. § 1127. Does it make sense to interpret this as a requirement that the defendant use the plaintiff's mark as a trademark? Can the statutory language do the work the authors require of it?

3. Should it matter that the defendant in *WhenU* is making money using the plaintiff's mark as a guide? Is "trading on the goodwill" associated with a trademark always illegal?

PROBLEM

Problem 5-6. Google, an Internet search engine, makes its money by selling targeted advertisements, so that people searching for, say, "cars" get ads for cars and not randomly selected ads. To accomplish this, Google lets its advertisers bid to run ads on particular "keywords." When users search for those terms on the Google Web page, the advertisers' ads appear alongside the search results in a separate box called "Sponsored Links."

The American Blind and Wallpaper Co. sells blinds and wallpaper. It has a registered trademark in the term "American Blind." It discovers that when a user types "american blind" into Google, its competitors' ads appear. Some of those competitors have chosen the phrase "american blind" as a keyword, while others have chosen general terms such as "blind." An ad also appears for the American Society for the Blind, which promotes the interests of blind people. American Blind sues Google, alleging that it is selling the American Blind trademark and therefore engaged in trademark infringement. Has Google made a trademark use of American Blind's mark?

2. Likelihood of Consumer Confusion

AMF Incorporated v. Sleekcraft Boats
United States Court of Appeals for the Ninth Circuit
599 F.2d 341 (9th Cir. 1979)

ANDERSON, Circuit Judge.

In this trademark infringement action, the district court, after a brief non-jury trial, found appellant AMF's trademark was valid, but not infringed, and denied AMF's request for injunctive relief.

AMF and appellee Nescher both manufacture recreational boats. AMF uses the mark Slickcraft, and Nescher uses Sleekcraft. The crux of this appeal is whether concurrent use of the two marks is likely to confuse the public. The district judge held that confusion was unlikely. We disagree and remand for entry of a limited injunction.

I. Facts

AMF's predecessor used the name Slickcraft Boat Company from 1954 to 1969 when it became a division of AMF. The mark Slickcraft was federally registered on April 1, 1969, and has been continuously used since then as a trademark for this line of recreational boats.

Slickcraft boats are distributed and advertised nationally. AMF has authorized over one hundred retail outlets to sell the Slickcraft line. For the years 1966–1974, promotional expenditures for the Slickcraft line averaged approximately $200,000 annually. Gross sales for the same period approached $50,000,000.

After several years in the boatbuilding business, appellee Nescher organized a sole proprietorship, Nescher Boats, in 1962. This venture failed in 1967. In late 1968 Nescher began anew and adopted the name Sleekcraft. Since then Sleekcraft has been the Nescher trademark. The name Sleekcraft was selected without knowledge of appellant's use. After AMF notified him of the alleged trademark infringement, Nescher adopted a distinctive logo and added the identifying phrase "Boats by Nescher" on plaques affixed to the boat and in much of its advertising. The Sleekcraft mark still appears alone on some of appellee's stationery, signs, trucks, and advertisements.

The Sleekcraft venture succeeded. Expenditures for promotion increased from $6,800 in 1970 to $126,000 in 1974. Gross sales rose from $331,000 in 1970 to over $6,000,000 in 1975. Like AMF, Nescher sells his boats through authorized local dealers.

Slickcraft boats are advertised primarily in magazines of general circulation. Nescher advertises primarily in publications for boat racing enthusiasts. Both parties exhibit their product line at boat shows, sometimes the same show....

IV. Likelihood of Confusion

When the goods produced by the alleged infringer compete for sales with those of the trademark owner, infringement usually will be found if the marks are sufficiently similar that confusion can be expected.[9] When the goods are related,[10] but not competitive, several other factors are added to the calculus. If the goods are totally unrelated, there can be no infringement because confusion is unlikely.

AMF contends these boat lines are competitive. Both lines are comprised of sporty, fiberglass boats often used for water skiing; the sizes of the boats are similar as are the prices. Nescher contends his boats are not competitive with Slickcraft boats because his are true high performance boats intended for racing enthusiasts.

The district court found that although there was some overlap in potential customers for the two product lines, the boats "appeal to separate sub-markets." Slickcraft boats are for general family recreation, and Sleekcraft boats are for persons who want high speed recreation; thus, the district court concluded, competition between the lines is negligible. Our research has led us to only one case in which a similarly fine distinction in markets has been recognized, Sleeper Lounge Co. v. Bell Manufacturing Co., 253 F.2d 720 (CA 9 1958). Yet, after careful review of all the exhibits introduced at trial, we are convinced the district court's finding was warranted by the evidence.

The Slickcraft line is designed for a variety of activities: fishing, water skiing, pleasure cruises, and sunbathing. The promotional literature emphasizes family fun. Sleekcraft boats are not for families. They are low-profile racing boats designed for racing, high speed cruises, and water skiing. Seating capacity and luxury are secondary. Unlike the Slickcraft line, handling capability is emphasized. The promotional literature projects an alluring, perhaps flashier, racing image; absent from the pictures are the small children prominently displayed in the Slickcraft brochures.

Even though both boats are designed for towing water skiers, only the highly skilled enthusiast would require the higher speeds the Sleekcraft promises. We therefore affirm the district court's finding that, despite the potential market overlap, the two lines are not competitive. Accordingly, we must consider all the relevant circumstances in assessing the likelihood of confusion. See Durox Co. v. Duron Paint Manufacturing Co., 320 F.2d 882, 885 (CA 4 1963).

V. Factors Relevant to Likelihood of Confusion

In determining whether confusion between related goods is likely, the following factors are relevant:[11]

1. strength of the mark;
2. proximity of the goods;
3. similarity of the marks;
4. evidence of actual confusion;
5. marketing channels used;
6. type of goods and the degree of care likely to be exercised by the purchaser;

---

9. The alleged infringer's intent in adopting the mark is weighed, both as probative evidence of the likelihood of confusion and as an equitable consideration.

10. Related goods are those "products which would be reasonably thought by the buying public to come from the same source if sold under the same mark." Standard Brands, Inc. v. Smidler, 151 F.2d 34, 37 (CA 2 1945). See Yale Electric Corp. v. Robertson, 26 F.2d 972 (CA 2 1928).

11. The list is not exhaustive. Other variables may come into play depending on the particular facts presented. Triumph Hosiery Mills, Inc. v. Triumph International Corp., 308 F.2d 196, 198 (CA 2 1962); Restatement of Torts §729, Comment a (1938).

7.   defendant's intent in selecting the mark; and
8.   likelihood of expansion of the product lines.

1. Strength of the Mark...

[W]e hold that Slickcraft is a suggestive mark when applied to boats.... Although appellant's mark is protectable and may have been strengthened by advertising,... it is a weak mark entitled to a restricted range of protection. Thus, only if the marks are quite similar, and the goods closely related, will infringement be found....

2. Proximity of the Goods

For related goods, the danger presented is that the public will mistakenly assume there is an association between the producers of the related goods, though no such association exists.... The more likely the public is to make such an association, the less similarity in the marks is requisite to a finding of likelihood of confusion.... Thus, less similarity between the marks will suffice when the goods are complementary,... the products are sold to the same class of purchasers,... or the goods are similar in use and function....

Although these product lines are non-competing, they are extremely close in use and function. In fact, their uses overlap. Both are for recreational boating on bays and lakes. Both are designed for water skiing and speedy cruises. Their functional features, for the most part, are also similar: fiberglass bodies, outboard motors, and open seating for a handful of people. Although the Sleekcraft boat is for higher speed recreation and its refinements support the market distinction the district court made, they are so closely related that a diminished standard of similarity must be applied when comparing the two marks....

3. Similarity of the Marks

The district court found that "the two marks are easily distinguishable in use either when written or spoken." Again, there is confusion among the cases as to whether review of this finding is subject to the clearly erroneous standard....

Similarity of the marks is tested on three levels: sight, sound, and meaning.... Each must be considered as they are encountered in the marketplace. Although similarity is measured by the marks as entities, similarities weigh more heavily than differences....

Standing alone the words Sleekcraft and Slickcraft are the same except for two inconspicuous letters in the middle of the first syllable.... To the eye, the words are similar.

In support of the district court's finding, Nescher points out that the distinctive logo on his boats and brochures negates the similarity of the words. We agree: the names appear dissimilar when viewed in conjunction with the logo, but the logo is often absent. The exhibits show that the word Sleekcraft is frequently found alone in trade journals, company stationery, and various advertisements.

Nescher also points out that the Slickcraft name is usually accompanied by the additional trademark AMF. As a result of this consistent use, Nescher argues, AMF has become the salient part of the mark indicative of the product's origin . . . .

Although Nescher is correct in asserting that use of a housemark can reduce the likelihood of confusion, . . . the effect is negligible here even though AMF is a well-known house name for recreational equipment. The exhibits show that the AMF mark is down-played in the brochures and advertisements; the letters AMF are smaller and skewed to one side. Throughout the promotional materials, the emphasis is on the Slickcraft name. Accordingly, we find that Slickcraft is the more conspicuous mark and serves to indicate the source of origin to the public . . . .

Sound is also important because reputation is often conveyed word-of-mouth. We recognize that the two sounds can be distinguishable, but the difference is only in a small part of one syllable. In G. D. Searle & Co. v. Chas. Pfizer & Co., 265 F.2d 385 (CA 7 1959), *cert. denied,* 361 U.S. 819 (1959), the court reversed the trial court's finding that Bonamine sounded "unlike" Dramamine, stating that: "Slight differences in the sound of trademarks will not protect the infringer." Id. at 387. The difference here is even slighter . . . .

Neither expert testimony nor survey evidence was introduced below to support the trial court's finding that the marks were easily distinguishable to the eye and the ear . . . . The district judge based his conclusion on a comparison of the marks. After making the same comparison, we are left with a definite and firm conviction that his conclusion is incorrect . . . .

The final criterion reinforces our conclusion. Closeness in meaning can itself substantiate a claim of similarity of trademarks. See, e.g., S. C. Johnson & Son, Inc. v. Drop Dead Co., 210 F. Supp. 816 (S.D. Cal. 1962), *aff'd,* 326 F.2d 87 (1963) (Pledge and Promise). Nescher contends the words are sharply different in meaning. This contention is not convincing; the words are virtual synonyms. Webster's New World Dictionary of the American Language 1371 (1966).

Despite the trial court's findings, we hold that the marks are quite similar on all three levels.

4. *Evidence of Actual Confusion*

Evidence that use of the two marks has already led to confusion is persuasive proof that future confusion is likely . . . . Proving actual confusion is difficult, however, . . . and the courts have often discounted such evidence because it was unclear or insubstantial . . . .

AMF introduced evidence that confusion had occurred both in the trade and in the mind of the buying public. A substantial showing of confusion among either group might have convinced the trial court that continued use would lead to further confusion . . . .

The district judge found that in light of the number of sales and the extent of the parties' advertising, the amount of past confusion was negligible. We cannot say this finding is clearly erroneous though we might have viewed the evidence more generously.

5. *Marketing Channels*

Convergent marketing channels increase the likelihood of confusion . . . . There is no evidence in the record that both lines were sold under the same roof except at boat shows; the normal marketing channels used by both AMF and Nescher are, however, parallel. Each sells through authorized retail dealers in diverse localities. The same sales methods are employed. The price ranges are almost identical. Each line is advertised extensively though different national magazines are used; the retail dealers also promote the lines, by participating in smaller boat shows and by advertising in local newspapers and classified telephone directories. Although different submarkets are involved, the general class of boat purchasers exposed to the products overlap.

6. *Type of Goods and Purchaser Care*

Both parties produce high quality, expensive goods. According to the findings of fact, the boats "are purchased only after thoughtful, careful evaluation of the product and the performance the purchaser expects."

In assessing the likelihood of confusion to the public, the standard used by the courts is the typical buyer exercising ordinary caution . . . . Although the wholly indifferent may be excluded, . . . the standard includes the ignorant and the credulous . . . . When the buyer has expertise in the field, a higher standard is proper though it will not preclude a finding that confusion is likely . . . . Similarly, when the goods are expensive, the buyer can be expected to exercise greater care in his purchases; again, though, confusion may still be likely . . . .

The parties vigorously dispute the validity of the trial court's finding on how discriminating the average buyer actually is. Although AMF presented expert testimony to the contrary, the court's finding is amply supported by the record. The care exercised by the typical purchaser, though it might virtually eliminate mistaken purchases, does not guarantee that confusion as to association or sponsorship is unlikely.

The district court also found that trademarks are unimportant to the average boat buyer. Common sense and the evidence indicate this is not the type of purchase made only on "general impressions." . . . This inattention to trade symbols does reduce the possibilities for confusion . . . .

The high quality of defendant's boats is also relevant in another way. The hallmark of a trademark owner's interest in preventing use of his mark on related goods is the threat such use poses to the reputation of his own goods . . . . When the alleged infringer's goods are of equal quality, there is little harm to the reputation earned by the trademarked goods. Yet this is no defense, for present quality is no assurance of continued quality . . . . The wrong inheres in involuntarily entrusting one's business reputation to another business . . . . AMF, of course, cannot control the quality of Sleekcraft boats . . . . [Indeed, e]quivalence in quality may actually contribute to the assumption of a common connection.

7. *Intent*

The district judge found that Nescher was unaware of appellant's use of the Slickcraft mark when he adopted the Sleekcraft name. There was no evidence that anyone attempted to palm off the latter boats for the former. And after notification of the purported infringement, Nescher designed a distinctive logo . . . . We agree with the district judge: appellee's good faith cannot be questioned.

When the alleged infringer knowingly adopts a mark similar to another's, reviewing courts presume that the defendant can accomplish his purpose: that is, that the public will be deceived . . . . Good faith is less probative of the likelihood of confusion, yet may be given considerable weight in fashioning a remedy.

8. Likelihood of Expansion

Inasmuch as a trademark owner is afforded greater protection against competing goods, a "strong possibility" that either party may expand his business to compete with the other will weigh in favor of finding that the present use is infringing . . . . When goods are closely related, any expansion is likely to result in direct competition . . . . The evidence shows that both parties are diversifying their model lines. The potential that one or both of the parties will enter the other's submarket with a competing model is strong.

Remedy

[A] limited mandatory injunction is warranted. Upon remand the district court should consider the above interests in structuring appropriate relief.

COMMENTS AND QUESTIONS

1. An interesting case on proximity of goods (factor 2) is Death Tobacco, Inc. v. Black Death USA, 31 U.S.P.Q.2d 1899, 1903 (C.D. Cal. 1993):

> In this case, defendant's ["Black Death"] vodka and plaintiff's ["Death"] cigarettes are related products. They are, to some extent, complementary in that smoking and drinking are related vices that are often undertaken together. They are also somewhat similar in use and function. Cigarettes and vodka both have a mood altering effect and both are used for recreation and relaxation . . . .
>
> The most important factor in the determination of relatedness, here, is that the two items are sold to the same class of purchasers. Defendant introduced evidence that Black Death vodka and Death cigarettes are sold in the same package stores in both San Francisco and Los Angeles. Several store owners display the two products adjacent to one another on the shelf.

2. For an older case listing factors to consider in determining likelihood of confusion, see Polaroid Corp. v. Polarad Electronics Corp., 287 F.2d 492, 495 (2d Cir.), *cert. denied*, 368 U.S. 820 (1961). The actual factors considered vary in each circuit, but in substance the different tests are similar.

3. Are all of these factors (however many there may in fact be) equally important? Strength of the mark may obviously be significant for validity purposes as well as for showing likelihood of confusion, but its use in an infringement proceeding presumes that the mark has already been determined valid. Must a plaintiff prove all the factors? Only some of them? Are there some that must be proven in all cases? Conversely, can proof of one factor ever be sufficient by itself?

4. Proof of actual confusion among consumers is obviously an important step towards showing a "likelihood of confusion." Such proof can sometimes be offered anecdotally, particularly where both products have already been in the market for a significant period of time. But confused consumers are difficult to find, and if the infringement is challenged early enough there may not be very many of them at all. In those circumstances, courts generally allow the results of consumer surveys as evidence of "actual" confusion. See, e.g., Union Carbide v. Ever-Ready, Inc., 531 F.2d 366, 387-388 (7th Cir. 1976); Mutual of Omaha Ins. Co. v. Novak, 836 F.2d 397, 400-401 (8th Cir. 1987). Because of the importance of actual confusion in proving likelihood of confusion, however, and because of the potential for abuse of the survey process, courts are relatively strict about the surveys they allow, routinely rejecting or discounting surveys that are improperly designed or ask ambiguous or leading questions.

5. Who must be likely to be confused? Obviously, the strictness with which this test is applied will determine the chances that a trademark plaintiff has of proving infringement. The consumer who is tested is the "reasonable purchaser" of the products at issue. This standard allows for a great deal of flexibility in testing marks used with different products. For example, the "reasonable purchaser" of fleets of commercial airplanes may be expected both to be more sophisticated and to pay more attention to the decision than the reasonable purchaser of pencils for home use. In light of this, should trademark law offer more protection to the owner of a trademark for pencils (and thus indirectly to the consumers of pencils) than to the owners of airplane trademarks?

A related question is how many consumers must be likely to be confused. The fact that the "reasonable consumer" is at issue might suggest that at least half of the consuming public must be confused in order to constitute trademark infringement. After all, if the median consumer is not "reasonable," who is? But courts have not been willing to require such a showing from plaintiffs. Instead, likelihood of confusion is regularly found if as few as 10 to 15 percent of the consumers surveyed were confused. Is it reasonable to test infringement on the basis of the reactions of a small minority of the population? If not, how much confusion should be required? See Mushroom Makers, Inc. v. R. G. Barry Corp., 580 F.2d 44, 47 (2d Cir. 1978) (testing the "likelihood that *an appreciable number of ordinarily prudent purchasers* will be misled, or indeed simply confused, as to the source of the goods in question.") (emphasis added).

6. What does it mean for two marks to be similar? The courts generally consider three kinds of similarity—sight, sound, and meaning. (Note that the second, and possibly the third, kind of similarity have little or no relevance to trade dress.) Similarity is generally tested by comparing the marks as a whole, rather than by dissecting them. This approach makes sense, doesn't it, since consumers are likely to pay attention to the whole mark in context? On the other hand, dissection is appropriate if the aim is to prevent trademark owners from exercising control over generic, functional, or disclaimed portions of a trademark or trade dress. The trademark owner shouldn't be able to point to similarities between parts of a composite mark that are generic.

Because the perception of the consumer when exposed to the whole mark in context is the linchpin of trademark infringement, a defendant's use of a similar or even identical trademark on similar products may be ameliorated by other differences between the mark and the packaging. For example, similarities in trade dress may not confuse consumers if the packages contain very different product names in large, obvious letters on the front. Disclaimers may also be effective in reducing confusion, if consumers notice them. If consumers would nonetheless be confused, however, efforts to ameliorate the effects of similar marks will not avoid trademark infringement.

If the plaintiff holds a family of related marks (such as the use of the "Mc" phrase in various McDonalds products), courts may be more willing to find similarity where the defendant copies the common element in the family.

7. What does the defendant's intent have to do with the likelihood that consumers will be confused by the two marks? Certainly, intentional copying is likely to be closer than accidental similarity, and therefore quite possibly more confusing, but isn't that adequately tested by the other factors that the court employs? Has the court slipped in a "fairness" factor to create what is in effect a presumption that counterfeiting (that is, intentional copying of a trademark) is illegal? Is such a presumption appropriate?

8. The fact that a trademark has been made incontestable precludes a defendant from challenging it on the basis of descriptiveness. See Park 'N Fly v. Dollar Park and Fly, discussed *supra*. An incontestable mark is not necessarily a *strong* mark for purposes of determining likelihood of confusion, however. In Petro Stopping Centers v. James River Petroleum Inc., 130 F.3d 88 (4th Cir. 1997), the Fourth Circuit affirmed a district court conclusion that plaintiff's Petro mark was merely descriptive and therefore weak, despite the fact that it was incontestable. The court noted that validity and infringement are separate inquiries and that incontestability applies only to the validity of a trademark. Accord Sports Authority Inc. v. Abercrombie & Fitch, 965 F. Supp. 925 (E.D. Mich. 1997).

Note on Other Types of Confusion

The traditional case for consumer confusion is that consumers will believe the infringer's product is the same as the trademark owner's product because of the similarity between the marks. Presumably, this confusion will allow the infringer to take sales away from the trademark owner by trading on the latter's goodwill. Although such confusion is the central problem at which trademark law is aimed, it is only one of many different ways in which consumers may be confused by different marks. Consider the following types of consumer confusion actionable under the Lanham Act.

Confusion as to Source. Two products bearing similar trademarks are likely to be confused only if both companies actually sell the same or similar products. Suppose, however, that an accused infringer uses a trademark on a group of products that the trademark owner does not sell. In that case, presumably, consumers cannot buy the infringer's products *instead of* the trademark owner's, because the trademark owner does not sell the products at all. For this reason, early courts sometimes found no infringement in situations where the plaintiff's and defendant's products did not directly compete. See, e.g., Borden Ice Cream v. Borden's Condensed Milk Co., 201 F. 510 (7th Cir. 1912) (use of identical trademark for different milk products does not constitute trademark infringement).

But courts have gradually come to recognize that identical trademarks can sometimes cause confusion *as to the source of the products* even in the absence of direct product competition. For example, "Prell" is a trademark for a brand of shampoo and conditioner. If another company uses the term "Prell" for its hair coloring product, consumers may well believe that they are buying a product sold by the same company that sells Prell shampoo, particularly given the proximity of the two goods. Cf. *Zazu Designs v. L'Oreal S.A.*, 979 F.2d 499 (7th Cir. 1992) (involving entry from one market into the other). While the infringer will not take sales of hair colors away from the trademark owner in this instance, the trademark owner may be hurt by this confusion in at least two ways. First, if the quality of the hair color is inferior, consumers may blame the maker of shampoo and stop buying Prell products altogether. Second, it is possible that the trademark owner may wish to expand into the market for hair colors. If it does, confusion between the products will almost certainly result if both parties use the Prell mark.

Confusion as to Sponsorship. Even in situations in which consumers will not believe that the trademark owner is the one selling the product, the use of a similar trademark may still confuse them by causing them to believe that the trademark owner is affiliated with or sponsors the infringer's products. For example, suppose that a company that sells soup uses the trade symbol of the United States Olympic Committee on its soup cans. Presumably, the USOC does not sell soup, and no reasonable consumer would be likely to conclude that she was in fact buying USOC soup. But consumers might well conclude that the infringer was somehow affiliated with the USOC, or that one group had agreed to sponsor the other. This confusion as to affiliation or sponsorship is actionable under the Lanham Act, assuming the other requirements for protecting a mark are met.[11]

In cases of confusion either as to source or as to sponsorship, the essential question remains whether consumers are likely to be confused by the similarity of the marks. Because of this, many of the same factors listed in *AMF* apply with equal force to these inquiries. For example, proximity between the products sold by the parties, while not required, does tend to increase both the likelihood of confusion as to source and the chance that the parties will eventually be in direct competition. A similar analysis of infringement can be conducted to determine whether any of these types of confusion is likely.

Initial Interest Confusion. What happens when confusion is dispelled before a product is ever purchased? For example, suppose that a competing fast-food restaurant put a large McDonald's sign by a highway exit. There is no McDonald's at the exit, so consumers won't end up thinking they are buying from one. However, they might be lured into the competitor's establishment once they have left the highway. Should this sort of deliberate effort to confuse consumers, but not at the point of purchase, be actionable? At least some courts have said yes. See *Mobil Oil Corp. v. Pegasus Petroleum Corp.*, 818 F.2d 254 (2d Cir. 1987); *Checkpoint Systems, Inc. v. Check Point Software Technologies, Inc.*, 269 F.3d 270 (3d Cir. 2001) (suggesting that whether the parties compete and the level of care exercised by consumers are relevant factors in determining initial interest confusion); McCarthy, Trademarks and Unfair Competition § 23:6.

11. Confusion as to affiliation or sponsorship is only expressly addressed in section 43(a) of the Lanham Act, which applies to unregistered marks. But at least one commentator has quite reasonably suggested that the protections afforded to unregistered marks under section 43(a) also apply to registered marks, whether under section 32 of the Lanham Act or under section 43(a). See Jay Dratler, Jr., Intellectual Property Law: Commercial, Creative, and Industrial Property § 10.01[1][i], at 10-5 to 10-6.

A similar problem has come up more recently on the Internet. Web pages frequently use "metatags," which are words on the page invisible to the user but which are read by computer search engines. Some companies have begun putting the names of their competitors' products (or even just popular trademarks such as "Playboy") into their metatags in an effort to draw unsuspecting consumers to their site. Is the use of a competitor's trademark in a metatag infringement? There are only a few reported decisions so far, but the cases seem to turn on whether there was a legitimate reason to make reference to the trademark on the Web page. Compare Brookfield Communications v. West Coast Entertainment Corp., 174 F.3d 1036 (9th Cir. 1999) (use of plaintiff's trademark in a Web site "metatag" was trademark infringement where it contributed to customer confusion) and People for the Ethical Treatment of Animals v. Doughney, 263 F.3d 359 (4th Cir. 2001) with Playboy Enterprises v. Welles, 279 F.3d 796 (9th Cir. 2002) (use of "Playmate of the Year" in a metatag to accurately describe defendant's resume was not illegal).

Doughney is a particularly interesting application of the idea of initial interest confusion. The plaintiff, People for the Ethical Treatment of Animals (PETA), is an advocacy group opposed to eating meat, wearing fur, and conducting research on animals. The defendant registered the Internet domain name peta.org, where he set up a page entitled "People Eating Tasty Animals" that was a parody of PETA and its goals. The Fourth Circuit found that using the domain name peta.org impermissibly caused initial-interest confusion, even though visitors to the site discovered almost immediately that it had no affiliation with PETA and even though Doughney was not competing with PETA in any commercial sense. Is this result correct? Or should initial-interest confusion be limited to cases in which the parties are direct competitors, as the Third Circuit held in *Checkpoint?* See Stacey L. Dogan & Mark A. Lemley, Trademarks and Search Costs on the Internet, 41 Hous. L. Rev. 777 (2004) (making this argument).

Commentators have been quite critical of the expansion of the doctrine, particularly in the Internet. See, e.g., Eric Goldman, Deregulating Relevancy in Internet Trademark Law, 54 Emory L.J. 507 (2005). They argue that attraction of initial interest is easy to reverse in the Internet context — the confused surfer simply clicks the "back" button on their Internet browser.

In Playboy Ents. v. Netscape Comms., 354 F.3d 1020 (9th Cir. 2004), the Ninth Circuit extended the doctrine of initial interest confusion to hold that a search engine was liable for either direct or contributory infringement (the court didn't specify which) when it encouraged competing pornography companies to run ads that appeared when users searched on the term "playboy." Concurring, Judge Berzon expressed significant concerns about the doctrine:

> I do not think it is reasonable to find initial interest confusion when a consumer is never confused as to source or affiliation, but instead knows, or should know, from the outset that a product or web link is not related to that of the trademark holder because the list produced by the search engine so informs him.
>
> There is a big difference between hijacking a customer to another website by making the customer think he or she is visiting the trademark holder's website (even if only briefly), which is what may be happening in this case when the banner advertisements are not labeled, and just distracting a potential customer with another *choice,* when it is clear that it is a choice. True, when the search engine list generated by the search for the trademark ensconced in a metatag comes up, an internet user might *choose* to visit westcoastvideo.com, the defendant's website in *Brookfield,* instead of the plaintiff's

moviebuff.com website, but such choices do not constitute trademark infringement off the internet, and I cannot understand why they should on the internet.

Playboy Ents. v. Netscape Comms., 354 F.3d 1020, 1034-35 (9th Cir. 2004) (Berzon, J., concurring). While *Playboy* applied the doctrine, even in the Fourth and Ninth Circuits courts have started to back away from broad application of initial interest confusion. See, e.g., Lamparello v. Falwell, 420 F.3d 309 (4th Cir. 2005); Bosley Med. Inst. v. Kremer, 403 F.3d 672 (9th Cir. 2005); Gibson Guitar Corp. v. Paul Reed Smith Guitars, 423 F.3d 539 (6th Cir. 2005) (all criticizing the broad reach of the doctrine and refusing to apply it).

Post-Sale Confusion. Suppose that consumers do not confuse two products but that, because of similarities in design, third parties who see the product from a distance (or stripped of identifying trademarks) will assume that the products are the same. This was the situation in Lois Sportswear v. Levi Strauss & Co., 799 F.2d 867 (2d Cir. 1986). There, Levi Strauss had registered not only its trade name and its jean labels, but also the pattern of stitching on the back pockets of its jeans. Lois Sportswear sold jeans with clearly different labels, but with an identical stitching pattern. The trial court found that there was no evidence of actual confusion by purchasers but that nonpurchasers, seeing the jeans "worn by a passer-by," would likely be confused. Should such "post-sale confusion" be actionable? The Second Circuit held that post-sale confusion constituted trademark infringement: "The confusion the Act seeks to prevent in this context is that a consumer seeing the familiar stitching pattern will associate the jeans with [Levi's] and that association will influence his buying decisions."

The district court also concluded that the Lois jeans were not of inferior quality. Given that, how is Levi Strauss & Co. hurt by any confusion that does arise? It will not lose sales, since customers themselves are not confused. Further, any confusion inures to the company's benefit, doesn't it? Whose interests are being protected by this decision?

Although some courts still require point of sale confusion, see, e.g., Gibson Guitar Corp. v. Paul Reed Smith Guitars, 423 F.3d 539 (6th Cir. 2005); Beneficial Corp. v. Beneficial Capital Corp., 529 F. Supp. 445 (S.D.N.Y. 1982) (holding that trademark law protects only against mistaken purchasing decisions); Nike, Inc. v. "Just Did It" Enterprises, 6 F.3d 1225 (7th Cir. 1993), many also recognize confusion after the sale of a product as actionable. In Mastercrafters Clock & Radio Co. v. Vacheron & Constantin-Le Coultre Watches, Inc., 221 F.2d 464 (2nd Cir. 1955), *cert. denied*, 350 U.S. 832 (1955), the junior user copied the distinctive appearance of the senior user's expensive "Atmos" clock. Judge Frank held that

> at least some customers would buy [the copier's] cheaper clock for the purpose of acquiring the prestige gained by displaying what many visitors at the customers' homes would regard as a prestigious article. [The copier's] wrong thus consisted of the fact that such a visitor would be likely to assume that the clock was an Atmos clock.... [T]he likelihood of such confusion suffices to render [the copier's] conduct actionable.

Id. at 466. This same problem arises in the sale of "knockoff" versions of high quality goods on street corners and flea markets. The purchaser of a "Rolex" branded watch for $25 appreciates that he or she is not purchasing an authentic product. See Rolex

Watch U.S.A., Inc. v. Canner, 645 F. Supp. 484 (S.D. Fla. 1986); United States v. Torkington, 812 F.2d 1347 (11th Cir. 1987). The Second Circuit has recognized, however, that

> [t]he creation of confusion in the post-sale context can be harmful in that if there are too many knockoffs in the market, sales of the originals may decline because the public is fearful that what they are purchasing may not be an original. Furthermore, the public may be deceived in the resale market if it requires expertise to distinguish between an original and a knockoff. Finally, the purchaser of an original is harmed by the widespread existence of knockoffs because the high value of originals, which derives in part from their scarcity, is lessened.

See Hermes Intern. v. Lederer de Paris Fifth Ave., Inc., 219 F.3d 104, 108 (2d Cir. 2000). See generally McCarthy, Trademarks and Unfair Competition §23:7.

Reverse Confusion. Ordinary trademark confusion occurs when the junior user trades on the reputation of the trademark owner, confusing the public into thinking that his goods are associated with those of the senior trademark owner. At times, however, a large company will adopt the mark of a smaller trademark owner. In this case, the danger is presumably not that the junior user will trade on the smaller company's goodwill. The risk is, instead, that the public will come to associate the mark not with its true owner but with the infringer (who has spent a great deal of money to advertise it). Several courts have made it clear that reverse confusion is trademark infringement, and that the relative size of the companies doesn't matter. Indeed, companies have periodically been forced to halt or even retract major advertising campaigns because of reverse confusion problems. See, e.g., Big O Tire Dealers, Inc. v. Goodyear Tire & Rubber Co., 561 F.2d 1365 (10th Cir. 1977); Sands, Taylor & Wood v. Quaker Oaks, 34 F.3d 1340 (7th Cir. 1994).

In a recent case, Judge Posner took a remarkably different approach. In Illinois High School Assn. v. GTE Vantage, 99 F.3d 244 (7th Cir. 1996), the plaintiffs were the owners of the trademark "March Madness" for their annual high school basketball tournament. After the media began using the term to describe the NCAA *college* basketball tournament, the NCAA and its various licensees adopted the term and used it in advertising their own products and services. The Illinois High School Association sued alleging reverse confusion — that their mark had been appropriated by the NCAA. The Seventh Circuit rejected this claim, creating a new trademark classification: the "dual use trademark." The court reasoned that where *the public,* not the defendant, had appropriated a trademark for use to describe another product or service, trademark law should not stand in the way.

Does this result make sense? Why isn't this an ordinary case of reverse confusion? Under the Seventh Circuit's opinion, presumably *both* IHSA and NCAA now have rights to the mark. Does this comport with the rationales for trademark protection? Who now has the right to sue infringers? To license the use of the mark?

A number of reverse confusion cases have come up in the context of plagiarism of works of authorship, where the author claims that the copier has engaged in "reverse passing off" — that is, selling the plaintiff's work as her own rather than vice versa. In Dastar Corp. v. Twentieth Century Fox Film Corp., 539 U.S. 23 (2003), however, the Supreme Court made it clear that Lanham Act §43(a) does not prevent the unaccredited copying of a work of authorship; any such rights must come from copyright law. For discussion, see Jane C. Ginsburg, The Right to Claim Authorship in U.S. Copyright and Trademark Laws, 41 Houston L. Rev. 263 (2004) and David Nimmer, The Moral Imperative Against Academic Plagiarism (Without a Moral Right Against Reverse Passing Off), 54 DePaul L. Rev. 1 (2004).

PROBLEMS

Problem 5-7. Bristol-Myers, a major pharmaceutical company, markets "Excedrin" pain reliever. Since 1968, B-M has marketed "Excedrin PM," which is a pain reliever that does not interfere with sleep. Excedrin PM tablets are sold in a solid blue box whose color fades from dark at the top to light at the bottom. The box contains the words "Excedrin PM" in large white letters across the top third of the box. In the bottom right-hand corner of the box is a depiction of two tablets labeled "PM." B-M also sells Excedrin PM capsules, which are packaged identically except that the background is green and the two capsules in the picture read "Excedrin PM." Both the mark Excedrin PM and the dress of both boxes are registered with the Trademark Office.

In 1991, McNeil Pharmaceuticals introduced "Tylenol PM," a pain reliever chemically identical to Excedrin PM. Tylenol PM tablets are sold in a solid green box whose color fades from dark at the top to light at the bottom. The box contains the words "Tylenol PM" in large white and yellow letters across the top third of the box. In the bottom right-hand corner of the box is a depiction of two tablets, one labeled "Tylenol" and the other labeled "PM." McNeil also sells Tylenol PM capsules, which are packaged identically except that the background is blue and the two capsules in the picture both read "Tylenol PM."

B-M sues McNeil, alleging that both its use of the term PM and its trade dress are likely to confuse consumers. Who should prevail?

Problem 5-8. Ivory Soap is sold for approximately $1.00 a bar. It is a heavily advertised brand name, which identifies itself as "99.44% pure." Whitewash Soap Co. makes and sells counterfeit Ivory Soap bars, which are packaged identically to Ivory (using the Ivory name and trade dress) and which contain chemically identical bars of soap. These counterfeit "Ivory" bars are sold for $.50 each. If a consumer buys the Whitewash soap thinking it is Ivory, how has he been injured? Hasn't he benefited? Is there any reason to protect Ivory's trademark against infringement at the expense of consumer welfare?

3. Dilution

Likelihood of confusion is the traditional benchmark of trademark infringement. Yet in some cases courts have found trademark infringement even where consumers were not likely to be confused as to source. The canonical case involved Kodak bicycles. Despite the fact — well recognized by the court — that consumers were not likely to assume a connection between Kodak, the famous maker of film and cameras, and the seller of bicycles, the court issued an injunction. It did so because it found that the use of the Kodak name on bicycles would harm Kodak, Inc., even in the absence of confusion as to source. See Eastman Photographic Materials Co. v. Kodak Cycle Co., 15 [British] R.P.C. 105 (1898). See generally Beverly W. Pattishall, *Dawning Acceptance of the Dilution Rationale for Trademark–Trade Identity Protection*, 74 Trademark Rep. 289, 289 n.2 (1984) (citing early trademark "dilution" cases).

What is the nature of this harm? The legislative history of the Federal Trademark Dilution Act of 1995, Pub. L. 104-98, codified at 15 U.S.C. § 1125(c), which added a new section 43(c) to the Lanham Act, contained this instructive summary:

> [This bill would] create a federal cause of action to protect famous marks from unauthorized users that attempt to trade upon the goodwill and established renown of such marks and, thereby, dilute their distinctive quality. The provision is intended to protect famous marks where the subsequent, unauthorized commercial use of such marks by others dilutes the distinctiveness of the mark. The bill defines the term "dilution" to mean "the lessening of the capacity of a famous mark to identify and distinguish goods or services regardless of the presence or absence of (a) competition between the parties, or (b) likelihood of confusion, mistakes, or deception." Thus, for example, the use of DUPONT shoes, BUICK aspirin, and KODAK pianos would be actionable under this legislation. The protection of marks from dilution differs from the protection accorded marks from trademark infringement. Dilution does not rely upon the standard test of infringement, that is, likelihood of confusion, deception or mistake. Rather, it applies when the unauthorized use of a famous mark reduces the public's perception that the mark signifies something unique, singular, or particular. As summarized in one decision:
>
> > Dilution is an injury that differs materially from that arising out of the orthodox confusion. Even in the absence of confusion, the potency of a mark may be debilitated by another's use. This is the essence of dilution. Confusion leads to immediate injury, while dilution is an infection, which if allowed to spread, will inevitably destroy the advertising value of the mark.
>
> Mortellito v. Nina of California, Inc., 335 F. Supp. 1288, 1296 (S.D.N.Y. 1972).

In 1996, section 43 of the Lanham Act was amended as follows:

> (c)(1) The owner of a famous mark shall be entitled, subject to the principles of equity and upon such terms as the court deems reasonable, to an injunction against another person's commercial use in commerce of a mark or trade name, if such use begins after the mark has become famous and causes dilution of the distinctive quality of the mark, and to obtain such other relief as is provided in this subsection. In determining whether a mark is distinctive and famous, a court may consider factors such as, but not limited to —
>
> > (A) the degree of inherent or acquired distinctiveness of the mark;
> > (B) the duration and extent of use of the mark in connection with the goods or services with which the mark is used;
> > (C) the duration and extent of advertising and publicity of the mark;
> > (D) the geographical extent of the trading area in which the mark is used;
> > (E) the channels of trade for the goods or services with which the mark is used;

(F) the degree of recognition of the mark in the trading areas and channels of trade used by the marks' owner and the person against whom the injunction is sought;

(G) the nature and extent of use of the same or similar marks by third parties; and

(H) whether the mark was registered under the Act of March 3, 1881, or the Act of February 20, 1905, or on the principal register.

(2) In an action brought under this subsection, the owner of the famous mark shall be entitled only to injunctive relief unless the person against whom the injunction is sought willfully intended to trade on the owner's reputation or to cause dilution of the famous mark. If such willful intent is proven, the owner of the famous mark shall also be entitled to the remedies set forth in sections 35(a) and 36 [i.e., 15 U.S.C. §§1117, 1118], subject to the discretion of the court and the principles of equity.

(3) The ownership by a person of a valid registration under the Act of March 3, 1881, or the Act of February 20, 1905, or on the principal register shall be a complete bar to an action against that person, with respect to that mark, that is brought by another person under the common law or a statute of a State and that seeks to prevent dilution of the distinctiveness of a mark, label, or form of advertisement.

(4) The following shall not be actionable under this section:

(A) Fair use of a famous mark by another person in comparative commercial advertising or promotion to identify the competing goods or services of the owner of the famous mark.

(B) Noncommercial use of a mark.

(C) All forms of news reporting and news commentary.

Federal Trademark Dilution Act of 1995, §3, P.L. 104-98, codified at 15 U.S.C. §1125(c).

The House Report, quoted in the introduction to this section, goes on to describe the effect of the act.

H.R. Rep. 104-374, 104th Cong., 1st Sess. (1995)

A federal dilution statute is necessary because famous marks ordinarily are used on a nationwide basis and dilution protection is currently only available on a patch-quilt system of protection, in that only approximately 25 states have laws that prohibit trademark dilution. Further, court decisions have been inconsistent and some courts are reluctant to grant nationwide injunctions for violation of state law where half of the states have no dilution law.[1] Protection for famous marks should not depend on whether the forum where suit is filed has a dilution statute. This simply encourages forum-shopping and increases the amount of litigation.

Moreover, the recently concluded Agreement on Trade-Related Aspects of Intellectual Property Rights, including Trade in Counterfeit Goods ("TRIPS") which was part of the Uruguay Round of the GATT agreement includes a provision designed to provide dilution protection to famous marks. Thus, enactment of this bill will be consistent with the terms of the agreement, as well as the Paris Convention, of which

1. Blue Ribbon Feed Co., Inc. v. Farmers Union Central Exchange, Inc., 731 F.2d 415, 422 (7th Cir. 1984); Deere & Co. v. MTD Products Inc., 34 U.S.P.Q. 2d 1706 (S.D.N.Y. 1995).

the U.S. is also a member. Passage of a federal dilution statute would also assist the executive branch in its bilateral and multilateral negotiations with other countries to secure greater protection for the famous marks owned by U.S. companies. Foreign countries are reluctant to change their laws to protect famous U.S. marks if the U.S. itself does not afford special protection for such marks.

It should be noted that as originally introduced, H.R. 1295 only applied to famous registered marks. However, based on testimony by the Patent and Trademark Office, Congresswoman Patricia Schroeder offered an amendment in the nature of a substitute to H.R. 1295, that was adopted by the Subcommittee, to include all famous marks within the scope of the bill. The Patent and Trademark Office made a compelling case that limiting the federal remedy against trademark dilution to those famous marks that are registered is not within the spirit of the United States' position as a leader setting the standards for strong worldwide protection of intellectual property. Such a limitation would undercut the United States position with our trading partners, which is that famous marks should be protected regardless of whether the marks are registered in the country where protection is sought.

The proposal adequately addresses legitimate First Amendment concerns espoused by the broadcasting industry and the media. The bill will not prohibit or threaten "noncommercial" expression, as that term has been defined by the courts. Nothing in this bill is intended to alter existing case law on the subject of what constitutes "commercial" speech. The bill includes specific language exempting from liability the "fair use" of a mark in the context of comparative commercial advertising or promotion as well as all forms of news reporting and news commentary. The latter provision which was added to [the act] as a result of an amendment offered by Congressman Moorhead that was adopted by the Committee, recognizes the heightened First Amendment protection afforded the news industry.

It is important to note that H.R. 1295 would not pre-empt existing state dilution statutes. State laws could continue to be applied in cases involving locally famous or distinctive marks.[2] Unlike patent and copyright laws, federal trademark law presently coexists with state trademark law, and it is to be expected that a federal dilution statute should similarly coexist with state dilution law. The ownership of valid federal registration would act as a complete bar to a dilution action brought under state law.

With respect to remedies, the bill limits the relief a court could award to an injunction unless the wrongdoer willfully intended to trade on the trademark owner's reputation or to cause dilution of the famous mark, in which case the remedies under sections 35(a) and 36 of the Trademark Act become available.

COMMENTS AND QUESTIONS

1. The rationale for protection against dilution was first articulated in a law review article by trademark lawyer Frank I. Schechter. See The Rational Basis of Trademark Protection, 40 Harv. L. Rev. 813 (1927). Schechter's proposal applied only to coined, fanciful, or arbitrary marks, only to situations in which the junior user's mark was identical to that of the senior user, and only to use of identical marks on noncompeting goods. Id. at 825-830. Courts and commentators have since expanded the doctrine to include marks that have acquired secondary meaning (see discussion in

2. See, e.g., Wedgewood Homes, Inc. v. Lund, 659 P.2d 377 (Or. 1983).

Howard Shire, Dilution Versus Deception—Are State Antidilution Laws an Appropriate Alternative to the Law of Infringement?, 77 Trademark Rep. 273, 275-278 (1987)), situations where the junior user's mark has substantial similarity to the senior user's mark, see, e.g., Mead Data Central, Inc. v. Toyota Motor Sales, U.S.A., Inc., 875 F.2d 1026 (2d Cir. 1989) (holding that Toyota's "LEXUS" mark for its line of luxury automobiles does not dilute Mead Data's "LEXIS" mark for computerized legal research services), and to use of similar marks on competing goods, see, e.g., Deere & Co. v. MTD Products, Inc., 41 F.3d 39 (2d Cir. 1994) (enjoining MTD, a manufacturer of tractors, from using an animated version of Deere's well-known leaping deer trademark in its television advertising campaign). For an overview of state dilution doctrines see Elliot Staffin, The Dilution Doctrine: Towards a Reconciliation with the Lanham Act, 6 Fordham Intell. Prop. Media & Ent. L.J. 105 (1995).

2. At least some commentators believe that antidilution law does not go far enough in protecting trademark owners. See Jerre B. Swann, Dilution Redefined for the Year 2002, 92 Trademark Reporter 585 (2002); Jerre B. Swann, Dilution Redefined for the Year 2000, 37 Houston L. Rev. 729 (2000); Jerre B. Swann & Theodore H. Davis, Jr., Dilution, An Idea Whose Time Has Gone: Brand Equity as Protectable Property, The New/Old Paradigm, 1 J. Intell. Prop. L. 219 (1994) (arguing for explicit property rights in brand equity, going beyond antidilution law). On the other hand, some commentators suggest the opposite. See Kenneth L. Port, The "Unnatural" Expansion of Trademark Rights: Is a Federal Dilution Statute Necessary?, 85 Trademark Rptr. 525 (1995); Megan Gray, Defending Against a Dilution Claim: A Practitioner's Guide, 4 Tex. Intell. Prop. L.J. 205 (1996).

Moseley v. V Secret Catalogue, Inc.
Supreme Court of the United States
123 S. Ct. 1115 (2003)

Justice STEVENS delivered the opinion of the Court.

In 1995 Congress amended §43 of the Trademark Act of 1946, 15 U.S.C. §1125, to provide a remedy for the "dilution of famous marks." 109 Stat. 985-986. That amendment, known as the Federal Trademark Dilution Act (FTDA), describes the factors that determine whether a mark is "distinctive and famous," and defines the term "dilution" as "the lessening of the capacity of a famous mark to identify and distinguish goods or services." The question we granted certiorari to decide is whether objective proof of actual injury to the economic value of a famous mark (as opposed to a presumption of harm arising from a subjective "likelihood of dilution" standard) is a requisite for relief under the FTDA.

I

Petitioners, Victor and Cathy Moseley, own and operate a retail store named "Victor's Little Secret" in a strip mall in Elizabethtown, Kentucky. They have no employees.

Respondents are affiliated corporations that own the VICTORIA'S SECRET trademark, and operate over 750 Victoria's Secret stores, two of which are in Louisville, Kentucky, a short drive from Elizabethtown. In 1998 they spent over $55 million advertising "the VICTORIA'S SECRET brand—one of moderately priced, high quality, attractively designed lingerie sold in a store setting designed to look like a wom[a]n's bedroom." They distribute 400 million copies of the Victoria's Secret catalog each year, including 39,000 in Elizabethtown. In 1998 their sales exceeded $1.5 billion.

In the February 12, 1998, edition of a weekly publication distributed to residents of the military installation at Fort Knox, Kentucky, petitioners advertised the "GRAND OPENING Just in time for Valentine's Day!" of their store "VICTOR'S SECRET" in nearby Elizabethtown. The ad featured "Intimate Lingerie *for every woman*"; "Romantic Lighting"; "Lycra Dresses"; "Pagers"; and "Adult Novelties/ Gifts." An army colonel, who saw the ad and was offended by what he perceived to be an attempt to use a reputable company's trademark to promote the sale of "unwholesome, tawdry merchandise," sent a copy to respondents. Their counsel then wrote to petitioners stating that their choice of the name "Victor's Secret" for a store selling lingerie was likely to cause confusion with the well-known victoria's secret mark and, in addition, was likely to "dilute the distinctiveness" of the mark. They requested the immediate discontinuance of the use of the name "and any variations thereof." In response, petitioners changed the name of their store to "Victor's Little Secret." Because that change did not satisfy respondents, they promptly filed this action in Federal District Court.

The complaint contained four separate claims: (1) for trademark infringement alleging that petitioners' use of their trade name was "likely to cause confusion and/or mistake in violation of 15 U.S.C. §1114(1)"; (2) for unfair competition alleging misrepresentation in violation of §1125(a); (3) for "federal dilution" in violation of the FTDA; and (4) for trademark infringement and unfair competition in violation of the common law of Kentucky. In the dilution count, the complaint alleged that petitioners' conduct was "likely to blur and erode the distinctiveness" and "tarnish the reputation" of the VICTORIA'S SECRET trademark.

After discovery the parties filed cross-motions for summary judgment. The record contained uncontradicted affidavits and deposition testimony describing the vast size of respondents' business, the value of the VICTORIA'S SECRET name, and descriptions of the items sold in the respective parties' stores. Respondents sell a "complete line of lingerie" and related items, each of which bears a VICTORIA'S SECRET label or tag. Petitioners sell a wide variety of items, including adult videos, "adult novelties," and lingerie. Victor Moseley stated in an affidavit that women's lingerie represented only about five per cent of their sales. In support of their motion for summary judgment, respondents submitted an affidavit by an expert in marketing who explained "the enormous value" of respondents' mark. Neither he, nor any other witness, expressed any opinion concerning the impact, if any, of petitioners' use of the name "Victor's Little Secret" on that value.

Finding that the record contained no evidence of actual confusion between the parties' marks, the District Court concluded that "no likelihood of confusion exists as a matter of law" and entered summary judgment for petitioners on the infringement and unfair competition claims. Civ. Action No. 3:98CV-395-S, 2000 WL 370525 (WD Ky., Feb. 9, 2000). With respect to the FTDA claim, however, the court ruled for respondents. [The district court found dilution by tarnishment, but not blurring].

The Court of Appeals for the Sixth Circuit affirmed. 259 F.3d 464 (2001)....

In reaching that conclusion the Court of Appeals expressly rejected the holding of the Fourth Circuit in *Ringling Bros.-Barnum & Bailey Combined Shows, Inc. v. Utah Div. of Travel Development,* 170 F.3d 449 (1999). In that case, which involved a claim that Utah's use on its license plates of the phrase "greatest *snow* on earth" was causing dilution of the "greatest *show* on earth," the court had concluded "that to establish dilution of a famous mark under the federal Act requires proof that (1) a defendant has made use of a junior mark sufficiently similar to the famous mark to evoke in a relevant universe of consumers a mental association of the two that (2) has caused (3) actual economic harm to the famous mark's economic value by lessening its former selling power as an advertising agent for its goods or services." *Id.,* at 461 (emphasis added). Because other Circuits have also expressed differing views about the "actual harm" issue, we granted certiorari to resolve the conflict. 535 U.S. 985, 122 S.Ct. 1536, 152 L.Ed.2d 463 (2002).

II

Traditional trademark infringement law is a part of the broader law of unfair competition, see Hanover Star Milling Co. v. Metcalf, 240 U.S. 403, 413, 36 S.Ct. 357, 60 L.Ed. 713 (1916), that has its sources in English common law, and was largely codified in the Trademark Act of 1946 (Lanham Act). See B. Pattishall, D. Hilliard, & J. Welch, Trademarks and Unfair Competition 2 (4th ed. 2000) ("The United States took the [trademark and unfair competition] law of England as its own"). That law broadly prohibits uses of trademarks, trade names, and trade dress that are likely to cause confusion about the source of a product or service. See 15 U.S.C. §§ 1114, 1125(a)(1)(A). Infringement law protects consumers from being misled by the use of infringing marks and also protects producers from unfair practices by an "imitating competitor." Qualitex Co. v. Jacobson Products Co., 514 U.S. 159, 163-164, 115 S.Ct. 1300, 131 L.Ed.2d 248 (1995).

Because respondents did not appeal the District Court's adverse judgement on counts 1, 2, and 4 of their complaint, we decide the case on the assumption that the Moseleys' use of the name "Victor's Little Secret" neither confused any consumers or potential consumers, nor was likely to do so. Moreover, the disposition of those counts also makes it appropriate to decide the case on the assumption that there was no significant competition between the adversaries in this case. Neither the absence of any likelihood of confusion nor the absence of competition, however, provides a defense to the statutory dilution claim alleged in count 3 of the complaint.

Unlike traditional infringement law, the prohibitions against trademark dilution are not the product of common-law development, and are not motivated by an interest in protecting consumers. The seminal discussion of dilution is found in Frank Schechter's 1927 law review article concluding "that the preservation of the uniqueness of a trademark should constitute the only rational basis for its protection." Rational Basis of Trademark Protection, 40 Harv. L.Rev. 813, 831. Schechter supported his conclusion by referring to a German case protecting the owner of the well-known trademark "Odol" for mouthwash from use on various noncompeting steel products. That case, and indeed the principal focus of the Schechter article, involved an established arbitrary mark that had been "added to rather than withdrawn from the human vocabulary" and an infringement that made use of the identical mark. *Id.*, at 829.[10] . . .

At least 25 States passed similar laws in the decades before the FTDA was enacted in 1995. See Restatement (Third) of Unfair Competition §25, Statutory Note (1995).

III

In 1988, when Congress adopted amendments to the Lanham Act, it gave consideration to an antidilution provision. During the hearings on the 1988 amendments, objections to that provision based on a concern that it might have applied to expression protected by the First Amendment were voiced and the provision was deleted from the amendments. H.R.Rep. No. 100-1028 (1988). The bill, H.R. 1295, 104th Cong., 1st Sess., that was introduced in the House in 1995, and ultimately enacted as the FTDA, included two exceptions designed to avoid those concerns: a provision allowing "fair use" of a registered mark in comparative advertising or promotion, and the provision that noncommercial use of a mark shall not constitute dilution. See 15 U.S.C. §1125(c)(4).

On July 19, 1995, the Subcommittee on Courts and Intellectual Property of the House Judiciary Committee held a 1-day hearing on H.R. 1295. No opposition to the bill was voiced at the hearing and, with one minor amendment that extended protection to unregistered as well as registered marks, the subcommittee endorsed the bill and it passed the House unanimously. The committee's report stated that the "purpose of H.R. 1295 is to protect famous trademarks from subsequent uses that blur the distinctiveness of the mark or tarnish or disparage it, even in the absence of a likelihood of confusion." H.R.Rep. No. 104-374, p. 1029 (1995), U.S.Code Cong. & Admin.News 1995, pp. 1029, 1030. As examples of dilution, it stated that "the use of DUPONT shoes, BUICK aspirin, and KODAK pianos would be actionable under this legislation." *Id.*, at 1030. In the Senate an identical bill, S. 1513, 104th Cong., 1st Sess., was introduced on December 29, 1995, and passed on the same day by voice vote without any hearings. In his explanation of the bill, Senator Hatch also stated that it was intended "to protect famous trademarks from subsequent uses that blur the distinctiveness of the mark or tarnish or disparage it," and referred to the Dupont Shoes, Buick aspirin, and Kodak piano examples, as well as to the Schechter law review article. 141 Cong. Rec. 38559-38561 (1995).

---

10. Schecter discussed this distinction at length: "The rule that arbitrary, coined or fanciful marks or names should be given a much broader degree of protection than symbols, words or phrases in common use would appear to be entirely sound. Such trademarks or tradenames as 'Blue Ribbon,' used, with or without registration, for all kinds of commodities or services, more than sixty times; 'Simplex' more than sixty times; 'Star,' as far back as 1898, nearly four hundred times; 'Anchor,' already registered over one hundred fifty times in 1898; 'Bull Dog,' over one hundred times by 1923; 'Gold Medal,' sixty-five times; '3-in-1' and '2-in-1,' seventy-nine times; 'Nox-all,' fifty times; 'Universal,' over thirty times; 'Lily White' over twenty times; — all these marks and names have, at this late date, very little distinctiveness in the public mind, and in most cases suggest merit, prominence or other qualities of goods or services in general, rather than the fact that the product or service, in connection with which the mark or name is used, emanates from a particular source. On the other hand, 'Rolls-Royce,' 'Aunt Jemima's,' 'Kodak,' 'Mazda,' 'Corona,' 'Nujol,' and 'Blue Goose,' are coined, arbitrary or fanciful words or phrases that have been added to rather than withdrawn from the human vocabulary by their owners, and have, from the very beginning, been associated in the public mind with a particular product, not with a variety of products, and have created in the public consciousness an impression or symbol of the excellence of the particular product in question." *Id.*, at 828-829.

IV

The Victoria's Secret mark is unquestionably valuable and petitioners have not challenged the conclusion that it qualifies as a "famous mark" within the meaning of the statute. Moreover, as we understand their submission, petitioners do not contend that the statutory protection is confined to identical uses of famous marks, or that the statute should be construed more narrowly in a case such as this. Even if the legislative history might lend some support to such a contention, it surely is not compelled by the statutory text.

The District Court's decision in this case rested on the conclusion that the name of petitioners' store "tarnished" the reputation of respondents' mark, and the Court of Appeals relied on both "tarnishment" and "blurring" to support its affirmance. Petitioners have not disputed the relevance of tarnishment, presumably because that concept was prominent in litigation brought under state antidilution statutes and because it was mentioned in the legislative history. Whether it is actually embraced by the statutory text, however, is another matter. Indeed, the contrast between the state statutes, which expressly refer to both "injury to business reputation" and to "dilution of the distinctive quality of a trade name or trademark," and the federal statute which refers only to the latter, arguably supports a narrower reading of the FTDA. See Klieger, Trademark Dilution: The Whittling Away of the Rational Basis for Trademark Protection, 58 U. Pitt. L.Rev. 789, 812-813, and n. 132 (1997).

The contrast between the state statutes and the federal statute, however, sheds light on the precise question that we must decide. For those state statutes, like several provisions in the federal Lanham Act, repeatedly refer to a "likelihood" of harm, rather than to a completed harm. The relevant text of the FTDA, quoted in full in note 1, *supra,* provides that "the owner of a famous mark" is entitled to injunctive relief against another person's commercial use of a mark or trade name if that use "*causes dilution* of the distinctive quality" of the famous mark. 15 U.S.C. § 1125(c)(1) (emphasis added). This text unambiguously requires a showing of actual dilution, rather than a likelihood of dilution.

This conclusion is fortified by the definition of the term "dilution" itself. That definition provides:

> "The term 'dilution' means the lessening of the capacity of a famous mark to identify and distinguish goods or services, regardless of the presence or absence of—
>> "(1) competition between the owner of the famous mark and other parties, or
>> "(2) likelihood of confusion, mistake, or deception."

§ 1127.

The contrast between the initial reference to an actual "lessening of the capacity" of the mark, and the later reference to a "likelihood of confusion, mistake, or deception" in the second caveat confirms the conclusion that actual dilution must be established.

Of course, that does not mean that the consequences of dilution, such as an actual loss of sales or profits, must also be proved. To the extent that language in the Fourth Circuit's opinion in the *Ringling Bros.* case suggests otherwise, see 170 F.3d, at 460-465, we disagree. We do agree, however, with that court's conclusion that, at least where the marks at issue are not identical, the mere fact that consumers mentally associate the junior user's mark with a famous mark is not sufficient to establish actionable dilution. As the facts of that case demonstrate, such mental association will not necessarily reduce the capacity of the famous mark to identify the goods of its owner, the statutory requirement for dilution under the FTDA. For even though Utah drivers may be reminded of the circus when they see a license plate referring to the "greatest *snow* on earth," it by no means follows that they will associate "the greatest show on earth" with skiing or snow sports, or associate it less strongly or exclusively with the circus. "Blurring" is not a necessary consequence of mental association. (Nor, for that matter, is "tarnishing.")

The record in this case establishes that an army officer who saw the advertisement of the opening of a store named "Victor's Secret" did make the mental association with "Victoria's Secret," but it also shows that he did not therefore form any different impression of the store that his wife and daughter had patronized. There is a complete absence of evidence of any lessening of the capacity of the Victoria's Secret mark to identify and distinguish goods or services sold in Victoria's Secret stores or advertised in its catalogs. The officer was offended by the ad, but it did not change his conception of Victoria's Secret. His offense was directed entirely at petitioners, not at respondents. Moreover, the expert retained by respondents had nothing to say about the impact of petitioners' name on the strength of respondents' mark.

Noting that consumer surveys and other means of demonstrating actual dilution are expensive and often unreliable, respondents and their *amici* argue that evidence of an actual "lessening of the capacity of a famous mark to identify and distinguish goods or services," § 1127, may be difficult to obtain. It may well be, however, that direct evidence of dilution such as consumer surveys will not be necessary if actual dilution can reliably be proven through circumstantial evidence — the obvious case is one where the junior and senior marks are identical. Whatever difficulties of proof may be entailed, they are not an acceptable reason for dispensing with proof of an essential element of a statutory violation. The evidence in the present record is not sufficient to support the summary judgment on the dilution count. The judgment is therefore reversed, and the case is remanded for further proceedings consistent with this opinion.

It is so ordered.

Justice KENNEDY, concurring.

As of this date, few courts have reviewed the statute we are considering, the Federal Trademark Dilution Act, 15 U.S.C. § 1125(c), and I agree with the Court that the evidentiary showing required by the statute can be clarified on remand. The conclusion that the VICTORIA'S SECRET mark is a famous mark has not been challenged throughout the litigation, *ante,* at 1120, 1124, and seems not to be in question. The remaining issue is what factors are to be considered to establish dilution.

For this inquiry, considerable attention should be given, in my view, to the word "capacity" in the statutory phrase that defines dilution as "the lessening of the capacity of a famous mark to identify and distinguish goods or services." 15 U.S.C. § 1127. When a competing mark is first adopted, there will be circumstances when the case can turn on the probable consequences its commercial use will have for the famous mark. In this respect, the word "capacity" imports into the dilution inquiry both the present and the potential power of the famous mark to identify and distinguish goods, and in some cases the fact that this power will be diminished could suffice to show dilution.

Capacity is defined as "the power or ability to hold, receive, or accommodate." Webster's Third New International Dictionary 330 (1961); see also Webster's New International Dictionary 396 (2d ed. 1949) ("Power of receiving, containing, or absorbing"); 2 Oxford English Dictionary 857 (2d ed. 1989) ("Ability to receive or contain; holding power"); American Heritage Dictionary 275 (4th ed. 2000) ("The ability to receive, hold, or absorb"). If a mark will erode or lessen the power of the famous mark to give customers the assurance of quality and the full satisfaction they have in knowing they have purchased goods bearing the famous mark, the elements of dilution may be established.

Diminishment of the famous mark's capacity can be shown by the probable consequences flowing from use or adoption of the competing mark. This analysis is confirmed by the statutory authorization to obtain injunctive relief. 15 U.S.C. §1125(c)(2). The essential role of injunctive relief is to "prevent future wrong, although no right has yet been violated." Swift & Co. v. United States, 276 U.S. 311, 326, 48 S.Ct. 311, 72 L.Ed. 587 (1928). Equity principles encourage those who are injured to assert their rights promptly. A holder of a famous mark threatened with diminishment of the mark's capacity to serve its purpose should not be forced to wait until the damage is done and the distinctiveness of the mark has been eroded.

In this case, the District Court found that petitioners' trademark had tarnished the VICTORIA'S SECRET mark. The Court of Appeals affirmed this conclusion and also found dilution by blurring. 259 F.3d 464, 477 (C.A.6 2001). The Court's opinion does not foreclose injunctive relief if respondents on remand present sufficient evidence of either blurring or tarnishment.

With these observations, I join the opinion of the Court.

COMMENTS AND QUESTIONS

1. The Court twice hints that a trademark owner may not have to show actual dilution if the plaintiff's and defendant's marks are identical. Why not? Is there any justification in the statute for such a distinction? In Savin Corp. v. Savin Group, 391 F.3d 439 (2d Cir. 2004), one of the few cases to find dilution since *Moseley*, the court presumed dilution from the use of identical marks.

If the marks are not identical, how similar must they be? Obviously the likelihood of confusion test cannot apply here. But there must be some degree of similarity, or consumers would not associate the defendant's mark with the plaintiff's. One court has held that the marks must be more similar for a finding of dilution than for a finding of consumer confusion. AutoZone, Inc. v. Tandy Corp., 373 F.3d 786 (6th Cir. 2004) ("Powerzone" not similar enough to dilute "AutoZone"). Does this make sense?

2. How can a famous trademark owner prove dilution under *Moseley*? Is it possible to design a survey to capture something so intangible as the "whittling away" of the significance of a mark? Does Justice Kennedy's concurrence provide a useful guide? How could a survey determine how unique the mark was before the diluting use began? The results of post-*Moseley* dilution litigation have not been encouraging to trademark owners. Defendants have won virtually all dilution cases decided since 2003.

3. The dilution statute normally provides only for injunctive relief, not damages. Is this consistent with the Court's decision to limit the statute to cases of actual dilution? If a trademark owner can't sue until dilution has already occurred, shouldn't they be able to recover damages for the harm they have suffered, not just prevent the harm from continuing?

Congress is considering reversing the result in *Moseley.* H.R. 683, pending at this writing, would restore the "likelihood of dilution" standard that predominated before the decision.

4. *Nature of Harm Cognizable for Dilution Purposes.* Dilution doctrine, as developed by the courts, comprises two principal types of harms: blurring and tarnishment. Nabisco, Inc. v. PF Brands, 191 F.3d 208 (2d Cir. 1999), illustrates the former: Nabisco's use of goldfish-shaped cheese crackers allegedly reduces the exclusive association that consumers have between the mark and Pepperidge Farm. Tarnishment arises where a junior user undermines the image that consumers hold of a famous mark by using the mark to advertise unsavory products. For example, the marketing of posters printed with the words "Enjoy Cocaine" featuring the same typeface and red and white color scheme as Coca-Cola's "Enjoy Coca-Cola" advertisements was found to tarnish Coca-Cola's famous mark. See Coca-Cola Co. v. Gemini Rising, Inc., 346 F. Supp. 1183 (E.D.N.Y. 1972); see also Dallas Cowboys Cheerleaders, Inc. v. Pussycat Cinema, Ltd., 604 F.2d 200 (2d Cir. 1979) (dilution to promote a pornographic movie by suggesting that Dallas Cowboys Cheerleaders were participants and to use actresses whose costumes resembled those of the Dallas Cowboys cheerleaders).

By contrast, using a mark to criticize the mark's owner is not tarnishment under the statute. Nor is it illegal to use the mark to display or refer to the plaintiff's own product, even if it is in a context the plaintiff might find repugnant. For example, in Wham-O, Inc. v. Paramount Pictures Corp., 286 F. Supp. 2d 1254 (C.D. Cal. 2003), the owner of the Slip 'N Slide trademark sued the makers of the film "Dickie Roberts: Former Child Star" over a scene in which the fictional Roberts injured himself by misusing a Slip 'N Slide water slide. The court held that the film's depiction of the product was "silly" but could not tarnish the plaintiff's mark. See also Caterpillar Inc. v. Walt Disney Co., 287 F. Supp. 2d 913 (N.D. Ill. 2003) (depicting plaintiff's bulldozers being used by a villain in a children's movie did not tarnish plaintiff's marks). Rather, tarnishment occurs only when the famous mark is used on the *defendant's* unsavory goods, causing the public to draw a connection between the plaintiff's goods and the defendant's.

Some also argue that dilution can occur through genericization of a famous mark, though no court has endorsed such a theory. See Elliott Staffin, The Dilution Doctrine: Towards Reconciliation with the Lanham Act, 6 Fordham Intell. Prop Media & Ent. L.J. 105, 139-42 (1995).

Does dilution by tarnishment survive the Supreme Court's opinion in *Moseley*?

5. *Fame.* The requirement of fame is designed as an important limitation on the reach of dilution protection. See Senate Judiciary Committee Report on S. 1883, S. Rep. No. 100-515, at 41-42 (Sept. 15, 1988) (noting that "Section 43(c) of the Act is to be applied selectively and is intended to provide protection only to those marks which are both truly distinctive and famous"); TCPIP Holding Co., Inc. v. Haar Communications, Inc., 244 F.3d 88 (2d Cir. 2001) ("It seems most unlikely that Congress intended to confer on marks that have enjoyed only brief fame in a small part of the country or among a small segment of the population, the power

to enjoin all other users throughout the nation in all realms of commerce."); see generally McCarthy, Trademarks and Unfair Competition § 24:112. Following the Act's passage, numerous courts uncritically applied the "fame" requirement in an effort to snare cyber-pirates. Thus courts found a host of obscure marks to be famous. See, e.g., Intermatic, Inc. v. Toeppen, 947 F. Supp. 1227 (N.D. Ill. 1996) ("Intermatic"); Teletech Customer Care Management, Inc. v. Tele-Tech Company, Inc., 977 F. Supp. 1407 (C.D. Cal. 1997) ("Teletech"); Panavision International v. Toeppen, 141 F. 3d 1316 (9th Cir. 1998) ("Panavision"); Archdiocese of St. Louis v. Internet Entertainment Group, 34 F. Supp. 2d 1145 (E.D. Mo. 1999) ("Papal Visit 1999," "Pastoral Visit," "1999 Papal Visit Commemorative Official Commemorative Items," and "Papal Visit 1999, St. Louis"). The passage of the Anticybersquatting Consumer Protection Act in 1999 substantially relieved the pressure to contort the FTDA by creating a more direct means of stopping cybersquatting, and courts have since applied the "fame" requirement more appropriately. See, e.g., Avery Dennison Corp. v. Sumpton, 189 F.3d 868 (9th Cir. 1999) (overturning a lower court decision finding "Avery" and "Dennison" to be famous trademarks).

Outside the domain name context, a few courts have significantly reduced the standard for fame by finding that a mark can be "famous" in a narrow product market. See Times Mirror Magazines v. Las Vegas Sporting News, 212 F.3d 157 (3d Cir. 2000), *cert. denied,* 531 U.S. 1071 (2001) (holding that "The Sporting News" is a famous mark); see also Syndicate Sales, Inc. v. Hampshire Paper Corp., 192 F.3d 633 (7th Cir. 1999) (noting that § 43(c)(1)(F) "indicates that fame may be constricted to a particular market"); Washington Speakers Bureau, Inc. v. Leading Authorities, Inc., 33 F. Supp. 2d 488 (E.D. Va. 1999) (same). Does this interpretation comport with the overall design and larger purposes of the FTDA? Should the Act be interpreted to provide such an extraordinary scope of exclusivity on the basis of relatively modest investments in building mark strength and relatively low consumer recognition of a mark? Cf. Advantage Rent-A-Car, Inc. v. Enterprise Rent-A-Car, Co., 238 F.3d 378 (5th Cir. 2001) (holding that Texas and Louisiana antidilution statutes require only distinctiveness and not fame). H.R. 683, pending in the 109th Congress at this writing, would strengthen the requirement of fame by abolishing the concept of niche fame.

6. *Dual Requirements of Distinctiveness and Fame?* In Nabisco, Inc. v. PF Brands, 191 F.3d 208 (2d Cir. 1999), the court suggested that the FTDA requires proof of both distinctiveness of the plaintiff's mark and fame in order to receive federal protection. It expanded upon this rationale in TCPIP Holding Co., Inc. v. Haar Communications, Inc., 244 F.3d 88 (2d Cir. 2001). TCPIP Holding Company operates a chain of stores selling children's clothing and accessories under the mark "The Children's Place." Haar Communications, Inc. registered the "thechildrensplace.-com" as a Web portal for children with the intention of providing links to a broad array of child-related products, services, and information. TCPIP sued for trademark infringement and dilution. While finding that the plaintiff was likely to prevail on its trademark infringement claim, the Second Circuit rejected the dilution claim on the ground that the FTDA requires proof of both distinctiveness and fame:

> The Dilution Act provides that "[t]he owner of a famous mark shall be entitled, subject to the principles of equity . . . , to an injunction against another person's commercial use in commerce of a mark or trade name, if such use begins after the mark has become famous *and causes dilution of the distinctive quality* of the mark." 15 U.S.C. §§ 1125(c)(1) (emphasis

added). Because the Act protects against the dilution of the mark's "distinctive quality," trademark owners seeking protection under the Act must establish that their marks possess a "distinctive quality" in order to state a claim for dilution. See Nabisco, Inc. v. PF Brands, Inc., 191 F.3d 208, 216 (2d Cir.1999) ("A mark that, notwithstanding its fame, has no distinctiveness is lacking the very attribute that the antidilution statute seeks to protect."). The Act furthermore invites courts to consider "the degree of inherent or acquired distinctiveness of the mark," 15 U.S.C. §§ 1125(c)(1)(A), among other listed factors.

The relevance of the "degree" of distinctiveness suggests that a bare minimum of distinctiveness is not adequate. If it were, the degree of distinctiveness would have no relevance.

TCPIP Holding Co., Inc. v. Haar Communications, Inc., 244 F.3d 88, 93 (2d Cir. 2001). The practical effect of this interpretation is to exclude descriptive marks from dilution protection. McDonald's, United Airlines, Ace Hardware, National Semiconductor, and American Airlines would not be eligible for federal dilution protection because they lack inherently distinctive marks (even though they would likely meet the "fame" requirement). Although he shares the Second Circuit's desire to cabin the dilution doctrine, McCarthy views the dual use of the terms "fame" and "distinctive" in this part of the statute as inartful drafting and counsels that the two terms be read as synonymous and not separate requirements. See McCarthy, Trademarks and Unfair Competition §§ 24:91.1-2; see also Times Mirror Magazines v. Las Vegas Sporting News, 212 F.3d 157 (3d Cir. 2000), *cert. denied,* 531 U.S. 1071 (2001) (rejecting the Second Circuit's interpretation). H.R. 683, pending in the 109th Congress at this writing, would explicitly permit dilution protection for famous descriptive marks.

On McCarthy's interpretation, does "distinctiveness" have any meaning in the statute? Is there good reason to deny special protection to descriptive marks such as United, Federal, National, and the like? Is there a greater need for companies in other fields to use those marks than there is to use an inherently distinctive mark?

7. *"Commercial Use in Commerce."* Another limitation on dilution protection (and another example of inartful drafting) is the requirement that the defendant make "commercial use in commerce" of the plaintiff's famous mark. 15 U.S.C. § 1125(c). Thus, mere registration of a trademark as a domain name does not constitute commercial use. As explained by one court,

> [w]hen a domain name is used only to indicate an address on the Internet, the domain name is not functioning as a trademark.... NSI's acceptance of domain name registrations is connected only with the names' technical function on the Internet to designate a set of computers.... Something more than the registration of the name is required before the use of a domain name is infringing.

Lockheed Martin Corp. v. Network Solutions, Inc., 985 F. Supp. 949, 956-57 (C.D. Cal. 1997), *aff'd,* 141 F.3d 1316 (9th Cir. 1998). The legislative history of the dilution statute made it clear that the "commercial use" language was intended to adopt the First Amendment test for commercial speech, limiting dilution law to the use of a mark to propose a commercial transaction.

In the interest of policing trademarks on the Internet, however, some courts have stretched the definition of "commercial use." See, e.g., Intermatic, Inc. v. Toeppen, 947 F. Supp. 1227 (N.D. Ill. 1996) (finding commercial use in the fact that the defendant was generally in the business of cybersquatting and ransoming domain names); Green Products Co. v. Independence Corn By-Products Co., 992 F. Supp. 1070 (N.D. Iowa 1997) (finding commercial use where a company registers its competitor's trademark as a domain name); cf. People for the Ethical Treatment of Animals v. Doughney, 263 F.3d 359 (4th Cir. 2001) (diverting traffic from a site through cybersquatting or linking to other commercial websites can constitute use of a mark "in connection with" goods or services); Planned Parenthood Federation of America, Inc. v. Bucci, 42 U.S.P.Q.2d 1430 (S.D.N.Y. 1997) (promoting books, disseminating information, and soliciting funds). Have the courts eviscerated the "commercial use" requirement?

8. *Referential Uses; Comparative Advertising.* It is a defense to a dilution claim that the famous mark was used in lawful comparative advertising. In Ty, Inc. v. Perryman, 306 F.3d 509 (7th Cir. 2002), the court held that it could also not dilute a mark to use it to refer to the trademark owner. Ty, the maker of Beanie Babies, sued an individual who accurately advertised second-hand Beanie Babies for resale. Judge Posner's opinion explained that there was no blurring or tarnishment of the Beanie Baby mark here, since the use of the mark was only to refer accurately to the plaintiff's own goods. The court acknowledged Ty's argument that Perryman was free riding on the fame of its mark, but said that "in that attenuated sense of free riding, almost everyone in business is free riding."

9. *Categories of Marks Protected Against Dilution.* As reflected in the *Nabisco* case through its protection of the product configuration of Pepperidge Farms' cracker, the FTDA can protect trade dress as well as more traditional trademarks. See also Sunbeam Prods. v. West Bend Co., 123 F.3d 246 (5th Cir. 1998). For criticism of the expansion of dilution, see Paul Heald, Exposing the Malign Application of the Federal Dilution Statute to Product Configurations, 5 J. Intell. Prop. L. 415 (1995).

10. *International Scope of Dilution Protection.* The Paris Convention, including Article 6[bis], is silent on the protection of trademarks against dilution. GATT-TRIPs signatory countries, however, must now provide some form of protection against dilution of a mark, at least if it is famous.

> Article 6[bis] of the Paris Convention (1967) shall apply, mutatis mutandis, to goods or services which are not similar to those in respect of which a trademark is registered, provided that use of that trademark in relation to those goods or services would indicate a connection between those goods and services and the owner of the registered trademark and provided that the interests of the owner of the registered trademark are likely to be damaged by such use.

Final Act Embodying the Results of the Uruguay Round of Multilateral Trade Negotiations, done at Marrakech, Morocco, April 15, 1994, Annex 1C: Agreement on Trade-Related Aspects of Intellectual Property Rights ("TRIPs"), at Article 16(3). For congressional approval, see Uruguay Rounds Agreements Act, Pub. L. No. 103-465, Dec. 8, 1994. Why can a mark not currently used in a country be protected against dilution in another country? Does it matter whether the mark is registered? Should foreign marks be protected against dilution in anticipation of their possible entry into the market? This provision has been used to target "trademark pirates": individuals who identify famous trademarks not yet used in their country and register those marks in anticipation of the famous trademark owner's expansion abroad.

Note on Dilution and "Search Theory"

How would you state the harm that comes with dilution in terms of the consumer search rationale described earlier in this chapter? As a consumer, is locating Kodak film in the store or picking out the Kodak film ad in a magazine more difficult if there are many unassociated products or ads that use the Kodak name? As an experiment, try walking down the "generic" aisle in a supermarket, where all the boxes are plain black and white, trying to pick out the macaroni among the spaghetti, rigatoni, and so on. Would distinctive labels — all brightly colored, with striking designs — make a difference? From a certain distance, to take a different example, it is difficult to pick out a single marathon runner from the crowd of other runners, even though — or because — each runner is dressed in day-glo warmups, running shoes, and so on. The point is simply this: if many products share a similar label, the fact that the label is striking when viewed in isolation is irrelevant. It is as difficult to discern a single blazing orange poppy in a field full of them as it is to identify a lone grey pebble out of a rock pile.

If, as suggested earlier, attention spans are truly taxed in the current era, with its barrage of information and images, dilution may be an even more important adjunct to traditional "consumer confusion" rationales for trademarks. Indeed, the harm that dilution seeks to address might best be described as a loss of consumer *attention* due to the proliferation of similar or identical symbols of trade. Trademark owners thus seek to protect the distinctiveness of their marks — first, to ensure a moment of unfettered attention, and then to indicate a unique connection between the mark and the seller's product. For an argument that prohibiting trademark dilution is consistent with the law's general focus on reducing consumer search costs, see Stacey L. Dogan & Mark A. Lemley, The Merchandising Right: Fragile Theory or Fait Accompli?, 54 Emory L.J. 461 (2005).

PROBLEM

Problem 5-9. Mead Data Central is the owner of the registered trademark LEXIS for a legal and business-related electronic database. The service is extremely well-known among lawyers, but not among the general public.

A decade after the first use of the Lexis mark, Toyota names its new luxury car the "Lexus." Mead sues for dilution of its Lexis mark. What result? On what facts does your decision depend?

---

4. Extension by Contract: Franchising and Merchandising

Trademarks and Organizational Forms: The Growth of Franchising. From its modern founding, trademark law has been at the service of emerging patterns in the organization of industry. In the beginning, that meant protecting emerging channels of trade in relatively local settings. Next came the great nationalization of the economy with the growth of large retail empires in the late nineteenth and early twentieth centuries. Today trademarks form an integral part of the variegated economic land-scape of the industrialized world.

Business historian Mira Wilkins has written:

> The legally-backed trade marks . . . became essential intangible assets, providing the basis for the rise of the modern enterprise . . . . The trade mark's fundamental contribution to the modern corporation was that it generated efficiency gains by creating for the firm the opportunity for large sales over long periods . . . . Without the trade mark, the introduction and acceptance by buyers of modern products, produced with economies of scale or scope, and marketed over long distances, would have been impossible . . . . The trade mark by reducing the costs of information led to efficiencies in production and distribution.

Mira Wilkins, The Neglected Intangible Asset: The Influence of the Trademark on the Rise of the Modern Corporation, 34 Bus. & Hist. 66, 87-88 (1992). Interestingly, Wilkins also points out the role that trademarks play in facilitating organizational diversity, or making possible various alternative forms of production. The clearest and most important recent example is franchising.

Trademarks are the "cornerstone of a franchise system." Susser v. Carvel Corp., 206 F. Supp. 636, 640 (S.D.N.Y. 1962), *aff'd*, 332 F.2d 505, *cert. granted*, 379 U.S. 885, *cert. dismissed*, 381 U.S. 125 (1965). The trademark of the franchisor is the identifiable symbol of continuity; it indicates the presence of the national brand at each location. Thus, whatever the precise nature of the franchise, the franchisor and franchisee are very likely to be parties to a trademark license agreement.

Here again we see the stretching of traditional theory. The individual source from which the national brand emanates is the franchisor. This is often a remote corporate entity, whereas one could argue that the franchisee is the real "source" (at the local level) of the goods. To maintain uniformity (and stay on the good side of the abandonment issue; see below), the franchisor almost invariably imposes certain contractual requirements on the franchisee. Yet it is still the franchisee, in the last instance, who actually runs the establishment where the goods are sold. See James M. Treece, Trademark Licensing and Vertical Restraints in Franchising Arrangements, 116 U. Pa. L. Rev. 435 (1968); Lynn M. Lopucki, Toward a Trademark-Based Liability System, 49 UCLA L. Rev. 1099 (2002) (suggesting that tort liability should follow the trademark rather than the franchise).

In economic terms, franchising is an interesting mix of contractual and integrated governance characteristics — a kind of hybrid organization in which the franchisee is neither an employee of the franchisor nor an arm's-length buyer of the franchisor's goods. See James A. Brickley & Frederick H. Dark, The Choice of Organizational Form: The Case of Franchising, 18 J. Fin. Econ. 401, 403-407 (1987); Gillian K. Hadfield, Problematic Relations: Franchising and the Law of Incomplete Contracts, 42 Stan. L. Rev. 927 (1990). The franchise trademark plays an interesting role in this relationship. It is one of the franchisor's great assets, one of the things it can sell to franchisees. Yet once a franchisee begins to use the trademark, he or she is in a position to harm the franchise's reputation by selling inferior quality goods. (Note that doing so hurts other franchisees as well.) Indeed, if many of the franchise's customers are travelers who come from afar and do not pass through often, a franchisee might be tempted to "free ride" off the quality investments of the franchisor and other franchisees by selling inferior goods. Many of the provisions in franchise agreements are directed at preventing such an outcome, e.g., agreements to purchase ingredients and other inputs from the franchisor, stipulations to frequent inspections, and profit-sharing arrangements. See Paul H. Rubin, The Theory of the Firm and the Structure of the Franchise Contract, 21 J. Law & Econ. 223 (1978). Cf. Patrick J. Kaufman & Francine Lafontaine, Costs of Control: The Source of Economic Rents for McDonald's Franchisees, 37 J.L. & Econ. 417 (1994) (finding substantial "rents" or profits left to the franchisee after the franchisor's "cut," indicating that incentive system in franchise contracts was working). In this complicated relationship, trademarks are a key strategic asset. Note also that in many states various aspects of franchise contracts — especially termination provisions — are closely regulated. See, e.g., Thomas M. Pitegoff, Franchise Relationship Laws: A Minefield for Franchisors, 45 Bus. Law 289 (1989).

What significance does the growth of franchising have for trademark doctrine? Should we be concerned about the disaggregation of trademarks from the goods they represent? See the discussion in the previous section on trademarks sold as valuable items in and of themselves. Does the selling of trademarks without goods attached to them make trademark owners less able to use the mark as a guarantee of quality? Alternatively, does it suggest that the trademark owner will invest even more in quality assurance, since the value of the mark is all it has to offer?

"Turning Logos into Profit Centers." Evidence is everywhere that a boom is under way in the licensing of trademarks. From sports team logos to university names to designer symbols, badges of affiliation and prestige are ever more common on products of all kinds. And, importantly, these badges come at a premium. As Robert Denicola has written,

> At any sporting goods store one can find plain, unadorned shirts, shorts, and jackets in assorted styles and colors. They are usually near the rear. Closer to the front are items apparently similar in all respects except one — they are prominently decorated with a variety of words and symbols. Some bear the names of athletic equipment manufacturers. Others display the names or insignia of professional sports teams, the name and seal of the state university, or the nickname and mascot of the local high school. They frequently cost significantly more than the items in the rear, yet they sell.

Robert C. Denicola, Institutional Publicity Rights: An Analysis of the Merchandising of Famous Trade Symbols, 62 N.C. L. Rev. 603 (1983).

Empirical evidence confirms the observation. A licensing industry newsletter placed the volume of licensing activity at $13.15 billion in 1994. See Licensing Letter (EPM Communications, Inc.), April 1, 1995, at p. 1. As for future growth prospects, the same newsletter quotes an industry source as a representative voice: "[O]ne of the trends in licensing last year [1994] was the tendency for corporations to examine the potential 'of anything marketable to turn logos into profit centers.'"

Turning logos into profit centers may make sense as a business strategy, but it poses problems for the legal system. The cases in this chapter highlight the fundamental problem: traditionally, trademarks were thought of as symbols representing products, rather than as products in and of themselves. Thus traditional trademark law protects a trademark only because, and only insofar as, it is emblematic of the goodwill behind a product. The mark itself is not the point; it is simply a vehicle to convey useful information regarding a product's quality, prestige, and so on. The trademark guides the consumer to the transaction; sale of the underlying product is the "profit center."

All this changes when the trademark becomes the *subject* of the transaction rather than an adjunct to it. In such a transaction, the mark does not represent the product: it *is* the product.

Consider for example the licensing of a sports team logo for use on a T-shirt or hat. The logo does not summarize information about the quality of the hat; it demonstrates loyalty to a team. Indeed, many if not most sports logos are licensed to a broad array of products, many of differing degrees of quality. The "high end" Red Sox cap ("just like the pros wear!") is a far different product—qua hat—than the cheap synthetic cap costing a few dollars and sold in discount stores. The same is true for sweatshirts and T-shirts emblazoned with the logos of college and professional teams. Similar examples can be drawn from other avenues of commerce. Thus, outdoor equipment companies with a certain consumer cachet have been known to lend their logos for application to sport-utility vehicles. Yet no one, or very few, can suppose that companies specializing in backpacks, long underwear, and hiking boots have suddenly taken up truck production.

In these cases, the consumer is buying an image. The trademark owner possesses rights in a symbol associated with certain qualities. Lending that symbol to diverse products connects those products to the feelings the symbol evokes. (The same thing happens outside commerce as well; consider the difference when the "stars and stripes" image is added to a plain object such as a flag or a tombstone.)

Trademark law has begun to recognize the importance of the logo as a product. For instance, a 1975 case held that unauthorized sales of team emblems violated the Lanham Act:

> The certain knowledge of the buyer that the source and origin of the trademark symbols were the plaintiffs satisfies the requirements of the Act. The argument that confusion must be as to source of the manufacture of the emblem itself is unpersuasive, where the trademark, originated by the team, is the triggering mechanism for the sale of the emblem.

Boston Professional Hockey Assn. v. Dallas Cap & Emblem Mfg., Inc., 510 F.2d 1004, 1012 (5th Cir. 1975) (permanent injunction), *cert. denied*, 423 U.S. 868, *reh'g denied*, 423 U.S. 991 (1975).

This holding seems to follow a felt sense that it would be improper to permit an anonymous third party to profit from the plaintiff's logo. However equitable it may appear, though, such a holding is arguably at odds with traditional trademark theory. How does the logo lower consumer search costs? If the logo does not summarize product attributes such as quality, why protect it? Tribunals such as the *Boston Professional Hockey* court have ignored these anomalies and protected logos against use by another.

Not all courts do so, however. Stacey Dogan and Mark Lemley review the cases and conclude that the majority refuse to protect trademarks on merchandise absent a showing of consumer confusion. Further, they point out that the Supreme Court has never recognized a merchandising right, and argue that the current Court is unlikely to do so, given its many recent decisions limiting trademark protection. Stacey L. Dogan & Mark A. Lemley, The Merchandising Right: Fragile Theory or Fait Accompli?, 54 Emory L.J. 461 (2005).

Assuming that Dogan and Lemley are right that a merchandising right doesn't fit well within existing trademark doctrine, what should happen if people have come to believe that only the Dallas Cowboys are permitted to sell Dallas Cowboy merchandise? Should the law expand over time to take account of changing consumer norms, or should it set the rules according to trademark principles and force consumer norms to adapt to those rules?

PROBLEM

Problem 5-10. A popular "cult" movie includes characters who have tattooed famous brand logos on various parts of their bodies. The practice quickly catches on, first among "avant-garde" artists, and then among college and high school students.

Post-Modern Concepts, Inc. (PMC), seeing an opportunity, opens a storefront tattoo parlor in a popular retail mall in the Midwest. Here PMC applies popular logos as tattoos. At first, customers are primarily interested in the ironic use of logos, as in the film; the logos are primarily used as a spoof or commentary on consumerism. One popular tattoo, for instance, includes the famous "Calvin Klein" wordmark and logo, drawn inside a red circle with a slash. Soon, however — perhaps predictably — the ironic intent of the original practice is lost as it becomes very chic to have a logo tattoo. Thus many people begin to request tattoos without the red slash and circle. The most popular trademark logo tattoos are "Rolls Royce," "Nike," the original "Calvin Klein," and "Harley-Davidson."

Success is instantaneous. In response to torrid demand, PMC sets up franchises all over the country.

Owners of the trademarks begin to notice PMC. They are concerned primarily about lost revenues, because PMC has never taken out any trademark licenses. But some trademark owners are also concerned that pictures of people with logo tattoos are beginning to appear in the press. Even some famous celebrities, such as popular musicians and sports stars, have tattoos in prominent places. When their picture appears in the newspaper, the logo — or part of it — does too. Other trademark owners are concerned that logo tattoos are being applied on parts of the body traditionally covered by clothing. These owners fear that when pictures of people with logo tattoos in these places appear — sometimes in press outlets not fit for family viewing — the trademarks are denigrated.

What hurdles will the trademark owners have to clear to enforce their rights? (Are these marks attached to a product? Also, PMC will undoubtedly argue that sports team fanatics have long painted their faces in team colors, or drawn the team logo on their faces; and that even before PMC, some permanent tattoos consisting of corporate logos were popular, such as "Chevy," "Red Sox Forever," or, perhaps most common, "Harley Davidson.") Are there any risks to letting the practice continue without any enforcement efforts? Does PMC, or its customers, have a defense not based in trademark law?

5. Domain Names and Cybersquatting

Since the early days of the Internet, people have communicated with each other using domain names, mnemonic devices that map to a particular Internet Protocol address. Individuals, organizations, and companies have registered domain names that correspond to their names, to their trademarks, or to their products. Thus, Microsoft can be found at www.microsoft.com. Internet domain names are registered on a first-come, first-served basis by a variety of registrars for a low price (less than $30 per domain name per year).

In the mid-1990s, a number of individuals registered domain names that corresponded to the name or trademark of someone else, often a large corporation. Many of these individuals registered the domain name not in order to use it themselves, but in order to sell the valuable domain name to the trademark owner for a price well in excess of $30. This practice is known as "cybersquatting." Others registered generic terms like business.com and resold them for a substantial profit. Still others may register a company's name for legitimate reasons: to criticize the company, or to advertise the availability of that company's products at a website.

Trademark owners have been trying for years to stop cybersquatting. Existing trademark law has not adapted terribly well to cybersquatting, and in recent years trademark owners have turned to new statutes and rules to help them in their fight. Courts have struggled to stop cybersquatting without punishing legitimate uses of domain names. In this section, we consider Internet domain name cases decided under two new legal regimes.

a. Anticybersquatting Consumer Protection Act

Trademark infringement and dilution are not well suited to preventing cybersquatting. While some uses of a domain name are infringing, most are not. Dilution reaches classic cases of cybersquatting—the registration of a domain name corresponding to someone else's trademark in order to extort money from the trademark owner—but only in the small number of cases in which the trademark in question is famous. This led to two problems. First, some instances of cybersquatting can't be punished under existing trademark laws. Second, courts sometimes stretched the requirements of the law in order to stop cybersquatting. For instance, many courts ignored or reduced the requirement of fame in order to apply dilution law against clear cases of cybersquatting.

In 1999, Congress sought to address both of these problems by passing a law directed specifically at cybersquatting. The Anticybersquatting Consumer Protection Act makes it illegal to register or use a domain name that corresponds to a trademark where the domain name registrant has no legitimate interest in using the name and acts in bad faith to deprive the trademark owner of the use of the name. 15 U.S.C. § 1125(d). Since the end of 1999, dozens of cases have been decided under the ACPA. The critical parts of the statute are the definitions of bad faith and legitimate interest, which are discussed in the cases that follow.

Shields v. Zuccarini
United States Court of Appeals for the Third Circuit
254 F.3d 476 (3d Cir. 2001)

ALDISERT, Circuit Judge.

John Zuccarini appeals from the district court's grant of summary judgment and award of statutory damages and attorneys' fees in favor of Joseph Shields under the new Anticybersquatting Consumer Protection Act ("ACPA" or "Act"). In this case of first impression in this court interpreting the ACPA, we must decide whether the district court erred in determining that registering domain names that are intentional misspellings of distinctive or famous names constitutes unlawful conduct under the Act. We must decide also whether the district court abused its discretion by assessing statutory damages of $10,000 per domain name. Finally, we must decide whether the court erred in awarding attorneys' fees in favor of Shields based on its determination that this case qualified as an "exceptional" case under the ACPA. We affirm the judgment of the district court . . . .

I

Shields, a graphic artist from Alto, Michigan, creates, exhibits and markets cartoons under the names "Joe Cartoon" and "The Joe Cartoon Co." His creations include the popular "Frog Blender," "Micro-Gerbil" and "Live and Let Dive" animations. Shields licenses his cartoons to others for display on T-shirts, coffee mugs and other items, many of which are sold at gift stores across the country. He has marketed his cartoons under the "Joe Cartoon" label for the past fifteen years.

On June 12, 1997, Shields registered the domain name joecartoon.com, and he has operated it as a web site ever since. Visitors to the site can download his animations and purchase Joe Cartoon merchandise. Since April 1998, when it won "shock site of the day" from Macromedia, Joe Cartoon's web traffic has increased exponentially, now averaging over 700,000 visits per month.

In November 1999, Zuccarini, an Andalusia, Pennsylvania "wholesaler" of Internet domain names, registered five world wide web variations on Shields's site: joescartoon.com, joecarton.com, joescartons.com, joescartoons.com and cartoonjoe.com. Zuccarini's sites featured advertisements for other sites and for credit card companies. Visitors were trapped or "mousetrapped" in the sites, which, in the jargon of the computer world, means that they were unable to exit without clicking on a succession of advertisements. Zuccarini received between ten and twenty-five cents from the advertisers for every click.

In December 1999, Shields sent "cease and desist" letters to Zuccarini regarding the infringing domain names. Zuccarini did not respond to the letters. Immediately after Shields filed this suit, Zuccarini changed the five sites to "political protest" pages and posted the following message on them:

> This is a page of POLITICAL PROTEST
> —Against the web site joecartoon.com—
> joecartoon.com is a web site that depicts the mutilation and killing of animals in a
> shockwave based cartoon format—many children are inticed [sic] to the web site, not

knowing what is really there and then encouraged to join in the mutilation and killing through use of the shockwave cartoon presented to them.

—Against the domain name policys [sic] of ICANN—
—Against the Cyberpiracy Consumer Protection Act—

As the owner of this domain name, I am being sued by joecartoon.com for $100,000 so he can use this domain to direct more kids to a web site that not only desensitizes children to killing animals, but makes it seem like great fun and games.

I will under no circumstances hand this domain name over to him so he can do that.
I hope that ICANN and Network Solutions will not assist him to attaining this goal.
—Thank You—

Shields v. Zuccarini, 89 F.Supp.2d 634, 635-636 (E.D.Pa.2000).

Shields's Complaint invoked the ACPA...

On March 22, 2000, the court entered a preliminary injunction in favor of Shields, which required Zuccarini to transfer the infringing domain names to Shields and to refrain from "using or abetting the use of" the infringing domain names or any other domain names substantially similar to Shields's marks....

On July 18, 2000, the district court entered an Order and Judgment awarding statutory damages in the amount of $10,000 for each infringing domain name and attorneys' fees and costs in the amount of $39,109.46....

III

On November 29, 1999, the ACPA became law, making it illegal for a person to register or to use with the "bad faith" intent to profit from an Internet domain name that is "identical or confusingly similar" to the distinctive or famous trademark or Internet domain name of another person or company. See 15 U.S.C. § 1125(d) (Supp.2000). The Act was intended to prevent "cybersquatting," an expression that has come to mean the bad faith, abusive registration and use of the distinctive trademarks of others as Internet domain names, with the intent to profit from the goodwill associated with those trademarks. See S.Rep. No. 106-140 (1999), 1999 WL 594571, at * 11-18. Under the ACPA, successful plaintiffs may recover statutory damages in an amount to be assessed by the district court in its discretion, from $1,000 to $100,000 per domain name. See 15 U.S.C. § 1117(d) (Supp. 2000). In addition, successful plaintiffs may recover attorneys' fees in "exceptional" cases. See id. at § 1117(a)....

A

To succeed on his ACPA claim, Shields was required to prove that (1) "Joe Cartoon" is a distinctive or famous mark entitled to protection; (2) Zuccarini's domain names are "identical or confusingly similar to" Shields's mark; and (3) Zuccarini registered the domain names with the bad faith intent to profit from them. See 15 U.S.C. § 1125(d)(1)(A); cf. Sporty's Farm L.L.C. v. Sportsman's Market Inc., 202 F.3d 489, 497-499 (2d Cir. 2000).[3]

3. Section 1125(d)(1)(A) of the Act provides in relevant part:

A person shall be liable in a civil action by the owner of a mark...if, without regard to the goods or services of the parties, that person

1

Under § 1125(d)(1)(A)(ii)(I) and (II), the district court first had to determine if "Joe Cartoon" is a "distinctive" or "famous" mark and, therefore, is entitled to protection under the Act. The following factors may be considered when making this inquiry:

> (A) the degree of inherent or acquired distinctiveness of the mark; (B) the duration and extent of use of the mark in connection with the goods or services with which the mark is used; (C) the duration and extent of advertising and publicity of the mark; (D) the geographical extent of the trading area in which the mark is used; (E) the channels of trade for the goods or services with which the mark is used; (F) the degree of recognition of the mark in the trading areas and channels of trade used by the marks' owner and the person against whom the injunction is sought; (G) the nature and extent of use of the same or similar marks by third parties.

15. U.S.C. § 1125(c)(1).

Shields runs the only "Joe Cartoon" operation in the nation and has done so for the past fifteen years. This suggests both the inherent and acquired distinctiveness of the "Joe Cartoon" name. In addition to using the "Joe Cartoon" name for fifteen years, Shields has used the domain name joecartoon.com as a web site since June 1997 to display his animations and sell products featuring his drawings. The longevity of his use suggests that "Joe Cartoon" has acquired some fame in the marketplace. The New York Times ran a page one story that quoted Shields and cited Joe Cartoon. See Andrew Pollack, Show Business Embraces the Web, But Cautiously, N.Y. Times, Nov. 9, 1999, at A1.

Joe Cartoon T-shirts have been sold across the country since at least the early 1990s, and its products appear on the web site of at least one nationally known retail chain, Spencer Gifts. Shields has advertised in an online humor magazine with a circulation of about 1.4 million. The Joe Cartoon web site receives in excess of 700,000 visits per month, bringing it wide publicity. According to Shields, word-of-mouth also generates considerable interest in the Joe Cartoon site. Shields trades nationwide in both real and virtual markets. The web site gives Joe Cartoon a global reach. Shields's cartoons and merchandise are marketed on the Internet, in gift shops and at tourist venues. The Joe Cartoon mark has won a huge following because of the work of Shields. In light of the above, we conclude that "Joe Cartoon" is distinctive, and, with 700,000 hits a month, the web site "joecartoon.com" qualifies as being famous. Therefore, the trademark and domain name are protected under the ACPA.

> (i) has a bad faith intent to profit from that mark, including a personal name which is protected as a mark under this section; and
> (ii) registers, traffics in, or uses a domain name that—
> (I) in the case of a mark that is distinctive at the time of registration of the domain name, is identical or confusingly similar to that mark;
> (II) in the case of a famous mark that is famous at the time of registration of the domain name, is identical or confusingly similar to or dilutive of that mark.

2

Under the Act, the next inquiry is whether Zuccarini's domain names are "identical or confusingly similar" to Shields's mark. The domain names — joescartoon.com, joecarton.com, joescartons.com, joescartoons.com and cartoonjoe.com — closely resemble "joecartoon.com," with a few additional or deleted letters, or, in the last domain name, by rearranging the order of the words. To divert Internet traffic to his sites, Zuccarini admits that he registers domain names, including the five at issue here, because they are likely misspellings of famous marks or personal names.[4] The strong similarity between these domain names and joecartoon.com persuades us that they are "confusingly similar." Shields also produced evidence of Internet users who were actually confused by Zuccarini's sites. See, e.g., Pltf's Exh. 22, at [4] (copy of an email stating, "I tried to look up you[r] website yesterday afternoon and a protest page came up. Will I have trouble entering the site at times because of this?").

On appeal, Zuccarini argues that registering domain names that are intentional misspellings of distinctive or famous names (or "typosquatting," his term for this kind of conduct) is not actionable under the ACPA. Zuccarini contends that the Act is intended to prevent "cybersquatting," which he defines as registering someone's famous name and trying to sell the domain name to them or registering it to prevent the famous person from using it themselves. This argument ignores the plain language of the statute and its stated purpose as discussed in the legislative history.

The statute covers the registration of domain names that are "identical" to distinctive or famous marks, but it also covers domain names that are "confusingly similar" to distinctive or famous marks. See 15 U.S.C. § 1125(d)(1)(A)(ii)(I), (II). A reasonable interpretation of conduct covered by the phrase "confusingly similar" is the intentional registration of domain names that are misspellings of distinctive or famous names, causing an Internet user who makes a slight spelling or typing error to reach an unintended site. The ACPA's legislative history contemplates such situations:

> [C]ybersquatters often register well-known marks to prey on consumer confusion by misusing the domain name to divert customers from the mark owner's site to the cybersquatter's own site, many of which are pornography sites that derive advertising revenue based on the number of visits, or "hits," the site receives. For example, the Committee was informed of a parent whose child *mistakenly typed in the domain name for 'dosney.com,' expecting to access the family-oriented content of the Walt Disney home page, only to end up staring at a screen of hardcore pornography because a cybersquatter had registered that domain name in anticipation that consumers would make that exact mistake.*

S. Rep. No. 106-140 (1999), 1999 WL 594571, at *15 (emphasis added).

Although Zuccarini's sites did not involve pornography, his intent was the same as that mentioned in the legislative history above — to register a domain name in anticipation that consumers would make a mistake, thereby increasing the number of hits his site would receive, and, consequently, the number of advertising dollars he would gain. We conclude that Zuccarini's conduct here is a classic example of a

---

4. Zuccarini testified that he was amazed to learn that "people mistype [domain names] as often as they do," and thus variants on likely search names would result in many unintended visitors to his sites. *Shields,* 89 F.Supp.2d at 640.

specific practice the ACPA was designed to prohibit. The district court properly found that the domain names he registered were "confusingly similar."

3

The final inquiry under the ACPA is whether Zuccarini acted with a bad faith intent to profit from Shields's distinctive and famous mark or whether his conduct falls under the safe harbor provision of the Act. Section 1125(d)(1)(B)(i) provides a non-exhaustive list of nine factors for us to consider when making this determination:

> (I) the trademark or other intellectual property rights of the person, if any, in the domain name; (II) the extent to which the domain name consists of the legal name of the person or a name that is otherwise commonly used to identify that person; (III) the person's prior use, if any, of the domain name in connection with the bona fide offering of any goods or services; (IV) the person's bona fide noncommercial or fair use of the mark in a site accessible under the domain name; (V) the person's intent to divert consumers from the mark owner's online location to a site accessible under the domain name that could harm the goodwill represented by the mark, either for commercial gain or with the intent to tarnish or disparage the mark, by creating a likelihood of confusion as to the source, sponsorship, affiliation, or endorsement of the site; (VI) the person's offer to transfer, sell, or otherwise assign the domain name to the mark owner or any third party for financial gain without having used, or having an intent to use, the domain name in the bona fide offering of any goods or services, or the person's prior conduct indicating a pattern of such conduct; (VII) the person's provision of material and misleading false contact information when applying for the registration of the domain name, the person's intentional failure to maintain accurate contact information, or the person's prior conduct indicating a pattern of such conduct; (VIII) the person's registration or acquisition of multiple domain names which the person knows are identical or confusingly similar to marks of others that are distinctive at the time of registration of such domain names, or dilutive of famous marks of others that are famous at the time of registration of such domain names, without regard to the goods or services of the parties; and (IX) the extent to which the mark incorporated in the person's domain name registration is or is not distinctive and famous within the meaning of subsection (c)(1) of this section.

Zuccarini's conduct satisfies a number of these factors. Zuccarini has never used the infringing domain names as trademarks or service marks; thus, he has no intellectual property rights in these domain names. See id. at (B)(i)(I). The domain names do not contain any variation of the legal name of Zuccarini, nor any other name commonly used to identify him. See id. at (B)(i)(II). Zuccarini has never used the infringing domain names in connection with the bona fide offering of goods or services. See id. at (B)(i)(III). He does not use these domain names for a non-commercial or "fair use" purpose. See id. at (B)(i)(IV). He deliberately maintains these domain names to divert consumers from Shields's web site. In so doing, he harms the goodwill associated with the mark. He does this either for commercial gain, or with the intent to tarnish or disparage Shields's mark by creating a likelihood of confusion. See id. at (B)(i)(V). He has knowingly registered thousands of Internet domain names that are identical to, or confusingly similar to, the distinctive marks of others, without the permission of the mark holders to do so. See id. at (B)(i)(VIII). We have already established that Shields's mark is distinctive and famous. See id. at (B)(i)(IX).

Zuccarini argues that his web sites were protected by the First Amendment because he was using them as self-described "protest pages." Therefore, he contends, his use falls under the safe harbor provision of § 1125(d)(1)(B)(ii), which states that "[b]ad faith intent . . . shall not be found in any case in which the court determines that the person believed and had reasonable grounds to believe that the use of the domain name was a fair use or otherwise lawful." The district court rejected this argument based on its conclusion that Zuccarini's claim of good faith and fair use was a "spurious explanation cooked up purely for this suit." *Shields*, 89 F.Supp.2d at 640. We agree.

Zuccarini used his "Joe Cartoon" web sites for purely commercial purposes before Shields filed this action. Zuccarini was on notice that his use of the domain names was considered unlawful when he received the cease and desist letters from Shields in December 1999. Zuccarini continued to use the infringing domain names for commercial purposes until Shields filed this lawsuit. Zuccarini testified that he put up the protest pages at 3:00 a.m. on February 1, 2000, just hours after being served with Shields's Complaint. Thus, by his own admission, Zuccarini submits that his alleged "fair use" of the offending domain names for "political protest" began only after Shields brought this action alleging a violation of the ACPA. We are aware of no authority providing that a defendant's "fair use" of a distinctive or famous mark only after the filing of a complaint alleging infringement can absolve that defendant of liability for his earlier unlawful activities. Indeed, were there such authority we think it would be contrary to the orderly enforcement of the trademark and copyright laws.

We conclude that the district court properly rejected Zuccarini's argument that his web sites were protected under the safe harbor provision. There was sufficient evidence for the district court to find that Zuccarini acted with a bad faith intent to profit when he registered and used the five domain names at issue here . . . .

IV

The Act provides for statutory damages for a violation of § 1125(d)(1) "in the amount of not less than $1,000 and not more than $100,000 per domain name, as the court considers just." 15 U.S.C. § 1117(d). Zuccarini argues that S 1117(d) does not apply to him because he registered the offending domain names before the ACPA became law. The district court held that Zuccarini's continued use of the domain names after November 29, 1999 subjects him to the statute's proscriptions and remedies. We agree with the teachings of Virtual Works, Inc. v. Volkswagen of America, Inc., 238 F.3d 264, 268 (4th Cir. 2001) ("A person who unlawfully registers, traffics in, or uses a domain name after the ACPA's date of enactment, November 29, 1999, can be liable for monetary damages . . . .") (emphasis added); and E. & J. Gallo Winery v. Spider Webs Ltd., 129 F.Supp.2d 1033, 1047-1048 (S.D.Tex. 2001) (holding that defendant who registered domain name in bad faith could be held liable for statutory damages even though registration was prior to enactment of the ACPA when defendant continued to use web site after the enactment of the Act).

In the alternative, Zuccarini argues that he only used the web site for sixty days after the passage of the ACPA and prior to this lawsuit being filed. He implies that, because he only used the web site for a short period of time, the district court's assessment of statutory damages was punitive in nature. Under the statute, the court has the discretion to award statutory damages that it "considers just" within a range from $1,000 to $100,000 per infringing domain name. See 15 U.S.C.

§ 1117(d). There is nothing in the statute that requires that the court consider the duration of the infringement when calculating statutory damages. We conclude that the district court properly exercised its discretion in awarding $10,000 for each infringing domain name....

The judgment and the award of statutory damages and attorneys' fees will be affirmed.

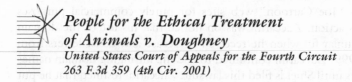

People for the Ethical Treatment of Animals v. Doughney
United States Court of Appeals for the Fourth Circuit
263 F.3d 359 (4th Cir. 2001)

GREGORY, Circuit Judge:

[The plaintiff, People for the Ethical Treatment of Animals (PETA), is an advocacy group opposed to eating meat, wearing fur, and conducting research on animals. The defendant registered the Internet domain name peta.org, where he set up a page entitled "People Eating Tasty Animals" that was a parody of PETA and its goals. In a portion of the opinion omitted here, the Fourth Circuit found that using the domain name peta.org impermissibly caused initial-interest confusion, and therefore found Doughney liable for trademark infringement.]

B. Anticybersquatting Consumer Protection Act

The district court found Doughney liable under the Anticybersquatting Consumer Protection Act ("ACPA"), 15 U.S.C. § 1125(d)(1)(A). To establish an ACPA violation, PETA was required to (1) prove that Doughney had a bad faith intent to profit from using the *peta.org* domain name, and (2) that the *peta.org* domain name is identical or confusingly similar to, or dilutive of, the distinctive and famous PETA Mark. 15 U.S.C. § 1125(d)(1)(A).

Doughney makes several arguments relating to the district court's ACPA holding: (1) that PETA did not plead an ACPA claim, but raised it for the first time in its motion for summary judgment; (2) that the ACPA, which became effective in 1999, cannot be applied retroactively to events that occurred in 1995 and 1996; (3) that Doughney did not seek to financially profit from his use of PETA's Mark; and (4) that Doughney acted in good faith.

None of Doughney's arguments are availing...we reject Doughney's first contention.

Doughney's second argument—that the ACPA may not be applied retroactively—also is unavailing. The ACPA expressly states that it "shall apply to all domain names registered before, on, or after the date of the enactment of this Act[.]" Pub.L. No. 106-113, § 3010, 113 Stat. 1536. See also Sporty's Farm L.L.C. v. Sportsman's Market, Inc., 202 F.3d 489, 496 (2d Cir. 2000) (same). Moreover, while the ACPA precludes the imposition of *damages* in cases in which domain names were registered, trafficked, or used before its enactment, Pub.L. No. 106-113, § 3010, 113 Stat. 1536 ("damages under subsection (a) or (d) of section 35 of the Trademark Act of 1946 (15 U.S.C. 1117),...shall not be available with respect to the registration, trafficking, or use of a domain name that occurs before the date of the enactment of this Act"), it does not preclude the imposition of *equitable remedies*. See also Virtual Works, Inc. v. Volkswagen of America, Inc., 238 F.3d 264, 268 (4th Cir. 2001). Here, the district court did not award PETA damages (nor did PETA request damages), but ordered Doughney to relinquish the domain name, transfer its registration to PETA, and limit his use of domain names to those that do not use PETA's Mark. *Doughney*, 113 F.Supp.2d at 922. Thus, the district court properly applied the ACPA to this case.

Doughney's third argument — that he did not seek to financially profit from registering a domain name using PETA's Mark — also offers him no relief. It is undisputed that Doughney made statements to the press and on his website recommending that PETA attempt to "settle" with him and "make him an offer." The undisputed evidence belies Doughney's argument.

Doughney's fourth argument — that he did not act in bad faith — also is unavailing. [The court recited the statutory factors listed in *Shields*]. In addition to listing these nine factors, the ACPA contains a safe harbor provision stating that bad faith intent "shall not be found in any case in which the court determines that the person believed and had reasonable grounds to believe that the use of the domain name was fair use or otherwise lawful." 15 U.S.C. § 1225(d)(1)(B)(ii).

The district court reviewed the factors listed in the statute and properly concluded that Doughney (I) had no intellectual property right in *peta.org;* (II) *peta.org* is not Doughney's name or a name otherwise used to identify Doughney; (III) Doughney had no prior use of *peta.org* in connection with the bona fide offering of any goods or services; (IV) Doughney used the PETA Mark in a commercial manner; (V) Doughney "clearly intended to confuse, mislead and divert internet users into accessing his web site which contained information antithetical and therefore harmful to the goodwill represented by the PETA Mark"; (VI) Doughney made statements on his web site and in the press recommending that PETA attempt to "settle" with him and "make him an offer"; (VII) Doughney made false statements when registering the domain name; and (VIII) Doughney registered other domain names that are identical or similar to the marks or names of other famous people and organizations. *People for the Ethical Treatment of Animals*, 113 F.Supp.2d at 920.

Doughney claims that the district court's later ruling denying PETA's motion for attorney fees triggers application of the ACPA's safe harbor provision. In that ruling, the district court stated that

> Doughney registered the domain name because he thought that he had a legitimate First Amendment right to express himself this way. The Court must consider Doughney's state of mind at the time he took the actions in question. Doughney thought he was within his First Amendment rights to create a parody of the plaintiff's organization.

People for the Ethical Treatment of Animals, Inc. v. Doughney, Civil Action No. 99-1336-A, Order at 4 (E.D.Va. Aug. 31, 2000). With its attorney's fee ruling, the district court did not find that Doughney "had reasonable grounds to believe" that his use of PETA's Mark was lawful. It held only that Doughney *thought it* to be lawful.

Moreover, a defendant "who acts even partially in bad faith in registering a domain name is not, as a matter of law, entitled to benefit from [the ACPA's] safe harbor provision." *Virtual Works, Inc.*, 238 F.3d at 270. Doughney knowingly provided false information to NSI upon registering the domain name, knew he was registering a domain name identical to PETA's Mark, and clearly intended to confuse

Internet users into accessing his website, instead of PETA's official website. Considering the evidence of Doughney's bad faith, the safe harbor provision can provide him no relief.

IV

For the foregoing reasons, the judgment of the district court is affirmed.

COMMENTS AND QUESTIONS

1. As noted above, a number of courts have rejected the *Doughney* court's conclusion (omitted here) that visitors to the peta.org site were confused, even briefly. Lamparello v. Falwell, 420 F.3d 309 (4th Cir. 2005); Bosley Med. Inst. v. Kremer, 403 F.3d 672 (9th Cir. 2005). Does it make sense to find a violation of the ACPA if there is no confusion? Does the ACPA do away with the requirement of proving infringement or dilution?

The statute applies only to domain names that are confusingly similar to ordinary trademarks, or that are likely to dilute a famous trademark. This is because trademark owners must demonstrate that they are entitled to control the domain names in question. But the infringement and dilution inquiries under the ACPA focus on the domain name itself, not how it is used. Does the fact that the ACPA imposes this requirement limit its effectiveness?

2. Is it fair to apply the ACPA retroactively, as *Shields* does, to anyone who registered a domain name before the effective date of the Act but who continues to use the name after 1999? What is the point of making injunctive relief retroactive and damages prospective if continued use of an existing domain name can justify a "prospective" claim for damages?

3. What constitutes a legitimate interest in a domain name? Consider the following possible uses:

- *The defendant has its own trademark that corresponds to the domain name.* Identical trademarks often coexist in different industries. Consider Delta, a well-known trademark for an airline, but also for other companies that make faucets, that provide dental insurance, and that even deal with telecommunications. If more than one company has a legitimate claim to a trademark, any of those companies can register the domain name. The ACPA does not require that the domain name be assigned to the "best" or largest trademark owner. Nonetheless, some courts have ruled for plaintiffs in cases in which they apparently concluded that a defendant's bad faith outweighed its legitimate rights in the name. For example, in Virtual Works, Inc. v. Volkswagen of America, 238 F.3d 264 (4th Cir. 2001), the court found Virtual Works (a legitimate trademark owner in its own right) to have violated the ACPA by registering vw.net. It concluded that Virtual's knowledge of likely confusion with Volkswagen's mark, and its discussion of the possibility of selling the domain name to Volkswagen, meant that it did not have a legitimate interest in the name despite its corporate name. Why does bad faith matter under the statute if the defendant is in fact a legitimate trademark owner?

- *The defendant is a legal reseller of the plaintiff's product.* A number of independent businesses sell trademarked products. They may be retailers who sell consumer products, independent service organizations who service or resell used cars, or the like. Can a company that services Porsches register "porscheservice.com"? See General Motors v. E-Publications, Inc., 2001 WL 1798648 (E.D. Mich. 2001) (legitimate parts seller can't refer to "GM" in a domain name); Toma's LLC v. Florida Hi Performance, AF 00968 (eResolution Sept. 10, 2001) (collecting decisions on both sides of the issue and ruling that advertising lawful resale is a legitimate use of a domain name).

- *The defendant wants to criticize or make fun of the plaintiff.* A number of cybersquatting cases involve suits against disaffected customers who set up criticism sites. For example, it is relatively common for an irate consumer to register [company name]sucks.com. It seems ludicrous to think that consumers viewing the site would be confused into thinking that it was owned or sponsored by the company. Nonetheless, while some cases have held that "sucks" sites are noninfringing, other decisions have granted the trademark owner the right to prevent the registration of such sites. Compare Ohio Art Co. v. Watts, No. 98 CV 7338, slip op. at 4 (N.D. Ohio June 23, 1998); BellSouth Corp. v. Internet Classified of Ohio, 1997 WL 33107251, No. 1:96-CV-0769-CC, slip op. at 29-30 (N.D. Ga. Nov. 12, 1997) (both finding infringement from the use of a "sucks" domain name) with Bally Total Fitness v. Faber, 29 F. Supp. 2d 1161 (C.D. Cal. 1998) (web page that used "ballysucks" in the URL was not dilution). Other defendants use the plaintiff's trademark in order to make fun of it. Consider the *PETA* case, above.

Should a bona fide desire to criticize a trademark owner be a legitimate reason to register a domain name? Should it matter whether the name consists merely of the trademark itself, or whether it contains an appendage like "sucks.com" that makes it unlikely anyone could confuse the site with the official trademark owner's site? If trademark owners can prevent the use of their mark in criticism, how can anyone criticize or make fun of the mark or its owner?

4. If a domain name is confusingly similar to a trademark, and the registrant has no legitimate interest in using the name, shouldn't that be enough? Why require proof that the defendant registered the domain name in bad faith? The answer seems to be that the ACPA is not intended to allocate all domain names to the proper owner. Rather, it is designed to single out a smaller class of cases in which the defendant's registration is clearly problematic.

How should a court decide whether a defendant has acted in bad faith? The statute provides a number of factors for the court to consider. Those factors are based on the characteristics of traditional "serial cybersquatters," many of whom have registered hundreds of different trademarks as domain names. Congress concluded that the registration of many different names, the use of false contact information, and demands for payment from trademark owners were all evidence of classic cybersquatting of the sort that the act was intended to prevent.

Was Doughney acting in bad faith? Suppose that he had come to you in 1998, before the ACPA was even adopted, to ask whether it was legal to register peta.org in order to parody PETA. Would you have told him he had a legitimate argument?

5. A variant on cybersquatting is "typosquatting." Typosquatters register common misspellings or variants on trademarked terms, hoping to draw consumers who type the wrong domain name into the URL. Should typosquatting be treated just as any other form of cybersquatting, because typosquatters can divert consumers away from their intended destination? Or is typosquatting different because the typosquatter is not creating confusion, but rather merely capitalizing on preexisting confusion? Courts have generally treated typosquatters just as they have cybersquatters. *Shields* is an obvious example. On the other hand, courts in other contexts have refused to hold a defendant liable for capitalizing on confusion it did not create. See Holiday Inns, Inc. v. 800 Reservation, Inc., 86 F.3d 619 (6th Cir. 1996) (finding that a company who registered a phone number that was a commonly misdialed variant of the Holiday Inns phone number and used it to sell hotel reservations wasn't liable for trademark infringement, since it explained to customers that it wasn't affiliated with Holiday Inns).

6. The ACPA is designed to protect trademarks against cybersquatters. But cybersquatters have also registered domain names that correspond to terms other than trademarks. In particular, cybersquatters have registered the names of various celebrities, such as Julia Roberts. Does the ACPA protect celebrities, just as it does trademark owners? Generally speaking, the answer is no. Personal names are generally not trademarks, and so not entitled to protection. There are exceptions, however — some celebrities, like O.J. Simpson, have registered their name as a trademark, and they are entitled to the same protection as any corporate trademark owner under the statute. 15 U.S.C. § 1125(d)(1)(A). And some UDRP decisions have ordered the transfer of personal names, despite the absence of any authorization to do so in the UDRP. See Roberts v. Boyd, No. D2000-0210 (WIPO May 29, 2000) (ordering transfer of juliaroberts.com). But see Turner v. Fahmi, No. D2002-0251 (WIPO July 4, 2002) (refusing to transfer tedturner.com); Springsteen v. Burgar, No. D2000-1532 (WIPO Jan. 25, 2001) (refusing to transfer brucespringsteen.com).

7. Cybersquatting is an international phenomenon, and one of the problems trademark owners face is how to find cybersquatters in order to sue them. The ACPA includes two ways for plaintiffs to get jurisdiction over cybersquatters. First, a trademark owner who finds a cybersquatter in the United States can sue him in any district with which the cybersquatter has minimum contacts. Second, the ACPA provides that if a trademark owner cannot find the cybersquatter, it can obtain in rem jurisdiction in the jurisdiction where the domain name "resides" by publishing legal notice of the suit and then filing suit "against" the domain name itself. 15 U.S.C. § 1125(d)(2).

In rem jurisdiction has traditionally been used against real property, which of course doesn't move. But a number of courts have held that domain names themselves aren't property. See, e.g., Network Solutions, Inc. v. Umbro Int'l, 529 S.E.2d 80 (Va. 2000). Is proceeding in rem against an intangible thing like a domain name consistent with due process? Or is it fundamentally unfair to the owner of the domain name, who is likely to receive no notice of the suit and be unable to defend the domain name? See Harrods Ltd. v. Sixty Internet Domain Names, 302 F.3d 214 (4th Cir. 2002); Porsche Cars North America v. Porsche.net, 302 F.3d 248 (4th Cir. 2002) (both upholding the constitutionality of the ACPA in rem provisions on the basis that domain names are property).

8. Because the ACPA — like other trademark doctrines — is designed to protect trademark owners, it does not create rights in generic terms. Nonetheless, generic domain names were very popular at the end of the 20th Century, and — because a

domain name is necessarily exclusive in the sense that an ordinary trademark isn't—names like pets.com and toys.com gained widespread recognition. Some courts have given protection to generic domain names on a theory of unfair competition. Cf. E-Cards v. King, No. C-99-3726 SC (N.D. Cal. Dec. 13, 1999) (denying preliminary injunction to E-Cards in suit against ecards.com; the plaintiff later won a jury verdict at trial). Interestingly, a German court has held that the very act of registering a generic domain name constitutes unfair competition, on the theory that the name will give the registrant an unfair advantage over those who cannot use the name.

b. The Uniform Dispute Resolution Procedure

Congress and the courts were not the only ones concerned with cybersquatting. In the latter half of the 1990s, Network Solutions Inc., the company that registered domain names in the global top-level domains (gTLDs), faced a series of lawsuits by trademark owners that sought to hold NSI liable for permitting cybersquatters to register their trademarks as domain names. To solve this problem, NSI drafted a series of dispute resolution policies to help it determine when to turn a domain name over to a trademark owner or to put the name "on hold." NSI's policies have generally provided that domain name registrants agreed to arbitrate disputes with NSI. They have not, however, mandated arbitration between trademark owners and domain name registrants.

When the Internet Corporation for Assigned Names and Numbers (ICANN) was formed in 1999, it sought to deal with the problem of cybersquatting in a comprehensive way outside the judicial system. ICANN established a compulsory private dispute resolution process for dealing with cases of cybersquatting and required all Internet registrars to require their customers to subscribe to the process. Under this process, known as the Uniform Dispute Resolution Procedure (UDRP), a trademark owner can file a complaint online with an approved dispute resolution provider for a small fee. The domain name owner is notified and has a short time to file a responsive pleading. Then the dispute panelist or panelists rule on the case. Unlike arbitration or other forms of private dispute resolution, there is no evidence, no oral presentation, and very little in the way of due process. However, the panelist's powers are limited to deciding whether or not to transfer the domain name in question to the complainant. If the loser objects to the decision, he or she has ten days to file suit in court challenging the result.

The UDRP is international in scope. The substantive rules governing UDRP cases do not rely expressly on any nation's trademark law. Rather, they are designed to cover only true cases of cybersquatting, leaving doubtful cases for resolution by the courts. The standard for proving cybersquatting under the UDRP is quite similar to the standard under the ACPA:

> This Paragraph sets forth the type of disputes for which you are required to submit to a mandatory administrative proceeding. These proceedings will be conducted before one of the administrative-dispute-resolution service providers listed at www.icann.org/udrp/approved-providers.htm (each, a "Provider").
>
> a. Applicable Disputes. You are required to submit to a mandatory administrative proceeding in the event that a third party (a "complainant") asserts to the applicable Provider, in compliance with the Rules of Procedure, that
>
> (i) your domain name is identical or confusingly similar to a trademark or service mark in which the complainant has rights; and

(ii) you have no rights or legitimate interests in respect of the domain name; and

(iii) your domain name has been registered and is being used in bad faith.

In the administrative proceeding, the complainant must prove that each of these three elements are present.

b. Evidence of Registration and Use in Bad Faith. For the purposes of Paragraph 4(a)(iii), the following circumstances, in particular but without limitation, if found by the Panel to be present, shall be evidence of the registration and use of a domain name in bad faith:

(i) circumstances indicating that you have registered or you have acquired the domain name primarily for the purpose of selling, renting, or otherwise transferring the domain name registration to the complainant who is the owner of the trademark or service mark or to a competitor of that complainant, for valuable consideration in excess of your documented out-of-pocket costs directly related to the domain name; or

(ii) you have registered the domain name in order to prevent the owner of the trademark or service mark from reflecting the mark in a corresponding domain name, provided that you have engaged in a pattern of such conduct; or

(iii) you have registered the domain name primarily for the purpose of disrupting the business of a competitor; or

(iv) by using the domain name, you have intentionally attempted to attract, for commercial gain, Internet users to your web site or other on-line location, by creating a likelihood of confusion with the complainant's mark as to the source, sponsorship, affiliation, or endorsement of your web site or location or of a product or service on your web site or location.

c. How to Demonstrate Your Rights to and Legitimate Interests in the Domain Name in Responding to a Complaint. When you receive a complaint, you should refer to Paragraph 5 of the Rules of Procedure in determining how your response should be prepared. Any of the following circumstances, in particular but without limitation, if found by the Panel to be proved based on its evaluation of all evidence presented, shall demonstrate your rights or legitimate interests to the domain name for purposes of Paragraph 4(a)(ii):

(i) before any notice to you of the dispute, your use of, or demonstrable preparations to use, the domain name or a name corresponding to the domain name in connection with a bona fide offering of goods or services; or

(ii) you (as an individual, business, or other organization) have been commonly known by the domain name, even if you have acquired no trademark or service mark rights; or

(iii) you are making a legitimate noncommercial or fair use of the domain name, without intent for commercial gain to misleadingly divert consumers or to tarnish the trademark or service mark at issue.

The fast and cheap nature of the UDRP process—and arguably the protrademark slant of its decisions—has attracted hordes of trademark owners. The UDRP went into effect at the end of 1999, around the same time as the ACPA. But while the ACPA had produced just over 50 reported decisions by the middle of 2002, there were UDRP decisions in cases involving over 7,000 domain names. The overwhelming majority of those panel decisions (around 80%) ruled for the trademark owner.

Individual UDRP decisions have garnered their share of controversy. Those controversies have centered on each of the four basic elements of a UDRP claim:

A legitimate trademark right. Ordinarily this is not a problem; almost all of the complainants in UDRP proceedings are trademark owners. But sometimes a complainant will not have a trademark in the name it seeks to control. When that happens, the proper result under the UDRP is to dismiss the case. But panelists and even courts have been persuaded to grant the complainant rights in words that it simply doesn't own as a trademark. For example, panelists have transferred the domain name sex.biz to a company claiming to have trademarked the term "sex," see Schatte v. Personal, FA0209000124756 (NAF Nov. 4, 2002); transferred the common French term merci.com to a company that claimed to own it as a trademark, August Storck KG v. Unimetal Sanayi ve Tic A.S., No. D2001-1125 (WIPO November 23, 2001); and transferred paint.biz to a company that claims to own a trademark in the word "paint," Valspar Sourcing, Inc. v. TIGRE, FA0204000112596 (NAF June 4, 2002). Cf. Archdiocese of St. Louis v. Internet Entertainment Group, 34 F. Supp. 2d 1145 (E.D. Mo. 1999) (concluding that "Papal Visit 1999" was a famous trademark entitled to dilution protection). By contrast, other panels have refused to transfer generic terms on the ground that they were not trademarks. For example, one panel refused to transfer the munich.biz domain name on the grounds that the city did not own the exclusive right to its name. Landeshauptstadt Muenchen v Beanmills, Inc., Case No. DBIZ2002-00147 (WIPO July 1, 2002). See also X/Open Company Limited v. Expeditious Investments, Case No. D2002-0294 (WIPO 2002) (refusing to transfer unix.com; a different panel decided the opposite way in a case involving unix.net).

An identical or confusingly similar domain name. Trademark owners are entitled to prevent the registration only of names that are identical to or confusingly similar to their trademarks. But some panelists have ordered the transfer of domain names that clearly fail to meet this standard. For example, a significant number of the cases involving domains of the type [trademark]sucks.com have been transferred to the trademark owner, even though no rational consumer would think that WalMart sponsored the domain name "walmartsucks.com." [Many other cases come to the opposite conclusion, however. For a discussion of the panel decisions on both sides, see Homer TLC v. GreenPeople, FA0508000550345 (NAF Oct. 25, 2005) (refusing to transfer home-depotsucks.com)]. And one panelist even held that the trademark "Tata," the name of a large Indian conglomerate, was infringed by the domain name "bodacioustatas.com." Tata Sons Limited v. D & V Enterprises, Case No. D2000-0479 (WIPO Aug. 18, 2000).

Absence of a legitimate right. Cybersquatting under the UDRP requires proof that the domain name registrant lacks any legitimate right to use the domain name. Trademark owners cannot use the UDRP to oust other trademark owners from the use of a mark; if both have a legitimate claim to use the name in commerce, the UDRP will not step in to decide who is "best" entitled to use the mark. But other claims to legitimate use are more controversial. Several panelists have considered whether a legitimate dealer in the plaintiff's products is entitled to use the plaintiff's trademark to advertise the presence of those products. See Toma's LLC v. Florida Hi-Performance, AF 0968 (eRes Sept. 10, 2001):

> Florida appears to be using Toma's trademarked domain name to sell Toma's products. The question is whether this is a legitimate use.
> A number of prior panel decisions have considered whether a distributor of a trademark owner's goods who uses the trademark as its domain name has made a bona fide use

of the trademark. Unfortunately, the decisions are not in agreement. Compare Kabushiki Kaisha Toshiba v. Distribution Purchasing and Logistics Corp., WIPO D2000-0464 (July 27, 2000); Hydraroll Ltd. v. Morgan Corp., NAF FA2000094108 (April 14, 2000); Mariah Boats, Inc. v. Shoreline Marina, NAF FA4000094392 (May 5, 2000) (all concluding that a distributor has no right to use the trademark of the distributed goods as a domain name) with Freni Brembo S.p.A. v. Webs We Weave, WIPO D2000-1717 (March 19, 2001); K&N Engineering v. Kinnor Servs., WIPO D2000-1077 (January 19, 2001); Weber-Stephen Prods. v. Armitage Hardware, WIPO D2000-0187 (May 11, 2000); Koninklijke Philips Electronics v. Cun Siang Wang, WIPO D2000-1778 (March 15, 2001); Hewlett-Packard Co. v. Napier, NAF FA3000094368 (April 28, 2000) (all concluding that a distributor *is* entitled to use the trademark of the distributed goods as a domain name). See also *Philips,* supra (noting three additional decisions on each side of the issue).

It is clear as a matter of trademark law that the purchaser of bona fide goods is entitled to resell those goods, and to truthfully advertise the goods bearing the trademark. Once the goods are sold into commerce, a trademark owner's rights in those goods are exhausted. See *Philips,* supra (citing authority). [3] Were it not so, distributors could not advertise their goods, car owners could not advertise their used cars, and so on. Since Toma's has not alleged that the "Fake Bake" goods sold on Florida's Web page are counterfeit or altered in any way, Florida is free to truthfully advertise that it is selling those products by using the trademark "Fake Bake." See generally Prestonettes, Inc. v. Coty, 264 U.S. 359, 368 (1924) ("When the mark is used in a way that does not deceive the public we see no such sanctity in the word as to prevent its being used to tell the truth.").

Here, though, Florida has done somewhat more than truthfully advertise the availability of bona fide "Fake Bake" products on its Web site. By using the domain name "fakebake.com," Toma's alleges that Florida has confused consumers into thinking that Toma's runs or sponsors the Web page. Toma's is particularly concerned about this because Florida also sells competing products on its page. Whether this exceeds Florida's rights as a reseller of legitimate goods is a difficult question under trademark law. In *Mariah Boats,* supra, the panel concluded that because Shoreline sold both Mariah's boats and other, competing boats, its use of the domain name mariahboats.net was likely to confuse consumers. By contrast, in both *Freni Brembo,* supra and *Weber-Stephens,* supra, the panels concluded that such a use was legitimate notwithstanding the fact that the respondent also sold other goods on its site. And in *Philips,* the panel concluded as follows:

> The Panel accepts that it is possible for a trader in branded goods to advertise and sell goods under the brand without using a domain name incorporating the brand name. Nevertheless such a domain name is of considerable assistance to a business selling the branded goods. The Panel considers that a person can have a legitimate interest in using a particular domain name even if it is not an essential requirement for him to do so.

Whether a bona fide reseller of trademarked goods is always entitled to use a domain name incorporating the trademark is an unsettled legal question. It depends in part on the detailed factual circumstances of the case — how the name is being used, what evidence of confusion there is, whether there is an agreement between the parties, and so on. None of that evidence has been presented to me in this case, other than Toma's allegation (unsupported by evidence) that consumers are being confused. Such a case is not appropriate for resolution under the UDRP, which was designed to deal only with a narrow class of abusive registrations by cybersquatters. The difficult and uncertain allocation of rights in

---

3. The "gray market" importation of bona fide goods from another country presents a more contentious issue, but there is no evidence of importation here.

this case should be decided by a court, not by a UDRP panel. *Accord Freni-Brembo; Weber-Stephen.* Thus, I cannot conclude that Florida's use here was necessarily illegitimate.

Compare Educational Testing Service v. TOEFL USA, Case No. D2002-0380 (WIPO July 11, 2002) (ordering transfer of toeflusa.com, a domain name owned by TOEFL USA as a trademark for services related to the TOEFL test).

As a general matter in trademark law, "nominative" or nontrademark use — the use of the plaintiff's mark to refer to the plaintiff's own products — is not infringement. But in the context of domain names, the defendant's registration of the plaintiff's mark as a domain name may be more likely to confuse consumers into thinking that the defendant is the actual source of the product. This depends on the nature of the name, though. A domain name like microsoft.com might suggest to consumers that the owner is Microsoft, not just an authorized dealer. By contrast, a name like porscheservice.com is more likely to accurately convey that the owner services Porsches, not that the owner is Porsche itself. But see Academy of Motion Picture Arts and Sciences v. Khan, FA0109000100142 (NAF Nov. 14, 2001) (ordering cancellation of the domain name oscarbetexchange.com for an online service in which people could bet on the outcome of the Oscars). Some trademark owners have even sued "fan sites" that register domain names in order to praise their subject.

Other disputed issues of legitimate use involve noncommercial uses of a name. For instance, a personal name or nickname may be a legitimate reason to register a domain name. An infamous case in this regard involves pokey.org, in which a 12-year-old nicknamed "Pokey" was threatened with suit by the owners of the "Gumby and Pokey" marks. See also Strick Corp. v. Strickland, No. FA #94801 (NAF July 3, 2000) (refusing to transfer strick.com, owned by an individual whose nickname was "Strick"). Proof of a legitimate nickname presumably provides a legitimate reason to register a domain name.[12] At the same time, nickname arguments are particularly susceptible to abuse. For example, one panelist held that the registrant of the domain oxforduniversity.com could not show a legitimate interest in the mark by changing his legal name to Oxford University. See University of Oxford v. Seagle, No. D2001-7046 (WIPO Aug. 14, 2001).

A final category of possible legitimate interests involves criticism of the trademark owner. Frustrated customers and others who want to criticize a company naturally want to post their criticisms at a domain name that includes the trademark of the company in question. Where the domain name in question is one — like walmart-sucks.com — that is unlikely to confuse consumers, the defendant clearly has a legitimate interest in using the mark. The question is more difficult if the critic registers only the trademark — say, walmart.com. Consumers may well be confused, at least initially, if walmart.com isn't run by Wal-Mart. See *Doughney, supra.* But even if the use will draw customers to the site because they are confused, arguably the critic has a legitimate interest in registering the domain name in order to attract the attention of those consumers. See Lucas Nursery & Landscaping Inc. v. Grosse, 359 F.3d 806 (6th Cir. 2004) (rejecting ACPA claim against dissatisfied customer who registered "lucas-nursery.com"); TMI, Inc. v. Maxwell, 368 F.3d 433 (5th Cir. 2004); Northland Ins.

---

12. It does not necessarily provide the registrant with the right to compete. For example, someone with the last name McDonald will be prevented from opening a hamburger stand using the name "McDonald's," because of the likelihood that consumers will be confused. See E. & J. Gallo Winery v. Gallo Cattle Co., 967 F.2d 1280 (9th Cir. 1992) (brother of Ernest & Julio Gallo could not use Gallo name to sell meat, because of the possibility of confusion).

Co. v. Blaylock, 115 F. Supp. 2d 1108 (D. Minn. 2000) (registration of company's trademark as a domain name in order to criticize the company was not infringing).

Bad faith. Like the ACPA, the UDRP identifies a number of factors evidentiary of bad faith. Registration of multiple names, demanding money to relinquish them, and providing false address information are all considered evidence of bad faith. Some panelists have gone farther, however, holding that a desire to criticize a trademark owner can itself be evidence of bad faith. See Reg Vardy Plc v. Wilkinson, Case No. D2001-0593 (WIPO July 3, 2001). This seems particularly perverse, given the discussion above of criticism as a legitimate reason to register a domain name. Compare Savannah College of Art & Design v. Houeix, No. CPR 0206 (CPR Feb. 19, 2002) (former employee could register scad.info in order to criticize the college).

On the other hand, at least one panelist has held that the mere registration of domain names does not constitute bad faith, leaving open the possibility that cybersquatters who do not ask for money may be entitled to keep the domain names they have registered even if they cannot show any legitimate interest in the name. See Ingram Micro v. Ingredients Among Modern Microwaves, No. D2002-0301 (WIPO May 15, 2002). But see Telstra Corporation Limited v. Nuclear Marshmallows, Case No. D2000-0003 (WIPO Feb. 18, 2000) (finding bad faith absent any "positive action" by registrant).

UDRP decisions have been roundly criticized as unduly favorable to trademark owners, particularly in the first two years of operation. See, e.g., Michael Geist, Fair. com?: An Examination of the Allegations of Systematic Unfairness in the ICANN UDRP, 27 Brook. J. Int'l. L. 903 (2002); Elizabeth G. Thornburg, Fast, Cheap and Out of Control: Lessons From the ICANN Dispute Resolution Process, 6 J. Sm. & Emerging Bus. L. 191 (2002). These critics focus both on the outcomes of the early cases, which overwhelmingly ordered transfers of domain names, and on deficiencies in the process by which the cases are decided. Because only the trademark owner gets to choose which dispute resolution company will decide the case, dispute resolution providers seem to compete for the approval of trademark owners. Eresolution, the one company whose decisions were roughly evenly divided between trademark owners and registrants, went out of business because trademark owners sent their work to NAF and WIPO instead. NAF and WIPO seem to assign cases to favored panelists, rather than randomly distributing them among eligible panelists. And NAF in particular has advertised its pro-trademark decisions in an effort to drum up more business. Notwithstanding these problems, panel opinions between 2003 and 2005 were actually substantially less skewed in favor of trademark owners than the earlier opinions.

Other concerns center on the short deadlines and the sketchy nature of the process required. Registrants have only a short time to respond to a UDRP complaint, and if they lose have only ten days to go to court to stop the transfer of their domain name. Many registrants are not represented by counsel, and so may have difficulty understanding the process and their obligations. As a result, a large percentage of registrants fail to answer a complaint, and relatively few registrants challenge the transfer of their domain names in court.

Froomkin has proposed a series of procedural reforms to improve the UDRP. A. Michael Froomkin, ICANN's "Uniform Dispute Resolution Policy": Causes and (Partial) Cures, 67 Brook. L. Rev. 605 (2002). Among other things, Froomkin suggests procedures for removing biased arbitrators; the creation of a substantive doctrine of "reverse domain name hijacking," in which domain name owners who are subject to abusive claims could recover money; and making it easier for respondents to meet the deadlines for filing a response and selecting an arbitration panel. Others have argued that an administrative appeal process would be a good idea. And it also seems reasonable to allow registrants to respond online, as eresolution once did.

Because of the limits of the UDRP process, courts to consider the issue have unanimously concluded that it is not "arbitration" subject to judicial deference under the Federal Arbitration Act. Rather, UDRP decisions are reviewed de novo in the courts. See, e.g., Sallen v. Corinthians Licenciamentos Ltd., 273 F.3d 14 (1st Cir. 2001); Parisi v. Netlearning, Inc., 139 F. Supp. 2d 745 (E.D. Va. 2001).

In 2001, ICANN permitted the creation of several new gTLDs, including .biz. The creation of these new gTLDs created new opportunities for cybersquatters. To deal with this problem, ICANN created a new dispute resolution process (called STOP) that included a "sunrise" registration period. Under this process, trademark owners get a chance to register their marks in the .biz domain before the registration is opened up to other registrants. Because some registrants falsely claimed to be trademark owners in order to take advantage of the sunrise process, a number of trademark owners have filed claims under the new STOP process. While the dispute resolution process is similar to the UDRP, it is not identical — and as a result the domain name dispute resolution procedure is no longer truly uniform.

6. Contributory Infringement

Both patent law and copyright law have well-developed doctrines of contributory infringement. Defendants are liable for contributory infringement if, although they did not themselves infringe the patent or copyright, they assisted or encouraged others to infringe. Liability for contributory infringement extends to the makers and vendors of machines on which infringements are performed, but only if the machines are not capable of a substantial non-infringing use.

Is there contributory infringement in trademark law? At least at first blush, it is hard to see what acts might be comparable to selling or making available a machine used in producing copies.

Nonetheless, a number of courts have held that contributory infringement of trademarks is illegal. For example, in Polo Ralph Lauren Corp. v. Chinatown Gift Shop, 855 F. Supp. 648 (S.D.N.Y. 1994), the District Court for the Southern District of New York held that a landlord who knew that a commercial tenant was engaging in trademark infringement by selling counterfeit goods and who did nothing to prevent the activity could be sued for contributory infringement.

Is this result defensible? Does it extend to newspapers that print advertisements by counterfeiters? To graphics and print shops that print ads? To those who sell furniture or office supplies to counterfeiters?

A related problem has arisen in the context of domain name registries. Should they be liable for registering domain names that infringe the trademarks of others? In Lockheed Martin Corp. v. Network Solutions, Inc., 194 F.3d 980 (9th Cir. 1999), the court held that Network Solutions, Inc., the exclusive domain name registry from 1991-1999, had no responsibility to screen domain names and bore "no affirmative duty to police the Internet in search of potentially infringing uses of domain names." The Anticybersquatting Consumer Protection Act in 1999 immunizes domain name registrars from monetary relief for registering a domain name that infringes trademark rights. See 15 U.S.C. §1114(2)(D); Lanham Act §32(2)(D). This section also provides a safe harbor for domain registries from liability for refusing to register, canceling, or transferring a domain name in furtherance of its dispute resolution policy.

A number of trademark owners have sued Google and other Internet search engines, alleging that their ads (which are targeted based on Internet keywords selected by the advertiser) infringe their trademarks. In GEICO v. Google, 2005 WL 1903128 (E.D. Va. Aug. 8, 2005), the district court rejected such a claim, ruling that the plaintiff could not demonstrate that the mere sale of a keyword confused consumers. The court left open the possibility that the advertisers themselves might be liable for infringement if the text of the ads were confusing, and that Google might be liable for contributory infringement if it encouraged such confusion. See also Stacey L. Dogan & Mark A. Lemley, Trademarks and Consumer Search Costs on the Internet, 41 Hous. L. Rev. 777 (2004) (arguing for this approach). Does it make sense to distinguish between ads that are likely to confuse consumers and those that aren't? Or is the mere use of a trademark as a keyword problematic even if no one will be confused by the resulting ad? Even if an advertiser is liable for running a confusing ad, is Google contributing to that infringement? How? What could Google do to avoid liability, short of terminating its entire advertising program?

7. False Advertising

Section 43(a) of the Lanham Act includes a specific prohibition on false or misleading advertising:

> (a) Any person who, on or in connection with any goods or services, or any container for goods, uses in commerce any word, term, name, symbol, or device, or any combination thereof, or any false designation of origin, false or misleading description of fact, or false or misleading representation of fact, which—
>
> (1)...
>
> (2) in commercial advertising or promotion, misrepresents the nature, characteristics, qualities, or geographic origin of his or her or another person's goods, services, or commercial activities,

shall be liable in a civil action by any person who believes that he or she is or is likely to be damaged by such act.

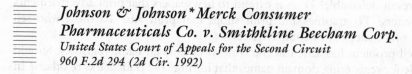

*Johnson & Johnson*Merck Consumer Pharmaceuticals Co. v. Smithkline Beecham Corp.*
United States Court of Appeals for the Second Circuit
960 F.2d 294 (2d Cir. 1992)

WALKER, Circuit Judge:

Johnson & Johnson*Merck Consumer Pharmaceuticals Company ("J&J* Merck") appeals from the final judgment of the United States District Court for the Southern District of New York, Honorable Miriam Goldman Cedarbaum, Judge, denying J&J*Merck injunctive relief and dismissing its complaint against defendants Smithkline Beecham Corporation ("Smithkline") and Jordan, McGrath, Case & Taylor, Inc. ("Jordan"). J&J*Merck, the manufacturer of MYLANTA, a nonprescription antacid, instituted this action in order to restrain Smithkline and Jordan, the manufacturer and advertiser of TUMS, respectively, a competing brand of antacid, from continuing to air certain television commercials that J&J*Merck claims are false and misleading in violation of §43(a) of the Lanham Act, 15 U.S.C. §1125(a), and §§349 and 350 of the New York General Business Law.

Because we agree with Judge Cedarbaum's legal analysis, and because we conclude that her factual findings were not clearly erroneous, we affirm the judgment of the district court.

Background

J&J*Merck manufactures and markets MYLANTA, a popular nonprescription antacid product that is used to reduce heartburn by dyspepsia sufferers across the nation. MYLANTA contains aluminum hydroxide and magnesium hydroxide as its acid neutralizing agents. Smithkline produces the antacid TUMS. While TUMS is also extremely popular among victims of gastric distress, its formula relies upon calcium carbonate, rather than aluminum or magnesium salts, in treating indigestion and related ailments.

In September of 1990, Smithkline and its advertising agency, Jordan, instituted a television advertising campaign that sought to promote TUMS by comparing the ingredients contained in TUMS to those contained in other leading nonprescription antacids. The first commercial in this media campaign was entitled "Ingredients." Ingredients shows a man and a woman seated at a lunch counter along with other patrons, all of whom are eating such marginally digestible items as grease-laden french fries and chili. A background announcer explains that this couple "ate the same lunch" and "got the same heartburn," but that "his antacid [TUMS] is very different." A series of competing products then appear on the screen — ROLAIDS, MAALOX, and MYLANTA. At the same time, the ingredients of each of these antacids are visually superimposed on the screen and listed by the announcer: ROLAIDS — "Aluminum Salt"; MAALOX — "Aluminum and Magnesium"; MYLANTA — "Aluminum and Magnesium." Immediately following this remedial menu, the ad continues with a close-up image of a roll of TUMS. The voice-over states that TUMS is "aluminum-free," and that "only TUMS helps wipe out heartburn and gives you calcium you need every day." After the woman is depicted pondering the virtues of a calcium-based antacid, the commercial ends with a visual and verbal statement of the advertisement's slogan: "Calcium rich, aluminum free TUMS."

The makers of ROLAIDS, MAALOX and MYLANTA found the Ingredients advertisement impossible to swallow. After it was first aired, they immediately complained to Smithkline that Ingredients falsely implied that their antacids were harmful. In early October, 1990, they filed formal protests with each of the three major television networks, objecting to the Ingredients broadcast on the grounds that it was false and misleading. Representatives of Smithkline and Jordan met with network officials to discuss the protests lodged by TUMS' competitors, at which time they advised the networks that they would revise the Ingredients commercial. Thereafter, Smithkline and Jordan voluntarily withdrew Ingredients from the air. Although necessary background, the Ingredients advertisement is not the subject of this action.

In late October, 1990, Smithkline and Jordan released a second version of the TUMS commercial, entitled "Ingredients-Revised." The difference between

Ingredients and Ingredients-Revised was that the second advertisement deleted all references to TUMS as an "aluminum-free" product. Rather, Ingredients-Revised emphasized the fact that TUMS contains calcium, ROLAIDS, MAALOX and MYLANTA did not, and that calcium is good for you. In addition to the full version of Ingredients-Revised, Smithkline and Jordan began showing a shortened one. This commercial singled out MYLANTA as the sole target of comparison, while keeping the rest of the commercial substantially the same.

... J&J*Merck sued in the district court to enjoin appellees from continuing to broadcast Ingredients-Revised and its shortened version, on the grounds that the commercials violated § 43(a) of the Lanham Act, 15 U.S.C § 1125(a), and §§ 349 and 350 of the New York General Business Law. Specifically, J&J*Merck alleged that both Ingredients-Revised, and the shortened version, falsely represented that: 1) occasional ingestion of TUMS, in the manner directed for antacid relief, results in nutritional benefit to the consumer; and 2) the magnesium and aluminum contained in MYLANTA are unsafe for human consumption.

Initially, J&J*Merck moved for a preliminary injunction. Instead of granting the motion, the district court consolidated the application with an expedited trial on the merits. After a five day bench trial, Judge Cedarbaum concluded that J&J*Merck had "not shown that the message that occasional Tums users will benefit from calcium is false or misleading." Furthermore, she found "that the challenged commercials do not communicate the message that aluminum or magnesium are harmful or unsafe." On the basis of this, Judge Cedarbaum concluded that J&J*Merck failed to demonstrate "that 'Ingredients-Revised' is either false or misleading." Accordingly, she dismissed both J&J*Merck's federal and state law claims. This appeal followed.

Discussion

On appeal, J&J*Merck has abandoned its claims relating to calcium and magnesium. Thus, the only issue before us now is whether the challenged commercials communicated a false or misleading message with respect to the safety of aluminum-based antacids. The district court explicitly found that "if the challenged commercials indicated that... aluminum [is] dangerous, they would be false and misleading."

J&J*Merck contends that, even though the content of the challenged commercials is literally true, Ingredients-Revised preys upon a publicly held misperception that the ingestion of aluminum causes Alzheimer's disease. According to J&J*Merck, the commercials accomplish this by repeatedly juxtaposing the absence of aluminum in TUMS with its presence in MYLANTA. In turn, this repetition supposedly links MYLANTA with an allegedly popularly held, yet unsubstantiated concern that aluminum is associated with Alzheimer's. Since the aluminum/Alzheimer's connection has not been scientifically established, J&J*Merck argues that Ingredients-Revised purposefully taps into a preexisting body of public misinformation in order to communicate the false and misleading message that aluminum-based antacids are harmful.

The gravamen of J&J*Merck's claim is that advertisers may be held liable for the knowing exploitation of public misperception. While this argument presents a novel theory of Lanham Act liability — one which we neither reject nor embrace — we note that, in any event, it would be unavailing in this case. Because J&J*Merck has failed to show that it has suffered any injury as a result of the challenged TUMS commercials, it cannot obtain relief under any theory of Lanham Act liability that is premised upon an implied falsehood . . . .

I. Liability for Implied Falsehoods . . .

The law governing false advertising claims under the Lanham Act is well settled in this circuit. In order to recover damages or obtain equitable relief, a plaintiff must show that either: 1) the challenged advertisement is literally false, or 2) while the advertisement is literally true it is nevertheless likely to mislead or confuse consumers. See McNeil-P.C.C., Inc. v. Bristol-Myers Squibb Co., 938 F.2d 1544, 1549 (2d Cir. 1991); Johnson & Johnson v. GAC Intl., Inc., 862 F.2d 975, 977 (2d Cir. 1988); Coca-Cola Co. v. Tropicana Products Inc., 690 F.2d 312, 317 (2d Cir. 1982); American Home Products Corp. v. Johnson & Johnson, 577 F.2d 160, 165 (2d Cir. 1978).

Where, as here, a plaintiff's theory of recovery is premised upon a claim of implied falsehood, a plaintiff must demonstrate, by extrinsic evidence, that the challenged commercials tend to mislead or confuse consumers. See GAC Intl., Inc., 862 F.2d at 977; American Home Products Corp., 577 F.2d at 165; Upjohn Co. v. American Home Products Corp., 598 F. Supp. 550, 556 (S.D.N.Y. 1984). It is not for the judge to determine, based solely upon his or her own intuitive reaction, whether the advertisement is deceptive. Rather, as we have reiterated in the past, "the question in such cases is — what does the person to whom the advertisement is addressed find to be the message?" American Home Products Corp., 577 F.2d at 166 (quoting American Brands, Inc. v. R. J. Reynolds Tobacco Co., 413 F. Supp. 1352, 1357 (S.D.N.Y. 1976)). That is, what does the public perceive the message to be?

The answer to this question is pivotal because, where the advertisement is literally true, it is often the only measure by which a court can determine whether a commercial's net communicative effect is misleading. Thus, the success of a plaintiff's implied falsity claim usually turns on the persuasiveness of a consumer survey. See Coca-Cola Co., 690 F.2d at 317 ("When the challenged advertisement is implicitly rather than explicitly false, its tendency to violate the Lanham Act by misleading, confusing or deceiving should be tested by public reaction."); American Home Products Corp., 577 F.2d at 165; Upjohn Co., supra.

Where a district court's factual findings regarding an advertisement's tendency to mislead or confuse consumers are based "on its analysis of . . . surveys [and] on the corroborating evidence of the market research experts," those findings are entitled to the clearly erroneous standard of review. American Home Products Corp., 577 F.2d at 167. J&J*Merck does not dispute this. However, in attempting to show that the district court's findings on this point were clearly erroneous, J&J*Merck argues that Judge Cedarbaum failed sufficiently to consider factors other than consumer survey evidence, such as: 1) the general "commercial context" or sea of information in which consumers are immersed; 2) the defendant's intent to harness public misperception; 3) the defendant's prior advertising history; and 4) the sophistication of the advertising audience. This argument causes us some discomfort.

Appellant's criticism of the district court's findings misconstrues the proper role of consumer survey evidence in the analysis of implied falsehood claims. Generally, before a court can determine the truth or falsity of an advertisement's message, it must first determine what message was actually conveyed to the viewing audience.

Consumer surveys supply such information. Id., 577 F.2d at 166. "Once the meaning to the target audience has been determined, the court, as the finder of fact, must then judge whether the evidence establishes that they were likely to be misled." Nikkal Indus., Ltd. v. Salton, Inc., 735 F. Supp. 1227, 1235 (S.D.N.Y. 1991) (citing McNeilab, Inc. v. American Home Products Corp., 501 F. Supp. 517, 525 (S.D.N.Y. 1980)).

Three of the factors listed by J&J*Merck, i.e., commercial context, defendant's prior advertising history, and sophistication of the advertising audience, only come into play, if at all, during the latter part of the court's analysis. In a particular case, these factors may shed some light on whether the challenged advertisement contributed to the meaning that was ultimately gleaned by the target audience. In other words, in determining whether an advertisement is likely to mislead or confuse, the district court may consider these factors after a plaintiff has established "that a not insubstantial number of consumers," Coca-Cola Co., 690 F.2d at 317, hold the false belief allegedly communicated in the ad.

Absent such a threshold showing, an implied falsehood claim must fail. This follows from the obvious fact that the injuries redressed in false advertising cases are the result of public deception. Thus, where the plaintiff cannot demonstrate that a statistically significant part of the commercial audience holds the false belief allegedly communicated by the challenged advertisement, the plaintiff cannot establish that it suffered any injury as a result of the advertisement's message. Without injury there can be no claim, regardless of commercial context, prior advertising history, or audience sophistication.

However, we have held that "where a plaintiff adequately demonstrates that a defendant has intentionally set out to deceive the public," and the defendant's "deliberate conduct" in this regard is of an "egregious nature," a presumption arises "that consumers are, in fact, being deceived." Resource Developers, Inc. v. Statue of Liberty—Ellis Island Foundation, Inc., 926 F.2d 134, 140 (2d Cir. 1991). This presumption which may be engendered by the expenditure "of substantial funds in an effort to deceive consumers and influence their purchasing decisions" relieves a plaintiff of the burden of producing consumer survey evidence that supports its claim. Id. (quoting U-Haul Intl., Inc. v. Jartran, Inc., 793 F.2d 1034, 1041 (9th Cir. 1986)). In such a case, once a plaintiff establishes deceptive intent, "the burden shifts to the defendant to demonstrate the absence of consumer confusion." Resource Developers, supra. J&J*Merck argues that this principle applies here. We are not persuaded.

The evidence of deceptive intent introduced at trial by J&J*Merck, which was controverted by defense witnesses, was comprised of certain conflicting Smithkline internal memoranda and related solely to the original Ingredients version of the TUMS commercial—the version that contained the "aluminum-free" slogan. No doubt, this evidence might have been used to establish Smithkline's deceptive intent in broadcasting the Ingredients commercial. However, as stated above, Smithkline withdrew Ingredients from the air, and that version of the commercial is not the subject of this action. In response to complaints lodged with Smithkline and the television networks by TUMS' competitors, the "aluminum-free" reference was dropped from the Ingredients ad. J&J*Merck introduced no additional evidence that the same "deceptive intent," which allegedly infected its production of the Ingredients commercial, informed the editing process that resulted in Ingredients-Revised. Judge Cedarbaum acknowledged that J&J*Merck "attempted to show that [a false message] was communicated through evidence of defendants' intent." Apparently, she found its attempt unavailing. In any event, given the indirect and controverted nature of the evidence regarding the intent behind the Ingredients-Revised advertisement, we are unwilling to extend the presumption of consumer confusion to this case.

II. The Bruno and Ridgway Survey

At trial, J&J*Merck introduced the results of a consumer survey that it had conducted in conjunction with Bruno and Ridgway Research Associates, a marketing firm that conducts between 200-300 such surveys each year. According to Mr. Joseph Ridgway, the firm's president and J&J*Merck's expert witness, the survey was designed to assess what messages are communicated to consumers by Ingredients-Revised. In order to gain this information, 150 male and 150 female adult nonprescription antacid users were shown the commercial and interviewed in eight different shopping malls.

The aspects of the survey that are relevant to this appeal are those which concern the commercial's message regarding aluminum. Questions 8a and 8b asked: "Aside from trying to get you to buy the product, what are the main ideas the commercial communicates to you?" and "What other ideas does the commercial communicate to you?" Out of the 300 people surveyed, 18 people generally responded that "other antacids contain ingredients that are bad/harmful to you," only six of which specifically commented that "aluminum is not good for you." Two other responses were listed in the survey's tally as "aluminum is bad for brain/causes alzheimer's," and lastly, one additional response was recorded under the heading "miscellaneous negative aluminum comments." Questions 14a-c were asked of the 220 respondents who recalled Ingredients-Revised stating that MAALOX and MYLANTA contained aluminum and magnesium. They respectively inquired:

14a — What, if anything, does the commercial communicate to you about the aluminum and magnesium in Maalox and Mylanta?
14b — What else, if anything, does the commercial communicate to you about the aluminum and magnesium in Maalox and Mylanta?
14c — Based on the commercial you just saw, how do you feel about taking a product for heartburn that contains aluminum and magnesium?

Of the 220 people who responded to these questions, 83 answered with a comment classified as "not good for you/harmful/detrimental to your health." Three answered that aluminum causes Alzheimer's disease and, in addition, 38 responded that these ingredients are not needed by the body.

By adding up all negative comments made about either aluminum or magnesium in response to any of the survey questions, Mr. Ridgway concluded that Ingredients-Revised communicated to 45% of those surveyed that aluminum is either unhealthful or not good for you.

Smithkline and Jordan called Dr. Yoram Wind, a professor of marketing, as their expert witness to testify on the nature of the messages communicated by Ingredients-Revised. Dr. Wind strongly criticized the Bruno and Ridgway study primarily on two grounds. First, he testified that, in his opinion, the survey was almost wholly comprised of leading questions. Second, he faulted the study for not taking into account the fact that respondents may have brought with them previously acquired information regarding calcium and aluminum, and for failing to adjust the survey accordingly.

In his opinion, the study should have contained a control group — people who were asked similar questions that sought to elicit their beliefs regarding the safety of antacid ingredients, but who were not shown the challenged commercials beforehand.

The evidentiary value of a survey's results rests upon the underlying objectivity of the survey itself. See Universal City Studios, Inc. v. Nintendo Co., Ltd., 746 F.2d 112, 118 (2d Cir. 1984). This objectivity, in turn, "depends upon many factors, such as whether [the survey] is properly 'filtered' to screen out those who got no message from the advertisement, whether the questions are directed to the real issues, and whether the questions are leading or suggestive." American Home Products Corp. v. Johnson & Johnson, 654 F. Supp. 568, 590 (S.D.N.Y. 1987); see also Weight Watchers Intl., Inc. v. Stouffer Corp., 744 F. Supp. 1259, 1272 (S.D.N.Y. 1990) (listing 7 criteria for survey trustworthiness).

Judge Cedarbaum's analysis of the Bruno and Ridgway survey was wholly in keeping with these principles. After hearing testimony from the parties' experts, and reviewing the results of the study itself, she specifically found that the study "did not show that 'Ingredients-Revised' communicates that the aluminum . . . in Mylanta is harmful or unsafe." In her view, the responses to questions 8a and 8b, which elicited only nine anti-aluminum reactions from the 300 people surveyed, were "the most persuasive evidence of the message communicated by 'Ingredients-Revised.'" She attributed this to the fact that questions 8a and 8b were "open-ended," and, therefore, more objective. . . .

The probative value of any given survey is a fact specific question that is uniquely contextual. While certain types of survey questions may be appropriate to discern the message of one advertisement, they may be completely inapposite with regard to another. Thus, "it [is] in the district court's province as trier of fact to weigh the evidence, and in particular the opinion research." American Home Products, 577 F.2d at 167. After reviewing the record in this case, we conclude that Judge Cedarbaum's evaluation of the survey questions is not clearly erroneous.

Conclusion

Based upon the literal message of the challenged commercials, and on the responses obtained from the consumer survey, the district court found that J&J*Merck failed to establish that commercials were either false or misleading. Upon review, we conclude that the district court's findings were not erroneous. Therefore, we affirm the district court's denial of injunctive relief, and its dismissal of J&J*Merck's complaint.

Affirmed.

COMMENTS AND QUESTIONS

1. What did Smithkline Beecham do wrong in its initial advertisement? Did it say anything that was untrue? Did it say that J&J's product was bad for consumers? The creation of often artificial brand distinctions is an integral part of product advertising. Is there anything wrong with creating such distinctions? Further, how should we test whether the "public" is deceived by a true statement? Should it be enough that 10 percent of the public is misled? 50 percent? Should the percentage required vary with the nature or severity of the deception? The Second Circuit has taken the

position that "[w]hether or not the statements made in the advertisements are literally true, section 43(a) of the Lanham Act encompasses more than blatant falsehoods. It embraces innuendo, indirect intimations, and ambiguous suggestions evidenced by the consuming public's misapprehension of the hard facts underlying an advertisement." Vidal Sassoon, Inc. v. Bristol-Myers Co., 661 F.2d 272 (2d Cir. 1981) (comparison ad that misled consumers as to the extent and methodology of the survey conducted was actionable under section 43(a)).

In some cases, unlike *SmithKline,* the advertisement is literally false on its face. In such cases, the courts tend to presume that consumers will be deceived by the statements, obviating the need for the plaintiff to prove that consumers are actually confused. See, e.g., Southland Sod Farms v. Stover Seed Co., 108 F.3d 1134 (9th Cir. 1997); Cashmere & Camel Hair Mfg. Inst. v. Saks Fifth Ave, 284 F.3d 402 (1st Cir. 2002) (misrepresentation as to the percentage of cashmere in a sweater was literally false, and so was presumed to have deceived consumers). The Third Circuit has gone so far as to hold that a completely unsubstantiated advertising claim is necessarily literally false, triggering the presumption. Novartis Consumer Health Inc. v. Johnson & Johnson-Merck Consumer Pharmaceuticals Co., 290 F.3d 578 (3d Cir. 2002).

2. Suppose that Smithkline Beecham had never mentioned its own product or compared the two, but had merely questioned the wisdom of consuming aluminum. Would J&J have a cause of action under section 43(a)? Until recently, the answer was no. Before the Trademark Law Reform Act of 1988, section 43(a) prohibited only false or misleading statements about the advertiser's own product. Disparagement of the plaintiff's product, while likely actionable under state law tort theories, did not violate the Lanham Act. See Bernard Food Indus. v. Dietene Co., 415 F.2d 1279 (7th Cir. 1969). The 1988 Amendments to section 43 clearly reverse this rule, at least as to statements made in "commercial advertising or promotion." Should federal law provide a cause of action for disparagement of goods or competitors? Johnson & Johnson also sued under state law, specifically §§ 349 and 350 of the New York General Business Law. These provisions cover, respectively, unfair trade practices in general and false advertising, and they are representative of state law false advertising statutes.

3. While advertisements comparing products or attacking one product often come from a competitor in the market, this is not necessarily the case. Consumers' rights groups like Consumer's Union (publisher of the popular magazine *Consumer Reports*) regularly compare competing products against each other. Government agencies do the same on occasion, and irate consumers have been known to vent their frustration by advertising the defects of a product they are particularly unhappy with. Do these comparisons or disparagements—which presumably are made by neutral third parties rather than competitors—fall within the scope of section 43(a)? The legislative history suggests that section 43(a) was not intended to reach "consumer or editorial comment" by a group like Consumer Reports. Cong. Rec. H10420-21 (daily ed. Oct. 19, 1988) (statement of Rep. Kastenmeier). The dilution provisions of section 43(c) expressly exempt news reporting and comparative advertising from the reach of that section.

4. What is "advertising"? Certainly television and print advertisements fall within the scope of the false advertising prohibition. But what about flyers handed out on the street? Sales presentations? Letters to a customer? The general rule is that the Lanham Act does not apply to isolated statements or correspondence to individuals. But in Seven-Up Co. v. Coca-Cola Co., 86 F.3d 1379 (5th Cir. 1996), the court concluded that a series of sales presentations made to bottlers qualified as advertising for purposes

of the Lanham Act. But see Sports Unlimited v. Lankford Ents., 275 F.3d 996 (10th Cir. 2002) (distribution of defamatory information to between two and seven people did not constitute "commercial advertising"); First Health Group Corp. v. BCE Emergis Corp., 269 F.3d 800 (7th Cir. 2001) (person-to-person pitches aren't advertising).

5. The false advertising prohibition in section 43(a) is the most obviously consumer-oriented of the provisions in the Lanham Act. Protecting consumers from false information about products obviously furthers one of the key goals of trademark law — to ensure that consumers have accurate information on which to make purchasing decisions. Given this, it is somewhat surprising that consumers do not have standing to sue for false advertising. Barrus v. Sylvania, 55 F.3d 468 (9th Cir. 1995) (no consumer standing under §43(a)); Serbin v. Ziebart Int'l Corp., 11 F.3d 1163 (3d Cir. 1993) (only parties with a "reasonable commercial interest" may bring suit under section 43(a)). Is this requirement inconsistent with the consumer protection rationale for section 43(a)? If not, is there another explanation? On the role of competitors as consumer champions, see B. Sanfield, Inc. v. Finlay Fine Jewelry, Inc., 857 F. Supp. 1241 (N.D. Ill. 1994) (finding that plaintiff, a competitor jeweler, stated viable cause of action under §43(a) and state unfair trade practices statute for defendant's misleading advertising concerning price discounting, despite fact that plaintiff's store, products, and prices are not mentioned in plaintiff's ads).

6. Is private litigation by competitors necessary to ensure accuracy in advertising? See Lillian R. BeVier, Competitor Suits for False Advertising Under Section 43(a) of the Lanham Act: A Puzzle in the Law of Deception, 78 Va. L. Rev. 1, 2-3 (1992) (describing most false advertising claims as "systematically trivial," and arguing that "competitors' incentives to sue are not correlated with the likelihood of consumer harm."). BeVier argues that for the most part the market will discipline producers whose ads are misleading.

Consider two other possible sources of advertising regulation: FTC complaints and "private advertising codes." See Arthur Best, Controlling False Advertising: A Comparative Study of Public Regulation, Industry Self-Policing, and Private Litigation, 20 Ga. L. Rev. 1 (1985) (comparing benefits of three different sources of ad regulation, and arguing for expansion of private litigation in various cases).

7. In Ross D. Petty, Supplanting Government Regulation with Competitor Lawsuits: The Case of Controlling False Advertising, 25 Ind. L. Rev. 351 (1991), the author describes industry-level self-regulation:

> Many industry trade associations have advertising codes, as do media and media associations. The National Advertising Division (NAD) of the Council of Better Business Bureaus has actively investigated advertising complaints since 1971. It is funded by dues paid to the Council of Better Business Bureaus by advertisers and advertising agencies. During 1983-85, forty-three percent of these complaints were made by competitors. If the NAD cannot resolve the complaint, the case can be appealed to the National Advertising Review Board. The Board is funded in the same manner as NAD, but has decided only forty-one cases that have been appealed to it out of the more than 2,000 investigated by NAD since 1971. In sixty-six percent of the cases, the Board upheld the NAD decision. In twenty percent of the cases, the Board reversed or modified the NAD decision, and fifteen percent of the cases were dismissed or withdrawn. Thus far, advertisers have always complied with a negative Board decision. NAD standards for reviewing advertising seem comparable to the FTC's standards. For example, in 1984, the NAD took formal action on 105 complaints. Eighty percent of these complaints questioned the adequacy of substantiation and eighty-three percent challenged misleading statements or depictions. In 1984, fifteen

of these cases involved explicit comparisons with rival offerings and nearly forty involved implicit comparisons (e.g., "the only lawn fertilizer there is"). Lastly, from 1973 through part of 1982, thirty percent of NAD cases dealt with companies in *Advertising Age's* 100 Leading National Advertisers. Thus, the NAD appears to be more willing than the FTC to deal with national advertisers and comparative advertising. The cost of complaining to NAD is comparable to complaining to the FTC and is lower than that of bringing a Lanham Act case. Like the Lanham Act courts, NAD acts quickly. It frequently resolves complaints within six months of receipt. NAD examines about one hundred complaints annually. Despite its lack of authority to issue binding orders, it obtains discontinuance or modification in about seventy-five percent of its cases with the remainder vindicating the challenged advertisement.

On the related topic of the post-1989 cause of action under § 43(a) for trade libel, and its pre-1989 common law antecedents, see McCarthy, Trademarks and Unfair Competition, § 27.09[1]-[3] (3d ed. 1992 & Supp. 1996). For more on "private" industry norms and practices supplanting formal, statutory law, see Robert P. Merges, Contracting into Liability Rules: Institutions for Facilitating Transactions in Intellectual Property Rights, 84 Cal. L. Rev. 1293 (1996); Robert Ellickson, Order Without Law (1991).

8. For the most part, plaintiffs in § 43(a) advertising cases seek injunctive relief. In some cases, however, they seek—and obtain—damages. See, e.g., U-Haul Int'l v. Jartran, Inc., 793 F.2d 1034 (9th Cir. 1986) (sustaining $40 million award of defendant's profits calculated with reference to defendant's advertising costs). Indeed, some courts will award some presumptive damages when claims are proven literally false. See Southland Sod Farms v. Stover Seed Co., 108 F.3d 1134, 1146 (9th Cir. 1997). Other courts require proof that the defendant has benefited from the false advertisements, however. See Logan v. Burgers Ozark Country Cured Hams, 263 F.3d 447 (5th Cir. 2001). At least one commentator has argued that damages ought to be more liberally awarded. See Arthur Best, Monetary Damages for False Advertising, 49 U. Pitt. L. Rev. 1 (1987) (advocating a presumption that ads cause economic injury, and that courts use a defendant's expenditures on the challenged ads as an approximate measure of the injury they caused plaintiff). Do we need enhanced damages remedies in these cases? Or does the nature of competition provide sufficient incentive for competitors to challenge false advertisements?

9. Suing a defendant for making false statements in the press raises obvious First Amendment concerns. Most of the cases involving independent testing agencies, as well as a number of the direct competitor disparagement cases, are actually litigated as First Amendment cases, in which the plaintiff must prove falsity plus actual malice if she is a public figure. See, e.g., *Bose, supra*, 466 U.S. at 485. We will discuss the First Amendment limits on a false advertising claim in more detail in section E.4 below, when we discuss the parody defense. For a more extensive treatment, see C. Edwin Baker, Advertising and the Democratic Process, 140 U. Pa. L. Rev. 2097 (1992).

E. DEFENSES

1. Genericness

For a term to serve the purpose of a trademark, it must point to a unique source. When a term refers instead to a general class of products, it is deemed "generic" and cannot serve as a trademark. "Kodak" is a source of products, a species; "film" is a class of products — a genus.

Generic terms are either "born generic," i.e., refused registration on the Principal Register because they are generic *ab initio,* or they become generic over time through a process called "genericide." For an example of the former, see Bellsouth Corp. v. Datanational Corp., 60 F.3d 1565 (Fed. Cir. 1995) ("Walking Fingers" logo for telephone yellow pages was generic). The case that follows discusses the latter issue.

The Murphy Door Bed Co., Inc. v. Interior Sleep Systems, Inc.

United States Court of Appeals for the Second Circuit
874 F.2d 95 (2d Cir. 1989)

MINER, Circuit Judge:

Defendants-appellants appeal from a judgment entered in the United States District Court for the Eastern District of New York [for]...plaintiff.... The judgment includes a permanent injunction enjoining all defendants from use of the Murphy name....

We hold that "Murphy bed" is a generic term, having been appropriated by the public to designate generally a type of bed. Consequently, defendants could not have infringed on plaintiff's trademark, alleged to be Murphy bed, and the district court erred in finding trademark infringement....

Background

At the turn of this century, William Lawrence Murphy invented and manufactured a bed that when not in use could be concealed in a wall closet. By using a counterbalancing mechanism, the bed could be lowered from or raised to a closet in a wall to which the bed is hinged. In 1918, the United States Patent Office granted Mr. Murphy a patent for a "pivot bed," which was substantially similar to the wall bed. Mr. Murphy incorporated in New York in 1925 as the Murphy Door Bed Company and began to sell the wall bed under the name of "Murphy bed." Since its inception, the Murphy Co. has used the words Murphy and Murphy bed as its trademark for concealed beds. Other manufacturers of wall beds generally describe their products as "wall beds," "concealed beds," "disappearing beds," "authentic adjustable hydraulic beds" and the like, but rarely as Murphy beds. In fact, at least twice, when independent companies marketed their products as Murphy beds, Murphy complained to them and, as a result, the companies refrained from further deliberate use of the term Murphy bed.

On March 23, 1981, and again on November 16, 1982, the Patent and Trademark Office ("PTO") denied the Murphy Co.'s application to register the Murphy bed trademark. The PTO examining attorney explained that the words "Murphy bed" had become generic and that the phrase Murphy bed was "merely descriptive of a characteristic of the goods." In August 1984, the Trademark Trial and Appeal Board ("TTAB") affirmed the denial of registration. See In re Murphy Door Bed Co., Inc., 223 U.S.P.Q. 1030 (TTAB 1984). The TTAB noted that "Murphy bed has for a long period of time been used by a substantial segment of the public as a generic term for a bed which folds into a wall or a closet." Id. at 1033.

In December 1981, [defendants] entered into a distributorship agreement with the Murphy Co. and became the exclusive distributor of the Murphy bed in . . . four Florida counties. . . . The agreement, in the form of a letter signed by both Murphy and Zarcone [on behalf of Interior Sleep Systems, Inc.], provided in part that:

5) — Whenever the Murphy name is used, it must be in capital letters and identified by the word trademark or TM adjacent to Murphy. Cabinets or other material not furnished by Murphy will not be promoted or sold as Murphy products. . . .

8) — Upon termination of this agreement Interior Sleep Systems, Inc. agrees to discontinue the use of the name "Murphy bed".

After learning of the TTAB's 1984 decision denying Murphy's application for trademark registration, [defendants formed a corporation with the term "Murphy Beds" in the title, and then filled certain orders with both beds of its own making and beds from the Murphy Company]. . . .

[The district] court ruled that the name Murphy was not generic because a "secondary meaning" had been attributed to it by the general public, and that the common law of trademark therefore protected the Murphy Co. . . . Thus, the court held that the burden was on the defendants to show abandonment of the Murphy trademark — i.e. that the trademark had lost its significance. . . . The court then found that defendants did not sustain their burden. . . .

Discussion . . .

A term or phrase is generic when it is commonly used to depict a genus or type of product, rather than a particular product. See Abercrombie & Fitch Co. v. Hunting World, Inc., 537 F.2d 4, 9 (2d Cir. 1976). When a term is generic, "trademark protection will be denied save for those markets where the term still has not become generic and a secondary meaning has been shown to continue." Id. at 10. We have held that "the burden is on plaintiff to prove that its mark is a valid trademark . . . [and] that its unregistered mark is not generic." Reese Publishing Co. v. Hampton Intl. Communications, Inc., 620 F.2d 7, 11 (2d Cir. 1980).

As the Murphy mark is unregistered, *Reese* suggests that the district court erred in shifting the burden of proof to the defendants. However, the words at issue in *Reese*, "Video Buyer's Guide," were of common use before the product developer applied them to his product, whereas here, the term Murphy bed was created for its purpose by the manufacturer and only thereafter was it adopted by the public as a matter of common use. See Gimix v. JS & A Group, Inc., 699 F.2d 901, 905 (7th Cir. 1983) (differentiating between term in common usage before application to product and coinage of term to suit product that is later expropriated). It was this genericness of an "invented" term that Learned Hand addressed when determining whether "aspirin," a coined word, had been so adopted by the lay public as to become generic. See Bayer Co. v. United Drug Co., 272 F. 505, 509 (S.D.N.Y. 1921); see also King-Seeley Thermos Co. v. Aladdin Indus., Inc., 321 F.2d 577, 579 (2d Cir. 1963) (wide-spread use of the word "thermos," despite having been invented by plaintiff for description of vacuum bottle, created genericness); DuPont Cellophane Co. v. Waxed Products Co., 85 F.2d 75, 81 (2d Cir.) (expropriation by public of word "cellophane" created genericness), *cert. denied*, 299 U.S. 601, 57 S. Ct. 194, 81 L. Ed. 443 (1936). We find this distinction important and hold that where the public is said to have expropriated a term established by a product developer, the burden is on the defendant to prove genericness. Thus, critical to a trial court's allocation of proof burdens is a determination of whether the term at issue is claimed to be generic by reason of common usage or by reason of expropriation. This presumption of nongenericness of a product name in the case of apparent public expropriation is justified by the commercial protection a developer of innovations deserves.

The Murphy Co. was the first to employ the word Murphy to describe a bed that could be folded into a wall closet. It is claimed that over time the public adopted, or, rather, expropriated, the term Murphy bed as a synonym for any bed that folds into a closet. Accordingly, the district court was correct in placing the burden of proof of genericness upon the defendants. We find, however, that [defendant] did indeed establish the genericness of the term Murphy bed.

The following factors combined lead us to conclude that [defendant] showed at trial that today the term Murphy bed, in the eyes of "a substantial majority of the public," King-Seeley, 321 F.2d at 579, refers to a species of bed that can fold into a wall enclosure. First, the decision of the PTO, and certainly the TTAB, is to be accorded great weight.... The district court explicitly rejected the decisions of the PTO and TTAB finding genericness, despite acknowledging their persuasive force. Second, the term Murphy bed is included in many dictionaries as a standard description of a wall-bed. See, e.g., Webster's Third New International Dictionary 1489 (1981). While dictionary definitions are not conclusive proof of a mark's generic nature, they are influential because they reflect the general public's perception of a mark's meaning and implication.... Third, [defendant] introduced as evidence numerous examples of newspaper and magazine use of the phrase Murphy bed to describe generally a type of bed. Again, such evidence is not proof positive, but it is a strong indication of the general public's perception that Murphy bed connotes something other than a bed manufactured by the Murphy Co....

In finding a lack of genericness, the district court was influenced by Murphy's efforts at policing its mark. The court noted with approval instances where Murphy complained to those who had used the term Murphy bed to describe beds not necessarily produced by Murphy.... However, when, as here, the mark has "entered the public domain beyond recall," policing is of no consequence to a resolution of whether a mark is generic. King-Seeley, 321 F.2d at 579....

We conclude that the defendants adequately demonstrated that the Murphy bed mark is generic....

COMMENTS AND QUESTIONS

1. *Preventing and Policing Genericide.* Companies often fight vigorously to prevent their trademarks from becoming generic through casual usage. Consider the strong message sent by Xerox Corporation in the accompanying advertisement (Figure 5-3).

Apart from general advertising such as this, firms also police the uses of their marks via lawsuits. See, e.g., Selchow & Righter Co. v. McGraw-Hill Book Co., 580 F.2d 25 (2d Cir. 1978) (granting Scrabble trademark holder preliminary injunction against publisher of "The Scrabble Dictionary," on grounds that publication would cause irreparable injury by possibly rendering trademark generic); Elliot Staffin, The

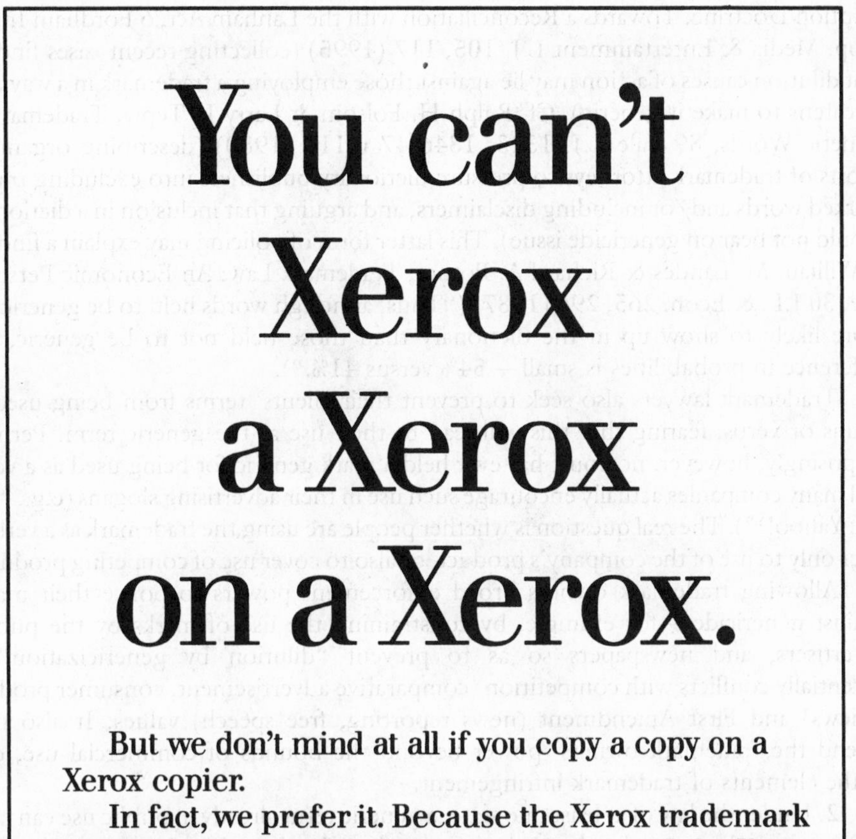
You can't Xerox a Xerox on a Xerox.
But we don't mind at all if you copy a copy on a Xerox copier.
In fact, we prefer it. Because the Xerox trademark should only identify products made by us. Like Xerox copiers, printers, fax machines, software and multi-function machines.
As a trademark, the term Xerox should always be used as an adjective, followed by a noun. And it is never used as a verb. Of course, helping us protect our trademark also helps you.
Because you'll continue to get what you're actually asking for.
And not an inferior copy.
THE DOCUMENT COMPANY
XEROX
THE DOCUMENT COMPANY XEROX Worldwide Sponsor
XEROX® The Document Company® and the stylized X are trademarks of XEROX CORPORATION. 36 USC 380.

FIGURE 5-3
Xerox advertisement.

Dilution Doctrine: Towards a Reconciliation with the Lanham Act, 6 Fordham Intell. Prop. Media & Entertainment L.J. 105, 117 (1995) (collecting recent cases finding that dilution causes of action may lie against those employing a trademark in a way that threatens to make it generic). Cf. Ralph H. Folsom & Larry L. Teply, Trademarked Generic Words, 89 Yale L.J. 1323, 1346-47 n.110 (1980) (describing organized efforts of trademark attorneys to pressure dictionary publishers into excluding trade-marked words and/or including disclaimers, and arguing that inclusion in a dictionary should not bear on genericide issue). This latter form of policing may explain a finding in William M. Landes & Richard A. Posner, Trademark Law: An Economic Perspective, 30 J.L. & Econ. 265, 296 (1987) ("Thus, although words held to be generic are more likely to show up in the dictionary than those held not to be generic, the difference in probabilities is small — 54% versus 41%.").

Trademark lawyers also seek to prevent their clients' terms from being used as nouns or verbs, fearing that this will lead to their use as the generic term. Perhaps surprisingly, however, no court has ever held a mark generic for being used as a verb, and many companies actually encourage such use in their advertising slogans (e.g., "Do you Yahoo!?"). The real question is whether people are using the trademark as a verb to refer only to use of the company's product, or also to cover use of competing products.

Allowing trademark owners broad enforcement powers to police their marks against genericide — for example, by constraining the use of marks by the public, advertisers, and newspapers so as to prevent "dilution by genericization" — potentially conflicts with competition (comparative advertisement, consumer product reviews) and First Amendment (news reporting, free speech) values. It also may extend the trademark owners' power beyond the bounds of commercial use, one of the elements of trademark infringement.

2. Under the law of at least one state, a generic term already in public use can later acquire secondary meaning by virtue of a product developer's unique use, thus warranting trademark protection. See, e.g., AntiDefamation League of B'Nai B'rith v. Arab Anti-Defamation League, 72 Misc. 2d 847, 855, 340 N.Y.S.2d 532, 534-44 (N.Y. Sup. Ct. 1972). Thus genericide may be reversible. Does this rule conflict with the federal trademark scheme?

By contrast, other courts have asked only whether a mark once became generic, reasoning that if it was generic that should deny protection even if it no longer is generic today. See Microsoft Corp. v. Lindows.com, Inc., 69 U.S.P.Q.2d 1863 (W.D. Wash. 2004). If consumers now associate a term such as "Windows" with a particular product, shouldn't the law validate that association?

3. In an omitted portion of the opinion, the court held that defendants breached their contractual obligation to refrain from using the term Murphy bed after termination of their distribution agreement with the Murphy Co., and that defendants engaged in unfair competition by passing off beds of their own manufacture as beds of the Murphy Co. Does it make sense to restrict the rights of the defendants — who had signed an agreement — to a greater degree than other competitors? Does this holding suggest that parties can "contract around" trademark genericness? What would be the effect of such agreements?

4. In Kellogg Co. v. National Biscuit Co., 305 U.S. 111, 116-18 (1938), Justice Brandeis set out a classic discussion of "genericide," though one infused also with elements of functionality and descriptiveness/secondary meaning:

> The plaintiff has no exclusive right to the use of the term "Shredded Wheat" as a
> trade name. For that is the generic term of the article, which describes it with a fair degree

of accuracy; and is the term by which the biscuit in pillow-shaped form is generally known by the public. Since the term is generic, the original maker of the product acquired no exclusive right to use it. As Kellogg Company had the right to make the article, it had, also, the right to use the term by which the public knows it.... Ever since 1894 the article has been known to the public as shredded wheat. For many years, there was no attempt to use the term "Shredded Wheat" as a trade-mark....

Moreover, the name "Shredded Wheat," as well as the product, the process and the machinery employed in making it, has been dedicated to the public. The basic patent for the product and for the process of making it, and many other patents for special machinery to be used in making the article, issued to Perky. In those patents the term "shredded" is repeatedly used as descriptive of the product. The basic patent expired October 15, 1912; the others soon after. Since during the life of the patents "Shredded Wheat" was the general designation of the patented product, there passed to the public upon the expiration of the patent, not only the right to make the article as it was made during the patent period, but also the right to apply thereto the name by which it had become known. As was said in Singer Mfg. Co. v. June Mfg. Co., 163 U.S. 169, 185:

> It equally follows from the cessation of the monopoly and the falling of the patented device into the domain of things public, that along with the public ownership of the device there must also necessarily pass to the public the generic designation of the thing which has arisen during the monopoly.... To say otherwise would be to hold that, although the public had acquired the device covered by the patent, yet the owner of the patent or the manufacturer of the patented thing had retained the designated name which was essentially necessary to vest the public with the full enjoyment of that which had become theirs by the disappearance of the monopoly.

It is contended that the plaintiff has the exclusive right to the name "Shredded Wheat," because those words acquired the "secondary meaning" of shredded wheat made at Niagara Falls by the plaintiff's predecessor. There is no basis here for applying the doctrine of secondary meaning. The evidence shows only that due to the long period in which the plaintiff or its predecessor was the only manufacturer of the product, many people have come to associate the product, and as a consequence the name by which the product is generally known, with the plaintiff's factory at Niagara Falls. But to establish a trade name in the term "shredded wheat" the plaintiff must show more than a subordinate meaning which applies to it. It must show that the primary significance of the term in the minds of the consuming public is not the product but the producer. This it has not done. The showing which it has made does not entitle it to the exclusive use of the term shredded wheat but merely entitles it to require that the defendant use reasonable care to inform the public of the source of its product.

The Court cites Singer Mfg. Co. v. June Mfg. Co., 163 U.S. 169, 185 (1896) for the proposition that once a patent on a device expires, the name by which that device has been sold also enters the public domain. Surely that is not always the case. Numerous products that once were patented are still sold under the same trademark. And this result — that patent expiration does not automatically end trademark protection — is also consistent with the purposes behind the two laws, which are very different. The key to *Singer*'s holding may lie in its expressed concern that "the owner of the patent... had retained the designated name which was essentially necessary to vest the public with the full enjoyment of" the product. Thus it is only when the name itself is "essentially necessary" to sales of the product — that is, when the name is generic — that trademark protection should not survive the expiration of a patent.

5. While most genericness cases involve word marks, trade dress and product configurations can also be generic. See, e.g., Kendall-Jackson Winery v. E. & J. Gallo Winery, 150 F.3d 1042 (9th Cir. 1998), where the court held that an autumnal grape

leaf featured on both plaintiff's and defendant's wine bottles was generic in the wine industry. See also Sunrise Jewelry Mfg. Corp. v. Fred, S.A. 175 F.3d 1322 (Fed. Cir. 1999) (holding that generic product configuration may be cancelled notwithstanding "incontestable" status; remanding for determination of genericness).

6. Lanham Act § 14 (15 U.S.C § 1064) offers some guidance on how to determine whether a mark has become generic. It provides that "[t]he primary significance of the registered mark to the relevant public rather than purchaser motivation shall be the test for determining whether the registered mark has become generic." This language was added to the statute by the Trademark Clarification Act of 1984, which was passed in order to reverse the holding of Anti-Monopoly, Inc. v. General Mills Fun Group, 684 F.2d 1316 (9th Cir. 1982), *cert. denied,* 459 U.S. 1227 (1983). That case held that a game called "Anti-Monopoly," which used rules and a game board very similar to Parker Brothers' "Monopoly" game, could be sold because "Monopoly" had become generic. The Ninth Circuit reasoned that most consumers did not buy the "Monopoly" game because it was made by Parker Brothers and because they liked the products made by this company. Yet only this motivation for buying, the court said, supports trademark rights in the game's name. Since consumers associated the term "Monopoly" with a single product (the game), *and not a single producer* (at least not one they could name), the court found the mark generic. The court apparently would have limited trademark rights only to those trademarks that point unmistakenly to a single, *known* source. This reasoning was quickly repudiated in the amendments to § 14 (15 U.S.C. § 1064) just described. Can you see what is wrong with the rationale in *Anti-Monopoly?*

Section 14 provides that a mark may become generic as to some, but not all, goods and that in that case it may be cancelled in part. Cancellation of a mark because it has become generic donates that mark to the public domain, enabling other competitors to use it for the same or similar products (such as Anti-Monopoly).

7. When a court declares a trademark generic, it destroys a right built up with considerable investment. Is this the same as instances where government decrees destroy private property; that is, should "just compensation" be paid for this "taking"? Does your answer depend on whether trademarks are viewed as property or instead as an unfair competition or consumer protection right? See Stephen L. Carter, Does It Matter Whether Intellectual Property Is Property?, 68 Chicago-Kent L. Rev. 715 (1993) (speculating about the desirability of such an arrangement); compare Ruckelshaus v. Monsanto Co., 467 U.S. 986 (1984) (government required to compensate trade secret owner when it compels disclosure of the secret).

The Second Circuit has periodically suggested that even generic marks are entitled to protection against some forms of unfair competition. See, e.g., Genesee Brewing Co. v. Stroh Brewing Co., 124 F.3d 137, 150 (2d Cir. 1997) (plaintiff's mark for "Honey Brown Ale" was generic, but defendant could still be liable if it did not use "every reasonable means to prevent confusion" as to the source of the products); Forschner Group Inc. v. Arrow Trading Co., 124 F.3d 402 (2d Cir. 1997); Home Builders Ass'n of Greater St. Louis v. L & L Exhibition Mgmt., Inc., 226 F.3d 944 (8th Cir. 2000) (holding that a generic mark that has acquired secondary meaning may require that other users take steps to avoid confusion). This seems contrary to the very purpose of the genericness doctrine.

8. Some commentators have noticed that trademarks have come to serve a more expressive role in addition to their original function of identifying a source for a commercial product. People make use of trademarks and taglines in conversation

without any attempt to identify a source, or even specifically connect the source to a commercial product. Such expressions have salience precisely because the owners of the mark have blanketed our culture with them. In this sense the communicative content of trademarks such as Barbie and Marlboro—the unique images and associations these terms call to mind—are an unintended offshoot of the advertiser's commercially-motivated saturation campaigns. Even so, these associations are very real.

This is not news to many owners of well-recognized marks. Many of them have instigated aggressive licensing campaigns to take advantage of their unique image in markets far beyond those they originally served. Thus logos such as "NBA" and "NFL," though they may have originally identified unique "brands" (high-quality sports leagues), now appear almost everywhere. Surely no one looks for the NBA logo on toy balls or teddy bears to assure herself of getting a well-made ball or toy. Rather, the buyer is seeking a piece of the NBA image.

Rochelle Dreyfuss questions whether trademark doctrine has kept up with these developments in popular culture. Rochelle C. Dreyfuss, Expressive Genericity: Trademarks as Language in the Pepsi Generation, 65 Notre Dame L. Rev. 397 (1990). She suggests a number of ways trademark law should be adjusted. Her primary suggestion is to extend traditional concepts of genericness to encompass the realm of expressive (as opposed to commercial, or what she calls "competitive") uses of trademarks. "[C]ourts," she writes,

> would first decide whether there is an expressive component to the challenged use and then consider how central the trademark is to the usage. If the mark is found to be rhetorically unique within its context, it would be considered expressively—but not necessarily competitively—generic, and the trademark owner would not be permitted to suppress its utilization in that context.... [T]he suggested analysis would permit courts to drop the fiction of consumer confusion, and instead focus on the primary concern, which is the expressive significance of the mark.

Dreyfuss, at 418-19.

9. *Generic.com.* The function of domain names as both unique cyberspace addresses and sources of goods and services has reversed the traditional dynamics of genericide. Many generic terms (e.g., escalator) begin their existence as product names of a particular manufacturer, but over time become identified with a particular good, thereby losing protection. By contrast, domain names (e.g., pets.com) begin as descriptive terms with a generic suffix (.com) but then build source identification and goodwill over time because of the uniqueness of the cyberspace address. Should courts protect this goodwill?

Note on Genericide, Language, and Policing Costs

It might seem that the notion of genericide does not make much sense. Why should a mark pass into the public domain because it has come to signify a category of products rather than a single product/source combination? If a trademark owner has "invented" a trademark, there is a clear sense in which the genericness doctrine takes away some rights that the creator was endowed with because of her creativity. Should trademark protection be permanent, or should we attempt a rough "calibration," along the lines of patent and copyright, to adjust the incentive to call forth the optimum (or at least desirable) level of this activity? Cf. William M. Landes & Richard A. Posner, Trademark Law: An Economic Perspective, 30 J.L. & Econ. 265, 270 (1987) (trademark protection offers a modest incentive to create new commercial terms).

Does it make sense that the very success of a trademark should be its undoing? In some ways, this is analogous to the argument that copyright protection on a software user interface should "lapse" when the interface becomes an industry standard. Cf. Lotus Devel. Corp. v. Borland Int'l, Inc., 49 F.3d 807 (1st Cir. 1995), *aff'd by equally divided court,* 516 U.S. 233 (1996). The unifying theme is the notion that the success of certain kinds of creative works is largely a result of the need for a single standard, rather than the creative brilliance of the work itself. Economists speak of these works as creating "network externalities": users benefit from the fact that others use the work. Widely used computer protocols, such as the TCP/IP standard allowing exchange of data over the Internet, are advantageous because they make a wide variety of content available to all users. Thus each new person who adopts a protocol benefits all current users of the protocol. See *infra,* Chapter 7, discussing these markets in detail.

Some courts and legal scholars point out that under such a dynamic, the success of works of this kind is attributable more to the *collective labor* of the users than to the labor of the work's creator. See, e.g., Lotus Devel. Corp. v. Borland Int'l, Inc., 49 F.3d 807, 821 (1st Cir. 1995) (Boudin, J., concurring) (accused infringer's product was "merely trying to give former Lotus users an option to exploit their own prior investment in learning.... [I]t is hard to see why customers who have learned the Lotus menu and devised macros for it should remain captives of Lotus because of an investment in learning made by the users and not by Lotus."). They argue that the efforts of the users in learning a new work, such as the computer protocol in the example above, rather than the efforts of the creator in designing the work, account for the success of these kinds of works. It follows from this argument that the claims of the work's creator must give way to those of the users in some cases. This logic is applied directly to generic trademarks in the following passage:

> Giving ownership in intellectual products that have come to serve as standards, such as West citations or generic trademarks, would not ordinarily leave "enough, and as good" [in the Lockean sense]. There may be room in the world for only one of a given type of thing, or a long-lived artifice may become a mode of communication. It is the nature of a standard that nothing "as good" is available. For these reasons, the [Lockean] proviso would be violated if the courts gave those who create standards in nonfungible goods a right to prevent people from utilizing them.

Wendy J. Gordon, A Property Right in Self-Expression: Equality and Individualism in the Natural Law of Intellectual Property, 102 Yale L.J. 1533, 1600 (1993) (footnote omitted). See also Justin Hughes, The Philosophy of Intellectual Property, 77 Geo. L.J. 287, 315-23 (1988) (suggesting that ownership of most intellectual products will easily satisfy the Lockean proviso, though "extraordinary ideas" like generic trademarks should remain open for all to use). A recent commentary makes explicit the *public's* contribution to strong trademarks. See Steven Wilf, Who Authors Trademarks?, 17 Cardozo Arts & Ent. L.J. 1 (1999).

Economists have observed that human languages exhibit network externalities. Professor Gordon's argument just above in effect extends the network externalities argument to individual words. When a word comes to describe a genus or class of goods, i.e., enters general use, it takes on some of the properties of the computer protocol described earlier. It becomes the standard shorthand descriptor for a complex, multi-attribute product type. Past a certain point, any alternative descriptor becomes clearly second-best. Requiring consumers and competitors to use this second-best descriptor entails costs. See Ralph H. Folsom & Larry L. Teply, Trademarked Generic Words, 89 Yale L.J. 1323, 1340-41 (1980). Though difficult to quantify, these costs may be substantial. See also Landes & Posner, *supra*, at 280, 292 (discussing costs of having to "describe around" a well-known term).

Consider the example of "plexiglas." The next best alternative to this well-known descriptor might be: "unbreakable clear plastic sheets, or window material." That is not only a mouthful; it is more expensive to advertise (because it is longer), harder to remember, and more subject to mistakes and confusion. Thus it is not hard to see why "plexiglas" became the preferred shorthand for it. Consequently, although the originator of this term might have put substantial effort into creating it and encouraging its use, there is a good argument that it has become a standard name (or descriptor)— and hence generic. But cf. Rohm & Haas Co. v. Polycast Technology Corp., 172 U.S.P.Q. 167 (D. Del. 1971) (enjoining defendant's use of Plexiglas mark). As another example, consider "Yo-Yo"; how would you describe this kind of toy without using the word *yo-yo?* Cf. Donald F. Duncan, Inc. v. Royal Tops Mfg. Co., 343 F.2d 655 (7th Cir. 1965) (holding *yo-yo* generic, besides being descriptive in Tagalog).

The Role of Policing Costs. Folsom and Teply argue that policing costs (reflected in ads such as the one by Xerox, above) are always wasted and therefore should not be considered in determining whether a trademark has become generic. See Folsom & Teply, *supra*, at 1354. Do you agree? Can you defend expenditures to keep a mark from becoming generic on the basis that they are attempts to provide alternative "standards" that will thereby retain the source-indicating function of threatened trademarks? If consumers accept the premise of the ads and faithfully refer to the "copier" instead of the "Xerox," are consumers harmed? And do such expenditures show that the cost of informing the public about alternative product descriptors is worth it to the firm, given its investment in its trademark? Cf. E. I du Pont de Nemours & Co. v. Yoshida Intl., Inc., 393 F. Supp. 502, 523-24 (E.D.N.Y. 1975) ("Teflon" not generic, in part because of du Pont's vigorous education and policing campaign). Of course, it may be impossible for a company to prevent genericide, for example, when a trademark achieves rapid and overwhelming acceptance. Then no amount of policing will work. Cf. Du Pont Cellophane Co. v. Waxed Prods. Co., 85 F.2d 75 (2d Cir.), *cert. denied*, 299 U.S. 601 (1936) ("Cellophane" generic despite vigorous policing efforts). Does such rapid acceptance indicate that the next best alternative descriptor is significantly less effective? Cf. Folsom & Teply, *supra*, at 1344 (noting that some alternative terms are better than others; comparing "lip balm" as alternative to "Chap Stick"; with "dextro-amphetamine sulphate" as alternative to "Dexadrine"); Comment, Trademarks and Generic Words: An Effect-on-Competition Test, 51 U. Chi. L. Rev. 868, 884-85 (1984) (advocating use of antitrust-like cross-elasticities of demand analysis to determine degree of substitutability between terms).

Can you see a less benign motive for expenditures to maintain the trademark status of a word that has become a widely used name for a product class? See Folsom & Teply, *supra*, at 1337 (suggesting two: (1) to maintain entry barriers to competition;

and (2) to obtain "free advertising" every time someone uses the trademarked word to refer to a product class). This perspective certainly points away from a *per se* rule that significant policing expenditures alone can preserve the trademark status of a term. And, of course, to the extent that firms attempt to police the noncommercial use of marks, there may be a significant effect on the free flow of information and ideas. The First Amendment, discussed below in the section on parody, comes into play in these situations.

2. Functionality

TrafFix Devices, Inc. v. Marketing Displays, Inc.
Supreme Court of the United States
532 U.S. 23 (2001)

Justice KENNEDY delivered the opinion of the Court.

Temporary road signs with warnings like "Road Work Ahead" or "Left Shoulder Closed" must withstand strong gusts of wind. An inventor named Robert Sarkisian obtained two utility patents for a mechanism built upon two springs (the dual-spring design) to keep these and other outdoor signs upright despite adverse wind conditions. The holder of the now-expired Sarkisian patents, respondent Marketing Displays, Inc. (MDI), established a successful business in the manufacture and sale of sign stands incorporating the patented feature. MDI's stands for road signs were recognizable to buyers and users (it says) because the dual-spring design was visible near the base of the sign.

This litigation followed after the patents expired and a competitor, TrafFix Devices, Inc., sold sign stands with a visible spring mechanism that looked like MDI's. MDI and TrafFix products looked alike because they were. When TrafFix started in business, it sent an MDI product abroad to have it reverse engineered, that is to say copied. Complicating matters, TrafFix marketed its sign stands under a name similar to MDI's. MDI used the name "WindMaster," while TrafFix, its new competitor, used "WindBuster." . . .

I

We are concerned with the trade dress question. The District Court ruled against MDI on its trade dress claim. 971 F.Supp. 262 (E.D.Mich. 1997). After determining that the one element of MDI's trade dress at issue was the dual-spring design, id., at 265, it held that "no reasonable trier of fact could determine that MDI has established secondary meaning" in its alleged trade dress, id., at 269. In other words, consumers did not associate the look of the dual-spring design with MDI. As a second, independent reason to grant summary judgment in favor of TrafFix, the District Court determined the dual-spring design was functional. . . .

The Court of Appeals for the Sixth Circuit reversed the trade dress ruling. 200 F.3d 929 (1999). . . . In its criticism of the District Court's ruling on the trade dress question, the Court of Appeals took note of a split among Courts of Appeals in various other Circuits on the issue whether the existence of an expired utility patent forecloses the possibility of the patentee's claiming trade dress protection in the product's design. 200 F.3d, at 939. Compare Sunbeam Products, Inc. v. West Bend Co., 123 F.3d 246 (C.A.5 1997) (holding that trade dress protection is not foreclosed), Thomas & Betts Corp. v. Panduit Corp., 138 F.3d 277 (C.A.7 1998) (same), and Midwest Industries, Inc. v. Karavan Trailers, Inc., 175 F.3d 1356 (C.A.Fed. 1999) (same), with Vornado Air Circulation Systems, Inc. v. Duracraft Corp., 58 F.3d 1498, 1500 (C.A.10 1995) ("Where a product configuration is a significant inventive component of an invention covered by a utility patent . . . it cannot receive trade dress protection"). To resolve the conflict, we granted certiorari.

II

It is well established that trade dress can be protected under federal law. The design or packaging of a product may acquire a distinctiveness which serves to identify the product with its manufacturer or source; and a design or package which acquires this secondary meaning, assuming other requisites are met, is a trade dress which may not be used in a manner likely to cause confusion as to the origin, sponsorship, or approval of the goods. In these respects protection for trade dress exists to promote competition. . . . Congress confirmed this statutory protection for trade dress by amending the Lanham Act to recognize the concept. Title 15 U.S.C. §1125(a)(3) (1994 ed., Supp. V) provides: "In a civil action for trade dress infringement under this chapter for trade dress not registered on the principal register, the person who asserts trade dress protection has the burden of proving that the matter sought to be protected is not functional." This burden of proof gives force to the well-established rule that trade dress protection may not be claimed for product features that are functional. *Qualitex,* supra, at 164-165; Two Pesos, Inc. v. Taco Cabana, Inc., 505 U.S. 763, 775 (1992). And in *Wal-Mart,* supra, we were careful to caution against misuse or over-extension of trade dress. We noted that "product design almost invariably serves purposes other than source identification." Id., at 213.

Trade dress protection must subsist with the recognition that in many instances there is no prohibition against copying goods and products. In general, unless an intellectual property right such as a patent or copyright protects an item, it will be subject to copying. As the Court has explained, copying is not always discouraged or disfavored by the laws which preserve our competitive economy. Bonito Boats, Inc. v. Thunder Craft Boats, Inc., 489 U.S. 141, 160 (1989). Allowing competitors to copy will have salutary effects in many instances. "Reverse engineering of chemical and mechanical articles in the public domain often leads to significant advances in technology." Ibid.

The principal question in this case is the effect of an expired patent on a claim of trade dress infringement. A prior patent, we conclude, has vital significance in resolving the trade dress claim. A utility patent is strong evidence that the features therein claimed are functional. If trade dress protection is sought for those features the strong evidence of functionality based on the previous patent adds great weight to the statutory presumption that features are deemed functional until proved otherwise by the party seeking trade dress protection. Where the expired patent claimed the features in question, one who seeks to establish trade dress protection must carry the heavy burden of showing that the feature is not functional, for instance by showing that it is merely an ornamental, incidental, or arbitrary aspect of the device.

In the case before us, the central advance claimed in the expired utility patents (the Sarkisian patents) is the dual-spring design; and the dual-spring design is the essential feature of the trade dress MDI now seeks to establish and to protect. The rule we have explained bars the trade dress claim, for MDI did not, and cannot, carry the burden of overcoming the strong evidentiary inference of functionality based on the disclosure of the dual-spring design in the claims of the expired patents.

The dual springs shown in the Sarkisian patents were well apart (at either end of a frame for holding a rectangular sign when one full side is the base) while the dual springs at issue here are close together (in a frame designed to hold a sign by one of its corners). As the District Court recognized, this makes little difference. The point is that the springs are necessary to the operation of the device. The fact that the springs in this very different-looking device fall within the claims of the patents is illustrated by MDI's own position in earlier litigation. In the late 1970's, MDI engaged in a long-running intellectual property battle with a company known as Winn-Proof. Although the precise claims of the Sarkisian patents cover sign stands with springs "spaced apart," U.S. Patent No. 3,646,696, col. 4; U.S. Patent No. 3,662,482, col. 4, the Winn-Proof sign stands (with springs much like the sign stands at issue here) were found to infringe the patents by the United States District Court for the District of Oregon, and the Court of Appeals for the Ninth Circuit affirmed the judgment. Sarkisian v. Winn-Proof Corp., 697 F.2d 1313 (1983). Although the Winn-Proof traffic sign stand (with dual springs close together) did not appear, then, to infringe the literal terms of the patent claims (which called for "spaced apart" springs), the Winn-Proof sign stand was found to infringe the patents under the doctrine of equivalents . . . In light of this past ruling — a ruling procured at MDI's own insistence — it must be concluded the products here at issue would have been covered by the claims of the expired patents.

The rationale for the rule that the disclosure of a feature in the claims of a utility patent constitutes strong evidence of functionality is well illustrated in this case. The dual-spring design serves the important purpose of keeping the sign upright even in heavy wind conditions; and, as confirmed by the statements in the expired patents, it does so in a unique and useful manner. As the specification of one of the patents recites, prior art "devices, in practice, will topple under the force of a strong wind." U.S. Patent No. 3,662,482, col. 1. The dual-spring design allows sign stands to resist toppling in strong winds. Using a dual-spring design rather than a single spring achieves important operational advantages. For example, the specifications of the patents note that the "use of a pair of springs . . . as opposed to the use of a single spring to support the frame structure prevents canting or twisting of the sign around a vertical axis," and that, if not prevented, twisting "may cause damage to the spring structure and may result in tipping of the device." U.S. Patent No. 3,646,696, col. 3. In the course of patent prosecution, it was said that "[t]he use of a pair of spring connections as opposed to a single spring connection . . . forms an important part of this combination" because it "forc[es] the sign frame to tip along the longitudinal axis of the elongated ground-engaging members." The dual-spring design affects the cost of the device as well; it was acknowledged that the device "could use three springs but this would unnecessarily increase the cost of the device." These statements made in the patent applications and in the course of procuring the patents demonstrate the functionality of the design. MDI does not assert that any of these representations are mistaken or inaccurate, and this is further strong evidence of the functionality of the dual-spring design.

III

In finding for MDI on the trade dress issue the Court of Appeals gave insufficient recognition to the importance of the expired utility patents, and their evidentiary significance, in establishing the functionality of the device. The error likely was caused by its misinterpretation of trade dress principles in other respects. As we have noted, even if there has been no previous utility patent the party asserting trade dress has the burden to establish the nonfunctionality of alleged trade dress features. MDI could not meet this burden. Discussing trademarks, we have said "'[i]n general terms, a product feature is functional,' and cannot serve as a trademark, 'if it is essential to the use or purpose of the article or if it affects the cost or quality of the article.'" *Qualitex,* 514 U.S., at 165 (quoting Inwood Laboratories, Inc. v. Ives Laboratories, Inc., 456 U.S. 844, 850, n. 10 (1982)). Expanding upon the meaning of this phrase, we have observed that a functional feature is one the "exclusive use of [which] would put competitors at a significant non-reputation-related disadvantage." 514 U.S. at 165. The Court of Appeals in the instant case seemed to interpret this language to mean that a necessary test for functionality is "whether the particular product configuration is a competitive necessity." 200 F.3d at 940. See also *Vornado,* 58 F.3d, at 1507 ("Functionality, by contrast, has been defined both by our circuit, and more recently by the Supreme Court, in terms of competitive need"). This was incorrect as a comprehensive definition. As explained in *Qualitex,* supra, and *Inwood,* supra, a feature is also functional when it is essential to the use or purpose of the device or when it affects the cost or quality of the device. The *Qualitex* decision did not purport to displace this traditional rule. Instead, it quoted the rule as *Inwood* had set it forth. It is proper to inquire into a "significant non-reputation-related disadvantage" in cases of aesthetic functionality, the question involved in *Qualitex.* Where the design is functional under the *Inwood* formulation there is no need to proceed further to consider if there is a competitive necessity for the feature. In *Qualitex,* by contrast, aesthetic functionality was the central question, there having been no indication that the green-gold color of the laundry press pad had any bearing on the use or purpose of the product or its cost or quality.

The Court has allowed trade dress protection to certain product features that are inherently distinctive. *Two Pesos,* 505 U.S., at 774. In *Two Pesos,* however, the Court at the outset made the explicit analytic assumption that the trade dress features in question (decorations and other features to evoke a Mexican theme in a restaurant) were not functional. Id., at 767, n. 6. The trade dress in those cases did not bar competitors from copying functional product design features. In the instant case, beyond serving the purpose of informing consumers that the sign stands are made by MDI (assuming it does so), the dual-spring design provides a unique and useful mechanism to resist the force of the wind. Functionality having been established, whether MDI's dual-spring design has acquired secondary meaning need not be considered.

There is no need, furthermore, to engage, as did the Court of Appeals, in speculation about other design possibilities, such as using three or four springs which might serve the same purpose. 200 F.3d, at 940. Here, the functionality of the spring design means that competitors need not explore whether other spring juxtapositions might be used. The dual-spring design is not an arbitrary flourish in the configuration of MDI's product; it is the reason the device works. Other designs need not be attempted.

Because the dual-spring design is functional, it is unnecessary for competitors to explore designs to hide the springs, say by using a box or framework to cover them, as suggested by the Court of Appeals. Ibid. The dual-spring design assures the user the device will work. If buyers are assured the product serves its purpose by seeing the operative mechanism that in itself serves an important market need. It would be at cross-purposes to those objectives, and something of a paradox, were we to require the manufacturer to conceal the very item the user seeks.

In a case where a manufacturer seeks to protect arbitrary, incidental, or ornamental aspects of features of a product found in the patent claims, such as arbitrary curves in the legs or an ornamental pattern painted on the springs, a different result might obtain. There the manufacturer could perhaps prove that those aspects do not serve a purpose within the terms of the utility patent. The inquiry into whether such features, asserted to be trade dress, are functional by reason of their inclusion in the claims of an expired utility patent could be aided by going beyond the claims and examining the patent and its prosecution history to see if the feature in question is shown as a useful part of the invention. No such claim is made here, however. MDI in essence seeks protection for the dual-spring design alone. The asserted trade dress consists simply of the dual-spring design, four legs, a base, an upright, and a sign. MDI has pointed to nothing arbitrary about the components of its device or the way they are assembled. The Lanham Act does not exist to reward manufacturers for their innovation in creating a particular device; that is the purpose of the patent law and its period of exclusivity. The Lanham Act, furthermore, does not protect trade dress in a functional design simply because an investment has been made to encourage the public to associate a particular functional feature with a single manufacturer or seller. The Court of Appeals erred in viewing MDI as possessing the right to exclude competitors from using a design identical to MDI's and to require those competitors to adopt a different design simply to avoid copying it. MDI cannot gain the exclusive right to produce sign stands using the dual-spring design by asserting that consumers associate it with the look of the invention itself. Whether a utility patent has expired or there has been no utility patent at all, a product design which has a particular appearance may be functional because it is "essential to the use or purpose of the article" or "affects the cost or quality of the article." *Inwood,* 456 U.S., at 850, n. 10.

TrafFix and some of its *amici* argue that the Patent Clause of the Constitution, Art. I, § 8, cl. 8, of its own force, prohibits the holder of an expired utility patent from claiming trade dress protection. We need not resolve this question. If, despite the rule that functional features may not be the subject of trade dress protection, a case arises in which trade dress becomes the practical equivalent of an expired utility patent, that will be time enough to consider the matter. The judgment of the Court of Appeals is reversed, and the case is remanded for further proceedings consistent with this opinion.

It is so ordered.

COMMENTS AND QUESTIONS

1. *Channeling Principles of Intellectual Property Law.* This case addresses the interplay of the patent and trademark domains. Functionality serves much the same role as Baker v. Selden, *supra,* Chapter 4.C.1, which channels works between the patent and copyright regimes. See generally Kellogg Co. v. National Biscuit Co., 305 U.S. 111, 122 (1938) (suggesting that lapse of patent on shredded wheat cereal,

together with lax policing of the mark and widespread use, obviate trademark status of term "Shredded Wheat"); cf. Trade-Mark Cases, 100 U.S. 82 (1879) (trademark statute cannot be predicated upon the patent and copyright clause of the Constitution). These cases bolster the primacy of the patent system as the means for protecting exclusive rights for functional features of products. Had MDI been able to protect its spring design for road signs as trade dress, it would have the equivalent of perpetual patent through the "back door" of trademark law. The social welfare loss of perpetual protection of functional features could be substantial indeed. In fact, the right to exclude others from making, using, selling, or importing a design during the patent term enables the patentee to build source identification in the marketplace for its distinctive (and functional) design. The rule articulated in *TrafFix* ensures that the patent quid pro quo — disclosure in exchange for protection of *limited* duration — continues to operate. Cf. Theodore H. Davis, Jr., Copying in the Shadow of the Constitution: The Rational Limits of Trade Dress Protection, 80 Minn. L. Rev. 595 (1996). But does this balance potentially come at cost in terms of consumer confusion? Should society's interest in free competition (after the expiration of a patent) trump some confusion as to the source of products?

2. Congress codified functionality as a basis to refuse registration, as well as a ground for opposition and cancellation and a defense to incontestability. Functional marks may appear on the Supplemental Register. See Pub. L. 105-330, Title II, §201(a)(4), Title III, §301, 112 Stat. 3070, codified at 15 U.S.C. §§1052, 1064, 1091, 1115 (1998).

3. *Definition of "Functional."* The Supreme Court states that a product feature is functional "if it is essential to the use or purpose of the article or if it affects the cost or quality of the article." As Landes and Posner suggest, in some sense every worthwhile design feature affects the "cost or quality" of the product. See William Landes and Richard Posner, Trademark Law: An Economic Perspective, 30 J. Law and Econ. 265, 297 (1987). But if this is true, what is left of trade dress law? Should all design attributes that succeed in differentiating products therefore be unprotectable? On the other hand, it is clear that a design choice need not be the only possible one in order to be functional. See Sportvision v. SportsMedia Tech Corp., 2005 WL 1869350 (N.D. Cal. Aug. 4, 2005) (holding that the color yellow was functional for an electronic line superimposed on a football field to indicate how far a team needed to travel to make a first down. While yellow was not the only possible color, it was easier than other colors for viewers to see).

Section 17 of the Restatement (Third) of Unfair Competition offers the following definition of "functional designs."

> A design is "functional" for purposes of the rule stated in §16 if the design affords benefits in the manufacturing, marketing, or use of the goods or services with which the design is used, apart from any benefits attributable to the design's significance as an indication of source, that are important to effective competition by others and that are not practically available through the use of alternative designs.
>
> Comment:
>
> a. [I]n determining whether a particular design is "functional" and therefore ineligible for protection as a trademark, the ultimate inquiry is whether a prohibition against copying will significantly hinder competition by others. . . .
>
> b. Functionality. A packaging or product feature is not functional merely because the feature serves a utilitarian purpose. The recognition of trademark rights is precluded only when the particular design affords benefits that are not practically available through

alternative designs. A bottle, for example, is a utilitarian element of the packaging for wine, but the design of a particular wine bottle is not functional under the rule of this Section unless the shape or other aspects of the bottle provide significant benefits that are not practically obtainable through the use of alternative bottle designs.

In general, a functional design is one that is costly to do without. Thus, the benefits afforded by a particular design do not themselves determine whether that design is functional; a design is functional only if those benefits cannot practically be duplicated through the use of other designs. The availability of alternative designs that satisfy the utilitarian requirements or that otherwise afford similar advantages is therefore decisive in determining functionality. If a particular design affords benefits that are superior to those of any practical alternative design, it is functional if the benefits are important to effective competition. . . .

The benefit conferred by a particular packaging or product design may take various forms. The benefit may lie, for example, in greater economy in manufacturing, shipping, or handling, in increased utility or durability, or in enhanced effectiveness or ease of use. (On aesthetic benefits, see Comment c.) If the benefit afforded by the design resides solely in its association with a particular source, however, the design is not functional; the extent to which others may use such a design is determined by the rules of this Chapter governing trademark infringement. See § 20.

Any evidence that relates to the advantages inherent in a particular design is relevant to the issue of functionality. Functionality may thus be inferred from evidence of advertising or other promotional efforts by the plaintiff that emphasize the benefits of the design. Benefits claimed in pursuit of a utility patent may be particularly persuasive evidence of functionality. Discussions of the design in textbooks or trade literature and engineering analysis of its utilitarian aspects are also relevant. The fact that other manufacturers familiar with the design have rejected it in favor of other designs is evidence that the design is not functional. . . .

How does this definition compare with the Supreme Court's definition? Are either of these tests practical?

One might well wonder after these inquiries whether designs can be protected directly without undermining the balance between free competition and innovation reflected in the patent law. Are designs so expensive that they need a special incentive system to call forth the optimal number? Or do design and manufacturing processes provide enough built-in "lead time" to compensate the creators of innovative designs? The prevalence of piracy in many parts of the world suggests that the answer to the last question may be no. See J.H. Reichman, Legal Hybrids Between the Patent and Copyright Paradigms, 94 Colum. L. Rev. 2432 (1994) (arguing that natural lead time has diminished in recent years). Note in this regard the growing expansion of design protection through *sui generis* regimes. See, e.g., 17 U.S.C. §§ 901-14 (semiconductors); 17 U.S.C. §§ 1301-32 (boat hull designs). Are these expansions of intellectual property protection merely power grabs by powerful industries or responses to gaps in the fabric of intellectual property law that hinder design innovation?

The Paris Convention for the Protection of Industrial Property obliges parties to grant protection to industrial designs and models, although this requirement can be met through copyright, trade dress, and unfair competition law protections. See Graeme B. Dinwoodie, William O. Hennessey & Shira Perlmutter, International Intellectual Property Law and Policy 348-71 (2001) (surveying design protection in the European Union); see also Mattel, Inc. v. MCA Records, 296 F.3d 894, 907-08 (9th Cir. 2002) (stating that the Paris Convention does not define specific minimum requirements for design protection but rather ensures "national treatment" of foreign nationals.)

4. *Separability of Functional and Distinctive Elements.* The fact that a product or feature has been patented does not necessarily mean that all of its aspects are functional. Many patented products have distinctive features that can be separated from the functional elements. What should courts do about designs that are partially distinctive and partially functional (a category which includes most, if not all, designs)? Should there be a threshold (say, 80% functional) below which defendants are free to copy the design? Or must copiers attempt to parse the functional from the nonfunctional? (Compare in this regard the copyright rules, which allow the copying of ideas but not their expression). Suppose the two cannot be separated? Ought there to be a "utilitarian function" exception in trademark law, along the lines of the useful article doctrine in copyright law?

We confronted this very problem with regard to the copyright/patent intersection in the useful article doctrine. See Chapter 4.C.1.b. Copyright law allows the manufacturer of useful articles to show that the expressive elements are physically or conceptually separable. As we saw, conceptual separability is not easily applied in particular cases. See, e.g., Brandir Int'l, Inc v. Cascade Pacific Lumber Co., 834 F.2d 1142 (2d Cir. 1987) (metal tubing sculpture used as a bike rack); Carol Barnhart Inc. v. Economy Cover Corp., 773 F.2d 411 (2d Cir. 1985) (upper torso sculpture sold as mannequins). The *TrafFix* case addresses the separability problem through the use of a legal presumption: "Where the expired patent claimed the features in question, one who seeks to establish trade dress protection must carry the heavy burden of showing that the feature is not functional, for instance by showing that it is merely an ornamental, incidental, or arbitrary aspect of the device." Is this standard any more determinative than the conceptual separability standard in copyright law?

5. *Aesthetic Functionality.* In *TrafFix*, the Supreme Court characterizes its prior decision in *Qualitex* (allowing protection of green-gold color for dry cleaning press pads) as centering on the question of "aesthetic functionality." In so doing, the Court confined its references to "competitive need" and putting competitors at "significant non-reputation-related disadvantage" in that case to the "aesthetic functionality" inquiry. Such concepts have no relevance to the analysis of utilitarian functionality. If a feature is found to be utilitarian, then it can be freely copied regardless of whether competitors have alternative means of competing effectively.

The concept of "aesthetic functionality" has troubled courts and commentators from its earliest recognition. See A. Samuel Oddi, The Functions of "Functionality" in Trademark Law, 22 Hous. L. Rev. 925, 963 (1985) (commenting that "[f]rom the outset, 'aesthetic functionality' has proved to be a most controversial and ill-defined concept.") Section 742 of the Restatement Torts (1938) captured the concept in the following manner: "[w]hen goods are bought largely for their aesthetic value, their features may be functional." The Restatement gave as an example a heart-shaped box for chocolates. Id. (comment a). Such a shape is no better at holding chocolates than a rectangular box (and hence cannot properly be characterized as functional in a utilitarian sense), but it is easy to see why the heart-shaped box would be particularly appealing to consumers, especially on Valentine's Day. The aesthetic functionality doctrine aims to ensure that such shapes are not monopolized (in the absence of patent or copyright protection).

The 9th Circuit addressed this doctrine in the 1950s in a case involving china designs. By that time, the Wallace China Company had developed a reputation for "overall" patterns of china in the commercial (hotel and restaurant) marketplace. Unlike other manufacturers, which used artistic designs to decorate the rim of their plates, Wallace china contained bold print designs, such as leaves and woven images, covering the entirety of their eating surface. Tepco appropriated these designs and sold them on plates made of lower quality china. In an action for unfair competition, the 9th Circuit held that Wallace could not enjoin Tepco's activities even though Wallace established secondary meaning. See Pagliero v. Wallace China Co., 198 F.2d 339 (9th Cir. 1952). The court concluded that the design constituted both an indication of sources and an aesthetically pleasing aspect of the product. "If the particular feature is an important ingredient in the commercial success of the product," to enjoin its use in the absence of a patent or copyright would undermine "the interest in free competition." Id. at 343.

Subsequent cases have criticized the *Pagliero* doctrine as unduly limiting the protection of trade dress. They have sought to balance the source identification benefit of trade dress protection for aesthetic features against the threat to free competition posed by such protection. They have relaxed the *Pagliero* standard (no trade dress protection if a design enhances the appeal of a product to consumers) by focusing on the extent to which trade dress protection forecloses alternative designs. See Wallace International Silversmiths, Inc. v. Godinger Silver Art Co., Inc., 916 F.2d 76 (2d Cir. 1990), *cert. denied* 499 U.S. 976 (1991); Restatement (Third) Unfair Competition § 17, cmt c (stating that a feature is aesthetically functional only if it "confers a significant benefit that cannot practically be duplicated by the use of alternative designs."); see generally Graeme B. Dinwoodie, The Death of Ontology: A Teleological Approach to Trademark Law, 84 Iowa L. Rev. 611, 692-93 (1999). But in *TrafFix*, the Supreme Court took a broader view of aesthetic functionality, asking only whether the inability to copy the design would put the defendant at a significant disadvantage for reasons not related to reputation.

Menell questions the retreat from the *Pagliero* standard, arguing that just as the utilitarian functionality doctrine properly channels protection between trademark and patent law, the aesthetic functionality doctrine should serve a comparable purpose with respect to the trademark/copyright intersection. See Peter S. Menell, Aesthetic Functionality and the Backdoor Copyright Problem (manuscript 2005) (arguing that the aesthetic functionality doctrine can most usefully be viewed as a means for preventing backdoor copyrights); Mitchell M. Wong, The Aesthetic Functionality Doctrine and the Law of Trade-Dress Protection, 83 Cornell L. Rev. 1116, 1157 (1998). These disputes do not arise nearly as frequently as utilitarian functionality disputes for a number of reasons, including the more limited market power typically associated with copyright protection and the ease with which copyright claims can be asserted (relative to patent protection).[13] Now that copyright protection is not subject to forfeiture as a result of failure to observe formalities, as occurred in *Pagliero*, resort to trademark for protection of aesthetic features has been rare. Relative to trademark concerns, the primacy of copyright law in the protection of creative expression rests on a comparable footing to the primacy of patent law in the protection of utilitarian

---

13. The *Pagliero* case, arising as it did prior to the effective elimination of the notice requirement under the 1976 Act, was undoubtedly brought as a trademark dispute because Wallace had failed to provide copyright notice upon the first publication of designs on its plates and hence forfeited copyright protection. Hence, trade dress protection served as an alternative potential theory for recovery. Had the plates been designed after 1978, Wallace would have been able to pursue a copyright cause of action even if notice had not been provided when the plates were "published."

features. Therefore, when a copyright has expired (or protection lost), the developer of an expressive work should not be able to gain protection for such expression through the backdoor of trademark law. Furthermore, engaging courts in the determination of whether copying an aesthetic design foreclosed competition may well be unworkable, see Kohler Co. v. Moen, Inc., 12 F.3d 632, 649 (7th Cir. 1993) (Cudahy, J. dissenting) (noting that "the attempt to categorize product features as 'essential' or 'nonessential' for competition is perplexing and ultimately vain"); see also McCarthy, Trademark and Unfair Competition ¶7.81, and improperly transmutes trademark law into design protection, see Krueger, Int'l Inc. v. Nightingale, 915 F.Supp. 595, 606 (S.D.N.Y. 1996) (criticizing aesthetic functionality doctrine on the grounds that it denied protection for design features "whose only sin was to delight the sense"); Keene Corp. v. Paraflex Indus., Inc., 653 F.2d 822, 825 (3d Cir. 1981) (observing that a narrow scope of trade-dress protection "provides a disincentive for development of imaginative and attractive design" and that "[t]he more appealing the design, the less protection it would receive" under a broad conception of the aesthetic functionality doctrine).

Doesn't *Krueger* have it backwards? Shouldn't the question be whether there is a reputation-related reason to protect designs?

Should our intellectual property system tolerate backdoor copyrights through trademark law? Why might courts be inclined to the view that utilitarian functionality represents a higher hurdle to overcome than aesthetic functionality (requiring proof of a "significant non-reputation-related disadvantage")? Could the problem be solved through remedies — e.g., allowing free competition in expressive designs so long as the copyist used reasonable indications of source (and possibly disclaimers) to minimize consumer confusion? Or should courts expressly balance trademark interests (avoiding consumer confusion) against the threat to free competition (where copyright does not afford protection)?

6. *Functional Marks.* While most functionality cases involve trade dress, the functionality doctrine also extends to word marks if they serve a function. Can you think of situations in which names, words, or other forms of traditional registered trademarks might be functional? How would a functional name differ from a generic name?

PROBLEMS

Problem 5-11. Ferrari is a world-famous maker of upscale sports cars. It limits the number of cars it produces, and each car costs approximately $200,000. Ferrari's cars have the same general features as normal cars — wheels, chassis, etc. — but they also have a distinctive look that is easily recognized. They are sleek and low to the ground, a fact that may make them accelerate more quickly and that makes them more attractive to look at.

Roberts sells a fiberglass kit that replicates the exterior features of a Ferrari, though not the engine or performance. When sued for trade dress infringement, Roberts defends on the grounds that the Ferrari design is functional. Is it?

Problem 5-12. Eighteen years ago, Spartan Laboratories invented and patented a new pain-relieving drug called asperol. During the term of the patent, Spartan retained the exclusive right to sell asperol, which it manufactured in bright orange capsules. After the Spartan patent expired, a number of other companies began making generic asperol. Each of these companies sells the generic asperol in the same bright orange capsules as Spartan. Although the orange capsules are not visible inside the manufacturers' boxes (which do not resemble each other), asperol is sold only by prescription, and pharmacists invariably remove the drug from its original packaging and repackage it in their own bottles. The result is that the consumer sees only the name "asperol" and the orange capsules, regardless of who makes the drug.

Spartan filed suit under section 43(a) of the Lanham Act, alleging that the other manufacturers had infringed its trade dress protection by coloring the capsules orange. At trial, Spartan proves that the color orange is protectable because it is distinctive and because, over the seventeen years of the patent, pharmacists and customers had come to equate orange with Spartan's asperol. The generic manufacturers defend the suit on the grounds that the color is functional. In support of this claim, they present survey evidence that patients, particularly elderly patients, may become upset if the color of the drug is changed, and may refuse to believe that the drug is in fact asperol. The generic manufacturers present further evidence that pharmacists rely in part on color in making sure that they have packaged and labelled drugs correctly. Is the color orange functional?

3. Abandonment

Major League Baseball Properties, Inc. v. Sed Non Olet Denarius, Ltd.
United States District Court for the Southern District of New York
817 F. Supp. 1103 (S.D.N.Y. 1993)

MOTLEY, District Judge.

I. Introduction

Plaintiffs, Major League Baseball Properties, Inc. ("Properties") and Los Angeles Dodgers, Inc. ("Los Angeles"), allege in their Amended Complaint that the conduct of the . . . defendants [corporate operators of three restaurants, and their principals] (hereinafter collectively "The Brooklyn Dodger") . . . constitute: a) an infringement upon the rights of plaintiffs' trademarks in violation of 15 U.S.C. §§ 1114 and 1117; b) a wrongful appropriation of plaintiffs' trademarks in violation of 15 U.S.C. § 1125; c) a violation of plaintiffs' common law trademark and property rights; d) a violation of plaintiffs' rights under the New York General Business Law § 368-d; e) unfair competition; and f) the intentional use by defendants of a counterfeit mark in violation of 15 U.S.C. § 1117(b).

Each of these six causes of action is alleged to flow from defendants' use of the words "The Brooklyn Dodger" as the name and servicemark of the restaurants which defendants have operated in Brooklyn, New York, beginning in March 1988. Plaintiffs initially sought permanent injunctive relief, an accounting of profits, the destruction of physical items containing the allegedly infringing marks, monetary damages, and attorneys' fees.

By their Answer and Amended Answer defendants denied any infringement of plaintiffs' alleged right to use a "Brooklyn Dodger" trademark. Defendants also pleaded the defenses of abandonment by plaintiffs of any "Brooklyn Dodgers" mark which plaintiffs may have owned at one time, as well as laches. The abandonment defense was premised upon the plaintiffs' failure to make any commercial or trademark use of the "Brooklyn Dodgers" name for at least 25 years after Los Angeles left Brooklyn in 1958. The laches defense was premised upon the fact that plaintiffs waited for more than a year and a half after learning of defendants' use of the allegedly infringing trademark before advising defendants of any alleged infringement. During this period defendants expended substantial resources and monies in establishing their restaurants in Brooklyn, New York. . . .

Finally, in their Amended Answer, defendants counterclaimed for the cancellation of various trademark registrations for "Brooklyn Dodgers" filed by plaintiffs after defendants' application to register the "Brooklyn Dodger" servicemark was filed on April 28, 1988. . . .

II. Findings of Facts . . .

Plaintiff Los Angeles is a corporation with offices and its principal place of business in Los Angeles, California. It is the owner of the Los Angeles Dodgers, a professional baseball team which, since 1958, has played baseball in Los Angeles, California under the name the "Los Angeles Dodgers." . . . Prior to 1958 the same professional baseball team played baseball in Brooklyn, New York and were known as the "Brooklyn Dodgers" or the "Dodgers."

In 1958, the team moved the site of its home games from Brooklyn to Los Angeles. . . . It pointedly changed its name to Los Angeles Dodgers, Inc. . . .

By agreement with the Major League Clubs, Properties has been granted the exclusive right to market, license, publish, publicize, promote nationally, and protect the trademarks owned by the Major League Clubs, including those owned by the Los Angeles Dodgers. . . .

. . . By 1991, retail sales of licensed Major League Baseball merchandise were in excess of $2 billion. [Plaintiffs estimate that of the total Dodgers merchandising revenue, only $9 million of 1991 sales carried the *Brooklyn* Dodgers name.]

. . . On March 17, 1988 SNOD began doing business as a restaurant under the name "The Brooklyn Dodger Sports Bar and Restaurant." . . .

It was the individual defendants' decision that their restaurants would emphasize the multiple themes of fun, sports and Brooklyn. Their intention was to create a nostalgic setting where Brooklynites could relax and reminisce about times gone by. . . .

They initially chose to name their establishment "Ebbets Field" after the former baseball park located in Brooklyn, New York in which a baseball team known as the "Brooklyn Dodgers" played baseball until October, 1957. . . .

[Defendants rejected the name "Ebbets Field" because of a conflict with a small restaurant elsewhere in New York.]

The defendants knew that the departure of the "Brooklyn Dodgers" in 1958 had been accompanied by monumental hard feelings in the Borough of Brooklyn. In fact the relocation was one of the most notorious abandonments in the history of sports.... At the time defendants selected their logo, they were aware that Los Angeles owned federal trademark registrations for the word "Dodgers."... However, However, at no time during their consideration of the "Brooklyn Dodger" name did the individual defendants have any reason to believe that "The Brooklyn Dodger" mark was being used by Los Angeles, and certainly not for restaurant or tavern services.... When considering the use of the "Brooklyn Dodger" mark, at no time was there any discussion among the individual defendants and Brian Boyle about trading on the goodwill of Los Angeles in Brooklyn.... Indeed, non-party witness Brian Boyle, a life-long Brooklyn resident, testified that, given the acrimonious abandonment of Brooklyn by Los Angeles, the idea of trading on Los Angeles' "goodwill" in Brooklyn is almost "laughable."

C. Defendants' Use of the Trademark

In connection with each of defendants' [three] "The Brooklyn Dodger" restaurants, defendants make and/or made prominent use of the "Dodger" name and the "Brooklyn Dodger" name, with the word "Dodger" in stylized script, in the color blue, and in blue script.

The defendants' composite design mark consisted of three words: "The," "Brooklyn" and "Dodger" are entwined with one another and with an impish character,... which was styled after the Charles Dickens' character, the "Artful Dodger" from the novel Oliver Twist, leaning against the "r" in "Dodger."... Defendants, however, make significant use of their logo without the cartoon character to promote their business, including on merchandise such as apparel, in advertisements, on their letterhead and as part of their servicemark.

Defendants' logo is similar to Los Angeles' trademarks. The name "Brooklyn Dodger" is similar to the name "Brooklyn Dodgers" as used by plaintiffs. The script used by the defendants in their logo is similar to that used in Los Angeles' trademarks. The color blue used by defendants is similar to the color blue used by and associated with Los Angeles' [team] in Brooklyn. The swash or tail of the word "Dodger" used by defendants is similar to that used in Los Angeles' trademarks in terms of style and length....

In selecting their logo, defendants intentionally sought to reproduce the Brooklyn Dodgers' trademarks. Indeed, the script for the defendants' logo was intentionally chosen by defendants to track the script used by the Brooklyn Dodgers....

D. Plaintiffs' Use of the Trademark...

Plaintiffs' use of the "Brooklyn Dodgers" mark was based upon its physical location, until October 1957, in Brooklyn, New York. However, in 1959, Los Angeles made prominent commercial use and reference to their Brooklyn heritage and trademarks in connection with the promotion of Roy Campanella Night, honoring the former Brooklyn Dodgers player and present employee.... Los Angeles made prominent use of their trademarks incorporating the word "Brooklyn" at their annual oldtimers games.... Oldtimers games are commercial baseball exhibitions at which former players are honored and perform so that older fans can recall the past and younger fans can learn about the history of the Club.

[The court describes extensive licensing and use of the Los Angeles Dodgers trademarks, including for food services and restaurants.]

While plaintiffs have from time to time made use of their former "Brooklyn Dodgers" mark occasionally and sporadically for historical retrospective[s] such as "Old Timer's Day" festivities, the documentary proof establishes that, following its departure from Brooklyn, Los Angeles' earliest licensing of the "Brooklyn Dodgers" mark occurred on April 6, 1981.

Between 1981 and March 17, 1988, the date of defendants' first use of their mark for restaurant and tavern services, plaintiffs used their "Brooklyn Dodgers" mark for a variety of purposes. Those uses were almost exclusively in the context of T-shirts, jackets, sportswear, sports paraphernalia and on various types of novelty items (i.e. on drinking mugs, cigarette lighters, pens, Christmas tree ornaments, wristbands, etc.).... However, none of these uses competes with defendants' use of the mark for restaurant and tavern services.

[The court describes several instances between 1981 and 1988 where businesses sought and received permission to use the "Brooklyn Dodgers" name and logo, including several that were restaurant-related.]...

III. Conclusions of Law

[The court concluded that plaintiffs had failed to prove a likelihood of confusion; for completeness, it then considered defenses.]

C. Defendants' Affirmative Defenses

1. Abandonment

Under the Lanham Act a federally registered trademark is considered abandoned if its "use has been discontinued with intent not to resume." Cerveceria Centroamericana, S.A. v. Cerveceria India, Inc., 892 F.2d 1021, 1023 (Fed. Cir. 1989) (quoting Lanham Act, 15 U.S.C. § 1127 (1988)).

Abandonment is defined in the Lanham Act: "A mark shall be deemed 'abandoned' — (a) When its use has been discontinued with intent not to resume. Intent not to resume may be inferred from circumstances." 15 U.S.C. § 1127 (1988).

... The burden of proving abandonment falls upon the party seeking cancellation of a registered mark because a certificate of registration is "'prima facie evidence of the validity of the registration' and continued use." *Cerveceria India,* 892 F.2d at 1023...; Exxon Corp. v. Humble Exploration Co., 695 F.2d 96, 99 (5th Cir.), reh'g denied, 701 F.2d 173 (5th Cir. 1983).... The party seeking cancellation must establish abandonment by a preponderance of the evidence. See *Cerveceria India,* 892 F.2d at 1023-24.

The Lanham Act provides that "[n]onuse for two consecutive years shall be prima facie abandonment." 15 U.S.C. § 1127.... Prima facie abandonment establishes a rebuttable presumption of abandonment.... The evidence presented at trial established that between 1958 and 1981 plaintiffs made no commercial trademark use whatsoever of any "Brooklyn Dodgers" mark. Minor changes in a trademark that do not affect the overall commercial impression of the mark do not constitute abandonment. . . . However, Los Angeles' change, in this case, from "Brooklyn Dodgers" to "Los Angeles Dodgers" was not minor; it involved an essential element affecting the public's perception of the mark and the team.[19] . . .

Plaintiffs argue that their "Dodgers" mark without a geographical reference — that is, "Dodgers" alone — is a protected use infringed by defendants actions. However, in this context, "Brooklyn" is more than a geographic designation or appendage to the word "Dodgers." The "Brooklyn Dodgers" was a non-transportable cultural institution separate from the "Los Angeles Dodgers" or the "Dodgers" who play in Los Angeles. It is not simply the "Dodgers," (and certainly not the "Los Angeles Dodgers"), that defendants seek to invoke in their restaurant; rather defendants specifically seek to recall the nostalgia of the cultural institution that was the "Brooklyn Dodgers." It was the "Brooklyn Dodgers" name that had acquired secondary meaning in New York in the early part of this century, prior to 1958. It was that cultural institution that Los Angeles abandoned.

. . . In this case, in order to maintain use of the mark, Los Angeles would have had to continue to use "Brooklyn Dodgers" as the name of its baseball team. Only in this way would the public continue to identify the name with the team. Defiance Button Mach. Co. v. C & C Metal Products Corp., 759 F.2d 1053, 1059 (2d Cir.), *cert. denied*, 474 U.S. 844 . . . (1985). . . .[20]

Rather than using the "Brooklyn Dodgers" mark in the ordinary course of trade, a more accurate description of Los Angeles' use of the mark, at least between 1958 and 1981, was given by its General Counsel in a 1985 letter to someone seeking to use it on a novelty item:

> Since the Dodgers moved to Los Angeles in 1958 the name "Brooklyn Dodgers" has been reserved strictly for use in conjunction with items of historical interest.

. . . Under the law, such warehousing is not permitted. . . . Rights in a trademark are lost when trademarks are "warehoused" as plaintiffs attempted to "warehouse" the "Brooklyn Dodgers" mark for more than two (2) decades. . . . Plaintiffs' failure to use the "Brooklyn Dodgers" trademark between 1958, when Los Angeles left Brooklyn, and 1981 constitutes abandonment of the trademark. . . .

19. While Los Angeles ceased commercial use of the trademark because of its relocation from Brooklyn to Los Angeles, its motive, justifiable or not, is irrelevant. *Stetson*, 955 F.2d at 851. Nevertheless, Los Angeles' nonuse in this case was voluntary. Courts have been more reluctant to find an absence of intent to resume where the trademark owner had an excusable reason for nonuse — that is, where nonuse was involuntary. See, e.g., Defiance Button Machine Co. v. C & C Metal Products Corp., 759 F.2d 1053, 1059 (2d Cir.), *cert. denied*, 474 U.S. 844, 106 S. Ct. 131, 88 L. Ed. 2d 108 (1985) (no abandonment where cessation of business was involuntary); American International Group, Inc. v. American International Airways, Inc., 726 F. Supp. 1470 (E.D. Pa. 1989) (where airline declared bankruptcy, remaining goodwill and lack of intent to abandon precluded finding abandonment).

20. The outcome in *Defiance*, finding no abandonment, is distinguishable from the case at hand because the court in *Defiance* held that the continued goodwill toward the button company after it stopped producing goods and the fact that the company intended to retain its trademark for some commercial use precluded a finding of abandonment. Similarly, the court in Schenley Industries, 441 F.2d 675 held that continued goodwill and lack of intent to abandon precluded finding of abandonment. In the unique facts of this case, however, plaintiffs have not succeeded in demonstrating that much goodwill in Brooklyn survived Los Angeles' move in 1957. But more importantly, while plaintiff in *Defiance* was at least able to demonstrate an intent not to abandon, plaintiffs here have not even demonstrated an intent not to abandon, much less the statutory requirement of intent to resume. A mark retains "residual" goodwill "if the proponent of a mark stops using it but demonstrates an intent to keep the mark alive for use in resumed business." Pan American World Airways, Inc. v. Panamerican School of Travel, Inc., 648 F.Supp. 1026, 1031 (S.D.N.Y.), *aff'd without op.*, 810 F.2d 1160 (2d Cir. 1986) (citing *Defiance*, supra). The claim to residual goodwill will not preclude a finding of abandonment where, as in this case, the owner unequivocally declares its intention to discontinue use.

Abandonment under the Lanham Act, however, requires both nonuse and intent not to resume use....

Once prima facie abandonment has been proven, the trademark registrants — in this case plaintiffs — must carry their burden of producing evidence that there was an intent to resume use of the trademark. *Cerveceria India*, 892 F.2d at 1025-26....

... Rather than merely proving that it did not intend to abandon its trademark, the trademark registrant must demonstrate that it intended to use or resume use. See *Exxon*, 695 F.2d at 99, 102-103 ("Stopping at an 'intent not to abandon' [rather than 'intent to resume'] tolerates an owner's protecting a mark with neither commercial use nor plans to resume commercial use. Such a license is not permitted by the Lanham Act")....

... Plaintiffs have in no way demonstrated their intent to resume commercial use of the "Brooklyn Dodgers" mark within two years after Los Angeles left Brooklyn in 1958 or at anytime within the ensuing quarter century....

2. *Resumption*

Having determined that plaintiffs abandoned the "Brooklyn Dodgers" mark, the next inquiry is to determine the effect of that abandonment, given that plaintiffs have recently resumed limited use of the trademark....

... [I]f plaintiffs have any interest in a "Brooklyn Dodgers" mark, that interest arose in 1981 when commercial use of the mark resumed after a twenty-three (23) year hiatus. Plaintiffs' preemptive rights in the "Brooklyn Dodgers" mark would extend only to the precise goods on or in connection with which the trademark was used since its resumption (i.e. clothing, jewelry, novelty items).

In other words, the fact that plaintiffs resumed use prior to defendants' use does not mean that plaintiffs may preclude defendants' use of the mark in their restaurant business in Brooklyn....

Accordingly, the court declines to enjoin defendants' very limited use of the "Brooklyn Dodger" mark by defendants for use in connection with its local restaurants directed toward older Brooklyn Dodgers fans in the Brooklyn community in the city of New York....

COMMENTS AND QUESTIONS

1. Would the result change if "The Brooklyn Dodger" restaurant began selling Brooklyn Dodger T-shirts? Other memorabilia?

2. Section 45 of the Lanham Act, on abandonment, was amended in 1994 (effective Jan. 1, 1996) to change the presumptive abandonment period from two to three years, in accordance with the Uruguay Round amendments. See GATT TRIPs, art. 19(1). This provision of the act was the U.S. version of article 19(1) of The Agreement on Trade-Related Aspects of Intellectual Property Rights, Including Trade in Counterfeit Goods, opened for signature Apr. 15, 1994, 33 I.L.M. 81.

Thus section 45 now reads in part:

A mark shall be deemed to be "abandoned" when either of the following occurs:

(1) When its use has been discontinued with intent not to resume such use. Intent not to resume may be inferred from circumstances. Nonuse for three consecutive years

shall be prima facie evidence of abandonment. "Use" of a mark means the bona fide use of such mark made in the ordinary course of trade, and not made merely to reserve a right in a mark.

(2) When any course of conduct of the owner, including acts of omission as commission, causes the mark to . . . lose its significance as a mark. . . .

Lanham Act § 45, 15 U.S.C. § 1127. The *Dawn Donuts* case and notes, below, deal with the § 45(2) type of abandonment.

3. If trademark rights are essentially about protecting consumers from confusion, why does the legal standard require anything other than a showing that the mark's meaning has faded sufficiently so that it can now be "reclaimed" by another? Why does the statute settle for a presumption; why not a per se rule, perhaps based on a longer period — say ten years? Cf. Note, The Song Is Over but the Melody Lingers On: Persistence of Goodwill and the Intent Factor in Trademark Abandonment, 56 Fordham L. Rev. 1003, 1006-07 (1988) ("[W]hen a trademark has persisting or residual goodwill, even after a period of nonuse, doubts should be resolved in favor of the trademark owner and against the competitor charging abandonment. . . . [Courts] should . . . make it as difficult as possible to find a trademark abandoned whenever goodwill in the mark persists."). What arguments can you think of in favor of this liberal rule? Against? Should there be an incentive to aggressively "recycle" marks that have shown some usefulness?

Do you think the association between the name "Brooklyn Dodgers" and the major league baseball team now located in Los Angeles had faded sufficiently for someone else to reclaim the name? Do the unusual facts in this case — especially the animosity between ex-Brooklyn Dodger fans and the Los Angeles Dodgers — suggest why a per se rule might not make sense in every case? Is the court right to assume the absence of "goodwill" here, where consumers recognize the Dodger mark but simply have bad associations with it? Or does the continual connection in the public's mind suggest an ongoing role for trademark law?

4. Are there reasons to limit the abandonment doctrine where it may affect names that clearly seem to "belong" to the trademark owner? In Kareem Abdul-Jabaar v. General Motors, 75 F.3d 1361 (9th Cir. 1996), the court rejected defendant's claim that basketball star Abdul-Jabaar had abandoned his birth name, Lew Alcindor. The court established a per se rule: "A proper name thus cannot be deemed abandoned throughout its possessor's life, despite his failure to use it. . . . "

5. *Securing Teams Names in Advance of Use.* Whereas Major League Baseball Properties, Inc. v. Sed Non Olet Denarius, Ltd. deals with rights in a team name after the team has left a city, a dispute involving the move of the Los Angeles Rams to St. Louis highlights some of the issues accompanying the birth of a mark following the relocation of an existing team to a new locale. On January 17, 1995, St. Louis Mayor Freeman Bosley and Georgia Frontiere, owner of the Los Angeles Rams Football Company, announced that the Los Angeles Rams football franchise intended to relocate to St. Louis, Missouri. By early February of 1995, the Rams organization had received over 72,000 personal seat license applications for the St. Louis stadium where the Rams would play. On February 22, 1995, Johnny Blastoff, a corporation in the business of creating and marketing cartoon characters, filed a trademark application in Wisconsin for the name of a fictional cartoon sports team named the "St. Louis Rams." On March 10, 1995, Blastoff filed two federal intent-to-use trademark

applications for the "St. Louis Rams" mark. One month later, the Rams' move to St. Louis was approved by the National Football League. Although a number of vendors began selling unlicensed merchandise bearing the mark "St. Louis Rams" in January of 1995, officially licensed vendors began using the "St. Louis Rams" mark in a wide variety of merchandise sales in April of 1995. In March 1997, Blastoff filed a declaratory judgment against the Rams football organization, the NFL, and others, alleging that he did not infringe the NFL defendants' trademark rights. In determining that the Rams first appropriated the mark through use as a designation of origin, the court placed emphasis on the nature of franchise relocations, see Indianapolis Colts, Inc. v. Metropolitan Baltimore Football Club Ltd., 34 F.3d 410, 413 (7th Cir. 1994) (finding a strong presumption of franchise owner priority in a team mark when a franchise relocates to a new city), and the role of "advertising brochures, catalogs, newspaper ads, and articles in newspapers and trade publications" and use on television and radio in establishing public identification of a mark with a product or service, see T.A.B. Systems v. Pactel Teletrac, 77 F.3d 1372, 1375 (Fed. Cir. 1996); In re Owens-Corning Fiberglas Corp., 774 F.2d 1116, 1125 (Fed. Cir. 1985). See Johnny Blastoff, Inc. v. Los Angeles Rams Football Co., 188 F.3d 427 (7th Cir. 1999), *cert. denied*, 528 U.S. 1188 (2000).

PROBLEMS

Problem 5-13. Until 1972 the Humble Oil & Refining Co. was one of the largest producers and sellers of gasoline in the world. In that year, after a merger with Esso, Humble decided to change its name to Exxon, an arbitrary mark it had selected for the purpose. The company invested an enormous amount of money advertising the new name and placing its new brand on all its products, services, and correspondence. However, the new Exxon also started a "brand maintenance program," under which it continued to make a few sales every year under the Humble name. It sent certain invoices on Humble letterhead, and bulk oil was sometimes sold in Humble barrels. Exxon continued this program throughout the 1970s.

In 1975, a new oil exploration company decides to call itself the Humble Oil Exploration Co. The new company picked the Humble name because they hoped to trade on residual goodwill left in the name, and because they believed Exxon had abandoned its old name. Can the new company use the Humble name?

Problem 5-14. IBM Corporation is a world-famous maker of computers and other business machines. For decades, people have informally referred to IBM as "Big Blue," both in conversation and in published articles and papers about the company. However, IBM does not itself use or claim the name "Big Blue." It has never registered the name Big Blue, its employees are forbidden by company policy to use the name in referring to the company, and at one point an exasperated company spokesperson is reported to have said "We're not Big Blue — please don't call us that. We want nothing to do with the name." Nonetheless, the nickname persists.

In 1994 a small repair shop that specializes in fixing IBM computers decided to call itself the "Big Blue Repair Shop." After BBRS started using the name, IBM sent it a cease-and-desist letter alleging that IBM owned the rights to the name Big Blue, since people used it to refer to IBM. Who has priority in use of the Big Blue mark?

a. Unsupervised Licenses

Dawn Donut Company, Inc. v. Hart's Food Stores, Inc.
United States Court of Appeals for the Second Circuit
267 F.2d 358 (2d Cir. 1959)

[Dawn Donut, a Michigan wholesaler of doughnut mix, owned federally registered trademarks Dawn and Dawn Donut. Dawn brought a trademark infringement suit against defendant Hart's Food Stores ("Hart"), which used the mark "Dawn" in connection with its sale of doughnuts and baked goods in stores in and around Rochester, N.Y. Hart filed a counterclaim to cancel plaintiff's registrations, on the ground that plaintiff had abandoned its trademarks due to inadequate quality control and supervision on the part of its licensees — the retail bakeries that sold doughnuts made from Dawn's mix. The district court denied Dawn's request for an injunction and dismissed Hart's counterclaim. This appeal followed. In an omitted (and oft-cited) portion of this opinion, Circuit Judge Lumbard held that plaintiff was not entitled to any relief under the Lanham Act because there was no present likelihood that plaintiff would expand its retail use of its trademarks into defendant's market area. A majority of the panel also found that there was no abandonment by plaintiff of its registrations, and that therefore defendant was not entitled to have plaintiff's registrations of trademarks cancelled.]

LUMBARD, Circuit Judge [dissenting in part]....

The final issue presented is raised by defendant's appeal from the dismissal of its counterclaim for cancellation of plaintiff's registration on the ground that the plaintiff failed to exercise the control required by the Lanham Act over the nature and quality of the goods sold by its licensees.

We are all agreed that the Lanham Act places an affirmative duty upon a licensor of a registered trademark to take reasonable measures to detect and prevent misleading uses of his mark by his licensees or suffer cancellation of his federal registration. The Act, 15 U.S.C.A. § 1064, provides that a trademark registration may be cancelled because the trademark has been "abandoned." And "abandoned" is defined in 15 U.S.C.A. § 1127 to include any act or omission by the registrant which causes the trademark to lose its significance as an indication of origin.

Prior to the passage of the Lanham Act many courts took the position that the licensing of a trademark separately from the business in connection with which it had been used worked an abandonment. Reddy Kilowatt, Inc. v. MidCarolina Electric Cooperative, Inc., 4 Cir., 1957, 240 F.2d 282, 289; American Broadcasting Co. v. Wahl Co., 2 Cir., 1941, 121 F.2d 412, 413; Everett O. Fisk & Co. v. Fisk Teachers' Agency, Inc., 8 Cir., 1924, 3 F.2d 7, 9. The theory of these cases was that:

A trade-mark is intended to identify the goods of the owner and to safeguard his good will. The designation if employed by a person other than the one whose business it serves to identify would be misleading. Consequently, "a right to the use of a trade-mark or a trade-name cannot be transferred in gross."

American Broadcasting Co. v. Wahl Co., supra, 121 F.2d at page 413.

Other courts were somewhat more liberal and held that a trademark could be licensed separately from the business in connection with which it had been used provided that the licensor retained control over the quality of the goods produced by the licensee. E. I. DuPont de Nemours & Co. v. Celanese Corporation of America, 1948, 167 F.2d 484, 35 CCPA 1061.... But even in the DuPont case the court was careful to point out that naked licensing, viz. the grant of licenses without the retention of control, was invalid. E. I. DuPont de Nemours & Co. v. Celanese Corporation of America, supra, 167 F.2d at page 489.

The Lanham Act clearly carries forward the view of these latter cases that controlled licensing does not work an abandonment of the licensor's registration, while a system of naked licensing does. 15 U.S.C.A. § 1055 provides:

> Where a registered mark or a mark sought to be registered is or may be used legitimately by related companies, such use shall inure to the benefit of the registrant or applicant for registration, and such use shall not affect the validity of such mark or of its registration, provided such mark is not used in such manner as to deceive the public....[14]

Without the requirement of control, the right of a trademark owner to license his mark separately from the business in connection with which it has been used would create the danger that products bearing the same trademark might be of diverse qualities. See American Broadcasting Co. v. Wahl Co., supra; Everett O. Fisk & Co. v. Fisk Teachers' Agency, Inc., supra. If the licensor is not compelled to take some reasonable steps to prevent misuses of his trademark in the hands of others the public will be deprived of its most effective protection against misleading uses of a trademark. The public is hardly in a position to uncover deceptive uses of a trademark before they occur and will be at best slow to detect them after they happen. Thus, unless the licensor exercises supervision and control over the operations of its licensees the risk that the public will be unwittingly deceived will be increased and this is precisely what the Act is in part designed to prevent. See Sen. Report No. 1333, 79th Cong., 2d Sess. (1946). Clearly the only effective way to protect the public where a trademark is used by licensees is to place on the licensor the affirmative duty of policing in a reasonable manner the activities of his licensees.

The critical question on these facts therefore is whether the plaintiff sufficiently policed and inspected its licensees' operations to guarantee the quality of the products they sold under its trademarks to the public. The trial court found that: "By reason of its contacts with its licensees, plaintiff exercised legitimate control over the nature and quality of the food products on which plaintiff's licensees used the trademark 'Dawn.' Plaintiff and its licensees are related companies within the meaning of Section 45 of the Trademark Act of 1946." It is the position of the majority of this court that the trial judge has the same leeway in determining what constitutes a reasonable degree of supervision and control over licensees under the facts and circumstances of the

---

14. Lanham Act § 45 now reads in relevant part: "The term 'related company' means any person whose use of a mark is controlled by the owner of the mark with respect to the nature and quality of the goods or services on or in connection with which the mark is used." 15 U.S.C. § 1127. The portion of 15 U.S.C. § 1055 (Lanham Act § 5) quoted just above in the case has not changed. — EDS.

particular case as he has on other questions of fact; and particularly because it is the defendant who has the burden of proof on this issue they hold the lower court's finding not clearly erroneous.

I dissent from the conclusion of the majority that the district court's findings are not clearly erroneous because while it is true that the trial judge must be given some discretion in determining what constitutes reasonable supervision of licensees under the Lanham Act, it is also true that an appellate court ought not to accept the conclusions of the district court unless they are supported by findings of sufficient facts. It seems to me that the only findings of the district judge regarding supervision are in such general and conclusory terms as to be meaningless. In the absence of supporting findings or of undisputed evidence in the record indicating the kind of supervision and inspection the plaintiff actually made of its licensees, it is impossible for us to pass upon whether there was such supervision as to satisfy the statute. There was evidence before the district court in the matter of supervision, and more detailed findings thereon should have been made.

Plaintiff's licensees fall into two classes: (1) those bakers with whom it made written contracts providing that the baker purchase exclusively plaintiff's mixes and requiring him to adhere to plaintiff's directions in using the mixes; and (2) those bakers whom plaintiff permitted to sell at retail under the 'Dawn' label doughnuts and other baked goods made from its mixes although there was no written agreement governing the quality of the food sold under the Dawn mark.[6]

The contracts that plaintiff did conclude, although they provided that the purchaser use the mix as directed and without adulteration, failed to provide for any system of inspection and control. Without such a system plaintiff could not know whether these bakers were adhering to its standards in using the mix or indeed whether they were selling only products made from Dawn mixes under the trademark "Dawn."

The absence, however, of an express contract right to inspect and supervise a licensee's operations does not mean that the plaintiff's method of licensing failed to comply with the requirements of the Lanham Act. Plaintiff may in fact have exercised control in spite of the absence of any express grant by licensees of the right to inspect and supervise.

The question then, with respect to both plaintiff's contract and non-contract licensees, is whether the plaintiff in fact exercised sufficient control.

Here the only evidence in the record relating to the actual supervision of licensees by plaintiff consists of the testimony of two of plaintiff's local sales representatives that they regularly visited their particular customers and the further testimony of one of them, Jesse Cohn, the plaintiff's New York representative, that "in many cases" he did

---

6. On cross-examination plaintiff's president conceded that during 1949 and 1950 the company in some instances, the number of which is not made clear by his testimony, distributed its advertising and packaging material to bakers with whom it had not reached any agreement relating to the quality of the goods sold in packages bearing the name "Dawn." It also appears from plaintiff's list of the 16 bakers who were operating as exclusive Dawn shops at the time of the trial that plaintiff's contract with 3 of these shops had expired and had not been renewed and that in the case of 2 other such shops the contract had been renewed only after a substantial period of time had elapsed since the expiration of the original agreement. The record indicates that these latter 2 bakers continued to operate under the name "Dawn" and purchase "Dawn" mixes during the period following the expiration of their respective franchise agreements with the plaintiff. Particularly damaging to plaintiff is the fact that one of the 2 bakers whose franchise contracts plaintiff allowed to lapse for a substantial period of time has also been permitted by plaintiff to sell doughnuts made from a mix other than plaintiff's in packaging labeled with plaintiff's trademark.

have an opportunity to inspect and observe the operations of his customers. The record does not indicate whether plaintiff's other sales representatives made any similar efforts to observe the operations of licensees.

Moreover, Cohn's testimony fails to make clear the nature of the inspection he made or how often he made one. His testimony indicates that his opportunity to observe a licensee's operations was limited to "those cases where I am able to get into the shop" and even casts some doubt on whether he actually had sufficient technical knowledge in the use of plaintiff's mix to make an adequate inspection of a licensee's operations.

The fact that it was Cohn who failed to report the defendant's use of the mark "Dawn" to the plaintiff casts still further doubt about the extent of the supervision Cohn exercised over the operations of plaintiff's New York licensees.

Thus I do not believe that we can fairly determine on this record whether plaintiff subjected its licensees to periodic and thorough inspections by trained personnel or whether its policing consisted only of chance, cursory examinations of licensees' operations by technically untrained salesmen. The latter system of inspection hardly constitutes a sufficient program of supervision to satisfy the requirements of the Act.

. . . I would direct the district court to order the cancellation of plaintiff's registrations if it should find that the plaintiff did not adequately police the operations of its licensees. . . .

The district court's denial of an injunction restraining defendant's use of the mark "Dawn" on baked and fried goods and its dismissal of defendant's [abandonment] counterclaim are affirmed.

COMMENTS AND QUESTIONS

1. Despite its age and the fact that it was a dissent, Judge Lumbard's discussion in *Dawn Donuts* regarding licensee supervision and abandonment is still the standard in the area. See McCarthy, Trademarks and Unfair Competition § 26.14 (3d ed. 1992 & Supp. 1996).

2. Trademark rights are regularly lost because of unsupervised licensing. See, e.g., Stanfield v. Osborne Indus., Inc., 52 F.3d 867 (10th Cir. 1995) (denying plaintiff's advertising-related trademark claim because rights were lost due to unsupervised license); Barcamerica Intern. USA Trust v. Tyfield Importers, Inc., 289 F. 2d 589 (9th Cir. 2002) (finding lack of express contractual right to inspect and supervise licensee's operation as well as no actual supervision).

3. The *Celanese* case cited in *Dawn Donuts*, E. I. du Pont de Nemours & Co. v. Celanese Corporation, 167 F.2d 484 (C.C.P.A. 1948), was one of the first cases to establish the legitimacy of trademark licensing over the objection that licensing necessarily entailed an abandonment. Obviously, the growth of franchising, character merchandising, and related practices depended on such a holding. Much modern business would be impossible if corporations could not expand their brand names in these ways. Cf. "Extension by Contract: Franchising and Merchandising," section D.4.

4. If a trademark owner wants to spend its own goodwill by licensing those who sell lower quality products, why not let it? The company would surely be permitted to reduce the quality of the products it makes itself.

b. The Rule Against Assignments in Gross

The rule that unsupervised or "naked" licenses can amount to abandonment has a logical corollary: that outright assignments of trademarks are invalid. This common law rule was codified in section 10 of the Lanham Act: "A registered mark or a mark for which application to register has been filed shall be assignable with the goodwill of the business in which the mark is used, or with that part of the goodwill of the business connected with the use of and symbolized by the mark." Lanham Act § 10, 15 U.S.C. § 1060.

Assignments of trademarks alone, without any underlying assets or "goodwill," are called "assignments in gross." They are invalid, see McCarthy, Trademarks and Unfair Competition § 18.01 (3d ed. 1992 & Supp. 1996), a result said to follow from the fact that trademarks are only repositories and symbols of goodwill, rather than true property rights. See American Steel Foundries v. Robertson, 269 U.S. 372, 380 (1926). That case reasoned that consumers, having come to rely on a trademark to identify the characteristics of a product as manufactured and sold by Firm *A*, would be harmed if *A* simply sold the mark outright to Firm *B* without transferring any of the employees or production machinery that Firm *A* had used to make its product. Id. By the same reasoning, courts invalidated trademarks when sold to another firm for use on a different product. See Filkins v. Blackman, 9 Fed. Cas. 50, 51 (No. 4786) (C.C.D. Conn. 1876) ("The right to the use of a trademark cannot be so enjoyed by an assignee that he shall have the right to affix the mark to goods differing in character or species from the article to which it was originally attached.").

As originally applied, the rule against assignments in gross was quite strict; in general, a firm was required to assign tangible assets along with the trademark. See, e.g., Pepsico, Inc. v. Grapette Co., 416 F.2d 285 (8th Cir. 1969) (assignment of trademark Peppy without underlying assets invalid). More recently, however, the traditional rule has been relaxed, partly in recognition of the increased frequency and importance of trademark-related transactions. The contemporary rule can be seen operating in cases involving assignment of "soft" trademark-related assets, such as customer lists, production formulas (as opposed to machinery), and even amorphous "goodwill." See, e.g., In re Roman Cleanser Co., 802 F.2d 207 (6th Cir. 1986) (validating transfer of trademark in satisfaction of security interest in it, together with formulas and customer lists); Money Store v. Harriscorp Finance, Inc., 689 F.2d 666 (7th Cir. 1982) (assignment of Money Store trademark by senior user for $1 not invalid; nominal recitation of "goodwill" in assignment contract, without transfer of any other assets, was enough). Cf. William M. Landes and Richard A. Posner, Trademark Law: An Economic Perspective, 30 J.L. & Econ. 265, 274-275 (1987) (arguing that the "assignment in gross" doctrine makes sense only in "final period" cases, where sellers of goods are leaving the market and hence do not care if consumers are disappointed by the low quality of the assignee's goods).

By contrast, Japan, for example, recognizes private property rights in the trademark itself. Therefore, assignments in gross are valid even if totally divorced from any goodwill. Trademark rights are also severable; they may be assigned by class, providing the goods of the remaining classification would not cause confusion with the goods of the class assigned. See generally Kazuko Matsuo, Trademarks, in 4 Doing Business in Japan (Zentaro Kitagawa ed., 1991).

The rule against assignments in gross makes sense from the point of view of protecting consumer associations between a mark and an underlying product. If the symbol changes hands, and is now used to "refer to" a different product, consumers might be confused. (Imagine if language experts decided to change the meaning of a common word, without telling anyone.) As Landes and Posner, *supra*, argue, the risk of confusion is greatest when the trademark assignor is leaving the market.

Of course, this rationale for the doctrine assumes that consumers cannot see the lower quality of the assignee's product when they look at the product. Moreover, the doctrine seems highly questionable as applied to cases where the assignee uses the mark on a completely different type of product altogether. In such a case, any consumers who retain an association between the mark and the old product will quickly grasp the changed circumstances when they see the new product.

With these points in mind, why not encourage the transfer of trademarks that have proven effective? Why restrict transfers to those accompanied by underlying assets? For cogent criticism of the rule against assignments in gross, see Allison Sell McDade, Note, Trading in Trademarks — Why the Anti-Assignment in Gross Doctrine Should Be Abolished When Trademarks Are Used as Collateral, 77 Tex. L. Rev. 465 (1998) (focusing on the use of trademarks as security interests); but see Mark A. Lemley, The Modern Lanham Act and the Death of Common Sense, 108 Yale L.J. 1687 (1999) (supporting the doctrine as consistent with the consumer-oriented focus of trademark law). What would happen to consumer expectations if a lending company foreclosed on a trademark?

Does (or should) the law similarly prohibit the original trademark owner from significantly decreasing the quality of his or her goods, or from changing the type of product to which the mark is attached? See 2 J. Thomas McCarthy, McCarthy on Trademarks and Unfair Competition § 17.09 (3d ed. 1992 & Supp. 1996) (citing cases and arguing that this would amount to deceit under the Lanham Act). The latter may be an issue for registered marks, where the classification of goods is important.

COMMENTS AND QUESTIONS

Article 21 of the General Agreements on Tariffs and Trade, which entered into force in the United States in 1995, provides that "the owner of a registered trademark shall have the right to assign his trademark with or without the transfer of the business to which the trademark belongs." Does this article require the United States to abolish the rule against assignment in gross?

4. Nontrademark (or Nominative) Use, Parody, and the First Amendment

Mattel, Inc. v. MCA Records
United States Court of Appeals for the Ninth Circuit
296 F.3d 894 (9th Cir. 2002)

KOZINSKI, Circuit Judge:

If this were a sci-fi melodrama, it might be called Speech-Zilla meets Trademark Kong.

I

Barbie was born in Germany in the 1950s as an adult collector's item. Over the years, Mattel transformed her from a doll that resembled a "German street walker," as she originally appeared, into a glamorous, long-legged blonde. Barbie has been labeled both the ideal American woman and a bimbo. She has survived attacks both psychic (from feminists critical of her fictitious figure) and physical (more than 500 professional makeovers). She remains a symbol of American girlhood, a public figure who graces the aisles of toy stores throughout the country and beyond. With Barbie, Mattel created not just a toy but a cultural icon.

With fame often comes unwanted attention. Aqua is a Danish band that has, as yet, only dreamed of attaining Barbie-like status. In 1997, Aqua produced the song *Barbie Girl* on the album *Aquarium*. In the song, one bandmember impersonates Barbie, singing in a high-pitched, doll-like voice; another bandmember, calling himself Ken, entices Barbie to "go party." (The lyrics are in the Appendix.) *Barbie Girl* singles sold well and, to Mattel's dismay, the song made it onto Top 40 music charts.

Mattel brought this lawsuit against the music companies who produced, marketed and sold *Barbie Girl*: MCA Records, Inc., Universal Music International Ltd....

Mattel appeals the district court's ruling that *Barbie Girl* is a parody of Barbie and a nominative fair use [and] that MCA's use of the term Barbie is not likely to confuse consumers as to Mattel's affiliation with *Barbie Girl* or dilute the Barbie mark...

III

A. A trademark is a word, phrase or symbol that is used to identify a manufacturer or sponsor of a good or the provider of a service. See New Kids on the Block v. News Am. Publ'g, Inc., 971 F.2d 302, 305 (9th Cir.1992). It's the owner's way of preventing others from duping consumers into buying a product they mistakenly believe is sponsored by the trademark owner. A trademark "inform[s] people that trademarked products come from the same source." Id. at 305 n. 2. Limited to this core purpose — avoiding confusion in the marketplace — a trademark owner's property rights play well with the First Amendment. "Whatever first amendment rights you may have in calling the brew you make in your bathtub 'Pepsi' are easily outweighed by the buyer's interest in not being fooled into buying it." [Alex Kozinski,] Trademarks Unplugged, 68 N.Y.U. L.Rev. 960, 973 (1993).

The problem arises when trademarks transcend their identifying purpose. Some trademarks enter our public discourse and become an integral part of our vocabulary. How else do you say that something's "the Rolls Royce of its class"? What else is a quick fix, but a Band-Aid? Does the average consumer know to ask for aspirin as "acetyl salicylic acid"? See Bayer Co. v. United Drug Co., 272 F. 505, 510 (S.D.N.Y.1921). Trademarks often fill in gaps in our vocabulary and add a contemporary flavor to our expressions. Once imbued with such expressive value, the trademark becomes a word in our language and assumes a role outside the bounds of trademark law.

Our likelihood-of-confusion test, see AMF Inc. v. Sleekcraft Boats, 599 F.2d 341, 348-49 (9th Cir.1979), generally strikes a comfortable balance between the trademark owner's property rights and the public's expressive interests. But when a trademark owner asserts a right to control how we express ourselves — when we'd find it difficult to describe the product any other way (as in the case of aspirin), or when the mark (like Rolls Royce) has taken on an expressive meaning apart from its source-identifying function — applying the traditional test fails to account for the full weight of the public's interest in free expression.

The First Amendment may offer little protection for a competitor who labels its commercial good with a confusingly similar mark, but "[t]rademark rights do not entitle the owner to quash an unauthorized use of the mark by another who is communicating ideas or expressing points of view." L.L. Bean, Inc. v. Drake Publishers, Inc., 811 F.2d 26, 29 (1st Cir.1987). Were we to ignore the expressive value that some marks assume, trademark rights would grow to encroach upon the zone protected by the First Amendment. See Yankee Publ'g, Inc. v. News Am. Publ'g, Inc., 809 F.Supp. 267, 276 (S.D.N.Y.1992) ("[W]hen unauthorized use of another's mark is part of a communicative message and not a source identifier, the First Amendment is implicated in opposition to the trademark right."). Simply put, the trademark owner does not have the right to control public discourse whenever the public imbues his mark with a meaning beyond its source-identifying function. See Anti-Monopoly, Inc. v. Gen. Mills Fun Group, 611 F.2d 296, 301 (9th Cir.1979) ("It is the source-denoting function which trademark laws protect, and nothing more.").

B. There is no doubt that MCA uses Mattel's mark: Barbie is one half of Barbie Girl. But Barbie Girl is the title of a song about Barbie and Ken, a reference that — at least today — can only be to Mattel's famous couple. We expect a title to describe the underlying work, not to identify the producer, and Barbie Girl does just that.

The Barbie Girl title presages a song about Barbie, or at least a girl like Barbie. The title conveys a message to consumers about what they can expect to discover in the song itself; it's a quick glimpse of Aqua's take on their own song. The lyrics confirm this: The female singer, who calls herself Barbie, is "a Barbie girl, in [her] Barbie world." She tells her male counterpart (named Ken), "Life in plastic, it's fantastic. You can brush my hair, undress me everywhere/Imagination, life is your creation." And off they go to "party." The song pokes fun at Barbie and the values that Aqua contends she represents. See Cliffs Notes, Inc. v. Bantam Doubleday Dell Publ'g Group, 886 F.2d 490, 495-96 (2d Cir.1989). The female singer explains, "I'm a blond bimbo girl, in a fantasy world/Dress me up, make it tight, I'm your dolly."

The song does not rely on the Barbie mark to poke fun at another subject but targets Barbie herself. See Campbell v. Acuff-Rose Music, Inc., 510 U.S. 569 (1994); see also Dr. Seuss Ents., L.P. v. Penguin Books USA, Inc., 109 F.3d 1394, 1400 (9th Cir.1997). This case is therefore distinguishable from *Dr. Seuss,* where we held that the book *The Cat NOT in the Hat!* borrowed Dr. Seuss's trademarks and lyrics to get attention rather than to mock *The Cat in the Hat!* The defendant's use of the Dr. Seuss trademarks and copyrighted works had "no critical bearing on the substance or style of" *The Cat in the Hat!,* and therefore could not claim First Amendment protection. Id. at 1401. *Dr. Seuss* recognized that, where an artistic work targets the original and does not merely borrow another's property to get attention, First Amendment interests weigh more heavily in the balance. See id. at 1400-02; see also Harley-Davidson, Inc. v. Grottanelli, 164 F.3d 806, 812-13 (2d Cir.1999) (a parodist whose expressive work aims its parodic commentary at a trademark is given considerable leeway, but a claimed parodic use that makes no comment on the mark is not a permitted trademark parody use).

The Second Circuit has held that "in general the [Lanham] Act should be construed to apply to artistic works only where the public interest in avoiding consumer confusion outweighs the public interest in free expression." Rogers v. Grimaldi, 875 F.2d 994, 999 (2d Cir.1989); see also Cliffs Notes, 886 F.2d at 494 (quoting Rogers, 875 F.2d at 999). Rogers considered a challenge by the actress Ginger Rogers to the film Ginger and Fred. The movie told the story of two Italian cabaret performers who made a living by imitating Ginger Rogers and Fred Astaire. Rogers argued that the film's title created the false impression that she was associated with it.

At first glance, Rogers certainly had a point. Ginger was her name, and Fred was her dancing partner. If a pair of dancing shoes had been labeled Ginger and Fred, a dancer might have suspected that Rogers was associated with the shoes (or at least one of them), just as Michael Jordan has endorsed Nike sneakers that claim to make you fly through the air. But Ginger and Fred was not a brand of shoe; it was the title of a movie and, for the reasons explained by the Second Circuit, deserved to be treated differently.

A title is designed to catch the eye and to promote the value of the underlying work. Consumers expect a title to communicate a message about the book or movie, but they do not expect it to identify the publisher or producer. See Application of Cooper, 254 F.2d 611, 615-16 (C.C.P.A.1958) (A "title . . . identifies a specific literary work, . . . and is not associated in the public mind with the . . . manufacturer." (internal quotation marks omitted)). If we see a painting titled "Campbell's Chicken Noodle Soup," we're unlikely to believe that Campbell's has branched into the art business. Nor, upon hearing Janis Joplin croon "Oh Lord, won't you buy me a Mercedes-Benz?," would we suspect that she and the carmaker had entered into a joint venture. A title tells us something about the underlying work but seldom speaks to its origin: Though consumers frequently look to the title of a work to determine what it is about, they do not regard titles of artistic works in the same way as the names of ordinary commercial products. Since consumers expect an ordinary product to be what the name says it is, we apply the Lanham Act with some rigor to prohibit names that misdescribe such goods. But most consumers are well aware that they cannot judge a book solely by its title any more than by its cover. Rogers, 875 F.2d at 1000 (citations omitted).

Rogers concluded that literary titles do not violate the Lanham Act "unless the title has no artistic relevance to the underlying work whatsoever, or, if it has some artistic relevance, unless the title explicitly misleads as to the source or the content of the work." Id. at 999 (footnote omitted). We agree with the Second Circuit's analysis and adopt the Rogers standard as our own.

Applying Rogers to our case, we conclude that MCA's use of Barbie is not an infringement of Mattel's trademark. Under the first prong of Rogers, the use of Barbie in the song title clearly is relevant to the underlying work, namely, the song itself. As noted, the song is about Barbie and the values Aqua claims she represents. The song title does not explicitly mislead as to the source of the work; it does not, explicitly or otherwise, suggest that it was produced by Mattel.

The only indication that Mattel might be associated with the song is the use of Barbie in the title; if this were enough to satisfy this prong of the Rogers test, it would render Rogers a nullity. We therefore agree with the district court that MCA was entitled to summary judgment on this ground. We need not consider whether the district court was correct in holding that MCA was also entitled to summary judgment because its use of Barbie was a nominative fair use.

IV

Mattel separately argues that, under the Federal Trademark Dilution Act ("FTDA"), MCA's song dilutes the Barbie mark in two ways: It diminishes the mark's capacity to identify and distinguish Mattel products, and tarnishes the mark because the song is inappropriate for young girls. See 15 U.S.C. § 1125(c)...

Originally a creature of state law, dilution received nationwide recognition in 1996 when Congress amended the Lanham Act by enacting the FTDA. The statute protects "[t]he owner of a famous mark...against another person's commercial use in commerce of a mark or trade name, if such use begins after the mark has become famous and causes dilution of the distinctive quality of the mark." 15 U.S.C. § 1125(c). Dilutive uses are prohibited unless they fall within one of the three statutory exemptions discussed below.... Barbie easily qualifies under the FTDA as a famous and distinctive mark, and reached this status long before MCA began to market the Barbie Girl song. The commercial success of Barbie Girl establishes beyond dispute that the Barbie mark satisfies each of these elements.

We are also satisfied that the song amounts to a "commercial use in commerce." Although this statutory language is ungainly, its meaning seems clear: It refers to a use of a famous and distinctive mark to sell goods other than those produced or authorized by the mark's owner. *Panavision,* 141 F.3d at 1324-25. That is precisely what MCA did with the Barbie mark: It created and sold to consumers in the marketplace commercial products (the Barbie Girl single and the *Aquarium* album) that bear the Barbie mark.

MCA's use of the mark is dilutive. MCA does not dispute that, while a reference to Barbie would previously have brought to mind only Mattel's doll, after the song's popular success, some consumers hearing Barbie's name will think of both the doll and the song, or perhaps of the song only. This is a classic blurring injury and is in no way diminished by the fact that the song itself refers back to Barbie the doll. To be dilutive, use of the mark need not bring to mind the junior user alone. The distinctiveness of the mark is diminished if the mark no longer brings to mind the senior user alone.[4]

We consider next the applicability of the FTDA's three statutory exemptions. These are uses that, though potentially dilutive, are nevertheless permitted: comparative advertising; news reporting and commentary; and noncommercial use. 15 U.S.C. § 1125(c)(4)(B). The first two exemptions clearly do not apply; only the exemption for noncommercial use need detain us.

A "noncommercial use" exemption, on its face, presents a bit of a conundrum because it seems at odds with the earlier requirement that the junior use be a "commercial use in commerce." If a use has to be commercial in order to be dilutive, how then can it also be noncommercial so as to satisfy the exception of section 1125(c)(4)(B)? If the term "commercial use" had the same meaning in both provisions, this would eliminate one of the three statutory exemptions defined by this subsection, because any use found to be dilutive would, of necessity, not be noncommercial.

Such a reading of the statute would also create a constitutional problem, because it would leave the FTDA with no First Amendment protection for dilutive speech other than comparative advertising and news reporting. This would be a serious problem because the primary (usually exclusive) remedy for dilution is an injunction.[5]

---

4. Because we find blurring, we need not consider whether the song also tarnished the Barbie mark.
5. The FTDA provides for both injunctive relief and damages, but the latter is only available if plaintiff can prove a willful intent to dilute. 15 U.S.C. § 1125(c)(2).

As noted above, tension with the First Amendment also exists in the trademark context, especially where the mark has assumed an expressive function beyond mere identification of a product or service. See supra; *New Kids on the Block,* 971 F.2d at 306-08. These concerns apply with greater force in the dilution context because dilution lacks two very significant limitations that reduce the tension between trademark law and the First Amendment.

First, depending on the strength and distinctiveness of the mark, trademark law grants relief only against uses that are likely to confuse. See 5 McCarthy § 30:3, at 30-8 to 30-11; Restatement § 35 cmt. c at 370. A trademark injunction is usually limited to uses within one industry or several related industries. Dilution law is the antithesis of trademark law in this respect, because it seeks to protect the mark from association in the public's mind with wholly unrelated goods and services. The more remote the good or service associated with the junior use, the more likely it is to cause dilution rather than trademark infringement. A dilution injunction, by contrast to a trademark injunction, will generally sweep across broad vistas of the economy.

Second, a trademark injunction, even a very broad one, is premised on the need to prevent consumer confusion. This consumer protection rationale — averting what is essentially a fraud on the consuming public — is wholly consistent with the theory of the First Amendment, which does not protect commercial fraud. Cent. Hudson Gas & Elec. v. Pub. Serv. Comm'n, 447 U.S. 557, 566 (1980); see Thompson v. W. States Med. Ctr., 122 S.Ct. 1497 (2002) (applying *Central Hudson*). Moreover, avoiding harm to consumers is an important interest that is independent of the senior user's interest in protecting its business.

Dilution, by contrast, does not require a showing of consumer confusion, 15 U.S.C. § 1127, and dilution injunctions therefore lack the built-in First Amendment compass of trademark injunctions. In addition, dilution law protects only the distinctiveness of the mark, which is inherently less weighty than the dual interest of protecting trademark owners and avoiding harm to consumers that is at the heart of every trademark claim.

Fortunately, the legislative history of the FTDA suggests an interpretation of the "noncommercial use" exemption that both solves our interpretive dilemma and diminishes some First Amendment concerns: "Noncommercial use" refers to a use that consists entirely of noncommercial, or fully constitutionally protected, speech. See 2 Jerome Gilson et al., Trademark Protection and Practice § 5.12[1][c][vi], at 5-240 (this exemption "is intended to prevent the courts from enjoining speech that has been recognized to be [fully] constitutionally protected," "such as parodies"). Where, as here, a statute's plain meaning "produces an absurd, and perhaps unconstitutional, result[, it is] entirely appropriate to consult all public materials, including the background of [the statute] and the legislative history of its adoption." Green v. Bock Laundry Mach. Co., 490 U.S. 504, 527 (1989) (Scalia, J., concurring).

The legislative history bearing on this issue is particularly persuasive. First, the FTDA's sponsors in both the House and the Senate were aware of the potential collision with the First Amendment if the statute authorized injunctions against protected speech. Upon introducing the counterpart bills, sponsors in each house explained that the proposed law "will not prohibit or threaten noncommercial expression, such as parody, satire, editorial and other forms of expression that are not a part of a commercial transaction." 141 Cong. Rec. S1930610, S19310 (daily ed.

Dec. 29, 1995) (statement of Sen. Hatch); 141 Cong. Rec. H14317-01, H14318 (daily ed. Dec. 12, 1995) (statement of Rep. Moorhead). The House Judiciary Committee agreed in its report on the FTDA. H.R.Rep. No. 104-374, at 4 (1995), reprinted in 1995 U.S.C.C.A.N. 1029, 1031 ("The bill will not prohibit or threaten, noncommercial expression, as that term has been defined by the courts.").[6]

The FTDA's section-by-section analysis presented in the House and Senate suggests that the bill's sponsors relied on the "noncommercial use" exemption to allay First Amendment concerns. H.R. Rep. No. 104-374, at 8, reprinted in 1995 U.S.C.C.A.N. 1029, 1035 (the exemption "expressly incorporates the concept of 'commercial' speech from the 'commercial speech' doctrine, and proscribes dilution actions that seek to enjoin use of famous marks in 'non-commercial' uses (such as consumer product reviews)"); 141 Cong. Rec. S19306-10, S19311 (daily ed. Dec. 29, 1995) (the exemption "is consistent with existing case law[, which] recognize[s] that the use of marks in certain forms of artistic and expressive speech is protected by the First Amendment"). At the request of one of the bill's sponsors, the section-by-section analysis was printed in the Congressional Record. 141 Cong. Rec. S19306-10, S19311 (daily ed. Dec. 29, 1995). Thus, we know that this interpretation of the exemption was before the Senate when the FTDA was passed, and that no senator rose to dispute it.

To determine whether Barbie Girl falls within this exemption, we look to our definition of commercial speech under our First Amendment caselaw. See H.R.Rep. No. 104-374, at 8, reprinted in 1995 U.S.C.C.A.N. 1029, 1035 (the exemption "expressly incorporates the concept of 'commercial' speech from the 'commercial speech' doctrine"); 141 Cong. Rec. S19306-10, S19311 (daily ed. Dec. 29, 1995) (the exemption "is consistent with existing [First Amendment] case law").

"Although the boundary between commercial and noncommercial speech has yet to be clearly delineated, the 'core notion of commercial speech' is that it 'does no more than propose a commercial transaction.'" Hoffman v. Capital Cities/ABC, Inc., 255 F.3d 1180, 1184 (9th Cir.2001) (quoting Bolger v. Youngs Drug Prods Corp., 463 U.S. 60, 66 (1983)). If speech is not "purely commercial" — that is, if it does more than propose a commercial transaction — then it is entitled to full First Amendment protection. Id. at 1185-86 (internal quotation marks omitted).

In *Hoffman,* a magazine published an article featuring digitally altered images from famous films. Computer artists modified shots of Dustin Hoffman, Cary Grant, Marilyn Monroe and others to put the actors in famous designers' spring fashions; a still of Hoffman from the movie "Tootsie" was altered so that he appeared to be wearing a Richard Tyler evening gown and Ralph Lauren heels. Hoffman, who had not given permission, sued under the Lanham Act and for violation of his right to publicity. Id. at 1183.

The article featuring the altered image clearly served a commercial purpose: "to draw attention to the for-profit magazine in which it appear[ed]" and to sell more copies. Id. at 1186. Nevertheless, we held that the article was fully protected under the First Amendment because it included protected expression: "humor" and "visual and verbal editorial comment on classic films and famous actors." Id. at 1185 (internal quotation marks omitted). Because its commercial purpose was "inextricably

---

6. Our interpretation of the noncommercial use exemption does not eliminate all tension between the FTDA and the First Amendment because the exemption does not apply to commercial speech, which enjoys "qualified but nonetheless substantial protection." Bolger, 463 U.S. 60, 68, 103 S.Ct. 2875 (applying Central Hudson Gas & Electric Corp. v. Pub. Serv. Comm'n, 447 U.S. 557(1980)). See also Thompson, 122 S.Ct. at 1503-04 (same). It is entirely possible that a dilution injunction against purely commercial speech would run afoul of the First Amendment. Because that question is not presented here, we do not address it.

entwined with [these] expressive elements," the article and accompanying photographs enjoyed full First Amendment protection. Id.

Hoffman controls: Barbie Girl is not purely commercial speech, and is therefore fully protected. To be sure, MCA used Barbie's name to sell copies of the song. However, as we've already observed, see pp. 10489-90 supra, the song also lampoons the Barbie image and comments humorously on the cultural values Aqua claims she represents. Use of the Barbie mark in the song Barbie Girl therefore falls within the noncommercial use exemption to the FTDA. For precisely the same reasons, use of the mark in the song's title is also exempted....

Affirmed.

Appendix

"Barbie Girl" by Aqua

- Hiya Barbie!
- Hi Ken!
- You wanna go for a ride?
- Sure, Ken!
- Jump in!
- Ha ha ha ha!

(CHORUS)

I'm a Barbie girl, in my Barbie world
Life in plastic, it's fantastic
You can brush my hair, undress me everywhere
Imagination, life is your creation
Come on Barbie, let's go party!

(CHORUS)

I'm a blonde bimbo girl, in a fantasy world
Dress me up, make it tight, I'm your dolly
You're my doll, rock and roll, feel the glamour in pink
Kiss me here, touch me there, hanky-panky
You can touch, you can play
If you say "I'm always yours," ooh ooh

(CHORUS)

(BRIDGE)

Come on, Barbie, let's go party, ah ah ah yeah

Come on, Barbie, let's go party, ooh ooh, ooh ooh
Come on, Barbie, let's go party, ah ah ah yeah
Come on, Barbie, let's go party, ooh ooh, ooh ooh

Make me walk, make me talk, do whatever you please
I can act like a star, I can beg on my knees
Come jump in, be my friend, let us do it again
Hit the town, fool around, let's go party
You can touch, you can play
You can say "I'm always yours"
You can touch, you can play
You can say "I'm always yours"

(BRIDGE)

(CHORUS ×2)

(BRIDGE)

- Oh, I'm having so much fun!
- Well, Barbie, we're just getting started!
- Oh, I love you Ken!

COMMENTS AND QUESTIONS

1. To similar effect is Mattel v. Walking Mountain Prods., 353 F.3d 792 (9th Cir. 2003), where the court held that an artist who posed Barbie dolls nude in photographs in which they were attacked by vintage household appliances was not liable for trademark infringement or dilution, since his use of the term Barbie accurately stated the content of his works, he was criticizing or parodying Barbie, and his use was not commercial use.

2. *Nontrademark or "Nominative" Use.* In New Kids on the Block v. News Am. Publ'g, Inc., 971 F.2d 302 (9th Cir. 1992), USA Today asked its readers: "Who's the best on the block?" This feature commented that "New Kids on the Block are pop's hottest group. Which of the five is your fave? Or are they a turn off? . . . Each call [to a 900 number] costs 50 cents. Results in Friday's Life Section." In finding that this use of the band's name did not infringe its trademark rights, Judge Kozinski articulated the contours and elements of the nominative use doctrine:

> [W]e may generalize a class of cases where the use of the trademark does not attempt to capitalize on consumer confusion or to appropriate the cachet of one product for a different one. Such *nominative use* of a mark — where the only word reasonably available to describe a particular thing is pressed into service — lies outside the strictures of trademark law: Because it does not implicate the source-identification function that is the purpose of trademark, it does not constitute unfair competition; such use is fair because it does not imply sponsorship or endorsement by the trademark holder. "When the mark is used in a way that does not deceive the public we see no such sanctity in the word as to prevent its being used to tell the truth." Prestonettes, Inc. v. Coty, 264 U.S. 359, 368 (1924) (Holmes, J.).

To be sure, this is not the classic fair use case where the defendant has used the plaintiff's mark to describe the defendant's *own* product. Here, the New Kids trademark is used to refer to the New Kids themselves. We therefore do not purport to alter the test applicable in the paradigmatic fair use case. If the defendant's use of the plaintiff's trademark refers to something other than the plaintiff's product, the traditional fair use inquiry will continue to govern. But, where the defendant uses a trademark to describe the plaintiff's product, rather than its own, we hold that a commercial user is entitled to a nominative fair use defense provided he meets the following three requirements: First, the product or service in question must be one not readily identifiable without use of the trademark; second, only so much of the mark or marks may be used as is reasonably necessary to identify the product or service; and third, the user must do nothing that would, in conjunction with the mark, suggest sponsorship or endorsement by the trademark holder.

New Kids on the Block v. News Am. Publ'g, Inc., 971 F.2d at 307-08.

How would the court rule if USA Today had used the band's logo in the feature? How would the court in *Mattel* have responded if MCA used the Barbie logo on the Aqua album cover? What if the Barbie logo had a red circle with a slash symbol ("no") through it? What if it merely had a Barbie look-alike on the cover?

3. How does the non-trademark use doctrine square with the merchandising rights in logos and university names that some courts have granted to trademark owners?

4. Terri Welles, Playboy Magazine's Playmate of the Year in 1981, created a Web site offering photographs of Welles (some for free, others for sale), membership in her photo club, and links to other commercial sites. Her Web site contained "playboy" and "playmate" in metatags, the phrase "Playmate of the Year 1981" on the masthead of the Web site and in banner advertisements, and the repeated use of the abbreviation "PMOY'81" as a watermark on the Web pages. How should each of these uses be evaluated under the *New Kids on the Block* test? See Playboy Enterprises, Inc. v. Welles, 279 F.3d 796 (9th Cir. 2002) (Welles was free to use Playboy's trademarks to accurately advertise her affiliation with the magazine, but could not go beyond that to trade on those marks).

5. *Non-Dilution Parody Analysis.* Parody has been an effective defense not only against claims grounded in state dilution laws, but also against trademark infringement per se. See, e.g., Mutual of Omaha Ins. Co. v. Novak, 836 F.2d 397 (8th Cir. 1987) (affirming district court's finding of likelihood of confusion between plaintiff's famous "Mutual of Omaha" mark and defendant's anti-nuclear T-shirts, bearing funny picture and "Mutant of Omaha" legend), *cert. denied,* 488 U.S. 933 (1988). But a parody that confuses consumers will not be immune from trademark infringement. Cliffs Notes, Inc. v. Bantam Doubleday Dell Publishing Group, Inc., 886 F.2d 490 (2d Cir. 1989). The court observed that

> A parody must convey two simultaneous — and contradictory — messages: that it is the original, but also that it is not the original and is instead a parody. To the extent that it does only the former but not the latter, it is not only a poor parody but also vulnerable under trademark law, since the customer will be confused.

Id. at 494.

Another case demonstrates the application of the likelihood of confusion to parody cases. In Hormel Foods Corp. v. Jim Henson Prods., Inc., 73 F.3d 497 (2d Cir. 1996), the court held that defendant's use of a character named "SPA'AM" in a movie was an acceptable parody of the plaintiff's famous registered mark "SPAM" for meat-related products. The opinion treats the case strictly under "likelihood of confusion" principles. Although no confusion is found to be likely, and the parody is permitted, the First Amendment is never mentioned. Should the First Amendment protect even confusing parodies? See Steven M. Perez, Confronting Biased Treatment of Trademark Parody Under the Lanham Act, 44 Emory L.J. 1451 (1995) (arguing that trademark-based "likelihood of confusion" analysis makes parody cases too unpredictable and inconsistent with free speech interests).

6. *Parody/Satire Distinction.* As we discussed in Chapter 4, courts distinguish between parody and satire in applying copyright's fair use doctrine, finding parody of a copyrighted work much more likely to fall within the scope of the doctrine than satirical treatment. The court in *Mattel* suggests that Congress intended to exclude "parody, satire, editorial and other forms of expression that are not a part of a commercial transaction" from the scope of the dilution cause of action. Should this more clear and broad exemption also apply to traditional (likelihood of confusion) trademark claims?

In Elvis Presley Enterprises v. Capece, 141 F.3d 188 (5th Cir. 1998), the Fifth Circuit imported the parody/satire distinction from copyright law into traditional trademark analysis. The court held that defendant's 1960s theme bar could not use the name "Velvet Elvis" because it infringed on the rights of Elvis Presley's estate. The owner of the bar claimed that he was engaged in a legitimate parody of the kitsch associated with certain aspects of the 1960s. The Fifth Circuit concluded that "parody is not a defense to trademark infringement," and that in any event the "Velvet Elvis" was engaged in satire and not parody because its statement did not require the use of the Elvis trademark.

Does it make sense to import the parody/satire distinction into trademark law? Aren't the purposes of the laws different? In any event, how likely is it that consumers will be confused by the use of the "Velvet Elvis" name?

PROBLEMS

Problem 5-15. Toho, Inc. is the owner of the copyrights in several Japanese "Godzilla" movies as well as the registered U.S. trademark "Godzilla." Toho granted a license to a major American movie company to use the name and the monster in its recent high-budget film and various related products and promotional efforts.

Capitalizing on the hype surrounding the new Godzilla film, Edgar publishes a book cataloguing the history of Godzilla in film and print. He titles his book "Godzilla" and places a picture of the monster on the cover. Toho sues for trademark infringement.

How should the court rule? Would it matter if Toho had licensed a different writer to produce an "authorized" history?

Problem 5-16. Anheuser-Busch sells beer under a variety of brand names, including Michelob. As part of a general trend of brand proliferation in the beer industry, Busch has recently produced sub-brands of Michelob, including Michelob Dry, Michelob Lite, and Michelob Ice.

After the waters near a Busch plant are polluted in an oil spill, Balducci runs the accompanying advertisement [Figure 5-4] for "Michelob Oily" on the back of a humor magazine. The ad identifies itself as a parody in micro-print along the side. The pictures in the ad are takeoffs from those in real Michelob commercials.

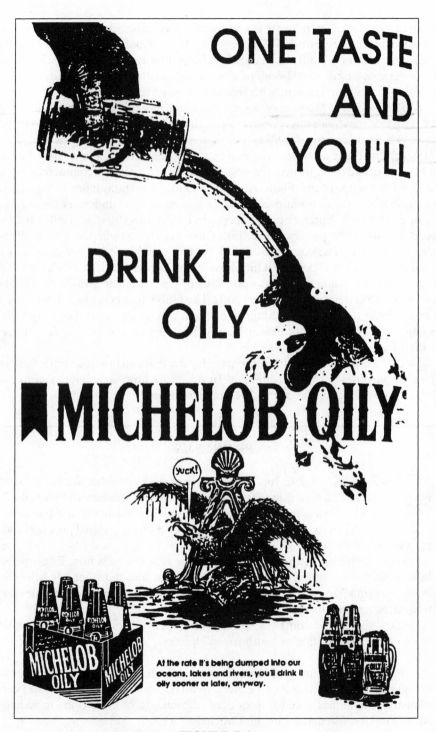

FIGURE 5-4
Parody of Michelob advertisement.

814

Anheuser-Busch sues Balducci for trademark infringement. At trial, Busch can show only a tiny percentage of consumers who thought Michelob Oily was a real product, but it proves that half the people surveyed believed Balducci should have to get permission from Busch to run the ad. Who should prevail?

Problem 5-17. The Coca-Cola Co. maintains an extremely strong and well-recognized trademark in the word "Coke" for its soft drink, and in the phrase "Enjoy Coke," particularly when used in connection with its red and white patterned logo. Gemini Rising, Inc., which distributes commercial posters, designs a poster with a logo identical to Coca-Cola's which reads "Enjoy Cocaine." After passersby who saw the poster complain to the local press, Coca-Cola brings suit against Gemini Rising. At trial, Coca-Cola offers evidence from a few members of the public who apparently believed that Coca-Cola had sponsored the posters, including one woman who threatened to organize a boycott of Coca-Cola products. What result?

Does your result change if Gemini Rising is a nonprofit political organization devoted to drug legalization? If it agrees to include a statement in medium-sized print at the bottom of the poster disclaiming any affiliation with the Coca-Cola Co.?

F. INTERNATIONAL ISSUES IN TRADEMARK

National borders are less of a barrier to economic exchange now than at almost any time in history. As economic activity continues its relentless drive toward worldwide scope, trademarks become even more important.[15] Just as the growth of the national economy in the United States set the stage for trademark law of national scope, the growth of the worldwide economy shows the importance of some sort of international trademark system.

U.S. Trademarks Abroad: The New Internationalization. Despite the obvious desirability of international trademark registration, the United States did not join an international convention that goes beyond the rather minimal protections afforded by the Paris Union until 2002. The Madrid Trademark Agreement (MTA) is the main international trademark convention.[16] The MTA does not itself protect any trademark rights; it simply facilitates trademark prosecution in member states. It is

---

15. Shari Caudron, Strike Up the Brand, Indus. Wk., Mar. 7, 1994, at 64 ("[I]n a world ruled by mass production and product standardization, the brand is the only remaining distinction among many products and services.").

16. Madrid Agreement Concerning the International Registration of Marks, April 14, 1891, 175 Consol. T.S. 57. This is an agreement under the Paris Convention for the Protection of Industrial Property, Stockholm Revision, done July 14, 1967, art. 2(1), 21 U.S.T. 1629, 1631, 828 U.N.T.S. 305, 313. There are approximately 40 members, including China and most of the European nations. See World Intellectual Property Org., World Intellectual Property Organization General Information 30 (1993). Various other international agreements exist, but for the most part they only cover relations between pairs of countries. See, e.g., 3 J. Thomas McCarthy, Trademarks and Unfair Competition § 29.10[2] (describing bilateral treaties signed by the U.S.).

effective in this regard, however: over 290,000 international Madrid registrations are now in force. On average, each international registration is extended to ten countries. Under the MTA, a trademark owner files a single registration with its domestic trademark office.[17] The owner may then extend this "basic registration" by filing, within six months of the original filing, an international application with the domestic trademark office designating other MTA member countries in which protection is sought.[18]

The U.S. recently joined the Madrid Protocol (MP), a corollary to the broader MTA. The MP considerably softens some of the harsher impacts of the MTA, especially on the United States.[19] Specifically, the MP (1) allows for filing in English or French, in the U.S. PTO or any other member country trademark office; (2) allows the filing of an international application based not only on a "basic" (or home country) registration but also on a basic *application* — a substantial advantage given the relatively stringent prosecution standards in the United States, which can lead to long gaps between applications and registrations, and especially now in light of intent-to-use registrations under the Lanham Act; and (3) substantially lessens the impact of a "central attack" provision, under which a trademark invalidated in its home country is invalid everywhere, by allowing an international registrant to convert a successfully attacked registration into a bundle of individual national registrations if the attack occurs during the first five years of the international registration.[20]

Although some problems remain,[21] the MP is a substantial step towards international harmonization. Indeed, some degree of harmonization would seem to be necessary, if only to keep up with trading partners. In Europe, for example, 1995 saw the advent of the Community Trademark Convention, a treaty providing for a single, Community Trade Mark (CTM) based on a single European trademark registration.[22] If further proof were needed, the CTC shows convincingly that the logic of

17. Madrid Agreement, art. 1(2), 828 U.N.T.S. at 393.
18. See arts. 3(1), 3[ter] (1), 828 U.N.T.S. at 395, 397; Art. 4(2), 828 U.N.T.S. at 399; Paris Convention, *supra*, Art. 4(C), 21 U.S.T. at 1632, 828 U.N.T.S. at 315 (six-month priority "grace period" for trademark registration filings).
19. Technically, members of the MP join the Madrid Union without signing the MTA. See MTA art. 1, 828 U.N.T.S. at 391 (states adhering to the Madrid Agreement "constitute a Special Union for the international registration of marks," known as the "Madrid Union."); Madrid Agreement Protocol, art. 1, at 9 (states adhering to the Madrid Agreement Protocol are also members of the Madrid Union).
20. MP, art. 9, at 43-45.
21. See, e.g., Allan Zelnick, The Madrid Protocol — Some Reflections, 82 Trademark Rep. 651 (1992). Zelnick argues that U.S. companies might still prefer to pursue foreign trademark rights by filing individual foreign applications because of the liberality of most foreign trademark systems regarding the classes of goods that can be identified with the mark in the application. (In many countries, a trademark application can specify *all* classes of goods — giving applicants substantially broader rights than under U.S. law.) According to Zelnick,

> [I]t seems likely that on the whole, most American trademark owners who file abroad today are unlikely to make use of the Madrid Protocol. The reason for this conclusion is that under our practice the specification of goods or services that will appear in domestic application, on which the international application will be based, must be narrowly drawn to the specific goods or services in respect of which the mark shall have been used in interstate or foreign commerce of the United States (in the case of applications based on use) or restricted to the particular items or services identified by their common trade names in respect of which the applicant shall have a bona fide intention to use the mark. Since the international application under the Madrid Protocol will have the same specification of goods or services as the United States application upon which it is based, we will have, in effect, transferred our domestic practice in this regard to United States trademark owners' international filings.

Id. at 652.
22. Council Regulation (EC) no. 40/94 of 20 December 1993 on the Community Trademark, OJ L 11/1. The CTM Regulation built on a 1980 Directive that substantially harmonized the national trademark laws of the EC member states. Primary features of the recently added CTM are:

trademark harmonization is compelling in this era of global trade and instantaneous communication.

Foreign Trademarks in the United States: Limited Internationalization. Even before U.S. adherence to an international trademark treaty, certain provisions of U.S. law made some slight concessions to the need for foreign trademark owners to register their marks in the United States. Primarily, section 44 of the Lanham Act allows foreign trademark owners to register their marks in the United States. Lanham Act §44, 15 U.S.C. §1126.

Although §44(d)(2) requires a statement of bona fide intent to use,[23] no subsequent proof of actual use in commerce in the United States is required. This allows foreign trademark owners to register their marks in the United States on terms somewhat more favorable than those extended to domestic applicants under the Lanham Act.

One court has held that the Lanham Act incorporates the substantive provisions of the Paris Convention governing trademarks and unfair competition, providing a cause of action for violation of those provisions. General Motors Corp. v. Lopez, 948 F. Supp. 684 (E.D. Mich. 1996).

Note on the "Gray Market"

United States law prohibits the importation of any goods bearing a registered United States trademark without the consent of the domestic trademark owner, even if the goods are genuine and even if they are imported by a licensed manufacturer (or the original trademark owner itself). 19 U.S.C. §1526. The so-called gray market or "parallel imports" problem is a troubling one for courts and commentators alike. The Supreme Court has described the problem as follows:

The gray market arises in any of three general contexts. The prototypical gray-market victim (case 1) is a domestic firm that purchases from an independent foreign firm the rights to register and use the latter's trademark as a United States trademark and to sell its foreign-manufactured products here. Especially where the foreign firm has already registered the trademark in the United States or where the product has already earned a reputation for quality, the right to use that trademark can be very valuable. If the foreign manufacturer could import the trademarked goods and distribute them here,

---

- Trademarks with renewable ten-year term, to be registered in the new European Trademark Office in Alicante, Spain.
- Marks valid in all member states but invalid everywhere if revoked.
- Substantive requirements for registration and grounds for opposition that would largely be familiar to American trademark practitioners, e.g., registration of nondistinctive marks upon proof of secondary meaning, and for a limited class of goods or services.
- The working languages are German, English, Spanish, French, and Italian. Applications can be filed in any Community language and are translated for free; opposition proceedings, however, are conducted in only one of the official languages, which must be selected at the time of filing.

See Eric P. Raciti, The Harmonization of Trademarks in the European Community: The Harmonization Directive and the Community Trademark, 78 J. Pat. & Trademark Off. Society 51 (1996).

23. This provision was added in 1988 (effective 1989) in response to the decision of Trademark Trial and Appeal Board's decision in Crocker National Bank v. Canadian Imperial Bank of Commerce, 223 U.S.P.Q. (BNA) 909 (TTAB 1984), which held valid a trademark registration application filed by a Canadian bank, despite the absence of an allegation of use anywhere in the world.

despite having sold the trademark to a domestic firm, the domestic firm would be forced into sharp intrabrand competition involving the very trademark it purchased. Similar intrabrand competition could arise if the foreign manufacturer markets its wares outside the United States, as is often the case, and a third party who purchases them abroad could legally import them. In either event, the parallel importation, if permitted to proceed, would create a gray market that could jeopardize the trademark holder's investment.

The second context (case 2) is a situation in which a domestic firm registers the United States trademark for goods that are manufactured abroad by an affiliated manufacturer. In its most common variation (case 2a), a foreign firm wishes to control distribution of its wares in this country by incorporating a subsidiary here. The subsidiary then registers under its own name (or the manufacturer assigns to the subsidiary's name) a United States trademark that is identical to its parent's foreign trademark. The parallel importation by a third party who buys the goods abroad (or conceivably even by the affiliated foreign manufacturer itself) creates a gray market. Two other variations on this theme occur when an American-based firm establishes abroad a manufacturing subsidiary corporation (case 2b) or its own unincorporated manufacturing division (case 2c) to produce its United States trademarked goods, and then imports them for domestic distribution. If the trademark holder or its foreign subsidiary sells the trademarked goods abroad, the parallel importation of the goods competes on the gray market with the holder's domestic sales.

In the third context (case 3), the domestic holder of a United States trademark authorizes an independent foreign manufacturer to use it. Usually the holder sells to the foreign manufacturer an exclusive right to use the trademark in a particular foreign location, but conditions the right on the foreign manufacturer's promise not to import its trademarked goods into the United States. Once again, if the foreign manufacturer or a third party imports into the United States, the foreign-manufactured goods will compete on the gray market with the holder's domestic goods.

K-Mart Corp. v. Cartier, Inc., 486 U.S. 281, 286-287 (1988).

What is wrong with importing gray market goods? In this case, unlike the typical trademark dispute, there is no question that the goods are genuine and are being sold as precisely what they are. Further, there is no question that they originated with the trademark owner (at least at some point). If a gray market transaction occurred entirely within the United States, would it constitute trademark infringement? If not, is there any reason to treat importation differently?

One obvious rationale for prohibiting gray market imports is that the United States trademark owner is losing the benefit of what it thought was an "exclusive" right to sell the trademarked goods in the United States. But does it really have such an exclusive right? Certainly, the right to use the Ford trademark on cars in the United States does not prevent bona fide purchasers of Ford cars from identifying them as such upon resale. And even if the United States trademark owner did obtain the exclusive rights to the first sale of goods in the United States from the manufacturer, isn't the problem here really one of breach of contract?

Another justification for prohibiting gray market goods is if they are of different quality than the goods normally sold in the United States. If so, consumers may be confused. Significantly, courts have held that gray market importation is legal unless the imported goods differ in some relevant way from the goods the trademark owner sold in the United States. American Circuit Breaker Corp. v. Oregon Breakers Inc.,

406 F.3d 577 (9th Cir. 2005) (circuit breakers that differed only in color from those sold in the United States could legally be imported after first sale abroad); SKF USA v. International Trade Comm'n, 2005 WL 2218460 (Fed. Cir. Sept. 14, 2005) (variation in goods could include variation in technical support provided, but in this case plaintiff did not uniformly provide support for its U.S. sales, and so could not complain of gray market imports without technical support).

Section 1526 of 19 U.S.C., which regulates gray market goods, may violate Article 2 of the Paris Convention because it treats United States trademark holders differently from foreign trademark holders, and United States infringers differently from foreign infringers. See Raimund Steiner & Robert Sabath, Intellectual Property and Trade Law Approaches to Gray Market Importation, and the Restructuring of Transnational Entities to Permit Blockage of Gray Goods in the United States, 15 Wm. Mitchell L. Rev. 433, 441 (1989) (after *K-Mart,* to obtain benefits of Section 526 of the Tariff Act of 1930 and the Lanham Act, companies must separate ownership of foreign and domestic trademarks); see generally Note, The Use of Copyright Law to Block the Importation of Gray-Market Goods: The Black and White of It All, 23 Loyola (LA) L. Rev. 645 (1990).

There is a debate in the economics literature about the wisdom of permitting gray market imports. The argument generally runs along these lines: those favoring restrictions on the gray market cite the importance of encouraging investments in brand quality by retailers, coupled with differing international product quality standards. In essence, the argument is that retailers will not engage in optimal expenditures to promote the product (advertising, clean showrooms, knowledgeable sales staff, etc.) if gray market imports will undercut the retailers' prices. On the opposite side, those who support the gray market say it undercuts blatant price discrimination. See Robert J. Staaf, International Price Discrimination and the Gray Market, 4 Intellectual Prop. J. 301 (1989) (gray market presents an opportunity for arbitrage, allowing competitors to circumvent attempted price discrimination and therefore benefiting consumers).

The Problem of Worldwide Famous Marks. Worldwide trademark protection is based on proof of either registration or use in each country in which protection is sought. In theory this creates a significant burden for companies with marks that are famous worldwide: they must register their mark in every country in the world or risk losing rights to that mark to a competing registrant. In fact a number of opportunistic individuals have registered famous marks such as "Coke" in countries where Coca-Cola does not yet do business.

In an effort to deal with this problem, courts around the world have increasingly been willing to recognize goodwill in a world-famous mark even if the mark is not actually used in the country. See Paris Convention art. 6bis. Courts have held that K-Mart was entitled to prevent a competing use in Jamaica (see Kmart Corp. v. Kay Mart Ltd., Suit No. C.L. 1993, K066) and that McDonald's was permitted to "take back" rights in South Africa to marks it had ceased using there during the world embargo of that country. Ironically, the United States seems less willing than other nations to act against domestic citizens who trade on the goodwill of foreign marks. See Buti v. Impressa Perosa S.R.L., 139 F.3d 98 (2d Cir. 1998); Person's Co. v. Christman, 900 F.2d 1565 (Fed. Cir. 1990).

G. REMEDIES

1. Injunctions

Like patent law and much of copyright law, trademark remedies are organized around a property rule. This means that infringers have no "right" to use the trademark upon payment of damages; trademark owners are entitled to injunctions against infringement as a matter of course. The application of a property rule makes sense in the trademark context because trademarks serve to protect a unique good — the plaintiff's business goodwill. Infringers who trade on or dilute (and thus appropriate or destroy) a plaintiff's goodwill cannot simply "buy back" that goodwill with money. Once it is dissipated, it is gone forever.

As is normal with property rule regimes, the principal remedy for trademark infringement is an injunction. Section 34(a) of the Lanham Act, 15 U.S.C. § 1116(a), gives courts the power to grant injunctions "to prevent the violation of any right of the registrant of a mark . . . or to prevent a violation under section 1125(a) of this title [Lanham Act § 43(a)]." Plaintiffs are entitled to injunctions as a matter of course once infringement is proven. See Champion Spark Plug Co. v. Sanders, 331 U.S. 125, 130-131 (1947). Injunctions may include prohibitions against the sale of goods or the use of a mark on goods or in advertising, or they may be affirmative injunctions requiring corrective advertising or the inclusion of disclaimers on products or advertising.

Trademark law is not a typical property rule, however. Owners of real property are entitled to sell it to whomever they wish. They may also let other people rent it for a fee. Trademark owners have no such unfettered rights to sell or license their trademarks. As discussed in section E above, both the sale and the licensing of trademarks are subject to significant legal restrictions. Having given trademark owners a property right to enable them to protect their goodwill, the government appears unwilling to allow trademark owners to do what they see fit with that right. Instead, restrictions on alienation of trademarks are designed to make sure that the trademark is in fact used to promote the goodwill of the associated business.

2. Damages

a. Infringer's Gain and Mark Owner's Loss

While injunctions are the fundamental Lanham Act remedy, they are not the only one. Damages are also sometimes available in trademark infringement cases. Section 35(a) of the Lanham Act, 15 U.S.C. § 1117(a), provides that plaintiffs are entitled:

> to recover (1) defendant's profits, (2) any damages sustained by the plaintiff, and (3) the costs of the action. . . . In assessing profits the plaintiff shall be required to prove defendant's sales only; defendant must prove all elements of cost or deduction claimed. In assessing damages the court may enter judgment . . . for any sum above the amount found as actual damages, not exceeding three times such amount.

Lindy Pen Company, Inc. v. Bic Pen Corporation
United States Court of Appeals for the Ninth Circuit
982 F.2d 1400 (9th Cir. 1993)

ROLL, District Judge:

This trademark infringement case brought by Lindy Pen Company, Inc. against Bic Pen Corporation returns to this court for the third time seeking review of the district court's damages order upon remand. For the reasons set forth below, we affirm the decision of the district court and deny Bic's cross-appeal.

Introduction

The origins of this case go back to the mid-1960s when Appellant Lindy Pen Company (Lindy) and Appellee Bic Pen Corporation (Bic) were competitors in the production and manufacture of ball point pens. Each marketed a fine-point tip for use by accountants and auditors. In 1965, before Lindy's trademark was perfected and registration issued, Bic used the mark "Auditor's" on its pen barrels. Lindy contacted Bic, making a claim to the mark, and Bic voluntarily agreed to stop using it. On September 20, 1966, United States Trademark Registration No. 815,488 was issued to Lindy for the word "Auditor's." Lindy Pen Co., Inc. v. Bic Pen Corp., 550 F. Supp. 1056, 1059 (C.D. Cal. 1982). Lindy perfected the registration by filing the appropriate documents pursuant to federal statute. Id.

Fourteen years later, Bic adopted the legend "Auditor's Fine Point" to describe a certain pen model. Prior to this use, Bic researched the term "Auditor's" and found that at least three other manufacturers employed a variation of the word "Auditor's" in their marketing materials. This investigation revealed that Lindy also used the term, but that Lindy exerted no proprietary interest over it in its advertising.

Lindy learned of Bic's renewed use of the mark and filed suit in 1980 alleging trademark infringement, unfair competition, breach of contract, and trademark dilution. The district court entered judgment in favor of Bic on all claims. The circuit court upheld the district court's ruling that Bic did not infringe in the major retail markets, but remanded the case to determine whether there was a likelihood of confusion in telephone order sales. Lindy Pen Co., Inc. v. Bic Pen Corp., 725 F.2d 1240 (9th Cir. 1984), cert. denied, 469 U.S. 1188, 105 S. Ct. 955, 83 L. Ed. 2d 962 (1985) (Lindy I). The court specifically found that "there is no evidence of actual confusion [and] that Bic adopted the designation 'Auditor's fine point' without intent to capitalize on Lindy's goodwill. . . ." Id. at 1246.

Upon remand, the district court determined that there was a likelihood of confusion in the telephone order market because the purchaser did not have the opportunity to view the pens at the time of purchase. The court went on to find, however, that the confusion could be cured upon receipt of the goods because the purchaser could return the product. The circuit court again disagreed and found that Lindy had established a likelihood of confusion in the telephone order market which post-sale inspection could not cure. Lindy Pen Co., Inc. v. Bic Pen Corp., 796 F.2d 254 (9th Cir. 1986) (Lindy II). In words which now comprise the nub of the current appeal, this court ordered that the case be remanded to district court "with instructions to enter an order enjoining Bic from using the word 'Auditor's' on or in connection with its pens. We additionally instruct the district court to order an accounting and to award damages and other relief as appropriate." Id. . . .

Accounting of Profits

Following the damages proceeding, the district court held that an accounting of profits was inappropriate because Bic's infringement was innocent and accomplished without intent to capitalize on Lindy's trade name. Section 35 of the Lanham Act, 15 U.S.C. § 1117(a), governs the award of monetary remedies in trademark infringement cases and provides for an award of defendant's profits, any damages sustained by the plaintiff, and the costs of the action.

The Supreme Court has indicated that an accounting of profits follows as a matter of course after infringement is found by a competitor. Hamilton-Brown Shoe Co. v. Wolf Bros. & Co., 240 U.S. 251, 259, 36 S. Ct. 269, 271, 60 L. Ed. 629, 634 (1916). Nonetheless, an accounting of profits is not automatic and must be granted in light of equitable considerations. Champion Spark Plug Co. v. Sanders, 331 U.S. 125, 131, 67 S. Ct. 1136, 1139, 91 L. Ed. 1386, 1391 (1947). Where trademark infringement is deliberate and willful, this court has found that a remedy no greater than an injunction "slights" the public. Playboy Enterprises, Inc. v. Baccarat Clothing Co., Inc., 692 F.2d 1272, 1274 (9th Cir. 1982). This standard applies, however, only in those cases where the infringement is "willfully calculated to exploit the advantage of an established mark." Id. The intent of the infringer is relevant evidence on the issue of awarding profits and damages and the amount. Maier Brewing Co., 390 F.2d at 123. When awarding profits, the court is cautioned that the "Plaintiff is not . . . entitled to a windfall." Bandag, Inc., 750 F.2d at 918.

Lindy posits that Bic's actions require an accounting as a remedy for willful infringement. The parties agree that as early as 1965, Bic had knowledge of Lindy's claim to trademark rights in the term "Auditor's" when Bic published a catalog advertising pens with the mark stamped on their barrels. Bic voluntarily suspended use of the mark after Lindy informed Bic of its claim to the term. Fourteen years after this exchange, Bic began using the legend "Auditor's Fine Point" on its extra-fine point pens, the subject of the current litigation.

Willful infringement carries a connotation of deliberate intent to deceive. Courts generally apply forceful labels such as "deliberate," "false," "misleading," or "fraudulent" to conduct that meets this standard. . . . Indeed, this court has cautioned that an accounting is proper only where the defendant is "attempting to gain the value of an established name of another." Maier Brewing Co., 390 F.2d at 123.

The present case simply does not involve willful infringement. Evidence taken at the damages proceeding shows that Lindy in general, and its Auditor's line in particular, was experiencing an overall business decline. The trial court also found that any knowledge that Bic may have had of Lindy's mark stemming from the 1965 interchange was conducted by outside counsel, thereby implying that Bic's knowledge of Lindy's interest was attenuated at best. Lindy Pen, 550 F. Supp. at 1059. The district court also determined that Lindy's mark was weak and that there was no evidence of actual confusion. Id. at 1058-60. Based on these facts, it was reasonable for the district court to conclude that Bic's actions were not "willfully calculated to exploit the advantage of an established mark," Bandag, Inc., 750 F.2d at 921, and that Bic's conduct did not rise to the level of willfulness which would have mandated an award.[4]

---

4. Even assuming that Bic could be termed a willful infringer, "willful infringement may support an award of profits to the plaintiff, but does not require one." Faberge, Inc. v. Saxony Products, Inc., 605 F.2d 426, 429 (9th Cir. 1979) (citing Maier Brewing Co., 390 F.2d at 121).

This circuit has announced a deterrence policy in response to trademark infringement and will grant an accounting of profits in those cases where infringement yields financial rewards. Playboy Enterprises, Inc. v. Baccarat Clothing Co., Inc., 692 F.2d 1272, 1274 (9th Cir. 1982) (plaintiff damaged through defendant's intentional use of counterfeit product labels). However, the policy considerations which clearly justified an award in *Playboy Enterprises* are not present in this case. Unlike the *Playboy Enterprises* plaintiff, Lindy's trademark was weak and Bic's infringement was unintentional. Moreover, Bic's major position in the pen industry makes it clear that it was not trading on Lindy's relatively obscure name. Accordingly, the policy considerations announced by *Playboy Enterprises* would be trivialized by insisting on an accounting in this case. See ALPO Petfoods Inc., 913 F.2d at 969 ("deterrence is too weak and too easily invoked a justification for the severe and often cumbersome remedy of a profits award...."). To award profits in this situation would amount to a punishment in violation of the Lanham Act which clearly stipulates that a remedy "shall constitute compensation not a penalty." 15 U.S.C. § 1117(a).

Award of Damages

15 U.S.C. § 1117(a) further provides for an award, subject to equitable principles, of "any damages sustained by the plaintiff...." A plaintiff must prove both the fact and the amount of damage. 2 J. Thomas McCarthy, Trademarks and Unfair Competition § 30:27, at 511 (2d ed. 1984). Damages are typically measured by any direct injury which a plaintiff can prove, as well as any lost profits which the plaintiff would have earned but for the infringement. Id. at 509. Because proof of actual damage is often difficult, a court may award damages based on defendant's profits on the theory of unjust enrichment. Id. at 511. See also Bandag, Inc., 750 F.2d at 918.

The district court gave Lindy the opportunity to prove its damages under both methods: actual damages in the form of its lost profits, or if that proved too difficult, through proof of Bic's unjust enrichment in the form of Bic's profits....

After final briefing, the court concluded that Lindy had failed to show any actual damage because it did not put forth sufficient proof of its lost profits. To establish damages under the lost profits method, a plaintiff must make a "prima facie showing of reasonably forecast profits." 2 J.T. McCarthy, Trademarks and Unfair Competition § 30:27, at 511 (2d ed. 1984). The court found Lindy's calculations to be irreparably flawed because Lindy did not isolate its own telephone order sales from total pen sales. Consequently, Lindy's calculations contained items in which no likelihood of confusion existed and therefore were inappropriately included. The district court reasoned that it had no rational basis upon which to estimate an award as to the infringing items and accordingly denied Lindy's request.

Trademark remedies are guided by tort law principles. Id. at 509 ("Plaintiff's damages should be measured by the tort standard under which the infringer-tortfeasor is liable for all injuries caused to plaintiff by the wrongful act...."). As a general rule, damages which result from a tort must be established with reasonable certainty. Dan B. Dobbs, Remedies § 3.3, at 151 (1973). The Supreme Court has held that "damages are not rendered uncertain because they cannot be calculated with absolute exactness," yet, a reasonable basis for computation must exist. Eastman

Kodak Co. v. Southern Photo Materials Co., 273 U.S. 359, 379, 47 S. Ct. 400, 405, 71 L. Ed. 684, 691 (1927)....

Lindy produced evidence of its total pen sales, as available, for the designated time period. Although it divided its sales into total sales and specific sales of Auditor's, it failed to further subdivide its data into the category of telephone order sales. Lindy was in the best position to identify its own sales, but declined to provide the court with any evidence of its loss caused by Bic's wrong doing. Although Lindy offers excuses for this deficiency, its explanations do not negate the fact that Lindy never furnished the court any reasonable estimate of its own sales. It would have been error for the district court to select an arbitrary percentage of total sales to represent the more narrow submarket of telephone sales. The court was correct, therefore, in finding that Lindy failed to sustain its burden of proving reasonably forecast profits.

Lindy also sought an accounting of Bic's profits based on a theory of unjust enrichment. Lindy assesses as error the district court's requirement that Lindy separate Bic's telephone order sales from total sales of goods. Lindy maintains that this requirement had the effect of shifting the burden of proving infringing sales from the infringer to the trademark holder. However, an accounting is intended to award profits only on sales that are attributable to the infringing conduct. The plaintiff has only the burden of establishing the defendant's gross profits from the infringing activity with reasonable certainty. Once the plaintiff demonstrates gross profits, they are presumed to be the result of the infringing activity. Mishawaka Rubber & Woolen Mfg. Co. v. S. S. Kresge Co., 316 U.S. 203, 206-07, 62 S. Ct. 1022, 1024-25, 86 L. Ed. 1381, 1385-86 (1942). The defendant thereafter bears the burden of showing which, if any, of its total sales are not attributable to the infringing activity, and, additionally, any permissible deductions for overhead. 15 U.S.C. § 1117(a).

Lindy failed to come forward with any evidence of sales of the Bic "Auditor's Fine Point" in the infringing market. Lindy instead brought forth proof of Bic's total sales. Lindy averred to the court that a division of Bic's sales into the telephone submarket "is impossible from Bic's records since Bic never separated its pens according to telephone sales...." To the contrary, Lindy had access through discovery to Bic's records from which a reasonable estimate could have been accomplished.

We find that Lindy's appeal on the question of actual damages fails due to a lack of proof at the damages proceeding....

COMMENTS AND QUESTIONS

1. How is the *public* "slighted" by a court's failure to award damages for willful infringement? Does it matter to the public whether an infringer pays damages to the trademark owner or not? Surely, any confusion that the public has suffered is not ameliorated because the trademark owner receives compensation. Why isn't an injunction, which the courts agree is a sufficient remedy in the case of "innocent infringement," sufficient in cases of intentional infringement as well?

2. Is an accounting for profits ever possible in a dilution case? If so, why? If the plaintiff is compensated for her losses caused by the dilution, does it make any sense to take away the defendant's profits in an unrelated market as well?

3. Who should bear the burden of proof in determining profits? See Mishawaka Rubber & Woolen Mfg. Co. v. S. S. Kresge Co., 316 U.S. 203, 206-07 (1942)

(defendant bears the burden of proving that not all profits associated with a sale should be awarded to plaintiff). The Court reasoned: "There may well be a windfall to the trade-mark owner where it is impossible to isolate the profits which are attributable to the use of the infringing mark. But to hold otherwise would give the windfall to the wrongdoer."

Suppose that a defendant has established significant goodwill of its own between the time it adopted a mark and the time it is found to have infringed. If the plaintiff is entitled to an injunction, does the defendant have any way of recovering the value of its investment in goodwill? Is it at least entitled to a setoff against damages for its monetary losses as a result of the injunction? Does it matter if the case is a dilution case and the defendant is in a separate market?

4. In general, many courts follow the holding in Champion Spark Plug v. Sanders, 331 U.S. 125, 130-31, 67 S. Ct. 1136, 1139, 91 L. Ed. 1386 (1947), that an accounting is appropriate only when fraud or palming off is present, and will grant an accounting of defendant's profits only if the defendant acted in bad faith. But in some circuits, lost profits may be awarded even in the absence of proof of willfulness. See, e.g., George Basch Co., Inc. v. Blue Coral, Inc., 968 F.2d 1532, 1540 (2d Cir. 1992) ("[I]n the absence of . . . a showing [of willfulness], a plaintiff is not foreclosed from receiving monetary relief. Upon proof of actual consumer confusion, a plaintiff may still obtain damages—which, in turn, may be inclusive of plaintiff's own lost profits."). For an interesting recent case that explores what constitutes "bad faith" for enhanced damages purposes, see Int'l Star Class Yacht Racing Assoc. v. Tommy Hilfiger U.S.A., Inc., 1999 WL 108739 (S.D. N.Y., Mar. 3, 1999) (on remand from Second Circuit, adhering to earlier opinion finding no bad faith in adoption of plaintiff's mark, despite very limited search of trademark databases). See also Roger D. Blair & Thomas F. Cotter, An Economic Analysis of Damage Rules in Intellectual Property Law, 39 Wm. & Mary L. Rev. 1585, 1691 (1998) (arguing that the law should require "some level of search activity" for a trademark defendant to escape a damage award). Cf. Banff, Ltd. v. Colberts, Inc., 996 F.2d 33, 35 (2d Cir.), *cert. denied,* 510 U.S. 1010 (1993).

5. On the availability of defendant's profits, see Minnesota Pet-Breeders, Inc. v. Schell & Kampeter, Inc., 843 F. Supp. 506, 512-13 (D. Minn. 1993), *aff'd on other grounds,* 41 F.3d 1242 (8th Cir. 1994):

> A number of courts have held that an accounting of defendant's profits in cases of trademark infringement or unfair competition is warranted if (1) the plaintiff sustained damage from the infringement, (2) the infringer is unjustly enriched, and (3) [it is] necessary to deter a willful infringer from doing so again. At least one court has held that "deterrence alone cannot justify such an award [of infringing defendant's profits]." [Alpo Petfoods, Inc. v. Ralston Purina Co., 913 F.2d 958, 969 (D.C. Cir. 1990).]

Note that in affirming the decision below, the Eighth Circuit added these qualifications: "[I]n this case, in which there was no actual competition between the trademarked and the infringing products, it is most likely that MPB would be entitled to injunctive but not monetary relief, and virtually inconceivable that only monetary relief would be appropriate." Id. at 1247-48.

6. For a detailed exposition of how to calculate the infringer's profits in an accounting, see Dennis S. Corgill, Measuring the Gains from Trademark Infringement, 65 Fordham L. Rev. 1909 (1997).

7. While the general rule is that damages in a trademark case are available only in cases of willful infringement, at least one court has held that damages will sometimes be permissible even in the absence of willful infringement. Quick Technologies v. Sage Group, 313 F.3d 338 (5th Cir. 2002).

b. Corrective Advertising

Big O Tire Dealers, Inc. v. The Goodyear Tire & Rubber Company
United States Court of Appeals for the Tenth Circuit
561 F.2d 1365 (10th Cir. 1977)

LEWIS, Chief Judge.

This civil action was brought by Big O Tire Dealers, Inc., ("Big O") asserting claims of unfair competition against The Goodyear Tire & Rubber Co. ("Goodyear") based upon false designation of origin under 15 U.S.C. § 1125(a) and common law trademark infringement. After a ten-day trial and three days of deliberation, the jury returned the following verdict:

> We the jury in the above entitled cause, upon our oath do say that we find the following as our verdict herein:
> Upon the claim of liability for trademark infringement we find for Big O Inc.
> Upon the claim of liability for false designation of origin we find for Goodyear.
> Upon the claim for trademark disparagement we find for Big O Inc.
> We find that plaintiff has proven special compensatory damages in the amount of $ None.
> We find that plaintiff has proven general compensatory damages in the amount of $2,800,000.
> We assess punitive or exemplary damages in the amount of $16,800,000.
> Dated September 4, 1975.

Filing a comprehensive post-trial opinion the United States District Court for the District of Colorado entered judgment on the jury's verdict, permanently enjoined Goodyear from infringing on Big O's trademark, and dismissed Goodyear's counterclaim for equitable relief. 408 F. Supp. 1219. Goodyear appeals that judgment.

Big O is a tire-buying organization which provides merchandising techniques, advertising concepts, operating systems, and other aids to approximately 200 independent retail tire dealers in 14 states who identify themselves to the public as Big O dealers. These dealers sell replacement tires using the Big O label on "private brand" tires. They also sell other companies' brands such as B. F. Goodrich and Michelin Tires. At the time of trial Big O's total net worth was approximately $200,000.

Goodyear is the world's largest tire manufacturer. In 1974 Goodyear's net sales totaled more than $5.25 billion and its net income after taxes surpassed $157 million. In the replacement market Goodyear sells through a nationwide network of company-owned stores, franchise dealers, and independent retailers.

In the fall of 1973 Big O decided to identify two of its lines of private brand tires as "Big O Big Foot 60" and "Big O Big Foot 70." These names were placed on the sidewall of the respective tires in raised white letters. The first interstate shipment of these tires occurred in February 1974. Big O dealers began selling these tires to the public in April 1974. Big O did not succeed in registering "Big Foot" as a trademark with the United States Patent and Trademark Office....

In July 1974 Goodyear decided to use the term "Bigfoot" in a nationwide advertising campaign to promote the sale of its new "Custom Polysteel Radial" tire. The name "Custom Polysteel Radial" was molded into the tire's sidewall. Goodyear employed a trademark search firm to conduct a search for "Bigfoot" in connection with tires and related products. This search did not uncover any conflicting trademarks. After this suit was filed Goodyear filed an application to register "Bigfoot" as a trademark for tires but withdrew it in 1975. Goodyear planned to launch its massive, nationwide "Bigfoot" advertising campaign on September 16, 1974.

On August 24, 1974, Goodyear first learned of Big O's "Big Foot" tires. Goodyear informed Big O's president, Norman Affleck, on August 26 of Goodyear's impending "Bigfoot" advertising campaign. Affleck was asked to give Goodyear a letter indicating Big O had no objection to this use of "Bigfoot." When Affleck replied he could not make this decision alone, it was suggested Affleck talk with John Kelley, Goodyear's vice-president for advertising.

Affleck called Kelley and requested more information on Goodyear's impending advertising campaign. A Goodyear employee visited Affleck on August 30 and showed him rough versions of the planned Goodyear "Bigfoot" commercials and other promotional materials. On September 10, Affleck and two Big O directors met in New Orleans, with Kelley and Goodyear's manager of consumer market planning to discuss the problem further. At this time the Big O representatives objected to Goodyear using "Bigfoot" in connection with tires because they believed any such use would severely damage Big O. They made it clear they were not interested in money in exchange for granting Goodyear the right to use the "Bigfoot" trademark, and asked Goodyear to wind down the campaign as soon as possible. Goodyear's response to this request was indefinite and uncertain.

During the trial several Goodyear employees conceded it was technically possible for Goodyear to have deleted the term "Bigfoot" from its television advertising as late as early September. However, on September 16, 1974, Goodyear launched its nationwide "Bigfoot" promotion on ABC's Monday Night Football telecast. By August 31, 1975, Goodyear had spent $9,690,029 on its massive, saturation campaign.

On September 17 [1974] Affleck wrote Kelley a letter setting forth his understanding of the New Orleans meeting that Goodyear would wind up its "Bigfoot" campaign as soon as possible. Kelley replied on September 20, denying any commitment to discontinue use of "Bigfoot" and declaring Goodyear intended to use "Bigfoot" as long as it continued to be a helpful advertising device.

On October 9 Kelley told Affleck he did not have the authority to make the final decision for Goodyear and suggested that Affleck call Charles Eaves, Goodyear's executive vice-president. On October 10 Affleck called Eaves and Eaves indicated the possibility of paying Big O for the use of the term "Bigfoot." When Affleck stated no interest in the possibility Eaves told him Goodyear wished to avoid litigation but that if Big O did sue, the case would be in litigation long enough that Goodyear might obtain all the benefits it desired from the term "Bigfoot."

This was the final communication between the parties until Big O filed suit on November 27, 1974. The district court denied Big O's request for a temporary restraining order and a preliminary injunction. After judgment was entered on the jury's verdict for Big O, Goodyear appealed to this court. Goodyear's allegations of error are discussed below....

IV

...Big O does not claim nor was any evidence presented showing Goodyear intended to trade on the goodwill of Big O or to palm off Goodyear products as being those of Big O. Instead, Big O contends Goodyear's use of Big O's trademark created a likelihood of confusion concerning the source of Big O's "Big Foot" tires.

The facts of this case are different from the usual trademark infringement case. As the trial judge stated, the usual trademark infringement case involves a claim by a plaintiff with a substantial investment in a well established trademark. The plaintiff would seek recovery for the loss of income resulting from a second user attempting to trade on the goodwill associated with that established mark by suggesting to the consuming public that his product comes from the same origin as the plaintiff's product. The instant case, however, involves reverse confusion wherein the infringer's use of plaintiff's mark results in confusion as to the origin of plaintiff's product. Only one reported decision involves the issue of reverse confusion. In Westward Coach Mfg. Co. v. Ford Motor Co., 7 Cir., 388 F.2d 627, *cert. denied,* 392 U.S. 927, 88 S. Ct. 2286, 20 L. Ed. 2d 1386, the court held reverse confusion is not actionable as a trademark infringement under Indiana law.

Consequently, Goodyear argues the second use of a trademark is not actionable if it merely creates a likelihood of confusion concerning the source of the first user's product. Since both parties agree Colorado law is controlling in this case, we must decide whether this so-called reverse confusion is actionable under Colorado law. To our knowledge, the Colorado courts have never considered whether a second use creating the likelihood of confusion about the source of the first user's products is actionable. However, the Colorado Court of Appeals in deciding a trade name infringement case involving an issue of first impression, cogently pointed out that the Colorado Supreme Court "has consistently recognized and followed a policy of protecting established trade names and preventing public confusion and the tendency has been to widen the scope of that protection." Wood v. Wood's Homes Inc., 33 Colo. App. 285, 519 P.2d 1212, 1215-16.

Using that language as a guiding light in divining what Colorado law is on this issue of first impression, we hold that the Colorado courts, if given the opportunity, would extend its common law trademark infringement actions to include reverse confusion situations. Such a rule would further Colorado's "policy of protecting trade names and preventing public confusion" as well as having "the tendency [of widening] the scope of that protection."

The district court very persuasively answered Goodyear's argument that liability for trademark infringement cannot be imposed without a showing that Goodyear intended to trade on the goodwill of Big O or to palm off Goodyear products as being those of Big O's when it said

> The logical consequence of accepting Goodyear's position would be the immunization from unfair competition liability of a company with a well established trade name and with the economic power to advertise extensively for a product name taken from a competitor. If the law is to limit recovery to passing off, anyone with adequate size and resources can adopt any trademark and develop a new meaning for that trademark as identification of the second user's products. The activities of Goodyear in this case are unquestionably unfair competition through an improper use of a trademark and that must be actionable.

408 F. Supp. at 1236.

Goodyear further argues there was no credible evidence from which the jury could have found a likelihood of reverse confusion. A review of the record demonstrates the lack of merit in this argument. Big O presented more than a dozen witnesses who testified to actual confusion as to the source of Big O's "Big Foot" tires after watching a Goodyear "Bigfoot" commercial. The jury could have reasonably inferred a likelihood of confusion from these witnesses' testimony of actual confusion. Moreover, two of Goodyear's executive officers, Kelley and Eaves, testified confusion was likely or even inevitable.

VII

Finally, Goodyear challenges the jury's verdict awarding Big O $2.8 million in compensatory damages and $16.8 million in punitive damages. . . .

Big O also asserts the evidence provided the jury with a reasonable basis for determining the amount of damages. Big O claims the only way it can be restored to the position it was in before Goodyear infringed its trademark is to conduct a corrective advertising campaign. Big O insists it should be compensated for the advertising expenses necessary to dispel the public confusion caused by Goodyear's infringement. Goodyear spent approximately $10 million on its "Bigfoot" advertising campaign. Thus, Big O advances two rationales in support of the $2.8 million award: (1) there were Big O Tire Dealers in 28 percent of the states (14 of 50) and 28 percent of $10 million equals the amount of the award; and (2) the Federal Trade Commission generally orders businesses who engage in misleading advertising to spend approximately 25 percent of their advertising budget on corrective advertising and this award is roughly 25 percent of the amount Goodyear spent infringing on Big O's trademark. The district court used the first rationale in denying Goodyear's motion to set the verdict aside. The second rationale was presented by Big O at oral argument. . . .

There is precedent for the recovery of corrective advertising expenses incurred by a plaintiff to counteract the public confusion resulting from a defendant's wrongful conduct. . . . Unlike the wronged parties in those cases Big O did not spend any money prior to trial in advertising to counteract the confusion from the Goodyear advertising. It is clear from the record Big O did not have the economic resources to conduct an advertising campaign sufficient to counteract Goodyear's $9,690,029 saturation advertising campaign. We are thus confronted with the question whether the law should apply differently to those who have the economic power to help themselves concurrently with the wrong than to those who must seek redress through the courts. Under the facts of this case we are convinced the answer must be no. Goodyear contends the recovery of advertising expenses should be limited to those actually incurred prior to trial. In this case the effect of such a rule would be to recognize that Big O has a right to the exclusive use of its trademark but has no remedy to be put in the position it was in prior to September 16, 1974, before Goodyear effectively usurped Big O's trademark. The impact of Goodyear's "Bigfoot" campaign was devastating. The infringing mark was seen repeatedly by millions of consumers. It is clear from the record that Goodyear deeply penetrated the public consciousness. Thus, Big O is entitled to recover a reasonable amount equivalent to that of a concurrent corrective advertising campaign.

As the district court pointed out, the jury's verdict of $2.8 million corresponds to 28 percent of the approximately $10 million Goodyear spent infringing Big O's mark. Big O has dealers in 14 states which equals 28 percent of the 50 states. Big O also points out the jury's award is close to 25 percent of the amount Goodyear spent infringing on Big O's mark. Big O emphasizes that the Federal Trade Commission often requires businesses who engage in misleading advertising to spend 25 percent of their advertising budget on corrective advertising.

Taking cognizance of these two alternative rationales for the jury's award for compensatory damages we are convinced the award is not capable of support as to any amount in excess of $678,302. As the district court implied in attempting to explain the jury's verdict, Big O is not entitled to the total amount Goodyear spent on its nationwide campaign since Big O only has dealers in 14 states, thus making it unnecessary for Big O to run a nationwide advertising campaign. Furthermore, implicit in the FTC's 25 percent rule in corrective advertising cases is the fact that dispelling confusion and deception in the consuming public's mind does not require a dollar-for-dollar expenditure. In keeping with "'(t)he constant tendency of the courts . . . to find some way in which damages can be awarded where a wrong has been done,'" we hold that the maximum amount which a jury could reasonably find necessary to place Big O in the position it was in before September 16, 1974, vis-a-vis its "Big Foot" trademark, is $678,302. We arrive at this amount by taking 28 percent of the $9,690,029 it was stipulated Goodyear spent on its "Bigfoot" campaign, and then reducing that figure by 75 percent in accordance with the FTC rule, since we agree with that agency's determination that a dollar-for-dollar expenditure for corrective advertising is unnecessary to dispel the effects of confusing and misleading advertising.

[The court also awarded $4 million in punitive damages].

COMMENTS AND QUESTIONS

1. How was Big O hurt by Goodyear's massive advertising campaign? If the products have the same name, and if Goodyear spends much more on advertising than Big O does, shouldn't sales of Big O Tires increase? One possible answer is that Big O feared that Goodyear's tires would be of lower quality than its own and that its public image would suffer as a result of being associated with Goodyear. But that doesn't seem to be the case here, since Goodyear was selling premium tires and Big O was selling off-brands. Can you envision any way in which Goodyear's advertising of the brand would reduce Big O's sales?

Is the public hurt by thinking that Big O is affiliated with Goodyear? How? Is Big O hurt? Should that matter?

2. In the *Big O Tires* case, the court ordered as a remedy that Goodyear pay $14 million in punitive damages plus $678,302 — what it would cost in corrective advertising to "undo" the "damage" Goodyear had done in connecting itself with the Big O name.

This punitive damage award raises an interesting point. Is the court justified in ordering compensation — whether in the form of corrective advertising or a cash payment — beyond the point where advertising would have benefited Big O? If Goodyear benefits more from the advertising than Big O would have — as is perhaps shown by the fact that it costs more to restore the status quo for Big O — does this establish that the trademark is better left in Goodyear's hands? What if Goodyear realizes that it can make better use of a trademark currently held by a competitor; should we allow Goodyear to appropriate the mark by paying damages as set by the court? (Note that this is an example of a "liability rule"; see the discussion on patent remedies in Chapter 3.) What would be the effects of such a rule?

Further, consider that *any* spending is really a windfall to Big O, since it did not have the resources to shell out anything like $4 million in advertising. Is there any consumer interest served by corrective advertising in a reverse confusion case? Won't it just confuse consumers more?

The idea of punitive damages for trademark infringement seems to deny the possibility of "efficient" trademark infringement. This in turn suggests that trademarks are property and that courts are willing to grant "specific performance" remedies in infringement cases. Does this view make sense?

3. How should damages be measured in cases of reverse confusion? Presumably, a company like Big O Tires might be able to prove that it lost sales (rather than gained them) as a result of Goodyear's use of its "Bigfoot" trademark. But is Big O also entitled to Goodyear's profits from infringement? Its "unjust enrichment"? Most courts would say yes, at least where Big O can prove intentional or willful infringement. But doing so can provide a significant windfall to Big O, which would never have made such profits in the absence of infringement.

What unjust profits has Goodyear gained *through infringement*? Must it turn over all its profits from the sale of "Bigfoot" tires, even though it is Goodyear and not Big O that has built a national reputation for the "Bigfoot" name? How about the increase in sales attributable to the use of the name "Bigfoot"? Or are Goodyear's profits from infringement limited to whatever sales it took away from Big O (presumably the same measure as Big O's lost profits)?

PROBLEM

Problem 5-18. A small Vermont company named STW (and its predecessors) has used the mark "Thirst-Aid" for soft drinks since 1921. STW has never sold its products widely, however, and in fact its sales are declining. In 1983, Quaker Oats adopted the slogan "Gatorade is Thirst Aid" for its popular Gatorade beverage. STW sued for trademark infringement. At trial, STW proves that Quaker Oats knew of STW's trademark, but its lawyers advised it that the Gatorade slogan made "fair use" of STW's "descriptive" mark. The trial court disagreed, finding that Quaker Oats had infringed STW's mark in bad faith.

During the period that the "Thirst-Aid" campaign ran, pretax profits on the sale of Gatorade came to $247 million on sales of $2.6 billion. Quaker Oats proves that Gatorade had approximately $475 million in sales before the ad campaign. It also demonstrates that STW had previously offered to license the "Thirst-Aid" mark to another company for 1/3 of 1% of sales, and that STW's total goodwill in 1984 was less than $100,000 and declining. What damages should be awarded in this case?

Note on the Trademark Counterfeiting Act of 1984

In 1984, Congress substantially increased the penalties for intentional copying of a trademark (called "counterfeiting").[24] 18 U.S.C. § 2320 makes it a felony to knowingly use a counterfeit mark in connection with the sale of goods or services. It provides for fines and imprisonment of the offenders, and it permits the destruction of goods bearing counterfeit marks (rather than simply the removal of the marks themselves). The act also added

- section 34(d) of the Lanham Act, 15 U.S.C. § 1116(d), which provides for the seizure of counterfeit goods and records of sale before trial upon an ex parte application;
- section 35(b) of the Lanham Act, 15 U.S.C. § 1117(b), which provides for the award of treble damages plus attorney fees and prejudgment interest against counterfeiters unless the court finds "extenuating circumstances";
- language in section 36 of the Lanham Act, 15 U.S.C. § 1118, concerning the right of the court to destroy counterfeit goods after trial.

An exception to the broad reach of the act applies to those who are authorized to use a trademark "at the time of the manufacture or production" of the goods. See U.S. v. Bohai Trading Co., Inc., 45 F.3d 577 (1st Cir. 1995) (upholding the constitutionality of this provision in the face of a challenge on the grounds of vagueness). Although the ex parte nature of the remedies has brought the 1984 Act under constitutional scrutiny, courts have generally upheld the legality of its remedies and procedures. See, e.g., United States v. McEvoy, 820 F.2d 1170 (11th Cir.), *cert denied*, 484 U.S. 902 (1987). But cf. Time Warner Entertainment Co., L.P. v. Does Nos. 1-2, 876 F. Supp. 407 (E.D.N.Y. 1994) (overturning on Fourth Amendment grounds proposed seizure order drafted by owners of copyrights and trademarks; court cited role of private investigator in conducting seizure and impoundment, failure to provide sufficient particularity for premises to be searched or articles to be seized, inclusion of private residence as site to be searched, and failure to sufficiently describe infringing materials).

COMMENTS AND QUESTIONS

1. The Trademark Counterfeiting Act of 1984 is clearly punitive in nature, providing for both criminal sanctions and treble damages. Is such a punitive approach warranted? If so, why not apply these remedies to all trademark infringements?

2. Are consumers hurt by counterfeiting, or is it only trademark owners who are being protected by the Lanham Act? What if the counterfeit goods were not in fact inferior to genuine goods with the same trademark? Have consumers suffered any injury in that case if they mistakenly buy counterfeit goods?

3. Should it be a defense to a counterfeiting claim that the reasonable consumer was aware that the goods were not genuine (for example, because they were purchased

---

24. Section 45 of the Lanham Act, 15 U.S.C. § 1127, defines a "counterfeit" as "a spurious mark which is identical with, or substantially indistinguishable from, a registered mark." While this definition does not contain an explicit intent requirement, courts are generally unwilling to find good faith use of an infringing mark to be "spurious."

from a street vendor without a certificate of authenticity for approximately 10 percent of the retail price of the genuine goods in stores)? Are consumers likely to be "confused" by such sales?

4. Judge Posner suggests one good reason for multiple damages in counterfeiting cases: deterrence. In holding that treble damages were an appropriate remedy even for innocent infringement, he argued:

> [T]he sale of counterfeit merchandise has become endemic — perhaps pandemic. Most of the infringing sellers are small retailers, such as K-Econo. Obtaining an injunction against each and every one of them would be infeasible. Trademark owners cannot hire investigators to shop every retail store in the nation. And even if they could and did, and obtained injunctions against all present violators, this would not stop the counterfeiting. Other infringers would spring up, and would continue infringing until enjoined.... Treble damages are a particularly suitable remedy in cases where surreptitious violations are possible, for in such cases simple damages (or profits) will underdeter; the violator will know that he won't be caught every time, and merely confiscating his profits in the cases in which he is caught will leave him with a net profit from infringement.... [T]he smaller the violator, the less likely he is to be caught, and the more needful, therefore, is a heavy punishment if he is caught.

Louis Vuitton S.A. v. Lee, 875 F.2d 584 (7th Cir. 1989). Is this rationale persuasive?

State Intellectual Property Law and Federal Preemption

As we saw in Chapter 2, state protection of intellectual property — whether rooted in statute or in the common law — has long been a feature of the legal landscape. But state intellectual property law extends far beyond trade secrets. Especially in recent years, state law has reached out to an ever-widening range of issues: from court decisions and legislation on publicity rights to "antidilution" statutes in the states (discussed in Chapter 5), "shrinkwrap" and "clickwrap" contracts purporting to protect data, and trespass concepts being used to limit access to websites. States have become a major force in the evolution of intellectual property law in the new technological age. As one commentator has observed:

> [T]echnology in this century has continually outpaced statutory law and litigants have repeatedly turned to judge-made law to protect important rights and large investments in the collection or creation of time-sensitive information and other commercially valuable content. It stands to reason that the faster a technology develops, the more rapidly it will surpass preexisting law, and the more prominent common law theories may become. It is not surprising, therefore, that as the Internet geometrically expands its speed, accessibility, and versatility — thereby vastly increasing the opportunities for economic free-riders to take, copy, and repackage information and information systems for profit — intellectual property owners again must consider the common law as a source of protection at the end of this century, much as it was at the beginning.

Bruce P. Keller, Condemned to Repeat the Past: The Reemergence of Misappropriation and Other Common Law Theories of Protection for Intellectual Property, 11 Harv. J.L. & Tech. 401, 428 (1998).

This chapter begins by surveying the most significant ways in which state law protects intellectual property (beyond trade secret and trademark law): misappropriation doctrine, contract law (shrinkwrap and related contract formation models), idea submissions (implied contract), right of publicity, state moral rights, and trespass to chattels. Just as different federal intellectual property regimes may conflict — e.g., expressive useful articles (patent-copyright) and distinctive product configurations (patent-trademark) — state intellectual property law may also clash with the federal intellectual property regime. Therefore, we turn in section B to the law of federal preemption and the doctrines that harmonize the overall system of protection. Preemption has taken on new importance in the current environment of state expansionism. The easiest cases on preemption are where state law attempts either to duplicate federal protection or to interfere with it; the Supremacy Clause of the Constitution nullifies such attempts. More difficult cases involve state laws that do not directly conflict with federal authority but instead address interstitial gaps within the federal regime. Courts must grapple with whether Congress intended to leave such gaps unfilled, thereby precluding state protection, or simply allowed state law to fill these voids. This logical structure — which involves a search for the intent behind uncovered cases — is what makes preemption such a challenging area. As you will see from the cases, doctrine in this area appears quite unstable, as it is not (yet) supported by an adequate conceptual framework.

A. STATE INTELLECTUAL PROPERTY LAW

1. The Tort of Misappropriation

International News Service v. Associated Press
Supreme Court of the United States
248 U.S. 215 (1918)

Justice PITNEY delivered the opinion of the court.

The parties are competitors in the gathering and distribution of news and its publication for profit in newspapers throughout the United States. The Associated Press, which was complainant in the District Court, is a cooperative organization, incorporated under the Membership Corporations Law of the State of New York, its members being individuals who are either proprietors or representatives of about 950 daily newspapers published in all parts of the United States.... Complainant gathers in all parts of the world, by means of various instrumentalities of its own, by exchange with its members, and by other appropriate means, news and intelligence of current and recent events of interest to newspaper readers and distributes it daily to its members for publication in their newspapers. The cost of the service, amounting approximately to $3,500,000 per annum, is assessed upon the members and becomes a part of their costs of operation, to be recouped, presumably with profit, through the publication of their several newspapers. Under complainant's by-laws each member agrees upon assuming membership that news received through complainant's service is received exclusively for publication in a particular newspaper, language, and place specified in the certificate of membership, that no other use of it shall be permitted, and that no member shall furnish or permit anyone in his employ or connected with his newspaper to furnish any of complainant's news in advance of publication to any person not a member. And each member is required to gather the local news of his district and supply it to the Associated Press and to no one else.

Defendant is a corporation organized under the laws of the State of New Jersey, whose business is the gathering and selling of news to its customers and clients, consisting of newspapers published throughout the United States, under contracts by which they pay certain amounts at stated times for defendant's service. It has wide-spread news-gathering agencies; the cost of its operations amounts, it is said, to more than $2,000,000 per annum; and it serves about 400 newspapers located in the various cities of the United States and abroad, a few of which are represented, also, in the membership of the Associated Press.

The parties are in the keenest competition between themselves in the distribution of news throughout the United States; and so, as a rule, are the newspapers that they serve, in their several districts.

Complainant in its bill, defendant in its answer, have set forth in almost identical terms the rather obvious circumstances and conditions under which their business is conducted. The value of the service, and of the news furnished, depends upon the promptness of transmission, as well as upon the accuracy and impartiality of the news; it being essential that the news be transmitted to members or subscribers as early or earlier than similar information can be furnished to competing newspapers by other news services, and that the news furnished by each agency shall not be furnished to newspapers which do not contribute to the expense of gathering it. And further, to quote from the answer:

> Prompt knowledge and publication of world-wide news is essential to the conduct of a modern newspaper, and by reason of the enormous expense incident to the gathering and distribution of such news, the only practical way in which a proprietor of a newspaper can obtain the same is, either through cooperation with a considerable number of other newspaper proprietors in the work of collecting and distributing such news, and the equitable division with them of the expenses thereof, or by the purchase of such news from some existing agency engaged in that business.

The bill was filed to restrain the pirating of complainant's news by defendant in three ways: First, by bribing employees of newspapers published by complainant's members to furnish Associated Press news to defendant before publication, for transmission by telegraph and telephone to defendant's clients for publication by them; Second, by inducing Associated Press members to violate its by-laws and permit defendant to obtain news before publication; and Third, by copying news from bulletin boards and from early editions of complainant's newspapers and selling this, either bodily or after rewriting it, to defendant's customers.

The District Court, upon consideration of the bill and answer, with voluminous affidavits on both sides, granted a preliminary injunction under the first and second heads; but refused at that stage to restrain the systematic practice admittedly pursued by defendant, of taking news bodily from the bulletin boards and early editions of complainant's newspapers and selling it as its own. The court expressed itself as satisfied that this practice amounted to unfair trade, but as the legal question was one of first impression it considered that the allowance of an injunction should await the outcome of an appeal. 240 Fed. Rep. 983, 996. Both parties having appealed, the Circuit Court of Appeals sustained the injunction order so far as it went, and upon complainant's appeal modified it and remanded the cause with directions to issue an injunction also against any bodily taking of the words or substance of complainant's news until its commercial value as news had passed away. 245 Fed. Rep. 244, 253. The present writ of certiorari was then allowed. 245 U.S. 644.

The only matter that has been argued before us is whether defendant may lawfully be restrained from appropriating news taken from bulletins issued by complainant or any of its members, or from newspapers published by them, for the purpose of selling it to defendant's clients. Complainant asserts that defendant's admitted course of conduct in this regard both violates complainant's property right in the news and constitutes unfair competition in business. And notwithstanding the case has proceeded only to the stage of a preliminary injunction, we have deemed it proper to consider the underlying questions, since they go to the very merits of the action and are presented upon facts that are not in dispute. As presented in argument, these questions are: 1. Whether there is any property in news; 2. Whether, if there be property in news collected for the purpose of being published, it survives the instant of its publication in the first newspaper to which it is communicated by the news-gatherer; and 3. Whether defendant's admitted course of conduct in appropriating for commercial use matter taken from bulletins or early editions of Associated Press publications constitutes unfair competition in trade.

The federal jurisdiction was invoked because of diversity of citizenship, not upon the ground that the suit arose under the copyright or other laws of the United States. Complainant's news matter is not copyrighted. It is said that it could not, in practice, be copyrighted, because of the large number of dispatches that are sent daily; and, according to complainant's contention, news is not within the operation of the copyright act. Defendant, while apparently conceding this, nevertheless invokes the analogies of the law of literary property and copyright, insisting as its principal contention that, assuming complainant has a right of property in its news, it can be maintained (unless the copyright act be complied with) only by being kept secret and confidential, and that upon the publication with complainant's consent of uncopyrighted news by any of complainant's members in a newspaper or upon a bulletin board, the right of property is lost, and the subsequent use of the news by the public or by defendant for any purpose whatever becomes lawful. . . .

In considering the general question of property in news matter, it is necessary to recognize its dual character, distinguishing between the substance of the information and the particular form or collocation of words in which the writer has communicated it.

No doubt news articles often possess a literary quality, and are the subject of literary property at the common law; nor do we question that such an article, as a literary production, is the subject of copyright by the terms of the act as it now stands. In an early case at the circuit Mr. Justice Thompson held in effect that a newspaper was not within the protection of the copyright acts of 1790 and 1802 (Clayton v. Stone, 2 Paine, 382; 5 Fed. Cas. No. 2872). But the present act is broader; it provides that the works for which copyright may be secured shall include "all the writings of an author," and specifically mentions "periodicals, including newspapers." Act of March 4, 1909, c. 320, §§ 4 and 5, 35 Stat. 1075, 1076. Evidently this admits to copyright a contribution to a newspaper, notwithstanding it also may convey news; and such is the practice of the copyright office, as the newspapers of the day bear witness. See Copyright Office Bulletin No. 15 (1917), pp. 7, 14, 16-17.

But the news element — the information respecting current events contained in the literary production — is not the creation of the writer, but is a report of matters that ordinarily are publici juris; it is the history of the day. It is not to be supposed that the framers of the Constitution, when they empowered Congress "to promote the progress of science and useful arts, by securing for limited times to authors and inventors the exclusive right to their respective writings and discoveries" (Const., Art I, § 8, par. 8), intended to confer upon one who might happen to be the first to report a historic event the exclusive right for any period to spread the knowledge of it.

We need spend no time, however, upon the general question of property in news matter at common law, or the application of the copyright act, since it seems to us the case must turn upon the question of unfair competition in business. And, in our opinion, this does not depend upon any general right of property analogous to the common-law right of the proprietor of an unpublished work to prevent its publication without his consent; nor is it foreclosed by showing that the benefits of the copyright act have been waived. We are dealing here not with restrictions upon publication but with the very facilities and processes of publication. The peculiar value of news is in the spreading of it while it is fresh; and it is evident that a valuable property interest in the news, as news, cannot be maintained by keeping it secret. Besides, except for matters improperly disclosed, or published in breach of trust or confidence, or in violation of law, none of which is involved in this branch of the case, the news of current events may be regarded as common property. What we are concerned with is the business of making it known to the world, in which both parties to the present suit are engaged. That business consists in maintaining a prompt, sure, steady, and reliable service designed to place the daily events of the world at the breakfast table of the millions at a price that, while of trifling moment to each reader, is sufficient in the aggregate to afford compensation for the cost of gathering and distributing it, with the added profit so necessary as an incentive to effective action in the commercial world. The service thus performed for newspaper readers is not only innocent but extremely useful in itself, and indubitably constitutes a legitimate business. The parties are competitors in this field; and, on fundamental principles, applicable here as elsewhere, when the rights or privileges of the one are liable to conflict with those of the other, each party is under a duty so to conduct its own business as not unnecessarily or unfairly to injure that of the other. Hitchman Coal & Coke Co. v. Mitchell, 245 U.S. 229, 254.

Obviously, the question of what is unfair competition in business must be determined with particular reference to the character and circumstances of the business. The question here is not so much the rights of either party as against the public but their rights as between themselves. See Morison v. Moat, 9 Hare, 241, 258. And although we may and do assume that neither party has any remaining property interest as against the public in uncopyrighted news matter after the moment of its first publication, it by no means follows that there is no remaining property interest in it as between themselves. For, to both of them alike, news matter, however little susceptible of ownership or dominion in the absolute sense, is stock in trade, to be gathered at the cost of enterprise, organization, skill, labor, and money, and to be distributed and sold to those who will pay money for it, as for any other merchandise. Regarding the news, therefore, as but the material out of which both parties are seeking to make profits at the same time and in the same field, we hardly can fail to recognize that for this purpose, and as between them, it must be regarded as quasi property, irrespective of the rights of either as against the public.

In order to sustain the jurisdiction of equity over the controversy, we need not affirm any general and absolute property in the news as such. The rule that a court of equity concerns itself only in the protection of property rights treats any civil right of a pecuniary nature as a property right (In re Sawyer, 124 U.S. 200, 210; In re Debs, 158 U.S. 564, 593); and the right to acquire property by honest labor or the conduct of a lawful business is as much entitled to protection as the right to guard property already acquired. Truax v. Raich, 239 U.S. 33, 37-38; Brennan v. United Hatters, 73 N.J.L. 729, 742; Barr v. Essex Trades Council, 53 N.J. Eq. 101. It is this right that furnishes the basis of the jurisdiction in the ordinary case of unfair competition....

Not only do the acquisition and transmission of news require elaborate organization and a large expenditure of money, skill, and effort; not only has it an exchange value to the gatherer, dependent chiefly upon its novelty and freshness, the regularity of the service, its reputed reliability and thoroughness, and its adaptability to the public needs; but also, as is evident, the news has an exchange value to one who can misappropriate it.

The peculiar features of the case arise from the fact that, while novelty and freshness form so important an element in the success of the business, the very processes of distribution and publication necessarily occupy a good deal of time. Complainant's service, as well as defendant's, is a daily service to daily newspapers; most of the foreign news reaches this country at the Atlantic seaboard, principally at the City of New York, and because of this, and of time differentials due to the earth's rotation, the distribution of news matter throughout the country is principally from east to west; and, since in speed the telegraph and telephone easily outstrip the rotation of the earth, it is a simple matter for defendant to take complainant's news from bulletins or early editions of complainant's members in the eastern cities and at the mere cost of telegraphic transmission cause it to be published in western papers issued at least as early as those served by complainant. Besides this, and irrespective of time differentials, irregularities in telegraphic transmission on different lines, and the normal consumption of time in printing and distributing the newspaper, result in permitting pirated news to be placed in the hands of defendant's readers sometimes simultaneously with the service of competing Associated Press papers, occasionally even earlier.

Defendant insists that when, with the sanction and approval of complainant, and as the result of the use of its news for the very purpose for which it is distributed, a portion of complainant's members communicate it to the general public by posting it upon bulletin boards so that all may read, or by issuing it to newspapers and distributing it indiscriminately, complainant no longer has the right to control the use to be made of it; that when it thus reaches the light of day it becomes the common possession of all to whom it is accessible; and that any purchaser of a newspaper has the right to communicate the intelligence which it contains to anybody and for any purpose, even for the purpose of selling it for profit to newspapers published for profit in competition with complainant's members.

The fault in the reasoning lies in applying as a test the right of the complainant as against the public, instead of considering the rights of complainant and defendant, competitors in business, as between themselves. The right of the purchaser of a single newspaper to spread knowledge of its contents gratuitously, for any legitimate purpose not unreasonably interfering with complainant's right to make merchandise of it, may be admitted; but to transmit that news for commercial use, in competition with complainant — which is what defendant has done and seeks to justify — is a very different matter. In doing this defendant, by its very act, admits that it is taking material that has been acquired by complainant as the result of organization and the expenditure of labor, skill, and money, and which is salable by complainant for money, and that defendant in appropriating it and selling it as its own is endeavoring to reap where it has not sown, and by disposing of it to newspapers that are competitors of complainant's members is appropriating to itself the harvest of those who have sown. Stripped of all disguises, the process amounts to an unauthorized interference with the normal operation of complainant's legitimate business precisely at the point where the profit is to be reaped, in order to divert a material portion of the profit from those who have earned it to those who have not; with special advantage to defendant in the competition because of the fact that it is not burdened with any part of the expense of gathering the news. The transaction speaks for itself, and a court of equity ought not to hesitate long in characterizing it as unfair competition in business.

The underlying principle is much the same as that which lies at the base of the equitable theory of consideration in the law of trusts — that he who has fairly paid the price should have the beneficial use of the property. Pom. Eq. Jur., §981. It is no answer to say that complainant spends its money for that which is too fugitive or evanescent to be the subject of property. That might, and for the purposes of the discussion we are assuming that it would, furnish an answer in a common-law controversy. But in a court of equity, where the question is one of unfair competition, if that which complainant has acquired fairly at substantial cost may be sold fairly at substantial profit, a competitor who is misappropriating it for the purpose of disposing of it to his own profit and to the disadvantage of complainant cannot be heard to say that it is too fugitive or evanescent to be regarded as property. It has all the attributes of property necessary for determining that a misappropriation of it by a competitor is unfair competition because contrary to good conscience.

The contention that the news is abandoned to the public for all purposes when published in the first newspaper is untenable. Abandonment is a question of intent, and the entire organization of the Associated Press negatives such a purpose. The cost of the service would be prohibitive if the reward were to be so limited. No single newspaper, no small group of newspapers, could sustain the expenditure. Indeed, it is one of the most obvious results of defendant's theory that, by permitting indiscriminate publication by anybody and everybody for purposes of profit in competition with the news-gatherer, it would render publication profitless, or so little profitable as in effect to cut off the service by rendering the cost prohibitive in comparison with the return. The practical needs and requirements of the business are reflected in complainant's by-laws which have been referred to. Their effect is that publication by each member must be deemed not by any means an abandonment of the news to the world for any and all purposes, but a publication for limited purposes; for the benefit of the readers of the bulletin or the newspaper as such; not for the purpose of making merchandise of it as news, with the result of depriving complainant's other members of their reasonable opportunity to obtain just returns for their expenditures.

It is to be observed that the view we adopt does not result in giving to complainant the right to monopolize either the gathering or the distribution of the news, or, without complying with the copyright act, to prevent the reproduction of its news articles; but only postpones participation by complainant's competitor in the processes of distribution and reproduction of news that it has not gathered, and only to the extent necessary to prevent that competitor from reaping the fruits of complainant's efforts and expenditure, to the partial exclusion of complainant, and in violation of the principle that underlies the maxim sic utere tuo, etc.

It is said that the elements of unfair competition are lacking because there is no attempt by defendant to palm off its goods as those of the complainant, characteristic of the most familiar, if not the most typical, cases of unfair competition. Howe Scale Co. v. Wyckoff, Seamans & Benedict, 198 U.S. 118, 140. But we cannot concede that the right to equitable relief is confined to that class of cases. In the present case the fraud upon complainant's rights is more direct and obvious. Regarding news matter as the mere material from which these two competing parties are endeavoring to make money, and treating it, therefore, as quasi property for the purposes of their business because they are both selling it as such, defendant's conduct differs from the ordinary case of unfair competition in trade principally in this that, instead of selling its own goods as those of complainant, it substitutes misappropriation in the place of misrepresentation, and sells complainant's goods as its own.

Besides the misappropriation, there are elements of imitation, of false pretense, in defendant's practices. The device of rewriting complainant's news articles, frequently resorted to, carries its own comment. The habitual failure to give credit to complainant for that which is taken is significant. Indeed, the entire system of appropriating complainant's news and transmitting it as a commercial product to defendant's clients and patrons amounts to a false representation to them and to their newspaper readers that the news transmitted is the result of defendant's own investigation in the field. But these elements, although accentuating the wrong, are not the essence of it. It is something more than the advantage of celebrity of which complainant is being deprived....

The decree of the Circuit Court of Appeals will be Affirmed.

HOLMES, J., concurring:

When an uncopyrighted combination of words is published there is no general right to forbid other people repeating them — in other words there is no property in the combination or in the thoughts or facts that the words express. Property, a creation of law, does not arise from value, although exchangeable — a matter of fact. Many exchangeable values may be destroyed intentionally without compensation. Property depends upon exclusion by law from interference, and a person is not excluded from using any combination of words merely because someone has used it before, even if it took labor and genius to make it. If a given person is to be prohibited from making the use of words that his neighbors are free to make some other ground must be found. One such ground is vaguely expressed in the phrase unfair trade. This means that the words are repeated by a competitor in business in such a way as to convey a misrepresentation that materially injures the person who first used them, by appropriating credit of some kind which the first user has earned. The ordinary case is a representation by device, appearance, or other indirection that the defendant's goods come from the plaintiff. But the only reason why it is actionable to make such a representation is that it tends to give the defendant an advantage in his competition with the plaintiff and that it is thought undesirable that an advantage should be gained in that way. Apart from that the defendant may use such unpatented devices and uncopyrighted combinations of words as he likes. The ordinary case, I say, is palming off the defendant's product as the plaintiff's, but the same evil may follow from the opposite falsehood — from saying, whether in words or by implication, that the plaintiff's product is the defendant's, and that, it seems to me, is what has happened here.

Fresh news is got only by enterprise and expense. To produce such news as it is produced by the defendant represents by implication that it has been acquired by the defendant's enterprise and at its expense. When it comes from one of the great news-collecting agencies like the Associated Press, the source generally is indicated, plainly importing that credit; and that such a representation is implied may be inferred with some confidence from the unwillingness of the defendant to give the credit and tell the truth. If the plaintiff produces the news at the same time that the defendant does, the defendant's presentation impliedly denies to the plaintiff the credit of collecting the facts and assumes that credit to the defendant. If the plaintiff is later in western cities it naturally will be supposed to have obtained its information from the defendant. The falsehood is a little more subtle, the injury a little more indirect, than in ordinary cases of unfair trade, but I think that the principle that condemns the one condemns the other. It is a question of how strong an infusion of fraud is necessary to turn a flavor into a poison. The dose seems to me strong enough here to need a remedy from the law. But as, in my view, the only ground of complaint that can be recognized without legislation is the implied misstatement, it can be corrected by stating the truth; and a suitable acknowledgment of the source is all that the plaintiff can require. I think that within the limits recognized by the decision of the Court the defendant should be enjoined from publishing news obtained from the Associated Press for __ hours after publication by the plaintiff unless it gives express credit to the Associated Press; the number of hours and the form of acknowledgment to be settled by the District Court....

BRANDEIS, J., dissenting:...

News is a report of recent occurrences. The business of the news agency is to gather systematically knowledge of such occurrences of interest and to distribute reports thereof. The Associated Press contended that knowledge so acquired is property, because it costs money and labor to produce and because it has value for which those who have it not are ready to pay; that it remains property and is entitled to protection as long as it has commercial value as news; and that to protect it effectively the defendant must be enjoined from making, or causing to be made, any gainful use of it while it retains such value. An essential element of individual property is the legal right to exclude others from enjoying it. If the property is private, the right of exclusion may be absolute; if the property is affected with a public interest, the right of exclusion is qualified. But the fact that a product of the mind has cost its producer money and labor, and has a value for which others are willing to pay, is not sufficient to ensure to it this legal attribute of property. The general rule of law is, that the noblest of human productions — knowledge, truths ascertained, conceptions, and ideas — become, after voluntary communication to others, free as the air to common use. Upon these incorporeal productions the attribute of property is continued after such communication only in certain classes of cases where public policy has seemed to demand it. These exceptions are confined to productions which, in some degree, involve creation, invention, or discovery. But by no means all such are endowed with this attribute of property. The creations which are recognized as property by the common law are literary, dramatic, musical, and other artistic creations; and these have also protection under the copyright statutes. The inventions and discoveries upon which this attribute of property is conferred only by statute, are the few comprised within the patent law. There are also many other cases in which courts interfere to prevent curtailment of plaintiff's enjoyment of incorporeal productions; and in which the right to relief is often called a property right, but is such only in a special sense. In those cases, the plaintiff has no absolute right to the protection of his production; he has merely the qualified right to be protected as against the defendant's acts, because of the special relation in which the latter stands or the wrongful method or means employed in acquiring the knowledge or the manner in which it is used. Protection of this character is afforded where the suit is based upon breach of contract or of trust or upon unfair competition.

The knowledge for which protection is sought in the case at bar is not of a kind upon which the law has heretofore conferred the attributes of property; nor is the manner of its acquisition or use nor the purpose to which it is applied, such as has heretofore been recognized as entitling a plaintiff to relief....

The means by which the International News Service obtains news gathered by the Associated Press is also clearly unobjectionable. It is taken from papers bought in the open market or from bulletins publicly posted. No breach of contract such as the court considered to exist in Hitchman Coal & Coke Co. v. Mitchell, 245 U.S. 229, 254; or of trust such as was present in Morison v. Moat, 9 Hare, 241; and neither fraud nor force, is involved. The manner of use is likewise unobjectionable. No reference is made by word or by act to the Associated Press, either in transmitting the news to sub-scribers or by them in publishing it in their papers. Neither the International News Service nor its subscribers is gaining or seeking to gain in its business a benefit from the reputation of the Associated Press. They are merely using its product without making compensation. See Bamforth v. Douglass Post Card & Machine Co., 158 Fed. Rep. 355; Tribune Co. of Chicago v. Associated Press, 116 Fed. Rep. 126. That, they have a legal right to do; because the product is not property, and they do not stand in any relation to the Associated Press, either of contract or of trust, which otherwise precludes such use. The argument is not advanced by characterizing such taking and use a misappropriation.

It is also suggested, that the fact that defendant does not refer to the Associated Press as the source of the news may furnish a basis for the relief. But the defendant and its subscribers, unlike members of the Associated Press, were under no contractual obligation to disclose the source of the news; and there is no rule of law requiring acknowledgment to be made where uncopyrighted matter is reproduced. The Inter-national News Service is said to mislead its subscribers into believing that the news transmitted was originally gathered by it and that they in turn mislead their readers. There is, in fact, no representation by either of any kind. Sources of information are sometimes given because required by contract; sometimes because naming the source gives authority to an otherwise incredible statement; and sometimes the source is named because the agency does not wish to take the responsibility itself of giving currency to the news. But no representation can properly be implied from omission to mention the source of information except that the International News Service is transmitting news which it believes to be credible.

... The great development of agencies now furnishing country-wide distribution of news, the vastness of our territory, and improvements in the means of transmitting intelligence, have made it possible for a news agency or newspapers to obtain, with-out paying compensation, the fruit of another's efforts and to use news so obtained gainfully in competition with the original collector. The injustice of such action is obvious. But to give relief against it would involve more than the application of existing rules of law to new facts. It would require the making of a new rule in analogy to existing ones. The unwritten law possesses capacity for growth; and has often satisfied new demands for justice by invoking analogies or by expanding a rule or principle. This process has been in the main wisely applied and should not be discontinued. Where the problem is relatively simple, as it is apt to be when private interests only are involved, it generally proves adequate. But with the increasing complexity of society, the public interest tends to become omnipresent; and the problems presented by new demands for justice cease to be simple. Then the creation or recognition by courts of a new private right may work serious injury to the general public, unless the boundaries of the right are definitely established and wisely guarded. In order to reconcile the new private right with the public interest, it may be necessary to prescribe limitations and rules for its enjoyment; and also to provide administrative machinery for enforcing the rules. It is largely for this reason that, in the effort to meet the many new demands for justice incident to a rapidly changing civilization, resort to legislation has latterly been had with increasing frequency.

The rule for which the plaintiff contends would effect an important extension of property rights and a corresponding curtailment of the free use of knowledge and of ideas; and the facts of this case admonish us of the danger involved in recognizing such a property right in news, without imposing upon news-gatherers corresponding obligations....

Courts are ill-equipped to make the investigations which should precede a determination of the limitations which should be set upon any property right in news or of the circumstances under which news gathered by a private agency should be deemed affected with a public interest. Courts would be powerless to prescribe the detailed regulations essential to full enjoyment of the rights conferred or to introduce the machinery required for enforcement of such regulations. Considerations such as these should lead us to decline to establish a new rule of law in the effort to redress a newly-disclosed wrong, although the propriety of some remedy appears to be clear.

COMMENTS AND QUESTIONS

1. Both the majority and the dissent seem to agree that there is no traditional intellectual property right in published news. Copyright law — the most likely candidate for protecting works of authorship — cannot protect AP's news. See Chapter 4. And the majority's reference to "unfair competition" suggests that its theory is grounded in a tort of wrongful appropriation. But the tort analogy runs into difficulty as well, since (as Justice Brandeis observes) it is hard to fault the conduct at issue in the case. (Note that the Supreme Court did not consider AP's claim that INS had bribed AP newspapers or employees, an allegation which if true seems much more likely to support a tort action.) Instead, the Court seems to settle into what it calls a theory of "quasi-property" that entitles its owner to protection only against direct competitors.

2. How should courts following *International News* determine whether two companies are direct competitors who are then subject to this quasi-property right? For example, suppose that a single newspaper (not a member of AP) had appropriated information from published AP reports. Does the single paper compete with AP in news gathering? Does it matter whether the paper competes directly with one of AP's members, or has its own city monopoly? Or should the test be whether AP could have expected licensing revenues from the paper, whether or not they are in competition? Cf. National Football League v. Delaware, 435 F. Supp. 1372 (D. Del. 1977) (NFL cannot prevent Delaware from implementing a lottery based on NFL games, because the lottery is a "collateral service" rather than one in competition with the NFL).

3. How long does this quasi-property right last? The Court suggests one answer: until AP has appropriated its news value as a return on its news-gathering activities. Justice Holmes's concurrence suggests that this would be a matter of hours. But is that necessarily true? What happens when different news media compete? Should there be different rules for CNN, scheduled television news, daily newspapers, and weekly and monthly news magazines? Is AP entitled to protect its news not only from appropriation by "immediate" news sources, but also from the weeklies and monthlies that seek to use its news reports?

4. What if INS or its newspapers had credited AP with reporting the story? In that case, AP would presumably have received some value for its news — credit for the "scoop." But its member papers would still have faced competition from nonmembers

that they would prefer to avoid. Should attribution be enough? Justice Holmes says yes, but neither the majority nor Justice Brandeis' dissent see this as the major issue (although the majority notes that INS's failure to attribute the news to AP "accentuates" the unfairness of its competition). Note that attribution should solve any Lanham Act problem that might otherwise arise, since there is no danger that INS or its member papers will be "passing off" AP stories as their own if the source of the story is clearly identified.

This fact points up a significant difference between the tort of misappropriation and trademark infringement. Trademark law is premised on harm to consumers stemming from a "likelihood of confusion." Absent consumer confusion, trademarks (and trade dress) are generally not protectable. *International News* cuts a much broader swath. If it is not based on consumer protection, what is the rationale for the tort of misappropriation? Is it incentive theory? If so, what exactly is the Court trying to encourage?

5. The *International News* decision has not fared well in the courts. Early decisions by lower courts attempted to construe the decision narrowly. Indeed, in 1929 Judge Learned Hand went so far as to say: "While it is of course true that the law ordinarily speaks in general terms, there are cases where the occasion is at once the justification for, and the limit of, what is decided. [*International News*] appears to us such an instance; we think that no more was covered than situations substantially similar to those then at bar." Cheney Bros. v. Doris Silk Corp., 35 F.2d 279 (2d Cir. 1929), *cert. denied*, 281 U.S. 728 (1930). Judge Hand went on to characterize the problems with the Court's decision as "insuperable," and to state that it "flagrantly conflict[ed]" with the federal statutory intellectual property laws. Id. More recently, Judge Posner has commented that he is "hard pressed to find a case in which a claim of misappropriation should have succeeded." See Richard A. Posner, Misappropriation: A Dirge, 40 Hous. L. Rev. 621, 632 (2003).

After Erie R. Co. v. Tompkins, 304 U.S. 64 (1938), abolished federal common law in diversity cases (the grounds on which *International News* was decided), it seemed to many that the *International News* misappropriation doctrine was dead. See, e.g., James Treece, Patent Policy and Pre-emption: The Stiffel and Compco Cases, 32 U. Chi. L. Rev. 80 (1964). But the doctrine has since reappeared in a number of cases. See, e.g., United States Golf Assn. v. St. Andrews Systems, 749 F.2d 1028 (3d Cir. 1984) (*International News* was based on direct competition between the parties; court refused to apply it absent such direct competition); Imax Corp. v. Cinema Technologies, Inc., 152 F.3d 1161 (9th Cir. 1998) (owner of movie projector equipment could recover on a common law misappropriation theory for the use of information disclosed in confidence, despite the fact that the plaintiff could not prevail on its trade secret claim); Ettore v. Philco Television Broadcasting Corp., 229 F.2d 481 (3d Cir. 1956), *cert. denied*, 351 U.S. 926 (1956) (producer of boxing match could recover damages from television station which broadcast the match without permission). In National Basketball Assn. v. Motorola, Inc., 105 F.3d 841 (2d Cir. 1997), the court held that *International News* survives today only where

(i) a plaintiff generates or gathers information at a cost; (ii) the information is time-sensitive; (iii) a defendant's use of the information constitutes free-riding on the plaintiff's efforts; (iv) the defendant is in direct competition with a product or service offered by the plaintiffs; and (v) the ability of other parties to free-ride on the efforts of the plaintiff or others would so reduce the incentive to produce the product that its existence or quality would be substantially threatened.

The court held that these limitations on the misappropriation doctrine made it qualitatively different from a cause of action for copyright infringement and, therefore, saved it from preemption by the copyright laws. We discuss copyright preemption *infra*. Cf. United States Golf Association v. Arroyo Software, 40 U.S.P.Q.2d 1840 (Cal. Super. Ct. 1996), *aff'd*, 69 Cal. App. 4th 607 (1999) (extending misappropriation doctrine to noncompetitors, finding that protection of USGA handicapping formula is necessary to "protect the basic incentive for the production of the idea or information involved").

6. Suppose that the plaintiff in Feist v. Rural Telephone Service, *supra* Chapter 4, had brought a claim for misappropriation and unfair competition rather than a copyright claim. Should the plaintiff prevail under *International News*? It certainly would be able to claim that it had invested substantial time and effort in putting together a telephone directory, and that defendant (a direct competitor) had merely copied that information as soon as it became public. Assuming this is enough to show unfair competition under *International News,* what does that fact suggest about the interaction between the copyright laws and the unfair competition doctrine? Does it make sense that the rule in *Feist* should coexist with unfair competition law?

7. One solution to the inconsistency between *Feist* and *International News* would be to create an explicit (presumably statutory) property right in factual compilations. Strong arguments in favor of such protection were made in the copyright context both before and after *Feist*. Furthermore, doesn't the Court have a valid point about the incentive to invest in a productive activity (collecting news) whose benefits will immediately be dissipated by imitators? Isn't that precisely the justification for the patent system?

On the other hand, what problems do you foresee with such a new federal intellectual property regime? Would it overwhelm patent and copyright as means of protecting intellectual property? What effect would it have on businesses? Would it encourage factual research and compilation, or just encourage monopolization of facts? On this subject, see also Leo J. Raskind, The Misappropriation Doctrine as a Competitive Norm of Intellectual Property Law, 75 Minn. L. Rev. 875, 876-77 (1991), which argues that:

> when courts hear patent, copyright, and trademark cases in which statutory protection is inappropriate, but nonetheless the conduct of a party is characterized as "chiseling," "piracy," "unethical," or the like, they should begin their analysis by considering the competitive relationship from which the claim originates. The clear legislative expression of a preference for competition contained in federal antitrust laws warrants this approach. Moreover, courts in these cases should recognize the Supreme Court's continued emphasis on the preemptive effect given federal legislation relating to competition. Ancillary doctrines that impinge on competition, such as misappropriation, should be invoked sparingly. From this perspective, courts should consider allegedly "inappropriate" conduct as an element of behavior in a competitive market context; within that framework, courts should view such conduct as an element of cost that a seller, as a competitor, considers when determining how much of a particular product to offer.

8. To what extent was the Court in *International News* merely enforcing the norms of commercial reasonableness in the newspaper industry? See Richard Epstein, International News Service v. Associated Press: Custom and Law as Sources of Property Rights in News, 78 Va. L. Rev. 85 (1992) (noting the different customary rules in the news trade under which "lifting" stories was wrong, but following up on another paper's "tips" to report the same story was acceptable; and concluding that, partially unwittingly, Justice Pitney ended up affirming these customary rules in his opinion); Douglas G. Baird, Common Law Intellectual Property and the Legacy of International News Service v. Associated Press, 50 U. Chi. L. Rev. 411 (1983). But see Stephen L. Carter, Custom, Adjudication, and Petrushevsky's Watch: Some Notes from the Intellectual Property Front, 78 Va. L. Rev. 129, 132 (1992) ("Even courts inclined to enforce private orderings might not be very good at anthropology. The judge, after all, is on the outside, looking in. Even assuming — and there is no reason to do so — that the parties tell the whole truth, it will not always be so easy for a court to discern an industry's customs.").

2. Protection by Contract

Inventors and creators can also use contract law to achieve some of the goals they might otherwise use intellectual property rights to obtain. Contracts are fundamentally intertwined with intellectual property, because rights owners use contracts to license those rights to others. But companies who are dissatisfied with the protections intellectual property affords may also seek to use contract law to create their own private intellectual property rights, or to modify or expand the rights the law gives them. In this section, we discuss a number of circumstances in which creators have sought to do just that.

When private parties seek the help of state courts to vary the rules of intellectual property, courts must decide whether such efforts are preempted by the federal intellectual property scheme. We discuss the possible preemption of contract claims in section B of this chapter.

ProCD, Inc. v. Zeidenberg
United States Court of Appeals for the Seventh Circuit
86 F.3d 1447 (7th Cir. 1996)

EASTERBROOK, Circuit Judge.

Must buyers of computer software obey the terms of shrinkwrap licenses? The district court held not, for two reasons: first, they are not contracts because the licenses are inside the box rather than printed on the outside; second, federal law forbids enforcement even if the licenses are contracts. 908 F. Supp. 640 (W.D. Wis. 1996). The parties and numerous amici curiae have briefed many other issues, but these are the only two that matter — and we disagree with the district judge's conclusion on each. Shrinkwrap licenses are enforceable unless their terms are objectionable on grounds applicable to contracts in general (for example, if they violate a rule of positive law, or if they are unconscionable). Because no one argues that the terms of the license at issue here are troublesome, we remand with instructions to enter judgment for the plaintiff.

I

ProCD, the plaintiff, has compiled information from more than 3,000 telephone directories into a computer database. We may assume that this database cannot be copyrighted, although it is more complex, contains more information (nine-digit zip codes and census industrial codes), is organized differently, and therefore is more original than the single alphabetical directory at issue in Feist Publications, Inc. v. Rural Telephone Service Co., 499 U.S. 340, 113 L. Ed. 2d 358, 111 S. Ct. 1282 (1991). See Paul J. Heald, The Vices of Originality, 1991 Sup. Ct. Rev. 143, 160-68. ProCD sells a version of the database, called SelectPhone (trademark), on CD-ROM discs. (CD-ROM means "compact disc — read only memory." The "shrinkwrap license" gets its name from the fact that retail software packages are covered in plastic or cellophane "shrinkwrap," and some vendors, though not ProCD, have written licenses that become effective as soon as the customer tears the wrapping from the package. Vendors prefer "end user license," but we use the more common term.) A proprietary method of compressing the data serves as effective encryption too. Customers decrypt and use the data with the aid of an application program that ProCD has written. This program, which is copyrighted, searches the database in response to users' criteria (such as "find all people named Tatum in Tennessee, plus all firms with 'Door Systems' in the corporate name"). The resulting lists (or, as ProCD prefers, "listings") can be read and manipulated by other software, such as word processing programs.

The database in SelectPhone (trademark) cost more than $10 million to compile and is expensive to keep current. It is much more valuable to some users than to others. The combination of names, addresses, and SIC codes enables manufacturers to compile lists of potential customers. Manufacturers and retailers pay high prices to specialized information intermediaries for such mailing lists; ProCD offers a potentially cheaper alternative. People with nothing to sell could use the database as a substitute for calling long distance information, or as a way to look up old friends who have moved to unknown towns, or just as a electronic substitute for the local phone book. ProCD decided to engage in price discrimination, selling its database to the general public for personal use at a low price (approximately $150 for the set of five discs) while selling information to the trade for a higher price. It has adopted some intermediate strategies too: access to the SelectPhone (trademark) database is available via the America Online service for the price America Online charges to its clients (approximately $3 per hour), but this service has been tailored to be useful only to the general public.

If ProCD had to recover all of its costs and make a profit by charging a single price — that is, if it could not charge more to commercial users than to the general public — it would have to raise the price substantially over $150. The ensuing reduction in sales would harm consumers who value the information at, say, $200. They get consumer surplus of $50 under the current arrangement but would cease to buy if the price rose substantially. If because of high elasticity of demand in the consumer segment of the market the only way to make a profit turned out to be a price attractive to commercial users alone, then all consumers would lose out — and so would the commercial clients, who would have to pay more for the listings because ProCD could not obtain any contribution toward costs from the consumer market.

To make price discrimination work, however, the seller must be able to control arbitrage. An air carrier sells tickets for less to vacationers than to business travelers, using advance purchase and Saturday-night-stay requirements to distinguish the categories. A producer of movies segments the market by time, releasing first to theaters, then to pay-per-view services, next to the videotape and laserdisc market, and finally to cable and commercial TV. Vendors of computer software have a harder task. Anyone can walk into a retail store and buy a box. Customers do not wear tags saying "commercial user" or "consumer user." Anyway, even a commercial-user-detector at the door would not work, because a consumer could buy the software and resell to a commercial user. That arbitrage would break down the price discrimination and drive up the minimum price at which ProCD would sell to anyone.

Instead of tinkering with the product and letting users sort themselves — for example, furnishing current data at a high price that would be attractive only to commercial customers, and two-year-old data at a low price — ProCD turned to the institution of contract. Every box containing its consumer product declares that the software comes with restrictions stated in an enclosed license. This license, which is encoded on the CD-ROM disks as well as printed in the manual, and which appears on a user's screen every time the software runs, limits use of the application program and listings to non-commercial purposes.

Matthew Zeidenberg bought a consumer package of SelectPhone (trademark) in 1994 from a retail outlet in Madison, Wisconsin, but decided to ignore the license. He formed Silken Mountain Web Services, Inc., to resell the information in the Select-Phone (trademark) database. The corporation makes the database available on the Internet to anyone willing to pay its price — which, needless to say, is less than ProCD charges its commercial customers. Zeidenberg has purchased two additional Select-Phone (trademark) packages, each with an updated version of the database, and made the latest information available over the World Wide Web, for a price, through his corporation. ProCD filed this suit seeking an injunction against further dissemination that exceeds the rights specified in the licenses (identical in each of the three packages Zeidenberg purchased). The district court held the licenses ineffectual because their terms do not appear on the outside of the packages. The court added that the second and third licenses stand no different from the first, even though they are identical, because they might have been different, and a purchaser does not agree to — and cannot be bound by — terms that were secret at the time of purchase. 908 F. Supp. at 654.

II

Following the district court, we treat the licenses as ordinary contracts accompanying the sale of products, and therefore as governed by the common law of contracts and the Uniform Commercial Code. Whether there are legal differences between "contracts" and "licenses" (which may matter under the copyright doctrine of first sale) is a subject for another day. See Microsoft Corp. v. Harmony Computers & Electronics, Inc., 846 F. Supp. 208 (E.D. N.Y. 1994). Zeidenberg does not argue that Silken Mountain Web Services is free of any restrictions that apply to Zeidenberg himself, because any effort to treat the two parties as distinct would put Silken Mountain behind the eight ball on ProCD's argument that copying the application program onto its hard disk violates the copyright laws. Zeidenberg does argue, and the district court held, that placing the package of software on the shelf is an "offer," which the customer "accepts" by paying the asking price and leaving the store with the goods. Peters v. State, 154 Wis. 111, 142 N.W. 181 (1913). In Wisconsin, as elsewhere, a contract includes only the terms on which the parties have agreed. One cannot agree to hidden terms, the judge concluded. So far, so good—but one of the terms to which Zeidenberg agreed by purchasing the software is that the transaction was subject to a license. Zeidenberg's position therefore must be that the printed terms on the outside of a box are the parties' contract—except for printed terms that refer to or incorporate other terms. But why would Wisconsin fetter the parties' choice in this way? Vendors can put the entire terms of a contract on the outside of a box only by using microscopic type, removing other information that buyers might find more useful (such as what the software does, and on which computers it works), or both. The "Read Me" file included with most software, describing system requirements and potential incompatibilities, may be equivalent to ten pages of type; warranties and license restrictions take still more space. Notice on the outside, terms on the inside, and a right to return the software for a refund if the terms are unacceptable (a right that the license expressly extends), may be a means of doing business valuable to buyers and sellers alike. See E. Allan Farnsworth, 1 Farnsworth on Contracts § .26 (1990); Restatement (2d) of Contracts § 211 comment a (1981) ("Standardization of agreements serves many of the same functions as standardization of goods and services; both are essential to a system of mass production and distribution. Scarce and costly time and skill can be devoted to a class of transactions rather than the details of individual transactions."). Doubtless a state could forbid the use of standard contracts in the software business, but we do not think that Wisconsin has done so.

Transactions in which the exchange of money precedes the communication of detailed terms are common. Consider the purchase of insurance. The buyer goes to an agent, who explains the essentials (amount of coverage, number of years) and remits the premium to the home office, which sends back a policy. On the district judge's understanding, the terms of the policy are irrelevant because the insured paid before receiving them. Yet the device of payment, often with a "binder" (so that the insurance takes effect immediately even though the home office reserves the right to withdraw coverage later), in advance of the policy, serves buyers' interests by accelerating effectiveness and reducing transactions costs. Or consider the purchase of an airline ticket. The traveler calls the carrier or an agent, is quoted a price, reserves a seat, pays, and gets a ticket, in that order. The ticket contains elaborate terms, which the traveler can reject by canceling the reservation. To use the ticket is to accept the terms, even terms that in retrospect are disadvantageous. See Carnival Cruise Lines, Inc. v. Shute, 499 U.S. 585 (1991); see also Vimar Seguros y Reaseguros, S.A. v. M/V Sky Reefer, 515 U.S. 528 (1995) (bills of lading). Just so with a ticket to a concert. The back of the ticket states that the patron promises not to record the concert; to attend is to agree. A theater that detects a violation will confiscate the tape and escort the violator to the exit. One could arrange things so that every concertgoer signs this promise before forking over money, but that cumbersome way of doing things not only would lengthen queues and raise prices but also would scotch the sale of tickets by phone or electronic data service.

Consumer goods work the same way. Someone who wants to buy a radio set visits a store, pays, and walks out with a box. Inside the box is a leaflet containing some terms, the most important of which usually is the warranty, read for the first time in the comfort of home. By Zeidenberg's lights, the warranty in the box is irrelevant; every consumer gets the standard warranty implied by the UCC in the event the contract is silent; yet so far as we are aware no state disregards warranties furnished with consumer products. Drugs come with a list of ingredients on the outside and an elaborate package insert on the inside. The package insert describes drug interactions, contraindications, and other vital information — but, if Zeidenberg is right, the purchaser need not read the package insert, because it is not part of the contract.

Next consider the software industry itself. Only a minority of sales take place over the counter, where there are boxes to peruse. A customer pay place an order by phone in response to a line item in a catalog or a review in a magazine. Much software is ordered over the Internet by purchasers who have never seen a box. Increasingly software arrives by wire. There is no box; there is only a stream of electrons, a collection of information that includes data, an application program, instructions, many limitations ("MegaPixel 3.14159 cannot be used with Byte-Pusher 2.718"), and the terms of sale. The user purchases a serial number, which activates the software's features. On Zeidenberg's arguments, these unboxed sales are unfettered by terms — so the seller has made a broad warranty and must pay consequential damages for any shortfalls in performance, two "promises" that if taken seriously would drive prices through the ceiling or return transactions to the horse-and-buggy age.

According to the district court, the UCC does not countenance the sequence of money now, terms later. (Wisconsin's version of the UCC does not differ from the Official Version in any material respect, so we use the regular numbering system. Wis. Stat. §402.201 corresponds to UCC §2-201, and other citations are easy to derive.) One of the court's reasons — that by proposing as part of the draft Article 2B a new UCC §2-2203 that would explicitly validate standard-form user licenses, the American Law Institute and the National Conference of Commissioners on Uniform Laws have conceded the invalidity of shrinkwrap licenses under current law, see 908 F. Supp. at 655-66 — depends on a faulty inference. To propose a change in a law's text is not necessarily to propose a change in the law's effect. New words may be designed to fortify the current rule with a more precise text that curtails uncertainty. To judge by the flux of law review articles discussing shrinkwrap licenses, uncertainty is much in need of reduction — although businesses seem to feel less uncertainty than do scholars, for only three cases (other than ours) touch on the subject, and none directly addresses it. See Step-Saver Data Systems, Inc. v. Wyse Technology, 939 F.2d 91 (3d Cir. 1991); Vault Corp. v. Quaid Software Ltd., 847 F.2d 255, 268-70 (5th Cir. 1988); Arizona Retail Systems, Inc. v. Software Link, Inc., 831 F. Supp. 759 (D. Ariz. 1993). As their titles suggest, these are not consumer transactions. Step-Saver is a battle-of-the-forms case, in which the parties exchange incompatible forms and a court must decide which prevails. See Northrop Corp. v. Litronic Industries, 29 F.3d 1173 (7th Cir. 1994) (Illinois law); Douglas G. Baird & Robert Weisberg, Rules, Standards, and the Battle of the Forms: A Reassessment of §2-207, 68 Va. L. Rev. 1217, 1227-31 (1982). Our case has only one form; UCC §2-207 is irrelevant. Vault holds that Louisiana's special shrinkwrap-license statute is preempted by federal law, a question to which we return. And Arizona Retail Systems did not reach the question, because the court found that the buyer knew the terms of the license before purchasing the software.

What then does the current version of the UCC have to say? We think that the place to start is §2-204(1): "A contract for sale of goods may be made in any manner sufficient to show agreement, including conduct by both parties which recognizes the existence of such a contract." A vendor, as master of the offer, may invite acceptance by conduct, and may propose limitations on the kind of conduct that constitutes acceptance. A buyer may accept by performing the acts the vendor proposes to treat as acceptance. And that is what happened. ProCD proposed a contract that a buyer would accept by using the software after having an opportunity to read the license at leisure. This Zeidenberg did. He had no choice, because the software splashed the license on the screen and would not let him proceed without indicating acceptance. So although the district judge was right to say that a contract can be, and often is, formed simply by paying the price and walking out of the store, the UCC permits contracts to be formed in other ways. ProCD proposed such a different way, and without protest Zeidenberg agreed. Ours is not a case in which a consumer opens a package to find an insert saying "you owe us an extra $10,000" and the seller files suit to collect. Any buyer finding such a demand can prevent formation of the contract by returning the package, as can any consumer who concludes that the terms of the license make the software worth less than the purchase price. Nothing in the UCC requires a seller to maximize the buyer's net gains.

Section 2-606, which defines "acceptance of goods", reinforces this understanding. A buyer accepts goods under §2-606(1)(b) when, after an opportunity to inspect, he fails to make an effective rejection under §2-602(1). ProCD extended an opportunity to reject if a buyer should find the license terms unsatisfactory; Zeidenberg inspected the package, tried out the software, learned of the license, and did not reject the goods. We refer to §2-606 only to show that the opportunity to return goods can be important; acceptance of an offer differs from acceptance of goods after delivery, see Gillen v. Atalanta Systems, Inc., 997 F.2d 280, 284 n.1 (7th Cir. 1993); but the UCC consistently permits the parties to structure their relations so that the buyer has a chance to make a final decision after a detailed review.

Some portions of the UCC impose additional requirements on the way parties agree on terms. A disclaimer of the implied warranty of merchantability must be "conspicuous." UCC §2-316(2), incorporating UCC §1-201(10). Promises to make firm offers, or to negate oral modifications, must be "separately signed." UCC §§2-205, 2-209(2). These special provisos reinforce the impression that, so far as the UCC is concerned, other terms may be as inconspicuous as the forum-selection clause on the back of the cruise ship ticket in Carnival Lines. Zeidenberg has not located any Wisconsin case—for that matter, any case in any state—holding that under the UCC the ordinary terms found in shrinkwrap licenses require any special prominence, or otherwise are to be undercut rather than enforced. In the end, the terms of the license are conceptually identical to the contents of the package. Just as no court would dream of saying that SelectPhone (trademark) must contain 3,100 phone books rather than 3,000, or must have data no more than 30 days old, or must sell for $100 rather than $150—although any of these changes would be welcomed by the customer, if all other things were held constant—so, we believe, Wisconsin would not let the buyer pick and choose among terms. Terms of use are no less a part of "the product" than are the size of the database and the speed with which the software compiles listings. Competition among vendors, not judicial revision of a package's contents, is how consumers are protected in a market economy. Digital Equipment Corp. v. Uniq Digital Technologies, Inc., 73 F.3d 756 (7th Cir. 1996). ProCD has rivals, which may elect to compete by offering superior software, monthly updates, improved terms of use, lower price, or a better compromise among these elements. As we stressed above, adjusting terms in buyers' favor might help Matthew Zeidenberg today (he already has the software) but would lead to a response, such as a higher price, that might make consumers as a whole worse off.

COMMENTS AND QUESTIONS

1. Courts are sharply split on the enforceability of shrinkwrap licenses. *ProCD* was the first court to enforce a shrinkwrap license; a number of courts before that time had refused to enforce them. See Step-Saver v. Wyse Technology, 939 F.3d 91 (3d Cir. 1991); Vault Corp. v. Quaid, Inc., 847 F.2d 255 (5th Cir. 1988); Arizona Retail Sys. v. Software Link, Inc., 831 F. Supp. 759 (D. Ariz. 1993); cf. Foresight Resource Corp. v. Pfortmiller, 719 F. Supp. 1006 (D. Kan. 1989) (dictum). Since *ProCD*, a majority of courts have enforced shrinkwrap licenses. See M.A. Mortenson Co. v. Timberline Software Corp., 998 P.2d 305 (Wash. 2000); Information Handling Servs. v. LRP, 2000 Copr. L. Decisions ¶28,177 (E.D. Pa. Sept. 20, 2000); Peerless Wall & Window Coverings v. Synchronics, 85 F. Supp. 2d 519 (W.D. Pa. 2000); Adobe v. One Stop Micro, 84 F. Supp. 2d 1086 (N.D. Cal. 2000); Davidson & Assocs. v. Jung, 422 F.3d 630 (8th Cir. 2005); Bowers v. Baystate Technologies, 320 F.3d 1317 (Fed. Cir. 2003).

A significant number of recent cases have refused to enforce shrinkwrap licenses, however. See Novell v. Network Trade Center, 25 F. Supp. 1218 (D. Utah 1997); Klocek v. Gateway, Inc., 104 F. Supp. 2d 1332 (D. Kan. 2000). Cf. Morgan Labs, Inc. v. Micro Data Base Sys., 41 U.S.P.Q.2d 1850 (N.D. Cal. 1997) (refusing to allow a shrinkwrap license to modify a prior signed contract). The reasoning of these cases differs. Some courts (*Arizona Retail,* for instance) follow the rationale of *Step-Saver,* holding that the shrinkwrap license should be treated as a "battle of the forms" under U.C.C. 2-207. Other courts (*Pfortmiller,* the district court in *Vault*) conclude that shrinkwrap licenses are unenforceable as "contracts of adhesion." And the Fifth Circuit in *Vault* relied on the federal intellectual property laws to preempt contrary provisions in shrinkwrap licenses.

The "contract of adhesion" rationale is suspect. Under traditional doctrines of contract law, a contract of adhesion is a standardized contract prepared entirely by one party and, due to the disparity in bargaining power between the parties, is effectively offered on a "take-it-or-leave-it" basis. Contrary to the holding of the district court in *Vault,* contracts of adhesion are generally enforceable unless certain other factors (such as "unconscionability") are present. According to the Restatement (Second) of Contracts, §211(3), "[w]here the other party has reason to believe that the party manifesting such assent would not do so if he knew that the writing contained a particular term, the term is not part of the agreement." Comment f states that "[r]eason to believe [that the adherent would not knowingly have signed] may be inferred from the fact that the term is bizarre or oppressive, from the fact that it eviscerates the nonstandard terms explicitly agreed to, or from the fact that it eliminates the dominant purpose of the transaction."

On the other hand, a "shrinkwrap license" is not an ordinary contract of adhesion. Unlike most form contracts, in this case the party bound by the contract does not even sign or otherwise manifest consent to the terms that bind her. Indeed, in many cases she may not be aware of them.

2. Why did shrinkwrap licenses develop in the software context? They were certainly a fixture of the mass-market software industry by the early 1980s. One answer

may be that shrinkwrap licenses serve (or once served) as a declaration of the vendor's rights. Generally, proprietary rights provisions of shrinkwrap licenses assert that the information contained in the accompanying computer software is confidential and proprietary to the vendor, and may not be copied or disclosed without the vendor's permission. Software vendors who needed proof that they were not in fact disclosing their trade secrets by selling copies to whoever wanted them created the legal fiction that they were really licensing rather than selling their software. Because the "license" contained provisions that required customers to keep the software confidential, the trade secrets contained therein could be protected under decisions like *Data General*.

This rationale has much less force today, when it is clear that patent and copyright law are available to protect software sold on the mass market. Unlike trade secrets, patent and copyright law operate against the world, and do not require proof of tortious conduct or of a confidential relationship (or indeed any relationship at all) between the parties.

3. Judge Easterbrook points out that certain special contract terms have special contract formation procedures and requirements under the U.C.C. Finding none for shrinkwrap contracts, he reverts to the general formation procedure under Article 2. But is the code section he relies on — section 2-204 — the right one? Other courts have characterized shrinkwrap licenses under U.C.C. 2-207 (dealing with the "battle of the forms") or U.C.C. 2-209 (dealing with modifications to an existing contract). The last provision seems particularly appropriate for a common class of shrinkwrap licenses, which give no indication of the license terms until after the consumer has purchased the product and left the store.

Is *ProCD* such a case? The box containing the ProCD disks did not disclose the license terms but did at least indicate that there *were* additional terms contained in the box. Should this make a difference in deciding whether to enforce the terms?

4. In Judge Easterbrook's view, ProCD's business model relied heavily on the ability to price discriminate between different classes of users.[1] Judge Easterbrook argues the benefits of this practice and of contracts to carry it out. Cf. Frank H. Easterbrook, Cyberspace and the Law of the Horse, 1996 U. Chi. Legal F. 205. Are there counterarguments or limits to the degree to which law ought to enable price discrimination? See Wendy J. Gordon, Intellectual Property as Price Discrimination: Implications for Contract, 73 Chi.-Kent L. Rev. 1367 (1998).

Note on "Clickwrap" Licenses and Electronic Commerce

ProCD and other shrinkwrap license cases have all been based on a rather traditional model of software distribution, in which software is sold off the shelf in a storefront, or delivered in a box via the mail. Increasingly, however, software is not distributed by these means at all. Rather, software (as well as other forms of digital information) is likely to be distributed over the Internet directly. This new distribution mechanism has a number of implications for the contract rules that govern the transactions.

---

1. In fact, there was no evidence introduced in the case that indicated ProCD would be willing to authorize a use such as Zeidenberg's at any price. The price discrimination rationale is a construct based on Judge Easterbrook's assumption about what a rational actor might do.

≡≡≡≡
≡≡≡≡ ***Specht v. Netscape Communications Corp.***
≡≡≡≡ *United States Court of Appeals for the Second Circuit*
≡≡≡≡ *306 F.3d 17 (2d Cir. 2002)*

SOTOMAYOR, Circuit Judge:

This is an appeal from a judgment of the Southern District of New York denying a motion by defendants-appellants Netscape Communications Corporation and its corporate parent, America Online, Inc. (collectively, "defendants" or "Netscape"), to compel arbitration and to stay court proceedings. In order to resolve the central question of arbitrability presented here, we must address issues of contract formation in cyberspace. Principally, we are asked to determine whether plaintiffs-appellees ("plaintiffs"), by acting upon defendants' invitation to download free software made available on defendants' webpage, agreed to be bound by the software's license terms (which included the arbitration clause at issue), even though plaintiffs could not have learned of the existence of those terms unless, prior to executing the download, they had scrolled down the webpage to a screen located below the download button. We agree with the district court that a reasonably prudent Internet user in circumstances such as these would not have known or learned of the existence of the license terms before responding to defendants' invitation to download the free software, and that defendants therefore did not provide reasonable notice of the license terms. In consequence, plaintiffs' bare act of downloading the software did not unambiguously manifest assent to the arbitration provision contained in the license terms....

We therefore affirm the district court's denial of defendants' motion to compel arbitration and to stay court proceedings.

Background

I. Facts

In three related putative class actions, plaintiffs alleged that, unknown to them, their use of SmartDownload transmitted to defendants private information about plaintiffs' downloading of files from the Internet, thereby effecting an electronic surveillance of their online activities in violation of two federal statutes, the Electronic Communications Privacy Act, 18 U.S.C. §§510 et seq., and the Computer Fraud and Abuse Act, 18 U.S.C. §1030.

Specifically, plaintiffs alleged that when they first used Netscape's Communicator — a software program that permits Internet browsing — the program created and stored on each of their computer hard drives a small text file known as a "cookie" that functioned "as a kind of electronic identification tag for future communications" between their computers and Netscape. Plaintiffs further alleged that when they installed SmartDownload — a separate software "plug-in" that served to enhance Communicator's browsing capabilities — SmartDownload created and stored on their computer hard drives another string of characters, known as a "Key," which similarly functioned as an identification tag in future communications with Netscape. According to the complaints in this case, each time a computer user employed Communicator to download a file from the Internet, SmartDownload "assume[d] from Communicator the task of downloading" the file and transmitted to Netscape the address of the file being downloaded together with the cookie created by Communicator and the Key created by SmartDownload. These processes, plaintiffs claim, constituted unlawful "eavesdropping" on users of Netscape's software products as well as on Internet websites from which users employing SmartDownload downloaded files.

In the time period relevant to this litigation, Netscape offered on its website various software programs, including Communicator and SmartDownload, which visitors to the site were invited to obtain free of charge. It is undisputed that five of the six named plaintiffs — Michael Fagan, John Gibson, Mark Gruber, Sean Kelly, and Sherry Weindorf — downloaded Communicator from the Netscape website. These plaintiffs acknowledge that when they proceeded to initiate installation of Communicator, they were automatically shown a scrollable text of that program's license agreement and were not permitted to complete the installation until they had clicked on a "Yes" button to indicate that they accepted all the license terms.[4] If a user attempted to install Communicator without clicking "Yes," the installation would be aborted. All five named user plaintiffs expressly agreed to Communicator's license terms by clicking "Yes." The Communicator license agreement that these plaintiffs saw made no mention of SmartDownload or other plug-in programs, and stated that "[t]hese terms apply to Netscape Communicator and Netscape Navigator" and that "all disputes relating to this Agreement (excepting any dispute relating to intellectual property rights)" are subject to "binding arbitration in Santa Clara County, California."

Although Communicator could be obtained independently of SmartDownload, all the named user plaintiffs, except Fagan, downloaded and installed Communicator in connection with downloading SmartDownload. Each of these plaintiffs allegedly arrived at a Netscape webpage captioned "SmartDownload Communicator" that urged them to "Download With Confidence Using SmartDownload!" At or near the bottom of the screen facing plaintiffs was the prompt "Start Download" and a tinted button labeled "Download." By clicking on the button, plaintiffs initiated the download of SmartDownload. Once that process was complete, SmartDownload, as its first plug-in task, permitted plaintiffs to proceed with downloading and installing Communicator, an operation that was accompanied by the clickwrap display of Communicator's license terms described above.

The signal difference between downloading Communicator and downloading SmartDownload was that no clickwrap presentation accompanied the latter operation. Instead, once plaintiffs Gibson, Gruber, Kelly, and Weindorf had clicked on the "Download" button located at or near the bottom of their screen, and the downloading of SmartDownload was complete, these plaintiffs encountered no further information about the plug-in program or the existence of license terms governing its use. The sole reference to SmartDownload's license terms on the "SmartDownload Communicator" webpage was located in text that would have become visible to plaintiffs only if they had scrolled down to the next screen.

---

4. This kind of online software license agreement has come to be known as "clickwrap" (by analogy to "shrinkwrap," used in the licensing of tangible forms of software sold in packages) because it "presents the user with a message on his or her computer screen, requiring that the user manifest his or her assent to the terms of the license agreement by clicking on an icon. The product cannot be obtained or used unless and until the icon is clicked." *Specht*, 150 F.Supp.2d at 593-94 (footnote omitted). Just as breaking the shrinkwrap seal and using the enclosed computer program after encountering notice of the existence of governing license terms has been deemed by some courts to constitute assent to those terms in the context of tangible software, see, e.g., ProCD, Inc. v. Zeidenberg, 86 F.3d 1447, 1451 (7th Cir.1996), so clicking on a webpage's clickwrap button after receiving notice of the existence of license terms has been held by some courts to manifest an Internet user's assent to terms governing the use of downloadable intangible software, see, e.g., Hotmail Corp. v. Van$ Money Pie Inc., 47 U.S.P.Q.2d 1020, 1025 (N.D.Cal.1998).

Had plaintiffs scrolled down instead of acting on defendants' invitation to click on the "Download" button, they would have encountered the following invitation: "Please review and agree to the terms of the *Netscape SmartDownload software license agreement* before downloading and using the software." Plaintiffs Gibson, Gruber, Kelly, and Weindorf averred in their affidavits that they never saw this reference to the SmartDownload license agreement when they clicked on the "Download" button. They also testified during depositions that they saw no reference to license terms when they clicked to download SmartDownload, although under questioning by defendants' counsel, some plaintiffs added that they could not "remember" or be "sure" whether the screen shots of the SmartDownload page attached to their affidavits reflected precisely what they had seen on their computer screens when they downloaded SmartDownload.[10]

In sum, plaintiffs Gibson, Gruber, Kelly, and Weindorf allege that the process of obtaining SmartDownload contrasted sharply with that of obtaining Communicator. Having selected SmartDownload, they were required neither to express unambiguous assent to that program's license agreement nor even to view the license terms or become aware of their existence before proceeding with the invited download of the free plug-in program. Moreover, once these plaintiffs had initiated the download, the existence of SmartDownload's license terms was not mentioned while the software was running or at any later point in plaintiffs' experience of the product.

Even for a user who, unlike plaintiffs, did happen to scroll down past the download button, SmartDownload's license terms would not have been immediately displayed in the manner of Communicator's clickwrapped terms. Instead, if such a user had seen the notice of SmartDownload's terms and then clicked on the underlined invitation to review and agree to the terms, a hypertext link would have taken the user to a separate webpage entitled "License & Support Agreements." The first paragraph on this page read, in pertinent part:

> The use of each Netscape software product is governed by a license agreement. You must read and agree to the license agreement terms BEFORE acquiring a product. Please click on the appropriate link below to review the current license agreement for the product of interest to you before acquisition. For products available for download, you must read and agree to the license agreement terms BEFORE you install the software. If you do not agree to the license terms, do not download, install or use the software.

Below this paragraph appeared a list of license agreements, the first of which was "*License Agreement for Netscape Navigator and Netscape Communicator Product Family* (Netscape Navigator, Netscape Communicator and Netscape SmartDownload)." If the user clicked on that link, he or she would be taken to yet another webpage that contained the full text of a license agreement that was identical in every respect to the Communicator license agreement except that it stated that its "terms apply to Netscape Communicator, Netscape Navigator, and Netscape Smart-Download." The license agreement granted the user a nonexclusive license to use and reproduce the software, subject to certain terms:

---

10. In the screen shot of the SmartDownload webpage attached to Weindorf's affidavit, the reference to license terms is partially visible, though almost illegible, at the bottom of the screen. In the screen shots attached to the affidavits of Gibson, Gruber, and Kelly, the reference to license terms is not visible.

BY CLICKING THE ACCEPTANCE BUTTON OR INSTALLING OR USING NETSCAPE COMMUNICATOR, NETSCAPE NAVIGATOR, OR NETSCAPE SMARTDOWNLOAD SOFTWARE (THE "PRODUCT"), THE INDIVIDUAL OR ENTITY LICENSING THE PRODUCT ("LICENSEE") IS CONSENTING TO BE BOUND BY AND IS BECOMING A PARTY TO THIS AGREEMENT. IF LICENSEE DOES NOT AGREE TO ALL OF THE TERMS OF THIS AGREEMENT, THE BUTTON INDICATING NON-ACCEPTANCE MUST BE SELECTED, AND LICENSEE MUST NOT INSTALL OR USE THE SOFTWARE.

Among the license terms was a provision requiring virtually all disputes relating to the agreement to be submitted to arbitration:

Unless otherwise agreed in writing, all disputes relating to this Agreement (excepting any dispute relating to intellectual property rights) shall be subject to final and binding arbitration in Santa Clara County, California, under the auspices of JAMS/End-Dispute, with the losing party paying all costs of arbitration....

II. Proceedings Below

In the district court, defendants moved to compel arbitration and to stay court proceedings pursuant to the Federal Arbitration Act ("FAA"), 9 U.S.C. §4, arguing that the disputes reflected in the complaints, like any other dispute relating to the SmartDownload license agreement, are subject to the arbitration clause contained in that agreement. Finding that Netscape's webpage, unlike typical examples of clickwrap, neither adequately alerted users to the existence of SmartDownload's license terms nor required users unambiguously to manifest assent to those terms as a condition of downloading the product, the court held that the user plaintiffs had not entered into the SmartDownload license agreement. Specht, 150 F.Supp.2d at 595-96.

[Defendants appealed]

Discussion . . .

III. Whether the User Plaintiffs Had Reasonable Notice of and Manifested Assent to the SmartDownload License Agreement

Whether governed by the common law or by Article 2 of the Uniform Commercial Code ("UCC"), a transaction, in order to be a contract, requires a manifestation of agreement between the parties. See Windsor Mills, Inc. v. Collins & Aikman Corp., 101 Cal.Rptr. 347, 350 (Cal. Ct. App.1972) ("[C]onsent to, or acceptance of, the arbitration provision [is] necessary to create an agreement to arbitrate."); see also Cal. Com.Code §2204(1) ("A contract for sale of goods may be made in any manner sufficient to show agreement, including conduct by both parties which recognizes the existence of such a contract.").[13] Mutual manifestation of assent,

---

13. The district court concluded that the SmartDownload transactions here should be governed by "California law as it relates to the sale of goods, including the Uniform Commercial Code in effect in California." *Specht*, 150 F.Supp.2d at 591. It is not obvious, however, that UCC Article 2 ("sales of goods")

whether by written or spoken word or by conduct, is the touchstone of contract. Binder v. Aetna Life Ins. Co., 89 Cal.Rptr.2d 540, 551 (Cal. Ct. App. 1999); cf. Restatement (Second) of Contracts § 19(2) (1981) ("The conduct of a party is not effective as a manifestation of his assent unless he intends to engage in the conduct and knows or has reason to know that the other party may infer from his conduct that he assents."). Although an onlooker observing the disputed transactions in this case would have seen each of the user plaintiffs click on the SmartDownload "Download" button, see Cedars Sinai Med. Ctr. v. Mid-West Nat'l Life Ins. Co., 118 F.Supp.2d 1002, 1008 (C.D. Cal. 2000) ("In California, a party's intent to contract is judged objectively, by the party's outward manifestation of consent."), a consumer's clicking on a download button does not communicate assent to contractual terms if the offer did not make clear to the consumer that clicking on the download button would signify assent to those terms, see Windsor Mills, 101 Cal.Rptr. at 351 ("[W]hen the offeree does not know that a proposal has been made to him this objective standard does not apply."). California's common law is clear that "an offeree, regardless of apparent manifestation of his consent, is not bound by inconspicuous contractual provisions of which he is unaware, contained in a document whose contractual nature is not obvious." Id.; see also Marin Storage & Trucking, Inc. v. Benco Contracting & Eng'g, Inc., 107 Cal.Rptr.2d 645, 651 (Cal. Ct. App. 2001) (same).

Arbitration agreements are no exception to the requirement of manifestation of assent. "This principle of knowing consent applies with particular force to provisions for arbitration." Windsor Mills, 101 Cal.Rptr. at 351....

---

applies to the licensing of software that is downloadable from the Internet. Cf. Advent Sys. Ltd. v. Unisys Corp., 925 F.2d 670, 675 (3d Cir.1991) ("The increasing frequency of computer products as subjects of commercial litigation has led to controversy over whether software is a 'good' or intellectual property. The [UCC] does not specifically mention software."); Lorin Brennan, Why Article 2 Cannot Apply to Software Transactions, PLI Patents, Copyrights, Trademarks, & Literary Property Course Handbook Series (Feb.-Mar. 2001) (demonstrating the trend in case law away from application of UCC provisions to software sales and licensing and toward application of intellectual property principles). There is no doubt that a sale of tangible goods over the Internet is governed by Article 2 of the UCC. See, e.g., Butler v. Beer Across Am., 83 F.Supp.2d 1261, 1263-64 & n.6 (N.D. Ala. 2000) (applying Article 2 to an Internet sale of bottles of beer). Some courts have also applied Article 2, occasionally with misgivings, to sales of off-the-shelf software in tangible, packaged formats. See, e.g., *ProCD*, 86 F.3d at 1450 ("[W]e treat the [database] licenses as ordinary contracts accompanying the sale of products, and therefore as governed by the common law of contracts and the Uniform Commercial Code. Whether there are legal differences between 'contracts' and 'licenses' (which may matter under the copyright doctrine of first sale) is a subject for another day."); i.LAN Sys., Inc. v. NetScout Serv. Level Corp., 183 F.Supp.2d 328, 332 (D. Mass. 2002) (stating, in the context of a dispute between business parties, that "Article 2 technically does not, and certainly will not in the future, govern software licenses, but for the time being, the Court will assume that it does").

Downloadable software, however, is scarcely a "tangible" good, and, in part because software may be obtained, copied, or transferred effortlessly at the stroke of a computer key, licensing of such Internet products has assumed a vast importance in recent years. Recognizing that "a body of law based on images of the sale of manufactured goods ill fits licenses and other transactions in computer information," the National Conference of Commissioners on Uniform State Laws has promulgated the Uniform Computer Information Transactions Act ("UCITA"), a code resembling UCC Article 2 in many respects but drafted to reflect emergent practices in the sale and licensing of computer information. UCITA, prefatory note (rev. ed. Aug. 23, 2001) (available at www.ucitaonline.com/ucita.html). UCITA—originally intended as a new Article 2B to supplement Articles 2 and 2A of the UCC but later proposed as an independent code—has been adopted by two states, Maryland and Virginia. See Md.Code Ann. Com. Law § § 22-101 et seq.; Va.Code Ann. § § 59.1-501.1 et seq.

We need not decide today whether UCC Article 2 applies to Internet transactions in downloadable products. The district court's analysis and the parties' arguments on appeal show that, for present purposes, there is no essential difference between UCC Article 2 and the common law of contracts. We therefore apply the common law, with exceptions as noted.

A. The Reasonably Prudent Offeree of Downloadable Software

Defendants argue that plaintiffs must be held to a standard of reasonable prudence and that, because notice of the existence of SmartDownload license terms was on the next scrollable screen, plaintiffs were on "inquiry notice" of those terms. We disagree with the proposition that a reasonably prudent offeree in plaintiffs' position would necessarily have known or learned of the existence of the SmartDownload license agreement prior to acting, so that plaintiffs may be held to have assented to that agreement with constructive notice of its terms. See Cal. Civ.Code § 1589 ("A voluntary acceptance of the benefit of a transaction is equivalent to a consent to all the obligations arising from it, so far as the facts are known, or ought to be known, to the person accepting."). It is true that "[a] party cannot avoid the terms of a contract on the ground that he or she failed to read it before signing." Marin Storage & Trucking, 107 Cal.Rptr.2d at 651. But courts are quick to add: "An exception to this general rule exists when the writing does not appear to be a contract and the terms are not called to the attention of the recipient. In such a case, no contract is formed with respect to the undisclosed term." Id.; cf. Cory v. Golden State Bank, 157 Cal. Rptr. 538, 541 (Cal. Ct. App. 1979) ("[T]he provision in question is effectively hidden from the view of money order purchasers until after the transactions are completed.... Under these circumstances, it must be concluded that the Bank's money order purchasers are not chargeable with either actual or constructive notice of the service charge provision, and therefore cannot be deemed to have consented to the provision as part of their transaction with the Bank."). ...

[R]eceipt of a physical document containing contract terms or notice thereof is frequently deemed, in the world of paper transactions, a sufficient circumstance to place the offeree on inquiry notice of those terms. "Every person who has actual notice of circumstances sufficient to put a prudent man upon inquiry as to a particular fact, has constructive notice of the fact itself in all cases in which, by prosecuting such inquiry, he might have learned such fact." Cal. Civ.Code § 19. These principles apply equally to the emergent world of online product delivery, pop-up screens, hyperlinked pages, clickwrap licensing, scrollable documents, and urgent admonitions to "Download Now!". What plaintiffs saw when they were being invited by defendants to download this fast, free plug-in called SmartDownload was a screen containing praise for the product and, at the very bottom of the screen, a "Download" button. Defendants argue that under the principles set forth in the cases cited above, a "fair and prudent person using ordinary care" would have been on inquiry notice of SmartDownload's license terms. *Shacket,* 651 F.Supp. at 690.

We are not persuaded that a reasonably prudent offeree in these circumstances would have known of the existence of license terms. Plaintiffs were responding to an offer that did not carry an immediately visible notice of the existence of license terms or require unambiguous manifestation of assent to those terms. Thus, plaintiffs' "apparent manifestation of ... consent" was to terms "contained in a document whose contractual nature [was] not obvious." *Windsor Mills,* 101 Cal.Rptr. at 351. Moreover, the fact that, given the position of the scroll bar on their computer screens, plaintiffs may have been aware that an unexplored portion of the Netscape webpage remained below the download button does not mean that they reasonably should have concluded that this portion contained a notice of license terms. In their deposition testimony, plaintiffs variously stated that they used the scroll bar "[o]nly if there is something that I feel I need to see that is on — that is off the page," or that the elevated position of the scroll bar suggested the presence of "mere [ ] formalities, standard lower banner links" or "that the page is bigger than what I can see." Plaintiffs testified, and defendants did not refute, that plaintiffs were in fact unaware that defendants intended to attach license terms to the use of SmartDownload.

We conclude that in circumstances such as these, where consumers are urged to download free software at the immediate click of a button, a reference to the existence of license terms on a submerged screen is not sufficient to place consumers on inquiry or constructive notice of those terms. The SmartDownload webpage screen was "printed in such a manner that it tended to conceal the fact that it was an express acceptance of [Netscape's] rules and regulations." *Larrus,* 266 P.2d at 147. Internet users may have, as defendants put it, "as much time as they need[ ]" to scroll through multiple screens on a webpage, but there is no reason to assume that viewers will scroll down to subsequent screens simply because screens are there. When products are "free" and users are invited to download them in the absence of reasonably conspicuous notice that they are about to bind themselves to contract terms, the transactional circumstances cannot be fully analogized to those in the paper world of arm's-length bargaining. In the next two sections, we discuss case law and other legal authorities that have addressed the circumstances of computer sales, software licensing, and online transacting. Those authorities tend strongly to support our conclusion that plaintiffs did not manifest assent to SmartDownload's license terms.

B. Shrinkwrap Licensing and Related Practices

Defendants cite certain well-known cases involving shrinkwrap licensing and related commercial practices in support of their contention that plaintiffs became bound by the SmartDownload license terms by virtue of inquiry notice. [The Court discusses *Hill, ProCD,* and *Mortenson.*]

These cases do not help defendants. To the extent that they hold that the purchaser of a computer or tangible software is contractually bound after failing to object to printed license terms provided with the product, *Hill* and *Brower* do not differ markedly from the cases involving traditional paper contracting discussed in the previous section. Insofar as the purchaser in *ProCD* was confronted with conspicuous, mandatory license terms every time he ran the software on his computer, that case actually undermines defendants' contention that downloading in the absence of conspicuous terms is an act that binds plaintiffs to those terms. In *Mortenson,* the full text of license terms was printed on each sealed diskette envelope inside the software box, printed again on the inside cover of the user manual, and notice of the terms appeared on the computer screen every time the purchaser executed the program. *Mortenson,* 970 P.2d at 806. In sum, the foregoing cases are clearly distinguishable from the facts of the present action.

C. Online Transactions

Cases in which courts have found contracts arising from Internet use do not assist defendants, because in those circumstances there was much clearer notice than in the present case that a user's act would manifest assent to contract terms. [The court discussed several clickwrap license cases, and also referred to UCITA as a relevant statement of analogous rules.]

After reviewing the California common law and other relevant legal authority, we conclude that under the circumstances here, plaintiffs' downloading of Smart-Download did not constitute acceptance of defendants' license terms. Reasonably conspicuous notice of the existence of contract terms and unambiguous manifestation of assent to those terms by consumers are essential if electronic bargaining is to have integrity and credibility. We hold that a reasonably prudent offeree in plaintiffs' position would not have known or learned, prior to acting on the invitation to download, of the reference to SmartDownload's license terms hidden below the "Download" button on the next screen. We affirm the district court's conclusion that the user plaintiffs, including Fagan, are not bound by the arbitration clause contained in those terms.

Conclusion

For the foregoing reasons, we affirm the district court's denial of defendants' motion to compel arbitration and to stay court proceedings.

COMMENTS AND QUESTIONS

1. Clickwrap and browsewrap licenses differ in significant respects from shrink-wrap licenses. First, bits sent over a modem seem less like traditional, tangible "goods" to which Article 2 has traditionally applied. This doesn't necessarily mean that the information in question is "licensed," of course. Software sold over the Internet is still distributed in the form of "copies," at least in a copyright law sense; it's just that those copies are created at the buyer's site, rather than being shipped there in physical form. Further, most of the indicia of sale in a mass-market transaction — the possession of one copy, with perpetual use rights in exchange for a single lump-sum payment — are still present in an online transaction. But even if the transaction looks more like a sale than a lease (or "license"), it is more and more difficult to apply Article 2 to such a deal, except perhaps by analogy.

Second, the nature of online sales gives vendors an opportunity to avoid the problems of shrinkwrap licenses entirely, by disclosing terms to buyers and obtaining agreement to those terms before the purchase is completed. It is relatively common for websites to require agreement to certain terms and conditions by asking the user to click "OK" before proceeding. Such an agreement, entered into before the vendor sells the software or other information, is certainly a "contract of adhesion." The buyer has no realistic choice, other than "take it or leave it." But it is not a shrinkwrap license. The buyer has the opportunity to read and agree to the terms before she pays her money. It is not surprising, therefore, that such "clickwrap" licenses are easier to enforce than their shrinkwrap counterparts. See Davidson & Assocs. v. Internet Gateway, 420 F.3d 630 (8th Cir. 2005); i.Lan Systems v. Netscout, 183 F. Supp. 328 (D. Mass. 2002); Caspi v. Microsoft Network LLC, 732 A.2d 528 (N.J. App. Div. 1999); Hotmail Corp. v. Van Money Pie, 47 U.S.P.Q.2d 1020 (N.D. Cal. 1998) (assuming such an agreement was enforceable, without discussing the issue).

2. Do consumers really benefit from being able to see the terms in advance, if they still have no choice regarding what those terms are? Is the real objection to shrinkwrap

licenses their immutable, monolithic character, and not how they are presented to the consumer? Cf. Robert P. Merges, Intellectual Property and the Costs of Commercial Exchange: A Review Essay, 93 Mich. L. Rev. 1570, 1612 (1995) (suggesting that shrinkwrap license terms should be preempted when there is sufficient uniformity in the industry that the terms in effect amount to "private legislation" by software vendors); Maureen A. O'Rourke, Copyright Preemption After the ProCD Case: A Market-Based Approach, 12 Berkeley Tech. L.J. 53 (1997).

3. Despite the obvious legal advantages to giving consumers terms to review and approve before entering into a contract, many websites (like the one in *Specht*) do not in fact require purchasers to read and agree to their terms. Not surprisingly, many courts have refused to enforce such unilateral statements as contracts. In addition to *Specht* and the cases it discusses, see Campbell v. General Dynamics, 407 F.3d 546 (1st Cir. 2005); Westcode, Inc. v. RBE Electronics, Inc., 2000 WL 124566 (E.D. Pa. Feb. 1, 2000). But a surprising number of courts have enforced such "browsewrap" agreements, often in the context of claims for invasion of property by electronic robots. See, e.g., Register.com v. Verio, Inc., 356 F.3d 393 (2d Cir. 2004); Ticketmaster Corp. v. Tickets.com, 2003 WL 21406289 (C.D. Cal. Mar. 7, 2003); Pollstar v. Gigmania, Ltd., 170 F. Supp. 2d 974 (E.D. Cal. 2000).

Can you think of any reason not to require a consumer to click "I agree"? What does such a policy say about the importance of the terms to the vendor? To the transaction?

If merely posting terms on a Web site creates a contract with anyone who visits the Web site, is there any way for a consumer to avoid entering into such a contract? What happens when companies post conflicting terms of use on their Web sites?

3. Idea Submissions

Courts have struggled over the years with whether and under what circumstances a person who supplies a valuable idea to another may obtain legal redress. See Lionel S. Sobel, The Law of Ideas, Revisited, 1 UCLA Ent. L. Rev. 9 (1994). On the one hand, copyright law expressly disavows protection for ideas, see § 102(b), and patent law imposes stringent requirements — examination, novelty, non-obviousness, utility, and disclosure — before affording protection. Therefore, it would seem incongruous for state law to protect ideas. On the other hand, we protect valuable information through state trade secret law. Should there be some role for state law to protect promises to pay for an idea that is not a trade secret? What if the idea is communicated without a formal contract being entered into? Without such protection, people with good ideas might lack the legal and economic security to exchange such information with those in a position to develop products, movies, and other valuable works. But if unpatented, nonsecret ideas are protectable, what is left of the public domain?

═════
═════
═════ *Nadel v. Play-by-Play Toys & Novelties, Inc.*
═════ United States Court of Appeals for the Second Circuit
═════ *208 F.3d 368 (2d Cir. 2000)*

SOTOMAYOR, Circuit Judge

Plaintiff-appellant Craig P. Nadel ("Nadel") brought this action against defendant-appellee Play-By-Play Toys & Novelties, Inc. ("Play-By-Play") for breach of contract, quasi contract, and unfair competition. The thrust of Nadel's complaint was that Play-By-Play took his idea for an upright, sound-emitting, spinning plush toy and that, contrary to industry custom, Play-By-Play used the idea in its "Tornado Taz" product without paying him compensation. . . .

Background

Nadel is a toy idea man. Toy companies regularly do business with independent inventors such as Nadel in order to develop and market new toy concepts as quickly as possible. To facilitate the exchange of ideas, the standard custom and practice in the toy industry calls for companies to treat the submission of an idea as confidential. If the company subsequently uses the disclosed idea, industry custom provides that the company shall compensate the inventor, unless, of course, the disclosed idea was already known to the company.

In 1996, Nadel developed the toy concept at issue in this case. He transplanted the "eccentric mechanism"[1] found in several hanging Halloween toys then on the market-such as "Spooky Skull" and "Shaking Mutant Pumpkin"—and placed the mechanism inside of a plush toy monkey skin to develop the prototype for a new table-top monkey toy. This plush toy figure sat upright, emitted sound, and spun when placed on a flat surface.

In October 1996, Nadel met with Neil Wasserman, an executive at Play-By-Play who was responsible for the development of its plush toy line. According to Nadel, he showed his prototype monkey toy to Wasserman, who expressed interest in adapting the concept to a non-moving, plush Tazmanian Devil toy that Play-By-Play was already producing under license from Warner Bros. Nadel contends that, consistent with industry custom, any ideas that he disclosed to Wasserman during their October 1996 meeting were subject to an agreement by Play-By-Play to keep such ideas confidential and to compensate Nadel in the event of their use.

Nadel claims that he sent his prototype monkey toy to Wasserman as a sample and awaited the "Taz skin" and voice tape, which Wasserman allegedly said he would send, so that Nadel could make a sample spinning/laughing Tazmanian Devil toy for Play-By-Play. Wasserman never provided Nadel with the Taz skin and voice tape, however, and denies ever having received the prototype monkey toy from Nadel.

Notwithstanding Wasserman's denial, his secretary, Melissa Rodriguez, testified that Nadel's prototype monkey toy remained in Wasserman's office for several months. According to Ms. Rodriguez, the monkey toy was usually kept in a glass cabinet behind Wasserman's desk, but she remembered that on one occasion she had seen it on a table in Wasserman's office. Despite Nadel's multiple requests, Wasserman did not return Nadel's prototype monkey toy until February 1997, after Play-by-Play introduced its "Tornado Taz" product at the New York Toy Fair.

The parties do not dispute that "Tornado Taz" has the same general characteristics as Nadel's prototype monkey toy. Like Nadel's toy, Tornado Taz is a plush toy that emits sounds (including "screaming," "laughing," "snarling," and "grunting"), sits upright, and spins by means of an internal eccentric vibration mechanism.

---

1. An eccentric mechanism typically consists of a housing containing a motor with an eccentric weight attached to the motor shaft. When the motor is activated, the motor rotates the weight centrifugally, causing the housing to shake or spin.

Nadel claims that, in violation of their alleged agreement, Play-By-Play used his idea without paying him compensation. Play-By-Play contends, however, that it independently developed the Tornado Taz product concept and that Nadel is therefore not entitled to any compensation. Specifically, Play-By-Play maintains that, as early as June or July of 1996, two of its officers — Wasserman and Slattery — met in Hong Kong and began discussing ways to create a spinning or vibrating Tazmanian Devil, including the possible use of an eccentric mechanism. Furthermore, Play-By-Play claims that in late September or early October 1996, it commissioned an outside manufacturing agent — Barter Trading of Hong Kong — to begin developing Tornado Taz.

Play-By-Play also argues that, even if it did use Nadel's idea to develop Tornado Taz, Nadel is not entitled to compensation because Nadel's concept was unoriginal and non-novel to the toy industry in October 1996. In support of this argument, Play-By-Play has submitted evidence of various toys, commercially available prior to October 1996, which used eccentric motors and allegedly contained the same characteristics as Nadel's prototype monkey toy. . . .

Discussion

I. *Nadel's Claims*

On January 21, 1999, the district court granted Play-By-Play's motion for summary judgment dismissing Nadel's claims for breach of contract, quasi contract, and unfair competition. Interpreting New York law, the district court stated that "a party is not entitled to recover for theft of an idea unless the idea is novel or original." Nadel v. Play By Play Toys & Novelties, Inc., 34 F. Supp. 2d 180, 184 (S.D.N.Y.1999). Applying that principle to Nadel's claims, the district court concluded that, even if the spinning toy concept were novel to Play-By-Play at the time Nadel made the disclosure to Wasserman in October 1996, Nadel's claims must nonetheless fail for lack of novelty or originality because "numerous toys containing the characteristics of [Nadel's] monkey were in existence prior to [ ] October 1996." Id. at 185. In essence, the district court interpreted New York law to require that, when a plaintiff claims that a defendant has either (1) misappropriated his idea (a "property-based claim") or (2) breached an express or implied-in-fact contract by using such idea (a "contract-based claim"), the idea at issue must be original or novel generally. See id. at 184 n. 1. Thus, according to the district court, a finding that an idea was novel as to Play By Play — i.e., novel to the buyer — cannot suffice to sustain any of Nadel's claims. See id.

On appeal, Nadel challenges the district court's conclusion that a showing of novelty to the buyer — i.e., that Nadel's idea was novel to Play-By-Play at the time of his October 1996 disclosure — cannot suffice to sustain his claims for breach of contract, quasi contract, and unfair competition under New York law. Nadel claims, moreover, that the record contains a genuine issue of material fact concerning whether his toy idea was novel to Play-By-Play at the time of his October 1996 disclosure to Wasserman and that the district court therefore erred in granting Play-By-Play's motion for summary judgment.

Nadel's factual allegations present a familiar submission-of-idea case: (1) the parties enter into a pre-disclosure confidentiality agreement; (2) the idea is subsequently disclosed to the prospective buyer; (3) there is no post-disclosure contract for payment based on use; and (4) plaintiff sues defendant for allegedly using the disclosed idea under either a contract-based or property-based theory. For the reasons that follow, we conclude that a finding of novelty as to Play-By-Play can suffice to provide consideration for Nadel's contract claims against Play-By-Play. Accordingly, because we also find that there exists a genuine issue of material fact as to whether Nadel's idea was novel to Play-By-Play at the time of his October 1996 disclosure, we vacate the district court's grant of summary judgment on Nadel's contract claims. With respect to Nadel's misappropriation claim, we similarly vacate the district court's grant of summary judgment and remand for further proceedings to determine whether Nadel's idea was original or novel generally.

A. Submission-of-Idea Cases Under New York Law

Our analysis begins with the New York Court of Appeals' most recent discussion of the law governing idea submission cases, Apfel v. Prudential-Bache Securities, Inc., 81 N.Y.2d 470, 600 N.Y.S.2d 433, 616 N.E.2d 1095 (1993). In *Apfel*, the Court of Appeals discussed the type of novelty an idea must have in order to sustain a contract-based or property-based claim for its uncompensated use. Specifically, *Apfel* clarified an important distinction between the requirement of "novelty to the buyer" for contract claims, on the one hand, and "originality" (or novelty generally) for misappropriation claims, on the other hand.

Under the facts of *Apfel*, the plaintiff disclosed his idea to the defendant pursuant to a confidentiality agreement and, subsequent to disclosure, entered into another agreement wherein the defendant agreed to pay a stipulated price for the idea's use. See id. at 474. The defendant used the idea but refused to pay plaintiff pursuant to the post-disclosure agreement on the asserted ground that "no contract existed between the parties because the sale agreement lacked consideration." Id. at 475. The defendant argued that an idea could not constitute legally sufficient consideration unless it was original or novel generally and that, because plaintiff's idea was not original or novel generally (it had been in the public domain at the time of the post-disclosure agreement), the idea provided insufficient consideration to support the parties' post-disclosure contract. See id. at 474-75.

In rejecting defendant's argument, the Court of Appeals held that there was sufficient consideration to support plaintiff's contract claim because the idea at issue had value to the defendant at the time the parties concluded their post-disclosure agreement. See id. at 476. The *Apfel* court noted that "traditional principles of contract law" provide that parties "are free to make their bargain, even if the consideration exchanged is grossly unequal or of dubious value," id. at 475, and that, so long as the "defendant received something of value" under the contract, the contract would not be void for lack of consideration, id. at 476. See also id. at 478 ("[T]he buyer knows what he or she is buying and has agreed that the idea has value, and the Court will not ordinarily go behind that determination.").

The *Apfel* court explicitly rejected defendant's contention that the court should carve out "an exception to traditional principles of contract law" for submission-of-idea cases by requiring that an idea must also be original or novel generally in order to constitute valid consideration. Id. at 477. In essence, the defendant sought to impose a requirement that an idea be novel in absolute terms, as opposed to only the defendant buyer, in order to constitute valid consideration for the bargain. In rejecting this argument, the *Apfel* court clarified the standards for both contract-based and property-based claims in submission-of-idea cases. That analysis guides our decision here.

The *Apfel* court first noted that "novelty as an element of an idea seller's claim" is a distinct element of proof with respect to both (1) "a claim based on a property theory" and (2) "a claim based on a contract theory." Id. at 477. The court then proceeded to discuss how the leading submission-of-idea case — Downey v. General Foods Corp., 31 N.Y.2d 56, 334 N.Y.S.2d 874, 286 N.E.2d 257 (1972) — treated novelty with respect to property-based and contract-based claims. First, the *Apfel* court explained that the plaintiff's property-based claims for misappropriation were dismissed in Downey because "the elements of novelty and originality [were] absent," i.e., the ideas were so common as to be unoriginal and known generally. *Apfel*, 81 N.Y.2d at 477 (quoting *Downey*, 31 N.Y.2d at 61) (alteration in original); accord *Downey*, 31 N.Y.2d at 61-62 (holding that the submitted idea — marketing Jell-O to children under the name "Mr. Wiggle" — was "lacking in novelty and originality" because the idea was merely the "use of a word ('wiggley' or 'wiggle') descriptive of the most obvious characteristic of Jell-O, with the prefix 'Mr.' added"). Second, the *Apfel* court explained that the plaintiff's contract claims in Downey had been dismissed on the separate ground that the "defendant possessed plaintiff's ideas prior to plaintiff's disclosure [and thus], the ideas could have no value to defendant and could not supply consideration for any agreement between the parties." *Apfel*, 81 N.Y.2d at 477; accord *Downey*, 31 N.Y.2d at 62 (finding that, where defendant had used the words "wiggles" and "wigglewam" in prior advertising, defendant could "rel [y] on its own previous experience" and "was free to make use of 'Mr. Wiggle' without being obligated to compensate the plaintiff").

By distinguishing between the two types of claims addressed in Downey and the different bases for rejecting each claim, the New York Court of Appeals clarified that the novelty requirement in submission-of-idea cases is different for misappropriation of property and breach of contract claims. . . .

Thus, the *Apfel* court refused to read *Downey* and "similar decisions" as requiring originality or novelty generally in all cases involving disclosure of ideas. See *Apfel*, 81 N.Y.2d at 476-77 ("These decisions do not support [the] contention that novelty [in absolute terms] is required in all cases involving disclosure of ideas."). Rather, the *Apfel* court clarified that the longstanding requirement that an idea have originality or general novelty in order to support a misappropriation claim does not apply to contract claims. See Oasis Music, Inc. v. 900 U.S.A., Inc., 161 Misc.2d 627, 614 N.Y.S.2d 878, 881 (1994) (noting that "the *Apfel* court did not repudiate the long line of cases requiring novelty in certain situations[,] . . . the *Apfel* court merely clarified that novelty is not required in all cases"). For contract-based claims in submission-of-idea cases, a showing of novelty to the buyer will supply sufficient consideration to support a contract.

Moreover, *Apfel* made clear that the "novelty to the buyer" standard is not limited to cases involving an express post-disclosure contract for payment based on an idea's use. The *Apfel* court explicitly discussed the pre-disclosure contract scenario present in the instant case, where "the buyer and seller contract for disclosure of the idea with payment based on use, but no separate post-disclosure contract for the use of the idea has been made." *Apfel*, 81 N.Y.2d at 477-78. In such a scenario, a seller might, as Nadel did here, bring an action against a buyer who allegedly used his ideas without payment, claiming both misappropriation of property and breach of an express or implied-in-fact contract.[2] Of course, the mere disclosure of an unoriginal idea to a defendant, to whom the idea is novel, will not automatically entitle a plaintiff to compensation upon the defendant's subsequent use of the idea. An implied-in-fact contract "requires such elements as consideration, mutual assent, legal capacity and legal subject matter." Id. (internal quotation marks omitted). The existence of novelty to the buyer only addresses the element of consideration necessary for the formation of the contract. Thus, apart from consideration, the formation of a contract implied-in-fact will depend on the presence of the other elements. The element of mutual assent, for example, must be inferred from the facts and circumstances of each case, including such factors as the specific conduct of the parties, industry custom, and course of dealing. Cf. id. at 94. The *Apfel* court recognized that these cases present courts with the difficult problem of determining "whether the idea the buyer was using was, in fact, the seller's." Id. at 478. Specifically, the court noted that, with respect to a misappropriation of property claim, it is difficult to "prove that the buyer obtained the idea from [the seller] and nowhere else." Id. With respect to a breach of contract claim, the court noted that it would be inequitable to enforce a contract if "it turns out upon disclosure that the buyer already possessed the idea." Id. The court then concluded that, with respect to these cases, "[a] showing of novelty, at least novelty as to the buyer" should address these problems.[3]

We note, moreover, that the "novelty to the buyer" standard comports with traditional principles of contract law. While an idea may be unoriginal or non-novel in a general sense, it may have substantial value to a particular buyer who is unaware of it and therefore willing to enter into contract to acquire and exploit it. See *Apfel,* 81 N.Y.2d at 475-76; Robert Unikel, Bridging the "Trade Secret" Gap: Protecting "Confidential Information" Not Rising to the Level of Trade Secrets, 29 Loy. U. Chi. L.J. 841, 877 n. 151 (1998) (noting that, if a valuable idea is already known to an industry but has not yet been acquired by a prospective buyer, one of two circumstances may exist: "(1) the person[ ] ha[s] not identified the potential value of the easily acquired information; or (2) the person[ ] ha[s] not identified the means, however easy or proper, for obtaining the valuable information"). As the *Apfel* court emphasized, "the buyer may reap benefits from such a contract in a number of ways—for instance, by not having to expend resources pursuing the idea through other channels or by having a profit-making idea implemented sooner rather than later." *Apfel,* 81 N.Y.2d at 478....

In contrast to contract-based claims, a misappropriation claim can only arise from the taking of an idea that is original or novel in absolute terms, because the law of property does not protect against the misappropriation or theft of that which

2. Where the pre-disclosure agreement is not an express contract to pay for the disclosed idea, an idea seller may argue in the alternative, cf. Watts v. Columbia Artists Mgmt. Inc., 188 A.D.2d 799, 591 N.Y.S.2d 234, 236 (3d Dep't 1992) (noting that express contract and implied-in-fact contract theories are mutually exclusive), that the parties entered into an agreement implied-in-fact. See Robbins v. Frank Cooper Assocs., 14 N.Y.2d 913, 252 N.Y.S.2d 318, 200 N.E.2d 860 (1964) (finding an implied-in-fact contract for the reasonable value of plaintiff's idea for a television program format where the parties had failed to reduce their agreement to writing). Unlike an express contract, an implied-in-fact contract arises "when the agreement and promise have simply not been expressed in words," but "a court may justifiably infer that the promise would have been explicitly made, had attention been drawn to it." Maas v. Cornell Univ., 94 N.Y.2d 87, 93-94, 699 N.Y.S.2d 716, 721 N.E.2d 966 (1999).

3. We note that this particular sentence could be read out of context to suggest that novelty to the buyer will alone support a misappropriation claim under New York law. However, nothing in *Apfel* otherwise suggests that the Court of Appeals meant to supplant the longstanding requirement that originality or novelty generally must be shown to support a misappropriation claim....

is free and available to all. See Murray v. National Broad. Co., 844 F.2d 988, 993 (2d Cir. 1988) ("Since...non-novel ideas are not protectible as property, they cannot be stolen."); cf. Ed Graham Prods., Inc. v. National Broad. Co., 75 Misc.2d 334, 347 N.Y.S.2d 766, 769 (1973) ("Ideas such as those presented by the plaintiff are in the public domain and may freely be used by anyone with impunity."); Educational Sales Programs, Inc. v. Dreyfus Corp., 65 Misc.2d 412, 317 N.Y.S.2d 840, 843 (1970) ("An idea is impalpable, intangible, incorporeal, yet it may be a stolen gem of great value, or mere dross of no value at all, depending on its novelty and uniqueness.").

Finally, although the legal requirements for contract-based claims and property-based claims are well-defined, we note that the determination of novelty in a given case is not always clear. Cf. AEB & Assocs. Design Group, Inc. v. Tonka Corp., 853 F. Supp. 724, 734 (S.D.N.Y.1994) ("In establishing an idea's originality, a plaintiff cannot rest on mere assertions, but must demonstrate some basis in fact for its claims."). The determination of whether an idea is original or novel depends upon several factors, including, inter alia, the idea's specificity or generality (is it a generic concept or one of specific application?), its commonality (how many people know of this idea?), its uniqueness (how different is this idea from generally known ideas?), and its commercial availability (how widespread is the idea's use in the industry?). Cf. Murray, 844 F.2d at 993 ("In assessing whether an idea is in the public domain, the central issue is the uniqueness of the creation."); AEB & Assocs., 853 F. Supp. at 734 ("[N]ovelty cannot be found where the idea consists of nothing more than a variation on a basic theme."); Educational Sales Programs, 317 N.Y.S.2d at 844 (noting that an idea "must show[ ] genuine novelty and invention, and not a merely clever or useful adaptation of existing knowledge" in order to be considered original or novel). Thus, for example, a once original or novel idea may become so widely disseminated over the course of time that it enters the body of common knowledge. When this occurs, the idea ceases to be novel or original. See, e.g., Murray, 844 F.2d at 989, 991-92 (affirming district court's finding that plaintiff's idea for a television sitcom about "Black American family life" was not novel or original because it "merely combined two ideas which had been circulating in the industry for a number of years — namely, the family situation comedy, which was a standard formula, and the casting of black actors in non-stereotypical roles," even though "the portrayal of a nonstereotypical black family on television was indeed a breakthrough").

Moreover, in assessing the interrelationship between originality and novelty to the buyer, we note that in some cases an idea may be so unoriginal or lacking in novelty that its obviousness bespeaks widespread and public knowledge of the idea, and such knowledge is therefore imputed to the buyer. See Soule, 195 N.Y.S. at 575 (noting that the idea that an increase in the wholesale price of groceries would result in an increase in profits was "not new" in 1922, "not original," and left the court "at a loss to understand how it could be deemed valuable"). In such cases, a court may conclude, as a matter of law, that the idea lacks both the originality necessary to support a misappropriation claim and the novelty to the buyer necessary to support a contract claim. See id. at 576 ("No person can by contract monopolize an idea that is common and general to the whole world."); Ed Graham Prods., 347 N.Y.S.2d at 769 ("[W]here plaintiff's idea is wholly lacking in novelty, no cause of action in contract or tort can stand....") (emphasis added); Educational Sales Programs, 317 N.Y.S.2d at 843-44 ("Nothing is bestowed if the facts of a 'secret' imparted in confidence are already the subject of general knowledge.").

In sum, we find that New York law in submission-of-idea cases is governed by the following principles: Contract-based claims require only a showing that the disclosed idea was novel to the buyer in order to find consideration.[4] Such claims involve a fact-specific inquiry that focuses on the perspective of the particular buyer. By contrast, misappropriation claims require that the idea at issue be original and novel in absolute terms. This is so because unoriginal, known ideas have no value as property and the law does not protect against the use of that which is free and available to all. Finally, an idea may be so unoriginal or lacking in novelty generally that, as a matter of law, the buyer is deemed to have knowledge of the idea. In such cases, neither a property-based nor a contract-based claim for uncompensated use of the idea may lie.

In light of New York's law governing submission-of-idea cases, we next consider whether Nadel's toy idea was original or novel in absolute terms so as to support his misappropriation claim and whether his idea was novel as to Play-By-Play so as to support his contract claims.

B. Nadel's Misappropriation Claim

This Court reviews a district court's grant of summary judgment de novo. See Quinn v. Green Tree Credit Corp., 159 F.3d 759, 764 (2d Cir. 1998). In so doing, we must consider the evidence in the record in a light most favorable to the non-movant, drawing all reasonable inferences in its favor. See id. at 764-65. Summary judgment is appropriate only where there is no genuine issue of material fact. See id.

In this case, the district court did not decide whether Nadel's idea — a plush toy that sits upright, emits sounds, and spins on a flat surface by means of an internal eccentric motor — was inherently lacking in originality. See *Nadel*, 34 F. Supp. 2d at 185 ("[We] need [not] reach the issue of whether combining elements of two commercially available toys to make another toy may be novel or is, as a matter of law, merely a 'clever adaptation of existing technology,' for Play-By-Play has demonstrated that plaintiff's idea was one which was already in use in the industry at the time that it was submitted"). We therefore remand this issue to the district court to determine whether Nadel's idea exhibited "genuine novelty or invention" or whether it was "a merely clever or useful adaptation of existing knowledge." Educational Sales Programs, 317 N.Y.S.2d at 844.

Moreover, insofar as the district court found that Nadel's idea lacked originality and novelty generally because similar toys were commercially available prior to October 1996, we believe that there remains a genuine issue of material fact on this point. While the record contains testimony of Play-By-Play's toy expert — Bert Reiner — in support of the finding that Nadel's product concept was already used in more than a dozen different plush toys prior to October 1996, the district court cited the "Giggle Bunny" toy as the only such example. See *Nadel*, 34 F. Supp. 2d at 185. Nadel disputes Reiner's contention and claims, furthermore, that the district court erroneously relied on an undated video depiction of the Giggle Bunny toy to conclude that upright, sound-emitting, spinning plush toys were commercially available prior to October 1996.

4. Of course, the mere formation of a contract in a submission-of-idea case does not necessarily mean that the contract has been breached by the defendant upon his use of the idea. In order to recover for breach of contract, a plaintiff must demonstrate some nexus or causal connection between his or her disclosure and the defendant's use of the idea, i.e., where there is an independent source for the idea used by the defendant, there may be no breach of contract, and the plaintiff's claim for recovery may not lie ....

With respect to the Giggle Bunny evidence, we agree with Nadel that the Giggle Bunny model depicted in the undated video exhibit is physically different from the earlier Giggle Bunny model known to be commercially available in 1994. Drawing all factual inferences in Nadel's favor, we cannot conclude as a matter of law that the upright, sound-emitting, spinning plush Giggle Bunny shown in the video exhibit was commercially available prior to October 1996, and we certainly cannot conclude based on this one exhibit that similar toys were in the public domain at that time.

Moreover, although we find highly probative Mr. Reiner's testimony that numerous toys with the same general characteristics of Nadel's toy idea were commercially available prior to October 1996, his testimony and related evidence are too ambiguous and incomplete to support a finding of unoriginality as a matter of law. Mr. Reiner's testimony fails to specify precisely which (if any) of the enumerated plush toys were designed to (1) sit upright, (2) on a flat surface, (3) emit sounds, and (4) spin or rotate (rather than simply vibrate like "Tickle Me Elmo," for example). Without this information, a reasonable finder of fact could discount Mr. Reiner's testimony as vague and inconclusive.

On remand, the district court is free to consider whether further discovery is warranted to determine whether Nadel's product concept was inherently original or whether it was novel to the industry prior to October 1996. A finding of unoriginality or lack of general novelty would, of course, preclude Nadel from bringing a misappropriation claim against Play-By-Play. Moreover, in evaluating the originality or general novelty of Nadel's idea in connection with his misappropriation claim, the district may consider whether the idea is so unoriginal that Play-By-Play should, as a matter of law, be deemed to have already possessed the idea, and dismiss Nadel's contract claims on that ground.

C. Nadel's Contract Claims

Mindful that, under New York law, a finding of novelty as to Play-By-Play will provide sufficient consideration to support Nadel's contract claims, we next consider whether the record exhibits a genuine issue of material fact on this point.

Reading the record in a light most favorable to Nadel, . . . we conclude that there exists a genuine issue of material fact as to whether Nadel's idea was, at the time he disclosed it to Wasserman in early October 1996, novel to Play-By-Play. Notably, the timing of Play-By-Play's development and release of Tornado Taz in relation to Nadel's October 1996 disclosure is, taken alone, highly probative. Moreover, although custom in the toy industry provides that a company shall promptly return all samples if it already possesses (or does not want to use) a disclosed idea, Play-By-Play in this case failed to return Nadel's prototype monkey toy for several months, despite Nadel's multiple requests for its return. According to Wasserman's secretary, Melissa Rodriguez, Nadel's sample was not returned until after the unveiling of "Tornado Taz" at the New York Toy Fair in February 1997. Ms. Rodriguez testified that from October 1996 through February 1997, Nadel's sample was usually kept in a glass cabinet behind Wasserman's desk, and on one occasion, she remembered seeing it on a table in Wasserman's office. These facts give rise to the reasonable inference that Play-By-Play may have used Nadel's prototype as a model for the development of Tornado Taz.

None of the evidence adduced by Play-By-Play compels a finding to the contrary on summary judgment. With regard to the discussions that Play-By-Play purportedly had in June or July of 1996 about possible ways to create a vibrating or spinning Tazmanian Devil toy, those conversations only lasted, according to Mr. Slattery, "a matter of five minutes." Play-By-Play may have "discussed the concept," as Mr. Slattery testified, but the record provides no evidence suggesting that, in June or July of 1996, Play-By-Play understood exactly how it could apply eccentric motor technology to make its Tazmanian Devil toy spin rather than, say, vibrate like Tickle Me Elmo. Similarly, although Play-By-Play asserts that it commissioned an outside manufacturing agent — Barter Trading of Hong Kong — to begin developing Tornado Taz in late September or early October of 1996, Play-By-Play admits that it can only "guess" the exact date. Play-By-Play cannot confirm that its commission of Barter Trading pre-dated Nadel's alleged disclosure to Wasserman on or about October 9, 1996. Nor has Play-By-Play produced any documents, technical or otherwise, relating to its purported business venture with Barter Trading or its independent development of a spinning Tornado Taz prior to October 1996. Based on this evidence, a jury could reasonably infer that Play-By-Play actually contacted Barter Trading, if at all, after learning of Nadel's product concept, and that Play-By-Play's development of Tornado Taz is attributable to Nadel's disclosure.

We therefore conclude that there exists a genuine issue of material fact as to whether Nadel's idea was, at the time he disclosed it to Wasserman in early October 1996, novel to Play-By-Play...

COMMENTS AND QUESTIONS

1. *Idea Submission Causes of Action.* As the *Nadel* case reflects, idea submission cases must be brought within some property, contract, or tort cause of action. Contract law has proven the most resilient basis for protecting ideas. *Nadel* discusses the availability of express and implied contract rights for the disclosure of an idea. In addition, some courts have protected ideas under the law of confidential relationships (and the related body of trade secret protection). See, e.g., Ralph Andrews Prods. v. Paramount Pictures, 222 Cal. App. 3d 676, 271 Cal. Rptr. 797 (1990) (trade secret suit alleging that a former employee stole an idea for a game show and solid it to Paramount Pictures). The misappropriation cause of action derives from property and tort principles. As noted in section 1, *supra*, it is quite narrow. Misappropriation and other property-type claims (e.g., conversion) no longer exist as to ideas for copyrightable works — stories, movies, songs. Some states jettisoned such protection expressly. See, e.g., *Desny* ("California does not now accord individual property type protection to abstract ideas"), excerpted below. State laws protecting such subject matter as of January 1, 1978, were preempted by the 1976 Copyright Act. See Nash v. CBS, Inc., 704 F. Supp. 823, 833-35 (N.D. Ill. 1989), *aff'd*, 899 F.2d 1537 (7th Cir. 1990) (rejecting claim that CBS had misappropriated ideas from plaintiff's book, *The Dillinger Report,* in an episode of the television drama "Simon and Simon"). We discuss the scope of copyright preemption more fully in section B, *infra*.

2. *Requirements.* Based upon the *Nadel* case, under what circumstances can ideas be protected? What elements must one who submits an idea to another prove in order to recover?

3. *Semantics.* The *Nadel* court uses words familiar from the study of patent and copyright law — originality, novelty, and obviousness. But do these intellectual property words of art have the same meaning in the law of idea submissions? What does the court mean by "original," "novel," and "obvious"? How do these meanings compare to the definitions from copyright and patent law? Does the court suggest that the "novelty" requirement for a misappropriation claim incorporates a broader conception of non-obviousness than patent law?

4. The novelty requirement was central to the Second Circuit's decision rejecting an idea submission claim in Murray v. National Broadcasting Co., 844 F.2d 988 (2d Cir. 1988), *cert. denied,* 488 U.S. 955 (1988). In that case, Murray (an NBC employee) submitted a two-page proposal to NBC for a show entitled "Father's Day" that would "combine humor with serious situations in a manner similar to that of the old Dick Van Dyke Show" but with a "Black perspective" and "a contemporary, urban setting." The proposal specifically identified Bill Cosby as the lead actor. NBC informed Murray that it was not interested in pursuing his proposal. When "The Cosby Show" aired four years later, starring Bill Cosby as Dr. Cliff Huxtable living in a contemporary urban setting with his lawyer-spouse and their five children, Murray sued. The court rejected Murray's claim for lack of novelty:

> [W]e believe, as a matter of law, that plaintiff's idea embodied in his "Father's Day" proposal was not novel because it merely represented an "adaptation of existing knowledge" and of "known ingredients" and therefore lacked "genuine novelty and invention." Educational Sales Programs, 317 N.Y.S.2d at 844.
>
> We recognize of course that even novel and original ideas to a greater or lesser extent combine elements that are themselves not novel. Originality does not exist in a vacuum. Nevertheless, where, as here, an idea consists in essence of nothing more than a variation on a basic theme — in this case, the family situation comedy — novelty cannot be found to exist. . . .
>
> Appellant would have us believe that by interpreting New York law as we do, we are in effect condoning the theft of ideas. On the contrary, ideas that reflect "genuine novelty and invention" are fully protected against unauthorized use. But those ideas that are not novel "are in the public domain and may freely be used by anyone with impunity." Since such non-novel ideas are not protectible as property, they cannot be stolen.

Id. at 992-93. How does this use of the term "novelty" compare to the patent law's definition?[2] Wouldn't a new combination of known elements be considered novel under the patent law's standards? Isn't the court really saying that the idea was obvious? How would this determination be made under the patent law's standards? Was there a suggestion or motivation to combine? Don't secondary considerations — especially commercial success — weigh against a finding of obviousness? Would the law of idea submissions be improved by adopting the patent law's framework for judging novelty and non-obviousness?

5. *The Novelty to the User Standard and Diffusion of Ideas.* Idea protection today is based upon an express or implied contract between an idea purveyor and a potential

---

2. It should be noted that Murray's claim of novelty would have surely failed under a patent law standard. Evidence at the trial revealed that Cosby himself had been quoted in a newspaper article 20 years earlier stating that it was his "dream" to appear in a situation comedy along the lines of "The Dick Van Dyke Show," but featuring an African-American family. Id. at 989. The court did not base its decision on this evidence, which would have gone to the question of whether NBC had independently developed the idea for the show.

user. Under *Apfel* and *Nadel,* the idea purveyor need only show that the idea was not known *to the defendant* in order to satisfy the novelty element. This element provides the consideration necessary for there to be a valid contract. The court notes, however, that "in some cases an idea may be so unoriginal or lacking in novelty that its obviousness bespeaks widespread and public knowledge of the idea, and such knowledge is therefore imputed to the buyer." Should protection for ideas turn on the sophistication of the buyer?

One justification for such a doctrine is to reward diffusion of ideas to those best situated to commercialize them. As we discussed in Chapter 1, economic historians and theorists find that diffusion of knowledge is a crucial element in the creation of social benefit flowing from innovation. This doctrine rewards not the inventor but those who have the insight and possibly the connections to diffuse (or market) information. Certainly a true inventor can benefit, because being first ensures that the idea will be novel to others. Should the intellectual property system reward diffusion of ideas in this way?

6. *Concreteness.* Some courts consider whether an idea is sufficiently "concrete" in determining liability based on breach of implied contract and confidential relationship causes of action, although not in express contract claims. In the words of one court, "[i]deas are the most intangible of property rights, and their lineage is uniquely difficult to trace. Paternity can be claimed in the most casual of ways, and once such a claim is lodged, definitive blood tests are notoriously lacking." Burten v. Milton Bradley Co., 592 F. Supp. 1021, 1031 (D.R.I. 1984), *rev'd on other grounds,* 763 F.2d 461 (1st Cir. 1985). In assessing a claim to protect the idea of organizing and sponsoring radio broadcasts of student talent shows, the court in Hamilton Nat'l Bank v. Belt, 210 F.2d 706 (D.C. Cir. 1953), noted that "[t]he law shies away from according protection to vagueness, and must do so especially in the realm of ideas with the obvious dangers of a contrary rule." Id. at 708. Does this amount to an indefinite claiming doctrine? Or is it more akin to a rejection under section 101 of the Patent Act for claiming only an abstract idea?

7. *Contract Analysis; Comparison to Trade Secret.* How does the contract analysis in *Nadel* compare to that in the trade secret/breach of confidence cases involving contracts, such as Smith v. Dravo Corp., discussed in Chapter 2? What (if any) differences are there in the relationship of the parties, the duties of the idea/information recipient, and the nature of the idea or information? In the remedies?

8. *Defenses: Waiver/Release.* Many companies and film production studios require that those who submit ideas sign waivers or releases from liability. The release used in the *Downey* case, 286 N.E.2d 257 (N.Y. Ct. App. 1972), stated:

> I submit this suggestion with the understanding, which is conclusively evidenced by my use and transmittal to you of this form, that this suggestion is not submitted to you in confidence, that no confidential relationship has been or will be established between us and that the use, if any, to be made of this suggestion by you and the compensation to be paid therefor, if any, if you use it, are matters resting solely in your discretion.

Suppose that the recipient used the idea without paying any compensation. Should such a clause be enforceable? Consider the argument put forth by Lon Sobel favoring enforcement in the context of script submissions: "If courts were to refuse to enforce releases, it is likely that unrepresented writers would be unable to get their material read at all — a consequence more harmful to aspiring writers than the possibility that releases will bar some of them from suing for the suspected theft of their ideas." See Lionel S. Sobel, The Law of Ideas, Revisited, 1 UCLA Entertainment L. Rev. 9, 91 (1994).

9. *Remedies.* What should be the remedy for the misappropriation of an idea or breach of a contract to pay for an idea? Assuming that Nadel prevails on his liability claims (misappropriation and contract) at trial, what should be the measure of damages? Assuming that Murray had carried his burden of proof on the novelty of "The Cosby Show," should he be entitled to all of the profits from the show?

Desny v. Wilder
Supreme Court of California
46 Cal. 2d 715, 299 P.2d 257, 110 U.S.P.Q. 433 (Cal. Sup. Ct. 1956)

SCHAUER, Justice.

Plaintiff appeals from a summary judgment rendered against him in this action to recover the reasonable value of a literary composition, or of an idea for a photoplay, a synopsis of which composition, embodying the idea, he asserts he submitted to defendants for sale, and which synopsis and idea, plaintiff alleges, were accepted and used by defendants in producing a photoplay . . .

[W]e have concluded, for reasons hereinafter stated, that the summary judgment in favor of defendants was erroneously granted and should be reversed . . .

[I]t appears from the present record that defendant [Billy] Wilder [a famous director] at the times here involved was employed by defendant Paramount Pictures Corporation . . . either as a writer, producer or director, or a combination of the three. In November, 1949, plaintiff telephoned Wilder's office. Wilder's secretary, who was also employed by Paramount, answered, and plaintiff stated that he wished to see Wilder. At the secretary's insistence that plaintiff explain his purpose, plaintiff 'told her about this fantastic unusual story . . . I told her that it was the life story of Floyd Collins who was trapped and made sensational news for two weeks . . . and I told her the plot.' . . . Two days later plaintiff, after preparing a three or four page outline of the story, telephoned Wilder's office a second time and told the secretary the synopsis was ready. The secretary requested plaintiff to read the synopsis to her over the telephone so that she could take it down in shorthand, and plaintiff did so . . . Plaintiff on his part told the secretary that defendants could use the story only if they paid him 'the reasonable value of it.' . . . She said that if Billy Wilder of Paramount uses the story, 'naturally we will pay you for it.' . . . Plaintiff's only subsequent contact with the secretary was a telephone call to her in July, 1950, to protest the alleged use of his composition and idea in a photoplay produced and exhibited by defendants. The photoplay, as hereinafter shown in some detail, closely parallels both plaintiff's synopsis and the historical material concerning the life and death of Floyd Collins.

Defendants concede, as they must, that "the act of disclosing an unprotectible idea, if that act is in fact the bargained for exchange for a promise, may be consideration to support the promise." They then add, "But once the idea is disclosed without the protection of a contract, the law says that anyone is free to use it. Therefore, subsequent use of the idea cannot constitute consideration so as to support a promise to pay for such use." And as to the effect of the evidence defendants argue that plaintiff "disclosed his material before . . . (defendants) did or could do anything to indicate their willingness or unwillingness to pay for the disclosure. The act of using the idea, from which appellant attempts to imply a promise to pay, came long after the disclosure." ...

Generally speaking, ideas are as free as the air ... But there can be circumstances when neither air nor ideas may be acquired without cost. The diver who goes deep in the sea, even as the pilot who ascends high in the troposphere, knows full well that for life itself he, or someone on his behalf, must arrange for air ... The theatrical producer likewise may be dependent for his business life on the procurement of ideas from other persons as well as the dressing up and portrayal of his self-conceptions; he may not find his own sufficient for survival. As counsel for the Writers Guild aptly say, ideas 'are not freely usable by the entertainment media until the latter are made aware of them.' The producer may think up the idea himself, dress it and portray it; or he may purchase either the conveyance of the idea alone or a manuscript embodying the idea in the author's concept of a literary vehicle giving it form, adaptation and expression. It cannot be doubted that some ideas are of value to a producer.

An idea is usually not regarded as property, because all sentient beings may conceive and evolve ideas throughout the gamut of their powers of cerebration and because our concept of property implies something which may be owned and possessed to the exclusion of all other persons. We quote as an accurate statement of the law in this respect the following language of Mr. Justice Brandeis, dissenting in International News Service v. Associated Press (1918), 248 U.S. 215, 250:

> An essential element of individual property is the legal right to exclude others from enjoying it. [...] But the fact that a product of the mind has cost its producer money and labor, and has a value for which others are willing to pay, is not sufficient to ensure to it this legal attribute of property. The general rule of law is, that the noblest of human productions, knowledge, truths ascertained, conceptions, and ideas become, after voluntary communication to others, free as the air to common use.

The principles above stated do not, however, lead to the conclusion that ideas cannot be a subject of contract. As Mr. Justice Traynor stated in his dissenting opinion in Stanley v. Columbia Broadcasting System (1950), 35 Cal. 2d 653, 674, 221 P.2d 73:

> The policy that precludes protection of an abstract idea by copyright does not prevent its protection by contract. Even though an idea is not property subject to exclusive ownership, its disclosure may be of substantial benefit to the person to whom it is disclosed. That disclosure may therefore be consideration for a promise to pay.... Even though the idea disclosed may be 'widely known and generally understood' (citation), it may be protected by an express contract providing that it will be paid for regardless of its lack of novelty.

The lawyer or doctor who applies specialized knowledge to a state of facts and gives advice for a fee is selling and conveying an idea. In doing that he is rendering a service. The lawyer and doctor have no property rights in their ideas, as such, but they do not ordinarily convey them without solicitation by client or patient. Usually the parties will expressly contract for the performance of and payment for such services, but, in the absence of an express contract, when the service is requested and rendered the law does not hesitate to infer or imply a promise to compensate for it. In other words the recovery may be based on contract either express or implied.

The person who can and does convey a valuable idea to a producer who commercially solicits the service or who voluntarily accepts it knowing that it is tendered for a price should likewise be entitled to recover. In so holding we do not fail to recognize that free-lance writers are not necessarily members of a learned profession and as such bound to the exalted standards to which doctors and lawyers are dedicated. So too we are not oblivious of the hazards with which producers of the class represented here by defendants and their related amici are confronted through the unsolicited submission of numerous scripts on public domain materials in which public materials the producers through their own initiative may well find nuclei for legitimately developing the "stupendous and colossal." The law, however, is dedicated to the proposition that for every wrong there is a remedy...To that end the law of implied contracts assumes particular importance in literary idea and property controversies...

[W]e conclude that conveyance of an idea can constitute valuable consideration and can be bargained for before it is disclosed to the proposed purchaser, but once it is conveyed, i.e., disclosed to him and he has grasped it, it is henceforth his own and he may work with it and use it as he sees fit. In the field of entertainment the producer may properly and validly agree that he will pay for the service of conveying to him ideas which are valuable and which he can put to profitable use. Furthermore, where an idea has been conveyed with the expectation by the purveyor that compensation will be paid if the idea is used, there is no reason why the producer who has been the beneficiary of the conveyance of such an idea, and who finds it valuable and is profiting by it, may not then for the first time, although he is not at that time under any legal obligation so to do, promise to pay a reasonable compensation for that idea, that is, for the past service of furnishing it to him and thus create a valid obligation...But, assuming legality of consideration, the idea purveyor cannot prevail in an action to recover compensation for an abstract idea unless (a) before or after disclosure he has obtained an express promise to pay, or (b) the circumstances preceding and attending disclosure, together with the conduct of the offeree acting with knowledge of the circumstances, show a promise of the type usually referred to as "implied" or "implied-in-fact."...

Such inferred or implied promise, if it is to be found at all, must be based on circumstances which were known to the producer at and preceding the time of disclosure of the idea to him and he must voluntarily accept the disclosure, knowing the conditions on which it is tendered...The idea man who blurts out his idea without having first made his bargain has no one but himself to blame for the loss of his bargaining power...So, if the plaintiff here is claiming only for the conveyance of the idea of making a dramatic production out of the life of Floyd Collins he must fail unless in conformity with the above stated rules he can establish a contract to pay.

From plaintiff's testimony, as epitomized above...it does not appear that a contract to pay for conveyance of the abstract photoplay idea had been made, or that the basis for inferring such a contract from subsequent related acts of the defendants had been established, at the time plaintiff disclosed his basic idea to the secretary. Defendants, consequently, were at that time and from then on free to use the abstract idea if they saw fit to engage in the necessary research and develop it to the point of a usable script. Whether defendants did that, or whether they actually accepted and used plaintiff's synopsis, is another question.

... Literary property which is protectible may be created out of unprotectible material such as historical events. It has been said (and does not appear to have been successfully challenged) that 'There are only thirty-six fundamental dramatic situations, various facets of which form the basis of all human drama.' (Georges Polti, 'The Thirty-Six Dramatic Situations'; see also, Henry Albert Phillips, 'The Universal Plot Catalog'; Eric Heath, 'Story Plotting Simplified.') It is manifest that authors must work with and from ideas or themes which basically are in the public domain ... Events from the life of Floyd Collins were avowedly the basic theme of plaintiff's story ...

[A]ny literary composition, conceivably, may possess value in someone's estimation and be the subject of contract ... Obviously the defendants here used someone's script in preparing and producing their photoplay. That script must have had value to them. As will be hereinafter shown, it closely resembles plaintiff's synopsis. Ergo, plaintiff's synopsis appears to be a valuable literary composition. Defendants had an unassailable right to have their own employees conduct the research into the Floyd Collins tragedy (an historical event in the public domain) and prepare a story based on those facts and to translate it into a script for the play. But equally unassailable (assuming the verity of the facts which plaintiff asserts) is plaintiff's position that defendants had no right except by purchase on the terms he offered to acquire and use the synopsis prepared by him.

[Affirmed in part, reversed in part, and remanded.]

CARTER, Justice [concurring in result] ...

When we consider the difference in economic and social backgrounds of those offering such merchandise for sale and those purchasing the same, we are met with the inescapable conclusion that it is the seller who stands in the inferior bargaining position. It should be borne in mind that producers are not easy to contact ... It should also be borne in mind that writers have no way of advertising their wares that, as is most graphically illustrated by the present opinion, no producer, publisher, or purchaser for radio or television, is going to buy a pig in a poke. And, when the writer, in an earnest endeavor to sell what he has written, conveys his idea or his different interpretation of an old idea, to such prospective purchaser, he has lost the result of his labor, definitely and irrevocably. And, in addition, there is no way in which he can protect himself. If he says to whomever he is permitted to see, or, as in this case, talk with over the telephone, "I won't tell you what my idea is until you promise to pay me for it," it takes no Sherlock Holmes to figure out what the answer will be! This case is a beautiful example of the practical difficulties besetting a writer with something to sell ...

I disagree with the statement in the majority opinion that: "The idea man who blurts out his idea without having first made his bargain has no one but himself to blame for the loss of his bargaining power." It seems to me that in the ordinary situation, when the so-called "idea man" has an opportunity to see, or talk with, the prospective purchaser, or someone in his employ, that it is at that time, without anything being said, known to both parties that the one is there to sell, and the other to buy. This is surely true of a department store when merchandise is displayed on the counter; it is understood by anyone entering the store that the merchandise so displayed is for sale. [I]t is completely unnecessary for the storekeeper, or anyone in his employ, to state to anyone entering the store that all articles there are for sale. I am at a loss to see why any different rules should apply when it is ideas for sale rather than the normal run of merchandise.

COMMENTS AND QUESTIONS

1. The concurring opinion makes an interesting point: "[W]hen the writer, in an earnest endeavor to sell what he has written, conveys his idea or his different interpretation of an old idea, to [a] prospective purchaser, he has lost the result of his labor." This general feature of the "market for information" was noted by Kenneth Arrow, in an article entitled Economic Welfare and the Allocation of Resources for Invention, in The Rate and Direction of Inventive Activity 615 (1962). Indeed, the phenomenon is sometimes called "Arrow's paradox of information." Without a property right, Arrow pointed out, the seller of information is in a pickle: if in trying to strike a deal she discloses the information, she has nothing left to sell, but if she does not disclose anything the buyer has no idea what is for sale. Arrow pointed out that when the information involves an invention, patents protect the seller so she can confidently offer her idea for sale. In other words, patents solve Arrow's Paradox. Does the opinion above prove that property rights are necessary to overcome the information paradox?

One problem with giving property rights to abstract ideas is that it can put the recipients of idea submissions in a bind. They are just as bound by Arrow's information paradox as plaintiffs — if they don't listen to the idea "pitch," they will never know whether the idea was worth paying for. The point of intellectual property protection in this situation is to encourage such idea submissions by making the plaintiff confident in her ability to protect her idea. But suppose that a defendant hears a "pitch" for an idea that is old, that she has already come up with herself, or that someone else has already pitched to her. If intellectual property protection extends to idea submissions, the recipient may be forced to pay for an idea she already has! Arguably, therefore, awarding property rights in idea submissions merely changes the nature of Arrow's paradox, rather than eliminating it entirely.

2. Wendy Gordon has explored the notion that a central problem of intellectual property law is to compensate creators of works who bestow benefits on those who follow, up to some socially justifiable point. In this analysis, the basic structure of intellectual property law is closely akin to the law of restitution, which seeks to determine when someone who bestows unbargained-for benefits deserves compensation. See, e.g., Wendy J. Gordon, On Owning Information: Intellectual Property and the Restitutionary Impulse, 78 Va. L. Rev. 149 (1992); Wendy J. Gordon, Of Harms and Benefits: Torts, Restitution and Intellectual Property, 21 J. Legal Stud. 449 (1992). On restitution generally, see Saul Levmore, Explaining Restitution, 71 Va. L. Rev. 65 (1985). Does this literature help explain the need for a remand in the *Desny* case? Recall that the purpose of the remand is to determine how much Desny's idea contributed to the profits of Wilder's film.

3. The holding in *Desny* has been applied and extended in a number of cases. For instance, in Blaustein v. Burton, 88 Cal. Rptr. 319, 9 Cal. App. 3d 161 (Cal. Ct. App. 1970), the plaintiff Blaustein orally "pitched" the idea of using Richard Burton and Elizabeth Taylor in a film version of Shakespeare's "Taming of the Shrew." Although there was little novel in the pitch, the court held that there were triable issues of fact concerning the enforceability of an implied contract for the idea. The case is notable in that the "pitch" was protectable even though it was never reduced to writing.

4. The oral presentation of ideas for movies, TV series, etc., is a well-recognized part of the entertainment industry. Recent cases show that it can be a dangerous

practice, however, for the "pitch-person." A number of recent cases find no liability for the uncompensated use of "pitched" ideas. See, e.g., Selby v. New Line Cinema Corp., 96 F. Supp. 2d 1053 (C.D. Cal. 2000) (refusing to enforce implied-in-fact contract for use of a movie idea, holding that enforcement of such a contract is preempted by the Copyright Act); Wrench LLC v. Taco Bell Corp., 51 F. Supp. 2d 840, 853 (W.D. Mich. 1999) (refusing to enforce implied-in-fact contract for use of feisty dog character in TV advertising, since enforcement of such a contract was preempted by the Copyright Act). But see Firoozye v. Earthlink Network, 153 F. Supp. 2d 1115 (N.D. Cal 2001) (finding implied-in-fact contract claims, in case involving ongoing relationship between computer programmer and customer, not preempted by the Copyright Act). The courts may be reacting to the flood of lawsuits (mostly meritless) against major movie studios in which plaintiffs allege that the idea for the movie was originally theirs.

PROBLEM

Problem 6-1. The Washington humorist and columnist Art Buchwald submitted an eight-page summary of a film idea entitled "It's a Crude, Crude World" to executives at Paramount. The summary described in some detail the storyline, which involved a third-world prince who came to America for "a state visit." Buchwald gave a two-page overview of the plot to Paramount Pictures. Paramount subsequently entered into a contract with Buchwald whereby Paramount bought the rights to Buchwald's story and concept, with the aim of making a movie, starring Eddie Murphy, to be called "King for a Day." Another person would write the actual script, but "King for a Day" was to be based on "It's a Crude, Crude World." Because of various production difficulties, Paramount abandoned the project in March 1985. In May 1986, Buchwald gave an option on his film idea to the Warner Brothers Studio. In the summer of 1987, Paramount began development of a similar script by Eddie Murphy. In this movie — which became the successful film *Coming to America* — Murphy portrayed an African prince who comes to America in search of a suitable woman to marry.

The two scripts have been contrasted as follows:

> In Buchwald's treatment, a rich, educated, arrogant, extravagant, despotic African potentate comes to America for a state visit. After being taken on a grand tour of the United States, the potentate arrives at the White House. A gaff[e] in remarks made by the President infuriates the African leader. His sexual desires are rebuffed by a black woman State Department officer assigned to him. She is requested by the President to continue to serve as the potentate's United States escort. While in the United States, the potentate is deposed, deserted by his entourage and left destitute. He ends up in the Washington ghetto, is stripped of his clothes, and befriended by a black lady. The potentate experiences a number of incidents in the ghetto, and obtains employment as a waiter. In order to avoid extradition, he marries the black lady who befriended him, becomes the emperor of the ghetto and lives happily ever after.
>
> In "Coming to America" the pampered prince of a mythical African Kingdom (Zamunda) wakes up on his 21st birthday to find that the day for his

prearranged marriage has arrived. Discovering his bride to be very subservient and being unhappy about that fact, he convinces his father to permit him to go to America for the ostensible purpose of sewing [sic] his "royal oats." In fact, the prince intends to go to America to find an independent woman to marry. The prince and his friend go to Queens, New York, where their property is stolen and they begin living in a slum area. The prince discovers his true love, Lisa, whose father — McDowell — operates a fast-food restaurant for whom the prince and his friend begin to work. The prince and Lisa fall in love, but when the King and Queen come to New York and it is disclosed who the prince is, Lisa rejects the prince's marriage invitation. The film ends with Lisa appearing in Zamunda, marrying the prince and apparently living happily ever after.

Because of *Coming to America*, Warner Brothers decided not to pursue Buchwald's story. Buchwald then sued Paramount. The contract between Buchwald and Paramount provided in pertinent part:

> "Work" means the aforementioned Material and includes all prior, present and future versions, adaptations and translations thereof (whether written by Author or by others), its theme, story, plot, characters and their names, its title or titles and subtitles, if any, . . . and each and every part of all thereof. "Work" does not include any material written or prepared by Purchaser or under Purchaser's Authority . . .
>
> ["Contingent consideration":]
>
> For the first theatrical motion picture (the "Picture"): If, but only if, a feature length theatrical motion picture shall be produced *based upon Author's Work*.

(Emphasis added.) How should the court rule?

COMMENTS AND QUESTIONS

1. If you were an independent scriptwriter, how would you protect your ideas while trying to market them? Under what circumstances would the career damage from obtaining a reputation for litigiousness be worth it?

2. If you were a film studio or television production company, how could you guard against suits by people like Buchwald, or by people who claim to have submitted ideas that no one in your company ever recalls seeing? Recall Sobel's argument favoring judicial enforcement of waivers and release clauses imposed as conditions for idea submissions. Under some cases habitual rejection and return of unsolicited ideas eliminates the prospect of liability. See, e.g., Davis v. General Foods Corp., 21 F. Supp. 445 (S.D.N.Y. 1937) (plaintiff's unsolicited recipe returned with form letter); Whitfield v. Lear, 751 F.2d 90 (2d Cir. 1984) (noting, in decision reversing summary judgment for defendant, plaintiff's evidence that it was customary in the television industry for a studio not desiring outside submissions to say so explicitly and to return scripts so submitted without opening them). Is it desirable for firms to routinely reject submissions, some of which may be worth pursuing, for fear of spurious lawsuits? Is this an example of nuisance lawsuits undermining an otherwise mutually beneficial market? Is it an example of the "market for lemons," where bad idea submissions (i.e., ones that lead to spurious suits) drive out the good ones? Is it enough that studios and production companies can rely on trusted middlepersons such as agents

to obtain ideas from proven submitters? See Julie Salamon, Bookshelf: Celluloid Immortals and Literary He-Men, Wall St. J., July 29, 1992, p. A7 col. 1:

> Whatever the outcome, the suit has made Hollywood acutely aware of where it gets its ideas. "Producers are going to have to be very careful what submissions they read, and that makes it harder in the way they conduct business," says producer Howard Rosenman, co-president of Sandollar Productions, adding, "I never accept any unsolicited material. Ever."

3. *The Hollywood Script Registry.* The Writer's Guild of America, West, developed a "Script Registry" to lower the risks to both aspiring writers and studios. Writers deposit a copy of a script they are going to submit with the Registry, which date-stamps it and stores it for five years. The Registry also operates an arbitration service for resolving disputes over writing credits.

PROBLEM

Problem 6-2. Lohr, an eccentric engineer, mails an idea she has had for a new invention to several engineering companies. The idea is mailed in a "double envelope." The outer envelope contains a confidentiality agreement, indicating that the contents of a second, sealed envelope are the property of Lohr and may not be used unless the user pays Lohr 10 percent of any profit that results. If the recipient does not agree to these terms, he is instructed to mail back the sealed, stamped, self-addressed envelope containing the idea. The "agreement" clearly provides that by opening the envelope, the recipient agrees to the terms.

Dupco receives the agreement and opens the inner envelope. The next day, Dupco discovers that Lohr has posted the complete text of her idea on the Internet with no confidentiality restriction. Can Dupco use Lohr's idea without compensating her? Does it matter how novel or original the idea was on the day Dupco opened the envelope?

---

4. The Right of Publicity

The right of publicity protects an individual's marketable image or persona. Although theoretically available to any individual, the right of publicity rarely arises outside of the celebrity realm. The right of publicity developed in response to the rise of mass advertising and the growing recognition that a celebrity's imprimatur on a product or even association of a product with a celebrity's persona enhances its appeal to consumers. This right affords individuals a property-type interest in the use of their name, likeness, photograph, portrait, voice, and other personal characteristics in connection with the marketing of products and services. Jurisdictions approach publicity rights in a variety of ways. Today, sixteen states recognize common law rights of publicity, and another fifteen states have codified the right of publicity in statute. Some states, like California, recognize both statutory and common law sources for this form of protection. New York's statutory privacy and publicity protections are embodied in a single statute.

The modern right of publicity reflects two distinct rationales – one grounded in privacy and the other in economic exploitation. The privacy branch can be traced back to an influential law review article published in the late 19th century. See Samuel D. Warren & Louis D. Brandeis, The Right to Privacy, 4 Harv. L. Rev. 193 (1890). Lamenting technological and cultural developments invading the private sphere, Warren and Brandeis advocated a right of privacy to forbid the publication of idle gossip and restore "propriety and dignity" to the press. Legislatures and courts gradually came to recognize this interest. New York led the way with its privacy law, enacted in 1903, banning the unauthorized use of "the name, portrait or picture of any living persons" for "advertising purposes, or for purposes of trade." The privacy right was quite limited in practice. Courts tended to view celebrities as inviting exploitation of their image and were reluctant to find the mere use of their image, even in advertising, to suggest endorsement. Even when liability was found, recovery was limited to the personal injury suffered, as opposed to the economic value to the advertiser.

As a result, celebrities pushed for stronger protection for the economic value of their image. In 1953, the Second Circuit found such a right in New York's common law, which it dubbed the "right of publicity." See Haelen Laboratories, Inc. v. Topps Chewing Gum, 202 F.2d 866 (2d Cir. 1953). In that case, Haelan had negotiated exclusive licenses from several Major League Baseball players authorizing the use of their images on baseball cards that it included with packs of gum. Topps sold its own gum with photographs of the same players. Although ruling that New York's statutory privacy law did not extend to such uses, the court ruled that "a man has a right in the publicity value of his photograph, i.e., the right to grant the exclusive privilege of publishing the picture, and that such a grant may validly be made 'in gross,' i.e., without an accompanying transfer of a business or of anything else" under New York's common law.[3] The concept was refined in a subsequent law review article by Melville Nimmer, The Right of Publicity, 19 Law & Contemp. Probs. 203 (1954). Beginning in the 1970s, a number of states enacted "publicity" statutes, which continue to evolve today. For a discussion of this history, see Stacey L. Dogan & Mark A. Lemley, What the Right of Publicity Can Learn From Trademark Law, 58 Stan. L. Rev. (2006) p. 1161.

Thus, a critical distinction to make among jurisdictions is the extent to which the right of publicity is recognized separately from the right of privacy. Perhaps the most substantial difference between "publicity" regimes oriented toward privacy and those oriented toward property is that, as a property right, a celebrity's interest is assignable and descendable. However, some jurisdictions place limits on the duration of publicity rights following a celebrity's death, and still others do not recognize the descendability of publicity rights at all.

As the home to a significant portion of the film, television, and sound recording industries, California has played a particularly important role in the development of

---

3. Ironically, Stephano v. News Group Publications, Inc., 64 N.Y.2d 174 (1984) held that *Haelan* had misinterpreted New York law, and that there was no right of publicity independent of the privacy protections in §§ 50 and 51 of the New York Civil Rights Law.

the right of publicity at both the statutory and jurisprudential levels. The main statutory provision, enacted in 1971, provides as follows:

Cal. Civ. Code § 3344. Unauthorized Commercial Use of Name, Voice, Signature, Photograph or Likeness

(a) Any person who knowingly uses another's name, voice, signature, photograph, or likeness, in any manner, on or in products, merchandise, or goods, or for purposes of advertising or selling, or soliciting purchases of, products, merchandise, goods or services, without such person's prior consent, or, in the case of a minor, the prior consent of his parent or legal guardian, shall be liable for any damages sustained by the person or persons injured as a result thereof. In addition, in any action brought under this section, the person who violated the section shall be liable to the injured party or parties in an amount equal to the greater of seven hundred fifty dollars ($750) or the actual damages suffered by him or her as a result of the unauthorized use, and any profits from the unauthorized use that are attributable to the use and are not taken into account in computing the actual damages. In establishing such profits, the injured party or parties are required to present proof only of the gross revenue attributable to such use, and the person who violated this section is required to prove his or her deductible expenses. Punitive damages may also be awarded to the injured party or parties. The prevailing party in any action under this section shall also be entitled to attorney's fees and costs.

(b) As used in this section, "photograph" means any photograph or photographic reproduction, still or moving, or any videotape or live television transmission, of any person, such that the person is readily identifiable . . .

(c) Where a photograph or likeness of an employee of the person using the photograph or likeness appearing in the advertisement or other publication prepared by or in behalf of the user is only incidental, and not essential, to the purpose of the publication in which it appears, there shall arise a rebuttable presumption affecting the burden of producing evidence that the failure to obtain the consent of the employee was not a knowing use of the employee's photograph or likeness.

(d) For purposes of this section, a use of a name, voice, signature, photograph, or likeness in connection with any news, public affairs, or sports broadcast or account, or any political campaign, shall not constitute a use for which consent is required under subdivision (a).

(e) The use of a name, voice, signature, photograph, or likeness in a commercial medium shall not constitute a use for which consent is required under subdivision (a) solely because the material containing such use is commercially sponsored or contains paid advertising. Rather it shall be a question of fact whether or not the use of the person's name, voice, signature, photograph, or likeness was so directly connected with the commercial sponsorship or with the paid advertising as to constitute a use for which consent is required under subdivision (a).

(f) Nothing in this section shall apply to the owners or employees of any medium used for advertising, including, but not limited to, newspapers, magazines, radio and television networks and stations, cable television systems, billboards, and transit ads, by whom any advertisement or solicitation in violation of this section is published or disseminated, unless it is established that such owners or employees had knowledge of the unauthorized use of the person's name, voice, signature, photograph, or likeness as prohibited by this section.

(g) The remedies provided for in this section are cumulative and shall be in addition to any others provided for by law.

In 1984, California provided for publicity rights in the persona of deceased celebrities. Those provisions have since been amended and are codified as Civil Code § 3344.1. Like the right of publicity held by a living person, § 3344.1(a) declares broadly that "[a]ny person who uses a deceased personality's name, voice, signature, photograph, or likeness, in any manner, on or in products, merchandise, or goods, or for purposes of advertising or selling, or soliciting purchases of, products, merchandise, goods, or services, without prior consent from the person or persons specified in subdivision (c), shall be liable for any damages sustained by the person or persons injured as a result thereof." The amount recoverable includes the greater of $750 or actual damages and any profits, as well as punitive damages, attorney fees, and costs. The statute provides that the post-mortem publicity right is freely transferable before or after the personality dies, by contract or by trust or will. § 33414.1(b). Consent to use the deceased personality's name, voice, photograph, etc., must be obtained from such a transferee or, if there is none, from certain described survivors of the personality. § 3341(c), (d). Any person claiming to be such a transferee or survivor must register the claim with the Secretary of State before recovering damages. § 3344.1(f). Drawing upon the duration of copyrights, the post-mortem right of publicity expires 70 years after the personality dies. § 3344.1(g). The statute contains exemptions for "news, public affairs, or sports broadcast or account, or any political campaign" purposes that are much broader than the exemptions that apply during the life of the celebrity.

Does allowing descendability and transfer of a publicity right make sense? Or is the right more personal to the person being depicted? Whose interests are served by preventing imitation after a singer's death?

Midler v. Ford Motor Co.
United States Court of Appeals for the Ninth Circuit
849 F.2d 460 (9th Cir. 1988)

NOONAN, Circuit Judge:

This case centers on the protectibility of the voice of a celebrated chanteuse from commercial exploitation without her consent. Ford Motor Company and its advertising agency, Young & Rubicam, Inc., in 1985 advertised the Ford Lincoln Mercury with a series of nineteen 30 or 60 second television commercials in what the agency called "The Yuppie Campaign." The aim was to make an emotional connection with Yuppies, bringing back memories of when they were in college. Different popular songs of the seventies were sung on each commercial. The agency tried to get "the original people," that is, the singers who had popularized the songs, to sing them. Failing in that endeavor in ten cases the agency had the songs sung by "sound alikes." Bette Midler, the plaintiff and appellant here, was done by a sound alike.

Midler is a nationally known actress and singer. She won a Grammy as early as 1973 as the Best New Artist of that year. Records made by her since then have gone Platinum and Gold. She was nominated in 1979 for an Academy award for Best Female Actress in The Rose, in which she portrayed a pop singer. Newsweek in its June 30, 1986 issue described her as an "outrageously original singer/comedian."

Time hailed her in its March 2, 1987 issue as "a legend" and "the most dynamic and poignant singer-actress of her time."

When Young & Rubicam was preparing the Yuppie Campaign it presented the commercial to its client by playing an edited version of Midler singing "Do You Want To Dance," taken from the 1973 Midler album, "The Divine Miss M." After the client accepted the idea and form of the commercial, the agency contacted Midler's manager, Jerry Edelstein. The conversation went as follows: "Hello, I am Craig Hazen from Young and Rubicam. I am calling you to find out if Bette Midler would be interested in doing...?" Edelstein: "Is it a commercial?" "Yes." "We are not interested."

Undeterred, Young & Rubicam sought out Ula Hedwig, whom it knew to have been one of "the Harlettes," a backup singer for Midler for ten years. Hedwig was told by Young & Rubicam that "they wanted someone who could sound like Bette Midler's recording of [Do You Want To Dance]." She was asked to make a "demo" tape of the song if she was interested. She made an a capella demo and got the job.

At the direction of Young & Rubicam, Hedwig then made a record for the commercial. The Midler record of "Do You Want To Dance" was first played to her. She was told to "sound as much as possible like the Bette Midler record," leaving out only a few "aahs" unsuitable for the commercial. Hedwig imitated Midler to the best of her ability.

After the commercial was aired Midler was told by "a number of people" that it "sounded exactly" like her record of "Do You Want To Dance." Hedwig was told by "many personal friends" that they thought it was Midler singing the commercial. Ken Fritz, a personal manager in the entertainment business not associated with Midler, declares by affidavit that he heard the commercial on more than one occasion and thought Midler was doing the singing.

Neither the name nor the picture of Midler was used in the commercial; Young & Rubicam had a license from the copyright holder to use the song. At issue in this case is only the protection of Midler's voice. The district court described the defendants' conduct as that "of the average thief." They decided, "If we can't buy it, we'll take it." The court nonetheless believed there was no legal principle preventing imitation of Midler's voice and so gave summary judgment for the defendants. Midler appeals.

The First Amendment protects much of what the media do in the reproduction of likenesses or sounds. A primary value is freedom of speech and press. Time, Inc. v. Hill, 385 U.S. 374, 388, 17 L. Ed. 2d 456, 87 S. Ct. 534 (1967). The purpose of the media's use of a person's identity is central. If the purpose is "informative or cultural" the use is immune; "if it serves no such function but merely exploits the individual portrayed, immunity will not be granted." Felcher and Rubin, "Privacy, Publicity and the Portrayal of Real People by the Media," 88 Yale L.J. 1577, 1596 (1979). Moreover, federal copyright law preempts much of the area. "Mere imitation of a recorded performance would not constitute a copyright infringement even where one performer deliberately sets out to simulate another's performance as exactly as possible." Notes of Committee on the Judiciary, 17 U.S.C.A. § 114(b). It is in the context of these First Amendment and federal copyright distinctions that we address the present appeal.

Nancy Sinatra once sued Goodyear Tire and Rubber Company on the basis of an advertising campaign by Young & Rubicam featuring "These Boots Are Made For Walkin'," a song closely identified with her; the female singers of the commercial were alleged to have imitated her voice and style and to have dressed and looked like her. The basis of Nancy Sinatra's complaint was unfair competition; she claimed that the song and the arrangement had acquired "a secondary meaning" which, under California law, was protectible. This court noted that the defendants "had paid a very substantial sum to the copyright proprietor to obtain the license for the use of the song and all of its arrangements." To give Sinatra damages for their use of the song would clash with federal copyright law. Summary judgment for the defendants was affirmed. Sinatra v. Goodyear Tire & Rubber Co., 435 F.2d 711, 717-718 (9th Cir. 1970), *cert. denied*, 402 U.S. 906, 28 L. Ed. 2d 646, 91 S. Ct. 1376 (1971). If Midler were claiming a secondary meaning to "Do You Want To Dance" or seeking to prevent the defendants from using that song, she would fail like Sinatra. But that is not this case. Midler does not seek damages for Ford's use of "Do You Want To Dance," and thus her claim is not preempted by federal copyright law. Copyright protects "original works of authorship fixed in any tangible medium of expression." 17 U.S.C. at 102(a). A voice is not copyrightable. The sounds are not "fixed." What is put forward as protectible here is more personal than any work of authorship.

Bert Lahr once sued Adell Chemical Co. for selling Lestoil by means of a commercial in which an imitation of Lahr's voice accompanied a cartoon of a duck. Lahr alleged that his style of vocal delivery was distinctive in pitch, accent, inflection, and sounds. The First Circuit held that Lahr had stated a cause of action for unfair competition, that it could be found "that defendant's conduct saturated plaintiff's audience, curtailing his market." Lahr v. Adell Chemical Co., 300 F.2d 256, 259 (1st Cir. 1962). That case is more like this one. But we do not find unfair competition here. One-minute commercials of the sort the defendants put on would not have saturated Midler's audience and curtailed her market. Midler did not do television commercials. The defendants were not in competition with her. See Halicki v. United Artists Communications, Inc., 812 F.2d 1213 (9th Cir. 1987).

California Civil Code section 3344 is also of no aid to Midler. The statute affords damages to a person injured by another who uses the person's "name, voice, signature, photograph or likeness, in any manner." The defendants did not use Midler's name or anything else whose use is prohibited by the statute. The voice they used was Hedwig's, not hers. The term "likeness" refers to a visual image not a vocal imitation. The statute, however, does not preclude Midler from pursuing any cause of action she may have at common law; the statute itself implies that such common law causes of action do exist because it says its remedies are merely "cumulative." Id. § 3344(g).

The companion statute protecting the use of a deceased person's name, voice, signature, photograph or likeness states that the rights it recognizes are "property rights." Id. § 990(b). By analogy the common law rights are also property rights. Appropriation of such common law rights is a tort in California. Motschenbacher v. R. J. Reynolds Tobacco Co., 498 F.2d 821 (9th Cir. 1974). In that case what the defendants used in their television commercial for Winston cigarettes was a photograph of a famous professional racing driver's racing car. The number of the car was changed and a wing-like device known as a "spoiler" was attached to the car; the car's features of white pinpointing, an oval medallion, and solid red coloring were retained. The driver, Lothar Motschenbacher, was in the car but his features were not visible. Some persons, viewing the commercial, correctly inferred that the car was his and that he was in the car and was therefore endorsing the product. The defendants were held to have invaded a "proprietary interest" of Motschenbacher in his own identity. Id. at 825.

Midler's case is different from Motschenbacher's. He and his car were physically used by the tobacco company's ad; he made part of his living out of giving commercial endorsements. But, as Judge Koelsch expressed it in Motschenbacher, California will recognize an injury from "an appropriation of the attributes of one's identity." Id. at 824. It was irrelevant that Motschenbacher could not be identified in the ad. The ad suggested that it was he. The ad did so by emphasizing signs or symbols associated with him. In the same way the defendants here used an imitation to convey the impression that Midler was singing for them.

Why did the defendants ask Midler to sing if her voice was not of value to them? Why did they studiously acquire the services of a sound-alike and instruct her to imitate Midler if Midler's voice was not of value to them? What they sought was an attribute of Midler's identity. Its value was what the market would have paid for Midler to have sung the commercial in person.

A voice is more distinctive and more personal than the automobile accouterments protected in Motschenbacher. A voice is as distinctive and personal as a face. The human voice is one of the most palpable ways identity is manifested. We are all aware that a friend is at once known by a few words on the phone. At a philosophical level it has been observed that with the sound of a voice, "the other stands before me." D. Ihde, Listening and Voice 77 (1976). A fortiori, these observations hold true of singing, especially singing by a singer of renown. The singer manifests herself in the song. To impersonate her voice is to pirate her identity. See W. Keeton, D. Dobbs, R. Keeton, D. Owen, Prosser & Keeton on Torts 852 (5th ed. 1984).

We need not and do not go so far as to hold that every imitation of a voice to advertise merchandise is actionable. We hold only that when a distinctive voice of a professional singer is widely known and is deliberately imitated in order to sell a product, the sellers have appropriated what is not theirs and have committed a tort in California. Midler has made a showing, sufficient to defeat summary judgment, that the defendants here for their own profit in selling their product did appropriate part of her identity.

Reversed and remanded for trial.

COMMENTS AND QUESTIONS

1. Young and Rubicam paid money to the owner of the copyright in the song "Do You Want to Dance," who permitted them to remake the song. Why wasn't that enough? Should Midler have the additional right to prevent imitation of her voice? Note that the owner of the copyright in sound recordings is often the producer or the studio, rather than the singer who made the original recording. Does it matter whether Young and Rubicam could have purchased the rights to the recording of Midler herself singing "Do You Want to Dance" from the studio? If an artist doesn't control the rights to her own recording, why should she be able to prevent imitations of that recording?

A related question is whether the right of publicity is subject to an implicit first sale defense. In Allison v. Vintage Sports Plaques, 136 F.3d 1443 (11th Cir. 1998), the Eleventh Circuit held that it was. The defendant had purchased authorized sports trading cards and framed them for resale. The court held that the defendant could lawfully resell the images of celebrities that he had lawfully purchased and that he was not impermissibly using the sports trading cards to sell the associated frames.

2. Midler, the imitator, and independent third parties all provided evidence that people hearing the Ford commercial were confused — they thought that Midler was the one singing. Is likelihood of confusion (the test for trademark infringement) the relevant question here? Could Midler prevail even if the attempt to imitate her was not very good, so that most people could tell the difference? If a disclaimer at the beginning of the ad had indicated that the song was an impersonation, rather than Midler herself?

A related question is whether the right of publicity applies only to celebrities. Historically, the answer has been no. The right of publicity is derived from the "commercial advantage" wing of the tort of invasion of privacy, and can be invoked by anyone whose name or likeness was appropriated by another for commercial advantage. See Restatement (2d) Torts § 652C. Does this lineage suggest that the "likelihood of confusion" test shouldn't limit the right of publicity?

White v. Samsung Electronics America, Inc.
United States Court of Appeals for the Ninth Circuit
989 F.2d 1512 (9th Cir. 1993)

Before Goodwin, Pregerson and Alarcon, Circuit Judges.

The panel has voted unanimously to deny the petition for rehearing...

KOZINSKI, Circuit Judge, with whom Circuit Judges O'SCANNLAIN and KLEINFELD join, dissenting from the order rejecting the suggestion for rehearing en banc.

I

Saddam Hussein wants to keep advertisers from using his picture in unflattering contexts. Clint Eastwood doesn't want tabloids to write about him. Rudolf Valentino's heirs want to control his film biography. The Girl Scouts don't want their image soiled by association with certain activities. George Lucas wants to keep Strategic Defense Initiative fans from calling it "Star Wars." Pepsico doesn't want singers to use the word "Pepsi" in their songs. Guy Lombardo wants an exclusive property right to ads that show big bands playing on New Year's Eve. Uri Geller thinks he should be paid for ads showing psychics bending metal through telekinesis. Paul Prudhomme, that household name, thinks the same about ads featuring corpulent bearded chefs. And scads of copyright holders see purple when their creations are made fun of.

Something very dangerous is going on here. Private property, including intellectual property, is essential to our way of life. It provides an incentive for investment and innovation; it stimulates the flourishing of our culture; it protects the moral entitlements of people to the fruits of their labors. But reducing too much to private property can be bad medicine. Private land, for instance, is far more useful if separated from other private land by public streets, roads and highways. Public parks, utility rights-of-way and sewers reduce the amount of land in private hands, but vastly enhance the value of the property that remains.

So too it is with intellectual property. Overprotecting intellectual property is as harmful as underprotecting it. Creativity is impossible without a rich public domain. Nothing today, likely nothing since we tamed fire, is genuinely new: Culture, like science and technology, grows by accretion, each new creator building on the works of those who came before. Overprotection stifles the very creative forces it's supposed to nurture.

The panel's opinion is a classic case of overprotection. Concerned about what it sees as a wrong done to Vanna White, the panel majority erects a property right of remarkable and dangerous breadth: Under the majority's opinion, it's now a tort for advertisers to remind the public of a celebrity. Not to use a celebrity's name, voice, signature or likeness; not to imply the celebrity endorses a product; but simply to evoke the celebrity's image in the public's mind. This Orwellian notion withdraws far more from the public domain than prudence and common sense allow. It conflicts with the Copyright Act and the Copyright Clause. It raises serious First Amendment problems. It's bad law, and it deserves a long, hard second look.

II

Samsung ran an ad campaign promoting its consumer electronics. Each ad depicted a Samsung product and a humorous prediction: One showed a raw steak with the caption "Revealed to be health food. 2010 A.D." Another showed [talk show host] Morton Downey, Jr. in front of an American flag with the caption "Presidential candidate. 2008 A.D." The ads were meant to convey — humorously — that Samsung products would still be in use twenty years from now.

The ad that spawned this litigation starred a robot dressed in a wig, gown and jewelry reminiscent of Vanna White's hair and dress; the robot was posed next to a Wheel-of-Fortune-like game board [see Figure 6-1]. The caption read "Longest-running game show. 2012 A.D." The gag here, I take it, was that Samsung would still be around when White had been replaced by a robot.

Perhaps failing to see the humor, White sued, alleging Samsung infringed her right of publicity by "appropriating" her "identity." Under California law, White has the exclusive right to use her name, likeness, signature and voice for commercial purposes. Cal. Civ. Code § 3344(a); Eastwood v. Superior Court, 149 Cal. App. 3d 409, 417, 198 Cal. Rptr. 342, 347 (1983). But Samsung didn't use her name, voice or signature, and it certainly didn't use her likeness. The ad just wouldn't have been funny had it depicted White or someone who resembled her — the whole joke was that the game show host(ess) was a robot, not a real person. No one seeing the ad could have thought this was supposed to be White in 2012.

The district judge quite reasonably held that, because Samsung didn't use White's name, likeness, voice or signature, it didn't violate her right of publicity. 971 F.2d at 1396-97. Not so, says the panel majority: The California right of publicity can't possibly be limited to name and likeness. If it were, the majority reasons, a "clever advertising strategist" could avoid using White's name or likeness but nevertheless remind people of her with impunity, "effectively eviscerat[ing]" her rights. To prevent this "evisceration," the panel majority holds that the right of publicity must extend beyond name and likeness, to any "appropriation" of White's "identity" — anything that "evoke[s]" her personality. Id. at 1398-99.

FIGURE 6-1
Samsung's robot of a game show hostess.

III

But what does "evisceration" mean in intellectual property law? Intellectual property rights aren't like some constitutional rights, absolute guarantees protected against all kinds of interference, subtle as well as blatant. They cast no penumbras, emit no emanations: The very point of intellectual property laws is that they protect only against certain specific kinds of appropriation. I can't publish unauthorized copies of, say, Presumed Innocent; I can't make a movie out of it. But I'm perfectly free to write a book about an idealistic young prosecutor on trial for a crime he didn't commit. So what if I got the idea from Presumed Innocent? So what if it reminds readers of the original? Have I "eviscerated" Scott Turow's intellectual property rights? Certainly not. All creators draw in part on the work of those who came before, referring to it, building on it, poking fun at it; we call this creativity, not piracy.

The majority isn't, in fact, preventing the "evisceration" of Vanna White's existing rights; it's creating a new and much broader property right, a right unknown in California law.[16] It's replacing the existing balance between the interests of the celebrity and those of the public by a different balance, one substantially more favorable to the celebrity. Instead of having an exclusive right in her name, likeness, signature or voice, every famous person now has an exclusive right to anything that reminds the viewer of her. After all, that's all Samsung did: It used an inanimate object to remind people of White, to "evoke [her identity]." 971 F.2d at 1399.[17]

Consider how sweeping this new right is. What is it about the ad that makes people think of White? It's not the robot's wig, clothes or jewelry; there must be ten million blond women (many of them quasi-famous) who wear dresses and jewelry like White's. It's that the robot is posed near the "Wheel of Fortune" game board. Remove the game board from the ad, and no one would think of Vanna White. But once you include the game board, anybody standing beside it—a brunette woman, a man wearing women's clothes, a monkey in a wig and gown—would evoke White's image, precisely the way the robot did. It's the "Wheel of Fortune" set, not the robot's face or dress or jewelry that evokes White's image. The panel is giving White an exclusive right not in what she looks like or who she is, but in what she does for a living.[18]

This is entirely the wrong place to strike the balance. Intellectual property rights aren't free: They're imposed at the expense of future creators and of the public at

16. In fact, in the one California case raising the issue, the three state Supreme Court Justices who discussed this theory expressed serious doubts about it. Guglielmi v. Spelling-Goldberg Prods., 25 Cal. 3d 860, 864 n.5, 160 Cal. Rptr. 352, 355 n.5, 603 P.2d 454, 457 n.5 (1979) (Bird, C. J., concurring) (expressing skepticism about finding a property right to a celebrity's "personality" because it is "difficult to discern any easily applied definition for this amorphous term"). Neither have we previously interpreted California law to cover pure "identity." Midler v. Ford Motor Co., 849 F.2d 460 (9th Cir. 1988), and Waits v. Frito-Lay, Inc., 978 F.2d 1093 (9th Cir. 1992), dealt with appropriation of a celebrity's voice. See id. at 1100-01 (imitation of singing style, rather than voice, doesn't violate the right of publicity). Motschenbacher v. R. J. Reynolds Tobacco Co., 498 F.2d 821 (9th Cir. 1974), stressed that, though the plaintiff's likeness wasn't directly recognizable by itself, the surrounding circumstances would have made viewers think the likeness was the plaintiff's. Id. at 827; see also Moore v. Regents of the Univ. of Cal., 51 Cal. 3d 120, 138, 271 Cal. Rptr. 146, 157, 793 P.2d 479, 490 (1990) (construing Motschenbacher as "hold[ing] that every person has a proprietary interest in his own likeness").

17. Some viewers might have inferred White was endorsing the product, but that's a different story. The right of publicity isn't aimed at or limited to false endorsements, Eastwood v. Superior Court, 149 Cal. App. 3d 409, 419-20, 198 Cal. Rptr. 342, 348 (1983); that's what the Lanham Act is for. Note also that the majority's rule applies even to advertisements that unintentionally remind people of someone. California law is crystal clear that the common-law right of publicity may be violated even by unintentional appropriations. Id. at 417 n.6, 198 Cal. Rptr. at 346 n.6; Fairfield v. American Photocopy Equipment Co., 138 Cal. App. 2d 82, 87, 291 P.2d 194 (1955).

18. Once the right of publicity is extended beyond specific physical characteristics, this will become a recurring problem: Outside name, likeness and voice, the things that most reliably remind the public of celebrities are the actions or roles they're famous for. A commercial with an astronaut setting foot on the moon would evoke the image of Neil Armstrong. Any masked man on horseback would remind people (over a certain age) of Clayton Moore. And any number of songs—"My Way," "Yellow Submarine," "Like a Virgin," "Beat It," "Michael, Row the Boat Ashore," to name only a few—instantly evoke an image of the person or group who made them famous, regardless of who is singing. See also Carlos V. Lozano, West Loses Lawsuit over Batman TV Commercial, L.A. Times, Jan. 18, 1990, at B3 (Adam West sues over Batman-like character in commercial); Nurmi v. Peterson, 10 U.S.P.Q.2d 1775, 1989 WL 407484 (C.D. Cal. 1989) (1950s TV movie hostess "Vampira" sues 1980s TV hostess "Elvira"); text accompanying notes 7-8 (lawsuits brought by Guy Lombardo, claiming big bands playing at New Year's Eve parties remind people of him, and by Uri Geller, claiming psychics who can bend metal remind people of him). Cf. Motschenbacher, where the claim was that viewers would think plaintiff was actually in the commercial, and not merely that the commercial reminded people of him.

large. Where would we be if Charles Lindbergh had an exclusive right in the concept of a heroic solo aviator? If Arthur Conan Doyle had gotten a copyright in the idea of the detective story, or Albert Einstein had patented the theory of relativity? If every author and celebrity had been given the right to keep people from mocking them or their work? Surely this would have made the world poorer, not richer, culturally as well as economically.

This is why intellectual property law is full of careful balances between what's set aside for the owner and what's left in the public domain for the rest of us: The relatively short life of patents; the longer, but finite, life of copyrights; copyright's idea-expression dichotomy; the fair use doctrine; the prohibition on copyrighting facts; the compulsory license of television broadcasts and musical compositions; federal preemption of overbroad state intellectual property laws; the nominative use doctrine in trademark law; the right to make soundalike recordings. All of these diminish an intellectual property owner's rights. All let the public use something created by someone else. But all are necessary to maintain a free environment in which creative genius can flourish.

The intellectual property right created by the panel here has none of these essential limitations: No fair use exception; no right to parody; no idea-expression dichotomy. It impoverishes the public domain, to the detriment of future creators and the public at large. Instead of well-defined, limited characteristics such as name, likeness or voice, advertisers will now have to cope with vague claims of "appropriation of identity," claims often made by people with a wholly exaggerated sense of their own fame and significance.... Future Vanna Whites might not get the chance to create their personae, because their employers may fear some celebrity will claim the persona is too similar to her own. The public will be robbed of parodies of celebrities, and our culture will be deprived of the valuable safety valve that parody and mockery create.

Moreover, consider the moral dimension, about which the panel majority seems to have gotten so exercised. Saying Samsung "appropriated" something of White's begs the question: Should White have the exclusive right to something as broad and amorphous as her "identity"? Samsung's ad didn't simply copy White's schtick — like all parody, it created something new. True, Samsung did it to make money, but White does whatever she does to make money, too; the majority talks of "the difference between fun and profit," 971 F.2d at 1401, but in the entertainment industry fun is profit. Why is Vanna White's right to exclusive for-profit use of her persona — a persona that might not even be her own creation, but that of a writer, director or producer — superior to Samsung's right to profit by creating its own inventions? Why should she have such absolute rights to control the conduct of others, unlimited by the idea-expression dichotomy or by the fair use doctrine?

To paraphrase only slightly Feist Publications, Inc. v. Rural Telephone Service Co., 111 S. Ct. 1282, 1289-90 (1991), it may seem unfair that much of the fruit of a creator's labor may be used by others without compensation. But this is not some unforeseen byproduct of our intellectual property system; it is the system's very essence. Intellectual property law assures authors the right to their original expression, but encourages others to build freely on the ideas that underlie it. This result is neither unfair nor unfortunate: It is the means by which intellectual property law advances the progress of science and art. We give authors certain exclusive rights, but in exchange we get a richer public domain. The majority ignores this wise teaching, and all of us are the poorer for it.

IV

The panel, however, does more than misinterpret California law: By refusing to recognize a parody exception to the right of publicity, the panel directly contradicts the federal Copyright Act. Samsung didn't merely parody Vanna White. It parodied Vanna White appearing in "Wheel of Fortune," a copyrighted television show, and parodies of copyrighted works are governed by federal copyright law.

Copyright law specifically gives the world at large the right to make "fair use" parodies, parodies that don't borrow too much of the original. Fisher v. Dees, 794 F.2d 432, 435 (9th Cir. 1986)...

The majority's decision decimates this federal scheme. It's impossible to parody a movie or a TV show without at the same time "evok[ing]" the "identit[ies]" of the actors. You can't have a mock Star Wars without a mock Luke Skywalker, Han Solo and Princess Leia, which in turn means a mock Mark Hamill, Harrison Ford and Carrie Fisher. You can't have a mock Batman commercial without a mock Batman, which means someone emulating the mannerisms of Adam West or Michael Keaton. See Carlos V. Lozano, West Loses Lawsuit over Batman TV Commercial, L.A. Times, Jan. 18, 1990, at B3 (describing Adam West's right of publicity lawsuit over a commercial produced under license from DC Comics, owner of the Batman copyright). The public's right to make a fair use parody and the copyright owner's right to license a derivative work are useless if the parodist is held hostage by every actor whose "identity" he might need to "appropriate."

Our court is in a unique position here. State courts are unlikely to be particularly sensitive to federal preemption, which, after all, is a matter of first concern to the federal courts. The Supreme Court is unlikely to consider the issue because the right of publicity seems so much a matter of state law. That leaves us. It's our responsibility to keep the right of publicity from taking away federally granted rights, either from the public at large or from a copyright owner. We must make sure state law doesn't give the Vanna Whites and Adam Wests of the world a veto over fair use parodies of the shows in which they appear, or over copyright holders' exclusive right to license derivative works of those shows. In a case where the copyright owner isn't even a party — where no one has the interests of copyright owners at heart — the majority creates a rule that greatly diminishes the rights of copyright holders in this circuit.

V

The majority's decision also conflicts with the federal copyright system in another, more insidious way. Under the dormant Copyright Clause, state intellectual property laws can stand only so long as they don't "prejudice the interests of other States." Goldstein v. California, 412 U.S. 546, 558, 93 S. Ct. 2303, 2310, 37 L. Ed. 2d 163 (1973). A state law criminalizing record piracy, for instance, is permissible because citizens of other states would "remain free to copy within their borders those works which may be protected elsewhere." Id. But the right of publicity isn't geographically limited. A right of publicity created by one state applies to conduct everywhere, so long as it involves a celebrity domiciled in that state. If a Wyoming resident creates an ad that features a California domiciliary's name or likeness, he'll be subject to California right of publicity law even if he's careful to keep the ad from being shown in California. See Acme Circus Operating Co. v. Kuperstock, 711 F.2d 1538, 1540 (11th Cir. 1983); Groucho Marx Prods. v. Day and Night Co., 689 F.2d 317, 320 (2d Cir. 1982); see also Factors Etc. v. Pro Arts, 652 F.2d 278, 281 (2d Cir. 1981).

The broader and more ill-defined one state's right of publicity, the more it interferes with the legitimate interests of other states. A limited right that applies to unauthorized use of name and likeness probably does not run afoul of the Copyright Clause, but the majority's protection of "identity" is quite another story. Under the majority's approach, any time anybody in the United States — even somebody who lives in a state with a very narrow right of publicity — creates an ad, he takes the risk that it might remind some segment of the public of somebody, perhaps somebody with only a local reputation, somebody the advertiser has never heard of. See note 17 supra (right of publicity is infringed by unintentional appropriations). So you made a commercial in Florida and one of the characters reminds Reno residents of their favorite local TV anchor (a California domiciliary)? Pay up.

This is an intolerable result, as it gives each state far too much control over artists in other states. No California statute, no California court has actually tried to reach this far. It is ironic that it is we who plant this kudzu in the fertile soil of our federal system.

VI

Finally, I can't see how giving White the power to keep others from evoking her image in the public's mind can be squared with the First Amendment. Where does White get this right to control our thoughts? The majority's creation goes way beyond the protection given a trademark or a copyrighted work, or a person's name or likeness. All those things control one particular way of expressing an idea, one way of referring to an object or a person. But not allowing any means of reminding people of someone? That's a speech restriction unparalleled in First Amendment law.

What's more, I doubt even a name-and-likeness-only right of publicity can stand without a parody exception. The First Amendment isn't just about religion or politics — it's also about protecting the free development of our national culture. Parody, humor, irreverence are all vital components of the marketplace of ideas. The last thing we need, the last thing the First Amendment will tolerate, is a law that lets public figures keep people from mocking them, or from "evok[ing]" their images in the mind of the public. 971 F.2d at 1399.

The majority dismisses the First Amendment issue out of hand because Samsung's ad was commercial speech. Id. at 1401 & n. 3. So what? Commercial speech may be less protected by the First Amendment than noncommercial speech, but less protected means protected nonetheless. Central Hudson Gas & Elec. Corp. v. Public Serv. Comm'n, 447 U.S. 557 (1980). And there are very good reasons for this. Commercial speech has a profound effect on our culture and our attitudes. Neutral-seeming ads influence people's social and political attitudes, and themselves arouse political controversy . . .

In our pop culture, where salesmanship must be entertaining and entertainment must sell, the line between the commercial and noncommercial has not merely blurred; it has disappeared. Is the Samsung parody any different from a parody on Saturday Night Live or in Spy Magazine? Both are equally profit-motivated. Both use a celebrity's identity to sell things — one to sell VCRs, the other to sell advertising. Both mock their subjects. Both try to make people laugh. Both add something, perhaps something worthwhile and memorable, perhaps not, to our culture. Both are things that the people being portrayed might dearly want to suppress...

VII

For better or worse, we are the Court of Appeals for the Hollywood Circuit. Millions of people toil in the shadow of the law we make, and much of their livelihood is made possible by the existence of intellectual property rights. But much of their livelihood — and much of the vibrancy of our culture — also depends on the existence of other intangible rights: The right to draw ideas from a rich and varied public domain, and the right to mock, for profit as well as fun, the cultural icons of our time.

In the name of avoiding the "evisceration" of a celebrity's rights in her image, the majority diminishes the rights of copyright holders and the public at large. In the name of fostering creativity, the majority suppresses it. Vanna White and those like her have been given something they never had before, and they've been given it at our expense. I cannot agree.

COMMENTS AND QUESTIONS

1. The Ninth Circuit at least partially reaffirmed its *White* decision in Wendt v. Host Intl., 125 F.3d 806 (9th Cir. 1997). That case also involved animatronic robots. These were representative of characters from the television show "Cheers" and were placed in licensed airport Cheers bars. The bars had obtained rights from the producers of the television show but not from the actors themselves. The court held that the actors retained publicity rights in their portrayal of the fictional characters under California law and remanded the case for an analysis of the similarities between the plaintiffs and the robots.

If the actors had brought a claim under California law based on the use of their images from the TV show itself, they would have lost. See Fleet v. CBS, 50 Cal. App. 4th 1911 (1996) (California publicity law preempted to the extent it imposes controls on the exploitation of name or likeness through distribution of a motion picture in which actor appeared); Page v. Something Weird Video, 960 F. Supp. 1438 (C.D. Cal. 1996) (First Amendment allows use of drawing of character in film to promote that film). Should the result be any different where a spin-off product is licensed?

2. Cases other than *White* discuss "celebrity attributes" besides appearance and voice. See, e.g., McFarland v. Miller, 14 F.3d 912 (3d Cir. 1994) (granting damages for use of appearance and name of "Our Gang" member Spanky McFarland in restaurant decor). Is there any natural boundary to the right of publicity? If mere "evocation of a celebrity" is required, which of the following could be protected by a right of publicity: (1) a characteristic walk or even running style; (2) a characteristic gesture,

such as Clint Eastwood's sneer, or Johnny Carson's musically accompanied golf swing, or Michael Jackson's "moonwalking" dance steps; (3) a "signature" joke, such as Henny Youngman's "take my wife — please," or Joan Rivers' "can we talk?" — notwithstanding that both jokes were well-known when they were adopted as "signatures"; (4) a style of chess opening that has become associated with a particular grand master; (5) a shot, move, or technique in sports that is closely identified with a particular athlete — e.g., Tiger Woods' fist pump.

On the other hand, is there any natural limit to Judge Kozinski's reasoning? Or does it suggest that the entire concept of a right of publicity is ill-considered? Is there some identifiable reason that Bette Midler's claim seems more plausible than Vanna White's? Should it matter that a consumer might be confused by the Midler imitation, but won't be by the robot?

3. Judge Kozinski emphasizes throughout his opinion that the right of publicity adds extra burdens to the creators of works that draw on celebrity attributes. For example, he states: "We must make sure state law doesn't give the Vanna Whites and Adam Wests of the world a veto over fair use parodies of the shows in which they appear, or over copyright holders' exclusive right to license derivative works of those shows." Judge Kozinski's point is that the right of publicity creates the need for an entirely new "layer" of transactions *on top of* the traditional copyright license. For example, in many of the "voice-alike" cases such as *Midler*, the defendant in the publicity action is a legitimate licensee of the copyright holder in the song that the defendant used. The right of publicity cases thus implicitly hold that the copyright license does not shield the licensee from liability for using the work, at least under some circumstances. It also means that, at least for copyrighted works assigned prior to the rapid growth of the right of publicity, creators of works can use the new right to extract some extra value. See Eben Shapiro, Rising Caution on Using Celebrity Images, N.Y. Times, Nov. 4, 1992, at D20.

4. Does the economic branch of the right of publicity serve utilitarian purposes? In Zacchini v. Scripps-Howard Broadcasting Co., 433 U.S. 562, 575 (1977) (finding that protection of a human cannonball performer's right of publicity against an unauthorized news broadcast of his exhibition did not violate the First Amendment), the Supreme Court commented that "protecting the proprietary interest of the individual in his act" serves in part "to encourage such entertainment," which is "closely analogous to the goals of patent and copyright law." Although it seems unlikely that failure to protect uses of a celebrity's image would discourage the pursuit of fame generally, see Diane Leenheer Zimmerman, Fitting Publicity Rights into Intellectual Property and Free Speech Theory: Sam, You Made the Pants Too Long!, 10 DePaul-LCA J. Art & Ent. L. & Pol'y 283, 306 (2000) (observing that "not a shred of empirical data exists to show that [celebrities] would invest less energy and talent in becoming famous" without a publicity right), the absence of protection could discourage some performers and interfere with licensing deals that might fund particular creative projects. Is *Zacchini* likely to be such a case? Is *White*? What would such a rationale say about the proper scope of the right of publicity doctrine?

A second utilitarian theory holds that failure to protect a celebrity's image could produce a congestion externality. Under this theory, the value of an image can be inefficiently depleted by oversaturation of the marketplace, which an exclusive right to exploit can prevent. See William M. Landes & Richard A. Posner, The Economic Structure of Intellectual Property Law 222-28 (2003); Mark Grady, A Positive Economic Theory of the Right of Publicity, 1 UCLA Entertainment L. Rev. 97 (1994).

Landes and Posner point to the Disney Corporation's self-imposed restraint on commercialization as a response to this problem: "To avoid overkill, Disney manages its character portfolio with care. It has hundreds of characters on its books, many of them just waiting to be called out of retirement . . . Disney practices good husbandry of its characters and extends the life of its brands by not overexposing them . . . They avoid debasing the currency." See B. Britt, "International Marketing: Disney's Global Goals," Marketing 22-26 (May 17, 1990). Landes and Posner suggest that a similar oversaturation can arise with regard to some rights of publicity (use of persona in advertising) and trademarks (justifying protection against dilution by blurring). Do you find this argument persuasive? See Stacey L. Dogan & Mark A. Lemley, What the Right of Publicity Can Learn from Trademark Law, 58 Stan. L. Rev. (2006) (disputing the congestion externality rationale on the grounds that it distorts the market by preventing the dissemination of truthful information); see also Dennis S. Karjala, Congestion Externalities and Extended Copyright Protection (manuscript 2005). If so, how far should it extend in constraining uses of another's image? What about parodies? satire? Does this theory apply to all celebrity images? If so, how do we account for nearly ubiquitous images, such as Mickey Mouse, Michael Jordan, and the Coca-Cola logo? For images that survive without any protection, such as Uncle Sam or the Statue of Liberty? Is it possible that for at least some images, more exposure means *higher* value? Should the law try to distinguish between such images and those for which overexposure is more clearly a problem — for instance, the Rolls Royce logo? See Mark A. Lemley, Ex Ante Versus Ex Post Justifications for Intellectual Property, 71 U. Chi. L. Rev. 129, 145 (2004) (arguing that the congestion externality may be limited to a narrow subset of images or works that become cultural icons).

Alternatively, is the right of publicity better understood under a Lockean rationale, a Kantian theory of personal autonomy, or a means of preventing of unjust enrichment? See Alice Haemmerli, Whose Who? The Case for a Kantian Right of Publicity, 49 Duke L.J. 383 (1999); Michael Madow, Private Ownership of Public Image: Popular Culture and Publicity Rights, 81 Cal. L. Rev. 127 (1993). What is the basis for this property right? Would a moral right based in personal autonomy justify the transferable property interest celebrities receive under the law today?

5. Consider Judge Kozinski's last argument, that the First Amendment limits the right of publicity. This point has some force. Certainly, newsworthy figures cannot prevent stories from being published about them by invoking the right of publicity. And this is true even though the paper may be using the name or likeness of the newsworthy figure to gain "commercial advantage" in the form of increased sales. In Montana v. San Jose Mercury News, 34 Cal. App. 4th 790, 40 Cal. Rptr. 2d 639 (1995), the court held that the *San Jose Mercury News* had a right to reprint its news story about and picture of football star Joe Montana on promotional posters "for the purpose of showing the quality and content of the newspaper."

For a case holding that there is a constitutionally grounded parody exception to the right of publicity — which might have applied in the *White* case as well — see Cardtoons v. Major League Baseball Players' Association, 95 F.3d 959 (10th Cir. 1996). See also Matthews v. Wozencraft, 15 F.3d 432 (5th Cir. 1994) (Texas right of publicity does not allow plaintiff to block a fictionalized biographical narrative, even though plaintiff can be identified from the book).

Comedy III Productions, Inc. v. Gary Saderup, Inc.
California Supreme Court
25 Cal. 4th 387, 21 P.3d 797, 106 Cal. Rptr. 2d 126 (2001)

MOSK, J.

A California statute grants the *right of publicity* to specified successors in interest of deceased celebrities, prohibiting any other person from using a celebrity's name, voice, signature, photograph, or likeness for commercial purposes without the consent of such successors. The United States Constitution prohibits the states from abridging, among other fundamental rights, freedom of speech. (U.S. Const., 1st and 14th Amends.) In the case at bar we resolve a conflict between these two provisions. The Court of Appeal concluded that the lithographs and silkscreened T-shirts in question here received no First Amendment protection simply because they were reproductions rather than original works of art. As will appear, this was error: reproductions are equally entitled to First Amendment protection. We formulate instead what is essentially a balancing test between the First Amendment and the right of publicity based on whether the work in question adds significant creative elements so as to be transformed into something more than a mere celebrity likeness or imitation. Applying this test to the present case, we conclude that there are no such creative elements here and that the right of publicity prevails. On this basis, we will affirm the judgment of the Court of Appeal . . .

II. Facts

Plaintiff Comedy III Productions, Inc. (hereafter Comedy III), brought this action against defendants Gary Saderup and Gary Saderup, Inc. (hereafter collectively Saderup), seeking damages and injunctive relief for violation of section 990 [the predecessor to § 3344.1] and related business torts. The parties waived the right to jury trial and the right to put on evidence, and submitted the case for decision on the following stipulated facts: Comedy III is the registered owner of all rights to the former comedy act known as The Three Stooges, who are deceased personalities within the meaning of the statute.

Saderup is an artist with over 25 years' experience in making charcoal drawings of celebrities. These drawings are used to create lithographic and silkscreen masters, which in turn are used to produce multiple reproductions in the form, respectively, of lithographic prints and silkscreened images on T-shirts. Saderup creates the original drawings and is actively involved in the ensuing lithographic and silkscreening processes. Without securing Comedy III's consent, Saderup sold lithographs and T-shirts bearing a likeness of The Three Stooges reproduced from a charcoal drawing he had made. These lithographs and T-shirts did not constitute an advertisement, endorsement, or sponsorship of any product.

Saderup's profits from the sale of unlicensed lithographs and T-shirts bearing a likeness of The Three Stooges was $75,000 and Comedy III's reasonable attorney fees were $150,000.

On these stipulated facts the court found for Comedy III and entered judgment against Saderup awarding damages of $75,000 and attorney fees of $150,000 plus costs. The court also issued a permanent injunction restraining Saderup from violating the statute by use of any likeness of The Three Stooges in lithographs, T-shirts, "or any other medium by which the [Saderup's] art work may be sold or marketed." The injunction further prohibited Saderup from "Creating, producing, reproducing, copying, distributing, selling or exhibiting any lithographs, prints, posters, t-shirts, buttons, or other goods, products or merchandise of any kind, bearing the photograph, image, face, symbols, trademarks, likeness, name, voice or signature of The Three Stooges or any of the individual members of The Three Stooges." The sole exception to this broad prohibition was Saderup's original charcoal drawing from which the reproductions at issue were made.

Saderup appealed. The Court of Appeal modified the judgment by striking the injunction. The court reasoned that Comedy III had not proved a likelihood of continued violation of the statute, and that the wording of the injunction was overbroad because it exceeded the terms of the statute and because it "could extend to matters and conduct protected by the First Amendment...."

The Court of Appeal affirmed the judgment as thus modified, however, upholding the award of damages, attorney fees, and costs. In so doing, it rejected Saderup's contentions that his conduct (1) did not violate the terms of the statute, and (2) in any event was protected by the constitutional guaranty of freedom of speech.

We granted review to address these two issues.

III. Discussion

A. The Statutory Issue

[The court held that the statute extends to the sale of products featuring a deceased personality's likeness as well as advertisements.]

B. The Constitutional Issue

Saderup next contends that enforcement of the judgment against him violates his right of free speech and expression under the First Amendment. He raises a difficult issue, which we address below.

The right of publicity is often invoked in the context of commercial speech when the appropriation of a celebrity likeness creates a false and misleading impression that the celebrity is endorsing a product. (See Waits v. Frito-Lay, Inc. (9th Cir. 1992) 978 F.2d 1093; Midler v. Ford Motor Co. (9th Cir. 1988) 849 F.2d 460.) Because the First Amendment does not protect false and misleading commercial speech (Central Hudson Gas & Elec. Corp. v. Public Serv. Com'n (1980) 447 U.S. 557, 563-564), and because even nonmisleading commercial speech is generally subject to somewhat lesser First Amendment protection (*Central Hudson*, at p. 566), the right of publicity may often trump the right of advertisers to make use of celebrity figures.

But the present case does not concern commercial speech. As the trial court found, Saderup's portraits of The Three Stooges are expressive works and not an advertisement for or endorsement of a product. Although his work was done for financial gain, "[t]he First Amendment is not limited to those who publish without charge.... [An expressive activity] does not lose its constitutional protection because it is undertaken for profit." (Guglielmi v. Spelling-Goldberg Productions (1979) 25 Cal. 3d 860, 868, 160 Cal. Rptr. 352, 603 P.2d 454 (conc. opn. of Bird, C.J.) (*Guglielmi*).)

The tension between the right of publicity and the First Amendment is highlighted by recalling the two distinct, commonly acknowledged purposes of the latter. First, "'to preserve an uninhibited marketplace of ideas' and to repel efforts to limit the "'uninhibited, robust and wide-open" debate on public issues.'" (*Guglielmi*, supra, 25 Cal. 3d at p. 866, 160 Cal. Rptr. 352, 603 P.2d 454.) Second, to foster a "fundamental respect for individual development and self-realization. The right to self-expression is inherent in any political system which respects individual dignity. Each speaker must be free of government restraint regardless of the nature or manner of the views expressed unless there is a compelling reason to the contrary." (Ibid., fn. omitted; see also Emerson, The System of Freedom of Expression (1970) pp. 6-7.)

The right of publicity has a potential for frustrating the fulfillment of both these purposes. Because celebrities take on public meaning, the appropriation of their likenesses may have important uses in uninhibited debate on public issues, particularly debates about culture and values. And because celebrities take on personal meanings to many individuals in the society, the creative appropriation of celebrity images can be an important avenue of individual expression. As one commentator has stated: "Entertainment and sports celebrities are the leading players in our Public Drama. We tell tales, both tall and cautionary, about them. We monitor their comings and goings, their missteps and heartbreaks. We copy their mannerisms, their styles, their modes of conversation and of consumption. Whether or not celebrities are 'the chief agents of moral change in the United States,' they certainly are widely used — far more than are institutionally anchored elites — to symbolize individual aspirations, group identities, and cultural values. Their images are thus important expressive and communicative resources: the peculiar, yet familiar idiom in which we conduct a fair portion of our cultural business and everyday conversation." (Madow, Private Ownership of Public Image: Popular Culture and Publicity Rights (1993) 81 Cal. L. Rev. 125, 128 (Madow, italics and fns. omitted.)

As Madow further points out, the very importance of celebrities in society means that the right of publicity has the potential of censoring significant expression by suppressing alternative versions of celebrity images that are iconoclastic, irreverent, or otherwise attempt to redefine the celebrity's meaning. (Madow, supra, 81 Cal. L. Rev. at pp. 143-145; see also Coombe, Author/izing the Celebrity: Publicity Rights, Postmodern Politics, and Unauthorized Genders (1992) 10 Cardozo Arts and Ent. L.J. 365, 377-388.) A majority of this court recognized as much in *Guglielmi*: "The right of publicity derived from public prominence does not confer a shield to ward off caricature, parody and satire. Rather, prominence invites creative comment." (*Guglielmi*, supra, 25 Cal. 3d at p. 869, 160 Cal. Rptr. 352, 603 P.2d 454.)

For similar reasons, speech about public figures is accorded heightened First Amendment protection in defamation law. As the United States Supreme Court held in Gertz v. Robert Welch, Inc. (1974) 418 U.S. 323, public figures may prevail in a libel action only if they prove that the defendant's defamatory statements were made with actual malice, i.e., actual knowledge of falsehood or reckless disregard for the truth, whereas private figures need prove only negligence. (Id. at pp. 328, 342, 344-45.) The rationale for such differential treatment is, first, that the public figure has greater access to the media and therefore greater opportunity to rebut defamatory statements, and second, that those who have become public figures have done so voluntarily and therefore "invite attention and comment." (Id. at pp. 344-345, 94 S.Ct. 2997.) Giving broad scope to the right of publicity has the potential of allowing a celebrity to accomplish through the vigorous exercise of that right the censorship of unflattering commentary that cannot be constitutionally accomplished through defamation actions.

Nor do Saderup's creations lose their constitutional protections because they are for purposes of entertaining rather than informing. As Chief Justice Bird stated in *Guglielmi*, invoking the dual purpose of the First Amendment: "Our courts have often observed that entertainment is entitled to the same constitutional protection as the exposition of ideas. That conclusion rests on two propositions. First, '[t]he line between informing and entertaining is too elusive for the protection of the basic right. Everyone is familiar with instances of propaganda through fiction. What is one man's amusement, teaches another doctrine." ' (*Guglielmi*, supra, 25 Cal. 3d at p. 867, 160 Cal. Rptr. 352, 603 P.2d 454, fn. omitted.) "Second, entertainment, as a mode of self-expression, is entitled to constitutional protection irrespective of its contribution to the marketplace of ideas. 'For expression is an integral part of the development of ideas, of mental exploration and of the affirmation of self. The power to realize his potentiality as a human being begins at this point and must extend at least this far if the whole nature of man is not to be thwarted." ' (Ibid.)

Nor does the fact that expression takes a form of nonverbal, visual representation remove it from the ambit of First Amendment protection. In Bery v. City of New York (2d Cir. 1996) 97 F.3d 689, the court overturned an ordinance requiring visual artists—painters, printers, photographers, sculptors, etc.—to obtain licenses to sell their work in public places, but exempted the vendors of books, newspapers or other written matter. As the court stated: "Both the [district] court and the City demonstrate an unduly restricted view of the First Amendment and of visual art itself. Such myopic vision not only overlooks case law central to First Amendment jurisprudence but fundamentally misperceives the essence of visual communication and artistic expression. Visual art is as wide ranging in its depiction of ideas, concepts and emotions as any book, treatise, pamphlet or other writing, and is similarly entitled to full First Amendment protection. . . . One cannot look at Winslow Homer's paintings on the Civil War without seeing, in his depictions of the boredom and hardship of the individual soldier, expressions of anti-war sentiments, the idea that war is not heroic." (Id. at p. 695.)

Moreover, the United States Supreme Court has made it clear that a work of art is protected by the First Amendment even if it conveys no discernable message: "[A] narrow, succinctly articulable message is not a condition of constitutional protection, which if confined to expressions conveying a 'particularized message,' [citation], would never reach the unquestionably shielded painting of Jackson Pollock, music of Arnold Schoenberg, or Jabberwocky verse of Lewis Carroll." (Hurley v. Irish-American Gay, Lesbian and Bisexual Group of Boston, Inc. (1995) 515 U.S. 557, 569.)

Nor does the fact that Saderup's art appears in large part on a less conventional avenue of communications, T-shirts, result in reduced First Amendment protection. As Judge Posner stated in the case of a defendant who sold T-shirts advocating the legalization of marijuana, "its T-shirts . . . are to [the seller] what the *New York Times* is to the Sulzbergers and the Ochs—the vehicle of her ideas and opinions." (Ayres v. City of Chicago (7th Cir. 1997) 125 F.3d 1010, 1017; see also Cohen v. California

(1971) 403 U.S. 15, [jacket with words "Fuck the Draft" on the back is protected speech].) First Amendment doctrine does not disfavor nontraditional media of expression.

But having recognized the high degree of First Amendment protection for non-commercial speech about celebrities, we need not conclude that all expression that trenches on the right of publicity receives such protection. The right of publicity, like copyright, protects a form of intellectual property that society deems to have some social utility. "Often considerable money, time and energy are needed to develop one's prominence in a particular field. Years of labor may be required before one's skill, reputation, notoriety or virtues are sufficiently developed to permit an economic return through some medium of commercial promotion. [Citations.] For some, the investment may eventually create considerable commercial value in one's identity." (*Lugosi*, supra, 25 Cal. 3d at pp. 834-835, 160 Cal. Rptr. 323, 603 P.2d 425 (dis. opn. of Bird, C. J.).)

The present case exemplifies this kind of creative labor. Moe and Jerome (Curly) Howard and Larry Fein fashioned personae collectively known as The Three Stooges, first in vaudeville and later in movie shorts, over a period extending from the 1920's to the 1940's. (See Fleming, The Three Stooges: Amalgamated Morons to American Icons (1999) pp. 10-46.) The three comic characters they created and whose names they shared — Larry, Moe, and Curly — possess a kind of mythic status in our culture. Their journey from ordinary vaudeville performers to the heights (or depths) of slapstick comic celebrity was long and arduous. (Ibid.) Their brand of physical humor — the nimble, comically stylized violence, the "nyuk-nyuks" and "whoop-whoop-whoops," eye-pokes, slaps and head conks (see, e.g., Three Little Pigskins (Columbia Pictures 1934), Hoi Polloi (Columbia Pictures 1935), A Gem of a Jam (Columbia Pictures 1943), Micro Phonies (Columbia Pictures 1945)) — created a distinct comedic trademark. Through their talent and labor, they joined the relatively small group of actors who constructed identifiable, recurrent comic personalities that they brought to the many parts they were scripted to play. "Groucho Marx just being Groucho Marx, with his moustache, cigar, slouch and leer, cannot be exploited by others. Red Skelton's variety of self-devised roles would appear to be protectible, as would the unique personal creations of Abbott and Costello, Laurel and Hardy and others of that genre. . . . '[W]e deal here with actors portraying themselves and developing their own characters.'" (*Lugosi*, supra, 25 Cal. 3d at pp. 825-826, 160 Cal. Rptr. 323, 603 P.2d 425 (conc. opn. of Mosk, J.).)

In sum, society may recognize, as the Legislature has done here, that a celebrity's heirs and assigns have a legitimate protectible interest in exploiting the value to be obtained from merchandising the celebrity's image, whether that interest be conceived as a kind of natural property right or as an incentive for encouraging creative work. (See 1 McCarthy, The Rights of Publicity and Privacy (2d ed. 2000) §§ 2.2-2.7, pp. 2-1 to 2-22 (McCarthy).) Although critics have questioned whether the right of publicity truly serves any social purpose, (see, e.g., Madow, supra, 81 Cal. L. Rev. at pp. 178-238), there is no question that the Legislature has a rational basis for permitting celebrities and their heirs to control the commercial exploitation of the celebrity's likeness.

Although surprisingly few courts have considered in any depth the means of reconciling the right of publicity and the First Amendment, we follow those that have in concluding that depictions of celebrities amounting to little more than the appropriation of the celebrity's economic value are not protected expression under the First Amendment. We begin with Zacchini v. Scripps-Howard Broadcasting Co.

(1977) 433 U.S. 562, 576, (*Zacchini*), the only United States Supreme Court case to directly address the right of publicity. Zacchini, the performer of a human cannonball act, sued a television station that had videotaped and broadcast his entire performance without his consent. The court held the First Amendment did not protect the television station against a right of publicity claim under Ohio common law. In explaining why the enforcement of the right of publicity in this case would not violate the First Amendment, the court stated: "'[T]he rationale for [protecting the right of publicity] is the straightforward one of preventing unjust enrichment by the theft of good will. No social purpose is served by having the defendant get free some aspect of the plaintiff that would have market value and for which he would normally pay.'" (Id. at p. 576.) The court also rejected the notion that federal copyright or patent law preempted this type of state law protection of intellectual property: "[Copyright and patent] laws perhaps regard the 'reward to the owner [as] a secondary consideration,' [citation], but they were 'intended definitely to grant valuable, enforceable rights' in order to afford greater encouragement to the production of works of benefit to the public. [Citation.] The Constitution does not prevent Ohio from making a similar choice here in deciding to protect the entertainer's incentive in order to encourage the production of this type of work." (Id. at p. 577.)

To be sure, *Zacchini* was not an ordinary right of publicity case: the defendant television station had appropriated the plaintiff's entire act, a species of common law copyright violation. Nonetheless, two principles enunciated in *Zacchini* apply to this case: (1) state law may validly safeguard forms of intellectual property not covered under federal copyright and patent law as a means of protecting the fruits of a performing artist's labor; and (2) the state's interest in preventing the outright misappropriation of such intellectual property by others is not automatically trumped by the interest in free expression or dissemination of information; rather, as in the case of defamation, the state law interest and the interest in free expression must be balanced, according to the relative importance of the interests at stake. (See Gertz v. Robert Welch, Inc., supra, 418 U.S. at pp. 347-350.)

Guglielmi adopted a similar balancing approach. The purported heir of Rudolph Valentino filed suit against the makers of a fictional film based on the latter's life. *Guglielmi* concluded that the First Amendment protection of entertainment superseded any right of publicity. This was in contrast to the companion *Lugosi* case, in which Chief Justice Bird concluded in her dissenting opinion that there may be an enforceable right of publicity that would prevent the merchandising of Count Dracula using the likeness of Bela Lugosi, with whom that role was identified. (*Lugosi*, supra, 25 Cal. 3d at pp. 848-849, 160 Cal. Rptr. 323, 603 P.2d 425 (dis. opn. of Bird, C. J.).) *Guglielmi* proposed a balancing test to distinguish protected from unprotected appropriation of celebrity likenesses: "an action for infringement of the right of publicity can be maintained only if the proprietary interests at issue clearly outweigh the value of free expression in this context." (*Guglielmi*, supra, 25 Cal. 3d at p. 871, 160 Cal. Rptr. 352, 603 P.2d 454.)

In Estate of Presley v. Russen (D.N.J. 1981) 513 F. Supp. 1339 (*Russen*), the court considered a New Jersey common law right of publicity claim by Elvis Presley's heirs against an impersonator who performed The Big El Show. The court implicitly used a balancing test similar to the one proposed in *Guglielmi*. Acknowledging that the First Amendment protects entertainment speech, the court nonetheless rejected that constitutional defense. "[E]ntertainment that is merely a copy or imitation, even if skillfully and accurately carried out, does not really have its own creative component and does not have a significant value as pure entertainment. As one authority has emphasized: 'The public interest in entertainment will support the sporadic, occasional and good-faith imitation of a famous person to achieve humor, to effect criticism or to season a particular episode, but it does not give a privilege to appropriate another's valuable attributes on a continuing basis as one's own without the consent of the other."' (*Russen*, supra, 513 F. Supp. at pp. 1359-1360.) Acknowledging also that the show had some informational value, preserving a live Elvis Presley act for posterity, the court nonetheless stated: "This recognition that defendant's production has some value does not diminish our conclusion that the primary purpose of defendant's activity is to appropriate the commercial value of the likeness of Elvis Presley." (Id. at p. 1360.)

On the other side of the equation, the court recognized that the Elvis impersonation, as in *Zacchini*, represented '"what may be the strongest case for a "right of publicity" — involving, not the appropriation of an entertainer's reputation to enhance the attractiveness of a commercial product, but the appropriation of the very activity by which the entertainer acquired his reputation in the first place."' (*Russen*, supra, 513 F. Supp. at p. 1361, quoting *Zacchini*, supra, 433 U.S. at p. 576.) Thus, in balancing the considerable right of publicity interests with the minimal expressive or informational value of the speech in question, the *Russen* court concluded that the Presley estate's request for injunctive relief would likely prevail on the merits. (*Russen*, at p. 1361; see also Factors Etc. Inc. v. Creative Card Co. (S.D.N.Y. 1977) 444 F. Supp. 279 [poster of Elvis Presley labeled "In Memory...1935-1977" did not possess sufficient newsworthiness to be eligible for First Amendment protection].)

In Groucho Marx Productions, Inc. v. Day & Night Co. (S.D.N.Y. 1981) 523 F. Supp. 485, reversed on other grounds (2d Cir. 1982) 689 F.2d 317, the court considered a right of publicity challenge to a new play featuring characters resembling the Marx Brothers. The court found in favor of the Marx Brothers' heirs, rejecting a First Amendment defense. In analyzing that defense, the court posed a dichotomy between "works...designed primarily to promote the dissemination of thoughts, ideas or information through news or fictionalization," which would receive First Amendment protection, and "use of the celebrity's name or likeness...largely for commercial purposes, such as the sale of merchandise," in which the right of publicity would prevail. (523 F. Supp. at p. 492.) In creating this dichotomy, the court did not appear to give due consideration to forms of creative expression protected by the First Amendment that cannot be categorized as ideas or information. Moreover, the court, borrowing from certain copyright cases, seemed to believe that the validity of the First Amendment defense turned on whether the play was a parody, without explaining why other forms of creative appropriation, such as using established characters in new theatrical works to advance various creative objectives, were not protected by the First Amendment. Nonetheless, the case is in line with *Zacchini*, *Guglielmi* and *Russen* in recognizing that certain forms of commercial exploitation of celebrities that violate the state law right of publicity do not receive First Amendment protection.

It is admittedly not a simple matter to develop a test that will unerringly distinguish between forms of artistic expression protected by the First Amendment and those that must give way to the right of publicity. Certainly, any such test must incorporate the principle that the right of publicity cannot, consistent with the First Amendment, be a right to control the celebrity's image by censoring disagreeable portrayals. Once the celebrity thrusts himself or herself forward into the limelight, the First Amendment dictates that the right to comment on, parody, lampoon, and make other expressive uses of the celebrity image must be given broad scope. The necessary implication of this observation is that the right of publicity is essentially an economic right. What the right of publicity holder possesses is not a right of censorship, but a right to prevent others from misappropriating the economic value generated by the celebrity's fame through the merchandising of the "name, voice, signature, photograph, or likeness" of the celebrity. (§ 990.)

Beyond this precept, how may courts distinguish between protected and unprotected expression? Some commentators have proposed importing the fair use defense from copyright law (17 U.S.C. § 107), which has the advantage of employing an established doctrine developed from a related area of the law. (See Barnett, First Amendment Limits on the Right of Publicity (1995) 30 Tort & Ins. L.J. 635, 650-657; Coyne, Toward a Modified Fair Use Defense in Right of Publicity Cases (1988) 29 Wm. & Mary L. Rev. 781, 812-820.) Others disagree, pointing to the murkiness of the fair use doctrine and arguing that the idea/expression dichotomy, rather than fair use, is the principal means of reconciling copyright protection and First Amendment rights. (2 McCarthy, supra, § 8.38, pp. 8-358 to 8-360; see also Kwall, The Right of Publicity vs. The First Amendment: A Property and Liability Rule Analysis (1994) 70 Ind. L.J. 47, 58, fn. 54.)

We conclude that a wholesale importation of the fair use doctrine into right of publicity law would not be advisable. At least two of the factors employed in the fair use test, "the nature of the copyrighted work" and "the amount and substantiality of the portion used" (17 U.S.C. § 107(2), (3)), seem particularly designed to be applied to the partial copying of works of authorship "fixed in [a] tangible medium of expression" (17 U.S.C. § 102); it is difficult to understand why these factors would be especially useful for determining whether the depiction of a celebrity likeness is protected by the First Amendment.

Nonetheless, the first fair use factor — "the purpose and character of the use" (17 U.S.C. § 107(1)) — does seem particularly pertinent to the task of reconciling the rights of free expression and publicity. As the Supreme Court has stated, the central purpose of the inquiry into this fair use factor "is to see, in Justice Story's words, whether the new work merely 'supersede[s] the objects' of the original creation [citations], or instead adds something new, with a further purpose or different character, altering the first with new expression, meaning, or message; it asks, in other words, whether and to what extent the new work is 'transformative.' Although such transformative use is not absolutely necessary for a finding of fair use, the goal of copyright, to promote science and the arts, is generally furthered by the creation of transformative works." (Campbell v. Acuff-Rose Music, Inc. (1994) 510 U.S. 569, 579, fn. omitted.)

This inquiry into whether a work is "transformative" appears to us to be necessarily at the heart of any judicial attempt to square the right of publicity with the First Amendment. As the above quotation suggests, both the First Amendment and copyright law have a common goal of encouragement of free expression and creativity, the former by protecting such expression from government interference, the latter by protecting the creative fruits of intellectual and artistic labor. (See 1 Nimmer on Copyright (2000 ed.) § 1.10, pp. 1-66.43 to 1-66.44 (Nimmer).) The right of publicity, at least theoretically, shares this goal with copyright law. (1 McCarthy, supra, § 2.6, pp. 2-14 to 2-19.) When artistic expression takes the form of a literal depiction or imitation of a celebrity for commercial gain,[9] directly trespassing on the right of publicity without adding significant expression beyond that trespass, the state law interest in protecting the fruits of artistic labor outweighs the expressive interests of the imitative artist. (See *Zacchini*, supra, 433 U.S. at pp. 575-576.)

On the other hand, when a work contains significant transformative elements, it is not only especially worthy of First Amendment protection, but it is also less likely to interfere with the economic interest protected by the right of publicity. As has been observed, works of parody or other distortions of the celebrity figure are not, from the celebrity fan's viewpoint, good substitutes for conventional depictions of the celebrity and therefore do not generally threaten markets for celebrity memorabilia that the right of publicity is designed to protect. (See Cardtoons, L.C. v. Major League Baseball Players Association (10th Cir. 1996) 95 F.3d 959, 974 (*Cardtoons*).) Accordingly, First Amendment protection of such works outweighs whatever interest the state may have in enforcing the right of publicity. The right-of-publicity holder continues to enforce the right to monopolize the production of conventional, more or less fungible, images of the celebrity.[10] *Cardtoons*, supra, 95 F.3d 959, cited by Saderup, is consistent with this "transformative" test. There, the court held that the First Amendment protected a company that produced trading cards caricaturing and parodying well-known major league baseball players against a claim brought under the Oklahoma right of publicity statute. The court concluded that "[t]he cards provide social commentary on public figures, major league baseball players, who are involved in a significant commercial enterprise, major league baseball," and that "[t]he cards are no less protected because they provide humorous rather than serious commentary." (*Cardtoons*, at p. 969.) The *Cardtoons* court weighed these First Amendment rights against what it concluded was the less-than-compelling interests advanced by the right of publicity outside the advertising context—especially in light of the reality that parody would not likely substantially impact the economic interests of celebrities—and found the cards to be a form of protected expression. (*Cardtoons*, at pp. 973-976.) While *Cardtoons* contained dicta calling into question the social value of the right of publicity, its conclusion that works parodying and caricaturing celebrities are protected by the First Amendment appears unassailable in light of the test articulated above.

9. Inquiry into the "purpose and character" of the work in copyright law also includes "whether such use is of a commercial nature or is for nonprofit educational purposes." (17 U.S.C. § 107(1).) It could be argued that reproduction of a celebrity likeness for noncommercial use-e.g., T-shirts of a recently deceased rock musician produced by a fan as a not-for-profit tribute-is a form of personal expression and therefore more worthy of First Amendment protection. This is an issue, however, that we need not decide in this case. It is undisputed that Saderup sold his reproductions for financial gain.

10. There is a fourth factor in the fair use test not yet mentioned, "the effect of the use upon the potential market for or value of the copyrighted work" (17 U.S.C. § 107(4)), that bears directly on this question. We do not believe, however, that consideration of this factor would usefully supplement the test articulated here. If it is determined that a work is worthy of First Amendment protection because added creative elements significantly transform the celebrity depiction, then independent inquiry into whether or not that work is cutting into the market for the celebrity's images-something that might be particularly difficult to ascertain in the right of publicity context (see Madow, supra, 81 Cal. L. Rev. at pp. 221-222)-appears to be irrelevant. Moreover, this "potential market" test has been criticized for circularity: it could be argued that if a defendant has capitalized in any way on a celebrity's image, he or she has found a potential market and therefore could be liable for such work. (See 4 Nimmer, supra, § 13.05[A][4] at pp. 13-183 to 13-184.) The "transformative" test elaborated in this opinion will, we conclude, protect the right-of-publicity holder's core interest in monopolizing the merchandising of celebrity images without unnecessarily impinging on the artists' right of free expression.

We emphasize that the transformative elements or creative contributions that require First Amendment protection are not confined to parody and can take many forms, from factual reporting (see, e.g., Rosemont Enterprises, Inc. v. Random House, Inc. (N.Y.Sup.Ct.1968) 58 Misc.2d 1, 294 N.Y.S.2d 122, 129, affd. mem. (1969) 32 A.D.2d 892, 301 N.Y.S.2d 948) to fictionalized portrayal (*Guglielmi*, supra, 25 Cal. 3d at pp. 871-872, 160 Cal. Rptr. 352, 603 P.2d 454; see also Parks v. LaFace Records (E.D.Mich.1999) 76 F. Supp. 2d 775, 779-782 [use of civil rights figure Rosa Parks in song title is protected expression]), from heavy-handed lampooning (see Hustler Magazine v. Falwell (1988) 485 U.S. 46) to subtle social criticism (see Coplans et al., Andy Warhol (1970) pp. 50-52 [explaining Warhol's celebrity portraits as a critique of the celebrity phenomenon]).

Another way of stating the inquiry is whether the celebrity likeness is one of the "raw materials" from which an original work is synthesized, or whether the depiction or imitation of the celebrity is the very sum and substance of the work in question. We ask, in other words, whether a product containing a celebrity's likeness is so transformed that it has become primarily the defendant's own expression rather than the celebrity's likeness. And when we use the word "expression," we mean expression of something other than the likeness of the celebrity.

We further emphasize that in determining whether the work is transformative, courts are not to be concerned with the quality of the artistic contribution — vulgar forms of expression fully qualify for First Amendment protection. (See, e.g., Hustler Magazine v. Falwell, supra, 485 U.S. 46, 99 L.Ed.2d 41; see also Campbell v. Acuff-Rose Music, Inc., supra, 510 U.S. at p. 582.) On the other hand, a literal depiction of a celebrity, even if accomplished with great skill, may still be subject to a right of publicity challenge. The inquiry is in a sense more quantitative than qualitative, asking whether the literal and imitative or the creative elements predominate in the work.

Furthermore, in determining whether a work is sufficiently transformative, courts may find useful a subsidiary inquiry, particularly in close cases: does the marketability and economic value of the challenged work derive primarily from the fame of the celebrity depicted? If this question is answered in the negative, then there would generally be no actionable right of publicity. When the value of the work comes principally from some source other than the fame of the celebrity — from the creativity, skill, and reputation of the artist — it may be presumed that sufficient transformative elements are present to warrant First Amendment protection. If the question is answered in the affirmative, however, it does not necessarily follow that the work is without First Amendment protection — it may still be a transformative work.

In sum, when an artist is faced with a right of publicity challenge to his or her work, he or she may raise as affirmative defense that the work is protected by the First Amendment inasmuch as it contains significant transformative elements or that the value of the work does not derive primarily from the celebrity's fame.

Turning to the present case, we note that the trial court, in ruling against Saderup, stated that "the commercial enterprise conducted by [Saderup] involves the sale of lithographs and T-shirts which are not original single works of art, and which are not protected by the First Amendment; the enterprise conducted by the [Saderup] was a commercial enterprise designed to generate profits solely from the use of the likeness of The Three Stooges which is the right of publicity...protected by section 990." Although not entirely clear, the trial court seemed to be holding that *reproductions* of celebrity images are categorically outside First Amendment protection. The Court of Appeal was more explicit in adopting this rationale: "Simply put, although the First Amendment protects speech that is sold, reproductions of an image, made to be sold for profit do not per se constitute speech." But this position has no basis in logic or authority. No one would claim that a published book, because it is one of many copies, receives less First Amendment protection than the original manuscript. It is true that the statute at issue here makes a distinction between a single and original work of fine art and a reproduction. (§ 990, subd. (n)(3).) Because the statute evidently aims at preventing the illicit merchandising of celebrity images, and because single original works of fine art are not forms of merchandising, the state has little if any interest in preventing the exhibition and sale of such works, and the First Amendment rights of the artist should therefore prevail. But the inverse — that a reproduction receives no First Amendment protection — is patently false: a reproduction of a celebrity image that, as explained above, contains significant creative elements is entitled to as much First Amendment protection as an original work of art. The trial court and the Court of Appeal therefore erred in this respect.

Rather, the inquiry is into whether Saderup's work is sufficiently transformative. Correctly anticipating this inquiry, he argues that all portraiture involves creative decisions, that therefore no portrait portrays a mere literal likeness, and that accordingly all portraiture, including reproductions, is protected by the First Amendment. We reject any such categorical position. Without denying that all portraiture involves the making of artistic choices, we find it equally undeniable, under the test formulated above, that when an artist's skill and talent is manifestly subordinated to the overall goal of creating a conventional portrait of a celebrity so as to commercially exploit his or her fame, then the artist's right of free expression is outweighed by the right of publicity. As is the case with fair use in the area of copyright law, an artist depicting a celebrity must contribute something more than a "merely trivial" variation, [but must create] something recognizably "his own" (L. Batlin & Son, Inc. v. Snyder (2d Cir. 1976) 536 F.2d 486, 490), in order to qualify for legal protection.

On the other hand, we do not hold that all reproductions of celebrity portraits are unprotected by the First Amendment. The silkscreens of Andy Warhol, for example, have as their subjects the images of such celebrities as Marilyn Monroe, Elizabeth Taylor, and Elvis Presley. Through distortion and the careful manipulation of context, Warhol was able to convey a message that went beyond the commercial exploitation of celebrity images and became a form of ironic social comment on the dehumanization of celebrity itself. (See Coplans et al., supra, at p. 52.) Such expression may well be entitled to First Amendment protection. Although the distinction between protected and unprotected expression will sometimes be subtle, it is no more so than other distinctions triers of fact are called on to make in First Amendment jurisprudence. (See, e.g., Miller v. California (1973) 413 U.S. 15, 24, [requiring determination, in the context of work alleged to be obscene, of "whether the work, taken as a whole, lacks serious literary, artistic, political, or scientific value"].)[12]

Turning to Saderup's work, we can discern no significant transformative or creative contribution. His undeniable skill is manifestly subordinated to the overall goal of

---

12. The novelist Don DeLillo gives this fictional account of an encounter with Warhol's reproductions of images of Mao Zedong: "He moved along and stood finally in a room filled with images of Chairman Mao. Photocopy Mao, silk-screen Mao, wallpaper Mao, synthetic-polymer Mao. A series of silk screens was installed over a broader surface of wallpaper serigraphs, the Chairman's face a pansy purple here, floating nearly free of its photographic source. Work that was unwitting of history appealed to [him]. He found it liberating. Had he ever realized the deeper meaning of Mao before he saw these pictures?" (DeLillo, Mao II (1991) p. 21.)

creating literal, conventional depictions of The Three Stooges so as to exploit their fame. Indeed, were we to decide that Saderup's depictions were protected by the First Amendment, we cannot perceive how the right of publicity would remain a viable right other than in cases of falsified celebrity endorsements.

Moreover, the marketability and economic value of Saderup's work derives primarily from the fame of the celebrities depicted. While that fact alone does not necessarily mean the work receives no First Amendment protection, we can perceive no transformative elements in Saderup's works that would require such protection.

Saderup argues that it would be incongruous and unjust to protect parodies and other distortions of celebrity figures but not wholesome, reverential portraits of such celebrities. The test we articulate today, however, does not express a value judgment or preference for one type of depiction over another. Rather, it reflects a recognition that the Legislature has granted to the heirs and assigns of celebrities the property right to exploit the celebrities' images, and that certain forms of expressive activity protected by the First Amendment fall outside the boundaries of that right. Stated another way, we are concerned not with whether conventional celebrity images should be produced but with who produces them and, more pertinently, who appropriates the value from their production. Thus, under section 990, if Saderup wishes to continue to depict The Three Stooges as he has done, he may do so only with the consent of the right-of-publicity holder.

IV. Disposition

The judgment of the Court of Appeal is affirmed.

Appendix

FIGURE 6-2
The Three Stooges T-shirt image.

COMMENTS AND QUESTIONS

1. The common law tort of right of publicity requires proof that the defendant used the plaintiff's name or likeness in "commercial advertising or promotion." There is no advertising use at issue in *Saderup*. Rather, *the product itself* is alleged to be illegal. Does the California statute go too far by extending the right of publicity from commercial advertising to reach the making of products?

2. Does the court's test adequately protect the freedom of expression? Is the court saying that art needs to go beyond literal interpretation and not be mass-marketed? Or possibly that the stature of the artist (e.g., Warhol) must be balanced against the popularity of the target? Would it be legal under the court's opinion to make and mass-market a statue of Bill Clinton?

3. The California Supreme Court's framework rests upon two relatively subjective pillars: fair use analysis and art criticism. Subsequent cases illustrate the unpredictability of this balancing framework. In a case that comes close to the conventional portrayal and direct commercial exploitation in *Saderup*, the majority of a divided Sixth Circuit panel ruled that a limited edition painting featuring Tiger Woods was immune from a right of publicity claim under Ohio common law on freedom of expression grounds. See ETW Corp. v. Jireh Pub., Inc., 332 F.3d 915 (6th Cir. 2003).

Just two years after its *Saderup* decision, the California Supreme Court revisited the right of publicity/First Amendment. Johnny and Edgar Winter, well-known musicians (their 1973 album "They Only Come Out at Night" featured the #1 hit instrumental "Frankenstein" and the top 15 single "Free Ride") with albino complexions and long white hair, brought a right of publicity action against defendant D.C. Comics for its publication of a comic book featuring the characters "Johnny and Edgar Autumn," half-worm, half-human creatures with pale faces and long white hair. The comic books series portrayed them as "vile, depraved, stupid, cowardly, subhuman individuals who engage in wanton acts of violence, murder and bestiality for pleasure and who should be killed." Applying the *Saderup* balancing framework, the California Supreme Court concluded that the expressive qualities of the work afforded D.C. Comics First Amendment protection from a right of publicity claim. Winter v. D.C. Comics, 30 Cal.4th 881, 134 Cal.Rptr.2d 634, 69 P.3d 473 (Cal. S.Ct. 2003). By contrast, the Missouri Supreme Court held that same year that another comic book publisher's use of a former professional hockey player's name and image as a metaphorical reference to tough-guy "enforcers" "was predominantly a ploy to sell comic books and related products rather than an artistic or literary expression" and hence "free speech must give way to the right of publicity." John Doe, a/k/a Tony Twist v. TCI Cablevision, 110 S.W.3d 363 (Mo. S. Ct. 2003).

Is there any other approach to balancing the tension between the right of publicity and the First Amendment? What about the use of disclaimers to the effect that a work of art was not authorized, sponsored, or endorsed by the target celebrity (or the holders of his or her post-mortem right of publicity)? If the right of publicity is predominantly an economic right, should society be deeply concerned about the economic plight of celebrities? Or is the right based on other interests, such as personal autonomy and identity? Do the facts in the *Winter* case suggest the need for a broader foundation for protection?

4. *Artistic Advertising.* In its March 1997 "Fabulous Hollywood Issue!," *Los Angeles Magazine* featured an article entitled "Grand Illusions," which used digitally-altered film stills to make it appear that the actors were wearing Spring 1997 fashions. The sixteen familiar scenes included movies and actors such as *Rear Window* (Grace Kelly and Jimmy Stewart), *The Seven Year Itch* (Marilyn Monroe), and *Thelma and Louise* (Susan Sarandon and Geena Davis). The feature also included an image from the film *Tootsie* in which Dustin Hoffman's head was digitally superimposed atop a picture of another model. The caption read "Dustin Hoffman isn't a drag in a butter-colored silk gown by Richard Tyler and Ralph Lauren heels." Hoffman sued, claiming violation of his right of publicity as well as Lanham Act and other state law claims. The Ninth Circuit reversed an award in excess of $3 million on the ground that *L.A. Magazine*'s appropriation of Hoffman's identity was "communicative" rather than "commercial" and hence was entitled to the highest level of protection under the First Amendment. See Hoffman v. Capital Cities/ABC, Inc., and L.A. Magazine, Inc., 255 F.3d 1180 (9th Cir. 2001). Can you square this result with the California Supreme Court's analytical framework in *Saderup?* Can freedom of expression and marketing be treated separately or are they inextricably intertwined? Will future Dustin Hoffmans abandon careers in acting (or otherwise alter their creative and commercial choices) as a result of this case? If not, doesn't this case suggest that we need not be so solicitous of right of publicity claimants? What other justifications might support celebrities in such cases?

PROBLEM

Problem 6-3. In the late 1980s, the New Kids on the Block were an enormously successful pop music group, especially among the younger teen market. Capitalizing on this success, the New Kids sold over 500 products or services bearing their trademarked name. Among those services were "900 numbers" that fans could call to learn more about the New Kids, or to talk to the New Kids themselves.

During the height of the New Kids craze, the newspaper *USA Today* conducted a telephone poll that allowed readers to "vote" for their favorite New Kid (or for "none of the above" if they did not like the band at all) by calling a *USA Today* 900 number. As a part of the poll, the paper included captioned pictures of each of the band members. Suppose that the New Kids on the Block sued *USA Today* for infringement of their right of publicity. Do they have a claim under California law?

---

5. State Moral Rights

Beginning in 1990, federal copyright law (17 U.S.C. § 106A) has provided limited protection for visual artists' moral rights. We discussed this protection in some detail in Chapter 4. In some states, this federal protection is augmented by separate state moral rights statutes. One recent summary describes the more expansive nature of state moral rights protection:

In New Mexico, any original work of art of recognized quality is protected from alteration or destruction if the work is in public view. Rhode Island protects all original works of visual art of any medium, with the exception of film. Connecticut provides an extensive list of protected media, including masters from which copies of an artistic work can be made. Some state laws permit recovery in a broader range of circumstances. For example, in California and Pennsylvania, no intentional defacement, alteration, or destruction is allowed regardless of the artist's ability to prove damage to honor or reputation, so long as the work is of recognized quality. Massachusetts goes further by adding a prohibition against acts of gross negligence.

Amy L. Landers, The Current State of Moral Rights Protection for Visual Artists in the United States, 15 Hastings Comm. & Ent. L.J. 165 (1992), citing 107 N.M. Stat. Ann. §13-4B-2(B), 3(B) (Michie 1991) (work of fine art is "any original work of visual or graphic art of any media ... of recognized quality."); R.I. Gen. Laws § 5-62-2(e) (Michie 1987); Conn. Gen. Stat. Ann. §42-116 § (2) (West Supp. 1991); Cal. Civ. Code §§987-990 (West Supp. 1991); Pa. Stat. Ann. title 73, §§2101-10 (1991); Mass. Gen. Laws Ann. ch. 231, §85S (West Supp. 1991). Most of these state statutes provide greater protection than does the federal statute. The state statutes generally provide more and broader rights to artists and restrict their alienability.

COMMENTS AND QUESTIONS

1. In one case construing Cal. Civ. Code §987, a California court held that a mural is a "painting" entitled to protection. Botello v. Shell Oil Co., 280 Cal. Rptr. 535, 229 Cal. App. 3d 1130 (Ct. App. 1991). Another court held that architectural plans are not works protected by the California Art Preservation Act, since they do not constitute "fine art." Robert H. Jacobs, Inc. v. Westoaks Realtors, Inc., 205 Cal. Rptr. 620, 159 Cal. App. 3d 637 (Ct. App. 1984).

2. The New York moral rights statute takes a different tack. It emphasizes the value of an artist's reputation, rather than the intrinsic value of the work itself. Under this statute, for example, display of a mutilated original art work (or a copy for limited-run works) is prohibited only if the artist's name is associated with it:

> [N]o person other than the artist or a person acting with the artist's consent shall knowingly display in a place accessible to the public or publish a work of fine art or limited edition multiple of not more than three hundred copies by that artist or a reproduction thereof in an altered, defaced, mutilated or modified form if the work is displayed, published or reproduced as being the work of the artist, or under circumstances under which it would reasonably be regarded as being the work of the artist, and damage to the artist's reputation is reasonably likely to result therefrom, except that this section shall not apply to sequential imagery such as that in motion pictures. . . .
>
> [T]he artist shall retain at all times the right to claim authorship, or, for just and valid reason, to disclaim authorship of such work. The right to claim authorship shall include the right of the artist to have his or her name appear on or in connection with such work as the artist. The right to disclaim authorship shall include the right of the artist to prevent his or her name from appearing on or in connection with such work as the artist. Just and valid reason for disclaiming authorship shall include that the work has been altered,

defaced, mutilated or modified other than by the artist, without the artist's consent, and damage to the artist's reputation is reasonably likely to result or has resulted therefrom.

N.Y. Arts and Cultural Affairs Law § 14.03 (1994). For interpretations of this statute, compare Morita v. Omni Publications Intern., Ltd., 741 F. Supp. 1107 (S.D.N.Y. 1990) (publication of photograph of Hiroshima juxtaposed with pro-nuclear slogan was not a mutilation under New York statute) with Wojnarowicz v. American Family Assn., 745 F. Supp. 130 (S.D.N.Y. 1990) (pamphlet containing photocopied fragments of only sexually explicit portions of artist's works, which was distributed in effort to stop public funding of National Endowment for the Arts, violated New York's Artists' Authorship Rights Act). See also Edward J. Damich, The New York Artists' Authorship Rights Act: A Comparative Critique, 84 Colum. L. Rev. 1733, 1735 (1984) ("The New York statute confines its protection of the integrity of an artwork to circumstances damaging to the artist's reputation, leaving the statute susceptible to the interpretation that it does not prohibit complete destruction of art works."). For an argument that the United States does not truly protect moral rights and is thus in violation of its Berne Convention obligations, see Brian T. McCartney, "Creepings" and "Glimmers" of the Moral Rights of Artists in American Copyright Law, 6 UCLA Ent. L. Rev. 35 (1998); but see Thomas F. Cotter, Pragmatism, Economics, and the Droit Moral, 76 N.C. L. Rev. 1 (1997) (agreeing, but arguing this is a good thing).

3. For a discussion of the preemption issues raised by state moral rights statutes and the Visual Artists Rights Act, see *infra* section B of this chapter.

4. *Resale Royalties ("Droit de Suite").* The "droit de suite" permits an artist to benefit from appreciation in the value of her works by entitling her to a percentage of all subsequent sales. California offers this benefit under the Artist's Resale Royalty Act, which stipulates that a 5 percent royalty shall be paid on each sale of a work. Puerto Rico also has implemented the droit de suite. See P.R. Laws Ann. title 31 (1991) (similar provision under the law of Puerto Rico). See also Larry S. Karp & Jeffrey M. Perloff, Legal Requirements That Artists Receive Resale Royalties, 13 Intl. Rev. L. & Econ. 163, 173 (1993) (arguing, on basis of formal model, that resale royalty mandated by law will almost surely cause more harm than good, and that better policy would be for government to help implement a transactional mechanism to make privately bargained resale royalty agreements more attractive).

6. Trespass to Chattels

eBay, Inc. v. Bidder's Edge, Inc.
United States District Court for the Northern District of California
100 F. Supp. 2d 1058 (N.D. Cal. 2000)

WHYTE, District Judge.

Plaintiff eBay, Inc.'s ("eBay") motion for preliminary injunction was heard by the court on April 14, 2000. The court has read the moving and responding papers and heard the argument of counsel. For the reasons set forth below, the court preliminarily enjoins defendant Bidder's Edge, Inc. ("BE") from accessing eBay's computer systems by use of any automated querying program without eBay's written authorization.

I. Background

eBay is an Internet-based, person-to-person trading site. eBay offers sellers the ability to list items for sale and prospective buyers the ability to search those listings and bid on items. The seller can set the terms and conditions of the auction. The item is sold to the highest bidder. The transaction is consummated directly between the buyer and seller without eBay's involvement. A potential purchaser looking for a particular item can access the eBay site and perform a key word search for relevant auctions and bidding status. eBay has also created category listings that identify items in over 2500 categories, such as antiques, computers, and dolls. Users may browse these category listing pages to identify items of interest.

Users of the eBay site must register and agree to the eBay User Agreement. Users agree to the seven page User Agreement by clicking on an "I Accept" button located at the end of the User Agreement. The current version of the User Agreement prohibits the use of "any robot, spider, other automatic device, or manual process to monitor or copy our web pages or the content contained herein without our prior expressed written permission." It is not clear that the version of the User Agreement in effect at the time BE began searching the eBay site prohibited such activity, or that BE ever agreed to comply with the User Agreement.

eBay currently has over 7 million registered users. Over 400,000 new items are added to the site every day. Every minute, 600 bids are placed on almost 3 million items. Users currently perform, on average, 10 million searches per day on eBay's database. Bidding for and sales of items are continuously ongoing in millions of separate auctions.

A software robot is a computer program which operates across the Internet to perform searching, copying and retrieving functions on the web sites of others. A software robot is capable of executing thousands of instructions per minute, far in excess of what a human can accomplish. Robots consume the processing and storage resources of a system, making that portion of the system's capacity unavailable to the system owner or other users. Consumption of sufficient system resources will slow the processing of the overall system and can overload the system such that it will malfunction or "crash." A severe malfunction can cause a loss of data and an interruption in services.

The eBay site employs "robot exclusion headers." A robot exclusion header is a message, sent to computers programmed to detect and respond to such headers, that eBay does not permit unauthorized robotic activity. Programmers who wish to comply with the Robot Exclusion Standard design their robots to read a particular data file, "robots.txt," and to comply with the control directives it contains . . .

eBay identifies robotic activity on its site by monitoring the number of incoming requests from each particular IP [(Internet Protocol)] address. Once eBay identifies an IP address believed to be involved in robotic activity, an investigation into the identity, origin and owner of the IP address may be made in order to determine if the activity is legitimate or authorized. If an investigation reveals unauthorized robotic activity, eBay may attempt to ignore ("block") any further requests from that IP address. Attempts to block requests from particular IP addresses are not always successful.

Organizations often install "proxy server" software on their computers. Proxy server software acts as a focal point for outgoing Internet requests. Proxy servers conserve system resources by directing all outgoing and incoming data traffic through a centralized portal ... Information requests sent through such proxy servers cannot easily be traced back to the originating IP address and can be used to circumvent attempts to block queries from the originating IP address. Blocking queries from innocent third party proxy servers is both inefficient, because it creates an endless game of hide-and-seek, and potentially counterproductive, as it runs a substantial risk of blocking requests from legitimate, desirable users who use that proxy server ...

BE does not host auctions. BE is an auction aggregation site designed to offer on-line auction buyers the ability to search for items across numerous on-line auctions without having to search each host site individually. As of March 2000, the BE web site contained information on more than five million items being auctioned on more than one hundred auction sites. BE also provides its users with additional auction-related services and information. The information available on the BE site is contained in a database of information that BE compiles through access to various auction sites such as eBay. When a user enters a search for a particular item at BE, BE searches its database and generates a list of every item in the database responsive to the search, organized by auction closing date and time. Rather than going to each host auction site one at a time, a user who goes to BE may conduct a single search to obtain information about that item on every auction site tracked by BE. It is important to include information regarding eBay auctions on the BE site because eBay is by far the biggest consumer to consumer on-line auction site ...

[The parties attempted to agree on terms for BE's access to the eBay site, but were unable to do so.] As a result, eBay attempted to block BE from accessing the eBay site; by the end of November, 1999, eBay had blocked a total of 169 IP addresses it believed BE was using to query eBay's system. BE elected to continue crawling eBay's site by using proxy servers to evade eBay's IP blocks.

Approximately 69% of the auction items contained in the BE database are from auctions hosted on eBay. BE estimates that it would lose one-third of its users if it ceased to cover the eBay auctions.

The parties agree that BE accessed the eBay site approximately 100,000 times a day. eBay alleges that BE activity constituted up to 1.53% of the number of requests received by eBay, and up to 1.10% of the total data transferred by eBay during certain periods in October and November of 1999. BE alleges that BE activity constituted no more than 1.11% of the requests received by eBay, and no more than 0.70% of the data transferred by eBay ... eBay has not alleged any specific incremental damages due to BE activity.

It appears that major Internet search engines, such as Yahoo!, Google, Excite and AltaVista, respect the Robot Exclusion Standard.[5]

---

5. BE appears to argue that this cannot be the case because searches performed on each of these search engines will return results that include eBay web pages. However, this does not establish that these sites do not respect robot exclusion headers. There are numerous ways in which search engines can obtain information in compliance with exclusion headers, including: obtaining consent, abiding by the robot.txt file guidelines, or manually searching the sites. BE did not present any evidence of any site ever complaining about the activities of any of these search engines.

eBay now moves for preliminary injunctive relief preventing BE from accessing the eBay computer system based on nine causes of action: trespass, false advertising, federal and state trademark dilution, computer fraud and abuse, unfair competition, misappropriation, interference with prospective economic advantage and unjust enrichment. However, eBay does not move, either independently or alternatively, for injunctive relief that is limited to restricting how BE can use data taken from the eBay site.[6] . . .

III. Analysis

A. *Balance of Harm*

According to eBay, the load on its servers resulting from BE's web crawlers represents between 1.11% and 1.53% of the total load on eBay's listing servers. eBay alleges both economic loss from BE's current activities and potential harm resulting from the total crawling of BE and others. In alleging economic harm, eBay's argument is that eBay has expended considerable time, effort and money to create its computer system, and that BE should have to pay for the portion of eBay's system BE uses. eBay attributes a pro rata portion of the costs of maintaining its entire system to the BE activity. However, eBay does not indicate that these expenses are incrementally incurred because of BE's activities, nor that any particular service disruption can be attributed to BE's activities. eBay provides no support for the proposition that the pro rata costs of obtaining an item represent the appropriate measure of damages for unauthorized use. In contrast, California law appears settled that the appropriate measure of damages is the actual harm inflicted by the conduct:

> Where the conduct complained of does not amount to a substantial interference with possession or the right thereto, but consists of intermeddling with or use of or damages to the personal property, the owner has a cause of action for trespass or case, and may recover only the actual damages suffered by reason of the impairment of the property or the loss of its use.

Zaslow v. Kroenert, 29 Cal. 2d 541, 551, 176 P.2d 1 (1946). Moreover, even if BE is inflicting incremental maintenance costs on eBay, potentially calculable monetary damages are not generally a proper foundation for a preliminary injunction. See e.g., Sampson v. Murray, 415 U.S. 61, 90, 94 S.Ct. 937, 39 L.Ed.2d 166 (1974). Nor does eBay appear to have made the required showing that this is the type of extraordinary case in which monetary damages may support equitable relief. . .

eBay's allegations of harm are based, in part, on the argument that BE's activities should be thought of as equivalent to sending in an army of 100,000 robots a day to check the prices in a competitor's store. This analogy, while graphic, appears inappropriate. Although an admittedly formalistic distinction, unauthorized robot intruders into a "brick and mortar" store would be committing a trespass to real property. There does not appear to be any doubt that the appropriate remedy for an ongoing

---

6. The bulk of eBay's moving papers and declarations address the alleged misuse of the eBay mark and the information BE obtains from the eBay computers. The court does not address the facts specific to these claims, nor the merits of these claims.

trespass to business premises would be a preliminary injunction. See e.g., State v. Carriker, 5 Ohio App.2d 255, 214 N.E.2d 809, 811-12 (1964) (interpreting Ohio criminal trespass law to cover a business invitee who, with no intention of making a purchase, uses the business premises of another for his own gain after his invitation has been revoked); General Petroleum Corp. v. Beilby, 213 Cal. 601, 605, 2 P.2d 797 (1931). More importantly, for the analogy to be accurate, the robots would have to make up less than two out of every one-hundred customers in the store, the robots would not interfere with the customers' shopping experience, nor would the robots even be seen by the customers. Under such circumstances, there is a legitimate claim that the robots would not pose any threat of irreparable harm. However, eBay's right to injunctive relief is also based upon a much stronger argument.

If BE's activity is allowed to continue unchecked, it would encourage other auction aggregators to engage in similar recursive searching of the eBay system such that eBay would suffer irreparable harm from reduced system performance, system unavailability, or data losses. BE does not appear to seriously contest that reduced system performance, system unavailability or data loss would inflict irreparable harm on eBay consisting of lost profits and lost customer goodwill. Harm resulting from lost profits and lost customer goodwill is irreparable because it is neither easily calculable, nor easily compensable and is therefore an appropriate basis for injunctive relief. See, e.g., People of California ex rel. Van De Kamp v. Tahoe Reg'l Planning Agency, 766 F.2d 1316, 1319 (9th Cir. 1985). Where, as here, the denial of preliminary injunctive relief would encourage an increase in the complained of activity, and such an increase would present a strong likelihood of irreparable harm, the plaintiff has at least established a possibility of irreparable harm.

In the patent infringement context, the Federal Circuit has held that a preliminary injunction may be based, at least in part, on the harm that would occur if a preliminary injunction were denied and infringers were thereby encouraged to infringe a patent during the course of the litigation. See Atlas Powder Co. v. Ireco Chemicals, 773 F.2d 1230, 1233 (Fed.Cir.1985). In the absence of preliminary injunctive relief, "infringers could become compulsory licensees for as long as the litigation lasts." Id. The Federal Circuit's reasoning is persuasive. "The very nature of the patent right is the right to exclude others...We hold that where validity and continuing infringement have been clearly established, as in this case, immediate irreparable harm is presumed. To hold otherwise would be contrary to the public policy underlying the patent laws." Smith Int'l, Inc. v. Hughes Tool Co., 718 F.2d 1573, 1581 (Fed. Cir. 1983) (footnotes omitted). Similarly fundamental to the concept of ownership of personal property is the right to exclude others. See Kaiser Aetna v. United States, 444 U.S. 164, 176, 100 S.Ct. 383, 62 L.Ed.2d 332 (1979) (characterizing "the right to exclude others" as "one of the most essential sticks in the bundle of rights that are commonly characterized as property"). If preliminary injunctive relief against an ongoing trespass to chattels were unavailable, a trespasser could take a compulsory license to use another's personal property for as long as the trespasser could perpetuate the litigation.

BE correctly observes that there is a dearth of authority supporting a preliminary injunction based on an ongoing to trespass to chattels. In contrast, it is black letter law in California that an injunction is an appropriate remedy for a continuing trespass to real property. See Allred v. Harris, 14 Cal. App. 4th 1386, 1390, 18 Cal. Rptr. 2d 530 (1993) (citing 5 B.E. Witkin, Summary of California Law, Torts §605 (9th ed.1988)). If eBay were a brick and mortar auction house with limited seating capacity, eBay would appear to be entitled to reserve those seats for potential bidders, to refuse entrance to individuals (or robots) with no intention of bidding on any of the items, and to seek preliminary injunctive relief against non-customer trespassers eBay was physically unable to exclude. The analytic difficulty is that a wrongdoer can commit an ongoing trespass of a computer system that is more akin to the traditional notion of a trespass to real property, than the traditional notion of a trespass to chattels, because even though it is ongoing, it will probably never amount to a conversion. The court concludes that under the circumstances present here, BE's ongoing violation of eBay's fundamental property right to exclude others from its computer system potentially causes sufficient irreparable harm to support a preliminary injunction...

If eBay's irreparable harm claim were premised solely on the potential harm caused by BE's current crawling activities, evidence that eBay had licensed others to crawl the eBay site would suggest that BE's activity would not result in irreparable harm to eBay. However, the gravamen of the alleged irreparable harm is that if BE is allowed to continue to crawl the eBay site, it may encourage frequent and unregulated crawling to the point that eBay's system will be irreparably harmed. There is no evidence that eBay has indiscriminately licensed all comers. Rather, it appears that eBay has carefully chosen to permit crawling by a limited number of aggregation sites that agree to abide by the terms of eBay's licensing agreement. "The existence of such a [limited] license, unlike a general license offered to all comers, does not demonstrate a decision to relinquish all control over the distribution of the product in exchange for a readily computable fee." Ty, Inc. v. GMA Accessories, Inc., 132 F.3d 1167, 1173 (7th Cir. 1997) (discussing presumption of irreparable harm in copyright infringement context). eBay's licensing activities appear directed toward limiting the amount and nature of crawling activity on the eBay site. Such licensing does not support the inference that carte blanche crawling of the eBay site would pose no threat of irreparable harm...

B. *Likelihood of Success...*

1. Trespass

Trespass to chattels "lies where an intentional interference with the possession of personal property has proximately cause injury." Thrifty-Tel v. Bezenek, 46 Cal. App. 4th 1559, 1566, 54 Cal. Rptr. 2d 468 (1996). Trespass to chattels "although seldom employed as a tort theory in California" was recently applied to cover the unauthorized use of long distance telephone lines. Id. Specifically, the court noted "the electronic signals generated by the [defendants'] activities were sufficiently tangible to support a trespass cause of action." Id. at n. 6. Thus, it appears likely that the electronic signals sent by BE to retrieve information from eBay's computer system are also sufficiently tangible to support a trespass cause of action.

In order to prevail on a claim for trespass based on accessing a computer system, the plaintiff must establish: (1) defendant intentionally and without authorization interfered with plaintiff's possessory interest in the computer system; and (2) defendant's unauthorized use proximately resulted in damage to plaintiff. See *Thrifty-Tel*, 46 Cal. App. 4th at 1566, 54 Cal. Rptr. 2d 468; see also Itano v. Colonial Yacht

Anchorage, 267 Cal. App. 2d 84, 90, 72 Cal. Rptr. 823 (1968) ("When conduct complained of consists of intermeddling with personal property 'the owner has a cause of action for trespass or case, and may recover only the actual damages suffered by reason of the impairment of the property or the loss of its use.'") (quoting Zaslow v. Kroenert, 29 Cal. 2d 541, 550, 176 P.2d 1 (1946)). Here, eBay has presented evidence sufficient to establish a strong likelihood of proving both prongs and ultimately prevailing on the merits of its trespass claim.

a. BE's Unauthorized Interference

eBay argues that BE's use was unauthorized and intentional. eBay is correct. BE does not dispute that it employed an automated computer program to connect with and search eBay's electronic database. BE admits that, because other auction aggregators were including eBay's auctions in their listing, it continued to "crawl" eBay's web site even after eBay demanded BE terminate such activity.

BE argues that it cannot trespass eBay's web site because the site is publicly accessible. BE's argument is unconvincing. eBay's servers are private property, conditional access to which eBay grants the public. eBay does not generally permit the type of automated access made by BE. In fact, eBay explicitly notifies automated visitors that their access is not permitted. "In general, California does recognize a trespass claim where the defendant exceeds the scope of the consent." Baugh v. CBS, Inc., 828 F. Supp. 745, 756 (N.D. Cal. 1993).

Even if BE's web crawlers were authorized to make individual queries of eBay's system, BE's web crawlers exceeded the scope of any such consent when they began acting like robots by making repeated queries. See City of Amsterdam v. Daniel Goldreyer, Ltd., 882 F. Supp. 1273, 1281 (E.D.N.Y. 1995) ("One who uses a chattel with the consent of another is subject to liability in trespass for any harm to the chattel which is caused by or occurs in the course of any use exceeding the consent, even though such use is not a conversion."). Moreover, eBay repeatedly and explicitly notified BE that its use of eBay's computer system was unauthorized. The entire reason BE directed its queries through proxy servers was to evade eBay's attempts to stop this unauthorized access. The court concludes that BE's activity is sufficiently outside of the scope of the use permitted by eBay that it is unauthorized for the purposes of establishing a trespass. See Civic Western Corp. v. Zila Industries, Inc., 66 Cal. App. 3d 1, 17, 135 Cal. Rptr. 915 (1977) ("It seems clear, however, that a trespass may occur if the party, entering pursuant to a limited consent . . . proceeds to exceed those limits . . .") (discussing trespass to real property).

eBay argues that BE interfered with eBay's possessory interest in its computer system. Although eBay appears unlikely to be able to show a substantial interference at this time, such a showing is not required. Conduct that does not amount to a substantial interference with possession, but which consists of intermeddling with or use of another's personal property, is sufficient to establish a cause of action for trespass to chattel. See *Thrifty-Tel*, 46 Cal. App. 4th at 1567, 54 Cal. Rptr. 2d 468 (distinguishing the tort from conversion). Although the court admits some uncertainty as to the precise level of possessory interference required to constitute an intermeddling, there does not appear to be any dispute that eBay can show that BE's conduct amounts to use of eBay's computer systems. Accordingly, eBay has made a strong showing that it is likely to prevail on the merits of its assertion that BE's use of eBay's computer system was an unauthorized and intentional interference with eBay's possessory interest.

b. Damage to eBay's Computer System

A trespasser is liable when the trespass diminishes the condition, quality or value of personal property. See CompuServe, Inc. v. Cyber Promotions, 962 F. Supp. 1015 (S.D. Ohio 1997). The quality or value of personal property may be "diminished even though it is not physically damaged by defendant's conduct." Id. at 1022. The Restatement offers the following explanation for the harm requirement:

The interest of a possessor of a chattel in its inviolability, unlike the similar interest of a possessor of land, is not given legal protection by an action for nominal damages for harmless intermeddlings with the chattel. In order that an actor who interferes with another's chattel may be liable, his conduct must affect some other and more important interest of the possessor. Therefore, one who intentionally intermeddles with another's chattel is subject to liability only if his intermeddling is harmful to the possessor's materially valuable interest in the physical condition, quality, or value of the chattel, or if the possessor is deprived of the use of the chattel for a substantial time, or some other legally protected interest of the possessor is affected . . . Sufficient legal protection of the possessor's interest in the mere inviolability of his chattel is afforded by his privilege to use reasonable force to protect his possession against even harmless interference.

Restatement (Second) of Torts § 218 cmt. e (1977).

eBay is likely to be able to demonstrate that BE's activities have diminished the quality or value of eBay's computer systems. BE's activities consume at least a portion of plaintiff's bandwidth and server capacity. Although there is some dispute as to the percentage of queries on eBay's site for which BE is responsible, BE admits that it sends some 80,000 to 100,000 requests to plaintiff's computer systems per day. Although eBay does not claim that this consumption has led to any physical damage to eBay's computer system, nor does eBay provide any evidence to support the claim that it may have lost revenues or customers based on this use, eBay's claim is that BE's use is appropriating eBay's personal property by using valuable bandwidth and capacity, and necessarily compromising eBay's ability to use that capacity for its own purposes. See *CompuServe*, 962 F. Supp. at 1022 ("any value [plaintiff] realizes from its computer equipment is wholly derived from the extent to which that equipment can serve its subscriber base.").

BE argues that its searches represent a negligible load on plaintiff's computer systems, and do not rise to the level of impairment to the condition or value of eBay's computer system required to constitute a trespass. However, it is undisputed that eBay's server and its capacity are personal property, and that BE's searches use a portion of this property. Even if, as BE argues, its searches use only a small amount of eBay's computer system capacity, BE has nonetheless deprived eBay of the ability to use that portion of its personal property for its own purposes. The law recognizes no such right to use another's personal property. Accordingly, BE's actions appear to have caused injury to eBay and appear likely to continue to cause injury to eBay. If the court were to hold otherwise, it would likely encourage other auction aggregators to crawl the eBay site, potentially to the point of denying effective access to eBay's customers. If preliminary injunctive relief were denied, and other aggregators began to crawl the eBay site, there appears to be little doubt that the load on eBay's computer system would qualify as a substantial impairment of condition or value. California law does not require eBay to wait for such a disaster before applying to this court for relief. The court concludes that eBay has made a strong showing that it is likely to prevail on the merits of its trespass claim, and that there is at least a possibility that it will suffer irreparable harm if preliminary injunctive relief is not granted. eBay is therefore entitled to preliminary injunctive relief...

COMMENTS AND QUESTIONS

1. The tort of trespass to chattels traditionally requires proof not just that the defendant "intermeddled" with a chattel, but that the defendant's use actually caused injury to the chattel or injured the owner by depriving it of the benefit of using the chattel. The chattel in this case is eBay's server, so the Restatement approach would seem to require proof that eBay's servers were damaged, or at least that eBay lost the use of its servers. But eBay could not prove either of those things. eBay's servers did not crash, nor did they need more frequent repair because Bidder's Edge sent a large number of requests to the servers. Nor was eBay deprived of the use of the server. eBay customers weren't kicked out, nor did they have their requests delayed, by reason of Bidder's Edge's use.

Judge Whyte's opinion points to two sorts of injury attributable to Bidder's Edge's conduct. First, Judge Whyte pointed out that when Bidder's Edge sent its flood of requests (normally in the middle of the night), it used just over 1% of eBay's capacity. The court worried that if dozens of other companies sent similar bulk requests at exactly the same time, eBay's server could crash. Second, Judge Whyte seemed to undo the injury requirement altogether by concluding that any use of the plaintiff's server was necessarily an interference with eBay's exclusive right of control, even if the use didn't prevent eBay from doing whatever it wanted with its server. Significantly, it is the second rationale that every subsequent court has pointed to in finding trespass to chattels even absent any possible injury to the chattel itself. [But note that it is no longer good law in California after the Intel v. Hamidi case discussed *infra*.]

Does *eBay*'s analysis of injury make sense? Or is the court really applying the rather different tort of trespass to land, which punishes intermeddling directly whether or not there is injury? See Dan L. Burk, The Trouble with Trespass, 4 J. Sm. & Emerging Bus. L. 27 (2000) (criticizing trespass cases for confusing the two doctrines).

2. Might eBay have anticompetitive motives for preventing Bidder's Edge from truthfully reporting the prices of various goods on different auction sites? If auction aggregators don't exist, it becomes harder for consumers to comparison-shop, and they may just buy from the auction site with the largest brand name — eBay. Law professors opposed to the *eBay* decision made this point on appeal:

> Established online merchants have a substantial incentive to use the doctrine of trespass to interfere with the flow of price and product information on the Internet. The district court's decision places the flow of information within their control, as it allows them to circumscribe the use of tools that gather information and present the results in a usable format. These tools are essential to make the process of gathering information both feasible and useful to the average consumer. Without information

about online alternatives, competition on the Internet will be reduced. Like eBay, many sites may have a motive to prevent such competition. eBay would rather not have a site like Bidder's Edge tell consumers that they can get the same product cheaper at a competing auction site. Similarly, online sellers of music, books, videos, toys and countless other sites that don't have the lowest prices won't want to participate in a price-comparison service. This is especially true of companies like eBay with large market shares, because they may believe that they could shut down a data aggregator altogether by choosing not to participate. Such companies will instead rely on their strong brand name to keep consumers coming to their site—and ignorant of the cheaper prices or better service that may be available elsewhere. The district court's opinion gives these companies the power to block such a price-comparison service altogether. While the promise of ecommerce is to improve consumer information and lower transaction costs, under a trespass theory many of those benefits will disappear.[4]

Is this a fair criticism? Should the doctrine of trespass to chattels take the plaintiff's motive into account? Note that some of the successful applications of trespass to chattels have been against price comparison sites. See Southwest Airlines Co. v. Farechase, Inc., 318 F. Supp. 2d 435 (N.D. Tex. 2004); American Airlines v. Farechase, Inc., No. 067-194022-02 (Tex. Dist. Ct. Mar. 8, 2003).

3. While the trespass to chattels cause of action has only recently been applied to the Internet, a number of recent decisions have used the tort to permit the owners of servers to prevent unwanted access. These cases fall into two basic categories. One set of cases—including *eBay*—involves a website owner's attempt to limit or prevent access to data the website owner has itself made open to the public.[5] In addition to *eBay*, courts in Register.com, Inc. v. Verio, Inc., 356 F.3d 393 (2d Cir. 2004) and Oyster Software, Inc. v. Forms Processing, 2001 WL 1736382 (N.D. Cal. Dec. 6, 2001) have held that a public website can prevent automated access, even to information (such as the WHOIS database in *Verio*) that the website is obligated to make available to the public. *Oyster Software* did not require any proof of burden on the server. Rather, the court followed the second theory in *eBay*, holding that any access to the server deprived the web site owner of its right of dominion. By contrast, the Second Circuit opinion in Register.com affirmed the district court's finding of trespass only because Verio's search "consumed a significant portion of the capacity of Register's computer systems," so that even though "Verio's robots alone would not incapacitate Register's systems . . . it was 'highly probable' that other Internet service providers would devise similar programs to access Register's data, and that the system would be overtaxed and would crash." 356 F.3d at 404.

The second set of cases involves not the collection of data from a website, but the sending of data to a mail server. The first applications of trespass to chattels to the Internet involved spam. See CompuServe, Inc. v. Cyber Promotions, Inc., 962 F. Supp. 1015 (S.D. Ohio 1997); America Online v. National Health Care Discount, Inc., 174 F. Supp. 2d 890 (N.D. Iowa 2001). In these cases, the court found that bulk

---

4. Consumers can obtain price information in the physical world by reading advertisements and catalogs or using existing comparison-shopping products like Consumer Reports. While in theory a store might be legally justified in excluding anyone who works for Consumer Reports from entering their store or buying their products, as a practical matter it would be difficult for them to do so. eBay, by contrast, has found in its trespass theory a perfect mechanism for preventing information-gathering.

5. The protection of information not open to the public—private or password-controlled servers—presents a different set of issues. The Computer Fraud and Abuse Act, 18 U.S.C. § 1030, precludes unauthorized access to a networked computer, and there is no need to apply trespass to preclude such conduct.

e-mail was overwhelming the servers of the plaintiff Internet service providers, and the court concluded that sending such bulk e-mail trespassed on the ISP servers. In Intel Corp. v. Hamidi, 114 Cal. Rptr. 2d 244 (Ct. App. 2001), rev'd (Cal. 2002), the court expanded the rationale of the spam cases by eliminating the requirement of burden on the servers. In that case, the court enjoined Hamidi, a former employee of Intel, from sending e-mails to current Intel employees complaining about the company. While there was clearly no harm to or burden on the servers, the court concluded that injury was not required, and in any event could be met by proof that Intel employees spent company time reading the e-mails.

After *Verio* and *Intel,* can you think of any electronic transmission to a public server that doesn't require advance permission?

The California courts have rejected trespass to chattels claims in the absence of evidence of actual harm to the chattel in question – the computer server. Ticketmaster Corp. v. Tickets.com, 2000 Copyright L. Decs. ¶28,146 (C.D. Cal. Aug. 10, 2000). In *Ticketmaster,* the court concluded that tickets.com was not trespassing on Ticketmaster's server by "scraping" the data from the Ticketmaster server, and then linking directly to places on the Ticketmaster site where users could buy tickets to various events.[6] The court was skeptical of the entire theory of trespass to chattels as applied to the Internet, but it distinguished *eBay* on the ground that Ticketmaster could not show actual harm to its servers. And in Intel Corp. v. Hamidi, 30 Cal.4th 1342, 71 P.3d 296, 1 Cal.Rptr.3d 32 (2003), the court limited the rationale of the spam trespass cases by requiring the plaintiff to prove damage to or dispossession of the chattel. In that case, the trial court enjoined Hamidi, a former employee of Intel, from sending e-mails to current Intel employees complaining about the company. The California Supreme Court reversed, noting that there was clearly no harm to or burden on the servers — they were not shut down and Intel's e-mail was not delayed. The court held that the requirement for injury could not be met merely by proof that Intel employees spent company time reading the e-mails.

> After reviewing the decisions analyzing unauthorized electronic contact with computer systems as potential trespasses to chattels, we conclude that under California law the tort does not encompass, and should not be extended to encompass, an electronic communication that neither damages the recipient computer system nor impairs its functioning. Such an electronic communication does not constitute an actionable trespass to personal property, i.e., the computer system, because it does not interfere with the possessor's use or possession of, or any other legally protected interest in, the personal property itself. The consequential economic damage Intel claims to have suffered, i.e., loss of productivity caused by employees reading and reacting to Hamidi's messages and company efforts to block the messages, is not an injury to the company's interest in its computers—which worked as intended and were unharmed by the communications—any more than the personal distress caused by reading an unpleasant letter would be an injury to the recipient's mailbox, or the loss of privacy caused by an intrusive telephone call would be an injury to the recipient's telephone equipment.
>
> Our conclusion does not rest on any special immunity for communications by electronic mail; we do not hold that messages transmitted through the Internet are exempt from the ordinary rules of tort liability. To the contrary, e-mail, like other forms of communication, may in some circumstances cause legally cognizable injury to

---

6. Presumably the linking claim would have to be based on some theory of contributory trespass, since it is Ticketmaster customers rather than tickets.com who are actually sending data to the Ticketmaster site.

the recipient or to third parties and may be actionable under various common law or statutory theories. Indeed, on facts somewhat similar to those here, a company or its employees might be able to plead causes of action for interference with prospective economic relations, see Guillory v. Godfrey 134 Cal. App. 2d 628, 630-632, 286 P.2d 474 (1955) (defendant berated customers and prospective customers of plaintiffs' cafe with disparaging and racist comments), interference with contract, see Blender v. Superior Court 55 Cal. App. 2d 24, 25-27, 130 P.2d 179 (1942) (defendant made false statements about plaintiff to his employer, resulting in plaintiff's discharge) or intentional infliction of emotional distress, see Kiseskey v. Carpenters' Trust for So. California 144 Cal.App.3d 222, 229-230, 192 Cal. Rptr. 492 (1983) (agents of defendant union threatened life, health, and family of employer if he did not sign agreement with union). And, of course, as with any other means of publication, third party subjects of e-mail communications may under appropriate facts make claims for defamation, publication of private facts, or other speech-based torts. See, e.g., Southridge Capital Management v. Lowry, 188 F.Supp.2d 388, 394-396 (S.D.N.Y.2002) (allegedly false statements in e-mail sent to several of plaintiff's clients support actions for defamation and interference with contract). Intel's claim fails not because e-mail transmitted through the Internet enjoys unique immunity, but because the trespass to chattels tort—unlike the causes of action just mentioned—may not, in California, be proved without evidence of an injury to the plaintiff's personal property or legal interest therein.

Nor does our holding affect the legal remedies of Internet service providers (ISP's) against senders of unsolicited commercial bulk e-mail (UCE), also known as "spam." See Ferguson v. Friendfinders, Inc., 94 Cal. App. 4th 1255, 1267, 115 Cal. Rptr. 2d 258 (2002). A series of federal district court decisions, beginning with CompuServe, Inc. v. Cyber Promotions, Inc. 962 F.Supp. 1015 (S.D. Ohio 1997), has approved the use of trespass to chattels as a theory of spammers' liability to ISP's, based upon evidence that the vast quantities of mail sent by spammers both overburdened the ISP's own computers and made the entire computer system harder to use for recipients, the ISP's customers. See id. at pp. 1022-1023. In those cases, discussed in greater detail below, the underlying complaint was that the extraordinary *quantity* of UCE impaired the computer system's functioning. In the present case, the claimed injury is located in the disruption or distraction caused to recipients by the *contents* of the e-mail messages, an injury entirely separate from, and not directly affecting, the possession or value of personal property.

The court went on to reject a policy argument that trespass law should be expanded by treating Web sites like real property:

> We discuss this debate among the amici curiae and academic writers only to note its existence and contours, not to attempt its resolution. Creating an absolute property right to exclude undesired communications from one's e-mail and Web servers might help force spammers to internalize the costs they impose on ISP's and their customers. But such a property rule might also create substantial new costs, to e-mail and e-commerce users and to society generally, in lost ease and openness of communication and in lost network benefits. In light of the unresolved controversy, we would be acting rashly to adopt a rule treating computer servers as real property for purposes of trespass law.

Justice Brown dissented. She took a broad view of Intel's property rights:

> Intel has invested millions of dollars to develop and maintain a computer system. It did this not to act as a public forum but to enhance the productivity of its employ-

ees. Kourosh Kenneth Hamidi sent as many as 200,000 e-mail messages to Intel employees. The time required to review and delete Hamidi's messages diverted employees from productive tasks and undermined the utility of the computer system. "There may ... be situations in which the value to the owner of a particular type of chattel may be impaired by dealing with it in a manner that does not affect its physical condition." Rest.2d Torts, §218, com. h, p. 422. This is such a case.

The majority repeatedly asserts that Intel objected to the hundreds of thousands of messages solely due to their content, and proposes that Intel seek relief by pleading content-based speech torts. This proposal misses the point that Intel's objection is directed not toward Hamidi's message but his use of Intel's property to display his message. Intel has not sought to prevent Hamidi from expressing his ideas on his Web site, through private mail (paper or electronic) to employees' homes, or through any other means like picketing or billboards. But as counsel for Intel explained during oral argument, the company objects to Hamidi's using Intel's property to advance his message.

Of course, Intel deserves an injunction even if its objections are based entirely on the e-mail's content. Intel is entitled, for example, to allow employees use of the Internet to check stock market tables or weather forecasts without incurring any concomitant obligation to allow access to pornographic Web sites.

4. As the *Intel* court suggests, many of the cases decided under the rubric of trespass to chattels could also be decided under other statutes. For example, many state laws now ban spam. See, e.g., Cal. Bus. & Prof. Code §17538.4. And many of the data cases, such as *eBay* and *Ticketmaster,* are really efforts to protect uncopyrightable data by creating a new form of intellectual property law. The intuitive case for protection of data compilations may motivate courts to rule for the plaintiffs on a trespass theory in order to effectively create a form of intellectual property for data protection. Does this result make sense? Or should the courts be careful to avoid undermining copyright law's limits? Note that copyright law may preempt state torts that create copyright-like rights within the constitutional field of copyright. See *Ticketmaster, supra,* in which the court held various state torts preempted by federal copyright law. For a fuller discussion of copyright preemption, see *infra* section B.

5. The broad application of trespass to the Internet effectively creates a consent standard for using the Internet. Websites and even e-mail recipients have an effective veto over the sending or collecting of data. Rather than simply deciding to make information open to the public or keep it private, a web site owner can declare that information is public but that particular specified individuals are barred from using it.

Understanding the trespass cases in this light puts the focus on how those declarations are made and communicated. In some cases, such as *eBay,* this is not much of an issue: it was quite clear to Bidder's Edge that eBay objected to its collection of data from the eBay website. But in other cases, courts have held that terms and conditions placed on a website—or even the use of an automated Robot Exclusion Header—constitute sufficient notice that a particular user is unwanted.

Does it make sense to give a website owner the right to post an announcement excluding certain users or anyone who would make certain uses? The alternative is for the website owner to rely on technical blocking alternatives. For example, after Ticketmaster lost its case, it reconfigured its website to block deep links from tickets.com. Should the courts stay out of such disputes and let the parties engage in their own technological competition? Or will the result be a wasteful technological "arms race"? David McGowan argues that website owners should have such a right. He doesn't worry that giving them such a right will interfere with the operation of the Internet because the vast majority of them will have incentives to permit most "trespasses" to their server by search engines, consumers, and others, although likely not by competitors. See David McGowan, Website Access: The Case for Consent, 35 Loy. U. Chi. L. Rev. 341 (2003).

6. The concept of trespass in cyberspace depends heavily on a conception of a web site or mail server as "property" from which, like land, the owner ought to have the right to exclude others. If a Website is like land, the trespass to land rules (which don't require proof of injury to the land for liability) seem like the ones to apply. But a number of commentators have challenged the metaphor of an Internet site as a "place" akin to land. See, e.g., Dan Hunter, Cyberspace as Place and the Tragedy of the Digital Anticommons, 91 Cal. L. Rev. 439 (2003); Mark A. Lemley, Place and Cyberspace, 91 Cal. L. Rev. 521 (2003); Maureen A. O'Rourke, Property Rights and Competition on the Internet: In Search of an Appropriate Analogy, 16 Berkeley Tech. L.J. 561 (2001). Hunter suggests that the metaphor is psychologically quite powerful, but troubling as a policy matter. Lemley argues that courts can and should focus on the ways the Internet is *not* like land. He also points out that treating the Internet like land does not necessarily mean that the owner should get an absolute right to exclude others. In the real world, the law gives many different sorts of rights of varying strengths to land owners. Similarly, a court might decide that an absolute right of exclusion is not the right legal analogy.

Are property principles relevant to the Internet at all? If so, how should a court decide which analogy is the right one to apply?

PROBLEM

Problem 6-4. Infoplex is a search engine. In order to run a comprehensive search engine, Infoplex must collect and index data from over 1 billion websites. In order to keep its search engine up-to-date, Infoplex must revisit those pages at least once a week, and often more frequently. To do this, Infoplex uses automated software robots called "spiders." These spiders crawl the web, automatically downloading all of the information on all of the websites it can find. Because of the practical impossibility of entering into contracts with the owners of all these websites — among other things, millions of new pages are added each year — Infoplex has configured its spiders to collect any data it can from any site that is open to the public.

Do any of the following plaintiffs have a claim against Infoplex for trespass to chattels based on its indexing of their sites?

   a)   Site A, which has a written "terms of use" statement precluding any collection of all or substantially all of its data by any for-profit company.

> b) Site B, which uses a Robot Exclusion Header in the source code of its web page. Such a header signals to robots that they are not welcome on the site. Not all robots recognize or respect such a header, though, and Infoplex has configured its robots to ignore such headers.
>
> c) Site C, an Internet service provider which has sent a letter to the general counsel of Infoplex demanding that Infoplex pay it $1,000 per month for the right to index any of its customers' pages.

B. FEDERAL PREEMPTION

1. Patent Preemption

Kewanee Oil Co. v. Bicron Corp.
Supreme Court of the United States
416 U.S. 470 (1974)

Mr. Chief Justice BURGER delivered the opinion of the Court.

We granted certiorari to resolve a question on which there is a conflict in the courts of appeals: whether state trade secret protection is pre-empted by operation of the federal patent law. In the instant case the Court of Appeals for the Sixth Circuit held that there was preemption. The Courts of Appeals for the Second, Fourth, Fifth, and Ninth Circuits have reached the opposite conclusion. . . .

Petitioner brought this diversity action in United States District Court for the Northern District of Ohio seeking injunctive relief and damages for the misappropriation of trade secrets. The district Court, applying Ohio trade secret law, granted a permanent injunction against the disclosure or use by respondents of 20 of the 40 claimed trade secrets until such time as the trade secrets had been released to the public, had otherwise generally become available to the public, or had been obtained by respondents from sources having the legal right to convey the information.

The Court of Appeals for the Sixth Circuit held that the findings of fact by the District Court were not clearly erroneous, and that it was evident from the record that the individual respondents appropriated to the benefit of Bicron secret information on processes obtained while they were employees at Harshaw. Further, the Court of Appeals held that the District Court properly applied Ohio law relating to trade secrets. Nevertheless, the Court of Appeals reversed the District Court, finding Ohio's trade secret law to be in conflict with the patent laws of the United States. The Court of Appeals reasoned that Ohio could not grant monopoly protection to processes and manufacturing techniques that were appropriate subjects for consideration under 35 U.S.C. § 101 for a federal patent but which had been in commercial use for over one year and so were no longer eligible for patent protection under 35 U.S.C. § 102(b).

We hold that Ohio's law of trade secrets is not preempted by the patent laws of the United States, and accordingly, we reverse. . . .

III

The first issue we deal with is whether the States are forbidden to act at all in the area of protection of the kinds of intellectual property which may make up the subject matter of trade secrets.

Article I, §8, cl. 8, of the Constitution grants to the Congress the power

> [t]o promote the Progress of Science and useful Arts, by securing for limited Times to Authors and Inventors the exclusive Right to their respective Writings and Discoveries....

In the 1972 Term, in Goldstein v. California, 412 U.S. 546 (1973), we held that the cl. 8 grant of power to Congress was not exclusive and that, at least in the case of writings, the States were not prohibited from encouraging and protecting the efforts of those within their borders by appropriate legislation. The States could, therefore, protect against the unauthorized rerecording for sale of performances fixed on records or tapes, even though those performances qualified as "writings" in the constitutional sense and Congress was empowered to legislate regarding such performances and could pre-empt the area if it chose to do so. This determination was premised on the great diversity of interests in our Nation — the essentially nonuniform character of the appreciation of intellectual achievements in the various States. Evidence for this came from patents granted by the States in the 18th century. 412 U.S., at 557.

Just as the States may exercise regulatory power over writings so may the States regulate with respect to discoveries. States may hold diverse viewpoints in protecting intellectual property relating to invention as they do in protecting the intellectual property relating to the subject matter of copyright. The only limitation on the States is that in regulating the area of patents and copyrights they do not conflict with the operation of the laws in this area passed by Congress, and it is to that more difficult question we now turn.

IV

The question of whether the trade secret law of Ohio is void under the Supremacy Clause involves a consideration of whether that law "stands as an obstacle to the accomplishment and execution of the full purposes and objectives of Congress." Hines v. Davidowitz, 312 U.S. 52, 67 (1941). See Florida Avocado Growers v. Paul, 373 U.S. 132, 141 (1963). We stated in Sears, Roebuck & Co. v. Stiffel Co., 376 U.S. 225, 229 (1964), that when state law touches upon the area of federal statutes enacted pursuant to constitutional authority, "it is 'familiar doctrine' that the federal policy 'may not be set at naught, or its benefits denied' by the state law. Sola Elec. Co v. Jefferson Elec. Co., 317 U.S. 173, 176 (1942). This is true, of course, even if the state law is enacted in the exercise of otherwise undoubted state power." ...

The stated objective of the Constitution in granting the power to Congress to legislate in the area of intellectual property is to "promote the Progress of Science and useful Arts." The patent laws promote this progress by offering a right of exclusion for a limited period as an incentive to inventors to risk the often enormous costs in terms of time, research, and development....

The maintenance of standards of commercial ethics and the encouragement of invention are the broadly stated policies behind trade secret law. "The necessity of good faith and honest, fair dealing, is the very life and spirit of the commercial world."...

As we noted earlier, trade secret law protects items which would not be proper subjects for consideration for patent protection under 35 U.S.C. § 101. As in the case of the recordings in Goldstein v. California, Congress, with respect to nonpatentable subject matter, "has drawn no balance; rather, it has left the area unattended, and no reason exists why the State should not be free to act." Goldstein v. California, supra, at 570 (footnote omitted).

Since no patent is available for a discovery, however useful, novel, and nonobvious, unless it falls within one of the express categories of patentable subject matter of 35 U.S.C. § 101, the holder of such a discovery would have no reason to apply for a patent whether trade secret protection existed or not. Abolition of trade secret protection would, therefore, not result in increased disclosure to the public of discoveries in the area of nonpatentable subject matter....

Congress has spoken in the area of those discoveries which fall within one of the categories of patentable subject matter of 35 U.S.C. § 101 and which are, therefore, of a nature that would be subject to consideration for a patent. Processes, machines, manufactures, compositions of matter, and improvements thereof, which meet the tests of utility, novelty, and nonobviousness are entitled to be patented, but those which do not, are not. The question remains whether those items which are proper subjects for consideration for a patent may also have available the alternative protection accorded by trade secret law.

Certainly the patent policy of encouraging invention is not disturbed by the existence of another form of incentive to invention. In this respect the two systems are not and never would be in conflict. Similarly, the policy that matter once in the public domain must remain in the public domain is not incompatible with the existence of trade secret protection. By definition a trade secret has not been placed in the public domain....

...Trade secret law will encourage invention in areas where patent law does not reach, and will prompt the independent innovator to proceed with the discovery and exploitation of his invention. Competition is fostered and the public is not deprived of the use of valuable, if not quite patentable, invention....

The final category of patentable subject matter to deal with is the clearly patentable invention, i.e., that invention which the owner believes to meet the standards of patentability. It is here that the federal interest in disclosure is at its peak....

Trade secret law provides far weaker protection in many respects than the patent law. While trade secret law does not forbid the discovery of the trade secret by fair and honest means, e.g., independent creation or reverse engineering, patent law operates "against the world," forbidding any use of the invention for whatever purpose for a significant length of time. The holder of a trade secret also takes a substantial risk that the secret will be passed on to his competitors, by theft or by breach of a confidential relationship, in a manner not easily susceptible of discovery or proof. Painton & Co. v. Bourns, Inc., 442 F.2d, at 224. Where patent law acts as a barrier, trade secret law functions relatively as a sieve. The possibility that an inventor who believes his invention meets the standards of patentability will sit back, rely on trade secret law, and after one year of use forfeit any right to patent protection, 35 U.S.C. § 102(b), is remote indeed.

Nor does society face much risk that scientific or technological progress will be impeded by the rare inventor with a patentable invention who chooses trade secret protection over patent protection. The ripeness-of-time concept of invention, developed from the study of the many independent multiple discoveries in history, predicts that if a particular individual had not made a particular discovery others would have, and in probably a relatively short period of time. If something is to be discovered at all very likely it will be discovered by more than one person....

... Trade secret law and patent law have co-existed in this country for over one hundred years. Each has its particular role to play, and the operation of one does not take away from the need for the other.... Congress, by its silence over these many years, has seen the wisdom of allowing the States to enforce trade secret protection. Until Congress takes affirmative action to the contrary, States should be free to grant protection to trade secrets....

Mr. Justice DOUGLAS, with whom Mr. Justice BRENNAN concurs, dissenting.

Today's decision is at war with the philosophy of Sears, Roebuck & Co. v. Stiffel Co., 376 U.S. 225, and Compco Corp. v. Day-Brite Lighting, Inc., 376 U.S. 234. Those cases involved patents — one of a pole lamp and one of fluorescent lighting fixtures — each of which was declared invalid. The lower courts held, however, that though the patents were invalid the sale of identical or confusingly similar products to the products of the patentees violated state unfair competition laws. We held that when an article is unprotected by a patent, state law may not forbid others to copy it, because every article not covered by a valid patent is in the public domain. Congress in the patent laws decided that where no patent existed, free competition should prevail; that where a patent is rightfully issued, the right to exclude others should obtain for no longer than 17 years, and that the States may not "under some other law, such as that forbidding unfair competition, give protection of a kind that clashes with the objectives of the federal patent laws," 376 U.S., at 231....

The conflict with the patent laws is obvious. The decision of Congress to adopt a patent system was based on the idea that there will be much more innovation if discoveries are disclosed and patented than there will be when everyone works in secret. Society thus fosters a free exchange of technological information at the cost of a limited 17-year monopoly....

A suit to redress theft of a trade secret is grounded in tort damages for breach of a contract — a historic remedy, Cataphote Corp. v. Hudson, 422 F.2d 1290. Damages for breach of a confidential relation are not pre-empted by this patent law, but an injunction against use is pre-empted because the patent law states the only monopoly over trade secrets that is enforceable by specific performance; and that monopoly exacts as a price full disclosure. A trade secret can be protected only by being kept secret. Damages for breach of a contract are one thing; an injunction barring disclosure does service for the protection accorded valid patents and is therefore pre-empted....

COMMENTS AND QUESTIONS

1. The rule set forth by Sears Roebuck & Co v. Stiffel Co., 376 U.S. 225 (1964) was fairly clear: Patent law reflects a compromise between the goal of

promoting innovation and the danger of condoning monopoly. Supplementing the scope of patent law may upset that balance, and is therefore prohibited. Supplementing *enforcement* of the federal intellectual property laws was condoned in dictum at the end of *Compco,* a companion case, and in Justice Harlan's concurrence in both cases. After these cases, state law served a very limited function in the scheme of intellectual property protection. States could work to further the goals of federal protection, but they had to work within the parameters set down by federal law.

The *Kewanee* opinion takes a remarkably different tack. Chief Justice Burger's opinion for the Court emphasizes only one of the two policies shaping the patent laws: the goal of promoting innovation. The opinion does not discuss the dangers intellectual property protection poses for free competition. As a result, the *Kewanee* Court finds no problem with trade secret protection that extends beyond the scope of the patent laws. Note that the Court seems to approve not only state laws that protect nonpatentable subject matter (an area in which it could be argued that the federal government has no interest),[7] but also the protection of inventions not patentable for some other reason (i.e., suppression, misuse, lack of novelty, or obviousness).

Is *Kewanee* reconcilable with *Sears?* The *Kewanee* court did not overrule *Sears* or *Compco;* indeed, it cited them in support of its holding. Thus state laws preventing copying were treated differently from trade secret laws after *Kewanee.* The latter, although broader in scope (they prevented far more than just outright copying of products), were permissible; the former were not. For an argument that *Kewanee* "closed the circle" on the open-ended preemption analysis of *Sears* and *Compco,* see Paul Goldstein, Kewanee Oil Co. v. Bicron Corp.: Notes on a Closing Circle, 1974 Sup. Ct. Rev. 81 (1974).

2. Is preemption a good idea? That depends on what you think of the balance the federal laws have struck. If you are more concerned about injury to competition by conferring monopoly rights on patentees, you are likely to favor the result in *Sears.* If on the other hand you think that innovation is underrewarded, it is reasonable to oppose federal preemption. One way to reconcile these cases may be to read *Sears* and *Compco* as expressing a federal policy in favor of reverse engineering of products in the public domain. If that is the overarching federal goal, it is logical to strike down the laws in *Sears* and *Compco* but not *Kewanee,* since trade secrets statutes (unlike the unfair competition laws we have discussed) generally allow reverse engineering. This result is also consistent with the reading of trade secret laws as merely an application of tort and contract principles.

3. Recall the Court's statement: "Where patent law acts as a barrier, trade secret law functions relatively as a sieve." Does this square with recent empirical evidence, such as the studies referred to in Chapters 1 and 2 that show trade secret protection to be the preferred form of protection in some industries? For more on the "election" of trade secret vs. patent protection, see the Note on "Electing" Trade Secrets or Patents, following the *Bonito Boats* case.

---

7. Even in this area, though, a federal interest may be discerned. If Congress has declared some subject matter unpatentable, that could reflect a federal determination that that matter is unworthy of protection, a determination that state law should not be allowed to upset.

Bonito Boats, Inc. v. Thunder Craft Boats, Inc.
Supreme Court of the United States
489 U.S. 141 (1989)

Justice O'CONNOR delivered the opinion of the Court....

I

In September 1976, petitioner Bonito Boats, Inc., a Florida corporation, developed a hull design for a fiberglass recreational boat which it marketed under the trade name Bonito Boat Model 5VBR. Designing the boat hull required substantial effort on the part of Bonito. A set of engineering drawings was prepared, from which a hardwood model was created. The hardwood model was then sprayed with fiberglass to create a mold, which then served to produce the finished fiberglass boats for sale. The 5VBR was placed on the market sometime in September 1976. There is no indication in the record that a patent application was ever filed for protection of the utilitarian or design aspects of the hull, or for the process by which the hull was manufactured. The 5VBR was favorably received by the boating public, and "a broad interstate market" developed for its sale.

In May 1983, after the Bonito 5VBR had been available to the public for over six years, the Florida Legislature enacted Fla. Stat. § 559.94 (1987). The statute makes "it . . . unlawful for any person to use the direct molding process to duplicate for the purpose of sale any manufactured vessel hull or component part of a vessel made by another without the written permission of that other person." § 559.94(2). The statute also makes it unlawful for a person to "knowingly sell a vessel hull or component part of a vessel duplicated in violation of subsection (2)." § 559.94(3). Damages, injunctive relief, and attorney's fees are made available to "any person who suffers injury or damage as the result of a violation" of the statute. § 559.94(4). The statute was made applicable to vessel hulls or component parts duplicated through the use of direct molding after July 1, 1983. § 559.94(5).

On December 21, 1984, Bonito filed this action in the Circuit Court of Orange County, Florida. The complaint alleged that respondent here, Thunder Craft Boats, Inc., a Tennessee corporation, had violated the Florida statute by using the direct molding process to duplicate the Bonito 5VBR fiberglass hull, and had knowingly sold such duplicates in violation of the Florida statute....

III

We believe that the Florida statute at issue in this case so substantially impedes the public use of the otherwise unprotected design and utilitarian ideas embodied in unpatented boat hulls as to run afoul of the teaching of our decisions in *Sears* and *Compco*. It is readily apparent that the Florida statute does not operate to prohibit "unfair competition" in the usual sense that the term is understood. The law of unfair competition has its roots in the common-law tort of deceit: its general concern is with protecting *consumers* from confusion as to source....

In contrast to the operation of unfair competition law, the Florida statute is aimed directly at preventing the exploitation of the design and utilitarian conceptions embodied in the product itself. The sparse legislative history surrounding its enactment indicates that it was intended to create an inducement for the improvement of boat hull designs. See Tr. of Meeting of Transportation Committee, Florida House of Representatives, May 3, 1983 ("There is no inducement for [a] quality boat man-ufacturer to improve these designs and secondly, if he does, it is immediately copied. This would prevent that and allow him recourse in circuit court"). To accomplish this goal, the Florida statute endows the original boat hull manufacturer with rights against the world, similar in scope and operation to the rights accorded a federal patentee. Like the patentee, the beneficiary of the Florida statute may prevent a competitor from "making" the product in what is evidently the most efficient manner available and from "selling" the product when it is produced in that fashion. Compare 35 U.S.C. § 154. The Florida scheme offers this protection for an unlimited number of years to all boat hulls and their component parts, without regard to their orna-mental or technological merit. Protection is available for subject matter for which patent protection has been denied or has expired, as well as for designs which have been freely revealed to the consuming public by their creators. That the Florida statute does not remove all means of reproduction and sale does not eliminate the conflict with the federal scheme. See *Kellogg*, 305 U.S., at 122. In essence, the Florida law prohibits the entire public from engaging in a form of reverse engineering of a product in the public domain. This is clearly one of the rights vested in the federal patent holder, but has never been a part of state protection under the law of unfair competi-tion or trade secrets. . . .

Moreover, as we noted in *Kewanee*, the competitive reality of reverse engineering may act as a spur to the inventor, creating an incentive to develop inventions that meet the rigorous requirements of patentability. 416 U.S., at 489-490. The Florida statute substantially reduces this competitive incentive, thus eroding the general rule of free competition upon which the attractiveness of the federal patent bargain depends. . . . The Florida statute is aimed directly at the promotion of intellectual crea-tion by substantially restricting the public's ability to exploit ideas that the patent system mandates shall be free for all to use. Like the interpretation of Illinois unfair competition law in *Sears* and *Compco*, the Florida statute represents a break with the tradition of peaceful coexistence between state market regulation and federal patent policy. The Florida law substantially restricts the public's ability to exploit an unpatented design in general circulation, raising the specter of state-created monopolies in a host of useful shapes and processes for which patent protection has been denied or is otherwise unob-tainable. It thus enters a field of regulation which the patent laws have reserved to Congress. The patent statute's careful balance between public right and private mono-poly to promote certain creative activity is a "scheme of federal regulation . . . so perva-sive as to make reasonable the inference that Congress left no room for the States to supplement it." Rice v. Santa Fe Elevator Corp., 331 U.S. 218, 230 (1947). . . .

COMMENTS AND QUESTIONS

1. Does this result make sense? It is certainly consistent with *Sears* and *Compco;* less so with *Kewanee. Bonito* can be reconciled with trade secrets statutes, though, if one accepts the reverse engineering rationale described above. Whatever one thinks of the *Bonito* result, the last sentence remains troubling. Characterizing patent law

as "pervasive federal regulation" suggests that it might preempt the field, automatically striking down all state laws that attempt to regulate intellectual property. If taken seriously, that approach would leave no room at all for state protection of inventions.

2. Note the similarity between the statutes struck down in *Sears* and *Bonito Boats*. In both cases, what was prohibited was the direct copying of a competitor's design. The *Bonito Boats* statute is more limited than that in *Sears*, since it prohibits only one particular method of copying. Nonetheless, the Court struck it down. Why should the courts be concerned about these comparatively narrow statutes when they allowed the far broader statute in *Kewanee* to pass muster?

3. The *Bonito Boats* decision has been roundly criticized. See., e.g., John S. Wiley, Jr., Bonito Boats: Uninformed but Mandatory Federal Innovation Policy, 1989 Sup. Ct. Rev. 283. Cf. Symposium, Product Simulation: A Right or a Wrong?, 64 Colum. L. Rev. 1178 (1964) (articles criticizing the analogous decisions in *Sears* and *Compco*).

A different form of criticism of the result in *Bonito Boats* was offered by Congress in 1998. As part of the Digital Millenium Copyright Act, Congress created a new federal intellectual property right protecting "original" boat hull designs. 17 U.S.C. § 1301 et seq. We discuss this statute in more detail in Chapter 4, *supra*. Can Congress lawfully accomplish here what the states cannot? Does *Bonito Boats* suggest some sort of constitutional limitation on any form of protection (state or federal) in this area?

4. Some lawyers have tried to apply *Bonito Boats* in contexts beyond the patent-like legislation actually considered in the case. This has not met with much success, however; like other broad preemption decisions before it, *Bonito Boats* has been hemmed in by subsequent qualifications. In Waits v. Frito-Lay, Inc., 978 F.2d 1093 (9th Cir. 1992), *cert. denied*, 506 U.S. 1080 (1993), a right of publicity case with facts remarkably similar to the *Midler* case reproduced above, the Ninth Circuit had the following to say about *Bonito Boats*:

> Bonito Boats involved a Florida statute giving perpetual patent-like protection to boat hull designs already on the market, a class of manufactured articles expressly excluded from federal patent protection. The Court ruled that the Florida statute was preempted by federal patent law because it directly conflicted with the comprehensive federal patent scheme. In reaching this conclusion, the Court cited its earlier decisions in Sears Roebuck & Co. v. Stiffel Co., 376 U.S. 225 (1964), and Compco Corp. v. Day-Brite Lighting, 376 U.S. 234 (1964), for the proposition that "publicly known design and utilitarian ideas which were unprotected by patent occupied much the same position as the subject matter of an expired patent," i.e., they are expressly unprotected. Bonito Boats, 489 U.S. at 152.
>
> The defendants seize upon this citation to Sears and Compco as a reaffirmation of the sweeping preemption principles for which these cases were once read to stand. They argue that Midler was wrongly decided because it ignores these two decisions, an omission that the defendants say indicates an erroneous assumption that Sears and Compco have been "relegated to the constitutional junkyard." Thus, the defendants go on to reason, earlier cases that rejected entertainers' challenges to imitations of their performances based on federal copyright preemption, were correctly decided because they relied on Sears and Compco. See Sinatra v. Goodyear Tire & Rubber Co., 435 F.2d 711, 716-18 (9th Cir. 1970), *cert. denied*, 402 U.S. 906 (1971); Booth v. Colgate-Palmolive Co., 362 F. Supp. 343, 348 (S.D.N.Y. 1973); Davis v. Trans World Airlines, 297 F. Supp. 1145, 1147 (C.D. Cal. 1969). This reasoning suffers from a number of flaws.

Bonito Boats itself cautions against reading Sears and Compco for a "broad pre-emptive principle" and cites subsequent Supreme Court decisions retreating from such a sweeping interpretation. "[T]he Patent and Copyright Clauses do not, by their own force or by negative implication, deprive the States of the power to adopt rules for the promotion of intellectual creation." Bonito Boats, 489 U.S. at 165 (citing, inter alia, Goldstein v. California, 412 U.S. 546, 552-61 (1973) and Kewanee Oil Co. v. Bicron Corp., 416 U.S. 470, 478-79 (1974)). Instead, the Court reaffirmed the right of states to "place limited regulations on the use of unpatented designs in order to prevent consumer confusion as to source." Id. Bonito Boats thus cannot be read as endorsing or resurrecting the broad reading of Compco and Sears urged by the defendants, under which Waits' state tort claim arguably would be preempted.

Moreover, the Court itself recognized the authority of states to protect entertainers' "right of publicity" in Zacchini v. Scripps-Howard Broadcasting Co., 433 U.S. 562 (1977). In Zacchini, the Court endorsed a state right-of-publicity law as in harmony with federal patent and copyright law, holding that an unconsented-to television news broadcast of a commercial entertainer's performance was not protected by the First Amendment. Id. at 573, 576-78. The cases Frito asserts were "rightly decided" all pre-date Zacchini and other Supreme Court precedent narrowing Sears' and Compco's sweeping preemption principles. In sum, our holding in Midler, upon which Waits' voice misappropriation claim rests, has not been eroded by subsequent authority.

Recent Federal Circuit cases have shown a willingness to permit state claims that tread on the area of patent law, particularly where the claims are made against patent owners rather than in an effort to create a new form of intellectual property protection. See, e.g., Dow Chemical v. Exxon Corp., 139 F.3d 1470 (Fed. Cir. 1998) (unfair competition claims by accused infringer based on alleged inequitable conduct not preempted); Univ. of Colorado Found. v. American Cyanamid, Inc., 196 F.3d 1366 (Fed. Cir. 1999) (state claims of fraud and unjust enrichment against patentee who stole invention from plaintiff not preempted). The court has, however, required that state claims that touch on areas of federal patent interest be judged by federal, not state, standards. See *Univ. of Colorado, supra*; Midwest Industries v. Karavan Trailers, 175 F.3d 1356 (Fed. Cir. 1999). Is this a reasonable compromise?

5. In a number of cases, restrictions on the content of licensing contracts are said to raise preemption issues. One might question whether contracts are equivalent to state statutes for preemption purposes. On at least one occasion, the Supreme Court has suggested that protection of unpatentable goods (which would be prohibited under state misappropriation statutes) is permissible under contract law. See, e.g., Aronson v. Quick Point Pencil, 440 U.S. 257 (1979). On the other hand, there are a variety of circumstances in which federal patent policy simply precludes the parties from contracting to the contrary. See, e.g., Brulotte v. Thys Co., 379 U.S. 29 (1964) (holding unenforceable patent licensed in agreement that extends beyond patent's then 17-year term); Lear, Inc. v. Adkins, 395 U.S. 653 (1969) (contracts estopping licensee from challenging the validity of a patent are void); Everex Systems Inc. v. Cadtrak Corp., 89 F.3d 673 (9th Cir. 1996) (federal policy precluding assignment of nonexclusive patent licenses prevailed over state doctrine permitting such assignments). Note that in some of these cases, such as *Everex*, federal preemption actually works to the benefit of the intellectual property owner.

Regardless of the particular limitations federal law imposes on licensing agreements, patent licensing in general is a question of state (not federal) law. The Federal

Circuit reaffirmed this in Gjerlov v. Schuyler Labs. Inc., 131 F.3d 1016 (Fed. Cir. 1997).

6. What is the prevailing rule regarding patent preemption? Can you articulate a workable guideline for distinguishing permissible from impermissible state laws?

Note on "Electing" Trade Secrets or Patents

The economic significance of the trade secret/patent tradeoff—which might be deemed the "doctrine of quasi-election"—is discussed in Friedman, Landes, and Posner, Some Economics of Trade Secret Law, 5 J. Econ. Persp. 61 (1991). The authors argue that trade secret law protects inventions that are not worth patenting, either because they are too trivial (and hence will fail 35 U.S.C. § 103) or because they are not worth the cost of patenting. Id. at 65-66. Moreover, they make an interesting point regarding the "race to invent/patent" literature: trade secret protection is better than patent protection in that it does not encourage over-rapid invention in a regime where the cost of invention is expected to fall over time. This takes some of the pressure off the racing/rent dissipation models that have become popular in the economic literature on patents in recent years. But there is more to the story than a simple election. As Paul Goldstein pointed out some time ago, it is anomalous to speak of state law "diverting" inventors away from patentable research; at the outset of the research, it is impossible to say whether it will be patentable. See Paul Goldstein, Kewanee Oil Co. v. Bicron Corp.: Notes on a Closing Circle, 1974 Sup. Ct. Rev. 81, 91-92 (1974).

This point bears heavily on one of the *Bonito Boats* court's justifications for its holding. According to the Supreme Court, state laws that protect the same subject matter as is protected by federal patent law have the potential to redirect inventive efforts away from potentially patentable research. Presumably this might "distort" research in directions contrary to those desired by Congress when it passed the Patent Act. As a consequence, it is implicit in the scheme of the federal patent system that state protection not get "too close," because this might divert research that would have been directed toward potentially patentable inventions. Here we refer to this as the "crowding out" rationale. The problem with this thinking is that it assumes that when researchers make decisions about where to invest they choose between two mutually exclusive types of research: potentially patentable and non-patentable. The Supreme Court envisions a world where inventors might "slack off," stopping their research efforts short of what is needed to achieve a patent, if they can get "equivalent" protection for less ambitious research. If such protection is available, they will invest in less ambitious, presumably "safer" projects; this will reduce the amount of "patent-oriented" research and cut into the volume of patents as well. This view of things also assumes that Congress has determined that the extra effort required to make something patentable makes that thing more valuable as well.

Are any of these assumptions true? Only the last one, perhaps; Congress does ask something significant of an inventor who would obtain a patent, whereas state protection schemes often require something more modest, as in *Bonito Boats*. As to the other assumptions, it appears that most researchers do research first and then consider various legal options for protecting it. One reason they do this is clear: *it is impossible in most cases to tell ex ante whether a particular project or project type is of the "patentable" variety or not.* Only when the researcher has the initial experimental results in hand does she turn to the patent attorney for guidance. Only at this point does she know whether she has something patentable or not. Although many researchers take patents into account in deciding which projects to pursue, it is not a major factor. Nor can it be, since patents are granted for results, not plans.

Seen in this light, state intellectual property laws are not a threat for the reasons identified by the Court. In fact, state protection might actually increase the amount of research firms take on — and ultimately the number of patents too! This is because state intellectual property laws may increase the overall incentives faced by the firm trying to decide whether to engage in a research project or not. If the incentive is greater, more research will be undertaken. And if more research is undertaken, more *patentable* research will result.

The preceding point can be restated as follows. The Supreme Court pictures an inventor faced with two urns, one marked "potentially patentable" and the other marked "state protection." The Court's point is that increasing the value of a ball picked from the second urn will decrease the number of balls chosen from the first urn. In fact, in some cases, and perhaps most cases, the inventor *does not know* which type of urn she is facing; it is as though she were presented with one large urn in which some of the balls are marked "patentable" and the others "state protection." Because she does not know at the outset of the game which type of ball she will pick, an increase in the value of *either type* of ball will increase her expected return and thus increase the number of balls she is willing to pick.

This is a more accurate representation of the choices facing many inventors.

Consequently one must recognize that state intellectual property laws may have an effect *opposite* of that stated by the Court: they may increase the number of "draws" from the undifferentiated urn marked "potential inventions," thereby increasing the number of patentable as well as state-protectable inventions.

This is not to say that the Court was wrong in its holding in *Bonito Boats*. On the contrary, there may be good reasons to guard against *over*-encouraging efforts to innovate. In addition, there will often be federalism-based concerns that are implicated by a system of state intellectual property laws. But the "crowding out" theory used by the Court may be wrong, at least in some cases, and therefore should not be used to justify a strong patent preemption position.

2. Copyright Preemption

ProCD, Inc. v. Zeidenberg
United States Court of Appeals for the Seventh Circuit
86 F.3d 1447 (7th Cir. 1996)

EASTERBROOK, Circuit Judge.

Must buyers of computer software obey the terms of shrinkwrap licenses? The district court held not, for two reasons: first, they are not contracts because the licenses are inside the box rather than printed on the outside; second, federal law forbids enforcement even if the licenses are contracts. 908 F. Supp. 640 (W.D. Wis. 1996). The parties and numerous amici curiae have briefed many other issues, but these are the only two that matter — and we disagree with the district judge's conclusion on each. Shrinkwrap licenses are enforceable unless their terms are objectionable on grounds applicable to contracts in general (for example, if they violate a rule of positive law, or if they are unconscionable). Because no one argues that the terms of the license at issue here are troublesome, we remand with instructions to enter judgment for the plaintiff.... [The facts of this case are reprinted in section A.3, supra. The Court concluded that shrinkwrap licenses were enforceable as a matter of contract law.]

III

The district court held that, even if Wisconsin treats shrinkwrap licenses as contracts, § 301(a) of the Copyright Act, 17 U.S.C. § 301(a), prevents their enforcement. 908 F. Supp. at 656-59. The relevant part of § 301(a) preempts any "legal or equitable rights [under state law] that are equivalent to any of the exclusive rights within the general scope of copyright as specified by section 106 in works of authorship that are fixed in a tangible medium of expression and come within the subject matter of copyright as specified by sections 102 and 103". ProCD's software and data are "fixed in a tangible medium of expression," and the district judge held that they are "within the subject matter of copyright." The latter conclusion is plainly right for the copyrighted application program, and the judge thought that the data likewise are "within the subject matter of copyright" even if, after Feist, they are not sufficiently original to be copyrighted. 908 F. Supp. at 656-57. Baltimore Orioles, Inc. v. Major League Baseball Players Ass'n, 805 F.2d 663, 676 (7th Cir. 1986), supports that conclusion, with which commentators agree. E.g., Paul Goldstein, III Copyright § 15.2.3 (2d ed. 1996); Melville B. Nimmer & David Nimmer, Nimmer on Copyright § 101[B] (1995); William F. Patry, II Copyright Law and Practice 1108-09 (1994). One function of § 301(a) is to prevent states from giving special protection to works of authorship that Congress has decided should be in the public domain, which it can accomplish only if "subject matter of copyright" includes all works of a type covered by sections 102 and 103, even if federal law does not afford protection to them. Cf. Bonito Boats, Inc. v. Thunder Craft Boats, Inc., 489 U.S. 141, 103 L. Ed. 2d 118, 109 S. Ct. 971 (1989) (same principle under patent laws).

But are rights created by contract "equivalent to any of the exclusive rights within the general scope of copyright"? Three courts of appeals have answered "no." National Car Rental Systems, Inc. v. Computer Associates International, Inc., 991 F.2d 426, 433 (8th Cir. 1993); Taquino v. Teledyne Monarch Rubber, 893 F.2d 1488, 1501 (5th Cir. 1990); Acorn Structures, Inc. v. Swantz, 846 F.2d 923, 926 (4th Cir. 1988). The district court disagreed with these decisions, 908 F. Supp. at 658, but we think them sound. Rights "equivalent to any of the exclusive rights within the general scope of copyright" are rights established by law — rights that restrict the options of persons who are strangers to the author. Copyright law forbids duplication, public performance, and so on, unless the person wishing to copy or perform the work gets permission; silence means a ban on copying. A copyright is a right against the world. Contracts, by contrast, generally affect only their parties; strangers may do as they please, so contracts do not create "exclusive rights." Someone who found a copy of SelectPhone (trademark) on the street would not be affected by the shrinkwrap license — though the federal copyright laws of their own force would limit the finder's ability to copy or transmit the application program.

Think for a moment about trade secrets. One common trade secret is a customer list. After *Feist*, a simple alphabetical list of a firm's customers, with address and telephone numbers, could not be protected by copyright. Yet Kewanee Oil Co. v. Bicron Corp., 416 U.S. 470, 40 L. Ed. 2d 315, 94 S. Ct. 1879 (1974), holds that contracts about trade secrets may be enforced — precisely because they do not affect strangers' ability to discover and use the information independently. If the amendment of § 301(a) in 1976 overruled *Kewanee* and abolished consensual protection of those trade secrets that cannot be copyrighted, no one has noticed — though abolition is a logical consequence of the district court's approach. Think, too, about everyday transactions in intellectual property. A customer visits a video store and rents a copy of Night of the Lepus. The customer's contract with the store limits use of the tape to home viewing and requires its return in two days. May the customer keep the tape, on the ground that § 301(a) makes the promise unenforceable?

A law student uses the LEXIS database, containing public-domain documents, under a contract limiting the results to educational endeavors; may the student resell his access to this database to a law firm from which LEXIS seeks to collect a much higher hourly rate? Suppose ProCD hires a firm to scour the nation for telephone directories, promising to pay $100 for each that ProCD does not already have. The firm locates 100 new directories, which it sends to ProCD with an invoice for $10,000. ProCD incorporates the directories into its database; does it have to pay the bill? Surely yes; Aronson v. Quick Point Pencil Co., 440 U.S. 257, 59 L. Ed. 2d 296, 99 S. Ct. 1096 (1979), holds that promises to pay for intellectual property may be enforced even though federal law (in *Aronson*, the patent law) offers no protection against third-party uses of that property. See also Kennedy v. Wright, 851 F.2d 963 (7th Cir. 1988). But these illustrations are what our case is about. ProCD offers software and data for two prices: one for personal use, a higher price for commercial use. Zeidenberg wants to use the data without paying the seller's price; if the law student and Quick Point Pencil Co. could not do that, neither can Zeidenberg.

Although Congress possesses power to preempt even the enforcement of contracts about intellectual property — or railroads, on which see Norfolk & Western Ry. v. Train Dispatchers, 499 U.S. 117 (1991) — courts usually read preemption clauses to leave private contracts unaffected. American Airlines, Inc. v. Wolens, 115 S. Ct. 817 (1995), provides a nice illustration. A federal statute preempts any state "law, rule, regulation, standard, or other provision . . . relating to rates, routes, or services of any air carrier." 49 U.S.C. App. § 1305(a)(1). Does such a law preempt the law of contracts — so that, for example, an air carrier need not honor a quoted price (or a contract to reduce the price by the value of frequent flyer miles)? The Court allowed that it is possible to read the statute that broadly but thought such an interpretation would make little sense. Terms and conditions offered by contract reflect private ordering, essential to the efficient functioning of markets. 115 S. Ct. at 824-25. Although some principles that carry the name of contract law are designed to defeat rather than implement consensual transactions, id. at 826 n.8, the rules that respect private choice are not preempted by a clause such as § 1305(a)(1). Section 301(a) plays a role similar to § 1301(a)(1): it prevents states from substituting their own regulatory systems for those of the national government. Just as § 301(a) does not itself interfere with private transactions in intellectual property, so it does not prevent states from respecting those transactions. Like the Supreme Court in *Wolens*, we think it prudent to refrain from adopting a rule that anything with the label "contract" is necessarily outside the preemption clause: the variations and possibilities are too numerous to foresee. National Car Rental likewise recognizes the possibility that some applications of the law of contract could interfere with the attainment of national objectives and therefore come within the domain of § 301(a). But general enforcement of shrinkwrap licenses of the kind before us does not create such interference.

Aronson emphasized that enforcement of the contract between Aronson and Quick Point Pencil Company would not withdraw any information from the public domain. That is equally true of the contract between ProCD and Zeidenberg. Everyone remains free to copy and disseminate all 3,000 telephone books that have been incorporated into ProCD's database. Anyone can add SIC codes and zip codes. ProCD's rivals have done so. Enforcement of the shrinkwrap license may even make information more readily available, by reducing the price ProCD charges to consumer buyers. To the extent licenses facilitate distribution of object code while concealing the source code (the point of a clause forbidding disassembly), they serve the same procompetitive functions as does the law of trade secrets. Rockwell Graphic Systems, Inc. v. DEV Industries, Inc., 925 F.2d 174, 180 (7th Cir. 1991). Licenses may have other benefits for consumers: many licenses permit users to make extra copies, to use the software on multiple computers, even to incorporate the software into the user's products. But whether a particular license is generous or restrictive, a simple two-party contract is not "equivalent to any of the exclusive rights within the general scope of copyright" and therefore may be enforced.

Reversed and remanded.

COMMENTS AND QUESTIONS

1. Section 301 is designed to prevent states from passing or enforcing laws "equivalent" to copyright. Whether a particular law is equivalent to copyright can be difficult to determine, however. Courts generally do not ask whether a body of law as a whole (contract or trade secrets, say) is equivalent to copyright. Rather, the question is whether the application of a state law to a particular factual circumstance would create a state-law right equivalent to copyright. Even *ProCD*, which takes a fairly categorical approach to section 301 preemption, is careful to note that it is not holding that section 301 will never preempt contract terms.

Nonetheless, Judge Easterbrook sees a categorical difference between enforcement of contract law and enforcement of other state laws. The court reasoned:

> Rights 'equivalent to any of the exclusive rights within the general scope of copyright' are rights established *by law* — rights that restrict the options of persons who are strangers to the author. Copyright law forbids duplication, public performance, and so on, unless the person wishing to copy or perform the work gets permission; silence means a ban on copying. A copyright is a right against the world. Contracts, by contrast, generally affect only their parties; strangers may do as they please, so contracts do not create "exclusive rights." Someone who found a copy of [the plaintiff's software product] on the street would not be affected by the shrinkwrap license . . .

Is this distinction between judicial enforcement of contracts and other state laws persuasive? Commentators have been skeptical. Radin and Wagner point out that the Legal Realist movement in the last century exploded the myth that contracts are purely "private" creatures; they depend on the legal system for their enforcement. Margaret Jane Radin & R. Polk Wagner, The Myth of Private Ordering: Rediscovering Legal Realism in Cyberspace, 73 Chi.-Kent L. Rev. 1295 (1998). And Lemley observes that "even truly 'private' contracts affect third parties who haven't agreed to the contract terms. Many contracts have significant negative externalities." Mark A. Lemley, Beyond Preemption: The Law and Policy of Intellectual Property Licensing, 87 Cal. L. Rev. 111 (1999). It is interesting to note that, whatever the validity of the general distinction, the court in *ProCD* applied it to validate a "shrinkwrap license," a peculiar form of contract that is drafted by the creator of the product and that purports to bind to its terms anyone who uses the product. While it is technically true in such a case that only "parties" to the contract are bound by it, anyone who has access to the product will automatically become such a party. Shrinkwrap licenses look less like contracts in the pure sense, and more like examples of private legislation. As Lemley argues:

> [T]he viability of the distinction between private contracts and public legislation is diminishing day by day. One of the main changes [*ProCD* and its progeny] would make in current law would be to render enforceable contract "terms" to which the parties did not agree in the classic sense, and indeed of which one party may be entirely unaware. [They] would also enable the enforcement of such contract terms "downstream" — that is, against whomever later acquires the software — despite the fact that a first sale under both patent and copyright law would free the purchaser from upstream contractual restrictions. Technology facilitates this change by allowing a vendor to interpose contract terms even in a downstream transaction that would not ordinarily be thought to demonstrate privity between the "contracting" parties.

Id.

2. Does copyright preemption of state contract law depend on the remedy asserted for breach of contract? Should federal law be more concerned if a party seeks by contract to bring the weapons of copyright law to bear? For example, the House Committee Notes to section 109(a) of the Copyright Act (the "first sale" doctrine) provide that a contract limiting a user's first sale rights may be enforceable by an action for breach of contract, but do not give rise to an action for copyright infringement. Contract remedies are ordinarily limited to expectation damages, rather than the consequential damages, statutory damages, attorney's fees and injunctions available under the copyright law. But suppose a license agreement provided that *resale* of a copyrighted work — permissible under copyright law — voids the entire license, rendering the reseller liable for copyright infringement. Should such a contract be preempted by copyright law? If not, is there anything left in practice of the first sale doctrine?

3. After *ProCD*, is there any set of circumstances in which section 301 will preempt a contract? Suppose that a contract provided that the buyer of a book could not make fair use of the book. Should such a contract term be enforced? Would section 301 stand in the way? Should the context of the contract (whether it was a shrinkwrap, whether the product was widely sold, whether the buyer was a consumer) matter?

4. While we have so far discussed copyright preemption of contract claims, copyright preemption also applies to a variety of state statutes and common law causes of action. Indeed, in *ProCD* itself, the district court held state tort claims for unfair competition, misappropriation, and violation of the Wisconsin Computer Crimes Act to be preempted, because all three claims were based on Zeidenberg's copying of uncopyrightable data. ProCD, Inc. v. Zeidenberg, 908 F. Supp. 640 (W.D. Wisc. 1996). The Seventh Circuit let these findings stand on appeal. The Eleventh Circuit has held claims for unfair competition and deceptive trade practices preempted where those claims were made in an effort to protect uncopyrightable databases. Lipscher v. LRP Publications, 266 F.3d 1305 (11th Cir. 2001). On copyright preemption of trade secret claims, see Computer Associates Int'l v. Altai, Inc., 982 F.2d 693, 715–17 (2d Cir. 1992) (copyright and trade secret claims could co-exist regarding the use of the same computer program, as long as misappropriation of trade secrets is based on a defendant's breach of a duty of trust or confidence). See also Automated Drawing Systems v. Integrated Network Services, 447 S.E.2d 109 (Ga. Ct. App. 1994) (copyright law did not preempt Georgia Computer Systems Protection Act, which prohibited the "misappropriation" of source code and provided tort damages).

5. *The "Additional Element" Test for Preemption Under Section 301.* Courts attempting to determine whether a right was "equivalent" to copyright have introduced a seemingly straightforward test for federal preemption of state laws in the general domain of copyright: the so-called extra element test. Under this test, if state law provides a cause of action that requires proof of at least one element in addition to those required for a copyright infringement claim, the state law survives. The test tries to state simply, in terms of required pleadings for an effective cause of action, the basic idea that a state can proscribe activities in the general domain of intellectual property as long as those activities are qualitatively different from the ones that are the subject of federal protection. Claims for breach of implied contract in the context of idea submissions (as seen in Desny v. Wilder) survive federal preemption as they turn "not upon the existence of a [copyright] ... but upon the implied promise to pay the reasonable value of the material disclosed." Grosso v. Miramax Film Corp., 383 F.3d 965, 968 (9th Cir. 2004) (quoting Landsberg v. Scrabble Crossword Game Players, Inc., 802 F.2d 1193, 1196 (9th Cir. 1986)).

As applied, the "additional element" test has an elastic quality. In National Car Rental System v. Computer Associates, 991 F.2d 426 (8th Cir. 1993), for example, the court held that a claim for breach of a license to use a computer program was not preempted under section 301 because the claim involved "use" of the program, and "use" of a computer program was not one of the rights conferred by the Copyright Act. If read broadly, so that an allegation of use *in addition to* copying of a copyrighted work survives preemption, *National Car Rental* doesn't leave much of section 301 preemption. Other examples of courts reaching to find an extra element include Taco Bell Corp. v. Wrench LLC, 256 F.3d 446 (6th Cir. 2001) (holding that the requirement of an "expectation of compensation" in a claim for implied-in-fact contract saved the claim from preemption).

6. *Supremacy Clause Preemption.* Section 301 is not all there is to copyright preemption. Congress enacted section 301 because it intended to "preempt the field" of copyright law, precluding states from passing laws that mimicked the federal statute. But even where Congress has not preempted the field, state statutes will still be preempted if they conflict with the specific mandate of federal

law, or if they stand as an obstacle to the purposes of a federal statute. In copyright law, such "conflicts" or "Supremacy Clause" preemption obviously occurs when a state law prevents the enforcement of a federal copyright. For example, if the state of California were to pass a law stating that the Copyright Act did not apply to citizens of California, the law would arguably survive section 301 preemption, since it contains an "extra element" (citizenship) and does not create a right equivalent to copyright. But it would surely be struck down by the courts as an interference with the federal scheme.

Not all interferences are so straightforward. In ASCAP v. Pataki, 930 F. Supp. 873 (S.D.N.Y. 1996), the court invalidated a state statute regulating the activities of performing rights societies such as ASCAP and BMI. The state statute required, among other things, that such groups provide owners of establishments performing music with written notice of an investigation within 72 hours after it is initiated, thus making it difficult for ASCAP and others to conduct "undercover" investigations for violations of the copyright laws. The court's opinion addressed only the issue of "conflict preemption, which occurs either where compliance with both federal and state regulations is a physical impossibility, or where state law stands as an obstacle to the accomplishment and execution of the full purposes and objectives of Congress." Id. The court found that the state statutory provisions "hinder the realization of the federal copyright scheme" for several reasons: the statute made it more difficult for copyright owners to enforce their rights; it effectively established a "statute of limitations" on copyright investigations that was shorter than the federal statute; and it gave copyright defendants a counterclaim that they could use to offset copyright damages. See also College Entrance Examination Board v. Pataki, 889 F. Supp. 554 (N.D.N.Y. 1995) (state law requiring disclosure of standardized test questions and answers preempted by Copyright Act because it conflicted with the rights of copyright owners to restrict distribution of copyrighted material).

Because the Copyright Act is an attempt by Congress to balance the interests of creators and users of intellectual property, state laws that give *too much* protection to copyrighted works may also interfere with the objectives of Congress. See, e.g., United States ex rel. Berge v. Trustees of the University of Alabama, 104 F.3d 1453 (4th Cir. 1997) ("The shadow actually cast by the Act's preemption is notably broader than the wing of its protection."). As the Supreme Court explained in Goldstein v. California, 412 U.S. 546 (1973), the question is whether Congress intended to place a particular work or use in the public domain:

> At any time Congress determines that a particular category of "writing" is worthy of national protection and the incidental expenses of federal administration, federal copyright protection may be authorized. Where the need for free and unrestricted distribution of a writing is thought to be required by the national interest, the Copyright Clause and the Commerce Clause would allow Congress to eschew all protection. In such cases, a conflict would develop if a State attempted to protect that which Congress intended to be free from restraint or to free that which Congress had protected. However, where Congress determines that neither federal protection nor freedom from restraint is required by the national interest, it is at liberty to stay its hand entirely. Since state protection would not then conflict with federal action, total relinquishment of the States' power to grant copyright protection cannot be inferred.

Although *Goldstein* was decided before the enactment of section 301, conflicts preemption is still a significant part of copyright law. Several cases have gone beyond section 301 to a general preemption analysis along the lines of the patent preemption cases such as *Kewanee* and *Bonito Boats.* In Associated Film Distribution Corp. v. Thornburgh, 520 F. Supp. 971 (E.D. Pa. 1981), *rev'd and remanded on other grounds*, 683 F.2d 808, 817 (3d Cir. 1982), *cert. denied*, 480 U.S. 933 (1982), the court held preempted a state statute regulating the procedure by which film exhibitors licensed major motion pictures from film distribution arms of the major movie companies. The court went beyond section 301, stating that the "more general question of conflict between the two statutory schemes under the supremacy clause is decisive." 520 F. Supp. at 993. See also Orson, Inc. v. Miramax Film Corp., 189 F.3d 377 (3d Cir. 1999) (en banc) (a Pennsylvania statute that prohibited an exclusive license of a first-run film for longer than six weeks was preempted by the Copyright Act); but see Allied Artists Pictures Corp. v. Rhodes, 496 F. Supp. 408 (S.D. Ohio 1980), *aff'd* and remanded in part, 679 F.2d 656, 665 (6th Cir. 1982) (reaching the opposite conclusion).

A number of cases have in fact preempted contract terms that conflict with federal patent and copyright policy. For example, contract terms that purport to extend a patent or copyright beyond its expiration have repeatedly been held unenforceable. On the distinction between "conflicts preemption" and copyright field preemption in the contract area, see David Nimmer et al., The Metamorphosis of Contract into Expand, 87 Cal. L. Rev. 17 (1999).

7. Is *ProCD* in fact a case of conflict between copyright and contract? The stage for such a conflict is set by the Supreme Court decision in *Feist*, which held that the telephone white pages at issue in *ProCD* were constitutionally ineligible for copyright protection. Does the fact that ProCD's shrinkwrap license gives if anything even greater protection than copyright would suggest there is a conflict here? In ProCD v. Zeidenberg, 908 F. Supp. 640 (W.D. Wisc. 1996), the district court reasoned in dictum that state contract law could not be used to prevent the copying of telephone white pages, which the Supreme Court had determined in *Feist* were uncopyrightable. The district court applied a constitutional copyright preemption analysis similar to the patent analysis undertaken in *Bonito Boats.* The Seventh Circuit reversed the district court's decision, concluding that under section 301 of the Copyright Act state contract law could not be preempted in this case. ProCD v. Zeidenberg, 86 F.3d 1447 (7th Cir. 1996). Judge Easterbrook's opinion did not mention Supremacy Clause preemption at all, even though the issue was briefed and was necessary to the decision, leaving the issue in some doubt.

Courts to consider the question of whether shrinkwrap licenses can be preempted when they conflict with federal policy are split. One court concluded that a state statute permitting enforcement of shrinkwrap licenses was invalid under the Supremacy Clause where the shrinkwrap license at issue forbad reverse engineering that was permissible under the Copyright Act. Vault Inc. v. Quaid Corp., 847 F.2d 255 (5th Cir. 1988). See also Bowers v. Baystate Technologies, 320 F.3d 1317, 1337 (Fed. Cir. 2003) (Dyk, J., dissenting) (endorsing this result). On the other hand, two more recent decisions have found no distinction between shrinkwrap licenses and negotiated contracts, and permitted copyright owners to block reverse engineering using a shrinkwrap provision. Davidson & Assocs. v. Jung, 422 F.3d 630 (8th Cir. 2005); Bowers v. Baystate Technologies, 320 F.3d 1317 (Fed. Cir. 2003).

If the federal copyright scheme intends unoriginal works like the telephone white pages to be in the public domain, is there any way to prevent the sort of use Zeidenberg made of this data? States presumably would lack the power to do so. There is some question whether Congress could avoid the constitutional limitation of originality in the Copyright Clause by enacting protection for databases under the Commerce Clause. In United States v. Moghadam, 175 F.3d 1269 (11th Cir. 1999), the court concluded that Congress could protect unfixed works under the Commerce Clause because it perceived no conflict between the anti-bootlegging statute at issue there and the policies of copyright law. But a Congressional effort to overrule *Feist* might be more problematic. See Paul J. Heald & Suzanna Sherry, Implied Limits on the Legislative Power: The Intellectual Property Clause as an Absolute Constraint on Congress, 2000 U. Ill. L. Rev. 1119.

PROBLEM

Problem 6-5. Presco is in the business of printing and selling academic journals to university libraries. When Presco receives a subscription request, it sends a form contract to the requestor. The requesting library is required to sign the contract before Presco will start the subscription. Presco's form contract provides in part that "Independent of and in addition to any provisions of state or federal law, the parties agree that Subscriber will not make, cause to be made, or allow to be made any copies of any Presco journals without the prior express written consent of Presco."

Is the contract provision preempted? Does it matter whether the subscriber's copying would constitute fair use under the copyright laws?

Note on Preemption of Right of Publicity Claims

Many applications of the right of publicity—such as the unauthorized use of a celebrity's name to endorse a product—easily survive federal preemption by the Copyright Act because enforcement of the right in such circumstances does not implicate a copyrighted work. Several contexts, however, strain the tests for federal preemption. Consider, for example, the implications of Bette Midler's publicity rights in her rendition of "Do You Want to Dance" for the owner of the underlying musical composition. Midler's right of publicity in her vocal rendition trumps other performers from imitating her version, yet section 114(b) of the Copyright Act expressly authorizes cover recordings that "imitate or simulate those in the copyrighted sound recording." While recognizing this implication, the Ninth Circuit in Midler v. Ford breezily concluded that Midler's assertion of the right of publicity was not preempted because "a voice is not copyrightable." 849 F.2d 462. Does such treatment adequately address the apparent conflict with federal copyright law?

By contrast, the Seventh Circuit found that federal copyright law preempted baseball players' assertions of state publicity rights in their images and game performances. See Baltimore Orioles, Inc. v. Major League Baseball Players Association, 805 F.2d 663 (7th Cir. 1986), *cert. denied*, 480 U.S. 941 (1987). Because each game was embodied in a copyrighted telecast and players uniformly assign their copyrights to their teams, the court reasoned that the game performances could not be the subject of independent state publicity rights. The Seventh Circuit has since construed this case narrowly, see Toney v. L'Oreal USA, Inc., 406 F.3d 905 (7th Cir. 2005) (holding that a model whose photograph was used in connection with packaging and promotion of a hair care product could recover under the right of publicity), and commentators question its logic, see Nimmer on Copyright, § 1.01 (criticizing the court's premise that a baseball game is a protected "work of authorship" under the Copyright Act). See generally Jennifer E. Rothman, Copyright Preemption and the Right of Publicity, 36 U.C. Davis L. Rev. 199 (2002) (arguing for a broad application of preemption doctrine).

A less controversial application of the preemption doctrine occurred in Fleet v. CBS, Inc., 50 Cal. App. 4th 1911, 58 Cal. Rptr. 2d 645 (1996), where actors in a copyrighted film alleged that the very exploitation of the copyrighted work infringed their state right of publicity. As the court recognized, "a party who does not hold the copyright in a performance captured on film cannot prevent the one who does from exploiting it by resort to state law." Id. at 1923. See also Ahn v. Midway Mfg. Co., 965 F. Supp. 1134 (N.D. Ill. 1997) (dismissing right of publicity claims brought by a ballerina and karate black belt who consented to having their movements digitally captured for use in developing characters for the copyrighted computer game "Mortal Kombat"). By contrast, a court allowed rock star Bret Michaels and actress Pamela Anderson Lee to block the unauthorized distribution over the Internet of a videotape showing them engaged in sex as a violation of their right of publicity. The basis of the right of publicity claim, however, was the use of their names, likenesses, and identities to advertise imminent distribution of the video, and not the distribution of the copyrighted tape itself. See Michaels v. Internet Entertainment Group, Inc., 5 F.Supp.2d 823, 837 (C.D. Cal. 1998).

Note on the Special Case of Moral Rights Preemption

As noted above, both state and federal statutes protect the moral rights of artists, and those rights can sometimes come into conflict. The Visual Artists Rights Act of 1990 added subsection (f) to § 301, the preemption provision of the Copyright Act:

17 U.S.C. § 301(f). Preemption Provision

(1) On [June 1, 1991], all legal and equitable rights that are equivalent to any of the rights conferred by 106A with respect to works of visual art to which the rights conferred by section 106A apply are governed exclusively by section 106A and section 113(d) and the provisions of this title relating to such sections. Thereafter, no person is entitled to any such right or equivalent right in any work of visual art under the common law or statutes of any State.

(2) Nothing in paragraph (1) annuls or limits any rights or remedies under the common law or statutes of any State with respect to—

(A) any cause of action from undertakings commenced before [June 1, 1991]

(B) activities violating legal or equitable rights that are not equivalent to any of the rights conferred by section 106A with respect to works of visual art;

(C) activities violating legal or equitable rights which extend beyond the life of the author.

COMMENTS AND QUESTIONS

1. Does the Copyright Act preempt the state moral rights statutes discussed earlier in this chapter? One case has indicated that the California moral rights statute, Cal. Civ. Code §987, "appears" to be preempted by this provision. Lubner v. City of Los Angeles, 53 Cal. Rptr. 2d 24 (Ct. App. 1996). The court did not need to reach the issue, however, since it held that negligently damaging a piece of art by losing control of a garbage truck which thereafter rolled into the art did not violate the California statute.

2. Does the Copyright Act preempt California's Artist Resale Royalty Act (droit de suite)? See Morseberg v. Balyon, 621 F.2d 972 (9th Cir.), *cert. denied*, 449 U.S. 983 (1980) (California Resale Royalty Act not preempted by 1909 Copyright Act; no opinion expressed on applicability of analysis to preemption under the 1976 Act).

3. Trademark Preemption

There has been surprisingly little litigation over the preemptive effect of the Lanham Act. Because the Lanham Act was passed under the aegis of the Commerce Clause of the Constitution, rather than the Patents and Copyrights Clause, the analysis developed in the remainder of this chapter is not directly relevant to the problem of trademark preemption. Instead, courts turn to other Commerce Clause cases, which focus on whether Congress intended to "preempt the field" of trademark law and on whether there is an actual or potential conflict between the state and federal statutes.

Two early cases held that the Lanham Act preempted all state trademark laws. Sargen & Co. v. Welco Feed Mfg., 195 F.2d 929 (8th Cir. 1952); Time, Inc. v. T.I.M.E. Inc., 123 F. Supp. 446 (S.D. Cal. 1954). But both cases are in disfavor, largely because Congress clearly intended in passing the Lanham Act to allow it to coexist with some state trademark laws. See 15 U.S.C. §§1065, 1115(b)(5) (both referring to the continued effect of state law). See Richard A. De Sevo, Antidilution Laws: The Unresolved Dilemma of Preemption Under the Lanham Act, 84 Trademark Rptr. 300, 301–04 (1994). However, at least some courts continue to hold that state statutes which are directed at the same types of conduct as the Lanham Act are preempted. See Three Blind Mice Designs Co. v. Cyrk Inc., 892 F. Supp. 303 (D. Mass. 1995) (state antidilution statute "wholly preempted" to the extent that it seeks to regulate directly competitive goods).

A more difficult question is presented where state trademark statutes may conflict with the general federal rule. Sometimes such conflicts are clear. For example, state laws that attempt to grant priority to a party other than the earliest federal registrant are surely preempted under 15 U.S.C. §1127, since the laws "interfere" with the rights granted federal registrants under the Act.

Note on the Preemption of State Antidilution Statutes

Preemption will also occur if the state law is at odds with the "purpose of Congress" in passing the Lanham Act. Depending on the purpose identified, state antidilution laws might fail this test, since they do not require a likelihood of consumer confusion. See De Sevo, *supra*, at 312–20. Courts and commentators have split on this issue. See David S. Welkowitz, Preemption, Extraterritoriality and the Problem of State Antidilution Laws, 67 Tul. L. Rev. 1, 4 (1992) (arguing that "injunctions like the one issued in Mead Data Central, Inc. v. Toyota Motor Sales, U.S.A., 702 F. Supp. 1031 (S.D.N.Y. 1988), *rev'd on other grounds*, 875 F.2d 1026 (2d Cir. 1989) exceed [constitutional] limits" under the Commerce Clause, though state antidilution statutes are not wholly preempted and in many cases injunctions based on state intellectual property law are enforceable under the Full Faith and Credit Clause). Cf. Milton W. Handler, Are the State Antidilution Laws Compatible with the National Protection of Trademarks?, 75 Trademark Rep. 269 (1985); Joseph P. Bauer, A Federal Law of Unfair Competition: What Should Be the Reach of Section 43(a) of the Lanham Act?, 31 UCLA L. Rev. 671 (1984); Charles Bunn, The National Law of Unfair Competition, 62 Harv. L. Rev. 987 (1949); Paul Heald, Comment, Unfair Competition and Federal Law: Constitutional Restraints on the Scope of State Law, 54 U. Chi. L. Rev. 1411 (1987).

The new federal antidilution law changes the preemption analysis, of course. Many people expected that such a law would preempt state antidilution statutes, replacing the patchwork of inconsistent protections with uniform national protection.[8] Instead, the new section 43(c)(3) of the Lanham Act charts a narrow course around state antidilution laws by allowing them to continue in force but granting owners of federally registered marks immunity from suit under such state laws. Unregistered marks, as well as those registered under state law, receive no such immunity. The legislative history explains:

> Under section 3 of the bill, a new Section 43(c)(3) of the Lanham Act would provide that ownership of a valid federal trademark registration is a complete bar to an action brought against the registrant under state dilution law. This section provides a further incentive for the federal registration of marks and recognizes that to permit a state to regulate the use of federally registered marks is inconsistent with the intent of the Lanham Act "to protect registered marks used in such commerce from interference by state, or territorial legislation." It is important to note that *the proposed federal dilution statute would not preempt state dilution laws*. Unlike patent and copyright laws, federal trademark law coexists with state trademark law, and it is to be expected that the federal dilution statute should similarly coexist with state dilution statutes.

Federal Trademark Dilution Act of 1995, House Report 104-374 (accompanying H.R. 1295), November 30, 1995 (emphasis added).

Thus parallel state antidilution claims will not wither away completely. Indeed, depending on whether state law affords more attractive remedies (note the "injunction only" rule for most cases under the federal dilution provision, § 43(c)(2)), parties may well continue to bring at least some antidilution cases under state law. Note that

---

8. Indeed, some saw little purpose for the federal law otherwise. See Kenneth L. Port, The "Unnatural" Expansion of Trademark Rights: Is a Federal Dilution Statute Necessary?, 18 Seton Hall Legis. J. 433 (1994).

for the most part, however, nationwide injunctions are more difficult to acquire under state antidilution statutes, making federal claims more likely if such an injunction is the trademark owner's major goal.

COMMENTS AND QUESTIONS

Is the lack of preemption troubling? Consider the following possibilities:

- A federal antidilution claim fails because the use is found to be a "fair" one (say, for news reporting purposes). A parallel action is brought under a state statute that contains no such limitation, and the court grants a nationwide injunction against dilution on the basis of the state statute.
- A nationally famous but unregistered product configuration is found to dilute a locally known product configuration under a state statute.

Do the state statutes at issue in these cases interfere with federal trademark policy?

7 *Protection of Computer Software*

The previous chapters have explored the principal modes of intellectual property protection. As we have seen, these statutory and common law regimes comprise a complex system of rights. In some circumstances, the law envisions overlapping layers of protection as in the case of design protection (covered simultaneously by copyright, design patent, and trade dress). In other circumstances, one mode of protection overrides or preempts others, as in the cases of the idea-expression dichotomy and the relationship of federal and state protections. With the advent of new technologies and the evolution of the law, the system of intellectual property protection has increasingly become integrated. The blurring of traditional doctrinal lines can be seen throughout the practice of intellectual property law. The major transactions and cases today affecting high technology firms typically involve complex technologies, multiple areas of intellectual property law, and antitrust issues. Nowhere is this pattern more evident than in the area of computer and information technologies.

In view of this reality, this chapter explores the manner in which the various modes of intellectual property protection have developed to protect computer technology. As essential background to this study, we explore the economics of computer markets.[1] We then work through the various modes of protection for computer technology. We begin with trade secret protection, which served as the principal mode of protection during the early and intermediate stages of the computer industry. With the advent and proliferation of microcomputers and mass-marketed software in the mid- to late 1970s, developers turned increasingly toward copyright protection, which has served as the principal mode of protection for software for the past two decades. Our focus then shifts to patents, which is now the principal battleground. We conclude the chapter with a look at alternative modes of protection for software, which enables us to explore the desirability and practicality of tailoring intellectual property protection to new technologies.

---

1. For background on computer technology, see Appendix B to the Statutory Supplement.

A. THE ECONOMICS OF COMPUTER MARKETS

Intellectual property protection for computer software has co-evolved with the underlying technology and platforms for software distribution. Therefore, it will be useful to have some general background about the industry, which has undergone continuous and dramatic change since its inception in the 1950s as a result of rapid technological advance. One of the goals of this chapter and Chapter 8 is to assess the role of intellectual property in this process and the likely effects — both beneficial and detrimental — on the future of software innovation and dissemination.

1. The Market for Computers and Computer Software

At the most basic level, consumers demand "computing services" to meet their data processing needs broadly defined — from accounting spreadsheets to ringtones. These needs can be satisfied completely by hardware, or they can be satisfied by general-purpose computer hardware equipped with appropriate application software. Thus hardware and software may be both substitutes and complements.[2]

Demand for Computer Services. The demand for computer services is driven by the great diversity of entities — businesses, government agencies, research institutions, individuals — with data processing, information, entertainment, and communication needs. These needs range from simple calculations to complex scientific applications and realistic video games. Consumers also differ in the variety of data processing tasks that they must accomplish. A medium-size business, for example, might have many data-processing tasks for which a computer might prove useful: handling the payroll, record keeping, word processing, and projecting business trends. By contrast, a manufacturing facility might simply need to regulate the temperature of a kiln. For many years, scientists required the most advanced computers to execute high-speed calculations and to maintain complex databases. With the development of high-resolution, online multiplayer video games, millions of teenagers today also desire such computing power.

Supply of Computer Services. The computer industry began in the 1950s as a hardware business. Early computers were quite limited, and much of the computing functionality was built directly into the architecture of the "main frame." With advances in general-purpose computing and the advent of more versatile programming languages, software development became a more significant part of the business. But with IBM and other leading mainframe manufacturers providing software to its hardware customers for no additional charge, there was little opportunity for a vibrant independent software sector to form. Even as late as the mid 1970s, most software firms produced custom programs for predominantly commercial customers. It was not until the proliferation of microcomputers in the 1980s that the software sector

2. In economic terms, goods are "substitutes" if purchase of one makes it unnecessary to buy the other. They are "complements" if they are traditionally purchased or used together.

took flight. Its role grew rapidly and by the early 1990s, Microsoft emerged as the dominant force in the computer industry. The development of the Internet and the World Wide Web further propelled the role of software. Open source software is now reshaping competition in the industry yet again.

The hardware sector of the computer industry consists of original equipment manufacturers, semiconductor chip manufacturers, and vendors. The vendors purchase computer components and chips from the other firms and assemble them into computer systems. A few major firms in the industry, like Hewlett-Packard and Apple, are involved in all aspects of hardware research and development as well as software. Other firms have significant shares of particular hardware markets but do not attempt to maintain a presence in vertically related markets (e.g., Dell in computers, Intel and Advanced Micro Devices in semiconductor chips).

The traditional software sector offers a wide variety of products and services. Its work includes the design of general operating systems, contract programming, the development of commercial application packages, and, increasingly, web-related programming. Many large hardware systems manufacturers develop operating and application software for their systems. There are also many small, independent firms that specialize in aspects of software services and product development. A few firms, such as Microsoft and Oracle (specializing in relational database business systems), hold dominant roles in important software markets. As with hardware, the largest software firms tend to be vertically integrated, selling operating systems, application programs, and services. Other software vendors concentrate on particular application program markets, such as video games. There is also a "middle ground" between operating systems and application programs: the market for client-server and network management software. Significant players in this market include Novell, Sun, and (increasingly) Microsoft, although the open source community (discussed below) is rapidly emerging as the leading force in this field.

During the past decade, information technology and software tools have increasingly shifted onto the Internet. Internet service providers, such as America Online and Verizon, now play a critical role in the computer marketplace, as do search companies (Google, Yahoo!) and makers of web servers and routers (Cisco). The emergence of the Internet has dramatically changed the marketplace for computer software and the way software is produced.

The traditional software supply chain has been built on a proprietary business model centered around protection of intellectual property. As we will see throughout this chapter, various modes of protection exist for software. Microsoft, for example, relies heavily on trade secret protection of its "crown jewels" (its source code), copyright protection of its products, trademarks on its flagship products (e.g., Windows), and patents for new software technologies. By protecting and successfully marketing its products, Microsoft and many other traditional software companies employ large hierarchically organized research staffs developing and supporting new generations of software.

The rise of network computing has led to the emergence of several new software models. As Sun Microsystems proclaimed during the dot-com boom, "the network is the computer." Due to the low marginal cost of distribution and marketing (web sites), a growing number of entrepreneurs provide software products and services as freeware (proprietary software that can be used without charge (but subject to license terms)) and shareware (proprietary software distributed based on an honor system, often with requests for voluntary payment and registration). Such offers can quickly expand adoption of such software and provide a vehicle for other forms of financial return, such as advertising.

The Internet has provided a fertile environment for a radically different software development and supply model: open source. The ethos of open source development draws upon the "free" and "open" environment in which software pioneers operated. As noted above, mainframe manufacturers generally distributed software for their products at no additional charge. And unlike commercial software for microcomputers, mainframe software was available in source code, thereby enabling programmers to study, tinker with, debug, and improve the software.

In an effort to restore the open culture of software development and to promote software reliability, quality, and rapid evolution, Richard Stallman, a well-known MIT artificial intelligence professor, proposed that programmers invert the logic of proprietary software. See Steven Weber, The Success of Open Source (2004). This model envisions software creation as a collaborative venture, with improvements coming from a decentralized team of users and innovators. See Eric S. Raymond, The Cathedral and the Bazaar (1999) (contrasting the traditional proprietary software development model (like a cathedral designed from the top down built by coordinated teams) with the open source process ("a great babbling bazaar of different agendas and approaches"). Under the "free software" development model, programmers must: (1) make source code available; (2) allow others to redistribute the software without licensing fees; and (3) allow others to modify the software and distribute the derivative works under these same conditions. In this way, "free" software can spread virally as others build upon and improve products.

Stallman formalized these principles in the General Public License (GPL), which is sponsored by the Free Software Foundation. In addition, the Open Source Foundation has promulgated a parallel set of principles. Various other entities have developed licensing variations that reflect these principles to varying degrees.

The open source model reinforces the relatively open system of protocols on which the Internet has been built and benefits from the ease of distribution and collaboration in this medium. Open source also provides a means for commercial firms to move away from the dominant shadow cast by Microsoft. As a result, open source software has become a significant force on the Internet, with over two-thirds of all active web sites using Apache's HTTP Server software. Other open source projects, such as the Linux operating system, Mozilla's Firefox browser, the PERL programming language, and the Sendmail e-mail transfer and management program, are widely used, and many of the leading commercial computer companies — such as Sun Microsystems, IBM, Hewlett-Packard, Dell, Oracle, and Intel — have committed to supporting open source technology to varying degrees. Several government agencies have also committed to use of open source software.

2. Market Failures

Public Goods Problem. As discussed in Chapter 1, all markets for goods embodying intellectual property exhibit an externality commonly referred to as the "public goods" problem. Public goods have two distinguishing features: (1) nonexcludability (it is difficult to prevent those who do not pay for the good from consuming it); and (2) nonrivalrous competition (additional consumers of the good do not deplete the supply of the good available to others). The result of the public goods problem is well recognized: the private market will undersupply public goods because producers cannot reap the marginal value of their investment in providing such goods.

Innovations in computer software are good (although not perfect) examples of a public good. Given the availability of low-cost copying, it is difficult to exclude non-purchasers from the benefits of innovative computer programs once they are made commercially available. Even programs written in object code and stored in ROM can be readily copied by inserting the chip into a laboratory device and downloading the information onto another chip or onto paper. Moreover, one person's use of the information does not detract from any other person's use of that same information, since the information can be copied without affecting the original. Since the authors and creators of computer software cannot reap the marginal value of their efforts, economic theory suggests that in the absence of other incentives to innovate they will undersupply technological advances in computer software.

The rise of open source, however, suggests that the market failure theory explains only part of the software development story, at least in the present technological environment. See Josh Lerner and Jean Tirole, The Economics of Technology Sharing: Open Source and Beyond, 19 J. Econ. Perspectives 99 (2005). The fact that thousands of programmers freely contribute to open source software suggests that traditional market motivations may not be needed for certain types of software development. Such programmers may benefit indirectly through enhanced career opportunities. There is also growing evidence that they derive psychic rewards from participating in the open source culture. The open source environment may also improve their productivity by enabling programmers to solve specific problems directly. In addition, open source provides a means for commercial ventures to develop successful service businesses supporting open source users (e.g., Red Hat) and for more mainstream computer companies to get out from under the control of monopolists in the software industry. See Robert P. Merges, A New Dynamism in the Public Domain, 71 U. Chi. L. Rev. 183 (2004). Network effects may also help in explaining the growth of open source.

Standardization and Network Externalities. The second principal market failure in the computer software market arises from the presence of network externalities. These exist in markets for products for which the utility or satisfaction that a consumer derives from the product increases with the number of other consumers of the product. The telephone is a classic example of a product with network externalities. The benefits to a person from owning a telephone are a function of the number of other people owning telephones connected to the same telephone network; the more people on the network, the more people each person can call and receive calls from. Another classic network externality flows from standardization. In this case, the value of learning a particular standard (say, how to use a certain word processing program) depends on how many other people use that standard. Consider the prevalence of a "normal" typewriter keyboard.[3] Because almost all English language typewriters feature the same keyboard configuration, commonly referred to as "QWERTY," typists need learn only one keyboard system. This standardization enhances worker mobility and the breadth of products available to those who use the QWERTY keyboard.

3. See Paul A. David, CLIO and the Economics of QWERTY, 75 Am. Econ. Rev. 332 (May 1985).

Network externalities also inhere to product standards that allow for the inter-changeability of complementary products. Examples of products for which this type of network externality is important are videocassette recorders, CD players, and computer operating systems. As discussed above, general-purpose computer operating systems allow consumers to use a variety of application software programs on the same system-unit hardware. The only requirement is that the application program be coded to work on the operating system embedded in the general computer system. Thus, the operating system serves as a "compatibility nexus" for a particular computer network. Application software producers will develop more programs for systems that are widely used; hardware producers will develop more configurations of disk drives, memory, and other features for popular operating systems. In general, the benefits of a larger computer operating system network include a wider variety of application software that can run on that operating system, lower search costs for consumers seeking particular application programs that run on that operating system, reduced retraining costs, greater labor mobility, and wider availability of compatible hardware configurations and peripherals. Similarly, the greater the extent to which computer-human interfaces for particular application programs, such as word processing or spreadsheets, are standardized, the easier it will be for computer users to utilize their skills in different working environments. Computer users also want the computer-human interface to be as easy to use as possible — i.e., "user-friendly." In part, user friendliness is a function of what the user has already learned: things that are familiar tend to be easier to use. More generally, user friendliness is determined by the extent to which the design is tailored to achieve human factor goals such as maximizing performance speed, minimizing learning time, minimizing rate of errors, and maximizing retention over time.

Standardization can occur by way of market processes, whereby subsequent entrants to a market adopt the standard of an existing firm. This has been referred to as "bandwagon standardization." In the computer industry, firms often foster this process by freely licensing the use of a standard and by publishing design and inter-operability specifications. Linux is an example of an open standard promoted by certain private companies, such as IBM and others.

Alternatively, standardization can occur through formal processes, such as government standard-setting and joint development of industry standards by a number of firms. This top-down approach to standard-setting has been common in the telecommunications industry, at least until recently. Regulatory agencies, in consultation with the affected companies, set standards for connecting peripheral devices (including computers) to telephone lines. Only devices that comply with these publicly available standards can operate over the telephone lines.

Finally, industries characterized by strong network externalities will sometimes develop "de facto" standards, even though neither the companies in the industry nor the government has made any effort to settle on a particular standard or ensure interoperability. An example is the Microsoft Windows operating system. Microsoft did not "open" its operating system to competitors. Nonetheless, once the Microsoft operating system reached a certain "critical mass" of users, its very size began to make it more attractive. More application programmers wrote programs to run on Windows because of the large user base; more users adopted the Microsoft system in part because of all the programs being written for it. In 2004, Windows ran on 90 percent of the personal computers in the world, despite the closed nature of the system and despite the presence of several arguably superior competing systems.

An important economic consideration in markets with significant network externalities is whether computer firms will have the correct incentives to develop or adopt compatible products, thereby enlarging existing networks. Economists have demonstrated that computer firms might prefer to adopt noncompatible product standards even though their adoption of compatible products would increase net social welfare.[4] The explanation for this behavior is that by adopting a compatible standard, a firm enlarges the size of a network that comprises both the adopter's product and its rivals' products. This will have the effect of increasing the desirability of the rivals' products to consumers, thereby reducing the adopter's market share (although of a larger market) relative to what it would have been had the firm adopted a noncompatible product standard. Although this effect by no means implies that subsequent market entrants will never adopt a compatible operating system or computer-user interface, it does suggest that incentives to adopt compatible standards might be suboptimal.

While a widely adopted product standard can offer important benefits to consumers and firms, it can also "trap" the industry in an obsolete or inferior standard. In essence, the installed base built upon the "old" standard — reflected in durable goods and human capital (training) specific to the old standard — can create an inertia that makes it much more difficult for any one producer to break away from the old standard by introducing a noncompatible product, even if the new standard offers a significant technological improvement over the current standard.[5] In this way, network externalities can retard innovation and slow or prevent adoption of improved product standards. This problem is particularly significant in industries that develop rapidly, such as the computer industry.

As an example of the danger of stagnation, investigators cite the persistence of the standard QWERTY typewriter keyboard despite the apparent availability of a better key configuration developed and patented by August Dvorak and W. L. Dealey in 1932. U.S. Navy studies and world typing speed records suggest that the Dvorak Simplex Keyboard is significantly more efficient than the QWERTY system.[6] Adoption of the better standard appears to have been effectively stymied by "switching costs" — the costs of converting or replacing QWERTY keyboards and retraining those who use them. Because of the fear that national standards would exacerbate the inertia problem, the National Bureau of Standards declined to set interface standards for computers in the early 1970s. Subsequent developments proved that a wise decision.

Nonetheless, standards are essential in many cases. Although the U.S. government did not adopt computer interface standards in the 1970s, the hardware industry had to develop de facto standards so that computers and peripheral devices could "talk" to each other. The software industry faced similar problems in ensuring compatibility between operating systems and applications programs in the 1980s, and the computer networking industry faces the same problem today. The problem goes away if one company controls the market for all related goods, as AT&T did in the

---

4. See Michael L. Katz & Carl Shapiro, Network Externalities, Competition, and Compatibility, 75 Am. Econ. Rev. 424, 435 (May 1985).

5. See Joseph Farrell & Garth Saloner, Standardization, Compatibility, and Innovation, 16 Rand J. Econ. 70 (1985).

6. However, evidence recently presented by two economists casts doubt on the technological superiority of the Dvorak keyboard. See S. J. Liebowitz & Stephen E. Margolis, Network Externality: An Uncommon Tragedy, 8 J. Econ. Persp. 133, 147 (1994).

telephone industry until 1984. If AT&T is the only company to sell both telephone services and telephones, there is unlikely to be a problem in ensuring that your telephone works with your phone line. But this form of standardization comes at the expense of competition. Controlling standards and compatibility in dynamic industries is one of the most vexing policy problems the computer industry faces. The problem is exacerbated by the availability of intellectual property protection for many of the potential standards.

COMMENTS AND QUESTIONS

1. What means exist — contractual, intellectual property, business strategy, and otherwise — to appropriate the value of innovation in computer software and hardware? Is the intellectual work embodied in computer technology any more "leaky" than other technologies? Does this leakiness justify special protection — beyond traditional copyright and patent protection — for computer technology, especially software, or can this technology fit neatly within existing structures?

2. Does legal protection for software run counter to reliability, quality, and rapid evolution? Would we get better software if all programmers (and companies) followed an open source model? Does your answer depend on the type of software? the market demand characteristics? other factors?

3. Which of the following types of firms have greater incentives to adopt an existing standard than to develop a new standard: large firms with significant name recognition in related product markets, or small firms with little name recognition? Why? See Joseph Farrell, Standardization and Intellectual Property, 30 Jurimetrics J. 35, 38 (1989).

4. Does intellectual property protection for products featuring network externalities increase or decrease the likelihood that firms will seek to develop proprietary standards? How does it affect the likelihood that inefficient standards will arise? that inertia can be broken? How does strong intellectual property protection affect the likelihood that formal standards — either by governmental bodies, private standard setting organizations, or consortia of leading firms — will be adopted? Should formal standardization be encouraged?

5. How does the presence of network externalities with regard to computer software affect the analysis of intellectual property protection? Of what relevance are network externalities to the scope of copyright protection? patent? trademark? Should the network externalities of operating systems be treated in the same manner as network externalities of user interfaces?

6. How should computer software be protected under the intellectual property laws? In particular, consider the implications of the following aspects of computer programming, suggested in Pamela Samuelson et al., A Manifesto Concerning the Legal Protection of Computer Programs, 94 Colum. L. Rev. 2308, 2315-16 (1994):

> Computer programs have a number of important characteristics that have been difficult for legal commentators and decisionmakers to perceive. First, the primary source of value in a program is its behavior, not its text. Second, program text and behavior are independent in the sense that a functionally indistinguishable imitation can be written by a programmer who has never seen the text of the original program. Third, programs are, in fact, machines (entities that bring about useful results, i.e. behavior) that have been constructed in a medium of text (source and object code). The engineering designs

embodied in programs could as easily be implemented in hardware as in software, and the user would be unable to distinguish between the two. Fourth, the industrial designs embodied in programs are typically incremental in character, the result of software engineering techniques and a large body of practical know-how.

Does the fact that we are concerned primarily with the behavior of a machine suggest that patent law is the appropriate vehicle for protecting computer software? Does it matter that virtually all of these "machines" are in fact constructed from text? Does the incremental nature of software development suggest the contrary, that patent law is not the appropriate vehicle of protection.

7. One of the themes of section 1 *supra* is that programmers focus on efficiency in achieving commercial or technical goals. Should improvements in efficiency be protectable as intellectual property? If so, what is the best way to accomplish this?

8. Paul Goldstein writes:

> Science and technology are centripetal, conducting toward a single optimal result. One water pump can be better than another water pump, and the rule of patent and trade secret is to direct investment toward such improvements. Literature and the arts are centrifugal, aiming at a wide variety of audiences with different tastes.... The aim of copyright is to direct investment toward abundant rather than efficient expression.

Paul Goldstein, Infringement of Copyright in Computer Programs, 47 U. Pitt. L. Rev. 1119, 1123 (1986). For all its parallels to literature and art, computer programming is at base engineering — it involves the construction of a machine. Does this fact suggest that copyright and trademark are not well-suited to protecting computer software?

9. Computer programs have a number of different elements — purpose, structure, user interfaces, program interfaces, code, etc. Further, as noted above, different people program in different ways. Do these factors suggest that different types of intellectual property protection might be appropriate for different parts of a computer program? Is it possible that different intellectual property laws could protect the *same* program elements? Do you see any problems with such an overlap? See J. H. Reichman, Legal Hybrids Between the Patent and Copyright Paradigms, 94 Colum. L. Rev. 2432 (1994).

B. TRADE SECRET PROTECTION

Computers are different from other creations subject to intellectual property protection. To a significant degree, they are different because they are "hybrids": part machine, part writing, and part artistry. Because of this, and because of the strangeness of the new technology (it is hard to imagine a "literary work" composed entirely of 1s and 0s), intellectual property law has struggled to protect the intellectual work embodied in computer programs.

The main contours of legal protection for computer technology have emerged in the past two decades. Copyright law protects computer software, and it is generally acknowledged that competitors may not make multiple copies of someone else's software without permission, with important exceptions and qualifications discussed later.

Patent law protects computer hardware and new processes or "structures" embodied in computer software. But in the late 1970s and early 1980s, things were very different. Whether copyright protected computer programs at all was an open question, debated by the federal Commission on New Technological Uses of Copyrighted Works (CONTU)[7] and not resolved until the enactment of the 1980 Amendments to the Copyright Act. During this period, and even after the 1980 Amendments, courts disagreed over such fundamental issues as the copyrightability of object code and whether copyright protected computer programs against nonliteral infringement.

At the same time, a similar debate was raging in the federal courts over the patentability of computer software. The Supreme Court had held that a computer program for converting numerical data from one coding format to another was unpatentable in Gottschalk v. Benson, 409 U.S. 63 (1972), and it did not reopen the question of whether software could be patented again until 1981. Even after 1981, the patentability of "pure" computer software was open to dispute. The issue was sufficiently clouded in 1994, for example, that the Federal Circuit took In re Alappat, 33 F.3d 1526 (Fed. Cir. 1994) (en banc) in a vain attempt to clarify the standards for issuing software patents. While software patents now appear to be here to stay, their precise scope remains unclear.

Against this backdrop, it was important for software vendors to establish their rights vis-à-vis users. One way for vendors to prevent users from freely copying their computer programs was to claim those programs as a trade secret. Trade secrecy was a particularly viable alternative in the early days of the computer industry because most computer software was customized and sold to particular customers through detailed licensing agreements. Trade secrets have also been important in the computer industry as a means of protecting projects under development from being taken by departing employees.

The general principles and doctrines of trade secret law were set out in Chapter 2. You should briefly review those materials at this time, taking particular note of the Uniform Trade Secrets Act. The materials below highlight the distinctive manner in which trade secret law has been applied to computer technology.

While it was established early on that the secret elements of a computer program could be protected from misappropriation, computer companies that wanted to protect their intellectual property still had a problem. It is a fundamental principle of trade secret law that the information protected must remain a secret. If the secret is disclosed to the public, protection is forever lost. Computer companies did not want to keep their hardware or software "secret" — they wanted to sell them! An influential early case held that they could do both.

Data General Corp. v. Digital Computer Controls, Inc.
Delaware Court of Chancery
297 A.2d 433 (Del. Ct. Chanc. 1971), aff'd, 297 A.2d 437 (Del. S. Ct. 1972)

MARVEL, Vice Chancellor:

Data General Corporation seeks an order preliminarily enjoining the defendant Digital Computer Controls from making use of claimed trade secrets allegedly contained in design drawings which accompany certain sales of plaintiff's Nova 1200

---

7. Final Report of the National Commission on New Technological Uses of Copyrighted Works (U.S. Govt. Printing Office 1978).

computer, defendant having acquired such a computer with accompanying drawings from one of plaintiff's customers. This is the decision of the Court on plaintiff's motion for a preliminary injunction as well as on defendants' cross-motion for summary judgment of dismissal of the present action.

The relevant facts thus far adduced are as follows: During the past several years plaintiff has developed and marketed successfully small general-purpose computers to which they have given the name Nova, a large part of plaintiff's research and development budget for the past several years having been allocated to the development of such small computers, which, the parties agree, are the only ones of their type presently being profitably marketed, there having been no device on the market comparable to plaintiff's machines until defendant's entry on the scene.

When a sale of a Nova 1200 computer is made, plaintiff makes available at no extra cost to those customers who wish to do their own maintenance the design or logic drawings of the device sold, in this case a Nova 1200 computer. Such type of maintenance has been found to be desired by some customers in order to avoid periods of unproductive delay while waiting for repairs to be made by plaintiff's trained personnel.

Design drawings made available to customers are furnished subject to the terms of a non-disclosure clause contained in a paper which accompanies a purchase agreement. Furthermore, all drawings bear a legend to the effect that they contain proprietary information of the plaintiff which is not to be used by a purchaser for manufacturing purposes. However, the Nova 1200 is not patented, and its design drawings have not been copyrighted.

In April, 1971, the defendant Ackley, the president of Digital, purchased for his company from a customer of Data General, namely Mini-Computer Systems, Inc., a Nova 1200 computer which the latter had earlier acquired from plaintiff. Although the exact circumstances surrounding such sale are not entirely clear on the present record, the corporate defendant's president, in consummating such purchase, acquired a set of design drawings of the Nova 1200, said drawings having been furnished to MiniComputer by the plaintiff in order to facilitate maintenance by the former of its Nova 1200's. The corporate defendant thereafter proceeded to use such design drawings as a pattern for the construction of a competing machine which it is now about to market, the corporate defendant's president conceding that only minor changes have been made in the basic design of plaintiff's device in the course of the development of defendant's comparable computer.

[The court held that plaintiff's trade secret claims are not preempted by the federal intellectual property or antitrust laws.]

II

Defendants' other arguments are accordingly based on the law of trade secrets. Thus, in order for the plaintiff to establish its right to relief here, it must demonstrate (1) the existence of a trade secret and that the corporate defendant has either (2) received the information within the confines of a confidential relationship and proposes to misuse the information in violation of such relationship, or (3) that the corporate defendant improperly received the information in question in such a manner that its confidential nature should have been known to it and that it nonetheless proposes to misuse such information.

Trade secrets have been defined in Restatement, Torts §757, comment b. as follows: "A trade secret may consist of any formula, pattern, device or compilation of information which is used in one's business, and which gives him an opportunity to obtain an advantage over competitors who do not know or use it."

Defendants insist, however, that plaintiff has not maintained that degree of secrecy which will preserve its right to relief, either by publicly selling an article alleged to contain a trade secret, or by failing to restrict access to the design drawings for its device, arguing that matters of common knowledge in an industry may not be claimed as trade secrets.

It has been recognized in similar cases that even though an unpatented article, device or machine has been sold to the public, and is therefore subject to examination and copying by anyone, the manner of making the article, device or machine may yet constitute a trade secret until such a copy has in fact been made, Schulenburg v. Signatrol, Inc., 33 Ill. 2d 379, 212 N.E.2d 865, and Tabor v. Hoffman, N.Y., 118 N.Y. 30, 23 N.E. 12.

Defendants contend, however, that the issuance by plaintiff of copies of design drawings to its customers was made without safeguards designed properly to maintain the secrecy requisite to the existence of a trade secret. In other words, it is contended that plaintiff's attempts to maintain secrecy merely consisted of (1) not giving copies of the design drawings to those customers who did not need them for maintenance of their computer, (2) obtaining agreements not to disclose the information from those customers who were given copies of the drawings, and (3) printing a legend on the drawings which contained the allegedly confidential information which identified the drawing as proprietary information, the use of which was restricted. Plaintiff argues, however, that disclosure of the design drawings to purchasers of the computer is necessary properly to maintain its device, that such disclosure was required by the very nature of the machine, and that reasonable steps were taken to preserve the secrecy of the material released. I conclude at this preliminary stage of the case that it cannot be held as a matter of law that such precautions were inadequate, a factual dispute as to the adequacy of such precautions having clearly been raised. Defendants' motion for summary judgment must accordingly be denied.

Finally, I am satisfied that if plaintiff were to prevail at final hearing, it would only be entitled to injunctive relief during that undetermined period of time which would be required for defendants substantially to reproduce the plaintiff's device without its accompanying drawings. Such period of time would, on the present record, vary according to the number of man hours devoted to so-called reverse engineering. Thus, the granting of a preliminary injunction at this juncture would grant plaintiff all the relief it might hope ultimately to obtain after final hearing. Compare Thomas C. Marshall, Inc. v. Holiday Inn, Inc., 40 Del. Ch. 77, 174 A.2d 27. In addition, while plaintiff has presented evidence of its attempts to preserve the secrecy of the alleged secrets it now seeks to have protected by injunction, it has not adduced sufficient evidence to establish that likelihood of ultimate success on final hearing which would entitle it to the issuance of a preliminary injunction.

On notice, an order may be presented denying defendants' motion for summary judgment as well as plaintiff's motion for a preliminary injunction, which order shall also include a provision that all drawings issued by the corporate defendant in connection with the sale of its Nova-type computer contain a restrictive legend of the type now set forth on plaintiff's drawings here in issue.

COMMENTS AND QUESTIONS

1. Even at a very early stage in the history of computer law, there was no serious question that computer hardware and software *could* be protected by trade secret law. See, e.g., University Computing Co. v. Lykes-Youngstown Corp., 504 F.2d 518 (5th Cir. 1974) (plaintiff's computer system was a protectable trade secret). Cases such as *Lykes-Youngstown* involved a fairly typical trade secret situation — people with access to a secret program still under development attempting to use or sell their own product based in part on the program. See also Cybertek Computer Products v. Whitfield, 203 U.S.P.Q. 1020 (Cal. Super. 1977).

But the contours of trade secret protection of computer programs are not always clear. For example, in Rivendell Forest Prods. v. Georgia-Pacific Corp., 28 F.3d 1042 (10th Cir. 1994), the Tenth Circuit reversed a district court opinion holding that Rivendell was not entitled to protect the basic concepts of organizing a price quote system against misappropriation by others. (The district court opinion is reported at 824 F. Supp. 961 (D. Colo. 1993).) The use of trade secret law to protect such abstract, high-level concepts is somewhat troubling. On the one hand, the district court is surely correct that if Rivendell's ideas are not in fact secret, Georgia-Pacific is free to copy them. On the other hand, the Court of Appeals points out that basic concepts are protectable under trade secret law in certain combinations, if those combinations are not generally known in the industry. On this issue, compare Integrated Cash Mgmt. Serv. v. Digital Transactions, Inc., 920 F.2d 171 (2d Cir. 1990) (finding the combination of publicly known utility programs into a specific arrangement is protectable as a trade secret) with Comprehensive Technologies, Inc. v. Software Artisans, Inc., 3 F.3d 730 (4th Cir. 1993) (combination of known elements did not qualify as a trade secret where the particular combination was logically required and common in the industry). Cf. Vermont Microsystems Inc. v. Autodesk Inc., 88 F.3d 142 (2d Cir. 1996) (holding that a shading algorithm could be protected as a trade secret even though other developers had independently created similar algorithms, because defendant derived value from using plaintiff's algorithm).

2. The plaintiff in *Data General* sold over 500 Nova computers to the general public. Each purchaser who requested one got a copy of the "confidential" design drawings. Why are these drawings still considered a secret? Does widespread disclosure compromise the secrecy claim at some point, even though all disclosures are made under an agreement of confidentiality? This issue arises frequently today in the software industry. As computers have become ubiquitous in business and quite common in the home, the numbers of "secret" programs in circulation may be counted in the millions rather than the hundreds. *Data General* implicitly concludes that even a relatively widespread disclosure to customers does not compromise the secrecy of the computer design. For cases addressing this issue in the context of computer software, compare Management Science of Am. v. Cyborg Sys., Inc., 1977-1 Trade Cas. (CCH) ¶61,472 (N.D. Ill. 1977) (holding that distribution of 600 copies of a program under a confidentiality agreement did not destroy secrecy) with Young Dental Mfg. Co. v. Q3 Special Prods., Inc., 891 F. Supp. 1345 (E.D. Mo. 1995) (characterizing as "completely frivolous" plaintiff's claim that its publicly sold software was a trade secret).

A closely related question involves attempts by the owners of information to "contract around" the requirement of secrecy. If the parties agree to treat a piece of information as secret, is the licensee bound not to use or disclose the information under contract principles regardless of whether or not it is in fact in the public domain? This issue is a recurring one in intellectual property law, and a problem that has never adequately been addressed.

3. Customers who buy a product on the open market are entitled to break it apart to see how it works. This process is called "reverse engineering" the product. Trade secret law does not protect owners against legitimate purchasers who discover the secret through reverse engineering. But does the possibility that a product might be reverse engineered foreclose *any* trade secret protection? At least one court has said no. In Data General Corp. v. Grumman Systems Support Corp., 825 F. Supp. 340, 359 (D. Mass. 1993), the court upheld a jury's verdict that Grumman had misappropriated trade secrets contained in object code form in Data General's computer program, despite the fact that many copies of the program had been sold on the open market. The court reasoned: "With the exception of those who lawfully licensed or unlawfully misappropriated MV/ADEX, Data General enjoyed the exclusive use of MV/ADEX. Even those who obtained MV/ADEX and were able to *use* MV/ADEX were unable to discover its trade secrets because MV/ADEX was distributed only in its object code form, which is essentially unintelligible to humans." The court noted that Data General took significant steps to preserve the secrecy of MV/ADEX, requiring that all users of the program sign licenses agreeing not to disclose the program to third parties. Under this decision, a defendant may have to prove that they had some sort of legitimate access to the plaintiff's information — for example, by demonstrating that they reverse-engineered it from a publicly available product — even though the product containing the secret is widely distributed.

The *Data General* cases suggest that reasonable efforts to protect the secrecy of an idea contained in a commercial product — such as locks, black boxes, or the use of unreadable code — may suffice to maintain trade secret protection even after the product itself is widely circulated. Does this result make sense? For a different approach, see Videotronics v. Bend Electronics, 564 F. Supp. 1471, 1476 (D. Nev. 1983) (holding that software cannot be a trade secret if it is publicly distributed and can be readily copied).

4. One way in which computer software cases may be distinguished from other trade secret cases has to do with the nature of what is being distributed. While a particular computer program may be widely distributed, in fact all that is sold to the consumer is a disk containing object code. Object code is virtually impossible for humans to read without machine assistance.[8] Because of this, computer software is in some sense unlike a physical product whose design is evident to the casual observer. Even after it is publicly distributed, object code is meaningless to the casual observer. Only a complex process of reverse engineering (sometimes called "disassembly" or "decompilation") can enable the user to decipher the source code that was originally written for the program.

Should it matter that a computer program is distributed only in object code form? Consider the following case, in which the defendant was accused of misappropriating a computer program in object code form:

---

8. It is possible to "reverse engineer" object code in some cases to create a kind of rough estimate of what must have been in the original source code. The process, however, is demanding and time consuming even for expert programmers. See, e.g., Andrew Johnson-Laird, Technical Demonstration of "Decompilation," paper presented at the Computer Law Association's 1993 International Computer Law Retreat, October 21-23, 1993, at 29.

The source code can and does qualify as a trade secret....

Whether the object code is a trade secret is a more difficult question.[7] Atkinson first contends that the object code cannot be a trade secret because it does not derive independent economic value from its secrecy, and therefore fails the first definitional requirement of a trade secret. This argument has no merit. Trandes generates most of its revenues by providing computer services....Armed with a copy of the object code, an individual would have the means to offer much the same engineering services as Trandes....

Atkinson next argues that the object code cannot be a trade secret because Trandes did not keep it secret....Atkinson asserts that the Tunnel System has been widely disclosed as a mass-marketed product and that its existence and its abilities are not secret. [The court concluded that the object code remained secret because it had only been distributed to two customers, and both of them signed licenses agreeing to keep the program a secret.]

Trandes Corp. v. Guy F. Atkinson Co., 996 F.2d 655, 663-64 (4th Cir. 1993). Consider the court's footnote. Can object code be a trade secret if it can easily be duplicated (whether or not the copier understands what he is copying)? Is the plaintiff in this case really trying to leverage copyright protection out of a trade secret claim? In both the *CTI* and *ICM* cases, cited above, the alleged trade secret at issue was not the source or object code of the computer program itself but certain high-level design features of the program (its "architecture"). Suppose that, rather than using what they had learned of the architecture of the program while employed by the company, ICM's former employees had copied the object code of the program altogether. (Leave aside for a moment questions of copyright infringement, and consider only the trade secret issue.) Would they be liable for misappropriating the trade secrets contained in the program architecture on the grounds that copying the program in its entirety necessarily copied the architecture? Or would the fact that the object code was publicly disclosed protect them from liability? How would the courts in *Trandes* and the *Data General* cases answer this question? Does the answer suggest a problem with relying on trade secrecy to protect computer programs?

5. While there once was a question as to the availability of copyright and/or patent protection for computer technology, there no longer is. In view of this fact, to what extent should hardware and software manufacturers rely on trade secret protection? Are there disadvantages to trade secret law as a means of protection for innovations in computer technology?

Section 301 of the Copyright Act of 1976, 17 U.S.C. §301, preempts state intellectual property rights that are equivalent to any of the rights granted by the federal copyright law. There has been a good deal of litigation in the computer industry over federal copyright preemption of state trade secret law. Most (but not all) have found no preemption. See, e.g., Computer Associates v. Altai, Inc., 982 F.2d 693 (2d Cir. 1992). If the trade secret laws are interpreted to preclude the copying of object code in the previous example, should they be preempted by copyright law?

---

7. This case presents an unusual set of facts. In the ordinary case, the owner of trade secret computer software will maintain the secrecy of the source code but freely distribute the object code. See, e.g., Q-Co Indus. v. Hoffman, 625 F. Supp. 608, 617 (S.D.N.Y. 1985) (program secret where source code secret, even though object code not secret). In such cases, the owner of the software cannot claim trade secret protection for the object code because its disclosure to the public destroyed its secrecy. In this case, however, Trandes maintained the secrecy of the source code and the object code, as we explain below.

PROBLEM

Problem 7-1. Microsoft Corp.'s Disk Operating System (MS-DOS) became the industry standard for personal computer operating systems. Microsoft sold several million copies of MS-DOS each year; there are estimated to be upwards of 25 million copies in circulation worldwide. Further, MS-DOS was normally "pre-installed" on many new personal computers, so that PC purchasers automatically purchased the operating system as well. Assuming that Microsoft has taken reasonable efforts to require both hardware manufacturers and end users to keep the program confidential, is Microsoft entitled to protect MS-DOS as a trade secret? What efforts would be "reasonable" under these circumstances?

Assume for the moment that Microsoft has succeeded in maintaining the secrecy of its computer operating system and has taken reasonable efforts to protect that secrecy. Are competing application programmers entitled to decompile the operating system to prepare application programs that will run on Microsoft's OS? Is there anything Microsoft can do that will prevent its competitors from legally reverse engineering its product?

C. COPYRIGHT LAW

Computer software, by its very nature as written work intended to serve utilitarian purposes, defies easy categorization within our intellectual property system. The copyright law has traditionally served as the principal source of legal protection for literary and artistic work, while the patent system and trade secret law have been the primary means for protecting utilitarian works. Faced with the difficult challenge of fitting computer and other new information technologies under the existing umbrella of intellectual property protection, Congress in 1974 established the National Commission on New Technological Uses of Copyrighted Works (CONTU) to study the implications of the new technologies and recommend revisions to federal intellectual property law. After conducting extensive hearings and receiving expert reports, a majority of the blue-ribbon panel of copyright authorities and interest group representatives comprising CONTU concluded in 1978 that the intellectual work embodied in computer software should be protected under copyright law, notwithstanding the fundamental principle that copyright cannot protect "any idea, procedure, process, system, method of operation, concept, principle, or discovery." 17 U.S.C. § 102(b). Congress adopted CONTU's recommendations in 1980, passing legislation almost identical to that suggested in the Final Report.

In light of the computer software industry's relative youth and anticipated rapid growth, CONTU's rough empirical judgment that copyright would best promote the invention, development, and diffusion of new and better software products was, by necessity, highly speculative. As CONTU recognized, it was impossible in 1978 to establish a precise line between copyrightable expression of computer programs and the uncopyrightable processes that they implement. Yet the location of this line — the idea/expression dichotomy — was critical to the rough cost-benefit analysis that guided CONTU's recommendation. Drawing the line too liberally in favor of copyright protection would bestow strong monopolies upon those who develop operating systems that become industry standards and upon the first to write programs performing specific applications and would thereby inhibit other creators from developing improved programs and computer systems. Drawing the line too conservatively would allow programmers' efforts to be copied easily, thus discouraging the creation of all but modest incremental advances. The wisdom of Congress's decision to bring computer programs within the scope of copyright law thus depends critically upon where the courts draw this line.

We begin with the first generation of copyright infringement suits under the 1980 Amendments; these cases focus on whether and to what extent literal copying of computer software violates copyright law. Because the coding of operating systems is critically important to hardware system compatibility, the principal economic effect of the first generation of cases was on competition among hardware manufacturers. The following section examines the second generation of copyright infringement suits, focusing on the extent to which nonliteral forms of copying constitute copyright infringement. We then turn to protection for operating elements of programs and program outputs.

1. The Scope of Software Copyright

a. Protection for Literal Elements of Program Code

The first copyright cases involving computer programs involved the simplest form of copyright infringement — direct copying of the program code. Because the copying in these early cases was "literal" — that is, the accused infringer took the actual text of the work, not merely the structure, organization, or output of the programs — there is no question of scope of protection or "substantial similarity." Rather, the disputed issue in these early cases was whether computer programs could be protected by copyright at all. Some early cases held that object codes or operating systems were not copyrightable. See Data Cash Systems v. JS & A Group, 480 F. Supp. 1063 (N.D. Ill. 1979). The rationale was that such works were not "communicative."

CONTU Commissioner John Hersey, in his dissent to CONTU's Final Report, highlighted the fact that computer "[p]rograms are profoundly different from the various forms of 'works of authorship' secured under the Constitution by copyright. Works of authorship have always been intended to be circulated to human beings and to be used by them — to be read, heard, or seen, for either pleasurable or practical ends. Computer programs, in their mature phase, are addressed to machines." Hersey argued that copyright protection should "not extend to a computer program in the form in which it is capable of being used to control computer operations."

Do you agree with Commissioner Hersey's critique of bringing computer software, in all its forms, within the copyright law? Does the Constitution limit copyright protection for "works of authorship" solely to works intended to be directly "read, heard, or seen" by humans? Should protection for computer software be limited in this way? Would Hersey's proposed limitation on the scope of copyright protection for software better balance the social interests affected by legal protection for software? What might justify CONTU's decision to accord software full inclusion within the copyright law, irrespective of whether it is directed toward operating machines or communicating with humans? For a detailed discussion of the problems raised by Commissioner Hersey in his dissent, see Pamela Samuelson, CONTU Revisited:

The Case Against Copyright Protection for Computer Programs in Machine-Readable Form, 1984 Duke L.J. 663 (1984).

Despite these arguments, it was well settled by the early 1980s that computer programs were copyrightable, just like any other literary work. See, e.g., Apple Computer v. Franklin Computer, 714 F.2d 1240 (3d Cir. 1983) (rejecting argument that operating system is not copyrightable on grounds that it is a process, system, or method of operation, or the idea/expression dichotomy; and observing that achieving compatibility with independently developed application programs "is a commercial and competitive objective which does not enter into the somewhat metaphysical issue of whether particular ideas and expressions have merged"); Williams Electronics v. Artic Int'l, 685 F.2d 870 (3d Cir. 1982). No cases since that time have held otherwise.

This argument has some surprising implications. One of the fundamental rules of computing is that anything that can be implemented in software can also, in principle, be implemented in hardware. Indeed, "software" itself is at base nothing more than a temporary (rather than permanent) way of ordering circuit switches in hardware. So if object code is copyrightable, why not more permanent instantiations in hardware — the microcode instructions embedded in computer chips, for example, or the layout of a circuit board itself?

Copyright law does sometimes protect the design elements of utilitarian objects, such as the sculpture/lamp of Mazer v. Stein, discussed in Chapter 4. While courts have had difficulty at times separating the artistic from the utilitarian, they have steadfastly refused to extend copyright protection to the utilitarian aspects of three-dimensional objects such as lamps, jewelry, and mannequins. Protecting object code and microcode raises many of the same issues. How should courts separate the unprotectable utilitarian aspects of a computer program from the protectable expressive aspects?

PROBLEM

Problem 7-2. Letni Corp. produces microprocessors, the central component of microcomputer systems. These "computers on a chip" are integrated circuits capable of performing arithmetic functions, manipulation of data, and other operations in response to object code instructions. Letni's 684 microprocessor currently commands a 70 percent market share of the microcomputer market. Letni's 684 chip utilizes a single "bus" or path for transferring data. It features 512 lines of microcode, each 21 bits long, which enables the microprocessor to interpret 133 assembly instructions.

CEN Corp., a competing microprocessor manufacturer, seeks to sell semiconductor chips capable of running the same software as Letni's 684 microprocessor. A prior patent cross-licensing agreement entitles CEN to duplicate Letni's 684 microarchitecture and hardware to the extent comprehended by the Letni patent. To increase processing speed and capability relative to Letni's 684 chip, CEN designed its U2 microprocessor with a dual bus system and a larger micro instruction word size (29 bits). The 1024 lines of microcode enable the microprocessor to interpret 156 assembly instructions; 552 of these lines carry out the instruction set of the Letni 684 microprocessor. The remaining lines perform other functions not carried out by the 684 chip. Because of the dual bus architecture and larger word size, CEN was able to write shorter and more efficient microcode.

Nonetheless, there were a number of similarities in the microcodes of the two microprocessors. Letni brought suit, alleging: (1) microprograms are computer programs and hence protected by the Copyright Act; and (2) CEN's U2 microcode infringed the copyright of Letni's 684 chip.

How should this case be resolved? Should microcode be copyrightable? If so, what should be the scope of protection? Can a hardware manufacturer obtain copyright protection for the configuration of the hardwiring of its computer? If so, under what theory? If your answer is different from that for microcode, what explains the distinction?

b. Protection for Non-literal Elements of Program Code

Apple v. Franklin established that exact copying of detailed computer code infringes the programmer's copyright in the code. This ruling served the important purpose of prohibiting outright piracy of computer programs. Soon thereafter, a second generation of computer software cases raised the question whether competitors may copy non-literal elements of a computer program, such as the program's underlying structure, sequence, or organization. As we saw in Chapter 4, the scope of copyright protection may extend beyond the literal elements of a work. But how far should these doctrines be extended in the context of computer programs?

The first major appellate test of these limits arose in Whelan Associates, Inc. v. Jaslow Dental Laboratory, Inc., 797 F.2d 1222 (3d Cir. 1986), *cert. denied* 479 U.S. 1031 (1987). The owner of a dental laboratory hired a custom software firm to develop a computer program that would organize the bookkeeping and administrative tasks of its business. Elaine Whelan, the principal programmer, interviewed employees about the operation of the laboratory and then developed a program to run on the laboratory's IBM Series One computer. Under the terms of an agreement, Whelan Associates retained the copyright in the program and agreed to use its best efforts to improve the program while Jaslow Laboratory agreed to use its best efforts to market the program. Rand Jaslow, an officer and shareholder of the laboratory, set out to create a version of the program that would run on other computer systems. Whelan sued for copyright infringement. At trial, the evidence showed that the Jaslow program did not literally copy Whelan's code, but there were overall structural similarities between the two programs. As a means of distinguishing protectable expression from unprotectable idea, the court reasoned:

> *[T]he purpose or function of a utilitarian work would be the work's idea, and everything that is not necessary to that purpose or function would be part of the expression of the idea.* Where there are many means of achieving the desired purpose, then the particular means chosen is not necessary to the purpose; hence, there is expression, not idea.

Id. at 1236 (emphasis in original). In applying this rule, the court defined the idea as "the efficient management of a dental laboratory," for which countless ways of expressing the idea would be possible. Drawing the idea/expression dichotomy at such a high level of abstraction implies an expansive scope of copyright protection. Furthermore, the court's conflation of merger analysis and the idea/expression dichotomy implicitly allows protection under copyright of procedures, processes, systems, and methods of operation, which are expressly excluded under § 102(b). Although the case did not directly address copyright protection for computer code establishing interoperability protocols for computer systems, the court's mode of analysis dramatically expanded the scope of copyright protection for computer programs. If everything below the general purpose of the program was protectable under copyright, then it would follow that particular protocols were protectable because there would be other ways of serving the general purpose of the program. Such a result would effectively bar competitors from developing interoperable programs and computer systems.

The *Whelan* test was roundly criticized by commentators. A Fifth Circuit decision handed down shortly thereafter declined to follow *Whelan*, finding that the similarities two programs used to facilitate cotton market transactions were dictated largely by standard practices in the cotton market (what the court called "externalities") which constitute unprotectable ideas. See Plains Cotton Cooperative Assoc. v. Goodpasture Computer Service, Inc., 807 F.2d 1256 (5th Cir. 1987). The court instead found persuasive the decision in Synercom Technology, Inc. v. University Computing Co., 462 F.Supp 1003 (N.D.Tex. 1978), in which Judge Higginbotham analogized the "input formats" of a computer program (the organization and configuration of information to be input into a computer) to the inherently utilitarian "figure-H" pattern of an automobile stick shift.

> Several different patterns may be imagined, some more convenient for the driver or easier to manufacture than others, but all representing possible configurations.... The pattern (analogous to the computer "format") may be expressed in several different ways: by a prose description in a driver's manual, through a diagram, photograph, or driver training film, or otherwise. Each of these expressions may presumably be protected through copyright. But the copyright protects copying of the particular expressions of the patterns, and does not prohibit another manufacturer from marketing a car using the same pattern. Use of the same pattern might be socially desirable, as it would reduce the retraining of drivers.

Id. at 1013. Five years later, the Second Circuit confronted copyright protection for nonliteral elements of program code.

═══ ### Computer Associates International v. Altai, Inc.
═══ *United States Court of Appeals for the Second Circuit*
═══ *982 F.2d 693 (2d Cir. 1992)*

WALKER, Circuit Judge:...

This appeal comes to us from the United States District Court for the Eastern District of New York, the Honorable George C. Pratt, Circuit Judge, sitting by designation. By Memorandum and Order entered August 12, 1991, Judge Pratt found that defendant Altai, Inc.'s ("Altai"), OSCAR 3.4 computer program had infringed plaintiff Computer Associates' ("CA"), copyrighted computer program entitled CA-SCHEDULER. Accordingly, the district court awarded CA $364,444 in actual damages and apportioned profits. Altai has abandoned its appeal from this award. With respect to CA's second claim for copyright infringement, Judge Pratt found that Altai's OSCAR 3.5 program was not substantially similar to a portion of CA-SCHEDULER called ADAPTER, and thus denied relief....

II. Facts...

The subject of this litigation originates with one of CA's marketed programs entitled CA-SCHEDULER. CA-SCHEDULER is a job scheduling program designed for IBM mainframe computers. Its primary functions are straightforward: to create a schedule specifying when the computer should run various tasks, and then to control the computer as it executes the schedule. CA-SCHEDULER contains a sub-program entitled ADAPTER, also developed by CA. ADAPTER is not an independently marketed product of CA; it is a wholly integrated component of CA-SCHEDULER and has no capacity for independent use.

Nevertheless, ADAPTER plays an extremely important role. It is an "operating system compatibility component," which means, roughly speaking, it serves as a translator. An "operating system" is itself a program that manages the resources of the computer allocating those resources to other programs as needed. The IBM System 370 family of computers, for which CA-SCHEDULER was created, is, depending upon the computer's size, designed to contain one of three operating systems: DOS/VSE, MVS, or CMS. As the district court noted, the general rule is that "a program written for one operating system, e.g., DOS/VSE, will not, without modification, run under another operating system such as MVS." Computer Assocs., 775 F. Supp. at 550. ADAPTER's function is to translate the language of a given program into the particular language that the computer's own operating system can understand....

A program like ADAPTER, which allows a computer user to change or use multiple operating systems while maintaining the same software, is highly desirable. It saves the user the costs, both in time and money, that otherwise would be expended in purchasing new programs, modifying existing systems to run them, and gaining familiarity with their operation. The benefits run both ways. The increased compatibility afforded by an ADAPTER-like component, and its resulting popularity among consumers, makes whatever software in which it is incorporated significantly more marketable.

Starting in 1982, Altai began marketing its own job scheduling program entitled ZEKE. The original version of ZEKE was designed for use in conjunction with a VSE operating system. By late 1983, in response to customer demand, Altai decided to rewrite ZEKE so that it could be run in conjunction with an MVS operating system.

[At that time, James P. Williams, then an employee of Altai and now its President, recruited Claude F. Arney, III, a long-standing friend and computer programmer who worked for CA, to assist Altai in designing an MVS version of ZEKE. Unknown to Williams, Arney was intimately familiar with CA's ADAPTER program and he took VSE and MVS source code versions of ADAPTER with him when he left CA to join Altai. Without disclosing his knowledge of ADAPTER, Arney persuaded Williams that the best way to modify ZEKE to run on an MVS operating system was to introduce a "common system interface" component, an approach that stemmed from Arney's familiarity with ADAPTER. Arney subsequently developed a component-program named OSCAR using the ADAPTER source code. Approximately 30% of the first generation of OSCAR was copied from CA's ADAPTER program. In mid 1988, CA discovered the copying from ADAPTER and brought this copyright infringement and trade secret action. Altai learned of the copying from the complaint.]

Upon advice of counsel, Williams initiated OSCAR's rewrite. The project's goal was to save as much of OSCAR 3.4 as legitimately could be used, and to excise those portions which had been copied from ADAPTER. Arney was entirely excluded from the process, and his copy of the ADAPTER code was locked away. Williams put eight other programmers on the project, none of whom had been involved in any way in the development of OSCAR 3.4. Williams provided the programmers with a description of the ZEKE operating system services so that they could rewrite the appropriate code. The rewrite project took about six months to complete and was finished in mid-November 1989. The resulting program was entitled OSCAR 3.5.

From that point on, Altai shipped only OSCAR 3.5 to its new customers....

Discussion

[The district court concluded that version 3.5 was not substantially similar to CA's ADAPTER.]

I. Copyright Infringement...

As a general matter, and to varying degrees, copyright protection extends beyond a literary work's strictly textual form to its non-literal components. As we have said, "[i]t is of course essential to any protection of literary property that the right cannot be limited literally to the text, else a plagiarist would escape by immaterial variations." Nichols v. Universal Pictures Co., 45 F.2d 119, 121 (2d Cir. 1930) (L. Hand, J.), *cert. denied*, 282 U.S. 902, 51 S. Ct. 216, 75 L. Ed. 795 (1931). Thus, where "the fundamental essence or structure of one work is duplicated in another," 3 Nimmer, § 13.03(A)[1], at 13-24, courts have found copyright infringement.... This black letter proposition is the springboard for our discussion.

A. Copyright Protection for the Non-literal Elements of Computer Programs

It is now well settled that the literal elements of computer programs, i.e., their source and object codes, are the subject of copyright protection.... Here, as noted earlier, Altai admits having copied approximately 30% of the OSCAR 3.4 program from CA's ADAPTER source code, and does not challenge the district court's related finding of infringement.

In this case, the hotly contested issues surround OSCAR 3.5. As recounted above, OSCAR 3.5 is the product of Altai's carefully orchestrated rewrite of OSCAR 3.4. After the purge, none of the ADAPTER source code remained in the 3.5 version; thus, Altai made sure that the literal elements of its revamped OSCAR program were no longer substantially similar to the literal elements of CA's ADAPTER.

According to CA, the district court erroneously concluded that Altai's OSCAR 3.5 was not substantially similar to its own ADAPTER program. CA argues that this occurred because the district court "committed legal error in analyzing [its] claims of copyright infringement by failing to find that copyright protects expression contained in the non-literal elements of computer software." We disagree.

CA argues that, despite Altai's rewrite of the OSCAR code, the resulting program remained substantially similar to the structure of its ADAPTER program. As discussed above, a program's structure includes its non-literal components such as general flow charts as well as the more specific organization of inter-modular relationships, parameter lists, and macros. In addition to these aspects, CA contends that OSCAR 3.5 is also substantially similar to ADAPTER with respect to the list of services that both ADAPTER and OSCAR obtain from their respective operating systems. We must decide whether and to what extent these elements of computer programs are protected by copyright law.

[The court agrees with Whelan v. Jaslow that the nonliteral elements of computer programs are entitled to copyright protection as literary works.]

1) Idea vs. Expression Dichotomy

It is a fundamental principle of copyright law that a copyright does not protect an idea, but only the expression of the idea. . . .

Drawing the line between idea and expression is a tricky business. Judge Learned Hand noted that "[n]obody has ever been able to fix that boundary, and nobody ever can," Nichols, 45 F.2d at 121. Thirty years later his convictions remained firm. "Obviously, no principle can be stated as to when an imitator has gone beyond copying the 'idea,' and has borrowed its 'expression,'" Judge Hand concluded. "Decisions must therefore inevitably be ad hoc." Peter Pan Fabrics, Inc. v. Martin Weiner Corp., 274 F.2d 487, 489 (2d Cir. 1960).

The essentially utilitarian nature of a computer program further complicates the task of distilling its idea from its expression. See SAS Inst., 605 F. Supp. at 829; cf. Englund, at 893. In order to describe both computational processes and abstract ideas, its content "combines creative and technical expression." See Spivack, at 755. The variations of expression found in purely creative compositions, as opposed to those contained in utilitarian works, are not directed towards practical application. For example, a narration of Humpty Dumpty's demise, which would clearly be a creative composition, does not serve the same ends as, say, a recipe for scrambled eggs — which is a more process oriented text. Thus, compared to aesthetic works, computer programs hover even more closely to the elusive boundary line described in § 102(b).

[The court reviewed the facts and holding of Baker v. Selden].

To the extent that an accounting text and a computer program are both "a set of statements or instructions . . . to bring about a certain result," 17 U.S.C. § 101, they are roughly analogous. In the former case, the processes are ultimately conducted by human agency; in the latter, by electronic means. In either case, as already stated, the processes themselves are not protectable. But the holding in *Baker* goes farther. The Court concluded that those aspects of a work, which "must necessarily be used as incident to" the idea, system or process that the work describes, are also not copyrightable. 101 U.S. at 104. Selden's ledger sheets, therefore, enjoyed no copyright protection because they were "necessary incidents to" the system of accounting that he described. Id. at 103. From this reasoning, we conclude that those elements of a computer program that are necessarily incidental to its function are similarly unprotectable.

While Baker v. Selden provides a sound analytical foundation, it offers scant guidance on how to separate idea or process from expression, and moreover, on how to further distinguish protectable expression from that expression which "must necessarily be used as incident to the work's underlying concept." In the context of computer programs, the Third Circuit's noted decision in *Whelan* has, thus far, been the most thoughtful attempt to accomplish these ends.

[The court quoted from the Whelan v. Jaslow decision.]

So far, in the courts, the *Whelan* rule has received a mixed reception. While some decisions have adopted its reasoning, see, e.g., Bull HN Info. Sys., Inc. v. American Express Bank, Ltd., 1990 Copyright Law Dec. (CCH) P 26,555 at 23,278 (S.D.N.Y. 1990); Dynamic Solutions, Inc. v. Planning & Control, Inc., 1987 Copyright Law Dec. (CCH) ¶26,062 at 20,912 (S.D.N.Y. 1987); Broderbund Software Inc. v. Unison World, Inc., 648 F. Supp. 1127, 1133 (N.D. Cal. 1986), others have rejected it. See Plains Cotton Co-op v. Goodpasture Computer Serv. Inc., 807 F.2d 1256, 1262 (5th Cir.), *cert. denied,* 484 U.S. 821 (1987); cf. Synercom Technology, Inc. v. University Computing Co., 462 F. Supp. 1003, 1014 (N.D. Tex. 1978) (concluding that order and sequence of data on computer input formats was idea not expression).

Whelan has fared even more poorly in the academic community, where its standard for distinguishing idea from expression has been widely criticized for being conceptually overbroad. See, e.g., Englund, at 881; Menell, at 1074, 1082; Kretschmer, at 837-39; Spivack, at 747-55; Thomas M. Gage, Note, Whelan Associates v. Jaslow Dental Laboratories: Copyright Protection for Computer Software Structure—What's the Purpose?, 1987 Wis. L. Rev. 859, 860-61 (1987). The leading commentator in the field has stated that, "[t]he crucial flaw in [Whelan's] reasoning is that it assumes that only one 'idea,' in copyright law terms, underlies any computer program, and that once a separable idea can be identified, everything else must be expression." 3 Nimmer § 13.03[F], at 13-62.34. This criticism focuses not upon the program's ultimate purpose but upon the reality of its structural design. As we have already noted, a computer program's ultimate function or purpose is the composite result of interacting subroutines. Since each subroutine is itself a program, and thus, may be said to have its own "idea," Whelan's general formulation that a program's overall purpose equates with the program's idea is descriptively inadequate.

Accordingly, we think that Judge Pratt wisely declined to follow *Whelan*. See *Computer Assocs.,* 775 F. Supp. at 558-60. In addition to noting the weakness in the *Whelan* definition of "program-idea," mentioned above, Judge Pratt found that *Whelan*'s synonymous use of the terms "structure, sequence, and organization," see *Whelan,* 797 F.2d at 1224 n.1, demonstrated a flawed understanding of a computer program's method of operation. See *Computer Assocs.,* 775 F. Supp. at 559-60 (discussing the distinction between a program's "static structure" and "dynamic structure"). Rightly, the district court found *Whelan*'s rationale suspect because it is so closely tied to what can now be seen — with the passage of time — as the opinion's somewhat outdated appreciation of computer science.

2) Substantial Similarity Test for Computer Program Structure: Abstraction-Filtration-Comparison

We think that Whelan's approach to separating idea from expression in computer programs relies too heavily on metaphysical distinctions and does not place enough emphasis on practical considerations. Cf. *Apple Computer,* 714 F.2d at 1253 (rejecting certain commercial constraints on programming as a helpful means of distinguishing idea from expression because they did "not enter into the somewhat metaphysical issue of whether particular ideas and expressions have merged"). As the cases that we shall discuss demonstrate, a satisfactory answer to this problem cannot be reached by resorting, *a priori*, to philosophical first principles.

As discussed herein, we think that district courts would be well-advised to undertake a three-step procedure, based on the abstractions test utilized by the district court, in order to determine whether the non-literal elements of two or more computer programs are substantially similar. This approach breaks no new ground; rather, it draws on such familiar copyright doctrines as merger, *scenes à faire,* and public domain. In taking this approach, however, we are cognizant that computer technology is a dynamic field which can quickly outpace judicial decisionmaking. Thus, in cases where the technology in question does not allow for a literal application of the procedure we outline below, our opinion should not be read to foreclose the district courts of our circuit from utilizing a modified version.

In ascertaining substantial similarity under this approach, a court would first break down the allegedly infringed program into its constituent structural parts. Then, by examining each of these parts for such things as incorporated ideas, expression that is necessarily incidental to those ideas, and elements that are taken from the public domain, a court would then be able to sift out all non-protectable material. Left with a kernel, or possibly kernels, of creative expression after following this process of elimination, the court's last step would be to compare this material with the structure of an allegedly infringing program. The result of this comparison will determine whether the protectable elements of the programs at issue are substantially similar so as to warrant a finding of infringement. It will be helpful to elaborate a bit further.

Step One: Abstraction

As the district court appreciated, see *Computer Assocs.,* 775 F. Supp. at 560, the theoretic framework for analyzing substantial similarity expounded by Learned Hand in the *Nichols* case is helpful in the present context. In *Nichols,* we enunciated what has now become known as the "abstractions" test for separating idea from expression:

> Upon any work . . . a great number of patterns of increasing generality will fit equally well, as more and more of the incident is left out. The last may perhaps be no more than the most general statement of what the [work] is about, and at times might consist only of its title; but there is a point in this series of abstractions where they are no longer protected, since otherwise the [author] could prevent the use of his "ideas," to which, apart from their expression, his property is never extended.

Nichols, 45 F.2d at 121.

While the abstractions test was originally applied in relation to literary works such as novels and plays, it is adaptable to computer programs. In contrast to the *Whelan* approach, the abstractions test "implicitly recognizes that any given work may consist of a mixture of numerous ideas and expressions." 3 Nimmer § 13.03[F] at 13-62.34-63.

As applied to computer programs, the abstractions test will comprise the first step in the examination for substantial similarity. Initially, in a manner that resembles reverse engineering on a theoretical plane, a court should dissect the allegedly copied program's structure and isolate each level of abstraction contained within it. This process begins with the code and ends with an articulation of the program's ultimate function. Along the way, it is necessary essentially to retrace and map each of the designer's steps — in the opposite order in which they were taken during the program's creation.

As an anatomical guide to this procedure, the following description is helpful:

> At the lowest level of abstraction, a computer program may be thought of in its entirety as a set of individual instructions organized into a hierarchy of modules. At a higher level of abstraction, the instructions in the lowest-level modules may be replaced conceptually by the functions of those modules. At progressively higher levels of abstraction, the functions of higher-level modules conceptually replace the implementations of those modules in terms of lower-level modules and instructions, until finally, one is left with nothing but the ultimate function of the program.... A program has structure at every level of abstraction at which it is viewed. At low levels of abstraction, a program's structure may be quite complex; at the highest level it is trivial.

Englund, at 897-98. Cf. *Spivack*, at 774.

Step Two: Filtration

Once the program's abstraction levels have been discovered, the substantial similarity inquiry moves from the conceptual to the concrete. Professor Nimmer suggests, and we endorse, a "successive filtering method" for separating protectable expression from non-protectable material. See generally 3 Nimmer § 13.03[F]. This process entails examining the structural components at each level of abstraction to determine whether their particular inclusion at that level was "idea" or was dictated by considerations of efficiency, so as to be necessarily incidental to that idea; required by factors external to the program itself; or taken from the public domain and hence is non-protectable expression. See also Kretschmer, at 844-45 (arguing that program features dictated by market externalities or efficiency concerns are unprotectable). The structure of any given program may reflect some, all, or none of these considerations. Each case requires its own fact specific investigation.

Strictly speaking, this filtration serves "the purpose of defining the scope of plaintiff's copyright." Brown Bag Software v. Symantec Corp., 960 F.2d 1465, 1475 (9th Cir.) (endorsing "analytic dissection" of computer programs in order to isolate protectable expression), *cert. denied*, 113 S. Ct. 198 (1992). By applying well developed doctrines of copyright law, it may ultimately leave behind a "core of protectable material." 3 Nimmer § 13.03[F](5), at 13-72. Further explication of this second step may be helpful.

(a) Elements Dictated by Efficiency

The portion of Baker v. Selden, discussed earlier, which denies copyright protection to expression necessarily incidental to the idea being expressed, appears to be the cornerstone for what has developed into the doctrine of merger. See Morrissey v. Proctor & Gamble Co., 379 F.2d 675, 678-79 (1st Cir. 1967) (relying on *Baker* for the proposition that expression embodying the rules of a sweepstakes contest was inseparable from the idea of the contest itself, and therefore were not protectable by copyright); see also *Digital Communications*, 659 F. Supp. at 457. The doctrine's underlying principle is that "[w]hen there is essentially only one way to express an idea, the idea and its expression are inseparable and copyright is no bar to copying that expression." Concrete Machinery Co. v. Classic Lawn Ornaments. Inc., 843 F.2d 600, 606 (1st Cir. 1988). Under these circumstances, the expression is said to have "merged" with the idea itself. In order not to confer a monopoly of the idea upon the copyright owner, such expression should not be protected. See Herbert Rosenthal Jewelry Corp. v. Kalpakian, 446 F.2d 738, 742 (9th Cir. 1971).

CONTU recognized the applicability of the merger doctrine to computer programs. In its report to Congress it stated that:

> [C]opyrighted language may be copied without infringing when there is but a limited number of ways to express a given idea. . . . In the computer context, this means that when specific instructions, even though previously copyrighted, are the only and essential means of accomplishing a given task, their later use by another will not amount to infringement.

CONTU Report at 20. While this statement directly concerns only the application of merger to program code, that is, the textual aspect of the program, it reasonably suggests that the doctrine fits comfortably within the general context of computer programs.

Furthermore, when one considers the fact that programmers generally strive to create programs "that meet the user's needs in the most efficient manner," Menell, at 1052, the applicability of the merger doctrine to computer programs becomes compelling. In the context of computer program design, the concept of efficiency is akin to deriving the most concise logical proof or formulating the most succinct mathematical computation. Thus, the more efficient a set of modules are, the more closely they approximate the idea or process embodied in that particular aspect of the program's structure.

While, hypothetically, there might be a myriad of ways in which a programmer may effectuate certain functions within a program — i.e., express the idea embodied in a given subroutine — efficiency concerns may so narrow the practical range of choice as to make only one or two forms of expression workable options. See 3 Nimmer § 13.03[F](2), at 13-63; see also *Whelan,* 797 F.2d at 1243 n.43 ("It is true that for certain tasks there are only a very limited number of file structures available, and in such cases the structures might not be copyrightable." . . .) Of course, not all program structure is informed by efficiency concerns. See Menell, at 1052 (besides efficiency, simplicity related to user accommodation has become a programming priority). It follows that, in order to determine whether the merger doctrine precludes copyright protection to an aspect of a program's structure that is so oriented, a court must inquire "whether the use of *this particular set of modules* is necessary efficiently to implement that part of the program's process" being implemented. Englund, at 902. If the answer is yes, then the expression represented by the programmer's choice of a specific module or group of modules has merged with their underlying idea and is unprotected. Id. at 902-03.

Another justification for linking structural economy with the application of the merger doctrine stems from a program's essentially utilitarian nature and the competitive forces that exist in the software marketplace. See Kretschmer, at 842. Working in tandem, these factors give rise to a problem of proof which merger helps to eliminate. Efficiency is an industry-wide goal. Since, as we have already noted, there may be only a limited number of efficient implementations for any given program task, it is quite possible that multiple programmers, working independently, will design the identical method employed in the allegedly infringed work. Of course, if this is the case, there is no copyright infringement. See *Roth Greeting Cards v. United Card Co.*, 429 F.2d 1106, 1110 (9th Cir. 1970); *Sheldon*, 81 F.2d at 54.

Under these circumstances, the fact that two programs contain the same efficient structure may as likely lead to an inference of independent creation as it does to one of copying. See 3 Nimmer § 13.03[F][2], at 13-65; cf. *Herbert Rosenthal Jewelry Corp.*, 446 F.2d at 741 (evidence of independent creation may stem from defendant's standing as a designer of previous similar works). Thus, since evidence of similarly efficient structure is not particularly probative of copying, it should be disregarded in the overall substantial similarity analysis. See 3 Nimmer § 13.03[F][2], at 13-65....

(b) *Elements Dictated by External Factors*

We have stated that where "it is virtually impossible to write about a particular historical era or fictional theme without employing certain 'stock' or standard literary devices," such expression is not copyrightable. *Hoehling v. Universal Studios, Inc.*, 618 F.2d 972, 979 (2d Cir.), *cert. denied*, 449 U.S. 1 841, 101 S. Ct. 42, 66 L. Ed. 2d 1 (1980)....

Professor Nimmer points out that "in many instances it is virtually impossible to write a program to perform particular functions in a specific computing environment without employing standard techniques." 3 Nimmer § 13.03[F][3], at 13-65. This is a result of the fact that a programmer's freedom of design choice is often circumscribed by extrinsic considerations such as (1) the mechanical specifications of the computer on which a particular program is intended to run; (2) compatibility requirements of other programs with which a program is designed to operate in conjunction; (3) computer manufacturers' design standards; (4) demands of the industry being serviced; and (5) widely accepted programming practices within the computer industry. Id. at 13-65-71....

(c) *Elements Taken from the Public Domain*

Closely related to the non-protectability of *scenes à faire*, is material found in the public domain. Such material is free for the taking and cannot be appropriated by a single author even though it is included in a copyrighted work. See *E. F. Johnson Co. v. Uniden Corp. of America*, 623 F. Supp. 1485, 1499 (D. Minn. 1985); see also *Sheldon*, 81 F.2d at 54. We see no reason to make an exception to this rule for elements of a computer program that have entered the public domain by virtue of freely accessible program exchanges and the like. See 3 Nimmer § 13.03[F][14]; see also *Brown Bag Software*, 960 F.2d at 1473 (affirming the district court's finding that "[p]laintiffs may not claim copyright protection of an...expression that is, if not standard, then commonplace in the computer software industry."). Thus, a court must also filter out this material from the allegedly infringed program before it makes the final inquiry in its substantial similarity analysis.

Step Three: Comparison

The third and final step of the test for substantial similarity that we believe appropriate for non-literal program components entails a comparison. Once a court has sifted out all elements of the allegedly infringed program which are "ideas" or are dictated by efficiency or external factors, or taken from the public domain, there may remain a core of protectable expression. In terms of a work's copyright value, this is the golden nugget. See *Brown Bag Software,* 960 F.2d at 1475. At this point, the court's substantial similarity inquiry focuses on whether the defendant copied any aspect of this protected expression, as well as an assessment of the copied portion's relative importance with respect to the plaintiff's overall program. See 3 Nimmer § 13.03[F][5]; *Data East USA,* 862 F.2d at 208 ("To determine whether similarities result from unprotectable expression, analytic dissection of similarities may be performed. If . . . all similarities in expression arise from use of common ideas, then no substantial similarity can be found.").

3) Policy Considerations

We are satisfied that the three step approach we have just outlined not only comports with, but advances the constitutional policies underlying the Copyright Act. Since any method that tries to distinguish idea from expression ultimately impacts on the scope of copyright protection afforded to a particular type of work, "the line [it draws] must be a pragmatic one, which also keeps in consideration 'the preservation of the balance between competition and protection. . . .'" *Apple Computer,* 714 F.2d at 1253 (citation omitted).

CA and some amici argue against the type of approach that we have set forth on the grounds that it will be a disincentive for future computer program research and development. At bottom, they claim that if programmers are not guaranteed broad copyright protection for their work, they will not invest the extensive time, energy and funds required to design and improve program structures. While they have a point, their argument cannot carry the day. The interest of the copyright law is not in simply conferring a monopoly on industrious persons, but in advancing the public welfare through rewarding artistic creativity, in a manner that permits the free use and development of non-protectable ideas and processes.

In this respect, our conclusion is informed by Justice Stewart's concise discussion of the principles that correctly govern the adaptation of the copyright law to new circumstances. In Twentieth Century Music Corp. v. Aiken, he wrote:

> The limited scope of the copyright holder's statutory monopoly, like the limited copyright duration required by the Constitution, reflects a balance of competing claims upon the public interest: Creative work is to be encouraged and rewarded, but private motivation must ultimately serve the cause of promoting broad public availability of literature, music, and the other arts. The immediate effect of our copyright law is to secure a fair return for an "author's" creative labor. But the ultimate aim is, by this incentive, to stimulate artistic creativity for the general public good. . . . When technological change has rendered its literal terms ambiguous, the Copyright Act must be construed in light of this basic purpose.

422 U.S. 151, 156 (1975)(citations and footnotes omitted).

Recently, the Supreme Court has emphatically reiterated that "[t]he primary objective of copyright is not to reward the labor of authors. . . ." Feist Publications,

Inc. v. Rural Tel. Serv. Co., 499 U.S. 340, 349 (1991) (emphasis added). While the *Feist* decision deals primarily with the copyrightability of purely factual compilations, its underlying tenets apply to much of the work involved in computer programming. *Feist* put to rest the "sweat of the brow" doctrine in copyright law. Id. at 359. The rationale of that doctrine "was that copyright was a reward for the hard work that went into compiling facts." Id. at 351. The Court flatly rejected this justification for extending copyright protection, noting that it "eschewed the most fundamental axiom of copyright law — that no one may copyright facts or ideas." Id.

Feist teaches that substantial effort alone cannot confer copyright status on an otherwise uncopyrightable work. As we have discussed, despite the fact that significant labor and expense often goes into computer program flow-charting and debugging, that process does not always result in inherently protectable expression. Thus, Feist implicitly undercuts the *Whelan* rationale, "which allow[ed] copyright protection beyond the literal computer code...[in order to] provide the proper incentive for programmers by protecting their most valuable efforts...." *Whelan*, 797 F.2d at 1237 (footnote omitted). We note that *Whelan* was decided prior to *Feist* when the "sweat of the brow" doctrine still had vitality. In view of the Supreme Court's recent holding, however, we must reject the legal basis of CA's disincentive argument.

Furthermore, we are unpersuaded that the test we approve today will lead to the dire consequences for the computer program industry that plaintiff and some amici predict. To the contrary, serious students of the industry have been highly critical of the sweeping scope of copyright protection engendered by the *Whelan* rule, in that it "enables first comers to 'lock up' basic programming techniques as implemented in programs to perform particular tasks." Menell, at 1087; see also Spivack, at 765 (*Whelan* "results in an inhibition of creation by virtue of the copyright owner's quasi-monopoly power").

To be frank, the exact contours of copyright protection for non-literal program structure are not completely clear. We trust that as future cases are decided, those limits will become better defined. Indeed, it may well be that the Copyright Act serves as a relatively weak barrier against public access to the theoretical interstices behind a program's source and object codes. This results from the hybrid nature of a computer program, which, while it is literary expression, is also a highly functional, utilitarian component in the larger process of computing.

Generally, we think that copyright registration — with its indiscriminating availability — is not ideally suited to deal with the highly dynamic technology of computer science. Thus far, many of the decisions in this area reflect the courts' attempt to fit the proverbial square peg in a round hole. The district court, see *Computer Assocs.*, 775 F. Supp. at 560, and at least one commentator have suggested that patent registration, with its exacting up-front novelty and non-obviousness requirements, might be the more appropriate rubric of protection for intellectual property of this kind. See Randell M. Whitmeyer, Comment, A Plea for Due Processes: Defining the Proper Scope of Patent Protection for Computer Software, 85 Nw. U.L. Rev. 1103, 1123-25 (1991); see also Lotus Dev. Corp. v. Borland Intl., Inc., 788 F. Supp. 78, 91 (D. Mass. 1992) (discussing the potentially supplemental relationship between patent and copyright protection in the context of computer programs). In any event, now that more than 12 years have passed since CONTU issued its final report, the resolution of this specific issue could benefit from further legislative investigation — perhaps a CONTU II.

In the meantime, Congress has made clear that computer programs are literary works entitled to copyright protection. Of course, we shall abide by these instructions, but in so doing we must not impair the overall integrity of copyright law. While incentive based arguments in favor of broad copyright protection are perhaps attractive from a pure policy perspective, see *Lotus Dev. Corp.*, 740 F. Supp. at 58, ultimately, they have a corrosive effect on certain fundamental tenets of copyright doctrine. If the test we have outlined results in narrowing the scope of protection, as we expect it will, that result flows from applying, in accordance with Congressional intent, long-standing principles of copyright law to computer programs. Of course, our decision is also informed by our concern that these fundamental principles remain undistorted.

B. The District Court Decision...

2) Evidentiary Analysis

The district court had to determine whether Altai's OSCAR 3.5 program was substantially similar to CA's ADAPTER. . . .

The district court took the first step in the analysis set forth in this opinion when it separated the program by levels of abstraction. The district court stated:

> As applied to computer software programs, this abstractions test would progress in order of "increasing generality" from object code, to source code, to parameter lists, to services required, to general outline. In discussing the particular similarities, therefore, we shall focus on these levels

Computer Assocs., 775 F. Supp. at 560. While the facts of a different case might require that a district court draw a more particularized blueprint of a program's overall structure, this description is a workable one for the case at hand.

Moving to the district court's evaluation of OSCAR 3.5's structural components, we agree with Judge Pratt's systematic exclusion of non-protectable expression. With respect to code, the district court observed that after the rewrite of OSCAR 3.4 to OSCAR 3.5, "there remained virtually no lines of code that were identical to ADAPTER." Id. at 561. Accordingly, the court found that the code "present[ed] no similarity at all." Id. at 562. Next, Judge Pratt addressed the issue of similarity between the two programs' parameter lists and macros. He concluded that, viewing the conflicting evidence most favorably to CA, it demonstrated that "only a few of the lists and macros were similar to protected elements in ADAPTER; the others were either in the public domain or dictated by the functional demands of the program." Id. As discussed above, functional elements and elements taken from the public domain do not qualify for copyright protection. With respect to the few remaining parameter lists and macros, the district court could reasonably conclude that they did not warrant a finding of infringement given their relative contribution to the overall program. See Warner Bros., Inc. v. American Broadcasting Cos., Inc., 720 F.2d 231, 242 (2d Cir. 1983) (discussing de minimis exception which allows for literal copying of a small and usually insignificant portion of the plaintiff's work); 3 Nimmer § 13.03[F][5], at 13-74. In any event, the district court reasonably found that, for lack of persuasive evidence, CA failed to meet its burden of proof on whether the macros and parameter lists at issue were substantially similar. See *Computer Assocs.*, 775 F. Supp. at 562.

The district court also found that the overlap exhibited between the list of services required for both ADAPTER and OSCAR 3.5 was "determined by the demands of the operating system and of the applications program to which it [was] to be linked through ADAPTER or OSCAR. . . ." Id. In other words, this aspect of the program's structure was dictated by the nature of other programs with which it was designed to interact and, thus, is not protected by copyright.

Finally, in his infringement analysis, Judge Pratt accorded no weight to the similarities between the two programs' organizational charts, "because [the charts were] so simple and obvious to anyone exposed to the operation of the program[s]." Id. CA argues that the district court's action in this regard "is not consistent with copyright law" — that "obvious" expression is protected, and that the district court erroneously failed to realize this. However, to say that elements of a work are "obvious," in the manner in which the district court used the word, is to say that they "follow naturally from the work's theme rather than from the author's creativity." 3 Nimmer § 13.03[F][3], at 1365. This is but one formulation of the *scenes à faire* doctrine, which we have already endorsed as a means of weeding out unprotectable expression. . . .

Since we accept Judge Pratt's factual conclusions and the results of his legal analysis, we affirm his dismissal of CA's copyright infringement claim based upon OSCAR 3.5. We emphasize that, like all copyright infringement cases, those that involve computer programs are highly fact specific. The amount of protection due structural elements, in any given case, will vary according to the protectable expression found to exist within the program at issue. . . .

COMMENTS AND QUESTIONS

1. How does the *Altai* court distinguish idea from expression? How would Baker v. Selden be decided under the *Altai* approach?

2. Is there a more appropriate way of distinguishing idea from expression in the context of application program code?

3. *Altai* seems to ask courts to analytically dissect a computer program to determine what is protectable and copied and compare this to the entire program, rather than to the protectable uncopied elements, to determine substantial similarity. Does this suggest that programs with relatively little protectable material can be freely copied?

4. Compare the manner in which *Altai* and Apple v. Franklin, discussed *supra*, treat the issue of achieving compatibility between computer programs. Recall that the Third Circuit observed in *Apple* that achieving compatibility with independently developed application programs "is a commercial and competitive objective which does not enter into the somewhat metaphysical issue of whether particular ideas and expressions have merged." *Apple* Computer v. Franklin Computer, 714 F.2d 1240, 1253 (3d Cir. 1983). Under the *Altai* test, could Franklin legitimately develop a computer program that achieves compatibility with the Apple operating system? If so, how would you advise Franklin to structure its development process so as to minimize the risk of copyright infringement?

5. The *Altai* case has been warmly praised by most commentators. See, e.g., David Bender, Computer Associates v. Altai: Rationality Prevails, Computer Law.,

Aug. 1992, at 1; Pamela Samuelson et al., The Nature of Copyright Analysis for Computer Program: Copyright Law Professors' Brief Amicus Curiae in Lotus v. Borland, 16 Hastings Comm. & Ent. L.J. 657 (1994) (brief of 24 copyright law professors endorsing *Altai*); Peter S. Menell, An Epitaph for Traditional Copyright Protection of Network Features of Computer Software, 43 Antitrust Bulletin 651 (1998). However, the decision has also been bitterly attacked by lawyers and scholars representing large computer companies. See Jack E. Brown, Analytical Dissection of Copyrighted Computer Software — Complicating the Simple and Confounding the Complex, 25 Ariz. St. L.J. 801 (1993) (representing Apple Computer); Anthony L. Clapes & Jennifer M. Daniels, Revenge of the Luddites: A Closer Look at Computer Associates v. Altai, Computer Lawyer, Nov. 1992, at 11 (lawyers for IBM). Professor Arthur Miller has expressed apprehension about the implications of the *Altai* approach:

> [A] court must employ considerable caution in excluding efficient or speedy program expression lest it undermine the effective protection of computer programs. For example, the mere fact that the expression is efficient should not, without more, bar protection for original authorship in the programming context any more than it does in prose work. An uncritical application of *Altai*'s language would penalize the most effective (and in some senses the most artistic) programmers

Arthur Miller, Copyright Protection for Computer Programs, Databases, and Computer-Generated Works: Is Anything New Since CONTU?, 106 Harv. L. Rev. 977, 1004-05 (1993). Do you agree with Miller's view of the scope of copyright protection for the efficient elements of program structure? Is protection of programming choices dictated by efficiency consistent with copyright cases outside the computer context?

6. The *Altai* test was rapidly adopted by most courts. Judicial convergence on the abstraction-filtration-comparison test has been so complete that every court to confront the issue since 1992 has chosen the *Altai* approach. See Bateman v. Mnemonics, Inc., 79 F.3d 1532, 1543-45 (11th Cir. 1996); Apple Computer v. Microsoft Corp., 35 F.3d 1435 (9th Cir. 1994); Engineering Dynamics, Inc. v. Structural Software, Inc., 26 F.3d 1335 (5th Cir. 1994); Kepner-Tregoe, Inc. v. Leadership Software, 12 F.3d 527 (5th Cir. 1994); Gates Rubber Co. v. Bando Chemical Indus., 9 F.3d 823 (10th Cir. 1993); Atari Games Corp. v. Nintendo, 975 F.2d 832 (Fed. Cir. 1992); CMAX/Cleveland, Inc. v. UCR, Inc., 804 F. Supp. 337 (M.D. Ga. 1992); Mark A. Lemley, Convergence in the Law of Software Copyright, 10 High Tech. L.J. 1 (1995). See also Brown Bag Software v. Symantec Corp., 960 F.2d 1465, 1475-76 (9th Cir. 1992) (endorsing "analytic dissection" of the elements of a computer program prior to the *Altai* decision). Cf. MiTek Holdings Inc. v. Arce Engineering Co., 89 F.3d 1548 (11th Cir. 1996) (upholding *Altai* approach, but allowing plaintiff to skip abstraction step by identifying in advance those elements it considers to be protectable and infringed). In addition, courts in Canada, the United Kingdom, and France have endorsed the *Altai* filtration analysis.

Not all of these courts have approached the abstraction-filtration-comparison analysis in precisely the same way, however. The Tenth Circuit decision in *Gates Rubber* is particularly notable for its elaboration of the test beyond the parameters of *Altai*. In that case, the court acknowledged that "[a]pplication of the abstractions test will necessarily vary from case-to-case and program-to-program. Given the complexity and ever-changing nature of computer technology, we decline to set forth any strict methodology for the abstraction of computer programs." Nonetheless, the court identified six levels of "generally declining abstraction": (1) the main purpose of the computer program, (2) the structure or architecture of a program, generally as represented in a flowchart, (3) "modules" that comprise particular program operations or types of stored data, (4) individual algorithms or data structures employed in each of the modules, (5) the source code that instructs the computer to carry out each necessary operation on each data structure, and (6) the object code that is actually read by the computer. The court used these levels of abstraction to facilitate its analysis of the program at issue.

The *Gates Rubber* court also gave further content to the filtration part of the *Altai* analysis. The court filtered out six unprotectable elements: ideas, the processes or methods of the computer program,[9] facts, material in the public domain, expression that has "merged" with an idea or process, and expression that is so standard or common as to be a "necessary incident" to an idea or process.[10] Finally, the court indicated that comparison of the protected elements of a program should be done on a case-by-case basis, with an eye toward determining whether a substantial portion of the protectable expression of the original work has been copied.

Is this analysis consistent with *Altai*? Is a court applying *Gates Rubber* likely to give more or less protection to a computer program than would the *Altai* court? Note that applying *Altai* does not mean that the nonliteral elements of computer programs get no protection at all. For example, in the Fifth Circuit, which has expressly adopted the *Gates Rubber* approach, one district court has concluded that a set of threshold values set by a computer manufacturer to indicate a failure threshold for replacement of a hard drive were copyrightable, since the business and engineering decisions that went into setting those values showed sufficient original expression to be protected. Compaq Computer v. Procom Technology, 908 F. Supp. 1409 (S.D. Tex. 1995). The court held that the merger and scenes à faire doctrines did not bar copyrightability, since the precise values Compaq chose were not necessary for competing manufacturers. See also Harbor Software Inc. v. Applied Systems Inc., 925 F. Supp. 1042 (S.D.N.Y. 1996) (finding nonliteral elements of a computer program protectable under the *Altai* test).

7. Note that while both *Whelan* and *Altai* are nominally about the scope of protection afforded computer programs under the copyright laws, both courts merge the analysis of infringement into the protectability analysis. This is particularly evident in *Altai*'s filtration analysis, where the comparison step involves identifying the similarities between the copyrighted program and the accused program.

---

9. Id. at 836-37. The court cited the legislative history accompanying section 102(b) of the Copyright Act, which clearly indicates that the actual processes or methods used in a computer program are not entitled to copyright protection. Id. The court noted that processes were most commonly found in the program architecture, module structure, and algorithms used. Id. at 837.

10. Id. This is the "scenes à faire" doctrine. The court held that such standard devices in computer programs include "those elements of a program that have been dictated by external factors," such as hardware and software standards, specifications, or compatibility requirements, customer design specifications, and basic industry practices. Id.

PROBLEMS

Problem 7-3. The Midwest Corn Cooperative Association (MCCA) is a nonprofit agricultural cooperative comprising approximately 15,000 corn farmers in the midwestern states. MCCA's purpose is to assist its members in the growing and marketing of corn. During the mid-1970s, MCCA developed a computer software system called CornMart that provided users with information regarding corn prices and availability, accounting services, and a means for electronically completing actual sales. The CornMart system was used by corn farmers and buyers through terminals connected by telephone lines to MCCA's large central computer.

CornMart was developed by a team of programmers that included George Cusher and Ben Fishman. In the early 1980s, after the program had been up and running for a number of years, Cusher and Fishman decided the system would have much greater usefulness if the CornMart program could be operated on personal computers, which were by then increasingly used by MCCA members. Cusher and Fishman took a complete source code version of Corn-Mart and supporting program documentation when they departed MCCA.

In 1985 Cusher and Fishman formed Cornware, Inc., and began work on a personal computer version of CornMart. By early 1986 they succeeded in writing a personal computer version of CornMart that they marketed as Cornware. Soon after they began marketing their new software, MCCA sued alleging copyright infringement.

Expert testimony at trial established that Cornware and CornMart were very similar in terms of functional specification, overall structure, and documentation. The main difference between the systems was that CornMart is designed to run on a mainframe computer and Cornware on a personal computer. There was, however, no literal copying of code. Cusher and Fishman testified that they had not copied or modified the CornMart program but instead had "drawn on their knowledge of the corn industry and expertise in computer programming and design gained over many years." Expert testimony at trial indicated that it would take significant effort to modify a program as complex as Corn Mart to run on a personal computer, perhaps more effort than rewriting the program entirely. Nonetheless, there remained significant structural similarities between the programs, although two experts testified that residual similarity between the programs reflected features of the corn market and that different programmers knowledgeable about the corn industry could conceivably make similar judgments. For example, both programs were designed to present the same information as is contained on a corn recap sheet, a standard form used for corn transactions.

How would you argue the case for MCCA? for Cornware?

Problem 7-4. Demento Corporation ("Demento") produces Demento II, the leading home video game system. The Demento II consists of a console, which runs game cartridges and attaches to a monitor or standard television, and a control device, which enables the user to manipulate characters or other images of particular games on the screen. Demento licenses a limited number of games per year.

To prevent unauthorized game producers from manufacturing games to run on the Demento II console, Demento developed a "lock out" device that governs access to the console. The console includes a "master chip" or "lock" that will run only game cartridges that contain an appropriate "slave chip" or "key." When a user inserts an authorized game cartridge into the Demento II, the slave chip transmits an arbitrary data stream "key" that is received by the master chip and unlocks the console, allowing the game to be played. The unlocking device is a sophisticated software program encoded in the master and slave chips. There are a multitude of different programs capable of generating the data stream that unlocks the console, although it would be almost impossible to decipher the "key" by trial and error. It would be like trying to find a needle in a haystack.

The rapid growth of the home video game market eroded the revenue base of Mutant Corporation ("Mutant"), a leading maker of arcade video games. Mutant sought to license the right to produce game cartridges for the Demento II system, but it was put off by the limits on the number of new games that Demento would license per year and the high cost of each license. Mutant decided instead to develop its own slave chip that would unlock the Demento II game cartridge.

Mutant engineers first chemically peeled the layers of the Demento II chips to allow microsopic examination of the circuitry. Through this means, they were able to decipher the object code. After that, they were able to construct the series of pulsating signals that unlocked the master chip. Mutant discovered that only a relatively small portion of the coded message was necessary to unlock the Demento II console. From this information, Mutant software designers built a slave chip that successfully unlocked the Demento II console and enabled Mutant's game cartridges to run. It included the entire coded message for fear that Demento might later alter new versions of the Demento II system in such a way as to make Mutant's cartridges inoperable on newer units seeking the fuller coded message. To differentiate its product, the Mutant slave program was written in a different computer language and employed a different microprocessor than the Demento II system. Since the Mutant microprocessor operated at a faster speed than the Demento II chip, the Mutant program included numerous pauses.

Despite these differences, Demento promptly sued Mutant for copyright infringement after Mutant introduced its first game cartridge for the Demento II system. What is the unprotectable idea of the Demento II system? What is the protectable expression? Is the lockout code functional? How would this case be resolved?

c. *Protection for Functional Elements and Protocols*

Many software copyright cases involve the taking of particular pieces of a plaintiff's program. Defendants in these cases may be those who attempted to "clone" the plaintiff's program, writing their own code to create a functionally equivalent program, see Lotus Dev. Corp. v. Paperback Software, 740 F. Supp. 37 (D. Mass. 1990), or those who simply attempt to reproduce the internal program interface necessary to make two programs compatible. Whether functional program elements could be protected under copyright law was tested in a series of cases involving the Lotus 1-2-3 spreadsheet.

Lotus Development Corp. v. Borland International
United States Court of Appeals for the First Circuit
49 F.3d 807 (1st Cir. 1995), aff'd by an equally divided court, 526 U.S. 233 (1996)

STAHL, Circuit Judge.

This appeal requires us to decide whether a computer menu command hierarchy is copyrightable subject matter. In particular, we must decide whether, as the district court held, plaintiff-appellee Lotus Development Corporation's copyright in Lotus 1-2-3, a computer spreadsheet program, was infringed by defendant-appellant Borland International, Inc., when Borland copied the Lotus 1-2-3 menu command hierarchy into its Quattro and Quattro Pro computer spreadsheet programs. See Lotus Dev. Corp. v. Borland Int'l, Inc., 788 F. Supp. 78 (D. Mass. 1992) ("Borland I"); Lotus Dev. Corp. v. Borland Int'l, Inc., 799 F. Supp. 203 (D. Mass. 1992) ("Borland II"); Lotus Dev. Corp. v. Borland Int'l, Inc., 831 F. Supp. 202 (D. Mass. 1993) ("Borland III"); Lotus Dev. Corp. v. Borland Int'l, Inc., 831 F. Supp. 223 (D. Mass. 1993) ("Borland IV").

I. Background

Lotus 1-2-3 is a spreadsheet program that enables users to perform accounting functions electronically on a computer. Users manipulate and control the program via a series of menu commands, such as "Copy," "Print," and "Quit." Users choose commands either by highlighting them on the screen or by typing their first letter. In all, Lotus 1-2-3 has 469 commands arranged into more than 50 menus and submenus. [See Figure 7-1.]

Lotus 1-2-3, like many computer programs, allows users to write what are called "macros." By writing a macro, a user can designate a series of command choices with a single macro keystroke. Then, to execute that series of commands in multiple parts of the spreadsheet, rather than typing the whole series each time, the user only needs to type the single pre-programmed macro keystroke, causing the program to recall and perform the designated series of commands automatically. Thus, Lotus 1-2-3 macros shorten the time needed to set up and operate the program.

Borland released its first Quattro program to the public in 1987, after Borland's engineers had labored over its development for nearly three years. Borland's objective was to develop a spreadsheet program far superior to existing programs, including Lotus 1-2-3. In Borland's words, "from the time of its initial release . . . Quattro included enormous innovations over competing spreadsheet products."

The district court found, and Borland does not now contest, that Borland included in its Quattro and Quattro Pro version 1.0 programs "a virtually identical copy of the entire 1-2-3 menu tree." Borland III, 831 F. Supp. at 212. In so doing, Borland did not copy any of Lotus's underlying computer code; it copied only the words and structure of Lotus's menu command hierarchy. Borland included the Lotus

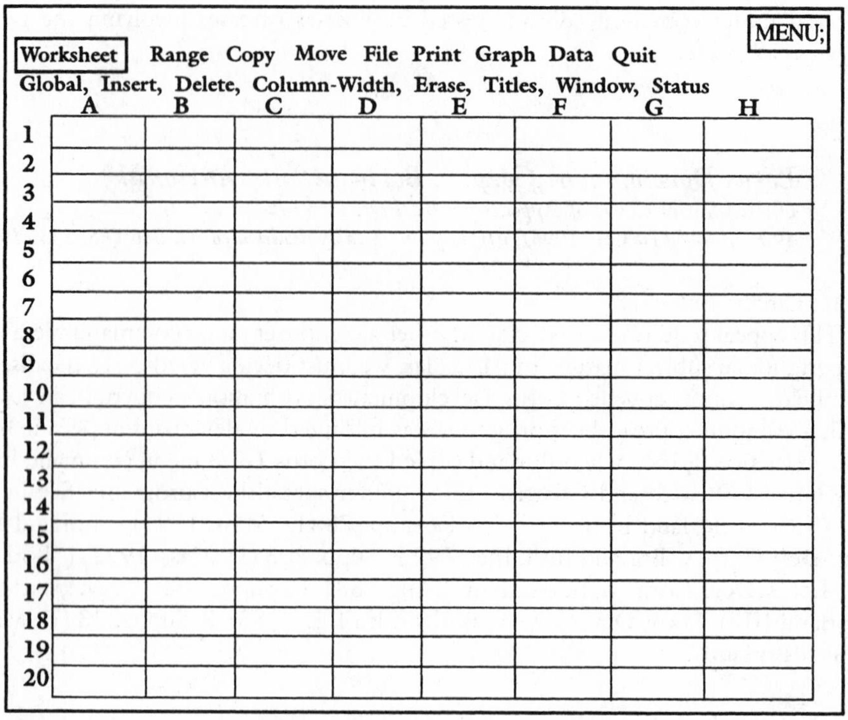

| Worksheet | Range Copy Move File Print Graph Data Quit | | | | | | | MENU; |

```
┌──────────────────────────────────────────────────────────────────────┐
│ ┌───────────┐                                              ┌──────┐    │
│ │ Worksheet │  Range  Copy  Move  File  Print  Graph  Data  Quit MENU; │
│ └───────────┘                                                          │
│ Global, Insert, Delete, Column-Width, Erase, Titles, Window, Status    │
│      A      B      C      D      E      F      G      H                 │
│  1                                                                     │
│  2                                                                     │
│  3                                                                     │
│  4                                                                     │
│  5                                                                     │
│  6                                                                     │
│  7                                                                     │
│  8                                                                     │
│  9                                                                     │
│ 10                                                                     │
│ 11                                                                     │
│ 12                                                                     │
│ 13                                                                     │
│ 14                                                                     │
│ 15                                                                     │
│ 16                                                                     │
│ 17                                                                     │
│ 18                                                                     │
│ 19                                                                     │
│ 20                                                                     │
└──────────────────────────────────────────────────────────────────────┘
```

FIGURE 7-1

Facsimile of a Lotus 1-2-3 screen display.

menu command hierarchy in its programs to make them compatible with Lotus 1-2-3 so that spreadsheet users who were already familiar with Lotus 1-2-3 would be able to switch to the Borland programs without having to learn new commands or rewrite their Lotus macros.

In its Quattro and Quattro Pro version 1.0 programs, Borland achieved compatibility with Lotus 1-2-3 by offering its users an alternate user interface, the "Lotus Emulation Interface." By activating the Emulation Interface, Borland users would see the Lotus menu commands on their screens and could interact with Quattro or Quattro Pro as if using Lotus 1-2-3, albeit with a slightly different looking screen and with many Borland options not available on Lotus 1-2-3. In effect, Borland allowed users to choose how they wanted to communicate with Borland's spreadsheet programs: either by using menu commands designed by Borland, or by using the commands and command structure used in Lotus 1-2-3 augmented by Borland-added commands.

Lotus filed this action against Borland in the District of Massachusetts on July 2, 1990, four days after a district court held that the Lotus 1-2-3 "menu structure, taken as a whole — including the choice of command terms [and] the structure and order of those terms," was protected expression covered by Lotus's copyrights. Lotus Dev. Corp. v. Paperback Software Int'l, 740 F. Supp. 37, 68, 70 (D. Mass. 1990) ("Paperback").[1] ...

———————

1. Judge Keeton presided over both the *Paperback* litigation and this case.

On July 31, 1992, the district court denied Borland's motion [for summary judgment] and granted Lotus's motion in part. The district court ruled that the Lotus menu command hierarchy was copyrightable expression because

[a] very satisfactory spreadsheet menu tree can be constructed using different commands and a different command structure from those of Lotus 1-2-3. In fact, Borland has constructed just such an alternate tree for use in Quattro Pro's native mode. Even if one holds the arrangement of menu commands constant, it is possible to generate literally millions of satisfactory menu trees by varying the menu commands employed.

Borland II, 799 F. Supp. at 217. The district court demonstrated this by offering alternate command words for the ten commands that appear in Lotus's main menu. Id. For example, the district court stated that "the 'Quit' command could be named 'Exit' without any other modifications," and that "the 'Copy' command could be called 'Clone,' 'Ditto,' 'Duplicate,' 'Imitate,' 'Mimic,' 'Replicate,' and 'Reproduce,' among others." Id. Because so many variations were possible, the district court concluded that the Lotus developers' choice and arrangement of command terms, reflected in the Lotus menu command hierarchy, constituted copyrightable expression.

In granting partial summary judgment to Lotus, the district court held that Borland had infringed Lotus's copyright in Lotus 1-2-3...Borland II, 799 F. Supp. at 223. The court nevertheless concluded that while the Quattro and Quattro Pro programs infringed Lotus's copyright, Borland had not copied the entire Lotus 1-2-3 user interface, as Lotus had contended....

Immediately following the district court's summary judgment decision, Borland removed the Lotus Emulation Interface from its products. Thereafter, Borland's spreadsheet programs no longer displayed the Lotus 1-2-3 menus to Borland users, and as a result Borland users could no longer communicate with Borland's programs as if they were using a more sophisticated version of Lotus 1-2-3. Nonetheless, Borland's programs continued to be partially compatible with Lotus 1-2-3, for Borland retained what it called the "Key Reader" in its Quattro Pro programs. Once turned on, the Key Reader allowed Borland's programs to understand and perform some Lotus 1-2-3 macros. With the Key Reader on, the Borland programs used Quattro Pro menus for display, interaction, and macro execution, except when they encountered a slash ("/") key in a macro (the starting key for any Lotus 1-2-3 macro), in which case they interpreted the macro as having been written for Lotus 1-2-3. Accordingly, people who wrote or purchased macros to shorten the time needed to perform an operation in Lotus 1-2-3 could still use those macros in Borland's programs. The district court permitted Lotus to file a supplemental complaint alleging that the Key Reader infringed its copyright.

The parties agreed to try the remaining liability issues without a jury. The district court held two trials, the Phase I trial covering all remaining issues raised in the original complaint (relating to the Emulation Interface) and the Phase II trial covering all issues raised in the supplemental complaint (relating to the Key Reader)....

In its Phase I trial decision, the district court found that "each of the Borland emulation interfaces contains a virtually identical copy of the 1-2-3 menu tree and that the 1-2-3 menu tree is capable of a wide variety of expression." Borland III, 831 F. Supp. at 218. The district court also rejected Borland's affirmative defenses of laches and estoppel. Id. at 218-23.

In its Phase II trial decision, the district court found that Borland's Key Reader file included "a virtually identical copy of the Lotus menu tree structure, but represented in a different form and with first letters of menu command names in place of the full menu command names." Borland IV, 831 F. Supp. at 228. In other words, Borland's programs no longer included the Lotus command terms, but only their first letters. The district court held that "the Lotus menu structure, organization, and first letters of the command names . . . constitute part of the protectable expression found in [Lotus 1-2-3]." Id. at 233. Accordingly, the district court held that with its Key Reader, Borland had infringed Lotus's copyright. Id. at 245. The district court also rejected Borland's affirmative defenses of waiver, laches, estoppel, and fair use. Id. at 235-45. The district court then entered a permanent injunction against Borland, id. at 245, from which Borland appeals. . . .

II. Discussion

On appeal, Borland does not dispute that it factually copied the words and arrangement of the Lotus menu command hierarchy. Rather, Borland argues that it "lawfully copied the unprotectable menus of Lotus 1-2-3." Borland contends that the Lotus menu command hierarchy is not copyrightable because it is a system, method of operation, process, or procedure foreclosed from protection by 17 U.S.C. § 102(b). Borland also raises a number of affirmative defenses.

A. Copyright Infringement Generally . . .

In this appeal, we are faced only with whether the Lotus menu command hierarchy is copyrightable subject matter in the first instance, for Borland concedes that Lotus has a valid copyright in Lotus 1-2-3 as a whole and admits to factually copying the Lotus menu command hierarchy. As a result, this appeal is in a very different posture from most copyright-infringement cases, for copyright infringement generally turns on whether the defendant has copied protected expression as a factual matter. Because of this different posture, most copyright-infringement cases provide only limited help to us in deciding this appeal. This is true even with respect to those copyright-infringement cases that deal with computers and computer software.

B. Matter of First Impression

Whether a computer menu command hierarchy constitutes copyrightable subject matter is a matter of first impression in this court. While some other courts appear to have touched on it briefly in dicta, see, e.g., Autoskill, Inc. v. National Educ. Support Sys., Inc., 994 F.2d 1476, 1495 n.23 (10th Cir.), *cert. denied*, 114 S. Ct. 307 (1993), we know of no cases that deal with the copyrightability of a menu command hierarchy standing on its own (i.e., without other elements of the user interface, such as screen displays, in issue). Thus we are navigating in uncharted waters.

Borland vigorously argues, however, that the Supreme Court charted our course more than 100 years ago when it decided Baker v. Selden, 101 U.S. 99 (1879). In Baker v. Selden, the Court held that Selden's copyright over the textbook in which he explained his new way to do accounting did not grant him a monopoly on the use of his accounting system. Borland argues: "The facts of Baker v. Selden, and even the arguments advanced by the parties in that case, are identical to those in this case. The only difference is that the 'user interface' of Selden's system was implemented by pen and paper rather than by computer." . . .

We do not think that Baker v. Selden is nearly as analogous to this appeal as Borland claims. Of course, Lotus 1-2-3 is a computer spreadsheet, and as such its grid of horizontal rows and vertical columns certainly resembles an accounting ledger or any other paper spreadsheet. Those grids, however, are not at issue in this appeal for, unlike Selden, Lotus does not claim to have a monopoly over its accounting system. Rather, this appeal involves Lotus's monopoly over the commands it uses to operate the computer. Accordingly, this appeal is not, as Borland contends, "identical" to Baker v. Selden.

C. *Altai*

Before we analyze whether the Lotus menu command hierarchy is a system, method of operation, process, or procedure, we first consider the applicability of the test the Second Circuit set forth in Computer Assoc. Int'l, Inc. v. Altai, Inc., 982 F.2d 693 (2d Cir. 1992). The Second Circuit designed its *Altai* test to deal with the fact that computer programs, copyrighted as "literary works," can be infringed by what is known as "nonliteral" copying, which is copying that is paraphrased or loosely paraphrased rather than word for word. See id. at 701 (citing nonliteral-copying cases); see also 3 Melville B. Nimmer & David Nimmer, Nimmer on Copyright §13.03[A][1] (1993). When faced with nonliteral-copying cases, courts must determine whether similarities are due merely to the fact that the two works share the same underlying idea or whether they instead indicate that the second author copied the first author's expression. The Second Circuit designed its *Altai* test to deal with this situation in the computer context, specifically with whether one computer program copied nonliteral expression from another program's code. . . .

In the instant appeal, we are not confronted with alleged nonliteral copying of computer code. Rather, we are faced with Borland's deliberate, literal copying of the Lotus menu command hierarchy. Thus, we must determine not whether nonliteral copying occurred in some amorphous sense, but rather whether the literal copying of the Lotus menu command hierarchy constitutes copyright infringement.

While the *Altai* test may provide a useful framework for assessing the alleged nonliteral copying of computer code, we find it to be of little help in assessing whether the literal copying of a menu command hierarchy constitutes copyright infringement. In fact, we think that the *Altai* test in this context may actually be misleading because, in instructing courts to abstract the various levels, it seems to encourage them to find a base level that includes copyrightable subject matter that, if literally copied, would make the copier liable for copyright infringement.[7] While that base (or literal) level would not be at issue in a nonliteral-copying case like *Altai,* it is precisely what is at issue in this appeal. We think that abstracting menu command hierarchies down to their individual word and menu levels and then filtering idea from expression at that stage, as both the *Altai* and the district court tests require, obscures the more fundamental question of whether a menu command hierarchy can be copyrighted at all. The initial inquiry should not be whether individual components of a menu command hierarchy are expressive, but rather whether the menu command hierarchy as a whole can be copyrighted. But see Gates Rubber Co. v. Bando Chem. Indus., Ltd., 9 F.3d 823 (10th Cir. 1993) (endorsing *Altai*'s abstraction-filtration-comparison test as a way of determining whether "menus and sorting criteria" are copyrightable).

D. The Lotus Menu Command Hierarchy: A "Method of Operation"

Borland argues that the Lotus menu command hierarchy is uncopyrightable because it is a system, method of operation, process, or procedure foreclosed from copyright protection by 17 U.S.C. § 102(b). Section 102(b) states: "In no case does copyright protection for an original work of authorship extend to any idea, procedure, process, system, method of operation, concept, principle, or discovery, regardless of the form in which it is described, explained, illustrated, or embodied in such work." Because we conclude that the Lotus menu command hierarchy is a method of operation, we do not consider whether it could also be a system, process, or procedure.

We think that "method of operation," as that term is used in § 102(b), refers to the means by which a person operates something, whether it be a car, a food processor, or a computer. Thus a text describing how to operate something would not extend copyright protection to the method of operation itself; other people would be free to employ that method and to describe it in their own words. Similarly, if a new method of operation is used rather than described, other people would still be free to employ or describe that method.

We hold that the Lotus menu command hierarchy is an uncopyrightable "method of operation." The Lotus menu command hierarchy provides the means by which users control and operate Lotus 1-2-3. If users wish to copy material, for example, they use the "Copy" command. If users wish to print material, they use the "Print" command. Users must use the command terms to tell the computer what to do. Without the menu command hierarchy, users would not be able to access and control, or indeed make use of, Lotus 1-2-3's functional capabilities.

The Lotus menu command hierarchy does not merely explain and present Lotus 1-2-3's functional capabilities to the user; it also serves as the method by which the program is operated and controlled. The Lotus menu command hierarchy is different from the Lotus long prompts, for the long prompts are not necessary to the operation

---

8. We recognize that *Altai* never states that every work contains a copyrightable "nugget" of protectable expression. Nonetheless, the implication is that for literal copying, "it is not necessary to determine the level of abstraction at which similarity ceases to consist of an 'expression of ideas,' because literal similarity by definition is always a similarity as to the expression of ideas." 3 Melville B. Nimmer & David Nimmer, Nimmer on Copyright § 13.03[A](2) (1993).

of the program; users could operate Lotus 1-2-3 even if there were no long prompts. The Lotus menu command hierarchy is also different from the Lotus screen displays, for users need not "use" any expressive aspects of the screen displays in order to operate Lotus 1-2-3; because the way the screens look has little bearing on how users control the program, the screen displays are not part of Lotus 1-2-3's "method of operation." The Lotus menu command hierarchy is also different from the underlying computer code, because while code is necessary for the program to work, its precise formulation is not. In other words, to offer the same capabilities as Lotus 1-2-3, Borland did not have to copy Lotus's underlying code (and indeed it did not); to allow users to operate its programs in substantially the same way, however, Borland had to copy the Lotus menu command hierarchy. Thus the Lotus 1-2-3 code is not an uncopyrightable "method of operation."

The district court held that the Lotus menu command hierarchy, with its specific choice and arrangement of command terms, constituted an "expression" of the "idea" of operating a computer program with commands arranged hierarchically into menus and submenus. *Borland II,* 799 F. Supp. at 216. Under the district court's reasoning, Lotus's decision to employ hierarchically arranged command terms to operate its program could not foreclose its competitors from also employing hierarchically arranged command terms to operate their programs, but it did foreclose them from employing the specific command terms and arrangement that Lotus had used. In effect, the district court limited Lotus 1-2-3's "method of operation" to an abstraction.

Accepting the district court's finding that the Lotus developers made some expressive choices in choosing and arranging the Lotus command terms, we nonetheless hold that that expression is not copyrightable because it is part of Lotus 1-2-3's "method of operation." We do not think that "methods of operation" are limited to abstractions; rather, they are the means by which a user operates something. If specific words are essential to operating something, then they are part of a "method of operation" and, as such, are unprotectable. This is so whether they must be highlighted, typed in, or even spoken, as computer programs no doubt will soon be controlled by spoken words.

The fact that Lotus developers could have designed the Lotus menu command hierarchy differently is immaterial to the question of whether it is a "method of operation." In other words, our initial inquiry is not whether the Lotus menu command hierarchy incorporates any expression. Rather, our initial inquiry is whether the Lotus menu command hierarchy is a "method of operation." Concluding, as we do, that users operate Lotus 1-2-3 by using the Lotus menu command hierarchy, and that the entire Lotus menu command hierarchy is essential to operating Lotus 1-2-3, we do not inquire further whether that method of operation could have been designed differently. The "expressive" choices of what to name the command terms and how to arrange them do not magically change the uncopyrightable menu command hierarchy into copyrightable subject matter.

Our holding that "methods of operation" are not limited to mere abstractions is bolstered by Baker v. Selden. In *Baker,* the Supreme Court explained that

> the teachings of science and the rules and methods of useful art have their final end in application and use; and this application and use are what the public derive from the publication of a book which teaches them.... The description of the art in a book, though entitled to the benefit of copyright, lays no foundation for an exclusive claim to the art

itself. The object of the one is explanation; the object of the other is use. The former may be secured by copyright. The latter can only be secured, if it can be secured at all, by letters-patent.

Baker v. Selden, 101 U.S. at 104-05. Lotus wrote its menu command hierarchy so that people could learn it and use it. Accordingly, it falls squarely within the prohibition on copyright protection established in Baker v. Selden and codified by Congress in § 102(b).

In many ways, the Lotus menu command hierarchy is like the buttons used to control, say, a video cassette recorder ("VCR"). A VCR is a machine that enables one to watch and record video tapes. Users operate VCRs by pressing a series of buttons that are typically labelled "Record, Play, Reverse, Fast Forward, Pause, Stop/Eject." That the buttons are arranged and labeled does not make them a "literary work," nor does it make them an "expression" of the abstract "method of operating" a VCR via a set of labeled buttons. Instead, the buttons are themselves the "method of operating" the VCR.

When a Lotus 1-2-3 user chooses a command, either by highlighting it on the screen or by typing its first letter, he or she effectively pushes a button. Highlighting the "Print" command on the screen, or typing the letter "P," is analogous to pressing a VCR button labeled "Play."

Just as one could not operate a buttonless VCR, it would be impossible to operate Lotus 1-2-3 without employing its menu command hierarchy. Thus the Lotus command terms are not equivalent to the labels on the VCR's buttons, but are instead equivalent to the buttons themselves. Unlike the labels on a VCR's buttons, which merely make operating a VCR easier by indicating the buttons' functions, the Lotus menu commands are essential to operating Lotus 1-2-3. Without the menu commands, there would be no way to "push" the Lotus buttons, as one could push unlabeled VCR buttons. While Lotus could probably have designed a user interface for which the command terms were mere labels, it did not do so here. Lotus 1-2-3 depends for its operation on use of the precise command terms that make up the Lotus menu command hierarchy. . . .

Computer programs, unlike VCRs, are copyrightable as "literary works." 17 U.S.C. § 102(a). Accordingly, one might argue, the "buttons" used to operate a computer program are not like the buttons used to operate a VCR, for they are not subject to a useful-article exception. The response, of course, is that the arrangement of buttons on a VCR would not be copyrightable even without a useful-article exception, because the buttons are an uncopyrightable "method of operation." Similarly, the "buttons" of a computer program are also an uncopyrightable "method of operation."

That the Lotus menu command hierarchy is a "method of operation" becomes clearer when one considers program compatibility. Under Lotus's theory, if a user uses several different programs, he or she must learn how to perform the same operation in a different way for each program used. For example, if the user wanted the computer to print material, then the user would have to learn not just one method of operating the computer such that it prints, but many different methods. We find this absurd. The fact that there may be many different ways to operate a computer program, or even many different ways to operate a computer program using a set of hierarchically arranged command terms, does not make the actual method of operation chosen copyrightable; it still functions as a method for operating the computer and as such is uncopyrightable.

Consider also that users employ the Lotus menu command hierarchy in writing macros. Under the district court's holding, if the user wrote a macro to shorten the time needed to perform a certain operation in Lotus 1-2-3, the user would be unable to use that macro to shorten the time needed to perform that same operation in another program. Rather, the user would have to rewrite his or her macro using that other program's menu command hierarchy. This is despite the fact that the macro is clearly the user's own work product. We think that forcing the user to cause the computer to perform the same operation in a different way ignores Congress's direction in § 102(b) that "methods of operation" are not copyrightable. That programs can offer users the ability to write macros in many different ways does not change the fact that, once written, the macro allows the user to perform an operation automatically. As the Lotus menu command hierarchy serves as the basis for Lotus 1-2-3 macros, the Lotus menu command hierarchy is a "method of operation." . . .

We also note that in most contexts, there is no need to "build" upon other people's expression, for the ideas conveyed by that expression can be conveyed by someone else without copying the first author's expression. In the context of methods of operation, however, "building" requires the use of the precise method of operation already employed; otherwise, "building" would require dismantling, too. Original developers are not the only people entitled to build on the methods of operation they create; anyone can. Thus, Borland may build on the method of operation that Lotus designed and may use the Lotus menu command hierarchy in doing so.

Our holding that methods of operation are not limited to abstractions goes against *Autoskill*, 994 F.2d at 1495 n.23, in which the Tenth Circuit rejected the defendant's argument that the keying procedure used in a computer program was an uncopyrightable "procedure" or "method of operation" under § 102(b). The program at issue, which was designed to test and train students with reading deficiencies, id. at 1481, required students to select responses to the program's queries "by pressing the 1, 2, or 3 keys." Id. at 1495 n.23. The Tenth Circuit held that, "for purposes of the preliminary injunction, . . . the record showed that [this] keying procedure reflected at least a minimal degree of creativity," as required by Feist for copyright protection. Id. As an initial matter, we question whether a programmer's decision to have users select a response by pressing the 1, 2, or 3 keys is original. More importantly, however, we fail to see how "a student selecting a response by pressing the 1, 2, or 3 keys," id., can be anything but an unprotectable method of operation.[13] . . .

Reversed.

BOUDIN, Circuit Judge, concurring.

The importance of this case, and a slightly different emphasis in my view of the underlying problem, prompt me to add a few words to the majority's tightly focused discussion.

I

Most of the law of copyright and the "tools" of analysis have developed in the context of literary works such as novels, plays, and films. In this milieu, the principal

14. The Ninth Circuit has also indicated in dicta that "menus, and keystrokes" may be copyrightable. Brown Bag Software v. Symantec Corp., 960 F.2d 1465, 1477 (9th Cir.), *cert. denied*, BB Asset Management, Inc. v. Symantec Corp., 121 L. Ed. 2d 141, 113 S. Ct. 198 (1992). In that case, however, the plaintiff did not show that the defendant had copied the plaintiff's menus or keystrokes, so the court was not directly faced with whether the menus or keystrokes constituted an unprotectable method of operation. Id.

problem — simply stated, if difficult to resolve — is to stimulate creative expression without unduly limiting access by others to the broader themes and concepts deployed by the author. The middle of the spectrum presents close cases; but a "mistake" in providing too much protection involves a small cost: subsequent authors treating the same themes must take a few more steps away from the original expression.

The problem presented by computer programs is fundamentally different in one respect. The computer program is a means for causing something to happen; it has a mechanical utility, an instrumental role, in accomplishing the world's work. Granting protection, in other words, can have some of the consequences of patent protection in limiting other people's ability to perform a task in the most efficient manner. Utility does not bar copyright (dictionaries may be copyrighted), but it alters the calculus.

Of course, the argument for protection is undiminished, perhaps even enhanced, by utility: if we want more of an intellectual product, a temporary monopoly for the creator provides incentives for others to create other, different items in this class. But the "cost" side of the equation may be different where one places a very high value on public access to a useful innovation that may be the most efficient means of performing a given task. Thus, the argument for extending protection may be the same; but the stakes on the other side are much higher.

It is no accident that patent protection has preconditions that copyright protection does not — notably, the requirements of novelty and non-obviousness — and that patents are granted for a shorter period than copyrights. This problem of utility has sometimes manifested itself in copyright cases, such as Baker v. Selden, 101 U.S. 99 (1879), and been dealt with through various formulations that limit copyright or create limited rights to copy. But the case law and doctrine addressed to utility in copyright have been brief detours in the general march of copyright law.

Requests for the protection of computer menus present the concern with fencing off access to the commons in an acute form. A new menu may be a creative work, but over time its importance may come to reside more in the investment that has been made by users in learning the menu and in building their own mini-programs — macros — in reliance upon the menu. Better typewriter keyboard layouts may exist, but the familiar QWERTY keyboard dominates the market because that is what everyone has learned to use. See P. David, CLIO and the Economics of QWERTY, 75 Am. Econ. Rev. 332 (1985). The QWERTY keyboard is nothing other than a menu of letters.

Thus, to assume that computer programs are just one more new means of expression, like a filmed play, may be quite wrong. The "form" — the written source code or the menu structure depicted on the screen — look hauntingly like the familiar stuff of copyright; but the "substance" probably has more to do with problems presented in patent law or, as already noted, in those rare cases where copyright law has confronted industrially useful expressions. Applying copyright law to computer programs is like assembling a jigsaw puzzle whose pieces do not quite fit. . . .

II

In this case, the raw facts are mostly, if not entirely, undisputed. Although the inferences to be drawn may be more debatable, it is very hard to see that Borland has shown any interest in the Lotus menu except as a fall-back option for those users already committed to it by prior experience or in order to run their own macros using

1-2-3 commands. At least for the amateur, accessing the Lotus menu in the Borland Quattro or Quattro Pro program takes some effort.

Put differently, it is unlikely that users who value the Lotus menu for its own sake — independent of any investment they have made themselves in learning Lotus' commands or creating macros dependent upon them — would choose the Borland program in order to secure access to the Lotus menu. Borland's success is due primarily to other features. Its rationale for deploying the Lotus menu bears the ring of truth.

Now, any use of the Lotus menu by Borland is a commercial use and deprives Lotus of a portion of its "reward," in the sense that an infringement claim if allowed would increase Lotus' profits. But this is circular reasoning: broadly speaking, every limitation on copyright or privileged use diminishes the reward of the original creator. Yet not every writing is copyrightable or every use an infringement. The provision of reward is one concern of copyright law, but it is not the only one. If it were, copyrights would be perpetual and there would be no exceptions.

The present case is an unattractive one for copyright protection of the menu. The menu commands (e.g., "print," "quit") are largely for standard procedures that Lotus did not invent and are common words that Lotus cannot monopolize. What is left is the particular combination and sub-grouping of commands in a pattern devised by Lotus. This arrangement may have a more appealing logic and ease of use than some other configurations; but there is a certain arbitrariness to many of the choices.

If Lotus is granted a monopoly on this pattern, users who have learned the command structure of Lotus 1-2-3 or devised their own macros are locked into Lotus, just as a typist who has learned the QWERTY keyboard would be the captive of anyone who had a monopoly on the production of such a keyboard. Apparently, for a period Lotus 1-2-3 has had such sway in the market that it has represented the de facto standard for electronic spreadsheet commands. So long as Lotus is the superior spreadsheet — either in quality or in price — there may be nothing wrong with this advantage.

But if a better spreadsheet comes along, it is hard to see why customers who have learned the Lotus menu and devised macros for it should remain captives of Lotus because of an investment in learning made by the users and not by Lotus. Lotus has already reaped a substantial reward for being first; assuming that the Borland program is now better, good reasons exist for freeing it to attract old Lotus customers: to enable the old customers to take advantage of a new advance, and to reward Borland in turn for making a better product. If Borland has not made a better product, then customers will remain with Lotus anyway.

Thus, for me the question is not whether Borland should prevail but on what basis. Various avenues might be traveled, but the main choices are between holding that the menu is not protectable by copyright and devising a new doctrine that Borland's use is privileged. No solution is perfect and no intermediate appellate court can make the final choice.

To call the menu a "method of operation" is, in the common use of those words, a defensible position. After all, the purpose of the menu is not to be admired as a work of literary or pictorial art. It is to transmit directions from the user to the computer, i.e., to operate the computer. The menu is also a "method" in the dictionary sense because it is a "planned way of doing something," an "order or system," and (aptly here) an "orderly or systematic arrangement, sequence or the like." Random House Webster's College Dictionary 853 (1991).

A different approach would be to say that Borland's use is privileged because, in the context already described, it is not seeking to appropriate the advances made by

Lotus' menu; rather, having provided an arguably more attractive menu of its own, Borland is merely trying to give former Lotus users an option to exploit their own prior investment in learning or in macros. The difference is that such a privileged use approach would not automatically protect Borland if it had simply copied the Lotus menu (using different codes), contributed nothing of its own, and resold Lotus under the Borland label.

The closest analogue in conventional copyright is the fair use doctrine. E.g., Harper & Row, Publishers, Inc. v. Nation Enters., 471 U.S. 539 (1985). Although invoked by Borland, it has largely been brushed aside in this case because the Supreme Court has said that it is "presumptively" unavailable where the use is a "commercial" one. See id. at 562. In my view, this is something less than a definitive answer; "presumptively" does not mean "always" and, in any event, the doctrine of fair use was created by the courts and can be adapted to new purposes.

But a privileged use doctrine would certainly involve problems of its own. It might more closely tailor the limits on copyright protection to the reasons for limiting that protection; but it would entail a host of administrative problems that would cause cost and delay, and would also reduce the ability of the industry to predict outcomes. Indeed, to the extent that Lotus' menu is an important standard in the industry, it might be argued that any use ought to be deemed privileged.

In sum, the majority's result persuades me and its formulation is as good, if not better, than any other that occurs to me now as within the reach of courts. Some solutions (e.g., a very short copyright period for menus) are not options at all for courts but might be for Congress. In all events, the choices are important ones of policy, not linguistics, and they should be made with the underlying considerations in view.

COMMENTS AND QUESTIONS

1. The issues raised by the Lotus decision remain unsettled. The Supreme Court granted certiorari in the case, but it deadlocked 4-4 and therefore affirmed the First Circuit but produced no precedential opinion. Lotus Dev. Corp. v. Borland Int'l, 516 U.S. 233 (1996). Note that the First Circuit concedes that its opinion is at odds with the Tenth Circuit's decision in Autoskill, Inc. v. National Educational Support Systems, 994 F.2d 1476, 1495 n.23 (10th Cir. 1993), *cert. denied*, 510 U.S. 916 (1993), and dictum from the Ninth Circuit's decision in Brown Bag Software v. Symantec Corp., 960 F.2d 1465, 1477 (9th Cir. 1992). It is also arguably inconsistent with the Fifth Circuit's decision in Engineering Dynamics, Inc. v. Structural Software Inc., 26 F.3d 1335 (5th Cir. 1994), which gave broad copyright protection to program interfaces and relied extensively on Judge Keeton's prior decision in Lotus v. Paperback.

On the other hand, the *Lotus* approach seems consistent with the legislative history of 17 U.S.C. § 102(b). The House Report accompanying that section states: "Section 102(b) is intended, among other things, to make clear that the expression adopted by the programmer is the copyrightable element in a computer program, and that the actual processes or methods embodied in the program are not within the scope of the copyright law." H.R. Rep. No. 1476, 94th Cong., 2d Sess. 57 (1976), reprinted in 1976 U.S.C.C.A.N. 5659, 5670.

Further, most commentators seem to agree with the ultimate result in *Lotus*, though many are troubled by the court's reasoning. See, e.g., Brief Amicus Curiae of Copyright Law Professors in Support of Respondent in Lotus v. Borland, No.

94-2003 (U.S. 1995) (brief submitted by 34 copyright professors urging affirmance for a variety of reasons); Mark A. Lemley, Convergence in the Law of Software Copyright?, 10 High Tech. L.J. 1 (1995); Glynn S. Lunney Jr., Lotus v. Borland: Copyright and Computer Programs, 70 Tulane L. Rev. 2397 (1996); Peter S. Menell, The Challenges of Reforming Intellectual Property Protection for Computer Software, 94 Colum. L. Rev. 2644, 2653 (1994). Finally, at least one other appellate court has followed the reasoning of *Lotus* in denying protection to computer systems as methods of operation. See MiTek Holdings v. Arce Engineering, 89 F.3d 1548 (11th Cir. 1996) (refusing to protect menu structures under section 102(b) because allowing such protection would infringe on the province of patent law). But see Mitel v. Iqtel, 124 F.3d 1366 (10th Cir. 1997), where the court explicitly rejected the *Lotus* approach, though it did find the communications protocols at issue there unprotectable under *Altai*.

2. Contrast the analysis of section 102(b) and Baker v. Selden in the *Lotus* case with that in Apple v. Franklin. Applying the *Lotus* court's logic, isn't object code a "method of operation" for instructing a CPU how to process information? If so, how can the copyrightability of computer programs be reconciled with the exclusion from copyright of "methods of operation"?

3. Note that the *Lotus* court chose to reject the filtration test adopted by the Second Circuit in Computer Associates v. Altai, and followed by most other circuits. Is the holding in *Lotus* really incompatible with *Altai*'s abstraction-filtration-comparison test? One possible way to reconcile *Lotus* and *Altai* is to treat *Lotus* not as a case involving "literal" copying of a menu structure, but as a case involving non-literal copying of the program itself. Lotus distributed object code which, in certain combinations, produces on a screen physical images that represent certain words and have certain effects. Borland did not copy that code; rather, it wrote its own code, which produced similar images and had the same effects as the Lotus code. The court could then have applied *Altai*'s abstraction-filtration-comparison analysis to this "non-literal" copying. This is in effect precisely what the *Lotus* court did — although when it reached the "filtration" step, it determined that the entire program was unprotectable at the menu command hierarchy level. See Mark A. Lemley, Convergence in the Law of Software Copyright?, 10 High Tech. L.J. 1 (1995).

Do the *Altai* and *Lotus* cases really present analogous issues? Is a "filtration" analysis that filters out the entire copyrighted work at one level of abstraction within the contemplation of the *Altai* court?

4. The *Lotus* court compares the 1-2-3 menu command hierarchy to the buttons on a VCR, which it says are uncopyrightable because they are methods of operation. But is this the right analogy? Lotus claimed that Borland had infringed its copyright by copying the structure, sequence, and organization of its menu hierarchy, not the words of the menus themselves. This "structure, sequence and organization" seems the equivalent of the location and arrangement of the VCR buttons, rather than the buttons themselves. Would an original way of arranging the buttons on a VCR be copyrightable? Should it be? Is the question one of originality and the idea-expression dichotomy, or is there some reason not to copyright the arrangement of buttons on a VCR even if it were original and expressive?

5. In the predecessor case to Lotus v. Borland, defendant Paperback Software argued that the command structure of the Lotus program is in essence a language. It can be used to create macros and run the spreadsheet. Is this a fair characterization? The district court attacked this analogy (which it called "strained") on two basic

grounds: that Lotus 1-2-3 was not in fact a language, and that languages could be copyrightable. The first conclusion is certainly debatable; one's definition of a language (or alternatively a "system") will go a long way toward answering this question. But consider the court's second premise — that languages might be copyrightable. Is copyrighting "natural" (i.e., human) languages consistent with what you know of copyright law? Is a language functional, or does it express? Does it matter whether the plaintiff is claiming the creation of certain words, on the one hand, or the entire system of grammar and interaction between words? On this point, compare Ronald Johnson & Allen Grogan, Copyright Protection for Command Driven Interfaces, 8:6 Computer Law. 1 (June 1991) (advocating copyright protection for languages) with Pamela Samuelson, How to Interpret the Lotus Decision; And How Not To, 33:11 Communications of the ACM 27 (Nov. 1990) and Elizabeth G. Lowry, Comment, Copyright Protection for Computer Languages: Creative Incentive or Technological Threat, 39 Emory L.J. 1293 (1990). Does the First Circuit's decision in Lotus v. Borland shed any light on this issue? Are languages "systems" or "methods of operation" which are denied protection under section 102(b)? The Australian case of Data Access Corp. v. Powerflex Services Pty., No. VG473 (Fed. Ct. Aus., Melbourne 1996), may prove instructive. That court held that even a single word serving a menu command function may be copyrightable, since it is a concatenation of letters. Copyright protection therefore clearly extended to the computer language at issue in that case, which consisted of 225 words. Is this result consistent with the goals of copyright?

6. Both the majority and Judge Boudin's concurrence express concern over the harm that would be done to users of Lotus 1-2-3 if they could not copy their macros for use on Quattro Pro. Certainly, users have invested time and effort in using Lotus 1-2-3, and this may make them reluctant to change spreadsheet programs, even if Borland's program really is superior. But is this the sort of problem with which the copyright law should be concerned? Why isn't a customer preference for a known, trusted product part of the reward that the copyright owner is entitled to reap?

Is the QWERTY keyboard protectable by copyright? Why or why not? Does the answer depend on timing — whether the QWERTY keyboard has just been introduced, or whether it is well established as the industry standard for typewriters? How would you apply section 102(b) in each case?

7. In his concurrence, Judge Boudin proposes (though he does not necessarily adopt) an alternative rationale for the decision: that Lotus's menu command hierarchy was copyrightable, but that Borland's use of the menu hierarchy in this case was privileged because it was done for the purpose of making the two programs compatible. The "privileged use" rationale is narrower than the majority's "method of operation" rule. Because the majority holds the menu command hierarchy entirely uncopyrightable, Lotus has no power to prevent anyone from copying its menu structure, regardless of the circumstances in which the copying occurs.

Suppose that, without taking any protected source or object code, a small company managed to acquire an early copy of Lotus 1-2-3. It replicated the menu command hierarchy and began to sell "clone" programs, which looked and worked just like Lotus 1-2-3. Cf. Lotus Dev. Corp. v. Paperback Software, 740 F. Supp. 37 (D. Mass. 1990), a prior decision by Judge Keeton which is apparently overruled by the First Circuit's decision. In such a case, the equities might seem to tilt towards Lotus more than they did in the *Borland* case. Nonetheless, under the majority's rule, the clone manufacturer has the right to make such copies. By contrast, the "privileged use" rationale might not protect the clone manufacturer from a copyright infringement suit. Does Judge Boudin's "privileged use" approach better address this circumstance? For a more detailed discussion of a privilege to copy certain aspects of a computer program in order to achieve compatibility, see *infra* section C.3 of this chapter. For an argument that the effort that goes into program design should be protected against misappropriation, though not by copyright law, see Pamela Samuelson et al., A Manifesto Concerning the Legal Protection of Computer Programs, 94 Colum. L. Rev. 2308 (1994).

d. *Protection for Program Outputs: User Interfaces*

Source and object code (as well as the non-literal elements of program structure) are protectable as literary works. Screen displays (the output of the computer program) are protectable as audiovisual or pictorial works. Given these two sets of doctrines, it might be considered evident that user interfaces — the screen displays that mediate the input to and output from a computer program — should also be protected by copyright.

Acknowledging this does not, however, tell us what *portions* of the user interface can be considered copyrightable expression. To a greater extent than other types of screen displays, user interfaces are dictated by industry and product market standards and functional considerations. As a result, assessing claims of copyright infringement is a challenging task.

The interface between the computer and the user consists of a variety of input/output devices, including a keyboard, pointing tools (such as a mouse, joystick, and interactive pen), disk drives, audio equipment, microphone, and screen displays. Copyright law excludes from protection such obviously functional works as keyboards, pointing objects, speaker systems, and other hardware devices. The courts have also found that data input formats, such as the order and size of data fields, are not protectable under copyright law. See e.g., Synercom Technology, Inc. v. University Computing Co., 462 F. Supp. 1003 (N.D. Tex. 1978); cf. Mitel, Inc. v. Iqtel, Inc., 124 F.3d 1366 (10th Cir. 1997). The visual images and text of screen displays may qualify as audiovisual, graphic, or literary works under copyright. See e.g., Stern Electronics, Inc. v. Kaufman, 669 F.2d 852 (2d Cir. 1982). Some early courts afforded broad protection to elements of a user interface. See e.g., Broderbund Software, Inc. v. Unison World, 648 F. Supp. 1127, 1134 (N.D. Cal. 1986) (holding that the choice of typeface on user screen display and choice of words such as "Choose a Font" as the title for a screen for producing cards, brochures, and other printing projects were examples of audiovisual displays "dictated primarily by artistic and aesthetic consideration, and not by utilitarian or mechanical ones"); Digital Communications Associates, Inc. v. Softklone Distributing Corp., 659 F. Supp. 449 (N.D. Ga. 1987) (finding that the arrangement of status screens and commands for a data communication program are protectable expression). Such works remain, however, subject to the originality requirement and functionality limitations of §102(b), the merger doctrine, and Baker v. Selden.

The Ninth Circuit addressed some of the limitations on copyright protection for audiovisual displays of computer programs in Data East USA, Inc. v. Epyx, Inc., 862 F.2d 204 (9th Cir. 1988), in which the manufacturer of a video game depicting a karate match sought to prevent another firm from marketing a competing game featuring many of the same audio and visual elements. Notwithstanding the many similarities between the two works, the court held that no infringement had occurred because the similarities flowed from "constraints inherent in the sport of karate itself" and "various constraints inherent in the use of [the particular type of] computer." Id. at 209. See also Interactive Network v. NTN Communications, 875 F. Supp 1398 (N.D. Cal. 1995), *aff'd* 57 F.3d 1083 (Fed. Cir. 1996) (finding that football video game was not infringed because similarities between works were based on the rules of football and the idea of an interactive prediction game). After filtering out the unprotectable ideas in the work, the court applied a standard of "virtual identity" in determining that the competing work did not infringe.

The most significant case to address the scope of copyright protection for network features of a computer-human interface is Apple Computer, Inc. v. Microsoft Corp., 799 F. Supp. 1006 (N.D. Cal. 1992), *aff'd in part, rev'd in part,* 35 F.2d 1435 (9th Cir. 1994), *cert. denied* 513 U.S. 1184 (1995), in which Apple Computer alleged that Microsoft's Windows operating system and Hewlett-Packard's NewWave operating system infringed Apple's copyrights in the desktop graphical user interface for its Macintosh computer system. The copyright issue was somewhat muddied by the existence of a licensing agreement authorizing the defendants to use aspects of Apple's graphical user interface. The court determined, however, that the licensing agreement was not a complete defense to the copyright claims and therefore undertook an analysis of the scope of copyright protection for a large range of audiovisual elements of computer screen displays.

In framing the analysis, the district court expressly recognized the importance of standardization to consumers and the cumulative nature of innovation to the scope of copyright protection. Apple Computer, Inc. v. Microsoft Corp., 709 F. Supp. 925, 930 (N.D. Cal. 1989); Apple Computer, Inc. v. Microsoft Corp., 717 F. Supp. 1428 (N.D. Cal. 1989); Apple Computer, Inc. v. Microsoft Corp., 759 F. Supp. 1444 (N.D. Cal. 1991). The court proceeded to determine those elements of the graphical user interface which were not protected on the grounds that they lacked originality or were not protectable under section 102(b), the doctrine of scenes à faire, or the merger of idea and expression, or due to the limited number of ways in which an idea could be expressed or the external constraints imposed by the computer system. The court found that all of the alleged similarities between Apple's works and Microsoft's Windows not authorized by the licensing agreement were either not protectable or subject to at least one of the limiting doctrines. As a result, the court applied the "virtual identity" standard in comparing the works as a whole and determined that no infringement had occurred. On appeal, the Ninth Circuit affirmed the district court's dissection of the work in question to determine which elements are protectable, its filtering out of unprotectable elements, and its application of the "virtual identity" standard in this context. Apple Computer, Inc. v. Microsoft Corp., 35 F.3d 1435 (9th Cir. 1994). The Ninth Circuit went on to note, however, that a copyright owner might be able to protect a user interface consisting of a unique combination of otherwise uncopyrightable elements.

The *Apple* litigation established that many elements of the desktop-based graphical user interface are in the public domain and that the originality requirement and functionality doctrines of copyright law substantially limit the protection afforded the desktop user interface. The Eleventh Circuit has since joined the Ninth Circuit in adopting the "virtual identity standard" for claims of software infringement in a computer-user interface based on a compilation of uncopyrightable elements. See

MiTek Holdings, Inc. v. ARCE Engineering Co., 89 F.3d 1548, 1558 (11th Cir. 1996).

COMMENTS AND QUESTIONS

1. What is the appropriate standard for testing infringement of the user interfaces of computer programs? The traditional test for evaluating copying has been the "substantial similarity" standard. *Apple*'s "virtual identicality" test comes from a number of cases, including the *Hoehling* case excerpted in Chapter 4, that deal with "thin" copyrights. Is it appropriate to vary the standard for infringement with the scope of copyright protection, or does that "double-count" the results of the filtration analysis?

2. In many non-literal software infringement cases, as well as in *Apple*, the vast majority of the copied work is determined to be uncopyrightable for one reason or another. If only a few "kernels" of copyrightable expression are mixed in with a large amount of public domain material, how should the courts test the "substantiality" of copying? If only 2 percent of a program is copyrightable, is taking that 2 percent the taking of a *de minimis* amount of the total program, or should it be considered the taking of the *entire* copyrighted work? Can the result possibly depend on whether uncopyrightable elements were also taken?

3. Consider the effect of dissection on copyright incentives. Does the fragmented protection that copyright affords to functional works such as computer programs adequately encourage creation of more programs? Does it encourage programmers to be more expressive when writing programs? If so, is that a desirable goal?

4. Is Lotus Dev. Corp. v. Borland Int'l, reprinted above, really a case about command structures or about user interfaces? If the court approached it as a user interface case, should it reach the same result? Would it? Cf. Mitel Inc. v. Iqtel Inc., 896 F. Supp. 1050 (D. Colo. 1995), where the court followed the First Circuit's decision in *Lotus* and, in addition, rejected the plaintiff's claim that its command codes were protectable as a user interface.

PROBLEM

Problem 7-5. Pueblo Graphics sells software that produces a variety of word processing fonts. (Fonts control the appearance of letters and numbers in typesetting). The software allows the fonts to appear on the screen in exactly the same way they will appear on the printed page. Pueblo's software contains a number of "classic" print fonts, such as Geneva, Helvetica, and Times Roman Bold. In addition, however, Pueblo has designed several new fonts that look significantly different from any fonts previously produced.

Pueblo would like to assert a copyright infringement suit against a competitor who has copied its fonts. The competitor wrote its own program to implement those fonts, so Pueblo wants to know if it can get copyright protection in the fonts themselves. How would you advise Pueblo? Is their claim appropriately characterized as one for a "user interface"? Why or why not?

2. Exclusive Rights in Computer Programs

a. *The Right to Make "Copies"*

Most of the cases considered in this chapter have involved the copying of all or part of a program by a company that sold a competing program. "Copying" has a potentially broader meaning in the computer context, however. The use of any modern computer program necessarily involves the creation of transitory copies in the RAM memory of a computer. Because "running" a computer program necessarily involves copying the program from long-term memory (ROM, CD-ROM, or disk) into the short-term memory of the computer, it is virtually impossible to use any computer program without making a "copy" of the program. However, that copy is not saved; it is erased whenever the application is exited or the computer is turned off. Whether such "copies" are fixed — and therefore potentially violate the copyright owner's rights in section 106(1) — has been a matter of some debate.

In 1993 the Ninth Circuit held that loading a computer program into RAM involved making a copy for purposes of section 106. MAI Systems Corp. v. Peak Computer, Inc., 991 F.2d 511, 518 (9th Cir. 1993). The court reasoned:

> The district court's grant of summary judgment on MAI's claims of copyright infringement reflects its conclusion that a 'copying' for purposes of copyright law occurs when a computer program is transferred from a permanent storage device to a computer's RAM. This conclusion is consistent with its finding, in granting the preliminary injunction, that: 'the loading of copyrighted computer software from a storage medium (hard disk, floppy disk, or read only memory) into the memory of a central processing unit ('CPU') causes a copy to be made. In the absence of ownership of the copyright or express permission by license, such acts constitute copyright infringement.' We find that this conclusion is supported by the record and by the law.
>
> Peak concedes that in maintaining its customer's computers, it uses MAI operating software 'to the extent that the repair and maintenance process necessarily involved turning on the computer to make sure it is functional and thereby running the operating system.' It is also uncontroverted that when the computer is turned on the operating system is loaded into the computer's RAM. As part of diagnosing a computer problem at the customer site, the Peak technician runs the computer's operating system software, allowing the technician to view the systems error log, which is part of the operating system, thereby enabling the technician to diagnose the problem.
>
> Peak argues that this loading of copyrighted software does not constitute a copyright violation because the 'copy' created in RAM is not 'fixed.' However, by showing that Peak loads the software into RAM and is then able to view the system error log and diagnose the problem with the computer, MAI has adequately shown that the representation created in the RAM is 'sufficiently permanent or stable to permit it to be perceived, reproduced, or otherwise communicated for a period of more than transitory duration.'
>
> After reviewing the record, we find no specific facts (and Peak points to none) which indicate that the copy created in the RAM is not fixed....
>
> We have found no case which specifically holds that the copying of software into RAM creates a 'copy' under the Copyright Act. However, it is generally accepted that the loading of software into a computer constitutes the creation of a copy under the Copyright Act. See, e.g., Vault Corp. v. Quaid Software, 847 F.2d 255, 260 (5th Cir. 1988) ('the act of loading a program from a medium of storage into a computer's memory creates a copy of the program'); 2 Nimmer on Copyright, §8.08 at 8-105 (1983)

('Inputting a computer program entails the preparation of a copy.'); Final Report of the National Commission on the New Technological Uses of Copyrighted Works, at 13 (1978) ('the placement of a work into a computer is the preparation of a copy'). We recognize that these authorities are somewhat troubling since they do not specify that a copy is created regardless of whether the software is loaded into the RAM, the hard disk or the read only memory ('ROM'). However, since we find that the copy created in the RAM can be 'perceived, reproduced, or otherwise communicated,' we hold that the loading of software into the RAM creates a copy under the Copyright Act.

COMMENTS AND QUESTIONS

1. If the Ninth Circuit is correct, every use of a computer or any computer program involves the making of at least one (and probably multiple) copies of the computer program. The fact that it is impossible to read or use a computer program without making such "copies" means that the application of the copyright laws to computer software is potentially far broader than to other types of works. While you can "use" a book without making a copy, under *MAI* any use of a computer program — even turning the computer on — necessarily implicates the copyright laws. But cf. NFLC, Inc. v. Devcom Mid-America, Inc., 45 F.3d 231 (7th Cir. 1995) (suggesting that the use of dumb terminals might not make a RAM copy).

Partly in recognition of this problem, Congress enacted section 117 of the Copyright Act in 1980. Section 117 makes it possible for users of computer programs to make certain copies without fear of liability. We discuss section 117 in detail in the next section of this chapter; for now, keep in mind that making a copy is not always a violation of the copyright laws.

Does the specific enactment by Congress of a statutory provision permitting a user to make copies "necessary" to run a program suggest that Congress agreed that RAM copies are "fixed" for copyright purposes?

2. *MAI* is arguably inconsistent with both prior case law and the intent of Congress. In Apple Computer v. Formula Intl., 594 F. Supp. 617, 621-22 (C.D. Cal. 1984), a case cited with apparent approval in *MAI,* the court indicated that copies stored in RAM, unlike those loaded in ROM, were only "temporary." And the House Report discussing the definition of "fixed" in section 101 of the Act had this to say on the subject of computer memory: "[T]he definition of 'fixation' would exclude from the concept purely evanescent or transient reproductions such as those projected briefly on a screen, shown electronically on a television or other cathode ray tube, or captured momentarily in the 'memory' of a computer." H.R. Rep. No. 94-1476, 94th Cong., 2d Sess. 52-53 (1976); cf. British Leland v. Armstrong, 1986 RPC 279 (U.K. House of Lords).

On the other hand, a number of courts have followed *MAI*'s lead in holding that copies loaded in RAM are fixed for purposes of copyright infringement. See Triad Systems v. Southeastern Express Co., 64 F.3d 1330 (9th Cir. 1995), *cert. denied,* 516 U.S. 1145 (1996); Stenograph LLC v. Bossard Assocs., 144 F.3d 96 (D.C. Cir. 1998); In re Independent Service Organizations Antitrust Litigation, 910 F. Supp. 1537, 1541 (D. Kan. 1995); ACS v. MAI, 845 F. Supp. 356 (E.D. Va. 1994).

Is RAM memory perceivable for "more than a transitory duration"? Is it more like movies projected on a screen, or like movies stored on videotape?

3. The Fourth Circuit takes a more narrow view of fixation relating to ISP activities:

[T]o establish direct liability under §§ 501 and 106 of the Act, something more must be shown than mere ownership of a machine used by others to make illegal copies. There must be actual infringing conduct with a nexus sufficiently close and causal to the illegal copying that one could conclude that the machine owner himself trespassed on the exclusive domain of the copyright owner. The Netcom [Religious Technology Center v. Netcom On-Line Communication Services, Inc., 907 F.Supp. 1361 (N.D.Cal.1995)] court described this nexus as requiring some aspect of volition or causation. 907 F.Supp. at 1370. Indeed, counsel for both parties agreed at oral argument that a copy machine owner who makes the machine available to the public to use for copying is not, without more, strictly liable under § 106 for illegal copying by a customer. The ISP in this case is an analogue to the owner of a traditional copying machine whose customers pay a fixed amount per copy and operate the machine themselves to make copies. When a customer duplicates an infringing work, the owner of the copy machine is not considered a direct infringer. Similarly, an ISP who owns an electronic facility that responds automatically to users' input is not a direct infringer. If the Copyright Act does not hold the owner of the copying machine liable as a direct infringer when its customer copies infringing material without knowledge of the owner, the ISP should not be found liable as a direct infringer when its facility is used by a subscriber to violate a copyright without intervening conduct of the ISP.

Moreover, in the context of the conduct typically engaged in by an ISP, construing the Copyright Act to require some aspect of volition and meaningful causation — as distinct from passive ownership and management of an electronic Internet facility — receives additional support from the Act's concept of "copying." A violation of § 106 requires copying or the making of copies. See 17 U.S.C. § 106(1), (3); id. § 102(a); Feist Publications, 499 U.S. at 361, 111 S.Ct. 1282. And the term "copies" refers to "material objects . . . in which a work is fixed." 17 U.S.C. § 101 ("Definitions") (emphasis added). A work is "fixed" in a medium when it is embodied in a copy "sufficiently permanent or stable to permit it to be perceived, reproduced, or otherwise communicated for a period of more than transitory duration." Id. (emphasis added). When an electronic infrastructure is designed and managed as a conduit of information and data that connects users over the Internet, the owner and manager of the conduit hardly "copies" the information and data in the sense that it fixes a copy in its system of more than transitory duration. Even if the information and data are "downloaded" onto the owner's RAM or other component as part of the transmission function, that downloading is a temporary, automatic response to the user's request, and the entire system functions solely to transmit the user's data to the Internet. Under such an arrangement, the ISP provides a system that automatically transmits users' material but is itself totally indifferent to the material's content. In this way, it functions as does a traditional telephone company when it transmits the contents of its users' conversations. While temporary electronic copies may be made in this transmission process, they would appear not to be "fixed" in the sense that they are "of more than transitory duration," and the ISP therefore would not be a "copier" to make it directly liable under the Copyright Act. With additional facts, of course, an ISP could become indirectly liable.

In concluding that an ISP has not itself fixed a copy in its system of more than transitory duration when it provides an Internet hosting service to its subscribers, we do not hold that a computer owner who downloads copyrighted software onto a computer cannot infringe the software's copyright. See, e.g., MAI Systems Corp. v. Peak Computer, Inc., 991 F.2d 511, 518-19 (9th Cir.1993). When the computer owner downloads copyrighted software, it possesses the software, which then functions in the service of the computer or its owner, and the copying is no longer of a transitory nature. See, e.g., Vault Corp. v. Quaid Software Ltd., 847 F.2d 255, 260 (5th Cir.1988). "Transitory duration" is thus both a qualitative and quantitative characterization. It is quantitative insofar as it describes the period during which the function occurs, and it is qualitative in the sense that it describes the status of transition. Thus, when the copyrighted software is downloaded onto the computer, because it may be used to serve the computer or the computer owner, it no longer remains transitory. This, however, is unlike an ISP, which provides a system that automatically receives a subscriber's infringing material and transmits it to the Internet at the instigation of the subscriber.

CoStar Group, Inc. v. LoopNet, Inc. 373 F.3d 544, 550-51 (4th Cir. 2004). Does the statute contemplate the meaning of "fixed" turning on the nature of the enterprise operating the computer or its knowledge? Doesn't the ISP fix the image through its design of its servers? Shouldn't the court address this issue directly by questioning *MAI*, since in that case the defendant did not download the software but merely booted up the computer?

4. Most commentators have been critical of the *MAI* decision. See, e.g., Anthony R. Reese, The Public Display Right: The Copyright Act's Neglected Solution to the Controversy Over RAM "Copies," 2001 U. Ill. L. Rev. 83; Niva Elkin-Koren, Copyright Law and Social Dialogue on the Information Superhighway: The Case Against Copyright Liability of Internet Providers, 13 Cardozo Arts & Ent. L.J. 345, 381-82 (1995); Michael E. Johnson, Note, The Uncertain Future of Computer Software Users' Rights in the Aftermath of MAI Systems, 44 Duke L.J. 327 (1994); Mark A. Lemley, Dealing with Overlapping Copyrights on the Internet, 22 U. Dayton L. Rev. 547 (1997); Jessica Litman, The Exclusive Right to Read, 13 Cardozo Arts & Ent. L.J. 29, 41-43 (1994). On the other hand, the Clinton administration enthusiastically endorsed it in its White Paper on copyright and the Internet. See NII Working Group on Intellectual Property, Intellectual Property and the National Information Infrastructure (August 1995).

b. Copies and Section 117

As noted above, making certain types of "copies" is an essential part of using computer programs. Because of this important difference between computer software and other types of copyrightable material, Congress followed the recommendations of CONTU and passed section 117. That section provides, in relevant part:

> Notwithstanding the provisions of section 106, it is not an infringement for the owner of a copy of a computer program to make or authorize the making of another copy or adaptation of that computer program provided:
> (1) that such a new copy or adaptation is created as an essential step in the utilization of the computer program in conjunction with a machine and that it is used in no other manner, or
> (2) that such new copy or adaptation is for archival purposes only and that all archival copies are destroyed in the event that continued possession of the computer program should cease to be rightful.

17 U.S.C. §117.

Section 117 has generated substantial controversy. We work through the principal complexities below.

i.   To whom does § 117 apply

By its terms, the statute authorizes copies or adaptations made by "the owner of a copy of a computer program" or persons she authorizes to make such copies. In its Final Report, CONTU indicated its intention "that *persons in rightful possession of copies of programs* be able to use them freely without fear of exposure to copyright liability." CONTU Final Report at 13 (emphasis added). The report seems to indicate that section 117 applies to anyone who has not obtained the program by improper means.

Without discussion or explanation, Congress changed "persons in rightful possession of copies" to "owners of copies." Does this mean that section 117 doesn't benefit "licensees" — a group which on some readings encompasses all users of a computer program? Should the phrase "owner of a copy of a computer program" be interpreted to refer to the rightful possessor of the physical copy, regardless of who retains the copyright in the program itself? This issue was addressed at length by the Federal Circuit in DSC Communications Corp. v. Pulse Communications Inc., 170 F.3d 1354 (Fed. Cir. 1999):

> The district court concluded that making copies of the POTS-DI software (in the resident memory of POTS cards) was an "essential step in the utilization" of the POTS-DI software and that there was no evidence that the RBOCs used the software in any other manner that would constitute infringement. Accordingly, under the district court's theory of the case there was no direct infringement (and thus no contributory infringement) if the RBOCs were section 117 "owners" of copies of the POTS-DI software.
>
> The district court then held that the RBOCs were "owners" of copies of the POTS-DI software because they obtained the software by making a single payment and obtaining a right to possession of the software for an unlimited period. Those attributes of the transaction, the court concluded, made the transaction a "sale."
>
> DSC challenges the district court's conclusion that, based on the terms of the purchase transactions between DSC and the RBOCs, the RBOCs were "owners" of copies of the POTS-DI software. In order to resolve that issue, we must determine what attributes are necessary to constitute ownership of copies of software in this context.
>
> Unfortunately, ownership is an imprecise concept, and the Copyright Act does not define the term. Nor is there much useful guidance to be obtained from either the legislative history of the statute or the cases that have construed it. The National Commission on New Technological Uses of Copyrighted Works ("CONTU") was created by Congress to recommend changes in the Copyright Act to accommodate advances in computer technology. In its final report, CONTU proposed a version of section 117 that is identical to the one that was ultimately enacted, except for a single change. The proposed CONTU version provided that "it is not an infringement for the rightful possessor of a copy of a computer program to make or authorize the making of another copy or adaptation of that program...." Final Report of the National Commission on New Technological Uses of Copyrighted Works, U.S. Dept. of Commerce, PB-282141, at 30 (July 31, 1978) (emphasis added). Congress, however, substituted the words "owner of a copy" in place of the words "rightful possessor of a copy." See Pub.L. No. 96-517, 96th Cong., 2d Sess. (1980). The legislative history does not explain the reason for the change, see H.R.Rep. No. 96-1307, 96th Cong., 2d Sess., pt. 1, at 23 (1980), but it is clear from the fact of the substitution of the term "owner" for "rightful possessor" that Congress must have meant to require more than "rightful possession" to trigger the section 117 defense.

In the leading case on section 117 ownership, the Ninth Circuit considered an agreement in which MAI, the owner of a software copyright, transferred copies of the copyrighted software to Peak under an agreement that imposed severe restrictions on Peak's rights with respect to those copies. See MAI Sys. Corp. v. Peak Computer, Inc., 991 F.2d 511 (9th Cir. 1995). The court held that Peak was not an "owner" of the copies of the software for purposes of section 117 and thus did not enjoy the right to copy conferred on owners by that statute. The Ninth Circuit stated that it reached the conclusion that Peak was not an owner because Peak had licensed the software from MAI. See id. at 518 n. 5. That explanation of the court's decision has been criticized for failing to recognize the distinction between ownership of a copyright, which can be licensed, and ownership of copies of the copyrighted software. See, e.g., 2 Melville B. Nimmer, Nimmer on Copyright ¶8.08[B][1], at 8-119 to 1-121 (3d ed. 1997). Plainly, a party who purchases copies of software from the copyright owner can hold a license under a copyright while still being an "owner" of a copy of the copyrighted software for purposes of section 117. We therefore do not adopt the Ninth Circuit's characterization of all licensees as non-owners. Nonetheless, the MAI case is instructive, because the agreement between MAI and Peak, like the agreements at issue in this case, imposed more severe restrictions on Peak's rights with respect to the software than would be imposed on a party who owned copies of software subject only to the rights of the copyright holder under the Copyright Act. And for that reason, it was proper to hold that Peak was not an "owner" of copies of the copyrighted software for purposes of section 117. See also Advanced Computer Servs. of Mich. v. MAI Sys. Corp., 845 F. Supp. 356, 367 (E.D.Va. 1994) ("MAI customers are not 'owners' of the copyrighted software; they possess only the limited rights set forth in their licensing agreements"). We therefore turn to the agreements between DSC and the RBOCs to determine whether those agreements establish that the RBOCs are section 117 "owners" of copies of the copyrighted POTS-DI software.

[The court concluded after a detailed analysis of the facts of the transaction that certain agreements in that case did not transfer ownership of the particular copies of the program at issue, but that other transactions did involve a transfer of ownership].

Id. at 1359-1362. It is worth noting that *DSC* involved a negotiated agreement between sophisticated parties — the buyers were the regional Bell operating companies. By contrast, in the mass-market context, most (though by no means all) cases that have considered the issue have concluded that mass-market, over-the-counter transfers of computer programs at retail stores constitute "sales of goods" subject to the Uniform Commercial Code, rather than license agreements.

The Second Circuit adopted a similarly flexible and fact-specific interpretation of "owner" in Krause v. Titleserv, Inc. 402 F.3d 119 (2d Cir. 2005). Krause had written several computer programs for Titleserv. Upon ending the working relationship, Krause left executable versions of several programs on Titleserv's file servers. Krause later objected to Titleserv's modification of the programs, which Titleserv defended under § 117. In order to effectuate Congress's purposes in enacting section 117, the court interpreted "owner of a copy of a computer program" broadly. It concluded that "formal title is not an absolute prerequisite for qualifying for § 117(a)'s affirmative defense," but merely a factor in determining "whether the party exercises sufficient incidents of ownership over a copy of the program to be sensibly considered the owner of the a[CE1] copy for purposes of § 117(a)."

In our view, Congress's decision to reject "rightful possessor" in favor of "owner" does not indicate an intention to limit the protection of the statute to those possessing formal title. The term "rightful possessor" is quite broad. Had that term been used, the authority granted by the statute would benefit a messenger delivering a program, a bailee, or countless others temporarily in lawful possession of a copy. Congress easily could have intended to reject so broad a category of beneficiaries without intending a narrow, formalistic definition of ownership dependent on title.

Several considerations militate against interpreting § 117(a) to require formal title in a program copy. First, whether a party possesses formal title will frequently be a matter of state law. See 2 Melville B. Nimmer & David Nimmer, Nimmer on Copyright § 8.08[B][1] (stating that copy ownership "arises presumably under state law"). The result would be to undermine some of the uniformity achieved by the Copyright Act. The same transaction might be deemed a sale under one state's law and a lease under another's. If § 117(a) required formal title, two software users, engaged in substantively identical transactions might find that one is liable for copyright infringement while the other is protected by § 117(a), depending solely on the state in which the conduct occurred. Such a result would contradict the Copyright Act's "express objective of creating national, uniform copyright law by broadly preempting state statutory and common-law copyright regulation." Cmty. for Creative Non-Violence v. Reid, 490 U.S. 730, 740 (1989); see also 17 U.S.C. § 301(a).

Second, it seems anomalous for a user whose degree of ownership of a copy is so complete that he may lawfully use it and keep it forever, or if so disposed, throw it in the trash, to be nonetheless unauthorized to fix it when it develops a bug, or to make an archival copy as backup security.

Based upon the circumstances surrounding the Titleserv's possession of the software, the court concluded that it was an "owner" of a copy for purposes of § 117(a):

Titleserv paid Krause substantial consideration to develop the programs for its sole benefit. Krause customized the software to serve Titleserv's operations. The copies were stored on a server owned by Titleserv. Krause never reserved the right to repossess the copies used by Titleserv and agreed that Titleserv had the right to continue to possess and use the programs forever, regardless whether its relationship with Krause terminated. Titleserv was similarly free to discard or destroy the copies any time it wished.

ii. § 117(a)(1): "essential step in the utilization of the computer program in conjunction with a machine and that it is used in no other manner."

This statutory phrase involves several complex elements: what constitutes "essential"? What if some parts of the software are not essential? What does it mean to use the software only in conjunction with a machine and "in no other manner"?

The basic intent of the section seems to be to allow copies like those made in *MAI v. Peak, supra,* which were necessary or naturally incident to the operation of the computer. Further, the language of section 117(1) extends not only to identical copies but to such "adaptation" of programs as is necessary to use them with a particular computer. In *Vault,* the Fifth Circuit held that such adaptations were not limited to those intended by the software vendor. Vault Corp. v. Quaid Software Ltd., 847 F.2d 255 (5th Cir. 1988). In that case, Vault sold a computer program called PROLOK, which was used to "copy-protect" software.[11] Quaid reverse

---

11. Copy protection, which was common during the early 1980s, refers to programs that attempt to prevent unauthorized copying of software by technical means.

engineered Vault's program in order to create its own program, called RAMKEY, which disabled PROLOK's copy protection. In the course of writing RAMKEY, Quaid loaded the PROLOK program into its computer memory. The court rejected Vault's claim of copyright infringement:

> In order to develop RAMKEY, Quaid analyzed Vault's program by copying it into its computer's memory. Vault contends that, by making this unauthorized copy, Quaid directly infringed upon Vault's copyright. The district court held that "Quaid's actions clearly fall within [the § 117(1)] exemption. The loading of [Vault's] program into the [memory] of a computer is an 'essential step in the utilization' of [Vault's] program. Therefore, Quaid has not infringed Vault's copyright by loading [Vault's program] into [its computer's memory]." Vault, 655 F. Supp. at 758.
>
> Section 117(1) permits an owner of a program to make a copy of that program provided that the copy "is created as an essential step in the utilization of the computer program in conjunction with a machine and that it is used in no other manner." Congress recognized that a computer program cannot be used unless it is first copied into a computer's memory, and thus provided the § 117(1) exception to permit copying for this essential purpose. See CONTU Report at 31. Vault contends that, due to the inclusion of the phrase "and that it is used in no other manner," this exception should be interpreted to permit only the copying of a computer program for the purpose of using it for *its intended purpose*. Because Quaid copied Vault's program into its computer's memory for the express purpose of devising a means of defeating its protective function, Vault contends that § 117(1) is not applicable.
>
> We decline to construe § 117(1) in this manner. Even though the copy of Vault's program made by Quaid was not used to prevent the copying of the program placed on the PROLOK diskette by one of Vault's customers (which is the purpose of Vault's program), and was, indeed, made for the express purpose of devising a means of defeating its protective function, the copy made by Quaid was "created as an essential step in the utilization" of Vault's program. Section 117(1) contains no language to suggest that the copy it permits must be employed for a use intended by the copyright owner, and, absent clear congressional guidance to the contrary, we refuse to read such limiting language into this exception. We therefore hold that Quaid did not infringe Vault's exclusive right to reproduce its program in copies under § 106(1).

The court in Krause v. Titleserv, *supra*, also read § 117(a)(1) expansively. Krause informed Titleserv that it was free to continue using the executable code as it existed on the day Krause left, but asserted that Titleserv had no right to modify the source code. Such a restriction would have severely limited the value of those programs, precluding Titleserv from adding new customers to the database, changing customer addresses, or fixing bugs. Employees of Titleserv subsequently circumvented the "lock" Krause had placed on the executable code, decompiled it back into source code, assigned variable names, and added comments. They also added capabilities, such as the ability to print checks and allow customers direct access to their records to make the software more responsive to the needs of the business. The court interpreted "essential to the utilization of the computer program" quite broadly to encompass all of these functions, rejecting Krause's narrow focus on the word "essential" to apply only to a modification without which the program could not function. See also Aymes v. Bonelli, 47 F.3d 23 (2d Cir. 1995) (finding that modifications to ensure their compatibility with successive generations of computer systems to be permissible under § 117(a)(1)).

The *Krause* court also concluded that Titleserv had not used the program in any "other manner," notwithstanding evidence showing that it had shared copies of the programs with its subsidiaries and granted two client banks dial-up access to the programs for a period of approximately one year. The court interpreted the statutory phrase "to depend on the type of use envisioned in the creation of the program." Since the "programs Krause designed for Titleserv were intended to run on Titleserv's file server, where they would be accessible to Titleserv's subsidiaries," the court did not consider these activities to be beyond the uses immunized under § 117(a)(1)).

One district court has taken a narrower view of 117 than the Second Circuit. See Madison River v. Business Mgmt. Software, 387 D.Supp.2d 521,537 (M.D.N.C. 2005) (concluding that defendant's copying of plaintiff's database was not an essential step in using the program, even though it made the use more effective). Another district court declined a motion to dismiss a copyright infringement action under circumstances similar to *Krause*, although the ruling leaves open the applicability of § 117 at trial, at which time the issue of ownership of a copy of the program in question can be fully litigated. See Logicom Inclusive v. WP Stewart & Co., 72 U.S.P.Q. 2D (BNA) 1632 (S.D.N.Y. 2004).

iii.   § 117(a)(2)

The right to make archival copies has not been controversial. It allows computer users to make "backup" copies of their programs, as well as to load the programs onto hard drives from a floppy disk. See Vault Corp. v. Quaid Software Ltd., 847 F.2d 255 (5th Cir. 1988); ProCD v. Zeidenberg, 908 F. Supp. 640 (W.D. Wisc. 1996), *rev'd on other grounds*, 86 F.3d 1447 (7th Cir. 1996). The language of the statute also contemplates that multiple archival copies may be made of a single computer program. It clearly does *not* allow users of a program to make copies and distribute them to others for commercial purposes. See, e.g., Allen Myland Inc. v. IBM, 746 F. Supp. 520 (E.D. Pa. 1990); Apple Computer v. Formula Int'l, 594 F. Supp. 617 (C.D. Cal. 1984).

iv.   Repair and Maintenance

The Digital Millennium Copyright Act, passed by Congress in 1998, reversed the holding of MAI v. Peak, but in a very narrow way. Rather than declare that section 117 applies to "rightful possessors" of a computer program, as CONTU recommended, the DMCA altered section 117 to create a particular exemption from liability for the owner or lessee of a computer "to make or authorize the making of a copy of a computer program if such copy is made solely by virtue of the activation of a machine that lawfully contains an authorized copy of the computer program, for purposes only of maintenance or repair." 17 U.S.C. § 117(c). However, this exception to copying liability is limited solely to "machine maintenance or repair." The copy made by activating the computer for such purpose must be "destroyed immediately after the maintenance or repair is completed," and aspects of the program that are not necessary for the machine to be activated may not be accessed or used ("other than to make such new copy by virtue of the activation of the machine"). The amendment defines "maintenance as "the servicing of the machine in order to make it work in accordance with its original specifications and any changes to those specifications authorized for that machine" and "repair" as "the restoring of the machine to the state of working in accordance with its original specifications and any changes to those specifications authorized for that machine." 17 U.S.C. § 117(d)(1), (2).

Does this solve the problem created by the *MAI* case? Or does it give legislative sanction to the idea that RAM copies are not permitted except in the narrow circumstances covered by the new act? One might read this new exemption as creating a negative implication that all other copies made by lessees or licensees were illegal even if they fell within sections 117(1) or (2). Thus this change may benefit independent service organizations that service computer hardware, but it does nothing for computer users in general.

The Federal Circuit has interpreted section 117(c)'s safe harbor flexibly to effectuate the purpose of permitting maintenance. See Storage Technology Corp. v. Custom Hardware Engineering, 421 F.3d 1307 (Fed. Cir. 2005). Storage Technology ("StorageTek") manufactures an automated cartridge library for massive amounts of computer data that makes use of software programs that are responsible for controlling and maintaining the library. StorageTek restricts access to the software through a password protection scheme called GetKey. Custom Hardware Engineering & Consulting, Inc., ("CHE") is an independent business that repairs data libraries manufactured by StorageTek. In order to diagnose problems with the libraries, CHE intercepts and interprets error messages produced by the maintenance code. To accomplish this task, CHE reverse engineered GetKey. StorageTek sued CHE for copyright infringement (relating to rebooting and reconfiguring its customers' computers) and violating the anticircumvention provisions of the DMCA, 17 U.S.C. § 1201(a), for circumventing the GetKey protection system.

The Federal Circuit rejected these claims on the grounds that CHE was authorized under section 117(c) to engage in these activities. Based on Congress's intent of providing a safe harbor for computer maintenance and repair activities, the court interpreted "maintenance" to include continuous monitoring of the StorageTek system for problems. Because CHE's copying fell within this safe harbor, the court determined that there could be no separate DMCA violation, relying upon its decision in Chamberlain Group, Inc. v. Skylink Technologies, Inc., 381 F.3d 1178, 1202-03 (Fed. Cir. 2004) (holding that Section 1201 "prohibits only forms of access that bear a reasonable relationship to the protections that the Copyright Act otherwise affords copyright owners" and therefore a copyright owner must prove that the circumvention of the technological measure either "infringes or facilitates infringing a right protected by the Copyright Act."). But cf. 321 Studios v. MGM, Inc., 307 F.Supp.2d 1085 (N.D.Cal. 2004) (holding that 321 Studios had violated the DMCA by selling software that allowed owners of a DVD to make a single backup copy, notwithstanding section 117).

Does the court stretch Section 117(c) beyond the express limits that Congress imposed? Dissenting in *StorageTek*, Judge Rader pointed to language in § 117(c)(2) specifically precluding a repair company from using copies of programs loaded into RAM upon powering-up that are "not necessary for that machine to be activated."

> Although at the time CHE reboots it repairs nothing, adjusts nothing, and checks nothing, CHE subsequently accesses and uses the maintenance code to send data packets that indicate the operation of the system. Therefore, CHE does not fall within the safe harbor of § 117(c)(2).

421 F.3d at 1322; see also PracticeWorks Inc. v. Professional Software Solutions of Illinois, 72 U.S.P.Q. 2D (BNA) 1691 (D. Md. 2004) (observing that "the loading of

the software is permissible under 117(c) only if it is for the purposes of repairing the machine into which it is loaded, not the software itself" and "applies to the repair and maintenance of the machine into which the program is loaded, not the repair and maintenance of a third party's machine" (emphasis in original)). How would you respond?

If Judge Rader is correct, should Congress revise § 117(c) to extend to software maintenance and repair? Is there a comparable social interest in maintaining a competitive market for both hardware and software maintenance? One way of thinking about the issue is to determine whether the copyright in a computer program should permit its owner to control the markets for hardware and software maintenance, or whether using the copyright to prevent competition in that market would expand copyright protection beyond its intended boundaries. See Pamela Samuelson, Modifying Copyrighted Software: Adjusting Copyright Doctrine to Accommodate a Technology, 28 Jurimetrics J. 179 (1988).

3. Fair Use and Derivative Works

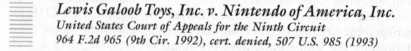

Lewis Galoob Toys, Inc. v. Nintendo of America, Inc.
United States Court of Appeals for the Ninth Circuit
964 F.2d 965 (9th Cir. 1992), cert. denied, 507 U.S. 985 (1993)

FARRIS, Circuit Judge:

Nintendo of America appeals the district court's judgment following a bench trial (1) declaring that Lewis Galoob Toys' Game Genie does not violate any Nintendo copyrights and dissolving a temporary injunction and (2) denying Nintendo's request for a permanent injunction enjoining Galoob from marketing the Game Genie. Lewis Galoob Toys, Inc. v. Nintendo of America, Inc., 780 F. Supp. 1283 (N.D. Cal. 1991). We have appellate jurisdiction pursuant to 15 U.S.C. § 1121 and 28 U.S.C. §§ 1291 and 1292(a)(1). We affirm.

Facts

The Nintendo Entertainment System is a home video game system marketed by Nintendo. To use the system, the player inserts a cartridge containing a video game that Nintendo produces or licenses others to produce. By pressing buttons and manipulating a control pad, the player controls one of the game's characters and progresses through the game. The games are protected as audiovisual works under 17 U.S.C. § 102(a)(6).

The Game Genie is a device manufactured by Galoob that allows the player to alter up to three features of a Nintendo game. For example, the Game Genie can increase the number of lives of the player's character, increase the speed at which the character moves, and allow the character to float above obstacles. The player controls the changes made by the Game Genie by entering codes provided by the Game Genie Programming Manual and Code Book. The player also can experiment with variations of these codes.

The Game Genie functions by blocking the value for a single data byte sent by the game cartridge to the central processing unit in the Nintendo Entertainment System and replacing it with a new value. If that value controls the character's strength, for example, then the character can be made invincible by increasing the value sufficiently. The Game Genie is inserted between a game cartridge and the Nintendo Entertainment System. The Game Genie does not alter the data that is stored in the game cartridge. Its effects are temporary.

Discussion . . .

2. *Fair Use*

"The doctrine of fair use allows a holder of the privilege to use copyrighted material in a reasonable manner without the consent of the copyright owner." Narell v. Freeman, 872 F.2d 907, 913 (9th Cir. 1989) (citations omitted). The district court concluded that, even if the audiovisual displays created by the Game Genie are derivative works, Galoob is not liable under 17 U.S.C. § 107 because the displays are a fair use of Nintendo's copyrighted displays. "Whether a use of copyrighted material is a 'fair use' is a mixed question of law and fact. If the district court found sufficient facts to evaluate each of the statutory factors, the appellate court may decide whether defendants may claim the fair use defense as a matter of law." Abend v. MCA, Inc., 863 F.2d 1465, 1468 (9th Cir. 1988), *aff'd sub nom.* Stewart v. Abend, 495 U.S. 207 (1990).

Section 107 codifies the fair use defense. . . .

Much of the parties' dispute regarding the fair use defense concerns the proper focus of the court's inquiry: (1) Galoob or (2) consumers who purchase and use the Game Genie. Nintendo's complaint does not allege direct infringement, nor did it try the case on that theory. The complaint, for example, alleges only that "Galoob's marketing advertising [sic], promoting and selling of Game Genie has and will contribute to the creation of infringing derivatives of Nintendo's copyrighted . . . . games." Contributory infringement is a form of third party liability. See Melville B. Nimmer & David Nimmer, 3 Nimmer on Copyright § 12.04[A]2, at 12-68 (1991). The district court properly focused on whether consumers who purchase and use the Game Genie would be infringing Nintendo's copyrights by creating (what are now assumed to be) derivative works.

Nintendo emphasizes that the district court ultimately addressed its direct infringement by authorization argument. The court concluded that, "because the Game Genie does not create a derivative work when used in conjunction with a copyrighted video game, *Galoob* does not 'authorize the use of a copyrighted work without the actual authority from the copyright owner.'" *Galoob*, 780 F. Supp. at 1298 (quoting *Sony*, 464 U.S. at 435 n.17). Although infringement by authorization is a form of direct infringement, this does not change the proper focus of our inquiry; a party cannot authorize another party to infringe a copyright unless the authorized conduct would itself be unlawful.

The district court concluded that "a family's use of a Game Genie for private home enjoyment must be characterized as a non-commercial, nonprofit activity." *Galoob*, 780 F. Supp. at 1293. Nintendo argues that Game Genie users are supplanting its commercially valuable right to make and sell derivative works. Nintendo's reliance on Harper & Row Publishers, Inc. v. Nation Enters., 471 U.S. 539, 562 (1985), is misplaced. The commercially valuable right at issue in Harper & Row was the right of first publication; Nation Enterprises intended to publish the copyrighted materials for profit. See id. at 562-63. See also *Sony,* 464 U.S. at 449 ("If the Betamax were used to make copies for a commercial or profit-making purpose, such use would presumptively be unfair."). Game Genie users are engaged in a non-profit activity. Their use of the Game Genie to create derivative works therefore is presumptively fair. See *Sony,* 464 U.S. at 449.

The district court also concluded that "the [Nintendo] works' published nature supports the fairness of the use." *Galoob,* 780 F. Supp. at 1293. Nintendo argues that it has not published the derivative works created by the Game Genie. This argument ignores the plain language of section 107: "the factors to be considered shall include...the nature of the copyrighted work." The argument also would make the fair use defense unavailable in all cases of derivative works, including "criticism, comment, news reporting, teaching..., scholarship, or research." 17 U.S.C. § 107. A commentary that incorporated large portions of *For Whom the Bell Tolls,* for example, would be undeserving of fair use protection because the incorporated portions would constitute an unpublished derivative work. This cannot be the law.

The district court further concluded that the amount of the portion used in relation to the copyrighted work as a whole "cannot assist Nintendo in overcoming the presumption of fair use." *Galoob,* 780 F. Supp. at 1293. The video tape recorders at issue in *Sony* allowed consumers to tape copyrighted works in their entirety. The Supreme Court nevertheless held that, "when one considers...that [video tape recording] merely enables a viewer to see such a work which he had been invited to witness in its entirety free of charge, the fact that the entire work is reproduced does not have its ordinary effect of militating against a finding of fair use." 464 U.S. 449 at 449-50 (citations omitted). Consumers are not invited to witness Nintendo's audiovisual displays free of charge, but, once they have paid to do so, the fact that the derivative works created by the Game Genie are comprised almost entirely of Nintendo's copyrighted displays does not militate against a finding of fair use.

Nintendo would distinguish *Sony* because it involved copying copyrighted works rather than creating derivative works based on those works. In other words, the consumers in *Sony* could lawfully copy the copyrighted works because they were invited to view those works free of charge. Game Genie users, in contrast, are not invited to view derivative works based on Nintendo's copyrighted works without first paying for that privilege. *Sony* cannot be read so narrowly. It is difficult to imagine that the Court would have reached a different conclusion if Betamax purchasers were skipping portions of copyrighted works or viewing denouements before climaxes. *Sony* recognizes that a party who distributes a copyrighted work cannot dictate how that work is to be enjoyed. Consumers may use a Betamax to view copyrighted works at a more convenient time. They similarly may use a Game Genie to enhance a Nintendo Game cartridge's audiovisual display in such a way as to make the experience more enjoyable.

"The fourth factor is the 'most important, and indeed, central fair use factor.'" Stewart, 495 U.S. at 238 (quoting 3 Nimmer on Copyright § 13.05[A], at 13-81). The district court concluded that "Nintendo has failed to show any harm to the present market for its copyrighted games and has failed to establish the reasonable likelihood of a potential market for slightly altered versions of the games at suit."

Galoob, 780 F. Supp. at 1295. Nintendo's main argument on appeal is that the test for market harm encompasses the potential market for derivative works. Because the Game Genie is used for a noncommercial purpose, the likelihood of future harm may not be presumed. See *Sony*, 464 U.S. at 451. Nintendo must show "by a preponderance of the evidence that some meaningful likelihood of future harm exists." Id.

Nintendo's argument is supported by case law. Although the Copyright Act requires a court to consider "the effect of the use upon the potential market for or value of the copyrighted work," 17 U.S.C. § 107(4) (emphasis added), we held in Abend that "although the motion picture will have no adverse effect on bookstore sales of the [underlying] novel — and may in fact have a beneficial effect — it is 'clear that [the film's producer] may not invoke the defense of fair use.'" 863 F.2d at 1482 (quoting 3 Nimmer on Copyright § 13.05[B], at 13-84). We explained: "'If the defendant's work adversely affects the value of any of the rights in the copyrighted work . . . the use is not fair even if the rights thus affected have not as yet been exercised by the plaintiff.'" Id. (quoting 3 Nimmer on Copyright § 13.05[B], at 13-84 to 13-85 (footnotes omitted)). The Supreme Court specifically affirmed our finding that the motion picture adaptation "impinged on the ability to market new versions of the story." *Stewart*, 495 U.S. at 238.

Still, Nintendo's argument is undermined by the facts. The district court considered the potential market for derivative works based on Nintendo game cartridges and found that: (1) "Nintendo has not, to date, issued or considered issuing altered versions of existing games," *Galoob*, 780 F. Supp. at 1295, and (2) Nintendo "has failed to show the reasonable likelihood of such a market." Id. The record supports the court's findings. According to Stephen Beck, Galoob's expert witness, junior or expert versions of existing Nintendo games would enjoy very little market interest because the original version of each game already has been designed to appeal to the largest number of consumers. Mr. Beck also testified that a new game must include new material or "the game player is going to feel very cheated and robbed, and [the] product will have a bad reputation and word of mouth will probably kill its sales." Howard Lincoln, Senior Vice President of Nintendo of America, acknowledged that Nintendo has no present plans to market such games.

The district court also noted that Nintendo's assertion that it may wish to re-release altered versions of its game cartridges is contradicted by its position in various other lawsuits:

> In those actions, Nintendo opposes antitrust claims by using the vagaries of the video game industry to rebut the impact and permanence of its market control, if any. Having indoctrinated this Court as to the fast pace and instability of the video game industry, Nintendo may not now, without any data, redefine that market in its request for the extraordinary remedy sought herein. . . . While board games may never die, good video games are mortal.

Galoob, 780 F. Supp. at 1295. The existence of this potential market cannot be presumed. See *Sony*, 464 U.S. at 451. See also Wright v. Warner Books, Inc., 953 F.2d 731, 739 (2d Cir. 1991) (affirming district court's finding of no reasonable likelihood of injury to alleged market because "plaintiff has offered no evidence that the project will go forward"). The fourth and most important fair use factor also favors Galoob.

Nintendo's most persuasive argument is that the creative nature of its audiovisual displays weighs against a finding of fair use. The Supreme Court has acknowledged that "fair use is more likely to be found in factual works than fictional works." *Stewart*, 495 U.S. at 237. This consideration weighs against a finding of fair use, but it is not dispositive. See *Sony*, 464 U.S. at 448 (fair use defense is an "equitable rule of reason"). The district court could properly conclude that Game Genie users are making a fair use of Nintendo's displays....

Affirmed.

COMMENTS AND QUESTIONS

1. Suppose that Nintendo had announced its intention to sell add-on devices to its entertainment system that would perform the same function as the Game Genie. Should this fact change the fair use analysis? Certainly, Nintendo could prove that it was losing money to Galoob in this example. Does this fact sway the fourth factor in Nintendo's favor? What if Nintendo could show the existence of a "market" for *licensing* devices like the Game Genie? Is Galoob depriving Nintendo of a royalty payment? On this issue, cf. American Geophysical Union v. Texaco, 37 F.3d 881 (2d Cir. 1994) (existence of "market" for royalty payments of copies undercuts fair use argument).

2. One of the reasons people play video games is for the challenge of overcoming obstacles. Does a device that gives a game character "super powers" or an infinite number of lives defeat this purpose? It is certainly possible that owners of the Game Genie will grow bored with video games more quickly.[12] If Nintendo can prove that the Game Genie will cut into its sales of games by making the games less challenging, should that affect its copyright argument?

Alternatively, does Nintendo have some "moral right" in the integrity of its audiovisual work that is being violated by Galoob's device?

3. A subsequent Ninth Circuit decision seems to find that any computer file that alters the operation of a computer game implicates the derivative work right. In Micro Star v. FormGen Inc., 154 F.3d 1107 (9th Cir. 1998), the defendant (Micro Star) distributed files that displayed new game levels of the plaintiff's "Duke Nukem" videogame when run in conjunction with the plaintiff's game. The files described the game levels in detail, but did not actually contain any of the plaintiff's code or graphics. They merely instructed the computer how to use images from the plaintiff's game. The Ninth Circuit held that the audiovisual displays created when the "Duke Nukem" software was run in conjunction with the defendant's files were recorded in permanent form in the defendant's files and therefore qualified as derivative works. The court's analysis, however, overlooks the fact that the defendant's files could not independently generate the audiovisual displays; they could generate the displays only when run in conjunction with the plaintiff's software. See Aaron Rubin, Are You Experienced? The Copyright Implications of Web Site Modification Technology, 89 Cal. L. Rev. 817, 834-37 (2001) (criticizing the *Micro Star* decision). Exactly what protected expression did the defendant copy? Isn't this really a case about contributory liability? If so, does Micro Star meet the requirements? Are users of its

---

12. Of course, this might induce them to buy more Nintendo games, rather than fewer. If that is the case, it is hard to see how Nintendo has been injured.

software (who presumably have legally acquired Formgen's software) direct infringers? Is the court's decision consistent with *Galoob*?

Note on Reverse Engineering

In view of the network externalities flowing from many software products and the cumulative nature of innovation in software, two critical issues in the computer industry are the manner by and extent to which competitors can develop compatible or interoperable programs, competitive programs, and enhancements of programs. Of particular relevance is the permissibility of reverse engineering. Reverse engineering is loosely defined as "starting with the known product and working backward to divine the process which aided in its development or manufacture." Kewanee Oil v. Bicron, 416 U.S. 470 (1974). It typically involves two phases: (1) disassembly or decompilation of the program in order to create human-readable source code that may be analyzed, and (2) using the results of this analysis to create a commercially viable program.

Most computer programs are distributed only in object code form, a fact that makes it difficult to discover the ideas and principles contained in a program without reverse engineering. In addition, some programs have special devices designed to prevent inspection. Nonetheless, reverse engineering of software, while difficult, is possible in some limited circumstances. There are a variety of approaches to reverse engineering of computer programs. One is referred to as "black box" testing. By systematically inputting instructions, a computer programmer can learn how the program processes information and can construct a program that produces the same outputs for given instructions. Because of the complexity of most programs today, however, this approach is often infeasible. A second method is to study the technical specifications and user manuals. These materials, however, often do not contain sufficient information to achieve full compatibility and interoperability. In order to feasibly achieve these objectives, it is often necessary to take a machine apart and directly observe its operation or, as is more common with regard to software, to decompile the program. After a program is understood, the competitor will use the results of the analysis to create a commercially viable program. The case and materials that follow consider the validity of such decompilation.

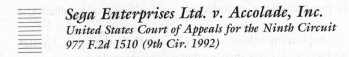

Sega Enterprises Ltd. v. Accolade, Inc.
United States Court of Appeals for the Ninth Circuit
977 F.2d 1510 (9th Cir. 1992)

REINHART, Circuit Judge: . . .

I. Background

Plaintiff-appellee Sega Enterprises, Ltd. ("Sega"), a Japanese corporation, and its subsidiary, Sega of America, develop and market video entertainment systems, including the "Genesis" console (distributed in Asia under the name "Mega-Drive") and video game cartridges. Defendant-appellant Accolade, Inc., is an independent developer, manufacturer, and marketer of computer entertainment software, including game cartridges that are compatible with the Genesis console, as well as game cartridges that are compatible with other computer systems.

Sega licenses its copyrighted computer code and its "SEGA" trademark to a number of independent developers of computer game software. Those licensees develop and sell Genesis-compatible video games in competition with Sega. Accolade is not and never has been a licensee of Sega. Prior to rendering its own games compatible with the Genesis console, Accolade explored the possibility of entering into a licensing agreement with Sega, but abandoned the effort because the agreement would have required that Sega be the exclusive manufacturer of all games produced by Accolade.

Accolade used a two-step process to render its video games compatible with the Genesis console. First, it "reverse engineered" Sega's video game programs in order to discover the requirements for compatibility with the Genesis console. As part of the reverse engineering process, Accolade transformed the machine-readable object code contained in commercially available copies of Sega's game cartridges into human-readable source code using a process called "disassembly" or "decompilation".[1] Accolade purchased a Genesis console and three Sega game cartridges, wired a decompiler into the console circuitry, and generated printouts of the resulting source code. Accolade engineers studied and annotated the printouts in order to identify areas of commonality among the three game programs. They then loaded the disassembled code back into a computer, and experimented to discover the interface specifications for the Genesis console by modifying the programs and studying the results. At the end of the reverse engineering process, Accolade created a development manual that incorporated the information it had discovered about the requirements for a Genesis-compatible game. According to the Accolade employees who created the manual, the manual contained only functional descriptions of the interface requirements and did not include any of Sega's code.

In the second stage, Accolade created its own games for the Genesis. According to Accolade, at this stage it did not copy Sega's programs, but relied only on the information concerning interface specifications for the Genesis that was contained in its development manual. Accolade maintains that with the exception of the interface specifications, none of the code in its own games is derived in any way from its examination of Sega's code. In 1990, Accolade released "Ishido", a game which it had originally developed and released for use with the Macintosh and IBM personal computer systems, for use with the Genesis console.

[With its Genesis III product, Sega included a new lockout code, called the TMSS system.]

Accolade learned of the impending release of the Genesis III in the United States in January, 1991, when the Genesis III was displayed at a consumer electronics show. When a demonstration at the consumer electronics show revealed that Accolade's "Ishido" game cartridges would not operate on the Genesis III, Accolade returned

---

2. Computer programs are written in specialized alphanumeric languages, or "source code". In order to operate a computer, source code must be translated into computer readable form, or "object code". Object code uses only two symbols, 0 and 1, in combinations which represent the alphanumeric characters of the source code. A program written in source code is translated into object code using a computer program called an "assembler" or "compiler", and then imprinted onto a silicon chip for commercial distribution. Devices called "disassemblers" or "decompilers" can reverse this process by "reading" the electronic signals for "0" and "1" that are produced while the program is being run, storing the resulting object code in computer memory, and translating the object code into source code. Both assembly and disassembly devices are commercially available, and both types of devices are widely used within the software industry.

to the drawing board. During the reverse engineering process, Accolade engineers had discovered a small segment of code—the TMSS initialization code—that was included in the "power-up" sequence of every Sega game, but that had no identifiable function. The games would operate on the original Genesis console even if the code segment was removed. Mike Lorenzen, the Accolade engineer with primary responsibility for reverse engineering the interface procedures for the Genesis console, sent a memo regarding the code segment to Alan Miller, his supervisor and the current president of Accolade, in which he noted that "it is possible that some future Sega peripheral device might require it for proper initialization."

In the second round of reverse engineering, Accolade engineers focused on the code segment identified by Lorenzen. After further study, Accolade added the code to its development manual in the form of a standard header file to be used in all games. The file contains approximately twenty to twenty-five bytes of data. Each of Accolade's games contains a total of 500,000 to 1,500,000 bytes. According to Accolade employees, the header file is the only portion of Sega's code that Accolade copied into its own game programs. In this appeal, Sega does not raise a separate claim of copyright infringement with respect to the header file.

In 1991, Accolade released five more games for use with the Genesis III, "Star Control", "Hardball!", "Onslaught", "Turrican", and "Mike Ditka Power Football." With the exception of "Mike Ditka Power Football", all of those games, like "Ishido", had originally been developed and marketed for use with other hardware systems. All contained the standard header file that included the TMSS initialization code....

[Sega sued Accolade for copyright infringement.]

III. Copyright Issues

Accolade raises four arguments in support of its position that disassembly of the object code in a copyrighted computer program does not constitute copyright infringement. First, it maintains that intermediate copying does not infringe the exclusive rights granted to copyright owners in section 106 of the Copyright Act unless the end product of the copying is substantially similar to the copyrighted work. Second, it argues that disassembly of object code in order to gain an understanding of the ideas and functional concepts embodied in the code is lawful under section 102(b) of the Act, which exempts ideas and functional concepts from copyright protection. Third, it suggests that disassembly is authorized by section 117 of the Act, which entitles the lawful owner of a copy of a computer program to load the program into a computer. Finally, Accolade contends that disassembly of object code in order to gain an understanding of the ideas and functional concepts embodied in the code is a fair use that is privileged by section 107 of the Act.

Neither the language of the Act nor the law of this circuit supports Accolade's first three arguments. Accolade's fourth argument, however, has merit....

A. *Intermediate Copying*

We have previously held that the Copyright Act does not distinguish between unauthorized copies of a copyrighted work on the basis of what stage of the alleged infringer's work the unauthorized copies represent. Walker v. University Books, 602

F.2d 859, 864 (9th Cir. 1979) ("[T]he fact that an allegedly infringing copy of a protected work may itself be only an inchoate representation of some final product to be marketed commercially does not in itself negate the possibility of infringement."). Our holding in *Walker* was based on the plain language of the Act. Section 106 grants to the copyright owner the exclusive rights "to reproduce the work in copies", "to prepare derivative works based upon the copyrighted work", and to authorize the preparation of copies and derivative works. 17 U.S.C. § 106 (1)-(2). Section 501 provides that "anyone who violates any of the exclusive rights of the copyright owner as provided by sections 106 through 118...is an infringer of the copyright." Id. § 501(a). On its face, that language unambiguously encompasses and proscribes "intermediate copying". *Walker,* 602 F.2d at 863-64.

In order to constitute a "copy" for purposes of the Act, the allegedly infringing work must be fixed in some tangible form, "from which the work can be perceived, reproduced, or otherwise communicated, either directly or with the aid of a machine or device." 17 U.S.C. § 101. The computer file generated by the disassembly program, the printouts of the disassembled code, and the computer files containing Accolade's modifications of the code that were generated during the reverse engineering process all satisfy that requirement. The intermediate copying done by Accolade therefore falls squarely within the category of acts that are prohibited by the statute.

Accolade points to a number of cases that it argues establish the lawfulness of intermediate copying. Most of the cases involved the alleged copying of books, scripts, or literary characters. See v. Durang, 711 F.2d 141 (9th Cir. 1983); Warner Bros. v. ABC, 654 F.2d 204 (2d Cir. 1981); Miller v. Universal City Studios, Inc., 650 F.2d 1365 (5th Cir. 1981). In each case, however, the eventual lawsuit alleged infringement only as to the final work of the defendants. We conclude that this group of cases does not alter or limit the holding of *Walker.*

The remaining cases cited by Accolade, like the case before us, involved intermediate copying of computer code as an initial step in the development of a competing product. Computer Assoc. Int'l v. Altai, Inc., [supra] (2d Cir. 1992) ("CAI"); NEC Corp. v. Intel Corp., 10 U.S.P.Q.2d (BNA) 1177 (N.D. Cal. 1989); E. F. Johnson Co. v. Uniden Corp., 623 F. Supp. 1485 (D. Minn. 1985). In each case, the court based its determination regarding infringement solely on the degree of similarity between the allegedly infringed work and the defendant's final product. A close reading of those cases, however, reveals that in none of them was the legality of the intermediate copying at issue. Sega cites an equal number of cases involving intermediate copying of copyrighted computer code to support its assertion that such copying is prohibited. Atari Games Corp. v. Nintendo of America, Inc., 18 U.S.P.Q.2d (BNA) 1935 (N.D. Cal. 1991); SAS Institute, Inc. v. S&H Computer Systems, Inc., 605 F. Supp. 816 (M.D. Tenn. 1985); S&H Computer Systems, Inc. v. SAS Institute, Inc., 568 F. Supp. 416 (M.D. Tenn. 1983); Hubco Data Products v. Management Assistance, Inc., 219 U.S.P.Q. (BNA) 450 (D. Idaho 1983). Again, however, it appears that the question of the lawfulness of intermediate copying was not raised in any of those cases.

In summary, the question whether intermediate copying of computer object code infringes the exclusive rights granted to the copyright owner in section 106 of the Copyright Act is a question of first impression. In light of the unambiguous language of the Act, we decline to depart from the rule set forth in *Walker* for copyrighted works generally. Accordingly, we hold that intermediate copying of computer object code may infringe the exclusive rights granted to the copyright owner in section

106 of the Copyright Act regardless of whether the end product of the copying also infringes those rights. If intermediate copying is permissible under the Act, authority for such copying must be found in one of the statutory provisions to which the rights granted in section 106 are subject.

B. The Idea/Expression Distinction

Accolade next contends that disassembly of computer object code does not violate the Copyright Act because it is necessary in order to gain access to the ideas and functional concepts embodied in the code, which are not protected by copyright. 17 U.S.C. § 102(b). Because humans cannot comprehend object code, it reasons, disassembly of a commercially available computer program into human-readable form should not be considered an infringement of the owner's copyright. Insofar as Accolade suggests that disassembly of object code is lawful *per se,* it seeks to overturn settled law.

Accolade's argument regarding access to ideas is, in essence, an argument that object code is not eligible for the full range of copyright protection. Although some scholarly authority supports that view, we have previously rejected it based on the language and legislative history of the Copyright Act. Johnson Controls, Inc. v. Phoenix Control Sys., Inc., 886 F.2d 1173, 1175 (9th Cir. 1989)....

Nor does a refusal to recognize a *per se* right to disassemble object code lead to an absurd result. The ideas and functional concepts underlying many types of computer programs, including word processing programs, spreadsheets, and video game displays, are readily discernible without the need for disassembly, because the operation of such programs is visible on the computer screen. The need to disassemble object code arises, if at all, only in connection with operations systems, system interface procedures, and other programs that are not visible to the user when operating — and then only when no alternative means of gaining an understanding of those ideas and functional concepts exists. In our view, consideration of the unique nature of computer object code thus is more appropriate as part of the case-by-case, equitable "fair use" analysis authorized by section 107 of the Act. See infra Part III(D). Accordingly, we reject Accolade's second argument.

C. Section 117

Section 117 of the Copyright Act allows the lawful owner of a copy of a computer program to copy or adapt the program if the new copy or adaptation "is created as an essential step in the utilization of the computer program in conjunction with a machine and . . . is used in no other manner." 17 U.S.C. § 117(1). Accolade contends that section 117 authorizes disassembly of the object code in a copyrighted computer program.

Section 117 was enacted on the recommendation of CONTU, which noted that "because the placement of any copyrighted work into a computer is the preparation of a copy [since the program is loaded into the computer's memory], the law should provide that persons in rightful possession of copies of programs be able to use them freely without fear of exposure to copyright liability." CONTU Report at 13. We think it is clear that Accolade's use went far beyond that contemplated by CONTU and authorized by section 117. Section 117 does not purport to protect a user who disassembles object code, converts it from assembly into source code, and makes printouts and photocopies of the refined source code version.

D. Fair Use

Accolade contends, finally, that its disassembly of copyrighted object code as a necessary step in its examination of the unprotected ideas and functional concepts embodied in the code is a fair use that is privileged by section 107 of the Act. Because, in the case before us, disassembly is the only means of gaining access to those unprotected aspects of the program, and because Accolade has a legitimate interest in gaining such access (in order to determine how to make its cartridges compatible with the Genesis console), we agree with Accolade. Where there is good reason for studying or examining the unprotected aspects of a copyrighted computer program, disassembly for purposes of such study or examination constitutes a fair use.

1

As a preliminary matter, we reject Sega's contention that the assertion of a fair use defense in connection with the disassembly of object code is precluded by statute. First, Sega argues that not only does section 117 of the Act *not* authorize disassembly of object code, but it also constitutes a legislative determination that any copying of a computer program *other* than that authorized by section 117 cannot be considered a fair use of that program under section 107. That argument verges on the frivolous. Each of the exclusive rights created by section 106 of the Copyright Act is expressly made subject to all of the limitations contained in sections 107 through 120. 17 U.S.C. § 106. Nothing in the language or the legislative history of section 117, or in the CONTU Report, suggests that section 117 was intended to preclude the assertion of a fair use defense with respect to uses of computer programs that are not covered by section 117, nor has section 107 been amended to exclude computer programs from its ambit.

Moreover, sections 107 and 117 serve entirely different functions. Section 117 defines a narrow category of copying that is lawful *per se*. 17 U.S.C. § 117. Section 107, by contrast, establishes a *defense* to an otherwise valid claim of copyright infringement. . . . Second, Sega maintains that the language and legislative history of section 906 of the Semiconductor Chip Protection Act of 1984 (SCPA) establish that Congress did not intend that disassembly of object code be considered a fair use. Section 906 of the SCPA authorizes the copying of the "mask work" on a silicon chip in the course of reverse engineering the chip. 17 U.S.C. § 906. The mask work in a standard ROM chip, such as those used in the Genesis console and in Genesis-compatible cartridges, is a physical representation of the computer program that is embedded in the chip. The zeros and ones of binary object code are represented in the circuitry of the mask work by open and closed switches. Sega contends that Congress's express authorization of copying in the particular circumstances set forth in section 906 constitutes a determination that other forms of copying of computer programs are prohibited.

The legislative history of the SCPA reveals, however, that Congress passed a separate statute to protect semiconductor chip products because it believed that semiconductor chips were intrinsically utilitarian articles that were not protected under the Copyright Act. H.R. Rep. No. 781, 98th Cong., 2d Sess. 8-10, reprinted in 1984 U.S.C.C.A.N. 5750, 5757-59. Accordingly, rather than amend the Copyright Act to extend traditional copyright protection to chips, it enacted "a sui generis form of protection, apart from and independent of the copyright laws." Id. at 10, 1984 U.S.C.C.A.N. at 5759. Because Congress did not believe that semiconductor chips were eligible for copyright protection in the first instance, the fact that it included an exception for reverse engineering of mask work in the SCPA says nothing about its intent with respect to the lawfulness of disassembly of computer programs under the Copyright Act. Nor is the fact that Congress did not contemporaneously amend the Copyright Act to permit disassembly significant, since it was focusing on the protection to be afforded to semiconductor chips. Here we are dealing not with an alleged violation of the SCPA, but with the copying of a computer program, which is governed by the Copyright Act. Moreover, Congress expressly stated that it did not intend to "limit, enlarge or otherwise affect the scope, duration, ownership or subsistence of copyright protection...in computer programs, data bases, or any other copyrightable works embodied in semiconductor chip products." Id. at 28, 1984 U.S.C.C.A.N. at 5777. Accordingly, Sega's second statutory argument also fails. We proceed to consider Accolade's fair use defense.

2 . . .

In determining that Accolade's disassembly of Sega's object code did not constitute a fair use, the district court treated the first and fourth statutory factors [of 17 U.S.C. § 107, reprinted above] as dispositive, and ignored the second factor entirely. Given the nature and characteristics of Accolade's direct use of the copied works, the ultimate use to which Accolade put the functional information it obtained, and the nature of the market for home video entertainment systems, we conclude that neither the first nor the fourth factor weighs in Sega's favor. In fact, we conclude that both factors support Accolade's fair use defense, as does the second factor, a factor which is important to the resolution of cases such as the one before us.

(a)

With respect to the first statutory factor, we observe initially that the fact that copying is for a commercial purpose weighs against a finding of fair use. *Harper & Row*, 471 U.S. at 562. However, the presumption of unfairness that arises in such cases can be rebutted by the characteristics of a particular commercial use. Hustler Magazine, Inc. v. Moral Majority, Inc., 796 F.2d 1148, 1152 (9th Cir. 1986); see also Maxtone-Graham v. Burtchaell, 803 F.2d 1253, 1262 (2d Cir. 1986), *cert. denied*, 481 U.S. 1059 (1987). Further "the commercial nature of a use is a matter of degree, not an absolute...." *Maxtone-Graham*, 803 F.2d at 1262.

Sega argues that because Accolade copied its object code in order to produce a competing product, the *Harper & Row* presumption applies and precludes a finding of fair use. That analysis is far too simple and ignores a number of important considerations. We must consider other aspects of "the purpose and character of the use" as well. As we have noted, the use at issue was an intermediate one only and thus any commercial "exploitation" was indirect or derivative.

The declarations of Accolade's employees indicate, and the district court found, that Accolade copied Sega's software solely in order to discover the functional requirements for compatibility with the Genesis console — aspects of Sega's programs that are not protected by copyright. 17 U.S.C. § 102(b). With respect to the video game programs contained in Accolade's game cartridges, there is no evidence in the record that Accolade sought to avoid performing its own creative work. Indeed, most of the games that Accolade released for use with the Genesis console were originally developed for other hardware systems. Moreover, with respect to the interface procedures for the Genesis console, Accolade did not seek to avoid paying a customarily charged fee for use of those procedures, nor did it simply copy Sega's code; rather, it wrote its own procedures based on what it had learned through disassembly. Taken together, these facts indicate that although Accolade's ultimate purpose was the release of Genesis-compatible games for sale, its direct purpose in copying Sega's code, and thus its direct use of the copyrighted material, was simply to study the functional requirements for Genesis compatibility so that it could modify existing games and make them usable with the Genesis console. Moreover, as we discuss below, no other method of studying those requirements was available to Accolade. On these facts, we conclude that Accolade copied Sega's code for a legitimate, essentially non-exploitative purpose, and that the commercial aspect of its use can best be described as of minimal significance.

We further note that we are free to consider the public benefit resulting from a particular use notwithstanding the fact that the alleged infringer may gain commercially. See *Hustler,* 796 F.2d at 1153 (quoting MCA, Inc. v. Wilson, 677 F.2d 180, 182 (2d Cir. 1981)). Public benefit need not be direct or tangible, but may arise because the challenged use serves a public interest. Id. In the case before us, Accolade's identification of the functional requirements for Genesis compatibility has led to an increase in the number of independently designed video game programs offered for use with the Genesis console. It is precisely this growth in creative expression, based on the dissemination of other creative works and the unprotected ideas contained in those works, that the Copyright Act was intended to promote. See Feist Publications, Inc. v. Rural Tel. Serv. Co., 111 S. Ct. 1282, 1290 (1991) (citing *Harper & Row,* 471 U.S. at 556-57). The fact that Genesis-compatible video games are not scholarly works, but works offered for sale on the market, does not alter our judgment in this regard. We conclude that given the purpose and character of Accolade's use of Sega's video game programs, the presumption of unfairness has been overcome and the first statutory factor weighs in favor of Accolade.

(b)

As applied, the fourth statutory factor, effect on the potential market for the copyrighted work, bears a close relationship to the "purpose and character" inquiry in that it, too, accommodates the distinction between the copying of works in order to make independent creative expression possible and the simple exploitation of another's creative efforts. We must, of course, inquire whether, "if [the challenged use] should become widespread, it would adversely affect the potential market for the copyrighted work," Sony Corp. v. Universal City Studios, 464 U.S. 417, 451 (1984), by diminishing potential sales, interfering with marketability, or usurping the market, *Hustler,* 796 F.2d at 1155-56. If the copying resulted in the latter effect, all other considerations might be irrelevant. The *Harper & Row* Court found a use that effectively usurped the market for the copyrighted work by supplanting that work to be dispositive. 471 U.S. at 567-69. However, the same consequences do not and could not attach to a use which simply enables the copier to enter the market for works of the same type as the copied work. Unlike the defendant in *Harper & Row,* which printed excerpts from President Ford's memoirs verbatim with the stated purpose of "scooping" a Time magazine review of the book, 471 U.S. at 562, Accolade did not attempt to "scoop" Sega's release of any particular game or games, but sought only to become a legitimate competitor in the field of Genesis-compatible video games. Within that market, it is the characteristics of the game program as experienced by the user that determine the program's commercial success. As we have noted, there is nothing in the record that suggests that Accolade copied any of those elements.

By facilitating the entry of a new competitor, the first lawful one that is not a Sega licensee, Accolade's disassembly of Sega's software undoubtedly "affected" the market for Genesis-compatible games in an indirect fashion. We note, however, that while no consumer except the most avid devotee of President Ford's regime might be expected to buy more than one version of the President's memoirs, video game users typically purchase more than one game. There is no basis for assuming that Accolade's "Ishido" has significantly affected the market for Sega's "Altered Beast", since a consumer might easily purchase both; nor does it seem unlikely that a consumer particularly interested in sports might purchase both Accolade's "Mike Ditka Power Football" and Sega's "Joe Montana Football", particularly if the games are, as Accolade contends, not substantially similar. In any event, an attempt to monopolize the market by making it impossible for others to compete runs counter to the statutory purpose of promoting creative expression and cannot constitute a strong equitable basis for resisting the invocation of the fair use doctrine. Thus, we conclude that the fourth statutory factor weighs in Accolade's, not Sega's, favor, notwithstanding the minor economic loss Sega may suffer.

(c)

The second statutory factor, the nature of the copyrighted work, reflects the fact that not all copyrighted works are entitled to the same level of protection. The protection established by the Copyright Act for original works of authorship does not extend to the ideas underlying a work or to the functional or factual aspects of the work. 17 U.S.C. § 102(b). To the extent that a work is functional or factual, it may be copied, Baker v. Selden, 101 U.S. 99, 102-04, (1879), as may those expressive elements of the work that "must necessarily be used as incident to" expression of the underlying ideas, functional concepts, or facts, id. at 104. Works of fiction receive greater protection than works that have strong factual elements, such as historical or biographical works, Maxtone-Graham, 803 F.2d at 1263 (citing Rosemont Enterprises, Inc. v. Random House, Inc., 366 F.2d 303, 307 (2d Cir. 1966), *cert. denied,* 385 U.S. 1009 (1967)), or works that have strong functional elements, such as accounting textbooks, *Baker,* 101 U.S. at 104. Works that are merely compilations of fact are copyrightable, but the copyright in such a work is "thin." *Feist Publications,* 111 S. Ct. at 1289.

Computer programs pose unique problems for the application of the "idea/ expression distinction" that determines the extent of copyright protection. To the extent that there are many possible ways of accomplishing a given task or fulfilling a particular market demand, the programmer's choice of program structure and design may be highly creative and idiosyncratic. However, computer programs are, in essence, utilitarian articles — articles that accomplish tasks. As such, they contain many logical, structural, and visual display elements that are dictated by the function to be performed, by considerations of efficiency, or by external factors such as compatibility requirements and industry demands. Computer Assoc. Int'l, Inc. v. Altai, Inc. In some circumstances, even the exact set of commands used by the programmer is deemed functional rather than creative for purposes of copyright. "When specific instructions, even though previously copyrighted, are the only and essential means of accomplishing a given task, their later use by another will not amount to infringement." CONTU Report at 20; see CAI, 23 U.S.P.Q. 2d at 1254.

Because of the hybrid nature of computer programs, there is no settled standard for identifying what is protected expression and what is unprotected idea in a case involving the alleged infringement of a copyright in computer software. We are in wholehearted agreement with the Second Circuit's recent observation that "thus far, many of the decisions in this area reflect the courts' attempt to fit the proverbial square peg in a round hole." CAI, 23 U.S.P.Q. 2d at 1257. In 1986, the Third Circuit attempted to resolve the dilemma by suggesting that the idea or function of a computer program is the idea of the program as a whole, and "everything that is not necessary to that purpose or function [is] part of the expression of that idea." Whelan Assoc., Inc. v. Jaslow Dental Laboratory, Inc., 797 F.2d 1222, 1236 (3d Cir. 1986) (emphasis omitted). The *Whelan* rule, however, has been widely — and soundly — criticized as simplistic and overbroad. See CAI, 23 U.S.P.Q. 2d at 1252 (citing cases, treatises, and articles). In reality, "a computer program's ultimate function or purpose is the composite result of interacting subroutines. Since each subroutine is itself a program, and thus, may be said to have its own 'idea,' *Whelan*'s general formulation . . . is descriptively inadequate." Id. For example, the computer program at issue in the case before us, a video game program, contains at least two such subroutines — the subroutine that allows the user to interact with the video game and the subroutine that allows the game cartridge to interact with the console. Under a test that breaks down a computer program into its component subroutines and sub-subroutines and then identifies the idea or core functional element of each, such as the test recently adopted by the Second Circuit in CAI, 23 U.S.P.Q.2d (BNA) at 1252-53, many aspects of the program are not protected by copyright. In our view, in light of the essentially utilitarian nature of computer programs, the Second Circuit's approach is an appropriate one.

Sega argues that even if many elements of its video game programs are properly characterized as functional and therefore not protected by copyright, Accolade copied protected expression. Sega is correct. The record makes clear that disassembly is wholesale copying. Because computer programs are also unique among copyrighted works in the form in which they are distributed for public use, however, Sega's observation does not bring us much closer to a resolution of the dispute.

The unprotected aspects of most functional works are readily accessible to the human eye. The systems described in accounting textbooks or the basic structural concepts embodied in architectural plans, to give two examples, can be easily copied without also copying any of the protected, expressive aspects of the original works. Computer programs, however, are typically distributed for public use in object code form, embedded in a silicon chip or on a floppy disk. For that reason, humans often cannot gain access to the unprotected ideas and functional concepts contained in object code without disassembling that code—i.e., making copies.[7] Atari Games v. Nintendo of America, 975 F.2d 832 (Fed. Cir. 1992).

Sega argues that the record does not establish that disassembly of its object code is the only available method for gaining access to the interface specifications for the Genesis console, and the district court agreed. An independent examination of the record reveals that Sega misstates its contents, and demonstrates that the district court committed clear error in this respect.

First, the record clearly establishes that humans cannot *read* object code. Sega makes much of Mike Lorenzen's statement that a reverse engineer can work directly from the zeros and ones of object code but "it's not as fun." In full, Lorenzen's statements establish only that the use of an electronic decompiler is not absolutely necessary. Trained programmers can disassemble object code by hand. Because even a trained programmer cannot possibly remember the millions of zeros and ones that make up a program, however, he must make a written or computerized copy of the disassembled code in order to keep track of his work. See generally Johnson-Laird, Technical Demonstration of "Decompilation", reprinted in Reverse Engineering: Legal and Business Strategies for Competitive Design in the 1990's 102 (Prentice Hall Law & Business ed. 1992). The relevant fact for purposes of Sega's copyright infringement claim and Accolade's fair use defense is that *translation* of a program from object code into source code cannot be accomplished without making copies of the code.

Second, the record provides no support for a conclusion that a viable alternative to disassembly exists. The district court found that Accolade could have avoided a copyright infringement claim by "peeling" the chips contained in Sega's games or in the Genesis console, as authorized by section 906 of the SCPA, 17 U.S.C. § 906. Even Sega's amici agree that this finding was clear error. The declaration of Dr. Harry Tredennick, an expert witness for Accolade, establishes that chip peeling yields only a physical diagram of the *object code* embedded in a ROM chip. It does not obviate the need to translate object code into source code. Atari Games Corp., slip op. at 22.

The district court also suggested that Accolade could have avoided a copyright infringement suit by programming in a "clean room". That finding too is clearly erroneous. A "clean room" is a procedure used in the computer industry in order to prevent direct copying of a competitor's code during the development of a

---

8. We do not intend to suggest that disassembly is always the only available means of access to those aspects of a computer program that are unprotected by copyright. As we noted in Part III(B), supra, in many cases the operation of a program is directly reflected on the screen display and therefore visible to the human eye. In those cases, it is likely that a reverse engineer would not need to examine the code in order to understand what the program does.

2. Title 35 U.S.C. § 100(b) provides:

> The term 'process' means process, art or method, and includes a new use of a known process, machine, manufacture, composition of matter, or material.

Title 35 U.S.C. § 101 provides:

> Whoever invents or discovers any new and useful process, machine, manufacture, or composition of matter, or any new and useful improvement thereof, may obtain a patent therefor, subject to the conditions and requirements of this title.

competing product. Programmers in clean rooms are provided only with the functional specifications for the desired program. As Dr. Tredennick explained, the use of a clean room would not have avoided the need for disassembly because disassembly was necessary in order to discover the functional specifications for a Genesis-compatible game.

In summary, the record clearly establishes that disassembly of the object code in Sega's video game cartridges was necessary in order to understand the functional requirements for Genesis compatibility. The interface procedures for the Genesis console are distributed for public use only in object code form, and are not visible to the user during operation of the video game program. Because object code cannot be read by humans, it must be disassembled, either by hand or by machine. Disassembly of object code necessarily entails copying. Those facts dictate our analysis of the second statutory fair use factor. If disassembly of copyrighted object code is *per se* an unfair use, the owner of the copyright gains a *de facto* monopoly over the functional aspects of his work — aspects that were expressly denied copyright protection by Congress. 17 U.S.C. § 102(b). In order to enjoy a lawful monopoly over the idea or functional principle underlying a work, the creator of the work must satisfy the more stringent standards imposed by the patent laws. Bonito Boats, Inc. v. Thunder Craft Boats, Inc., 489 U.S. 141, 159-64 (1989). Sega does not hold a patent on the Genesis console. Because Sega's video game programs contain unprotected aspects that cannot be examined without copying, we afford them a lower degree of protection than more traditional literary works. See CAI, 23 U.S.P.Q.2d at 1257. In light of all the considerations discussed above, we conclude that the second statutory factor also weighs in favor of Accolade.

(d)

As to the third statutory factor, Accolade disassembled entire programs written by Sega. Accordingly, the third factor weighs against Accolade. The fact that an entire work was copied does not, however, preclude a finding of fair use. *Sony Corp.,* 464 U.S. at 449-50; *Hustler,* 795 F.2d at 1155 ("*Sony Corp.* teaches us that the copying of an entire work does not preclude fair use per se."). In fact, where the ultimate (as opposed to direct) use is as limited as it was here, the factor is of very little weight. Cf. Wright v. Warner Books, Inc., 953 F.2d 731, 738 (2d Cir. 1991).

(e)

In summary, careful analysis of the purpose and characteristics of Accolade's use of Sega's video game programs, the nature of the computer programs involved, and the nature of the market for video game cartridges yields the conclusion that the first, second, and fourth statutory fair use factors weigh in favor of Accolade, while only the third weighs in favor of Sega, and even then only slightly. Accordingly, Accolade clearly has by far the better case on the fair use issue.

We are not unaware of the fact that to those used to considering copyright issues in more traditional contexts, our result may seem incongruous at first blush. To oversimplify, the record establishes that Accolade, a commercial competitor of Sega, engaged in wholesale copying of Sega's copyrighted code as a preliminary step in the development of a competing product. However, the key to this case is that we are dealing with computer software, a relatively unexplored area in the world of copyright law. We must avoid the temptation of trying to force "the proverbial square peg into a round hole." CAI, 23 U.S.P.Q.2d at 1257.

In determining whether a challenged use of copyrighted material is fair, a court must keep in mind the public policy underlying the Copyright Act. "'The immediate effect of our copyright law is to secure a fair return for an author's creative labor. But the ultimate aim is, by this incentive, to stimulate artistic creativity for the general public good.'" *Sony Corp.*, 464 U.S. at 432 (quoting Twentieth Century Music Corp. v. Aiken, 422 U.S. 151, 156 (1975)). When technological change has rendered an aspect or application of the Copyright Act ambiguous, "'the Copyright Act must be construed in light of this basic purpose.'" Id. As discussed above, the fact that computer programs are distributed for public use in object code form often precludes public access to the ideas and functional concepts contained in those programs, and thus confers on the copyright owner a de facto monopoly over those ideas and functional concepts. That result defeats the fundamental purpose of the Copyright Act — to encourage the production of original works by protecting the expressive elements of those works while leaving the ideas, facts, and functional concepts in the public domain for others to build on. *Feist Publications*, 111 S. Ct. at 1290.

Sega argues that the considerable time, effort, and money that went into development of the Genesis and Genesis-compatible video games militate against a finding of fair use. Borrowing from antitrust principles, Sega attempts to label Accolade a "free rider" on its product development efforts. In *Feist Publications*, however, the Court unequivocally rejected the "sweat of the brow" rationale for copyright protection. 111 S. Ct. at 1290-95. Under the Copyright Act, if a work is largely functional, it receives only weak protection. "This result is neither unfair nor unfortunate. It is the means by which copyright advances the progress of science and art." Id. at 1290; see also id. at 1292 ("In truth, 'it is just such wasted effort that the proscription against the copyright of ideas and facts . . . [is] designed to prevent.'") (quoting Rosemont Enterprises, Inc. v. Random House, Inc., 366 F.2d 303, 310 (2d Cir. 1966), *cert. denied*, 385 U.S. 1009 (1967)); CAI, 23 U.S.P.Q.2d at 1257. Here, while the work may not be largely functional, it incorporates functional elements which do not merit protection. The equitable considerations involved weigh on the side of public access. Accordingly, we reject Sega's argument.

(f)

We conclude that where disassembly is the only way to gain access to the ideas and functional elements embodied in a copyrighted computer program and where there is a legitimate reason for seeking such access, disassembly is a fair use of the copyrighted work, as a matter of law. Our conclusion does not, of course, insulate Accolade from a claim of copyright infringement with respect to its finished products. Sega has reserved the right to raise such a claim, and it may do so on remand.

COMMENTS AND QUESTIONS

1. The Sega decision has been widely followed. See DSC Communications v. DGI Technologies, 81 F.3d 597, 601 (5th Cir. 1996); Bateman v. Mnemonics, Inc., 79 F.3d 1532, 1539 n.18 (11th Cir. 1995); Lotus Dev. Corp. v. Borland Int'l,

49 F.3d 807, 817-18 (1st Cir. 1995) (Boudin, J., concurring); Atari Games Corp. v. Nintendo of America, 975 F.2d 832, 843-44 (Fed. Cir. 1992); Sega, Inc. v. Accolade, 977 F.2d 1510, 1527-28 (9th Cir. 1992); Vault v. Quaid, 847 F.2d 255, 270 (5th Cir. 1988); Mitel Inc. v. Iqtel Inc., 896 F. Supp. 1050 (D. Colo. 1995), *aff'd* 124 F.3d 1366 (10th Cir. 1997); cf. DSC Communications Corp. v. Pulse Communications, Inc., 170 F.3d 1354 (Fed. Cir. 1999) (acknowledging that reverse engineering could be a fair use, but holding that it was not always fair). Most commentators have also endorsed a right to reverse engineer, at least in some circumstances. See, e.g., Pamela Samuelson & Suzanne Scotchmer, The Law and Economics of Reverse Engineering, 111 Yale L.J. 1575 (2002); Peter S. Menell, An Epitaph for Traditional Copyright Protection of Network Features of Computer Software, 43 Antitrust Bulletin 651 (1998); Jonathan Band & Masanobu Katoh, Interfaces on Trial (1995); Julie Cohen, Reverse Engineering and the Rise of Electronic Vigilantism: Intellectual Property Implications of "Lock-Out" Technologies, 68 S. Cal. L. Rev. 1091 (1995); Lawrence D. Graham & Richard O. Zerbe Jr., Economically Efficient Treatment of Computer Software: Reverse Engineering, Protection, and Disclosure, 22 Rutgers Comp. & Tech. L.J. 61 (1996); Dennis S. Karjala, Copyright Protection of Computer Documents, Reverse Engineering, and Professor Miller, 19 U. Dayton L. Rev. 975, 1016-18 (1994); David A. Rice, Sega and Beyond: A Beacon for Fair Use Analysis . . . At Least as Far as It Goes, 19 U. Dayton L. Rev. 1131, 1168 (1994); LaST Frontier Conference Report on Copyright Protection of Computer Software, 30 Jurimetrics J. 15, 24-25 (1989) (conference consensus statement).

2. Do you agree with the *Sega* court's analysis of the fourth fair use factor? Hasn't Sega been injured because it can no longer control who produces games for its machine? Won't it lose the ability to charge a fee to game writers? If so, is this injury the result of copyright infringement or of Accolade's independent contributions to its games?

3. In Sony Computer Entertainment, Inc. v. Connectix Corp., 203 F.3d 596 (9th Cir.), *cert. denied*, 531 U.S. 871 (2000), the Ninth Circuit took the *Sega* doctrine a significant step further. This case concerned Sony's successful Playstation game platform. Unlike the situation in *Sega* — where Accolade had reverse engineered the Sega platform for purposes of manufacturing games that could be run on Sega machines — Connectix reverse engineered the interoperability specifications for Sony's platform in order to enable games to be run on the Apple iMac microcomputer platform. In essence, Connectix enabled the porting of Playstation games from the Sony console to another platform. In this way, Connectix directly competed with Sony in platform technology. Should this difference matter under the fair use test? See Phil Weiser, Networks Unplugged: Towards a Model of Compatibility Regulation Between Information Platforms, Cornell University Library, arxiv, Computer Science abstract cs, CY/0109070, Sept. 24, 2001, *http://arxiv.org/abs/cs.CY/0109070* (arguing that *Sony* is a significant step beyond *Sega*); cf. Pamela Samuelson & Suzanne Scotchmer, The Law and Economics of Reverse Engineering, 111 Yale L.J. 1575, 1621-26 (2002) (distinguishing between incentives to develop games within a platform and incentives to innovate new platforms).

4. Is standardization a valid rationale for copying? Professor Menell suggests that even arbitrary elements of user interfaces that become de facto industry standards should lose copyright protection in order to foster network externalities. Peter Menell, An Analysis of the Scope of Copyright Protection for Computer Programs, 41 Stan. L. Rev. 1045, 1101 (1989). He argues that the high premium that consumers place on widely learned standards makes compatibility with such standards

critical to entering the marketplace. In effect, what began as an arbitrary element has become functional due to widespread consumer learning.

On the other hand, not every case requires standardization. The *Sega* court is careful to note that intermediate copying qualifies as fair use only if it is necessary to produce an interoperable program. At least one court has rejected a claim of fair use on the grounds that the copyrighted material taken was not needed for interoperability. Compaq Computer v. Procom Technology, 908 F. Supp. 1409 (S.D. Tex. 1995).

Should a copyright owner be deprived of protection merely because its product has been adopted as a standard? Will this affect incentives to develop computer programs?

5. Is the fair use doctrine an appropriate means of encouraging compatibility or standardization? Does accommodating those policy goals require stretching the doctrine too far? Some commentators have suggested that encouraging compatibility is best accomplished by legislating a right to reverse engineer, analogous to the right provided in the Semiconductor Chip Protection Act of 1984. See, e.g., Timothy S. Teter, Merger and the Machines: An Analysis of the Pro-Compatibility Trend in Computer Software Copyright Cases, 45 Stan. L. Rev. 1061, 1089-97 (1993).

6. *Antitrust/Misuse Constraints.* Is Sega entitled to control not only the sales of its copyrighted software-hardware package, but also video games that will run on the Sega system? The court's decision obliquely suggests that there are antitrust (or possibly copyright misuse) problems with giving Sega such power: "an attempt to monopolize the market by making it impossible for others to compete runs counter to the statutory purpose of promoting creative expression and cannot constitute a strong equitable basis for resisting the invocation of the fair use doctrine." See also Alcatel USA Inc. v. DGI Technologies, Inc., 166 F.3d 772 (5th Cir. 1999) (holding that it was copyright misuse to attempt to prevent a competitor from testing the compatibility of its cards with the plaintiff's technology). Is this concern properly addressed in the fair use doctrine? Should antitrust or copyright misuse provide the frameworks for resolving these disputes? See Julie E. Cohen, Reverse Engineering and the Rise of Electronic Vigilantism: Intellectual Property Implications of "Lock-Out" Technologies, 68 S. Cal. L. Rev. 1091 (1995) (suggesting that such a misuse-based approach may be warranted).

7. At the time *Sega* was decided, reverse engineering a computer program was a difficult and time-consuming task. If reverse engineering were simple and flawless, should it still be legal? For an argument that "easy" reverse engineering would be problematic, see Pamela Samuelson, Randall Davis, Mitchell D. Kapor, & Jerome H. Reichman, A Manifesto Concerning the Legal Protection of Computer Programs, 94 Colum. L. Rev. 2308 (1994).

8. *Patent and Trade Secret.* Patent and trade secrets laws treat reverse engineering very differently. Reverse engineering a patented product necessarily constitutes infringement if it involves making, using, or selling the patented invention. There is no statutory defense for reverse engineering. By contrast, the Uniform Trade Secrets Act expressly provides that reverse engineering a commercially available product is a legitimate means of discovering a trade secret. What explains the different treatment of reverse engineering among patent, copyright, and trade secret laws?

9. *Contract.* Should software manufacturers be able to prohibit reverse engineering through contractual terms in shrinkwrap licenses? Should such provisions be preempted by federal copyright law? Compare Vault Corp. v. Quaid Software Ltd., 847 F.2d 255 (5th Cir. 1988) (finding standard-form contractual restriction on

reverse engineering to be preempted) with Davidson & Assocs. v. Jung, 422 F.3d 630 (8th Cir. 2005) (enforcing standard-form contractual restriction on reverse engineering); Bowers v. Baystate Technologies, Inc., 302 F.3d 1334 (Fed. Cir. 2002) (same). See David Nimmer, Elliot Brown, Gary N. Frischling, the Metamorphosis of Contract into Expand, 87 Cal. L. Rev. 17 (1999); Mark A. Lemley, Beyond Preemption: The Federal Law and Policy of Intellectual Property Licensing, 87 Cal. L. Rev. 111 (1999); Dennis J. Karjala, Federal Preemption of Shrinkwrap and On-Line Licenses, 22 U. Dayton L. Rev. 511 (1997).

4. Open Source Licensing

Although disdaining proprietary control of software innovation, the open source movement relies on copyright protection as the vehicle for weaning the software industry off its proprietary business models. Thus, the General Public License (GPL) and other open source licenses integrate copyright concepts, such as the derivative works, in encouraging a culture of free software.

Brian W. Carver, Share and Share Alike:
Understanding and Enforcing Open Source and
Free Software Licenses
20 Berkeley Tech. L.J. 443 (2005)

. . .

II. UNDERSTANDING OPEN SOURCE AND
FREE SOFTWARE LICENSES

A. *Comparing the Free Software Definition and the Open Source Definition*

Richard Stallman and the FSF have taken a stand against the "open source" label because in their view it minimizes the most important part of the concept: the user's freedom. Consequently, what the FSF considers a "free software" license is based on the Free Software Definition, while what the OSI [Open Source Initiative] considers an "open source" license is based on the Open Source Definition.

The Free Software Definition is largely comprised of the four freedoms:

0. The freedom to run the program, for any purpose.
1. The freedom to study how the program works, and adapt it to your needs. Access to the source code is a precondition for this.
2. The freedom to redistribute copies so you can help your neighbor.
3. The freedom to improve the program, and release your improvements to the public, so that the whole community benefits. Access to the source code is a precondition for this.

In contrast, the Open Source Definition contains ten criteria, requiring, among other things, free redistribution rights, access to source code, permission to modify and distribute modifications, and forbidding discrimination against persons, groups, or fields of endeavor. The first criterion reads:

1. Free Redistribution

The license shall not restrict any party from selling or giving away the software as a component of an aggregate software distribution containing programs from several different sources. The license shall not require a royalty or other fee for such sale. This is indicative of the style of the other nine criteria. In general, the Open Source Definition is more detailed than the Free Software Definition and arguably provides clearer direction to those hoping to write a software license about what features such a license should and should not contain.

Software licensed under the GPL, such as the Linux kernel, satisfies both definitions and hence may accurately be called both "free software" and "open source" software. Indeed, the vast majority of Free Software and Open Source licenses satisfy both definitions, however, this is not always the case.

B. *Classifying Software Licenses: Copyleft and GPL-Compatibility*

There are several features of a software license that can be useful in classifying it. First, a license may be a proprietary license, a free software license, or an open source license. Secondly, within free software and open source licenses, the license might also be a copyleft, or reciprocal, license. A copyleft license requires that any derivative work made from the copyleft-licensed work be itself licensed under the same copyleft license, preserving all the same rights and responsibilities the original licensee had to downstream licensees. Third, it is useful to classify licenses according to whether they are compatible with the GPL, since it is the most widely-used free software license. A license is "compatible" with the GPL if that license allows licensees to combine a module that was released under that license with a GPL-covered module to make one larger program. Note that neither the Free Software Definition nor the Open Source Definition requires a license to be either a copyleft license or to be GPL-compatible in order to qualify as a free software or open source license, respectively. Access to source code and the freedoms to use and modify that code are distinct issues, separable from copyleft status or GPL-compatibility....

III. THE GNU GENERAL PUBLIC LICENSE

A. *Key License Terms*

The Preamble to the GPL is indicative of its intended audience. While the GPL is a legal document and a copyright license, it is not written primarily with a legal audience in mind. Instead, Stallman intended software developers to use this license on their software and for software users to think about software freedoms. Consequently, the GPL is written in clear language that anyone can understand. The Preamble concisely explains the FSF's philosophy and the key terms of the license, which are specified in detail later.

1. Copyleft

The GPL's Preamble explains the notion of copyleft without introducing the term. Instead the Preamble states that if you distribute copies of a program licensed under the GPL, "you must give the recipients all the rights that you have." . . . Copyleft licenses, such as the GPL, prevent someone from turning free software into proprietary software. The copyleft condition of the GPL appears formally in Section 2(b).

2. Fees

The GPL's Preamble mentions twice that charging a fee for the distribution of GPL-covered software is not only allowed but a guaranteed right. This point is often confused with the fact that the GPL forbids requiring that any third party who receives the software pay a license fee. A license fee would violate Section 6 of the GPL, which reads in relevant part:

> Each time you redistribute the Program (or any work based on the Program), the recipient automatically receives a license from the original licensor to copy, distribute or modify the Program subject to these terms and conditions. You may not impose any further restrictions on the recipients' exercise of the rights granted herein. To require a royalty or license fee of those third parties would be to impose a further restriction on them. In contrast, nothing in the GPL prevents someone from charging fees for service, warranty protection, or indemnification of software licensed under the GPL.

3. Source Code

The ability to access, modify, and distribute source code is at the heart of the GPL. The Preamble explains that the license guarantees "that you receive source code or can get it if you want it" and that "you can change the software or use pieces of it in new free programs." Those who distribute GPL-covered software must ensure that recipients "receive or can get the source code." . . .

5. Patents

Finally, the Preamble flags a threat to the free software community: patents. The GPL requires that those who would distribute software under the GPL not bind themselves to a patent license that would "not permit royalty-free redistribution of the Program by all those who receive copies directly or indirectly through you . . . ." If a developer of GPL-covered software had a patent license that was not available to those who might receive the software, then such recipients would become patent infringers. To avoid this problem the GPL requires such developers either to ensure that the patent holder permits royalty-free redistribution by all or to simply refrain from distributing patented software under the GPL.

B. *The GPL and Derivative Works*

Determining when a piece of software is a "derivative work" of a GPL-covered work is critical to the licensing regime, particularly considering the fact that the GPL requires that derivative works that are distributed must also be licensed under the GPL. The lack of certainty regarding what counts as a derivative work is only partially the fault of the GPL. In general, what constitutes a derivative work within software is not sufficiently clear from either the statutes or the case law. . . .

The nature of software is one of the reasons why defining derivative works for software is difficult. Most software is made up of numerous individual programs and files that work together to produce the results the end-user experiences. Software programs also make use of "libraries." A library is a collection of subprograms used to develop software. "Libraries are distinguished from executables in that they are not independent programs; rather, they are 'helper' code that provides services to some other independent program." A library may be either statically or dynamically linked to the independent programs that it helps.

Static linking involves embedding the library in the independent program when it is compiled. A dynamically linked library is not incorporated into a program, rather it exists both in its own place on a computer's hard drive and in its own memory space while in use. Multiple programs might even be dynamically linked to the same library and communicate with it while it sits in a single memory space. Consequently, some have argued that a proprietary program could dynamically link to a library licensed under the GPL, or a program licensed under the GPL could dynamically link to a proprietary library, and in neither case would a "derivative work" of the GPL-covered work be created, since the two programs retain distinct existences and only tenuous connections.

This is not the view of the FSF, however. The FSF argues that even dynamic linking creates a derivative work, and has written a different license, the GNU Lesser General Public License (LGPL), for those who wish to permit such proprietary linking. However, the FSF does not recommend the use of the LGPL, as the FSF ultimately sees the LGPL as encouraging an unhealthy reliance on proprietary software, and it would prefer that the needed proprietary software, which would be linked, be rewritten, and licensed under a free software license. . . .

C. *Software Patents and Free Software*

Many within the free software community now believe that software patents are the greatest threat to open source and free software. A report by Open Source Risk Management found that 283 granted patents could potentially be used against the Linux kernel. However, none of these patents have been tested in court, and there is no indication that the Linux kernel is any more susceptible to such claims than other large software projects. . . .

IV. ENFORCEMENT OF THE GPL

Historically, the GPL has been primarily enforced through private negotiation and settlement agreements. This process has been successful thus far because most alleged violators have apparently been eager to correct any defects in their compliance. These extra-judicial resolutions have generally satisfied the goals of the copyright holders enforcing the GPL, but left the question of the GPL's enforceability in court largely unanswered. In 2004, however, two cases began to illuminate how courts will enforce the GPL.

A. *Private GPL Enforcement by the Free Software Foundation*

The only party who can license a work under the GPL is the copyright holder. Consequently, the only party who can seek a remedy for violation of the GPL as applied to that work is the copyright holder. There is a misconception that because the FSF wrote the GPL that it is therefore the enforcement police for all software licensed under the GPL. While the FSF is often willing to help, and asks to be informed of GPL violations, ultimately the copyright holder is responsible for enforcement. The FSF does hold the copyrights to many widely-used free software programs, particularly those within its GNU project. For those programs, the FSF does handle enforcement, primarily through its pro bono general counsel.

The FSF engaged in an interesting public example of private enforcement of the GPL in 2003. Linksys was distributing its popular WRT54G wireless router with GPL-covered software, but was not providing source code. Initial discussions with Linksys were not moving fast enough to satisfy some developers of the Linux kernel. The FSF stepped in to negotiate with Linksys' new owner, Cisco Systems, who eventually complied with the license and now provides the GPL-covered source code from the Linksys website for not only the WRT54G, but also for dozens of Linksys devices using GPL-covered software. Consequently, this router has become particularly popular among hackers, who have even developed projects to install more full-featured GNU/Linux installations on the router.

B. *Private GPL Enforcement by Other Copyright Holders*

The FSF is not the only copyright holder who licenses software under the GPL and who has had significant success enforcing the GPL privately. One of the more prominent examples of such private enforcement is provided by Harald Welte, who would later be the plaintiff in [a]s Munich court decision enforcing the GPL. Welte is the head member of the Netfilter/iptables core software development team. The netfilter/iptables software acts as a firewall for GNU/Linux systems and is in widespread use, especially in hardware routers. The netfilter software is itself a part of the Linux kernel and since its source code is freely available on the web, it becomes a natural choice for inclusion by many device manufacturers. Problems arise when these device manufacturers do not understand or ignore the licensing terms under which the software is provided. Manufacturers that distribute modified versions of the software are required by the GPL to provide both the full text of the GPL license itself, as well as the source code to the modified software they are distributing. Several manufacturers have failed to do this, and Harald Welte has sought to enforce the terms of the GPL against them.

Welte's first success came in February 2004, when he announced an out-of-court settlement with Allnet GmbH. Allnet was offering two routers, both including software developed by the netfilter/iptables project. However, Allnet did not fulfill the obligations of the GPL regarding the netfilter/iptables software because it did not make any source code offering or include the terms of the GPL with its products. In the settlement, Allnet agreed to adhere to all clauses of the license and to inform its customers about their respective rights and obligations under the GPL. Allnet also agreed to refrain from offering any new netfilter/iptables based products without adhering to the GPL. Finally, Allnet made "a significant donation" to the FSF Europe and to the Foundation of a Free Information Infrastructure.

Buoyed by this success, Welte pursued similar successful GPL-enforcement actions against Fujitsu-Siemens, ASUS, and Securepoint before coming across a violator of the GPL that was reluctant to comply with the license's terms: Sitecom Germany GmbH.

C. *Judicial Enforcement of the GPL: The Munich District Court Decisions*

The first clear resolution of the GPL's enforceability came this year from a German district court in Munich hearing Welte's case against Sitecom. The court held that the GPL was a valid and enforceable copyright license. . . .

D. *Court Discussion of the GPL in the United States: SCO v. IBM*

Though version 2 of the GPL was released in 1991, no U.S. court has ruled on its enforceability in the ensuing fourteen years. Indeed, the GPL has only been mentioned in passing in U.S. courts. However, in March of 2003, Caldera Systems, Inc., doing business as The SCO Group, filed a complaint against IBM in Utah state court, alleging misappropriation of trade secrets, unfair competition, interference with contract, and breach of contract. SCO claimed that while it was working with IBM on "Project Monterey," a project to develop enterprise-class UNIX systems on Intel processor-based platforms, IBM allegedly misappropriated SCO's proprietary knowledge. IBM's misappropriation was allegedly in the form of contributions of SCO's property to GNU/Linux systems in order to destroy the value of UNIX systems, and thereby, of SCO's business. . . .

While many were skeptical of SCO's claims from the beginning, conspiracy theorists felt vindicated when it was revealed that, just over two months after SCO filed suit against IBM, Microsoft bought a UNIX license from SCO for between $10-20 million. Some speculated that this UNIX license was merely a means for Microsoft to pay for someone else to blunt the competition it was increasingly seeing from GNU/Linux.

...SCO's dispute with IBM may yet provide a U.S. ruling on the GPL. Finally, a Version 3 of the GPL is in the works, and it also will likely address these issues while continuing to encourage software users and developers to share and share alike.

COMMENTS AND QUESTIONS

1. Many of the same issues posed earlier in this chapter about the enforceability of shrinkwrap and other mass-market licenses apply to the GPL and other open source licenses. The GPL assumes that the first developer of GPL software can bind all subsequent users of the software to the terms of the GPL, even those with whom the developer has never had any dealings. Under what theory or theories might such a developer be able to bind subsequent users to GPL terms? The GPL also assumes that downstream users and modifiers of GPL software will also be bound by subsequent versions of the GPL, should it be revised. Under what theory or theories might this be possible? See Christian H. Nadan, Open Source Licensing: Virus or Virtue?, 10 Tex. Intellectual Prop. L.J. 349 (2002); Margaret Jane Radin, Humans, Computers, and Binding Commitment, 75 Ind. L.J. 1125 (2000); Ira Heffan, Copyleft: Licensing Collaborative Works, 49 Stan. L. Rev. 1487 (1997).

2. Some legal questions about the GPL and other open source licenses are more specific to these licenses. For example, the GPL and many open source licenses disclaim all warranties for the software. Do you agree with open source developers that the law should allow them to disclaim all liability for errors? A licensing lawyer for Microsoft has argued that Microsoft and other developers of software need the same freedom from warranty liability. See Robert Gomulkiewicz, How Copyleft Uses License Rights To Succeed in the Open Source Revolution and the Implications for Article 2B, 36 Houston L. Rev. 179 (1999). What if any reasons might there be to distinguish between warranty disclaimers by GPL developers and those by Microsoft Corp.?

3. Microsoft officials have sometimes been very critical of open source initiatives and in particular of the GPL. See Prepared Text of Remarks by Craig Mundie, Microsoft Senior Vice President, The Commercial Software Model, The New York University Stern School of Business, May 3, 2001 *http://www.microsoft.com/presspass/exec/craig/05-03sharedsource.mspx*. Among other things, Mundie states: "The GPL mandates that any software that incorporates source code already licensed under the GPL will itself become subject to the GPL. When the resulting software product is distributed, its creator must make the entire source code base freely available to everyone, at no additional charge. This viral aspect of the GPL poses a threat to the intellectual property of any organization making use of it. It also fundamentally undermines the independent commercial software sector because it effectively makes it impossible to distribute software on a basis where recipients pay for the product rather than just the cost of distribution." Do you agree?

4. Several legal commentators have explored a range of interesting implications of open source development of computer programs. See, e.g., Yochai Benkler, Coase's Penguin, or Linux and the Nature of the Firm, Yale L.J. 369 (2002); David McGowan, Legal Implications of Open Source Software, 2001 U. Ill. L. Rev. 241 (2001); Stephen McJohn, The Paradoxes of Free Software, 9 Geo. Mason L. Rev. 25 (2000); Lawrence Lessig, The Limits in Open Code: Regulatory Standards and the Future of the Net, 14 Berkeley Tech. L.J. 759 (1999).

5. According to freshmeat.net and SourceForge.net, the leading online repositories of open source software, nearly 70 percent of open source programs use the GPL. Does this indicate network effects? the superiority of the GPL specification for most collaborative open source ventures?

6. Is the GPL an enforceable contract? What evidence of assent is there? Is there privity between all the users of open source software?

7. How should the concept of derivative work be applied in the open source context? Is it an issue of contract interpretation or an application of copyright's rather imprecise definition? Would the Free Software Foundation be better off defining the term as part of the license agreement? Would such an effort be preempted by the Copyright Act? As noted above, the courts are split on whether "shrinkwrap licenses" can be enforced where they vary the rules of copyright law.

8. If you were advising a company about whether to adopt open source products for its internal use, what would be the principal advantages? disadvantages? How would your advice change if your client was a software maker that would be integrating open source components into its software products? What are the risks of adopting open source software?

9. *GPL v.3.* The Free Software Foundation recently announced that it will oversee a participatory process bringing together "thousands of organisations, software developers, and software users from around the globe during 2006" to revise the GPL, the most widely adopted open source license. A draft revision is at http://gplv3.fsf.org/draft. Coming at a time at which major for-profit companies are becoming active in the open source community could generate some interesting tensions, politics, and dynamics. It seems likely that the scope of the "viral" feature of the GPL could be controversial. A second question is whether those who modify GPL code but do not distribute it — i.e., use the code for internal purposes only — should be required to make such modifications publicly available. Several major Internet companies use Linux in this way and profit handsomely, yet do not have to share their "enhancements" to the extent that they use Linux code internally. Other companies use open source software for selling software-related services that do not entail the distribution of code. Free software purists, such as Richard Stallman might well seek to expand the GPL to cover these activities, while more commercially-oriented users will likely resist such changes. How should this play out? A tightening of the GPL could reduce the spread of Linux, the most important GPL software program, while enhancing the viability of other software products offering "open source" characteristics, but more flexibility for commercial exploitation, such as Sun's Solaris operating system. Which approach is most like to promote competition and benefit consumers in the long run?

D. PATENT PROTECTION

1. Is Software Patentable Subject Matter?

With the growth of computer technology came an early crop of patent applications. In the 1950s and early 1960s, the Patent Office met these with a uniform response: whatever software is, it is definitely *not* patentable subject matter.

Some programmers persisted in their efforts to have software recognized with the traditional badge of technical achievement, an issued patent. Some even obtained patents, but not for software per se. Overall, the early unwillingness of the PTO and courts to grant patents on software set the stage for the next two decades of legal protection for computer software, bringing trade secrets and copyright to the forefront.

On several notable occasions, early programmer/applicants brought their fight to patent software to the Supreme Court. These early cases — and indeed the vast majority of software patent cases to date — focused on the question of whether an "invention" consisting of the use of a mathematical algorithm could be patentable subject matter under 35 U.S.C. §101. Gottschalk v. Benson, 409 U.S. 63 (1972), said no. In a broadly worded majority opinion Justice Douglas equated "mathematical algorithms," such as the numerical conversion software claimed in the case, with laws of nature and naturally occurring products. *Benson*'s emphasis on "algorithms" was a departure from prior case law. See Pamela Samuelson, *Benson* Revisited: The Case Against Patent Protection for Algorithms and Other Computer Program-Related Inventions, 39 Emory L.J. 1025, 1042-1043 (1990). A subsequent case, Diamond v. Diehr, 450 U.S. 175 (1981), seemed to open the door to some software-related inventions. That case held that a rubber-curing process that included a software component did present patentable subject matter. Even so, the language of "algo-rithms," and the spectre of *Benson*, cast a shadow over efforts to obtain patents for computer software for many years. The issue was not conclusively resolved until the mid- to late-1990s.

In the meantime, the author of the authoritative patent law treatise, Donald Chisum, criticized the *Benson* decision and called for it to be overruled. The result in the case, he argued, "stemmed from an antipatent judicial bias that cannot be reconciled with the basic elements of the patent system established by Congress." Donald Chisum, The Future of Software Protection: The Patentability of Algorithms, 47 U. Pitt. L. Rev. 959, 961 (1986). Professor Chisum stated that the "awkward distinctions and seemingly irreconcilable results of the case law since *Benson* ... are the product of the analytical and normative weakness of *Benson* itself." Id., at 961-962. Chisum believed there were strong policy reasons to favor the patentability of computer algorithms.

For an updated version of many of these arguments, see Richard S. Gruner, Intangible Inventions: Patentable Subject Matter for an Information Age, 35 Loy. L.A. L. Rev. 355 (2002).

Note on "Floppy Disk" Claims

Some inventors obtained software patents during the 1980s and early 1990s, but the issue was clouded with doubt. For a discussion of this history, and how software patentees evaded the rules using what the authors call "the doctrine of the magic words," see Julie E. Cohen & Mark A. Lemley, Patent Scope and Innovation in the Software Industry, 89 Cal. L. Rev. 1 (2001). Antipathy toward pure software claims began to change in 1995, when IBM appealed the PTO's rejection of a claim to "software contained on a floppy disk" to the Federal Circuit. See In re Beauregard,

53 F.3d 1583 (Fed. Cir. 1995). While the appeal was pending, the PTO decided not to oppose the claim, and promised the court that it would shortly issue new examining guidelines for software patents. After several false starts, the PTO did issue final Examination Guidelines for Computer-Implemented Inventions in January of 1996 (reprinted at 17 J. Marshall J. of Computer & Info. L. 311 (1998)). These guidelines direct examiners as follows:

> The subject matter courts have found to be outside the four statutory categories of invention is limited to abstract ideas, laws of nature and natural phenomena. While this is easily stated, determining whether an applicant is seeking to patent an abstract idea, a law of nature or a natural phenomenon has proven to be challenging. These three exclusions recognize that subject matter that is not a *practical application or use* of an idea, a law of nature or a natural phenomenon is not patentable....
>
> Where certain types of descriptive material, such as music, literature, art, photographs and mere arrangements or compilations of facts or data, are merely stored so as to be read or outputted by a computer without creating any functional interrelationship, either as part of the stored data or as part of the computing processes performed by the computer, then such descriptive material alone does not impart functionality either to the data as so structured, or to the computer. Such "descriptive material" is not a process, machine, manufacture or composition of matter.
>
> The policy that precludes the patenting of non-functional descriptive material would be easily frustrated if the same descriptive material could be patented when claimed as an article of manufacture. For example, music is commonly sold to consumers in the format of a compact disc. In such cases, the known compact disc acts as nothing more than a carrier for non-functional descriptive material. The purely non-functional descriptive material cannot alone provide the practical application for the manufacture.
>
> Office personnel should be prudent in applying the foregoing guidance. Non-functional descriptive material may be claimed in combination with other functional descriptive material on a computer-readable medium to provide the necessary functional and structural interrelationship to satisfy the requirements of § 101. The presence of the claimed non-functional descriptive material is not necessarily determinative of non-statutory subject matter. For example, a computer that recognizes a particular grouping of musical notes read from memory and upon recognizing that particular sequence, causes another defined series of notes to be played, defines a functional interrelationship among that data and the computing processes performed when utilizing that data, and as such is statutory because it implements a statutory process....
>
> Office personnel must treat each claim as a whole. The mere fact that a hardware element is recited in a claim does not necessarily limit the claim to a specific machine or manufacture. If a product claim encompasses *any and every* computer implementation of a process, when read in light of the specification, it should be examined on the basis of the underlying process. Such a claim can be recognized as it will:
>
> - define the physical characteristics of a computer or computer component exclusively as functions or steps to be performed on or by a computer, and
> - encompass *any and every* product in the stated class (e.g., computer, computer-readable memory) *configured in any manner* to perform that process....
>
> A claim that requires one or more acts to be performed defines a process. However, not all processes are statutory under § 101. To be statutory, a claimed computer-related process must either: (1) result in a physical transformation outside the computer for which a practical application in the technological arts is either disclosed in the specification or would have been known to a skilled artisan (discussed in (i) below), or (2) be limited by the language in the claim to a practical application within the technological arts

(discussed in (ii) below). The claimed practical application must be a further limitation upon the claimed subject matter if the process is confined to the internal operations of the computer. If a physical transformation occurs outside the computer, it is not necessary to claim the practical application. A disclosure that permits a skilled artisan to practice the claimed invention, i.e., to put it to a practical use, is sufficient. On the other hand, it is necessary to claim the practical application if there is no physical transformation or if the process merely manipulates concepts or converts one set of numbers into another. . . .

There is always some form of physical transformation within a computer because a computer acts on signals and transforms them during its operation and changes the state of its components during the execution of a process. Even though such a physical transformation occurs within a computer, such activity is not determinative of whether the process is statutory because such transformation alone does not distinguish a statutory computer process from a non-statutory computer process. What is determinative is not how the computer performs the process, but what the computer does to achieve a practical application.

A process that merely manipulates an abstract idea or performs a purely mathematical algorithm is non-statutory despite the fact that it might inherently have some usefulness. For such subject matter to be statutory, the claimed process must be limited to a practical application of the abstract idea or mathematical algorithm in the technological arts. For example, a computer process that simply calculates a mathematical algorithm that models noise is non-statutory. However, a claimed process for digitally filtering noise employing the mathematical algorithm is statutory. . . .

If the "acts" of a claimed process manipulate only numbers, abstract concepts or ideas, or signals representing any of the foregoing, the acts are not being applied to appropriate subject matter. Thus, a process consisting solely of mathematical operations, i.e., converting one set of numbers into another set of numbers, does not manipulate appropriate subject matter and thus cannot constitute a statutory process.

COMMENTS AND QUESTIONS

1. For more on claims to information residing on storage media (sometimes called "floppy disk" claims), see Symposium: "Article of Manufacture" Patent Claims for Computer Instruction, 17 J. Marshall J. Computer & Info. L. 5 et seq. (1998).

2. Do the guidelines' provisions on patentable subject matter affect the scope of the patents that result? One possible explanation for the formal distinctions drawn in the guidelines is that they are designed to allow narrow patent claims while excluding broad ones. For example, a claim directed to a computer program itself is not statutory, but a claim for a program implemented on a particular medium is statutory. Does this distinction imply that the claim is infringed only by the use of the same program *recorded on the same medium*?

3. Despite the ongoing struggle to define the contours of algorithm patentability in the United States, the world trade community in 1994 adopted an expansive patent subject matter provision as part of the Trade-Related Aspects of Intellectual Property (TRIPs) accord in the "Uruguay Round" of agreements under the General Agreement on Tariffs and Trade (GATT). See General Agreement on Tariffs and Trade: Multilateral Trade Negotiations Final Act Embodying the Results of the Uruguay Round of Trade Negotiations, Done at Marrakesh, April 15, 1994, 33 International Legal Materials 1125 (1994), at Article 27, ¶1:

Subject to the provisions of paragraphs 2 and 3 [which do not deal with software], patents shall be available for any inventions, whether products or processes, in all fields of technology, provided that they are new, involve an inventive step and are capable of industrial application.... [P]atents shall be available and patent rights enjoyable without discrimination as to the place of invention, the field of technology and whether products are imported or locally produced.

This language suggests agreement upon a more or less uniform rule worldwide favoring patents for at least some types of software, although there is no evidence that this was intended by the drafters of Article 27. Notwithstanding Article 27, however, most European courts have taken the position that "pure" software is not patentable. But efforts to change the European rule against pure software patents foundered in 2005. Article 52 of the European Patent Convention has increasingly been interpreted in ways similar to the old U.S. rule, permitting software inventions to be patented if they are claimed as part of a larger machine.

The changes begun in the *Beauregard* case came fully to fruition three years later, when the Federal Circuit decided the following case.

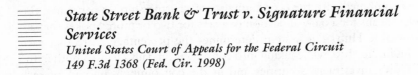

State Street Bank & Trust v. Signature Financial Services
United States Court of Appeals for the Federal Circuit
149 F.3d 1368 (Fed. Cir. 1998)

RICH, Circuit Judge.

Signature Financial Group, Inc. (Signature) appeals from the decision of the United States District Court for the District of Massachusetts granting a motion for summary judgment in favor of State Street Bank & Trust Co. (State Street), finding U.S. Patent No. 5,193,056 (the '056 patent) invalid on the ground that the claimed subject matter is not encompassed by 35 U.S.C. §101 (1994). See State Street Bank & Trust Co. v. Signature Financial Group, Inc., 927 F. Supp. 502, 38 USPQ2d 1530 (D.Mass. 1996). We reverse and remand because we conclude that the patent claims are directed to statutory subject matter.

Background

Signature is the assignee of the '056 patent which is entitled "Data Processing System for Hub and Spoke Financial Services Configuration." The '056 patent issued to Signature on March 9, 1993, naming R. Todd Boes as the inventor. The '056 patent is generally directed to a data processing system (the system) for implementing an investment structure which was developed for use in Signature's business as an administrator and accounting agent for mutual funds. In essence, the system, identified by the proprietary name Hub and Spoke, facilitates a structure whereby mutual funds (Spokes) pool their assets in an investment portfolio (Hub) organized as a partnership. This investment configuration provides the administrator of a mutual fund with the advantageous combination of economies of scale in administering investments coupled with the tax advantages of a partnership.

State Street and Signature are both in the business of acting as custodians and accounting agents for multi-tiered partnership fund financial services. State Street negotiated with Signature for a license to use its patented data processing system described and claimed in the '056 patent. When negotiations broke down, State Street brought a declaratory judgment action asserting invalidity, unenforceability, and non-infringement in Massachusetts district court, and then filed a motion for partial summary judgment of patent invalidity for failure to claim statutory subject matter under sec. 101. The motion was granted and this appeal followed.

Discussion

... The patented invention relates generally to a system that allows an administrator to monitor and record the financial information flow and make all calculations necessary for maintaining a partner fund financial services configuration. As previously mentioned, a partner fund financial services configuration essentially allows several mutual funds, or "Spokes," to pool their investment funds into a single portfolio, or "Hub," allowing for consolidation of, inter alia, the costs of administering the fund combined with the tax advantages of a partnership. In particular, this system provides means for a daily allocation of assets for two or more Spokes that are invested in the same Hub. The system determines the percentage share that each Spoke maintains in the Hub, while taking into consideration daily changes both in the value of the Hub's investment securities and in the concomitant amount of each Spoke's assets.

In determining daily changes, the system also allows for the allocation among the Spokes of the Hub's daily income, expenses, and net realized and unrealized gain or loss, calculating each day's total investments based on the concept of a book capital account. This enables the determination of a true asset value of each Spoke and accurate calculation of allocation ratios between or among the Spokes. The system additionally tracks all the relevant data determined on a daily basis for the Hub and each Spoke, so that aggregate year end income, expenses, and capital gain or loss can be determined for accounting and for tax purposes for the Hub and, as a result, for each publicly traded Spoke.

It is essential that these calculations are quickly and accurately performed. In large part this is required because each Spoke sells shares to the public and the price of those shares is substantially based on the Spoke's percentage interest in the portfolio. In some instances, a mutual fund administrator is required to calculate the value of the shares to the nearest penny within as little as an hour and a half after the market closes. Given the complexity of the calculations, a computer or equivalent device is a virtual necessity to perform the task. ...

The district court began its analysis by construing the claims to be directed to a process, with each "means" clause merely representing a step in that process. However, "machine" claims having "means" clauses may only be reasonably viewed as process claims if there is no supporting structure in the written description that corresponds to the claimed "means" elements. See In re Alappat, 33 F.3d 1526, 1540-41, 31 USPQ2d 1545, 1554 (Fed. Cir. 1994) (en banc). This is not the case now before us.

When independent claim 1 is properly construed in accordance with section 112, para. 6, it is directed to a machine, as demonstrated below, where representative claim 1 is set forth, the subject matter in brackets stating the structure the written description discloses as corresponding to the respective "means" recited in the claims.

1. A data processing system for managing a financial services configuration of a portfolio established as a partnership, each partner being one of a plurality of funds, comprising:

   (a)   computer processor means [a personal computer including a CPU] for processing data;

   (b)   storage means [a data disk] for storing data on a storage medium;

   (c)   first means [an arithmetic logic circuit configured to prepare the data disk to magnetically store selected data] for initializing the storage medium;

   (d)   second means [an arithmetic logic circuit configured to retrieve information from a specific file, calculate incremental increases or decreases based on specific input, allocate the results on a percentage basis, and store the output in a separate file] for processing data regarding assets in the portfolio and each of the funds from a previous day and data regarding increases or decreases in each of the funds, [sic, funds'] assets and for allocating the percentage share that each fund holds in the portfolio;

   (e)   third means [an arithmetic logic circuit configured to retrieve information from a specific file, calculate incremental increases and decreases based on specific input, allocate the results on a percentage basis and store the output in a separate file] for processing data regarding daily incremental income, expenses, and net realized gain or loss for the portfolio and for allocating such data among each fund;

   (f)   fourth means [an arithmetic logic circuit configured to retrieve information from a specific file, calculate incremental increases and decreases based on specific input, allocate the results on a percentage basis and store the output in a separate file] for processing data regarding daily net unrealized gain or loss for the portfolio and for allocating such data among each fund; and

   (g)   fifth means [an arithmetic logic circuit configured to retrieve information from specific files, calculate that information on an aggregate basis and store the output in a separate file] for processing data regarding aggregate year-end income, expenses, and capital gain or loss for the portfolio and each of the funds.

Each claim component, recited as a "means" plus its function, is to be read, of course, pursuant to sec. 112, para. 6, as inclusive of the "equivalents" of the structures disclosed in the written description portion of the specification. Thus, claim 1, properly construed, claims a machine, namely, a data processing system for managing a financial services configuration of a portfolio established as a partnership, which machine is made up of, at the very least, the specific structures disclosed in the written description and corresponding to the means-plus-function elements (a)-(g) recited in the claim. A "machine" is proper statutory subject matter under § 101. We note that, for the purposes of a § 101 analysis, it is of little relevance whether claim 1 is directed to a "machine" or a "process," as long as it falls within at least one of the four enumerated categories of patentable subject matter, "machine" and "process" being such categories.

This does not end our analysis, however, because the court concluded that the claimed subject matter fell into one of two alternative judicially-created exceptions to statutory subject matter. The court refers to the first exception as the "mathematical algorithm" exception and the second exception as the "business method" exception. . . .

The repetitive use of the expansive term "any" in § 101 shows Congress's intent not to place any restrictions on the subject matter for which a patent may be obtained beyond those specifically recited in § 101. Indeed, the Supreme Court has acknowledged that Congress intended § 101 to extend to "anything under the sun that is made by man." Diamond v. Chakrabarty, 447 U.S. 303, 309 (1980); see also Diamond v. Diehr, 450 U.S. 175, 182 (1981). Thus, it is improper to read limitations into § 101 on the subject matter that may be patented where the legislative history indicates that Congress clearly did not intend such limitations. See *Chakrabarty*, 447 U.S. at 308 ("We have also cautioned that courts 'should not read into the patent laws limitations and conditions which the legislature has not expressed.'" (citations omitted)).

The "Mathematical Algorithm" Exception

The Supreme Court has identified three categories of subject matter that are unpatentable, namely "laws of nature, natural phenomena, and abstract ideas." *Diehr*, 450 U.S. at 185. Of particular relevance to this case, the Court has held that mathematical algorithms are not patentable subject matter to the extent that they are merely abstract ideas. See *Diehr*, 450 U.S. 175; Parker v. Flook, 437 U.S. 584 (1978); Gottschalk v. Benson, 409 U.S. 63 (1972). In *Diehr*, the Court explained that certain types of mathematical subject matter, standing alone, represent nothing more than abstract ideas until reduced to some type of practical application, i.e., "a useful, concrete and tangible result." *Alappat*, 33 F.3d at 1544.[3]

Unpatentable mathematical algorithms are identifiable by showing they are merely abstract ideas constituting disembodied concepts or truths that are not "useful." From a practical standpoint, this means that to be patentable an algorithm must be applied in a "useful" way. In *Alappat*, we held that data, transformed by a machine through a series of mathematical calculations to produce a smooth waveform display on a rasterizer monitor, constituted a practical application of an abstract idea (a mathematical algorithm, formula, or calculation), because it produced "a useful, concrete and tangible result" — the smooth waveform.

Similarly, in Arrhythmia Research Technology Inc. v. Corazonix Corp., 958 F.2d 1053 (Fed. Cir. 1992), we held that the transformation of electrocardiograph signals from a patient's heartbeat by a machine through a series of mathematical calculations constituted a practical application of an abstract idea (a mathematical algorithm, formula, or calculation), because it corresponded to a useful, concrete or tangible thing — the condition of a patient's heart.

---

4. "To Promote the Progress of . . . Useful Arts," Report of the President's Commission on the Patent System (1966). This has come to be known as the mathematical algorithm exception. This designation has led to some confusion, especially given the *Freeman-Walter-Abele* analysis. By keeping in mind that the mathematical algorithm is unpatentable only to the extent that it represents an abstract idea, this confusion may be ameliorated.

Today, we hold that the transformation of data, representing discrete dollar amounts, by a machine through a series of mathematical calculations into a final share price, constitutes a practical application of a mathematical algorithm, formula, or calculation, because it produces "a useful, concrete and tangible result"—a final share price momentarily fixed for recording and reporting purposes and even accepted and relied upon by regulatory authorities and in subsequent trades.

The district court erred by applying the *Freeman-Walter-Abele* test to determine whether the claimed subject matter was an unpatentable abstract idea....

After *Diehr* and *Chakrabarty,* the *Freeman-Walter-Abele* test has little, if any, applicability to determining the presence of statutory subject matter. As we pointed out in *Alappat,* 33 F.3d at 1543, application of the test could be misleading, because a process, machine, manufacture, or composition of matter employing a law of nature, natural phenomenon, or abstract idea is patentable subject matter even though a law of nature, natural phenomenon, or abstract idea would not, by itself, be entitled to such protection. The test determines the presence of, for example, an algorithm. Under *Benson,* this may have been a sufficient indicium of nonstatutory subject matter. However, after *Diehr* and *Alappat,* the mere fact that a claimed invention involves inputting numbers, calculating numbers, outputting numbers, and storing numbers, in and of itself, would not render it nonstatutory subject matter, unless, of course, its operation does not produce a "useful, concrete and tangible result." *Alappat,* 33 F.3d at 1544....

The question of whether a claim encompasses statutory subject matter should not focus on which of the four categories of subject matter a claim is directed to—process, machine, manufacture, or composition of matter—but rather on the essential characteristics of the subject matter, in particular, its practical utility. Section 101 specifies that statutory subject matter must also satisfy the other "conditions and requirements" of Title 35, including novelty, nonobviousness, and adequacy of disclosure and notice. See In re Warmerdam, 33 F.3d 1354, 1359 (Fed. Cir. 1994). For purpose of our analysis, as noted above, claim 1 is directed to a machine programmed with the Hub and Spoke software and admittedly produces a "useful, concrete, and tangible result." *Alappat,* 33 F.3d at 1544. This renders it statutory subject matter, even if the useful result is expressed in numbers, such as price, profit, percentage, cost, or loss.

COMMENTS AND QUESTIONS

1. Any thought that the new test announced in *State Street* was an aberration, or that it could easily be reconciled with prior precedent, was put to rest in AT&T v. Excel Communications, 172 F.3d 1352 (Fed. Cir. 1999). In that case the court held that method claims to a method for "generating a message record for an interexchange call" and recording who the call should be billed to, are patentable subject matter. The court applied *State Street*'s "useful, concrete and tangible result" test, concluding that the generation of billing records was clearly useful. The court specifically rejected the argument that a patentable software claim must have physical structure associated with it, noting that physical transformation is only one of several possible ways to bring about a useful result. In sustaining the patent, the court laid to rest much of the history of the mathematical algorithm exception:

A mathematical formula alone, sometimes referred to as a mathematical algorithm, viewed in the abstract, is considered unpatentable subject matter. Courts have used the terms "mathematical algorithm," "mathematical formula," and "mathematical equation," to describe types of nonstatutory mathematical subject matter without explaining whether the terms are interchangeable or different. Even assuming the words connote the same concept, there is considerable question as to exactly what the concept encompasses. See, e.g., *Diehr*, 450 U.S. at 186 n. 9 ("The term 'algorithm' is subject to a variety of definitions . . . [Petitioner's] definition is significantly broader than the definition this Court employed in *Benson* and *Flook*."); accord In re Schrader, 22 F.3d 290, 293 n. 5 (Fed. Cir. 1994).

This court recently pointed out that any step-by-step process, be it electronic, chemical, or mechanical, involves an "algorithm" in the broad sense of the term. Because § 101 includes processes as a category of patentable subject matter, the judicially-defined proscription against patenting of a "mathematical algorithm," to the extent such a proscription still exists, is narrowly limited to mathematical algorithms in the abstract. See also *Benson*, 409 U.S. at 65 (describing a mathematical algorithm as a "procedure for solving a given type of mathematical problem").

Since the process of manipulation of numbers is a fundamental part of computer technology, we have had to reexamine the rules that govern the patentability of such technology. The sea-changes in both law and technology stand as a testament to the ability of law to adapt to new and innovative concepts, while remaining true to basic principles. In an earlier era, the PTO published guidelines essentially rejecting the notion that computer programs were patentable. As the technology progressed, our predecessor court disagreed, and, overturning some of the earlier limiting principles regarding § 101, announced more expansive principles formulated with computer technology in mind. In our recent decision in State Street, this court discarded the so-called "business method" exception and reassessed the "mathematical algorithm" exception, see 149 F.3d at 1373-77, both judicially-created "exceptions" to the statutory categories of § 101. As this brief review suggests, this court (and its predecessor) has struggled to make our understanding of the scope of § 101 responsive to the needs of the modern world. . . .

The State Street formulation, that a mathematical algorithm may be an integral part of patentable subject matter such as a machine or process if the claimed invention as a whole is applied in a "useful" manner, follows the approach taken by this court en banc in In re Alappat, 33 F.3d 1526, 31 USPQ2d 1545 (Fed. Cir. 1994). . . . [T]he *Alappat* inquiry simply requires an examination of the contested claims to see if the claimed subject matter as a whole is a disembodied mathematical concept representing nothing more than a "law of nature" or an "abstract idea," or if the mathematical concept has been reduced to some practical application rendering it "useful." Id. at 1544. In *Alappat*, we held that more than an abstract idea was claimed because the claimed invention as a whole was directed toward forming a specific machine that produced the useful, concrete, and tangible result of a smooth waveform display. See id. at 1544.

In both *Alappat* and *State Street*, the claim was for a machine that achieved certain results. In the case before us, because Excel does not own or operate the facilities over which its calls are placed, AT&T did not charge Excel with infringement of its apparatus claims, but limited its infringement charge to the specified method or process claims. Whether stated implicitly or explicitly, we consider the scope of § 101 to be the same regardless of the form — machine or process — in which a particular claim is drafted.

Id at 23-24.

2. Related to the record-generation claim in *AT&T*, but potentially even more far-reaching, is a new type of patent claim to "propagated signals." These claims are

directed to "a manufactured transient phenomenon, such as an electrical, optical, or acoustical signal." Jeffrey R. Kuester et al., A New Frontier in Patents: Patent Claims to Propagated Signals, 17 J. Marshall J. Comp. & Info. L. 75 (1998). The claim would be directly infringed by the sending of data covered by the claim over computer wires, telephone lines, or the airwaves.

After *State Street,* is there any reason such a claim would not be patentable? Does it produce a useful, concrete, and tangible result?

3. The PTO revised its Examination Guidelines in late 1998 in response to the decision in *State Street,* adding several "training examples" dealing with business, artificial intelligence, and mathematical processing claims. The examples show a reluctance on the part of the PTO to abandon the physical transformation approach that characterized the case law through *Alappat.*

4. The Patent Office Board of Appeals and Interferences decided a case in 2005 that once again tested the limits of patentable subject matter. The case, In re Lundgren, 76 U.S.P.Q. 2D (BNA) 1385 (Bd. Pat. App & Interf. Apr 20, 2005), rejected a "technological arts" test for section 101. The claim in *Lundgren* was to a "method for compensating a manager" so as to prevent collusion among firms in an industry with a small number of competitors, i.e., an oligopoly. The basic idea is to tie managerial compensation to a performance measure based on comparison to the profitability of other firms in the industry, thereby reducing the gains from collusion. (This works because it sets up a "zero sum game" among managers in the industry, under which collusive bargaining will usually fail.) The invention was a pure business method, not one implemented in a computer or using any other technology. The majority opinion reversed an examiner's final rejection, stating:

> Our determination is that there is currently no judicially recognized separate "technological arts" test to determine patent eligible subject matter under § 101. We decline to propose to create one. Therefore, it is apparent that the examiner's rejection can not be sustained.

76 U.S.P.Q. 2D (BNA), at 1388. A lengthy concurrence and dissent, and a separate lengthy dissent, take issue with various features of the majority opinion. From the dissent:

> I would affirm the rejection, therefore, not because it is directed to a method of doing business, but rather, because the process as claimed is not tied to any known science or technology.... There is no science or technology associated with the claimed invention.... I find it ludicrous, however, to think that the writers of the Constitution would have found the idea of providing compensation to an executive, as claimed, to be something that would qualify for a patent.... If the majority simply wants to take comfort in the idea that all categories of nonstatutory subject matter have been established, and no new categories will be considered, then I disagree. The majority's position that essentially anything that can be claimed as a process is entitled to a patent under 35 U.S.C. § 101 opens the floodgate for patents on essentially any activity which can be pursued by human beings without regard to whether those activities have anything to do with the traditional sciences or whether they enhance the technological arts in any manner..... Unfortunately, the federal judiciary cannot get jurisdiction of this issue unless someone takes the issue to it. The majority has ensured that, at least in this case and probably the foreseeable future, the entity best capable of deciding the constitutionality question will not get a chance to consider it.... I cannot be concerned that an affirmance of the examiner's

rejection may imply that many other previously issued patents should not have been granted. It cannot possibly be good public policy to continue to issue invalid patents just to be consistent with the past.

76 U.S.P.Q. 2D (BNA), at 1389. Administrative Judge Smith goes on to say, "I only hope that this decision will open a public discourse on th[is] topic ....," id., at 1389 — a hope quite likely to come true in coming years.

5. As *Lundgren* shows, *State Street* is widely viewed as a seminal case. There is now a small mountain of commentary on the case, its holding, and particularly what it portends for the patent system. This note captures only a few of the highlights from this growing literature.

Most academics disagreed with the breadth of *State Street*. See, e.g., Rochelle Cooper Dreyfuss, Are Business Method Patents Bad for Business?, 16 Santa Clara Computer & High Tech. L.J. 263 (2000); Alan Durham, "Useful Arts" in the Information Age, 1999 B.Y.U. L. Rev. 1419; Leo Raskind, The State Street Bank Decision: The Bad Business of Unlimited Patent Protection for Methods of Doing Business, 10 Fordham Intellectual Prop. Media & Ent. L.J. 61 (1999). Others foresaw a flood of new patent applications predicted to reveal latent structural weaknesses in the patent system. Mark D. Janis, Inter Partes Patent Reexamination, 10 Fordham Intellectual Prop. Media & Ent. L.J. 481 (2000); Robert P. Merges, As Many as Six Impossible Patents before Breakfast: Property Rights for Business Concepts and Patent System Reform, 14 Berkeley Tech. L.J. 577 (1999).

Still others saw *State Street* as an opportunity to examine the foundations of patentable subject matter doctrine in the United States. These scholars divide into two camps: (1) those who see *State Street* as wrong and use it to define with greater precision inherent notions of traditional patentable inventions — i.e., "technology"; and (2) those who see *State Street* as a watershed in our expanding conception of what constitutes an economically valuable contribution.

The first group — the "hard technology" school — believes that "traditional" patentable subject matter can be defined and cordoned off from other subject matter such as business methods. Some simply assume this; for example, Professor Rochelle Dreyfuss asserts:

> The costs of business method patents are very high. The benefits, at least the traditional benefits, are low. The ratio is terrible. The case for patents on business methods is simply not there, at least not in general.... State Street now provides the opportunity to tie up ... knowledge for the future, to privatize it, and prevent it from leaking out to all users.

Dreyfuss, *supra*, 16 Santa Clara Computer & High Tech. L.J. at 276-277 (footnote omitted).

Professor John R. Thomas takes a similar tack but is much more explicit about how to divide patentable from unpatentable subject matter. The key for Thomas is that patents should protect only "technology." John R. Thomas, The Patenting of the Liberal Professions, 40 B.C. L. Rev. 1139 (1999). By clearly defining "technology," Thomas argues, we will have a defensible criterion by which to distinguish what is patentable from what is not. His definition is as follows:

> [T]echnological activities are concerned with the production or transformation of artifacts through the systematic manipulation of physical forces. Bounded by interaction with the external environment, technological activities expend resources and

knowledge in order to fabricate or modify products, or to develop procedural systems for so doing. Furthermore, technology presents a form of rational and systematic knowledge, oriented towards efficiency and capable of being assessed through objective criteria.

Id., at 1142. The "hard technologists" can also point to comparative law sources for support: the European Patent Office has long relied on notions such as "industrial use" and "technical effect" to resolve various doctrinal issues. See Note, Brian P. Biddinger, Limiting the Business Method Patent: A Comparison and Proposed Alignment of European, Japanese and United States Patent Law, 69 Fordham L. Rev. 2523 (2001).

On the other side of the debate stands the "soft technology" school. These scholars observe that our economy is increasingly driven by "information." As more of the value in the economy is contributed by software, marketing schemes, and internet trade and communication, they argue that the patent system must adjust. The core of their critique of the "hard technology" school is this: since patents were adopted in this country to cover the *most valuable* assets of the founding era, our era's patent system must be expanded to cover what is of value *today*. See, e.g., Richard S. Gruner, Intangible Inventions: Patentable Subject Matter for an Information Age, 35 Loy. L.A. L. Rev. 355 (2002). See also Note, Erik S. Maurer, An Economic Justification for a Broad Interpretation of Patentable Subject Matter, 95 Nw. U. L. Rev. 1057, 1058 (2001) (arguing that the "wealth-generating characteristics of innovation fundamentally justify a broad interpretation of patentable subject matter").

Thus, one way to view the debate is to ask whether "technology"—the patentable category created, but of course in no way fixed at the nation's founding—is still of paramount importance to the economy, or is for other reasons still deserving of unique legal status. If technology really is special, then the "hard technologists" have a case. Perhaps it is special because it still has distinctive qualities that require property rights to call it forth in the socially optimal amount. Perhaps it is special because it is easier to define and delimit than other value-adding components of the economy. Or perhaps the critics are right—the era is over when plows, axes, guns, and harvesters (and their "hard technology" descendants) make a special contribution to the economy.

This debate is not purely esoteric. A 2001 Patent Office case, for example, explicitly relied on the notion of the "technological arts" in affirming a rejection of claims to a method of creating a chart to represent the value of an asset and plotting points on the chart. After pointing out that the claims were not limited to a computer environment—and hence would cover hand-drawn charts—the Board of Appeals and Interferences noted:

> The phrase "technological arts" has been created to offer another view of the term "useful arts." The Constitution of the United States authorizes and empowers the government to issue patents only for inventions which promote the progress [of science and] the useful arts. We find that the invention before us, as disclosed and claimed, does not promote the progress of science and the useful arts, and does not fall within the definition of technological arts. The abstract idea which forms the heart of the invention before us does not become a technological art merely by the recitation in the claim of "transforming physical media into a chart" [sic, drawing or creating a chart] and "physically plotting a point on said chart."

Ex Parte Bowman, 61 U.S.P.Q.2d 1669, 1671 (Bd. Pat. App. & Interferences June 12, 2001).

The second practical application of the debate over proper subject matter pertains to the "first inventor" defense of 35 U.S.C. §273. This section makes an exception to general U.S. law in permitting a first inventor who does not file an application to continue using technology claimed in another inventor's later patent application (see Chapter 3). Section 273(a)(3) limits the defense to inventions claiming a "method of doing or conducting business," so the definition of this category will have immediate practical impact. If the "hard technology" school succeeds in establishing a notion of "traditional" patentable subject matter, perhaps this can be used to define (negatively, as it were) what constitutes non-traditional "business methods."

As a practical matter, software is now quite clearly patentable in the United States. See, e.g., Netscape Communications Corp. v. Konrad, 295 F.3d 1315 (Fed. Cir. 2002) (§101 not an issue for software invention patent infringement case). This is true also of business methods that happen to be implemented in software form. See, e.g., Ex Parte Baker, 2002 WL 465345 (Bd. Pat. App & Interf. 2002) (reversing rejection of application reciting claims to a method for choosing an optimized portfolio of equity stocks). But even though §101 is not a real barrier, software patents must still meet the other criteria for validity.

2. Examination and Validity of Software Patents

A valid patent must pass four other tests in examination in the Patent Office: it must be useful (35 U.S.C. §101), novel (35 U.S.C. §102), nonobvious (35 U.S.C. §103), and clearly described in the patent application (35 U.S.C. §112 ¶1). Utility is not a particularly strict test, and most computer programs that work will meet it easily. But each of the other steps in the patent examination process presents unique issues in the software context.

Since the advent of software patents, many programmers have been concerned that too many patents were issuing on programs that were in fact well known. Defenders of the patent system would often respond by arguing that the requirements of patentability were designed to weed out well-known inventions, and that the patent system would be able to sort out the old from the new. Another concern was that the software industry was simply not accustomed to dealing with patents, and the cost of doing so would be a significant burden on an industry that has been highly innovative in its brief history.

Software patents have become routine features on the landscape of the industry. Because of this, infringement litigation involving software inventions is growing. The reported cases are important not only for establishing the application of basic patent principles in the software industry, but also as a window into the older debate on the effect of software patents on the software industry.

a. Novelty and Statutory Bars

Netscape Communications Corp. v. Konrad, 295 F.3d 1315 (Fed. Cir. 2002) established two important patent principles in the software industry: (1) even informal software demonstrations may constitute "public use" of a software invention under 35 U.S.C. § 102(b) (and presumably by extension also § 102(a)'s "known or used" requirement); and (2) offers to allow others to use software even in a quasi-commercial setting may trigger the § 102(b) on-sale bar.

The *Netscape* case involved several patents issued to Konrad on systems that allow a computer user to access and search a database residing on a remote computer. Konrad began working as a staff scientist for the Lawrence Berkeley Laboratory in 1977, where he studied how an individual computer workstation user could obtain services from a remote computer. Beginning in 1990, Konrad successfully tested a remote database access system. The Federal Circuit affirmed a district court holding that a 1991 demonstration of the Konrad database constituted public use under § 102(b):

> Konrad argues that the district court erred in determining that his 1991 demonstration of . . . [a] remote database . . . to University of California computing personnel was an invalidating public use. He maintains that the invention disclosure [form] he submitted to the Lawrence Berkeley Laboratory patent department in October of 1990, established an expectation of confidentiality from [computing personnel] Roth and Peters. Netscape responds that the district court correctly determined that Konrad failed to establish a genuine issue of material fact that the demonstrated prototype was not in public use, and was prior art for the purpose of evaluating the validity of the patents in suit.
>
> We agree with Netscape. Konrad did not show that Roth or Peters were ever made aware of any requirement of confidentiality or even apprised of the invention disclosure forms that he submitted to the Lawrence Berkeley Laboratory patent department. He also did not make any discernable effort to inform the 1991 demonstration attendees of the requirement of confidentiality, or otherwise indicate to them that they would owe him a duty of confidentiality." . . . Lack of a confidentiality agreement is significant here because Roth and Peters were computer personnel who could easily demonstrate the invention to others.

295 F.3d at 1321.

The second issue in *Netscape* is the application of the on-sale bar. Here the facts are somewhat unusual. Konrad offered the system to another research laboratory in exchange for four months' salary. The court found this a commercial offer for sale, despite Konrad's assertion that both his main employer and the potential customer received Department of Energy funding, and hence they were not "separate commercial entities" for purposes of the on-sale bar. The court rejected this argument as well, reasoning that the shared funding source did not mean that the two labs were controlled by same entity. Id. at 1324.

COMMENTS AND QUESTIONS

1. *Guide to Action?* What does the *Netscape* case tell you about strategies for invalidating software patents? If *Netscape* is any indication, at least some programmers can be unsophisticated about patent issues. How will this affect efforts to invalidate software patents?

2. *Online Prior Art.* Although not directly concerned with "online" prior art, the case of In re Klopfenstein, 380 F.3d 1345 (Fed. Cir. 2004) signals a clear direction for

such cases in the future. That case involved the public display of graphical overhead slides and other data for two days as part of a "poster session" at a scientific meeting. The court held that the nonsecret display of the poster was a "printed publication" under 35 U.S.C. § 102(b). This suggests strongly the commonsense rule that the term "printed publication" will apply to information posted on the World Wide Web or other widely accessible networks, even if the posting is for a relatively brief period.

b. Nonobviousness

Lockwood v. American Airlines, Inc.
United States Court of Appeals for the Federal Circuit
107 F.3d 1565 (Fed. Cir. 1997)

LOURIE, Circuit Judge.

Lawrence B. Lockwood appeals from the final judgment of the United States District Court for the Southern District of California granting summary judgment in favor of American Airlines, Inc. In that summary judgment, the court held that (1) U.S. Patent Re. 32,115, U.S. Patent 4,567,359, and U.S. Patent 5,309,355 were not infringed by American's SABREvision reservation system, and that (2) the '355 patent and the asserted claims of the '359 patent were invalid under 35 U.S.C. § 102 and 35 U.S.C. § 103, respectively. Lockwood v. American Airlines, Inc., 834 F. Supp. 1246, 28 USPQ2d 1114 (S.D.Cal. 1993), *req. for reconsideration denied,* 847 F. Supp. 777 (S.D. Cal. 1994) (holding the '115 and '359 patents not infringed); Lockwood v. American Airlines, Inc., 877 F. Supp. 500 (S.D. Cal. 1994) (holding the asserted claims of the '355 patent invalid and not infringed); Lockwood v. American Airlines, Inc., 37 USPQ2d 1534, 1995 WL 822659 (S.D. Cal. 1995) (holding the '359 patent invalid). Because the district court correctly determined that there were no genuine issues of material fact in dispute and that American was entitled to judgment as a matter of law, we affirm.

Background

The pertinent facts are not in dispute. Lockwood owns the '115, '355, and '359 patents, all of which relate to automated interactive sales terminals that provide sales presentations to customers and allow the customers to order goods and services. Lockwood sued American asserting that American's SABREvision airline reservation system infringed all three patents. SABREvision is used by travel agents to access schedule and fare information, to book itineraries, and to retrieve photographs of places of interest, including hotels, restaurants, and cruises, for display to consumers. It improves upon American's SABRE reservation system, which originated in the 1960s and which cannot display photographs....

The '359 patent discloses a system of multiple interactive self-service terminals that provide audio-visual sales presentations and dispense goods and services from multiple institutions. Claim 1, the only independent claim, reads in pertinent part:

A system for automatically dispensing information, goods, and services for a plurality of institutions in a particular industry, comprising:

> ...
>> at least one customer sales and information terminal...
>> ...
>> said sales and information terminal including:
>> audio-visual means for interaction with a customer, comprising:
>> means for storing a sequence of audio and video information to be selectively transmitted
> to a customer;
>> means for transmitting a selected sequence of said stored information to the customer;
>> customer operated input means for gathering information from a customer....

The district court held that SABREvision did not infringe the '359 patent because it lacked the "audio-visual means" and "customer operated input means." The court also held the '359 patent invalid because it would have been obvious in light of the original SABRE system in combination with the self-service terminal disclosed in U.S. Patent 4,359,631, which issued in 1982 and was subsequently reissued as the '115 patent....

Discussion ...

A. Validity

The district court held that the asserted claims of the '359 patent would have been obvious in light of the prior art '631 patent and the original SABRE system. A determination of obviousness under 35 U.S.C. § 103 is a legal conclusion involving factual inquiries. Uniroyal, Inc. v. Rudkin-Wiley Corp., 837 F.2d 1044, 1050, 1438 (Fed. Cir. 1988). Lockwood argues that the subject matter of the '359 claims would not have been obvious and that the district court impermissibly drew adverse factual inferences in concluding that the patent was invalid. Lockwood first argues that the district court erred in concluding that the SABRE system qualified as prior art.

American submitted an affidavit averring that the SABRE system was introduced to the public in 1962, had over one thousand connected sales desks by 1965, and was connected to the reservation systems for most of the other airlines by 1970. Lockwood does not dispute these facts, but argues that because "critical aspects" of the SABRE system were not accessible to the public, it could not have been prior art. American's expert conceded that the essential algorithms of the SABRE software were proprietary and confidential and that those aspects of the system that were readily apparent to the public would not have been sufficient to enable one skilled in the art to duplicate the system. However, American responds that the public need not have access to the "inner workings" of a device for it to be considered "in public use" or "used by others" within the meaning of the statute.

We agree with American that those aspects of the original SABRE system relied on by the district court are prior art to the '359 patent. The district court held that SABRE, which made and confirmed reservations with multiple institutions (e.g., airlines, hotels, and car rental agencies), combined with the terminal of the '631 patent rendered the asserted claims of the '359 patent obvious. The terminal of the '631 patent admittedly lacked this "multiple institution" feature. It is undisputed, however, that the public was aware that SABRE possessed this capability and that the public had been using SABRE to make travel reservations from independent travel agencies prior to Lockwood's date of invention.

If a device was "known or used by others" in this country before the date of invention or if it was "in public use" in this country more than one year before the date of application, it qualifies as prior art. See 35 U.S.C. § 102(a) and (b) (1994). Lockwood attempts to preclude summary judgment by pointing to record testimony that one skilled in the art would not be able to build and practice the claimed invention without access to the secret aspects of SABRE. However, it is the claims that define a patented invention. See Constant v. Advanced Micro-Devices, Inc., 848 F.2d 1560, 1571 (Fed. Cir. 1988). As we have concluded earlier in this opinion, American's public use of the high-level aspects of the SABRE system was enough to place the claimed features of the '359 patent in the public's possession. See In re Epstein, 32 F.3d 1559, 1567-68 (Fed. Cir. 1994) ("Beyond this 'in public use or on sale' finding, there is no requirement for an enablement-type inquiry."). Lockwood cannot negate this by evidence showing that other, unclaimed aspects of the SABRE system were not publicly available. Moreover, the '359 patent itself does not disclose the level of detail that Lockwood would have us require of the prior art. For these reasons, Lockwood fails to show a genuine issue of material fact precluding summary judgment.

Lockwood further argues that even if the SABRE system is effective prior art, the combination of that system and the '631 patent would not have yielded the invention of the '359 patent. The terminal in the claims of the '359 patent includes a number of means-plus-function limitations, subject to 35 U.S.C. § 112, ¶6, including "means for gathering information from a customer" and "means for storing a sequence of audio and video information to be selectively transmitted to a customer." Means-plus-function clauses are construed "as limited to the corresponding structure[s] disclosed in the specification and equivalents thereof." In re Donaldson Co., 16 F.3d 1189, 1195 (Fed. Cir. 1994) (en banc); see 35 U.S.C. § 112, ¶6 (1994). Lockwood argues that the structures disclosed in the '359 patent differ substantially from the terminal disclosed in the '631 patent and that, at the very least, his expert's declaration raised genuine issues of material fact sufficient to preclude summary judgment.

We do not agree. We believe that American has met its burden, even in light of the presumption of patent validity, to show that the means limitations relating to the terminal in the claims of the '359 patent appear in the '631 specification. Lockwood has failed to respond by setting forth specific facts that would raise a genuine issue for trial. Specifically, Lockwood has not alleged that the '631 disclosure lacks the structures disclosed in the '359 patent specification or their equivalents. As the district court noted, Lockwood's expert, Dr. Tuthill, relied on structures that are not mentioned in either the '631 or the '359 patents. For example, Tuthill states that the claimed invention differs from the '631 patent because the terminal described in the '631 patent uses a "backward-chaining" system to solve problems while the '359 patent uses a "forward-chaining" system. Yet neither the '359 nor the '631 patents mentions backward- or forward-chaining. Nor does the '359 specification describe any hardware or software structure as being limited to any particular problem-solving technique. In addition, Lockwood argues that the hardware and software disclosed in the two patents are not equivalent to each other. However, the '359 patent claims the hardware and software in broad terms, and the patents both describe similar computer controlled self-service terminals employing video disk players that store and retrieve audio-visual information. For example, with regard to the "means for controlling said storage and transmitting means," Lockwood's expert avers that the "structure described in the '359 patent which corresponds to this means is the processor unit and the application program which the processor executes." Yet, the only software descriptions in the '359 patent consist of high level exemplary functional flowcharts. Lockwood's arguments and his expert's statements are thus conclusory. They fail to identify which structures in the '359 patent are thought to be missing from the '631 patent disclosure. Accordingly, we agree with the district court that Lockwood's and his expert's declarations have not adequately responded to American's motion by raising genuine issues of material fact, and we therefore conclude that the district court properly held the asserted claims of the '359 patent to have been obvious as a matter of law....

AFFIRMED.

COMMENTS AND QUESTIONS

1. The court holds that the SABRE system was "known or used by others" before the critical date of the '359 patent. Is affirmatively secret use sufficient to constitute prior art? Has the public really "known or used" the invention if they are prevented from seeing how it works or duplicating it? Does it matter whether American protected the SABRE system as a trade secret during the 1960s?

2. The court notes that Lockwood has not disclosed much more structural detail in his patent specification than the prior art '631 patent and therefore cannot claim to have differentiated his invention from that prior patent. Does this holding improperly conflate the enablement requirement and the test for obviousness?

3. Should programmers be able to patent new implementations of old solutions, for example the coding of a shorter version of a well-known computer algorithm (such as a "sort" routine) in a new computer language? Cf. Northern Telecom, Inc. v. Datapoint Corp., 908 F.2d 931, 941-42 (Fed. Cir. 1990) ("[T]he conversion of a complete thought...into a language a machine understands is necessarily a mere clerical function to a skilled programmer." (quoting In re Sherwood, 613 F.2d 809, 817 n.6 (C.C.P.A. 1980)).) For a discussion of why the courts should be wary of overly broad interpretation of patent claims in the software area, see Julie E. Cohen and Mark A. Lemley, Patent Scope and Innovation in the Software Industry, 89 Cal. L. Rev. 1 (2001).

4. Software and business method patents have been criticized for ignoring much of the most relevant prior art.[13] Knowledgeable observers of the patent system offer at least three explanations for these concerns. First, because software was not clearly patentable subject matter until the mid-1990s, the Patent Office was slow to hire

___

[13]. It is not clear that software patents are worse than others in this regard. See John R. Allison & Mark A. Lemley, Who's Patenting What? An Empirical Exploration of Patent Prosecution, 53 Vand. L. Rev. 2099 (2000) (comprehensive study finding that software patents cite more non-patent prior art than average).

patent examiners qualified in computer software or related fields. During the 1980s and the early part of the 1990s, the flood of software patent applications was handled largely by people operating outside their area of expertise. Abundant evidence indicates that the Patent Office has issued software patents on a number of applications during this period that did not meet the standard tests of novelty and nonobviousness. For anecdotes discussing some of the more extreme examples, see Simson L. Garfinkel, Patently Absurd, Wired, July 1994, at 104.

Second, for similar reasons, the PTO's classification system is not equipped to handle software patents. As a result, software patents tend to be classified according to the field in which the software will ultimately be used, rather than according to the nature of the software invention. This makes it much harder for examiners to find what prior art does exist. This problem too could be solved, either by reclassification or by an increased emphasis on computer search systems.

The final problem with prior art in the software field is more enduring: prior art in this particular industry may be difficult or, in some cases, impossible to find. As Julie Cohen explains:

> [I]n the field of computers and computer programs, much that qualifies as prior art lies outside the areas in which the PTO has traditionally looked — previously issued patents and previous scholarly publications. Many new developments in computer programming are not documented in scholarly publications at all. Some are simply incorporated into products and placed on the market; others are discussed only in textbooks or user manuals that are not available to examiners on line. In an area that relies so heavily on published, "official" prior art, a rejection based on "common industry knowledge" that does not appear in the scholarly literature is unlikely. Particularly where the examiner lacks a computer science background, highly relevant prior art may simply be missed. In the case of the multimedia data retrieval patent granted to Compton's New Media,[14] industry criticism prompted the PTO to reexamine the patent and ultimately to reject it because it did not represent a novel and nonobvious advance over existing technology. However, it would be inefficient, and probably impracticable, to reexamine every computer program-related patent, and the PTO is unlikely to do so.

Julie E. Cohen, Reverse Engineering and the Rise of Electronic Vigilantism: Intellectual Property Implications of "Lock-Out" Technologies, 68 S. Cal. L. Rev. 1091, 1179 (1995). The current PTO Examination Guidelines for Computer-Related Inventions direct that examiners must "conduct a thorough search of the prior art," but they give no indication of how to accomplish that task.

5. The first case involving the obviousness of a software-implemented invention is, perhaps surprisingly, a Supreme Court case from the 1970s. In Dann v. Johnston, 425 U.S. 219 (1976), the Court held a patent on a "machine system for automatic record-keeping of bank checks and deposits" invalid for obviousness. The Court took a rather broad view of obviousness in the computer industry, focusing on whether analogous systems to the patentee's had been implemented in computer before, rather than analyzing the precise differences between the patentee's program and the prior

---

14. Compton's received a patent on an application filed in the late 1980s which purported to cover basic use of hypertext. When Compton's announced the existence of the patent at a computer trade show, and offered everyone in the multimedia industry a chance to license it, public outrage prompted the PTO to initiate a reexamination proceeding sua sponte. The PTO rejected the application after considering a number of prior art references submitted by third parties. — EDS.

art programs. The clear implication of the opinion is that if a reasonably skilled programmer could produce a program analogous to the patented one, and if there was motivation in the prior art to do so, the patented program is obvious.

On the other hand, the Federal Circuit in In re Zurko, 111 F.3d 887 (Fed. Cir. 1997), held that a patented invention was nonobvious even though each of the elements of the invention could be found in the prior art, where the prior art did not identify the problem to be solved. This offers another outlet for patentees: they can focus their invention on solving a previously unknown problem rather than on the particular elements of their solution.

≡ *Amazon.com, Inc, v. barnesandnoble.com, inc.*
≡ **United States Court of Appeals for the Federal Circuit**
≡ **239 F.3d 1343 (Fed. Cir. 2001)**

CLEVENGER, Circuit Judge.

This is a patent infringement suit brought by Amazon.com, Inc. ("Amazon") against barnesandnoble.com, inc., and barnesandnoble.com LLC (together, "BN"). Amazon moved for a preliminary injunction to prohibit BN's use of a feature of its web site called "Express Lane." BN resisted the preliminary injunction on several grounds, including that its Express Lane feature did not infringe the claims of Amazon's patent, and that substantial questions exist as to the validity of Amazon's patent. The United States District Court for the Western District of Washington rejected BN's contentions [and granted a preliminary injunction.] BN brings its timely appeal from the order entering the preliminary injunction.

After careful review of the district court's opinion, the record, and the arguments advanced by the parties, we conclude that BN has mounted a substantial challenge to the validity of the patent in suit. Because Amazon is not entitled to preliminary injunctive relief under these circumstances, we vacate the order of the district court that set the preliminary injunction in place and remand the case for further proceedings.

I

This case involves United States Patent No. 5,960,411 ("the '411 patent"), which issued on September 28, 1999, and is assigned to Amazon....

The '411 patent describes a method and system in which a consumer can complete a purchase order for an item via an electronic network using only a "single action," such as the click of a computer mouse button on the client computer system. Amazon developed the patent to cope with what it considered to be frustrations presented by what is known as the "shopping cart model" purchase system for electronic commerce purchasing events. In previous incarnations of the shopping cart model, a purchaser using a client computer system (such as a personal computer executing a web browser program) could select an item from an electronic catalog, typically by clicking on an "Add to Shopping Cart" icon, thereby placing the item in the "virtual" shopping cart. Other items from the catalog could be added to the shopping cart in the same manner. When the shopper completed the selecting process, the electronic commercial event would move to the check-out counter, so to speak. Then, information regarding the purchaser's identity, billing and shipping addresses, and credit payment method would be inserted into the transactional information base by the soon-to-be purchaser. Finally, the purchaser would "click" on a button displayed on the screen or somehow issue a command to execute the completed order, and the server computer system would verify and store the information concerning the transaction.

As is evident from the foregoing, an electronic commerce purchaser using the shopping cart model is required to perform several actions before achieving the ultimate goal of the placed order. The '411 patent sought to reduce the number of actions required from a consumer to effect a placed order.... How, one may ask, is the number of purchaser interactions reduced? The answer is that the number of purchaser interactions is reduced because the purchaser has previously visited the seller's web site and has previously entered into the database of the seller all of the required billing and shipping information that is needed to effect a sales transaction. Thereafter, when the purchaser visits the seller's web site and wishes to purchase a product from that site, the patent specifies that only a single action is necessary to place the order for the item. In the words of the written description, "once the description of an item is displayed, the purchaser need only take a single action to place the order to purchase that item." Col. 3, ll. 64-66.

The '411 patent has 26 claims, 4 of which are independent.... Although there are significant differences among the various independent and dependent claims in issue, for purposes of this appeal we may initially direct our primary focus on the "single action" limitation that is included in each claim. This focus is appropriate because BN's appeal attacks the injunction on the grounds that either its accused method does not infringe the "single action" limitation present in all of the claims, that the "single action" feature of the patent is invalid, or both.

We set forth below [claim 1], with emphasis added to highlight the disputed claim terms:

> 1. A method of placing an order for an item comprising: under control of a client system, displaying information identifying the item; and *in response to only a single action being performed*, sending a request to order the item along with an identifier of a purchaser of the item to a server system; under control of *a single-action ordering component* of the server system, receiving the request; retrieving additional information previously stored for the purchaser identified by the identifier in the received request; and generating an order to purchase the requested item for the purchaser identified by the identifier in the received request using the retrieved additional information; and *fulfilling the generated order* to complete purchase of the item whereby the item is ordered without using a *shopping cart ordering model*.

The district court interpreted the key "single action" claim limitation, which appears in each of the pertinent claims, to mean:

> The term "single action" is not defined by the patent specification.... As a result, the term "single action" as used in the '411 patent appears to refer to one action (such as clicking a mouse button) that a user takes to purchase an item once the following information is displayed to the user: (1) a description of the item; and (2) a description of the single action the user must take to complete a purchase order for that item.

With this interpretation of the key claim limitation in hand, the district court turned to BN's accused ordering system. BN's short-cut ordering system, called "Express

Lane," like the system contemplated by the patent, contains previously entered billing and shipping information for the customer. In one implementation, after a person is presented with BN's initial web page (referred to as the "menu page" or "home page"), the person can click on an icon on the menu page to get to what is called the "product page." BN's product page displays an image and a description of the selected product, and also presents the person with a description of a single action that can be taken to complete a purchase order for the item. If the single action described is taken, for example by a mouse click, the person will have effected a purchase order using BN's Express Lane feature. . . .

[The court concluded that under a proper claim interpretation, Amazon had made the showing that it is likely to succeed at trial on its infringement case. The court then turned to the issue of patent validity.]

The district court considered, but ultimately rejected, the potentially invalidating impact of several prior art references cited by BN. [W]e find that the district court committed clear error by misreading the factual content of the prior art references cited by BN and by failing to recognize that BN had raised a substantial question of invalidity of the asserted claims in view of these prior art references.

Validity challenges during preliminary injunction proceedings can be successful, that is, they may raise substantial questions of invalidity, on evidence that would not suffice to support a judgment of invalidity at trial. See, e.g., Helifix Ltd. v. Blok-Lok, Ltd., 208 F.3d 1339, 1352 (Fed. Cir. 2000). The test for invalidity at trial is by evidence that is clear and convincing. WMS Gaming, Inc. v. Int'l Game Tech., 184 F.3d 1339, 1355 (Fed. Cir. 1999). To succeed with a summary judgment motion of invalidity, for example, the movant must demonstrate a lack of genuine dispute about material facts and show that the facts not in dispute are clear and convincing in demonstrating invalidity. In resisting a preliminary injunction, however, one need not make out a case of actual invalidity. Vulnerability is the issue at the preliminary injunction stage, while validity is the issue at trial. The showing of a substantial question as to invalidity thus requires less proof than the clear and convincing showing necessary to establish invalidity itself. That this is so is plain from our cases. . . .

When the heft of the asserted prior art is assessed in light of the correct legal standards, we conclude that BN has mounted a serious challenge to the validity of Amazon's patent. We hasten to add, however, that this conclusion only undermines the prerequisite for entry of a preliminary injunction. Our decision today on the validity issue in no way resolves the ultimate question of invalidity. That is a matter for resolution at trial. It remains to be learned whether there are other references that may be cited against the patent, and it surely remains to be learned whether any shortcomings in BN's initial preliminary validity challenge will be magnified or dissipated at trial. All we hold, in the meantime, is that BN cast enough doubt on the validity of the '411 patent to avoid a preliminary injunction, and that the validity issue should be resolved finally at trial.

One of the references cited by BN was the "CompuServe Trend System." The undisputed evidence indicates that in the mid-1990s, CompuServe offered a service called "Trend" whereby CompuServe subscribers could obtain stock charts for a surcharge of 50 cents per chart. Before the district court, BN argued that this system anticipated claim 11 of the '411 patent. The district court failed to recognize the substantial question of invalidity raised by BN in citing the CompuServe Trend reference, in that this system appears to have used "single action ordering technology" within the scope of the claims in the '411 patent.

First, the district court dismissed the significance of this system partly on the basis that "[t]he CompuServe system was not a world wide web application." This distinction is irrelevant, since none of the claims mention either the Internet or the World Wide Web (with the possible exception of dependent claim 15, which mentions HTML, a program commonly associated with both the Internet and the World Wide Web). Moreover, the '411 patent specification explicitly notes that "[o]ne skilled in the art would appreciate that the single-action ordering techniques can be used in various environments other than the Internet." Col. 6, ll. 22-24.

More importantly, one of the screen shots in the record (reproduced [in Figure 7-2]) indicates that with the CompuServe Trend system, once the "item" to be purchased (*i.e.*, a stock chart) has been displayed (by typing in a valid stock symbol), only a single action (*i.e.*, a single mouse click on the button labeled "Chart ($.50)") is required to obtain immediate electronic delivery (*i.e.*, "fulfillment") of the item. Once the button labeled "Chart ($.50)" was activated by a purchaser, an electronic version of the requested stock chart would be transmitted to the purchaser and displayed on the purchaser's computer screen, and an automatic process to charge the purchaser's account 50 cents for the transaction would be initiated. In terms of the language of claims 2 and 11 in the CompuServe Trend system, the item to be ordered is "displayed" when the screen echoes back the characters of the stock symbol typed in by the purchaser before clicking on the ordering button.

The evidence before us indicates that the billing process for the electronic stock chart would not actually commence until the client system sent a message to the server system indicating that the electronic stock chart had been received at the client system.

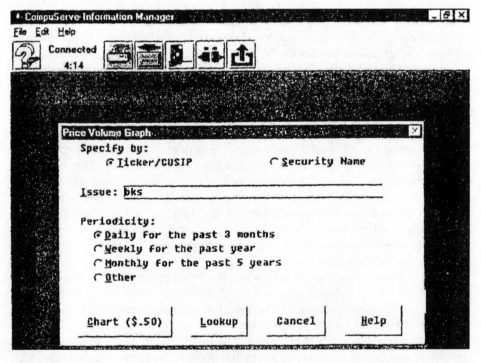

FIGURE 7-2
The CompuServe Prior Art.

In its brief, Amazon argues that this feature of the CompuServe Trend system amounts to an additional "confirmation step necessary to complete the ordering process," and that the CompuServe Trend system therefore does not use "single action" technology within the scope of the claims in the '411 patent. However, all of the claims only require sending a *request* to order an item in response to performance of only a single action. In the CompuServe Trend system, this requirement is satisfied when a purchaser performs the single action of "clicking" on the button labeled "Chart ($.50)." The claims do not require that the billing process for the item must also be initiated in response to performance of the single action. Furthermore, in the CompuServe Trend system, the "action" of sending a message from the client system to the server system confirming successful reception of the electronic stock chart is performed automatically, without user intervention.

At oral argument, Amazon's counsel articulated three differences between the CompuServe Trend system and the claimed invention. First, Amazon's counsel repeated the district court's reasoning, and asserted that the CompuServe Trend system is not on the Internet or the World Wide Web. As mentioned above, the '411 patent specification indicates that this distinction is irrelevant.

Second, Amazon's counsel claimed that the CompuServe Trend system was different from the claims of the '411 patent because it required a user to "log in" at the beginning of each session, and therefore would not send the claimed "identifier" along with a request to purchase each item. However, claim 11 does not require transmission of an identifier along with a request to order an item. This requirement is found only in claims 1, 6, and 9, and their respective dependent claims.

On its face, the CompuServe Trend reference does not mention transmission of the claimed identifier along with a request to purchase each item. Nor does the evidence in the record at this stage indicate that the CompuServe Trend system transmitted such an identifier. BN has therefore not demonstrated that the Compu-Serve Trend reference anticipates the asserted claims of the '411 patent requiring transmission of such an identifier with the degree of precision necessary to obtain summary judgment on this point. However, as noted above, validity challenges during preliminary injunction proceedings can be successful on evidence that would not suffice to support a judgment of invalidity at trial. See *Helifix,* 208 F.3d at 1352. The record in this case is simply not yet developed to the point where a determination can be made whether the CompuServe Trend system transmits the claimed identifier along with a request to order an item, or whether this limitation is obvious in view of the prior art. For example, United States Patent No. 5,708,780 ("the '780 patent") (a reference cited by BN which is discussed more fully below), describes "forwarding a service request from the client to the server and appending a session identification (SID) to the request and to subsequent service requests from the client to the server within a session of requests." See '780 patent, col. 3, ll. 12-16.

Moreover, the '411 patent specification itself dismisses the distinction between ordering systems in which an identifier is transmitted along with each request to order an item, and systems in which a user logs in once at the beginning of each session. See '411 patent at col. 10, ll. 6-10 ("[T]he purchaser can be alternatively identified by a unique customer identifier that is provided by the customer when the customer initiates access to the server system and sent to the server system with each message.").

The final distinction drawn by Amazon's counsel between the claimed invention and the CompuServe Trend system was that—according to Amazon—the *only* reason that a purchaser would "call up" the screen would be to actually order an electronic stock chart, and that therefore an earlier action taken by a purchaser to invoke the screen should count as an extra purchaser action. According to this argument, the CompuServe Trend system would not meet the "single action" limitation because at least two actions would need to be taken to order an item: one action to invoke the ordering screen, and a second action to click on the ordering button. However, as the screen shot plainly indicates, a purchaser could use the display screen for purposes other than to order an electronic stock chart (*e.g.*, to "Lookup" a stock symbol). Furthermore, to the extent that Amazon argues that the CompuServe Trend fails to meet the "single action" limitation due to the "click" necessary to activate the stock chart ordering screen in the first place, Amazon also admits that BN's Express Lane feature fails to meet the same limitation because of the "click" required to proceed from a menu page to a product page when using the Express Lane feature.

As the CompuServe Trend stock chart ordering screen indicates, we note that once a purchaser types in a valid stock symbol, the screen displays both "information identifying the item" (*i.e.*, the stock symbol identifying the desired electronic stock chart) *and* an indication of the "single action" to be performed to order the identified item (*i.e.*, clicking on the button labeled "Chart ($.50)"). Therefore, the substantial question of invalidity raised by the CompuServe Trend reference is the same regardless of whether one considers claims explicitly requiring that both of these pieces of information be displayed (*i.e.*, claims 2 and 11) or claims requiring that only the "information identifying the item" be displayed (*i.e.*, claims 1, 6, and 9).

In view of the above, we conclude that the district court erred in failing to recognize that the CompuServe Trend reference raises a substantial question of invalidity. Whether the CompuServe Trend reference either anticipates and/or renders obvious the claimed invention in view of the knowledge of one of ordinary skill in the relevant art is a matter for decision at trial.

[The court also considered several other pieces of prior art that raised substantial questions of validity.]

Conclusion

While it appears on the record before us that Amazon has carried its burden with respect to demonstrating the likelihood of success on infringement, it is also true that BN has raised substantial questions as to the validity of the '411 patent. For that reason, we must conclude that the necessary prerequisites for entry of a preliminary injunction are presently lacking. We therefore vacate the preliminary injunction and remand the case for further proceedings.

Vacated and remanded.

COMMENTS AND QUESTIONS

1. The district court's preliminary injunction was issued on December 1, 1999 — i.e., right in the heart of the Christmas gift-buying season. See Amazon.com v. barnesandnoble.com, 73 F. Supp. 2d 1228 (W.D. Wash. 1999), *vacated and*

remanded, 239 F.3d 1343 (Fed. Cir. 2001). At the time, the case caused quite a stir, with one popular commentator claiming: "When 21st-century historians look back at the breakdown of the United States patent system, they will see a turning point in the case of Jeff Bezos and Amazon.com and their special invention: 'The patented One Click® feature,' as Bezos calls it." James Gleick, Patently Absurd, N.Y. Times Magazine, Mar. 12, 2000, at 44.

One year after the Federal Circuit vacated the preliminary injunction, the parties settled the case without disclosing the terms of their agreement. After the settlement, BN's Express Checkout option still required two clicks, with the second click confirming the order.

2. What is the status of the One-Click patent after the decision? Given the procedural posture of the case—it was only the preliminary injunction stage—did the court actually conclude that the patent was invalid?

c. Enablement of Software Inventions

35 U.S.C. §112, ¶1

The specification shall contain a written description of the invention, and of the manner and process of making and using it, in such full, clear, concise and exact terms as to enable any person skilled in the art to which it pertains, or with which it is most nearly connected, to make and use the same, and shall set forth the best mode contemplated by the inventor of carrying out his invention.

In the early days of software patenting, the primary issue under section 112, the disclosure provision of the patent code, was whether a patent applicant must deposit source code to meet the enablement (i.e., "how to make and use") requirement of section 112, ¶1. Two cases defined the terms of this debate. In White Consolidated Industries, Inc. v. Vega Servo-Control, Inc., 713 F.2d 788 (Fed. Cir. 1983), the Federal Circuit invalidated a patent for a machine tool control system that was run by a computer program. Part of the invention was a programming language translator designed to convert an input program into machine language so that the system could then execute the program. The patent specification identified an example of a translator program, the so-called SPLIT program, which was a trade secret of Sundstrand, a company that later became the plaintiff White Consolidated. When the application was filed, the SPLIT program was available exclusively from Sundstrand.

The defendant, Vega, asserted patent invalidity as a defense to the infringement claim. Specifically, Vega argued that the SPLIT program was the only suitable translator program, and merely identifying it was not sufficient to meet the standards of section 112. White claimed that widely available equivalent translators could also be used and that in any event the specification described the characteristics of the necessary translator program in terms that enabled a programmer of ordinary skill to create it from scratch.

The court held that the program translator was an integral part of the invention and that mere identification of it was not sufficient to discharge the applicant's duty under section 112. The court seemed concerned that maintaining the translator program as a trade secret would allow White to extend the patent beyond the statutory term.

In Northern Telecom, Inc. v. Datapoint Corp., 908 F.2d 931 (Fed. Cir.), *cert. denied*, 498 U.S. 920 (1990), the Federal Circuit once again confronted the enablement requirement in the context of an invention containing a computer program. In *Northern Telecom*, the court reversed the district court's holding of invalidity on enablement grounds. Here the patent claimed an improved method of entering, verifying, and storing (or "batching") data with a special data entry terminal. The district court invalidated certain claims of the patent on the grounds that they were inadequately disclosed under section 112.

In a decision reversing this aspect of the district court's ruling, the Federal Circuit held that when claims pertain to a computer program that implements a claimed device or method, the enablement requirement varies according to the nature of the claimed invention as well as the role and complexity of the computer program needed to implement it. Under the facts in this case, the core of the claimed invention was the combination of components or steps rather than the details of the program the applicant actually used. The court noted expert testimony that various programs could be used to implement the invention, and that it would be "relatively straightforward [in light of the specification] for a skilled computer programmer to design a program to carry out the claimed invention." Id. at 941-42. The court continued:

> The computer language is not a conjuration of some black art, it is simply a highly structured language.... [T]he conversion of a complete thought (as expressed in English and mathematics, i.e. the known input, the desired output, the mathematical expressions needed and the methods of using those expressions) into a language a machine understands is necessarily a mere clerical function to a skilled programmer.

Id. The court refused to state categorically that a program listing would never have to be disclosed, however, especially in cases such as *White*, where it was estimated that it would take two years for a skilled programmer to produce a working program of the type called for by the specification.

While the courts have given software patentees broad leeway under the enablement doctrine to write broad functional claims without disclosing source code or implementation details, those broad claims may turn out to be vulnerable under a different doctrine: written description. The 2005 case of Lizardtech, Inc. v. Earth Resource Mapping, Inc., 424 F.3d 1336 (Fed. Cir. 2005), demonstrates the risks. In that case, the patentee Lizardtech had sued defendant Earth Resource Mapping (ERM) for infringement of Lizardtech's patents on software for compressing and displaying digital images. The software invention at issue was for a particular type of "discrete wavelet transform," or DWT, a mathematical tool for compressing and displaying digital images. The claimed DWT method was said to solve a problem in the prior art encountered when digital images were broken into subparts and then reassembled into "tiles." The edges of tiles were distorted in the final image because of the way the data was stored and reassembled. Lizardtech solved the problem by changing the way the DWT operated, thereby creating what it called a "seamless" DWT image. ERM argued, however, that Lizardtech's broad generic claim to DWT transforms (claim 21 in its patent) was invalid under the "written description" requirement of section 112.

> The trouble with allowing claim 21 to cover all ways of performing DWT-based compression processes that lead to a seamless DWT is that there is no support for such a broad

claim in the specification. The specification provides only a single way of creating a seamless DWT, which is by maintaining updated sums of DWT coefficients. There is no evidence that the specification contemplates a more generic way of creating a seamless array of DWT coefficients.

424 F.3d 1336, at 1345.

For this reason, the court found the patent invalid. The court also rooted its holding squarely in the "possession test" of section 112, noting in particular that claim 21 was fully enabled under section 112. 424 F.3d at 1346. Thus *Lizardtech* brings to the software field the same (disputed) issue that has concerned the biotechnology industry, viz., the precise scope and effect of an aggressive "written description" requirement.

COMMENTS AND QUESTIONS

1. Failure to disclose the specifics of a program that would take two years to re-create renders a patent invalid under section 112; what if the program would take one year to re-create? How about six months? One month?

2. The rule that actual program code need not be disclosed in a patent specification is now well accepted. See, e.g., Ex Parte Tsu-Chang Lee, 2000 WL 33741050 (Bd. Pat. App & Interf. Nov 30, 2000), at 3:

> The examiner maintains that the specification fails to disclose the program in some form and therefore, the claim is not supported by the specification.... [T]he examiner maintains that the skilled artisan would not only have to be skilled in multiple arts, but that a solution implementing multiple disciplines would require entirely too much experimentation to implement the claimed invention. We disagree with the examiner that the solution would require undue experimentation. Appellant argues... that the specification is enabling and that the actual computer program code is not necessary to enable the claimed invention. We agree with appellant.

3. Even where disclosure is adequate under the enablement standard, it can run into problems under the "best mode" aspect of section 112. An inventor runs afoul of the best mode requirement if she intentionally withholds from the patent specification information that discloses the best means of which she is aware for practicing her invention. Does the best mode requirement compel the disclosure of source code, on the theory that the actual way in which the programmer has implemented an invention must be the "best" way known to the inventor? See In re Hayes Microcomputer Products, Inc. Patent Litigation, 982 F.2d 1527 (Fed. Cir. 1992) (maker of Hayes modems was not required to disclose secret settings used on those modems in patent application, because settings were arbitrarily chosen).

In Fonar v. General Elec. Co., 107 F.3d 1543 (Fed. Cir. 1997), the Federal Circuit not only concluded that source code need not be disclosed but decided that even flowcharts were unnecessary—a broad description of what the software did was sufficient:

> As a general rule, where software constitutes part of a best mode of carrying out an invention, description of such a best mode is satisfied by a disclosure of the functions of the software. This is because, normally, writing code for such software is within the skill

of the art, not requiring undue experimentation, once its functions have been disclosed. It is well established that what is within the skill of the art need not be disclosed to satisfy the best mode requirement as long as that mode is described. Stating the functions of the best mode software satisfies that description test. We have so held previously and we so hold today. See In re Hayes Microcomputer Prods., Inc. Patent Litigation, 982 F.2d 1527, 1537-38 (Fed. Cir. 1992); In re Sherwood, 613 F.2d 809, 816-17 (CCPA 1980). Thus, flow charts or source code listings are not a requirement for adequately disclosing the functions of software. See *Sherwood,* 613 F.2d at 816-17. Here, substantial evidence supports a finding that the software functions were disclosed sufficiently to satisfy the best mode requirement. See *Hayes,* 982 F.2d at 1537 (stating that there was no best mode violation where the specification failed to disclose a firmware listing or flow charts, but did disclose sufficient detail to allow one skilled in the art to develop a firmware listing for implementing the invention).

Id. at 1549.

Most software patent cases have taken a similar position. In two cases decided in 1997, the Federal Circuit concluded that the patentees satisfied the written description and best mode requirements for inventions partially implemented in software even though they did not use the terms "computer" or "software" anywhere in the specification! See Robotic Vision Systems, Inc. v. View Engineering, Inc., 112 F.3d 1163 (Fed. Cir. 1997) (it would be "plainly apparent" to one of ordinary skill in the art that software would be required, and it was not necessary to further disclose the nature of the code actually used); In re Dossel, 115 F.3d 942 (Fed. Cir. 1997) (means-plus-function claim which includes a "means for reconstructing data" is obviously implemented in software and so disclosure of that fact is not required; software program itself did not have to be disclosed). For a recent summary of practitioner issues, see William F. Heinze, A Risk-Balancing Approach to Best Mode Disclosure in Software Patent Applications, 84 J. Pat. & Trademark Off. Soc'y 40 (2002).

Would Lizardtech have been better off not disclosing how they performed the DWT at all? Wouldn't a broad description of the DWT suffice under the earlier Federal Circuit cases?

3. Infringement

≡≡≡ *British Telecommunications Plc. v. Prodigy*
Communications Corp.
United States District Court for the Southern District of New York
217 F. Supp. 2d 399 (S.D.N.Y. 2002)

MCMAHON, J.

Plaintiff British Telecommunications ("BT") asserts that Defendant Prodigy Communications Corp. ("Prodigy"), through its business activities as an Internet Service Provider ("ISP"), directly infringes claims 3, 5, 6, and 7 (the "Asserted Claims") of U.S. Patent No. 4,873,662 (the "Sargent Patent" or "'662 Patent"). BT also alleges that Prodigy induces and contributes to infringement by Prodigy subscribers who infringe the Sargent patent by accessing the Internet through the Prodigy service.

The Court has already construed the claims of the patent in its *Markman* Opinion. British Telecommunications PLC v. Prodigy Communications Corp., 189 F. Supp.2d 101 (S.D.N.Y.2002) ("Markman Op.")....[Claim 3 of the patent at issue (the Sargent patent) reads as follows, with key terms highlighted:

3. A digital information storage, retrieval and display system comprising:

> *a central computer means* in which *plural blocks of information* are stored at respectively corresponding locations each of which locations is designated by a predetermined address therein by means of which a block can be selected, each of said blocks comprising *a first portion containing information for display and a second portion containing information not for display* but including *the complete address* for each of plural other blocks of information;

plural remote terminal means, each including (a) modem means for effecting input/output digital data communications with said central computer means via the telephone lines of a telephone network, (b) local memory means for locally storing digital data representing at least the first portion of the selected block of information received via said modem means from the central computer and for processing digital data, (c) display means for visually displaying such a locally stored first portion of a block of information and (d) keypad means connected to communicate data to at least said local memory means for manual entry of keyed digital data; and further memory means being provided as part of said local memory means at each of said remote terminal means for receiving and storing said second portion of the selected block of information in response to the selection of the block and when its respective first portion is transmitted thereto, said local memory means utilizing keyed digital data of less extent than any one of said complete addresses for another block of information but nevertheless uniquely indicative of one of the complete addresses contained in said second portion of the block of information which contains the first portion then being displayed for selectively accessing said further memory means and for supplying data to be transmitted by said modem means and indicative of the complete address of the next block of information which is to be retrieved and utilized for display purposes.]

For the reasons stated below, I find that as a matter of law, no jury could find that Prodigy infringes the Sargent patent, nor that Prodigy contributes to infringement of the Sargent patent, nor actively induces others to infringe that patent. I therefore grant Prodigy's motion for summary judgment....

BT argues that Prodigy infringes the Sargent patent through its business activities as an Internet Service Provider. BT contends that Prodigy's web servers provide access to information in a manner that literally infringes the Sargent patent. BT also alleges that the Internet infringes the Sargent patent and that Prodigy facilitates infringement by its subscribers by providing them with access to the Internet. BT contends that Prodigy contributorily infringes or actively induces the infringement of the Sargent patent by providing the necessary software and encouraging its subscribers to access pages of information from Web servers maintained by third parties. Therefore, BT argues, even if Prodigy's servers do not infringe the Sargent patent as a matter of law, summary judgment should be denied because Prodigy infringes the '662 patent by making and using infringing remote terminals....

Prodigy is an Internet Service Provider ("ISP") that has supplied its customers with access to the Internet since October, 1996. Prodigy's services include dial-up access, and broadband DSL access service....

When a user dials in to the Prodigy system, the user's computer is assigned an IP address dynamically by Prodigy. An IP address is a unique binary number assigned to an interface connection of a computer to the Internet and is used by the other computers to send packets using the IP protocol to this computer. The member can then use the Web browser to retrieve Web pages from the Prodigy Web site or from other Web servers connected to the Internet....

BT contends that:

1. Each Web server on the Internet is a "central computer" as defined in the Sargent patent because each Web server has its own centralized data store.
2. HTML files qualify as "blocks of information" either literally or under the doctrine of equivalents.
3. Each URL address is a "complete address" within the meaning of the Court's construction of the term, either literally or under the doctrine of equivalents.

BT makes other arguments but it is not necessary to reach them because its failure to raise any disputed issue of fact as to these points means that Prodigy is entitled to judgment as a matter of law.

I. The Internet Does Not Infringe the Sargent Patent

A. The Internet Has No "Central Computer"

1. Literal Infringement

A "central computer" in the Sargent patent is:

a single device, in one location. It is referred to as "central" because it is connected to numerous physically separate stations, called "remote terminals," by the telephone lines of a telephone network. So there is one computer, connected to many remote terminals. The central computer means in this patent thus serves as the hub of a digital information storage, retrieval and display system — and all of the remote terminals connect to it.

The central computer stores information. The central computer contains a "main store." In the context of this patent, the main store is a mass information storage or memory device. An example of a main store is a magnetic disk, which is a rotating circular plate having a magnetizable surface on which information may be stored as a pattern of polarized spots on concentric recording tracks.

The central computer contains an information database, which is "centralized" in the sense that all of the remote users can access it by accessing the central computer.

Markman Op. at 112.

BT asserts that, while the Internet is made up of an enormous number of computers, each individual web server is a central computer in one location (i.e. central relative to the remote terminals within the meaning of the patent).

The cornerstone of this argument is BT's assertion that a central computer is not limited to a single computer as a matter of law.... The patent claims as construed clearly provide that the central computer is one device, in one location. Just as a circle has but one center, hub-and-spoke networks have only a single hub. There may be other circles with other centers, just as there may be other hub-and-spoke networks with other central computers or hubs. But each system (network) of the type claimed in the Sargent patent can have only one central computer. Therefore, viewing the Internet as a system (as BT asks me to do), it does not literally infringe the Sargent patent, because it contains no such central computer.

The central computer in the Sargent patent also contains an information database, which is "centralized" in the sense that all remote users can access it by accessing the central computer. [*British Telecomm I,* 189 F. Supp.2d 109, 112.] Prodigy argues that the Internet has no such "centralized data store," but rather, contains an extremely diffuse data storage architecture. According to Prodigy's expert, this distribution is the essence of the Internet, because it allows users to access information stored throughout a global network of networks of computers and storage devices that are loosely linked through adherence to an open group of protocols and standards adopted by the Internet community.

BT does not dispute that this is the case. It responds, however, that its argument is not that the Internet has a centralized data store, but that each Web server (or central computer, as BT would call it) has its own centralized data store because it has a main storage device for storing HTML files.

However, the claims of the '662 patent state that the centralized data store contains *all* of the blocks of information accessible by the remote terminals. [*British Telecomm. I,* 189 F. Supp.2d 109, 122.] BT does not dispute that there is no centralized data store for the Internet that contains all of the data remote users might care to access.

The Internet is a network of computers intertwined with each other in order to allow users around the world to exchange information. The whole purpose of the Internet is for the *sources* of information to be in many places rather than centralized. Any user can retrieve information that is stored on a Web server in any physical location, as long as that server is connected to the Web. For example, a Prodigy user does not have to rely on Prodigy to gather information from multiple sources and put it on its own server in order for a Prodigy customer to have access to the information. Rather, when Prodigy users connect to the Internet through Prodigy's system, users can access blocks of information located on remote systems, e.g. a computer in Alaska, rather than Prodigy's own computers in New York. This "network of networks" or "system of systems" allows users access to information from a variety of sources, in any location.

BT cites the general rule that addition to an accused apparatus of one or more features than the claim requires does not preclude a finding of infringement. But what BT characterizes as "additions" are fundamental differences in the nature of the claim elements. As it did during the *Markman* hearing, BT would have me exclude the word "central" from the construction of "central computer." The Court expressly rejected BT's interpretation by ruling that the Sargent patent claims a central computer that is a single computer with a centralized database. Because the Internet is not a computer network consisting of a centralized computer that stores all of the data accessible by remote terminals, Web servers on the Internet cannot literally infringe the '662 patent.

2. Doctrine of Equivalents

BT may not rely on the doctrine of equivalents to withstand Prodigy's motion for summary judgment. The doctrine of equivalents requires insubstantial differences between the patented invention and the accused product, as determined on an element-by-element basis. *Warner-Jenkinson,* 520 U.S. at 29-30. Whether the differences are insubstantial can be determined by the function-way-result test, which requires that the claimed and accused elements "perform substantially the same function in substantially the same way to obtain the same result." Id. at 40. In this case, the Internet fails the function-way-result test, because a central computer containing all of the data accessible by a remote terminal user operates in substantially different ways from the Internet.

A central computer, as claimed in the Sargent patent, has a one-to-one, hub and spoke relationship with numerous physically separate stations called remote terminals. The remote terminals are connected to the central computer by the telephone lines of a telephone network. All of the remote terminals connect to the central computer that has one centralized main store for storing information. As a result, a remote terminal in the Sargent patent does not identify in its communication protocol a computer with which it would like to communicate, for it communicates with only one central computer. In contrast, a computer operating on the Internet must at all times identify a specific computer with which it seeks to communicate, because not all of the remote users are connected to a single central source of information. If all Internet users were connected to Prodigy's Web server alone, BT's argument might have some force; but all Internet users are not connected to Prodigy. Instead, Internet users have access to multiple Web servers on the Internet that store information in various locations.

Indeed, the Internet is the very antithesis of a digital information storage system having a central computer. The opposite of a claim limitation cannot be considered its equivalent. Moore U.S.A. Inc. v. Standard Register Co., 229 F.3d 1091, 1106 (Fed. Cir. 2000). In *Moore,* the Federal Circuit held that a claim limitation reciting that an adhesive "extend[s] the majority of the lengths of said longitudinal marginal portions," could not be infringed either literally or under the doctrine of equivalents by a product having adhesive extending only a minority of a marginal portion (i.e. less than 50%) — even if it was almost a majority (e.g., 47.8%) Id. According to the Federal Circuit, "it would defy logic to conclude that a minority — the very antithesis of a majority — could be insubstantially different from a claim limitation requiring a majority." Id.

BT cannot dispute that any user throughout the world can access information stored in any of the millions of computers connected to the Internet. In contrast, the Sargent patent claims revolve around a central computer — a single device, in one location, with one main data store. The Internet, is, in short, an entirely different beast from the system described in the Sargent patent. Consequently, the Internet does not infringe the Sargent patent either literally or under the doctrine of equivalents. Prodigy is therefore entitled to summary judgment as a matter of law....

C. The Internet Does Not Contain Blocks of Information as Required by the Sargent Patent

1. Literal Infringement

A key distinction between the '662 patent and the Prodigy Internet Service is the requirement of the '662 patent that information be stored as blocks of information. These blocks have special characteristics:

Each block has a first portion and a second portion. These portions are separable, contiguous and co-stored sub-units. That means that the portions are stored together, and they are stored next to each other, yet they can be separated from each other. A block of information may contain very limited programming information, the purpose of which is to reduce the complexity of keying required to communicate with the central computer.

The two parts of a block of information are the first portion and the second portion. The first portion of a block of information is intended for visual display on a remote terminal. The second portion is information not intended for display. The second portion contains the complete address for each of the other blocks of information referenced in the first portion. So if a block of information is referenced in the first portion, then the complete address for that block of information will be in the second portion. The second portion may also contain other information as well, such as information to influence the display or to reduce the complexity of keying required to communicate with the central computer. But it never contains information intended for display.

[*British Telecomm. I,* 189 F. Supp.2d 109, 116.]

Unlike the blocks of information required by the '662 patent, HTML code, which is the primary language of the World Wide Web and of the Prodigy Internet Service, does not use blocks. HTML code does not separate displayed information into a first sub-unit, and non-displayed information in a contiguous, separable second sub-unit. Rather, HTML code contains information to be displayed intermingled with other information concerning formatting and linking, such as URLs and anchors.

In HTML, hypertext references include each URL link adjacent to each phrase or image for display. For example, in the code,

<script>document.write(HTMLCacheArray[34]<><A href = '*http:// www.msnbc.com/ modules/exports/ct — prodigy.asp?/news/736921.asp*' target = '*— top*'> Yahoo! profits meet forecasts</A><TD>

the information for display is shown in bold, and the URL is shown in italics. The first linking tag (which is referred to in HTML as an "anchor tag," and identified with the "A"), is the information always for display associated with each URL. This information appears on the screen. The user can mouse this link, and click to activate it. The URL for each such component of displayed information is set forth in the HTML code immediately prior to the information for display. Consequently, URLs associated with information for display are not located contiguously and separably with a sub-unit of information not for display.

2. Doctrine of Equivalents

BT cannot rely on the doctrine of equivalents to withstand Prodigy's motion for summary judgment.

a. Application of the Doctrine of Equivalents is Barred with Respect to "Blocks of Information" Because the Applicant Made Unmistakable Assertions to Avoid the Prior Art.

The doctrine of prosecution history estoppel prevents a patent owner from relying upon the doctrine of equivalents when the patent applicant relinquishes coverage of subject matter during the prosecution of the patent, either by amendment or argument. A patentee cannot invoke the doctrine of equivalents to "embrace a structure that was specifically excluded from the claims." Dolly, Inc. v. Spalding & Evenflo Cos., *Inc.*, 16 F.3d 394, 400 (Fed. Cir.1994).

During the prosecution of the Sargent patent, a new limitation was added to the phrase "blocks of information" in order to distinguish the Sargent patent from [several prior art] references. To distinguish [one such reference, for example, the Applicant cited to several of its narrowed claim limitations:

> For example, there is absolutely no suggestion anywhere in [the reference] that blocks of stored data should include a first portion containing information for display and a second portion containing information not for display but including the complete address for each of plural other blocks of information. Nor does [the reference] teach . . . manual entry of keyed digital data of less extent than any one complete address but nevertheless uniquely indicative of one of the complete addresses contained in the second portion of the block of information (which contains the first portion then being displayed).

Because BT relied on narrowing amendments to overcome prior art, equivalents are unavailable with respect to these claim limitations. *Warner Jenkinson*, 520 U.S. at 33. Thus, BT is barred from asserting that Prodigy's Internet service meets the above claim elements under the doctrine of equivalents. . . .

II.   *Contributory Infringement and Active Inducement*

Because the Internet itself does not infringe the Sargent patent, Prodigy can not be liable for contributory infringement or active inducement for providing its users with access to the Internet. I therefore need not address BT's arguments concerning contributory infringement and active inducement in any detail. . . .

Note on Software Patent Scope

The trend in software infringement cases seems to be towards interpreting the claims narrowly. Consider for example Wiener v. NEC Electronics, Inc., 102 F.3d 534 (Fed. Cir. 1996). In that case the Federal Circuit upheld the district court's finding of noninfringement under the doctrine of equivalents because there were substantial differences between the patent's requirement that a computer program "call on" columns of data one byte at a time and the defendant's product, where the columns alleged to be equivalent were not in the data matrix and therefore were not called upon to read data. The court rejected the "conclusory" declaration of plaintiff's expert that the two processes were identical. In General Electric v. Nintendo of America, 179 F.3d 1350 (Fed. Cir. 1999), the court held that Nintendo's video game systems did not infringe GE's television switch patents because the patents, written in means-plus-function format, did not disclose a function for the switches identical to Nintendo's function. In Digital Biometrics, Inc. v. Identix, Inc., 149 F.3d 1335 (Fed. Cir. 1998), the court narrowly construed a patent claim to "image arrays" storing a two-dimensional slice of video data that were merged into a "composite array" storing a fingerprint image. The court held that the defendant's systems, which constructed the composite array directly rather than by using two-dimensional slices, did not create "image arrays" within the meaning of the claims. See also SuperGuide Corp. v. DirecTV Enterprises, Inc., 211 F. Supp.2d 725 (W.D.N.C. 2002) (interactive TV program guide patent not infringed, in part because of differences between analog signal modulation of claimed system and pure digital signals used by accused system).

Do these decisions make sense in light of what we know about the validity of software patents? Do they render software patents ineffective because of the fast-changing nature of the technology?

One possible explanation for the narrow construction given software patents in infringement cases to date is that most of those patents have claim elements written in means-plus-function format. A growing number of software patent claims are drafted in means-plus-function language largely to meet the "structure" requirements the courts have imposed on patentees under section 101. How should the "corresponding structure or equivalents thereof" in the specification be interpreted for infringement purposes? In particular, suppose that a patentee, to come within the dictates of section 101, claims a "means for processing data" of a particular type. In the specification, the patentee discloses the use of an IBM-compatible personal computer with a Pentium microprocessor. Is this claim infringed by a defendant who runs the same software on a 486 microprocessor? an Apple computer using a Motorola microprocessor? a Sun Microsystems SPARC workstation? a Cray supercomputer? a cash register containing a microprocessor?

The same question arises when the claims and disclosed structure relate strictly to software, rather than the hardware that runs it. For example, Wang Laboratories, Inc. v. America Online, Inc., 197 F.3d 1377 (Fed. Cir. 1999), centered on a patent from the 1980s-era "videotex" industry, a short-lived predecessor to the Internet in which customers subscribed to individual "content provider" clearinghouses via dedicated telephone lines. The patent at issue included the key term "frame," which is roughly equivalent to an Internet-era "page" of data. But the patent specification consistently referred to "frames" as including only characters. The patentee, however, argued that the normal technical meaning of frames included "bit-mapped" content, which (unlike character pages) can include graphics. Under this definition, the patentee had an argument that all Internet service providers (ISPs), such as defendant America Online, infringed the patent.

The court construed the claim term "frame" to be limited to "character-based protocols," even though the ordinary meaning of the term could arguably be applied to "bit-mapped protocols." 197 F.3d at 1382. Although the court noted that the only system described and enabled in the specification used a character-based protocol, it also noted that a skilled programmer would not have understood bit-mapped protocols to be included. In addition, the court pointed out that during prosecution the patentee distinguished the "pel [picture element] level" from the "character level," and presented the invention as involving a character-based system. 197 F.3d at 1384. The court concluded by stating that the question of what structures are disclosed, and what their equivalents are under § 112 ¶6, is a highly fact-intensive inquiry.

On the question of infringement under the doctrine of equivalents, the court made these instructive observations leading up to its holding:

The defendants argue that Wang's inability to implement bit-mapped technology should not be rewarded with a judgment of equivalency after others later succeeded. The evidence in the record before us supports the district court's ruling of non-equivalency under § 112 ¶6. The expert testimony was to the general effect that although these protocols are today interchangeable as modules when placed in self-contained "black boxes" and inserted into the appropriate software, they function in accordance with markedly different principles, have greatly different capabilities, and generally do not operate in substantially the same way. The '669 inventors testified that Wang had discontinued development of the bit-mapped module in this system based, at least in part, on technologic difficulties. Witnesses also testified that Wang chose to work with [a] character-based protocol because it had already achieved popularity and was less complex than the bit-mapped...system, although Wang knew that [its system] had limited graphics capabilities. Wang's statement in the prosecution of the parent application that [a] reference to the [bit-mapped] system was not applicable because it did not describe a character-based system weighs heavily against Wang's entitlement to reach this system under § 112 ¶6. Thus the specification, the prosecution history, and the testimony of witnesses support the conclusion that the bit-mapped protocol is not an equivalent under § 112 ¶6. The district court's summary judgment that there is not literal infringement is affirmed.

197 F.3d 1377, at 1385. For a similar case involving videotex-era patents asserted against internet-era businesses, see British Telecommunications Plc. v. Prodigy Communications Corp., 189 F. Supp. 2d 101 (S.D.N.Y.2002) (narrowly construing claim terms such as "central computer means"), later opinion British Telecommunications Plc. v. Prodigy Communications Corp., 217 F. Supp.2d 399 (S.D.N.Y. 2002) (finding no infringement).

In many ways, the Federal Circuit has been struggling with similar issues for some time. Compare Texas Instruments, Inc. v. International Trade Commn., 805 F.2d 1558 (Fed. Cir. 1986) (modern calculators did not infringe TI's original patent on the integrated circuit) with Hughes Aircraft Co. v. United States, 717 F.2d 1351 (Fed. Cir. 1983) (patent on means for controlling satellites from the ground via telemetry was infringed by use of on-board microprocessors to control satellites). For a case applying the doctrine of equivalents in a software context, see Safe Flight Instrument Corp. v. Sundstrand Data Control, Inc., 706 F. Supp. 1146 (D. Del. 1989), aff'd, 899 F.2d 1228 (Fed. Cir. 1990).

Significant light was shed on this important question in Alpex Corp. v. Nintendo of America, 102 F.3d 1214 (Fed. Cir. 1996). There the court reversed a jury verdict of infringement in a case involving an old video-game patent asserted against modern technology. Differences in the nature of the devices doomed the infringement claim. See also WMS Gaming, Inc. v. International Game Technology, 184 F.3d 1339 (Fed. Cir. 1999) (newer generation of computer-controlled slot machine did not infringe patent on older generation technology).

In summary, both in general and with respect to § 112 ¶6 claims, the Federal Circuit has tended to construe software patents narrowly. This is a salutary development, according to Julie Cohen and Mark Lemley. One aspect of this is a narrow application of the doctrine of equivalents:

[W]e argue that in light of the special nature of innovation within the software industry, courts should apply the doctrine of equivalents narrowly in infringement cases....The test of equivalence is the known interchangeability of claimed and accused elements at the time of (alleged) infringement. A number of factors unique to software and the software industry—a culture of reuse and incremental improvement, a lack of reliance on systems of formal documentation used in other fields, the short effective life of software innovations, and the inherent plasticity of code—severely complicate post hoc assessments of the "known interchangeability" of software elements. A standard for equivalence of code elements that ignores these factors risks stifling legitimate, successful efforts to design around existing software patents. To avoid this danger, courts should construe software claims narrowly, and should refuse a finding of equivalence if the accused element is "interchangeable" with prior art that should have narrowed the original patent, or if the accused improvement is too many generations removed from the original invention.

Julie E. Cohen and Mark A. Lemley, Patent Scope and Innovation in the Software Industry, 89 Cal. L. Rev. 1, 3-4 (2001).

In addition to the doctrine of equivalents, patent infringement analysis also requires application of the "reverse doctrine of equivalents." As the name suggests, the reverse doctrine of equivalents operates to protect patent defendants whose products fall within the literal scope of patent claims, but which are in fact substantially different from the patented invention. This doctrine may be of particular importance in the software industry because the application of computer technology has resulted in major changes in the way in which certain physical operations are performed. The continuing vitality of the reverse doctrine of equivalents was questioned in dictum in Tate Access Floors, Inc. v. Interface Architectural Resources, Inc., 279 F.3d 1357 (Fed. Cir. 2002). But the Federal Circuit has since reaffirmed the doctrine.

E. SUI GENERIS PROTECTION OF COMPUTER TECHNOLOGY

As the previous sections suggest, computer technology does not fit neatly within any of the categories of intellectual property law. Different aspects of computers and computer programs are eligible for protection as trade secrets or by means of patents, copyright, or trademarks. Further, many important works—particularly computer programs—receive overlapping protection from several sources.

None of these sources is a perfect "fit" with computer software. Patent, copyright, trade secret, and trademark law all protect some aspects of computer programs, but what they protect is not necessarily what we ideally want to encourage. To address this problem, numerous commentators have suggested that we need a new intellectual property statute, one directed specifically at computer technology and responsive to the particular needs of the industry. In 1984, Congress responded in part to these concerns by passing the Semiconductor Chip Protection Act ("SCPA"), 17 U.S.C. §§ 901-914. That act provided specific protection to semiconductor "mask works." Its purpose and history were described by the Federal Circuit in Brooktree Corp. v. Advanced Micro Devices, 977 F.2d 1555 (Fed. Cir. 1992):

> The Semiconductor Chip Protection Act of 1984, Pub. L. 98-620, Title III, 98 Stat. 3347, codified at 17 U.S.C. §§ 901-914, arose from concerns that existing intellectual property laws did not provide adequate protection of proprietary rights in semiconductor chips that had been designed to perform a particular function. The act, enacted after

extensive congressional consideration and hearings over several years, adopted relevant aspects of existing intellectual property law, but for the most part created a new law, specifically adapted to the protection of design layouts of semiconductor chips.

Chip design layouts embody the selection and configuration of electrical components and connections in order to achieve the desired electronic functions. The electrical elements are configured in three dimensions, and are built up in layers by means of a series of "masks" whereby, using photographic depositing and etching techniques, layers of metallic, insulating, and semiconductor material are deposited in the desired pattern on a wafer of silicon. This set of masks is called a "mask work," and is part of the semiconductor chip product. The statute defines a mask work as:

a series of related images, however fixed or encoded

(A) having or representing the predetermined, three dimensional pattern of metallic, insulating, or semiconductor material present or removed from the layers of a semiconductor chip product; and

(B) in which series the relation of the images to one another is that each image has the pattern of the surface of one form of a semiconductor chip product.

17 U.S.C. §901(a)(2). The semiconductor chip product in turn is defined as: the final or intermediate form of any product —

(A) having two or more layers of metallic, insulating, or semiconductor material, deposited or otherwise placed on, or etched away or otherwise removed from, a piece of semiconductor material in accordance with a predetermined pattern; and

(B) intended to perform electronic circuitry functions.

17 U.S.C. §901(a)(1).

The design of a satisfactory chip layout may require extensive effort and be extremely time consuming, particularly as new and improved electronic capabilities are sought to be created. A new semiconductor chip may incur large research and development costs, yet after the layout is imprinted in the mask work and the chip is available in commerce, it can be copied at a fraction of the cost to the originator. Thus there was concern that widespread copying of new chip layouts would have adverse effects on innovative advances in semiconductor technology...

The patent system alone was deemed not to provide the desired scope of protection of mask works. Although electronic circuitry and electronic components are within the statutory subject matter of patentable invention, see 35 U.S.C. § 101, and some original circuitry may be patentable if it also meets the requirements of the Patent Act, as is illustrated in this case, Congress sought more expeditious protection against copying of original circuit layouts, whether or not they met the criteria of patentable invention. Senate Report at 8....

The Semiconductor Chip Protection Act provides for the grant of certain exclusive rights to owners of registered mask works, including the exclusive right "to reproduce the mask work by optical, electronic, or any other means", and the exclusive right "to import or distribute a semiconductor chip product in which the mask work is embodied". 17 U.S.C. §905. Mask works that are not "original", or that consist of "designs that are staple, commonplace, or familiar in the semiconductor industry, or variations of such designs, combined in a way that, considered as a whole, is not original", are excluded from protection. 17 U.S.C. §902(b). Protection is also not extended to any "idea, procedure, process, system, method of operation, concept, principle, or discovery, regardless of the form in which it is described, explained, illustrated or embodied" in the mask work. 17 U.S.C. §902(c). The sponsors and supporters of this legislation foresaw that there would be areas of uncertainty in application of this new law to particular situations, and referred to "gray areas" wherein factual situations could arise that would not have easy answers. Those areas are emphasized by both parties in the assignments of error on this appeal....

Litigation under the SCPA is extremely rare. The second reported decision to address the SCPA in the statute's two decades on the books was handed down in 2005. It illustrates similarities between infringement analysis under the SCPA and copyright as well as the application of the SCPA's reverse engineering defense.

Altera Corp. v. Clear Logic, Inc.
United States Court of Appeals for the Ninth Circuit
424 F.3d 1079 (9th Cir. 2005)

HUG, Circuit Judge.

. . .

I. Overview

Filling the gap between copyright law and patent law, the SCPA aims to protect the substantial investment of innovative firms in creating the semiconductor chips that are "at the vortex" of the modern information age. H.R.Rep. No. 98-781, at 2 (1984). These chips operate microwave ovens, televisions, computers, robots, Xray machines, and countless other now indispensable apparatuses. Each chip carries its own blueprint. Pirate firms can strip the layers of a semiconductor chip and replicate the design at a cost substantially lower than the original firm's investment. Id.

Altera and Clear Logic are competitors in the semiconductor industry. Altera manufactures programmable logic devices ("PLDs"), which are chips that can be programmed to perform various logic functions. A customer uses Altera's MAX+-PLUS II software to program the PLD to perform the desired function.[1] The software helps to route the functions through the thousands of transistors that make up the PLD, ideally achieving the maximum functionality for the particular function desired. Because the PLD can be programmed and reprogrammed, the customer, working with Altera, can continue to work with the PLD and the software until the PLD meets the customer's exact needs. This process can take months.

Clear Logic manufactures a different type of chip: Application-Specific Integrated Circuits ("ASICs"). These chips are designed to perform one specific function and cannot be programmed by the customer. They use less power, are smaller and, for a customer with a large order, are often cheaper. Customers will sometimes start with PLDs and switch to ASICs once they have determined exactly what they need the chips to do. Traditionally, a company that converts from PLDs to ASICs must again start from a high level of description and work toward the end product, the ASIC. This can take a few months and there is a substantial risk that even after the initial attempt, the first chip will not work and more time and money will have to be invested in perfecting the product. Clear Logic works from a different business model. When customers program Altera devices, using the Altera software, a file called a bitstream is

---

1. Altera sells chips to companies that use those chips to perform logic functions in devices they produce, not to individual consumers. For example, a company that manufactures printers might purchase PLDs from Altera to perform the functions necessary to operate the printer.

generated. Clear Logic asks customers to send the bitstream to Clear Logic, and Clear Logic uses the bitstream to create an ASIC for the customer. Clear Logic only produces ASICs that are compatible with Altera chips. The laser process Clear Logic uses to create chips with the bitstream allows for a turnaround time of just a few weeks, and rarely produces an incompatible chip.

Faced with the loss of millions of dollars in business, Altera has challenged Clear Logic's business model. In the district court, Altera argued that Clear Logic infringed its rights under the SCPA by copying the layout design of its registered mask works for three families of chip products. Clear Logic denied the infringement and asserted an affirmative defense of reverse engineering. The jury returned a verdict in favor of Altera on the infringement claim....

II. The Semiconductor Chip Design Process and the Semiconductor Chip Protection Act

Clear Logic and Altera define the stages and terms relating to the chip design process differently. According to Altera, the layout is "the physical arrangement of the components on the chip." The architecture is comprised of "the components and the structures that are physically arranged within the chip." Clear Logic argues that the architecture is essentially a block diagram showing the basic arrangement of the chip. From this conceptual plan, the designer creates floor plans that show the arrangement of functional modules, focusing on how the designer will group major components. The floor plan and the architecture are both at high levels of abstraction. The designer next creates an electrical schematic, which is a two-dimensional abstract drawing. After this, a layout designer creates a three-dimensional layout design which includes the specific placement of all of the elements of the chip and is used to make the glass masks that are printed onto the chip.

Despite their disagreement over the specific definitions of industry terms and the specific steps in the process of chip design, Altera and Clear Logic agree that chip design starts with a high level idea and moves toward the placement of individual transistors on a chip in several layers. Ultimately, the schematics and floor plans are used to develop the specific placement of every transistor that will eventually go on the chip. Glass disks are etched with the pattern for each layer of the chip, and these glass disks, called masks, are printed onto the semiconductor chip, one layer at a time, by photolithography. Generally, there are eight to twelve layers to the chip, each of which requires a separate mask. The series of all of the masks is the mask work.

Reverse engineering has long been an accepted practice in the semiconductor chip industry. By photographing and chemically dissolving each layer of the chip, a second company can recreate the entire mask work for any chip. The process allows legitimate analysis of chips to spur innovation and improvement on existing designs, but also makes direct copying of chips feasible. At the behest of the semiconductor industry, Congress sought to protect this important facet of the American economy, and indeed of modern American society, while preserving the validity of reverse engineering as an appropriate form of competition.

After extensive debates, Congress passed the Semiconductor Chip Protection Act in 1985, creating sui generis protection for mask works embodied in semiconductor chip products. The Act borrows heavily from copyright law, and was initially proposed as an extension of existing copyright protection. S. Rep. 98-425 at 9; S.Rep. No. 98-425 at 12-13 (1984) (analogizing mask works to technical drawings or audiovisual works subject to copyright protection). In analogizing semiconductor chips to traditional areas of copyright law, the legislative history notes that, just as a plagiarist who copies only one chapter of a book may be held liable for infringement, a person may be liable for copying a part of a mask work if it is a qualitatively important portion that results in substantial similarity. S.Rep. No. 98-425 at 16-18. The realities of the business world, however, would purportedly make it impractical for a copyist to copy a part of a chip but create the rest of the chip independently. Id. Where chips appeared to be similar, Congress assumed that admitting expert testimony to assist in determining whether subtle changes in a mask work layout were significant would resolve the problem of distinguishing a copy from a legitimate reverse engineering attempt in most cases. Id. . . . .

The SCPA grants the owner of a mask work the exclusive rights to reproduce the mask work and to "import or distribute a semiconductor chip product in which the mask work is embodied." 17 U.S.C. §905. The Act does not, however, extend protection "to any idea, procedure, process, system, method of operation, concept, principle, or discovery, regardless of the form in which it is described, explained, illustrated, or embodied in such work." 17 U.S.C. §902(c).

Before trial, the district court granted in part Altera's summary judgment motion regarding the scope of the SCPA. The court held that "Altera's layout is fixed in the final chip product: the placement of the components and their interconnection lines on the actual chip are created using masks which are the physical embodiment of the layout design chosen by Altera design engineers. Thus Altera's layout design is more than a mere idea. It is the blueprint for the layout of the semiconductor chip." The court concluded that "the Act is broad enough to cover the type of claims made by Altera," but left for the jury the factual question of whether Altera had proven infringement. Relying on Brooktree Corp. v. Advanced Micro Devices, Inc., 977 F.2d 1555 (Fed.Cir.1992), the only case to date to examine the SCPA, the court explained that "copying groupings of transistors and interconnection lines may constitute a violation of the Act." The ruling on Altera's motion for summary judgment involved a purely legal issue which we review de novo.

In *Brooktree*, the plaintiff alleged that the defendant company had copied its ten-transistor SRAM cell. 977 F.2d at 1563. The court stated that "If the copied portion [of the mask work] is qualitatively important, the finder of fact may properly find substantial similarity under copyright law and under the Semiconductor Chip Protection Act," even if other portions of the chip were not copied. Id. at 1564 (citations omitted). The district court had appropriately allowed the jury to determine whether the copying of the layout of a cell within the chip was an infringement of the act. Id. at 1565.

Altera asserts that the placement of groupings of transistors on the chip was copied, and does not specifically address the layout of the transistors within those groupings. Clear Logic argues that the placement of the groupings is a system or an idea and is not entitled to protection under the SCPA. We reject this contention; the boundaries and organization of these groupings are more than conceptual. As both John Reed and John Turner testified, these groupings are physically present in the mask work. Commentators have suggested analyzing the levels of abstraction in the production of a computer program or a mask work to identify the distinction between ideas and expression, and the degree of similarity, in these formats. See 4 Nimmer on Copyrights § 13.03 (2005) (comparing the analysis of broad ideas, plots, structure, dialogue, or sequence of events in a novel or play to the levels of abstraction in creating a computer program); Copyright Protection for Semiconductor Chips: Hearings on H.R. 1028 Before the House Subcomm. on Courts, Civil Liberties, and the Administration of Justice of the Comm. on the Judiciary, 98th Cong. 316-21 (1983) (letter and article submitted by Eric W. Petraske, patent attorney) (identifying ideas from electrical information, geometric information about component placement, size, shape, circuit design within the mask level). Our own cases have suggested such an approach is appropriate. See Data East USA, Inc. v. Epyx, Inc., 862 F.2d 204, 207-09 (9th Cir.1988). Under this approach, we identify the broad idea behind the design and assess each successive step in the design process, identifying the point at which the idea becomes protectable expression. See Computer Assocs. Int'l, Inc. v. Altai, Inc., 982 F.2d 693, 706-12 (2d Cir.1992) (as amended) (explaining the abstraction-filtration-comparison test for computer programs).

In considering the chip design process, we recognize, as do the parties, that with each step, the ideas become more concrete until they are finally expressed in the layout of the transistors in the mask work. The customer's idea is at the highest level of abstraction, and the schematics and floor plans convey more concrete ideas, designating how the chip may be structured or organized. These drawings are preliminary sketches that would not be protected under traditional copyright principles. Copyright Protection for Semiconductor Chips: Hearings on H.R. 1028 Before the House Subcomm. on Courts, Civil Liberties, and the Administration of Justice of the Comm. on the Judiciary, 98th Cong. 116-17 (1983) (statement of Dorothy Schrader, Copyright Office). It is not until the level of the mask work, the piece of the process that Congress chose to protect, that there is an expression of that idea. Those ideas that are physically expressed in the mask work are subject to protection under the SCPA.

The district court correctly determined that the organization of the groupings is physically a part of the mask work. The mask work is structured according to the groupings that Altera highlighted and the district court correctly allowed the jury to determine whether those similarities constituted an infringement of the act. Unlike the outline of an article or the chapters in a book, these groupings physically dictate where certain functions will occur on a chip and describe the interaction of parts of the chip. The placement of logic groupings in a mask work is not an abstract concept; it is embodied in the chip and affects the chip's performance and efficiency as well as the chip's timing. In accordance with the Federal Circuit's holding in Brooktree, it is the province of the jury to determine whether those aspects of the mask work are material, and whether the similarity between the mask works is substantial. 977 F.2d at 1570. The district court did not err in finding that the organization of the groupings of logic functions on Altera's mask works, and the interconnections between them, was protected under the SCPA. The arrangement of the transistors within those blocks is also entitled to protection under the SCPA, but, as in Brooktree, the jury determines whether the similarities are more important than the differences. Id.

III. Reverse Engineering

The SCPA specifically protects the right of

(1) a person to reproduce the mask work solely for the purpose of teaching, analyzing, or evaluating the concepts or techniques embodied in the mask work or the circuitry, logic flow, or organization of components used in the mask work; or

(2) a person who performs the analysis or evaluation described in paragraph (1) to incorporate the results of such conduct in an original mask work which is made to be distributed.

17 U.S.C. §906. This reverse engineering provision explicitly protects industry practices and encourages innovation. The second mask work must not be "substantially identical to the original," and as long as there is evidence of "substantial toil and investment" in creating the second mask work, rather than "mere plagiarism," the second chip will not "infringe the original chip, even if the layout of the two chips is, in substantial part, similar." Brooktree, 977 F.2d at 1566(quoting Mathias-Leahy Memorandum at S12,917). The legislative history, relying on the testimony of industry representatives, indicated that most cases would probably present clear cut evidence of direct copying or of innovation and that in cases falling into the gray area between outright copying and complete originality, the courts should consider the presence or absence of a paper trail by the second firm. See id. (summarizing relevant statements from the legislative history discussing reverse engineering and paper trails). A firm that simply copied another firm's mask work would have no evidence of its own investment and labor, whereas a legitimate reverse engineering job would require a trail of paper work documenting the analysis of the original chip as well as the development of an independent design. Id. The district court in this case instructed the jury that this was a factor to be considered in determining whether Clear Logic had proven a valid reverse engineering defense.

In Brooktree, the Federal Circuit analyzed the defendant's paper trail, but held that "the sheer volume of paper" was not dispositive. Id. at 1569. The trail in that case was susceptible to the interpretation that the defendant copied the chip. The court held that the jury was entitled to weigh the evidence and consider the differences in the chips as well as the similarities, and could find the similarities sufficient to invalidate the reverse engineering defense. Id.

In this case, Clear Logic challenges the district court's formulation of one of the reverse engineering jury instructions. . . .

The district court instructed the jury that

A defendant does not infringe the mask works of a plaintiff if the defendant can prove by a preponderance of the evidence that it has engaged in legitimate reverse engineering. A question on the verdict form asks you to indicate whether you find that the defendant has proved that it has — that it engaged in legitimate reverse engineering. To establish a legitimate reverse engineering defense, Clear Logic must prove by a preponderance of the evidence, the following requirements:

Number 1. That Clear Logic reproduced Altera's mask works solely for the purpose of teaching, substantially analyzing, or substantially evaluating non-protectable concepts or techniques embodied in Altera's mask works; and,

2. That Clear Logic did not use or copy any of Altera's protectable expression in the Clear Logic chips, but rather used only the non-protectable concepts or techniques embodied in Altera's mask works in developing clear logic's own chips, which contain an original layout design.

(Tr. 1478-79) (emphasis added). The judge continued to instruct the jury that the law actually permits a competitor to reproduce the mask work if that reproduction is done solely as a step in a process of creating its own original mask work. A mask work is original as long as it exhibits any minimal degree of creativity in any aspect of the mask work layout. The requisite level of creativity for originality is extremely low; even a slight amount will suffice.

> Notice that I use the term "legitimate" reverse engineering. In the process of reverse engineering, the competitor is allowed to photograph and reproduce a registered mask work. If the competitor uses this information to reproduce a substantially identical mask work, as for example a virtual photocopy, the competitor has not engaged in legitimate reverse engineering.
>
> Legitimate reverse engineering would start with photographing and reproducing the mask work, but the photograph and reproduction is used for the purpose of study and analysis, and then the competitor must combine this study and analysis with its own engineers' engineering efforts to yield its own original mask work.
>
> Because the new mask work is designed to be interchangeable with the registered mask, it may be substantially similar to the registered mask work.

(Tr. 1481-82).

The question on the verdict form pertinent to the defense of reverse engineering stated:

> Did Clear Logic prove that it reproduced Altera's mask works solely for the purpose of analysis to develop its own original layout design, namely one which is not substantially identical to components or groupings of components or interconnections on the corresponding Altera mask work which are material and original?
> *Answer as to each mask works which is the subject of this action. A "Yes." answer is a finding that the defense of legitimate reverse engineering was proved. A "No." answer is a finding that the defense of legitimate reverse engineering was not proved.*

The jury responded "No" with respect to each mask work, finding that Clear Logic had not established a reverse engineering defense for any of the chip families.

The SCPA's reverse engineering provision allows copying the entire mask work. It does not distinguish between the protectable and non-protectable elements of the chip as long as the copying is for the purpose of teaching, evaluating, or analyzing the chip. Although the product created from that analysis must be original, the process of studying the chip is not limited to copying ideas or concepts. The district court's instructions initially define "legitimate reverse engineering" to allow copying and analyzing only "non-protectable concepts or techniques." This is an incorrect statement of the law.

The district judge's instructions in this case were extremely thorough; he made a real effort to make the instructions on this complex subject understandable to the jury. The instructions on the reverse engineering defense extend for several pages of the transcript. The later instructions correctly state the law applicable to the defense of reverse engineering and remedy the initial incorrect statement. They explain that it is permissible to reproduce "a registered mask work" as a step in the process of creating an original chip, so long as the purpose of reproducing the chip is appropriate. The district judge correctly explained that only minimal ingenuity is necessary for a second chip to qualify as original, and he emphasized the distinction between substantial identity and substantial similarity. As the court in Brooktree discussed, the similarities in a chip may outweigh the differences. 977 F.2d at 1570. In this case, there was sufficient testimony about both the similarities and the differences in the mask works to allow a jury to determine whether the Clear Logic mask works were original as defined under the Act. The incorrect statement in the instructions was not prejudicial error.

As our opinion in In re Asbestos Cases stated:

> Prejudicial error results when looking to the instructions as a whole, the substance of the applicable law was [not] fairly and correctly covered. The instructions and interrogatories must fairly present the issues to the jury. If the issues are fairly presented, the district court has broad discretion regarding the precise wording.

847 F.2d 523, 524-25 (9th Cir. 1988) (internal citations and quotation marks omitted, alteration in original). In the case at hand, the interrogatory (or question) presented to the jury was clear and concise and correctly stated the law. It is significant that this was the statement of the law to which the jury particularly responded. We hold that, viewing the instructions and the interrogatory as a whole, the legal issue was fairly presented to the jury. . . .

RYMER, Circuit Judge, concurring:

I agree that the district court's instructions did not constitute reversible error on the record presented to us for review. As the district court held, the selection, combination and arrangement of electrical components in the mask works (which it called the "mask work layout design"), including the placement, orientation, and interrelationship of groupings of transistors and interconnection lines are protectible. However, I write . . . to emphasize that our decision today does not mean that the SCPA protects chip functionality or provides protection to chips in any way synonymous with the protections provided through patent law. To the extent our opinion or the district court's rulings pertain to functionality, it is in the context of a case where the experts differed with each other (and even occasionally with themselves) about what the layout design amounts to, and what the architecture of the semiconductor chips at issue is in relation to the organization of components, groupings of transistors, and interconnecting lines. Based on the state of the evidence, I cannot say that the district court erred in allowing the jury to consider the entirety of the mask works for both chips including not only the design layout of the transistors, but also how these transistors came together on the chip to form components and how both chips placed interconnecting lines between those components and groups of transistors. . . .

COMMENTS AND QUESTIONS

1. One of the most remarkable aspects of the SCPA has been the dearth of litigation. Congressional testimony prior to the SCPA's enactment indicated that piracy was rampant. A number of explanations have been offered for low litigation rate. First, most "piracy" prior to the act may have been due to uncertainty about the scope of mask work protection under the copyright law. The SCPA clarified the law, thereby leading firms to conform their behavior to the requirements of the act. Second, the level of piracy alluded to during the hearings may have been exaggerated. Alternatively, advances in chip design in the mid to late 1980s may have alleviated piracy problems. Highly integrated modern mask works cannot be quickly and accurately duplicated without using a fabrication process virtually identical to that of the original manufacturer. Since such processes are highly proprietary, it is unlikely that

pirates could duplicate complex, high-density chip designs without costly and time consuming reverse engineering. Ron Laurie, The First Year's Experience Under the Chip Act or "Where Are the Pirates Now That We Need Them?" 3 Computer Lawyer 11, 21 (Feb. 1986).

2. Given the importance of the reverse engineering exception to SCPA liability, are you persuaded that the district court's error in instructing the jury in Altera was harmless? Would a jury reading those instructions really understand that copying of even the protectable elements of the mask work is permissible if done through reverse engineering, so long as the defendant's work as a whole is not "substantially identical" to the plaintiff's?

3. *Database Protection.* One other form of sui generis protection has taken root outside the United States. The European Community adopted database protection in 1996. See Directive 96/9/EC of the European Parliament and of the Council of 11 March 1996 on the legal protection of databases. This new form of protection has proven to be controversial. See Stephen M. Maurer, P. Bernt Hugenholtz, and Harlan J. Onsrud, 'Europe's Database Experiment', 294 Science 789 (Oct. 26, 2001); James Boyle: A Natural Experiment; Thomas Hazlett: Changes Merit Symmetric Scrutiny; Richard Epstein: Why Rule Out Copyright Protection for All Databases? Financial Times (Nov. 22, 2004). We have seen repeated efforts to adopt such legislation in the United States over the past dozen years, but it has struggled to gain traction.

4. How well does the sui generis model of the SCPA work in the software context? As the principal software copyright cases have highlighted, the highly utilitarian nature of computer programming poses substantial problems for judges seeking to maintain the integrity of the copyright law. In the *Altai* case, Judge Walker commented

> Generally, we think that copyright registration — with its indiscriminating availability — is not ideally suited to deal with the highly dynamic technology of computer science. Thus far, many of the decisions in this area reflect the courts' attempt to fit the proverbial square peg in a round hole. The district court, see Computer Assocs., 775 F. Supp. at 560, and at least one commentator have suggested that patent registration, with its exacting up-front novelty and non-obviousness requirements, might be the more appropriate rubric of protection for intellectual property of this kind. See Randell M. Whitmeyer, Comment, A Plea for Due Processes: Defining the Proper Scope of Patent Protection for Computer Software, 85 Nw. U. L. Rev. 1103, 1123-25 (1991); see also Lotus Dev. Corp. v. Borland Int'l, Inc., 788 F. Supp. 78, 91 (D. Mass. 1992) (discussing the potentially supplemental relationship between patent and copyright protection in the context of computer programs). In any event, now that more than 12 years have passed since CONTU issued its final report, the resolution of this specific issue could benefit from further legislative investigation — perhaps a CONTU II.

Judge Boudin, in his thoughtful concurrence in *Lotus,* expressly acknowledged courts' limitations in confronting the challenges posed by legal protection for computer software. He concluded by noting that "Some solutions (e.g., a very short copyright period for menus) are not options at all for courts but might be for Congress. In all events, the choices are important ones of policy, not linguistics, and they should be made with the underlying considerations in view."

This section considers alternatives and assesses the potential for the development of a sui generis regime for the protection of computer software.

Peter S. Menell, Tailoring Legal Protection for Computer Software
39 Stan. L. Rev. 1329 (1987)

. . .

B. Tailoring Legal Protection for Operating Systems

Part of the reason for copyright's inability to promote economic efficiency in the provision of computer products is that the public goods and network externality problems suggest conflicting modes of legal protection. Public goods problems are alleviated by expanding legal protection for intellectual work. External benefits from networks are promoted by facilitating access to a standard. Thus, the difficult policy question is how to promote standardization while at the same time encouraging continuing innovation (along the entire spectrum from software to hardware). By closely tailoring legal protection to reward desired innovation while permitting reasonable access to industry standards, it is possible to reach a satisfactory accommodation of these apparently conflicting objectives.

In theory, patent law is more appropriate than copyright for protecting the intellectual work contained in computer operating systems. A patent protects new and useful processes and machines. Given the interchangeability of hardware and software, it seems logical to protect computer operating systems and dedicated computers that embody a particular operating system with the same form of legal protection. Because patent law protects ideas, those who create patentable operating systems could be better assured of appropriating a substantial portion of the benefits of their efforts.

. . . [T]he importance of network externalities flowing from widespread access to a common mini- and microcomputer operating system suggests that legal protection should be hard to come by and relatively short in duration.

To encourage innovation in operating system technology, Congress should consider creating a hybrid form of patent protection specifically tailored to accommodate the market failures endemic to the provision of computer operating systems.[190] As with traditional patent law, the standard for protection should be novelty, nonobviousness, and usefulness; dominant firms (or anyone else) should not be able to "lock up" an industry standard simply by expressing it in a unique way.

To be feasible, the modified form of patent protection for computer operating systems should be based on a timely examination of patent applications. And given the rapid pace of technological change in the computer field and the interest in promoting access to industry standards, patent protection for operating systems should be shorter in duration than traditional patent protection.

---

190. Congress has recently followed a sui generis approach in designing legal protection for semiconductor chips. . . . Other commentators have also urged Congress to create a hybrid form of legal protection for computer software. See Davidson, supra note 63, at 673-82; Galbi, Proposal for New Legislation to Protect Computer Programming, 17 J. Copyright Soc'y 280 (1970); Karjala, supra note 63, at 61-81; Samuelson, Creating a New Kind of Intellectual Property: Applying the Lessons of the Chip Law to Computer Programs, 70 Minn. L. Rev. 471, 507, 529-31 (1985); Stern, The Case of the Purloined Object Code: Can It Be Solved? (Part 2), Byte, Oct. 1982, at 210, 222.

In order to promote continued innovation in widely used operating systems, the operating system patent code should, like the Semiconductor Chip Protection Act, permit some limited form of reverse engineering. And like traditional patent law, the hybrid code should allow consumers to buy a ROM chip or other device containing a patented operating system and modify it for sale to a third person.

Because traditional patent law affords absolute protection, however, it would inhibit realization of network externalities from operating systems satisfying the above subject matter requirements. In order to facilitate realization of network externalities, therefore, the hybrid patent code should contain a flexible compulsory licensing provision. Such a provision would promote access to an industry standard while assuring rewards to the creator of an innovative and socially valuable operating system. It would also limit the ability of dominant firms in the industry to engage in anticompetitive practices. . . .

In light of the strong network externalities flowing from compatibility, computer operating systems serve as "essential facilities" in computer hardware markets. Unless a firm can get onto the network, its products will be at a great disadvantage relative to those that can run the vast stock of application programs designed for the industry standard. The operating system royalty rate per use should be set so as to compensate true innovators for the cost of building a useful "highway" for the market plus a fair profit (adjusted for the risk of failure). For high volume products, these rates would probably be low. In the microcomputer market, for example, the rate would probably be less than one dollar for access to the major operating systems (assuming that the Apple and IBM operating systems merited hybrid patent protection at all.)

A patent code for operating systems based on the above outline strikes a preferable balance of the conflicting policy concerns raised by computer operating systems. By providing solid protection for truly innovative and useful operating systems, the code would reward innovation in operating systems. The limits on this regime of protection — moderate duration, reverse engineering, adaptation — and the provision for compulsory licensing would promote access to operating systems that emerge as industry standards, wide diffusion of computer products, and innovation in hardware products. The code would also avoid wasteful expenditure of resources on efforts to emulate an industry standard.

The proposed operating system code would probably entail somewhat higher administrative costs than the current system. Patent examinations, though streamlined, would be significantly more expensive than the cost of copyright registration. Moreover, compulsory licensing proceedings, as well as the cost of monitoring use of protected operating systems, would add to the expense of the system. If the royalty rates were low (as the microcomputer example indicates), however, members of the industry could be expected to cooperate in ensuring that patent owners were properly compensated.

C. Tailoring Legal Protection for Application Programs

As with legal protection for operating systems, Congress should consider creating a special form of legal protection for application programs. Given the importance of improving existing programs as a primary mode of technological innovation and the presence of some network externalities, legal protection should be significantly shorter in duration than traditional copyright protection. The relatively short commercial life of most application programs indicates that legal protection should be correspondingly short.

The regime for protecting application programs should also allow for reverse engineering. In designing legal protection for semiconductor chips, Congress recognized the importance of reverse engineering in enabling researchers to advance a field in which innovations are cumulative. A limited reverse engineering provision in the application software code would similarly promote the advancement of application software technology.

Congress should also consider the desirability of a limited form of compulsory licensing of application packages. In order to realize the benefits of network externalities and to promote creativity in the integration of software programs, it would seem worthwhile to allow limited access to application programs, particularly those that emerge as industry standards. This could be achieved without dulling primary creative incentives by delaying the availability of compulsory licensing for a limited period to allow the creator of the program to reap the rewards of commercial success.

COMMENTS AND QUESTIONS

1. Does Menell's proposal give too little protection for computer software? Since the mid 1980s, lawyers for IBM have argued vehemently for affording broad protection under copyright for non-literal elements of program code (structure, sequence, and organization) and compatibility elements. See Anthony L. Clapes, Patrick Lynch, and Mark R. Steinberg, Silicon Epics and Binary Bards: Determining the Proper Scope of Copyright Protection for Computer Programs, 34 UCLA L. Rev. 1493 (1987). Their basic argument is that

> [c]omputer programs are works of authorship in which the range and variety of expression are broad and deep. These works of authorship exhibit all the attributes of literary works of a kind with which the general public and the copyright laws are already quite conversant. Although written in languages that are unfamiliar to mose, they are not qualitiatively different from other literary works in ways that should affect the scope of legal protection available to their detailed design elements.

34 UCLA L. Rev. at 1583. They analogize computer programming to the writing of poetry and the composing of opera. Is this a more persuasive analogy than that between computer programming and engineering and other endeavors traditionally protected by patent law? Does it justify broad copyright protection for software? Is it an adequate basis for discouraging the study of *sui generis* regimes?

2. A multidisciplinary group of legal scholars and computer technologists has proposed a relatively brief period (one to five years) of automatic anticloning protection for program behavior and industrial designs of programs in order to appropriately reward software development. See Pamela Samuelson, Randall Davis, Mitchell D. Kapor, & J.H. Reichman, A Manifesto Concerning the Legal Protection of Computer Programs, 94 Colum. L. Rev. 2308 (1994). Such a system would augment existing intellectual property protection. The authors also propose that a registration system be developed by which innovative components of programs, such as command hierarchies and internal interfaces, would be licensed in a manner similar to the man-

agement of copyrights by copyright collectives. More recently, Professor Larry Lessig has advocated a reduced term for copyright protection. See Lawrence Lessig, The Future of Ideas: The Fate of the Commons in a Connected World (2002).

3. In 1991, the European Community adopted the Council Directive on the Legal Protection of Computer Programs. Council Directive 91/250, 1991 O.J. (L 122) 42. Article 1.2 provides that the Directive protects the "expression" of any form of a computer program, while "[i]deas and principles which underlie any element of a computer program, including those which underlie its interfaces, are not protected." Article 5.3 entitles users of programs "to observe, study or test the functioning of the program in order to determine the ideas and principles of the program, if he does so while performing any of the acts of loading, displaying, running, transmitting or storing the program . . ." Article 6 (Decompilation) provides:

> 1. The authorization of the rightholder shall not be required where reproduction of the code and translation of its form . . . are indispensable to obtain information necessary to achieve the interoperability of an independently created computer program with other programs, provided that the following conditions are met:
>
>   (a) these acts are performed by the licensee or by another person having a right to use a copy of a program, or on their behalf by a person authorized to do so;
>   (b) the information necessary to achieve interoperability has not previously been available to the persons referred to in subparagraph (a); and
>   (c) these acts are confined to the parts of the original program which are necessary to achieve interoperability.
>
> 2. The provisions of paragraph 1 shall not permit the information obtained through its application:
>
>   (I)   to be used for goals other than to achieve the interoperability of the independently created computer program;
>   (II) to be given to others, except when necessary for the interoperability of the independently created computer program; or
>   (III) to be used for the development, production or marketing of a computer program substantially similar in its expression or for any other act which infringes copyright. . . .

Is the EC Directive more or less restrictive than U.S. law in enabling a competitor to copy non-literal elements of computer programs? to reverse engineer computer programs? to achieve operating system compatibility? user interface compatibility? See Leo Raskind, Protecting Computer Software in the European Economic Community: The Innovative New Direction, 18 Brooklyn J. Int'l L. 729, 745-46 (1992); Thomas Vinje, The Development of Interoperable Products Under the EC Software Directive, 8 Computer Lawyer 1 (Nov. 1991); Huet and Ginsburg, Computer Programs in Europe: A Comparative Analysis of the EC Software Directive, 30 Colum. J. Transnational L. 327, 362-63 (1992).

4. *The Political Economy of Intellectual Property Reform.*

> In view of the substantial vesting of economic interests that has already occurred around the existing legal regime for computer software, it is difficult to imagine Congress giving serious attention to a comprehensive reform effort without a crying need. After many years of confusion, the existing legal regime appears to be muddling through. In the view of most commentators, the . . . *Altai* and *Sega* decisions have correctly resolved two of the

three major problems that have plagued copyright protection for software. Patent protection for software may prove to be problematic, although the extent of problems to date has been modest. . . .

Any attempt at *sui generis* reform would face not only the private interests vested in federal copyright law, but also the inertia flowing from the hard-won battle to establish an international copyright regime for computer software . . . [T]he United States government devoted substantial effort over the past decade to browbeating most of the developed nations into following its path. Neither the U.S. government nor the many entities desiring uniform protection for their products across national borders are interested in starting a new fight.

In this political climate it would be more efficacious to address the individual problems most amenable to reform than to attempt a comprehensive, *sui generis* scheme. The two most important targets would appear to be concerns about patent protection and impediments to the nascent field of multi-media software products.

Peter S. Menell, The Challenges of Reforming Intellectual Property Protection for Computer Software, 94 Colum. L. Rev. 2644, 2651-54 (1994). Has the window of opportunity for addressing software reform — copyright, patent, contract, or otherwise — passed? What proposals would be both desirable and politically feasible? One explanation for the lack of political moment may be the inherent adaptability of copyright and patent to at least some aspects of technological change. Is this explanation — that courts are "muddling through" using existing laws — the correct one?

8

Intellectual Property and Competition Policy

A. PRINCIPLES OF ANTITRUST LAW

1. The Scope of Antitrust Law

Antitrust law protects competition and the competitive process by preventing certain types of conduct that threaten a free market. For example, antitrust law prohibits competitors from agreeing on the price they will charge. It prohibits certain "predatory" practices designed to exclude competitors from the market, and it places certain limits on the behavior of firms with market power.

Competition is good for a variety of reasons. Basic economics teaches that firms in competition will produce more and charge lower prices than monopolists. Monopolists not only take money away from consumers by raising prices, but they impose a "deadweight loss" on society by reducing their output below the level that consumers would be willing to purchase at a competitive price. Monopoly has other problems as well. It inherently reduces consumer choice, and monopolists will usually have fewer incentives to innovate than do competitive firms.

The first United States antitrust law was passed in 1890. The Sherman Act, as the statute is called, was a reaction to populist pressure on Congress to do something about the "trusts" that had come to dominate the American business landscape. The Sherman Act granted broad powers to government to break up trusts and other conspiracies in restraint of trade. The Sherman Act is still in force today; it provides in part:

> **Section 1 [15 U.S.C. § 1].** Every contract, combination in the form of trust or otherwise, or conspiracy, in restraint of trade or commerce among the several States, or with a foreign nation, is declared to be illegal....
> **Section 2 [15 U.S.C. § 2].** Every person who shall monopolize, or attempt to monopolize, or combine or conspire with any other person or persons, to monopolize any part of the trade or commerce among the several States, or with foreign nations, shall be deemed guilty of a felony....

By their terms, these laws are broad indeed. In some sense, *every* contract necessarily restrains trade, since it forecloses options that were once open. Under the literal terms of section 1, employment contracts and sales contracts might be considered antitrust violations.

In practice, courts quickly gave the Sherman Act a more restrictive interpretation. The Supreme Court read section 1 as prohibiting only *unreasonable* restraints of trade. Much antitrust jurisprudence in the past century has attempted to delineate reasonable and unreasonable restraints of trade.

The courts also limited the reach of section 2 to ensure that successful businesses would not be punished for their success. Because the antitrust laws provide for felony criminal punishments, private treble-damage actions and injunctions, and plaintiffs' attorney fees, the possibility of overdeterring legitimate business conduct is a real concern. To avoid this problem, courts have distinguished between merely possessing a monopoly and actively acquiring or maintaining a monopoly through anticompetitive conduct. Section 2 prohibits only the latter.

Congress has never amended the core sections of the Sherman Act. In 1914, however, in response to what it perceived as lax judicial enforcement of the antitrust laws, Congress passed the Clayton Act. The Clayton Act contains a number of specific provisions that served to clarify the Sherman Act. For example, section 3 of the Clayton Act enumerates and prohibits certain types of agreements that are also included within the general scope of Sherman Act section 1.

The Clayton Act was amended in 1950 to prohibit certain types of mergers. Section 7 of the Clayton Act prohibits mergers and acquisitions (of a corporation, stock, or assets) where "the effect of such acquisition may be substantially to lessen competition or to tend to create a monopoly" in "any line of commerce . . . in any section of the country." Thus the antitrust laws prevent businesses from acting anticompetitively in three principal ways.

Monopolization. It is not illegal to have a monopoly. However, monopolists and firms gaining market power are subject to greater scrutiny for anticompetitive conduct and can violate the law even if they act unilaterally. A monopolist violates section 2 if it has "market power" (defined as the power to raise prices or exclude competition in a relevant market) and engages in anticompetitive conduct designed to acquire, maintain, or extend that power. Over time, courts have identified several anticompetitive practices (and the circumstances in which they are actionable); we will discuss several in the balance of this chapter. A company may also be guilty of "attempted monopolization" if it intends to monopolize a market, engages in anticompetitive conduct, and has a "dangerous probability of successful monopolization." See, e.g., Spectrum Sports v. McQuillen, 506 U.S. 447 (1993).

To find that a defendant has "monopolized" a market, the court must first decide what the relevant market is. Specifically, the court must identify a product or set of products and a geographic region within which most consumers make their purchases. Controlling such a market will allow a monopolist to raise prices without losing customers to competitors from outside the market. The definition of a relevant market and the analysis of power in that market are both extremely complex questions. A good deal of legal and economic work has gone into the attempt to define exactly what market power is. For a more detailed discussion of this issue, see United States

Department of Justice and Federal Trade Commission, Revised Merger Guidelines § 1 (April 2, 1992).

Agreements. Courts have identified two basic types of agreements that may be in restraint of trade — agreements among competitors (called "horizontal restraints") and agreements between buyers and sellers (called "vertical restraints").[1] Vertical restraints are generally less threatening to competition than horizontal restraints. With the exception of vertical minimum price fixing, they are generally judged under the "rule of reason." Under the rule of reason, courts balance the anticompetitive harms of a restraint against its procompetitive benefits. Only those restraints that produce harms significantly in excess of benefits to competition are deemed unreasonable.

Horizontal restraints are more troubling because they may allow the participants to create a cartel, which can then behave anticompetitively, much as a monopolist would. At first, most agreements between competitors were deemed illegal "per se," without any necessity for a weighing of harms and benefits to competition. Today the Supreme Court has retreated from that position, recognizing that certain agreements among competitors may be efficient and procompetitive. Most horizontal restraints are now judged under the rule of reason. Only certain forms of "naked" agreements to fix prices or divide territories remain illegal per se.

Mergers. Section 7 of the Clayton Act is intended to catch monopolies "in their incipiency." Because of this, the standard for proving the existence of market power is significantly lower in the case of a merger than in a section 2 monopolization claim. If a merger would substantially lessen competition in a relevant market, the courts may prohibit it. As with most other areas of antitrust law, however, courts are usually willing to consider a merger's potential benefits to competition in deciding whether it should be prevented.

Most merger enforcement takes place at the Antitrust Division of the United States Department of Justice and at the Federal Trade Commission. The Hart-Scott-Rodino Act of 1976 requires companies to file a statement with the division and the FTC and to seek approval for mergers over a certain size. If the division or the FTC feels that a merger will restrain competition, it can challenge the merger in court before the merger takes place. 15 U.S.C. § 7A.

Note on Antitrust Theory

The antitrust laws clearly reflect some congressional judgment that "big is bad." But *why* it is bad has been the subject of considerable debate during the past hundred years. At least three different rationales have been advanced to explain the antitrust laws. The first view may be called the "populist" view of antitrust. Populists take the position that big is *intrinsically* bad. It may be bad for a variety of reasons: because it concentrates wealth, because it reduces product diversity, or because it concentrates political power, for example. But regardless of the reason, populists would use the

---

1. Actually, the term "vertical restraints" refers to a whole class of transactions between companies in a vertical relationship in the chain of distribution, including dealers, franchisors, distributors, resellers, etc.

antitrust laws as a weapon against all monopolies, and against oligopolies as well. Populists can find substantial support for their position in the legislative history of the Sherman Act and in the political climate in the late nineteenth century. For an elaboration of the populist view, see Walter Adams, The Bigness Complex (1986); Victor H. Kramer, The Supreme Court and Tying Arrangements: Antitrust as History, 69 Minn. L. Rev. 1013 (1985). However, the populist view of antitrust is out of vogue today because it ignores economic justifications for corporate growth and is inconsistent with the more recent trend toward permitting mergers and agreements that are justified on efficiency grounds.

A second group sees antitrust as designed to protect small businesses from being driven out of the market. This view is related to the populist view, since both are likely to vigorously oppose monopolies and growing concentration in an industry. But they differ in the instrumental value served by preventing monopoly. Advocates of this view tend to view antitrust as an "unfair competition" statute and to apply the antitrust laws to a number of practices that are bad for competitors but good for consumers. For example, small business advocates are likely to be skeptical of price cuts by large national companies, which they fear will "squeeze out" local independents. The legislative history of the Clayton Act and subsequent antitrust laws lends some support to this view, but it too is inconsistent with the recent trend of antitrust decisions. Indeed, a fundamental maxim of modern antitrust law is that it "protects competition, not competitors." Brown Shoe Co. v. United States, 370 U.S. 294 (1962).

A third theory of the antitrust laws — and the theory that clearly dominates modern antitrust analysis — is the "economic" or "social welfare" model. This approach sees the purpose of the antitrust laws as promoting social welfare by ensuring that markets work freely and without interference. Antitrust law is aimed at practices that corrupt the market or usurp its function. On this view, antitrust is particularly concerned with preventing horizontal restraints of trade. While there is relatively little in the history of the antitrust laws that suggests Congress was primarily concerned with economics in 1890,[2] this approach has gained great currency with courts and scholars in the past 25 years. See, e.g., Robert H. Bork, The Antitrust Paradox: A Policy at War with Itself (1978); Richard Posner, Antitrust Law: An Economic Perspective (1976). Most antitrust experts now use economic effects as the basis for their analysis, although they continue to disagree vigorously over *how* those effects are to be analyzed.

2. Intellectual Property and Antitrust Law

Intellectual property law grants certain exclusive rights to the creators, inventors, or discoverers of certain intangible but valuable assets. In previous chapters, we have discussed the nature of these rights at length. There is substantial disagreement as to whether intellectual property rights are "property" in the ordinary sense of the term. But whatever they are, intellectual property rights give their owner the right to exclude others from using or copying his invention or creation.

---

2. But cf. F. M. Scherer, Efficiency, Fairness, and the Early Contributions of Economists to the Antitrust Debate, 29 Washburn L.J. 243, 250-51 (1990) (arguing that the views of mainstream economists‧ in 1890 were consistent with the goals of the Sherman Act).

Traditionally, the conventional wisdom was that the antitrust laws and the intellectual property laws are in conflict. See, e.g., Stephen Calkins, Patent Law: The Impact of the 1988 Patent Misuse Reform Act and Noerr-Pennington Doctrine on Misuse Defenses and Antitrust Counterclaims, 38 Drake L. Rev. 175, 176 n.1 (1989) (noting this conflict). Baldly stated, the conflict arises because the intellectual property laws grant "monopolies" to inventors, while the goal of the antitrust laws is to prevent or restrict monopoly.

However, scholars are increasingly taking the position that the two laws are not in conflict at all. Rather, they are complementary efforts to promote an efficient marketplace and long-run, dynamic competition through innovation. Paul Goldstein was an early proponent of the idea that the intellectual property laws were designed to promote competition. See Paul Goldstein, The Competitive Mandate: From Sears to Lear, 59 Cal. L. Rev. 971 (1971). So too was Ward Bowman.

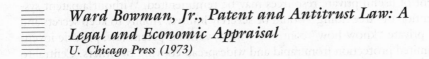

Ward Bowman, Jr., Patent and Antitrust Law: A Legal and Economic Appraisal
U. Chicago Press (1973)

. . .

Antitrust law and patent law are frequently viewed as standing in diametric opposition. How can there be compatibility between anti-trust law, which promotes competition, and patent law, which promotes monopoly? In terms of the economic goals sought, the supposed opposition between these laws is lacking. Both antitrust law and patent law have a common central economic goal: *to maximize wealth by producing what consumers want at the lowest cost.* In serving this common goal, reconciliation between patent and antitrust law involves serious problems of assessing effects, but not conflicting purposes. Antitrust law does not demand competition under all circumstances. Quite properly, it permits monopoly when monopoly makes for greater output than would the alternative of an artificially fragmented (inefficient) industry. The patent monopoly fits directly into this scheme insofar as its central aim is achieved. It is designed to provide something which consumers value and which they could not have at all or have as abundantly were no patent protection afforded.

Antitrust Law

Under the Sherman Act agreements in restraint of trade (section 1) and monopolization or an attempt to monopolize (section 2) are condemned as illegal. The economic rationale for this condemnation is that monopoly makes it possible to restrict output and raise prices so that consumers pay more for and get less of the things they want most. The output restriction made possible by monopoly idles, or transfers to less urgent uses, those resources, which but for the monopolistic restriction would be more efficiently employed. Conversely, a competitive market process allocates scarce resources to those uses the community values most highly. Insofar as the antitrust laws are successful, they promote a market-oriented, profit-incentive process unimpeded by artificial roadblocks to efficiency. Such is the rationale of market competition and the antitrust laws that support it.

Patent Law

Patent law, thought by some to be an exception to a general rule in favor of competition, shares with antitrust law its central purpose — efficiently providing those things consumers value. But the means are different. Patent law pursues this goal by encouraging the invention of new and better products. Invention, like other forms of productive activity, is not costless. Those who undertake it, therefore, must be rewarded. And so elusive a commodity as an idea which qualifies as invention is peculiarly susceptible to being freely appropriated by others. A patent is a legal device to insure that there can be a property right in certain ideas. Thus the temporary right of a patentee to exclude others is a means of preventing "free riding" so that the employment of useful private resources may be remunerated. Without a patent system, prevention of "free riding" would be severely limited. Ability to keep secrets and to enforce private "know-how" contracts would, without patent law, provide inventors very limited protection from rapid and widespread copying by others. Central to the economic justification of a patent system is the presumption that without the patent right, too few resources would be devoted to invention.

The "exclusive right to make, use and vend the invention or discovery," which Congress has long granted patentees, is thus a legal monopoly exempt from the more general proscription of trade restraints and monopolization under early common law and more recent antitrust statutes.

Goal Evaluation

The goal of both antitrust law and patent law is to maximize allocative efficiency (making what consumers want) and productive efficiency (making these goods with the fewest scarce resources). In achieving this goal under either antitrust or patent law the detriment to be avoided is output restriction. This may arise from monopolization which diverts production from more urgent to less urgent use or from legal rules requiring inefficient methods of production. The evil then may be viewed as net output restriction after efficiency increases are accounted for. Both antitrust and patent law seek output expansion not output restriction. Competition deserves support insofar as it brings about this result. And so it is with patents. The temporary monopoly afforded by a patent, once a particular invention has come into being, will have all the output-restrictive disabilities of any monopoly. The argument for patents is that without this temporary monopoly there would be insufficient profit incentives to produce the invention, and that because an invention is profitable only if consumers are willing to pay what the patentee charges, the consumers are therefore better off than they would be without the invention, even if they are charged "monopoly" prices. If this is so, a trade-off (some monopoly restraint for greater output in the long run) is in the interest of socially desirable resource allocation. An appraisal of alleged conflicts between antitrust law and patent law depends upon understanding the role of profits in providing the incentive for undertaking efficient production of those things consumers value . . . .

As a public policy tool perfect competition does provide useful insights into the equilibrating forces of a competitive process unencumbered by either external or internal impediments to the flow of resources responding to price and profit signals. Monopoly and agreements in restraint of trade are such impediments, and, as has been stressed, they are condemned for this reason. This does not mean, however, and the antitrust laws have never interpreted to mean, that every departure from the perfect competitive model becomes illegal. Rather, these laws, when consistently and rationally applied, view competition as a process designed to work for the benefit of consumers. They take the world as it is and evaluate available alternatives. And the allocation of resources cannot be fully evaluated unless productive efficiencies (cost savings or better products) are assessed along with restraints. Stressing perfect competition as an ideal involves the danger of underplaying the importance of productive efficiencies: savings arising from large size and efficiencies arising from innovation in manufacturing and from company organization and management, as well as from dissemination of knowledge through sales effort and advertising. Employee relations and myriad other activities also are sources of productive efficiency. None of these commonplace examples would exist if competition were perfect. These examples, moreover, are not output restricting. Imperfect competition is not monopoly. Dealing with imperfections is what competition as a process rather than a static model is all about . . . .

Efficiency and Patents

If the temporary monopoly afforded to inventors is to be justified as on balance desirable, its existence and the uses to which it is put must face up to the kind of restraint/efficiency assessment applied under antitrust law. Analytically, the combination of two competing patents raises the same questions as the merger of two competing companies; both combine competing assets. That the productive efficiencies predictable from a patent pool may seem *de minimis* does not change the nature of the analysis. It may, however, seem more analogous to the antitrust cartel rule than to the rule on mergers. And, of course, jointly fixing the price to be charged for competing patents is an even more obvious analogy.

The problem of assessing permissible output restriction under patent law is, in one important sense, different from that under antitrust law. The legal propriety of a basic patent monopoly has to be recognized. Consequently, evaluating whether certain patent licensing practices should be sanctioned will involve the proper scope of the legal monopoly. Is more being monopolized than what the patent grants, or is the practice merely maximizing the reward attributable to the competitive advantage afforded by a patent? . . . [M]any cases of alleged patent misuse—many, indeed, which are said to violate antitrust law—involve vertical contract relationships between patentee and licensee in which no such extension of scope is explicable. Many such contracts between patentees and licensees, moreover, lead to productive efficiencies unattainable if the practices are prohibited. Given a legal patent monopoly, it is important not only to appraise whether the monopoly has been broadened—scope extension—but also to insure that prohibiting alleged misuses does not leave the holder of the patent a permissible alternative which, although less profitable, is more restrictive than the alleged misuse. The "misuse" may actually be the more efficient use of the patent from the social as well as the private point of view. It pays monopolists as well as competitors to be efficient, as has been stressed....

Growth makes an economy richer in successive years. Achieving continuously more efficient resource allocation is not only compatible with growth but is an essential aspect of it. But economic growth is more often associated with a decision to produce more and newer products for tomorrow by foregoing consumption today. This may be done by voluntary savings and investment or by a centralized decision about what should be growing. In neither case, however, is there any compelling reason to believe that tomorrow is necessarily more important than today. And this is as true for that unique commodity information, the subject matter of patent law, as it is for other unpatentable commodities. Thus, even though the patent legislation looks to the promotion of the progress of science and useful arts, and the antitrust law prohibits artificial barriers to the entry of new products or new production methods, neither need favor or disfavor investment over current consumption, and neither need favor particular forms of investment. The allocative function which antitrust law should support and patent law should supplement is market motivation, making it profitable to produce the wide variety of things that people choose. Free exercise of choice is not limited to choices among goods, or choices between goods and services, or choices between either of these and leisure. Timing — the choice between having now and having later — is also involved.

The patent law is often described as a special inducement to growth. But, as has been indicated, the patent law, like antitrust law, does not foster growth for growth's sake. Rather, it is designed to make possible those new products and new production methods which customers deem worth paying for. Promoting science and the useful arts ... is not to be viewed as a reasonably exact facsimile of a crop-support program for the over-production of invention. If the patent system works as it is designed to work, growth can be expected to be fostered. Scarce resources are diverted from current consumption to better products for tomorrow. Growth under a patent system is thus not a separate goal competing with consumer choice. It is the result of it....

COMMENTS AND QUESTIONS

1. In the first half of this century, Joseph Schumpeter argued that monopoly, not competition, is most conducive to innovation. In his view, competition will lead to a focus on short-run marginal cost, to the exclusion of (long-run efficient) capital investments in research and development. In the absence of intellectual property protection, the result of competition is insufficient innovation. If that is true (and we have certainly assumed that it is for much of this book), what does it say about the wisdom of applying the antitrust laws to intellectual property? Is it a good idea to enforce competition in high-technology industries? Or should the goal of antitrust law be modified somehow in this context?

2. Recent advances in economic thought have complicated the innovation picture. In particular, a significant paper by Richard Levin et al. forcefully demonstrates that different industries depend on different types of appropriability mechanisms. Richard Levin, Alvin Klevorick, Richard Nelson & Sidney Winter, Appropriating the Returns from Industrial Research and Development, 3 Brookings Papers on

Economic Activity 783 (1987). The fact that appropriability varies from industry to industry significantly complicates the simple economic model of patents that Bowman sets forth. Further, work by David Teece and others suggests that firms within an industry will adapt strategically to the appropriability environment of that industry by taking advantage of what Teece calls "co-specific assets." David J. Teece, Profiting from Technological Innovation: Implications for Integration, Collaboration, Licensing and Public Policy, 15 Res. Pol. 285 (1986). The upshot of this recent work is that the economic model of patent incentives is substantially more complicated than it appeared to be a few years ago.

3. Consider carefully what Bowman is arguing. He asserts that intellectual property and antitrust are not ends in themselves but ways toward a larger goal of wealth maximization. As a matter of policy, he is probably correct.

Bowman's conclusion that the patent laws and the antitrust laws do not have conflicting purposes can be misleading, however. Bowman is best understood as saying that both patent law and antitrust law are tools to be used in promoting wealth maximization. They are designed to be used together to achieve a certain result. However, they strive toward that result in ways that are often in tension. Antitrust law seeks to maximize wealth by preventing monopoly, while patent law may in some cases encourage monopoly.

It is too facile to say that the patent laws give a monopoly to the patentee. Most patents cover a product or products that are only a small subset of a *product market*. It is therefore possible to have a perfectly competitive industry composed of patent holders, if each holds a patent on one of several possible products in a market. Only where an innovation creates an entirely new market, or represents a quantum advance in an old one, is the patent likely to confer an economic monopoly directly.

There is room, therefore, for the patent and antitrust laws to coexist in the service of long-run, dynamic efficiencies. In this scheme, patents constitute an exception to the reach of the antitrust laws, and antitrust constrains what a patentee can do with its patent. Efficient wealth maximization requires that a line be drawn between conduct which is permissible and that which is impermissible. Drawing that line is the subject of this chapter.

B. MONOPOLIZATION

1. Defining the Market

The first step in any antitrust inquiry is to define the relevant product and geographic markets. Monopolization does not occur in a vacuum. To be liable, an antitrust defendant must have monopolized *something, somewhere*. Market definition and market power analysis are well developed in antitrust law; we will not repeat the general issues here.

There are, however, some market definition issues that are specific to intellectual property law. The first of these is the long-standing presumption that intellectual property rights confer market power. The Supreme Court has reasoned that, since a patent confers the right to exclude competitors *from practicing the patented invention,* it must confer monopoly power over that invention. Thus the Court has on several occasions presumed market power from the existence of patents or copyrights. E.g., Jefferson Parish Hosp. Dist. v. Hyde, 466 U.S. 2 (1984); United States v. Loew's, Inc., 371 U.S. 38 (1962); see also Digidyne v. Data General Corp., 734 F.2d 1336 (9th Cir. 1984), *cert. denied*, 473 U.S. 908 (1985). In Independent Ink v. Illinois Tool Works, 396 F.3d 1342 (Fed. Cir. 2005), *cert. granted*, 125 S. Ct. 2937 (2005), the Federal Circuit held that it was bound by this prior dicta and that it would presume that all intellectual property rights conferred market power.

This presumption that intellectual property rights confer market power has little basis in fact. Patents grant the right to exclude in a tightly defined technological domain. In most cases, this does not translate into what an economist would call a "monopoly," since the technological domain is rarely coextensive with an economic product market. For example, consider a typical mechanical patent, say on a new type of muffler for lawn mowers. This typical patent does not confer the right to exclude others from making lawn mowers, of course; nor even from making most types of mufflers for lawn mowers. Only the new muffler is covered. Whether one considers the market for lawn mowers, for lawn mower components, or even lawn mower mufflers, this patent does not confer a monopoly. *Only* in the very limited product area defined by this new type of muffler and its close technological substitutes does the patent holder have exclusionary power.[3] Most patents are like that. Occasionally, however, a patent is granted on a "pioneering" or "basic" invention — e.g., the light-bulb, or the laser, or the communication satellite, or recombinant DNA techniques. In these rare instances, the patent may confer exclusionary power over an entire market and hence qualify as a true economic monopoly.

Commentators have been virtually unanimous in their criticism of the market power presumption. See, e.g., Herbert Hovenkamp et al., IP and Antitrust ch. 4; Philip Areeda & Louis Kaplow, Antitrust Analysis 441 (1987); Herbert Hovenkamp, Economics and Federal Antitrust Law § 8.3, at 219 (1985); William Montgomery, Note, The Presumption of Economic Power for Patented and Copyrighted Products in Tying Arrangements, 85 Colum. L. Rev. 1140, 1156 (1985). Moreover, the United States Antitrust Guidelines for the Licensing and Acquisition of Intellectual Property ("Intellectual Property Guidelines"), issued jointly by the Federal Trade Commission and the Antitrust Division of the Department of Justice, specifically reject the market power presumption:

> Market power is the ability profitably to maintain prices above, or output below, competitive levels for a significant period of time. The Agencies will not presume that a patent, copyright or trade secret necessarily confers market power upon its owner. Although the intellectual property right confers the power to exclude with respect to the *specific* product, process, or work in question, there will often be sufficient actual or potential close substitutes for such product, process or work to prevent the exercise of market power.

Intellectual Property Guidelines § 2.2 (1995). Legislation is periodically introduced in Congress to abolish the market power presumption. See, e.g., H.R. 2674, 104th Cong., 1st Sess. (1995). But the fact that the Supreme Court has agreed to

---

3. It is sometimes asserted that a patent is merely a property right, as opposed to a monopoly, and that the exclusionary power of a patent should not be deemed any more forceful than property rights in one's house. This is misleading, since unlike a house, or any piece of tangible (real or personal) property, a patent grants its owner the right to exclude others from doing certain things with their own physical property. For example, my ownership of a house does not ordinarily limit what you may do with your own house, but my "ownership" of a patented invention will prevent you from making or using the same invention, even if you developed it independently of me.

consider the issue may make these efforts moot; most commentators expect the Supreme Court to hold that an intellectual property right does not necessarily confer market power.

Innovation presents other, more complex problems in determining market power. See, for example, David J. Teece, Raymond Hartman & Will Mitchell, Assessing Market Power in Regimes of Rapid Technological Change, 2 Indus. & Corp. Change 3, 7 (1993) ("The more innovative the new product or process, the greater the conventional market power will appear, because price changes will have little or no influence on demand for a truly innovative product."); id. at 29 (proposing to measure market power by performance attributes, not price changes). See also Kenneth Burchfiel, Patent Misuse and Antitrust Reform: "Blessed Be the Tie," 4 Harv J.L. & Tech. 1 (1991) (suggesting simpler tests for determining market power, including: (1) whether a royalty was paid by licensees; (2) whether accused infringers chose to keep infringing during litigation; and (3) the profit derived from the patented product); Jesse Markham, Concentration: A Stimulus or Retardant to Innovation?, in Industrial Concentration: The New Learning 247 (Harvey Goldschmid, H. Michael Mann & J. Fred Watson eds. 1974) (summarizing empirical tests of the Schumpeterian hypothesis).

The Antitrust Division takes the position that markets may need to be defined differently in the intellectual property context. In addition to evaluating the effect of intellectual property rights on markets for goods, the division has defined two additional types of markets:

3.2.2 Technology Markets

Technology markets consist of intellectual property that is licensed (the "licensed technology") and its close substitutes — that is, the technologies or goods that are close enough substitutes significantly to constrain the exercise of market power with respect to the intellectual property that is licensed.[18] When rights to intellectual property are marketed separately from the products in which they are used, the Agencies may rely on technology markets to analyze the competitive effects of a licensing arrangement.

To identify a technology's close substitutes and thus to delineate the relevant technology market, the Agencies will, if the data permit, identify the smallest group of technologies and goods over which a hypothetical monopolist of those technologies and goods likely would exercise market power — for example, by imposing a small but significant and nontransitory price increase. The Agencies recognize that technology often is licensed in ways that are not readily quantifiable in monetary terms. In such circumstances, the Agencies will delineate the relevant market by identifying other technologies and goods which buyers would substitute at a cost comparable to that of using the licensed technology.

In assessing the competitive significance of current and likely potential participants in a technology market, the Agencies will take into account all relevant evidence. When market share data are available and accurately reflect the competitive significance of market participants, the Agencies will include market share data in this assessment.

---

18. For example, the owner of a process for producing a particular good may be constrained in its conduct with respect to that process not only by other processes for making that good, but also by other goods that compete with the downstream good and by the processes used to produce those other goods.

The Agencies also will seek evidence of buyers' and market participants' assessments of the competitive significance of technology market participants.... When market share data or other indicia of market power are not available, and it appears that competing technologies are comparably efficient, the Agencies will assign each technology the same market share. For new technologies, the Agencies generally will use the best available information to estimate market acceptance over a two-year period, beginning with commercial introduction.

3.2.3 Research and Development: Innovation Markets

...A licensing arrangement may have competitive effects on innovation that cannot be adequately addressed through the analysis of goods or technology markets. For example, the arrangement may affect the development of goods that do not yet exist. Alternatively, the arrangement may affect the development of new or improved goods or processes in geographic markets where there is not actual or likely potential competition in the relevant goods.

An innovation market consists of the research and development directed to particular new or improved goods or processes, and the close substitutes for that research and development. The close substitutes are research and development efforts, technologies, and goods that significantly constrain the exercise of market power with respect to the relevant research and development, for example by limiting the ability and incentive of a hypothetical monopolist to retard the pace of research and development. The Agencies will delineate an innovation market only when the capabilities to engage in the relevant research and development can be associated with specialized assets or characteristics of specific firms.

In assessing the competitive significance of current and likely potential participants in an innovation market, the Agencies will take into account all relevant evidence. When market share data are available and accurately reflect the competitive significance of market participants, the Agencies will include market share data in this assessment. The Agencies also will seek evidence of buyers' and market participants' assessments of the competitive significance of innovation market participants.... The Agencies may base the market shares of participants in an innovation market on their shares of identifiable assets or characteristics upon which innovation depends, on shares of research and development expenditures, or on shares of a related product. When entities have comparable capabilities and incentives to pursue research and development that is a close substitute for the research and development activities of the parties to a licensing arrangement, the Agencies may assign equal market shares to such entities.

Intellectual Property Guidelines §§ 3.2.2, 3.2.3 (citations omitted).

COMMENTS AND QUESTIONS

1. Does defining separate technology and innovation markets make sense? Another way of approaching this question is to ask whether these market definitions capture any anticompetitive conduct that could not be challenged under a traditional product market definition. One possible answer is that alleging a market for innovation allows the Antitrust Division to challenge mergers or anticompetitive conduct earlier than it could in product markets, since it need not wait until an actual product market is affected. Whether this is good or bad depends on your confidence in the ability of courts to define innovation markets and assess anticompetitive effects accurately.

But it is also possible that technology and innovation markets allow the antitrust laws to reach conduct that could not be attacked under traditional product market definition. For example, a dominant company in a product market may be able to merge with or drive out a competitor whose current market share is insignificant, but who has the best technology and is therefore well placed to challenge the dominant firm. If nothing else, technology markets may make market power analysis more forward-looking, by requiring the parties to focus on likely future market power rather than just current market shares. At the same time, technology and innovation markets may be used by the government to attack mergers that do not present product market overlaps at all.

2. How can we define a market for "research and development"? The Intellectual Property Guidelines suggest several possible ways of determining market shares, including arbitrarily assigning each participant an equal share. Is this reasonable? Is R&D expenditure a fair measure? One problem with the guidelines' assumption that innovation monopolists can profitably retard or restrict innovation is that it depends on a fairly static model of monopoly. Monopolists who cut back on research and development may reap the rewards in the short term but are likely to be left behind as new companies "enter the market" for research and development. Can an argument be made the other way—that the real danger is that a dominant firm will extend its product monopoly by spending excessively on research and development?

PROBLEM[4]

Problem 8-1. Two companies that specialize in advanced metallurgy agree to cross-license future patents relating to the development of a new component for aircraft jet turbines. Innovation in the development of the component requires the capability to work with very high tensile strength materials for jet engines. Aspects of the licensing arrangement raise the possibility that competition in research and development of this and related components will be lessened. The Antitrust Division is considering whether to define an innovation market in which to evaluate the competitive effects of the arrangement. What factors should it take into account in making this decision?

---

2. Intellectual Property and Anticompetitive Conduct

It is not enough to define a market and determine that an intellectual property owner is a monopolist. The reason is that monopoly alone is not illegal. Rather, it must be combined with some form of anticompetitive conduct. Assuming that the intellectual property owner has market power in a particular case, what conduct will suffice? It is in this context that the "conflict" between intellectual property and the antitrust laws arises most starkly. Certainly, antitrust law must make some accommodation for monopolies that result from intellectual property if those intellectual property rights are to have any meaning. In particular, it is evident that merely

---

4. This problem is adapted from Example 3 in the Intellectual Property Guidelines.

possessing and enforcing a patent, copyright, trademark, or trade secret against competitors ought not to violate the antitrust laws:

> Although the conflict between the patent and the antitrust laws has long been thought troublesome, it is in fact even more deep-seated than is generally perceived. Consider a patentee that intends to employ a particular restriction, practice, or strategy in exploiting its patent. Limiting the analysis to the antitrust issues, which is the intended scope of this Article, one might initially conclude that the practice should be held permissible only if it does not violate the antitrust laws.
>
> Not long after the passage of the Sherman Act, however, courts realized that this approach is too facile. A practice is typically deemed to violate the antitrust laws because it is anticompetitive. But the very purpose of a patent grant is to reward the patentee by limiting competition, in full recognition that monopolistic evils are the price society will pay. Generalizing from this principle, one could reverse the initial conclusion, arguing that any action by a patentee in violation of the antitrust laws is privileged under the patent statute.
>
> Courts subsequently recognized that this conclusion was also too simplistic, because the patent statute was plainly not intended to bestow upon each patentee carte blanche in all its endeavors. For example, a patentee who negotiates a favorable royalty by holding a prospective licensee at gunpoint clearly will not be relieved from the proscriptions of either criminal or contract law. The question is whether one should view antitrust law any differently. At a minimum, it seems clear that a firm having one otherwise insignificant patent may not freely engage in price-fixing, mergers, predatory pricing, or anything else it wishes solely on that account. The statutory limitation of the patent grant to seventeen years illustrates the position, now generally accepted by commentators, that the reward for inventive activity should not be unbounded. This realization, however, does not indicate whether any or all practices in violation of the antitrust laws are out of bounds.

Louis Kaplow, The Patent-Antitrust Intersection: A Reappraisal, 97 Harv. L. Rev. 1815 (1984).

If we start from the proposition that it cannot be illegal to possess a patent (or other intellectual property right), what role remains for antitrust law? Two types of factual situations may still present antitrust problems. First, the patentee may have acquired his patent illegally (for example, through fraud). Second, a patentee may use his patent to acquire or maintain power beyond that granted by the intellectual property right itself. In both situations, antitrust courts are properly concerned with the effects that anticompetitive use of the patent may have. See generally Beal Corp. Liquidating Trust v. Valleylab, Inc., 927 F. Supp. 1350 (D. Colo. 1996) (patent law does not preclude antitrust action regarding patent license, where effect of license is to expand the scope of the patent).

a. Enforcing Intellectual Property Rights

Walker Process Equipment, Inc. v. Food Machinery & Chemical Corp.
Supreme Court of the United States
382 U.S. 172 (1965)

Mr. J. CLARK delivered the opinion of the Court.

The question before us is whether the maintenance and enforcement of a patent obtained by fraud on the Patent Office may be the basis of an action under § 2 of the

Sherman Act.... [Food Machinery sued Walker for patent infringement, and Walker counterclaimed for violations of the antitrust laws.] Walker then amended its counter-claim to charge that Food Machinery had "illegally monopolized interstate and for-eign commerce by fraudulently and in bad faith obtaining and maintaining...its patent...well knowing that it had no basis for...a patent." It alleged fraud on the basis that Food Machinery had sworn before the Patent Office that it neither knew nor believed that its invention had been in public use in the United States for more than one year prior to filing its patent application when, in fact, Food Machinery was a party to prior use within such time. [The District Court dismissed Walker's counterclaim, and the Court of Appeals affirmed.]

... The gist of Walker's claim is that since Food Machinery obtained its patent by fraud it cannot enjoy the limited exception to the prohibitions of § 2 of the Sherman Act, but must answer under that section and § 4 of the Clayton Act in treble damages to those injured by any monopolistic action taken under the fraudulent patent claim. Nor can the interest in protecting patentees from "innumerable vexatious suits" be used to frustrate the assertion of rights conferred by the antitrust laws. It must be remembered that we deal only with a special class of patents, i.e. those procured by intentional fraud.

Under the decisions of this Court a person sued for infringement may challenge the validity of the patent on various grounds, including fraudulent procurement. In fact, one need not await the filing of a threatened suit by the patentee; the validity of the patent may be tested under the Declaratory Judgment Act.... At the same time, we have recognized that an injured party may attack the misuse of patent rights. To permit recovery of treble damages for the fraudulent procurement of the patent coupled with violations of § 2 accords with these long-recognized procedures. It would also promote the purposes so well expressed in Precision Instrument [Mfg. Co. v. Automotive Maintenance Machinery, 324 U.S. 806 (1945)] at 816:

> A patent by its very nature is affected with a public interest.... [It] is an exception to the general rule against monopolies and to the right to access to a free and open market. The far-reaching social and economic consequences of a patent, therefore, give the public a paramount interest in seeing that patent monopolies spring from backgrounds free from fraud or other inequitable conduct and that such monopolies are kept within their legitimate scope.

III

Walker's counterclaim alleged that Food Machinery obtained the patent by knowingly and willfully misrepresenting facts to the Patent Office. Proof of this asser-tion would be sufficient to strip Food Machinery of its exemption from the antitrust laws. By the same token, Food Machinery's good faith would furnish a complete defense. This includes an honest mistake as to the effect of prior installation upon patentability — so-called "technical fraud."

To establish monopolization or attempt to monopolize a part of trade or com-merce under § 2 of the Sherman Act, it would then be necessary to appraise the exclusionary power of the illegal patent claim in terms of the relevant market for the product involved. Without a definition of that market there is no way to measure Food Machinery's ability to lessen or destroy competition. It may be that the device — knee-action swing diffusers — used in sewage treatment systems does not comprise a relevant market. There may be effective substitutes for the device which do not infringe the patent. This is a matter of proof, as is the amount of damages suffered by Walker.

[The Court remanded the case with instructions to consider the antitrust claim on its merits.]

Brunswick Corporation v. Riegel Textile Corporation

United States Court of Appeals for the Seventh Circuit
752 F.2d 261 (7th Cir. 1984)

POSNER, Circuit Judge.

Brunswick Corporation appeals from the dismissal, on the pleadings, of its antitrust suit against Riegel Textile Corporation. 578 F. Supp. 893 (N.D. Ill. 1983). The appeal requires us to consider aspects of the relationship between patent and antitrust law. The complaint alleges that in 1967 Brunswick invented a new process for making "antistatic yarn," which is used to make garments worn in hospital operating rooms and other areas where there are volatile gases that could be ignited by static electricity. Brunswick, which is not itself a textile manufacturer, disclosed its invention to Riegel, which is. Riegel promised to keep the invention secret. In April 1970 Brunswick applied for a patent on the new process and in August Riegel did likewise — in breach of its agreement with Brunswick. (Riegel denies that this was a breach, but as Brunswick has been given no chance to substantiate the allegations of its complaint we must treat them as true for purposes of this appeal.) Without considering Brunswick's application the Patent Office issued a patent to Riegel in 1972. The Patent Office discovered the Brunswick application in 1973, and in 1975 instituted a patent-interference proceeding to determine priority of invention between Riegel and Brunswick. See 35 U.S.C. § 135; 1 Rosenberg, Patent Law Fundamentals § 10.02 (2d ed. 1984). That proceeding was still pending before the Patent Office when Brunswick brought this lawsuit in 1982, but since the Patent Office has held that although Brunswick indeed invented the process first, its patent application was invalid. Brunswick has challenged this ruling in another lawsuit in the Northern District of Illinois, and it has also sued Riegel in an Illinois state court for unfair competition.

Brunswick's complaint in this case is that by procuring a patent by fraud and then defending the patent's validity groundlessly in the patent-interference proceeding, Riegel monopolized in the production of antistatic yarn in violation of section 2 of the Sherman Act, 15 U.S.C. § 2. The complaint also describes Riegel's misconduct as an attempt to monopolize and as a conspiracy (with Riegel's agents and employees) to monopolize, which are also forbidden by section 2, except that a conspiracy between a corporation and its employees is not actionable under antitrust law. University Life Ins. Co. v. Unimarc Ltd., 699 F.2d 846, 852 (7th Cir. 1983). The district court dismissed the suit on alternative grounds: the complaint fails to state an antitrust cause of action; the suit is barred by the antitrust statute of limitations.

Getting a patent by means of fraud on the Patent Office can, but does not always, violate section 2 of the Sherman Act. See, e.g., Walker Process Equipment, Inc. v. Food Machinery & Chem. Corp., 382 U.S. 172 (1965); United States v. Singer Mfg. Co., 374 U.S. 174, 196-97 (1963); American Cyanamid Co. v. FTC, 363 F.2d 757, 770-71 (6th Cir. 1966); see generally 3 Areeda & Turner, Antitrust Law

¶¶707a-b, d, f (1978). A patent entitles the patentee to prevent others from making or selling the patented product or, as here, using the patented production process, and he may be able to use this legal right to restrict competition. If antistatic yarn cannot be produced efficiently other than by using Riegel's patented process, Riegel may be able to exclude competition in the sale of such yarn, in which event it may have "monopoly power" — the "power to control prices or exclude competition." United States v. E. I. du Pont de Nemours & Co., 351 U.S. 377, 391-92 (1956) (plurality opinion). And to create (or attempt to create, or conspire to create) monopoly power by improper means is to monopolize (or attempt to monopolize, or conspire to monopolize) within the meaning of section 2 of the Sherman Act. See, e.g., United States v. Grinnell Corp., 384 U.S. 563, 570-71 (1966).

But "may" is not "does"; and for a patent fraud actually to create or threaten to create monopoly power, and hence violate section 2, three conditions must be satisfied besides proof that the defendant obtained a patent by fraud:

1. The patent must dominate a real market. See Walker Process Equipment, Inc. v. Food Machinery & Chem. Corp., supra, 382 U.S. at 177-78; American Hoist & Derrick Co. v. Sowa & Sons, Inc., 725 F.2d 1350, 1366-67 (Fed. Cir. 1984); Handgards, Inc. v. Ethicon, Inc., 601 F.2d 986, 993 n.13 (9th Cir. 1979). Although the Patent Office will not issue a patent on an invention that has no apparent utility, the invention need not have any commercial value at all (other products or processes may be superior substitutes), and it certainly need not have enough value to enable the patentee to drive all or most substitutes from the market. If a patent has no significant impact in the marketplace, the circumstances of its issuance cannot have any antitrust significance.

2. The invention sought to be patented must not be patentable. If the invention is patentable, it does not matter from an antitrust standpoint what skullduggery the defendant may have used to get the patent issued or transferred to him. The power over price that patent rights confer is lawful, and is no greater than it otherwise would be just because the person exercising the rights is not the one entitled by law to do so. The distinction between a fraud that leads the Patent Office to issue a patent on an unpatentable invention (as in a case where the patent applicant concealed from the Patent Office the fact that the invention already was in the public domain) and one that merely operates to take the patent opportunity away from the real inventor (who but for the fraud would have gotten a valid patent that would have yielded him a royalty measured by the monopoly power that the patent conferred) is supported by analogy to cases holding that fraud on the Patent Office, to be actionable as patent fraud, must be material in the sense that the patent would not have been issued but for the misconduct. See, e.g., E. I. du Pont de Nemours & Co. v. Berkley & Co., 620 F.2d 1247, 1274 (8th Cir. 1980); Norton v. Curtiss, 433 F.2d 779, 794 (C.C.P.A. 1970). Equally, for a fraud to be material in an antitrust sense the plaintiff must show that but for the fraud no patent would have been issued to anyone. If a patent would have been issued to someone, the fraud could but have diverted market power from the one who had the right to possess and exploit it to someone else.

3. The patent must have some colorable validity, conferred for example by the patentee's efforts to enforce it by bringing patent-infringement suits. Indeed, some formulations of the antitrust offense of patent fraud make it seem that the offense is not the fraudulent procuring of a patent in circumstances that create monopoly power but the bringing of groundless suits for patent infringement. See, e.g., Handgards, Inc. v. Ethicon, Inc., supra, 601 F.2d at 993 and n.13. This metamorphosis is natural

because most patent-antitrust claims are asserted as counter-claims to patent-infringement suits and because the abusive prosecution of such suits could violate the antitrust laws even if the patent had not been obtained by fraud. See, e.g., id. at 994. But enforcement actions are not a sine qua non of monopolizing by patent fraud. Since a patent known to the trade to be invalid will not discourage competitors from making the patented product or using the patented process, and so will not confer monopoly power, suing an infringer is some evidence that the patent has (or at least the patentee is seeking to clothe it with) some colorable validity that might deter competitors. But it is not indispensable evidence; the concern of section 2 is with the exclusion of competition, not with the particular means of exclusion. Indeed, one might argue that just by virtue of being issued, a patent would have some apparent validity and that no more should be necessary. But this would go too far the other way. Since patents are issued in ex parte proceedings, and since by hypothesis the patent applicant had to use fraud to persuade the Patent Office to issue the patent, the patent might not fool anybody in the defendant's market.

Let us see whether these three conditions are satisfied by the complaint in this case. Right off the bat there is a problem with condition (1), as Brunswick's complaint does not allege that the business of making and selling antistatic yarn is an economically meaningful market. However, facts are pleaded that allow inference that antistatic yarn probably does not have good substitutes, in which event its sole producer could maintain its price significantly above the cost of production and sale.

There is a serious problem, however, with condition (2). Far from alleging that the process for making antistatic yarn that Riegel patented is not patentable, the complaint alleges that it is. Brunswick's only objection is to the patentee's identity; it thinks that it rather than Riegel should be the patentee. But as we have already suggested, to say that a patent should have been issued because the invention covered by it is patentable, but should have been issued to a different person and would have been but for fraud (the breach of the promise to Brunswick not to disclose its invention), is to say in effect that the patentee stole the patent from its rightful owner; and stealing a valid patent is not at all the same thing, from an antitrust standpoint, as obtaining an invalid patent. Until unmasked in an infringement or cancellation[5] or other proceeding, a patent on an unpatentable invention may create a monopoly by discouraging (through litigation or other means) others from making the patented product, just as a valid patent may, but the monopoly that such a patent creates is illegal, and hence actionable under antitrust law. The theft of a perfectly valid patent, in contrast, creates no monopoly power; it merely shifts a lawful monopoly into different hands. This has no antitrust significance, although it hurts the lawful owner of the monopoly power.

The purpose of the antitrust laws as it is understood in the modern cases is to preserve the health of the competitive process — which means, so far as a case such as this is concerned, to discourage practices that make it hard for consumers to buy at competitive prices — rather than to promote the welfare of particular competitors. This point was implicit in the famous dictum of Brown Shoe Co. v. United States, 370 U.S. 294, 320 (1962), that antitrust law (the Court was speaking of section 7 of the Clayton Act, but the point has been understood to be general) is concerned "with the protection of *competition,* not *competitors*" (emphasis in original), and has been

---

5. There is no procedure for "cancellation" of a patent. Judge Posner was apparently referring to "reexamination" of the patent by the PTO at the request of a third party. — EDS.

repeated with growing emphasis in recent years by this and other courts. See, e.g., Sutliff, Inc. v. Donovan Cos., 727 F.2d 648, 655 (7th Cir. 1984), and cases cited here. True, competitors as well as consumers still have standing to complain about antitrust violations, but that is because competitors are thought to be effective (maybe indispensable) surrogates for the many consumers who do not realize they are the victims of monopolistic practices, or if they do may lack incentives to bring suit because the harm to an individual consumer may be tiny even though the aggregate harm is immense. See Landes, Optimal Sanctions for Antitrust Violations, 50 U. Chi. L. Rev. 652, 671-72 (1983); cf. In re Industrial Gas Antitrust Litigations, 681 F.2d 514, 520 (7th Cir. 1982). If no consumer interest can be discerned even remotely in a suit brought by a competitor — if, as here, a victory for the competitor can confer no benefit, certain or probable, present or future, on consumers — a court is entitled to question whether a violation of antitrust law is being charged. See generally Easterbrook, The Limits of Antitrust, 63 Tex. L. Rev. 1, 33-39 (1984). If injury to a competitor, caused by wrongful conduct, were enough to bring the antitrust laws into play, the whole state tort law of unfair competition would be absorbed into federal antitrust law; it has not been: "unfair competition, as such, does not violate the antitrust laws." Sutliff, Inc. v. Donovan Cos., supra, 727 F.2d at 655; see also Car Carriers, Inc. v. Ford Motor Co., 745 F.2d 1101, 1107-08 (7th Cir. 1984); Havoco of America, Ltd. v. Shell Oil Co., 626 F.2d 549, 544-59 (7th Cir. 1980).

We cannot find the consumer interest in this case. Brunswick is complaining not because Riegel is gouging the consumer by charging a monopoly price for antistatic yarn, but because Riegel took away a monopoly that rightfully belonged to Brunswick as the real inventor. It is true that when Riegel got its patent Brunswick's patent application was still pending. But there is no suggestion that any other competitors were in the picture. If Riegel had not committed the alleged fraud, Brunswick would have had the whole field to itself — would have had the monopoly of antistatic yarn that it accuses Riegel of having stolen. This would be true even if Brunswick had not tried to get a patent on its process for making antistatic yarn. It still could, and as a rational profit-maximizer presumably would, have tried to license the invention as a trade secret; and a trade secret known only to, and licensed by, one firm may create as much monopoly power as a patent (more, even, if the secret can be kept for more than 17 years).

The nature of the remedy sought in an antitrust case is often, as here, an important clue to the soundness of the antitrust claim. Brunswick is asking, as a main part of the remedy, for an order transferring ownership of the patent to Riegel to itself. There is no contention that in asking for this Brunswick is motivated by altruism. It wants to make as much money as it can from the patent — as much as Riegel made, or, if possible, even more. There is nothing discreditable in this ambition but we do not see how consumers can benefit from its achievement. The nature of the remedy sought shows that Brunswick, far from contesting the propriety of a patent monopoly of antistatic yarn, makes that propriety the very foundation for the judicial relief that it seeks.

It makes no difference that Brunswick, which as we said is not a textile manufacturer, says that if it owned the patent it would license production to several manufacturers. There would then be more manufacturers of antistatic yarn than there are today, but there would not be more competition if the "competitors" were constrained by the terms of the patent license to charge the monopoly price. And they would be. As rational profit-maximizer Brunswick would charge its licensees a royalty designed to extract from them all the monopoly profits that the patent made possible; and the licensees would raise their prices to consumers to cover the royalty expense. The price to the consumer would be the same as it is, today, with Riegel the only seller in the market.

It is not a purpose of antitrust law to confer patents or to resolve disputes between rival applicants for a patent. From the standpoint of antitrust law, concerned as it is with consumer welfare, it is a matter of indifference whether Riegel or Brunswick exploits a monopoly of antistatic yarn. Cf. *Products Liability Ins. Agency, Inc. v. Crum & Forster Ins. Cos.,* 682 F.2d 660, 665 (7th Cir. 1982). Indeed, if anything, competitive pricing is more likely if Brunswick loses this suit that if it wins it. If Brunswick is confident that Riegel's patent is invalid, it can go into the anti-static-yarn business itself, with little fear of being held liable for patent infringement; and by entering, it will inject some competition into that market for the first time. Brunswick argues that it could not induce textile manufacturers to produce antistatic yarn under license from it since they would fear that Riegel would sue them, however baselessly, for patent infringement. But when a patentee (or, as in this case, a patent applicant) licenses his patent to other firms, he typically agrees to indemnify them for any costs incurred in patent-infringement suits brought against them. Brunswick, a large corporation, can afford to indemnify its licensees and would promise to do so if it really believed that it and not Riegel was the lawful owner of the patent....

The third condition for a patent fraud to violate section 2 of the Sherman Act — that the defendant has made efforts to give the color of validity to his patent on an unpatentable invention — cannot be met in a case where the plaintiff himself asserts that the underlying invention is patentable. This reinforces our conclusion that the complaint states no antitrust cause of action.

COMMENTS AND QUESTIONS

1. The Supreme Court's decision in *Walker Process* has given rise to a whole series of antitrust claims (such as *Brunswick*) that are based on fraudulent procurement of a patent. Indeed, these antitrust suits have come to be referred to as "*Walker Process* claims."

However, the idea that fraudulent procurement of a patent itself violates the antitrust laws is highly misleading. As the last section of the Court's opinion makes clear, the only effect of the defendant's fraud was that it lost the protection of the patent laws. The Court still required that Walker prove all the substantive elements of a section 2 violation (market power, anticompetitive conduct, and, for attempt claims, intent to monopolize). Indeed, Justice Harlan concurred precisely in order to emphasize that point:

> We hold today that a treble-damage action for monopolization...may be maintained under §4 of the Clayton Act if two conditions are satisfied: (1) the relevant patent is shown to have been produced by knowing and willful fraud...; and (2) all the elements otherwise necessary to establish a §2 monopolization charge are proved. Conversely, such a private cause of action would *not* be made out if the plaintiff...failed to prove the elements of a §2 charge even though he has established actual fraud in the procurement of the patent and the defendant's knowledge of that fraud.

382 U.S. at 179 (Harlan, J., concurring). Thus *Walker Process* is limited to removing the cloak of protection afforded by patent law to patents themselves, where the patent in question was obtained by fraud.

2. While proof of fraud on the PTO doesn't itself violate the antitrust laws, can it be used as evidence of anticompetitive conduct to support a monopolization claim? See 1 Herbert Hovenkamp et al., IP and Antitrust § 11.4d (arguing that it can in appropriate cases). Does it matter whether the patentee ever enforces the patent? Whether it threatens to do so? Could the mere existence of a patent deter potential competitors?

3. *Brunswick* suggests that a *Walker Process* claim can never be maintained in the case of an ownership dispute, because the plaintiff in such a case doesn't object to the existence of a patent right, merely its allocation to the patentee. 752 F.2d at 265. While Judge Posner is right to question whether there is really an anticompetitive effect here, he is arguably wrong to suggest there can never be harm to competition in such a case. Antitrust law should not always be indifferent to who owns patent rights; in some cases the ownership of a patent by one party could cement its control of the market, while ownership of the patent by a different party might enable real competition in the market. For example, consider a market dominated by an existing patented technology. If a new, competing technology is developed, the market structure may look very different depending on whether it is the existing patent owner or a new competitor who patents the new technology. The focus of the courts should be on the market effect of the fraud in the context of the facts of any given case.

4. The antitrust status of patents obtained by fraud is, of course, closely related to the definition of such fraud. We discussed this problem at some length in Chapter 3, when we considered the patent defense of inequitable conduct. Does it make sense to punish fraudulent patents both in the patent law and through antitrust law?

Note on Antitrust Petitioning Immunity

Even if a patent is lawfully acquired, using it as an anticompetitive weapon might violate the antitrust laws. After all, baseless infringement suits on a valid patent are just as troubling as infringement suits on a baseless patent. Several courts have held that "anticompetitive litigation," standing alone, can suffice to violate the antitrust laws. See, e.g., Handgards, Inc. v. Ethicon, Inc., 601 F.2d 986, 996 (9th Cir. 1979), *cert. denied*, 444 U.S. 1025 (1980); CVD, Inc. v. Raytheon Co., 769 F.2d 842, 851 (1st Cir. 1985), *cert. denied*, 475 U.S. 1016 (1986). Still other courts have held that litigation may constitute anticompetitive conduct sufficient to satisfy one element of the monopolization test. See Herbert Hovenkamp et al., IP and Antitrust §§ 11.3-4 (collecting cases).

There is a significant constitutional limitation on antitrust claims or counterclaims that are based on the filing of an infringement suit. Before considering the merits of an antitrust counterclaim, courts must first decide whether the patentee is immune from antitrust liability under the *Noerr-Pennington* doctrine. That doctrine protects antitrust defendants from liability for "petitioning the government." The Supreme Court has consistently held that filing a lawsuit or an action before an administrative agency is "petitioning" and therefore is presumptively entitled to immunity from antitrust suit. There is an exception to antitrust immunity for "sham litigation," however. If a lawsuit is a sham, rather than a "genuine effort . . . to influence" the decision maker, it is not entitled to antitrust immunity, and the counterclaim will be evaluated on the merits.

Allied Tube & Conduit Corp. v. Indian Head, Inc., 486 U.S. 492 (1988). Not surprisingly, there has been heated debate over the precise scope of the "sham exception." The Supreme Court visited this issue in the 1993 case of Professional Real Estate Investors v. Columbia Pictures, Inc., 508 U.S. 49 (1993). In that case, Columbia Pictures had brought a copyright infringement suit against PREI based on PREI's performance of copyrighted movies in guests' hotel rooms, and PREI counterclaimed on the grounds that Columbia had conspired to monopolize the market and restrain trade. Columbia lost its copyright case on summary judgment. But the district court held that PREI was not entitled to pursue its antitrust claim because Columbia's copyright suit, though unsuccessful, was not a sham.

The Supreme Court affirmed the application of antitrust immunity to PREI's counterclaim. In doing so, it set out a new, two-part test to determine whether a lawsuit is a sham. "First, the lawsuit must be objectively baseless in the sense that no reasonable litigant could realistically expect success on the merits." *PREI, supra.* Second, "the court should focus on whether the baseless lawsuit conceals an attempt to interfere directly with the business relationships of a competitor through the use of governmental *process* — as opposed to the *outcome* of that process — as an anticompetitive weapon." Id. (emphasis in original). These are commonly referred to as the "objectively baseless" and "subjectively baseless" tests, respectively. Only if the suit is a sham under this definition will the court proceed to consider the substantive elements of an antitrust violation. The effect of this new test is to make it extremely difficult to prevail in an antitrust counterclaim based on wrongful enforcement of an intellectual property right.

Interestingly, the Federal Circuit has held that *Noerr-Pennington* immunity does not apply to allegations of fraudulent patent procurement under *Walker Process,* but only to allegations of sham litigation. Nobelpharma AB v. Implant Innovations, Inc., 141 F.3d 1059 (Fed. Cir. 1998).

b. Other Anticompetitive Conduct

Data General Corp. v. Grumman Systems Support Corp.
United States Court of Appeals for the First Circuit
36 F.3d 1147 (1st Cir. 1994)

STAHL, Circuit Judge.

While this case raises numerous issues touching on copyright law, Grumman's most intriguing argument — presented below as both a defense and a counterclaim — is that DG illegally maintained its monopoly in the market for service of DG computers by unilaterally refusing to license ADEX to Grumman and other competitors. The antitrust claims are intriguing because they present a curious conflict, namely, whether (and to what extent) the antitrust laws, in the absence of any statutory exemption, must tolerate short-term harm to the competitive process when such harm is caused by the otherwise lawful exercise of an economically potent "monopoly" in a copyrighted work.

After a careful analysis, we affirm [a jury verdict for Data General on the question of copyright infringement] on all but one relatively minor issue concerning the calculation of damages.

I. Background

DG and Grumman are competitors in the market for service of computers manufactured by DG, and the present litigation stems from the evolving nature of their competitive relationship. DG not only designs and manufactures computers, but also offers a line of products and services for the maintenance and repair of DG computers. Although DG has no more than a 5% share of the highly competitive "primary market" for mini-computers, DG occupies approximately 90% of the "aftermarket" for service of DG computers. As a group, various "third party maintainers" ("TPMs") earn roughly 7% of the service revenues; Grumman is the leading TPM with approximately 3% of the available service business. The remaining equipment owners (typically large companies in the high technology industry) generally maintain their own computers and peripherals, although they occasionally need outside service on a "time and materials" basis....

From 1976 until some point in the mid-1980s, DG affirmatively encouraged the growth of TPMs with relatively liberal policies concerning TPM access to service tools. DG sold or licensed diagnostics directly to TPMs, and allowed TPMs to use diagnostics sold or licensed to DG equipment owners. DG did not restrict access by TPMs to spare parts manufactured by DG or other manufacturers. DG allowed (or at least tolerated) requests by TPMs for DG's repair depot to fix malfunctioning circuit boards, the heart of a computer's central processing unit ("CPU"). DG sold at least some schematics and other documentation to TPMs. DG also sold TPMs engineering change order kits. And finally, DG training classes were open to TPM field engineers. Grumman suggests that DG's liberal policies were beneficial to DG because increased capacity (and perhaps competition) in the service aftermarket would be a selling point for DG equipment.

3. Increased Restrictions

In the mid-1980s, DG altered its strategy. With the goal of maximizing revenues from its service business, DG began to refuse to provide many service tools directly to TPMs. DG would not allow TPMs to use the DG repair depot, nor would it permit TPMs to purchase schematics, documentation, "change order" kits, or certain spare parts. DG no longer allowed TPM technicians to attend DG training classes. Finally, DG developed and severely restricted the licensing of ADEX, a new software diagnostic for its MV computers. The MV series was at once DG's most advanced computer hardware and an increasingly important source of sales and service revenue for DG.

A number of items unavailable to TPMs directly from DG were either available to all equipment owners (even customers of TPMs) from DG, or were available to TPMs from sources other than DG. For example, DG depot service, change order kits, and at least some documentation were available to all equipment owners. There is also evidence that Grumman had its own repair depot and that Grumman could make use of repair depots run by other service organizations (sometimes called "fourth party maintainers"). Likewise, there is evidence that TPMs could purchase at least some spare parts from sources other than DG.

The situation was different with respect to ADEX. DG service technicians would use ADEX [a computer diagnostic program developed by Data General] in performing service for DG equipment owners. DG would also license ADEX for the exclusive use of the in-house technicians of equipment owners who perform most of their own service. However, DG would not license ADEX to its own service customers or to the customers of TPMs. Nor was ADEX available to TPMs from sources other than DG. At least two other diagnostics designed to service DG's MV computers may have become available as early as 1989, but no fully functional substitute was available when this case was tried in 1992.

Grumman found various ways to skirt DG's ADEX restrictions. Some former DG employees, in violation of their employment agreements, brought copies of ADEX when they joined Grumman. In addition, DG field engineers often stored copies of ADEX at the work sites of their service customers, who were bound to preserve the confidentiality of any DG proprietary information in their possession. Although DG service customers had an obligation to return copies of ADEX to DG should they cancel their service agreement and switch to a TPM, few customers did so. It is essentially undisputed that Grumman technicians used and duplicated copies of ADEX left behind by DG field engineers. There is also uncontroverted evidence that Grumman actually acquired copies of ADEX in this manner in order to maintain libraries of diagnostics so that Grumman technicians could freely duplicate and use any copy of ADEX to service any of Grumman's customers with DG's MV computers.

C. The Present Litigation

In 1988, DG filed suit against Grumman in the United States District Court for the District of Massachusetts.... In one count, DG alleged that Grumman's use and duplication of ADEX infringed DG's ADEX copyrights, and requested injunctive relief, 17 U.S.C. §502 (1988), as well as actual damages and profits, 17 U.S.C. §504(b) (1988). In another count, DG alleged that Grumman had violated Massachusetts trade secrets law by misappropriating copies of ADEX in violation of confidentiality agreements binding on former DG employees and DG service customers....

b. Antitrust Defenses

Grumman claimed that DG could not maintain its infringement action because DG had used its ADEX copyrights to violate Sections 1 and 2 of the Sherman Antitrust Act, 15 U.S.C. §§1 and 2 (1988 & Supp. IV 1992). Specifically, Grumman charged that DG misused its copyrights by (1) tying the availability of ADEX to a consumer's agreement either to purchase DG support services (a "positive tie") or not to purchase support services from TPMs (a "negative tie"), and (2) willfully maintaining its monopoly in the support services aftermarket by imposing the alleged tie-in and refusing to deal with TPMs....

In rejecting Grumman's motion for reconsideration of the grant of summary judgment on the monopolization claim, the district court also directly addressed Grumman's contention that DG's refusal to license ADEX to TPMs constitutes exclusionary conduct. The court stated that DG's refusal to license ADEX to TPMs was not exclusionary because "DG offers to the public a license to use MV/ADEX on any computer owned by the customer," and therefore DG "'did not withhold from one member of the public a service offered to the rest[.]'" *Grumman III*, slip op. at 5 (citing Olympia Equip. Leasing Co. v. Western Union Tel. Co., 797 F.2d 370, 377 (7th Cir. 1986), *cert. denied*, 480 U.S. 934, 94 L. Ed. 2d 765, 107 S. Ct. 1574 (1987)). . . .

B. Grumman's Antitrust Counterclaims . . .

2. Monopolization . . .

a. Unilateral Refusals to Deal

Because a monopolization claim does not require proof of concerted activity, even the unilateral actions of a monopolist can constitute exclusionary conduct. See 15 U.S.C. §2 (referring to "every person who shall monopolize . . . or combine or conspire with any other person . . . to monopolize"); Moore v. Jas. H. Matthews & Co., 473 F.2d 328, 332 (9th Cir. 1973) (observing that "section 2 is not limited to concerted activity"). Thus, a monopolist's unilateral refusal to deal with its competitors (as long as the refusal harms the competitive process) may constitute prima facie evidence of exclusionary conduct in the context of a Section 2 claim. See Kodak, 112 S. Ct. at 2091 n.32 (citing Aspen Skiing Co. v. Aspen Highlands Skiing Corp., 472 U.S. 585, 602-05, 86 L. Ed. 2d 467, 105 S. Ct. 2847 (1985)). A monopolist may nevertheless rebut such evidence by establishing a valid business justification for its conduct. See Kodak, 112 S. Ct. at 2091 n.32 (suggesting that monopolist may rebut an inference of exclusionary conduct by establishing "legitimate competitive reasons for the refusal"); Aspen Skiing, 472 U.S. at 608 (suggesting that sufficient evidence of harm to consumers and competitors triggers further inquiry as to whether the monopolist has "persuaded the jury that its [harmful] conduct was justified by [a] normal business purpose"). In general, a business justification is valid if it relates directly or indirectly to the enhancement of consumer welfare. Thus, pursuit of efficiency and quality control might be legitimate competitive reasons for an otherwise exclusionary refusal to deal, while the desire to maintain a monopoly market share or thwart the entry of competitors would not. See Kodak, 112 S. Ct. at 2091 (discussing the validity and sufficiency of various business justifications); Aspen Skiing, 472 U.S. at 608-11 (same); see generally 7 Areeda & Turner ¶1504, at 377-83; 9 Areeda & Turner ¶¶1713, 1716-17, at 148-61, 185-239. In essence, a unilateral refusal to deal is prima facie exclusionary if there is evidence of harm to the competitive process; a valid business justification requires proof of countervailing benefits to the competitive process.

Despite the theoretical possibility, there have been relatively few cases in which a unilateral refusal to deal has formed the basis of a successful Section 2 claim. Several of the cases commonly cited for a supposed duty to deal were actually cases of joint conduct in which some competitors joined to frustrate others. See Associated Press v. United States, 326 U.S. 1, 89 L. Ed. 2013, 65 S. Ct. 1416 (1945); United States v. Terminal R.R. Ass'n, 224 U.S. 383, 56 L. Ed. 810, 32 S. Ct. 507 (1912). Prior to Aspen Skiing, the case that probably came closest to condemning a true unilateral refusal to deal was Otter Tail Power Co. v. United States, 410 U.S. 366, 35 L. Ed. 2d 359, 93 S. Ct. 1022 (1973), which condemned the refusal of a wholesale power supplier either to sell wholesale power to municipal systems or to "wheel power" when Otter Tail's retail franchises expired and local municipalities sought to supplant Otter Tail's local distributors. The case not only involved a capital-intensive public utility facility—which could not effectively be duplicated and occupied a distinct separate market—but the Supreme Court laid considerable emphasis on "supported" findings in the district court "that Otter Tail's refusals to sell at wholesale or to wheel were solely to prevent municipal power systems from eroding its monopolistic position." 410 U.S. at 378.

In Aspen Skiing, the Court criticized a monopolist's unilateral refusal to deal in a very different situation, casting serious doubt on the proposition that the Court has adopted any single rule or formula for determining when a unilateral refusal to deal is unlawful. In that case, an "all-Aspen" ski ticket—valid at any mountain in Aspen—had been developed and jointly marketed when the three (later four) ski areas in Aspen were owned by independent entities. 472 U.S. at 589. Some time after Aspen Skiing Company ("Ski Co.") came into control of three of the four ski areas, Ski Co. refused to continue a joint agreement with Aspen Highlands Skiing Corp. ("Highlands"), the owner of the fourth area. Id. at 592-93. Although there was no "essential facility" involved, the Court found that it was exclusionary for Ski Co., as a monopolist, to refuse to continue a presumably efficient "pattern of distribution that had originated in a competitive market and had persisted for several years." Id. at 603.

It is not entirely clear whether the Court in Aspen Skiing merely intended to create a category of refusal-to-deal cases different from the essential facilities category or whether the Court was inviting the application of more general principles of antitrust analysis to unilateral refusals to deal. We follow the parties' lead in assuming that Grumman need not tailor its argument to a preexisting "category" of unilateral refusals to deal.

b. Unilateral Refusals to License

DG attempts to undermine Grumman's monopolization claim by proposing a powerful irrebuttable presumption: a unilateral refusal to license a copyright can never constitute exclusionary conduct. We agree that some type of presumption is in order, but reach that conclusion only after an exhaustive inquiry touching on the general character of presumptions, the role of market analysis in the copyright context, existing responses to the tension between the antitrust and patent laws, the nature of the rights extended by the copyright laws, and our duty to harmonize two conflicting statutes.

(1) The Propriety of a Presumption

We begin our analysis with two observations. First, DG's rule of law could be characterized as either an empirical assumption or a policy preference. For example, if we were convinced that refusals to license a copyright always have a net positive effect on the competitive process, we might adopt a presumption to this effect in order to preclude wasteful litigation about a known fact. On the other hand, if we were convinced that the rights enumerated in the Copyright Act should take precedence over the responsibilities set forth in the Sherman Act, regardless of the realities of the market, we might adopt a blanket rule of preference. DG's argument contains elements of both archetypal categories of presumptions.

Second, we note that the phrase "competitive process" may need some refinement in order to evaluate either an empirical assumption or a policy presumption concerning the desirability of unilateral refusals to license a copyright. Antitrust law generally seeks to punish and prevent harm to consumers in particular markets, with a focus on relatively specific time periods. See, e.g., Jefferson Parish, 466 U.S. at 18 (holding that "any inquiry into the validity of a tying arrangement must focus on the market or markets in which the two products are sold, for that is where the anticompetitive forcing has its impact"). Thus, in determining whether conduct is exclusionary in the context of a monopolization claim, we ordinarily focus on harm to the competitive process in the relevant market and time period. See generally 3 Areeda & Turner ¶517-28, at 346-88, ¶¶533-36, at 406-431. Confining the competitive process in this way assists courts in deciding particular disputes based primarily on case-specific adjudicative facts rather than generally-applicable "legislative" facts or assumptions. The use and protection of copyrights also affects the "competitive process," but it may not be appropriate to judge the effect of the use of a copyright by looking only at one market or one time period. We now consider what appears to be an empirical proclamation from DG: "The refusal to make one's innovation available to rivals . . . is pro-competitive conduct." As support, DG cites Grinnell, 384 U.S. at 570-71, in which the Court held that willful maintenance of monopoly does not include "growth or development as a consequence of a superior product." It is not the superiority of a work that allows the author to exclude others, however, but rather the limited monopoly granted by copyright law. Moreover, one reason why the Copyright Act fosters investment and innovation is that it may allow the author to earn monopoly profits by licensing the copyright to others or reserving the copyright for the author's exclusive use. See Sony Corp. of Am. v. Universal City Studios, Inc., 464 U.S. 417, 429 (1984) (explaining that the limited copyright monopoly "is intended to motivate the creative activity of authors and inventors by the provision of a special reward"). Thus, at least in a particular market and for a particular period of time, the Copyright Act tolerates behavior that may harm both consumers and competitors. Cf. SCM Corp. v. Xerox Corp., 645 F.2d 1195, 1203 (2d Cir. 1981) ("The primary purpose of the antitrust laws — to preserve competition — can be frustrated, albeit temporarily, by a holder's exercise of the patent's inherent exclusionary power during its term."), *cert. denied,* 455 U.S. 1016 (1982).

DG does not in fact argue that consumers are better off in the short term because of the inability of TPMs to license ADEX. Instead, DG suggests that allowing copyright owners to exclude others from the use of their works creates incentives which ultimately work to the benefit of consumers in the DG service aftermarket as well as to the benefit of consumers generally. In other words, DG seeks to justify any immediate harm to consumers by pointing to countervailing long-term benefits. Certainly, a monopolist's refusal to license others to use a commercially successful patented idea is likely to have more profound anti-competitive consequences than a refusal to allow others to duplicate the copyrighted expression of an unpatented idea (although such differences may become less pronounced if copyright law becomes increasingly protective of intellectual property such as computer software). But by no means is a monopolist's refusal to license a copyright entirely "procompetitive" within the ordinary economic framework of the Sherman Act. Accordingly, it may be inappropriate to adopt an empirical assumption that simply ignores harm to the competitive process caused by a monopolist's unilateral refusal to license a copyright. Even if it is clear that exclusive use of a copyright can have anti-competitive consequences, some type of presumption may nevertheless be appropriate as a matter of either antitrust law or copyright law.

(2) Antitrust Law and the Accommodation of Patent Rights

Antitrust law is somewhat instructive. Although creation and protection of original works of authorship may be a national pastime, the Sherman Act does not explicitly exempt such activity from antitrust scrutiny and courts should be wary of creating implied exemptions. See Square D Co. v. Niagara Frontier Tariff Bureau, Inc., 476 U.S. 409, 421 (1986) ("Exemptions from the antitrust laws are strictly construed and strongly disfavored."); cf. Flood v. Kuhn, 407 U.S. 258 (1972) (holding that the longstanding judicially created exemption of professional baseball from the Sherman Act is an established "aberration" in which Congress has acquiesced). The Supreme Court has suggested that an otherwise reasonable yet anti-competitive use of a copyright should not "be deemed a per se violation of the Sherman Act," Broadcast Music, Inc. v. CBS, Inc., 441 U.S. 1, 19 (1979), but a monopolistic refusal to license might still violate the rule of reason, see Rural Tel. Serv. Co. v. Feist Publications, Inc., 957 F.2d 765, 767-69 (10th Cir.) (analyzing reasonableness of monopolist's unilateral refusal to license copyrighted telephone listings to a competing distributor of telephone directories), *cert. denied*, 113 S. Ct. 490 (1992).[63] Should an antitrust plaintiff be allowed to demonstrate the anti-competitive effects of a monopolist's unilateral refusal to grant a copyright license? Would the monopolist then have to justify its refusal to license by introducing evidence that the protection of the copyright laws enabled the author to create a work which advances consumer welfare?

The courts appear to have partly settled an analogous conflict between the patent laws and the antitrust laws, treating the former as creating an implied limited exception to the latter. In Simpson v. Union Oil Co., 377 U.S. 13, 24 (1964), the Supreme Court stated that "the patent laws which give a 17-year monopoly on 'making, using, or selling the invention' are in pari materia with the antitrust laws and modify them pro tanto." Similarly, we have suggested that the exercise of patent rights is a "legitimate means" by which a firm may maintain its monopoly power. Barry Wright, 724 F.2d at 230. Other courts have specifically held that a monopolist's unilateral refusal to license a patent is ordinarily not properly viewed as exclusionary conduct. See Miller Insituform, 830 F.2d at 609 ("A patent holder who lawfully acquires a patent cannot be held liable under Section 2 of the Sherman Act for maintaining the monopoly

---

63. It is in any event well settled that concerted and contractual behavior that threatens competition is not immune from antitrust inquiry simply because it involves the exercise of copyright privileges. See, e.g., Kodak, 112 S. Ct. at 2089 n.29 ("The Court has held many times that power gained through some natural and legal advantage such as a patent, copyright, or business acumen can give rise to liability if 'a seller exploits his dominant position in one market to expand his empire into the next.'") (quoting Times-Picayune Publishing Co. v. United States, 345 U.S. 594, 611 (1953) (tying case)); United States v. Paramount Pictures, Inc., 334 U.S. 131, 143 (1948) (holding that horizontal conspiracy to engage in price-fixing in copyright licenses is illegal per se); id. at 159 (holding that block-booking of motion pictures — "a refusal to license one or more copyrights unless another copyright is accepted" — is an illegal tying arrangement); Straus v. American Publishers' Ass'n, 231 U.S. 222, 234 (1913) ("No more than the patent statute was the copyright act intended to authorize agreements in unlawful restraint of trade ...."); Digidyne Corp. v. Data General Corp., 734 F.2d 1336 (9th Cir. 1984) (affirming finding of illegal tie between copyrighted software and computer hardware), *cert. denied*, 473 U.S. 908 (1985); cf. Miller Insituform, Inc. v. Insituform of N. Am., Inc., 830 F.2d 606, 608-09 & n.4 (6th Cir. 1987) (describing ways in which patent holder may violate the antitrust laws), *cert. denied*, 484 U.S. 1064 (1988); United States v. Westinghouse Elec. Corp., 648 F.2d 642, 646-47 (9th Cir. 1981) (same).

power he lawfully acquired by refusing to license the patent to others."); Westinghouse, 648 F.2d at 647 (finding no antitrust violation because "Westinghouse has done no more than to license some of its patents and refuse to license others"); SCM Corp., 645 F.2d at 1206 (holding that "where a patent has been lawfully acquired, subsequent conduct permissible under the patent laws cannot trigger any liability under the antitrust laws"); see also 3 Areeda & Turner ¶704, at 114 ("The patent is itself a government grant of monopoly and is therefore an exception to usual antitrust rules."). This exception is inoperable if the patent was unlawfully "acquired." SCM Corp., 645 F.2d at 1208-09 (analyzing legality of Xerox's acquisition of plain-paper copier patent); see generally 3 Areeda & Turner ¶705-707, at 117-45 (discussing effect of patent acquisition, internal development of patents, and improprieties in patent procurement on applicability of antitrust laws).

The "patent exception" is largely a means of resolving conflicting rights and responsibilities, i.e., a policy presumption. See, e.g., Miller Insituform, 830 F.2d at 609 (declaring summarily that "there is no adverse effect on competition since, as a patent monopolist, [the patent holder] had [the] exclusive right to manufacture, use, and sell his invention"). At the same time, the exception is grounded in an empirical assumption that exposing patent activity to wider antitrust scrutiny would weaken the incentives underlying the patent system, thereby depriving consumers of beneficial products. See, e.g., SCM Corp., 645 F.2d at 1209 (holding that imposition of antitrust liability for an arguably unreasonable refusal to license a lawfully acquired patent "would severely trample upon the incentives provided by our patent laws and thus undermine the entire patent system").

(3) Copyright Law

Copyright law provides further guidance. The Copyright Act expressly grants to a copyright owner the exclusive right to distribute the protected work by "transfer of ownership, or by rental, lease, or lending." 17 U.S.C. §106. Consequently, "the owner of the copyright, if [it] pleases, may refrain from vending or licensing and content [itself] with simply exercising the right to exclude others from using [its] property." Fox Film Corp. v. Doyal, 286 U.S. 123, 127 (1932). See also Stewart v. Abend, 495 U.S. 207, 229 (1990). We may also venture to infer that, in passing the Copyright Act, Congress itself made an empirical assumption that allowing copyright holders to collect license fees and exclude others from using their works creates a system of incentives that promotes consumer welfare in the long term by encouraging investment in the creation of desirable artistic and functional works of expression. See Feist Publications, Inc. v. Rural Tel. Serv. Co., 499 U.S. 340 (1991) ("The primary objective of a copyright is not to reward the labor of authors, but 'to promote the Progress of Science and useful Arts.'") (quoting U.S. Const. art. I. §8, cl. 8); Sony Corp., 464 U.S. at 429 (discussing goals and incentives of copyright protection); Twentieth Century Music Corp. v. Aiken, 422 U.S. 151, 156 (1975) ("The immediate effect of our copyright law is to secure a fair return for an 'author's' creative labor. But the ultimate aim is, by this incentive, to stimulate artistic creativity for the general public good."). We cannot require antitrust defendants to prove and reprove the merits of this legislative assumption in every case where a refusal to license a copyrighted work comes under attack. Nevertheless, although "nothing in the copyright statutes would prevent an author from hoarding all of his works during the term of the copyright," Stewart, 495 U.S. at 228-29 (emphasis added), the Copyright Act does not explicitly purport to limit the scope of the Sherman Act. And, if the Copyright Act is silent on the subject generally, the silence is particularly acute in cases where a monopolist harms consumers in the monopolized market by refusing to license a copyrighted work to competitors.

We acknowledge that Congress has not been entirely silent on the relationship between antitrust and intellectual property laws. Congress amended the patent laws in 1988 to provide that "no patent owner otherwise entitled to relief for infringement . . . of a patent shall be denied relief or deemed guilty of misuse or illegal extension of the patent right by reason of [the patent owner's] refusal to license or use any rights to the patent." 35 U.S.C. § 271(d) (1988). Section 271(d) clearly prevents an infringer from using a patent misuse defense when the patent owner has unilaterally refused a license, and may even herald the prohibition of all antitrust claims and counterclaims premised on a refusal to license a patent. See Richard Calkins, Patent Law: The Impact of the 1988 Patent Misuse Reform Act and Noerr-Pennington Doctrine on Misuse Defenses and Antitrust Counterclaims, 38 Drake L. Rev. 192-97 (1988-89). Nevertheless, while Section 271(d) is indicative of congressional "policy" on the need for antitrust law to accommodate intellectual property law, Congress did not similarly amend the Copyright Act.

(4) Harmonizing the Sherman Act and the Copyright Act

Since neither the Sherman Act nor the Copyright Act works a partial repeal of the other, and since implied repeals are disfavored, e.g., Watt v. Alaska, 451 U.S. 259, 267 (1981), we must harmonize the two as best we can, id., mindful of the legislative and judicial approaches to similar conflicts created by the patent laws. We must not lose sight of the need to preserve the economic incentives fueled by the Copyright Act, but neither may we ignore the tension between the two very different policies embodied in the Copyright Act and the Sherman Act, both designed ultimately to improve the welfare of consumers in our free market system. Drawing on our discussion above, we hold that while exclusionary conduct can include a monopolist's unilateral refusal to license a copyright, an author's desire to exclude others from use of its copyrighted work is a presumptively valid business justification for any immediate harm to consumers.[64]

c. DG's Refusal to License ADEX to non-CMOs

Having arrived at the applicable legal standards, we may resolve Grumman's principal allegation of exclusionary conduct. Although there may be a genuine factual dispute about the effect on DG equipment owners of DG's refusal to license ADEX to TPMs, DG's desire to exercise its rights under the Copyright Act is a presumptively valid business justification . . . .

Grumman attempts to analogize this case to Aspen Skiing by focusing on the fact that DG once encouraged firms to enter the DG service aftermarket by allowing

---

64. Wary of undermining the Sherman Act, however, we do not hold that an antitrust plaintiff can never rebut this presumption, for there may be rare cases in which imposing antitrust liability is unlikely to frustrate the objectives of the Copyright Act.

liberal access to service tools, but no longer does so. The analytical framework of *Aspen Skiing* cannot function in these circumstances, however, because we are unable to view DG's market practices in both competitive and noncompetitive conditions. While TPMs have made inroads in the market for service of DG computers, DG has always been a monopolist in that market, and competitive conditions have never prevailed. Therefore, it would not be "appropriate to infer" from DG's change of heart that its former policies "satisfy consumer demand in free competitive markets." *Aspen Skiing*, 472 U.S. at 603.

Nor does it appear that Grumman would be able at trial to overcome the presumption on any other theory. There is no evidence that DG acquired its ADEX copyrights in any unlawful manner; indeed, the record suggests that DG developed all its software internally. Cf. 3 Areeda & Turner ¶706, at 127-28 (arguing that although an internally developed patent may be as exclusionary as one acquired from outside a firm, labeling the former as exclusionary would "discourage progressiveness by monopolists"). And, while there is evidence that DG knew that developing a "proprietary position" in the area of diagnostic software would help to maintain its monopoly in the aftermarket for service of DG computers, there is also evidence that DG set out to create a state-of-the-art diagnostic that would help to improve the quality of DG service. Cf. id. ¶706, at 128-29 (suggesting that "nearly all commercial research rests on a mixture of motivations" and that a search for an overriding "antisocial" motivation would be unilluminating). In fact, there is clearly some evidence that ADEX is a significant benefit to owners of DG's MV computers. ADEX is a better product than any other diagnostic for MV computers. The use of ADEX appears to have increased the efficiency and reduced the cost of service because technicians can locate problems more quickly and, through the use of the software's "remote assistance" capability, can arrive at customer sites having determined ahead of time what replacement parts are necessary. In addition to the possibility of lower prices occasioned by such gains in efficiency, ADEX also promises to lower prices through gains in effectiveness. For example, customers may save the cost of replacing expensive hardware components because the use of advanced diagnostics increases the possibility that technicians can locate a problem and repair the component.

COMMENTS AND QUESTIONS

1. Consider the court's conclusion that a refusal to license a copyrighted work is a presumptively valid business justification. How does this square with the rationales behind the copyright law? Is it inconsistent with the ultimate goal of copyright — to promote the *dissemination* of works of authorship to the general public?

2. Does it make a difference that this case arises under copyright and not under patent law? The *Data General* court notes that Congress amended the Patent Act in 1988 to provide that refusing to license a patent does not constitute patent misuse (though the statute is silent on whether refusal to license can violate the antitrust laws). Are there different policies in the patent and copyright laws that justify treating the two differently? In particular, consider whether the fact that the information contained in a patent is disclosed to the world makes a patentee's refusal to license less troubling than the same action by a copyright owner.

3. Some recent cases have cast doubt on the strength of the presumption that an intellectual property owner is always justified in unilaterally refusing to license its patents. In Kodak v. Image Tech. Servs., 125 F.3d 1195 (9th Cir. 1997), the court faced an appeal from a jury verdict that Kodak had tied the purchase of replacement parts for its copiers to the purchase of service from Kodak, in an effort to drive independent service organizations out of business. (A prior Supreme Court ruling had accepted the legal theory of monopolization by tying in this case.) One of Kodak's arguments on appeal was that because some of its spare parts were patented, it was free to refuse to license those patents to anyone it chose. The Ninth Circuit rejected that argument, reasoning that while Kodak had a presumptively legitimate business justification to refuse to license its patents, that presumption was rebutted in this case because the patent-related justifications were pretextual. The court relied on the fact that Kodak refused to sell all its parts to ISOs, not just the patented ones, and the testimony of Kodak's own marketing manager that patents "never crossed his mind." Rather, the court found that Kodak's intent was to exclude a competitor in the service market.

If Kodak did indeed have patents covering its parts, why should its purpose in refusing to license them matter?

4. The Federal Circuit dealt with the legality of unilateral refusals to license patent rights in two cases at the tail end of the twentieth century. The first case is Intergraph Corp. v. Intel Corp., 195 F.3d 1346 (Fed. Cir. 1999). Intergraph alleged that Intel's failure to continue to supply it with access to proprietary information, chips, and technical support constituted a refusal to deal that violated section 2. The court rejected this claim. It began by observing that

> it is well established that in the absence of any purpose to create or maintain a monopoly, the Sherman act does not restrict the long-recognized right of a trader or manufacturer engaged in an entirely private business, freely to exercise his own independent discretion as to parties with whom he will deal.

Id. at 1358. The court acknowledged that refusals to deal "may raise antitrust concerns when the refusal is directed against competition and the purpose is to create, maintain, or enlarge a monopoly." Id. Nonetheless, it found no section 2 violation based on Intel's refusal to deal, both because it concluded that Intel and Intergraph did not compete at all, and because it saw the fact that Intergraph had sued Intel as a valid reason for Intel to cease giving Intergraph preferential treatment. The court made no specific reference to Intel's patents as a factor in this decision, despite the district court's determination that Intel had used its patents to restrain trade, and that its patent rights did not immunize it from antitrust liability. See Intergraph Corp. v. Intel Corp., 3 F. Supp. 2d 1255, 1279 (N.D. Ala. 1998).

Only three months later, the Federal Circuit took a somewhat different approach to unilateral refusals to license intellectual property in In re Independent Service Organizations Antitrust Litigation, 203 F.3d 1322 (Fed. Cir. 2000) (*Xerox*). In that case, a group of independent service organizations (ISOs) that provided service for Xerox copiers sued Xerox for violating the antitrust laws because Xerox refused to sell parts to them or their customers. Xerox designed its policy so that it sold parts only to end-users that serviced the machines themselves, or to end-users who hired Xerox to perform service. The effect of the Xerox policy was to drive the ISOs out of the business of servicing Xerox copiers, and to reserve that business exclusively to Xerox.

Xerox counterclaimed for patent and copyright infringement, arguing that it had patents on a number of its parts and copyrights on its service drawings that the ISOs had infringed. Xerox also argued that it could not be held liable for violating the antitrust laws if all it did was unilaterally refuse to sell patented or copyrighted products to the ISOs, regardless of the purpose or effect of that refusal.

The Federal Circuit agreed with Xerox. The court asserted that there was "no reported case in which a court has imposed antitrust liability for a unilateral refusal to sell or license a patent," though in fact *Kodak* had done exactly that. Id. at 1326. The court held that a patentee's right to refuse to license its intellectual property right was limited only in certain circumstances: where the patent was obtained through fraud, where a lawsuit to enforce the patent was a sham, or where the patent holder uses his "statutory right to refuse to sell patented parts to gain a monopoly in a market *beyond the scope of the patent.*" Id. at 1327.

In the case before it, the court held that Xerox had not sought to extend its patents beyond the scope of the statutory grant. It noted that patents themselves could cover more than one market, and it held without explanation that Xerox's parts patents entitled it to control the market for service of Xerox copiers as well. And it refused to inquire into Xerox's motivation for refusing to license its parts patents. Thus, the Federal Circuit created a per se rule of legality, in accord with earlier statements from both the Second and Sixth Circuits.[6]

The *Xerox* court also considered Xerox's refusal to license its copyrights. In doing so, it applied Tenth Circuit law. In the absence of any precedent from the Tenth Circuit, the court adopted the First Circuit's approach in *Data General,* and rejected the Ninth Circuit's approach in *Image Technical.*

5. An interesting development in the patent context is C.R. Bard v. M3 Systems, 157 F.3d 1340 (Fed. Cir. 1998), where the Federal Circuit (over a vigorous dissent by Judge Newman) held that deliberate acts to create incompatibility between a patented system and a competitor's product could be the sort of predatory conduct that was actionable under the antitrust laws. For a discussion of the circumstances in which design changes can be predatory, see Herbert Hovenkamp et al., IP and Antitrust ch. 12.

Note on a Monopolist's Duty to Disclose

Antitrust law sometimes imposes burdens on monopolists that it does not impose on ordinary competitors. One of the most contentious issues in antitrust law is the duty of monopolists to give competitors access to their facilities. In limited circumstances, courts have been willing to require monopolists to provide such access on nondiscriminatory terms. For example, the Supreme Court has held that a group of railroads that together owned the central railroad switching yard in St. Louis, Missouri could not deny access to that yard to nonmembers. United States v. Terminal Ry., 212 U.S. 1 (1912). That case reasoned that because it was the only rail yard

---

6. See Miller Insituform Inc. v. Insituform of North Am., 830 F.2d 606, 609 (6th Cir. 1987) ("A patent holder who lawfully acquires a patent cannot be held liable under Section 2 of the Sherman Act for maintaining the monopoly power he lawfully acquired by refusing to license the patent to others."); SCM Corp. v. Xerox Corp., 645 F.2d 1195, 1206 (2d Cir. 1981) ("where a patent has been lawfully acquired, subsequent conduct permissible under the patent laws cannot trigger any liability under the antitrust laws.").

serving St. Louis, and because of the importance of the city as a hub, the rail yard was an "essential facility" from which its owners could not exclude others to the detriment of competition. A more recent case held that the owner of ski resorts with local market power could not discontinue a multi-area ski lift ticket it had shared with a smaller competitor in the absence of a legitimate business justification, even though the Court was unwilling to describe the lift ticket as an "essential facility." Aspen Skiing Co. v. Aspen Highlands Skiing Corp., 472 U.S. 585 (1985).

With these limited exceptions, however, the general rule is that businesses — even monopolists — do not have to give their competitors access to their facilities. And the Supreme Court seems unlikely to read these precedents broadly. In its most recent treatment of the issue, the Court engaged in a bit of revisionist history, distancing itself from the essential facilities doctrine and claiming that the Court had "never recognized such a doctrine." Verizon Communications v. Law Offices of Curtis V. Trinko, 540 U.S. __, 124 S. Ct. 872 (2004). The Court described *Aspen Skiing* as existing "at or near the outer boundary of §2 liability."

In Berkey Photo, Inc. v. Eastman Kodak Co., 603 F.2d 263 (2d Cir. 1979), *cert. denied*, 444 U.S. 1093 (1980), the court rejected Berkey's argument that Kodak owed it a duty to predisclose its forthcoming products, so that Berkey could prepare compatible products. The court recognized that Kodak's size and advance knowledge of its own plans gave it an advantage over Berkey. But, the court said,

> a large firm does not violate section 2 simply by reaping the competitive rewards attributable to its efficient size, nor does an integrated business offend the Sherman Act whenever one of its departments benefits from association with a division possessing a monopoly in its own market. So long as we allow a firm to compete in several fields, we must expect it to seek the competitive advantages of its broad-based activity....
>
> Berkey postulates that Kodak had a duty to disclose limited types of information to certain competitors under specific circumstances. But it is difficult to comprehend how a major corporation, accustomed though it is to making business decisions with antitrust considerations in mind, could possess the omniscience to anticipate all the instances in which a jury might one day in the future retrospectively conclude that predisclosure was warranted. And it is equally difficult to discern workable guidelines that a court might set forth to aid the firm's decision. For example, how detailed must the information conveyed be? And how far must research have progressed before it is "ripe" for disclosure? These inherent uncertainties would have an inevitable chilling effect on innovation. They go far, we believe, towards explaining why no court has ever imposed the duty Berkey seeks to create here.

Id.

The already difficult line between these two strands of cases is further complicated when the subject matter at issue is intellectual property. An intellectual property right is an exclusive right. It would be stripped of much of its value if antitrust law were to compel intellectual property owners to share it with their competitors. But there may be circumstances in which access to a patent or copyright has become an "essential facility" for competitors in the same or in a downstream market. In those circumstances, courts must face a direct conflict between the rights granted by intellectual property law and the prohibitions of antitrust law. Cf. David Scheffman, The

Application of Raising Rivals' Costs Theory to Antitrust, 37 Antitrust Bull. 187 (1992) (suggesting that raising the cost of a key input to a rival creates a problem similar to the problem of denying access to an essential facility).

In Intergraph Corp. v. Intel Corp., 195 F.3d 1346 (Fed. Cir. 1999), discussed above, Intergraph sued Intel after Intel cut off its supply of microprocessors and proprietary information. When Intergraph ultimately sued Intel for patent infringement, it also made a variety of antitrust claims based on Intel's efforts to cut off the flow of technology to Intergraph.[7]

Among Intergraph's claims was an essential facilities argument. Intergraph argued that access to Intel's chips and technical know-how was vital to its business, and that Intel should be compelled to license its patents and trade secrets to Intergraph on reasonable and nondiscriminatory terms. The district court granted a preliminary injunction, finding that Intel's intellectual property rights related to its chip architecture were indeed essential facilities. Intergraph Corp. v. Intel Corp., 3 F. Supp. 2d 1255 (N.D. Ala. 1998).

The Federal Circuit reversed. On the essential facilities issue, the court reviewed the doctrine in detail and concluded that an essential facilities claim could not be made out unless the owner of the essential facility and the antitrust plaintiff competed in a market that required access to the facility. The court noted that the gravamen of an essential facilities claim has always been an attempt to use control of such a facility to gain an unfair competitive advantage in a downstream market in which the defendant and the plaintiff competed.[8] Because it held that Intergraph and Intel did not compete at all, the court concluded that Intergraph could not possibly make out an essential facilities claim.[9] Cf. Aldridge v. Microsoft Corp., 995 F. Supp. 728 (S.D. Tex. 1998) (holding that Microsoft's Windows 95 operating system was potentially subject to an essential facilities claim but that the portions of the program at issue in this case were not, in fact, essential facilities).

As we shall see in the next section, antitrust law normally is intensely skeptical of agreements and exchanges of information between competitors. The plaintiff in *Berkey Photo* seems to be asking that the court *require* that competitors share information about their new product specifications. Should the court be concerned about helping facilitate a cartel between the parties in such an "essential facilities" case?

---

7. Most of those antitrust claims are based on Intel's efforts to link the two sets of intellectual property rights together. As such, they involve conditional rather than pure unilateral refusals to deal.

8. Id. at 1357 ("the essential facility theory is not an invitation to demand access to the property or privileges of another, on pain of antitrust penalties and compulsion; thus the courts have required anticompetitive action by a monopolist that is intended to 'eliminate competition in the downstream market.'").

9. Id. ("A non-competitor's asserted need for a manufacturer's business information does not convert the withholding of that information into an antitrust violation."). Cf. Multivideo Labs v. Intel Corp., 2000 WL 502866 (S.D.N.Y. April 27, 2000) (monopoly leveraging claim fails where the parties are not competitors).

The court was arguably incorrect to conclude that the parties were not in the same market. While Intergraph sold its products in a market downstream from Intel's, Intergraph's *intellectual property rights,* the assertion of which triggered the dispute, were in the same technology market as Intel's primary line of business (microprocessors). The court's failure to recognize this doesn't affect the essential facilities analysis, however, because any such competition would exist in the market for the essential facility itself, not the downstream market Intel was allegedly trying to control.

C. AGREEMENTS TO RESTRAIN TRADE

1. Vertical Restraints

a. *Tying Arrangements*

In a tying arrangement, the seller forces the buyer to buy a product he does not want (the tied product) in order to get a product he does want (the tying product). According to antitrust cases, sellers with market power in the tying product can leverage that monopoly into the tied product market by forcing captive buyers to buy the tied product exclusively from them. Whether this strategy is economically viable has been the subject of heated debate, as we discuss below. Tying arrangements, like joint ventures, have proven difficult to pigeonhole for antitrust purposes. They have been attacked under section 1 of the Sherman Act as involving a conspiracy,[10] under section 2 of the Sherman Act as involving monopolization of the tied product market, and under section 3 of the Clayton Act as requiring an exclusive dealing arrangement.[11]

Tying is one of the theories most commonly asserted against intellectual property owners. A common antitrust complaint is that a patentee, who has some power in the (patented) tying product by virtue of her intellectual property right, is requiring buyers to purchase unpatented tied products as well.

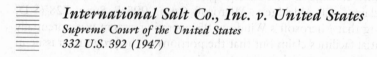

International Salt Co., Inc. v. United States
Supreme Court of the United States
332 U.S. 392 (1947)

Mr. Justice JACKSON delivered the opinion of the Court.

The Government brought this civil action to enjoin the International Salt Company, appellant here, from carrying out provisions of the leases of its patented machines to the effect that lessees would use therein only International's salt products. The restriction is alleged to violate § 1 of the Sherman Act, and § 3 of the Clayton Act. Upon appellant's answer and admissions of fact, the Government moved for summary judgment under Rule 56 of the Rules of Civil Procedure, upon the ground that no issue as to a material fact was presented and that, on the admissions, judgment followed as a matter of law. Neither party submitted affidavits. Judgment was granted and appeal was taken directly to this Court.

It was established by pleadings or admissions that the International Salt Company is engaged in interstate commerce in salt, of which it is the country's largest producer for industrial uses. It also owns patents on two machines for utilization of salt products. One, the "Lixator," dissolves rock salt into a brine used in various industrial processes. The other, the "Saltomat," injects salt, in tablet form, into canned products

---

10. The presumed "conspiracy" here is between the coercing seller and the coerced buyer. This is hardly the sort of collusion one normally thinks of as subject to section 1. But see Perma-Life Mufflers, Inc. v. International Parts Co., 392 U.S. 134 (1968) (defendant may be guilty of conspiracy with plaintiff he coerces under section 1).

11. Only some tying arrangements fall within the scope of this last provision. A tying arrangement can be either exclusive or nonexclusive, depending on whether it requires buyers to buy *all* of their supply of the tied product from the seller or only a specified amount of it.

during the canning process. The principal distribution of each of these machines is under leases which, among other things, require the lessees to purchase from appellant all unpatented salt and salt tablets consumed in the leased machines.

Appellant had outstanding 790 leases of an equal number of "Lixators," all of which leases were on appellant's standard form containing the tying clause and other standard provisions; of 50 other leases which somewhat varied the terms, all but 4 contained the tying clause. It also had in effect 73 leases of 96 "Saltomats," all containing the restrictive clause. In 1944, appellant sold approximately 119,000 tons of salt, for about $500,000, for use in these machines.

The appellant's patents confer a limited monopoly of the invention they reward. From them appellant derives a right to restrain others from making, vending or using the patented machines. But the patents confer no right to restrain use of, or trade in, unpatented salt. By contracting to close this market for salt against competition, International has engaged in a restraint of trade for which its patents afford no immunity from the antitrust laws. Morton Salt Co. v. G. S. Suppiger Co., 314 U.S. 488; Mercoid Corp. v. Mid-Continent Investment Co., 320 U.S. 661; Mercoid Corp. v. Minneapolis-Honeywell Co., 320 U.S. 680.

Appellant contends, however, that summary judgment was unauthorized because it precluded trial of alleged issues of fact as to whether the restraint was unreasonable within the Sherman Act or substantially lessened competition or tended to create a monopoly in salt within the Clayton Act. We think the admitted facts left no genuine issue. Not only is price-fixing unreasonable, per se, United States v. Socony-Vacuum Oil Co., 310 U.S. 150; United States v. Trenton Potteries Co., 273 U.S. 392, but also it is unreasonable, per se, to foreclose competitors from any substantial market. Fashion Originators Guild v. Federal Trade Commission, 114 F.2d 80, *affirmed*, 312 U.S. 457. The volume of business affected by these contracts cannot be said to be insignificant or insubstantial and the tendency of the arrangement to accomplishment of monopoly seems obvious. Under the law, agreements are forbidden which "tend to create a monopoly," and it is immaterial that the tendency is a creeping one rather than one that proceeds at full gallop; nor does the law await arrival at the goal before condemning the direction of the movement.

Appellant contends, however, that the "Lixator" contracts are saved from unreasonableness and from the tendency to monopoly because they provided that if any competitor offered salt of equal grade at a lower price, the lessee should be free to buy in the open market, unless appellant would furnish the salt at an equal price; and the "Saltomat" agreements provided that the lessee was entitled to the benefit of any general price reduction in lessor's salt tablets. The "Lixator" provision does, of course, afford a measure of protection to the lessee, but it does not avoid the stifling effect of the agreement on competition. The appellant had at all times a priority on the business at equal prices. A competitor would have to undercut appellant's price to have any hope of capturing the market, while appellant could hold that market by merely meeting competition. We do not think this concession relieves the contract of being a restraint of trade, albeit a less harsh one than would result in the absence of such a provision. The "Saltomat" provision obviously has no effect of legal significance since it gives the lessee nothing more than a right to buy appellant's salt tablets at appellant's going price. All purchases must in any event be of appellant's product. . . .

United States v. Microsoft Corp., 147 F.3d 935 (D.C. Cir. 1998). In a much-reported decision, the D.C. Circuit considered the legality of Microsoft's bundling its Internet Explorer browser into its Windows 95 operating system.[12] The fundamental facts of the case are that Microsoft had originally sold Windows 95 and Internet Explorer separately but proceeded through various upgrades to link the two more closely together, selling Windows 95 only in conjunction with Internet Explorer (though not vice versa). The Justice Department charged that this violated a 1994 consent decree. For purposes of interpreting that consent decree, the key question was whether the two bundled programs constituted a permissible "integrated product." The district court issued an injunction precluding Microsoft from offering Windows 95 only on the condition that the user also take Internet Explorer.

The D.C. Circuit reversed. The majority held that Microsoft's actions did not violate the consent decree. Two of the judges went beyond that question, however, and addressed the propriety of a tying theory in far-reaching dictum:

> The point of the test is twofold and may be illustrated by its application to the paradigm case of the Novell complaint and the subsequent release of Windows 95. First, "integration" suggests a degree of unity, something beyond merely placing disks in the same box. If an OEM or end user (referred to generally as "the purchaser") could buy separate products and combine them himself to produce the "integrated product," then the integration looks like a sham. If Microsoft had simply placed the disks for Windows 3.11 and MS-DOS in one package and covered it with a single license agreement, it would have offered purchasers nothing they could not get by buying the separate products and combining them on their own.[11]
>
> Windows 95, by contrast, unites the two functionalities in a way that purchasers could not; it is not simply a graphical user interface running on top of MS-DOS. Windows 95 is integrated in the sense that the two functionalities — DOS and graphical interface — do not exist separately: the code that is required to produce one also produces the other. Of course one can imagine that code being sold on two different disks, one containing all the code necessary for an operating system, the other with all the code necessary for a graphical interface. But as the code in the two would largely overlap, it would be odd to speak of either containing a discrete functionality. Rather, each would represent a disabled version of Windows 95. The customer could then "repair" each by installing them both on a single computer, but in such a case it would not be meaningful to speak of the customer "combining" two products. Windows 95 is an example of what Professor Areeda calls "physical or technological interlinkage that the customer cannot perform." X Areeda, Antitrust Law § 1746b at 227, 228 (1996).
>
> So the combination offered by the manufacturer must be different from what the purchaser could create from the separate products on his own. The second point is that it must also be better in some respect; there should be some technological value to integration. Manufacturers can stick products together in ways that purchasers cannot

---

12. The inclusion of the browser in the later Windows 98 operating system is being litigated in a separately filed case discussed below.

11. The same analysis would apply to peripherals. If, for example, Microsoft tried to bundle its mouse with the operating system, it would have to show that the mouse/operating system package worked better if combined by Microsoft than it would if combined by OEMs. This is quite different from showing that the mouse works better with the operating system than other mice do. See X Areeda, Elhauge & Hovenkamp, Antitrust Law ¶1746b. Problems seem unlikely to arise with peripherals, because their physical existence makes it easier to identify the act of combination. It seems unlikely that a plausible claim could be made that a mouse and an operating system were integrated in the sense that neither could be said to exist separately. An operating system used with a different mouse does not seem like a different product. But Windows 95 without IE's code will not boot, and adding a rival browser will not fix this. If the add/remove utility is run to hide the IE 4 technologies, Windows 95 reverts to an earlier version, OEM service release ("OSR") 2.0.

without the link serving any purpose but an anticompetitive one. The concept of integration should exclude a case where the manufacturer has done nothing more than to metaphorically "bolt" two products together, as would be true if Windows 95 were artificially rigged to crash if IEXPLORE.EXE were deleted. Cf. ILC Peripherals Leasing Corp. v. International Business Machines Corp., 448 F. Supp. 228, 233 (N.D. Cal. 1978) ("If IBM had simply bolted a disk pack or data module into a drive and sold the two items as a unit for a single price, the 'aggregation' would clearly have been an illegal tying arrangement.") aff'd per curiam sub nom. Memorex Corp. v. International Business Machines Corp., 636 F.2d 1188 (9th Cir. 1980); X Areeda, Elhauge & Hovenkamp, Antitrust Law ¶1746 at 227 (discussing literal bolting). Thus if there is no suggestion that the product is superior to the purchaser's combination in some respect, it cannot be deemed integrated.[12]

It might seem difficult to put the two elements discussed above together. If purchasers cannot combine the two functionalities to make Windows 95, it might seem that there is nothing to test Windows 95 against in search of the required superiority. But purchasers can combine the functionalities in their stand-alone incarnations. They can install MS-DOS and Windows 3.11. The test for the integration of Windows 95 then comes down to the question of whether its integrated design offers benefits when compared to a purchaser's combination of corresponding stand-alone functionalities. The decree's evident embrace of Windows 95 as a permissible single product can be taken as manifesting the parties' agreement that it met this test.

The short answer is thus that integration may be considered genuine if it is beneficial when compared to a purchaser combination. But we do not propose that in making this inquiry the court should embark on product design assessment. In antitrust law, from which this whole proceeding springs, the courts have recognized the limits of their institutional competence and have on that ground rejected theories of "technological tying." A court's evaluation of a claim of integration must be narrow and deferential. As the Fifth Circuit put it, "[S]uch a violation must be limited to those instances where the technological factor tying the hardware to the software has been designed for the purpose of tying the products, rather than to achieve some technologically beneficial result. Any other conclusion would enmesh the courts in a technical inquiry into the justifiability of product innovations." Response of Carolina, Inc. v. Leasco Response, Inc., 537 F.2d 1307, 1330 (5th Cir. 1976)....

We emphasize that this analysis does not require a court to find that an integrated product is superior to its stand-alone rivals. See ILC Peripherals Leasing Corp. v. International Business Machines Corp., 458 F. Supp. 423, 439 (N.D. Cal. 1978) ("Where there is a difference of opinion as to the advantages of two alternatives which can both be defended from an engineering standpoint, the court will not allow itself to be enmeshed 'in a technical inquiry into the justifiability of product innovations.'") (quoting *Leasco*, 537 F.2d at 1330), *aff'd per curiam sub nom.* Memorex Corp. v. IBM Corp., 636 F.2d 1188 (9th Cir. 1980). We do not read §IV(E)(i) to "put[] judges and juries in the unwelcome position of designing computers." IX Areeda, Antitrust Law ¶1700j at 15 (1991). The question is not whether the integration is a net plus but merely whether there is a plausible claim that it brings some advantage. Whether or not this is the appropriate test for antitrust law generally, we believe it is the only sensible reading of §IV(E)(i).

Judge Wald dissented from this opinion. She suggested a somewhat less deferential interpretation of the consent decree — and of tying law generally: "I think the prohibition [in the consent decree on tying] and the proviso [allowing integrated products]

---

12. Thus of course we agree with the separate opinion that "commingling of code...alone is not sufficient evidence of true integration." Commingling for an anticompetitive purpose (or for no purpose at all) is what we refer to as "bolting."

could reasonably be construed to state that Microsoft may offer an 'integrated' product to OEMs under one license only if the integrated product achieves synergies great enough to justify Microsoft's extension of its monopoly to an otherwise distinct market." Id. Judge Wald would balance the productive efficiencies against the harm to competition in determining whether an antitrust violation had occurred.

COMMENTS AND QUESTIONS

1. Is the deference shown by the majority in *Microsoft* appropriate? Under Judge Williams' opinion, how likely is it that any "integrated" product will ever be held to be a tie? Does the opinion have implications beyond the integration of software products, to cover any sort of synergistic combination of products?

Surely there is some value to embedding Internet Explorer in Windows 95; it is easier for novice computer users to access the Internet without having to obtain separate software to do it. Is this value enough to outweigh the elimination of competition in the browser market (which will almost certainly result from Microsoft's bundling)?

The D.C. Circuit revisited the issue three years later in the Windows 98 antitrust case. United States v. Microsoft, 253 F.3d 34 (D.C. Cir. 2001). In *Microsoft*, the D.C. Circuit treated Microsoft's efforts to bundle the browser into the operating system under both tying and monopolization law. Interestingly, it reached different results in those two analyses. Most of what was alleged to be unlawful tying was independently condemned as monopolization under section 2, without any finding of tying law's "separate products" — an approach advocated by Areeda & Hovenkamp, Antitrust Law ¶777 (2d ed.), when the defendant is a monopolist in the "tying product." The court did hold that the rule of reason, not the per se rule, was appropriate for the limited case of technological ties involving computer operating systems. However, the fact that the court found section 2 liability for bundling made most of its subsequent tying discussion superfluous on the basic question of the legality or illegality of specific practices. Indeed, the government abandoned its tying claims on remand, since it had won essentially the same arguments in the monopolization context.

2. In *International Salt*, the patentee allegedly tied the sale of its patented machine to the sale of a basic staple commodity, salt. What is wrong with that? The theory of leveraging is that the defendant will use its power in the tying market to obtain power in the tied product market (in this case, the market for salt). This power is considered to be beyond the scope of the patent grant, since it restrains trade in a separate market. But was there really any danger that International Salt would gain market power over such a basic commodity as salt? If not, is there any reason to be troubled by such a tying arrangement? See Joel Dirlam & Alfred Kahn, Fair Competition: The Law and Economics of Antitrust Policy 97 (1954) (criticizing *International Salt* for failing to inquire into market power).

On this point, the Supreme Court is sharply divided. In Jefferson Parish Hosp. Dist. v. Hyde, 466 U.S. 2 (1984), the Court split 5-4 on the standard to be used to evaluate tying cases. The majority concluded that tying was illegal "per se," but only in circumstances in which the defendant had market power in the tying product. This creates a "hybrid" rule somewhere between the "per se" rule and the rule of reason. The four dissenters would have required proof of three separate elements in tying cases: (a) a demonstration that there were really two separate products at stake;

(b) proof of market power in the tying product; and (c) a dangerous probability of acquiring power in the tied product. Even then, the dissent would treat tying arrangements under the rule of reason. Decisions subsequent to *Jefferson Parish* have left some question as to whether the majority rule in that case still stands. See Eastman Kodak Co. v. Image Technical Services, 504 U.S. 451 (1992) (applying the rule of reason in a Section 2 monopolization-by-tying case).

3. Consider the last paragraph of the court's opinion in *International Salt*. Is there really a tie here? Can't the "bound" parties get out of the contract if they find a better deal elsewhere? The court suggests that this "right of first refusal" is anticompetitive, but might it not serve the laudable (or at least inoffensive) purpose of keeping International Salt informed about the competitive market price? On this point, see Mark Grady & Jay Alexander, Patent Law and Rent Dissipation, 78 Va. L. Rev. 305 (1992). But see Aaron S. Edlin, Do Guaranteed-Low-Price Policies Guarantee High Prices, and Can Antitrust Rise to the Challenge?, 111 Harv. L. Rev. 528 (1997).

4. Intellectual property tying arrangements are not always between patented machines and staple products. In Automatic Radio Mfg. Co. v. Hazeltine Research, 339 U.S. 827 (1950), Hazeltine Research, a patent holding company, had licensed a group of 570 patents and 200 patent applications in a single block. Automatic Radio accused Hazeltine of "tying" desirable patents to undesirable patents. The Court disagreed:

> [P]etitioner urges that this case "is identical in principle" with the "Tie-in" cases. It is contended that the licensing provision requiring royalty payments of a percentage of the sales of the licensee's products constitutes a misuse of patents because it ties in a payment on unpatented goods. Particular reliance is placed on language from United States v. U.S. Gypsum Co., 333 U.S. 364, 389, 400. That case was a prosecution under the Sherman Act for an alleged conspiracy of Gypsum and its licensees to extend the monopoly of certain patents and to eliminate competition by fixing prices on patented and unpatented gypsum board. The license provisions based royalties on all sales of gypsum board, both patented and unpatented. It was held that the license provisions, together with evidence of an understanding that only patented board would be sold, showed a conspiracy to restrict the production of unpatented products which was an invalid extension of the area of the patent monopoly. 333 U.S. at 397. There is no indication here of conspiracy to restrict production of unpatented or any goods to effectuate a monopoly, and thus the Gypsum case does not aid petitioner. That which is condemned as against public policy by the "Tie-in" cases is the extension of the monopoly of the patent to create another monopoly or restraint of competition — a restraint not countenanced by the patent grant. See, e.g., Mercoid Corp. v. Mid-Continent Investment Co., 320 U.S. 661, 665-666; Morton Salt Co. v. Suppiger Co., 314 U.S. 488; Ethyl Gasoline Corp. v. United States, 309 U.S. 436, 456. The principle of those cases cannot be contorted to circumscribe the instant situation. This royalty provision does not create another monopoly; it creates no restraint of competition beyond the legitimate grant of the patent. The right to a patent includes the right to market the use of the patent at a reasonable return. See 46 Stat. 376, 35 U. S. C. §40; Hartford-Empire Co. v. United States, 323 U.S. 386, 417, 324 U.S. 570, 574.
>
> The licensing agreement in issue was characterized by the District Court as essentially a grant by Hazeltine to petitioner of a privilege to use any patent or future development of Hazeltine in consideration of the payment of royalties. Payment for the privilege is required regardless of use of the patents. The royalty provision of the licensing agreement was sustained by the District Court and the Court of Appeals on the theory that it was a convenient mode of operation designed by the parties to avoid the necessity

of determining whether each type of petitioner's product embodies any of the numerous Hazeltine patents. 77 F. Supp. at 496. The Court of Appeals reasoned that since it would not be unlawful to agree to pay a fixed sum for the privilege to use patents, it was not unlawful to provide a variable consideration measured by a percentage of the licensee's sales for the same privilege. 176 F.2d at 804. Numerous District Courts which have had occasion to pass on the question have reached the same result on similar grounds, and we are of like opinion.

Id. Can the bundling of patent licenses ever constitute actionable tying of products under the antitrust laws? Cf. LePage's, Inc. v. 3M, 324 F.3d 141 (3d Cir. 2003) (discounted bundling can be exclusionary). Does it matter whether the patents confer actual market power? Whether the licenses are exclusive or nonexclusive?

Automatic Radio was significantly limited by the Supreme Court in Zenith Radio Corp. v. Hazeltine Research, Inc., 395 U.S. 100 (1969). *Zenith* concludes that charging a single rate for a group of licenses is allowable because the patentee can charge anything it wants for licenses, but that compelling the purchase of other patented or unpatented products is unlawful. In other words, Hazeltine can charge Zenith $1,000 to use patent *A,* but Hazeltine cannot force Zenith to use patents *A* and *B* together, even if it only charges $1,000 for the set. Is this distinction logical? The Federal Circuit seemed to reject this distinction in U.S. Philips Corp. v. International Trade Comm'n, 424 F.3d 1179 (Fed. Cir. 2005). There, the court held that Philips's policy of charging a single license price for the rights to all its patents covering the manufacture of compact discs, both those essential to make CDs and those that were nonessential, was not patent misuse. The court reasoned that because the buyer was free to use alternative technologies, and because there was no evidence that the price for just the essential patents would have been lower, the buyer was not harmed by the restriction. It also pointed to substantial procompetitive benefits associated with package patent licensing: reducing transactions costs and avoiding unfair surprise to buyers. Finally, it noted the difficulty of drawing lines between essential and nonessential patents, lines that are unnecessary under its opinion.

Are there dangers to competition in allowing the bundling of essential and nonessential patents? How likely is it that a customer will adopt a competing nonessential technology if they have already gotten a "free" license to the patentee's technology?

Even if package licensing is not strictly "tying" of multiple products, is it nonetheless objectionable? See George Stigler, United States v. Loew's, Inc.: A Note on Block Booking, 1963 Sup. Ct. Rev. 152 (discussing package licensing as a means of price discrimination and/or cross-subsidization).

5. Several companies have alleged that after they asserted their intellectual property rights against Intel (either by filing suit or by entering into license negotiations), Intel threatened to "cut them off" from the supply of its crucial technology unless they dropped their suits. In effect the allegation is that Intel is using its power in the market for microprocessors to obtain royalty-free licenses to everyone else's intellectual property. Is this a reciprocal tying arrangement of a sort, between Intel's technology and the intellectual property asserted against it? If so, does it violate the antitrust laws? See Intergraph Corp. v. Intel Corp., 195 F.3d 1346 (Fed. Cir. 1999) (no); In re Intel Corp., FTC Dock. 9288 (complaint filed June 8, 1998) (yes).

Intel and the FTC entered into a consent decree in 1999. Under that decree Intel agreed not to terminate its supply of chips or information to a buyer merely because that buyer sued it for intellectual property infringement. Intel retained the right to terminate supply for a number of legitimate business reasons, however. Further, the consent decree provided that Intel could cut off any purchaser who sued it for infringement if that purchaser sought to enjoin the sale of Intel's core products. In essence, therefore, Intel retained the power to demand a compulsory license of intellectual property rights from the companies with which it does business.

Is the consent decree an appropriate compromise between the positions of the parties? Is it likely to have procompetitive effects?

6. Another common "tie" involving intellectual property is between patented machines and "nonstaple" products. (A nonstaple product is a product that is useful only in connection with the machine to which it is tied. Thus certain specially made replacement parts are nonstaple products.) Does a patent on a machine confer the right to a monopoly over replacement parts usable only on that machine? Congress resolved this issue in the patent misuse context in 1952. After a number of Supreme Court decisions had found patentees guilty of patent misuse for tying their patent machines to nonstaple products, Congress explicitly provided in 35 U.S.C. §271(d) that tying of nonstaple products did not constitute misuse. See Dawson Chemical v. Rohm & Haas Co., 448 U.S. 176, 209 (1980) (reviewing cases and legislative history).

In the antitrust context, the tying of nonstaple goods and services remains a major issue. A 1992 Supreme Court case (not involving intellectual property) held that an "aftermarket" for parts or service for a single brand of machines could be a relevant antitrust market. Eastman Kodak Co. v. Image Technical Servs. Inc., 504 U.S. 451 (1992). A large number of post-*Kodak* cases turn on whether the plaintiff can characterize the tied product as a genuinely separate product market. See, e.g., Tricom Inc. v. Electronic Data Sys., 902 F. Supp. 741 (E.D. Mich. 1995) (conditioning of a software lease agreement on the purchase of CPU time from the software dealer was actionable tying arrangement); Advanced Computer Servs. v. MAI Systems, 845 F. Supp. 356 (E.D. Va. 1994) (rejecting aftermarket antitrust claim because there was no separate market for MAI computer systems).

7. Is there any reason to treat intellectual property owners differently from companies whose power in aftermarkets is "inherent" in their product? The answer may depend both on whether you believe that intellectual property law justifies certain restraints on trade in order to encourage innovation, and on whether you think there is a real danger that power in an equipment market could profitably be leveraged into aftermarkets. Economists have debated the validity of "leveraging" theory for some time. Both Robert Bork and Richard Posner have advanced the idea that leveraging is economically irrational. See Robert Bork, The Antitrust Paradox: A Policy at War with Itself (1978):

> The fallacy of the cases on tying arrangements may be shown through a hypothetical example based on the facts of *International Salt*. Suppose that a food canner is just willing to pay $100 for a one-year lease of a salt-dispensing machine. How is it possible for the lessor of the machine to make him pay $100 and, in addition, require him to take all his salt from the lessor? If the requirement is necessary, the lessee is giving up something he values. If the lessor has charged the full value of the machine, he cannot then charge still more in the form of coercion to take what amounts to a requirements contract for salt. That is double counting of monopoly power. The tying arrangement, whatever else it may accomplish, is obviously not a means of gaining two monopoly profits from a single monopoly.
>
> The argument is identical with respect to reciprocity. In *Consolidated Foods* the Supreme Court struck down Consolidated's acquisition of Gentry, a small manufacturer

of dehydrated onion and garlic, on the theory that Consolidated, a large food processor, might condition its purchases from food suppliers upon their willingness to purchase from Gentry. There is, however, no way in which that practice could be anti-competitive. Suppose that both Consolidated and Gentry buy and sell in fully competitive markets. Consolidated will then have no market power with which to force its suppliers to purchase from Gentry. The suppliers can turn to other customers. Suppose, however, that all markets involved display large elements of market power. Consolidated, before its acquisition of Gentry, may be presumed to have negotiated the best price it could from its suppliers. The acquisition of Gentry does not alter Consolidated's purchasing power, but the attempt after the acquisition to force Gentry's onions and garlic on suppliers as a condition of continued purchases by Consolidated is merely a way of demanding a still lower price from the suppliers. If the tactic works, Consolidated merely learns that it was paying too high a price to begin with: it would be better off renegotiating prices than cramming unwanted onions and garlic down its suppliers' throats. Nor is the theory improved by assuming that Consolidated understands all this and agrees to pay suppliers more in return for purchases from Gentry as a tactic of monopolizing the onion market. Rival sellers in Gentry's market can respond in a variety of ways, the most obvious being a price cut that just matches the price increase laid out by Consolidated. In that kind of price war, Consolidated has no advantages.

But perhaps the most concise devastation of the law's tie-in theory was provoked by the Supreme Court's 1962 *Loew's* decision. Distributors of pre-1948 copyrighted motion-picture feature films for television exhibition engaged in the practice of block booking their films to television stations. The stations had to take entire groups of films and could not pick and choose particular films from the proffered packages. The Court majority held, of course, that the practice violated Section 1 of the Sherman Act. The opinion's economic analysis of block booking was confined to a recitation of Justice Frankfurter's by then clearly indefensible dictum on the purpose of tying arrangements.

The inadequacy of the Supreme Court's theory of tying was well stated by George Stigler in a critique of the *Loew's* opinion:

> Consider the following simple example. One film, Justice Goldberg cited 'Gone With the Wind', is worth $10,000 to the buyer, while a second film, the Justice cited 'Getting Gertie's Garter', is worthless to him. The seller could sell the one for $10,000 and throw away the second, for no matter what its cost, bygones are forever bygones. Instead the seller compels the buyer to take both. But surely he can obtain no more than $10,000, since by hypothesis this is the value of both films to the buyer. Why not, in short, use his monopoly power directly on the desirable film? It seems no more sensible, on this logic, to blockbook the two films than it would be to compel the exhibitor to buy 'Gone With the Wind' and seven Ouija boards, again for $10,000.

To these queries the law has no answers. Indeed, it has so far successfully managed to ignore the existence of the queries.

Id. at 373-74; see also Richard Posner, Antitrust Law: An Economic Perspective (2d ed. 2001). Cf. Melissa Hamilton, Software Tying Arrangements Under the Antitrust Laws: A More Flexible Approach, 71 Denv. L. Rev. 607 (1994) (making the rather different argument that tying is efficient in the software context because it allows programmers to continue to upgrade their products). The logic of Judge Bork's argument is appealing. But is it correct? Louis Kaplow has offered a powerful critique of the Bork-Posner theory. Louis Kaplow, Extension of Monopoly Power Through Leverage, 85 Colum. L. Rev. 515 (1985). There, he argues that

> There are a number of deficiencies in the analysis of recent commentators who have attempted to proclaim the death of leverage theory. The basic mistake in their central

thesis is that antitrust law should be indifferent to the exploitation of monopoly power because extant power is a fixed sum and thus will result in the same damage regardless of how it is deployed. Although of some superficial appeal, it can readily be demonstrated that their analysis is strongly counterintuitive. Consider the case of a terrorist on the loose with one stick of dynamite. The fixed sum thesis posits that since the power is fixed — that is, the terrorist has one and only one dynamite stick — we should be indifferent to where the dynamite is placed. It is all too obvious, however, that the potential damage resulting from power in this context, as well as in virtually any other we can imagine, is overwhelmingly dependent upon how it may be used....

The position of these commentators [Bork, Posner, others] can be given more meaning only by developing it more extensively. The most reasonable interpretation, I believe, is that their two categories represent an implicit attempt to distinguish between short-run and long-run phenomena — or, to use more technical language, between static and dynamic models. Profit-maximizing practices are meant to refer roughly to those actions that can have fairly direct and immediate effects, while monopoly extension refers to behavior designed to have implications on the magnitude of profits and welfare loss in the future. The prototypical example of a profit maximization device is a pricing decision by a firm with market power, a decision which can be implemented rather quickly. By contrast, practices designed to affect the market share and elasticity of market demand might be labeled monopoly extension devices. These practices do not increase short-run profits, and might even decrease them. The firm's motivation is to change the structural conditions it faces in the future in order that it may receive greater profits in the future.

Id. at 516, 523-24. Professor Kaplow goes on to point out a number of reasons why leveraging may be possible in the real world. His reasons include the dynamic nature of markets, in which firms may give up short-run profits to maximize long-run gains, and inherent market imperfections that may allow monopolists to gain limited power in new markets through leveraging. See also George Stigler, United States v. Loew's, Inc.: A Note on Block Booking, 1963 Sup. Ct. Rev. 152 (explaining price discrimination potential inherent in tying).

Despite scholarly criticism, and to the surprise of many, the Eleventh Circuit held in 1999 that block-booking of television shows was illegal per se as a tying arrangement. MCA Television Ltd. v. Public Interest Corp., 171 F.3d 1265 (11th Cir. 1999).

8. Some forms of intellectual property may confer or promote limited power over price by their very nature. Trademarks, for example, are designed to differentiate products in the minds of consumers. The result of product differentiation is that producers have a form of constrained monopoly power, even in otherwise competitive industries. (For example, users of Colgate toothpaste may put up with a price increase rather than buy a competitor's product, because of "brand loyalty" generated by trademarks.) In the 1970s, the FTC investigated leaders in the packaged breakfast cereal industry, operating on the theory that brand names give each cereal a certain degree of market power. For a discussion of trademarks, brand loyalty, and market power, see Glynn S. Lunney, Jr., Trademark Monopolies, 48 Emory L.J. 367 (1999).

PROBLEMS

Problem 8-2. Advanced Micro Devices (AMD) and Intel both make microprocessor chips. Chip designs tend to travel in waves, or "generations," which reflect newer technology and which successively become the industry standard. The old generation of chips were known as 286 chips. They were replaced by the next generation, the 386. Intel has since developed three new generations of chips, the 486, the Pentium, and the Pentium Pro. In 1991, AMD filed suit against Intel, alleging that Intel was intimidating chip consumers into buying Intel's rather than AMD's 386 chips by threatening to withhold 486 chips from any firm that defects to AMD. The suit claims that this is an illegal tying arrangement.

How should such a "temporal tying arrangement" be treated under the antitrust laws? Is leveraging a problem here? Does it matter if Intel obtains a patent on the 486?

Problem 8-3. Tastee Turkee (TT) is a national fast-food franchise that specializes in roast turkey meals. TT does not operate any of its own restaurants. However, it has a series of requirements that franchisees must meet in order to use the Tastee Turkee name. In addition to requirements regarding color scheme, logo, building, and products, TT requires that its franchisees purchase all of their packaging materials, food, and kitchen equipment from TT itself or from approved TT vendors.

Cantor is a franchisee who believes that he can obtain food and packaging material more cheaply from his own sources. When TT refuses to allow him to do so, he brings suit under the antitrust laws. Cantor's theory is that TT has tied the "Tastee Turkee" trademark and image to the purchase of various staple goods.

Is this a "tying arrangement"? If so, does it violate the antitrust laws?

b. Licensing Restrictions

One of the most difficult antitrust issues surrounding intellectual property rights concerns the limitations placed on the licensing of intellectual property. Intellectual property owners may wish to license their rights to others for a number of reasons. The owners may be ill-equipped to make the protected product themselves; they may want a revenue stream without having to invest in producing and selling the product; they may wish to reserve one geographic or product market to themselves, while allowing others to exploit the intellectual property right elsewhere; or they may simply feel that broad dissemination of their product will redound to their benefit (for example, because there is value in having their product become an industry standard). Economic theory encourages licensing because it allows the market to transfer use of the intellectual property right to the most productive user of that right.

There is no antitrust problem with licensing per se. Indeed, the Antitrust Division's Intellectual Property Guidelines take the position that licensing is essentially procompetitive:

> Intellectual property typically is one component among many in a production process and derives value from its combination with complementary factors. Complementary factors of production include manufacturing and distribution facilities, workforces, and other items of intellectual property. The owner of intellectual property has to arrange for its combination with other necessary factors to realize its commercial value. Often, the owner finds it most efficient to contract with others for these factors, to sell rights to

the intellectual property, or to enter into a joint venture arrangement for its development, rather than supplying these complementary factors itself.

Licensing, cross-licensing, or otherwise transferring intellectual property (hereinafter "licensing") can facilitate integration of the licensed property with complementary factors of production. This integration can lead to more efficient exploitation of the intellectual property, benefiting consumers through the reduction of costs and the introduction of new products. Such arrangements increase the value of intellectual property to consumers and to the developers of the technology. By potentially increasing the expected returns from intellectual property, licensing also can increase the incentive for its creation and thus promote greater investment in research and development....

Field-of-use, territorial and other limitations on intellectual property licenses may serve procompetitive ends by allowing the licensor to exploit its property as efficiently and effectively as possible. These various forms of exclusivity can be used to give a licensee an incentive to invest in the commercialization and distribution of products embodying the licensed intellectual property and to develop additional applications for the licensed property. The restrictions may do so, for example, by protecting the licensee against free-riding on the licensee's investments by other licensees or by the licensor. They may also promote the licensor's incentive to license, for example, by protecting the licensor from competition in the licensor's own technology in a market niche that it prefers to keep to itself. These benefits of licensing restrictions apply to patent, copyright, and trade secret licenses, and to know-how agreements.

Intellectual Property Guidelines § 2.3. For many of these reasons, the division's prior set of guidelines governing intellectual property licensing (promulgated by the Reagan Administration in 1988 and contained in the International Guidelines) took a very "hands-off" approach to intellectual property licensing transactions. That has changed with the current guidelines, however. While the new guidelines recognize the procompetitive benefits of licensing, they also identify some licensing arrangements that are cause for antitrust concern:

> [A]ntitrust concerns may arise when a licensing arrangement harms competition among entities that would have been actual or likely potential competitors in a relevant market in the absence of a license (entities in a "horizontal relationship"). A restraint in a licensing arrangement may harm such competition, for example, if it facilitates market division or price fixing. In addition, license restrictions with respect to one market may harm such competition in another market by anticompetitively foreclosing access to, or significantly raising the price of, an important input, or by facilitating coordination to increase price or reduce output.

Id., § 3.1.

COMMENTS AND QUESTIONS

1. Consider the Antitrust Division's new policy statement — that it is concerned about agreements that reduce competition that would have existed but for the license. On the one hand, this policy can be (and has been) criticized as being too aggressive. By declaring that intellectual property is like any other form of property, the guidelines narrowly circumscribe the terms of licensing agreements. The effect may be to prevent some efficient licensing agreements from being signed, even

though the procompetitive effects of the agreement as a whole would outweigh its negative effects on preexisting competition.

On the other hand, perhaps the policy can be criticized from the other side for not being aggressive enough. Why should the antitrust laws be concerned only with protecting competition that would have existed absent the license? If licensing is favored because it is procompetitive, doesn't antitrust law have some obligation to see that those procompetitive benefits that result from the license itself are not lost because of restrictive license provisions?

One way of resolving this debate is to distinguish between the effects inherent to the license and the effects that result from restrictive provisions ancillary to the license. For example, an agreement to license a patent may promote competition in the market for the patented product, but the agreement may also contain a term dividing geographic territories in the sale of unpatented products or setting resale prices. To the extent that these restrictive terms can be severed from the balance of the license, antitrust law may be able to challenge them without threatening the license itself.

2. The guidelines also establish that most licensing agreements will be treated under the rule of reason, rather than the per se rule. Intellectual Property Guidelines § 3.4. Does this provision help to alleviate any concern that the guidelines are too aggressive in their treatment of intellectual property licensing?

Note on Intellectual Property and Exclusive Dealing

Section 3 of the Clayton Act prohibits some "exclusive dealing arrangements" under the rule of reason.[13] Exclusive dealing refers to agreements that prevent the agreeing party from buying or selling the goods of a competitor. For example, a company with a significant share of the market may attempt to drive its competitors out of business by requiring that all its customers buy exclusively from it. If the customers consider the exclusive dealer to be their most important source of supply, they will probably agree to purchase all their goods from that dealer. The result may be that the exclusive dealer expands its market power into an actual monopoly. One example of such a case is Fashion Originators Guild of America v. FTC, 312 U.S. 457, 466-67 (1941), where the Court applied section 3 of the Clayton Act to bar fashion designers with a collective 40% share of the market from insisting that any store that bought from the designers not buy from their competitors.

On the other hand, exclusive dealing arrangements may have procompetitive justifications. Agreements to buy everything you need from (or sell everything you produce to) one source may significantly reduce transactions costs and allow smaller companies to take advantage of discounts resulting from economies of scale in purchasing. The problem for antitrust law is to distinguish exclusive dealing arrangements that restrict competition from those that merely serve to make business transactions easier. To draw this line, antitrust law requires a showing of "substantial foreclosure" of competition to make out a section 3 violation.

---

13. Strictly speaking, section 3 of the Clayton Act applies only to transactions for the sale of goods. However, courts have interpreted sections 1 and 2 of the Sherman Act (which do extend to service transactions) to cover most of the same exclusive dealing arrangements as Clayton Act § 3.

Intellectual property licenses often contain exclusivity provisions. Such provisions can bind either the licensor or the licensee (or conceivably both). The licensor might agree to grant an "exclusive license" for his patent. An exclusive license gives only the licensee the right to work the patent. The licensor cannot license anyone else to work the patent and in some circumstances gives up the right to work it himself. On the other end, a license may bind the licensee not to use or sell products other than the patentee's product. In either case, the contract expressly requires that one party deal exclusively with the other. The Antitrust Division's Intellectual Property Guidelines take the position that exclusive licenses do not raise antitrust concerns unless the parties to the agreement (or other potential licensees) would have been actual or potential competitors absent the exclusivity provision. See Intellectual Property Guidelines § 4.1.2. Does this approach make sense? Would it allow the agencies to challenge agreements like those in *Fashion Originators Guild?*

Exclusive dealing practices need not be express. Certain types of licensing provisions have the practical effect of coercing the licensee into using only the licensor's product. One example is the nonmetered royalty. In a recent complaint, the Antitrust Division charged that Microsoft had dominated the market for installed computer operating systems by requiring that computer manufacturers pay royalties to Microsoft on each computer shipped, regardless of whether the computer had a Microsoft operating system installed. Because the manufacturers had to pay for Microsoft's operating system anyway, they were extremely unlikely to install a competitor's operating system, since if they did they would have had to pay twice for an operating system. United States v. Microsoft, Inc., 159 F.R.D. 318 (D.D.C.), *rev'd on other grounds,* 56 F.3d 1448 (D.C. Cir. 1995). When Microsoft settled the division's complaint, it agreed to stop the use of nonmetered royalties.

Yet another Microsoft case offers some interesting thoughts on the role of intellectual property law in justifying licensing restrictions. In United States v. Microsoft Corp., 253 F.3d 34 (D.C. Cir. 2001), the case brought against Windows 98, Microsoft defended its allegedly anticompetitive licensing restrictions in part on the ground that its license terms "merely highlight and expressly state the rights that Microsoft already enjoys under federal copyright law" and therefore could not violate the antitrust laws. The court resoundingly rejected this argument, noting that the copyright laws did not give Microsoft unlimited power over its software. It wrote:

> Microsoft argues that the license restrictions are legally justified because, in imposing them, Microsoft is simply "exercising its rights as the holder of valid copyrights."
>
> Microsoft's primary copyright argument borders upon the frivolous. The company claims an absolute and unfettered right to use its intellectual property as it wishes: "[I]f intellectual property rights have been lawfully acquired," it says, then "their subsequent exercise cannot give rise to antitrust liability." That is no more correct than the proposition that use of one's personal property, such as a baseball bat, cannot give rise to tort liability. As the Federal Circuit succinctly stated: "Intellectual property rights do not confer a privilege to violate the antitrust laws." In re Indep. Serv. Orgs. Antitrust Litig., 203 F.3d 1322, 1325 (Fed. Cir. 2000).

Id. at 62-63. Is this approach consistent with the general principles of antitrust and intellectual property law we have discussed so far?

PROBLEM

Problem 8-4. Microsoft Corporation succeeds in registering a trademark in the name "Windows" to describe a graphical user interface on a computer operating system. Microsoft agrees to license the "Windows" trademark to certain companies who wish to indicate that their application programs are "Windows-compatible." However, Microsoft requires as a condition of the license that the companies sell only application programs that run exclusively on Microsoft Windows operating systems. Has Microsoft violated the antitrust laws? What role should market power play in the analysis? Does the Windows trademark itself confer market power?

Note on Other Common Licensing Restrictions

Besides exclusivity, a number of other common features of intellectual property licenses raise potential antitrust issues. We consider several here.

Grantback Clauses. Intellectual property owners sometimes require as a condition of the license that the licensee "grant back" to the licensor the rights to use any improvements the licensee makes in the technology. Some grantback clauses simply require that the licensee agree not to sue the licensor for infringing on any improvement patents that result from the licensee's work. Others are more strict, however, requiring the assignment or exclusive license of the rights to any improvement.

One reason licensors use grantback clauses is obvious — they don't want to be held "hostage" by their licensees or lose the right to practice improvements on their invention. By retaining the right to use any improvements the licensee makes, the original intellectual property owner can ensure that she stays current in her technology. Grantback clauses may also help to avoid the "blocking patents" problem, by ensuring that the senior intellectual property owner always has the dominant right.

The precise nature of the grantback clause may differ from case to case. Some patentees require that their licensee *assign* any improvement patents back to the licensee. In this instance, the grantback clause avoids the blocking patents problem entirely, since the original patentee ends up owning both the original and the improvement patents. By contrast, if the original patentee requires only that the improver "license back" the right to use the improvement, the grantback clause looks more like an ex ante cross-licensing solution to a prospective blocking patents problem.

On the other hand, grantback clauses may be used to stifle competition. Grantbacks allow a dominant industry player to license her intellectual property freely, without worrying that her licensees may develop an improved process for manufacturing her product and therefore compete with her. Indeed, such licenses may be an effective way of *preventing* serious competition, since they allow the dominant player to "capture" potential competitors and take advantage of their ideas. Further, because licensees no longer have the exclusive (or perhaps any) right to their improvements, they may be less willing to invest in research and development of such improvements.

Traditionally, most grantback clauses have not been considered to violate the antitrust laws. See Transparent-Wrap Machine Corp. v. Stokes & Smith, 329 U.S. 637, 648 (1947) (grantback clauses not illegal per se because they can serve valuable procompetitive purposes).[14] The Antitrust Division's current Intellectual Property Guidelines, however, conclude that "[g]rantbacks may adversely affect competition . . . if they substantially reduce the licensee's incentives to engage in research and development and thereby limit rivalry in innovation markets." In such circumstances, federal agencies will weigh any harm to competition against "offsetting procompetitive effects, such as (1) promoting the dissemination of licensees' improvements to the licensed technology, (2) increasing the licensors' incentives to disseminate the licensed technology, or (3) otherwise increasing competition and output in a relevant technology or innovation market." Intellectual Property Guidelines, §5.6. The guidelines encourage the use of nonexclusive grantback clauses.

Is there reason to treat patent grantbacks differently from copyright grantbacks, because of the difference between the "blocking patents" rule and the derivative works rule in copyright?

Field-of-use Restrictions. Alluded to earlier, field-of-use restrictions in intellectual property licenses grant licensees the right to work a patent in only a limited area. The area may be limited either geographically or by product market. Such restrictions may enable a patentee to exploit his technology efficiently in several markets while maintaining the exclusivity granted him by the patent laws. But field-of-use restrictions may also be an efficient means of enforcing a territorial market division agreement between horizontal competitors. Horizontal market division can have the effect of creating mini-monopolies in what would otherwise be a competitive market. If the patent licensed is a sham, or is only a minor part of the business of the licensor and licensee, it is possible that the patent license is really a "front" for such a market division scheme. Territorial market division outside the scope of an intellectual property right is illegal per se under the rule of United States v. Topco Associates, Inc., 405 U.S. 596 (1972).

Are there legitimate reasons for some field-of-use restrictions that inhere in the nature of the intellectual property laws? Consider, for example, a geographic market division scheme that grants each licensee an exclusive territory in a different country. Because of the national nature of patents, this means that each licensee receives the exclusive right to work the patent in one country (although not the exclusive worldwide right to sell the technology). Is there anything wrong with such a scheme? See United States v. Westinghouse Elec. Corp., 648 F.2d 642 (9th Cir. 1981) (approving such an arrangement); 35 U.S.C. §261 (permitting territorially restricted patent licenses).

Extensions of Patent Term. A license may seek to extend the exclusive rights of the patent laws beyond the old 17-year (or the new 20-year) patent term, either by compelling the continued payment of royalties or by imposing collateral obligations (such as exclusive dealing arrangements or grantback clauses) that do not terminate when the patent expires.[15] Such agreements are illegal per se. See Scheiber v. Dolby Labs, 293 F.3d 1014 (7th Cir. 2002).

14. A notable exception is United Shoe Machinery Co. v. United States, 258 U.S. 451 (1922).
15. This is primarily an issue in patent law, because patents have a shorter term than copyrights.

Whether extensions of the patent term by contract should be of concern to anyone but the contracting parties themselves has been a matter of some debate. Compare Robert Merges, Reflections on Current Legislation Affecting Patent Misuse, 70 J. Pat. & Trademk. Off. Society 793, 801-02 (1988) (patent term extensions are anticompetitive) with Mark A. Lemley, Comment, The Economic Irrationality of the Patent Misuse Doctrine, 78 Calif. L. Rev. 1599, 1630 (1990) (agreements to extend patent term are not anticompetitive). Is the issue here the same that we have been considering repeatedly throughout this book — the right of private parties to change the scope of the intellectual property laws by contract? Or do the antitrust laws put a different spin on the debate? Judge Posner's opinion in *Scheiber* criticized the economic rationale for prohibiting extensions beyond patent term. Constrained by the precedent of *Brulotte*, however, the court found an extension to be per se misuse.

Finally, note that while patents and copyrights have fixed dates on which they expire, trademarks and trade secrets do not. Licensees of multiple intellectual property rights (i.e., both patents and trade secrets) sometimes allege that the intellectual property owner is illegally extending the patent term by continuing to enforce her trade secret rights after her patents have expired. The Justice Department made such a claim in United States v. Pilkington plc, Civ. No. 94-345-TUC-WDB (D. Ariz. filed May 25, 1994). Assuming that the licensor still has valid trade secrets, is there anything wrong with continuing such a "group license"? See Christianson v. Colt Indus. Operating Corp., 870 F.2d 1292, 1303 (7th Cir. 1989), *cert. denied*, 493 U.S. 822 (1989) (trade secrets may extend beyond the expiration of a patent on the same technology, provided that section 112 of the Patent Act did not require the patentee to disclose the secrets in the patent itself).

2. Horizontal Restraints

a. Pooling and Cross-Licensing

Cartels — loosely defined as agreements among competitors to restrict output and raise prices — are illegal under section 1 of the Sherman Act, 15 U.S.C. § 1. When competitors have agreed to set prices or output levels or divide markets, the Sherman Act may be violated without regard to proof of market power or effect. This is the "per se" rule. An early Supreme Court case concluded that it may not be appropriate to subject price agreements involving patented products to the per se rule. United States v. General Electric Co., 272 U.S. 476 (1926). But subsequent decisions have made horizontal agreements setting the price of products illegal in most circumstances. See, e.g., United States v. Line Material Co., 333 U.S. 287 (1948); Herbert Hovenkamp et al., IP and Antitrust ch. 31 (*General Electric* no longer good law, but price agreements should be treated under the rule of reason).

The economic problem with *General Electric* is straightforward. Because cartels price above the competitive level and thus produce excess profits, each cartel member has an incentive to "cheat" by expanding its output to capture a larger percentage of the profits. If all members cheat, though, output will increase back to the competitive level, and prices will fall. Cartels, therefore, are inherently unstable.

The only way to run a cartel effectively over the long run is to employ some sort of "policing" mechanism to prevent cheating. The difficulty with most such mechanisms is that they must be secret, since cartels are illegal (at least in the United States). The

General Electric decision has the effect of providing the perfect policing mechanism for cartel enforcers. By patenting a product and licensing it to its competitors, a company can control the prices at which its competitors sell under *General Electric*. What is more, since such control is legal, the patent holder can enforce the cartel in court, bringing breach of contract actions against any competitor who tries to cut prices! As a result, *General Electric* would appear to allow a patent holder in a competitive or oligopolistic industry to convert its control over one product into effective control over an entire market.

Is there a good rationale for giving a patentee control over licensee pricing? The Court suggests that the patentee may wish to avoid losing profits to its licensee. This rationale makes sense where the patentee and the licensee are likely to compete directly, but not otherwise. Does it suggest that the courts should treat exclusive licenses differently from nonexclusive licenses? Is there an economic reason to be less concerned about one or the other?

Note on Patent Pools

A common form of horizontal agreement in the intellectual property context is a patent pool. Patent pools are formed when two or more intellectual property owners agree to combine their intellectual property rights under common management. At first glance, this may seem to be another form of cartel. Just as agreeing on the price of a product will restrict competition, combining to agree on the price to be charged for intellectual property licenses or on the terms and conditions of those licenses will restrict competition in the market for uses of the patented technology. On this theory, horizontal agreements among intellectual property owners are simply another way of avoiding competition, and should be illegal per se.

One problem with the per se approach is that not all patent pooling is anti-competitive. Indeed, an argument can be made that patent pooling is more desirable than controls on the licensing of a single patent. The argument focuses on the problem of "blocking patents" discussed in Chapter 3. Where one company patents an original invention and a second company (possibly but not necessarily a licensee) patents an improvement to that invention, the law leaves the two companies in a rather peculiar situation. In order to make the most efficient use of the invention, each company wants to use both the original invention and its improvement. But alone, neither company can do so without infringing the other's patent. This is the blocking patents problem. One obvious solution is for the companies to cross-license their patents. This allows both companies to use the invention efficiently, benefiting society and preserving some measure of competition. Why should such an arrangement be subject to greater antitrust scrutiny than the sort of cartel facilitation apparent in *General Electric*?

The problem of "blocking patents" is a common one in the patent-antitrust literature. In addition to the examples cited by the Court, see Eleanor M. Fox & Lawrence A. Sullivan, Antitrust 353-54 (1989); Robert P. Merges, Intellectual Property Rights and Bargaining Breakdown: The Case of Blocking Patents, 62 Tenn. L. Rev. 75 (1994). Blocking patents can take one of two forms. First, two or more companies may have valid patents on different steps of a process in which each step is necessary to make the end product. This can be thought of as "pure" technical necessity. Without cross-licensing, neither company would be able to produce the product at all. In this circumstance, the economic case against applying antitrust law is fairly clear: preventing patent pooling will reduce rather than increase market participation. In this circumstance, at least some courts have been willing to disregard *Line Material* and permit pools of blocking patents. For example, in Carpet Seaming Tape Licensing Corp. v. Best Seam, Inc., 616 F.2d 1133 (9th Cir. 1980), *cert. denied*, 464 U.S. 818 (1983), the court held:

> A well-recognized legitimate purpose for a pooling agreement is exchange of blocking patents. The trial court...demonstrates a misunderstanding as to the nature of the relationship which can give rise to a blocking situation. The trial judge found that the Winkler and Clymin patents did not block the Burgess patents, but ignored the possibility that the Burgess patents may well have blocked the Winkler and Clymin patents despite well-established law that patents on basic processes and products may block patents on improvements to those products and processes. As the Supreme Court stated in Standard Oil Co. v. United States, 283 U.S. at 172, n.5:
>
> > "This is often the case where patents covering improvements of a basic process, owned by one manufacturer, are granted to another. A patent may be rendered quite useless, or 'blocked,' by another unexpired patent which covers a vitally related feature of the manufacturing process. Unless some agreement can be reached, the parties are hampered and exposed to litigation. And, frequently, the cost of litigation to a patentee is greater than the value of a patent for a minor improvement."
>
> . . .

The economics underlying this attitude toward the accumulation of improvement patents is ably and succinctly stated by Professor Ward S. Bowman:

> "If...one patent was subservient to the other, and improvement patent unusable without infringing the basic patent, then combining or pooling them eliminates no user alternative. In terms of possible trade restraint, this case is indistinguishable from a vertical merger. The two patents combined...could not restrict output or raise price any more than if the two were exploited separately."

W. Bowman, Patent and Antitrust Law at 201 (1973).

See also Clorox Co. v. Sterling Winthrop, 932 F. Supp. 469 (E.D.N.Y. 1996) (agreement settling trademark dispute over the terms Lysol and Pine-Sol by limiting the product markets each could sell to was not anticompetitive, since owner of Pine-Sol can compete with Lysol using other marks).

A more common problem involves a number of patents whose validity is open to question. This may be thought of as "practical technical necessity." Rather than entering into prohibitively expensive and protracted litigation over the validity of their respective patents, can firms in this situation settle their differences by pooling their patents (at least one of which is presumably valid)? Standard Oil Co. v. United States, 283 U.S. 163 (1931), seems to suggest that such an agreement would be treated under the rule of reason. Of course, that does not mean that the agreement will always be upheld. If the companies with blocking patents together would comprise only a small part of the industry, their agreement would presumably survive rule of reason scrutiny. But if the new technology encompassed by the patents would dominate a market, even *Standard Oil* presents an antitrust problem. Is this a desirable outcome? On the one hand, allowing the agreement would get the product to market more quickly and inexpensively. Further, because each firm in our hypothetical situation holds an arguably valid patent, one firm is likely to end up with a monopoly in the product anyway (the firm whose patent is upheld). On the other hand, allowing competitors to resolve their differences by cross-licensing their patents could foreclose any incentive for those firms to compete in the future. Instead of a monopoly patent holder being faced with several competitors trying to unseat it, the patent cartel would face no foreseeable competition. Whether you believe the rule of reason or outright legality should govern this situation depends on your assessment of the relative costs of each approach.

Does it matter for antitrust analysis whether the parties to the pool are in the same market? Is the existence of the pool presumptive evidence that they are? Market definition problems have plagued patent pooling cases. See Roger B. Andewelt, Analysis of Patent Pools Under the Antitrust Laws, 53 Antitrust L.J. 611 (1984).

Can an argument be made in favor of patent pooling on the grounds that it reduces transactions costs? Consider the following excerpt:

> Patent pools arise because it makes sense for firms in industries characterized by crossing and conflicting IPRs [intellectual property rights] to "institutionalize" technology exchange. In many cases, pools are creatures of necessity. For example, where different firms hold patents on the basic building blocks of the industry's products, they will have to cross-license to produce at all. This was the case, for example, with the aircraft industry in the early days of the twentieth century, and with sewing machines. Even where no single patent or set of patents is essential, however, firms in an industry often find that they engage in such frequent negotiations that a regularized institution with formal rules, or even general guidelines, is helpful in reducing transaction costs. An example of a pool such as this is the one formed by the early shoe machinery industry. The economic literature on institutions explains this quite well; to use one popular metaphor, the "repeat-play" nature of an institution makes it easier to reach agreement on any particular issue, because disparities tend to balance out over many transactions....
>
> All patent pools share one fundamental characteristic: they provide a regularized transactional mechanism in place of the statutory property rule baseline which requires an individual bargain for each transaction. But my review of particular pools shows that this general structure has encompassed a diversity of organizational forms....
>
> Many patent pools are in essence contracts. Firms agree to consolidate patents and license them collectively. The royalties from licensing the patents, and sometimes from sales of products made by one or more parties to the pooling agreement, are divided according to a contractual formula.... In this simple example, the contract integrated numerous transactions that would otherwise have been negotiated separately. And, most importantly for present purposes, it translated the contribution of each major patent holder into a precise percentage of the royalty stream.

Robert P. Merges, Contracting into Liability Rules: Intellectual Property Rights and Collective Rights Organizations, 84 Cal. L. Rev. 1293 (1996).

Does a similar argument justify copyright pools? Why or why not?

Because patent pools can serve both pro- and anticompetitive purposes, they are best evaluated under the rule of reason. In several recent Business Review Letters involving patent pools in the electronics industry, the Department of Justice has given some guidance on how to minimize the anticompetitive risks of patent pools. Patent pools are more likely to be legal if they cover only patents that are strictly necessary to the use of a technology, if they hire independent experts to assess technical necessity, and if they do not attempt to control the output or pricing decisions of the licensees.

But in U.S. Philips Corp. v. Int'l Trade Comm'n, 424 F.3d. 1179 (Fed. Cir. 2005), the Federal Circuit seemed to reject this distinction as unnecessary. Should pools be allowed to include technologies that compete with one another, rather than just ones that are complements?

PROBLEM

Problem 8-5. Discworld sues the two largest makers of compact discs, alleging that their patented processes infringe its pioneer patent on compact disc technology. After years of discovery, the parties settle the suit by agreeing that each company will license all its patents to the other two, forming a "pool." The settlement agreement contains a "grantback" clause, under which each party agrees to license to pool members any improvements in its technology. The parties to the pool also agree that they will license the collective patents to outside companies, but only if those outside companies agree to use only the technology developed and patented by members of the pool.

NextDisc, Inc., a new entrant into the compact disc market, is approached by the pool and asked to take a license. NextDisc considers the requested royalty to be outrageously high. Rather than taking a license, NextDisc sues the pool members for violating the antitrust laws. What result?

---

b. Patent Settlements

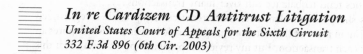

In re Cardizem CD Antitrust Litigation
United States Court of Appeals for the Sixth Circuit
332 F.3d 896 (6th Cir. 2003)

OBERDORFER, District Judge.

This antitrust case arises out of an agreement entered into by the defendants, Hoescht Marion Roussel, Inc. ("HMR"), the manufacturer of the prescription drug Cardizem CD, and Andrx Pharmaceuticals, Inc. ("Andrx"), then a potential manufacturer of a generic version of that drug. The agreement provided, in essence, that Andrx, in exchange for quarterly payments of $10 million, would refrain from marketing its generic version of Cardizem CD even after it had received FDA approval (the "Agreement"). The plaintiffs are direct and indirect purchasers of Cardizem CD who filed complaints challenging the Agreement as a violation of federal and state antitrust laws. After denying the defendants' motions to dismiss, see In re Cardizem CD Antitrust Litigation, 105 F.Supp.2d 618 (E.D.Mich.2000) (*"Dist.Ct.Op. I"*) and granting the plaintiffs' motions for partial summary judgment, id., 105 F.Supp.2d 682 (E.D.Mich.2000) (*"Dist.Ct.Op. II"*), the district court certified two questions for interlocutory appeal:
* * *

(2)...In determining whether Plaintiffs' motions for partial judgment were properly granted, whether the Defendants' September 24, 1997 Agreement constitutes a restraint of trade that is illegal *per se* under section 1 of the Sherman Antitrust Act, 15 U.S.C. § 1, and under the corresponding state antitrust laws at issue in this litigation. JA 607. Our answers, explained more fully herein, are as follows:

* * *

Answer to Second Certified Question: Yes. The Agreement whereby HMR paid Andrx $40 million per year not to enter the United States market for Cardizem CD and its generic equivalents is a horizontal market allocation agreement and, as such, is *per se* illegal under the Sherman Act and under the corresponding state antitrust laws.

Accordingly, the district court properly granted summary judgment for the plaintiffs on the issue of whether the Agreement was *per se* illegal.

I. BACKGROUND

As the district court has set forth a complete outline of the relevant statutory framework, *see Dist. Ct. Op. I,* at 627-29; *Dist. Ct. Op. II,* at 685-86, facts, *see Dist. Ct. Op. I,* at 629-32; *Dist. Ct. Op. II,* at 686-89, and procedural history, *see Dist. Ct. Op. I,* at 632-33; *Dist. Ct. Op. II,* at 689-90, we repeat here only what is necessary to our analysis of the issues on appeal.

A. Statutory Framework

In 1984, Congress enacted the Hatch-Waxman Amendments, *see* Drug Price Competition & Patent Term Restoration Act of 1984, Pub. L. No. 98-417, 98 Stat. 1585 (1984), to the Federal Food, Drug, and Cosmetic Act, 21 U.S.C. §§ 301-399. Those amendments permit a potential generic.[1] manufacturer of a patented pioneer drug to file an abbreviated application for approval with the Food and Drug Administration ("FDA") (known as an Abbreviated New Drug Application ("ANDA")). *See* 21 U.S.C. § 355(j)(1). Instead of submitting new safety and efficacy studies, an ANDA may rely on the FDA's prior determination, made in the course of approving an earlier "pioneer" drug, that the active ingredients of the proposed new drug are safe and effective. *Id.* § 355(j)(2)(A). Every ANDA must include a "certification that, in the opinion of the applicant and to the best of his knowledge, the proposed generic drug does not infringe any patent listed with the FDA as covering the pioneer drug." *Id.* § 355(j)(2)(A)(vii). That certification can take several forms. Relevant here is the so-called "paragraph IV certification" whereby the applicant certifies that any such patent "is invalid or will not be infringed by the manufacture, use, or sale of the new drug for which the application is submitted." *Id.* § 355(j)(2)(A)(vii)(IV). An applicant filing a paragraph IV certification must give notice to the patent-holder, *id.* § 355(j)(2)(B); the patent-holder then has forty-five days to file a patent infringement action against the applicant. *Id.* § 355(j)(5)(B)(iii). If the patent-holder files suit, a thirty-month stay goes into effect, meaning that unless before that time the court hearing the patent infringement case

---

1. A "generic" drug contains the same active ingredients but not necessarily the same inactive ingredients as a "pioneer" drug sold under a brand name. United States v. Generix Drug Corp., 460 U.S. 453, 454-55 (1983).

finds that the patent is invalid or not infringed, the FDA cannot approve the generic drug before the expiration of that thirty-month period. *Id.* § 355(j)(5)(B)(iii)(I). In order to encourage generic entry, and to compensate for the thirty-month protective period accorded the patent holder, the first generic manufacturer to submit an ANDA with a paragraph IV certification receives a 180-day period of exclusive marketing rights, during which time the FDA will not approve subsequent ANDA applications. *Id.* § 355(j)(5)(B)(iv). The 180-day period of exclusivity begins either (1) when the first ANDA applicant begins commercial marketing of its generic drug (the marketing trigger) or (2) when there is a court decision ruling that the patent is invalid or not infringed (the court decision trigger), whichever is earlier. *Id.*

B. Facts

Unless otherwise noted, the following facts are undisputed. HMR manufactures and markets Cardizem CD, a brand-name prescription drug which is used for the treatment of angina and hypertension and for the prevention of heart attacks and strokes. The active ingredient in Cardizem CD is diltiazem hydrochloride, which is delivered to the user through a controlled-release system that requires only one dose per day. HMR's patent for diltiazem hydrochloride expired in November 1992.

On September 22, 1995, Andrx filed an ANDA with the FDA seeking approval to manufacture and sell a generic form of Cardizem CD. On December 30, 1995, Andrx filed a paragraph IV certification stating that its generic product did not infringe any of the patents listed with the FDA as covering Cardizem CD. Andrx was the first potential generic manufacturer of Cardizem CD to file an ANDA with a paragraph IV certification, entitling it to the 180-day exclusivity period once it received FDA approval.

In November 1995, the United States patent office issued Carderm Capital, L.P. ("Carderm") U.S. Patent No. 5,470,584 ("'584 patent"), for Cardizem CD's "dissolution profile," which Carderm licensed to HMR. The dissolution profile claimed by the '584 patent was for 0-45% of the total diltiazem to be released within 18 hours ("45%-18 patent").

In January 1996, HMR and Carderm filed a patent infringement suit against Andrx in the United States District Court for the Southern District of Florida, asserting that the generic version of Cardizem CD that Andrx proposed would infringe the '584 patent. The complaint sought neither damages nor a preliminary injunction. However, filing that complaint automatically triggered the thirty-month waiting period during which the FDA could not approve Andrx's ANDA and Andrx could not market its generic product. In February 1996, Andrx brought antitrust and unfair competition counterclaims against HMR. In April 1996, Andrx amended its ANDA to specify that the dissolution profile for its generic product was not less than 55% of total diltiazem released within 18 hours ("55%-18 generic"). HMR nonetheless continued to pursue its patent infringement litigation against Andrx in defense of its 45%-18 patent. On June 2, 1997, Andrx represented to the patent court that it intended to market its generic product as soon as it received FDA approval.

On September 15, 1997, the FDA tentatively approved Andrx's ANDA, indicating that it would be finally approved as soon as it was eligible, either upon expiration of the thirty-month waiting period in early July 1998, or earlier if the court in the patent infringement action ruled that the '584 patent was not infringed.

Nine days later, on September 24, 1997, HMR and Andrx entered into the Agreement. It provided that Andrx would not market a bioequivalent or generic version of Cardizem CD in the United States until the earliest of: (1) Andrx obtaining a favorable, final and unappealable determination in the patent infringement case; (2) HMR and Andrx entering into a license agreement; *or* (3) HMR entering into a license agreement with a third party. Andrx also agreed to dismiss its antitrust and unfair competition counterclaims, to diligently prosecute its ANDA, and to not "relinquish or otherwise compromise any right accruing thereunder or pertaining thereto," including its 180-day period of exclusivity. In exchange, HMR agreed to make interim payments to Andrx in the amount of $40 million per year, payable quarterly, beginning on the date Andrx received final FDA approval.[3] HMR further agreed to pay Andrx $100 million per year, less whatever interim payments had been made, once: (1) there was a final and unappealable determination that the patent was not infringed; (2) HMR dismissed the patent infringement case; or (3) there was a final and unappealable determination that did not determine the issues of the patent's validity, enforcement, or infringement, and HMR failed to refile its patent infringement action. HMR also agreed that it would not seek preliminary injunctive relief in the ongoing patent infringement litigation.

On July 8, 1998, the statutory thirty-month waiting period expired. On July 9, 1998, the FDA issued its final approval of Andrx's ANDA. Pursuant to the Agreement, HMR began making quarterly payments of $10 million to Andrx, and Andrx did not bring its generic product to market.

On September 11, 1998, Andrx, in a supplement to its previously filed ANDA, sought approval for a reformulated generic version of Cardizem CD. Andrx informed HMR that it had reformulated its product; it also urged HMR to reconsider its infringement claims. On February 3, 1999, Andrx certified to HMR that its reformulated product did not infringe the '584 patent.

On June 9, 1999, the FDA approved Andrx's reformulated product. That same day, HMR and Andrx entered into a stipulation settling the patent infringement case and terminating the Agreement. At the time of settlement, HMR paid Andrx a final sum of $50.7 million, bringing its total payments to $89.83 million. On June 23, 1999, Andrx began to market its product under the trademark Cartia XT, and its 180-day period of marketing exclusivity began to run. Since its release, Cartia XT has sold for a much lower price than Cardizem CD and has captured a substantial portion of the market.

* * *

For all of the plaintiffs, the foundation for their claims is the allegation that but for the Agreement, specifically the payment of $40 million per year, Andrx would have brought its generic product to market once it received FDA approval and at a lower price than the patented Cardizem CD sold by HMR. They further allege that the Agreement protected HMR from competition from both Andrx and other potential generic competitors because Andrx's delayed market entry postponed the start of its 180-day exclusivity period, which it had agreed not to relinquish or transfer. The Sherman Act Class Plaintiffs and the Individual Sherman Act Plaintiffs bring claims under the federal antitrust laws, specifically section 1 of the Sherman Act, 15 U.S.C.

---

3. The payments were scheduled to end on the earliest of: (1) a final and unappealable order or judgment in the patent infringement case; (2) if HMR notified Andrx that it intended to enter into a license agreement with a third party, the earlier of: (a) the expiration date of the required notice period or (b) the date Andrx effected its first commercial sale of the Andrx product; or (3) if Andrx exercised its option to acquire a license from HMR, the date the license agreement became effective.

§ 1; they seek treble damages under section 4 of the Clayton Act, 15 U.S.C. § 4. The State Law Class Plaintiffs bring claims under various state antitrust laws....

II. DISCUSSION

As we believe that our answer to the second question sheds light on our consideration of the first, we address first whether the Agreement was a *per se* illegal restraint of trade before considering whether the plaintiffs adequately alleged antitrust injury.

A. *Per Se Illegal Restraint of Trade*

1. Relevant Antitrust Law

Section 1 of the Sherman Act provides that "Every contract, combination in the form of trust or otherwise, or conspiracy, in restraint of trade or commerce among the several States, or with foreign nations, is declared to be illegal...." 15 U.S.C. § 1. Read "literally," section 1 "prohibits *every* agreement in restraint of trade." Arizona v. Maricopa Cty. Medical Soc., 457 U.S. 332, 342 (1982). However, the Supreme Court has long recognized that Congress intended to outlaw only "unreasonable" restraints. State Oil Co. v. Khan, 522 U.S. 3, 10 (1997); *Maricopa Cty.*, 457 U.S. at 342-43 (citing United States v. Joint Traffic Ass'n, 171 U.S. 505 (1898)). Most restraints are evaluated using a "rule of reason." *State Oil*, 522 U.S. at 10, 118 S.Ct. 275. Under this approach, the "finder of fact must decide whether the questioned practice imposes an unreasonable restraint on competition, taking into account a variety of factors, including specific information about the relevant business, its condition before and after the restraint was imposed, and the restraint's history, nature, and effect." *Id.* (citing *Maricopa Cty.*, 457 U.S. at 343 & n. 13).

Other restraints, however, "are deemed unlawful *per se*" because they "have such predictable and pernicious anticompetitive effect, and such limited potential for procompetitive benefit." *Id.* (citing Northern Pacific Ry. Co. v. United States, 356 U.S. 1, 5 (1958)). "*Per se* treatment is appropriate '[o]nce experience with a particular kind of restraint enables the Court to predict with confidence that the rule of reason will condemn it.'" *Id.* (quoting *Maricopa Cty.*, 457 U.S. at 344); see also Broadcast Music, Inc. v. Columbia Broadcasting System, Inc., 441 U.S. 1, 19-20 (1979) (a *per se* rule is applied when "the practice facially appears to be one that would always or almost always tend to restrict competition and decrease output."). The *per se* approach thus applies a "conclusive presumption" of illegality to certain types of agreements, *Maricopa Cty.*, 457 U.S. at 344; where it applies, no consideration is given to the intent behind the restraint, to any claimed pro-competitive justifications, or to the restraint's actual effect on competition.[11] National College Athletic Ass'n

---

11. The risk that the application of a *per se* rule will lead to the condemnation of an agreement that a rule of reason analysis would permit has been recognized and tolerated as a necessary cost of this approach. *See, e.g., Maricopa Cty.*, 457 U.S. at 344 ("As in every rule of general application, the match between the presumed and the actual is imperfect. For the sake of business certainty and litigation efficiency, we have tolerated the invalidation of some agreements that a full-blown inquiry might have proved to be reasonable.")...

("NCAA") v. Board of Regents, 468 U.S. 85, 100, 104 S.Ct. 2948, 82 L.Ed.2d 70 (1984). As explained by the Supreme Court, "[t]he probability that anticompetitive consequences will result from a practice and the severity of those consequences must be balanced against its procompetitive consequences. Cases that do not fit the generalization may arise, but a *per se* rule reflects the judgment that such cases are not sufficiently common or important to justify the time and expense necessary to identify them." Continental T.V., Inc. v. GTE Sylvania Inc., 433 U.S. 36, 50 n. 6 (1977).

The Supreme Court has identified certain types of restraints as subject to the *per se* rule. The classic examples are naked, horizontal restraints pertaining to prices or territories. *See, e.g., NCAA,* 468 U.S. at 100 ("Horizontal price fixing and output limitation are ordinarily condemned as a matter of law under an 'illegal *per se*' approach because the probability that these practices are anticompetitive is so high."); Copperweld Corp. v. Independence Tube Corp., 467 U.S. 752, 768 (1984) ("Certain agreements, such as horizontal price fixing and market allocation, are thought so inherently anticompetitive that each is illegal *per se* without inquiry into the harm it has actually caused."); United States v. Topco Assocs., 405 U.S. 596, 608 (1972) ("One of the classic examples of a *per se* violation of § 1 is an agreement between competitors at the same level of the market structure to allocate territories in order to minimize competition. Such concerted action is usually termed a 'horizontal' restraint, in contradistinction to combinations of persons at different levels of the market structure, *e.g.,* manufacturers and distributors, which are termed 'vertical' restraints. This Court has reiterated time and time again that horizontal territorial limitations . . . are naked restraints of trade with no purpose except stifling of competition. Such limitations are *per se* violations of the Sherman Act." (internal citations omitted)); *Northern Pacific Ry.,* 356 U.S. at 5 ("Among the practices which the courts have heretofore deemed to be unlawful in and of themselves are price fixing, division of markets, group boycotts, and tying arrangements." (internal citations omitted)).

2. Application

In answering the question whether the Agreement here was *per se* illegal, the following facts are undisputed and dispositive. The Agreement guaranteed to HMR that its only potential competitor at that time, Andrx, would, for the price of $10 million per quarter, refrain from marketing its generic version of Cardizem CD even after it had obtained FDA approval, protecting HMR's exclusive access to the market for Cardizem CD throughout the United States until the occurrence of one of the end dates contemplated by the Agreement. (In fact, Andrx and HMR terminated the Agreement and the payments in June 1999, before any of the specified end dates occurred.) In the interim, however, from July 1998 through June 1999, Andrx kept its generic product off the market and HMR paid Andrx $89.83 million. By delaying Andrx's entry into the market, the Agreement also delayed the entry of other generic competitors, who could not enter until the expiration of Andrx's 180-day period of marketing exclusivity, which Andrx had agreed not to relinquish or transfer.[12] There

---

12. As the district court for the Eastern District of New York recently observed, in distinguishing the district court's opinion in the present case (*Cardizem II*),

> By agreeing both not to end the underlying patent dispute and not to market a generic drug product in the relevant domestic market, Andrx . . . effectively precluded or seriously delayed both the

is simply no escaping the conclusion that the Agreement, all of its other conditions and provisions notwithstanding, was, at its core, a horizontal agreement to eliminate competition in the market for Cardizem CD throughout the entire United States, a classic example of a *per se* illegal restraint of trade.

None of the defendants' attempts to avoid *per se* treatment is persuasive. As explained in greater detail in the district court's opinion, *see Dist. Ct. Op. II*, at 700-705, the Agreement cannot be fairly characterized as merely an attempt to enforce patent rights or an interim settlement of the patent litigation. As the plaintiffs point out, it is one thing to take advantage of a monopoly that naturally arises from a patent, but another thing altogether to bolster the patent's effectiveness in inhibiting competitors by paying the only potential competitor $40 million per year to stay out of the market. Nor does the fact that this is a "novel" area of law preclude *per se* treatment, *see Maricopa Cty.*, 457 U.S. at 349. To the contrary, the Supreme Court has held that " '[w]hatever may be its peculiar problems and characteristics, the Sherman Act, so far as price-fixing agreements are concerned, establishes one uniform rule applicable to all industries alike.'" *Id.* at 349 (quoting United States v. Socony-Vacuum Oil Co., 310 U.S. 150, 222 (1940)). We see no reason not to apply that rule here, especially when the record does not support the defendants' claim that the district court made "errors" in its analysis. Finally, the defendants' claims that the Agreement lacked anticompetitive effects and had procompetitive benefits are simply irrelevant. *See, e.g., Maricopa Cty.*, 457 U.S. at 351. To reiterate, the virtue/vice of the *per se* rule is that it allows courts to presume that certain behaviors as a class are anticompetitive without expending judicial resources to evaluate the actual anticompetitive effects or procompetitive justifications in a particular case. As the Supreme Court explained in *Maricopa County*:

> The respondents' principal argument is that the *per se* rule is inapplicable because their agreements are alleged to have procompetitive justifications. The argument indicates a misunderstanding of the *per se* concept. The anticompetitive potential inherent in all price-fixing agreements justifies their facial invalidation even if procompetitive justifications are offered for some. Those claims of enhanced competition are so unlikely to prove significant in any particular case that we adhere to the rule of law that is justified in its general application.

457 U.S. at 351. Thus, the law is clear that once it is decided that a restraint is subject to *per se* analysis, the claimed lack of any actual anticompetitive effects or presence of procompetitive effects is irrelevant. Of course, our holding here does not resolve the issues of causation and damages, both of which will have to be proved before the plaintiffs can succeed on their claim for treble damages under the Clayton Act.

. . .

[patent] court decision and the commercial marketing trigger of the 180-day exclusivity period. As a result, any future [generic] filers were delayed in coming to market....

In re Ciprofloxacin Hydrochloride Antitrust Litigation, 261 F.Supp.2d 188, 242 (E.D.N.Y. 2003); *see also In re Tamoxifen Citrate Antitrust Litigation*, MDL No. 1408, 2003 WL 21196817, at *9 (E.D.N.Y. May 13, 2003).

≣≣≣ **Valley Drug Co. v. Geneva**
≣≣≣ **Pharmaceuticals, Inc.,**
≣≣≣ *United States Court of Appeals for the Eleventh Circuit*
≣≣≣ *344 F.3d 1294 (11th Cir. 2003)*

ANDERSON, Circuit Judge:

This case comes to us on interlocutory appeal from the district court's order granting plaintiffs' motion for partial summary judgment. The issue with which we are presented is whether the district court properly determined that two agreements among the defendants were *per se* violations of § 1 of the Sherman Act, 15 U.S.C. § 1. Because we conclude that the district court incorrectly applied the law, the order below will be reversed and the case remanded for further proceedings consistent with this opinion.

I. BACKGROUND

This is a private antitrust lawsuit, or rather numerous private antitrust lawsuits, brought against three pharmaceuticals manufacturers. The various cases asserting antitrust injury from the actions of the defendants have been consolidated by the Judicial Panel on Multidistrict Litigation in the Southern District of Florida. Plaintiffs basically assert that two agreements, one between defendant-appellant Abbott Laboratories ("Abbott") and defendant-appellant Geneva Pharmaceuticals ("Geneva") and another between Abbott and defendant Zenith Goldline Pharmaceuticals ("Zenith"), entered into in 1998, violated the Sherman Act's prohibition against contracts in restraint of trade. Abbott, a manufacturer of the pioneer drug Hytrin, entered separate agreements with generic manufacturers Zenith and Geneva while those companies were pursuing FDA approval of generic versions of Hytrin and while embroiled in patent litigation with those companies. Because the facts of this case took place against a complex regulatory background, we will first describe the relevant regulatory context, then the facts of the case as they appear in the current procedural posture, and then the relevant procedural history.

A. Regulatory Framework

[see *In re Cardizem CD* for a discussion of the Hatch-Waxman regulatory scheme]

B. Factual History

1. Zenith's and Geneva's ANDAs

Abbott manufactures Hytrin, a brand-name drug with the active ingredient dihydrate terazosin hydrochloride. Hytrin is used to treat hypertension and enlarged prostate, and has been a very successful product for Abbott. Abbott obtained FDA approval of its NDA for Hytrin in 1987 and has held a number of patents related to terazosin hydrochloride over the years. Its first patent, issued in 1977, covered the basic terazosin hydrochloride compound. See Abbott Labs. v. Novopharm Ltd., Nos. 96-C-611 & 95-C-6657, 1996 WL 131498, *1 (N.D.Ill. March 15, 1996). That patent has since expired, though Abbott has been issued other patents for various crystalline forms of the compound and various methods of using and preparing the compound.

Geneva filed four ANDAs based on Hytrin between 1993 and 1996, each time making paragraph IV certifications with respect to Abbott's listed patents. Abbott brought infringement suits under 35 U.S.C. § 271(e), invoking the 30-month stay of FDA approval of Geneva's ANDAs. On April 29, 1996, Geneva filed two additional ANDAs based on Hytrin, one for a capsule form of terazosin hydrochloride and one for a tablet form, making paragraph IV certifications with respect to the relevant patents. Within forty-five days of receiving notice of Geneva's paragraph IV certifications, Abbott filed an infringement suit based on the submission of the tablet ANDA, asserting that the Geneva's tablet terazosin hydrochloride product infringed Abbott's Patent No. 5,504,207 ("the '207 patent"). In the suit, Geneva admitted infringement but contested the patent's validity. Apparently through oversight, Abbott failed to file an infringement suit based on the submission of the capsule ANDA. FDA consideration of that ANDA, therefore, proceeded unhindered, and Geneva's capsule ANDA was approved in March of 1998. When Abbott was notified of this approval, it began efforts to amend its complaint to allege that Geneva's terazosin hydrochloride capsule infringed the '207 patent.

Zenith, meanwhile, filed an ANDA for a terazosin hydrochloride drug in June of 1994, making a paragraph IV certification with respect to Abbott's Hytrin patents. Abbott was issued a patent claiming forms of terazosin hydrochloride on May 2, 1995, Patent No. 5,412,095 ("the '095 patent"), and another on April 2, 1996, the '207 patent. Abbott timely filed this patent information with the FDA, which then required Zenith to amend its ANDA to make a certification with regard to these newly listed patents. Zenith resisted, hoping to avoid the 30-month stay of approval and the 180-day delay of approval based on Geneva's earlier-filed ANDAs. Instead of certifying, Zenith brought suit against Abbott in an attempt to force Abbott to delist its '095 and '207 patents, relieving Zenith of the obligation to certify with respect to those patents, and seeking a declaration that its terazosin hydrochloride drug did not infringe those patents. Zenith's suit alleged that Abbott listed the '095 and '207 patents knowing that the patents did not claim Hytrin or a method of using Hytrin. Abbott counterclaimed for infringement. Zenith's attempt to obtain a preliminary injunction against the listing of the patents was unsuccessful. See Zenith Labs. v. Abbott Labs., No. 96-1661, 1997 U.S. Dist. LEXIS 23954 (D.N.J. Oct. 1, 1997). Zenith appealed the denial to the Federal Circuit.

2. The Agreements

The agreements challenged by the plaintiffs arose against this backdrop. On March 31, 1998, Abbott and Zenith entered an agreement dismissing Zenith's delisting claims and Abbott's infringement counterclaims ("Zenith Agreement"). In the Zenith Agreement, Zenith acknowledged the validity of each of Abbott's patents claiming terazosin hydrochloride and admitted that any terazosin hydrochloride product Zenith might market would infringe these patents. Zenith agreed not to sell or distribute any pharmaceutical product containing any form of terazosin hydrochloride until someone else introduced a generic terazosin hydrochloride product first or until Abbott's Patent No. 4,215,532 ("the '532 patent") expired. Zenith agreed not to sell or transfer its rights under any ANDA application relating to a terazosin hydrochloride drug, not to aid any other person in gaining FDA approval of a terazosin hydrochloride drug, and not to aid any other person in opposing or invalidating any of Abbott's patents claiming terazosin. In return, Abbott agreed to a payment schedule according to which Abbott would pay Zenith $3 million up front, $3 million after three months, and $6 million every three months thereafter until March 1, 2000, or until the Agreement terminated by its own terms. If another generic manufacturer introduced a terazosin hydrochloride drug and obtained a 180-day exclusivity period, Abbott's payments would be halved until the period expired. Abbott agreed not to sue Zenith for infringement if it entered the market consistent with the Agreement.

Abbott entered an agreement with Geneva on April 1, 1998 ("Geneva Agreement"). According to the Geneva Agreement, Geneva agreed not to sell or distribute any pharmaceutical product containing any form of terazosin hydrochloride until either Abbott's '532 patent expired, someone else introduced a generic terazosin hydrochloride drug, or Geneva obtained a court judgment that its terazosin tablets and capsules did not infringe the '207 patent or that the patent was invalid. This latter condition required a final judgment from which no further appeal could be taken, including petition for certiorari to the Supreme Court. Geneva agreed not to transfer or sell its rights under its ANDAs, including its right to the 180-day exclusivity period. Geneva also agreed to oppose any subsequent ANDA applicant's attempt to seek approval of its application based on Geneva's failure to satisfy the then-existing successful defense requirement and to join and support any attempt by Abbott to seek an extension of the 30-month stay of FDA approval on Geneva's tablet ANDA. In return, Abbott agreed to pay Geneva $4.5 million each month until either someone else brought a generic terazosin hydrochloride product to market or Abbott won a favorable decision in the district court on its infringement claim. If Geneva won in district court, Abbott's $4.5 million monthly payments would go into escrow pending resolution of the appeal, with the escrowed funds going to the party prevailing on appeal. Abbott reserved the right to terminate its payments after February 8, 2000, if no other generic terazosin hydrochloride product had been marketed as of that date. If Abbott exercised this right, it would execute a release in Geneva's favor of any claims of infringement based on the '207 patent.

3. Termination of the Agreements

The district court hearing Abbott's infringement suit against Geneva handed down its decision on September 1, 1998. Abbott Labs. v. Geneva Pharms., Inc., Nos. 96-C-3331, 96-C-5868, & 97-C-7587, 1998 WL 566884 (N.D.Ill. Sept. 1, 1998). The court held the '207 patent invalid because the crystalline form of terazosin hydrochloride claimed in the patent was on sale in the United States more than one year before Abbott applied for the patent, *see* 35 U.S.C. § 102(b). 1998 WL 566884 at *7. Abbott appealed to the Federal Circuit, which affirmed on July 1, 1999. 182 F.3d 1315 (Fed.Cir.1999). Abbott's petition for certiorari was denied on January 10, 2000. 528 U.S. 1078 (2000).

The Agreements did not terminate on their own terms, however. The parties terminated the Agreements on August 13, 1999, apparently in response to an FTC investigation of those Agreements. The FTC action resulted in a consent settlement. *See Matter of Abbott Labs.*, No. C-3945, 2000 WL 681848 (F.T.C. May 22, 2000), *also available at* http://www.ftc.gov/os/2000/05/c3945.do.htm.

. . .

D. *The Order Granting Summary Judgment*

The December 13, 2000, Order granting plaintiffs' motion for partial summary judgment concluded that the Agreements were *per se* violations of §1 of the Sherman Act. The Order characterized the Agreements as geographic market allocation Agreements between horizontal competitors, essentially allocating the entire United States market for terazosin drugs to Abbott, who shared its monopoly profits with the other cartel members during the life of the Agreements.

The court found that on the eve of the Agreements,

> both Geneva and Zenith were poised to market generic versions of Hytrin in the United States. Geneva received final FDA approval for its generic capsule in March subject to "validation," and the 30-month stay on its generic tablet proposal was set to expire in October. Zenith declared that it was ready to market a generic tablet upon receipt of a favorable decision from the Federal Circuit and final FDA approval.

In re Terazosin Hydrochloride Antitrust Litig., 164 F.Supp.2d 1340, 1345-46 (S.D.Fla.2000). Despite being poised to enter the market, however, "Geneva and Zenith forswore competing with Abbott in the United States market for terazosin hydrochloride drugs. . . ." *Id.* at 1348-49.

The court identified four elements of the Agreement with Geneva that were anticompetitive: (1) Geneva's promise not to market its terazosin capsule until the Agreement terminated; (2) Geneva's promise not to market its terazosin tablet until the Agreement terminated; (3) Geneva's promise not to sell its rights in its capsule and tablet ANDAs until the Agreement terminated; and (4) Geneva's promise to aid Abbott in opposing any attempt by other ANDA applicants to enter the market before the Agreement terminated. The court identified three anticompetitive elements of the Agreement with Zenith: (1) Zenith's agreement to dismiss its delisting suit; (2) Zenith's promise not to aid any other entity's challenge to the validity of Abbott's terazosin patents; and (3) Zenith's promise not to market a generic terazosin product until the Agreement terminated. The essence of the Agreements, the court concluded, was to "dissuade[ ] Geneva and Zenith from marketing the first generic terazosin hydrochloride drugs in the United States for an indefinite period [and] eliminat[e] the risk that either drug maker would sell or purchase the right to introduce such drugs in the interim. . . ." *Id.* at 1349.

Despite holding the Agreements *per se* unlawful, the court nonetheless entertained, and rejected, the defendants' arguments that the Agreements were either procompetitive or benign. Defendants argued that the Agreements eliminated the "substantial legal and financial risks" that accompany market entry while patent disputes remain unresolved. This justification was rejected for three reasons. First, the Agreement with Geneva was unnecessary to avoid these risks, because Geneva's unilateral decision to forego entry would achieve the same result. Second, the Agreement with Geneva did not resolve the patent litigation; "in fact, it tended to prolong that dispute to Abbott's advantage." *Id.* at 1350. The litigation was prolonged by the provisions extending the Agreement until the patent dispute was finally resolved, including Supreme Court review, and by Geneva's promise to aid Abbott in any motion seeking to extend the 30-month stay of FDA approval of Geneva's tablet ANDA. Finally, the court rejected the argument that the provision of the Agreement permitting Geneva to enter the market if Abbott elected to suspend its payments was pro-competitive. Although the court may have been willing to "infer that this clause was a catalyst for competition if Geneva paid Abbott for it, [ ] the suggestion that Abbott handsomely paid Geneva to spur competition in its own lucrative domestic market for terazosin hydrochloride products is patently unreasonable." *Id.* at 1351. The court did not analyze any potentially efficiency-enhancing effects of the Agreement with Zenith, concluding that the Agreement "would indefinitely postpone Zenith's entry into the United States market and would permit competition only once Abbott lost its exclusive market." *Id.*

Defendants argued that because the Agreement with Geneva was analogous to an interim patent settlement and the Agreement with Zenith terminated the litigation between Zenith and Abbott, the court should treat the Agreements as patent litigation settlements. The court rejected this argument as well, concluding that the Agreement with Geneva did not resolve Abbott's infringement suit and that the termination of litigation between Zenith and Abbott "was part of a larger scheme to restrain the domestic sale of generic terazosin hydrochloride products." *Id.* at 1353. Even if the Agreements were patent litigation settlements, the court held, such settlements were not immune from *per se* analysis. *Id.*

. . .

III. DISCUSSION

. . .

The district court focused on the agreements by Geneva and Zenith not to enter the market with FDA-approved (or approval pending) generic terazosin drugs, holding that this exclusionary effect of the Agreements constituted an allocation of the market between horizontal competitors and that the Agreements were therefore *per se* illegal. We begin our discussion by addressing this exclusionary effect of the Agreements.

A.

An agreement between competitors to allocate markets is, as the district court noted, clearly anticompetitive. Such an agreement has the obvious tendency to diminish output and raise prices. When a firm pays its only potential competitor not to compete in return for a share of the profits that firm can obtain by being a monopolist, competition is reduced. See, e.g., Palmer v. BRG of Georgia, Inc., 498 U.S. 46, 49-50 (1990) (per curiam) (agreements not to compete within certain territorial limits are obviously anticompetitive); United States v. Topco Assocs., 405 U.S. 596, 608 (1972) (observing that an agreement between competitors to allocate territories is a "classic example" of a *per se* violation of the Sherman Act, with no purpose other than reducing competition).

If this case merely involved one firm making monthly payments to potential competitors in return for their exiting or refraining from entering the market, we would readily affirm the district court's order. This is not such a case, however, because one of the parties owned a patent. For reasons explained below, and in light of the fact that all parties seem to agree that the ANDAs that were the subjects of the infringement suits infringed Abbott's '207 patent, we reject the district court's characterization of the instant Agreements as illegal *per se*. We believe any such characterization is premature without further analysis of the kind suggested in this opinion. Because the market allocation characterization was central to the district court's conclusion that the Agreements in their entireties are *per se* violations of §1 of the Sherman Act, we will reverse the grant of summary judgment.

A patent grants its owner the lawful right to exclude others. *See* 35 U.S.C. §§ 271(a) (defining infringement) & 283 (providing injunctive relief for infringement); Dawson Chem. Co. v. Rohm & Haas Co., 448 U.S. 176, 215 (1980) ("[T]he essence of a patent grant is the right to exclude others from profiting by the patented invention."). This exclusionary right is granted to allow the patentee to exploit whatever degree of market power it might gain thereby as an incentive to induce investment in innovation and the public disclosure of inventions. Bonito Boats, Inc., v. Thunder Craft Boats, Inc., 489 U.S. 141, 150-51(1989); United States v. Studiengesellschaft Kohle, m.b.H., 670 F.2d 1122, 1127 (D.C.Cir.1981). The exclusionary right cannot be exploited in every way–patentees cannot pool their patents and fix the prices at which licensees will sell the patented article, for example, *see* United States v. New Wrinkle, Inc., 342 U.S. 371 (1952)—but a patentee can choose to exclude everyone from producing the patented article or can choose to be the sole supplier itself, see, e.g., In re Indep. Serv. Orgs. Antitrust Litig., 203 F.3d 1322, 1328 (Fed.Cir.2000); SCM Corp. v. Xerox Corp., 645 F.2d 1195, 1209 (2d Cir.1981); or grant exclusive territorial licenses carving up the United States among its licensees, see 35 U.S.C. § 261. Within reason, patentees can also subdivide markets in ways other than territorial, such as by customer class. Gen. Talking Pictures Corp. v. Western Elec. Co., 304 U.S. 175 *aff'd on reh'g,* 305 U.S. 124 (1938) (approving a license restricting the licensee's sales to non-commercial customers). Such arrangements undoubtedly tend to result in lower production and higher prices of the patented article than if competition were unrestrained, but these anticompetitive tendencies do not render them in violation of the Sherman Act.

The above discussion illustrates the point that a patentee's allocation of territories is not always the kind of territorial market allocation that triggers antitrust liability, and this is so because the patent gives its owner a lawful exclusionary right. In characterizing the Agreements as territorial market allocations agreements, the district court did not consider that the '207 patent gave Abbott the right to exclude others from making, using, or selling anhydrous terazosin hydrochloride until October of 2014, when it is due to expire. To the extent that Zenith and Geneva agreed not to market admittedly infringing products before the '207 patent expired or was held invalid, the market allocation characterization is inappropriate.

Zenith's agreement not to market an infringing generic terazosin hydrochloride drug terminated, by its own terms, when either another generic manufacturer marketed a terazosin product and any exclusivity period expired or Abbott's '532 patent expired in February of 2000. The effect of the Zenith Agreement on the production of Zenith's infringing terazosin product appears to be no broader than the potential exclusionary effect of the '207 patent, and was actually narrower to the extent it permitted Zenith to market its drug before the '207 patent expired. Geneva's agreement not to market an infringing terazosin product terminated at the earliest of (1) a final, unappealable judgment holding the '207 patent invalid; (2) the marketing of a terazosin product by another generic manufacturer; or (3) the expiration of the '532 patent. The effect of the Geneva Agreement on the production of Geneva's infringing generic terazosin product may have been no broader than the potential exclusionary effect of the '207 patent. The '207 patent may have allowed Abbott to obtain preliminary injunctive relief or a stay of an adverse judgment pending appeal, which also would have prevented Geneva from marketing its terazosin hydrochloride products during this period.

With respect to the foregoing aspects of the two Agreements' exclusionary effects, which were the foundation of the district court's characterization of the Agreements as market allocation agreements, these are at the heart of the patent right and cannot trigger the *per se* label.[18] Unlike some kinds of agreements that are *per se* illegal whether engaged in by patentees or anyone else, such as tying or price-fixing, the exclusion of infringing competition is the essence of the patent grant. As one court has concluded, "when patents are involved . . . the exclusionary effect of the patent must be considered before making any determination as to whether the alleged restraint is *per se* illegal." *In re Ciprofloxacin Hydrochloride Antitrust Litig.*, 261 F.Supp.2d 188, 249 (E.D.N.Y.2003). Because the district court failed to consider the exclusionary power of Abbott's patent in its antitrust analysis, its rationale was flawed and its conclusion that these Agreements constitute *per se* violations of the antitrust laws must be reversed.

While our holding at this early stage of the litigation is appropriately narrow, the decision below and the arguments of the parties invite our discussion of several matters that promise to be relevant on remand. We first discuss appellees' argument that the antitrust analysis need not consider Abbott's patent rights because the '207 patent was declared invalid. We then discuss appellees' argument that the lawful right of exclusion does not include the right to pay competitors not to produce infringing products. Finally, we offer several observations with respect to the framework to be developed on remand for deciding the appropriate antitrust analysis.

B.

The individual Sherman Act plaintiffs-appellees argue that because the '207 patent was declared invalid after the Agreements were entered into, Abbott never had any patent rights and our antitrust analysis need not consider the '207 patent. We reject the appellees' argument that the agreements by Geneva and Zenith not to produce infringing products are subject to *per se* condemnation and treble-damages liability merely because the '207 patent was subsequently declared invalid. We begin with the proposition that the reasonableness of agreements under the antitrust laws

18. Appellees argue that the Agreements have broader exclusionary tendencies in that they also prohibited the marketing of non-infringing terazosin products, prohibited Geneva from marketing infringing products beyond the date a district court held the '207 patent invalid, and prohibited Geneva from waiving its 180-day exclusivity period. As we explain below, these prohibitions may be beyond the scope of Abbott's lawful right to exclude and, if so, would expose appellants to antitrust liability for any actual exclusionary effects resulting from these provisions that appellees can prove at the causation and damages stages of litigation. Our point thus far is that appellees have failed to prove that appellants should face *per se* antitrust liability for treble damages for the failure of Zenith and Geneva to market admittedly infringing products when no court had declared Abbott's patent invalid or unenforceable at the time of the Agreements.

are to be judged at the time the agreements are entered into. Polk Bros. v. Forest City Enters., 776 F.2d 185, 189 (7th Cir.1985); *SCM Corp.*, 645 F.2d at 1207. At the time the Agreements were entered into, no court had declared Abbott's '207 patent invalid, and the appellees have advanced no argument other than mere invalidity. We hold that the mere subsequent invalidity of the patent does not render the patent irrelevant to the appropriate antitrust analysis.

The right of exclusion conferred by a patent has been characterized as a defense to an antitrust claim, see Walker Process Equip., Inc., v. Food Mach. & Chem. Corp., 382 U.S. 172, 179 (1965) (Harlan, J., concurring), or as a limited exception to the general rule that markets should be free from barriers to competition, *see id.*, 382 U.S. at 176; United States v. Line Material Co., 333 U.S. 287, 309 (1948)*id.*, 333 U.S. at 310. Appellees' argument implies that this defense is unavailable with respect to the Agreements because the '207 patent was subsequently held invalid.

The only time the Supreme Court has addressed the circumstances under which the patent immunity from antitrust liability can be pierced, it held that the antitrust claimant must prove that the patentee enforced a patent with the knowledge that the patent was procured by fraud on the Patent Office. *Walker Process*, 382 U.S. at 177 (considering a monopolization claim against a patentee for suing to enforce a patent allegedly procured by fraud). Good faith procurement furnishes a complete defense to the antitrust claim. *Id.* Justice Harlan's concurrence explained that the effect of anti-trust liability on the incentives for innovation and disclosure created by the patent regime must be taken into account when a court considers whether a patentee is stripped of its immunity from the antitrust laws:

> It is well also to recognize the rationale underlying this decision, aimed of course at achieving a suitable accommodation in this area between the differing policies of the patent and antitrust laws. To hold, as we do, that private suits may be instituted under § 4 of the Clayton Act to recover damages for Sherman Act monopolization knowingly practiced under the guise of a patent procured by deliberate fraud, cannot well be thought to impinge upon the policy of the patent laws to encourage inventions and their disclosure. Hence, as to this class of improper patent monopolies, antitrust remedies should be allowed room for full play. On the other hand, to hold, as we do not, that private antitrust suits might also reach monopolies practiced under patents that for one reason or another may turn out to be voidable under one or more of the numerous technicalities attending the issuance of a patent, might well chill the disclosure of inven-tions through the obtaining of a patent because of fear of the vexations or punitive consequences of treble-damage suits. Hence, this private antitrust remedy should not be deemed to reach § 2 monopolies carried on under a nonfraudulently procured patent.

382 U.S. at 179-80. This approach is commonly used by courts considering the intersection of patent law and antitrust law. See, *e.g.*, *SCM Corp.*, 645 F.2d at 1203-06 (considering antitrust claims against a patentee for acquiring and refusing to license various patents); Handgards, Inc., v. Ethicon, Inc., 601 F.2d 986, 992-93 (9th Cir.1979) (considering a monopolization claim against a patentee for asserting a patent it allegedly knew to be invalid). A suitable accommodation between antitrust law's free competition requirement and the patent regime's incentive system is required by the complementary objectives of the two:

> It is commonly said . . . that the patent and antitrust laws necessarily clash . . . . At the same time, the two regimes seek the same object: the welfare of the public . . . . [A]ntitrust law

forbids certain agreements tending to restrict output and elevate prices and profits above the competitive level. Patent law also serves the interests of consumers by protecting invention against prompt imitation in order to encourage more innovation than would otherwise occur.

H. Hovenkamp, Antitrust Law: An Analysis of Antitrust Principles and Their Application, ¶1780a (1999) ("Hovenkamp") (citations and quotations omitted). *See also* D. Crane, Exit Payments in Settlement of Patent Infringement Lawsuits: Antitrust Rules and Economic Implications, 54 Fla. L. Rev. 747, 748 n. 1 (2002) ("It is generally recognized that antitrust and patent law, although polar opposites in their treatment of monopolies, share common objectives."); Zenith Elecs. Corp. v. Exzec, Inc., 182 F.3d 1340, 1352 (Fed.Cir.1999) ("The patent and antitrust laws are complementary in purpose in that they each promote innovation and competition....") (citation omitted).

Employing this approach, we conclude that exposing settling parties to antitrust liability for the exclusionary effects of a settlement reasonably within the scope of the patent merely because the patent is subsequently declared invalid would undermine the patent incentives. Patent litigation is too complex and the results too uncertain for parties to accurately forecast whether enforcing the exclusionary right through settlement will expose them to treble damages if the patent immunity were destroyed by the mere invalidity of the patent. This uncertainty, coupled with a treble damages penalty, would tend to discourage settlement of any validity challenges except those that the patentee is certain to win at trial and the infringer is certain to lose. By restricting settlement options, which would effectively increase the cost of patent enforcement, the proposed rule would impair the incentives for disclosure and innovation. *See Ciprofloxacin*, 261 F.Supp.2d at 256 (expressing concern for the effect of settlement-restricting antitrust liability rules on the incentives for research and development); *cf. Walker Process*, 382 U.S. at 180 (Harlan, J., concurring) (permitting antitrust liability based on a showing that mere invalidity "might well chill the disclosure of inventions through the obtaining of a patent because of fear of the vexations or punitive consequences of treble-damages suits").

There may be circumstances under which the unreasonableness of a settlement agreement regarding a subsequently-invalidated or unenforceable patent would be sufficiently apparent that antitrust liability would not undermine the encouragement of genuine invention and disclosure. *Cf. Walker Process*, 382 U.S. at 179-80 (Harlan, J., concurring) (noting that exposing patentees to antitrust liability for the assertion of a patent known to have been procured by fraud "cannot well be thought to impinge upon the policy of the patent laws to encourage inventions and their disclosure"). In this regard we note that some lower courts have extended *Walker Process* to permit antitrust claims against patentees for the anticompetitive effects of infringement lawsuits when the antitrust claimant proves that the patentee knew that the patent was invalid, *Handgards*, 601 F.2d at 994-96, or knew that the patent was not infringed, Loctite Corp. v. Ultraseal Ltd., 781 F.2d 861, 876-77 (Fed.Cir.1985), *overruled on other grounds by* Nobelpharma AB v. Implant Innovations, Inc., 141 F.3d 1059 (Fed. Cir.1998). To the extent that the appellees have demonstrated nothing more than subsequent invalidity, we hold that this alone is insufficient to render the patent's potential exclusionary effects irrelevant to the antitrust analysis.

C.

The class action plaintiffs-appellees argue that the patent right does not include the right to pay infringers, with the implication that any exclusion resulting from payment rather than judicial enforcement is not protected from *per se* antitrust liability by the patent laws. Our discussion with regard to the important role played by settlement in the enforcement of patent rights leads us to reject this argument as well. Appellees have not explained why a monetary payment as part of a patent litigation settlement should be flatly prohibited as a *per se* violation, particularly where the alleged infringer has not yet caused the patentee any harm and the patentee does not have a damages claim to bargain with. *See Ciprofloxacin*, 261 F.Supp.2d at 251-52 (discussing the asymmetries of litigation risk created by Hatch-Waxman and rejecting the argument that payments from the patentee to the infringer are subject to *per se* antitrust analysis); Crane, 54 Fla. L. Rev. at 774 (discussing the dynamics of settlement before the alleged infringer has entered the market).

We cannot conclude that the exclusionary effects of the Agreements not to enter the market were necessarily greater than the exclusionary effects of the '207 patent merely because Abbott paid Geneva and Zenith in return for their respective agreements. If Abbott had a lawful right to exclude competitors, it is not obvious that competition was limited more than that lawful degree by paying potential competitors for their exit. The failure to produce the competing terazosin drug, rather than the payment of money, is the exclusionary effect, and litigation is a much more costly mechanism to achieve exclusion, both to the parties and to the public, than is settlement. *See Aro Corp.*, 531 F.2d at 1372. To hold that an ostensibly reasonable settlement of patent litigation gives rise to *per se* antitrust liability if it involves any payment by the patentee would obviously chill such settlements, thereby increasing the cost of patent enforcement and decreasing the value of patent protection generally. We are not persuaded that such a *per se* rule would be an appropriate accommodation of the competing policies of the patent and antitrust laws.

It may be that the size of the payment to refrain from competing, sometimes called a "reverse payment" or an "exit payment," raises the suspicion that the parties lacked faith in the validity of the patent,[22] particularly when those payments are non-refundable in the event that the patentee prevails on the infringement claim (as a bond posted as part of a preliminary injunction would be). However, in the instant case and given the state of the current record, it is difficult to infer from the size of the payments alone that the infringement suits lacked merit. We do not know, for example, what lost profits Abbott expected from generic competition or what profits Geneva and Zenith expected to gain from entry, the risk of the defendants' inability to satisfy a judgment, or the litigation costs each side expected to save from settlement. We do not know how much of the payment might have been in exchange for provisions of the Agreements other than Zenith's and Geneva's acknowledgment of validity. Without these facts we cannot confidently draw the conclusion, merely from the size of the payments, that there were no genuine disputes over the validity of the patent. Given the asymmetries of risk and large profits at stake, even a patentee confident in the validity of its patent might pay a potential infringer a substantial sum in settlement.

---

22. For example, the size of the payments might be evidence supporting a claim that the patentee knew that the patent was procured by fraud, or knew that the patent was invalid, or that there was no objective basis to believe the patent was valid.

See, e.g., Ciprofloxacin Hydrochloride, 261 F.Supp.2d at 196, 234 (settlement agreements under which a patentee paid an infringer $49.1 million to acknowledge the validity of a patent, and at least $398 million for the infringer to remain off the market, though the patent was subsequently approved by the PTO on reexamination and unsuccessfully challenged in court three times).

Another possibly suspicious characteristic of the payments to Geneva is their structure (*i.e.,* tying their duration to the length of the litigation), which, as the district court noted, may have given Geneva an incentive to delay resolution of the infringement suit. But if the payments were in furtherance of the seemingly reasonable purpose of compensating Geneva for any lost profits during the course of litigation (much like a bond posted as part of a preliminary injunction), it is difficult to imagine how else to structure the payments but by tying them to the length of the litigation.

We recognize that the Sixth Circuit appeared to take the opposite view in *In re Cardizem CD Antitrust Litig.,* 332 F.3d 896, 908 (6th Cir. 2003) ("[I]t is one thing to take advantage of a monopoly that naturally arises from a patent, but another thing altogether to bolster the patent's effectiveness in inhibiting competition by paying the only potential competitor $40 million per year to stay out of the market."). When the exclusionary power of a patent is implicated, however, the antitrust analysis cannot ignore the scope of the patent exclusion. "[T]he protection of the patent laws and the coverage of the antitrust laws are not separate issues." *Studiengesellschaft,* 670 F.2d at 1128. As the above discussion indicates, we do not think that a payment from the patentee to the alleged infringer should be automatically condemned under the antitrust laws, nor do we think that the evidence regarding the exit payments in this case allows a confident conclusion to be drawn at this stage of the litigation that the exclusionary effect of the Agreements were bolstered by the exit payments to a degree that exceeds the potential exclusionary power of the patent.

Because these matters were given scant attention in the district court, considering the court's primary conclusion that the Agreements were market division agreements like those in *Palmer* and *Topco,* these questions have not been explored. We simply note them here to explain our conclusion that the presence of an exit payment as part of the settlement does not alone demonstrate that the Agreements had obvious anticompetitive tendencies above and beyond Abbott's potential exclusionary rights under the '207 patent.

D.

Our discussion so far has been limited to the exclusionary effects of the instant Agreements to delay entrance into the market, the subsequent invalidation of the patent, and the mere fact of exit payments. To the extent that these or other effects of the Agreements are within the scope of the exclusionary potential of the patent, such effects are not subject to *per se* antitrust condemnation.[27] From the preceding discussion, we conclude that deciding the antitrust implications of these exclusionary

---

27. Application of rule of reason analysis is similarly inappropriate, as the anticompetitive effects of exclusion cannot be seriously debated. Rule of reason and *per se* analysis are both aimed at assessing the anticompetitive effects of particular conduct; what is required here is an analysis of the extent to which antitrust liability might undermine the encouragement of innovation and disclosure, or the extent to which the patent laws prevent antitrust liability for such exclusionary effects.

effects requires an analysis of the effects of antitrust liability on the innovation and disclosure incentives created by the patent regime, with the aim of "achieving a suitable accommodation between the differing policies." *Walker Process*, 382 U.S. at 179, 86 S.Ct. at 351 (Harlan, J., concurring).

As alluded to earlier, however, *see supra* at 1306 n. 18, the instant Agreements are not confined to matters involving restrictions on infringing products, exit payments, and a subsequent court decision declaring the patent invalid. Appellees also challenge other provisions of the Agreements, such as those prohibiting the marketing of "any" generic terazosin product; Geneva's agreement not to waive its 180-day exclusivity period; and Geneva's agreement not to come to market until a final, unappealable judgment of invalidity rather than on a district court judgment. These arguments require consideration of the scope of the exclusionary potential of the patent, the extent to which these provisions of the Agreements exceed that scope, and the anticompetitive effects thereof. Because these considerations require a different analytic framework than that applied by the district court and advocated by the parties on appeal, it is appropriate to remand to the district court to apply an appropriate framework derived from this decision, and the arguments of the parties and the facts developed on remand. To aid in the development of an appropriate framework, we offer the following observations.

We recognize the patent exception to antitrust liability, but also recognize that the exception is limited by the terms of the patent and the statutory rights granted the patentee. "[T]he precise terms of the grant define the limits of a patentee's monopoly and the area in which the patentee is freed from competition of price, service, quality or otherwise." *Line Material*, 333 U.S. at 300. *See also New Wrinkle*, 342 U.S. at 378 ("Patents give no protection from the prohibitions of the Sherman Act to [price fixing] when the licenses are used, as here, in the scheme to restrain."); United States v. Masonite Corp., 316 U.S. 265, 277 (1942) ("The owner of a patent cannot extend his statutory grant by contract or agreement. A patent affords no immunity for a monopoly not fairly or plainly within the grant.").

The appropriate analysis on remand will likely require an identification of the protection afforded by the patents and the relevant law and consideration of the extent to which the Agreements reflect a reasonable implementation of these. Appellants, for example, contend that certain provisions of the Geneva Agreement are analogous to a consensual preliminary injunction and stay of judgment pending appeal. To evaluate this claim, the provisions of this Agreement should be compared to the protections afforded by the preliminary injunction and stay mechanisms and considered in light of the likelihood of Abbott's obtaining such protections. *Cf.* Hovenkamp at ¶2046 ("some care must be taken to ensure that . . . the settlement . . . is not more anticompetitive than a likely outcome of the litigation").

Any provisions of the Agreements found to have effects beyond the exclusionary effects of Abbott's patent may then be subject to traditional antitrust analysis to assess their probable anticompetitive effects in order to determine whether those provisions violate §1 of the Sherman Act. Standard Oil Co., Ind., v. United States, 283 U.S. 163, 175 (1931) (recognizing that cross-licensing agreements and a division of royalties could be used to monopolize or fix prices and examining the evidence to assess the potential for anticompetitive effects as a result of the agreements at issue). The appropriate analysis of the market effects of such provisions may depend on the nature of the provision challenged. It may be that some challenged provisions are so obviously anticompetitive that they can be condemned as illegal on the evidence so far adduced, or it may be that the tendencies of some challenged provisions cannot be confidently predicted without further inquiry. . . .

IV. CONCLUSION

The district court's order granting partial summary judgment is REVERSED and this case is REMANDED for further proceedings consistent with this opinion.

COMMENTS AND QUESTIONS

1. Recent years have seen a flurry of cases involving allegedly anticompetitive agreements between pharmaceutical patent owners and generics. The FTC has settled several high-profile cases with consent decrees precluding agreements to keep generic competitors out of the market. See, e.g., In re Geneva Pharm., Dkt. No. C-3946 (May 22, 2000); In re Hoechst Marion Roussel Inc., Dkt. No. 9293 (May 8, 2001). And several private suits are in litigation. In addition to the cases cited above, see Andrx Pharmaceuticals v. Biovail, 256 F.3d 799 (D.C. Cir. 2001); In re Ciprofloxacin Hydrochloride, 363 F. Supp. 2d 514 (E.D.N.Y. 2005). For a full discussion, see Herbert Hovenkamp et al., IP and Antitrust ch. 7e; David A. Balto, Pharmaceutical Patent Settlements: The Antitrust Risks, 55 Food & Drug L.J. 321 (2000); Thomas F. Cotter, Antitrust Implications of Patent Settlements Involving Reverse Payments: Defending a Rebuttable Presumption of Illegality in Light of Some Recent Scholarship, 71 Antitrust L.J. 1069 (2004).

2. Settlement of ongoing patent cases has obvious social benefits, both because it saves the costs of litigation and because it may eliminate uncertainty that could prevent both the patentee and the accused infringer from fully investing in innovation. Further, it is possible that the outcome of the patent suit would be an injunction preventing the defendant from entering the market for the life of the patent. Given this, what is wrong with settling a case by having the defendant stipulate that it will not enter the market? Should such settlements be permissible, at least in cases in which the patentee has a reasonable argument that its patent is valid and infringed?

Hovenkamp et al. argue otherwise. They observe that

> the parties have every incentive to enter a collusive settlement whether or not the underlying patents are enforceable. In some cases the payments made by the branded manufacturer to the generic infringement defendant exceeded the market value of the latter's anticipated entry into the market; so the generic producer would be better off accepting the payment than it would be in entering. Further, ordinarily no one reviews patent settlements. The suspicious transaction in such cases is not the litigation over the enforceability of the branded drug's patents, but the payment of money from the branded manufacturer to the incipient generic manufacturer in exchange for the latter's promise to stay out of the market. Given the considerable danger of competitive harm, that payment can rightfully be regarded as patent misuse, or perhaps even as an unlawful restraint of trade.

Herbert Hovenkamp et al., IP and Antitrust § 33.9. See also Carl Shapiro, Antitrust Limits to Patent Settlements, *http://www.law.berkeley.edu/institutes/bclt/pubs/wp/*

501.pdf. Shapiro argues that we need not inquire into the substantive merits of the underlying patent case, because the fact that the patentee is willing to pay money to keep a competitor out of the market tells us that the patentee's case is not strong. And Lemley and Shapiro point out that all patents are probabilistic rights, since so many are found invalid or not infringed if they are litigated. Mark A. Lemley & Carl Shapiro, Probabilistic Patents, 19 J. Econ. Persp. 75 (2005). They point out that the PTO engages in only a cursory look at the hundreds of thousands of applications it receives every year, and grants patents to the overwhelming majority of such applications. When patents are enforced, their validity is almost always at issue, and 46 percent of the patents litigated to judgment are determined to be invalid.

Do Lemley and Shapiro go too far? Wouldn't the logic of their argument suggest that a patentee could not pay an infringer to stay out of the market even if the parties were 90 percent certain that the patent was valid? Note that Hovenkamp et al suggest that a payment sufficient to compensate for the attorneys fees that the patent owner would have paid to go to trial shouldn't be illegal. Further, they approve the practice of "delayed entry" settlements, in which the generic delays entry into the market for less than the full patent term, as long as such a settlement is not accompanied by a payment. Herbert Hovenkamp et al., Anticompetitive Settlement of Intellectual Property Disputes, 87 Minn. L. Rev. 1719 (2003).

3. The Eleventh Circuit went even further than its *Valley Labs* opinion in reversing the FTC's decision in Schering Plough Corp. v. FTC, 402 F.3d 1056 (11th Cir. 2005). There, the court concluded that "neither the rule of reason nor the per se analysis is appropriate in this context." Id. at 1065. This impliedly suggests that exclusion payments are per se legal in the Eleventh Circuit, though the court does not expressly hold that. Instead, the court constructs a three-part test in which it looks at "(1) the scope of the exclusionary potential of the patent; (2) the extent to which the agreements exceed that scope; and (3) the resulting anticompetitive effects." Id. at 1066. Stated that way, the Eleventh Circuit's legal test does not seem dissimilar to the FTC's rule of reason approach. But the court diverged sharply from the FTC in engaging in an all-but-conclusive presumption that an issued patent was valid. Because it presumed that a patent that had not yet been invalidated was necessarily valid, at least unless the patentee's infringement lawsuits were shams, the court found no expansion beyond the proper legal scope of the patent.

Does the *Schering* approach make sense? While patents enjoy a strong legal presumption of validity, it is not an irrebuttable one. Shouldn't antitrust plaintiffs be entitled to adduce evidence tending to suggest that the patent would likely have been held invalid? And wouldn't a large payment from the patentee to the accused infringer be strong evidence of probable invalidity, and therefore that consumers are, on expectation, losing competition they might otherwise have enjoyed.

The *Schering* case is complicated by the fact that the patentee's payment not only settled a validity dispute but also purchased another drug from the payee. To be evidence of invalidity in this context, a payment must clearly exceed the fair market value of the drug being purchased. The evidence before the ALJ on this point was disputed.

While *Schering* seems to suggest that exclusion payments will always be legal, *Valley Labs* does not hold that. Rather, it seems to open the door to a fact-based analysis of the underlying merits of the patent suit. Indeed, that is exactly what happened on remand. The district court conducted a detailed analysis of the likely outcome of the patent litigation and found that the patent owner's arguments were "weak and unlikely to result in a District Court finding that the '207 patent was valid." In re Terazosin Hydrochloride Antitrust Litig., 352 F. Supp. 2d 1279, 1306 (S.D. Fla. 2005). As a result, the district court concluded that the patent could not shield the patentee's exclusion payment from antitrust liability and went on to find that the settlement agreement including the payment was per se unlawful.

Does it make sense to put district courts to the detailed reconstruction of hypothetical litigation outcomes? Does the fact that the former opponents are now on the same side create any problems of proof in such a case?

4. Paying a defendant to stay out of the market is not the only way a pharmaceutical company might seek to discourage generic competition. A related set of cases involves acquisitions by the owners of expiring pharmaceutical patents of generic companies that intend to compete to produce the drug. In Eon Labs Mfg. v. Watson Pharmaceuticals, 164 F. Supp. 2d 350 (S.D.N.Y. 2001), the court concluded that while pharmaceutical settlements of the type at issue in *Cardizem* were illegal per se, an antitrust plaintiff could not maintain a section 1 claim in a case in which the patentee purchased the original ANDA applicant's right to enter the generic market, and later did itself enter the generic market. Such conduct should be evaluated under section 7 of the Clayton Act, governing mergers and acquisitions. By contrast, in Eli Lilly & Co. v. American Cyanamid, 2001 WL 30191 (S.D. Ind. Jan. 8, 2001), the court held that a market division claim alleging that the holder of an expiring pharmaceutical patent conspired with likely generics to outsource production to them, thus co-opting potential competition, survived a motion to dismiss.

A second disputed practice is known as "evergreening." In this case, patentees obtain multiple patents and assert that they cover a drug of interest. Under Hatch-Waxman, each patent they listed on the Orange Book entitled them to a separate 30-month stay, even if the patent claim was weak or frivolous. As a result, patentees were able to obtain successive injunctions against generic entry even after their core patent expired or was invalidated. While generics challenged the practice, the filing of lawsuits is generally protected under the *Noerr-Pennington* doctrine, and so those challenges have failed for the most part. In re Ciprofloxacin Hydrochloride, 363 F. Supp. 2d 514 (E.D.N.Y. 2005).

Regulatory and legislative changes effective in 2004 deal effectively with the problem of multiple 30-month stays, both by giving a generic ANDA applicant sued for patent infringement the right to assert a counterclaim challenging the listing of information in the Orange Book, 21 U.S.C. § 355(j)(5)(C)(2), and by limiting patentees to a single 30-month stay for any given drug, regardless of the number of patents listed as covering that drug. Applications for FDA Approval to Market a New Drug, 68 Fed. Reg. 36,676-77, 21 C.F.R. pt. 314 (June 18, 2003).

5. Pharmaceutical patent owners have engaged in a second form of evergreening, one that might be described as "product-hopping." Product-hopping pharmaceutical companies faced with the possibility of generic competition once a patent expires or is held invalid sometimes make trivial alterations to their approved drugs, get FDA approval for those trivial alterations, and then replace the old product with the new product. For example, a patentee might switch from selling a drug in capsule form to selling the same formulation of the same drug in tablet form. While the change won't matter much to consumers, it can be sufficient to require a generic company to start the ANDA filing process over from scratch, delaying the date of generic entry and triggering an entirely new round of patent litigation. Because the

patented pharmaceutical is now being sold only in the new tablet formulation, the generic company will be unable to rely on generic substitution to sell its ANDA-approved capsules. A number of antitrust challenges to product-hopping are pending as of this writing.

How should the courts evaluate product-hopping? A pharmaceutical patent owner has no legal duty either to help its generic competitors or to continue selling a particular product. Patent owners may argue with some justification, therefore, that they cannot be held liable for stopping the sale of a product. At the same time, product-hopping seems clearly to be an effort to game the rather intricate FDA rules to anticompetitive effect. The patentee is making a product change with no technological benefit solely in order to delay competition. See Herbert Hovenkamp et al., IP and Antitrust § 12.5 (suggesting that such a change could qualify as a predatory product change).

The product-hopping problem could be solved if the FDA Act permitted generic substitution across formulations. In the absence of such a statutory change, antitrust cases will continue to arise.

6. In Asahi Glass v. Pentech Pharmaceuticals, 289 F. Supp. 2d 986 (N.D. Ill. 2003), Judge Posner (sitting by designation) approved a settlement between a pharmaceutical company and a generic under unusual circumstances. The patentee Glaxo had sued Apotex, and Judge Posner had held the patent valid but not infringed, meaning that Apotex could enter the market. Glaxo had also sued Pentech, which made the same product as Apotex and therefore presumably also would not infringe. The parties to the second suit settled with an agreement that Pentech would enter the market as soon as Apotex did, but would exit the market if and when Apotex exited. Glaxo did not pay Pentech to stay out of the market unless and until Apotex entered but did supply it with the product it would sell when it entered.

This agreement seems designed to discourage entry by Apotex by facing it immediately with generic competition, denying it the benefit of 180-day generic exclusivity. Further, the fact that Glaxo was supplying Pentech suggests that its planned entry may be below cost. Nonetheless, Judge Posner approved the agreement. The court held that "the general policy of the law is to favor . . . settlement of patent infringement suits." It found nothing suspicious about this settlement and worried about the risk of antitrust liability chilling settlements. In so doing, the court set the antitrust bar for settlements quite high, concluding that they were "legal unless a neutral observer would reasonably think that the patent was almost certain" to be invalid or not infringed. Id. at 993. Judge Posner concluded that that standard was not met in the *Pentech* suit, even though he had previously held that the patent did not cover the drug in question.

The situation presented in *Asahi* is different and less obviously anticompetitive than in the exclusion payment cases. As Judge Posner noted,

> There is a difference between the reverse-payment case and other forms of settlement. In a reverse-payment case, the settlement leaves the competitive situation unchanged from before the defendant tried to enter the market. In this case, in contrast, the settlement led to increased competition, first in Puerto Rico and now throughout the United States.

Id. at 994. Is Judge Posner right? Or does a rule of virtually per se legality in this situation go too far?

c. *Industry Standardization*

Broadcast Music, Inc., et al. v. Columbia Broadcasting System, Inc.
Supreme Court of the United States
441 U.S. 1 (1979)

Mr. Justice WHITE delivered the opinion of the Court.

This case involves an action under the antitrust and copyright laws brought by respondent Columbia Broadcasting System, Inc. (CBS), against petitioners, American Society of Composers, Authors and Publishers (ASCAP) and Broadcast Music, Inc. (BMI), and their members and affiliates. The basic question presented is whether the issuance by ASCAP and BMI to CBS of blanket licenses to copyrighted musical compositions at fees negotiated by them is price fixing per se unlawful under the antitrust laws....

BMI, a nonprofit corporation owned by members of the broadcasting industry, was organized in 1939, is affiliated with or represents some 10,000 publishing companies and 20,000 authors and composers, and operates in much the same manner as ASCAP. Almost every domestic copyrighted composition is in the repertory either of ASCAP, with a total of three million compositions, or of BMI, with one million.

Both organizations operate primarily through blanket licenses, which give the licensees the right to perform any and all of the compositions owned by the members or affiliates as often as the licensees desire for a stated term. Fees for blanket licenses are ordinarily a percentage of total revenues or a flat dollar amount, and do not directly depend on the amount or type of music used. Radio and television broadcasters are the largest users of music, and almost all of them hold blanket licenses from both ASCAP and BMI. Until this litigation, CBS held blanket licenses from both organizations for its television network on a continuous basis since the late 1940's and had never attempted to secure any other form of license from either ASCAP or any of its members. Id., at 752-754.

The complaint filed by CBS charged various violations of the Sherman Act and the copyright laws. CBS argued that ASCAP and BMI are unlawful monopolies and that the blanket license is illegal price fixing, an unlawful tying arrangement, a concerted refusal to deal, and a misuse of copyrights. The District Court, though denying summary judgment to certain defendants, ruled that the practice did not fall within the *per se* rule. 337 F. Supp. 394, 398 (SDNY 1972). After an 8-week trial, limited to the issue of liability, the court dismissed the complaint, rejecting again the claim that the blanket license was price fixing and a per se violation of § 1 of the Sherman Act, and holding that since direct negotiation with individual copyright owners is available and feasible there is no undue restraint of trade, illegal tying, misuse of copyrights, or monopolization. 400 F. Supp., at 781-783....

[The Court of Appeals reversed.]

A

As a preliminary matter, we are mindful that the Court of Appeals' holding would appear to be quite difficult to contain. If, as the court held, there is a *per se* antitrust violation whenever ASCAP issues a blanket license to a television network for a single fee, why would it not also be automatically illegal for ASCAP to negotiate and issue blanket licenses to individual radio or television stations or to other users who perform copyrighted music for profit? Likewise, if the present network licenses issued through ASCAP on behalf of its members are *per se* violations, why would it not be equally illegal for the members to authorize ASCAP to issue licenses establishing various categories of uses that a network might have for copyrighted music and setting a standard fee for each described use?

Although the Court of Appeals apparently thought the blanket license could be saved in some or even many applications, it seems to us that the *per se* rule does not accommodate itself to such flexibility and that the observations of the Court of Appeals with respect to remedy tend to impeach the *per se* basis for the holding of liability.

CBS would prefer that ASCAP be authorized, indeed directed, to make all its compositions available at standard per-use rates within negotiated categories of use. 400 F. Supp., at 747 n. 7.[28] But if this in itself or in conjunction with blanket licensing constitutes illegal price fixing by copyright owners, CBS urges that an injunction issue forbidding ASCAP to issue any blanket license or to negotiate any fee except on behalf of an individual member for the use of his own copyrighted work or works.[29] Thus, we are called upon to determine that blanket licensing is unlawful across the board. We are quite sure, however, that the *per se* rule does not require any such holding.

B

In the first place, the line of commerce allegedly being restrained, the performing rights to copyrighted music, exists at all only because of the copyright laws. Those who would use copyrighted music in public performances must secure consent from the copyright owner or be liable at least for the statutory damages for each infringement and, if the conduct is willful and for the purpose of financial gain, to criminal penalties. Furthermore, nothing in the Copyright Act of 1976 indicates in the slightest that Congress intended to weaken the rights of copyright owners to control the public performance of musical compositions. Quite the contrary is true. Although the copyright laws confer no rights on copyright owners to fix prices among themselves or otherwise to violate the antitrust laws, we would not expect that any market arrangements reasonably necessary to effectuate the rights that are granted would be deemed a *per se* violation of the Sherman Act. Otherwise, the commerce anticipated by the Copyright Act and protected against restraint by the Sherman Act would not exist at all or would exist only as a pale reminder of what Congress envisioned.

C

More generally, in characterizing this conduct under the *per se* rule, our inquiry must focus on whether the effect and, here because it tends to show effect, see United

28. Surely, if ASCAP abandoned the issuance of all licenses and confined its activities to policing the market and suing infringers, it could hardly be said that member copyright owners would be in violation of the antitrust laws by not having a common agent issue per-use licenses. Under the copyright laws, those who publicly perform copyrighted music have the burden of obtaining prior consent. Cf. Zenith Radio Corp. v. Hazeltine Research, Inc., 395 U.S., at 139-140.

29. In its complaint, CBS alleged that it would be "wholly impracticable" for it to obtain individual licenses directly from the composers and publishing houses, but it now says that it would be willing to do exactly that if ASCAP were enjoined from granting blanket licenses to CBS or its competitors in the network television business.

States v. United States Gypsum Co., 438 U.S. 422, 436 n. 13 (1978), the purpose of the practice are to threaten the proper operation of our predominantly free-market economy — that is, whether the practice facially appears to be one that would always or almost always tend to restrict competition and decrease output, and in what portion of the market, or instead one designed to "increase economic efficiency and render markets more, rather than less, competitive." Id., at 441 n.16; see National Society of Professional Engineers v. United States, 435 U.S., at 688; Continental T.V., Inc. v. GTE Sylvania Inc., 433 U.S., at 50 n.16; Northern Pac. R. Co. v. United States, 356 U.S., at 4.

The blanket license, as we see it, is not a "naked restrain[t] of trade with no purpose except stifling of competition," White Motor Co. v. United States, 372 U.S. 253, 263 (1963), but rather accompanies the integration of sales, monitoring, and enforcement against unauthorized copyright use. See L. Sullivan, Handbook of the Law of Antitrust § 59, p. 154 (1977). As we have already indicated, ASCAP and the blanket license developed together out of the practical situation in the marketplace: thousands of users, thousands of copyright owners, and millions of compositions. Most users want unplanned, rapid, and indemnified access to any and all of the repertory of compositions, and the owners want a reliable method of collecting for the use of their copyrights. Individual sales transactions in this industry are quite expensive, as would be individual monitoring and enforcement, especially in light of the resources of single composers. Indeed, as both the Court of Appeals and CBS recognize, the costs are prohibitive for licenses with individual radio stations, nightclubs, and restaurants, 562 F.2d, at 140 n.26, and it was in that milieu that the blanket license arose.

A middleman with a blanket license was an obvious necessity if the thousands of individual negotiations, a virtual impossibility, were to be avoided. Also, individual fees for the use of individual compositions would presuppose an intricate schedule of fees and uses, as well as a difficult and expensive reporting problem for the user and policing task for the copyright owner. Historically, the market for public-performance rights organized itself largely around the single-fee blanket license, which gave unlimited access to the repertory and reliable protection against infringement. When ASCAP's major and user-created competitor, BMI, came on the scene, it also turned to the blanket license.

With the advent of radio and television networks, market conditions changed, and the necessity for and advantages of a blanket license for those users may be far less obvious than is the case when the potential users are individual television or radio stations, or the thousands of other individuals and organizations performing copyrighted compositions in public. But even for television network licenses, ASCAP reduces costs absolutely by creating a blanket license that is sold only a few, instead of thousands, of times, and that obviates the need for closely monitoring the networks to see that they do not use more than they pay for. ASCAP also provides the necessary resources for blanket sales and enforcement, resources unavailable to the vast majority of composers and publishing houses. Moreover, a bulk license of some type is a necessary consequence of the integration necessary to achieve these efficiencies, and a necessary consequence of an aggregate license is that its price must be established.

Mr. Justice STEVENS, dissenting.

The Court holds that ASCAP's blanket license is not a species of price fixing categorically forbidden by the Sherman Act. I agree with that holding. The Court remands the cases to the Court of Appeals, leaving open the question whether the blanket license as employed by ASCAP and BMI is unlawful under a rule-of-reason inquiry. I think that question is properly before us now and should be answered affirmatively . . . .

COMMENTS AND QUESTIONS

1. There is no "technical necessity" for the copyright collective in *Broadcast Music* in the same sense that there is for a cross-license of blocking patents. It is certainly possible in theory to require each copyright holder to sell licenses independently, and to punish cooperation in the market for copyright licenses in the same way that section 1 punishes other cooperation among competitors. But the result would be that most licenses currently sold would not be sold, because of prohibitive transaction costs. "Economic necessity" therefore requires that we exempt such sales from the per se rule. The Court concludes that the rule of reason applies, and remands the case.

The transactions cost rationale seems to require that copyright collectives have a certain degree of market power. After all, if the industry were composed of thousands of small "collectives," transactions costs would still be a significant barrier to copyright licenses. In fact, the "efficient scale" of a copyright collective would seem to be the entire industry. Given this fact, should antitrust courts be less concerned about the market power of copyright collectives? Or, on the contrary, should it subject the industry to pervasive regulation of the sort found in other "natural monopoly" contexts?

2. The Court reasons that since the "market" in question was created by the copyright collectives in the first place, they ought to be protected from per se antitrust scrutiny. But is it fair to say that ASCAP and BMI "created" the market? Certainly, they facilitate the granting of copyright permissions, although they do so at the expense of competition among copyright owners in the sale of permissions. It may be that, absent large organizations such as BMI with significant bargaining power, most musicians would trade away any performance right royalty in exchange for the public exposure that comes from having their songs played on the radio. If this is true, BMI skews the ordinary market distribution in favor of copyright owners, with the result that radio stations pay a higher price.

There are two possible responses to this argument. First, the transactions cost problem discussed above may be so great that absent a copyright collective, artists and broadcasters might never connect at all. As a result, broadcasters would be unable to play songs by artists who would like to have their songs played. Second, BMI may actually restore equity to the bargaining situation by providing countervailing power to match the power of the broadcasters. Individual copyright owners might feel they have no choice but to give up their royalties; they are likely to be in a much better bargaining position as part of a large collective organization.

3. The music industry is not the only one to have developed collectives of copyright owners. In the publishing industry, an organization called the Copyright Clearance Center (CCC) has taken a leading role in granting permissions and collecting fees from people who photocopy pages out of books or periodicals. The CCC's standard fee for photocopying permissions is fairly high — 25 cents per page. (Thus the royalty for copying a 200-page book is $50. This price does not include any reproduction cost.) While the CCC is in much the same situation as BMI or ASCAP, the people with whom it is bargaining (individuals, corporations, and universities) are generally smaller than broadcasters and have less bargaining power. Does this fact make CCC's existence as the dominant publishing collective troublesome from an antitrust perspective?

Note on Intellectual Property and Standard-setting Organizations

In industries characterized by large standardization externalities — that is, where the market naturally tends towards a single product or type of product — group standard-setting organizations can serve valuable procompetitive purposes. They may allow a number of companies to make and sell competing products that are compatible with each other, thus encouraging more flexible use of the products by consumers and preventing one firm from dominating the market with a de facto standard. See Mark A. Lemley, Antitrust and the Internet Standardization Problem, 28 Conn. L. Rev. 1041 (1996). Unfortunately, standard-setting organizations are sometimes subject to "capture" by a particular participant. In the context of high-technology markets, the most likely means of capturing a standard is by the strategic use of intellectual property rights.[16]

For example, in 1992, the Video Electronics Standards Association (VESA) adopted a computer hardware standard called the VL-Bus standard, which governs the transmission of information between a computer's CPU and its peripheral devices. Each of the members voting to adopt the standard, including Dell Computer Corporation, was required by VESA rules to affirm that it did not own any patent rights that covered the VL-Bus standard.[17] Dell's representative did in fact make such a statement. Nonetheless, Dell asserted a patent against other VESA members for using the VL-Bus standard eight months later, after the VL-Bus standard had been widely adopted. By working to adopt as a group standard a technology Dell allegedly knew was proprietary,[18] Dell could obtain the help of its competitors in establishing a standard that it would ultimately be able to control.

The competitive harms of this form of capture are relatively clear. Not only does the capturing party end up with exclusive control over the market standard, converting a group standard-setting process into a de facto one, but the capturing party can use the group standard to achieve a dominant position it could not have attained in an open standards competition. Had Dell announced up front that the standards it was backing were proprietary, it is unlikely that the affected industry would have chosen those standards. At the very least, those standards would have faced stiffer competition than they did.

The most likely avenue of antitrust attack against such capture does not involve section 1 at all but rather is an attempted monopolization claim under section 2. Attempted monopolization has three elements — intent to monopolize, anticompetitive conduct in furtherance of that intent, and a dangerous probability of success. Spectrum Sports v. McQuillen, 506 U.S. 970 (1993). Assuming the failure to disclose relevant intellectual property rights was intentional and not an oversight, the first

---

16. In more traditional products markets, capture of standard-setting organizations sometimes takes the form of controlling the organization and using it to vote down a competitor's standard. This is what apparently happened in Allied Tube & Conduit Corp. v. Indian Head, Inc., 486 U.S. 492 (1988). That approach will not have the same effect in industries such as the software market unless the capturing party has intellectual property rights in the dominant standard. Absent such rights, competitors will alter their products to comply with the winning standard. This may require some time and expense, but the expense will be well worth it to competitors if the lock-in effects of the standard are significant.

17. Many standard-setting organizations, including the American National Standards Institute (ANSI) and Semiconductor Equipment and Materials International (SEMI) have similar rules.

18. Whether Dell in fact knew this is a matter of some dispute. In her dissent to the Federal Trade Commission's proposed consent decree involving Dell, Commissioner Mary Azcuenaga claimed that there was "no evidence to support such a finding of intentional conduct."

element should be easy to satisfy. Efforts to capture an industry standard in any given case would constitute anticompetitive conduct precisely in the situation where those efforts are likely to lead to monopolization — that is, where the standard being set is one that will likely dominate the industry. Market power may be the necessary result of patent enforcement in some cases, while in others the patent owner's control over the market stems from a failure of information in the market, a failure that the patent owner herself has induced. While the fact that the antitrust defendant does own intellectual property rights governing the technology suggests some caution in applying the antitrust laws, the mere possession of an intellectual property right will not protect its owner from a charge of dominating a market by extending the scope of that right. In the *Dell* case, the FTC entered into a consent decree in which Dell agreed not to assert its intellectual property rights in the VL-Bus. See In re Dell Computer Corp., 121 FTC 616 (FTC 1995).

But proof of abuse of the standard-setting process may not always be sufficient. In Rambus v. Infineon Techs. AG, 318 F.3d 1081 (Fed. Cir. 2003), the plaintiff had attended JEDEC standard-setting meetings and then redrafted its patent applications to cover what JEDEC was doing, without disclosing the existence of those applications to the standard-setting group. The district court found that Rambus had engaged in fraud, but the Federal Circuit reversed. The court found that the JEDEC policy "did not address a member's intentions to file future patent applications. . . . JEDEC could have drafted a patent policy broad enough to capture a member's failed attempts to mine a disclosed specification for broader undisclosed claims. It could have. It simply did not." Id. at 1102. As a result, the court found, Rambus had not breached any duty of disclosure to JEDEC. Judge Prost dissented, citing abundant evidence that Rambus knew of the disclosure obligation and skirted it in order to catch the industry unawares.

The *Rambus* and *Dell* situations involve efforts to abuse the standard-setting process by capturing a return greater than a patentee's invention would justify. Of course, not all patents covering standards will necessarily be anticompetitive. While one approach to standards is to require them to be intellectual property-free (the approach of the Internet Engineering Task Force (IETF), at least until recently), intellectual property can coexist with procompetitive standard-setting. For example, the American National Standards Institute and other groups do not require that an intellectual property owner give up any claim to a standard, but merely that they license their intellectual property rights on a reasonable basis. Other examples of reasonable and even procompetitive uses of intellectual property in the standard-setting context are possible. It is only in that subset of cases where the patent is used as a competitive weapon that concerns about market control are implicated.[19] For a detailed discussion of the antitrust issues that arise in standard-setting, see Mark A. Lemley, Intellectual Property Rights and Standard-Setting Organizations, 90 Calif. L. Rev. 1889 (2002).

---

19. In rare cases, a rule *precluding* patents on standards might be found to be anticompetitive. In In re Am. Society of Sanitary Eng., 106 FTC 324, 328-29 (1985), the FTC alleged that a standard-setting organization could not refuse to consider revising its standards to include a new product solely on the grounds that that product was patented. The case was settled by consent decree. It is significant that the standard in question was inclusive rather than exclusive, and so allowing the complainant's product to be included would not have restricted the rights of other members to make use of other technology covered by the standard. Nonetheless, the case should serve as a caution for rules such as IETF's requiring that participants relinquish their intellectual property rights.

D. MERGERS AND ACQUISITIONS

≡ *SCM Corporation v. Xerox Corporation*
United States Court of Appeals for the Second Circuit
645 F.2d 1195 (2d Cir. 1981)

MESKILL, Circuit Judge:

The plaintiff, SCM Corporation (SCM), appeals from an order entered in the United States District Court for the District of Connecticut, Jon O. Newman, Judge, dismissing its claim for monetary damages asserted in this private antitrust action for injuries sustained as a result of alleged exclusionary acts committed by the defendant, Xerox Corporation (Xerox), in violation of §§ 1 and 2 of the Sherman Act, 15 U.S.C. §§ 1, 2 (1976), and § 7 of the Clayton Act, 15 U.S.C. § 18 (1976). The trebled amount of damages calculated by the jury on this claim totalled $111.3 million. The principal anticompetitive acts alleged by SCM concerned patent acquisitions made by Xerox. SCM averred that Xerox's acquisition of certain patents and subsequent refusal to license those patents excluded SCM from competing effectively in a relevant product market and submarket dominated by Xerox products that embraced the patented art. Judge Newman ruled below that a need to accommodate the antitrust and patent laws precluded damage liability predicated upon Xerox's refusal to license its patents; however, he left open the possibility of granting the plaintiff equitable relief. Judge Newman certified his order for interlocutory review pursuant to 28 U.S.C. § 1292(b) (1976) and, as developed below, we exercised our discretion under that section to accept this appeal. Without commenting upon Judge Newman's remedial theory, we affirm the denial of monetary damages in connection with SCM's exclusionary claim based upon our determination that none of Xerox's patent-related conduct, the only conduct alleged by SCM to have caused it any harm, contributed to any antitrust violation . . . .

SCM has argued that Xerox's acquisition of its patents and subsequent exercise of the exclusionary power in them violated the antitrust laws and injured SCM. Xerox contends that its acquisition of the patents was lawful and its decision not to license its patents for plain-paper copying constituted a lawful exercise of patent power. Xerox does not dispute that it achieved monopoly power in the relevant market and submarket by 1969, but contends that an examination of the circumstances under which this feat was accomplished reveals the lawfulness of the monopoly it attained. Our analysis commences with a review of the relationship between the patent and antitrust laws . . . .

II

The law is unsettled concerning the effect under the antitrust laws, if any, that the evolution of a patent monopoly into an economic monopoly might have upon a patent holder's right to exercise the exclusionary power ordinarily inherent in a patent. Indeed, implicit in Judge Newman's decision below is a deep concern over the uncertain antitrust law implications just such an event might have had in this case. His thoughtful analysis of the relationship between the patent and antitrust laws led him to conclude that "the need to accommodate the patent laws with the antitrust laws precludes the imposition of damage liability . . . for a unilateral refusal to license valid patents." 463 F. Supp. at 1012-13. Judge Newman opined that whether or not Xerox's refusal to license the patents it acquired under the 1956 agreement transgressed any provisions of the antitrust laws, monetary damage liability could not be imposed upon Xerox without seriously undermining the patent system. The district court's thesis rests on the assumption that despite the lawfulness of a patent's acquisition, "[i]n some circumstances, [a] refusal to license may be considered a § 2 violation." 463 F. Supp. at 1012.

SCM has contended that a unilateral refusal to license a patent should be treated like any other refusal to deal by a monopolist, see generally Otter Tail Power Co. v. United States, 410 U.S. 366 (1973); Lorain Journal Co. v. United States, 342 U.S. 143 (1951); Eastman Kodak Co. v. Southern Photo Materials Co., 273 U.S. 359 (1927), where the patent has afforded its holder monopoly power over an economic market. While, as SCM suggests, a concerted refusal to license patents is no less unlawful than other concerted refusals to deal, in such cases the patent holder abuses his patent by attempting to enlarge his monopoly beyond the scope of the patent granted him. See, e.g., Zenith Radio Corp. v. Hazeltine Research, Inc., supra, 305 U.S. at 118-19; United States v. Singer Manufacturing Co., 374 U.S. 174, 192-97 (1963); United States v. Line Material Co., 333 U.S. 287, 314-15 (1948); Hartford-Empire Co. v. United States, 323 U.S. 386, 406-07 (1945); United States v. Masonite Corp., 316 U.S. 265, 277 (1942). Where a patent holder, however, merely exercises his "right to exclude others from making, using, or selling the invention," 35 U.S.C. § 154 (1976), by refusing unilaterally to license his patent for its seventeen-year term, see, e.g., Bement v. National Harrow Co., 186 U.S. 70, 88-90 (1902), such conduct is expressly permitted by the patent laws. "The heart of [the patentee's] legal monopoly is the right to invoke the State's power to prevent others from utilizing his discovery without his consent." Zenith Radio Corp. v. Hazeltine Research, Inc., supra, 395 U.S. at 135 (citing Crown Die & Tool Co. v. Nye Tool & Machine Works, 261 U.S. 24 (1923); Continental Paper Bag Co. v. Eastern Paper Bag Co., 210 U.S. 405 (1908)). Simply stated, a patent holder is permitted to maintain his patent monopoly through conduct permissible under the patent laws.

No court has ever held that the antitrust laws require a patent holder to forfeit the exclusionary power inherent in his patent the instant his patent monopoly affords him monopoly power over a relevant product market. In *Alcoa* this Court never questioned the legality of the economic monopoly Alcoa maintained by virtue of the two successive patents it had acquired. United States v. Aluminum Co. of America, supra, 148 F.2d at 422, 430. Indeed, Judge Learned Hand termed Alcoa's economic monopoly during the terms of those patents "lawful." 148 F.2d at 430. We do not interpret Judge Wyzanski's decision in United States v. United Shoe Machinery Corp., 110 F. Supp. 295 (D. Mass. 1953), *aff'd per curiam*, 347 U.S. 521 (1954), as supporting SCM's argument to the contrary. In *United Shoe*, the primary vehicle found to have been employed by United Shoe in achieving and maintaining its monopoly was its lease-only system of distributing its machines. 110 F. Supp. at 344. The patent acquisitions scrutinized by Judge Wyzanski occurred after United Shoe possessed substantial market power and were not "one of the principal factors... enabling [United Shoe] to achieve and hold its share of the market." 110 F. Supp. at 312. Thus, contrary to appellant's contention, the *United Shoe* case stands in stark contrast to the one at bar where the patents were acquired prior to the appearance of the relevant product market and where the patents themselves afforded Xerox the power to achieve eventual market dominance.

In *Alcoa* Judge Learned Hand stated that the "successful competitor, having been urged to compete, must not be turned upon when he wins." 148 F.2d at 430. And while that statement was made in regard to a hypothetical situation where only one of a group of competitors ultimately survives, it at least indicates a concern Judge Hand had for preserving those economic incentives that provide the primary impetus for competition. Subsequently, the Supreme Court in United States v. Grinnell Corp., 384 U.S. 563 (1966), amplified this consideration when it set forth the elements of a § 2 violation as follows:

> The offense of monopoly under § 2 of the Sherman Act has two elements: (1) the possession of monopoly power in the relevant market and (2) the willful acquisition or maintenance of that power *as distinguished from growth or development as a consequence of a superior product, business acumen, or historic accident.*

Id. at 570-71 (emphasis added).

Thus, in Berkey Photo, Inc. v. Eastman Kodak Co., 603 F.2d 263, 275 (2d Cir. 1979), *cert. denied*, 444 U.S. 1093 (1980), this Court stated that "[w]e tolerate the existence of monopoly power . . . only insofar as necessary to preserve competitive incentives and to be fair to the firm that has attained its position innocently." In United States v. Griffith, 334 U.S. 100 (1948), the Supreme Court declared, however, that "the use of monopoly power, however lawfully acquired, to foreclose competition, to gain a competitive advantage, or to destroy a competitor, is unlawful." Id. at 107. Echoing the same consideration in *Berkey*, Judge Kaufman stated that while "[t]he mere possession of monopoly power does not *ipso facto* condemn a market participant . . . , the firm must refrain at all times from conduct directed at smothering competition." Berkey Photo, Inc. v. Eastman Kodak Co., supra, 603 F.2d at 275.

The tension between the objectives of preserving economic incentives to enhance competition while at the same time trying to contain the power a successful competitor acquires is heightened tremendously when the patent laws come into play. As the facts of this case demonstrate, the acquisition of a patent can create the potential for tremendous market power.

III

Patent *acquisitions* are not immune from the antitrust laws. Surely, a § 2 violation will have occurred where, for example, the dominant competitor in a market acquires a patent covering a substantial share of the same market that he knows when added to his existing share will afford him monopoly power. See generally Kobe, Inc. v. Dempsey Pump Co., 198 F.2d 416 (10th Cir.), *cert. denied*, 344 U.S. 837 (1952); United States v. Besser Manufacturing Co., 96 F. Supp. 304, 310-11 (E.D. Mich. 1951), *aff'd*, 343 U.S. 444 (1952). That the asset acquired is a patent is irrelevant; in such a case the patented invention already has been commercialized successfully, and the magnitude of the transgression of the antitrust laws' proscription against willful aggregations of market power outweighs substantially the negative effect that the elimination of that class of purchasers for commercialized patents places upon the patent system.

The patent system would be seriously undermined, however, were the threat of potential antitrust liability to attach upon the acquisition of a patent at a time prior to the existence of the relevant market and, even more disconcerting, at a time prior to the commercialization of the patented art. As SCM itself admits, the procurement of a patent by the inventor will not violate §2 even where it is likely that the patent monopoly will evolve into an economic monopoly; yet SCM would deny the same reward to anyone but the patentee.[9]

If the antitrust laws were interpreted to proscribe the natural evolution of a patent monopoly into an economic monopoly, then Judge Newman's concern would be well founded. If the threat of treble damage liability for refusing to license were imbedded in the minds of potential patent holders as a likely prospect incident to every successful commercial exploitation of a patented invention, the efficacy of the economic incentives afforded by our patent system might be severely diminished.

Nevertheless, it is especially clear that the economic incentives provided by the patent laws were intended to benefit only those persons who lawfully acquire the rights granted under our patent system. Cf. Walker Process Equipment, Inc. v. Food Machinery & Chemical Corp., 382 U.S. 172 (1965) (patent obtained by fraud on Patent Office). Where a patent in the first instance has been lawfully acquired, a patent holder ordinarily should be allowed to exercise his patent's exclusionary power even after achieving commercial success; to allow the imposition of treble damages based on what a reviewing court might later consider, with the benefit of hindsight, to be too much success would seriously threaten the integrity of the patent system. Where, however, the acquisition itself is unlawful, the subsequent exercise of the ordinarily lawful exclusionary power inherent in the patent would be a continuing wrong, a continuing unlawful exclusion of potential competitors.

Without passing upon the validity of Judge Newman's theory to preclude antitrust damage liability in all cases where the injury is predicated upon a patent holder's refusal to license, we hold that where a patent has been lawfully acquired, subsequent conduct permissible under the patent laws cannot trigger any liability under the antitrust laws. This holding, we believe, strikes an adequate balance between the patent and antitrust laws. Therefore, to determine whether Xerox incurred any antitrust damage liability to SCM in 1969, our inquiry must now shift to determining whether the acquisition of the Carlson and Battelle patents pursuant to the 1956 agreement violated either the Sherman Act or the Clayton Act. Because the essence of a patent is the monopoly or exclusionary power it confers upon the holder, analyzing the lawfulness of the acquisition of a patent necessitates that we primarily focus upon the circumstances of the acquiring party and the status of the relevant product and geographic markets at the time of acquisition.

9. Notwithstanding that "[t]he law . . . recognizes that [a patentee] may assign to another his patent, in whole or in part, and may license others to practice his invention." Zenith Radio Corp. v. Hazeltine Research, Inc., supra, 395 U.S. at 135, SCM argues that a distinction should be made between the exploitation of a patent by an inventor and an investor. We assume, therefore, that had Chester Carlson possessed the resources to commercialize xerography to the same extent as did Xerox, SCM would not have challenged a refusal by the inventor to license his patents as violative of the antitrust laws. Investors, however, play a key role, if not an indispensable one today, in both the inventive process and commercialization of inventions. And it is fair to say, we think, that the contribution of the investor in both the funding of research that leads to inventions and the promotion that necessarily must follow to achieve successful commercialization is of comparable value. See generally Picard v. United Aircraft Corp., 128 F.2d 632, 642 (2d Cir.), cert. denied, 317 U.S. 651 (1942) (Frank, J., concurring). In either case, the ultimate intended beneficiary of the patent laws — the public — is equally benefitted. See generally Mannington Mills, Inc. v. Congoleum Industries, Inc., 610 F.2d 1059, 1070-71 (3d Cir. 1979); United States v. Parker-Rust-Proof Co., 61 F. Supp. 805, 808 (E.D. Mich. 1945); In re Anthony, 414 F.2d 1383, 1398 (C.C.P.A. 1969); In re Herr, 377 F.2d 610, 619 (C.C.P.A. 1967). Since Xerox participated financially in both the inventive process by funding research at Battelle and the subsequent commercialization of xerography by bringing the first plain-paper copier to the market, we see little reason to deny Xerox the full benefit of the patents it acquired on this basis alone.

IV

Section 2 of the Sherman Act

Turning to the facts of this case, the patents about which we are concerned were acquired in 1956, four years prior to the production of the 914, Xerox's first automatic plain-paper copier, and at least eight years prior to the appearance of the relevant market and submarket. In 1956 Xerox had achieved success in the commercialization of xerography but not in the field of automatic plain-paper copying. There is evidence in the record, however, that Xerox in 1956 valued a non-exclusive license in xerography at $70 million and that key personnel at Xerox believed that they already possessed the necessary technology in 1956 to produce a plain-paper copier. Notwithstanding their optimistic forecasts, however, the confidence of the Xerox organization was still tempered by the risks of producing a new, technologically-sophisticated product line. Thus, in 1958 Xerox considered the possibility of having IBM manufacture and market the 914. But before Xerox's management made a final decision on the matter, IBM informed them that it was not interested in manufacturing or marketing the 914 or another model, the 813, having concluded that both were a bad business risk. While SCM argues that IBM's turndown was directed exclusively at the models 914 and 813 and did not amount to a rejection of plain-paper copying entirely, at the very least the episode demonstrates that as of 1958 the achievement of commercial success in plain-paper copying was not a foregone conclusion.

It was also in 1956 that Xerox acquired non-exclusive licenses from Horizons Corporation, another research organization, covering a small number of xerographic patents.[11] Additionally, Xerox began to cultivate relationships with its international family of companies in 1956. But SCM has not argued that either the Horizons patents or the international agreements caused it any injury. Rather, SCM contends that these facts constitute proof of Xerox's willful acquisition of monopoly power over the relevant market and submarket that came into being, according to the jury, between eight and thirteen years later. Likewise, other aspects of the 1956 agreement with Battelle, such as the promise by Battelle to transfer to Xerox all know-how it developed and patents it obtained in the future were claimed to be additional evidence of Xerox's willful acquisition of its market dominance. But the promise to transfer all xerographic know-how developed and patents obtained in the future was conditioned on Xerox's promise to contribute at least $25,000 a year for research that would help develop the know-how and patents. There appears to be little distinction, if any, between patents obtained under a contract with a research organization and patents generated internally by a company, see P. Areeda & D. Turner, Antitrust Law: An Analysis of Antitrust Principles and Their Application ¶704e (1978), and ordinarily there is no limitation on a company's freedom to generate its own patents. See generally Automatic Radio Manufacturing Co. v. Hazeltine Research, Inc., 339 U.S. 827, 834 (1950). The jury's specific finding that by 1969 Xerox had not obtained any patents primarily for the purpose of blocking the development and marketing of competitive products laid to rest any suspicion that either Xerox's internal R & D program or its R & D work subcontracted to Battelle was driven principally by

---

11. In 1960 Horizons Corporation agreed to assign these patents to Xerox.

anticompetitive animus.[12] In any event, none of Xerox's conduct other than the acquisition of the Carlson and Battelle Patents under the 1956 agreement caused SCM any harm. But even more important, all of the events described occurred between eight and thirteen years prior to the appearance of the relevant product market and submarket defined by SCM.

In scrutinizing acquisitions of patents under §2 of the Sherman Act, the focus should be upon the market power that will be conferred by the patent *in relation* to the market position then occupied by the acquiring party. We agree with Professors Areeda and Turner that whether limitations should be imposed on the patent rights of an acquiring party should be dictated by the extent of the power already possessed by that party in the relevant market into which the products embodying the patented art enter. See Areeda & Turner, supra, at ¶819. Therefore, that Xerox acquired the patents in this case four years prior to the production of the first plain-paper copier and at least eight years prior to the appearance of the relevant product market and submarket over which those patents eventually afforded it monopoly power would seem to dispose entirely of SCM's 1969 exclusion claim under §2.
SCM argues, however, that

> [t]o uphold the jury's verdicts in this case, this Court need hold only that an agreement to purchase patents that eliminate an existing potential for competition in a reasonably foreseeable economic market can be found to be unreasonable if it (a) results in the acquisition of persistent, substantial, real-world economic monopoly power and (b) imposes a restraint on competition that is greater than reasonably necessary to induce the purchaser to develop and market the product involved.

SCM's proposition is that even prior to the commercialization of the patented invention and prior to the appearance of the relevant market over which Xerox eventually achieved monopoly power a §2 violation occurred. SCM suggests that the antitrust laws impose a limitation on the extent of the rights in a patent a purchaser may acquire, and that in some instances a patent with its inherent exclusionary power may not be transferred *in toto*. The limitation that SCM would impose, however, turns not upon the market position of the acquiring party, but rather, upon the potential for commercial success a particular patent may hold. Thus, SCM argues that a purchaser of a patent is entitled only to the rights in a patent reasonably necessary to induce his investment to commercialize the patent. Presumably, under SCM's proposed rule, where the commercial success of a patented invention virtually is guaranteed, no person other than the inventor can hold exclusive rights in the patent, at least where it is foreseeable that the products generated under the patent will create their own relevant product market.

SCM contends that the test it proposes represents the appropriate rule of reason analysis to be employed in patent acquisition cases. By introducing the concept of foreseeability, SCM seeks to escape an unfavorable disposition of its case that it apparently feared might be based upon the absence of the relevant product market and submarket at the time of the patent acquisitions in 1956. Implicit in the jury's findings was that it was reasonably foreseeable in 1956 that the agreement with Battelle would permit Xerox to obtain monopoly power in a relevant product

---

12. SCM also argues that the two-year covenants not to compete imposed by Xerox on its employees up until 1972 also evidence Xerox's willful acquisition of monopoly power. We find this argument wholly without merit.

market.[13] While sufficient evidence was presented by Xerox to support a contrary finding, we are unable to hold, as a matter of law, that no rational jury could find that a reasonably foreseeable effect of the 1956 agreement was the eventual acquisition by Xerox of monopoly power in a relevant market. But notwithstanding the jury's implicit finding, we conclude that, under the facts presented here, the policies of the patent laws preclude the imposition of antitrust liability.

It is undisputed that the first automatic plain-paper copier was not produced by Xerox until four years after the 1956 agreement was executed. Additionally, while Xerox concedes that its plain-paper copiers eventually formed an independent relevant product market, SCM has not challenged on appeal the jury's finding that this event did not occur until some time after 1964, eight years after the agreement. Furthermore, Xerox contributed in a very substantial way to the development of an automatic plain-paper copier by investing in research and development not only after 1956 but also for almost a decade before the agreement. Moreover, the party from whom Xerox purchased the patent under the 1956 agreement was not a potential competitor. We believe that, under the circumstances presented here, to impose antitrust liability upon Xerox would severely trample upon the incentives provided by our patent laws and thus undermine the entire patent system. Therefore, irrespective of the jury's implicit finding that Xerox's commercial success was reasonably foreseeable in 1956, Xerox was lawfully entitled to purchase the patents it did pursuant to the agreement it made with Battelle that year.

With respect to Xerox's subsequent unilateral refusal to license the Carlson and Battelle patents, which we have held were lawfully acquired, that conduct was permissible under the patent laws and, therefore, did not give rise to any liability under § 2 . . . .

Section 7 of the Clayton Act

Section 7 of the Clayton Act proscribes a corporation from acquiring the whole or any part of the assets of another corporation where "the effect of such acquisition may be substantially to lessen competition, or to tend to create a monopoly [in any line of commerce]," 15 U.S.C. § 18 (1976). Since a patent is a form of property, see generally Transparent-Wrap Machine Corp. v. Stokes & Smith Co., 329 U.S. 637, 643 (1947), and thus an asset, there seems little reason to exempt patent acquisitions from scrutiny under this provision. See United States v. Lever Brothers Co., 216 F. Supp. 887, 889 (S.D.N.Y. 1963); see generally L. Sullivan, Handbook of the Law of Antitrust, § 180 (1977); 16a J. von Kalinowski, Business Organizations: Antitrust Laws and Trade Regulation § 16.05 (1980); Kessler & Stern, Competition, Contract, and Vertical Integration, 69 Yale L.J. 1, 75-78 (1959).

Section 7 principally was designed to curtail the anti-competitive consequences of corporate acquisitions in their "incipiency." Brown Shoe Co. v. United States, 370 U.S. 294, 317 (1962). Thus, the analysis ordinarily employed in determining the lawfulness of a corporate acquisition under § 7 is prospective in nature. The jury

---

13. This finding was implicit in the jury's affirmative answer to question 20: "Was the probable effect of Xerox's acquisition of patents pursuant to the 1956 Xerox-Battelle agreement, when the agreement was made, substantially to lessen competition or to tend to create a monopoly in any relevant market or submarket that you have found to exist?" In explaining this question, Judge Newman instructed the jurors to determine "whether the acquisition at the time it was made was reasonably probable to have the proscribed effect in the reasonably foreseeable future."

found that the probable effect of the 1956 Xerox-Battelle agreement was substantially to lessen competition or to tend to create a monopoly in the relevant product market and submarket that appeared between eight and thirteen years later. The jury additionally found that, as of 1964 and 1969, the probable effect of Xerox's continued holding of patents acquired pursuant to the 1956 Xerox-Battelle agreement was substantially to lessen competition or to tend to create a monopoly. We conclude that neither of these findings can stand as a matter of law.

While the Supreme Court in Brown Shoe Co. v. United States, supra, 370 U.S. at 323, stated that the language contained in §7 is indicative that Congress' "concern was with probabilities, not certainties," the speculative aspect of this antitrust law was intended to allow courts to appreciate immediately the potential consequences that a particular acquisition might have upon an *existing* line of commerce. Thus in *Brown Shoe*, the Supreme Court stated:

> Because §7 of the Clayton Act prohibits any merger which may substantially lessen competition "in *any* line of commerce," (emphasis supplied) it is necessary to examine the effects of a merger in each such economically significant submarket to determine if there is a reasonable probability that the merger will substantially lessen competition.

370 U.S. at 325. The existing market provides the framework in which the probability and extent of an adverse impact upon competition may be measured. In the case at bar it would have been impossible to examine the effects of the Xerox-Battelle agreement upon the relevant product market and submarket in 1956 because those markets did not come into being until sometime between 1964 and 1969. The jury was instructed that it should include in its considerations whether the appearance of the relevant market and submarket and Xerox's domination of those markets was reasonably foreseeable in 1956.

Judge Newman concisely stated below that "[s]ection 7 is concerned with undue concentrations of power and the anti-competitive effects of permitting one entity with market power to strengthen its position by acquisition." 463 F. Supp. at 1001-02. Where, as here, it is conceded that the relevant product market and submarket did not exist until eight years following the patent acquisitions and that Xerox possessed no power whatsoever in even the inchoate market and submarket until 1960 when it introduced its 914 copier, as a matter of law the 1956 agreement did not violate §7 at the time it was made. Finally, our decision regarding SCM's foreseeability argument under §§1 and 2 of the Sherman Act disposes of its argument propounded under §7 along those lines.

SCM argues alternatively that the Supreme Court's decisions in *du Pont-GM*, supra, and United States v. ITT Continental Baking Co., 420 U.S. 223 (1975), require that we affirm the jury's second finding under §7 that Xerox's continued "holding" of the patents it obtained under the Xerox-Battelle agreement in 1956 violated §7 as of 1969. In *du Pont-GM*, the government commenced its action approximately thirty years after du Pont had purchased a twenty-three percent stock interest in General Motors Corporation, and alleged that du Pont had used its stock ownership to attain a "commanding position as General Motors' supplier of automotive finishes and fabrics," 353 U.S. at 588-89. There, the Court held that

> any acquisition by one corporation of all or any part of the stock of another corporation, competitor or not, is within the reach of the section *whenever the reasonable likelihood*

appears that the acquisition will result in a restraint of commerce or in the creation of a monopoly in any line of commerce.

353 U.S. at 592 (emphasis added). Subsequently, in *ITT Continental Baking,* the Court reaffirmed, albeit in dictum, its holding in *du Pont-GM* that the term acquisition as it is employed in §7 comprehends both the initial "acquiring" and subsequent "retaining" of the stock of another corporation. 420 U.S. at 241-42. Relying on these cases, SCM would have us hold that Xerox's acquisition of the Carlson and Battelle patents in 1956 became actionable under §7 as soon as the monopoly afforded by the patents burgeoned into an economic monopoly. Whatever the meaning that may be ascribed to §7 in other contexts, the patent laws circumscribe the scope of the provision here. Where a corporation's acquisition of a patent is not violative of §7, as was the case here, its subsequent holding of the patent cannot later be deemed violative of this section. Where a company has acquired patents lawfully, it must be entitled to hold them free from the threat of antitrust liability for the seventeen years that the patent laws provide. To hold otherwise would unduly trespass upon the policies that underlie the patent law system. The restraint placed upon competition is temporally limited by the term of the patents, and must, in deference to the patent system, be tolerated throughout the duration of the patent grants....

COMMENTS AND QUESTIONS

1. Automatic Radio Mfg. v. Hazeltine Research, Inc., 339 U.S. 827, 834 (1950), established the proposition that "[t]he mere accumulation of patents, no matter how many, is not in and of itself illegal." However, the strength of that case was diluted by the amendment of the Clayton Act the same year. Section 7 of the Clayton Act prohibits the acquisition of any company, stock, *or asset* where such acquisition is likely to restrain competition. Courts have viewed patents and copyrights as "assets" subject to the provisions of that section. *SCM* does so; see also United States v. Lever Bros., 216 F. Supp. 887, 889 (S.D.N.Y. 1963); United States v. Columbia Pictures Corp., 189 F. Supp. 153, 182-83 (S.D.N.Y. 1960). SCM recognizes that in some circumstances section 7 would prohibit patent acquisitions (or mergers between companies holding patents in the same market), even though in-house development of the same intellectual property would be perfectly legal.

Most patents are not commercialized by the original inventor.[20] Companies that purchase a patent outright are subject to the Clayton Act; companies that develop a number of patents "in-house" are not. Is there more reason to be concerned about acquisitions than about internal growth? If not, why the differential treatment? One reason may have to do with the nature of the antitrust remedy. It is much easier to enjoin a proposed purchase or merger, or to order divestiture of a recently acquired asset, than it is to fashion a remedy against a company that has developed and patented its own research.

---

20. In some fields such as biotechnology, however, "bundled" patenting of groups of similar inventions is becoming increasingly common. For example, the designer of a synthetic analog to a naturally occurring chemical may patent a whole series of similar analogs to prevent imitation by competitors. Large companies and universities (which have major research departments) are more likely to engage in bundled patenting than are small inventors. To the extent that these patents are commercialized (and they often are not), they will normally be commercialized by the same firm.

More common than outright purchase of patent rights is the long-term licensing of those rights by inventors to companies that will commercialize the invention. Do such licenses constitute the "acquisition" of an asset within the meaning of section 7? The law is not clear. Should licenses fall within the scope of the Clayton Act? Does it matter whether the license is exclusive or nonexclusive? The Antitrust Division's Intellectual Property Guidelines provide that exclusive licenses will be treated like acquisitions under section 7. Intellectual Property Guidelines § 5.7.

PROBLEMS

Problem 8-6. In the summer of 1991, Borland International announced its plan to buy Ashton-Tate Corp. Both companies were major players in the database software market for personal computers. Together, they would presumably become the dominant player in that market, reducing competition. However, the merged company also planned to enter the database software market for minicomputers, a market dominated at that time by Oracle Corp. Should antitrust law permit the expected increase in competition in that market to offset the damage to competition in the PC market?

The Antitrust Division expressed concern that the combined intellectual property libraries of Borland and Ashton-Tate would enable the combined company to prevent further entry into the database market. In particular, the division was concerned that large potential entrants like Lotus Corp. would be deterred by the threat of intellectual property litigation. How would you evaluate a proposal by the division, as a condition of permitting the merger, to require the new company to give up its right to sue new entrants for copyright infringement?

Problem 8-7. Megaplex, Inc., is a major petrochemical manufacturer. Concerned that Newco is "gaining" on it in the market for new gasoline additives through greater research prowess, Megaplex approaches several solo inventors who hold patents in the field and "buys" the rights to the patents from the inventors. Armed with a patent portfolio that Newco will have difficulty "inventing around," Megaplex manages to maintain its market share in spite of Newco's research efforts. Has Megaplex violated the antitrust laws?

Table of Cases

Principal cases are in italics.

Table of Statutes, Regulations, and Treaties

Index